Algebra and Trigonometry

5th Edition

Algebra and Trigonometry

5th Edition

MERVIN L. KEEDY
Purdue University

MARVIN L. BITTINGER
Indiana University—Purdue University at Indianapolis

ADDISON-WESLEY PUBLISHING COMPANY

Reading, Massachusetts • Menlo Park, California • New York
Don Mills, Ontario • Wokingham, England • Amsterdam • Bonn
Sydney • Singapore • Tokyo • Madrid • San Juan

Sponsoring Editor	*Elizabeth Burr*
Production Supervisor	*Jack Casteel*
Design, Editorial, and Production Services	*Quadrata, Inc.*
Art Consultant	*Loretta Bailey*
Illustrator	*Textbook Art Associates; Hardlines*
Manufacturing Supervisor	*Roy Logan*
Cover Designer	*Marshall Henrichs*

PHOTO CREDITS

1, © David R. Frazier Photolibrary/Photo Researchers, Inc. **65**, St. Louis Regional Committee and Growth Association **131**, © Stan Goldblatt/Photo Researchers, Inc. **205**, Ken Karp **275**, Ulrike Welsch 1981/Stock, Boston **340 (left)**, AP/Wide World Photos **340 (right)**, Rick Haston **347**, © 1982 Peter Fink/ Photo Researchers, Inc. **431**, F. B. Grunzweig/Photo Researchers, Inc. **493**, Jean-Claude LeJeune/Stock, Boston **551**, W. B. Finch/Stock, Boston **625**, William Clark/U.S. Department of the Interior **661**, AP/Wirephoto **721**, AP/Wide World Photos

Library of Congress Cataloging-in-Publication Data

Keedy, Mervin Laverne.

Algebra and trigonometry: a functions approach / by Mervin L. Keedy and Marvin L. Bittinger.—5th ed.

p. cm.

Rev. ed. of: Algebra & trigonometry. 4th ed. c1986.

Includes index.

ISBN 0-201-14995-8

1. Algebra. 2. Trigonometry. I. Bittinger, Marvin L. II. Keedy, Mervin Laverne. Algebra & trigonometry. III. Title. QA154.2.K43 1989

512'.13—dc20 89-31219

ABCDEFGHIJ-MU-954321089

This volume is part of the Bittinger precalculus series for college algebra and trigonometry courses. Other texts in the series are the following.

BITTINGER/BEECHER

Algebra and Trigonometry covers precalculus algebra and trigonometry with an emphasis on functions. Trigonometric functions are presented first through a discussion of right-triangle trigonometry.

CONTENTS

Basic Concepts of Algebra. Equations, Inequalities, and Problem Solving. Relations, Functions, and Transformations. Linear and Quadratic Functions and Inequalities. Polynomial and Rational Functions. Exponential and Logarithmic Functions. The Trigonometric Functions. Trigonometric Identities, Inverse Functions, and Equations. Triangles, Vectors, and Applications. Systems of Equations and Matrices. Equations of Second Degree and Their Graphs. Sequences, Series, and Probability.

College Algebra covers the basic topics of college-level precalculus algebra.

CONTENTS

Basic Concepts of Algebra. Equations, Inequalities, and Problem Solving. Relations, Functions, and Transformations. Linear and Quadratic Functions and Inequalities. Polynomial and Rational Functions. Exponential and Logarithmic Functions. Systems of Equations and Matrices. Equations of Second Degree and Their Graphs. Sequences, Series, and Probability.

Trigonometry introduces the trigonometric functions and their properties, beginning with the trigonometry of the right triangle.

CONTENTS

Relations, Functions, and Transformations. The Trigonometric Functions. Trigonometric Identities, Inverse Functions, and Equations. Triangles, Vectors, and Applications. Imaginary and Complex Numbers. Exponential and Logarithmic Functions.

Preface

This text covers college-level algebra and trigonometry and is appropriate for a one- or two-term course in precalculus mathematics. Its approach is more intuitive and interactive than most precalculus texts, and is designed to help students achieve a greater understanding of sophisticated mathematical concepts than they might be able to achieve with other texts. Our goal is to provide as much support and help for students as we can, in order to ease the difficult transition into their first college-level mathematics course. At the same time, we want to give students solid preparation for calculus and maintain the appropriate topical coverage and level of presentation for a precalculus course.

CONTENT FEATURES

Algebra Review

- A lack of basic algebra skills is a common cause for student struggles in precalculus and calculus. This text contains a careful and detailed review of intermediate algebra topics in Chapters 1 and 2. (See pp. 1–129.)

Graphing and Functions

- Graphing and functions are both introduced in Chapter 3. We present graphing as a means of representing relations or correspondences visually, and use it throughout the text to give students an intuitive and visual understanding of the material they are studying. For example, whenever a new type of function is introduced, we immediately show its graph. (See pp. 136, 290, and 357.)
- Functions are presented in Chapter 3 after correspondences, relations, and

graphs have been introduced. We include a detailed presentation of topics that will be important in calculus: the algebra of functions, the composition of functions. (See pp. 145 and 181.)

Modeling, Problem Solving, and Applications

- Throughout the text, we try to motivate the material by providing meaningful real-life applications of the mathematics. We also include many problem-solving sections. (See pp. 155, 214, 315, and 499.)
- Since solving applied problems is one of the most difficult and important areas of algebra, we present a problem-solving algorithm in Section 2.3, and use this algorithm consistently throughout the text. The algorithm provides students with a starting point, and allows them to focus on the mathematics needed to solve the problem. To give students the background they need to translate problems themselves, we discuss extensively the "familiarization" and "translation" steps of the problem-solving process. (See pp. 79 and 557.)
- In Chapter 4, we include a section on mathematical models, which describes the characteristics of a mathematical model and discusses when a model will work, and, more important, when it will not work. (See pp. 235–242.)

Trigonometry

- We begin the trigonometry material with a discussion of distances on the unit circle in Section 6.1, to give the student a thorough understanding of where points like π and $-\pi/3$ are located. We then develop trigonometric functions of the real numbers on the unit circle in Sections 6.2 and 6.3, before introducing angles, radians, and degrees in Section 6.4. Trigonometric functions of the right triangle are introduced in Section 6.5 and applied in Section 8.1. (See pp. 348–399 and 494.)
- In Chapter 7, we spend an entire section discussing and practicing techniques for proving trigonometric identities. We present two different methods for proving identities and discuss the merits of each one. (See pp. 461–466.)

Calculus Preparation

- Throughout the text, we give careful attention to the presentation of concepts that will become important in later calculus courses. For example, in the introduction to relations and functions, we emphasize the idea of domain, discussing it three times: first in the context of relations, then with graphs of relations, and finally with functions. In addition, the material on polynomials and rational functions includes an intuitive and graphical introduction to the idea of limit. (See pp. 134, 142, 151, and 688.)

Exponential and Logarithmic Functions

- The material on exponential and logarithmic functions is introduced in the context of inverse functions, with careful emphasis placed on the relationship between exponents and logarithms. (See pp. 276 and 296.)

WHAT'S NEW IN THE FIFTH EDITION?

We have rewritten many key topics in response to user feedback, and have made significant improvements to several chapters. Detailed information about the changes made to this material is available in the form of a *Conversion Guide.*

Please ask your local Addison-Wesley sales representative for more information. Following is a list of the major organizational changes in this revision.

- Many applications and exercises have been added throughout the text.
- Sets and set-builder notation are now introduced in Chapter 1.
- The introduction to complex numbers is now covered in Chapter 2, and complex solutions of quadratic equations are now introduced in this chapter. The material on polar notation and DeMoivre's theorem has been moved into the third trigonometry chapter.
- Chapter 3, "Relations, Functions, and Transformations," has been extensively rewritten in response to user feedback and now includes many new applications as well as material on the algebra of functions and the composition of functions.
- The material on systems of equations and matrices has been combined and moved after the trigonometry chapters. The introduction to polynomial functions and some material on graphs of polynomials has been moved to the end of the chapter on linear and quadratic functions. This new organization provides a logical and unbroken flow of the functions material through the trigonometric functions.
- The material on circles from the chapter on conic sections has been moved earlier in the text.
- The material on exponential and logarithmic functions has been rewritten to provide a clearer understanding of the relationship between the exponential and the logarithmic functions. Many applications have also been added to this material.
- The introduction to trigonometric functions has been reorganized so that it flows more logically, and more time is spent introducing students to distances on the unit circle.
- The material on trigonometric identities has been reorganized and includes the material on cofunction identities. We now also include two methods for proving trigonometric identities with a discussion of the relative advantages and disadvantages of each method.

PEDAGOGICAL FEATURES

Intuitive and Interactive Presentation

- Whenever possible, we base the presentation of a concept on students' prior experience. For example, the introduction to relations and functions is based on correspondences that students are familiar with from daily life. (See pp. 132 and 145.)
- Throughout each section, students are directed to do exercises in the margin; these have answers at the back of the text. Margin exercises involve students interactively in the development of the material and provide a built-in study guide. Many instructors use the margin exercises for in-class reinforcement of the material. (See p. 552.)

Section Objectives

- Objectives for each section are stated in the margin at the beginning of the section, and keyed with a symbol like **2**. This symbol is used throughout the text to key sections of the exposition, exercises, and answers at the back of the text to specific objectives, allowing students to easily find appropriate material for review. (See p. 206.)

Variety of Exercises

- In addition to regular computational and applied exercises, most exercise sets contain two levels of challenge exercises. *Synthesis exercises* require students to synthesize objectives from several sections and help to develop their critical thinking skills. *Challenge exercises* go beyond the section objectives and are designed to challenge the brightest students. (See pp. 216 and 224–226.)

Tests and Review

- Answers to review sections and chapter tests are at the back of the text, together with section references so that students can easily find the correct material to restudy if they miss an exercise. (See p. A-1.)

ACKNOWLEDGMENTS

The authors wish to express their appreciation to the many people who helped with the development of this book, particularly John K. Baumgart and Donna DeSpain for their precise proofreading and checking of the manuscript.

In addition, we wish to thank the following professors for their thorough reviewing and feedback:

Susan Addington, *Harvard University*
Leonard Andrusaitis, *University of Lowell*
Bruce Bemis, *Westminster College*
Barbara Buchalter, *University of Nebraska—Omaha*
Earl Carpenter, *Bob Jones University*
Henry Cohen, *University of Pittsburgh*
Chris Ennis, *Carleton College*
Carol J. Flakus, *Lower Columbia College*
Lenore Frank, *SUNY—Stonybrook*
Julie Guelich, *Normandale Community College*
Scott Higinbotham, *Middlesex Community College*
Glen Jacobs, *Greenville Tech*
Don Jessup, *University of Nevada—Reno*
Jo Lane, *Eastern Kentucky University*
Joy Mark, *University of Connecticut*
Rosalie Mahony, *College of San Mateo*
John A. Schumaker, *Rockford College*
Sharon Sledge, *San Jacinto College Central Campus*
Dorothy Sulock, *University of North Carolina at Asheville*
Ray Treadway, *Sue Bennett College*
Jeanette Tyson, *Valencia Community College*
John Weerts, *Triton College.*

We would also like to thank particularly the members of the mathematics department of Georgia State University, who contributed greatly to this revision.

M.L.K.
M.L.B.

SUPPLEMENTS FOR THE INSTRUCTOR

Addison-Wesley is committed to providing the best possible service and support for your classroom needs. If you have any questions about the extensive supplements package that accompanies this text for you or your students, please feel free to call 1-800-227-5210.

Instructor's Resource Guide

This includes the following:

- Six forms of each chapter test and six final examinations with answers;
- Transparency masters;
- BASIC computer programs for each chapter designed to give students an idea of the power and utility of the computer in solving exercises encountered in this text;
- User notes for *Cactusplot*, one of Addison-Wesley's function-graphing software packages (see below for more information), and suggestions for integrating *Cactusplot* into your course;
- Videotape indexes cross-referenced to each section of the text.

Instructor's Solutions Manual by Judith A. Penna and John K. Baumgart

This contains worked-out solutions to every exercise in the text.

Printed Test Bank by Betty P. Givan

This is a collection of over 3000 multiple-choice test items organized by text section and objective.

Videotapes

Using the blackboard and manipulative aids, Professor John K. Baumgart of North Park College gives a careful, detailed, and polished presentation of the text material in 50 videotapes ranging from 30 minutes to 45 minutes in length. Professor Baumgart's use of the blackboard and step-by-step discussion brings the concepts of algebra and trigonometry to life and provides ideal review or supplementary coverage for classroom lectures.

Transparency Masters

This is a booklet of 120 black-line masters for key rules, definitions, theorems, proofs, and figures taken from the text. Twenty of these are available on four-color acetates for qualifying adopters.

SUPPLEMENT FOR THE STUDENT

Student's Solutions Manual by Judith A. Penna

This contains completely worked-out solutions to the odd-numbered exercises and answers to the even-numbered exercises in the text.

SOFTWARE SUPPLEMENTS

Testing Software

AWTest. Addison-Wesley's random-number test-generating system is available for the Apple II Series with this text. Using *AWTest*, you can generate up to 99 variations of any particular test with a few keystrokes. You can also choose test items by number from a bank of over 320 multiple-choice algorithm-driven test items or request tests to be printed out in chapter-test format.

AWTestEdit. This is a computerized test item bank containing over 3000 multiple-choice test items for the IBM PC. The program also allows you to edit existing test items and enter your own as easily as using a word processor. You can create tests with both multiple-choice and open-ended questions, save them, and create multiple versions of each test by scrambling question order and multiple-choice distractors.

Utilities Software

Cactusplot: Addison-Wesley Student Version. This is an easy-to-use software utilities package for the Apple II series and IBM PC by John Losse of Scottsdale Community College (IBM adaptation by David Yunker). *Cactusplot* can perform many mathematical operations, such as graphing functions, creating tables, solving equations, and finding the area under curves. Suggestions for integrating *Cactusplot* into your course are included in the Instructor's Manual for this text. A site license to the Student Version of *Cactusplot* is available from Addison-Wesley and is free to qualifying adopters.

The Student Version of *Cactusplot* is available only from Addison-Wesley. A professional version of *Cactusplot* with many enhanced capabilities is available directly from

John Losse
The Cactusplot Company
4712 E. Osborn
Phoenix, AZ 85281
(602) 945-1667.

Related Addison-Wesley Software Titles

Demana/Waits Master Grapher and *3D Grapher Software.* These two utilities packages are available for the IBM PC, MacIntosh, and Apple II series computers. *Master Grapher* allows you to graph and manipulate functions in two dimensions. You can change function parameters, rotate axes, overlay one graph with another, and perform a variety of transformations. The *3D Grapher* graphs complex functions in three dimensions. You can zoom in on the function, change the resolution of its graph, rotate the axis of the graph horizontally and vertically, change its scale, and perform other operations designed to help students visualize graphics in three dimensions more clearly. Contact your Addison-Wesley sales representative for more information.

Contents

5 EXPONENTIAL AND LOGARITHMIC FUNCTIONS 275

6 THE TRIGONOMETRIC OR CIRCULAR FUNCTIONS 347

7 TRIGONOMETRIC IDENTITIES, INVERSE FUNCTIONS, AND EQUATIONS 431

10 CONIC SECTIONS 625

11 POLYNOMIAL AND RATIONAL FUNCTIONS 661

12 SEQUENCES, SERIES, AND COMBINATORICS 721

TABLES

Algebra and Trigonometry

5th Edition

Basic Concepts of Algebra

1

The square-root formula in the problem below illustrates a basic concept of algebra. This chapter is a study of the basic concepts of algebra. We assume that you have already studied intermediate algebra. That being the case, you should find most of this material merely a review. We cover the properties of the real-number system and various kinds of algebraic expressions and manipulations of them—for example, how to add them, multiply them, factor them, and so on.

If your study of algebra is recent, you might be able to skip this chapter, or at least most of it. To determine whether that is the case, you can work through the test at the end of the chapter. If you answer 75% to 85% of the questions correctly, then it might be wise for you to go on to Chapter 2.

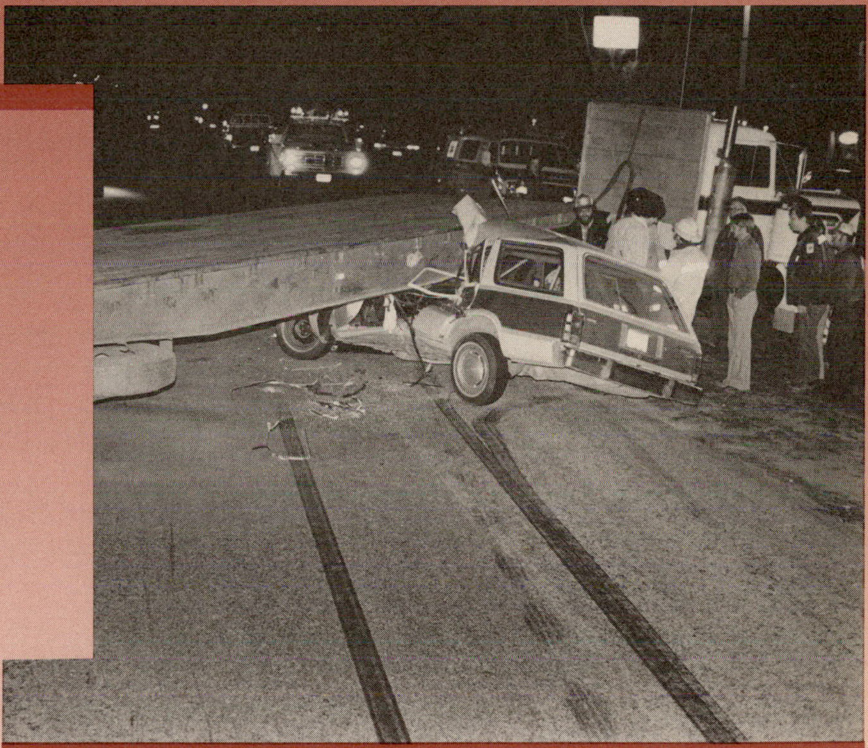

FEATURE PROBLEM

Speed of a speeding car. How do police determine the speed of a car that has skidded? The formula

$$r = 2\sqrt{5L}$$

can be used to approximate the speed r, in miles per hour, of a car that has left a skid mark of length L, in feet. What was the speed of a car that left skid marks 306 ft long?

THE MATHEMATICS

We substitute 306 for L in the formula $r = 2\sqrt{5L}$ and use a calculator to approximate r:

$$r = 2\sqrt{5(306)} = 2\sqrt{1530}$$
$$\approx 2(39.115) = 78.23.$$

OBJECTIVES

You should be able to:

1 Identify various kinds of real numbers.

2 Add, subtract, multiply, and divide positive and negative real numbers.

All answers to the margin exercises throughout the book are located in the answer section at the back of the book.

Consider the numbers

$$1, \frac{3}{4}, -6, 0, 19, -\frac{8}{7}.$$

1. Which are natural numbers?

2. Which are whole numbers?

3. Which are integers?

4. Which are rational numbers?

1.1 The Real-Number System

1 Real Numbers

There are various kinds of numbers. The set of numbers most used in elementary algebra is the set of **real numbers.** Later, we will consider a more comprehensive set of numbers called the **complex numbers.** The real numbers are often shown in one-to-one correspondence with the points of a line, as follows.

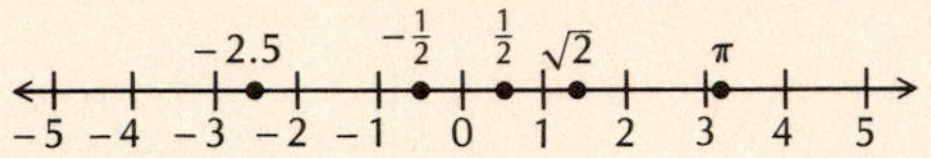

The positive numbers are shown to the right of zero and the negative numbers to the left. Zero itself is neither positive nor negative.

There are many ways to name sets. For example, the set containing the numbers 2, 3, 4, and 5 can be denoted {2, 3, 4, 5}. This method of naming sets is called the **roster method.**

There are several subsets of the real-number system. Some important subsets are as follows.

> Natural Numbers. **Those numbers used for counting: {1, 2, 3, . . . }.**
>
> Whole Numbers. **The natural numbers and 0: {0, 1, 2, 3, . . . }.**
>
> Integers. **The whole numbers and their opposites:**
>
> **{. . . , −3, −2, −1, 0, 1, 2, 3, . . . }.**
>
> Rational Numbers. **The integers and all quotients of integers (excluding division by 0): for example, $\frac{4}{5}$, $-\frac{4}{7}$, $\frac{9}{1}$, 6, −4, 0, $\frac{78}{-5}$, $-\frac{2}{3}$ (can also be named $\frac{-2}{3}$ or $\frac{2}{-3}$).**

We can describe the set of rational numbers precisely using what is known as **set-builder notation.** To do so, we abbreviate the sentence

"The set of all quotients a/b such that a and b are integers and $b \neq 0$."

as

$$\{a/b \mid a \text{ and } b \text{ are integers and } b \neq 0\}.$$

We will consider set-builder notation more extensively when we study equations and inequalities.

DO EXERCISES 1–4 (IN THE MARGIN).

Any real number that is not rational is called **irrational.** The rational numbers and the irrational numbers can be described in several ways.

> **The *rational numbers* are:**
>
> **1. Those numbers that can be named with fractional notation a/b, where a and b are integers and $b \neq 0$ (definition);**
>
> **2. Those numbers for which decimal notation either ends or repeats.**

EXAMPLES All of these are rational numbers.

1. $\frac{5}{16} = 0.3125$ — Ending (terminating) decimal
2. $-\frac{8}{7} = -1.142857142857\ldots = -1.\overline{142857}$ — Repeating decimal. The bar indicates the repeating part.
3. $\frac{3}{11} = 0.2727\ldots = 0.\overline{27}$

The *irrational numbers* are:

1. Those numbers that are not rational (definition);
2. Those real numbers that cannot be named with fractional notation a/b, where a and b are integers and $b \neq 0$.

There are many irrational numbers. For example, $\sqrt{2}$ is irrational. We can find rational numbers a/b for which $(a/b)^2$ is close to 2, but we cannot find such a number a/b for which $(a/b)^2$ is *exactly* 2.

Unless a whole number is a perfect square, its square root is irrational. For example, $\sqrt{9}$ and $\sqrt{25}$ are rational, but all of the following are irrational:

$$\sqrt{3}, \quad -\sqrt{14}, \quad \sqrt{45}.$$

There are also many irrational numbers that cannot be obtained by taking square roots. The number π is an example.* Decimal notation for π does not end and does not repeat.

EXAMPLES All of these are irrational numbers.

4. $\pi = 3.1415926535\ldots$ — There is no repeating block of digits.
5. $-1.101001000100001000001\ldots$ — Though there is a pattern, there is no repeating block of digits.
6. $\sqrt{45} = 6.70820393\ldots$ — There is no repeating block of digits.
7. $\sqrt[3]{2} = 1.25992105\ldots$ — There is no repeating block of digits.

The following figure shows the relationship between the various kinds of numbers.

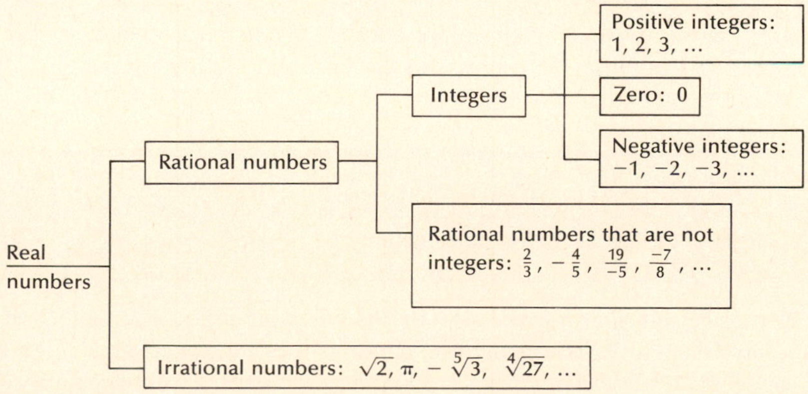

DO EXERCISES 5–10.

In arithmetic, we use numbers, performing calculations to obtain certain answers. In algebra, we use arithmetic symbolism, but we also use symbols to represent unknown numbers. We do calculations and manipulations of symbols

*$\frac{22}{7}$, 3.14, and $\frac{355}{113}$ are only rational approximations to the irrational number π.

Which of the following are rational? Which are irrational?

5. $\frac{-4}{5}$

6. $-\sqrt{64}$

7. 7.42

8. 0.47474747. . .
(There is a repeating block of digits.)

9. 2.57340046631. . .
(There is no repeating block of digits.)

10. $\sqrt{27}$

Add.

11. $-5 + (-7)$

12. $-1.2 + (-3.5)$

13. $-\frac{6}{5} + \frac{2}{5}$

14. $0.5 + (-0.7)$

15. $8 + (-3)$

16. $\frac{14}{3} + \left(-\frac{14}{3}\right)$

on the basis of the properties of numbers, which we review now. Algebra is thus an extension of arithmetic and a more powerful tool for solving problems.

2 Operations on the Real Numbers

Addition

Let us review how the definition of addition is extended to include the negative numbers.

> **1. The sum of two positive real numbers is a positive real number. The sum of two negative real numbers is a negative real number.**
>
> **2. The sum of a positive and a negative number can be either positive or negative, depending on which number is farther from 0 on a number line. If the numbers are the same distance from 0, then their sum is 0.**

EXAMPLES Add.

8. $-5 + (-6) = -11$

9. $8.6 + (-4.2) = 4.4$

10. $-5 + 3 = -2$

11. $\pi + (-\pi) = 0$

12. $8 + (-5) = 3$

13. $-\sqrt{3} + (-4\sqrt{3}) = -5\sqrt{3}$

14. $-\frac{9}{5} + \frac{3}{5} = -\frac{6}{5}$

15. $-\frac{5}{6} + \left(-\frac{7}{8}\right) = -\frac{20}{24} + \left(-\frac{21}{24}\right) = -\frac{41}{24}$

DO EXERCISES 11–16.

PROPERTIES OF REAL NUMBERS UNDER ADDITION. In solving equations and doing other kinds of work in algebra, we manipulate algebraic symbols in various ways, such as collecting like terms. For example, instead of

$$4x + 7x,$$

we might write

$$11x,$$

knowing that the two expressions represent the same number no matter what x represents. In that sense, the expressions $4x + 7x$ and $11x$ are **equivalent.** We define equivalent expressions in general.

> DEFINITION
>
> **If two expressions represent the same number for *any* meaningful replacement, then the expressions are said to be *equivalent*.**

A replacement for an expression that is not meaningful is a number that when substituted for a variable in the expression does not give us a real number. For example, the number 2 is not a sensible replacement for y in the expression

$$\frac{8 + y}{y - 2}$$

because it gives a denominator of 0, and we know that we cannot divide by 0.

We now list for review the fundamental properties of real numbers under addition. These are properties on which algebraic manipulations are based, especially when symbols for unknown numbers are used. These properties allow us to find equivalent expressions.

Commutative Law. For any real numbers a and b,

$$a + b = b + a.$$

(The *order* in which numbers are added does not affect the sum.)

Associative Law. For any real numbers a, b, and c,

$$a + (b + c) = (a + b) + c.$$

(When *only* additions are involved, parentheses for grouping purposes may be placed as we please without affecting the sum.)

Identity. There exists a unique real number 0 such that for any real number a,

$$a + 0 = 0 + a = a.$$

Inverses. For any real number a, there exists a unique real number, denoted $-a$, called the *additive inverse*, or *opposite*, for which

$$-a + a = a + (-a) = 0.$$

ADDITIVE INVERSES, OR OPPOSITES. Concerning additive inverses, or opposites, we caution you about one of the most misunderstood, or confusing, ideas in elementary algebra: It is common to read an expression such as $-x$ as "negative x." This can be confusing, because $-x$ may be positive, negative, or zero, depending on the value of x. The symbol $-$, used in this way, indicates an **additive inverse**, or **opposite**. The same symbol may also indicate a negative number, as in -5, or it may indicate subtraction, as in $3 - x$.

CAUTION! An initial $-$ sign, as in $-x$ or $-(x^2 + 3x - 2)$, should always be interpreted as meaning "the additive inverse of" or "the opposite of," *not* "the negative of"! The entire expression may be positive, negative, or zero, depending on the value of the part of the expression that follows the $-$ sign.

Taking the additive inverse or opposite is sometimes called "changing the sign."

EXAMPLES Find each of the following.

16. $-x$, when $x = 5$ $\quad -(5) = -5$ $\quad$ The opposite of 5 is negative 5.

17. $-x$, when $x = -3$ $\quad -(-3) = 3$ $\quad$ The additive inverse of negative 3 is 3.

18. $-x$, when $x = 0$ $\quad -(0) = 0$ $\quad$ The opposite of 0 is 0.

19. $-(x^2 + 4x + 2)$, when $x = 1$ $\quad -(1^2 + 4 \cdot 1 + 2) = -(7) = -7$

$-(x^2 + 4x + 2)$, when $x = -1$ $\quad -[(-1)^2 + 4(-1) + 2] = -(-1) = 1$ ■

We can easily show that $-1 \cdot x = -x$ and $-(-x) = x$ for any number x. That is, multiplying a number x by negative 1 produces the opposite of x, and the opposite of the opposite of a number is the number itself. We list this as a theorem.

THEOREM 1

For any real number a,

$$-1 \cdot a = -a \quad \text{and} \quad -(-a) = a.$$

(Multiplying a number by -1 produces its opposite, and the opposite of the opposite of a number is the number itself.)

Find $-x$ and $-1 \cdot x$ when:

17. $x = 6$.

18. $x = -8$.

19. $x = -3.4$.

20. Find $-(x^2 - 3x)$ when $x = 2$ and $x = -1$.

Multiply.

21. $4 \cdot (-6)$

22. $-\frac{7}{5} \cdot \left(-\frac{3}{5}\right)$

23. $(-2) \cdot (-3) \cdot (-4) \cdot (-6)$

DO EXERCISES 17–20.

Multiplication

Assuming that multiplication of nonnegative real numbers poses no problem, let us review how the definition of multiplication is extended to include the negative numbers.

> **1. The product of two positive numbers is positive. The product of two negative numbers is positive. In other words, if the signs of two numbers are the same, then the product is positive.**
>
> **2. The product of a positive number and a negative number is negative. In other words, if the signs of two numbers are different, then the product is negative.**

EXAMPLES Multiply.

20. $3 \cdot (-4) = -12$

21. $1.5 \cdot (-3.8) = -5.7$

22. $-5 \cdot (-4) = 20$

23. $-\frac{2}{3} \cdot \left(-\frac{4}{5}\right) = \frac{8}{15}$

24. $-3 \cdot (-2) \cdot (-4) = -24$

DO EXERCISES 21–23.

PROPERTIES OF REAL NUMBERS UNDER MULTIPLICATION. We now list the properties of the real numbers under multiplication. Again, these are properties that allow us to find equivalent expressions.

> **Commutative Law.** For any real numbers a and b,
>
> $$ab = ba.$$
>
> (The *order* in which numbers are multiplied does not affect the product.)
>
> **Associative Law.** For any real numbers a, b, and c,
>
> $$a(bc) = (ab)c.$$
>
> (When *only* multiplications are involved, parentheses for grouping purposes may be placed as we please without affecting the product.)
>
> **Identity.** There exists a unique real number 1 such that for any real number a,
>
> $$a \cdot 1 = 1 \cdot a = a.$$
>
> (Multiplying any number by 1 gives that same number.)
>
> **Inverses.** For each nonzero real number a, there exists a unique number, denoted $1/a$ or a^{-1}, called the *multiplicative inverse*, or *reciprocal*, for which
>
> $$a\left(\frac{1}{a}\right) = \frac{1}{a}(a) = 1$$

EXAMPLES

25. The multiplicative inverse, or reciprocal, of 2 is $\frac{1}{2}$, because $2\left(\frac{1}{2}\right) = 1$.

26. The multiplicative inverse of $-\frac{2}{3}$ is $-\frac{3}{2}$, because $\left(-\frac{2}{3}\right)\left(-\frac{3}{2}\right) = 1$.

27. The reciprocal of 0.16 is 6.25, because $(0.16)(6.25) = 1$.

There is a very special property that connects addition and multiplication as follows.

Distributive Law. **For any real numbers a, b, and c,**

$$a(b + c) = ab + ac.$$

(This is also called the distributive law of multiplication over addition.)

The expression $ab + ac$ means $(a \cdot b) + (a \cdot c)$. By agreement, we can omit parentheses around multiplications. According to this agreement, multiplications are to be done before additions or subtractions.

Any number system having the preceding properties for addition and for multiplication is called a **field.** Thus we refer to these properties as the **field properties.**

Many other properties important in algebraic manipulations can be proved from the field properties. We list some of these as theorems.

Subtraction

Subtraction is the operation opposite to addition, as given in the following definition.

DEFINITION **Subtraction**

For any real numbers a and b,

$$a - b = c \quad \text{if and only if} \quad b + c = a.$$

($a - b$ is the number that when added to b gives a.)

In any field, we actually subtract by adding the opposite of the number being subtracted, according to the following theorem.

THEOREM 2*

For any real numbers a and b,

$$a - b = a + (-b).$$

Theorem 2 follows immediately from the definitions of subtraction and additive inverse. It says that to subtract a number, we can add its additive inverse, or opposite.

EXAMPLES Subtract.

28. $8 - 5 = 8 + (-5) = 3$

29. $3 - 7 = 3 + (-7) = -4$

30. $8.6 - (-2.3) = 8.6 + 2.3 = 10.9$

31. $-15 - (-5) = -15 + 5 = -10$

32. $10 - (-4) = 10 + 4 = 14$

33. $\frac{2}{3} - \frac{5}{6} = \frac{4}{6} + \left(-\frac{5}{6}\right) = -\frac{1}{6}$

DO EXERCISES 24–27.

*Theorem 2 is often used as a *definition* of subtraction. The definition of subtraction used here is more general, since it does not depend on the existence of inverses. Our definition is valid in the system of natural numbers, for example, where Theorem 2 would not even make sense since additive inverses do not exist.

Subtract.

24. $2.5 - 1.2$

25. $12 - (-5)$

26. $-\frac{8}{5} - \frac{3}{5}$

27. $-20 - (-7)$

Divide.

28. $\dfrac{-20}{-5}$

29. $\dfrac{4.5}{-1.5}$

30. $-\dfrac{4}{5} \div \dfrac{3}{10}$

31. $-\dfrac{5}{6} \div \left(-\dfrac{5}{12}\right)$

Multiplication is distributive over subtraction in the real-number system, as the following theorem states.

> **THEOREM 3 The Distributive Law**
>
> **For any real numbers a, b, and c,**
>
> $$a(b - c) = ab - ac.$$
>
> **(This is the distributive law of multiplication over subtraction.)**

Theorem 3 follows easily from Theorem 2 and the other distributive law. We will often use the term "distributive law" when we mean either law, since one is easily proven from the other and subtraction can always be expressed as an addition.

Division

Division is the operation opposite to multiplication, as given in the following definition.

> **DEFINITION Division**
>
> **For any number a and any nonzero number b,**
>
> $$a \div b = c \quad \text{if and only if} \quad b \cdot c = a.$$
>
> **($a \div b$ is the number that when multiplied by b gives a.)**

In any field, we usually divide by multiplying by a reciprocal, according to the following theorem.

> **THEOREM 4**
>
> **For any real number a and any nonzero number b,**
>
> $$a \div b = a\left(\frac{1}{b}\right).$$

The definition of division parallels the one for subtraction, and Theorems 2 and 4 are also parallel.

EXAMPLES Divide by multiplying by a reciprocal.

34. $\frac{3}{4} \div \left(-\frac{2}{3}\right) = \frac{3}{4}\left(-\frac{3}{2}\right) = -\frac{9}{8}$

35. $-\frac{6}{7} \div \left(-\frac{3}{5}\right) = -\frac{6}{7}\left(-\frac{5}{3}\right) = \frac{30}{21} = \frac{10}{7}$

From Theorem 4, it follows easily that the quotient of two negative numbers is positive and that the quotient of a positive number and a negative number is negative.

DO EXERCISES 28–31.

Order

The order of the real numbers is shown intuitively by a number line.

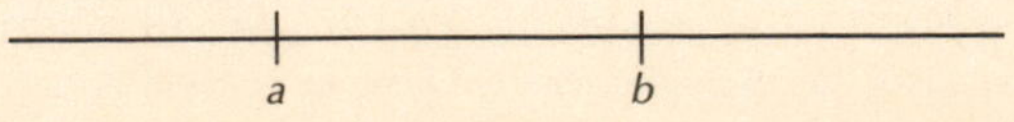

If a number a is pictured to the left of a number b, then *a is less than b* $(a < b)$. In this case, if we subtract a from b, then the answer will be a positive number. This idea motivates the definition of "less than."

DEFINITION

For any real numbers a and b, $a < b$ if and only if $b - a$ is positive.

EXAMPLES Verify each inequality using the definition of less than.

36. $2 < 8$ $8 - 2 = 6$ and 6 is positive.

37. $-4 < 9$ $9 - (-4) = 13$ and 13 is positive.

38. $-7 < -5$ $-5 - (-7) = 2$ and 2 is positive. ■

The symbol for "greater than" $(>)$ is defined in terms of "less than" $(<)$, as follows.

DEFINITION

For any real numbers a and b, $a > b$ is defined to mean $b < a$.

Thus to show that $a > b$, we can show that $a - b$ is positive.

The symbol $a \leq b$, read "a is less than or equal to b," is true when either $a < b$ is true or $a = b$ is true.

DO EXERCISES 32–35.

The Use of Calculators

We assume that you own a calculator and will use it while studying this book. A scientific calculator with a power key, exponential and logarithmic keys, and trigonometric keys will be the most beneficial to you. You will note that certain exercises are designed for use of a calculator. They are indicated by the symbol ▦. Certain examples and discussions are similarly marked.

Keep in mind that there can be differences in answers found on the calculator because of rounding-error differences. For example, suppose you were asked to approximate $\sqrt{18}$, precise to seven decimal places. On a certain kind of calculator with an eight-digit readout, the display would show

$$\sqrt{18} \approx 4.2426406$$

On another kind of calculator with a ten-digit readout, the display would show

$$\sqrt{18} \approx 4.242640687$$

The answer would be found by rounding back to the seventh decimal place and would be given as

$$\sqrt{18} = 4.2426407$$

Thus there can be variance in the seventh decimal place.

In this case, if two-place accuracy were desired, then

$$\sqrt{18} \approx 4.24,$$

and there is no discrepancy. Keep in mind that $\sqrt{18}$ is an exact symbol for the positive square root of 18, whereas the decimal values given above are approximations.

Determine whether each of the following is true.

32. $-3.4 < -3.8$

33. $-3 \leq 5$

34. $234 > -56$

35. $\dfrac{2}{3} \leq 0.\overline{3}$

EXERCISE SET 1.1

1 Consider the numbers

$$-6,\ 0,\ 3,\ -\tfrac{1}{2},\ \sqrt{3},\ -2,\ -\sqrt{7},\ \sqrt[3]{2},\ \tfrac{5}{8},\ 14,\ -\tfrac{9}{4},\ 8.53,\ 9\tfrac{1}{2}.$$

1. Which are natural numbers?
2. Which are whole numbers?
3. Which are irrational numbers?
4. Which are rational numbers?
5. Which are integers?
6. Which are real numbers?

Which of the following are rational? Which are irrational?

7. $-\frac{6}{5}$
8. $-\frac{3}{7}$
9. -9.032
10. 3.14
11. $4.\overline{516}$
12. $-7.\overline{32}$
13. 4.303003000300003. . . (No repeating block of digits)
14. 6.414114111411114. . . (No repeating block of digits)
15. $\sqrt{6}$
16. $\sqrt{7}$
17. $-\sqrt{14}$
18. $-\sqrt{12}$
19. $\sqrt{49}$
20. $-\sqrt{16}$
21. $\sqrt[3]{5}$
22. $\sqrt[4]{10}$

2 Find $-x$ and $-1 \cdot x$, when:

23. $x = -7$.
24. $x = -\frac{10}{3}$.
25. $x = 57$.
26. $x = \frac{13}{14}$.

Find $-(x^2 - 5x + 3)$, when:

27. $x = 12$.
28. $x = -8$.

Find $-(7 - y)$, when:

29. $y = -9$.
30. $y = 19$.

Compute.

31. $-3.1 + (-7.2)$
32. $-735 + 319$
33. $\frac{9}{2} + \left(-\frac{3}{5}\right)$
34. $-6 + (-4) + (-10)$
35. $-7(-4)$
36. $-\frac{8}{3}\left(-\frac{9}{2}\right)$
37. $(-8.2) \times 6$
38. $-6(-2)(-4)$
39. $-7(-2)(-3)(-5)$
40. $(-7.1)(-2.3)$
41. $-\frac{14}{3}\left(-\frac{17}{5}\right)\left(-\frac{21}{2}\right)$
42. $-\frac{13}{4}\left(-\frac{16}{5}\right)\left(\frac{23}{2}\right)$
43. $\frac{-20}{-4}$
44. $\frac{49}{-7}$
45. $\frac{-10}{70}$
46. $\frac{-40}{8}$
47. $\frac{2}{7} \div \left(-\frac{14}{3}\right)$
48. $-\frac{3}{5} \div \left(-\frac{6}{7}\right)$
49. $-\frac{10}{3} \div \left(-\frac{2}{15}\right)$
50. $-\frac{12}{5} \div (-0.3)$
51. $11 - 15$
52. $-12 - 17$
53. $12 - (-6)$
54. $-13 - (-4)$
55. $15.8 - 27.4$
56. $-19.04 - 15.76$
57. $-\frac{21}{4} - \left(-\frac{7}{8}\right)$
58. $\frac{2}{3} - \left(-\frac{17}{4}\right)$

SYNTHESIS

Calculate. Round to six decimal places. The symbol 🖩 indicates an exercise meant to be done with a calculator. Recall that $(1.4)^2$ means $(1.4)(1.4)$ and that $(2.1)^3$ means $(2.1)(2.1)(2.1)$.

59. 🖩 a) $(1.4)^2$
$(1.41)^2$
$(1.414)^2$
$(1.4142)^2$
$(1.41421)^2$
b) What number does the sequence of numbers 1.4, 1.41, 1.414, 1.4142, and so on, seem to approach?

60. 🖩 a) $(2.1)^3$
$(2.15)^3$
$(2.154)^3$
$(2.1544)^3$
$(2.15443)^3$
b) What number does the sequence of numbers 2.1, 2.15, 2.154, 2.1544, and so on, seem to approach?

What property is illustrated by each sentence?

61. $k + 0 = k$
62. $ax = xa$
63. $-1(x + y) = (-1x) + (-1y)$
64. $4 + (t + 6) = (4 + t) + 6$
65. $c + d = d + c$
66. $-67 \cdot 1 = -67$
67. $4(xy) = (4x)y$
68. $5(a + t) = 5a + 5t$
69. $y\left(\frac{1}{y}\right) = 1, y \neq 0$
70. $-x + x = 0$

71. $a + (b + c) = a + (c + b)$

72. $a(b + c) = (b + c)a$

73. $a(b + c) = a(c + b)$

74. $ab + ac = a(b + c)$

75. Show that subtraction is not commutative. That is, find real numbers a and b such that $a - b \neq b - a$.

76. Show that division is not commutative.

77. Show that division is not associative.

78. Show that subtraction is not associative.

79. Which is a better approximation to π: 3.14, 22/7, or 3927/1250?

80. At what decimal place does 22/7 differ from π?

To convert from repeating decimal notation to fractional notation, consider an example such as 8.97656565. . . or $8.97\overline{65}$. Let $n = 8.97\overline{65} = 8.976565. \ldots$ Then

$$\begin{aligned} 10{,}000n &= 89765.6565\ldots \\ 100n &= 897.6565\ldots \\ \hline 9{,}900n &= 88{,}868 \\ n &= \frac{88{,}868}{9900}. \end{aligned}$$

Convert to fractional notation.

81. $0.\overline{9}$

82. $3.\overline{74}$

83. $18.3\overline{245}$

84. $12.347\overline{652}$

CHALLENGE

85. Prove that for any real numbers $(b + c)a = ba + ca$.

86. Prove that any positive number is greater than 0.

1.2 Exponential, Scientific, and Absolute-Value Notation

1 Integers as Exponents

When an integer greater than 1 is used as an **exponent**, the integer tells us the number of times that the base is used as a factor. For example, 5^3 means $5 \cdot 5 \cdot 5$. An exponent of 1 does not change the meaning of an expression. For example, $(-3)^1 = -3$. When 0 occurs as the exponent of a nonzero expression, we agree that the expression is equal to 1. For example, $37^0 = 1$.

DEFINITION

For any real number a, $a^1 = a$.
For any nonzero real number a, $a^0 = 1$.

We will see later why 0 is not allowed as an exponential base.

CAUTION! When a negative sign occurs in exponential notation, a certain caution is in order. For example, $(-4)^2$ means that -4 is to be raised to the second power. Hence, $(-4)^2 = (-4)(-4) = 16$. On the other hand, -4^2 represents the opposite of 4^2. Thus, $-4^2 = -16$. It may help to think of $-x^2$ as $-1 \cdot x^2$, according to Theorem 1.

OBJECTIVES

You should be able to:

1. Simplify expressions with integer exponents.
2. Convert between decimal and scientific notation, and solve problems using scientific notation.
3. Carry out a calculation using the rules for order of operations.
4. Simplify expressions involving absolute value, leaving as little as possible inside the absolute-value signs.

Rename with exponents.

1. $8 \cdot 8 \cdot 8 \cdot 8$

2. xxx

3. $4y \cdot 4y \cdot 4y \cdot 4y$

Rename without exponents.

4. 3^4

5. $(5x)^4$

6. $(-5)^4$

7. -5^4

8. $(3x)^0$

Simplify.

9. $(5y)^2$

10. $(-2x)^3$

11. Rename $1/4^3$ using a negative exponent.

12. Rename 10^{-4} without using a negative exponent.

13. Write three other symbols for 4^{-3}.

Multiply and simplify.

14. $8^{-3} \cdot 8^7$

15. $y^7 y^{-2}$

16. $(9x^4)(-2x^7)$

17. $(-3x^{-4})(25x^{-10})$

18. $(5x^{-3}y^4)(-2x^{-9}y^{-2})$

19. $(4x^{-2}y^4)(15x^2y^{-3})$

DO EXERCISES 1–10.

Negative integers as exponents are defined as follows.

DEFINITION

If n is any positive integer, then a^{-n} means $1/a^n$ for $a \neq 0$. In other words, a^n and a^{-n} are reciprocals of each other.

EXAMPLES

1. $\dfrac{1}{5^2} = 5^{-2}$

2. $7^{-3} = \dfrac{1}{7^3}$

3. $5^{-4} = \dfrac{1}{5^4} = \dfrac{1}{5 \cdot 5 \cdot 5 \cdot 5} = \dfrac{1}{625}$

DO EXERCISES 11–13.

Properties of Exponents

Multiplication

Let us consider an example involving multiplication:

$$b^5 \cdot b^{-2} = (b \cdot b \cdot b \cdot b \cdot b) \cdot \frac{1}{b \cdot b} = \frac{b \cdot b}{b \cdot b} \cdot (b \cdot b \cdot b) = 1 \cdot (b \cdot b \cdot b) = b^3.$$

Note that the result can be obtained by adding the exponents. This is true in general.

THEOREM 5 The Product Rule

For any number a and any integers m and n,

$$a^m \cdot a^n = a^{m+n}.$$

We can use Theorem 5 to find equivalent expressions for products of exponential expressions with the same base. We will usually express our final answer with a positive exponent wher possible.

EXAMPLES Multiply and simplify.

4. $x^4 \cdot x^{-2} = x^{4+(-2)} = x^2$

5. $5^4 \cdot 5^6 = 5^{10}$

6. $c^{-3} \cdot c^{-2} = c^{-5} = \dfrac{1}{c^5}$

7. $a^4 \cdot a^3 = a^7$

DO EXERCISES 14–19.

Division

Let us consider an example involving division:

$$\frac{8^5}{8^3} = 8^5 \cdot \frac{1}{8^3} = 8^5 \cdot 8^{-3} = 8^{5+(-3)} = 8^{5-3} = 8^2.$$

Note that we could also have obtained this result by subtracting the exponents. Here is another example:

$$\frac{7^{-2}}{7^3} = 7^{-2} \cdot 7^{-3} = 7^{-2+(-3)}$$

$$= 7^{-2-3} = 7^{-5} = \frac{1}{7^5}.$$

Again, the result could be obtained by subtracting the exponents. This is true in general.

THEOREM 6 The Quotient Rule

For any nonzero number a and any integers m and n,

$$a^m/a^n = a^{m-n}.$$

EXAMPLES Divide and simplify.

8. $\frac{9^{-2}}{9^5} = 9^{-2-5} = 9^{-7} = \frac{1}{9^7}$

9. $\frac{7^{-4}}{7^{-5}} = 7^{-4-(-5)} = 7^1 = 7$

10. $\frac{x^{10}}{x^8} = x^{10-8} = x^2$

11. $\frac{y^{-5}}{y^{-8}} = y^{-5-(-8)} = y^3$

We can use the quotient rule to show why a^0 is not defined when $a = 0$. Consider the following:

$$a^0 = a^{3-3} = \frac{a^3}{a^3}.$$

If a were 0, we would then have 0/0, which is meaningless.

DO EXERCISES 20–26.

Raising Powers to Powers

Consider this example:

$$(5^2)^4 = 5^2 \cdot 5^2 \cdot 5^2 \cdot 5^2 = 5^8.$$

We can obtain the same result by multiplying the exponents. This is true in general.

THEOREM 7 The Power Rule

For any number a and any integers m and n,

$$(a^m)^n = a^{mn}.$$

When an expression inside parentheses is raised to a power, the inside expression is the base. For example, $(3a)^2 = (3a)(3a) = 9a^2$. We can evaluate the power $(3a)^2$ by raising 3 to the power 2 and a to the power 2. A similar thing happens to quotients.

Divide and simplify.

20. $\frac{4^8}{4^5}$

21. $\frac{5^4}{5^{-2}}$

22. $\frac{10^{-5}}{10^9}$

23. $\frac{9^{-8}}{9^{-2}}$

24. $\frac{y^6}{y^{-5}}$

25. $\frac{10y^2}{2y^3}$

26. $\frac{42x^7y^6}{-21y^{-3}x^{10}}$

Simplify.

27. $(3^7)^7$

28. $(8^2)^{-7}$

29. $(y^4)^{-7}$

30. $(2xy)^3$

31. $(4x^{-2}y^7)^2$

32. $\left(\frac{3x^4y^2}{z^5}\right)^{-3}$

33. $\left(\frac{10x^{-4}y^7z^{-2}}{5x^6y^{-8}z^{-3}}\right)^3$

THEOREM 8 Product or Quotient to a Power

For any real numbers a and b and any integer n (provided $ab \neq 0$ when $n < 0$),

$$(ab)^n = a^n b^n.$$

For any real numbers a and b, $b \neq 0$, and any integer n,

$$\left(\frac{a}{b}\right)^n = \frac{a^n}{b^n}.$$

EXAMPLES Simplify. Write answers with positive exponents.

12. $(8^{-2})^3 = 8^{-2 \cdot 3} = 8^{-6} = \frac{1}{8^6}$

13. $(x^{-5})^4 = x^{-20} = \frac{1}{x^{20}}$

14. $(3x^2y^{-2})^3 = 3^3(x^2)^3(y^{-2})^3 = 3^3x^6y^{-6} = 27x^6y^{-6} = \frac{27x^6}{y^6}$

15. $(5x^3y^{-5}z^2)^4 = 5^4x^{12}y^{-20}z^8 = 625x^{12}y^{-20}z^8 = \frac{625x^{12}z^8}{y^{20}}$

16. $\left(\frac{12x^2y^{-3}}{3y^8z^{-5}}\right)^{-4} = \left(\frac{4x^2}{y^{11}z^{-5}}\right)^{-4} = \frac{4^{-4}(x^2)^{-4}}{(y^{11})^{-4}(z^{-5})^{-4}} = \frac{x^{-8}}{256y^{-44}z^{20}} = \frac{y^{44}}{256x^8z^{20}}$ ■

CAUTION! When raising a product such as $8x^2y^{-2}$ to a power, don't forget to raise *all* the factors to the power. For example,

$$(8x^2y^{-2})^3 = 8^3(x^2)^3(y^{-2})^3.$$

DO EXERCISES 27–33.

2 Scientific Notation

Scientific notation is particularly useful for naming very large or very small numbers. It also has uses in our later study of logarithms. The following are examples of scientific notation:

$$7.8 \times 10^{13} \quad \text{and} \quad 5.64 \times 10^{-8}.$$

DEFINITION

Scientific notation **for a number is an expression of the type**

$$N \times 10^n,$$

where 1 is less than or equal to N and N is less than 10 ($1 \leq N < 10$) and N is expressed in decimal notation. The expression 10^n is considered to be scientific notation when $N = 1$.

We keep in mind that positive exponents correspond to large numbers and negative exponents correspond to small numbers. We can convert to scientific notation by multiplying by 1, choosing a name like $10^k \cdot 10^{-k}$ for the number 1.

EXAMPLE 17 The population of the United States is about 243,000,000. Convert this number to scientific notation.

Solution We want to move the decimal point 8 places, between the 2 and the 4, so we choose $10^{-8} \times 10^8$ as a name for 1. Then we multiply:

$$\begin{aligned} 243{,}000{,}000 &= 243{,}000{,}000 \times 10^{-8} \times 10^8 \quad \text{Multiplying by 1} \\ &= 2.43 \times 10^8. \end{aligned}$$

With practice, you should do these conversions mentally as much as possible.

EXAMPLE 18 The mass of a hydrogen atom is

$$0.0000000000000000000000017 \text{ gram.}$$

Convert this number to scientific notation.

Solution We make this conversion mentally. We want the decimal point to be positioned between the 1 and the 7. We count the number of moves of the decimal point: It is 24. Since the number to be converted is small, the exponent is negative. Thus,

$$0.0000000000000000000000017 = 1.7 \times 10^{-24}.$$

EXAMPLES Convert to decimal notation.

19. $6.043 \times 10^5 = 604{,}300$

20. $4.7 \times 10^{-8} = 0.000000047$

DO EXERCISES 34–39.

On a calculator, a number like 370,000,000 might be expressed using notation like "3.7 E 8", or with a space, simply as "3.7 8". This is the way the calculator would show a very large or small number using scientific notation.

EXAMPLES Convert to decimal notation.

21. 4.23 E −5 = 0.0000423

22. 7.31 E 12 = 7,310,000,000,000

DO EXERCISES 40 AND 41.

Scientific notation is often used in problem solving.

EXAMPLE 23 Alpha Centauri is the star, apart from the sun, that is closest to the earth. It is about 4.3 light-years from the earth. One *light-year* is the distance that light travels in one year and is about 5.88×10^{12} miles. How many miles is it from earth to Alpha Centauri? Express your answer in scientific notation.

Solution The distance from the earth to Alpha Centauri is the number of light-years times the number of miles that light travels in one year and is given by

$$\begin{aligned} 4.3 \times (5.88 \times 10^{12}) &= (4.3 \times 5.88) \times 10^{12} \\ &= 25.284 \times 10^{12} \\ &= (2.5284 \times 10^1) \times 10^{12} \\ &= 2.5284 \times 10^{13}. \end{aligned}$$

In Example 23, you may have been tempted to quit when you obtained

Convert to scientific notation.

34. 465,000

35. 3789

36. 0.000145

37. 0.00000000067

Convert to decimal notation.

38. 4.67×10^{-5}

39. 7.894×10^{12}

Convert to decimal notation.

40. 8.166 E 9

41. 1.103 E −6

42. Find scientific notation for the number of seconds in one year. Use 365 days for a year.

25.284×10^{12}. That would be a correct answer numerically, but it is not in scientific notation, since 25.284 is not a number between 1 and 10.

DO EXERCISE 42.

3 Order of Operations

What does $3 + 2 \cdot 7^2$ mean? If we add 3 and 2, to get 5, and then multiply by 7^2, which is 49, we get 245. If we multiply 2 times 49 and then add 3, we get 101. Clearly, both results are not correct. To determine which procedure to use, mathematicians have agreed on the following rules for **order of operations.**

Rules for Order of Operations

1. **Do all calculations within grouping symbols before operations outside.**
2. **Evaluate all exponential expressions.**
3. **Do all multiplications and divisions in order from left to right.**
4. **Do all additions and subtractions in order from left to right.**

EXAMPLE 24 Calculate: $8 + 2(4 - 9)^2$.

Solution

$$\begin{aligned} 8 + 2(4 - 9)^2 &= 8 + 2(-5)^2 && \text{Working within parentheses first} \\ &= 8 + 2(25) && \text{Simplifying } (-5)^2 \\ &= 8 + 50 && \text{Multiplying} \\ &= 58 && \text{Adding} \end{aligned}$$

In addition to the common grouping symbols such as parentheses ( ), brackets [ ], and braces { }, a fraction bar may act as a grouping symbol.

EXAMPLE 25 Calculate:

$$\frac{14(11 - 2) + 8 \cdot 6}{5^2 + 2^3}.$$

Solution An equivalent expression using brackets as grouping symbols is

$$[14(11 - 2) + 8 \cdot 6] \div [5^2 + 2^3].$$

What this shows, in effect, is that first we do the calculations in the numerator and in the denominator, and then we divide the results:

$$\frac{14(11 - 2) + 8 \cdot 6}{5^2 + 2^3} = \frac{14(9) + 8 \cdot 6}{25 + 8} = \frac{126 + 48}{33} = \frac{174}{33}.$$

It turns out that the conditions in rules (3) and (4) regarding "in order from left to right" can be ignored if we convert all subtractions to additions of opposites and all divisions to products by reciprocals. Then the commutative and the associative laws allow us to proceed from either left or right. For example,

$$10 \cdot 5 - 12 \div 3 = 10 \cdot 5 + (-12) \cdot \tfrac{1}{3} = 50 + (-4) = 46.$$

The use of order of operations is especially relevant to those programming a computer—a skill that is more and more important in today's society.

DO EXERCISES 43–45.

4 Absolute Value

Informally, we say that the **absolute value** of a number is its distance from 0 on a number line. The absolute value of a number a is denoted $|a|$.

EXAMPLES Simplify.

26. $|-7|$ The distance of -7 from 0 is 7, so $|-7| = 7$.

27. $|5|$ The distance of 5 from 0 is 5, so $|5| = 5$.

28. $|0|$ The distance of 0 from 0 is 0, so $|0| = 0$.

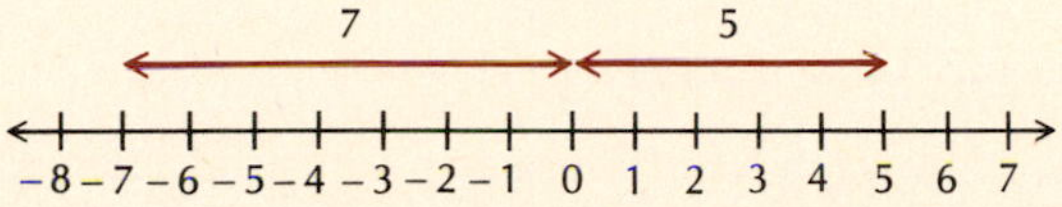

We can use the notation of additive inverse, or opposite, to give a formal definition of absolute value. The absolute value of a nonnegative number is that number itself. The absolute value of a negative number is its opposite. We define this as follows.

> DEFINITION Absolute Value
>
> **For any real number a,**
>
> $$|a| = a \text{ if } a \geq 0$$
>
> **(if a number is nonnegative, then its absolute value is the number itself)**
>
> **and**
>
> $$|a| = -a \text{ if } a < 0$$
>
> **(if a number is negative, then its absolute value is its opposite).**

The absolute value of a number x is the number x itself, if x is not negative, and the opposite of x, $-x$, if x is negative. A common source of confusion occurs if you interpret $-x$ as something negative, rather than "the additive inverse of x," or "the opposite of x."

DO EXERCISES 46–49.

We now consider certain properties involving absolute value and use them to simplify certain expressions. In that way, we can find equivalent expressions. Consider, for example, the absolute value of the product $(-3)5$. Now

$$|-3 \cdot 5| = |-15| = 15 \quad \text{and} \quad |-3| \cdot |5| = 3 \cdot 5 = 15,$$
$$\text{so} \quad |-3 \cdot 5| = |-3| \cdot |5|.$$

We note that the absolute value of a product is the product of the absolute values.

Similarly, the absolute value of a quotient is the quotient of the absolute values. We can check this as follows:

$$\left|\frac{25}{-5}\right| = |-5| = 5 \quad \text{and} \quad \frac{|25|}{|-5|} = \frac{25}{5} = 5, \quad \text{so} \quad \left|\frac{25}{-5}\right| = \frac{|25|}{|-5|}.$$

Calculate.

43. **a)** $3 \cdot 5^2 + 4$

b) $3 \cdot (5^2 + 4)$

44. $\left(\dfrac{(3+2)^2 - 3 + 2^2 + 1}{2^3 + 5^0}\right)^3$

45. **a)** Calculate $20 \div 4 - 3 \cdot 6$ using the rules for order of operations.

b) Do the same calculation by first converting the division to a multiplication and the subtraction to an addition.

Simplify.

46. $|2|$

47. $|\sqrt{3}|$

48. $|-11.3|$

49. $\left|-\frac{3}{4}\right|$

Simplify.

50. $|(-5)(-4)|$

51. $|-5| \cdot |-4|$

52. $\frac{|-20|}{|-5|}$

53. $\left|\frac{-20}{-5}\right|$

Simplify.

54. $|-6ab|$

55. $|x^8|$

56. $|10m^2n^3|$

57. $\left|\frac{-2x^3}{y^2}\right|$

For any real numbers a and b, $|ab| = |a||b|$.

For any real numbers a and b, $b \neq 0$, $\left|\frac{a}{b}\right| = \frac{|a|}{|b|}$.

Suppose that we want to simplify the absolute value of a number raised to an even power, such as $|a^2|$. No matter what the value of a, a^2 is nonnegative. Thus, $|a^2| = a^2$ using the definition of absolute value. Similarly, $|a^n| = a^n$, when n is even. That is, the absolute value of any even power is the even power.

The absolute value of the opposite of a number is the same as the absolute value of the number. We can prove this using the definition of absolute value. Suppose a is positive. Then $-a$ is negative. Thus,

$$|a| = a \quad \text{and} \quad |-a| = -(-a) = a,$$

by Theorem 1. Now suppose a is negative or zero. Then $|a| = -a$ by the definition of absolute value and also $|-a| = -a$ since $-a$ is positive. Another way to think of the property $|a| = |-a|$ is that the distances of a and $-a$ from 0 are the same.

EXAMPLES Simplify.

29. $|(-3)^2| = |9| = 9$ and $(-3)^2 = 9$, so $|(-3)^2| = (-3)^2$.

30. $|-3| = 3$ and $|3| = 3$, so $|-3| = |3|$.

DO EXERCISES 50–53.

Theorem 9 summarizes the properties of absolute value that we have used. Each can be proven using the definition of absolute value.

THEOREM 9

For any real numbers a and b and any nonzero number c:

1. $|ab| = |a| \cdot |b|$;
2. $\left|\frac{a}{c}\right| = \frac{|a|}{|c|}$;
3. $|a^n| = a^n$, **if n is an even integer;**
4. $|-a| = |a|$.

We can use Theorem 9 to find equivalent expressions.

EXAMPLES Simplify, leaving as little as possible inside the absolute-value signs.

31. $|3x| = |3| \cdot |x| = 3|x|$

32. $|x^2| = x^2$

33. $|x^2y^3| = |x^2 \cdot y^2 \cdot y| = |x^2| \cdot |y^2| \cdot |y| = x^2y^2|y|$

34. $\left|\frac{x^2}{y}\right| = \frac{|x^2|}{|y|} = \frac{x^2}{|y|}$

35. $|-3x| = 3|x|$

DO EXERCISES 54–57.

EXERCISE SET 1.2

1 Simplify.

1. $2^3 \cdot 2^{-4}$
2. $3^4 \cdot 3^{-5}$
3. $b^2 \cdot b^{-2}$
4. $c^3 \cdot c^{-3}$
5. $4^2 \cdot 4^{-5} \cdot 4^6$
6. $5^2 \cdot 5^{-4} \cdot 5^5$
7. $2x^3 \cdot 3x^2$
8. $3y^4 \cdot 4y^3$
9. $(5a^2b)(3a^{-3}b^4)$
10. $(4xy^2)(3x^{-4}y^5)$
11. $(2x)^3(3x)^2$
12. $(4y)^2(3y)^3$
13. $(6x^5y^{-2}z^3)(-3x^2y^3z^{-2})$
14. $(5x^4y^{-3}z^2)(-2x^2y^4z^{-1})$
15. $\dfrac{b^{40}}{b^{37}}$
16. $\dfrac{a^{39}}{a^{32}}$
17. $\dfrac{x^2y^{-2}}{x^{-1}y}$
18. $\dfrac{x^3y^{-3}}{x^{-1}y^2}$
19. $\dfrac{9a^2}{(-3a)^2}$
20. $\dfrac{16y^2}{(-4y)^2}$
21. $\dfrac{24a^5b^3}{8a^4b}$
22. $\dfrac{30x^6y^4}{5x^3y^2}$
23. $\dfrac{12x^2y^3z^{-2}}{21xy^2z^3}$
24. $\dfrac{15x^3y^4z^{-3}}{45xyz^5}$
25. $(2ab^2)^3$
26. $(4xy^3)^2$
27. $(-2x^3)^4$
28. $(-3x^2)^4$
29. $-(2x^3)^4$
30. $-(3x^2)^4$
31. $(6a^2b^3c)^2$
32. $(5x^3y^2z)^2$
33. $(-5c^{-1}d^{-2})^{-2}$
34. $(-4x^{-1}z^{-2})^{-2}$
35. $\dfrac{4^{-2} + 2^{-4}}{8^{-1}}$
36. $\dfrac{3^{-2} + 2^{-3}}{7^{-1}}$
37. $\dfrac{(-2)^4 + (-4)^2}{(-1)^8}$
38. $\dfrac{(-3)^2 + (-2)^4}{(-1)^6}$
39. $\dfrac{(3a^2b^{-2}c^4)^3}{(2a^{-1}b^2c^{-3})^2}$
40. $\dfrac{(2a^3b^{-3}c^3)^3}{(3a^{-1}b^{-3}c^{-5})^2}$
41. $\dfrac{6^{-2}x^{-3}y^2}{3^{-3}x^{-4}y}$
42. $\dfrac{5^{-2}x^{-4}y^3}{2^{-3}x^{-5}y}$
43. $\left(\dfrac{24a^{10}b^{-8}c^7}{3a^6b^{-3}c^5}\right)^5$
44. $\left(\dfrac{125p^{12}q^{-14}r^{22}}{25p^8q^6r^{-15}}\right)^{-4}$

Find $-x^2$ and $(-x)^2$, when:

45. $x = 5.$
46. $x = -7.$
47. $x = -1.08.$
48. $x = \sqrt{3}.$

2 Convert to scientific notation.

49. 58,000,000
50. 27,000
51. 365,000
52. 3645
53. 0.0000027
54. 0.0000658
55. 0.027
56. 0.0038
57. The mass of an electron is 0.000000000000000000000000000911 gram.
58. The distance from the earth to the sun is 93,000,000 miles.

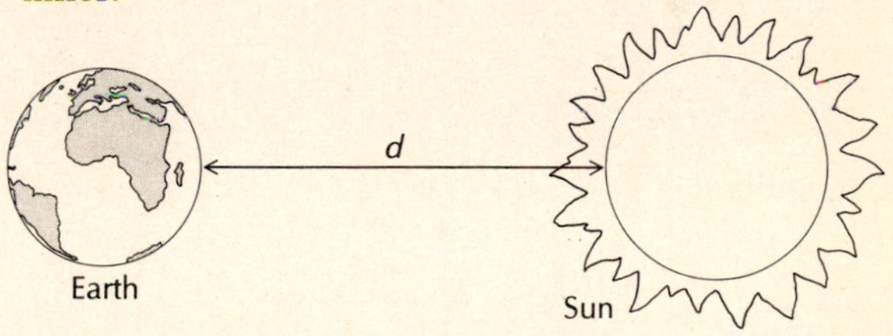

59. The distance from the sun to Pluto is 3,664,000,000 miles.
60. An oxygen atom is about 0.000000001 times the size of a drop of water.

Convert to decimal notation.

61. 4×10^5
62. 5×10^{-4}
63. 6.2×10^{-3}
64. 7.8×10^6
65. 7.69×10^{12}
66. 8.54×10^{-7}
67. 5.67 E −7
68. 1.314 E 12
69. Light travels 9.46×10^{12} kilometers in one year.
70. The wavelength of a certain red light is 6.6×10^{-5}.
71. 7.69 E −8
72. 8.603 E −10
73. 2.567 E 8
74. 1.113 E 11

Solve. Write answers using scientific notation.

75. The average discharge of water from the mouth of the Amazon river is 4,200,000 cubic feet per second. How much water is discharged in one hour?
76. Americans drink 3 million gallons of orange juice in one day. How much orange juice is consumed by Americans in one year? Use 365 days for 1 year.
77. A *nanosecond* is one billionth of a second. Find scientific notation for 1 nanosecond.
78. The average distance from the earth to the sun is 9.3×10^7 miles. About how far does the earth travel in a yearly orbit about the sun? (*Hint:* Assume a circular orbit.)

79. How far, in miles, does light travel in 13 weeks?

80. A certain thin plastic sheet is used in many applications of building and landscaping. The sheet is packaged in rolls that are 1 m wide and 30 m long. The thickness of the sheet is 0.8 mm. Find the volume of plastic in a roll.

3 Calculate.

81. $3 \cdot 2 + 4 \cdot 2^2 - 6(3 - 1)$

82. $3[(2 + 4 \cdot 2^2) - 6(3 - 1)]$

83. $\dfrac{4(8 - 6)^2 + 4 \cdot 3 - 2 \cdot 8}{3^1 + 19^0}$

84. $\dfrac{[4(8 - 6)^2 + 4](3 - 2 \cdot 8)}{2^2(2^3 + 5)}$

85. $16 \div 4 \cdot 4 \div 2 \cdot 256$

86. $2^6 \cdot 2^{-3} \div 2^{10} \div 2^{-8}$

87. $\left[\dfrac{5^2}{8} + 5(5) - \dfrac{5^3}{12}\right] - \left[\dfrac{(-4)^2}{8} + 5(-4) - \dfrac{(-4)^3}{12}\right]$

88. $\left[\dfrac{2^2}{8} + \dfrac{2}{2} - \dfrac{2^3}{12}\right] - \left[\dfrac{(-1)^2}{8} - \dfrac{1}{2} - \dfrac{(-1)^3}{12}\right]$

4 Simplify.

89. $|12|$

90. $|-2.56|$

91. $|-47|$

92. $|0|$

93. $|-7a|$

94. $|-10mn|$

95. $|-8x^6|$

96. $|5x^4y^8|$

97. $|9xy|$

98. $|y^4|$

99. $|3a^2b|$

100. $\left|\dfrac{4a}{b^2}\right|$

SYNTHESIS

Simplify. Assume that all exponents are integers.

101. $(x^t \cdot x^{3t})^2$

102. $(x^y \cdot x^{-y})^3$

103. $(t^{a+x} \cdot t^{x-a})^4$

104. $(m^{x-y} \cdot m^{3y-x})^t$

105. $(x^a y^b \cdot x^b y^a)^c$

106. $(m^{x-b} \cdot n^{x+b})^x (m^b n^{-b})^x$

107. $\left[\dfrac{(3x^a y^b)^3}{(-3x^a y^b)^2}\right]^2$

108. $\left[\left(\dfrac{x^r}{y^t}\right)^2 \left(\dfrac{x^{2r}}{y^{4t}}\right)^{-2}\right]^{-3}$

The formula

$$M = P\left[\frac{\frac{i}{12}\left(1 + \frac{i}{12}\right)^n}{\left(1 + \frac{i}{12}\right)^n - 1}\right]$$

gives the monthly mortgage payment M on a home loan of P dollars at interest rate i, where n is the total number of payments (12 times the number of years).

109. 🖩 The cost of a house is \$92,000. The down payment is \$14,000. The interest rate is $10\frac{3}{4}\%$. The loan period is 25 years. What is the monthly payment?

110. 🖩 Repeat Exercise 109 for loan periods of 20 years and 30 years.

Find the error(s) in each of the following. Explain why each is an error. Then find the correct answer.

111. $x^4(x^3)^2 = x^9$

112. $\dfrac{x^4y^{-7}}{x^{-2}y^5} = \dfrac{x^2}{y^2}$

113. $(2x^{-4}y^6z^3)^3 = 6x^{-1}y^3z^6$

1.3 Addition and Subtraction of Algebraic Expressions

OBJECTIVES

You should be able to:

1. Determine the degree of each term of a polynomial and the degree of the polynomial.
2. Add polynomials and other algebraic expressions.
3. Find the additive inverse, or opposite, of an algebraic expression.
4. Subtract polynomials and other algebraic expressions.

1 Polynomials

Expressions like the following are called **polynomials in one variable:**

$$-7x + 5, \qquad 3y^3 - 5y^2 + 7y - 4, \qquad 0, \qquad -5t^4, \qquad x^5 - 9.$$

A **variable** is a symbol that can represent different numbers. Letters are generally used for variables. For example, if a is used to represent your age, then a is a variable. Letters used to represent numbers are not always variables, however. For example, if we choose to represent the distance to the moon by the letter d, then in that context d is not a variable, but a **constant.**

> DEFINITION
>
> A *polynomial in one variable* is any expression of the type
>
> $$a_nx^n + a_{n-1}x^{n-1} + \cdots + a_2x^2 + a_1x + a_0,$$
>
> where n is a nonnegative integer and $a_n, \ldots, a_0$ are real numbers, called *coefficients*. Some or all of the coefficients may be 0. Each of the parts separated by plus signs is called a *term*.

The following are *not* polynomials:

$$(1)\ x^2 + 3x + \frac{2}{x}; \qquad (2)\ 2x + \sqrt{x}; \qquad (3)\ \frac{x^3 + 4}{x - 7}.$$

Expression (1) is not a polynomial because $2/x = 2x^{-1}$ and -1 is a negative integer. Expression (2) is not a polynomial because $\sqrt{x} = x^{1/2}$ and $\frac{1}{2}$ is not an integer. Expression (3) is not a polynomial because it represents a division that cannot be expressed as an equivalent polynomial.

The question might arise whether an expression such as

$$8x^3 - 6x^2 + 7x - 5$$

is a polynomial. It is indeed, because it is equivalent to

$$8x^3 + (-6x^2) + 7x + (-5).$$

Note that coefficients can be negative.

Expressions like the following are called **polynomials in several variables:**

$$5x^2y^3 + 17x^2y - 2, \qquad 14a^2b, \qquad \pi r^2 + 2\pi rh.$$

EXAMPLE 1 Find the terms and the coefficients of the polynomial

$$5x^3y - 7xy^2 + 2.$$

Solution The terms are

$$5x^3y, \qquad -7xy^2, \quad \text{and} \quad 2.$$

The coefficients of the terms are 5, -7, and 2. ■

The **degree of a term** is the sum of the exponents of the variables. For example, the degree of the term $-10x^3y$ is 4. The degree of a nonzero constant

Determine the degree of each term and the degree of the polynomial.

1. $x^8 - 7x^6 + 2x^4 - 3x^9 + 2$

2. $8y^4 - 7xy^3 + 6x^2y^3 - 9x^5y - 1$

Combine similar terms.

3. $5x^3y^2 - 2x^2y^3 + 4x^3y^2$

4. $3xy^2 - 4x^2y + 4xy^2 + 2x^2y$

5. $5x^4\sqrt{y} - 2x^4\sqrt{y} + 2$

term, such as the number -10 by itself, is 0. We can think of -10 as $-10x^0$. The **degree of a nonzero polynomial** is the degree of the term of highest degree.

The polynomial consisting only of the number 0 is a special case. Mathematicians agree that it has *no* degree either as a term or as a polynomial. This is because we can express 0 as $0 = 0x^5 = 0x^{13}$, and so on, using any exponent we wish.

EXAMPLE 2 In the polynomial $5x^3y - 7xy^2 + 2$, find the degrees of the terms and the degree of the polynomial.

Solution The degrees of the terms are 4, 3, and 0. The polynomial is of degree 4.

A polynomial with just one term is called a **monomial.** If there are just two terms, a polynomial is called a **binomial.** If there are just three terms, it is called a **trinomial.** A polynomial in one variable of second degree is also called a **quadratic polynomial.**

DO EXERCISES 1 AND 2.

CAUTION! In many applications, lower-case and capital letters are used to represent different numbers, as in

$$R + r.$$

In copying expressions, be careful not to use a capital letter if a lower-case letter is given.

2 Addition

Much of the algebraic manipulation that we do with polynomials can also be done with algebraic expressions that are not polynomials. Here are some examples.

EXAMPLES

3. $3\sqrt{x} + 4y$ — This is a sum.

4. $\dfrac{3x^2 + 2}{x - 1}$ — This is a quotient.

5. $4x^{1/2} - 5y^{3/2}$ — This is a difference.

If two terms of an expression have the same letters raised to the same powers, then the terms are called **like,** or **similar.** Similar terms can be "combined" using the distributive laws.

EXAMPLES

6. $3x^2 - 4y + 2x^2 = 3x^2 + 2x^2 - 4y$ — Rearranging using the commutative and the associative laws

$= (3 + 2)x^2 - 4y$ — Using a distributive law

$= 5x^2 - 4y$

7. $4x^{1/2}y + 7x^{1/2}y = 11x^{1/2}y$

8. $-2x^2\sqrt{y^3} + 5x^2\sqrt{y^3} = 3x^2\sqrt{y^3}$

DO EXERCISES 3–5.

We can find the sum of two polynomials by writing a plus sign between them and then combining similar terms. Ordinarily, this can be done mentally.

EXAMPLE 9 Add $-3x^3 + 2x - 4$ and $4x^3 + 3x^2 + 2$.

Solution

$$(-3x^3 + 2x - 4) + (4x^3 + 3x^2 + 2) = x^3 + 3x^2 + 2x - 2 \quad \blacksquare$$

DO EXERCISES 6 AND 7.

3 Additive Inverses, or Opposites

We can find the additive inverse, or opposite, of an expression by placing an inverse sign before the expression. Another equivalent expression can be found using the following theorem.

> **THEOREM 10**
>
> **The additive inverse, or opposite, of a polynomial can be found by replacing every term by its opposite.**

EXAMPLE 10 Find two equivalent expressions for the opposite, or additive inverse, of $-3xy^2 + 4x^2y - 5x - 3$.

Solution One expression is

$$-(-3xy^2 + 4x^2y - 5x - 3).$$

Another equivalent expression is

$$3xy^2 - 4x^2y + 5x + 3. \quad \blacksquare$$

EXAMPLE 11 Find two equivalent expressions for the additive inverse, or opposite, of $7ab^2 - 6ab - 4b + 8$.

Solution One expression is

$$-(7ab^2 - 6ab - 4b + 8).$$

Another equivalent expression is

$$-7ab^2 + 6ab + 4b - 8. \quad \blacksquare$$

The preceding examples may bring to mind the following rule: To remove parentheses preceded by an additive-inverse, or opposite, sign, change the sign of every term inside the parentheses.

DO EXERCISES 8 AND 9.

4 Subtraction

By Theorem 2, we can subtract by adding the opposite of the number being subtracted. Thus to subtract one polynomial or other algebraic expression from another, we add the opposite of the expression being subtracted. We change the sign of each term of the expression to be subtracted and then add. In simple cases, this can be done mentally.

Add.

6. $(3x^3 + 4x^2 - 7x - 2) + \left(-7x^3 - 2x^2 + 3x + \frac{1}{2}\right)$

7. $(5p^2q^4 - 2p^2q^2 - 3q) + (-6pq^2 + 3p^2q^2 + 5)$

Find two equivalent expressions for the opposite, or additive inverse, of each expression.

8. $5x^2t^2 - 4xy^2t - 3xt + 6x - 5$

9. $-3x^2y + 5xy - 7x + 4y + 2$

Subtract.

10. $(5xy^4 - 7xy^2 + 4x^2 - 3) - (-3xy^4 + 2xy^2 - 2y + 4)$

11. $(5x^2y - 7x^3y^2 - x^2y^2 + 4y) - (2x^2y + 2x^3y^2 - 5x^2y^3 - 5y)$

EXAMPLE 12 Subtract:

$$(-9x^5 - x^3 + 2x^2 + 4) - (2x^5 - x^4 + 4x^3 - 3x^2).$$

Solution

$$(-9x^5 - x^3 + 2x^2 + 4) - (2x^5 - x^4 + 4x^3 - 3x^2)$$
$$= (-9x^5 - x^3 + 2x^2 + 4) + [-(2x^5 - x^4 + 4x^3 - 3x^2)]$$ Adding the opposite
$$= (-9x^5 - x^3 + 2x^2 + 4) + (-2x^5 + x^4 - 4x^3 + 3x^2)$$
$$= -11x^5 + x^4 - 5x^3 + 5x^2 + 4$$

EXAMPLE 13 Subtract:

$$(4x^2y - 6x^3y^2 + x^2y^2 - 5y) - (4x^2y + x^3y^2 + 3x^2y^3 + 6y).$$

Solution

$$(4x^2y - 6x^3y^2 + x^2y^2 - 5y) - (4x^2y + x^3y^2 + 3x^2y^3 + 6y)$$
$$= -7x^3y^2 - 3x^2y^3 + x^2y^2 - 11y.$$

DO EXERCISES 10 AND 11.

EXERCISE SET 1.3

1 Determine the degree of each term and the degree of the polynomial.

1. $-11x^4 - x^3 + x^2 + 3x - 9$
2. $t^3 - 3t^2 + t + 1$
3. $y^3 + 2y^6 + x^2y^4 - 8$
4. $u^2 + 3v^5 - u^3v^4 - 7$
5. $a^5 + 4a^2b^4 + 6ab + 4a - 3$
6. $8p^6 + 2p^4t^4 - 7p^3t + 5p^2 - 14$

2 Add.

7. $(5x^2y - 2xy^2 + 3xy - 5) + (-2x^2y - 3xy^2 + 4xy + 7)$
8. $(6x^2y - 3xy^2 + 5xy - 3) + (-4x^2y - 4xy^2 + 3xy + 8)$
9. $(-3pq^2 - 5p^2q + 4pq + 3) + (-7pq^2 + 3pq - 4p + 2q)$
10. $(-5pq^2 - 3p^2q + 6pq + 5) + (-4pq^2 + 5pq - 6p + 4q)$
11. $(2x + 3y + z - 7) + (4x - 2y - z + 8) + (-3x + y - 2z - 4)$
12. $(2x^2 + 12xy - 11) + (6x^2 - 2x + 4) + (-x^2 - y - 2)$
13. $\left(7x\sqrt{y} - 3y\sqrt{x} + \frac{1}{5}\right) + \left(-2x\sqrt{y} - y\sqrt{x} - \frac{3}{5}\right)$
14. $\left(10x\sqrt{y} - 4y\sqrt{x} + \frac{4}{3}\right) + \left(-3x\sqrt{y} - y\sqrt{x} - \frac{1}{3}\right)$

3 Find two equivalent expressions for the opposite, or additive inverse, of each expression.

15. $5x^3 - 7x^2 + 3x - 6$
16. $-4y^4 + 7y^2 - 2y - 1$

4 Subtract.

17. $(3x^2 - 2x - x^3 + 2) - (5x^2 - 8x - x^3 + 4)$
18. $(5x^2 + 4xy - 3y^2 + 2) - (9x^2 - 4xy + 2y^2 - 1)$
19. $(4a - 2b - c + 3d) - (-2a + 3b + c - d)$
20. $(5a - 3b - c + 4d) - (-3a + 5b + c - 2d)$
21. $(x^4 - 3x^2 + 4x) - (3x^3 + x^2 - 5x + 3)$
22. $(2x^4 - 5x^2 + 7x) - (5x^3 + 2x^2 - 3x + 5)$
23. $(7x\sqrt{y} - 4y\sqrt{x} + 7.5) - (-2x\sqrt{y} - y\sqrt{x} - 1.6)$
24. $\left(10x\sqrt{y} - 4y\sqrt{x} + \frac{4}{3}\right) - \left(-3x\sqrt{y} + y\sqrt{x} - \frac{1}{3}\right)$

SYNTHESIS

Simplify.

25. ▦ $(0.565p^2q - 2.167pq^2 + 16.02pq - 17.1) + (-1.612p^2q - 0.312pq^2 - 7.141pq - 87.044)$
26. ▦ $(5003.2xy^{-2} + 3102.4\sqrt{xy} - 5280) - (2143.6xy^{-2} + 6153.8xy - 4141\sqrt{xy} + 4979.12)$

1.4 Multiplication of Algebraic Expressions

OBJECTIVE

You should be able to:

1 **Multiply any two polynomials, striving for speed and accuracy. Whenever possible, you should write only the answer. In particular, you should be able to:**

a) Square a binomial, writing only the answer.
b) Multiply the sum and difference of the same two expressions, writing only the answer.
c) Multiply any two binomials, writing only the answer.
d) Cube a binomial.

1 Multiplication of Any Two Polynomials

Multiplication of polynomials is based on the distributive laws. For example,

$$(x - 3)(x + y + 5) = (x - 3)x + (x - 3)y + (x - 3)5 \quad \text{Using the distributive law}$$
$$= x \cdot x - 3x + xy - 3y + 5x - 3 \cdot 5$$
$$= x^2 + 2x + xy - 3y - 15.$$

What we have done is to multiply each term of one polynomial by every term of the other and then add the results. We can also do such a multiplication using columns, as follows.

EXAMPLE 1 Multiply: $(4x^4y - 7x^2y + 3y)(2y - 3x^2y)$.

Solution

$$\begin{array}{rl} 4x^4y - 7x^2y + 3y & \\ 2y - 3x^2y & \\ \hline -12x^6y^2 + 21x^4y^2 - 9x^2y^2 \qquad\qquad & \text{Multiplying by } -3x^2y \\ 8x^4y^2 - 14x^2y^2 + 6y^2 & \text{Multiplying by } 2y \\ \hline -12x^6y^2 + 29x^4y^2 - 23x^2y^2 + 6y^2 & \text{Adding} \end{array}$$

DO EXERCISES 1 AND 2.

Multiply.

1. $(3x^2y - 2xy + 3y)(xy + 2y)$

2. $(p^2q + 2pq + 2q)(2p^2q - pq + q)$

The following methods allow us to multiply binomials more efficiently.

Products of Two Binomials

We can find a product of two binomials mentally. We multiply the First terms, then the Outside terms, then the Inside terms, then the Last terms (this procedure is sometimes abbreviated **FOIL**), and then add the results. This procedure also works for multiplying algebraic expressions that are not polynomials.

EXAMPLES Multiply.

F O I L

2. $(3xy + 2x)(x^2 + 2xy^2) = 3x^3y + 6x^2y^3 + 2x^3 + 4x^2y^2$

3. $(x + \sqrt{2})(y - \sqrt{2}) = xy - \sqrt{2}x + \sqrt{2}y - 2$

4. $(2x - \sqrt{3})(y + 2) = 2xy + 4x - \sqrt{3}y - 2\sqrt{3}$

5. $(2x + 3y)(x - 4y) = 2x^2 - 5xy - 12y^2$

DO EXERCISES 3–5.

Multiply.

3. $(2xy + 3x)(x^2 - 2)$

4. $(3x - 2y)(5x + 3y)$

5. $(2x + \sqrt{2})(3y - \sqrt{2})$

Squares of Binomials

Using FOIL to multiply a binomial $A + B$ by itself, we obtain the following:

$$(A + B)^2 = A^2 + AB + BA + B^2 = A^2 + 2AB + B^2$$

and

$$(A - B)^2 = A^2 - 2AB + B^2.$$

Multiply.

6. $(4x - 5y)^2$

7. $(2y^2 + 6x^2y)^2$

Multiply.

8. $(4x + 7)(4x - 7)$

9. $(5x^2y + 2y)(5x^2y - 2y)$

10. $(4y^2 + \sqrt{3})(4y^2 - \sqrt{3})$

11. $(2x + 3 + 5y)(2x + 3 - 5y)$

12. $(-2x^3y^2 + 5t)(2x^3y^2 + 5t)$

This gives us a way to square a binomial that is faster than FOIL. We square the first term, add twice the product of the terms, and then add the square of the second term.

EXAMPLES Multiply.

6. $(2x + 9y^2)^2 = (2x)^2 + 2(2x)(9y^2) + (9y^2)^2$
$= 4x^2 + 36xy^2 + 81y^4$

7. $(3x^2 - 5xy^2)^2 = (3x^2)^2 - 2(3x^2)(5xy^2) + (-5xy^2)^2$
$= 9x^4 - 30x^3y^2 + 25x^2y^4$

The second term is $-5xy^2$, so twice the product of the terms is $-2(3x^2)(5xy^2)$.

CAUTION! The square of a sum is *not* the sum of the squares; that is,

$$(A + B)^2 \neq A^2 + B^2.$$

DO EXERCISES 6 AND 7.

Products of Sums and Differences

We can also use FOIL to find the product of a sum and a difference of the same two expressions:

$$\begin{aligned}(A + B)(A - B) &= A^2 - AB + AB - B^2\\ &= A^2 - B^2.\end{aligned}$$

The product of a sum and a difference of the same two terms is the difference of their squares. Thus to find such a product, we square the first term, square the second term, and write a minus sign between the results:

$$\mathbf{(A + B)(A - B) = A^2 - B^2.}$$

EXAMPLES Multiply.

8. $(y + 5)(y - 5) = y^2 - 5^2$
$= y^2 - 25$

9. $(3x + 2)(3x - 2) = (3x)^2 - 2^2$
$= 9x^2 - 4$

10. $(2xy^2 + 3x)(2xy^2 - 3x) = (2xy^2)^2 - (3x)^2$
$= 4x^2y^4 - 9x^2$

11. $(5x + \sqrt{2})(5x - \sqrt{2}) = (5x)^2 - (\sqrt{2})^2$
$= 25x^2 - 2$

12. $(5y + 4 + 3x)(5y + 4 - 3x) = (5y + 4)^2 - (3x)^2$
$= 25y^2 + 40y + 16 - 9x^2$

13. $(3xy^2 + 4y)(-3xy^2 + 4y) = (4y + 3xy^2)(4y - 3xy^2)$
$= (4y)^2 - (3xy^2)^2$
$= 16y^2 - 9x^2y^4$

DO EXERCISES 8–12.

Cubing Binomials

The following multiplication gives another result to be remembered:

$$\begin{aligned}(A + B)^3 &= (A + B)(A + B)^2 \\ &= (A + B)(A^2 + 2AB + B^2) \\ &= (A + B)A^2 + (A + B)2AB + (A + B)B^2 \\ &= A^3 + A^2B + 2A^2B + 2AB^2 + AB^2 + B^3 \\ &= A^3 + 3A^2B + 3AB^2 + B^3.\end{aligned}$$

The result to be remembered is as follows:

$$\mathbf{(A + B)^3 = A^3 + 3A^2B + 3AB^2 + B^3.}$$

EXAMPLES Multiply.

14. $(x + 2)^3 = x^3 + 3x^2(2) + 3x(2)^2 + 2^3$
$= x^3 + 6x^2 + 12x + 8$

15. $(x - 2)^3 = [x + (-2)]^3$
$= x^3 + 3x^2(-2) + 3x(-2)^2 + (-2)^3$
$= x^3 - 6x^2 + 12x - 8$

16. $(5m^2 - 4n^3)^3 = (5m^2)^3 + 3(5m^2)^2(-4n^3) + 3(5m^2)(-4n^3)^2 + (-4n^3)^3$
$= 125m^6 - 300m^4n^3 + 240m^2n^6 - 64n^9$ ■

Note in Examples 15 and 16 that a separate formula for $(A - B)^3$ need not be memorized. We can think of $(A - B)^3$ as $[A + (-B)]^3$.

DO EXERCISES 13–16.

In the following exercises you should do mentally as much of the calculating as you can. If possible, write only the answer. Work for speed with accuracy.

Multiply.

13. $(x + 1)^3$

14. $(x - 1)^3$

15. $(t^2 - 3b)^3$

16. $(2a^3 - 5b^2)^3$

EXERCISE SET 1.4

1 Multiply.

1. $(2x^2 + 4x + 16)(3x - 4)$

2. $(3y^2 - 3y + 9)(2y + 3)$

3. $(4a^2b - 2ab + 3b^2)(ab - 2b + 1)$

4. $(2x^2 + y^2 - 2xy)(x^2 - 2y^2 - xy)$

5. $(a - b)(a^2 + ab + b^2)$

6. $(t + 1)(t^2 - t + 1)$

7. $(2x + 3y)(2x + y)$

8. $(2a - 3b)(2a - b)$

9. $\left(4x^2 - \frac{1}{2}y\right)\left(3x + \frac{1}{4}y\right)$

10. $\left(2y^3 + \frac{1}{5}x\right)\left(3y - \frac{1}{4}x\right)$

11. $(2p^2q^3 - r^2)(5pq - 2r)$

12. $(3y^2 - 2)(3y - x)$

13. $(2x + 3y)^2$

14. $(5x + 2y)^2$

15. $(2x^2 - 3y)^2$

16. $(4x^2 - 5y)^2$

17. $(2x^3 + 3y^2)^2$

18. $(5x^3 + 2y^2)^2$

19. $\left(\frac{1}{2}x^2 - \frac{3}{5}y\right)^2$

20. $\left(\frac{1}{4}x^2 - \frac{2}{3}y\right)^2$

21. $(0.5x + 0.7y^2)^2$

22. $(0.3x + 0.8y^2)^2$

23. $(3x - 2y)(3x + 2y)$

24. $(3x + 5y)(3x - 5y)$

25. $(x^2 + yz)(x^2 - yz)$

26. $(2x^2 + 5xy)(2x^2 - 5xy)$

27. $(3x^2 - \sqrt{2})(3x^2 + \sqrt{2})$

28. $(5x^2 - \sqrt{3})(5x^2 + \sqrt{3})$

29. $(2x + 3y + 4)(2x + 3y - 4)$

30. $(5x + 2y + 3)(5x + 2y - 3)$

31. $(x^2 + 3y + y^2)(x^2 + 3y - y^2)$
32. $(2x^2 + y + y^2)(2x^2 + y - y^2)$
33. $(x + 1)(x - 1)(x^2 + 1)$
34. $(y - 2)(y + 2)(y^2 + 4)$
35. $(2x + y)(2x - y)(4x^2 + y^2)$
36. $(5x + y)(5x - y)(25x^2 + y^2)$
37. 🖩 $(0.051x + 0.04y)^2$
38. 🖩 $(1.032x - 2.512y)^2$
39. 🖩 $(37.86x + 1.42)(65.03x - 27.4)$
40. 🖩 $(3.601x - 17.5)(47.105x + 31.23)$
41. $(y + 5)^3$
42. $(t - 7)^3$
43. $(m^2 - 2n)^3$
44. $(3t^2 + 4)^3$
45. $(\sqrt{2}x^2 - y^2)(\sqrt{2}x - 2y)$
46. $(\sqrt{3}y^2 - 2)(\sqrt{3}y - x)$

SYNTHESIS

Multiply. Assume that all exponents are natural numbers.

47. $(a^n + b^n)(a^n - b^n)$
48. $(t^a + 4)(t^a - 7)$
49. $(x^m - t^n)^3$
50. $y^3z^n(y^{3n}z^3 - 4yz^{2n})$
51. $(x - 1)(x^2 + x + 1)(x^3 + 1)$
52. $(a^n + b^n)^2$
53. $[(2x - 1)^2 - 1]^2$
54. $[(a + b)(a - b)][5 - (a + b)][5 + (a + b)]$
55. $(x^{a-b})^{a+b}$
56. $(t^{m+n})^{m+n} \cdot (t^{m-n})^{m-n}$
57. $(a + b + c)^2$
58. $(a + b + c)^3$
59. $(a + b)^4$
60. $(x - y)(x^4 + x^3y + x^2y^2 + xy^3 + y^4)$
61. $(m + t)(m^4 - m^3t + m^2t^2 - mt^3 + t^4)$
62. $(a - b)(a^7 + a^6b + a^5b^2 + a^4b^3 + a^3b^4 + a^2b^5 + ab^6 + b^7)$

Find the error(s) in each of the following. Explain why each is an error. Then find the correct answer.

63. $(3a + b)^2 = 3a^2 + b^2$
64. $(2x - 3y)(2x - 3y) = 4x^2 - 9y^2$
65. $2x(x + 3) + 4(x^2 - 3) = 2x^2 + 3x + 4x^2 - 3 \quad (1)$
$ = 6x^2 + x \quad (2)$
66. $(2a - 3b)(3a + 2b) = 6a^2 - 6b^2 \quad (1)$
$ = a^2 - b^2 \quad (2)$

CHALLENGE

67. Multiply. Assume that n is a natural number.

$$(x - y)(x^{n-1} + x^{n-2}y + x^{n-3}y^2 + \cdots + x^3y^{n-4} + x^2y^{n-3} + xy^{n-2} + y^{n-1})$$

1.5 Factoring

OBJECTIVE

You should be able to:

1 Determine the kind of factoring to try when an expression is to be factored. Then factor expressions:

a) by removing a common factor;
b) that are differences of squares;
c) that are trinomials;
d) that are trinomial squares;
e) that are sums or differences of cubes.

1 Factoring Polynomials

To **factor** a polynomial, we do the reverse of multiplying; that is, we find an equivalent expression that is a product. Factoring is an important algebraic skill.

Terms with Common Factors

When an expression is to be factored, we should always look first for a possible factor that is common to all terms. We then "factor it out" using the distributive laws. We usually look for a constant with the largest absolute value and variables with the largest exponent.

EXAMPLE 1 Factor: $4x^2 + 8$.

Solution

$$4x^2 + 8 = 4 \cdot x^2 + 4 \cdot 2 = 4(x^2 + 2)$$

Note that $4x^2$ and 8 are *terms* of the expression. The number 4 is a common factor of each term, so we factor it out. The expression $4 \cdot x^2 + 4 \cdot 2$ is not a correct answer. Although each term is factored, the entire expression has not been factored, that is, expressed as a product. The expression $4(x^2 + 2)$ is an equivalent expression that is a product. ■

EXAMPLES Factor.

2. $12x^2y - 20x^3y = 4x^2y \cdot 3 - 4x^2y \cdot 5x = 4x^2y(3 - 5x)$
3. $7x\sqrt{y} + 14x^2\sqrt{y} - 21\sqrt{y} = 7\sqrt{y}(x + 2x^2 - 3)$
4. $(a - b)(x + 5) + (a - b)(x - y^2) = (a - b)[(x + 5) + (x - y^2)]$
$\qquad = (a - b)(2x + 5 - y^2)$ ■

In some polynomials, pairs of terms have a common factor that can be removed, as in the following examples. This process is called **factoring by grouping**, and uses the distributive laws repeatedly.

EXAMPLES Factor.

5. $y^3 + 3y^2 - 5y - 15 = y^2(y + 3) - 5(y + 3)$
$\qquad = (y + 3)(y^2 - 5)$
6. $ax^2 + ay + bx^2 + by = a(x^2 + y) + b(x^2 + y)$
$\qquad = (a + b)(x^2 + y)$ ■

DO EXERCISES 1–3.

Differences of Squares

Recall that $(A + B)(A - B) = A^2 - B^2$. We can use this equation in reverse to factor an expression that is a *difference of two squares*.

EXAMPLES Factor.

7. $x^2 - 9 = (x + 3)(x - 3)$
8. $y^2 - 2 = (y + \sqrt{2})(y - \sqrt{2})$
9. $9a^2 - 16x^4 = (3a)^2 - (4x^2)^2 = (3a + 4x^2)(3a - 4x^2)$
10. $9y^4 - 9x^4 = 9(y^4 - x^4)$ **Remove the common factor first.**
$\qquad = 9(y^2 + x^2)(y^2 - x^2)$
$\qquad = 9(y^2 + x^2)(y + x)(y - x)$ ■

DO EXERCISES 4–7.

Factoring Trinomials

Some trinomials can be factored into two binomials. To do this, we factor by trial and error and check by multiplying using the **FOIL** equation.

EXAMPLE 11 Factor: $x^2 + 7x + 12$.

Solution We look for factors of 12 whose sum is 7. Since the constant term 12 is positive, its factors are either both negative or both positive. We want the sum of the factors to be 7, which is positive, so we consider only the positive factors

Factor.

1. $20x^3y + 12x^2y$

2. $(p + q)(x + 2) + (p + q)(x + y)$

3. $4x^3 + 20x^2 - 3x - 15$

Factor.

4. $x^2 - 16$

5. $25y^4 - 16x^2$

6. $2y^4 - 32x^4$

7. $x^2 - 3$

Factor.

8. $x^2 + 6x + 5$

9. $x^2 + 5x - 14$

10. $w^4 - 7w^2 + 10$

11. $3x^2 + 5x + 2$

12. $6x^4y^6 - 9x^2y^3 - 60$

13. $t^2 - t + 5$

of 12. By trial, we determine the factors to be 3 and 4 and the factorization to be

$$(x + 4)(x + 3).$$

EXAMPLE 12 Factor: $x^4 + 3x^2 - 10$.

Solution We can think of this polynomial mentally as $u^2 + 3u - 10$, where we have mentally substituted u for x^2. The constant term is negative this time, so any pair of factors must have one positive number and one negative number. The coefficient of the middle term is positive, so we look for pairs of factors for which the positive number has the larger absolute value. By trial, we determine the factors to be 5 and -2 and the factorization to be

$$u^2 + 3u - 10 = (u + 5)(u - 2).$$

Then substituting x^2 for u, we obtain the factorization of the original trinomial:

$$(x^2 + 5)(x^2 - 2).$$

EXAMPLE 13 Factor: $3x^2 - 10x - 8$.

Solution

Method 1. We look for binomials $ax + b$ and $cx + d$ for which the product of the first terms is $3x^2$. The product of the last terms must be -8. When we multiply the inside terms, then the outside terms, and add, we must have $-10x$. By trial, we determine the factorization to be

$$(3x + 2)(x - 4).$$

Method 2. We multiply the leading coefficient 3 and the constant -8: $3(-8) = -24$. Then we try to factor -24 so that the sum of the factors is -10. By trial, we find these factors to be -12 and 2. We then write the middle term $-10x$ as a sum using -12 and 2. That is, we split the middle term as follows:

$$-10x = -12x + 2x.$$

Now we factor by grouping:

$$\begin{aligned} 3x^2 - 10x - 8 &= 3x^2 - 12x + 2x - 8 \\ &= 3x(x - 4) + 2(x - 4) \\ &= (3x + 2)(x - 4) \end{aligned}$$

CAUTION! Keep in mind that any factoring you do can and should be checked by multiplying. It is an easy check and can help you avoid many mistakes.

Not all polynomials can be factored into polynomials with integer, rational, or real coefficients. An example is $x^2 - x - 7$. There are no real factors of -7 whose sum is -1. In such a case, we say that the polynomial is "not factorable."

DO EXERCISES 8–13.

Trinomial Squares

Certain trinomials are squares of binomials. You can use trial and error to factor such trinomials, but it is more efficient to make use of the following rules, which reverse the rules for squaring binomials. You should recall that

$$A^2 + 2AB + B^2 = (A + B)^2 \quad \text{and} \quad A^2 - 2AB + B^2 = (A - B)^2.$$

We can use these equations to factor trinomials that are squares. To factor a trinomial, you should check to see if it is a square. For this to be the case, two of the terms must be squares and the other term must be twice the product of the square roots, or the additive inverse of that product.

EXAMPLES Factor.

14. $x^2 - 10x + 25 = (x - 5)^2$

15. $16y^2 + 56y + 49 = (4y + 7)^2$

16. $-4y^2 - 144y^8 + 48y^5 = -4y^2(1 + 36y^6 - 12y^3)$ We first removed the common factor.

$$= -4y^2(1 - 12y^3 + 36y^6)$$
$$= -4y^2(1 - 6y^3)^2$$

DO EXERCISES 14–16.

Sums or Differences of Cubes

We can use the following equations to factor a sum or a difference of two cubes:

$$A^3 + B^3 = (A + B)(A^2 - AB + B^2),$$
$$A^3 - B^3 = (A - B)(A^2 + AB + B^2).$$

Check them by multiplying the right-hand sides.

EXAMPLE 17 Factor: $x^3 - 27$.

Solution We have

$$x^3 - 27 = x^3 - 3^3.$$

In one set of parentheses, we write the cube root of the first expression x, then we write a minus sign, and then the cube root of the second expression 27. This gives us $x - 3$.

$$(x - 3)(\qquad\qquad)$$

To get the next factor, we think of $x - 3$ and do the following.

1. Square the first expression: x^2.
2. Multiply the expressions and then change the sign: $3x$.
3. Square the second expression: 9.

$$(x - 3)(x^2 + 3x + 9)$$

Note: We cannot factor $x^2 + 3x + 9$ as a product of polynomials with real coefficients. (It is not a trinomial square nor can it be factored by trial and error.)

DO EXERCISES 17 AND 18.

EXAMPLE 18 Factor: $125x^3 + y^3$.

Solution We have

$$125x^3 + y^3 = (5x)^3 + y^3.$$

Factor.

14. $9y^2 - 30y + 25$

15. $16x^2 + 72xy + 81y^2$

16. $-12x^4y^2 + 60x^2y^5 - 75y^8$

Factor.

17. $x^3 - 8$

18. $64 - t^3$

In one set of parentheses, we write the cube root of the first expression, then a plus sign, and then the cube root of the second expression.

$$(5x + y)(\qquad\qquad)$$

To get the next factor, we think of $5x + y$ and do the following.

1. Square the first expression: $(5x)^2$ or $25x^2$.
2. Multiply the expressions and then change the sign: $-5xy$.
3. Square the second expression: y^2.

$$(5x + y)(25x^2 - 5xy + y^2)$$

DO EXERCISES 19 AND 20.

Factor.

19. $27x^3 + y^3$

20. $8m^3 + 125t^3$

EXAMPLE 19 Factor: $16x^7y + 54xy^7$.

Solution We first look for a common factor.

$$\begin{aligned}16x^7y + 54xy^7 &= 2xy(8x^6 + 27y^6)\\ &= 2xy[(2x^2)^3 + (3y^2)^3]\\ &= 2xy(2x^2 + 3y^2)(4x^4 - 6x^2y^2 + 9y^4)\end{aligned}$$

DO EXERCISE 21.

21. Factor: $128y^7 - 250x^6y$.

EXAMPLE 20 Factor: $a^6 - b^6$.

Solution We can express this polynomial as a difference of squares:

$$(a^3)^2 - (b^3)^2.$$

We factor as follows:

$$(a^3 + b^3)(a^3 - b^3).$$

One factor is a sum of cubes, and the other is a difference of cubes. We factor them:

$$(a + b)(a^2 - ab + b^2)(a - b)(a^2 + ab + b^2).$$

The factoring is complete.

In Example 20, had we thought of factoring first as a difference of cubes, we would have had

$$\begin{aligned}(a^2)^3 - (b^2)^3 &= (a^2 - b^2)(a^4 + a^2b^2 + b^4)\\ &= (a + b)(a - b)(a^4 + a^2b^2 + b^4).\end{aligned}$$

In this case, we have missed some factors; $a^4 + a^2b^2 + b^4$ can be factored as $(a^2 - ab + b^2)(a^2 + ab + b^2)$, but we probably would not have known to do such factoring.

DO EXERCISE 22.

22. Factor: $p^6 - 64$.

Remember the following about factoring sums or differences of squares and cubes.

Sum of cubes:	$A^3 + B^3 = (A + B)(A^2 - AB + B^2)$
Difference of cubes:	$A^3 - B^3 = (A - B)(A^2 + AB + B^2)$
Difference of squares:	$A^2 - B^2 = (A + B)(A - B)$
Sum of squares:	$A^2 + B^2$ cannot be factored using real-number coefficients.

EXERCISE SET 1.5

1 Factor.

1. $p^2 + 6p + 8$
2. $w^2 - 7w + 10$
3. $n^2 + n - 56$
4. $y^2 + 5y - 14$
5. $y^4 - 4y^2 - 21$
6. $m^4 - m^2 - 90$
7. $18a^2b - 15ab^2$
8. $4x^2y + 12xy^2$
9. $a(b - 2) + c(b - 2)$
10. $a(x^2 - 3) - 2(x^2 - 3)$
11. $x^3 + 3x^2 + 6x + 18$
12. $3x^3 + x^2 - 18x - 6$
13. $y^3 - 3y^2 - 4y + 12$
14. $p^3 - 2p^2 - 9p + 18$
15. $9x^2 - 25$
16. $16x^2 - 9$
17. $4xy^4 - 4xz^2$
18. $5xy^4 - 5xz^4$
19. $y^2 - 6y + 9$
20. $x^2 + 8x + 16$
21. $1 - 8x + 16x^2$
22. $1 + 10x + 25x^2$
23. $4x^2 - 5$
24. $16x^2 - 7$
25. $x^2y^2 - 14xy + 49$
26. $x^2y^2 - 16xy + 64$
27. $4ax^2 + 20ax - 56a$
28. $21x^2y + 2xy - 8y$
29. $a^2 + 2ab + b^2 - c^2$
30. $x^2 - 2xy + y^2 - z^2$
31. $x^2 + 2xy + y^2 - a^2 - 2ab - b^2$
    [*Hint:* Factor $x^2 + 2xy + y^2$ and $-1(a^2 + 2ab + b^2)$.]
32. $r^2 + 2rs + s^2 - t^2 + 2tv - v^2$
33. $5y^4 - 80x^4$
34. $6y^4 - 96x^4$
35. $x^3 + 8$
36. $y^3 - 64$
37. $3x^3 - \frac{3}{8}$
38. $5y^3 + \frac{5}{27}$
39. $x^3 + 0.001$
40. $y^3 - 0.125$
41. $3z^3 - 24$
42. $4t^3 + 108$
43. $a^6 - t^6$
44. $64m^6 + y^6$
45. $16a^7b + 54ab^7$
46. $24a^2x^4 - 375a^8x$
47. $x^2 - 17.6$
48. $x^2 - 8.03$
49. $37x^2 - 14.5y^2$
    (*Hint:* First remove the common factor 37.)
50. $1.96x^2 - 17.4y^2$
    (*Hint:* First remove the common factor 1.96.)

SYNTHESIS

Factor.

51. $(x + h)^3 - x^3$
52. $(x + 0.01)^2 - x^2$
53. $y^4 - 84 + 5y^2$
54. $11x^2 + x^4 - 80$
55. $y^2 - \frac{8}{49} + \frac{2}{7}y$
56. $x^2 + \frac{3}{5}x - \frac{4}{25}$
57. $t^2 - 0.27 + 0.6t$
58. $0.4m - 0.05 + m^2$

Factor. Assume that variables in exponents represent natural numbers.

59. $x^{2n} + 5x^n - 24$
60. $4x^{2n} - 4x^n - 3$
61. $x^2 + ax + bx + ab$
62. $bdy^2 + ady + bcy + ac$
63. $\frac{1}{4}t^2 - \frac{2}{5}t + \frac{4}{25}$
64. $\frac{4}{27}r^2 + \frac{5}{9}rs + \frac{1}{12}s^2 - \frac{1}{3}rs$
65. $25y^{2m} - (x^{2n} - 2x^n + 1)$
66. $4x^{4a} + 12x^{2a} + 10x^{2a} + 30$
67. $3x^{3n} - 24y^{3m}$
68. $x^{6a} - t^{3b}$
69. $(y - 1)^4 - (y - 1)^2$
70. $x^6 - 2x^5 + x^4 - x^2 + 2x - 1$

Express each of the following in the form $A(x + B)$.

71. $5x - 9$
72. $\frac{2}{3}x - 7$

73. a) Multiply: $(x^2 - x + 1)(x^3 + x^2 - 1)$.
 b) Factor: $x^5 + x - 1$.

1.6 Fractional Expressions

OBJECTIVES

You should be able to:

1 Determine meaningful replacements in fractional expressions.

2 Simplify fractional expressions.

3 Multiply or divide fractional expressions, and simplify.

4 Add or subtract fractional expressions, and simplify.

5 Simplify complex fractional expressions.

Determine the meaningful replacements.

1. $\dfrac{x^2 - 9}{x - 3}$

2. $\dfrac{x^3 - xy^2}{x^2 + 7x + 12}$

1 Replacements in Fractional Expressions

Expressions like the following are called **fractional expressions** or **rational expressions:**

$$\frac{8}{5}, \quad \frac{x^2 - 9}{x - 3}, \quad \frac{3x^2 + 5\sqrt{x} - 2}{x^2 - y^2}, \quad \frac{x - 3}{x^2 - x - 2}.$$

Fractional expressions represent division. Certain substitutions are not meaningful in such expressions. Since division by zero is not defined, any number that makes a denominator zero is not a meaningful replacement. For example, 3 is not a meaningful replacement in

$$\frac{x^2 - 9}{x - 3}$$

because the denominator $x - 3$ is 0 when x is replaced by 3. All real numbers other than 3 are meaningful replacements. As another example, consider

$$\frac{x - 3}{x^2 - x - 2}.$$

To determine the meaningful replacements, we can first factor the denominator:

$$\frac{x - 3}{x^2 - x - 2} = \frac{x - 3}{(x + 1)(x - 2)}.$$

The factor $x + 1$ is 0 when $x = -1$. The factor $x - 2$ is 0 when $x = 2$. Thus, -1 and 2 are not meaningful replacements. All real numbers except -1 and 2 are meaningful replacements.

DO EXERCISES 1 AND 2.

Multiplication and Division

To multiply two fractional expressions, we multiply their numerators and also their denominators. By Theorem 4, when we divide, we multiply by the reciprocal of the divisor.

EXAMPLE 1 Multiply:

$$\frac{x + 3}{y - 4} \cdot \frac{x^3}{y + 5}.$$

Solution

$$\frac{x+3}{y-4}\cdot\frac{x^3}{y+5}=\frac{(x+3)x^3}{(y-4)(y+5)}$$

EXAMPLE 2 Divide:

$$\frac{x-2}{x+1}\div\frac{x+5}{x-3}.$$

Solution

$$\frac{x-2}{x+1}\div\frac{x+5}{x-3}=\frac{x-2}{x+1}\cdot\frac{x-3}{x+5}\quad\text{Multiplying by the reciprocal}$$

$$=\frac{(x-2)(x-3)}{(x+1)(x+5)}\quad\text{Multiplying}$$

In Example 2, we could go on and finish multiplying in the numerator and the denominator, but we choose not to do so because we want to simplify, if possible. It also eases addition, subtraction, and equation solving if we do not carry out the multiplication.

DO EXERCISES 3 AND 4.

2 Simplifying

The basis for simplifying fractional expressions lies in the fact that certain expressions have a value of 1 for all meaningful replacements. Such expressions have the same numerator and denominator.* Here are some examples:

$$\frac{x-2}{x-2}=1,\qquad\frac{3x^2-4x+2}{3x^2-4x+2}=1,\qquad\frac{4x-5}{4x-5}=1.$$

When we multiply by such an expression, we obtain an equivalent expression. This means that the new expression will name the same number as the first for all meaningful replacements. The set of meaningful replacements may not be the same for the two expressions.

EXAMPLE 3 Multiply:

$$\frac{y+4}{y-3}\cdot\frac{y-2}{y-2}.$$

Solution

$$\frac{y+4}{y-3}\cdot\frac{y-2}{y-2}=\frac{(y+4)(y-2)}{(y-3)(y-2)}$$

$$=\frac{y^2+2y-8}{y^2-5y+6}$$

The expressions $(y+4)/(y-3)$ and $(y^2+2y-8)/(y^2-5y+6)$ are equivalent. That is, they name the same number for all meaningful replacements. The only replacement that is not meaningful in $(y+4)/(y-3)$ is 3. For the expression $(y^2+2y-8)/(y^2-5y+6)$, or $(y^2+2y-8)/[(y-2)(y-3)]$, the replacements that are not meaningful are 2 and 3.

DO EXERCISE 5.

*By Theorem 4, $a \div a = a/a = a(1/a)$, and since a and $1/a$ are reciprocals, their product is 1.

3. Multiply:

$$\frac{x+y}{2x^2-1}\cdot\frac{x+y}{7x}.$$

4. Divide:

$$\frac{x-2}{x+2}\div\frac{x+2}{x+4}.$$

5. Multiply

$$\frac{x+2}{x-5}\text{ by }\frac{x+3}{x+3}$$

to obtain an equivalent expression. Name the meaningful replacements for the two expressions.

Simplify. Name the meaningful replacements in the original and the simplified expressions.

6. $\dfrac{6x^2 + 4x}{2x^2 + 4x}$

7. $\dfrac{y^2 + 3y + 2}{y^2 - 1}$

Simplification can be accomplished if we reverse the procedure in the above example; that is, we try to factor the fractional expression in such a way that one of the factors is equal to 1 and then we "remove" that factor. In this way, we obtain an expression that is simpler, or less complicated, than the original, but equivalent to it.

EXAMPLE 4 Simplify:

$$\frac{15x^3y^2}{20x^2y}.$$

Solution

$$\frac{15x^3y^2}{20x^2y} = \frac{(5x^2y)3xy}{(5x^2y)4} \qquad \text{Factoring the numerator and the denominator}$$

$$= \frac{5x^2y}{5x^2y} \cdot \frac{3xy}{4} \qquad \text{Factoring the expression}$$

$$= \frac{3xy}{4} \qquad \text{"Removing" a factor of 1} \qquad \blacksquare$$

In Example 4, note that in the original expression, neither x nor y can be replaced by 0. That is, 0 is not a meaningful replacement. In the simplified expression, however, all replacements are meaningful. The expressions are equivalent for all meaningful replacements.

EXAMPLE 5 Simplify:

$$\frac{x^2 - 1}{2x^2 - x - 1}.$$

Solution

$$\frac{x^2 - 1}{2x^2 - x - 1} = \frac{(x - 1)(x + 1)}{(2x + 1)(x - 1)}$$

$$= \frac{x - 1}{x - 1} \cdot \frac{x + 1}{2x + 1}$$

$$= \frac{x + 1}{2x + 1} \qquad \blacksquare$$

In the original expression in Example 5, the meaningful replacements are all real numbers except 1 and $-\frac{1}{2}$. In the simplified expression, all real numbers except $-\frac{1}{2}$ are meaningful replacements. The expressions are equivalent.

DO EXERCISES 6 AND 7.

3 Multiplying, Dividing, and Simplifying

EXAMPLES

6. Multiply and simplify.

$$\frac{x + 2}{x - 2} \cdot \frac{x^2 - 4}{x^2 + x - 2} = \frac{(x + 2)(x^2 - 4)}{(x - 2)(x^2 + x - 2)} \qquad \text{Multiplying}$$

$$= \frac{(x + 2)(x + 2)(x - 2)}{(x - 2)(x + 2)(x - 1)} \qquad \text{Factoring}$$

$$= \frac{(x + 2)(x - 2)}{(x + 2)(x - 2)} \cdot \frac{x + 2}{x - 1}$$

$$= \frac{x + 2}{x - 1} \qquad \text{"Removing" a factor of 1}$$

7. Divide and simplify.

$$\frac{a^2 - 1}{a + 1} \div \frac{a^2 - 2a + 1}{a + 1} = \frac{a^2 - 1}{a + 1} \cdot \frac{a + 1}{a^2 - 2a + 1}$$

$$= \frac{(a + 1)(a - 1)(a + 1)}{(a + 1)(a - 1)(a - 1)}$$

$$= \frac{a + 1}{a - 1} \qquad \blacksquare$$

DO EXERCISES 8 AND 9.

4 **Addition and Subtraction**

When fractional expressions have the same denominator, we can add or subtract them by adding or subtracting the numerators and retaining the common denominator. If the denominators are not the same, we then find equivalent expressions with the same denominator and add. If one denominator is the opposite of another, we can find a common denominator by multiplying by $-1/-1$.

EXAMPLE 8 Add:

$$\frac{3x^2 + 4x - 8}{x^2 + y^2} + \frac{-5x^2 + 5x + 7}{x^2 + y^2}.$$

Solution

$$\frac{3x^2 + 4x - 8}{x^2 + y^2} + \frac{-5x^2 + 5x + 7}{x^2 + y^2} = \frac{-2x^2 + 9x - 1}{x^2 + y^2} \qquad \blacksquare$$

In the following example, one denominator is the additive inverse of the other.

EXAMPLE 9 Add:

$$\frac{3x^2 + 4}{x - y} + \frac{5x^2 - 11}{y - x}.$$

Solution

$$\frac{3x^2 + 4}{x - y} + \frac{5x^2 - 11}{y - x} = \frac{3x^2 + 4}{x - y} + \frac{-1}{-1} \cdot \frac{5x^2 - 11}{y - x}$$

We multiply by 1 using $-1/-1$ to convert the second denominator to its opposite.

$$= \frac{3x^2 + 4}{x - y} + \frac{-1(5x^2 - 11)}{-1(y - x)}$$

$$= \frac{3x^2 + 4}{x - y} + \frac{11 - 5x^2}{x - y} \qquad -1(y - x) = -y + x = x - y$$

$$= \frac{-2x^2 + 15}{x - y} \qquad \blacksquare$$

DO EXERCISES 10 AND 11.

When denominators are different, but not opposites, we find a common

8. Multiply and simplify:

$$\frac{x^2 - 2xy + y^2}{x + y} \cdot \frac{3x + 3y}{x^2 - y^2}.$$

9. Divide and simplify:

$$\frac{a^2 - b^2}{ab} \div \frac{a^2 - 2ab + b^2}{2a^2b^2}.$$

Add.

10. $\dfrac{2x^2 + 5x - 9}{x - 5} + \dfrac{x^2 - x + 11}{x - 5}$

11. $\dfrac{3x^2 + 4}{x - 5} + \dfrac{x^2 - 7}{5 - x}$

12. Add:

$$\frac{x^2 - 4xy + 4y^2}{2x^2 - 3xy + y^2} + \frac{x + 4y}{2x - 2y}.$$

denominator by factoring the denominators. Then we multiply each term by 1 in such a way as to get the common denominator in each expression.

EXAMPLE 10 Add:

$$\frac{1}{2x} + \frac{5x}{x^2 - 1} + \frac{3}{x + 1}.$$

Solution We first find the **least common multiple** (LCM) of the denominators, also referred to as the **least common denominator.** To find the LCM, we first factor each denominator:

$$2x = 2x,$$
$$x^2 - 1 = (x + 1)(x - 1),$$
$$x + 1 = x + 1.$$

Then we consider how often each factor occurs in each factorization. We make up a product using each factor the greatest number of times that it occurs in each factorization. We use 2 as a factor once, x as a factor once, $x + 1$ as a factor once even though it occurs in two of the factorizations, and $x - 1$ as a factor once. The LCM is $2x(x + 1)(x - 1)$. Now we multiply each fractional expression by 1 in such a way as to get the LCM:

$$\frac{1}{2x} + \frac{5x}{x^2 - 1} + \frac{3}{x + 1}$$

$$= \frac{1}{2x} \cdot \frac{(x + 1)(x - 1)}{(x + 1)(x - 1)} + \frac{5x}{(x + 1)(x - 1)} \cdot \frac{2x}{2x} + \frac{3}{x + 1} \cdot \frac{2x(x - 1)}{2x(x - 1)}$$

$$= \frac{1(x + 1)(x - 1)}{2x(x + 1)(x - 1)} + \frac{5x(2x)}{(x + 1)(x - 1)(2x)} + \frac{3(2x)(x - 1)}{(x + 1)(2x)(x - 1)}$$

$$= \frac{(x + 1)(x - 1) + 10x^2 + 6x(x - 1)}{2x(x + 1)(x - 1)}$$

$$= \frac{17x^2 - 6x - 1}{2x(x + 1)(x - 1)}, \quad \text{or} \quad \frac{17x^2 - 6x - 1}{2x^3 - 2x}.$$ ■

DO EXERCISE 12.

EXAMPLE 11 Subtract:

$$\frac{x}{x^2 + 5x + 6} - \frac{2}{x^2 + 3x + 2}.$$

Solution

$$\frac{x}{x^2 + 5x + 6} - \frac{2}{x^2 + 3x + 2}$$

$$= \frac{x}{(x + 2)(x + 3)} - \frac{2}{(x + 1)(x + 2)}$$

The denominators have been factored in order to determine the LCM. We use each factor the greatest number of times that it occurs in each factorization. The LCM is $(x + 1)(x + 2)(x + 3)$.

$$= \frac{x}{(x+2)(x+3)} \cdot \frac{x+1}{x+1} - \frac{2}{(x+1)(x+2)} \cdot \frac{x+3}{x+3}$$

$$= \frac{x(x+1) - [2(x+3)]}{(x+1)(x+2)(x+3)}$$

We use the color brackets here to make sure that we subtract the *entire* numerator, not just part of it.

$$= \frac{x^2 + x - [2x + 6]}{(x+1)(x+2)(x+3)}$$

$$= \frac{x^2 + x - 2x - 6}{(x+1)(x+2)(x+3)}$$

$$= \frac{x^2 - x - 6}{(x+1)(x+2)(x+3)}$$

$$= \frac{(x-3)(x+2)}{(x+1)(x+2)(x+3)}$$

$$= \frac{x-3}{(x+1)(x+3)}$$

Always simplify at the end if possible.

We could multiply out the denominator, but it will be convenient when we solve fractional equations to have the denominators factored.

DO EXERCISE 13.

CAUTION! When subtracting one fractional expression from another, subtract numerators:

$$\frac{A}{C} - \frac{B}{C} = \frac{A - (B)}{C}.$$

Always be sure to subtract the *entire* numerator B and not just part of it. The use of parentheses or brackets helps. See Example 11.

5 Complex Fractional Expressions

A **complex fractional expression** has a fractional expression within its numerator or denominator or both. To simplify such an expression, we can use either of two methods, which we shall exemplify as follows.

To simplify a complex fractional expression:

Method 1. Find the LCM of all the denominators *within* the complex fractional expression. Then multiply by 1 using that LCM as the numerator and the denominator of that expression for 1.

Method 2. First add or subtract, if necessary, to get a single fractional expression in both numerator and denominator. Then divide by multiplying by the reciprocal of the denominator.

EXAMPLE 12 Simplify:

$$\frac{x + \frac{1}{5}}{x - \frac{1}{3}}.$$

13. Subtract:

$$\frac{x}{x^2 + 11x + 30} - \frac{5}{x^2 + 9x + 20}.$$

Solution

Method 1. The denominators within the complex fractional expression are 3 and 5. The LCM is $3 \cdot 5$, or 15. We multiply by 1 using 15/15:

$$\frac{x+\frac{1}{5}}{x-\frac{1}{3}} = \left(\frac{x+\frac{1}{5}}{x-\frac{1}{3}}\right)\frac{15}{15} = \frac{\left(x+\frac{1}{5}\right)15}{\left(x-\frac{1}{3}\right)15} = \frac{15x+\frac{1}{5}\cdot 15}{15x-\frac{1}{3}\cdot 15} = \frac{15x+3}{15x-5}.$$

Method 2. We carry out the addition in the numerator and the subtraction in the denominator separately to obtain a single fractional expression for both the numerator and the denominator. Then we divide:

$$\frac{x+\frac{1}{5}}{x-\frac{1}{3}} = \frac{x\cdot\frac{5}{5}+\frac{1}{5}}{x\cdot\frac{3}{3}-\frac{1}{3}}$$

$$= \frac{\frac{5x+1}{5}}{\frac{3x-1}{3}}$$ Now we have a single fractional expression for both numerator and denominator.

$$= \frac{5x+1}{5}\cdot\frac{3}{3x-1}$$ Here we divided by multiplying by the reciprocal of the denominator.

$$= \frac{15x+3}{15x-5}$$

EXAMPLE 13 Simplify:

$$\frac{a^{-3}-b^{-3}}{a^{-1}-b^{-1}}.$$

Solution We first note that

$$\frac{a^{-3}-b^{-3}}{a^{-1}-b^{-1}} = \frac{\frac{1}{a^3}-\frac{1}{b^3}}{\frac{1}{a}-\frac{1}{b}}.$$

Method 1. The denominators within the complex fractional expression are a, b, a^3, and b^3. The LCM of these expressions is a^3b^3. We multiply by 1 using a^3b^3/a^3b^3.

$$\frac{a^{-3}-b^{-3}}{a^{-1}-b^{-1}} = \frac{\frac{1}{a^3}-\frac{1}{b^3}}{\frac{1}{a}-\frac{1}{b}} = \frac{\frac{1}{a^3}-\frac{1}{b^3}}{\frac{1}{a}-\frac{1}{b}}\cdot\frac{a^3b^3}{a^3b^3} = \frac{\left(\frac{1}{a^3}-\frac{1}{b^3}\right)a^3b^3}{\left(\frac{1}{a}-\frac{1}{b}\right)a^3b^3}$$

$$= \frac{\frac{1}{a^3}(a^3b^3)-\frac{1}{b^3}(a^3b^3)}{\frac{1}{a}(a^3b^3)-\frac{1}{b}(a^3b^3)} = \frac{b^3-a^3}{a^2b^3-a^3b^2}$$

$$= \frac{(b-a)(b^2+ab+a^2)}{a^2b^2(b-a)}$$

$$= \frac{b^2+ab+a^2}{a^2b^2}$$

Method 2. We carry out the subtractions in the numerator and the denominator separately to obtain a single fractional expression for both the numerator and the denominator. Then we divide.

$$\frac{a^{-3}-b^{-3}}{a^{-1}-b^{-1}}=\frac{\frac{1}{a^3}-\frac{1}{b^3}}{\frac{1}{a}-\frac{1}{b}}=\frac{\frac{1}{a^3}\cdot\frac{b^3}{b^3}-\frac{1}{b^3}\cdot\frac{a^3}{a^3}}{\frac{1}{a}\cdot\frac{b}{b}-\frac{1}{b}\cdot\frac{a}{a}}=\frac{\frac{b^3}{a^3b^3}-\frac{a^3}{a^3b^3}}{\frac{b}{ab}-\frac{a}{ab}}=\frac{\frac{b^3-a^3}{a^3b^3}}{\frac{b-a}{ab}}$$

$$=\frac{b^3-a^3}{a^3b^3}\cdot\frac{ab}{b-a}=\frac{(b-a)(b^2+ab+a^2)ab}{ab(a^2b^2)(b-a)}$$

$$=\frac{(b-a)ab}{(b-a)ab}\cdot\frac{b^2+ab+a^2}{a^2b^2}=\frac{b^2+ab+a^2}{a^2b^2}$$

DO EXERCISES 14 AND 15.

Simplify.

14. $\dfrac{1+\frac{x}{a}}{a-\frac{x^2}{a}}$

15. $\dfrac{\frac{1}{a}+\frac{1}{b}}{\frac{1}{a^3}+\frac{1}{b^3}}$

EXERCISE SET 1.6

1 Determine the meaningful replacements.

1. $\dfrac{3x-3}{x(x-1)}$
2. $\dfrac{(x^2-4)(x+1)}{(x+2)(x^2-1)}$
3. $\dfrac{7x^2-28x+28}{(x^2-4)(x^2+3x-10)}$
4. $\dfrac{7x^2+11x-6}{x(x^2-x-6)}$

2 Simplify. Then determine replacements that are meaningful in the simplified expression.

5. $\dfrac{3x-3}{x(x-1)}$
6. $\dfrac{(x^2-4)(x+1)}{(x+2)(x^2-1)}$
7. $\dfrac{7x^2-28x+28}{(x^2-4)(x^2+3x-10)}$
8. $\dfrac{7x^2+11x-6}{x(x^2-x-6)}$
9. $\dfrac{25x^2y^2}{10xy^2}$
10. $\dfrac{x^2-4}{x^2+5x+6}$
11. $\dfrac{x^2-3x+2}{x^2+x-2}$
12. $\dfrac{a^2+2a+4}{(a^3-8)(a+5)}$

3 Multiply or divide, and simplify.

13. $\dfrac{x^2-y^2}{(x-y)^2}\cdot\dfrac{1}{x+y}$
14. $\dfrac{r-s}{r+s}\cdot\dfrac{r^2-s^2}{(r-s)^2}$
15. $\dfrac{x^2-2x-35}{2x^3-3x^2}\cdot\dfrac{4x^3-9x}{7x-49}$
16. $\dfrac{x^2+2x-35}{3x^3-2x^2}\cdot\dfrac{9x^3-4x}{7x+49}$
17. $\dfrac{a^2-a-6}{a^2-7a+12}\cdot\dfrac{a^2-2a-8}{a^2-3a-10}$
18. $\dfrac{a^2-a-12}{a^2-6a+8}\cdot\dfrac{a^2+a-6}{a^2-2a-24}$
19. $\dfrac{m^2-n^2}{r+s}\div\dfrac{m-n}{r+s}$
20. $\dfrac{a^2-b^2}{x-y}\div\dfrac{a+b}{x-y}$
21. $\dfrac{3x+12}{2x-8}\div\dfrac{(x+4)^2}{(x-4)^2}$
22. $\dfrac{a^2-a-2}{a^2-a-6}\div\dfrac{a^2-2a}{2a+a^2}$
23. $\dfrac{x^2-y^2}{x^3-y^3}\cdot\dfrac{x^2+xy+y^2}{x^2+2xy+y^2}$
24. $\dfrac{c^3+8}{c^2-4}\div\dfrac{c^2-2c+4}{c^2-4c+4}$
25. $\dfrac{(x-y)^2-z^2}{(x+y)^2-z^2}\div\dfrac{x-y+z}{x+y-z}$
26. $\dfrac{(a+b)^2-9}{(a-b)^2-9}\cdot\dfrac{a-b-3}{a+b+3}$

4 Add or subtract, and simplify.

27. $\dfrac{3}{2a+3}+\dfrac{2a}{2a+3}$
28. $\dfrac{a-3b}{a+b}+\dfrac{a+5b}{a+b}$
29. $\dfrac{y}{y-1}+\dfrac{2}{1-y}$
30. $\dfrac{a}{a-b}+\dfrac{b}{b-a}$
31. $\dfrac{x}{2x-3y}-\dfrac{y}{3y-2x}$
32. $\dfrac{3a}{3a-2b}-\dfrac{2a}{2b-3a}$
33. $\dfrac{3}{x+2}+\dfrac{2}{x^2-4}$
34. $\dfrac{5}{a-3}-\dfrac{2}{a^2-9}$
35. $\dfrac{y}{y^2-y-20}+\dfrac{2}{y+4}$
36. $\dfrac{6}{y^2+6y+9}-\dfrac{5}{y+3}$

37. $\dfrac{3}{x+y} + \dfrac{x-5y}{x^2-y^2}$

38. $\dfrac{a^2+1}{a^2-1} - \dfrac{a-1}{a+1}$

39. $\dfrac{9x+2}{3x^2-2x-8} + \dfrac{7}{3x^2+x-4}$

40. $\dfrac{3y}{y^2-7y+10} - \dfrac{2y}{y^2-8y+15}$

41. $\dfrac{5a}{a-b} + \dfrac{ab}{a^2-b^2} + \dfrac{4b}{a+b}$

42. $\dfrac{6a}{a-b} - \dfrac{3b}{b-a} + \dfrac{5}{a^2-b^2}$

43. $\dfrac{7}{x+2} - \dfrac{x+8}{4-x^2} + \dfrac{3x-2}{4-4x+x^2}$

44. $\dfrac{6}{x+3} - \dfrac{x+4}{9-x^2} + \dfrac{2x-3}{9-6x+x^2}$

45. $\dfrac{1}{x+1} - \dfrac{x}{x-2} + \dfrac{x^2+2}{x^2-x-2}$

46. $\dfrac{x-1}{x-2} - \dfrac{x+1}{x+2} + \dfrac{x-6}{x^2-4}$

5 Simplify.

47. $\dfrac{\dfrac{x^2-y^2}{xy}}{\dfrac{x-y}{y}}$

48. $\dfrac{\dfrac{a-b}{b}}{\dfrac{a^2-b^2}{ab}}$

49. $\dfrac{a-a^{-1}}{a+a^{-1}}$

50. $\dfrac{a-\dfrac{a}{b}}{b-\dfrac{b}{a}}$

51. $\dfrac{c+\dfrac{8}{c^2}}{1+\dfrac{2}{c}}$

52. $\dfrac{x^{-1}+y^{-1}}{x^{-3}+y^{-3}}$

53. $\dfrac{x^2+xy+y^2}{\dfrac{x^2}{y}-\dfrac{y^2}{x}}$

54. $\dfrac{\dfrac{a^2}{b}+\dfrac{b^2}{a}}{a^2-ab+b^2}$

55. $\dfrac{\dfrac{x}{y}-\dfrac{y}{x}}{\dfrac{1}{y}+\dfrac{1}{x}}$

56. $\dfrac{\dfrac{a}{b}-\dfrac{b}{a}}{\dfrac{1}{a}-\dfrac{1}{b}}$

57. $\dfrac{x^2y^{-2}-y^2x^{-2}}{xy^{-1}+yx^{-1}}$

58. $\dfrac{a^2b^{-2}-b^2a^{-2}}{ab^{-1}-ba^{-1}}$

59. $\dfrac{\dfrac{a}{1-a}+\dfrac{1+a}{a}}{\dfrac{1-a}{a}+\dfrac{a}{1+a}}$

60. $\dfrac{\dfrac{1-x}{x}+\dfrac{x}{1+x}}{\dfrac{1+x}{x}+\dfrac{x}{1-x}}$

61. $\dfrac{\dfrac{1}{a^2}+\dfrac{2}{ab}+\dfrac{1}{b^2}}{\dfrac{1}{a^2}-\dfrac{1}{b^2}}$

62. $\dfrac{\dfrac{1}{x^2}-\dfrac{1}{y^2}}{\dfrac{1}{x^2}-\dfrac{2}{xy}+\dfrac{1}{y^2}}$

SYNTHESIS

Simplify.

63. $\dfrac{(x+h)^2-x^2}{h}$

64. $\dfrac{\dfrac{1}{x+h}-\dfrac{1}{x}}{h}$

65. $\dfrac{(x+h)^3-x^3}{h}$

66. $\dfrac{\dfrac{1}{(x+h)^2}-\dfrac{1}{x^2}}{h}$

67. $\left[\dfrac{\dfrac{x+1}{x-1}+1}{\dfrac{x+1}{x-1}-1}\right]^5$

68. $1+\cfrac{1}{1+\cfrac{1}{1+\cfrac{1}{1+\cfrac{1}{x}}}}$

Find the error(s) in each of the following. Explain why each is an error. Then find the correct answer.

69. $\dfrac{a}{b} \div \left(\dfrac{a}{3}+\dfrac{b}{4}\right) = \dfrac{a}{b}\cdot\left(\dfrac{3}{a}+\dfrac{4}{b}\right)$ (1)

$= \dfrac{a}{b}\cdot\left(\dfrac{3b+4a}{ab}\right)$ (2)

$= \dfrac{a(3b+4a)}{ab^2}$ (3)

$= \dfrac{4a+3b}{b}$ (4)

70. $\dfrac{5x}{2y}+\dfrac{3y}{4x} = \dfrac{5x^2}{4xy}+\dfrac{3y^2}{4xy}$ (1)

$= \dfrac{5x^2+3y^2}{4xy}$ (2)

$= \dfrac{5x+3y^2}{4y}$ (3)

$= \dfrac{5x+3y}{4}$ (4)

Add and simplify.

71. $\frac{n(n+1)(n+2)}{2 \cdot 3} + \frac{(n+1)(n+2)}{2}$

72. $\frac{n(n+1)(n+2)(n+3)}{2 \cdot 3 \cdot 4} + \frac{(n+1)(n+2)(n+3)}{2 \cdot 3}$

1.7 Radical Notation

OBJECTIVES

You should be able to:

1 Determine meaningful replacements in fractional expressions.

2 Simplify fractional expressions.

3 Multiply or divide fractional expressions, and simplify.

4 Add or subtract fractional expressions, and simplify.

5 Simplify complex fractional expressions.

A number a is said to be a **square root** of c if $a^2 = c$. Thus, -3 is a square root of 9 because $(-3)^2 = 9$. Similarly, 3 is also a square root of 9 because $3^2 = 9$. A number a is said to be an ***n*th root** of c if $a^n = c$. For example, 5 is a third root (called a **cube root**) of 125 because $5^3 = 125$. The number 125 has no other real-number cube root. Any real number has only one real-number cube root.

The symbol $\sqrt{a}$ denotes the nonnegative square root of the number a. The symbol $\sqrt[3]{a}$ denotes the real-number cube root of a, and $\sqrt[n]{a}$ denotes the nth root of a, that is, a number whose nth power is a. The symbol $\sqrt[n]{\ }$ is called a **radical,** and the symbol under the radical is called the **radicand.** The number n (which is omitted when it is 2) is called the **index.** Examples of nth roots are

$$\sqrt[3]{125} = 5 \quad \text{and} \quad -\sqrt[4]{16} = -2.$$

Odd and Even Roots

Any positive real number has two square roots, one positive and one negative. The same is true for fourth roots, or roots of any even index. The positive root is called the **principal root.** When a radical such as $\sqrt{4}$ or $\sqrt[4]{18}$ is used, it is understood to represent the principal (nonnegative) root. To denote a nonpositive root, we use $-\sqrt{4}$, $-\sqrt[4]{18}$, and so on.

> DEFINITION
>
> **A radical expression $\sqrt[n]{a}$, where n is even, represents the *principal* (nonnegative) nth root of a. The nonpositive root is denoted $-\sqrt[n]{a}$.**

> CAUTION! Again, keep in mind that when the index is an even number E, then $\sqrt[E]{x}$ *never* represents a negative number. For example, $\sqrt{9}$ represents 3, and *not* -3!

1 Meaningful Replacements

Since negative numbers do not have even roots in the system of real numbers, any replacement that makes a radicand negative when the index is even is not meaningful. For example, $\sqrt{-1}$ and $\sqrt[4]{-8.37}$ do not represent real numbers.

Every real number, positive, negative, or zero, has just one cube root, and the same is true for any odd root. Thus, $\sqrt[n]{a}$, where n is odd, represents the only nth root of a. In this case, all real numbers are meaningful replacements in the radicand.

EXAMPLE 1 Determine whether 0 and 3 are meaningful replacements in $\sqrt{5x - 4}$.

Determine whether the numbers are meaningful replacements in the given expression.

1. $\sqrt{x-2}$; 5, −1

2. $\sqrt{x+3}$; −7, −3

3. $\sqrt{x^2}$; 18, $-\frac{1}{2}$

4. $\sqrt{x^2+1}$; −23, 47.2

Simplify.

5. $\sqrt{(x+2)^2}$

6. $\sqrt{x^2(y-2)^2}$

7. $\sqrt[4]{(x+2)^4}$

8. $\sqrt{x^2+8x+16}$

9. $\sqrt[3]{(-4xy)^3}$

Solution We substitute 0 for x in the radicand $5x - 4$:

$$5(0) - 4 = 0 - 4 = -4.$$

Since the radicand is negative, 0 is not a meaningful replacement. We substitute 3 for x in $5x - 4$:

$$5(3) - 4 = 15 - 4 = 11.$$

Since the radicand is not negative, 3 is a meaningful replacement.

DO EXERCISES 1–4.

2 Simplifying Radical Expressions

Consider the expression $\sqrt{(-3)^2}$. This is equivalent to $\sqrt{9}$, which simplifies to 3. Similarly, $\sqrt{3^2} = 3$. This illustrates an important general principle for simplifying radicals of even index.

THEOREM 11

For any radicand R, $\sqrt{R^2} = |R|$. Similarly, for any even index n, $\sqrt[n]{R^n} = |R|$.

EXAMPLES

2. $\sqrt{x^2} = |x|$
3. $\sqrt{x^2 - 2ax + a^2} = \sqrt{(x-a)^2} = |x - a|$
4. $\sqrt{x^2y^6} = \sqrt{(xy^3)^2} = |xy^3| = |y^2xy| = y^2|xy|$

If an index is odd, no absolute-value signs are necessary, because there is only one root and it has the same sign as the radicand.

THEOREM 12

For any radicand R and any odd index n, $\sqrt[n]{R^n} = R$.

EXAMPLE 5 Simplify: $\sqrt[3]{(15ab)^3}$.

Solution

$$\sqrt[3]{(15ab)^3} = 15ab$$

EXAMPLE 6 Simplify: $\sqrt[5]{(-3xy^2)^5}$.

Solution

$$\sqrt[5]{(-3xy^2)^5} = -3xy^2$$

DO EXERCISES 5–9.

The next property enables us to multiply radicals. We illustrate it with an example.

EXAMPLE 7 Compare $\sqrt{4} \cdot \sqrt{9}$ and $\sqrt{4 \cdot 9}$.

Solution

$\sqrt{4} \cdot \sqrt{9} = 2 \cdot 3 = 6$ and $\sqrt{4 \cdot 9} = \sqrt{36} = 6$, so $\sqrt{4} \cdot \sqrt{9} = \sqrt{4 \cdot 9}$

THEOREM 13

For any nonnegative real numbers a and b and any index n,

$$\sqrt[n]{a} \cdot \sqrt[n]{b} = \sqrt[n]{a \cdot b}.$$

EXAMPLES Multiply.

8. $\sqrt{3} \cdot \sqrt{5} = \sqrt{3 \cdot 5} = \sqrt{15}$
9. $\sqrt{x+2} \cdot \sqrt{x-2} = \sqrt{(x+2)(x-2)} = \sqrt{x^2-4}$
10. $\sqrt[3]{4} \cdot \sqrt[3]{5} = \sqrt[3]{4 \cdot 5} = \sqrt[3]{20}$

DO EXERCISES 10–12.

Theorem 13 also enables us to simplify radical expressions. The idea is to factor the radicand, obtaining factors that are perfect nth powers.

EXAMPLES Simplify.

11. $\sqrt{50} = \sqrt{25 \cdot 2} = \sqrt{25} \cdot \sqrt{2} = 5\sqrt{2}$
12. $\sqrt{5x^2} = \sqrt{x^2 \cdot 5} = \sqrt{x^2} \cdot \sqrt{5} = |x|\sqrt{5}$
13. $\sqrt[3]{32} = \sqrt[3]{8 \cdot 4} = \sqrt[3]{8} \cdot \sqrt[3]{4} = 2\sqrt[3]{4}$
14. $\sqrt{216x^5y^3} = \sqrt{36 \cdot 6 \cdot x^4 \cdot x \cdot y^2 \cdot y} = |6x^2y|\sqrt{6xy} = 6x^2|y|\sqrt{6xy}$
15. $\sqrt{2x^2-4x+2} = \sqrt{2(x-1)^2} = |x-1|\sqrt{2}$

DO EXERCISES 13–17.

The next property of radicals involves division.

THEOREM 14

For any nonnegative number a and any positive number b, and any index n,

$$\sqrt[n]{\frac{a}{b}} = \frac{\sqrt[n]{a}}{\sqrt[n]{b}}.$$

This property can be used to divide and to simplify radical expressions.

EXAMPLES Simplify.

16. $\sqrt{16x^3y^{-4}} = \sqrt{\dfrac{16x^3}{y^4}} = \dfrac{\sqrt{16x^3}}{\sqrt{y^4}} = \dfrac{\sqrt{16x^2 \cdot x}}{\sqrt{y^4}} = \dfrac{4|x|\sqrt{x}}{y^2}$
17. $\sqrt[3]{\dfrac{27y^5}{343x^3}} = \dfrac{\sqrt[3]{27y^5}}{\sqrt[3]{343x^3}} = \dfrac{\sqrt[3]{27y^3 \cdot y^2}}{\sqrt[3]{343x^3}} = \dfrac{3y\sqrt[3]{y^2}}{7x}$
18. $\dfrac{18\sqrt{72}}{6\sqrt{6}} = 3\sqrt{\dfrac{72}{6}} = 3\sqrt{12} = 3\sqrt{4 \cdot 3} = 3 \cdot 2\sqrt{3} = 6\sqrt{3}$
19. $\dfrac{\sqrt[3]{32}}{\sqrt[3]{2}} = \sqrt[3]{\dfrac{32}{2}} = \sqrt[3]{16} = \sqrt[3]{8 \cdot 2} = 2\sqrt[3]{2}$

DO EXERCISES 18–24.

Another fundamental principle of radicals involves an exponent under the radical. We illustrate with an example.

Multiply.

10. $\sqrt{19} \cdot \sqrt{7}$

11. $\sqrt{x+2y} \cdot \sqrt{x-2y}$

12. $\sqrt[4]{27} \cdot \sqrt[4]{3}$

Simplify.

13. $\sqrt{300}$

14. $\sqrt{36y^2}$

15. $\sqrt{2x^2+4x+2}$

16. $\sqrt[3]{16}$

17. $\sqrt[3]{(a+b)^4}$

Simplify.

18. $\sqrt{\dfrac{49}{64}}$

19. $\sqrt{\dfrac{25}{y^2}}$

20. $\sqrt[4]{\dfrac{16}{81}}$

21. $\sqrt[3]{\dfrac{7}{125}}$

22. $\dfrac{\sqrt{75}}{\sqrt{3}}$

23. $\dfrac{\sqrt{2x^3}}{\sqrt{50x}}$

24. $\dfrac{\sqrt[3]{24x^3y}}{\sqrt[3]{3y^4}}$

Simplify.

25. $\sqrt[3]{27^{10}}$

26. $(\sqrt{3})^8$

Simplify.

27. $7\sqrt{5} + 3\sqrt{5} - 8\sqrt{20}$

28. $5\sqrt[3]{16y^4} + 7\sqrt[3]{2y}$

29. $(\sqrt{3} - 5\sqrt{2})(2\sqrt{3} + \sqrt{2})$

EXAMPLE 20 Compare $\sqrt{3^4}$ and $(\sqrt{3})^4$.

Solution We have

$\sqrt{3^4} = \sqrt{(3^2)^2} = 3^2 = 9$ and $(\sqrt{3})^4 = \sqrt{3} \cdot \sqrt{3} \cdot \sqrt{3} \cdot \sqrt{3} = 3 \cdot 3 = 9$.

Thus, $\sqrt{3^4} = (\sqrt{3})^4$.

The general principle is given in Theorem 15.

> **THEOREM 15**
>
> **For any nonnegative number a and any index n and any natural number m,**
>
> $$\sqrt[n]{a^m} = (\sqrt[n]{a})^m.$$

Theorem 15 sometimes facilitates radical simplification.

EXAMPLES Simplify.

21. $\sqrt[3]{8^5} = (\sqrt[3]{8})^5 = 2^5 = 32$

22. $(\sqrt{2})^6 = \sqrt{2^6} = \sqrt{(2^3)^2} = 2^3 = 8$

DO EXERCISES 25 AND 26.

Various calculations with radicals can be carried out using the properties of radicals and the properties of numbers, such as the distributive property. The following examples illustrate.

EXAMPLES Simplify.

23. $3\sqrt{8} - 5\sqrt{2} = 3\sqrt{4 \cdot 2} - 5\sqrt{2} = 3 \cdot 2\sqrt{2} - 5\sqrt{2} = 6\sqrt{2} - 5\sqrt{2}$
$= (6 - 5)\sqrt{2}$ **Here we use a distributive law.**
$= \sqrt{2}$

24. $(4\sqrt{3} + \sqrt{2})(\sqrt{3} - 5\sqrt{2}) = 4(\sqrt{3})^2 - 20\sqrt{3}\sqrt{2} + \sqrt{2}\sqrt{3} - 5(\sqrt{2})^2$
$= 4 \cdot 3 - 20\sqrt{6} + \sqrt{6} - 5 \cdot 2$
$= 12 - 19\sqrt{6} - 10$
$= 2 - 19\sqrt{6}$

DO EXERCISES 27–29.

3 Applications

EXAMPLE 25 *Speed of a skidding car.* How do police determine the speed of a car that has skidded? The formula

$$r = 2\sqrt{5L}$$

can be used to approximate the speed r, in miles per hour, of a car that has left a skid mark of length L, in feet. What was the speed of a car that left skid marks 306 ft long?

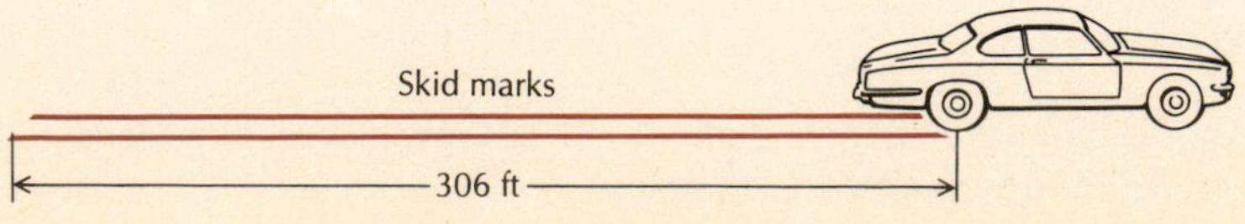

Solution We substitute 306 for L in the formula $r = 2\sqrt{5L}$ and use a calculator to approximate r:

$$r = 2\sqrt{5(306)} = 2\sqrt{1530}$$
$$\approx 2(39.115) \quad \text{Using a calculator}$$
$$= 78.23.$$

The speed of the car was about 78.23 mph.

DO EXERCISE 30.

30. What was the speed of a car that left skid marks 70 ft long?

EXAMPLE 26 *The Pythagorean theorem.* By the **Pythagorean theorem,** in right triangles, $c^2 = a^2 + b^2$ and $c = \sqrt{a^2 + b^2}$, where a and b are the lengths of the legs and c is the length of the hypotenuse. A surveyor is trying to measure a certain distance PR across a pond. Poles are placed at points P, Q, and R. The distances that the surveyor was able to measure are shown in the figure. What is the approximate distance from P to R?

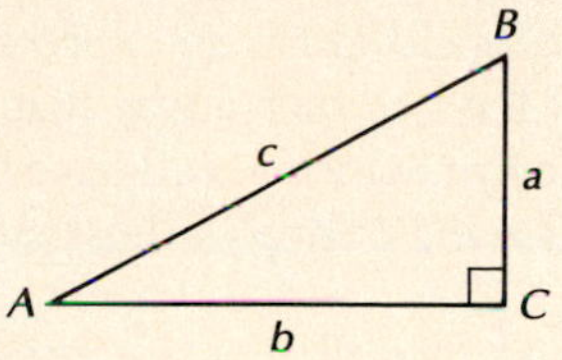

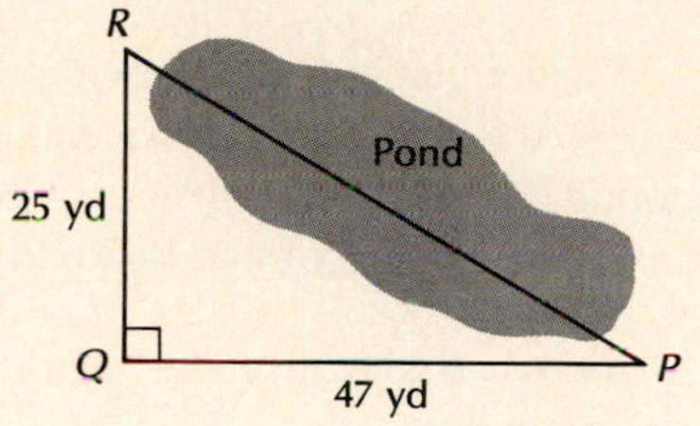

Solution We have the lengths of two of the legs given. We find PR, the hypotenuse, as follows:

$$PR^2 = RQ^2 + PQ^2$$

so

$$PR = \sqrt{RQ^2 + PQ^2}$$
$$PR = \sqrt{25^2 + 47^2}$$
$$PR = \sqrt{625 + 2209}$$
$$= \sqrt{2834} \approx 53.24.$$

Thus the distance across the pond is about 53.24 yd.

DO EXERCISE 31.

31. How long must a wire be to reach from the top of a 14-m telephone pole to a point on the ground 7 m from the foot of the pole?

4 Rationalizing Denominators or Numerators

Fractional expressions are often considered simpler when the denominator is free of radicals. Thus in simplifying, we generally remove the radicals in a denominator. This is called **rationalizing the denominator,** and it can be done by multiplying by 1 in such a way as to obtain a perfect power in the denominator. On occasion, we prefer to rationalize the numerator. This is often the case in calculus. In either case, we can accomplish the rationalization by multiplying by 1, as in the following examples.

EXAMPLES Simplify.

27. $\sqrt{\frac{1}{2}} = \sqrt{\frac{1}{2} \cdot \frac{2}{2}} = \sqrt{\frac{2}{4}} = \frac{\sqrt{2}}{\sqrt{4}} = \frac{\sqrt{2}}{2}$ We multiply by $\frac{2}{2}$ so that we have a perfect square in the denominator.

28. $\sqrt[3]{\frac{7}{9}} = \sqrt[3]{\frac{7}{9} \cdot \frac{3}{3}} = \sqrt[3]{\frac{21}{27}} = \frac{\sqrt[3]{21}}{\sqrt[3]{27}} = \frac{\sqrt[3]{21}}{3}$ We multiply by $\frac{3}{3}$ so that we have a perfect cube in the denominator.

Rationalize the denominator. Assume that all letters represent positive numbers.

32. $\dfrac{1}{\sqrt{3}-\sqrt{5}}$

33. $\dfrac{\sqrt{x}-5}{\sqrt{x}+2}$

29. $\dfrac{\sqrt{7}}{\sqrt{5}} = \dfrac{\sqrt{7}}{\sqrt{5}} \cdot \dfrac{\sqrt{5}}{\sqrt{5}} = \dfrac{\sqrt{35}}{\sqrt{25}} = \dfrac{\sqrt{35}}{5}$

30. $\dfrac{\sqrt{2a}}{\sqrt{5b}} = \dfrac{\sqrt{2a}}{\sqrt{5b}} \cdot \dfrac{\sqrt{5b}}{\sqrt{5b}} = \dfrac{\sqrt{10ab}}{\sqrt{(5b)^2}}$

$= \dfrac{\sqrt{10ab}}{|5b|} = \dfrac{\sqrt{10ab}}{5b}$

The absolute-value sign in the denominator is not necessary since $\sqrt{5b}$ would not exist at the outset unless $b > 0$.

31. $\dfrac{\sqrt[3]{54x^3}}{\sqrt[3]{4y^5}} = \sqrt[3]{\dfrac{54x^3}{4y^5} \cdot \dfrac{2y}{2y}}$

$= \sqrt[3]{\dfrac{27x^3 \cdot 4y}{8y^6}}$

$= \dfrac{\sqrt[3]{27x^3} \cdot \sqrt[3]{4y}}{\sqrt[3]{8y^6}}$

$= \dfrac{3x \cdot \sqrt[3]{4y}}{2y^2}$

When a numerator or a denominator to be rationalized has an expression like $a + \sqrt{b}$ or $\sqrt{c} - \sqrt{d}$, we choose the symbol for 1 in such a way that the denominator becomes a difference of squares. The symbol for 1 will have two terms in its numerator and denominator. The following examples illustrate.

EXAMPLES Rationalize the denominator. Assume that all letters represent positive numbers.

32. $\dfrac{1}{\sqrt{2}+\sqrt{3}} = \dfrac{1}{\sqrt{2}+\sqrt{3}} \cdot \dfrac{\sqrt{2}-\sqrt{3}}{\sqrt{2}-\sqrt{3}}$

The number $\sqrt{2} - \sqrt{3}$ is called the *conjugate* of $\sqrt{2} + \sqrt{3}$. It is found by changing the middle sign. We use the conjugate to form the symbol for 1.

$= \dfrac{\sqrt{2}-\sqrt{3}}{(\sqrt{2}+\sqrt{3})(\sqrt{2}-\sqrt{3})}$

$= \dfrac{\sqrt{2}-\sqrt{3}}{(\sqrt{2})^2-(\sqrt{3})^2}$

$= \dfrac{\sqrt{2}-\sqrt{3}}{2-3} = \dfrac{\sqrt{2}-\sqrt{3}}{-1}$

$= \sqrt{3} - \sqrt{2}$

33. $\dfrac{\sqrt{x}+\sqrt{y}}{\sqrt{x}-\sqrt{y}} = \dfrac{\sqrt{x}+\sqrt{y}}{\sqrt{x}-\sqrt{y}} \cdot \dfrac{\sqrt{x}+\sqrt{y}}{\sqrt{x}+\sqrt{y}}$

The conjugate of $\sqrt{x} - \sqrt{y}$ is $\sqrt{x} + \sqrt{y}$.

$= \dfrac{(\sqrt{x}+\sqrt{y})^2}{(\sqrt{x})^2-(\sqrt{y})^2}$

$= \dfrac{x + 2\sqrt{xy} + y}{x - y}$

EXAMPLES Rationalize the numerator. Assume that all letters represent positive numbers and that all radicands are positive.

34. $\dfrac{1-\sqrt{2}}{5} = \dfrac{1-\sqrt{2}}{5} \cdot \dfrac{1+\sqrt{2}}{1+\sqrt{2}}$

The conjugate of $1 - \sqrt{2}$ is $1 + \sqrt{2}$.

$= \dfrac{(1-\sqrt{2})(1+\sqrt{2})}{5(1+\sqrt{2})} = \dfrac{1-2}{5(1+\sqrt{2})}$

$= \dfrac{-1}{5+5\sqrt{2}}$

35. $$\frac{\sqrt{x+h}-\sqrt{x}}{h} = \frac{\sqrt{x+h}-\sqrt{x}}{h}\cdot\frac{\sqrt{x+h}+\sqrt{x}}{\sqrt{x+h}+\sqrt{x}}$$
$$= \frac{(x+h)-x}{h(\sqrt{x+h}+\sqrt{x})} = \frac{h}{h(\sqrt{x+h}+\sqrt{x})}$$
$$= \frac{1}{\sqrt{x+h}+\sqrt{x}}$$

The simplification in Example 35 is often needed in calculus.

DO EXERCISES 32–35. (EXERCISES 32 AND 33 ARE ON THE PRECEDING PAGE.)

Rationalize the numerator. Assume that all letters represent positive numbers.

34. $\dfrac{\sqrt{a+2}-\sqrt{a}}{2}$

35. $\dfrac{\sqrt{x}-\sqrt{5}}{\sqrt{x}+\sqrt{5}}$

EXERCISE SET 1.7

1 Determine whether the given numbers are sensible replacements in the expression.

1. $\sqrt{x-3}$; $-2, 5$
2. $\sqrt{2x-5}$; $3, 2$
3. $\sqrt{3-4x}$; $-1, 1$
4. $\sqrt{x^2+3}$; $0, 4.3$
5. $\sqrt{1-x^2}$; $1, 3$
6. $\sqrt{x^2+2x+1}$; $-3, 4$
7. $\sqrt[3]{2x+7}$; $-4, 5$
8. $\sqrt[4]{3-5x}$; $1, 2$

2 Simplify.

9. $\sqrt{(-11)^2}$
10. $\sqrt{(-1)^2}$
11. $\sqrt{16x^2}$
12. $\sqrt{36t^2}$
13. $\sqrt{(b+1)^2}$
14. $\sqrt{(2c-3)^2}$
15. $\sqrt[3]{-27x^3}$
16. $\sqrt[3]{-8y^3}$
17. $\sqrt{x^2-4x+4}$
18. $\sqrt{y^2+16y+64}$
19. $\sqrt[5]{32}$
20. $\sqrt[5]{-32}$
21. $\sqrt{180}$
22. $\sqrt{48}$
23. $\sqrt[3]{54}$
24. $\sqrt[3]{135}$
25. $\sqrt{128c^2d^4}$
26. $\sqrt{162c^4d^6}$
27. $\sqrt{3}\cdot\sqrt{6}$
28. $\sqrt{6}\cdot\sqrt{8}$

Simplify, assuming that all letters represent positive numbers and that all radicands are positive. Thus no absolute-value signs will be needed.

29. $\sqrt{2x^3y}\sqrt{12xy}$
30. $\sqrt{3y^4z}\sqrt{20z}$
31. $\sqrt[3]{3x^2y}\sqrt[3]{36x}$
32. $\sqrt[5]{8x^3y^4}\sqrt[5]{4x^4y}$
33. $\sqrt[3]{2(x+4)}\sqrt[3]{4(x+4)^4}$
34. $\sqrt[3]{4(x+1)^2}\sqrt[3]{18(x+1)^2}$
35. $\dfrac{\sqrt{21ab^2}}{\sqrt{3ab}}$
36. $\dfrac{\sqrt{128ab^2}}{\sqrt{16a^2b}}$
37. $\dfrac{\sqrt[3]{40m}}{\sqrt[3]{5m}}$
38. $\dfrac{\sqrt{40xy}}{\sqrt{8x}}$
39. $\dfrac{\sqrt[3]{3x^2}}{\sqrt[3]{24x^5}}$
40. $\dfrac{\sqrt[3]{40xy^3}}{\sqrt[3]{8x}}$
41. $\dfrac{\sqrt{a^2-b^2}}{\sqrt{a-b}}$
42. $\dfrac{\sqrt{x^3-y^3}}{\sqrt{x-y}}$
43. $\sqrt{\dfrac{9a^2}{8b}}$
44. $\sqrt{\dfrac{5b^2}{12a}}$
45. $\sqrt[3]{\dfrac{2x^2 2y^3}{25z^4}}$
46. $\sqrt[3]{\dfrac{24x^3y}{3y^4}}$
47. $\dfrac{(\sqrt[3]{32x^4y})^2}{(\sqrt[3]{xy})^2}$
48. $\dfrac{(\sqrt[3]{16x^2y})^2}{(\sqrt[3]{xy})^2}$
49. $\dfrac{3\sqrt{a^2b^2}\sqrt{4xy}}{2\sqrt{a^{-1}b^{-2}}\sqrt{9x^{-3}y^{-1}}}$
50. $\dfrac{4\sqrt{xy^2}\sqrt{9ab}}{3\sqrt{x^{-1}y^{-2}}\sqrt{16a^{-5}b^{-1}}}$
51. $9\sqrt{50}+6\sqrt{2}$
52. $11\sqrt{27}-4\sqrt{3}$
53. $8\sqrt{2}-6\sqrt{20}-5\sqrt{8}$
54. $\sqrt{12}-\sqrt{27}+\sqrt{75}$
55. $2\sqrt[3]{8x^2}+5\sqrt[3]{27x^2}-3\sqrt[3]{x^3}$
56. $5a\sqrt{(a+b)^3}-2ab\sqrt{a+b}-3b\sqrt{(a+b)^3}$
57. $3\sqrt{3y^2}-\dfrac{y\sqrt{48}}{\sqrt{2}}+\sqrt{\dfrac{12}{4y^{-2}}}$
58. $\sqrt[3]{x^5}-\dfrac{2\sqrt[3]{x}}{\sqrt[3]{x^{-1}}}+\sqrt[3]{\dfrac{8}{x^{-5}}}$
59. $(\sqrt{3}-\sqrt{2})(\sqrt{3}+\sqrt{2})$
60. $(\sqrt{8}+2\sqrt{5})(\sqrt{8}-2\sqrt{5})$
61. $(1+\sqrt{3})^2$
62. $(\sqrt{2}-5)^2$
63. $(\sqrt{t}-x)^2$
64. $\left(\sqrt{a}+\dfrac{1}{\sqrt{a}}\right)^2$
65. $5\sqrt{7}+\dfrac{35}{\sqrt{7}}$
66. $(\sqrt{a^2b}+3\sqrt{y})(2a\sqrt{b}-\sqrt{y})$
67. $(\sqrt{x+3}-\sqrt{3})(\sqrt{x+3}+\sqrt{3})$
68. $(\sqrt{x+h}-\sqrt{x})(\sqrt{x+h}+\sqrt{x})$

3

69. In Example 25, what was the speed of a car that left skid marks 90 ft long?

70. In Example 25, what was the speed of a car that left skid marks 110 ft long?

71. ▦ *Pendulums.* The period T of a pendulum is the time it takes to make a move from one side to the other and back. A formula for the period is

$$T = 2\pi\sqrt{\frac{L}{32}},$$

where T is in seconds and L, the length of the pendulum, is in feet. Find the periods of pendulums of lengths 2 ft, 8 ft, 64 ft, and 100 ft. Use 3.14 for π.

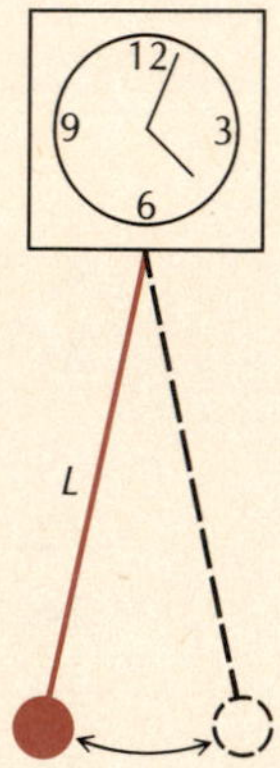

72. A slow-pitch softball diamond is actually a square 65 ft on a side. How far is it from home to second base?

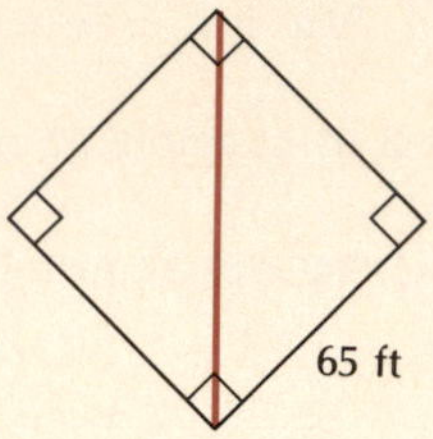

73. An airplane is flying at an altitude of 3700 ft. The slanted distance directly to the airport is 14,200 ft. How far horizontally is the airplane from the airport?

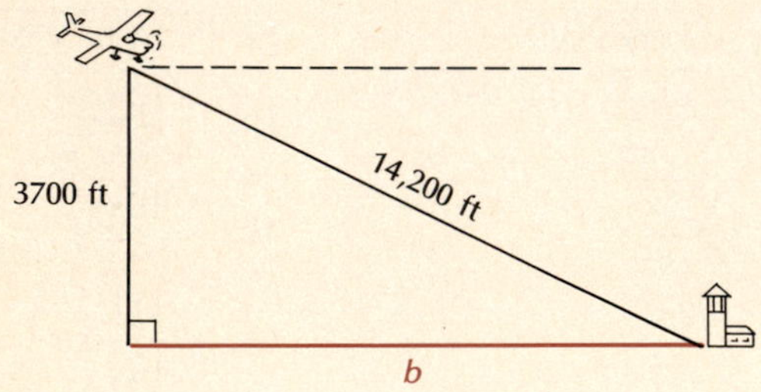

74. During the summer heat, a 2-mi bridge expands 2 ft in length. Assuming that the bulge occurs straight up the middle, estimate the height of the bulge. (The answer may surprise you. In reality, bridges are built with expansion joints to control such buckling.)

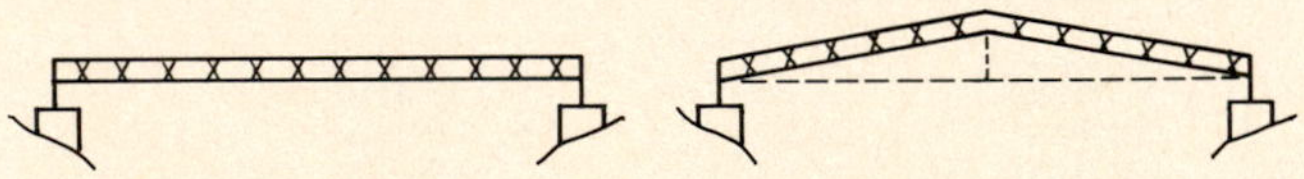

An *equilateral* triangle is shown at the right.

75. Find an expression for its height h in terms of a.

76. Find an expression for its area A in terms of a.

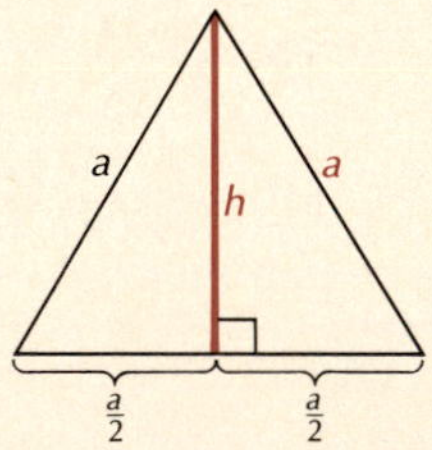

77. Figure $ABCD$ is a square. Find the length of $\overline{AC}$.

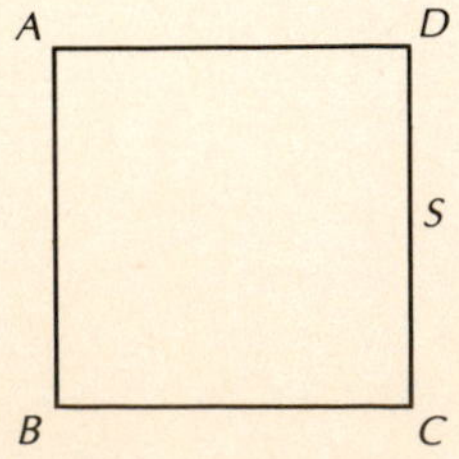

78. An isosceles right triangle has two sides of length s. Find a formula for the length of the third side.

79. The diagonal of a square has length $8\sqrt{2}$. Find the length of a side of the square.

80. The area of square $PQRS$ is 100 ft^2 (square feet), and A, B, C, and D are the midpoints of the sides. Find the area of square $ABCD$.

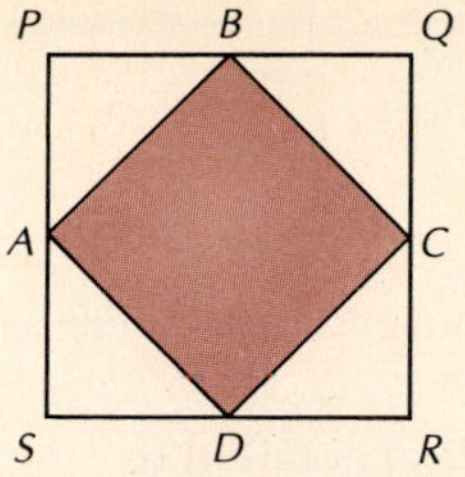

4 Rationalize the denominator. Assume that all letters represent positive numbers and that all radicands are positive.

81. $\dfrac{6}{3+\sqrt{5}}$

82. $\dfrac{2}{\sqrt{3}-1}$

83. $\sqrt[3]{\dfrac{16}{9}}$

84. $\dfrac{\sqrt[3]{3}}{\sqrt[3]{6}}$

85. $\dfrac{4\sqrt{x}-3\sqrt{xy}}{2\sqrt{x}+5\sqrt{y}}$

86. $\dfrac{5\sqrt{x}+2\sqrt{xy}}{3\sqrt{x}-2\sqrt{y}}$

Rationalize the numerator. Assume that all letters represent positive numbers and that all radicands are positive.

87. $\dfrac{\sqrt{2}+\sqrt{5a}}{6}$

88. $\dfrac{\sqrt{3}+\sqrt{5y}}{4}$

89. $\dfrac{\sqrt{x+1}+1}{\sqrt{x+1}-1}$

90. $\dfrac{\sqrt{x+4}-2}{\sqrt{x+4}+2}$

91. $\dfrac{\sqrt{a+3}-\sqrt{3}}{3}$

92. $\dfrac{\sqrt{a+h}-\sqrt{a}}{h}$

SYNTHESIS

Simplify, assuming that all letters represent positive numbers.

93. $\sqrt{8.2x^3y}\sqrt{12.5xy}$

94. $\sqrt{0.012y^4z}\sqrt{1.305z}$

95. $\sqrt{\dfrac{6.03a^2}{17.13b}}$

96. $\sqrt{\dfrac{3.2b^2}{82.1a}}$

Simplify.

97. $\sqrt{1+x^2}+\dfrac{1}{\sqrt{1+x^2}}$

98. $\sqrt{1-x^2}-\dfrac{x^2}{2\sqrt{1-x^2}}$

99. Show that $\sqrt{a+b}=\sqrt{a}+\sqrt{b}$ is false for positive real numbers a and b by finding two positive numbers a and b for which $\sqrt{a+b}\neq\sqrt{a}+\sqrt{b}$.

100. Show that $(\sqrt{5+\sqrt{24}})^2=(\sqrt{2}+\sqrt{3})^2$.

101. *Water flow from a faucet.* As water flows at a velocity v_0 from a faucet with internal diameter d, the width w of the stream at a distance h below the outlet is given by

$$w = d\sqrt{\frac{v_0}{\sqrt{v_0^2+19.6h}}}.$$

Find the width of a stream 0.1 m below a faucet of internal diameter 0.03 m if the water is flowing at a rate of 0.6 m/sec.

102. Select the smallest positive number from the following list:

$$10\sqrt{26}-51,\quad 51-10\sqrt{26},\quad 18-5\sqrt{13},$$
$$3\sqrt{11}-10,\quad 10-3\sqrt{11}.$$

103. *Escape velocity.* The *escape velocity* V_0 of a projectile is the initial velocity needed in order for it to escape the gravitational pull of the planet. Escape velocity is given in meters per second by

$$V_0 = \sqrt{\frac{2GM}{R}},$$

where G is the universal gravitational constant given by $G = 6.672 \times 10^{-11}$ N-m^2/kg^2, M is the mass of the planet, in kilograms, and R is its radius, in meters, where N is newtons in kg-m/sec^2.

a) The mass of the earth is 5.97×10^{24} kg, and its radius is 6.37×10^6 m. What velocity is necessary for a rocket to escape the gravitational pull of the earth?

b) The mass of Mars is 6.27×10^{23} kg, and its radius is 3.3917×10^6 m. What velocity is necessary for a rocket to escape the gravitational pull of Mars?

OBJECTIVES

You should be able to:

1. Convert from exponential notation to radical notation.
2. Convert from radical notation to exponential notation.
3. Simplify expressions using the properties of rational exponents and the arithmetic of rational numbers.
4. Simplify certain expressions to expressions containing a single radical.
5. Factor and simplify expressions involving negative and fractional exponents.

Convert to radical notation and simplify.

1. $n^{3/2}$

2. $y^{-6/7}$

3. $32^{4/5}$

4. $64^{-2/3}$

Convert to exponential notation and simplify.

5. $\sqrt[3]{(5ab)^4}$

6. $\sqrt[4]{16^3}$

7. $\sqrt[6]{a^4}$

8. $\sqrt{\sqrt[3]{4}}$

9. $\sqrt{5^3}\sqrt[3]{5}$

1.8 Rational Exponents

We are motivated to define *fractional exponents* so that the same rules, or laws, hold for them as for integer exponents. For example, if the laws of exponents are to hold, we must have

$$a^{1/2} \cdot a^{1/2} = a^{1/2 + 1/2} = a^1 = a.$$

Thus we are led to define $a^{1/2}$ to mean $\sqrt{a}$. Similarly, $a^{1/n}$ would mean $\sqrt[n]{a}$. Again, if the usual laws of exponents are to hold, we must have

$$(a^{1/n})^m = (a^m)^{1/n} = a^{m/n}.$$

Thus we are led to define $a^{m/n}$ to mean $(\sqrt[n]{a})^m$, or $\sqrt[n]{a^m}$.

DEFINITION

An expression $a^{m/n}$, where a is positive and m and n are natural numbers, is defined to mean $(\sqrt[n]{a})^m$. An expression $a^{-m/n}$ is defined to mean $1/a^{m/n}$.

Note that in this definition we require a to be positive. Thus in manipulations with fractional exponents we assume that all letters represent positive numbers and that all radicands are positive. No absolute-value signs need be used.

Once the definition of rational exponents is made, the question arises whether the usual laws of exponents actually do hold. We will not prove it here, but the answer is that they do. Thus we can simplify or otherwise manipulate expressions containing rational exponents using those laws and the usual arithmetic of rational numbers.

1 Converting to Radical Notation

EXAMPLES Convert to radical notation and simplify, if possible.

1. $m^{2/3} = \sqrt[3]{m^2}$
2. $t^{-1/2} = \dfrac{1}{t^{1/2}} = \dfrac{1}{\sqrt{t}}$, or $\dfrac{\sqrt{t}}{t}$
3. $64^{5/2} = (64^{1/2})^5 = (\sqrt{64})^5 = 8^5 = 32{,}768$
4. $16^{-3/4} = \dfrac{1}{16^{3/4}} = \dfrac{1}{(16^{1/4})^3} = \dfrac{1}{(\sqrt[4]{16})^3} = \dfrac{1}{2^3} = \dfrac{1}{8}$

DO EXERCISES 1–4.

2 Converting to Exponential Notation

EXAMPLES Convert to exponential notation and simplify.

5. $(\sqrt[4]{7xy})^5 = (7xy)^{5/4}$
6. $\sqrt[3]{8^4} = 8^{4/3} = (8^{1/3})^4 = 2^4 = 16$
7. $\sqrt[6]{x^3} = x^{3/6} = x^{1/2}$, or $\sqrt{x}$
8. $\sqrt[6]{4} = 4^{1/6} = (2^2)^{1/6} = 2^{2/6} = 2^{1/3}$, or $\sqrt[3]{2}$
9. $\sqrt[3]{\sqrt{7}} = \sqrt[3]{7^{1/2}} = (7^{1/2})^{1/3} = 7^{1/6}$, or $\sqrt[6]{7}$
10. $\sqrt{6^3}\sqrt[3]{6} = 6^{1/2} \cdot 6^{1/3} = 6^{1/2 + 1/3} = 6^{5/6}$, or $\sqrt[6]{6^5}$

DO EXERCISES 5–9 ON THE PRECEDING PAGE.

3 Simplifying Expressions with Rational Exponents

EXAMPLES Simplify and then write radical notation.

11. $x^{5/6} \cdot x^{2/3} = x^{5/6 + 2/3}$ Adding exponents

$= x^{9/6} = x^{3/2} = \sqrt{x^3}$

$= x\sqrt{x}$

12. $(a^5)^{-2/3} = a^{-10/3}$ Multiplying exponents

$= \dfrac{1}{a^{10/3}} = \dfrac{1}{\sqrt[3]{a^{10}}}$

$= \dfrac{1}{\sqrt[3]{a^9 \cdot a}} = \dfrac{1}{a^3 \cdot \sqrt[3]{a}}$

13. $(5^{1/3} - 5^{-5/3}) \cdot 5^{1/3} = 5^{1/3} \cdot 5^{1/3} - 5^{-5/3} \cdot 5^{1/3}$ Using a distributive law

$= 5^{2/3} - 5^{-4/3} = \sqrt[3]{5^2} - \dfrac{1}{\sqrt[3]{5^4}}$

$= \sqrt[3]{25} - \dfrac{1}{5\sqrt[3]{5}} = \sqrt[3]{25} - \dfrac{1}{5\sqrt[3]{5}} \cdot \dfrac{\sqrt[3]{25}}{\sqrt[3]{25}}$

$= \sqrt[3]{25} - \dfrac{\sqrt[3]{25}}{25} = \dfrac{24}{25}\sqrt[3]{25}$

DO EXERCISES 10–12.

Simplify and then write radical notation.

10. $a^{3/4} \cdot a^{1/2}$

11. $(x^{-3})^{2/5}$

12. $(2^{1/4} + 2^{-3/4}) \cdot 2^{1/2}$

4 Writing Single Radicals

In certain expressions containing radicals or fractional exponents, it is possible to simplify in such a way that there is a single radical.

EXAMPLES Write an expression containing a single radical.

14. $a^{1/2}b^{-1/2}c^{5/6} = a^{3/6}b^{-3/6}c^{5/6}$

$= (a^3b^{-3}c^5)^{1/6} = \sqrt[6]{a^3b^{-3}c^5}$

15. $\dfrac{a^{1/4}b^{3/8}}{a^{1/2}b^{1/8}} = a^{-1/4}b^{1/4} = (a^{-1}b)^{1/4} = \sqrt[4]{a^{-1}b} = \sqrt[4]{\dfrac{b}{a}}$

16. $\sqrt[4]{7}\sqrt{3} = 7^{1/4} \cdot 3^{1/2} = 7^{1/4} \cdot 3^{2/4}$

$= (7 \cdot 3^2)^{1/4} = \sqrt[4]{63}$

17. $\dfrac{\sqrt[4]{(x+2)^3}\sqrt[5]{x+2}}{\sqrt{x+2}} = \dfrac{(x+2)^{3/4}(x+2)^{1/5}}{(x+2)^{1/2}}$

$= (x+2)^{3/4 + 1/5 - 1/2}$

$= (x+2)^{9/20} = \sqrt[20]{(x+2)^9}$

DO EXERCISES 13–15.

Write an expression containing a single radical.

13. $\sqrt[3]{5} \cdot \sqrt{2}$

14. $x^{2/3}y^{1/2}z^{5/6}$

15. $\dfrac{\sqrt[4]{(x+y)^3}}{\sqrt{x+y}}$

EXAMPLE 18 *Road pavement signs.* In a psychological study, pavement signs were found to be most readable by a driver when the letters in the sign are of length L, given by

$$L = \frac{0.000169d^{2.27}}{h},$$

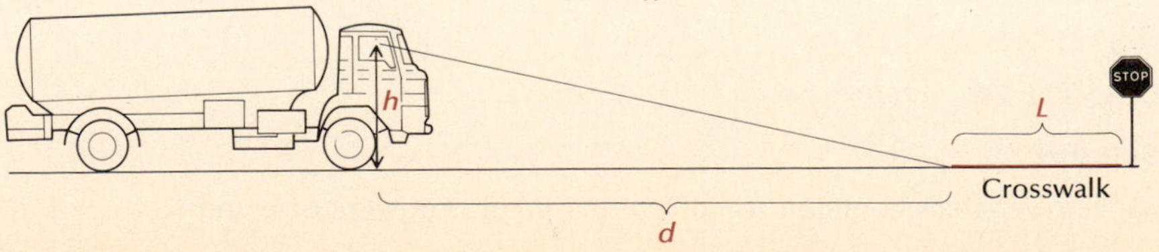

16. Find the length L of the letters in a road pavement sign when $h = 5$ ft and $d = 205$ ft.

where d is the distance from the car to the lettering and h is the height of the eye above the road. All units are in feet. Find L from a view from the window of a truck for which $h = 7$ ft and $d = 175$ ft.

Solution We substitute 7 for h and 175 for d in the formula for L, and calculate its value using a calculator with a y^x key:

$$L = \frac{0.000169d^{2.27}}{h} = \frac{0.000169(175)^{2.27}}{7} \approx 2.98 \text{ ft.}$$

DO EXERCISE 16.

5 Factoring Expressions with Rational Exponents in Calculus

In calculus, it is often important to be able to factor expressions involving rational exponents. Before considering such factoring, let us look again at the kind of factoring we have done in Section 1.5. Consider the expression

$$a^2b^6 + a^4b^3.$$

After factoring, we have

$$a^2b^3(b^3 + a^2).$$

We can use the following procedure to do the factoring:

a) We decide which constants and variables are common in the terms. In this case, there are no common constants (other than -1 and 1), but the common variables are powers of a and b.

b) The largest common factor involving a is the smallest exponent of the a-factors. The largest common factor involving b is the smallest exponent of the b-factors. Thus, a^2 and b^3 make up the largest common factor.

c) Determine the factors that remain in each term:

$$a^2b^3 \text{ factored out of } a^2b^6 \text{ leaves } a^{2-2}b^{6-3} = a^0b^3 = b^3;$$
$$a^2b^3 \text{ factored out of } a^4b^3 \text{ leaves } a^{4-2}b^{3-3} = a^2b^0 = a^2.$$

d) Thus, $a^2b^6 + a^4b^3 = a^2b^3 \cdot b^3 + a^2b^3 \cdot a^2 = a^2b^3(b^3 + a^2)$.

We just factored an expression involving nonnegative exponents, but the process does not change with negative exponents.

EXAMPLE 19 Factor: $a^{-3}b^5 + a^4b^{-2}$.

Solution

a) The variables common to both terms involve powers of a and b.

b) a^{-3} has a smaller exponent than a^4.
b^{-2} has a smaller exponent than b^5.

c) $a^{-3}b^{-2}$ factored out of $a^{-3}b^5$ leaves $a^{-3-(-3)}b^{5-(-2)} = a^0b^7 = b^7$.
$a^{-3}b^{-2}$ factored out of a^4b^{-2} leaves $a^{4-(-3)}b^{-2-(-2)} = a^7b^0 = a^7$.

d) Thus,

$$a^{-3}b^5 + a^4b^{-2} = a^{-3}b^{-2}(b^7 + a^7) = \frac{b^7 + a^7}{a^3b^2}.$$

EXAMPLE 20 Factor: $3a^{1/2}b^{-3/4} - a^{-1/2}b^{1/4}$.

Solution

a) The variables common to both terms involve powers of a and b.

b) $a^{-1/2}$ has a smaller exponent than $a^{1/2}$.
$b^{-3/4}$ has a smaller exponent than $b^{1/4}$.

c) $a^{-1/2}b^{-3/4}$ factored out of $3a^{1/2}b^{-3/4}$ leaves
$3a^{1/2-(-1/2)}b^{-3/4-(-3/4)} = 3a^1b^0 = 3a$.

$a^{-1/2}b^{-3/4}$ factored out of $a^{-1/2}b^{1/4}$ leaves
$a^{-1/2-(-1/2)}b^{1/4-(-3/4)} = a^0b^1 = b$.

d) Thus,

$$3a^{1/2}b^{-3/4} - a^{-1/2}b^{1/4} = a^{-1/2}b^{-3/4}(3a - b) = \frac{3a - b}{a^{1/2}b^{3/4}}.$$

DO EXERCISES 17 AND 18.

EXAMPLE 21 Factor and simplify:

$$\frac{3x^2(2x-1)^{1/2} - x^3(\frac{1}{2})(2x-1)^{-1/2}(2)}{[(2x-1)^{1/2}]^2}.$$

Solution

$$\frac{3x^2(2x-1)^{1/2} - x^3\left(\frac{1}{2}\right)(2x-1)^{-1/2}(2)}{[(2x-1)^{1/2}]^2}$$

$$= \frac{3x^2(2x-1)^{1/2} - x^3(2x-1)^{-1/2}}{2x-1} \quad \text{Simplifying}$$

$$= \frac{x^2(2x-1)^{-1/2}[3(2x-1) - x]}{2x-1} \quad \text{Factoring the numerator}$$

$$= \frac{x^2(2x-1)^{-1/2}[6x - 3 - x]}{2x-1}$$

$$= \frac{x^2(5x-3)}{(2x-1)^{1/2}(2x-1)}$$

$$= \frac{x^2(5x-3)}{(2x-1)^{3/2}}$$

DO EXERCISE 19.

Factor and simplify.

17. $p^8q^{-5} - p^{-6}q^7$

18. $5x^{2/3}y^{-1/4} + 4x^{-1/3}y^{1/2}$

19. Factor and simplify.

$$\frac{3x^2(2x+5)^{1/2} - x^3\left(\frac{1}{2}\right)(2x+5)^{-1/2}(2)}{[(2x+5)^{1/2}]^2}.$$

EXERCISE SET 1.8

1 Convert to radical notation and simplify.

1. $x^{3/4}$
2. $y^{2/5}$
3. $16^{3/4}$
4. $4^{7/2}$
5. $125^{-1/3}$
6. $32^{-4/5}$
7. $a^{5/4}b^{-3/4}$
8. $x^{2/5}y^{-1/5}$

2 Convert to exponential notation and simplify.

9. $\sqrt[3]{20^2}$
10. $\sqrt[5]{17^3}$
11. $(\sqrt[4]{13})^5$
12. $(\sqrt[5]{12})^4$
13. $\sqrt[3]{\sqrt{11}}$
14. $\sqrt[3]{\sqrt[4]{7}}$
15. $\sqrt{5}\sqrt[3]{5}$
16. $\sqrt[3]{2}\sqrt{2}$
17. $\sqrt[5]{32^2}$
18. $\sqrt[3]{64^{-2}}$
19. $\sqrt[3]{8y^6}$
20. $\sqrt[5]{32c^{10}d^{15}}$
21. $\sqrt[3]{a^2 + b^2}$
22. $\sqrt[4]{a^3 - b^3}$
23. $\sqrt[3]{27a^3b^9}$
24. $\sqrt[4]{81x^8y^8}$
25. $\sqrt[6]{\dfrac{m^{12}n^{24}}{64}}$
26. $\sqrt[8]{\dfrac{m^{16}n^{24}}{2^8}}$

3 Simplify and then write radical notation, unless inappropriate.

27. $(2a^{3/2})(4a^{1/2})$

28. $(3a^{5/6})(8a^{2/3})$

29. $\left(\dfrac{x^6}{9b^{-4}}\right)^{-1/2}$

30. $\left(\dfrac{x^{2/3}}{4y^{-2}}\right)^{-1/2}$

31. $\dfrac{x^{2/3}y^{5/6}}{x^{-1/3}y^{1/2}}$

32. $\dfrac{a^{1/2}b^{5/8}}{a^{1/4}b^{3/8}}$

4 Write an expression containing a single radical and simplify.

33. $\sqrt[3]{6}\sqrt{2}$

34. $\sqrt{2}\sqrt[4]{8}$

35. $\sqrt[4]{xy}\sqrt[3]{x^2y}$

36. $\sqrt[3]{ab^2}\sqrt{ab}$

37. $\sqrt[3]{a^4\sqrt{a^3}}$

38. $\sqrt{a^3\sqrt[3]{a^2}}$

39. $\dfrac{\sqrt{(a+x)^3}\sqrt[3]{(a+x)^2}}{\sqrt[4]{a+x}}$

40. $\dfrac{\sqrt[4]{(x+y)^2}\sqrt[3]{(x+y)}}{\sqrt{(x+y)^3}}$

Simplify. Round to three decimal places. (*Note:* Since $x^{1/4} = (x^{1/2})^{1/2}$, you can take a fourth root by taking a square root, and then the square root of the result. Or you can find decimal notation for the exponent, obtaining $x^{0.25}$, and use the power key, x^y. Remember, answers can vary depending on the type and the readout of your calculator.)

41. $(\sqrt[4]{13})^5$

42. $\sqrt[4]{17^3}$

43. $12.3^{3/2}$

44. $1.345^{5/2}$

45. $105.6^{3/4}$

46. $7.14^{5/4}$

Using Example 18, find the length L of the letters in the road-pavement sign given the values of h and d.

47. $h = 4$ ft, $d = 180$ ft

48. $h = 4$ ft, $d = 100$ ft

49. $h = 4$ ft, $d = 200$ ft

50. $h = 4$ ft, $d = 300$ ft

Industrial psychology. In most working situations, the more times a task is performed, the less time it takes to do the task. In a certain situation, industrial psychologists discovered that

$$T = 34x^{-0.41},$$

where T is the number of hours of labor required for the xth unit to be produced.

51. Find T when $x = 1, 6, 8, 10, 32$, and 64.

52. Find T when $x = 2, 3, 5, 9, 11, 19$, and 100.

5 Factor and simplify.

53. $a^{-2}b^5 - a^3b^{-5}$

54. $p^8q^{-2} + p^{-5}q^4$

55. $5a^{2/3}b^{-1/2} + 2a^{-1/3}b^{1/2}$

56. $p^{4/5}q^{-2} + 2p^{-1/5}q^2$

57. $x^{-1/3}y^{3/4} - x^{2/3}y^{-1/4}$

58. $4a^{1/2}b^{-3/4} - 6a^{-1/2}b^{1/4}$

59. $(2x-3)^{-3}(x+1)^{5/4} + (2x-3)^{-2}(x+1)^{1/4}$

60. $2x(5x+3)^{2/3} + 3x^2(5x+3)^{-1/3}$

61. $2(x+1)^{1/2}(3x+4)^{-3/4} - 10(x+1)^{-1/2}(3x+4)^{1/4}$

62. $-4(2x-5)^{-3}(3x+1)^{1/3} + 8(2x-5)^{-2}(3x+1)^{-2/3}$

63. $3(x^2+1)^3 + 3(3x-5)(x^2+1)^2(2x)$

64. $3x^2(x-1)^4 + x^3(4)(x-1)^3$

65. $\dfrac{x^3(2x) - (x^2+1)(3x^2)}{x^6}$

66. $\dfrac{1}{2}x^{-1/2}(x^3-4) + x^{1/2}(3x^2)$

67. $\dfrac{x^2(x^2+1)^{-1/2}(x) - (2x)(x^2+1)^{1/2}}{x^4}$

68. $\dfrac{(x-1)^{1/2} - (x+1)(\frac{1}{2})(x-1)^{-1/2}}{x-1}$

SYNTHESIS

Simplify.

69. $(\sqrt{a^{\sqrt{a}}})^{\sqrt{a}}$

70. $(2a^3b^{5/4}c^{1/7})^4 \div (54a^{-2}b^{2/3}c^{6/5})^{-1/3}$

1.9 Handling Dimension Symbols

We now learn to manipulate dimension symbols. We make calculations, simplifications, and changes of unit. The algebraic skills we have reviewed up to now are quite useful for this. The following table contains abbreviations for some dimension symbols.

Dimension Symbol	Unit
m	meter
cm	centimeter (0.01 m)
km	kilometer (1000 m)
g	gram
cg	centigram (0.01 g)
kg	kilogram (1000 g)
s or sec	second
h or hr	hour
L	liter
mL	milliliter (0.001 L)

SPEED

Speed is often determined by measuring a distance and a time and then dividing the distance by the time (this is **average speed**):

$$\text{Speed} = \frac{\text{Distance}}{\text{Time}}.$$

If a distance is measured in kilometers and the time required to travel that distance is measured in hours, the speed will be computed in *kilometers per hour* (km/h*). For example, if a car travels 100 km in 2 hr, the average speed is

$$\frac{100 \text{ km}}{2 \text{ hr}}, \quad \text{or } 50\,\frac{\text{km}}{\text{hr}} \quad \text{or } 50 \text{ km/h}.$$

DO EXERCISES 1 AND 2.

1 Dimension Symbols

The symbol 100 km/2 hr makes it look as though we are dividing 100 km by 2 hr. It may be argued that we cannot divide 100 km by 2 hr (we can only divide 100 by 2). Nevertheless, it is convenient to treat dimension symbols such as *kilometers, hours, feet, seconds,* and *pounds* as though they were numerals or variables, because correct results can thus be obtained mechanically. Compare, for example,

$$\frac{100x}{2y} = \frac{100}{2} \cdot \frac{x}{y} = 50\,\frac{x}{y}$$

with

$$\frac{100 \text{ km}}{2 \text{ hr}} = \frac{100}{2} \cdot \frac{\text{km}}{\text{hr}} = 50\,\frac{\text{km}}{\text{hr}}.$$

The analogy holds in other situations, as shown in the following examples.

*The standard abbreviation for kilometers per hour is km/h and for meters per second is m/s. When the abbreviations for hour and second stand alone, "hr" and "sec" are used instead of "h" and "s."

OBJECTIVES

You should be able to:

1 **Perform a given calculation involving dimension symbols and simplify, if possible, without making any unit changes.**

2 **Perform a given change of dimension symbols, using substitution or "multiplying by 1."**

What is the speed in m/s?

1. 186,000 m, 10 sec

2. 8 m, 16 sec

Add the measures.

3. 45 ft, 17 ft

4. $\frac{3}{4}$ kg, $\frac{2}{5}$ kg

5. $70\ \frac{\text{cm}}{\text{sec}}$, $35\ \frac{\text{cm}}{\text{sec}}$

Perform these calculations and simplify if possible. Do not make any unit changes.

6. $36\ \text{ft} \cdot \frac{1\ \text{yd}}{3\ \text{ft}}$

7. $5\ \text{lb} \cdot \frac{16\ \text{oz}}{1\ \text{lb}}$

8. $\frac{4\ \text{kg}}{5\ \text{ft}} \cdot \frac{7\ \text{ft}}{8\ \text{kg}}$

9. $\frac{5\ \text{in.} \cdot 9\ \text{lb/hr}}{4\ \text{hr}}$

10. $\frac{10\ \text{lb}}{7\ \text{m}} \cdot \frac{14\ \text{lb}}{5\ \text{m}}$

EXAMPLE 1 Compare

$$3\ \text{ft} + 2\ \text{ft} = (3 + 2)\ \text{ft} = 5\ \text{ft}$$

with

$$3x + 2x = (3 + 2)x = 5x.$$

This looks like a distributive law in use.

DO EXERCISES 3–5.

EXAMPLE 2 Compare

$$4\ \text{m} \cdot 3\ \text{m} = 3 \cdot 4 \cdot \text{m} \cdot \text{m} = 12\ \text{m}^2\ (\text{sq m})$$

with

$$4x \cdot 3x = 4 \cdot 3 \cdot x \cdot x = 12x^2.$$

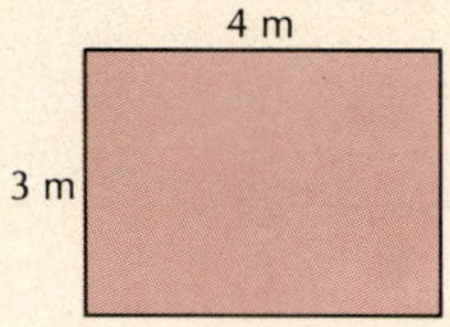

EXAMPLE 3 Compare

$$5\ \text{men} \cdot 8\ \text{hr} = 5 \cdot 8 \cdot \text{man-hr} = 40\ \text{man-hr}$$

with

$$5x \cdot 8y = 5 \cdot 8 \cdot x \cdot y = 40xy.$$

In each of the examples above, dimension symbols are treated as though they were variables or numerals, and as though a symbol such as 3 m represents a product 3 times m. A symbol like km/h is treated as though it represents a division of km by hr (*kilometers* by *hours*). Any two measures can be "multiplied" or "divided."

EXAMPLE 4 Perform this calculation and simplify, if possible. Do not make any unit changes.

$$7\ \text{hr} \cdot \frac{5\ \text{mi}}{8\ \text{hr}}$$

Solution We have

$$7\ \text{hr} \cdot \frac{5\ \text{mi}}{8\ \text{hr}} = \frac{7 \cdot 5}{8} \cdot \frac{\text{hr}}{\text{hr}} \cdot \text{mi} = 4.375\ \text{hr}.$$

We treated hr/hr as a symbol for 1.

DO EXERCISES 6–10.

2 Changes of Unit

Changes of unit can be accomplished by substitutions.

EXAMPLE 5 Change to inches: 25 yd.

Solution

$$
\begin{aligned}
25 \text{ yd} &= 25 \cdot 1 \text{ yd} \\
&= 25 \cdot 3 \text{ ft} \qquad \text{Substituting 3 ft for 1 yd} \\
&= 25 \cdot 3 \cdot 1 \text{ ft} \\
&= 25 \cdot 3 \cdot 12 \text{ in.} \qquad \text{Substituting 12 in. for 1 ft} \\
&= 900 \text{ in.}
\end{aligned}
$$

DO EXERCISES 11–13.

The notion of "multiplying by 1" can also be used to change units.

EXAMPLE 6 Change 72 in. to yd.

Solution

$$72 \text{ in.} = 72 \text{ in.} \cdot \frac{1 \text{ ft.}}{12 \text{ in.}} \cdot \frac{1 \text{ yd}}{3 \text{ ft}}$$

Each of these is equal to 1.

$$
\begin{aligned}
&= \frac{72}{12 \cdot 3} \cdot \frac{\text{in.}}{\text{in.}} \cdot \frac{\text{ft}}{\text{ft}} \\
&= 2 \text{ yd}
\end{aligned}
$$

In Example 6, we first used the following symbol for 1:

$$\frac{1 \text{ ft}}{12 \text{ in.}}$$

← "ft" in the numerator is the unit we are changing *to*.
← "in." in the denominator is the unit we are changing *from*.

In the final multiplication, we were converting from ft to yd, so 1 yd was in the numerator and 3 ft was in the denominator.

DO EXERCISES 14–16.

EXAMPLE 7 Change $60 \frac{\text{km}}{\text{hr}}$ to $\frac{\text{m}}{\text{sec}}$.

Solution

$$
\begin{aligned}
60 \frac{\text{km}}{\text{hr}} &= 60 \frac{\text{km}}{\text{hr}} \cdot \frac{1000 \text{ m}}{1 \text{ km}} \cdot \frac{1 \text{ hr}}{60 \text{ min}} \cdot \frac{1 \text{ min}}{60 \text{ sec}} \\
&= \frac{60 \cdot 1000}{60 \cdot 60} \cdot \frac{\text{km}}{\text{km}} \cdot \frac{\text{hr}}{\text{hr}} \cdot \frac{\text{min}}{\text{min}} \cdot \frac{\text{m}}{\text{sec}} \\
&= 16.67 \frac{\text{m}}{\text{sec}}, \quad \text{or } 16.67 \text{ m/s.}
\end{aligned}
$$

EXAMPLE 8 Change $55 \frac{\text{mi}}{\text{hr}}$ to $\frac{\text{ft}}{\text{sec}}$.

Solution

$$
\begin{aligned}
55 \frac{\text{mi}}{\text{hr}} &= 55 \frac{\text{mi}}{\text{hr}} \cdot \frac{5280 \text{ ft}}{1 \text{ mi}} \cdot \frac{1 \text{ hr}}{60 \text{ min}} \cdot \frac{1 \text{ min}}{60 \text{ sec}} \\
&= \frac{55 \cdot 5280}{60 \cdot 60} \cdot \frac{\text{mi}}{\text{mi}} \cdot \frac{\text{hr}}{\text{hr}} \cdot \frac{\text{min}}{\text{min}} \cdot \frac{\text{ft}}{\text{sec}} \\
&= 80 \frac{2}{3} \frac{\text{ft}}{\text{sec}}
\end{aligned}
$$

DO EXERCISES 17–20.

Perform the following changes of unit. Use substitution.

11. 34 yd; change to in.

12. 11 mi; change to ft

13. 5 hr; change to sec

Perform the following changes of unit. Use multiplying by 1.

14. 720 in.; change to yd

15. 36,960 m; change to km

16. 360,000 sec; change to hr

Perform the following changes of unit. Use multiplying by 1.

17. $120 \frac{\text{mi}}{\text{hr}}$; change to $\frac{\text{ft}}{\text{sec}}$

18. 3600 cm^2; change to m^2

19. $50 \frac{\text{kg}}{\text{L}}$; change to $\frac{\text{g}}{\text{cm}^3}$ (*Hint*: 1 L = 1000 cm^3.)

20. $\frac{\$72}{\text{day}}$; change to $\frac{¢}{\text{hr}}$

EXERCISE SET 1.9

1 Perform the calculations and simplify if possible. Do not make any unit changes.

1. $36 \text{ ft} \cdot \dfrac{1 \text{ yd}}{3 \text{ ft}}$

2. $6 \text{ lb} \cdot \dfrac{16 \text{ oz}}{1 \text{ lb}}$

3. $6 \text{ kg} \cdot 8 \dfrac{\text{hr}}{\text{kg}}$

4. $9 \dfrac{\text{km}}{\text{hr}} \cdot 3 \text{ hr}$

5. $3 \text{ cm} \cdot \dfrac{2 \text{ g}}{2 \text{ cm}}$

6. $\dfrac{9 \text{ km}}{3 \text{ days}} \cdot 6 \text{ days}$

7. $6 \text{ m} + 2 \text{ m}$

8. $10 \text{ tons} + 6 \text{ tons}$

9. $5 \text{ ft}^3 + 7 \text{ ft}^3$

10. $10 \text{ yd}^3 + 17 \text{ yd}^3$

11. $\dfrac{3 \text{ kg}}{5 \text{ m}} \cdot \dfrac{7 \text{ kg}}{6 \text{ m}}$

12. $3 \text{ acres} \times 60 \dfrac{1}{\text{acre}}$

13. $\dfrac{2000 \text{ lb} \cdot (6 \text{ mi/hr})^2}{100 \text{ ft}}$

14. $\dfrac{7 \text{ m} \cdot 8 \text{ kg/sec}}{4 \text{ sec}}$

15. $\dfrac{6 \text{ cm}^2 \cdot 5 \text{ cm/sec}}{2 \text{ sec}^2/\text{cm}^2 \cdot 2 \dfrac{1}{\text{kg}}}$

16. $\dfrac{320 \text{ lb} \cdot (5 \text{ ft/sec})^2}{2 \cdot 32 \dfrac{\text{ft}}{\text{sec}^2}}$

2 Perform the following changes of unit, using substitution or multiplying by 1.

17. 72 in.; change to ft

18. 17 hr; change to min

19. 2 days; change to sec

20. 360 sec; change to hr

21. $60 \dfrac{\text{kg}}{\text{m}}$; change to $\dfrac{\text{g}}{\text{cm}}$

22. $44 \dfrac{\text{ft}}{\text{sec}}$; change to $\dfrac{\text{mi}}{\text{hr}}$

23. 216 m^2; change to cm^2

24. $60 \dfrac{\text{lb}}{\text{ft}^3}$; change to $\dfrac{\text{ton}}{\text{yd}^3}$

25. $\dfrac{\$36}{\text{day}}$; change to $\dfrac{¢}{\text{hr}}$

26. 1440 man-hr; change to man-days

27. $1.73 \dfrac{\text{mL}}{\text{sec}}$; change to $\dfrac{\text{L}}{\text{hr}}$
(*Hint*: 1 liter = 1 L = 1000 mL = 1000 milliliters.)

28. $1800 \dfrac{\text{g}}{\text{L}}$; change to $\dfrac{\text{cg}}{\text{mL}}$

29. $186{,}000 \dfrac{\text{mi}}{\text{sec}}$ (speed of light); change to $\dfrac{\text{mi}}{\text{yr}}$
(Let 365 days = 1 yr.)

30. $1100 \dfrac{\text{ft}}{\text{sec}}$ (speed of sound); change to $\dfrac{\text{mi}}{\text{yr}}$
(Let 365 days = 1 yr.)

Use Table 6 at the back of the book to do the following unit changes.

31. ▦ $89.2 \dfrac{\text{ft}}{\text{sec}}$; change to $\dfrac{\text{m}}{\text{min}}$

32. ▦ 1013 yd^3; change to m^3

33. ▦ 640 mi^2; change to km^2

34. ▦ $312.2 \dfrac{\text{kg}}{\text{m}}$; change to $\dfrac{\text{lb}}{\text{ft}}$

35. If a steel rod 2 cm long weighs 5 g, how much does a rod of the same type weigh whose length is 3 cm? 5 m?

36. In Exercise 35, how long is a rod that weighs 4.3 g? 20 cg?

In chemistry, 1 *mole* of a substance is that mass of the substance, in grams, equal to its molecular weight. For example, 1 mole of oxygen is 32 *grams* because its molecular weight is 32, and 1 mole of neon is 20.2 grams.

Convert to grams.

37. 50 moles of oxygen

38. 44 moles of neon

Convert to moles.

39. 303 grams of neon

40. 377.6 grams of oxygen

SYNTHESIS

41. *Einstein's equation of relativity* states that
$$E = mc^2,$$
where E is the energy emitted from a mass m and c is the speed of light. An atom bomb is exploded. It contains 5000 g of radioactive uranium. The speed of light is 2.9979×10^8 m/sec. Find the amount of energy created from the explosion.

42. Density is mass divided by volume, $D = M/V$. The density of potassium is 0.86 g/mL. What is the mass of 18 cm^3 of potassium? (*Hint:* 1 mL = 1 cm^3.)

43. 1 nanosecond = 10^{-9} sec. Estimate how far light travels in 1 nanosecond.

SUMMARY AND REVIEW: CHAPTER 1

TERMS TO KNOW

Real number
Whole number
Integer
Rational number
Irrational number
Equivalent expressions
Commutative
Associative
Identity
Additive inverse
Opposite
Multiplicative inverse
Reciprocal
Distributive
Exponent
Scientific notation
Order of operations
Absolute value
Polynomial
Term
Degree
Binomial
Trinomial
FOIL
Factoring
Factoring by grouping
Difference of squares
Trinomial square
Sum or difference of cubes
Fractional expression
Least common multiple
Complex fractional expression
Radical
Radicand
Index
Principal root
Pythagorean theorem
Rationalizing numerators or denominators
Rational exponent

REVIEW EXERCISES

The following review exercises are for practice. Answers are at the back of the book. If you miss an exercise, restudy the section indicated alongside the answer.

Consider the numbers

$$-43.89,\ 12,\ -3,\ -\tfrac{1}{5},\ \sqrt{7},\ \sqrt[3]{10},\ -1,\ -\tfrac{4}{3},\ 7\tfrac{2}{3},\ -19,\ 31,\ 0.$$

1. Which are integers?
2. Which are natural numbers?
3. Which are rational numbers?
4. Which are real numbers?
5. Which are irrational numbers?
6. Which are whole numbers?

Compute.

7. $15 + (-19)$
8. $-12 + |-4|$
9. $-2.5 + (-2.5)$
10. $22 - (-8)$
11. $\dfrac{18}{-3}$
12. $(-17)(-9)$
13. $-10(20)(-5)(-3)$
14. $-\dfrac{15}{16} + \dfrac{3}{4}$
15. $\dfrac{5}{12} - \left(-\dfrac{7}{8}\right)$
16. $5^3 - [2(4^2 - 3^2 - 6)]^3$
17. $\dfrac{3^4 - (6 - 7)^4}{2^3 - 2^4}$

Convert to decimal notation.

18. 3.261×10^6
19. 4.1×10^{-4}
20. 2.77 E 8
21. 1.009 E −4

Convert to scientific notation.

22. 0.01432
23. 43,210

Simplify. Write answers using positive exponents where possible.

24. $(7a^2b^4)(-2a^{-4}b^3)$
25. $\dfrac{54x^6y^{-4}z^2}{9x^{-3}y^2z^{-4}}$
26. $\sqrt[4]{81}$
27. $\sqrt[5]{-32}$
28. $\dfrac{b - a^{-1}}{a - b^{-1}}$
29. $\dfrac{\dfrac{x^2}{y} + \dfrac{y^2}{x}}{y^2 - xy + x^2}$
30. $(\sqrt{3} - \sqrt{7})(\sqrt{3} + \sqrt{7})$
31. $(5x^2 - \sqrt{2})^2$
32. $8\sqrt{5} + \dfrac{25}{\sqrt{5}}$
33. $(x + t)(x^2 - xt + t^2)$
34. $(5a + 4b)^3$
35. $(5xy^4 - 7xy^2 + 4x^2 - 3) - (-3xy^4 + 2xy^2 - 2y + 4)$

Factor.

36. $x^3 + 2x^2 - 3x - 6$
37. $12a^3 - 27ab^4$
38. $24x + 144 + x^2$
39. $9x^3 + 35x^2 - 4x$
40. $8x^3 - 1$
41. $27x^6 + 125y^6$

Write an expression containing a single radical.

42. $\sqrt{y^5}\sqrt[3]{y^2}$
43. $\dfrac{\sqrt{(a+b)^3}\sqrt[3]{a+b}}{\sqrt[6]{(a+b)^7}}$

44. Convert to radical notation:
$$b^{7/5}.$$

45. Convert to exponential notation and simplify:
$$\sqrt[8]{\frac{m^{32}n^{16}}{3^8}}.$$

46. Divide and simplify:
$$\frac{3x^2 - 12}{x^2 + 4x + 4} \div \frac{x - 2}{x + 2}.$$

47. Subtract and simplify:
$$\frac{x}{x^2 + 9x + 20} - \frac{4}{x^2 + 7x + 12}.$$

48. Rationalize the numerator:
$$\frac{\sqrt{x} - \sqrt{y}}{\sqrt{x} + \sqrt{y}}.$$

49. Rationalize the denominator:
$$\frac{\sqrt{x} - \sqrt{y}}{\sqrt{x} + \sqrt{y}}.$$

Factor and simplify.

50. $2x^{1/2}y^{-3/4} - 3x^{-1/2}y^{1/4}$
51. $(x - 2)^{-3/4}(3x + 5)^{5/2} + (x - 2)^{1/4}(3x + 5)^{3/2}$

52. How long is a guy wire reaching from the top of a 17-ft pole to a point on the ground 8 ft from the pole?
53. Change $10\dfrac{\text{km}}{\text{h}}$ to $\dfrac{\text{m}}{\text{min}}$.

SYNTHESIS

What property is illustrated by each sentence?

54. $t + (-t) = 0$
55. $8(a + b) = 8a + 8b$
56. $-3(ab) = (-3a)b$
57. $tx = xt$

Multiply. Assume that all exponents are integers.

58. $(x^n + 10)(x^n - 4)$
59. $(t^a + t^{-a})^2$
60. $(y^b - z^c)(y^b + z^c)$
61. $(a^n - b^m)^3$

Factor.

62. $y^{2n} + 16y^n + 64$
63. $x^{2t} - 3x^t - 28$
64. $m^{6n} - m^{3n}$

65. Simplify:
$$\frac{\dfrac{2^{n+1}x^{n+1}}{(n+1)^5}}{\dfrac{2^n x^n}{n^5}}.$$

66. Subtract and simplify:
$$\frac{(n-1)(n-2)(n-3)}{2 \cdot 3} - \frac{n(n-1)(n-2)(n-3)}{2 \cdot 3 \cdot 4}.$$

TEST: CHAPTER 1

Consider the numbers

$$-14,\ 23.77,\ -5,\ -\tfrac{4}{7},\ \sqrt{8},\ -\sqrt[3]{11},\ 56\tfrac{1}{4},\ \tfrac{9}{2},\ 0,\ 233.$$

1. Which are whole numbers?
2. Which are irrational numbers?
3. Which are real numbers?
4. Which are rational numbers?
5. Which are natural numbers?
6. Which are integers?

Compute.

7. $-7 + |-7|$
8. $-7.4 + 9.4$
9. $(-6)(-2)$
10. $3 - (-5)$
11. $\dfrac{15}{-5}$
12. $\dfrac{27 \div 3^2 - 3^4 \cdot 3^2}{5(7 - 16) + 4 \cdot 10}$

Convert to decimal notation.

13. 2.834×10^{-3}

14. 4.7×10^{2}

15. 4.45 E 9

16. 4.45 E −5

Convert to scientific notation.

17. 0.000816

18. 480.57

Simplify.

19. $(4x^4y^{-2})(-3x^5y^{-4})$

20. $\dfrac{24\,pq^8r^{-4}}{36p^{-3}q^{-6}r^5}$

21. $\sqrt[3]{-27}$

22. $\sqrt[4]{625}$

23. $(\sqrt{8} - \sqrt{2})(\sqrt{8} + \sqrt{2})$

24. $\dfrac{\dfrac{x}{y^2} + \dfrac{y}{x^2}}{\dfrac{y^2 - yx + x^2}{y}}$

25. $(5a^2 - 2b)^2$

26. $(2x^2y - 3xy + y^2 - 4) - (-6x^2y + 2xy - y + 8)$

27. $(4y - 3)^3$

28. Write an expression containing a single radical:

$$\frac{\sqrt[4]{(c + d)^3} \cdot \sqrt{c + d}}{\sqrt[5]{(c + d)^4}}.$$

29. Convert to radical notation:

$$t^{2/7}.$$

Factor.

30. $16x^2 + 48x + 36$

31. $t^3 - 343$

32. $m^5 - 9m^3n^2$

33. $12p^4 + 9p^2 - 30$

34. Divide and simplify:

$$\frac{x^2 - 10x + 25}{x^2 + 10x + 25} \div \frac{x^2 - 25}{(x + 5)^3}.$$

35. Subtract and simplify:

$$\frac{x}{x^2 - 4x - 12} - \frac{9}{x^2 - 36}.$$

36. Rationalize the denominator:

$$\frac{7 - \sqrt{x}}{7 + \sqrt{x}}.$$

37. Multiply:

$$(x^t + x^{-t})^3.$$

38. Rationalize the numerator:

$$\frac{7 - \sqrt{x}}{7 + \sqrt{x}}.$$

39. Factor:

$$a^{-2/3}b^{8/5} - a^{4/3}b^{3/5}.$$

40. A baseball diamond is actually a square 90 ft on a side. A catcher fields a bunt along the third-base line 12 ft from home plate. How far would the catcher have to throw the ball to first base?

41. Factor and simplify:

$$\frac{3x^2(2x + 3)^{1/2} - x^3\left(\frac{1}{2}\right)(2x + 3)^{-1/2}(2)}{[(2x + 3)^{1/2}]^2}.$$

42. Change $1200 \dfrac{\text{m}}{\text{min}}$ to $\dfrac{\text{km}}{\text{hr}}$.

SYNTHESIS

43. Factor:

$$x^{16} - 16.$$

44. Simplify:

$$\left| \frac{\dfrac{(-1)^{n+2}(x - 2)^{n+1}}{(n + 1)2^{n+1}}}{\dfrac{(-1)^{n+1}(x - 2)^n}{n \cdot 2^n}} \right|.$$

Equations, Inequalities, and Problem Solving

2

In this chapter, we continue the review that we began in Chapter 1. In particular, we review basic equation-solving techniques. We also study the numbers that today bear the name *imaginary*. Those numbers are square roots of negative numbers. We will then allow addition of real numbers and imaginary numbers, thereby forming a system of numbers known as the *complex numbers*. The use of the complex numbers will allow us to find complete solutions to quadratic equations. Complex numbers have many practical applications, though these applications are too involved to be considered in this book.

We also begin to consider the payoff that results from the ability to solve equations: *problem solving*. From this introduction to problem solving, we will go on to consider problem solving many times in later chapters.

FEATURE PROBLEM

An object is dropped from the top of the Gateway Arch, which is 195 m high, in St. Louis. How long does it take for the object to reach the ground?

THE MATHEMATICS

When an object is dropped or thrown downward, the distance s, in meters, that it falls in t seconds is given by the formula

$$s = 4.9t^2 + v_0t,$$

where v_0 is the initial velocity. Since the object was dropped, its initial velocity was 0. We substitute 0 for v_0 and 195 for s and solve for t.

2.1 Solving Equations and Inequalities

OBJECTIVES

You should be able to:

1. Solve simple equations using the addition and multiplication principles and the principle of zero products.
2. Solve simple inequalities, and write set-builder notation for a set.
3. Use inequalities to find meaningful replacements in radical expressions.

Solve (find the solution set) by trial and error.

1. $x - 2 = 7$

2. $y^2 + y = 0$

1 Solving Equations

DEFINITION

A *solution of an equation* is any number that makes the equation true when that number is substituted for the variable.

The number 3 is a solution of $5x = 15$ because $5(3) = 15$ is true. The number -4 is not a solution of $5x = 15$, because $5(-4) = 15$ is false.

DEFINITION

The set of all solutions of an equation is called its *solution set*. When we find all the solutions of an equation (find its solution set), we say that we have solved the equation.

The number 3 is the only solution of $5x = 15$, so the solution set consists of the number 3 and is denoted $\{3\}$. As another example, consider

$$y^2 - y = 0.$$

The number 0 is a solution, and so is the number 1. There are no other solutions, so the solution set is $\{0, 1\}$.

DO EXERCISES 1 AND 2.

Equation-Solving Principles

The Addition and Multiplication Principles

Two simple principles allow us to solve many equations. The first of these is as follows:

The Addition Principle

For any real numbers a, b, and c, if an equation $a = b$ is true, then $a + c = b + c$ is true.

This principle and the next are actually very easy theorems. Suppose that $a = b$ is true. Then a and b are the same number. If we add c to this number, the result is $a + c$. It is also $b + c$. Similarly, if we multiply by c, the result is ac. It is also bc.

The second principle is similar to the first.

The Multiplication Principle

For any real numbers a, b, and c, if an equation $a = b$ is true, then $ac = bc$ is true.

Note that these principles also cover "subtracting on both sides" and "dividing on both sides," because subtracting c can be accomplished by adding

$-c$, and dividing by c can be accomplished by multiplying by $1/c$, provided that $c \neq 0$.

Now let us use the principles to solve some equations.

EXAMPLE 1 Solve: $3x + 4 = 15$.

Solution

$$3x + 4 = 15$$

$$3x + 4 + (-4) = 15 + (-4) \quad \text{Using the addition principle; adding } -4$$

$$3x = 11 \quad \text{Simplifying}$$

$$\tfrac{1}{3} \cdot 3x = \tfrac{1}{3} \cdot 11 \quad \text{Using the multiplication principle; multiplying by } \tfrac{1}{3}$$

$$x = \tfrac{11}{3}$$

Check:

$3x + 4$	$= 15$	
$3 \cdot \frac{11}{3} + 4$	15	Substituting $\frac{11}{3}$ for x
$11 + 4$		
15		

The solution is $\frac{11}{3}$. The solution set is $\{\frac{11}{3}\}$. ∎

EXAMPLE 2 Solve: $3(7 - 2x) = 14 - 8(x - 1)$.

Solution

$$3(7 - 2x) = 14 - 8(x - 1)$$

$$21 - 6x = 14 - 8x + 8 \quad \text{Multiplying, using the distributive laws, to remove parentheses}$$

$$21 - 6x = 22 - 8x \quad \text{Collecting like terms}$$

$$8x - 6x = 22 - 21 \quad \text{Adding } -21 \text{ and also } 8x$$

$$2x = 1 \quad \text{Collecting like terms}$$

$$x = \tfrac{1}{2} \quad \text{Multiplying by } \tfrac{1}{2}$$

Check:

$3(7 - 2x)$	$= 14 - 8(x - 1)$
$3(7 - 2 \cdot \frac{1}{2})$	$14 - 8(\frac{1}{2} - 1)$
$3(7 - 1)$	$14 - 8(-\frac{1}{2})$
$3 \cdot 6$	$14 + 4$
18	18

The solution is $\frac{1}{2}$. The solution set is $\{\frac{1}{2}\}$. ∎

EXAMPLE 3 Solve: $x + 3 = x$.

Solution We have

$$x + 3 = x$$

$$-x + x + 3 = -x + x \quad \text{Adding } -x$$

$$3 = 0. \quad \text{Collecting like terms}$$

We get a false equation. No replacement for x will make the equation true. Since there are no solutions, the solution set is the **empty set,** denoted $\emptyset$. ∎

EXAMPLE 4 Solve: $x + 8 = 8 + x$.

Solve.

3. $9x - 4 = 8$

4. $-4x + 2 + 5x = 3x - 15$

5. $3(y - 1) - 1 = 2 - 5(y + 5)$

6. $x - 7 = x$

7. $y + 6 = 6 + y$

Solution We have

$$\begin{aligned} x + 8 &= 8 + x \\ -x + x + 8 &= -x + 8 + x \quad \text{Adding } -x \\ 8 &= 8. \end{aligned}$$

We get a true equation. Any replacement for x will make the equation true. Thus the solution set is the entire set of real numbers. We also know this by the commutative law of addition.

An equation that is true for all meaningful replacements of the variable is called an **identity**. The equation $x + 8 = 8 + x$ of Example 4 is an identity. The equation $3(7 - 2x) = 14 - 8(x - 1)$ of Example 2 is *not* an identity, because there is only one replacement for which it is true, even though all real numbers are meaningful replacements.

DO EXERCISES 3–7.

The Principle of Zero Products

A third principle for solving equations is called the **principle of zero products**, as follows.

> **The Principle of Zero Products**
>
> **For any numbers a and b, if $ab = 0$, then $a = 0$ or $b = 0$; and if $a = 0$ or $b = 0$, then $ab = 0$.**

We can abbreviate the principle of zero products by saying "$ab = 0$ if and only if $a = 0$ or $b = 0$."

To solve an equation using this principle, we must have a 0 on one side of the equation and a product on the other. We then obtain the solutions by setting the factors equal to 0 separately.

EXAMPLE 5 Solve: $x^2 + x - 12 = 0$.

Solution We have

$$\begin{aligned} x^2 + x - 12 &= 0 \\ (x + 4)(x - 3) &= 0 \quad \text{Factoring} \\ x + 4 = 0 \quad \text{or} \quad x - 3 &= 0 \quad \text{Using the principle of zero products} \\ x = -4 \quad \text{or} \quad x &= 3. \end{aligned}$$

The solutions are -4 and 3. The solution set is $\{-4, 3\}$.

EXAMPLE 6 Solve: $2x^3 - x^2 = 3x$.

Solution We have

$$\begin{aligned} 2x^3 - x^2 &= 3x \\ 2x^3 - x^2 - 3x &= 0 \quad \text{Addition principle} \\ x(2x^2 - x - 3) &= 0 \quad \text{Factoring} \\ x(2x - 3)(x + 1) &= 0 \\ x = 0 \quad \text{or} \quad 2x - 3 = 0 \quad \text{or} \quad x + 1 &= 0 \quad \text{Principle of zero products} \\ x = 0 \quad \text{or} \quad x = \tfrac{3}{2} \quad \text{or} \quad x &= -1. \end{aligned}$$

The solution set is $\{0, \frac{3}{2}, -1\}$.

EXAMPLE 7 Solve: $x^3 + 5x^2 - 4x - 20 = 0$.

Solution We first factor by grouping:

$$x^3 + 5x^2 - 4x - 20 = 0$$
$$x^2(x + 5) - 4(x + 5) = 0$$
$$(x^2 - 4)(x + 5) = 0$$
$$(x - 2)(x + 2)(x + 5) = 0$$

$x - 2 = 0$ or $x + 2 = 0$ or $x + 5 = 0$ Principle of zero products

$x = 2$ or $x = -2$ or $x = -5$.

The solution set is $\{2, -2, -5\}$.

DO EXERCISES 8–13.

Not all equations with four terms can be solved using factoring by grouping. An example is $x^3 - 2x^2 - 2x - 3 = 0$.

2 Solving Inequalities

Principles for solving inequalities are similar to those for solving equations. We can add the same number on both sides of an inequality. We can also multiply on both sides by the same nonzero number, but if that number is negative, we must reverse the inequality sign.

The Addition Principle For Inequalities

For any real numbers a, b, and c, if $a < b$ is true, then $a + c < b + c$ is true.

The Multiplication Principle for Inequalities

1. **For any real numbers a, b, and c, if $a < b$ and $c > 0$ are true, then $ac < bc$ is true.**
2. **For any real numbers a, b, and c, if $a < b$ and $c < 0$ are true, then $ac > bc$ is true.**

Similar statements hold when $<$ is replaced by $\leq$.

EXAMPLE 8 Solve: $3x < 11 - 2x$.

Solution We have

$$3x < 11 - 2x$$
$$3x + 2x < 11 \quad \text{Adding } 2x$$
$$5x < 11 \quad \text{Collecting like terms}$$
$$x < \tfrac{11}{5}. \quad \text{Multiplying by } \tfrac{1}{5}$$

Any number less than $\frac{11}{5}$ is a solution. The solution set is the set of all x such that $x < \frac{11}{5}$. We abbreviate this using **set-builder notation** as follows:

$$\left\{x \middle| x < \tfrac{11}{5}\right\}.$$

Solve.

8. $(x - 7)(2x + 3) = 0$

9. $x^2 - x = 20$

10. $x^2 = 5x$

11. $25x^2 + 10x + 1 = 0$

12. $3x^3 - 11x^2 = 4x$

13. $5x^3 + x^2 - 5x - 1 = 0$

Solve.

14. $5x > 12 - 3x$

15. $17 - 5y \leq 8y - 5$

16. $12x - 6 < 10x + 4$

Write set-builder notation for the set.

17. The set of all x such that $x > \frac{5}{2}$

18. The set of all y such that $y \geq -7$

19. The set of all x such that $x^2 = 5$

Determine the meaningful replacements in the expression.

20. $\sqrt{x - 2}$

21. $\sqrt{x + 3}$

22. $\sqrt{22 - 4x}$

We can make a graph of the solution set, as follows:

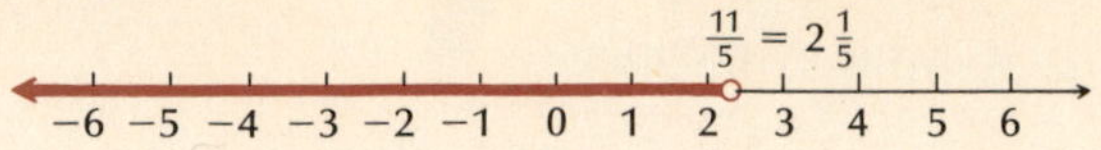

A **graph** is a drawing that represents the solution set of an equation.

EXAMPLE 9 Solve: $16 - 7y \geq 10y - 4$.

Solution We have

$$16 - 7y \geq 10y - 4$$
$$-16 + 16 - 7y \geq -16 + 10y - 4 \qquad \text{Adding } -16$$
$$-7y \geq 10y - 20 \qquad \text{Simplifying}$$
$$-10y - 7y \geq -10y + 10y - 20 \qquad \text{Adding } -10y$$
$$-17y \geq -20 \qquad \text{Simplifying}$$
$$y \leq \frac{20}{17}. \qquad \text{Multiplying by } -\tfrac{1}{17} \text{ and reversing the inequality sign}$$

Any number less than or equal to $\frac{20}{17}$ is a solution. Thus, the solution set is $\{y | y \leq \frac{20}{17}\}$.

DO EXERCISES 14–19.

3 Finding Meaningful Replacements In Radical Expressions

We can solve an inequality to find the *meaningful replacements* in a radical expression.

EXAMPLE 10 Determine the meaningful replacements in $\sqrt{5x - 4}$.

Solution The meaningful replacements are those values of x for which $\sqrt{5x - 4}$ is a real number. Such replacements are those that make the radicand nonnegative, that is, numbers x for which

$$5x - 4 \geq 0$$
$$5x \geq 4$$
$$x \geq \frac{4}{5}.$$

Thus the meaningful replacements are any numbers x for which $x \geq \frac{4}{5}$. These form the set $\{x | x \geq \frac{4}{5}\}$.

DO EXERCISES 20–22.

EXERCISE SET 2.1

1 Solve.

1. $4x + 12 = 60$
2. $2y - 11 = 37$
3. $4 + \frac{1}{2}x = 1$
4. $4.1 - 0.2y = 1.3$
5. $y + 1 = 2y - 7$
6. $5 - 4x = x - 13$
7. $5x - 2 + 3x = 2x + 6 - 4x$
8. $5x - 17 - 2x = 6x - 1 - x$
9. $1.9x - 7.8 + 5.3x = 3.0 + 1.8x$
10. $2.2y - 5 + 4.5y = 1.7y - 20$
11. $7(3x + 6) = 11 - (x + 2)$
12. $4(5y + 3) = 3(2y - 5)$
13. $2x - (5 + 7x) = 4 - [x - (2x + 3)]$
14. $y - (9y - 8) = [5 - 2y - 3(2y - 3)] + 29$

15. $(2x - 3)(3x - 2) = 0$
16. $(5x - 2)(2x + 3) = 0$
17. $x(x - 1)(x + 2) = 0$
18. $x(x + 2)(x - 3) = 0$
19. $3x^2 + x - 2 = 0$
20. $10x^2 - 16x + 6 = 0$
21. $(x - 1)(x + 1) = 5(x - 1)$
22. $6(y - 3) = (y - 3)(y - 2)$
23. $x[4(x - 2) - 5(x - 1)] = 2$
24. $14[(x - 4) - \frac{1}{14}(x + 2)] = (x + 2)(x - 4)$
25. $(3x^2 - 7x - 20)(2x - 5) = 0$
26. $(8x + 11)(12x^2 - 5x - 2) = 0$
27. $16x^3 = x$
28. $9x^3 = x$
29. $2x^2 = 6x$
30. $18x + 9x^2 = 0$
31. $3y^3 - 5y^2 - 2y = 0$
32. $3t^3 + 2t = 5t^2$
33. $(2x - 3)(3x + 2)(x - 1) = 0$
34. $(y - 4)(4y + 12)(2y + 1) = 0$
35. $(2 - 4y)(y^2 + 3y) = 0$
36. $(y^2 - 9)(y^2 - 36) = 0$
37. $x + 4 = 8 + x$
38. $x - 7 = 9 + x$
39. $7x^3 + x^2 - 7x - 1 = 0$
40. $3x^3 + x^2 - 12x - 4 = 0$
41. $y^3 + 2y^2 - y - 2 = 0$
42. $t^3 + t^2 - 25t - 25 = 0$
43. $11 + x = x + 11$
44. $0 \cdot x = 0$

2 Solve.

45. $x + 6 < 5x - 6$
46. $3 - x < 4x + 7$
47. $3x - 3 + 2x \geq 1 - 7x - 9$
48. $5y - 5 + y \leq 2 - 6y - 8$
49. $14 - 5y \leq 8y - 8$
50. $8x - 7 < 6x + 3$
51. $-\frac{3}{4}x \geq -\frac{5}{8} + \frac{2}{3}x$
52. $-\frac{5}{6}x \leq \frac{3}{4} + \frac{8}{3}x$
53. $4x(x - 2) < 2(2x - 1)(x - 3)$
54. $(x + 1)(x + 2) > x(x + 1)$

Write set-builder notation for the set.

55. The set of all x such that $x > 2.5$
56. The set of all y such that $y \leq -7$
57. The set of all t such that $2t^2 = 10$
58. The set of all m such that $m^3 + 3 = m^2 - 2$

3 Determine the meaningful replacements.

59. $\sqrt{x - 3}$
60. $\sqrt{2x - 5}$
61. $\sqrt{3 - 4x}$
62. $\sqrt{x^2 + 3}$

SYNTHESIS

Solve.

63. ▦ $2.905x - 3.214 + 6.789x = 3.012 + 1.805x$
64. ▦ $(13.14x + 17.152)(15.15 - 7.616x) = 0$
65. ▦ $3.12x^2 - 6.715x = 0$
66. ▦ $9.25x^2 + 18.03x = 0$
67. ▦ $1.52(6.51x + 7.3) < 11.2 - (7.2x + 13.52)$
68. ▦ $4.73(5.16y + 3.62) \geq 3.005(2.75y - 6.31)$

CHALLENGE

Solve.

69. $(x + 1)^3 = (x - 1)^3 + 26$
70. $(x - 2)^3 = x^3 - 2$
71. $(x^2 - x - 20)(x^2 - 25) = 0$
72. $(x^3 - 3x^2 - 16x + 48)(x^3 + x^2 - 9x - 9) = 0$

2.2 Fractional Equations

1 Equivalent Equations

DEFINITION

Equations that have the same solution set are called *equivalent equations*.

OBJECTIVES

You should be able to:

1. Determine whether equations are equivalent.
2. Solve fractional equations.

Determine whether the equations of each pair are equivalent.

1. $3x = 7$,
   $-15x = -35$

2. $x = 7$,
   $5x = 35$

3. $x = -5$,
   $x^2 = 25$

4. $x = -5$,
   $x + 1 = -4$

5. $\dfrac{(x-2)(x+8)}{x-2} = x + 8$,
   $x + 8 = x + 8$

6. $3x^2 = 5x$,
   $3x = 5$

7. Explain how you would derive $2x + 5x^2 = 5x^2$ from $2x = 0$.

8. Explain how you would derive $2x = 0$ from $2x + 5x^2 = 5x^2$.

EXAMPLE 1 Determine whether the equations $3x = 6$, $3x + 5 = 11$, and $-12x = -24$ are equivalent.

Solution

$3x = 6$	$3x + 5 = 11$	$-12x = -24$
Solution set: $\{2\}$	Solution set: $\{2\}$	Solution set: $\{2\}$

Each equation has the same solution set as the others, $\{2\}$. Thus all three equations are equivalent. ■

EXAMPLE 2 Determine whether the equations $3x = 4x$ and $3/x = 4/x$ are equivalent.

Solution

$3x = 4x$	$\dfrac{3}{x} = \dfrac{4}{x}$
Solution set: $\{0\}$	The solution set is $\emptyset$, the empty set (no solution, since division by 0 is not defined).

The empty set $\emptyset$ and the set containing the number 0, $\{0\}$, are not the same set. There are no elements in $\emptyset$. There is one element in $\{0\}$, the number 0. The solution sets are *not* the same, so the equations are *not* equivalent. ■

EXAMPLE 3 Determine whether the equations $x = 1$ and $x^2 = x$ are equivalent.

Solution

$x = 1$	$x^2 = x$
Solution set: $\{1\}$	Solution set: $\{0, 1\}$

The solution sets are *not* the same, so the equations are *not* equivalent. ■

DO EXERCISES 1–6.

Let us examine our equation-solving principles from the standpoint of equivalent equations. Consider the following example illustrating the addition principle:

$$x + 5 = 9$$
$$x + 5 + (-5) = 9 + (-5) \quad \text{Adding } -5 \text{ on both sides}$$
$$x = 4.$$

In this example, the original equation and the last equation have exactly the same solutions. They are equivalent. Whenever the steps in an argument are reversible, the equations are equivalent. When we use the addition principle, an equivalent equation is obtained unless an expression is added having replacements that are not meaningful. To see that, note that we start with an equation $a = b$ and obtain $a + c = b + c$. By adding $-c$, we can always obtain $a = b$ again, so the steps are reversible.

DO EXERCISES 7 AND 8.

The use of the addition principle produces an equation equivalent to the original, unless an expression is added having replacements that are not meaningful. Checking by substituting is therefore not necessary except to detect errors in solving.

Now let us consider the multiplication principle, which says that if $a = b$ is true, then $ac = bc$ is true. Does the multiplication principle yield equivalent equations? In other words, are the steps reversible? If the number c by which we multiply is not 0, then we can reverse the step by multiplying by $1/c$; but if c is 0, then $1/c$ does not exist, and the step is not reversible.

The use of the multiplication principle produces an equation equivalent to the original, if we multiply by a nonzero number.

Now let us look at the principle of zero products. According to this principle, if we start with an equation $ab = 0$, we obtain $a = 0$ or $b = 0$. Also, if we start with $a = 0$ or $b = 0$, we obtain $ab = 0$. Thus when we use this principle, that step is reversible, so we have equivalent equations.

The use of the principle of zero products yields the solutions of the original equation. Checking by substituting is not necessary except to detect errors in carrying out the algebra.

2 Fractional Equations

Let us now consider equations containing fractional expressions. These are called **fractional equations.** Finding the solutions of such equations often involves multiplying by expressions with variables, as in the following example. We multiply by the LCM of all the denominators.

EXAMPLE 4 Solve:

$$\frac{x-3}{x-7} = \frac{4}{x-7}.$$

Solution The LCM of the denominators is $x - 7$. We multiply by the LCM:

$$\frac{x-3}{x-7} = \frac{4}{x-7}$$

$$(x-7)\cdot\frac{x-3}{x-7} = (x-7)\cdot\frac{4}{x-7} \qquad \text{Multiplying by } x-7$$

$$x - 3 = 4 \qquad \text{Simplifying}$$

$$x = 7.$$

The possible solution is 7. We check:

Check:

$\frac{x-3}{x-7}$	$= \frac{4}{x-7}$
$\frac{7-3}{7-7}$	$\frac{4}{7-7}$
$\frac{4}{0}$	$\frac{4}{0}$

Since division by 0 is undefined, 7 is not a solution. The equation has no solutions. The solution set is $\emptyset$. ■

In Example 4, we did not obtain equivalent equations because the multiplication principle gives equivalent equations only when we are multiplying by nonzero numbers.

9. Solve. Don't forget to check.

$$\frac{x+4}{x+5} = \frac{-1}{x+5}$$

When we use the multiplication principle and multiply (or divide) by an expression with a variable, we may not obtain equivalent equations. We must check possible solutions by substituting in the original equation.

DO EXERCISE 9.

EXAMPLE 5 Solve:

$$\frac{x^2}{x-3} = \frac{9}{x-3}.$$

Solution The LCM of the denominators is $x - 3$. We multiply by the LCM:

$$\frac{x^2}{x-3} = \frac{9}{x-3}$$

$$(x-3)\cdot\frac{x^2}{x-3} = (x-3)\cdot\frac{9}{x-3} \quad \text{Multiplying by } x-3$$

$$x^2 = 9 \quad \text{Simplifying}$$

$$x^2 - 9 = 0$$

$$(x+3)(x-3) = 0$$

$$x + 3 = 0 \quad \text{or} \quad x - 3 = 0 \quad \text{Principle of zero products}$$

$$x = -3 \quad \text{or} \quad x = 3.$$

The possible solutions are 3 and -3. We must check, since we have multiplied by an expression with a variable.

For 3:

$$\frac{x^2}{x-3} = \frac{9}{x-3}$$

$\frac{3^2}{3-3}$	$\frac{9}{3-3}$
$\frac{9}{0}$	$\frac{9}{0}$

3 does not check.

For -3:

$$\frac{x^2}{x-3} = \frac{9}{x-3}$$

$\frac{(-3)^2}{-3-3}$	$\frac{9}{-3-3}$
$-\frac{9}{6}$	$-\frac{9}{6}$

-3 checks.

Thus the solution set is $\{-3\}$.

We can actually determine the nonmeaningful replacements before we start solving by noting when the denominators of the original equation are 0.

Solve. Don't forget to check.

10. $\frac{y^2}{y+4} = \frac{16}{y+4}$

11. $\frac{x^2}{x-5} = \frac{36}{x-5}$

DO EXERCISES 10 AND 11.

The general procedure for solving fractional equations involves multiplying on both sides by the LCM of all the denominators. This procedure is called **clearing of fractions.**

EXAMPLE 6 Solve:

$$\frac{14}{x+2} - \frac{1}{x-4} = 1.$$

Solution We note at the outset that -2 and 4 are *not* meaningful replacements.

We multiply by the LCM of all the denominators: $(x + 2)(x - 4)$.

$$\frac{14}{x+2} - \frac{1}{x-4} = 1$$

$$(x + 2)(x - 4) \cdot \left[\frac{14}{x+2} - \frac{1}{x-4}\right] = (x + 2)(x - 4) \cdot 1$$

$$(x + 2)(x - 4) \cdot \frac{14}{x+2} - (x + 2)(x - 4) \cdot \frac{1}{x-4} = (x + 2)(x - 4) \cdot 1$$

Using the distributive law

$$14(x - 4) - (x + 2) = (x + 2)(x - 4)$$

Simplifying

$$14x - 56 - x - 2 = x^2 - 2x - 8$$

$$13x - 58 = x^2 - 2x - 8$$

$$0 = x^2 - 15x + 50$$

$$0 = (x - 10)(x - 5)$$

$$x - 10 = 0 \quad \text{or} \quad x - 5 = 0$$

Principle of zero products

$$x = 10 \quad \text{or} \quad x = 5$$

The possible solutions are 10 and 5. These check, so the solution set is $\{10, 5\}$.

DO EXERCISE 12.

EXAMPLE 7 Solve:

$$\frac{12x}{x-4} - \frac{3x^2}{x+4} = \frac{384}{x^2 - 16}.$$

Solution We note at the outset that 4 and -4 are *not* meaningful replacements. We first find the LCM by factoring the denominators:

$$\frac{12x}{x-4} - \frac{3x^2}{x+4} = \frac{384}{(x-4)(x+4)}.$$

The LCM is $(x - 4)(x + 4)$, or $x^2 - 16$. We multiply by the LCM. After clearing of fractions, we factor by grouping:

$$(x - 4)(x + 4)\left[\frac{12x}{x-4} - \frac{3x^2}{x+4}\right] = (x - 4)(x + 4) \cdot \frac{384}{x^2 - 16}$$

$$(x - 4)(x + 4) \cdot \frac{12x}{x-4} - (x - 4)(x + 4) \cdot \frac{3x^2}{x+4} = (x - 4)(x + 4) \cdot \frac{384}{x^2 - 16}$$

Using the distributive law

$$12x(x + 4) - 3x^2(x - 4) = 384$$

Simplifying

$$12x^2 + 48x - 3x^3 + 12x^2 = 384$$

$$-3x^3 + 24x^2 + 48x - 384 = 0$$

$$3x^3 - 24x^2 - 48x + 384 = 0$$

Multiplying by -1 to ease the factoring

$$3(x^3 - 8x^2 - 16x + 128) = 0$$

Factoring out the common factor 3

12. Solve:

$$\frac{4}{x+5} + \frac{1}{x-5} = \frac{1}{x^2 - 25}.$$

13. Solve:

$$\frac{7x-12}{x-3}-\frac{x^2}{x+3}=\frac{54}{x^2-9}.$$

and

$$x^3-8x^2-16x+128=0 \quad \text{Multiplying by } \tfrac{1}{3}$$
$$x^2(x-8)-16(x-8)=0 \quad \text{Factoring by grouping}$$
$$(x^2-16)(x-8)=0$$
$$(x-4)(x+4)(x-8)=0$$
$$x-4=0 \quad \text{or} \quad x+4=0 \quad \text{or} \quad x-8=0$$
$$x=x \quad \text{or} \quad x=-4 \quad \text{or} \quad x=8.$$

The number 8 checks, but the numbers -4 and 4 do not. The solution is 8. The solution set is $\{8\}$. ■

DO EXERCISE 13.

EXERCISE SET 2.2

1 Determine whether the equations of the pair are equivalent.

1. $3x+5=12$, $3x=7$

2. $x^2=-7x$, $x=-7$

3. $x=3$, $x^2=9$

4. $2y+1=-3$, $8y+4=-12$

5. $\dfrac{(x-3)(x+9)}{(x-3)}=x+9$, $x+9=x+9$

6. $x^2+x-20=0$, $x^2-25=0$

2 Solve.

7. $\dfrac{1}{4}+\dfrac{1}{5}=\dfrac{1}{t}$

8. $\dfrac{1}{3}-\dfrac{5}{6}=\dfrac{1}{x}$

9. $\dfrac{3}{x-8}=\dfrac{x-5}{x-8}$

10. $\dfrac{23}{y}=\dfrac{-5}{y}$

11. $\dfrac{x+2}{4}-\dfrac{x-1}{5}=15$

12. $\dfrac{t+1}{3}-\dfrac{t-1}{2}=1$

13. $x+\dfrac{6}{x}=5$

14. $x=\dfrac{12}{x}+1$

15. $\dfrac{x+2}{2}+\dfrac{3x+1}{5}=\dfrac{x-2}{4}$

16. $\dfrac{2x-1}{3}-\dfrac{x-2}{5}=\dfrac{x}{2}$

17. $\dfrac{1}{2}+\dfrac{2}{x}=\dfrac{1}{3}+\dfrac{3}{x}$

18. $\dfrac{1}{t}+\dfrac{1}{2t}+\dfrac{1}{3t}=5$

19. $\dfrac{4}{x^2-1}-\dfrac{2}{x-1}=\dfrac{3}{x+1}$

20. $\dfrac{3y+5}{y^2+5y}+\dfrac{y+4}{y+5}=\dfrac{y+1}{y}$

21. $\dfrac{1}{2t}-\dfrac{2}{5t}=\dfrac{1}{10t}-3$

22. $\dfrac{3}{m+2}+\dfrac{2}{m-2}=\dfrac{4m-4}{m^2-4}$

23. $1-\dfrac{3}{x}=\dfrac{40}{x^2}$

24. $1-\dfrac{15}{y^2}=\dfrac{2}{y}$

25. $\dfrac{11-t^2}{3t^2-5t+2}=\dfrac{2t+3}{3t-2}-\dfrac{t-3}{t-1}$

26. $\dfrac{1}{3y^2-10y+3}=\dfrac{6y}{9y^2-1}+\dfrac{2}{1-3y}$

27. $\dfrac{7x}{x-3}-\dfrac{21}{x}+11=\dfrac{63}{x^2-3x}$

28. $\dfrac{3x}{x-5}-\dfrac{15}{x+5}=\dfrac{150}{x^2-25}$

29. ▦ $\dfrac{2.315}{y}-\dfrac{12.6}{17.4}=\dfrac{6.71}{7}+0.763$

30. ▦ $\dfrac{6.034}{x}-43.17=\dfrac{0.793}{x}+18.15$

31. $\dfrac{2x^2}{x-3}+\dfrac{4x-6}{x+3}=\dfrac{108}{x^2-9}$

32. $\dfrac{3x^2}{x+2}+\dfrac{48}{x^2-4}=\dfrac{2x+8}{x-2}$

33. $\dfrac{24}{x^2 - 2x + 4} = \dfrac{3x}{x + 2} + \dfrac{72}{x^3 + 8}$

34. $\dfrac{90}{x^2 - 3x + 9} - \dfrac{5x}{x + 3} = \dfrac{405}{x^3 + 27}$

35. $\dfrac{5}{x - 1} + \dfrac{9}{x^2 + x + 1} = \dfrac{15}{x^3 - 1}$

36. $\dfrac{7}{x + 2} + \dfrac{5}{x^2 - 2x + 4} = \dfrac{84}{x^3 + 8}$

37. $\dfrac{7}{x - 9} - \dfrac{7}{x} = \dfrac{63}{x^2 - 9x}$

38. $\dfrac{26}{x + 13} - \dfrac{14}{x + 7} = \dfrac{12x}{x^2 + 20x + 91}$

39. $\dfrac{(x - 3)^2}{x - 3} = x - 3$

40. $\dfrac{x^2 + 6x - 16}{x - 2} = x + 8$

41. $\dfrac{x^3 + 8}{x + 2} = x^2 - 2x + 4$

42. $\dfrac{x + 8}{x - 2} = \dfrac{8 + x}{-2 + x}$

SYNTHESIS

43. Determine whether each equation is equivalent to the one that follows:

$$x^2 - x - 20 = x^2 - 25 \quad (1)$$
$$(x - 5)(x + 4) = (x - 5)(x + 5) \quad (2)$$
$$x + 4 = x + 5 \quad (3)$$
$$4 = 5. \quad (4)$$

Equations that are true for all meaningful replacements of the variables are called **identities.** (See Section 2.1.) Determine whether each of the following equations is an identity.

44. $\dfrac{x^2 + 6x - 16}{x - 2} = x + 8$

45. $x + 4 = 4 + x$

46. $(x - 1)(x^2 + x + 1) = x^3 - 1$

47. $\dfrac{x^3 + 8}{x^2 - 4} = \dfrac{x^2 - 2x + 4}{x - 2}$

48. $(x + 7)^2 = x^2 + 49$

49. $\sqrt{x^2 - 16} = x - 4$

CHALLENGE

50. Solve: $\dfrac{x + 3}{x + 2} - \dfrac{x + 4}{x + 3} = \dfrac{x + 5}{x + 4} - \dfrac{x + 6}{x + 5}$.

2.3 Formulas and Problem Solving

OBJECTIVES

You should be able to:

1 Solve a formula for a given variable.

2 Use the problem-solving strategy to solve applied problems.

1 Formulas

A formula is a "recipe" for doing a calculation. An example is $A = \pi rs + \pi r^2$, which gives the area A of a cone in terms of the slant height s and radius of the base r.

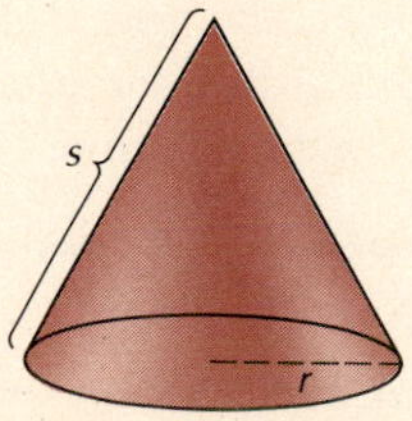

Suppose we want to find the slant height s when the area A and the radius r are known. Our knowledge of equations allows us to get s alone on one side or, as we say, "solve the formula for s."

EXAMPLE 1 Solve $A = \pi rs + \pi r^2$ for s.

1. Solve

$$C = \tfrac{5}{9}(F - 32)$$

for F. (This is a formula for Celsius, or Centigrade, temperature in terms of Fahrenheit temperature F.)

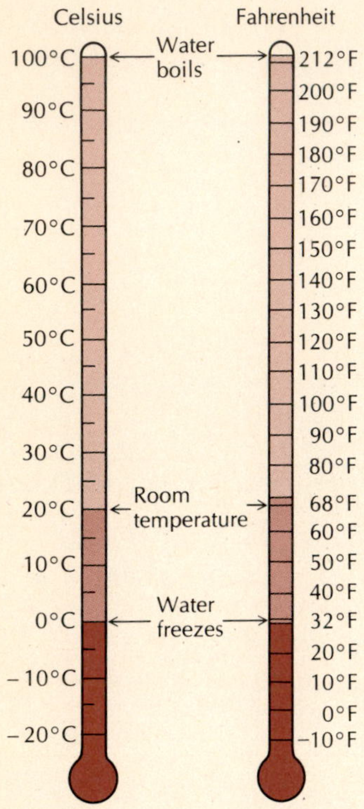

2. Solve

$$\frac{1}{R} = \frac{1}{r_1} + \frac{1}{r_2}$$

for r_2.

Solution We have

$$A = \pi rs + \pi r^2$$
$$A - \pi r^2 = \pi rs$$
$$\frac{A - \pi r^2}{\pi r} = s. \quad \text{Multiplying by } \frac{1}{\pi r}$$

An equivalent expression for s is

$$\frac{A}{\pi r} - r.$$

EXAMPLE 2 Solve

$$\frac{1}{R} = \frac{1}{r_1} + \frac{1}{r_2}$$

for R. (This is a formula from electricity.)

Resistance in parallel:

$$\frac{1}{R} = \frac{1}{r_1} + \frac{1}{r_2}$$

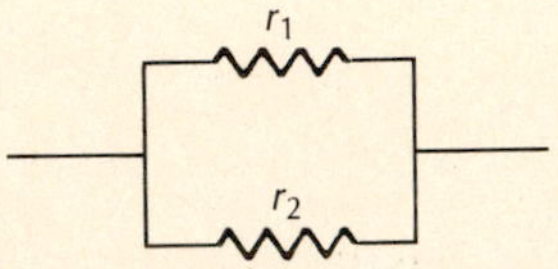

Solution We first multiply by the LCM, which is Rr_1r_2:

$$Rr_1r_2 \cdot \frac{1}{R} = Rr_1r_2 \cdot \left(\frac{1}{r_1} + \frac{1}{r_2}\right)$$
$$Rr_1r_2 \cdot \frac{1}{R} = Rr_1r_2 \cdot \frac{1}{r_1} + Rr_1r_2 \cdot \frac{1}{r_2} \quad \text{Using a distributive law}$$
$$r_1r_2 = Rr_2 + Rr_1 \quad \text{Simplifying}$$
$$r_1r_2 = (r_2 + r_1)R \quad \text{Factoring out the common factor } R$$
$$\frac{r_1r_2}{r_2 + r_1} = R. \quad \text{Multiplying by } \frac{1}{r_2 + r_1}$$

DO EXERCISES 1 AND 2.

2 Problem Solving

By an **applied problem,** we mean a problem in which mathematical techniques are used to answer some question. Problems like this may be posed orally. They may come about in the course of a conversation, or they can be hatched within the mind of one person. Thus to call them "word problems" or "story problems" is misleading.

There is no one rule that will enable us to solve applied problems, because they are of many different kinds. We can, however, describe an overall, or general, strategy. The idea is to translate the problem situation to mathematical language and then calculate to find a solution.

The Five-Step Problem-Solving Process

1. ***Familiarize*** **yourself with the problem situation. If the problem is presented to you in written words, then, of course, this means to read carefully. Some or all of the following can also be helpful.**
   a) **Make a drawing, if it makes sense to do so. It is difficult to overemphasize the importance of this!**
   b) **Make a written list of the known facts and a list of what you wish to find out.**

c) Organize the information in a chart or table.
d) Assign certain variables to represent unknown quantities.
e) Find further information. Look up a formula or consult a reference book or an expert in the field.
f) Guess or estimate the answer.

2. *Translate* the problem situation to mathematical language or symbolism. For most of the problems you will encounter in algebra, this means to write one or more equations, but sometimes an inequality or some other mathematical symbolism may be appropriate.
3. *Carry out* some kind of mathematical manipulation. This means to use your mathematical knowledge to find a possible solution. In algebra, this usually means to solve an equation or system of equations.
4. *Check* to see whether your possible solution actually fits the problem situation and is thus really a solution of the problem. You may be able to solve an equation, but the solution(s) of the equation may or may not be solution(s) of the original problem.
5. *State* the answer clearly.

Whether you use all or part of the process depends on the difficulty of the problem. You may indeed guess a possible solution and check it in the problem. In this text, you will generally need to do more than that. We may sometimes illustrate all of the process, and sometimes not. In general, the more difficult the problem, the more of the process you will need to use.

Problems stated in textbooks are of necessity somewhat contrived. Problems that you encounter in nonclassroom situations will almost invariably contain insufficient information to obtain a firm, or exact, answer. They may also contain a good deal of extraneous, useless information.

Let us consider a problem.

EXAMPLE 3 Assume that a hydrogen atom is a sphere. Find its volume.

Solution

1. *Familiarize.* This problem is typical of those encountered in real life. There is insufficient information to obtain an answer. To get an answer, we might do the following:

   a) Try to look up the answer in a reference book or simply call a chemist, or
   b) Look up a formula for finding the volume of a sphere. On the basis of the variables in the formula, look up certain information.

   Let us suppose we did the latter. We look up a formula for the volume of a sphere (see the table of geometric formulas at the back of the book) and find it to be

   $$V = \tfrac{4}{3}\pi r^3,$$

   where V = the volume and r = the radius of the sphere. In a reference book, we find that the diameter of a hydrogen atom is 0.0000000001 m. Now we know that $r = d/2$, so we can indeed complete the problem.

2. *Translate.*

   $$V = \tfrac{4}{3}\pi r^3 \quad \text{and} \quad r = d/2.$$

3. *Carry out.* We substitute the value for d and calculate r. Then we compute V.

   $$r = d/2 = 0.0000000001/2 = 5.0 \times 10^{-11},$$
   $$V = \tfrac{4}{3}\pi r^3 \approx \tfrac{4}{3}(3.14)(5.0 \times 10^{-11})^3 \approx 5.233 \times 10^{-31}$$

3. What percent of 76 is 13.68?

4. 24% of what is 13.68?

5. 36% of 75 is what?

4. *Check.* In a problem such as this, simply recalculating is a sufficient check.
5. *State.* The volume of a hydrogen atom is about 5.233×10^{-31} cubic meters (m^3).

> ***Calculator note:*** **Many calculators have a π key, which gives a better approximation than 3.14. If such a key were used, another approximation in the problem above might be 5.236×10^{-31}.**

In the following examples, we illustrate how a problem situation can be translated to mathematical language. In certain simple situations, the translation is easy because certain words translate directly to mathematical symbols. Note that the word "is" translates to an equals sign, the word "what" translates to a variable, and the word "of" translates to a multiplication sign.

EXAMPLE 4 What percent of 84 is 11.76?

Solution

1. *Familiarize.* We let $x =$ the percent. The percent symbol, %, will have to be added to the answer.
2. *Translate.*

$$\underbrace{\text{What percent}}_{x\%}\ \underbrace{\text{of}}_{\cdot}\ \underbrace{84}_{84}\ \underbrace{\text{is}}_{=}\ \underbrace{11.76?}_{11.76}$$

3. *Carry out.* We solve the equation:

$$x \cdot 0.01 \cdot 84 = 11.76 \qquad \text{"\%" means " } \cdot 0.01\text{"}$$
$$x \cdot 0.84 = 11.76$$
$$x = \frac{11.76}{0.84} = 14.$$

4. *Check.* $14\% \cdot 84 = 0.14 \cdot 84 = 11.76$
5. *State.* The answer is 14%.

EXAMPLE 5 14% of what is 11.76?

Solution

1. *Familiarize.* We let $y =$ the number of which a percentage is being taken.
2. *Translate.*

$$\underbrace{14\%}_{14\%}\ \underbrace{\text{of}}_{\cdot}\ \underbrace{\text{what}}_{y}\ \underbrace{\text{is}}_{=}\ \underbrace{11.76?}_{11.76}$$

3. *Carry out.* We solve the equation:

$$0.14 \cdot y = 11.76$$
$$y = \frac{11.76}{0.14} = 84.$$

4. *Check.* $14\% \cdot 84 = 0.14 \cdot 84 = 11.76$
5. *State.* The answer is 84.

DO EXERCISES 3–5.

In the remainder of this section, we consider applied problems of various types. Although there is no rule for solving applied problems because they can be so different, it does help somewhat to consider a few different types of problems. *The best way to learn to solve applied problems is to solve a lot of them and to use the five steps that we have given you.*

Problems Involving Compound Interest

EXAMPLE 6 An investment is made at 11%, compounded annually. It grows to \$1443 at the end of one year. How much was originally invested?

Solution

1. *Familiarize.* We first restate the situation as follows:

   The invested amount *plus* the interest is \$1443.

   We might guess an answer, say, \$1200. We take \$1200 plus 11% of \$1200, and get \$1200 + \$132, or \$1332. We see that our guess is too small. We could continue our guessing, but we want to use our algebra as a tool. Though we have not found the answer, we are much more familiar with the problem.

   We let $x =$ the amount originally invested. Then we translate.

2. *Translate.* Since the interest is 11% of the amount invested, we have the following, which translates directly:

$$\underbrace{\text{Invested amount}}_{x}\ \underbrace{\text{plus}}_{+}\ \underbrace{\text{11\% of invested amount}}_{11\% \cdot x}\ \underbrace{\text{is}}_{=}\ \underbrace{\$1443}_{\$1443}.$$

3. *Carry out.* We solve the equation:

$$x + 11\%x = 1443$$
$$x + 0.11x = 1443$$
$$(1 + 0.11)x = 1443$$
$$1.11x = 1443$$
$$x = \frac{1443}{1.11} = 1300.$$

4. *Check.* We check in much the same manner that we made a guess when familiarizing:

   \$1300 + 11% of \$1300 = \$1300 + \$143, or \$1443.

5. *State.* The amount originally invested was \$1300. ■

DO EXERCISE 6.

Let us now consider an investment over a period longer than one year. If we invest P dollars at an interest rate i, compounded annually, we will have an amount in the account at the end of a year that we will call A_1. Now, $A_1 = P + Pi$, or

$$A_1 = P(1 + i).$$

Going into the second year, we have $P(1 + i)$ dollars. By the end of the second year, we will have A_2 dollars, given by

$$A_2 = A_1 \cdot (1 + i).$$

6. An investment is made at 7%, compounded annually. It grows to \$2782 at the end of one year. How much was originally invested?

7. Suppose that \$1000 is invested at 12.5%, compounded annually. How much will be in the account at the end of 8 years?

But $A_1 = P(1 + i)$, so $A_2 = P(1 + i)(1 + i)$, or

$$A_2 = P(1 + i)^2.$$

Similarly, the amount A_3 in the account at the of three years is given by

$$A_3 = P(1 + i)^3,$$

and so on. In general, the following applies.

THEOREM 1

If principal P is invested at an interest rate i, compounded annually, in t years it will grow to an amount A given by

$$A = P(1 + i)^t.$$

EXAMPLE 7 Suppose that \$1000 is invested at 12%, compounded annually. What amount will be in the account at the end of 10 years?

Solution We do not need the entire problem-solving strategy. We have a formula. We arrive at the solution by substituting into the formula. Using the equation $A = P(1 + i)^t$, we get

$$A = 1000(1 + 0.12)^{10} = 1000(1.12)^{10}$$
$$\approx 3105.85. \quad \text{Using the } y^x \text{ key}$$

The answer is \$3105.85.

DO EXERCISE 7.

Interest may be compounded more often than once a year. Suppose it is compounded four times a year, or *quarterly*. The formula derived above can be altered to apply. We consider one fourth of a year to be an *interest period*. The *rate of interest* for such a period is then $i/4$. The number of periods will be four times the number of years. This is shown in the following diagram.

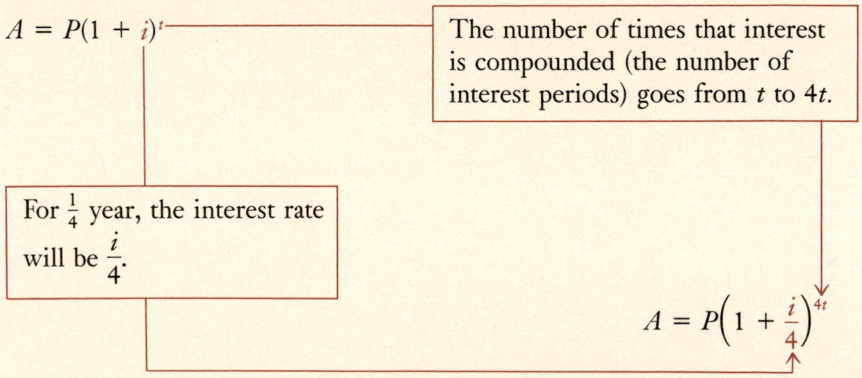

$A = P(1 + i)^t$
The number of times that interest is compounded (the number of interest periods) goes from t to $4t$.
For $\frac{1}{4}$ year, the interest rate will be $\frac{i}{4}$.
$A = P\left(1 + \frac{i}{4}\right)^{4t}$

Now suppose that the number of interest periods per year is something other than 4, say n. Using the reasoning illustrated above, we obtain a general formula.

THEOREM 2

If principal P is invested at an interest rate i, compounded n times per year, in t years it will grow to an amount A given by

$$A = P\left(1 + \frac{i}{n}\right)^{nt}.$$

When problems involving compound interest are translated to mathematical language, the preceding formula is almost always used.

EXAMPLE 8 Suppose that \$1000 is invested at 12%, compounded quarterly. How much will be in the account at the end of 10 years?

Solution In this case, $n = 4$ and $t = 10$. We substitute into the formula:

$$A = P\left(1 + \frac{i}{n}\right)^{nt} = 1000\left(1 + \frac{0.12}{4}\right)^{4 \cdot 10}$$

$$= 1000(1.03)^{40}$$

$$\approx 3262.04. \quad \text{Using the } y^x \text{ key}$$

The answer is \$3262.04.

DO EXERCISE 8.

8. Suppose that \$1000 is invested at 12.5%, compounded semiannually ($n = 2$). How much will be in the account at the end of 8 years?

Problems Involving Area

EXAMPLE 9 The radius of a circular swimming pool is 10 ft. A sidewalk of uniform width is constructed around the outside and has an area of 44π ft^2. How wide is the sidewalk?

Solution

1. *Familiarize.* First make a drawing:

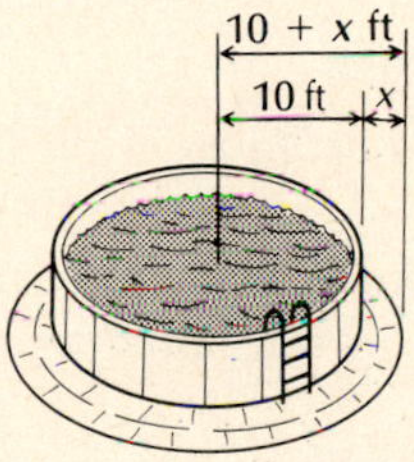

Let x represent the width of the walk. Then, recalling that a formula for the area of a circle is $A = \pi r^2$, we have

Area of pool $= \pi \cdot 10^2 = 100\pi$;

Area of sidewalk plus pool $= \pi \cdot (10 + x)^2 = \pi(100 + 20x + x^2)$.

2. *Translate.* The translation is as follows:

$$\underbrace{\text{Area of sidewalk plus pool}} \; \underbrace{-} \; \underbrace{\text{Area of pool}} = \underbrace{\text{Area of sidewalk}}$$

$$\pi(100 + 20x + x^2) - 100\pi = 44\pi.$$

3. *Carry out.* We solve the equation:

$$100 + 20x + x^2 - 100 = 44 \quad \text{Multiplying by } 1/\pi$$

$$x^2 + 20x = 44$$

$$x^2 + 20x - 44 = 0$$

$$(x + 22)(x - 2) = 0$$

$$x = -22 \quad \text{or} \quad x = 2.$$

4. *Check.* We know that -22 ft is not a solution of the original problem because width must be positive. The number 2 checks; that is, when the sidewalk is 2 ft wide, the area of the pool plus sidewalk is $\pi \cdot 12^2$, or 144π ft^2.

9. A rectangular garden is 60 ft by 80 ft. Part of the garden is torn up to install a sidewalk of uniform width around the garden. The new area of the garden is one sixth of the old area. How wide is the sidewalk?

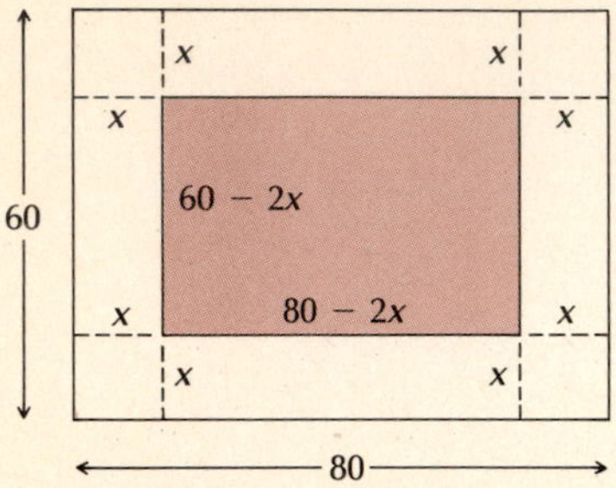

5. *State.* The width of the sidewalk is 2 ft.

DO EXERCISE 9.

Problems Involving Motion

For problems that deal with distance, time, and speed, we almost always need to recall the **distance formula**, or something equivalent to it.

> **Distance = Rate (or speed) times Time or $d = rt$, where d = distance, r = speed, and t = time.**

If you memorize the equation $d = rt$, you can easily obtain either of the two equivalent equations, $r = d/t$ or $t = d/r$, as needed.

When translating these problems to mathematical language, it is often helpful to look for some quantity that is constant in the problem. For example, two cars may travel for the same length of time, or two boats may travel the same distance. Such facts often provide the basis for setting up an equation.

EXAMPLE 10 A boat travels 246 km downstream in the same time that it takes to travel 180 km upstream. The speed of the current in the stream is 5.5 km/h. Find the speed of the boat in still water.

Solution

1. *Familiarize.* We first make a drawing and lay out the known facts and any other pertinent information. The time to go downstream is the same as the time to go upstream. We call it t. We know that the speed of the current is 5.5 km/h, but we do not know the speed of the boat. Let's call it r.

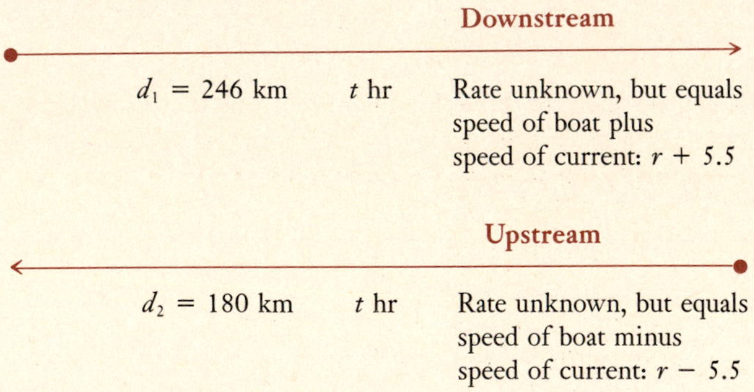

We can also organize the data in a table.

	Distance	Speed	Time
Downstream	246	$r + 5.5$	t
Upstream	180	$r - 5.5$	t

2. *Translate.* The table suggests that we use the formula $t = d/r$, because we can get two different expressions for t. We get two equations:

$$t = \frac{246}{r + 5.5} \quad \text{and} \quad t = \frac{180}{r - 5.5}.$$

Thus,

$$\frac{246}{r + 5.5} = \frac{180}{r - 5.5}.$$

3. *Carry out.* Solving for r, we get 35.5 km/h.
4. *Check.* We leave the check to the student.
5. *State.* The speed of the boat in still water is 35.5 km/h. ■

DO EXERCISE 10.

In the following example, although the distance is the same in both directions, the key to the translation lies in an additional piece of given information.

EXAMPLE 11 The speed of a boat in still water is 10 mph. It travels 24 mi upstream and 24 mi downstream in a total time of 5 hr. What is the speed of the current?

Solution

1. *Familiarize.* We first make a drawing and write out the pertinent information. We know that the speed of the boat is 10 mph, but we do not know the speed of the current. Let's call it c.

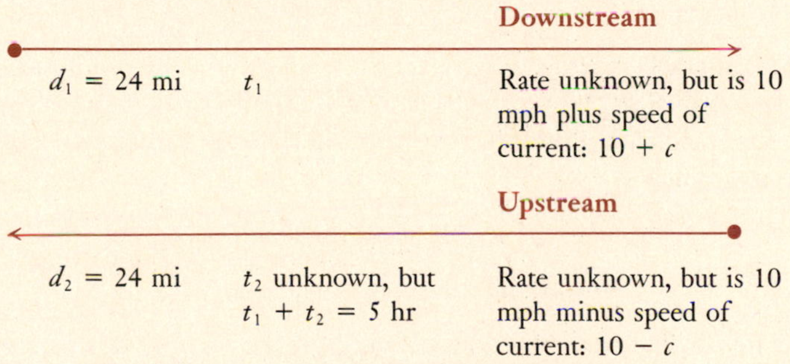

We can also organize the information in a table.

	Distance	Speed	Time
Downstream	24	$10 + c$	t_1
Upstream	24	$10 - c$	t_2

2. *Translate.* This time, the basis of our translation is the equation involving the times:

$$t_1 + t_2 = 5.$$

We use $t = d/r$ with the information from the table. This leads us to

$$\frac{d_1}{r_1} + \frac{d_2}{r_2} = 5$$

and then

$$\frac{24}{10 + c} + \frac{24}{10 - c} = 5,$$

where c is the speed of the current.

3. *Carry out.* Solving for c, we get $c = -2$ or $c = 2$.

10. An airplane flies 1062 km with the wind in the same time that it takes to fly 738 km against the wind. The speed of the plane in still air is 200 km/h. Find the speed of the wind.

4. *Check.* Since speed cannot be negative in this problem, -2 cannot be a solution. But 2 checks.

5. *State.* The speed of the current is 2 mph.

DO EXERCISES 11 AND 12.

Problems Involving Work

EXAMPLE 12 Typist A can do a certain job in 3 hr. Typist B can do the same job in 5 hr. How long would it take both, working together, to do the same job?

Solution

1. *Familiarize.* We list the facts:

A can do the typing job in 3 hr;

B can do the same typing job in 5 hr.

We want to know how long it will take them working together. Let's let t be that number of hours. We first consider two common, *but incorrect*, approaches to the problem.

a) One *incorrect* approach is to simply add the two times:

$$3 \text{ hr} + 5 \text{ hr} = 8 \text{ hr}.$$

This cannot be correct since either A or B can do the typing job alone in less than 8 hr.

b) Another *incorrect* approach is to split up the job so that each does half the job. Then

A does $\frac{1}{2}$ the job in $\frac{1}{2}(3 \text{ hr})$, or 1.5 hr, and

B does $\frac{1}{2}$ the job in $\frac{1}{2}(5 \text{ hr})$, or 2.5 hr.

But this would leave A idle for 1 hr, and they would not be working *together* the entire time. We can at least see that, by working together, A and B can finish the job in a time somewhere between 1.5 hr and 2.5 hr.

Let's consider how much of the job is done in 1 hr, 2 hr, 3 hr, and so on. Since A can do the whole typing job in 3 hr, A can do $\frac{1}{3}$ of it in 1 hr. Since B can do the whole job in 5 hr, B can do $\frac{1}{5}$ of it in 1 hr. Working together, they can do

$$\frac{1}{3} + \frac{1}{5} = \frac{8}{15} \text{ in 1 hr.} \qquad (1)$$

In 2 hr, A can do $2(\frac{1}{3})$ of the job and B can do $2(\frac{1}{5})$ of the job. Working together, they can do

$$2\left(\frac{1}{3}\right) + 2\left(\frac{1}{5}\right) = \frac{16}{15} \quad \text{or} \quad 1\frac{1}{15} \text{ in 2 hr.} \qquad (2)$$

But $1\frac{1}{15}$ would represent doing more than 1 job.

We let t = the number of hours required for A and B, working together, to do the job.

2. *Translate.* From equations (1) and (2), we see that the time we want is some number t for which

$$t\left(\frac{1}{3}\right) + t\left(\frac{1}{5}\right) = 1, \quad \text{or} \quad \frac{t}{3} + \frac{t}{5} = 1.$$

11. A train leaves a station and travels north at a speed of 75 km/h. Two hours later, a second train leaves on a parallel track traveling north at a speed of 125 km/h. How far from the station will the second train overtake the first train?

12. A train leaves Oldtown and travels 300 km to Newtown. On the return trip, it travels 10 km/h faster. The total time for the round trip was 11 hr. How fast did the train travel on each part of the trip?

3. *Carry out.* We solve the equation:

$$\frac{t}{3} + \frac{t}{5} = 1$$

$$15\left(\frac{t}{3}\right) + 15\left(\frac{t}{5}\right) = 15 \cdot 1 \quad \text{Multiplying by the LCM, 15}$$

$$5t + 3t = 15$$

$$8t = 15$$

$$t = \frac{15}{8}, \text{ or } 1\frac{7}{8} \text{ hr.}$$

4. *Check.* Consider $1\frac{7}{8}$ hr. If A works $1\frac{7}{8}$ hr, then A will do $\left(\frac{1}{3}\right)\left(\frac{15}{8}\right)$, or $\frac{5}{8}$ of the job. Then B can do $\left(\frac{1}{5}\right)\left(\frac{15}{8}\right)$, or $\frac{3}{8}$ of the job. Together they will do $\frac{5}{8} + \frac{3}{8}$, or 1 complete job.

   We have another partial check in noting from the familiarization that the entire job can be done between 1.5 hr, the time it takes A to do half the job, and 3 hr, the time it takes A alone.

5. *State.* It will take $1\frac{7}{8}$ hr for A and B to do the job together.

DO EXERCISE 13.

EXAMPLE 13 It takes Red 9 hr longer to build a wall than it takes Mort. If they work together, they can build the wall in 20 hr. How long would it take each, working alone, to build the wall?

Solution Let $t =$ the amount of time it takes Mort working alone. Then $t + 9 =$ the amount of time it takes Red working alone.

Thus Mort can do $1/t$ of the work in 1 hr and Red can do $1/(t + 9)$ of it in 1 hr. We know that working together, they can do the job in 20 hr. In 20 hr, Mort does $20(1/t)$ of the job and Red does $20(1/(t + 9))$ of the job. If we add these fractional parts, we get the entire job, represented by 1. This gives us the following equation:

$$20\left(\frac{1}{t}\right) + 20\left(\frac{1}{t + 9}\right) = 1.$$

Solving the equation, we get $t = -5$ or $t = 36$. Since negative time has no meaning in the problem, -5 is not a solution of the original problem. The number 36 checks in the original problem. Thus it would take Mort 36 hr and Red 45 hr.

DO EXERCISE 14.

13. A can mow a lawn in 4 hr, while B can mow the same lawn in 5 hr. How long will it take them to mow the lawn if they work together?

14. Fran and Helen work together and get a certain job completed in 4 hr. It would take Helen 6 hr longer, working alone, to do the job than it would Fran. How long would each need to do the job working alone?

EXERCISE SET 2.3

1 Solve the formula for the indicated letter.

1. Solve $P = 2l + 2w$ for w. (Perimeter of a rectangle)
2. Solve $C = 2\pi r$ for r. (Circumference of a circle)
3. Solve $A = \frac{1}{2}bh$ for b. (Area of a triangle)
4. Solve $A = \pi r^2$ for π. (Area of a circle)
5. Solve $d = rt$ for r. (Distance formula)
6. Solve $F = ma$ for a. (Force = Mass times Acceleration)
7. Solve $E = IR$ for I.
8. Solve $F = \frac{km_1m_2}{d^2}$ for m_2.
9. Solve $\frac{P_1V_1}{T_1} = \frac{P_2V_2}{T_2}$ for T_1.
10. Solve $\frac{P_1V_1}{T_1} = \frac{P_2V_2}{T_2}$ for V_2.

11. Solve $S = \dfrac{H}{m(v_1 - v_2)}$ for v_1.

12. Solve $S = \dfrac{H}{m(v_1 - v_2)}$ for v_2.

13. Solve $\dfrac{1}{F} = \dfrac{1}{m} + \dfrac{1}{p}$ for p.

14. Solve $\dfrac{1}{F} = \dfrac{1}{m} + \dfrac{1}{p}$ for F.

Solve for x.

15. $(x + a)(x - b) = x^2 + 5$

16. $(c + d)x + (c - d)x = c^2$

17. $10(a + x) = 8(a - x)$

18. $4(a + b + x) + 3(a + b - x) = 8a$

2 Problem Solving

19. 79.2 is what percent of 180?

20. 6% of what number is 480?

21. What percent of 28 is 1.68?

22. What is 7% of 45.3?

23. A person gets an 11% raise that is \$1595. What was the old salary? the new salary?

24. A person gets a 12% raise that is \$2520. What was the old salary? the new salary?

25. An investment is made at 8%, compounded annually. It grows to \$702 at the end of one year. How much was originally invested?

26. An investment is made at 9%, compounded annually. It grows to \$926.50 at the end of one year. How much was originally invested?

27. In triangle ABC, angle B is five times as large as angle A. The measure of angle C is 2° less than that of angle A. Find the measures of the angles. (*Hint:* The sum of the angle measures is 180°.)

28. In triangle ABC, the angle B is twice as large as angle A. Angle C measures 20° more than angle A. Find the measures of the angles.

29. The perimeter of a rectangle is 322 m. The length is 25 m more than the width. Find the dimensions.

30. The length of a rectangle is twice the width. The perimeter is 39 m. Find the dimensions.

31. A student's scores on three tests are 87%, 64%, and 78%. What must the student score on the fourth test so that the average will be 80%?

32. A student's scores on three tests are 74%, 55%, and 68%. What must the student score on the fourth test so that the average will be 70%?

33. An open box is made from a 10-cm by 20-cm piece of tin by cutting a square from each corner and folding up the edges. The area of the resulting base is 96 cm^2. What is the length of the sides of the squares?

34. The frame of a picture is 28 cm by 32 cm outside and is of uniform width. What is the width of the frame if 192 cm^2 of the picture shows?

35. After a 2% increase, the population of a city is 826,200. What was the former population?

36. After a 3% increase, the population of a city is 741,600. What was the former population?

37. A boat travels 50 km downstream in the same time that it takes to go 30 km upstream. The speed of the stream is 3 km/h. Find the speed of the boat in still water.

38. A boat travels 50 km downstream in the same time that it takes to go 30 km upstream. The speed of the boat in still water is 16 km/h. Find the speed of the stream.

39. The speed of train A is 12 mph slower than the speed of train B. Train A travels 230 mi in the same time that it takes train B to travel 290 mi. Find the speed of each train.

40. The speed of a passenger train is 14 mph faster than the speed of a freight train. The passenger train travels 400 mi in the same time that it takes the freight train to travel 330 mi. Find the speed of each train.

41. An airplane leaves Chicago at a speed of 475 mph. Twenty minutes later, a plane leaves Cleveland, which is 350 mi from Chicago, at a speed of 500 mph. When they meet, how far are they from Cleveland? (*Hint:* It is usually best to make the units consistent. That is, consider 20 min as $\frac{1}{3}$ hr.)

42. A private airplane leaves an airport and flies due east at a speed of 180 km/h. Two hours later, a jet leaves the same airport and flies due east at a speed of 900 km/h. How far from the airport will the jet overtake the private plane?

43. A can do a certain job in 3 hr, B can do the same job in 5 hr, and C can do the same job in 7 hr. How long would the job take with all three working together?

44. Pipe A can fill a tank in 4 hr, pipe B can fill it in 10 hr, and pipe C can fill it in 12 hr. The pipes are connected to the same tank. How long does it take to fill the tank if all three are running together?

45. A can do a certain job, working alone, in 3.15 hr. Working with B, A can do the same job in 2.09 hr. How long would it take B, working alone, to do the job?

46. At a factory, smokestack A pollutes the air 2.13 times as fast as smokestack B. When both stacks operate together, they yield a certain amount of pollution in 16.3 hr. Find the amount of time that it would take each to yield the same amount of pollution if it operated alone.

47. Suppose that \$1000 is invested at $13\frac{3}{4}\%$. How much is in the account at the end of one year, if interest is compounded (a) annually? (b) semiannually? (c) quarterly? (d) daily (use 365 days per yr)? (e) hourly?

48. Suppose that \$1000 is invested at 15.5%. How much is in the account at the end of five years, if interest is compounded (a) annually? (b) semiannually? (c) quarterly? (d) daily (use 365 days per yr)? (e) hourly?

SYNTHESIS

49. A car is driven 144 mi. If it had gone 4 mph faster, it could have made the trip in $\frac{1}{2}$ hr less time. What was the speed? (*Average speed* is defined as total distance divided by total time.)

50. A car is driven 280 mi. If it had gone 5 mph faster, it could have made the trip in 1 hr less time. What was the speed?

51. A student drove 3 hr on a freeway at a speed of 55 mph and then drove 10 mi in the city at 35 mph. What was the average speed?

52. For the first 100 km of a 200-km trip, a student drove at a speed of 40 km/h. For the second half of the trip, the student drove at a speed of 60 km/h. What was the average speed for the entire trip? (It is not 50 km/h.)

53. A driver drove half the distance of a trip at a speed of 40 mph. At what speed would the driver have to drive for the rest of the distance so that the average speed for the entire trip would be 45 mph and the trip would be completed in 1 hr?

54. At what time after 4:00 will the minute hand and the hour hand of a clock first be in the same position?

55. At what time after 10:30 will the hands of a clock first be perpendicular?

56. Three trucks, A, B, C, working together, can move a load of sand in t hours. When working alone, it takes A 1 extra hour to move the sand; B, 6 extra hours; and C, t extra hours. Find t.

57. If you are married and filing a joint tax return, and your taxable income is between \$29,751 and \$71,900, your federal taxes are \$4462 plus 28% of the amount of taxable income that exceeds \$29,751. Suppose under these conditions that you pay federal taxes of \$8711. What was your taxable income that year?

58. If you are single, and your taxable income is between \$17,851 and \$43,150, your federal taxes are \$2677 plus 28% of the amount of taxable income that exceeds \$17,851. Suppose under these conditions that you pay federal taxes of \$6786. What was your taxable income that year?

CHALLENGE

59. An airplane is flying from Los Angeles to Hawaii at a speed of 750 mph with a tailwind of 50 mph. The distance, in statute miles, from Los Angeles to Honolulu is 2574 mi.

   a) Find the point at which it takes the same amount of time to go back to Los Angeles as it does to go on to Honolulu.

   b) After traveling 1187 mi, the pilot determines that it is necessary to make an emergency landing. Would it require less time to continue to Honolulu or to return to Los Angeles?

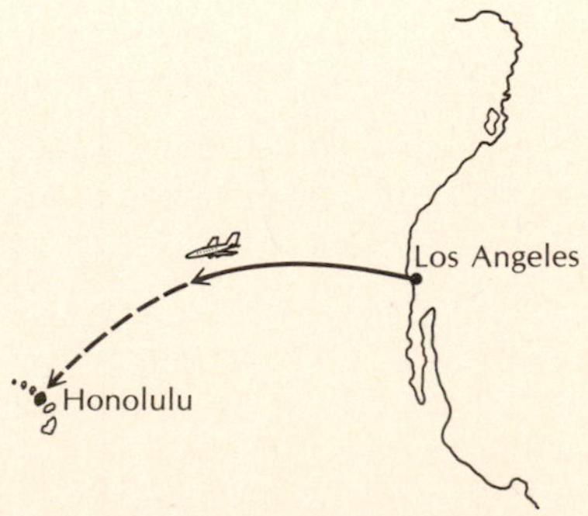

60. A commuter drives to work at a speed of 45 mph and arrives one minute early. At 40 mph, the commuter would arrive one minute late. How far is it to work?

61. Suppose that b is 20% more than a, c is 25% more than b, and d is $k\%$ less than c. Find k such that $a = d$.

62. A loan of \$1000 is taken at 11%, compounded quarterly. What is the equivalent interest rate if it were compounded only annually?

63. Suppose your father gathers the family together and gives half of all the money in his pocket to your mother. Then he gives one fourth of what is left to your sister, and one third of what is left after that to your brother. He then gives you half of what is left, which happens to be \$2. How much was in his pocket at the outset?

64. A student walks into a bakery and says to the owner, "I will buy half of all the pies in the store, plus half a pie." The sale is then made. Another student comes into the store and makes the same statement and purchase. Then so does a third student. The owner then has exactly one pie left. How many pies did the owner have in the store to begin with?

2.4 The Complex Numbers

OBJECTIVES

You should be able to:

1. Express imaginary numbers (that is, square roots of negative numbers) in terms of i, and simplify.
2. Add, subtract, and multiply complex numbers, expressing the answer as $a + bi$. Also, factor sums of squares.
3. Determine whether a complex number is a solution of an equation.
4. Use the fact that for equality of complex numbers, the real parts must be the same and the imaginary parts must be the same.
5. Find the conjugate of a complex number and divide complex numbers.
6. Find the reciprocal of a complex number and express it in the form $a + bi$.
7. Solve linear equations with complex-number coefficients.

1 Imaginary Numbers

Since numbers do not have square roots in the system of real numbers, certain equations such as $x^2 = -1$ have no solutions. A new kind of number, called *imaginary*, was invented so that negative numbers would have square roots and certain equations would have solutions. These numbers were devised, starting with an imaginary unit named i, with the agreement that $i^2 = -1$ and $i = \sqrt{-1}$.

> DEFINITION
>
> **The number i is defined such that $i = \sqrt{-1}$ and $i^2 = -1$.**

All other imaginary numbers can then be expressed as a product of i and a nonzero real number.

> DEFINITION
>
> **An *imaginary number* is a number that can be named bi, where b is some nonzero real number.**

In the complex-number system, when $a > 0, \sqrt{-a} = \sqrt{-1} \cdot \sqrt{a}$. We can use this property to express certain roots in terms of i.

EXAMPLES Express in terms of i.

1. $\sqrt{-7} = \sqrt{-1 \cdot 7} = \sqrt{-1} \cdot \sqrt{7} = i\sqrt{7}$, or $\sqrt{7}i$
2. $\sqrt{-16} = \sqrt{-1 \cdot 16} = \sqrt{-1} \cdot \sqrt{16} = i \cdot 4 = 4i$
3. $-\sqrt{-13} = -\sqrt{-1 \cdot 13} = -\sqrt{-1} \cdot \sqrt{13} = -i\sqrt{13}$, or $-\sqrt{13}i$
4. $-\sqrt{-64} = -\sqrt{-1 \cdot 64} = -\sqrt{-1} \cdot \sqrt{64} = -i \cdot 8 = -8i$
5. $\sqrt{-48} = \sqrt{-1 \cdot 48} = \sqrt{-1} \cdot \sqrt{48}$
   $= i \cdot \sqrt{48} = i \cdot 4\sqrt{3} = 4\sqrt{3}i = 4i\sqrt{3}$

It is also to be understood that the imaginary numbers obey the familiar laws of real numbers, such as the commutative and the associative laws.

EXAMPLE 6 Simplify: $\sqrt{-3}\sqrt{-7}$.

Solution *Important:* We first express the two imaginary numbers in terms of i:

$$\sqrt{-3}\sqrt{-7} = i\sqrt{3} \cdot i\sqrt{7}.$$

Now, rearranging and combining, we have

$$\begin{aligned} i^2\sqrt{3}\sqrt{7} &= -1 \cdot \sqrt{21} \\ &= -\sqrt{21}. \end{aligned}$$

Had we not expressed the imaginary numbers in terms of i at the outset, we would have obtained $\sqrt{21}$ instead of $-\sqrt{21}$. ■

CAUTION! All imaginary numbers must be expressed in terms of i before simplifying.

DO EXERCISES 1–5.

EXAMPLE 7 Simplify: $-\sqrt{20}/\sqrt{-5}$.

Solution

$$\begin{aligned} \frac{-\sqrt{20}}{\sqrt{-5}} &= \frac{-\sqrt{20}}{i\sqrt{5}} \cdot \frac{i}{i} \\ &= \frac{-i\sqrt{20}}{i^2\sqrt{5}} \\ &= \frac{-i}{-1}\sqrt{\frac{20}{5}} = i\sqrt{4} \\ &= 2i \end{aligned}$$ ■

EXAMPLE 8 Simplify: $\sqrt{-9} + \sqrt{-25}$.

Solution

$$\begin{aligned} \sqrt{-9} + \sqrt{-25} &= i\sqrt{9} + i\sqrt{25} \\ &= 3i + 5i = (3 + 5)i \\ &= 8i \end{aligned}$$ ■

DO EXERCISES 6–10.

Powers of i

We can simplify powers of i using the fact that $i^2 = -1$ and expressing the given power of i in terms of i^2. Consider the following:

$$\begin{aligned} &i, \\ &i^2 = -1, \\ &i^3 = i^2 \cdot i = (-1)i = -i, \\ &i^4 = (i^2)^2 = (-1)^2 = 1, \\ &i^5 = i^4 \cdot i = (i^2)^2 \cdot i = (-1)^2 \cdot i = i, \\ &i^6 = (i^2)^3 = (-1)^3 = -1. \end{aligned}$$

Note that the powers of i cycle themselves through the values of i, -1, $-i$, and 1.

EXAMPLES Simplify.

9. $i^{37} = i^{36} \cdot i = (i^2)^{18} \cdot i = (-1)^{18} \cdot i = 1 \cdot i = i$
10. $i^{58} = (i^2)^{29} = (-1)^{29} = -1$

Express in terms of i.

1. $\sqrt{-6}$
2. $-\sqrt{-10}$
3. $\sqrt{-4}$
4. $-\sqrt{-25}$
5. Simplify: $\sqrt{-5}\sqrt{-2}$.

Simplify.

6. $\dfrac{\sqrt{-22}}{\sqrt{-2}}$
7. $\dfrac{\sqrt{-21}}{\sqrt{3}}$
8. $\sqrt{-16} + \sqrt{-9}$
9. $\sqrt{-25} - \sqrt{-4}$
10. $\sqrt{-17} + \sqrt{-9}$

Simplify.

11. i^{25}

12. i^{18}

13. i^{31}

11. $i^{75} = i^{74} \cdot i = (i^2)^{37} \cdot i = -1 \cdot i = -i$

12. $i^{80} = (i^2)^{40} = (-1)^{40} = 1$

DO EXERCISES 11–13.

2 Complex Numbers

The equation $x^2 + 1 = 0$ has no solution in the real-number system, but it has the imaginary solutions i and $-i$ in the complex-number system. There are still rather simple-looking equations that do not have either real or imaginary solutions. For example, $x^2 - 2x + 2 = 0$ does not. If we allow sums of real and imaginary numbers, however, this equation and many others have solutions, as we will show.

In order that more equations will have solutions, we invent a new system of numbers called the **system of complex numbers.*** To form the system of complex numbers, we take the imaginary numbers and the real numbers and all the possible sums of real and imaginary numbers. These are examples of complex numbers:

$$7 - 4i, \qquad -\pi + 9i, \qquad 37, \qquad i\sqrt{8}.$$

DEFINITION

A *complex number* is any number that can be named $a + bi$, where a and b are any real numbers. Note that a and b can both be 0.

For the complex number $a + bi$, we say that the **real part** is a and the **imaginary part** is bi.

We also agree that the familiar properties of real numbers (the *field properties*) hold for complex numbers.† We list them here.

Commutative Law.	**Addition and multiplication are commutative.**
Associative Law.	**Addition and multiplication are associative.**
Distributive Law.	**Multiplication is distributive over addition and also over subtraction.**
Identities.	**The additive identity is $0 + 0i$, or 0. The multiplicative identity is $1 + 0i$, or 1.**
Additive Inverses.	**Every complex number $a + bi$ has the additive inverse $-a - bi$.**
Multiplicative Inverses.	**Every nonzero complex number has a multiplicative inverse, or reciprocal.**

The complex-number system is an extension of the real-number system.‡ Any number $a + bi$, where a and b are real numbers, is a complex number. The number b can be 0, in which case we have $a + 0i$, which simplifies to the real

*You may wonder why we do not invent a system of imaginary numbers. That would not make sense because products of imaginary numbers are not necessarily imaginary. Consider, for example, $i^2 = -1$.

†In a more rigorous treatment, addition and multiplication are defined for complex numbers. It is then proved that the field properties hold.

‡It is important to keep in mind some comparisons between numbers that have real-number roots and those that have complex-number roots, which are not real. For example, $\sqrt{-48}$ is a complex number that is not a real number because we are taking the square root of a negative number. *But,* $\sqrt[3]{-125}$ is a real number because we are taking the cube root of a negative number and *any* real number has a cube root that is a real number.

number a. Thus the complex numbers include all the real numbers. They also include all the imaginary numbers, because any imaginary number bi is equal to $0 + bi$.

Calculations

Since the field properties hold in the system of complex numbers, calculations are much the same as those for real numbers. The primary difference is that one must remember that $i^2 = -1$.

EXAMPLE 13 Simplify: $(8 + 6i) + (3 + 2i)$.

Solution

$$(8 + 6i) + (3 + 2i) = 8 + 3 + 6i + 2i$$
$$= 11 + 8i \qquad \blacksquare$$

EXAMPLE 14 Simplify: $(1 + 2i)(1 + 3i)$.

Solution

$$(1 + 2i)(1 + 3i) = 1 + 3i + 2i + 6i^2$$
$$= 1 - 6 + 2i + 3i \qquad i^2 = -1$$
$$= -5 + 5i \qquad \blacksquare$$

EXAMPLE 15 Simplify: $(3 + 2i) - (5 - 2i)$.

Solution

$$(3 + 2i) - (5 - 2i) = 3 + 2i - 5 + 2i$$
$$= 3 - 5 + 2i + 2i$$
$$= -2 + 4i \qquad \blacksquare$$

DO EXERCISES 14–19.

In the system of real numbers, a sum of two squares cannot be factored. In the system of complex numbers, a sum of squares is always factorable.

EXAMPLE 16 Factor: $x^2 + y^2$.

Solution

$$x^2 + y^2 = (x + yi)(x - yi)$$

A check by multiplying will show that this is correct. $\qquad \blacksquare$

DO EXERCISES 20 AND 21.

3 Solutions of Equations

In the system of complex numbers, a great many equations have solutions. In fact, any equation $P(x) = 0$, where $P(x)$ is a nonconstant polynomial, has a solution.*

EXAMPLE 17 Determine whether $1 + i$ is a solution of $x^2 - 2x + 2 = 0$.

*See Chapter 11 for more about this.

Simplify.

14. $(8 - i) + (4 + 2i)$

15. $(9 + 2i) - (4 + 3i)$

16. $(2 + 4i)(3 + i)$

17. $(4 + 5i) + (4 - 5i)$

18. $2i(4 + 3i)$

19. $(5 + 6i) - (5 + 3i)$

Factor.

20. $x^2 + 4$

21. $9 + y^2$

22. Determine whether $1 - i$ is a solution of $x^2 - 2x + 2 = 0$.

23. Given that

$$3x + 1 + (y + 2)i = 2x + 2yi,$$

find x and y.

Find the conjugate.

24. $7 + 2i$

25. $6 - 4i$

26. $-5i$

27. $3i$

28. -3

29. 8

Solution

$$\begin{array}{r|l}
\multicolumn{2}{c}{x^2 - 2x + 2 = 0} \\
\hline
(1 + i)^2 - 2(1 + i) + 2 & 0 \quad \text{Substituting} \\
1 + 2i + i^2 - 2 - 2i + 2 & \\
1 + 2i - 1 - 2 - 2i + 2 & \\
0 &
\end{array}$$

The number $1 + i$ is a solution.

DO EXERCISE 22.

4 Equality for Complex Numbers

An equation $a + bi = c + di$ will be true if and only if the real parts are the same and the imaginary parts are the same. In other words, we have the following:

$$a + bi = c + di \quad \text{if and only if} \quad a = c \text{ and } b = d.$$

EXAMPLE 18 Suppose that $3x + yi = 5x + 1 + 2i$. Find x and y.

Solution We equate the real parts: $3x = 5x + 1$. Solving this equation, we obtain $x = -\frac{1}{2}$. We then equate the imaginary parts: $yi = 2i$. Thus, $y = 2$.

DO EXERCISE 23.

5 Conjugates and Division

We define the *conjugate* of a complex number as follows.

> **DEFINITION**
>
> **The *conjugate* of a complex number $a + bi$ is $a - bi$, and the *conjugate* of a complex number $a - bi$ is $a + bi$.**

We illustrate:

The conjugate of $3 + 4i$ is $3 - 4i$.
The conjugate of $5 - 7i$ is $5 + 7i$.
The conjugate of $5i$ is $-5i$.
The conjugate of 6 is 6.

DO EXERCISES 24–29.

Division

Fractional notation is useful for division of complex numbers. We also use the notion of conjugates.

EXAMPLE 19 Divide $4 + 5i$ by $1 + 4i$.

Solution We write fractional notation and then multiply by 1:

$$\frac{4+5i}{1+4i} = \frac{4+5i}{1+4i} \cdot \frac{1-4i}{1-4i} \qquad \text{Note that } 1-4i \text{ is the conjugate of the divisor.}$$
$$= \frac{(4+5i)(1-4i)}{1^2 - 4^2 i^2}$$
$$= \frac{4 - 11i - 20i^2}{1+16}$$
$$= \frac{24-11i}{17} \qquad i^2 = -1$$
$$= \frac{24}{17} - \frac{11}{17}i.$$

The procedure shown in Example 19 allows us always to find a quotient of two numbers and express it in the form $a + bi$. This is true because the product of a number and its conjugate is always a real number, which gives us a real-number denominator:

$$(a - bi)(a + bi) = a^2 - (bi)^2$$
$$= a^2 - b^2 i^2$$
$$= a^2 - b^2(-1)$$
$$= a^2 + b^2.$$

DO EXERCISES 30 AND 31.

Divide.

30. $\frac{1+3i}{3+2i}$

31. $\frac{2+i}{3-2i}$

6 Reciprocals

We can find the reciprocal, or multiplicative inverse, of a complex number by division. The reciprocal of a complex number $a + bi$ is $1/(a + bi)$.

EXAMPLE 20 Find the reciprocal of $2 - 3i$ and express it in the form $a + bi$.

Solution

a) The reciprocal of $2 - 3i$ is $1/(2 - 3i)$.
b) We can express it in the form $a + bi$ as follows:

$$\frac{1}{2-3i} = \frac{1}{2-3i} \cdot \frac{2+3i}{2+3i}$$
$$= \frac{2+3i}{2^2 - 3^2 i^2}$$
$$= \frac{2+3i}{4+9}$$
$$= \frac{2}{13} + \frac{3}{13}i.$$

DO EXERCISE 32.

32. Find the reciprocal of $3 + 4i$ and express it in the form $a + bi$.

7 Linear Equations

Linear equations with complex coefficients are solved in the same way as equations with real-number coefficients. The steps used in solving depend on the field properties in each case.

EXAMPLE 21 Solve: $3ix + 4 - 5i = (1 + x)x + 2i$.

33. Solve: $3 - 4i + 2ix = 3i - (1 - i)x$.

Solution

$$3ix + 4 - 5i = (1 + x)x + 2i$$

$$3ix - (1 + i)x = 2i - (4 - 5i) \quad \text{Adding } -(1+i)x \text{ and } -(4-5i)$$

$$(-1 + 2i)x = -4 + 7i \quad \text{Simplifying}$$

$$x = \frac{-4 + 7i}{-1 + 2i} \quad \text{Dividing}$$

$$x = \frac{-4 + 7i}{-1 + 2i} \cdot \frac{-1 - 2i}{-1 - 2i} \quad \text{Simplifying}$$

$$x = \frac{18 + i}{5} = \frac{18}{5} + \frac{1}{5}i$$

DO EXERCISE 33.

EXERCISE SET 2.4

1 Express in terms of i.

1. $\sqrt{-15}$
2. $\sqrt{-17}$
3. $\sqrt{-81}$
4. $\sqrt{-25}$
5. $-\sqrt{-12}$
6. $-\sqrt{-20}$

Simplify. Leave answers in terms of i in Exercises 7–10.

7. $\sqrt{-16} + \sqrt{-25}$
8. $\sqrt{-36} - \sqrt{-4}$
9. $\sqrt{-7} - \sqrt{-10}$
10. $\sqrt{-5} + \sqrt{-7}$
11. $\sqrt{-5}\sqrt{-11}$
12. $\sqrt{-7}\sqrt{-8}$
13. $-\sqrt{-4}\sqrt{-5}$
14. $-\sqrt{-9}\sqrt{-7}$
15. $\dfrac{-\sqrt{5}}{\sqrt{-2}}$
16. $\dfrac{\sqrt{-7}}{-\sqrt{5}}$
17. $\dfrac{\sqrt{-9}}{-\sqrt{4}}$
18. $\dfrac{-\sqrt{25}}{\sqrt{-16}}$
19. $\dfrac{-\sqrt{-36}}{\sqrt{-9}}$
20. $\dfrac{\sqrt{-25}}{-\sqrt{-16}}$

Simplify.

21. i^{18}
22. i^{14}
23. i^{15}
24. i^{16}
25. i^{39}
26. i^{40}
27. i^{46}
28. i^{72}

2 Simplify.

29. $(2 + 3i) + (4 + 2i)$
30. $(5 - 2i) + (6 + 3i)$
31. $(4 + 3i) + (4 - 3i)$
32. $(2 + 3i) + (-2 - 3i)$
33. $(8 + 11i) - (6 + 7i)$
34. $(9 - 5i) - (4 + 2i)$
35. $2i - (4 + 3i)$
36. $3i - (5 + 2i)$
37. $(1 + 2i)(1 + 3i)$
38. $(1 + 4i)(1 - 3i)$
39. $(1 + 2i)(1 - 3i)$
40. $(2 + 3i)(2 - 3i)$
41. $3i(4 + 2i)$
42. $5i(3 - 4i)$
43. $(2 + 3i)^2$
44. $(3 - 2i)^2$

3 Factor.

45. $4x^2 + 25y^2$
46. $16a^2 + 49b^2$
47. Determine whether $1 + 2i$ is a solution of $x^2 - 2x + 5 = 0$.
48. Determine whether $1 - 2i$ is a solution of $x^2 - 2x + 5 = 0$.

4 Solve for x and y.

49. $4x + 7i = -6 + yi$
50. $-4 + (x + y)i = 2x - 5y + 5i$

5 Simplify. Write the answer in the form $a + bi$.

51. $\dfrac{4 + 3i}{1 - i}$
52. $\dfrac{2 - 3i}{5 - 4i}$
53. $\dfrac{\sqrt{2} + i}{\sqrt{2} - i}$
54. $\dfrac{\sqrt{3} + i}{\sqrt{3} - i}$
55. $\dfrac{3 + 2i}{i}$
56. $\dfrac{2 + 3i}{i}$
57. $\dfrac{i}{2 + i}$
58. $\dfrac{3}{5 - 11i}$
59. $\dfrac{1 - i}{(1 + i)^2}$
60. $\dfrac{1 + i}{(1 - i)^2}$
61. $\dfrac{3 - 4i}{(2 + i)(3 - 2i)}$
62. $\dfrac{(4 - i)(5 + i)}{(6 - 5i)(7 - 2i)}$
63. $\dfrac{1 + i}{1 - i} \cdot \dfrac{2 - i}{1 - i}$
64. $\dfrac{1 - i}{1 + i} \cdot \dfrac{2 + i}{1 + i}$
65. $\dfrac{3 + 2i}{1 - i} + \dfrac{6 + 2i}{1 - i}$
66. $\dfrac{4 - 2i}{1 + i} + \dfrac{2 - 5i}{1 + i}$

6 Find the reciprocal and express it in the form $a + bi$.

67. $4 + 3i$
68. $4 - 3i$
69. $5 - 2i$
70. $2 + 5i$
71. i
72. $-i$
73. $-4i$
74. $5i$

7 Solve.

75. $(3 + i)x + i = 5i$
76. $(2 + i)x - i = 5 + i$
77. $2ix + 5 - 4i = (2 + 3i)x - 2i$
78. $5ix + 3 + 2i = (3 - 2i)x + 3i$
79. $(1 + 2i)x + 3 - 2i = 4 - 5i + 3ix$
80. $(1 - 2i)x + 2 - 3i = 5 - 4i + 2x$
81. $(5 + i)x + 1 - 3i = (2 - 3i)x + 2 - i$
82. $(5 - i)x + 2 - 3i = (3 - 2i)x + 3 - i$

SYNTHESIS

83. Show that the general rule for radicals, in real numbers, $\sqrt{a \cdot b} = \sqrt{a} \cdot \sqrt{b}$, does not hold for complex numbers.
84. Show that the general rule for radicals, in real numbers, $\sqrt{a/b} = \sqrt{a}/\sqrt{b}$, does not hold for complex numbers.

We can use a single letter for a complex number. For example, we could shorten $a + bi$ to z. To denote the conjugate of a number, we use a bar. The conjugate of z is $\bar{z}$. Or, the conjugate of $a + bi$ is $\overline{a + bi}$. Of course, by the definition of conjugates, $\overline{a + bi} = a - bi$ and $\overline{a - bi} = a + bi$. Using this notation, prove the following properties of conjugates.

85. For any complex number z, $z \cdot \bar{z}$ is a real number.
86. For any complex number z, $z + \bar{z}$ is a real number.
87. For any complex numbers z and w, $\overline{z + w} = \bar{z} + \bar{w}$.
88. For any complex numbers z and w, $\overline{z \cdot w} = \bar{z} \cdot \bar{w}$.
89. For any complex number z, $\overline{z^n} = \bar{z}^n$, where n is a natural number.
90. If z is a real number, then $\bar{z} = z$.
91. Using the properties proved in Exercises 85–90, find a polynomial in $\bar{z}$ that is the conjugate of $3z^5 - 4z^2 + 3z - 5$.
92. Solve: $z + 6\bar{z} = 7$.
93. Solve: $5z - 4\bar{z} = 7 + 8i$.
94. Let $z = a + bi$. Find $\frac{1}{2}(z + \bar{z})$.
95. Let $z = a + bi$. Find $\frac{1}{2}(\bar{z} - z)$.

CHALLENGE

96. Solve $z^2 = -2i$. (*Hint:* Let $z = a + bi$.)
97. Solve $z^2/4 = i$. (*Hint:* Let $z = a + bi$.)
98. Let $z = a + bi$. Find a general expression for $1/z$.
99. Let $z = a + bi$ and $w = c + di$. Find a general expression for w/z.

2.5 Quadratic Equations

OBJECTIVES

You should be able to:

1 Solve quadratic equations by:
 a) taking the square root on both sides;
 b) using completing the square;
 c) using the quadratic formula.

2 Use the discriminant to determine the nature of the solutions of a given quadratic equation.

3 Write a quadratic equation having specified solutions.

1 Solving Quadratic Equations

DEFINITION

A *quadratic equation* is an equation equivalent to one of the form

$$ax^2 + bx + c = 0,$$

where the numbers a, b, and c are real numbers, called *coefficients*, and $a \neq 0$.

The **standard form of a quadratic equation** is

$$ax^2 + bx + c = 0,$$

where the coefficients a, b, and c are real numbers. The coefficients b and c might be 0, but the **leading coefficient** a cannot be. If it were, the polynomial

Solve for x.

1. $x^2 = 3$

2. $5x^2 = 0$

3. $3x^2 = \pi$

4. $mx^2 = n$

5. Solve: $2x^2 + 1 = 0$.

Solve.

6. $(x + 4)^2 = 7$

7. $(x - 5)^2 = 3$

8. $(x + 5)^2 = 4$

would not be quadratic, that is, of second degree. You have solved such equations by factoring in Section 2.1. We now consider other methods.

Solving $(x + h)^2 = k$

First consider the equation $x^2 = k$, where k is positive. Since a positive real number has two square roots, the solutions of this equation are $\sqrt{k}$ and $-\sqrt{k}$. We will often abbreviate this by saying that the solutions are $\pm\sqrt{k}$. If $k = 0$, there is just one solution, 0. If k is negative, there are two nonreal complex solutions.

EXAMPLE 1 Solve: $5x^2 = 15$.

Solution We have

$$5x^2 = 15$$
$$x^2 = 3 \qquad \text{Multiplying by } \tfrac{1}{5}$$
$$x = \sqrt{3} \quad \text{or} \quad x = -\sqrt{3}. \qquad \text{Taking square roots}$$

These numbers check, so the solutions are $\sqrt{3}$ and $-\sqrt{3}$. The solution set is $\{\sqrt{3}, -\sqrt{3}\}$, or $\{\pm\sqrt{3}\}$. ■

DO EXERCISES 1–4.

Sometimes we get solutions that are not real numbers but complex numbers.

EXAMPLE 2 Solve: $4x^2 + 9 = 0$.

Solution We have

$$4x^2 + 9 = 0$$
$$x^2 = -\tfrac{9}{4} \qquad \text{Adding } -9 \text{ and multiplying by } \tfrac{1}{4}$$
$$x = \sqrt{-\tfrac{9}{4}} \quad \text{or} \quad x = -\sqrt{-\tfrac{9}{4}} \qquad \text{Taking square roots}$$
$$x = \tfrac{3}{2}i \quad \text{or} \quad x = -\tfrac{3}{2}i. \qquad \text{Simplifying}$$

The numbers $\frac{3}{2}i$ and $-\frac{3}{2}i$ check, so they are solutions. Thus the solution set is $\{\frac{3}{2}i, -\frac{3}{2}i\}$ or $\{\pm\frac{3}{2}i\}$. ■

DO EXERCISE 5.

EXAMPLE 3 Solve: $(x + 5)^2 = 3$.

Solution

$$(x + 5)^2 = 3$$
$$x + 5 = \pm\sqrt{3} \qquad \text{Taking square roots}$$
$$x = -5 \pm \sqrt{3}.$$

The solution set is $\{-5 - \sqrt{3}, -5 + \sqrt{3}\}$, or $\{-5 \pm \sqrt{3}\}$. ■

DO EXERCISES 6–8.

Completing the Square

If the equation is not in the form $(x + h)^2 = k$, we can put it into that form by *completing the square*.

EXAMPLE 4 Solve by completing the square: $x^2 - 6x - 12 = 0$.

Solution We consider first the x^2- and x-terms:

$$x^2 - 6x \qquad - 12 = 0.$$

We construct a trinomial square. First note that $x^2 - 6x + 9$ is a perfect square: $(x - 3)^2$. So if we add 9 to $x^2 - 6x$, we will have a perfect square. To get an equivalent equation, however, we must subtract 9 as well. This is the same as adding $9 - 9$, or 0. Then we proceed as follows:

$$\begin{aligned} x^2 - 6x + 9 - 9 - 12 &= 0 \\ (x-3)^2 - 21 &= 0 \\ (x-3)^2 &= 21 \\ x - 3 &= \pm\sqrt{21} \\ x &= 3 \pm \sqrt{21}. \end{aligned}$$

The solution set is $\{3 + \sqrt{21}, 3 - \sqrt{21}\}$, or $\{3 \pm \sqrt{21}\}$.

The following is the general procedure we call *completing the square*.

To solve $ax^2 + bx + c = 0$, by *completing the square*:

1. **If $a \neq 1$, multiply both sides of the equation by $1/a$ to get the equation in the form $x^2 + Bx + C = 0$.**
2. **To complete the square on $x^2 + Bx$, take half the coefficient of x and square it. Then add and subtract that number: $(B/2)^2$.**
3. **Take the square roots and solve for x.**

EXAMPLE 5 Solve by completing the square: $x^2 + 3x - 5 = 0$.

Solution

1. Since $a = 1$, no multiplication by $1/a$ is necessary to get the equation in the proper form: $x^2 + 3x - 5 = 0$.
2. To complete the square on $x^2 + 3x$, we take half the coefficients of x and square it. Then we add and subtract the number: $(\frac{1}{2} \cdot 3)^2$, or $\frac{9}{4}$:

$$\begin{aligned} x^2 + 3x + \tfrac{9}{4} - \tfrac{9}{4} - 5 &= 0 \\ x^2 + 3x + \tfrac{9}{4} - \tfrac{29}{4} &= 0 \\ \left(x + \tfrac{3}{2}\right)^2 - \tfrac{29}{4} &= 0 \\ \left(x + \tfrac{3}{2}\right)^2 &= \tfrac{29}{4}. \end{aligned}$$

3. We now take the square roots and solve for x:

$$\begin{aligned} x + \frac{3}{2} &= \pm\sqrt{\frac{29}{4}} = \pm\frac{\sqrt{29}}{2} \\ x &= -\frac{3}{2} \pm \frac{\sqrt{29}}{2} = \frac{-3 \pm \sqrt{29}}{2}. \end{aligned}$$

The solutions are

$$\frac{-3 + \sqrt{29}}{2} \quad \text{and} \quad \frac{-3 - \sqrt{29}}{2}, \quad \text{or} \quad \frac{-3 \pm \sqrt{29}}{2}.$$

The solution set is

$$\left\{\frac{-3 \pm \sqrt{29}}{2}\right\}.$$

Find the term that completes the square; then fill in the second expression.

9. $x^2 + 4x +$ ______ $= (\qquad)^2$

10. $x^2 - 6x +$ ______ $= (\qquad)^2$

11. $x^2 + 5x +$ ______ $= (\qquad)^2$

12. $x^2 - 7x +$ ______ $= (\qquad)^2$

13. $x^2 + \frac{3}{4}x +$ ______ $= (\qquad)^2$

14. $x^2 - x +$ ______ $= (\qquad)^2$

Solve by completing the square.

15. $x^2 + 4x - 3 = 0$

16. $x^2 - 6x + 8 = 0$

17. $x^2 - 5x + 6 = 0$

Solve by completing the square.

18. $2x^2 + 2x - 3 = 0$

19. $4x^2 + 3x - 1 = 0$

DO EXERCISES 9–17.

EXAMPLE 6 Solve by completing the square: $2x^2 - 3x - 1 = 0$.

Solution

1. Since $a \neq 1$, we multiply on both sides by $1/a$, which is $\frac{1}{2}$:

$$2x^2 - 3x - 1 = 0$$
$$x^2 - \tfrac{3}{2}x - \tfrac{1}{2} = 0. \qquad \text{Multiplying by } \tfrac{1}{2}$$

2. To complete the square on $x^2 - \frac{3}{2}x$, we take half the coefficient of x and square it. Then we add and subtract that number: $[\frac{1}{2}(-\frac{3}{2})]^2$, or $\frac{9}{16}$:

$$x^2 - \tfrac{3}{2}x + \tfrac{9}{16} - \tfrac{9}{16} - \tfrac{1}{2} = 0$$
$$\left(x - \tfrac{3}{4}\right)^2 - \tfrac{17}{16} = 0$$
$$\left(x - \tfrac{3}{4}\right)^2 = \tfrac{17}{16}.$$

3. We now take the square roots and solve for x:

$$x - \frac{3}{4} = \pm\sqrt{\frac{17}{16}} = \pm\frac{\sqrt{17}}{4}$$
$$x = \frac{3}{4} \pm \frac{\sqrt{17}}{4} = \frac{3 \pm \sqrt{17}}{4}.$$

The solution set is

$$\left\{\frac{3 + \sqrt{17}}{4}, \frac{3 - \sqrt{17}}{4}\right\}, \quad \text{or} \quad \left\{\frac{3 \pm \sqrt{17}}{4}\right\}.$$

DO EXERCISES 18 AND 19.

The Quadratic Formula

We studied completing the square for two reasons. The most important is that it is a useful tool in other places in mathematics. The second reason is that it can be used to derive a general formula for solving quadratic equations, called the **quadratic formula**. We consider the standard form of the quadratic equation $ax^2 + bx + c = 0$, with unspecified coefficients, solving as we have in the preceding examples. We assume that $a > 0$. If $a < 0$, we can first multiply on both sides by -1. Let's solve by completing the square:

$$ax^2 + bx + c = 0$$
$$x^2 + \frac{b}{a}x + \frac{c}{a} = 0 \qquad \text{Multiplying by } \frac{1}{a}$$

Half of b/a is $\frac{1}{2} \cdot b/a$, or $b/2a$. The square is $b^2/4a^2$. Thus we add and subtract $b^2/4a^2$:

$$x^2 + \frac{b}{a}x + \frac{b^2}{4a^2} - \frac{b^2}{4a^2} + \frac{c}{a} = 0$$
$$x^2 + \frac{b}{a}x + \frac{b^2}{4a^2} = \frac{b^2}{4a^2} - \frac{c}{a}$$
$$\left(x + \frac{b}{2a}\right)^2 = \frac{b^2}{4a^2} - \frac{4ac}{4a^2}$$
$$\left(x + \frac{b}{2a}\right)^2 = \frac{b^2 - 4ac}{4a^2}$$
$$x + \frac{b}{2a} = \sqrt{\frac{b^2 - 4ac}{4a^2}} \quad \text{or} \quad x + \frac{b}{2a} = -\sqrt{\frac{b^2 - 4ac}{4a^2}}$$

Since $a > 0$, then $\sqrt{4a^2} = 2|a| = 2a$, so

$$x + \frac{b}{2a} = \frac{\sqrt{b^2 - 4ac}}{2a} \quad \text{or} \quad x + \frac{b}{2a} = -\frac{\sqrt{b^2 - 4ac}}{2a}.$$

Thus,

$$x = -\frac{b}{2a} + \frac{\sqrt{b^2 - 4ac}}{2a} \quad \text{or} \quad x = -\frac{b}{2a} - \frac{\sqrt{b^2 - 4ac}}{2a}.$$

The solution set is

$$\left\{\frac{-b \pm \sqrt{b^2 - 4ac}}{2a}\right\}.$$

This gives us the quadratic formula.

THEOREM 3 The Quadratic Formula

The solutions of a quadratic equation $ax^2 + bx + c = 0$ are given by

$$x = \frac{-b \pm \sqrt{b^2 - 4ac}}{2a}.$$

When using the quadratic formula, it is helpful to first find the standard form so that the coefficients a, b, and c can be determined.

EXAMPLE 7 Solve $3x^2 + 2x - 7$. Find exact and approximate solutions. Round to the nearest hundredth.

Solution We first find the standard form and determine a, b, and c:

$$3x^2 + 2x - 7 = 0;$$
$$a = 3, \quad b = 2, \quad c = -7.$$

We then use the quadratic formula:

$$x = \frac{-b \pm \sqrt{b^2 - 4ac}}{2a} = \frac{-2 \pm \sqrt{2^2 - 4 \cdot 3 \cdot (-7)}}{2 \cdot 3}$$

$$x = \frac{-2 \pm \sqrt{4 + 84}}{6}$$ **To prevent careless mistakes, it helps to write out *all* the steps.**

$$x = \frac{-2 \pm \sqrt{88}}{6} = \frac{-2 \pm \sqrt{4 \cdot 22}}{6}$$

$$x = \frac{-2 \pm 2\sqrt{22}}{6} = \frac{2(-1 \pm \sqrt{22})}{2 \cdot 3} = \frac{-1 \pm \sqrt{22}}{3}.$$

The solution set is $\left\{\frac{-1 \pm \sqrt{22}}{3}\right\}$.

Should such irrational solutions arise in an applied problem, we can find approximations using a calculator:

$$\frac{-1 + \sqrt{22}}{3} \approx \frac{-1 + 4.6904}{3}$$

$$\approx \frac{3.6904}{3} \approx 1.23;$$ **Rounding to the nearest hundredth**

$$\frac{-1 - \sqrt{22}}{3} \approx \frac{-1 - 4.6904}{3}$$

$$\approx \frac{-5.6904}{3} \approx -1.90.$$ **Rounding to the nearest hundredth**

20. a) Solve $2x^2 + 7x = 4$ by factoring.
b) Solve the equation in part (a) using the quadratic formula. Compare your answers.

Solve using the quadratic formula.

21. $5x^2 - 8x = 3$

22. $x^2 - x + 2 = 0$

The set of approximate solutions is $\{1.23, -1.90\}$.

EXAMPLE 8 Solve: $x^2 + x + 1 = 0$.

Solution

$$x^2 + x + 1 = 0$$
$$a = 1, \quad b = 1, \quad c = 1.$$
$$x = \frac{-b \pm \sqrt{b^2 - 4ac}}{2a}$$
$$x = \frac{-1 \pm \sqrt{1^2 - 4 \cdot 1 \cdot 1}}{2 \cdot 1} = \frac{-1 \pm \sqrt{1 - 4}}{2}$$
$$x = \frac{-1 \pm \sqrt{-3}}{2}$$
$$x = \frac{-1 \pm i\sqrt{3}}{2}$$

The solutions of a quadratic equation can *always* be found using the quadratic formula. Solutions that are irrational are difficult to find by factoring. A general strategy for solving quadratic equations is as follows.

1. **Try factoring.**
2. **If factoring seems difficult, use the quadratic formula. It *always works*!**

DO EXERCISES 20–22.

2 The Discriminant

From the quadratic formula, we know that the solutions x_1 and x_2 of a quadratic equation are given by

$$x_1 = \frac{-b + \sqrt{b^2 - 4ac}}{2a} \quad \text{and} \quad x_2 = \frac{-b - \sqrt{b^2 - 4ac}}{2a}.$$

The expression $b^2 - 4ac$ shows the nature of the solutions. This expression is called the **discriminant.** If it is 0, then it doesn't matter whether we choose the plus or the minus sign in the formula: Hence there is just one solution. If the discriminant is positive, there will be two real solutions. If it is negative, we will be taking the square root of a negative number; hence there will be two nonreal solutions, and they will be complex conjugates.

THEOREM 4

If $b^2 - 4ac = 0$, then $ax^2 + bx + c = 0$ has just one real-number solution.

If $b^2 - 4ac > 0$, then $ax^2 + bx + c = 0$ has two different real-number solutions.

If $b^2 - 4ac < 0$, then $ax^2 + bx + c = 0$ has two different nonreal-number solutions (complex conjugates).

EXAMPLE 9 Determine the nature of the solutions of $9x^2 - 12x + 4 = 0$.

Solution We have

$$a = 9, \quad b = -12, \quad \text{and} \quad c = 4.$$

We compute the discriminant:

$$\begin{aligned} b^2 - 4ac &= (-12)^2 - 4 \cdot 9 \cdot 4 \\ &= 144 - 144 = 0. \end{aligned}$$

Since the discriminant is 0, there is just one solution and it is a real number.

EXAMPLE 10 Determine the nature of the solutions of $x^2 + 5x + 8 = 0$.

Solution We have

$$a = 1, \quad b = 5, \quad \text{and} \quad c = 8.$$

We compute the discriminant:

$$\begin{aligned} b^2 - 4ac &= 5^2 - 4 \cdot 1 \cdot 8 \\ &= 25 - 32 = -7. \end{aligned}$$

Since the discriminant is negative, there are two nonreal solutions.

EXAMPLE 11 Determine the nature of the solutions of $x^2 + 5x + 6 = 0$.

Solution

$$a = 1, \quad b = 5, \quad \text{and} \quad c = 6$$
$$b^2 - 4ac = 5^2 - 4 \cdot 1 \cdot 6 = 1$$

Since the discriminant is positive, there are two solutions and they are real numbers.

DO EXERCISES 23–25.

Determine the nature of the solutions without solving.

23. $x^2 + 5x - 3 = 0$

24. $9x^2 - 6x + 1 = 0$

25. $3x^2 - 2x + 1 = 0$

3 Writing Equations from Solutions

We know by the principle of zero products that $(x - 2)(x + 3) = 0$ has solutions 2 and -3. If we know the solutions of an equation, we can write the equation.

EXAMPLE 12 Find a quadratic equation whose solutions are 3 and $-\frac{2}{5}$.

Solution

$$x = 3 \quad \text{or} \quad x = -\tfrac{2}{5}$$

$$x - 3 = 0 \quad \text{or} \quad x + \tfrac{2}{5} = 0 \qquad \text{Getting the 0's on one side}$$

$$(x - 3)\left(x + \tfrac{2}{5}\right) = 0 \qquad \text{Principle of zero products (multiplying)}$$

$$x^2 + \tfrac{2}{5}x - 3x - 3 \cdot \tfrac{2}{5} = 0 \qquad \text{Using FOIL}$$

$$x^2 - \tfrac{13}{5}x - \tfrac{6}{5} = 0 \qquad \text{Collecting like terms}$$

$$5x^2 - 13x - 6 = 0 \qquad \text{Multiplying by 5}$$

EXAMPLE 13 Write a quadratic equation whose solutions are $\sqrt{3}$ and $-2\sqrt{3}$.

Find a quadratic equation having the following solutions.

26. -4 and $\frac{5}{3}$

27. $-2\sqrt{2}$ and $\sqrt{2}$

28. $5i$ and $-5i$

Solution

$$x = \sqrt{3} \quad \text{or} \quad x = -2\sqrt{3}$$

$$x - \sqrt{3} = 0 \quad \text{or} \quad x + 2\sqrt{3} = 0 \qquad \text{Getting the 0's on one side}$$

$$(x - \sqrt{3})(x + 2\sqrt{3}) = 0 \qquad \text{Principle of zero products}$$

$$x^2 + 2\sqrt{3}x - \sqrt{3}x - 2(\sqrt{3})^2 = 0 \qquad \text{Using FOIL}$$

$$x^2 + \sqrt{3}x - 6 = 0 \qquad \text{Collecting like terms}$$

EXAMPLE 14 Write a quadratic equation whose solutions are $2i$ and $-2i$.

Solution

$$x = 2i \quad \text{or} \quad x = -2i$$

$$x - 2i = 0 \quad \text{or} \quad x + 2i = 0 \qquad \text{Getting the 0's on one side}$$

$$(x - 2i)(x + 2i) = 0 \qquad \text{Principle of zero products (multiplying)}$$

$$x^2 + 2ix - 2ix = x^2 - (2i)^2 = 0 \qquad \text{Using FOIL}$$

$$x^2 + 4 = 0$$

DO EXERCISES 26–28.

EXERCISE SET 2.5

1 Solve for x.

1. $3x^2 = 27$
2. $5x^2 = 80$
3. $x^2 = -1$
4. $x^2 = -4$
5. $4x^2 = 20$
6. $3x^2 = 21$
7. $10x^2 = 0$
8. $9x^2 = 0$
9. $2x^2 - 3 = 0$
10. $3x^2 - 7 = 0$
11. $2x^2 + 14 = 0$
12. $3x^2 + 15 = 0$
13. $ax^2 = b$
14. $\pi x^2 = k$
15. $(x - 7)^2 = 5$
16. $(x + 3)^2 = 2$
17. $\frac{4}{9}x^2 - 1 = 0$
18. $\frac{16}{25}x^2 - 1 = 0$
19. $(x - h)^2 - 1 = a$
20. $y = a(x - h)^2 + k$

1 Solve by completing the square. (It is important to practice this method because we will need it later.)

21. $x^2 + 6x + 4 = 0$
22. $x^2 - 6x - 4 = 0$
23. $y^2 + 7y - 30 = 0$
24. $y^2 - 7y - 30 = 0$
25. $5x^2 - 4x - 2 = 0$
26. $12y^2 - 14y + 3 = 0$
27. $2x^2 + 7x - 15 = 0$
28. $9x^2 - 30x = -25$

1 Solve using the quadratic formula.

29. $x^2 + 4x = 5$
30. $x^2 = 2x + 15$
31. $2y^2 - 3y - 2 = 0$
32. $5m^2 + 3m - 2 = 0$
33. $3t^2 + 8t + 3 = 0$
34. $3u^2 = 18u - 6$
35. $3 + u^2 = 12u$
36. $40 + 30p + 5p^2 = 0$
37. $x^2 - x + 1 = 0$
38. $x^2 + x + 2 = 0$
39. $x^2 + 13 = 4x$
40. $2x + 1 = -5x^2$
41. $5x^2 = 13x + 17$
42. $x^2 = \frac{5}{3}x + \frac{4}{3}$
43. $0.03 + 0.08v = v^2$
44. $3x^2 - 2x - 5\frac{1}{3} = 0$
45. $\dfrac{1}{x} + \dfrac{1}{x + 3} = 7$
46. $\dfrac{1}{x - 2} - \dfrac{1}{x + 2} = 5$
47. $1 + \dfrac{x + 5}{(x + 1)^2} - \dfrac{2}{x + 1} = 0$
48. $\dfrac{x^2}{x + 11} + 1 = 0$

2 Determine the nature of the solutions.

49. $x^2 - 6x + 9 = 0$
50. $x^2 + 10x + 25 = 0$
51. $x^2 + 7 = 0$
52. $x^2 + 2 = 0$
53. $x^2 - 2 = 0$
54. $x^2 - 5 = 0$
55. $4x^2 - 12x + 9 = 0$
56. $4x^2 + 8x - 5 = 0$
57. $x^2 - 2x + 4 = 0$
58. $x^2 + 3x + 4 = 0$
59. $9t^2 - 3t = 0$
60. $4m^2 + 7m = 0$
61. $y^2 = \frac{1}{2}y + \frac{3}{5}$
62. $y^2 + \frac{9}{4} = 4y$
63. $4x^2 - 4\sqrt{3}x + 3 = 0$
64. $6y^2 - 2\sqrt{3}y - 1 = 0$

3 Write a quadratic equation with the given solutions.

65. $-11, 9$
66. $-4, 4$
67. 7, only solution
68. $-\frac{2}{3}$, only solution
69. $-\frac{2}{5}, \frac{6}{5}$
70. $-\frac{1}{4}, -\frac{1}{2}$
71. $\frac{c}{2}, \frac{d}{2}$
72. $\frac{k}{3}, \frac{m}{4}$
73. $\sqrt{2}, 3\sqrt{2}$
74. $-\sqrt{3}, 2\sqrt{3}$
75. $3i, -3i$
76. $4i, -4i$

SYNTHESIS

Solve.

77. $x^2 - 0.75x - 0.5 = 0$
78. $5.33x^2 - 8.23x - 3.24 = 0$

Solve using any method. (In general, try factoring first. Then use the quadratic formula if factoring is not possible.)

79. $x + \frac{1}{x} = \frac{13}{6}$
80. $\frac{3}{x} + \frac{x}{3} = \frac{5}{2}$
81. $t^2 + 0.2t - 0.3 = 0$
82. $p^2 + 0.3p - 0.2 = 0$
83. $x^2 + x - \sqrt{2} = 0$
84. $x^2 - x - \sqrt{3} = 0$
85. $x^2 + \sqrt{5}x - \sqrt{3} = 0$
86. $2x^2 + \sqrt{3}x - \pi = 0$
87. $\sqrt{2}x^2 - \sqrt{3}x - \sqrt{5} = 0$
88. $\sqrt{2}x^2 + 5x + \sqrt{2} = 0$
89. $(2t - 3)^2 + 17t = 15$
90. $2y^2 - (y + 2)(y - 3) = 12$
91. $(x + 3)(x - 2) = 2(x + 11)$
92. $9t(t + 2) - 3t(t - 2) = 2(t + 4)(t + 6)$
93. $2x^2 + (x - 4)^2 = 5x(x - 4) + 24$
94. $(c + 2)^2 + (c - 2)(c + 2) = 44 + (c - 2)^2$

CHALLENGE

95. Prove each of the following.
    a) The sum of the solutions of $ax^2 + bx + c = 0$ is $-b/a$.
    b) The product of the solutions of $ax^2 + bx + c = 0$ is c/a.

For each equation under the given condition, (a) find k and (b) find the other solution.

96. $kx^2 - 17x + 33 = 0$; one solution is 3
97. $kx^2 - 2x + k = 0$; one solution is -3
98. $x^2 - kx + 2 = 0$; one solution is $1 + i$
99. $x^2 - (6 + 3i)x + k = 0$; one solution is 3

100. Find k for which $kx^2 - 4x + (2k - 1) = 0$ and the product of the solution is 3.

101. Find a quadratic equation for which the sum of the solutions is $\sqrt{3}$ and the product is 8.

102. Write a quadratic equation having the given numbers as solutions.
    a) $\frac{2 + \sqrt{3}}{2}, \frac{2 - \sqrt{3}}{2}$
    b) $\frac{g}{h}, -\frac{h}{g}$
    c) $2 - 5i, 2 + 5i$

103. Find h and k, where $3x^2 - hx + 4k = 0$, the sum of the solutions is -12, and the product of the solutions is 20.

104. One solution of the equation

$$p(q - r)y^2 + q(r - p)y + r(p - q) = 0$$

is 2. Find the other.

105. The sum of the squares of the solutions of

$$x^2 + 2kx - 5 = 0$$

is 26. Find the absolute value of k.

106. Prove that the solutions of $ax^2 + bx + c = 0$ are the reciprocals of the solutions of the equation

$$cx^2 + bx + a = 0, \quad c \neq 0, \quad a \neq 0.$$

2.6 Formulas and Problem Solving

OBJECTIVES

You should be able to:

1. Solve a formula for a given letter.
2. Solve applied problems involving quadratic equations.

1. Solve $V = \frac{1}{3}\pi r^2 h$ for r.

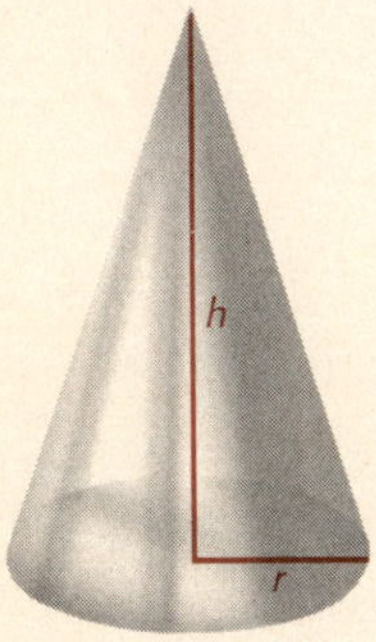

2. Solve $S = 16t^2 + v_0 t$ for t.

1 Formulas

To solve a formula for a certain letter, we use the principles of equation solving that we have developed until we have an equation with the letter alone on one side. In most formulas, the letters represent nonnegative numbers, so we generally need not use absolute values when taking principal square roots.

EXAMPLE 1 Solve $V = \pi r^2 h$ for r.

Solution

$$V = \pi r^2 h$$
$$\frac{V}{\pi h} = r^2$$
$$\sqrt{\frac{V}{\pi h}} = r$$

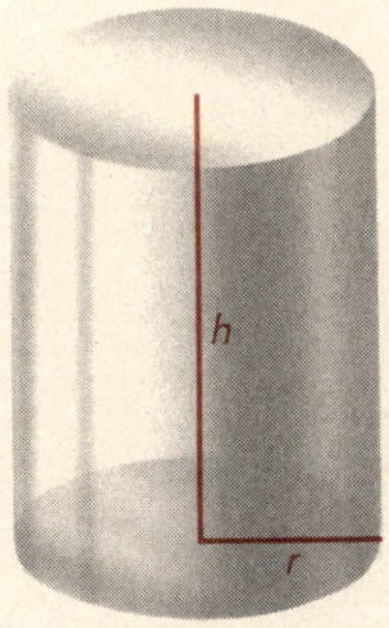

DO EXERCISE 1.

EXAMPLE 2 Solve $A = \pi rs + \pi r^2$ for r.

Solution We have

$$\pi r^2 + \pi rs - A = 0.$$

Then $a = \pi$, $b = \pi s$, and $c = -A$, and we use the quadratic formula:

$$r = \frac{-b \pm \sqrt{b^2 - 4ac}}{2a}$$
$$r = \frac{-\pi s \pm \sqrt{(\pi s)^2 - 4 \cdot \pi \cdot (-A)}}{2\pi}$$
$$r = \frac{-\pi s \pm \sqrt{\pi^2 s^2 + 4\pi A}}{2\pi},$$

or just

$$r = \frac{-\pi s + \sqrt{\pi^2 s^2 + 4\pi A}}{2\pi},$$

since the negative square root would result in a negative solution.

DO EXERCISE 2.

2 Problem Solving

EXAMPLE 3 *Compound interest.* An investment of \$2560 is made at interest rate i, compounded annually. In 2 years, it grows to \$3240. What is the interest rate?

Solution We substitute 2560 for P, 3240 for A, and 2 for t in the formula $A = P(1 + i)^t$, and solve for i:

$$\begin{aligned} A &= P(1+i)^t \\ 3240 &= 2560(1+i)^2 \\ \frac{3240}{2560} &= (1+i)^2 \\ \pm\frac{9}{8} &= 1 + i \qquad \text{Taking the square roots} \\ -1 + \frac{9}{8} &= i \quad \text{or} \quad -1 - \frac{9}{8} = i \\ \frac{1}{8} &= i \quad \text{or} \quad -\frac{17}{8} = i. \end{aligned}$$

Since the interest rate cannot be negative, $i = \frac{1}{8} = 0.125 = 12.5\%$.

DO EXERCISE 3.

3. An investment of $2560 is made at interest rate i, compounded annually. In 2 years, it grows to $3610. What is the interest rate?

EXAMPLE 4 A ladder 10 ft long leans against a wall. The bottom of the ladder is 6 ft from the wall. The bottom of the ladder is then pulled out 3 ft farther. How much does the top end move down the wall?

Solution

1. *Familiarize.* The first thing to do is to make a drawing and label it with both the known and unknown data. The dashed line in the figure shows the ladder in its original position, with the lower end 6 ft from the wall.

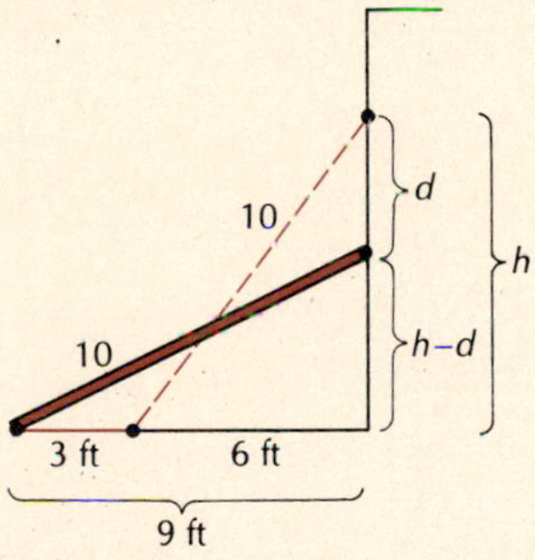

2. *Translate.* We see that there are right triangles in the figure. This is a clue that we may wish to use the Pythagorean theorem. We use that theorem and begin to write equations. From the taller triangle, we get

$$h^2 + 6^2 = 10^2. \tag{1}$$

From the other triangle, we get

$$(h - d)^2 + 9^2 = 10^2. \tag{2}$$

3. *Carry out.* We solve equation (1) for h and get $h = 8$ ft. Thus we know that $h = 8$ in equation (2), so we have

$$(8 - d)^2 + 9^2 = 10^2,$$

or

$$d^2 - 16d + 45 = 0.$$

Using the quadratic formula, we get

$$d = \frac{16 \pm \sqrt{76}}{2} = \frac{16 \pm 2\sqrt{19}}{2} = 8 \pm \sqrt{19}.$$

4.,5. *Check and State.* The length $8 + \sqrt{19}$ is not a solution since it exceeds the original length. The number $8 - \sqrt{19} \approx 8 - 4.359 = 3.641$

4. A 13-ft ladder leans against a wall. The bottom of the ladder is 5 ft from the wall. The bottom is then pulled out 4 ft farther. How much does the top end move down the wall?

5. a) An object is dropped from the top of the Statue of Liberty, which is 92 m tall. How long does it take for the object to reach the ground?

b) An object is thrown downward from the statue at an initial velocity of 40 m/sec. How long does it take for the object to reach the ground?

c) How far will an object fall in 1 sec if it is thrown downward from the statue at an initial velocity of 40 m/sec?

checks and is the solution. Therefore, the top of the ladder moves down 3.641 ft when the bottom is moved out 3 ft. ■

DO EXERCISE 4.

We use the following information in Example 5.

When an object is dropped or thrown downward, the distance, in meters, that it falls in t seconds is given by the following formula:

$$s = 4.9t^2 + v_0 t.$$

In this formula, v_0 is the initial velocity, in meters per second.

The distance, in feet, that the object falls in t seconds is given by

$$s = 16t^2 + v_0 t,$$

where v_0 is the initial velocity in feet per second.

EXAMPLE 5

a) An object is dropped from the top of the Gateway Arch, which is 195 m high, in St. Louis. How long does it take for the object to reach the ground?

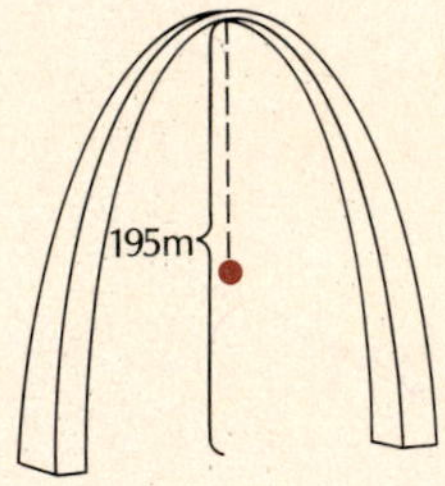

Solution Since the object was *dropped*, its initial velocity was 0. We substitute 0 for v_0 and 195 for s and then solve for t:

$$195 = 4.9t^2 + 0 \cdot t$$
$$195 = 4.9t^2$$
$$t^2 = 39.8$$
$$t = \sqrt{39.8} \approx 6.31. \quad \text{Using the square root key}$$

Thus it takes about 6.31 sec for the object to reach the ground.

b) An object is thrown downward from the arch at an initial velocity of 16 m/sec. How long does it take to reach the ground?

Solution We substitute 195 for s and 16 for v_0 and solve for t:

$$195 = 4.9t^2 + 16t$$
$$0 = 4.9t^2 + 16t - 195.$$

Using the quadratic formula and a calculator, we obtain

$$t \approx -8.15 \quad \text{or} \quad t \approx 4.88.$$

The negative answer is meaningless in this problem, so the answer is about 4.88 sec.

c) How far will an object fall in 3 sec if it is thrown downward from the arch at an initial velocity of 16 m/sec?

Solution We substitute 16 for v_0 and 3 for t and solve for s:

$$s = 4.9t^2 + v_0t = 4.9(3)^2 + 16 \cdot 3 = 92.1.$$

Thus the object falls 92.1 m in 3 sec.

DO EXERCISE 5 ON THE PRECEDING PAGE.

EXERCISE SET 2.6

1 Solve the formula for the given letter. Assume that all letters represent positive numbers.

1. $F = \dfrac{kM_1M_2}{d^2}$, for d
2. $E = mc^2$, for c
3. $S = \dfrac{1}{2}at^2$, for t
4. $S = 4\pi r^2$, for r
5. $s = -16t^2 + v_0t$, for t
6. $A = 2\pi r^2 + 3\pi rh$, for r
7. $d = \dfrac{n^2 - 3n}{2}$, for n
8. $\sqrt{2}t^2 + 3k = \pi t$, for t
9. $A = P(1 + i)^2$, for i
10. $A = P\left(1 + \dfrac{i}{2}\right)^2$, for i

2 Problem Solving

What is the interest rate if interest is compounded annually?

11. \$5120 grows to \$7220 in 2 years
12. \$1000 grows to \$1210 in 2 years
13. \$8000 grows to \$9856.80 in 2 years
14. \$1000 grows to \$1271.26 in 2 years

The number of diagonals, d, of a polygon of n sides is given by

$$d = \frac{n^2 - 3n}{2}.$$

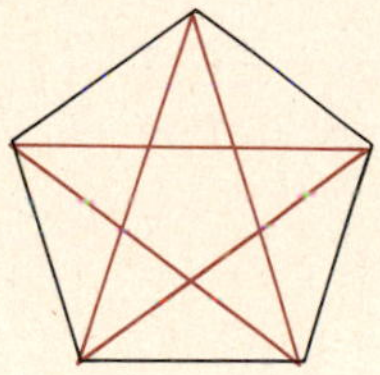

15. A polygon has 27 diagonals. How many sides does it have?
16. A polygon has 44 diagonals. How many sides does it have?
17. A ladder 25 ft long leans against a wall. The bottom of the ladder is 7 ft from the wall. The bottom of the ladder is then pulled out 2 ft farther. How much does the top end move down the wall?
18. A 15-ft ladder leans against a wall. The bottom of the ladder is 4 ft from the wall. The bottom is then pulled out 3 ft farther. How much does the top end move down the wall?
19. A ladder 10 ft long leans against a wall. The bottom of the ladder is 6 ft from the wall. How much would the lower end of the ladder have to be pulled away so that the top end would be pulled down the same amount?
20. A ladder 13 ft long leans against a wall. The bottom of the ladder is 5 ft from the wall. How much would the lower end of the ladder have to be pulled away so that the top end would be pulled down the same amount?
21. The area of a triangle is 18 cm^2. The base is 3 cm longer than the height. Find the height.
22. A baseball diamond is a square 90 ft on a side. How far is it directly from second base to home?
23. Trains A and B leave the same city at right angles at the same time. Train B travels 5 mph faster than train A. After 2 hr, they are 50 mi apart. Find the speed of each train.
24. Trains A and B leave the same city at right angles at the same time. Train A travels 14 km/h faster than train B. After 5 hr, they are 130 km apart. Find the speed of each train.

For Exercises 25 and 26, use the formula $s = 4.9t^2 + v_0 t$.

25. **a)** An object is dropped 75 m from an airplane. How long does it take for the object to reach the ground?
b) An object is thrown downward from the plane at an initial velocity of 30 m/sec. How long does it take for the object to reach the ground?
c) How far will an object fall in 2 sec if it is thrown downward at an initial velocity of 30 m/sec?

26. **a)** An object is dropped 500 m from an airplane. How long does it take for the object to reach the ground?
b) An object is thrown downward from the plane at an initial velocity of 30 m/sec. How long does it take for the object to reach the ground?
c) How far will an object fall in 5 sec if it is thrown downward at an initial velocity of 30 m/sec?

27. The diagonal of a square is 1.341 cm longer than a side. Find the length of the side.

28. The hypotenuse of a right triangle is 8.312 cm long. The sum of the lengths of the legs is 10.23 cm. Find the lengths of the legs.

29. A rectangular garden is 60 ft by 80 ft. Part of the garden is torn up to install a sidewalk of uniform width around the garden. The new area of the garden is $\frac{2}{3}$ of the old area. How wide is the sidewalk?

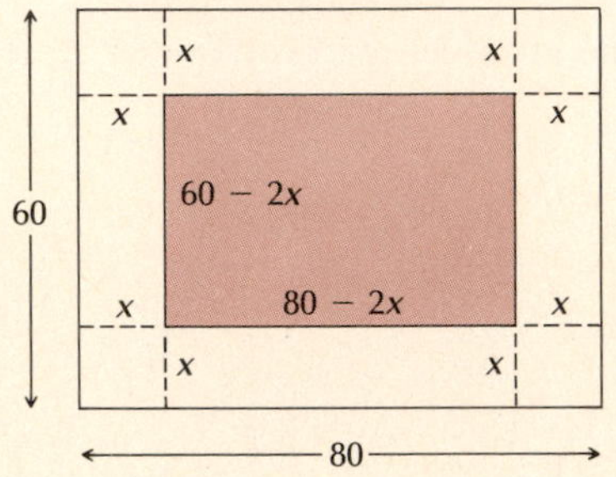

30. The speed of a boat in still water is 20 mph. It travels 24 mi upstream and 24 mi downstream in a total time of 10 hr. What is the speed of the current?

31. An open box is made from a 10-cm by 20-cm piece of tin by cutting a square from each corner and folding up the edges. The area of the resulting base is 90 cm^2. What is the length of the sides of the squares?

32. The frame of a picture is 28 cm by 32 cm outside and is of uniform width. What is the width of the frame if 200 cm^2 of the picture shows?

33. During the first part of a trip, a car travels 50 mi at a certain speed. It travels 80 mi on the second part of a trip at a speed of 10 mph slower. The total time for the trip is 2 hr. Find the speed of the car on each part of the trip.

34. A car travels 120 mi at a certain speed. If the speed had been 10 mph faster, the trip could have been made in 2 hr less time. Find the speed.

35. A boat travels 12 mi upstream and 12 mi back. The time required for the round trip is 2 hr. The speed of the stream is 3 mph. Find the speed of the boat in still water.

36. Working together, two people can do a job in 4 hr. Person A takes 5 hr longer, working alone, than person B alone. How long would it take person B to do the job?

The following formula will be helpful in Exercises 37–40:

$$T = c \cdot N.$$

Total cost = (Cost per item) · (Number of items)
Total cost = (Cost per person) · (Number of persons)

37. A group of students share equally in the \$140 cost of a boat. At the last minute, 3 students drop out, and this raises the share of each remaining student \$15. How many students were in the group at the outset?

38. An investor bought a group of lots for \$8400. All but 4 of the lots were sold for \$8400. The selling price for each lot was \$350 greater than the cost. How many lots were bought?

39. An investor buys some stock for \$720. If each share had cost \$15 less, 4 more shares could have been bought for the same \$720. How many shares of stock were bought?

40. A sorority is going to spend \$112 for a party. When 14 new pledges join the sorority, this reduces each student's cost by \$4. How much did it cost each student before?

SYNTHESIS

Solve for x.

41. $kx^2 + (3 - 2k)x - 6 = 0$

42. $x^2 - 2x + kx + 1 = kx^2$

43. $(m + n)^2x^2 + (m + n)x = 2$

44. Solve $x^2 - 3xy - 4y^2 = 0$ (a) for x; (b) for y.

45. For interest compounded annually, what is the interest rate when \$9826 grows to \$13,704 in 3 years?

46. An equilateral triangle is inscribed in a circle whose circumference is 6π. Find the area of the triangle.

47. Two equilateral triangles have sides of respective lengths a_1 and a_2. Find the length of a side a_3 of a third equilateral triangle whose area is the sum of the areas of the two triangles.

48. The sides of triangle A are 25 ft, 25 ft, and 30 ft. The sides of triangle B are 25 ft, 25 ft, and 40 ft. Which triangle has the greatest area?

CHALLENGE

49. A rectangle of 12-cm^2 area is inscribed in the right triangle ABC as shown in the figure. What are its dimensions?

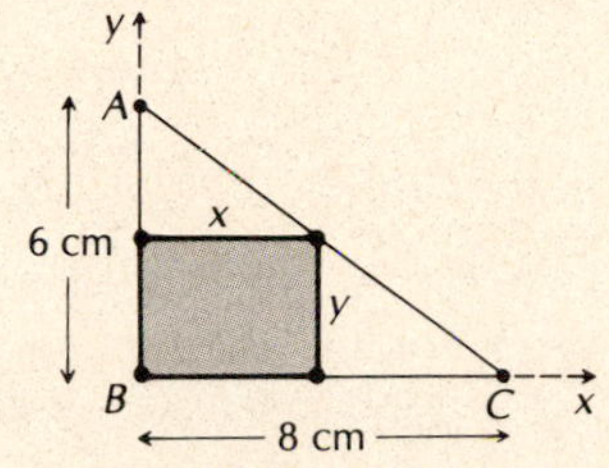

50. The world record for free-fall to the earth by a woman without a parachute is 175 ft and is held by Kitty O'Neill. (She fell into a bed of foam padding.) Approximately how long did the fall take?

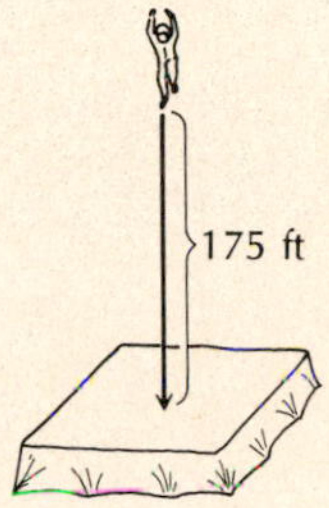

2.7 Radical Equations

1 Solving Radical Equations

A **radical equation** is an equation in which variables occur in one or more radicands. An example is the equation $\sqrt{2x-5} - \sqrt{x-3} = 1$. To solve such equations, we need a new principle.

THEOREM 5 The Principle of Powers

For any positive number n, if an equation $a = b$ is true, then $a^n = b^n$ is true.

It is important to check when using the principle of powers. This principle may *not* produce equivalent equations. For example, when we square both sides of an equation, the new equation may have solutions that the first equation does not. Consider, for example, the equation

$$x = 3.$$

This equation has just one solution, the number 3. When we square both sides, we get

$$x^2 = 9,$$

which has two solutions, 3 and -3. Thus the equations $x = 3$ and $x^2 = 9$ are *not* equivalent.

As another example, consider $\sqrt{x} = -3$. At the outset, we should note that this equation has no real-number solution because, by definition, $\sqrt{x}$ must be a *nonnegative* number. Suppose, though, that we try to solve by squaring both sides. We would get $(\sqrt{x})^2 = (-3)^2$, or $x = 9$. The number 9 does not check.

OBJECTIVE

You should be able to:

1. Solve radical equations using the principle of powers.

Solve. Don't forget to check!

1. $\sqrt{2x} = -5$

2. $x - 1 = \sqrt{x + 5}$

3. Solve. Don't forget to check.

$$\sqrt[4]{3x - 1} = 2$$

CAUTION! When using the principle of powers, it is imperative to check possible solutions in the original equation.

EXAMPLE 1 Solve: $x - 5 = \sqrt{x + 7}$.

Solution

$$x - 5 = \sqrt{x + 7}$$

$$(x - 5)^2 = (\sqrt{x + 7})^2 \quad \text{Principle of powers; squaring both sides}$$

$$x^2 - 10x + 25 = x + 7$$

$$x^2 - 11x + 18 = 0$$

$$(x - 9)(x - 2) = 0$$

$$x = 9 \quad \text{or} \quad x = 2$$

The possible solutions are 9 and 2. We check, as follows. For 9, we substitute 9 for x on each side of the equation and simplify each side separately:

$x - 5$	$= \sqrt{x + 7}$
$9 - 5$	$\sqrt{9 + 7}$
4	4

Since the results, 4, are the same, 9 checks. Thus it is a solution.

For 2, we substitute 2 for x on each side and simplify each side separately:

$x - 5$	$= \sqrt{x + 7}$
$2 - 5$	$\sqrt{2 + 7}$
-3	3

Since the results are not the same, 2 is not a solution. The only solution is 9. Thus the solution set is $\{9\}$.

DO EXERCISES 1 AND 2.

EXAMPLE 2 Solve: $\sqrt[3]{4x^2 + 1} = 5$.

Solution

$$\sqrt[3]{4x^2 + 1} = 5$$

$$(\sqrt[3]{4x^2 + 1})^3 = 5^3 \quad \text{Principle of powers; cubing both sides}$$

$$4x^2 + 1 = 125$$

$$4x^2 = 124$$

$$x^2 = 31$$

$$x = \pm\sqrt{31}$$

Both $\sqrt{31}$ and $-\sqrt{31}$ check. The solution set is $\{\pm\sqrt{31}\}$.

DO EXERCISE 3.

Equations with Two Radical Terms

A general strategy for solving equations with two radical terms is as follows.

1. Isolate one of the radical terms.
2. Use the principle of powers.
3. If a radical term remains, perform steps (1) and (2) again.
4. Check possible solutions.

EXAMPLE 3 Solve: $\sqrt{x-3} + \sqrt{x+5} = 4$.

Solution

$$\sqrt{x-3} + \sqrt{x+5} = 4$$

$$\sqrt{x-3} = 4 - \sqrt{x+5}$$ Adding $-\sqrt{x+5}$, which isolates one of the radical terms

$$(\sqrt{x-3})^2 = (4 - \sqrt{x+5})^2$$ Principle of powers; squaring both sides

Here we are squaring the binomial. We square 4, then we subtract twice the product of 4 and $\sqrt{x+5}$, and then we add the square of $\sqrt{x+5}$. Recall that $(A - B)^2 = A^2 - 2AB + B^2$.

$$x - 3 = 16 - 8\sqrt{x+5} + (x + 5)$$

$$-3 = 21 - 8\sqrt{x+5}$$ Adding $-x$ and collecting like terms

$$-24 = -8\sqrt{x+5}$$ Isolating the remaining radical term

$$3 = \sqrt{x+5}$$

$$3^2 = (\sqrt{x+5})^2$$ Squaring

$$9 = x + 5$$

$$4 = x$$

The number 4 checks and is the solution. The solution set is $\{4\}$.

CAUTION! A common error in solving equations like

$$\sqrt{x-3} + \sqrt{x+5} = 4$$

is to square the left side, obtaining $(x - 3) + (x + 5)$. That is wrong because the square of a sum is *not* the sum of the squares.

Example: $\sqrt{9} + \sqrt{16} = 7$, but $9 + 16 = 25$, not 7^2.

DO EXERCISE 4.

EXAMPLE 4 Solve: $\sqrt{2x-5} = 1 + \sqrt{x-3}$.

4. Solve:

$$\sqrt{x} - \sqrt{x-5} = 1.$$

Solution

$$\sqrt{2x-5} = 1 + \sqrt{x-3}$$

$$(\sqrt{2x-5})^2 = (1+\sqrt{x-3})^2 \quad \text{One radical is already isolated; we square both sides.}$$

$$2x - 5 = 1 + 2\sqrt{x-3} + (x-3)$$

$$x - 3 = 2\sqrt{x-3} \quad \text{Isolating the remaining radical term}$$

$$(x-3)^2 = (2\sqrt{x-3})^2 \quad \text{Squaring both sides}$$

$$x^2 - 6x + 9 = 4(x-3)$$

$$x^2 - 6x + 9 = 4x - 12$$

$$x^2 - 10x + 21 = 0$$

$$(x-7)(x-3) = 0 \quad \text{Factoring}$$

$$x = 7 \quad \text{or} \quad x = 3 \quad \text{Using the principle of zero products}$$

The numbers 7 and 3 check and are the solutions. Thus the solution set is $\{3, 7\}$. ■

DO EXERCISE 5.

5. Solve:

$$\sqrt{3x+1} = 1 + \sqrt{x+4}.$$

EXAMPLE 5 Solve

$$A = \sqrt{1 + \frac{a^2}{b^2}}$$

for a. Assume that the variables represent positive numbers.

Solution

$$A = \sqrt{1 + \frac{a^2}{b^2}}$$

$$A^2 = 1 + \frac{a^2}{b^2}$$

$$b^2A^2 = b^2 + a^2$$

$$b^2A^2 - b^2 = a^2$$

$$\sqrt{b^2A^2 - b^2} = a$$

$$\sqrt{b^2(A^2-1)} = a$$

$$b\sqrt{A^2-1} = a$$ ■

DO EXERCISE 6.

6. Solve

$$P = \sqrt{\frac{m^2-1}{m^2}}$$

for m. Assume that the variables represent positive numbers.

EXERCISE SET 2.7

1 Solve. Don't forget to check!

1. $\sqrt{3x-4} = 1$
2. $\sqrt[3]{2x+1} = -5$
3. $\sqrt[4]{x^2-1} = 1$
4. $\sqrt{m+1} - 5 = 8$
5. $\sqrt{y-1} + 4 = 0$
6. $5 + \sqrt{3x^2 + \pi} = 0$
7. $\sqrt{x-3} + \sqrt{x+5} = 4$ (*Hint:* You are squaring a binomial.)
8. $\sqrt{x} - \sqrt{x-5} = 1$
9. $\sqrt{3x-5} + \sqrt{2x+3} + 1 = 0$
10. $\sqrt{2m-3} = \sqrt{m+7} - 2$
11. $\sqrt[3]{6x+9} + 8 = 5$
12. $\sqrt[5]{3x+4} = 2$
13. $\sqrt{6x+7} = x + 2$
14. $\sqrt{6x+7} - \sqrt{3x+3} = 1$
15. $\sqrt{20-x} = \sqrt{9-x} + 3$
16. $\sqrt{n+2} + \sqrt{3n+4} = 2$
17. $\sqrt{x} - \sqrt{3x-3} = 1$
18. $\sqrt{2x+1} - \sqrt{x} = 1$
19. $\sqrt{2y-5} - \sqrt{y-3} = 1$
20. $\sqrt{4p-5} - \sqrt{p-2} = 3$
21. 🖩 $\sqrt{7.35x + 8.051} = 0.345x + 0.067$
22. 🖩 $\sqrt{1.213x + 9.333} = 5.343x + 2.312$

23. $x^{1/3} = -2$

24. $t^{1/5} = 2$

25. $t^{1/4} = 3$

26. $m^{1/2} = -7$

27. $8 = \dfrac{1}{\sqrt{x}}$

28. $3 = \dfrac{1}{\sqrt{y}}$

29. $\sqrt[3]{m} = -5$

30. $\sqrt[4]{t} = -5$

For Exercises 31 and 32, assume that the variables represent positive numbers.

31. Solve $T = 2\pi\sqrt{L/g}$ for L; for g.

32. Solve $H = \sqrt{c^2 + d^2}$ for c.

Distance to the horizon. The formula $V = 1.2\sqrt{h}$ can be used to approximate the distance V, in miles, that a person can see to the horizon from a height h, in feet.

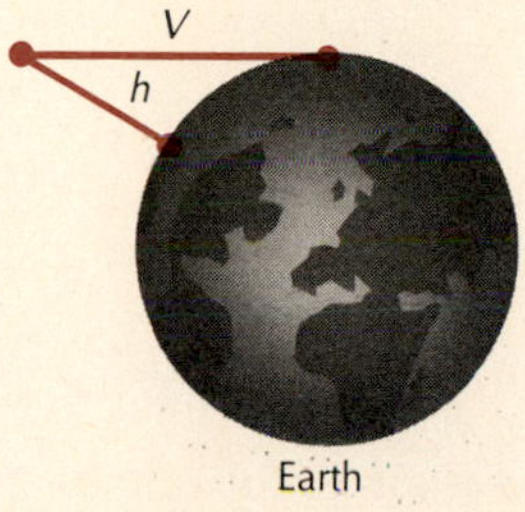

33. How far can you see to the horizon through an airplane window at a height of 30,000 ft?

34. How far can a sailor see to the horizon from the top of a 72-ft mast?

35. A person can see 144 mi to the horizon from an airplane window. How high is the airplane?

36. A sailor can see 11 mi to the horizon from the top of a mast. How high is the mast?

SYNTHESIS

Solve.

37. $(x - 5)^{2/3} = 2$

38. $(x - 3)^{2/3} = 2$

39. $\dfrac{x + \sqrt{x + 1}}{x - \sqrt{x + 1}} = \dfrac{5}{11}$

40. $\sqrt{\sqrt{x + 25} - \sqrt{x}} = 5$

41. $\sqrt{x + 2} - \sqrt{x - 2} = \sqrt{2x}$

42. $2\sqrt{x + 3} = \sqrt{x} + \sqrt{x + 8}$

43. $\sqrt[4]{x + 2} = \sqrt{3x + 1}$

44. $\sqrt[3]{2x - 1} = \sqrt[6]{x + 1}$

45. $\dfrac{14}{3 + \sqrt{7 + x}} - \dfrac{\sqrt{7 + x}}{2} = 0$

46. $\dfrac{\sqrt{10 + x}}{4} = \dfrac{6}{2 + \sqrt{10 + x}}$

47. $\sqrt{3x + 1} - \sqrt{2x} = \dfrac{5}{\sqrt{3x + 1}}$

48. $\sqrt{3 + x} + \sqrt{x} = \dfrac{5}{\sqrt{x}}$

49. $\sqrt{15 + \sqrt{2x + 80}} = 5$

50. $\sqrt{x + 5} + 1 = \dfrac{6}{\sqrt{x + 5}}$

2.8 Equations Reducible to Quadratic

OBJECTIVES

You should be able to:

1. Solve equations that are reducible to quadratic.
2. Solve applied problems involving equations reducible to quadratic.

1 Solving Equations Reducible to Quadratic

Certain equations that are not really quadratic can be thought of in such a way that they can be solved as quadratic. For example,

$$x + 3\sqrt{x} - 10 = 0$$

$$\downarrow \quad \downarrow \quad \downarrow$$

$$(\sqrt{x})^2 + 3\sqrt{x} - 10 = 0 \qquad \text{Thinking of } x \text{ as } (\sqrt{x})^2$$

$$\downarrow \quad \downarrow \quad \downarrow$$

$$u^2 + 3u - 10 = 0 \qquad \text{To make this clearer, write } u \text{ instead of } \sqrt{x}.$$

1. a) Solve: $x + \sqrt{x} - 12 = 0$.

 b) Can you think of another procedure for solving this equation? See Section 2.7. Which procedure seemed easier?

We can solve the equation $u^2 + 3u - 10 = 0$ by factoring or by using the quadratic formula. After that, we can find x by remembering that $\sqrt{x} = u$. Equations that can be solved in this way are said to be **reducible to quadratic**.

> **To solve equations reducible to quadratic, we first make a substitution, solve for the new variable, and then solve for the original variable.**

EXAMPLE 1 Solve: $x + 3\sqrt{x} - 10 = 0$.

Solution Let $u = \sqrt{x}$. Then we solve the equation resulting from substituting u for $\sqrt{x}$:

$$x + 3\sqrt{x} - 10 = 0$$
$$u^2 + 3u - 10 = 0$$
$$(u + 5)(u - 2) = 0$$
$$u = -5 \quad \text{or} \quad u = 2.$$

We have solved for u. Now we solve for x. We substitute $\sqrt{x}$ for u and solve these equations:

$$\sqrt{x} = -5 \quad \text{or} \quad \sqrt{x} = 2$$
$$\text{(no solution)} \quad \text{or} \quad x = 4.$$

Check: $\dfrac{x + 3\sqrt{x} - 10 = 0}{\begin{array}{c|c} 4 + 3\sqrt{4} - 10 & 0 \\ 4 + 6 - 10 & \\ 0 & \end{array}}$

The solution is 4. The solution set is $\{4\}$. ■

DO EXERCISE 1.

EXAMPLE 2 Solve: $x^4 - 6x^2 + 7 = 0$.

Solution Let $u = x^2$. Then we solve the equation resulting from substituting u for x^2. We have

$$x^4 - 6x^2 + 7 = 0$$
$$(x^2)^2 - 6x^2 + 7 = 0$$
$$u^2 - 6u + 7 = 0;$$
$$a = 1, \quad b = -6, \quad c = 7;$$

$$u = \frac{-b \pm \sqrt{b^2 - 4ac}}{2a} = \frac{-(-6) \pm \sqrt{(-6)^2 - 4 \cdot 1 \cdot 7}}{2 \cdot 1}$$
$$u = \frac{6 \pm \sqrt{8}}{2} = \frac{2 \cdot 3 \pm 2\sqrt{2}}{2 \cdot 1}$$
$$u = 3 \pm \sqrt{2}.$$

We have solved for u. Now we solve for x. We substitute x^2 for u and solve for x:

$$x^2 = 3 + \sqrt{2} \qquad \text{or} \quad x^2 = 3 - \sqrt{2}$$
$$x = \pm\sqrt{3 + \sqrt{2}} \quad \text{or} \quad x = \pm\sqrt{3 - \sqrt{2}}.$$

Thus the solution set is

$$\{\sqrt{3+\sqrt{2}},\ -\sqrt{3+\sqrt{2}},\ \sqrt{3-\sqrt{2}},\ -\sqrt{3-\sqrt{2}}\}.$$

DO EXERCISE 2.

EXAMPLE 3 Solve: $(x^2 - x)^2 - 14(x^2 - x) + 24 = 0$.

Solution Let $u = x^2 - x$. Then we solve the equation resulting from substituting u for $x^2 - x$:

$$u^2 - 14u + 24 = 0$$
$$(u - 12)(u - 2) = 0$$
$$u = 12 \quad \text{or} \quad u = 2.$$

We have solved for u. Now we solve for x. We substitute $x^2 - x$ for u and solve:

$$x^2 - x = 12 \quad \text{or} \quad x^2 - x = 2$$
$$x^2 - x - 12 = 0 \quad \text{or} \quad x^2 - x - 2 = 0$$
$$(x - 4)(x + 3) = 0 \quad \text{or} \quad (x - 2)(x + 1) = 0$$
$$x = 4 \quad \text{or} \quad x = -3 \quad \text{or} \quad x = 2 \quad \text{or} \quad x = -1.$$

The solutions are 4, -3, 2, -1. The solution set is $\{4, -3, 2, -1\}$.

DO EXERCISE 3.

EXAMPLE 4 Solve: $t^{2/5} - t^{1/5} - 2 = 0$.

Solution Let $u = t^{1/5}$. Then we solve the equation resulting from substituting u for $t^{1/5}$:

$$(t^{1/5})^2 - t^{1/5} - 2 = 0$$
$$u^2 - u - 2 = 0$$
$$(u - 2)(u + 1) = 0$$
$$u = 2 \quad \text{or} \quad u = -1.$$

Now we substitute $t^{1/5}$ for u and solve:

$$t^{1/5} = 2 \quad \text{or} \quad t^{1/5} = -1$$
$$t = 32 \quad \text{or} \quad t = -1.$$

Principle of powers; raising to the fifth power

The solutions are 32 and -1. The solution set is $\{-1, 32\}$.

DO EXERCISE 4.

EXAMPLE 5 Solve: $x^4 + 3x^2 - 4 = 0$.

Solution Let $u = x^2$. Then we solve the equation resulting from substituting u for x^2:

$$u^2 + 3u - 4 = 0$$
$$(u + 4)(u - 1) = 0$$
$$u = -4 \quad \text{or} \quad u = 1.$$

2. Solve: $2x^4 - 10x^2 + 11 = 0$.

3. Solve: $(x^2 - 1)^2 - (x^2 - 1) - 2 = 0$.

4. Solve: $t^{2/3} - 3t^{1/3} - 10 = 0$.

5. Solve: $x^4 + 5x^2 - 36 = 0$.

CAUTION! Remember that you are solving for x, *not* u.

Now we substitute x^2 for u and solve for x:

$$x^2 = -4 \quad \text{or} \quad x^2 = 1$$
$$x = \pm 2i \quad \text{or} \quad x = \pm 1.$$

The solutions are $2i$, $-2i$, 1, and -1. The solution set is $\{2i, -2i, 1, -1\}$.

DO EXERCISE 5.

2 Problem Solving

EXAMPLE 6 *A well problem.* An object is dropped into a well. Two seconds later the sound of the splash is heard at the top. The speed of sound is 1100 ft/sec. How deep is the well?

Solution

1. *Familiarize.* We first make a drawing and label it with known and unknown information. We can picture the situation as follows. We have a well s feet deep. The time it takes for the object to fall to the bottom of the well can be represented by t_1. The time it takes for the sound to get back to the top of the well can be represented by t_2. The total amount of time is the sum of t_1 and t_2. This gives us the equation

$$t_1 + t_2 = 2. \tag{1}$$

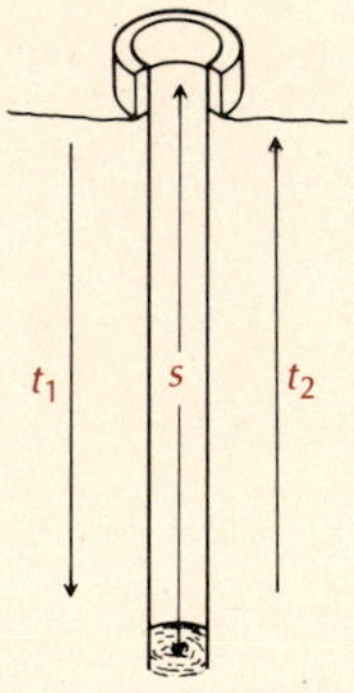

2. *Translate.* Now can we find any relationship between the two times and the distance s? Often in problem solving you may need to look up related formulas in a physics book, an encyclopedia, or maybe another mathematics book. It turns out that the formula for falling objects, which we considered in Section 2.6, becomes

$$s = 16t^2 + v_0 t,$$

when the distance is given in feet. Since the object is dropped, $v_0 = 0$. The time t_1 that it takes for the object to reach the bottom of the well can be found as follows:

$$s = 16t_1^2, \quad \text{or} \quad t_1 = \frac{\sqrt{s}}{4}. \tag{2}$$

We now have an expression for t_1. What about t_2? To find how long it takes for the sound to get back to the top of the well, we can use the formula $d = rt$, which we considered in Section 2.3. We want to find s, so

$d = s$. The speed r is the speed of sound. (It is given, but in another problem-solving situation, you might have needed to look it up.) It is 1100 ft/sec. Then

$$s = 1100 \cdot t_2, \quad \text{or} \quad t_2 = \frac{s}{1100}. \tag{3}$$

We have an expression for t_1 in equation (2) and an expression for t_2 in equation (3) both in terms of s. We substitute these into equation (1) and obtain

$$t_1 + t_2 = 2, \quad \text{or} \quad \frac{\sqrt{s}}{4} + \frac{s}{1100} = 2. \tag{4}$$

3. *Carry out.* We solve equation (4) for s. Multiplying by 1100, we get

$$275\sqrt{s} + s = 2200, \quad \text{or} \quad s + 275\sqrt{s} - 2200 = 0.$$

This equation is reducible to quadratic with $u = \sqrt{s}$. Substituting, we get

$$u^2 + 275u - 2200 = 0.$$

Using the quadratic formula, we can solve for u:

$$u = \frac{-275 + \sqrt{275^2 + 8800}}{2} \qquad \text{We want the positive solution.}$$

$$= \frac{-275 + \sqrt{84{,}425}}{2}$$

$$\approx \frac{-275 + 290.56}{2} \qquad \text{Use your calculator and approximate.}$$

$$u \approx \frac{15.56}{2} = 7.78.$$

Thus, $u \approx 7.78 \approx \sqrt{s}$, so $s \approx 60.5284$.

4. *Check.* We check by substituting 60.53 for s back through the related equations.

5. *State.* The well is about 60.53 ft deep.

6. An object is dropped into a well. In 5 sec, the sound of the splash reaches the top of the well. If we assume that the speed of sound is 1100 ft/sec, how deep is the well?

DO EXERCISE 6.

EXERCISE SET 2.8

1 Solve.

1. $x - 10\sqrt{x} + 9 = 0$
2. $2x - 9\sqrt{x} + 4 = 0$
3. $x^4 - 10x^2 + 25 = 0$
4. $x^4 - 3x^2 + 2 = 0$
5. $t^{2/3} + t^{1/3} - 6 = 0$
6. $w^{2/3} - 2w^{1/3} - 8 = 0$
7. $z^{1/2} = z^{1/4} + 2$
8. $6 = m^{1/3} - m^{1/6}$
9. $(x^2 - 6x)^2 - 2(x^2 - 6x) - 35 = 0$
10. $(1 + \sqrt{x})^2 + (1 + \sqrt{x}) - 6 = 0$
11. $(y^2 - 5y)^2 + (y^2 - 5y) - 12 = 0$
12. $(2t^2 + t)^2 - 4(2t^2 + t) + 3 = 0$
13. $w^4 - 4w^2 - 2 = 0$
14. $t^4 - 5t^2 + 5 = 0$
15. $x^{-2} - x^{-1} - 6 = 0$
16. $4x^{-2} - x^{-1} - 5 = 0$
17. $2x^{-2} + x^{-1} = 1$
18. $10 - 9m^{-1} = m^{-2}$
19. $x^4 - 24x^2 - 25 = 0$
20. $x^4 - 5x^2 - 36 = 0$
21. $\left(\frac{x^2 - 2}{x}\right)^2 - 7\left(\frac{x^2 - 2}{x}\right) - 18 = 0$
22. $\left(\frac{x^2 + 1}{x}\right)^2 - 8\left(\frac{x^2 + 1}{x}\right) + 15 = 0$

23. $\dfrac{x}{x-1} - 6\sqrt{\dfrac{x}{x-1}} - 40 = 0$

24. $\dfrac{2x+1}{x} - 30 = 7\sqrt{\dfrac{2x+1}{x}}$ $\left(\textit{Hint: } u = \sqrt{\dfrac{2x+1}{x}}.\right)$

25. $5\left(\dfrac{x+2}{x-2}\right)^2 = 3\left(\dfrac{x+2}{x-2}\right) + 2$

26. $\left(\dfrac{x+1}{x+3}\right)^2 + \left(\dfrac{x+1}{x+3}\right) - 6 = 0$

27. $\left(\dfrac{x^2-1}{x}\right)^2 - \left(\dfrac{x^2-1}{x}\right) - 2 = 0$

28. $\left(\dfrac{x+8}{x-8}\right)^2 - 6 = 5\left(\dfrac{x+8}{x-8}\right)$

2 Problem Solving

29. A stone is dropped from a cliff. In 3 sec, the sound of the stone striking the ground reaches the top of the cliff. If we assume that the speed of sound is 1100 ft/sec, how high is the cliff?

30. A stone is dropped from a cliff. In 4 sec, the sound of the stone striking the ground reaches the top of the cliff. If we assume that the speed of sound is 1100 ft/sec, how high is the cliff?

31. At the beginning of the year, \$2000 is deposited in a bank at a certain interest rate. At the beginning of the next year, \$1200 is deposited in another bank at the same interest rate. At the beginning of the third year, there is a total of \$3573.80 in both accounts. Interest is compounded annually. What is the interest rate?

32. At the beginning of the year, \$3500 is deposited in a bank at a certain interest rate. At the beginning of the next year, \$4000 is deposited in another bank at the same interest rate. At the beginning of the third year, there is a total of \$8518.35 in both accounts. Interest is compounded annually. What is the interest rate?

SYNTHESIS

Solve. Check possible solutions by substituting into the original equation.

33. $6.75x = \sqrt{35x} + 5.36$

34. $\pi x^4 - \sqrt{99.3} = \pi^2 x^2$

Solve.

35. $9x^{3/2} - 8 = x^3$

36. $\sqrt[3]{2x+3} = \sqrt[6]{2x+3}$

37. $\sqrt{x-3} - \sqrt[4]{x-3} = 2$

38. $a^3 - 26a^{3/2} - 27 = 0$

39. $x^6 - 28x^3 + 27 = 0$

40. $x^6 + 7x^3 = 8$

41. $(x^2 - 5x - 2)^2 - 5(x^2 - 5x - 2) + 4 = 0$

42. $x^{1/2} + \dfrac{1}{x^{1/2}} = \dfrac{13}{6}$

43. $\left(y + \dfrac{2}{y}\right)^2 + 3y + \dfrac{6}{y} = 4$

44. $x^2 + 3x + 1 - \sqrt{x^2 + 3x + 1} = 8$

CHALLENGE

45. Solve: $\dfrac{2x+1}{x} = 3 + 7\sqrt{\dfrac{2x+1}{x}}$.

46. At the beginning of a year, \$2000 is deposited. Six months later, \$3000 is deposited in another account at the same interest rate. At the beginning of the next year, there is \$956.80 more in the second account than in the first account. If the interest is compounded semiannually, what is the interest rate?

OBJECTIVE

You should be able to:

1 Find equations of variation and solve applied problems involving variation.

2.9 Variation

1 Types of Variation

Direct Variation

There are many situations that yield linear equations like $y = kx$, where k is some positive constant. Note that as x increases, y increases. In such a situation, we say that we have **direct variation**, and k is called the **variation constant**. Usually only positive values of x and y are considered.

DEFINITION

If two variables x and y are related as in the equation $y = kx$, where k is a positive constant, we say that "y varies directly as x," or that "y is directly proportional to x."

For example, the circumference C of a circle varies directly as its diameter D: $C = \pi D$. A car moves at a constant speed of 65 mph. The distance d that it travels varies directly as the time t: $d = 65t$.

EXAMPLE 1 Find an equation of variation in which y varies directly as x, and $y = 5.6$ when $x = 8$.

Solution We know that $y = kx$, so $5.6 = k \cdot 8$ and $0.7 = k$. Thus the equation of variation is $y = 0.7x$.

DO EXERCISE 1.

EXAMPLE 2 *A spring problem. Hooke's law* states that the distance d that an elastic object such as a spring is stretched by placing a certain weight on it varies directly as the weight w of the object. If the distance is 40 cm when the weight is 3 kg, what is the distance stretched when a 2-kg weight is attached?

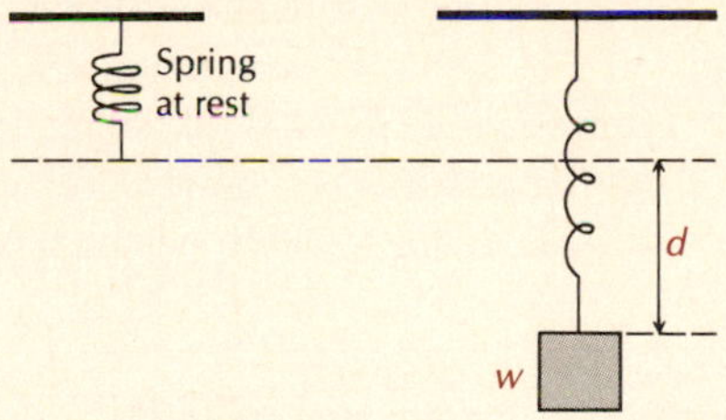

Solution The equation of variation is

$$d = kw.$$

We first find k using the fact that $d = 40$ cm when $w = 3$ kg:

$$40 = k \cdot 3$$
$$\frac{40}{3} = k.$$

Then the equation of variation becomes

$$d = \frac{40}{3}w.$$

We then substitute 2 kg for w and compute d:

$$d = \frac{40}{3} \cdot 2 = \frac{80}{3} \approx 26.7 \text{ cm}.$$

Thus a 2-kg weight will stretch the spring about 26.7 cm.

DO EXERCISES 2–4.

Inverse Variation

There are also situations that yield equations of the type $y = k/x$, where k is a positive constant. Note that as positive values of x increase, y decreases. In such

1. Find an equation of variation in which y varies directly as x, and $y = 32$ when $x = 0.2$.

2. Under the conditions of Example 2, what weight would be required to stretch the spring 60 cm?

3. *Ohm's law* states that the voltage V in an electric circuit varies directly as the number of amperes I of electric current in the circuit. If the voltage is 10 volts when the current is 3 amperes, what is the voltage when the current is 15 amperes?

4. The amount of garbage G produced in the United States varies directly as the number of people N who produce the garbage. It is known that 50 tons of garbage is produced by 200 people in 1 year. The population of San Francisco is 705,000. How much garbage is produced by San Francisco in 1 year?

a situation, we say that we have **inverse variation**, and k is called the **variation constant**. In applications, we are generally considering only positive values of x and y.

DEFINITION

If two variables x and y are related as in the equation $y = k/x$, where k is a positive constant, we say that "y varies inversely as x," or that "y is inversely proportional to x."

EXAMPLE 3 Find an equation of variation in which y varies inversely as x, and $y = 5.6$ when $x = 8$.

Solution We know that $y = k/x$, so $5.6 = k/8$ and $k = 44.8$. Thus the equation of variation is $y = 44.8/x$.

CAUTION! Keep in mind that an answer to an example like Examples 1 and 3 is an *equation*. The value of k is *not* the answer!

DO EXERCISE 5.

EXAMPLE 4 *Stocks and gold.* Certain economists theorize that stock prices are inversely proportional to the price of gold. That is, when the price of gold goes up, the prices of stock go down; and when the price of gold goes down, the prices of stock go up. Let us assume that the Dow-Jones Industrial Average, D, an index of the overall price of stock, is inversely proportional to the price of gold, G, in dollars per ounce. One day the Dow-Jones Industrial Average was 1846 and the price of gold was \$462 per ounce. What will the Dow-Jones Average be if the price of gold drops to \$440 per ounce?

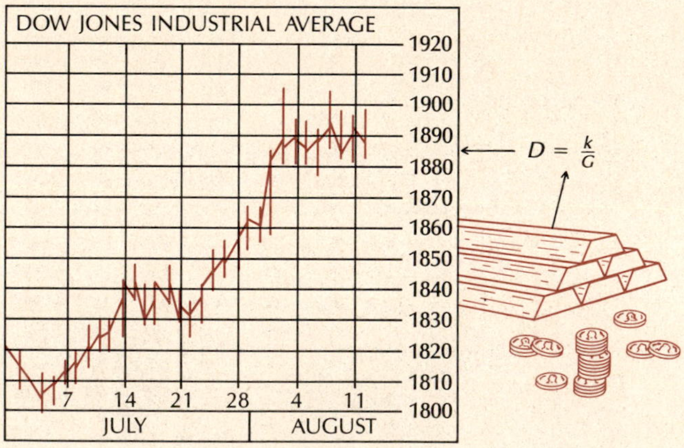

Solution The equation of variation is

$$D = \frac{k}{G}.$$

We first find k using the fact that $D = 1846$ when $G = \$462$:

$$1846 = \frac{k}{462}$$

$$k = 1846(462) = 852{,}852.$$

Then the equation of variation becomes

$$D = \frac{852{,}852}{G}.$$

We substitute \$440 for G and compute D:

$$D = \frac{852{,}852}{440} = 1938.3.$$

Thus the Dow-Jones Average will be 1938.3 when the price of gold is \$440 per ounce. ■

DO EXERCISE 6.

Other Kinds of Variation

There are many other kinds of variation.

> **DEFINITION**
>
> y varies *directly* as the square of x if there is some positive constant k such that $y = kx^2$.

For example, the area A of a circle varies directly as the square of the radius r: $A = \pi r^2$.

DO EXERCISE 7.

> **DEFINITION**
>
> y varies *inversely* as the square of x if there is some positive constant k such that $y = k/x^2$.

For example, the weight W of a body varies inversely as the square of the distance d from the center of the earth: $W = k/d^2$.

DO EXERCISE 8.

> **DEFINITION**
>
> y varies *jointly* as x and z if there is some positive constant k such that $y = kxz$.

For example, consider the equation for the area of a triangle: $A = \frac{1}{2}bh$. The area varies jointly as b and h. The variation constant is $\frac{1}{2}$. Joint variation implies a product of variables.

DO EXERCISE 9.

Several kinds of variation can occur together. For example, if

$$y = k \cdot xz^3/w^2,$$

then y varies jointly as x and the cube of z and inversely as the square of w.

DO EXERCISE 10.

5. Find an equation of variation in which y varies inversely as x, and $y = 32$ when $x = 0.2$.

6. The time t required to drive a fixed distance varies inversely as the speed r. It takes 5 hr at 60 km/h to drive a fixed distance. How long would it take to drive the fixed distance at 40 km/h?

7. Find an equation of variation in which y varies directly as the square of x, and $y = 12$ when $x = 2$.

8. Find an equation of variation in which y varies inversely as the square of x, and $y = \frac{1}{4}$ when $x = 6$.

9. Find an equation of variation in which y varies jointly as x and z, and $y = 42$ when $x = 2$ and $z = 3$.

10. Find an equation of variation in which y varies jointly as x and z and inversely as the square of w, and $y = 105$ when $x = 3$, $z = 20$, and $w = 2$.

11. The distance S that an object falls from some point above the ground varies directly as the square of the time t that it falls. If the object falls 4 ft in 0.5 sec, how long will it take the object to fall 64 ft?

EXAMPLE 5 *The volume of a tree trunk.* The volume of wood V in a tree trunk varies jointly as the height h and the square of the girth g (girth is distance around). If the volume is 7750 ft^3 when the height is 100 ft and the girth is 5 ft, what is the height when the volume is 37,975 ft^3 and the girth is 7 ft?

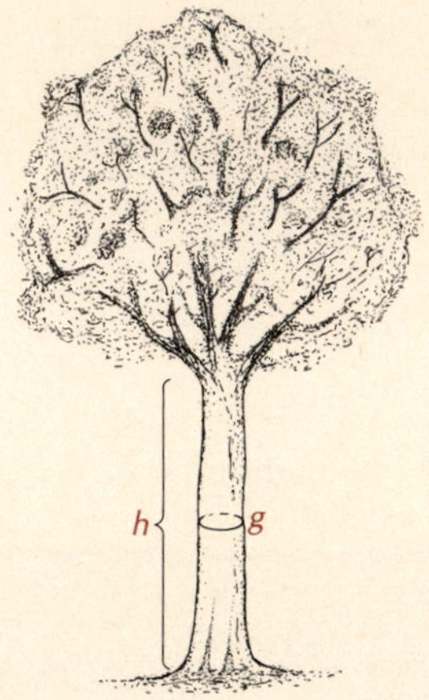

Solution The equation of variation is

$$V = khg^2.$$

We first find k using the fact that $V = 7750$ ft^3 when $h = 100$ ft and $g = 5$ ft:

$$7750 = k \cdot 100 \cdot 5^2$$
$$3.1 = k.$$

Then the equation of variation becomes

$$V = 3.1hg^2.$$

We substitute 37,975 ft^3 for V and 7 ft for g and solve for h:

$$37{,}975 = 3.1 \cdot h \cdot 7^2$$
$$h = 250 \text{ ft}.$$

DO EXERCISE 11.

EXAMPLE 6 *The weight of an astronaut.* The weight W of an object varies inversely as the square of the distance d from the object to the center of the earth. At sea level (4000 mi from the center of the earth), an astronaut weighs 200 lb. Find his weight when he is 100 mi above the surface of the earth and the spacecraft is not in motion.

Solution The equation of variation is

$$W = \frac{k}{d^2}.$$

We first find k using the fact that $W = 200$ lb when $d = 4000$ mi:

$$200 = \frac{k}{(4000)^2}$$
$$200 = \frac{k}{16{,}000{,}000}$$
$$3{,}200{,}000{,}000 = k.$$

Then the equation of variation becomes

$$W = \frac{3{,}200{,}000{,}000}{d^2}.$$

We add 100 mi to 4000 mi and substitute 4100 for d, and compute W:

$$W = \frac{3{,}200{,}000{,}000}{(4100)^2} = \frac{3{,}200{,}000{,}000}{16{,}810{,}000} = 190 \text{ lb} \quad \text{(to the nearest pound).}$$

Thus the weight of the astronaut 100 mi above the surface of the earth is 190 lb.

DO EXERCISE 12.

12. a) How much will the 200-lb astronaut weigh when he is 1000 mi from earth?
b) How far is he from earth when his weight is reduced to 50 lb?

EXERCISE SET 2.9

1 Find an equation of variation for the given situation.

1. y varies directly as x, and $y = 0.6$ when $x = 0.4$
2. y varies directly as x, and $y = 125$ when $x = 32$
3. y varies inversely as x, and $y = 125$ when $x = 32$
4. y varies inversely as x, and $y = 0.4$ when $x = 0.8$
5. y varies directly as x, and $y = 8.6$ when $x = 1.6$
6. y varies inversely as x, and $y = 5.4$ when $x = 3.8$
7. y varies inversely as the square of x, and $y = 0.15$ when $x = 0.1$
8. y varies jointly as x and z, and $y = 56$ when $x = 7$ and $z = 10$
9. y varies jointly as x and z and inversely as w, and $y = \frac{3}{2}$ when $x = 2$, $z = 3$, and $w = 4$
10. y varies jointly as x and the square of z, and $y = 105$ when $x = 14$ and $z = 5$
11. y varies jointly as x and z and inversely as the square of w, and $y = \frac{12}{5}$ when $x = 16$, $z = 3$, and $w = 5$
12. y varies jointly as x and z and inversely as the product of w and p, and $y = \frac{3}{28}$ when $x = 3$, $z = 10$, $w = 7$, and $p = 8$
13. Suppose that y varies directly as x and x is doubled. What is the effect on y?
14. Suppose that y varies inversely as x and x is tripled. What is the effect on y?
15. Suppose that y varies inversely as the square of x and x is multiplied by n. What is the effect on y?
16. Suppose that y varies directly as the square of x and x is multiplied by n. What is the effect on y?
17. The amount of pollution A entering the atmosphere varies directly as the amount of people N living in an area. If 60,000 people cause 42,600 tons of pollutants, how many tons entered the atmosphere in a city with a population of 750,000?
18. The volume V of a given mass of gas varies directly as the temperature T and inversely as the pressure P. If $V = 231 \text{ in}^3$ when $T = 420°$ and $P = 20 \text{ lb/in}^2$, what is the volume when $T = 300°$ and $P = 15 \text{ lb/in}^2$?
19. *The strength of a beam.* The safe load (the amount it supports without breaking) L of a beam varies jointly as its width w and the square of its height h and inversely as its length l. If the width and the height are doubled at the same time that the length is halved, what is the effect on L?

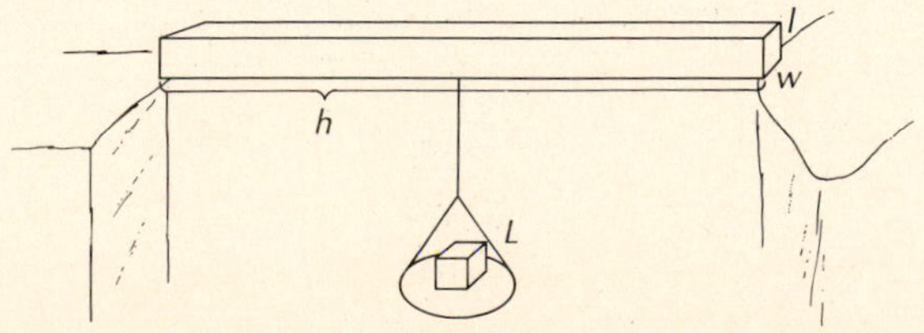

20. For a chord $\overline{PQ}$ through a fixed point A in a circle, the length of $\overline{PA}$ is inversely proportional to the length of $\overline{AQ}$. If the length of $\overline{PA} = 64$ when the length of $\overline{AQ} = 16$, what is the length of $\overline{AQ}$ when the length of $\overline{PA} = 4$?

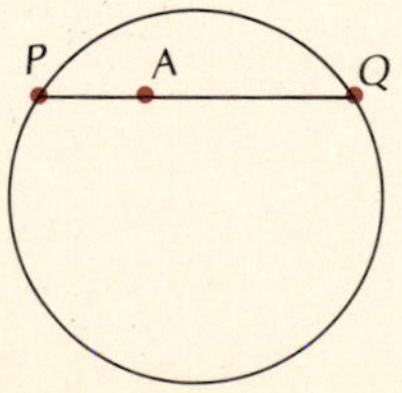

21. *A sighting problem.* The distance d that one can see to the horizon varies directly as the square root of the height above sea level. If a person 19.5 m above sea level can see 28.97 km, how high above sea level must one be in order to see 54.32 km?

22. *Electrical resistance.* At a fixed temperature and chemical composition, the resistance of a wire varies directly as the length l and inversely as the square of the diameter d. If the resistance of a certain kind of wire is 0.112 ohm when the diameter is 0.254 cm and the length is 15.24 m, what is the resistance of a wire whose length is 608.7 m and whose diameter is 0.478 cm?

23. *Area of a cube.* The area of a cube varies directly as the square of the length of a side. If a cube has an area of 168.54 m^2 when the length of a side is 5.3 m, what will the area be when the length of a side is 10.2 m?

24. *Intensity of light.* The intensity of light I from a light bulb varies inversely as the square of the distance d from the bulb. Suppose that I is 90 W/m^2 when the distance is 5 m. Find the intensity at a distance of 10 m.

25. *Earned run average.* A pitcher's earned run average A varies directly as the number of earned runs R allowed and inversely as the number of innings pitched I. In a recent year a pitcher had an earned run average of 2.92. He gave up 85 earned runs in 262 innings. How many earned runs would he have given up had he pitched 300 innings? Round to the nearest whole number.

26. *Volume of a gas.* The volume V of a given mass of a gas varies directly as the temperature T and inversely as the pressure P. If $V = 231$ cm^3 when $T = 42°$ and $P = 20$ kg/cm^2, what is the volume when $T = 30°$ and $P = 15$ kg/cm^2?

SYNTHESIS

27. Show that if p varies directly as q, then q varies directly as p.

28. Show that if u varies inversely as v, then v varies inversely as u, and $1/u$ varies directly as v.

29. The area of a circle varies directly as the square of the length of a diameter. What is the variation constant?

30. P varies directly as the square of t. How does t vary in relationship to P?

CHALLENGE

31. *The gravity model in sociology.* It has been determined that the average number of telephone calls in a day, N, between two cities is directly proportional to the populations P_1 and P_2 of the cities and inversely proportional to the square of the distance d between the cities. That is,

$$N = \frac{kP_1P_2}{d^2}.$$

This theory is known as the *gravity model* because the equation is similar to Newton's theory of gravity. Use a calculator to find solutions to these problems.

a) The population of Indianapolis is about 744,624, the population of Cincinnati is about 452,524, and the distance between the cities is 174 km. The average number of daily phone calls between the two cities is 11,153. Find the value k and write the equation of variation.

b) The population of Detroit is about 1,511,482 and it is 446 km from Indianapolis. Find the average number of daily phone calls between Detroit and Indianapolis.

c) The average number of daily phone calls between Indianapolis and New York City is 4270 and the population of New York City is about 7,895,563. Find the distance between Indianapolis and New York City.

d) Why is this model not appropriate for adjoining cities such as Minneapolis and St. Paul? (Sociologists say that as communication between two cities increases, the cities tend to merge.)

SUMMARY AND REVIEW: CHAPTER 2

TERMS TO KNOW

Solution
Solution set
Addition principle
Multiplication principle
Identity
Principle of zero products
Meaningful replacements
Equivalent equations
Fractional equations
Clearing of fractions
Compound interest
Distance formula
Imaginary numbers
Complex numbers
Conjugate
Quadratic equation
Completing the square
Quadratic formula
Discriminant
Radical equation
Reducible to quadratic
Direct variation
Inverse variation
Variation constant
Joint variation

REVIEW EXERCISES

Simplify. Leave answers in terms of i.

1. $-\sqrt{-40}$
2. $\sqrt{-12} \cdot \sqrt{-20}$

Simplify.

3. $(2 - 2i)(3 + 4i)$
4. $(3 - 5i) - (2 - i)$
5. $(6 + 2i) + (-4 - 3i)$
6. $\dfrac{2 - 3i}{1 - 3i}$
7. Determine whether $1 - 3i$ is a solution of $x^2 - 2x - 10 = 0$.
8. Find the reciprocal of $6 - 7i$ and express it in the form $a + bi$.
9. Solve for x and y: $4x + 2i = 8 - (2 + y)i$.

Solve.

10. $(6 - i)x + 4 - 8i = 3 - 3i + 2ix$
11. $\dfrac{3x + 2}{7} - \dfrac{5x}{3} = \dfrac{32}{21}$
12. $(p - 3)(3p + 2)(p + 2) = 0$
13. $3x^2 + 2x - 8 = 0$
14. $r^2 - 2r + 10 = 0$
15. $x^3 + 5x^2 - 4x - 20 = 0$
16. $\dfrac{5}{2x + 3} + \dfrac{1}{x - 6} = 0$
17. $y^4 - 3y^2 + 1 = 0$
18. $x = 2\sqrt{x} - 1$
19. $(x^2 - 1)^2 - (x^2 - 1) - 2 = 0$
20. $t^{2/3} - 10 = 3t^{1/3}$
21. $\sqrt{x - 1} - \sqrt{x - 4} = 1$
22. $\sqrt{5x + 1} - 1 = \sqrt{3x}$
23. Solve by completing the square: $y^2 - 6y = 16$. Show your work.
24. $y^2 - 3y = 18$
25. $3[x - 5(4 + 2x)] = 7x - 10(3x - 2)$
26. $(z^2 - 1) + z = 14 - z$
27. $(x - 2)(x + 3) + 4 = 0$
28. $14 - 4y < 22$
29. $(x - 5)(x + 5) \geq (x - 5)(x + 4)$
30. Determine the meaningful replacements in the radical expression $\sqrt{24 - 6x}$.

Determine the nature of the solutions of the equation.

31. $4y^2 + 5y + 10 = 0$
32. $3x^2 + 2x - 1 = 0$
33. Write a quadratic equation whose solutions are -3 and $\frac{1}{2}$.
34. Find an equation having the solutions $1 - 2i$, $1 + 2i$.
35. Solve $v = \sqrt{2gh}$ for h.
36. Solve $\dfrac{1}{a} + \dfrac{1}{b} = \dfrac{1}{t}$ for t.

Solve.

37. A student scores 73% and 79% on two tests. If the third test counts as though it were two tests, what score must the student make on the third test so the average will be 85%?

38. A can mow a lawn in 4 hr. B can mow it in 2 hr. How long would it take if they worked together?

39. A can mow a lawn in 3 hr. Working together, A and B can mow the lawn in 1 hr. How long would it take B, working alone, to mow the lawn?

40. 18 is 30% of what?

41. What is 9 percent of 50?

42. Two trains leave the same city at right angles. The first train travels at a speed of 60 km/h. In 1 hr, the trains are 100 km apart. How fast is the second train traveling?

43. In a right triangle, the perimeter is 40 and the sum of the squares of the sides is 578. Find the lengths of the sides.

44. A boat travels 2 km upstream and 2 km downstream. The total time for both parts of the trip is 1 hr. The speed of the stream is 2 km/h. What is the speed of the boat in still water? Round to the nearest tenth.

Solve.

45. Find an equation of variation in which y varies inversely as the square of x, and $y = 0.005$ when $x = 10$.

46. Find an equation of variation in which T varies directly as the square of x and inversely as p, and $T = 0.01$ when $x = 6$ and $p = 20$.

47. It is theorized that the dividends paid on utilities stocks are inversely proportional to the prime (interest) rate. Recently, the dividends D on the stock of Indianapolis Power and Light were $2.09 per share and the prime rate was 19%. The prime rate, R, dropped to 17.5%. What dividends were then paid?

48. For a body falling freely from rest, the distance (s ft) that the body falls varies directly as the square of the time (t sec). Given that $s = 64$ when $t = 2$, find a formula for s in terms of t. How long will it take the body to fall 900 ft?

Determine whether the equations of the pair are equivalent.

49. $x = 5$,
$x^2 = 25$

50. $x - 7 = \dfrac{x^2 - 49}{x + 7}$,
$x - 7 = x - 7$

SYNTHESIS

51. Suppose a, b, c, and d are nonzero real numbers such that c and d are solutions of

$$x^2 + ax + b = 0$$

and a and b are solutions of

$$x^2 + cx + d = 0.$$

Find $a + b + c + d$.

52. Determine whether

$$x - 7 = \frac{x^2 - 49}{x + 7}$$

is an identity.

53. The area of a circle varies directly as the square of its circumference. Write an equation of variation and find the variation constant.

54. Solve:

$$\sqrt{\sqrt{\sqrt{x}}} = 2.$$

55. Determine whether

$$\frac{x + 1}{x + 2} = \frac{1}{2}$$

is an identity.

56. Solve:

$$2x - 3y = 7 + 7i,$$
$$3x + 2y = 4 - 9i.$$

TEST: CHAPTER 2

Simplify.

1. i^{83}
2. $-\sqrt{-3}\sqrt{-16}$
3. $(7 - i)(2 + 4i)$
4. $(1 + 6i) - (-3 - 8i)$
5. $\dfrac{5 - 2i}{2 - 3i}$
6. $(1 - i)(1 + i)$
7. Find the reciprocal of $4 + 2i$ and express it in the form $a + bi$.
8. Solve for x and y:

$$3x + 4 - i = 10 + (2 + y)i.$$

Solve.

9. $(2y - 9)(y + 4)(y - 5) = 0$
10. $x - 7\sqrt{x} + 6 = 0$
11. $2t^2 - t - 21 = 0$
12. $4t^2 - 3t - 5 = 0$
13. $\dfrac{8}{3x - 5} = \dfrac{4}{x + 3}$
14. $5x^2 - 4x + 12 = 0$
15. $\sqrt{y - 11} = \sqrt{y + 10} - 3$
16. $\dfrac{2x - 3}{3} = \dfrac{x + 4}{4} - 2$
17. $(a + 3)(a - 7) - 11 = 0$
18. $22 - 8y < 38$
19. $x^3 - 7x^2 - x + 7 = 0$
20. $3x - 4(x + 6) = 2[x - 3(4 - x)]$
21. 18 is what percent of 9?
22. Grain flows through spout A five times faster than through spout B. When grain flows through both spouts, a grain bin is filled in 4 hr. How many hours would it take to fill the grain bin if grain flows through spout B alone?
23. The speed of train A is 8 mph slower than the speed of train B. Train A travels 144 mi in the same time that it takes train B to travel 240 mi. Find the speed of train A.
24. Solve by completing the square. Show your work.

$$3x^2 - 12x - 6 = 0$$

25. Determine the nature of the solutions of

$$5z^2 + 8z - 4 = 0.$$

26. Write a quadratic equation whose solutions are $5i$, $-5i$.
27. Solve $\dfrac{W_1}{S_1T_1} = \dfrac{W_2}{S_2T_2}$ for T_2.
28. The hypotenuse of a triangle is 50 ft. One leg is 10 ft longer than the other. What are the lengths of the legs?
29. Find an equation of variation in which y varies jointly as x and w and inversely as the square of z, and $y = 10$ when $x = \frac{4}{3}$, $w = 5$, and $z = 2$.
30. The length l of rectangles of fixed area is inversely proportional to the width w. Suppose that the length is 64 cm when the width is 3 cm. Find the length when the width is 12 cm.
31. Determine whether these equations are equivalent.

$$2x + 3 = -4,$$
$$3x + 3 = x - 4$$

32. Determine the meaningful replacements in the radical expression $\sqrt{10-5x}$.

SYNTHESIS

Solve.

33. $x = \dfrac{1}{1 + x}$
34. $\sqrt{7x} - \sqrt{3x} = 7 - 3$

Relations, Functions, and Transformations

3

This chapter begins by considering the concept of a *relation*. A relation is defined as any set of ordered pairs. The most common relations that we will consider are solution sets of equations in two variables, since those solution sets are sets of ordered pairs.

A *function* is defined to be a special kind of relation, and probably the most important in mathematics. For this reason, we will consider a function from many standpoints, for example, as an input–output machine and as a mapping. We also consider graphs of functions and many of their applications.

When we draw a picture that represents the solution set of a relation or a function, we say that we have *graphed* the function. We will also consider alterations of the graph called *transformations*.

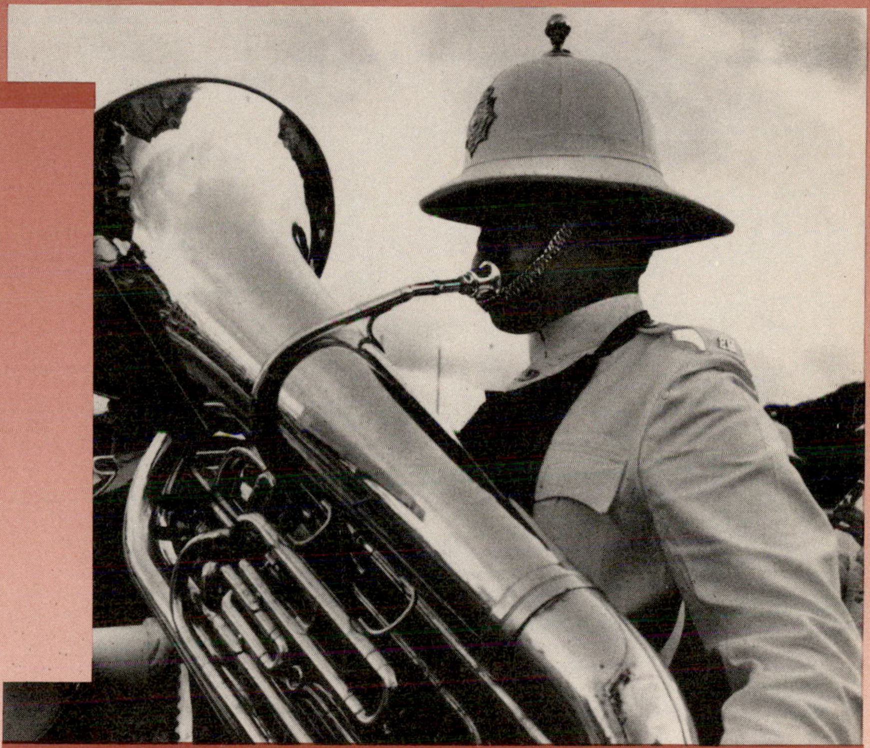

FEATURE PROBLEM

Speed of sound in air. The speed of sound S in air is a function of the temperature T, in degrees Fahrenheit, and is given by

$$S(T) = 1087.7\sqrt{\frac{5T + 2457}{2457}},$$

where S is in feet per second. Find the speed of sound in air when the temperature is 0° and when it is 70°.

THE MATHEMATICS

We find the function values as follows. A calculator with a square root key would be helpful.

$$S(0) = 1087.7\sqrt{\frac{5(0) + 2457}{2457}}$$
$$= 1087.7 \text{ ft/sec},$$

$$S(70) = 1087.7\sqrt{\frac{5(70) + 2457}{2457}}$$
$$= 1162.6 \text{ ft/sec}$$

3.1 Correspondences, Relations, and Ordered Pairs

OBJECTIVES

1. Given a correspondence, write it as a set of ordered pairs.
2. Find Cartesian products of small sets.
3. Indicate or find certain relations.
4. Find the domain and the range of a relation.

Write the correspondence as a set of ordered pairs.

1.

Total Revenue of Sears, Roebuck and Co.	
Year	**Revenue (in millions)**
1983 ⟶	\$35,883
1984 ⟶	38,828
1985 ⟶	40,715
1987 ⟶	48,440

2.

U.S. Senator–State Correspondence	
U.S. Senator	**State**
DeConcini ⟶	Arizona
McCain ⟶	
Mack ⟶	Florida
Graham ⟶	
Dixon ⟶	Illinois
Simon ⟶	
Hatfield ⟶	Oregon
Packwood ⟶	

1 Correspondences

Mathematics often concerns itself with how things *relate* or *correspond.* Below, for example, are four correspondences. We will make frequent reference to them.

Cost of First-Class Postage (First Ounce) at the End of the Year	
Year	**Cost (in cents)**
1964 ⟶	5
1974 ⟶	10
1978 ⟶	15
1983 ⟶	20
1984 ⟶	
1986 ⟶	22
1989 ⟶	25

State–U.S. Senator Correspondence	
State	**U.S. Senator**
Arizona ⟶	DeConcini
	McCain
Florida ⟶	Mack
	Graham
Illinois ⟶	Dixon
	Simon
Oregon ⟶	Hatfield
	Packwood

Cubing	
Number	**Cube**
−3 ⟶	−27
−2 ⟶	−8
−1 ⟶	−1
0 ⟶	0
1 ⟶	1
2 ⟶	8
3 ⟶	27

Squaring	
Number	**Square**
−3 ⟶	9
3 ⟶	
−2 ⟶	4
2 ⟶	
−1 ⟶	1
1 ⟶	
0 ⟶	0

Rather than draw arrows each time to show a correspondence, we can use ordered pairs. For example, instead of the correspondence

$$-3 \longrightarrow -27, \quad \text{we write} \quad (-3, -27).$$

Then we can gather all the ordered pairs into a set.

EXAMPLE 1 Write each of the preceding correspondences as a set of ordered pairs.

Solution

First-class postage = {(1964, 5), (1974, 10), (1978, 15), (1983, 20), (1984, 20), (1986, 22), (1989, 25)}

U.S. Senator = {(Arizona, DeConcini), (Arizona, McCain), (Florida, Mack), (Florida, Graham), (Illinois, Dixon), (Illinois, Simon), (Oregon, Hatfield), (Oregon, Packwood)}

Cubing = $\{(-3, -27), (-2, -8), (-1, -1), (0, 0), (1, 1), (2, 8), (3, 27)\}$

Squaring = $\{(-3, 9), (-2, 4), (-1, 1), (0, 0), (1, 1), (2, 4), (3, 9)\}$ ■

DO EXERCISES 1 AND 2.

2 Cartesian Products

It is from the preceding ideas that we make the following definitions. These become the basis for much of our work in this chapter.

Consider the following sets:

$$A = \{1, 2, 3\} \quad \text{and} \quad B = \{a, b\}.$$

From these sets, we form a set of ordered pairs, choosing the first elements from set A and the second elements from set B:

$$\{(1, a), (1, b), (2, a), (2, b), (3, a), (3, b)\}.$$

The set of all ordered pairs formed in this way is called the **Cartesian product** of sets A and B and is denoted $A \times B$. We read $A \times B$ as "A cross B," or as "the Cartesian product of A and B." We agree that

$$(a, b) = (c, d) \quad \textbf{if and only if} \quad a = b \textbf{ and } c = d.$$

In (a, b), we say that a is the **first member** and b is the **second member**. Thus, two ordered pairs are equal if and only if the first members are equal and the second members are equal. Thus,

$$\left(\tfrac{1}{2}, 89\%\right) = (0.5, 0.89) \quad \text{but} \quad (1, a) \neq (a, 1).$$

In general, $A \times B$ does not produce the same set as $B \times A$. For example, for the above sets, $A \times B$ is not the same set as $B \times A$.

> DEFINITION
>
> **The *Cartesian product* of two sets A and B, denoted $A \times B$, is the set of all ordered pairs having the first members from set A and the second members from set B.**

EXAMPLE 2 Find the Cartesian product $B \times A$, where $A = \{1, 2, 3\}$ and $B = \{a, b\}$.

Solution First members come from set B and second members come from set A. Thus,

$$B \times A = \{(a, 1), (a, 2), (a, 3), (b, 1), (b, 2), (b, 3)\}.$$

> CAUTION! When forming Cartesian products, be sure that in each pair of $A \times B$, the first member comes from A and the second member comes from B.

DO EXERCISE 3.

The two sets used to find a Cartesian product may be the same.

EXAMPLE 3 Find the Cartesian product $Q \times Q$, where $Q = \{2, 3, 4, 5\}$.

Solution The Cartesian product $Q \times Q$ is as follows:

$$\begin{aligned} Q \times Q = \{&(2, 2), (2, 3), (2, 4), (2, 5),\\ &(3, 2), (3, 3), (3, 4), (3, 5),\\ &(4, 2), (4, 3), (4, 4), (4, 5),\\ &(5, 2), (5, 3), (5, 4), (5, 5)\}. \end{aligned}$$

DO EXERCISES 4 AND 5.

3. Let $A = \{d, e, f\}$ and $B = \{1, 2\}$.
   a) List all the ordered pairs in $A \times B$.
   b) List all the ordered pairs in $B \times A$.

4. Find the Cartesian product $W \times W$, where $W = \{1, 2, 3, 4\}$.

5. Find the Cartesian product $M \times M$, where $M = \{-2, -1, 0, 5, 7\}$.

3 Relations

Suppose we select certain ordered pairs out of a Cartesian product. Such a set of ordered pairs is called a **relation**.

EXAMPLE 4 In the relation of Example 3, list the set of all ordered pairs in which the first member is the same as the second. We can call this relation *equals* (=).

Solution The relation *equals* is as follows:

$$\{(2, 2), (3, 3), (4, 4), (5, 5)\}.$$

EXAMPLE 5 In the relation of Example 3, list the set of all ordered pairs in which the first member is less than the second. This set of ordered pairs is the relation *less than* (<).

Solution The relation *less than* is as follows:

$$\{(2, 3), (2, 4), (2, 5), (3, 4), (3, 5), (4, 5)\}.$$

DO EXERCISES 6 AND 7.

6. In the Cartesian product of Margin Exercise 4, list all the ordered pairs in which the first member is the same as the second. This is the relation equals (=).

7. In the Cartesian product of Margin Exercise 4, list all the ordered pairs in which the first member is greater than the second. This set of ordered pairs is the relation >.

DEFINITION

A *relation* from a set A to a set B is *any* set of ordered pairs in $A \times B$.

Any time we select a set of ordered pairs from a Cartesian product, we have created a relation. This is true even when we have not followed any particular guideline.

EXAMPLE 6 Let A = the set of states consisting of Arizona, Florida, Illinois, and Oregon, and let B = the set of U.S. Senators from these states. The following is a relation from A to B:

U.S. Senator = {(Arizona, DeConcini), (Arizona, McCain), (Florida, Mack), (Florida, Graham), (Illinois, Dixon), (Illinois, Simon), (Oregon, Hatfield), (Oregon, Packwood)}.

Are there any ordered pairs of $A \times B$ that are not in the relation?

Solution There are ordered pairs of $A \times B$ not in this relation. For example, the pair (Arizona, Dixon) is in $A \times B$, but it is not in the relation, since Dixon is not a U.S. Senator from Arizona.

DO EXERCISE 8.

8. Let $A = \{1, 2, 3\}$ and $B = \{a, b\}$. The Cartesian product $A \times B$ is as follows: $\{(1, a), (1, b), (2, a), (2, b), (3, a), (3, b)\}$. The set $\{(2, b), (1, a)\}$ is an example of a relation in $A \times B$. Make up a relation of your own that is different from the given one.

4 Domain and Range

DEFINITION

The set of all first members of ordered pairs in a relation is called the *domain*. The set of all second members is called the *range*.

EXAMPLE 7 List the domain and the range of the relation in Example 4.

Solution The relation *equals* is as follows:

$$\{(2, 2), (3, 3), (4, 4), (5, 5)\};$$
$$\text{Domain} = \{2, 3, 4, 5\}; \quad \text{Range} = \{2, 3, 4, 5\}.$$

EXAMPLE 8 List the domain and the range of the relation in Example 5.

Solution The relation *less than* is as follows:

$$\{(2, 3), (2, 4), (2, 5), (3, 4), (3, 5), (4, 5)\};$$
$$\text{Domain} = \{2, 3, 4\}; \quad \text{Range} = \{3, 4, 5\}.$$

EXAMPLE 9 List the domain and the range of the relation in Example 6.

Solution The relation is as follows:

U.S. Senator = {(Arizona, DeConcini), (Arizona, McCain), (Florida, Mack), (Florida, Graham), (Illinois, Dixon), (Illinois, Simon), (Oregon, Hatfield), (Oregon, Packwood)}

Domain = {Arizona, Florida, Illinois, Oregon};

Range = {DeConcini, McCain, Mack, Graham, Dixon, Simon, Hatfield, Packwood}.

DO EXERCISES 9–12.

9. List the domain and the range of the relation in Margin Exercise 1.

10. List the domain and the range of the relation in Margin Exercise 2.

11. List the domain and the range of the relation in Margin Exercise 7.

12. List the domain and the range of the relation

$$\{(2, 2), (2, 3), (-4, 5), (-6, 7)\}.$$

EXERCISE SET 3.1

1 Write the correspondence as a set of ordered pairs (relation).

1. Sports Teams

City		Team
New York	→	Mets
	→	Giants
Atlanta	→	Braves
	→	Falcons
Houston	→	Astros
	→	Oilers
San Diego	→	Padres
	→	Chargers

(New York, Atlanta, Houston and San Diego each have arrows to the two teams listed beside and below them.)

2. Population of the USSR

Year		Population (in millions)
1959	→	209
1969	→	231
1979	→	255
1989	→	282
1999	→	312

3. Absolute Value

Number, x		$\lvert x \rvert$
-3	→	3
3	→	3
-2	→	2
2	→	2
-1	→	1
1	→	1
0	→	0

4. Reciprocal

Number, x		$1/x$
-16	→	-0.0625
$3/4$	→	$4/3$
-2	→	-0.5
1	→	1
0.875	→	$8/7$
10	→	0.1
5^3	→	5^{-3}

2 Find the Cartesian product.

5. $A \times B$, where $A = \{0, 2, 4, 5\}$ and $B = \{a, b, c\}$

6. $A \times B$, where $A = \{\text{h, q, t, w}\}$ and $B = \{1, 4, 7\}$

7. $B \times C$, where $B = \{x, y, z\}$ and $C = \{1, 2\}$

8. $B \times C$, where $B = \{5, 7, 10\}$ and $C = \{a, z\}$

9. $D \times D$, where $D = \{5, 6, 7, 8\}$

10. $E \times E$, where $E = \{-2, 0, 2, 4\}$

11. Let $A = \{0, 2\}$ and $B = \{a, b, c\}$. Find the following Cartesian products.
    a) $A \times B$
    b) $B \times A$
    c) $A \times A$
    d) $B \times B$

12. Let $A = \{1, 3, 5, 9\}$ and $B = \{d, e, f\}$. Find the following Cartesian products.
    a) $A \times B$
    b) $B \times A$
    c) $A \times A$
    d) $B \times B$

3 Consider the Cartesian product $E \times E$, where $E = \{-7, -3, 1, 2, 5\}$. List the ordered pairs in each of the following relations.

13. $<$ (less than)
14. $>$ (greater than)
15. $\leq$ (less than or equal to)
16. $\geq$ (greater than or equal to)
17. $=$ (equal)
18. $\neq$ (not equal)
19. Make up a relation of your own not like those in Exercises 1–18.
20. Make up a relation of your own not like those in Exercises 1–19.

4 List the domain and the range.

21. $\{(5, 2), (6, 4), (8, 6)\}$
22. $\{(7, 1), (8, 2), (9, 5)\}$
23. $\{(6, 0), (7, 5), (8, 5), (-4, -7)\}$
24. $\{(8, 2), (10, -10), (6, 3), (-2, 5)\}$
25. $\{(8, 1), (-8, 1), (5, 1), (-3, 1)\}$
26. $\{(6, 2), (-5, 2), (0, 2), (-3, 2)\}$
27. $\{(5, -6)\}$
28. $\{(-7, 4)\}$

List the domain and the range of the relation in each of the following.

29. Exercise 1
30. Exercise 2
31. Exercise 3
32. Exercise 4
33. Exercise 13
34. Exercise 14
35. Exercise 15
36. Exercise 16

37. a) List all the ordered pairs in $C \times C$, where
$$C = \{-1, 0, 1, 2\}.$$
    b) Consider the relation $\{(0, 0), (1, 1), (0, 1), (1, 2)\}$. Circle all the ordered pairs in $C \times C$ that are in this relation.
    c) List the domain and the range of this relation.

38. a) List all the ordered pairs in $D \times D$, where
$$D = \{-1, 1, 3, 5\}.$$
    b) Consider the relation $\{(-1, 1), (1, 1), (-1, 3), (1, 3)\}$. Circle all the ordered pairs in $D \times D$ that are in this relation.
    c) List the domain and the range of this relation.

SYNTHESIS

39. Consider $Q \times Q$, where $Q = \{2, 3, 4, 5\}$. Find
$$\{(x, y) | y > x + 1\}.$$

40. Consider $E \times E$, where $E = \{-1, 0, 1, 2\}$. Find
$$\{(x, y) | x^2 = y^2\}.$$

3.2 Graphs of Equations

OBJECTIVES

You should be able to:

1 Graph ordered pairs and simple relations.

2 Determine whether an ordered pair of numbers is a solution of an equation with two variables.

3 Graph certain relations.

4 Given the graph of a relation, describe the domain and the range.

1 Graphing Points and Relations

We are most interested in relations involving $R \times R$, where R is the set of real numbers. Now R is an infinite set: It contains more members than we could ever list. Thus relations involving R may be infinite and therefore cannot be indicated by listing the ordered pairs one at a time. We usually indicate such relations with some sort of picture in $R \times R$. This kind of picture, or drawing, is called a **graph**.

On a number line, each point corresponds to a number. On a plane, each point corresponds to an ordered pair of numbers from $R \times R$. The idea of using two perpendicular real-number lines, called **axes**, to identify points is commonly attributed to the great French mathematician René Descartes (1596–1650). To represent $R \times R$, we draw an x-axis and a y-axis perpendicular to each other. The variable x is most commonly represented on the horizontal axis, and the variable y is most commonly represented on the vertical axis. Their intersection is called the **origin** and is labeled O (capital letter O). The arrows show the positive directions. This method of showing $R \times R$ is called the **Cartesian coordinate system.**

The first member of an ordered pair is called the **first coordinate,** or the **x-coordinate,** or the **abscissa.** The second member is called the **second coordinate,** or the **y-coordinate,** or the **ordinate.** Together these are called the **coordinates of a point.**

The axes divide the plane into four regions called **quadrants,** indicated by the Roman numerals numbered counterclockwise from the upper right (see the figure in Example 1).

To **graph** a relation, we plot the points that correspond to the ordered pairs in the relation.

EXAMPLE 1 Graph the relation

$$\{(4, 3), (-3, 5), (-4, -2), (3, -4), (0, 0), (-3, 0), (0, 4)\}.$$

Solution The origin O has coordinates (0, 0). For the ordered pair $(-3, 5)$, the x-coordinate tells us to move from the origin 3 units to the left of the y-axis. The y-coordinate tells us to move from the origin 5 units up from the x-axis.

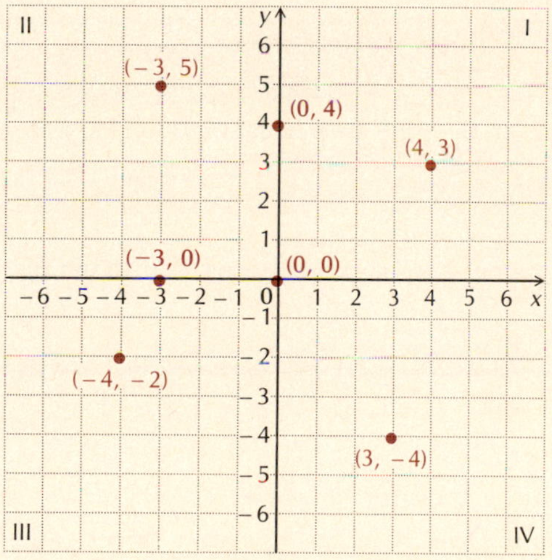

DO EXERCISE 1.

2 Solutions of Equations

If an equation has two variables, its solutions are ordered pairs of numbers. Suppose the variables in the equation are x and y. A **solution** is an ordered pair of numbers such that when the first coordinate is substituted for x and the second coordinate is substituted for y, the result is a true equation. Note that substitutions are usually made in alphabetical order.

EXAMPLE 2 Determine whether the following ordered pairs are solutions of the equation $y = 3x - 1$: $(-1, -4)$ and $(7, 5)$.

1. a) Graph the relation
$\{(3, 2), (-5, -2), (-4, 3), (-2, 0), (0, 4)\}$.

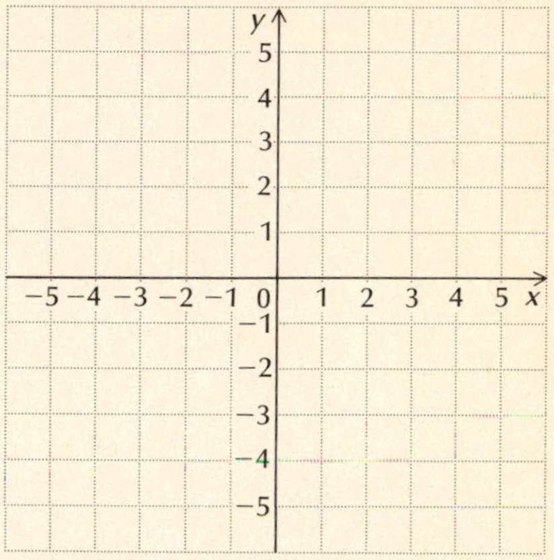

b) Find the domain and the range of the relation.

2. Determine whether (1, 7) is a solution of $y = 2x + 5$.

3. Determine whether $(-1, 4)$ is a solution of $y = 2x + 5$.

4. Determine whether $(-2, 5)$ is a solution of $y = x^2$.

5. Determine whether $(4, -5)$ is a solution of $x^3 - y^2 = 39$.

Solution We have

$$\begin{array}{c|l} y & = 3x - 1 \\ \hline -4 & 3(-1) - 1 \\ & -3 - 1 \\ & -4 \end{array}$$

We substitute -1 for x and -4 for y (alphabetical order of variables).

The equation $-4 = -4$ is true, so $(-1, -4)$ is a solution.

Now

$$\begin{array}{c|l} y & = 3x - 1 \\ \hline 5 & 3 \cdot 7 - 1 \\ & 21 - 1 \\ & 20 \end{array}$$

We substitute.

The equation $5 = 20$ is false, so (7, 5) is not a solution.

DO EXERCISES 2–5.

3 Graphs of Equations

The equation $y = 3x - 1$ considered in Example 2 actually has an infinite number of solutions. Rather than attempt to list all the solutions, we will use a graph as a convenient representation of all the solutions.

> DEFINITION
>
> **The solutions of an equation in two variables are ordered pairs and thus constitute a relation. To *graph* an equation is to make a drawing that represents its solutions.**

Following are some general suggestions for graphing an equation or relation.

> **Graphing suggestions**
>
> a) **Use graph paper.**
> b) **Label axes with symbols for the variables.**
> c) **Use arrows on the axes to indicate positive directions.**
> d) **Scale the axes; that is, mark numbers on the axes.**
> e) **Calculate solutions and list the ordered pairs in a table.**
> f) **Plot solutions, look for patterns, and complete the graph. When finished, label the graph with the equation or the relation being graphed.**

EXAMPLE 3 Graph: $y = 3x - 1$.

Solution We find some ordered pairs that are solutions. To find an ordered pair, we choose *any* number that is a meaningful replacement for x and then determine y. For example, if we choose 2 for x, then $y = 3(2) - 1$, or 5. We have found the solution (2, 5). We continue making choices for x and finding the corresponding values for y. We make some negative choices for x, as well as positive ones. We keep track of the solutions in a table.

x	y	(x, y)
0	-1	$(0, -1)$
1	2	$(1, 2)$
2	5	$(2, 5)$
-1	-4	$(-1, -4)$
-2	-7	$(-2, -7)$

(1) Select values for x.
(2) Compute values for y.

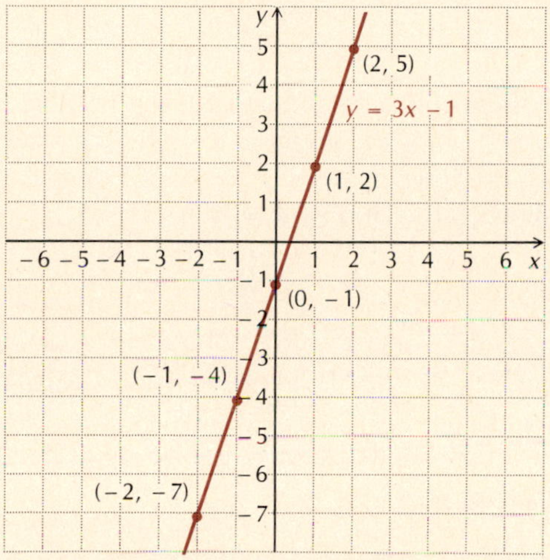

The table gives us the ordered pairs $(0, -1)$, $(1, 2)$, $(2, 5)$, and so on. Next, we plot these points. If we could calculate enough of them, they would make a solid line. We can draw the line with a ruler and label it $y = 3x - 1$.

Note that the equation $y = 3x - 1$ has an infinite (unending) set of solutions. The *graph of the equation* is a drawing that represents the relation that is made up of its solutions. Thus the relation consists of all pairs (x, y) such that $y = 3x - 1$ is true. That is, $\{(x, y) | y = 3x - 1\}$.

DO EXERCISE 6.

EXAMPLE 4 Graph: $y = x^2 - 5$.

Solution We select numbers for x and find the corresponding values for y. The table gives us the ordered pairs $(0, -5)$, $(-1, -4)$, and so on.

x	y	(x, y)
0	-5	$(0, -5)$
-1	-4	$(-1, -4)$
1	-4	$(1, -4)$
-2	-1	$(-2, -1)$
2	-1	$(2, -1)$
-3	4	$(-3, 4)$
3	4	$(3, 4)$

(1) Select values for x.
(2) Compute values for y.

6. Graph $y = -3x + 1$.

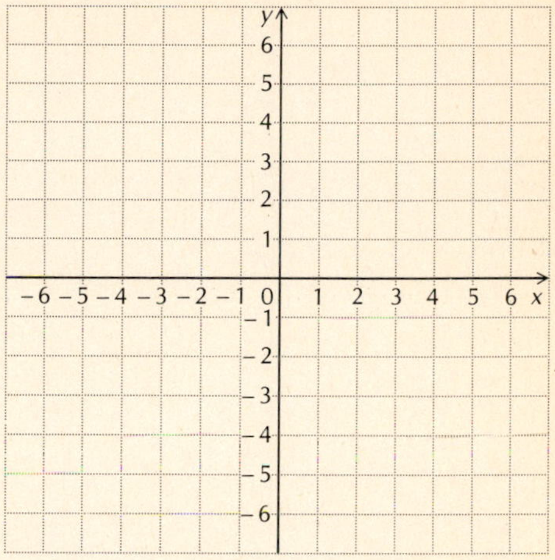

7. Graph $y = 3 - x^2$.

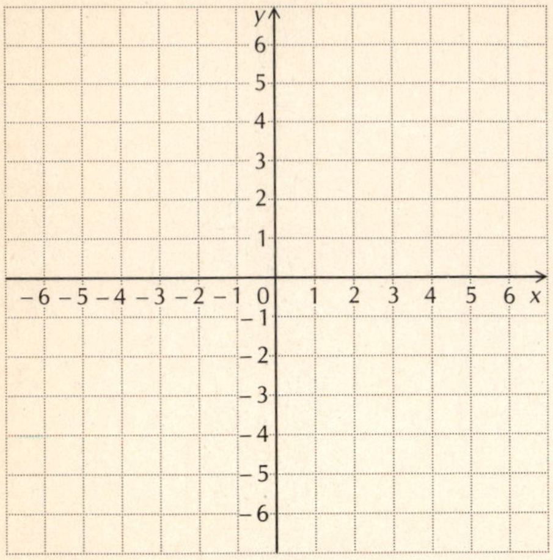

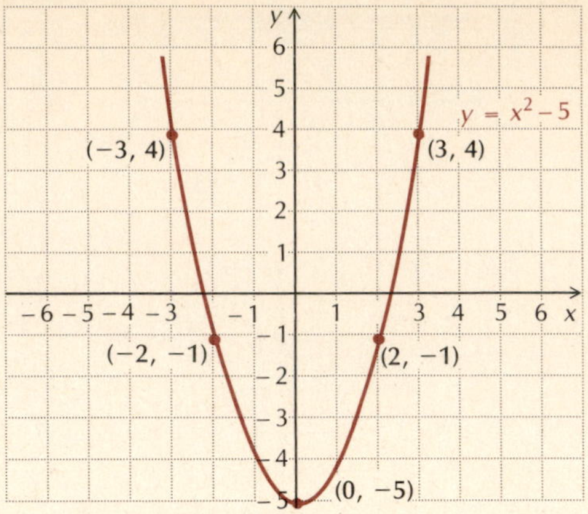

Next, we plot these points. We note that as the absolute value of x increases, $x^2 - 5$ also increases. Thus the graph is a curve that rises gradually on both sides of the y-axis, as shown above.

This graph shows the relation $\{(x, y) | y = x^2 - 5\}$. ■

DO EXERCISE 7.

EXAMPLE 5 Graph: $x = y^2 + 1$.

Solution Since x is expressed in terms of y, we select numbers for y and then find the corresponding values for x.

x	y	(x, y)
1	0	(1, 0)
2	−1	(2, −1)
2	1	(2, 1)
5	−2	(5, −2)
5	2	(5, 2)

① Select values for y.
② Compute values for x.

We must remember that x is the first coordinate and y is the second coordinate. Thus the table gives us the ordered pairs (1, 0), (2, −1), (2, 1), and so on. We note that as the absolute value of y gradually increases, the value of x also gradually increases. Thus the graph is a curve that stretches farther and farther to the right as it gets farther from the x-axis.

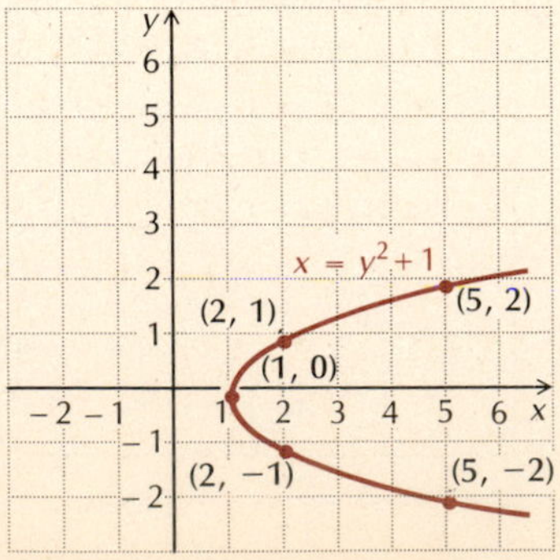

This graph shows the relation $\{(x, y) | x = y^2 + 1\}$.

DO EXERCISE 8.

You can always use a calculator to find as many values as you wish. This can be especially helpful when you are uncertain about the shape of a graph. There are also many computer software packages and calculators that will create graphs.

EXAMPLE 6 Graph: $xy = 12$.

Solution We first find numbers that satisfy the equation. To do so, it helps to solve for y (that is, $y = 12/x$) or solve for x (that is, $x = 12/y$). Then we make substitutions.

x	1	−1	2	−2	3	−3	4	−4	6	−6	12	−12
y	12	−12	6	−6	4	−4	3	−3	2	−2	1	−1

We plot these points and connect them. Note that since neither x nor y can be 0, the graph does not cross either axis. As the absolute value of x gets small, the absolute value of y must get large, and vice versa. Thus the graph consists of two curves, as follows.

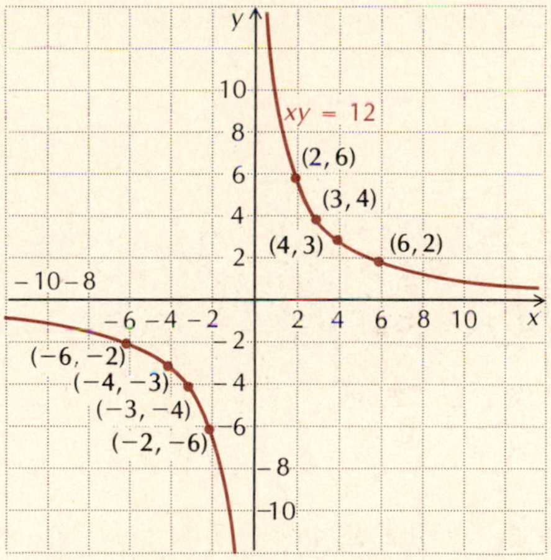

The graph shows the relation $\{(x, y) | xy = 12\}$.

DO EXERCISE 9.

EXAMPLE 7 Graph: $y = |x|$.

Solution We find numbers that satisfy the equation. For example, for $x = -3$, $y = |-3| = 3$. For $x = 2$, $y = |2| = 2$, and for $x = 0$, $y = |0| = 0$.

x	0	1	−1	2	−2	3	−3	4	−4
y	0	1	1	2	2	3	3	4	4

8. Graph $x = y^2 - 5$. (*Hint:* Select values for y and then find the corresponding values of x. When you plot, be sure to find x (horizontally) first.) Compare it with the graph of Example 4.

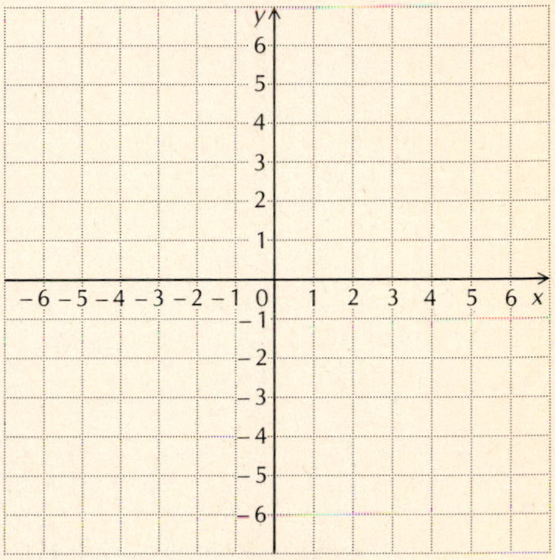

9. Graph $xy = 1$.

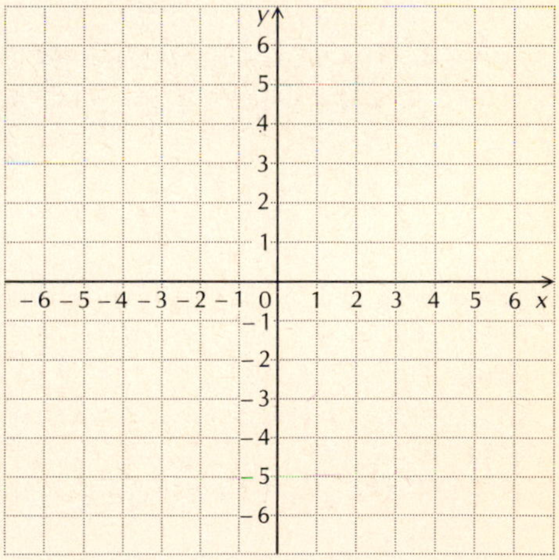

10. Graph the equation $x = |y|$. Compare it with the graph of Example 7.

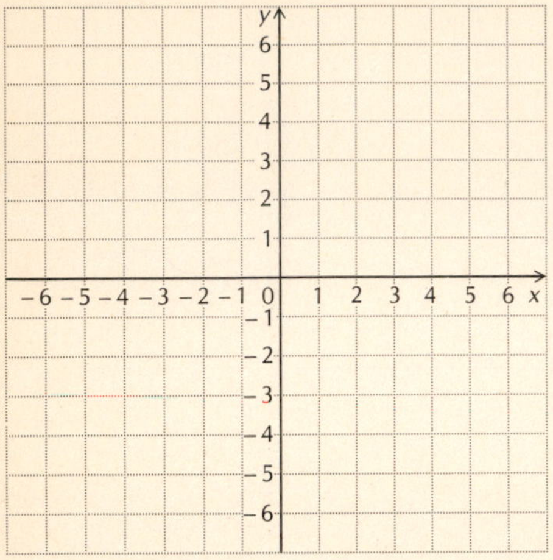

On each diagram in Exercises 11 and 12:

a) shade (on the x-axis) the domain, and

b) shade (on the y-axis) the range.

11.

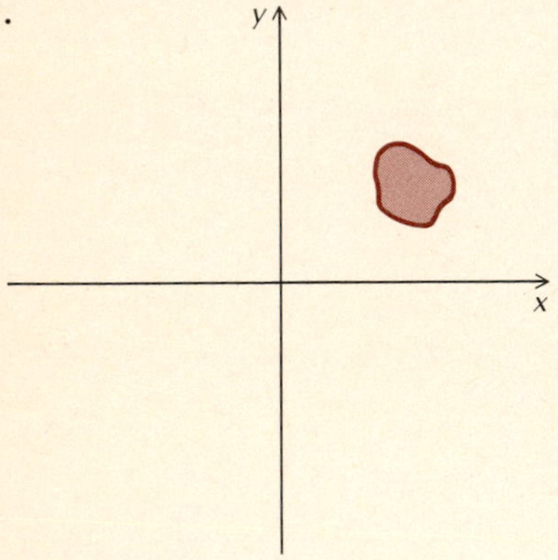

12.

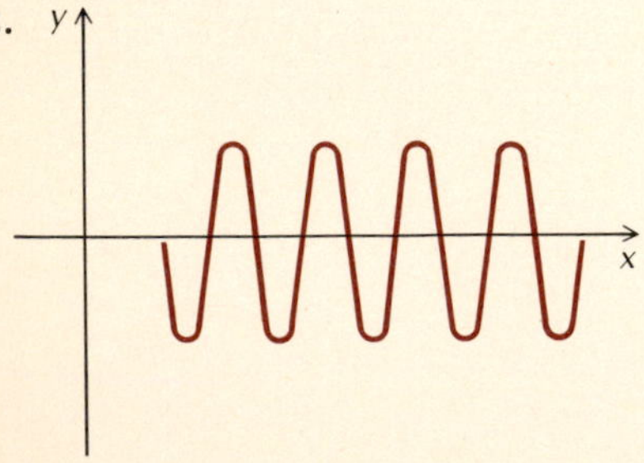

We plot these points and connect them. Note that as we get farther from the origin, to the left or the right, the absolute value of x increases. Thus the graph is a V-shaped curve that rises to the left and the right of the y-axis. It actually consists of parts of two straight lines, as shown here.

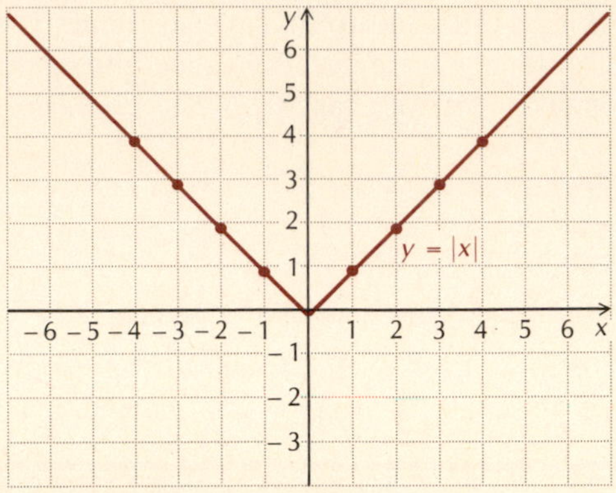

The graph shows the relation $\{(x, y) | y = |x|\}$. ■

DO EXERCISE 10.

4 Domains and Ranges

Recall that the **domain** of a relation is the set of all first coordinates and the **range** is the set of all second coordinates.

EXAMPLE 8 For the relation in part (a) of the figure, locate or shade the domain on the x-axis and the range on the y-axis.

Solution

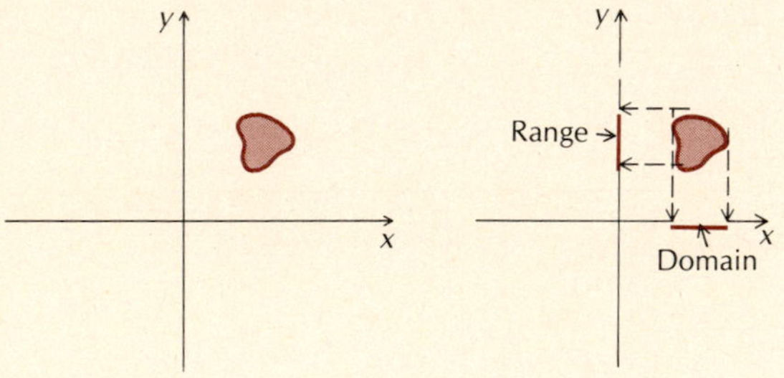

(a) (b)

From a point (x, y) on the graph, we find x by drawing or thinking of a vertical line back to the x-axis. To locate the domain, we want all such numbers on the x-axis. For this relation, we determine the outermost left and right extremities of the relation, if possible, and draw vertical lines back to the x-axis. The domain consists of the numbers where the lines cross the x-axis and all numbers in between.

From a point (x, y) on the graph, we find y by drawing or thinking of a horizontal line back to the y-axis. To locate the range, we want all such numbers on the y-axis. For this relation, we determine the outermost upper or lower extremities of the relation, and draw horizontal lines from them to the y-axis. The range consists of the numbers where the lines cross the y-axis and all numbers in between. The domain and the range are shown in part (b) of the figure. ■

EXAMPLE 9 Graph $x = y^2 + 1$. Locate the domain on the x-axis and the range on the y-axis.

Solution The graph is shown in part (a) of the figure. It is the relation that we considered in Example 5. The left extremity of the graph yields the number 1 on the x-axis. There is no right extremity. The domain is the set of all real numbers greater than or equal to 1. The graph extends up and to the right, and down and to the right. There are no upper–lower extremities. The range consists of the entire set of real numbers. The domain and the range are shown in part (b) of the figure.

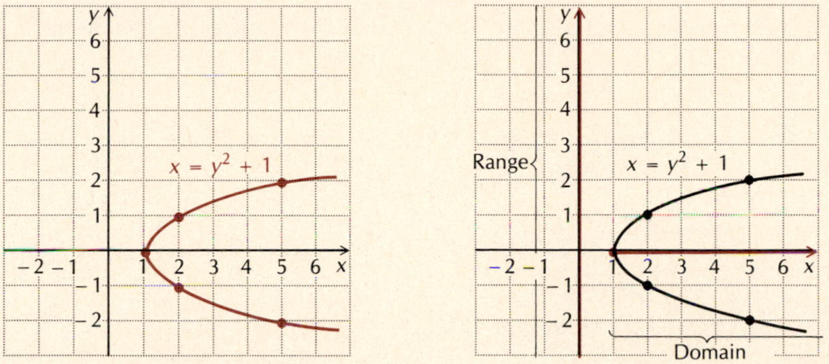

(a) (b)

DO EXERCISES 11–13. (EXERCISES 11 AND 12 ARE ON THE PRECEDING PAGE.)

13. Graph $y = x^2$. Then shade and describe the domain on the x-axis and the range on the y-axis.

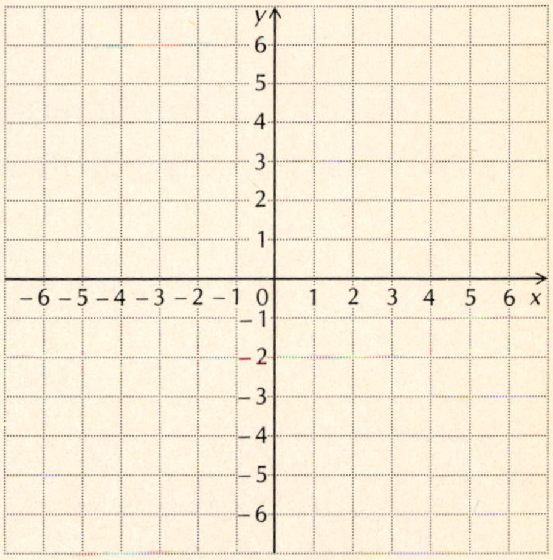

EXERCISE SET 3.2

1 Graph the relation.

1. $\{(3, 0), (4, 2), (5, 4), (6, 6)\}$
2. $\{(1, 1), (2, 3), (3, 5), (4, 7)\}$
3. $\{(3, -4), (3, -3), (3, -2), (3, -1), (3, 0)\}$
4. $\{(-2, 1), (-2, 2), (-2, 3), (-2, 4), (-2, 5)\}$
5. $\{(4, 3), (4, 2), (3, 2), (3, 3), (5, 2), (5, 3)\}$
6. $\{(2, -2), (3, -2), (2, -3), (3, -3), (2, -4), (3, -4)\}$
7. $\{(-1, 1), (-2, 1), (-2, 2), (-3, 1), (-3, 2), (-3, 3)\}$
8. $\{(-1, -1), (-1, -2), (-1, -3), (-2, -2), (-2, -3), (-3, -3)\}$

2 Determine whether the given ordered pairs are solutions of the given equation.

9. $(1, -1), (0, 3); y = 2x - 3$
10. $(2, 5), (-2, -7); y = 3x - 1$
11. $(3, 4), (-3, 5); 3s + t = 4$
12. $(2, 3), (-5, -15); 2p + q = 5$
13. $(3, 5), (-2, -15); 4x - y = 7$
14. $(2, 7), (-1, 8); 5x - y = 3$
15. $\left(0, \frac{3}{5}\right), \left(-\frac{1}{2}, -\frac{4}{5}\right); 2a + 5b = 3$
16. $\left(0, \frac{3}{2}\right), \left(\frac{2}{3}, 1\right); 3f + 4g = 6$
17. $(2, -1), (-0.75, 2.75); 4r + 3s = 5$
18. $(2, -4), (2.4, -5); 5w + 2z = 2$
19. $(3, 2), (22, 31); -3x + 2y = -4$
20. $(1, 2), (-40, 14); 2x - 5y = -6$

3 Graph.

21. $y = x$
22. $y = 2x$
23. $y = -2x$
24. $y = -\frac{1}{2}x$
25. $y = x + 3$
26. $y = x - 2$
27. $y = 3x - 2$
28. $y = -4x + 1$
29. $y = x^2$
30. $y = -x^2$, or $-1 \cdot x^2$
31. $y = x^2 + 2$
32. $y = x^2 - 2$
33. $x = y^2 + 2$
34. $x = y^2 - 2$
35. $y = |x + 1|$
36. $y = |x - 1|$
37. $x = |y + 1|$
38. $x = |y - 1|$
39. $xy = 10$
40. $xy = -18$
41. $y = 1/x$
42. $y = -2/x$
43. $y = 1/x^2$
44. $y = x^3$
45. $y = \sqrt{x}$
46. $y = x^{1/3}$
47. $y = 8 - x^2$
48. $x = 4 - y^2$

Graph and compare.

49. $y = x^2 + 1, y = (-x)^2 + 1$
50. $y = x^2 - 2, y = 2 - x^2$

4

51. Graph a relation as follows.
   a) Using a compass, draw a circle with radius 2, centered at the point (4, 3).
   b) Shade (on the x-axis) the domain. Describe the domain.
   c) Shade (on the y-axis) the range. Describe the range.

52. Graph a relation as follows.
   a) Draw a triangle with vertices at (1, 1), (4, 2), and (3, 6).
   b) Shade (on the x-axis) the domain. Describe the domain.
   c) Shade (on the y-axis) the range. Describe the range.

Graphs of the following were created in the indicated exercises. Describe the domain and the range of the relation.

53. $y = -x^2$ (Exercise 30)
54. $y = x - 2$ (Exercise 26)
55. $x = |y + 1|$ (Exercise 37)
56. $y = 1/x$ (Exercise 41)
57. $y = \sqrt{x}$ (Exercise 45)
58. $y = x^{1/3}$ (Exercise 46)
59. $y = 8 - x^2$ (Exercise 47)
60. $x = 4 - y^2$ (Exercise 48)

SYNTHESIS

Graph each relation. All relations are in $R \times R$, where R is the set of real numbers.

	First Coordinate	*Second Coordinate*
61.	Any real number	3
62.	-3	Any real number
63.	Any real number	1 more than the first coordinate
64.	Any real number	1 less than the first coordinate
65.	Any real number	Twice the first coordinate
66.	Any real number	Half the first coordinate
67.	Any real number	The square of the first coordinate
68.	The second coordinate squared	Any real number

Graph the following equations. The definition of absolute value in Section 1.2 may be helpful for some of the exercises.

69. $y = |x| + x$
70. $y = x|x|$
71. $y = |x^2 - 4|$
72. $y = x^{2/3}$
73. $|y| = x + 1$
74. $y = |x^3|$
75. $|y| = |x|$
76. $|x| + |y| = 0$
77. $|xy| = 1$
78. Graph the relation $\{(x, y) |\ 1 < x < 4 \text{ and } -3 < y < -1\}$.

COMPUTER-CALCULATOR EXERCISES

Use a computer software package or a graphing calculator to graph each of the following equations. If neither is available, compute lots of solutions using a calculator.

79. $y = \frac{1}{3}x^3 - x + \frac{2}{3}$
80. $y = -\frac{1}{3}x^3 + \frac{1}{2}x^2 + 2x - 1$
81. $y = 1 + \sqrt{2 - x}$
82. $y = 1 + \dfrac{2}{x + 1}$

CHALLENGE

83. Graph: $|x| + |y| = 1$.
84. Graph the relation $\{(x, y) \,|\, |x| \leq 1 \text{ and } |y| \leq 2\}$.

3.3 Functions

OBJECTIVES

You should be able to:

1. Determine whether a correspondence is a function.
2. Determine whether a relation is a function.
3. Using the vertical-line test, determine whether the graph of a relation is the graph of a function.
4. Use a set of ordered pairs, a graph, or a formula to find function values.
5. Find the domain of a function given by a formula.

One of the most important concepts in mathematics is a *function*. It is so important that we are going to consider it from several viewpoints. If you have trouble with one viewpoint of a function, keep reading, as the next may give you the needed understanding.

1 Functions as Correspondences

We looked at correspondences in Section 3.2. A *function* is a certain kind of correspondence from one set to another. For example:

To each person in a math class	There corresponds	His or her grade
To each automobile in a state	There corresponds	Its license plate number
To each triangle	There corresponds	Its area

In each example, the first set is called the **domain**. The second set is called the **range**. Given a member of the domain, there is *exactly one* member of the range to which it corresponds. This kind of correspondence is called a **function**.

> DEFINITION
>
> **A *function* is a correspondence between a first set, called a *domain*, and a second set, called a *range*, such that to any member of the domain, there corresponds *exactly one* member of the range.**

EXAMPLE 1 Each of the following correspondences was considered in Section 3.1. Which are functions?

a) Cost of First-Class Postage (First Ounce) at the End of the Year

Year (domain)	Cost (in cents) (range)
1964 →	5
1974 →	10
1978 →	15
1983 →	20
1984 →	20
1986 →	22
1989 →	25

b) State–U.S. Senator Correspondence

State (domain)	U.S. Senator (range)
Arizona →	DeConcini
Arizona →	McCain
Florida →	Mack
Florida →	Graham
Illinois →	Dixon
Illinois →	Simon
Oregon →	Hatfield
Oregon →	Packwood

c) Cubing

Number (domain)	Cube (range)
−3 →	−27
−2 →	−8
−1 →	−1
0 →	0
1 →	1
2 →	8
3 →	27

d) Squaring

Number (domain)	Square (range)
−3 →	9
3 →	9
−2 →	4
2 →	4
−1 →	1
1 →	1
0 →	0

Which of the following correspondences are functions?

1.

Total Revenue of Sears, Roebuck and Co.	
Year (domain)	**Revenue (in millions) (range)**
1983 →	$35,883
1984 →	38,828
1985 →	40,715
1987 →	48,440

2.

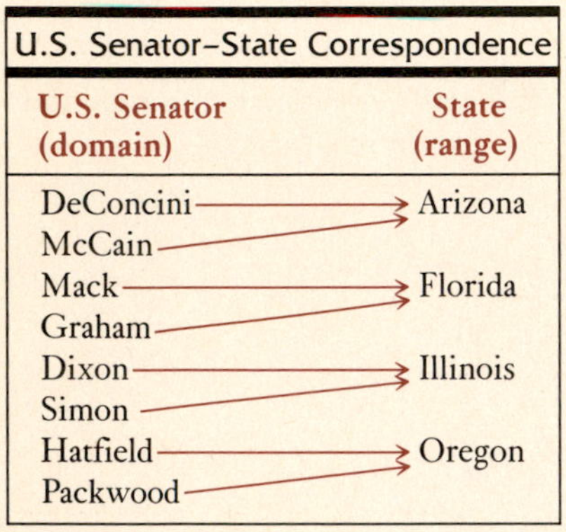
U.S. Senator–State Correspondence
U.S. Senator (domain)
State (range)
DeConcini
McCain
Mack
Graham
Dixon
Simon
Hatfield
Packwood
Arizona
Florida
Illinois
Oregon

3.

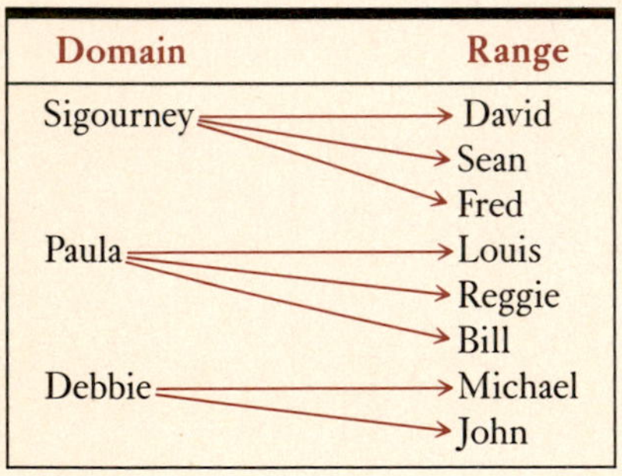
Domain
Range
Sigourney
Paula
Debbie
David
Sean
Fred
Louis
Reggie
Bill
Michael
John

4.

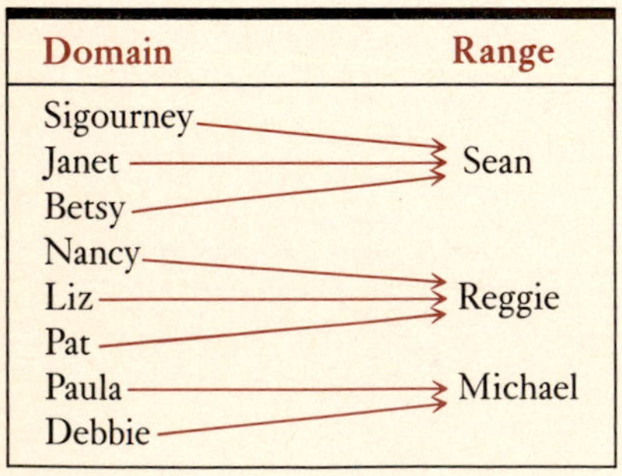
Domain
Range
Sigourney
Janet
Betsy
Nancy
Liz
Pat
Paula
Debbie
Sean
Reggie
Michael

5. Which of the following relations are functions?

$A = \{(9, 0), (3, 8), (5, 8), (9, -1)\}$

$B = \{(0, t), (9, e), (-2, q), (-5, b)\}$

$C = \{(-3, 5), (7, -2), (4, -6)\}$

$D = \{(0, 1), (1, 0), (-1, 1)\}$

$E = \{(7, -7), (-7, -7)\}$

Solution

a) This is a function. Each member of the domain corresponds to exactly one member of the range, even though both 1983 and 1984 correspond to 20. There is one and only one member of the range, 20, that corresponds to 1983. Similarly, there is one and only one member of the range, 20, that corresponds to 1984.

b) This is *not* a function. There are two members of the range, *DeConcini* and *McCain*, that correspond to *Arizona*. There are other instances that show that this is not a function, but one case is all it takes.

c) This is a function.

d) This is a function. ■

DO EXERCISES 1–4.

2 Functions as Relations

In Section 3.1, we saw that we can write a correspondence as a set of ordered pairs, or a relation. Then how do we know when a relation is a function? Let us consider the correspondences in Example 1 and write them as relations.

a) First-class postage = {(1964, 5), (1974, 10), (1978, 15), (1983, 20), (1984, 20), (1986, 22), (1989, 25)}.

b) U.S. Senator = {(Arizona, DeConcini), (Arizona, McCain), (Florida, Mack), (Florida, Graham), (Illinois, Dixon), (Illinois, Simon), (Oregon, Hatfield), (Oregon, Packwood)}

c) Cubing $= \{(-3, -27), (-2, -8), (-1, -1), (0, 0), (1, 1), (2, 8), (3, 27)\}$

d) Squaring $= \{(-3, 9), (-2, 4), (-1, 1), (0, 0), (1, 1), (2, 4), (3, 9)\}$

In terms of ordered pairs, we know that (b) is not a function because it has at least two ordered pairs, (Arizona, DeConcini) and (Arizona, McCain), with the same first coordinates but different second coordinates. This does not happen with (a), (c), and (d). Thus we know that (a), (c), and (d) are functions, but (b) is not.

> **DEFINITION**
>
> **A *function* is a relation in which no two ordered pairs have the same first coordinate but different second coordinates.**

EXAMPLE 2 Determine whether the relation

$$A = \{(2, 3), (5, 9), (1, 0), (10, -2)\}$$

is a function.

Solution The relation A is a function because no two ordered pairs have the same first coordinate and a different second coordinate. ■

EXAMPLE 3 Determine whether the relation

$$B = \{(4, 5), (-3, 2), (4, 0), (-1, 9)\}$$

is a function.

Solution The relation B is *not* a function because the ordered pairs (4, 5) and (4, 0) have the same first coordinate and a different second coordinate. ■

DO EXERCISE 5.

3 The Vertical-Line Test

Suppose that a relation has two ordered pairs with the same first coordinate, but different second coordinates. The graphs of these two ordered pairs would be points on the same vertical line. This gives a method to test whether a graph of a relation is the graph of a function.

> **The Vertical-Line Test**
>
> **If no vertical line intersects a graph at more than one point, then the graph is the graph of a function.**

When applying the vertical-line test, we, in effect, try to make it fail. That is, we try to find a vertical line that meets the graph more than once. If we do, then the relation is not a function. If we do not, then the relation is a function.

EXAMPLE 4 Which of the following are graphs of functions?

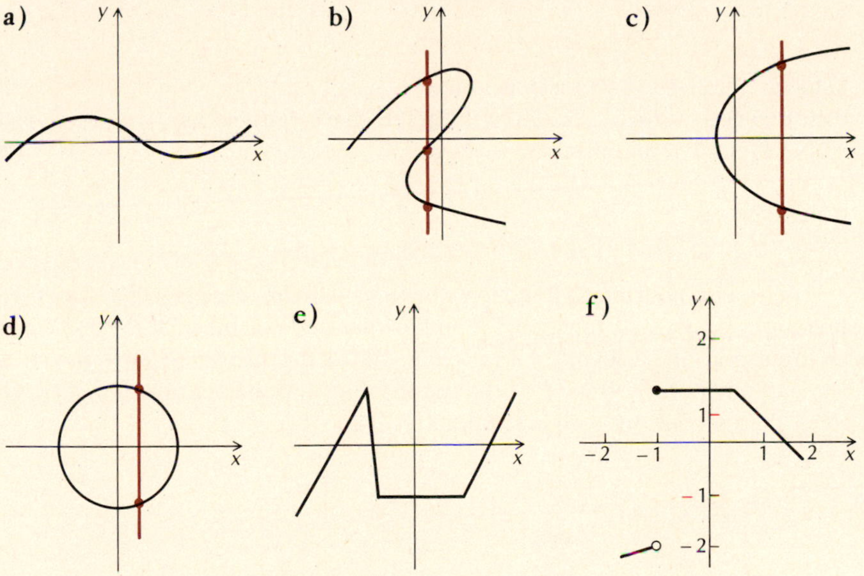

In graph (f), the solid dot shows that $(-1, 1)$ belongs to the graph. The open circle shows that $(-1, -2)$ does *not* belong to the graph.

Solution Graphs (a), (e), and (f) are graphs of functions. Graphs (b), (c), and (d) fail the vertical-line test. In graph (b), the points $(-1, 0)$, $(-1, 3)$, and $(-1, -2)$ are all in the relation and so it is not a function. Also, in (c) and (d), we can find a vertical line that crosses the graph more than once.

DO EXERCISE 6.

4 Function Notation; Functions as Machines

Functions are often named by letters, such as f and g. A function f is thus a set of ordered pairs. If we represent the first coordinate of a pair by x, then we can represent the second coordinate, y, by $f(x)$. The ordered pair is then $(x, f(x))$. The function has been named f. We call the input x and its output $f(x)$. This is read "f of x," or "f at x," or "the value of f at x." Note that $f(x)$ does not mean "f times x."

EXAMPLE 5 Given the function $g = \{(1, 4), (2, 3), (3, 2), (4, -8), (5, 2)\}$, find $g(1)$, $g(2)$, $g(3)$, $g(4)$, and $g(5)$. What is the domain? What is the range?

6. Which of the following are graphs of functions?

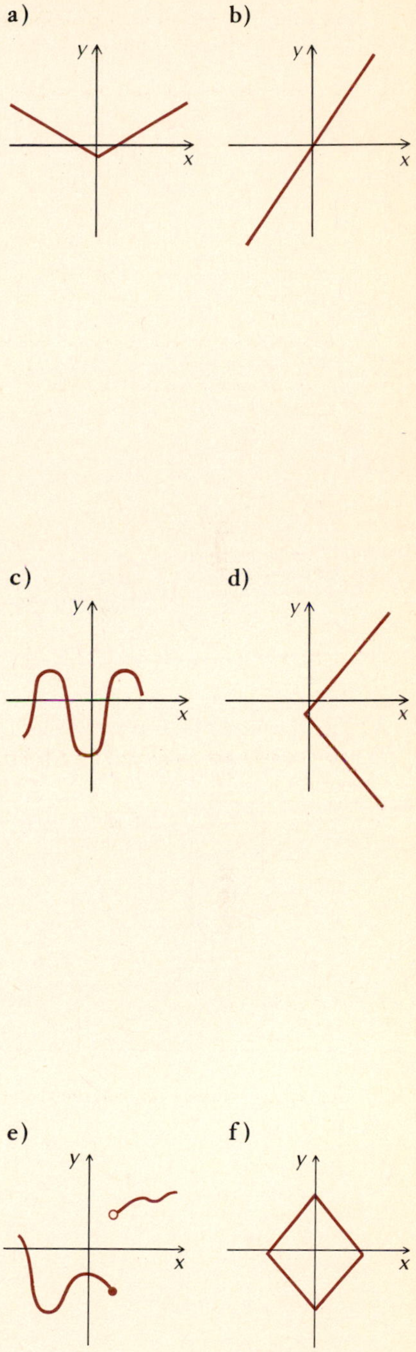

7. Given the function

$$f = \{(2, 4), (4, 4), (3, 3), (1, -5)\},$$

find $f(1), f(2), f(3)$, and $f(4)$. What is the domain? What is the range?

Solution Since $(1, 4)$ is in the function, $g(1) = 4$. Since $(2, 3)$ is in the function, $g(2) = 3$. Similarly, $g(3) = 2, g(4) = -8$, and $g(5) = 2$. The domain is $\{1, 2, 3, 4, 5\}$. The range is $\{-8, 2, 3, 4\}$. ■

DO EXERCISE 7.

EXAMPLE 6 The function f is graphed below. Use the graph to estimate $f(-2), f(0)$, and $f(3)$.

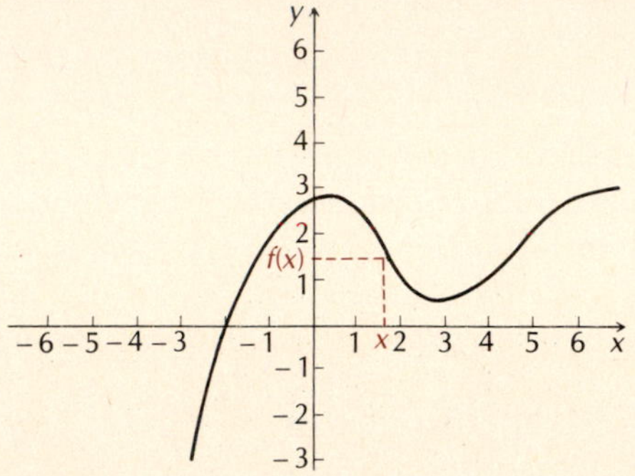

Solution To find function values given a graph, we locate x on the x-axis, and then find $f(x)$ on the y-axis. From the graph above, we see that $f(-2) = 0$, that $f(0)$ is about 2.7, and that $f(3)$ is about 0.6. ■

8. The function f is graphed below. Use the graph to estimate $f(-3), f(0), f(2)$, and $f(4)$.

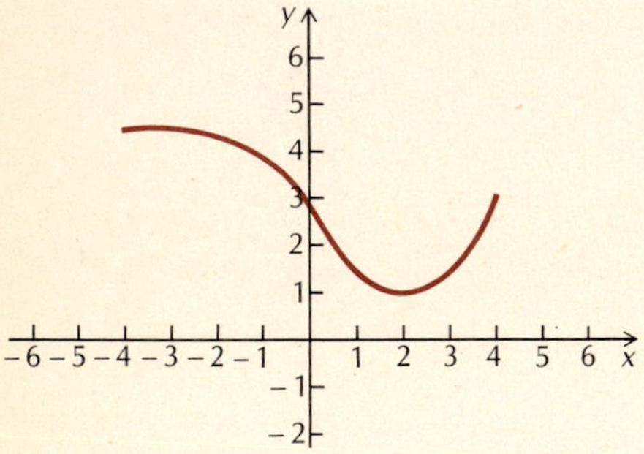

DO EXERCISE 8.

To better understand function notation, it helps to imagine a **function machine.** Consider a function f. Think of putting a member of the domain x, an **input,** into the machine. The machine knows the correspondence and gives you a member of the range $f(x)$, the **output.** Since we have a function, for each input there is one and only one output.

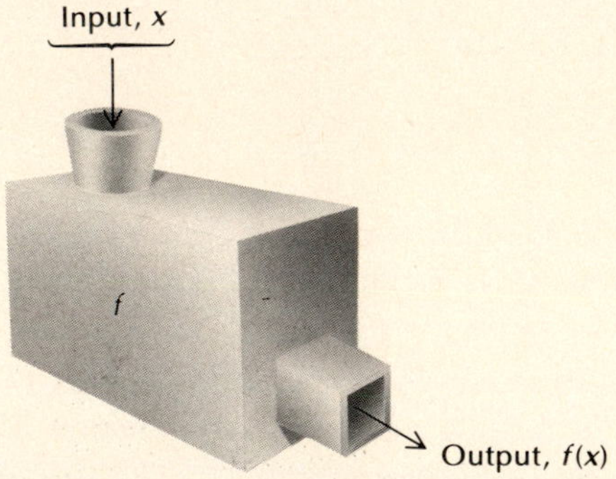

Often functions with real-number domains and ranges can be described by formulas or equations. For example, $f(x) = 3x - 4$ describes the function that takes an input x, multiplies it by 3, and then subtracts 4. To find the output $f(-1)$, we take the input -1, multiply it by 3 to get -3, and then subtract 4 to get -7:

$$f(-1) = 3(-1) - 4$$
$$= -7.$$

Sometimes, instead of writing $f(x) = 3x - 4$, we might write $y = 3x - 4$, where it is understood that the value of y, the **dependent variable,** is calculated after first choosing a value for x, the **independent variable.** To understand why $f(x)$ notation is so useful, consider these two statements, which have the same meaning:

a) If $f(x) = 3x - 4$, then $f(-1) = -7$.

b) If $y = 3x - 4$, then the value of y is -7 when x is -1.

Note that the notation used in (a) is more concise.

We find function values by making substitutions for the variables. When a substitution is not meaningful, we say that the function value *does not exist.*

EXAMPLE 7 For the function f given by $f(z) = 2z^2 - z + 3$, find $f(0)$, $f(-7)$, $f(5a)$, and $f(a - 4)$.

Solution One way to find function values when a formula is given is to think of the formula as follows:

$$f(\quad) = 2(\quad)^2 - (\quad) + 3.$$

Whatever goes in the blank on the left between parentheses goes in the blanks on the right between parentheses:

$$\begin{aligned} f(0) &= 2 \cdot (0)^2 - 0 + 3 = 0 - 0 + 3 = 3; \\ f(-7) &= 2(-7)^2 - (-7) + 3 = 2 \cdot 49 + 7 + 3 = 108; \\ f(5a) &= 2(5a)^2 - 5a + 3 = 2 \cdot 25a^2 - 5a + 3 = 50a^2 - 5a + 3; \\ f(a-4) &= 2(a-4)^2 - (a-4) + 3 = 2(a^2 - 8a + 16) - a + 4 + 3 \\ &= 2a^2 - 17a + 39. \end{aligned}$$

Note above that whether we write $f(z) = 2z^2 - z + 3$ or $f(x) = 2x^2 - x + 3$ or $f(\quad) = 2(\quad)^2 - (\quad) + 3$, we still have $f(-7) = 108$. Thus our independent variable can be regarded as a *dummy* variable. The letter chosen for the dummy variable is not as important as the algebraic manipulation to which it is subjected.

DO EXERCISES 9 AND 10.

EXAMPLE 8 For the function f given by $f(x) = 1/x$, find $f(2)$, $f(-\frac{1}{4})$, and $f(0)$, if possible.

Solution

$$f(2) = \frac{1}{2}, \qquad f\left(-\frac{1}{4}\right) = \frac{1}{-\frac{1}{4}} = -4,$$

$f(0)$ is not defined, so $f(0)$ does not exist.

DO EXERCISE 11.

Sometimes all the outputs of a function are the same. In such a case, we have what is called a **constant function.**

EXAMPLE 9 For the constant function $g(x) = 3$, find $g(5)$, $g(-7)$, and $g(24.96)$.

Solution All the outputs are the same number, 3. Thus,

$$g(5) = 3, \qquad g(-7) = 3, \quad \text{and} \quad g(24.96) = 3.$$

Note that each of the ordered pairs $(5, 3)$, $(-7, 3)$, and $(24.96, 3)$ are in the constant function g. The range contains only one number, 3.

DO EXERCISE 12.

9. For the function f given by
$$f(x) = 2x - 3,$$
find $f(-1)$, $f(0)$, $f(5.6)$, and $f(10)$.

10. For the function f given by
$$f(x) = x^2 + 5x - 1,$$
find each of the following.
   a) $f(0)$
   b) $f(1)$
   c) $f(-1)$
   d) $f(2a)$
   e) $f(a + 1)$

11. For the function f given by $f(x) = \sqrt{x}$, find $f(16)$, $f(3)$, and $f(-4)$, if possible.

12. For the function G given by $G(x) = 7$, find $G(0)$, $G(-3)$, and $G(\frac{1}{2})$. What is the range?

For each function f given as follows, construct and simplify the expression

$$\frac{f(a+h)-f(a)}{h}.$$

13. $f(x) = 3x^2 + 1$

14. $f(x) = x^2 - x$

In calculus, it is a valuable skill to be able to simplify expressions like

$$\frac{f(a+h)-f(a)}{h}.$$

EXAMPLE 10 For the function f given by $f(x) = 2x^2 - 3$, construct and simplify the expression

$$\frac{f(a+h)-f(a)}{h}.$$

Solution We first find $f(a+h)$ and $f(a)$:

$$f(a+h) = 2(a+h)^2 - 3; \qquad f(a) = 2a^2 - 3.$$

Then

$$\frac{f(a+h)-f(a)}{h} = \frac{[2(a+h)^2 - 3]-[2a^2 - 3]}{h}.$$

Now we simplify:

$$\frac{f(a+h)-f(a)}{h} = \frac{2a^2 + 4ah + 2h^2 - 3 - 2a^2 + 3}{h}$$

$$= \frac{4ah + 2h^2}{h} = 4a + 2h.$$

DO EXERCISE 13.

EXAMPLE 11 For the function f given by $f(x) = x^3 + x$, construct and simplify the expression

$$\frac{f(a+h)-f(a)}{h}.$$

Solution We first find $f(a+h)$ and $f(a)$:

$$f(a+h) = (a+h)^3 + (a+h) \quad \text{Replacing each occurrence of } x \text{ by } a+h$$

$$= a^3 + 3a^2h + 3ah^2 + h^3 + a + h;$$

$$f(a) = a^3 + a.$$

Then

$$\frac{f(a+h)-f(a)}{h} = \frac{[a^3 + 3a^2h + 3ah^2 + h^3 + a + h] - [a^3 + a]}{h}.$$

Now we simplify:

$$\frac{f(a+h)-f(a)}{h} = \frac{a^3 + 3a^2h + 3ah^2 + h^3 + a + h - a^3 - a}{h}$$

$$= \frac{3a^2h + 3ah^2 + h^3 + h}{h} = \frac{h(3a^2 + 3ah + h^2 + 1)}{h}$$

$$= 3a^2 + 3ah + h^2 + 1.$$

DO EXERCISE 14.

Some comment is in order regarding the notation $f(x)$ and f. When we refer to a function f, we are referring to a set of ordered pairs. The notation $f(x)$ refers to the second coordinate of an ordered pair that has x as its first coordinate. Nevertheless, it is a fact of life in the literature of mathematics to be careless about this notation, referring instead to "the function $f(x)$" or "the function $f(x) = x^2$."

5 Finding Domains of Functions

When a function f in $R \times R$ is given by a formula, the domain is understood to be the set of all real numbers that are meaningful replacements for x. For example, consider the function f given by $f(x) = 1/x$. The number 0 is not a meaningful replacement, because division by 0 is not possible. All other real numbers are meaningful replacements, so the domain of f consists of all nonzero real numbers, $\{x|x \neq 0\}$. Consider the function g given by $g(x) = \sqrt{x}$. The negative numbers are not meaningful replacements. Thus the domain of g consists of all nonnegative numbers, $\{x|x \geq 0\}$.

EXAMPLE 12 Find the domain of the function g given by

$$g(x) = \frac{x}{x^2 + 2x - 3}.$$

Solution The formula is meaningful so long as a replacement for x does not make the denominator 0. To find those replacements that do make the denominator 0, we solve $x^2 + 2x - 3 = 0$:

$$x^2 + 2x - 3 = 0$$
$$(x - 1)(x + 3) = 0$$
$$x - 1 = 0 \quad \text{or} \quad x + 3 = 0$$
$$x = 1 \quad \text{or} \quad x = -3.$$

Thus the domain consists of the set of all real numbers except 1 and -3. We can name this set $\{x|x \neq 1 \text{ and } x \neq -3\}$. ■

EXAMPLE 13 Find the domain of the function f given by

$$f(x) = \frac{\sqrt{5x - 3}}{x^2 + 1}.$$

Solution Let us first consider the numerator, $\sqrt{5x - 3}$. The formula is meaningful so long as replacements make the radicand nonnegative (no negative number has a real square root). Thus to find the domain, we solve the inequality $5x - 3 \geq 0$:

$$5x - 3 \geq 0$$
$$5x \geq 3$$
$$x \geq \tfrac{3}{5}.$$

Now let us consider the denominator. Since x^2 is nonnegative, $x^2 + 1$ is always positive. Thus the denominator is always nonzero, and there are no restrictions on numbers that can be substituted into the denominator. The domain of the function is $\{x|x \geq \frac{3}{5}\}$. ■

EXAMPLE 14 Find the domain of the function t given by $t(x) = x^3 + |x|$.

Solution There are no restrictions on the numbers that we can substitute into this formula. We can cube any real number, we can take the absolute value of any real number, and we can add the results. Thus the domain is the entire set of real numbers. ■

DO EXERCISES 15–17.

Find the domain of the function.

15. $f(x) = \dfrac{x + 1}{3x^2 + 10x + 8}$

16. $g(x) = \dfrac{\sqrt{10x + 25}}{x^2 + 3}$

17. $p(x) = x^3 - 4x^2 + 2x + 8$

EXERCISE SET 3.3

1 Determine whether each of the correspondences in Exercises 1–8 is a function.

1. Sports Teams

City (Domain)	Team (Range)
New York	→ Mets
New York	→ Giants
Atlanta	→ Braves
Atlanta	→ Falcons
Houston	→ Astros
Houston	→ Oilers
San Diego	→ Padres
San Diego	→ Chargers

2. Population of the USSR

Year (Domain)	Population (in millions) (Range)
1959	→ 209
1969	→ 231
1979	→ 255
1989	→ 282
1999	→ 312

3. Absolute Value

Number, x (Domain)	$\lvert x\rvert$ (Range)
-3	→ 3
3	→ 3
-2	→ 2
2	→ 2
-1	→ 1
1	→ 1
0	→ 0

4. Reciprocal

Number, x (Domain)	$1/x$ (Range)
-16	→ -0.0625
$3/4$	→ $4/3$
-2	→ -0.5
1	→ 1
0.875	→ $8/7$
10	→ 0.1
5^3	→ 5^{-3}

2 Determine whether each of the following relations is a function.

5. $K = \{(1, 2), (2, 3), (3, 4), (4, 1)\}$
6. $L = \{(-3, 5), (0, 0), (-5, 3), (-3, -5)\}$
7. $T = \{(-4, -4), (-1, -6), (-4, 4), (-6, -1)\}$
8. $V = \{(2, -9), (4, 2), (0, 5), (4, -9)\}$
9. $B = \{(6, -6), (-2, 2), (0, 0), (2, -2), (-6, 6)\}$
10. $F = \{(p, r), (q, r), (a, r), (z, m), (q, t), (q, w)\}$

3 Determine whether each of the following is a graph of a function. An open circle indicates that the point does not belong on the graph.

11.

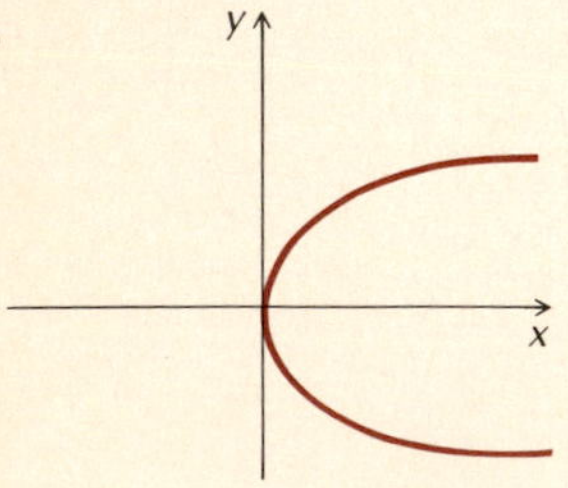

12.

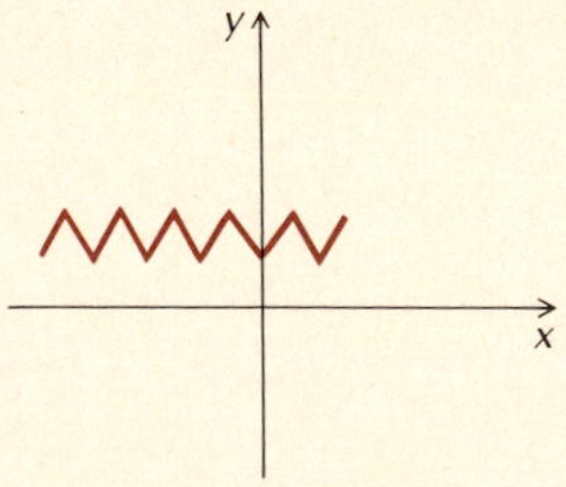

13.

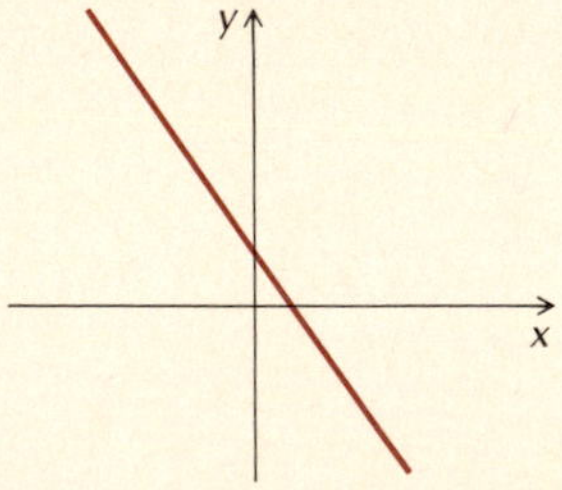

14.

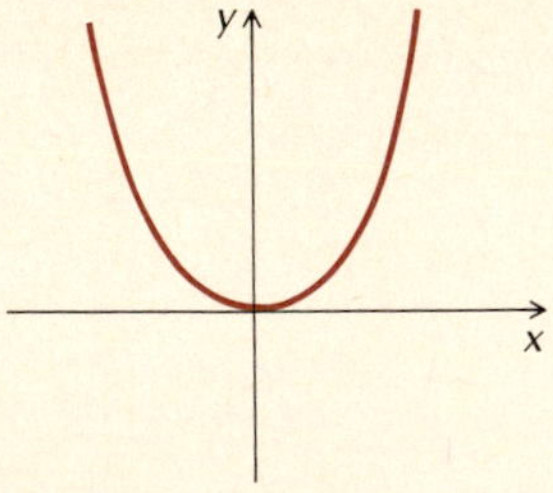

15.

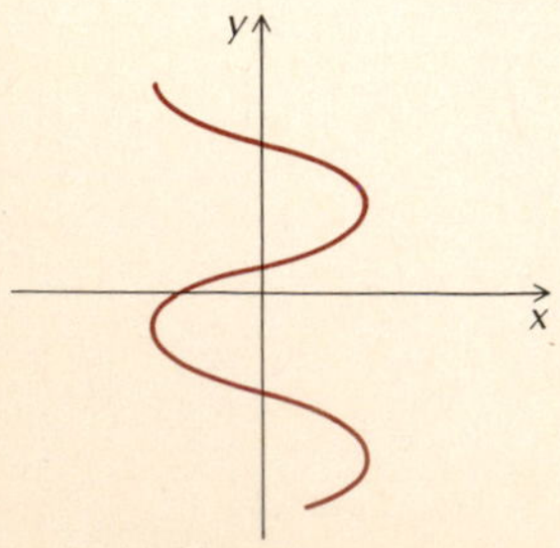

16.

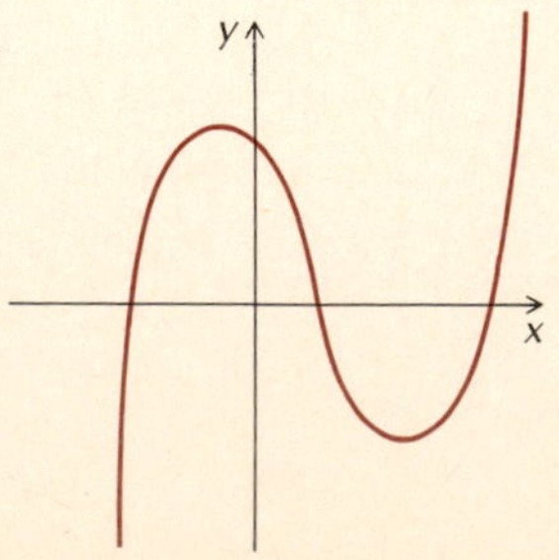

17.

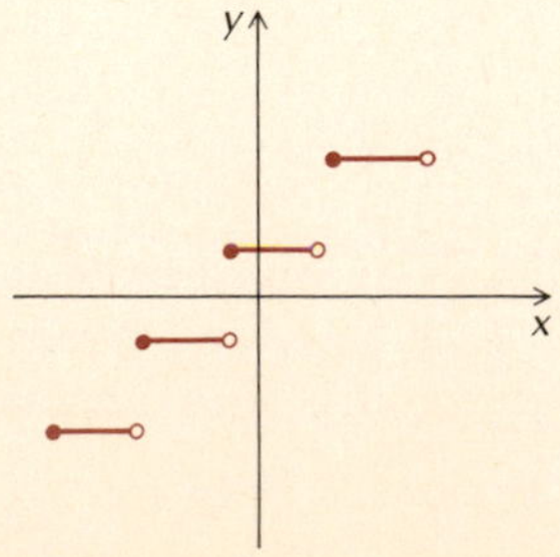

18.

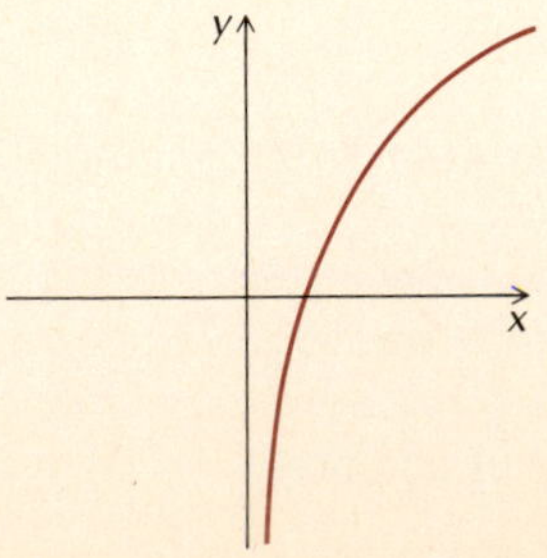

4 In Exercises 19–26, functions are given by formulas. Find the indicated function values and construct and simplify expressions as

$$\frac{[f(a+h)-f(a)]}{h}.$$

19. Given $f(x) = 5x^2 + 4x$, find:
a) $f(0)$; b) $f(-1)$; c) $f(3)$; d) $f(t)$; e) $f(t-1)$; f) $\dfrac{f(a+h)-f(a)}{h}$.

20. Given $g(x) = 3x^2 - 2x + 1$, find:
a) $g(0)$; b) $g(-1)$; c) $g(3)$; d) $g(t)$; e) $g(a+h)$; f) $\dfrac{g(a+h)-g(a)}{h}$.

21. Given $f(x) = 2|x| + 3x$, find:
a) $f(1)$; b) $f(-2)$; c) $f(-4)$; d) $f(2y)$; e) $f(a+h)$; f) $\dfrac{f(a+h)-f(a)}{h}$.

22. Given $g(x) = x^3 - 2x$, find:
a) $g(1)$; b) $g(-2)$; c) $g(-4)$; d) $g(3y)$; e) $g(2+h)$; f) $\dfrac{g(2+h)-g(2)}{h}$.

23. Given $f(x) = 4.3x^2 - 1.4x$, find:
a) $f(1.034)$; b) $f(-3.441)$; c) $f(27.35)$; d) $f(-16.31)$.

24. Given $g(x) = \sqrt{2.2|x| + 3.5}$, find:
a) $g(17.3)$; b) $g(-64.2)$; c) $g(0.095)$; d) $g(-6.33)$.

25. Given $f(x) = \dfrac{x^2 - x - 2}{2x^2 - 5x - 3}$, find:
a) $f(0)$; b) $f(4)$; c) $f(-1)$; d) $f(3)$; e) $f(2-h)$; f) $f(a+b)$.

26. Given $s(x) = \sqrt{\dfrac{3x-4}{2x+5}}$, find:
a) $s(10)$; b) $s(2)$; c) $s(1)$; d) $s(-1)$; e) $s(3+h)$; f) $s(a-b)$.

27. Given $f(x) = x^2$, find
$$\frac{f(a+h)-f(a)}{h}.$$

28. Given $f(x) = x^3$, find
$$\frac{f(a+h)-f(a)}{h}.$$

29. Given $g(x) = x + \sqrt{x^2 - 1}$, find $f(0)$, $f(2)$, $f(10)$.

30. Given $g(x) = x/\sqrt{1 - x^2}$, find $g(0)$, $g(3)$, $g\left(\frac{1}{2}\right)$.

31. Given the function
$$f = \{(-1, 2), (-3, 4), (5, -6), (7, 9)\},$$
find $f(-1)$, $f(7)$, $f(5)$, and $f(-3)$. What is the domain? What is the range?

32. Given the function
$$g = \{(2, -3), (-4, 5), (6, -7), (8, -7)\},$$
find $g(8)$, $g(6)$, $g(-4)$, and $g(2)$. What is the domain? What is the range?

33. From this graph, estimate $g(-2)$, $g(-3)$, $g(0)$, and $g(2)$.

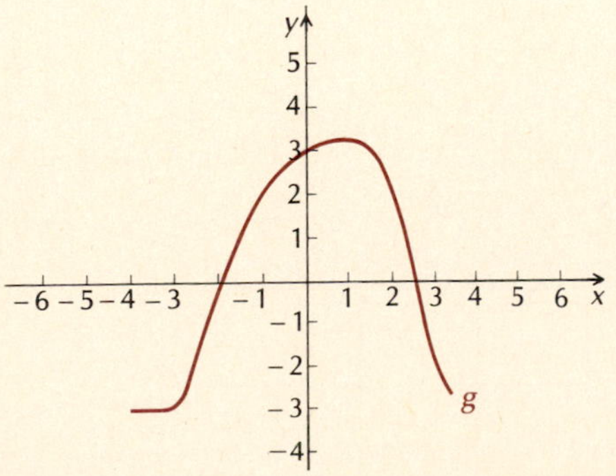

34. From this graph, estimate $h(-2)$, $h(0)$, $h(3)$, and $h(-3)$.

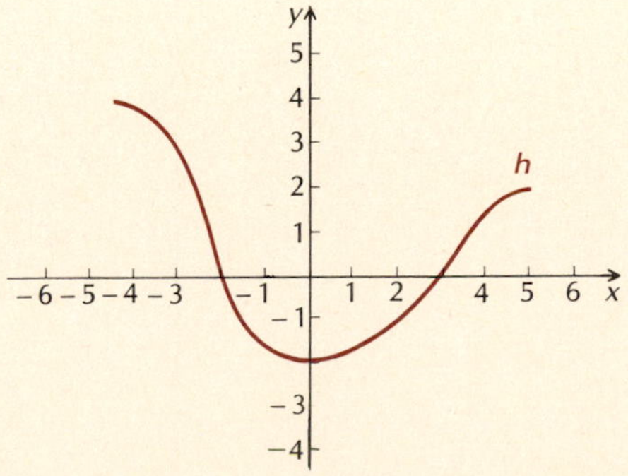

5 In Exercises 39–50, functions are given by formulas. Find the domain of each function.

35. $f(x) = 7x + 4$

36. $f(x) = |3x - 2|$

37. $f(x) = 4 - \dfrac{2}{x}$

38. $f(x) = \sqrt{x - 3}$

39. $f(x) = \sqrt{7x + 4}$

40. $f(x) = \dfrac{1}{9 - x^2}$

41. $f(x) = \dfrac{1}{x^2 - 4}$

42. $f(x) = \dfrac{2x + 6}{x^3 - 4x}$

43. $f(x) = \dfrac{4x^3 + 4}{4x^2 - 5x - 6}$

44. $f(x) = \dfrac{x^3 + 8}{x^2 - 4}$

45. $f(x) = \dfrac{4x^3 + 4}{x(x + 2)(x - 1)}$

46. $f(x) = x^3 - x^2 + x - 2$

SYNTHESIS

47. A complex-valued function f is defined as follows: $f(z) = z^2 - 4z + i$. Find $f(3 + i)$.

48. A complex-valued function g is defined as follows: $g(z) = 2z^2 + z - 2i$. Find $g(2 - i)$.

For each function, construct and simplify $\dfrac{f(x + h) - f(x)}{h}$.

49. $f(x) = \dfrac{1}{x}$

50. $f(x) = \dfrac{1}{x^2}$

51. $f(x) = \sqrt{x}$ (*Hint:* Rationalize the numerator.)

Find the domain.

52. $f(x) = \dfrac{\sqrt{x}}{2x^2 - 3x - 5}$

53. $f(x) = \dfrac{\sqrt{x + 3}}{x^2 - x - 2}$

54. $f(x) = \sqrt{\dfrac{x + 1}{x + |x|}}$

55. $f(x) = \sqrt{x^2 + 1}$

56. Determine whether the relation $\{(x, y) | xy = 0\}$ is a function.

OBJECTIVES

You should be able to:

1 Graph simple functions given by formulas.

2 Given a formula for a function involved in an application, find related function values.

3 Given an applied problem, find a function formula related to the situation.

3.4 Graphs and Applications of Functions

1 Graphs of Functions

Most of the functions that we study in this course and in calculus are given by formulas. We can graph them in much the same way as we did equations. We consider various inputs x, and then compute outputs $f(x)$. Then we plot the ordered pairs $(x, f(x))$, look for patterns, and complete the graph. As you learn more about functions and graphs, you will refine your skills. For example, you will know the shape of many graphs from their formulas, and you will know how to derive one graph from another.

EXAMPLE 1 *Absolute-value function.* Graph the function f given by $f(x) = |x| - 3$. We will often just say "Graph $f(x) = |x| - 3$."

Solution We compute some function values:

$$f(-3) = |-3| - 3 = 3 - 3 = 0; \quad f(1) = |1| - 3 = 1 - 3 = -2;$$
$$f(-2) = |-2| - 3 = 2 - 3 = -1; \quad f(2) = |2| - 3 = 2 - 3 = -1;$$
$$f(-1) = |-1| - 3 = 1 - 3 = -2; \quad f(3) = |3| - 3 = 3 - 3 = 0.$$
$$f(0) = |0| - 3 = 0 - 3 = -3;$$

x	$f(x)$	$(x, f(x))$
−3	0	(−3, 0)
−2	−1	(−2, −1)
−1	−2	(−1, −2)
0	−3	(0, −3)
1	−2	(1, −2)
2	−1	(2, −1)
3	0	(3, 0)

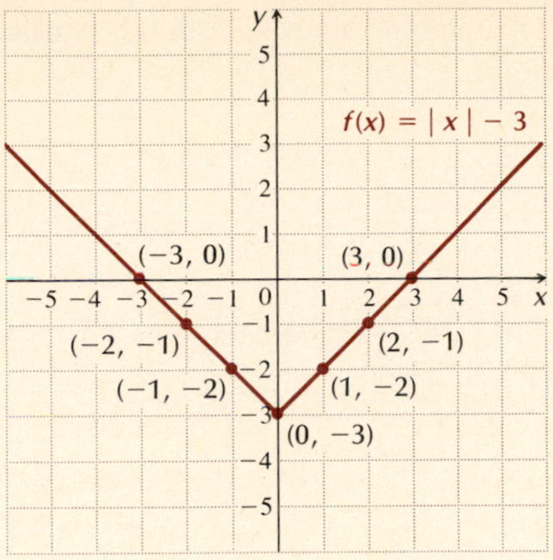

We plot these points, look for a pattern, and sketch the graph. Note that as we get farther from 0, the absolute value increases. Thus the graph rises to the left and the right of the y-axis. ■

DO EXERCISES 1 AND 2.

2 Applications of Functions

There are many applications that involve functions. For example, the volume V of a sphere with radius r is given by the function

$$V = \tfrac{4}{3}\pi r^3.$$

We say that "V is a function of r." We can express the formula using function notation as

$$V(r) = \tfrac{4}{3}\pi r^3.$$

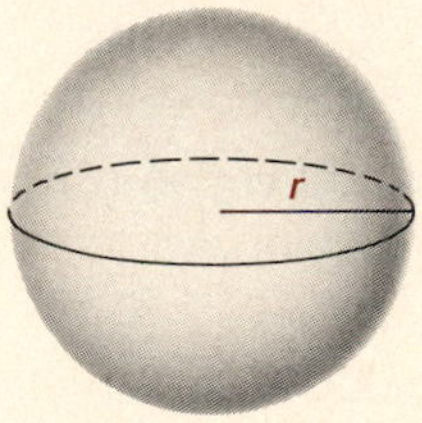

The radius of a bowling ball is 4.08 in. To approximate its volume, we find $V(4.08)$, using 3.14 for π. A calculator would be helpful:

$$V(4.08) = \tfrac{4}{3}\pi(4.08)^3 \approx 284.3 \text{ in}^3.$$

DO EXERCISE 3.

Let us consider another application.

EXAMPLE 2 *Speed of sound in air.* The speed of sound S in air is a function of the temperature T, in degrees Fahrenheit, and is given by

$$S(T) = 1087.7\sqrt{\frac{5T + 2457}{2457}},$$

where S is in feet per second. Find the speed of sound in air when the temperature is 0°, 32°, and 70°.

Graph.

1. $f(x) = |x - 2|$

2. $g(x) = -3$ (a constant function)

3. The surface area S of a sphere is a function of the radius r, given by

$$S(r) = 4\pi r^2.$$

The radius of a bowling ball is 4.08 in. Find its surface area.

4. Find the speed of sound in air when the temperature is $-10°$, $55°$, and $90°$.

Solution We find the function values as follows. A calculator with a square root key would be helpful.

$$S(0) = 1087.7\sqrt{\frac{5(0) + 2457}{2457}} = 1087.7 \text{ ft/sec};$$

$$S(32) = 1087.7\sqrt{\frac{5(32) + 2457}{2457}} = 1087.7\sqrt{\frac{160 + 2457}{2457}} = 1087.7\sqrt{\frac{2617}{2457}}$$
$$= 1087.7\sqrt{1.065120065} \approx 1087.7(1.032046542) \approx 1122.6 \text{ ft/sec};$$

$$S(70) = 1087.7\sqrt{\frac{5(70) + 2457}{2457}} = 1087.7\sqrt{\frac{350 + 2457}{2457}} = 1087.7\sqrt{\frac{2807}{2457}}$$
$$= 1087.7\sqrt{1.142450142} \approx 1087.7(1.068854593) \approx 1162.6 \text{ ft/sec}$$

CAUTION! When using a calculator to do exercises like this, do not stop to round as you make successive calculations. Answers at the back of the text generally will have been found by not stopping to round, even though an example may not show as many decimal places as Example 2 does.

DO EXERCISE 4.

3 Finding Formulas for Functions

In calculus, many problems are solved that seek to find the *maximum* or *minimum* values of a function. These occur in so-called *maximum–minimum* problems. The task is to find the inputs to a function for which it has its maximum or minimum output.

The critical skill in solving this type of problem is to find a formula for the function. Typically, we are trying to represent one variable in terms of another variable even though there may be many variables involved in the problem. This task involves the *Familiarize* and *Translate* skills of our five steps for problem solving. Let us consider some examples.

EXAMPLE 3 *Maximizing area.* A hobby store has 20 ft of fencing to fence off a rectangular area for an electric train in one corner of its display room. The sides up against the wall require no fence.

a) Express the area A as a function of the length x of one side of the rectangle.
b) Complete a table of function values and sketch a graph of the function. Then estimate the length and the width that will yield the maximum area. Estimate the maximum area.

Solution

a) Intuitively (meaning on the basis of your previous mathematical experience), you might think that any dimensions whose sum is 20 ft will give the same area. We will see that this is not true. We first draw a picture. We let $x =$ the length of one side and $y =$ the length of the other. Then since the sum of the lengths must be 20 ft, we have

$$x + y = 20, \quad \text{or} \quad y = 20 - x.$$

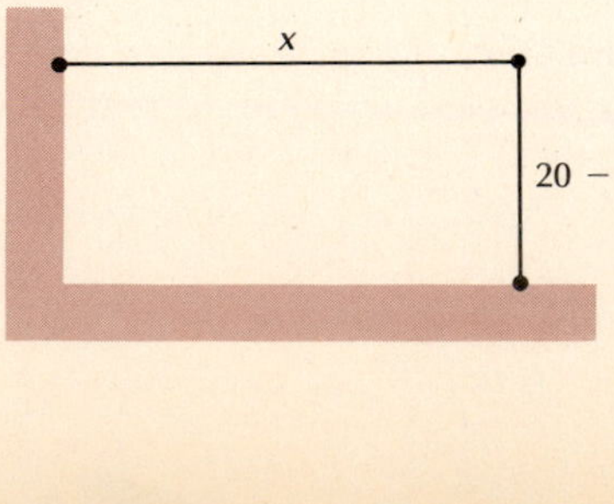

Then the area is given by

$$A = xy$$
$$A = x(20 - x) \quad \text{Substituting } 20 - x \text{ for } y$$
$$A = 20x - x^2.$$

We have thus expressed A as a function of x: $A(x) = 20x - x^2$. This function gives the area A of the rectangular region as a function of the length x of one side.

b) In the next chapter, we will introduce algebraic skills that will allow us to solve this problem exactly. You will learn other skills for an exact solution in calculus. For now, let us complete a table of function values and estimate the dimensions that will yield the maximum area.

x	$y = 20 - x$	$A(x) = x(20 - x)$
0	20	0
4	16	64
6.5	13.5	87.75
8	12	96
10	10	100
12	8	96
13.2	6.8	89.76
20	0	0

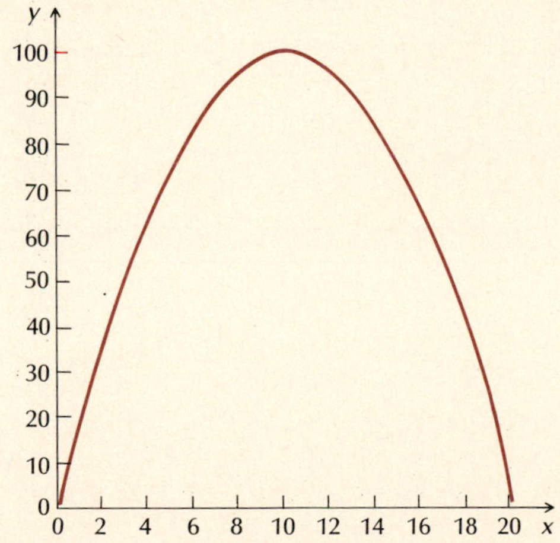

The problem restricts the domain of the function to values of x such that $0 < x < 20$. Otherwise, the lengths would not be positive.

From the table, we can estimate that the maximum area might be 100, but since we cannot substitute all possible values, we do not know for sure that the maximum value occurs at $x = 10$ and $y = 10$. We sketch the graph as further evidence, but we do not know for certain. We estimate that the maximum area of 100 ft^2 occurs when the dimensions are 10 ft by 10 ft.

DO EXERCISE 5.

5. *Maximizing area.* A store has 40 ft of fencing to fence off a rectangular area in one corner of its display room. The sides up against the wall require no fence.

a) Express the area A as a function of the length x of one side of the rectangle.

b) Complete a table of function values and sketch a graph of the function. Then estimate the maximum area and the length and the width that will yield the maximum area.

6. A hot-air balloon is released and rises from the ground at the rate of 120 ft/min. The balloon is tracked from a rangefinder at point P, which is 400 ft from the release point Q of the balloon. Let $d =$ the distance from the balloon to the rangefinder at point P and $t =$ the time, in minutes, after which the balloon is released. Express d as a function of t.

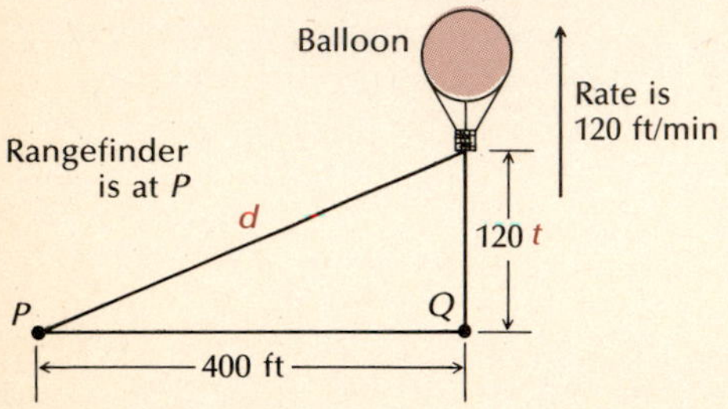

EXAMPLE 4 Two cars leave the same intersection at right angles to each other. One travels at a speed of 55 mph and the other at 65 mph. Express the distance d, in miles, between the cars as a function of the time t, in hours.

Solution We let $d =$ the distance, in miles, between the cars and $t =$ the time, in hours. Then we make a drawing, as shown below.

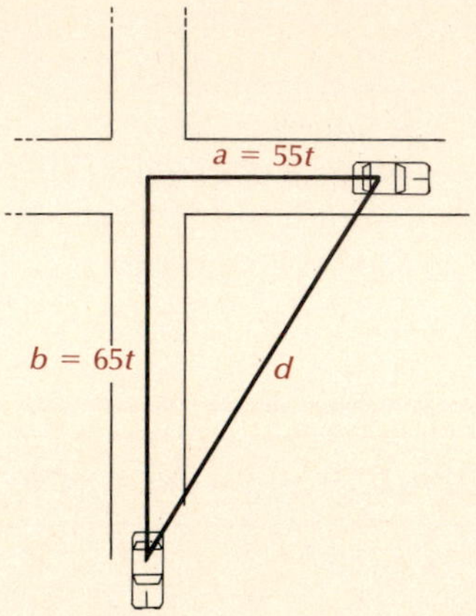

The first car travels at 55 mph. So after t hours, it has traveled $55t$ miles. Similarly, in time t, the second car has traveled $65t$ miles. We can then apply the Pythagorean theorem:

$$d^2 = a^2 + b^2$$
$$= (55t)^2 + (65t)^2.$$

Thus,

$$d = \sqrt{(55t)^2 + (65t)^2} = \sqrt{3025t^2 + 4225t^2}$$
$$d = \sqrt{7250t^2}$$
$$d = \sqrt{7250}t, \quad \text{or} \quad 5\sqrt{290}t.$$

DO EXERCISE 6.

EXERCISE SET 3.4

1 Graph.

1. $f(x) = |x| + 2$
2. $f(x) = 2 - |x|$
3. $g(x) = 4 - x^2$
4. $f(x) = |x + 2|$
5. $f(x) = \frac{2}{x}$
6. $f(x) = \frac{-3}{x}$
7. $f(x) = \frac{1}{2}|x|$
8. $g(x) = x^2 - 4$
9. $g(x) = 3$ (constant function)
10. $g(x) = -2$ (constant function)
11. $f(x) = |x - 1|$
12. $g(x) = 2|x| - 3$
13. $f(x) = |x| + x$
14. $g(x) = |x| - x$
15. $f(x) = \sqrt{x} + 3$
16. $g(x) = \sqrt{x} - 5$

2

17. *Median age of women at first marriage.* Our society is marrying at a later age. The median age of women at first marriage can be approximated by the function

$$A(t) = 0.08t + 19.7,$$

where $A(t) =$ the median age of women at first marriage the tth year after 1950. Thus, $A(0)$ is the median age of women at first marriage in the year 1950, $A(30)$ is the median age in 1980, and so on.

a) Find $A(0)$, $A(1)$, $A(10)$, $A(30)$, and $A(40)$.

b) What will be the median age of women at first marriage in 1996?

18. *Average price of a movie ticket.* The average price of a movie ticket can be estimated by the function

$$P(y) = 0.1522y - 298.592,$$

where y = the year and $P(y)$ = the average price, in dollars. The price is lower than what might be expected due to senior citizen's discounts, children's prices, and special volume discounts. Thus, $V(1991)$ is the average price of a movie ticket in 1991.

a) Use the function to predict the average price of a ticket in 1991, 1995, and 2000.
b) When will the average price of a movie ticket be $6.00?

19. *Number of diagonals of a polygon.* The number D of diagonals of a polygon is a function of the number n of sides and is given by

$$D(n) = \frac{n^2 - 3n}{2}.$$

Find the number of diagonals of a pentagon (5 sides), an octagon (8 sides), a decagon (10 sides), and a dodecagon (12 sides).

20. *Number of games in a sports league.* The number N of games in a sports league is a function of the number of teams n, where each team plays every other team twice, and is given by

$$N(n) = n(n - 1).$$

a) Find the number of games in a women's softball league containing 6 teams where each team plays every other team twice.
b) Find the number of games in a Big-10 basketball season. There are 10 teams and each team plays every other team twice.

21. *Boiling point and elevation.* The elevation E, in meters, above sea level at which the boiling point of water is T degrees Celsius is given by

$$E(T) = 1000(100 - T) + 580(100 - T)^2.$$

a) At what elevation is the boiling point 99.5°?
b) At what elevation is the boiling point 100°?

22. *Territorial area of an animal.* The territorial area of an animal is defined to be its defended, or exclusive, region. For example, a lion has a certain region over which it is ruler. It has been shown that the territorial region T, in acres, of predatory animals is a function of body weight w, in pounds, as given by

$$T(w) = w^{1.31}.$$

Find the territorial area of animals whose body weights are 0.5 lb, 10 lb, 20 lb, 40 lb, 100 lb, and 200 lb.

3

23. The length L of a rectangle is 4 m longer than the width W.

a) Express the area A of the rectangle as a function of the length L.
b) Express the area A of the rectangle as a function of the width W.

24. The base b of a triangle is 3 less than twice the height h.

a) Express the area A of the triangle as a function of the height h.
b) Express the area A of the triangle as a function of the base b.

25. A rectangle of dimensions x by y has a perimeter of 34 ft. Express the area A as a function of x.

26. A rectangle of dimensions x by y has a perimeter of 48 m. Express the area A as a function of x.

27. A rectangle of dimensions x by y is inscribed in a circle of radius 8 ft. Express the area A as a function of x.

28. A rectangle of dimensions x by y is inscribed in a circle of diameter 20 yd. Express the area A as a function of x.

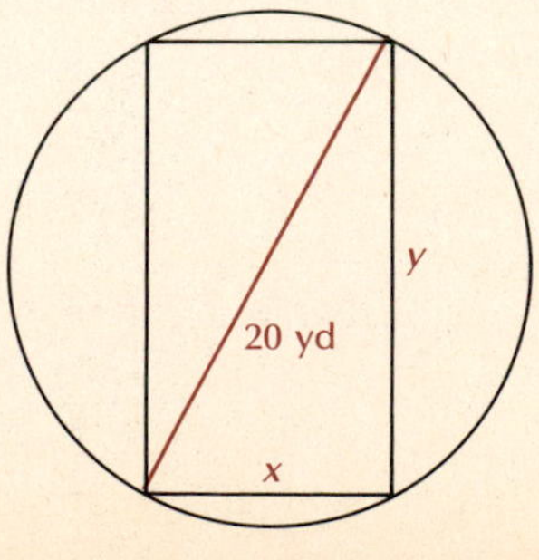

29. *Compound interest.* Suppose that \$1000 is invested in a savings account at interest rate i, compounded annually, for 4 years. Express the amount A in the account as a function of i.

30. *Compound interest.* Suppose that \$1000 is invested in a savings account at interest rate 8%, compounded annually. Express the amount A in the account as a function of the time t.

31. *Maximizing volume.* From a thin piece of cardboard 10 in. by 10 in., square corners are cut out so that the sides can be folded up to make a box.
   a) Express the volume V as a function of the length x of a side of the square base.
   b) Complete a table of function values and sketch a graph of the function. Then estimate the maximum volume and the dimensions that will yield the maximum volume.

32. *Maximizing area.* A store has 30 ft of fencing to fence off a rectangular area in one corner of its display room. The sides up against the wall require no fence.
   a) Express the area A as a function of the length x of one side of the rectangle.
   b) Complete a table of function values and sketch a graph of the function. Then estimate the maximum area and the length and the width that will yield the maximum area.

33. A container firm is designing an open-top rectangular box, with a square base, that will hold 108 in^3 (see the figure). Let x = the length of a side of the base. Express the surface area as a function of x.

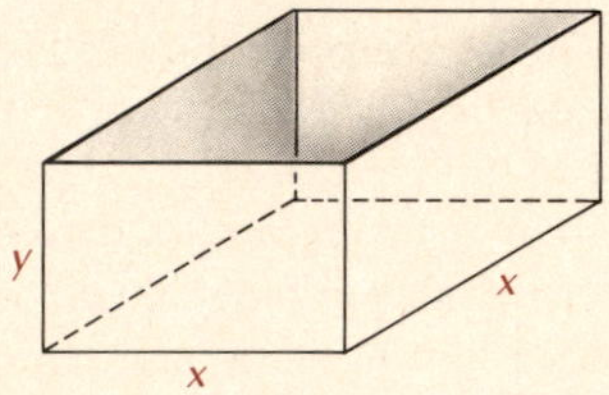

34. *Cost.* A rectangular box with a volume of 320 ft^3 is to be constructed with a square base and top (see the figure). The cost per square foot for the bottom is \$1.50, for the top is \$1, and for the sides is \$2.50. Express the cost C as a function of x.

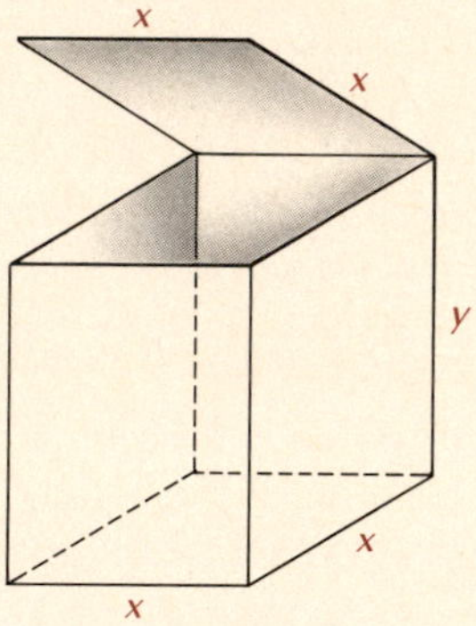

35. *Golf distance finder.* A device used in golf to estimate the distance d, in yards, to a hole measures the size s, in inches, that the 7-ft pin *appears* to be in a viewfinder. Express the distance d as a function of s.

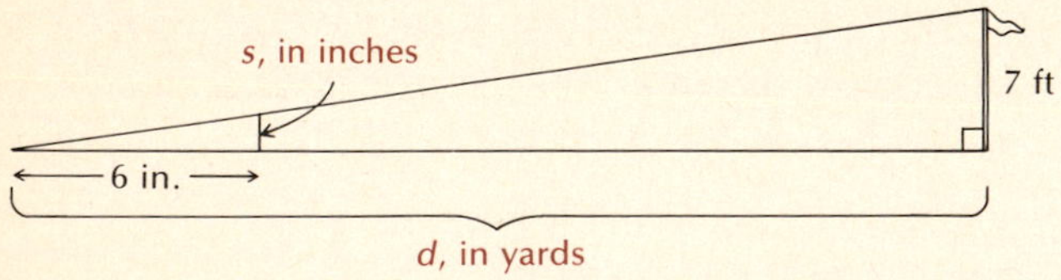

36. *Norman window.* A Norman window is a rectangle with a semicircle on top. Suppose the perimeter of a particular Norman window is to be 24 ft. Express the area A as a function of the radius of the circle x.

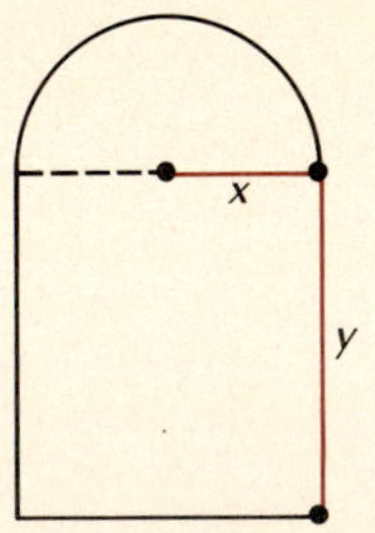

A Norman window.

37. An airplane is flying at an altitude of 3700 ft. The slanted distance directly to the airport is d ft. Express the horizontal distance h as a function of d.

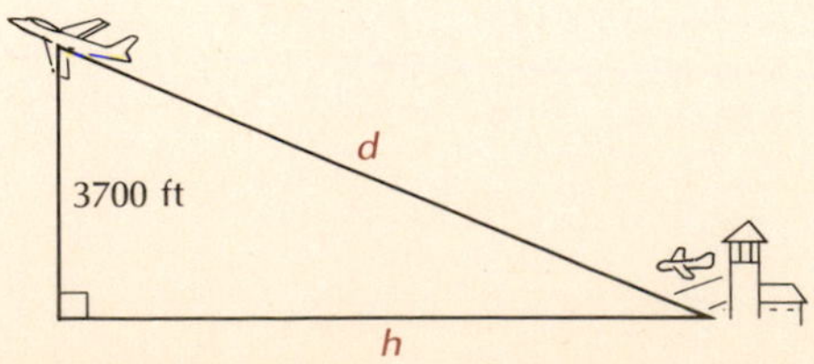

38. A gas tank has ends that are hemispheres with radius r and a cylindrical center with the same radius and a height or length of 6 ft.

a) Express the volume as a function of r.

b) Express the surface area S of the tank as a function of r.

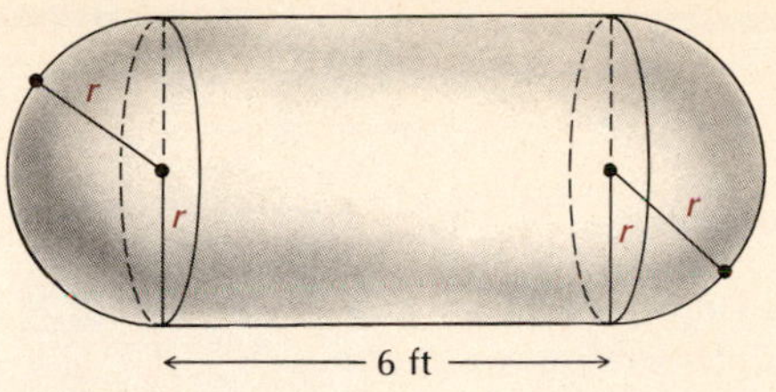

39. Express the area of an equilateral triangle as a function of the length of a side a.

40. A softball diamond is a square 65 ft on a side. A catcher fields a ball x ft up the third-base line. Express the distance D of a throw to first base as a function of x.

SYNTHESIS

41. A 24-in. piece of string is cut into two pieces. One piece is used to form a circle and the other to form a square. Express the sum of the areas S as a function of the length x cut to form the circle.

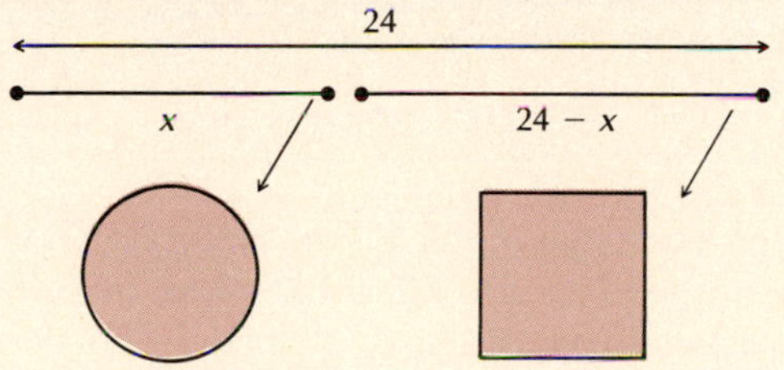

42. A power line is to be constructed from a power station at point A (see the figure) to an island at point C, which is directly 1 mi out in the water from a point B on the shore. Point B is 4 mi downshore from the power station at A. It costs \$5000 per mile to lay the power line under water and \$3000 per mile to lay the line under ground. At what point S downshore from A should the line come to the shore in order to minimize cost? Note that S could very well be A or B. Let $x =$ the distance from B to S. Express the cost C of laying the line as a function of x.

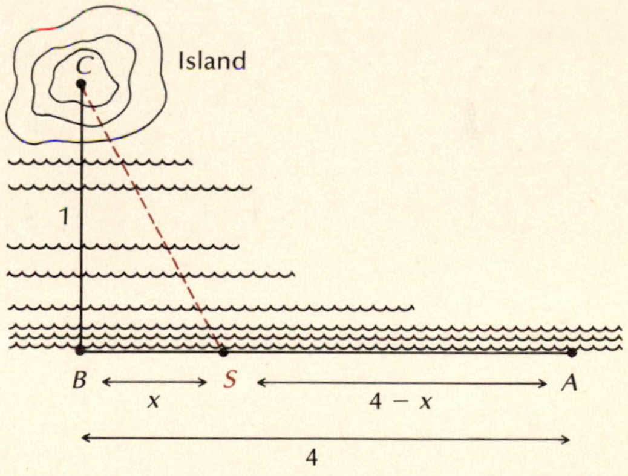

43. A right circular cylinder of height h and radius r is inscribed in a right circular cone of height 10 ft and base with a radius of 6 ft.

a) Express the height h as a function of r.

b) Express the volume V of the cylinder as a function of r.

c) Express the volume V of the cylinder as a function of h.

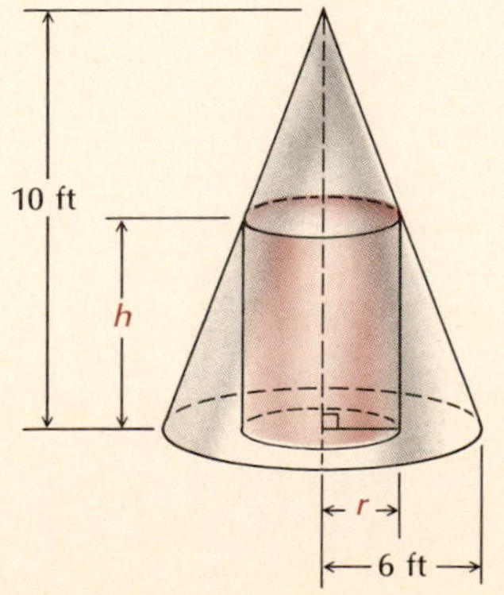

3.5 Symmetry

OBJECTIVES

You should be able to:

1. Given an equation defining a relation, determine whether a graph is symmetric with respect to a coordinate axis.
2. Given an equation defining a relation, determine whether a graph is symmetric with respect to the origin.
3. Given the graph of a function or a formula, determine whether the function is even, odd, or neither even nor odd.

1 Symmetry with Respect to a Line

In the figure, points P and P_1 are said to be **symmetric with respect to line l.** They are the same distance from l, and the segment $\overline{PP_1}$ is perpendicular to l.

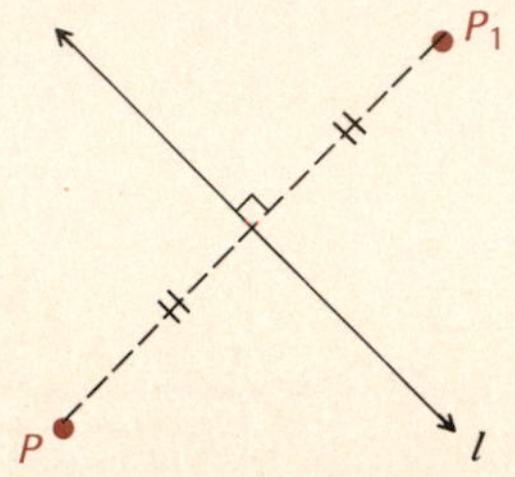

DEFINITION

Two points P and P_1 are *symmetric with respect to a line* l if and only if l is the perpendicular bisector of the segment $\overline{PP_1}$. The line l is known as the *line of symmetry*.

We also say that the two points P and P_1 are **reflections** of each other across the line. The line is therefore also known as a **line of reflection.**

Now consider a set of points (geometric figure) as shown in the color curve below. This figure is said to be symmetric with respect to the line l, because if you pick any point Q in the set, you can find another point Q_1 in the set such that Q and Q_1 are symmetric with respect to l.

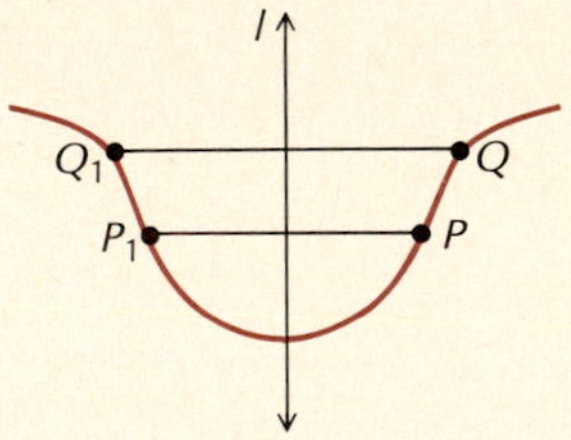

DEFINITION

A figure, or set of points, is *symmetric with respect to a line* l if and only if for each point Q in the set there exists another point Q_1 in the set for which Q and Q_1 are symmetric with respect to line l.

Imagine picking up the preceding figure and flipping it about the line l. Points P and P_1 would be interchanged. Points Q and Q_1 would be interchanged. These are, then, pairs of symmetric points. The entire figure would look exactly like it did before it was flipped. This means that *each* point of the figure is symmetric with *some* point of the figure. Thus the figure is symmetric with respect to the line. A point and its reflection are known as **images** of each other. Thus P_1 is the image of P, for example. The line l is known as an **axis of symmetry.**

Symmetry with Respect to the Axes

There are special and interesting kinds of symmetry in which a coordinate axis is a line of symmetry. The following example shows figures that are symmetric with respect to an axis and a figure that is not.

EXAMPLE 1 In graph (a), flipping the figure about the y-axis would not change the figure. In graph (b), flipping the graph about the x-axis would not change the figure. In graph (c), flipping about either axis would change the figure.

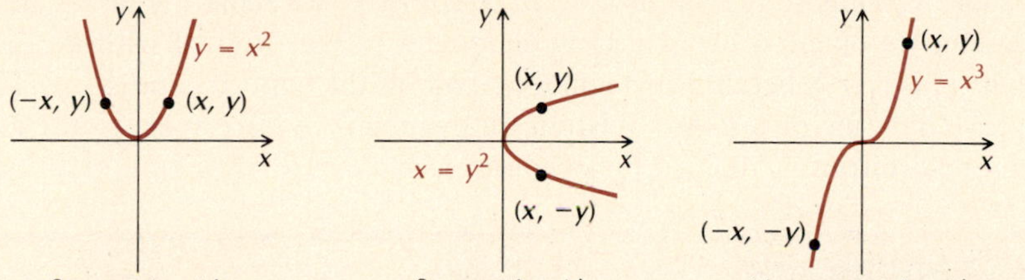

Symmetric with respect to y-axis (a) | Symmetric with respect to x-axis (b) | Not symmetric with respect to either axis (c)

Let us consider a figure like graph (a), symmetric with respect to the y-axis. For every point of the figure, there is another point that is the same distance across the y-axis. The first coordinates of such a pair of points are additive inverses of each other.

EXAMPLE 2 In the relation $y = x^2$, there are points (2, 4) and (−2, 4). The first coordinates, 2 and −2, are additive inverses of each other, whereas the second coordinates are the same. For every point of the figure (x, y), there is another point $(-x, y)$.

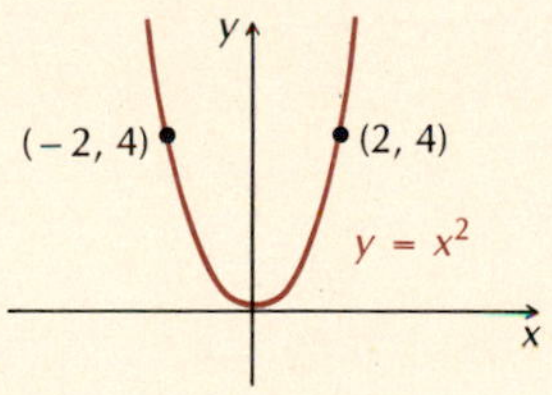

DO EXERCISE 1.

Let us consider a figure symmetric with respect to the x-axis, like (b) in Example 1. For every point of such a figure there is another point the same distance across the x-axis. The second coordinates of such a pair of points are additive inverses of each other.

EXAMPLE 3 In the relationship $x = y^2$ there are points (4, 2) and (4, −2). The second coordinates, 2 and −2, are additive inverses, or opposites, of each other, while the first coordinates are the same. For every point of the figure (x, y), there is another point $(x, -y)$.

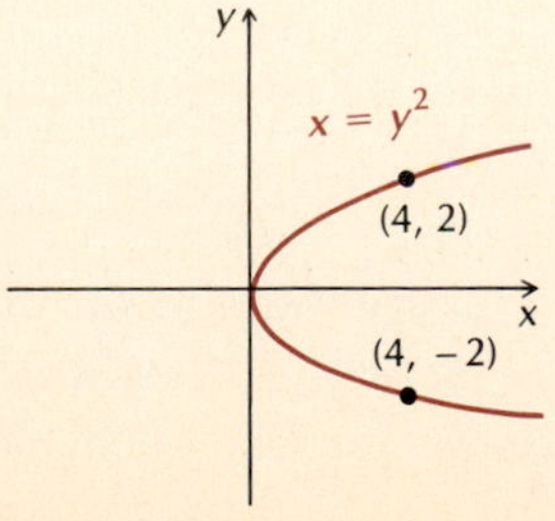

1. a) Plot the point (3, 2). Let the y-axis be a line of symmetry. Plot the point symmetric to (3, 2). What are its coordinates?

   b) Plot the point (−4, −5). Let the y-axis be a line of reflection. Plot the image of (−4, −5). What are its coordinates?

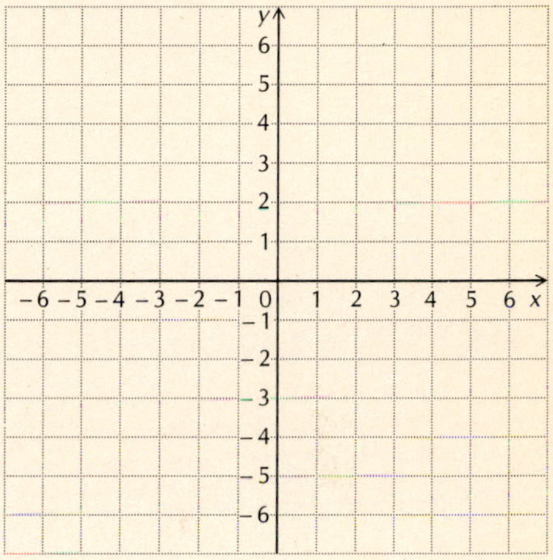

2. Let the x-axis be a line of symmetry.
   a) Plot the point (4, 3). Plot the point symmetric to it. What are its coordinates?
   b) Plot the point (3, −5). Plot its image after reflection across the x-axis. What are its coordinates?

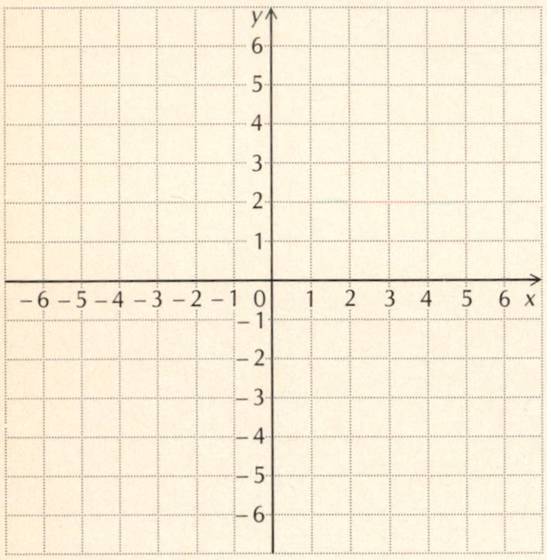

DO EXERCISE 2.

Let us consider a figure that is symmetric with respect to the y-axis, as in Example 2. Suppose it is defined by an equation. If in this equation we replace x by $-x$, we obtain a new equation equivalent to the original with the same graph. This is true because any number x gives us the same y-value as its additive inverse, $-x$.

Let us consider a figure that is symmetric with respect to the x-axis, as in Example 3. Suppose it is defined by an equation. If we replace y by $-y$ in the equation, we obtain a new equation equivalent to the original with the same graph. This is true because any number y gives us the same x-value as $-y$. Thus we have a means of testing a relation for symmetry with respect to the axes, when that relation is defined by an equation.

DEFINITION

When a relation is defined by an equation:

1. **If replacing x by $-x$ produces an equivalent equation, then the graph is symmetric with respect to the y-axis.**
2. **If replacing y by $-y$ produces an equivalent equation, then the graph is symmetric with respect to the x-axis.**

EXAMPLE 4 Test $y = x^2 + 2$ for symmetry with respect to the y-axis.

Solution

a) Replace x by $-x$.

$$y = x^2 + 2 \qquad (1)$$
$$\downarrow$$
$$y = (-x)^2 + 2$$

b) Simplify, if possible.

$$y = (-x)^2 + 2 = x^2 + 2 \qquad (2)$$

c) Is the resulting equation (2) equivalent to the original (1)?

Since the answer is yes, the graph is symmetric with respect to the y-axis. ■

EXAMPLE 5 Test $y = x^2 + 2$ for symmetry with respect to the x-axis.

Solution

a) Replace y by $-y$.

$$y = x^2 + 2 \qquad (1)$$
$$\downarrow$$
$$-y = x^2 + 2$$

b) Simplify, if possible.

The equation is simplified for the most part, although we could multiply on both sides by −1, obtaining

$$y = -x^2 - 2. \qquad (2)$$

c) Is the resulting equation (2) equivalent to the original (1)?

This answer may be obvious to you and is no. To be sure, one might use some trial-and-error reasoning, as follows. Suppose we substitute 0 for x in

Eq. (1). Then

$$y = 0^2 + 2 = 2,$$

so $(0, 2)$ is a solution of Eq. (1). In order for Eqs. (1) and (2) to be equivalent, $(0, 2)$ must also be a solution of Eq. (2). We substitute to find out:

$$
\begin{array}{c|c}
\multicolumn{2}{c}{y = -x^2 - 2} \\
\hline
2 & -0^2 - 2 \\
 & -2
\end{array}
$$

Thus $(0, 2)$ is not a solution of Eq. (2); hence the equations are not equivalent and the graph is not symmetric with respect to the x-axis. ■

EXAMPLE 6 Test $a^2 + b^4 - 5 = 0$ for symmetry with respect to the a-axis. The a-axis is ordinarily the horizontal axis.

Solution

a) Replace b by $-b$.

$$a^2 + b^4 - 5 = 0 \tag{1}$$
$$\downarrow$$
$$a^2 + (-b)^4 - 5 = 0$$

b) Simplify, if possible.

$$a^2 + (-b)^4 - 5 = a^2 + b^4 - 5 = 0 \tag{2}$$

c) Is the resulting equation equivalent to the first?

Since the answer is yes, the graph is symmetric with respect to the a-axis. ■

DO EXERCISES 3–8.

2 Symmetry with Respect to a Point

Two points are **symmetric with respect to a point** when they are situated as shown in the following figure. That is, the points are the same distance from that point, and all three points are on a line.

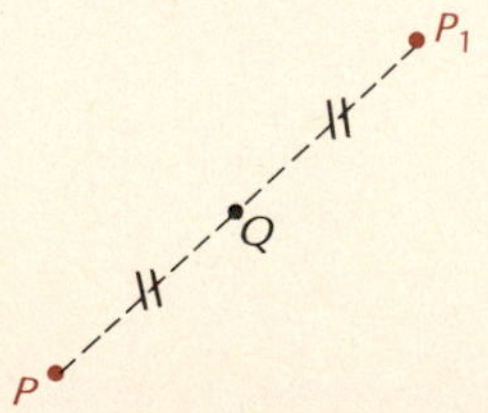

> **DEFINITION**
>
> Two points P and P_1 are *symmetric with respect to a point Q* if and only if Q (the point of symmetry) is the midpoint of segment $\overline{PP_1}$.

A *set* of points is symmetric with respect to a point if for each point P in the set there exists another point P_1 in the set, such that P and P_1 are symmetric with respect to the point. This is illustrated below. Imagine sticking a pin in this figure at O and then rotating the figure 180°. Points P and P_1 would be interchanged. Points Q and Q_1 would be interchanged. These are pairs of symmetric points. The entire figure would look exactly as it did before it was

Test for symmetry with respect to the coordinate axes.

3. $y = x^2 - 3$

4. $y^2 = x^3$

5. $x^4 = y^2 + 2$

6. $3y^2 + 4x^2 = 12$

7. $a + 3b = 5$

8. $2p^3 + 4q^3 = 1$

9. Draw coordinate axes. Let the origin be a point of symmetry.
   a) Plot the point (3, 2). Plot the point symmetric to it. What are its coordinates?
   b) Plot the point (−4, 3). Plot the point symmetric to it. What are its coordinates?
   c) Plot the point (−5, −7). Plot the point symmetric to it. What are its coordinates?

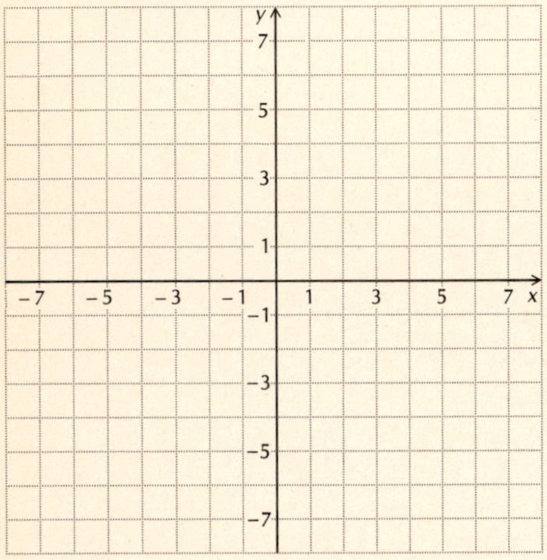

rotated. This means that for *each* point P of the figure, there is a point P_1 of the figure such that the points P and P_1 are symmetric with respect to O. Thus the figure is symmetric with respect to the point O.

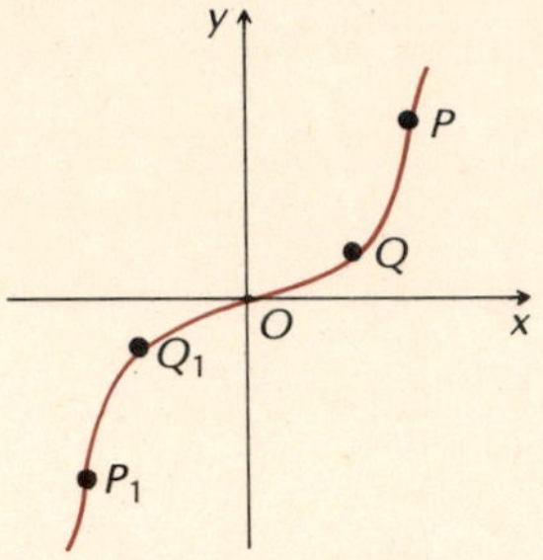

DEFINITION

A set of points is *symmetric with respect to a point B* if and only if for every point P in the set there exists another point P_1 in the set for which P and P_1 are symmetric with respect to B.

Equivalently, a set of points is symmetric with respect to a point B if when the set is rotated 180°, the resulting figure coincides with the original.

Symmetry with Respect to the Origin

A special kind of symmetry with respect to a point is symmetry with respect to the origin.

EXAMPLE 7 In graphs (a) and (b), rotating the figure 180° about the origin would not change the figure. In graph (c), such a rotation would change the figure.

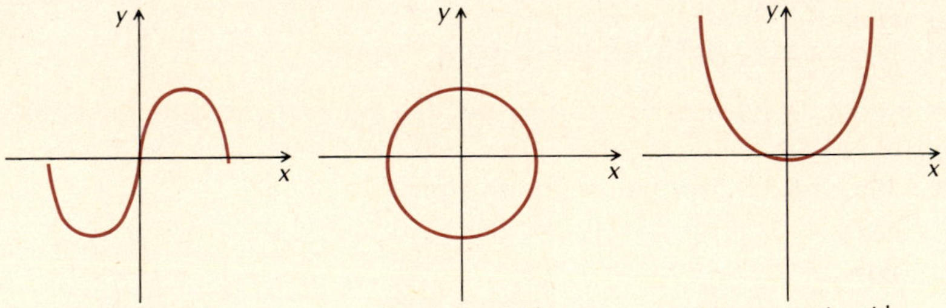

Symmetric with respect to origin
(a)

Symmetric with respect to origin
(b)

Not symmetric with respect to origin
(c)

Let us consider figures like (a) and (b) above, symmetric with respect to the origin. For every point of the figure, there is another point that is the same distance across the origin. The first coordinates of such pairs are additive inverses, or opposites, of each other, and the second coordinates are additive inverses, or opposites, of each other.

DO EXERCISE 9.

EXAMPLE 8 The relation $y = x^3$ is symmetric with respect to the origin. In this relation are the points (2, 8) and (−2, −8). The first coordinates are additive inverses of each other. The second coordinates are additive inverses of each other. For every point of the figure (x, y), there is another point $(-x, -y)$.

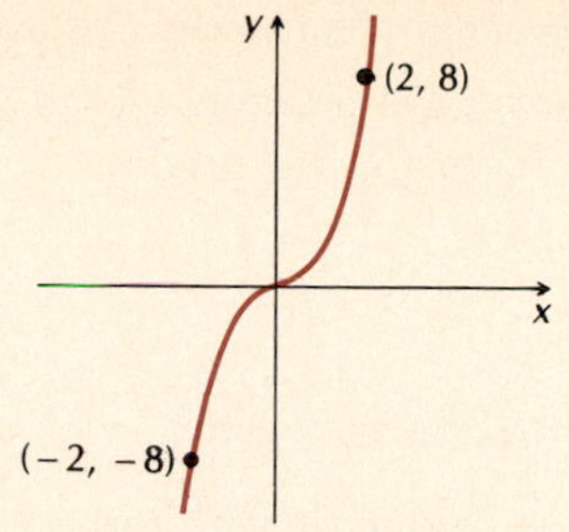

In a relation that is symmetric with respect to the origin, as in Example 8, if we replace x by $-x$ and y by $-y$, we obtain a new equation, but we will get the same figure. This is true because whenever a point (x, y) is in the relation, the point $(-x, -y)$ is also in the relation. This gives us a means of testing a relation for symmetry with respect to the origin when it is defined by an equation.

DEFINITION

When a relation is defined by an equation, if replacing x by $-x$ and replacing y by $-y$ produces an equivalent equation, then the graph is symmetric with respect to the origin.

EXAMPLE 9 Test $x^2 = y^2 + 2$ for symmetry with respect to the origin.

Solution

a) Replace x by $-x$ and y by $-y$.

$$x^2 = y^2 + 2$$
$$\downarrow \qquad \downarrow$$
$$(-x)^2 = (-y)^2 + 2$$

b) Simplify, if possible.

Since $(-x)^2 = x^2$ and $(-y)^2 = y^2$, then $(-x)^2 = (-y)^2 + 2$ simplifies to

$$x^2 = y^2 + 2.$$

c) Is the resulting equation equivalent to the original?

Since the answer is yes, the graph is symmetric with respect to the origin.

EXAMPLE 10 Test $2a + 3b = 8$ for symmetry with respect to the origin.

Solution

a) Replace a by $-a$ and b by $-b$.

$$2a \quad + \quad 3b \quad = 8$$
$$\downarrow \qquad \downarrow$$
$$2(-a) + 3(-b) = 8$$

b) Simplify, if possible.

We have

$$2(-a) + 3(-b) = -2a - 3b,$$

so

$$-(2a + 3b) = 8 \quad \text{and} \quad 2a + 3b = -8.$$

Test for symmetry with respect to the origin.

10. $x^2 + 3y^2 = 4$

11. $x = y$ (*Hint:* After you substitute, multiply on both sides by -1.)

12. $x = -y$

13. $xy = 5$

14. $ab = -5$

15. $u = |v|$

16. Determine whether the function is even.

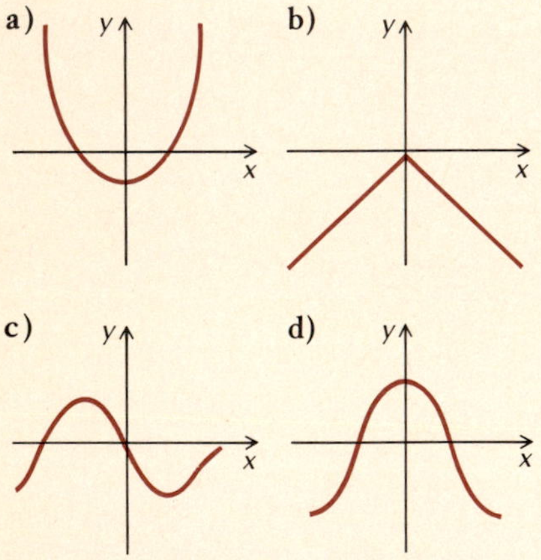

17. Determine whether the function is even.

a) $f(x) = x^2 + 3x$
b) $f(x) = |x|$
c) $f(x) = 3x^2 - x^4$
d) $f(x) = 2x^2 + 1$

c) Is the resulting equation equivalent to the original?

The equation $2a + 3b = -8$ is *not* equivalent to $2a + 3b = 8$, so the graph is not symmetric with respect to the origin. ■

DO EXERCISES 10–15.

3 Even and Odd Functions

If the graph of a function is symmetric with respect to the y-axis, then it is an **even function.** A function will be symmetric to the y-axis if in its equation we can replace x by $-x$ and obtain an equivalent equation. Thus if we have a function given by $y = f(x)$, then $y = f(-x)$ will give the same function if the function is even. In other words, an even function is one for which $f(x) = f(-x)$ for all x in its domain.

> DEFINITION
>
> **A function f is an *even function* if $f(x) = f(-x)$ for all x in the domain of f.**

EXAMPLE 11 Determine whether the function $f(x) = x^2 + 1$ is even.

Solution

a) Find $f(-x)$ and simplify.

$$f(-x) = (-x)^2 + 1 = x^2 + 1$$

b) Compare $f(x)$ and $f(-x)$.

Since $f(x) = f(-x)$ for all x in the domain, f is an even function.

Note that the graph is symmetric with respect to the y-axis.

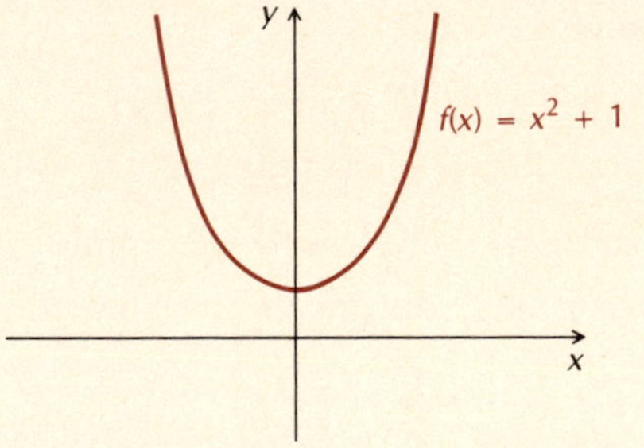

■

EXAMPLE 12 Determine whether the function $f(x) = x^2 + 8x^3$ is even.

Solution

a) Find $f(-x)$ and simplify.

$$f(-x) = (-x)^2 + 8(-x)^2 = x^2 - 8x^3$$

b) Compare $f(x)$ and $f(-x)$.

Since $f(x)$ and $f(-x)$ are not the same for all x in the domain, f is not an even function. ■

DO EXERCISES 16 AND 17.

If the graph of a function is symmetric with respect to the origin, then it is an **odd function.** A function will be symmetric with respect to the origin if in its equation we can replace x by $-x$ and y by $-y$ and obtain an equivalent equation. Thus if we have a function given by $y = f(x)$, then replacing y by $-y$ and $f(x)$ by $f(-x)$ yields the equation $-y = f(-x)$, which is equivalent to $y = f(x)$, if f is an odd function. In other words, an odd function is one for which $f(-x) = -f(x)$ for all x in the domain.

> DEFINITION
>
> **A function f is an *odd function* when $f(-x) = -f(x)$ for all x in the domain of f.**

EXAMPLE 13 Determine whether $f(x) = x^3$ is an odd function.

Solution

a) Find $f(-x)$ and $-f(x)$ and simplify.

$$f(-x) = (-x)^3 = -x^3,$$
$$-f(x) = -x^3$$

b) Compare $f(-x)$ and $-f(x)$.

Since $f(-x) = -f(x)$ for all x in the domain, f is odd.

Note that the graph is symmetric with respect to the origin.

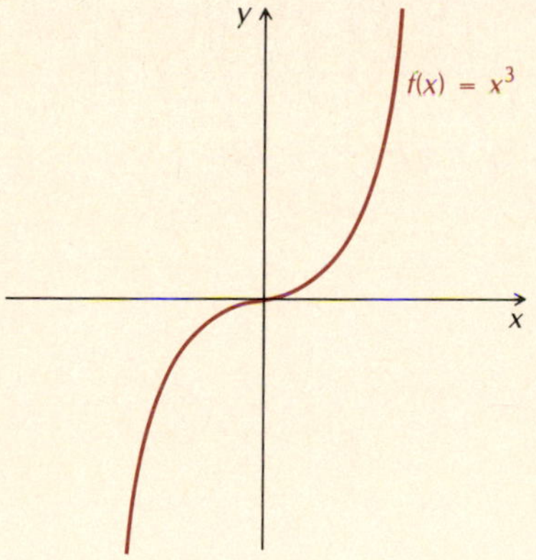

EXAMPLE 14 Determine whether $f(x) = x^2 - 4x^3$ is even, odd, or neither even nor odd.

Solution

a) Find $f(-x)$ and $-f(x)$ and simplify.

$$f(x) = x^2 - 4x^3,$$
$$f(-x) = (-x)^2 - 4(-x)^3 = x^2 + 4x^3,$$
$$-f(x) = -x^2 + 4x^3$$

b) Compare $f(x)$ and $f(-x)$ to determine whether f is even.

Since $f(x)$ and $f(-x)$ are *not* the same for all x in the domain, f is *not* even.

c) Compare $f(-x)$ and $-f(x)$ to determine whether f is odd.

Since $f(-x)$ and $-f(x)$ are *not* the same for all x in the domain, f is *not* odd. Thus f is *neither* even nor odd.

DO EXERCISES 18 AND 19.

18. Determine whether the function is even, odd, or neither even nor odd.

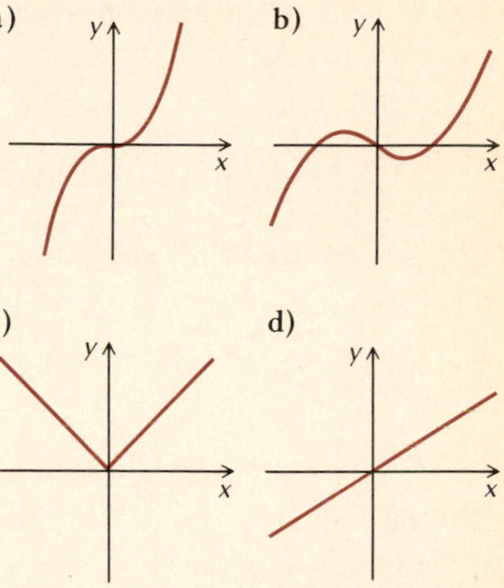

19. Determine whether the function is even, odd, or neither even nor odd.

a) $f(x) = x^3 + 2$

b) $f(x) = x^4 - x^6$

c) $f(x) = x^3 + x$

d) $f(x) = 3x^2 + 3x^5$

e) $f(x) = x^2 - \dfrac{1}{x}$

EXERCISE SET 3.5

1 Test for symmetry with respect to the coordinate axes and the origin.

1. $3y = x^2 + 4$
2. $5y = 2x^2 - 3$
3. $y^3 = 2x^2$
4. $3y^3 = 4x^2$
5. $2x^4 + 3 = y^2$
6. $3y^2 = 2x^4 - 5$
7. $2y^2 = 5x^2 + 12$
8. $3x^2 - 2y^2 = 7$
9. $2x - 5 = 3y$
10. $5y = 4x + 5$
11. $3b^3 = 4a^3 + 2$
12. $p^3 - 4q^3 = 12$

2 Test for symmetry with respect to the origin.

13. $3x^2 - 2y^2 = 3$
14. $5y^2 = -7x^2 + 4$
15. $5x - 5y = 0$
16. $3x = 3y$
17. $3x + 3y = 0$
18. $7x = -7y$
19. $3x = \frac{5}{y}$
20. $3y = \frac{7}{x}$
21. $y = |2x|$
22. $3x = |y|$
23. $3a^2 + 4a = 2b$
24. $5v = 7u^2 - 2u$
25. $3x = 4y$
26. $6x + 7y = 0$
27. $xy = 12$
28. $y = -\frac{4}{x}$

3

29. Determine whether the function is even, odd, or neither even nor odd.

a)

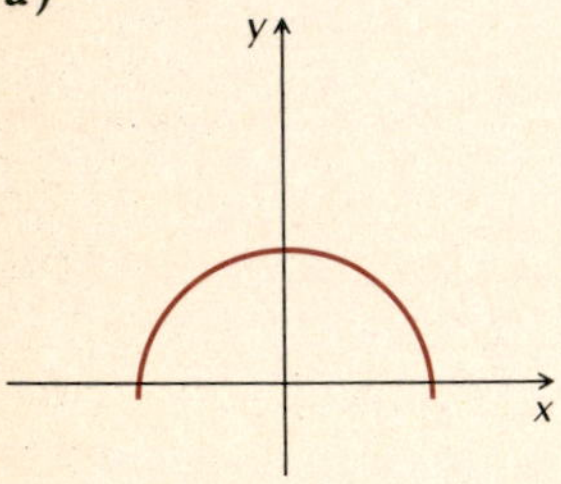

b)

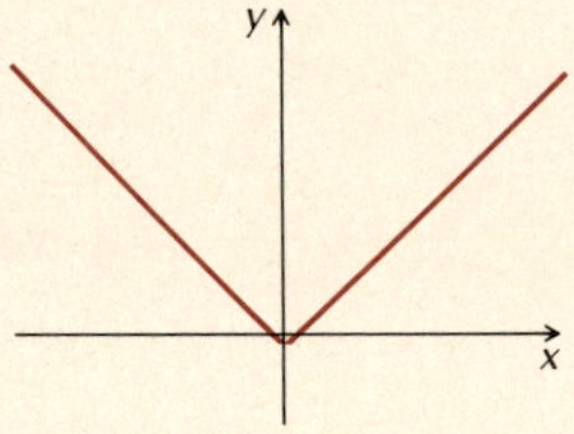

c)

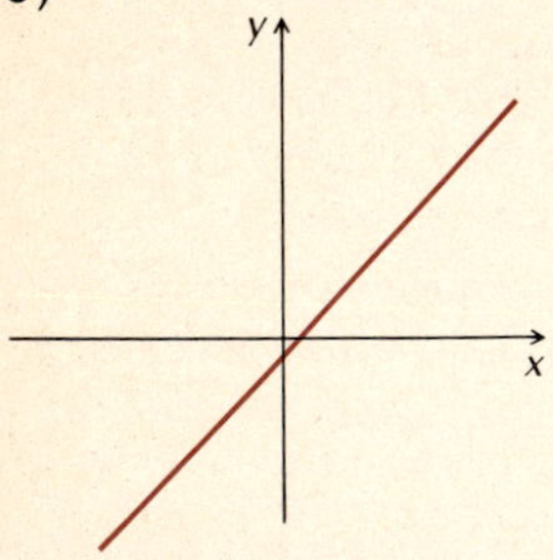

d)

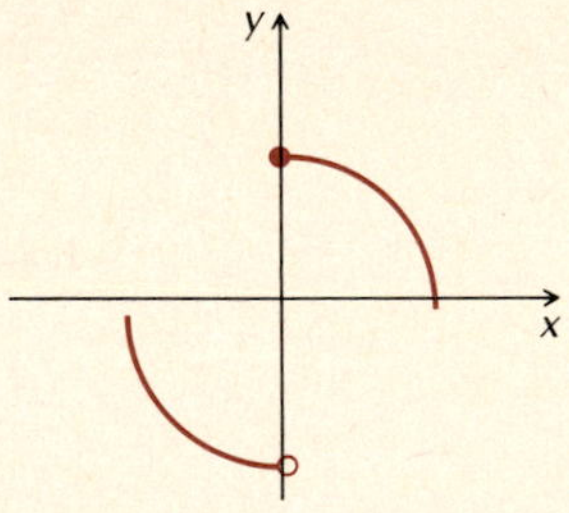

30. Determine whether the function is even, odd, or neither even nor odd.

a)

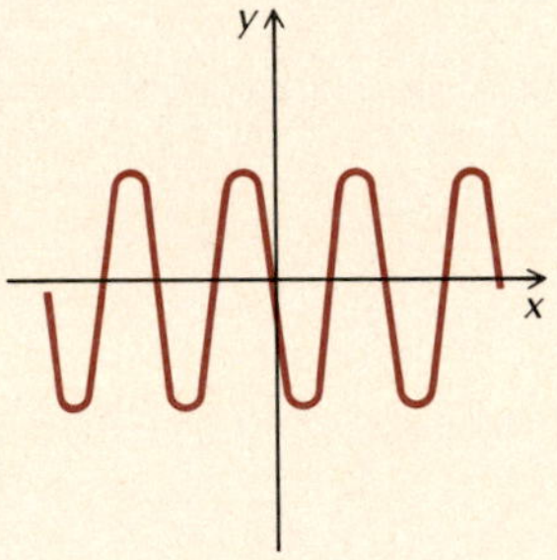

b)

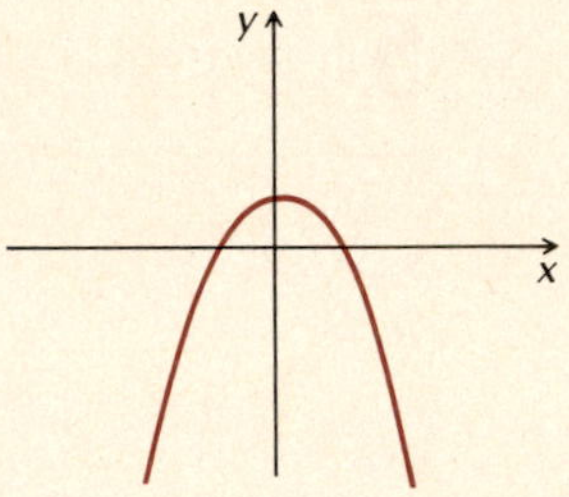

c)

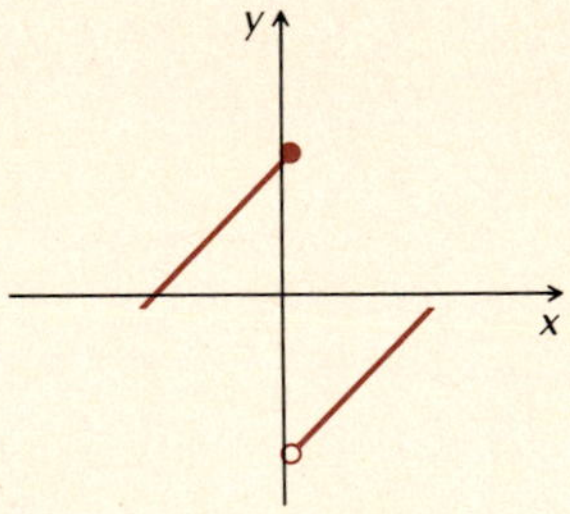

d)

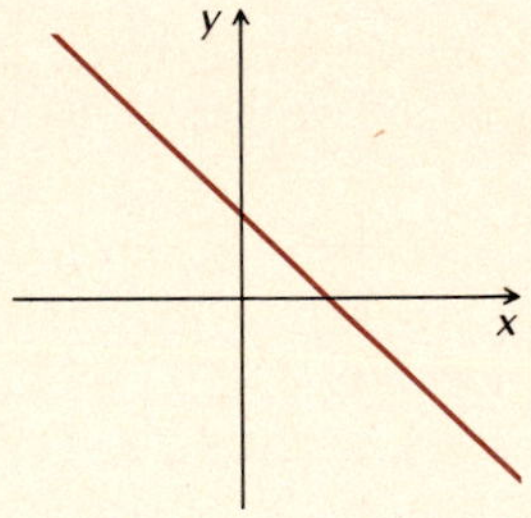

Determine whether the function is even, odd, or neither even nor odd.

31. $f(x) = 2x^2 + 4x$
32. $f(x) = -3x^3 + 2x$
33. $f(x) = 3x^4 - 4x^2$
34. $f(x) = 5x^2 + 2x^4 - 1$
35. $f(x) = 7x^3 + 4x - 2$
36. $f(x) = 4x$
37. $f(x) = |3x|$
38. $f(x) = x^{24}$
39. $f(x) = x^{17}$
40. $f(x) = x + \frac{1}{x}$
41. $f(x) = x - |x|$
42. $f(x) = \sqrt{x}$
43. $f(x) = \sqrt[3]{x}$
44. $f(x) = 7$
45. $f(x) = 0$
46. $f(x) = \sqrt[3]{x - 2}$
47. $f(x) = \sqrt{x^2 + 1}$
48. $f(x) = \frac{x^2 + 1}{x^3 - x}$

SYNTHESIS

Consider the figure on the right for Exercises 49–52.

49. Graph the reflection across the x-axis.

50. Graph the reflection across the y-axis.

51. Graph the reflection across the line $y = x$.

52. Graph the figure formed by reflecting each point through the origin.

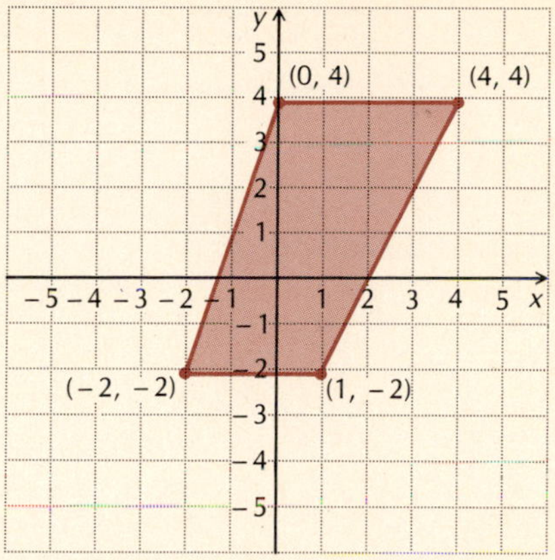

CHALLENGE

53. Consider symmetries with respect to the x-axis and the origin. Prove that symmetry with respect to any two of these implies symmetry with respect to the other.

3.6 Some Special Classes of Functions

OBJECTIVES

You should be able to:

1. Write interval notation for certain sets.
2. Given the graph of a function, determine whether it is periodic, and if it is periodic, determine its period.
3. Given the graph of a function, determine whether it is continuous over a specified interval, and indicate discontinuities.
4. Given the graph of a function, determine whether it is increasing, decreasing, or neither increasing nor decreasing, and find the equilibrium point of a pair of supply and demand functions.
5. Graph functions defined piecewise.

In this section, we consider some special classes of functions, such as *periodic*, *continuous*, *increasing*, *decreasing*, and those defined *piecewise*. Before doing so, however, we need to define a new kind of notation for certain sets of real numbers. This is called **interval notation**.

1 Interval Notation

The set of real numbers corresponds to the set of points on a line.

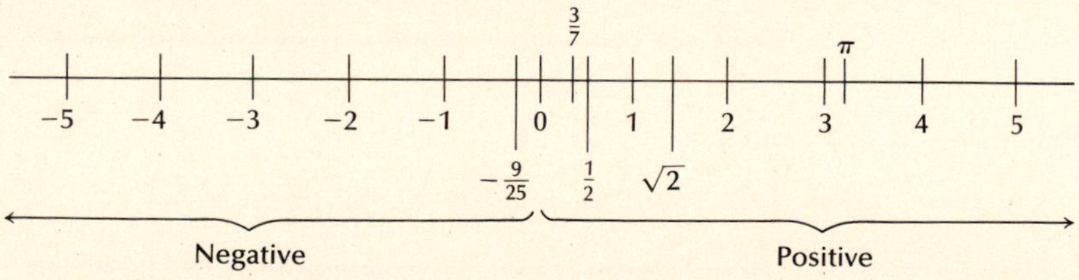

For real numbers a and b such that $a < b$ (a is to the left of b on a number line), we define the **open interval** (a, b) to be the set of numbers between, but not including, a and b. That is,

$$(a, b) = \text{the set of all numbers } x \text{ such that } a < x < b$$
$$= \{x | a < x < b\}.$$

The graph of (a, b) is shown above. The points a and b are called **endpoints**. The open circles and the parentheses indicate that the endpoints a and b are not included in the interval. Be careful not to confuse this notation with that of an ordered pair. The context of the writing should make the meaning clear. If not, we might say "the interval $(-2, 3)$." When we mean an ordered pair, we might say "the pair $(-2, 3)$."

1. Write interval notation for the graph.

a)
b)

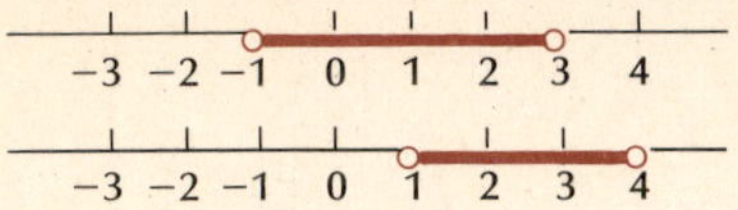

2. Write interval notation for the set.

a) $\{x|-2 < x < 3\}$

b) $\{x|0 < x < 1\}$

c) $\left\{x|-\frac{1}{4} < x < \sqrt{2}\right\}$

3. Write interval notation for the graph.

a)
b)
c)
d)

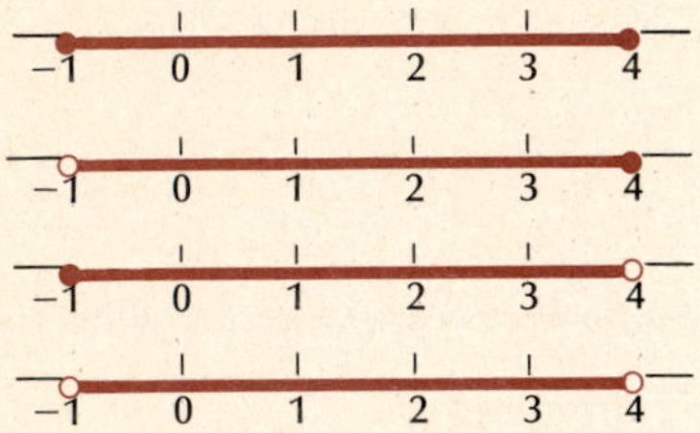

4. Write interval notation for the set.

a) $\{x|4 \leq x \leq 5\frac{1}{2}\}$

b) $\{x|-3 < x \leq 0\}$

c) $\{x|-\frac{1}{2} \leq x < \frac{1}{2}\}$

d) $\{x|-\pi < x < \pi\}$

5. Write interval notation for the graph.

a)
b)
c)
d)

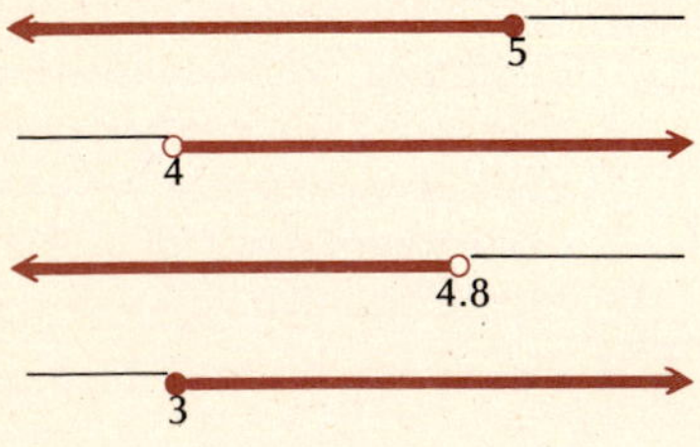

6. Write interval notation for the set.

a) $\{x|x \geq 8\}$

b) $\{x|x < -7\}$

c) $\{x|x > 10\}$

d) $\{x|x \leq 0.78\}$

DO EXERCISES 1 AND 2.

The **closed interval** $[a, b]$ is the set of numbers between and including a and b. That is,

$$[a, b] = \text{the set of all numbers } x \text{ such that } a \leq x \leq b$$
$$= \{x|a \leq x \leq b\}.$$

The graph of $[a, b]$ is shown above. The solid circles and the brackets indicate that a and b are included.

There are two kinds of **half-open intervals,** defined as follows:

$$(a, b] = \text{the set of all numbers } x \text{ such that } a < x \leq b$$
$$= \{x|a < x \leq b\}.$$

The open circle and the parenthesis indicate that a is not included. The solid circle and the bracket indicate that b is included. Also,

$$[a, b) = \text{the set of all numbers } x \text{ such that } a \leq x < b$$
$$= \{x|a \leq x < b\}.$$

The solid circle and the bracket indicate that a is included. The open circle and the parenthesis indicate that b is not included.

DO EXERCISES 3 AND 4.

Some intervals are of unlimited extent in one or both directions. In such cases, we use the infinity symbol ∞. For example,

$$[a, \infty) = \text{the set of all numbers } x \text{ such that } x \geq a.$$

Note that ∞ is not a number.

$$(a, \infty) = \text{the set of all numbers } x \text{ such that } x > a.$$

$$(-\infty, b] = \text{the set of all numbers } x \text{ such that } x \leq b.$$

$$(-\infty, b) = \text{the set of all numbers } x \text{ such that } x < b.$$

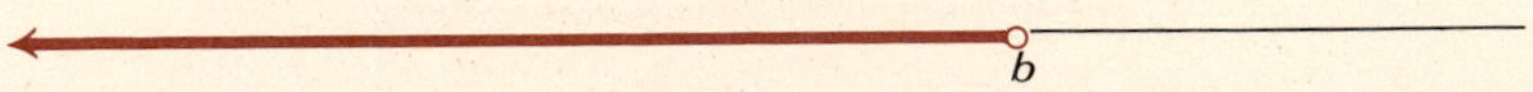

We can name the entire set of real numbers using $(-\infty, \infty)$.

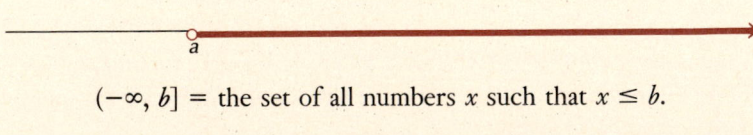

DO EXERCISES 5 AND 6.

Any point in an interval that is not an endpoint is an **interior point.**

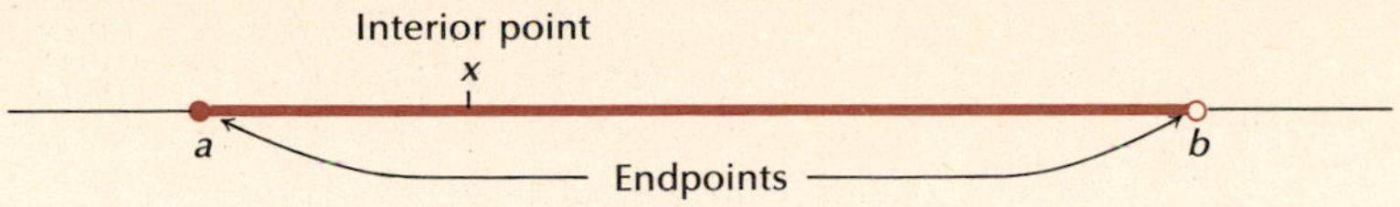

Note that all the points in an open interval are interior points.

Note that interval notation gives us an alternative notation for the solution set of an inequality. For example, we might say that the solution set of $x + 2 < 7$ is the set $\{x | x < 5\}$ or the interval $(-\infty, 5)$.

2 Periodic Functions

Certain functions with a repeating pattern are called **periodic.** Here are some examples.

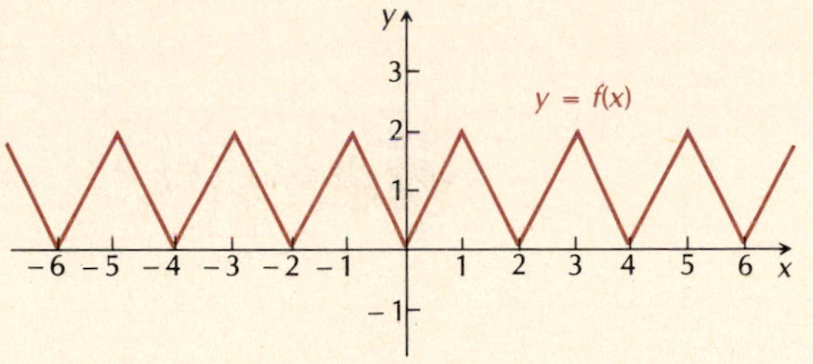

(a)

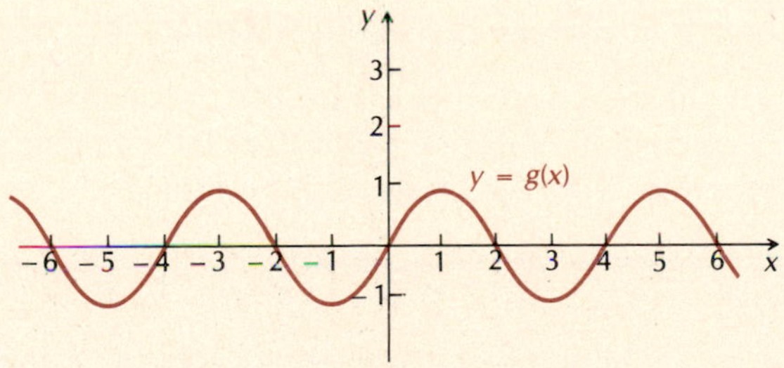

(b)

The function values of the function f repeat themselves every two units as we move from left to right. In other words, for any x, we have $f(x) = f(x + 2)$. To see this another way, think of the part of the graph between 0 and 2 and note that the rest of the graph consists of copies of it. In terms of translations, if we translate the graph 2 units to the left or right, the original graph will be obtained.

In the function g, the function values repeat themselves every 4 units. Hence $g(x) = g(x + 4)$ for any x, and if the graph is translated 4 units to the left or right, it will coincide with itself. Or, think of the part of the graph between 0 and 4. The rest of the graph consists of copies of it.

We say that f has a *period* of 2 and that g has a *period* of 4.

> DEFINITION
>
> **If a function f has the property that $f(x + p) = f(x)$ whenever x and $x + p$ are in the domain, where p is a positive constant, then f is said to be *periodic*. The smallest positive number p (if there is one) for which $f(x + p) = f(x)$ for all x is called the *period* for the function.**

7. For the function f whose graph was just considered, how does $f(x)$ compare with $f(x + 4)$? with $f(x + 6)$?

8. a) Determine whether this function is periodic.

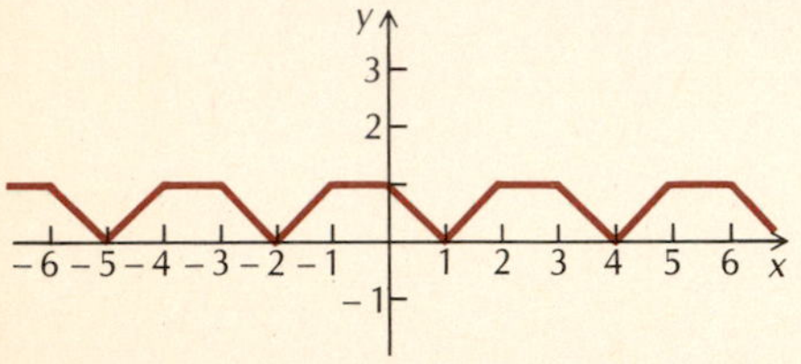

b) If so, what is the period?

9. For the function t, where $t(x) = 3$, how does $t(x)$ compare with $t(x + 1)$? By the definition, is t periodic? If so, does it have a period?

10. Is this function continuous

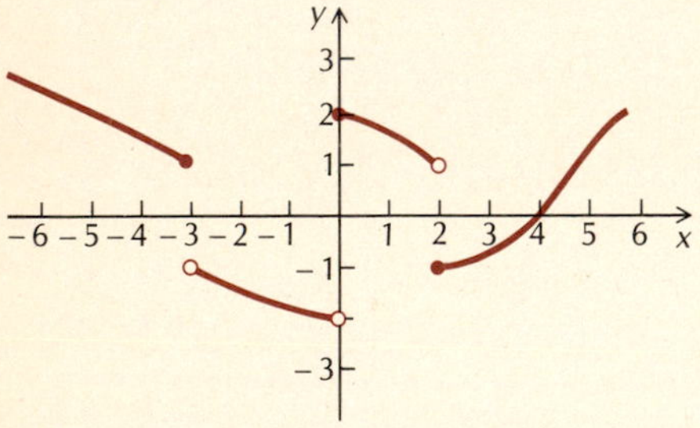

a) on the interval $(-1, 1)$?
b) on the interval $[-2, -1]$?
c) on the interval $(1, 2]$?
d) on the interval $(1, 2)$?
e) on the interval $(-5, 0)$?

11. Where are the discontinuities of the function above?

The period p can be thought of as the length of the shortest recurring interval.

DO EXERCISES 7–9.

3 Continuous Functions

Some functions have graphs that are continuous curves, without breaks or holes in them. Such functions are called **continuous functions.**

EXAMPLE 1 The function f below is continuous because it has no breaks, jumps, or holes in it. The function g has *discontinuities* where $x = -2$ and $x = 5$. The function g is continuous on the interval $[-2, 5)$ or on any interval contained therein. It is also continuous on other intervals, but it is not continuous on an interval such as $[-3, 1]$ or $(3, 7)$.

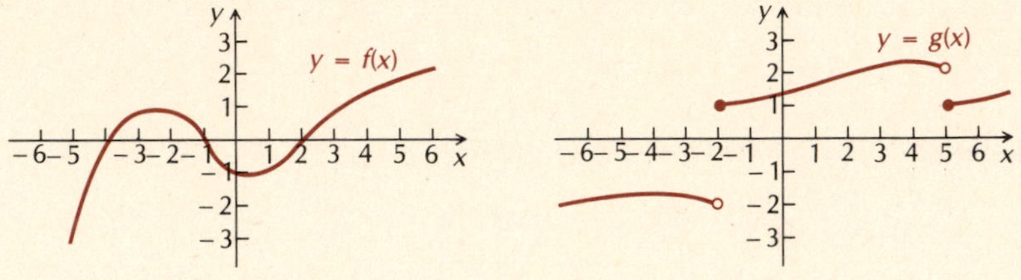

(a) (b)

DO EXERCISES 10 AND 11.

4 Increasing and Decreasing Functions

If the graph of a function rises from left to right, it is said to be an **increasing function.** If the graph of a function drops from left to right, it is said to be a **decreasing function.** This can be stated more formally.

> **DEFINITION**
>
> 1. A function f is an *increasing* function when for all a and b in the domain of f, if $a < b$, then $f(a) < f(b)$.
> 2. A function f is a *decreasing* function when for all a and b in the domain of f, if $a < b$, then $f(a) > f(b)$.

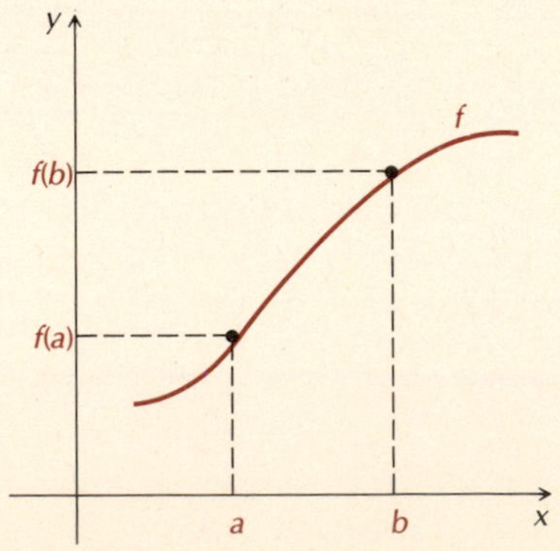

An increasing function
If $a < b$, then $f(a) < f(b)$.

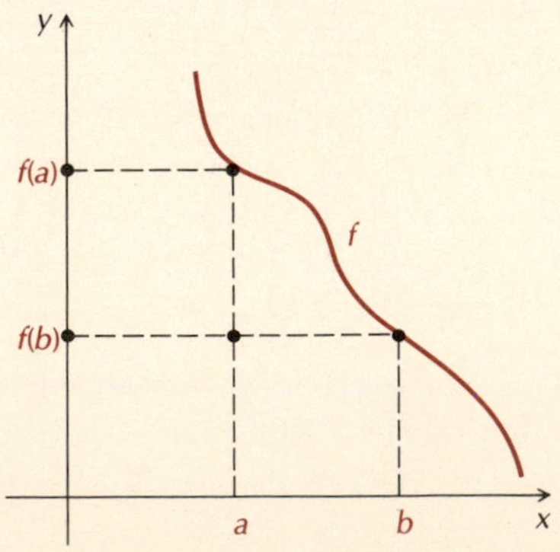

A decreasing function
If $a < b$, then $f(a) > f(b)$.

EXAMPLES Determine whether the function is increasing, decreasing, or neither.

2.

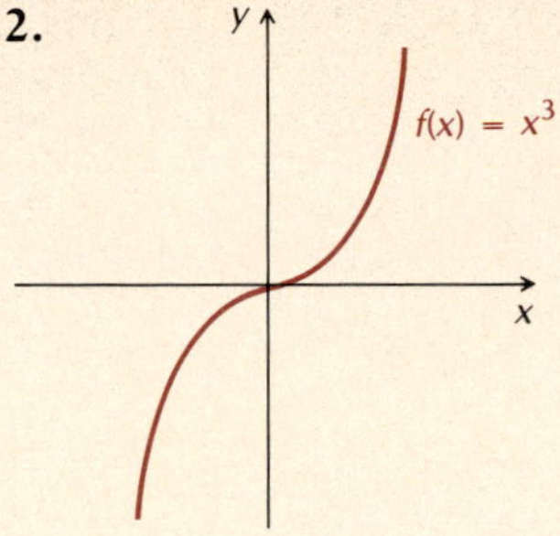

f is increasing.

3.

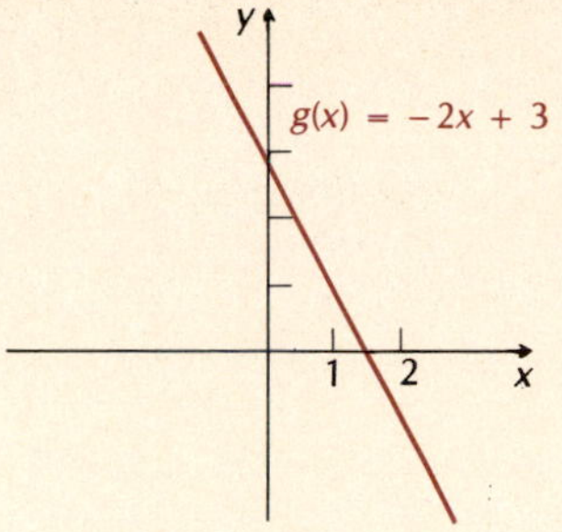

g is decreasing.

4.

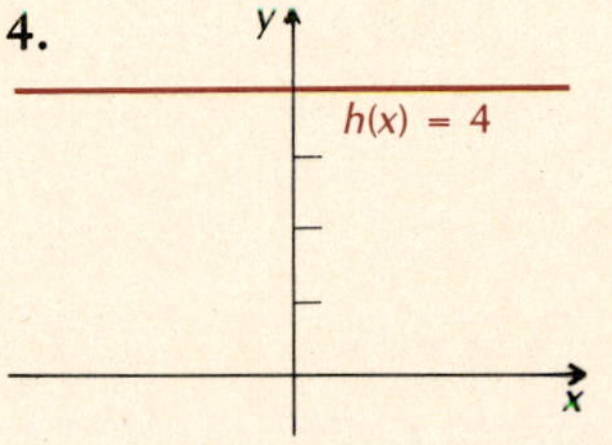

h is neither increasing nor decreasing—it is a constant function.

5.

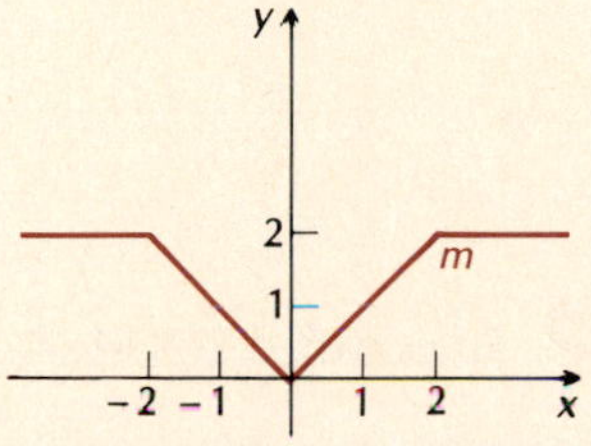

m is neither increasing nor decreasing.

In Example 5, the function m is neither increasing nor decreasing on the entire real line. But it is increasing on the interval $[0, 2]$ and decreasing on the interval $[-2, 0]$.

DO EXERCISES 12 AND 13.

EXAMPLE 6 *Supply and demand.* In economics, a **demand function,** D, is such that $D(x)$ is the price per unit of an item when x units are demanded by the consumer. A **supply function,** S, is such that $S(x)$ is the price per unit of an item at which the seller is willing to supply x units of a product to the consumer. The point of intersection of a supply-and-demand curve (x_E, p_E) is called the **equilibrium point.** The equilibrium price p_E is the point at which the amount x_E that the seller willingly supplies is the same as the amount that the consumer willingly demands. Find the equilibrium point for the demand and supply functions

$$D(x) = (x - 6)^2 \quad \text{and} \quad S(x) = x^2 + x + 10.$$

Solution To find the equilibrium point, we set $D(x) = S(x)$ and solve:

$$\begin{aligned} (x - 6)^2 &= x^2 + x + 10 \\ x^2 - 12x + 36 &= x^2 + x + 10 \\ -12x + 36 &= x + 10 \\ -13x &= -26 \\ x &= \frac{-26}{-13} \\ x &= 2. \end{aligned}$$

Thus, $x_E = 2$ (units). To find p_E, we substitute x_E into either $D(x)$ or $S(x)$. We use $D(x)$. Then

$$p_E = D(x_E) = D(2) = (2 - 6)^2 = (-4)^2 = \$16.$$

12. Determine whether the function is increasing, decreasing, or neither increasing nor decreasing.

a)

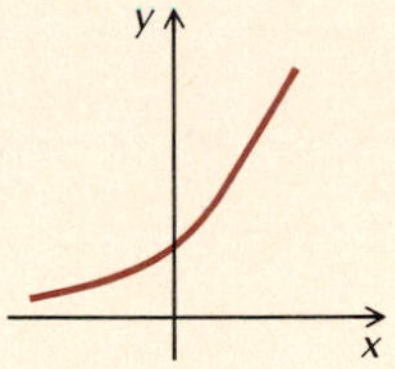

b)

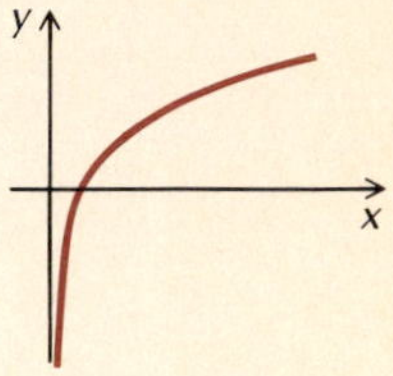

c)

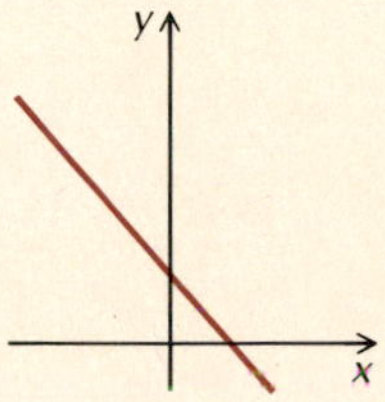

d)

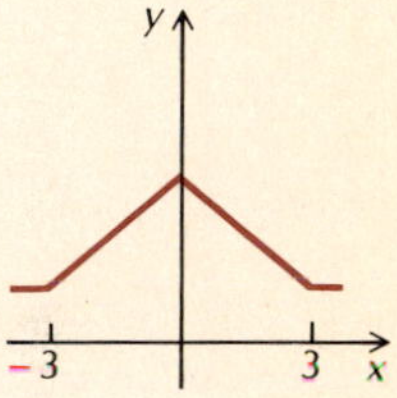

13. For the function in Margin Exercise 12(d), find an interval on which the function is (a) increasing; (b) decreasing.

14. Given $D(x) = (x - 5)^2$ and $S(x) = x^2 + x + 3$, find the equilibrium point.

Thus the equilibrium price is \$16 per unit and the equilibrium point is (2, \$16).

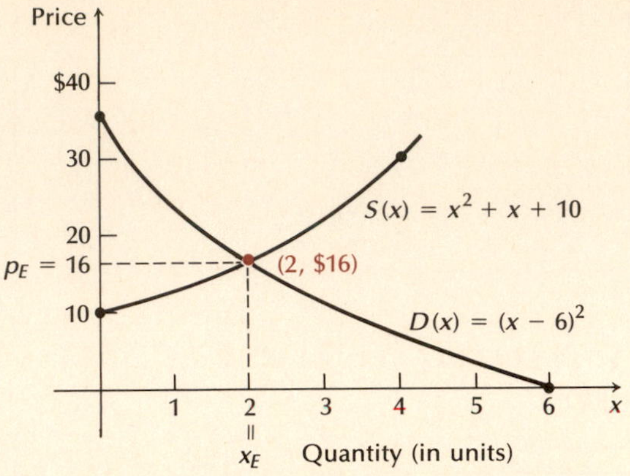

DO EXERCISE 14.

5 Functions Defined Piecewise

Sometimes functions are defined **piecewise**. That is, we have different output formulas for different parts of the domain.

EXAMPLE 7 Graph the function defined as follows:

$$f(x) = \begin{cases} 4, & \text{for } x \leq 0, \\ 4 - x^2, & \text{for } 0 < x \leq 2, \\ 2x - 6, & \text{for } x > 2. \end{cases}$$

(This means that for any input x less than or equal to 0, the output is 4.)

(This means that for any input x greater than 0 and less than or equal to 2, the output is $4 - x^2$.)

(This means that for any input x greater than 2, the output is $2x - 6$.)

Solution See the following graph.

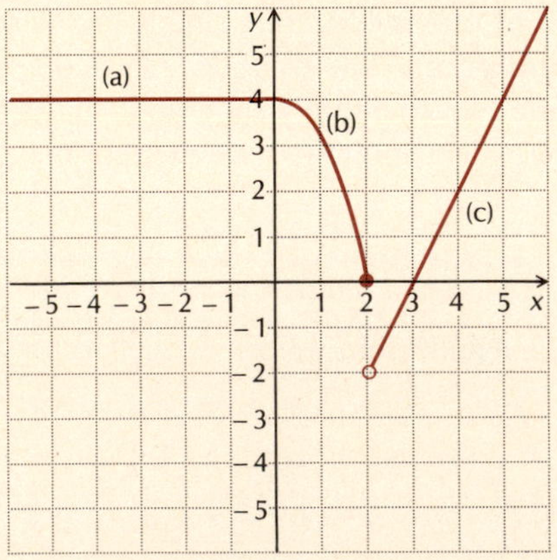

a) We graph $f(x) = 4$ for inputs less than or equal to 0 (that is, $x \leq 0$). Note that $f(x) = 4$ *only* for numbers x on the interval $(\infty, 0]$.

b) We graph $f(x) = 4 - x^2$ for inputs greater than 0 and less than or equal to 2 (that is, $0 < x \leq 2$). Note that $f(x) = 4 - x^2$ *only* on the interval $(0, 2]$.

c) We graph $f(x) = 2x - 6$ for inputs greater than 2 (that is, $x > 2$). Note that $f(x) = 2x - 6$ *only* for numbers x on the interval $(2, \infty)$.

DO EXERCISE 15.

EXAMPLE 8 Graph the function f defined as follows:

$$f(x) = \begin{cases} \dfrac{x^2 - 4}{x + 2}, & \text{for } x \neq 2, \\ 3, & \text{for } x = -2. \end{cases}$$

Solution When $x \neq -2$, the denominator of $(x^2 - 4)/(x + 2)$ is nonzero, so we can simplify:

$$\frac{x^2 - 4}{x + 2} = \frac{(x + 2)(x - 2)}{x + 2}$$
$$= x - 2.$$

Thus,

$$f(x) = x - 2 \quad \text{for } x \neq -2.$$

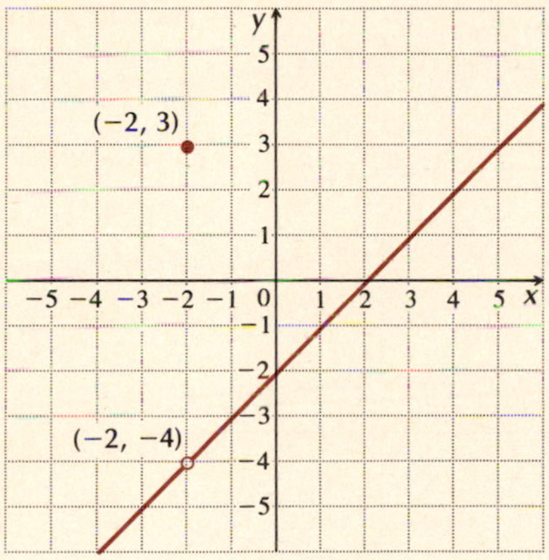

The graph of this part of the function consists of a line with a hole at the point $(-2, -4)$ indicated by the open circle. At $x = -2$, $f(-2) = 3$, so the point $(-2, 3)$ is plotted above $(-2, -4)$. ■

DO EXERCISE 16.

EXAMPLE 9 *The greatest integer function.* A function with importance in calculus and computer programming is the **greatest integer function** f, given by

$$f(x) = \text{INT}(x)$$

and defined as follows:

> **DEFINITION**
>
> **INT(x) = the greatest integer less than or equal to x.**

For example, INT(4.5) = 4, INT(−1) = −1, and INT(−3.8) = −4. In some texts, the notation $[x]$ is used for INT(x).

The greatest integer function can also be defined by a piecewise function

15. Graph the function defined as

$$f(x) = \begin{cases} x + 3, & \text{for } x \leq -2, \\ 1, & \text{for } -2 < x \leq 3, \\ x^2 - 10, & \text{for } 3 < x. \end{cases}$$

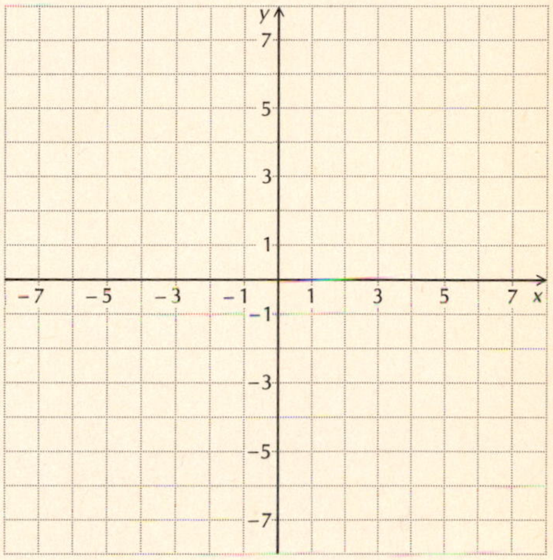

16. Graph the function f defined as

$$f(x) = \begin{cases} \dfrac{x^2 - 4}{x - 2}, & \text{for } x \neq 2, \\ -1, & \text{for } x = 2. \end{cases}$$

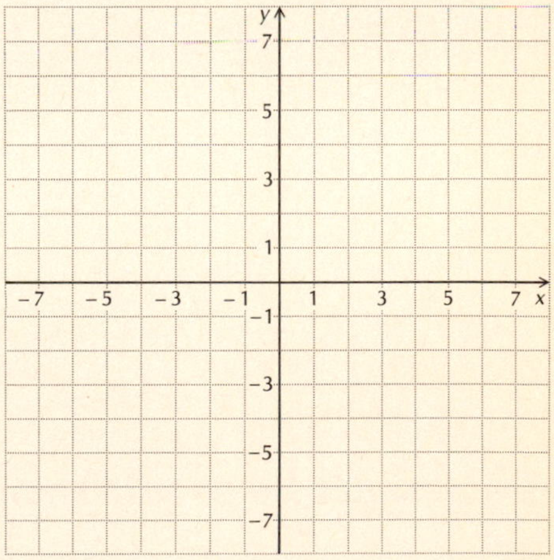

17. a) Find INT(7.3), INT(−2), INT(1.986734199), INT(3), INT(−2.6), and INT(−0.9).

b) Graph $f(x) = \text{INT}(x) + 1$.

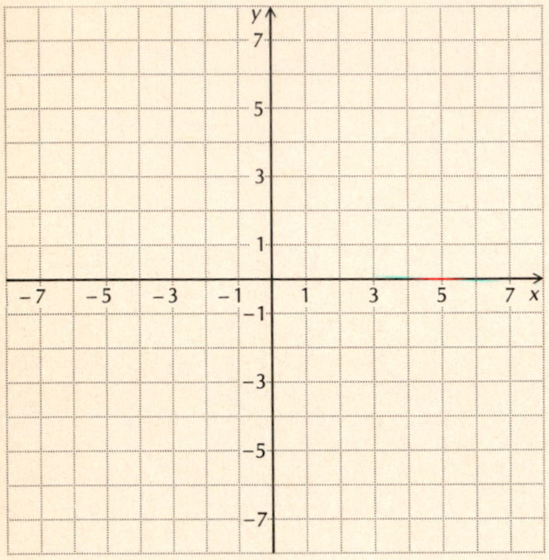

c) Graph $f(x) = \text{INT}(x + 1)$.

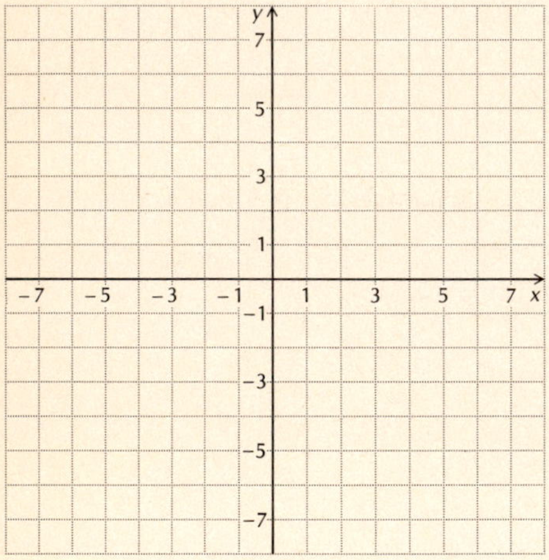

with an infinite number of statements:

$$f(x) = \text{INT}(x) = \begin{cases} \quad\vdots \\ -3 & \text{if } -3 \le x < -2 \\ -2 & \text{if } -2 \le x < -1 \\ -1 & \text{if } -1 \le x < 0 \\ 0 & \text{if } 0 \le x < 1 \\ 1 & \text{if } 1 \le x < 2 \\ 2 & \text{if } 2 \le x < 3 \\ 3 & \text{if } 3 \le x < 4 \\ \quad\vdots \end{cases}$$

Graph the greatest integer function.

Solution

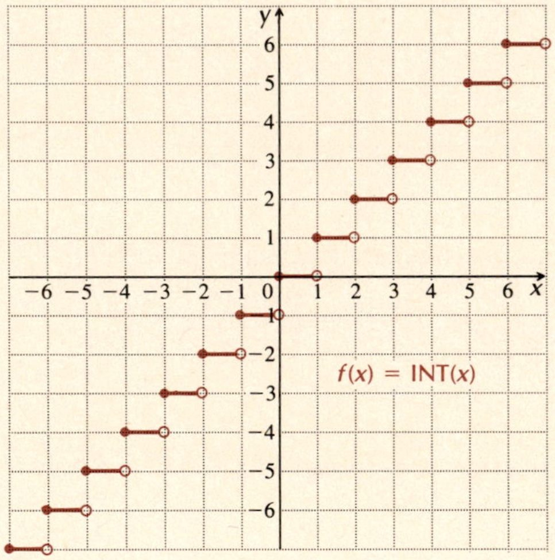

DO EXERCISE 17.

EXERCISE SET 3.6

1 Write interval notation for the graph.

1. 

2. 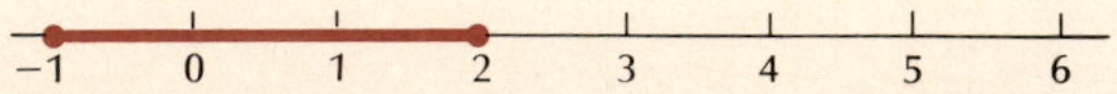

3. 

4. 

5. 

6. 

7. 

8. 

Write interval notation for the set.

9. $\{x|-3 \leq x \leq 3\}$ **10.** $\{x|-4 < x < 4\}$ **11.** $\{x|-14 \leq x < -11\}$ **12.** $\{x|6 < x \leq 20\}$

13. $\{x|x \leq -4\}$ **14.** $\{x|x > -5\}$ **15.** $\{x|x < 3.8\}$ **16.** $\{x|x \geq \sqrt{3}\}$

2

17. Determine whether the function is periodic.

a)

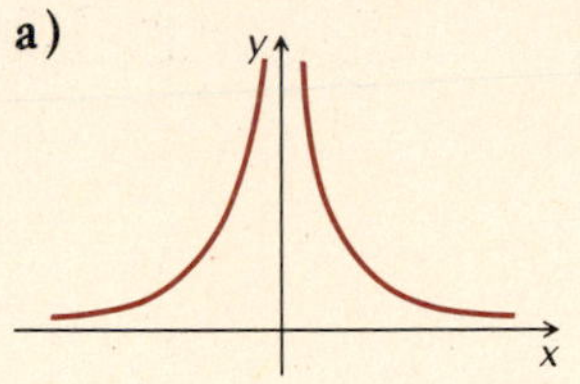

b)

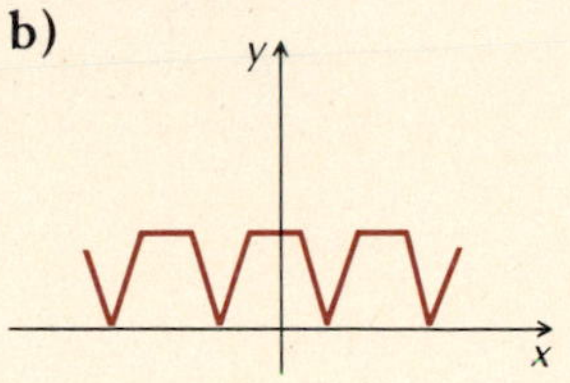

c)

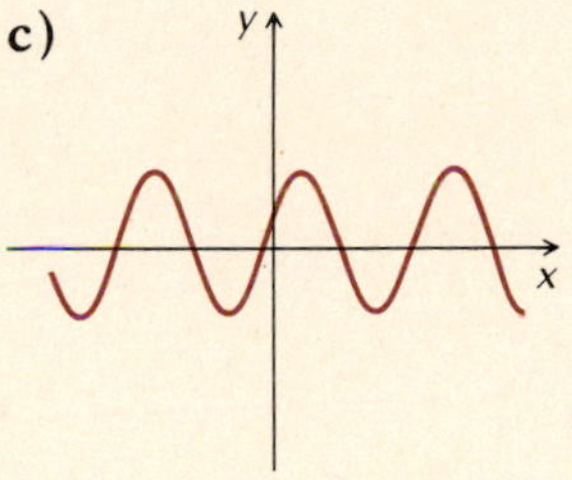

d)

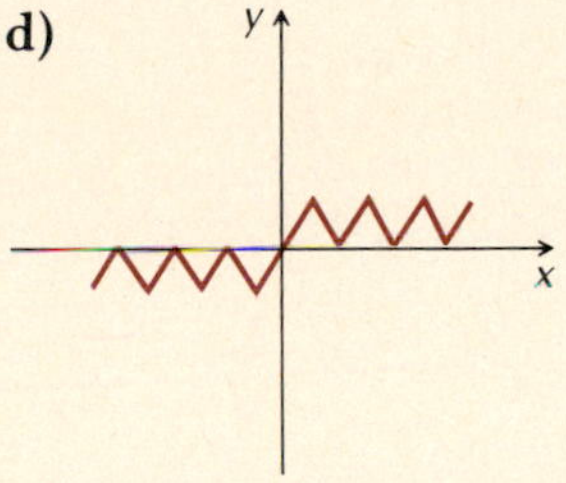

18. Determine whether the function is periodic.

a)

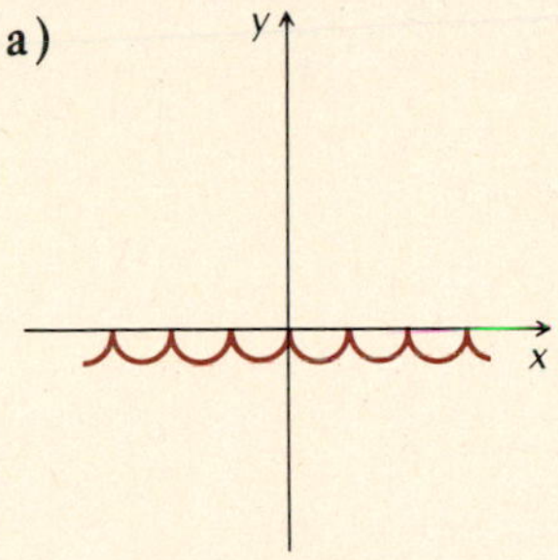

b)

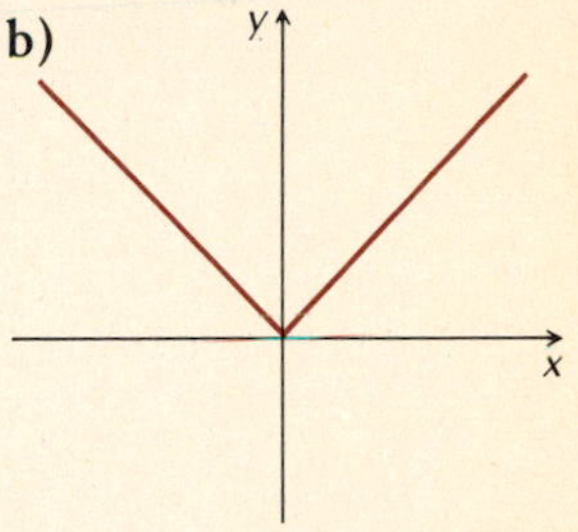

c)

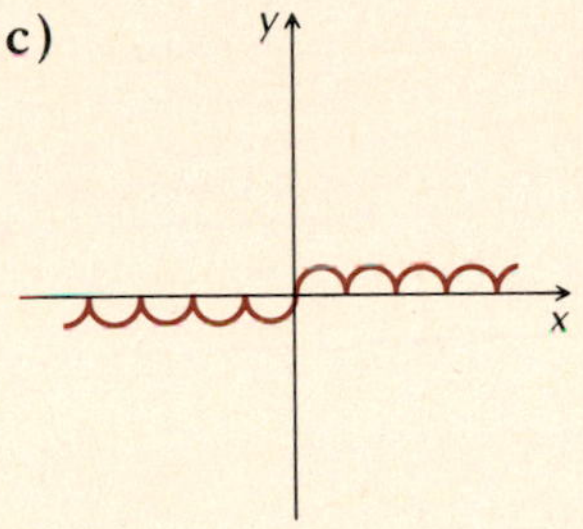

d)

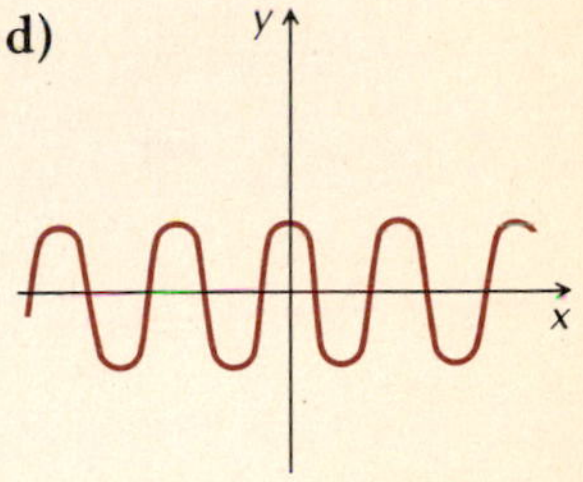

19. What is the period of this function?

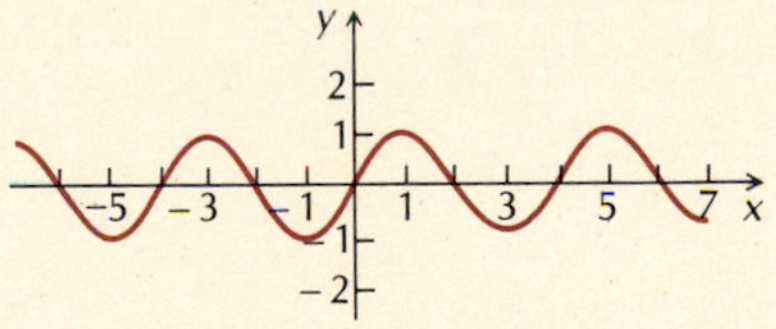

20. What is the period of this function?

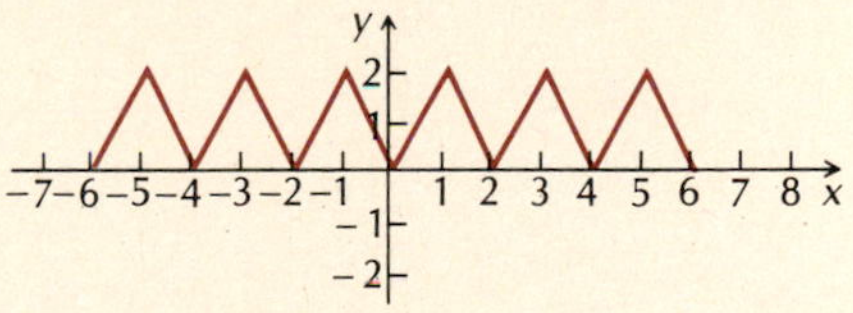

3

21. Is this function continuous

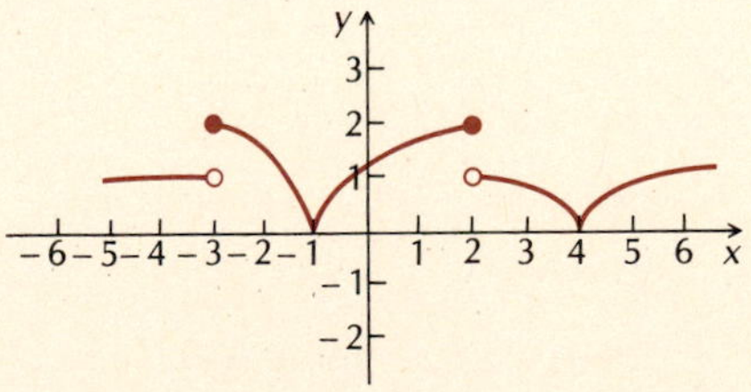

a) on the interval [0, 2]?
b) on the interval (−2, 0)?
c) on the interval [0, 3)?
d) on the interval [−3, −1]?
e) on the interval (−3, −1]?

22. Is this function continuous

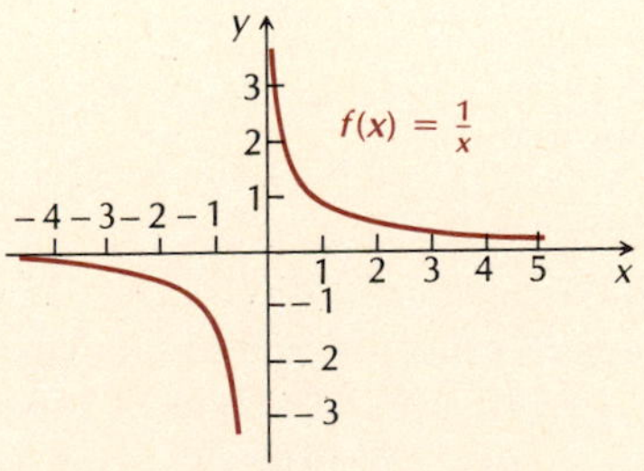

a) on the interval [−3, −1]?
b) on the interval (1, 4)?
c) on the interval [−1, 1]?
d) on the interval [−2, 4]?
e) on the interval (0, 1]?

23. Where are the discontinuities of the function in Exercise 21?

24. Where are the discontinuities of the function in Exercise 22?

4

25. Determine whether the function is increasing, decreasing, or neither increasing nor decreasing.

a) 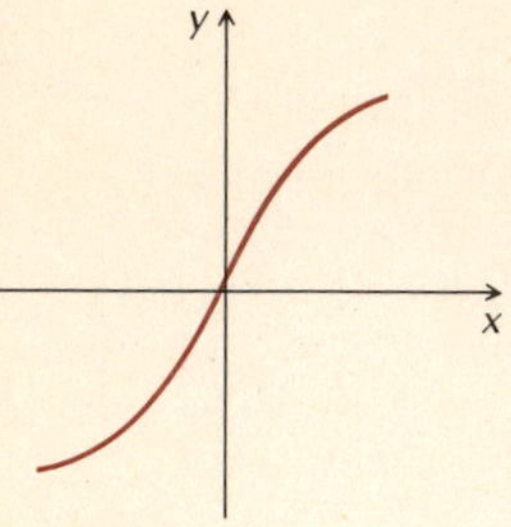

b) 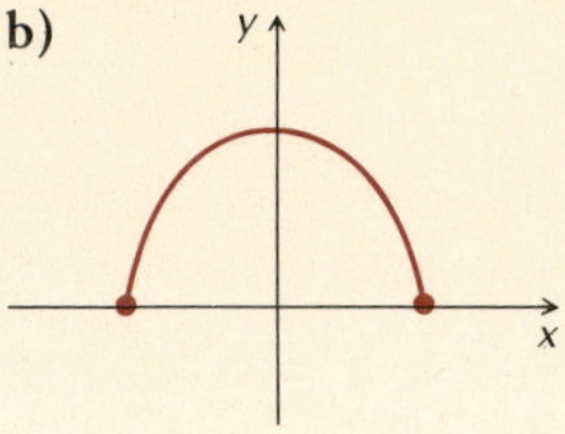

c) 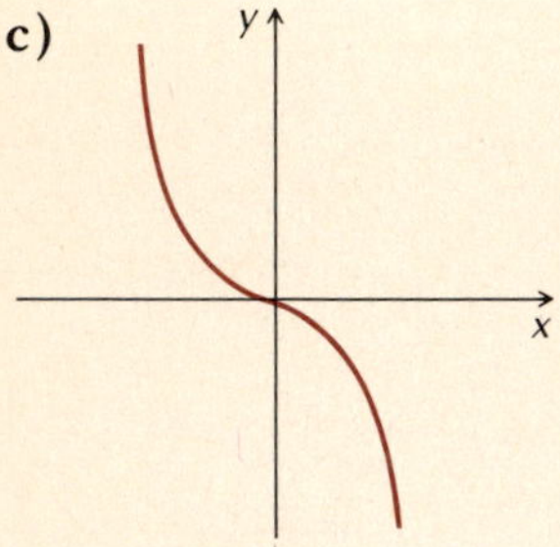

d) 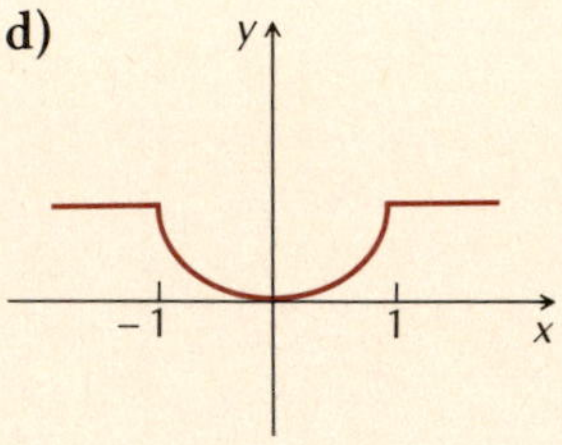

26. Determine whether the function is increasing, decreasing, or neither increasing nor decreasing.

a) 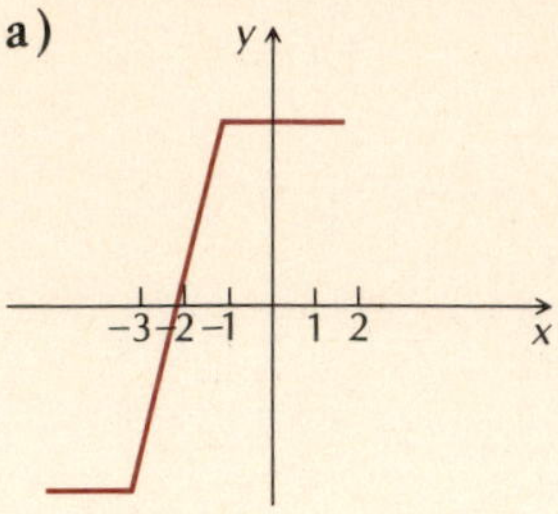

b) 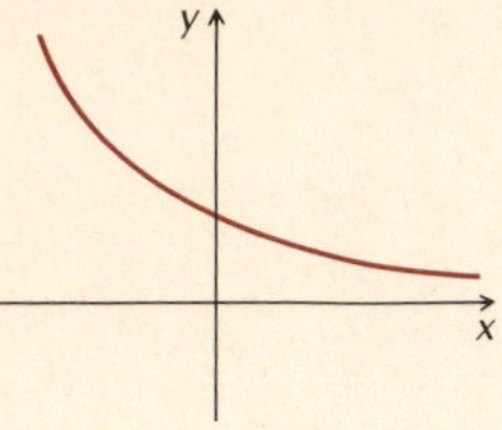

c) 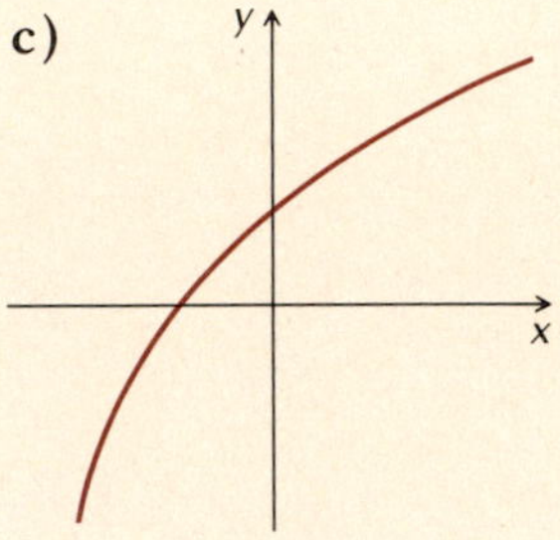

d) 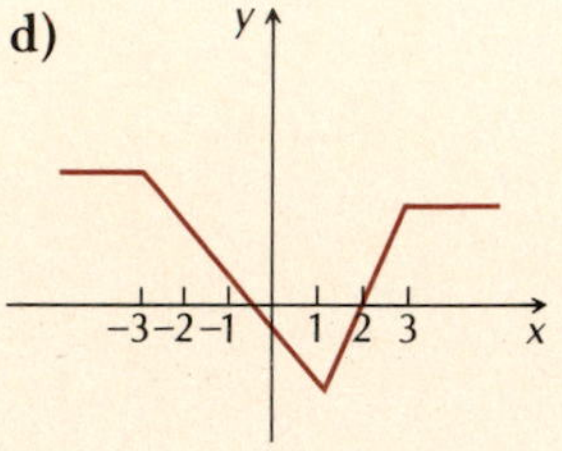

Find the equilibrium point for each of the following demand and supply functions.

27. $D(x) = -2x + 8,\ S(x) = x + 2$

28. $D(x) = -\frac{5}{6}x + 10,\ S(x) = \frac{1}{2}x + 2$

29. $D(x) = (x - 3)^2,\ S(x) = x^2 + 2x + 1$

30. $D(x) = (x - 4)^2,\ S(x) = x^2 + 2x + 6$

31. $D(x) = (x - 4)^2,\ S(x) = x^2$

32. $D(x) = (x - 6)^2,\ S(x) = x^2$

5 Graph.

33. $f(x) = \begin{cases} 1, & \text{for } x < 0, \\ -1, & \text{for } x \geq 0 \end{cases}$

34. $f(x) = \begin{cases} 2, & \text{for } x \text{ an integer}, \\ -2, & \text{for } x \text{ not an integer} \end{cases}$

35. $f(x) = \begin{cases} 3, & \text{for } x \leq -3, \\ |x|, & \text{for } -3 < x \leq 3, \\ -3, & \text{for } x > 3 \end{cases}$

36. $f(x) = \begin{cases} -2x - 6, & \text{for } x \leq -2, \\ 2 - x^2, & \text{for } -2 < x < 2, \\ 2x - 6, & \text{for } x \geq 2 \end{cases}$

37. $f(x) = \begin{cases} \dfrac{x^2 - 1}{x - 1}, & \text{for } x \neq 1, \\ -2, & \text{for } x = 1 \end{cases}$

38. $f(x) = \begin{cases} \dfrac{x^2 - 9}{x + 3}, & \text{for } x \neq -3, \\ 4, & \text{for } x = -3 \end{cases}$

39. *The postage function.* Postage rates are as follows: 25 cents for the first ounce plus 20 cents for each additional ounce or fraction thereof. Thus if x is the weight of a letter in ounces, then $p(x)$ is the cost of mailing the letter, where

$$p(x) = \begin{cases} 25 \text{ cents}, & \text{if } 0 < x \leq 1, \\ 45 \text{ cents}, & \text{if } 1 < x \leq 2, \\ 65 \text{ cents}, & \text{if } 2 < x \leq 3, \end{cases}$$

and so on, up to 12 ounces, after which postal cost also depends on distance. Graph this function for x such that $0 < x \leq 12$.

40. Graph

$$f(x) = \begin{cases} 3 + x, & \text{for } x \leq 0, \\ \sqrt{x}, & \text{for } 0 < x < 4, \\ x^2 - 4x - 1, & \text{for } x \geq 4. \end{cases}$$

Graph.

41. $f(x) = \text{INT}(x - 2)$

42. $f(x) = \text{INT}(x) - 1$

43. $f(x) = \text{INT}(x) + 2$

44. $f(x) = \text{INT}(x - 1)$

SYNTHESIS

45. For the function in Exercise 25(d), find an interval on which the function is (a) increasing; (b) decreasing.

46. For the function in Exercise 26(d), find an interval on which the function is (a) increasing; (b) decreasing.

47. Graph each function. Then determine whether it is increasing, decreasing, or neither increasing nor decreasing.

a) $f(x) = 3x + 4$ b) $f(x) = -3x + 4$
c) $f(x) = x^2 + 1$ d) $f(x) = -2$
e) $f(x) = x^3 + 1$ f) $f(x) = |x|$

48. Graph each function. Then determine whether it is increasing, decreasing, or neither increasing nor decreasing.

a) $f(x) = -2x - 3$ b) $f(x) = 2x - 3$
c) $f(x) = 3x^2$ d) $f(x) = \sqrt{3}$
e) $f(x) = |x| + 2$ f) $f(x) = x^3 - 2$

Graph.

49. $f(x) = \dfrac{|x|}{x}$

50. $g(x) = \dfrac{x}{|x|}$

51. If $\text{INT}(x) = 4$, what are the possible inputs for x?

52. If $\text{INT}(x) = -3$, what are the possible inputs for x?

53. *Bowling average.* A person's bowling average is defined as follows: *Bowling average* $= \text{INT}(p/n)$, where p = the total number of pins and n = the total number of games bowled. Find the following bowling averages.

a) 547 pins in 3 games
b) 4621 pins in 27 games

54. *Decimal place function.* We define a *decimal place function* f_n as follows: Given a number x, find decimal notation for x. Then $f_n(x)$ = the digit in the nth decimal place. For example,

$$f_4(3/5) = f_4(0.600000\ldots) = 0,$$
$$f_4(17/21) = f_4(0.8095238\ldots) = 5,$$
$$f_4(-89/56) = f_4(-1.5892857\ldots) = 2, \text{ and so on.}$$

a) Find

$$f_4(7/8),\ f_4(17/23),\ f_4(-49/37), \text{ and } f_4(\sqrt{11}).$$

b) Determine the domain and the range of f_4.
c) Find $f_4(\pi) + f_5(\pi) + f_6(\pi) + f_7(\pi)$.

CHALLENGE

55. Graph the equation $\text{INT}(y) = \text{INT}(x)$. Is this the graph of a function?

3.7 The Algebra of Functions

OBJECTIVES

You should be able to:

1 **Compute function values for the sum, difference, product, and quotient of two functions, and determine their domains.**

2 **Find the composition of two functions f and g, giving formulas for $f \circ g(x)$ and $g \circ f(x)$, find the domains of the composition, and decompose a function as a composition of two functions.**

Next we look at methods of combining two functions to obtain a new function.

1 Sums, Differences, Products, and Quotients of Functions

Consider two functions f and g given as follows:

$$f(x) = x^2 - 5 \quad \text{and} \quad g(x) = x + 7.$$

Suppose we have an input 2. This number is in the domain of both functions. That is, it can be substituted into each formula. Let us find the outputs $f(2)$ and $g(2)$:

$$f(2) = 2^2 - 5 = 4 - 5 = -1,$$
$$g(2) = 2 + 7 = 9.$$

We can then add the outputs: $f(2) + g(2) = -1 + 9 = 8$. We start with an input 2 and we end up with exactly one output number, 8. Doing this for any input x that is in the domain of *both* functions creates a new function $f + g$, called the **sum** of f and g, described by $f(x) + g(x)$.

Suppose we subtract the function values: $f(2) - g(2) = -1 - 9 = -10$. Doing this for any input x that is in the domain of *both* functions creates a new function $f - g$, called the **difference** of f and g, described by $f(x) - g(x)$.

Suppose we multiply the function values: $f(2) \cdot g(2) = (-1)9 = -9$. Doing this for any input x that is in the domain of *both* functions creates a new function fg, called the **product** of f and g, described by $f(x)g(x)$.

Suppose we divide the function values: $f(2)/g(2) = -1/9 = -\frac{1}{9}$. Doing this for any input x that is in the domain of *both* functions and for which $g(x)$ is not zero creates a new function f/g, called the **quotient** of f and g, described by $f(x)/g(x)$.

DEFINITION

From any functions f and g, we can form new functions defined as follows, assuming that x is in the domain of f and the domain of g.

1. **The *sum* of f and g:** $(f + g)(x) = f(x) + g(x)$.
2. **The *difference* of f and g:** $(f - g)(x) = f(x) - g(x)$.
3. **The *product* of f and g:** $fg(x) = f(x)g(x)$.
4. **The *quotient* of f and g:** $(f/g)(x) = f(x)/g(x)$, **where** $g(x) \neq 0$.

EXAMPLES Given f and g described by $f(x) = 8 - x$ and $g(x) = \sqrt{2x + 3}$.

1. Find $(f + g)(5)$ and $(f + g)(-4)$.

$$(f + g)(5) = f(5) + g(5) = (8 - 5) + (\sqrt{2(5) + 3})$$
$$= 3 + \sqrt{13},$$
$$(f + g)(-4) = f(-4) + g(-4) = [8 - (-4)] + \sqrt{2(-4) + 3}$$
$$= 12 + \sqrt{-5},$$ **which does not exist as a real number.**

Note that $(f + g)(-4)$ does not exist because -4 is *not* in the domain of g.

2. Find $(f - g)(7)$ and $(f - g)(-10)$.

$$(f - g)(7) = f(7) - g(7) = (8 - 7) - \sqrt{2(7) - 3} = 1 - \sqrt{17}$$

$(f - g)(-10)$ does not exist because -10 is not in the domain of g since it produces a negative radicand.

3. Find $fg(7)$ and $gg(7)$.

$$fg(7) = f(7)g(7) = (8 - 7)(\sqrt{2(7) + 3}) = \sqrt{17},$$
$$gg(7) = g(7)g(7) = [g(7)]^2 = [\sqrt{17}]^2 = 17$$

4. Find $(f/g)(5)$, $(g/f)(5)$, and $(f/g)(-\frac{3}{2})$.

$$(f/g)(5) = \frac{f(5)}{g(5)} = \frac{8 - 5}{\sqrt{2(5) + 3}} = \frac{3}{\sqrt{13}}, \quad \text{or} \quad \frac{3\sqrt{13}}{13},$$

$$(g/f)(5) = \frac{g(5)}{f(5)} = \frac{\sqrt{13}}{3}$$

$$(f/g)\left(-\frac{3}{2}\right) = \frac{f\left(-\frac{3}{2}\right)}{g\left(-\frac{3}{2}\right)} = \frac{8 - \left(-\frac{3}{2}\right)}{\sqrt{2\left(-\frac{3}{2}\right) + 3}} = \frac{\frac{19}{2}}{0}, \quad \text{which does not exist.}$$

5. Find the domain of $f + g$, $f - g$, fg, and f/g.

For the function f above, the domain is the set of all real numbers. For the function g, the domain is the set of all real numbers for which $2x + 3 \geq 0$, which is the set $\{x|x \geq -3/2\}$ or the interval $[-3/2, \infty)$. The domain of $f + g, f - g$, and fg is therefore the interval $[-3/2, \infty)$. The domain of f/g must exclude any input that makes the denominator 0, namely, $-3/2$. Thus the domain of f/g is the interval $(-3/2, \infty)$. ■

DO EXERCISE 1.

It makes sense that if we were to be doing lots of computations with sums, differences, products, and quotients of functions f and g, we might want to find general formulas.

EXAMPLES Given f and g described by $f(x) = x^2 - 3$ and $g(x) = x + 4$, find the following.

6. $(f + g)(x) = f(x) + g(x) = (x^2 - 3) + (x + 4) = x^2 + x + 1$

7. $(f - g)(x) = f(x) - g(x) = (x^2 - 3) - (x + 4) = x^2 - x - 7$

8. $fg(x) = f(x)g(x) = (x^2 - 3)(x + 4) = x^3 + 4x^2 - 3x - 12$

9. $gg(x) = [g(x)]^2 = (x + 4)^2 = x^2 + 8x + 16$

10. $(f/g)(x) = \dfrac{f(x)}{g(x)} = \dfrac{x^2 - 3}{x + 4}$, where $x \neq -4$.

11. Find the domain of $f + g, f - g, fg$, and f/g.

The domain of f is the set of all real numbers. The domain of g is the set of all real numbers. The domain of $f + g, f - g$, and fg is the set of numbers in the intersection of the two domains—that is, the set of numbers in both domains, which is again the set of real numbers. For f/g, we must exclude -4 since $g(-4) = 0$. Thus the domain of f/g is $\{x|x$ is a real number and $x \neq -4\}$. ■

If a function is described by a polynomial, such as $f(x) = x^2 - 3$ and $g(x) = x + 4$, we say that it is a **polynomial function**. Note in Examples 6–10 that the sum, difference, and product of polynomial functions are also polynomial functions, but the quotient may not be. The quotient of two polynomial functions is called a **rational function**.

DO EXERCISE 2.

Let us now consider an economic application.

EXAMPLE 12 *Total cost, revenue, and profit.* In economics, we are frequently concerned with functions defined as follows:

Total cost $= C(x) =$ the total cost of producing x units of a product (usually considered in some time period);

Total revenue $= R(x) =$ the total revenue from the sale of x units of a product;

Total profit $= P(x) =$ the total profit from the production and sale of x units of a product $= R(x) - C(x)$.

Given

$$R(x) = 40x - 0.1x^2,$$
$$C(x) = 2x^3 - 12x^2 + 40x + 10,$$

1. Given $f(x) = 1/(x - 2)$ and $g(x) = x^2 - 25$, find each of the following, if possible.

a) $(f + g)(5)$ and $(f + g)(2)$

b) $(f - g)(7)$ and $(f - g)(-3)$

c) $fg(-3)$ and $gg(1)$

d) $(f/g)(1)$, $(g/f)(5)$, $(f/g)(5)$, and $(f/g)(-5)$

e) Find the domains of $f, g, f + g, f - g, fg$, and f/g.

2. Given $f(x) = x^2 + 3$ and $g(x) = x^2 - 9$, find each of the following.

a) $(f + g)(x)$

b) $(f - g)(x)$

c) $fg(x)$

d) $(f/g)(x)$

e) $ff(x)$

f) the domains of $f + g, f - g, fg$, and f/g

3. Given

$$R(x) = 50x - 0.5x^2,$$

and

$$C(x) = 10x + 3,$$

find each of the following.

a) $P(x)$

b) $R(40)$, $C(40)$, $P(40)$

find each of the following.

a) $P(x)$

b) $R(2)$, $C(2)$, and $P(2)$

Solution

a) $P(x) = R(x) - C(x)$
$= (40x - 0.1x^2) - (2x^3 - 12x^2 + 40x + 10)$
$= -2x^3 + 11.9x^2 - 10$

b) $R(2) = 40(2) - 0.1(2)^2 = 80 - 0.4 = \$79.60$
(the total revenue from the sale of 2 units);
$C(2) = 2(2)^3 - 12(2)^2 + 40(2) + 10 = \$58$
(the total cost of producing 2 units);
$P(2) = R(2) - C(2) = \$79.60 - \$58 = \$21.60$
(the total profit from the production and sale of 2 units) ■

DO EXERCISE 3.

2 The Composition of Functions

In the real world, functions frequently arise in which some variable depends on the choice of a third variable. For instance, the number of employees hired by a firm may depend on the firm's profits, which may in turn depend on the number of items the firm produces. Functions like this are called **composite functions.**

There is a function g that gives a correspondence between women's shoe sizes in the United States and those in Italy. The function is given by $g(x) = 2(x + 12)$, where x is a shoe size in the United States and $g(x)$ is a shoe size in Italy. For example, a shoe size of 4 in the United States corresponds to a shoe size of $g(4) = 2(4 + 12)$, or 32, in Italy.

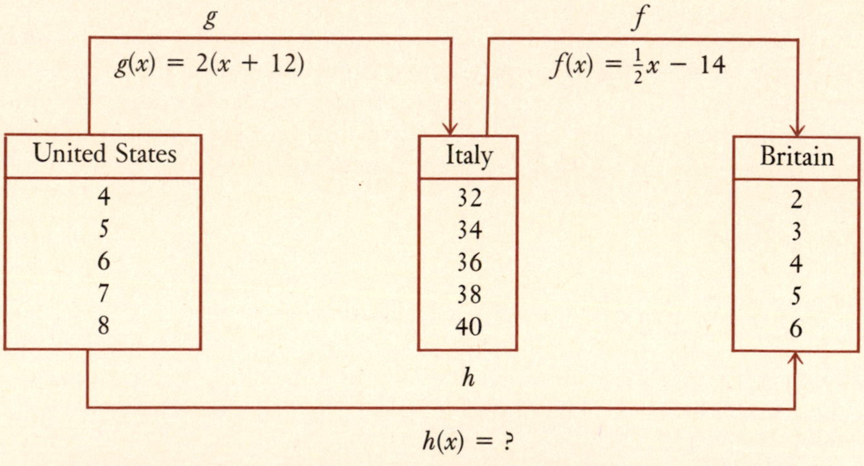

There is also a function that gives a correspondence between women's shoe sizes in Italy and those in Britain. The function is given by $f(x) = \frac{1}{2}x - 14$, where x is a shoe size in Italy and $f(x)$ is the corresponding shoe size in Britain. For example, a shoe size of 32 in Italy corresponds to a shoe size of $f(32) = \frac{1}{2}(32) - 14$, or 2, in Britain.

It seems reasonable to assume that a shoe size of 4 in the United States corresponds to a shoe size of 2 in Britain and that there is a function h that describes this correspondence. Can we find a formula for h? If we look at the tables, we might guess that such a formula is $h(x) = x - 2$, and that is indeed correct. For more complicated formulas, however, we would need to do some algebra.

A shoe size x in the United States corresponds to a shoe size $g(x)$ in Italy, where

$$g(x) = 2(x + 12).$$

Now $2(x + 12)$ is a shoe size in Italy. If we replace x in $f(x)$ by $2(x + 12)$, we can find the corresponding shoe size in Britain:

$$f(g(x)) = \tfrac{1}{2}[2(x + 12)] - 14 = \tfrac{1}{2}[2x + 24] - 14$$
$$= x + 12 - 14 = x - 2.$$

This gives a formula for h: $h(x) = x - 2$. Thus a shoe size of 4 in the United States corresponds to a shoe size of $h(4) = 4 - 2$, or 2, in Britain. The function h is called the **composition** of f and g and is symbolized $f \circ g$.

DEFINITION

The *composite* function $f \circ g$, the *composition* of f and g, is defined as

$$f \circ g(x) = f(g(x)).$$

We can visualize the composition of functions as follows. To find $f \circ g(x)$, we substitute $g(x)$ for x in $f(x)$.

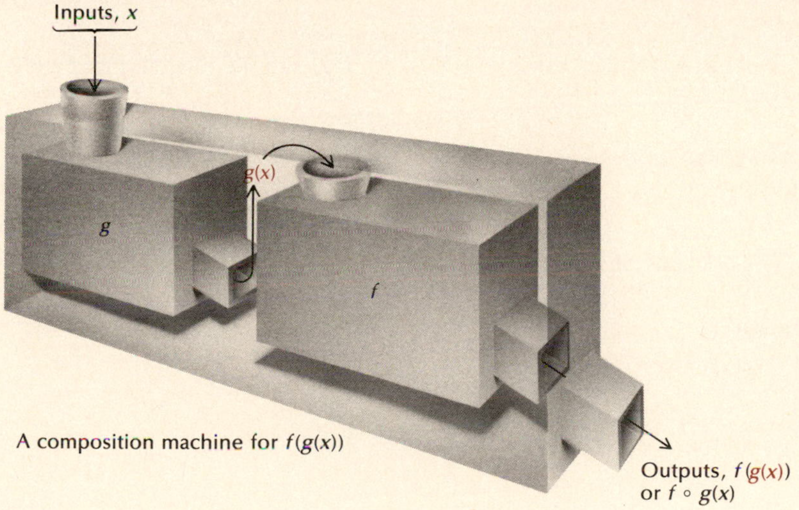

A composition machine for $f(g(x))$

EXAMPLE 13 Given $f(x) = 3x$ and $g(x) = 1 + x^2$, find each of the following.

a) $f \circ g(5)$ and $g \circ f(5)$

b) $f \circ g(x)$ and $g \circ f(x)$

Solution Consider each function separately:

$$f(x) = 3x \quad \text{This function multiplies each input by 3.}$$

and

$$g(x) = 1 + x^2. \quad \text{This function adds 1 to the square of each input.}$$

a) To find $f \circ g(5)$, we first find $g(5)$ by substituting into the formula for g: We square 5 and add it to 1, to get 26. Then we substitute the result, 26, into the formula for f: We multiply 26 by 3.

$$f \circ g(5) = f(g(5))$$
$$= f(1 + 5^2) = f(26)$$
$$= 3(26) = 78$$

To find $g \circ f(5)$, we first find $f(5)$ by substituting into the formula for f: We multiply 5 by 3, to get 15. Then we substitute the result, 15, into the formula for g: We square 15 and add 1.

$$g \circ f(5) = g(f(5)) = g(3 \cdot 5) = g(15)$$
$$= 1 + 15^2$$
$$= 1 + 225 = 226$$

b) $f \circ g$ first does what g does (adds 1 to the square) and then does what f does (multiplies by 3). We find $f \circ g(x)$ by substituting $g(x)$ for x:

$$f \circ g(x) = f(g(x)) = f(1 + x^2) \quad \text{Substituting } 1 + x^2 \text{ for } x$$
$$= 3(1 + x^2) = 3 + 3x^2.$$

$g \circ f$ first does what f does (multiplies by 3) and then does what g does (adds 1 to the square). We find $g \circ f(x)$ by substituting $f(x)$ for x:

$$g \circ f(x) = g(f(x)) = g(3x) \quad \text{Substituting } 3x \text{ for } x$$
$$= 1 + (3x)^2$$
$$= 1 + 9x^2.$$

DO EXERCISE 4.

4. For functions f and g given by $f(x) = x + 5$ and $g(x) = x^2 - 1$, find each of the following.

a) $f \circ g(-3)$ and $g \circ f(-3)$

b) $f \circ g(x)$ and $g \circ f(x)$

Note in Example 13 that $f \circ g(5) \neq g \circ f(5)$ and, in general, that $f \circ g(x) \neq g \circ f(x)$.

EXAMPLE 14 Given $f(x) = \sqrt{1 - x}$ and $g(x) = 2x - 3$, find each of the following.

a) $f \circ g(x)$ and $g \circ f(x)$

b) The domains of f, g, $f \circ g$, and $g \circ f$

Solution

a) $f \circ g(x) = f(g(x)) = f(2x - 3)$
$= \sqrt{1 - (2x - 3)} = \sqrt{4 - 2x}$,
$g \circ f(x) = g(f(x)) = g(\sqrt{1 - x})$
$= 2(\sqrt{1 - x}) - 3 = 2\sqrt{1 - x} - 3$

b) The domain of f is the set of x such that $1 - x \geq 0$, which is $\{x \mid x \leq 1\}$, or the interval $(-\infty, 1]$. The domain of g is the set of all real numbers. To find the domain of $f \circ g$, we must consider that the outputs for g must be acceptable as inputs for f. In other words, if any number $g(x)$ is not in the domain of f, then x is not in the domain of $f \circ g$. The domain of $f \circ g$ is thus the set of all real numbers x such that $4 - 2x \geq 0$, or $x \leq 2$. The domain is $\{x \mid x \leq 2\}$ or the interval $(-\infty, 2]$.

Since the domain of g is the set of all real numbers, any number that is in the domain of f is in the domain of $g \circ f$. The domain of $g \circ f$ is thus the set of real numbers x such that $1 - x \geq 0$, which is $\{x \mid x \leq 1\}$ or the interval $(-\infty, 1]$.

DO EXERCISE 5.

5. For functions f and g given by $f(x) = \sqrt{x + 3}$ and $g(x) = x^2$, find each of the following.

a) $f \circ g(x)$ and $g \circ f(x)$

b) The domains of f, g, and $f \circ g$

EXAMPLE 15 *Speed of sound in air.* In Section 3.4, we learned that the speed of sound S in air is a function of the temperature F, in degrees Fahrenheit, and is given by

$$S(F) = 1087.7\sqrt{\frac{5F + 2457}{2457}},$$

where S is in feet per second. Suppose we wanted a function that would tell us the speed of sound in air when the Celsius temperature is given. The following function can be used to first convert the Celsius temperature C to the corresponding Fahrenheit temperature F:

$$F(C) = \tfrac{9}{5}C + 32.$$

Find a formula for $S \circ F(C)$ that can be used to find the speed of sound in air when the temperature is in degrees Celsius. Then find $S \circ F(20°)$.

Solution

$$S \circ F(C) = S(F(C)) = 1087.7\sqrt{\frac{5\left(\frac{9}{5}C + 32\right) + 2457}{2457}} = 1087.7\sqrt{\frac{9C + 2617}{2457}},$$

$$S \circ F(20°) = 1087.7\sqrt{\frac{9(20) + 2617}{2457}} \approx 1160.5 \text{ ft/sec.}$$

DO EXERCISE 6.

It is important in calculus to be able to recognize how a function can be expressed as a composition. In this way, we are "decomposing" the function.

EXAMPLE 16 Find $f(x)$ and $g(x)$ such that $h(x) = f \circ g(x)$:

$$h(x) = (2x - 3)^5.$$

Solution This is $2x - 3$ to the 5th power. Two functions that can be used for the composition are $f(x) = x^5$ and $g(x) = 2x - 3$. We can check by forming the composition:

$$h(x) = f \circ g(x) = f(g(x))$$
$$= f(2x - 3) = (2x - 3)^5.$$

This is the most "obvious" answer to the question. There can be other less obvious answers. For example, if

$$f(x) = (x + 7)^5 \quad \text{and} \quad g(x) = 2x - 10,$$

then

$$h(x) = f \circ g(x) = f(g(x)) = f(2x - 10) = [(2x - 10) + 7]^5 = (2x - 3)^5.$$

DO EXERCISE 7.

6. The function K given by

$$K(C) = C + 273$$

converts the Celsius temperature C to degrees in Kelvin units. The function C given by

$$C(F) = \tfrac{5}{9}(F - 32)$$

converts Fahrenheit temperature F to Celsius temperature C.

a) Find $K \circ C(F)$ and explain its meaning and usage.

b) Find the Kelvin temperature when the temperature is $-13°F$.

7. Find $f(x)$ and $g(x)$ such that $h(x) = f \circ g(x)$. Answers may vary, but try to select the most obvious.

a) $h(x) = \sqrt[3]{x^2 + 1}$

b) $h(x) = \dfrac{1}{(x + 5)^4}$

EXERCISE SET 3.7

1 For each of the following functions,

a) find $(f + g)(x)$, $(f - g)(x)$, $fg(x)$, $ff(x)$, $(f/g)(x)$, $(g/f)(x)$, $f \circ g(x)$, and $g \circ f(x)$.

b) find the domains of f, g, $f + g$, $f - g$, fg, ff, f/g, g/f, $f \circ g$, and $g \circ f$.

1. $f(x) = x - 3$, $g(x) = x + 4$
2. $f(x) = x^2 - 1$, $g(x) = 2x + 5$
3. $f(x) = x^3$, $g(x) = 2x^2 + 9x - 3$
4. $f(x) = x^2$, $g(x) = \sqrt{x}$

1, **2** Let $f(x) = x^2 - 4$ and $g(x) = 2x + 5$. Find each of the following.

5. $(f - g)(3)$
6. $(f + g)(-1)$
7. $(f - g)(x)$
8. $(f + g)(x)$
9. $fg(3)$
10. $(f/g)(-1)$
11. $(g/f)(-2)$
12. $(f/g)(-2.5)$

13. $fg(x)$

14. $(f/g)(x)$

15. $(g/f)(x)$

16. $ff(x)$

17. $f \circ g(x)$

18. $f \circ f(x)$

19. $g \circ g(x)$

20. $g \circ f(x)$

21. Given

$$R(x) = 60x - 0.4x^2,$$
$$C(x) = 3x + 13,$$

find each of the following.

a) $P(x)$

b) $R(20)$, $C(20)$, $P(20)$

22. Given

$$R(x) = 15x,$$
$$C(x) = 0.001x^2 + 1.2x + 60,$$

find each of the following.

a) $P(x)$

b) $R(100)$, $C(100)$, $P(100)$

2 Find $f \circ g(x)$ and $g \circ f(x)$.

23. $f(x) = \frac{4}{5}x$, $g(x) = \frac{5}{4}x$

24. $f(x) = x + 3$, $g(x) = x - 3$

25. $f(x) = 3x - 7$, $g(x) = \frac{x + 7}{3}$

26. $f(x) = \frac{2}{3}x - \frac{4}{5}$, $g(x) = 1.5x + 1.2$

27. $f(x) = x^3 - 1$, $g(x) = \sqrt[3]{x + 1}$

28. $f(x) = \sqrt[5]{x + 2}$, $g(x) = x^5 - 2$

29. $f(x) = \sqrt{x + 5}$, $g(x) = x^2 - 5$

30. $f(x) = x^4$, $g(x) = \sqrt[4]{x}$

31. $f(x) = \frac{1 - x}{x}$, $g(x) = \frac{1}{1 + x}$

32. $f(x) = \frac{x^2 - 1}{x^2 + 1}$, $g(x) = \frac{3x - 4}{5x - 2}$

33. $f(x) = -6$, $g(x) = 12$

34. $f(x) = 20$, $g(x) = 0.2x + 1$

2 In Exercises 35–46, find $f(x)$ and $g(x)$ such that $h(x) = f \circ g(x)$. Answers may vary, but try to select the most obvious answer.

35. $h(x) = (4 - 3x)^5$

36. $h(x) = \sqrt[3]{x^2 - 8}$

37. $h(x) = \frac{1}{(x - 1)^4}$

38. $h(x) = \frac{1}{\sqrt{3x + 7}}$

39. $h(x) = \frac{x^3 - 1}{x^3 + 1}$

40. $h(x) = |9x^2 - 4|$

41. $h(x) = \left(\frac{2 + x^3}{2 - x^3}\right)^6$

42. $h(x) = (\sqrt{x} - 3)^4$

43. $h(x) = \sqrt{\frac{x - 5}{x + 2}}$

44. $h(x) = \sqrt{1 + \sqrt{1 + x}}$

45. $h(x) = (x + 3)^5 + (x + 3)^4 + (x + 3)^3$

46. $h(x) = 4(x - 1)^{2/3} + 5 - (x + 3)^2 + 4(x + 3)$

47. An airplane is 300 ft from the control tower at the end of the runway. It takes off at a speed of 250 mph.

a) Let a be the distance that the plane travels down the runway. Find a formula for a in terms of the time t that the plane travels. That is, find an expression for $a(t)$.

b) Let P be the distance of the plane from the control tower. Find a formula for P in terms of the distance a. That is, find an expression for $P(a)$.

c) Find $(P \circ a)(t)$. Explain the meaning of this function.

48. A stone is thrown into a pond. A circular ripple is spreading over the pond in such a way that the radius is increasing at the rate of 3 ft/sec.

a) Find a function $r(t)$ for the radius in terms of the time t.

b) Find a function $A(r)$ for the area of the ripple in terms of the radius r.

c) Find $(A \circ r)(t)$. Explain the meaning of this function.

SYNTHESIS

In Exercises 49–52, graph the equation. Then graph $y = (f + g)(x)$ by adding second coordinates.

49. $f(x) = x^2$, $g(x) = -2x + 3$

50. $f(x) = 2/x$, $g(x) = x$

51. $f(x) = \sqrt{x}$, $g(x) = 1 - x^2$

52. $f(x) = 5 - x^2$, $g(x) = x^2 - 5$

53. Consider $f(x) = 3x + b$ and $g(x) = 2x - 1$. Find b such that $f \circ g(x) = g \circ f(x)$ for all real numbers x.

54. Consider $f(x) = 3x - 4$ and $g(x) = mx + b$. Find m and b such that $f \circ g(x) = g \circ f(x) = x$ for all real numbers x.

55. For $f(x) = 1/(1 - x)$, find $f \circ f(x)$ and $f \circ f \circ f(x)$.

56. Prove that the composition of two even functions is even.

57. Prove that the sum of two even functions is even.

58. Prove that the product of two odd functions is even.

59. Prove that the composition of two odd functions is odd.

60. Prove that the product of an even function and an odd function is odd.

61. Prove that if f and g are increasing functions, then $f \circ g$ is increasing and $f + g$ is increasing.

62. Prove that if f is *any* function, then the function E defined by

$$E(x) = \frac{f(x) + f(-x)}{2}$$

is even.

63. Prove that if f is *any* function, then the function O defined by

$$O(x) = \frac{f(x) - f(-x)}{2}$$

is odd.

64. Consider the functions E and O of Exercises 72 and 73. Prove that

$$f(x) = E(x) + O(x),$$

which shows that every function can be expressed as the sum of an even function and an odd function.

65. Graph $f(x) = |x| + \text{INT}(x)$.

3.8 Transformations of Relations and Functions

OBJECTIVE

You should be able to:

1 Given the graph of a function or a relation, graph its transformation under translations, reflections, and shrinkings.

1 Transformations

Vertical Translations

Given a relation or a function, we can find various ways of altering it to obtain another relation or function. Such an alteration is called a **transformation.** If such an alteration consists merely of moving the graph without changing its shape or orientation, the transformation is called a **translation.**

EXAMPLE 1 Consider functions of the type $f(x) = x^2 + a$. Using the same set of axes, sketch and compare the graphs of

$$f(x) = x^2,$$
$$f(x) = x^2 + 1, \quad \text{and}$$
$$f(x) = x^2 - 3.$$

Solution The graphs are shown at the right. Note that the graph of $f(x) = x^2 + 1$ has the same shape as the graph of $f(x) = x^2$, but is moved upward a distance of 1 unit. Each function value, or output, is increased by 1 unit. The graph of $f(x) = x^2 - 3$ has the same shape as the graph of $f(x) = x^2$, but is moved downward 3 units. Each function value, or output, is decreased by 3 units.

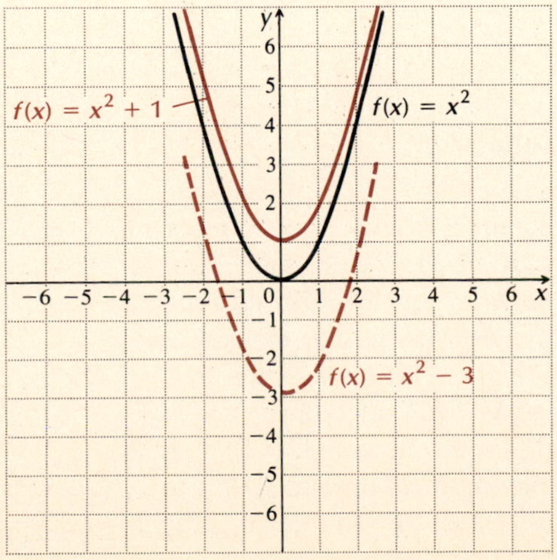

1. Graph

$$f(x) = |x|,$$
$$f(x) = |x| - 2, \quad \text{and}$$
$$f(x) = |x| + 3$$

using the same set of axes.

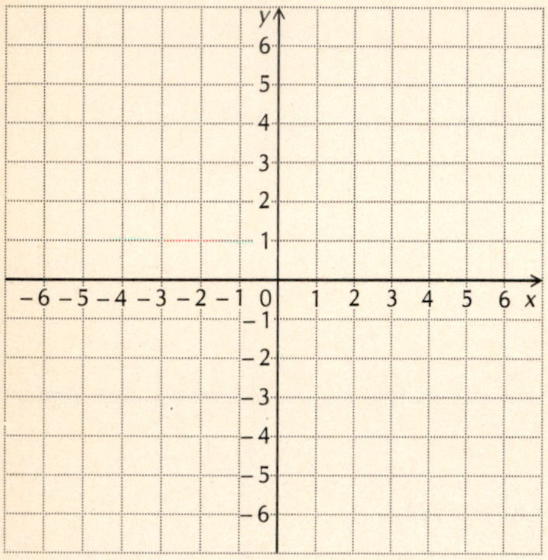

Consider any equation of a function $y = f(x)$. Adding a constant a to produce $y = f(x) + a$ changes each function value by the same amount, a. Thus it produces no change in the shape of the graph, but merely translates it upward if the constant a is positive and downward if the constant a is negative.

> **THEOREM 1F** Vertical Translations of Functions
>
> **The graph of $y = f(x) + a$ is a translation upward $|a|$ units if $a > 0$ and is a translation downward $|a|$ units if $a < 0$.**

DO EXERCISE 1.

Now let us consider translations of relations in general—for example, $y = x^2 + 1$. This equation, or relation, is equivalent to $y - 1 = x^2$. Thus the translation amounts to replacing y in the equation $y = x^2$ by $y - 1$. In this case, the constant a is 1 (positive) and the translation is upward. Similarly, if in an equation we replace y by $y + 3$, this is the same as replacing it by $y - (-3)$. In this case, the constant a is -3 (negative) and the translation is downward. If we replace y by $y - 5$, the constant a is 5 (positive) and the translation is upward.

We can extend Theorem 1 to relations as follows.

> **THEOREM 1R** Vertical Translations of Relations
>
> **In an equation of a relation, replacing y by $y - a$, where a is constant, translates the graph vertically $|a|$ units. If $a > 0$, the translation is upward. If $a < 0$, the translation is downward.**

EXAMPLE 2 Shown at the right is the graph of the relation $y^2 = x$. Sketch a graph of $(y - 1)^2 = x$ and $(y + 3)^2 = x$.

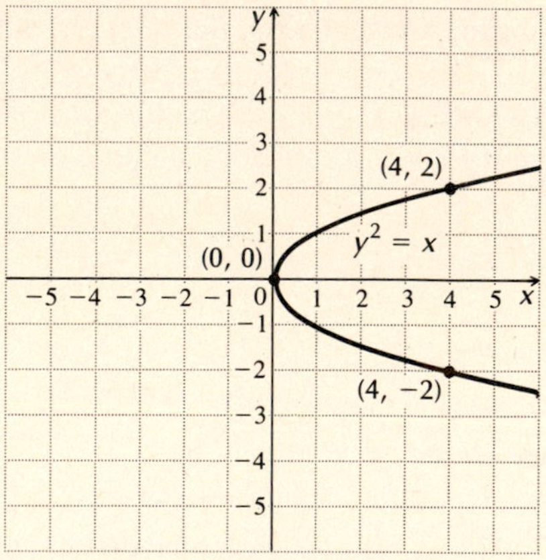

Solution

The equation $(y - 1)^2 = x$ is obtained from $y^2 = x$ by replacing y by $y - 1$. We are subtracting a positive number. This tells us that the translation is 1 unit upward. We can achieve the graph by looking for key points, such as (0, 0), (4, 2), and (4, −2), and moving them upward 1 unit to (0, 1), (4, 3), and (4, −1). The graph is shown here.

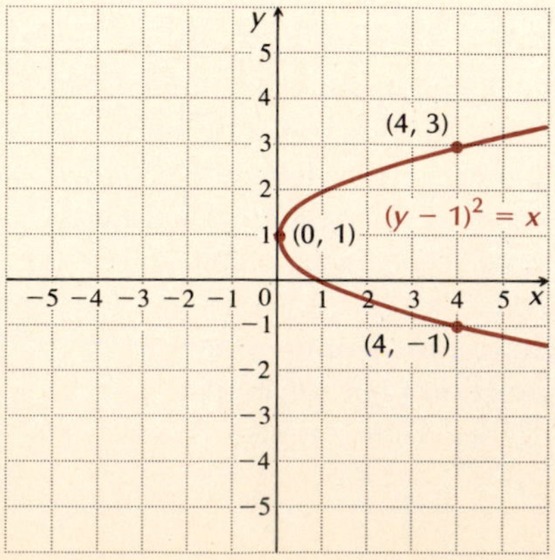

The equation $(y + 3)^2 = x$ is obtained from $y^2 = x$ by replacing y by $y + 3$, or $y - (-3)$. We are subtracting a negative number. This tells us that the translation is 3 units downward. We can achieve the graph by looking for key points such as (0, 0), (4, 2), and (4, −2), and moving them downward 3 units to (0, −3), (4, −1), and (4, −5). The graph is shown here.

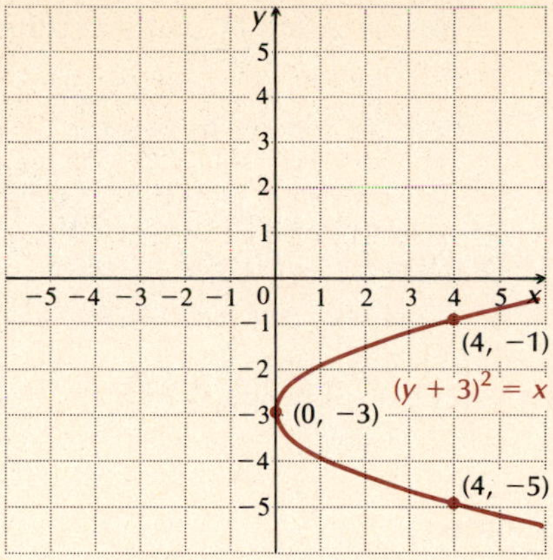

DO EXERCISE 2.

Horizontal Translations

We now consider horizontal translations.

EXAMPLE 3 Consider functions of the type $f(x) = (x + a)^2$. Using the same set of axes, sketch and compare the graphs of

$$f(x) = x^2,$$
$$f(x) = (x - 3)^2, \quad \text{and}$$
$$f(x) = (x + 4)^2.$$

Solution The graphs are shown below. The graph of $f(x) = (x - 3)^2$ is a horizontal translation of the graph of $f(x) = x^2$, 3 units to the right. The graph of $f(x) = (x + 4)^2$ is a horizontal translation of the graph of $f(x) = x^2$, 4 units to the left.

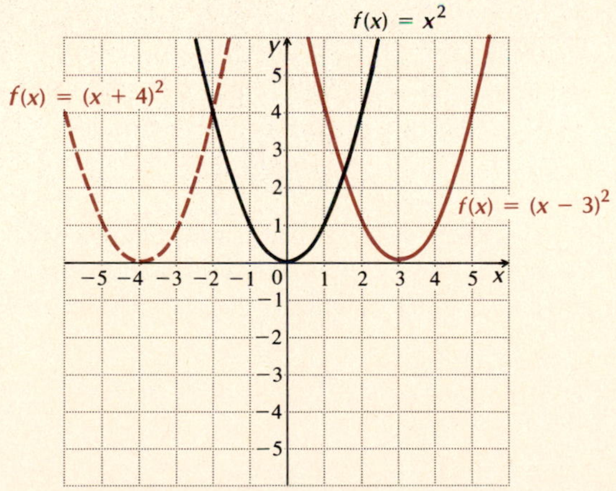

THEOREM 2F Horizontal Translations of Functions

The graph of $y = f(x - a)$ is a translation to the right $|a|$ units if $a > 0$ and is a translation to the left $|a|$ units if $a < 0$.

DO EXERCISE 3.

2. Graph

$$|y| = x,$$
$$|y - 3| = x, \quad \text{and}$$
$$|y + 1| = x$$

using the same set of axes.

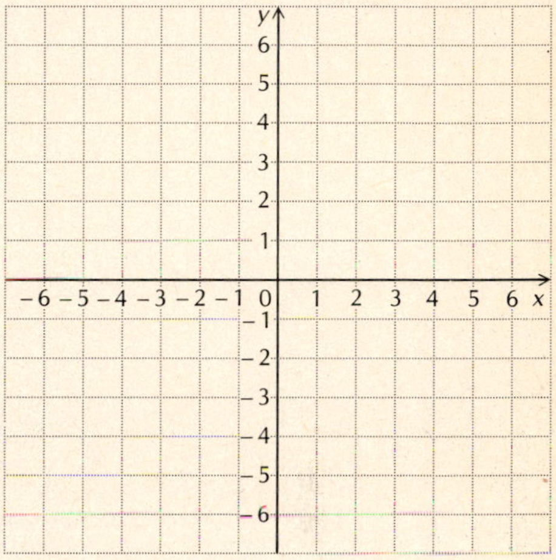

3. Graph

$$f(x) = |x|,$$
$$f(x) = |x - 2|, \quad \text{and}$$
$$f(x) = |x + 3|$$

using the same set of axes.

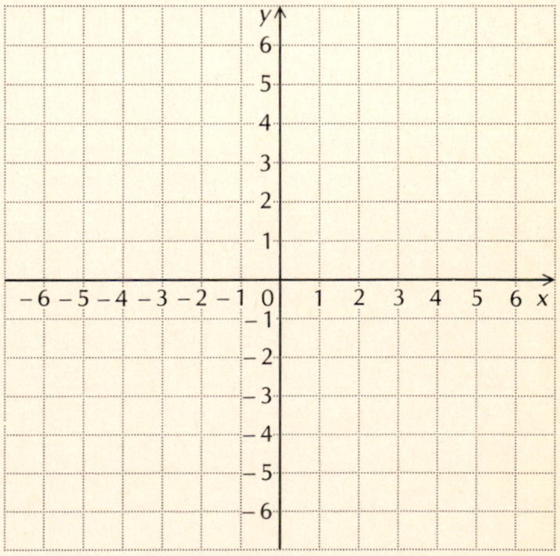

4. Graph

$$|y| = x,$$
$$|y| = x + 5, \quad \text{and}$$
$$|y| = x - 3$$

using the same set of axes.

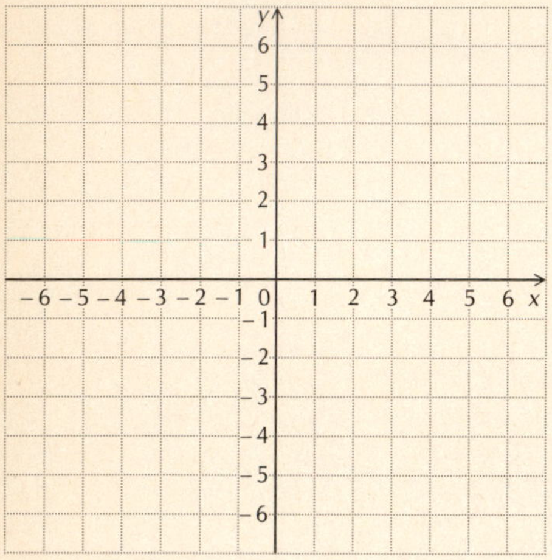

5. Graph the circle

$$(x + 3)^2 + (y - 2)^2 = 1$$

by translating the graph of the circle

$$x^2 + y^2 = 1.$$

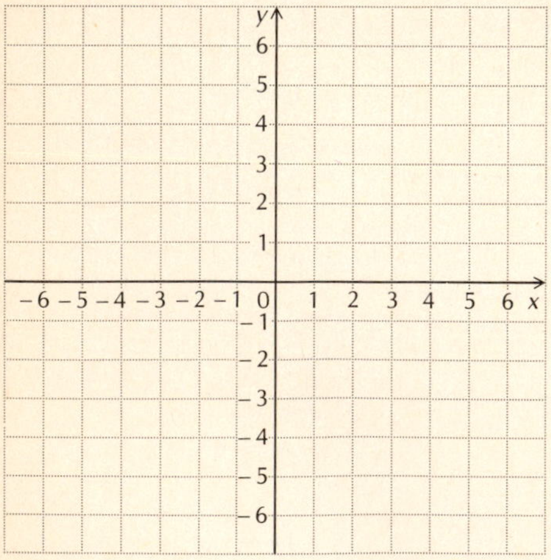

A similar result holds for relations.

THEOREM 2R Horizontal Translations of Relations

In an equation of a relation, replacing x by $x - b$, where b is a constant, translates the graph horizontally $|b|$ units. If $b > 0$, the translation is to the right. If $b < 0$, the translation is to the left.

EXAMPLE 4 The graph of $y^2 = x$ is shown in Example 2. Sketch a graph of $y^2 = x - 1$ and $y^2 = x + 4$.

Solution

The graph of $y^2 = x - 1$ is obtained from the graph of $y^2 = x$ by replacing x by $x - 1$. Since 1 is positive, the translation is 1 unit to the right. The graph is as follows.

The graph of $y^2 = x + 4$ is obtained from the graph of $y^2 = x$ by replacing x by $x + 4$, or $x - (-4)$. Since -4 is negative, the translation is 4 units to the left. The graph is as follows.

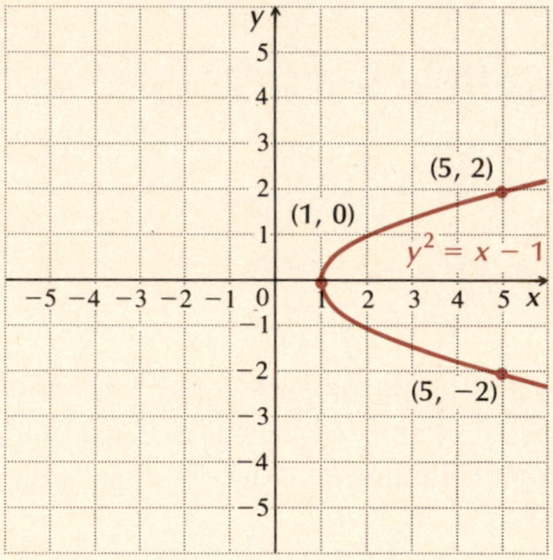

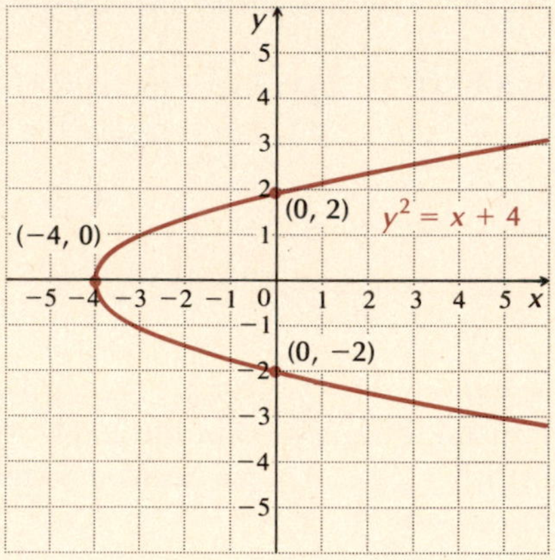

EXAMPLE 5 A circle centered at the origin with radius of length 1 has an equation $x^2 + y^2 = 1$. Sketch a graph of the equation

$$(x - 1)^2 + (y + 2)^2 = 1.$$

Solution In $(x - 1)^2 + (y + 2)^2 = 1$, x has been replaced by $x - 1$ and y has been replaced by $y + 2$, or $y - (-2)$. We translate the circle, $x^2 + y^2 = 1$, 1 unit to the right and 2 units downward, so that the center is at the point $(1, -2)$.

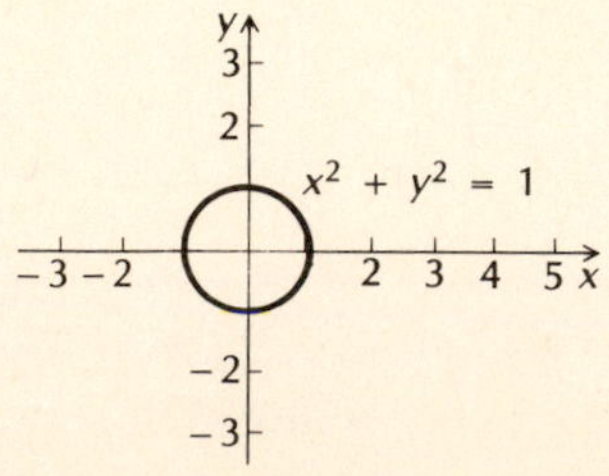

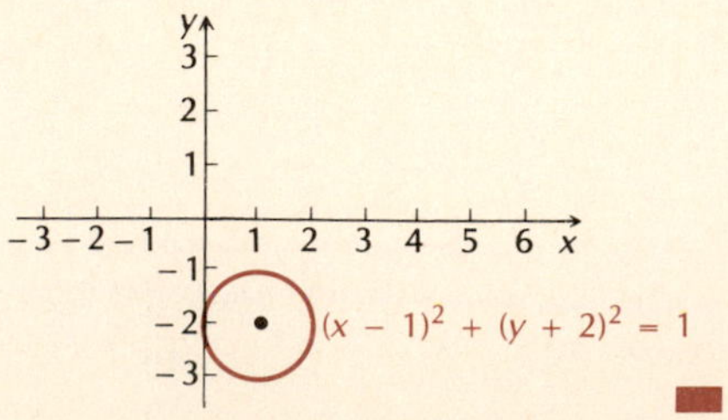

DO EXERCISES 4 AND 5.

Vertical Stretchings and Shrinkings

Let us now consider functions of the type $y = cf(x)$, where c is some constant.

EXAMPLE 6 Using the same set of axes, sketch and compare the graphs of

$$f(x) = |x|, \quad f(x) = 2|x|, \quad \text{and} \quad f(x) = \tfrac{1}{2}|x|.$$

Solution

x	$2\|x\|$	$\frac{1}{2}\|x\|$
-3	6	$\frac{3}{2}$
-2	4	1
0	0	0
2	4	1
3	6	$\frac{3}{2}$

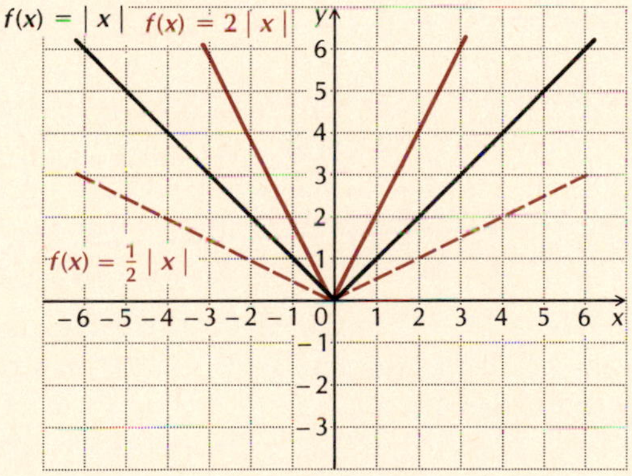

The graph of $f(x) = 2|x|$ looks like that of $f(x) = |x|$, but every function value, or output, is doubled, so the graph is *stretched* in a vertical direction. The graph of $f(x) = \frac{1}{2}|x|$ is flattened, or *shrunk*, in a vertical direction since every output, or function value, is halved.

Consider any function f given by $y = f(x)$. Multiplying on the right by any constant c greater than 1 to obtain $y = cf(x)$ will *stretch* the graph away from the horizontal axis. If the constant c is between 0 and 1, then the graph will be flattened or *shrunk* vertically toward the horizontal axis.

DO EXERCISE 6.

When we multiply by a negative constant, the graph is *reflected*, or flipped, across the x-axis as well as being stretched or shrunk.

EXAMPLE 7 Using the same set of axes, sketch and compare the graphs of

$$f(x) = |x|, \quad f(x) = -2|x|, \quad \text{and} \quad f(x) = -\tfrac{1}{2}|x|.$$

6. Using the same set of axes, graph

$$f(x) = x^2,$$
$$f(x) = 2x^2, \quad \text{and}$$
$$f(x) = 0.8x^2.$$

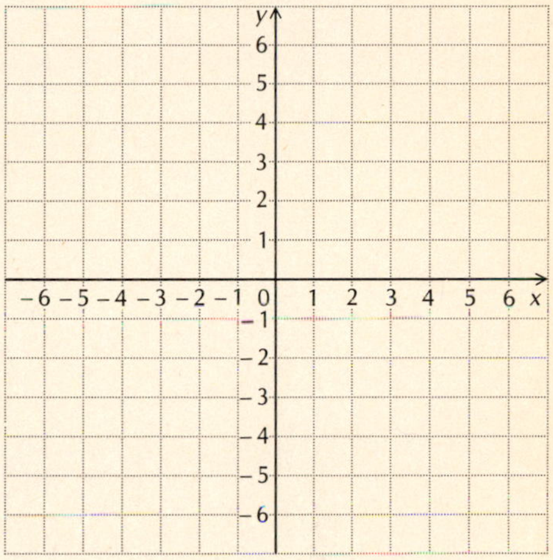

7. Using the same set of axes, graph

$$f(x) = -x^2,$$
$$f(x) = -2x^2, \quad \text{and}$$
$$f(x) = -0.8x^2.$$

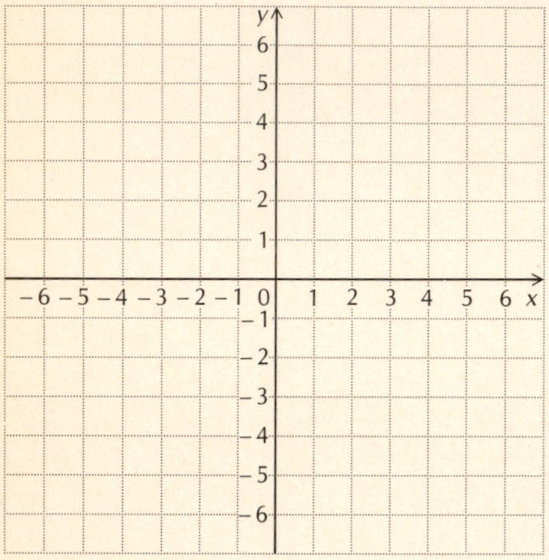

8. Using the same set of axes, graph

$$f(x) = x^2 - 4 \quad \text{and}$$
$$f(x) = -(x^2 - 4) = 4 - x^2.$$

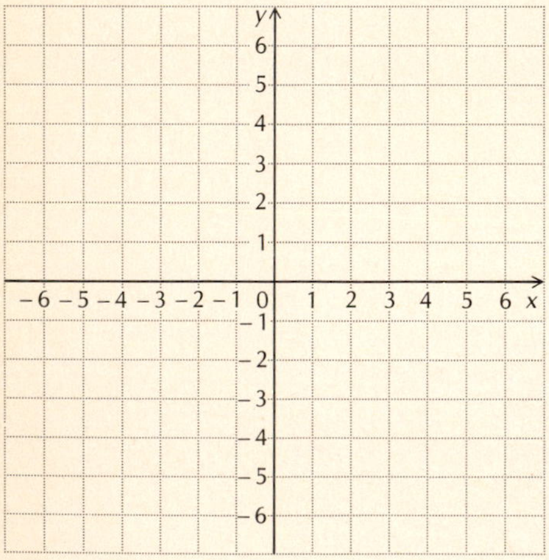

Solution The graph of $f(x) = -2|x|$ is obtained from the graph of $f(x) = |x|$ by stretching each output by a factor of 2 and reflecting across the x-axis. The graph of $f(x) = -\frac{1}{2}|x|$ is obtained by shrinking each output by a factor of $\frac{1}{2}$ and reflecting across the x-axis.

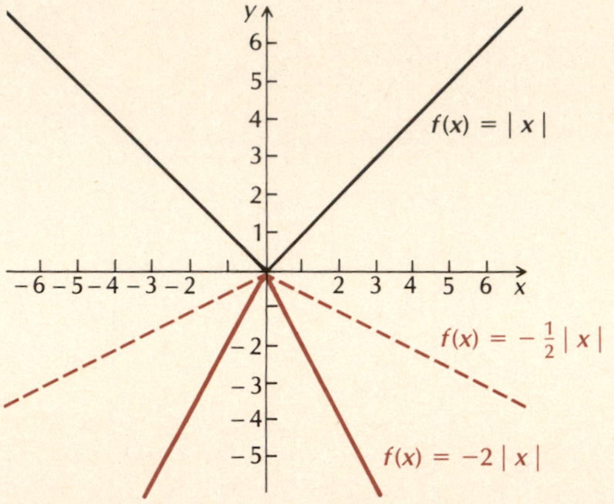

> **THEOREM 3 Vertical Stretchings and Shrinkings of Functions**
>
> **The graph of the function $y = cf(x)$ is obtained from the graph of $y = f(x)$ by**
>
> **a) stretching vertically if $|c| > 1$,**
>
> **b) shrinking vertically if $|c| < 1$, and**
>
> **c) reflecting across the x-axis if $c < 0$.**
>
> **The new graph can be obtained by multiplying the y-coordinate of each ordered-pair solution of $y = f(x)$ by c.**

Note that multiplying by -1 has the effect of reflecting the graph without stretching or shrinking. That is, the graph of $y = -f(x)$ is a **reflection** of the graph of $y = f(x)$ across the x-axis.

DO EXERCISES 7 AND 8.

Horizontal Stretchings and Shrinkings

The constant c in the equation $y = cf(x)$ has the effect of vertically stretching or shrinking the graph of $y = f(x)$, with a possible reflection across the x-axis if $c < 0$. The constant c in the equation $y = f(cx)$ has the effect of horizontally stretching or shrinking the graph of $y = f(x)$, with a possible reflection across the y-axis if $c < 0$.

> **Theorem 4 Horizontal Stretchings and Shrinkings of Functions**
>
> **The graph of the function $y = f(cx)$ is obtained from the graph of $y = f(x)$ by**
>
> **a) shrinking horizontally if $|c| > 1$,**
>
> **b) stretching horizontally if $|c| < 1$, and**
>
> **c) reflecting across the y-axis if $c < 0$.**
>
> **The new graph can be obtained by dividing the x-coordinate of each ordered-pair solution of $y = f(x)$ by c.**

EXAMPLE 8 Below is a graph of $y = f(x)$ for some function f. No formula for f is given. Sketch a graph of $y = f(2x)$, $y = f\left(\frac{1}{2}x\right)$, and $y = f\left(-\frac{1}{2}x\right)$.

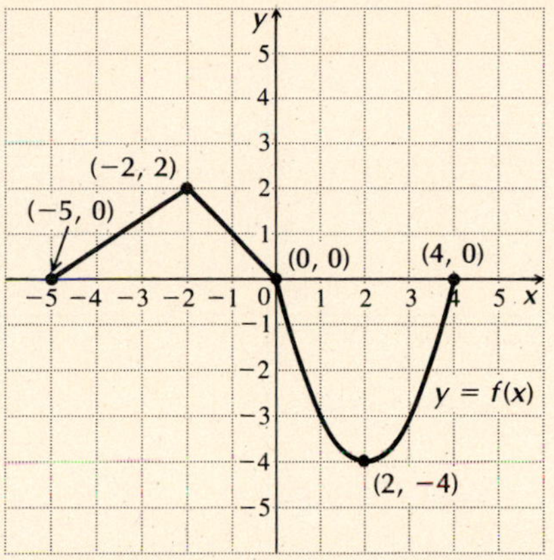

Solution Since $|2| > 1$, the graph of $y = f(2x)$ is a horizontal shrinking of the graph of $y = f(x)$. We can consider the key points $(-5, 0)$, $(-2, 2)$, $(0, 0)$, $(2, -4)$, and $(4, 0)$. The transformation divides each x-coordinate by 2 to obtain the key points $(-2.5, 0)$, $(-1, 2)$, $(0, 0)$, $(1, -4)$, and $(2, 0)$ of the graph of $y = f(2x)$. The graph is as follows.

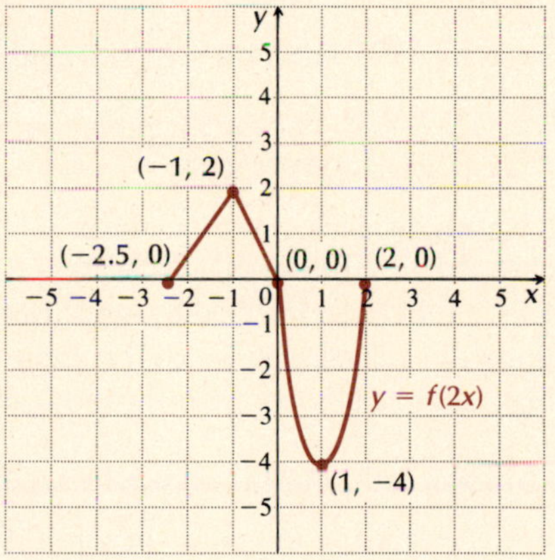

Since $\left|\frac{1}{2}\right| < 1$, the graph of $y = f\left(\frac{1}{2}x\right)$ is a horizontal stretching of the graph of $y = f(x)$. We can consider the key points $(-5, 0)$, $(-2, 2)$, $(0, 0)$, $(2, -4)$, and $(4, 0)$. The transformation divides each x-coordinate by $1/2$ (which is the same as multiplying by 2) to obtain the key points $(-10, 0)$, $(-4, 2)$, $(0, 0)$, $(4, -4)$, and $(8, 0)$ of the graph of $y = f(2x)$. The graph is as follows.

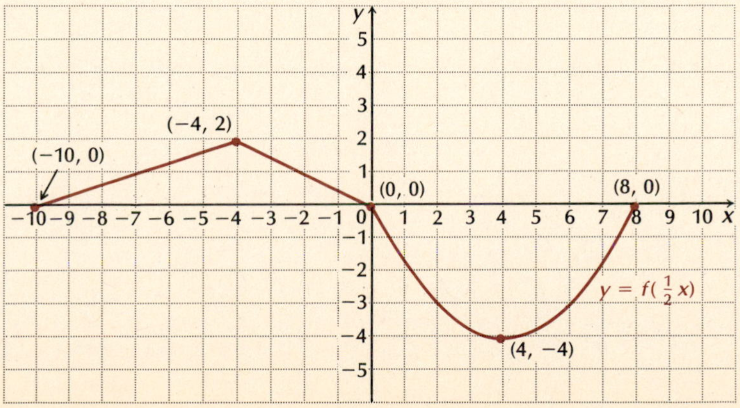

Below is a graph of $y = t(x)$. No formula for t is given. Sketch a graph of each of the following using graph paper.

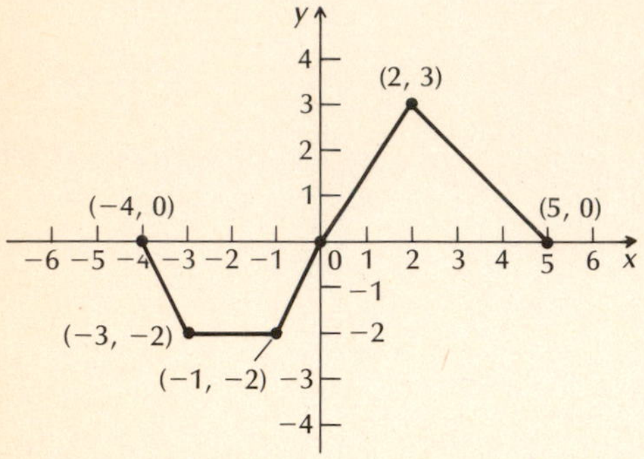

9. $y = \frac{1}{2}t(x)$

10. $y = -2t(x)$

11. $y = t(2x)$

12. $y = t\left(-\frac{1}{2}x\right)$

13. Use the graph of $y = t(x)$ used for Margin Exercises 9–12. Graph $y = -3t(x + 1) - 4$.

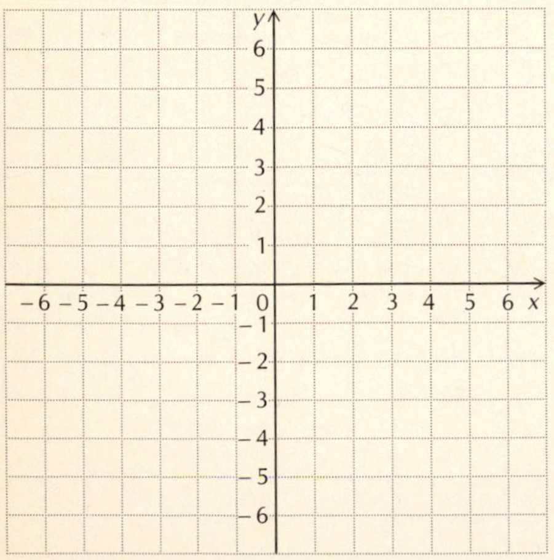

The graph of $y = f\left(-\frac{1}{2}x\right)$ can be obtained by reflecting the graph of $y = f\left(\frac{1}{2}x\right)$ across the y-axis and is as follows.

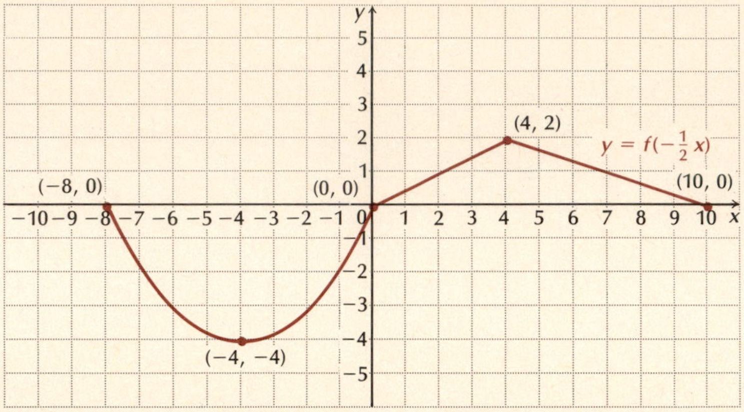

DO EXERCISES 9–12.

Combined Transformations

Suppose we knew the graph of a function $y = f(x)$ and wanted to graph the function

$$y = af(x + c) + b.$$

From the preceding, we might anticipate that there is a vertical as well as a horizontal translation, together with a vertical as well as a horizontal stretching or shrinking and perhaps a reflection across the x-axis. But, in what order do we proceed? By not thinking this out clearly, we could obtain the wrong graph. An easy way to keep things straight is to work from the inside out, as follows:

To graph $y = af(x + c) + b$ using the graph of $y = f(x)$:

a) Obtain the graph of $y = f(x + c)$ by translating horizontally.

b) Obtain the graph of $y = af(x + c)$ by vertically stretching or shrinking the graph of $y = f(x + c)$ with a possible reflection across the x-axis.

c) Obtain the graph of $y = af(x + c) + b$ from the graph of $y = af(x + c)$ by translating vertically.

EXAMPLE 9 Use the graph of $y = f(x)$ given in Example 8. Sketch a graph of

$$y = -2f(x - 3) + 1.$$

Solution

$y = f(x)$

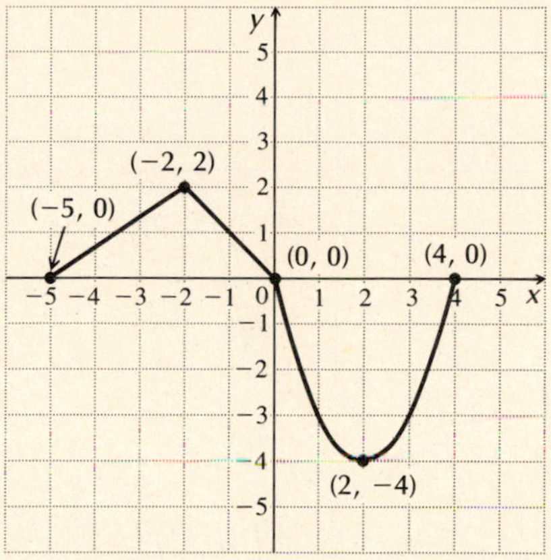

$y = f(x - 3)$

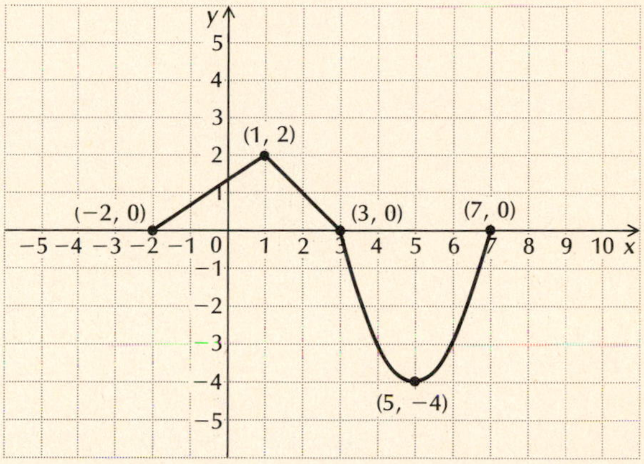

$y = -2f(x - 3)$

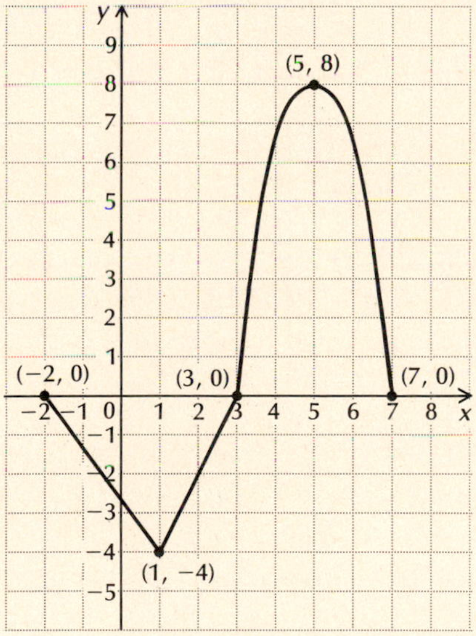

$y = -2f(x - 3) + 1$

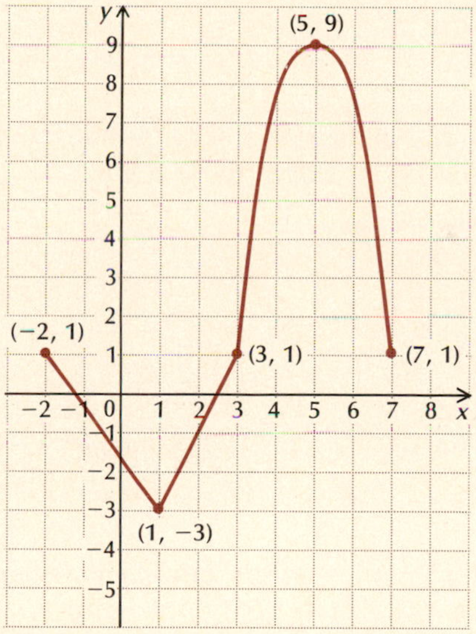

DO EXERCISE 13 ON THE PRECEDING PAGE.

EXERCISE SET 3.8

1 Sketch the graph by transforming the graph of $f(x) = |x|$.

1. $f(x) = |x| - 3$

2. $f(x) = 2 + |x|$

3. $f(x) = |x - 1|$

4. $f(x) = |x + 2|$

5. $f(x) = -4|x|$

6. $f(x) = 3|x|$

7. $f(x) = \frac{1}{3}|x|$

8. $f(x) = -\frac{1}{4}|x|$

9. $f(x) = |2x|$

10. $f(x) = \left|\frac{x}{3}\right|$

11. $f(x) = |x - 2| + 3$

12. $f(x) = 2|x + 1| - 3$

13. $f(x) = -3|x - 2|$

14. $f(x) = \frac{1}{3}|x + 2| + 1$

Here is a graph of $y = f(x)$. No formula will be given for this function. In Exercises 15–34, sketch the graph by transforming this one.

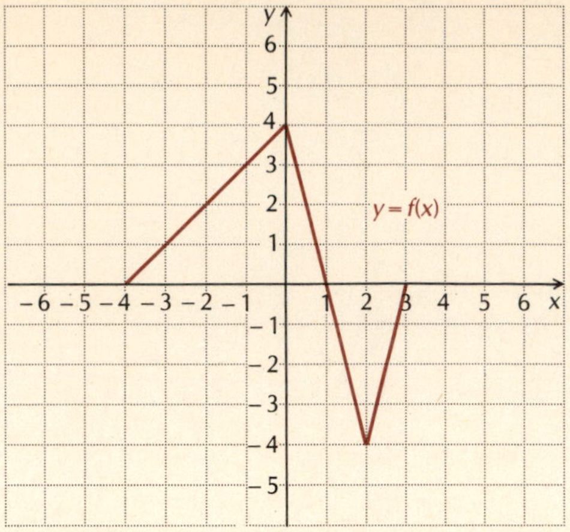

15. $y = 2 + f(x)$
16. $y + 1 = f(x)$
17. $y = f(x - 1)$
18. $y = f(x + 2)$
19. $\dfrac{y}{-2} = f(x)$
20. $y = 3f(x)$
21. $y = \frac{1}{3}f(x)$
22. $y = -\frac{1}{2}f(x)$
23. $y = f(2x)$
24. $y = f(3x)$
25. $y = f(-2x)$
26. $y = f(-3x)$
27. $y = f\left(\dfrac{x}{-2}\right)$
28. $y = f\left(\frac{1}{3}x\right)$
29. $y = f(x - 2) + 3$
30. $y = -3f(x - 2)$
31. $y = 2 \cdot f(x + 1) - 2$
32. $y = \frac{1}{2}f(x + 2) - 1$
33. $y = -\frac{1}{2}f(x - 3) + 2$
34. $y = -3f(x - 1) + 4$

Here is a graph of $y = f(x)$. No formula will be given for this function. In Exercises 35–38, sketch the graph by transforming this one.

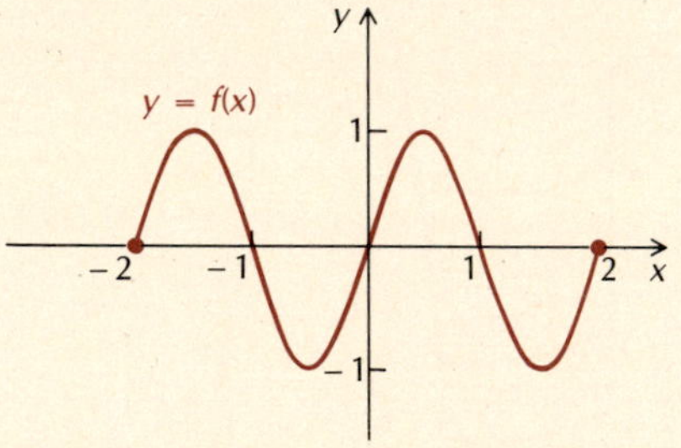

35. $y = -2f(x + 1) - 1$
36. $y = 3f(x + 2) + 1$
37. $y = \frac{5}{2}f(x - 3) - 2$
38. $y = -\frac{2}{3}f(x - 4) + 3$

The graph of the relation $x^2 + y^2 = 1$ is the circle with radius 1 centered at (0, 0) shown in Example 5. Sketch a graph of each of the following relations.

39. $(x - 1)^2 + (x + 3)^2 = 1$
40. $(x + 2)^2 + (x + 3)^2 = 1$
41. $x^2 + (y - 2)^2 = 1$
42. $(x + 3)^2 + y^2 = 1$

The graph of the relation $|x| + |y| = 1$ is shown below. Sketch a graph of each of the following relations.

43. $|x| + |y + 3| = 1$
44. $|x - 2| + |y| = 1$
45. $|x - 4| + |y| = 1$
46. $|x - 2| + |y + 4| = 1$

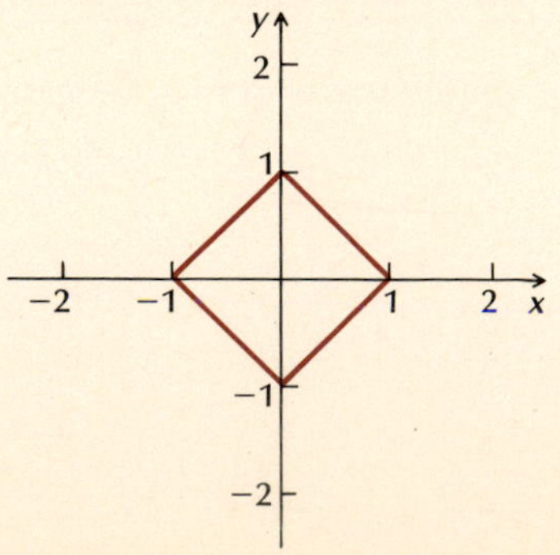

SYNTHESIS

Given the graph of $y = f(x)$ for Exercises 35–38, graph each of the following.

47. $y = -\sqrt{2}f(x + 1.8)$

48. $y = \frac{\sqrt{3}}{2} \cdot f(x - 2.5) - 5.3$

CHALLENGE

Given the graph of $y = f(x)$ for Exercises 35–38, graph each of the following.

49. $\frac{y}{3} = f\left(2x + \frac{1}{2}\right)$

50. $y = -4 \cdot f(5x + 10)$

SUMMARY AND REVIEW: CHAPTER 3

TERMS TO KNOW

Correspondence
Cartesian product
Relation
Domain
Range
Graph
Coordinate
Solution
Function as a correspondence
Function as a relation
Vertical-line test
Function as a machine
Input
Output
Constant function
Function as a mapping
Symmetry with respect to the axes
Symmetry with respect to the origin
Even function
Odd function
Interval notation
Periodic function
Continuous function
Increasing function
Decreasing function
Equilibrium point
Greatest integer function
Composition of functions
Function of two variables
Transformation
Translation
Stretching
Shrinking
Reflection

REVIEW EXERCISES

1. Find the Cartesian product $G \times G$, where $\{1, 3, 5, 7\}$.

Consider the relation $\{(3, 1), (5, 3), (7, 7), (3, 5)\}$ for Exercises 2–4.

2. Determine whether the relation is a function.
3. Find the domain.
4. Find the range.

Graph.

5. $x = |y|$
6. $y = (x + 1)^2$
7. $g(x) = |x| - 2$
8. $f(x) = \sqrt{x}$
9. $f(x) = \sqrt{x - 2}$
10. $f(x) = 2\sqrt{x + 3}$
11. $f(x) = \frac{1}{2}\sqrt{x - 1} + 2$
12. $f(x) = \text{INT}(x)$
13. $f(x) = \text{INT}(x) - 3$
14. $f(x) = \text{INT}(x - 3)$

Consider the following for Exercises 15–17.

a) $y = 7$
b) $x^2 + y^2 = 4$
c) $x^3 = y^3 - y$
d) $y^2 = x^2 + 3$
e) $x + y = 3$
f) $x = 3$
g) $y = x^2$
h) $y = x^3$

15. Which are symmetric with respect to the x-axis?
16. Which are symmetric with respect to the y-axis?
17. Which are symmetric with respect to the origin?
18. Given $R(x) = 120x - 0.5x^2$ and $C(x) = 15x + 6$, find $P(x)$.
19. Find the equilibrium point for the supply and demand functions

$$D(x) = (x - 7)^2$$

and

$$S(x) = x^2 + x + 4.$$

20. Which of the following are graphs of functions?

a)

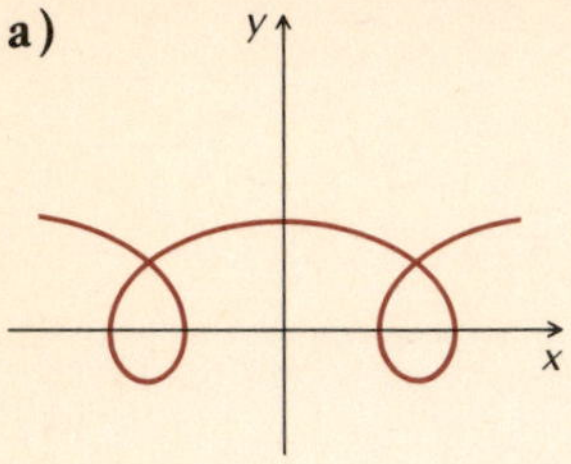

b)

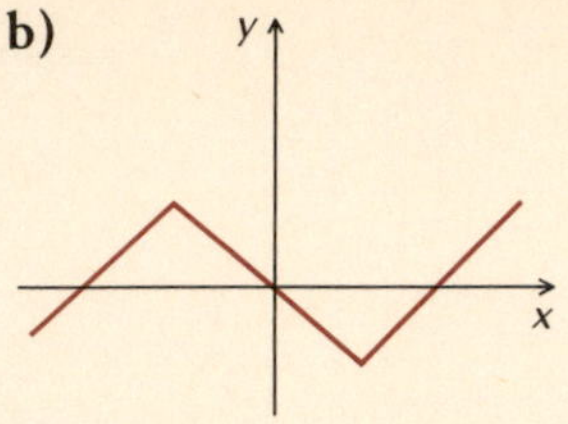

Use $f(x) = x^2 - x - 3$ for Exercises 21–23. Find:

21. $f(0)$

22. $f(-3)$

23. $\dfrac{f(a+h) - f(a)}{h}$

Use $g(x) = 2\sqrt{x-1}$ for Exercises 24–26. Find:

24. $g(1)$

25. $g(5)$

26. $g(a+2)$

Find the domain.

27. $f(x) = \sqrt{7 - 3x}$

28. $f(x) = \dfrac{1}{x^2 - 6x + 5}$

In Exercises 29 and 30, find

a) $(f+g)(x)$, $(f-g)(x)$, $fg(x)$, $(f/g)(x)$, $f \circ g(x)$, and $g \circ f(x)$.

b) the domain of f, g, $f+g$, $f-g$, fg, f/g, $f \circ g$, and $g \circ f$.

29. $f(x) = \dfrac{4}{x^2}$; $g(x) = 3 - 2x$

30. $f(x) = 3x^2 + 4x$; $g(x) = 2x - 1$

Here is a graph of $f(x) = \sqrt{9 - x^2}$. Sketch the graph of each of the following.

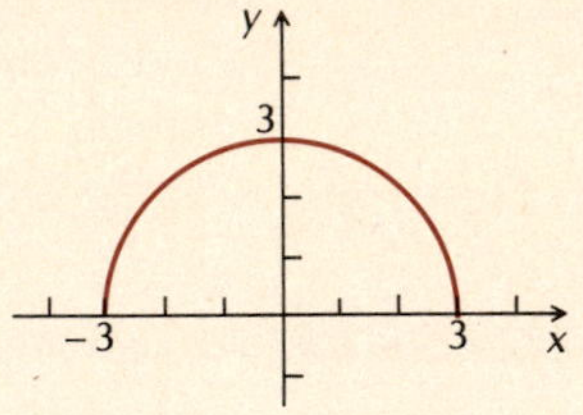

31. $y = 1 + f(x)$

32. $y = \frac{1}{2}f(x)$

33. $y = f(x+1)$

Use the following for Exercises 34–36.

a)

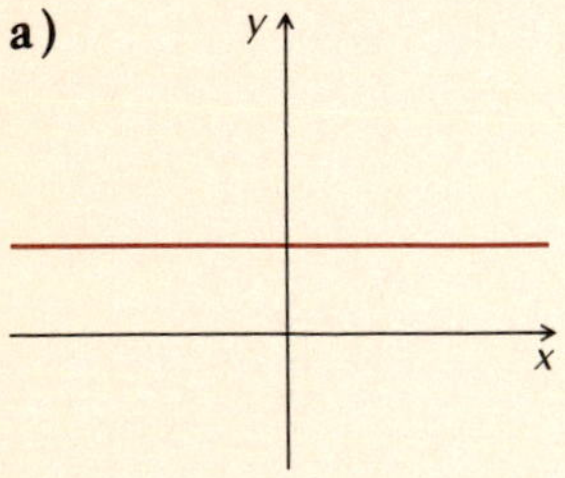

b)

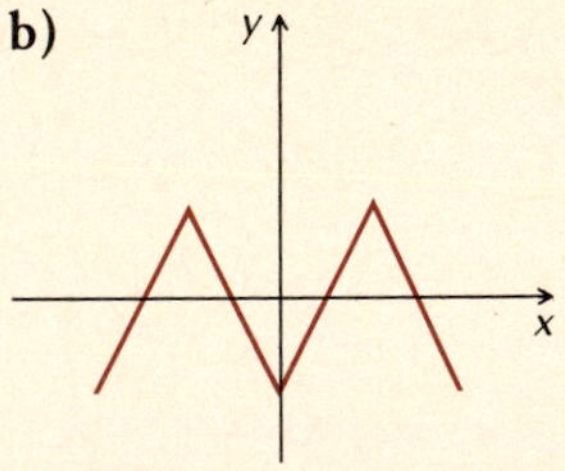

c) $f(x) = 3x^2 - 2$

d) $f(x) = x + 3$

e) $f(x) = 3x^3$

f) $f(x) = x^5 - x^3$

34. Which are even?

35. Which are odd?

36. Which are neither even nor odd?

37. Which of the following functions are periodic?

a)

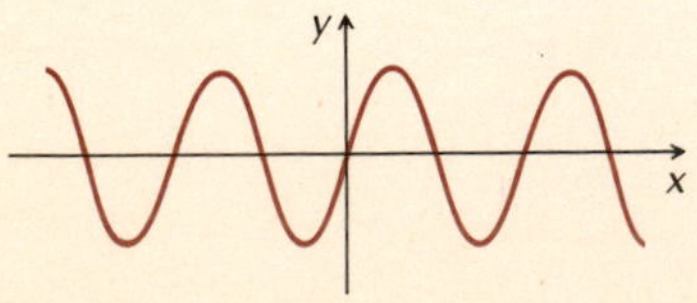

b)

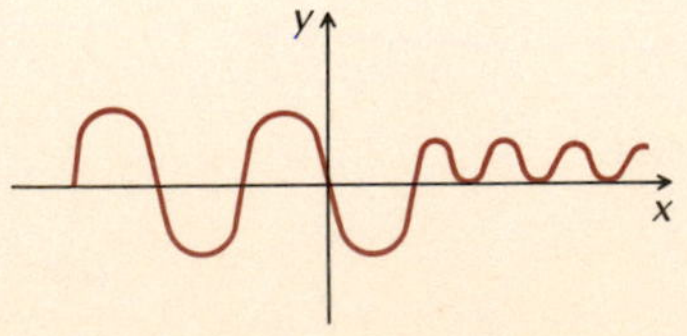

c)

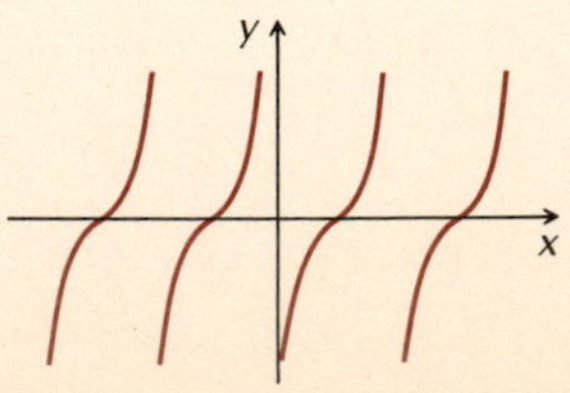

38. What is the period of this function?

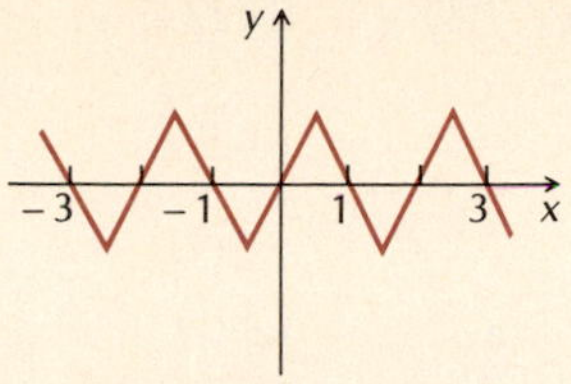

39. Is this function continuous in the interval (a) $[-3, -1]$? (b) $-1, 1$?

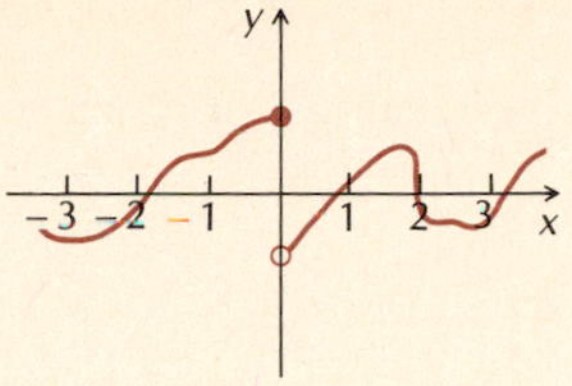

Use the following graphs of functions for Exercises 40–42.

a)

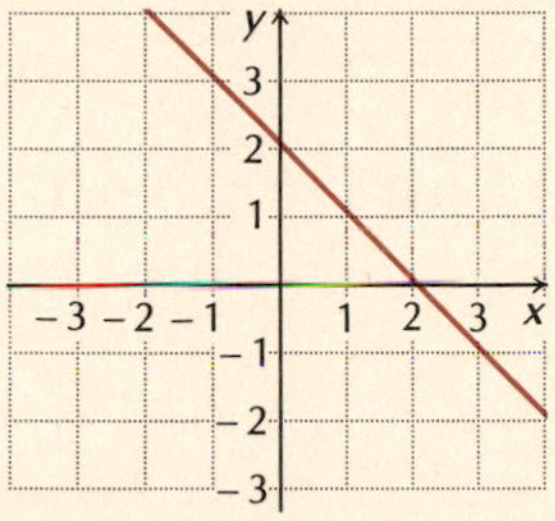

b)

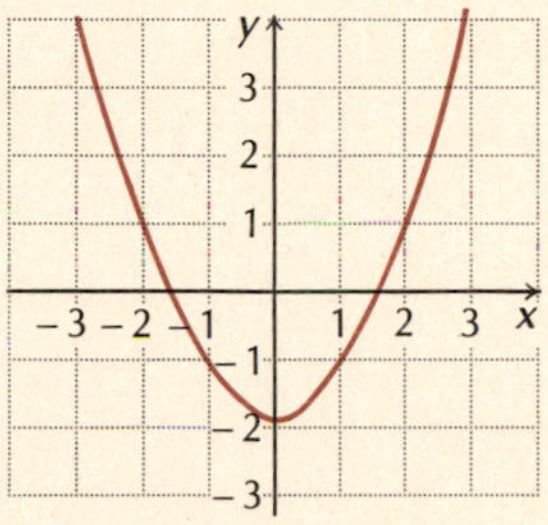

c)

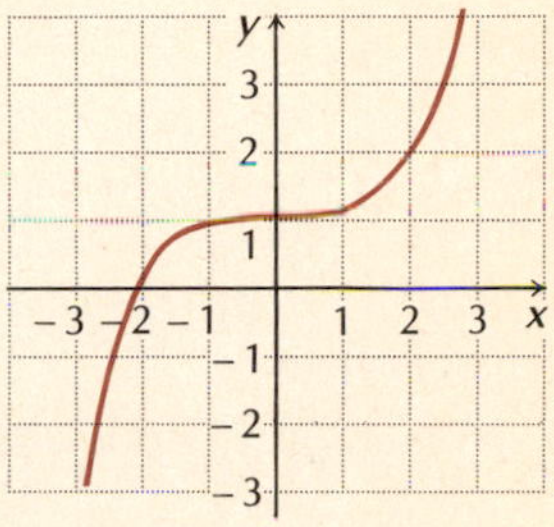

40. Which are increasing?

41. Which are decreasing?

42. Which are neither increasing nor decreasing?

Write interval notation for the set.

43. $\{x \mid -\pi \leq x \leq 2\pi\}$

44. $\{x \mid 0 < x \leq 1\}$

45. $\{x \mid x < 14\}$

Graph.

46. $f(x) = \begin{cases} x^2 + 2, & \text{for } x < 0, \\ x^3, & \text{for } 0 \leq x < 2, \\ -4x + 5, & \text{for } x \geq 2 \end{cases}$

47. $f(x) = \begin{cases} \dfrac{x^2 - 1}{x + 1}, & \text{for } x \neq -1, \\ 3, & \text{for } x = -1 \end{cases}$

48. Two cars leave the same intersection at right angles to each other. One travels at a speed of 50 mph and the other at 55 mph. Express the distance d, in miles, between the cars as a function of time t, in hours.

49. A right circular cylinder with radius r and height h is inscribed in a sphere with radius a. Express the volume of the cylinder as a function of a.

50. Find $f(x)$ and $g(x)$ such that $h(x) = f \circ g(x)$.

a) $h(x) = \sqrt{5x + 2}$

b) $h(x) = \dfrac{x^3 + 1}{x^3 - 1}$

SYNTHESIS

Find the domain.

51. $f(x) = (x - 9x^{-1})^{-1}$

52. $f(x) = \dfrac{\sqrt{1 - x}}{x - |x|}$

53. Graph several functions of the type $y = |f(x)|$. Describe a procedure, involving transformations, for graphing such functions.

54. Graph: $|x - y| = 1$.

TEST: CHAPTER 3

1. Find the Cartesian product $A \times B$, where $A = \{1, 3, 7\}$ and $B = \{3, 6\}$.

Consider the relation $\{(2, 7), (-2, -7), (7, -2), (0, 2)\}$ for Questions 2–4.

2. Determine whether the relation is a function.

3. Find the range.

4. Find the domain.

Graph.

5. $f(x) = (x - 2)^2$

6. $x = |y + 1|$

7. $f(x) = 2\ \mathrm{INT}(x)$

Consider the following relations for Questions 8 and 9.

a) $2y - \frac{5}{x} = 0$

b) $x = -5$

c) $y = x^2 - 2$

d) $x^3 - x = y^3$

e) $x^2 - 1 = y^2$

f) $3y = |x|$

8. Which are symmetric with respect to the origin?

9. Which are symmetric with respect to the x-axis?

10. Find the domain of the function f given by

$$\frac{1}{16 - x^2}.$$

Consider the functions f and g given by $f(x) = 2x - 1$ and $g(x) = x^2 + 6$ for Questions 11 and 12.

11. Find $(f + g)(x)$, $(f - g)(x)$, $fg(x)$, $(f/g)(x)$, $f \circ g(x)$, and $g \circ f(x)$.

12. Find the domain of f, g, $f + g$, $f - g$, fg, f/g, $f \circ g$, and $g \circ f$.

13. Which of the following are graphs of functions?

a)

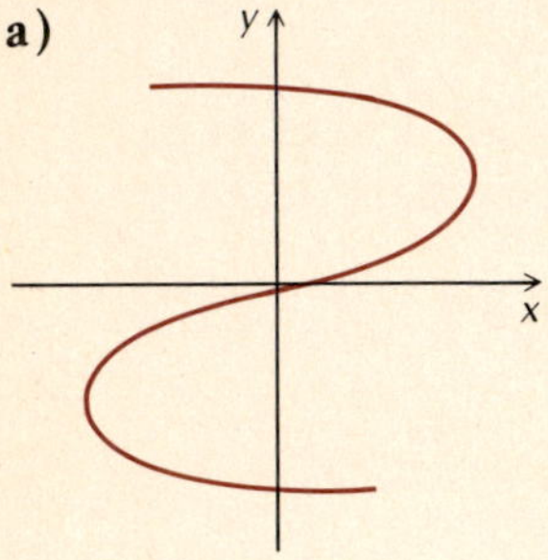

b)

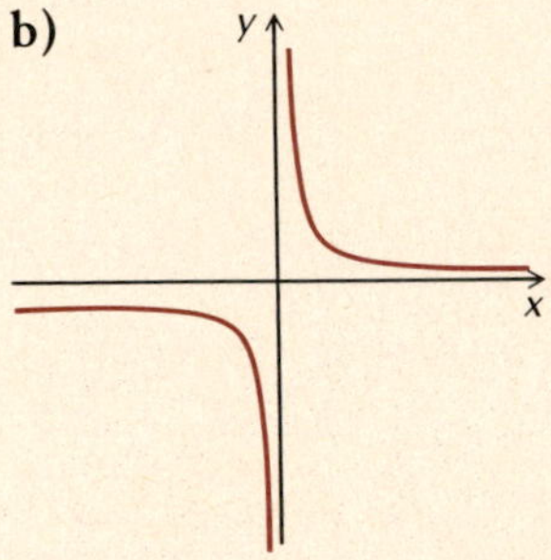

Use $g(x) = x^2 - x + 2$ for Questions 14–17. Find:

14. $g(-1)$

15. $g(0)$

16. $g(a - 1)$

17. $\frac{g(a + h) - g(a)}{h}$

18. Given

$$R(x) = x^2 + 110x + 60$$

and

$$C(x) = 1.1x^2 + 10x + 80,$$

find $P(x)$.

19. Find the equilibrium point for the supply and demand functions

$$D(x) = x^2 - 11x + 51$$

and

$$S(x) = x^2 + 18.$$

20. Here is a graph of $y = f(x)$. Sketch the graph of each of the following.

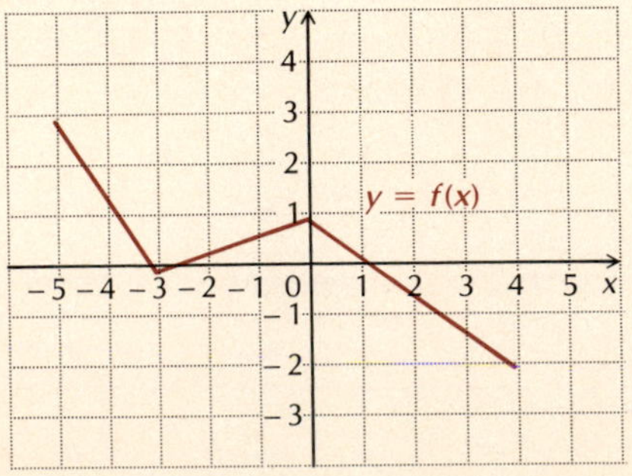

a) $y = f(x - 1)$

b) $y = f(2x)$

c) $y = 3 + f(x)$

Use the following for Questions 21 and 22.

a)

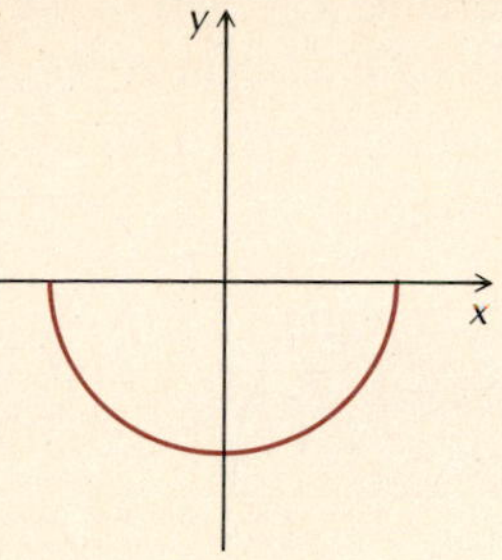

b)

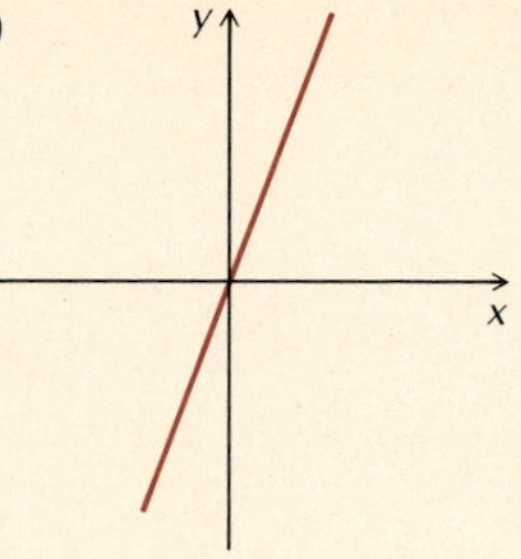

c) $f(x) = x^2 - 9$

d) $f(x) = x^3 - 2x + 4$

e) $f(x) = |x|$

f) $f(x) = x^7 - x^5$

21. Which are even?

22. Which are neither even nor odd?

Write interval notation for the set.

23. $\{x| -7 < x < 2\}$

24. $\{x|x \geq -2\}$

25. Which of the following functions are periodic?

a)

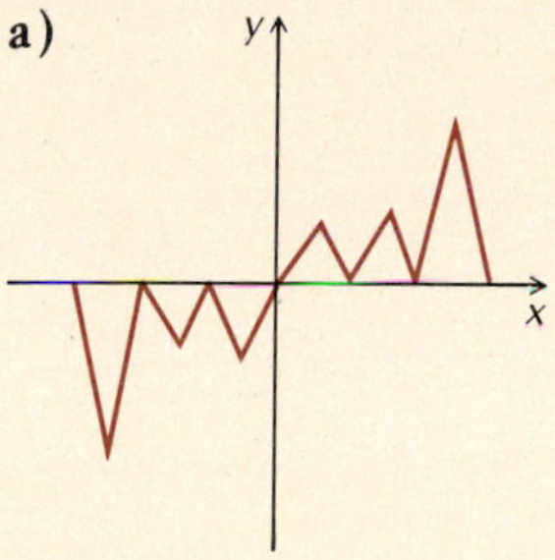

b)

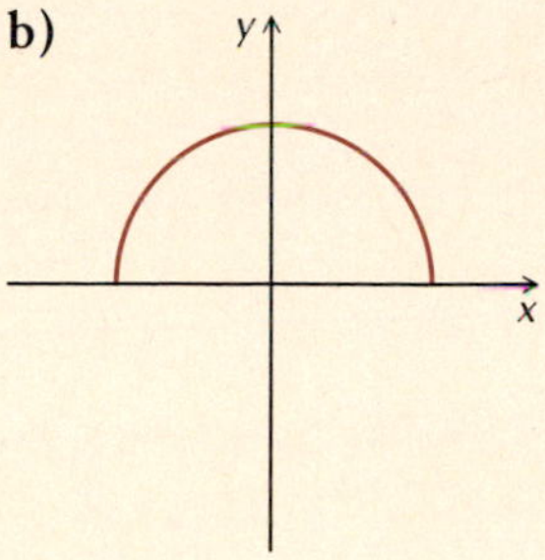

c)

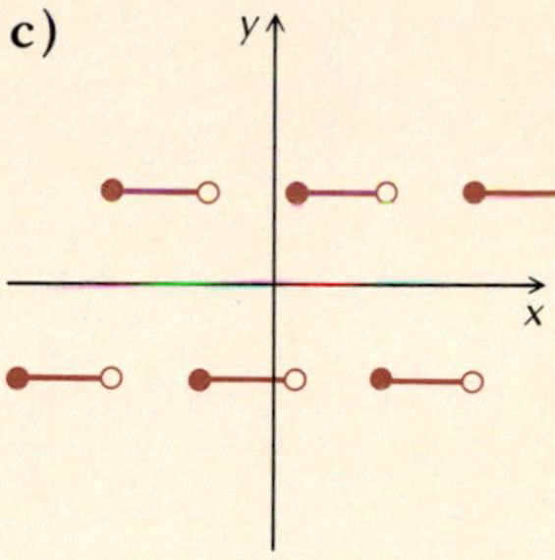

26. What is the period of this function?

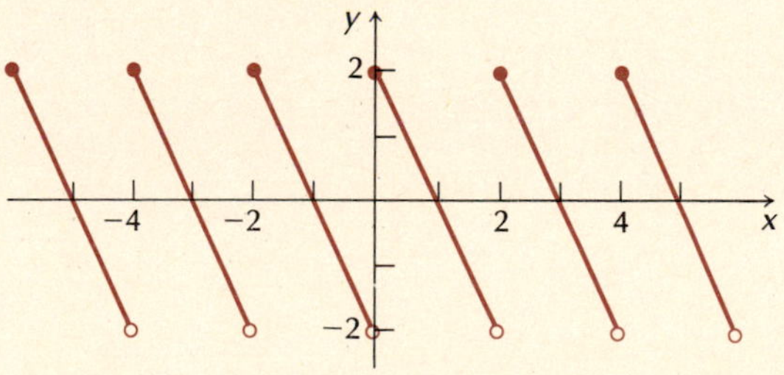

27. Is this function continuous in the interval (a) $[-1, 1]$? (b) $(-2, -1)$?

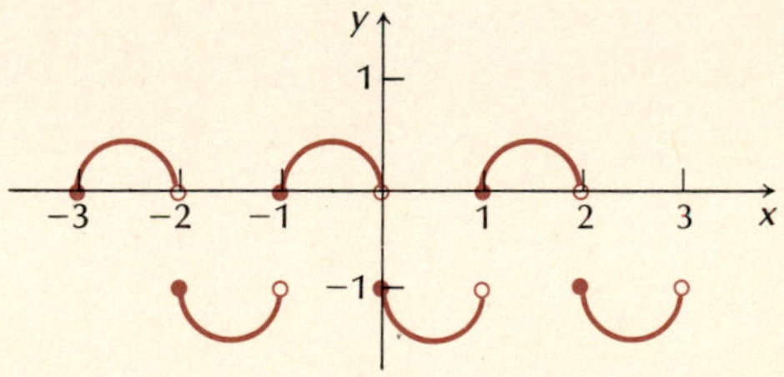

Use the following for Questions 28 and 29.

a)

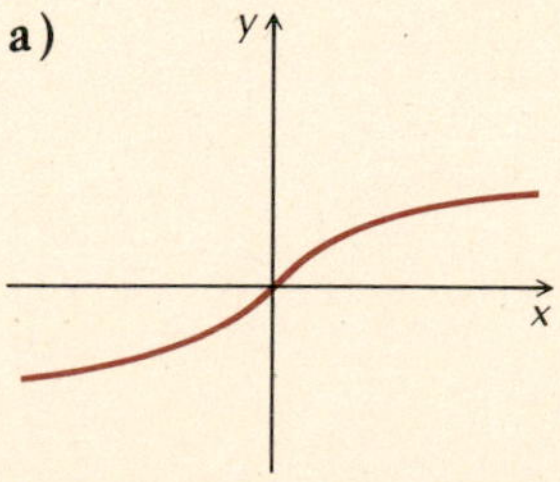

b)

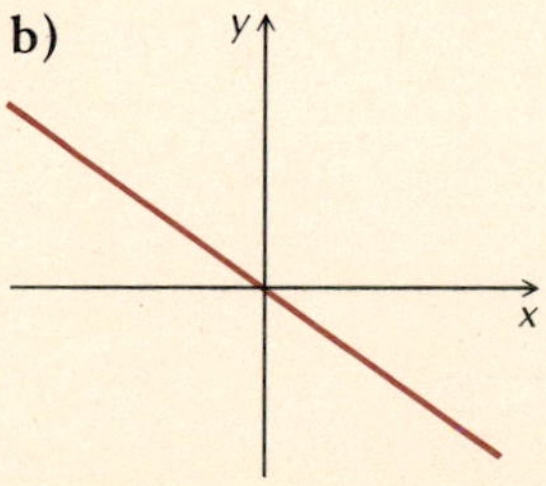

c)

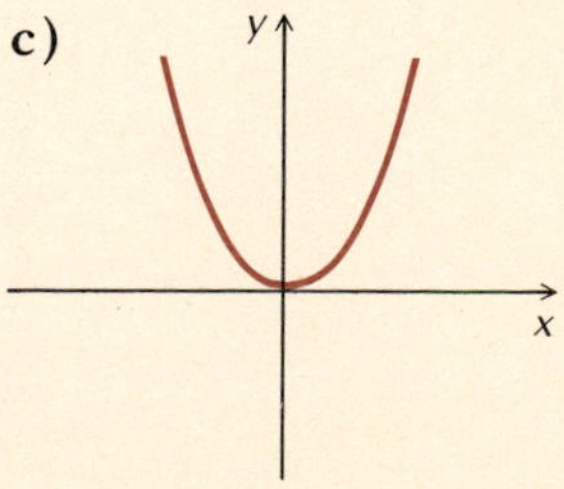

28. Which are decreasing?

29. Which are neither increasing nor decreasing?

30. Graph.

$$f(x) = \begin{cases} x^3, & \text{for } x < -2, \\ |x|, & \text{for } -2 \leq x < 2, \\ \sqrt{x-1}, & \text{for } x \geq 2. \end{cases}$$

31. A rectangle is inscribed in a semicircle of radius 2 as shown in the figure. The variable x = half of the length. Express the area of the rectangle as a function of x.

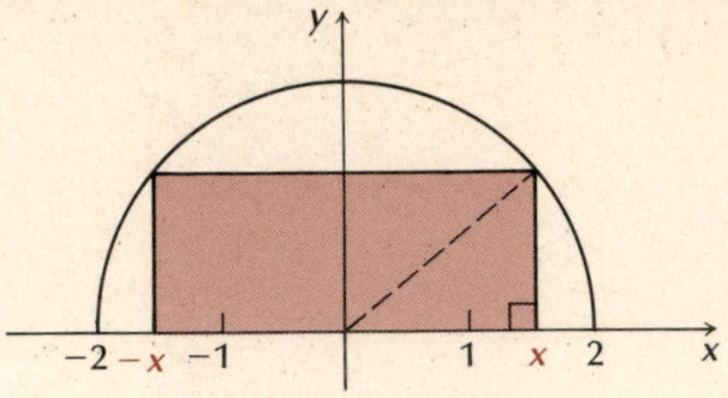

32. Find $f(x)$ and $g(x)$ such that $h(x) = f \circ g(x)$.

a) $h(x) = \dfrac{1}{\sqrt{7x + 2}}$

b) $h(x) = 4(5x - 1)^2 + 9$

SYNTHESIS

33. Prove that the sum of two odd functions is odd.

Linear, Quadratic, and Polynomial Functions

4

The most important kind of elementary function is the *polynomial function.* A polynomial function is a function that can be described by a polynomial expression. For example, a *linear function* is a polynomial function that can be described by an equation of the type $f(x) = mx + b$. A *quadratic function* is a polynomial function that can be described by an equation of the type $f(x) = ax^2 + bx + c,\ c \neq 0$. Another example of a polynomial function is given in the problem above. In this chapter, we study these kinds of functions in some detail, looking for their properties and using them in problem solving.

We will also consider linear, quadratic, and rational inequalities as well as some functions that involve the use of absolute value.

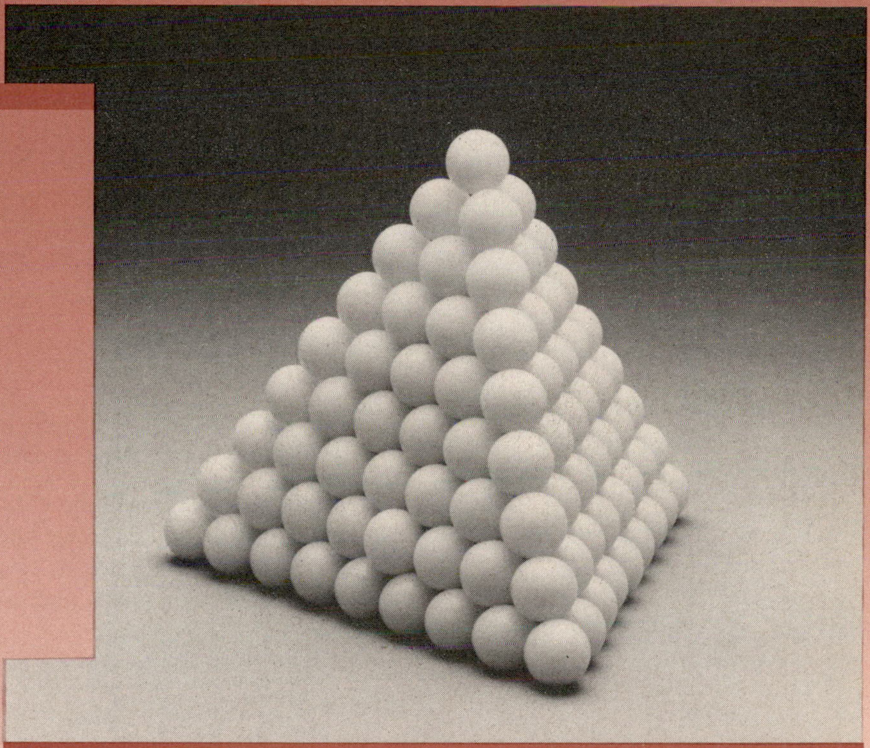

FEATURE PROBLEM

How many spheres, such as marbles or balls or oranges, are in the pile above? You might see such a question in connection with a guessing contest in a store. If the layers are equilateral triangles and all the spheres are the same size, there is a polynomial function that can be used to tell the count. The number N is given by the polynomial function

$$N(x) = \frac{1}{6}x^3 + \frac{1}{2}x^2 + \frac{1}{3}x,$$

where x is the number of layers and $N(x)$ is the total number of spheres. Thus, to win the contest, count the number of layers and substitute the value into the formula. Find the number of spheres in a pile with 9 layers.

THE MATHEMATICS

We find the number by substituting 9 for x in the polynomial:

$$\begin{aligned} N(9) &= \frac{1}{6}(9)^3 + \frac{1}{2}(9)^2 + \frac{1}{3}(9) \\ &= \frac{1}{6}(729) + \frac{1}{2}(81) + 3 \\ &= 121.5 + 40.5 + 3 = 165. \end{aligned}$$

When there are 9 layers, there will be 165 spheres.

OBJECTIVES

You should be able to:

1. Determine whether a given equation is linear.
2. Graph linear equations.
3. Find the slope, if it exists, of the line containing two given points.
4. Given the slope and the coordinates of one point on the line, find an equation of the line.
5. Given the coordinates of two points, find an equation of the line containing them.
6. Given an equation of a line, find its slope, if it exists, and its y-intercept, and graph linear equations using slope and y-intercept.
7. Solve problems involving applications of linear functions.

1. Determine whether each equation is linear.
   a) $3x = 2y + 4$
   b) $7y = 11$
   c) $5y^2x = 13$
   d) $x = \frac{4}{y}$

4.1 Lines and Linear Functions

1 Linear Equations

An equation of the type $Ax + By = C$ is called a **linear equation** because its graph is a straight line. In this case, A, B, and C are constants, but A and B cannot both be 0. Any equation that is equivalent to one of this form has a straight-line graph.

EXAMPLE 1 Determine whether each equation is linear.

a) $5x + 8y - 3 = 0$
b) $3x^2 - 4y + 5 = 0$
c) $4x + 3xy = 25$

Solution

a) The equation $5x + 8y - 3 = 0$ is linear because it is equivalent to $5x + 8y = 3$. Here, $A = 5$, $B = 8$, and $C = 3$.
b) The equation $3x^2 - 4y + 5 = 0$ is not linear because x is squared.
c) The equation $4x + 3xy = 25$ is not linear because the product xy occurs.

DO EXERCISE 1.

2 Graphs of Linear Equations

Since two points determine a line, we can graph a linear equation by finding two of its points. Then we draw a line through those points.

A third point should always be used as a check. The easiest points to find are often the intercepts (the points at which the line crosses the axes).

EXAMPLE 2 Graph: $4x + 5y = 20$.

Solution We set $x = 0$ and find that $y = 4$. Thus $(0, 4)$ is a point of the graph (the y-intercept).

We set $y = 0$ and find that $x = 5$. Thus $(5, 0)$ is a point of the graph (the x-intercept). The graph is shown below. A third point can be used as a check. We substitute any point, say -2, for x, and solve for y:

$$\begin{aligned} 4(-2) + 5y &= 20 \\ -8 + 5y &= 20 \\ 5y &= 28, \\ y &= 5\tfrac{3}{5}. \end{aligned}$$

This gives a check point: $(-2, 5\frac{3}{5})$.

x	y	
0	4	←y-intercept
5	0	←x-intercept
-2	$5\frac{3}{5}$	

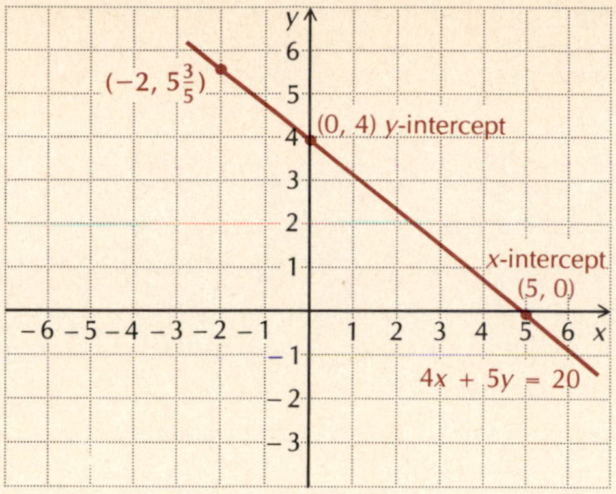

DO EXERCISES 2 AND 3.

If a graph, such as that of $y = 7x$, goes through the origin, then it has only one intercept, and other points will be needed for graphing.

If an equation has a missing variable ($A = 0$ or $B = 0$), then its graph is parallel to one of the axes. Vertical lines are parallel to the y-axis and are of the form $x = c$. Horizontal lines are parallel to the x-axis and are of the form $y = c$.

EXAMPLE 3 Graph (a) $y = 3$ and (b) $x = -2$.

Solution

a) Graph $y = 3$.

The graph is shown below. All y-coordinates are 3.

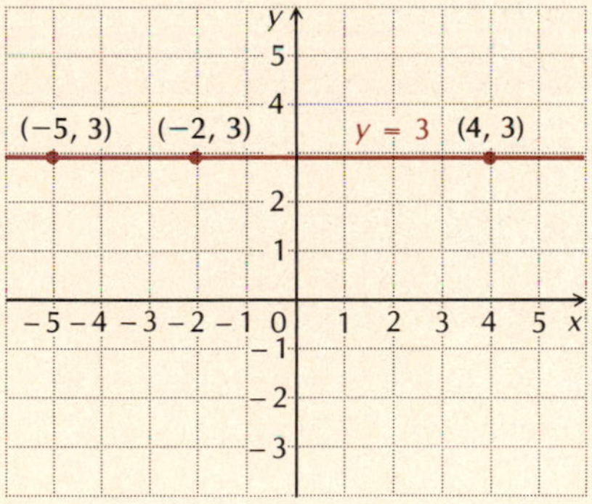

b) Graph $x = -2$.

The graph is shown below. All x-coordinates are -2.

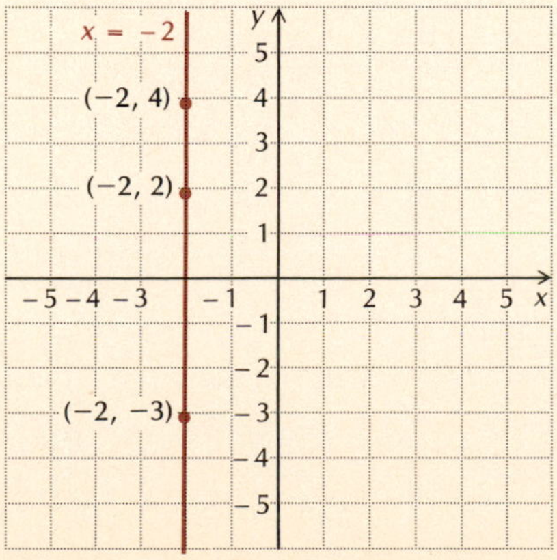

2. Graph $6x - 4y = 12$.

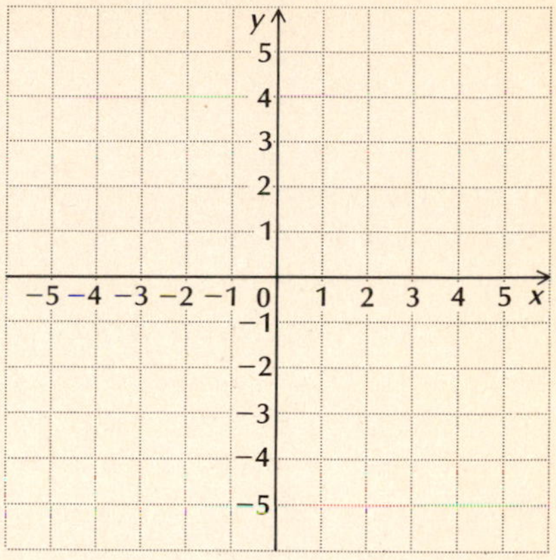

3. Graph $3x + 2y = 6$.

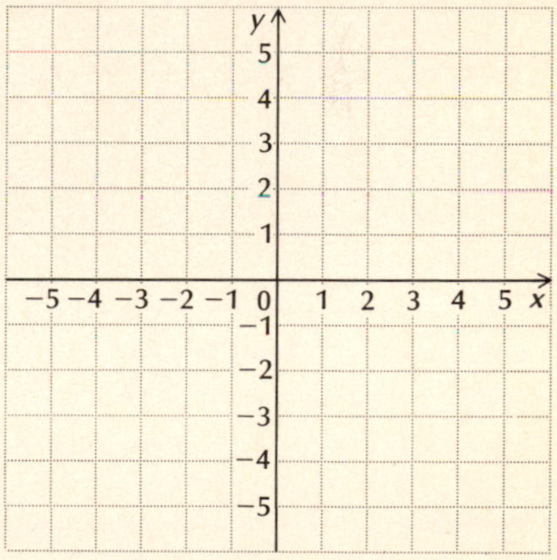

Graph.

4. $x = 4$

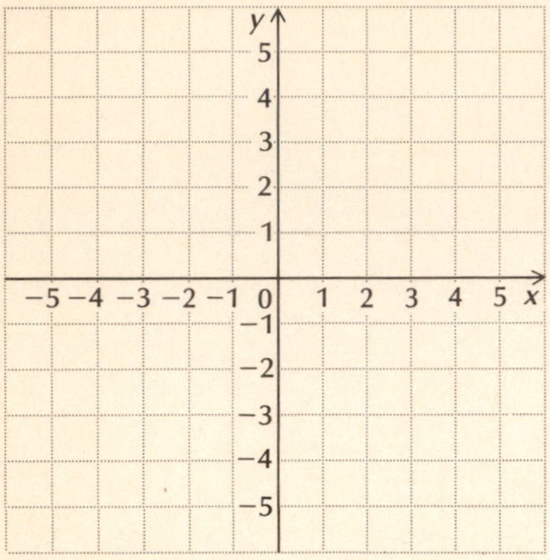

5. $y = -3$

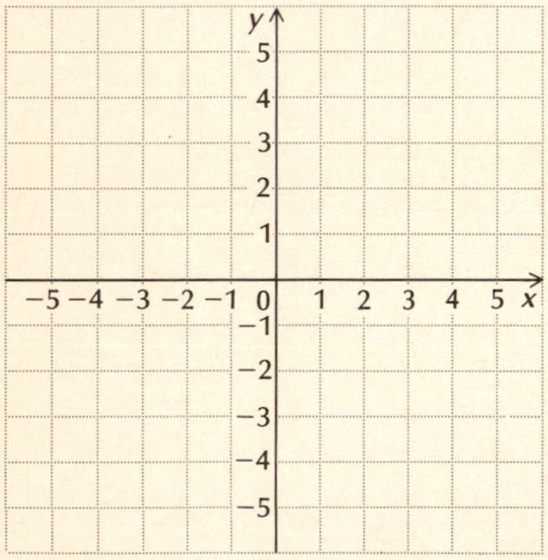

DO EXERCISES 4 AND 5.

3 Slope

The graph of a linear equation may slant upward or downward, or be horizontal or vertical. Let us see how this relates to equations.

Suppose that points P_1 and P_2 with coordinates (x_1, y_1) and (x_2, y_2) are two different points on a line not parallel to an axis. Consider a right triangle, as shown, with legs parallel to the axes. The point P_3 with coordinates (x_2, y_1) is the third vertex of a triangle. As we move from P_1 to P_2, y changes from y_1 to y_2. The change in y is $y_2 - y_1$. Similarly, the change in x is $x_2 - x_1$. The ratio of these changes is called the **slope**.

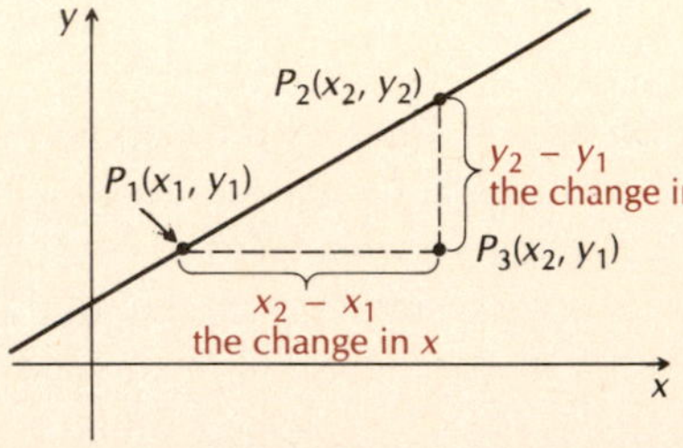

DEFINITION

The *slope* m of a line containing two points (x_1, y_1) and (x_2, y_2) is defined by

$$m = \frac{y_2 - y_1}{x_2 - x_1}, \text{ where } x_2 \neq x_1.$$

Note that when $x_2 = x_1$, $x_2 - x_1 = 0$, and the slope is not defined.

EXAMPLE 4 Graph the line through the points (1, 2) and (3, 6) and find its slope.

Solution Let us call the slope m and let (1, 2) be (x_1, y_1) and (3, 6) be (x_2, y_2). Applying the definition, we obtain

$$m = \frac{y_2 - y_1}{x_2 - x_1} = \frac{6 - 2}{3 - 1} = \frac{4}{2} = 2.$$

Note that we can also use the points in the opposite order, so long as we are consistent. We get the same slope:

$$m = \frac{y_2 - y_1}{x_2 - x_1} = \frac{2 - 6}{1 - 3} = \frac{-4}{-2} = 2.$$

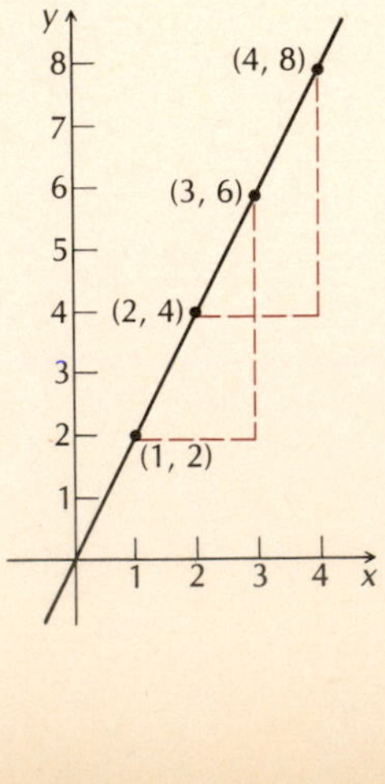

From Example 4, we see that it does not matter in which order we choose the points, so long as we take differences in the same order. From Example 4, we can also see that it does not matter which two points of a line we choose to determine the slope. No matter what points we choose, we get the same number for the slope. For example, if we choose (2, 4) and (4, 8), we get

$$\frac{8-4}{4-2} = \frac{4}{2} = 2$$

for the slope.

DO EXERCISES 6–11.

If a line slants upward from left to right, it has a positive slope, and the larger the slope is, the steeper the line. If a line slants downward from left to right, the change in x and the change in y are of opposite signs, so the line has a negative slope. The larger the absolute value of the slope, the steeper the line.

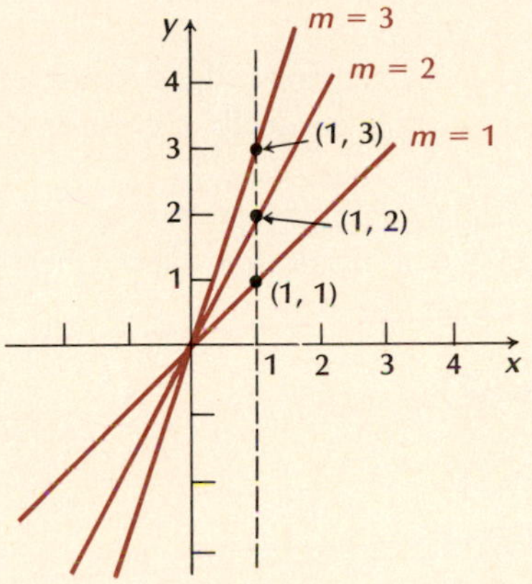

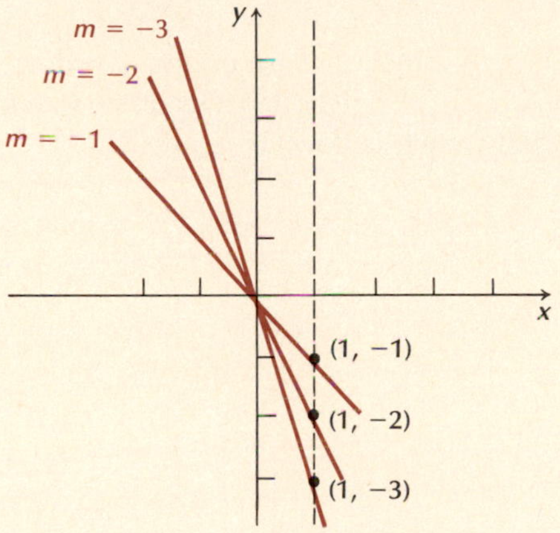

The following summarizes the results for horizontal and vertical lines.

a) If a line is horizontal, the change in y for any two points is 0. Thus a horizontal line has zero slope.
b) If a line is vertical, the change in x for any two points is 0. Thus the slope is not defined, because we cannot divide by zero.

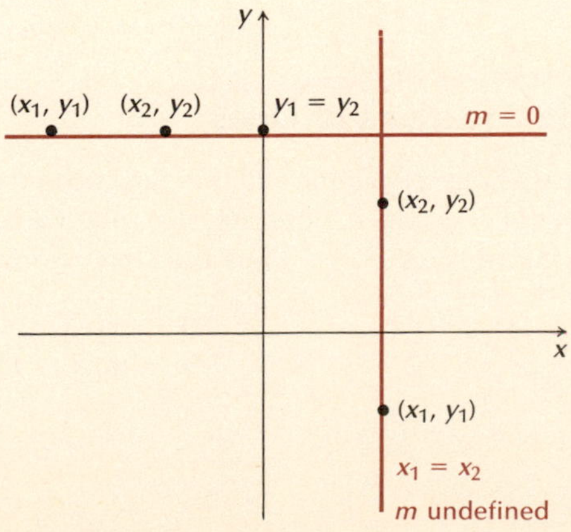

DO EXERCISES 12–14.

Use graph paper. Graph the line through the points and find its slope.

6. (1, 3) and (2, 5)

7. (3, 7) and (5, 3)

8. (1, 1) and (2, 3)

9. (3, 5) and (2, −1)

10. (−1, −1) and (2, −4)

11. (0, 2) and (3, 1)

Find the slope, if it exists, of the line containing these points.

12. (4, 6) and (−2, 6)

13. (−3, 5) and (−3, 7)

14. (9, 0) and (6, 0)

15. Find the slope, or pitch, of the right side of this roof.

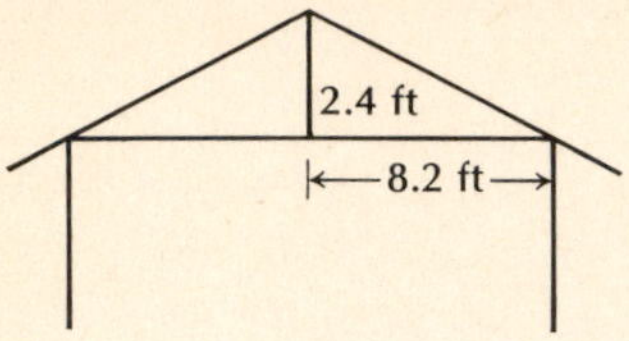

Applications of Slope

Slope has many real-world applications. For example, numbers like 2%, 3%, and 10% are often used to represent the **grade** of a road. Such a number is meant to tell how steep a road up a hill or mountain is. For example, a 5% grade means that for every horizontal distance of 100 ft, the road rises 5 ft, if a vehicle is going upward; and −5% means that the road is dropping 5 ft for every 100 ft, if the vehicle is going downward. The concept of grade is also relevant in cardiology when a person runs on a treadmill. A physician may change the slope or grade to measure its effects on heartbeat.

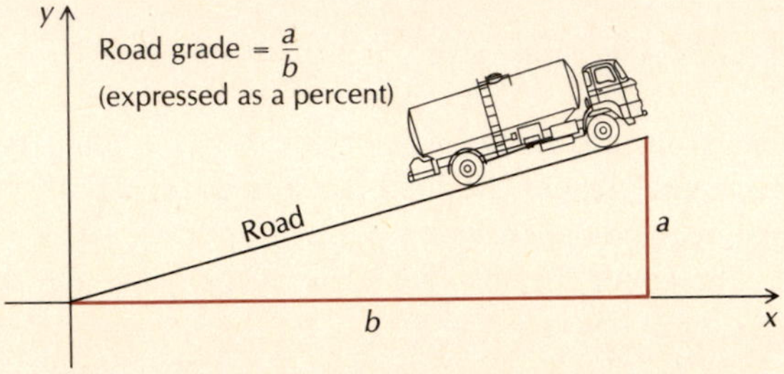

Another example occurs in hydrology. When a river flows, the strength or force of the river depends on how much the river falls vertically compared to how much it flows horizontally.

EXAMPLE 5 A treadmill 5 ft long is raised 0.6 ft vertically when a heart arrhythmia occurs. What is the grade of the treadmill?

Solution The grade is positive for treadmill tests and is the slope of the line on which the patient is jogging. Thus,

$$m = \frac{0.6 \text{ ft}}{5 \text{ ft}} = 12\%.$$

DO EXERCISE 15.

4 Point–Slope Equations of Lines

Suppose that we have a nonvertical line and that the coordinates of one point P_1 are (x_1, y_1). We think of P_1 as fixed. Suppose, also, that we have a movable point P on the line with coordinates (x, y). Thus the slope would be given by

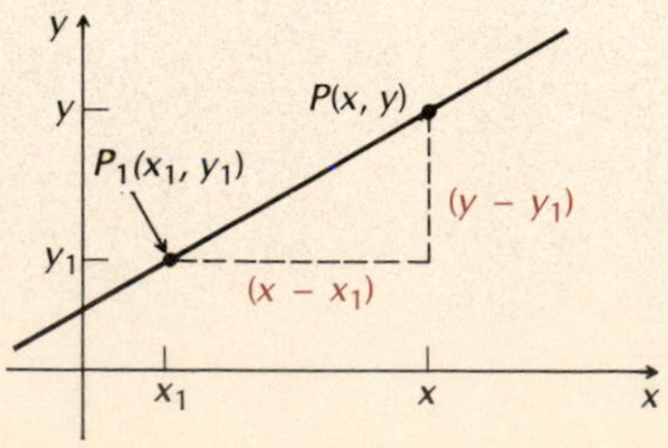

$$\frac{y - y_1}{x - x_1} = m. \quad (1)$$

Note that this is true only when (x, y) is a point different from (x_1, y_1). If we use the multiplication principle, we get*

$$(y - y_1) = m(x - x_1). \quad \textit{Point–slope equation} \quad (2)$$

Equation (2) will be true even if $(x, y) = (x_1, y_1)$. Equation (2) is called the **point–slope equation** of a line. Thus if we know the slope of a line and the coordinates of a point on the line, we can find an equation of the line.

EXAMPLE 6 Find an equation of the line containing the point $\left(\frac{1}{2}, -1\right)$ with slope 5.

Solution If we substitute in $(y - y_1) = m(x - x_1)$, we get $y - (-1) = 5(x - \frac{1}{2})$, which simplifies to

$$y + 1 = 5\left(x - \frac{1}{2}\right),$$

or

$$y = 5x - \frac{5}{2} - 1,$$

or

$$y = 5x - \frac{7}{2}.$$

DO EXERCISES 16–18.

5 Two-Point Equations of Lines

Suppose that a nonvertical line contains the points $P_1(x_1, y_1)$ and $P_2(x_2, y_2)$. The slope of the line is

$$\frac{y_2 - y_1}{x_2 - x_1}.$$

If we substitute $(y_2 - y_1)/(x_2 - x_1)$ for m in the point–slope equation,

$$y - y_1 = m(x - x_1),$$

we have†

$$y - y_1 = \frac{y_2 - y_1}{x_2 - x_1}(x - x_1). \quad \textit{Two-point equation}$$

This is known as the **two-point equation** of a line. Note that either of the two given points can be called P_1 or P_2 and (x, y) is any point on the line.

EXAMPLE 7 Find an equation of the line containing the points (2, 3) and (1, −4).

*Hint: It may be easier to remember Eq. (1), since it relates to the slope formula. We can easily get Eq. (2) from it.

†Hint: It may be easier to remember

$$\frac{y - y_1}{x - x_1} = \frac{y_2 - y_1}{x_2 - x_1},$$

interpreting each side as slope.

Find an equation of the line.

16. Containing the point $\left(-2, \frac{1}{4}\right)$ with slope -3

17. With y-intercept $(0, -9)$ and slope $\frac{1}{4}$

18. With x-intercept $(5, 0)$ and slope $-\frac{1}{2}$

Find an equation of the line containing the given pair of points.

19. (1, 4) and (3, −2)

20. (3, −6) and (0, 4)

Solution If we take (2, 3) as P_1 and (1, −4) as P_2 and use the two-point equation, we get

$$y - 3 = \frac{-4 - 3}{1 - 2}(x - 2),$$

which simplifies to $y = 7x - 11$.

DO EXERCISES 19 AND 20.

6 Slope–Intercept Equations of Lines

Suppose that a nonvertical line has slope m and y-intercept $(0, b)$. We sometimes, for brevity, refer to the number b as the y-intercept. Let us substitute m and $(0, b)$ in the point–slope equation. We get

$$y - y_1 = m(x - x_1)$$
$$y - b = m(x - 0).$$

This simplifies to

$$y = mx + b. \qquad \textit{Slope–intercept equation}$$

This is called the **slope–intercept equation** of a line. The advantage of such an equation is that we can read the slope m and the y-intercept b from the equation.

Find the slope and the y-intercept.

21. $y = -7x + 11$

22. $y = -4$

EXAMPLE 8 Find the slope and the y-intercept of $y = 5x - \frac{1}{4}$.

Solution

$$y = 5x - \frac{1}{4}$$

Slope: 5 $\qquad$ y-intercept: $-\frac{1}{4}$

EXAMPLE 9 Find the slope and the y-intercept of $y = 8$.

Solution We can rewrite this equation as $y = 0x + 8$. We then see that the slope is 0 and the y-intercept is 8. The graph of the equation is a horizontal line.

DO EXERCISES 21 AND 22.

To find the slope–intercept equation of a line, when given another equation, we solve for y.

EXAMPLE 10 Find the slope and the y-intercept of the line having the equation $3x - 6y - 7 = 0$.

Solution We solve for y, obtaining $y = \frac{1}{2}x - \frac{7}{6}$. Thus the slope is $\frac{1}{2}$ and the y-intercept is $-\frac{7}{6}$.

If a line is vertical, it has no slope. Thus it has no slope–intercept equation. Such a line does have a simple equation, however. All vertical lines have equations $x = c$, where c is some constant.

DO EXERCISE 23.

We can graph linear equations using the slope and the y-intercept.

EXAMPLE 11 Graph: $y = -\frac{2}{3}x + 1$.

Solution First we plot the y-intercept (0, 1). We think of the slope as $-2/3$. Starting at the y-intercept and using the slope, we find another point by moving 2 units down (since the numerator is *negative* and corresponds to the change in y) and 3 units right (since the denominator is *positive* and corresponds to the change in x). We get to a new point, $(3, -1)$. In a similar manner, we can find other points such as $(6, -3)$ by moving on from $(3, -1)$.

If we think of the slope as $2/-3$, then we start again at the y-intercept, (0, 1), and find another point by moving 2 units up (since the numerator is *positive* and corresponds to the change in y) and 3 units left (since the denominator is *negative* and corresponds to the change in x). We get to another point on the line, $(-3, 3)$.

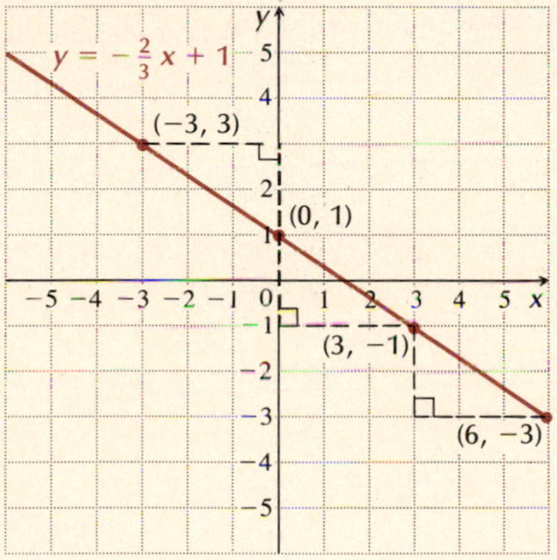

EXAMPLE 12 Graph: $y = \frac{2}{5}x + 4$.

Solution First we plot the y-intercept, (0, 4). We then consider the slope, $\frac{2}{5}$. Starting at the y-intercept and using the slope, we find another point by moving 2 units up (since the numerator is *positive* and corresponds to the change in y) and 5 units right (since the denominator is *positive* and corresponds to the change in x). We get to a new point, (5, 6).

By thinking of the slope as $-2/-5$, we can find another point, $(-5, 2)$.

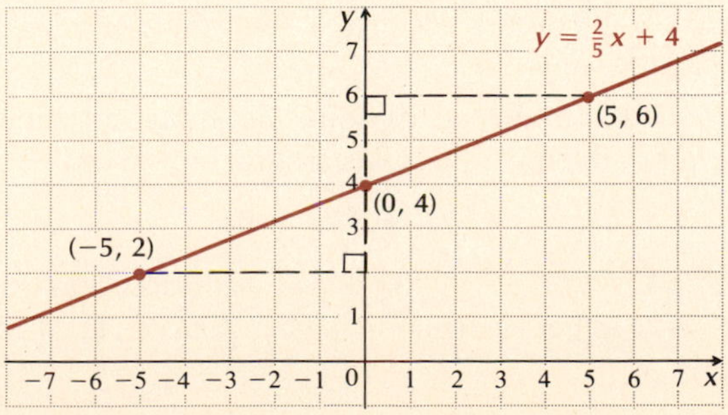

DO EXERCISES 24 AND 25.

23. a) Find the slope–intercept equation of the line whose equation is

$$-2x + 3y - 6 = 0.$$

b) Find the slope and the y-intercept of this line.

Graph.

24. $y = \frac{3}{5}x + 2$

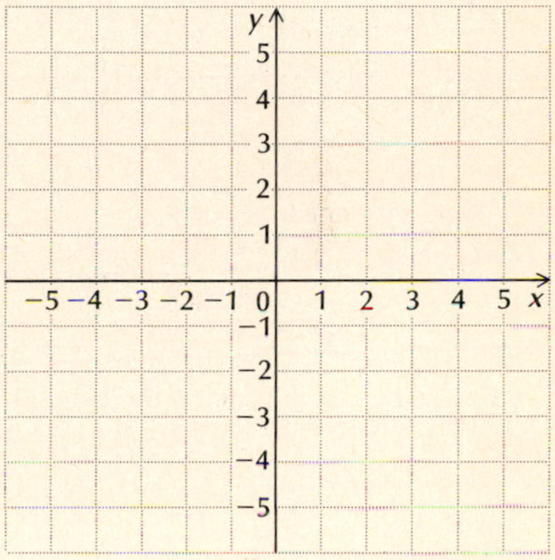

25. $y = -\frac{3}{5}x - 1$

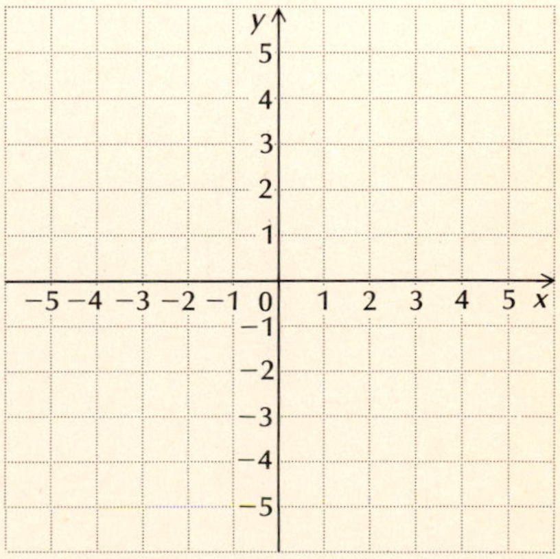

26. *Pressure at sea depth.* The pressure P, given by

$$P(d) = 1 + \frac{1}{33}d,$$

gives the pressure, in atmospheres (atm), at a depth d, in feet, in the sea.

a) Find $P(0)$, $P(5)$, $P(10)$, $P(33)$, and $P(200)$.

b) Graph P.

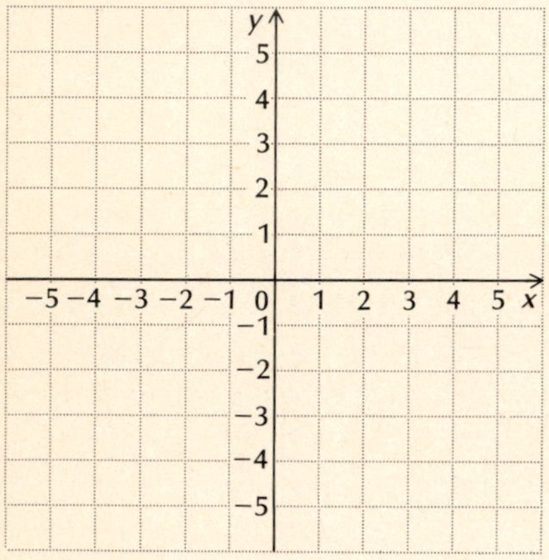

c) Discuss the domain of the function.

7 Linear Functions and Applications

Any nonvertical straight line is the graph of a function. Such a function is called a *linear function.* Any nonvertical line has an equation $y = mx + b$. Thus a function f is a **linear function** if and only if it has an equation $f(x) = mx + b$. If the slope m is zero, then the equation simplifies to $f(x) = b$. A function like this is called a **constant function.** If the slope m is positive, the function is *increasing.* If the slope is negative, the function is *decreasing.*

Let us consider an application of a linear function.

EXAMPLE 13 *Stopping distance on glare ice.* The stopping distance D (at some fixed speed) of regular tires on glare ice is a function of the air temperature F, in degrees Fahrenheit. This function D is estimated by

$$D(F) = 2F + 115,$$

where $D(F) =$ the stopping distance, in feet, when the air temperature is F, in degrees Fahrenheit.

a) Find $D(0°)$, $D(-20°)$, $D(10°)$, and $D(32°)$.
b) Graph F.
c) Explain why the domain of the function should be restricted to the interval $[-57.5°, 32°]$.

Solution

a) We find the function values by substitution:

$$\begin{aligned} D(0°) &= 2(0) + 115 \\ &= 115 \text{ ft}, \\ D(-20°) &= 2(-20) + 115 \\ &= -40 + 115 \\ &= 75 \text{ ft}, \\ D(10°) &= 2(10) + 115 \\ &= 135 \text{ ft}, \\ D(32°) &= 2(32) + 115 \\ &= 179 \text{ ft}. \end{aligned}$$

b) The graph of the function is as follows. Note that this is an increasing function.

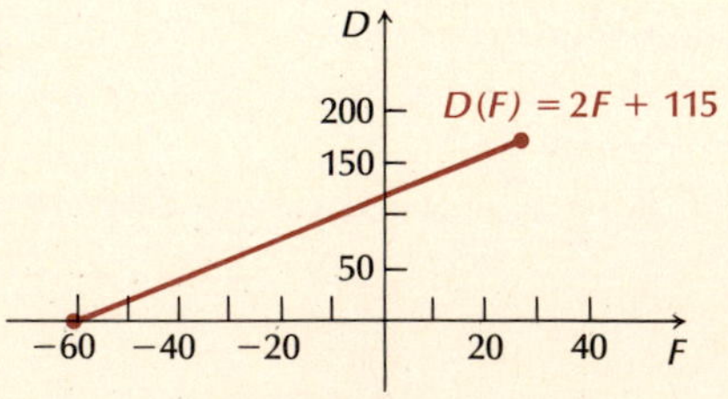

c) The function has certain restrictions. Stopping distance can only be nonnegative. Thus the function is meaningful only for values of x for which

$$2F + 115 \geq 0.$$

Solving, we get $F \geq -57.5°$. Now we have ice only when $F \leq 32°$. This gives us another restriction. The domain is thus the interval $[-57.5°, 32°]$. ■

DO EXERCISE 26.

EXERCISE SET 4.1

1

1. Determine whether the equation is linear.

a) $3y = 2x - 5$ b) $5x + 3 = 4y$ c) $3y = x^2 + 2$ d) $y = 3$

e) $xy = 5$ f) $3x^2 + 2y = 4$ g) $3x + \frac{1}{y} = 4$ h) $5x - 2 = 4y$

2. Determine whether the equation is linear.

a) $5y = 3x - 4$ b) $3x + 5 = 7y$ c) $4y = 3x^2 - 4$ d) $4x - \frac{2}{y} = 3$

e) $2xy = 4$ f) $5x^2 + 3y = -4$ g) $x = -4$ h) $6x - 7 = 3y$

2 Graph.

3. $8x - 3y = 24$
4. $5x - 10y = 50$
5. $3x + 12 = 4y$
6. $4x - 20 = 5y$
7. $y = -2$
8. $2y - 3 = 9$
9. $5x + 2 = 17$
10. $19 = 5 - 2x$

3 Find the slope of the line containing the given points.

11. $(6, 2)$ and $(-2, 1)$
12. $(-2, 1)$ and $(-4, -2)$
13. $(2, -4)$ and $(4, -3)$
14. $(5, -3)$ and $(-5, 8)$
15. $(\pi, 5)$ and $(\pi, 4)$
16. $(\sqrt{2}, -4)$ and $(\pi, -4)$
17. $(\sqrt{2}, 13)$ and $(\pi, 13)$
18. $(-8, \sqrt{2})$ and $(-8, \pi)$

Find the road grade and an equation giving the height y as a function of the horizontal distance x.

19.

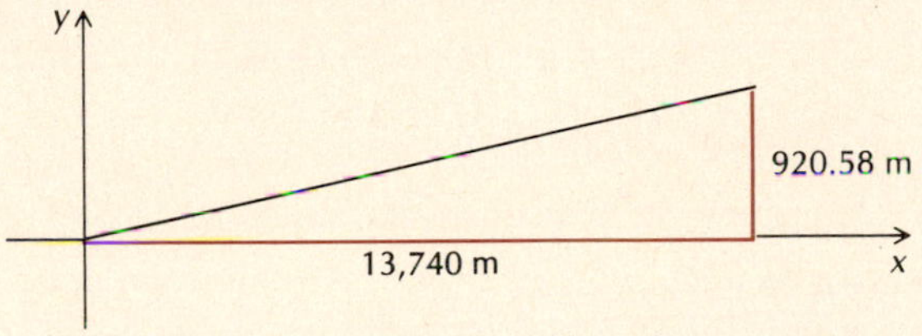

20.

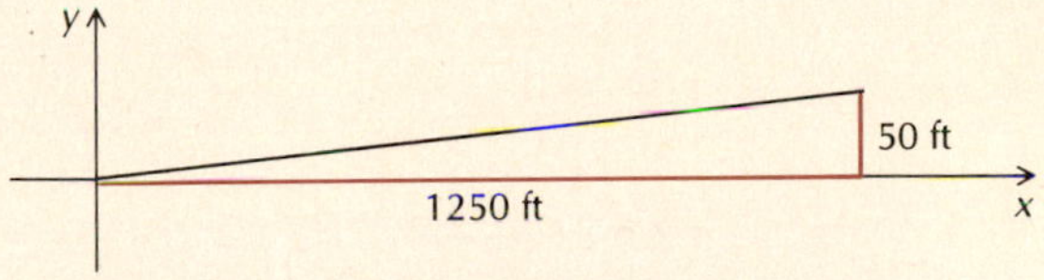

21. A road drops 158.4 ft vertically for every 5280 ft horizontally. What is the grade of the road?
22. A river drops 55.71 ft vertically for every 1238 ft horizontally. What is the slope of the river?
23. A treadmill is 5 ft long and is set at an 8% grade when a heart arrhythmia occurs. How high vertically is the end of the treadmill?
24. A river flows at a slope of 0.12. How many feet does it fall vertically for every 250 ft horizontally?

4 Find the equation of the line.

25. Through $(3, 2)$ with $m = 4$
26. Through $(4, 7)$ with $m = -2$
27. With y-intercept -5 and $m = 2$
28. With y-intercept π and $m = \frac{1}{4}$
29. Through $(-4, 7)$ with $m = -\frac{2}{3}$
30. Through $(-3, -5)$ with $m = \frac{3}{4}$
31. Through $(5, -8)$ with $m = 0$
32. Through $(5, -8)$ with m undefined

5 Find the equation of the line.

33. Containing $(1, 4)$ and $(5, 6)$
34. Containing $(-2, 0)$ and $(2, 3)$
35. Containing $(-2, 5)$ and $(-4, -7)$
36. Containing $\left(\frac{2}{3}, -\frac{4}{5}\right)$ and $\left(-\frac{1}{2}, 8.2\right)$
37. Containing $(3, 6)$ and $(-2, 6)$
38. Containing $\left(-\frac{3}{8}, 0\right)$ and $\left(-\frac{3}{8}, \frac{8}{3}\right)$

6 Find the slope and the y-intercept of the line.

39. $y = 2x + 3$
40. $y = 6 - x$
41. $2y = -6x + 10$
42. $-3y = -12x + 9$
43. $3x - 4y = 12$
44. $5x + 2y = -7$
45. $3y + 10 = 0$
46. $y = 7$

Graph.

47. $y = -\frac{3}{2}x$
48. $y = \frac{2}{3}x$
49. $y = -\frac{5}{2}x - 2$
50. $y = -\frac{5}{3}x + 3$
51. $y = \frac{1}{2}x + 1$
52. $y = \frac{1}{3}x - 1$
53. $y = \frac{4}{3} - \frac{1}{3}x$
54. $y = -\frac{1}{4}x - \frac{1}{2}$

▦ Find the equation of the line.

55. Through $(3.014, -2.563)$ with slope 3.516

56. Through $(-173.4, -17.6)$ with slope -0.00014

57. Through the points $(1.103, 2.443)$ and $(8.114, 11.012)$

58. Through the points $(473.78, 910.2)$ and $(993.55, 171.43)$

7

59. *Temperature and depth in the earth.* The function T given by
$$T(d) = 10d + 20$$
can be used to determine the temperature T, in degrees Celsius, at a depth d, in kilometers, inside the earth.
a) Find $T(5\text{ km})$, $T(20\text{ km})$, and $T(1000\text{ km})$.
b) Graph T.
c) The radius of the earth is about 5600 km. Use this fact to determine the domain of this function.

60. *Toll road charges.* The function C given by
$$C(d) = 0.027d + 0.32$$
can be used to estimate the cost C, in dollars, of driving a car d miles on the Indiana toll road.
a) Find $C(0)$, $C(10)$, $C(100)$, and $C(153)$.
b) Graph C.
c) The entire distance of the toll road is 153 mi. Use this fact to determine the domain of the function.

61. *Tail length of a snake.* It has been found in a study that the total length L, and the tail length T, both in millimeters, of females of the snake species *Lampropeltis Polyzona* are related by the linear function
$$T(L) = 0.143L - 1.18.$$
a) Find $T(0\text{ mm})$, $T(50\text{ mm})$, and $T(80\text{ mm})$.
b) Graph T.
c) Determine the domain of the function.

62. *Straight-line depreciation.* A company buys an office machine for \$5200 on January 1 of a given year. The machine is expected to last 8 years, at the end of which time its *trade-in,* or *salvage, value* will be \$1100. If the company figures the decline or depreciation in value to be the same each year, then the salvage value V, after t years, $0 \le t \le 8$, is given by the linear function
$$V(t) = \$5200 - \$512.50t.$$
a) Find $V(0)$, $V(1)$, $V(2)$, $V(3)$, and $V(8)$.
b) Graph V.
c) Determine the domain of the function.

63. *Biology: Spread of an organism.* A certain kind of organism is released over an area of 2 mi^2. It grows and spreads over more area. The area covered by the organism after time t is given by the linear function
$$A(t) = 1.1t + 2,$$
where $A(t)$ = the area covered, in square miles, after time t, in years.
a) Find $A(0)$, $A(1)$, $A(4)$, and $A(10)$.
b) Graph $A(t)$.
c) Why should the domain be restricted to the interval $[0, \infty)$?

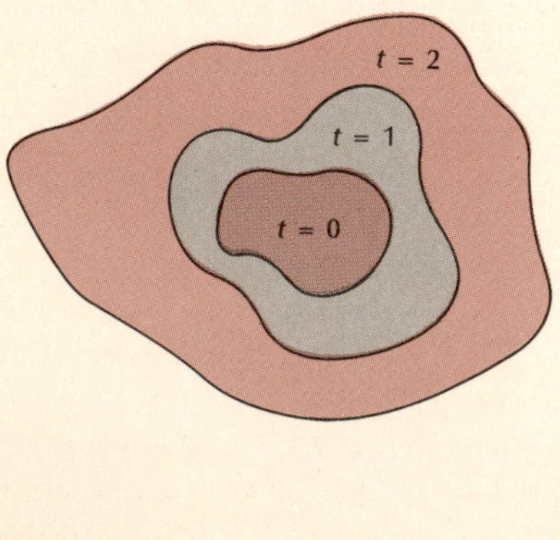

64. *Anthropology: Estimating heights.* An anthropologist can use certain linear functions to estimate the height of a male or female, given the length of certain bones. A *humerus* is the bone from the elbow to the shoulder. Let x = the length of the humerus in centimeters. Then the height, in centimeters, of a male with a humerus of length x is given by
$$M(x) = 2.89x + 70.64.$$
The height, in centimeters, of a female with a humerus of length x is given by
$$F(x) = 2.75x + 71.48.$$
A 45-cm humerus was uncovered in a ruins.
a) If we assume that it was from a male, how tall was he?
b) If we assume that it was from a female, how tall was she?
c) Discuss the domain of the function.

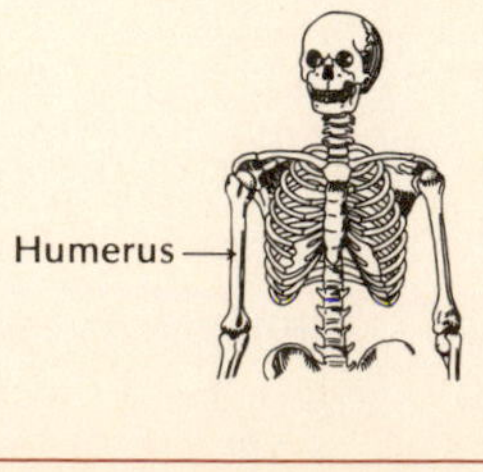

SYNTHESIS

Suppose that f is a linear function. Then $f(x) = mx + b$. Find a formula for $f(x)$ given each of the following.

65. $f(3x) = 3f(x)$

66. $f(kx) = kf(x)$, for some number k, $k \neq 1$

67. $f(x + 2) = f(x) + 2$

68. $[f(x)]^2$ is linear

Suppose that f is a linear function. Then $f(x) = mx + b,\ b \neq 0$. Determine whether each of the following is true or false.

69. $f(c + d) = f(c) + f(d)$

70. $f(cd) = f(c)f(d)$

71. $f(kx) = kf(x)$

72. $f(c - d) = f(c) - f(d)$

73. f is increasing, if $m > 0$

74. f is decreasing, if $m < 0$

75. Determine whether these three points are on a line. [*Hint:* Compare the slopes of $\overline{AB}$ and $\overline{BC}$. ($\overline{AB}$ refers to the segment from A to B.)]

$$A(9, 4),\quad B(-1, 2),\quad C(4, 3)$$

76. Determine whether these three points are on a line. (See the hint for Exercise 75.)

$$A(-1, -1),\quad B(2, 2),\quad C(-3, -4)$$

77. Use graph paper. Plot the points $A(0, 0)$, $B(8, 2)$, $C(11, 6)$, and $D(3, 4)$. Draw $\overline{AB}$, $\overline{BC}$, $\overline{CD}$, and $\overline{DA}$. Find the slopes of these four segments. Compare the slopes of $\overline{AB}$ and $\overline{CD}$. Compare the slopes of $\overline{BC}$ and $\overline{DA}$. (Figure $ABCD$ is a parallelogram and its opposite sides are parallel.)

78. Use graph paper. Plot the points $E(-2, -5)$, $F(2, -2)$, $G(7, -2)$, and $H(3, -5)$. Draw $\overline{EF}$, $\overline{FG}$, $\overline{GH}$, $\overline{HE}$, $\overline{EG}$, and $\overline{FH}$. Compare the slopes of $\overline{EG}$ and $\overline{FH}$. (Figure $EFGH$ is a rhombus and its diagonals are perpendicular.)

79. *Fahrenheit temperature as a function of Celsius temperature.* Fahrenheit temperature F is a linear function of Celsius (or Centigrade) temperature C. When C is 0, F is 32. When C is 100, F is 212. Use these data to express F as a linear function of C.

80. *Celsius temperature as a function of Fahrenheit temperature.* Celsius (Centigrade) temperature C is a linear function of Fahrenheit temperature F. When F is 32, C is 0. When F is 212, C is 100. Use these data to express C as a linear function of F.

81. Suppose that P is a nonconstant linear function of Q. Show that Q is a linear function of P.

82. Suppose that y is directly proportional to x. Show that y is a linear function of x.

4.2 Parallel and Perpendicular Lines; The Distance Formula; Circles

OBJECTIVES

You should be able to:

1. **Given equations of two lines, tell whether the lines are parallel, perpendicular, or neither.**
2. **Given an equation of a line and the coordinates of a point, find equations of lines parallel or perpendicular to that line and containing the given point.**
3. **Given the coordinates of two points, find the distance between them and determine whether three points with given coordinates are vertices of a right triangle.**
4. **Given the coordinates of the endpoints of a segment, find the coordinates of its midpoint.**
5. **Given the center and radius of a circle, find an equation for the circle. Given an equation of a circle, complete the square if necessary, and find the center and radius. Given the center and a point through which a circle passes, find an equation of the circle.**

1 Parallel and Perpendicular Lines

If two lines are vertical, then they are parallel. Thus equations such as $x = c_1$ and $x = c_2$ (where c_1 and c_2 are unequal constants) have graphs that are *parallel lines.* Now we consider nonvertical lines. In order that such lines be parallel, they must have the same slope but different y-intercepts. Thus equations such as $y = mx + b_1$ and $y = mx + b_2$, $b_1 \neq b_2$, have graphs that are *parallel lines.*

> THEOREM 1
>
> **Vertical lines are *parallel*. Nonvertical lines are *parallel* if and only if they have the same slope and different y-intercepts.**

If two equations are equivalent, then they represent the same line. Thus if two lines have the same slope and the same y-intercept, then they are not really two different lines. They are the same line. In such a case, we sometimes speak of **coincident lines.**

If one line is vertical and the other is horizontal, such as $x = c_1$ and $y = c_2$, then they are perpendicular. Otherwise, how can we tell whether two lines are perpendicular? Let's look at the figure at the top of the next page.

Consider a line $\overleftrightarrow{AB}$ as shown, with slope a/b. Then think of rotating the figure 90° to get a line perpendicular to $\overleftrightarrow{AB}$. For the new line, the change in y

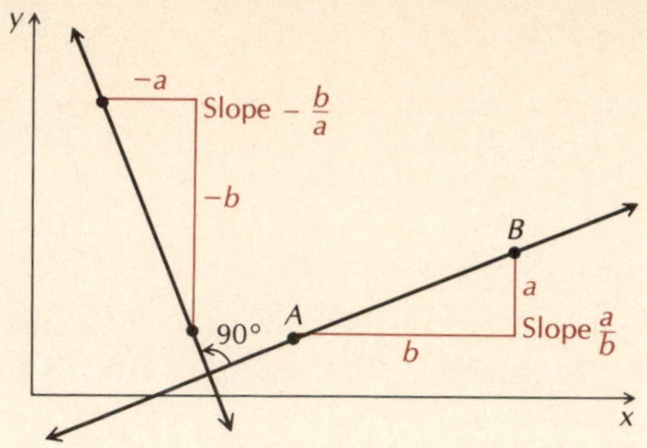

and the change in x are interchanged, but the change in x is now the additive inverse of what was the change in y. Thus the slope of the new line is $-b/a$. Let us multiply the slopes:

$$\frac{a}{b}\left(-\frac{b}{a}\right) = -1.$$

This is the condition under which lines will be *perpendicular*.

> **THEOREM 2**
>
> **Two lines with slopes m_1 and m_2 are *perpendicular* if and only if $m_1 m_2 = -1$.**

If one line has slope m_1, the slope m_2 of a line perpendicular to it is $-1/m_1$.

DO EXERCISES 1 AND 2.

EXAMPLE 1 Determine whether the pair of lines is parallel, perpendicular, or neither.

a) $y + 2 = 5x,\ 5y + x = -15$

b) $2y + 4x = 8,\ 5 + 2x = -y$

c) $2x + 1 = y,\ y + 3x = 4$

Solution

a) $y + 2 = 5x,\ 5y + x = -15$

We solve each equation for y:

$$y = 5x - 2, \qquad y = -\tfrac{1}{5}x - 3.$$

The slopes are 5 and $-\frac{1}{5}$. Their product is -1, so the lines are perpendicular.

b) $2y + 4x = 8,\ 5 + 2x = -y$

By solving for y, we get $y = -2x + 4$ and $y = -2x - 5$ and can determine that $m_1 = -2$ and $m_2 = -2$. Since the y-intercepts, -4 and -5, are different, the lines are parallel.

c) $2x + 1 = y,\ y + 3x = 4$

By solving for y, we determine that $m_1 = 2$ and $m_2 = -3$, so the lines are neither parallel nor perpendicular. ■

DO EXERCISES 3–5.

In each situation, slopes of the lines are given. Determine whether the lines are perpendicular.

1.

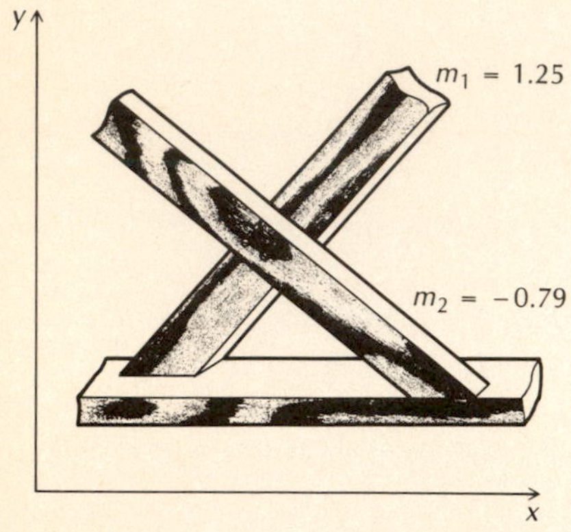

2.

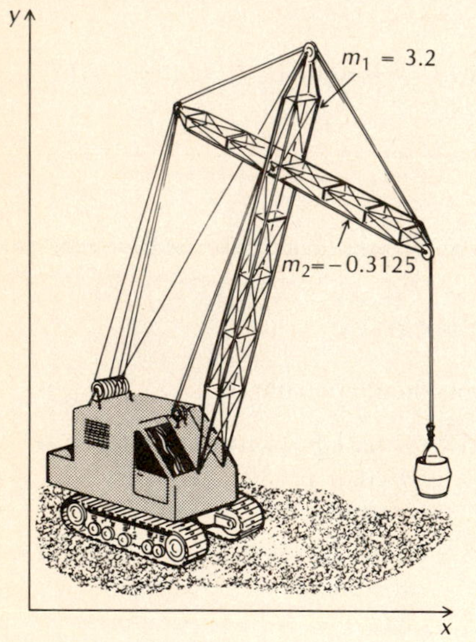

Determine whether the pair of lines is parallel, perpendicular, or neither.

3. $2y - x = 2,\ y + 2x = 4$

4. $3y = 2x + 15,\ 2y = 3x = 10$

5. $5y = 3 - 4x,\ 8x + 10y = 1$

2 Parallel or Perpendicular Lines Through a Point

EXAMPLE 2 Write equations of the lines parallel and perpendicular to the line $4y - x = 20$ and containing the point $(2, -3)$.

Solution We first solve for y: $y = \frac{1}{4}x + 5$, so the slope is $\frac{1}{4}$.

The line parallel to the given line will have slope $\frac{1}{4}$. Then we use the point–slope equation with slope $\frac{1}{4}$ and containing the point $(2, -3)$:

$$y - y_1 = m(x - x_1)$$
$$y - (-3) = \tfrac{1}{4}(x - 2).$$

This simplifies to $y = \frac{1}{4}x - \frac{7}{2}$. The slope of the perpendicular line is -4.

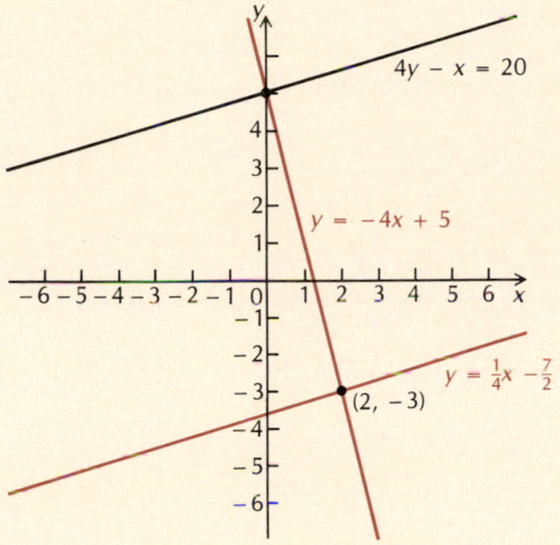

Now we use the point–slope equation to write an equation with slope -4 and containing the point $(2, -3)$:

$$y - y_1 = m(x - x_1)$$
$$y - (-3) = -4(x - 2).$$

This simplifies to $y = -4x + 5$.

DO EXERCISE 6.

Horizontal and Vertical Lines

If a line is vertical, it has no slope. Every **vertical line** has an equation $x = c$, where c is a constant. Any line parallel to a vertical line must also be vertical, hence must also have an equation $x = k$, where k is a constant and $k \neq c$.

A line will be perpendicular to a vertical line if and only if that line is horizontal. Every **horizontal line** has zero slope and has an equation $y = c$, where c is a constant. Thus if a line is vertical, it is simple to determine whether another line is parallel to it or perpendicular to it.

EXAMPLE 3 Find equations of the lines parallel and perpendicular to the line $x = 4$ and containing the point $(-2, 3)$.

Solution The line $x = 4$ is vertical, so any line parallel to it must be vertical. The line we seek has one x-coordinate, which is -2, so all x-coordinates on the line must be -2. The equation is $x = -2$. The line perpendicular to $x = 4$ must be horizontal and has one y-coordinate, which is 3, so all y-coordinates must be 3. The equation is $y = 3$.

6. Find equations of the lines parallel and perpendicular to the line $4 - y = 2x$ and containing the point $(3, 4)$.

7. Find equations of the lines parallel and perpendicular to the line $x = -3$ and containing the point $(5, -4)$.

8. Find equations of the lines parallel and perpendicular to the line $y = \frac{1}{2}$ and containing the point $\left(-6, \frac{11}{2}\right)$.

If a line is horizontal, it has an equation $y = b$. A line parallel to it will also be horizontal and have an equation $y = k$, where $b \neq k$. A line perpendicular to it will be vertical and have an equation $x = a$.

DO EXERCISES 7 AND 8.

3 The Distance Formula

We develop a formula for finding the *distance between two points* whose coordinates are known. Suppose that the points are on a horizontal line, thus having the same second coordinate. We can find the distance between them by subtracting their first coordinates. This difference may be negative, depending on the order in which we subtract. So to make sure we get a positive number, we take the absolute value of this difference. The distance between two points on a horizontal line (x_1, y) and (x_2, y) is thus $|x_2 - x_1|$. Similarly, the distance between two points on a vertical line (x, y_1) and (x, y_2) is $|y_2 - y_1|$.

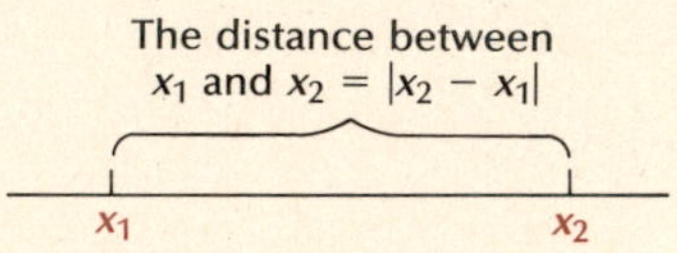

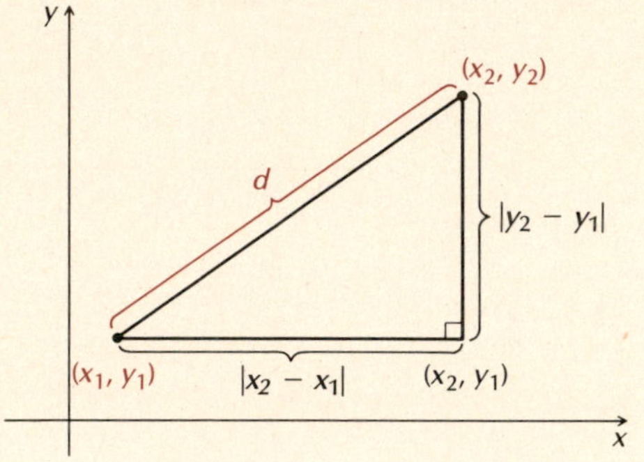

Now consider any two points (x_1, y_1) and (x_2, y_2) not on a horizontal or vertical line. These points are vertices of a right triangle, as shown. The other vertex is (x_2, y_1). The legs of this triangle have the lengths $|x_2 - x_1|$ and $|y_2 - y_1|$. Now by the Pythagorean theorem, we obtain a relation between the length of the hypotenuse d and the lengths of the legs:

$$d^2 = |x_2 - x_1|^2 + |y_2 - y_1|^2.$$

We may now dispense with the absolute-value signs because squares of numbers are never negative. Thus we have

$$d^2 = (x_2 - x_1)^2 + (y_2 - y_1)^2.$$

By taking the square root, we obtain the distance between two points:

THEOREM 3 The Distance Formula

The distance between any two points (x_1, y_1) and (x_2, y_2) is given by

$$d = \sqrt{(x_1 - x_2)^2 + (y_1 - y_2)^2}.$$

Although we derived the distance formula by considering two points not on a horizontal or a vertical line, the formula holds for *any* two points. The subtraction of the x-coordinates can be done in any order, as can the subtraction of the y-coordinates.

EXAMPLE 4 Find the distance between the points $A(-2, 2)$ and $B(4, -3)$ on the islands in the figure.

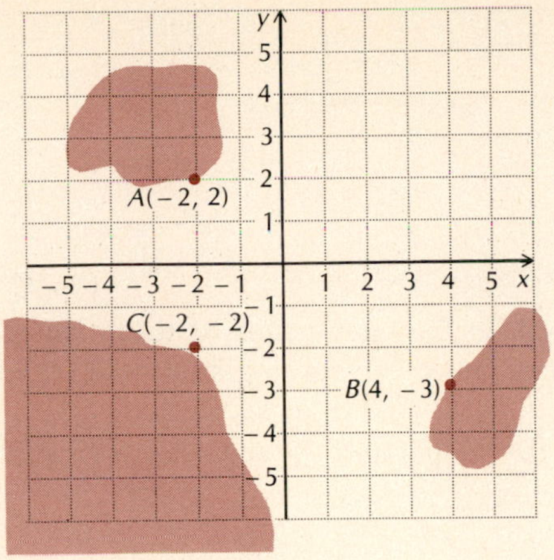

Solution We have

$$\begin{aligned} d &= \sqrt{[4-(-2)]^2 + (-3-2)^2} \\ &= \sqrt{(6)^2 + (-5)^2} \\ &= \sqrt{36 + 25} \\ &= \sqrt{61}. \end{aligned}$$

An approximation for this distance is 7.8.

DO EXERCISES 9–12.

We can use the distance formula to determine whether three points are vertices of a right triangle.

EXAMPLE 5 Determine whether the points $A(-2, 2)$, $B(4, -3)$ and $C(-2, -3)$ are vertices of a right triangle.

Solution First we find the squares of the distances between the points:

$$\begin{aligned} d_1^2 &= [4-(-2)]^2 + (-3-2)^2 \\ &= (6)^2 + (-5)^2 = 61, \\ d_2^2 &= [-2-(-2)]^2 + (-3-2)^2 \\ &= (0)^2 + (-5)^2 = 25, \\ d_3^2 &= (-2-4)^2 + [-3-(-3)]^2 \\ &= (-6)^2 + (0)^2 = 36. \end{aligned}$$

Since $d_2^2 + d_3^2 = d_1^2$, the points are vertices of a right triangle.

The points of Example 4 are not the vertices of a right triangle, since no sum of any two squares is another.

DO EXERCISES 13 AND 14.

4 Midpoints of Segments

The distance formula can be used to verify or derive a formula for finding the coordinates of the *midpoint* of a segment when the coordinates of its endpoints are known. We will not derive this formula but simply state it.

Find the distance between the pair of points.

9. $(-5, 3)$ and $(2, -7)$

10. $(3, 3)$ and $(-3, -3)$

11. $(9, -5)$ and $(9, 11)$

12. $(0, \pi)$ and $(8, \pi)$

Determine whether the points are vertices of a right triangle.

13. $(11, 1)$, $(6, 6)$, $(2, 2)$

14. $(10, -4)$, $(3, 5)$ $(0, 0)$

Find the midpoints of the segments having endpoints as given.

15. $(-2, 1)$ and $(5, -6)$

16. $(9, -6)$ and $(9, -4)$

THEOREM 4 The Midpoint Formula

If the endpoints of a segment are (x_1, y_1) and (x_2, y_2), then the coordinates of the midpoint are

$$\left(\frac{x_1 + x_2}{2}, \frac{y_1 + y_2}{2}\right).$$

Note that we obtain the coordinates of the midpoint by averaging the coordinates of the endpoints. This is an easy way to remember this formula.

EXAMPLE 6 Find the midpoint of the segment in Example 4 with endpoints $A(-2, 2)$ and $B(4, -3)$.

Solution Using the midpoint formula, we obtain

$$\left(\frac{-2 + 4}{2}, \frac{2 + (-3)}{2}\right), \quad \text{or} \quad \left(1, -\frac{1}{2}\right).$$

DO EXERCISES 15 AND 16.

5 Circles

A **circle** is a set of points in a plane that are a fixed distance r, called the **radius,** from a fixed point (h, k), called the **center.** If a point (x, y) is on the circle, then by the definition of a circle and the distance formula, it must follow that

$$r = \sqrt{(x - h)^2 + (y - k)^2}, \quad \text{or}$$
$$r^2 = (x - h)^2 + (y - k)^2.$$

The equation, in standard form, of a circle with center (h, k) and radius r is

$$(x - h)^2 + (y - k)^2 = r^2.$$

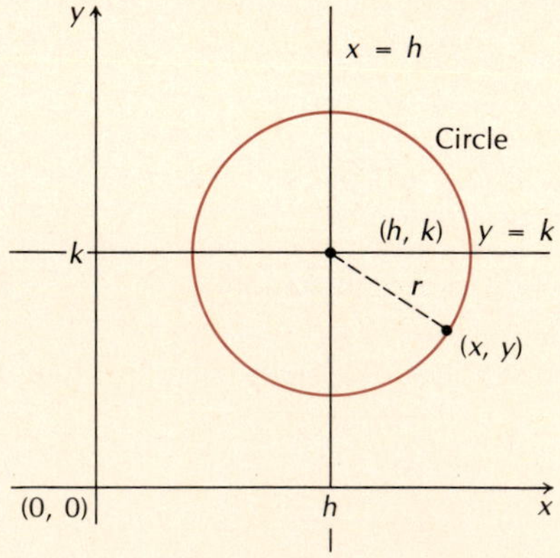

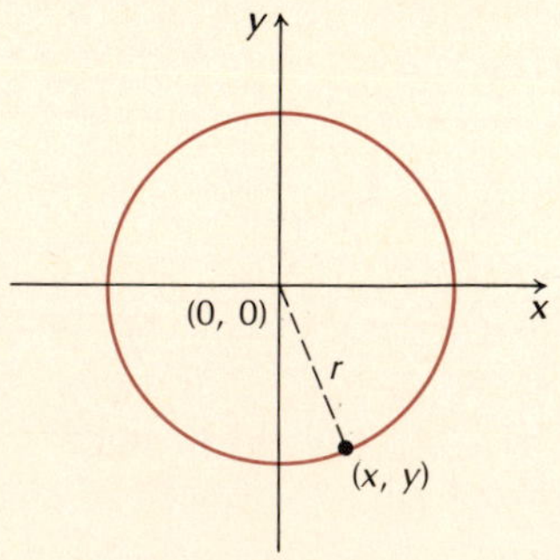

If $h = 0$ and $k = 0$, then the circle is centered at the origin. If a circle centered at the origin has radius $r = 1$, the circle is called a **unit circle.**

EXAMPLE 7 Find an equation of the circle having center $(4, -5)$ and radius 6.

Solution Using the standard form, we obtain

$$(x - 4)^2 + [y - (-5)]^2 = 6^2,$$

or

$$(x - 4)^2 + (y + 5)^2 = 36.$$

DO EXERCISES 17 AND 18.

EXAMPLE 8 Find the center and the radius of $(x - 2)^2 + (y + 3)^2 = 16$. Then graph the circle.

Solution We can first write standard form: $(x - 2)^2 + (y + 3)^2 = 4^2$. Then the center is $(2, -3)$ and the radius is 4. Now the graph is easy to draw, as shown, using a compass.

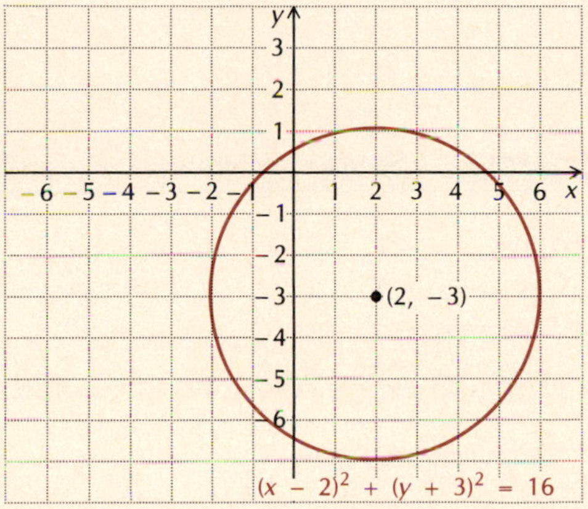

DO EXERCISE 19.

Completing the square allows us to find the standard form for the equation of a circle.

EXAMPLE 9 Find the center and the radius of the circle

$$x^2 + y^2 + 8x - 2y + 15 = 0.$$

Solution We complete the square twice to get standard form:

$$(x^2 + 8x + \qquad) + (y^2 - 2y + \qquad) = -15.$$

We take half the coefficient of the x-term and square it, obtaining 16. We add $16 - 16$ in the first parentheses. Similarly, we add $1 - 1$ in the second parentheses:

$$(x^2 + 8x + 16 - 16) + (y^2 - 2y + 1 - 1) = -15.$$

Next we do some rearranging and factoring:

$$(x^2 + 8x + 16) + (y^2 - 2y + 1) - 16 - 1 = -15$$

$$(x + 4)^2 + (y - 1)^2 = 2. \quad \text{This is standard form.}$$

The center is $(-4, 1)$ and the radius is $\sqrt{2}$.

DO EXERCISE 20.

17. Find an equation of the circle centered at the origin, (0, 0), and having radius $2\sqrt{5}$.

18. Find an equation of the circle having center $(-3, 7)$ and radius 5.

19. Find the center and the radius of

$$(x + 1)^2 + (y - 3)^2 = 4.$$

Then graph the circle.

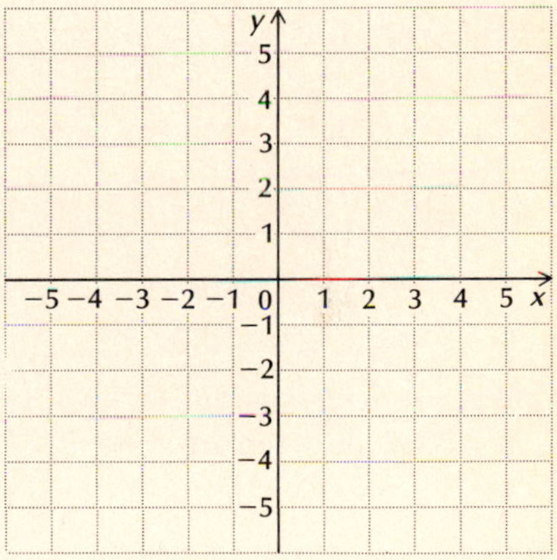

20. Find the center and the radius of the circle

$$x^2 + y^2 - 14x + 4y - 11 = 0.$$

21. Find an equation of the circle with center $(-1, 4)$ that passes through $(3, -1)$.

EXAMPLE 10 Find an equation of a circle with center $(-2, -3)$ that passes through the point $(1, 1)$.

Solution Since $(-2, -3)$ is the center, we have

$$(x + 2)^2 + (y + 3)^2 = r^2.$$

The circle passes through $(1, 1)$. We find r by substituting 1 for x and 1 for y in the above equation:

$$\begin{aligned}(1 + 2)^2 + (1 + 3)^2 &= r^2\\ 9 + 16 &= r^2\\ 25 &= r^2\\ 5 &= r.\end{aligned}$$

Then $(x + 2)^2 + (y + 3)^2 = 25$ is an equation of the circle.

DO EXERCISE 21.

EXERCISE SET 4.2

1 Determine whether the lines are parallel, perpendicular, or neither.

1. $2x - 5y = -3,$ $2x + 5y = 4$
2. $x + 2y = 5,$ $2x + 4y = 8$
3. $y = 4x - 5,$ $4y = 8 - x$
4. $y = -x + 7,$ $y = x + 3$

2 Find equations of the lines parallel and perpendicular to the given line and containing the given point.

5. $(0, 3)$, $3x - y = 7$
6. $(-4, -5)$, $2x + y = -4$
7. $(3, 8)$, $x = 2$
8. $(3, -3)$, $x = -1$
9. $(-2, -3)$, $y = 4$
10. $(-3, 2)$, $y = -3$
11. $(-3, -5)$, $5x - 2y = 4$
12. $(3, -2)$, $3x + 4y = 5$
13. $(0, 3)$, $x = 1$
14. $(-2, -2)$, $x = 3$
15. $(-3, -7)$, $y = 2$
16. $(4, -5)$, $y = -1$
17. ▦ Find an equation of the line parallel to the one given, and containing the given point.
 $4.323x - 7.071y = 16.61$, $(-2.603, 1.818)$
18. ▦ Find an equation of the line containing the given point and perpendicular to the given line.
 $6.232x + 4.001y = 4.881$, $(3.149, -2.908)$

3 Find the distance between the pair of points.

19. $(-3, -2)$ and $(1, 1)$
20. $(5, 9)$ and $(-1, 6)$
21. $(0, -7)$ and $(3, -4)$
22. $(2, 2)$ and $(-2, -2)$
23. $(a, -3)$ and $(2a, 5)$
24. $(5, 2k)$ and $(-3, k)$
25. $(0, 0)$ and (a, b)
26. $(\sqrt{2}, \sqrt{3})$ and $(0, 0)$
27. $(\sqrt{a}, \sqrt{b})$ and $(-\sqrt{a}, \sqrt{b})$
28. $(c - d, c + d)$ and $(c + d, d - c)$
29. ▦ $(7.3482, -3.0991)$ and $(18.9431, -17.9054)$
30. ▦ $(-25.414, 175.31)$ and $(275.34, -95.144)$

Determine whether the points are vertices of a right triangle.

31. $(9, 6)$, $(-1, 2)$, and $(1, -3)$
32. $(-5, -8)$, $(1, 6)$, and $(5, -4)$

4 Find the midpoints of the segments having the following endpoints.

33. $(-4, 7)$ and $(3, -9)$
34. $(4, 5)$ and $(6, -7)$
35. (a, b) and $(a, -b)$
36. $(-c, d)$ and (c, d)
37. ▦ $(-3.895, 8.1212)$ and $(2.998, -8.6677)$
38. ▦ $(4.1112, 6.9898)$ and $(5.1928, 6.9143)$

5 Find the center and the radius of the circle. Then graph the circle.

39. $x^2 + y^2 = 36$
40. $x^2 + y^2 = 25$
41. $x^2 + y^2 = 3$

42. $x^2 + y^2 = 2$

43. $(x + 1)^2 + (y + 3)^2 = 4$

44. $(x - 2)^2 + (y + 3)^2 = 1$

Find the center and the radius of the circle.

45. $(x - 8)^2 + (y + 3)^2 = 40$

46. $(x + 5)^2 + (y - 1)^2 = 75$

47. $(x - 3)^2 + y^2 = \frac{1}{25}$

48. $x^2 + (y - 1)^2 = \frac{1}{4}$

49. $x^2 + y^2 + 8x - 6y - 15 = 0$

50. $x^2 + y^2 + 25x + 10y + 12 = 0$

51. $x^2 + y^2 + 6x = 0$

52. $x^2 + y^2 - 4x = 0$

53. $x^2 + y^2 + 8x = 84$

54. $x^2 + y^2 - 75 = 10y$

55. $x^2 + y^2 + 21x + 33y + 17 = 0$

56. $x^2 + y^2 - 7x + 3y - 10 = 0$

57. ▦ $x^2 + y^2 + 8.246x - 6.348y - 74.35 = 0$

58. ▦ $x^2 + y^2 + 25.074x + 10.004y + 12.054 = 0$

59. $9x^2 + 9y^2 = 1$

60. $16x^2 + 16y^2 = 1$

Find an equation of the circle satisfying the given conditions.

61. Center (0, 0), passing through (−3, 4)

62. Center (3, −2), passing through (11, −2)

63. Center (−4, 1), passing through (−2, 5)

64. Center (−3, −3), passing through (1.8, 2.6)

SYNTHESIS

65. Find equations of the lines containing the point (4, −2) parallel and perpendicular to the line containing (−1, 4) and (2, −3).

66. Find equations of the lines containing the point (−1, 3) parallel and perpendicular to the line containing (3, −5) and (−2, −7).

67. Find the point on the x-axis that is equidistant from the points (1, 3) and (8, 4).

68. Find the point on the y-axis that is equidistant from the points (−2, 0) and (4, 6).

69. Find k so that the line containing $(-3, k)$ and (4, 8) is parallel to the line containing (6, 4) and (2, −5).

70. Find k so that the line containing $(-3, k)$ and (4, 8) is perpendicular to the line containing (6, 4) and (2, −5).

71. Find an equation of the perpendicular bisector of the line segment with endpoints (−1, 3) and (−6, 7).

72. Find an equation of the perpendicular bisector of the line segment with endpoints (1, 0) and (−7, −12).

Find an equation of a circle satisfying the given conditions.

73. Center (2, 4) and tangent (touching at one point) to the x-axis

74. Center (−3, −2) and tangent to the y-axis

75. The endpoints of a diameter are (5, −3) and (−3, 7).

76. The endpoints of a diameter are (9, 8) and (−4, −3)

77. Center (−8, 5) with a circumference of 10π units

78. Center (−3, −8) with an area of 36π square units

79. Find the center and the radius of the circle that is inscribed in the square with vertices $A(8, 4)$, $B(3, 9)$, $C(8, 14)$, and $D(13, 9)$.

80. A ferris wheel has a radius of 25.3 ft. Assuming that the center is 30.8 ft off the ground and that the origin is below the center as in the drawing below, find an equation of the outside of the ferris wheel.

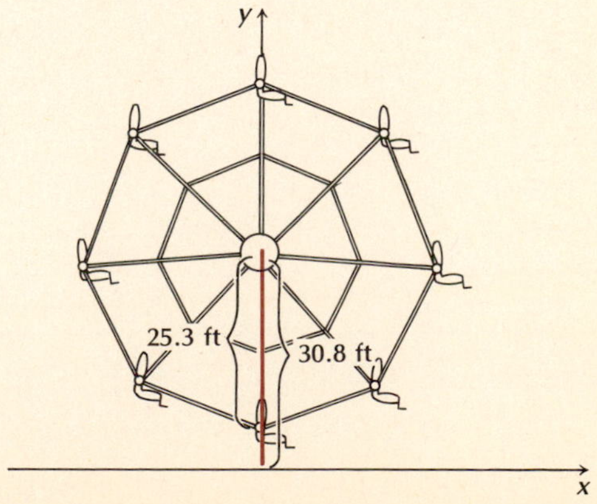

81. A circular swimming pool has a radius of 8.4 ft. Assuming that the pool is at the origin of a coordinate system, what is the equation of the circle?

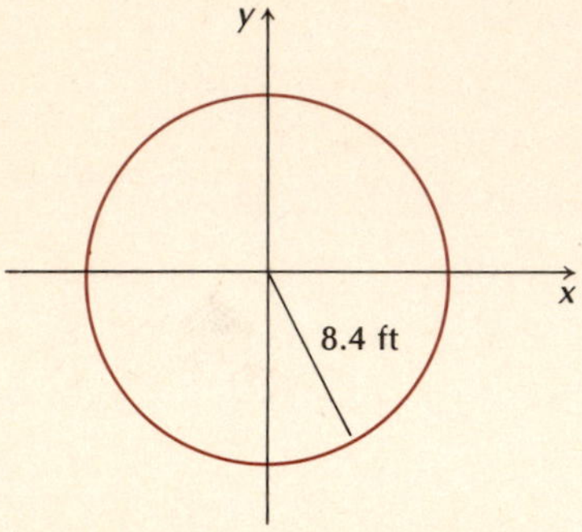

82. ▦ A swimming pool is being constructed in the corner of a lot as shown. In laying out the pool, the contractor wishes to know the distances a_1 and a_2. Find them.

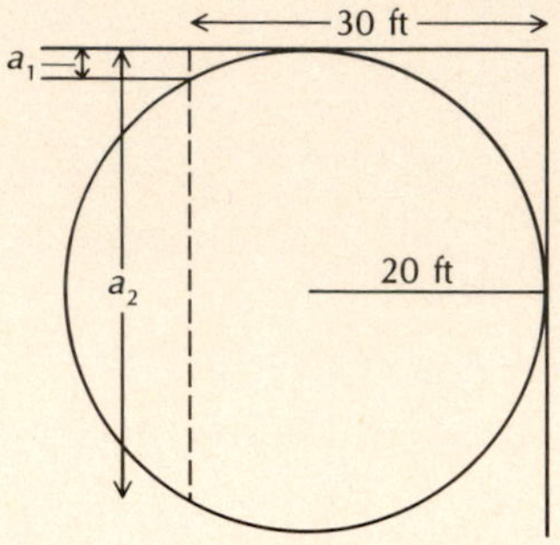

83. Show that the following equation is an equation of a circle with center (h, k) and radius r.

$$\begin{vmatrix} x - h & -(y - k) \\ y - k & x - h \end{vmatrix} = r^2$$

84. a) Graph $x^2 + y^2 = 4$. Is this relation a function?
b) Solve $x^2 + y^2 = 4$ for y.
c) Graph $y = \sqrt{4 - x^2}$ and determine whether it is a function. Find the domain and the range.
d) Graph $y = -\sqrt{4 - x^2}$ and determine whether it is a function. Find the domain and the range.

Determine whether each of the following lies on the unit circle $x^2 + y^2 = 1$.

85. $(-1, 0)$

86. $\left(\frac{\sqrt{3}}{2}, -\frac{1}{2}\right)$

87. ▦ $(0.838670568, -0.544639035)$

88. $\left(\frac{\pi}{3}, \frac{3}{\pi}\right)$

CHALLENGE

89. Consider any right triangle with base b and height h, situated as shown. Show that the midpoint of the hypotenuse P is equidistant from the three vertices of the triangle.

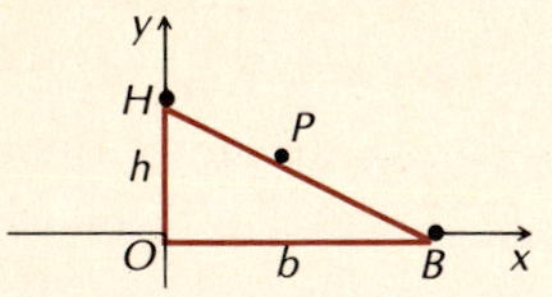

90. Consider any quadrilateral situated as shown. Show that the segments joining the midpoints of the sides, in order as shown, form a parallelogram.

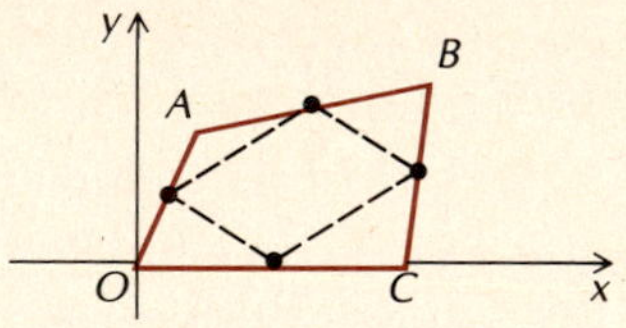

91. Prove that the distance formula holds when two points are on either a vertical line or a horizontal line.

92. Prove that $\angle ABC$ is a right angle. Assume that point B is on the circle whose radius is a and whose center is at the origin. (*Hint:* Use slopes and an equation of the circle.)

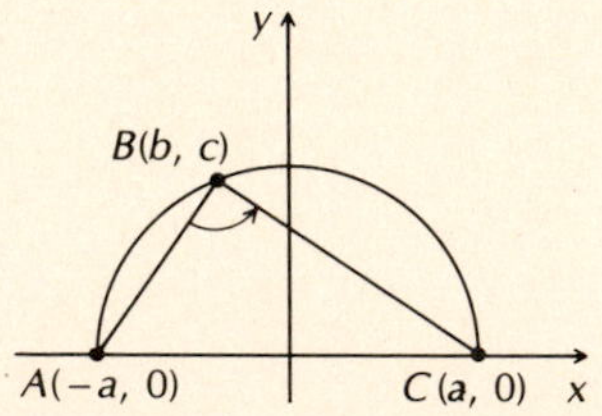

4.3 Quadratic Functions

OBJECTIVES

You should be able to:

1. Given a quadratic function, find the vertex of its graph, the line of symmetry, and the maximum or minimum value.
2. Graph a quadratic function.
3. Find the x-intercepts of the graph of a quadratic function.

1 Properties of Quadratic Functions

If a function can be described by a second-degree polynomial, then it is called **quadratic.** The following is a more precise definition.

> **DEFINITION**
>
> A *quadratic function* is a function that can be described as follows:
>
> $$f(x) = ax^2 + bx + c, \quad \text{where } a \neq 0.$$

In this definition, we insist that $a \neq 0$; otherwise the polynomial would not be of degree two. One or both of the constants b and c can be 0.

Consider $f(x) = x^2$. This is an even function, so the y-axis is a line of symmetry. The graph opens upward, as shown below. If we multiply by a constant a to get $f(x) = ax^2$, we obtain a vertical stretching or shrinking, and a reflection if a is negative. Examples of this are shown in the figure. Bear in mind that if a is negative, the graph opens downward.

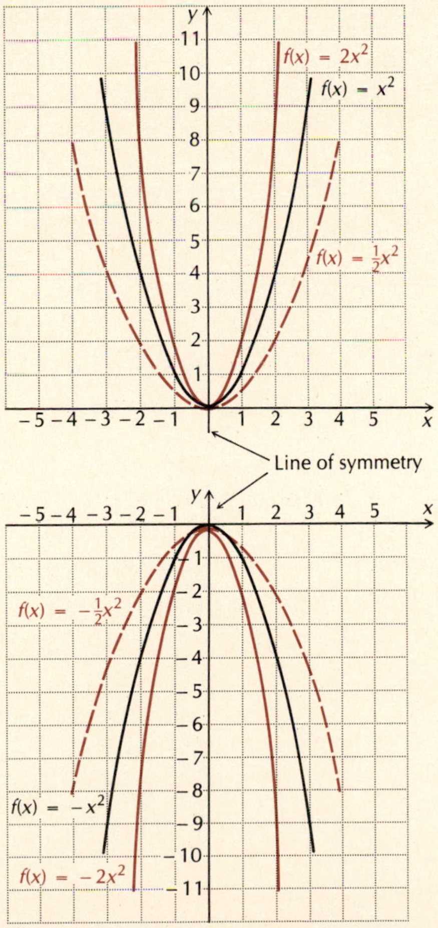

Graphs of quadratic functions. Parabolas.

Graphs of quadratic functions are called **parabolas.** The highest or lowest point at which the graph turns is called the **vertex.** In each parabola shown, the

Use graph paper for Margin Exercises 1–4.

1. a) Graph $f(x) = 2x^2$.
   b) Does the graph open upward or does it open downward?
   c) What is the line of symmetry?
   d) What is the minimum value of the function?
   e) What is the vertex?

2. a) Graph $f(x) = -0.4x^2$.
   b) Does the graph open upward or does it open downward?
   c) What is the line of symmetry?
   d) What is the maximum value of the function?
   e) What is the vertex?

3. a) Graph $f(x) = 3x^2$.
   b) Use the graph in (a) to graph $f(x) = 3(x - 2)^2$.
   c) What is the vertex of the graph in (b)?
   d) What is the line of symmetry of the graph in (b)?
   e) What is the minimum value?
   f) Does the graph open upward or downward?
   g) Is the graph of $f(x) = 3(x - 2)^2$ a horizontal translation to the left or to the right?

4. a) Graph $f(x) = -3x^2$.
   b) Use the graph in (a) to graph
   $$f(x) = -3(x + 2)^2 = -3[x - (-2)]^2.$$
   c) What is the vertex of the graph in (b)?
   d) What is the line of symmetry of the graph in (b)?
   e) What is the maximum value?
   f) Does the graph open upward or downward?
   g) Is the graph of $f(x) = -3(x + 2)^2$ a horizontal translation to the left or to the right?

point (0, 0) is the *vertex* and the line $x = 0$, or the y-axis, is the **line of symmetry.**

DO EXERCISES 1 AND 2.

Let us consider $f(x) = a(x - h)^2$. We have replaced x by $x - h$ in ax^2, and will therefore obtain a horizontal translation. The translation will be to the right if h is positive and will be to the left if h is negative. The examples in the figures below illustrate translations.

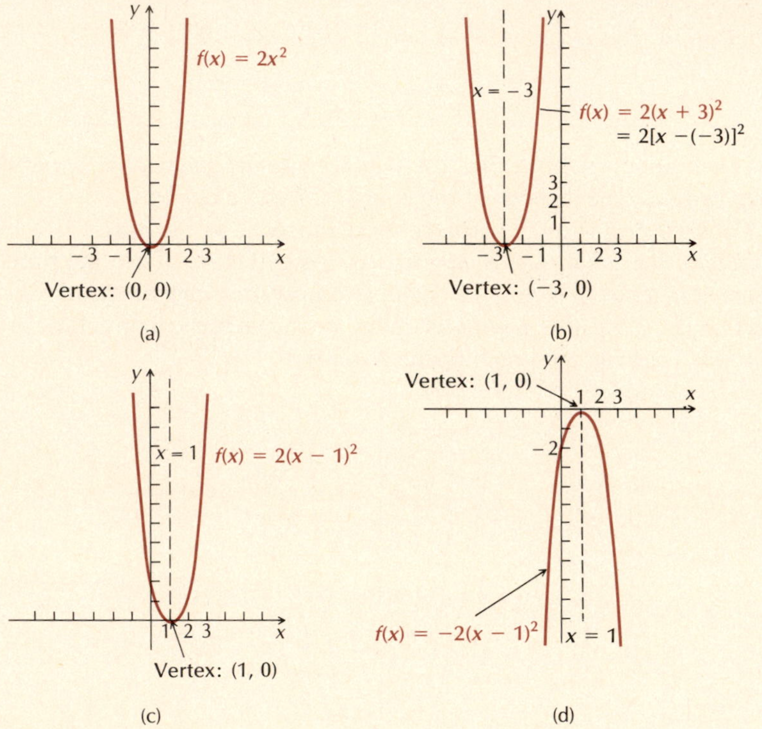

DO EXERCISES 3 AND 4.

Now consider $f(x) = a(x - h)^2 + k$, or $f(x) - k = a(x - h)^2$. We have replaced $f(x)$ by $f(x) - k$ in the equation $f(x) = a(x - h)^2$. Thus we have a translation. If k is positive, the translation is upward. If k is negative, the translation is downward. Consider these examples. Note that the vertex has been moved off the x-axis.

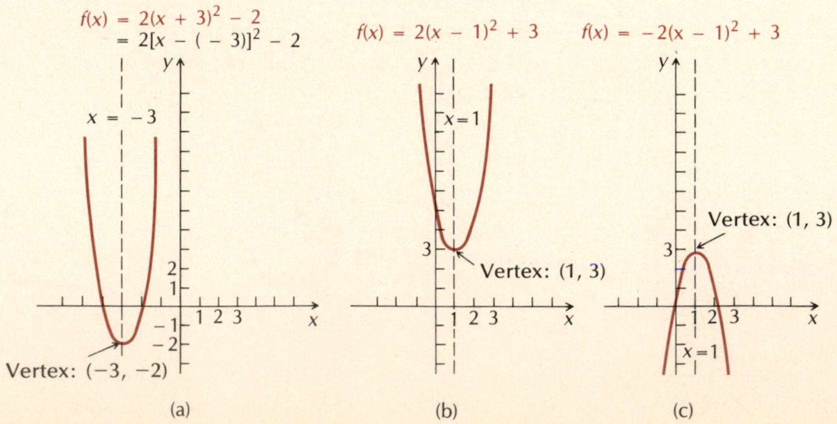

The graph of $f(x) = a(x - h)^2 + k$:

a) opens upward if $a > 0$, downward if $a < 0$;
b) has (h, k) as a vertex;
c) has $x = h$ as a line of symmetry;
d) has k as a minimum value (output) if $a > 0$, has k as a maximum value (output) if $a < 0$.

Thus without graphing, we can determine a lot of information about a function described by $f(x) = a(x - h)^2 + k$. The following table is an example.

Function	$f(x) = 3(x - \frac{1}{4})^2 - 2$ $= 3(x - \frac{1}{4})^2 + (-2)$	$g(x) = -3(x + 5)^2 + 7$ $= -3[x - (-5)]^2 + 7$
a) What is the vertex?	$(\frac{1}{4}, -2)$	$(-5, 7)$
b) What is the line of symmetry?	$x = \frac{1}{4}$	$x = -5$
c) Is there a maximum? What is it?	No: graph extends upward; $3 > 0$.	Yes, 7: graph extends downward; $-3 < 0$.
d) Is there a minimum? What is it?	Yes, -2: graph extends upward; $3 > 0$	No: graph extends downward; $-3 < 0$.

Note that the vertex (h, k) is used to find the maximum or minimum. The maximum or minimum is the number k, *not* the ordered pair (h, k).

DO EXERCISES 5–10.

Now let us consider a quadratic function $f(x) = ax^2 + bx + c$. Note that it is not in the form $f(x) = a(x - h)^2 + k$. We can put it into that form by *completing the square.*

EXAMPLE 1 Use completing the square to put the function

$$f(x) = x^2 - 6x + 4$$

into the form $f(x) = a(x - h)^2 + k$.

Solution We consider first the x^2- and x-terms:

$$f(x) = x^2 - 6x \qquad + 4.$$

We construct a trinomial square. To do so, we take half the coefficient of x and square it. The number is $(-6/2)^2$, or 9. We now add that number to complete the square. We also must subtract it to get an equivalent equation. We can think of this simply as adding $9 - 9$, which is 0.

$$\begin{aligned} f(x) &= x^2 - 6x + 9 - 9 + 4 \\ &= x^2 - 6x + 9 - 5 \end{aligned}$$

We now factor the trinomial square $x^2 - 6x + 9$ and we are finished:

$$f(x) = (x - 3)^2 - 5.$$

DO EXERCISE 11.

If the coefficient of x^2 is not 1, a preliminary step is needed.

Answer the following questions in Margin Exercises 5–10.

a) What is the vertex?
b) What is the line of symmetry?
c) Is there a maximum? What is it?
d) Is there a minimum? What is it?

Use graph paper for Margin Exercises 5 and 6.

5. Graph $f(x) = 3(x - 2)^2 + 4$.

6. Graph
$$\begin{aligned} f(x) &= -3(x + 2)^2 - 1 \\ &= -3[x - (-2)]^2 + (-1). \end{aligned}$$

Without graphing, answer the above questions for each function.

7. $f(x) = (x - 5)^2 + \pi$

8. $f(x) = -3(x - 5)^2$

9. $f(x) = 2\left(x + \frac{1}{4}\right)^2 - 6$

10. $f(x) = -\frac{1}{4}(x + 9)^2 + 3$

11. Use completing the square to put the function
$$f(x) = x^2 - 4x + 7$$
into the form
$$f(x) = a(x - h)^2 + k.$$

12. Use completing the square to put the function

$$f(x) = 3x^2 + 24x + 10$$

into the form

$$f(x) = a(x - h)^2 + k.$$

EXAMPLE 2 Use completing the square to put the function

$$f(x) = 2x^2 + 12x - 1$$

into the form $f(x) = a(x - h)^2 + k$.

Solution Again we consider the x^2- and x-terms, but we begin by factoring out the x^2-coefficient, as follows:

$$f(x) = 2(x^2 + 6x) - 1.$$

Next we proceed as before, *inside* the parentheses. We take half the coefficient of x and square it. That number is $(6/2)^2$, or 9. We add $9 - 9$, but do it *inside* the parentheses:

$$f(x) = 2(x^2 + 6x + 9 - 9) - 1.$$

We now have an extra, unwanted term inside the parentheses, so we use the distributive law and multiplication to get the $(-2 \cdot 9)$ outside, as follows:

$$\begin{aligned} f(x) &= 2(x^2 + 6x + 9) - 2 \cdot 9 - 1 \\ &= 2(x^2 + 6x + 9) - 18 - 1 \\ &= 2(x + 3)^2 - 19. \end{aligned}$$

DO EXERCISE 12.

In many situations, we want to be able to read off the vertex (h, k) directly from $f(x) = ax^2 + bx + c$ without repeatedly completing the square. We look for a formula to compute h and k. We proceed in a manner quite similar to what we did when we proved the quadratic formula.

THEOREM 5

The vertex of the graph of the quadratic function $f(x) = ax^2 + bx + c$ is

$$\left(-\frac{b}{2a}, -\frac{b^2 - 4ac}{4a}\right).$$

Proof. We proceed in a manner similar to Example 2. We begin by factoring out the x^2-coefficient, as follows:

$$f(x) = a\left(x^2 + \frac{b}{a}x\right) + c.$$

Next, we proceed *inside* the parentheses. We take half the coefficient of x and square it. That number is $(b/2a)^2$. We add $(b/2a)^2 - (b/2a)^2$, but do it *inside* the parentheses:

$$f(x) = a\left(x^2 + \frac{b}{a}x + \left(\frac{b}{2a}\right)^2 - \left(\frac{b}{2a}\right)^2\right) + c.$$

We now have an extra, unwanted term inside the parentheses, so we use the distributive law and multiplication to simplify, as follows:

$$\begin{aligned} f(x) &= a\left(x^2 + \frac{b}{a}x + \left(\frac{b}{2a}\right)^2\right) - a\left(\frac{b}{2a}\right)^2 + c \\ &= a\left(x^2 + \frac{b}{a}x + \left(\frac{b}{2a}\right)^2\right) - \frac{b^2}{4a} + c \\ &= a\left(x^2 + \frac{b}{a}x + \left(\frac{b}{2a}\right)^2\right) - \frac{b^2 - 4ac}{4a} = a\left(x + \frac{b}{2a}\right)^2 - \frac{b^2 - 4ac}{4a} \\ &= a\left(x - \left(-\frac{b}{2a}\right)\right)^2 + \left(-\frac{b^2 - 4ac}{4a}\right). \end{aligned}$$

The vertex is

$$\left(-\frac{b}{2a}, -\frac{b^2 - 4ac}{4a}\right).$$

Note that the numerator for the second coordinate of the vertex contains the discriminant, used to solve quadratic equations. We can use all of the formula for the vertex, or we can find $-b/2a$ and substitute into the original formula for the function to find the second coordinate, which turns out to be the maximum or minimum value.

EXAMPLE 3 For the function $f(x) = 4x^2 + 12x + 7$, (a) find the vertex and the line of symmetry and (b) determine whether there is a maximum or minimum value and find that value.

Solution

a) We note that $a = 4$, $b = 12$, and $c = 7$. The first coordinate of the vertex is

$$-\frac{b}{2a} = -\frac{12}{2(4)} = -\frac{3}{2}.$$

We find the second coordinate of the vertex as follows:

$$-\frac{b^2 - 4ac}{4a} = -\frac{12^2 - 4(4)(7)}{4(4)}$$
$$= -2.$$

We can also substitute $-\frac{3}{2}$ into the original function formula:

$$f\left(-\frac{3}{2}\right) = 4\left(-\frac{3}{2}\right)^2 + 12\left(-\frac{3}{2}\right) + 7$$
$$= -2.$$

The vertex is $\left(-\frac{3}{2}, -2\right)$. The line of symmetry goes through this point, hence is the line $x = -\frac{3}{2}$.

b) Since the coefficient of x^2 is positive, the parabola opens upward. Thus we have a minimum value. The value is -2. ■

DO EXERCISE 13.

EXAMPLE 4 For the function $f(x) = -2x^2 + 10x - 7$, find the vertex and the maximum or minimum value.

Solution We note that $a = -2$, $b = 10$, and $c = -7$. The first coordinate of the vertex is

$$-\frac{b}{2a} = -\frac{10}{2(-2)} = \frac{5}{2}.$$

We find the second coordinate by substituting $\frac{5}{2}$ into the original function formula:

$$f\left(\frac{5}{2}\right) = -2\left(\frac{5}{2}\right)^2 + 10\left(\frac{5}{2}\right) - 7$$
$$= \frac{11}{2}.$$

The vertex is $\left(\frac{5}{2}, \frac{11}{2}\right)$. The maximum value is $\frac{11}{2}$, since the coefficient of x^2 is negative. ■

DO EXERCISE 14.

13. For the function

$$f(x) = 4x^2 - 12x - 5,$$

a) find the vertex and the line of symmetry;
b) determine whether there is a maximum or minimum value and find that value.

14. For the function

$$f(x) = -3x^2 - 18x + 7,$$

find the vertex and the maximum or minimum value.

Graph.

15. $f(x) = x^2 - 6x + 4$

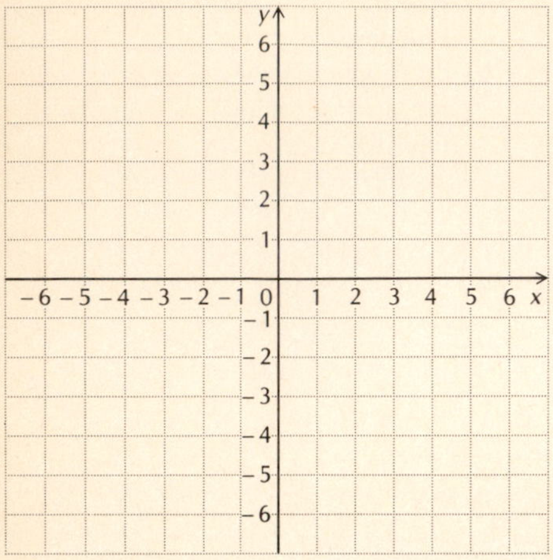

16. $f(x) = -4x^2 + 12x - 5$

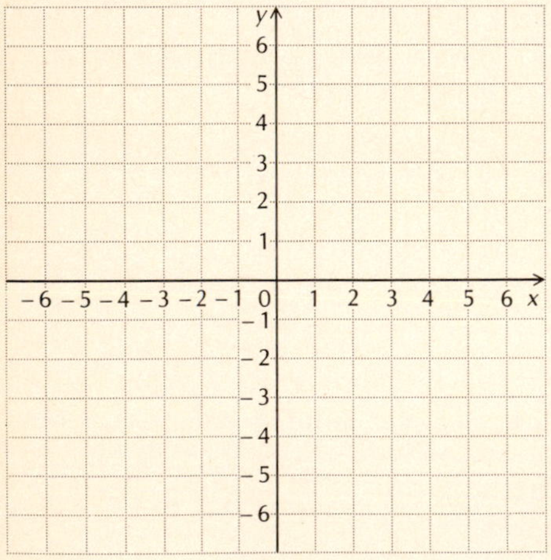

2 Graphing Quadratic Functions

We know that the graph of any quadratic function is a *parabola*. If the x^2-coefficient is positive, the parabola opens upward. If the x^2-coefficient is negative, the graph opens downward. We also know how to find the vertex and the line of symmetry. We can compute function values on each side of the vertex. We can also find other values by reflecting across the line of symmetry.

EXAMPLE 5 Graph: $f(x) = -2x^2 + 10x - 7$.

Solution We note that the coefficient of x^2 is negative. Therefore, the parabola opens downward. We find the vertex and the line of symmetry. From Example 4, we know that the vertex is $\left(\frac{5}{2}, \frac{11}{2}\right)$. Thus the line of symmetry is $x = \frac{5}{2}$. We then plot several input–output pairs. We start with $x = 2$, since it is near the vertex and is an integer, making computations easy:

$$f(2) = -2(2)^2 + 10(2) - 7 = 5.$$

This gives us the pair (2, 5). By reflecting this pair across the line of symmetry, we also get the pair (3, 5). We then consider $x = 1$. Since $f(1) = 1$, we get another pair (1, 1) and by reflecting, we get the pair (4, 1). Finding the y-intercept by computing $f(0)$ is easy to do. We get $f(0) = -7$. This gives us another pair $(0, -7)$. We then plot the ordered pairs and complete the graph.

x	$f(x)$	
$\frac{5}{2}$	$\frac{11}{2}$	←Vertex
2	5	
3	5	
1	1	
4	1	
0	−7	← y-intercept

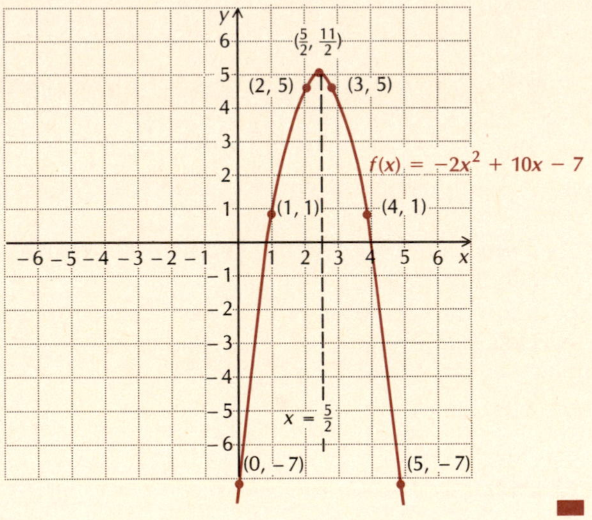

To graph a quadratic function:

1. **Note whether the x^2-coefficient is positive or negative and thus determine whether the curve opens upward or downward.**
2. **Find the vertex and the line of symmetry.**
3. **Find several other input–output pairs. Use reflection across the line of symmetry to find other pairs. Compute $f(0)$ to determine the y-intercept.**

DO EXERCISES 15 AND 16.

3 *x*-Intercepts

The points at which a graph crosses the x-axis are called its ***x*-intercepts.** These are the points at which $f(x) = 0$. Thus the x-values at the intercepts are the solutions of the equation $f(x) = 0$. These are also called **roots**, or **zeros**, of the function.

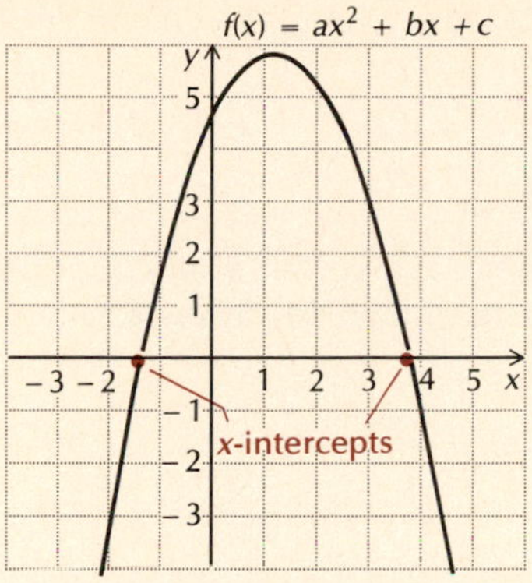

In order to find the x-intercepts of a quadratic function $f(x) = ax^2 + bx + c$, we solve the equation $f(x) = 0$, or

$$ax^2 + bx + c = 0.$$

EXAMPLE 6 Find the x-intercepts of the graph of $f(x) = x^2 - 2x - 2$.

Solution To find the x-intercepts, we solve the equation $f(x) = 0$. In this case, we solve

$$x^2 - 2x - 2 = 0.$$

Since the equation is difficult to factor, we use the quadratic formula and get $x = 1 \pm \sqrt{3}$. Thus the x-intercepts are $(1 - \sqrt{3}, 0)$ and $(1 + \sqrt{3}, 0)$. For plotting, we approximate, to get $(-0.7, 0)$ and $(2.7, 0)$. We sometimes refer to the x-coordinates as intercepts.

Note that the x-intercepts could also be used when graphing quadratic functions, but they are not essential.

DO EXERCISE 17.

The **discriminant**, $b^2 - 4ac$, tells us how many real-number solutions the equation $ax^2 + bx + c = 0$ has, so it also indicates how many intercepts there are. Compare.

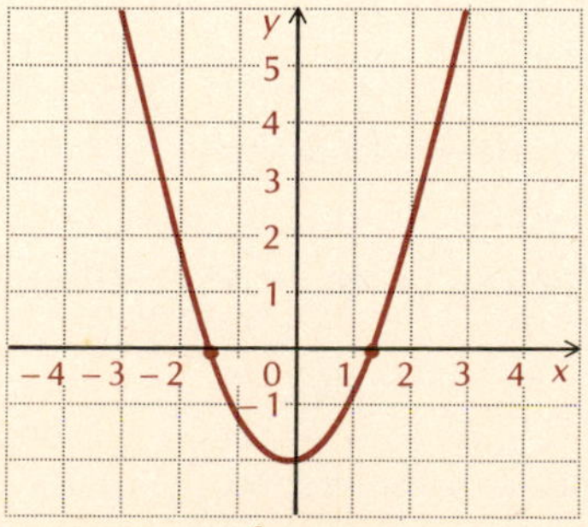

$y = ax^2 + bx + c$
$b^2 - 4ac > 0$
Two real solutions
Two x-intercepts

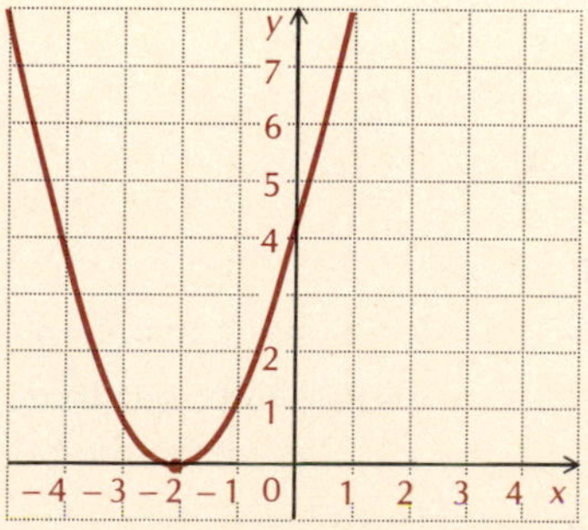

$y = ax^2 + bx + c$
$b^2 - 4ac = 0$
One real solution
One x-intercept

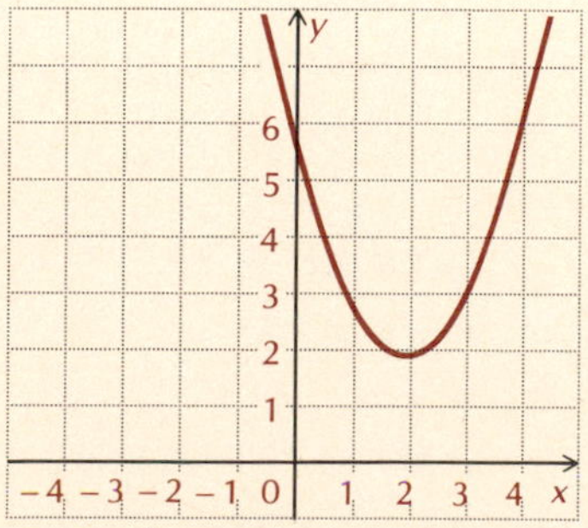

$y = ax^2 + bx + c$
$b^2 - 4ac < 0$
Two nonreal, complex solutions
No x-intercepts

DO EXERCISES 18–20.

17. Find the x-intercepts of $f(x) = x^2 - 2x - 5$.

Find the x-intercepts, if they exist.

18. $f(x) = x^2 - 2x - 3$

19. $f(x) = x^2 + 8x + 16$

20. $f(x) = -2x^2 - 4x - 3$

EXERCISE SET 4.3

1 For each of the following functions, (a) find the vertex; (b) find the line of symmetry; and (c) determine whether there is a maximum or minimum function value and find that value.

1. $f(x) = x^2$
2. $f(x) = -5x^2$
3. $f(x) = -2(x - 9)^2$
4. $f(x) = 5(x - 7)^2$
5. $f(x) = 2(x - 1)^2 - 4$
6. $f(x) = -(x + 4)^2 - 3$

For each of the following functions, (a) use completing the square to put each equation into the form $f(x) = a(x - h)^2 + k$; (b) find the vertex; and (c) determine whether there is a maximum or minimum function value and find that value.

7. $f(x) = -x^2 + 2x + 3$
8. $f(x) = -x^2 + 8x - 7$
9. $f(x) = x^2 + 3x$
10. $f(x) = x^2 - 9x$
11. $f(x) = -\frac{3}{4}x^2 + 6x$
12. $f(x) = \frac{3}{2}x^2 + 3x$
13. $f(x) = 3x^2 + x - 4$
14. $f(x) = -2x^2 + x - 1$

Find the maximum or minimum value of the function.

15. $f(x) = -5(x + 2)^2$
16. $f(x) = 0.4(x - 3)^2$
17. $f(x) = 8(x - 1)^2 + 5$
18. $f(x) = -6(x - 4)^2 - 11$
19. $f(x) = -4x^2 + x - 13$
20. $f(x) = \frac{2}{3}x^2 + 0.1x + 3.9$
21. $f(x) = \frac{1}{2}x^2 - \frac{2}{5}x - \frac{67}{100}$
22. $f(x) = -31.8x^2 + 12.3x - 17.2$
23. $g(x) = -\$120{,}000x^2 + \$430{,}000x - \$240{,}000$
24. $g(x) = \sqrt{8}x^2 - \pi x + \sqrt{5}$

2 Graph.

25. $f(x) = -x^2 + 2x + 3$
26. $f(x) = x^2 - 3x - 4$
27. $f(x) = x^2 - 8x + 19$
28. $f(x) = -x^2 - 8x - 17$
29. $f(x) = -\frac{1}{2}x^2 - 3x + \frac{1}{2}$
30. $f(x) = 2x^2 - 4x - 2$
31. $f(x) = 3x^2 - 24x + 50$
32. $f(x) = -2x^2 + 2x + 1$

3 Find the x-intercepts.

33. $f(x) = -x^2 + 2x + 3$
34. $f(x) = x^2 - 3x - 4$
35. $f(x) = x^2 - 8x + 5$
36. $f(x) = -x^2 - 3x - 3$
37. $f(x) = -5x^2 + 6x - 5$
38. $f(x) = 2x^2 + x - 5$

SYNTHESIS

Find an equation of the type $f(x) = a(x - h)^2 + k$ for each of the following.

39. $f(x) = ax^2 + bx + c$
40. $f(x) = 3x^2 + mx + m^2$

Graph.

41. $f(x) = |x^2 - 1|$ (*Hint:* Consider two cases,
$$x^2 - 1 \geq 0 \quad \text{and} \quad x^2 - 1 < 0;$$
or graph $y = x^2 - 1$ and then reflect negative values across the x-axis.)
42. $f(x) = |3 - 2x - x^2|$

Find the maximum or minimum value of the function.

43. ▦ $f(x) = 2.31x^2 - 3.135x - 5.89$
44. ▦ $f(x) = -18.8x^2 + 7.92x + 6.18$
45. Find a such that
$$f(x) = ax^2 + 3x - 8$$
has a minimum value at $x = -2$.
46. Find b such that
$$f(x) = -4x^2 + bx + 3$$
has a maximum value of 50.
47. Find c such that
$$f(x) = -0.2x^2 - 3x + c$$
has a maximum value of -225.
48. Find a quadratic function that has $(4, -5)$ as a vertex and contains the point $(-3, 1)$.
49. Find the vertex of the quadratic function
$$f(x) = qx^2 - 2x + q.$$
50. Find the line of symmetry of the graph of
$$f(x) = -3x^2 + px - 7.$$
51. Find c such that
$$f(x) = x^2 - 6x + c$$
has a vertex on the x-axis.
52. Find b such that
$$f(x) = -0.1x^2 + bx - 5$$
has a vertex on the y-axis.

4.4 Mathematical Models

OBJECTIVES

You should be able to:

1. Given a situation that is described by a linear function, find that function from two sets of data and then use that function to make predictions.
2. Given a situation described by a quadratic function, use that function to make predictions, including finding a maximum or minimum value of the function.

What Is a Mathematical Model?

When the essential parts of a problem are described in mathematical language, we say that we have a **mathematical model**. For example, the arithmetic of the natural numbers constitutes a mathematical model for situations in which counting is the essential ingredient. Situations in which algebra and calculus can be brought to bear often require the use of equations and functions. Typically in calculus, there is concern with the way in which a change in one variable affects a change in another.

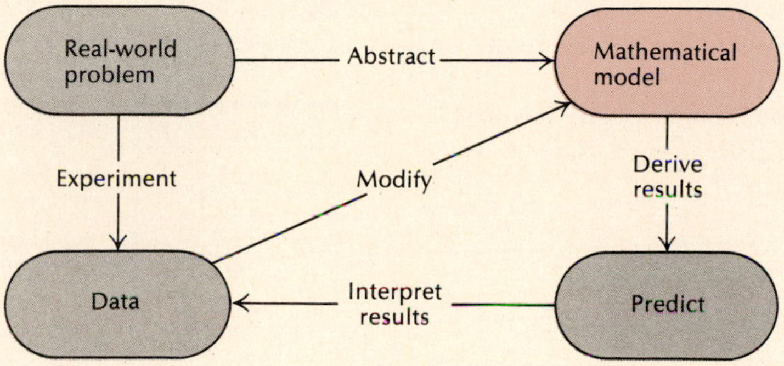

Mathematical models are abstracted from real-world situations (see the figure). Procedures within the mathematical model then give results that allow one to predict what will happen in that real-world situation. To the extent that these predictions are inaccurate or the results of experimentation do not conform to the model, the model is in need of modification.

The figure above indicates that mathematical modeling is an ongoing, possibly everchanging, process. This is often the case. For example, finding a mathematical model that will enable accurate prediction of population growth is not a simple problem. Surely any population model one might devise will need to be altered as relevant information is acquired.

Although models can reveal worthwhile information, one must always be cautious using them. An interesting case in point is a study* showing that world records in *any* running race can be modeled by a linear function. In particular, for the mile run,

$$R = -0.00582x + 15.3476,$$

where R is the world record in minutes and x is the year. Roger Bannister shocked the world in 1954 by breaking the 4-min mile. Had people been aware of this model, they would not have been shocked, for when we substitute 1954 for x, we get

$$R = -0.00582(1954) + 15.3476 = 3.97532 \approx 3{:}58.5.$$

The actual record was 3:59.4. Although for 40 to 50 years this model will continue to be worthwhile in predicting the world record in the mile run, we see that we can't get meaningful answers to some questions. For example, we could use the model to find when the 1-min mile will be broken. We set $R = 1$ and solve for x:

$$1 = -0.00582x + 15.3476$$
$$2465 = x.$$

*H. W. Ryder, H. J. Carr, and P. Herget, "Future Performance in Footracing," *Scientific American*, 234 (June 1976): 109–119.

Most track people would assure us that the 1-min mile is beyond human capability. In fact, at the time of this writing, experienced runners think it will never reach 3:40.0, the current world record being 3:48.8. Going to an even further extreme, we see that the model predicts that the 0-min mile will be run in the year 2637. In conclusion, one must be careful in the use of any model.

1 Linear Functions as Mathematical Models

As a result of gathering data, as in the case of the running problem, or other experimenting, we often acquire data that indicate that a function of some sort would be a good mathematical model.

How do we know when a linear function fits a situation?

EXAMPLE 1 Suppose that we are interested in discovering a function that can be used to predict a person's SAT score on the basis of our knowledge of their IQ. Plot the data and determine whether a linear function fits the data.

Solution These factual data have been gathered on people who have taken the SAT test.

IQ Scores, I	SAT Scores, S
124	1000
146	1250
122	1110
99	690
104	980

We make a graph with an I-axis (horizontal) and an S-axis (vertical) and plot these data. We see that a representative line can be drawn through the data points, and while there is not a close fit, we might use a linear function in this situation. If this were a carefully done scientific piece of research, then thousands of data points would be used and more advanced statistical analysis would tell us how confident we might be of using a line to represent the data.

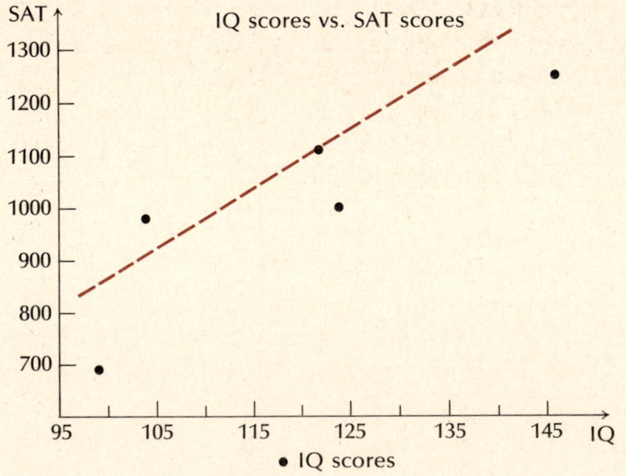

EXAMPLE 2 Wind friction, or *resistance*, increases with speed. Here are some measurements made in a wind tunnel. Plot the data and determine whether a linear function will give an approximate fit.

Velocity, km/h	10	21	34	40	45	52
Force of resistance, kg	3	4.2	6.2	7.1	15.1	29.0

Solution We make a graph with a V- (velocity) and an F- (force) axis and plot the data. They do not lie on a straight line, even approximately. Therefore, we cannot use a linear function in this situation. It turns out that these data fit a linear function on the interval containing smaller values—in this case, about [0, 38]—and a quadratic function on the interval (38, ∞). Thus a piecewise function can be used as a model.

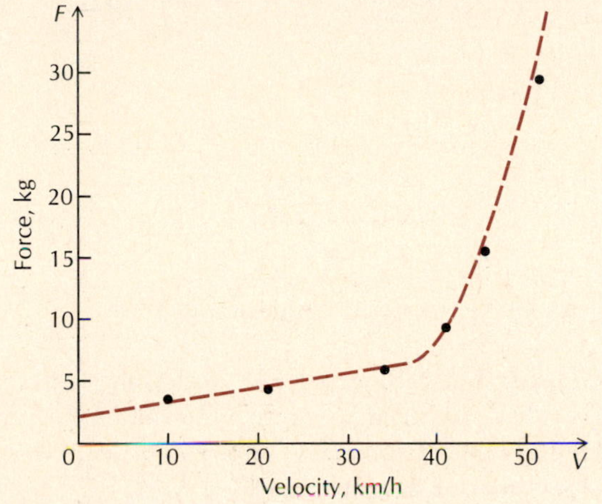

We will consider many kinds of models in this text. For now, we will consider linear and quadratic models.

EXAMPLE 3 It is known that a person with an IQ of 124 made an SAT score of 1000, and another person with an IQ of 146 made an SAT score of 1250. Assume that a linear function fits the data.

a) Find a linear function that fits the data.
b) Use the function to predict the SAT score of a person with an IQ of 110; of 130.

Solution

a) We can use our five steps for problem solving.

1. *Familiarize.* We know from the statement of the problem that a linear function or equation fits the data. We let I = the IQ score and S = the SAT score. Thus the function is of the form

$$S = mI + b.$$

The wording of the problem gives us two ordered pairs: (124, 1000) and (146, 1250).

2. *Translate.* To find the equation, we use the two known ordered pairs, (124, 1000) and (146, 1250). We call these **data points.** We use the two-point equation and substitute the data points:

$$S - S_1 = \frac{S_2 - S_1}{I_2 - I_1}(I - I_1)$$

$$S - 1000 = \frac{1250 - 1000}{146 - 124}(I - 124). \quad \text{Substituting}$$

3. *Carry out.* We simplify to the following function:

$$S = \frac{125}{11}I - \frac{4500}{11}.$$

4. *Check.* We can check by going over our work a second time. We can also substitute the value $I = 124$. We do get $S = 1000$.

1. A class of college students wanted to determine a function from which they could predict a final exam score S from a midterm test score d. To do this, they checked with two students who took the class before. One student scored 70 on the midterm and 75 on the final. Another scored 85 on the midterm and 89 on the final.
   a) Assuming that S is a linear function of d, express S in terms of d.
   b) Use the equation obtained in (a) to predict a student's final exam score, given that an 81 was scored on the midterm.

5. *State.* A linear function that fits the two data points is

$$S = \frac{125}{11}I - \frac{4500}{11}.$$

b) Using the formula $S = \frac{125}{11}I - \frac{4500}{11}$, we find S when $I = 110$:

$$S = \frac{125}{11}(110) - \frac{4500}{11} \approx 841.$$

When $I = 130$,

$$S = \frac{125}{11}(130) - \frac{4500}{11} \approx 1068.$$

DO EXERCISE 1.

Many applications are modeled by linear functions.

EXAMPLE 4 *Business: Total cost.* Raggs, Ltd., a clothing firm, has *fixed costs* of \$10,000 per year. These costs, such as rent and maintenance, must be paid no matter how much the company produces. To produce x units of a certain kind of suit, it costs \$20 per unit in addition to the fixed costs. That is, the *variable costs* for producing x of these units is $20x$ dollars. These are costs that are directly related to production, such as material, wages, and fuel. Then the *total cost*, $C(x)$, of producing x suits in a year is given by a function C:

$$C(x) = (\text{Variable costs}) + (\text{Fixed costs}) = 20x + 10{,}000.$$

a) Graph the variable-cost, the fixed-cost, and the total-cost functions.
b) What is the total cost of producing 100 suits? 400 suits?
c) How much more does it cost to produce 400 suits than 100 suits?

Solution

a) The variable-cost and the fixed-cost functions appear in the first graph below. The total-cost function is shown in the second graph. From a practical standpoint, the domains of these functions are nonnegative integers 0, 1, 2, 3, and so on, since it does not make sense to make a negative number of suits or a fractional number of suits. Nevertheless, it is common practice to draw the graphs as though the domains were the entire set of nonnegative real numbers.

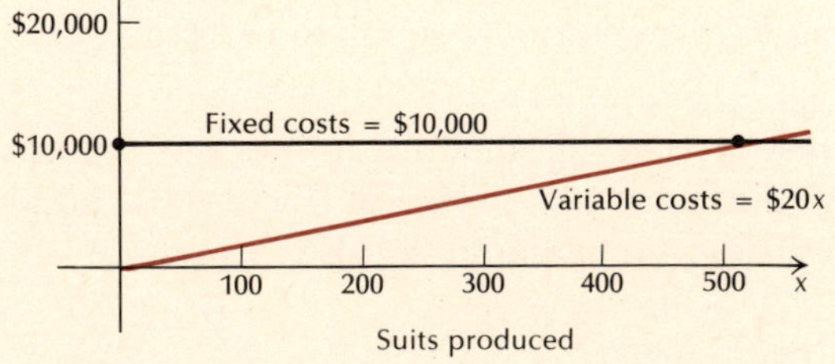

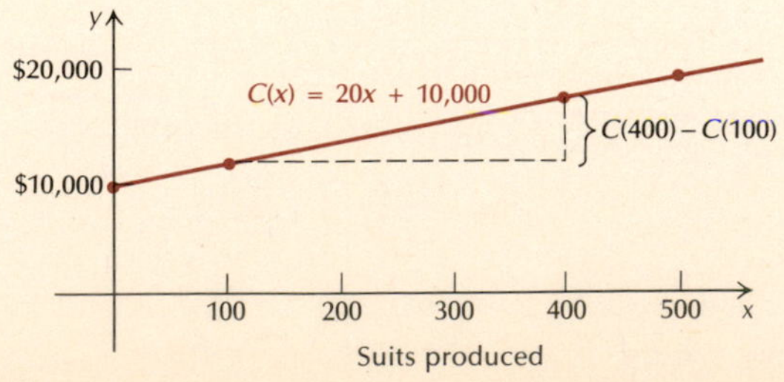

b) The total cost of producing 100 suits is

$$C(100) = 20 \cdot 100 + 10{,}000 = \$12{,}000.$$

The total cost of producing 400 suits is

$$C(400) = 20 \cdot 400 + 10{,}000 = \$18{,}000.$$

c) The extra cost of producing 400 suits rather than 100 suits is given by

$$C(400) - C(100) = \$18{,}000 - \$12{,}000 = \$6000.$$

DO EXERCISE 2.

EXAMPLE 5 *Business: Profit-and-loss analysis.* In reference to Example 4, Raggs, Ltd., determines that its total revenue from the sale of x suits is \$80 per suit. That is, the total revenue $R(x)$ is given by the function

$$R(x) = 80x.$$

a) Graph $R(x)$ and $C(x)$ using the same set of axes.
b) The total profit $P(x)$ is given by a function P:

$$P(x) = (\text{Total revenue}) - (\text{Total costs}) = R(x) - C(x).$$

Determine $P(x)$ and draw its graph using the same set of axes.
c) The company will *break even* at that value of x for which $P(x) = 0$ (that is, no profit and no loss). This is where $R(x) = C(x)$. Find the break-even value of x.

Solution

a) The graphs of $R(x) = 80x$ and $C(x) = 20x + 10{,}000$ are shown here. When $C(x)$ is above $R(x)$, a loss will occur. This is shown by the color-shaded region. When $R(x)$ is above $C(x)$, a gain will occur. This is shown by the gray-shaded region.

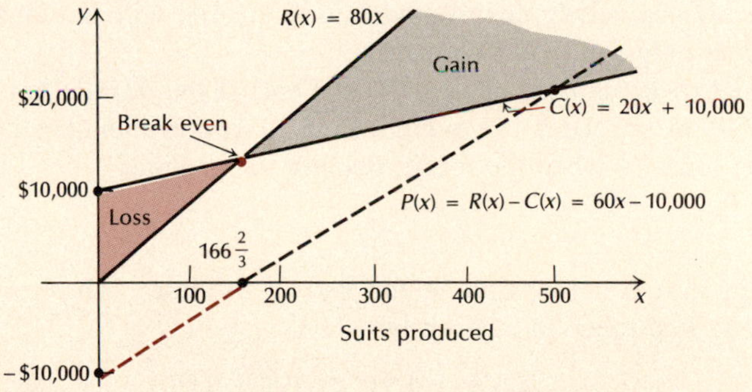

b) We see that

$$P(x) = R(x) - C(x) = 80x - (20x + 10{,}000) = 60x - 10{,}000.$$

The graph of $P(x)$ is shown by the dashed line. The color dashed line shows a "negative" profit, or loss. The black dashed line shows a "positive" profit, or gain.

c) To find the break-even value, we solve $R(x) = C(x)$:

$$R(x) = C(x)$$
$$80x = 20x + 10{,}000$$
$$60x = 10{,}000$$
$$x = 166\tfrac{2}{3}.$$

2. Rework Example 4, given that variable costs $= 30x$, fixed costs $= \$15{,}000$, and total costs $= C(x) = 30x + 15{,}000$.

3. Rework Example 5, given that

$$C(x) = 30x + 15{,}000$$

and

$$R(x) = 90x.$$

How do we interpret the fractional answer, since it is not possible to produce $\frac{2}{3}$ of a suit? We simply round to 167. Estimates of break-even points are usually sufficient because companies want to operate well away from break-even points in order to maximize profit. ■

DO EXERCISE 3.

2 Quadratic Functions as Mathematical Models

EXAMPLE 6 *A projectile problem.* When an object such as a bullet or a ball is shot or thrown upward with an initial velocity v_0, its height is given, approximately, by a quadratic function:

$$s(t) = -4.9t^2 + v_0t + h.$$

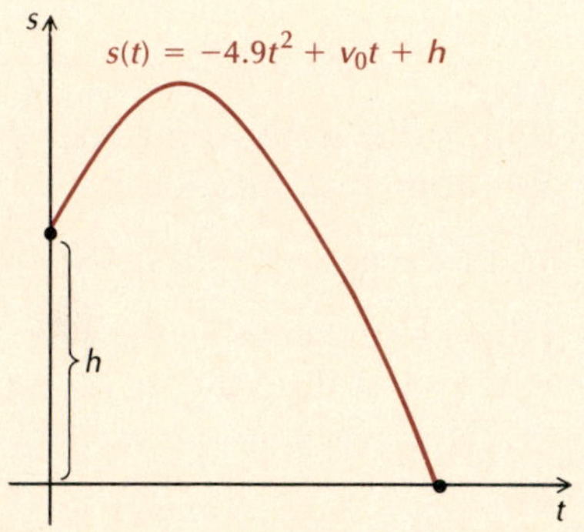

In this function, h is the starting height in meters, s is the actual height (also in meters), and t is the time from projection in seconds.

This model is constructed from theoretical principles, rather than experiment. It is based on the assumption that there is no air resistance, and that the force of gravity pulling the object earthward is constant. Neither of these conditions exists precisely, so this model (as in the case with most mathematical models) gives only approximate results.

A model rocket is fired upward. At the end of the burn, it has an upward velocity of 49 m/sec and is 155 m high. Find (a) its maximum height and when it is attained and (b) when the rocket reaches the ground.

Solution

a) We will start counting time at the end of the burn. Thus, $v_0 = 49$ and $h = 155$. The function is given by

$$s(t) = -4.9t^2 + 49t + 155.$$

We first find the vertex. The first coordinate of the vertex is

$$-\frac{b}{2a} = -\frac{49}{2(-4.9)} = 5.$$

We find the second coordinate by substituting:

$$s(5) = -4.9(5)^2 + 49(5) + 155 = 277.5.$$

The vertex of the graph is the point (5, 277.5) and the graph is shown below. Since the coefficient of t^2 is negative, we know that a maximum height is reached and is 277.5 m, attained 5 sec after the end of the burn.

b) To find when the rocket reaches the ground, we set $s(t) = 0$, and solve for t. That is, we find the t-intercept, root, or zero of the function. We do this most conveniently using the quadratic formula:

$$-4.9t^2 + 49t + 155 = 0$$
$$a = -4.9, \quad b = 49, \quad c = 155$$
$$t = \frac{-b \pm \sqrt{b^2 - 4ac}}{2a}$$
$$t = \frac{-49 \pm \sqrt{49^2 - 4(-4.9)(155)}}{2(-4.9)}$$
$$t \approx 12.525.$$

The rocket will reach the ground about 12.525 sec after the end of the burn.

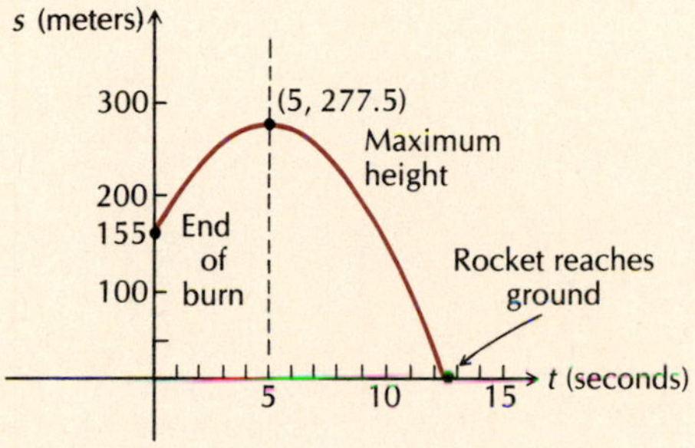

DO EXERCISE 4.

We will fit a quadratic function to a set of data in Chapter 9. We now solve a maximum problem that we began in Example 4 of Section 3.4.

EXAMPLE 7 *Maximizing area.* A hobby store has 20 ft of fencing to fence off a rectangular area for an electric train in one corner of its display room. The sides up against the wall require no fence. Find the maximum area that can be enclosed and the dimensions of the maximum area.

Solution

1., 2. *Familiarize* and *Translate.* We first draw a picture. We let x = the length of one side and y = the length of the other. Since the sum of the lengths must be 20 ft, we have

$$x + y = 20 \quad \text{and} \quad y = 20 - x.$$

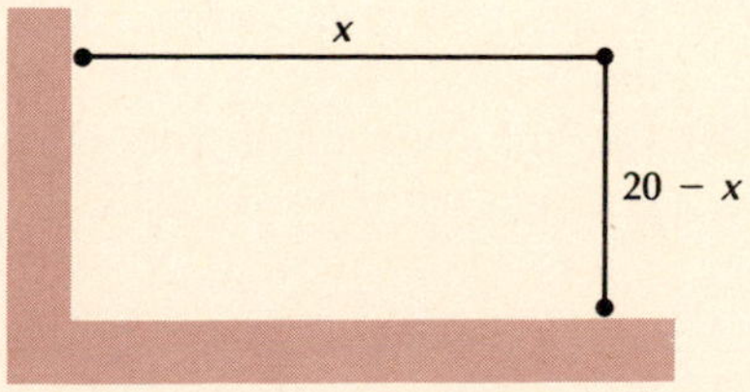

Then the area is given by

$$A = xy$$
$$A = x(20 - x) = 20x - x^2. \quad \text{Substituting } 20 - x \text{ for } y$$

We have thus expressed A as a function of x: $A(x) = 20x - x^2$.

3. *Carry out.* We want to maximize this function over the interval (0, 20). We first find the vertex of the function. The first coordinate of the vertex

4. A ball is thrown upward from the top of a cliff that is 12 m high, at a velocity of 2.8 m/sec. Find:
   a) its maximum height and when it is attained;
   b) when the ball reaches the ground.

5. *Maximizing area.* A store has 40 ft of fencing to fence off a rectangular area in one corner of its display room. The sides up against the wall require no fence. Find the maximum area and the length and the width that will yield the maximum area.

is

$$-\frac{b}{2a} = -\frac{20}{2(-1)} = 10.$$

The second coordinate is found by substituting:

$$A(10) = 20(10) - (10)^2 = 200 - 100 = 100.$$

The vertex is (10, 100). Since the leading coefficient of $A(x) = 20x - x^2$, -1, is negative, we know that the function has a maximum value. That maximum value is 100 when $x = 10$.

4. *Check.* We check our work by going over it another time. We can also check by examining the graph and estimating the maximum value.

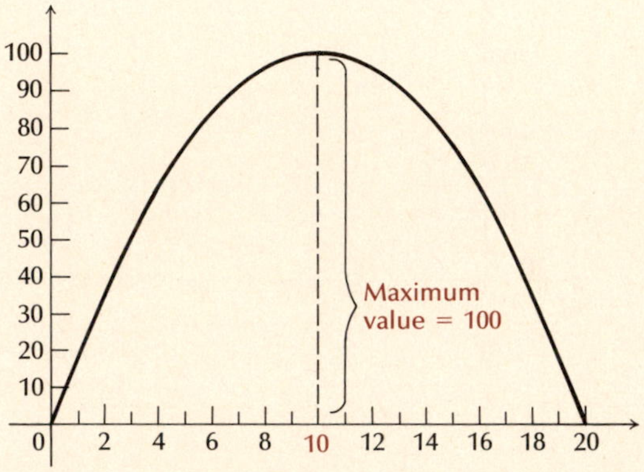

5. *State.* The maximum area of 100 ft^2 occurs when $x = 10$ ft. Thus the dimensions of 10 ft by 10 ft will yield the maximum area.

DO EXERCISE 5.

EXERCISE SET 4.4

1

1. *Life expectancy of females in the United States.* In 1950, the life expectancy of females was 72 years. In 1970, it was 75 years. Let E represent life expectancy and t the number of years since 1950. ($t = 0$ gives 1950 and $t = 10$ gives 1960.)
   a) Assuming that E is a linear function of t, express E in terms of t. [*Hint:* Data points are (0, 72) and (20, 75).]
   b) Use the equation in part (a) to predict the life expectancy of females in 1993; in 1997.

2. *Life expectancy of males in the United States.* In 1950, the life expectancy of males was 65 years. In 1970, it was 68 years. Let E represent life expectancy and t the number of years since 1950.
   a) Assuming that E is a linear function of t, express E in terms of t.
   b) Use the equation in part (a) to predict the life expectancy of males in 1993; in 1997.

3. *World record in the 100-meter dash.* In 1920, the world record for the 100-m dash was 10.43 sec. In 1970, it was 9.93 sec. Let R represent the record in the 100-m dash and t the number of years since 1920.
   a) Assuming that R is a linear function of t, express R in terms of t.
   b) Use the function in part (a) to predict the record in 1996; in 2000.
   c) According to this model, in what year will the record be 9.0 sec?

4. *Natural gas demand.* In 1950, natural gas demand in the United States was 19 quadrillion BTU. In 1960, the demand was 21 quadrillion BTU. Let D represent the demand for natural gas t years after 1950.
   a) Assuming that D is a linear function of t, express D in terms of t.
   b) Use the equation in part (a) to predict the natural gas demand in 1995; in 2000.

5. *Women in the military.* In 1971, the percentage of the total active military force that was women was 1.6%. In 1987, it was 10.2%. Let P = the percentage of women in the active duty force and t = the number of years since 1971. ($t = 0$ gives 1971 and $t = 16$ gives 1987.)
   a) Assuming that P is a linear function of t, express P in terms of t.
   b) Use the equation in part (a) to predict the percentage of women in the active duty force in 1997; in 2010.

6. *1-800-toll free numbers.* In 1983, there were 260,000 toll-free listings. In 1987, there were 535,000. Let N = the number of toll-free listings and t = the number of years since 1983.
   a) Assuming that N is a linear function of t, express N in terms of t.
   b) Use the equation in part (a) to predict the number of toll-free listings in 1998; in 2020.

7. *Business: Profit-and-loss analysis.* A ski manufacturer is planning a new line of skis. For the first year, the fixed costs for setting up the new production line are \$22,500. The variable costs for producing each pair of skis are estimated at \$40. The sales department projects that 3000 pairs can be sold during the first year at a price of \$85 per pair.
   a) Formulate a function $C(x)$ for the total cost of producing x pairs of skis.
   b) Formulate a function $R(x)$ for the total revenue from the sale of x pairs of skis.
   c) Formulate a function $P(x)$ for the total profit from the production and sale of x pairs of skis.
   d) What profit or loss will the company realize if the expected sales of 3000 pairs occurs?
   e) How many pairs must the company sell in order to break even?

8. *Business: Profit-and-loss analysis.* Boxowitz, Inc., a computer firm, is planning to sell a new minicalculator. For the first year, the fixed costs for setting up the new production line are \$100,000. The variable costs for producing each calculator are estimated at \$20. The sales department projects that 150,000 calculators can be sold during the first year at a price of \$45 each.
   a) Formulate a function $C(x)$ for the total cost of producing x calculators.
   b) Formulate a function $R(x)$ for the total revenue from the sale of x calculators.
   c) Formulate a function $P(x)$ for the total profit from the production and sale of x calculators.
   d) What profit or loss will the company realize if the expected sales of 150,000 calculators occurs?
   e) How many calculators must the firm sell in order to break even?

2

9. Of all the numbers whose sum is 50, find the two that have the maximum product. That is, find the maximum value of $Q = xy$, where $x + y = 50$.

10. Of all the numbers whose difference is 16, find the two that have the minimum product.

11. The sum of the base and the height of a triangle is 20 cm. Find the dimensions for which the area is a maximum.

12. The sum of the base and the height of a parallelogram is 138 yd. Find the dimensions for which the area is a maximum.

13. A rancher wants to build a rectangular fence next to a river, using 120 yd of fencing. What dimensions of the rectangle will maximize the area? What is the maximum area? Note that the rancher need not fence in the side next to the river.

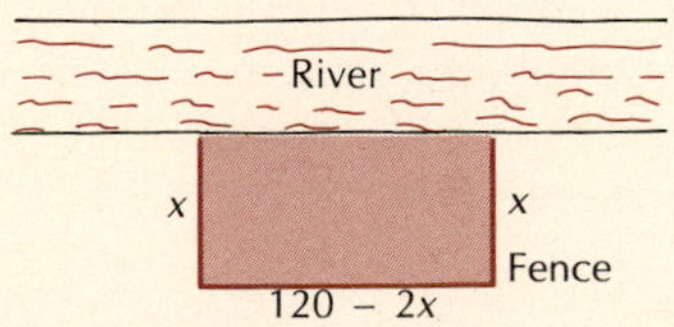

14. A rancher wants to enclose two congruent rectangular areas near a river, one for sheep and one for cattle. There is 240 yd of fencing available. What is the largest total area that can be enclosed?

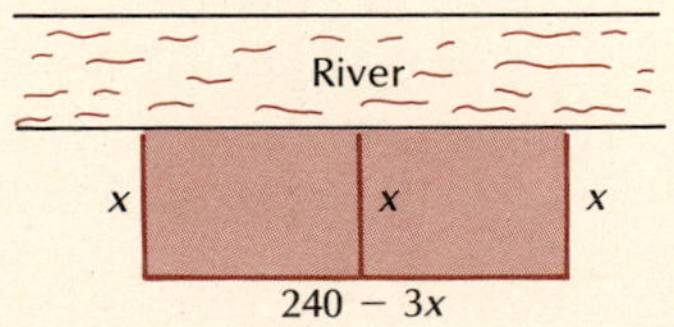

15. A carpenter is building a rectangular room with a fixed perimeter of 54 ft. What are the dimensions of the largest room that can be built? What is its area?

16. Of all rectangles that have a perimeter of 34 ft, find the dimensions of the one with the largest area. What is its area?

Business: Maximizing profit. Find the maximum profit and the number of units that must be produced and sold in order to yield the maximum profit.

17. $R(x) = 50x - 0.5x^2$, $C(x) = 4x + 10$

18. $R(x) = 50x - 0.5x^2$, $C(x) = 10x + 3$

19. $R(x) = 2x$, $C(x) = 0.01x^2 + 0.6x + 30$

20. $R(x) = 5x$, $C(x) = 0.001x^2 + 1.2x + 60$

21. Raggs, Ltd., a clothing firm, determines that in order to sell x suits, its price per suit must be

$$p = D(x) = 150 - 0.5x.$$

It also determines that its total cost of producing x suits is given by

$$C(x) = 4000 + 0.25x^2.$$

a) Total revenue = $R(x) = x\,D(x)$. Find the total revenue $R(x)$.
b) Find the total profit $P(x)$.
c) How many suits must the company produce and sell in order to maximize profit?
d) What is the maximum profit?
e) What price per suit must be charged in order to make this maximum profit?

22. An appliance firm is marketing a new refrigerator. It determines that in order to sell x refrigerators, its price per refrigerator must be

$$p = D(x) = 280 - 0.4x.$$

It also determines that its total cost of producing x refrigerators is given by

$$C(x) = 5000 + 0.6x^2.$$

a) Total revenue = $R(x) = x\,D(x)$. Find the total revenue $R(x)$.
b) Find the total profit $P(x)$.
c) How many refrigerators must the company produce and sell in order to maximize profit?
d) What is the maximum profit?
e) What price per refrigerator must be charged in order to make this maximum profit?

23. A rocket is fired upward. At the end of the burn, it has an upward velocity of 147 m/sec and is 560 m high. Find (a) its maximum height and when it is attained and (b) when the rocket reaches the ground.

24. A rocket is fired upward. At the end of the burn, it has an upward velocity of 245 m/sec and is 1240 m high. Find (a) its maximum height and when it is attained and (b) when the rocket reaches the ground.

25. *Norman window.* A Norman window is a rectangle with a semicircle on top. Suppose that the perimeter of a particular Norman window is to be 24 ft. What should its dimensions be in order to allow the maximum amount of light to enter through the window?

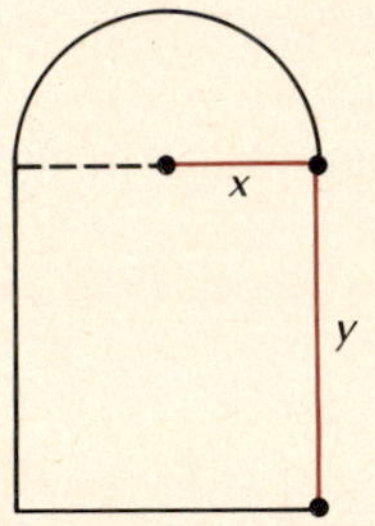

A Norman window.

SYNTHESIS

26. A university is trying to determine what price to charge for football tickets. At a price of \$6 per ticket, it averages 70,000 people per game. For every increase of \$1, it loses 10,000 people from the average number. Every person at the game spends an average of \$1.50 on concessions. What price per ticket should be charged in order to maximize revenue? How many people will attend at that price?

27. Suppose that you are the owner of a 30-unit motel. All units are occupied when you charge \$20 a day per unit. For every increase of x dollars in the daily rate, there are x units vacant. Each occupied room costs \$2 per day to service and maintain. What should you charge per unit in order to maximize profit?

28. An apple farm yields an average of 30 bushels of apples per tree when 20 trees are planted on an acre of ground. Each time 1 more tree is planted per acre, the yield decreases 1 bushel per tree due to the extra congestion. How many trees should be planted in order to get the highest yield?

29. When a theater owner charges \$3 for admission, there is an average attendance of 100 people. For every \$0.10 increase in admission, there is a loss of 1 customer from the average. What admission should be charged in order to maximize revenue?

CHALLENGE

30. A 24-in. piece of string is cut into two pieces. One piece is used to form a circle and the other to form a square. Express the sum of the areas S as a function of the length x cut to form the circle.

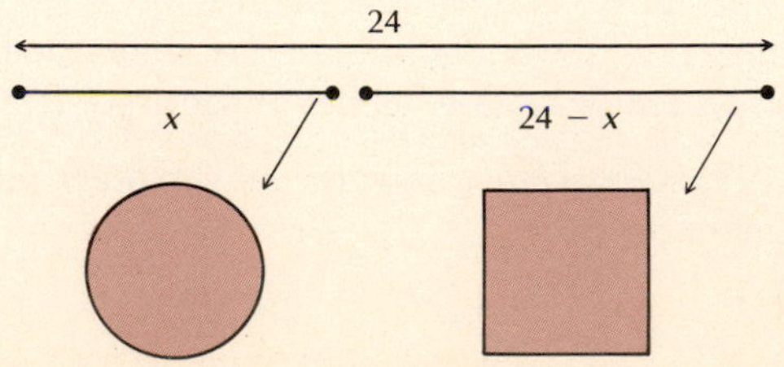

31. Find the dimensions and the area of the largest rectangle that can be inscribed as shown in a right triangle ABC whose sides have lengths 9 cm, 12 cm, and 15 cm.

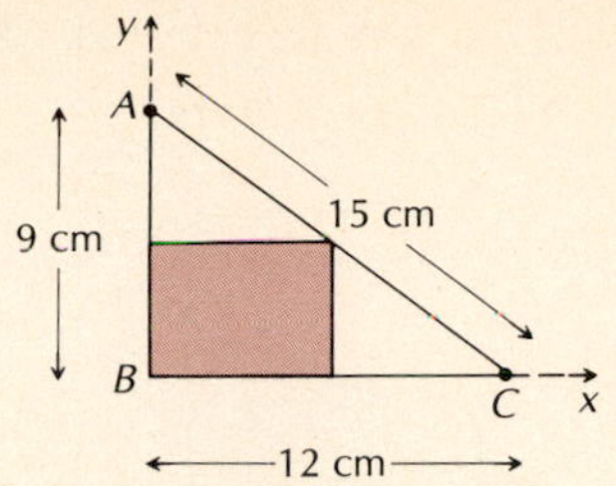

4.5 Sets, Sentences, and Inequalities

OBJECTIVES

You should be able to:

1 **Find the union and the intersection of sets like $\{1, 2, 3, 4\}$ and $\{3, 4, 5\}$, and graph the union and the intersection of sets like $\{x|x > 1\}$ and $\{x|x \leq 3\}$.**

2 **Solve conjunctions and disjunctions of inequalities.**

1 Intersections and Unions

The **intersection** of two sets consists of those elements common to both sets. Intersection is illustrated in the figure on the left. Note in the figure on the right that the intersection of two sets may be the empty set. The intersection of sets A and B is indicated as $A \cap B$.

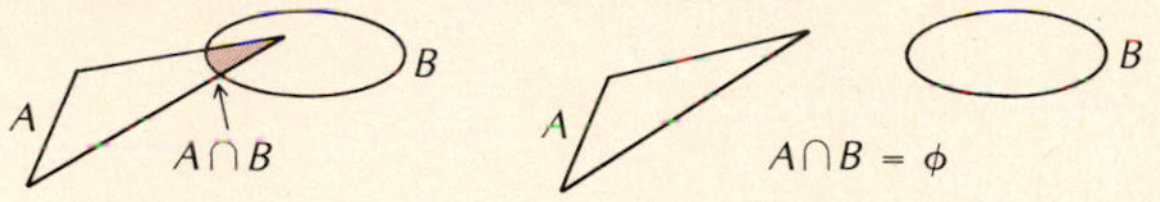

The intersection of sets $\{1, 3, 5, 7, 9\}$ and $\{2, 4, 5, 7, 11\}$ is

$$\{1, 3, 5, 7, 9\} \cap \{2, 4, 5, 7, 11\} = \{5, 7\}.$$

Graphs are set diagrams. They represent the solution set of an equation or inequality. We can find intersections of solution sets using graphs. In the following example, we find the intersection of the set of all x such that 3 is less than x and the set of all x less than or equal to 5. These sets are indicated, and the symbolism is read, as follows:

$\{x|3 < x\}$ "the set of all x such that 3 is less than x."
$\{x|x \leq 5\}$ "the set of all x such that x is less than or equal to 5."

EXAMPLE 1 Find the graph of $\{x|3 < x\} \cap \{x|x \leq 5\}$.

Solution We first graph the two solution sets separately and then find the intersection.

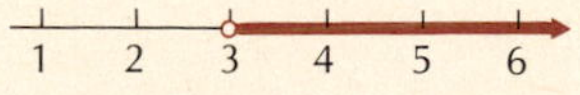

$\{x \mid 3 < x\}$

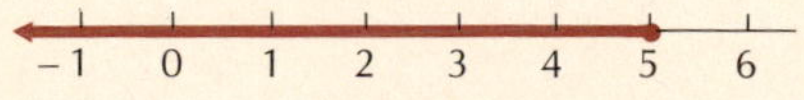

$\{x \mid x \leq 5\}$

The open circle at 3 indicates that 3 is not in the solution set. The solid circle at 5 indicates that 5 is in the solution set. The intersection is as follows.

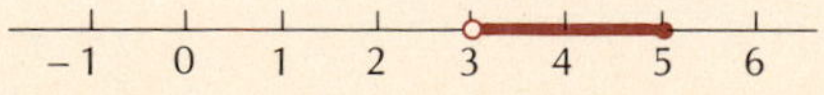

$\{x \mid 3 < x\} \cap \{x \mid x \leq 5\}$, or, in interval notation $(3, 5]$

Find the intersection.

1. $\{-3, -4, 2, 3, 4\} \cap \{1, 4, -3, 8, 9, 11\}$

2. $\{2, b, c, d, e\} \cap \{1, 2, d, e, f, g\}$

Graph.

3. $\{x|1 < x\} \cap \{x|x \le 3\}$

-6 -5 -4 -3 -2 -1 0 1 2 3 4 5 6

4. $\{x|-4 < x\} \cap \{x|x < -1\}$

-6 -5 -4 -3 -2 -1 0 1 2 3 4 5 6

5. $\{x|0 \le x\} \cap \{x|2 \le x\}$

-6 -5 -4 -3 -2 -1 0 1 2 3 4 5 6

6. $\{x|1 \le x\} \cap \{x|x \le -2\}$

-6 -5 -4 -3 -2 -1 0 1 2 3 4 5 6

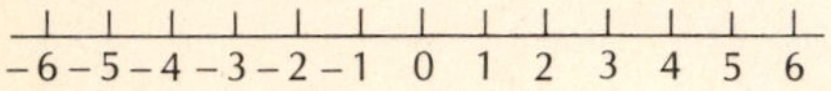

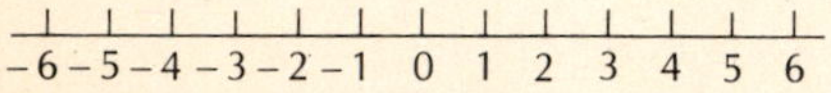

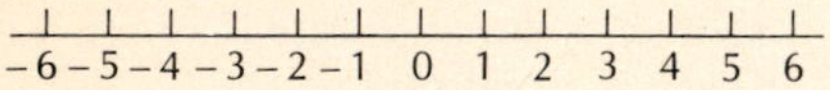

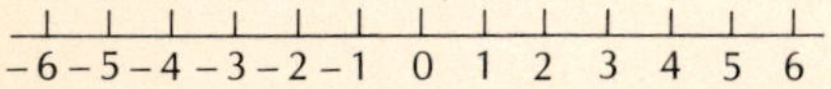

Find the union.

7. $\{1, 2\} \cup \{2, 3, 4, 5\}$

8. $\{-3, -4, 2, 3, 4\} \cup \{1, 4, -3, 8, 9, 11\}$

9. $\{2, a, b, c\} \cup \{1, 2, c, d, e\}$

Graph.

10. $\{x|x < -3\} \cup \{x|x \ge 1\}$

-6 -5 -4 -3 -2 -1 0 1 2 3 4 5 6

11. $\{x|x \le -3\} \cup \{x|x \le 3\}$

-6 -5 -4 -3 -2 -1 0 1 2 3 4 5 6

12. $\{x|x > 0\} \cup \{x|x < 1\}$

-6 -5 -4 -3 -2 -1 0 1 2 3 4 5 6

13. $\left\{x\left|x > \dfrac{1}{2}\right.\right\} \cup \left\{x\left|x = \dfrac{1}{2}\right.\right\}$

-6 -5 -4 -3 -2 -1 0 1 2 3 4 5 6

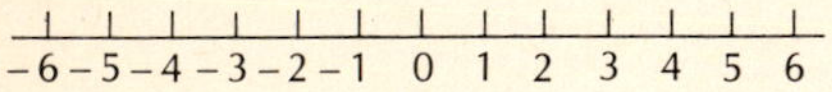

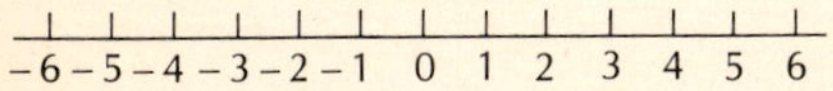

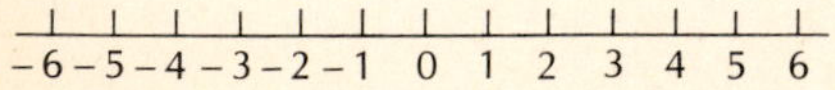

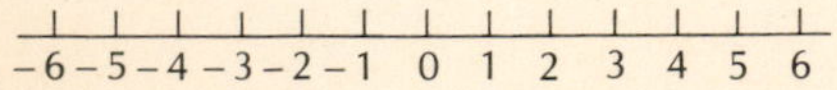

EXAMPLE 2 Graph: $\{x|-3 \le x\} \cap \{x|-1 \le x\}$.

Solution Again, we graph the solution sets separately and then find the intersection.

$\{x|-3 \le x\} \cap \{x|-1 \le x\}$, or $[-1, \infty)$

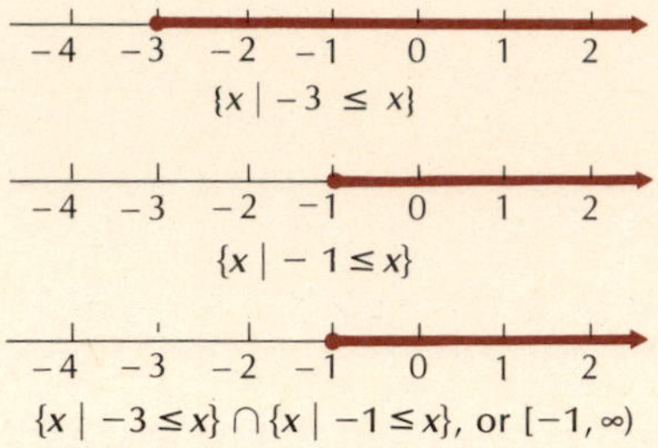

DO EXERCISES 1–6.

The **union** of two sets consists of the members that are in one or both of the sets. Union is illustrated in the following figure. The union of sets A and B is indicated as $A \cup B$.

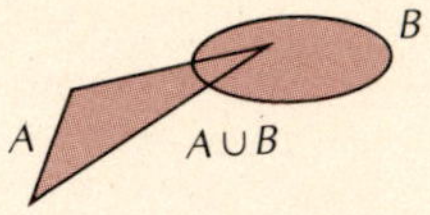

The union of sets $\{1, 3, 5, 7, 9\}$ and $\{2, 4, 5, 7, 11\}$ is

$$\{1, 3, 5, 7, 9\} \cup \{2, 4, 5, 7, 11\} = \{1, 2, 3, 4, 5, 7, 9, 11\}.$$

In the following examples, we find unions of solution sets. Note that we graph the individual sets separately and then combine them.

EXAMPLE 3 Graph: $\{x|x \ge -1\} \cup \{x|x < 2\}$.

Solution

$\{x|x \ge -1\} \cup \{x|x < 2\}$
= the set of all real numbers, or $(-\infty, \infty)$

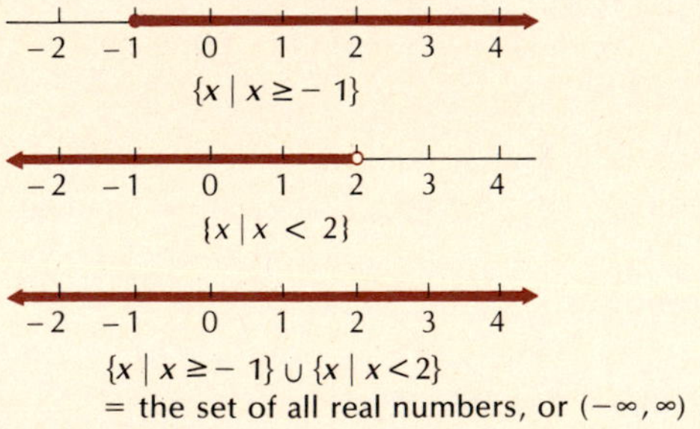

EXAMPLE 4 Graph: $\{x|x \le -2\} \cup \{x|x > 1\}$.

Solution

$\{x|x \le -2\} \cup \{x|x > 1\}$, or $(-\infty, -2] \cup (1, \infty)$

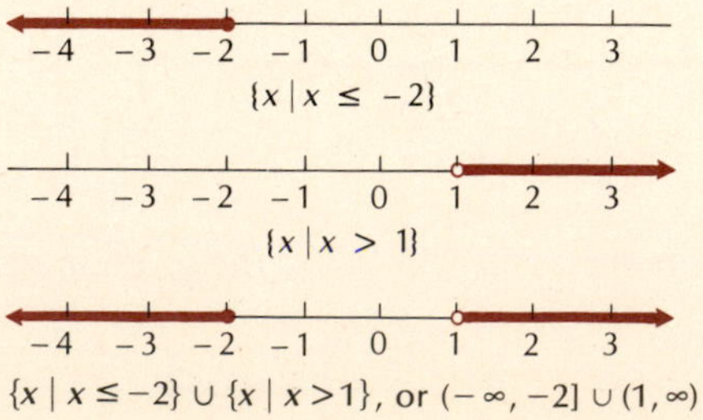

DO EXERCISES 7–13.

2 Compound Sentences

When two sentences are joined by the word *and,* a compound sentence is formed. Such a sentence is called a **conjunction.** (The word conjunction used in this way is a logical term; the meaning is not the same as in ordinary grammar.) A conjunction of two sentences is true when both parts are true. Thus the solution set is the intersection of the solution sets of the parts. Consider, for example, the conjunction

$$-2 \leq x \text{ and } x < 1.$$

Any number that makes this sentence true must make both parts true. We can graph the sentence by graphing the parts and then finding their intersection.

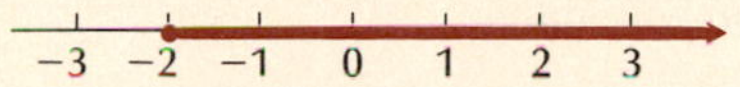

$-2 \leq x$

Any number in this set makes the first part true.

(a)

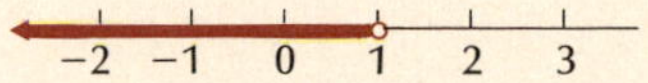

$x < 1$

Any number in this set makes the second part true.

(b)

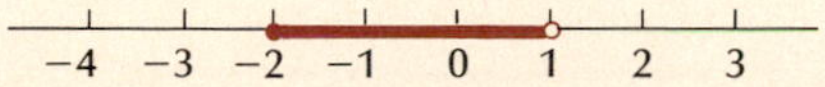

Any number in this set (the intersection) makes both parts true ($-2 \leq x$ *and* $x < 1$ are both true).

(c)

We often abbreviate certain conjunctions of inequalities. In this case,

$$-2 \leq x \text{ and } x < 1 \quad \text{is abbreviated} \quad -2 \leq x < 1.$$

The latter is read "-2 is less than or equal to x and x is less than 1," or "-2 is less than or equal to x is less than 1." Thus,

$$\{x|-2 \leq x\} \cap \{x|x < 1\} = \{x|-2 \leq x \text{ and } x < 1\}$$
$$= \{x|-2 \leq x < 1\}.$$

The word *and* corresponds to set *intersection.*

DO EXERCISES 14–20.

EXAMPLE 5 Solve and graph: $-3 < 2x + 5 < 7$.

Solution

Method 1

$-3 < 2x + 5$	*and*	$2x + 5 < 7$	Rewriting using *and*
$-8 < 2x$	*and*	$2x < 2$	Adding -5
$-4 < x$	*and*	$x < 1$	Multiplying by $\frac{1}{2}$

Method 2

$-3 < 2x + 5 < 7$	
$-8 < 2x < 2$	Adding -5
$-4 < x < 1$	Multiplying by $\frac{1}{2}$

Abbreviate the conjunction.

14. $4 < x$ *and* $x < 8$

15. $-3 \leq x$ *and* $x < 0$

16. $-5 \leq x$ *and* $x \leq -2$

17. $-1 < x$ *and* $x \leq -\frac{1}{4}$

Rewrite, using the word *and.*

18. $-\frac{1}{2} < x < 1$

19. $-\frac{17}{3} \leq x < -2$

20. $\frac{19}{4} \leq x \leq \frac{37}{6}$

Solve. Then graph.

21. $-4 < 3x - 2 \leq 10$

22. $-2 < \frac{3 - x}{4} < 2$

23. $\frac{2}{3} \leq 1 - 2x \leq \frac{5}{3}$

The solution set is

$$\{x \mid -4 < x\} \cap \{x \mid x < 1\}, \quad \text{or } \{x \mid -4 < x < 1\}, \quad \text{or } (-4, 1).$$

The graph is as follows.

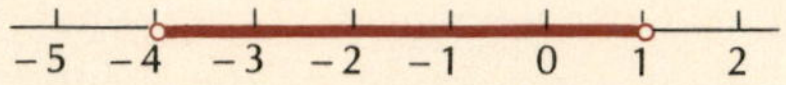

EXAMPLE 6 Solve and graph:

$$-4 < \frac{5 - 3x}{2} \leq 5.$$

Solution We have the following:

$$-8 < 5 - 3x \leq 10 \quad \text{Multiplying by 2}$$

$$-13 < -3x \leq 5 \quad \text{Adding } -5$$

$$\tfrac{13}{3} > x \geq -\tfrac{5}{3} \quad \text{Multiplying by } -\tfrac{1}{3}$$

$$-\tfrac{5}{3} \leq x < \tfrac{13}{3} \quad x \geq -\tfrac{5}{3} \text{ means } -\tfrac{5}{3} \leq x, \text{ and } \tfrac{13}{3} > x \text{ means } x < \tfrac{13}{3}$$

The solution set is $\{x \mid -\frac{5}{3} \leq x < \frac{13}{3}\}$, or $[-\frac{5}{3}, \frac{13}{3})$.

The graph is as follows.

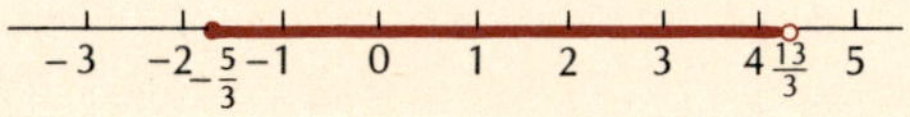

DO EXERCISES 21–23.

When two sentences are joined by the word *or*, a compound sentence is formed. Such a sentence is called a **disjunction**. A disjunction of two sentences is true when either part is true. It is also true when both parts are true. The solution set of a disjunction is thus the union of the solution sets of the parts. Consider the disjunction

$$x < -2 \text{ or } x > \tfrac{1}{4}.$$

Any number that makes either or both of the parts true makes the disjunction true.

We can graph the sentence by graphing the two parts and then finding their union. The word *or* corresponds to the set *union*.

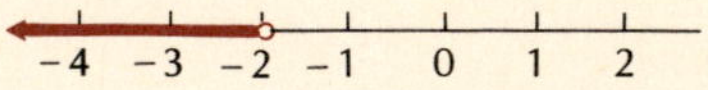

$x < -2$

Any number in this set makes the first part true.

(a)

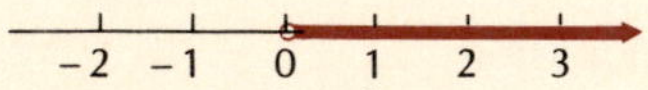

$x > \frac{1}{4}$

Any number in this set makes the second part true.

(b)

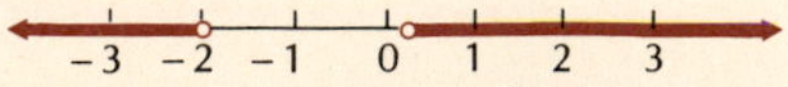

$x < -2$ *or* $x > \frac{1}{4}$

$\{x \mid x < -2\} \cup \{x \mid x > \frac{1}{4}\}$

Any number in this set makes one or both of the parts true.

(c)

CAUTION! There is no compact way to abbreviate disjunctions of inequalities, ordinarily. *Be careful about this*! For example, if you try to abbreviate $-3 < x$ *or* $x < 4$ as $-3 < x < 4$, you will be *wrong*, because $-3 < x < 4$ is an abbreviation for the conjunction $-3 < x$ *and* $x < 4$. To state this another way, you can write $\{x|-2 \leq x\} \cap \{x|x < 1\}$ as $\{x|-2 \leq x < 1\}$ without the word "and," but you cannot write $\{x|x < -2\} \cup \{x|x > \frac{1}{4}\}$ without the word "or" or without the union symbol $\cup$. Note that $3x \leq 15$ can be written as an abbreviation for $3x < 15$ *or* $3x = 15$.

EXAMPLE 7 Solve $2x - 5 < -7$ *or* $2x - 5 > 7$. Then graph.

Solution We have

$$\begin{aligned} 2x - 5 < -7 \quad &\text{or} \quad 2x - 5 > 7 \\ 2x < -2 \quad &\text{or} \quad 2x > 12 \qquad \text{Adding 5} \\ x < -1 \quad &\text{or} \quad x > 6. \qquad \text{Multiplying by } \tfrac{1}{2} \end{aligned}$$

The solution set is $\{x|x < -1$ *or* $x > 6\}$, which is the union $\{x|x < -1\} \cup \{x|x > 6\}$ or $(-\infty, -1) \cup (6, \infty)$. The graph is as follows.

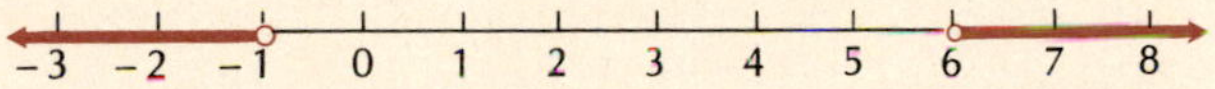

EXAMPLE 8 Solve $\frac{4 - 3x}{2} < -1$ *or* $\frac{4 - 3x}{2} \geq 1$. Then graph.

Solution We have

$$\begin{aligned} \frac{4 - 3x}{2} < -1 \quad &\textit{or} \quad \frac{4 - 3x}{2} \geq 1 \\ 4 - 3x < -2 \quad &\textit{or} \quad 4 - 3x \geq 2 \qquad \text{Multiplying by 2} \\ 4 < -2 + 3x \quad &\textit{or} \quad 4 \geq 2 + 3x \qquad \text{Adding } 3x \\ 6 < 3x \quad &\textit{or} \quad 2 \geq 3x \\ 2 < x \quad &\textit{or} \quad \frac{2}{3} \geq x. \end{aligned}$$

The solution set is

$$\{x|2 < x \textit{ or } \tfrac{2}{3} \geq x\},$$

which is the union

$$\{x|2 < x\} \cup \{x|\tfrac{2}{3} \geq x\}, \quad \textit{or } (2, \infty) \cup (-\infty, \tfrac{2}{3}].$$

The graph is as follows.

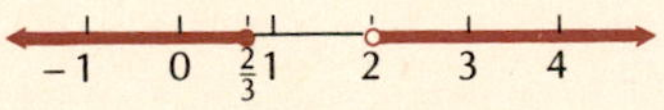

DO EXERCISES 24–27.

Solve. Then graph.

24. $x + 4 < -3$ *or* $x + 4 > 3$

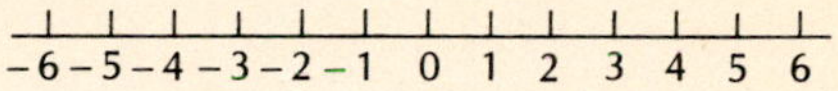

25. $2x - 3 \leq -5$ *or* $2x - 3 > 5$

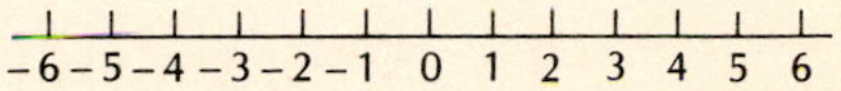

26. $4 - 3x \leq -1$ *or* $4 - 3x \geq 1$

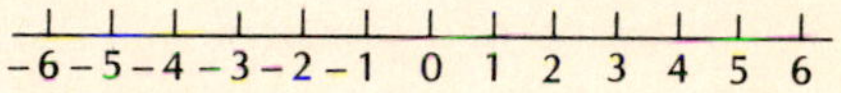

27. $\frac{4x + 5}{3} < -2$ *or* $\frac{4x + 5}{3} \geq 2$

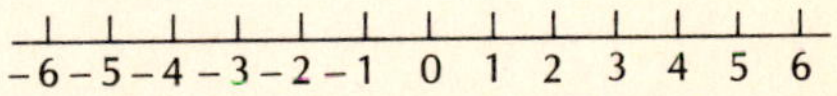

EXERCISE SET 4.5

1 Find the union or the intersection.

1. $\{3, 4, 5, 8, 10\} \cap \{1, 2, 3, 4, 5, 6, 7\}$
2. $\{3, 4, 5, 8, 10\} \cup \{1, 2, 3, 4, 5, 6, 7\}$
3. $\{0, 2, 4, 6, 8\} \cup \{4, 6, 9\}$
4. $\{0, 2, 4, 6, 8\} \cap \{4, 6, 9\}$
5. $\{a, b, c\} \cap \{c, d\}$
6. $\{a, b, c\} \cup \{c, d\}$

Graph.

7. $\{x|7 \leq x\} \cup \{x|x < 9\}$
8. $\{x|-\frac{1}{2} \leq x\} \cup \{x|x < \frac{1}{2}\}$
9. $\{x|-\frac{1}{2} \leq x\} \cap \{x|x < \frac{1}{2}\}$
10. $\{x|x > \frac{1}{4}\} \cap \{x|1 \geq x\}$
11. $\{x|x < -\pi\} \cup \{x|x > \pi\}$
12. $\{x|-\pi \leq x\} \cap \{x|x < \pi\}$
13. $\{x|x < -7\} \cup \{x|x = -7\}$
14. $\{x|x > \frac{1}{2}\} \cup \{x|x = \frac{1}{2}\}$
15. $\{x|x \geq 5\} \cap \{x|x \leq -3\}$
16. $\{x|x \geq -3\} \cup \{x|x \leq 0\}$

2 Solve. Use interval notation for solution sets where appropriate.

17. $-2 \leq x + 1 < 4$
18. $-3 < x + 2 \leq 5$
19. $5 \leq x - 3 \leq 7$
20. $-1 < x - 4 < 7$
21. $-3 \leq x + 4 \leq -3$
22. $-5 < x + 2 < -5$
23. $-2 < 2x + 1 < 5$
24. $-3 \leq 5x + 1 \leq 3$
25. $-4 \leq 6 - 2x < 4$
26. $-3 < 1 - 2x \leq 3$
27. $-5 < \frac{1}{2}(3x + 1) \leq 7$
28. $\frac{2}{3} \leq -\frac{4}{5}(x - 3) < 1$
29. $3x \leq -6$ *or* $x - 1 > 0$
30. $2x < 8$ *or* $x + 3 \geq 1$
31. $2x + 3 \leq -4$ *or* $2x + 3 \geq 4$
32. $3x - 1 < -5$ *or* $3x - 1 > 5$
33. $2x - 20 < -0.8$ *or* $2x - 20 > 0.8$
34. $5x + 11 \leq -4$ *or* $5x + 11 \geq 4$
35. $x + 14 \leq -\frac{1}{4}$ *or* $x + 14 \geq \frac{1}{4}$
36. $x - 9 < -\frac{1}{2}$ *or* $x - 9 > \frac{1}{2}$
37. ▦ The length of a rectangle is 15.23 cm. What widths will give a perimeter greater than 40.23 cm and less than 137.8 cm?
38. ▦ The height of a triangle is 15 m. What lengths of the base will keep the area less than or equal to 305.4 m^2 (and, of course, positive)?
39. To get an A in a course, a student's average must be greater than or equal to 90%. It will, of course, be less than or equal to 100%. On the first three tests, a student scored 83%, 87%, and 93%. What scores on the fourth test will produce an A? Is an A possible?
40. In Exercise 39, suppose that the scores on the first three tests are 75%, 70%, and 83%. What scores on the fourth test will produce an A? Is an A possible?
41. *Temperatures of liquids.* The function $C(F) = \frac{5}{9}(F - 32)$ can be used to convert Fahrenheit temperatures F to Celsius temperatures C. Copper is a liquid for Celsius temperatures C such that $1083° \leq C < 2580°$. Find such an inequality for the corresponding Fahrenheit temperatures.
42. *Pressure at sea depth.* The pressure P, given by
$$P(d) = 1 + \frac{1}{33}d,$$
gives the pressure, in atmospheres (atm), at a depth d, in feet, in the sea. Find the depths d for which the pressure P satisfies $2 \leq P \leq 10$.
43. *Toll road charges.* The function C, given by
$$C(d) = 0.027d + 0.32,$$
can be used to estimate the cost C, in dollars, of driving a car d miles on the Indiana toll road. Find the distances d if the cost of driving on the toll road is between \$2 and \$4.
44. *Straight-line depreciation.* A company buys an office machine for \$5200 on January 1 of a given year. The machine is expected to last 8 years, at the end of which time its *trade-in*, or *salvage, value* will be \$1100. If the company figures the decline or depreciation in value to be the same each year, then the salvage value V, after t years, $0 \leq t \leq 8$, is given by the linear function
$$V(t) = \$5200 - \$512.50t.$$
For what time period is the value of the machine between \$2000 and \$4000?

SYNTHESIS

Solve.

45. $x \leq 3x - 2 \leq 2 - x$
46. $2x \leq 5 - 7x < 7 + x$
47. $(x + 1)^2 > x(x - 3)$

48. $(x + 4)(x - 5) > (x + 1)(x - 7)$

49. $(x + 1)^2 \leq (x + 2)^2 \leq (x + 3)^2$

50. $(x - 1)(x + 1) < (x + 1)^2 \leq (x - 3)^2$

Find the domain of the function.

51. $f(x) = \dfrac{\sqrt{x + 2}}{\sqrt{x - 2}}$

52. $f(x) = \dfrac{\sqrt{3 - x}}{\sqrt{x + 5}}$

4.6 Equations and Inequalities with Absolute Value

OBJECTIVE

You should be able to:

1 Solve and graph equations and inequalities with absolute value.

1 Recall the definition of absolute value:

$$|x| = \begin{cases} x, & \text{if } x \geq 0, \\ -x, & \text{if } x < 0. \end{cases}$$

An informal way of thinking of absolute value is that it is the distance from 0 of a number on a number line. For example, $|4|$ is 4 because 4 is 4 units from 0; $|-5|$ is 5 because -5 is 5 units from 0. This idea is helpful in solving *equations* and *inequalities with absolute value*.

EXAMPLE 1 Solve: $|x| = 3$.

Solution To solve, we look for all numbers x whose distance from 0 is 3. There are two of them, so there are two solutions, 3 and -3. The solution set is $\{-3, 3\}$. The graph is as follows.

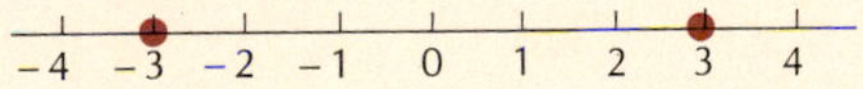

DO EXERCISES 1 AND 2.

EXAMPLE 2 Solve: $|x| < 3$.

Solution This time, we look for all numbers x whose distance from 0 is less than 3. These are the numbers between -3 and 3. The solution set and its graph are as follows.

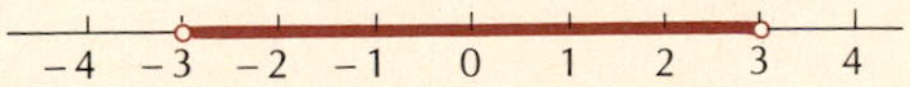

$\{x \ -3 < x < 3\}$, or $(-3, 3)$

DO EXERCISES 3 AND 4.

EXAMPLE 3 Solve: $|x| \geq 3$.

Solution This time, we look for all numbers x whose distance from 0 is 3 or greater. The solution set and its graph are as follows.

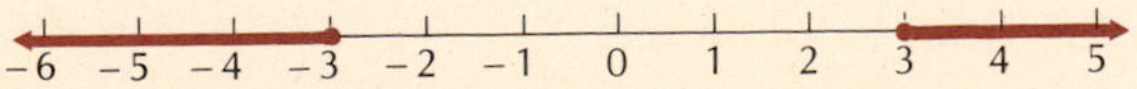

$\{x \mid x \leq -3 \text{ or } x \geq 3\}$, or $(-\infty, -3] \cup [3, \infty)$

Solve and graph.

1. $|x| = 5$

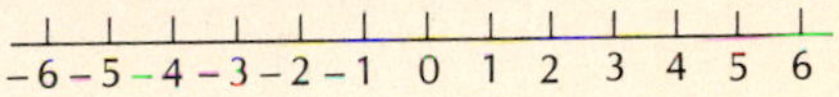

2. $|x| = \dfrac{1}{4}$

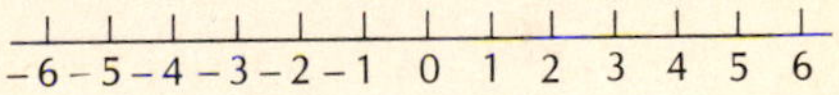

Solve and graph.

3. $|x| < 5$

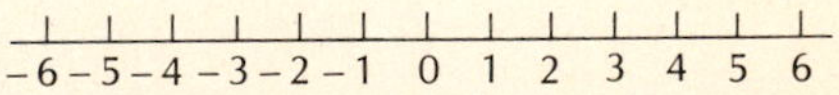

4. $|x| \leq \dfrac{1}{4}$

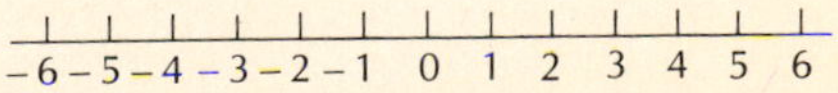

Solve and graph.

5. $|x| \geq 5$

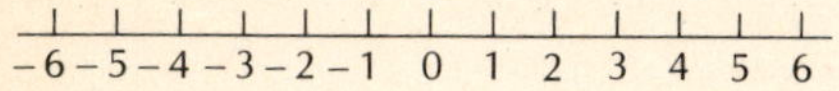

DO EXERCISES 5 AND 6.

The results of the above examples can be generalized as follows.

> **For any $a > 0$:**
> **i) $|x| = a$ is equivalent to $x = -a$ or $x = a$.**
> **ii) $|x| < a$ is equivalent to $-a < x < a$.**
> **iii) $|x| > a$ is equivalent to $x < -a$ or $x > a$.**

Similar statements hold true for $|x| \leq a$ and $|x| \geq a$.

EXAMPLE 4 Solve: $|x - 2| = 3$.

Solution Note that this is a translation of $|x| = 3$, two units to the right. We first graph $|x| = 3$.

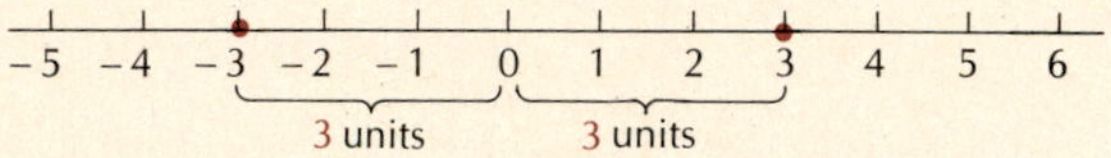

This consists of the numbers that are a distance of 3 from 0. We now translate.

6. $|x| > \frac{1}{4}$

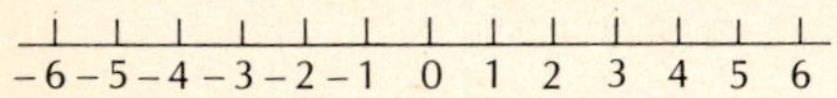

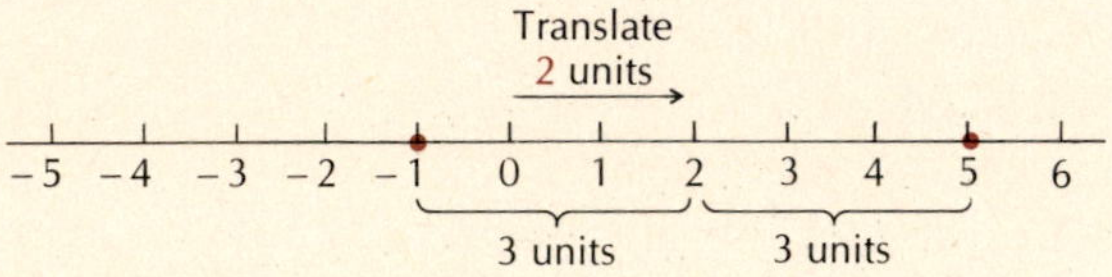

This solution set consists of the numbers that are a distance of 3 from 2 (note that 2 is where 0 went in the translation). The solutions of $|x - 2| = 3$ are -1 and 5. The solution set is $\{-1, 5\}$.

The results of Example 4 can be generalized as follows.

> **For any x, $|x - a|$ is the distance between a and x.**

EXAMPLE 5

a) $|x - 5|$ is the distance between 5 and x.
b) $|x + 7|$ is the distance between -7 and x [because $x + 7 = x - (-7)$].

Here are some further examples of solving inequalities. In Example 6, we use three methods. The first two provide understanding. The third is the most efficient for general solving.

EXAMPLE 6 Solve: $|x + 2| < 3$.

Solution

Method 1. An equivalent inequality is $|x - (-2)| < 3$. We translate the graph of $|x| < 3$ to the left 2 units.

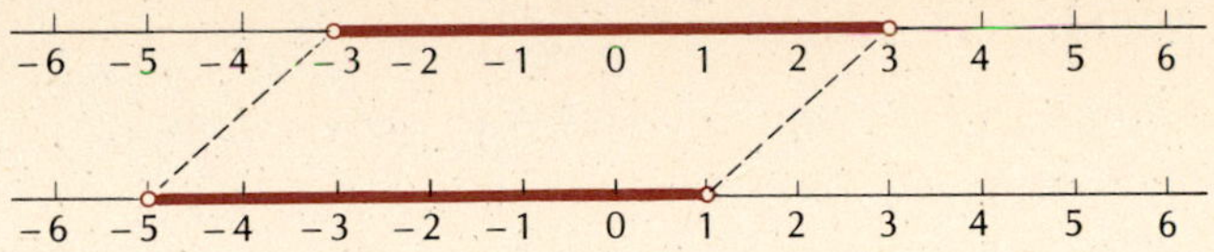

The solution set is $\{x \mid -5 < x < 1\}$, *or* $(-5, 1)$.

Method 2. The solutions are those numbers x whose distance from -2 is less than 3. Thus to find the solutions graphically, we locate -2. Then we locate those numbers that are less than 3 units to the left and less than 3 units to the right. Thus the solution set is $\{x \mid -5 < x < 1\}$, or $(-5, 1)$.

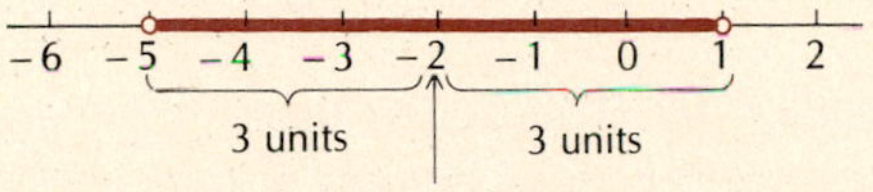

Method 3. We use property (ii), replacing x by $x + 2$:

$$|x + 2| < 3$$
$$-3 < x + 2 < 3$$
$$-5 < x < 1. \quad \text{Adding } -2$$

The solution set is $\{x \mid -5 < x < 1\}$, *or* $(-5, 1)$.

DO EXERCISES 7–9.

The sentences in the next examples are more complicated. Although we could continue using translations and graphs, this actually gets more difficult than if we use properties (i)–(iii).

EXAMPLE 7 Solve: $|2x + 3| = 1$.

Solution We have

$$2x + 3 = -1 \quad or \quad 2x + 3 = 1 \qquad \text{Property (i)}$$
$$2x = -4 \quad or \quad 2x = -2 \qquad \text{Adding } -3$$
$$x = -2 \quad or \quad x = -1. \qquad \text{Multiplying by } \tfrac{1}{2}$$

The solution set is $\{-2, -1\}$.

EXAMPLE 8 Solve: $|2x + 3| \le 1$.

Solution We have

$$-1 \le 2x + 3 \le 1 \qquad \text{Property (ii)}$$
$$-4 \le 2x \le -2 \qquad \text{Adding } -3$$
$$-2 \le x \le -1. \qquad \text{Multiplying by } \tfrac{1}{2}$$

The solution set is $\{x \mid -2 \le x \le -1\}$, or $[-2, -1]$.

EXAMPLE 9 Solve: $|3 - 4x| > 2$.

Solve. Use all three methods.

7. $|x + 1| = 4$

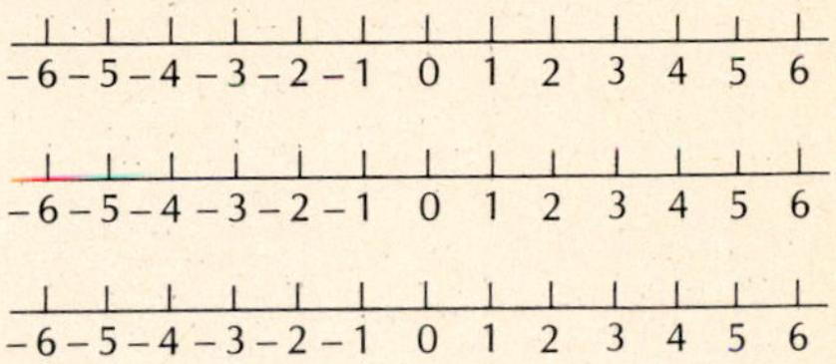

8. $|x + 7| < 2$

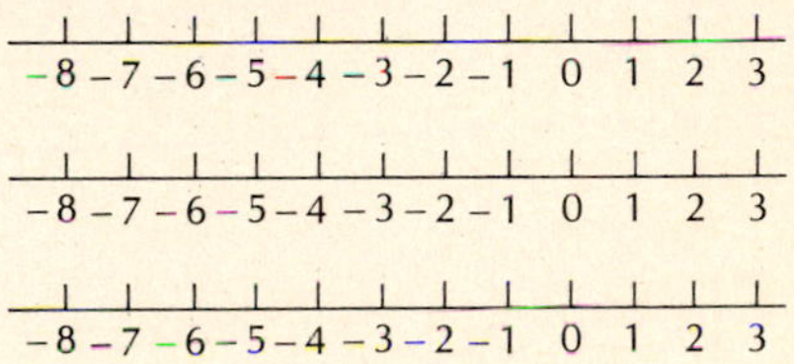

9. $|x - 3| > 2$

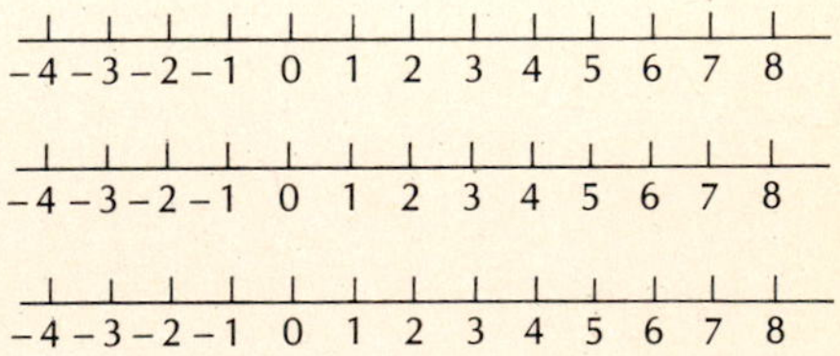

Solve.

10. $|3x - 4| = 7$

11. $\left|5x + \frac{1}{2}\right| < 1$

12. $|3x - 4| > 7$

Solve.

13. $|2x - 3| = |x + 5|$

14. $|x - 5| = |x + 8|$

Solution We have

$$\begin{array}{lcll} 3 - 4x < -2 & or & 3 - 4x > 2 & \text{Property (iii)} \\ -4x < -5 & or & -4x > -1 & \text{Adding } -3 \\ x > \frac{5}{4} & or & x < \frac{1}{4} & \text{Multiplying by } -\frac{1}{4} \end{array}$$

Note that the inequality signs must also be reversed in the preceding step. The solution set is $\{x | x < \frac{1}{4} \text{ or } x > \frac{5}{4}\}$, or $(-\infty, \frac{1}{4}) \cup (\frac{5}{4}, \infty)$.

DO EXERCISES 10–12.

EXAMPLE 10 Solve: $|5x - 3| = |x + 4|$.

Solution Consider $|a| = |b|$. This asserts that a and b are the same distance from 0. Thus they are either the same number or opposites of each other. Since $|5x - 3| = |x + 4|$, either $5x - 3 = x + 4$ *or* $5x - 3 = -(x + 4)$. We now solve each equation separately:

$$\begin{array}{lcl} 5x - 3 = x + 4 & or & 5x - 3 = -(x + 4) \\ 4x = 7 & or & 5x - 3 = -x - 4 \\ x = \frac{7}{4} & or & 6x = -1 \\ x = \frac{7}{4} & or & x = -\frac{1}{6}. \end{array}$$

The solution set is $\{-\frac{1}{6}, \frac{7}{4}\}$.

EXAMPLE 11 Solve: $|x - 3| = |x + 10|$.

Solution We have

$$\begin{array}{lcl} x - 3 = x + 10 & or & x - 3 = -(x + 10) \\ -3 = 10 & or & x - 3 = -x - 10 \\ -3 = 10 & or & 2x = -7 \\ -3 = 10 & or & x = -\frac{7}{2}. \end{array}$$

The first equation has no solution. The second equation has solution $-\frac{7}{2}$. The solution set is $\emptyset \cup \{-\frac{7}{2}\}$, or $\{-\frac{7}{2}\}$.

DO EXERCISES 13 AND 14.

EXAMPLE 12 Solve: $|2x - 3| \le 4 + 5x$.

Solution We have

$$\begin{array}{rl} |2x - 3| \le 4 + 5x & \\ -(4 + 5x) \le 2x - 3 \le 4 + 5x & \text{Property (ii)} \\ -4 - 5x \le 2x - 3 \le 4 + 5x & \\ -4 \le 7x - 3 \le 4 + 10x. & \end{array}$$

Note that we have a variable remaining in two parts of the inequality. Any addition, to get the variable alone, must be done to all three parts and will reintroduce the variable on the left side. To deal with this, we write the conjunction, solve each inequality separately, and then find the intersection of the solution sets:

$$-4 \le 7x - 3 \quad \textit{and} \quad 7x - 3 \le 4 + 10x$$
$$-1 \le 7x \quad \textit{and} \quad -3 \le 4 + 3x$$
$$-\tfrac{1}{7} \le x \quad \textit{and} \quad -7 \le 3x$$
$$-\tfrac{1}{7} \le x \quad \textit{and} \quad -\tfrac{7}{3} \le x.$$

The solution set of $-\frac{1}{7} \le x$ is the interval $[-\frac{1}{7}, \infty)$. The solution set of $-\frac{7}{3} \le x$ is $[-\frac{7}{3}, \infty)$. The solution set of the conjunction is the intersection of the intervals:

$$\left[-\tfrac{1}{7}, \infty\right) \cap \left[-\tfrac{7}{3}, \infty\right) = \left[-\tfrac{1}{7}, \infty\right).$$

DO EXERCISES 15 AND 16.

Solve.

15. $|5 - 6x| \le 10 + 7x$

16. $|5 - 6x| \ge 10 + 7x$

EXERCISE SET 4.6

1 Solve and graph. Use interval notation for the solution set where appropriate.

1. $|x| = 7$
2. $|x| = \pi$
3. $|x| < 7$
4. $|x| \le \pi$
5. $|x| \ge \pi$
6. $|x| > 7$

Solve. Use three methods. Use interval notation for the solution set where appropriate.

7. $|x - 1| = 4$
8. $|x - 7| = 5$
9. $|x + 8| < 9$
10. $|x + 6| \le 10$
11. $|x + 8| \ge 9$
12. $|x + 6| > 10$
13. $|x - \frac{1}{4}| < \frac{1}{2}$
14. $|x - 0.5| \le 0.2$

Solve. Use any method. Use interval notation for the solution set where appropriate.

15. $|3x| = 1$
16. $|5x| = 4$
17. $|3x + 2| = 1$
18. $|7x - 4| = 8$
19. $|3x| < 1$
20. $|5x| \le 4$
21. $|2x + 3| \le 9$
22. $|2x + 3| < 13$
23. $|x - 5| > 0.1$
24. $|x - 7| \ge 0.4$
25. $\left|x + \frac{2}{3}\right| \le \frac{5}{3}$
26. $\left|x + \frac{3}{4}\right| < \frac{1}{4}$
27. $|6 - 4x| \le 8$
28. $|5 - 2x| > 10$
29. $\left|\frac{2x + 1}{3}\right| > 5$
30. $\left|\frac{3x + 2}{4}\right| \le 5$
31. $\left|\frac{13}{4} + 2x\right| > \frac{1}{4}$
32. $\left|\frac{5}{6} + 3x\right| < \frac{7}{6}$
33. $\left|\frac{3 - 4x}{2}\right| \le \frac{3}{4}$
34. $\left|\frac{2x - 1}{3}\right| \ge \frac{5}{6}$
35. $|x| = -3$
36. $|x| < -3$
37. $|2x - 4| < -5$
38. $|3x + 5| < 0$
39. ▦ $|x + 17.217| > 5.0012$
40. ▦ $|x - 2.0245| < 0.1011$
41. ▦ $|-2.1437x + 7.8814| \ge 9.1132$
42. ▦ $|3.0147x - 8.9912| \le 6.0243$
43. $|2x - 8| = |x + 3|$
44. $|x - 7| = |3x + 4|$
45. $\left|\frac{2x + 3}{6}\right| = \left|\frac{4 - 5x}{8}\right|$
46. $\left|\frac{2}{3} - \frac{5}{6}x\right| = \left|\frac{3}{4} + \frac{3}{5}x\right|$
47. $|x - 2| = x - 2$
48. $|4x - 5| = x + 1$
49. $|7x - 2| = x + 5$
50. $|x - 3| = 3x - 8$
51. $|3x - 4| \le 2x + 1$
52. $|5x + 8| \le 4 - 3x$
53. $|3x - 4| \ge 2x + 1$
54. $|5x + 8| \ge 4 - 3x$

SYNTHESIS

Solve.

55. $|4x - 5| = |x| + 1$
56. $|2x + 3| = |x| + 8$
57. $||x| - 1| = 3$
58. $|x + 2| > x$
59. $|x + 2| \le |x - 5|$
60. $|3x - 1| > 5x - 2$
61. $|x| + |x - 1| < 10$
62. $|x| - |x - 3| < 7$
63. $|x - 3| + |2x + 5| > 6$
64. $|p - 4| + |p + 4| < 8$

CHALLENGE

65. Solve: $|x - 3| + |2x + 5| + |3x - 1| = 12.$
66. Solve for x: $|x - x_0| < \delta.$

Prove the following for any real numbers a and b.

67. $-|a| \le a \le |a|$

68. $|a + b| \le |a| + |b|$ (the triangle inequality)

69. Show that if $|a| < \frac{\epsilon}{2}$ and $|b| < \frac{\epsilon}{2}$, then $|a + b| < \epsilon$.

70. a) Prove that

$$\left|x - \frac{a+b}{2}\right| < \frac{b-a}{2}$$

is equivalent to $a < x < b$.

Use graphs or the result of part (a) to find an inequality with absolute value for each of the following.

b) $-5 < x < 5$ **c)** $-6 < x < 6$ **d)** $-1 < x < 7$ **e)** $-5 < x < 1$

71. Use absolute value to prove that the number halfway between a and b is $(a + b)/2$.

72. Solve for $f(x)$: $|f(x) - L| < \epsilon$.

4.7 Quadratic and Rational Inequalities

OBJECTIVES

You should be able to:

1. Solve quadratic and other polynomial inequalities.
2. Solve rational inequalities.

1 Quadratic and Other Polynomial Inequalities

Inequalities such as the following are called **quadratic inequalities:**

$$x^2 + 3x - 10 < 0, \qquad 5x^2 - 3x + 2 \ge 0.$$

In each case, we have a polynomial of degree 2 on the left. We will consider the solving of such inequalities three ways. The first two provide understanding and the last yields the fastest method.

The first method for solving quadratic inequalities is by considering the graph of the related function in the plane.

EXAMPLE 1 Solve: $x^2 + 3x - 10 > 0$.

Solution Consider the function $f(x) = x^2 + 3x - 10$ and its graph. Its graph opens upward since the leading coefficient is positive.

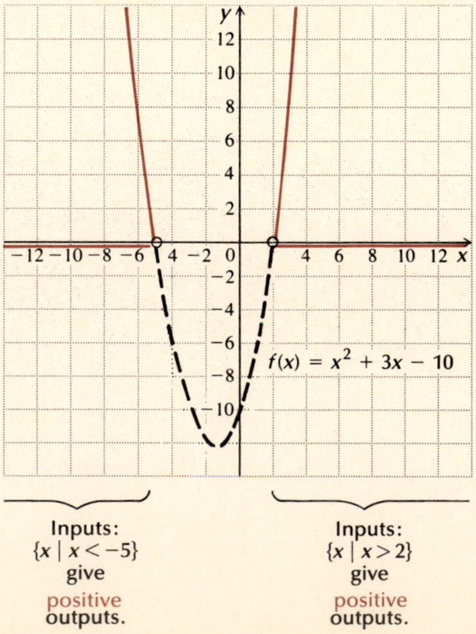

Function values will be positive to the left and right of the intercepts, as shown. Then the solution set of the inequality is

$$\{x \mid x < -5 \text{ or } x > 2\}, \quad \text{or} \quad (-\infty, -5) \cup (2, \infty).$$ ▪

DO EXERCISE 1.

We can solve any inequality by considering a graph of the related function and finding intercepts, as in Example 1. In some cases, we may need to use the quadratic formula to find the intercepts.

EXAMPLE 2 Solve: $x^2 + 3x - 10 < 0$.

Solution Looking at the graph or at least visualizing it mentally tells us that the function is negative for those inputs x such that x is between -5 and 2. That is, the solution set is

$$\{x \mid -5 < x < 2\}, \quad \text{or} \quad (-5, 2).$$

Inputs:
$\{x \mid -5 < x < 2\}$
give
negative
outputs.

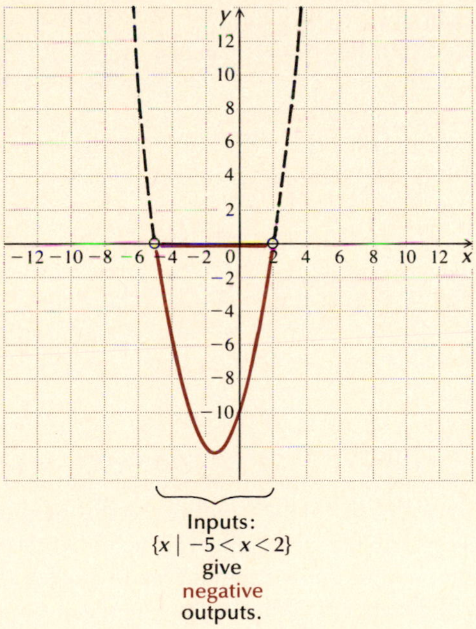

DO EXERCISE 2.

When an inequality contains $\leq$ or $\geq$, the x-values of the intercepts need to be included. For example, the solution set of $x^2 + 3x - 10 \geq 0$ is

$$\{x \mid x \leq -5 \text{ or } x \geq 2\}, \quad \text{or} \quad (-\infty, -5] \cup [2, \infty).$$

DO EXERCISE 3.

Let us now consider another method of solving inequalities that works for any polynomial that is factored into a product of first-degree polynomials.

EXAMPLE 3 Solve: $x^2 + 3x - 10 < 0$.

1. Solve by graphing:
$$x^2 + 2x - 3 > 0.$$

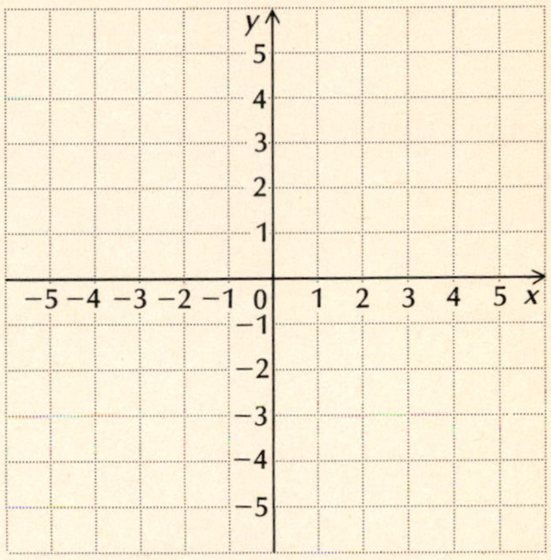

2. Solve by graphing:
$$x^2 + 2x - 3 < 0.$$

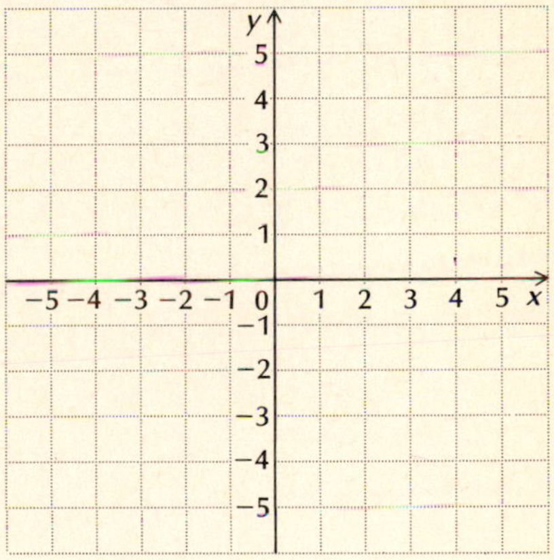

3. Solve by graphing:
$$x^2 + 2x - 3 \leq 0.$$

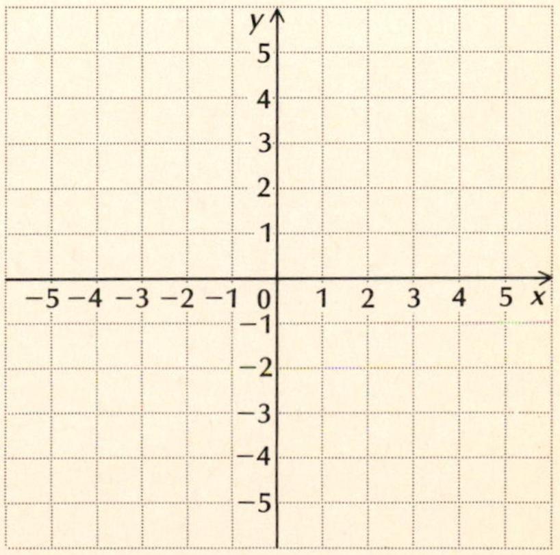

Solve. Use the method of Example 3.

4. $x^2 + 3x > 4$

Solution We factor the inequality, obtaining $(x + 5)(x - 2) < 0$. The solutions of $(x + 5)(x - 2) = 0$ are -5 and 2. They are not solutions of the inequality, but they divide the real-number line in a natural way, pictured as follows. When we graphed the function as in Examples 1–3, these intercepts also provided a division of the number line. The product $(x + 5)(x - 2)$ is positive or negative for values other than -5 and 2, depending on the signs of the factors $x + 5$ and $x - 2$. We can determine this efficiently with a diagram as follows.

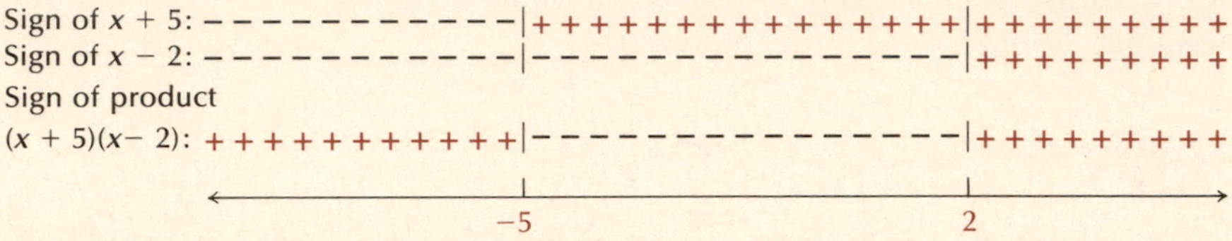
Sign of x + 5: − − − − − − − − − − − |+ + + + + + + + + + + + + + + + |+ + + + + + + + + +
Sign of x − 2: − − − − − − − − − − − |− − − − − − − − − − − − − − − − |+ + + + + + + + + +
Sign of product
(x + 5)(x− 2): + + + + + + + + + + + |− − − − − − − − − − − − − − − − |+ + + + + + + + + +
−5
2

To set up the diagram, we solve $x + 5 > 0$. We get $x > -5$. Thus, $x + 5$ is positive for all numbers to the right of -5. We indicate that with the + signs. Accordingly, $x + 5 < 0$ for all numbers to the left of -5. We indicate that with the − signs.

Similarly, we solve $x - 2 > 0$ and get $x > 2$. Thus, $x - 2$ is positive for all numbers to the right of 2. Accordingly, $x - 2 < 0$ for all numbers to the left of 2. We indicate this with the + and − signs.

Finally, we determine the signs of the product by using the rules for multiplication. In order for the product $(x + 5)(x - 2)$ to be negative, one factor must be positive and the other negative. In the table, this situation occurs only when $-5 < x < 2$. The solution set of the inequality is $\{x | -5 < x < 2\}$, or $(-5, 2)$. ■

5. $x^2 + 3x \leq 4$

Note in Example 4 that setting up a diagram in this manner also tells us the solutions of $(x + 5)(x - 2) > 0$. That solution set is $\{x | x < -5 \text{ or } x > 2\}$, or $(-\infty, -5) \cup (2, \infty)$.

DO EXERCISES 4 AND 5.

We now consider our final method for solving quadratic inequalities. The preceding method provides the understanding for this method. In Example 4, we see that the intercepts divide the number line up into intervals. If a particular function has a positive output for one number in an interval, it will be positive for all the numbers in the interval. Thus we can merely make a test substitution in each interval in order to solve the inequality.

EXAMPLE 4 Solve: $x^2 + 3x - 10 < 0$.

Solution We set the polynomial equal to 0 and solve. The solutions of $x^2 + 3x - 10 = 0$, or $(x + 5)(x - 2) = 0$, are -5 and 2. Then we locate them on a number line as follows. Note that the numbers divide the number line into three intervals A, B, and C.

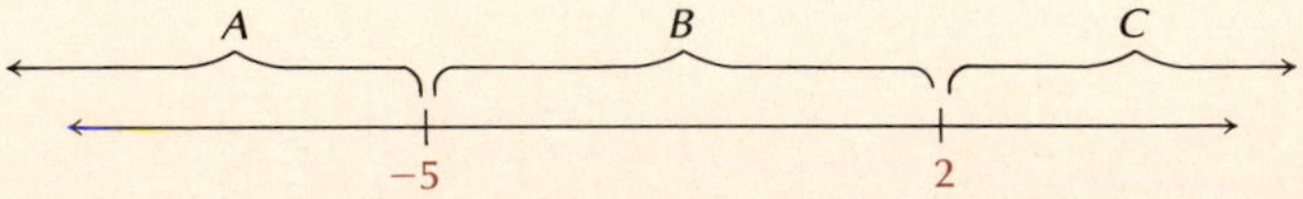

We pick a test number in interval A, say -7, and substitute it for x in the function $f(x) = x^2 + 3x - 10$:

$$f(-7) = (-7)^2 + 3(-7) - 10$$
$$= 49 - 21 - 10 = 18.$$

Note that $f(-7) > 0$, so the function will be positive for any number in interval A. Next we try a test number in interval B, say 1, and find $f(1)$:

$$f(1) = 1^2 + 3(1) - 10 = -6.$$

Note that $f(1) < 0$, so the function will be negative for any number in interval B. Next we try a test number in interval C, say 4, and find $f(4)$:

$$\begin{aligned} f(4) &= 4^2 + 3(4) - 10 \\ &= 16 + 12 - 10 = 18. \end{aligned}$$

Note that $f(4) > 0$, so the function will be positive for any number in interval C. We are looking for numbers x for which $x^2 + 3x - 10 < 0$. Thus any number x in interval B is a solution. If the inequality had been $\leq$ or $\geq$, we would also have needed both of the intercepts, -5 and 2. The solution set is $\{x | -5 < x < 2\}$. ■

Let us review the last method.

To solve a quadratic inequality:

1. **Get 0 on one side, set the polynomial on the other side equal to 0, and solve to find the intercepts.**
2. **Use the numbers found in step (1) to divide the number line into intervals.**
3. **Substitute a number from each interval into the related function. If the value is positive, then the function will be positive for all numbers in the interval. If the value is negative, then the function will be negative for all numbers in the interval.**
4. **Select the intervals for which the inequality is satisfied and write set-builder notation or interval notation for the solution set. Include the intercepts in the solution sets if the inequality sign is $\leq$ or $\geq$.**

DO EXERCISES 6 AND 7.

EXAMPLE 5 Solve: $7x(x + 3)(x - 2) \geq 0$.

Solution The solutions of $7x(x + 3)(x - 2) = 0$ are -3, 0, and 2. They divide the real-number line into four intervals, as follows.

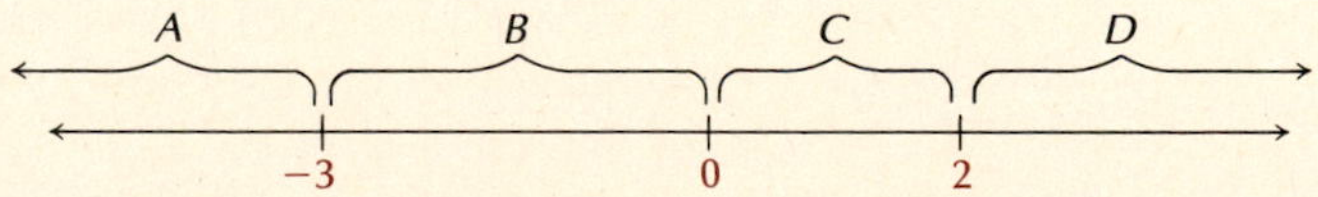

We try a test number in each interval:

A: Test -5, $f(-5) = 7(-5)(-5 + 3)(-5 - 2) = -490$;
B: Test -2, $f(-2) = 7(-2)(-2 + 3)(-2 - 2) = 56$;
C: Test 1, $f(1) = 7(1)(1 + 3)(1 - 2) = -28$;
D: Test 3, $f(3) = 7(3)(3 + 3)(3 - 2) = 126$.

Function values are positive in intervals B and D. Since the inequality symbol is $\geq$, we will need to include the intercepts. The solution set of the inequality is

$$\{x | -3 \leq x \leq 0 \text{ or } 2 \leq x\}, \quad \text{or} \quad [-3, 0] \cup [2, \infty).$$ ■

DO EXERCISE 8.

Solve. Use the method of Example 4.

6. $x^2 + 3x > 4$

7. $x^2 + 3x \leq 4$

8. Solve: $4x(x + 1)(x - 1) < 0$.

2 Rational Inequalities

We adapt the preceding method when an inequality involves rational expressions. We call these **rational inequalities.**

EXAMPLE 6 Solve: $\dfrac{x-3}{x+4} \geq 2.$

Solution We write the related equation by changing the $\geq$ symbol to $=$:

$$\frac{x-3}{x+4} = 2.$$

Then we solve this related equation. We multiply on both sides of the equation by the LCM, which is $x + 4$:

$$\begin{aligned} (x+4)\cdot\frac{x-3}{x+4} &= (x+4)\cdot 2 \\ x - 3 &= 2x + 8 \\ -11 &= x. \end{aligned}$$

In the case of rational inequalities, we also need to determine those replacements that are not meaningful. These are those that make the denominator 0. We set the denominator equal to 0 and solve:

$$\begin{aligned} x + 4 &= 0 \\ x &= -4. \end{aligned}$$

Now we use the numbers -11 and -4 to divide the number line into intervals, as follows:

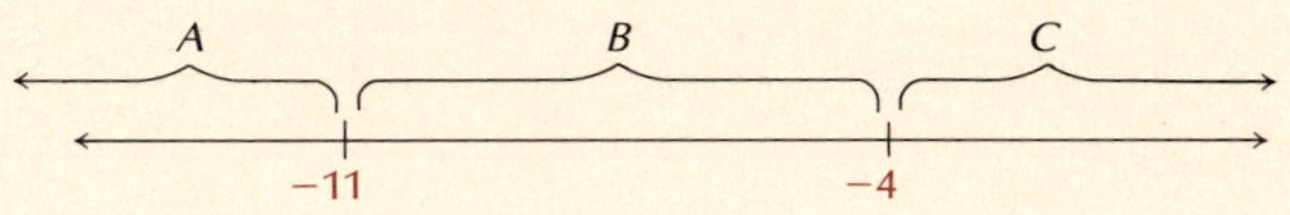

We try test numbers in each interval. We see if each satisfies the original inequality.

A: Test -15,

$$\begin{array}{c|c} \dfrac{x-3}{x+4} & \geq 2 \\ \hline \dfrac{-15-3}{-15+4} & 2 \\ \dfrac{-18}{-11} & \\ \dfrac{18}{11} & \end{array}$$

Since $\frac{18}{11} \geq 2$ is a false inequality, the number -15 is not a solution of the inequality, so the interval A is not part of the solution set.

B: Test -8,

$$\begin{array}{c|c} \dfrac{x-3}{x+4} & \geq 2 \\ \hline \dfrac{-8-3}{-8+4} & 2 \\ \dfrac{-11}{-4} & \\ \dfrac{11}{4} & \end{array}$$

Since $\frac{11}{4} \geq 2$ is a true inequality, the number -8 is a solution of the inequality, so the interval B is part of the solution set.

C: Test 1, $\frac{x-3}{x+4} \geq 2$

$\frac{1-3}{1+4}$	2
$\frac{-2}{5}$	
$-\frac{2}{5}$	

Since $-\frac{2}{5} \geq 2$ is a false inequality, the number 1 is not a solution of the inequality, so the interval C is not part of the solution set.

The solution set includes the interval B. The number -11 is also included, since the inequality symbol is $\geq$ and -11 is a solution of the related equation. The number -4 is not included, since it is not a meaningful replacement. Thus the solution set of the original inequality is

$$\{x \mid -11 \leq x < -4\}, \quad \text{or} \quad [-11, -4).$$

To solve a rational inequality:

1. **Change the inequality symbol to an equals sign and solve the related equation.**
2. **Find the replacements that are not meaningful.**
3. **Use the numbers found in steps (1) and (2) to divide the number line into intervals.**
4. **Substitute a number from each interval into the inequality. If the number is a solution, then the interval to which it belongs is part of the solution set.**
5. **Select the intervals for which the inequality is satisfied and write set-builder notation for the solution set. If the inequality symbol is $\leq$ or $\geq$, then the solutions to step (1) should also be included in the solution set.**

DO EXERCISES 9 AND 10.

Solve.

9. $\frac{x+1}{x-2} \geq 3$

10. $\frac{x}{x-5} < 2$

EXERCISE SET 4.7

1 Solve.

1. $(x+5)(x-3) > 0$
2. $(x+4)(x-1) > 0$
3. $(x-1)(x+2) \leq 0$
4. $(x+5)(x-3) \leq 0$
5. $x^2 + x - 2 < 0$
6. $x^2 - x - 2 < 0$
7. $x^2 \geq 1$
8. $x^2 < 25$
9. $9 - x^2 \leq 0$
10. $4 - x^2 \geq 0$
11. $x^2 - 2x + 1 \geq 0$
12. $x^2 + 6x + 9 < 0$
13. $x^2 + 8 < 6x$
14. $x^2 - 12 > 4x$
15. $4x^2 + 7x < 15$
16. $4x^2 + 7x \geq 15$
17. $2x^2 + x > 5$
18. $2x^2 + x \leq 2$
19. $3x(x+2)(x-2) < 0$
20. $5x(x+1)(x-1) > 0$
21. $(x+3)(x-2)(x+1) > 0$
22. $(x-1)(x+2)(x-4) < 0$
23. $(x+3)(x+2)(x-1) < 0$
24. $(x-2)(x-3)(x+1) < 0$

2 Solve.

25. $\frac{1}{4-x} < 0$
26. $\frac{-4}{2x+5} > 0$
27. $3 < \frac{1}{x}$
28. $\frac{1}{x} \leq 5$
29. $\frac{3x+2}{x-3} > 0$
30. $\frac{5-2x}{4x+3} < 0$

31. $\dfrac{x+2}{x} \le 0$

32. $\dfrac{x}{x-3} \ge 0$

33. $\dfrac{x+1}{2x-3} \ge 1$

34. $\dfrac{x-1}{x-2} \ge 3$

35. $\dfrac{x+1}{x+2} \le 3$

36. $\dfrac{x+1}{2x-3} \le 1$

37. $\dfrac{x-6}{x} > 1$

38. $\dfrac{x}{x+3} > -1$

39. $(x+1)(x-2) > (x+3)^2$

40. $(x-4)(x+3) > (x-1)^2$

41. $x^3 - x^2 > 0$

42. $x^3 - 4x > 0$

43. $x + \dfrac{4}{x} > 4$

44. $\dfrac{1}{x^2} \le \dfrac{1}{x^3}$

45. $\dfrac{1}{x^3} \le \dfrac{1}{x^2}$

46. $x + \dfrac{1}{x} > 2$

47. $\dfrac{2 + x - x^2}{x^2 + 5x + 6} < 0$

48. $\dfrac{4}{x^2} - 1 > 0$

SYNTHESIS

Solve.

49. $x^4 - 2x^2 \le 0$

50. $x^4 - 3x^2 > 0$

51. $\left|\dfrac{x+3}{x-4}\right| < 2$

52. $|x^2 - 5| = 5 - x^2$

53. $(7 - x)^{-2} < 0$

54. $(1 - x)^3 > 0$

55. $\left|1 + \dfrac{1}{x}\right| < 3$

56. $(x+5)^{-2} > 0$

57. $|x|^2 - 4|x| + 4 \ge 9$

58. $2|x|^2 - |x| + 2 \le 5$

59. $\left|2 - \dfrac{1}{x}\right| \le 2 + \left|\dfrac{1}{x}\right|$

60. $\dfrac{(x-2)^2(x-3)^3(x+1)}{(x+2)(4-x)} \ge 0$

61. $|x^2 + 3x - 1| < 3$ (*Hint:* Solve $-3 < x^2 + 3x - 1 < 3$.)

62. $|1 + 5x - x^2| \ge 5$ (*Hint:* Solve the disjunction $1 + 5x - x^2 \le -5$ or $1 + 5x - x^2 \ge 5$.)

63. The base of a triangle is 4 cm greater than the height. Find the possible heights h such that the area of the triangle will be greater than 10 cm^2.

64. The length of a rectangle is 3 m greater than the width. Find the possible widths w such that the area of the rectangle will be greater than 15 m^2.

65. *Total profit.* A company determines that its total-profit function is given by
$$P(x) = -3x^2 + 630x - 6000.$$
   a) A company makes a profit for those nonnegative values of x for which $P(x) > 0$. Find the values of x for which the company makes a profit.
   b) A company loses money for those nonnegative values of x for which $P(x) < 0$. Find the values of x for which the company loses money.

66. *Height of a thrown object.* The function
$$s(t) = -16t^2 + 32t + 1920$$
gives the height s of an object thrown from a cliff 1920 ft high, after time t seconds.
   a) For what times is the height greater than 1920 ft?
   b) For what times is the height less than 640 ft?

67. *Number of handshakes.* There are n people in a room. The number N of possible handshakes by the people is given by the function
$$N(n) = \frac{n(n-1)}{2}.$$
For what number of people n is $78 \le N \le 1225$?

68. *Number of diagonals.* A polygon with n sides has D diagonals, where D is given by the function
$$D(n) = \frac{n(n-3)}{2}.$$
For what number of sides n is $35 \le D \le 740$?

69. A company has the following total-cost and total-revenue functions to use in producing and selling x units of a certain product (for reference, see p. 000):
$$R(x) = 50x - x^2, \quad C(x) = 5x + 350.$$
   a) Find the break-even values.
   b) Find the values of x that produce a profit.
   c) Find the values of x that result in a loss.

70. A company has the following total-cost and total-revenue functions to use in producing and selling x units of a certain product:
$$R(x) = 80x - x^2, \quad C(x) = 10x + 600.$$
   a) Find the break-even values.
   b) Find the values of x that produce a profit.
   c) Find the values of x that result in a loss.

71. Find the numbers k for which the quadratic equation $x^2 + kx + 1 = 0$ has (a) two real-number solutions and (b) no real-number solution.

72. Find the numbers k for which the quadratic equation $2x^2 - kx + 1 = 0$ has (a) two real-number solutions and (b) no real-number solution.

Find the domain of the function.

73. $f(x) = \sqrt{1 - x^2}$

74. $f(x) = \dfrac{1}{\sqrt{1 - x^2}}$

75. $g(x) = \sqrt{x^2 + 2x - 3}$

76. $g(x) = \sqrt{4 - x^2}$

COMPUTER–CALCULATOR EXERCISES

There are many computer software packages and calculators that graph functions and solve equations. For each of the following functions, graph the function and find solutions of $f(x) = 0$. These numbers are often called roots, or zeros. Then solve the inequalities $f(x) < 0$ and $f(x) > 0$.

77. $f(x) = x^3 - 2x^2 - 5x + 6$

78. $f(x) = \frac{1}{3}x^3 - x + \frac{2}{3}$

79. $f(x) = x + \frac{1}{x}$

80. $f(x) = x - \sqrt{x}, x \geq 0$

81. $f(x) = x^4 - 4x^3 - x^2 + 16x - 12$

82. $f(x) = \dfrac{x^3 + x^2 - 2x}{x^2 + x - 6}$

4.8 Polynomial Functions

OBJECTIVES

You should be able to:

1 Sketch a graph of a polynomial function.

2 Solve problems involving polynomial functions.

Any function that can be described by a polynomial in one variable such as

$$f(x) = 5x^7 + 3x^6 - 4x^2 - 5$$

is called a **polynomial function.** Here is a formal definition.

DEFINITION

A *polynomial function* f is given by

$$f(x) = a_nx^n + a_{n-1}x^{n-1} + \cdots + a_2x^2 + a_1x + a_0,$$

where n is a nonnegative integer and $a_n, a_{n-1}, \ldots, a_2, a_1, a_0$ are real numbers, called the *coefficients* of the polynomial. The first nonzero coefficient is assumed to be a_n and is called the *leading coefficient*. The *degree* of the polynomial function is n.

1 Graphs of Polynomial Functions

Graphs of first-degree linear polynomial functions are lines; graphs of second-degree quadratic functions are parabolas. We now consider polynomials of higher degree.

EXAMPLE 1 Sketch graphs of each of the functions

$$f(x) = x^2 \quad \text{and} \quad g(x) = x^3$$

using the same set of axes. Find the domain and the range of each function.

Solution Since $f(-x) = (-x)^2 = x^2 = f(x)$, we know that f is an even function, which means that it is symmetric with respect to the y-axis. Thus any time we have a point (a, b) on the graph, we know that $(-a, b)$ is also on the graph. Since $g(-x) = (-x)^3 = -x^3 = -g(x)$, we know that g is an odd function, which means that its graph is symmetric with respect to the origin. Thus any time we have a point (a, b) on the graph, we know that $(-a, -b)$ is also on the graph. We compute some function values and use them and the symmetry ideas to find others. Then we complete the graphs.

x	$f(x) = x^2$	$g(x) = x^3$
-2	4	-8
-1	1	-1
$-\frac{1}{2}$	$\frac{1}{4}$	$-\frac{1}{8}$
0	0	0
$\frac{1}{2}$	$\frac{1}{4}$	$\frac{1}{8}$
1	1	1
2	4	8

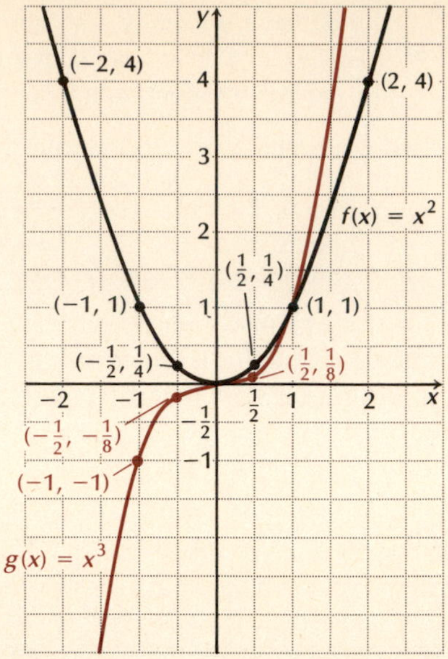

The domain of each function is the set of all real numbers. The range of f is the interval $[0, \infty)$. The range of g is the set of all real numbers, or the interval $(-\infty, \infty)$. ■

DO EXERCISES 1 AND 2.

Some graphs of $f(x) = ax^n$, where n is *even*, are shown below. Note that $g(x) = x^4$ has the same general shape as $f(x) = x^2$, but the larger the power of n, the closer the graph gets to the x-axis for values of x in the interval $[-1, 1]$. For other values of x, the graphs get steeper. When a is negative, the graphs are reflected across the x-axis. Each function is even, has the y-axis as a line of symmetry, and passes through the origin.

Sketch a graph of each of the following functions.

1. $f(x) = (x - 1)^3$

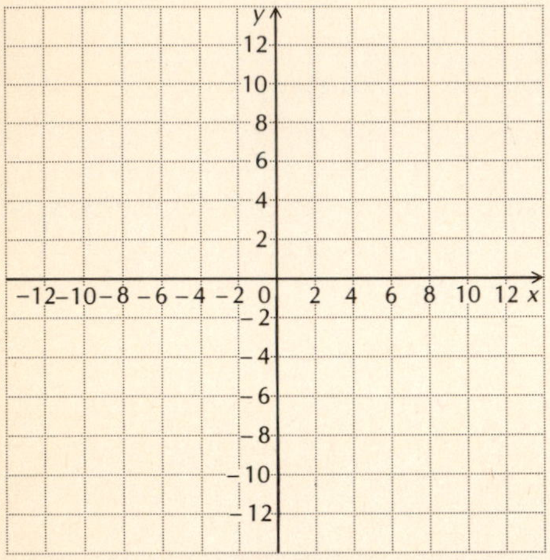

2. $f(x) = -\frac{1}{2}x^3 + 2$

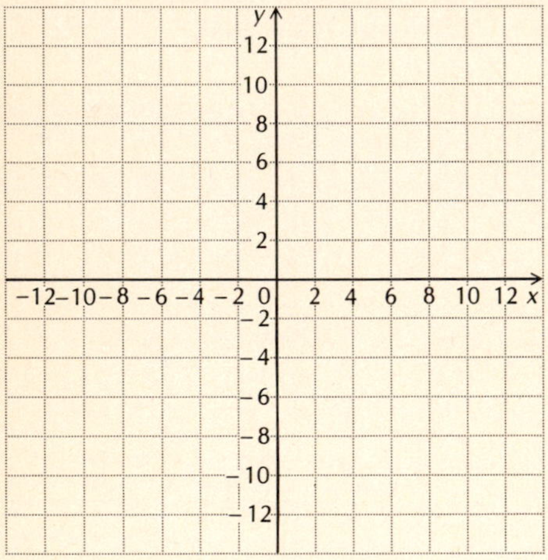

x	x^2	x^3	x^4	x^5
0	0	0	0	0
1	1	1	1	1
1.5	≈ 2.3	≈ 3.4	≈ 5.0	≈ 7.6
0.5	≈ 0.3	≈ 0.13	0.06	0.03
2	4	8	16	32
2.5	≈ 6.3	15.6	39.1	97.7

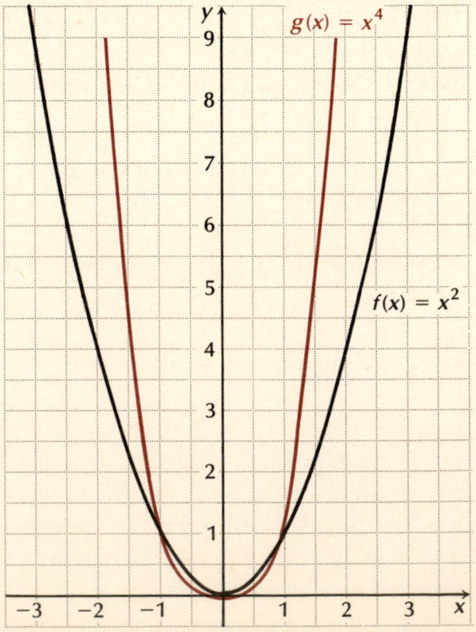

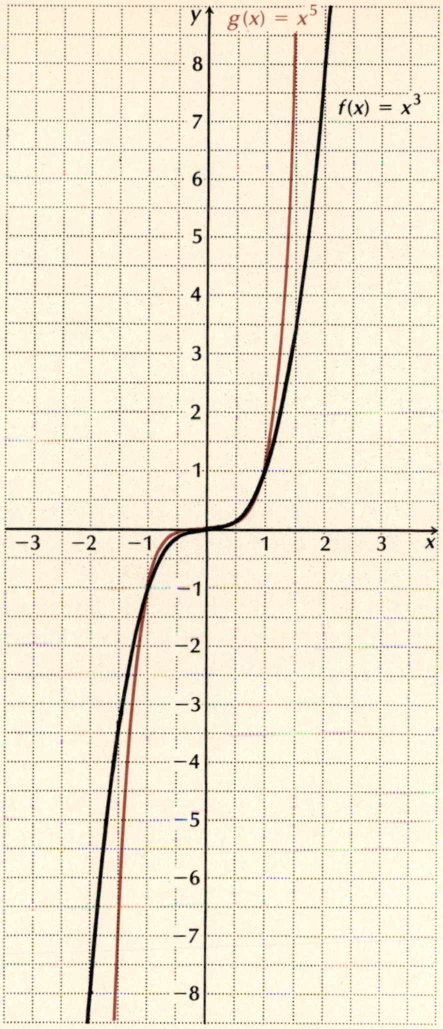

The graphs of $f(x) = ax^n$, where n is *odd*, are shown above. Note that $g(x) = x^5$ has the same general shape as $f(x) = x^3$, but the larger the power of n, the closer the graph gets to the x-axis for values of x in the interval $[-1, 1]$. For other values of x, the graphs get steeper. When a is negative, the graphs are reflected across the x-axis. Each function is odd, has the origin as a point of symmetry, and passes through the origin.

DO EXERCISES 3 AND 4.

The following is a theorem about polynomials that can be helpful in graphing.

THEOREM 6

The domain of any polynomial function is the set of real numbers. The graph of every polynomial function is a continuous function.

We say that c is a **zero** of f, if $f(c) = 0$. We also call c a **root** of the equation $f(x) = 0$, or simply a **root of the polynomial.** We know by Theorem 6 that every real number can be used as an input and that the graph of any polynomial function can be drawn without lifting the pencil from the paper. Suppose that

Sketch a graph of each of the following functions.

3. $f(x) = x^4$

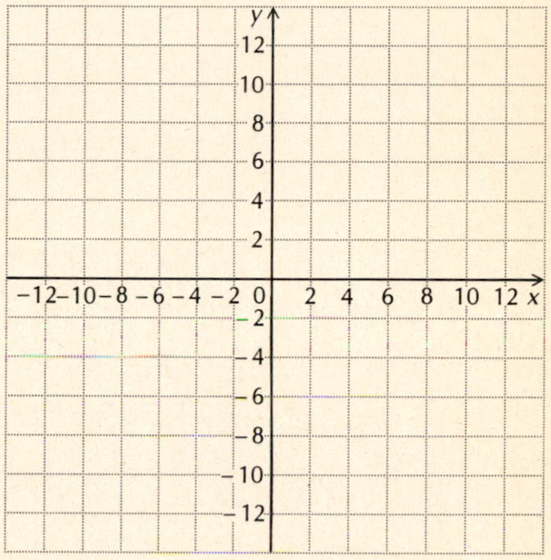

4. $f(x) = -\frac{1}{2}x^5$

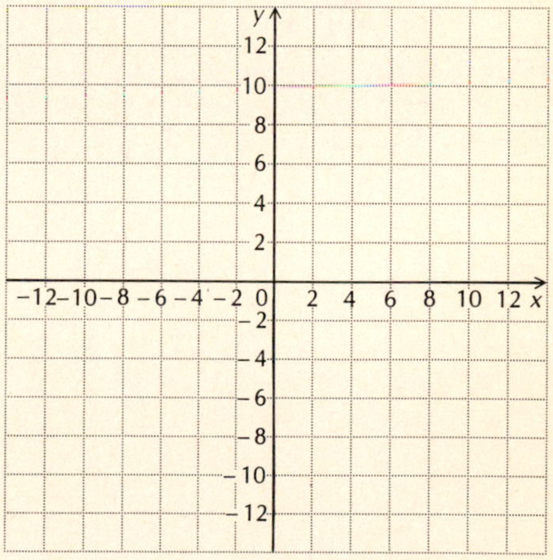

we have two function values $f(a)$ and $f(b)$ that are of opposite sign. Since we have a continuous function and we must be able to draw a curve from $f(a)$ to $f(b)$ without lifting the pencil from the paper, it follows that the line must cross the x-axis somewhere between a and b. Thus, f must have a zero, or root, somewhere between a and b. This fact can be helpful in graphing and in a later chapter when we spend more time finding zeros of polynomial functions.

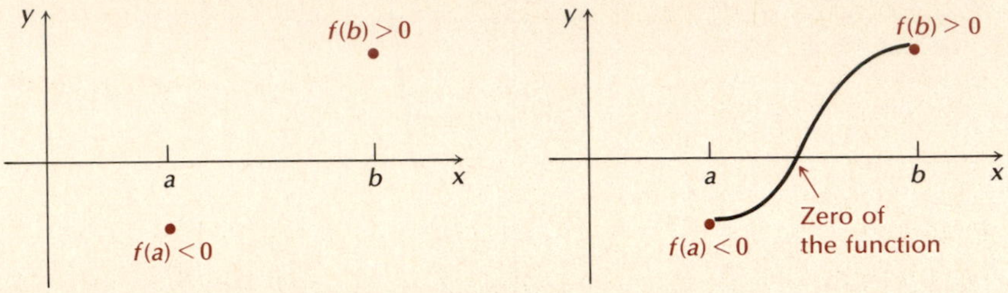

Suppose that we have determined all the real-number zeros of a polynomial, which often is no small task. Then between successive zeros, the values of the polynomial are either all positive or all negative. If there were a sign change, then there would be an additional zero. Try drawing a continuous curve between successive zeros. You assume that you have found all the zeros, so the curve cannot cross the x-axis. Thus values between zeros are either all positive or all negative. By using *test values* between successive zeros, we know the sign of the function on the entire interval. We used this idea when solving inequalities in Section 4.7.

Let us now sketch some more complicated graphs. We will be considering polynomials for which it is relatively easy to find the roots. In a later chapter, we will consider other polynomials for which finding roots and sketching graphs is more complicated.

EXAMPLE 2 Sketch a graph of the polynomial function f given by

$$f(x) = 2x^3 + x^2 - 8x - 4.$$

Solution We first try to factor the function in order to find the zeros. In this case, we can use factoring by grouping:

$$\begin{aligned} f(x) &= 2x^3 + x^2 - 8x - 4 \\ &= x^2(2x + 1) - 4(2x + 1) \\ &= (x^2 - 4)(2x + 1) \\ &= (x - 2)(x + 2)(2x + 1). \end{aligned}$$

We see that the solutions of the equation $f(x) = 0$ are -2, -0.5, and 2. These are also the zeros of the polynomial function. The zeros divide the number line into four intervals: $(-\infty, -2)$, $(-2, -0.5)$, $(-0.5, 2)$, and $(2, \infty)$.

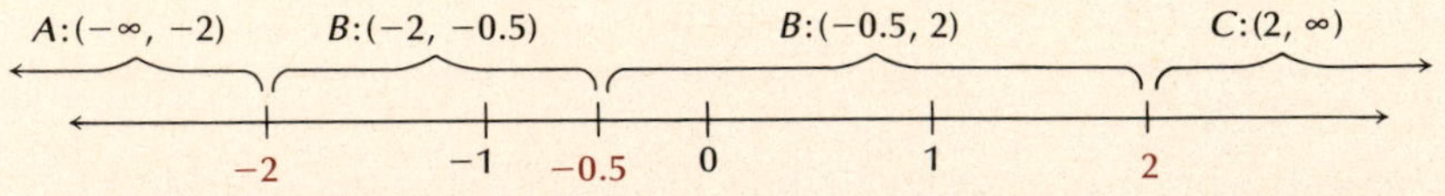

We try test numbers, as follows, in each interval. These also give us function values that can be used for graphing.

A: Test -3, $f(-3) = 2(-3)^3 + (-3)^2 - 8(-3) - 4 = -25;$

B: Test -1, $f(-1) = 2(-1)^3 + (-1)^2 - 8(-1) - 4 = 3;$

C: Test 1, $f(1) = 2(1)^3 + (1)^2 - 8(1) - 4 = -9;$

D: Test 3, $f(3) = 2(3)^3 + (3)^2 - 8(3) - 4 = 35.$

Since $f(-3) = -25$, the function values are all negative and all the points on the graph lie below the x-axis on the interval $(-\infty, -2)$. Since $f(-1) = 3$, the function values are all positive and all the points on the graph lie above the x-axis on the interval $(-2, -0.5)$. Since $f(1) = -9$, the function values are all negative and all the points on the graph lie below the x-axis on the interval $(-0.5, 2)$. Since $f(3) = 35$, the function values are all positive and all the points on the graph lie above the x-axis on the interval $(2, \infty)$. We summarize this information in the following table.

Interval	$(-\infty, -2)$	$(-2, -0.5)$	$(-0.5, 2)$	$(2, \infty)$
Test value	$f(-3) = -25$	$f(-1) = 3$	$f(1) = -9$	$f(3) = 35$
Sign of $f(x)$	Negative	Positive	Negative	Positive
Location of points on graph	Below	Above	Below	Above

Using these results and calculating additional function values, we complete the graph as follows.

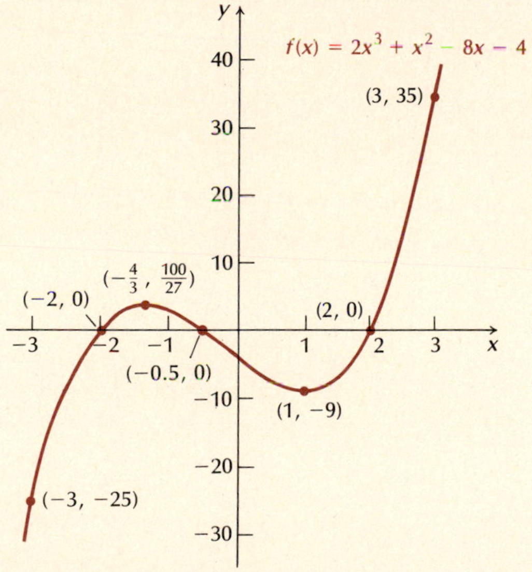

The points $(-\frac{4}{3}, \frac{100}{27})$ and $(1, -9)$ of the preceding graph are called **turning points.** They are points at which the graph changes from increasing to decreasing or from decreasing to increasing. In calculus, you will learn skills that will allow you to find these points easily. The following tip may also help you graph a polynomial function.

THEOREM 7

A polynomial function of degree n has at most $n - 1$ turning points and at most n zeros.

DO EXERCISE 5.

EXAMPLE 3 Sketch the graph of the polynomial function f given by

$$f(x) = x^4 - 4x^2 + 3.$$

5. Sketch a graph of

$$f(x) = x^3 + 3x^2 - x - 3.$$

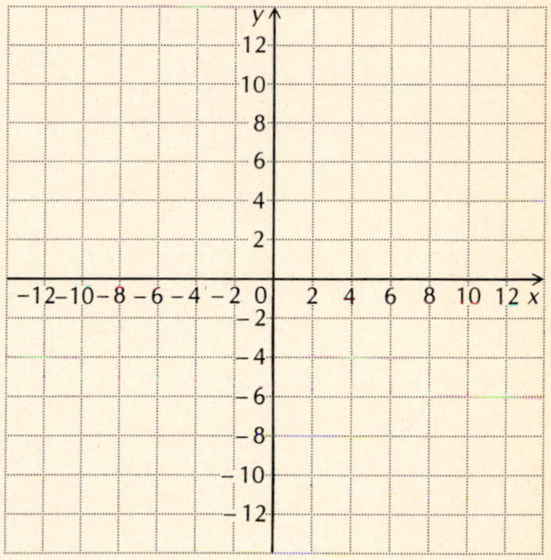

6. Sketch a graph of

$$f(x) = x^4 - 10x^2 + 9.$$

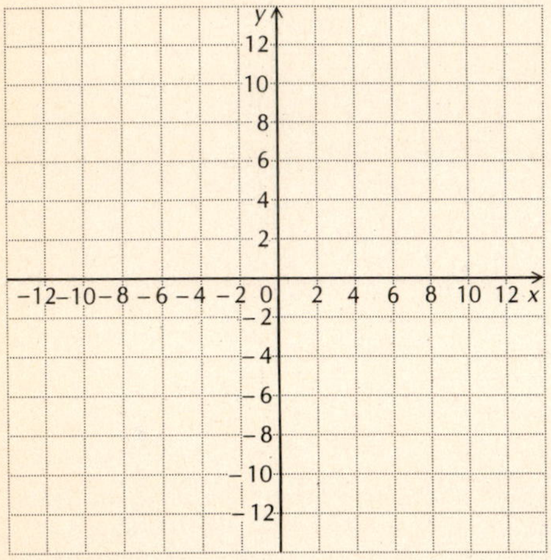

Solution We first factor the function as follows:

$$\begin{aligned} f(x) &= x^4 - 4x^3 + 3 \\ &= (x^2 - 3)(x^2 - 1) \\ &= (x + \sqrt{3})(x - \sqrt{3})(x + 1)(x - 1). \end{aligned}$$

The zeros of the function are $-\sqrt{3}$, -1, 1, and $\sqrt{3}$.

These zeros divide the real-number line into five open intervals listed in the following table. We try a test value in each interval and determine the sign of the function for values of x in the interval. The results are summarized in the following table.

Interval	$(-\infty, -\sqrt{3})$	$(-\sqrt{3}, -1)$	$(-1, 1)$	$(1, \sqrt{3})$	$(3, \infty)$
Test value	−2	−1.5	0	1.5	2
Sign of $f(x)$	3	−0.9375	3	−0.9375	3
Location of points on graph	Positive	Negative	Positive	Negative	Positive

We use the information in the table, calculate some extra function values, if needed, plot points, and sketch the graph, as shown below.

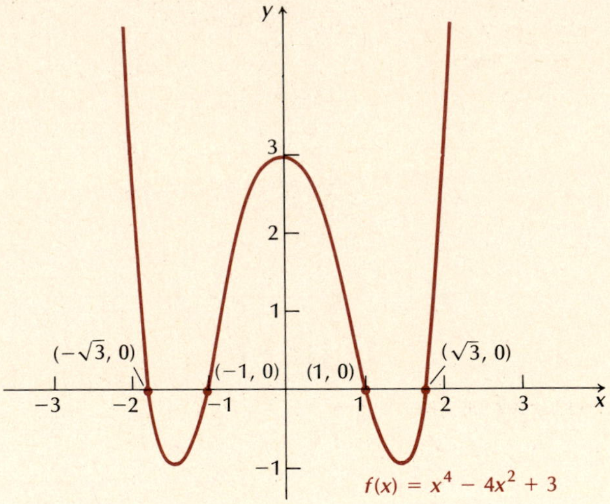

DO EXERCISE 6.

2 Applications of Polynomial Functions

We now consider an application of polynomial functions.

EXAMPLE 4 How many spheres, such as marbles or balls or oranges, are in the pile below? You might see such a question in connection with a guessing contest in a store.

If the layers are equilateral triangles and all the spheres are the same size, there is a polynomial function that can be used to tell the count. The number N is given by the polynomial function

$$N(x) = \frac{1}{6}x^3 + \frac{1}{2}x^2 + \frac{1}{3}x,$$

where x is the number of layers and $N(x)$ is the total number of spheres. Thus to win the contest, count the number of layers and substitute the answer into the formula.

a) Find the number of spheres in a pile with 9 layers.
b) Sketch a graph of the function over the interval $[0, \infty)$.

Solution

a) We will find the number by substituting 9 for x in the polynomial:

$$\begin{aligned} N(9) &= \frac{1}{6}(9)^3 + \frac{1}{2}(9)^2 + \frac{1}{3}(9) \\ &= \frac{1}{6}(729) + \frac{1}{2}(81) + 3 \\ &= 121.5 + 40.5 + 3 \\ &= 165. \end{aligned}$$

When there are 9 layers, there will be 165 spheres.

b) The following is a graph of the function for x-values in the interval $[0, \infty)$. Note that it is an increasing function over that interval, which is reasonable since the more layers there are, the more spheres there are in the pile.

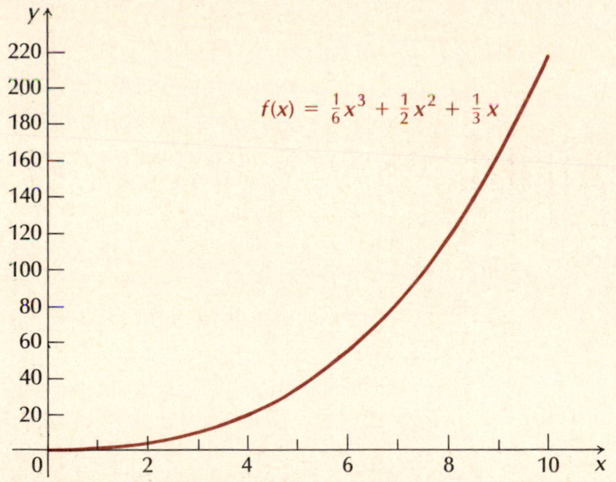

DO EXERCISE 7.

7. Referring to Example 4, find the number of spheres in a pile with 13 layers; with 25 layers.

EXERCISE SET 4.8

1 Sketch a graph of the polynomial function.

1. $f(x) = \frac{1}{3}x^6$
2. $f(x) = -\frac{2}{3}x^3$
3. $f(x) = -0.6x^5$
4. $f(x) = \frac{4}{5}x^4 - 3$
5. $f(x) = (x + 1)^5 - 4$
6. $g(x) = -(x - 2)^3 + 1$
7. $f(x) = \frac{1}{4}(x + 1)^4$
8. $f(x) = -0.7(x + 4)^3$
9. $f(x) = (x + 3)(x - 2)(x + 1)$
10. $f(x) = (x - 1)(x + 2)(x - 4)$
11. $f(x) = 9x^2 - x^4$
12. $f(x) = 4x^2 - x^4$

13. $f(x) = x^4 - x^3$
14. $f(x) = x^4 - 4x^2$
15. $f(x) = x^3 - 4x$
16. $f(x) = 25x - x^3$
17. $f(x) = x^3 + x^2 - 2x$
18. $f(x) = -x^3 - x^2 + 6x$
19. $f(x) = x^4 - 9x^2 + 20$
20. $f(x) = x^4 - 3x^2 + 2$
21. $f(x) = x^3 - 3x^2 - 4x + 12$
22. $f(x) = x^3 - 2x^2 - 9x + 18$
23. $f(x) = -x^4 - 3x^3 - 3x^2$
24. $f(x) = -3x^4 - x^3 + 2x^2$
25. $f(x) = x(x - 2)(x + 1)(x + 3)$
26. $f(x) = x(x - 1)(x - 2)(x - 3)$

2

27. *Beam deflection.* A beam rests at two points P and Q and has a concentrated load applied to the center of the beam, as shown in the figure. Let y denote the deflection of the beam at a distance of x units from the left end of the beam. The deflection depends on the elasticity of the board, the load, and other physical characteristics. Suppose under certain conditions that y is given by the polynomial function

$$y = \frac{1}{13}x^3 - \frac{1}{14}x.$$

a) Find the deflection of the beam at distances of 1 unit, 2 units, 3 units, 6 units, 8 units, 9 units, and 10 units.
b) Sketch a graph of the function for those values of x for which $y \geq 0$.

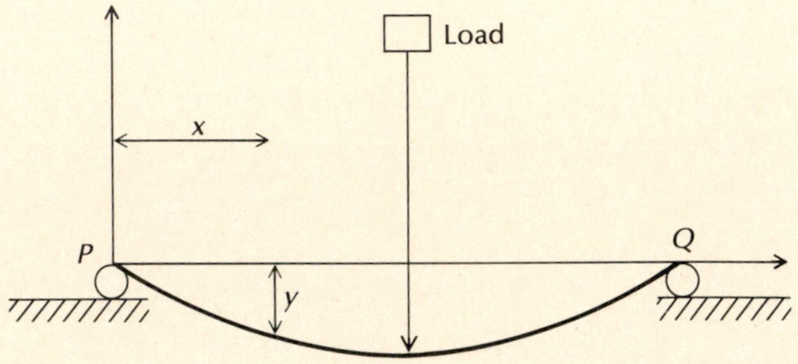

28. *Medical dosage.* The function

$$N(t) = -0.046t^3 + 2.08t + 2$$

gives the bodily concentration in parts per million of a certain dosage of medication after time t, in hours.

a) Find the concentration after 1 hr, 2 hr, 3 hr, 4 hr, 5 hr, 6 hr, and 7 hr.
b) Using a calculator, find several other function values and sketch a graph of the function over the interval (0, 7).

29. *Weight threshold.* In a medical study done by Alwin Shemesh in 1976, it was found that *threshold weight* W, defined as that weight above which the risk of mortality rises astronomically, is given as a function of height h, by

$$W(x) = (h/12.3)^3,$$

where W is in pounds and height h is in inches.

a) What is the threshold weight of a person who is 5 ft 7 in.? 5 ft 10 in.?
b) One of the authors of this book is 6 ft 1 in. tall and weighs 209 lb. Does he need to watch his weight?

30. *Windmill power.* Under certain conditions, the power P, in watts, generated by a windmill when the speed of the wind is V, in miles per hour, is given by

$$P = 0.015V^3.$$

a) Find the power generated by a wind of speed 8 mph, 10 mph, 15 mph.
b) How fast would the wind have to blow in order to generate 120 watts of power?

31. *Volume.* From a thin piece of cardboard 10 in. by 10 in., square corners are cut out so that the sides can be folded up to make a box.

a) Express the volume V as a function of the length x of a side of the square base.
b) Sketch a graph of the function using the entire set of real numbers as domain.
c) Over what intervals is the function positive?

32. *Volume.* From a thin piece of cardboard 8 in. by 8 in., square corners are cut out so that the sides can be folded up to make a box.

a) Express the volume V as a function of the length x of a side of the square base.
b) Sketch a graph of the function using the entire set of real numbers as domain.
c) Over what intervals is the function positive?

SYNTHESIS

33. Make a graph with interval $[-1, 1]$ and a large scale on the x-axis, perhaps two inches.

a) Sketch graphs of each of the following functions:

$$f(x) = x, \quad f(x) = x^3, \quad f(x) = x^5, \quad f(x) = x^7.$$

b) Look for as many patterns as you can in the graphs. Without actually graphing, explain the possible shape and location of the graph of $f(x) = x^{13}$.

34. Make a graph with interval $[-1, 1]$ and a large scale on the x-axis, perhaps two inches.

a) Sketch graphs of each of the following functions.

$$f(x) = x^2, \quad f(x) = x^4, \quad f(x) = x^6, \quad f(x) = x^8.$$

b) Look for as many patterns as you can in the graphs. Without actually graphing, explain the possible shape and location of the graph of $f(x) = x^{18}$.

35. Which of the functions in Exercises 1–26 is an even function?

36. Which of the functions in Exercises 1–26 is an odd function?

37. Under what conditions is a polynomial function of real variables an even function?

38. Under what conditions is a polynomial function of real variables an odd function?

COMPUTER–CALCULATOR EXERCISES

There are many computer software packages and calculators that will graph polynomials and solve equations. Using a computer or calculator, graph the following functions, and approximate their zeros, or roots.

39. $f(x) = x^3 + 4x^2 + x - 6$

40. $f(x) = x^3 + x^2 - 6x + 6$

41. $f(x) = -x^4 + x^3 + 4x^2 - 2x - 4$

42. $f(x) = x^5 - x^4 - 9x^3 + 3x^2 + 18x$

43. Consider the function in Exercise 28.

a) Graph the function.
b) Approximate all of its zeros.
c) After what time will the concentration be 0?

SUMMARY AND REVIEW: CHAPTER 4

TERMS TO KNOW

Linear equation	Circle	Union
Slope	Radius	Compound sentence
Point–slope equation	Center	Conjunction
Two-point equation	Quadratic function	Disjunction
Slope–intercept equation	Parabola	Equations with absolute value
Linear function	Vertex	Inequalities with absolute value
Parallel lines	Line of symmetry	Quadratic inequality
Perpendicular lines	x-intercept	Rational inequality
Distance formula	Mathematical model	Polynomial function
Midpoint formula	Intersection	Zero of a function

REVIEW EXERCISES

1. Graph $5y - 2x = 10$.

2. Find the slope and the y-intercept of the line

$$-2x - y = 7.$$

3. Graph $y = -\frac{2}{3}x - 4$ using the slope and the y-intercept.

4. Find the slope of the line containing the points $(7, -2)$ and $(1, 4)$.

5. Find an equation of the line through $(-2, -1)$ with $m = 3$.

6. Find an equation of the line containing $(4, 1)$ and $(-2, -1)$.

7. Find the distance between (3, 7) and (−2, 4).

8. Find the midpoint of the segment with endpoints (3, 7) and (−2, 4).

Given the point (1, −1) and the line $2x + 3y = 4$:

9. Find an equation of the line containing the given point and parallel to the given one.

10. Find an equation of the line containing the given point and perpendicular to the given line.

Determine whether the lines are parallel, perpendicular, or neither.

11. $3x - 2y = 8$, $6x - 4y = 2$

12. $y - 2x = 4$, $2y - 3x = -7$

13. $y = \frac{3}{2}x + 7$, $y = -\frac{2}{3}x - 4$

14. Find an equation of the circle with center (−2, 6) and radius $\sqrt{13}$.

15. Find the center and the radius of the circle

$$(x + 1)^2 + (y - 3)^2 = \frac{9}{4}.$$

Then graph the circle.

16. Find the center and the radius of the circle

$$x^2 + y^2 + 4y + 21 = 10x.$$

17. Find an equation of the circle having its center at (3, 4) and passing through the origin.

18. Find an equation of the circle having a diameter with endpoints (−3, 5) and (7, 3).

For the functions in Questions 19 and 20:

a) use completing the square to put each equation into the form $f(x) = a(x - h)^2 + k$;
b) find the vertex;
c) find the line of symmetry; and
d) determine whether the second coordinate of the vertex is a maximum or minimum, and find the maximum or minimum.

19. $f(x) = 3x^2 + 6x + 1$

20. $f(x) = -2x^2 - 3x + 6$

21. Graph $f(x) = 3x^2 + 6x + 1$.

22. Find the x-intercepts of $f(x) = -x^2 - x - 1$.

23. Find $\{4, 5, 8, 12, 13\} \cap \{3, 5, 7, 9, 11\}$.

24. Find $\{4, 5, 8, 12, 13\} \cup \{3, 5, 7, 9, 11\}$.

25. Graph $\{x|-2 < x\} \cap \{x|x \leq 3\}$ on a line.

26. Graph $\{x|x < -2\} \cup \{x|x > 2\}$ on a line.

Solve.

27. $-3x + 2 \leq -4$ *and* $x - 1 \leq 3$

28. $|x - 6| < 5$

29. $|-3x - 2| > 2$

30. $|\frac{1}{2}x + 4| \leq 6$

31. $|2x + 5| = 9$

32. $x^2 - 9 < 0$

33. $2x^2 - 3x - 2 > 0$

34. $(1 - x)(x + 4)(x - 2) < 0$

35. $\dfrac{x - 2}{x + 3} < 4$

36. *Temperature as a linear function of cricket chirps.* It has been shown experimentally that the chirps of crickets can be used to estimate the temperature. When crickets chirp 40 times per minute, the temperature is 50°F. When crickets chirp 120 times per minute, the temperature is 70°F. Assume that a linear function fits the data.

a) Find a linear function that fits the data.
b) Use the function to find the temperature when crickets chirp 76 times per minute; 100 times per minute.
c) At what temperature do crickets stop chirping? Discuss the domain of the function.

37. The sum of the length and the width of a rectangle is 40. Find the dimensions for which the area is a maximum.

38. A company has the following total-cost and total-revenue functions to use in producing and selling x units of a product:

$$C(x) = 15x + 200, \qquad R(x) = 100 + x^2.$$

a) Find the break-even values.
b) Find the values of x that produce a profit.
c) Find the values of x that result in a loss.

Sketch a graph of the polynomial function.

39. $f(x) = x^3 + 3x^2 - 2x - 6$

40. $f(x) = x^4 - 3x^3 + 2x^2$

SYNTHESIS

Solve.

41. $4x \leq 5x + 2 < 4 - x$

42. $\left|1 - \frac{1}{x^2}\right| < 3$

43. $(x - 2)^{-3} < 0$

44. Under what conditions is the square of a linear function a quadratic function?

Find the domain of the function.

45. $f(x) = \sqrt{1 - |3x - 2|}$

46. $g(x) = \frac{1}{\sqrt{5 - |7x + 2|}}$

47. There are n teams in a sports league. If each team plays each other team once, then the total number of games is given by

$$N = \frac{n(n - 1)}{2}.$$

For what values of n is $N \geq 105$?

TEST: CHAPTER 4

1. Find the slope and the y-intercept of the line $10 - 7y = 3x$.

2. Find an equation of the line through $(-2, 2)$ with $m = 5$.

3. Find an equation of the line containing $(2, -3)$ and $(5, 3)$.

4. Find the distance between $(4, -5)$ and $(6, -2)$.

5. Find the midpoint of the segment with endpoints $(4, -9)$ and $(7, 6)$.

6. Determine whether these lines are parallel, perpendicular, or neither.

$$2x - 3y = -9,$$
$$3x + 2y = 6$$

7. Graph $y = \frac{3}{2}x - 1$ using the slope and the y-intercept.

8. Find an equation of the line containing the given point and parallel to the given line:

$$(2, -4);\ y = \tfrac{1}{3}x - 6.$$

9. Find an equation of the circle with center $(-2, 4)$ and radius 4.

10. Find the center and the radius of the circle

$$x^2 + y^2 - 2x + 6y + 5 = 0.$$

For the functions in Questions 11 and 12:

a) use completing the square to put each equation into the form $f(x) = a(x - h)^2 + k$;
b) find the vertex; and
c) determine whether there is a maximum or minimum function value and find that value.

11. $f(x) = 5x^2 - 10x + 3$

12. $f(x) = -4x^2 + 3x - 1$

13. Graph $f(x) = 5x^2 - 10x + 3$.

14. Find the x-intercepts of $f(x) = 3x^2 - x - 1$.

15. Find $\{3, 4, 7, 8, 11, 12\} \cup \{2, 4, 6, 8, 10\}$.

16. Graph $\{x|x \leq \frac{3}{2}\} \cap \{x|x > \frac{1}{2}\}$.

Solve.

17. $x - 3 > -8$ *and* $-2x + 3 \geq 9$

18. $|-4x + 3| \geq 9$

19. $|x - 3| < 2$

20. $|6x - 3| = 15$

21. $x^2 - 8x + 12 > 0$

22. $8x^2 + 10x - 3 < 0$

23. $\frac{x - 4}{2x + 3} < 1$

24. Find two numbers whose sum is -20 and whose product is a maximum.

25. To produce and sell x units of a certain product, a company has the following total-cost and total-revenue functions:

$$R(x) = 65x - x^2, \qquad C(x) = 25x + 300.$$

a) Find the break-even values.
b) Find the values of x that produce a profit.
c) Find the values of x that result in a loss.

26. *Items in supermarkets.* The number of things that we can buy in a supermarket has increased considerably in recent years. In 1981, the average number of items in a supermarket was 12,877. In 1985, the average was 17,469. Assume that a linear function fits the data.

a) Find a linear function that fits the data.
b) Use the function to find the average number of items in a supermarket in 1995; in 2001.
c) Discuss the domain of the function.

Sketch a graph of the polynomial function.

27. $f(x) = x^4 - 5x^2$

28. $f(x) = x^4 - 5x^2 + 6$

SYNTHESIS

29. Suppose that an object is thrown upward with an initial velocity of 80 ft/sec from a height of 224 ft. Its height after t seconds is a function s given by

$$s(t) = -16t^2 + 80t + 224.$$

a) Find its maximum height and when the object attains it.
b) Find when the object reaches the ground.
c) Over what interval of time is the height greater than 320 ft?

30. Solve: $\left|2 + \frac{1}{x}\right| < 6.$

31. Find the domain of $f(x) = \sqrt{x^2 + 3x - 10}.$

Exponential and Logarithmic Functions

5

In this chapter, we will consider two kinds of functions, closely related. The first kind, called *exponential functions,* are functions defined by using variables for the exponents. Such functions have many applications—one, for example, is to problems of population growth.

Imagine putting a function machine in reverse. If it works, it will represent what we call the *inverse* of the original function. Functions that are inverses of each other are thus closely related. The inverses of the exponential functions, called *logarithmic functions,* or *logarithm functions,* are also important in many applications.

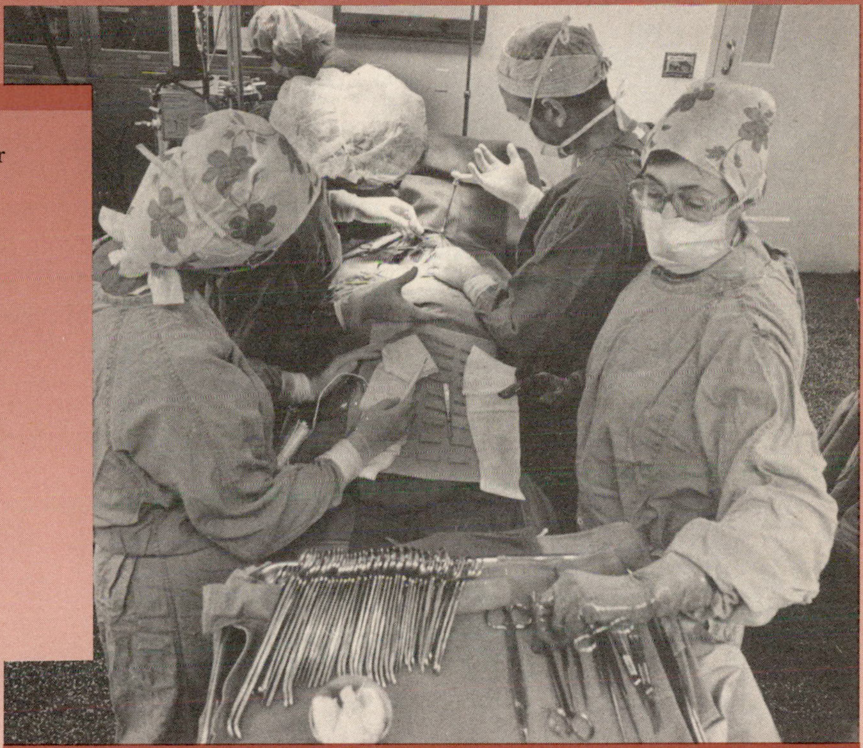

THE MATHEMATICS

Medical application: Heart transplants. In 1967, Dr. Christian Barnard, of South Africa, staggered the world by performing the first heart transplant. There was only 1 such transplant that year. In 1987, there were 1418 heart transplants. Assuming that the number N of heart transplants during a given year can be modeled by an exponential-growth function, find a formula for this function.

FEATURE PROBLEM

An exponential growth function that models this situation is given by

$$N(t) = e^{0.363t},$$

where t = the number of years since 1967.

5.1 Inverses of Relations and Functions

OBJECTIVES

You should be able to:

1 Given a relation that is a set of ordered pairs, find its inverse relation. Given an equation defining a relation, write an equation of the inverse relation.

2 Given a relation that is a set of ordered pairs, graph the relation and its inverse. Given a graph of a relation, sketch a graph of its inverse, or given an equation defining a relation, graph it and then graph its inverse.

3 Given an equation defining a relation, determine whether the graph is symmetric with respect to the line $y = x$.

4 Given a relation, determine whether it is one-to-one and thus has an inverse that is a function. If the function is one-to-one, find a formula for the inverse. Graph a function and its inverse using the same set of axes.

5 Determine whether one function is the inverse of another, use the property $x = f \circ f^{-1}(x)$ to find the inverse of a function, and simplify expressions of the type $f \circ f^{-1}(x)$ and $f^{-1} \circ f(x)$.

1. Find the inverse of the relation Q given by

$$Q = \{(-1, 4), (2, 5), (0, -3), (5, 1)\}.$$

2. Write an equation of the inverse of each relation.

a) $y = 3x + 2$

b) $y = x$

c) $x^2 + 3y^2 = 4$

d) $y = 5x^2 + 2$

e) $y^2 = 4x - 5$

f) $xy = 5$

3. Consider the relation P given by $P = \{(5, 0), (3, -2), (0, -3), (4, -4)\}$.

a) Draw the graph of the relation P in black.

b) Find the inverse of P and draw its graph in color.

1 Inverses of Relations

Consider the relation r given as follows:

$$r = \{(2, 4), (-1, 3), (-2, 0)\}.$$

Suppose we *interchange* the first and second coordinates. The relation we obtain is called the **inverse** of the relation r. It is given as follows:

$$\text{Inverse of } r = \{(4, 2), (3, -1), (0, -2)\}.$$

DO EXERCISE 1.

If a relation is defined by an equation, then an equation of the inverse can be found by interchanging the variables. The solutions of the second equation will be the same as those of the first equation, except that the first and second coordinates will be interchanged in each ordered pair.

When a relation is defined by an equation, interchanging x and y produces an equation of the *inverse* relation.

EXAMPLE 1 Find an equation of the inverse of the relation $y = x^2 - 5$.

Solution We interchange x and y and obtain $x = y^2 - 5$. This is an equation of the inverse relation.

DO EXERCISE 2.

2 Graphs of Inverse Relations

EXAMPLE 2 Consider the relation r given by

$$r = \{(2, 4), (-1, 3), (-2, 0)\}.$$

Draw the graph of the relation in black. Find the inverse relation and draw its graph in color.

Solution The relation r is shown on the graph in black. The inverse of the relation is $\{(4, 2), (3, -1), (0, -2)\}$ and is shown in color.

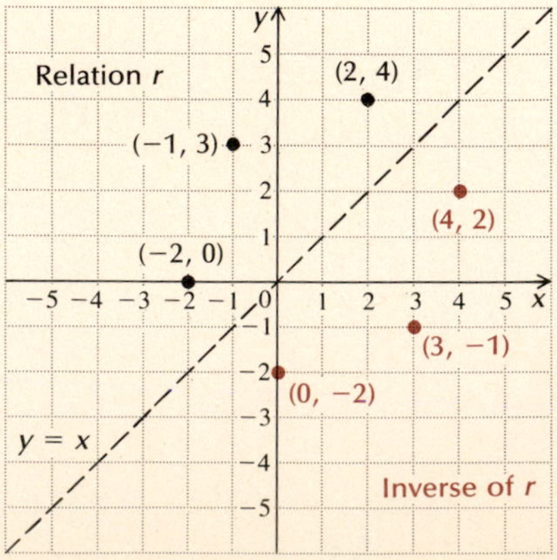

DO EXERCISE 3 ON THE PRECEDING PAGE.

Compare the relation and its inverse in Example 2. Note that the inverse can be obtained by reflecting each ordered pair across the line $y = x$. Interchanging x and y in an equation to find the inverse has the effect of reflecting each ordered pair in the original relation across the line $y = x$. Thus the graphs of a relation and its inverse are always reflections of each other across the line $y = x$. (This assumes that the same scale is used on both axes.)

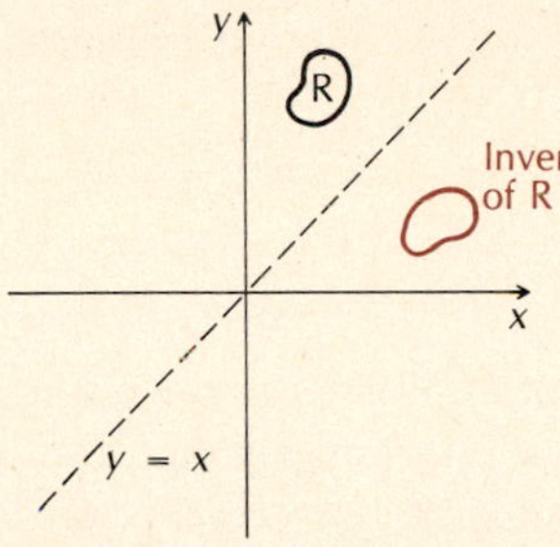

EXAMPLES In each case, a relation is shown in black. Graph the inverse.

Solution The graph of each inverse is shown in color.

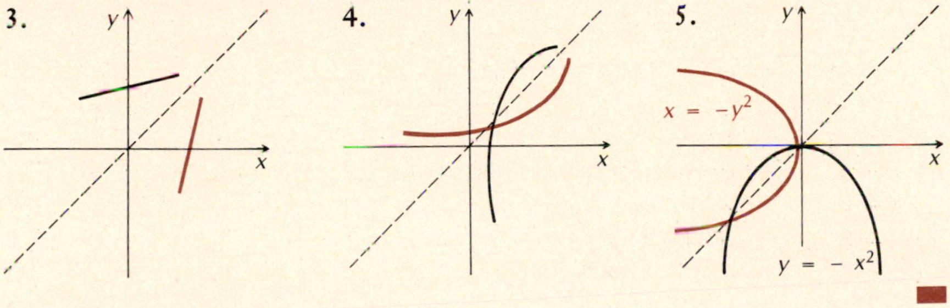

DO EXERCISES 4 AND 5.

3 Symmetry with Respect to the Line $y = x$

It can happen that a relation is its own inverse; that is, when x and y are interchanged or the relation is reflected across the line $y = x$, there is no change. Such a relation is symmetric with respect to the line $y = x$. The following are two examples of relations symmetric with respect to the line $y = x$.

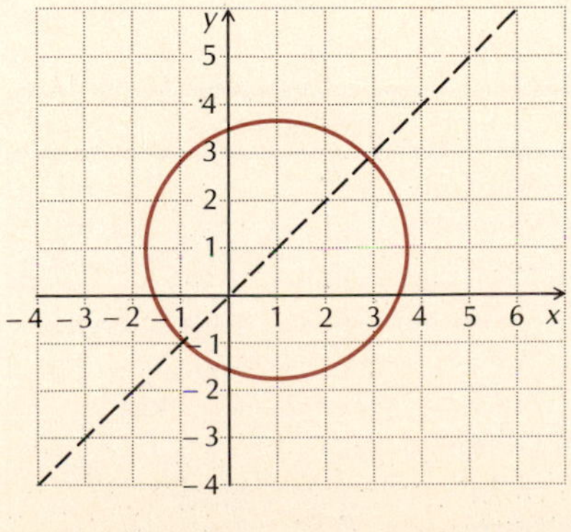

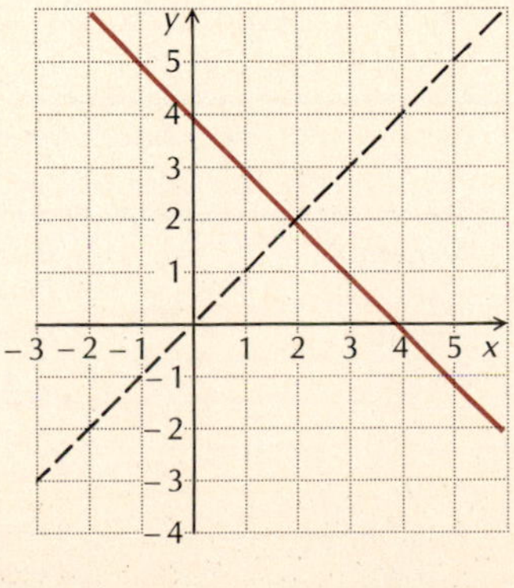

None of the relations shown in black in Examples 3–5 is symmetric with respect to the line $y = x$. If a relation is symmetric with respect to that line, that relation is its own inverse.

4. a) On a piece of thin paper, draw the coordinate axes.
   b) Draw the line $y = x$.
   c) Draw a relation as shown here.

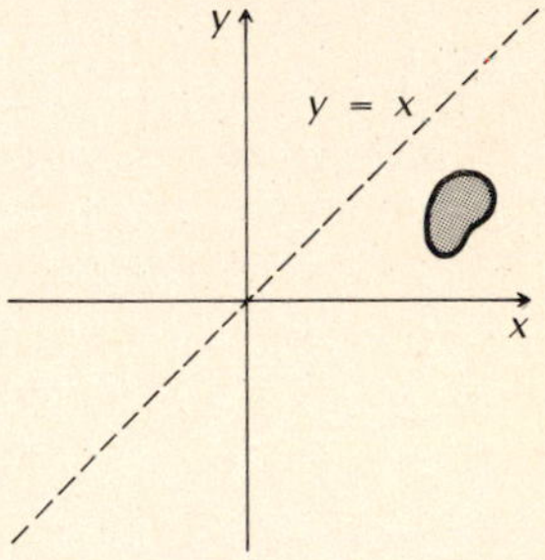

   d) Flip the paper over (to interchange the x-axis and the y-axis). Look through the paper. How do the graphs of the relation and its inverse compare?

5. Graph the inverse of each relation by reflecting across the line $y = x$.
   a)

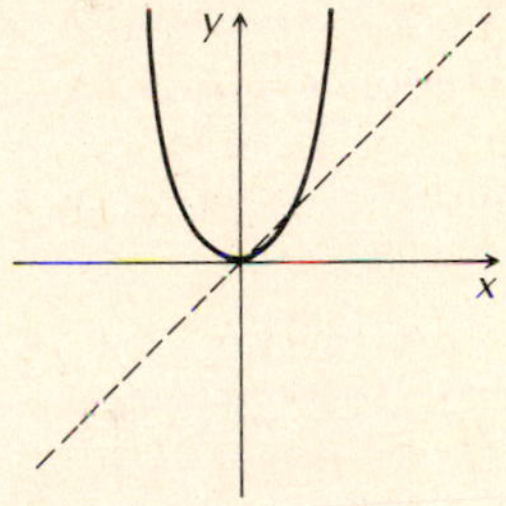

   b)

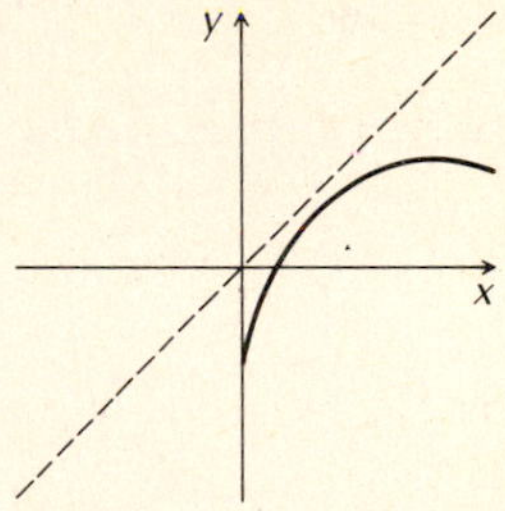

   c) Graph $y = 4 - x^2$. Then by reflecting across the line $y = x$, graph its inverse.

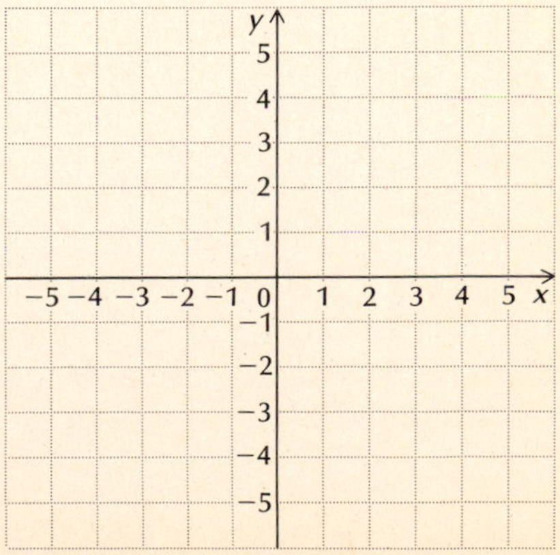

Test for symmetry with respect to the line $y = x$.

6. $y = -x$

7. $x + y = 4$

8. $xy = 3$

9. $y = |x|$

10. $3x^2 + 3y^2 = 4$

11. $|x| = |y|$

12. $y = x^3$

13. $x - y = 4$

For a relation defined by an equation: If interchanging x and y produces an equivalent equation, the relation is its own inverse, and the graph of the equation is symmetric with respect to the line $y = x$.

EXAMPLE 6 Test $3x + 3y = 5$ for symmetry with respect to the line $y = x$.

Solution

a) We interchange x and y. This amounts to replacing each occurrence of x by y and each y by x:

$$\begin{array}{ccccc} 3x & + & 3y & = & 5 \\ \downarrow & & \downarrow & & \\ 3y & + & 3x & = & 5. \end{array}$$

b) Is the resulting equation equivalent to the original? The commutative law of addition guarantees that the resulting equation is equivalent to the original. Thus the graph is symmetric with respect to the line $y = x$. ■

EXAMPLE 7 Test $y = x^2$ for symmetry with respect to the line $y = x$.

Solution

a) We interchange x and y:

$$\begin{array}{ccc} y & = & x^2 \\ \downarrow & & \downarrow \\ x & = & y^2. \end{array}$$

b) Is the resulting equation equivalent to the original? Note that (2, 4) is a solution of the original equation $y = x^2$, but it is not a solution of the resulting equation $x = y^2$.

$$\begin{array}{c|c} y & x^2 \\ \hline 4 & 2^2 \\ & 4 \end{array} \qquad \begin{array}{c|c} x & y^2 \\ \hline 2 & 4^2 \\ & 16 \end{array}$$

Thus the equations are not equivalent, so the graph of $y = x^2$ is *not* symmetric with respect to the line $y = x$. ■

DO EXERCISES 6–13.

4 Inverses and One-to-One Functions

Let us consider the following two functions. They are relations. Let us think of them as correspondences.

Cost of a 60-second Commercial During the Super Bowl

Domain (Set of Inputs)		Range (Set of Outputs)
1967	→	\$80,000
1970	→	\$200,000
1977	→	\$324,000
1981	→	\$550,000
1983	→	\$800,000
1985	→	\$1,100,000
1988	→	\$1,350,000

U.S. Senators

Domain (Set of Inputs)		Range (Set of Outputs)
Cranston	→	California
Wilson	→	California
Mack	→	Florida
Graham	→	Florida
Simon	→	Illinois
Dixon	→	Illinois
D'Amato	→	New York
Moynihan	→	New York

Suppose we reverse the arrows. We obtain what is called the *inverse correspondence*, or *relation*. Now are these new correspondences functions?

Cost of a 60-second Commercial During the Super Bowl		
Domain (Set of Inputs)		**Range (Set of Outputs)**
1967	←	\$80,000
1970	←	\$200,000
1977	←	\$324,000
1981	←	\$550,000
1983	←	\$800,000
1985	←	\$1,100,000
1988	←	\$1,350,000

U.S. Senators		
Domain (Set of Inputs)		**Range (Set of Outputs)**
Cranston	←	California
Wilson	←	California
Mack	←	Florida
Graham	←	Florida
Simon	←	Illinois
Dixon	←	Illinois
D'Amato	←	New York
Moynihan	←	New York

We see that the inverse of the first correspondence is a function, but the inverse of the second correspondence is not a function.

Recall that for each input, a function provides exactly one output. However, nothing in our definition of function precludes having the same output for two or more different inputs. That is, it is possible with a function for different inputs to correspond to the same output in the range. When this possibility is excluded, the inverse is also a function.

In the Super Bowl function, different inputs have different outputs. Thus this function is what is called a **one-to-one function.** In the U.S. Senator function, the input Simon has the output Illinois, and the input Dixon also has the output Illinois. Thus this is not a one-to-one function.

DEFINITION

A function f is *one-to-one* if different inputs have different outputs. That is, if $a \neq b$, then $f(a) \neq f(b)$.

A function f is *one-to-one* if when the outputs are the same, the inputs are the same. That is, if $f(a) = f(b)$, then $a = b$.

If a function is one-to-one, then its inverse correspondence is a function.

DO EXERCISES 14 AND 15.

Let us consider this idea from the standpoint of a relation as a set of ordered pairs.

EXAMPLE 8 Consider the relation G given by

$$G = \{(1, 3), (2, 4), (6, 3), (7, 7)\}.$$

a) Determine whether G is a function.
b) Find the inverse of G.
c) Determine whether the inverse of G is a function.

Solution

a) G is a function because no two ordered pairs have the same first coordinates but different second coordinates.
b) The inverse of $G = \{(3, 1), (4, 2), (3, 6), (7, 7)\}$.
c) The inverse of G is *not* a function because there are two ordered

For each function, find its inverse correspondence and determine whether that inverse is a function.

14.

Women's Dress Sizes	
Domain (United States)	**Range (France)**
6	38
8	40
10	42
12	44
14	46
16	48
18	50

15.

Sports Teams	
Domain	**Range**
Lakers	
Dodgers	Los Angeles
Rams	
Knickerbockers	
Yankees	New York
Giants	

pairs—namely, (3, 1) and (3, 6)—that have the same first coordinates but different second coordinates. Thus, G is not one-to-one, so the inverse of G is not a function. ■

DO EXERCISES 16 AND 17.

How can we tell graphically whether a function is one-to-one and hence has an inverse that is a function? The graph below shows a function, in black, and its inverse, in color. To determine whether the inverse is a function, we can apply the vertical-line test to its graph. By reflecting each such vertical line back across the line $y = x$, we obtain an equivalent **horizontal-line test** for the original function.

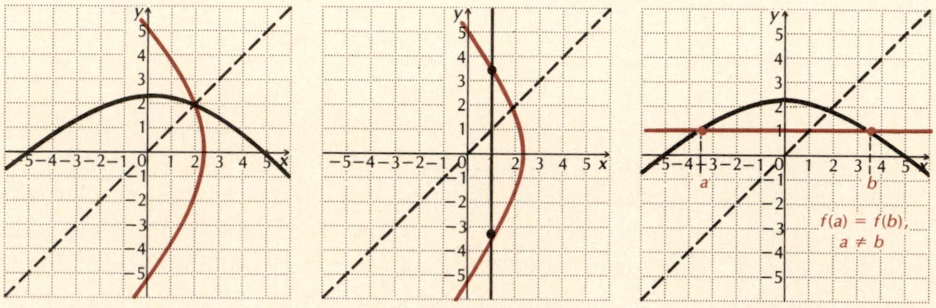

The inverse is not a function.

The function is not one-to-one.

The Horizontal-Line Test

A function is one-to-one and has an inverse that is a function if there is no horizontal line that crosses the graph more than once.

Thus a graph is that of a function if no vertical line crosses the graph more than once. A function has an inverse that is also a function if no horizontal line crosses the graph of the original function more than once.

EXAMPLE 9 Determine whether the function $f(x) = x^2$ is one-to-one and thus has an inverse that is also a function.

Solution The graph of $f(x) = x^2$ is shown below. Note that there are many horizontal lines that cross the graph more than once. In particular, the line $y = 4$ crosses the graph more than once. If you note where the line crosses, the first coordinates are -2 and 2. Although these are different inputs, they have the same outputs. That is, $-2 \neq 2$, but

$$f(-2) = (-2)^2 = 4 = 2^2 = f(2).$$

Thus the function is not one-to-one and does not have an inverse that is a function.

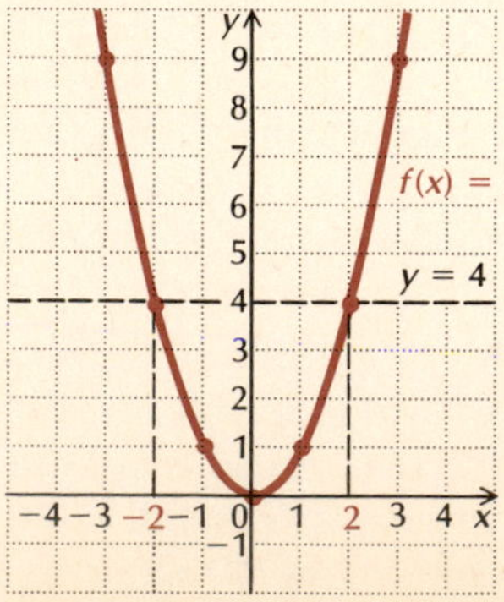

■

For each relation in Exercises 16 and 17,

a) determine whether the relation is a function;

b) find the inverse of the relation;

c) determine whether the inverse is a function.

16. $G = \{(5, -4), (9, 2), (8, -6), (10, -4)\}$

17. $H = \{(2, 3), (9, -1), (7, 4), (4, -3), (10, 12)\}$

Consider the Super Bowl function again. Note that if we start with the input 1985, we get \$1,100,000 as an output. Then consider the inverse relation. If we start with the input \$1,100,000, we get 1985 as the output. Thus the inverse takes us back to where we started. This does not happen with the U.S. Senator function. If we start with Cranston as an input, we get California as an output. If we go to the inverse relation and start with California, we have two possibilities for outputs, Cranston and Wilson. The U.S. Senator function is not one-to-one and its inverse is not a function.

DO EXERCISES 18–21.

If the inverse of a function f is also a function, it can be named f^{-1} (read "f-inverse").

CAUTION! The -1 in f^{-1} is not an exponent!

Suppose that a function is described by a formula. If it has an inverse that is a function, how do we find a formula for its inverse?

EXAMPLE 10 Consider $f(x) = x + 3$.

a) Determine whether the function is one-to-one.
b) If it is one-to-one, find a formula for $f^{-1}(x)$.

Solution

a) The graph of $f(x) = x + 3$ is shown below. It passes the horizontal-line test, so it is one-to-one. Thus its inverse is a function.

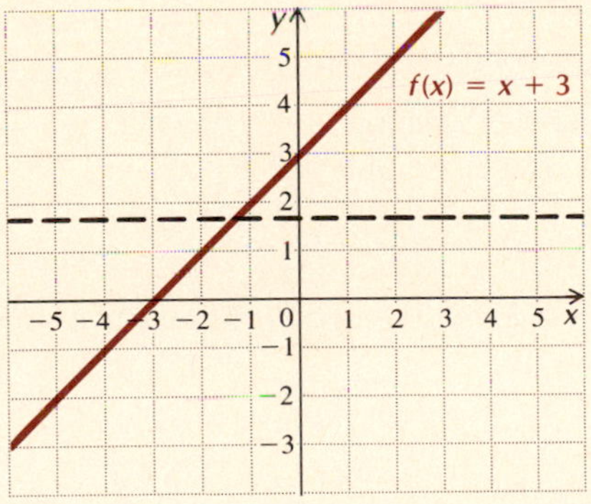

b) To find a formula, think of this function as $y = x + 3$. Normally, you are given x-values, or inputs, and asked to find y-values, or outputs. Suppose you start with $x = 5$. Then $y = 5 + 3 = 8$. If you are given the output 8 and asked to find the input from which it came, you could set $y = 8$ in the equation and solve for x:

$$8 = x + 3$$
$$5 = x.$$

Suppose the output is 13 and you want the input from which it came. Set $y = 13$ and solve for x:

$$13 = x + 3$$
$$10 = x.$$

In general, the inverse has the effect of interchanging the domain and the range. If we start with an output y and solve the equation for x, we get the

Determine whether the function is one-to-one and has an inverse that is also a function.

18. $f(x) = 4 - x$

19. $f(x) = x^2 - 1$

20. $f(x) = |x| - 3$

21. Which of the following have inverses that are functions?

a)

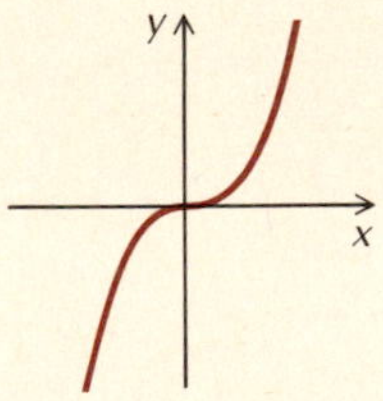

b)

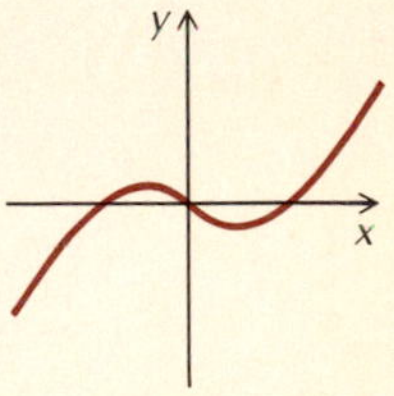

c)

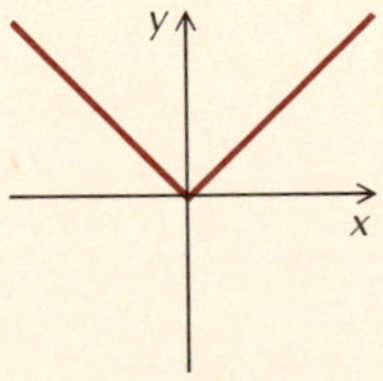

d)

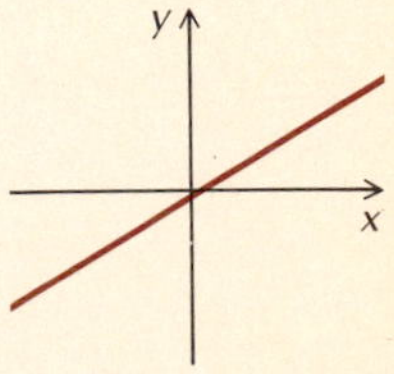

input from which y came:

$$y = x + 3$$
$$y - 3 = x.$$

The inputs of the inverse are what we substitute into the function to get outputs, and x is used for that variable. The outputs are represented by y. Thus we can interchange x and y, which gives us the inverse, represented by

$$(x) - 3 = [y], \quad \text{or } y = x - 3.$$

Or, using $f^{-1}(x)$ for y, we finally obtain a formula for the inverse:

$$f^{-1}(x) = x - 3.$$

Note in Example 10 that f maps any x onto $x + 3$. (This function adds 3 to each number in its domain.) Its inverse, f^{-1}, maps any number x onto $x - 3$. (This function subtracts 3 from each member of its domain.) Thus the function and its inverse reverse each other.

To Find a Formula for the Inverse of a Function

If a function is one-to-one, a formula for its inverse can be found as follows:

1. **Replace $f(x)$ by y.**
2. **Solve the equation for x.**
3. **Interchange x and y.**
4. **Replace y by $f^{-1}(x)$.**

EXAMPLE 11 Consider $g(x) = 2x - 3$.

a) Determine whether the function is one-to-one.
b) If it is one-to-one, find a formula for $g^{-1}(x)$.

Solution

a) The graph of $g(x) = 2x - 3$ is shown below. It passes the horizontal-line test and is one-to-one.

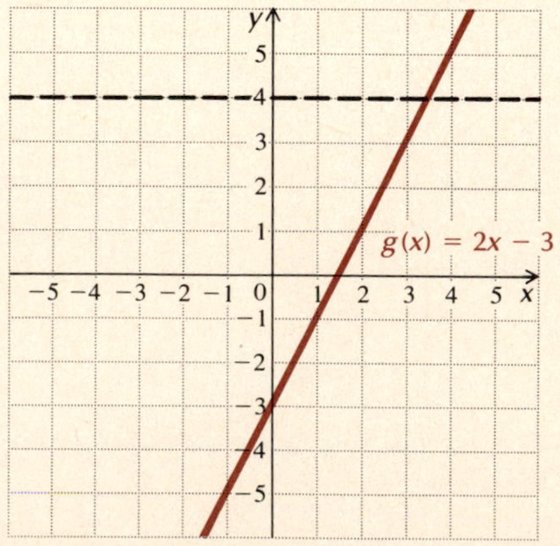

b) We first replace $g(x)$ by y:

$$y = 2x - 3.$$

We then solve the equation for x:

$$y + 3 = 2x$$
$$\frac{y + 3}{2} = x.$$

Next we interchange x and y:

$$\frac{x + 3}{2} = y.$$

We then replace y by $g^{-1}(x)$:

$$g^{-1}(x) = \frac{x + 3}{2}.$$

DO EXERCISES 22 AND 23.

Let us consider inverses of functions in terms of function machines. Suppose that the function f programmed into a machine has an inverse that is also a function. Suppose then that the function machine has a reverse switch. The machine is programmed to do the inverse mapping g^{-1} when the switch is thrown. Inputs then enter at the opposite end and the entire process is reversed.

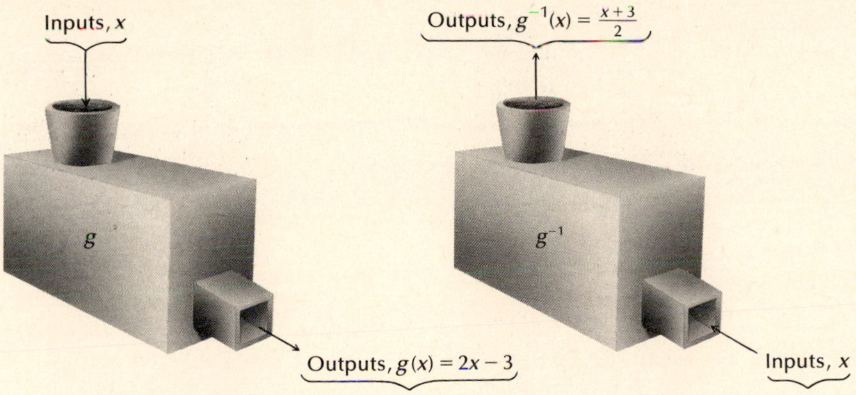

For example, the function $g(x) = 2x - 3$ has as its inverse

$$g^{-1}(x) = \frac{x + 3}{2},$$

from Example 11. Consider the input 5. Now

$$g(5) = 2(5) - 3 = 10 - 3 = 7.$$

The output is 7. Now we use 7 for the input in the inverse:

$$g^{-1}(7) = \frac{7 + 3}{2} = \frac{10}{2} = 5.$$

The function g takes 5 to 7. The inverse function g^{-1} takes the number 7 back to 5.

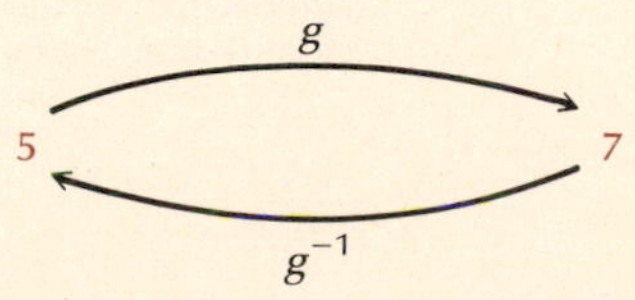

How do the graphs of a function and its inverse compare?

EXAMPLE 12 Graph $g(x) = 2x - 3$ and $g^{-1}(x) = (x + 3)/2$ using the same set of axes. Then compare.

Given each function,

a) Determine whether it is one-to-one.

b) If it is one-to-one, find a formula for the inverse.

22. $f(x) = 3 - x$

23. $g(x) = 3x - 2$

24. Graph

$$g(x) = 3x - 2 \quad \text{and} \quad g^{-1}(x) = \frac{x + 2}{3}$$

using the same set of axes.

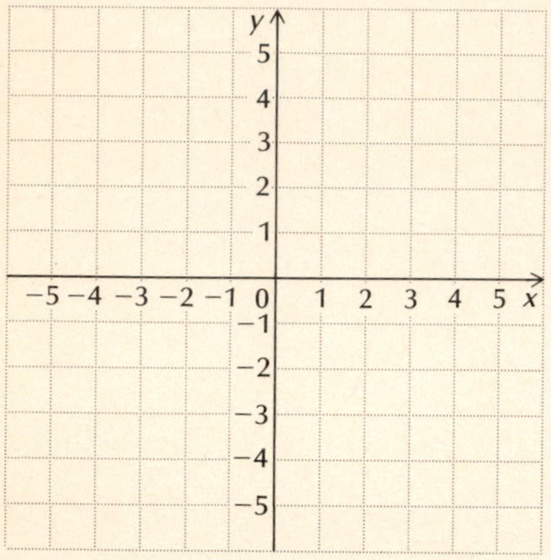

Solution The graph of each function is shown below.

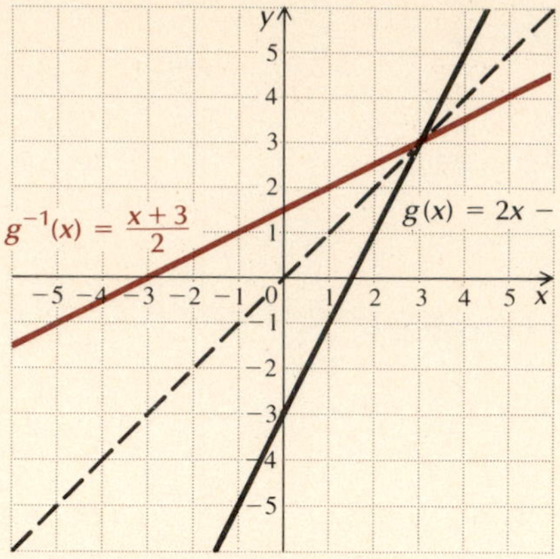

Note that the graph of g^{-1} can be obtained by reflecting the graph of g across the line $y = x$. That is, if you graph $g(x) = 2x - 3$ and $y = x$ and fold the paper along the line $y = x$, the graph of

$$g^{-1}(x) = \frac{x + 3}{2}$$

will be the result of "flipping" the graph of $g(x) = 2x - 3$ across the line.

The graph of f^{-1} is a reflection of the graph of f across the line $y = x$.

DO EXERCISE 24.

EXAMPLE 13 Consider $f(x) = x^3 + 2$.

a) Determine whether the function is one-to-one.
b) If it is one-to-one, find a formula for its inverse.
c) Graph the inverse.

Solution

a) The graph of $f(x) = x^3 + 2$ is shown. It passes the horizontal-line test, and thus has an inverse.
b) We replace $f(x)$ by y:

$$y = x^3 + 2.$$

We solve for x:

$$y - 2 = x^3$$
$$\sqrt[3]{y - 2} = x.$$

Since a function has only one cube root, we can solve for x.

Next we interchange x and y:

$$\sqrt[3]{x - 2} = y.$$

We then replace y by $f^{-1}(x)$:

$$f^{-1}(x) = \sqrt[3]{x - 2}.$$

c) We find the graph by flipping the graph of $f(x) = x^3 + 2$ over the line $y = x$, interchanging x- and y-coordinates. We could also find the graph by substituting into $f^{-1}(x) = \sqrt[3]{x - 2}$ to find function values. The two graphs are shown using the same set of axes.

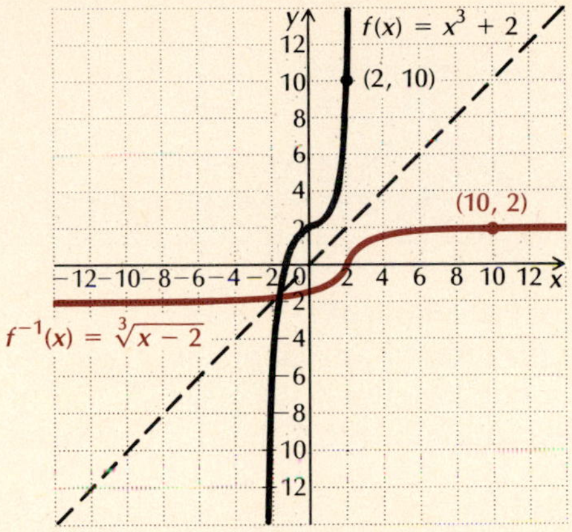

DO EXERCISE 25.

25. Consider $f(x) = x^3 + 1$.

a) Determine whether the function is one-to-one.

b) If it is one-to-one, find a formula for its inverse.

c) Graph the function and its inverse using the same set of axes.

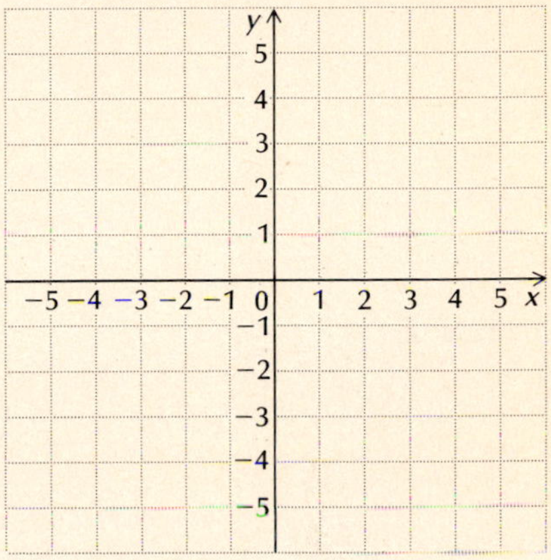

Now let us consider a situation where the inverse of a function is *not* a function. Consider

$$f(x) = x^2.$$

The graph is shown below. It does not pass the horizontal-line test. Thus it is not one-to-one and does not have an inverse that is a function. Nevertheless, suppose we reflect the graph across the line $y = x$. We see again that the inverse is not a function since the graph (in color) does not pass the vertical-line test.

Suppose we had not known this and had tried to find a formula for the inverse as follows:

$$y = x^2 \qquad \textbf{Replacing } f(x) \textbf{ by } y$$
$$\pm\sqrt{y} = x.$$

We cannot solve for x and get only one value, since most real numbers have two square roots. If we interchange x and y, we get the inverse relation

$$\pm\sqrt{x} = y.$$

This is not the equation of a function. An input of, say, 4 yields two outputs, -2 and 2.

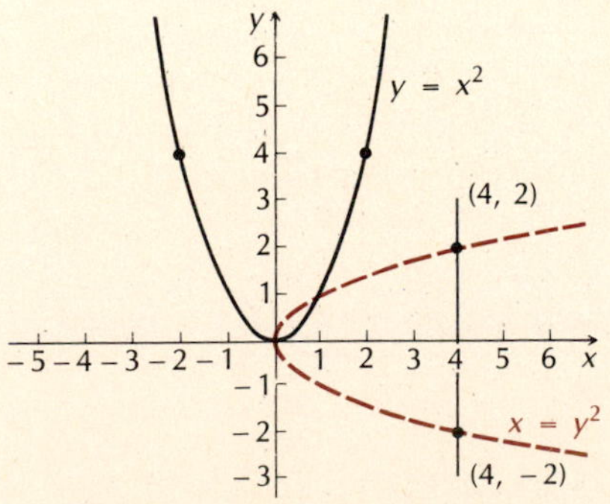

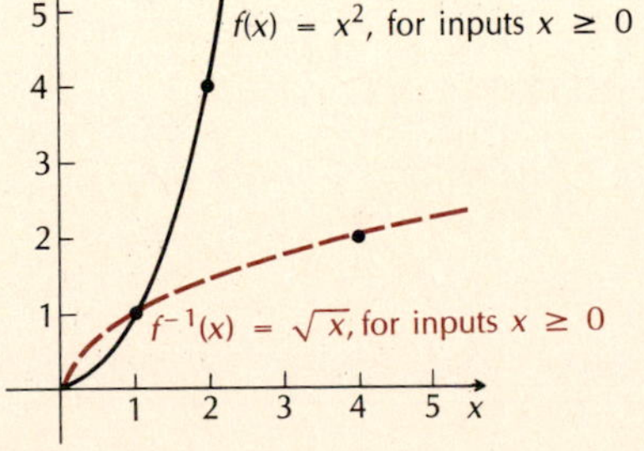

In such cases, it is often convenient to consider "part" of the function by restricting the domain of $f(x) = x^2$ to nonnegative numbers. Then its inverse is a function. See the graphs of $f(x) = x^2$, $x \geq 0$, and $f^{-1}(x) = \sqrt{x}$, $x \geq 0$ above.

DO EXERCISE 26.

26. Given $f(x) = x^2 - 4$, $x \geq 0$, find $f^{-1}(x)$.

27. Let

$$f(x) = \frac{4}{x} - 3.$$

Show that

$$f^{-1}(x) = \frac{4}{x+3}.$$

28. Simplify $f(f^{-1}(1992))$ and $f^{-1}(f(-23{,}456))$.

5 Inverse Functions and Composition

THEOREM 1

If a function f is one-to-one and hence has an inverse f^{-1} that is a function, then

a) $f^{-1} \circ f(x) = f^{-1}(f(x)) = x$, for each x in the domain of f, and
b) $f \circ f^{-1}(x) = f(f^{-1}(x)) = x$, for each x in the domain of f^{-1}.

Proof. Suppose x is in the domain of f. Then $f(x) = y$, for some number y in the range of f. Then the ordered pair (x, y) is in f, and by definition of f^{-1}, the pair (y, x) is in f^{-1}. It follows that $f^{-1}(y) = x$. Then by substituting $f(x)$ for y, we get $f^{-1}(f(x)) = x$. A similar proof shows the second part.

The first condition asserts that if you start with an input x for the function f and find its output, then the inverse will take the output back to x. The second condition asserts that if you start with an input x for the function f^{-1}, the inverse takes it to its output, and if you apply the original function to that output, you will get back to x.

EXAMPLE 14 Let $f(x) = 5x + 1$. Show that $f^{-1}(x) = (x - 1)/5$.

Solution We find $f^{-1} \circ f(x)$ and $f \circ f^{-1}(x)$ and check to see that each is x.

a) $$f^{-1} \circ f(x) = f^{-1}(f(x)) = f^{-1}(5x + 1) = \frac{(5x+1) - 1}{5} = \frac{5x}{5} = x$$

b) $$f \circ f^{-1}(x) = f(f^{-1}(x)) = f\left(\frac{x-1}{5}\right) = 5\left(\frac{x-1}{5}\right) + 1 = x - 1 + 1 = x$$

EXAMPLE 15 Simplify $g^{-1}(g(283))$ and $g(g^{-1}(-12{,}045))$.

Solution Assuming that 283 is in the domain of g, we have

$$g^{-1}(g(283)) = 283.$$

Assuming that $-12{,}045$ is in the domain of g^{-1}, we have

$$g(g^{-1}(-12{,}045)) = -12{,}045.$$

DO EXERCISES 27 AND 28.

EXERCISE SET 5.1

1 Find the inverse of each relation.

1. $\{(0, 1), (5, 6), (-2, -4)\}$
2. $\{(-1, 3), (2, 5), (-3, 5), (2, 0)\}$
3. $\{(7, 8), (-2, 8), (3, -4), (8, -8)\}$
4. $\{(-1, -1), (-3, 4)\}$

Write an equation of the inverse relation.

5. $y = 4x - 5$
6. $y = 3x + 5$
7. $x^2 - 3y^2 = 3$
8. $2x^2 + 5y^2 = 4$
9. $y = 3x^2 + 2$
10. $y = 5x^2 - 4$
11. $xy = 7$
12. $xy = -5$

2

13. Graph $y = x^2 + 1$. Then by reflection across the line $y = x$, graph its inverse.

14. Graph $x = y^2 - 3$. Then by reflection across the line $y = x$, graph its inverse.

15. Graph $x = |y|$. Then by reflection across the line $y = x$, graph its inverse.

16. Graph $y = |x|$. Then by reflection across the line $y = x$, graph its inverse.

3 Test for symmetry with respect to the line $y = x$.

17. $3x + 2y = 4$
18. $5x - 2y = 7$
19. $4x + 4y = 5$
20. $5x + 5y = -1$
21. $xy = 10$
22. $xy = -12$
23. $3x = \dfrac{4}{y}$
24. $4y = \dfrac{5}{x}$
25. $y = |2x|$
26. $3x = |2y|$
27. $4x^2 + 4y^2 = 3$
28. $3x^2 + 3y^2 = 5$

4 Determine whether the function is one-to-one.

29. $f(x) = 5x - 8$
30. $f(x) = 3 - 2x$
31. $f(x) = x^2 - 7$
32. $f(x) = 1 - x^2$
33. $g(x) = 3$
34. $g(x) = -0.8$
35. $g(x) = |x|$
36. $h(x) = |x| - 2$
37. $f(x) = |x + 1|$
38. $f(x) = |x - 3|$
39. $g(x) = \dfrac{-4}{x}$
40. $h(x) = \dfrac{1}{x}$

Given each function, (a) determine whether it is one-to-one and (b) if it is one-to-one, find a formula for the inverse.

41. $f(x) = x + 4$
42. $f(x) = x + 5$
43. $f(x) = 5 - x$
44. $f(x) = 7 - x$
45. $g(x) = x - 3$
46. $g(x) = x - 10$
47. $f(x) = 2x$
48. $f(x) = 5x$
49. $g(x) = 2x + 5$
50. $g(x) = 5x + 8$
51. $h(x) = \dfrac{4}{x + 7}$
52. $h(x) = \dfrac{1}{x - 6}$
53. $f(x) = \dfrac{1}{x}$
54. $f(x) = -\dfrac{4}{x}$
55. $f(x) = \dfrac{2x + 3}{4}$
56. $f(x) = \dfrac{3x - 5}{4}$
57. $g(x) = \dfrac{x + 4}{x - 3}$
58. $g(x) = \dfrac{5x - 3}{2x + 1}$
59. $f(x) = x^3 - 1$
60. $f(x) = x^3 + 7$
61. $G(x) = (x - 4)^3$
62. $G(x) = (x + 5)^3$
63. $f(x) = \sqrt[3]{x}$
64. $f(x) = \sqrt[3]{x - 8}$
65. $f(x) = 4x^2 + 3,\ x > 3$
66. $f(x) = 5x^2 - 2,\ x > 2$
67. $f(x) = \sqrt{x + 1}$
68. $g(x) = \sqrt{2x - 3}$

69. Which of the following have inverses that are functions?

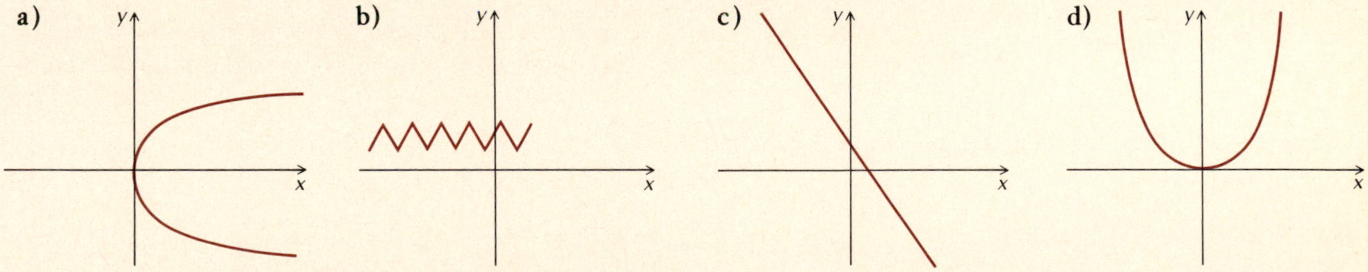

70. Which of the following have inverses that are functions?

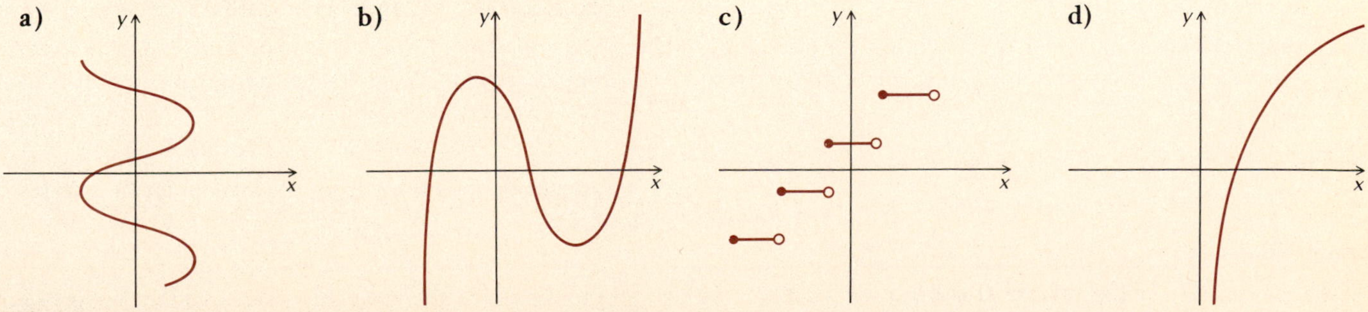

Graph the function and its inverse using the same set of axes.

71. $f(x) = \frac{1}{2}x - 4$
72. $g(x) = x + 4$
73. $f(x) = x^3$
74. $f(x) = x^3 - 1$

75. $f(x) = \sqrt{x - 3}$

76. $f(x) = \sqrt{2x + 5}$

77. $g(x) = \frac{1}{x}$

78. $f(x) = -\frac{2}{x}$

79. $f(x) = 3 - x^2, x \geq 0$

80. $f(x) = x^2 - 1, x \leq 0$

5

81. Let $f(x) = \frac{7}{8}x$. Show that $f^{-1}(x) = \frac{8}{7}x$.

82. Let $f(x) = (x + 5)/4$. Show that $f^{-1}(x) = 4x - 5$.

83. Let $f(x) = (1 - x)/x$. Show that $f^{-1}(x) = 1/(x + 1)$.

84. Let $f(x) = x^3 - 4$. Show that $f^{-1}(x) = \sqrt[3]{x + 4}$.

85. Let $f(x) = 35x - 173$. Simplify

$$f^{-1}(f(3)) \quad \text{and} \quad f(f^{-1}(-125)).$$

86. Let $g(x) = (15 - 173x)/3$. Simplify

$$g^{-1}(g(5)) \quad \text{and} \quad g(g^{-1}(-12)).$$

87. Let $f(x) = x^3 + 2$. Simplify

$$f^{-1}(f(12{,}053)) \quad \text{and} \quad f(f^{-1}(-17{,}243)).$$

88. Let $g(x) = x^3 - 486$. Simplify

$$g^{-1}(g(489)) \quad \text{and} \quad g(g^{-1}(-17{,}422)).$$

89. *Women's dress sizes in the United States and France.* Sizes of clothing and shoes are not the same numbers in different countries. For example, a size-6 dress in the United States is size 38 in France. A function that will convert dress sizes in the United States to those in France is

$$f(x) = x + 32.$$

a) Find the dress sizes in France that correspond to sizes 8, 10, 14, and 18 in the United States.
b) Determine whether this function has an inverse that is a function. If so, find a formula for the inverse.
c) Use the inverse function to find dress sizes in the United States that correspond to sizes 40, 42, 46, and 50 in France.

90. *Women's dress sizes in the United States and Italy.* A size-6 dress in the United States is size 36 in Italy. A function that will convert dress sizes in the United States to those in Italy is

$$g(x) = 2(x + 12).$$

a) Find the dress sizes in Italy that correspond to sizes 8, 10, 14, and 18 in the United States.
b) Determine whether this function has an inverse that is a function. If so, find a formula for the inverse.
c) Use the inverse function to find dress sizes in the United States that correspond to sizes 40, 44, 52, and 60 in Italy.

SYNTHESIS

91. Does the constant function $f(x) = 5$ have an inverse that is a function? If so, find a formula. If not, explain why.

92. An organization determines that the cost per person of chartering a bus is given by the function

$$C(x) = \frac{100 + 5x}{x},$$

where x = the number of people in the group and $C(x)$ is in dollars. Determine $C^{-1}(x)$ and explain this inverse function.

Determine whether the functions are inverses of each other.

93. $f(x) = \frac{2}{3}, g(x) = \frac{3}{2}$

94. $f(x) = \sqrt[5]{x}, g(x) = x^5$

95. $f(x) = \sqrt[4]{x}, x \geq 0, g(x) = x^4$

96. $f(x) = \frac{2x - 5}{4x + 7}, g(x) = \frac{7x - 4}{5x + 2}$

97. Find three examples of functions that are their own inverses. That is, $f = f^{-1}$.

98. Prove that if a function is increasing, then it is one-to-one and hence has an inverse that is a function.

99. Graph the equation $y = 1/x^2$. Then test for symmetry with respect to the x-axis, the y-axis, the origin, and the line $y = x$.

100. The following formulas for the conversion between Fahrenheit and Celsius temperatures have been considered several times in the text:

$$C = \tfrac{5}{9}(F - 32), \qquad F = \tfrac{9}{5}C + 32.$$

Discuss the functions from the standpoint of inverses.

CHALLENGE

Graph the equation and its inverse. Then test for symmetry with respect to the x-axis, the y-axis, the origin, and the line $y = x$.

101. $|x| - |y| = 1$

102. $y = \frac{|x|}{x}$

5.2 Exponential Functions

OBJECTIVES

You should be able to:

1. Graph exponential equations and functions.
2. Graph exponential equations when x and y have been interchanged.
3. Solve problems involving applications of exponential functions and their graphs.

The following graph shows the years in which the cost of first-class postage was raised. A curve drawn along the graph would approximate the graph of an *exponential function*. We now consider such graphs. We will also study graphs and properties of *logarithmic functions*, which are inverses of the exponential functions. You will see that these new functions are rich in applications.

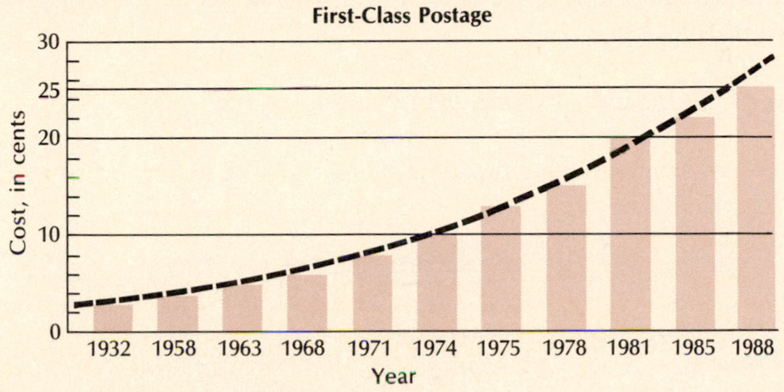

1 Graphing Exponential Functions

In Chapter 1, we gave meaning to exponential expressions with rational-number exponents such as

$$5^{1/4}, \quad 3^{-3/4}, \quad 7^{2.34}, \quad 8^{1.73}.$$

For example, $5^{1.73}$, or $5^{173/100}$, means to raise 5 to the 173rd power and then take the positive 100th root. We now give meaning to expressions with irrational exponents, such as

$$5^{\sqrt{3}}, \quad 7^{\pi}, \quad 9^{-\sqrt{2}}.$$

Consider $5^{\sqrt{3}}$. Let us think of rational numbers r close to $\sqrt{3}$ and look at 5^r. As r gets closer to $\sqrt{3}$, 5^r gets closer to some real number.

r closes in on $\sqrt{3}$.	5^r closes in on some real number p.
r	5^r
$1 < \sqrt{3} < 2$	$5 = 5^1 < p < 5^2 = 25$
$1.7 < \sqrt{3} < 1.8$	$15.426 = 5^{1.7} < p < 5^{1.8} = 18.119$
$1.73 < \sqrt{3} < 1.74$	$16.189 = 5^{1.73} < p < 5^{1.74} = 16.452$
$1.732 < \sqrt{3} < 1.733$	$16.241 = 5^{1.732} < p < 5^{1.733} = 16.267$

As r closes in on $\sqrt{3}$, 5^r closes in on some real number p. We define $5^{\sqrt{3}}$ to be the number p to seven decimal places:

$$5^{\sqrt{3}} \approx 16.2424508.$$

Any positive irrational exponent can be defined in a similar way. Negative irrational exponents are then defined in the same way as negative integer exponents. Thus the expression a^x has meaning for any real number x. The usual laws of exponents still hold, but we will not prove that here. We now define exponential functions.

DEFINITION

The function $f(x) = a^x$, where x is real and a is a positive constant different from 1, is called the *exponential function*, base a.

1. Graph: $y = f(x) = 3^x$.

a) Complete this table of solutions.

x	y, or $f(x)$
0	
1	
2	
3	
−1	
−2	
−3	

b) Plot the points from the table and connect them with a smooth curve.

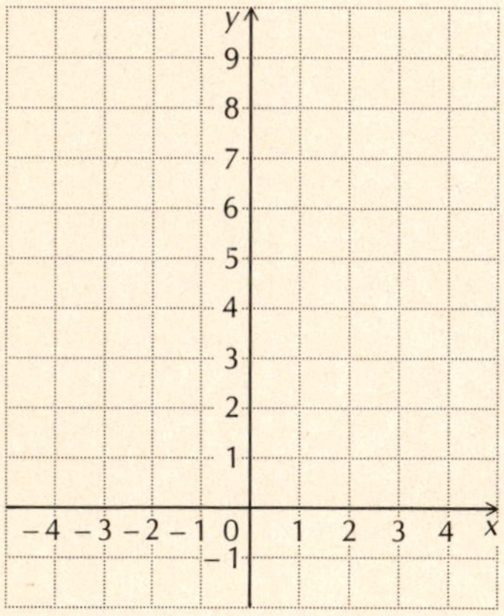

We restrict the **base** a to be positive to avoid the possibility of taking even roots of negative numbers—for example, the square root of -1, $(-1)^{1/2}$, which is not a real number. We restrict the base from being 1, because $f(x) = 1^x = 1$, and this function does not have an inverse. We will see that all other exponential functions, with the base restrictions, do have inverses.

The following are examples of exponential functions:

$$f(x) = 2^x, \qquad f(x) = \left(\tfrac{1}{2}\right)^x, \qquad f(x) = (0.04)^x.$$

Note that, in contrast to polynomial functions like $f(x) = x^2$ and $f(x) = x^3$, the variable in an exponential function is in the *exponent*. Let us consider graphs of exponential functions.

EXAMPLE 1 Graph the exponential function $y = f(x) = 2^x$.

Solution We compute some function values, thinking of y as $f(x)$, and list the results in a table.

$f(0) = 2^0 = 1,$

$f(1) = 2^1 = 2,$

$f(2) = 2^2 = 4$

$f(3) = 2^3 = 8,$

$f(-1) = 2^{-1} = \dfrac{1}{2^1} = \dfrac{1}{2},$

$f(-2) = 2^{-2} = \dfrac{1}{2^2} = \dfrac{1}{4},$

$f(-3) = 2^{-3} = \dfrac{1}{2^3} = \dfrac{1}{8}$

x	y, or $f(x)$
0	1
1	2
2	4,
3	8
−1	$\frac{1}{2}$
−2	$\frac{1}{4}$
−3	$\frac{1}{8}$

Next, we plot these points and connect them with a smooth curve. Be sure to plot enough points to determine how steeply the curve rises.

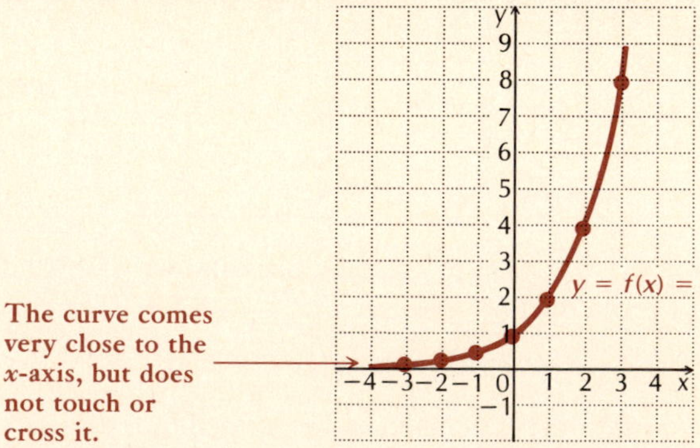

Note that as x increases, the function values increase indefinitely. As x decreases, the function values decrease, getting very close to 0. The x-axis is an **asymptote**, meaning that the curve comes very close to, but never touches the axis. ■

DO EXERCISE 1.

EXAMPLE 2 Graph the exponential function $y = f(x) = \left(\tfrac{1}{2}\right)^x$.

Solution We compute some function values, thinking of y as $f(x)$, and list the results in a table. Before we do this, note that

$$y = f(x) = \left(\tfrac{1}{2}\right)^x = (2^{-1})^x = 2^{-x}.$$

$f(0) = 2^{-0} = 1,$

$f(1) = 2^{-1} = \frac{1}{2^1} = \frac{1}{2},$

$f(2) = 2^{-2} = \frac{1}{2^2} = \frac{1}{4},$

$f(3) = 2^{-3} = \frac{1}{2^3} = \frac{1}{8},$

$f(-1) = 2^{-(-1)} = 2^1 = 2,$

$f(-2) = 2^{-(-2)} = 2^2 = 4,$

$f(-3) = 2^{-(-3)} = 2^3 = 8$

x	y, or $f(x)$
0	1
1	$\frac{1}{2}$
2	$\frac{1}{4}$
3	$\frac{1}{8}$
-1	2
-2	4
-3	8

We plot these points and draw the curve. Note that this graph is a reflection of the graph in Example 1 across the y-axis.

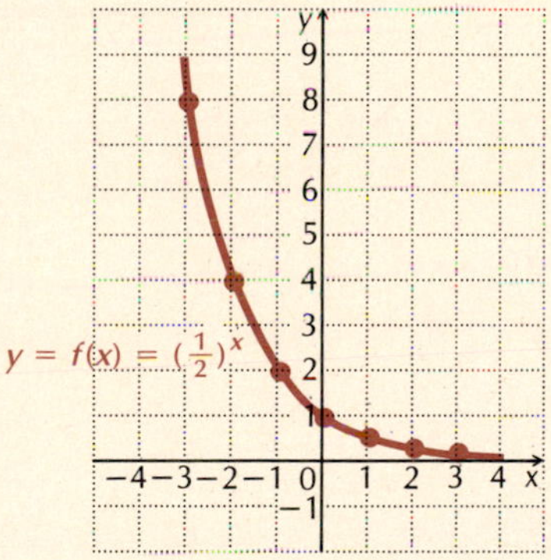

DO EXERCISE 2.

The preceding examples illustrate exponential functions with various bases. Let us list some of these characteristics. Keep in mind that the definition of an exponential function, $f(x) = a^x$, requires that the base be positive and different from 1. When $a = 1$, the graph is the line $y = 1$, which is not one-to-one and thus does not have an inverse.

THEOREM 2

1. **When $a > 1$, the function f given by $f(x) = a^x$ is a continuous, increasing, one-to-one function. The greater the value of a, the faster the function increases.**
2. **When $0 < a < 1$, the function f given by $f(x) = a^x$ is a continuous, decreasing, one-to-one function. The greater the value of a, the more slowly the function decreases.**

The y-intercept of every exponential function is the point (0, 1). The domain of each function is the set of real numbers. The range is the set of all nonnegative real numbers.

2. Graph: $y = \left(\frac{1}{3}\right)^x$.

a) Complete this table of solutions.

x	y, or $f(x)$
0	
1	
2	
3	
-1	
-2	
-3	

b) Plot the points from the table and connect them with a smooth curve.

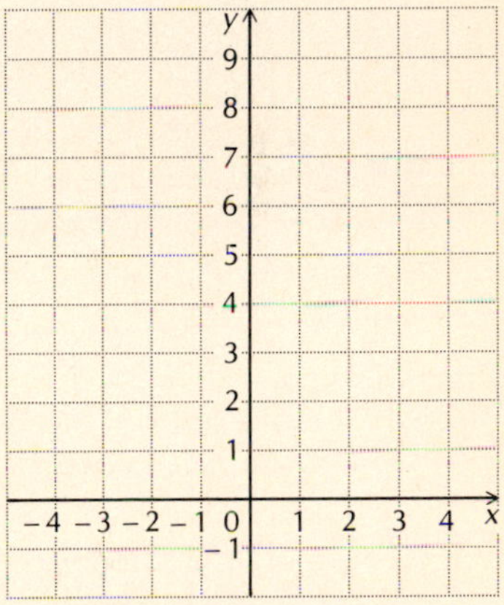

Graph.

3. $f(x) = 4^x$

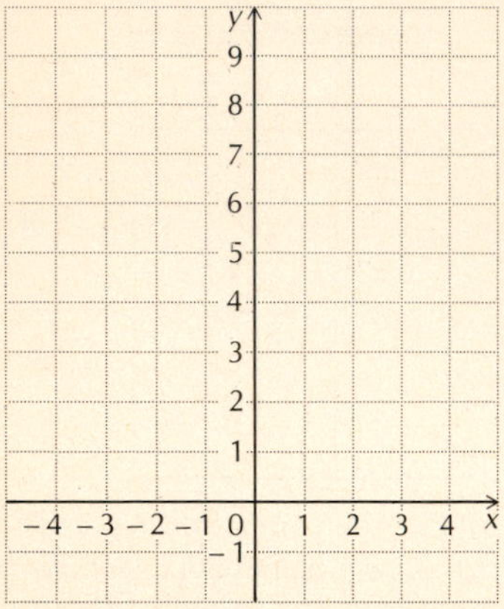

4. $f(x) = \left(\frac{1}{4}\right)^x$

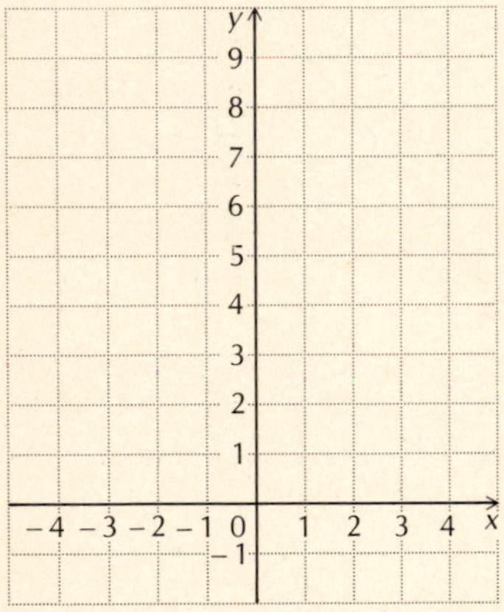

5. Graph: $y = 2^{x+2}$.

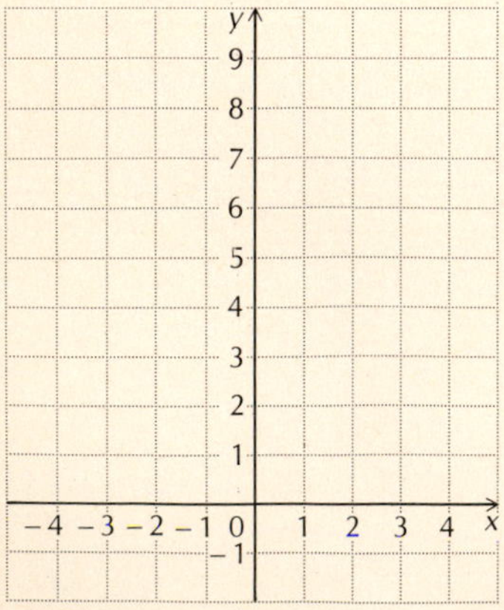

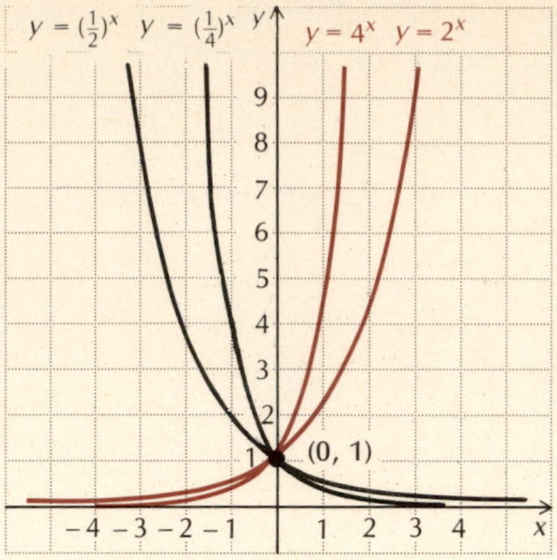

DO EXERCISES 3 AND 4.

EXAMPLE 3 Graph $y = f(x) = 2^{x-2}$.

Solution We construct a table of values. Then we plot the points and connect them with a smooth curve. Be sure to note that $x - 2$ is the *exponent*.

$f(0) = 2^{0-2} = 2^{-2} = \frac{1}{2^2} = \frac{1}{4},$

$f(1) = 2^{1-2} = 2^{-1} = \frac{1}{2^1} = \frac{1}{2},$

$f(2) = 2^{2-2} = 2^0 = 1,$

$f(3) = 2^{3-2} = 2^1 = 2,$

$f(4) = 2^{4-2} = 2^2 = 4,$

$f(-1) = 2^{-1-2} = 2^{-3} = \frac{1}{2^3} = \frac{1}{8},$

$f(-2) = 2^{-2-2} = \frac{1}{2^{-4}} = \frac{1}{2^4} = \frac{1}{16}$

x	y, or $f(x)$
0	$\frac{1}{4}$
1	$\frac{1}{2}$
2	1
3	2
4	4
-1	$\frac{1}{8}$
-2	$\frac{1}{16}$

The graph is a translation of the graph of $f(x) = 2^x$, to the right 2 units.

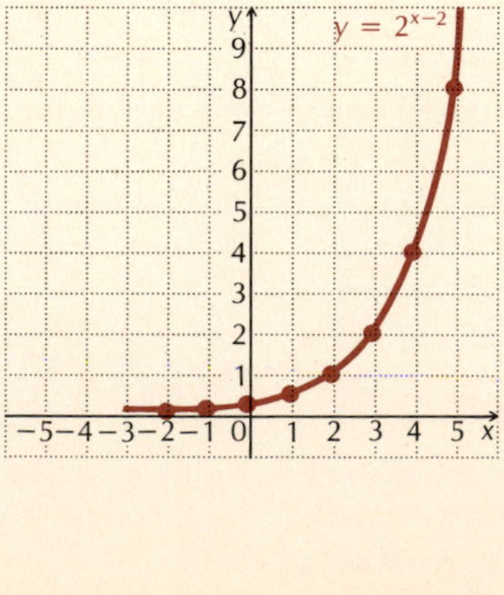

DO EXERCISE 5.

2 Graphs of Inverses of the Exponential Functions

We have noted that every exponential function $a > 0$ and $a \neq 1$ is one-to-one and therefore has an inverse that is a function. In the next section, we will consider these functions and give them names. For now, we draw graphs of these inverses by interchanging x and y.

EXAMPLE 4 Graph: $x = 2^y$.

Solution The inverse of $y = 2^x$ is $x = 2^y$. Note that x is alone on one side of the equation. We can find ordered pairs that are solutions more easily by picking values for y and then computing the x-values.

For $y = 0$, $x = 2^0 = 1$.
For $y = 1$, $x = 2^1 = 2$.
For $y = 2$, $x = 2^2 = 4$.
For $y = 3$, $x = 2^3 = 8$.
For $y = -1$, $x = 2^{-1} = \frac{1}{2^1} = \frac{1}{2}$.
For $y = -2$, $x = 2^{-2} = \frac{1}{2^2} = \frac{1}{4}$.
For $y = -3$, $x = 2^{-3} = \frac{1}{2^3} = \frac{1}{8}$.

x	y
1	0
2	1
4	2
8	3
$\frac{1}{2}$	-1
$\frac{1}{4}$	-2
$\frac{1}{8}$	-3

(1) Pick values for y.
(2) Compute values for x.

We plot the points and connect them with a smooth curve. Note that the curve does not touch or cross the y-axis.

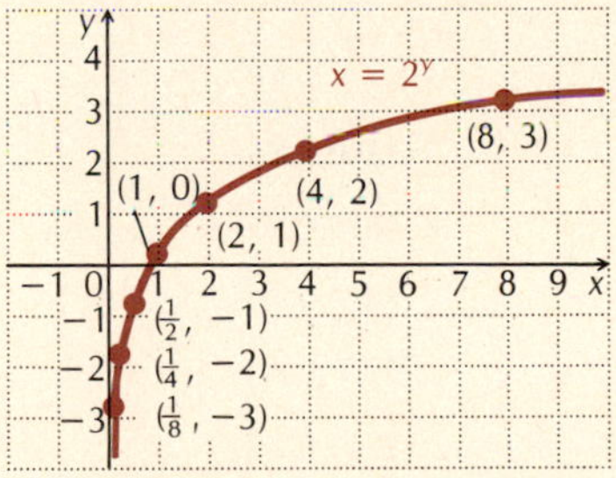

Note too that this curve looks just like the graph of $y = 2^x$, except that it is reflected across the line $y = x$, as we would expect for an inverse.

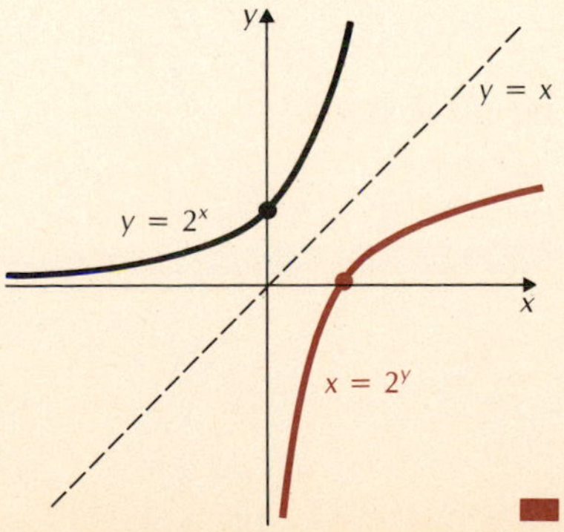

6. Graph: $x = 3^y$.

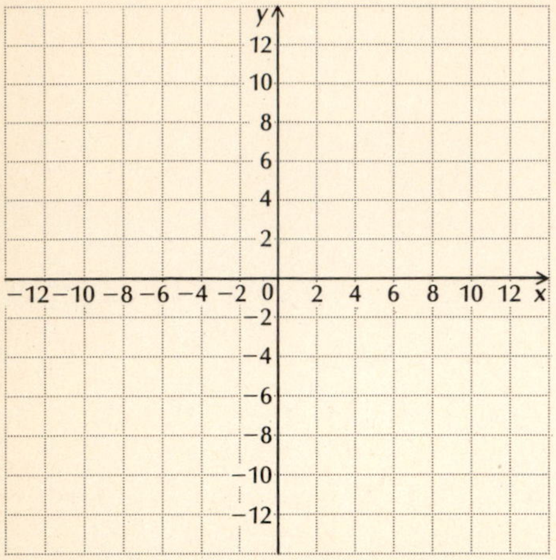

DO EXERCISE 6.

3 Applications of Exponential Functions

EXAMPLE 5 *Interest compounded annually.* The amount of money A that a principal P will be worth after t years at interest rate i, compounded annually, is given by the formula

$$A = P(1 + i)^t.$$

Suppose that \$100,000 is invested at 8% interest, compounded annually.

a) Find a function for the amount in the account after t years.
b) Find the amount of money in the account at $t = 0$, $t = 4$, $t = 8$, and $t = 10$.
c) Graph the function.

Solution

a) If $P = \$100{,}000$ and $i = 8\% = 0.08$, we can substitute these values and form the following function:

$$A(t) = \$100{,}000(1 + 0.08)^t = \$100{,}000(1.08)^t.$$

b) To find the function values, a calculator with a power key might be helpful.

$$A(0) = \$100{,}000(1.08)^0 = \$100{,}000(1) = \$100{,}000;$$
$$A(4) \approx \$100{,}000(1.08)^4 = \$100{,}000(1.36048896) \approx \$136{,}048.90;$$
$$A(8) \approx \$100{,}000(1.08)^8 = \$100{,}000(1.85093021) \approx \$185{,}093.02;$$
$$A(10) \approx \$100{,}000(1.08)^{10} = \$100{,}000(2.158924997) \approx \$215{,}892.50$$

c) We use the function values computed in part (b) and others if we wish, and draw the graph as follows. Note that the axes are scaled differently because of the large numbers and that t is restricted to nonnegative values, because negative time values have no meaning.

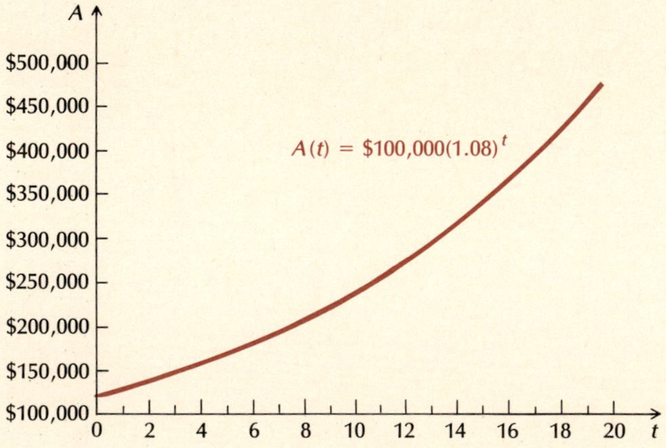

DO EXERCISE 7.

7. Suppose that \$80,000 is invested at 8% interest, compounded annually.

a) Find a function for the amount in the account after t years.

b) Find the amount of money in the account at $t = 0$, $t = 4$, $t = 8$, and $t = 10$.

c) Graph the function.

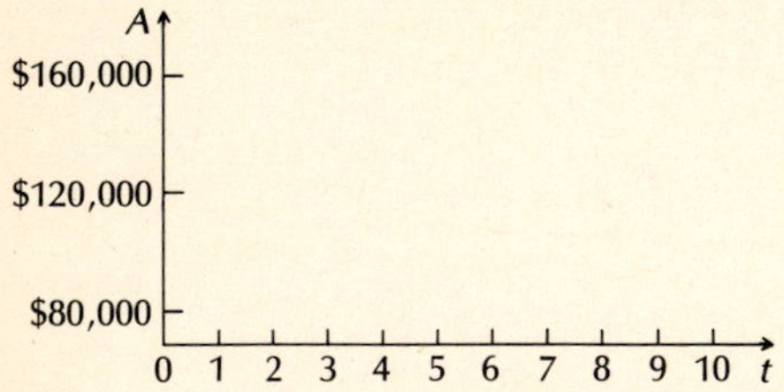

EXERCISE SET 5.2

1 Graph.

1. $y = f(x) = 2^x$
2. $y = f(x) = 3^x$
3. $y = 5^x$
4. $y = 6^x$
5. $y = 2^{x+1}$
6. $y = 2^{x-1}$
7. $y = 3^{x-2}$
8. $y = 3^{x+2}$
9. $y = 2^x - 3$
10. $y = 2^x + 1$
11. $y = 5^{x+3}$
12. $y = 6^{x-4}$

13. $y = \left(\frac{1}{2}\right)^x$ 14. $y = \left(\frac{1}{3}\right)^x$ 15. $y = \left(\frac{1}{5}\right)^x$ 16. $y = \left(\frac{1}{4}\right)^x$

17. $y = 2^{2x-1}$ 18. $y = 3^{4-x}$ 19. $y = 2^{x-1} - 3$ 20. $y = 2^{x+3} - 4$

2 Graph.

21. $x = 2^y$ 22. $x = 6^y$ 23. $x = \left(\frac{1}{2}\right)^y$ 24. $x = \left(\frac{1}{3}\right)^y$

25. $x = 5^y$ 26. $x = 3^y$ 27. $x = \left(\frac{2}{3}\right)^y$ 28. $x = \left(\frac{4}{3}\right)^y$

Graph both equations using the same set of axes.

29. $y = 2^x, x = 2^y$ 30. $y = 3^x, x = 3^y$ 31. $y = \left(\frac{1}{2}\right)^x, x = \left(\frac{1}{2}\right)^y$ 32. $y = \left(\frac{1}{4}\right)^x, x = \left(\frac{1}{4}\right)^y$

3

33. *Compact discs.* The number of compact discs purchased each year is increasing exponentially. The number N purchased, in millions, is given by

$$N(t) = 7.5(6)^{0.5t},$$

where $t = 0$ corresponds to 1985, $t = 1$ corresponds to 1986, and so on, t being the number of years after 1985.

a) Find the number of compact discs sold in 1986, 1987, 1990, 1995, and 2000.
b) Graph the function.

34. *Growth of bacteria* Escherichi coli. The bacteria *Escherichi coli* is commonly found in the bladder of human beings. Suppose 3000 of the bacteria are present at time $t = 0$. Then t minutes later, the number of bacteria present will be

$$N(t) = 3000(2)^{t/20}.$$

a) How many bacteria will be present after 10 min? 20 min? 30 min? 40 min? 60 min?
b) Graph the function.

35. *Interest compounded annually.* Suppose that $50,000 is invested at 9% interest, compounded annually.

a) Find a function for the amount in the account after t years.
b) Find the amount of money in the account at $t = 0$, $t = 4$, $t = 8$, and $t = 10$.
c) Graph the function.

36. *Recycling aluminum cans.* It is known that $\frac{1}{4}$ of all aluminum cans distributed will be recycled each year. A beverage company distributes 250,000 cans. The number still in use after time t, in years, is given by the function

$$N(t) = 250{,}000\left(\frac{1}{4}\right)^t.$$

a) How many cans are still in use after 0 years? 1 year? 4 years? 10 years?
b) Graph the function.

37. *Salvage value.* An office machine is purchased for $5200. Its value each year is about 75% of the value the preceding year. Its value after t years is given by the exponential function

$$V(t) = \$5200(0.75)^t.$$

a) Find the value of the machine after 0 years, 1 year, 2 years, 5 years, and 10 years.
b) Graph the function.

38. *Turkey consumption.* The amount of turkey consumed by each person in this country is increasing exponentially. Assuming $t = 0$ corresponds to 1937, the amount of turkey, in pounds per person, consumed t years after 1937 is given by the function

$$N(t) = 2.3(3)^{0.033t}.$$

a) How much turkey was consumed per person in 1940, 1950, 1980, and 1985?
b) How much will be consumed per person in 2007?
c) Graph the function.

SYNTHESIS

39. Approximate each of the following to six decimal places.

a) 7^3 b) $7^{3.1}$ c) $7^{3.14}$ d) $7^{3.141}$ e) $7^{3.1415}$ f) $7^{3.14159}$

Determine which of the two numbers is larger.

40. 7^π or π^7 41. $\pi^{3.2}$ or $\pi^{2.3}$ 42. $\sqrt{4^3}$ or $4^{\sqrt{3}}$

Graph. You will find a calculator with a power key $\boxed{y^x}$ most helpful.

43. $f(x) = (2.7)^x$ 44. $f(x) = (5.8)^x$ 45. $g(x) = (0.745)^x$ 46. $g(x) = (0.8)^x$

Graph.

47. $y = 2^x + 2^{-x}$ 48. $y = \left(\frac{1}{2}\right)^x - 1$ 49. $y = 3^x + 3^{-x}$ 50. $f(x) = 2^{|x|}$

51. $y = 2^{-(x-1)}$ 52. $y = |2^x - 1|$ 53. $y = |2^x - 2|$ 54. $g(x) = 2^{-|x|}$

Graph both equations using the same set of axes.

55. $y = 3^{-(x-1)}$, $x = 3^{-(y-1)}$

56. $y = 1^x$, $x = 1^y$

Solve graphically.

57. $2^x > 1$

58. $3^x \leq 1$

59. *Typing speed.* A person studies typing one semester in college. After he has studied for t hours, his speed, in words per minute, is given by

$$S(t) = 200[1 - (0.86)^t].$$

a) What is the speed of the typist after studying for 10 hr? 20 hr? 40 hr? 100 hr?

b) Graph the function.

CHALLENGE

Graph.

60. $y = 2^{-x^2}$

61. $y = 3^{-(x+1)^2}$

62. $y = |2^{x^2} - 8|$

OBJECTIVES

You should be able to:

1. Graph logarithmic functions.
2. Convert from exponential equations to logarithmic equations and from logarithmic equations to exponential equations.
3. Solve certain logarithmic equations.

5.3 Logarithmic Functions

We now consider a kind of function called a *logarithm function*, or *logarithmic function*. Such functions have many applications to problem solving.

1 Graphs of Logarithmic Functions

Consider the exponential function $f(x) = 2^x$. Does this function have an inverse that is a function? We see from the graph that this function is one-to-one and does have an inverse f^{-1} that is a function.

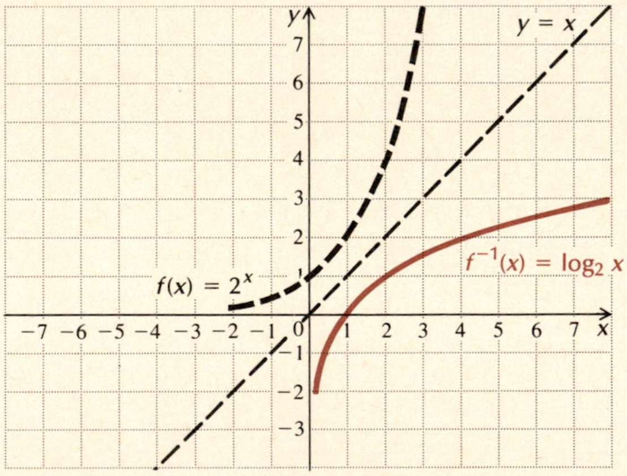

Consider the input 3 and the original function

$$f(x) = 2^x.$$

Then

$$f(3) = 2^3 = 8.$$

This also tells us that $f^{-1}(8) = 3$. If we did not know this, then to find $f^{-1}(8)$, we would be looking for an x such that

$$8 = 2^x.$$

Now we can probably reason that x is 3, but suppose we wanted to find $f^{-1}(13)$. The input for the inverse function is the number 13. We would be looking for an x such that

$$13 = 2^x.$$

Since $2^3 = 8$ and $2^4 = 16$, it seems reasonable that the number x that we are seeking is somewhere between 3 and 4. We know that the inverse exists, but we do not have a way to name $f^{-1}(13)$ as yet. Mathematicians have invented a name for the *inverse* of $f(x) = 2^x$. It is the *logarithm function, base 2,* denoted

$$f^{-1}(x) = \log_2 x.$$

We read $\log_2 x$ as "the logarithm, base 2, of x." Now $\log_2 8$ is the power to which we raise 2 to get 8. Thus, $\log_2 8 = 3$. Similarly, $\log_2 13$ is the power to which we raise 2 to get 13. We have no simpler way to write this for now.

For any exponential function $f(x) = a^x$, the inverse is called a **logarithmic function, base a.** The graph of the inverse can, of course, be obtained by reflecting the graph of $y = a^x$ across the line $y = x$, to obtain $x = a^y$, by interchanging x and y. Then $x = a^y$ is equivalent to $y = \log_a x$.

The inverse of $f(x) = a^x$ is given by

$$f^{-1}(x) = \log_a x.$$

We read $\log_a x$ as the "logarithm, base a, of x."

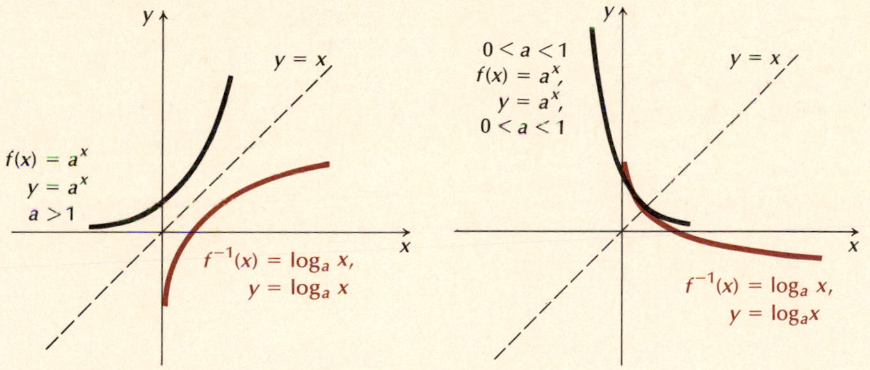

To be consistent with our previous use of function notation, we should probably write $f^{-1}(x) = \log_a (x)$, but unless the parentheses are actually needed for clarity, they are ordinarily omitted. Almost always, we use a number a that is greater than 1 for a logarithmic base.

DEFINITION Logarithms

We define $y = \log_a x$ as that number y such that $x = a^y$, where $x > 0$ and a is a positive constant other than 1.

It is helpful in dealing with logarithmic functions to remember that the logarithm of a number is an *exponent*. It is the exponent y in $x = a^y$. You might also think to yourself, "the logarithm, base a, of a number x is the power to which a must be raised in order to get x."

A logarithm is an exponent.

The following is a comparison of exponential and logarithmic functions.

1. Graph $y = f(x) = \log_3 x$. Determine the domain and the range.

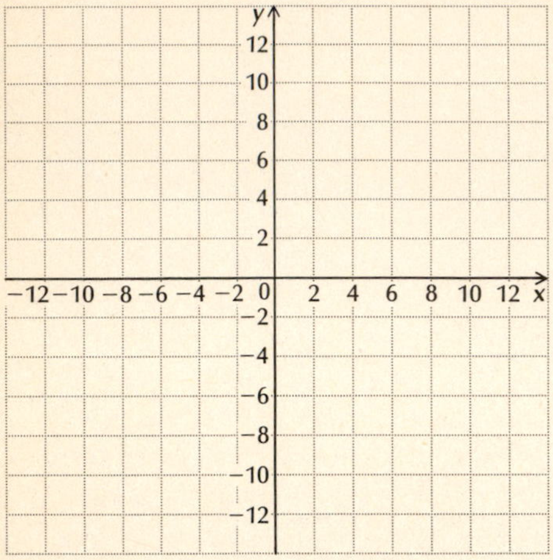

Exponential Function	Logarithmic Function
$y = a^x$ $f(x) = a^x$ $a > 0, a \neq 1$ Domain = The set of real numbers Range = The set of positive real numbers	$x = a^y$ $f^{-1}(x) = \log_a x$ $a > 0, a \neq 1$ Range = The set of real numbers Domain = The set of positive real numbers

Why do we exclude 1 from being a logarithmic base? If we included it, we would be considering $x = 1^y = 1$. The graph of this equation is a vertical line and is not a function. It does not pass the vertical-line test. We also exclude it from being a logarithmic base, because we exclude it from being an exponential base.

EXAMPLE 1 Graph: $y = f(x) = \log_5 x$. Determine the domain and the range.

Solution The equation $y = \log_5 x$ is equivalent to $5^y = x$. We can find ordered pairs that are solutions by picking values for y and computing the x-values.

For $y = 0$, $x = 5^0 = 1$.
For $y = 1$, $x = 5^1 = 5$.
For $y = 2$, $x = 5^2 = 25$.
For $y = 3$, $x = 5^3 = 125$.
For $y = -1$, $x = 5^{-1} = \frac{1}{5}$.
For $y = -2$, $x = 5^{-2} = \frac{1}{25}$.

x, or 5^y	y
1	0
5	1
25	2
125	3
$\frac{1}{5}$	−1
$\frac{1}{25}$	−2

(1) Select y.
(2) Compute x.

We plot the ordered pairs and connect them with a smooth curve. The graph of $y = 5^x$ has been shown only for reference.

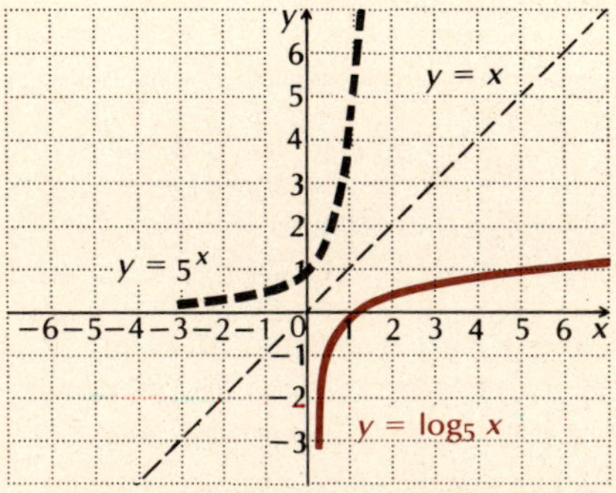

The graph of $f(x) = \log_a x$, for any a, has the x-intercept $(1, 0)$. The domain is the set of positive real numbers. The range is the set of all real numbers.

DO EXERCISE 1.

2 Converting Between Exponential and Logarithmic Equations

We use the definition of logarithms to convert from exponential to logarithmic equations.

> $y = \log_a x$ is equivalent to $a^y = x$.

Be sure to memorize this relationship! It is probably the most important definition in the chapter. Many times this definition will be a justification for a proof or a procedure that we are considering.

EXAMPLES Convert each of the following to a logarithmic equation.

2. $8 = 2^x \rightarrow x = \log_2 8$ — The exponent is the logarithm. The base remains the base.
3. $y^{-1} = 4 \rightarrow -1 = \log_y 4$
4. $a^b = c \rightarrow b = \log_a c$

DO EXERCISES 2–5.

We also use the definition of logarithms to convert from logarithmic to exponential equations.

EXAMPLES Convert each of the following to an exponential equation.

5. $y = \log_3 5 \rightarrow 3^y = 5$ — The logarithm is the exponent. The base does not change.
6. $-2 = \log_a 7 \rightarrow a^{-2} = 7$
7. $a = \log_b d \rightarrow b^a = d$

DO EXERCISES 6–9.

3 Solving Certain Logarithmic Equations

Certain equations involving logarithms can be solved by first converting to exponential equations. We will solve more complicated equations later.

EXAMPLE 8 Solve: $\log_3 x = -2$.

Solution

$$\log_3 x = -2$$
$$3^{-2} = x \qquad \text{Converting to an exponential equation}$$
$$\tfrac{1}{9} = x \qquad \text{Computing } 3^{-2}$$

Check: $\log_3 \frac{1}{9}$ is the exponent to which we raise 3 to get $\frac{1}{9}$. Since $3^{-2} = \frac{1}{9}$, we know that $\frac{1}{9}$ checks and is the solution. The solution set is $\{\frac{1}{9}\}$.

EXAMPLE 9 Solve: $\log_x 25 = 2$.

Solution

$$\log_x 25 = 2$$
$$x^2 = 25 \qquad \text{Converting to an exponential equation}$$
$$x = 5 \quad \text{or} \quad x = -5$$

Convert to a logarithmic equation.

2. $6^0 = 1$
3. $10^{-3} = 0.001$
4. $16^{0.25} = 2$
5. $m^T = P$

Convert to an exponential equation.

6. $\log_2 32 = 5$
7. $\log_{10} 1000 = 3$
8. $\log_a Q = 7$
9. $\log_t M = x$

Solve.

10. $\log_{10} x = 4$

11. $\log_x 81 = 4$

12. $\log_2 x = -2$

Check: $\log_5 25 = 2$ because $5^2 = 25$. Thus, 5 is a solution. Since all logarithmic bases must be positive, $\log_{-5} 25$ is not defined. Therefore, -5 is not a solution. Logarithmic bases must be positive because logarithms are defined in terms of exponential functions, which are defined only for positive bases. ■

DO EXERCISES 10–12.

Solving an equation like $\log_b a = x$ amounts to finding the logarithm, base b, of the number a. You have done this before in graphing logarithmic functions. To think of finding logarithms as solving equations may help in some cases.

EXAMPLE 10 Find $\log_{10} 10{,}000$.

Solution

Method 1. Let $\log_{10} 10{,}000 = x$. Then

$$10^x = 10{,}000 \qquad \text{Converting to an exponential equation}$$
$$10^x = 10^4$$
$$x = 4. \qquad \text{The exponents are the same.}$$

Therefore, $\log_{10} 10{,}000 = 4$.

Method 2. Think of the meaning of $\log_{10} 10{,}000$. It is the exponent to which you raise 10 to get 10,000. That exponent is 4. Therefore, $\log_{10} 10{,}000 = 4$. ■

EXAMPLE 11 Find $\log_{10} 0.01$.

Solution

Method 1. Let $\log_{10} 0.01 = x$. Then

$$10^x = 0.01$$
$$10^x = \frac{1}{100}$$
$$10^x = 10^{-2}$$
$$x = -2.$$

Therefore, $\log_{10} 0.01 = -2$.

Method 2. We know that $\log_{10} 0.01$ is the exponent to which you raise 10 to get 0.01. Since $0.01 = 1/100 = 1/10^2$, it follows that the exponent is -2. Therefore, $\log_{10} 0.01 = -2$. ■

Here are some other examples. Think out mentally how they can be found and compare them.

$$\log_{10} 1000 = 3, \qquad \log_{64} 64 = 1,$$
$$\log_{10} 100 = 2, \qquad \log_8 64 = 2,$$
$$\log_{10} 10 = 1, \qquad \log_4 64 = 3,$$
$$\log_{10} 1 = 0, \qquad \log_2 64 = 6.$$
$$\log_{10} 0.1 = -1,$$
$$\log_{10} 0.01 = -2;$$

EXAMPLE 12 Find $\log_6 1$.

Solution

Method 1. Let $\log_6 1 = x$. Then

$$6^x = 1 \quad \text{Converting to an exponential equation}$$
$$6^x = 6^0 \quad \text{Renaming 1 as } 6^0$$
$$x = 0.$$

Therefore, $\log_6 1 = 0$.

Method 2. We know that $\log_6 1$ is the exponent to which 6 is raised to get 1. That exponent is 0. Therefore, $\log_6 1 = 0$. ■

DO EXERCISES 13–15.

Example 12 illustrates an important property of logarithms.

THEOREM 3

For any base a, $a > 0$, $a \neq 1$,

$$\log_a 1 = 0.$$

The logarithm, base a, of 1 is always 0.

The proof follows from the fact that $a^0 = 1$. This is equivalent to the logarithmic equation $\log_a 1 = 0$.

Another property follows similarly. We know that $a^1 = a$ for any real number a. In particular, this holds for any positive number a. This is equivalent to the logarithmic equation $\log_a a = 1$.

THEOREM 4

For any base a, $a > 0$, $a \neq 1$,

$$\log_a a = 1.$$

DO EXERCISES 16–19.

Find each of the following.

13. $\log_{10} 100{,}000$

14. $\log_{10} 0.0001$

15. $\log_7 1$

Simplify.

16. $\log_7 1$

17. $\log_4 4$

18. $\log_m m$

19. $\log_m 1$

EXERCISE SET 5.3

1 Graph.

1. $y = \log_3 x$
2. $y = \log_6 x$
3. $y = \log_{10} x$
4. $y = \log_2 x$
5. $f(x) = \log_4 x$
6. $f(x) = \log_5 x$
7. $f(x) = \log_{1/2} x$
8. $f(x) = \log_{2.5} x$
9. $f(x) = \log_2 (x + 3)$
10. $f(x) = \log_3 (x - 2)$

Graph each function using the same set of axes.

11. $f(x) = 3^x$, $f^{-1}(x) = \log_3 x$
12. $f(x) = 4^x$, $f^{-1}(x) = \log_4 x$

2 Convert to a logarithmic equation.

13. $10^3 = 1000$
14. $10^2 = 100$
15. $5^{-3} = \frac{1}{125}$
16. $4^{-5} = \frac{1}{1024}$
17. $8^{1/3} = 2$
18. $16^{1/4} = 2$
19. $10^{0.3010} = 2$
20. $10^{0.4771} = 3$
21. $e^3 = t$
22. $p^k = 3$
23. $Q^t = x$
24. $p^m = V$
25. $e^3 = 20.0855$
26. $e^2 = 7.3891$
27. $e^{-1} = 0.3679$
28. $e^{-6} = 0.00247$

Convert to an exponential equation.

29. $t = \log_4 7$
30. $h = \log_6 29$
31. $\log_2 32 = 5$
32. $\log_5 5 = 1$
33. $\log_{10} 0.1 = -1$
34. $\log_{10} 0.01 = -2$
35. $\log_{10} 7 = 0.845$
36. $\log_{10} 3 = 0.4771$
37. $\log_e 30 = 3.4012$
38. $\log_e 10 = 2.3036$
39. $\log_t Q = k$
40. $\log_m P = a$
41. $\log_e 0.38 = -0.9676$
42. $\log_e 0.906 = -0.0987$
43. $\log_r M = -x$
44. $\log_c W = -w$

3 Solve.

45. $\log_{10} x = 3$
46. $\log_2 16 = x$
47. $\log_3 3 = x$
48. $\log_5 \frac{1}{25} = x$
49. $\log_3 x = 2$
50. $\log_4 x = 3$
51. $\log_x 16 = 2$
52. $\log_x 64 = 3$
53. $\log_2 x = -1$
54. $\log_3 x = -2$
55. $\log_8 x = \frac{1}{3}$
56. $\log_{32} x = \frac{1}{5}$

Find each of the following.

57. $\log_{10} 1000$
58. $\log_{10} 10{,}000{,}000$
59. $\log_{10} 0.1$
60. $\log_{10} 0.001$
61. $\log_{10} 1$
62. $\log_{10} 10$
63. $\log_5 625$
64. $\log_2 64$
65. $\log_5 \frac{1}{25}$
66. $\log_2 \frac{1}{16}$
67. $\log_3 1$
68. $\log_8 8$
69. $\log_e 1$
70. $\log_e e$
71. $\log_{81} 9$
72. $\log_8 2$
73. $\log_e e^5$
74. $\log_e e^{-2}$
75. $\log_{10} 10^m$
76. $\log_5 5^t$

SYNTHESIS

77. Graph both equations using the same set of axes.

$$y = \left(\frac{2}{3}\right)^x, \; y = \log_{2/3} x$$

Graph.

78. $y = |\log_2 x|$
79. $y = \log_2 |x|$

Solve.

80. $|\log_3 x| = 3$
81. $\log_{125} x = \frac{2}{3}$
82. $\log_\pi \pi^4 = x$
83. $\log_{\sqrt{5}} x = -3$
84. $\log_b b^{2x^2} = x$
85. $\log_4 (3x - 2) = 2$
86. $\log_8 (2x - 3) = -1$
87. $\log_x \sqrt[5]{36} = \frac{1}{10}$
88. $\log_{10} (x^2 + 21x) = 2$

Simplify.

89. $\log_{1/4} \frac{1}{64}$
90. $\log_{81} 3 \cdot \log_3 81$
91. $\log_{10} (\log_4 (\log_3 81))$
92. $\log_2 (\log_2 (\log_4 256))$
93. $\log_{\sqrt{3}} \frac{1}{81}$
94. $\log_{1/5} 25$

What is the domain of the function?

95. $f(x) = 3^x$
96. $f(x) = \log_{10} x$
97. $f(x) = \log_a x^2$
98. $f(x) = \log_4 x^3$
99. $f(x) = \log_{10} (3x - 4)$
100. $f(x) = \log_5 |x|$
101. $f(x) = \log_6 (x^2 - 9)$

Solve by graphing.

102. $\log_2 x < 0$
103. $\log_2 x \geq 4$
104. $\log_2 (x - 3) \leq 5$

5.4 Properties of Logarithmic Functions

OBJECTIVES

You should be able to:

1. Express the logarithm of a product as a sum of logarithms, and conversely.
2. Express the logarithm of a power as a product.
3. Express the logarithm of a quotient as a difference of logarithms, and conversely.
4. Convert from logarithms of products, quotients, and powers to expressions in terms of individual logarithms, and conversely.
5. Simplify expressions of the type $\log_a a^x$ and $a^{\log_a x}$.

Logarithmic functions are important in many applications and in more advanced mathematics. We now establish some basic properties that are fundamental to the use of the functions.

1 Logarithms of Products

One of the interesting properties special to logarithmic functions is the following.

PROPERTY 1 The Product Rule

For any positive numbers M and N,

$$\log_a M \cdot N = \log_a M + \log_a N.$$

(The logarithm of a product is the sum of the logarithms of the factors. The number a can be any logarithmic base.)

Proof. Let $\log_a M = x$ and $\log_a N = y$. Converting to exponential equations, we have $a^x = M$ and $a^y = N$. Then we multiply the latter two equations, to obtain

$$M \cdot N = a^x \cdot a^y$$
$$= a^{x+y}.$$

Converting back to a logarithmic equation, we get

$$\log_a M \cdot N = x + y.$$

Remembering what x and y represent, it follows that

$$\log_a M \cdot N = \log_a M + \log_a N,$$

which was to be shown.

The use of the rule $a^x \cdot a^y = a^{x+y}$, which is the product rule for exponents, gives rise to the name product rule for logarithms. Recall that a logarithm is an exponent. To find the product of exponential expressions with the same base, we add the exponents. To find the logarithm of a product, which is an exponent, we add logarithms of the factors, which is a sum of exponents.

EXAMPLE 1 Express as a sum of logarithms: $\log_3 (9 \cdot 27)$.

Solution

$$\log_3 (9 \cdot 27) = \log_3 9 + \log_3 27 \quad \text{By Property 1}$$
$$= 2 + 3 = 5$$

EXAMPLE 2 Express as a single logarithm: $\log_{10} 0.01 + \log_{10} 3947$.

Solution

$$\log_{10} 0.01 + \log_{10} 3947 = \log_{10} (0.01 \times 3947) \quad \text{By Property 1}$$
$$= \log_{10} 39.47$$

Express as a sum of logarithms.

1. $\log_5 (25 \cdot 5)$

2. $\log_b PQ$

Express as a single logarithm.

3. $\log_3 7 + \log_3 5$

4. $\log_a C + \log_a A + \log_a B + \log_a I + \log_a N$

Express as a product.

5. $\log_7 4^5$

6. $\log_a \sqrt{5}$

CAUTION! The logarithm of a product is *not* the product of the logarithms—that is,

$$\log_a MN \neq (\log_a M)(\log_a N).$$

DO EXERCISES 1–4.

2 Logarithms of Powers

The second basic property is as follows.

PROPERTY 2 The Power Rule

For any positive number M and any real number p,

$$\log_a M^p = p \cdot \log_a M.$$

(The logarithm of a power of M is the exponent times the logarithm of M. The number a can be any logarithmic base.)

Proof. Let $x = \log_a M$. Then we convert to an exponential equation, to get $a^x = M$. Raising both sides to the pth power, we obtain

$$(a^x)^p = M^p, \quad or \quad a^{xp} = M^p.$$

Converting back to a logarithmic equation, we get

$$\log_a M^p = xp.$$

But $x = \log_a M$, so

$$\log_a M^p = (\log_a M)p = p \cdot \log_a M,$$

which was to be shown.

We call this the power rule for logarithms because of the analogous power rule for exponents.

EXAMPLES Express as a product.

3. $\log_a 11^{-3} = -3 \log_a 11$ By Property 2

4. $\log_a \sqrt[4]{7} = \log_a 7^{1/4}$ Writing exponential notation

$= \frac{1}{4} \log_a 7$ By Property 2

DO EXERCISES 5 AND 6.

3 Logarithms of Quotients

Here is the third basic property.

PROPERTY 3 The Quotient Rule

For any positive numbers M and N,

$$\log_a \frac{M}{N} = \log_a M - \log_a N.$$

(The logarithm of a quotient is the logarithm of the numerator minus the logarithm of the denominator. The number a can be any logarithmic base.)

Proof. The proof makes use of Property 1 and Property 2:

$$\log_a \frac{M}{N} = \log_a MN^{-1}$$
$$= \log_a M + \log_a N^{-1} \quad \text{Property 1}$$
$$= \log_a M + (-1)\log_a N \quad \text{Property 2}$$
$$= \log_a M - \log_a N.$$

EXAMPLE 5 Express as a difference of logarithms: $\log_t \frac{8}{Q}$.

Solution

$$\log_t \frac{8}{Q} = \log_t 8 - \log_t Q \quad \text{By Property 3}$$

EXAMPLE 6 Express as a single logarithm: $\log_b 54 - \log_b 27$.

Solution

$$\log_b 54 - \log_b 27 = \log_b \frac{54}{27} \quad \text{By Property 3}$$
$$= \log_b 2$$

EXAMPLE 7 Express as a single logarithm: $\log_{10} 10{,}000 - \log_{10} 100$.

Solution

$$\log_{10} 10{,}000 - \log_{10} 100 = \log_{10} \frac{10{,}000}{100} = \log_{10} 100$$

CAUTION! The logarithm of a quotient is *not* the quotient of the logarithms. That is,

$$\log_a \frac{M}{N} \neq \frac{(\log_a M)}{(\log_a N)}.$$

DO EXERCISES 7 AND 8.

4 Using the Properties Together

EXAMPLES Express in terms of logarithms.

8. $$\log_a \frac{p^3q^2}{z^4} = \log_a (p^3q^2) - \log_a z^4 \quad \text{Using Property 3}$$
$$= \log_a p^3 + \log_a q^2 - \log_a z^4 \quad \text{Using Property 1}$$
$$= 3\log_a p + 2\log_a q - 4\log_a z \quad \text{Using Property 2}$$

9. $$\log_a \sqrt[4]{\frac{ab}{c^3}} = \log_a \left(\frac{ab}{c^3}\right)^{1/4} \quad \text{Writing exponential notation}$$
$$= \frac{1}{4} \cdot \log_a \frac{ab}{c^3} \quad \text{Using Property 2}$$
$$= \frac{1}{4}(\log_a ab - \log_a c^3) \quad \text{Using Property 3}$$
$$= \frac{1}{4}(\log_a a + \log_a b - 3\log_a c) \quad \text{Using Properties 1 and 2}$$
$$= \frac{1}{4}(1 + \log_a b - 3\log_a c) \quad \log_a a = 1$$

7. Express as a difference of logarithms:

$$\log_b \frac{P}{Q}.$$

8. Express as a single logarithm:

$$\log_2 x - \log_2 25.$$

Express in terms of logarithms.

9. $\log_a \sqrt{\dfrac{z^3}{xy}}$

10. $\log_a \dfrac{x^2}{y^3 a}$

11. $\log_m \dfrac{a^3 b^4}{m^5 n^9}$

Express as a single logarithm.

12. $5 \log_a x - \log_a y + \frac{1}{4} \log_a z$

13. $\log_a \dfrac{\sqrt{x}}{b} - \log_a \sqrt{bx}$

10. $\log_b \dfrac{ab^5}{m^3 n^4}$

$= \log_b ab^5 - \log_b m^3 n^4$ Using Property 3

$= (\log_b a + \log_b b^5) - (\log_b m^3 + \log_b n^4)$ Using Property 1

$= \log_b a + \log_b b^5 - \log_b m^3 - \log_b n^4$ Removing parentheses

$= \log_b a + 5 \log_b b - 3 \log_b m - 4 \log_b n$ Using Property 2

$= \log_b a + 5 - 3 \log_b m - 4 \log_b n$ $\log_b b = 1$ ■

DO EXERCISES 9–11.

EXAMPLES

11. $\frac{1}{2} \log_a x - 7 \log_a y + \log_a z = \log_a x^{1/2} - \log_a y^7 + \log_a z$ Using Property 2

$= \log_a \dfrac{\sqrt{x}}{y^7} + \log_a z$ Using Property 3

$= \log_a \dfrac{z\sqrt{x}}{y^7}$ Using Property 1

12. $\log_a \dfrac{b}{\sqrt{x}} + \log_a \sqrt{bx}$

$= \log_a b - \log_a \sqrt{x} + \log_a \sqrt{bx}$ Using Property 3

$= \log_a b - \frac{1}{2} \log_a x + \frac{1}{2} \log_a (bx)$ Using Property 2

$= \log_a b - \frac{1}{2} \log_a x + \frac{1}{2} (\log_a b + \log_a x)$ Using Property 1

$= \log_a b - \frac{1}{2} \log_a x + \frac{1}{2} \log_a b + \frac{1}{2} \log_a x$

$= \frac{3}{2} \log_a b$ Collecting like terms

$= \log_a b^{3/2}$ Using Property 2

Example 12 could also be done as follows:

$$\log_a \frac{b}{\sqrt{x}} + \log_a \sqrt{bx} = \log_a \frac{b}{\sqrt{x}} \sqrt{bx} = \log_a b\sqrt{b} = \log_a b^{3/2}$$ ■

DO EXERCISES 12 AND 13.

EXAMPLES Given

$$\log_a 2 = 0.301,$$
$$\log_a 3 = 0.477,$$

find each of the following.

13. $\log_a 6$ $\log_a 6 = \log_a (2 \cdot 3) = \log_a 2 + \log_a 3$ Property 1
$= 0.301 + 0.477$
$= 0.778$

14. $\log_a \frac{2}{3}$ $\log_a \frac{2}{3} = \log_a 2 - \log_a 3$ Property 3
$= 0.301 - 0.477$
$= -0.176$

15. $\log_a 81$ $\log_a 81 = \log_a 3^4 = 4 \log_a 3$ Property 2
$= 4(0.477)$
$= 1.908$

16. $\log_a \frac{1}{3}$ $\quad \log_a \frac{1}{3} = \log_a 1 - \log_a 3$ Property 3
$= 0 - 0.477$
$= -0.477$

17. $\log_a \sqrt{a}$ $\quad \log_a \sqrt{a} = \log_a a^{1/2}$
$= \frac{1}{2} \log_a a$ Property 2
$= \frac{1}{2} \cdot 1$
$= \frac{1}{2}$

18. $\log_a 5$ No way to find using these properties and the given information.
$(\log_a 5 \neq \log_a 2 + \log_a 3)$

19. $\dfrac{\log_a 3}{\log_a 2}$ $\quad \dfrac{\log_a 3}{\log_a 2} \approx \dfrac{0.477}{0.301} \approx 1.58.$
We simply divided, not using any of the properties.

DO EXERCISES 14–21.

5 Simplifying Expressions $\log_a a^x$ and $a^{\log_a x}$

We have two final properties to consider.

PROPERTY 4

For any base a and any real number x,

$$\log_a a^x = x.$$

The logarithm, base a, of a to a power is the power.

Proof. The proof involves Property 2 and the fact that $\log_a a = 1$:

$$\log_a a^x = x(\log_a a) \quad \text{Using Property 2}$$
$$= x \cdot 1 \quad \text{Using } \log_a a = 1$$
$$= x.$$

If you forget Property 4, you can apply Property 2 and the fact that $\log_a a = 1$.

EXAMPLES Simplify.

20. $\log_a a^7 = 7$
21. $\log_{10} 10^{8.6} = 8.6$
22. $\log_e e^{-t} = -t$

DO EXERCISES 22–24.

PROPERTY 5

For any base a and any positive real number x,

$$a^{\log_a x} = x.$$

The number a raised to the power $\log_a x$ is x.

Given

$\log_a 2 = 0.301,$
$\log_a 5 = 0.699,$

find each of the following.

14. $\log_a 4$

15. $\log_a 10$

16. $\log_a \frac{2}{5}$

17. $\log_a \frac{5}{2}$

18. $\log_a \frac{1}{5}$

19. $\log_a \sqrt{a^3}$

20. $\log_a 5a$

21. $\log_a 16$

Simplify.

22. $\log_2 2^8$

23. $\log_{10} 10^{4.3}$

24. $\log_e e^{23}$

Simplify.

25. $4^{\log_4 3}$

26. $7^{\log_7 x}$

27. $b^{\log_b 42}$

Proof. The proof follows directly from the definition of logarithms. Let

$$M = \log_a x.$$

Then $a^M = x$ from the definition of logarithms. But $M = \log_a x$, and if we substitute $\log_a x$ for M, we obtain the desired result:

$$a^{\log_a x} = x.$$

EXAMPLES Simplify.

23. $a^{\log_a 3} = 3$
24. $2^{\log_2 5} = 5$
25. $10^{\log_{10} t} = t$

DO EXERCISES 25–27.

EXERCISE SET 5.4

1 Express as a sum of logarithms.

1. $\log_2 (64 \cdot 8)$
2. $\log_3 (81 \cdot 27)$
3. $\log_4 (32 \cdot 64)$
4. $\log_5 (125 \cdot 25)$
5. $\log_c QP$
6. $\log_t 9Y$

Express as a single logarithm.

7. $\log_b 8 + \log_b 90$
8. $\log_a 75 + \log_a 2$
9. $\log_c P + \log_c Q$
10. $\log_e M + \log_e T$

2 Express as a product.

11. $\log_a x^4$
12. $\log_b t^3$
13. $\log_c y^5$
14. $\log_{10} y^8$
15. $\log_b Q^{-6}$
16. $\log_c K^{-6}$

3 Express as a difference of logarithms.

17. $\log_a \dfrac{76}{13}$
18. $\log_t \dfrac{M}{8}$
19. $\log_b \dfrac{5}{4}$
20. $\log_a \dfrac{x}{y}$

Express as a single logarithm.

21. $\log_a 18 - \log_a 5$
22. $\log_b 54 - \log_b 6$

4 Express in terms of logarithms.

23. $\log_a x^3y^2z$
24. $\log_a 6xy^5z^4$
25. $\log_b \dfrac{x^2y}{b^3}$
26. $\log_b \dfrac{p^2q^5}{m^4b^9}$
27. $\log_c \sqrt[3]{\dfrac{x^4}{y^3z^2}}$
28. $\log_a \sqrt{\dfrac{x^6}{p^5q^8}}$
29. $\log_a \sqrt[4]{\dfrac{m^8n^{12}}{a^3b^5}}$
30. $\log_a \sqrt{\dfrac{a^6b^8}{a^2b^5}}$

Express as a single logarithm and simplify, if possible.

31. $\frac{2}{5} \log_a x - \frac{1}{3} \log_a y$
32. $\frac{1}{2} \log_a x + 4 \log_a y - 3 \log_a x$
33. $\log_a 2x + 3 (\log_a x - \log_a y)$
34. $\log_a x^2 - 2 \log_a \sqrt{x}$
35. $\log_a \dfrac{a}{\sqrt{x}} - \log_a \sqrt{ax}$
36. $\log_a (x^2 - 4) - \log_a (x - 2)$

Given $\log_b 3 = 0.5283$ and $\log_b 5 = 0.7740$, find each of the following.

37. $\log_b 15$
38. $\log_b \frac{5}{3}$
39. $\log_b \frac{3}{5}$
40. $\log_b \frac{1}{5}$
41. $\log_b \frac{1}{3}$
42. $\log_b \sqrt{b}$
43. $\log_b \sqrt{b^3}$
44. $\log_b 5b$
45. $\log_b 3b$
46. $\log_b 9$
47. $\log_b 25$
48. $\log_b 75$

5 Simplify.

49. $\log_t t^{11}$
50. $\log_p p^3$
51. $\log_e e^{|x-4|}$
52. $\log_Q Q^{\sqrt{5}}$
53. $3^{\log_3 4x}$
54. $5^{\log_5 (4x-3)}$
55. $a^{\log_a Q}$
56. $t^{\log_t e^5}$

Solve for x.

57. $a^{\log_a x} = 15$

58. $5^{\log_5 8} = 2x$

59. $\log_e e^x = -7$

60. $\log_a a^x = 2.7$

SYNTHESIS

Determine whether each of the following is false.

61. $\dfrac{\log_a M}{\log_a N} = \log_a M - \log_a N$

62. $\dfrac{\log_a M}{\log_a N} = \log_a \dfrac{M}{N}$

63. $\dfrac{\log_a M}{c} = \log_a M^{1/c}$

64. $\log_N (M \cdot N)^x = x \log_N M + x$

65. $\log_a 2x = 2 \log_a x$

66. $\log_a 2x = \log_a 2 + \log_a x$

67. $\log_a (M + N) = \log_a M + \log_a N$

68. $\log_a x^3 = 3 \log_a x$

69. $\log a - \log b = \log \left(\dfrac{a}{b}\right)$

70. $\log a - \log b = \dfrac{\log a}{\log b}$

Express as a single logarithm and simplify, if possible.

71. $\log_a (x^8 - y^8) - \log_a (x^2 + y^2)$

72. $\log_a (x + y) + \log_a (x^2 - xy + y^2)$

Express as a sum or difference of logarithms.

73. $\log_a \sqrt{4 - x^2}$

74. $\log_a \dfrac{x - y}{\sqrt{x^2 - y^2}}$

75. If $\log_a x = 2$, $\log_a y = 3$, and $\log_a z = 4$, what is

$$\log_a \frac{\sqrt[3]{x^2 z}}{\sqrt[3]{y^2 z^{-2}}}?$$

Solve.

76. $(x - 4) \cdot \log_a a^x = x$

77. $\log_a 3x = \log_a 3 + \log_a x$

78. $\log_\pi \pi^{2x+3} = 4$

79. $3^{\log_3 (8x - 4)} = 5$

80. $4^{2 \log_4 x} = 7$

81. $8^{2 \log_8 x + \log_8 x} = 27$

82. $\log_a x^2 = 2 \log_a x$

83. $\log_b \dfrac{5}{x + 2} = \log_b 5 - \log_b (x + 2)$

CHALLENGE

84. If $\log_a x = 2$, what is $\log_a \left(\dfrac{1}{x}\right)$?

85. If $\log_a x = 2$, what is $\log_{1/a} x$?

Prove the following for any base a and any positive number x.

86. $\log_a \left(\dfrac{1}{x}\right) = -\log_a x$

87. $\log_a \left(\dfrac{x + \sqrt{x^2 - 5}}{5}\right) = -\log_a (x - \sqrt{x^2 - 5})$

88. $\log_a \left(\dfrac{1}{x}\right) = \log_{1/a} x$

89. $\log_{a^{1/n}} x = \log_a x^n$

5.5 Finding Logarithmic Function Values on a Calculator

OBJECTIVES

You should be able to:

1. Find logarithms and antilogarithms, base 10, on a calculator.
2. Find logarithms and antilogarithms, base e, on a calculator.
3. Use the change-of-base formula to find logarithms to bases other than e or 10.

Any positive number different from 1 can be used as the base of a logarithmic function. However, some numbers are easier to use than others, and there are logarithmic bases that fit into certain applications more naturally than others. Base-10 logarithms are called **common logarithms**. They are useful because they are the same base as our *commonly* used decimal system for naming numbers. Before calculators became so widely available, common logarithms

Using a calculator, find each of the following.

1. log 98,021,544

2. log 0.000617

3. log 53.1

4. log 0.05357

were extensively used in calculations. In fact, that is why logarithms were invented.

Another logarithmic base that is used a great deal today is, strangely enough, an irrational number. This number is named e and is about 2.7182818. Logarithms, base e, are called **natural logarithms**. We will consider e and natural logarithms later in this section. We first consider common logarithms.

1 Common Logarithms on a Calculator

Before the invention of calculators, tables were developed in order to find common logarithms. It is faster to use calculators and they are now quite inexpensive to purchase, so here we find common logarithms using calculators.

The abbreviation log, with no base written, is used for logarithms, base 10, or common logarithms. Thus,

$$\log 29 \quad \text{means} \quad \log_{10} 29.$$

On scientific calculators, the key for common logarithms is usually marked LOG. To find the common logarithm of a number, enter that number and press the LOG key. It is important that you read the instructions for your particular calculator to be sure that you are carrying out the steps correctly.

EXAMPLE 1 Find log 64,577.

Solution We enter 64,577 and then press the LOG key. We find that

$$\log 64,577 \approx 4.8101. \qquad \text{Rounded to four decimal places} \qquad \blacksquare$$

Keep in mind that 4.8101 is the exponent used with a base of 10 to get 64,577. That is, $10^{4.8101} \approx 64,577$.

EXAMPLE 2 Find log 0.0000239.

Solution We enter 0.0000239 and then press the LOG key. We find that

$$\log 0.0000239 \approx -4.6216. \qquad \text{Rounded to four decimal places} \qquad \blacksquare$$

DO EXERCISES 1–4.

The inverse of a logarithmic function is, of course, an exponential function. The inverse of finding a logarithm is also called finding an **antilogarithm**. To find an antilogarithm, we *exponentiate*:

$$f(100) = \log 100 = 2,$$
$$f^{-1}(2) = \text{antilog } 2 = 10^2 = 100.$$

Generally, there is no key on a calculator marked "antilog." It is up to you to know that to find the inverse, or antilogarithm, you must use the 10^x key, if there is one. If there is no such key, then you must raise 10 to the x power using an exponential key.

EXAMPLE 3 Find antilog 2.1792.

Solution The problem gives the exponent, 2.1792.

a) *Using the 10^x key.* We enter 2.1792 and then press the 10^x key. We find that

$$\text{antilog } 2.1792 = 10^{2.1792} \approx 151.078.$$

b) *Using an exponential key, y^x.* We enter 10 and also 2.1792. Then, pressing keys in the order appropriate for your particular calculator (you must read the instructions), you will find that

$$\text{antilog } 2.1792 = 10^{2.1792} \approx 151.078.$$

EXAMPLE 4 Find antilog (-3.261194).

Solution

$$\text{antilog } (-3.261194) = 10^{-3.261194} \approx 0.00054803$$

DO EXERCISES 5–8.

Using a calculator, find each of the following.

5. antilog 6.1053

6. antilog 0.001256

7. antilog (-4.52)

8. antilog (-0.144567)

2 The Base *e* and Natural Logarithms on a Calculator

The compound-interest formula, which we considered in Chapter 1, is

$$A = P\left(1 + \frac{i}{n}\right)^{nt},$$

where A is the amount that an initial investment P will be worth after t years at interest rate i, compounded n times per year. Suppose that \$1 is an initial investment at 100% interest for 1 year (no bank would pay this). The above formula becomes a function A defined in terms of the number of compounding periods n:

$$A(n) = \left(1 + \frac{1}{n}\right)^n.$$

Let us find some function values. We round to six decimal places. We use a calculator with a power key y^x.

n	$A(n) = \left(1 + \frac{1}{n}\right)^n$
1 (compounded annually)	\$2.00
2 (compounded semiannually)	\$2.25
3	\$2.370370
4 (compounded quarterly)	\$2.441406
5	\$2.488320
100	\$2.704814
365 (compounded daily)	\$2.714567
8760 (compounded hourly)	\$2.718121

The numbers in this table get closer and closer to a very important number in mathematics, called e. The number e occurs in a great many applications. It may seem like a strange one to use as a logarithmic base, because it is an irrational number. Its decimal representation does not terminate or repeat:

$$e \approx 2.7182818284\ldots$$

Logarithms to the base e are called **natural logarithms**.

The abbreviation "ln" is generally used with natural logarithms. Thus

$$\ln 53 \quad \text{means} \quad \log_e 53.$$

On scientific calculators, the key for the natural logarithmic function is marked LN.

EXAMPLE 5 Find ln 4568.

Solution We enter 4568 and then press the LN key. We find that

$$\ln 4568 \approx 8.4268. \quad \text{Rounded to four decimal places}$$

EXAMPLE 6 Find ln 2.

Solution

$$\ln 2 \approx 0.6931$$

EXAMPLE 7 Find ln 0.0005142.

Solution We enter 0.0005142 and then press the LN key. We find that

$$\ln 0.0005142 \approx -7.5729.$$

DO EXERCISES 9–12.

To find the antilogarithm, base e, we use the e^x key, if there is one. If not, we use a power key y^x and an approximation for e, say, 2.71828.

EXAMPLE 8 Find $\text{antilog}_e\ 2.1792$.

Solution The problem gives the exponent.

a) *Using the e^x key.* We enter 2.1792 and then press the e^x key. We find that

$$\text{antilog}_e\ 2.1792 = e^{2.1792} \approx 8.8392.$$

b) *Using an exponential key, y^x.* We enter 2.1792 and also an approximate value of e, say, 2.71828. Then, pressing keys in the order appropriate for your calculator (you must read the instructions), you will find that

$$\text{antilog}_e\ 2.1792 = e^{2.1792} \approx 8.8392.$$

EXAMPLE 9 Find $\text{antilog}_e\ (-6.5399)$.

Solution

$$\text{antilog}_e\ (-6.5399) = e^{-6.5399} \approx 0.001445$$

DO EXERCISES 13–16.

3 Changing Logarithmic Bases

Most calculators give the values of both common logarithms and natural logarithms. To find a logarithm with some other base, we can use the following conversion formula.

Using a calculator, find each of the following.

9. ln 85,122

10. ln 0.001127

11. ln 0.39

12. ln 1544.923

Using a calculator, find each of the following.

13. $\text{antilog}_e\ 7.1485$

14. $\text{antilog}_e\ 0.0123$

15. $\text{antilog}_e\ (-1.11)$

16. $\text{antilog}_e\ (-6.86678)$

THEOREM 5 Change-of-Base Formula

For any logarithmic bases a and b, and any positive number M,

$$\log_b M = \frac{\log_a M}{\log_a b}.$$

Proof. Let $x = \log_b M$. Then, writing an equivalent exponential equation, we have $b^x = M$. Next we take the logarithmic base a on both sides. This gives us

$$\log_a b^x = \log_a M.$$

By Property 2,

$$x \log_a b = \log_a M,$$

and solving for x, we obtain

$$x = \frac{\log_a M}{\log_a b}.$$

But, $x = \log_b M$, so we have

$$\log_b M = \frac{\log_a M}{\log_a b}$$

which is the change-of-base formula.

EXAMPLE 10 Find $\log_5 8$ using common logarithms.

Solution Let $a = 10$, $b = 5$, and $M = 8$. Then substitute into the change-of-base formula:

$$\log_5 8 = \frac{\log_{10} 8}{\log_{10} 5} \quad \text{Substituting}$$

$$\approx \frac{0.9031}{0.6990} \quad \text{When using your calculator, you need not round before dividing.}$$

$$\approx 1.2920.$$

To check, we use a calculator with a power key y^x to verify that

$$5^{1.2920} \approx 8.$$

DO EXERCISE 17.

We can also use base e for a conversion.

EXAMPLE 11 Find $\log_4 31$ using natural logarithms.

Solution Substituting e for a, 4 for b, and 31 for M, we have

$$\log_4 31 = \frac{\log_e 31}{\log_e 4} \quad \text{Using the change-of-base formula}$$

$$= \frac{\ln 31}{\ln 4}$$

$$\approx \frac{3.4340}{1.3863}$$

$$\approx 2.4771.$$

DO EXERCISE 18.

17. Find $\log_6 8$ using common logarithms.

18. Find $\log_3 546$ using natural logarithms.

EXERCISE SET 5.5

1 Use a calculator to find the following common logarithms and antilogarithms.

1. log 3
2. log 7
3. log 8
4. log 13
5. log 2.34
6. log 3.07
7. log 65
8. log 84
9. log 62.4
10. log 10.8
11. log 532
12. log 196
13. log 13,400
14. log 93,100
15. log 0.57
16. log 0.69
17. log 0.052
18. log 0.387
19. log 0.009808
20. log 0.0005123
21. antilog 3
22. antilog 5
23. antilog 2.7
24. antilog 14.8
25. antilog 0.477133
26. antilog 0.06532
27. antilog (−0.5465)
28. antilog (−0.3404)
29. $10^{-2.9523}$
30. $10^{4.8982}$
31. $\log (5.621 \times 10^5)$
32. $\log (4.625 \times 10^{-12})$
33. log (−4.923)
34. log (−7.891)
35. $10^{(3.8146 \times 10^{-3})}$
36. $10^{(1.6773 \times 10^{-2})}$

2 Find the following logarithms and antilogarithms, base e, using a calculator.

37. ln 3
38. ln 2
39. ln 8
40. ln 13
41. ln 82
42. ln 50
43. ln 8365
44. ln 809.3
45. ln 0.0059
46. ln 0.00037
47. $\text{antilog}_e\ 3.6052$
48. $\text{antilog}_e\ 4.9312$
49. $\text{antilog}_e\ (-6.0751)$
50. $\text{antilog}_e\ (-2.3001)$
51. $\text{antilog}_e\ 0.00567$
52. $\text{antilog}_e\ 0.01111$
53. $\text{antilog}_e\ 34$
54. $\text{antilog}_e\ 56$
55. $\ln (8.041 \times 10^{28})$
56. $\ln (2.031 \times 10^{12})$
57. $\ln (5.043 \times 10^{-14})$
58. $\ln (3.051 \times 10^{-51})$
59. $\text{antilog}_e\ 7.4012$
60. $\text{antilog}_e\ 6.3058$
61. $e^{1.0312}$
62. $e^{-6.3783}$
63. $e^{-12.832}$
64. $e^{17.814}$

3 Find the following logarithms using the change-of-base formula.

65. $\log_4 100$
66. $\log_3 20$
67. $\log_2 12$
68. $\log_5 40$
69. $\log_{100} 0.3$
70. $\log_{200} 50$
71. $\log_{0.5} 7$
72. $\log_{0.1} 2$
73. $\log_3 0.3$
74. $\log_2 0.06$
75. $\log_\pi 100$
76. $\log_\pi 25$

SYNTHESIS

Verify each of the following.

77. $\ln x = 2.3026 \log x$
78. $\log x = 0.4343 \ln x$
79. Using function values obtained on a calculator, plot points and draw a precise graph of $y = f(x) = 10^x$.
80. Using function values obtained on a calculator, plot points and draw a precise graph of $y = g(x) = e^x$.
81. Using function values obtained on a calculator and values obtained in Exercise 79, plot points and draw a precise graph of $y = f^{-1}(x) = \log x$.
82. Using function values obtained on a calculator and values obtained in Exercise 80, plot points and draw a precise graph of $y = g^{-1}(x) = \ln x$.

Simplify.

83. $\dfrac{\log_5 8}{\log_5 2}$
84. $\dfrac{\log_3 64}{\log_3 16}$

Solve for x.

85. $\log 872x = 5.3442$
86. $\log 43x^2 = 8.0166$
87. $\log 784 + \log x = \log 2322$
88. $\dfrac{2.34}{\ln x} = \dfrac{57}{4.03}$

Use the change-of-base formula to derive each of the following formulas.

89. $\log e = \dfrac{1}{\ln 10}$
90. $\log M = \dfrac{\ln M}{\ln 10}$
91. $\log_b M = \dfrac{1}{\log_M b}$
92. $\ln M = \dfrac{\log M}{\log e}$
93. $\log_a (\log_a x) = \log_a (\log_b x) - \log_a (\log_b a)$

94. Given $f(x) = (1 + x)^{1/x}$, find $f(1)$, $f(0.5)$, $f(0.2)$, $f(0.1)$, $f(0.01)$, and $f(0.001)$ to six decimal places. This sequence of numbers approaches the number e.

95. Given $f(x) = t^{1/(t-1)}$, find $f(0.5)$, $f(0.9)$, $f(0.99)$, $f(0.999)$, and $f(0.9999)$ to six decimal places. This sequence of numbers approaches the number e.

96. Which is larger, e^{π} or π^{e}?

97. Which is larger, $e^{\sqrt{\pi}}$ or $\sqrt{e^{\pi}}$?

98. In some textbooks and computer applications, log x is used to represent $\log_e x$. Discuss some methods you might use to discover what the base actually is.

5.6 Graphs: Base *e* and Applications

OBJECTIVES

You should be able to:

1. Graph exponential functions, base e.
2. Graph natural logarithmic functions.
3. Sketch graphs of functions in applications.

Exponential and logarithmic functions, base e, are two of the most valuable functions that we study in mathematics. Because of their importance in many applications, it is helpful to study their graphs.

1 Exponential Functions, Base *e*

Graphs of $f(x) = e^{kx}$ and $f(x) = e^{-kx}$

EXAMPLE 1 Graph $f(x) = e^x$ and $f(x) = e^{-x}$.

Solution We use a calculator with an e^x key to find approximate values of e^x and e^{-x}. Using these values, we can draw the graphs of the functions.

x	e^x	e^{-x}
0	1	1
1	2.7	0.4
2	7.4	0.1
−1	0.4	2.7
−2	0.1	7.4

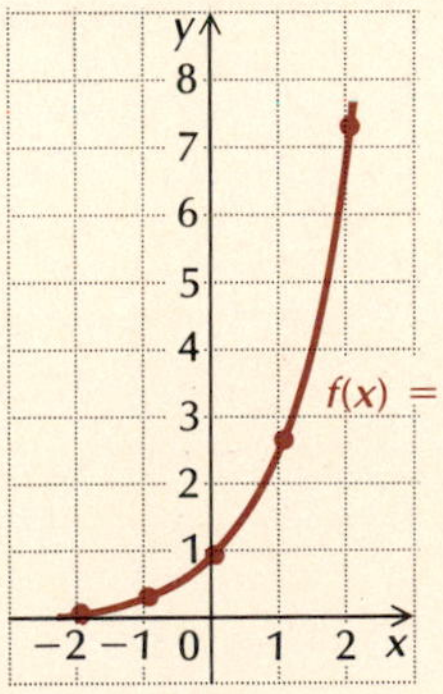

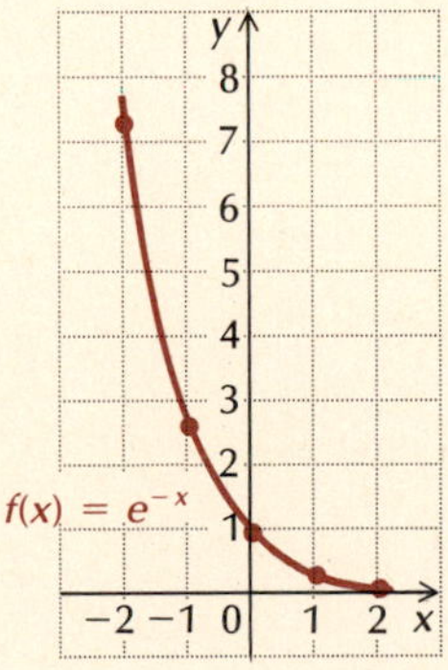

Note that the graph of e^{-x} is a reflection of the graph of e^x across the y-axis.

EXAMPLE 2 Graph $f(x) = e^{-0.5x}$.

Solution We find some solutions with a calculator, plot them, and then draw the graph. For example, $f(2) = e^{-0.5(2)} = e^{-1} \approx 0.4$.

x	$e^{-0.5x}$
0	1
1	0.6
2	0.4
3	0.2
-1	1.6
-2	2.7
-3	4.5

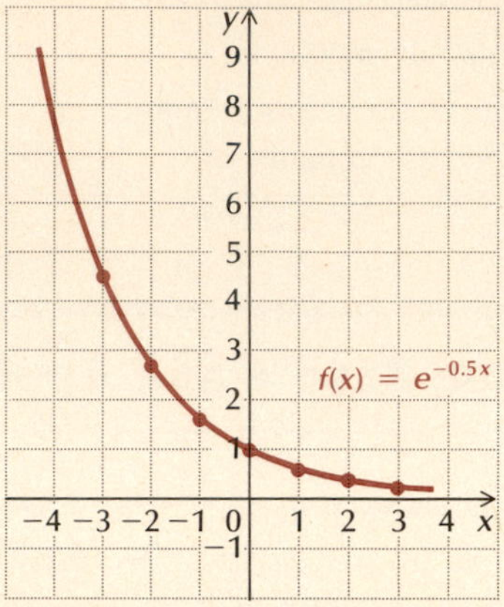

DO EXERCISES 1–3.

Graph.

1. $f(x) = e^{2x}$

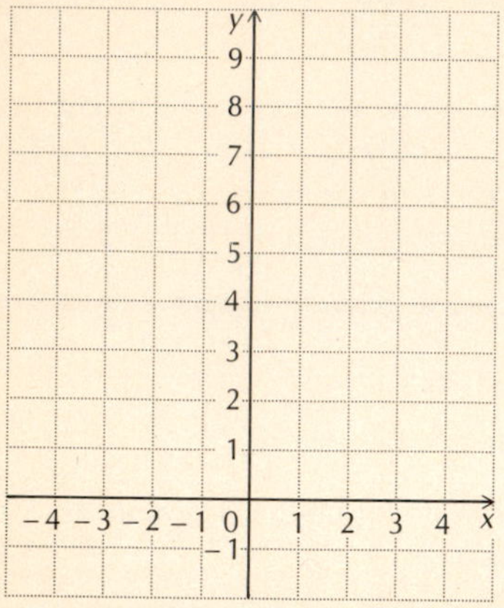

2. $f(x) = e^{-2x}$

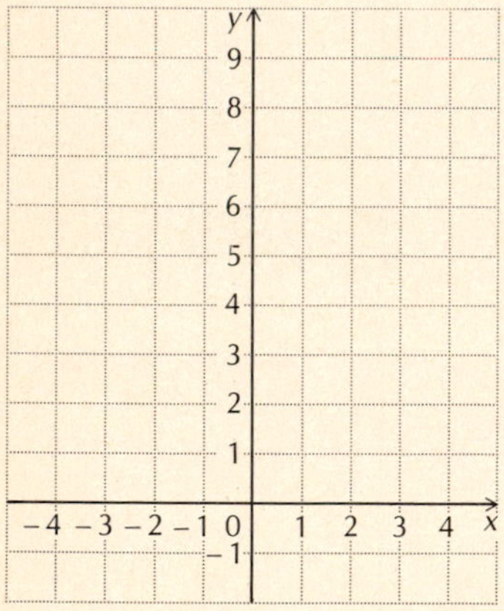

3. $f(x) = e^{0.2x}$

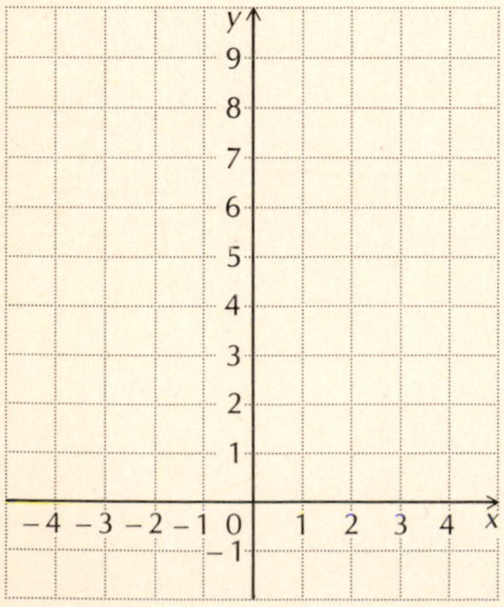

Graphs of $f(x) = 1 - e^{-kx}$

Functions of the type $f(x) = 1 - e^{-kx}$ are also important.

EXAMPLE 3 Graph $f(x) = 1 - e^{-2x}$ for nonnegative values of x.

Solution We obtain these values using a calculator with an e^x key. For example, when $x = 1$,

$$
\begin{aligned}
f(1) &= 1 - e^{-2(1)} && \text{Substituting} \\
&= 1 - e^{-2} \\
&\approx 1 - 0.135335 && \text{Using a calculator} \\
&\approx 0.86.
\end{aligned}
$$

x	e^{-2x}	$1 - e^{-2x}$
0	1	0
$\frac{1}{2}$	0.367879	0.63
1	0.135335	0.86
2	0.018316	0.98
3	0.002479	0.998

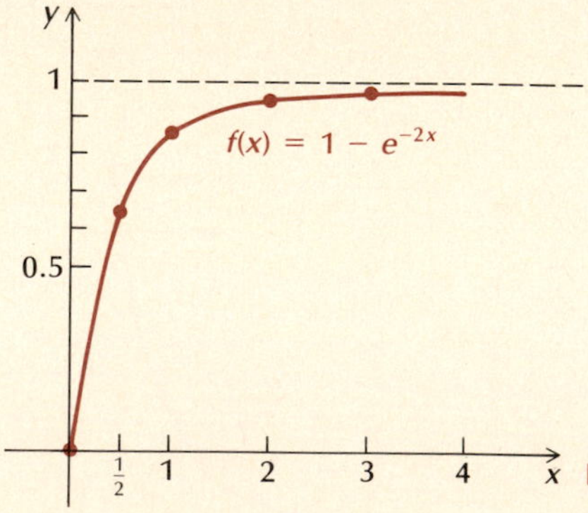

In general, the graph of $f(x) = 1 - e^{-kx}$, for $k > 0$, increases from 0 and approaches 1 as x gets larger.

DO EXERCISE 4 ON THE FOLLOWING PAGE.

2 Natural Logarithmic Functions

EXAMPLE 4 Graph $g(x) = \ln x$.

Solution There are two ways in which we might obtain the graph of $y = g(x) = \ln x$. One is by writing its equivalent equation, $x = e^y$.

We select values of y and use a calculator to find the corresponding values of e^y. We then plot points, remembering that x still is the first coordinate.

x, or e^y	y
0.1	−2
0.4	−1
1	0
2.7	1
7.4	2
20	3

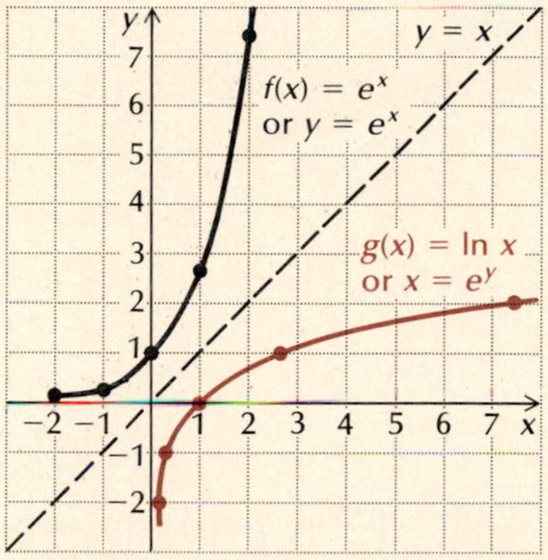

Note that f and g are inverses of each other. That is, the graph of $y = \ln x$ is a reflection across the line $y = x$ of the graph of $y = e^x$.

The second method of graphing $y = \ln x$ is to use a calculator. For example, $\ln 2 = 0.6931 \approx 0.7$.

DO EXERCISE 5.

EXAMPLE 5 Graph $f(x) = \ln(x + 3)$.

Solution We find some solutions with a calculator, plot them, and then draw the graph. When $x = 2$, $y = \ln(2 + 3) = \ln 5 \approx 1.6$.

x	y, or $\ln(x + 3)$
0	1.1
1	1.4
2	1.6
3	1.8
4	1.9
−1	0.7
−2	0
−2.5	−0.7

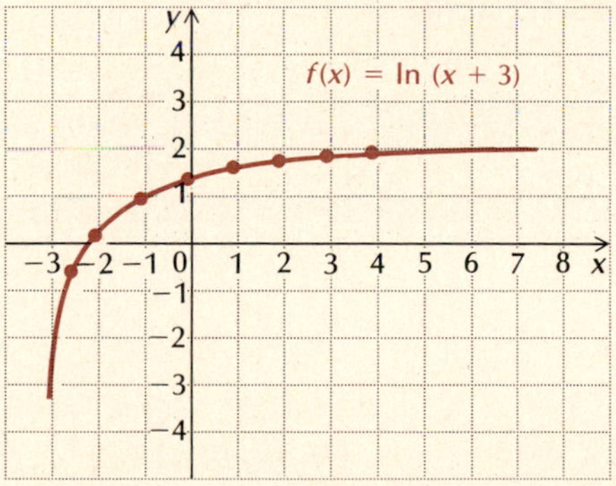

Note that the graph of $y = \ln(x + 3)$ is a horizontal translation (three units to the left) of the graph of $y = \ln x$.

DO EXERCISE 6.

3 Graphs in Applications

EXAMPLE 6 *Business.* A company begins a radio advertising campaign in New York City to market a new product. The percentage of the target market that buys a product is normally a function of the length of the advertising campaign. The estimated percentage is given by

$$f(t) = 1 - e^{-0.04t},$$

4. Graph $f(x) = 1 - e^{-x}$ for nonnegative values of x.

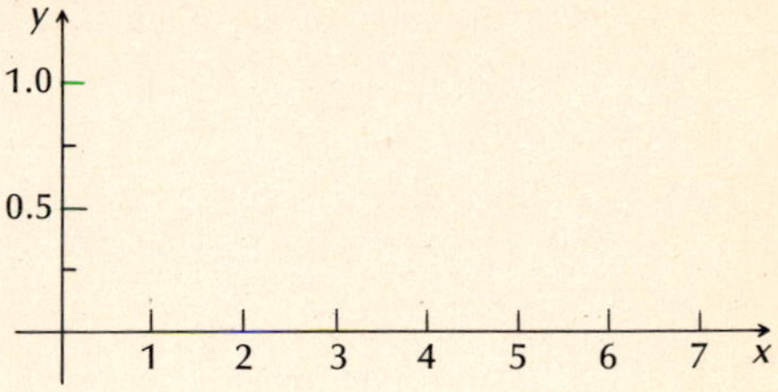

5. Graph $f(x) = 2 \ln x$.

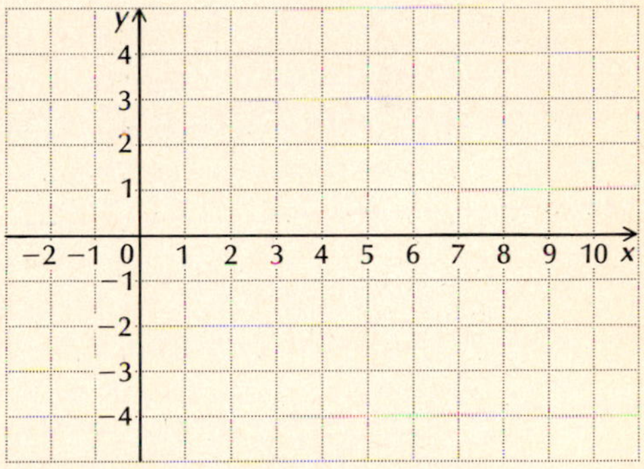

6. Graph $f(x) = \ln(x - 1)$.

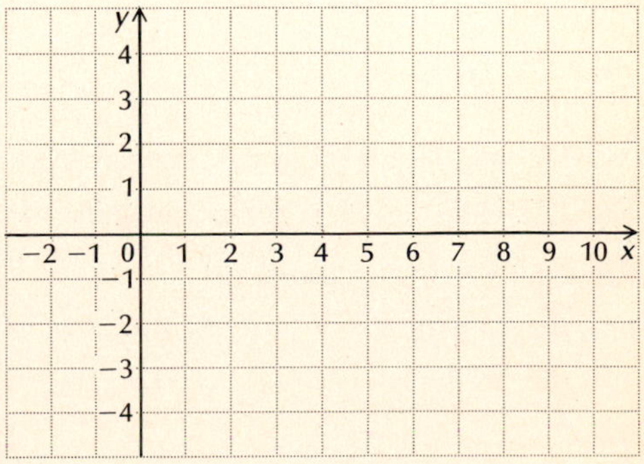

7. The value of a stock is given by

$$V(t) = \$45(1 - e^{-0.8t}) + \$15,$$

where V is the value of the stock after time t, in months.

a) Find $V(6)$.

b) Find other function values, and sketch a graph of the function.

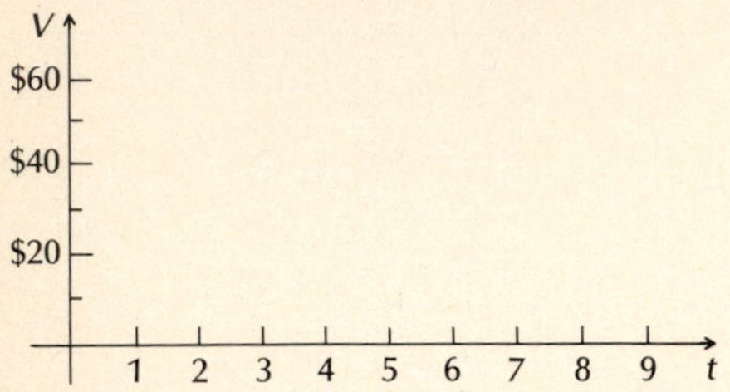

where t = the number of days of the campaign.

a) Find $f(25)$, the percentage of the target market that has bought the product after a 25-day advertising campaign.

b) Sketch a graph of the function.

Solution

a) We evaluate $f(t)$ when $t = 25$:

$$\begin{aligned} f(t) &= 1 - e^{-0.04t} \\ f(25) &= 1 - e^{-0.04(25)} \quad \text{Substituting} \\ &= 1 - e^{-1} \\ &\approx 1 - 0.367879 \\ &\approx 0.632121 \\ &\approx 63.2\%. \end{aligned}$$

b) We find other solutions, plot them, and then sketch the graph.

t	$f(t)$
25	63.2%
50	86.5%
75	95.0%
100	98.2%

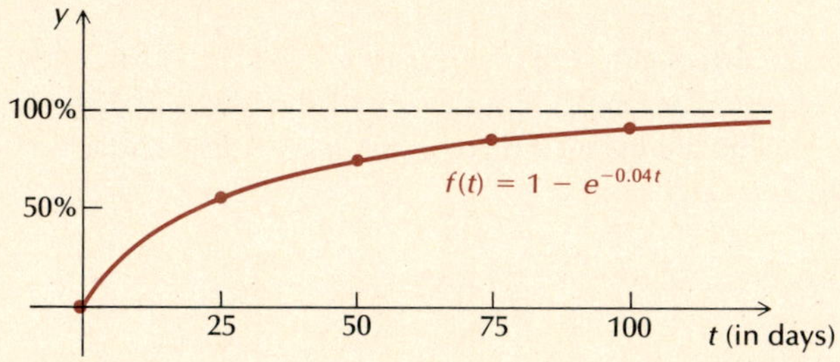

The function increases from 0 (0%) to 1 (100%). The longer the advertising campaign, the larger the percentage of the market that has bought the product.

DO EXERCISE 7.

EXERCISE SET 5.6

1 Graph using a calculator.

1. $f(x) = 2e^x$
2. $f(x) = 0.5e^x$
3. $f(x) = e^{(1/2)x}$
4. $f(x) = e^{-0.3x}$
5. $f(x) = e^{x+1}$
6. $f(x) = e^{-x+1}$
7. $f(x) = e^{2x} + 1$
8. $f(x) = e^x - 2$
9. $f(x) = 1 - e^{-0.01x}$, for nonnegative values of x
10. $f(x) = 1 - e^{-3x}$, for nonnegative values of x
11. $f(x) = 2(1 - e^{-x})$, for nonnegative values of x
12. $f(x) = \frac{1}{2}(1 - e^{-2x})$, for nonnegative values of x

2 Graph using a calculator.

13. $f(x) = 4 \ln x$
14. $f(x) = 3 \ln x$
15. $f(x) = \frac{1}{2} \ln x$
16. $f(x) = 0.2 \ln x$
17. $f(x) = \ln (x - 2)$
18. $f(x) = \ln (x + 1)$
19. $f(x) = 2 - \ln x$
20. $f(x) = (\ln x) - 4$

3

21. *Biomedical: Acceptance of a new medicine.* The percentage P of doctors who accept a new medicine is given by

$$P(t) = 1 - e^{-0.2t},$$

where t = time, in months.

a) Find $P(1)$, $P(4)$, $P(6)$, and $P(12)$.

b) Sketch a graph of the function.

22. *Psychology: Hullian learning model.* A typist learns to type W words per minute after t weeks of practice, where W is given by

$$W(t) = 100(1 - e^{-0.3t}).$$

a) Find $W(1)$, $W(5)$, $W(8)$, and $W(10)$.

b) Sketch a graph of the function.

23. *Business: Growth of a stock.* The value of a stock is given by

$$V(t) = \$58(1 - e^{-1.1t}) + \$20,$$

where V is the value of the stock after t months.

a) Find $V(1)$, $V(2)$, $V(4)$, $V(6)$, and $V(12)$.
b) Sketch a graph of the function.

24. *Business.* A toy company begins a television advertising campaign in Houston to market a new product. The television station uses the following function to estimate the percentage of the target market that buys the toy after t days of the campaign:

$$P(t) = 1 - e^{-0.03t}.$$

a) Find $P(10)$, $P(30)$, $P(60)$, and $P(120)$.
b) Sketch a graph of the function.

SYNTHESIS

Graph.

25. $g(x) = e^{|x|}$

26. $f(x) = \ln |x|$

27. $f(x) = |\ln x|$

28. $g(x) = |\ln (x - 1)|$

29. $f(x) = \dfrac{e^x + e^{-x}}{2}$

30. $f(x) = \dfrac{e^x - e^{-x}}{2}$

31. *Spread of a rumor.* In a college with a student population of 800, a group of 6 students spread the rumor, "Men go for women who study calculus, and women go for men who study calculus!" The number of people who have heard the rumor after t minutes is given by

$$N(t) = \frac{4800}{6 + 794e^{-0.4t}}.$$

a) Find $N(3)$, $N(5)$, $N(10)$, and $N(15)$.
b) Sketch the graph of the function.

32. *Spread of an epidemic.* In a town whose total population is 2000, the disease *Rottenich* creates an epidemic. The initial number of people infected is 10. The number of people infected after time t, in weeks, is given by

$$P(t) = \frac{20{,}000}{10 + 1990e^{-6t}}.$$

a) Find $P(0.2)$, $P(0.5)$, $P(0.8)$, $P(1)$, $P(1.5)$, and $P(2)$.
b) Sketch the graph of the function.

COMPUTER–CALCULATOR EXERCISES

Use a computer software graphing package or graphing calculator. For each of the following:

a) Graph the function.
b) Estimate the zeros.
c) Estimate the maximum and the minimum values.

33. $f(x) = x^2e^{-x}$

34. $f(x) = e^{-x^2}$

35. $f(x) = x^2 \ln x$

36. $f(x) = \dfrac{\ln x}{x^2}$

5.7 Solving Exponential and Logarithmic Equations

OBJECTIVES

You should be able to:

1 Solve exponential equations.
2 Solve logarithmic equations.

1 Solving Exponential Equations

Equations with variables in exponents, such as $3^x = 20$ and $2^{5x} = 64$, are called **exponential equations**. Sometimes, as is the case with $2^{5x} = 64$, we can write each side as a power of the same number:

$$2^{5x} = 2^6.$$

Then the exponents are the same, and we can set them equal and solve:

$$5x = 6$$
$$x = \tfrac{6}{5}.$$

Solve.

1. $5^{2x} = 25$

2. $4^{4x-3} = 64$

We use the following property.

THEOREM 6

For any $a > 0$, $a \neq 1$,

$$a^x = a^y$$

is equivalent to $x = y$.

Proof. The theorem follows from the fact that $f(x) = a^x$ is a one-to-one function. If $a^x = a^y$ is true, then $f(x) = f(y)$. Then since f is one-to-one (see the definition in Section 5.1), it follows that $x = y$. Conversely, if $x = y$, it follows that $a^x = a^y$, since we are raising a to the same power.

EXAMPLE 1 Solve: $2^{3x-7} = 32$.

Solution Note that $32 = 2^5$. Thus we can write each side as a power of the same number:

$$2^{3x-7} = 2^5.$$

Since the base is the same, 2, the exponents must be the same. Thus,

$$\begin{aligned} 3x - 7 &= 5 \\ 3x &= 12 \\ x &= 4. \end{aligned}$$

Check:

2^{3x-7}	$=$ 32
$2^{3(4)-7}$	32
2^{12-7}	
2^5	
32	

The solution is 4. The solution set is $\{4\}$.

DO EXERCISES 1 AND 2.

When it does not seem possible to write each side as a power of the same base, we can take the common or natural logarithm on each side and then use Property 2.

EXAMPLE 2 Solve: $3^x = 20$.

Solution

$$\begin{aligned} \log 3^x &= \log 20 && \text{Taking the common logarithm on both sides} \\ x \log 3 &= \log 20 && \text{Property 2} \\ x &= \frac{\log 20}{\log 3} && \text{Solving for } x \end{aligned}$$

CAUTION! This is not $\log 20 - \log 3$!

This is an exact answer. We cannot simplify further, but we can approximate using a calculator:

$$x = \frac{\log 20}{\log 3} \approx \frac{1.3010}{0.4771} \approx 2.7268.$$

You can check this answer by finding $3^{2.7268}$ using a y^x key on a calculator. ■

DO EXERCISE 3.

3. Solve: $7^x = 20$.

If the base is e, we can take the logarithm with e as the base. This will ease our work.

EXAMPLE 3 Solve: $e^{0.08t} = 2500$.

Solution We take the natural logarithm on both sides:

$$\ln e^{0.08t} = \ln 2500 \quad \text{Taking ln on both sides}$$

$$0.08t = \ln 2500 \quad \text{Here we use Property 4: } \log_a a^x = x.$$

$$t = \frac{\ln 2500}{0.08}.$$

We can approximate using a calculator:

$$t = \frac{\ln 2500}{0.08} \approx \frac{7.8240}{0.08} \approx 97.8.$$

■

DO EXERCISE 4.

4. Solve: $e^{0.3t} = 80$.

EXAMPLE 4 Solve $\dfrac{e^x + e^{-x}}{2} = t$ for x.

Solution Note that we are to solve for x. However, we have more than one term with x in the exponent. To get a single expression with x in the exponent, we do the following:

$$e^x + e^{-x} = 2t \quad \text{Multiplying by 2}$$

$$e^x + \frac{1}{e^x} = 2t \quad \text{Rewriting with a positive exponent}$$

$$e^{2x} + 1 = 2te^x \quad \text{Multiplying on both sides by } e^x$$

$$(e^x)^2 - 2t \cdot e^x + 1 = 0.$$

This equation is reducible to quadratic, with $u = e^x$. The coefficients of the reduced quadratic equation are $a = 1$, $b = -2t$, and $c = 1$. Using the quadratic formula, we obtain

$$e^x = \frac{2t \pm \sqrt{4t^2 - 4}}{2} = t \pm \sqrt{t^2 - 1}.$$

5. Solve $\frac{e^x - e^{-x}}{2} = t$ for x.

6. Solve: $\log_2 x = 3$.

7. Solve: $\log_4 (8x - 6) = 3$.

We can now take the natural logarithm on both sides:

$$\ln e^x = \ln (t \pm \sqrt{t^2 - 1})$$
$$x = \ln (t \pm \sqrt{t^2 - 1}). \quad \text{Using Property 4}$$

DO EXERCISE 5.

2 Solving Logarithmic Equations

Equations containing logarithmic expressions are called **logarithmic equations.** We solved some logarithmic equations in Section 5.3. We did so by converting to an equivalent exponential equation.

EXAMPLE 5 Solve: $\log_2 x = 4$.

Solution We obtain an equivalent exponential expression:

$$x = 2^4$$
$$x = 16.$$

The solution is 16. The solution set is $\{16\}$.

DO EXERCISE 6.

To solve logarithmic equations, we first try to obtain a single logarithmic expression on one side and then write an equivalent exponential equation.

EXAMPLE 6 Solve: $\log_3 (5x + 7) = 2$.

Solution We already have a single logarithmic expression, so we write an equivalent exponential equation:

$$5x + 7 = 3^2 \quad \text{Writing an equivalent exponential equation}$$
$$5x + 7 = 9$$
$$5x = 2$$
$$x = \tfrac{2}{5}.$$

Check:

$\log_3 (5x + 7)$	$= 2$
$\log_3 \left(5 \cdot \frac{2}{5} + 7\right)$	2
$\log_3 (2 + 7)$	
$\log_3 9$	
2	

The solution is $\frac{2}{5}$. The solution set is $\{\frac{2}{5}\}$.

DO EXERCISE 7.

EXAMPLE 7 Solve: $\log x + \log (x + 3) = 1$.

Solution We have common logarithms here. Writing in the base 10's will help us understand the problem:

$$\log_{10} x + \log_{10} (x + 3) = 1$$

$$\log_{10} [x(x + 3)] = 1 \quad \text{Using Property 1 to obtain a single logarithm}$$

$$x(x + 3) = 10^1 \quad \text{Writing an equivalent exponential equation}$$

$$x^2 + 3x = 10$$

$$x^2 + 3x - 10 = 0$$

$$(x - 2)(x + 5) = 0 \quad \text{Factoring}$$

$$x - 2 = 0 \quad \text{or} \quad x + 5 = 0 \quad \text{Principle of zero products}$$

$$x = 2 \quad \text{or} \quad x = -5$$

Check: For 2:

$\log x + \log (x + 3) = 1$	
$\log 2 + \log (2 + 3)$	1
$\log 2 + \log 5$	
$\log 10$	
1	

For -5:

$\log x + \log (x + 3) = 1$	
$\log (-5) + \log (-5 + 3)$	1

The number -5 is not a solution because negative numbers do not have logarithms. The solution is 2. The solution set is $\{2\}$.

DO EXERCISE 8.

8. Solve: $\log x + \log (x - 3) = 1$.

EXERCISE SET 5.7

1 Solve.

1. $2^x = 32$
2. $4^x = 256$
3. $3^x = 81$
4. $5^x = 625$
5. $2^{2x} = 8$
6. $4^{5x} = 32$
7. $3^{7x} = 27$
8. $5^{3x} = 125$
9. $2^x = 33$
10. $2^x = 20$
11. $2^x = 40$
12. $2^x = 19$
13. $5^{4x-7} = 125$
14. $4^{3x+5} = 16$
15. $3^{x^2+4x} = \frac{1}{27}$
16. $27 = 3^{5x} \cdot 9^{x^2}$
17. $84^x = 70$
18. $28^x = 10$
19. $e^t = 1000$
20. $e^t = 100$
21. $e^{-t} = 0.3$
22. $e^{-t} = 0.04$
23. $e^{-0.03t} = 0.08$
24. $e^{0.09t} = 4$
25. $3^x = 2^{x-1}$
26. $5^{x+2} = 4^{x-1}$
27. $(3.9)^x = 48$
28. $(5.6)^x = 100$
29. $250 - (1.87)^x = 0$
30. $4805 - (21.3)^t = 0$
31. $4^{2x} = 8^{3x-4}$
32. $25^{3x-2} = 625^{2x+7}$
33. $\dfrac{e^x - e^{-x}}{t} = 5$
34. $e^x + e^{-x} = 5$
35. $\dfrac{e^x + e^{-x}}{e^x - e^{-x}} = t$
36. $\dfrac{5^x - 5^{-x}}{5^x + 5^{-x}} = t$

2 Solve for x.

37. $\log_5 x = 4$
38. $\log_2 x = 2$
39. $\log_5 x = -3$
40. $\log_{25} x = \frac{1}{2}$
41. $\log x = 2$
42. $\log x = 1$
43. $\log x = -3$
44. $\log x = -4$
45. $\ln x = 1$
46. $\ln x = 3$
47. $\ln x = -2$
48. $\ln x = -1$

49. $\log_5 (8 - 7x) = 3$
50. $\log_2 (10 + 3x) = 5$
51. $\log x + \log (x - 9) = 1$
52. $\log x + \log (x + 9) = 1$
53. $\log x - \log (x + 3) = -1$
54. $\log (x + 9) - \log x = 1$
55. $\log_2 (x + 1) + \log_2 (x - 1) = 3$
56. $\log_8 (x + 1) - \log_8 (x) = 2$
57. $\log_8 (x + 1) - \log_8 x = \log_8 4$
58. $\log (2x + 1) - \log (x - 2) = 1$
59. $\log_4 (x + 3) + \log_4 (x - 3) = 2$
60. $\log_5 (x + 4) + \log_5 (x - 4) = 2$
61. $\log \sqrt[4]{x} = \sqrt{\log x}$
62. $\log \sqrt[3]{x} = \sqrt{\log x}$
63. $\log_5 \sqrt{x^2 + 1} = 1$
64. $\log \sqrt{x} = \sqrt{\log x}$
65. $\log x^2 = (\log x)^2$
66. $(\log_3 x)^2 - \log_3 x^2 = 3$
67. $\log_3 (\log_4 x) = 0$
68. $\log (\log x) = 2$

SYNTHESIS

Solve.

69. $(\log_a x)^{-1} = \log_a x^{-1}$
70. $|\log_5 x| = 2$
71. $\log_7 \sqrt{x^2 - 9} = 1$
72. $x \log \frac{1}{6} = \log 6$
73. $\log (\log x) = 3$
74. $2^{x^2 + 4x} = \frac{1}{8}$
75. $\log_3 |x| = 2$
76. $\log_5 |x| = 3$
77. $\log x^{\log x} = 4$
78. $\log \sqrt{2x} = \sqrt{\log 2x}$
79. $\log_a a^{x^2} + 5x = 24$
80. $x^{\log x} = \dfrac{x^3}{100}$
81. $x^{\log_{10} x} = \dfrac{x^{-4}}{1000}$
82. $5^{2x} - 9 \cdot 5^x + 14 = 0$
83. $(32^{x-2})(64^{x+1}) = 16^{2x-3}$
84. $49^{x+2} = 5140 + 49^x$
85. $4^{3x} - 4^{3x-1} = 48$
86. $x^{\log x} = 100x$
87. $\dfrac{(e^{3x+1})^2}{e^4} = e^{10x}$
88. $\dfrac{\sqrt{(e^{2x} \cdot e^{-5x})^{-4}}}{e^x \div e^{-x}} = e^7$

Solve for t.

89. $P = P_0 e^{kt}$
90. $P = P_0 e^{-kt}$
91. $T = T_0 + (T_1 - T_0)e^{-kt}$
92. Solve for n. Use $\log_V$.

$$PV^n = c$$

93. Solve for Q.

$$\log_a Q = \frac{1}{3} \log_a y + b$$

CHALLENGE

94. Given that $2^y = 16^{x-3}$ and $3^{y+2} = 27^x$, find the value of $x + y$.
95. If $x = (\log_{125} 5)^{\log_5 125}$, what is the value of $\log_3 x$?

Solve.

96. $|\log_5 x| + 3 \log_5 |x| = 4$
97. $|\log_a x| = \log_a |x|$
98. $8x^{0.3} - 8x^{-0.3} = 63$
99. $(0.5)^x < \frac{4}{5}$
100. Given that

$$\begin{aligned}\log_2 [\log_3 (\log_4 x)] &= \log_3 [\log_2 (\log_4 y)]\\ &= \log_4 [\log_3 (\log_2 z)]\\ &= 0,\end{aligned}$$

find $x + y + z$.

101. If $2 \log_3 (x - 2y) = \log_3 x + \log_3 y$, find x/y.

102. Find the ordered pair (x, y) for which
$$4^{\log_{16} 27} = 2^x 3^y.$$

103. Suppose that
$$a = \log_8 225$$
and
$$b = \log_2 15.$$
Express a as a function of b.

5.8 Applications of Exponential and Logarithmic Functions

OBJECTIVE

You should be able to:

1 Solve problems involving applications of exponential and logarithmic equations.

1 In this section, we consider applications of exponential and logarithmic functions that involve bases other than e.

EXAMPLE 1 *Interest compounded annually.* The amount A that principal P will be worth after t years at interest rate i, compounded annually, is given by the formula $A = P(1 + i)^t$. Suppose that \$100,000 is invested at 8% interest, compounded annually.

a) Express A as a function of t. (See Example 5 of Section 5.2.)
b) After what amount of time will there be \$500,000 in the account?
c) Let $T =$ the amount of time it takes for the \$100,000 to double itself. T is called the *doubling time.* Find the doubling time.

Solution

a) We have
$$A(t) = \$100{,}000(1.08)^t:$$

b) We set $A(t) = \$500{,}000$ and solve for t:

$$500{,}000 = 100{,}000(1.08)^t$$
$$\frac{500{,}000}{100{,}000} = (1.08)^t$$
$$5 = (1.08)^t$$
$$\log 5 = \log (1.08)^t \quad \text{Taking the common logarithm on both sides}$$
$$\log 5 = t \log 1.08 \quad \text{Property 2}$$
$$\frac{\log 5}{\log 1.08} = t.$$

We simplify further by using a calculator and approximating the logarithms:
$$t = \frac{\log 5}{\log 1.08} \approx \frac{0.69897}{0.03342} \approx 20.9.$$

It will take about 20.9 years for the \$100,000 to grow to \$500,000.

Calculator note. When you are doing a calculation like this on your calculator, it is not necessary to stop and round the approximate values of the logarithms. Just find the numbers and divide. Answers will be found that way in the exercises. You may notice some slight variation in the last one or two decimal places if you round as you go. You might check with your instructor regarding such variation.

1. Suppose that \$80,000 is invested at 7% interest, compounded annually.
   a) Express the amount A as a function of time t, in years.
   b) After what amount of time will there be \$240,000 in the account?
   c) Find the doubling time.

c) To find the doubling time, we set $A(t) = \$200{,}000$ and $t = T$ and solve for T:

$$200{,}000 = 100{,}000(1.08)^T$$
$$2 = (1.08)^T$$
$$\log 2 = \log (1.08)^T \quad \text{Taking the common logarithm on both sides}$$
$$\log 2 = T \log 1.08 \quad \text{Property 2}$$
$$T = \frac{\log 2}{\log 1.08} \approx \frac{0.30103}{0.03342} \approx 9.0.$$

The doubling time is about 9 years.

DO EXERCISE 1.

EXAMPLE 2 *Forgetting.* Here is a mathematical model from psychology. A group of people take a test and make an average score of A. After a time t, in months, they take an equivalent form of the same test. At that time, the average score is $S(t)$, given by

$$S(t) = A - B \log (t + 1), \qquad t \geq 0.$$

The model is appropriate only over the interval $[0, 10^{A/B} - 1]$. Students in a zoology class took a final exam and then took equivalent forms of the exam at monthly intervals thereafter. The average scores, $S(t)$, were found to be given by the function

$$S(t) = 80 - 62 \log (t + 1), \qquad t \geq 0.$$

a) What was the average score when they took the test originally? after 1 month? after 9 months? after 1 year?
b) Graph the function for values of t such that $t \geq 0$.
c) After what amount of time will the average score be 20%?

Solution

a) $S(0) = 80 - 62 \log (0 + 1) = 80 - 62(0) = 80\%$; Original score
$S(1) = 80 - 62 \log (1 + 1) \approx 80 - 62(0.3010) = 80 - 18.662 \approx 61\%$;
$S(9) = 80 - 62 \log (9 + 1) = 80 - 62(1) = 18\%$;
$S(12) = 80 - 62 \log 13 \approx 80 - 62(1.1139) \approx 80 - 69.0618 \approx 11\%$

b) Using the values computed in part (a) and any others we want, we can sketch the graph as follows:

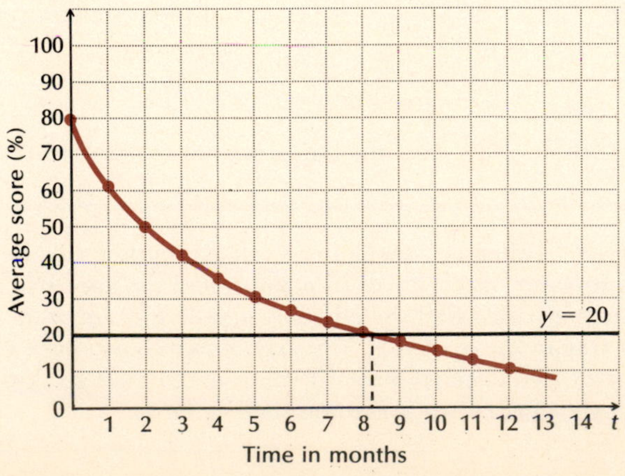

c) The time t at which the average score will be 20% is found by setting $S(t) = 20$ and solving for t. Note visually that this is where the line $y = 20$ crosses the graph.

$$20 = 80 - 62 \log (t + 1)$$
$$-60 = -62 \log (t + 1)$$
$$0.96774 \approx \log (t + 1)$$
$$10^{0.96774} \approx t + 1 \qquad \text{Using the definition of logarithms or taking the antilog: } 10^{0.96774} \approx 9.3$$
$$9.3 \approx t + 1$$
$$8.3 \approx t$$

After about 8.3 months, the average score will be 20%.

DO EXERCISE 2.

EXAMPLE 3 *Chemistry: pH of substances.* In chemistry, the **pH** of a substance is defined as follows:

$$\text{pH} = -\log [H^+],$$

where H^+ is the hydrogen ion concentration in moles per liter.

a) The hydrogen ion concentration of a common brand of mouthwash is 6.3×10^{-7} moles per liter. Find the pH.

b) The pH of a common hair rinse is 2.9. Find the hydrogen ion concentration.

Solution

a) To find the pH of the mouthwash, we substitute 6.3×10^{-7} for $[H^+]$ in the formula for pH:

$$\text{pH} = -\log [H^+] = -\log(6.3 \times 10^{-7}) = -[\log 6.3 + \log 10^{-7}]$$
$$= -[\log 6.3 + (-7)] \approx -[0.7993 + (-7)] = -[-6.2007] \approx 6.2.$$

The pH of the mouthwash is about 6.2.

b) To find the hydrogen ion concentration of the hair rinse, we substitute 2.9 for pH in the formula and solve for $[H^+]$:

$$2.9 = -\log [H^+]$$
$$-2.9 = \log [H^+]$$
$$10^{-2.9} = [H^+] \qquad \text{Using the definition of logarithms or taking the antilogarithm: } 10^{-2.9} \approx 0.0013$$
$$0.0013 \approx [H^+]$$
$$1.3 \times 10^{-3} \approx [H^+].$$

The hydrogen ion concentration of the hair rinse is 1.3×10^{-3}.

DO EXERCISE 3.

EXAMPLE 4 *Earthquake magnitude.* The magnitude R (measured on the Richter scale) of an earthquake of intensity I is defined as

$$R = \log \frac{I}{I_0},$$

where I_0 is a minimum intensity used for comparison. We can think of I_0 as a

2. *Business advertising.* A model for advertising response is given by

$$N(a) = 5000 + 200 \log a, \qquad a \geq 1,$$

where $N(a)$ = the number of units sold and a = the amount spent on advertising, in thousands of dollars.

a) How many units were sold after spending \$1000 ($a = 1$) on advertising?

b) How many units were sold after spending \$7000?

c) Graph the function.

d) How much would have to be spent in order to sell 5140 units?

3. **a)** The hydrogen ion concentration of milk is 2.3×10^{-6}. Find the pH.

b) The pH of vinegar is 5.8. Find the hydrogen ion concentration.

4. The Mexico City earthquake of 1978 had an intensity of $10^{7.85} \cdot I_0$. What was its magnitude on the Richter scale?

5. a) Find the loudness, in decibels, of the sound of an automobile for which the intensity I is $3{,}100{,}000 \cdot I_0$.
   b) Find the loudness of sound pain having an intensity $10^{14} \cdot I_0$.

threshold intensity that is the weakest earthquake that can be recorded on a seismograph. If one earthquake is 10 times as intense as another, its magnitude on the Richter scale is 1 higher. If one earthquake is 100 times as intense as another, its magnitude on the Richter scale is 2 higher, and so on. Thus an earthquake whose magnitude is 7 on the Richter scale is 10 times as intense as an earthquake whose magnitude is 6. Earthquakes can be interpreted as multiples of the minimum intensity I_0.

The devastating earthquake in Armenia had an intensity of $10^{6.9} \cdot I_0$. What was its magnitude on the Richter scale?

Solution We substitute into the formula:

$$R = \log \frac{10^{6.9} I_0}{I_0} = \log 10^{6.9} = 6.9.$$

The magnitude of the earthquake was 6.9 on the Richter scale.

DO EXERCISE 4.

EXAMPLE 5 *Loudness of sound.* The sensation of loudness of sound is not proportional to the energy intensity, but rather is a logarithmic function. **Loudness L,** in Bels (after Alexander Graham Bell), of a sound of intensity I is defined to be

$$L = \log \frac{I}{I_0},$$

where I_0 is the minimum intensity detectable by the human ear (such as the tick of a watch at 20 ft under quiet conditions). If a sound is 10 times as intense as another, its loudness is 1 Bel greater. If a sound is 100 times as intense as another, it is louder by 2 Bels, and so on. The Bel is a large unit, so a subunit, a **decibel,** is generally used. For L in decibels, the formula is

$$L = 10 \log \frac{I}{I_0}.$$

a) Find the loudness, in decibels, of the sound in a radio studio for which the intensity I is $199 \cdot I_0$.
b) Find the loudness of a jet aircraft for which the intensity is $10^{12} \cdot I_0$.

Solution In each case, we substitute into the formula.

a) $$L = 10 \log \frac{199 I_0}{I_0} = 10 \log 199 \approx 10(2.2989) \approx 23 \text{ decibels}$$

b) $$L = 10 \log \frac{10^{12} I_0}{I_0} = 10 \log 10^{12} = 10(12) = 120 \text{ decibels}$$

DO EXERCISE 5.

EXERCISE SET 5.8

1 Solve.

1. *Compact discs.* The number of compact discs purchased each year is increasing exponentially. The number N, in millions, purchased is given by

$$N(t) = 7.5(6)^{0.5t},$$

where $t = 0$ corresponds to 1985, $t = 1$ corresponds to 1986, and so on (t being the number of years after 1985).

a) After what amount of time will 1 billion compact discs be sold in a year?
b) What is the doubling time on the sale of compact discs?

2. *Growth of bacteria* Escherichi coli. The bacteria *Escherichi coli* is commonly found in the bladder of human beings. Suppose that 3000 of the bacteria is present at time $t = 0$. Then t minutes later, the number of bacteria present will be

$$N(t) = 3000(2)^{t/20}.$$

a) After what amount of time will there be 60,000 bacteria?
b) If the bladder is not emptied and 100,000,000 bacteria accumulate, a bladder infection can occur. What amount of time would have to pass before a possible bladder infection would occur?
c) What is the doubling time?

3. *Interest compounded annually.* Suppose that \$50,000 is invested at 9% interest, compounded annually. After time t, in years, it grows to an amount A.

a) Express A as a function of t.
b) After what amount of time will there be \$450, 000 in the account?
c) Find the doubling time.

4. *Recycling aluminum cans.* It is known that $\frac{1}{4}$ of all aluminum cans distributed will be recycled each year. A beverage company distributes 250,000 cans. The number still in use after time t, in years, is given by the function

$$N(t) = 250{,}000\left(\frac{1}{4}\right)^{t}.$$

a) After what year will 60,000 cans still be in use?
b) After what amount of time will only 10 cans still be in use?

5. *Salvage value.* An office machine is purchased for \$5200. Its value each year is about 80% of the value the preceding year. Its value after t years is given by the exponential function

$$V(t) = \$5200(0.8)^{t}.$$

a) After what amount of time will the salvage value be \$1200?
b) After what amount of time will the salvage value be half of its original value? This is known as the **half-life.**

6. *Turkey consumption.* The amount of turkey consumed by each person in this country is increasing exponentially. Assuming $t = 0$ corresponds to 1937, the amount of turkey, in pounds per person, consumed t years after 1937 is given by the function

$$N(y) = 2.3(3)^{0.033t}.$$

a) After what amount of time will each person consume 20 lb of turkey?
b) What is the doubling time on the consumption of turkey?

7. *Psychology: Forgetting.* Students in an English class took a final exam. They took equivalent forms of the exam in monthly intervals thereafter. The average score $S(t)$, in percent, after t months was found to be given by

$$S(t) = 68 - 20 \log (t + 1), \quad t \geq 0.$$

a) What was the average score when they initially took the test, $t = 0$?
b) What was the average score after 4 months? after 24 months?
c) Graph the function.
d) After what time t was the average score 50?

8. *Psychology: Forgetting.* Students in an accounting class took a final exam. They took equivalent forms of the exam in monthly intervals thereafter. The average score $S(t)$, in percent, after t months was found to be given by

$$S(t) = 78 - 15 \log (t + 1), \quad t \geq 0.$$

a) What was the average score when they initially took the test, $t = 0$?
b) What was the average score after 4 months? after 24 months?
c) Graph the function.
d) After what time t was the average score 30?

9. *Business: Advertising.* A model for advertising response is given by

$$N(a) = 1000 + 200 \log a, \quad a \geq 1,$$

where $N(a)$ = the number of units sold and a = the amount spent on advertising, in thousands of dollars.

a) How many units were sold after spending \$1000 ($a = 1$) on advertising?
b) How many units were sold after spending \$5000?
c) Graph the function.
d) How much would have to be spent in order to sell 1276 units?

10. *Business: Advertising.* A model for advertising response is given by

$$N(a) = 2000 + 500 \log a, \quad a \geq 1,$$

where $N(a)$ = the number of units sold and a = the amount spent on advertising, in thousands of dollars.

a) How many units were sold after spending \$1000 ($a = 1$) on advertising?
b) How many units were sold after spending \$8000?
c) Graph the function.
d) How much would have to be spent in order to sell 5000 units?

Consider the pH formula for Exercises 11–18.

Find the pH of each substance given the hydrogen ion concentration.

11. Pineapple juice; $[H^+] = 1.6 \times 10^{-4}$
12. A common brand of insect repellent; $[H^+] = 4.0 \times 10^{-8}$
13. Tomato; $[H^+] = 6.3 \times 10^{-5}$
14. Egg; $[H^+] = 1.6 \times 10^{-8}$

Find the hydrogen ion concentration of each substance given the pH.

15. Rainwater; pH = 5.4
16. Water; pH = 7
17. Wine; pH = 4.8
18. Orange juice; pH = 3.2
19. The San Francisco earthquake of 1906 had an intensity of $10^{8.25} \cdot I_0$. What was its magnitude on the Richter scale?
20. In 1986, there was an earthquake near Cleveland, Ohio. It had an intensity of $10^5 \cdot I_0$. What was its magnitude on the Richter scale?
21. Find the loudness, in decibels, of the sound in a library that is 2510 times as intense as the minimum intensity I_0.
22. Find the loudness, in decibels, of the sound of a dishwasher that is 2,500,000 times as intense as the minimum intensity I_0.
23. Find the loudness, in decibels, of conversational speech having an intensity that is 10^6 times as intense as I_0.
24. Find the loudness, in decibels, of the sound of a heavy truck having an intensity 10^9 times as intense as I_0.

SYNTHESIS

25. *Typing speed.* A person is studying typing one semester in college. After t hours, the speed of the typist, in words per minute, is given by

$$S(t) = 200[1 - (0.86)^t].$$

a) When will the typist's speed be 100 words per minute?
b) After the course is completed, the typist's speed is 150 words per minute. How many hours of studying occurred in the course?

26. *Compound interest.* The amount A that principal P will be worth after t years at interest rate i, compounded n times per year, is given by the formula

$$A = P\left(1 + \frac{i}{n}\right)^{nt}.$$

Suppose that \$100,000 is invested at 8% interest, compounded quarterly.

a) Express A as a function of t.
b) How much time will it take for the original investment to grow to \$1 million?
c) What is the doubling time?

27. Solve the earthquake-magnitude formula for I.
28. Solve the loudness-of-sound formula for I.
29. Solve the pH formula for $[H^+]$.
30. *Loudness of sound.* Two sounds have intensities I_1 and I_2, respectively.

a) Show that the difference in the loudness of the sounds can be expressed as

$$L_2 - L_1 = 10 \log \frac{I_2}{I_1}.$$

b) Find a formula for the sum of the loudness of the two sounds.

31. *Apparent magnitude of stars.* The **apparent magnitude** of a star is the measure of its brightness. In ancient times, the brightest star was of magnitude 1 and those just visible to the naked eye were of magnitude 6. In more recent times, with the advent of improved measuring devices, stars of negative magnitude and of magnitude as great as 25 have been observed. Two stars with *apparent magnitudes* M_1 and M_2 and *apparent brightness* I_1 and I_2 are related by

$$M_2 - M_1 = 2.5 \log \frac{I_1}{I_2}.$$

a) The star Sirius, which is actually made up of more than one star, has an apparent magnitude of -1.5, and was the brightest star known as of this writing. The star closest to the earth is Alpha Centauri and has an apparent magnitude of -0.3. Find the ratio of the intensity of Sirius to the intensity of Alpha Centauri.

b) The star Luyten has an apparent magnitude of 12.6. Find the ratio of the intensity of Sirius to the intensity of Luyten.

c) Two stars have a difference in apparent magnitude of 5. How many times brighter is one over the other?

32. *Absolute magnitude of stars.* When a star's brightness is compared to that of the sun, we obtain what is called the star's **absolute magnitude.** The **luminosity** of a star is the measure of the total amount of energy radiated. The sun's luminosity is 3.9×10^{26} watts. The absolute magnitude M of a star is, then, a function of its luminosity as given by the function

$$M = 4.75 - 2.5 \log \left(\frac{L}{3.9 \times 10^{26}}\right).$$

a) The star Luyten has an absolute magnitude of 14.9. Find its luminosity.

b) The star Sirius has an absolute magnitude of 1.4. Find its luminosity.

c) Solve the formula for L.

33. *Allometric law.* Biologists speak of an *allometric law,* or *allometry,* between two variables x and y when there exist constants b and k such that $y = bx^k$. When such a relationship exists, it can be shown that $Y = \log y$ is a linear function of $X = \log x$, and this is usually discovered from data by considering graphs of logarithms of data, rather than the data themselves.

a) Show that Y is a linear function of X.

b) In an experiment on baboons, skull length y, in millimeters, was related to face length x, in millimeters, by allometry. When $x = 31.0$ mm, $y = 78.5$; and when $x = 144.25$ mm, $y = 122$. Find a linear relationship between $Y = \log x$ and $X = \log y$.

c) From your answer to part (b), express y as a function of x.

d) Use the answer to part (c) to predict the skull length of a baboon with a face length of 131.0 mm.

5.9 Applications of Exponential and Logarithmic Functions, Base *e*

OBJECTIVE

You should be able to:

1 Solve problems involving applications of exponential and logarithmic functions, base *e*.

1 Exponential and logarithmic functions, base e, are rich in application to many fields such as psychology, business, sociology, and science. We now consider many of these applications. A calculator with logarithmic and power keys would be most helpful for this section.

EXAMPLE 1 *Psychology: Walking speed.* In a study by psychologists Bornstein and Bornstein, it was found that the average walking speed R of a person living in a city of population P, in thousands, is given by the function

$$R(P) = 0.37 \ln P + 0.05,$$

where R is in feet per second.

a) The population of Albuquerque, New Mexico, is 290,000. Find the average walking speed of people living in Albuquerque.

1. The population of certain cities is given below. Find the walking speed of people in each city.

   a) Los Angeles, California: 3,097,000

   b) Key West, Florida: 29,600

   c) Seattle, Washington: 531,000

b) Graph the function.

Solution

a) We substitute 290 for P, since P is in thousands:

$$R(290) = 0.37 \ln 290 + 0.05 \quad \text{Substituting}$$
$$\approx 0.37(5.6699) + 0.05 \quad \text{Finding the natural logarithm on a calculator}$$
$$\approx 2.1 \text{ ft/sec.}$$

The average walking speed of people living in Albuquerque is 2.1 ft/sec.

b) We find several function values using a calculator and sketch the graph, as follows. Note that the axes have to be scaled differently because inputs are very large and outputs are small by comparison.

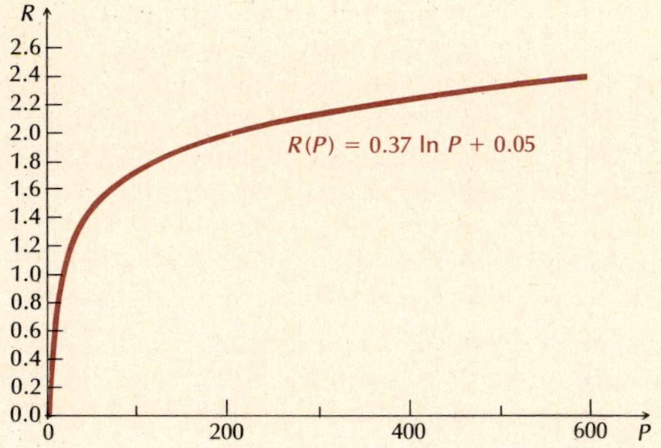

DO EXERCISE 1.

Population Growth

The equation

$$P(t) = P_0 e^{kt}$$

is an effective model of many kinds of population growth, whether it be a population of people or a population of money. In this equation, P_0 is the number of people at time 0, P is the population after time t, and k is often called the **exponential growth rate**. The graph of such an equation is shown here:

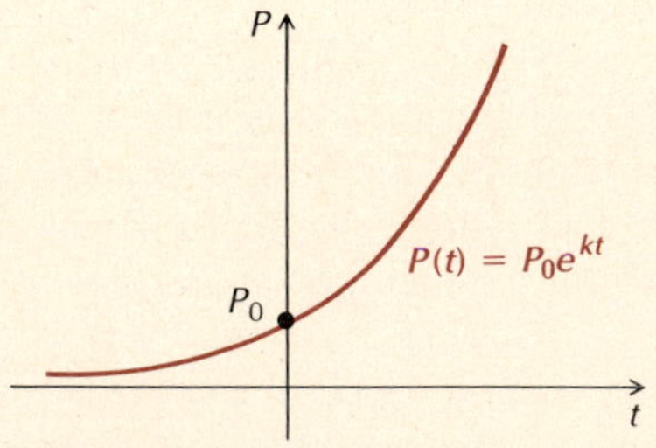

EXAMPLE 2 *Growth of the United States.* In 1985, the population of the United States was 234 million and the exponential growth rate was 0.8% per year.

a) Find the exponential growth function.

b) What will the population be in 1996? in 2000?

c) Graph the exponential growth function.

Solution

a) At $t = 0$ (1985), the population was 234 million. We substitute 234 for P_0 and 0.8%, or 0.008, for k to obtain the exponential growth function

$$P(t) = 234e^{0.008t}.$$

b) In 1996, $t = 11$; that is, 11 years have passed. To find the population in 1996, we substitute 11 for t:

$$\begin{aligned} P(11) &= 234e^{0.008(11)} && \text{Substituting 11 for } t \\ &= 234e^{0.088} \\ &\approx 234(1.0920) && \text{Finding } e^{0.088} \text{ using a calculator} \\ &\approx 256. \end{aligned}$$

The population of the United States in 1996 will be about 256 million.

Calculator note. **When you are doing calculations like this on your calculator, you need not round to a number like 1.0920 before multiplying by 234. Just carry out the multiplication. Answers will be found that way in the exercises. You may notice some variance in the last one or two decimal places if you round as you go. You might check with your instructor regarding such variation.**

In 2000, $t = 15$, that is, 15 years have passed. To find the population in 2000, we substitute 15 for t:

$$\begin{aligned} P(15) &= 234e^{0.008(15)} && \text{Substituting 15 for } t \\ &= 234e^{0.12} \\ &\approx 234(1.1275) && \text{Finding } e^{0.12} \text{ using a calculator} \\ &\approx 264 \text{ million.} \end{aligned}$$

The population of the United States in 2000 will be about 264 million.

c) We find other function values and sketch the graph as follows.

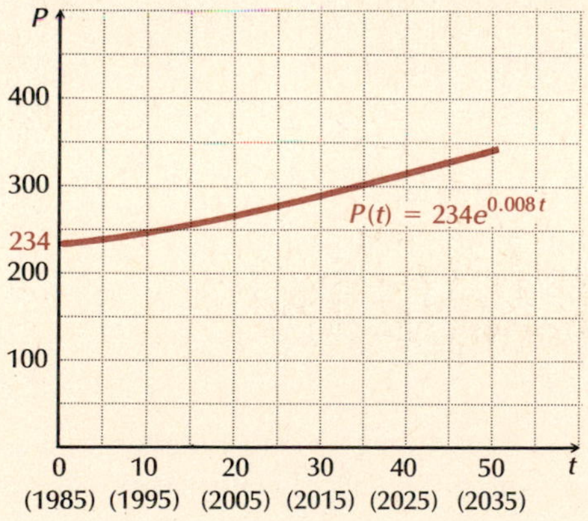

DO EXERCISE 2.

EXAMPLE 3 *The cost of a first-class postage stamp.* The cost of a first-class postage stamp became 3 cents in 1932, and the exponential growth rate of the cost was 3.8% per year. The exponential growth function for the cost is found by substituting 3 for P_0 and 3.8%, or 0.038, for k to obtain the function

$$P(t) = 3e^{0.038t}.$$

2. What will the population of the United States be in 1998? in 2020?

3. a) What will the cost of a first-class stamp be in 1995?
b) When will a first-class stamp cost $2?

a) The cost of first-class postage increased to 25 cents in 1988. Use the function to predict the cost in 1988 and compare.
b) What will the cost of a first-class stamp be in 2000?
c) When will the cost of a first-class postage stamp be $1.00?

Solution

a) At $t = 0$ (1932), the cost of a stamp was 3 cents. Since 1988 is 56 years from 1932, in 1988, we have $t = 56$. To find the cost in 1988, according to the function, we substitute 56 for t:

$$\begin{aligned} P(56) &= 3e^{0.038(56)} && \text{Substituting 56 for } t \\ &= 3e^{2.128} \\ &\approx 3(8.3981) && \text{Finding } e^{2.128} \text{ using a calculator} \\ &\approx 25.2. \end{aligned}$$

The function seems to be a fairly accurate predictor of the cost of first-class postage. The actual amount is 25 cents.

b) In the year 2000, $t = 68$, that is, 68 years have passed. To find the cost in 2000, we substitute 68 for t:

$$\begin{aligned} P(68) &= 3e^{0.038(68)} && \text{Substituting 68 for } t \\ &= 3e^{2.584} \\ &\approx 3(13.2500) && \text{Finding } e^{2.584} \text{ using a calculator} \\ &\approx 40. \end{aligned}$$

The cost of a first-class stamp in the year 2000 will be about 40 cents.

c) To find when the cost will be $1.00, we substitute 100 for $P(t)$ and solve for t:

$$\begin{aligned} 100 &= 3e^{0.038t} \\ \frac{100}{3} &= e^{0.038t} \\ \ln \frac{100}{3} &= \ln e^{0.038t} && \text{Taking the natural logarithm on both sides} \\ \ln 33.3333 &\approx 0.038t && \text{Using Property 3 and Property 4, recalling that } \ln e^{0.038t} = \log_e e^{0.038t} \\ 3.5066 &\approx 0.038t \\ 92 &\approx t. \end{aligned}$$

In 92 years from 1932, or in 2024, the cost of first-class postage is predicted to be $1.00. ■

DO EXERCISE 3.

In order to fit an exponential growth function to a situation, we have to determine P_0 and k from given data. Then we can make predictions.

EXAMPLE 4 *Medical: Heart transplants.* In 1967, Dr. Christian Barnard, of South Africa, staggered the world by performing the first heart transplant. There was 1 transplant in 1967. In 1987, there were 1418 such transplants. The numbers of transplants through these years are summarized in the table shown at the left.

Year	Number of Heart Transplants
1967	1
1971	13
1975	23
1980	36
1985	719
1987	1418

Source: The 1987 National Heart Transplant Study and the International Society for Heart Transplantation.

a) Graph the function. Do the table and the graph seem to be of an exponential function?
b) Find an exponential growth function that fits the data.
c) Use the function to predict the number of heart transplants in 1995.

Solution

a) We use the data from the table and sketch the following graph. We draw a curve as close to the data as possible. It is very close to the graph of an exponential function.

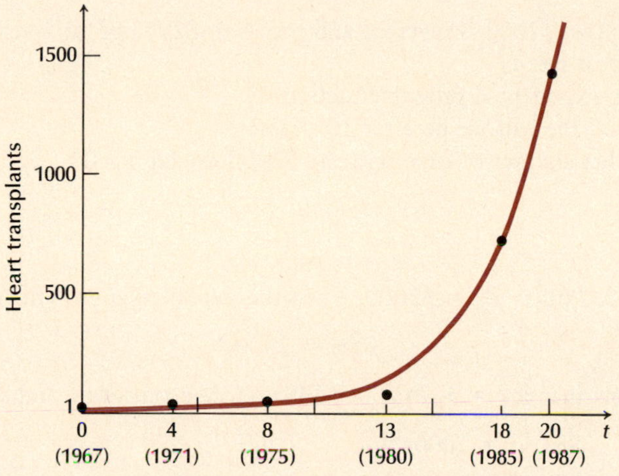

Considering the rapid growth and comparing the graph to other exponential functions we have considered, the assumption of exponential growth seems reasonable. A more thorough analysis would require more advanced mathematics.

b) The exponential growth function is

$$N(t) = N_0 e^{kt}.$$

We assume that $t = 0$ corresponds to 1967 and that $N(0) = N_0 e^{k(0)} = N_0 = 1$. Then the growth function is

$$N(t) = e^{kt}.$$

To find k, we can use the fact that at $t = 20$ (1987), the number of transplants was 1418. Then we substitute and solve for k as follows:

$$1418 = e^{k(20)} \quad \text{Substituting}$$
$$1418 = e^{20k}$$
$$\ln 1418 = \ln e^{20k} \quad \text{Taking the natural logarithm on both sides}$$
$$\ln 1418 = 20k \quad \text{Remember, } \log_a a^k = k, \text{ so } \ln e^{20k} = 20k, \text{ since } \log e^{20k} = \log_e e^{20k}$$
$$\frac{\ln 1418}{20} = k$$
$$\frac{7.2570}{20} \approx k$$
$$0.363 \approx k.$$

Thus the exponential growth function is

$$N(t) = e^{0.363t}.$$

c) The year 1995 is 28 years from 1967. We let $t = 28$ and find $N(28)$:

$$N(28) = e^{0.363(28)} \approx 25{,}952.$$

Thus, according to the exponential growth function, there will be about 25,952 heart transplants in 1995.

DO EXERCISE 4.

4. Predict the number of heart transplants in 2010.

5. a) Suppose that \$10,000 is invested and grows to \$14,049.48 in 4 years. What is the interest rate k, assuming interest is compounded continuously?
   b) Find the exponential growth function.
   c) What will the balance be after 10 years?
   d) After what amount of time will the \$10,000 double itself?

EXAMPLE 5 *Business: Interest compounded continuously.* Suppose that an amount P_0 is invested in a savings account at interest rate k, compounded continuously. The balance $P(t)$, after t years, is given by the exponential function

$$P(t) = P_0 e^{kt}.$$

a) Suppose that \$2000 is invested and grows to \$2983.65 in 5 years. What is the interest rate k?
b) Find the exponential growth function.
c) What will the balance be after 10 years?
d) After what amount of time will the \$2000 double itself?

Solution

a) At $t = 0$, $P(0) = P_0 = \$2000$. Thus the exponential growth function is

$$P(t) = 2000e^{kt}.$$

We know that at $t = 5$, $P(5) = \$2983.65$. We substitute and solve for k:

$$2983.65 = 2000e^{k(5)}$$

$$2983.65 = 2000e^{5k}$$

$$\frac{2983.65}{2000} = e^{5k} \qquad \text{Dividing both sides by 2000}$$

$$1.491825 = e^{5k}$$

$$\ln 1.491825 = \ln e^{5k} \qquad \text{Taking the natural logarithm on both sides}$$

$$0.4 \approx 5k \qquad \text{Finding ln 1.491825 on a calculator and simplifying ln } e^{5k}$$

$$\frac{0.4}{5} \approx k$$

$$0.08 \approx k.$$

The interest rate is about 0.08, or 8%.

b) The exponential growth function is

$$P(t) = 2000e^{0.08t}.$$

c) The balance after 10 years is

$$P(10) = 2000e^{0.08(10)} = 2000e^{0.8}$$
$$\approx 2000(2.2255410) \approx \$4451.08.$$

d) To find the doubling time T, we set $P(T) = \$4000$ and solve for T:

$$4000 = 2000e^{0.08T}$$

$$2 = e^{0.08T}$$

$$\ln 2 = \ln e^{0.08T}$$

$$\ln 2 = 0.08T$$

$$\frac{\ln 2}{0.08} = T$$

$$\frac{0.693147}{0.08} \approx T$$

$$8.7 \approx T.$$

Thus the original investment of \$2000 will double itself in about 8.7 years. ■

DO EXERCISE 5.

We can find a general expression relating the growth rate k and the doubling time T by solving the following equation:

$$2P_0 = P_0 e^{kT} \quad \text{Substituting } 2P_0 \text{ for } P$$
$$2 = e^{kT} \quad \text{Multiplying by } 1/P_0$$
$$\ln 2 = \ln e^{kT}$$
$$\ln 2 = kT.$$

THEOREM 7

The growth rate k and the doubling time T are related by

$$kT = \ln 2 \approx 0.693147,$$

or

$$k = \frac{\ln 2}{T} \approx \frac{0.693147}{T}$$

and

$$T = \frac{\ln 2}{k} \approx \frac{0.693147}{k}.$$

Note that this relationship between k and T does not depend on P_0.

EXAMPLE 6 The population of the world is now doubling every 24.8 years. What is its exponential growth rate?

Solution

$$k = \frac{\ln 2}{T} \approx \frac{0.693147}{24.8} \approx 0.028 = 2.8\%$$

The growth rate of the world is about 2.8% per year. ■

DO EXERCISES 6–8.

Exponential Decay

The function

$$P(t) = P_0 e^{-kt}$$

is an effective model of the decline, or decay, of a population. An example is the decay of a radioactive substance. Here P_0 is the amount of the substance at time $t = 0$, P is the amount of the substance left after time t, k is a positive constant that depends on the situation. The constant k is called the **decay rate.**

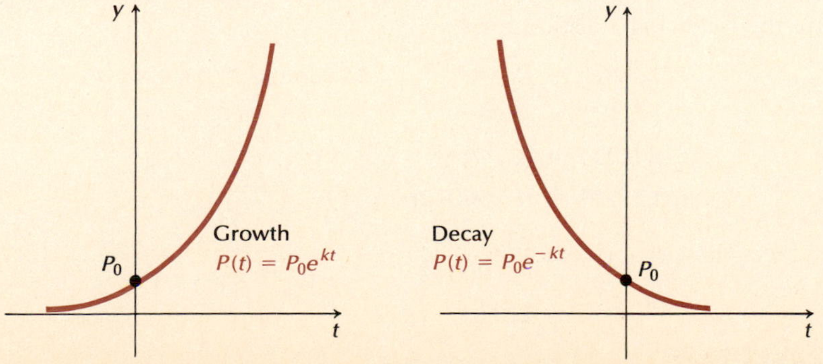

6. The exponential growth rate of the Bahamas is about 4.1%. What is the doubling time of the population?

7. The doubling time of the population of Guam is about 13.9 years. What is the exponential growth rate?

8. A financial institution advertises that it will double your money in 6.8 years. What is the interest rate, assuming interest is compounded continuously?

The **half-life** of bismuth is 5 days. This means that half of an amount of bismuth will cease to be radioactive in 5 days. The effect of half-life is shown in the graph below. The exponential function gets close to 0, but never reaches 0, as t gets larger. Thus, in theory, a radioactive substance never completely decays.

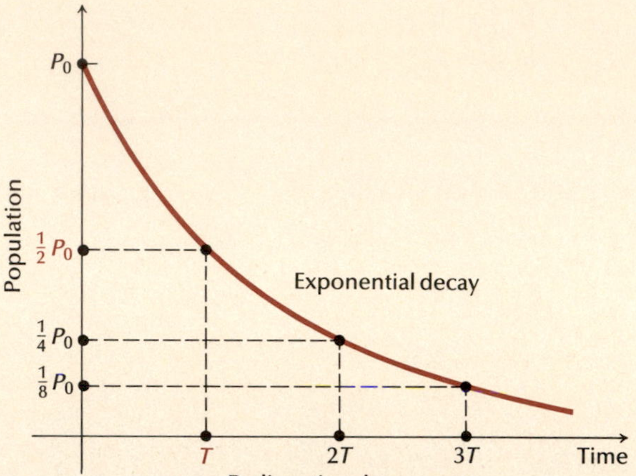

Radioactive decay curve

EXAMPLE 7 *Carbon dating.* The radioactive element carbon-14 has a half-life of 5750 years. The percentage of carbon-14 present in the remains of living matter can be used to determine their age. Archeologists found that the linen wrapping from one of the Dead Sea Scrolls had lost 22.3% of its carbon-14. How old was the linen wrapping?

Solution We first find k. To do so, we use the concept of half-life. When $t = 5750$ (half-life), P will be half of P_0. We substitute $\frac{1}{2}P_0$ for P and 5570 for t and solve for k. Then

$$\frac{1}{2}P_0 = P_0e^{-k(5750)}, \quad \text{or} \quad \frac{1}{2} = e^{-5750k}.$$

We take the natural logarithm on both sides:

$$\ln\frac{1}{2} = \ln e^{-5750k} = -5750k.$$

Then

$$k = \frac{\ln 0.5}{-5750} \approx \frac{-0.6931}{-5750} \approx 0.00012.$$

Now we have the function

$$P(t) = P_0e^{-0.00012t}.$$

(*Note:* This equation can be used for any subsequent carbon-dating problem.) If the linen wrapping has lost 22.3% of its carbon-14 from an initial amount P_0, then $77.7\%(P_0)$ is the amount present. To find the age t of the wrapping, we solve the following equation for t:

$$\begin{aligned} 77.7\%P_0 &= P_0e^{-0.00012t} && \text{Substituting } 77.7\%P_0 \text{ for } P \\ 0.777 &= e^{-0.00012t} \\ \ln 0.777 &= \ln e^{-0.00012t} \\ -0.25231 &\approx -0.00012t \\ t &= \frac{-0.25231}{-0.00012} \\ t &\approx 2103. \end{aligned}$$

How can scientists determine that an animal bone has lost 30% of its carbon-14? The assumption is that the percentage of carbon-14 in the atmosphere and in living plants and animals is the same. When a plant or an animal dies, the amount of carbon-14 decays exponentially. The scientist burns the animal bone and uses a Geiger counter to determine the percentage of the smoke that is carbon-14. It is the amount that this varies from the percentage in the atmosphere that tells how much carbon-14 has been lost.

The process of carbon-14 dating was developed by the American chemist Willard E. Libby in 1952. It is known that the radioactivity in a living plant is 16 disintegrations per gram per minute. Since the half-life of carbon-14 is 5750 years old, an object with an activity of 8 disintegrations per gram per minute is 5750 years old, one with an activity of 4 disintegrations per gram per minute is 11,500 years old, and so on. Carbon-14 dating can be used to measure the age of objects from 30,000 to 40,000 years old. Beyond such an age, it is too difficult to measure the radioactivity and some other method would have to be used.

Carbon-14 was indeed used to find the age of the Dead Sea Scrolls. It was used recently to refute the authenticity of the Shroud of Turin, presumed to have covered the body of Christ.

Thus the linen wrapping on the Dead Sea Scrolls is about 2103 years old. ■

DO EXERCISE 9.

9. How old is a skeleton that has lost 80% of its carbon-14?

EXERCISE SET 5.9

1 Various cities and their populations are given below. Find the walking speed of people in each city. (See Example 1.)

1. New York, New York: 7,900,000
2. Pittsburgh, Pennsylvania: 853,000
3. Rome, New York: 50,400
4. Reno, Nevada: 106,000

5. *Cost of a Hershey bar.* The cost of a Hershey chocolate bar in 1962 was 5 cents and was increasing at an exponential growth rate of 9.7%.
   a) Find an exponential function describing the growth of the cost of a Hershey bar.
   b) What will a Hershey bar cost in 1990? in 2000?
   c) When will a Hershey bar cost $5?
   d) Graph the function.
   e) What is the doubling time of the cost of a Hershey bar?

6. *World population growth.* The population of the world passed 5.0 billion in 1987. The exponential growth rate was 2.8% per year.
   a) Find the exponential growth function.
   b) Predict the population of the world in 1996; in 2000.
   c) When will the world population be 6.0 billion?
   d) Graph the function.

7. *Consumer price index.* The consumer price index is often in the news. It compares the cost of goods and services over various years, where 1967 is used as a base (P_0). Although rates vary for different years, an average exponential growth rate is 6%.
   a) Find the exponential function for the consumer price index.
   b) Goods and services that cost $100 in 1967 will cost how much in 1995?
   c) Goods and services that cost $100 in 1967 will cost how much in 2000?
   d) Graph the function.

8. *Population growth of rabbits.* Under ideal conditions, a population increase of rabbits has an exponential growth rate of 11.7% per day. Suppose one starts with a population of 100 rabbits.
   a) Find an exponential function describing the growth of the population of rabbits after t days.
   b) What will the population of rabbits be after 7 days?
   c) Graph the function.
   d) What is the doubling time of the population of rabbits?

9. *Cost of a 60-second commercial during the Super Bowl.* Past data on the cost of a 60-second commercial during the Super Bowl are given in the table below.

Year	1967	1970	1977	1981	1983	1985	1988
Cost of 60-sec TV Commercial During the Super Bowl	$80,000	$200,000	$324,000	$550,000	$800,000	$1,100,000	$1,350,000

   a) Make a graph of the data and analyze the table. Does it appear that we can fit an exponential function to the data?
   b) Find an exponential growth function for the cost of a Super Bowl commercial. Assume that C_0 = $80 thousand. That is, at $t = 0$ (1967), C = $80 (in thousands). Find k using the data point $C(21)$ = $1350 thousand. That is, in 1988, $1350 thousand, or $1,350,000, was the cost of a 60-second commercial.
   c) What will the cost of a 60-second commercial be in 1995?
   d) When will the cost of a 60-second commercial be $3,000,000?
   e) What is the doubling time for the cost of a 60-second Super Bowl commercial?

10. *Cost of a double-dip ice cream cone.* In 1970, the cost of a double-dip ice cream cone was 52 cents. In 1978, it was 66 cents. Assuming the exponential model:

a) Find the value k ($P_0 = 52$). Write the exponential growth function.
b) Estimate the cost of a cone in 1994.
c) After what period of time will the cost of a cone be twice that of 1978?
d) When will the cost of a cone be $3?

11. *Interest compounded continuously.* Suppose that P_0 is invested in a savings account in which interest is compounded continuously at 9% per year. That is, the balance $P(t)$ after time t, in years, is

$$P(t) = P_0 e^{kt}.$$

a) Find the exponential function for $P(t)$ in terms of P_0 and 0.09.
b) Suppose that $5000 is invested. What is the balance after 1 year? after 2 years?
c) When will an investment of $5000 double itself?

12. *Interest compounded continuously.* Suppose that P_0 is invested in a savings account in which interest is compounded continuously at 10% per year. That is, the balance $P(t)$ after time t, in years, is

$$P(t) = P_0 e^{kt}.$$

a) Find the solution of the equation in terms of P_0 and 0.10.
b) Suppose that $35,000 is invested. What is the balance after 1 year? after 2 years?
c) When will an investment of $35,000 double itself?

13. *Ecology: Population growth.* The growth rate of the population of Mexico is 3.5% per year (one of the highest in the world). What is the doubling time?

14. *Population growth.* The growth rate of the population in Europe west of the USSR is 1% per year. What is the doubling time?

15. *Annual interest rate.* A bank advertises that it compounds interest continuously and that it will double your money in 7 years. What is its annual interest rate?

16. *Annual interest rate.* A bank advertises that it compounds interest continuously and that it will double your money in 5.4 years. What is its annual interest rate?

17. *Value of a Van Gogh painting.* The Van Gogh painting *Irises* sold for $84,000 in 1947, but was sold again for $53,900,000 in 1987. Assuming that the growth in the value V of the painting was exponential:

a) Find the value k and determine the exponential growth function assuming $P_0 = 84{,}000$.
b) Estimate the value of the painting in 2007.
c) What is the doubling time for the value of the painting?
d) After what amount of time will the value of the painting be $1 billion?

Van Gogh's *Irises,* a 28-by-32-inch oil on canvas.

18. *Exponential growth of the value of a baseball card.* The collecting of baseball cards and other memorabilia has become a profitable hobby. The card shown here contains a photograph of Eddie Murray in his rookie season of 1978. The value of that card in 1983 was $7.75. Its value in 1987 was $27.00. The value of the card has increased so much because Murray has turned out to be such an outstanding player. Assume that the value of the card has grown exponentially.

a) Find the value k and determine the exponential growth function, assuming $V_0 = 7.75$.
b) Estimate the value of the card in 1995; in 2000.
c) What is the doubling time for the value of the card?
d) After what amount of time will the value of the card be $2000?

19. *Coal demand.* The growth rate of the demand for coal in the world is 4% per year. When will the demand be double that of 1990?

20. *Oil demand.* The growth rate of the demand for oil in the United States is 10% per year. When will the demand be double that of 1990?

21. *Population growth.* The population of Los Angeles, California, was 2,812,000 in 1970. In 1984, it was 3,097,000. Assuming that growth was exponential:

    a) Find the value of k ($P_0 = 2{,}812{,}000$). Write the function.
    b) Estimate the population of Los Angeles in 1996.

22. *Population growth.* The population of San Antonio, Texas, was 786,000 in 1980. In 1984, it was 843,000. Assuming that growth was exponential:

    a) Find the value of k ($P_0 = 786{,}000$). Write the function.
    b) Estimate the population of San Antonio in 2000.

23. A mummy discovered in the pyramid Khufu in Egypt has lost 46% of its carbon-14. Determine its age.

24. The statue of Zeus at Olympia in Greece is one of the Seven Wonders of the World. It is made of gold and ivory. The ivory was found to have lost 35% of its carbon-14. Determine the age of the statue.

25. The half-life of polonium is 3 minutes. What is its decay rate?

26. The half-life of lead is 22 years. What is its decay rate?

27. The decay rate of iodine-131 is 9.6% per day. What is its half-life?

28. The decay rate of krypton-85 is 6.3% per year. What is its half-life?

29. *Weight loss.* The initial weight of a starving animal is W_0. Its weight W after t days is given by

$$W(t) = W_0 e^{-0.007t}.$$

    a) What percentage of its weight does it lose each day?
    b) What percentage of its initial weight remains after 30 days?

30. *Satellite power.* The power supply of a satellite is a radioisotope. The power output P, in watts, decreases at a rate proportional to the amount present. P is given by

$$P(t) = 60e^{-0.006t},$$

where t = the time in days.

    a) How much power will be available after 365 days?
    b) What is the half-life of the power supply?
    c) The satellite's equipment cannot operate on fewer than 10 watts of power. How long can the satellite stay in operation?
    d) How much power did the satellite have to begin with?

31. *Atmospheric pressure.* Atmospheric pressure P at altitude a is given by

$$P = P_0 e^{-0.00005a},$$

where P_0 = the pressure at sea level. Assume that $P_0 = 14.7 \text{ lb/in}^2$ (pounds per square inch).

    a) Find the pressure at an altitude of 2000 ft.
    b) Find the pressure at the top of Mt. Shasta in California, which is 14,162 ft above sea level.
    c) At what altitude is the pressure 1.47 lb/in^2?
    d) Blood will boil when atmospheric pressure drops below 0.39 lb/in^2. At what altitude, in an unpressurized vehicle, will a pilot's blood boil?

32. *Salvage value.* A business estimates that the salvage value V of a piece of machinery after t years is given by

$$V(t) = \$46{,}000e^{-t}.$$

    a) What did the machinery cost initially?
    b) What is the salvage value after 2 years?

SYNTHESIS

33. *The Beer–Lambert law.* A beam of light enters a medium such as water or smog with initial intensity I_0. Its intensity decreases depending on the thickness (or concentration) of the medium. The intensity I at a depth (or concentration) of x units is given by

$$I = I_0 e^{-\mu x}.$$

The constant μ (the Greek letter "mu") is called the **coefficient of absorption**, and it varies with the medium. For sea water, $\mu = 1.4$.

    a) What percentage of light intensity I_0 remains at a depth of sea water that is 1 m? 3 m? 5 m? 50 m?
    b) Plant life cannot exist below 10 m. What percentage of I_0 remains at 10 m?

34. *Present value.* Following the birth of a child, a parent wants to make an initial investment P_0 that will grow to \$50,000 for the child's education at age 18. Interest is compounded continuously at 8%. What should the initial investment be? Such an amount is called the **present value** of \$50,000 18 years from now.

35. *Velocity of a rocket.* The theory of rocket flight shows that the velocity of a rocket when its propellant is burned to depletion is expressed by the equation

$$v = c \ln R,$$

where v = the velocity gained by the rocket during launch, c = the exhaust velocity of the engine, and R = the mass ratio of the rocket = (Takeoff weight)/(Burnout weight). Solve the formula for R.

36. *Electricity.* The formula

$$i = \frac{V}{R}[1 - e^{-(R/L)t}]$$

occurs in the theory of electricity. Solve for t.

37. *Supply and demand.* The supply and demand for the sale of stereos by a sound company are given by

$$S(x) = e^x, \qquad D(x) = 163{,}000e^{-x},$$

where $S(x)$ = the price at which the company is willing to supply x stereos and $D(x)$ = the demand price for a quantity of x stereos. Find the equilibrium point. (For reference, see Section 3.6.)

38. In reference to Exercise 31, explain how you might use a barometer, or some other device for measuring atmospheric pressure, to find the height of the Empire State Building.

39. *Newton's Law of Cooling.* An object which has a different temperature other than its surroundings will either cool down or heat up to that of its surroundings. Suppose a body that has a temperature T_1 is placed in surroundings with temperature T_0. The body will either cool or warm to temperature $T(t)$ after time t, in minutes, where

$$T(t) = T_0 + |T_1 - T_0|e^{-kt}.$$

A cup of coffee whose temperature is 105°F is placed in a freezer whose temperature is 32°F. After 5 minutes its temperature is 70°. What will its temperature be after 10 minutes?

CHALLENGE

40. *When was the murder committed?* The police discover the body of a math professor. Critical to solving the crime is determining when the murder was committed. The police call the coroner, who arrives at 12:00 PM. The coroner immediately takes the temperature of the body and finds it to be 94.6°. The coroner takes the temperature 1 hour later and finds it to be 93.4°. The temperature of the room is 70°. When was the murder committed? Use Newton's Law of Cooling, Exercise 39.

SUMMARY AND REVIEW: CHAPTER 5

TERMS TO KNOW

Inverse relation	Exponential function	Natural logarithm
One-to-one function	Logarithmic function	Antilogarithm
Inverse function	Base	Exponential equation
Horizontal-line test	Common logarithm	Logarithmic equation

REVIEW EXERCISES

1. Find the inverse of the relation H given by

$$H = \{(-4, 5), (2, -3), (1, 7), (8, 8), (5, -4)\}.$$

Write an equation of the inverse.

2. $y = 3x^2 + 2x - 1$

3. $y = \sqrt{x + 2}$

4. Which of the following relations have graphs that are symmetric with respect to the line $y = x$?

a) $y = 7$ b) $x^2 + y^2 = 4$ c) $x^3 = y^3 - y$ d) $y^2 = x^2 + 3$

e) $x + y = 3$ f) $x = 3$ g) $y = x^2$ h) $y = x^3$

5. Which of the following have inverses that are functions?

a)
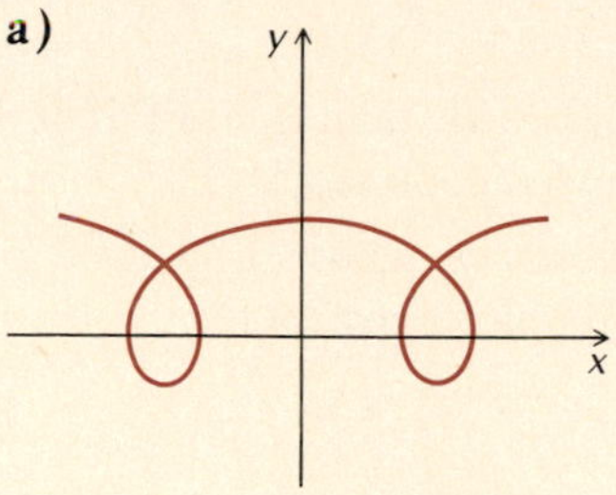

b)
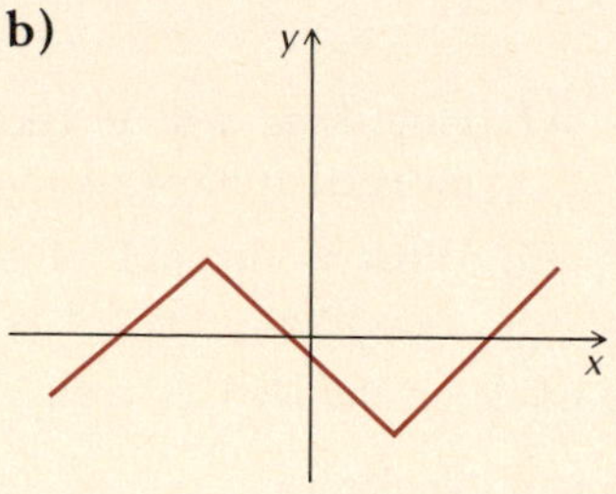

c)
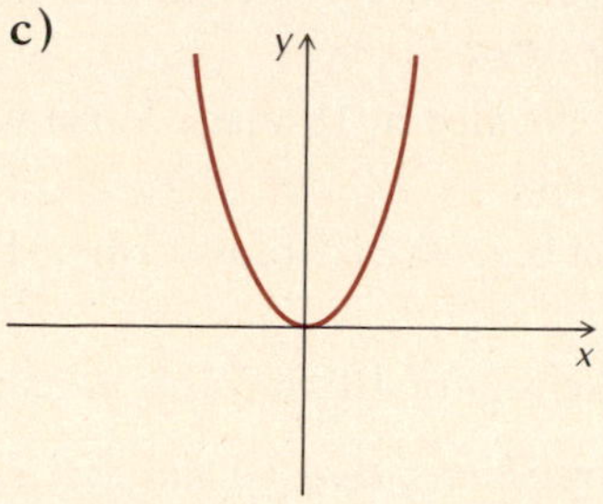

d)
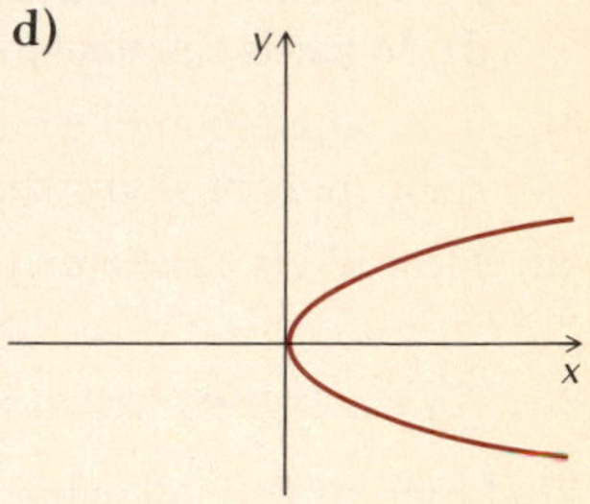

Find a formula for $f^{-1}(x)$.

6. $f(x) = \dfrac{\sqrt{x}}{2} + 2$

7. $f(x) = x^3 + 8$

8. Find $f(f^{-1}(a))$:
$$f(x) = x^3 + 2.$$

9. Find $h^{-1}(h(t))$:
$$h(x) = x^{17} + x^{65}.$$

Graph.

10. $y = \log_2 (x - 1)$

11. $y = \left(\frac{1}{2}\right)^x$

12. $f(x) = 3(1 - e^{-x})$, for nonnegative values of x

13. $f(x) = \ln (x - 4)$

14. Find $\log_3 10$ using common logarithms.

15. Find $\log_6 2$ using natural logarithms.

16. Write an exponential equation equivalent to $\log_8 \frac{1}{4} = -\frac{2}{3}$.

17. Write a logarithmic equation equivalent to $7^{2.3} = x$.

18. Write an equivalent expression containing a single logarithm:
$$\tfrac{1}{2} \log_b a + \tfrac{3}{2} \log_b c - 4 \log_b d.$$

19. Express in terms of logarithms of M and N:
$$\log \sqrt[3]{M^2/N}.$$

Given that $\log_a 2 = 0.301$, $\log_a 3 = 0.477$, and $\log_a 7 = 0.845$, find each of the following.

20. $\log_a 18$ 21. $\log_a \frac{7}{2}$ 22. $\log_a \frac{1}{4}$ 23. $\log_a \sqrt{3}$

Simplify.

24. $\log_{12} 12^{x^2 + 1}$

25. $\log_8 8^{\sqrt{9}}$

Solve.

26. $\log_x 64 = 3$ 27. $\log_{16} 4 = x$ 28. $\log_5 125 = x$

29. $3^{1-x} = 9^{2x}$ 30. $e^x = 80$ 31. $\log x^2 = \log x$

32. $\log (x^2 - 1) - \log (x - 1) = 1$ 33. $\log 2 + 2 \log x = \log (5x + 3)$ 34. $\log_2 (x - 1) + \log_2 (x + 1) = 3$

35. How many years will it take an investment of \$1000 to double if interest is compounded annually at 13%?

36. What is the loudness, in decibels, of a sound whose intensity is $1000I_0$?

37. The half-life of a radioactive substance is 15 days. How much of a 25-gram sample will remain radioactive after 30 days?

38. *Forgetting.* In an art class, students were tested at the end of the course on a final exam. They were tested again after 6 months. The forgetting formula was determined to be
$$S(t) = 82 - 38 \log (t + 1),$$
where t is the time, in months, after taking the first test.

a) What was the average score when they initially took the test, $t = 0$?

b) What was the average score after 6 months?

c) After what time was the average score 54?

39. *The cost of a prime-rib dinner.* The average cost C of a prime-rib dinner was \$4.65 in 1962. In 1986, it was \$15.81. Assume that the growth followed the exponential growth function.

a) Find k and write the exponential growth function.
b) How much will a prime-rib dinner cost in 2010?
c) When will the average cost of a prime-rib dinner be \$20?
d) What is the doubling time?

40. The population of a city doubled in 18 years. What was the exponential growth rate?

41. How long will it take \$7600 to double itself if it is invested at 8.6%, compounded continuously?

42. How old is a skeleton that has lost 27% of its carbon-14?

43. What is the pH of a substance whose hydrogen ion concentration is 3.8×10^{-7} moles per liter?

44. An earthquake has an intensity of $10^8 I_0$. What is its magnitude on the Richter scale?

Find each of the following logarithms, base 10 and base e, using a calculator.

45. $\log 0.00216$
46. $\log 1{,}342{,}000$
47. $\log (2.037 \times 10^{-5})$
48. $\ln 87{,}380$
49. $\ln 0.00002776$
50. $\ln (4.369 \times 10^{10})$

Find each of the following antilogarithms, base 10 and base e, using a calculator.

51. antilog 3.0287
52. $10^{-5.4632}$
53. antilog 0.0008162
54. $\text{antilog}_e\ 6.5692$
55. $e^{11.637}$
56. $\text{antilog}_e\ (-4.089)$

SYNTHESIS

Solve.

57. $|\log_4 x| = 3$

58. $\log x = \ln x$

Graph.

59. $y = |\log_3 x|$

60. $y = |e^x - 4|$

Find the domain.

61. $f(x) = \dfrac{1}{\sqrt{5 \ln x - 6}}$

62. $f(x) = \dfrac{8}{e^{4x} - 10}$

TEST: CHAPTER 5

1. Find the inverse of the relation H given by

$$H = \{(3, -5), (6, -3), (8, -3), (-5, 3), (1.3, -2.7)\}.$$

2. Write an equation of the inverse of the relation $y = |x|$.

3. Which of the following relations have graphs that are symmetric with respect to the line $y = x$?

a) $2y - \dfrac{5}{x} = 0$
b) $x = -3$
c) $y = 2 - x$
d) $x^3 - x = y^3$
e) $x^2 - 1 = y^2$
f) $3y = |x|$

4. Which of the following have inverses that are functions?

a)
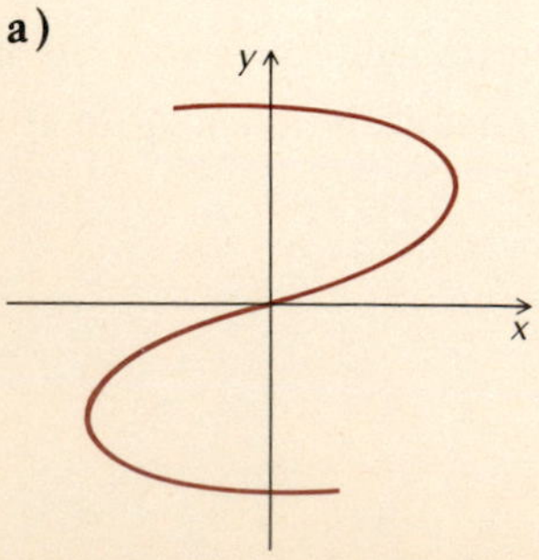

b)
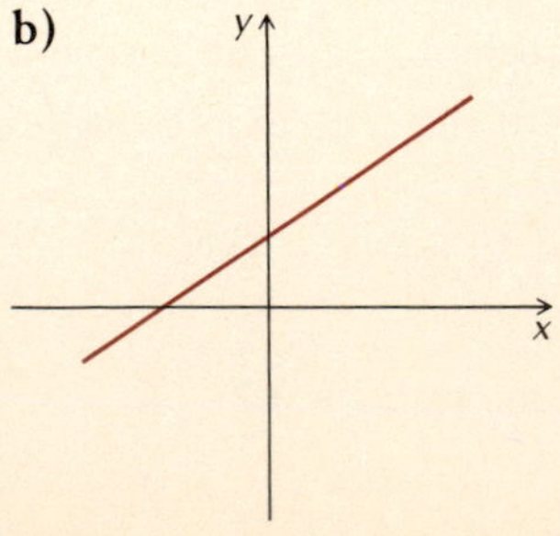

c)
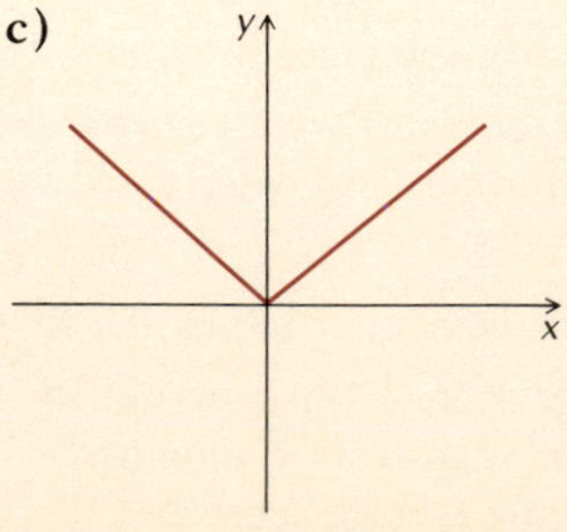

d)
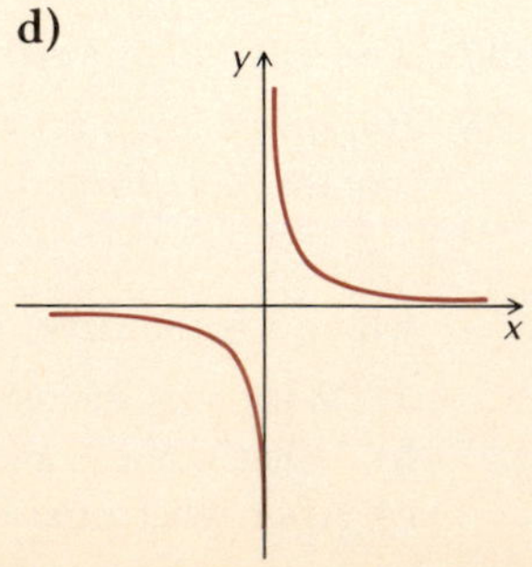

5. Find a formula for $f^{-1}(x)$:

$$f(x) = \sqrt{x - 6}.$$

6. Find $h(h^{-1}(3))$:

$$h(x) = \frac{-21x + 3}{20}.$$

7. Graph: $y = \log_3 x$.

8. Graph: $f(x) = e^{x-3}$.

9. Find $\log_5 23$ using natural logarithms.

10. Write an exponential equation equivalent to $\log_{\sqrt{3}} 9 = 4$.

11. Write a logarithmic equation equivalent to $x^5 = 0.03125$.

12. Simplify $6^{\log_6 3x}$.

13. Write an equivalent expression containing a single logarithm:

$$3 \log_c x - 4 \log_c y + \tfrac{1}{2} \log_c z.$$

14. Express in terms of logarithms of x and r:

$$\log \sqrt[4]{wr^3}.$$

Given that $\log_a 2 = 0.301$, $\log_a 5 = 0.699$, and $\log_a 6 = 0.778$, find each of the following.

15. $\log_a 3$

16. $\log_a 50$

17. $\log_a \sqrt[3]{5}$

18. Solve for x: $\log_b b^{2x^2} = x$.

Solve.

19. $\log_4 x = 2$

20. $4^{2x-1} - 3 = 61$

21. $\log_2 x + \log_2 (x - 2) = 3$

22. $e^{-x} = 0.2$

Find using a calculator.

23. $\log 0.005243$

24. $\ln 3.86$

25. $10^{2.0755}$

26. $\text{antilog}_e\, 2.037$

27. $\log 87{,}200$

28. $\text{antilog}\,(-2.003)$

29. $\ln 0.00000029$

30. $\log (-14.68)$

31. How many years will it take an investment of \$5000 to double if interest is compounded annually at 12%?

32. What is the loudness, in decibels, of a sound whose intensity is $50{,}000I_0$?

33. The population of a city was 80,000 in 1970 and 100,000 in 1980. Estimate the population in 2000.

34. *Walking speed.* The average walking speed R of people living in a city of population P, in thousands, is given by

$$R = 0.37 \ln P + 0.05,$$

where R is in feet per second.

a) The population of Akron, Ohio, is 660,000. Find the average walking speed.

b) A city's population has an average walking speed of 2.3 ft/sec. Find the population.

35. *Population of the USSR.* The population of the USSR was 209 million in 1959, and the exponential growth rate was 1% per year.

a) Write an exponential function describing the growth of the population of the USSR.

b) What will the population be in 1998? in 2020?

c) When will the population be 300 million?

d) What is the doubling time?

36. The population of a city doubled in 30 years. What was the exponential growth rate?

37. How long will it take an investment to double itself if it is invested at 8.6%, compounded continuously?

38. How old is an animal bone that has lost 38% of its carbon-14?

39. What is the loudness, in decibels, of a sound whose intensity is $230I_0$?

40. The hydrogen ion concentration of water is 1.8×10^{-9}. What is the pH?

SYNTHESIS

41. True or false:

$$\log_a (3x^5) = 15 \log_a x.$$

42. Find the domain:

$$f(x) = \log_3 (\ln x).$$

43. Solve: $5^{\sqrt{x}} = 625$.

The Trigonometric or Circular Functions

6

In this chapter, we consider an important class of functions called *trigonometric functions*, or *circular functions*. Historically, these functions arose from a study of triangles, hence the name "trigonometric." We will first develop the six basic circular functions from a unit circle (radius = 1), centered at the origin, which allows us to "see" the values of sin s and cos s, where s is a real-number arc length on the circle. Then we will consider trigonometric functions of angles or rotations on any circle centered at the origin, with both radian and degree measure. We will examine the characteristics of these functions, developing many relationships between them.

FEATURE PROBLEM

The rays of the sun over the top of a building cast a 40-ft shadow and form an angle of 60° with the ground. Find the height h of the building.

THE MATHEMATICS

We can use the tangent function to solve the problem. We have

$$\tan 60° = \frac{h}{40}.$$

Knowing that $\tan 60° = \sqrt{3}$, we can solve for h.

OBJECTIVES

You should be able to:

1 Given a real number that is an integral number of halves, fourths, or sixths of π, find on the unit circle the point determined by it. Given a point on the unit circle that is an integral number of fourths, eighths, or twelfths of the distance around the circle, find the real numbers between -2π and 2π that determine that point.

2 For the unit circle, state the coordinates of the intercepts and the midpoints of the arcs in each quadrant, and given the coordinates of a point on the unit circle, find its reflection across the u-axis, the v-axis, and the origin.

1. How far will a point travel if it goes:

a) $\frac{1}{4}$ of the way around the unit circle?

b) $\frac{3}{8}$ of the way around the unit circle?

c) $\frac{3}{4}$ of the way around the unit circle?

6.1 The Unit Circle

We refer to a circle centered at the origin with radius 1 as a **unit circle.** Such a circle has an equation

$$u^2 + v^2 = 1.$$

We use u and v for the variables here because we wish to reserve x and y for later purposes.

To develop the trigonometric, or circular, functions, we use the unit circle. The inputs for the functions that we will define are to be numbers that are associated with points on a unit circle, and the outputs are to be numbers associated with the coordinates of points on the circle.

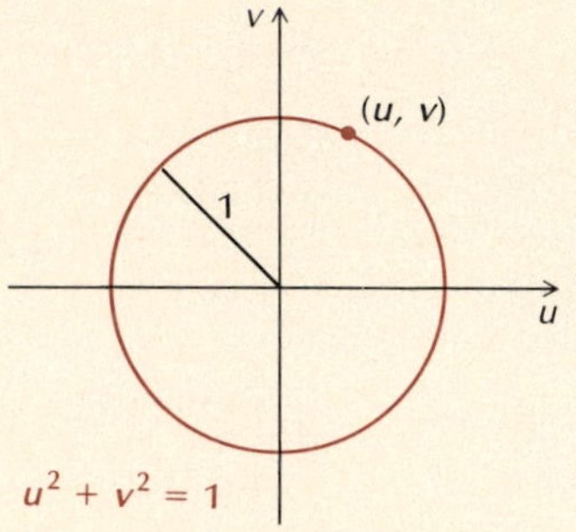

1 Distances on the Unit Circle

The circumference of a circle of radius r is $2\pi r$. Thus for the unit circle, where $r = 1$ the circumference is 2π. If a point starts at A and travels counterclockwise around the circle, it will travel a distance of 2π. If it travels halfway around the circle, it will travel a distance of π.

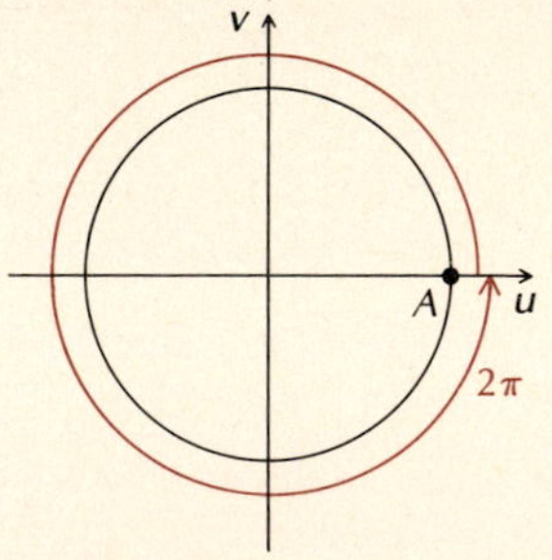

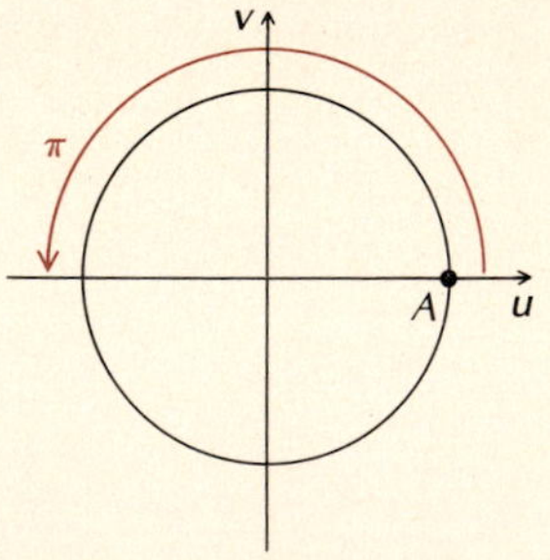

EXAMPLE 1 How far will a point travel if it goes $\frac{1}{12}$ of the way around the unit circle?

Solution The distance will be $\frac{1}{12}$ of the total distance around the circle, or $\frac{1}{12} \cdot 2\pi$, which is $\pi/6$.

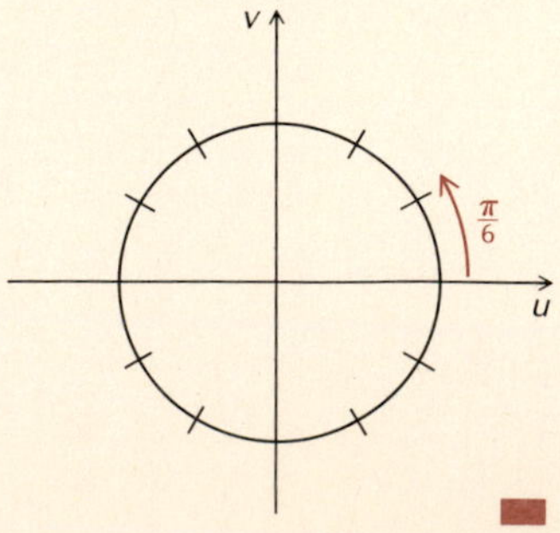

DO EXERCISE 1.

If a point travels $\frac{1}{8}$ of the way around the circle, it will travel a distance of $\frac{1}{8} \cdot 2\pi$, or $\pi/4$. Note that the stopping point is $\frac{1}{4}$ of the way from A to C and $\frac{1}{2}$ of the way from A to B.

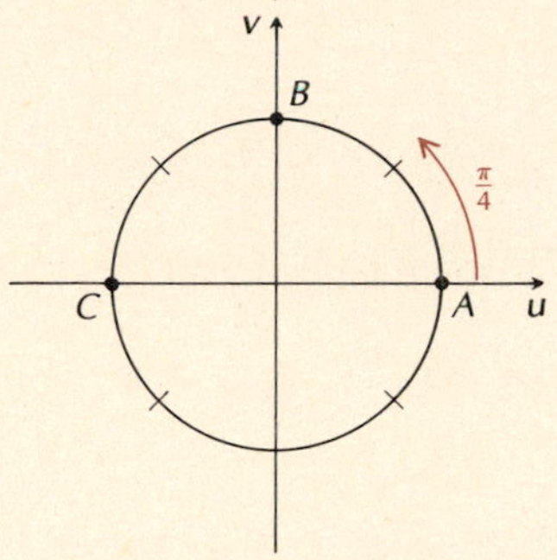

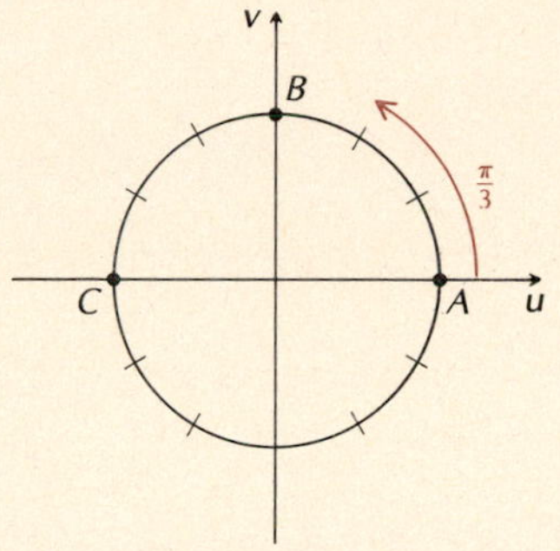

If a point travels $\frac{1}{6}$ of the way around the circle ($\frac{2}{12}$ of the way, as shown), it will travel a distance of $\frac{1}{6} \cdot 2\pi$, or $\pi/3$. Note that the stopping point is $\frac{1}{3}$ of the way from A to C and $\frac{2}{3}$ of the way from A to B.

EXAMPLE 2 How far will a point move on the unit circle, counterclockwise, going from point A to point Q?

Solution If the point went to C, it would move a distance of π. It goes $\frac{3}{4}$ of that distance to get to Q. Hence it goes a distance of $3\pi/4$ to get to Q.

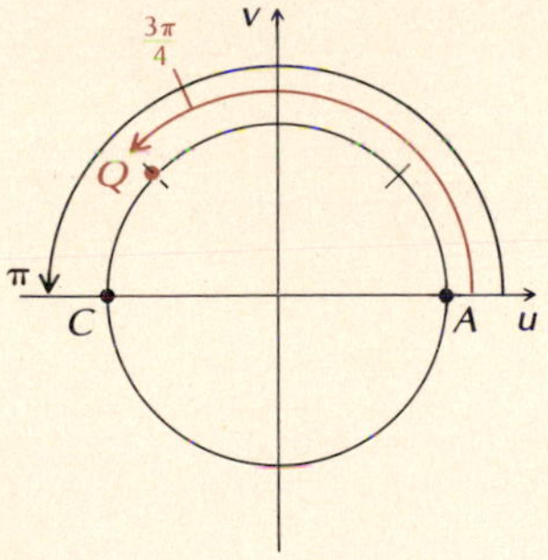

EXAMPLE 3 How far will a point move on the unit circle, counterclockwise, going from point A to point G?

Solution If the point went to C, it would move a distance of π. It goes $\frac{5}{6}$ of that distance to G. Hence it goes a distance of $5\pi/6$ to get to G.

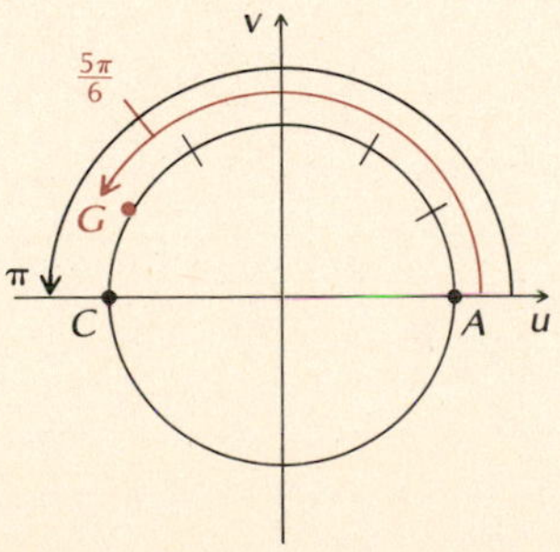

DO EXERCISES 2 AND 3.

2. How far will a point move on the unit circle, counterclockwise, going from point A to:
   a) point M?
   b) point N?
   c) point P?

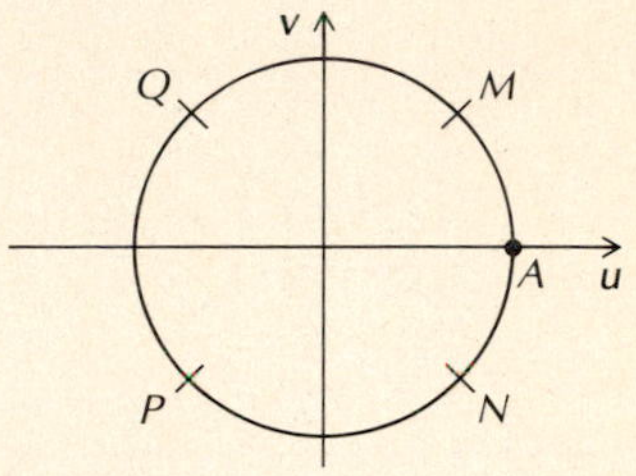

3. How far will a point move on the unit circle, counterclockwise, in going from point A to:
   a) point F?
   b) point H?
   c) point J?

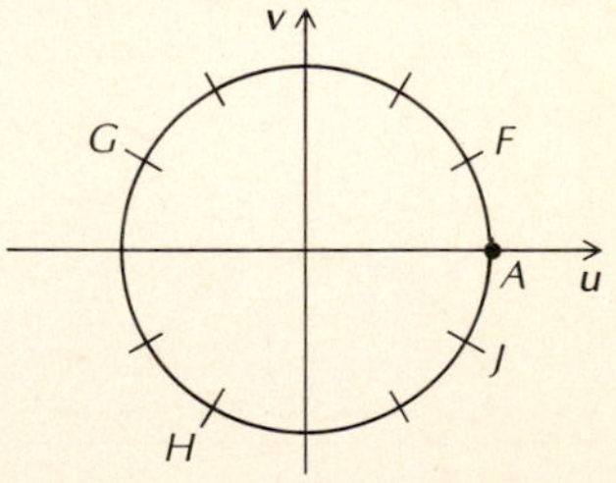

4. On this unit circle, mark the points determined by each of the following.
   a) $\frac{\pi}{4}$
   b) $\frac{7}{4}\pi$
   c) $\frac{9}{4}\pi$
   d) $\frac{13}{4}\pi$

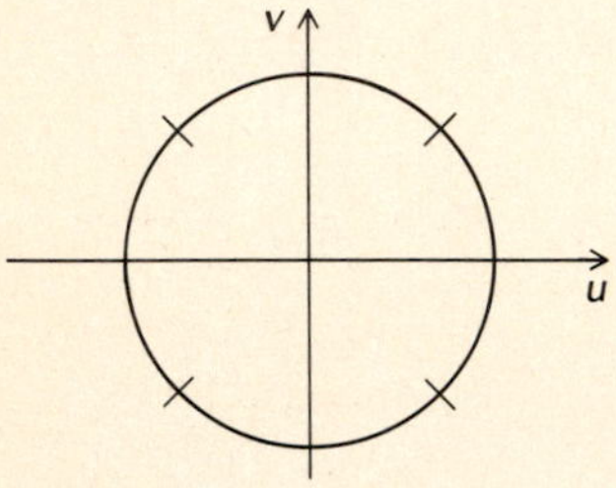

5. On this unit circle, mark the points determined by each of the following.
   a) $\frac{\pi}{6}$
   b) $\frac{7}{6}\pi$
   c) $\frac{13}{6}\pi$
   d) $\frac{25}{6}\pi$

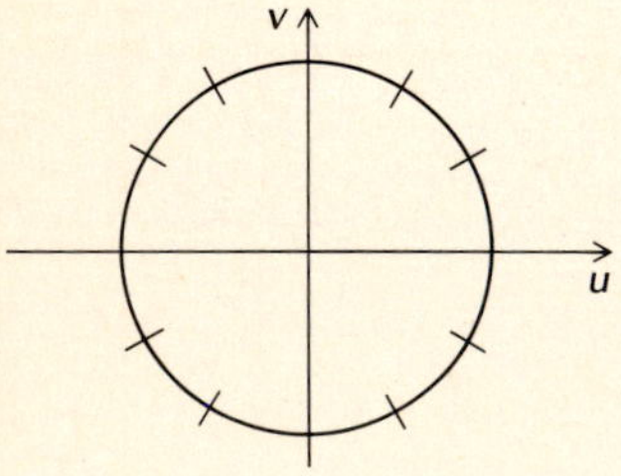

A point may travel completely around the circle and then continue. For example, if it goes around once and then continues $\frac{1}{4}$ of the way around, it will have traveled a distance of $2\pi + \frac{1}{4} \cdot 2\pi$, or $5\pi/2$.

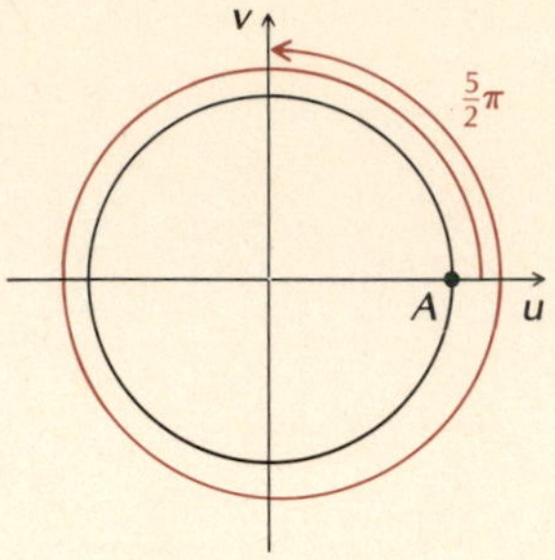

Any positive number thus determines a point on the unit circle. For the number 35, for example, we start at A and travel counterclockwise a distance of 35. The point at which we stop is the point "determined" by the number 35.

EXAMPLE 4 On the unit circle, mark the point determined by $\frac{5\pi}{4}$.

Solution From A to B, the distance is π, or $\left(\frac{4}{4}\right)\pi$, so we need to go another distance of $\pi/4$.

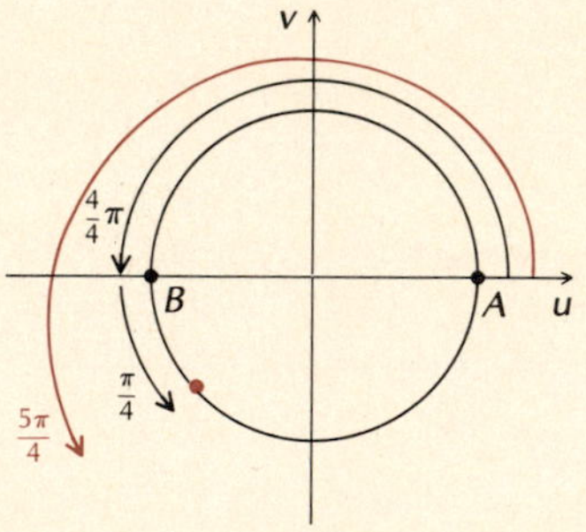

DO EXERCISES 4 AND 5.

Negative numbers also determine points on the unit circle. For a negative number, we move clockwise around the circle. Points for $-\pi/4$ and $-3\pi/2$ are shown. Any negative number thus determines a point on the unit circle.

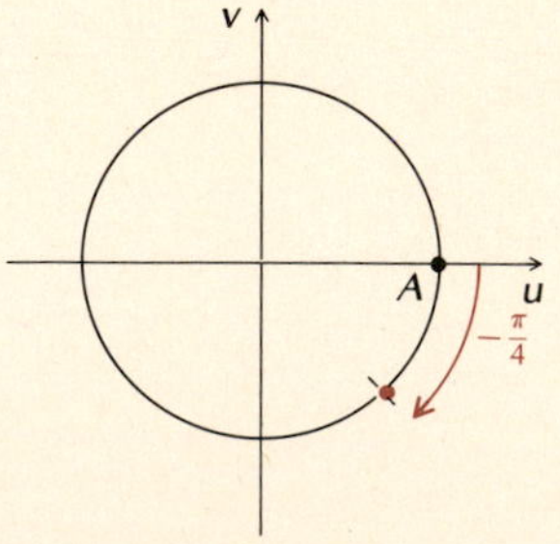

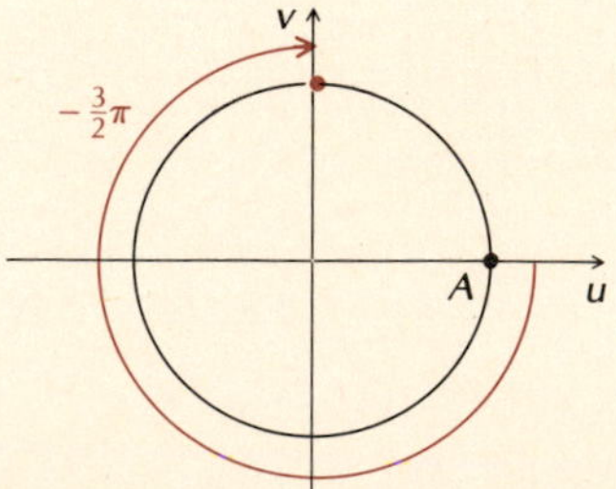

The number 0 determines the point A. Hence there is a point on the unit circle for every real number.

EXAMPLE 5 On the unit circle, mark the point determined by $\frac{-7\pi}{6}$.

Solution From A to B, the distance is $-\pi$, or $-\frac{6}{6}\pi$, so we need to go another distance of $\pi/6$, clockwise.

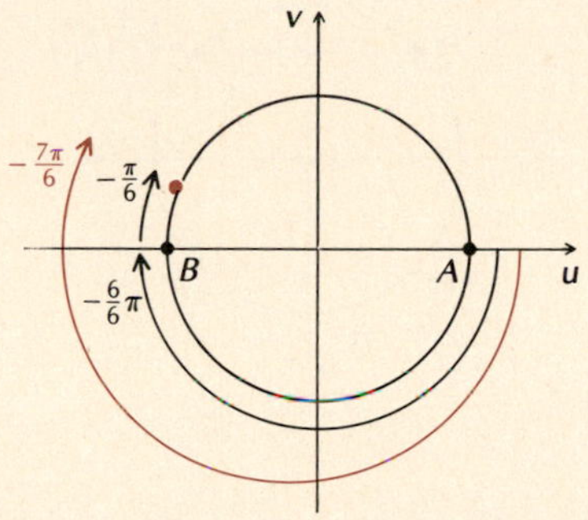

DO EXERCISE 6.

6. On this unit circle, mark the points determined by each of the following.

a) $-\frac{\pi}{2}$

b) $-\frac{3}{4}\pi$

c) $-\frac{7}{4}\pi$

d) -2π

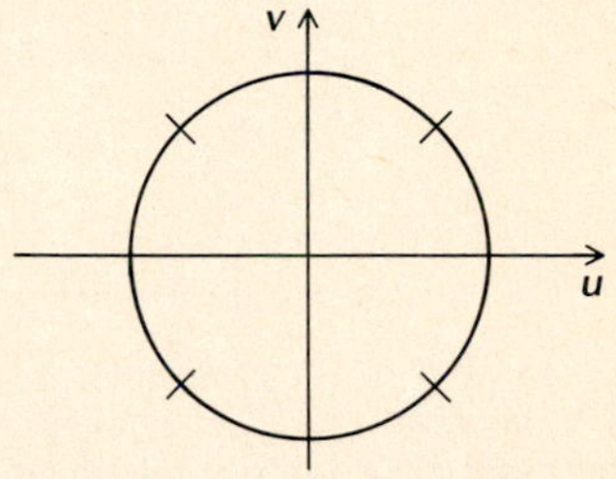

2 Special Points

The intercepts (points of intersection with the axes) of the unit circle A, B, C, and D have coordinates as shown. Let us find the coordinates of point E, which is the intersection of the circle with the line $u = v$. Note that this point divides the arc in the first quadrant in half.

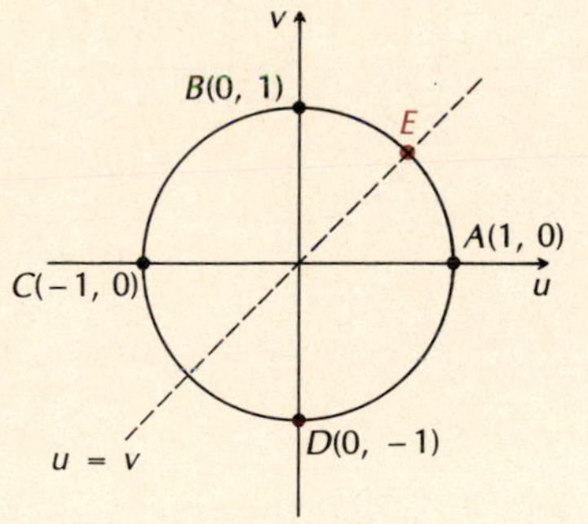

Since for any point (u, v) on the circle, $u^2 + v^2 = 1$, and for point E we have $u = v$, we know that the coordinates E satisfy $u^2 + u^2 = 1$ (substituting u for v in the equation of the circle). Let us solve this equation:

$$u^2 + u^2 = 1, \text{ or } 2u^2 = 1$$

$$u^2 = \frac{1}{2}$$

$$u = \sqrt{\frac{1}{2}}, \text{ or } \frac{\sqrt{1}}{\sqrt{2}}, \text{ or } \frac{1}{\sqrt{2}}.$$

Now

$$\frac{1}{\sqrt{2}} = \frac{1}{\sqrt{2}} \cdot \frac{\sqrt{2}}{\sqrt{2}} = \frac{\sqrt{2}}{2}.$$

Thus the coordinates of point E are $(\sqrt{2}/2, \sqrt{2}/2)$.

A unit circle is symmetric with respect to the u-axis, the v-axis, and the origin. We can use the coordinates of one point to find coordinates of reflections.

7. The point $\left(\frac{3}{5}, -\frac{4}{5}\right)$ is on the unit circle. Find the coordinates of its reflection across:

a) the u-axis;

b) the v-axis;

c) the origin.

d) Are these points on the circle? Why?

8. The point $(-\sqrt{35}/6, -1/6)$ is on the unit circle. Find the coordinates of its reflection across:

a) the u-axis;

b) the v-axis;

c) the origin.

d) Are these points on the circle? Why?

9. The point M, as shown, has coordinates $(\sqrt{3}/2, 1/2)$. Find the coordinates of the points N, P, and R, which are reflections of M.

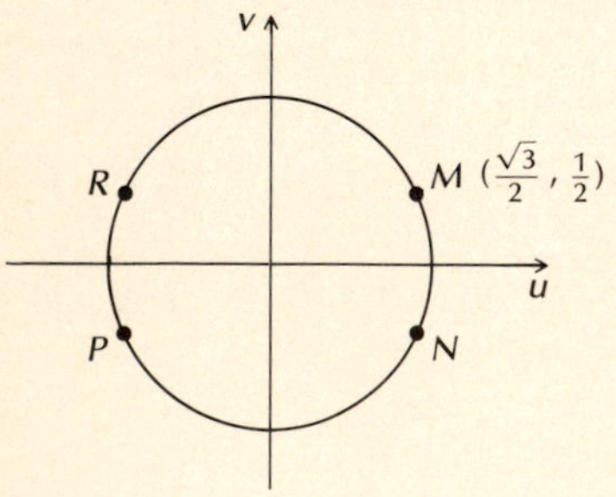

EXAMPLE 6 The point E on the unit circle, as shown below, has coordinates $(\sqrt{2}/2, \sqrt{2}/2)$. Find the coordinates of points F, G, and H, which are the reflections of E.

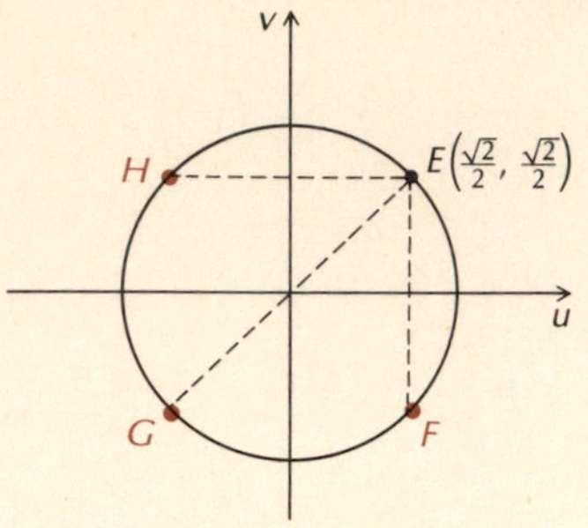

Solution

a) Point F is the reflection of point E across the u-axis. Hence its coordinates are $(\sqrt{2}/2, -\sqrt{2}/2)$.

b) Point G is the reflection of point E across the origin. Hence its coordinates are $(-\sqrt{2}/2, -\sqrt{2}/2)$.

c) Point H is the reflection of point E across the v-axis. Hence its coordinates are $(-\sqrt{2}/2, \sqrt{2}/2)$.

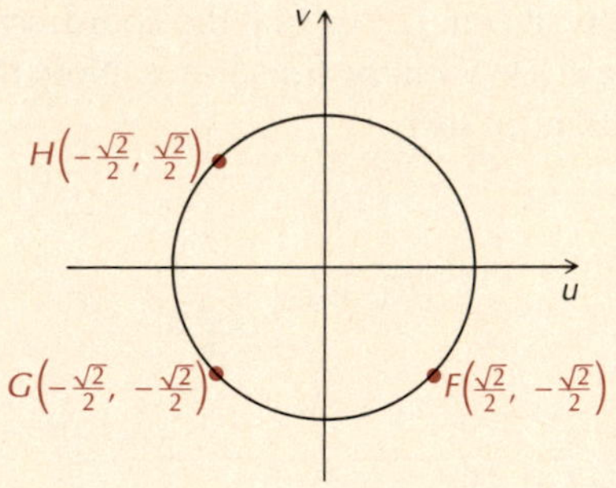

DO EXERCISES 7 AND 8.

EXAMPLE 7 The point $(1/2, \sqrt{3}/2)$ is on the unit circle. Find its reflection across (a) the u-axis, (b) the v-axis, and (c) the origin.

Solution

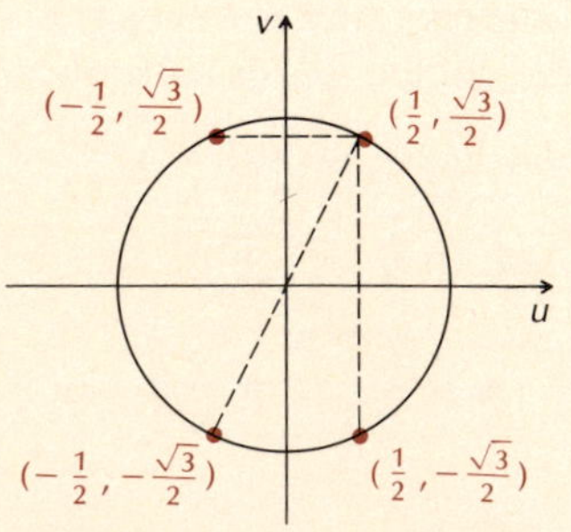

a) The reflection across the u-axis is $(1/2, -\sqrt{3}/2)$.

b) The reflection across the v-axis is $(-1/2, \sqrt{3}/2)$.

c) The reflection across the origin is $(-1/2, -\sqrt{3}/2)$.

By symmetry, all these points are on the circle.

DO EXERCISE 9.

EXERCISE SET 6.1

1 For each of Exercises 1–6, sketch a unit circle such as the one shown below.

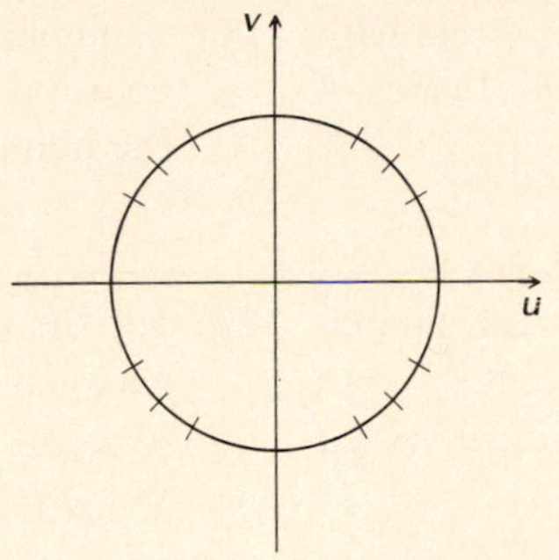

1. Mark the points determined by (a) $\frac{\pi}{4}$, (b) $\frac{3}{2}\pi$, (c) $\frac{3}{4}\pi$, (d) π, (e) $\frac{11}{4}\pi$, (f) $\frac{17}{4}\pi$.
2. Mark the points determined by (a) $\frac{\pi}{2}$, (b) $\frac{5}{4}\pi$, (c) 2π, (d) $\frac{9}{4}\pi$, (e) $\frac{13}{4}\pi$, and (f) $\frac{23}{4}\pi$.
3. Mark the points determined by (a) $\frac{\pi}{6}$, (b) $\frac{2}{3}\pi$, (c) $\frac{7}{6}\pi$, (d) $\frac{10}{6}\pi$, (e) $\frac{14}{6}\pi$, and (f) $\frac{23}{4}\pi$.
4. Mark the points determined by (a) $\frac{\pi}{3}$, (b) $\frac{5}{6}\pi$, (c) $\frac{11}{6}\pi$, (d) $\frac{13}{6}\pi$, (e) $\frac{23}{6}\pi$, and (f) $\frac{33}{6}\pi$.
5. Mark the points determined by (a) $-\frac{\pi}{2}$, (b) $-\frac{3}{4}\pi$, (c) $-\frac{5}{6}\pi$, (d) $-\frac{5}{2}\pi$, (e) $-\frac{17}{6}\pi$, and (f) $-\frac{9}{4}\pi$.
6. Mark the points determined by (a) $-\frac{3}{2}\pi$, (b) $-\frac{\pi}{3}$, (c) $-\frac{7}{6}\pi$, (d) $-\frac{13}{6}\pi$, (e) $-\frac{19}{6}\pi$, and (f) $-\frac{11}{4}\pi$.
7. Find two real numbers between -2π and 2π that determine each of these points.

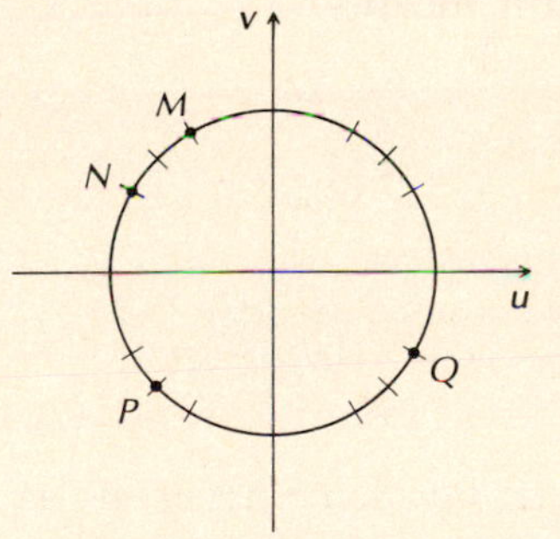

8. Find two real numbers between -2π and 2π that determine each of these points.

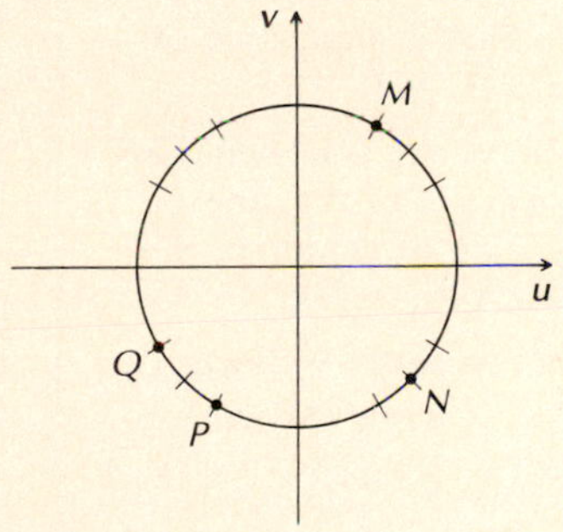

2

9. The point $(-0.375, 0.927)$ is on the unit circle. Find the coordinates of its reflection across (a) the u-axis, (b) the v-axis, and (c) the origin. (d) Are these points on the unit circle? Why?
10. The point $(0.625, -0.781)$ is on the unit circle. Find the coordinates of its reflection across (a) the u-axis, (b) the v-axis, and (c) the origin. (d) Are these points on the unit circle. Why?

The following points are on the unit circle. Find the coordinates of their reflections across (a) the u-axis, (b) the v-axis, and (c) the origin.

11. $\left(-\frac{\sqrt{2}}{2}, \frac{\sqrt{2}}{2}\right)$
12. $\left(-\frac{3}{4}, \frac{\sqrt{7}}{4}\right)$
13. $\left(\frac{2}{3}, \frac{\sqrt{5}}{3}\right)$
14. $\left(-\frac{\sqrt{3}}{2}, -\frac{1}{2}\right)$
15. Find the coordinates of these points on the unit circle.

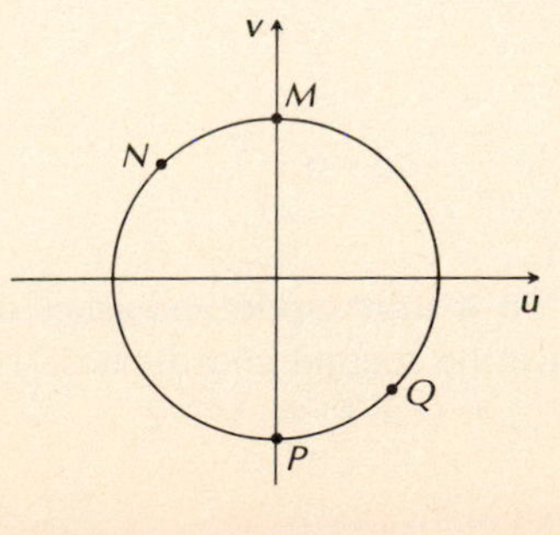

16. Find the coordinates of these points on the unit circle.

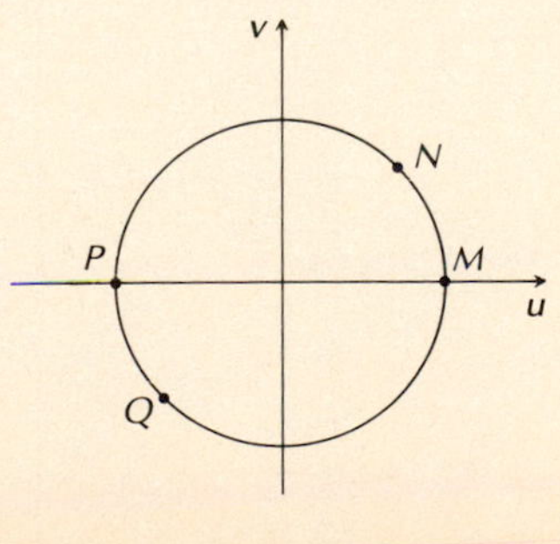

17. The number $\pi/6$ determines a point on the unit circle with coordinates $(\sqrt{3}/2, 1/2)$. What are the coordinates of the point determined by $-\pi/6$?

18. The number $\pi/3$ determines a point on the unit circle with coordinates $(1/2, \sqrt{3}/2)$. What are the coordinates of the point determined by $-\pi/3$?

19. A number α determines a point on the unit circle with coordinates $(3/4, -\sqrt{7}/4)$. What are the coordinates of the point determined by $-\alpha$?

20. A number β determines a point on the unit circle with coordinates $(-2/3, \sqrt{5}/3)$. What are the coordinates of the point determined by $-\beta$?

SYNTHESIS

21. For the points given in Exercise 7, find a real number between 2π and 4π that determines each point.

22. For the points given in Exercise 8, find a real number between -2π and -4π that determines each point.

23. A point on the unit circle has u-coordinate $-1/3$. What is its v-coordinate?

24. A point on the unit circle has u-coordinate $-\sqrt{21}/5$. What is its v-coordinate?

25. A point on the unit circle has u-coordinate 0.25671. What is its v-coordinate?

26. A point on the unit circle has v-coordinate -0.88041. What is its u-coordinate?

6.2 The Sine and the Cosine Functions

OBJECTIVES

You should be able to:

1 Determine $\sin s$ and $\cos s$ for any real number s that is a multiple of $\pi/4$, $\pi/6$, or $\pi/3$.

2 Sketch a graph of the sine function. State the properties of the sine function and state or complete the following identities.

$$\sin s = \sin (s + 2k\pi),\quad k \text{ any integer},$$
$$\sin (-s) = -\sin s,$$
$$\sin (s \pm \pi) = -\sin s,$$
$$\sin (\pi - s) = \sin s$$

3 Sketch a graph of the cosine function. State the properties of the cosine function and state or complete the following identities.

$$\cos s = \cos (s + 2k\pi),\quad k \text{ any integer},$$
$$\cos (-s) = \cos s,$$
$$\cos (s \pm \pi) = -\cos s,$$
$$\cos (\pi - s) = -\cos s$$

1 Definitions and Function Values

DEFINITION The Sine Function

Every real number s determines a point T on a unit circle. To the real number s, we assign the second coordinate of point T as a function value. The function thus defined is called the *sine function*. Function values are denoted by "sine s" or more briefly, "sin s."

We could use the more general function notation $\sin(s)$, but the parentheses are ordinarily omitted unless the input s has more than one term.

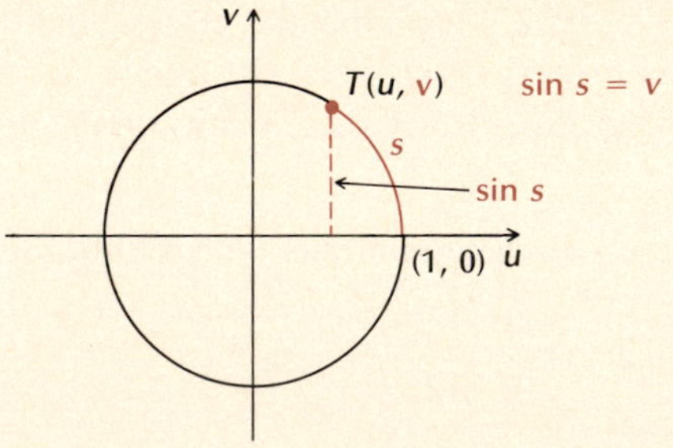

Let us find some function values.

EXAMPLE 1 Find $\sin 0$, $\sin \dfrac{\pi}{2}$, $\sin \dfrac{\pi}{4}$, and $\sin \dfrac{3\pi}{4}$.

Solution We locate the points on a unit circle and label the coordinates. The values of the sine function are the second coordinates. Thus, $\sin 0 = 0$, $\sin (\pi/2) = 1$, $\sin (\pi/4) = \sqrt{2}/2$, $\sin (3\pi/4) = \sqrt{2}/2$.

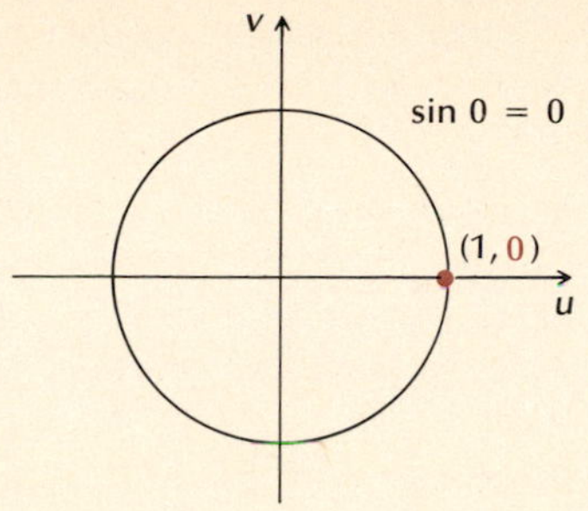

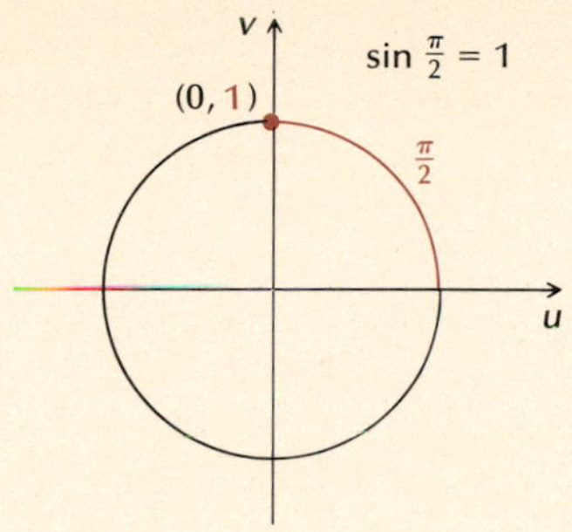

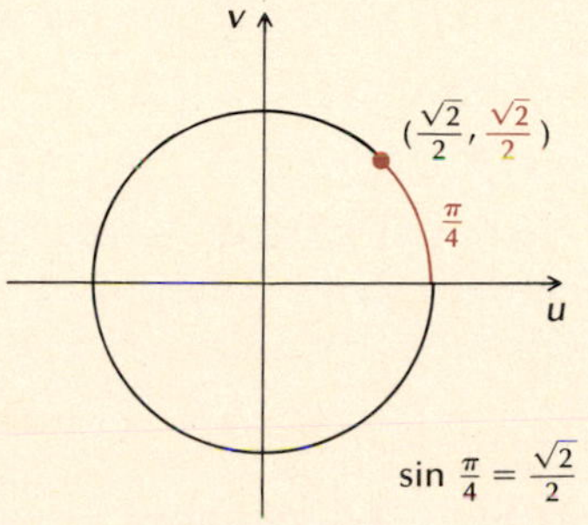

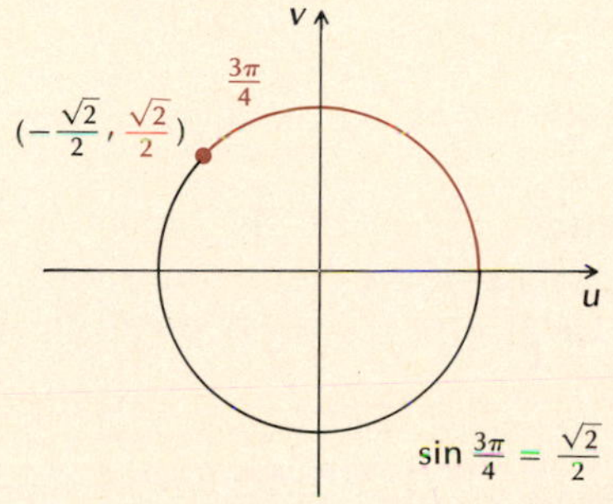

We will verify later that the point $\pi/6$ on the unit circle has coordinates $(\sqrt{3}/2, 1/2)$, and the point $\pi/3$ has coordinates $(1/2, \sqrt{3}/2)$. These facts allow us to find other values of the sine function.

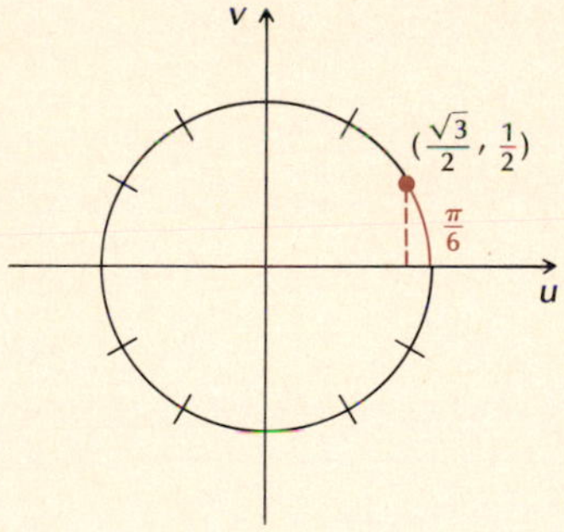

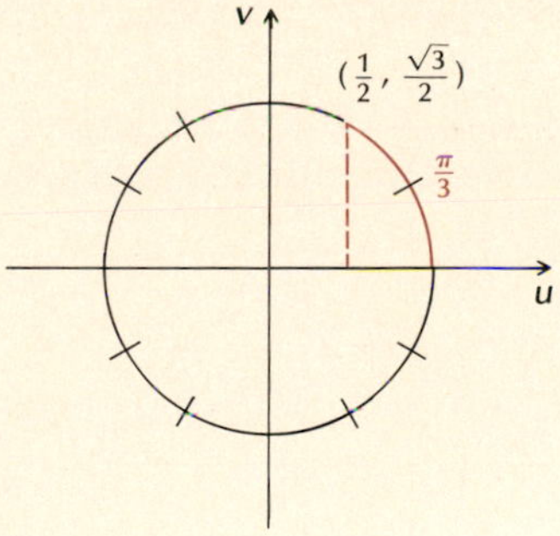

EXAMPLE 2 Find $\sin \dfrac{\pi}{6}$ and $\sin \dfrac{\pi}{3}$.

Solution From the coordinates given in the preceding figures, we have

$$\sin \frac{\pi}{6} = \frac{1}{2} \quad \text{and} \quad \sin \frac{\pi}{3} = \frac{\sqrt{3}}{2}.$$

DO EXERCISE 1.

DEFINITION The Cosine Function

Every real number s determines a point T on a unit circle. To the real number s, we assign the first coordinate of point T as a function value. The function thus defined is called the *cosine function*. Function values are denoted by "cosine s" or more briefly, "cos s."

1. Using a unit circle, find each of the following.

   a) $\sin \pi$ b) $\sin \dfrac{3}{2}\pi$

   c) $\sin \dfrac{\pi}{4}$ d) $\sin \left(-\dfrac{\pi}{4}\right)$

   e) $\sin \dfrac{2\pi}{3}$ f) $\sin \left(-\dfrac{\pi}{6}\right)$

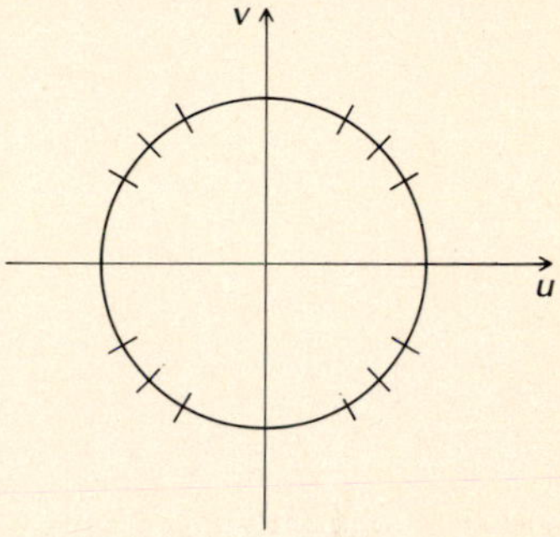

2. Using a unit circle, find each of the following.

   a) $\cos \pi$ b) $\cos \frac{3}{2}\pi$

   c) $\cos \frac{\pi}{4}$ d) $\cos \left(-\frac{\pi}{4}\right)$

   e) $\cos \frac{5\pi}{6}$ f) $\cos \left(-\frac{\pi}{3}\right)$

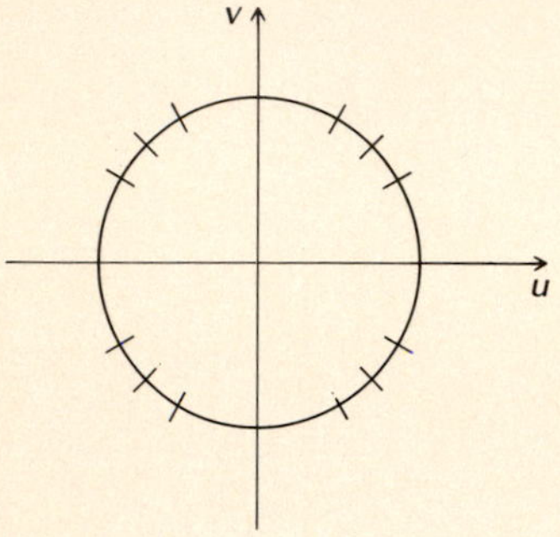

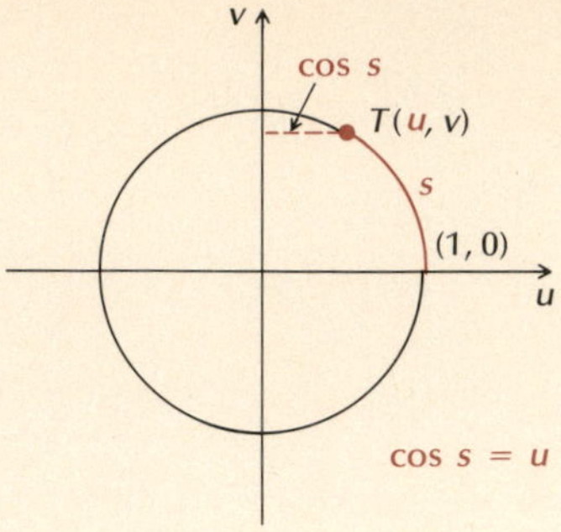

EXAMPLE 3 Find $\cos 0$, $\cos \frac{\pi}{2}$, $\cos \frac{\pi}{4}$, and $\cos \frac{3\pi}{4}$.

Solution We locate the points on a unit circle and label the coordinates. The values of the cosine function are the first coordinates. Thus, $\cos 0 = 1$, $\cos (\pi/2) = 0$, $\cos (\pi/4) = \sqrt{2}/2$, and $\cos (3\pi/4) = -\sqrt{2}/2$.

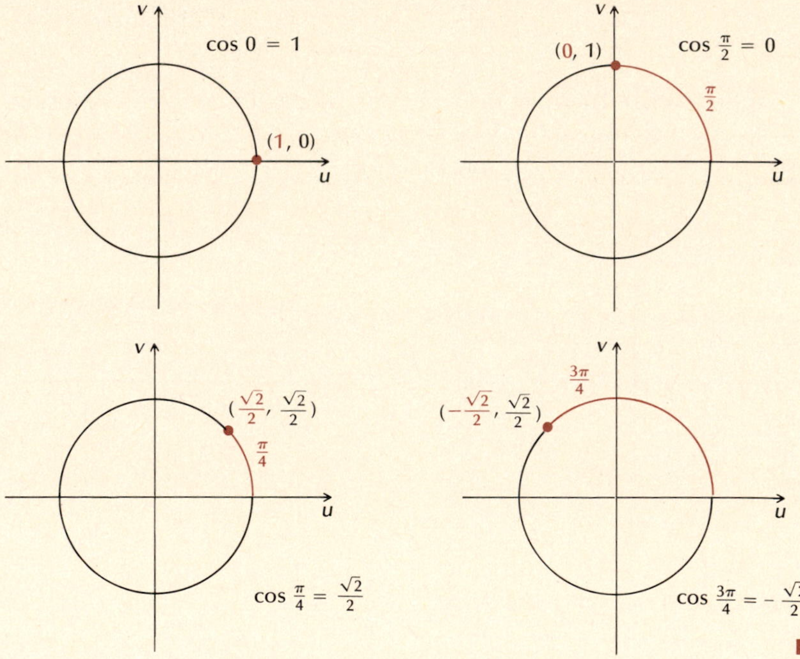

The point $\pi/6$ on the unit circle has coordinates $(\sqrt{3}/2, 1/2)$, and the point $\pi/3$ has coordinates $(1/2, \sqrt{3}/2)$. These facts allow us to find other values of the cosine function.

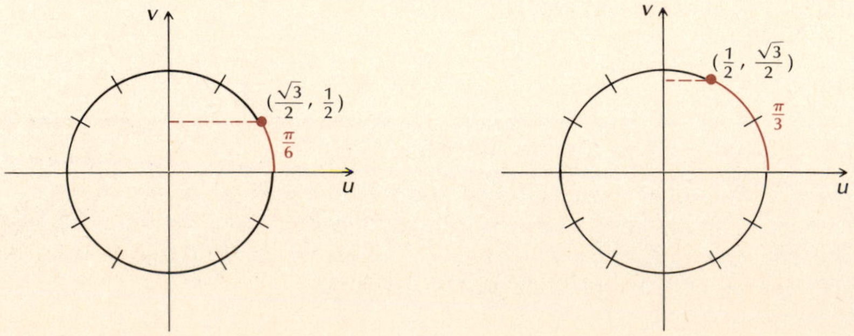

EXAMPLE 4 Find $\cos \frac{\pi}{6}$ and $\cos \frac{\pi}{3}$.

Solution From the coordinates given in the preceding figures, we have

$$\cos \frac{\pi}{6} = \frac{\sqrt{3}}{2} \quad \text{and} \quad \cos \frac{\pi}{3} = \frac{1}{2}.$$

DO EXERCISE 2 ON THE PRECEDING PAGE.

2 Properties of the Sine Function

Let us consider making a graph of the sine function. There are two ways we might make such a graph. The first way is to take the function values we have found in the preceding examples and margin exercises, make a table of values, plot points, look for patterns, and complete the graph. Below is a table of function values.

s	0	$\frac{\pi}{6}$	$\frac{\pi}{4}$	$\frac{\pi}{3}$	$\frac{\pi}{2}$	$\frac{3\pi}{4}$	π	$\frac{5\pi}{4}$	$\frac{3\pi}{2}$	2π
$\sin s$	0	$\frac{1}{2}$	$\frac{\sqrt{2}}{2}$	$\frac{\sqrt{3}}{2}$	1	$\frac{\sqrt{2}}{2}$	0	$-\frac{\sqrt{2}}{2}$	-1	0

s	$-\frac{\pi}{6}$	$-\frac{\pi}{4}$	$-\frac{\pi}{3}$	$-\frac{\pi}{2}$	$-\frac{3\pi}{4}$	$-\pi$	$-\frac{5\pi}{4}$	$-\frac{3\pi}{2}$	-2π
$\sin s$	$-\frac{1}{2}$	$-\frac{\sqrt{2}}{2}$	$-\frac{\sqrt{3}}{2}$	-1	$-\frac{\sqrt{2}}{2}$	0	$\frac{\sqrt{2}}{2}$	1	0

We find decimal approximations of the expressions involving square roots. Then we plot points, look for patterns, and connect them with a curve. The graph is as follows. Note that on the s-axis, π appears at about 3.14, and $\pi/2$ appears at about 1.57.

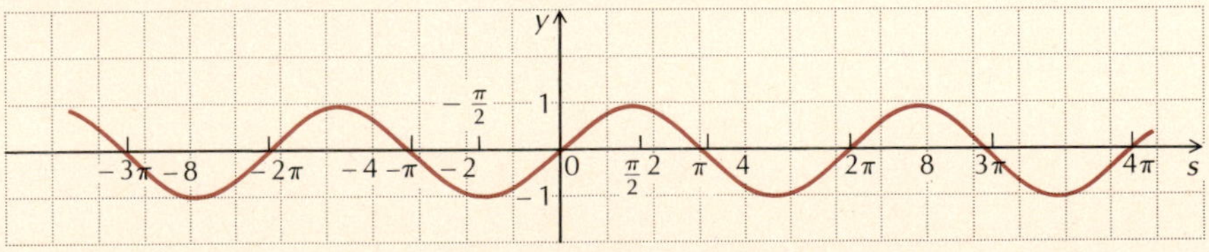
The sine function

The second way to construct a graph is by considering a unit circle. Then from the unit circle, we transfer vertical distances with a compass. You are asked to draw such a graph in Margin Exercise 3.

DO EXERCISE 3.

Note in the graph of the sine function that function values increase from 0 at 0 to 1 at $\pi/2$, then decrease to 0 at π, decrease further to -1 at $3\pi/2$, and then increase to 0 at 2π. A similar pattern follows for negative inputs. Moreover, the function is periodic, with period 2π, and the change in function values is smooth.

3. Construct a graph of the sine function. From the unit circle, transfer vertical distances with a compass.

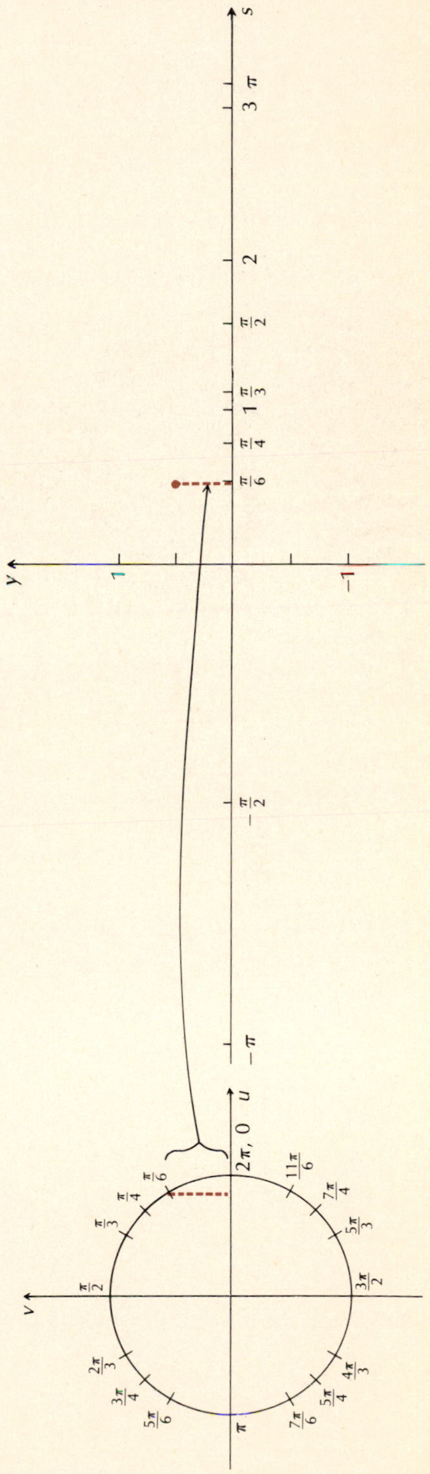

4. Is the sine function periodic? If so, what is its period?

5. Is the sine function even? Is it odd?

6. Does the sine function appear to be continuous?

7. What is the domain of the sine function?

8. What is the range of the sine function?

The following is a device that could be constructed to draw a graph of the sine function. A light shines horizontally, casting a shadow of the ball fastened to the wheel at P. The wheel rotates at a constant speed and the paper on which the shadow falls moves at a constant speed. The shadow of the ball traces a sine graph.

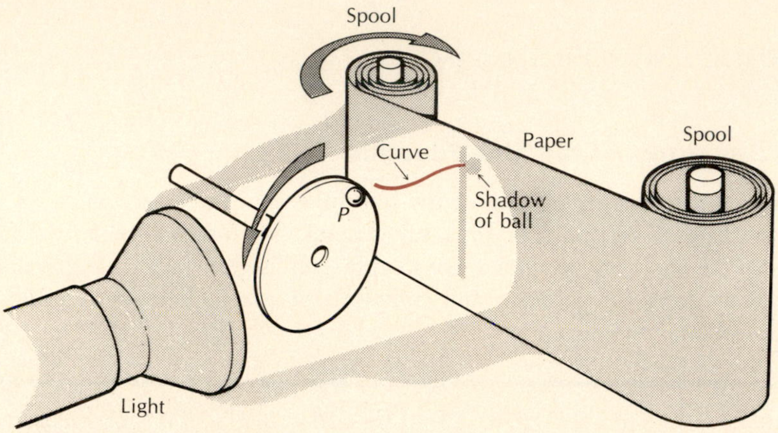

DO EXERCISES 4–8.

The unit circle can be used to verify some properties of the sine function that were conjectured in Margin Exercises 4–8.

Let us verify that the period is 2π. Consider any real number s and the point T that it determines on a unit circle. Now we increase s by 2π. The point for $s + 2\pi$ is again the point T. Hence for any real number s,

$$\sin s = \sin (s + 2\pi).$$

There is no number smaller than 2π for which this is true. Thus the period is 2π.

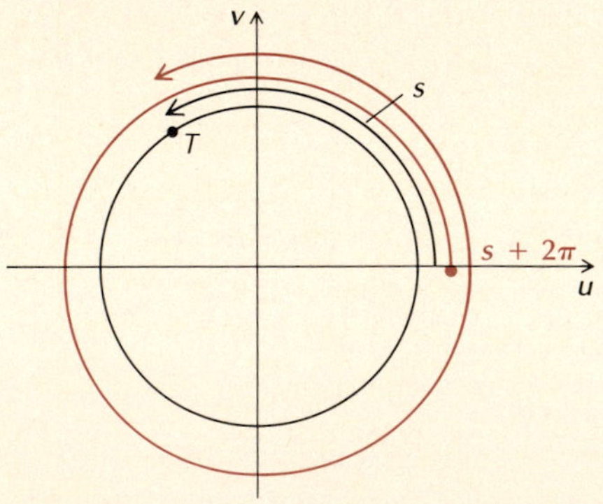

Some Identities

The statement $\sin s = \sin (s + 2\pi)$ holds for any real number s. Equations that hold for all meaningful replacements for the variable are known as **identities**. Thus we have shown that $\sin s = \sin (s + 2\pi)$ is an identity.

Actually, we know that $\sin s = \sin [s + k(2\pi)]$, or $\sin s = \sin (s + 2k\pi)$ *for any integer k*. For example, if $k = 1$, we have $\sin s = \sin (s + 2\pi)$. For $k = 2$, we have $\sin s = \sin (s + 4\pi)$; the point for $s + 4\pi$ is found by going around the unit circle from s, two full circuits. For $k = 3$, we have $\sin s = \sin (s + 6\pi)$, corresponding to three full circuits. For $k = -5$, we have $\sin s = \sin (s - 10\pi)$, corresponding to 5 full circuits, *but in the clockwise direction.*

Consider any real number s and its additive inverse, or opposite, $-s$. These numbers determine points on a unit circle that are symmetric with respect to the u-axis. Hence their second coordinates are additive inverses of each other. Thus we know that for any number s, $\sin(-s) = -\sin s$, so we have shown another identity as well as showing that the sine function is odd.

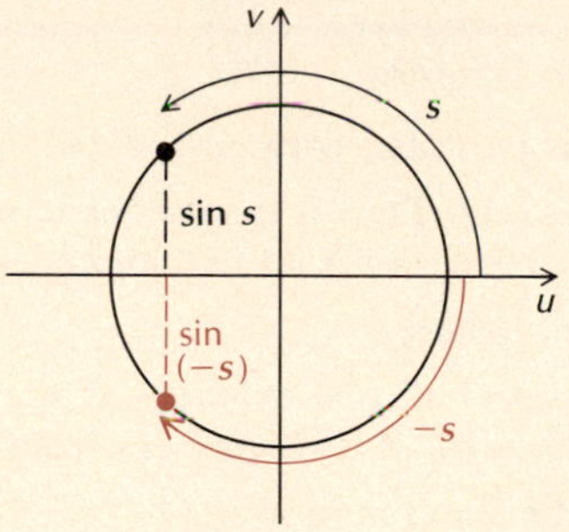

DO EXERCISE 9.

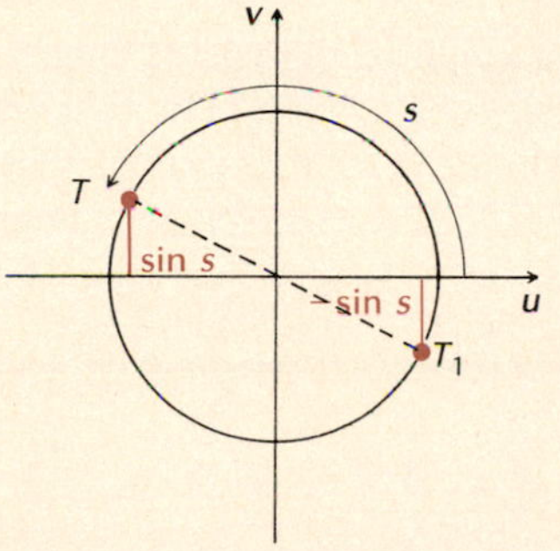

Consider any real number s and the point T that it determines on a unit circle. Then $s + \pi$ and $s - \pi$ both determine a point T_1 symmetric with T with respect to the origin. Thus the sines of $s + \pi$ and $s - \pi$ are the same, and are the additive inverses of $\sin s$. This gives us another important identity and property.

For any real number s, $\sin(s + \pi) = -\sin s$ and $\sin(s - \pi) = -\sin s$.

Let us add to our list a property concerning $\sin(\pi - s)$. Since $\pi - s$ is the additive inverse of $s - \pi$, it follows that $\sin(\pi - s) = \sin[-(s - \pi)]$. Since the sine function is odd, the latter is equal to $-\sin(s - \pi)$. We know that $\sin(s - \pi) = -\sin s$, so $-\sin(s - \pi) = \sin s$. Thus our last identity, or property, is

$$\sin(\pi - s) = \sin s.$$

We have proven the following identities concerning the sine function.

Identities for the Sine Function

$$\sin s = \sin(s + 2k\pi),\ k \text{ any integer},$$
$$\sin(-s) = -\sin s,$$
$$\sin(s \pm \pi) = -\sin s$$
$$\sin(\pi - s) = \sin s$$

The **amplitude** of a periodic function is defined to be one-half the difference between its maximum and minimum function values. It is always

9. A real number s determines the point T, as shown.
   a) Find the point for $s + \pi$. What are its coordinates?
   b) Find the point for $s - \pi$. What are its coordinates?

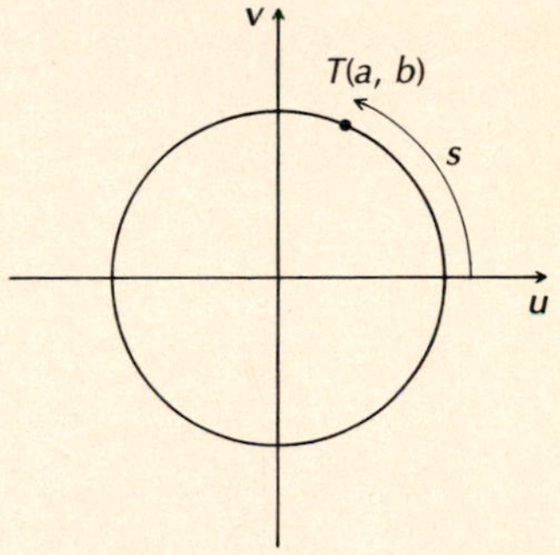

10. Construct a graph of the cosine function. From the unit circle, transfer horizontal distances with a compass.

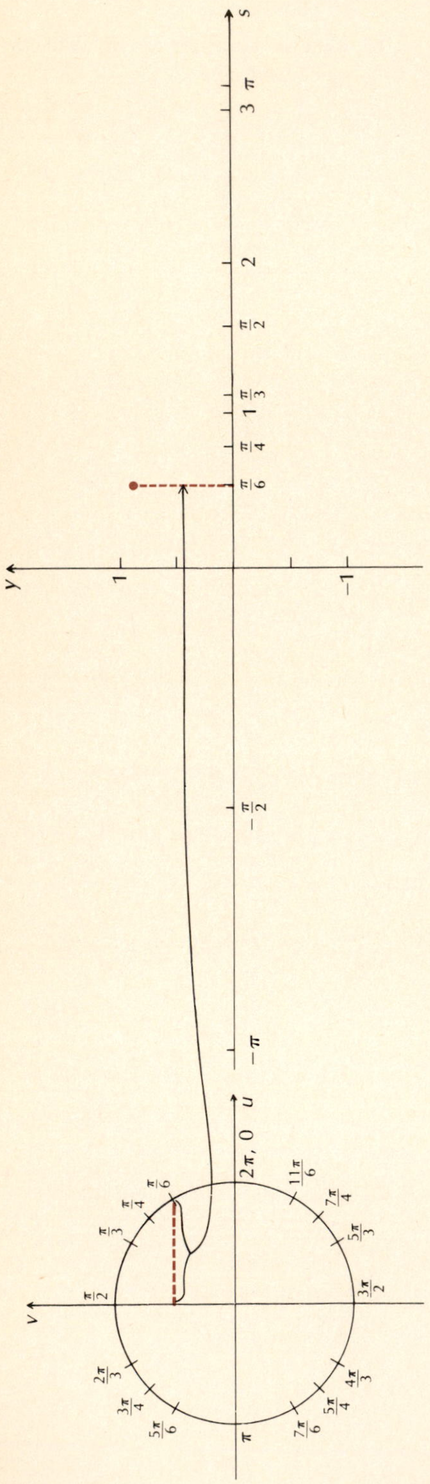

positive. We can see from either the graph or a unit circle that the maximum value of the sine function is 1, whereas the minimum value is -1. Thus,

$$\text{The amplitude of the sine function} = \tfrac{1}{2}[1 - (-1)] = 1.$$

The following is a summary of properties of the sine function.

Properties of the Sine Function

1. **The sine function is periodic, with period 2π.**
2. **The sine function is odd. Thus it is symmetric with respect to the origin and $\sin(-s) = -\sin s$ for all real numbers s.**
3. **It is continuous everywhere.**
4. **The domain is the set of all real numbers. The range is the interval $[-1, 1]$, that is, the set of all real numbers from -1 to 1.**
5. **The amplitude of the sine function is 1.**

3 Properties of the Cosine Function

Let us consider making a graph of the cosine function. There are again two ways in which we can make such a graph. The first way is to take the function values that we have found in the preceding examples and margin exercises, make a table of values, plot points, look for patterns, and complete the graph. Below is a table of function values.

s	0	$\frac{\pi}{6}$	$\frac{\pi}{4}$	$\frac{\pi}{3}$	$\frac{\pi}{2}$	$\frac{3\pi}{4}$	π	$\frac{5\pi}{4}$	$\frac{3\pi}{2}$	2π
$\cos s$	1	$\frac{\sqrt{3}}{2}$	$\frac{\sqrt{2}}{2}$	$\frac{1}{2}$	0	$-\frac{\sqrt{2}}{2}$	-1	$-\frac{\sqrt{2}}{2}$	0	1

s	$-\frac{\pi}{6}$	$-\frac{\pi}{4}$	$-\frac{\pi}{3}$	$-\frac{\pi}{2}$	$-\frac{3\pi}{4}$	$-\pi$	$-\frac{5\pi}{4}$	$-\frac{3\pi}{2}$	-2π
$\cos s$	$\frac{\sqrt{3}}{2}$	$\frac{\sqrt{2}}{2}$	$\frac{1}{2}$	0	$-\frac{\sqrt{2}}{2}$	-1	$-\frac{\sqrt{2}}{2}$	0	1

We find decimal approximations of the expressions involving square roots. Then we plot points, look for patterns, and connect them with a curve. The graph is as follows.

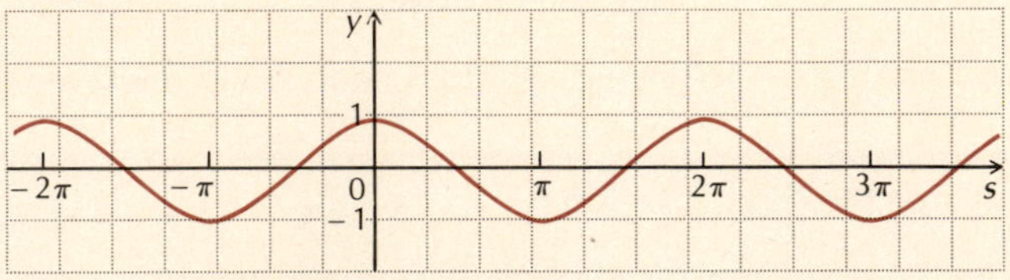

The cosine function

The second way to construct a graph is by considering a unit circle. Then from the unit circle, we transfer horizontal distances with a compass. You are asked to draw such a graph in Margin Exercise 10.

DO EXERCISE 10.

Note in the graph of the cosine function that function values start at 1, when $s = 0$, and decrease to 0 at $\pi/2$. They decrease further to -1 at π, and then begin to increase, reaching the value 1 at 2π. A similar pattern follows for negative inputs. Moreover, the function is periodic, with period 2π, and the change in function values is smooth.

DO EXERCISES 11–15.

The unit circle can be used to verify some properties of the cosine function that were conjectured in Margin Exercises 11–15.

Let us verify that the period is 2π. Consider any real number s and the point T that it determines on a unit circle. Now we increase s by 2π. The point for $s + 2\pi$ is again the point T. Hence for any real number s,

$$\cos s = \cos (s + 2\pi).$$

There is no number smaller than 2π for which this is true. Thus the period is 2π.

Some Identities

The unit circle can be used to verify that $\cos s = \cos (s + 2k\pi)$ holds for any real number s and any integer k.

Consider any real number s and its additive inverse, or opposite, $-s$. They determine points symmetric with respect to the u-axis, hence they have the same first coordinates. Thus we know that for any number s, $\cos (-s) = \cos s$, so we have proved another identity as well as showing that the cosine function is even.

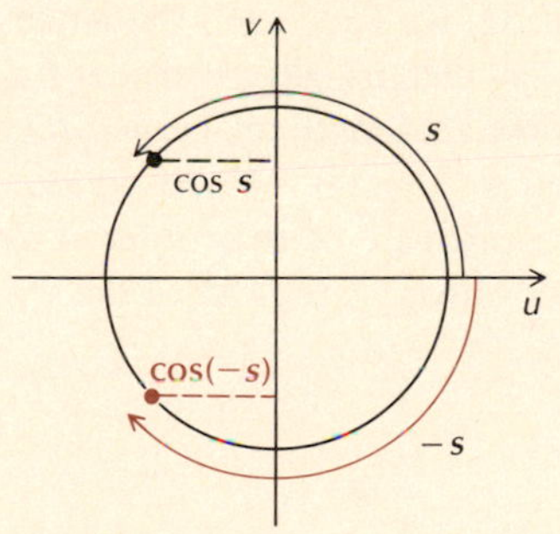

Consider any real number s and the point that it determines on a unit circle. Then $s + \pi$ and $s - \pi$ both determine points that are symmetric with respect to the origin. Thus their cosines are the same, and are the additive inverses of $\cos s$. This gives us another important identity and property.

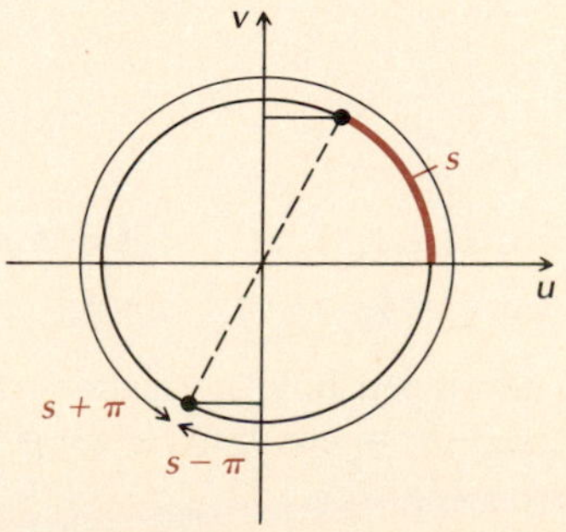

For any real number s,

$$\cos (s + \pi) = -\cos s \quad \text{and} \quad \cos (s - \pi) = -\cos s.$$

11. Is the cosine function periodic? If so, what is its period?

12. Is the cosine function even? Is it odd?

13. Does the cosine function appear to be continuous?

14. What is the domain of the cosine function?

15. What is the range of the cosine function?

16. Complete each of the following in reference to the figure.

$\cos y =$ ________.

$\sin y =$ ________.

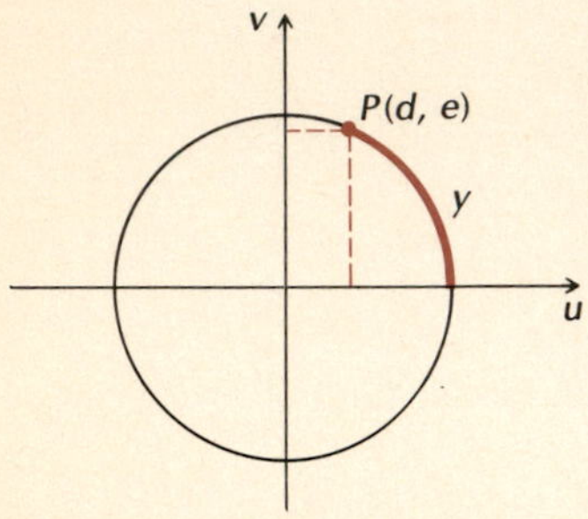

Let us consider $\cos(\pi - s)$. Since $\pi - s$ is the additive inverse of $s - \pi$, it follows that $\cos(\pi - s) = \cos[-(s - \pi)]$. Since the cosine function is even, the latter is equal to $\cos(s - \pi)$. By the result proved above, we know that $\cos(s - \pi) = -\cos s$. Hence we have $\cos(\pi - s) = -\cos s$. We now list the identities that we have developed for the cosine function.

Identities for the Cosine Function

$$\cos s = \cos(s + 2k\pi), \quad k \text{ any integer},$$
$$\cos(-s) = \cos s,$$
$$\cos(s \pm \pi) = -\cos s,$$
$$\cos(\pi - s) = -\cos s$$

The following is a summary of the properties of the cosine function.

Properties of the Cosine Function

1. **The cosine function is periodic, with period 2π.**
2. **The cosine function is even. Thus it is symmetric with respect to the y-axis, and $\cos(-s) = \cos s$ for all real numbers s.**
3. **It is continuous everywhere.**
4. **The domain is the set of all real numbers. The range is the interval $[-1, 1]$, that is, the set of all real numbers from -1 to 1.**
5. **The amplitude of the cosine function is 1.**

For your understanding, we have used the letter s for arc length and have avoided the letters x and y, which usually represent first and second coordinates. In fact, we can represent the arc length on a unit circle by any variable, such as s, t, x, or θ (the Greek letter "theta"). When we do, the point determines an ordered pair. The first coordinate of that ordered pair is the *cosine* of the arc length, and the second coordinate is the *sine* of the arc length. This is illustrated in the following figures.

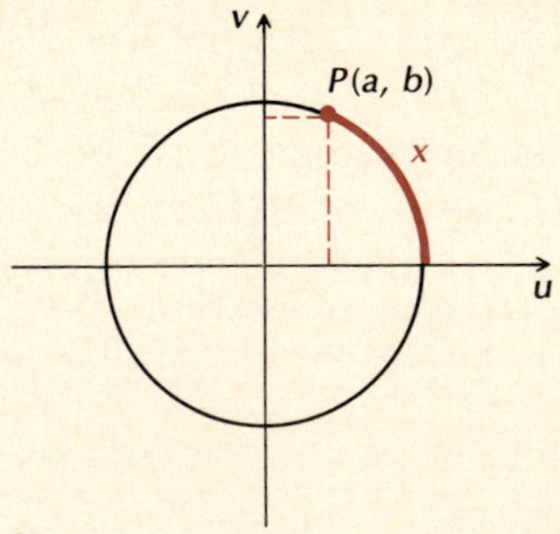

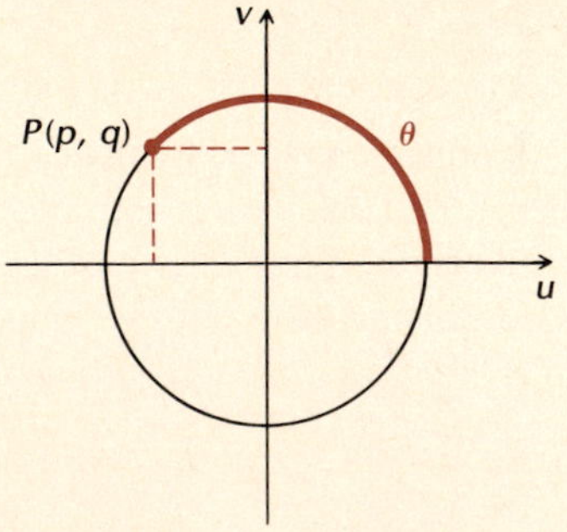

$\cos x = a,$ $\quad\cos\theta = p,$

$\sin x = b;$ $\quad\sin\theta = q.$

The identities we have developed hold no matter what symbols are used for variables. For example, $\cos(-s) = \cos s$, $\cos(-x) = \cos x$, $\cos(-\theta) = \cos\theta$, and $\cos(-y) = \cos y$.

DO EXERCISE 16.

EXERCISE SET 6.2

1 Find the function value.

1. $\sin \frac{\pi}{4}$
2. $\sin \pi$
3. $\sin \frac{\pi}{6}$
4. $\sin \frac{5}{4}\pi$
5. $\sin \frac{3}{2}\pi$
6. $\sin \frac{\pi}{2}$
7. $\sin \frac{\pi}{3}$
8. $\sin \frac{3}{4}\pi$
9. $\sin \frac{7}{4}\pi$
10. $\sin 2\pi$
11. $\sin \frac{5}{6}\pi$
12. $\sin \left(-\frac{\pi}{4}\right)$
13. $\sin (-\pi)$
14. $\sin \left(-\frac{\pi}{2}\right)$
15. $\sin \left(-\frac{3}{4}\pi\right)$
16. $\sin \left(-\frac{7}{4}\pi\right)$
17. $\sin \left(-\frac{5}{4}\pi\right)$
18. $\sin \left(-\frac{3}{2}\pi\right)$
19. $\sin \left(-\frac{\pi}{3}\right)$
20. $\sin \left(-2\pi\right)$
21. $\sin \left(-\frac{2}{3}\pi\right)$
22. $\sin \left(-\frac{\pi}{6}\right)$
23. $\cos \frac{\pi}{4}$
24. $\cos \pi$
25. $\cos \frac{5}{4}\pi$
26. $\cos \frac{\pi}{6}$
27. $\cos \frac{\pi}{2}$
28. $\cos \frac{\pi}{3}$
29. $\cos \frac{3}{4}\pi$
30. $\cos \frac{7}{4}\pi$
31. $\cos \frac{3}{2}\pi$
32. $\cos \frac{5}{6}\pi$
33. $\cos \left(-\frac{\pi}{4}\right)$
34. $\cos (-\pi)$
35. $\cos 2\pi$
36. $\cos \left(-\frac{\pi}{2}\right)$
37. $\cos \left(-\frac{3}{4}\pi\right)$
38. $\cos \left(-\frac{5}{4}\pi\right)$
39. $\cos \left(-\frac{3}{2}\pi\right)$
40. $\cos \left(-\frac{5}{3}\pi\right)$
41. $\cos \left(-\frac{\pi}{6}\right)$
42. $\cos \left(-\frac{7}{4}\pi\right)$
43. $\cos \left(-\frac{2}{3}\pi\right)$
44. $\sin \left(-\frac{2}{3}\pi\right)$

2, **3** Complete.

45. $\cos (-x) =$ ________
46. $\sin (-x) =$ ________
47. $\sin (x + \pi) =$ ________
48. $\sin (x - \pi) =$ ________
49. $\cos (\pi - x) =$ ________
50. $\sin (\pi - x) =$ ________
51. $\cos (x + 2k\pi) =$ ________
52. $\sin (x + 2k\pi) =$ ________
53. $\cos (x - \pi) =$ ________
54. $\cos (x + \pi) =$ ________

55. a) Sketch a graph of $y = \sin x$.
    b) By reflecting the graph in part (a), sketch a graph of $y = \sin (-x)$.
    c) By reflecting the graph in part (a), sketch a graph of $y = -\sin x$.
    d) How do the graphs in parts (b) and (c) compare?

56. a) Sketch a graph of $y = \cos x$.
    b) By reflecting the graph in part (a), sketch a graph of $y = \cos (-x)$.
    c) By reflecting the graph in part (a), sketch a graph of $y = -\cos x$.
    d) How do the graphs in parts (a) and (b) compare?

57. a) Sketch a graph of $y = \sin x$.
    b) By translating, sketch a graph of $y = \sin (x + \pi)$.
    c) By reflecting the graph of part (a), sketch a graph of $y = -\sin x$.
    d) How do the graphs of parts (b) and (c) compare?

58. a) Sketch a graph of $y = \sin x$.
    b) By translating, sketch a graph of $y = \sin (x - \pi)$.
    c) By reflecting the graph of part (a), sketch a graph of $y = -\sin x$.
    d) How do the graphs of parts (b) and (c) compare?

59. a) Sketch a graph of $y = \cos x$.
    b) By translating, sketch a graph of $y = \cos (x + \pi)$.
    c) By reflecting the graph of part (a), sketch a graph of $y = -\cos x$.
    d) How do the graphs of parts (b) and (c) compare?

60. a) Sketch a graph of $y = \cos x$.
    b) By translating, sketch a graph of $y = \cos (x - \pi)$.
    c) By reflecting the graph of part (a), sketch a graph of $y = -\cos x$.
    d) How do the graphs of parts (b) and (c) compare?

SYNTHESIS

61. For which numbers is:

a) $\sin x = 1$? b) $\sin x = -1$?

62. For which numbers is:

a) $\cos x = 1$? b) $\cos x = -1$?

63. Solve for x:

$$\sin x = 0.$$

64. Solve for x:

$$\cos x = 0.$$

65. Find $f \circ g$ and $g \circ f$, where $f(x) = x^2 + 2x$ and $g(x) = \cos x$.

66. Calculate each of the following. Do not use the trigonometric function keys.

a) $\sin \frac{\pi}{4}$ b) $\sin \frac{\pi}{3}$

67. Calculate each of the following. Do not use the trigonometric function keys.

a) $\cos \frac{\pi}{6}$ b) $\cos \frac{\pi}{4}$

68. Show that the sine function does not have the *linearity* property, i.e., that it is not true that

$$\sin (x + y) = \sin x + \sin y$$

for all numbers x and y.

69. Graph: $y = 3 \sin x$.

70. Graph: $y = \sin x + \cos x$.

What are the domain, the range, the period, and the amplitude?

71. $y = \sin^2 x$

72. $y = |\cos x| + 1$

What is the domain?

73. $f(x) = \sqrt{\cos x}$

74. $g(x) = \dfrac{1}{\sin x}$

75. $f(x) = \dfrac{\sin x}{\cos x}$

76. $y = \log (\sin x)$

77. Consider $(\sin x)/x$, when x is between 0 and $\pi/2$. What can you say about this function? As x approaches 0, this function approaches a limit. What is it?

78. Solve $\sin x < \cos x$ in the interval $[-\pi, \pi]$.

6.3 The Other Basic Circular Functions

OBJECTIVES

You should be able to:

1 **Define the tangent, the cotangent, the secant, and the cosecant functions and find function values (when they exist) for multiples of $\pi/6$, $\pi/4$, and $\pi/3$.**

2 **Sketch graphs of the tangent, the cotangent, the secant, and the cosecant functions and describe properties of the functions.**

3 **State and prove basic and Pythagorean identities and use them to find function values given a function value of s and the quadrant in which s lies.**

4 **Prove simple trigonometric identities.**

There are six basic circular, or trigonometric, functions. In this section, we define the remaining four. Many other related functions can be obtained from the basic ones by adding them, multiplying them, dividing them, translating them, and so on.

1 Definitions and Function Values

There are four other circular functions, all defined in terms of the sine and the cosine functions. Their definitions are as follows. We again consider a point T on a unit circle determined by a real number s.

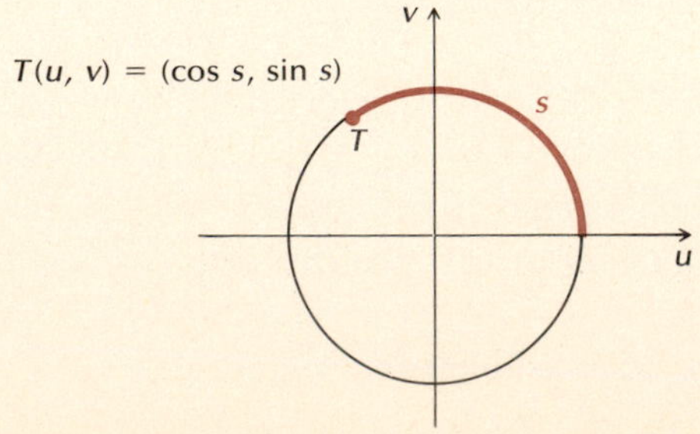

$\cos s = u,$
$\sin s = v.$

DEFINITIONS

The tangent function: $\tan s = \dfrac{\text{second coordinate}}{\text{first coordinate}} = \dfrac{\sin s}{\cos s} = \dfrac{v}{u}$;

The cotangent function: $\cot s = \dfrac{\text{first coordinate}}{\text{second coordinate}} = \dfrac{\cos s}{\sin s} = \dfrac{u}{v}$;

The secant function: $\sec s = \dfrac{1}{\text{first coordinate}} = \dfrac{1}{\cos s} = \dfrac{1}{u}$;

The cosecant function: $\csc s = \dfrac{1}{\text{second coordinate}} = \dfrac{1}{\sin s} = \dfrac{1}{v}$

Let us find some function values.

EXAMPLE 1 Find tan 0, cot 0, sec 0, and csc 0.

Solution We locate the point 0 on a unit circle and note that it determines a point with coordinates (1, 0). Then

$$\tan 0 = \frac{\text{second coordinate}}{\text{first coordinate}} = \frac{\sin 0}{\cos 0} = \frac{0}{1} = 0;$$

$$\cot 0 = \frac{\text{first coordinate}}{\text{second coordinate}} = \frac{\cos 0}{\sin 0}$$

$$= \frac{1}{0}, \text{ which is not defined;}$$

$$\sec 0 = \frac{1}{\text{first coordinate}} = \frac{1}{\cos 0} = \frac{1}{1} = 1;$$

$$\csc 0 = \frac{1}{\text{second coordinate}} = \frac{1}{\sin 0}$$

$$= \frac{1}{0}, \text{ which is not defined.}$$

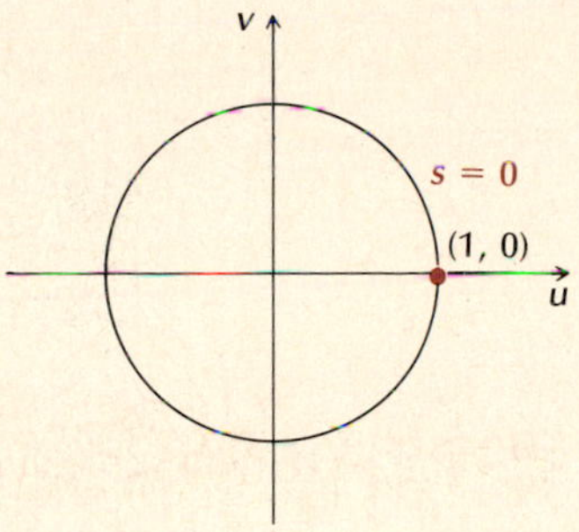

DO EXERCISE 1.

EXAMPLE 2 Find $\tan\frac{\pi}{4}$, $\cot\frac{\pi}{4}$, $\sec\frac{\pi}{4}$, and $\csc\frac{\pi}{4}$.

Solution We locate $\pi/4$ on a unit circle and recall from Section 6.2 that it determines a point with coordinates $(\sqrt{2}/2, \sqrt{2}/2)$. Then

$$\tan\frac{\pi}{4} = \frac{\sin\frac{\pi}{4}}{\cos\frac{\pi}{4}} = \frac{\frac{\sqrt{2}}{2}}{\frac{\sqrt{2}}{2}} = 1;$$

$$\cot\frac{\pi}{4} = \frac{\cos\frac{\pi}{4}}{\sin\frac{\pi}{4}} = \frac{\frac{\sqrt{2}}{2}}{\frac{\sqrt{2}}{2}} = 1;$$

$$\sec\frac{\pi}{4} = \frac{1}{\cos\frac{\pi}{4}} = \frac{1}{\frac{\sqrt{2}}{2}} = \frac{2}{\sqrt{2}} = \frac{2\sqrt{2}}{2} = \sqrt{2};$$

$$\csc\frac{\pi}{4} = \frac{1}{\sin\frac{\pi}{4}} = \frac{1}{\frac{\sqrt{2}}{2}} = \frac{2}{\sqrt{2}} = \sqrt{2}.$$

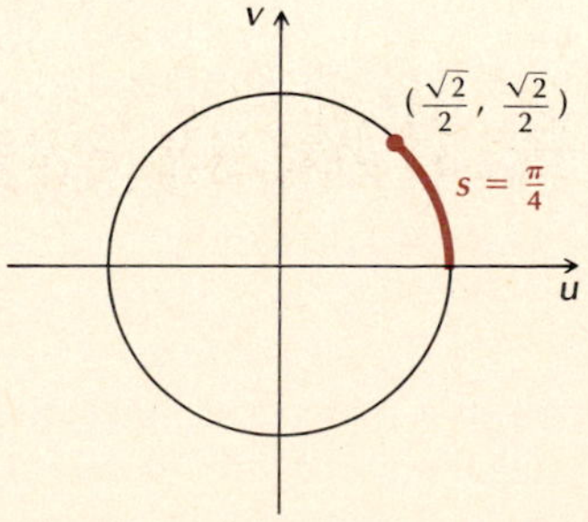

DO EXERCISE 2.

1. Find $\tan\pi$, $\cot\pi$, $\sec\pi$, and $\csc\pi$.

2. Find $\tan\frac{3\pi}{4}$, $\cot\frac{3\pi}{4}$, $\sec\frac{3\pi}{4}$, and $\csc\frac{3\pi}{4}$.

3. Find $\tan \frac{\pi}{6}$, $\cot \frac{\pi}{6}$, $\sec \frac{\pi}{6}$, and $\csc \frac{\pi}{6}$.

4. Find $\tan\left(-\frac{\pi}{3}\right)$, $\cot\left(-\frac{\pi}{3}\right)$, $\sec\left(-\frac{\pi}{3}\right)$, and $\csc\left(-\frac{\pi}{3}\right)$.

5. Find $\tan \frac{\pi}{2}$, $\tan\left(-\frac{\pi}{2}\right)$, and $\tan\left(-\frac{3\pi}{2}\right)$.

EXAMPLE 3 Find $\tan \frac{\pi}{3}$, $\cot \frac{\pi}{3}$, $\sec \frac{\pi}{3}$, and $\csc \frac{\pi}{3}$.

Solution We locate $\pi/3$ on a unit circle and recall from Section 6.2 that it determines a point with coordinates $(1/2, \sqrt{3}/2)$.

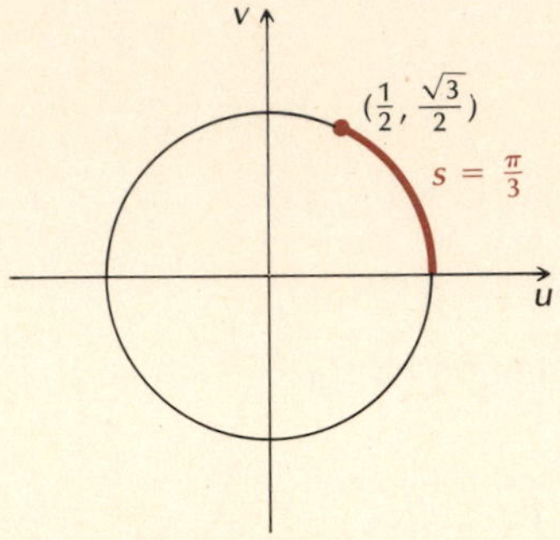

Then:

$$\tan \frac{\pi}{3} = \frac{\sin \frac{\pi}{3}}{\cos \frac{\pi}{3}} = \frac{\frac{\sqrt{3}}{2}}{\frac{1}{2}} = \sqrt{3};$$

$$\cot \frac{\pi}{3} = \frac{\cos \frac{\pi}{3}}{\sin \frac{\pi}{3}} = \frac{\frac{1}{2}}{\frac{\sqrt{3}}{2}} = \frac{1}{\sqrt{3}} = \frac{\sqrt{3}}{3};$$

$$\sec \frac{\pi}{3} = \frac{1}{\cos \frac{\pi}{3}} = \frac{1}{\frac{1}{2}} = 2;$$

$$\csc \frac{\pi}{3} = \frac{1}{\sin \frac{\pi}{3}} = \frac{1}{\frac{\sqrt{3}}{2}} = \frac{2}{\sqrt{3}} = \frac{2\sqrt{3}}{3}.$$

DO EXERCISES 3–5.

2 Properties of the Tangent, Cotangent, Secant, and Cosecant Functions

The Tangent Function

Note the following:

$$\tan \frac{\pi}{2} = \frac{\sin \frac{\pi}{2}}{\cos \frac{\pi}{2}} = \frac{1}{0},$$

which is undefined, since division by 0 is not possible. The tangent function is undefined for any number whose cosine is 0. Thus it is undefined for $(\pi/2) + k\pi$, where k is any integer. In the first quadrant, the function values are positive. In the second quadrant, however, the sine and cosine values have opposite signs, so the tangent values are negative. Tangent values are also negative in the fourth quadrant, and they are positive in the third quadrant.

A graph of the tangent function is shown below. It can be constructed using the values we have found in the preceding examples and margin exercises and by computing other values as needed. Note that the function value is 0 when $x = 0$, and the values increase as x increases toward $\pi/2$. As we approach $\pi/2$, the denominator becomes very small, so the tangent values become very large. In fact, they increase without bound. The dashed vertical line $x = \pi/2$ is called an **asymptote**. The graph gets closer and closer to the asymptote as x gets closer to $\pi/2$ with values that are smaller than $\pi/2$, but it never crosses the line. Each of the vertical dashed lines is an asymptote.

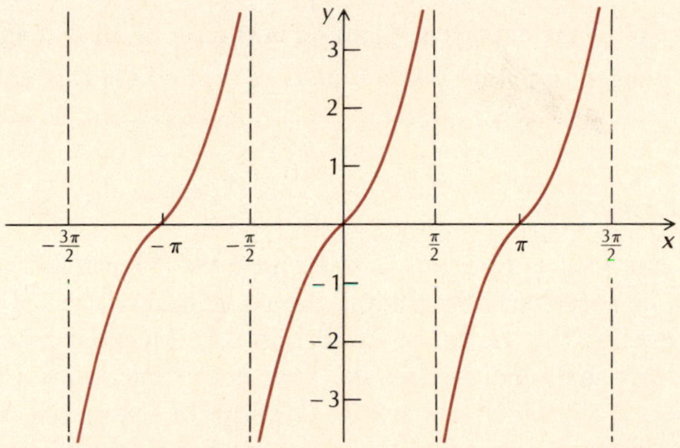

The tangent function

DO EXERCISES 6–11.

The following is a summary of the properties of the tangent function.

Properties of the Tangent Function

1. **The tangent function is periodic, with period π.**
2. **The domain of the tangent function is the set of all real numbers except $(\pi/2) + k\pi$, where k is any integer. Tan x is not defined where $\cos x = 0$.**
3. **The range of the tangent function is the set of all real numbers.**
4. **The tangent function is continuous except where it is not defined.**

The Cotangent Function

The graph of the cotangent function is shown below. The cotangent of a number is the quotient of its cosine by its sine. Thus the cotangent function is undefined for any number whose sine is 0.

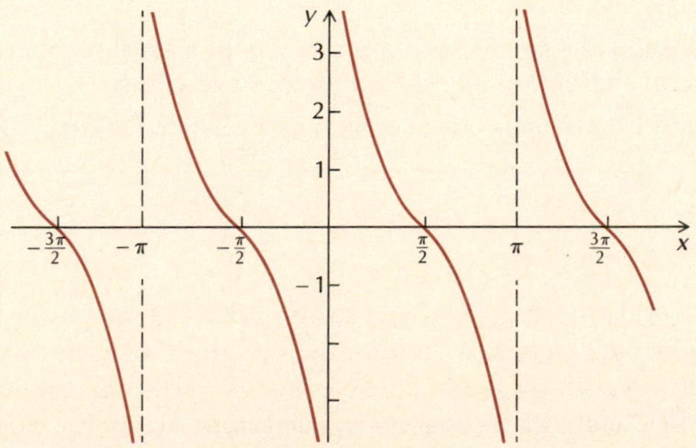

The cotangent function

DO EXERCISES 12–18.

6. Find three real numbers other than $\pi/2$ at which the tangent function is undefined.

7. Graph the tangent function.

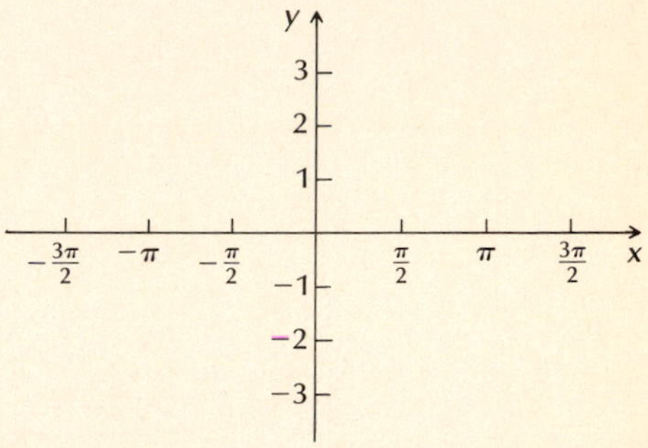

8. What is the period of the tangent function?

9. What is the domain of the tangent function?

10. What is the range of the tangent function?

11. Does the tangent function appear to be even? Is it odd?

12. Find three real numbers at which the cotangent function is undefined.

13. Graph the cotangent function.

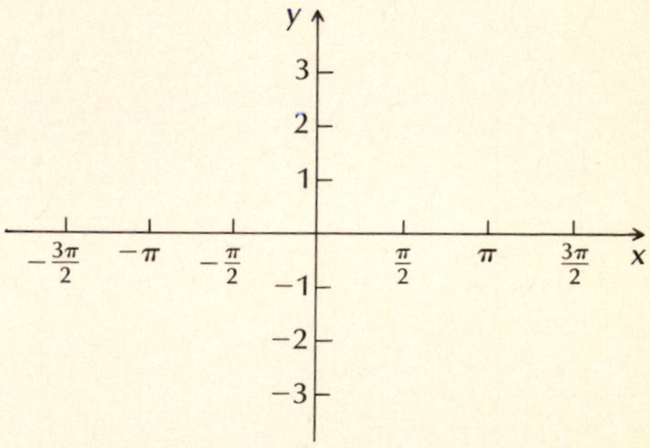

14. What is the period of the cotangent function?

15. What is the domain of the cotangent function?

16. What is the range of the cotangent function?

17. In which quadrants are the function values of the cotangent function positive? negative?

18. Does the cotangent function appear to be even? Is it odd?

Properties of the Cotangent Function

1. **The cotangent function is periodic, with period π.**
2. **The domain of the cotangent function is the set of all real numbers except $k\pi$, where k is any integer. Cot x is not defined where $\sin x = 0$.**
3. **The range of the cotangent function is the set of all real numbers.**
4. **The cotangent function is continuous except where it is not defined.**

The Secant Function

The secant and cosine functions are reciprocals. The graph of the secant function can be constructed by finding the reciprocal of each of the values of the cosine function that we found in Section 6.2. Thus the functions will be positive together and negative together. The secant function is not defined for those numbers x for which $\cos x = 0$. A graph of the secant function is as follows. The cosine function is shown for reference by the dashed curve, since these functions are reciprocals.

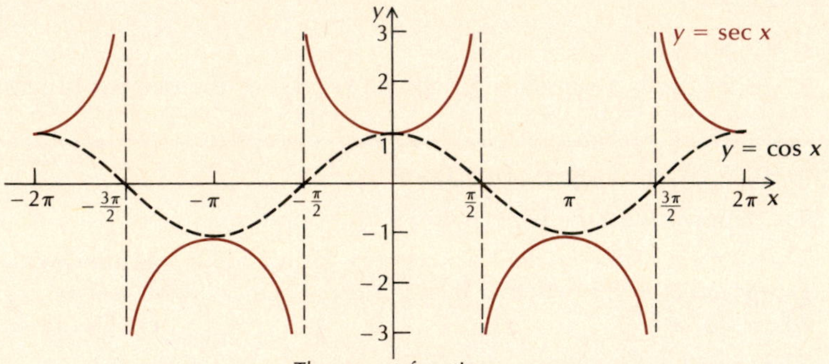

The secant function

The following is a summary of the properties of the secant function.

Properties of the Secant Function

1. **The secant function is periodic, with period 2π.**
2. **The domain of the secant function is the set of all real numbers except $(\pi/2) + k\pi$, where k is any integer. Sec x is not defined where $\cos x = 0$.**
3. **The range of the secant function consists of all real numbers 1 and greater, in addition to all real numbers -1 and less.**
4. **The secant function is continuous wherever it is defined.**

The Cosecant Function

The cosecant and sine functions are reciprocals. The graph of the cosecant function can be constructed by finding the reciprocal of each of the values of the sine function that we found in Section 6.2. Thus the functions will be positive together and negative together. The secant function is not defined for those numbers x for which $\sin x = 0$. A graph of the cosecant function is as follows. The sine function is shown by the dashed curve, since these functions are reciprocals.

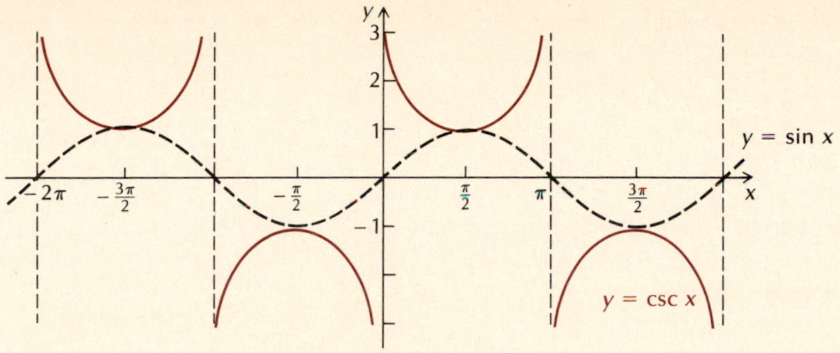

The cosecant function

The following is a summary of the properties of the cosecant function.

Properties of the Cosecant Function

1. **The cosecant function is periodic, with period 2π.**
2. **The domain of the cosecant function is the set of all real numbers except $k\pi$, where k is any integer. Csc x is not defined where $\sin x = 0$.**
3. **The range of the cosecant function consists of all real numbers 1 and greater, in addition to all real numbers -1 and less.**
4. **The cosecant function is continuous wherever it is defined.**

DO EXERCISES 19–28.

Signs of the Functions

Suppose a real number s determines a point on the unit circle in the first quadrant. Since both coordinates are positive there, the function values for all the circular functions will be positive. In the second quadrant, the first coordinate is negative and the second coordinate is positive. Thus the sine function has positive values and the cosine function has negative values. The secant, being the reciprocal of the cosine, also has negative values, and the cosecant, being the reciprocal of the sine, has positive values. The tangent and the cotangent both have negative values in the second quadrant.

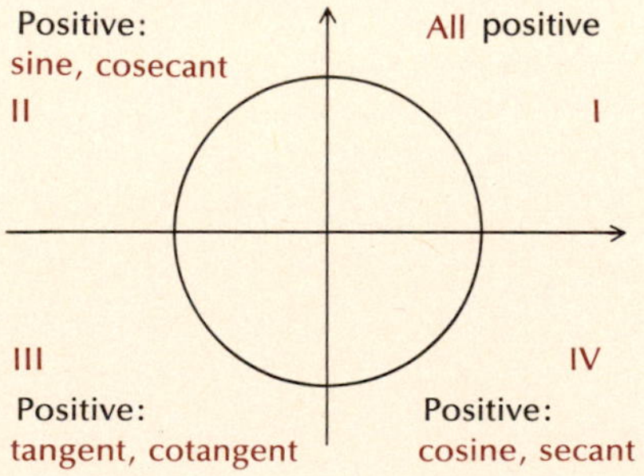

DO EXERCISE 29.

The preceding shows the quadrants in which the circular function values are positive. Although the diagram need not be memorized, because the signs can be readily determined by reference to a unit circle, it can be helpful to remember the diagram quickly by thinking of the memory device ALLSIN-TANCOS. This means, moving counterclockwise, that ALL functions are

19. a) Find three numbers at which the secant function is undefined.
    b) Find three numbers at which the cosecant function is undefined.

20. Graph the secant function.

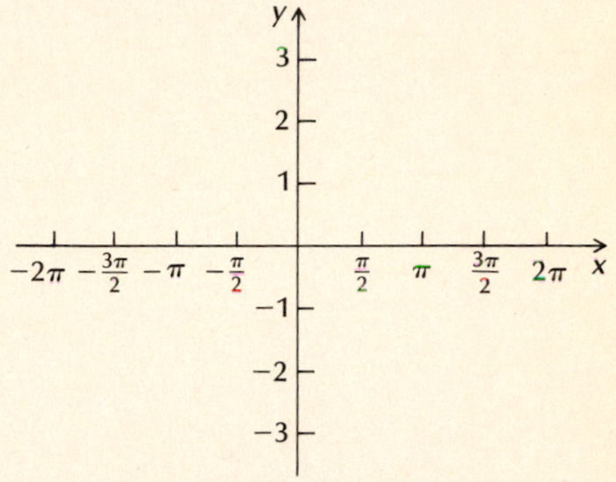

21. Graph the cosecant function.

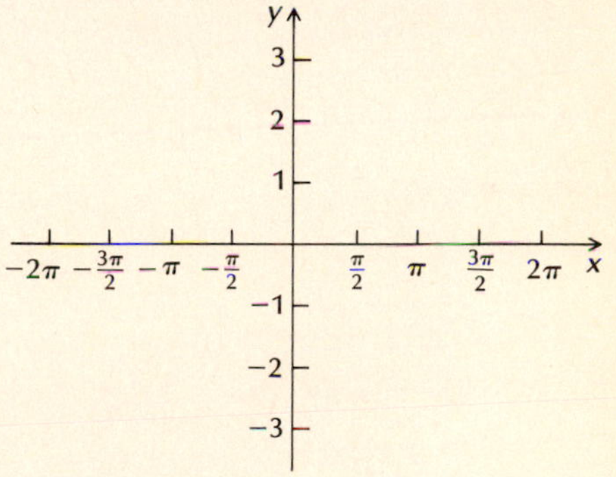

22. What is the period of the secant function? the cosecant function?

23. What is the domain of the secant function? the cosecant function?

24. What is the range of the secant function? the cosecant function?

25. In which quadrants are the function values of the secant function positive? negative?

26. In which quadrants are the function values of the cosecant function positive? negative?

27. Does the secant function appear to be even? odd?

28. Does the cosecant function appear to be even? odd?

29. What are the signs of the six circular functions in each of the four quadrants?

30. Prove the identity $\tan s = \dfrac{1}{\cot s}$.

positive in the first quadrant, the SIN function is positive in the second quadrant, the TAN function is positive in the third quadrant, and the COS function is positive in the fourth quadrant. Reciprocals can also be thought out from this memory device, since, for example, the cotangent is positive where the tangent is positive.

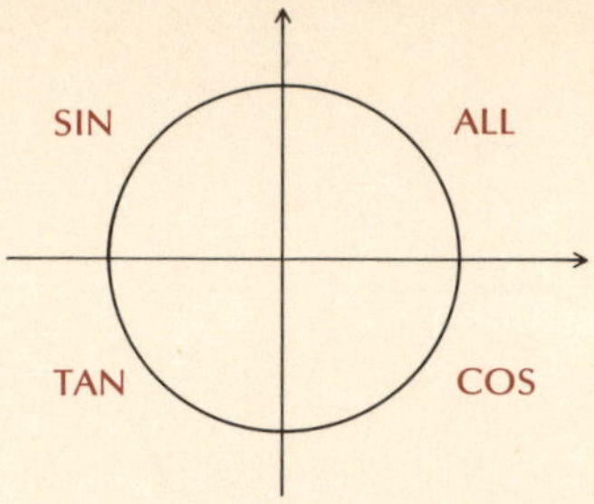

ALLSINTANCOS

3 The Functions Interrelated

The definitions of the four trigonometric functions introduced in this section give us eight basic identities.

Basic Identities

$$\sin s = \frac{1}{\csc s} \qquad \csc s = \frac{1}{\sin s} \qquad \tan s = \frac{\sin s}{\cos s}$$

$$\cos s = \frac{1}{\sec s} \qquad \sec s = \frac{1}{\cos s} \qquad \cot s = \frac{\cos s}{\sin s}$$

$$\tan s = \frac{1}{\cot s} \qquad \cot s = \frac{1}{\tan s}$$

Some of these have not been proved, but are straightforward.

EXAMPLE 4 Prove the identity $\cot s = \dfrac{1}{\tan s}$.

Proof

$$\cot s = \frac{\cos s}{\sin s} = \frac{1}{\dfrac{\sin s}{\cos s}} = \frac{1}{\tan s}$$ ■

DO EXERCISE 30.

31. Prove the identity $\cos s = \dfrac{1}{\sec s}$.

EXAMPLE 5 Prove the identity $\sin s = \dfrac{1}{\csc s}$.

Proof

$$\sin s = \frac{1}{\dfrac{1}{\sin s}} = \frac{1}{\csc s}$$ ■

DO EXERCISE 31.

The Pythagorean Identities

Here we consider three other identities that are fundamental to a study of trigonometry. They are called the **Pythagorean identities**. Recall that the equation of a unit circle in a uv-plane is

$$u^2 + v^2 = 1.$$

For any point on a unit circle, the coordinates u and v satisfy this equation. Suppose a real number s determines a point T on a unit circle, with coordinates (u, v), or $(\cos s, \sin s)$. Then $u = \cos s$ and $v = \sin s$.

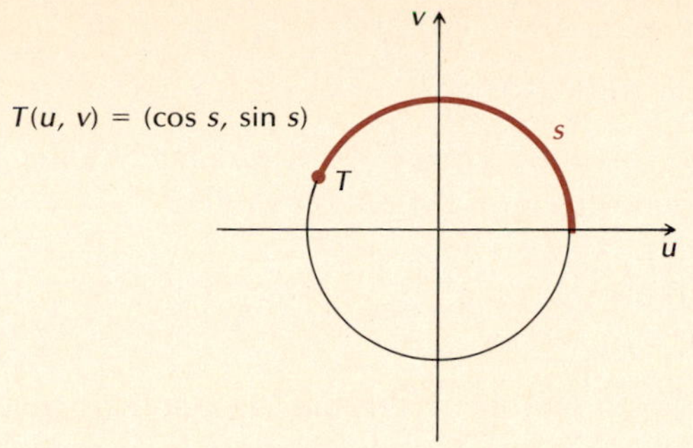

Substituting cos s for u and sin s for v in the equation of the unit circle gives us the identity

$$(\cos s)^2 + (\sin s)^2 = 1,$$

which can be expressed as

$$\sin^2 s + \cos^2 s = 1.$$

It is conventional in trigonometry to use the notation $\sin^2 s$ rather than $(\sin s)^2$—there are fewer symbols to write. This identity relates the sine and the cosine of any real number s. It is an important Pythagorean identity. We now develop another. We divide on both sides of the preceding identity by $\sin^2 s$:

$$\frac{\sin^2 s}{\sin^2 s} + \frac{\cos^2 s}{\sin^2 s} = \frac{1}{\sin^2 s}.$$

Simplifying, we get

$$1 + \cot^2 s = \csc^2 s.$$

This equation is true for any replacement of s by a real number for which $\sin^2 s \neq 0$, since we divided by $\sin^2 s$. But the numbers for which $\sin^2 s = 0$ (or $\sin s = 0$) are exactly the ones for which the cotangent and the cosecant functions are undefined. Hence our new equation holds for all numbers s for which cot s and csc s are defined, and is thus an identity.

DO EXERCISE 32.

32. Prove the identity $1 + \tan^2 s = \sec^2 s$.

The third Pythagorean identity, obtained by dividing on both sides of the first Pythagorean identity by $\cos^2 s$, is

$$1 + \tan^2 s = \sec^2 s.$$

The Pythagorean Identities

$$\sin^2 s + \cos^2 s = 1,$$
$$1 + \cot^2 s = \csc^2 s,$$
$$1 + \tan^2 s = \sec^2 s$$

The Pythagorean identities should be memorized. Certain variations of them, although most useful, need not be memorized, because they are so easily obtained from these three.

EXAMPLE 6 Derive identities that give $\cos^2 x$ and $\cos x$ in terms of $\sin x$.

Solution We begin with $\sin^2 x + \cos^2 x = 1$, and solve it for $\cos^2 x$. We have $\cos^2 x = 1 - \sin^2 x$. This gives us $\cos^2 x$ in terms of $\sin x$. Then taking square

33. Derive an identity that gives $\sin^2 x$ in terms of $\cos x$.

34. Derive an identity that gives $\sin x$ in terms of $\cos x$.

35. Given that $\sin s = -\frac{3}{8}$ and that s is in the third quadrant, find the other function values.

roots, we have

$$\cos x = \pm\sqrt{1 - \sin^2 x},$$

with the understanding that the sign must be determined by the quadrant in which the point determined by x on the unit circle lies. ■

DO EXERCISES 33 AND 34.

Using identities, we can find the rest of the trigonometric function values from one value if we know the quadrant in which s lies on a unit circle.

EXAMPLE 7 Given that $\cos s = \frac{1}{5}$ and that s is in the fourth quadrant, find the other function values.

Solution From Margin Exercise 34, we know that

$$\sin s = \pm\sqrt{1 - \cos^2 s},$$

where the sign of the radical depends on the quadrant in which s lies. Since s is in the fourth quadrant, the sine function is negative. We can then solve for $\sin s$ by substituting $\frac{1}{5}$ for $\cos s$ in the preceding identity:

$$\sin s = -\sqrt{1 - \cos^2 s} = -\sqrt{1 - \left(\frac{1}{5}\right)^2} = -\sqrt{1 - \frac{1}{25}}$$
$$= -\sqrt{\frac{24}{25}} = -\frac{\sqrt{24}}{5} = -\frac{2\sqrt{6}}{5}.$$

Now that we have $\sin s$ and $\cos s$, we can find the other function values:

$$\tan s = \frac{\sin s}{\cos s} = \frac{-\frac{2\sqrt{6}}{5}}{\frac{1}{5}} = -2\sqrt{6};$$

$$\cot s = \frac{1}{\tan s} = \frac{1}{-2\sqrt{6}} = -\frac{\sqrt{6}}{12};$$

$$\sec s = \frac{1}{\cos s} = \frac{1}{\frac{1}{5}} = 5;$$

$$\csc s = \frac{1}{\sin s} = \frac{1}{-\frac{2\sqrt{6}}{5}} = -\frac{5}{2\sqrt{6}} = -\frac{5\sqrt{6}}{12}.$$ ■

DO EXERCISE 35.

EXAMPLE 8 Given that $\cot s = 2$ and that s is in the third quadrant, find the other function values.

Solution Solving the identity $1 + \cot^2 s = \csc^2 s$ for $\csc s$, we get

$$\csc s = \pm\sqrt{1 + \cot^2 s},$$

where the sign of the radical depends on the quadrant in which s lies. Since s is in the third quadrant, $\csc s = 1/\sin s$, and the sine function is negative in the third quadrant, it follows that the cosecant function is negative in the third quadrant. Thus,

$$\csc s = -\sqrt{1 + \cot^2 s} = -\sqrt{1 + 2^2} = -\sqrt{5}.$$

From the identity $\sin s = 1/\csc s$, we can find $\sin s$ as follows:

$$\sin s = \frac{1}{\csc s} = \frac{1}{-\sqrt{5}}$$
$$= -\frac{\sqrt{5}}{5}.$$

Since the cosine function is negative in the third quadrant, we can find $\cos s$ as follows:

$$\cos s = -\sqrt{1 - \sin^2 s} = -\sqrt{1 - \left(-\frac{\sqrt{5}}{5}\right)^2}$$
$$= -\sqrt{1 - \frac{5}{25}} = -\sqrt{\frac{4}{5}}$$
$$= -\frac{2\sqrt{5}}{5}.$$

Finally, we get the tangent as the reciprocal of the cotangent:

$$\tan s = \frac{1}{\cot s}$$
$$= \frac{1}{2}.$$

DO EXERCISE 36.

36. Given that $\tan s = -3$ and that s is in the second quadrant, find the other function values.

4 Some Other Identities

The tangent function appears to be odd. Let us investigate by finding $\tan(-x)$. By definition of the tangent function,

$$\tan(-x) = \frac{\sin(-x)}{\cos(-x)}.$$

But we know that $\sin(-x) = -\sin x$ and $\cos(-x) = \cos x$. Therefore,

$$\tan(-x) = \frac{-\sin x}{\cos x}$$
$$= -\tan x.$$

Thus the tangent function is indeed odd. Similarly, it can be shown that the cotangent function is odd, the secant function is even, and the cosecant function is odd. Each such demonstration gives us an identity.

By using the previously developed identities involving the sine and the cosine functions, we can develop a number of other identities.

EXAMPLE 9 Prove the identity $\tan(x + \pi) = \tan x$.

Proof

$$\tan(x + \pi) = \frac{\sin(x + \pi)}{\cos(x + \pi)}$$ By definition of the tangent function

$$= \frac{-\sin x}{-\cos x}$$ Using $\sin(x + \pi) = -\sin x$ and $\cos(x + \pi) = -\cos x$

$$= \frac{\sin x}{\cos x}$$
$$= \tan x$$

37. Prove the identity

$$\cot(-x) = -\cot x.$$

38. Prove the identity

$$\tan(x - \pi) = \tan x.$$

39. Prove the identity

$$\csc(\pi - x) = \csc x.$$

EXAMPLE 10 Prove the identity $\sec(\pi - x) = -\sec x$.

Proof

$$\sec(\pi - x) = \frac{1}{\cos(\pi - x)} \quad \text{By definition of the secant function}$$

$$= \frac{1}{-\cos x} \quad \text{Using } \cos(\pi - x) = -\cos x$$

$$= -\frac{1}{\cos x}$$

$$= -\sec x$$

EXAMPLE 11 Prove the identity $\csc(x - \pi) = -\csc x$.

Proof

$$\csc(x - \pi) = \frac{1}{\sin(x - \pi)} \quad \text{By definition of the cosecant function}$$

$$= \frac{1}{-\sin x} \quad \text{Using } \sin(x - \pi) = -\sin x$$

$$= -\csc x$$

DO EXERCISES 37–39.

EXERCISE SET 6.3

1 Find the function values.

1. $\cot \frac{\pi}{4}$
2. $\tan\left(-\frac{\pi}{4}\right)$
3. $\tan \frac{\pi}{6}$
4. $\cot \frac{5\pi}{6}$
5. $\sec \frac{\pi}{4}$
6. $\csc \frac{3\pi}{4}$
7. $\tan \frac{3\pi}{2}$
8. $\cot \pi$
9. $\tan \frac{2\pi}{3}$
10. $\cot\left(-\frac{2\pi}{3}\right)$
11. $\sec\left(-\frac{7\pi}{4}\right)$
12. $\csc\left(-\frac{4\pi}{3}\right)$
13. $\tan \frac{5\pi}{6}$
14. $\cot \frac{7\pi}{6}$

2

15. Of the six circular functions, which are even?
16. Of the six circular functions, which are odd?
17. Which of the six circular functions have period 2π?
18. Which of the six circular functions have period π?
19. In which quadrants is the tangent function positive? negative?
20. In which quadrants is the cotangent function positive? negative?
21. In which quadrants is the secant function positive? negative?
22. In which quadrants is the cosecant function positive? negative?

3

23. Complete this table of approximate function values. Do not use trigonometric function keys. Round to five decimal places.

	$\pi/16$	$\pi/8$	$\pi/6$	$\pi/4$	$3\pi/8$	$7\pi/16$
sin	0.19509	0.38268			0.92388	0.98079
cos	0.98079	0.92388			0.38268	0.19509
tan						
cot						
sec						
csc						

24. Complete this table of approximate function values. Do not use trigonometric function keys. Round to five decimal places.

	$-\pi/16$	$-\pi/8$	$-\pi/6$	$-\pi/4$	$-\pi/3$
sin	-0.19509	-0.38268	-0.50000	-0.70711	-0.86603
cos	0.98079	0.92388	0.86603	0.70711	0.50000
tan					
cot					
sec					
csc					

Find the other function values, given each set of conditions.

25. $\cos s = \frac{1}{3}$, s in the first quadrant
26. $\sin s = \frac{2}{3}$, s in the first quadrant
27. $\tan s = 3$, s in the third quadrant
28. $\cot s = 4$, s in the third quadrant
29. $\sec s = -\frac{5}{3}$, s in the second quadrant
30. $\csc s = -\frac{5}{4}$, s in the fourth quadrant
31. $\sin s = -\frac{2}{5}$, s in the third quadrant
32. $\cos s = -\frac{1}{6}$, s in the third quadrant

4 Prove the following identities.

33. $\sec(-x) = \sec x$
34. $\csc(-x) = -\csc x$
35. $\cot(x + \pi) = \cot x$
36. $\cot(x - \pi) = \cot x$
37. $\sec(x + \pi) = -\sec x$
38. $\tan(\pi - x) = -\tan x$

SYNTHESIS

39. Verify the identity $\sec(x - \pi) = -\sec x$ graphically.
40. Verify the identity $\tan(x + \pi) = \tan x$ graphically.
41. Describe how the graphs of the tangent and the cotangent functions are related.
42. Describe how the graphs of the secant and the cosecant functions are related.
43. Which pairs of circular functions have the same zeros? (A "zero" of a function is an input that produces an output of 0.)
44. Describe how the asymptotes of the tangent, cotangent, secant, and cosecant functions are related to the inputs that produce outputs of 0.

Graph.

45. $f(x) = |\tan x|$
46. $f(x) = |\sin x|$
47. $g(x) = \sin |x|$
48. $f(x) = |\cos x|$

CHALLENGE

Solve graphically.

49. $\cos x \leq \sec x$
50. $\sin x > \csc x$

6.4 Angles and Rotations

OBJECTIVES

You should be able to:

1 **Given the measure of an angle or rotation in degrees, tell in which quadrant the terminal side lies; find two positive angles and two negative angles that are coterminal with a given angle; find the complement and the supplement of a given angle; and convert between angle measure in degrees, minutes, seconds, and decimal measure.**

(continued)

1 Angles, Rotations, and Degree Measure

An *angle* is a familiar figure.

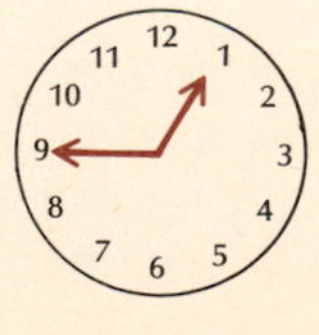

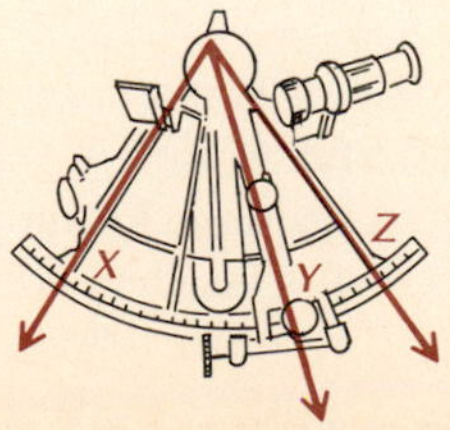

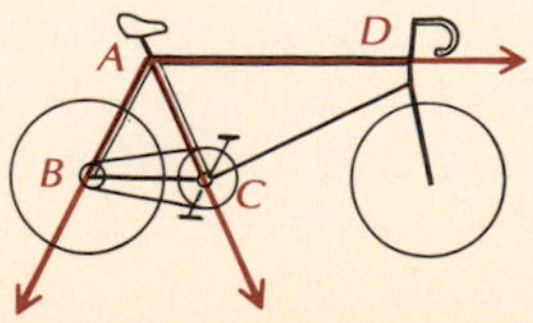

2 Convert between radian measure and degree measure.

3 Find the length of an arc of a circle, given the measure of its central angle and the length of a radius. Also, find the measure of a central angle of a circle, given the length of its arc and the length of a radius.

4 Convert between linear and angular speed.

5 Find total distance or total angle, when speed, radius, and time are given.

1. a) Draw $\angle ABC$.

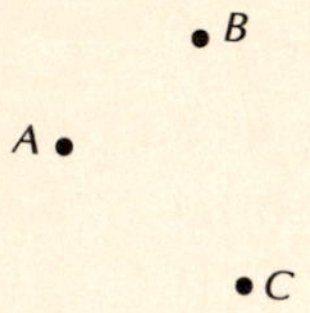

b) Draw $\angle BAC$.

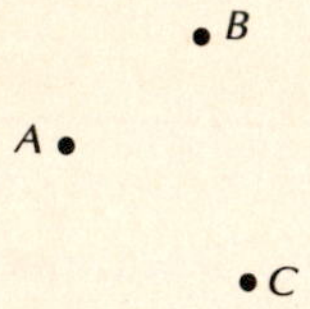

Let us review the definition of angle from our study of geometry. Below is a line determined by two points A and B. If we consider the points A and B and all points between A and B, we have the **line segment** AB, denoted $\overline{AB}$. If we consider the line segment AB and all points C such that B is between A and C, then we have the **ray** AB, denoted $\overrightarrow{AB}$. The point A is called the **endpoint** of the ray. Note that $\overrightarrow{AB}$ and $\overrightarrow{BA}$ are not the same rays, but their union is the line AB.

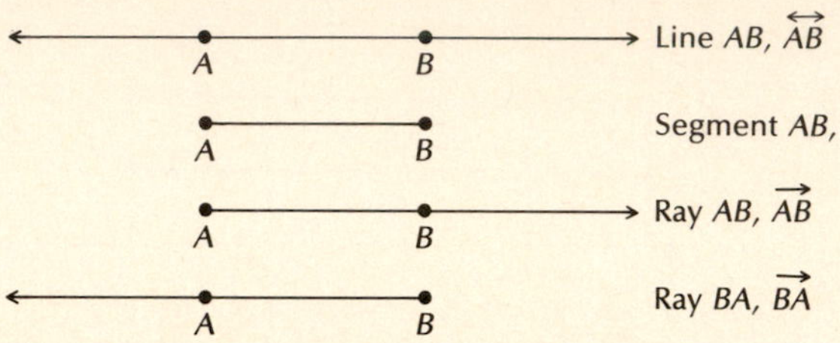

An **angle** is the union of rays with a common endpoint. Each ray is a **side** of the angle. If B is the common endpoint, it is called the **vertex**. If A is on one ray and C is on the other, and B is the vertex, we can name the angle as $\angle ABC$, where it is understood that the vertex is the middle letter.

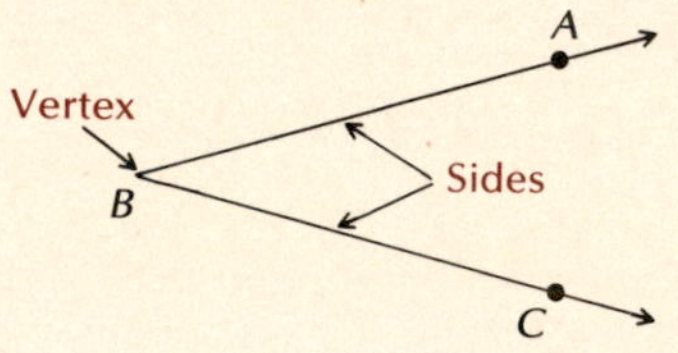

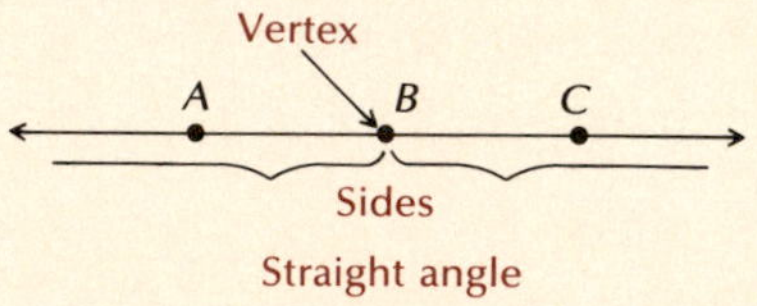

If the two rays that make up an angle are on the same line, we say that the angle is a *straight angle*.

DO EXERCISE 1.

In trigonometry, we often think of an angle as a **rotation.** To do so, think of locating a ray along the positive x-axis with its vertex at the origin. This ray is called the **initial side** of an angle. Though we leave that ray fixed, think of making a copy of it and rotating it. A rotation counterclockwise is a **positive rotation**, and a rotation clockwise is a **negative rotation**. The ray at the end of the rotation is called the **terminal side** of the angle. An angle so formed is said to be in **standard position.**

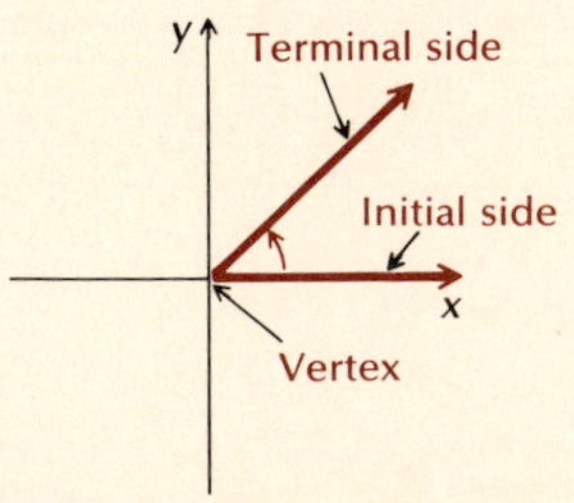

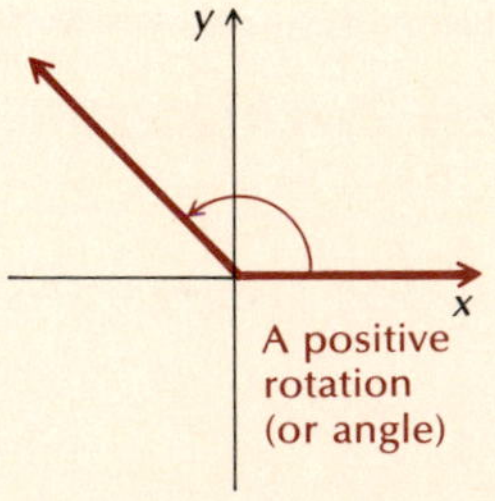

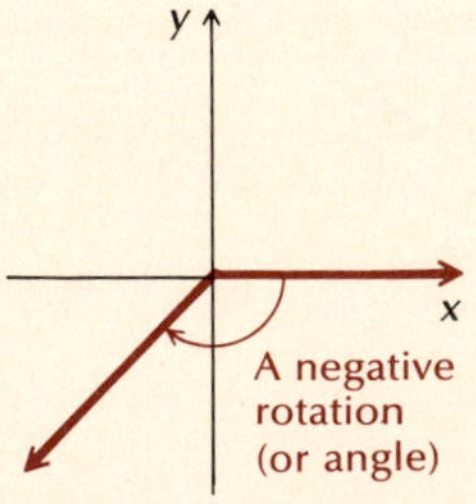

It is common to refer to an angle or rotation by Greek letters such as α (alpha), β (beta), γ (gamma), θ (theta), and ϕ (phi). A rotation need not stop after it goes around to the initial side. It can continue for more and more revolutions. If two rotations have the same terminal side, they are said to be **coterminal**. The following are two examples of coterminal angles.

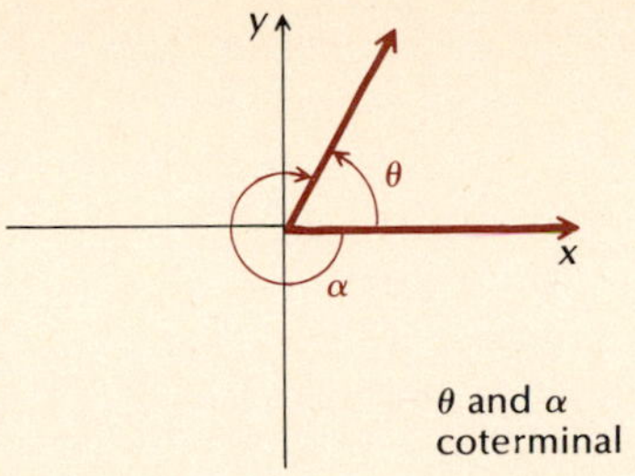

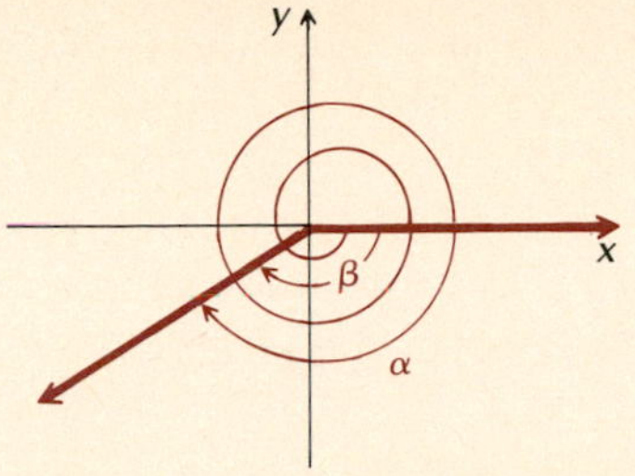

The measure of an angle or rotation may be given in **degrees**. The Babylonians developed the idea of dividing the circumference of a circle into 360 equal parts. Then one complete positive revolution or rotation has a measure of 360 *degrees*, which we denote with a degree symbol as 360°. One half of a revolution has a measure of 180°, one fourth of a revolution has a measure of 90°, and so on. We also speak of an angle of measure 60°, 135°, 415°, and so on. Measures of negative rotations can be −38°, −270°, −750°, and so on.

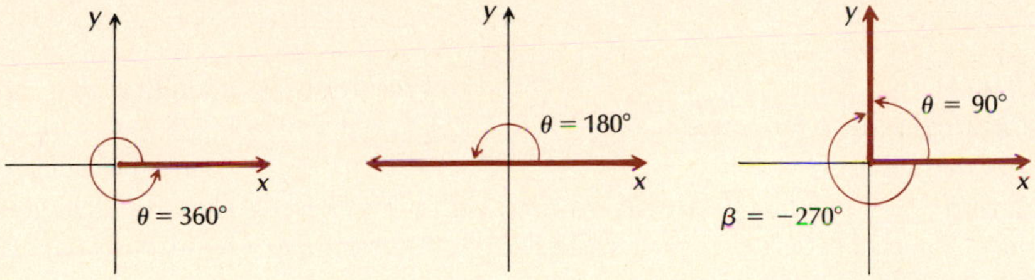

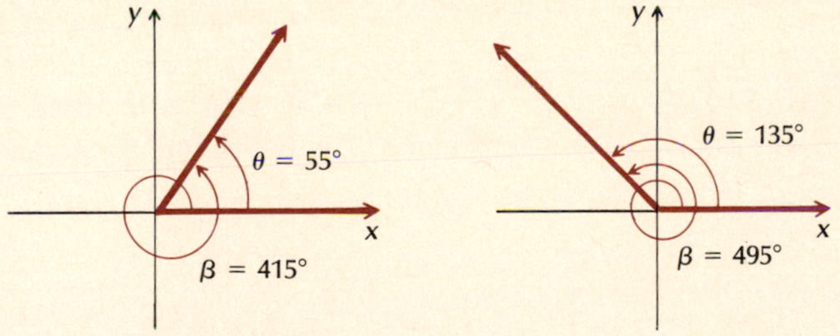

When the measure of an angle is greater than 360° or less than −360°, the rotating ray has gone through more than one complete revolution. For example, 415° will have the same terminal side as 55°. Thus the terminal side will be in the first quadrant. Note that angles of measure 415° and 55° are coterminal. We will often speak of an angle by its measure, rather than the more cumbersome terminology "a angle whose measure is . . . ." Thus we might say "the 45° angle," rather than "the angle whose measure is 45°."

EXAMPLES In which quadrant does the terminal side of each angle lie?

1. 53°

   The terminal side lies in the *first* quadrant.

2. −126°

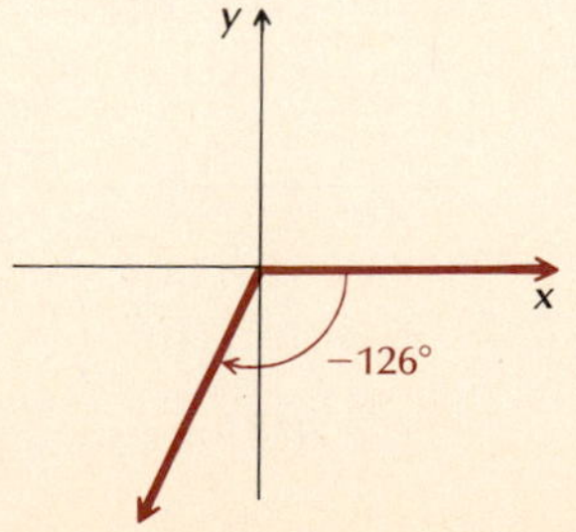

The terminal side lies in the *third* quadrant.

3. 460°

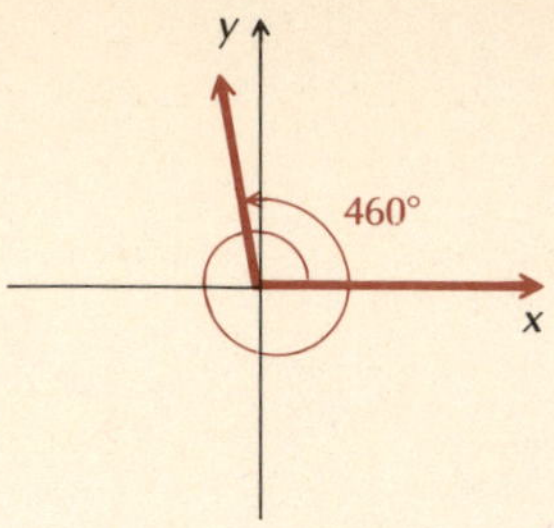

The terminal side lies in the *second* quadrant.

4. 253°

The terminal side lies in the *third* quadrant.

5. −373°

The terminal side lies in the *fourth* quadrant. ■

DO EXERCISES 2 AND 3.

EXAMPLE 6 Find two positive angles and two negative angles that are coterminal with a 30° angle.

Solution Consider a 30° angle or rotation. If we have a rotation that goes around another 360°, we get another angle, coterminal with 30°. If we go around twice or 720°, we get a second angle coterminal with a 30° angle. Since

$$30° + 360° = 390° \quad \text{and} \quad 30° + 720° = 750°,$$

we know that 390° and 750° are two positive angles that are coterminal with a 30° angle. Many other answers are possible by adding multiples of 360°. Similarly, we get the negative coterminal angles by adding −360° and −720°:

$$30° + (-360°) = -330° \quad \text{and} \quad 30° + (-720°) = -690°. \quad ■$$

DO EXERCISE 4.

An angle whose measure is 90° is called a **right angle.** An angle whose measure is between 0° and 90° is called an **acute angle.** An angle whose measure is between 90° and 180° is called an **obtuse angle.** A **straight angle** is an angle whose measure is 180°.

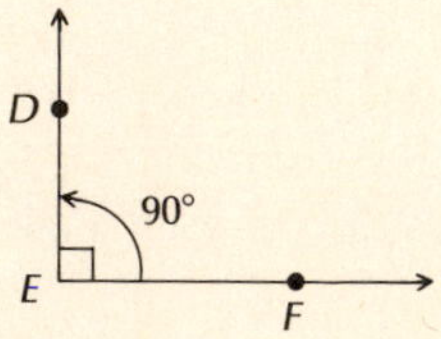

$\angle DEF$ is a right angle.

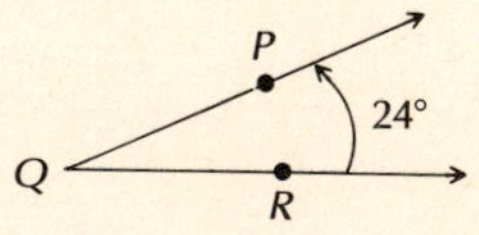

$\angle PQR$ is an acute angle.

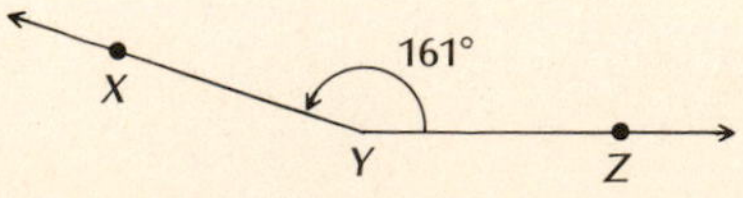

$\angle XYZ$ is an obtuse angle.

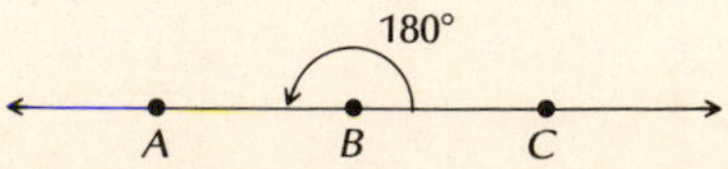

$\angle ABC$ is a straight angle.

2. In which quadrant does the terminal side of each angle lie?
   a) 47°
   b) 212°
   c) −43°
   d) −145°
   e) 365°
   f) −365°
   g) 740°

3. How many degrees are there in:
   a) one revolution?
   b) one half of a revolution?
   c) one fourth of a revolution?
   d) one eighth of a revolution?
   e) one sixth of a revolution?
   f) one twelfth of a revolution?

4. Find two positive angles and two negative angles that are coterminal with a 45° angle.

Subunits of degrees are often used. One **minute** is denoted $1'$ and is such that 60 minutes $= 60' = 1°$. One **second** is denoted $1''$ and is such that 60 seconds $= 60'' = 1'$. Then

$$34°42'28'' = 34 \text{ degrees, } 42 \text{ minutes, } 28 \text{ seconds} = 34° + 42' + 28''.$$

DO EXERCISES 5 AND 6.

Two acute angles are **complementary** if their sum is 90°. Thus, if θ is acute, then θ and $90° - \theta$ are complementary. If the sum of two positive angles is 180°, then the angles are **supplementary**.

EXAMPLE 7 Find the complement and the supplement of $56°23'16''$.

Solution

$$\begin{array}{rr} 90° = & 89°59'60'' \\ & -56°33'16'' \\ \hline & 33°36'44'' \end{array} \qquad \begin{array}{rr} 180° = & 179°59'60'' \\ & -56°23'16'' \\ \hline & 123°36'44'' \end{array}$$

Thus the complement of $56°23'16''$ is $33°36'44''$ and the supplement is $123°36'44''$. ■

DO EXERCISE 7.

EXAMPLE 8 Find the complement and the supplement of 87.46°.

Solution

$$90° - 87.46° = 2.54°,$$
$$180° - 87.46° = 92.54°.$$

Thus the complement of 87.46° is 2.54° and the supplement is 92.54°. ■

DO EXERCISE 8.

When working with a calculator, we will need to know how to convert a measure like $34°41'52''$ to decimal parts of degrees and conversely.

EXAMPLE 9 Convert $34°41'52''$ to degrees and decimal parts of degrees. Round to two decimal places.

Solution Using $1' = \left(\frac{1}{60}\right)^\circ$ and $1'' = \left(\frac{1}{3600}\right)^\circ$, we have

$$\begin{aligned} 34°41'52'' &= 34° + 41' + 52'' \\ &= 34° + \left(\tfrac{41}{60}\right)^\circ + \left(\tfrac{52}{3600}\right)^\circ \\ &\approx 34° + 0.68 + 0.01 \\ &= 34.69°. \end{aligned}$$ ■

EXAMPLE 10 Convert 16.35° to degrees and minutes.

Solution We have

$$16.35° = 16° + 0.35 \times 1°.$$

5. Classify each of the following angles as right, acute, obtuse, or straight.

a)

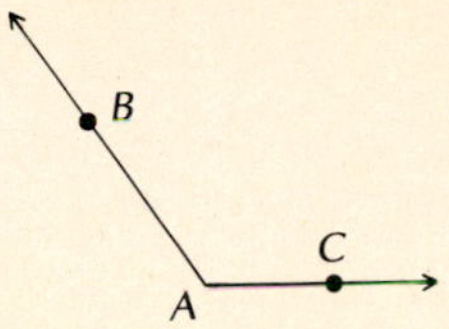

b)

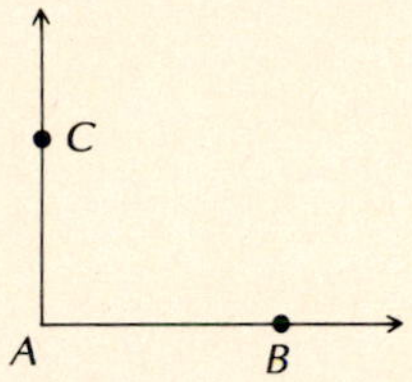

c)

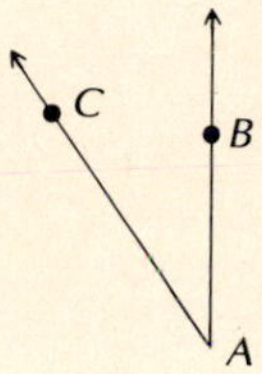

d)

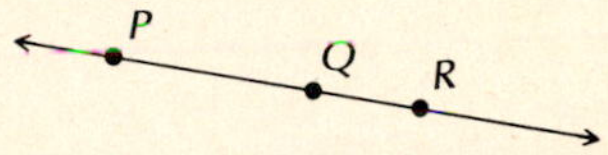

6. Complete.
   a) $1' = (\_\_\_\_)°$
   b) $1'' = (\_\_\_\_)°$
   c) $56°23'11'' =$
   $(\_\_\_\_)° + (\_\_\_\_)' + (\_\_\_\_)''$

7. Find the complement and the supplement of $85°12'38''$.

8. Find the complement and the supplement of 34.89°.

9. Convert 67°13′16″ to degrees and decimal parts of degrees. Round to four decimal places.

10. Convert 37.45° to degrees and minutes.

Now substituting 60′ for 1°, we have

$$0.35 \times 1° = 0.35 \times 60' = 21'.$$

Thus,

$$16.35° = 16°21'.$$

DO EXERCISES 9 AND 10.

2 Radian Measure

Degree measure is a common unit of angle measure in many everyday applications and in such fields as navigation and surveying. But in many scientific fields and in mathematics and calculus, there is another commonly used unit of measure called the **radian.**

Consider a circle with its center at the origin and radius of length 1 (a *unit circle*). Suppose we measure an arc, moving counterclockwise, of length 1, and mark a point T on the circle. If we draw a ray from the origin through T, we have formed an angle. The measure of that angle is 1 **radian.** The word radian comes from the word *radius*. Thus, measuring 1 radius along the circumference of the circle determines an angle whose measure is 1 *radian*. One radian is about 57.3°.

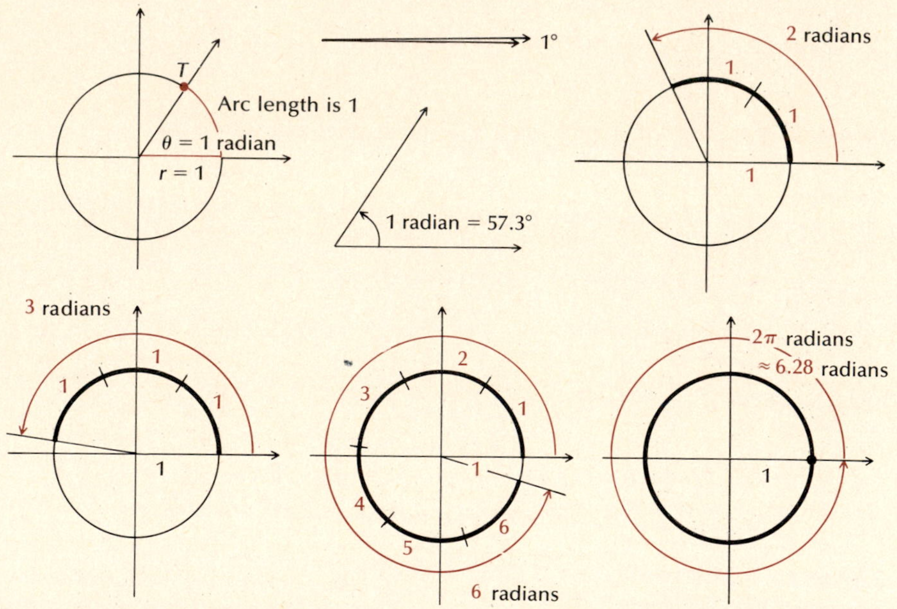

Angles that measure 2 radians and 3 radians are also shown above. When we make a complete (counterclockwise) revolution, the terminal side coincides with the initial side on the positive x-axis. We then have an angle whose measure is 2π radians, or about 6.28 radians, which is the circumference of the circle:

$$2\pi r = 2\pi(1) = 2\pi.$$

Thus a rotation of 360° (1 revolution) has a measure of 2π radians. Half of a revolution is a rotation of 180°, or π radians. A quarter revolution is a rotation of 90°, or $\pi/2$ radians, and so on.

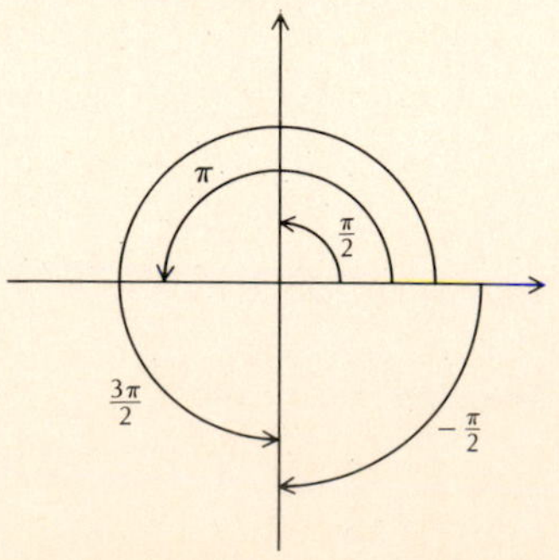

To convert between degrees and radians, we first note that

$$360° = 2\pi \text{ radians}.$$

It follows that

$$\mathbf{180° = \pi \text{ radians}.}$$

To make conversions, we use the notion of "multiplying by one" (see Section 1.9), noting the following:

$$\frac{1 \text{ revolution}}{1 \text{ revolution}} = \frac{\pi \text{ radians}}{180°} = \frac{180°}{\pi \text{ radians}} = 1.$$

When a rotation is given in radians, the word "radians" is optional and is most often omitted. Thus, if no unit is given for a rotation, the rotation is understood to be in radians.

EXAMPLE 11 Convert 60° to radians.

Solution We have

$$\begin{aligned} 60° &= 60° \cdot \frac{\pi \text{ radians}}{180°} \\ &= \frac{60°}{180°}\pi \text{ radians} \\ &= \frac{\pi}{3} \text{ radians}, \quad \text{or } \frac{\pi}{3} \end{aligned}$$

Using 3.14 for π, we find that $\pi/3$ radians is about 1.047 radians.

EXAMPLE 12 Convert $\frac{3\pi}{4}$ radians to degrees.

Solution

$$\begin{aligned} \frac{3\pi}{4} \text{ radians} &= \frac{3\pi}{4} \text{ radians} \cdot \frac{180°}{\pi \text{ radians}} \\ &= \frac{3\pi}{4\pi} \cdot 180° = \frac{3}{4} \cdot 180° \\ &= 135° \end{aligned}$$

EXAMPLE 13 Convert 1 radian to degrees.

Solution

$$\begin{aligned} 1 \text{ radian} &= 1 \text{ radian} \cdot \frac{180°}{\pi \text{ radians}} \\ &= \frac{180°}{\pi} \approx \frac{180°}{3.14} \\ &\approx 57.3° \end{aligned}$$

DO EXERCISES 11–14.

11. Convert to radian measure. Leave answers in terms of π.
    a) 225°
    b) 315°
    c) −720°

12. Convert to radian measure. Do not leave answers in terms of π (use 3.14 for π).
    a) $72\frac{1}{2}°$ (a safe working angle for ladders)

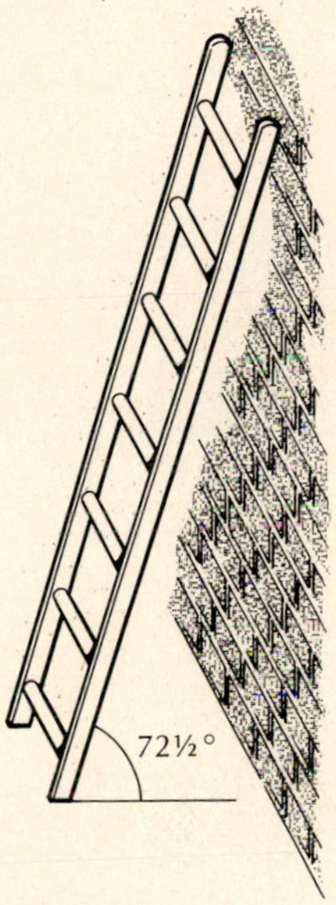

    b) 300°
    c) −315°

13. Convert to degree measure.
    a) $\frac{4\pi}{3}$
    b) $\frac{5\pi}{2}$
    c) $-\frac{4\pi}{5}$

14. In which quadrant does the terminal side of each angle lie?
    a) $\frac{5\pi}{4}$
    b) $\frac{17\pi}{8}$
    c) $-\frac{\pi}{15}$
    d) 37.3π

3 Arc Length and Central Angles

Radian measure can be determined using a circle other than a unit circle. In the following figure, a unit circle is shown along with another circle. The angle shown is a **central angle** of both circles; hence the arcs that it intercepts have their lengths in the same ratio as the radii of the circles. The radii of the circles are r and 1.

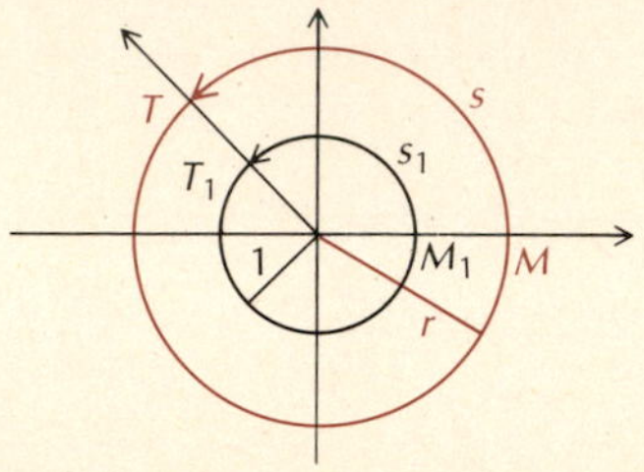

The corresponding arc lengths are MT and M_1T_1, or more simply, s and s_1. We therefore have the proportion

$$\frac{s}{r} = \frac{s_1}{1}.$$

Now s_1 is the radian measure of the rotation in question. It is more common to use a Greek letter, such as θ, for the measure of an angle or rotation. We commonly use the letter s for arc length. Adopting this convention, we rewrite the proportion above as $\theta = s/r$. In any circle, arc length, central angle, and length of radius are related in this fashion. Or, in general, the following is true.

The radian measure θ of a rotation is the ratio of the distance s traveled by a point at a radius r from the center of rotation, to the length of the radius r:

$$\theta = \frac{s}{r}.$$

EXAMPLE 14 Find the length of an arc of a circle of 5-cm radius associated with a central angle of $\pi/3$ radians.

Solution We have

$$\theta = \frac{s}{r}, \quad \text{or} \quad s = r\theta.$$

Therefore, $s = 5 \cdot \pi/3$ cm, or, if we use 3.14 for π, about 5.23 cm. ■

EXAMPLE 15 Find the measure of a rotation in radians where a point 2 m from the center of rotation travels 4 m.

Solution

$$\theta = \frac{s}{r} = \frac{4\text{ m}}{2\text{ m}} = 2 \qquad \text{The unit is understood to be radians.} \qquad ■$$

A look at Examples 14 and 15 will show why the word radian is most often omitted. In Example 15, we have the division 4 m/2 m, which simplifies to the number 2, since m/m = 1. From this point of view, it would seem preferable to omit the word radians. In Example 14, had we used the word radians all the way through, our answer would have come out to be 5.23 cm-radians. It is a distance we seek; hence we know the unit should be centimeters. Thus we must omit the word radians. Since a measure in radians is simply a number, it is usually preferable to omit the word radians.

CAUTION! In using the formula $\theta = s/r$, you must make sure that θ is in radians and that s and r are expressed in the same unit.

DO EXERCISES 15–17.

4 Angular Speed

Speed is defined to be distance traveled per unit of time. Similarly, **angular speed** is defined to be amount of rotation per unit of time. For example, we might speak of the angular speed of a wheel as 150 revolutions per minute or the angular speed of the earth as 2π radians per day. The Greek letter ω (omega) is generally used for angular speed. Thus angular speed is defined as

$$\omega = \frac{\theta}{t}.$$

Relating Linear and Angular Speed

For many applications, it is important to know a relationship between *angular speed* and *linear speed.* For example, we might wish to find the linear speed of a point on the earth, knowing its angular speed. Or, we might wish to know the linear speed of an earth satellite, knowing its angular speed. To develop the relationship we seek, we recall the relation between angle and distance from the preceding section: $\theta = s/r$. This is equivalent to

$$s = r\theta.$$

We divide by time, t, to obtain

$$\frac{s}{t} = r\frac{\theta}{t}$$

$$\downarrow \qquad \downarrow$$

$$v \qquad \omega$$

Now s/t is linear speed v, and θ/t is angular speed ω. Thus we have the relation we seek.

The *linear speed* v of a point a distance r from the center of rotation is given by

$$v = r\omega,$$

where ω is the *angular speed* in radians per unit time.

In deriving this formula, we used the equation $s = r\theta$, in which the units for s and r must be the same and θ must be in radians. So, for our new formula $v = r\omega$, the units of distance for v and r must be the same, ω must be in radians per unit of time, and the units of time must be the same for v and ω.

15. Find the length of an arc of a circle with 10-cm radius associated with a central angle of measure $11\pi/6$. (Use 3.14 for π.)

16. Find the radian measure of a rotation where a point 2.5 cm from the center of rotation travels 15 cm.

17. Find the radian measure of a rotation where a point 24 in. from the center of rotation travels 3 ft.

18. A wheel with 12-cm diameter is rotating at a speed of 10 revolutions per second. What is the velocity of a point on the rim?

EXAMPLE 16 An earth satellite in circular orbit 1200 km high makes one complete revolution every 90 min. What is its linear speed? Use 6400 km for the length of a radius of the earth.

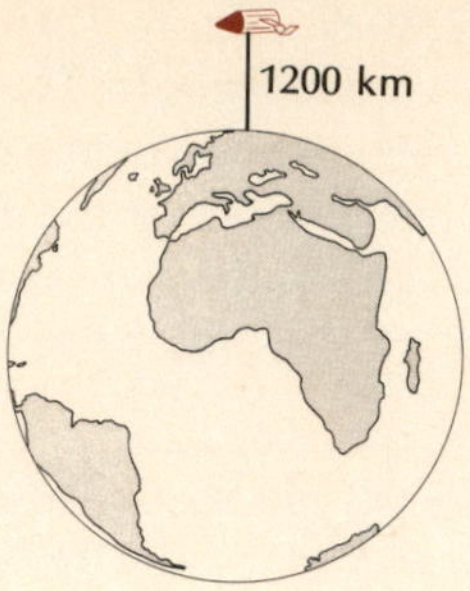

Solution We will use the formula

$$v = r\omega;$$

thus we will need to know r and ω:

$$r = 6400 \text{ km} + 1200 \text{ km} \quad \text{Radius of earth plus height of satellite}$$
$$= 7600 \text{ km},$$
$$\omega = \frac{2\pi \text{ radians}}{90 \text{ min}} = \frac{\pi}{45 \text{ min}}. \quad \text{We have, as usual, omitted the word radians.}$$

Now, using $v = r\omega$, we have

$$v = 7600 \text{ km} \cdot \frac{\pi}{45 \text{ min}} = \frac{7600\pi}{45} \cdot \frac{\text{km}}{\text{min}}.$$

Using 3.14 for π, we obtain $v = 530\frac{\text{km}}{\text{min}}$.*

DO EXERCISE 18.

EXAMPLE 17 An anchor is being hoisted at a rate of 2 ft/sec, the chain being wound around a capstan with a 1.8-yd diameter. What is the angular speed of the capstan?

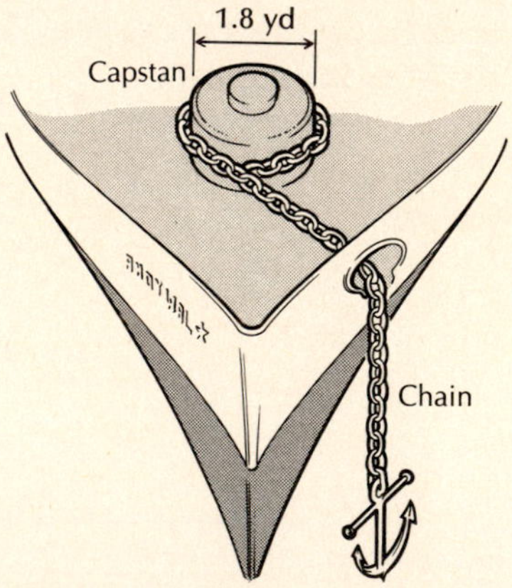

Solution We will use the formula $\omega = v/r$, taking care to use the proper units. Since v is given in feet per second, we need r in feet. Then ω will be in radians per second:

$$r = \frac{1.8}{2} \text{ yd} \cdot \frac{3 \text{ ft}}{\text{yd}} = 2.7 \text{ ft},$$
$$\omega = \frac{v}{r} = \frac{2}{2.7} = 0.741 \text{ radian/sec}.$$

*It may be appropriate at this point to review Section 1.9 on handling units.

> CAUTION! In applying the formula $v = r\omega$, we must be sure that the distance units for v and r are the same and that ω is in radians per unit of time. The units of time must be the same for v and ω.

DO EXERCISES 19–21.

5 Total Distance and Total Angle

The formulas $\theta = s/r$ and $v = r\omega$ can be used in combination to find distances and angles in various situations involving rotational motion.

EXAMPLE 18 A car is traveling at a speed of 45 mph. Its tires have a 20-in. radius. Find the angle through which a wheel turns in 5 sec.

Solution Recall that $\omega = \theta/t$, or $\theta = \omega t$. Thus we can find θ if we know ω and t. To find ω, we use $v = r\omega$. For convenience, we will first convert 45 mph to ft/sec:

$$v = 45\,\frac{\text{mi}}{\text{hr}} \cdot \frac{1\text{ hr}}{60\text{ min}} \cdot \frac{1\text{ min}}{60\text{ sec}} \cdot \frac{5280\text{ ft}}{1\text{ mi}}$$
$$= 66\,\frac{\text{ft}}{\text{sec}}.$$

Now $r = 20$ in. We will convert to ft, since v is in ft/sec:

$$r = 20\text{ in.} \cdot \frac{1\text{ ft}}{12\text{ in.}}$$
$$= \frac{20}{12}\text{ ft} = \frac{5}{3}\text{ ft}.$$

Using $v = r\omega$, we have

$$66\,\frac{\text{ft}}{\text{sec}} = \frac{5}{3}\text{ ft} \cdot \omega, \quad \text{so} \quad \omega = 39.6\,\frac{\text{radians}}{\text{sec}}.$$

Then

$$\theta = \omega t = 39.6\,\frac{\text{radians}}{\text{sec}} \cdot 5\text{ sec} = 198\text{ radians}.$$

DO EXERCISE 22.

19. In using $v = r\omega$, if v is given in centimeters per second, what must be the units for r and ω?

20. In using $v = r\omega$, if ω is given in radians per year and r is in kilometers, what must be the units for v?

21. The old oaken bucket is being raised at a rate of 3 ft/sec. The radius of the drum is 10 in. What is the angular speed of the handle?

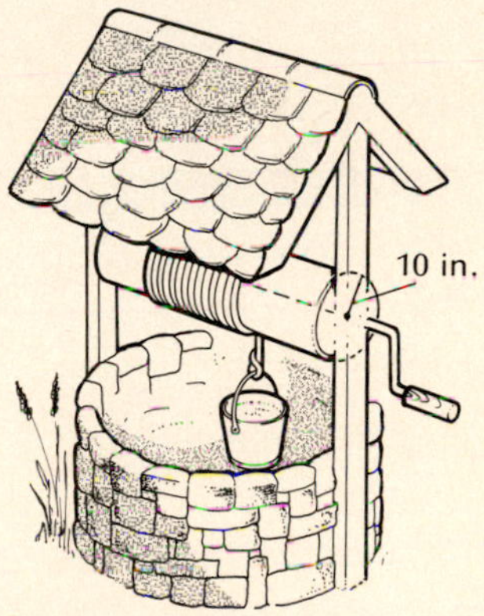

22. The diameter of a wheel of a car is 30 in. When the car is traveling at a speed of 60 mph (88 ft/sec), how many revolutions does the wheel make in 1 sec?

EXERCISE SET 6.4

1, 2 For angles of the following measures, state in which quadrant the terminal side lies.

1. $34°$
2. $320°$
3. $-120°$
4. $175°$
5. $\frac{\pi}{3}$
6. $-\frac{3\pi}{4}$
7. $\frac{11\pi}{4}$
8. $\frac{19\pi}{4}$

Find two positive angles and two negative angles that are coterminal with the given angle.

9. $58°$
10. $320°$
11. $-120°$
12. $-215°$

Find the complement and the supplement.

13. $57°23'$
14. $47°38'$
15. $73°45'11''$
16. $12°03'14''$
17. $67.31°$
18. $13.68°$
19. $11.2344°$
20. $85.06312°$

Convert to degrees and minutes.

21. $8.6°$ **22.** $47.8°$ **23.** $72.25°$ **24.** $11.75°$
25. $46.38°$ **26.** $85.21°$ **27.** $67.84°$ **28.** $38.48°$

Convert to degrees, minutes, and seconds.

29. $87.3456°$ **30.** $11.0256°$ **31.** $48.02498°$ **32.** $27.899461°$

Convert to degrees and decimal parts of degrees. Round to two decimal places.

33. $9°45'$ **34.** $52°15'$ **35.** $35°50'$ **36.** $64°40'$
37. $80°33'$ **38.** $27°19'$ **39.** $3°02'$ **40.** $10°08'$

Convert to degrees and decimal parts of degrees. Round to four decimal places.

41. $19°47'23''$ **42.** $49°38'46''$ **43.** $31°57'55''$ **44.** $76°11'34''$

2 Convert to radian measure. Leave answers in terms of π.

45. $30°$ **46.** $15°$ **47.** $60°$ **48.** $200°$
49. $75°$ **50.** $300°$ **51.** $37.71°$ **52.** $12.73°$
53. $214.6°$ **54.** $73.87°$

Convert to radian measure. Do not leave answers in terms of π. Use 3.14 for π.

55. $120°$ **56.** $240°$ **57.** $320°$ **58.** $75°$
59. $200°$ **60.** $300°$ **61.** $117.8°$ **62.** $231.2°$
63. $1.354°$ **64.** $327.9°$

Convert to degree measure.

65. 1 radian **66.** 2 radians **67.** 8π **68.** -12π
69. $\frac{3}{4}\pi$ **70.** $\frac{5}{4}\pi$ **71.** 1.303 **72.** 2.347
73. 0.7532π **74.** -1.205π

75. Certain positive angles are marked here in degrees. Find the corresponding radian measures.

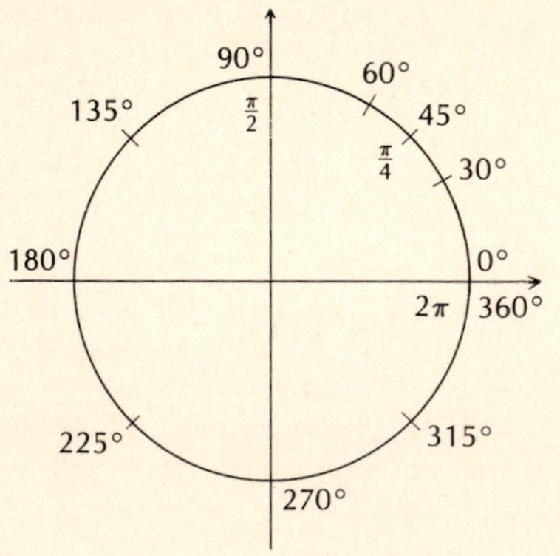

76. Certain negative angles are marked here in degrees. Find the corresponding radian measures.

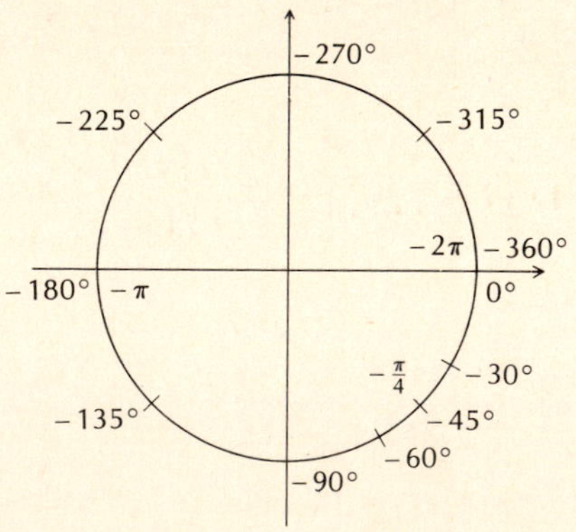

3

77. In a circle with 120-cm radius, an arc 132 cm long subtends a central angle of how many radians? how many degrees, to the nearest degree?

78. In a circle with 200-cm radius, an arc 65 cm long subtends a central angle of how many radians? how many degrees, to the nearest degree?

79. Through how many radians does the minute hand of a clock rotate in 50 min?

80. A wheel on a car has a 14-in. radius. Through what angle (in radians) does the wheel turn while the car travels 1 mi?

81. In a circle with 10-m radius, how long is an arc associated with a central angle of 1.6 radians?

82. In a circle with 5-m radius, how long is an arc associated with a central angle of 2.1 radians?

4

83. A flywheel is rotating at 7 radians/sec. It has a 15-cm diameter. What is the linear speed of a point on its rim, in cm/min?

84. A wheel is rotating at 3 radians/sec. The wheel has a 30-cm radius. What is the linear speed of a point on its rim, in m/min?

85. A $33\frac{1}{3}$-rpm record has a radius of 15 cm. What is the linear velocity of a point on the rim, in cm/sec?

86. A 45-rpm record has a radius of 8.7 cm. What is the linear velocity of a point on the rim, in cm/sec?

87. The earth has a 4000-mi radius and rotates one revolution every 24 hr. What is the linear speed of a point on the equator, in mph?

88. The earth is 93,000,000 miles from the sun and traverses its orbit, which is nearly circular, every 365.25 days. What is the linear velocity of the earth in its orbit, in mph?

89. A wheel has a 32-cm diameter. The speed of a point on its rim is 11 m/s. What is its angular speed?

90. A horse on a merry-go-round is 7 m from the center and travels at a speed of 10 km/h. What is its angular speed?

91. *Determining the speed of a river.* A water wheel has a 10-ft radius. To get a good approximation of the speed of the river, you count the revolutions of the wheel and find that it makes 14 revolutions per minute. What is the speed of the river?

92. *Determining the speed of a river.* A water wheel has a 10-ft radius. To get a good approximation of the speed of the river, you count the revolutions of the wheel and find that it makes 16 rpm. What is the speed of the river?

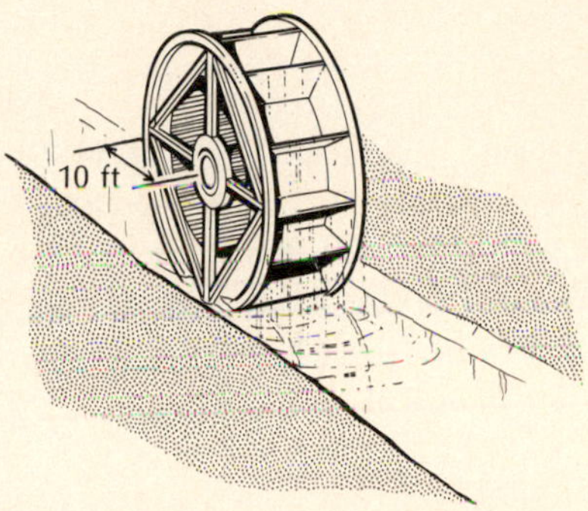

5

93. The wheels of a bicycle have a 24-in. diameter. When the bicycle is being ridden so that the wheels make 12 rpm, how far will the bike travel in 1 min?

94. The wheels of a car have a 15-in. radius. When the car is being driven so that the wheels make 10 revolutions/sec, how far will the car travel in 1 min?

95. A car is traveling at a speed of 30 mph. Its wheels have a 14-in. radius. Find the angle through which a wheel rotates in 10 sec.

96. A car is traveling at a speed of 40 mph. Its wheels have a 15-in. radius. Find the angle through which a wheel rotates in 12 sec.

SYNTHESIS

97. The **grad** is a unit of angle measure similar to a degree. A right angle has a measure of 100 grads. Convert each of the following to grads.

a) 48°
b) 153°
c) $\pi/8$ radians
d) $5\pi/7$ radians

98. A **mil** is a unit of angle measure. A right angle has a measure of 1600 mils. Convert each of the following to degrees, minutes, and seconds.

a) 100 mils
b) 350 mils

99. On the earth, one degree of latitude is how many kilometers? how many miles? (Assume that the radius of the earth is 6400 km, or 4000 mi, approximately.)

100. One minute of latitude on the earth is equivalent to one *nautical mile*. Find the circumference and the radius of the earth in nautical miles.

101. An astronaut on the moon observes the earth, about 240,000 mi away. The diameter of the earth is about 8000 mi. Find the angle α.

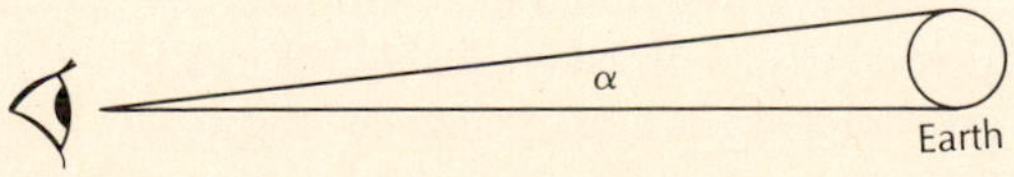

102. The circumference of the earth was computed by Eratosthenes (276–195 B.C.). He knew the distance from Aswan to Alexandria to be about 500 mi. From each town he observed the sun at noon, finding the angular difference to be 7°12′ (7 degrees, 12 minutes). Do Eratosthenes' calculation.

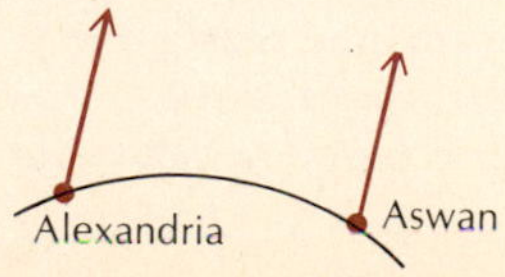

103. Two pulleys, 50 cm and 30 cm in diameter, respectively, are connected by a belt. The larger pulley makes 12 revolutions/min. Find the angular speed of the smaller pulley, in radians/sec.

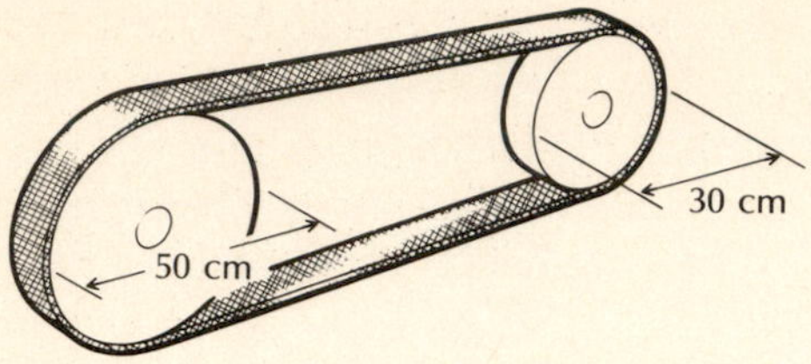

104. One gear wheel turns another, the teeth being on the rims. The wheels have 40-cm and 50-cm radii, and the smaller wheel rotates at 20 rpm. Find the angular speed of the larger wheel, in radians/sec.

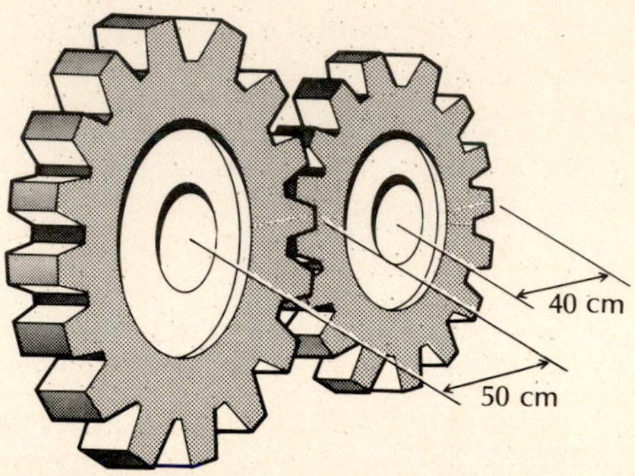

105. An airplane engine is idling at 800 rpm. When the throttle is opened, it takes 4.3 sec for the speed to come up to 2500 rpm. What was the angular acceleration (a) in rpm/sec? (b) in radians/sec/sec?

106. The linear speed of an airplane, flying at low level over the sea, is 175 knots (nautical miles per hour). It accelerates to 325 knots in 12 sec.

a) What was its linear acceleration in knots/sec?

b) What was its angular acceleration in radians/sec/sec? (*Hint:* One nautical mile is equivalent to one minute of latitude.)

CHALLENGE

107. What is the angle between the hands of a clock at 7:45?

108. At what time between noon and 1:00 P.M. are the hands of a clock perpendicular?

109. The diameter of the earth at the equator has been measured to be 7926.4 statute miles. Use this and the result of Exercise 100 to find the number of statute miles in a nautical mile.

110. To find the distance between two points on the earth when their latitude and longitude are known, we can use a plane right triangle for an excellent approximation if the points are not too far apart. Point A is at latitude $38°27'30''$ N, longitude $82°57'15''$ W; and point B is at latitude $38°28'45''$ N, longitude $82°56'30''$ W. Find the distance from A to B in nautical miles (one minute of latitude is one nautical mile).

6.5 Trigonometric Functions Involving Angles or Rotations

OBJECTIVES

You should be able to:

1 Find trigonometric function values of angles measured in both degrees and radians. State the definitions of the trigonometric functions, given a point $P(x, y)$ on any circle. Given an ordered pair on the terminal side of an angle θ, or given the equation of the line that is the terminal side, find the six trigonometric function values for the angle.

(*continued*)

1 Function Values of Angles Measured in Degrees

The domains of the circular, or trigonometric, functions are sets of real numbers. We can also consider the domains to be real numbers that represent degree measures. In this context, the number considered is a real-number multiple of a unit called a degree. For example, 47° is really 47 × 1°. If a rotation is in radians, the numbers are the same as those we have already considered. For example, the sine of $\pi/2$ radians is the same as $\sin(\pi/2)$. If a rotation is given in degrees, we can merely change to radians to determine the function values. For example,

$$\sin 30° = \sin \frac{\pi}{6}.$$

The diagrams below show unit circles marked in both radians and degrees. You should memorize values in the first quadrant and be able to reason the others.

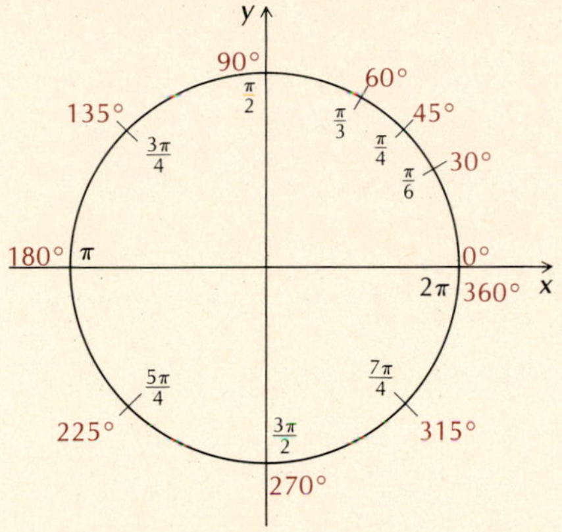

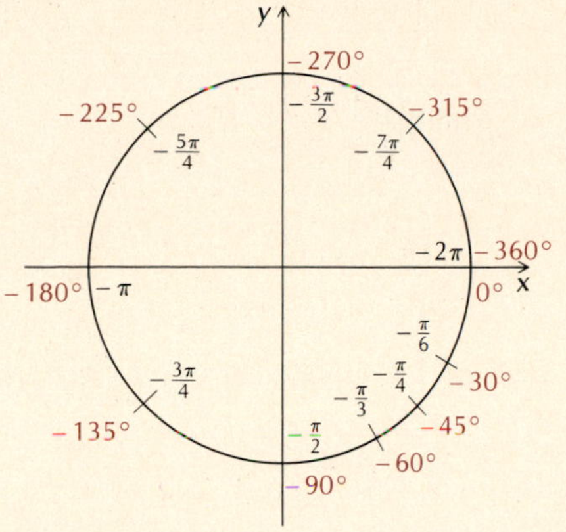

EXAMPLES Find the following function values.

1. sin 90°

From the diagram, we have $90° = \pi/2$. Hence

$$\sin 90° = \sin \frac{\pi}{2} = 1.$$

2. tan 135°

$$\tan 135° = \tan \frac{3\pi}{4} = -1$$

3. cos (−210°)

$$\cos(-210°) = \cos\left(-\frac{7\pi}{6}\right) = -\frac{\sqrt{3}}{2}$$

DO EXERCISES 1–5.

When we defined the trigonometric functions, we considered a point on a *unit circle*. Now we extend the definition to a point P on *any circle* of radius r. Let us consider a circle of radius r other than a unit circle, as shown below.

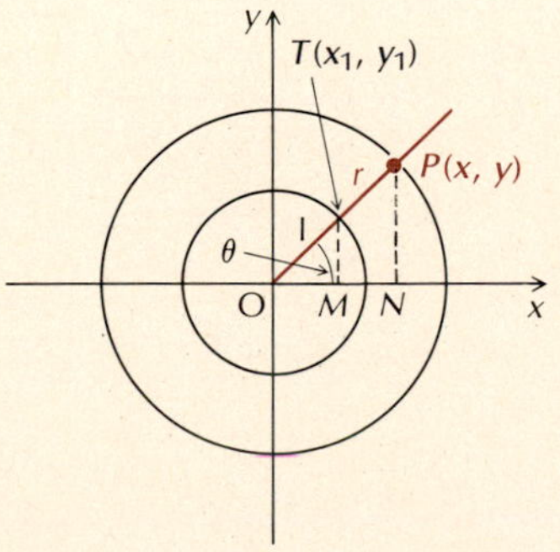

$$OT = 1,$$
$$OP = r$$

Also shown is a unit circle and a rotation θ. The triangles OMT and ONP are similar, being right triangles with the same acute angle. Thus we know that certain sides have the same ratio. For example,

$$\frac{MT}{OT} = \frac{NP}{OP}.$$

2 Determine the signs of the trigonometric function values for a rotation measured in degrees θ in any specified quadrant.

3 State function values (when defined) for any angle whose terminal side lies on an axis.

4 State the definitions of the trigonometric function values for any acute angle of a right triangle in terms of the adjacent side, the opposite side, and the hypotenuse. Find the trigonometric function values when the sides of a right triangle are known. Find the function values for any acute angle of 30°, 45°, or 60°.

5 Given the function value for an angle and the quadrant in which the terminal side lies, find the other five function values.

Find the following.

1. sin 180°

2. tan 0°

3. cos 45°

4. cot 45°

5. sin (−135°)

But $MT = y_1$ and $OT = 1$; $NP = y$ and $OP = r$. Thus we have

$$\frac{y_1}{1} = \frac{y}{r}.$$

Since T is a point on a unit circle, we know that $y_1 = \sin\theta$. Thus, $\sin\theta = y/r$.

Similarly,

$$\frac{OM}{OT} = \frac{ON}{OP}$$
$$= \frac{x_1}{1} = \frac{x}{r}.$$

Because T is a point on a unit circle, we know that $x_1 = \cos\theta$. Thus, $\cos\theta = x/r$.

It also follows that

$$\frac{MT}{OM} = \frac{NP}{ON}$$
$$= \frac{y_1}{x_1} = \frac{y}{x}.$$

Since T is on a unit circle, we know that $y_1/x_1 = \tan\theta$. Thus, $\tan\theta = y/x$.

The point P, which is a point other than the vertex on the terminal side of the angle, can be anywhere on a circle of radius r. Its coordinates may be positive, negative, or zero, depending on the quadrant in which the terminal side lies. The length of the radius, which is also the hypotenuse of what we call a **reference triangle**, is always considered positive. The angle is always measured from the *positive* half of the x-axis. Regardless of the location of point P, we have

$$\sin\theta = \frac{y}{r}, \qquad \cos\theta = \frac{x}{r}, \quad \text{and} \quad \tan\theta = \frac{y}{x}.$$

The following figure shows angles whose terminal sides lie in quadrants II, III, and IV. The reference triangle for the angle θ is triangle ONP, where $\overline{PN}$ is the perpendicular from P to the x-axis.

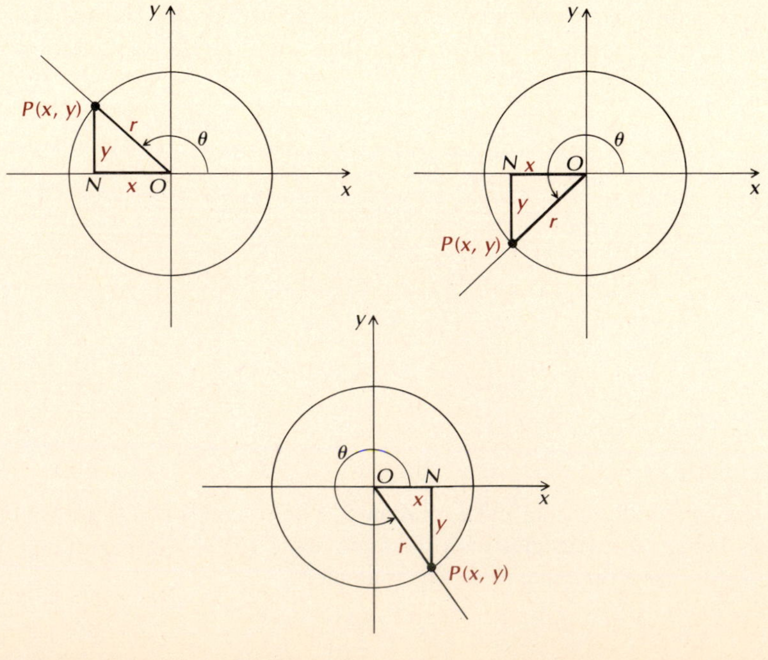

DEFINITION Trigonometric Functions: Any Circle

Suppose that $P(x, y)$ is any point on the terminal side of any angle θ in standard position, and r is the radius, or distance, from the origin to $P(x, y)$. Then the trigonometric functions are defined as follows:

$$\sin\theta = \frac{\text{second coordinate}}{\text{radius}} = \frac{y}{r}, \qquad \csc\theta = \frac{\text{radius}}{\text{second coordinate}} = \frac{r}{y},$$

$$\cos\theta = \frac{\text{first coordinate}}{\text{radius}} = \frac{x}{r}, \qquad \sec\theta = \frac{\text{radius}}{\text{first coordinate}} = \frac{r}{x},$$

$$\tan\theta = \frac{\text{second coordinate}}{\text{first coordinate}} = \frac{y}{x}, \qquad \cot\theta = \frac{\text{first coordinate}}{\text{second coordinate}} = \frac{x}{y}.$$

Given any point $P(x, y)$ other than the vertex on the terminal side of an angle θ in standard position, we can find the length r using the distance formula:

$$\begin{aligned} r &= \sqrt{(x-0)^2 + (y-0)^2} \\ &= \sqrt{x^2 + y^2}. \end{aligned}$$

EXAMPLE 4 Find the six trigonometric function values for the angle shown.

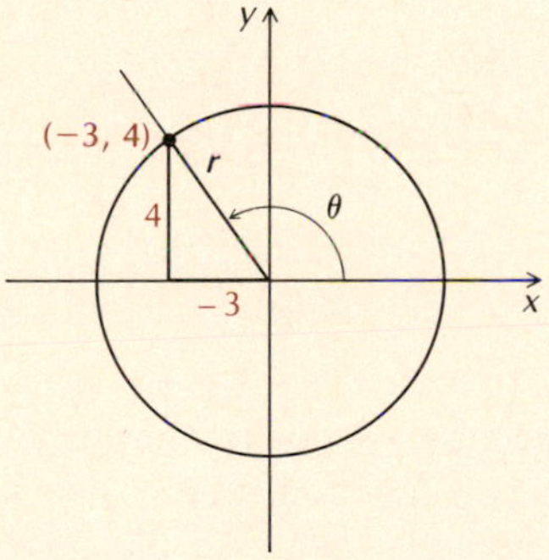

Solution We first determine r, the distance from the origin to the point $(-3, 4)$:

$$\begin{aligned} r &= \sqrt{x^2 + y^2} \\ r &= \sqrt{(-3)^2 + 4^2} \qquad \text{Substituting } -3 \text{ for } x \text{ and } 4 \text{ for } y \\ r &= \sqrt{9 + 16} \\ r &= \sqrt{25}, \quad \text{or} \quad 5. \end{aligned}$$

Using the definitions of the trigonometric functions, we can now find the function values for θ. We substitute -3 for x, 4 for y, and 5 for r:

$$\sin\theta = \frac{y}{r} = \frac{4}{5} = 0.8, \qquad \csc\theta = \frac{r}{y} = \frac{5}{4} = 1.25,$$

$$\cos\theta = \frac{x}{r} = \frac{-3}{5} = -0.6, \qquad \sec\theta = \frac{r}{x} = \frac{5}{-3} \approx -1.67,$$

$$\tan\theta = \frac{y}{x} = \frac{4}{-3} \approx -1.33, \qquad \cot\theta = \frac{x}{y} = \frac{-3}{4} = -0.75.$$

EXAMPLE 5 Find the six trigonometric function values for the angle shown in the following graph.

6. Find the six trigonometric function values for the angle shown here.

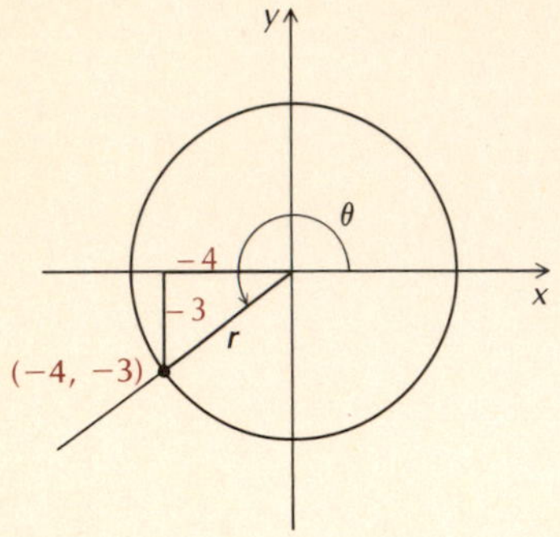

7. Find the six trigonometric function values for the angle shown here.

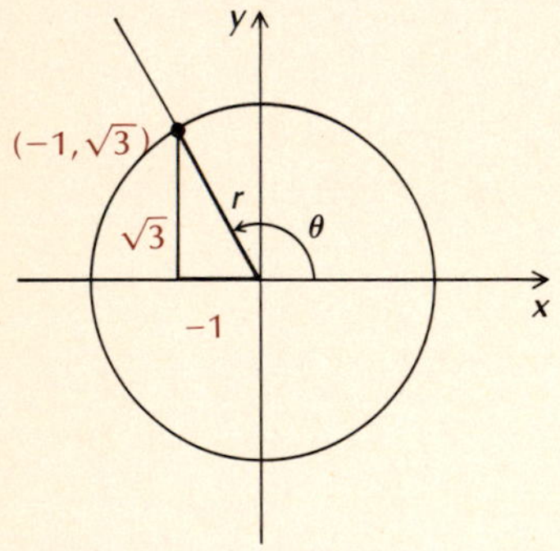

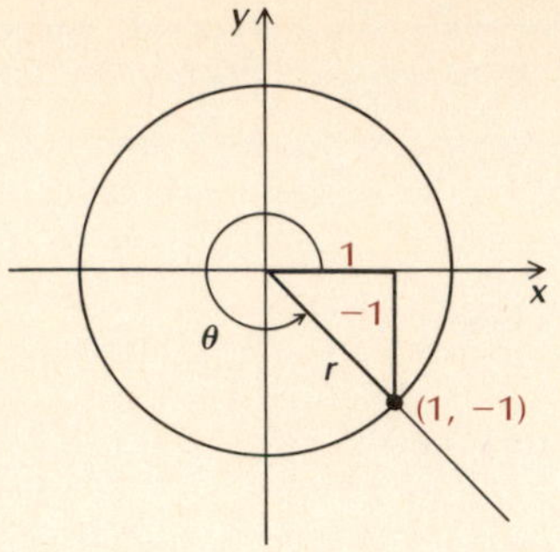

Solution We first determine r, the distance from the origin to the point $(1, -1)$:

$$r = \sqrt{x^2 + y^2}$$

$$r = \sqrt{1^2 + (-1)^2} \qquad \text{Substituting 1 for } x \text{ and } -1 \text{ for } y$$

$$r = \sqrt{1 + 1}$$

$$r = \sqrt{2}.$$

Substituting 1 for x, -1 for y, and $\sqrt{2}$ for r, we find that the trigonometric function values of θ are

$$\sin\theta = \frac{y}{r} = \frac{-1}{\sqrt{2}} = -\frac{\sqrt{2}}{2}, \qquad \csc\theta = \frac{r}{y} = \frac{\sqrt{2}}{-1} = -\sqrt{2},$$

$$\cos\theta = \frac{x}{r} = \frac{1}{\sqrt{2}} = \frac{\sqrt{2}}{2}, \qquad \sec\theta = \frac{r}{x} = \frac{\sqrt{2}}{1} = \sqrt{2},$$

$$\tan\theta = \frac{y}{x} = \frac{-1}{1} = -1, \qquad \cot\theta = \frac{x}{y} = \frac{1}{-1} = -1.$$

DO EXERCISES 6 AND 7.

Any point other than the origin on the terminal side of an angle can be used to determine the trigonometric function values. The function values are the same regardless of which point is used. We illustrate this in the following example.

EXAMPLE 6 The terminal side of angle θ in standard position lies on the line $x + 2y = 0$. If the terminal side is in quadrant II, find $\sin\theta$, $\cos\theta$, and $\tan\theta$.

Solution First we draw the graph of $x + 2y = 0$ and determine a second quadrant solution of the equation. Using $(-4, 2)$, we then determine r:

$$r = \sqrt{(-4)^2 + 2^2} = \sqrt{20} = 2\sqrt{5}.$$

Then using

$$x = -4,\ y = 2, \text{ and } r = 2\sqrt{5},$$

we find that

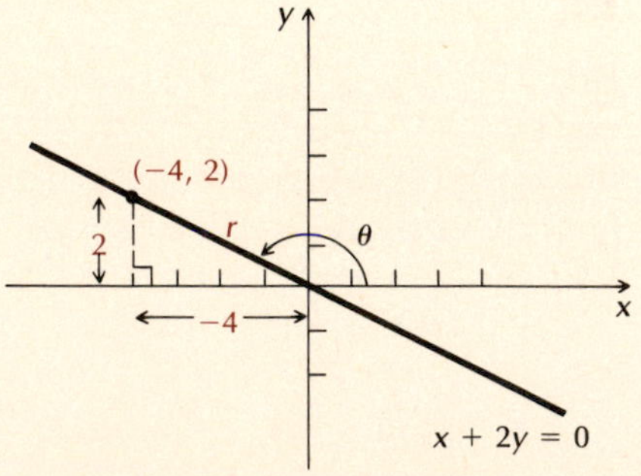

$$\sin\theta = \frac{2}{2\sqrt{5}} = \frac{\sqrt{5}}{5}, \qquad \cos\theta = \frac{-4}{2\sqrt{5}} = -\frac{2\sqrt{5}}{5}, \qquad \tan\theta = \frac{2}{-4} = -\frac{1}{2}.$$

Would we get the same function values for θ if we had chosen a different point on the terminal side? The point $(-8, 4)$ is also a solution of the equation that is in the second quadrant and on the terminal side of θ.

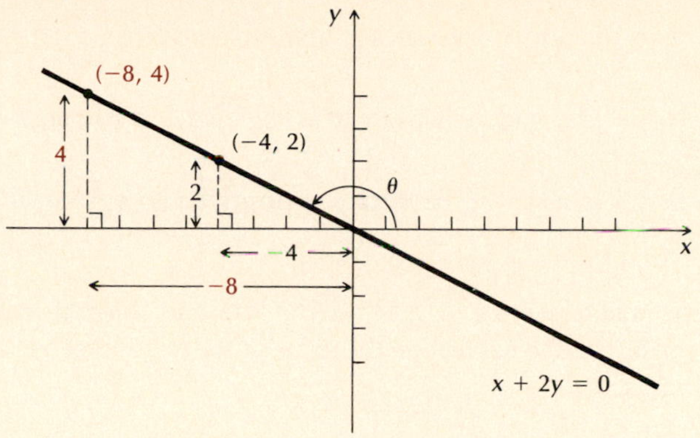

We determine r and find the function values:

$$r = \sqrt{(-8)^2 + 4^2} = \sqrt{80} = 4\sqrt{5}.$$

Then

$$\sin\theta = \frac{4}{4\sqrt{5}} = \frac{\sqrt{5}}{5},$$
$$\cos\theta = \frac{-8}{4\sqrt{5}} = -\frac{2\sqrt{5}}{5},$$
$$\tan\theta = \frac{4}{-8} = -\frac{1}{2}.$$

We see that the function values are the same as what we had obtained earlier.

Any point other than the origin on the terminal side of an angle can be used to determine the trigonometric function values.

The trigonometric function values of θ depend only on the size of the angle, not on the choice of the point on the terminal side.

DO EXERCISE 8.

2 Signs of the Functions

Function values of the generalized trigonometric functions can be positive, negative, or zero, depending on where the terminal side of the angle lies. In the first quadrant, all function values are positive because x, y, and r are all positive. In the second quadrant, first coordinates are negative and second coordinates are positive.

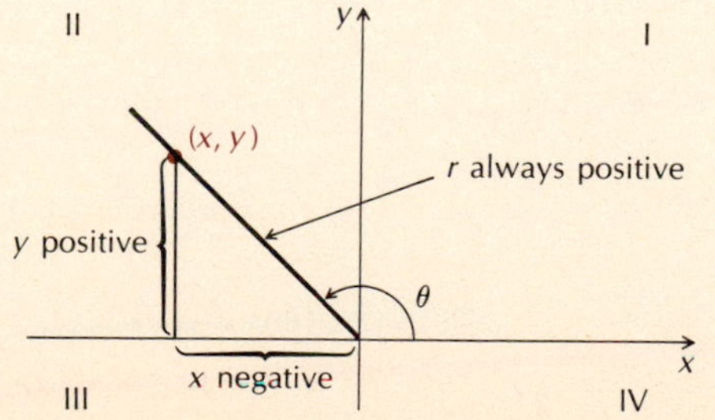

8. Suppose that θ is an angle whose terminal side lies in quadrant III and has equation $y - 2x = 0$. Find $\sin\theta$, $\cos\theta$, and $\tan\theta$.

9. Complete the following table with the appropriate signs of the trigonometric functions of θ when the terminal side of angle θ lies in each of the four quadrants.

	I	II	III	IV
$\sin\theta$				
$\cos\theta$				
$\tan\theta$				
$\cot\theta$				
$\sec\theta$				
$\csc\theta$				

Give the signs of the six trigonometric function values for each of the following rotations.

10. $-30°$

11. $160°$

12. $343°$

13. Find the trigonometric function values for 270°.

Thus if θ terminates in quadrant II,

$$\sin\theta = \frac{y}{r} \text{ is positive, because } y \text{ and } r \text{ are positive;}$$

$$\cos\theta = \frac{x}{r} \text{ is negative, because } x \text{ is negative and } r \text{ is positive;}$$

$$\tan\theta = \frac{y}{x} \text{ is negative, because } y \text{ is positive and } x \text{ is negative;}$$

and so on.

The same mental device ALLSINTANCOS and diagram considered in Section 6.3 can be used to determine the signs of the trigonometric functions in various quadrants.

DO EXERCISE 9.

EXAMPLE 7 Give the signs of the six trigonometric function values for a rotation of 225°.

Solution

$$180° < 225° < 270°, \quad \text{so } P(x, y) \text{ is in the third quadrant.}$$

The tangent and the cotangent are positive, and the other four function values are negative. ■

DO EXERCISES 10–12.

3 Terminal Side on an Axis

Now let us suppose that the terminal side of an angle falls on one of the axes. In that case, one of the coordinates is zero. The definitions of the functions still apply, but in some cases, functions will not be defined because a denominator will be 0.

EXAMPLE 8 Find the trigonometric function values for 180°.

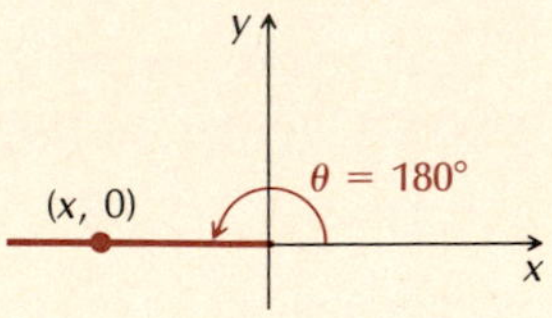

Solution Let $(x, 0)$ represent any point, other than the vertex, on the terminal side of a 180° angle in standard position. We note that the first coordinate is negative, that the second coordinate is 0, and that x and r have the same absolute value (r being always positive). Thus we have

$$\sin 180° = \frac{0}{r} = 0,$$

$$\cos 180° = \frac{x}{r} = -1, \qquad \text{Since } |x| = |r|, \text{ but } x \text{ and } r \text{ have opposite signs}$$

$$\tan 180° = \frac{y}{x} = \frac{0}{x} = 0,$$

$$\cot 180° = \frac{x}{y} = \frac{x}{0}, \qquad \text{Thus, cot 180° is undefined.}$$

$$\sec 180° = \frac{r}{x} = -1, \qquad \text{The reciprocal of cos 180°}$$

$$\csc 180° = \frac{r}{0}. \qquad \text{Thus, csc 180° is undefined.}$$ ■

DO EXERCISE 13.

4 Function Values for Acute Angles

We are now ready to consider the trigonometry of right triangles. Historically, this is where trigonometry began. We will develop trigonometric function values of acute angles. These are angles that have a measure greater than 0° and less than 90°. Consider the right $\triangle ONP$ shown at the left below. Note that $\angle O$ has measure θ. Side PN is called the **opposite** side because it is opposite $\angle O$, ON is called the **adjacent** side because it is adjacent to $\angle O$, and side OP is called the **hypotenuse.**

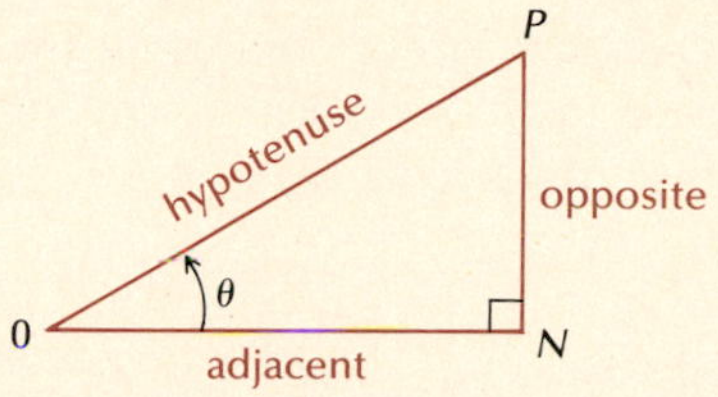

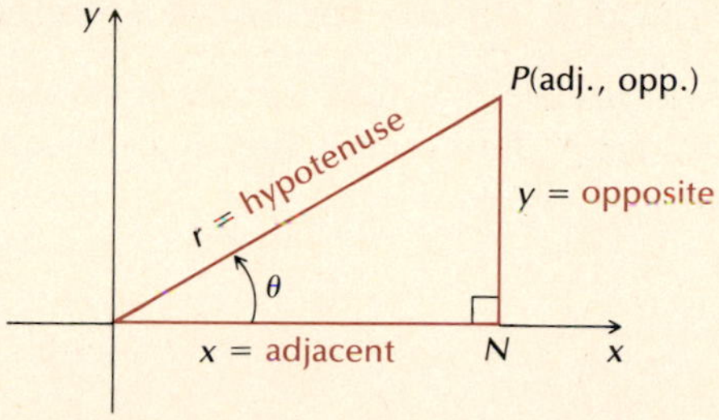

In the figure at the right above, we have moved $\triangle ONP$ onto a coordinate system in such a way that point O is at the origin, with its adjacent side on the x-axis. Then we can express $P(x, y)$ as P(adjacent, opposite) and the radius = hypotenuse. That is, x = adjacent, y = opposite, and r = hypotenuse.

This leads us to the following definition of trigonometric function values of acute angles θ.

DEFINITION Trigonometric Function Values of an Acute Angle θ

$$\sin\theta = \frac{\text{opposite}}{\text{hypotenuse}},\qquad \csc\theta = \frac{\text{hypotenuse}}{\text{opposite}},$$
$$\cos\theta = \frac{\text{adjacent}}{\text{hypotenuse}},\qquad \sec\theta = \frac{\text{hypotenuse}}{\text{adjacent}},$$
$$\tan\theta = \frac{\text{opposite}}{\text{adjacent}},\qquad \cot\theta = \frac{\text{adjacent}}{\text{opposite}}$$

EXAMPLE 9 In this triangle, find each of the following:

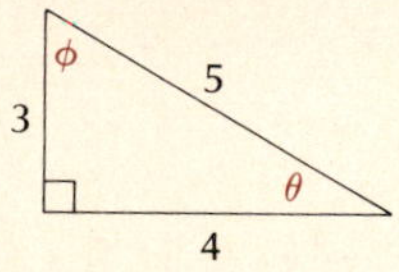

a) The trigonometric function values for θ

b) The trigonometric function values for ϕ (the Greek letter "phi")

Solution

$$\sin\theta = \frac{\text{side opposite } \theta}{\text{hypotenuse}} = \frac{3}{5},\qquad \sin\phi = \frac{\text{side opposite } \phi}{\text{hypotenuse}} = \frac{4}{5},$$
$$\cos\theta = \frac{\text{side adjacent } \theta}{\text{hypotenuse}} = \frac{4}{5},\qquad \cos\phi = \frac{\text{side adjacent } \phi}{\text{hypotenuse}} = \frac{3}{5},$$
$$\tan\theta = \frac{\text{side opposite } \theta}{\text{side adjacent } \theta} = \frac{3}{4},\qquad \tan\phi = \frac{\text{side opposite } \phi}{\text{side adjacent } \phi} = \frac{4}{3},$$
$$\cot\theta = \frac{\text{side adjacent } \theta}{\text{side opposite } \theta} = \frac{4}{3},\qquad \cot\phi = \frac{\text{side adjacent } \phi}{\text{side opposite } \phi} = \frac{3}{4},$$
$$\sec\theta = \frac{\text{hypotenuse}}{\text{side adjacent } \theta} = \frac{5}{4},\qquad \sec\phi = \frac{\text{hypotenuse}}{\text{side adjacent } \phi} = \frac{5}{3},$$
$$\csc\theta = \frac{\text{hypotenuse}}{\text{side opposite } \theta} = \frac{5}{3},\qquad \csc\phi = \frac{\text{hypotenuse}}{\text{side opposite } \phi} = \frac{5}{4}$$

DO EXERCISE 14.

14. Find the following.

a) $\sin\theta$, $\cos\theta$, $\tan\theta$, $\cot\theta$, $\sec\theta$, $\csc\theta$

b) $\sin\phi$, $\cos\phi$, $\tan\phi$, $\cot\phi$, $\sec\phi$, $\csc\phi$

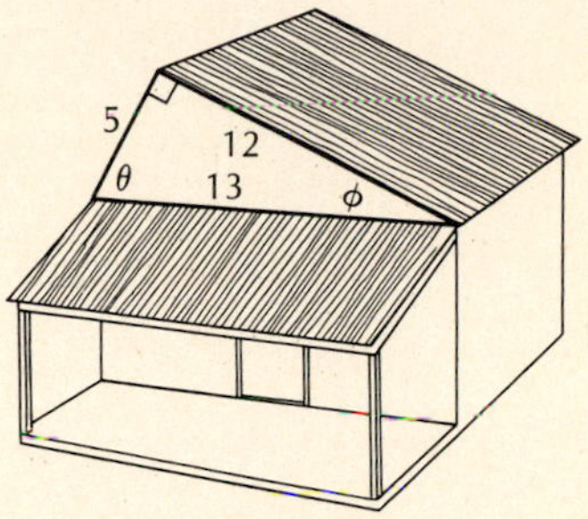

In Example 9, we note that the value of cot θ, $\frac{4}{3}$, is the reciprocal of $\frac{3}{4}$, the value of tan θ. Likewise, we see the same reciprocal relationship between the values of sec θ and cos θ and between the values of csc θ and sin θ. For any angle, the cotangent, secant, and cosecant function values are the respective reciprocals of the tangent, cosine, and sine function values.

$$\cot\theta = \frac{1}{\tan\theta}, \qquad \sec\theta = \frac{1}{\cos\theta}, \qquad \csc\theta = \frac{1}{\sin\theta}$$

Function Values for Some Special Angles

We can determine the function values for certain angles using our knowledge of geometry. First, recall the Pythagorean theorem. It says that in any right triangle, $a^2 + b^2 = c^2$, where c is the length of the hypotenuse.

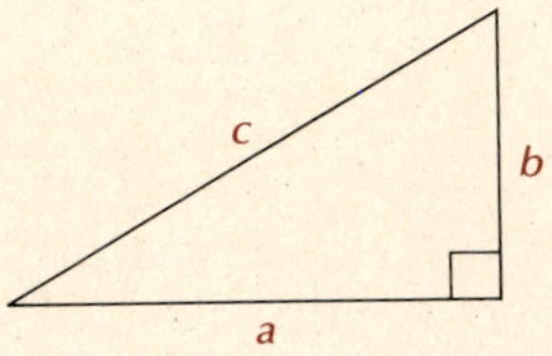

A right triangle with a 45° angle actually has two 45° angles. Thus the triangle is isosceles, and the legs are the same length. Let us consider such a triangle whose legs have length 1. Then its hypotenuse has length c:

$$1^2 + 1^2 = c^2, \quad \text{or} \quad c^2 = 2, \quad \text{or} \quad c = \sqrt{2}.$$

Such a triangle is shown below. From this diagram, we can easily determine the trigonometric function values for 45°, or $\pi/4$.

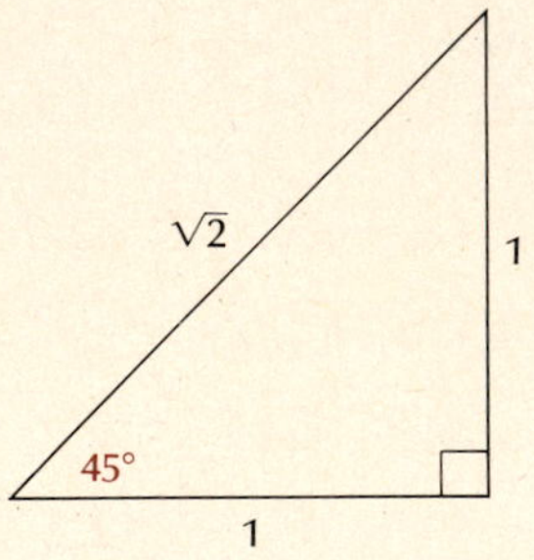

$$\sin 45° = \frac{\text{opposite}}{\text{hypotenuse}} = \frac{1}{\sqrt{2}} = \frac{\sqrt{2}}{2} \approx 0.707,$$

$$\cos 45° = \frac{\text{adjacent}}{\text{hypotenuse}} = \frac{1}{\sqrt{2}} = \frac{\sqrt{2}}{2} \approx 0.707,$$

$$\tan 45° = \frac{\text{opposite}}{\text{adjacent}} = \frac{1}{1} = 1,$$

$$\cot 45° = \frac{\text{adjacent}}{\text{opposite}} = \frac{1}{1} = 1,$$

$$\sec 45° = \frac{\text{hypotenuse}}{\text{adjacent}} = \frac{\sqrt{2}}{1} = \sqrt{2} \approx 1.414,$$

$$\csc 45° = \frac{\text{hypotenuse}}{\text{opposite}} = \frac{\sqrt{2}}{1} = \sqrt{2} \approx 1.414$$

In a similar way, we can determine function values for 30° and 60°, or $\pi/6$ and $\pi/3$. A right triangle with 30° and 60° acute angles is half of an equilateral triangle, as shown in the following diagram. Thus if we choose an equilateral triangle whose sides have length 2 and take half of it, we obtain a right triangle that has a hypotenuse of length 2 and a leg of length 1. The other leg has length a, which can be found using the Pythagorean theorem as follows:

$$a^2 + 1^2 = 2^2$$
$$a^2 = 3$$
$$a = \sqrt{3}.$$

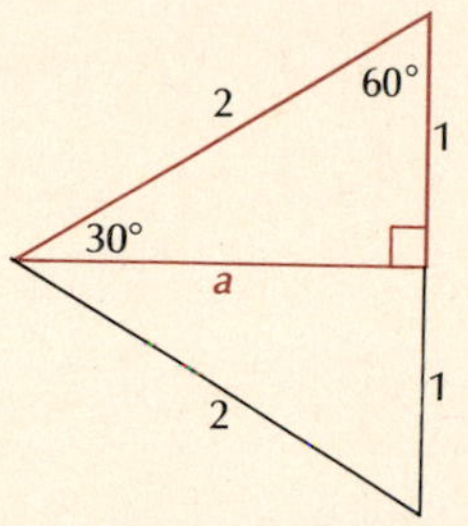

We can now determine function values for 30° and 60°.

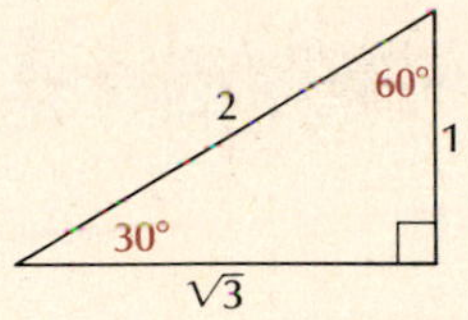

$$\sin 30° = \frac{1}{2}, \qquad \sin 60° = \frac{\sqrt{3}}{2},$$
$$\cos 30° = \frac{\sqrt{3}}{2}, \qquad \cos 60° = \frac{1}{2},$$
$$\tan 30° = \frac{1}{\sqrt{3}} = \frac{\sqrt{3}}{3}, \qquad \tan 60° = \frac{\sqrt{3}}{1} = \sqrt{3},$$
$$\cot 30° = \frac{\sqrt{3}}{1} = \sqrt{3}, \qquad \cot 60° = \frac{1}{\sqrt{3}} = \frac{\sqrt{3}}{3},$$
$$\sec 30° = \frac{2}{\sqrt{3}} = \frac{2\sqrt{3}}{3}, \qquad \sec 60° = \frac{2}{1} = 2,$$
$$\csc 30° = \frac{2}{1} = 2, \qquad \csc 60° = \frac{2}{\sqrt{3}} = \frac{2\sqrt{3}}{3}$$

Since we will often use the function values for 45°, 30°, and 60°, they should be memorized. It is sufficient to learn only those for the sine, cosine, and tangent, since the others are their reciprocals.

$$\sin 45° = \frac{\sqrt{2}}{2}, \qquad \cos 45° = \frac{\sqrt{2}}{2}, \qquad \tan 45° = 1,$$
$$\sin 30° = \frac{1}{2}, \qquad \cos 30° = \frac{\sqrt{3}}{2} \qquad \tan 30° = \frac{\sqrt{3}}{3},$$
$$\sin 60° = \frac{\sqrt{3}}{2} \qquad \cos 60° = \frac{1}{2} \qquad \tan 60° = \sqrt{3}$$

We can now use what we have learned about trigonometric functions of special angles to solve problems. We will consider such applications in much greater detail in Section 8.1.

15. A baseball diamond is really a square 90 ft on a side. If a line is drawn from third base to first base, then a right triangle QPR is formed, where $\angle QPR$ is 45°. Use the sine function to find the length PQ from third base to first base.

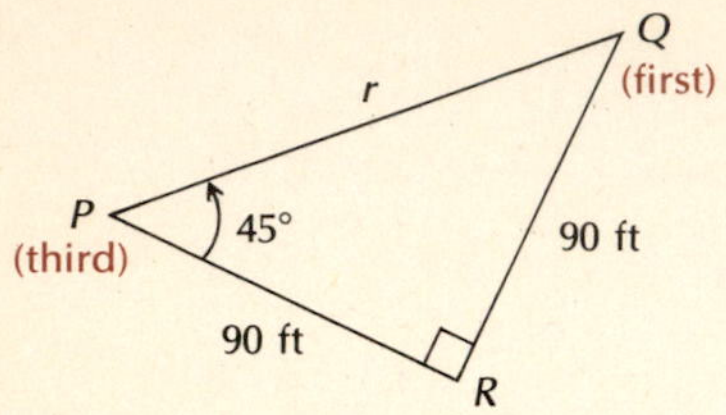

EXAMPLE 10 The rays of the sun over the top of a building cast a 40-ft shadow and form an angle of 60° with the ground. Find the height h of the building.

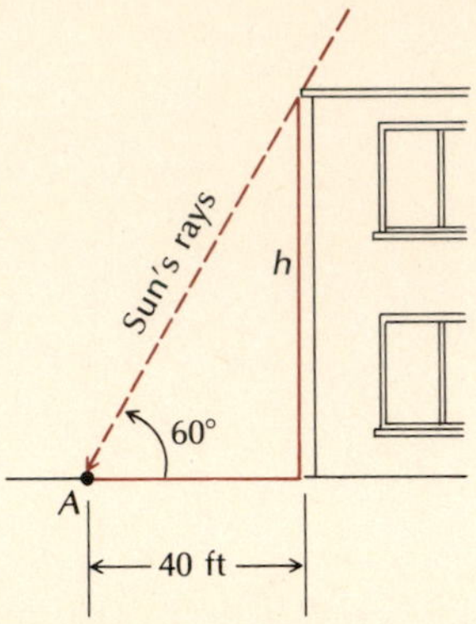

Solution We will typically refer to an angle A and its measure by the same notation. Thus, $A = 60°$. We know the side adjacent to $\angle A$ and also the measure of $\angle A$. Since we want the length of the opposite side, we can use the tangent ratio or the cotangent ratio. Here we use the tangent:

$$\tan 60° = \frac{\text{opposite}}{\text{adjacent}} = \frac{h}{40}$$

$$\sqrt{3} = \frac{h}{40} \quad \textbf{Substituting}$$

$$40\sqrt{3} = h$$

$$69.3 \text{ ft} \approx h.$$

The height of the building is about 69.3 ft. ■

DO EXERCISE 15.

5 The Six Trigonometric Functions Related

The six basic circular, or trigonometric, functions are related in a number of interesting ways. We have already seen that to be the case as we studied identities. In this section, we will use diagrams. When we know one of the function values for an angle, we can find the other five if we know the quadrant in which the terminal side lies. The idea is to sketch a reference triangle in the appropriate quadrant, use the Pythagorean theorem as needed to find the lengths of its sides, and then read off the ratios of the sides.

EXAMPLE 11 Given that $\tan\theta = -\frac{2}{3}$ and that θ is in the second quadrant, find the other function values.

Solution We first sketch a second-quadrant triangle. Since $\tan\theta = -\frac{2}{3}$, we make the legs of lengths 2 and 3. The hypotenuse must then have length $\sqrt{13}$. Now we can read off the appropriate ratios:

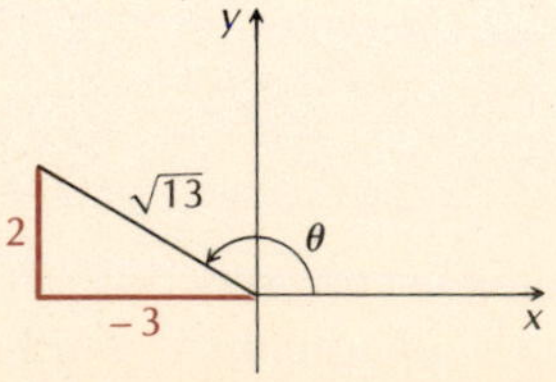

$$\sin\theta = \frac{2}{\sqrt{13}}, \qquad \csc\theta = \frac{\sqrt{13}}{2},$$

$$\cos\theta = -\frac{3}{\sqrt{13}}, \qquad \sec\theta = -\frac{\sqrt{13}}{3},$$

$$\tan\theta = -\frac{2}{3}, \qquad \cot\theta = -\frac{3}{2}.$$

■

EXAMPLE 12 Given that $\cot \theta = 2$ and θ is in the third quadrant, find $\sin \theta$, $\cos \theta$, and $\tan \theta$.

Solution Again, we sketch a reference triangle and label the sides, then read off the appropriate ratios:

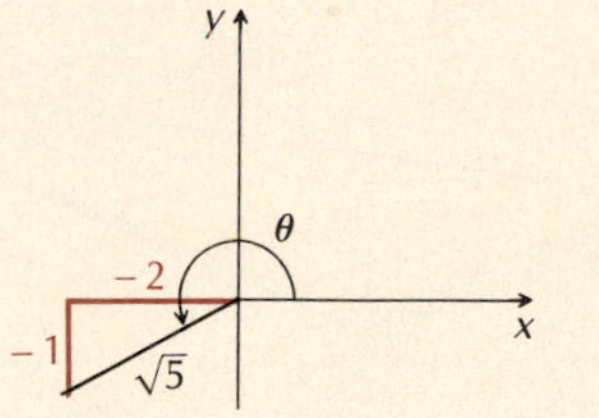

$$\sin \theta = -\frac{1}{\sqrt{5}},$$
$$\cos \theta = -\frac{2}{\sqrt{5}},$$
$$\tan \theta = \frac{1}{2}.$$

DO EXERCISES 16 AND 17.

16. Given $\cos \theta = \frac{3}{4}$ and that the terminal side is in quadrant IV, find the other function values.

17. Given $\cot \theta = -3$ and that the terminal side is in quadrant II, find the other function values.

EXERCISE SET 6.5

1 Find each of the following, if they exist.

1. $\cos 180°$
2. $\cot 0°$
3. $\sin 45°$
4. $\tan 45°$
5. $\cos(-135°)$
6. $\sin 150°$
7. $\cot(-60°)$
8. $\tan(-120°)$

Find the six trigonometric function values for the angle θ.

9.

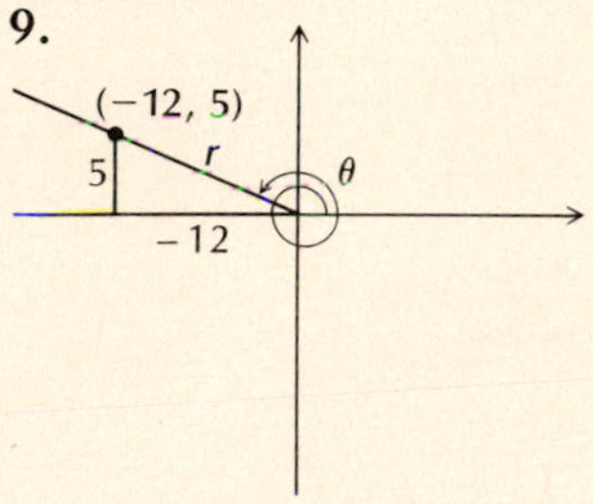

10.

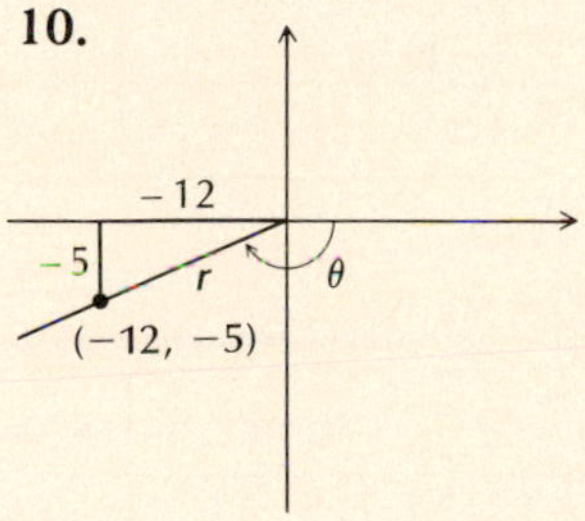

11.

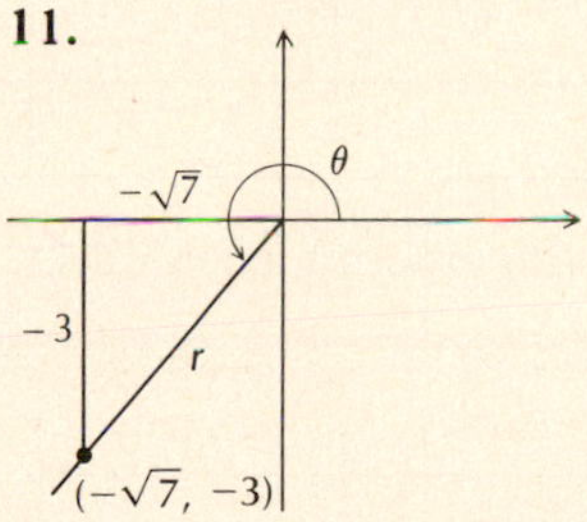

12.

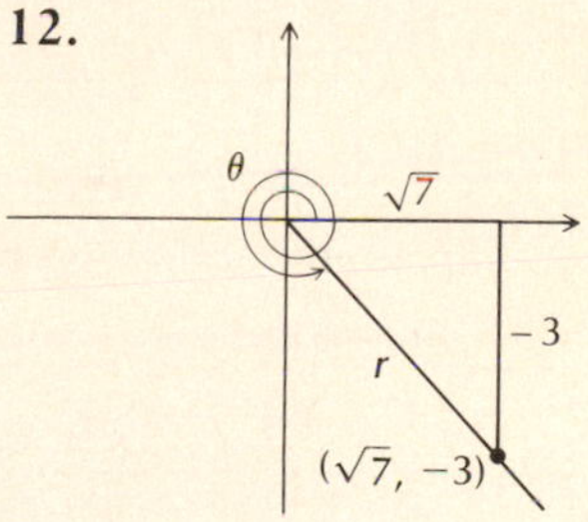

The terminal side of angle θ in standard position lies on the given line in the given quadrant. Find $\sin \theta$, $\cos \theta$, and $\tan \theta$.

13. $2x + 3y = 0$; quadrant IV
14. $4x + y = 0$; quadrant II
15. $5x - 4y = 0$; quadrant I
16. $y = 0.8x$; quadrant III

2 Find the signs of the six trigonometric function values for the given rotation.

17. $319°$
18. $-57°$
19. $-620°$
20. $194°$
21. $-215°$
22. $290°$
23. $91°$
24. $-272°$

3 Find the trigonometric function values for the given angle.

25. $90°$
26. $360°$
27. $-180°$
28. $-270°$

4 Find the six trigonometric function values for the specified angle.

29.

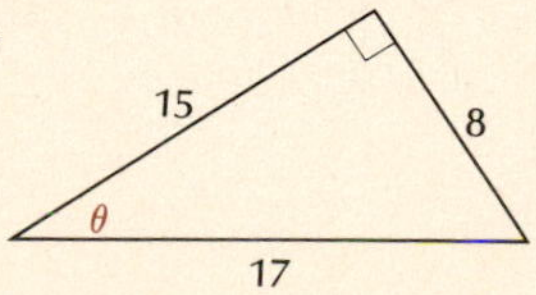

30.

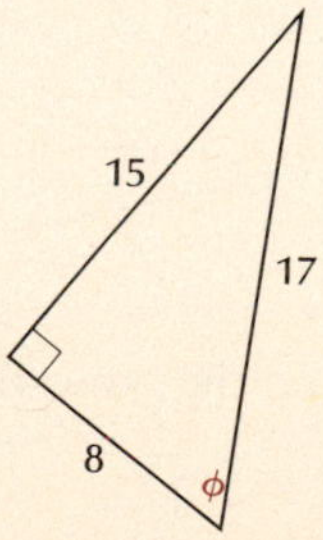

31.

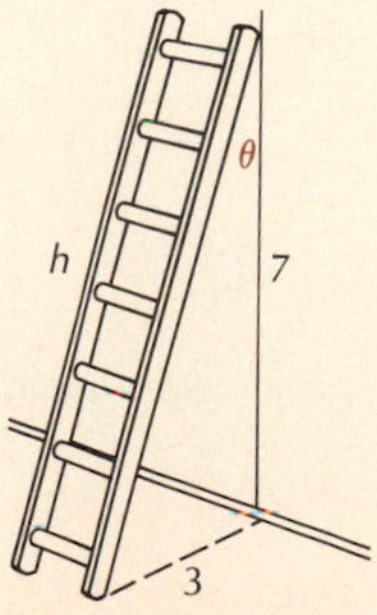

32.

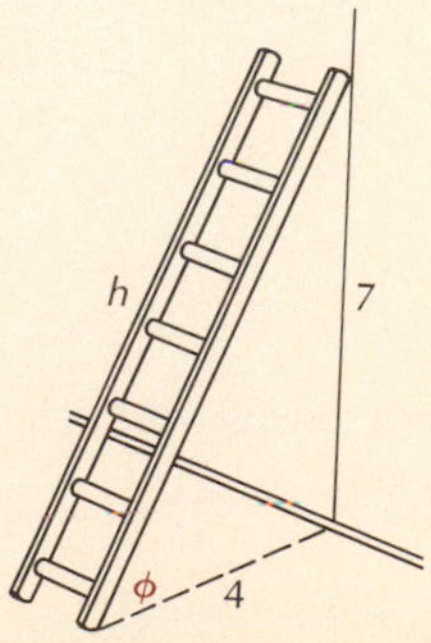

33.

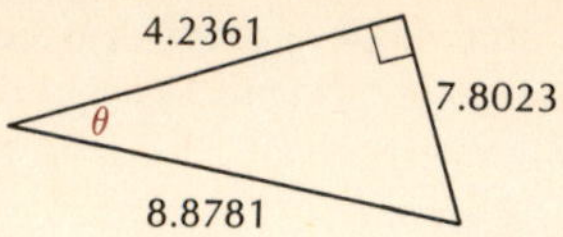

34.

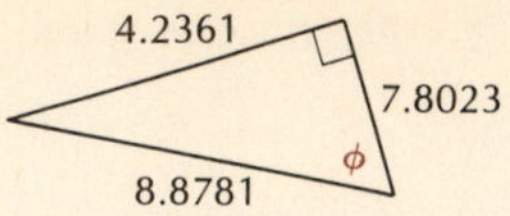

35.

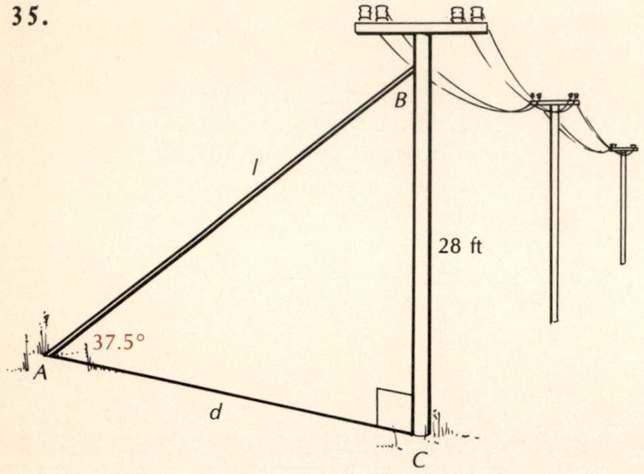

36.

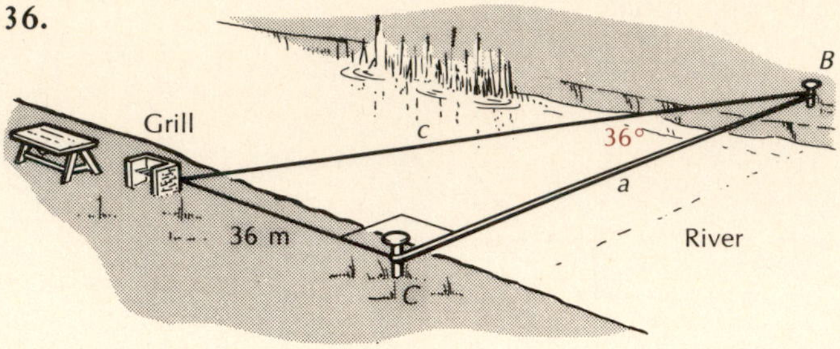

37. Complete the following table with exact function values.

θ	sine	cosine	tangent	cotangent	secant	cosecant
45°						
30°						
60°						

38. Complete the following table with exact function values.

θ	sine	cosine	tangent	cotangent	secant	cosecant
−45°						
−30°						
−60°						

39. Find the distance a across the river.

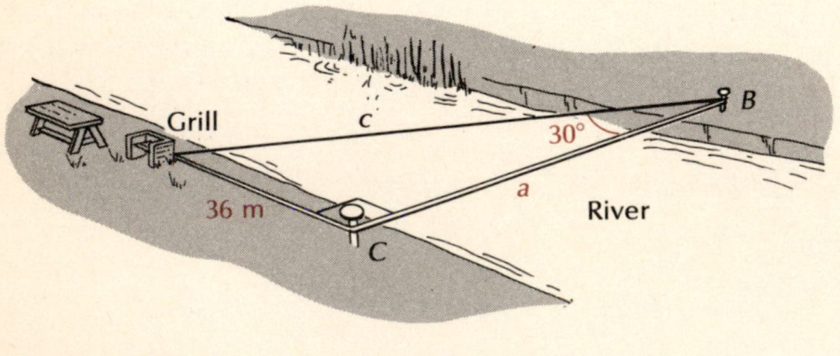

40. Find the length L from point A to the top of the pole.

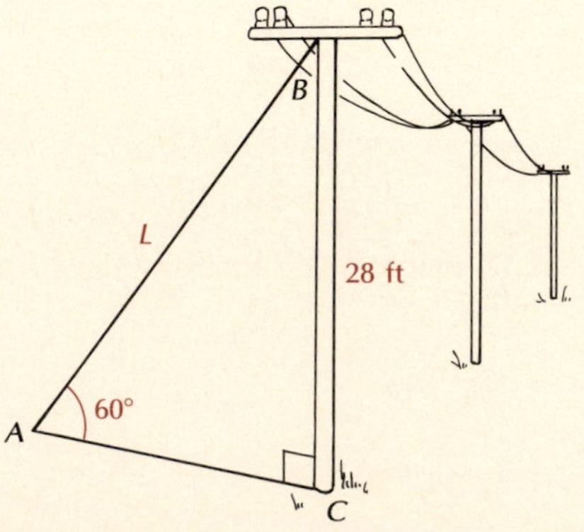

5 In Exercises 41–48, a function value and a quadrant are specified. Find the other five function values.

41. $\sin\theta = -\frac{1}{3}$; III

42. $\sin\theta = -\frac{1}{5}$; IV

43. $\cos\theta = \frac{3}{5}$; IV

44. $\cos\theta = -\frac{4}{5}$; II

45. $\cot\theta = -2$; IV

46. $\tan\theta = 5$; III

47. $\sin\theta = \frac{1}{3}$; II

48. $\cos\theta = \frac{4}{5}$; IV

In Exercises 49–52, assume that θ is an acute angle. A function value is specified. Find the other five function values.

49. $\sin\theta = \frac{24}{25}$

50. $\cos\theta = 0.7$

51. $\tan\phi = 2$

52. $\sec\phi = \sqrt{17}$

SYNTHESIS

53. This diagram shows a piston of a steam engine, driving a drive wheel of a locomotive. The radius of the drive wheel (from the center of the wheel to the pin P) is R, and the length of the rod is L. Suppose that the drive wheel is rotating at a speed of ω radians per second (so that $\theta = \omega t$). Show that the distance of pin Q from 0, the center of the wheel, is a function of time given by

$$x = \sqrt{L^2 - R^2 \sin^2 \omega t} + R\cos\omega t.$$

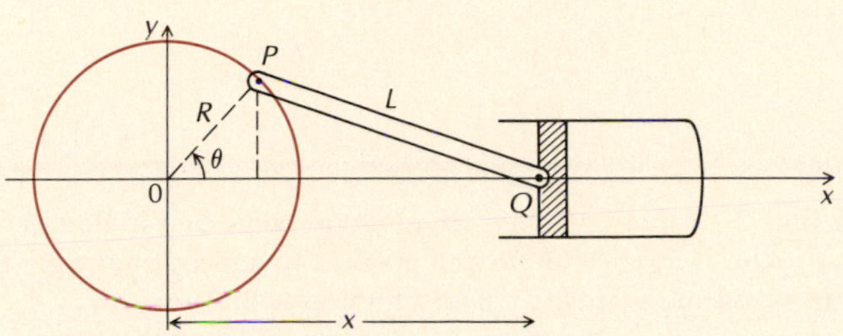

54. One of the motivations for developing trigonometry with a unit circle is that you can actually "see" $\sin\theta$ and $\cos\theta$ on the circle. Note in the figure that $AP = \sin\theta$ and $OA = \cos\theta$. It turns out that you can also "see" the other four trigonometric functions. Prove each of the following.

a) $BD = \tan\theta$
b) $OD = \sec\theta$
c) $OE = \csc\theta$
d) $CE = \cot\theta$

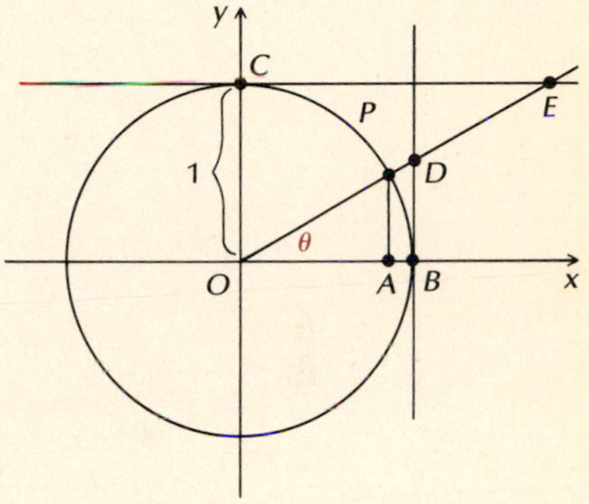

6.6 Finding Trigonometric Function Values

OBJECTIVES

You should be able to:

1 **Find the six trigonometric function values for angles that are multiples of 30° ($\pi/6$), 45° ($\pi/4$), and 60° ($\pi/3$).**

2 **Using a calculator, find the six trigonometric function values for any angle, measured in either radians or degrees. Given a trigonometric function value of θ, find θ given an interval in which θ lies, providing the answer in either radians or degrees.**

1 Function Values for Special Angles

We are now able to determine trigonometric function values for many other angles, whether they are measured in radians or degrees. If the terminal side of an angle falls on one of the axes, the function values are 0 or 1 or -1, or are undefined. Thus we can determine function values for any multiple of 90°, or $\pi/2$. We can also determine the function values for any angle whose terminal side makes a 30° ($\pi/6$), 45° ($\pi/4$), or 60° ($\pi/3$) angle with the x-axis.

Consider, for example, an angle of 150°. The terminal side makes a 30° angle with the x-axis, since $180° - 150° = 30°$. As the figure below shows, $\triangle ONR$ is congruent to $\triangle ON'R'$; thus the ratios of the sides of the two triangles are the same except perhaps for sign. We could determine the function values directly from $\triangle ONR$, but this is not necessary. If we remember that in quadrant II the sine is positive and the cosine and the tangent are negative, we can simply use the values of 30°, prefixing the appropriate sign. The $\triangle ONR$ is called a **reference triangle** and its acute angle is called a **reference angle**. The reference angle for a rotation, or angle, is the acute angle formed by the terminal side and the x-axis.

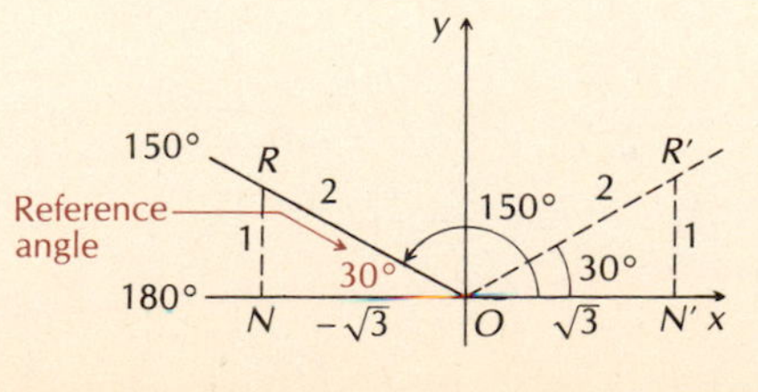

1. Find the trigonometric function values for 120°.

If we considered the same angle measured in radians, $5\pi/6$, then we would think of an arc length of $s = 5\pi/6$ on a unit circle. The number s measured on a unit circle determines a point T. Then we would consider a **reference number**, which is the shortest positive arc length s' between the point T and the x-axis. In the case of $5\pi/6$, the reference number is $\pi/6$, since $\pi - 5\pi/6 = \pi/6$.

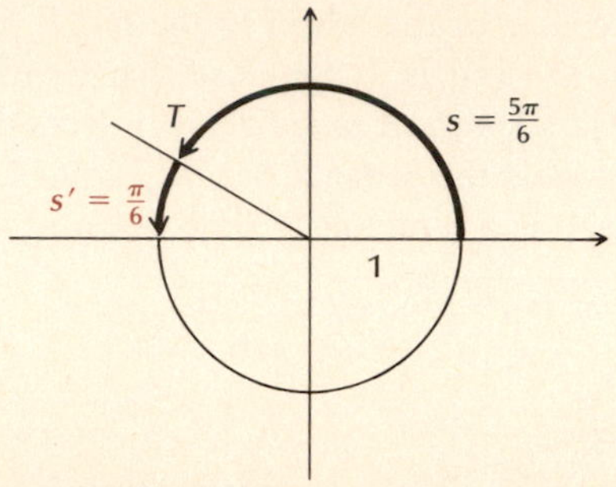

In general, to find the function values of an angle measured in degrees, we find them for the reference angle and prefix the appropriate sign. If the angle is measured in radians, we find function values for the reference number and prefix the appropriate sign. A reference angle formed by the terminal side and the x-axis is always positive and acute. A reference number, being the shortest positive arc length between the terminal side and the x-axis, is always positive.

EXAMPLE 1 Find the trigonometric function values for 225°.

Solution We draw a figure showing the terminal side of a 225° angle. The reference angle is $225° - 180°$, or 45°.

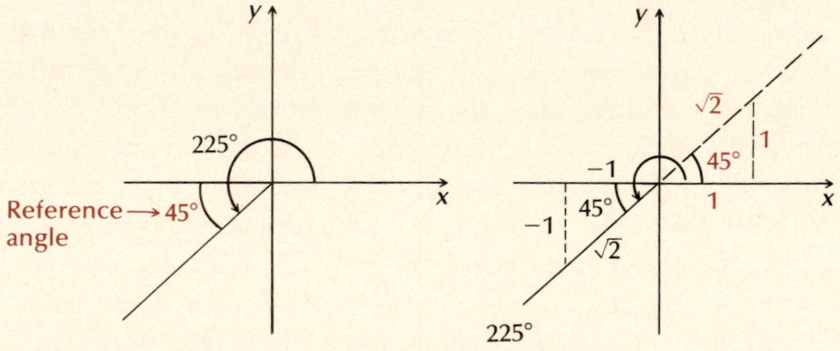

We recall from Section 6.5 that $\sin 45° = \sqrt{2}/2$, $\cos 45° = \sqrt{2}/2$, and $\tan 45° = 1$. We also note that in the third quadrant, the sine and the cosine are negative and the tangent is positive. We can easily determine the other three function values—the cotangent, the secant, and the cosecant—by remembering that they are the respective reciprocals of the sine, cosine, and tangent function values. Thus we have

$$\sin 225° = -\frac{\sqrt{2}}{2}, \qquad \csc 225° = -\frac{2}{\sqrt{2}}, \text{ or } -\sqrt{2},$$

$$\cos 225° = -\frac{\sqrt{2}}{2}, \qquad \sec 225° = -\frac{2}{\sqrt{2}}, \text{ or } -\sqrt{2},$$

$$\tan 225° = 1, \qquad \cot 225° = 1.$$

DO EXERCISE 1.

EXAMPLE 2 Find the trigonometric function values for 660°.

Solution We find the multiple of 180° nearest 660°:

$$180° \times 2 = 360°$$
$$180° \times 3 = 540°$$

and ← 660°

$$180° \times 4 = 720°.$$

The nearest multiple is 720°. The difference between 720° and 660° is 60°. This gives us the reference angle.

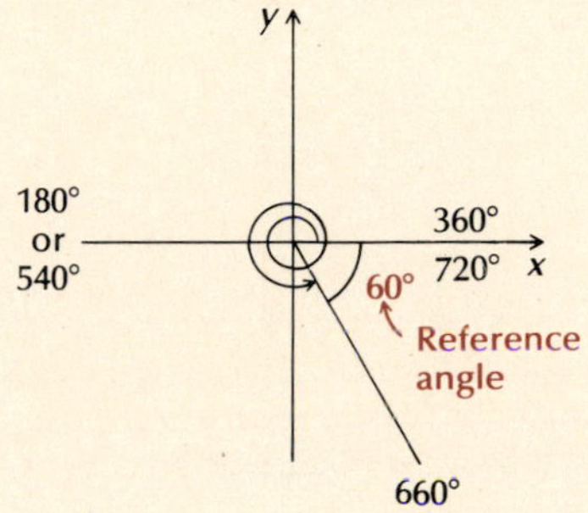

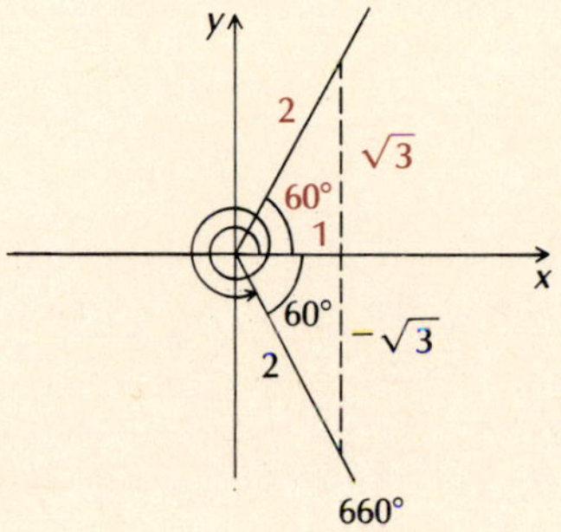

We recall that $\sin 60° = \sqrt{3}/2$, $\cos 60° = 1/2$, and $\tan 60° = \sqrt{3}$. In the fourth quadrant, the cosine is positive and the sine and tangent are negative. Thus we have

$$\sin 660° = -\frac{\sqrt{3}}{2}, \qquad \csc 660° = -\frac{2}{\sqrt{3}}, \text{ or } -\frac{2\sqrt{3}}{3},$$
$$\cos 660° = \frac{1}{2}, \qquad \sec 660° = \frac{2}{1} = 2,$$
$$\tan 660° = -\sqrt{3}, \qquad \cot 660° = -\frac{1}{\sqrt{3}}, \text{ or } -\frac{\sqrt{3}}{3}.$$

DO EXERCISE 2.

2. Find the trigonometric function values for 570°.

We can use the same procedure for negative rotations.

EXAMPLE 3 Find the sine, the cosine, and the tangent of −1050°.

Solution We find the multiple of −180° nearest −1050°:

$$-180° \times 5 = -900° \quad \text{and} \quad -180° \times 6 = -1080°.$$

The nearest multiple is −1080°. The difference between −1050° and −1080° is 30°. This gives us the reference angle.

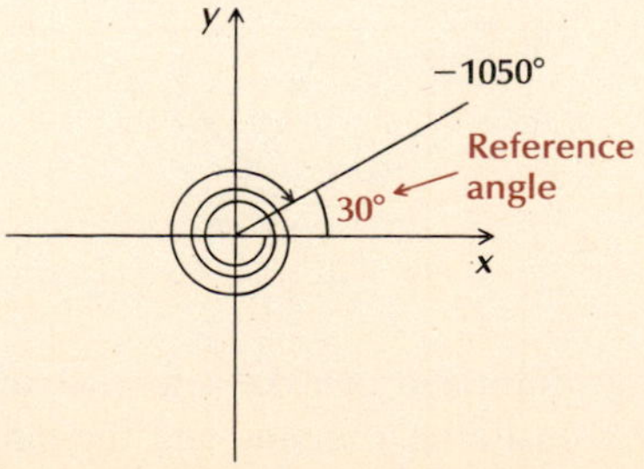

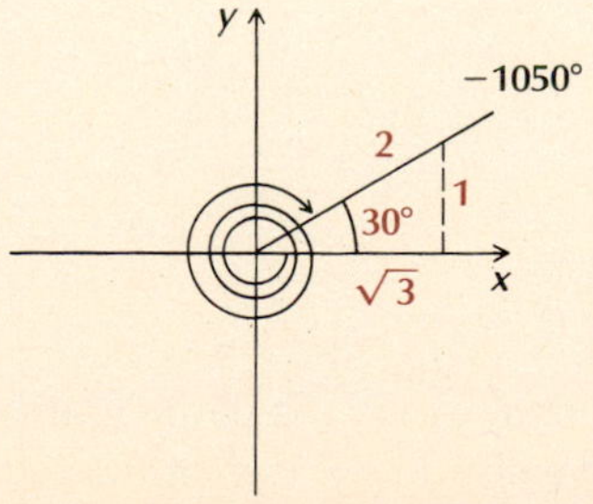

3. Find the trigonometric function values for $-945°$.

All trigonometric function values are positive in the first quadrant, and $\sin 30° = 1/2$, $\cos 30° = \sqrt{3}/2$, and $\tan 30° = \sqrt{3}/3$. Thus we have

$$\sin(-1050°) = \frac{1}{2}, \qquad \cos(-1050°) = \frac{\sqrt{3}}{2}, \quad \text{and} \quad \tan(-1050°) = \frac{\sqrt{3}}{3}.$$

DO EXERCISE 3.

EXAMPLE 4 Find the trigonometric function values of $\frac{7\pi}{6}$.

Solution Since we have radian measure, we consider a unit circle. To find $7\pi/6$ on a unit circle, we subtract as many multiples of π as we can. Since

$$\frac{7\pi}{6} = \frac{6\pi}{6} + \frac{\pi}{6} = \pi + \frac{\pi}{6},$$

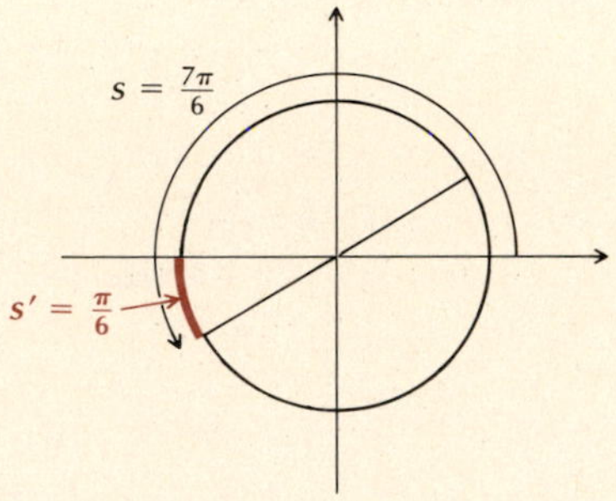

we see that $7\pi/6$ is in the third quadrant and that the reference number is $\pi/6$. We also know that in the third quadrant, the sine and the cosine are negative and the tangent is positive. Thus we have

4. Find the trigonometric function values for $\frac{17\pi}{6}$.

$$\sin\frac{7\pi}{6} = -\frac{1}{2}, \qquad \csc\frac{7\pi}{6} = -2,$$
$$\cos\frac{7\pi}{6} = -\frac{\sqrt{3}}{2}, \qquad \sec\frac{7\pi}{6} = -\frac{2\sqrt{3}}{3},$$
$$\tan\frac{7\pi}{6} = \frac{\sqrt{3}}{3}, \qquad \cot\frac{7\pi}{6} = \sqrt{3}.$$

DO EXERCISE 4.

EXAMPLE 5 Find the trigonometric function values of $-\frac{13\pi}{3}$.

Solution Since we have radian measure, we consider a unit circle. To find $-13\pi/3$ on a unit circle, we subtract as many multiples of π as we can. Since

$$-\frac{13\pi}{3} = -\frac{12\pi}{3} - \frac{\pi}{3} = -\left(4\pi + \frac{\pi}{3}\right),$$

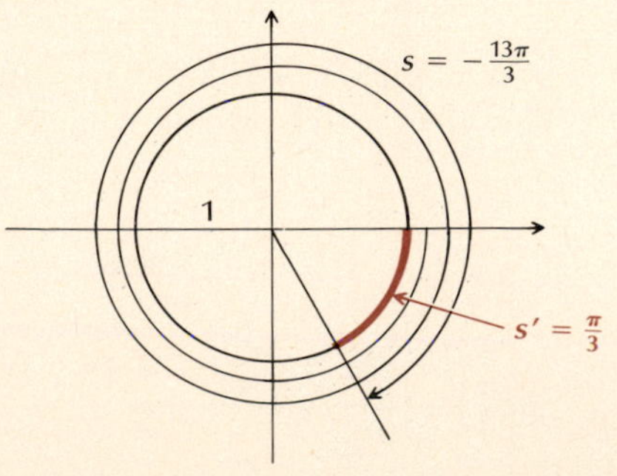

we see that $-13\pi/3$ is in the fourth quadrant and that the reference number is $\pi/3$. We also know that in the fourth quadrant, the sine and the tangent are negative and the cosine is positive. Thus we have

$$\sin\left(-\frac{13\pi}{3}\right) = -\frac{\sqrt{3}}{2}, \qquad \csc\left(-\frac{13\pi}{3}\right) = -\frac{2\sqrt{3}}{3},$$
$$\cos\left(-\frac{13\pi}{3}\right) = \frac{1}{2}, \qquad \sec\left(-\frac{13\pi}{6}\right) = 2,$$
$$\tan\left(-\frac{13\pi}{3}\right) = -\sqrt{3}, \qquad \cot\left(-\frac{13\pi}{3}\right) = -\frac{\sqrt{3}}{3}.$$

DO EXERCISES 5 AND 6.

5. Find the trigonometric function values for $-\frac{22\pi}{3}$.

6. Find the trigonometric function values for $\frac{29\pi}{4}$.

2 Finding Trigonometric Function Values with a Calculator

How do we find function values when the angle or arc length is not one of the special values 0, $\pi/6$, $\pi/2$, $\pi/3$, and $\pi/2$; or 0°, 30°, 45°, 60°, and 90°? Before the advent of scientific calculators, tables were used to find such values. These were prepared by mathematicians using advanced procedures for approximation. Table 3, at the back of the book, is such a table. It contains function values for degrees in the interval [0°, 90°] and for radians in the interval $[0, \pi/2]$.

With the advent of the scientific calculator, the need for tables has lessened. The advantage of using a calculator is that no reference angle or reference number need be considered. The main concern is whether the calculator is operating in degree mode or radian mode. You will have to know which mode. Be sure to read your instruction book carefully. One quick check to keep in mind is that

$$\sin 90° = 1, \quad \text{or} \quad \sin\frac{\pi}{2} = 1.$$

If you try sin 90 and get 0.893996664, you know that you are in radian mode. But if you try sin 90 and get 1, you know that you are in degree mode.

When finding values on a calculator, remember that different calculators round to different decimal places, so there may be variance in answers found on your calculator. Also keep in mind the difference between an exact answer and an approximation. For example,

$$\sin 60° = \frac{\sqrt{3}}{2}. \qquad \textbf{This is exact!}$$

But on a calculator, you might get an answer like

$$\sin 60° = 0.866025404. \qquad \textbf{This is approximate!}$$

Generally, to ease notation, we will not use an approximation symbol, ≈ .

Scientific calculators usually provide values of the sine, the cosine, and the tangent functions only. You can find values of the cotangent, the secant, and the cosecant by taking reciprocals of the tangent, the cosine, and the sine function, respectively. Since calculators are different, we suggest that you carefully read the directions for the calculator you are using.

EXAMPLE 6 Find tan 81°.

Solution We check to be sure that the calculator is in degree mode. Then we enter 81 and press the TAN key. We find that

$$\tan 81° = 6.3138.$$

We rounded the answer to four decimal places.

Find each of the following to six decimal places.

7. $\sin 229°$

8. $\tan 22.6°$

9. $\cos(-1007°)$

10. $\csc(143.567°)$

11. $\cot(-789.23°)$

12. $\sec(209.056778°)$

In each of the following cases, a wrong answer was found using a calculator. Explain the incorrect procedure used.

13. $\sin \frac{21\pi}{19} = 0.060565745$

14. $\tan(-405) = -1$

15. $\cos \frac{16\pi}{17} = 0.99866872$

Find each of the following to four decimal places.

16. $\cos 61°54'$

17. $\tan 347°02'38''$

18. $\sec 55°13'43''$

EXAMPLE 7 Find $\cos 715.8234°$.

Solution We enter 715.8234 and press the COS key. We need not consider a reference angle. We find that

$$\cos 715.8234° = 0.997344305.$$

This time we rounded to nine decimal places. ■

EXAMPLE 8 Find $\sin(-8.0113°)$. Round to four decimal places.

Solution We enter -8.0113 and press the SIN key. Again, we do not need to consider a reference angle. We find that

$$\sin(-8.0013°) = -0.1392.$$ ■

EXAMPLE 9 Find $\sec 43°$. Round to nine decimal places.

Solution The secant function value can be found by taking the reciprocal of the cosine function value. We enter 43 and press the COS key:

$$\cos 43° = 0.731353702.$$

We then press the reciprocal key $1/x$, or divide 1 by 0.731353702. You need not round before taking the reciprocal. We find that

$$\sec 43° = \frac{1}{\cos 43°} = 1.367327461.$$ ■

EXAMPLE 10 Find $\cot(-389°)$. Round to nine decimal places.

Solution The cotangent function is the reciprocal of the tangent. We enter -389, press the TAN key, and then press the reciprocal key:

$$\cot(-389°) = \frac{1}{\tan(-389°)} = -1.804047755.$$ ■

DO EXERCISES 7–15.

When the angle measure is given in degrees, minutes, and seconds, we first convert to degrees and decimal parts of degrees.

EXAMPLE 11 Find $\sin 38°25'57''$. Round to four decimal places.

Solution We first convert to degrees and decimal parts of degrees. You can do this on your calculator. You need not round values before adding. We get

$$\begin{aligned} 38°25'57'' &= 38° + \left(\frac{25}{60}\right)^{\circ} + \left(\frac{57}{3600}\right)^{\circ} \\ &= 38° + (0.41666667)° + (0.01583333)° \\ &= 38.4325°. \end{aligned}$$

We then press the SIN key and get

$$\sin 38°25'57'' = 0.6216.$$ ■

DO EXERCISES 16–18.

Using a calculator, we can find function values of real numbers, or radians, directly without changing to degrees. Most scientific calculators have both degree and radian modes. When directly finding function values of real numbers or radian measures, we must set the calculator in radian mode.

EXAMPLE 12 Find $\cos \frac{2\pi}{5}$. Round to four decimal places.

Solution Using the π key, we get

$$\frac{2\pi}{5} = 1.256637062.$$

Again, you need not round off before continuing. (If you do not have a π key, then use 3.14 or 3.1416, or some other approximation as directed by your instructor.) Then remembering to use the radian mode, we press the COS key. We find that

$$\cos \frac{2\pi}{5} = 0.3090.$$

EXAMPLE 13 Find $\tan\left(-\frac{\pi}{7}\right)$. Round to six decimal places.

Solution

$$\tan\left(-\frac{\pi}{7}\right) = \tan(-0.448798951) = -0.481575.$$

EXAMPLE 14 Find sin 13,523.86. Round to six decimal places.

Solution We enter 13,523.86 and press the SIN key. The calculator must be in radian mode. We find that

$$\sin 13{,}523.86 = 0.641440.$$

EXAMPLE 15 Find csc 429. Round to six decimal places.

Solution The cosecant function value can be found by taking the reciprocal of the sine function value. Remembering to use radian mode, we enter 429, press the SIN key, and then press the reciprocal key. We find that

$$\csc 429 = \frac{1}{\sin 429} = \frac{1}{0.985141084} = 1.015083.$$

DO EXERCISES 19–24.

Sometimes we need to work backwards and find an angle that has a specified function value. There are many such values unless restrictions to a quadrant or an interval are specified. For example, since

$$\sin 30° = \sin 150° = \sin 390° = \sin(-210°) = \tfrac{1}{2},$$

we see that there are many solutions, in degrees, to the equation

$$\sin\theta = \tfrac{1}{2}.$$

In fact, there are infinitely many solutions. Similarly, there are infinitely many solutions in radians.

We can use a calculator in reverse—for example, to find an angle when its sine is given—but the answers are restricted to certain intervals. That is, if we

Find the function value. Round to five decimal places.

19. $\cos \frac{11\pi}{13}$

20. $\tan \frac{\pi}{9}$

21. sin 21

22. cot 1026

23. $\cos\left(-\frac{3\pi}{11}\right)$

24. sec (−90,234.567)

Find the acute angle θ, to the nearest minute, for each of the following.

25. $\tan \theta = 0.4621$

26. $\sin \theta = 0.7660$

27. $\cos \theta = 0.7009$

Find θ in the interval indicated.

28. $\sin \theta = -0.2363$, $[-90°, 90°]$

29. $\cos \theta = -0.5712$, $[0, \pi]$

30. $\cos \theta = -0.5712$, $[0°, 180°]$

31. $\tan \theta = 124.84$, $(-\pi/2, \pi/2)$

ask a calculator to find the angle θ for which

$$\sin \theta = \tfrac{1}{2},$$

the calculator will give us the answer in degrees only in the interval $[-90°, 90°]$, or in radians only in the interval $[-\pi/2, \pi/2]$.

To use a calculator in reverse—for example, to find an angle whose sine is given—we first enter the sine value. Then we press the key marked either SIN^{-1} (on some calculators) or ARCSIN (on other calculators). On still others, two keys must be pressed in sequence: ARC and then SIN, INV and then SIN, or f^{-1} and then SIN. Be sure to read the instructions for the calculator you are using.

EXAMPLE 16 Find the acute angle θ, to the nearest minute, for which $\cos \theta = 0.5417$.

Solution We check to be certain that the calculator is in degree mode. We enter 0.5417 and press the COS^{-1} key. We find that

$$\theta = \cos^{-1} 0.5417 = 57.2° = 57°12'.$$

Thus, $\theta = 57°12'$. ■

DO EXERCISES 25–27.

Inverse functions for the sine, the cosine, and the tangent functions are defined and developed in Section 7.6. The restrictions are given as follows.

The Inverse Function Key

1. SIN^{-1} gives values of θ in degrees in the interval $[-90°, 90°]$ and in radians in the interval $[-\pi/2, \pi/2]$.

2. COS^{-1} gives values of θ in degrees in the interval $[0°, 180°]$ and in radians in the interval $[0, \pi]$.

3. TAN^{-1} gives values of θ in degrees in the interval $(-90°, 90°)$ and in radians in the interval $(-\pi/2, \pi/2)$.

EXAMPLES Find the value of θ satisfying the given condition on the specified interval.

17. $\sin \theta = -0.8789$, θ in $[-90°, 90°]$

Using the SIN^{-1} key in degree mode, we find that $\theta = -61.51°$.

18. $\tan \theta = -6.2051$, θ in $(-\pi/2, \pi/2)$

Using the TAN^{-1} key in radian mode, we find that $\theta = -1.4110$. ■

DO EXERCISES 28–31.

To find function values in other intervals, we must consider the reference angle.

EXAMPLE 19 Given that $\sin \theta = 0.2812$, find θ between 90° and 180°.

Solution We first find the reference angle. We enter 0.2812 and press the SIN^{-1} key. The reference angle is 16.33°, or 16°20'.

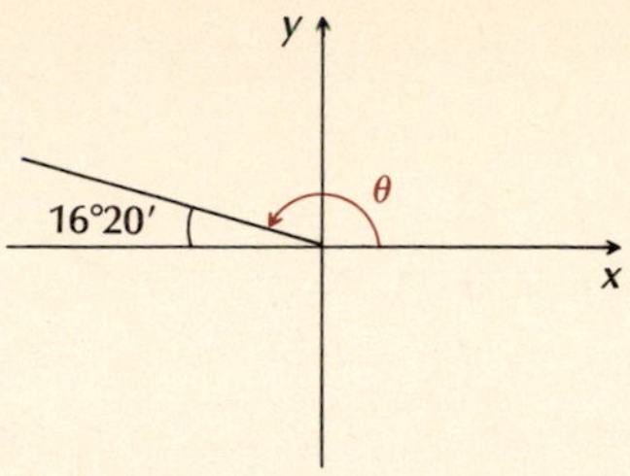

We find the angle θ by subtracting 16.33° from 180°:

$$180° - 16.33° = 163.67°, \quad \text{or} \quad 163°40'.$$

Thus, $\theta = 163°40'$.

EXAMPLE 20 Given that $\tan\theta = -6.2051$, find θ between 270° and 360°.

Solution We find the reference angle, ignoring the fact that $\tan\theta$ is negative. We enter 6.2051 and press the TAN^{-1} key. The reference angle is 80.85°, or 80°51′.

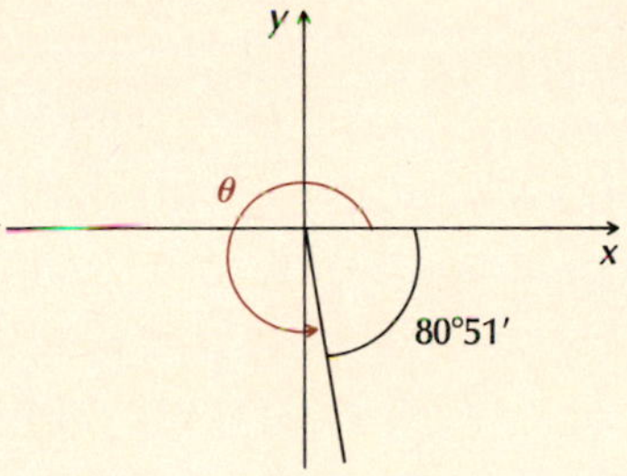

We find the angle θ by subtracting 80.85° from 360°:

$$360° - 80.85° = 279.15°, \quad \text{or} \quad 279°09'.$$

Thus, $\theta = 279°09'$.

DO EXERCISES 32–35.

Find θ in the interval indicated.

32. $\cos\theta = 0.5712$, (270°, 360°)

33. $\tan\theta = -2.4778$, (90°, 180°)

34. $\sin\theta = -0.2363$, (180°, 270°)

35. $\sec\theta = -1.2020$, (180°, 270°)

EXERCISE SET 6.6

1 Find each of the following, if they exist.

1. sec 315°	2. csc 315°	3. sin 150°	4. cos 150°
5. cot 570°	6. tan 570°	7. csc 270°	8. sin (−450°)
9. cot (−225°)	10. sec (−225°)	11. sin 1050°	12. cos 675°
13. tan (−135°)	14. sin (−135°)	15. sec 1125°	16. csc 1125°
17. sin 495°	18. cot 330°	19. cos 5220°	20. sin (−7560°)
21. $\sin\frac{13\pi}{4}$	22. $\cos\frac{19\pi}{4}$	23. $\tan\frac{11\pi}{3}$	24. $\sec\frac{11\pi}{3}$

25. Find the six trigonometric function values for $23\pi/6$.

26. Find the six trigonometric function values for $-29\pi/3$.

27. Find the six trigonometric function values for $-19\pi/3$.

28. Find the six trigonometric function values for $37\pi/6$.

29. Find the six trigonometric function values for $31\pi/3$.

30. Find the six trigonometric function values for $-25\pi/6$.

31. Find the six trigonometric function values for −750°.

32. Find the six trigonometric function values for 1590°.

2 Use a calculator to find each of the following function values.

33. $\cos 18°$
34. $\sin 37°$
35. $\tan 2.6°$
36. $\cos 34.8°$
37. $\sin 62°20'$
38. $\tan 55°30'$
39. $\cos 15°35'$
40. $\sin 40°55'$
41. $\csc 29°11'36''$
42. $\cot 54°44'53''$
43. $\sec 10°31'42''$
44. $\csc 45°15'23''$
45. $\sin 561.2344°$
46. $\tan(-234.567°)$
47. $\cot(-900.23°)$
48. $\sec(0.2045°)$
49. $\tan 295°14'$
50. $\cos 230°53'$
51. $\sec 146.9°$
52. $\sin 98.4°$
53. $\sin 756°25'$
54. $\cot 820°40'$
55. $\cos(-1000.85°)$
56. $\tan(-1086.15°)$
57. $\sin 37$
58. $\cos 810$
59. $\cos(-10)$
60. $\sin(-4)$
61. $\tan 5\pi$
62. $\cot 28\pi$
63. $\cot 1000$
64. $\tan 13$
65. $\sec\left(-\frac{\pi}{5}\right)$
66. $\cot\left(-\frac{7\pi}{5}\right)$
67. $\sec\frac{10\pi}{7}$
68. $\cot\frac{2\pi}{7}$
69. $\cos(-13\pi)$
70. $\sin(-10\pi)$
71. $\tan 2.5$
72. $\cot 5.2$

Use a calculator to find the acute angle θ that is a solution of the equation. Give the answer in degrees and decimal parts of degrees, as well as in radians.

73. $\sin\theta = 0.5125$
74. $\cos\theta = 0.8241$
75. $\cos\theta = 0.6512$
76. $\sin\theta = 0.8220$
77. $\tan\theta = 7.425$
78. $\tan\theta = 3.163$
79. $\csc\theta = 6.277$
80. $\sec\theta = 1.175$

Find θ in the interval indicated.

81. $\sin\theta = -0.9956$, $(270°, 360°)$
82. $\sin\theta = 0.4313$, $(90°, 180°)$
83. $\cos\theta = -0.9388$, $(180°, 270°)$
84. $\cos\theta = -0.0990$, $(90°, 180°)$
85. $\tan\theta = 0.2460$, $(180°, 270°)$
86. $\tan\theta = -3.0541$, $(270°, 360°)$
87. $\sec\theta = -1.0485$, $(90°, 180°)$
88. $\csc\theta = 1.0480$, $(0°, 90°)$
89. $\sin\theta = 0.5766$, $(\pi/2, \pi)$
90. $\sin\theta = -0.8873$, $(3\pi/2, 2\pi)$
91. $\cos\theta = -0.211499$, $(\pi/2, \pi)$
92. $\cos\theta = -0.9388$, $(\pi, 3\pi/2)$
93. $\tan\theta = -19.0541$, $(3\pi/2, 2\pi)$
94. $\tan\theta = 32.246$, $(\pi, 3\pi/2)$
95. $\csc\theta = 11.21048$, $(0, \pi/2)$
96. $\sec\theta = -4.23785$, $(\pi/2, \pi)$

SYNTHESIS

97. The valve cap on a bicycle wheel is 24.5 in. from the center of the wheel. From the position shown, the wheel starts rolling. After the wheel has turned 390°, how far above the ground is the valve cap? Assume that the outer radius of the tire is 26 in.

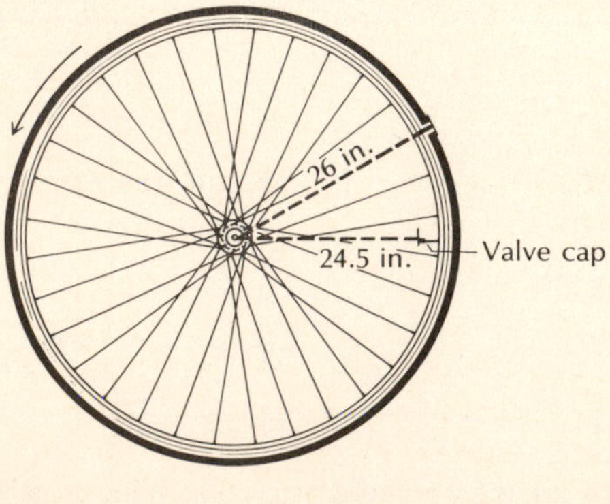

98. The seats of a ferris wheel are 35 ft from the center of the wheel. When you board the wheel, you are 5 ft above the ground. After you have rotated through an angle of 765°, how far above the ground are you?

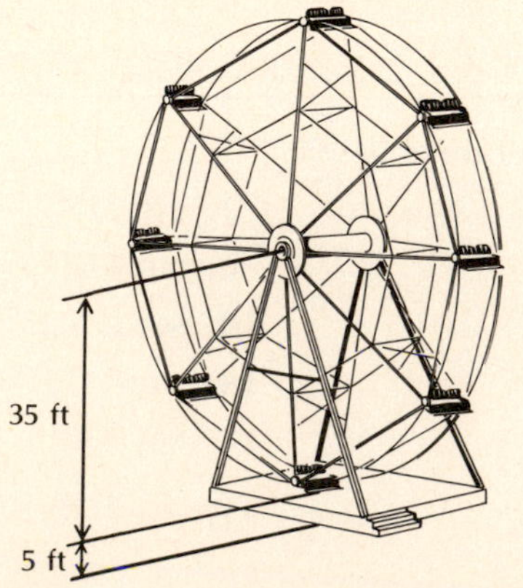

99. How big must an angle be in order that it first differs from its tangent in the fourth decimal place?

100. How big must an angle be in order that it first differs from its tangent in the third decimal place?

101. Angles are measured in degrees, minutes ($60' = 1°$), and seconds ($60'' = 1'$). Convert radian measure of $\pi/7$ to degrees, minutes, and seconds.

102. Convert $61°38'22''$ to degrees and decimal parts of degrees.

103. The formula

$$\sin x = x - \frac{x^3}{6} + \frac{x^5}{120}$$

gives an approximation for sine values when x is in radians. Calculate sin 0.5 and compare your answer with Table 3.

104. Using values from the table, graph the tangent function between $-90°$ and $90°$. What is the domain of the tangent function?

105. Use the fact that $\sin\theta \approx \theta$ when θ is small to calculate the diameter of the sun. The sun is 93 million miles from the earth and the angle it subtends at the earth's surface is about $31'59''$.

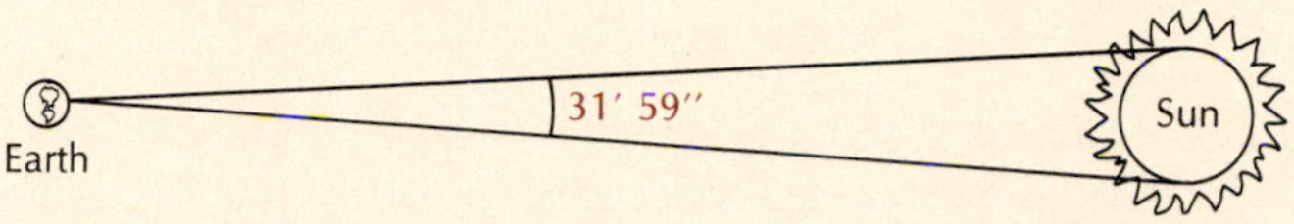

106. *Snell's law.* When light passes from one substance to another, rays are bent, depending on the speed of light in those substances. For example, light traveling in air is bent as it enters water. Angle i is called the **angle of incidence,** and angle r is called the **angle of refraction.** Snell's law states that $\sin i/\sin r$ is constant. The constant is called the **index of refraction.** The index of refraction of a certain crystal is 1.52. Light strikes it, making an angle of incidence of 27°. What is the angle of refraction?

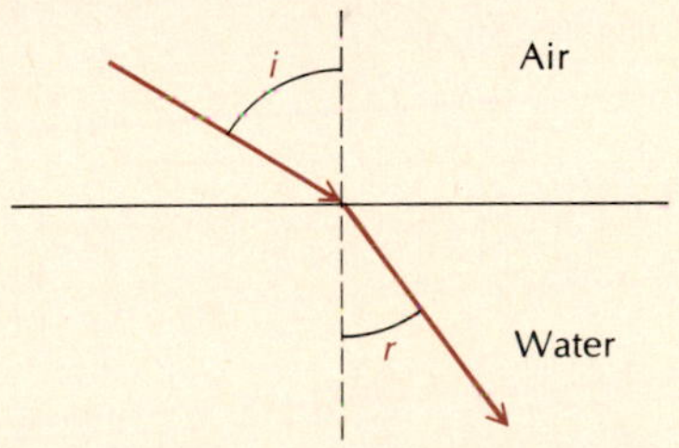

CHALLENGE

107. *Projectile motion.* An object is thrown into the air at a velocity V, in meters per second, at an angle θ. After t seconds, the object is at a horizontal distance x and a vertical distance y from its release point, where

$$x = (V\cos\theta)t$$

and

$$y = -\tfrac{1}{2}gt^2 + (V\sin\theta)t + h,$$

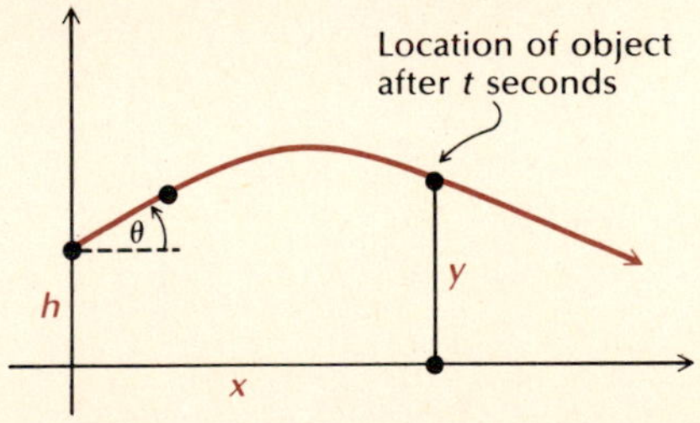

and where h = the height at which the shot is released and g = the acceleration due to gravity (air resistance is neglected).

a) At what time will the object hit the ground?
b) At what horizontal distance will the object hit the ground?
c) In August, 1975, Marianne Adam of East Germany put a shot 21.6 m. Assume that she was 1.6 m tall and that $\theta = 44°$ and $g = 9.8\text{ m/sec}^2$. What was the velocity of the shot when she released it?
d) All other things being equal, what advantage does a tall shotputter have over a shorter one?
e) When the units are given in feet and seconds, $g = 32\text{ ft/sec}^2$. A softball player hits a softball at a height of 4.5 ft and an angle of 20°. The ball travels for a home run just over an 8-ft fence that is 300 ft from home plate. How long does it take the ball to get to the fence? What was its velocity when hit?

6.7 Graphs of Transformed Sine and Cosine Functions

OBJECTIVES

You should be able to:

1 Sketch graphs of
$$y = B + A \sin (Cx - D)$$
and
$$y = B + A \cos (Cx - D)$$
for various values of the constants A, B, C, and D.

2 For functions like these, determine the amplitude, the period, and the phase shift.

3 By addition of ordinates, graph sums of functions.

1. Sketch a graph of $y = -2 + \cos x$.

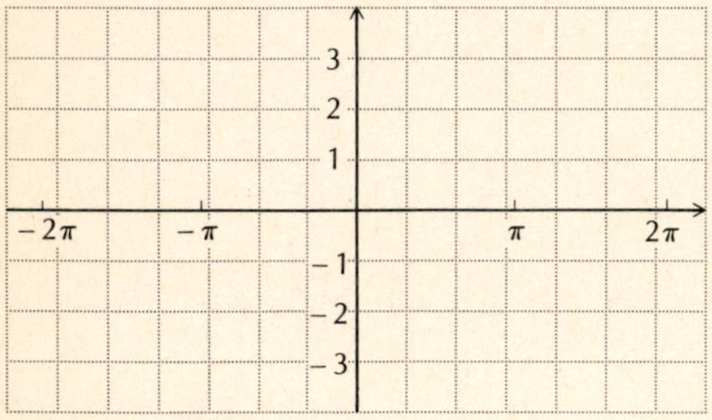

1, 2 Variations of Basic Graphs

In Sections 6.2 and 6.3, we graphed the six circular, or trigonometric, functions. It might be helpful to review those graphs and have their shapes well in mind because in this section, we will consider variations of these graphs. In particular, we are interested in graphs in the form

$$y = B + A \sin (Cx - D)$$

and

$$y = B + A \cos (Cx - D),$$

where A, B, C, and D are constants, some of which may be 0. These constants have the effect of translating, stretching, reflecting, or shrinking the basic graphs. (It might be helpful also to review Section 3.6.) Let us first examine the effect of each constant individually. Then we will consider the combined effect of more than one of the constants.

We first consider the effect of the constant B.

EXAMPLE 1 Sketch a graph of $y = 3 + \sin x$.

Solution The graph of $y = 3 + \sin x$ is a translation of the graph of $y = \sin x$ up 3 units. One way to sketch the graph is to first consider $y = \sin x$ on an interval of length 2π, say, $[0, 2\pi]$. The zeros of the function and the maximum and minimum values can be considered key points. These are

$$(0, 0), \quad \left(\frac{\pi}{2}, 1\right), \quad (\pi, 0), \quad \left(\frac{3\pi}{2}, -1\right), \quad (2\pi, 0).$$

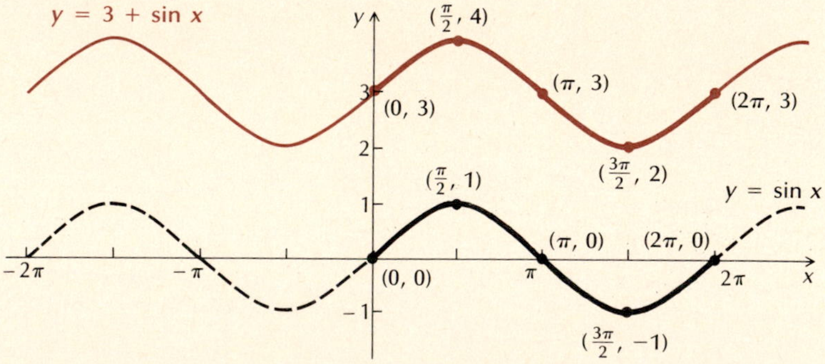

These key points are transformed up 3 units to obtain the key points of the graph of $y = 3 + \sin x$. These are

$$(0, 3), \quad \left(\frac{\pi}{2}, 4\right), \quad (\pi, 3), \quad \left(\frac{3\pi}{2}, 2\right), \quad (2\pi, 3).$$

The graph of $y = 3 + \sin x$ can be sketched over the interval $[0, 2\pi]$ and extended to obtain the rest of the graph by repeating the graph over intervals of length 2π. ■

DO EXERCISE 1.

Next we consider the effect of the constant A.

EXAMPLE 2 Sketch a graph of $y = 2 \sin x$. What is the amplitude?

Solution The constant 2 in $y = 2 \sin x$ has the effect of stretching the graph of $y = \sin x$ vertically by a factor of 2 units. The function values of $y = \sin x$ are such that $-1 \leq \sin x \leq 1$, and $|\sin x| \leq 1$. The function values of $y = 2 \sin x$ are such that $-2 \leq 2 \sin x \leq 2$, or $|2 \sin x| \leq 2$. The maximum value of $y = 2 \sin x$ is 2, and the minimum value is -2. Thus the amplitude is $\frac{1}{2}[2 - (-2)]$, or 2.

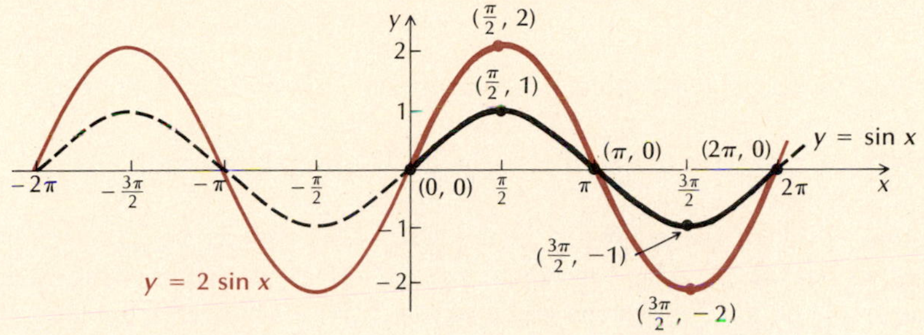

We draw the graph of $y = \sin x$ and consider its key points

$$(0, 0), \quad \left(\frac{\pi}{2}, 1\right), \quad (\pi, 0), \quad \left(\frac{3\pi}{2}, -1\right), \quad (2\pi, 0)$$

over the interval $[0, 2\pi]$.

The second coordinates are then multiplied by 2 to obtain the key points of $y = 2 \sin x$. These are

$$(0, 0), \quad \left(\frac{\pi}{2}, 2\right), \quad (\pi, 0), \quad \left(\frac{3\pi}{2}, -2\right), \quad (2\pi, 0).$$

We plot these points and sketch the graph over the interval $[0, 2\pi]$. Then we repeat this part of the graph over other intervals of length 2π. ■

If the constant A in $y = A \sin x$ is negative, there will also be a reflection across the x-axis. If the absolute value of A is less than 1, then there will be a vertical shrinking. In general, we have the following.

> DEFINITION
>
> **The *amplitude* of $y = B + A \sin (Cx - D)$ is $|A|$.**

EXAMPLE 3 Sketch a graph of $y = -\frac{1}{2} \sin x$.

Solution The amplitude of the graph is $|-\frac{1}{2}|$, or $\frac{1}{2}$. The graph of $y = -\frac{1}{2} \sin x$ is a vertical shrinking and a reflection of the graph of $y = \sin x$. In graphing, the key points of $y = \sin x$,

$$(0, 0), \quad \left(\frac{\pi}{2}, 1\right), \quad (\pi, 0), \quad \left(\frac{3\pi}{2}, -1\right), \quad (2\pi, 0),$$

are transformed to

$$(0, 0), \quad \left(\frac{\pi}{2}, -\frac{1}{2}\right), \quad (\pi, 0), \quad \left(\frac{3\pi}{2}, \frac{1}{2}\right), \quad (2\pi, 0).$$

2. Sketch a graph of $y = -2 \cos x$. What is the amplitude?

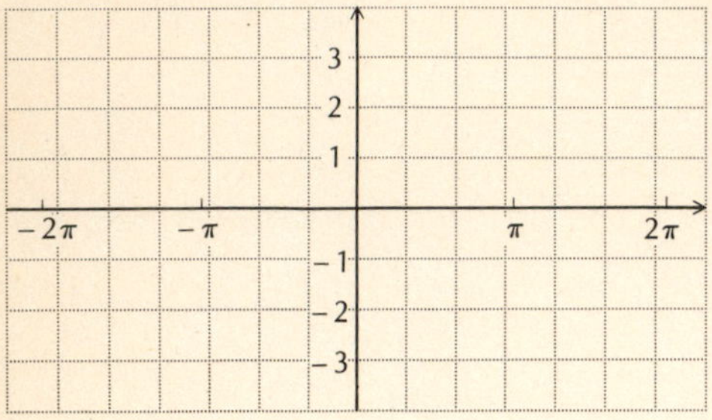

3. Sketch a graph of $y = \frac{2}{3} \cos x$. What is the amplitude?

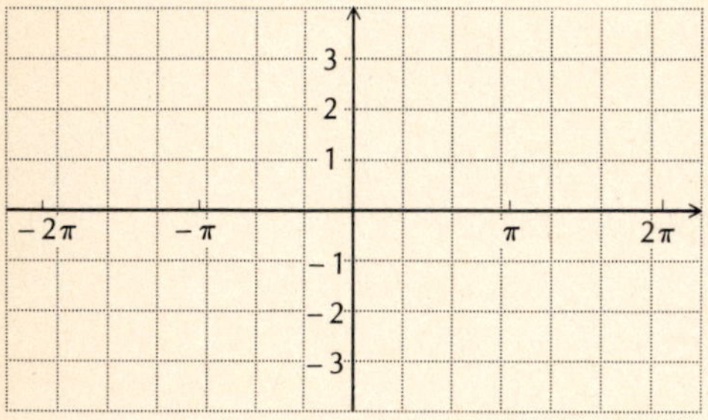

4. Sketch a graph of $y = \cos \frac{1}{2}x$. What is the period?

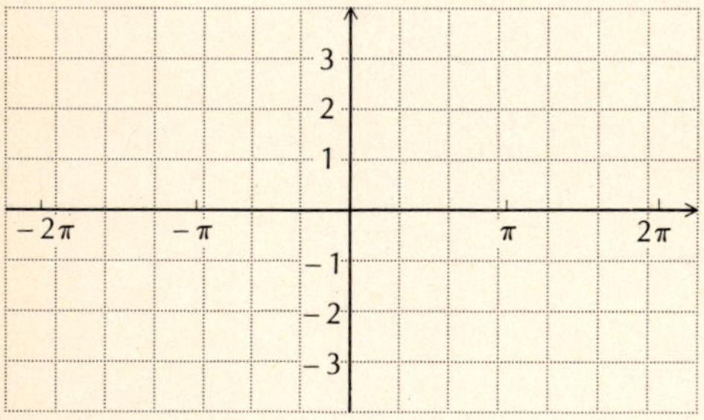

5. Sketch a graph of $y = \sin(-2x)$. What is the period?

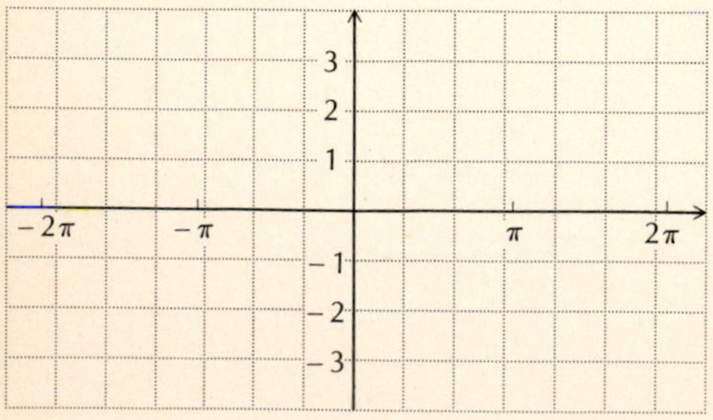

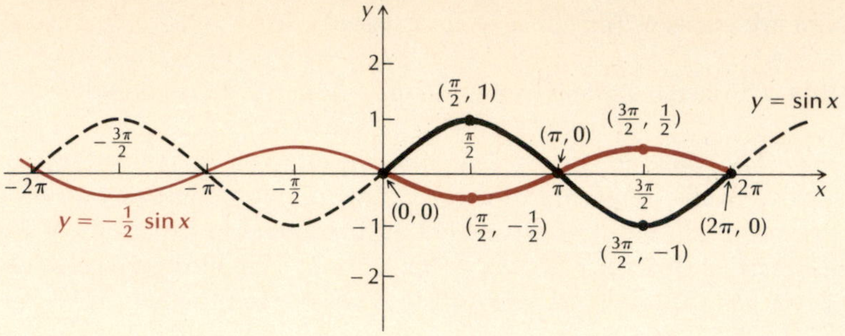

DO EXERCISES 2 AND 3.

Now we consider the effect of the constant C in $y = B + A \sin(Cx - D)$.

EXAMPLE 4 Sketch a graph of $y = \sin 2x$. What is the period?

Solution Each of the graphs in Examples 1–3, as well as the graph of $y = \sin x$, has period 2π. The constant C has the effect of changing the period. Recall from Section 3.8 that the graph of $y = f(2x)$ is obtained from the graph of $y = f(x)$ by shrinking the graph horizontally by a factor of 2. The new graph is obtained by dividing the first coordinate of each ordered-pair solution of $y = f(x)$ by 2. The key points of $y = \sin x$ are

$$(0, 0), \quad \left(\frac{\pi}{2}, 1\right), \quad (\pi, 0), \quad \left(\frac{3\pi}{2}, -1\right), \quad (2\pi, 0).$$

These are transformed to the key points of $y = \sin 2x$, which are

$$(0, 0), \quad \left(\frac{\pi}{4}, 1\right), \quad \left(\frac{\pi}{2}, 0\right), \quad \left(\frac{3\pi}{4}, -1\right), \quad (\pi, 0).$$

We plot these key points and sketch in the graph over the shortened interval $[0, \pi]$, which is of length π. Then we repeat the graph over other intervals of length π.

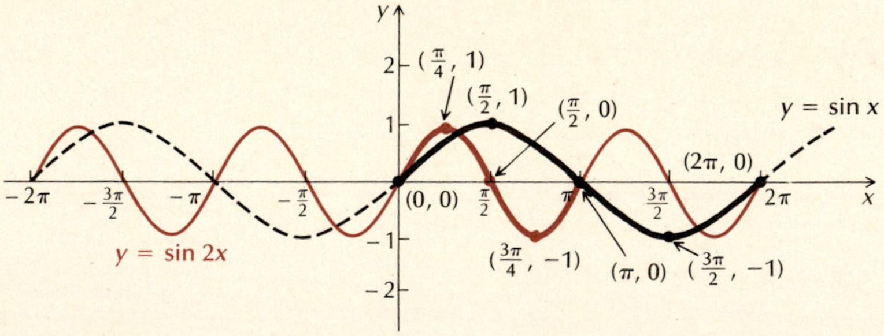

> **DEFINITION**
> The *period* of the graph of $y = B + A \sin(Cx - D)$ is $\left|\frac{2\pi}{C}\right|$.

DO EXERCISES 4 AND 5.

Now we examine the effect of the constant D in $y = B + A \sin(Cx - D)$.

EXAMPLE 5 Sketch a graph of $y = \sin\left(x - \frac{\pi}{2}\right)$.

Solution The graph of $y = \sin(x - \pi/2)$ is a translation of the graph of $y = \sin x$ to the right $\pi/2$ units. The key points of $y = \sin x$,

$$(0, 0), \quad \left(\frac{\pi}{2}, 1\right), \quad (\pi, 0), \quad \left(\frac{3\pi}{2}, -1\right), \quad (2\pi, 0),$$

are transformed by adding $\pi/2$ to each of the first coordinates to obtain the following key points of $y = \sin(x - \pi/2)$:

$$\left(\frac{\pi}{2}, 0\right), \quad (\pi, 1), \quad \left(\frac{3\pi}{2}, 0\right), \quad (2\pi, -1), \quad \left(\frac{5\pi}{2}, 0\right).$$

We plot these key points and sketch the curve over the interval $[\pi/2, 5\pi/2]$. Then we repeat the graph over other intervals of length 2π.

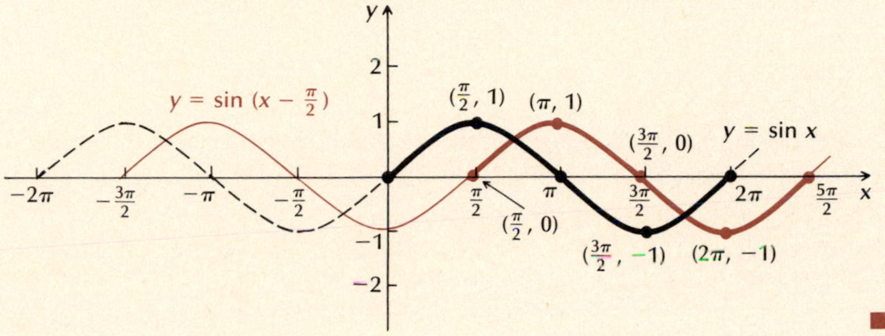

DO EXERCISE 6.

Now we consider combined transformations of graphs.

EXAMPLE 6 Sketch a graph of $y = \sin(2x - \pi)$. What is the period?

Solution The graph of $y = \sin(2x - \pi)$ is the same as the graph of $y = \sin[2(x - \pi/2)]$. The $\pi/2$ translates the graph of $y = \sin 2x$ a distance of $\pi/2$ to the right. (See Example 5.) The 2 factored out of the parentheses shrinks the period by half, making the period $|2\pi/2|$, or π.

Thus to form this graph, we first graph $y = \sin 2x$, as in Example 4. The key points of $y = \sin 2x$ are

$$(0, 0), \quad \left(\frac{\pi}{4}, 1\right), \quad \left(\frac{\pi}{2}, 0\right), \quad \left(\frac{3\pi}{4}, -1\right), \quad (\pi, 0).$$

We use them to obtain the key points of $y = \sin[2(x - \pi/2)]$, which are obtained, in manner similar to Example 5, by adding $\pi/2$ to each first coordinate:

$$\left(\frac{\pi}{2}, 0\right), \quad \left(\frac{3\pi}{4}, 1\right), \quad (\pi, 0), \quad \left(\frac{5\pi}{4}, -1\right), \quad \left(\frac{3\pi}{2}, 0\right).$$

We draw one period of the graph over the interval $[\pi/2, 3\pi/2]$, which has length π. Then we repeat the graph over other intervals of length π.

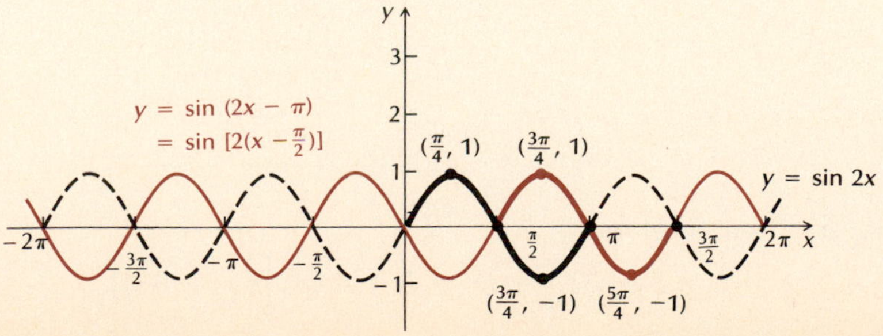

6. Sketch a graph of $y = \sin(x - \pi/2)$.

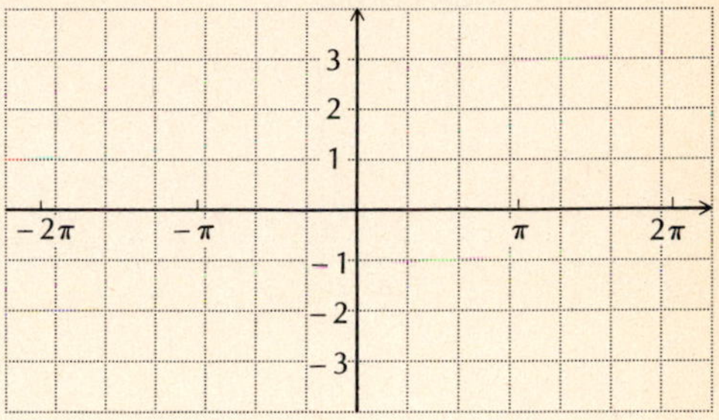

7. Sketch a graph of $y = \cos(2x + \pi)$. What is the period?

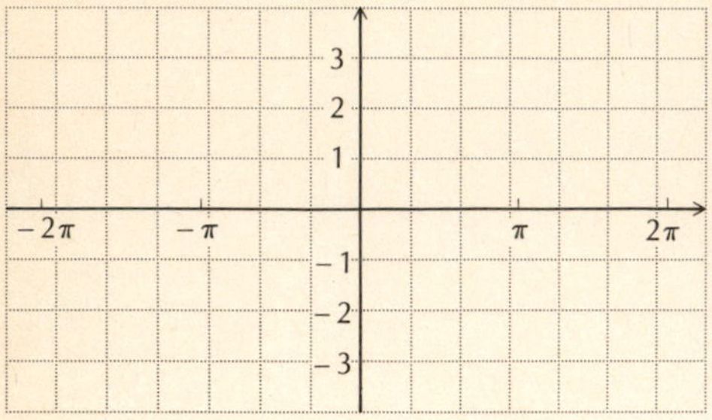

DEFINITION

For the graph of

$$y = B + A \sin(Cx - D) = B + A \sin\left[C\left(x - \frac{D}{C}\right)\right],$$

the quantity D/C translates the graph to the right if D/C is positive and to the left if D/C is negative, and is called the *phase shift*. Be sure to do the horizontal stretching or shrinking based on the constant C *before* the translation based on the phase shift D/C.

In Example 6, the phase shift is $\pi/2$. This means that the graph of $y = \sin 2x$ has been shifted to the right a distance of $\pi/2$ units.

DO EXERCISE 7.

Let us summarize the effects of the constants. We carry out the procedures in the order listed.

To graph

(1) Stretches the graph horizontally by a factor of $|C|$ if $|C| < 1$. Shrinks the graph horizontally by a factor of $|C|$ if $|C| > 1$. If $C < 0$, the graph is also reflected across the y-axis. The period is $|2\pi/C|$.

(2) Stretches the graph vertically by a factor of $|A|$ if $|A| > 1$. Shrinks the graph vertically by a factor of $|A|$ if $|A| < 1$. If $A < 0$, the graph is also reflected across the x-axis. The *amplitude* of the graph is $|A|$.

$$y = B + A \sin(Cx - D) = B + A \sin\left[C\left(x - \frac{D}{C}\right)\right].$$

(3) Translates the graph to the right $|D/C|$ units if $D/C > 0$ or to the left $|D/C|$ units if D/C is negative. The *phase shift* is D/C.

(4) Translates the graph $|B|$ units up if $B > 0$ or $|B|$ units down if $B < 0$.

Similar statements hold for

$$y = B + A \cos(Cx - D) = B + A \cos\left[C\left(x - \frac{D}{C}\right)\right].$$

EXAMPLE 7 Sketch a graph of $y = 3 \sin (2x + \pi/2)$. Find the amplitude, the period, and the phase shift.

Solution We first note that

$$y = 3 \sin\left(2x + \frac{\pi}{2}\right) = 3 \sin\left[2\left(x - \left(-\frac{\pi}{4}\right)\right)\right].$$

Then

$$\text{Amplitude} = |A| = |3| = 3,$$

$$\text{Period} = \left|\frac{2\pi}{C}\right| = \left|\frac{2\pi}{2}\right| = \pi,$$

$$\text{Phase shift} = \frac{D}{C} = -\frac{\pi}{4}.$$

To create the final graph, we sketch graphs of each of the following equations in sequence:

1. $y = \sin 2x$,
2. $y = 3 \sin 2x$
3. $y = 3 \sin\left[2\left(x - \left(-\frac{\pi}{4}\right)\right)\right]$.

Start with $y = \sin x$.

1. Then graph $y = \sin 2x$. (See Example 4.)

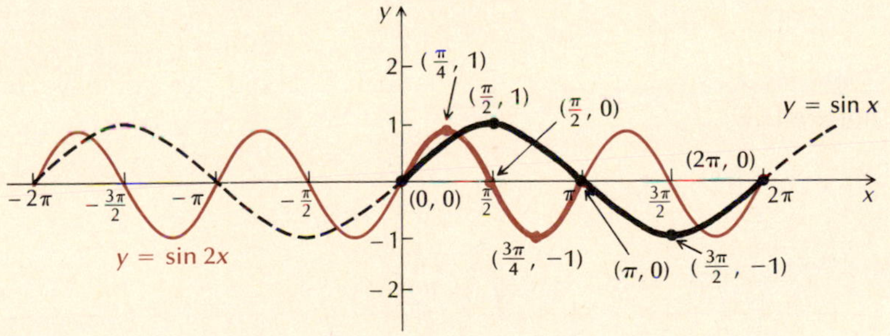

2. Then graph $y = 3 \sin 2x$ by stretching the graph vertically by a factor of 3.
3. Then graph $y = 3 \sin [2(x - (-\pi/4))]$ by translating the graph of $y = 3 \sin 2x$ to the left $\pi/4$ units.

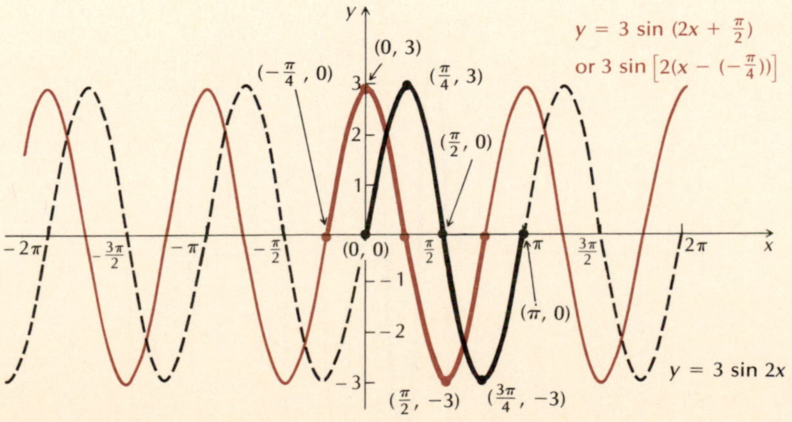

DO EXERCISE 8.

8. Sketch a graph of

$$y = 3 \cos (2x - \pi/2).$$

What is the amplitude? the period? the phase shift?

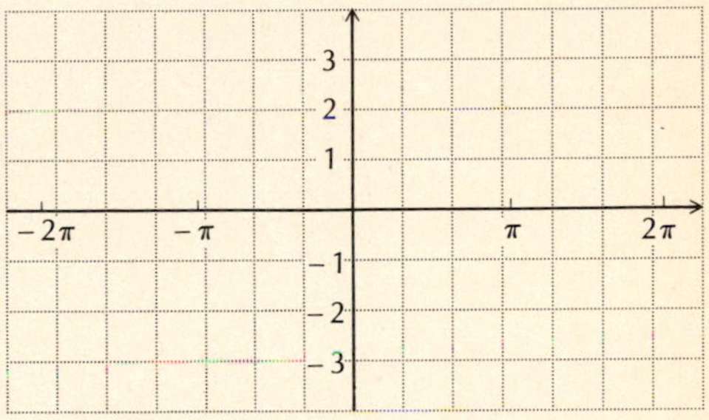

EXAMPLE 8 Sketch a graph of $y = -1 + 3\cos(2\pi x)$. Find the amplitude, the period, and the phase shift.

Solution First we note the following:

$$\text{Amplitude} = |A| = |3| = 3,$$

$$\text{Period} = \left|\frac{2\pi}{C}\right| = \left|\frac{2\pi}{2\pi}\right| = |1| = 1,$$

$$\text{Phase shift} = \frac{D}{C} = \frac{0}{2\pi} = 0.$$

There is no phase shift in this problem since the constant $D = 0$. Thus there is no horizontal translation.

To create the final graph, we graph each of the following equations in sequence:

1. $y = \cos(2\pi x)$,
2. $y = 3\cos(2\pi x)$,
3. $y = -1 + 3\cos(2\pi x)$.

Start with $y = \cos x$.

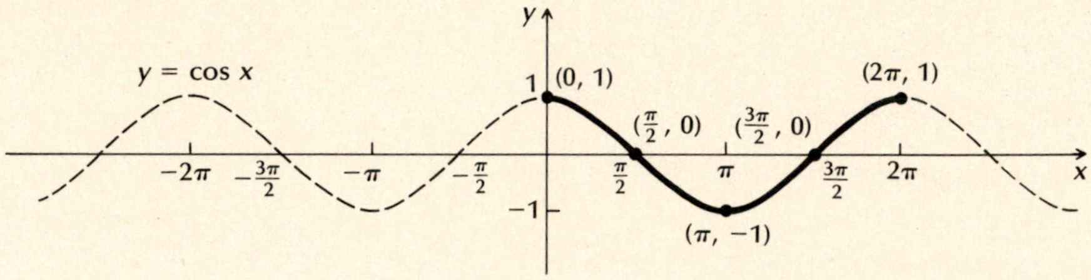

1. Then graph $y = \cos(2\pi x)$. The period is 1 and the graph is shrunk horizontally.

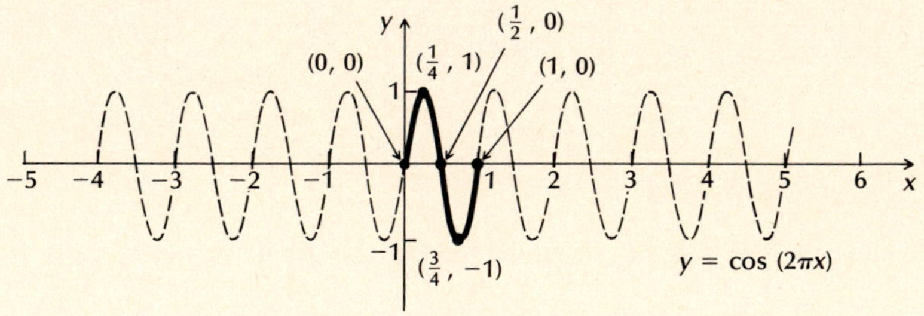

2. Then graph $y = 3\cos(2\pi x)$. The amplitude is 3. The graph is stretched vertically by a factor of 3.

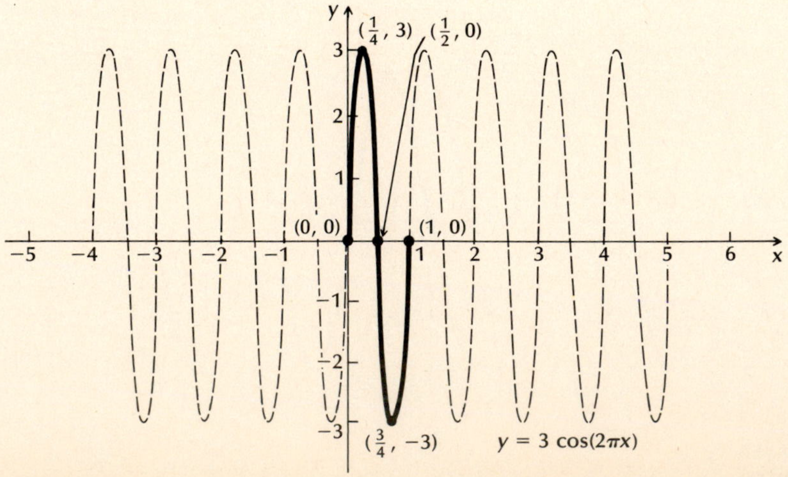

3. Then graph $y = -1 + 3\cos(2\pi x)$. The graph is translated down 1 unit.

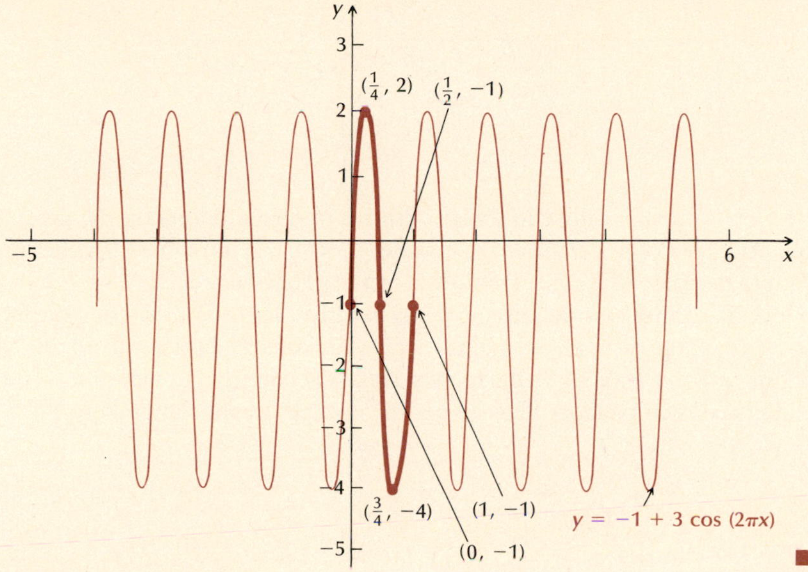

9. Sketch a graph of $y = 2 - 2\sin(\pi x)$. What is the amplitude? the period? the phase shift?

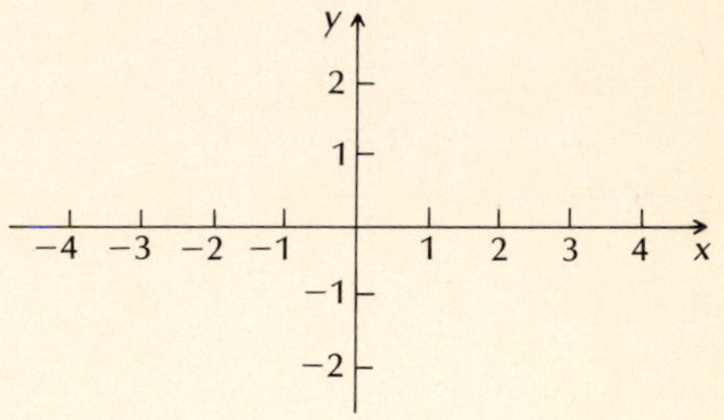

DO EXERCISE 9.

The oscilloscope shown here is an electronic device that draws graphs like we have done in the preceding examples. These graphs are often called **sinusoidal**. By manipulating the controls, we can change such things as the amplitude, the period, and the phase shift. The oscilloscope has many applications, and the trigonometric functions play a major role in many of them.

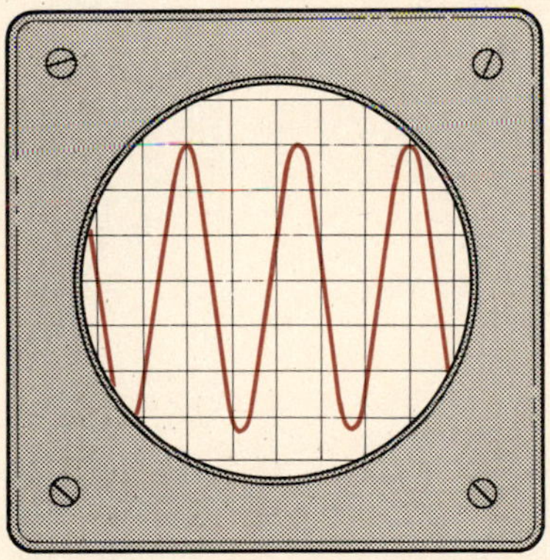

An oscilloscope display, showing a sinusoidal (sine-shaped) curve.

3 Graphs of Sums: Addition of Ordinates

A function that is a sum of two functions can often be graphed by a method called **addition of ordinates**. We graph the two functions separately and then add the second coordinates (called **ordinates**) graphically. A compass may be helpful.

EXAMPLE 9 Graph: $y = 2\sin x + \sin 2x$.

Solution We graph $y = 2\sin x$ and $y = \sin 2x$ using the same axes.

10. By addition of ordinates, sketch a graph of $y = \cos x + \sin 2x$.

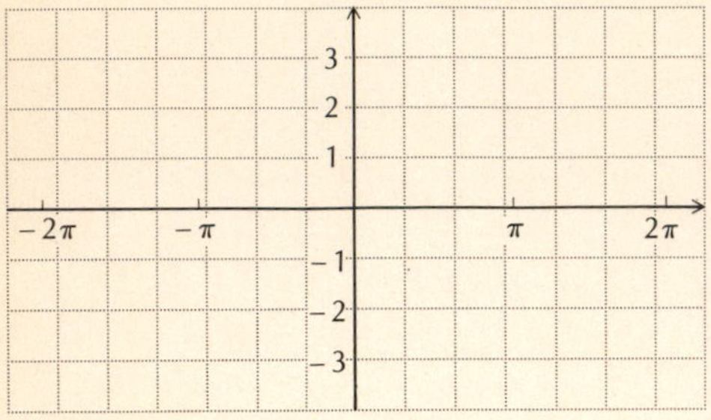

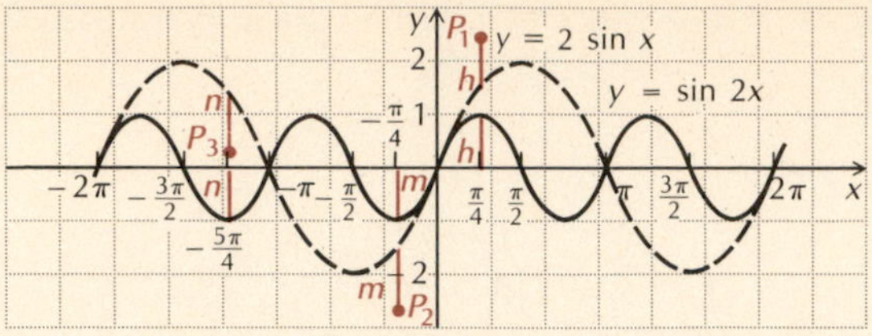

Now we graphically add some ordinates to obtain points on the graph we seek. At $x = \pi/4$, we transfer the distance h, which is the value of $\sin 2x$, up to add it to the value of $2 \sin x$. Point P_1 is on the graph we seek. At $x = -\pi/4$, we do a similar thing, but the distance m is negative. Point P_2 is on our graph. At $x = -5\pi/4$, the distance n is negative, so we in effect subtract it from the value of $2 \sin x$. Point P_3 is on the graph we seek. We continue to plot points in this fashion and then connect them to get the desired graph, shown here.

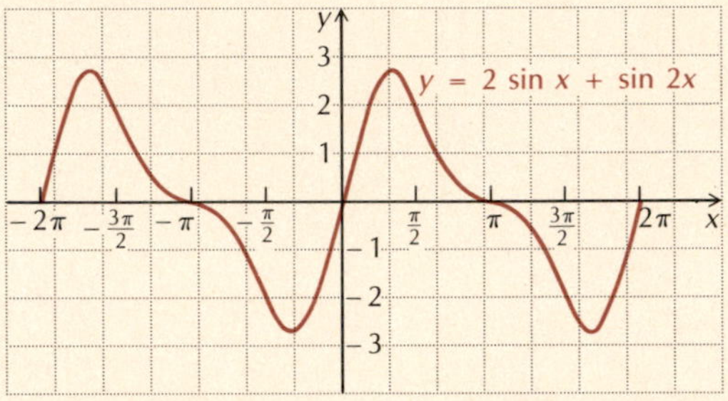

DO EXERCISE 10.

A "sawtooth" function, such as the one shown on this oscilloscope, has numerous applications—for example, in television circuits. This sawtooth function can be approximated extremely well by adding, electronically, several sine and cosine functions. The graph in Example 9 could be a first step in such an approximation.

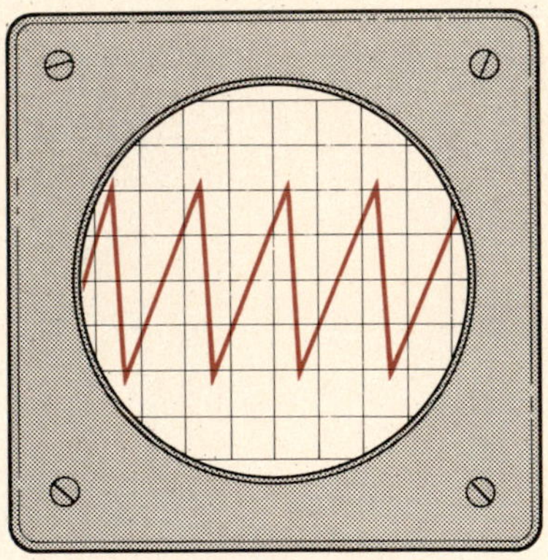

An oscilloscope displaying a sawtooth function. The function has been synthesized using sines and cosines.

EXERCISE SET 6.7

1, **2** Use graph paper to sketch the graph of the function. Determine the amplitude, the period, and the phase shift.

1. $y = 2 + \sin x$
2. $y = -3 + \cos x$
3. $y = \frac{1}{2} \sin x$
4. $y = \frac{1}{3} \cos x$
5. $y = 2 \cos x$
6. $y = 3 \cos x$
7. $y = -\frac{1}{2} \cos x$
8. $y = -2 \sin x$
9. $y = \cos 2x$
10. $y = \cos 3x$
11. $y = \cos (-2x)$
12. $y = \cos (-3x)$
13. $y = \sin \frac{1}{2}x$
14. $y = \cos \frac{1}{3}x$
15. $y = \sin \left(-\frac{1}{2}x\right)$
16. $y = \cos \left(-\frac{1}{3}x\right)$

17. $y = \cos(2x - \pi)$
18. $y = \sin(2x + \pi)$
19. $y = 2\cos\left(\frac{1}{2}x - \frac{\pi}{2}\right)$
20. $y = 4\sin\left(\frac{1}{4}x + \frac{\pi}{8}\right)$
21. $y = -3\cos(4x - \pi)$
22. $y = -3\sin\left(2x + \frac{\pi}{2}\right)$
23. $y = 2 + 3\cos(\pi x - 3)$
24. $y = 5 - 2\cos\left(\frac{\pi}{2}x + \frac{\pi}{2}\right)$

2 Determine the amplitude, the period, and the phase shift.

25. $y = 3\cos\left(3x - \frac{\pi}{2}\right)$
26. $y = 4\sin\left(4x - \frac{\pi}{3}\right)$
27. $y = -5\cos\left(4x + \frac{\pi}{3}\right)$
28. $y = -4\sin\left(5x + \frac{\pi}{2}\right)$
29. $y = \frac{1}{2}\sin(2\pi x + \pi)$
30. $y = -\frac{1}{4}\cos(\pi x - 4)$

3 Graph.

31. $y = 2\cos x + \cos 2x$
32. $y = 3\cos x + \cos 3x$
33. $y = \sin x + \cos 2x$
34. $y = 2\sin x + \cos 2x$
35. $y = \sin x - \cos x$
36. $y = 3\cos x - \sin x$
37. $y = 3\cos x + \sin 2x$
38. $y = 3\sin x - \cos 2x$

SYNTHESIS

39. *Temperature during an illness.* The temperature T of a patient during a 12-day illness is given by

$$T(t) = 101.6° + 3\sin\left(\frac{\pi}{8}t\right).$$

a) Sketch a graph of the function over the interval $[0, 12]$.

b) What are the maximum and the minimum temperatures during the illness?

40. *Periodic sales.* A company in a northern climate has sales of skis as given by

$$S(t) = 10\left(1 - \cos\frac{\pi}{6}t\right),$$

where t = time, in months ($t = 0$ corresponds to July 1), and $S(t)$ is in thousands of dollars.

a) Sketch a graph of the function over a 12-month interval $[0, 12]$.

b) What is the period of the function?

c) What is the minimum amount of sales and when does it occur?

d) What is the maximum amount of sales and when does it occur?

41. *Satellite location.* A satellite circles the earth in such a way that it is y miles from the equator (north or south, height not considered) t minutes after its launch, where

$$y = 3000\left[\cos\frac{\pi}{45}(t - 10)\right].$$

a) Sketch a graph of the function.

b) What are the amplitude, the period, and the phase shift?

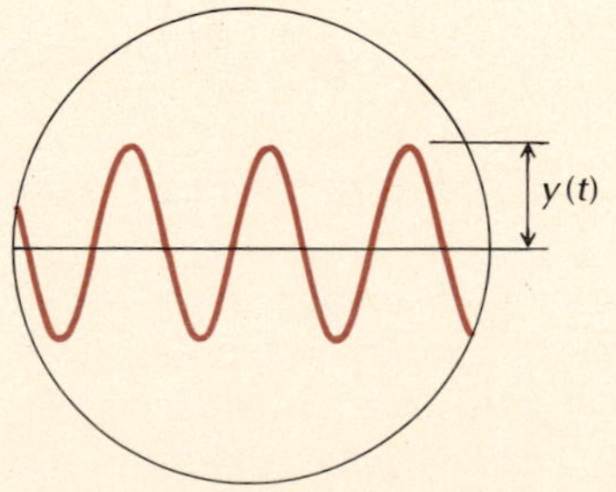

42. *Water wave.* The cross-section of a water wave is given by

$$y = 3\sin\left(\frac{\pi}{4}x + \frac{\pi}{4}\right),$$

where y is the vertical height of the water wave and x is the distance from the origin to the wave.

a) Sketch a graph of the function.

b) What are the amplitude, the period, and the phase shift?

43. Find a sine function that has amplitude 8.6, period 5, and phase shift -11.

44. Find a cosine function that has amplitude 6.7, period π, and phase shift $\frac{4}{7}$.

45. Find a cosine function that has amplitude 16, period $\pi/3$, and phase shift $-2/\pi$.

46. Find a sine function that has amplitude 34, period $\pi/5$, and phase shift $-8/\pi$.

Graph.

47. $y = |4 \sin x|$

48. $y = |\cos 3x|$

CHALLENGE

Graph using addition of ordinates. You might want to use a calculator to find some function values to plot.

49. $y = x + \sin x$

50. $y = \cos x - x$

OBJECTIVES

You should be able to:

1 Sketch graphs that are transformations of the tangent, cotangent, secant, and cosecant functions.

2 Sketch graphs of functions found by multiplying trigonometric functions by other functions.

6.8 Other Graphs of Trigonometric Functions

1 Transformations of Tangent, Cotangent, Secant, and Cosecant

The transformational techniques that we learned in Section 6.7 for graphing sine and cosine can also be applied to the other trigonometric functions.

EXAMPLE 1 Sketch a graph of $y = \frac{1}{2} \tan \left(x - \frac{\pi}{4}\right)$.

Solution The graph of $y = \tan x$ is shown below. Recall that it has period π and is defined for all real numbers except $\pi/2 + 2k\pi$.

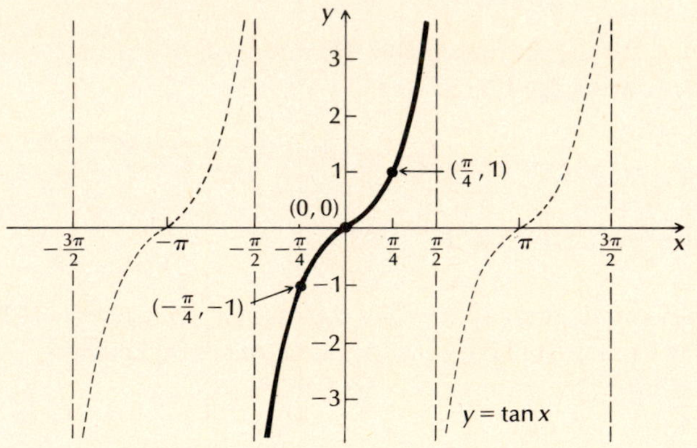

Let us consider the following ordered pairs as key points in the interval $(-\pi/2, \pi/2)$. Note that the function is not defined at the endpoints, so we indicate the following as key points,

$$\left(-\frac{\pi}{2}, \text{undefined}\right), \quad \left(-\frac{\pi}{4}, -1\right), \quad (0, 0), \quad \left(\frac{\pi}{4}, 1\right), \quad \left(\frac{\pi}{2}, \text{undefined}\right),$$

keeping in mind that "undefined" means that the first coordinate is not in the domain.

There is no amplitude since the function has no maximum or minimum value. Nevertheless, the constant $\frac{1}{2}$ shrinks each function value of $y = \tan x$ by a factor of $\frac{1}{2}$. Thus the graph of $y = \frac{1}{2} \tan x$ is obtained from the graph of $y = \tan x$ by a vertical shrinking. The key points are transformed by multiplying each second coordinate by $\frac{1}{2}$:

$$\left(-\frac{\pi}{2}, \text{undefined}\right), \quad \left(-\frac{\pi}{4}, -\frac{1}{2}\right), \quad (0, 0), \quad \left(\frac{\pi}{4}, \frac{1}{2}\right), \quad \left(\frac{\pi}{2}, \text{undefined}\right).$$

We plot these key points and sketch the curve over the interval $(-\pi/2, \pi/2)$.

We can use a calculator to find other function values, if desired. Then we repeat the graph over other intervals of length π.

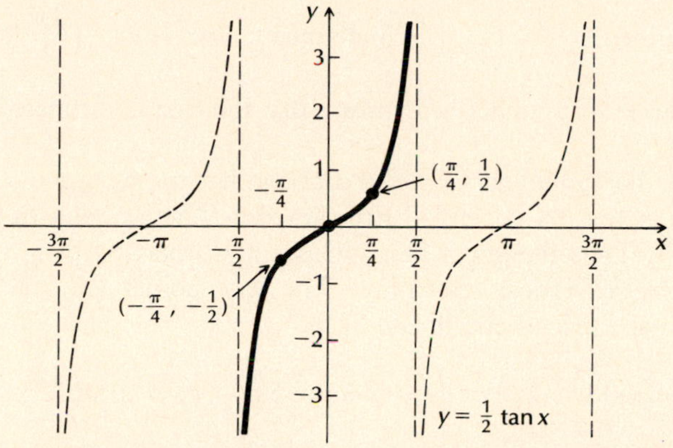

Next, we obtain the graph of $y = \frac{1}{2}\tan(x - \pi/4)$ by translating the graph of $y = \frac{1}{2}\tan x$ to the right $\pi/4$ units.

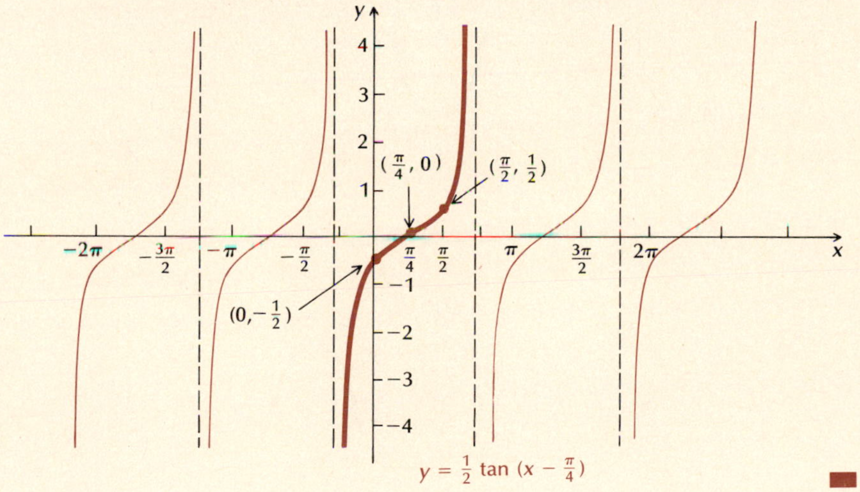

DO EXERCISE 1.

EXAMPLE 2 Sketch the graph of $y = 3\sec\left(x + \dfrac{\pi}{2}\right)$.

Solution The graph of $y = \sec x$ is shown below. Recall that it has period 2π and is defined for all real numbers except $\pi/2 + 2k\pi$.

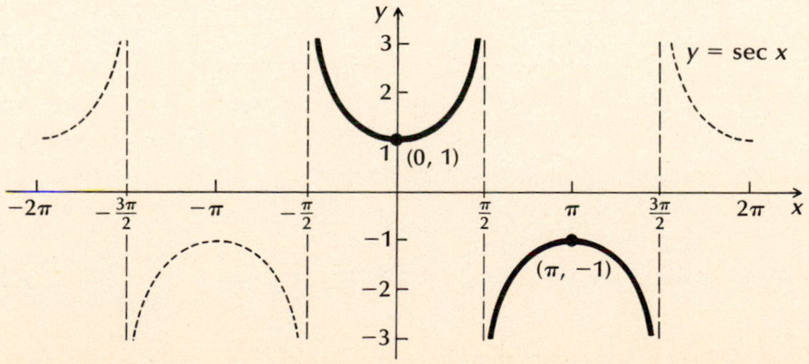

1. Sketch a graph of

$$y = 2\cot\left(x + \frac{\pi}{4}\right).$$

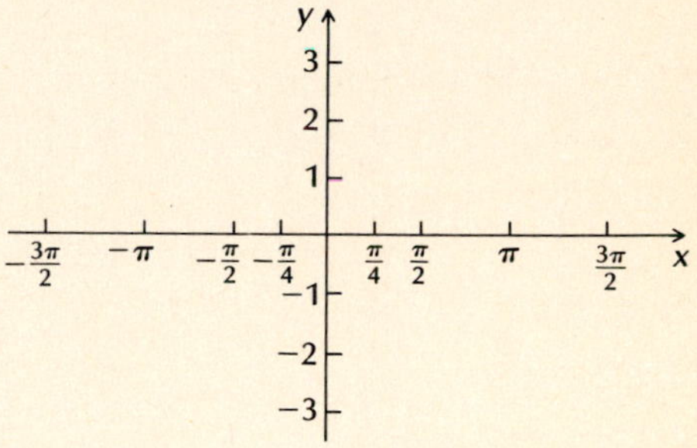

2. Sketch a graph of

$$y = -\frac{2}{3}\csc\left(x - \frac{\pi}{2}\right).$$

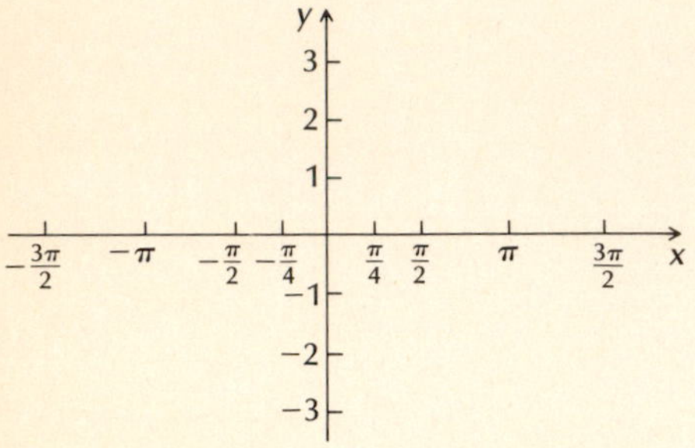

Let us consider the following ordered pairs as key points in the interval $(-\pi/2, 3\pi/2)$. Note that the function is not defined at the endpoints or at $\pi/2$, so we indicate the following as key points,

$$\left(-\frac{\pi}{2}, \text{undefined}\right), \quad (0, 1), \quad \left(\frac{\pi}{2}, \text{undefined}\right), \quad (\pi, -1), \quad \left(\frac{3\pi}{2}, \text{undefined}\right),$$

keeping in mind that "undefined" means that the first coordinate is not in the domain.

There is no amplitude since the function has no maximum or minimum value. Nevertheless, the constant 3 stretches each function value of $y = \sec x$ by a factor of 3. Thus the graph of $y = 3 \sec x$ is obtained from the graph of $y = \sec x$ by a vertical stretching. The key points are transformed by multiplying each second coordinate by 3:

$$\left(-\frac{\pi}{2}, \text{undefined}\right), \quad (0, 3), \quad \left(\frac{\pi}{2}, \text{undefined}\right), \quad (\pi, -3), \quad \left(\frac{3\pi}{2}, \text{undefined}\right).$$

We plot these key points and sketch the curve over the interval $(-\pi/2, 3\pi/2)$. We can use a calculator to find other function values, if desired. Then we repeat the graph over other intervals of length 2π.

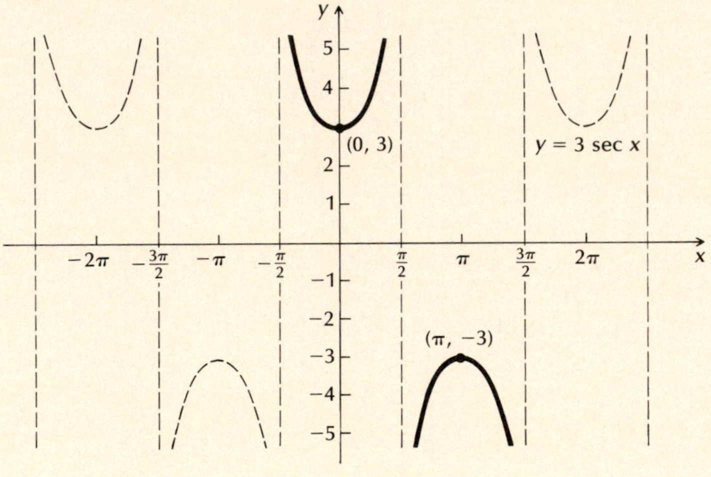

Next, we obtain the graph of $y = 3 \sec [x - (-\pi/2)]$ by translating the graph of $y = 3 \sec x$ to the left $\pi/2$ units.

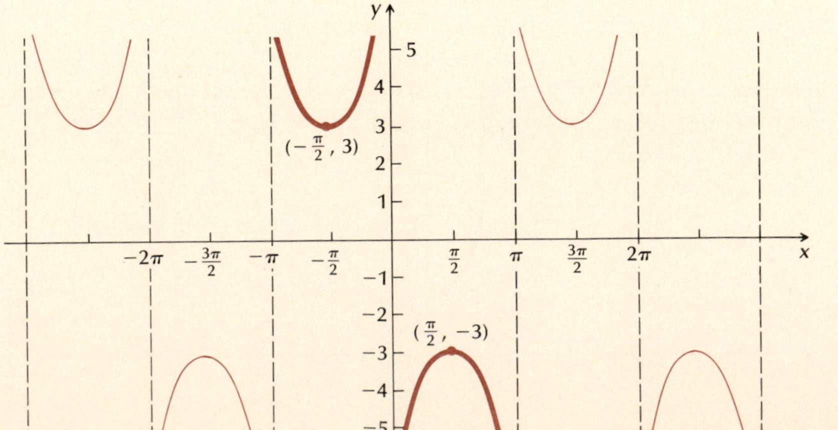

DO EXERCISE 2.

2 Multiplication of Ordinates

Suppose that a weight is attached to a spring and the spring is stretched and put into motion. The weight oscillates up and down. If we could assume falsely that the weight will bob up and down forever, then its height h after time t in seconds might be approximated by a function like

$$h(t) = 5 + 2 \sin (6\pi t).$$

Over a short time period, this might be a valid model, but experience tells us that eventually the spring will come to rest. A more appropriate model is provided by the following example, which illustrates the **theory of damped oscillations.**

EXAMPLE 3 Sketch a graph of $f(x) = e^{-x/2} \sin x$.

Solution The function f is the product of two functions g and h, where

$$g(x) = e^{-x/2} \quad \text{and} \quad h(x) = \sin x.$$

Thus, to find function values, we can **multiply ordinates.** Let us do some more analysis before graphing. Note that for any real number x,

$$-1 \le \sin x \le 1.$$

Recall from Chapter 5 that all values of the exponential function are positive. Thus we can multiply by $e^{-x/2}$ and obtain the inequality

$$-e^{-x/2} \le e^{-x/2} \sin x \le e^{-x/2}.$$

The direction of the inequality symbols does not change since $e^{-x/2} > 0$. This also tells us that the original function crosses the x-axis only at values for which $\sin x = 0$. These are the numbers $k\pi$, for any integer k.

The inequality tells us that the function f is constrained between the graphs of $y = -e^{-x/2}$ and $y = e^{-x/2}$. We start by graphing these functions using dashed lines. We also know that $f(x) = 0$ when $x = k\pi$, k an integer. We mark these points on the graph. Then we use a calculator and compute other function values. The graph is as follows:

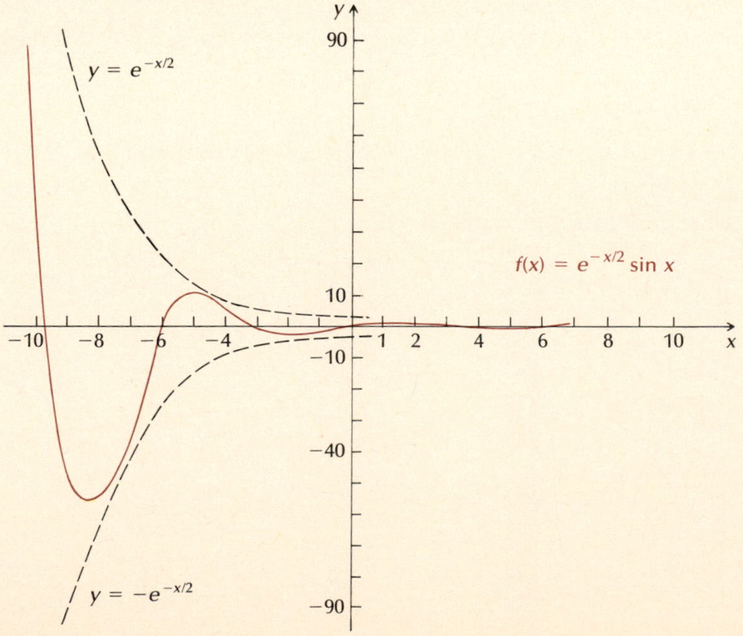

DO EXERCISE 3.

3. Sketch a graph of

$$f(x) = \tfrac{1}{2}x^2 \sin x.$$

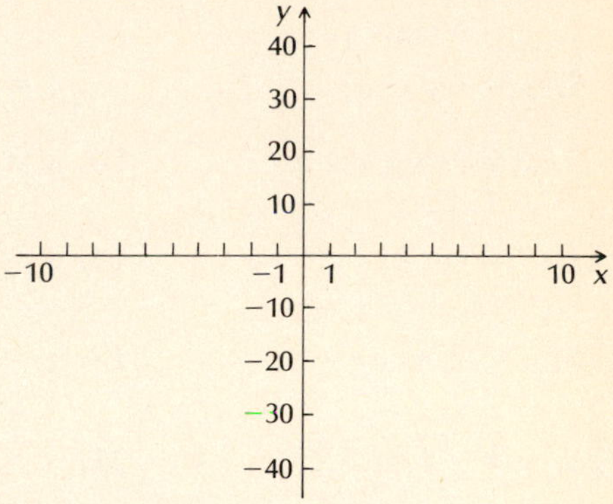

EXERCISE SET 6.8

1 Sketch the graph.

1. $y = -\tan x$
2. $y = -\cot x$
3. $y = -\csc x$
4. $y = -\sec x$
5. $y = \sec(-x)$
6. $y = \csc(-x)$
7. $y = \cot(-x)$
8. $y = \tan(-x)$
9. $y = -2 + \cot x$
10. $y = -3 + \sec x$
11. $y = -\frac{3}{2}\csc x$
12. $y = 0.25 \cot x + 3$
13. $y = \cot 2x$
14. $y = 2 \tan \frac{1}{2}x$
15. $y = \sec\left(-\frac{1}{2}x\right)$
16. $y = \csc\left(-\frac{1}{3}x\right)$
17. $y = 2 \sec(x - \pi)$
18. $y = 3 \csc(x + \pi)$
19. $y = 4 \tan\left(\frac{1}{4}x + \frac{\pi}{8}\right)$
20. $y = 2 \sec\left(\frac{1}{2}x - \frac{\pi}{2}\right)$
21. $y = -3 \cot\left(2x + \frac{\pi}{2}\right)$
22. $y = -3 \tan(4x - \pi)$
23. $y = 4 \sec(2x - \pi)$
24. $y = 2 \csc\left(\frac{1}{2}x - \frac{3\pi}{4}\right)$

2 Sketch the graph of each of the following functions.

25. $f(x) = e^{-x/2} \cos x$
26. $f(x) = e^{-0.4x} \sin x$
27. $f(x) = 0.6x^2 \cos x$
28. $f(x) = e^{-x/4} \sin x$
29. $f(x) = x \sin x$
30. $f(x) = |x| \cos x$
31. $f(x) = 2^{-x} \sin x$
32. $f(x) = 2^{-x} \cos x$

SYNTHESIS

Sketch the graph of each of the following functions.

33. $f(x) = |\tan x|$
34. $f(x) = |\csc x|$
35. $f(x) = (\ln x)(\sin x)$
36. $f(x) = \sin^2 x$
37. $f(x) = \sec^2 x$
38. $f(x) = (\tan x)(\csc x)$
39. *Damped oscillations.* Suppose the motion of a spring is given by

$$d(t) = 6e^{-0.8t} \cos(6\pi t) + 4,$$

where d is the distance, in inches, of a weight from the point at which the spring is attached to a ceiling, after t seconds.

a) Sketch the graph over the interval [0, 10].
b) How far do you think the spring is from the ceiling when the spring stops bobbing?

40. *Temperature during an illness.* A patient's temperature during an illness is given by

$$T(t) = 98.6° + 2e^{-0.6t} \sin t,$$

where T is the temperature after time t, in hours.

a) Sketch the graph over the interval [0, 12].
b) What is the patient's normal temperature after the illness?

41. *Rotating beacon.* A police car is parked 10 ft from a wall. On top of the car is a beacon rotating in such a way that the light is at a distance $d(t)$ from point Q after t seconds, where

$$d(t) = 10 \tan(2\pi t).$$

When d is positive, the light is pointing north of Q, and when d is negative, the light is pointing south of Q.

a) Sketch a graph of the function over the interval [0, 2].
b) Explain the meaning of the values of t for which the function is undefined.

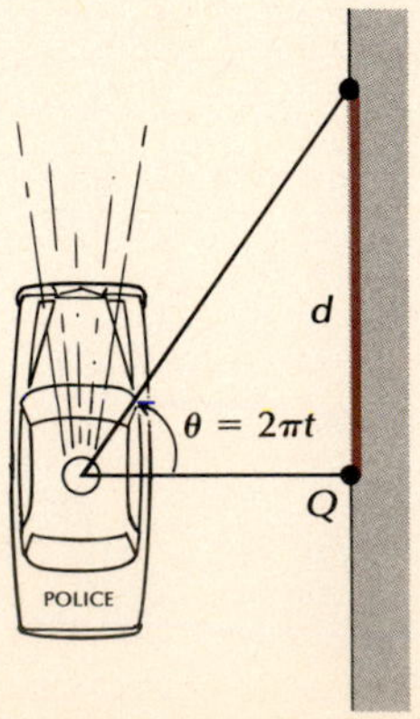

COMPUTER–CALCULATOR EXERCISES

Use a computer software package or graphing calculator to graph each of the following over the given interval and approximate the zeros.

42. $f(x) = \dfrac{\sin x}{x}$; $[-12, 12]$

43. $f(x) = \dfrac{\sin^2 x}{x}$; $[-4, 4]$

44. $f(x) = \ln|\sec x + \tan x|$; $[-12, 12]$

45. $f(x) = \dfrac{\cos x - 1}{x}$; $[-12, 12]$

SUMMARY AND REVIEW: CHAPTER 6

TERMS TO KNOW

Unit circle
Sine function
Cosine function
Identity
Period
Amplitude
Phase shift
Tangent function
Cotangent function
Secant function
Cosecant function
Basic identities
Pythagorean identities
Angle
Vertex
Rotation
Initial side
Terminal side
Standard position
Coterminal angles
Degrees
Minutes
Seconds
Right angle
Acute angle
Obtuse angle
Straight angle
Complementary angles
Supplementary angles
Radians
Arc length
Central angle
Linear speed
Angular speed
Reference angle
Reference triangle
Addition of ordinates
Multiplication of ordinates

REVIEW EXERCISES

1. The point $\left(\frac{3}{5}, -\frac{4}{5}\right)$ is on a unit circle. Find the coordinates of its reflections across the x-axis, the y-axis, and the origin.

2. On a unit circle, mark and label the points determined by $7\pi/6$, $-3\pi/4$, $-\pi/3$, and $9\pi/4$.

3. In which quadrants are the signs of the sine and the tangent the same?

4. Complete the following table. Give exact values. Do not use a calculator.

	$\pi/6$	$\pi/4$	$\pi/2$	$2\pi/3$	$-\pi$	$-5\pi/4$
$\sin x$						
$\cos x$						
$\cot x$						
$\csc x$						

For angles of the following measures, state in which quadrant the terminal side lies, convert to radian measure in terms of π, and convert to radian measure not in terms of π. Use 3.14 for π.

5. $87°$

6. $145°$

7. $-30°$

Convert to degree measure.

8. $\dfrac{3\pi}{2}$

9. 3

10. Find two positive angles and two negative angles coterminal with a 65° angle.

11. Find the length of an arc of a circle, given a central angle of $\pi/4$ and a radius of 7 cm.

12. An arc 18 m long on a circle of radius 8 m subtends an angle of how many radians? how many degrees, to the nearest degree?

13. A phonograph record revolves at 45 rpm. What is the linear velocity, in cm/min, of a point 4 cm from the center?

14. An automobile wheel has a diameter of 26 in. If the car travels at a speed of 30 mph, what is the angular velocity, in radians/hr, of a point on the edge of the wheel?

15. Find the six trigonometric function values for the angle θ as shown.

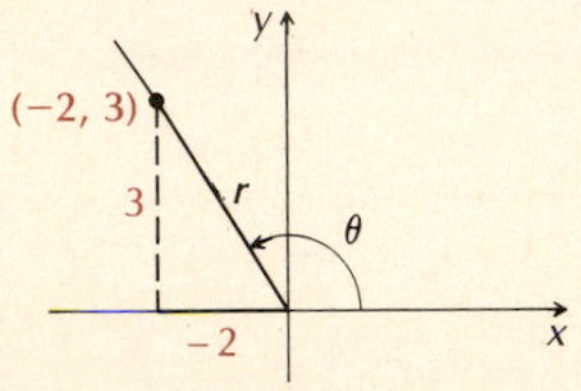

16. Find the six trigonometric ratios for the specified angle.

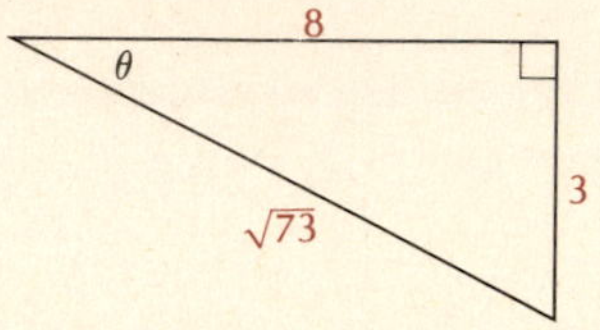

Find the following if they exist. Do not use a calculator.

17. $\sin 495°$
18. $\tan (-315°)$
19. $\cot 210°$
20. $\cos 150°$
21. $\sec (-270°)$
22. $\csc 600°$

23. Given that $\tan \theta = 2/\sqrt{5}$ and the terminal side is in quadrant III, find the other five function values.

24. Convert 22.20° to degrees and minutes.

25. Convert 47°33′27″ to degrees and decimal parts of degrees.

Use a calculator to find each of the following.

26. $\sin (-190°40')$
27. $\cos 0.1134$
28. $\sec 83°18'24''$

Use a calculator to find θ in degrees and minutes in the interval indicated.

29. $\tan \theta = 0.1228$, $(180°, 270°)$
30. $\cos \theta = -0.2672$, $(90°, 180°)$

Use a calculator to find the acute angle θ in radians that is a solution of the equation.

31. $\sin \theta = 0.2756$
32. $\tan \theta = 6.197$

Use a calculator to find each of the following function values.

33. $\sin 24$
34. $\csc\left(-\frac{\pi}{5}\right)$
35. $\tan 5.2256$
36. $\cot 16\pi$
37. $\sec 14.3$
38. $\cos\left(-\frac{3\pi}{7}\right)$

39. Sketch a graph of $y = \sin x$.

40. Sketch a graph of $y = \cot x$.

41. Sketch a graph of $y = \csc x$.

42. What are the period and the range of the sine function?

43. What are the amplitude and the domain of the sine function?

44. What are the period and the range of the cosecant function?

45. Sketch a graph of $y = \sin\left(x + \frac{\pi}{2}\right)$.

46. Sketch a graph of $y = 3 + \cos\left(x - \frac{\pi}{4}\right)$.

47. What is the phase shift of the function in Exercise 52?

48. What is the period of the function in Exercise 52?

49. Sketch a graph of $y = 3\cos x + \sin x$ for values of x between 0 and 2π.

Sketch the graph.

50. $y = 2\tan\left(x - \frac{\pi}{2}\right)$
51. $f(x) = e^{-0.7x}\cos x$

SYNTHESIS

52. Does $5\sin x = 7$ have a solution for x? Why or why not?

53. For what values of x in $(0, \pi/2]$ is $\sin x < x$ true?

54. Graph $y = 3\sin(x/2)$ and determine the domain, the range, and the period.

55. Find the domain of $y = \log(\cos x)$.

56. Given that $\sin x = 0.6144$ is in quadrant II, find the other basic circular function values.

TEST: CHAPTER 6

1. The point $\left(-\frac{4}{5}, -\frac{2}{5}\right)$ is on a unit circle. Find the coordinates of its reflection across the origin, the x-axis, and the y-axis.

2. On the unit circle, mark and label the points determined by $5\pi/6$, $-7\pi/2$, and $13\pi/4$.

3. Sketch a graph of $y = \cos x$.

4. What is the range of the cosine function?

5. What is the period of the cosine function?

6. What is the domain of the sine function?

7. What is the amplitude of the sine function?

8. Complete the following table.

	$\pi/2$	$-4\pi/2$	$7\pi/2$
$\sin x$			
$\cos x$			

9. Sketch a graph of the secant function.

10. What is the period of the secant function?

11. What is the domain of the secant function?

12. In which quadrants are the signs of the sine and the secant different?

For angles of the following measures, state in which quadrant the terminal side lies, convert to radian measure in terms of π, and convert to radian measure not in terms of π. Use 3.14 for π.

13. $-150°$

14. $117°$

15. $225°$

Convert to degree measure.

16. $\frac{7\pi}{4}$ radians

17. 2 radians

18. Find the length of an arc of a circle, given a central angle of $\pi/3$ and a radius of 10 cm.

19. An arc 8 in. long on a circle of radius 16 in. subtends an angle of how many radians? how many degrees, to the nearest degree?

20. A rock on a 4-ft string is rotated at 80 rpm. What is its linear speed, in ft/min?

21. A pulley belt runs, uncrossed, around two pulleys of radii 10 cm and 4 cm, respectively. A point on the belt travels at a rate of 50 m/sec. Find the angular speed, in radians/sec of the smaller pulley.

22. Find the six trigonometric function values for the angle θ shown.

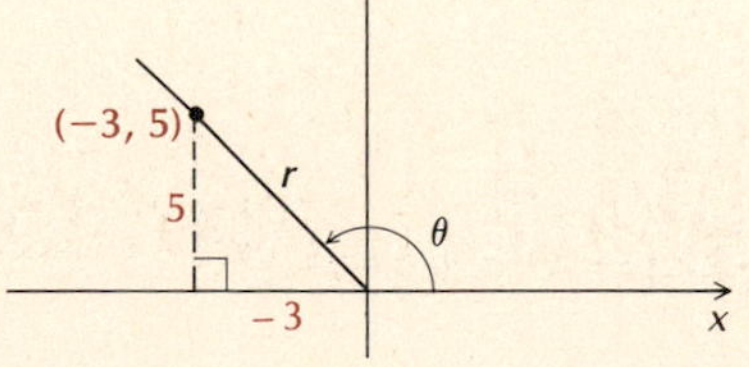

Find the following exactly. Do *not* use a calculator.

23. $\tan 495°$

24. $\sin(-150°)$

25. $\cos 300°$

26. Given that $\sin\theta = \frac{1}{4}$ and that the terminal side is in quadrant II, find the other five function values.

27. Convert 19.25° to degrees and minutes.

28. Convert 72°48′ to degrees and decimal parts of degrees.

Use a calculator to find each of the following.

29. $\cos(-213°48'13'')$

30. $\tan 1.228$

31. $\sin(-34\pi/19)$

32. $\sec 38°17'$

33. $\csc 3.4128$

34. $\cot(3472°)$

Use a calculator to find the acute angle θ that is a solution of the equation. Give the answer in degrees and decimal parts of degrees, as well as in radians.

35. $\sin \theta = 0.8714$

36. $\cot \theta = 8$

Use a calculator to find θ in the interval indicated.

37. $\sin \theta = 0.4305$, $(90°, 180°)$

38. $\tan \theta = -11.412$, $(-\pi/2, \pi/2)$

39. Sketch a graph of $y = 2 \cos (x - \pi/4)$.

40. What is the period of the function in Question 39?

41. What is the phase shift of the function in Question 39?

42. Sketch the graph of $y = 2 \cos x - \sin x$ for values between 0 and 2π.

SYNTHESIS

43. Given that $\cos (\pi/32) = 0.99518$, find each of the following.

a) $\sin (15\pi/32)$
b) $\sin (17\pi/32)$

44. What is the angle between the hands of a clock at 9:25?

Trigonometric Identities, Inverse Functions, and Equations

7

There are a number of relationships between the trigonometric functions, given by identities, that are important in several respects. They are important in algebraic and trigonometric manipulation and in both practical problems and theoretical developments. A large part of this chapter is devoted to those identities. Among the manipulations in which the identities are used is the solving of *trigonometric equations*. A trigonometric equation is an equation with some trigonometric function in it. You will learn to solve such equations in this chapter.

We have not yet considered the inverses of the trigonometric functions beyond the use of calculators in reverse. We now look at those inverses. The chapter is an important one, both for problem solving and as a foundation for further work in mathematics.

FEATURE PROBLEM

In a circus, a guy wire is attached to the top of a 30-ft pole. Another wire is used for the performers to walk up to the tight wire, 10 ft above the ground. If the wires are secured to the ground 40 ft from the base of the pole, how large is the angle between the wires?

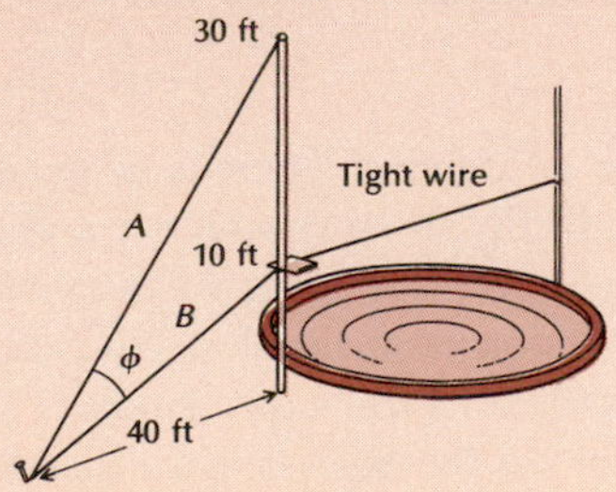

THE MATHEMATICS

A general formula for finding the angle from one line to another is

$$\tan \phi = \frac{m_2 - m_1}{1 + m_2 m_1},$$

where m_1 and m_2 are the slopes of the lines.

When solving a problem, a drawing is often helpful. The slope of line A is $\frac{30}{40}$, or 0.75, and the slope of line B is $\frac{10}{40}$, or 0.25. We substitute 0.75 for m_2 and 0.25 for m_1 and solve for ϕ.

7.1 Trigonometric Manipulations and Identities

OBJECTIVES

You should be able to:

1. Multiply and factor expressions containing trigonometric expressions.
2. Simplify and manipulate expressions containing trigonometric expressions.
3. Simplify and manipulate radical expressions containing trigonometric expressions, including rationalizing numerators or denominators.

Trigonometric expressions such as $\sin 2x$ or $\tan (x - \pi)$ represent real numbers, in the same way that purely algebraic expressions do. Thus the laws of real numbers hold whether we are working with algebraic expressions or trigonometric expressions, or combinations thereof. We can factor, simplify, and manipulate expressions in much the same way that we manipulate purely algebraic expressions.

The laws of real numbers can be combined with the fundamental identities considered in Chapter 6 to further enhance our ability to manipulate trigonometric expressions. The following is a list of these fundamental identities.

$$\sin x = \frac{1}{\csc x}, \quad \csc x = \frac{1}{\sin x}, \quad \tan x = \frac{\sin x}{\cos x},$$

$$\cos x = \frac{1}{\sec x}, \quad \sec x = \frac{1}{\cos x}, \quad \cot x = \frac{\cos x}{\sin x},$$

$$\tan x = \frac{1}{\cot x}, \quad \cot x = \frac{1}{\tan x},$$

$$\sin^2 x + \cos^2 x = 1, \quad 1 + \tan^2 x = \sec^2 x, \quad 1 + \cot^2 x = \csc^2 x,$$

$$\sin (-x) = -\sin x, \quad \csc (-x) = -\csc x,$$

$$\cos (-x) = \cos x, \quad \sec (-x) = \sec x,$$

$$\tan (-x) = -\tan x, \quad \cot (-x) = -\cot x$$

With these identities in mind, let us now consider an example.

1 Multiplying and Factoring

EXAMPLE 1 Multiply and simplify: $\cos y (\tan y - \sec y)$.

Solution

$$\cos y (\tan y - \sec y) = \cos y \tan y - \cos y \sec y$$

The multiplication has now been accomplished, but we can simplify:

$$\begin{aligned} \cos y \tan y - \cos y \sec y &= \cos y \frac{\sin y}{\cos y} - \cos y \frac{1}{\cos y} \\ &= \sin y - 1. \end{aligned}$$

Recall that an **identity** is an equation that is true for all meaningful replacements of the variables. Thus we have proved in Example 1 that

$$\cos y (\tan y - \sec y) = \sin y - 1$$

is an identity. There is no general rule for doing a simplification that creates an identity as in Example 1 and no specific result, but it is often helpful to put everything in terms of sines and cosines. For the most part, the main goal of this section is to develop your skill by the practice you receive in manipulating trigonometric expressions. This kind of simplifying is very important for the rest of the trigonometry that you will study in this chapter and in other courses that you will take in mathematics and science.

EXAMPLE 2 Factor and simplify: $\sin^2 x \cos^2 x + \cos^4 x$.

Solution

$$\sin^2 x \cos^2 x + \cos^4 x = \cos^2 x (\sin^2 x + \cos^2 x)$$

The factoring has been accomplished, but we can now simplify, using the Pythagorean identity $\sin^2 x + \cos^2 x = 1$:

$$\cos^2 x (\sin^2 x + \cos^2 x) = \cos^2 x.$$

DO EXERCISES 1 AND 2.

1. Multiply and simplify: $\sin x (\cot x + \csc x)$.

2. Factor and simplify: $\sin^3 x + \sin x \cos^2 x$.

2 Simplifying

EXAMPLE 3 Simplify: $\dfrac{\cot(-\theta)}{\csc(-\theta)}$.

Solution

$$\frac{\cot(-\theta)}{\csc(-\theta)} = \frac{\dfrac{\cos(-\theta)}{\sin(-\theta)}}{\dfrac{1}{\sin(-\theta)}} = \frac{\cos(-\theta)}{\sin(-\theta)} \cdot \sin(-\theta) = \cos(-\theta) = \cos\theta$$

EXAMPLE 4 Simplify: $\dfrac{\sin x - \sin x \cos x}{\sin x + \sin x \tan x}$.

Solution

$$\frac{\sin x - \sin x \cos x}{\sin x + \sin x \tan x} = \frac{\sin x (1 - \cos x)}{\sin x (1 + \tan x)} \quad \text{Factoring}$$

$$= \frac{1 - \cos x}{1 + \tan x} \quad \text{Simplifying}$$

EXAMPLE 5 Subtract and simplify: $\dfrac{2}{\sin x - \cos x} - \dfrac{3}{\sin x + \cos x}$.

Solution

$$\frac{2}{\sin x - \cos x} - \frac{3}{\sin x + \cos x}$$

$$= \frac{2}{\sin x - \cos x} \cdot \frac{\sin x + \cos x}{\sin x + \cos x} - \frac{3}{\sin x + \cos x} \cdot \frac{\sin x - \cos x}{\sin x - \cos x}$$

$$= \frac{2(\sin x + \cos x) - 3(\sin x - \cos x)}{\sin^2 x - \cos^2 x}$$

$$= \frac{-\sin x + 5\cos x}{\sin^2 x - \cos^2 x}$$

EXAMPLE 6 Add and simplify: $\dfrac{2}{\tan^3 x - 2\tan^2 x} + \dfrac{2}{\tan x - 2}$.

Solution

$$\frac{2}{\tan^3 x - 2\tan^2 x} + \frac{2}{\tan x - 2} = \frac{2}{\tan^2 x (\tan x - 2)} + \frac{2}{\tan x - 2} \cdot \frac{\tan^2 x}{\tan^2 x}$$

$$= \frac{2 + 2\tan^2 x}{\tan^2 x (\tan x - 2)} = \frac{2(1 + \tan^2 x)}{\tan^2 x (\tan x - 2)}$$

Now we use the Pythagorean identity $1 + \tan^2 x = \sec^2 x$ in the numerator and obtain for that numerator

$$2 \sec^2 x, \quad \text{or} \quad 2\frac{1}{\cos^2 x}.$$

Thus we have

$$\frac{2}{\cos^2 x \dfrac{\sin^2 x}{\cos^2 x}(\tan x - 2)},$$

and finally

$$\frac{2}{\sin^2 x(\tan x - 2)}.$$

DO EXERCISES 3–6.

3 Radical Expressions

When radicals occur, the use of absolute value is sometimes necessary, but it can be difficult to determine when to use it. In the examples and exercises that follow, we will assume that all radicals are nonnegative.

EXAMPLE 7 Multiply and simplify: $\sqrt{\sin^3 x \cos x} \cdot \sqrt{\cos x}$.

Solution

$$\begin{aligned}\sqrt{\sin^3 x \cos x} \cdot \sqrt{\cos x} &= \sqrt{\sin^3 x \cos^2 x}\\ &= \sqrt{\sin^2 x \cos^2 x \sin x}\\ &= \sin x \cos x \sqrt{\sin x}\end{aligned}$$

EXAMPLE 8 Rationalize the denominator: $\sqrt{\dfrac{2}{\tan x}}$.

Solution

$$\sqrt{\frac{2}{\tan x}} = \sqrt{\frac{2}{\tan x} \cdot \frac{\tan x}{\tan x}} = \sqrt{\frac{2 \tan x}{\tan^2 x}} = \frac{\sqrt{2 \tan x}}{\tan x}$$

DO EXERCISES 7 AND 8.

EXAMPLE 9 Express $\sin \theta$ in terms of $\tan \theta$.

Solution We know that $1 + \tan^2 \theta = \sec^2 \theta$ and that $\sec \theta = 1/\cos \theta$. Then

$$1 + \tan^2 \theta = \frac{1}{\cos^2 \theta}.$$

Since $\sin^2 \theta + \cos^2 \theta = 1$, it follows that $\cos^2 \theta = 1 - \sin^2 \theta$. Thus,

$$1 + \tan^2 \theta = \frac{1}{1 - \sin^2 \theta},$$

or

$$1 - \sin^2 \theta = \frac{1}{1 + \tan^2 \theta}.$$

Simplify.

3. $\dfrac{\tan(-\theta)}{\sin(-\theta)}$

4. $\dfrac{\cos x + \sin x \cos x}{\cos x - \cos x \cot x}$

Add and simplify.

5. $\dfrac{2}{\sin x - \cos x} + \dfrac{3}{\sin x + \cos x}$

6. $\dfrac{2}{\cot^3 x - 2\cot^2 x} + \dfrac{2}{\cot x - 2}$

7. Multiply and simplify:

$$\sqrt{\tan x \sin^2 x} \cdot \sqrt{\tan x \sin x}.$$

8. Rationalize the numerator:

$$\sqrt{\frac{\cos x}{3}}.$$

Then we solve for $\sin\theta$ as follows. We have

$$-\sin^2\theta = -1 + \frac{1}{1+\tan^2\theta}$$

and

$$\sin^2\theta = 1 - \frac{1}{1+\tan^2\theta},$$

so

$$\sin^2\theta = 1 - \frac{1}{1+\tan^2\theta} = \frac{\tan^2\theta + 1}{\tan^2\theta + 1} - \frac{1}{1+\tan^2\theta} = \frac{\tan^2\theta}{1+\tan^2\theta}.$$

Then

$$\sin\theta = \frac{\pm\tan\theta}{\sqrt{1+\tan^2\theta}},$$

because the sign depends on the quadrant in which θ lies. ■

DO EXERCISE 9.

Often in calculus, a substitution such as the following is a useful manipulation.

EXAMPLE 10 Use the substitution $x = a\tan\theta$ to express $\sqrt{a^2+x^2}$ as a trigonometric function without radicals. Assume that $a > 0$ and that $-\pi/2 < \theta < \pi/2$. Then find $\sin\theta$ and $\cos\theta$.

Solution Substituting $a\tan\theta$ for x, we get

$$\begin{aligned}\sqrt{a^2+x^2} &= \sqrt{a^2+(a\tan\theta)^2} = \sqrt{a^2+a^2\tan^2\theta}\\ &= \sqrt{a^2(1+\tan^2\theta)} = \sqrt{a^2\sec^2\theta}\\ &= \sqrt{a^2}\sqrt{\sec^2\theta} = |a|\cdot|\sec\theta| = a\sec\theta.\end{aligned}$$

The latter equality follows from the fact that $a > 0$ and that $\sec\theta > 0$ for values of θ for which $-\pi/2 < \theta < \pi/2$.

We can express the preceding result as

$$\sec\theta = \frac{\sqrt{a^2+x^2}}{a}.$$

In terms of right triangles, we know that $\sec\theta$ is hypotenuse/adjacent, where θ is one of the acute angles. Now interpreting the hypotenuse to be $\sqrt{a^2+x^2}$ and the adjacent side to be a, we can find the length opposite θ to be x using the Pythagorean theorem.

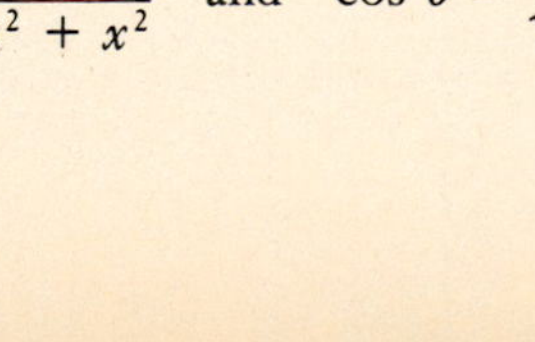

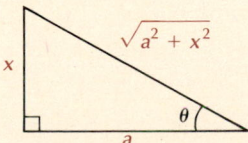

Then from the right triangle we see that

$$\sin\theta = \frac{x}{\sqrt{a^2+x^2}} \quad\text{and}\quad \cos\theta = \frac{a}{\sqrt{a^2+x^2}}.$$ ■

DO EXERCISE 10.

9. Express $\cos\theta$ in terms of $\tan\theta$.

10. Use the substitution $x = 2\sin\theta$ to express $\sqrt{4-x^2}$ as a trigonometric function without radicals. Assume that $-\pi/2 \le \theta \le \pi/2$. Then find $\tan\theta$ and $\cos\theta$.

EXERCISE SET 7.1

1 Multiply and simplify.

1. $(\sin x - \cos x)(\sin x + \cos x)$
2. $(\tan y - \cot y)(\tan y + \cot y)$
3. $\tan x\,(\cos x - \csc x)$
4. $\cot x\,(\sin x + \sec x)$
5. $\cos y \sin y\,(\sec y + \csc y)$
6. $\tan y \sin y\,(\cot y - \csc y)$
7. $(\sin x + \cos x)(\csc x - \sec x)$
8. $(\sin x + \cos x)(\sec x + \csc x)$
9. $(\sin y - \cos y)^2$
10. $(\sin y + \cos y)^2$
11. $(1 + \tan x)^2$
12. $(1 + \cot x)^2$
13. $(\sin y - \csc y)^2$
14. $(\cos y + \sec y)^2$
15. $(\cos x - \sec x)(\cos^2 x + \sec^2 x + 1)$
16. $(\sin x + \csc x)(\sin^2 x + \csc^2 x - 1)$
17. $(\cot x - \tan x)(\cot^2 x + 1 + \tan^2 x)$
18. $(\cot y + \tan y)(\cot^2 y - 1 + \tan^2 y)$
19. $(1 - \sin x)(1 + \sin x)$
20. $(1 + \cos x)(1 - \cos x)$

Factor and simplify.

21. $\sin x \cos x + \cos^2 x$
22. $\sec x \csc x - \csc^2 x$
23. $\sin^2 \theta - \cos^2 \theta$
24. $\tan^2 \theta - \cot^2 \theta$
25. $\tan x + \sin(\pi - x)$
26. $\cot x - \cos(\pi - x)$
27. $\sin^4 x - \cos^4 x$
28. $\tan^4 x - \sec^4 x$
29. $3\cot^2 y + 6\cot y + 3$
30. $4\sin^2 y + 8\sin y + 4$
31. $\csc^4 x + 4\csc^2 x - 5$
32. $-8 + \tan^4 x - 2\tan^2 x$
33. $\sin^3 y + 27$
34. $1 - 125\tan^3 y$
35. $\sin^3 v - \csc^3 v$
36. $\cos^3 u - \sec^3 u$

2 Simplify.

37. $\dfrac{\sin^2 x \cos x}{\cos^2 x \sin x}$
38. $\dfrac{\cos^2 x \sin x}{\sin^2 x \cos x}$
39. $\dfrac{4\sin x \cos^3 x}{18\sin^2 x \cos x}$
40. $\dfrac{30\sin^3 x \cos x}{6\cos^2 x \sin x}$
41. $\dfrac{\cos^2 x - 2\cos x + 1}{\cos x - 1}$
42. $\dfrac{\sin^2 x + 2\sin x + 1}{\sin x + 1}$
43. $\dfrac{\cos^2 \alpha - 1}{\cos \alpha - 1}$
44. $\dfrac{\sin^2 \alpha - 1}{\sin \alpha + 1}$
45. $\dfrac{4\tan x \sec x + 2\sec x}{6\sin x \sec x + 2\sec x}$
46. $\dfrac{6\tan x \sin x - 3\sin x}{9\sin^2 x + 3\sin x}$
47. $\dfrac{\csc(-\theta)}{\cot(-\theta)}$
48. $\tan(-\beta) + \cot(-\beta)$
49. $\dfrac{\sin^4 x - \cos^4 x}{\sin^2 x - \cos^2 x}$
50. $\dfrac{\sec^4 x - \tan^4 x}{\sec^2 x + \tan^2 x}$
51. $\dfrac{2\sin^2 x}{\cos^3 x} \cdot \left(\dfrac{\cos x}{2\sin x}\right)^2$
52. $\dfrac{4\cos^3 x}{\sin^2 x} \cdot \left(\dfrac{\sin x}{4\cos x}\right)^2$
53. $\dfrac{3\sin x}{\cos^2 x} \cdot \dfrac{\cos^2 x + \cos x \sin x}{\cos^2 x - \sin^2 x}$
54. $\dfrac{5\cos x}{\sin^2 x} \cdot \dfrac{\sin^2 x - \sin x \cos x}{\sin^2 x - \cos^2 x}$
55. $\dfrac{\tan^2 \gamma}{\sec \gamma} \div \dfrac{3\tan^3 \gamma}{\sec \gamma}$
56. $\dfrac{\cot^3 \phi}{\csc \phi} \div \dfrac{4\cot^2 \phi}{\csc \phi}$
57. $\dfrac{1}{\sin^2 y - \cos^2 y} - \dfrac{2}{\cos y + \sin y}$
58. $\dfrac{3}{\cos y - \sin y} - \dfrac{2}{\sin^2 y - \cos^2 y}$
59. $\left(\dfrac{\sin x}{\cos x}\right)^2 - \dfrac{1}{\cos^2 x}$
60. $\left(\dfrac{\cot x}{\csc x}\right)^2 + \dfrac{1}{\csc^2 x}$

61. $\dfrac{\sin^2 x - 9}{2\cos x + 1} \cdot \dfrac{10\cos x + 5}{3\sin x + 9}$

62. $\dfrac{9\cos^2 x - 25}{2\cos x - 2} \cdot \dfrac{\cos^2 x - 1}{6\cos x - 10}$

3 Simplify. Assume that expressions in radicands are nonnegative.

63. $\sqrt{\sin^2 x \cos x} \cdot \sqrt{\cos x}$

64. $\sqrt{\cos^2 x \sin x} \cdot \sqrt{\sin x}$

65. $\sqrt{\sin^3 y} + \sqrt{\sin y \cos^2 y}$

66. $\sqrt{\cos y \sin^2 y} - \sqrt{\cos^3 y}$

67. $\sqrt{\sin^2 x + 2\cos x \sin x + \cos^2 x}$

68. $\sqrt{\tan^2 x - 2\tan x \sin x + \sin^2 x}$

69. $(1 - \sqrt{\sin y})(\sqrt{\sin y} + 1)$

70. $(2 - \sqrt{\tan y})(\sqrt{\tan y} + 2)$

71. $\sqrt{\sin x}\,(\sqrt{2}\sin x + \sqrt{\sin x \cos x})$

72. $\sqrt{\cos x}\,(\sqrt{2}\cos x + \sqrt{\sin x \cos x})$

Rationalize the denominator.

73. $\sqrt{\dfrac{\sin x}{\cos x}}$

74. $\sqrt{\dfrac{\cos x}{\sin x}}$

75. $\sqrt{\dfrac{\sin x}{\cot x}}$

76. $\sqrt{\dfrac{\cos x}{\tan x}}$

77. $\sqrt{\dfrac{\cos^2 x}{2\sin^2 x}}$

78. $\sqrt{\dfrac{\sin^2 x}{3\cos^2 x}}$

79. $\sqrt{\dfrac{1 + \sin x}{1 - \sin x}}$

80. $\sqrt{\dfrac{1 - \cos x}{1 + \cos x}}$

Rationalize the numerator.

81. $\sqrt{\dfrac{\sin x}{\cos x}}$

82. $\sqrt{\dfrac{\cos x}{\sin x}}$

83. $\sqrt{\dfrac{\sin x}{\cot x}}$

84. $\sqrt{\dfrac{\cos x}{\tan x}}$

85. $\sqrt{\dfrac{\cos^2 x}{2\sin^2 x}}$

86. $\sqrt{\dfrac{\sin^2 x}{3\cos^2 x}}$

87. $\sqrt{\dfrac{1 + \sin x}{1 - \sin x}}$

88. $\sqrt{\dfrac{1 - \cos x}{1 + \cos x}}$

Express each of the six basic trigonometric functions in terms of the given function.

89. $\sin\theta$

90. $\cos\theta$

91. $\tan\theta$

92. $\cot\theta$

93. $\sec\theta$

94. $\csc\theta$

Use the given substitution to express the given radical expression as a trigonometric function without radicals. Assume that $a > 0$ and $0 < \theta < \pi/2$. Then find expressions for the indicated trigonometric functions.

95. Let $x = a\sin\theta$ in $\sqrt{a^2 - x^2}$. Then find $\cos\theta$ and $\tan\theta$.

96. Let $x = 2\tan\theta$ in $\sqrt{4 + x^2}$. Then find $\sin\theta$ and $\cos\theta$.

97. Let $x = 3\sec\theta$ in $\sqrt{x^2 - 9}$. Then find $\sin\theta$ and $\cos\theta$.

98. Let $x = a\sec\theta$ in $\sqrt{a\sec\theta}$. Then find $\sin\theta$ and $\cos\theta$.

Use the given substitution to express the given radical expression as a trigonometric function without radicals. Assume that $0 < \theta < \pi/2$.

99. Let $x = \sin\theta$ in $\dfrac{x^2}{\sqrt{1 - x^2}}$.

100. Let $x = 4\sec\theta$ in $\dfrac{\sqrt{x^2 - 16}}{x^2}$.

SYNTHESIS

Show that each of the following is *not* an identity by finding a replacement or replacements for which each side of the equation does not name the same number.

101. $(\sin x + \cos x)^2 = \sin^2 x + \cos^2 x$

102. $\sqrt{\sin^2\theta} = \sin\theta$

103. $\dfrac{\sin x}{\sin y} = \dfrac{x}{y}$

104. $\cos x = \sqrt{1 - \sin^2 x}$

105. $\cos(\alpha + \beta) = \cos\alpha + \cos\beta$

106. $\sin(-\theta) = \sin\theta$

107. $\cos(2\theta) = 2\cos\theta$

108. $\tan^2\theta + \cot^2\theta = 1$

109. $\ln(\sin\theta) = \sin(\ln\theta)$

7.2 Sum and Difference Formulas

OBJECTIVES

You should be able to:

1. Use the sum and difference identities to find function values.
2. Simplify expressions using the sum and difference identities.
3. Find angles between lines, knowing their slopes.

1 Difference Identities

We now develop some important identities involving sums or differences of two numbers (or angles), first an identity for the cosine of the difference of two numbers. We use the Greek letters α (alpha) and β (beta) for these numbers. Let us consider a real number α in the interval $[\pi/2, \pi]$ and a real number β in the interval $[0, \pi/2]$. These determine points A and B on the unit circle as shown. The arc length s is $\alpha - \beta$, and we know that $0 \leq s \leq \pi$. Recall that the coordinates of A are $(\cos \alpha, \sin \alpha)$, and the coordinates of B are $(\cos \beta, \sin \beta)$.

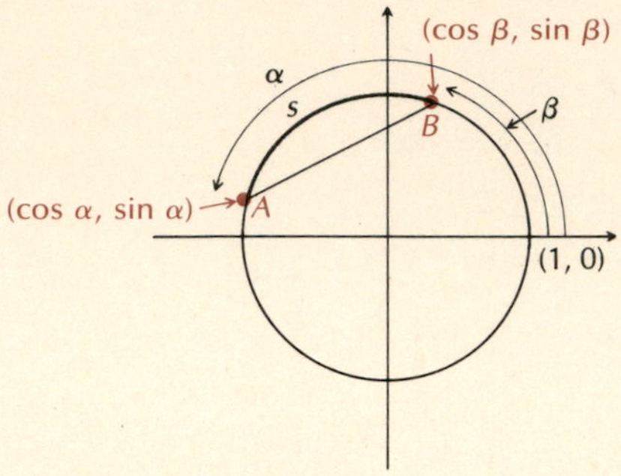

Using the distance formula, we can write an expression for the square of the distance AB:

$$AB^2 = (\cos \alpha - \cos \beta)^2 + (\sin \alpha - \sin \beta)^2.$$

This can be simplified as follows:

$$\begin{aligned} AB^2 &= \cos^2 \alpha - 2 \cos \alpha \cos \beta + \cos^2 \beta + \sin^2 \alpha - 2 \sin \alpha \sin \beta + \sin^2 \beta \\ &= (\sin^2 \alpha + \cos^2 \alpha) + (\sin^2 \beta + \cos^2 \beta) - 2(\cos \alpha \cos \beta + \sin \alpha \sin \beta) \\ &= 2 - 2(\cos \alpha \cos \beta + \sin \alpha \sin \beta). \end{aligned}$$

Now let us imagine rotating the circle above so that point B is at $(1, 0)$. The coordinates of point A are now $(\cos s, \sin s)$. The distance AB has not changed.

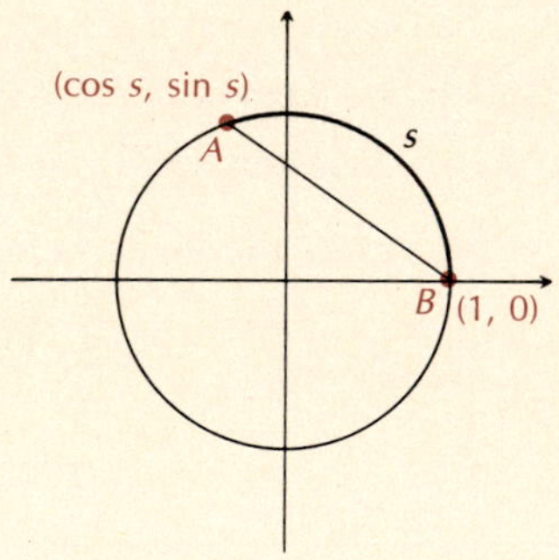

Again we use the distance formula to write an expression for the square of AB:

$$AB^2 = (\cos s - 1)^2 + (\sin s - 0)^2.$$

This simplifies as follows:

$$\begin{aligned} AB^2 &= \cos^2 s - 2 \cos s + 1 + \sin^2 s \\ &= (\sin^2 s + \cos^2 s) + 1 - 2 \cos s \\ &= 2 - 2 \cos s. \end{aligned}$$

Equating our two expressions for AB^2 and simplifying, we obtain

$$2 - 2(\cos \alpha \cos \beta + \sin \alpha \sin \beta) = 2 - 2 \cos s.$$

Solving this equation for $\cos s$, we obtain

$$\cos s = \cos \alpha \cos \beta + \sin \alpha \sin \beta. \tag{1}$$

But $s = \alpha - \beta$, so we have the equation

$$\cos (\alpha - \beta) = \cos \alpha \cos \beta + \sin \alpha \sin \beta. \tag{2}$$

This formula holds for any numbers α and β for which $\alpha - \beta$ is the length of the shortest arc from A to B; in other words, when $0 \leq \alpha - \beta \leq \pi$.

EXAMPLE 1 Find $\cos\left(\frac{3\pi}{4} - \frac{\pi}{3}\right)$.

Solution

$$\begin{aligned}\cos\left(\frac{3\pi}{4} - \frac{\pi}{3}\right) &= \cos\frac{3\pi}{4}\cos\frac{\pi}{3} + \sin\frac{3\pi}{4}\sin\frac{\pi}{3}\\ &= -\frac{\sqrt{2}}{2}\cdot\frac{1}{2} + \frac{\sqrt{2}}{2}\cdot\frac{\sqrt{3}}{2}\\ &= \frac{\sqrt{2}}{4}(\sqrt{3} - 1), \quad \text{or} \quad \frac{\sqrt{6} - \sqrt{2}}{4}\end{aligned}$$

EXAMPLE 2 Find $\cos 15°$.

Solution

$$\begin{aligned}\cos 15° &= \cos(45° - 30°)\\ &= \cos 45° \cos 30° + \sin 45° \sin 30°\\ &= \frac{\sqrt{2}}{2}\cdot\frac{\sqrt{3}}{2} + \frac{\sqrt{2}}{2}\cdot\frac{1}{2}\\ &= \frac{\sqrt{2}}{4}(\sqrt{3} + 1), \quad \text{or} \quad \frac{\sqrt{6} + \sqrt{2}}{4}\end{aligned}$$

DO EXERCISES 1 AND 2.

Formula (1) above holds when s is the length of the shortest arc from A to B. Given any real numbers α and β, the length of the shortest arc from A to B is not always $\alpha - \beta$. However, $\cos s$ is always equal to $\cos(\alpha - \beta)$. We will not prove this, but the proof is similar to the preceding.

Formula (2) holds for all real numbers α and β. That formula is thus the identity we sought.

$$\boldsymbol{\cos(\alpha - \beta) = \cos\alpha\cos\beta + \sin\alpha\sin\beta}$$

2 Simplification

EXAMPLE 3 Simplify:

$$\sin\left(-\frac{5\pi}{2}\right)\sin\frac{\pi}{2} + \cos\frac{\pi}{2}\cos\left(-\frac{5\pi}{2}\right).$$

Solution The expression is equal to

$$\cos\frac{\pi}{2}\cos\left(-\frac{5\pi}{2}\right) + \sin\frac{\pi}{2}\sin\left(-\frac{5\pi}{2}\right).$$

By the identity above, we can simplify to

$$\cos\left[\frac{\pi}{2} - \left(-\frac{5\pi}{2}\right)\right], \quad \text{or} \quad \cos\frac{6\pi}{2}, \quad \text{or} \quad \cos 3\pi,$$

which is -1.

DO EXERCISES 3–5.

1. Find $\cos\left(\frac{\pi}{2} - \frac{\pi}{6}\right)$.

2. Find $\cos 105°$ as $\cos(150° - 45°)$.

Simplify.

3. $\sin\frac{\pi}{3}\sin\left(-\frac{\pi}{4}\right) + \cos\left(-\frac{\pi}{4}\right)\cos\frac{\pi}{3}$

4. $\cos 37° \cos 12° + \sin 12° \sin 37°$

5. $\cos\alpha\cos(-\beta) + \sin\alpha\sin(-\beta)$

6. Find $\sin\left(\frac{\pi}{4} + \frac{\pi}{3}\right)$.

7. Simplify:

$\sin \alpha \cos(-\beta) + \cos \alpha \sin(-\beta)$.

1, 2 Other Formulas

The other sum and difference formulas we seek follow easily from the one we have just derived. Let us consider $\cos(\alpha + \beta)$. This is equal to $\cos[\alpha - (-\beta)]$, and by the identity above, we have

$$\cos(\alpha + \beta) = \cos[\alpha - (-\beta)] = \cos \alpha \cos(-\beta) + \sin \alpha \sin(-\beta).$$

But $\cos(-\beta) = \cos \beta$ and $\sin(-\beta) = -\sin \beta$, so the identity we seek is the following:

$$\boldsymbol{\cos(\alpha + \beta) = \cos \alpha \cos \beta - \sin \alpha \sin \beta.}$$

To develop an identity for the sine of a sum, we first need two intermediary identities. Let us consider $\cos(\pi/2 - \theta)$. We can use the identity for the cosine of a difference to simplify as follows:

$$\begin{aligned}\cos\left(\frac{\pi}{2} - \theta\right) &= \cos \frac{\pi}{2} \cos \theta + \sin \frac{\pi}{2} \sin \theta \\ &= 0 \cdot \cos \theta + 1 \cdot \sin \theta = \sin \theta.\end{aligned}$$

Thus we have developed the identity $\cos(\pi/2 - \theta) = \sin \theta$, which we can apply in reverse as

$$\sin \theta = \cos\left(\frac{\pi}{2} - \theta\right). \tag{1}$$

This identity holds for any real number θ. From it we can obtain a similar identity for the sine function. Let α be any real number. Then replace θ in $\sin \theta = \cos(\pi/2 - \theta)$ by $\pi/2 - \alpha$. This gives us

$$\sin\left(\frac{\pi}{2} - \alpha\right) = \cos\left[\frac{\pi}{2} - \left(\frac{\pi}{2} - \alpha\right)\right] = \cos \alpha,$$

which yields the identity

$$\sin\left(\frac{\pi}{2} - \alpha\right) = \cos \alpha. \tag{2}$$

This identity holds for any real number α.

Using identities (1) and (2) and the identity for the cosine of a difference, we now obtain an identity for the sine of a sum. We start with identity **(1)** and substitute $\alpha + \beta$ for θ, obtaining

$$\sin(\alpha + \beta) = \cos\left[\frac{\pi}{2} - (\alpha + \beta)\right].$$

Now, using the identity for the cosine of a difference, we get

$$\begin{aligned}\sin(\alpha + \beta) &= \cos\left[\frac{\pi}{2} - (\alpha + \beta)\right] \\ &= \cos\left[\left(\frac{\pi}{2} - \alpha\right) - \beta\right] \\ &= \cos\left(\frac{\pi}{2} - \alpha\right) \cos \beta + \sin\left(\frac{\pi}{2} - \alpha\right) \sin \beta \\ &= \sin \alpha \cos \beta + \cos \alpha \sin \beta.\end{aligned}$$

Using identities (1) and (2)

Thus the identity we seek is

$$\boldsymbol{\sin(\alpha + \beta) = \sin \alpha \cos \beta + \cos \alpha \sin \beta.}$$

DO EXERCISES 6 AND 7.

To find a formula for the sine of a difference, we can use the identity just derived, substituting $-\beta$ for β. We obtain

$$\sin(\alpha - \beta) = \sin\alpha\cos\beta - \cos\alpha\sin\beta.$$

A formula for the tangent of a sum can be derived as follows, using identities already established:

$$\begin{aligned}
\tan(\alpha+\beta) &= \frac{\sin(\alpha+\beta)}{\cos(\alpha+\beta)}\\
&= \frac{\sin\alpha\cos\beta + \cos\alpha\sin\beta}{\cos\alpha\cos\beta - \sin\alpha\sin\beta} \cdot \frac{\dfrac{1}{\cos\alpha\cos\beta}}{\dfrac{1}{\cos\alpha\cos\beta}}\\
&= \frac{\dfrac{\sin\alpha\cos\beta}{\cos\alpha\cos\beta} + \dfrac{\cos\alpha\sin\beta}{\cos\alpha\cos\beta}}{\dfrac{\cos\alpha\cos\beta}{\cos\alpha\cos\beta} - \dfrac{\sin\alpha\sin\beta}{\cos\alpha\cos\beta}}\\
&= \frac{\dfrac{\sin\alpha}{\cos\alpha} + \dfrac{\sin\beta}{\cos\beta}}{1 - \dfrac{\sin\alpha\sin\beta}{\cos\alpha\cos\beta}}\\
&= \frac{\tan\alpha + \tan\beta}{1 - \tan\alpha\tan\beta}.
\end{aligned}$$

Similarly, a formula for the tangent of a difference can be established. Following is a list of **sum and difference formulas.*** These should be memorized.

$$\cos(\alpha \mp \beta) = \cos\alpha\cos\beta \pm \sin\alpha\sin\beta,$$
$$\sin(\alpha \mp \beta) = \sin\alpha\cos\beta \mp \cos\alpha\sin\beta,$$
$$\tan(\alpha \mp \beta) = \frac{\tan\alpha \mp \tan\beta}{1 \pm \tan\alpha\tan\beta}$$

DO EXERCISE 8.

8. Derive the formula for $\tan(\alpha - \beta)$.

The identities involving sines and tangents can be used in the same way as those involving cosines in the earlier examples.

EXAMPLE 4 Find tan 15°.

Solution

$$\begin{aligned}
\tan 15^\circ &= \tan(45^\circ - 30^\circ)\\
&= \frac{\tan 45^\circ - \tan 30^\circ}{1 + \tan 45^\circ \tan 30^\circ}\\
&= \frac{1 - \sqrt{3}/3}{1 + \sqrt{3}/3}\\
&= \frac{3 - \sqrt{3}}{3 + \sqrt{3}}
\end{aligned}$$

*There are six identities here, half of them obtained by using the signs shown in color.

9. Find tan 75° as tan (45° + 30°).

10. Simplify:

$$\sin\frac{\pi}{2}\cos\frac{\pi}{3} - \sin\frac{\pi}{3}\cos\frac{\pi}{2}.$$

11. Assume that $\cos\alpha = \frac{3}{4}$ and $\cos\beta = \frac{1}{4}$ and that α and β are between 0 and $\pi/2$. Then evaluate $\cos(\alpha - \beta)$.

EXAMPLE 5 Simplify: $\sin\frac{\pi}{3}\cos\pi + \sin\pi\cos\frac{\pi}{3}$.

Solution The expression is equal to

$$\sin\frac{\pi}{3}\cos\pi + \cos\frac{\pi}{3}\sin\pi.$$

Thus by the fourth identity in the list above, we can simplify to

$$\sin\left(\frac{\pi}{3} + \pi\right), \quad \text{or} \quad \sin\frac{4\pi}{3}, \quad \text{or} \ -\frac{\sqrt{3}}{2}.$$

DO EXERCISES 9 AND 10.

EXAMPLE 6 Assume that $\sin u = \frac{2}{3}$ and $\sin v = \frac{1}{3}$ and that u and v are between 0 and $\pi/2$. Then evaluate $\sin(u + v)$.

Solution Using the identity for the sine of a sum, we have

$$\begin{aligned}\sin(u + v) &= \sin u\cos v + \cos u\sin v\\ &= \tfrac{2}{3}\cos v + \tfrac{1}{3}\cos u.\end{aligned}$$

To finish, we need to know the values of $\cos v$ and $\cos u$. We first find $\cos v$. We know that $\sin^2 v + \cos^2 v = 1$. Solving for $\cos v$, we get

$$\cos v = \sqrt{1 - \sin^2 v}.$$

We know that the radical is positive because v is between 0 and $\pi/2$. Then substituting $\frac{1}{3}$ for $\sin v$, we get

$$\cos v = \sqrt{1 - \left(\frac{1}{3}\right)^2} = \sqrt{1 - \frac{1}{9}} = \sqrt{\frac{8}{9}} = \frac{\sqrt{8}}{3} = \frac{2\sqrt{2}}{3}.$$

Similarly,

$$\cos u = \sqrt{1 - \sin^2 u} = \sqrt{1 - \left(\frac{2}{3}\right)^2} = \sqrt{1 - \frac{4}{9}} = \sqrt{\frac{5}{9}} = \frac{\sqrt{5}}{3}.$$

Substituting these values gives us

$$\sin(u + v) = \frac{2}{3}\cdot\frac{2\sqrt{2}}{3} + \frac{1}{3}\cdot\frac{\sqrt{5}}{3} = 4\frac{\sqrt{2}}{9} + \frac{\sqrt{5}}{9} = \frac{4\sqrt{2} + \sqrt{5}}{9}.$$

DO EXERCISE 11.

3 Angles Between Lines

Recall that a nonvertical line has an equation $y = mx + b$, where m is the slope. Note that such a line makes an angle with the positive half of the x-axis whose tangent is the slope of the line, as in $\tan\theta = m$.

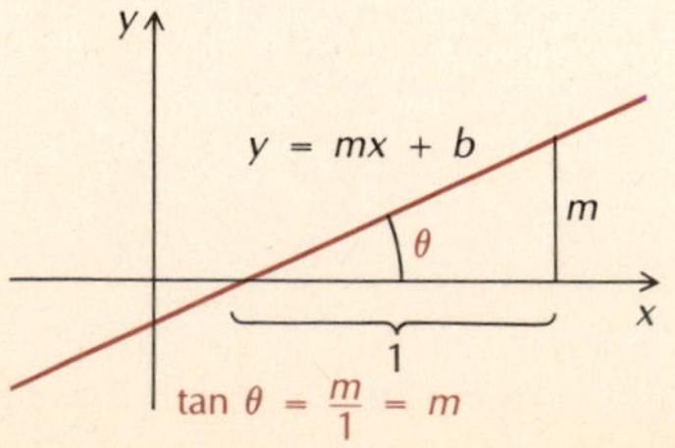

One of the identities just developed gives us an easy way to find an angle formed by two lines. We consider the case in which neither line is vertical. Thus we will consider two lines with equations:

$$l_1: y = m_1x + b_1$$

or

$$l_2: y = m_2x + b_2.$$

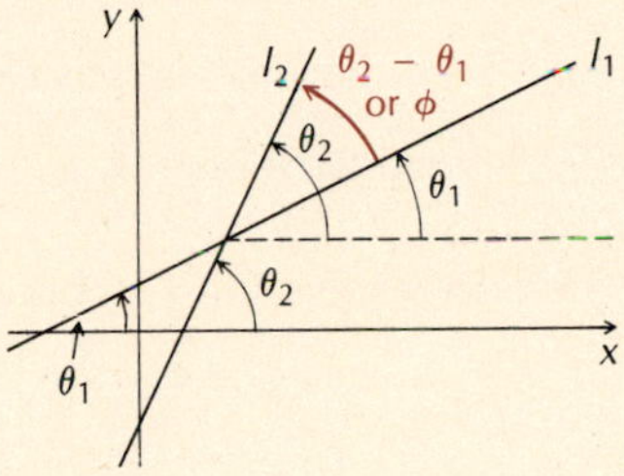

The slopes m_1 and m_2 are the tangents of the angles θ_1 and θ_2 that the lines form with the positive half of the x-axis. Thus we have $m_1 = \tan\theta_1$ and $m_2 = \tan\theta_2$. We want to find the measure of the smallest angle through which l_1 can be rotated in the positive direction in order to get to l_2. In this case, the angle is $\theta_2 - \theta_1$, or ϕ. We proceed as follows:

$$\tan\phi = \tan(\theta_2 - \theta_1) = \frac{\tan\theta_2 - \tan\theta_1}{1 + \tan\theta_2 \tan\theta_1} = \frac{m_2 - m_1}{1 + m_2m_1}.$$

Suppose we had taken the lines in the reverse order. Then the smallest *positive* angle from l_1 to l_2 would be ϕ, as shown here. Note that $\tan\theta_1 = m_1$.

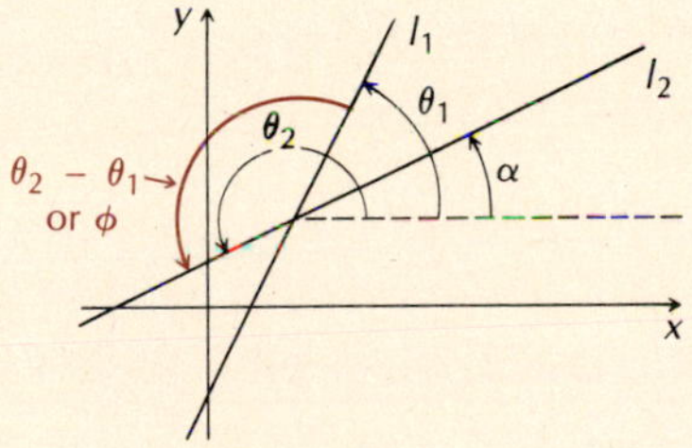

Since $\theta_2 = \alpha + 180°$, $\tan\theta_2 = \tan\alpha$ and $\tan\theta_2 = m_2$. Thus the formula for $\tan\phi$ derived above holds. In the first case, ϕ is an acute angle, so $\tan\phi$ will be positive. In the second case, ϕ is obtuse, so $\tan\phi$ will be negative. Thus we have a general formula for finding the angle from one line to another.

THEOREM 1

The smallest positive angle ϕ from a line l_1 to a line l_2 (where neither line is vertical) can be obtained from

$$\tan\phi = \frac{m_2 - m_1}{1 + m_2m_1}$$

where m_1 and m_2 are the slopes of l_1 and l_2, respectively. If $\tan\phi$ is positive, then ϕ is acute. If $\tan\phi$ is negative, then ϕ is obtuse.

EXAMPLE 7 Two lines have the following slopes: l_1 has slope $\sqrt{3}/3$ and l_2 has slope $\sqrt{3}$. How large is the smallest positive angle from l_1 to l_2?

Solution We have $m_1 = \sqrt{3}/3$ and $m_2 = \sqrt{3}$. Then

$$\tan\phi = \frac{\sqrt{3} - \dfrac{\sqrt{3}}{3}}{1 + \sqrt{3}\cdot\dfrac{\sqrt{3}}{3}} = \frac{\dfrac{3\sqrt{3} - \sqrt{3}}{3}}{\dfrac{3 + 3}{3}} = \frac{2\sqrt{3}}{6} = \frac{\sqrt{3}}{3}.$$

12. Find the smallest positive angle from l_1 to l_2. Using graph paper, sketch a graph and show this angle.

$$l_1: 3y = \sqrt{3}x + 6,$$
$$l_2: y + 5 = \sqrt{3}x.$$

13. Find the smallest positive angle from l_1 to l_2. Using graph paper, sketch a graph and show this angle.

$$l_1: y = \sqrt{3}x - 4,$$
$$l_2: 3y = \sqrt{3}x + 3.$$

14. A loading ramp rises 4 ft with a run of 10 ft. A wire from the top of a building is attached to the ground 8 ft from the building. Find the angle θ that the wire makes with the ramp.

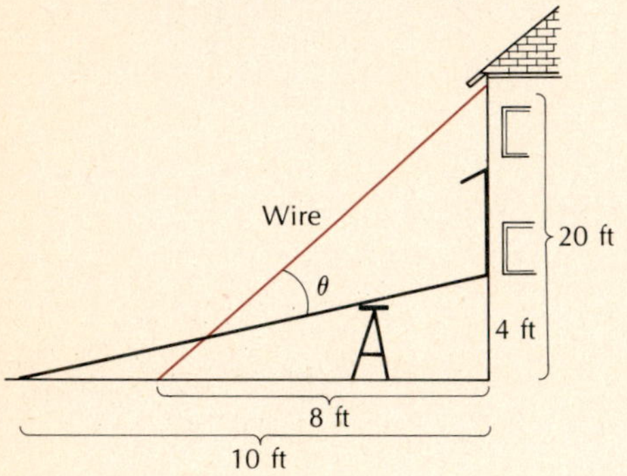

Since $\tan \phi$ is positive, ϕ is acute. Thus,

$$\phi = \frac{\pi}{6}, \quad \text{or } 30°.$$

EXAMPLE 8 Find the smallest positive angle from l_1 to l_2:

$$l_1: 2y = x + 3, \qquad l_2: 3y + 4x + 6 = 0.$$

Solution We first solve for y to obtain the slopes:

$$l_1: y = \tfrac{1}{2}x + \tfrac{3}{2}, \quad \text{so} \quad m_1 = \tfrac{1}{2};$$
$$l_2: y = -\tfrac{4}{3}x - 2, \quad \text{so} \quad m_2 = -\tfrac{4}{3}.$$

Then we have

$$\tan \phi = \frac{-\frac{4}{3} - \frac{1}{2}}{1 + \left(-\frac{4}{3}\right)\left(\frac{1}{2}\right)} = \frac{\frac{-8-3}{6}}{\frac{6-4}{6}}$$
$$= -\frac{11}{2} = -5.5.$$

Since $\tan \phi$ is negative, we know that ϕ is obtuse. From a calculator, we find that an angle whose tangent is 5.5 is approximately 79.7°, or 79°42′. This is a reference angle. We obtain $\phi = 180° - 79°42' = 179°60' - 79°42'$, or

$$\phi = 100°18'.$$

DO EXERCISES 12–14.

EXERCISE SET 7.2

1 Use the sum and difference identities to evaluate.

1. $\sin 75°$ (*Hint:* $75° = 45° + 30°$.)
2. $\cos 75°$
3. $\sin 15°$ (*Hint:* $15° = 45° - 30°$.)
4. $\cos 15°$
5. $\sin 105°$ (*Hint:* Use the results of Exercises 1 and 2.)
6. $\cos 105°$
7. $\tan 75°$
8. $\cot 15°$

2 Simplify or evaluate.

9. $\sin 37° \cos 22° + \cos 37° \sin 22°$
10. $\cos 37° \cos 22° - \sin 22° \sin 37°$
11. $\dfrac{\tan 20° + \tan 32°}{1 - \tan 20° \tan 32°}$
12. $\dfrac{\tan 35° - \tan 12°}{1 + \tan 35° \tan 12°}$

Assume that $\sin u = \frac{3}{5}$ and $\sin v = \frac{4}{5}$ and that u and v are between 0 and $\pi/2$. Then evaluate.

13. $\sin (u + v)$
14. $\sin (u - v)$
15. $\cos (u + v)$
16. $\cos (u - v)$
17. $\tan (u + v)$
18. $\tan (u - v)$

In Exercises 19 and 20, assume that $\sin \theta = 0.6249$ and $\cos \phi = 0.1102$ and that θ and ϕ are both first-quadrant angles. Evaluate.

19. $\sin (\theta + \phi)$
20. $\cos (\theta + \phi)$

Simplify.

21. $\sin (\alpha + \beta) + \sin (\alpha - \beta)$
22. $\sin (\alpha + \beta) - \sin (\alpha - \beta)$
23. $\cos (\alpha + \beta) + \cos (\alpha - \beta)$
24. $\cos (\alpha + \beta) - \cos (\alpha - \beta)$
25. $\cos (u + v) \cos v + \sin (u + v) \sin v$
26. $\sin (u - v) \cos v + \cos (u - v) \sin v$

3 Find the angle from l_1 to l_2.

27. l_1: $3y = \sqrt{3}x + 2$, l_2: $3y + \sqrt{3}x = -3$

28. l_1: $3y = \sqrt{3}x + 3$, l_2: $y = \sqrt{3}x + 2$

29. l_1: $2x = 3 - 2y$, l_2: $x + y = 5$

30. l_1: $2x = 3 + 2y$, l_2: $x - y = 5$

31. l_1: $2x - 5y + 1 = 0$, l_2: $3x + y - 7 = 0$

32. l_1: $2x + y - 4 = 0$, l_2: $y - 2x + 5 = 0$

33. l_1: $y = 3$, l_2: $x + y = 5$

34. l_1: $y = 5$, l_2: $x - y = 2$

In each figure, find the angle from l_1 to l_2.

35. Two pieces of wood are attached to a base in such a way that one has slope 1.25 and the other has slope -0.79.

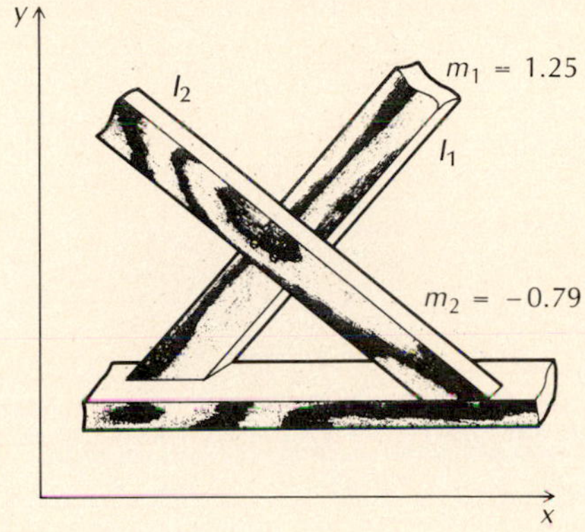

36. Two arms of a crane intersect in such a way that one has slope 3.2 and the other has slope -0.38.

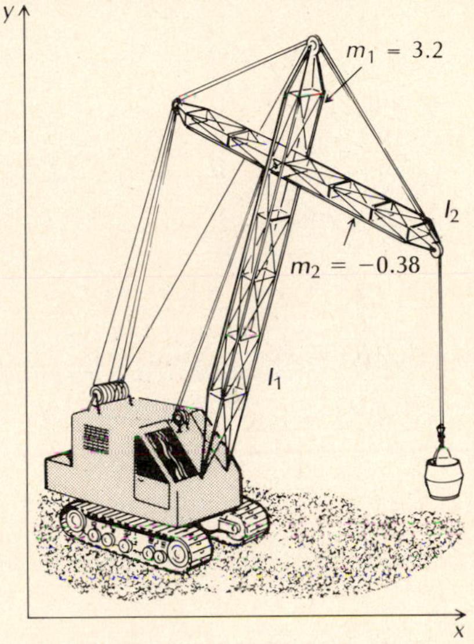

SYNTHESIS

37. Find an identity for $\sin 2\theta$. (*Hint:* $2\theta = \theta + \theta$.)

38. Find an identity for $\cos 2\theta$. (*Hint:* $2\theta = \theta + \theta$.)

39. Derive an identity for $\cot(\alpha + \beta)$ in terms of $\cot \alpha$ and $\cot \beta$.

40. Derive an identity for $\cot(\alpha - \beta)$ in terms of $\cot \alpha$ and $\cot \beta$.

Derive each of the following identities.

41. $\sin\left(x + \frac{3\pi}{2}\right) = -\cos x$

42. $\cos\left(x - \frac{3\pi}{2}\right) = -\sin x$

43. $\cos\left(x + \frac{3\pi}{2}\right) = \sin x$

44. $\sin\left(x - \frac{3\pi}{2}\right) = \cos x$

45. $\tan\left(x + \frac{\pi}{4}\right) = \frac{1 + \tan x}{1 - \tan x}$

46. $\tan\left(x - \frac{\pi}{4}\right) = \frac{\tan x - 1}{\tan x + 1}$

47. $\frac{\sin(\alpha + \beta)}{\cos(\alpha - \beta)} = \frac{\tan \alpha + \tan \beta}{1 + \tan \alpha \tan \beta}$

48. $\frac{\sin(\alpha + \beta)}{\sin(\alpha - \beta)} = \frac{\tan \alpha + \tan \beta}{\tan \alpha - \tan \beta}$

49. $\sin(\alpha + \beta) + \sin(\alpha - \beta) = 2 \sin \alpha \cos \beta$

50. $\cos(\alpha + \beta) \cdot \cos(\alpha - \beta) = \cos^2 \alpha - \sin^2 \beta$

Find the slope of line l_1, where m_2 is the slope of line l_2 and ϕ is the smallest positive angle from l_1 to l_2.

51. $m_2 = \frac{4}{3}$, $\phi = 45°$

52. $m_2 = \frac{2}{3}$, ϕ is the smallest angle for which l_1 has slope $\frac{5}{2}$

53. Line l_1 contains the points $(-2, 4)$ and $(5, -1)$. Find the slope of line l_2 such that the angle from l_1 to l_2 is 45°.

54. Line l_1 contains the points $(3, -1)$ and $(-4, 2)$. Find the slope of line l_2 such that the angle from l_1 to l_2 is $-45°$.

55. ▦ Line l_1 contains the points $(-2.123, 3.899)$ and $(-4.892, -0.9012)$. Line l_2 contains $(0, -3.814)$ and $(5.925, 4.013)$. Find the smallest positive angle from l_1 to l_2.

56. Line l_1 contains the points $(-3, 7)$ and $(-3, -2)$. Line l_2 contains $(0, -4)$ and $(2, 6)$. Find the smallest positive angle from l_1 to l_2.

57. In a circus, a guy wire A is attached to the top of a 30-ft pole. Wire B is used for performers to walk up to the tight wire, 10 ft above the ground. Find the angle ϕ between the wires if they are attached to the ground 40 ft from the pole.

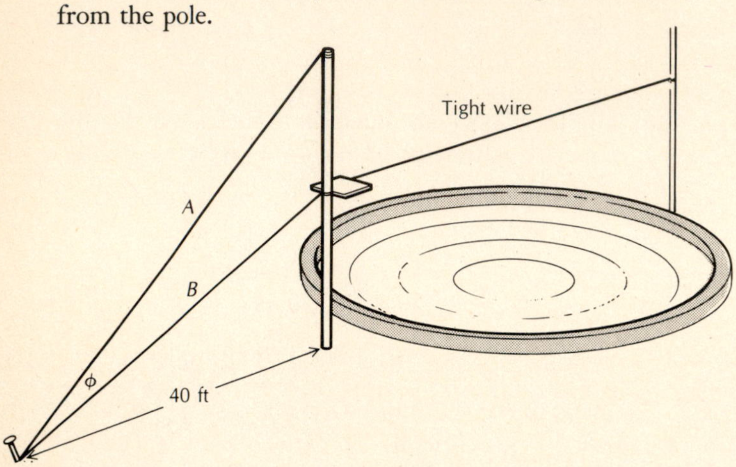

58. Given that $f(x) = \sin x$, show that
$$\frac{f(x+h) - f(x)}{h} = \sin x \left(\frac{\cos h - 1}{h}\right) + \cos x \left(\frac{\sin h}{h}\right).$$

59. Given that $f(x) = \cos x$, show that
$$\frac{f(x+h) - f(x)}{h} = \cos x \left(\frac{\cos h - 1}{h}\right) - \sin x \left(\frac{\sin h}{h}\right).$$

60. Given that $f(x) = \tan x$, show that
$$\frac{f(x+h) - f(x)}{h} = \frac{\sec^2 x}{1 - \tan x \tan h}\left(\frac{\tan h}{h}\right).$$

CHALLENGE

61. Find an identity for $\cos(\alpha + \beta)$ involving only cosines.

62. Find an identity for $\sin(\alpha + \beta + \gamma)$.

7.3 Cofunction and Related Identities

OBJECTIVES

You should be able to:

1 State the cofunction identities for the sine and the cosine, and derive cofunction identities for the other functions. Use the cofunction identities and a given function value for an acute angle to find the function value for its complement.

2 Apply reduction formulas to find identities.

3 Given an expression
$$c \sin ax + d \cos ax,$$
find an equivalent expression $A \sin(ax + b)$; and given an equation
$$y = c \sin ax + d \cos ax,$$
find an equation of the type $y = A \sin(ax + b)$ and then graph the equation.

Prove each of the following identities.

1. $\cot\left(\frac{\pi}{2} - \theta\right) = \tan\theta$

2. $\sec\left(\frac{\pi}{2} - \theta\right) = \csc\theta$

1 Cofunction Identities

In Section 7.2, we developed the following two identities, which we used to prove two other important identities:
$$\cos\left(\frac{\pi}{2} - \theta\right) = \sin\theta \quad\text{and}\quad \sin\left(\frac{\pi}{2} - \theta\right) = \cos\theta.$$

We can now use these to prove another related identity:
$$\tan\left(\frac{\pi}{2} - \theta\right) = \frac{\sin\left(\frac{x}{2} - \theta\right)}{\cos\left(\frac{\pi}{2} - \theta\right)} = \frac{\cos\theta}{\sin\theta} = \cot\theta.$$

DO EXERCISES 1–3. (EXERCISE 3 IS ON THE FOLLOWING PAGE.)

These identities turn out to be very important unto themselves. Each yields a conversion to a cofunction. For this reason, we call them **cofunction identities.** They are summarized below.

Cofunction Identities

$$\sin\left(\frac{\pi}{2} - \theta\right) = \cos\theta, \quad \tan\left(\frac{\pi}{2} - \theta\right) = \cot\theta, \quad \sec\left(\frac{\pi}{2} - \theta\right) = \csc\theta,$$
$$\cos\left(\frac{\pi}{2} - \theta\right) = \sin\theta, \quad \cot\left(\frac{\pi}{2} - \theta\right) = \tan\theta, \quad \csc\left(\frac{\pi}{2} - \theta\right) = \sec\theta$$

These identities hold for all real numbers, and thus, for all degree measures, but if we restrict θ to values such that $0° < \theta < 90°$, then we have a special application to the acute angles of a right triangle.

In a right triangle, the acute angles are complementary, since the sum of all three angle measures is 180° and the right angle accounts for 90° of this total. Thus if one acute angle of a right triangle is θ, the other is $90° - \theta$, or $\pi/2 - \theta$. Note that the sine of $\angle A$ is also the cosine of $\angle B$, its complement:

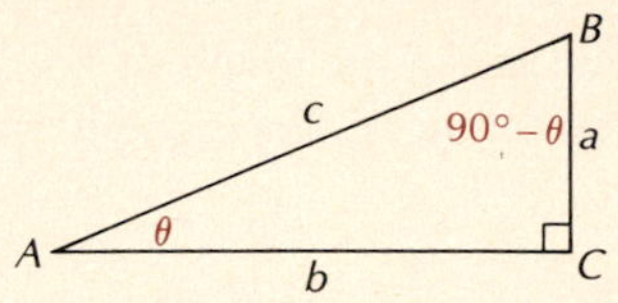

Similarly, the tangent of $\angle A$ is the cotangent of its complement, and the secant of $\angle A$ is the cosecant of its complement.

These pairs of functions are called **cofunctions**. The name **cosine** originally meant the sine of the complement, **cotangent** meant the tangent of the complement, and **cosecant** meant the secant of the complement.

EXAMPLE 1 Given that

$$\sin 18° = 0.3090, \quad \cos 18° = 0.9511,$$
$$\tan 18° = 0.3249, \quad \cot 18° = 3.078,$$
$$\sec 18° = 1.051, \quad \csc 18° = 3.236,$$

find the six function values for 72°.

Solution Since 72° and 18° are complements, we have $\sin 72° = \cos 18°$, and so on, and the function values are

$$\sin 72° = 0.9511, \quad \cos 72° = 0.3090,$$
$$\tan 72° = 3.078, \quad \cot 72° = 0.3249,$$
$$\sec 72° = 3.236, \quad \csc 72° = 1.051.$$

DO EXERCISE 4.

Other Cofunction Identities

Suppose we wanted to derive an identity for

$$\sin\left(\theta + \frac{\pi}{2}\right).$$

We have at least two ways to prove this identity. One way to do so is to use the identity for the sine of a sum developed in Section 7.2:

$$\begin{aligned}\sin\left(\theta + \frac{\pi}{2}\right) &= \sin\theta\cos\frac{\pi}{2} + \cos\theta\sin\frac{\pi}{2}\\ &= \sin\theta \cdot 0 + \cos\theta \cdot 1\\ &= \cos\theta.\end{aligned}$$

Another proof of this identity can be found by considering the left side in terms of x in the form of a function, as follows:

$$g(x) = \sin\left(x + \frac{\pi}{2}\right) = \sin\left[x - \left(-\frac{\pi}{2}\right)\right].$$

3. $\csc\left(\frac{\pi}{2} - \theta\right) = \sec\theta$

4. Given that

$$\sin 75° = 0.9659, \quad \cos 75° = 0.2588,$$
$$\tan 75° = 3.732, \quad \cot 75° = 0.2679,$$
$$\sec 75° = 3.864, \quad \csc 75° = 1.035,$$

find the six function values for the complement of 75°.

5. a) Graph $y = \cos x$.
   b) Translate to obtain a graph of
   $$y = \cos\left(x - \frac{\pi}{2}\right).$$
   c) Graph $y = \sin x$.
   d) How do the graphs of parts (b) and (c) compare?
   e) Write the identity thus established.

6. Prove the identity
   $$\cos\left(x - \frac{\pi}{2}\right) = \sin x$$
   using the formula for the cosine of a difference.

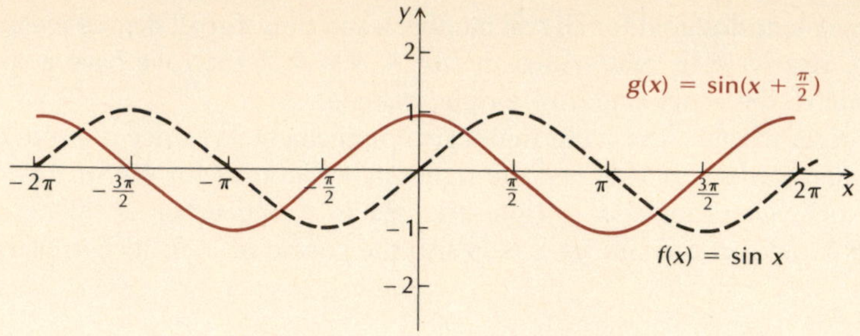

The graph of g is found by translating the graph of $f(x) = \sin x$ to the left a distance of $\pi/2$ units. The translation is also a graph of the cosine function. Thus, we obtain the identity

$$\sin\left(x + \frac{\pi}{2}\right) = \cos x.$$

DO EXERCISE 5.

In Margin Exercise 5, you gave a graphical proof of the following identity:

$$\cos\left(x - \frac{\pi}{2}\right) = \sin x.$$

We can also give a proof using the identity $\cos(-x) = \cos x$ and a cofunction identity:

$$\cos\left(x - \frac{\pi}{2}\right) = \cos\left[-\left(\frac{\pi}{2} - x\right)\right] = \cos\left(\frac{\pi}{2} - x\right) = \sin x.$$

Yet another proof can be given using the formula for the cosine of a difference.

DO EXERCISE 6.

Consider the following graph. The graph of $f(x) = \sin x$ has been translated to the right a distance of $\pi/2$, to obtain the graph of $g(x) = \sin(x - \pi/2)$. The latter is a reflection of the cosine function across the x-axis. In other words, it is a graph of $g(x) = -\cos x$. We thus obtain the following identity:

$$\sin\left(x - \frac{\pi}{2}\right) = -\cos x.$$

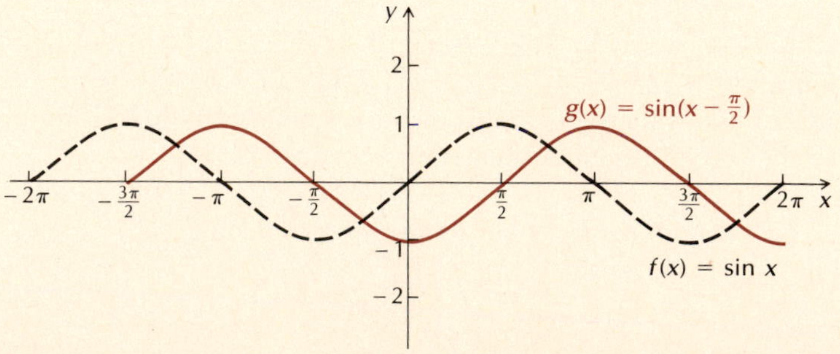

We have at least two other ways to prove this identity. One way to do so is to use the identity $\sin(-x) = -\sin x$ and a cofunction identity:

$$\sin\left(x - \frac{\pi}{2}\right) = \sin\left[-\left(\frac{\pi}{2} - x\right)\right] = -\sin\left(\frac{\pi}{2} - x\right) = -\cos x.$$

Yet another proof can be given using the formula for the sine of a difference.

DO EXERCISE 7.

By using the formula for the cosine of a sum or a graphical method, we can prove the following identity:

$$\cos\left(x + \frac{\pi}{2}\right) = -\sin x.$$

DO EXERCISE 8.

We now have established four more cofunction identities. We give those involving sine and cosine in the following list. Two of them are obtained by taking the top signs, the other two by taking the bottom signs.

Cofunction Identities for Sine and Cosine

$$\sin\left(x \pm \frac{\pi}{2}\right) = \pm\cos x, \qquad \cos\left(x \pm \frac{\pi}{2}\right) = \mp\sin x.$$

These should be learned. A good memory device is to think of the "s" in "sine" as "same," meaning that the signs stay the same. For cosine, think of "c" for "change," meaning that the signs change. In all cases, there is a change to a cofunction. Cofunction identities for the other functions can be obtained from these easily, using the definitions of the other basic circular functions.

EXAMPLE 2 Find an identity for $\tan(x + \pi/2)$.

Solution By definition of the tangent function,

$$\tan\left(x + \frac{\pi}{2}\right) = \frac{\sin\left(x + \frac{\pi}{2}\right)}{\cos\left(x + \frac{\pi}{2}\right)}.$$

Using a cofunction identity, we obtain

$$\frac{\sin\left(x + \frac{\pi}{2}\right)}{\cos\left(x + \frac{\pi}{2}\right)} = \frac{\cos x}{-\sin x} = -\frac{\cos x}{\sin x} = -\cot x.$$

Thus the identity we seek is

$$\tan\left(x + \frac{\pi}{2}\right) = -\cot x.$$

DO EXERCISE 9.

EXAMPLE 3 Find an identity for $\sec(x - \pi/2)$.

Solution By definition of the secant function,

$$\sec\left(x - \frac{\pi}{2}\right) = \frac{1}{\cos\left(x - \frac{\pi}{2}\right)}.$$

7. Prove the identity

$$\sin\left(x - \frac{\pi}{2}\right) = -\cos x$$

using the formula for the sine of a difference.

8. Prove the identity

$$\cos\left(x + \frac{\pi}{2}\right) = -\sin x$$

a) by graphing. Explain your proof.

b) by using the formula for the cosine of a sum.

9. Find an identity for $\cot\left(x + \frac{\pi}{2}\right)$.

10. Find an identity for $\csc\left(x - \frac{\pi}{2}\right)$.

Using a cofunction identity, we obtain

$$\frac{1}{\cos\left(x - \frac{\pi}{2}\right)} = \frac{1}{\sin x} = \csc x.$$

Thus the identity we seek is

$$\sec\left(x - \frac{\pi}{2}\right) = \csc x.$$

DO EXERCISE 10.

2 Reduction Formulas

In Margin Exercises 11 and 12, you are asked to prove the following identities using the formulas for the sine and cosine of a sum:

$$\sin(x + \pi) = -\sin x \quad \text{and} \quad \tan(\theta + 270°) = -\cot\theta.$$

Prove each of the following identities.

11. $\sin(x + \pi) = -\sin x$

12. $\tan(\theta + 270°) = -\cot\theta$

DO EXERCISES 11 AND 12.

We can go on from here and develop a set of identities for such expressions as $\sin(x \pm \pi)$ or $\tan(x \pm 3\pi/2)$ or $\cos(\pm 90°n \pm \theta)$. In the exercises, you will be asked to prove many identities for expressions like these. All of these fall into a class of identities called **reduction formulas.** Rather than have you memorize them all, we are going to consider a procedure for simplifying such identities quickly.

Suppose that f is any of the six basic trigonometric functions: sine, cosine, tangent, cotangent, secant, or cosecant. Then

$$f\left(\frac{\pi}{2}n \pm \theta\right) = \pm g(\theta),$$

where g is either the same function or the cofunction of f.

If n is even, then g is f. If n is odd, then g is the cofunction of f.

The $\pm$ signs do not necessarily correspond. To determine which sign to use, consider a value of θ, but think of it being in the first quadrant. Then determine the quadrant in which $(\pi/2)n \pm \theta$ lies. Then determine the sign of $f[(\pi/2)n \pm \theta]$ in that quadrant. Attach that sign to g on the right side. The result is an identity that holds for all θ.

EXAMPLE 4 Find an identity for $\tan(\theta + 270°)$.

Solution We first express

$$\tan(\theta + 270°)$$

as

$$\tan(\theta + 270°) = \tan(90° \cdot 3 + \theta).$$

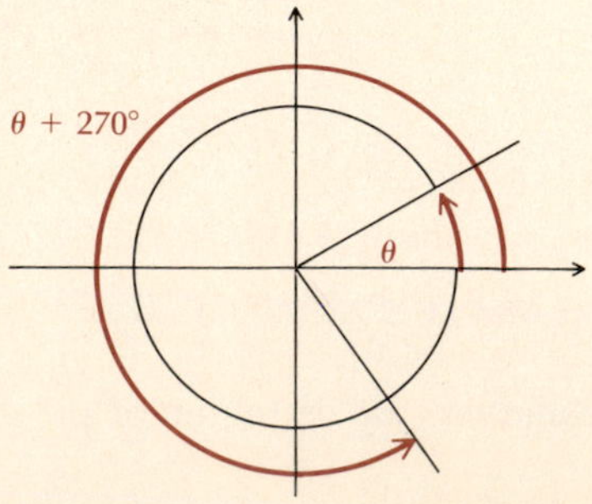

Since $n = 3$ and 3 is odd, we know that

$$\tan(\theta + 270°) = \pm\cot\theta.$$

Suppose that θ is in the first quadrant.
Then $\theta + 270°$ is in the fourth quadrant. Since the tangent function is negative

in the fourth quadrant, we use a minus sign. Thus,

$$\tan(\theta + 270^\circ) = -\cot\theta.$$

This identity holds for any degree measure. ■

DO EXERCISES 13–15.

3 An Identity for $c \sin ax + d \cos ax$

An expression $c \sin ax + d \cos ax$, where a, c, and d are constants, is equivalent to an expression $A \sin(ax + b)$, where A and b are also constants. To prove this, we look for the numbers A and b. First, let us consider the numbers $c/\sqrt{c^2 + d^2}$ and $d/\sqrt{c^2 + d^2}$. These numbers are the coordinates of a point on the unit circle because

$$\left(\frac{c}{\sqrt{c^2 + d^2}}\right)^2 + \left(\frac{d}{\sqrt{c^2 + d^2}}\right)^2 = 1.$$

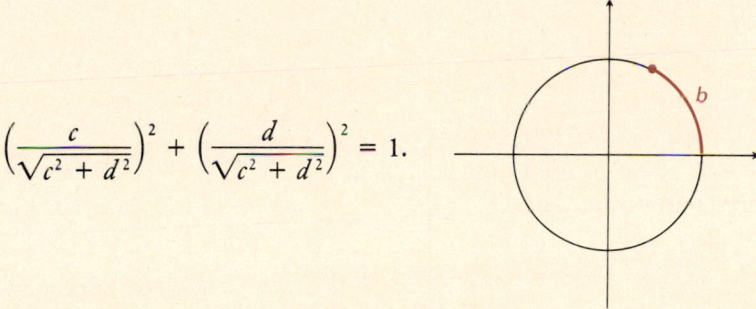

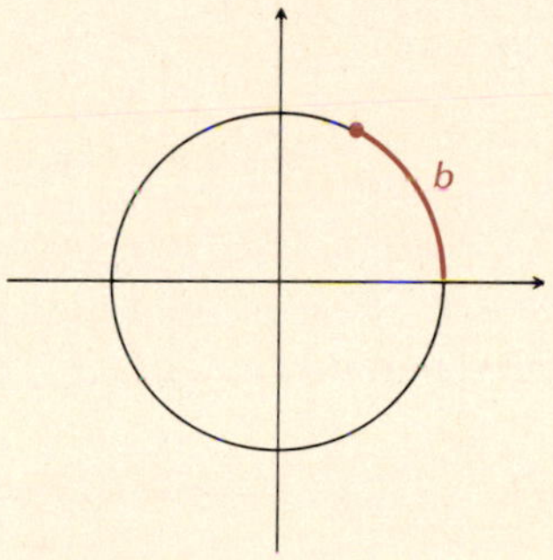

Thus there is a number b for which these numbers are the cosine and the sine, respectively. In other words,

$$\frac{c}{\sqrt{c^2 + d^2}} = \cos b \quad \text{and} \quad \frac{d}{\sqrt{c^2 + d^2}} = \sin b,$$

for some number b. Now we will consider the expression $c \sin ax + d \cos ax$, and multiply it by 1, as follows:

$$\begin{aligned} &\frac{\sqrt{c^2 + d^2}}{\sqrt{c^2 + d^2}} [c \sin ax + d \cos ax] \\ &\quad = \sqrt{c^2 + d^2}\left[(\sin ax) \cdot \frac{c}{\sqrt{c^2 + d^2}} + (\cos ax)\frac{d}{\sqrt{c^2 + d^2}}\right] \\ &\quad = \sqrt{c^2 + d^2}[\sin ax \cos b + \cos ax \sin b]. \end{aligned}$$

Now, by the identity for the sine of a sum, we obtain

$$\sqrt{c^2 + d^2} \sin(ax + b).$$

Thus we have an identity as follows.

THEOREM 2

For any numbers a, c, and d,

$$c \sin ax + d \cos ax = A \sin(ax + b),$$

where $A = \sqrt{c^2 + d^2}$ and b is a number whose cosine is c/A and whose sine is d/A.

EXAMPLE 5 Find an expression equivalent to $\sqrt{3} \sin x + \cos x$ that involves only the sine function.

Find an identity for each of the following.

13. $\tan(\theta + 180^\circ)$

14. $\sin(270^\circ - \theta)$

15. $\cos(3\pi - x)$

Find an equivalent expression involving only the sine function.

16. $\sin 2x + \cos 2x$

17. $12 \sin x - 5 \cos x$

Solution

$$A = \sqrt{c^2 + d^2} = \sqrt{3 + 1} = 2, \qquad \cos b = \frac{\sqrt{3}}{2}, \quad \text{and} \quad \sin b = \frac{1}{2}.$$

Thus we can use 30°, or $\pi/6$, for b. Therefore,

$$\sqrt{3} \sin x + \cos x = 2 \sin\left(x + \frac{\pi}{6}\right).$$

EXAMPLE 6 Find an expression equivalent to $\sqrt{3} \sin (\pi/4)t + \cos (\pi/4)t$ involving only the sine function.

Solution In this case, $c = \sqrt{3}$, $d = 1$, and $a = \pi/4$:

$$A = \sqrt{3 + 1} = 2, \qquad \cos b = \frac{\sqrt{3}}{2}, \quad \text{and} \quad \sin b = \frac{1}{2}.$$

Thus we have

$$\sqrt{3} \sin \frac{\pi}{4}t + \cos \frac{\pi}{4}t = 2 \sin\left(\frac{\pi}{4}t + \frac{\pi}{6}\right).$$

EXAMPLE 7 Find an expression equivalent to $3 \sin 2x - 4 \cos 2x$ involving only the sine function.

Solution In this case, $c = 3$, $d = -4$, and $a = 2$:

$$A = \sqrt{3^2 + (-4)^2} = 5, \qquad \cos b = \frac{3}{5}, \quad \text{and} \quad \sin b = -\frac{4}{5}.$$

Thus we have $3 \sin 2x - 4 \cos 2x = 5 \sin (2x + b)$, where b is a number whose cosine is $\frac{3}{5}$ and whose sine is $-\frac{4}{5}$.

DO EXERCISES 16 AND 17.

We can use the expression $y = A \sin (ax + b)$ for graphing purposes.

EXAMPLE 8 For

$$y = \sin x - \cos x,$$

find an equation in the form $y = A \sin (ax + b)$. Then graph.

Solution We first transform the right-hand side to an expression involving only the sine function. Since $c = 1$ and $d = -1$, we have

$$A = \sqrt{1^2 + (-1)^2} = \sqrt{2}, \qquad \cos b = \frac{1}{\sqrt{2}}, \quad \text{and} \quad \sin b = -\frac{1}{\sqrt{2}}.$$

So $b = -\pi/4$ and we have $y = \sqrt{2} \sin (x - \pi/4)$. This can be graphed by stretching the sine graph vertically and translating it $\pi/4$ units to the right.

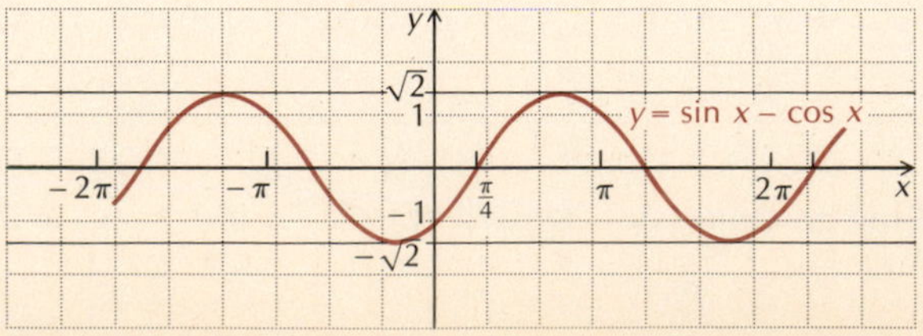

Skill in making graphs like the one in Example 10 is important in such fields as engineering, electricity, and physics.

DO EXERCISE 18.

18. Graph: $y = \sqrt{3}\sin x + \cos x$.

EXERCISE SET 7.3

1

1. Given that

$$\sin 65° = 0.9063, \quad \cos 65° = 0.4226,$$
$$\tan 65° = 2.145, \quad \cot 65° = 0.4663,$$
$$\sec 65° = 2.366, \quad \csc 65° = 1.103,$$

find the six function values for 58°.

2. Given that

$$\sin 32° = 0.5299, \quad \cos 32° = 0.8480,$$
$$\tan 32° = 0.6249, \quad \cot 32° = 1.600,$$
$$\sec 32° = 1.179, \quad \csc 32° = 1.887,$$

find the six function values for 58°.

3. Given that

$$\sin 22.5° = \frac{\sqrt{2 - \sqrt{2}}}{2}$$

and

$$\cos 22.5° = \frac{\sqrt{2 + \sqrt{2}}}{2},$$

find:

a) the other four function values for 22.5°;
b) the six function values for 67.5°.

4. Given that

$$\sin \frac{\pi}{12} = \frac{\sqrt{2 - \sqrt{3}}}{2}$$

and

$$\cos \frac{\pi}{12} = \frac{\sqrt{2 + \sqrt{3}}}{2},$$

find:

a) the other four function values for $\pi/12$;
b) the six function values for $11\pi/12$.

2

5. Find and prove an identity for $\tan(x - \pi/2)$.
6. Find and prove an identity for $\cot(x - \pi/2)$.
7. Find and prove an identity for $\csc(x + \pi/2)$.
8. Find and prove an identity for $\sec(x + \pi/2)$.
9. Find and prove an identity for $\sin(x - \pi)$.
10. Find and prove an identity for $\cos(\theta - 270°)$.

Find an identity for each of the following.

11. $\tan(\theta - 270°)$
12. $\sin(\theta - 270°)$
13. $\sin(\theta + 450°)$
14. $\cos(\theta - 450°)$
15. $\sec(\pi - \theta)$
16. $\csc(\pi + \theta)$
17. $\tan(x - 4\pi)$
18. $\sec(x - 10\pi)$
19. $\cos\left(\frac{9\pi}{2} + x\right)$
20. $\cot\left(\frac{9\pi}{2} - x\right)$
21. $\csc(540° - \theta)$
22. $\tan(540° + \theta)$
23. $\cot\left(x - \frac{3\pi}{2}\right)$
24. $\sin\left(x + \frac{11\pi}{2}\right)$

3 Find an equivalent expression involving the sine function only.

25. $\sin 2x + \sqrt{3}\cos 2x$
26. $\sqrt{3}\sin 3x - \cos 3x$
27. $4\sin x + 3\cos x$
28. $4\sin 2x - 3\cos 2x$
29. $6.75\sin 0.374x + 4.08\cos 0.374x$
30. $97.81\sin 0.8081x - 4.89\cos 0.8081x$

Find an equation of the form $y = A\sin(ax + b)$. Then graph.

31. $y = \sin 2x - \cos 2x$
32. $y = \sin x + \sqrt{3}\cos x$

SYNTHESIS

33. Given that $\sin\theta = \frac{1}{3}$ and that the terminal side is in quadrant II.

a) Find the other function values for θ.
b) Find the six function values for $\pi/2 - \theta$.
c) Find the six function values for $\pi + \theta$.
d) Find the six function values for $\pi - \theta$.
e) Find the six function values for $2\pi - \theta$.

34. Given that $\cos\theta = \frac{4}{5}$ and that the terminal side is in quadrant IV.

a) Find the other function values for θ.
b) Find the six function values for $\pi/2 + \theta$.
c) Find the six function values for $\pi + \theta$.
d) Find the six function values for $\pi - \theta$.
e) Find the six function values for $2\pi - \theta$.

35. Given that $\sin 27° = 0.45399$.

a) Find the other function values for 27°.
b) Find the six function values for 63°.

36. Given that $\cot 54° = 0.72654$.

a) Find the other function values for 54°.
b) Find the six function values for 36°.

37. Given that $\cos 38° = 0.78801$, find the six function values for 128°.

38. Given that $\tan 73° = 3.27085$, find the six function values for 343°.

Simplify.

39. $\sin\left(\frac{\pi}{2} - x\right)[\sec x - \cos x]$

40. $\cos\left(\frac{\pi}{2} - x\right)[\csc x - \sin x]$

41. $\sin x \cos y - \cos\left(x + \frac{\pi}{2}\right)\tan y$

42. $\cos\left(x - \frac{\pi}{2}\right)\tan y + \sin x \cot y$

43. $\cos(\pi - x) + \cot x \sin\left(x - \frac{\pi}{2}\right)$

44. $\sin(\pi - x) - \tan x \cos\left(\frac{\pi}{2} - x\right)$

45. $\dfrac{\cos^2 x - 1}{\sin\left(\frac{\pi}{2} - x\right) - 1}$

46. $\dfrac{\sin^2 x - 1}{\cos\left(\frac{\pi}{2} - x\right) + 1}$

47. $\dfrac{\sin x - \cos\left(x - \frac{\pi}{2}\right)\cos x}{-\sin x - \cos\left(x - \frac{\pi}{2}\right)\tan x}$

48. $\dfrac{\cos x - \sin\left(\frac{\pi}{2} - x\right)\sin x}{\cos x - \cos(\pi - x)\tan x}$

49. $\dfrac{\cos^2 x + 2\sin\left(x - \frac{\pi}{2}\right) + 1}{\sin\left(\frac{\pi}{2} - x\right) - 1}$

50. $\dfrac{\sin^2 x - 2\cos\left(x - \frac{\pi}{2}\right) + 1}{\cos\left(\frac{\pi}{2} - x\right) - 1}$

51. $\dfrac{\sin^2 y \cos\left(y + \frac{\pi}{2}\right)}{\cos^2 y \cos\left(\frac{\pi}{2} - y\right)}$

52. $\dfrac{\cos^2 y \sin\left(y + \frac{\pi}{2}\right)}{\sin^2 y \sin\left(\frac{\pi}{2} - y\right)}$

OBJECTIVES

You should be able to:

1. Use the double-angle identities to find function values of twice an angle when one function value is known for that angle.
2. Use the half-angle identities to find the function values of half an angle when one function value is known for that angle.
3. Simplify certain trigonometric expressions using the half-angle and the double-angle formulas. Also, derive some identities by simplifying.

7.4 Some Important Identities

Two important classes of trigonometric identities are known as the *double-angle identities* and the *half-angle identities*.

1 Double-Angle Identities

To develop these identities, we will use the sum formulas from the preceding section. We first develop a formula for $\sin 2\theta$. Recall that

$$\sin(\alpha + \beta) = \sin\alpha\cos\beta + \cos\alpha\sin\beta.$$

We will consider a number θ and substitute it for both α and β in this identity. We obtain

$$\begin{aligned}\sin(\theta + \theta) &= \sin 2\theta \\ &= \sin\theta\cos\theta + \cos\theta\sin\theta \\ &= 2\sin\theta\cos\theta.\end{aligned}$$

The identity we seek is

$$\boldsymbol{\sin 2\theta = 2\sin\theta\cos\theta.}$$

EXAMPLE 1 If $\sin\theta = \frac{3}{8}$ and θ is in the first quadrant, what is $\sin 2\theta$?

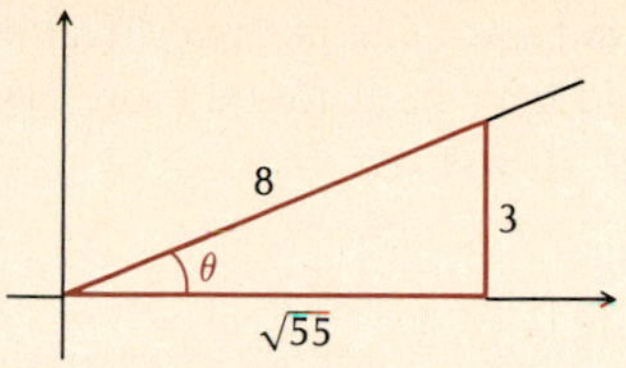

Solution From the figure, we see that $\cos\theta = \sqrt{55}/8$. Thus,

$$\sin 2\theta = 2\sin\theta\cos\theta = 2\cdot\frac{3}{8}\cdot\frac{\sqrt{55}}{8} = \frac{3\sqrt{55}}{32}.$$

DO EXERCISE 1.

1. Given that $\sin\theta = \frac{3}{5}$ and that θ is in the first quadrant, what is $\sin 2\theta$?

Double-angle identities for the cosine and tangent functions can be derived in much the same way as the identity above:

$$\cos(\alpha+\beta) = \cos\alpha\cos\beta - \sin\alpha\sin\beta$$

$$\cos 2\theta = \cos(\theta+\theta) = \cos\theta\cos\theta - \sin\theta\sin\theta$$

$$= \cos^2\theta - \sin^2\theta$$

$$\boldsymbol{\cos 2\theta = \cos^2\theta - \sin^2\theta.}$$

Now we derive an identity for $\tan 2\theta$:

$$\tan(\alpha+\beta) = \frac{\tan\alpha+\tan\beta}{1-\tan\alpha\tan\beta}$$

$$\tan 2\theta = \tan(\theta+\theta) = \frac{\tan\theta+\tan\theta}{1-\tan\theta\tan\theta} = \frac{2\tan\theta}{1-\tan^2\theta}$$

$$\boldsymbol{\tan 2\theta = \frac{2\tan\theta}{1-\tan^2\theta}.}$$

EXAMPLE 2 Given that $\tan\theta = -\frac{3}{4}$ and θ is in the second quadrant, find $\sin 2\theta$, $\cos 2\theta$, $\tan 2\theta$, and the quadrant in which 2θ lies.

Solution By drawing a diagram as shown, we find that

$$\sin\theta = \tfrac{3}{5} \text{ and } \cos\theta = -\tfrac{4}{5}.$$

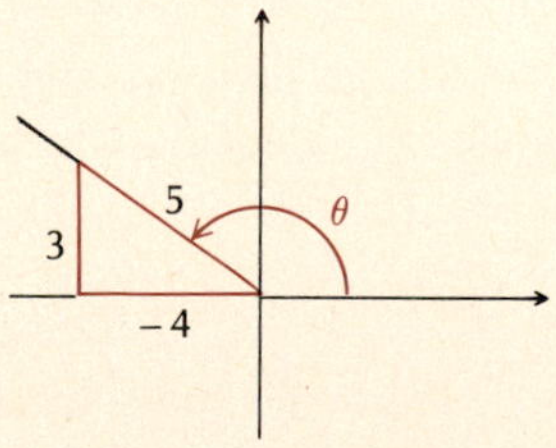

Now,

$$\sin 2\theta = 2\sin\theta\cos\theta$$

$$= 2\cdot\frac{3}{5}\cdot\left(-\frac{4}{5}\right) = -\frac{24}{25};$$

$$\cos 2\theta = \cos^2\theta - \sin^2\theta = \left(-\frac{4}{5}\right)^2 - \left(\frac{3}{5}\right)^2$$

$$= \frac{16}{25} - \frac{9}{25} = \frac{7}{25};$$

$$\tan 2\theta = \frac{2\tan\theta}{1-\tan^2\theta} = \frac{2\cdot\left(-\frac{3}{4}\right)}{1-\left(-\frac{3}{4}\right)^2}$$

$$= \frac{-\frac{3}{2}}{1-\frac{9}{16}} = -\frac{24}{7}.$$

2. Given that $\cos\theta = -\frac{5}{13}$ and that θ is in the third quadrant, find $\sin 2\theta$, $\cos 2\theta$, and $\tan 2\theta$. Also, determine the quadrant in which 2θ lies.

Since $\sin 2\theta$ is negative and $\cos 2\theta$ is positive, we know that 2θ is in quadrant IV. Note that $\tan 2\theta$ could have been found more easily in this case by simply dividing the values of $\sin 2\theta$ and $\cos 2\theta$. ■

DO EXERCISE 2.

Two other useful identities for $\cos 2\theta$ can be derived easily, as follows:

$$\begin{aligned}\cos 2\theta &= \cos^2\theta - \sin^2\theta \\ &= (1 - \sin^2\theta) - \sin^2\theta \qquad \text{Using } \sin^2\theta + \cos^2\theta = 1 \\ &= 1 - 2\sin^2\theta.\end{aligned}$$

Similarly,

$$\begin{aligned}\cos 2\theta &= \cos^2\theta - \sin^2\theta \\ &= \cos^2\theta - (1 - \cos^2\theta) \\ &= 2\cos^2\theta - 1.\end{aligned}$$

Solving these two identities for $\sin^2\theta$ and $\cos^2\theta$, respectively, we obtain two more identities that are often useful. Following is a list of **double-angle identities.** It should be memorized.

$$\begin{aligned}\sin 2\theta &= 2\sin\theta\cos\theta; \\ \cos 2\theta &= \cos^2\theta - \sin^2\theta \\ &= 1 - 2\sin^2\theta \\ &= 2\cos^2\theta - 1; \\ \tan 2\theta &= \frac{2\tan\theta}{1 - \tan^2\theta}; \\ \sin^2\theta &= \frac{1 - \cos 2\theta}{2}; \\ \cos^2\theta &= \frac{1 + \cos 2\theta}{2}\end{aligned}$$

Using division and the last two identities, we can easily deduce the following identity, which is also often useful:

$$\tan^2\theta = \frac{1 - \cos 2\theta}{1 + \cos 2\theta}.$$

From the basic identities (in the lists to be memorized), others can be obtained.

EXAMPLE 3 Find a formula for $\sin 3\theta$ in terms of function values of θ.

Solution

$$\begin{aligned}\sin 3\theta &= \sin(2\theta + \theta) \\ &= \sin 2\theta\cos\theta + \cos 2\theta\sin\theta \\ &= (2\sin\theta\cos\theta)\cos\theta + (2\cos^2\theta - 1)\sin\theta \\ &= 2\sin\theta\cos^2\theta + 2\sin\theta\cos^2\theta - \sin\theta \\ &= 4\sin\theta\cos^2\theta - \sin\theta\end{aligned}$$ ■

EXAMPLE 4 Find a formula for $\cos^3 x$ in terms of function values of x or $2x$, raised only to the first power.

Solution

$$\cos^3 x = \cos^2 x \cos x = \frac{1 + \cos 2x}{2} \cos x$$

DO EXERCISES 3 AND 4.

2 Half-Angle Identities

To develop these identities, we use three of the previously developed ones. As shown below, we take square roots and replace θ by $\phi/2$:

$$\sin^2 \theta = \frac{1 - \cos 2\theta}{2} \longrightarrow \left|\sin \frac{\phi}{2}\right| = \sqrt{\frac{1 - \cos \phi}{2}},$$

$$\cos^2 \theta = \frac{1 + \cos 2\theta}{2} \longrightarrow \left|\cos \frac{\phi}{2}\right| = \sqrt{\frac{1 + \cos \phi}{2}},$$

$$\tan^2 \theta = \frac{1 - \cos 2\theta}{1 + \cos 2\theta} \longrightarrow \left|\tan \frac{\phi}{2}\right| = \sqrt{\frac{1 - \cos \phi}{1 + \cos \phi}}.$$

The half-angle formulas are those on the right above. We can eliminate the absolute-value signs by introducing $\pm$ signs, with the understanding that our use of + or − depends on the quadrant in which the angle $\phi/2$ lies. We thus obtain these formulas in the following form.

$$\sin \frac{\phi}{2} = \pm\sqrt{\frac{1 - \cos \phi}{2}},$$

$$\cos \frac{\phi}{2} = \pm\sqrt{\frac{1 + \cos \phi}{2}},$$

$$\tan \frac{\phi}{2} = \pm\sqrt{\frac{1 - \cos \phi}{1 + \cos \phi}}$$

These formulas should be memorized.

There are two other formulas for $\tan (\phi/2)$ that are often useful. They can be obtained as follows:

$$\left|\tan \frac{\phi}{2}\right| = \sqrt{\frac{1 - \cos \phi}{1 + \cos \phi}} = \sqrt{\frac{1 - \cos \phi}{1 + \cos \phi} \cdot \frac{1 + \cos \phi}{1 + \cos \phi}}$$

$$= \sqrt{\frac{1 - \cos^2 \phi}{(1 + \cos \phi)^2}} = \sqrt{\frac{\sin^2 \phi}{(1 + \cos \phi)^2}}$$

$$= \frac{|\sin \phi|}{|1 + \cos \phi|}.$$

Now $1 + \cos \phi$ cannot be negative because $\cos \phi$ is never less than -1. Thus the absolute-value signs are not necessary in the denominator. As the following graph shows, $\tan (\phi/2)$ and $\sin \phi$ have the same sign for all ϕ for which $\tan (\phi/2)$ is defined.

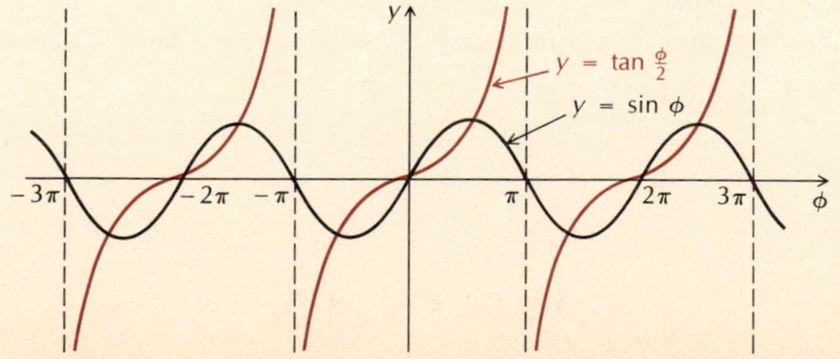

3. Find a formula for $\cos 3\theta$ in terms of function values of θ.

4. Find a formula for $\sin^3 x$ in terms of function values of x or $2x$, raised only to the first power.

5. Find $\cos 15°$.

6. Find $\tan \frac{\pi}{12}$.

Thus we can dispense with the other absolute-value signs, and obtain the formula we seek. A second formula can be obtained in a similar way.

$$\tan \frac{\phi}{2} = \frac{\sin \phi}{1 + \cos \phi}, \qquad \tan \frac{\phi}{2} = \frac{1 - \cos \phi}{\sin \phi}$$

These formulas have the advantage that they give the sign of $\tan (\phi/2)$ directly.

EXAMPLE 5 Find $\sin 15°$.

Solution

$$\sin 15° = \sin \frac{30°}{2} = \pm \sqrt{\frac{1 - \cos 30°}{2}}$$
$$= \pm \sqrt{\frac{1 - (\sqrt{3}/2)}{2}} = \pm \sqrt{\frac{2 - \sqrt{3}}{4}}$$
$$= \frac{\sqrt{2 - \sqrt{3}}}{2}$$

The expression is positive, because 15° is in the first quadrant. ■

EXAMPLE 6 Find $\tan \frac{\pi}{8}$.

Solution

$$\tan \frac{\pi}{8} = \tan \frac{\frac{\pi}{4}}{2} = \frac{\sin \frac{\pi}{4}}{1 + \cos \frac{\pi}{4}} = \frac{\frac{\sqrt{2}}{2}}{1 + \frac{\sqrt{2}}{2}}$$
$$= \frac{\sqrt{2}}{2 + \sqrt{2}}$$
$$= \sqrt{2} - 1 \qquad \text{Rationalizing the denominator}$$ ■

DO EXERCISES 5 AND 6.

3 Simplification

Many simplifications of trigonometric expressions are possible through the use of the identities that we have developed.

EXAMPLE 7 Simplify:

$$\frac{1 - \cos 2x}{4 \sin x \cos x}.$$

Solution We search the list of identities, looking for some substitution that will simplify the expression. In this case, one might note that the denominator is $2(2 \sin x \cos x)$ and thus find a simplification, since $2 \sin x \cos x = \sin 2x$. This gives us

$$\frac{1 - \cos 2x}{2 \sin 2x}.$$

Now

$$\frac{1 - \cos \phi}{\sin \phi} = \tan \frac{\phi}{2}.$$

Using this, we obtain

$$\frac{1}{2} \cdot \frac{1 - \cos 2x}{\sin 2x} = \frac{1}{2} \tan x.$$

In many cases, we can derive an identity by simplifying an expression, as in the following example.

EXAMPLE 8 By simplifying, derive an identity:

$$\frac{\sin x \cos x}{\frac{1}{2} \cos 2x}.$$

Solution We can obtain $2 \sin x \cos x$ in the numerator by multiplying the expression by $\frac{2}{2}$:

$$\frac{2 \sin x \cos x}{2 \cdot \frac{1}{2} \cos 2x} = \frac{\sin 2x}{\cos 2x} = \tan 2x.$$

We have thus derived the identity

$$\frac{\sin x \cos x}{\frac{1}{2} \cos 2x} = \tan 2x.$$

DO EXERCISES 7 AND 8.

7. Simplify:

$$\frac{2(\tan x + \tan^3 x)}{1 - \tan^4 x}.$$

8. By simplifying, derive an identity:

$$\frac{\frac{1}{2}(\cos^2 x - \sin^2 x)}{\sin x \cos x}.$$

EXERCISE SET 7.4

1 Find $\sin 2\theta$, $\cos 2\theta$, $\tan 2\theta$, and the quadrant in which 2θ lies.

1. $\sin \theta = \frac{4}{5}$; θ in quadrant I
2. $\sin \theta = \frac{5}{13}$; θ in quadrant I
3. $\cos \theta = -\frac{4}{5}$; θ in quadrant III
4. $\cos \theta = -\frac{3}{5}$; θ in quadrant III
5. $\tan \theta = \frac{4}{3}$; θ in quadrant III
6. $\tan \theta = \frac{3}{4}$; θ in quadrant III
7. Find a formula for $\sin 4\theta$ in terms of function values of θ.
8. Find a formula for $\cos 4\theta$ in terms of function values of θ.
9. Find a formula for $\sin^4 \theta$ in terms of function values of θ or 2θ or 4θ, raised only to the first power.
10. Find a formula for $\cos^4 \theta$ in terms of function values of θ or 2θ or 4θ, raised only to the first power.

2 Find each of the following without using a calculator.

11. $\sin 75°$ (*Hint:* $75 = 150/2$.)
12. $\cos 75°$
13. $\tan 75°$
14. $\tan 67.5°$ (*Hint:* $67.5 = 135/2$.)
15. $\sin \frac{5\pi}{8}$
16. $\cos \frac{5\pi}{8}$
17. $\sin 112.5°$
18. $\cos 22.5°$
19. $\cos \frac{\pi}{8}$
20. $\tan 15°$
21. $\sin 22.5°$
22. $\tan 112.5°$

1, **2** ▦ Given that $\sin \theta = 0.3416$ and that θ is in the first quadrant, find each of the following.

23. $\sin 2\theta$
24. $\cos 2\theta$
25. $\sin 4\theta$
26. $\cos 4\theta$
27. $\sin \frac{\theta}{2}$
28. $\cos \frac{\theta}{2}$

3 Simplify.

29. $\dfrac{\sin 2x}{2 \sin x}$
30. $\dfrac{\sin 2x}{2 \cos x}$
31. $1 - 2 \sin^2 \frac{x}{2}$
32. $2 \cos^2 \frac{x}{2} - 1$

33. $2 \sin \frac{x}{2} \cos \frac{x}{2}$

34. $2 \sin 2x \cos 2x$

35. $\cos^2 \frac{x}{2} - \sin^2 \frac{x}{2}$

36. $\cos^4 x - \sin^4 x$

37. $(\sin x + \cos x)^2 - \sin 2x$

38. $(\sin x - \cos x)^2 + \sin 2x$

39. $2 \sin^2 \frac{x}{2} + \cos x$

40. $2 \cos^2 \frac{x}{2} - \cos x$

41. $(-4 \cos x \sin x + 2 \cos 2x)^2 + (2 \cos 2x + 4 \sin x \cos x)^2$

42. $(-4 \cos 2x + 8 \cos x \sin x)^2 + (8 \sin x \cos x + 4 \cos 2x)^2$

43. $2 \sin x \cos^3 x + 2 \sin^3 x \cos x$

44. $2 \sin x \cos^3 x - 2 \sin^3 x \cos x$

By simplifying, derive an identity.

45. $(\sin x + \cos x)^2$

46. $\cos^4 x - \sin^4 x$

47. $\frac{2 \cot x}{\cot^2 x - 1}$

48. $\frac{2 - \sec^2 x}{\sec^2 x}$

49. $2 \sin^2 2x + \cos 4x$

50. $\frac{1 + \sin 2x + \cos 2x}{1 + \sin 2x - \cos 2x}$

SYNTHESIS

Find $\sin \theta$, $\cos \theta$, and $\tan \theta$ under the given conditions.

51. $\sin 2\theta = \frac{1}{5}, \frac{\pi}{2} \leq 2\theta \leq \pi$

52. $\cos 2\theta = \frac{7}{12}, \frac{3\pi}{2} \leq 2\theta \leq 2\pi$

53. $\tan \frac{\theta}{2} = \frac{1}{4}, \frac{3\pi}{2} \leq \theta \leq 2\pi$

54. $\tan \frac{\theta}{2} = -\frac{5}{3}, \pi < \theta \leq \frac{3\pi}{2}$

55. *Nautical mile. Latitude* is used to measure North–South location on the earth between the equator and the poles. For example, the north end of Chicago has latitude 42°N. (See the figure.) In Great Britain, the **nautical mile** is defined as the length of a minute of arc of the earth's radius. Since the earth is flattened at the poles, a British nautical mile varies with latitude. In fact, it is given, in feet, by the function

$$N(\phi) = 6066 - 31 \cos 2\phi,$$

where ϕ is the latitude in degrees.

a) What is the length of a British nautical mile at the north end of Chicago?
b) What is the length of a British nautical mile at the North Pole?
c) Express $N(\phi)$ in terms of $\cos \phi$ only. That is, eliminate the double angle.

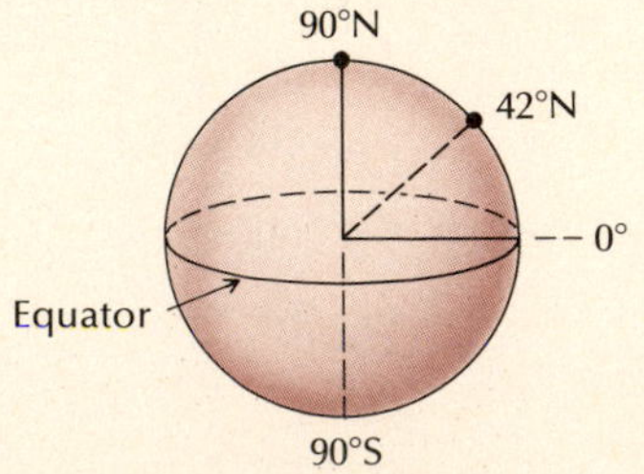

56. *Acceleration due to gravity.* The acceleration due to gravity, often denoted by g in a formula such as $S = \frac{1}{2} gt^2$, means the distance that an object falls in time t. It has to do with the physics of motion near the earth's surface and is usually considered constant. In fact, however, g is not constant, but varies slightly with latitude. If ϕ stands for latitude, in degrees, g is given with good approximation by the formula

$$g = 9.78049(1 + 0.005288 \sin^2 \phi - 0.000006 \sin^2 2\phi),$$

where g is measured in m/sec^2 at sea level.

a) Chicago has latitude 42°N. Find g.
b) Philadelphia has latitude 40°N. Find g.
c) Express g in terms of $\sin \phi$ only. That is, eliminate the double angle.
d) Where on earth is g greatest? least?

CHALLENGE

57. Graph: $f(x) = \cos^2 x - \sin^2 x$.

58. Graph: $f(x) = |\sin x \cos x|$.

7.5 Proving Trigonometric Identities

OBJECTIVES

You should be able to:

1. Prove identities using other identities.
2. Prove the sum–product identities and use them to prove other identities.

Basic Identities

We have proved a great many identities in the last chapter and in this chapter. Here we consider the proving of identities more intensely. You will be amazed how many trigonometric identities we can prove, and how the proving of some enables the proving of others. The proofs we considered in Sections 7.1–7.4 are good examples.

A first step in learning to prove identities is to have at hand as many identities as possible that have already been learned. The following list contains many such identities. While it is not necessary to learn them all, the more you learn, the better your identity-proving skill will be.

There are many "tricks" to minimize the number of identities you need to learn. For example, knowing identities for $\sin(-x)$ and $\cos(-x)$ might eliminate the need to learn an identity for $\tan(-x)$ because you can divide $\sin(-x)$ by $\cos(-x)$, and simplify. As another example, the reduction formulas, covered in Section 7.3, can eliminate the need to memorize all the identities involving $\sin(x \pm \pi/2)$. Another aid to remembering an identity is to consider a unit circle or a graph. You may be able to recall, for example, that $\sin(x + \pi/2) = \cos x$ by thinking of translating the graph of $y = \sin x$ to the left $\pi/2$ units.

The following is a list of important identities.

Basic Identities

$$\sin x = \frac{1}{\csc x}, \qquad \csc x = \frac{1}{\sin x}, \qquad \tan x = \frac{\sin x}{\cos x},$$

$$\cos x = \frac{1}{\sec x}, \qquad \sec x = \frac{1}{\cos x}, \qquad \cot x = \frac{\cos x}{\sin x},$$

$$\tan x = \frac{1}{\cot x}, \qquad \cot x = \frac{1}{\tan x},$$

$$\sin(-x) = -\sin x, \qquad \cos(-x) = \cos x, \qquad \tan(-x) = -\tan x,$$

$$\csc(-x) = -\csc x, \qquad \sec(-x) = \sec x, \qquad \cot(-x) = -\cot x$$

Pythagorean Identities

$$\sin^2 x + \cos^2 x = 1,$$

$$1 + \tan^2 x = \sec^2 x,$$

$$1 + \cot^2 x = \csc^2 x$$

Cofunction Identities

$$\sin\left(x \pm \frac{\pi}{2}\right) = \pm\cos x, \qquad \sin\left(\frac{\pi}{2} \pm x\right) = \cos x,$$

$$\cos\left(x \pm \frac{\pi}{2}\right) = \mp\sin x, \qquad \cos\left(\frac{\pi}{2} \pm x\right) = \mp\sin x$$

Sum and Difference Identities

$$\sin(\alpha \pm \beta) = \sin\alpha\cos\beta \pm \cos\alpha\sin\beta,$$
$$\cos(\alpha \pm \beta) = \cos\alpha\cos\beta \mp \sin\alpha\sin\beta,$$
$$\tan(\alpha \pm \beta) = \frac{\tan\alpha \pm \tan\beta}{1 \mp \tan\alpha\tan\beta}$$

Double-Angle Identities

$$\sin 2x = 2\sin x\cos x,$$
$$\cos 2x = \cos^2 x - \sin^2 x = 1 - 2\sin^2 x$$
$$= 2\cos^2 x - 1,$$
$$\tan 2x = \frac{2\tan x}{1 - \tan^2 x},$$
$$\sin^2 x = \frac{1 - \cos 2x}{2},$$
$$\cos^2 x = \frac{1 + \cos 2x}{2},$$
$$\tan^2 x = \frac{1 - \cos 2x}{1 + \cos 2x}$$

Half-Angle Identities

$$\sin\frac{x}{2} = \pm\sqrt{\frac{1 - \cos x}{2}},$$
$$\cos\frac{x}{2} = \pm\sqrt{\frac{1 + \cos x}{2}},$$
$$\tan\frac{x}{2} = \pm\sqrt{\frac{1 - \cos x}{1 + \cos x}} = \frac{\sin x}{1 + \cos x}$$
$$= \frac{1 - \cos x}{\sin x}$$

1 Proving Identities

The Logic of Proving Identities

The method of attack in proving identities proceeds somewhat as follows. We use any law or theorem regarding real numbers and all identities that have been proven to prove a new identity. We will outline two methods for proving identities.

Method 1. *Start with either the left or the right side of an identity and deduce the other side.* For example, suppose you are trying to prove that the equation $P = Q$ is an identity. You might try to produce a string of statements like the following, which start at P and end with Q:

$$\begin{aligned} P &= S_1 \\ &= S_2 \\ &\ \ \vdots \\ &= Q. \end{aligned}$$

Method 2. *Work with each side separately until you deduce the same expression.* For example, suppose you are trying to prove that $P = Q$ is an identity. You might be able to produce two strings of statements like the following, each ending with the same statement S.

$$\begin{aligned} P &= S_1 \\ &= S_2 \\ &\;\;\vdots \\ &= S. \end{aligned} \qquad \begin{aligned} Q &= P_1 \\ &= P_2 \\ &\;\;\vdots \\ &= S. \end{aligned}$$

The number of steps in each string might be different, but in each case the result is S.

Examples 1–3 illustrate Method 1. Examples 4 and 5 illustrate Method 2.

EXAMPLE 1 Prove the identity

$$\frac{\sec t - 1}{t \sec t} = \frac{1 - \cos t}{t}.$$

Proof. We use Method 1, starting with the left side and deducing the right side. We note that the left side involves sec t, whereas the right side involves cos t, so it might be wise to make use of an identity that involves these two expressions. That basic identity is $\sec t = 1/\cos t$.

$$\frac{\sec t - 1}{t \sec t} = \frac{\frac{1}{\cos t} - 1}{t\frac{1}{\cos t}} \qquad \text{Substituting } \frac{1}{\cos t} \text{ for } \sec t$$

$$= \frac{\left(\frac{1}{\cos t} - 1\right)}{\frac{t}{\cos t}} \cdot \frac{\cos t}{\cos t} \qquad \text{Multiplying by 1 in order to simplify the complex fractional expression}$$

$$= \frac{\left(\frac{1}{\cos t} - 1\right)\cos t}{\left(\frac{t}{\cos t}\right)\cos t}$$

$$= \frac{1 - \cos t}{t} \qquad \text{Multiplying and simplifying}$$

We started with the left side and deduced the right side, so the proof is complete.

DO EXERCISES 1 AND 2.

EXAMPLE 2 Prove the identity

$$1 + \sin 2\theta = (\sin\theta + \cos\theta)^2.$$

Proof. We again use Method 1. This time we start with the right side and deduce the left side.

$$(\sin\theta + \cos\theta)^2 = \sin^2\theta + 2\sin\theta\cos\theta + \cos^2\theta \qquad \text{Squaring}$$

$$= 1 + 2\sin\theta\cos\theta \qquad \text{Recalling the identity } \sin^2\theta + \cos^2\theta = 1 \text{ and substituting}$$

$$= 1 + \sin 2\theta \qquad \text{Recalling the identity } \sin 2\theta = 2\sin\theta\cos\theta \text{ and substituting}$$

Prove each of the following identities.

1. $\dfrac{\csc t - 1}{t \csc t} = \dfrac{1 - \sin t}{t}$

2. $\sec\theta \csc\theta = \tan\theta + \cot\theta$

We started with the right side and deduced the left side, so the proof is complete. ■

DO EXERCISES 3 AND 4.

EXAMPLE 3 Prove the identity

$$\frac{\sin\theta - \cos\theta}{\sin\theta + \cos\theta} = -\frac{\cos 2\theta}{1 + \sin 2\theta}.$$

Proof. We start with the left side and deduce the right side. In the first step, we multiply by 1, where the symbol for 1 is formed from the conjugate of the numerator of the original expression.

$$\frac{\sin\theta - \cos\theta}{\sin\theta + \cos\theta} = \frac{\sin\theta - \cos\theta}{\sin\theta + \cos\theta}\cdot\frac{\sin\theta + \cos\theta}{\sin\theta + \cos\theta}$$ Multiplying by 1

$$= \frac{(\sin\theta - \cos\theta)(\sin\theta + \cos\theta)}{(\sin\theta + \cos\theta)(\sin\theta + \cos\theta)}$$

$$= \frac{\sin^2\theta - \cos^2\theta}{(\sin\theta + \cos\theta)^2}$$

$$= \frac{\sin^2\theta - \cos^2\theta}{1 + \sin 2\theta}$$ Substituting, using the identity of Example 2

$$= \frac{-(\cos^2\theta - \sin^2\theta)}{1 + \sin 2\theta}$$

$$= -\frac{\cos 2\theta}{1 + \sin 2\theta}$$ Recalling the identity $\cos^2\theta - \sin^2\theta = \cos 2\theta$ and substituting

We started with the left side and deduced the right side, so the proof is complete. ■

DO EXERCISE 5.

EXAMPLE 4 Prove the identity

$$\tan^2 x - \sin^2 x = \sin^2 x \tan^2 x.$$

Proof. For this proof, we are going to work with each side separately using Method 2. We try to deduce the same expression. In practice, when carrying out this method of proof, you might work on one side for awhile, then work on the other side separately, and then go back to the other side. That is, you bounce back and forth until you arrive at the same expression. Let us start with the left side:

$$\tan^2 x - \sin^2 x = \frac{\sin^2 x}{\cos^2 x} - \sin^2 x$$ Recalling the identity $\tan x = \frac{\sin x}{\cos x}$ and substituting

$$= \frac{\sin^2 x}{\cos^2 x} - \sin^2 x\cdot\frac{\cos^2 x}{\cos^2 x}$$ Multiplying by 1 in order to subtract

$$= \frac{\sin^2 x - \sin^2 x\cos^2 x}{\cos^2 x}$$ Carrying out the subtraction

$$= \frac{\sin^2 x(1 - \cos^2 x)}{\cos^2 x}$$ Factoring

$$= \frac{\sin^2 x\sin^2 x}{\cos^2 x}$$ Recalling the identity $\sin^2 x + \cos^2 x = 1$ or $1 - \cos^2 x = \sin^2 x$ and substituting

$$= \frac{\sin^4 x}{\cos^2 x}.$$

Prove each of the following identities.

3. $(\sin\theta - \cos\theta)^2 = 1 - \sin 2\theta$

4. $(\sec u - \tan u)(1 + \sin u) = \cos u$

5. Prove the identity

$$\frac{\sin t}{1 + \cos t} = \frac{1 - \cos t}{\sin t}.$$

At this point, we stop and work with the right side of the original identity and try to end with the same expression that we ended with on the left side:

$$\sin^2 x \tan^2 x = \sin^2 x \frac{\sin^2 x}{\cos^2 x} \qquad \text{Recalling the identity } \tan x = \frac{\sin x}{\cos x} \text{ and substituting}$$

$$= \frac{\sin^4 x}{\cos^2 x}.$$

From each side we have deduced the same expression, so the proof is complete.

DO EXERCISE 6.

EXAMPLE 5 Prove the identity

$$\frac{\sin 2\theta}{\sin \theta} - \frac{\cos 2\theta}{\cos \theta} = \sec \theta.$$

Proof. We are again using Method 2, beginning with the left side:

$$\frac{\sin 2\theta}{\sin \theta} - \frac{\cos 2\theta}{\cos \theta} = \frac{2 \sin \theta \cos \theta}{\sin \theta} - \frac{\cos^2 \theta - \sin^2 \theta}{\cos \theta} \qquad \text{Using the identities } \sin 2\theta = 2 \sin \theta \cos \theta \text{ and } \cos 2\theta = \cos^2 \theta - \sin^2 \theta \text{ and substituting}$$

$$= 2 \cos \theta - \frac{\cos^2 \theta - \sin^2 \theta}{\cos \theta} \qquad \text{Simplifying}$$

$$= \frac{2 \cos^2 \theta}{\cos \theta} - \frac{\cos^2 \theta - \sin^2 \theta}{\cos \theta} \qquad \text{Multiplying } 2 \cos \theta \text{ by 1, or } \cos \theta/\cos \theta$$

$$= \frac{2 \cos^2 \theta - \cos^2 \theta + \sin^2 \theta}{\cos \theta} \qquad \text{Carrying out the subtraction}$$

$$= \frac{\cos^2 \theta + \sin^2 \theta}{\cos \theta}$$

$$= \frac{1}{\cos \theta}. \qquad \text{Recalling the identity } \sin^2 x + \cos^2 x = 1$$

At this point, we stop and work with the right side of the original identity and try to end with the same expression:

$$\sec \theta = \frac{1}{\cos \theta}. \qquad \text{Recalling a basic identity}$$

From each side we have deduced the same expression, so the proof is complete.

DO EXERCISE 7.

Hints for Proving Identities

1. Use Methods 1 or 2 previously outlined.
2. Work with the more complex side first.
3. Do the algebraic manipulations, such as adding, subtracting, multiplying, or factoring.
4. Multiplying by 1 can often be helpful when fractional expressions are involved.
5. Converting all expressions to sines and cosines is often helpful.
6. Try something! Put your pencil to work and get involved. You will be amazed at how often this leads to success.

6. Prove the identity

$$\cot^2 x - \cos^2 x = \cos^2 x \cot^2 x.$$

7. Prove the identity

$$\frac{\sin 2\theta + \sin \theta}{\cos 2\theta + \cos \theta + 1} = \tan \theta.$$

8. Prove the first identity in the list of sum–product identities. *Hint:* Start with the right side and use the sine of a sum and the sine of a difference.

2 Sum–Product Identities

On occasion, it is convenient to convert a product of trigonometric expressions to a sum, or the reverse. The following identities are useful in this connection. Proofs are left as exercises.

$$\sin u \cdot \cos v = \frac{1}{2}[\sin(u+v) + \sin(u-v)],$$

$$\cos u \cdot \sin v = \frac{1}{2}[\sin(u+v) - \sin(u-v)],$$

$$\cos u \cdot \cos v = \frac{1}{2}[\cos(u-v) + \cos(u+v)],$$

$$\sin u \cdot \sin v = \frac{1}{2}[\cos(u-v) - \cos(u+v)],$$

$$\sin x + \sin y = 2 \sin\frac{x+y}{2} \cos\frac{x-y}{2},$$

$$\sin x - \sin y = 2 \cos\frac{x+y}{2} \sin\frac{x-y}{2},$$

$$\cos y + \cos x = 2 \cos\frac{x+y}{2} \cos\frac{x-y}{2},$$

$$\cos y - \cos x = 2 \sin\frac{x+y}{2} \sin\frac{x-y}{2}$$

DO EXERCISE 8.

EXERCISE SET 7.5

1 Prove each of the following identities.

1. $\csc x - \cos x \cot x = \sin x$
2. $\sec x - \sin x \tan x = \cos x$
3. $\dfrac{1+\cos\theta}{\sin\theta} + \dfrac{\sin\theta}{\cos\theta} = \dfrac{\cos\theta + 1}{\sin\theta\cos\theta}$
4. $\dfrac{1}{\sin\theta\cos\theta} - \dfrac{\cos\theta}{\sin\theta} = \dfrac{\sin\theta\cos\theta}{1-\sin^2\theta}$
5. $\dfrac{1-\sin x}{\cos x} = \dfrac{\cos x}{1+\sin x}$
6. $\dfrac{1-\cos x}{\sin x} = \dfrac{\sin x}{1+\cos x}$
7. $\dfrac{1+\tan\theta}{1+\cot\theta} = \dfrac{\sec\theta}{\csc\theta}$
8. $\dfrac{\cot\theta - 1}{1-\tan\theta} = \dfrac{\csc\theta}{\sec\theta}$
9. $\dfrac{\sin x + \cos x}{\sec x + \csc x} = \dfrac{\sin x}{\sec x}$
10. $\dfrac{\sin x - \cos x}{\sec x - \csc x} = \dfrac{\cos x}{\csc x}$
11. $\dfrac{1+\tan\theta}{1-\tan\theta} + \dfrac{1+\cot\theta}{1-\cot\theta} = 0$
12. $\dfrac{\cos^2\theta + \cot\theta}{\cos^2\theta - \cot\theta} = \dfrac{\cos^2\theta\tan\theta + 1}{\cos^2\theta\tan\theta - 1}$
13. $\dfrac{1+\cos 2\theta}{\sin 2\theta} = \cot\theta$
14. $\dfrac{2\tan\theta}{1+\tan^2\theta} = \sin 2\theta$
15. $\sec 2\theta = \dfrac{\sec^2\theta}{2-\sec^2\theta}$
16. $\cot 2\theta = \dfrac{\cot^2\theta - 1}{2\cot\theta}$
17. $\dfrac{\sin(\alpha+\beta)}{\cos\alpha\cos\beta} = \tan\alpha + \tan\beta$
18. $\dfrac{\cos(\alpha-\beta)}{\cos\alpha\sin\beta} = \tan\alpha + \cot\beta$
19. $1 - \cos 5\theta\cos 3\theta - \sin 5\theta\sin 3\theta = 2\sin^2\theta$
20. $2\sin\theta\cos^3\theta + 2\sin^3\theta\cos\theta = \sin 2\theta$
21. $\dfrac{\tan\theta + \sin\theta}{2\tan\theta} = \cos^2\dfrac{\theta}{2}$
22. $\dfrac{\tan\theta - \sin\theta}{2\tan\theta} = \sin^2\dfrac{\theta}{2}$

23. $\cos^4 x - \sin^4 x = \cos 2x$

24. $\dfrac{\cos^4 x - \sin^4 x}{1 - \tan^4 x} = \cos^4 x$

25. $\dfrac{\tan 3\theta - \tan \theta}{1 + \tan 3\theta \tan \theta} = \dfrac{2 \tan \theta}{1 - \tan^2 \theta}$

26. $\left(\dfrac{1 + \tan \theta}{1 - \tan \theta}\right)^2 = \dfrac{1 + \sin 2\theta}{1 - \sin 2\theta}$

27. $\dfrac{\cos^3 x - \sin^3 x}{\cos x - \sin x} = \dfrac{2 + \sin 2x}{2}$

28. $\dfrac{\sin^3 t + \cos^3 t}{\sin t + \cos t} = \dfrac{2 - \sin 2t}{2}$

29. $\sin (\alpha + \beta) \sin (\alpha - \beta) = \sin^2 \alpha - \sin^2 \beta$

30. $\cos (\alpha + \beta) \cos (\alpha - \beta) = \cos^2 \alpha - \sin^2 \beta$

31. $\cos (\alpha + \beta) + \cos (\alpha - \beta) = 2 \cos \alpha \cos \beta$

32. $\sin (\alpha + \beta) + \sin (\alpha - \beta) = 2 \sin \alpha \cos \beta$

33. $\sin^2 x - \cos^2 x = 1 - 2 \cos^2 x$

34. $\cos^2 x\,(1 - \sec^2 x) = -\sin^2 x$

35. $\sin^2 \theta = \cos^2 \theta\,(\sec^2 \theta - 1)$

36. $\tan \theta\,(\tan \theta + \cot \theta) = \sec^2 \theta$

37. $\dfrac{\tan x}{\sec x - \cos x} = \dfrac{\sec x}{\tan x}$

38. $\dfrac{\tan x + \sin x}{1 + \sec x} = \sin x$

39. $\dfrac{\cos \theta + \sin \theta}{\cos \theta} = 1 + \tan \theta$

40. $\dfrac{1 + \tan^2 \theta}{\csc \theta} = \tan^2 \theta$

41. $\dfrac{\tan^2 x}{1 + \tan^2 x} = \sin^2 x$

42. $\dfrac{1 + \cos^2 x}{\sin^2 x} = 2 \csc^2 x - 1$

43. $\dfrac{\cot x + \tan x}{\csc x} = \sec x$

44. $\dfrac{\csc x - \sin x}{\cot x} = \cos x$

45. $\dfrac{\tan \theta + \cot \theta}{\csc \theta} = \sec \theta$

46. $\dfrac{\csc \theta - \sin \theta}{\cos^2 \theta} = \csc \theta$

47. $\dfrac{\sin x}{1 + \cos x} + \dfrac{1 + \cos x}{\sin x} = 2 \csc x$

48. $\dfrac{1 + \sin x}{1 - \sin x} + \dfrac{\sin x - 1}{1 + \sin x} = 4 \sec x \tan x$

49. $\cos \theta\,(1 + \csc \theta) - \cot \theta = \cos \theta$

50. $\cos^2 \theta \cot^2 \theta = \cot^2 \theta - \cos^2 \theta$

51. $\dfrac{\tan x + \cot x}{\sec x + \csc x} = \dfrac{1}{\cos x + \sin x}$

52. $\dfrac{\cos^2 x - 1}{1 - \sec^2 x} = \dfrac{1}{\tan^2 x + 1}$

53. $(\sec \theta - \tan \theta)\,(1 + \csc \theta) = \cot \theta$

54. $\tan \theta - \cot \theta = (\sec \theta - \csc \theta)\,(\sin \theta + \cos \theta)$

55. $(\sec x + \tan x)\,(1 - \sin x) = \cos x$

56. $\csc x - \cot x = \dfrac{1}{\csc x + \cot x}$

57. $\cos^2 \theta - \sin^2 \theta = \cos^4 \theta - \sin^4 \theta$

58. $2 \sin^2 \theta \cos^2 \theta + \cos^4 \theta = 1 - \sin^4 \theta$

59. $\dfrac{\cos \theta}{1 - \cos \theta} = \dfrac{1 + \sec \theta}{\tan^2 \theta}$

60. $\dfrac{\cot \theta}{\csc \theta - 1} = \dfrac{\csc \theta + 1}{\cot \theta}$

61. $\dfrac{\sin x - \cos x}{\cos^2 x} = \dfrac{\tan^2 x - 1}{\sin x + \cos x}$

62. $\dfrac{1 + \sin x}{1 - \sin x} = (\sec x + \tan x)^2$

63. $1 + \sec^4 \theta = \tan^4 \theta + 2 \sec^2 \theta$

64. $\sec^4 \theta - \tan^2 \theta = \tan^4 \theta + \sec^2 \theta$

65. $\dfrac{1 + \tan x}{1 - \tan x} = \dfrac{\cot x + 1}{\cot x - 1}$

66. $1 + \sec x = \csc x\,(\sin x + \tan x)$

67. $\dfrac{\sin \theta \tan \theta}{\sin \theta + \tan \theta} = \dfrac{\tan \theta - \sin \theta}{\tan \theta \sin \theta}$

68. $\dfrac{\sin^3 \theta - \cos^3 \theta}{\sin \theta - \cos \theta} = \sin \theta \cos \theta + 1$

69. $\dfrac{\cos x + 1}{\sin x} + \dfrac{\sin x}{\cos x + 1} = \dfrac{2}{\sin x}$

70. $2 \sec^2 x + \dfrac{\csc x}{1 - \csc x} = \dfrac{\csc x}{\csc x + 1}$

71. $\sec \theta \csc \theta + \dfrac{\cot \theta}{\tan \theta - 1} = \dfrac{\tan \theta}{1 - \cot \theta} - 1$

72. $\cos \theta + \dfrac{\sin \theta}{\cot \theta - 1} = \dfrac{\cos \theta}{1 - \tan \theta} - \sin \theta$

73. $\cot x + \csc x = \dfrac{\sin x}{1 - \cos x}$

74. $\dfrac{\cos x + \cot x}{1 + \csc x} = \cos x$

75. $\tan \theta + \cot \theta = \dfrac{1}{\cot \theta \sin^2 \theta}$

76. $\sec^2 \theta - \csc^2 \theta = \dfrac{\tan \theta - \cot \theta}{\sin \theta \cos \theta}$

77. $2 \cos^2 x - 1 = \cos^4 x - \sin^4 x$

78. $\sec^4 x - 4 \tan^2 x = (1 - \tan^2 x)^2$

79. $(\cos x + \sin x)\,(1 - \sin x \cos x) = \cos^3 x + \sin^3 x$

80. $\dfrac{\tan^2 x + \sec^2 x}{\sec^4 x} = 1 - \sin^4 x$

2

81. Prove the second, third, and fourth of the sum–product identities using the sum and difference formulas for sine and cosine.

82. Prove the last four of the sum–product identities. (*Hint:* Use the first four sum–product identities.)

Use the sum–product identities to find identities for each of the following.

83. $\sin 3\theta - \sin 5\theta$
84. $\sin 7x - \sin 4x$
85. $\sin 8\theta + \sin 5\theta$
86. $\cos \theta - \cos 7\theta$
87. $\sin 7u \sin 5u$
88. $2 \sin 7\theta \cos 3\theta$
89. $7 \cos \theta \sin 7\theta$
90. $\cos 2t \sin t$
91. $\cos 55° \sin 25°$
92. $7 \cos 5\theta \cos 7\theta$

Use the sum–product identities to prove each of the following.

93. $\sin 4\theta + \sin 6\theta = \cot \theta (\cos 4\theta - \cos 6\theta)$

94. $\tan 2x (\cos x + \cos 3x) = \sin x + \sin 3x$

95. $\cot 4x (\sin x + \sin 4x + \sin 7x) = \cos x + \cos 4x + \cos 7x$

96. $\tan \dfrac{x + y}{2} = \dfrac{\sin x + \sin y}{\cos x + \cos y}$

97. $\cot \dfrac{x + y}{2} = \dfrac{\sin y - \sin x}{\cos x - \cos y}$

98. $\tan \dfrac{\theta + \phi}{2} \tan \dfrac{\theta - \phi}{2} = \dfrac{\cos \theta - \cos \phi}{\cos \theta + \cos \phi}$

99. $\tan \dfrac{\theta + \phi}{2} (\sin \theta - \sin \phi) = \tan \dfrac{\theta - \phi}{2} (\sin \theta + \sin \phi)$

100. $\sin 2\theta + \sin 4\theta + \sin 6\theta = 4 \cos \theta \cos 2\theta \sin 3\theta$

SYNTHESIS

Prove each of the following identities.

101. $\ln |\sec x| = -\ln |\cos x|$

102. $\ln |\tan x| = -\ln |\cot x|$

103. $\ln |\tan x| = \ln |\sin x| - \ln |\cos x|$

104. $\ln |\csc x| = -\ln |\sin x|$

105. $\ln e^{\sin t} = \sin t$

106. $e^{\ln |\cos t|} = |\cos t|$

107. $\ln |\csc \theta - \cot \theta| = -\ln |\csc \theta + \cot \theta|$

108. $\ln |\sec \theta + \tan \theta| = -\ln |\sec \theta - \tan \theta|$

109. Show that $\log (\cos x - \sin x) + \log (\cos x + \sin x) = \log \cos 2x$.

110. The following equation occurs in the study of mechanics:
$$\sin \theta = \frac{I_1 \cos \phi}{\sqrt{(I_1 \cos \phi)^2 + (I_2 \sin \phi)^2}}.$$
It can happen that $I_1 = I_2$. Assuming that this happens, simplify the equation.

111. In the theory of alternating current, the following equation occurs:
$$R = \frac{1}{\omega C (\tan \theta + \tan \phi)}.$$
Show that this equation is equivalent to
$$R = \frac{\cos \theta \cos \phi}{\omega C \sin (\theta + \phi)}.$$

CHALLENGE

112. In electrical theory the following equations occur:
$$E_1 = \sqrt{2}E_t \cos \left(\theta + \frac{\pi}{P}\right), \quad E_2 = \sqrt{2}E_t \cos \left(\theta - \frac{\pi}{P}\right).$$
Assuming that these equations hold, show that
$$\frac{E_1 + E_2}{2} = \sqrt{2}E_t \cos \theta \cos \frac{\pi}{P} \quad \text{and} \quad \frac{E_1 - E_2}{2} = -\sqrt{2}E_t \sin \theta \sin \frac{\pi}{P}.$$

7.6 Inverses of the Trigonometric Functions

OBJECTIVE

You should be able to:

1 Find values of the inverse trigonometric functions.

In this section, we develop inverse trigonometric functions. It may be helpful for you to review the material on inverse functions in Section 5.1.

Following are the graphs of the sine, cosine, tangent, and cotangent functions. Do these functions have inverses that are functions? They do if these functions are one-to-one, which means that they pass the horizontal-line test.

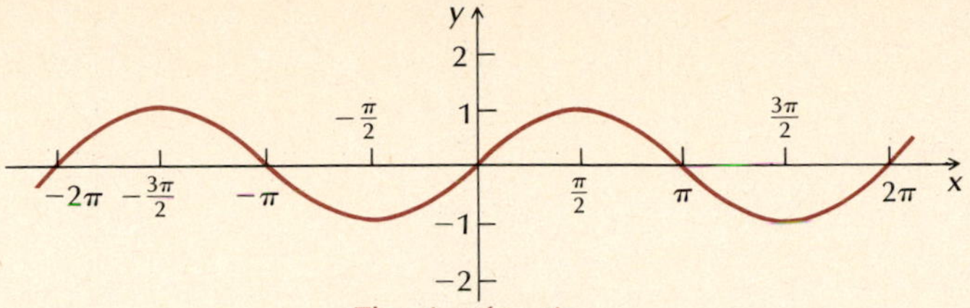

The sine function

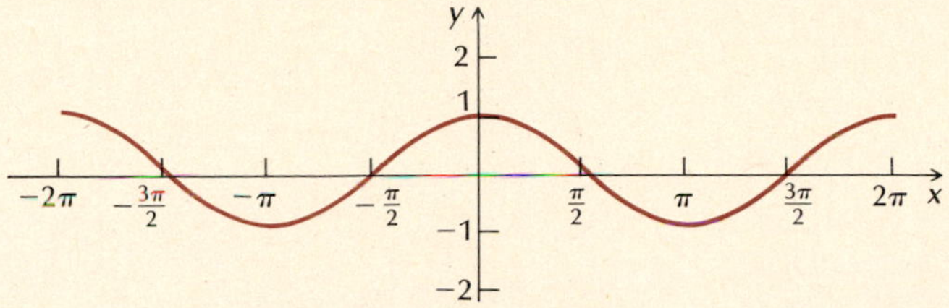

The cosine function

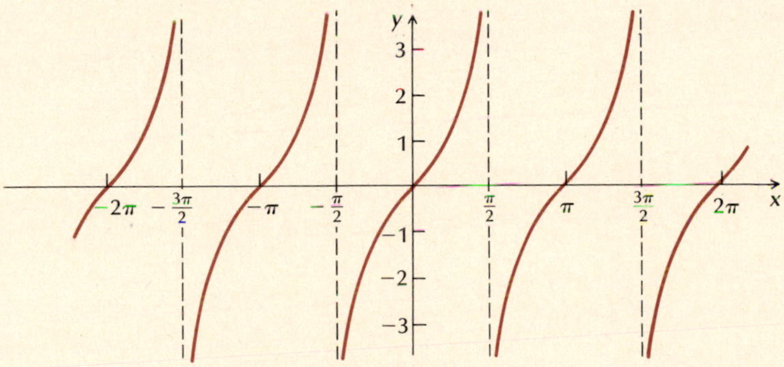

The tangent function

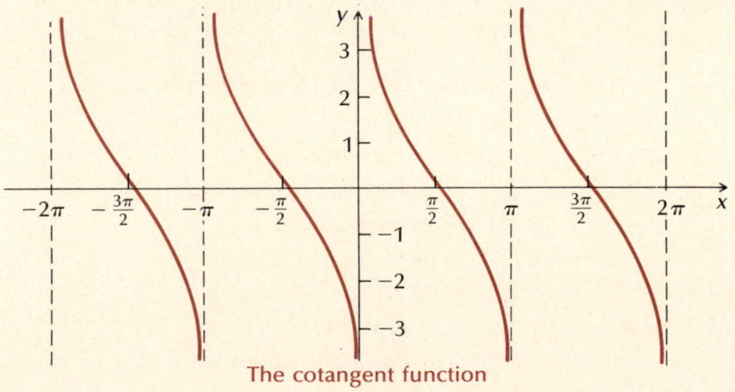

The cotangent function

Note that each function has a horizontal line that crosses the graph more than once. Thus each fails the horizontal-line test. Therefore, none of them has an inverse that is a function.

Recall that to obtain the inverse of any relation, we interchange the first and second members of each ordered pair in the relation. If a relation is defined in terms of, say, x and y, interchanging x and y produces an equation of the inverse relation. The graphs of an equation and its inverse are reflections of each other across the line $y = x$. Let us examine the inverses of each of the four trigonometric functions graphed above. The graphs are as follows.

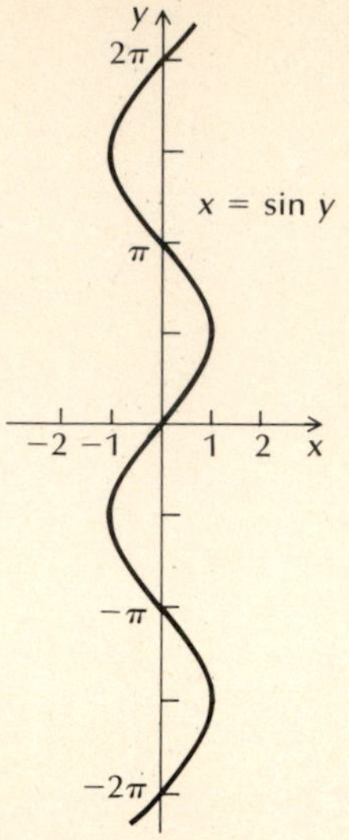

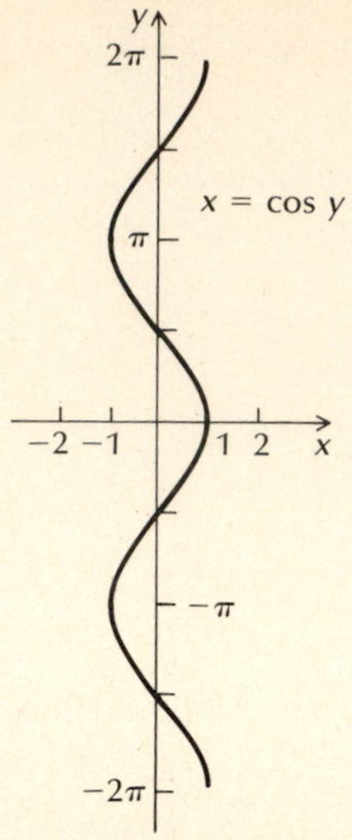

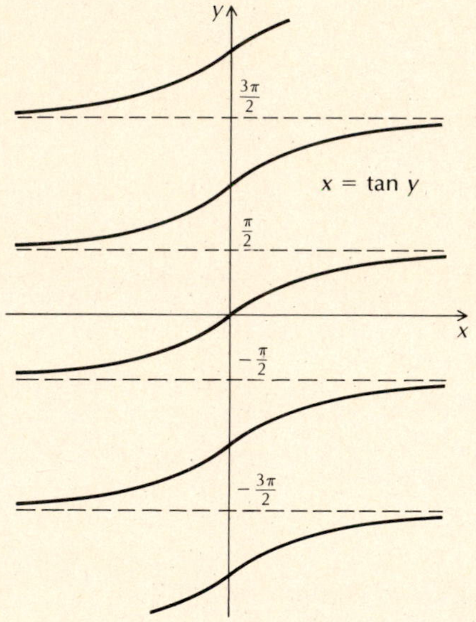

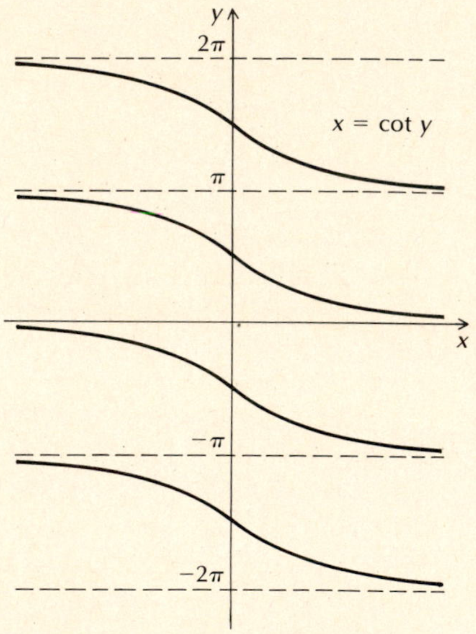

We can check again to see whether these are graphs of functions by using the vertical-line test. In each case, there is a vertical line that crosses the graph more than once, so each fails to be a function.

Let us consider specifically the graph of the inverse of $y = \sin x$, which is $x = \sin y$. Consider the input $x = 1/2$. On the graph of the inverse of the sine function, we draw a vertical line at $x = 1/2$, as shown. It intersects the graph at points whose y-value is such that $1/2 = \sin y$. Some numbers whose sine is 1/2 are $\pi/6$, $5\pi/6$, $-7\pi/6$, and so on. From the graph, we see that $\pi/6$ plus any multiple of 2π is such a number. Also, $5\pi/6$ plus any multiple of 2π is such a number. The complete set of values is given by $\pi/6 + 2k\pi$, k an integer, and $5\pi/6 + 2k\pi$, k an integer. Indeed, we have many points that cross the vertical line and verify that the inverse of the sine function is not a function.

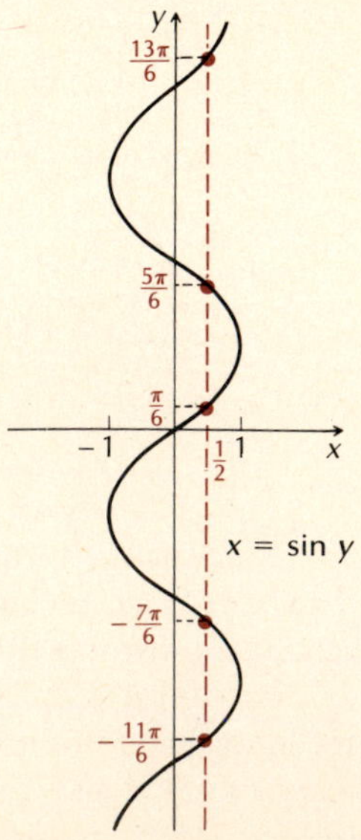

1 Restricting Ranges to Define Inverse Functions

Recall that a function like $f(x) = x^2$ did not have an inverse that was a function, but by restricting the domain of f to nonnegative numbers, we have a new squaring function that has an inverse, $f^{-1}(x) = \sqrt{x}$, $x \geq 0$. This is equivalent to restricting the range of the inverse relation to exclude ordered pairs that contain negative numbers.

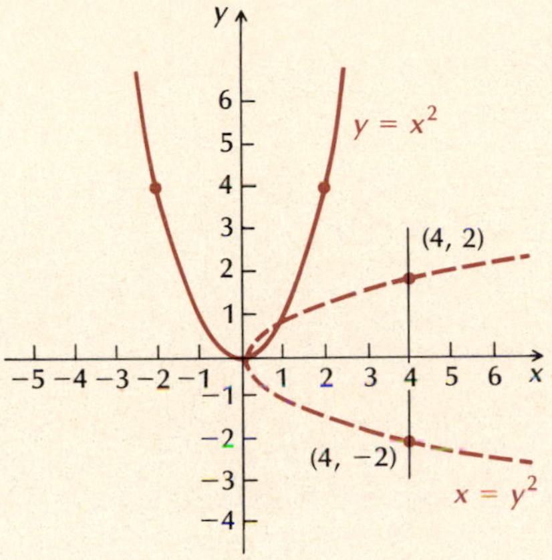

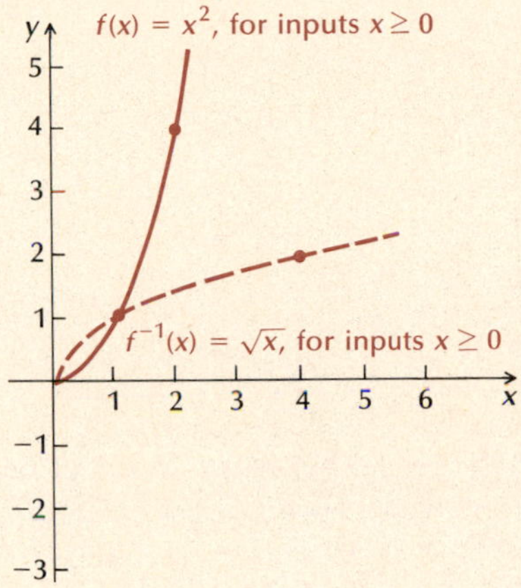

In a similar manner, we can define new trigonometric functions whose inverses are functions. We do this by restricting the ranges of the inverses. This can be done in many ways, but we choose restrictions that are fairly standard in mathematics. For the inverse sine function, we choose a range close to the origin that allows all inputs in the interval $[-1, 1]$ to have function values in the interval $[-\pi/2, \pi/2]$. For the inverse cosine function, we choose a range close to the origin that allows all inputs in the interval $[-1, 1]$ to have function values in the interval $[0, \pi]$. For the inverse tangent function, the domain is the set of all real numbers and the range is the interval $(-\pi/2, \pi/2)$. For the inverse cotangent function, the domain is the set of all real numbers and the range is the interval $(0, \pi)$.

We denote the inverse trigonometric functions as follows:

The Inverse Sine Function

$$y = \sin^{-1} x, \quad \text{or} \quad y = \arcsin x, \quad \text{where } x = \sin y.$$

The domain $= [-1, 1]$, and the range $= [-\pi/2, \pi/2]$.

The notation arcsin x arises because it is the length of an arc on the unit circle for which the sine is x. The notation $\sin^{-1} x$ is *not* exponential notation. It does not mean $1/\sin x$! Either of the two kinds of notation above can be read "the inverse sine of x" or "the arc sine of x" or "the number (or angle) whose sine is x." Notation is chosen similarly for the inverse of the other trigonometric functions: $\cos^{-1} x$, or $\arccos x$, $\tan^{-1} x$, or $\arctan x$, and so on. The following diagrams show the restricted ranges for the inverse trigonometric functions on a unit circle. The restricted ranges should be memorized. The missing endpoints indicate inputs that are not in the domain of the original function.

In each of Margin Exercises 1–4, the graph shown is an inverse of a trigonometric function. In each case, shade the range so the result is the restricted inverse that is a function.

1.

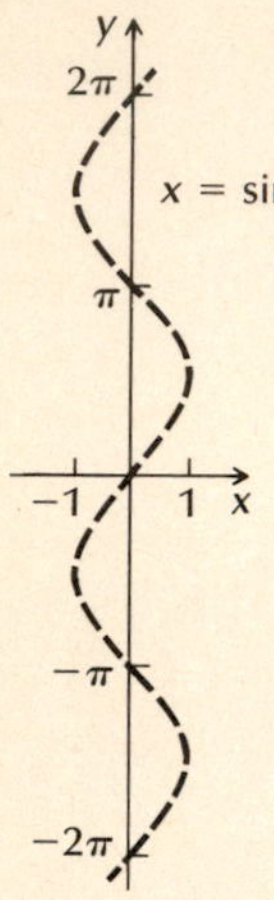

2.

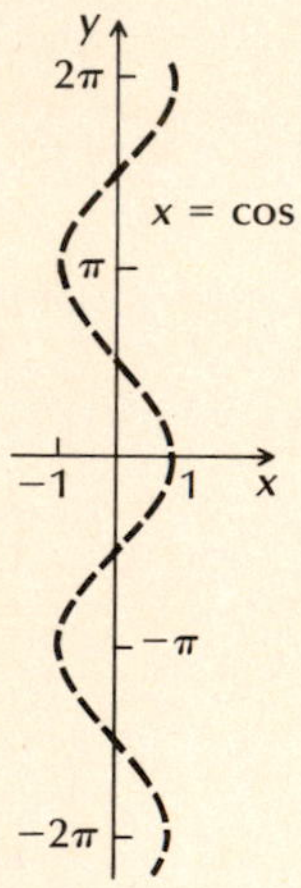

3.

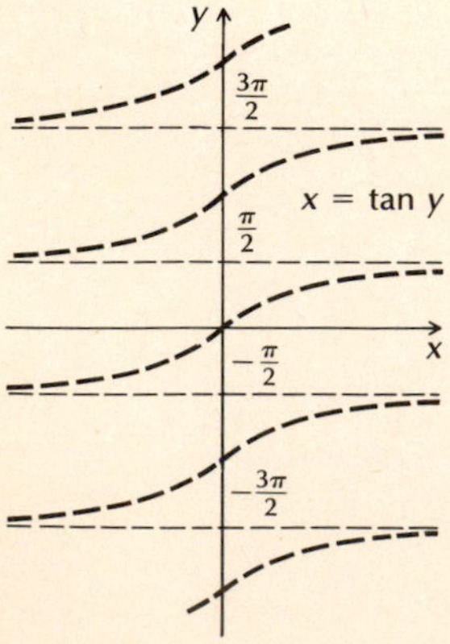

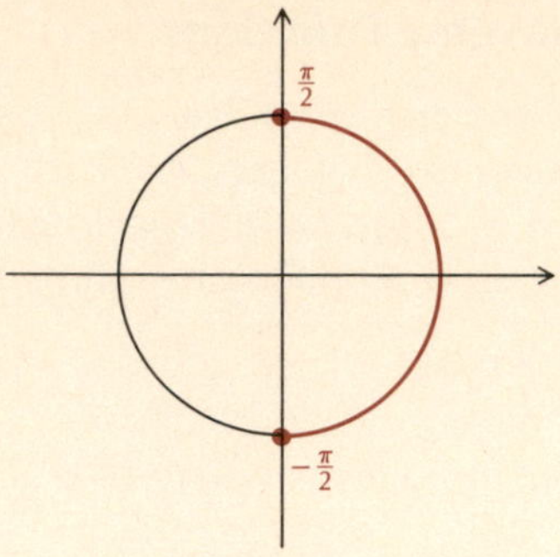

arcsine

Range $\left[-\frac{\pi}{2}, \frac{\pi}{2}\right]$

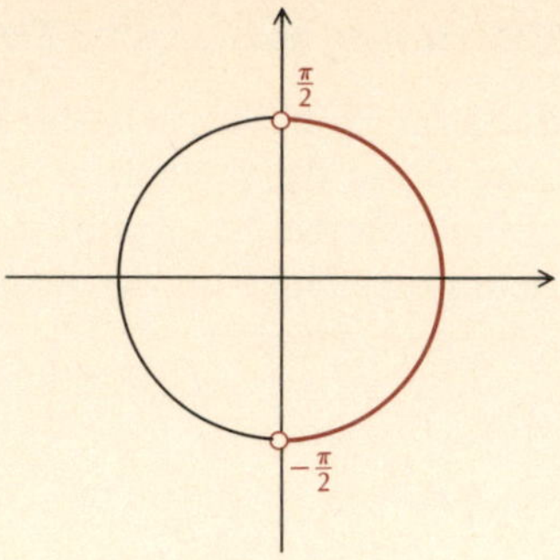

arctangent

Range $\left(-\frac{\pi}{2}, \frac{\pi}{2}\right)$

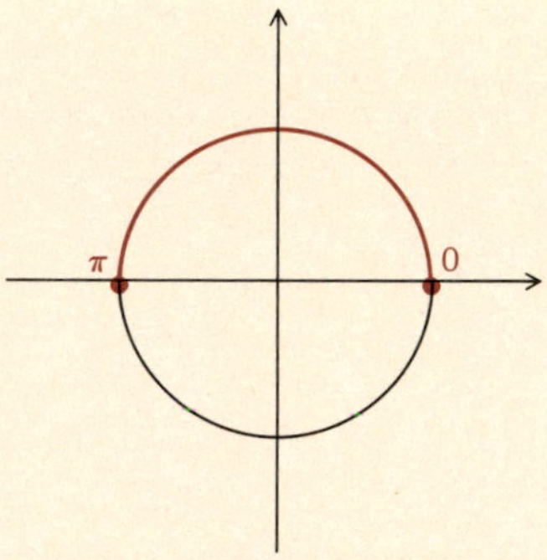

arccosine

Range $[0, \pi]$

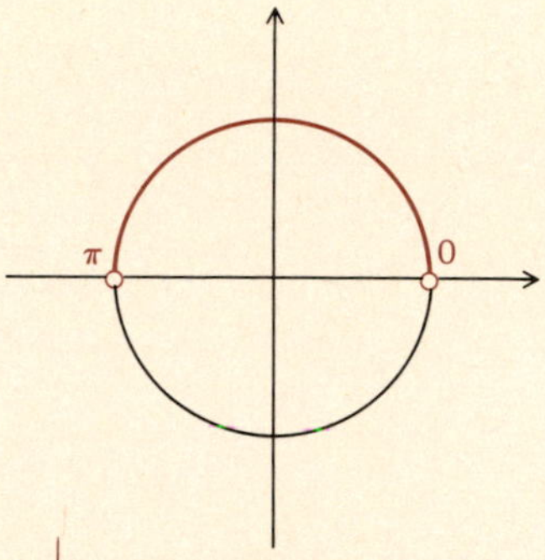

arccotangent

Range $(0, \pi)$

The graphs of the inverse trigonometric functions are as follows:

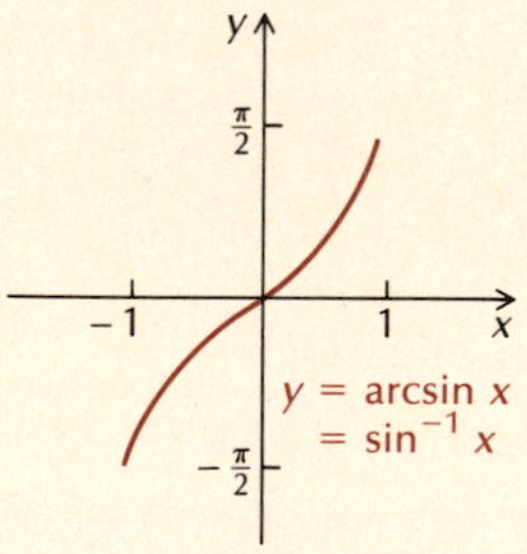

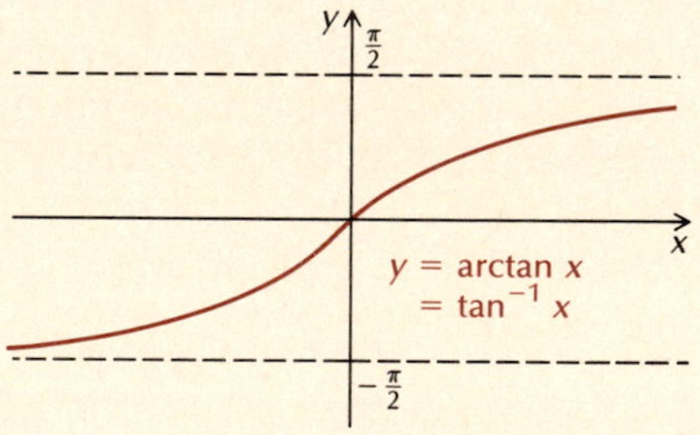

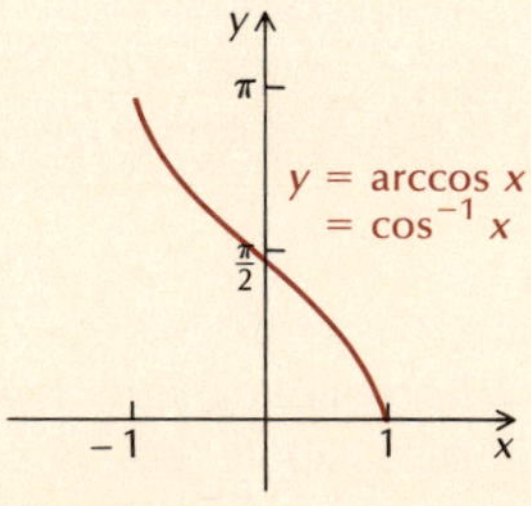

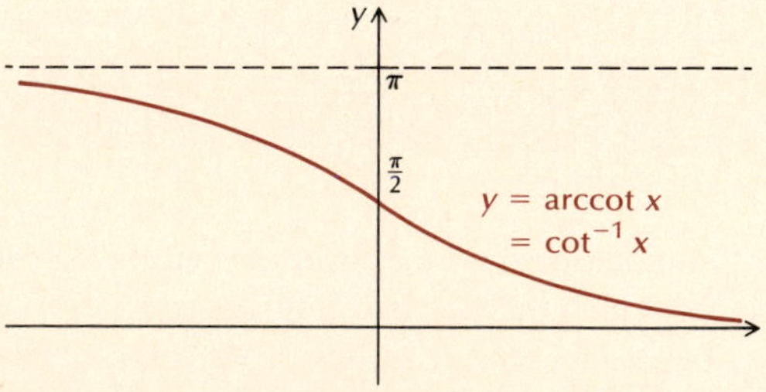

DO EXERCISES 1–4. (EXERCISE 4 IS ON THE FOLLOWING PAGE.)

Now let us find some function values.

EXAMPLE 1 Find $\sin^{-1} \frac{1}{2}$.

Solution In the restricted range, as shown in the figure, the only number whose sine is $\frac{1}{2}$ is $\pi/6$. Hence $\sin^{-1}\frac{1}{2} = \pi/6$. In degrees, we have $\sin^{-1}\frac{1}{2} = 30°$.

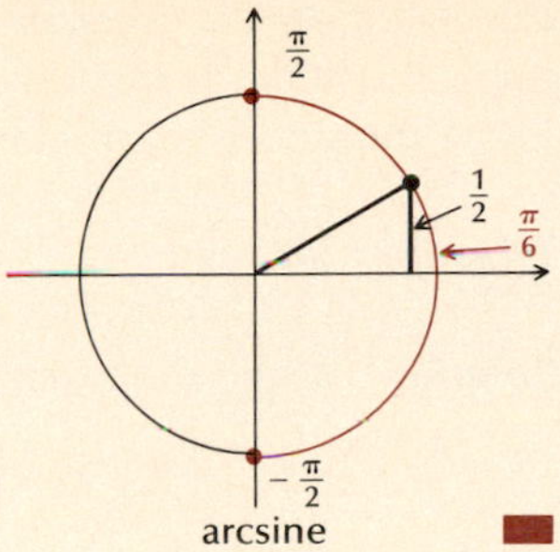

EXAMPLE 2 Find $\arcsin\dfrac{\sqrt{2}}{2}$.

Solution In the restricted range, as shown in the figure, the only number whose sine is $\sqrt{2}/2$ is $\pi/4$. Hence $\arcsin(\sqrt{2}/2) = \pi/4$, or 45°.

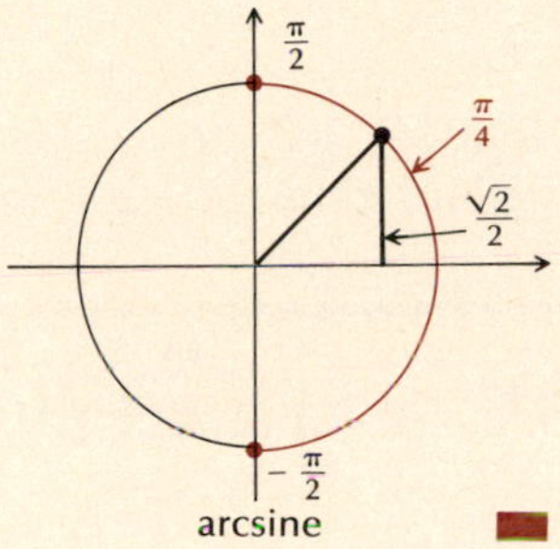

EXAMPLE 3 Find $\cos^{-1}\left(-\frac{1}{2}\right)$.

Solution The only number whose cosine is $-\frac{1}{2}$ in the restricted range $[0, \pi]$ is $2\pi/3$. Hence $\cos^{-1}(-\frac{1}{2}) = 2\pi/3$, or 120°.

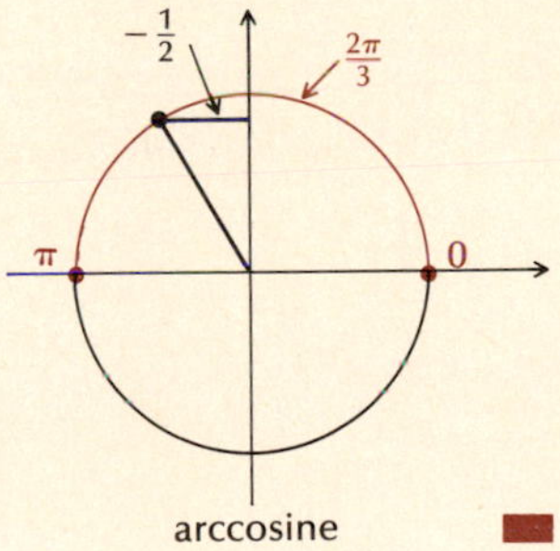

EXAMPLE 4 Find $\tan^{-1} 1$.

Solution The only number in the restricted range $(-\pi/2, \pi/2)$ whose tangent is 1 is $\pi/4$, or 45°.

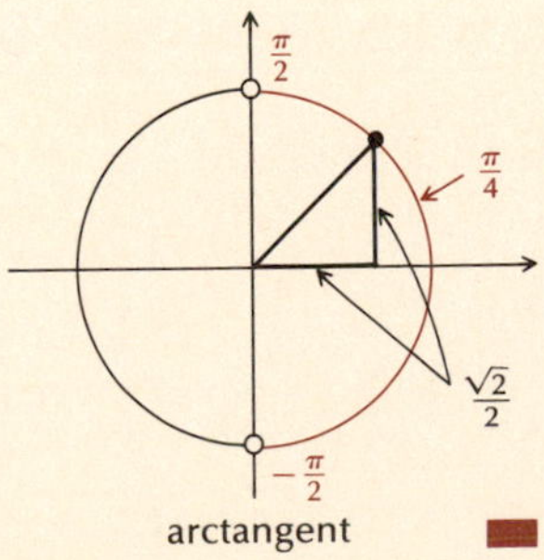

DO EXERCISES 5–8.

We can also use a calculator to find inverse trigonometric function values. Some calculators give inverse function values in both radians and degrees. Some give values only in degrees. The key strokes involved in finding inverse function values vary with the calculator. Be sure to read the instructions for the

4.

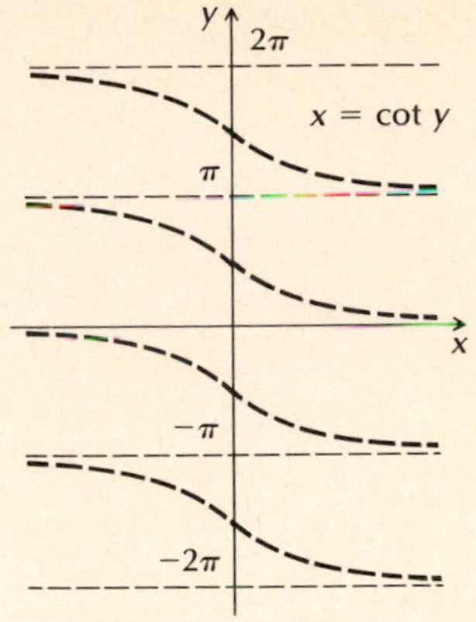

Find each of the following.

5. $\arcsin\dfrac{\sqrt{3}}{2}$

6. $\cos^{-1}\left(-\dfrac{\sqrt{2}}{2}\right)$

7. $\operatorname{arccot}(-1)$

8. $\tan^{-1}(-1)$

Find each of the following using a calculator.

9. $\cos^{-1} 0.63254$

10. arcsin (-0.10203)

11. arctan 1.34568

12. $\tan^{-1} (-22.467)$

calculator you are using, or try a simple value such as $\sin^{-1} 1$. If you get 90, you know the values are in degrees. If you get 1.570796327, which is about $\pi/2$, you know the values are in radians.

EXAMPLE 5 Find $\cos^{-1} (-0.925678)$ in degrees, using a calculator.

Solution Using the inverse trigonometric keys on a calculator, we find that

$$\cos^{-1} (-0.925678) = 157.77°, \quad \text{or } 157°46'.$$

DO EXERCISES 9–12.

The following is a summary of the domains and the ranges of the trigonometric functions together with a summary of the domains and the ranges of the inverse trigonometric functions. For completeness, we have included the arcsecant and the arccosecant, though there is a lack of uniformity on their definitions in the mathematical literature.

Function	Domain	Range	Inverse Function	Domain	Range
sin	All reals, $(-\infty, \infty)$	$[-1, 1]$	$\sin^{-1}$	$[-1, 1]$	$[-\pi/2, \pi/2]$
cos	All reals, $(-\infty, \infty)$	$[-1, 1]$	$\cos^{-1}$	$[-1, 1]$	$[0, \pi]$
tan	All reals except $k\pi/2$, k odd	All reals, $(-\infty, \infty)$	$\tan^{-1}$	All reals, $(-\infty, \infty)$	$\left(-\frac{\pi}{2}, \frac{\pi}{2}\right)$
cot	All reals except $k\pi$	All reals, $(-\infty, \infty)$	$\cot^{-1}$	All reals, $(-\infty, \infty)$	$(0, \pi)$
sec	All reals except $k\pi/2$, k odd	$(-\infty, -1] \cup [1, \infty)$	$\sec^{-1}$	$(-\infty, -1] \cup [1, \infty)$	$\left[0, \frac{\pi}{2}\right) \cup \left[\pi, \frac{3\pi}{2}\right)$
csc	All reals except $k\pi$	$(-\infty, -1] \cup [1, \infty)$	$\csc^{-1}$	$(-\infty, -1] \cup [1, \infty)$	$\left(0, \frac{\pi}{2}\right] \cup \left(-\pi, -\frac{\pi}{2}\right]$

EXERCISE SET 7.6

1 Find each of the following without using a calculator.

1. arcsin $\frac{\sqrt{2}}{2}$
2. arcsin $\frac{\sqrt{3}}{2}$
3. $\cos^{-1} \frac{\sqrt{2}}{2}$
4. $\cos^{-1} \frac{\sqrt{3}}{2}$
5. $\sin^{-1} \left(-\frac{\sqrt{2}}{2}\right)$
6. $\sin^{-1} \left(-\frac{\sqrt{3}}{2}\right)$
7. arccos $\left(-\frac{\sqrt{2}}{2}\right)$
8. arccos $\left(-\frac{\sqrt{3}}{2}\right)$
9. arctan $\sqrt{3}$
10. arctan $\frac{\sqrt{3}}{3}$
11. $\cot^{-1} 1$
12. $\cot^{-1} \sqrt{3}$
13. arctan $\left(-\frac{\sqrt{3}}{3}\right)$
14. arctan $(-\sqrt{3})$
15. arccot (-1)
16. arccot $(-\sqrt{3})$
17. arcsec 1
18. arcsec 2
19. $\csc^{-1} 1$
20. $\csc^{-1} 2$
21. arcsin $\left(-\frac{\sqrt{2}}{2}\right)$
22. arcsin $\frac{1}{2}$
23. $\cos^{-1} \frac{1}{2}$
24. arccos $\frac{\sqrt{2}}{2}$
25. arcsin $\left(-\frac{\sqrt{3}}{2}\right)$
26. $\sin^{-1} \left(-\frac{1}{2}\right)$
27. $\cos^{-1} \left(-\frac{\sqrt{2}}{2}\right)$
28. $\cos^{-1} \left(-\frac{\sqrt{3}}{2}\right)$

29. $\tan^{-1}\left(-\frac{\sqrt{3}}{3}\right)$
30. $\tan^{-1}(-\sqrt{3})$
31. $\text{arccot}\left(-\frac{\sqrt{3}}{3}\right)$
32. $\text{arccot}(-\sqrt{3})$

Use a calculator to find each of the following in degrees.

33. arcsin 0.3907
34. arcsin 0.9613
35. $\sin^{-1}(-0.619867)$
36. $\sin^{-1}(-0.867314)$
37. arccos 0.7990
38. arccos 0.9265
39. $\cos^{-1}(-0.981028)$
40. $\cos^{-1}(-0.271568)$
41. $\tan^{-1} 0.3673$
42. $\tan^{-1} 1.091$
43. $\cot^{-1} 1.265$
44. $\cot^{-1} 0.4770$
45. $\sec^{-1} 1.1677$
46. $\sec^{-1}(-1.4402)$
47. arccsc (−6.2774)
48. arccsc 1.11123
49. arcsin 0.2334
50. arcsin 0.4514
51. $\sin^{-1}(-0.6361)$
52. $\sin^{-1}(-0.8192)$
53. arcsin (−0.8886)
54. arccos (−0.2935)
55. $\tan^{-1}(-0.4087)$
56. $\tan^{-1}(-0.2410)$
57. $\cot^{-1}(-5.936)$
58. $\cot^{-1}(-1.319)$
59. $\cot^{-1}(-23)$
60. $\tan^{-1}(158)$

SYNTHESIS

61. Graph the function $y = \sec^{-1} x$.
62. Graph the function $y = \csc^{-1} x$.

Show that each of the following is *not* an identity.

63. $\sin^{-1} x = (\sin x)^{-1}$
64. $\cos^{-1} x = (\cos x)^{-1}$
65. $\tan^{-1} x = (\tan x)^{-1}$
66. $\cot^{-1} x = (\cot x)^{-1}$
67. $\tan^{-1} x = \dfrac{\sin^{-1} x}{\cos^{-1} x}$
68. $(\cos^{-1} x)^2 + (\sin^{-1} x)^2 = 1$

69. A guy wire is attached to the top of a 50-ft pole and stretched to a point that is b ft from the bottom of the pole. Show that the angle of inclination θ of the wire to the top is given by

$$\theta = \tan^{-1}\left(\frac{50}{b}\right).$$

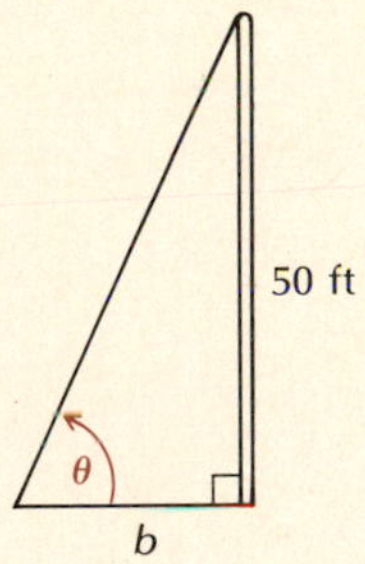

70. An airplane at an altitude of 2000 ft is flying toward an island. The straight-line distance from the airplane to the island is h ft. Show that the angle of depression θ is given by

$$\theta = \sin^{-1}\left(\frac{2000}{h}\right).$$

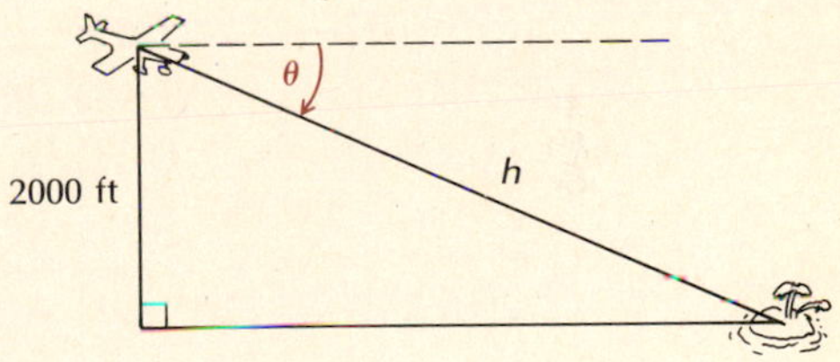

71. a) Use your calculator to approximate the following expression:

$$16\tan^{-1}\frac{1}{5} - 4\tan^{-1}\frac{1}{239}.$$

b) What number does this expression seem to approximate?

7.7 Composition of Trigonometric Functions and Their Inverses

OBJECTIVES

You should be able to:

1 Simplify expressions such as $\sin(\sin^{-1} x)$ and $\sin^{-1}(\sin x)$.

2 Simplify expressions involving compositions such as $\sin\left(\cos^{-1}\frac{1}{2}\right)$, without using a calculator or a table, and simplify expressions such as sin arctan (a/b) by drawing a triangle or triangles and reading off appropriate ratios.

1 Immediate Simplification

Various compositions of trigonometric functions and their inverses often arise in practice. For example, we might want to try to simplify expressions such as

$$\sin(\sin^{-1} x) \quad \text{or} \quad \sin\left(\text{arccot}\,\frac{x}{2}\right).$$

In the expression on the left, we are finding "the sine of a number whose sine is

Find each of the following.

1. $\sin\left(\sin^{-1}\frac{1}{2}\right)$

2. $\tan\left(\tan^{-1}1\right)$

3. $\cos\left(\arccos\frac{\sqrt{2}}{2}\right)$

4. $\sin\left[\sin^{-1}(-5.7)\right]$

Find each of the following.

5. $\cos^{-1}\left(\cos\frac{2\pi}{3}\right)$

6. $\arctan\left(\tan\frac{3\pi}{4}\right)$

7. $\sin^{-1}\left(\sin\frac{\pi}{6}\right)$

8. $\arccos\left[\cos\left(-\frac{2\pi}{3}\right)\right]$

x." Recall from Section 5.1 that if a function f has an inverse that is also a function, then

$$f(f^{-1}(x)) = x, \quad \text{for all } x \text{ in the domain of } f^{-1},$$

and

$$f^{-1}(f(x)) = x, \quad \text{for all } x \text{ in the domain of } f.$$

Thus, if $f(x) = \sin x$ and $f^{-1}(x) = \sin^{-1} x$, then $\sin(\sin^{-1} x) = x$, if x is in the domain of $\sin^{-1}$, which is any number in the interval $[-1, 1]$. Similar results hold for the other trigonometric functions. These yield the following theorem.

THEOREM 3

a) $\sin(\sin^{-1} x) = x$, for any x in the domain of $\sin^{-1}$.

b) $\cos(\cos^{-1} x) = x$, for any x in the domain of $\cos^{-1}$.

c) $\tan(\tan^{-1} x) = x$, for any x in the domain of $\tan^{-1}$.

d) $\cot(\cot^{-1} x) = x$, for any x in the domain of $\cot^{-1}$.

e) $\sec(\sec^{-1} x) = x$, for any x in the domain of $\sec^{-1}$.

f) $\csc(\csc^{-1} x) = x$, for any x in the domain of $\csc^{-1}$.

EXAMPLE 1 Find $\cos\left(\cos^{-1}\frac{\sqrt{3}}{2}\right)$.

Solution Since $\sqrt{3}/2$ is in the interval $[-1, 1]$, it follows that

$$\cos\left(\cos^{-1}\frac{\sqrt{3}}{2}\right) = \frac{\sqrt{3}}{2}. \quad ■$$

EXAMPLE 2 Find $\sin(\sin^{-1} 12.3)$.

Solution Since 12.3 is not in the interval $[-1, 1]$, we cannot evaluate this expression. Thinking one step at a time, we know that there is no number whose sine is 12.3. Since we cannot find $\sin^{-1} 12.3$, we cannot evaluate the expression. ■

DO EXERCISES 1–4.

Now let us consider an expression like $\sin^{-1}(\sin x)$. We might also suspect that this is equal to x for any x, but this is not true unless x is in the range of the $\sin^{-1}$ function. Note that in order to define $\sin^{-1}$, we had to restrict the domain of the sine function. In so doing, we restricted the range of the inverse sine function.

EXAMPLE 3 Find $\arcsin\left(\sin\frac{3\pi}{4}\right)$.

Solution We first find $\sin(3\pi/4)$. It is $\sqrt{2}/2$. Next we find $\arcsin(\sqrt{2}/2)$. It is $\pi/4$. So

$$\arcsin\left(\sin\frac{3\pi}{4}\right) = \frac{\pi}{4}. \quad ■$$

DO EXERCISES 5–8.

For any x in the range of the arcsine function, $[-\pi/2, \pi/2]$, we do have $\arcsin(\sin x) = x$, or $\sin^{-1}(\sin x) = x$. Similar conditions hold for the other functions.

THEOREM 4

The following are true for any x in the range of the inverse function.

a) $\sin^{-1}(\sin x) = x$, for any x in the range of $\sin^{-1}$.
b) $\cos^{-1}(\cos x) = x$, for any x in the range of $\cos^{-1}$.
c) $\tan^{-1}(\tan x) = x$, for any x in the range of $\tan^{-1}$.
d) $\cot^{-1}(\cot x) = x$, for any x in the range of $\cot^{-1}$.
e) $\sec^{-1}(\sec x) = x$, for any x in the range of $\sec^{-1}$.
f) $\csc^{-1}(\csc x) = x$, for any x in the range of $\csc^{-1}$.

2 Simplifying Other Compositions

Now we find some other function composition.

EXAMPLE 4 Find $\sin[\arctan(-1)]$.

Solution We first find $\arctan(-1)$. It is $-\pi/4$. Now we find the sine of this, which is $\sin(-\pi/4) = -\sqrt{2}/2$, so $\sin[\arctan(-1)] = -\sqrt{2}/2$.

EXAMPLE 5 Find $\cos^{-1}\left(\sin \frac{\pi}{2}\right)$.

Solution We first find $\sin(\pi/2)$. It is 1. Now we find $\cos^{-1} 1$. It is 0, so $\cos^{-1}[\sin(\pi/2)] = 0$.

DO EXERCISES 9–13.

Now let us consider

$$\cos\left(\arcsin \tfrac{3}{5}\right).$$

Without using a calculator, we cannot find $\arcsin \frac{3}{5}$. However, we can still evaluate the entire expression without using a calculator. We are looking for an angle θ such that $\arcsin \frac{3}{5} = \theta$, or $\sin\theta = \frac{3}{5}$. In a case like this, we sketch a triangle, as shown below. The angle θ in this triangle is an angle whose sine is $\frac{3}{5}$ (it is $\arcsin \frac{3}{5}$). We wish to find the cosine of this angle. Since the triangle is a right triangle, we can find the length of the base, b. It is 4. Thus we know that $\cos\theta = b/5$, or $\frac{4}{5}$. Therefore,

$$\cos\left(\arcsin \tfrac{3}{5}\right) = \tfrac{4}{5}.$$

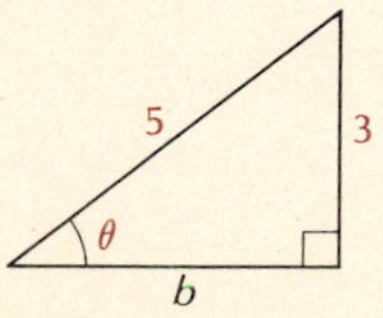

EXAMPLE 6 Find $\sin\left(\operatorname{arccot} \frac{x}{2}\right)$.

Solution We draw a right triangle whose sides have lengths x and 2, so that $\cot\theta = x/2$. If x is negative, we get the triangle in standard position shown on the left in the following figure. If x is positive, we get the triangle shown on the right.

Find each of the following.

9. $\sin(\arctan 1)$

10. $\cos\left(\arcsin \frac{1}{2}\right)$

11. $\cos^{-1}\left(\sin \frac{\pi}{6}\right)$

12. $\sin^{-1}\left(\tan \frac{\pi}{4}\right)$

13. $\cos(\sin^{-1} 0)$

Find each of the following.

14. $\cos\left(\arctan \frac{b}{3}\right)$

15. $\sin(\arctan 2)$
$\left(Hint: 2 = \frac{2}{1}.\right)$

16. $\tan(\arcsin t)$

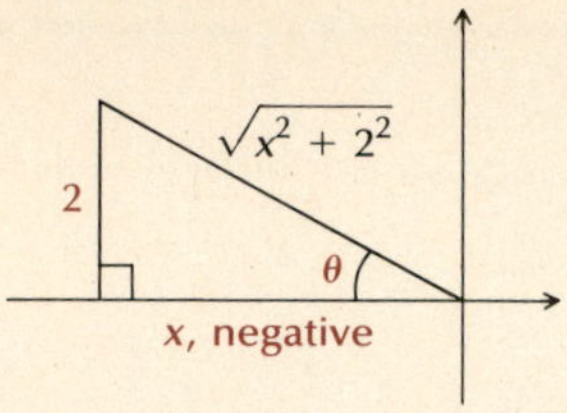

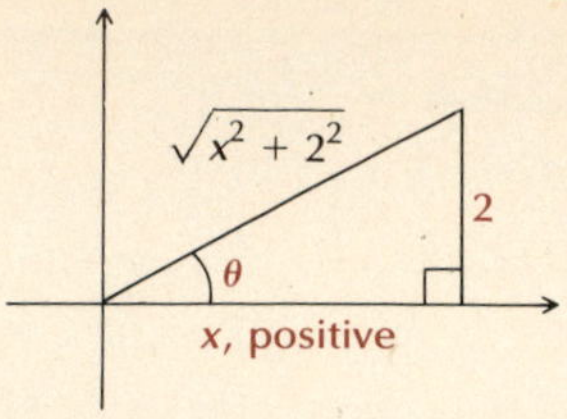

We find the length of the hypotenuse and then read off the sine ratio. In either case, we get

$$\sin\left(\operatorname{arccot} \frac{x}{2}\right) = \frac{2}{\sqrt{x^2 + 2^2}}, \quad \text{or} \quad \frac{2}{\sqrt{x^2 + 4}}.$$

EXAMPLE 7 Find $\cos(\arctan p)$.

Solution We draw two right triangles, two of whose sides have lengths p and 1, so that $\tan \theta = p/1$.

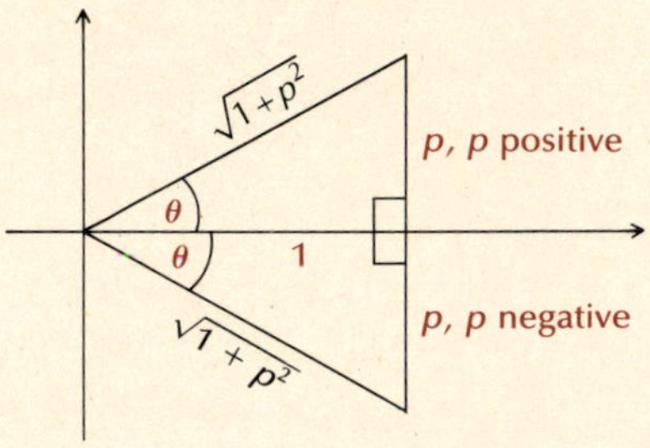

We find the length of the other side and then read off the cosine ratio. In either case, we get

$$\cos(\arctan p) = \frac{1}{\sqrt{1 + p^2}}.$$

The idea in Examples 6 and 7 is to first sketch a triangle, or triangles, two of whose sides have the appropriate ratio. Then use the Pythagorean theorem to find the length of the third side and read off the desired ratio.

DO EXERCISES 14–16.

In some cases, the use of certain identities is needed to evaluate expressions.

EXAMPLE 8 Evaluate $\sin\left(\sin^{-1} \frac{1}{2} + \cos^{-1} \frac{4}{5}\right)$.

Solution To simplify the use of an identity, we make substitutions. Let

$$u = \sin^{-1} \frac{1}{2} \quad \text{and} \quad v = \cos^{-1} \frac{4}{5}.$$

Then we have $\sin(u + v)$. By a sum formula, this is equivalent to

$$\sin u \cos v + \cos u \sin v.$$

Now we make the reverse substitutions and obtain

$$\sin\left(\sin^{-1} \tfrac{1}{2}\right) \cdot \cos\left(\cos^{-1} \tfrac{4}{5}\right) + \cos\left(\sin^{-1} \tfrac{1}{2}\right) \cdot \sin\left(\cos^{-1} \tfrac{4}{5}\right).$$

This immediately simplifies to

$$\tfrac{1}{2} \cdot \tfrac{4}{5} + \cos\left(\sin^{-1} \tfrac{1}{2}\right) \cdot \sin\left(\cos^{-1} \tfrac{4}{5}\right).$$

Now $\cos(\sin^{-1}\frac{1}{2})$ readily simplifies to $\sqrt{3}/2$. To find $\sin(\cos^{-1}\frac{4}{5})$, we need a triangle. We set up the triangle so that $\cos^{-1}\frac{4}{5} = \theta$, or $\cos\theta = \frac{4}{5}$. Then we find $\sin\theta$.

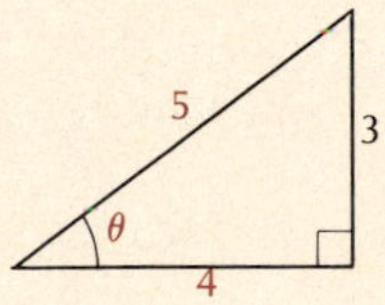

We see that $\sin(\cos^{-1}\frac{4}{5}) = \frac{3}{5}$. Our expression is now simplified to

$$\frac{1}{2}\cdot\frac{4}{5} + \frac{\sqrt{3}}{2}\cdot\frac{3}{5}, \quad \text{or} \quad \frac{4+3\sqrt{3}}{10}.$$

Thus,

$$\sin\left(\sin^{-1}\frac{1}{2} + \cos^{-1}\frac{4}{5}\right) = \frac{4+3\sqrt{3}}{10}.$$

DO EXERCISES 17 AND 18.

Evaluate.

17. $\cos\left(\sin^{-1}\frac{\sqrt{3}}{2} - \cos^{-1}\frac{1}{2}\right)$

18. $\tan\left(\frac{1}{2}\arcsin\frac{3}{5}\right)$

$\left(\textit{Hint:}\text{ Let } \arcsin\frac{3}{5} = u \text{ and use a half-angle formula.}\right)$

EXERCISE SET 7.7

1 Evaluate or simplify.

1. $\sin(\arcsin 0.3)$
2. $\cos(\arccos 0.2)$
3. $\tan[\tan^{-1}(-4.2)]$
4. $\cot[\cot^{-1}(-1.5)]$
5. $\arcsin\left(\sin\frac{2\pi}{3}\right)$
6. $\arccos\left(\cos\frac{3\pi}{2}\right)$
7. $\sin^{-1}\left[\sin\left(-\frac{3\pi}{4}\right)\right]$
8. $\cos^{-1}\left[\cos\left(-\frac{\pi}{4}\right)\right]$
9. $\sin^{-1}\left(\sin\frac{\pi}{5}\right)$
10. $\cos^{-1}\left(\cos\frac{\pi}{7}\right)$
11. $\tan^{-1}\left(\tan\frac{2\pi}{3}\right)$
12. $\cot^{-1}\left(\cot\frac{2\pi}{3}\right)$

2

13. $\sin(\arctan\sqrt{3})$
14. $\sin\left(\arctan\frac{\sqrt{3}}{3}\right)$
15. $\cos\left(\arcsin\frac{\sqrt{3}}{2}\right)$
16. $\cos\left(\arcsin\frac{\sqrt{2}}{2}\right)$
17. $\tan\left(\cos^{-1}\frac{\sqrt{2}}{2}\right)$
18. $\tan\left(\cos^{-1}\frac{\sqrt{3}}{2}\right)$
19. $\cos^{-1}\left(\sin\frac{\pi}{3}\right)$
20. $\cos^{-1}(\sin\pi)$
21. $\arcsin\left(\cos\frac{\pi}{6}\right)$
22. $\arcsin\left(\cos\frac{\pi}{4}\right)$
23. $\sin^{-1}\left(\tan\frac{\pi}{4}\right)$
24. $\sin^{-1}\left[\tan\left(-\frac{\pi}{4}\right)\right]$
25. $\sin\left(\arctan\frac{x}{2}\right)$
26. $\sin\left(\arctan\frac{a}{3}\right)$
27. $\tan\left(\cos^{-1}\frac{3}{x}\right)$
28. $\cot\left(\sin^{-1}\frac{5}{y}\right)$
29. $\cot\left(\sin^{-1}\frac{a}{b}\right)$
30. $\tan\left(\cos^{-1}\frac{p}{q}\right)$
31. $\cos\left(\tan^{-1}\frac{\sqrt{2}}{3}\right)$
32. $\cos\left(\tan^{-1}\frac{\sqrt{3}}{4}\right)$
33. $\tan(\arcsin 0.1)$
34. $\tan(\arcsin 0.2)$
35. $\cot[\cos^{-1}(-0.2)]$
36. $\cot[\cos^{-1}(-0.3)]$
37. $\sin(\operatorname{arccot} y)$
38. $\sin(\operatorname{arccot} x)$
39. $\cos(\arctan t)$
40. $\sin(\arctan t)$
41. $\cot(\sin^{-1} y)$
42. $\tan(\cos^{-1} y)$
43. $\sin(\cos^{-1} x)$
44. $\cos(\sin^{-1} x)$
45. $\tan\left(\frac{1}{2}\arcsin\frac{4}{5}\right)$

46. $\tan\left(\frac{1}{2}\arcsin\frac{1}{2}\right)$

47. $\cos\left(\frac{1}{2}\arcsin\frac{1}{2}\right)$

48. $\cos\left(\frac{1}{2}\arcsin\frac{\sqrt{3}}{2}\right)$

49. $\sin\left(2\cos^{-1}\frac{3}{5}\right)$

50. $\sin\left(2\cos^{-1}\frac{1}{2}\right)$

51. $\cos\left(2\sin^{-1}\frac{5}{13}\right)$

52. $\cos\left(2\cos^{-1}\frac{4}{5}\right)$

53. $\sin\left(\sin^{-1}\frac{1}{2}+\cos^{-1}\frac{3}{5}\right)$

54. $\sin\left(\sin^{-1}\frac{1}{2}-\cos^{-1}\frac{4}{5}\right)$

55. $\cos\left(\sin^{-1}\frac{\sqrt{2}}{2}+\cos^{-1}\frac{3}{5}\right)$

56. $\cos\left(\sin^{-1}\frac{4}{5}-\cos^{-1}\frac{1}{2}\right)$

57. $\sin(\sin^{-1}x+\cos^{-1}y)$

58. $\sin(\sin^{-1}x-\cos^{-1}y)$

59. $\cos(\sin^{-1}x+\cos^{-1}y)$

60. $\cos(\sin^{-1}x-\cos^{-1}y)$

61. ▦ $\sin(\sin^{-1}0.6032+\cos^{-1}0.4621)$

62. ▦ $\cos(\sin^{-1}0.7325-\cos^{-1}0.4838)$

63. $\tan\left(\arcsin\frac{p}{\sqrt{p^2+9}}\right)$

64. $\csc\left(\tan^{-1}\frac{\sqrt{25-p^2}}{p}\right)$

SYNTHESIS

65. Suppose that $\theta = \sin^{-1} x$. Find expressions in terms of x for $\sin\theta$, $\cos\theta$, $\tan\theta$, $\cot\theta$, $\sec\theta$, and $\csc\theta$.

66. Suppose that $\theta = \arccos x$. Find expressions in terms of x for $\sin\theta$, $\cos\theta$, $\tan\theta$, $\cot\theta$, $\sec\theta$, and $\csc\theta$.

67. Suppose that $\theta = \tan^{-1} x$. Find expressions in terms of x for $\sin\theta$, $\cos\theta$, $\tan\theta$, $\cot\theta$, $\sec\theta$, and $\csc\theta$.

68. Suppose that $\theta = \operatorname{arccot} x$. Find expressions in terms of x for $\sin\theta$, $\cos\theta$, $\tan\theta$, $\cot\theta$, $\sec\theta$, and $\csc\theta$.

Prove each of the following identities.

69. $\sin^{-1}x+\cos^{-1}x=\frac{\pi}{2}$

70. $\tan^{-1}x+\cot^{-1}x=\frac{\pi}{2}$

71. $\sin^{-1}x=\tan^{-1}\frac{x}{\sqrt{1-x^2}}$

72. $\tan^{-1}x=\sin^{-1}\frac{x}{\sqrt{x^2+1}}$

73. For $x \geq 0$,
$$\arcsin x = \arccos\sqrt{1-x^2}.$$

74. For $x \geq 0$,
$$\arccos x = \arctan\frac{\sqrt{1-x^2}}{x}.$$

75. An observer's eye is at a point A, looking at a mural of height h, with the bottom of the mural y feet above the eye. The eye is x feet from the wall. Write an expression for θ in terms of x, y, and h.

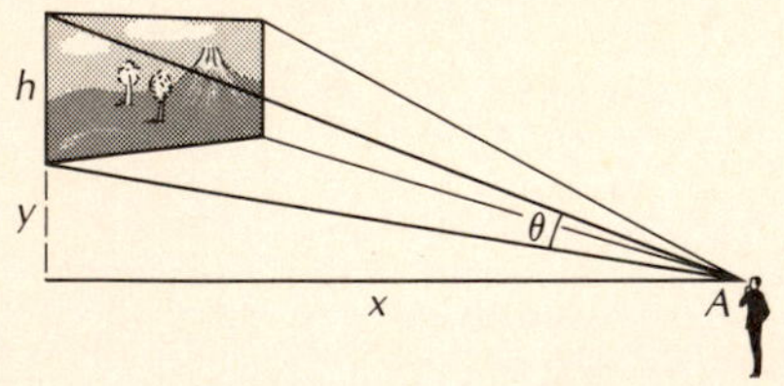

76. ▦ Evaluate the expression given in Exercise 75 when $x = 20$ ft, $y = 7$ ft, and $h = 25$ ft.

7.8 Trigonometric Equations

OBJECTIVES

You should be able to:

1 Solve simple trigonometric equations not requiring the use of calculators, finding all solutions, or all solutions in $[0, 2\pi)$.

(continued)

1 Simple Equations

When an equation contains a trigonometric expression with a variable, such as $\sin x$, it is called a **trigonometric equation**. We have worked with many trigonometric equations that are identities. Now we consider equations that may not be identities. To solve such an equation, we find all replacements for the variable that make the equation true.

EXAMPLE 1 Solve: $2 \sin x = 1$.

Solution We first solve for $\sin x$:

$$\sin x = \tfrac{1}{2}.$$

Now we note that the solutions are those numbers having a sine of $\frac{1}{2}$. We look for them. The unit circle is helpful.

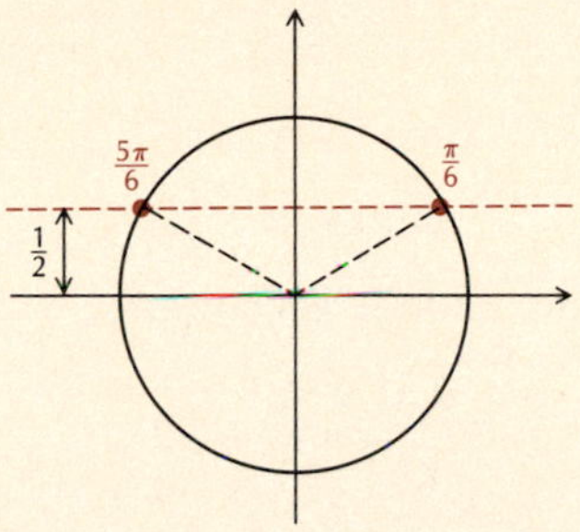

There are just two points on it for which the sine is $\frac{1}{2}$, as shown. They are the points for $\pi/6$ and $5\pi/6$. These numbers, plus any multiple of 2π, are the solutions

$$\frac{\pi}{6} + 2k\pi \quad \text{or} \quad \frac{5\pi}{6} + 2k\pi,$$

where k is any integer. In degrees, the solutions are

$$30° + k \cdot 360° \quad \text{or} \quad 150° + k \cdot 360°,$$

where k is any integer.

Note that what we did in Example 1 was comparable to solving $x = \sin^{-1} \frac{1}{2}$, only now we get *all* the numbers whose sines are $\frac{1}{2}$, not just $\pi/6$.

DO EXERCISE 1.

EXAMPLE 2 Solve: $4 \cos^2 x = 1$.

Solution We first solve for $\cos x$:

$$\cos^2 x = \tfrac{1}{4}$$
$$\cos x = \pm\tfrac{1}{2}.$$

Now we use the unit circle to find those numbers having a cosine of $\frac{1}{2}$ or $-\frac{1}{2}$. The solutions are $\pi/3$, $2\pi/3$, $4\pi/3$, $5\pi/3$, plus any multiple of 2π. In degrees, the solutions are 60°, 120°, 240°, 300°, plus any multiple of 360°.

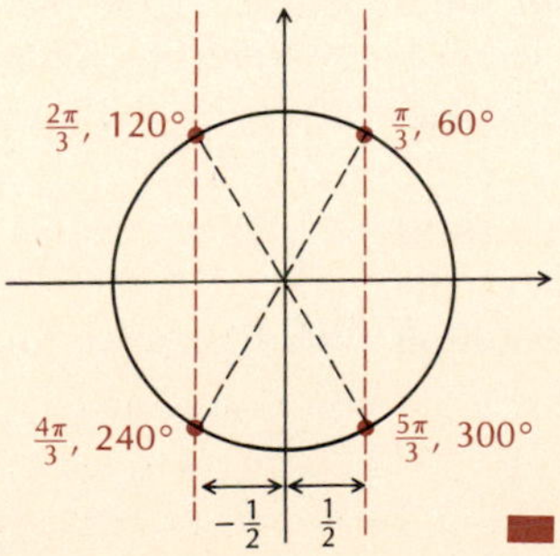

DO EXERCISE 2.

2 **Solve simple trigonometric equations requiring the use of calculators, finding all solutions, or all solutions in [0°, 360°).**

3 **Solve trigonometric equations by factoring or using the quadratic formula.**

1. Solve. Give answers in both degrees and radians.

$$2 \cos x = 1$$

2. Find all solutions in $[0, 2\pi)$. Give answers in both degrees and radians.

$$4 \sin^2 x = 1$$

3. Find all solutions in $[0, 2\pi)$. Leave answers in terms of π.

$$2 \cos 2x = 1$$

In solving trigonometric equations as in most practical applications, it is usually sufficient to find just the solutions from 0 to 2π, or from 0° to 360°. We then remember that any multiple of 2π, or 360°, can be added to obtain the rest of the solutions.

The following example illustrates that when we look for solutions in the interval $[0, 2\pi)$, or $[0°, 360°)$, we must be cautious.

EXAMPLE 3 Solve $2 \sin 2x = -1$ in the interval $[0, 2\pi)$.

Solution We first solve for $\sin 2x$:

$$2 \sin 2x = -1$$
$$\sin 2x = -\tfrac{1}{2}.$$

We are looking for solutions x to the equation for which

$$0 \le x < 2\pi.$$

Multiplying by 2, we get

$$0 \le 2x < 4\pi,$$

which is the interval we consider to solve $\sin 2x = \frac{1}{2}$.

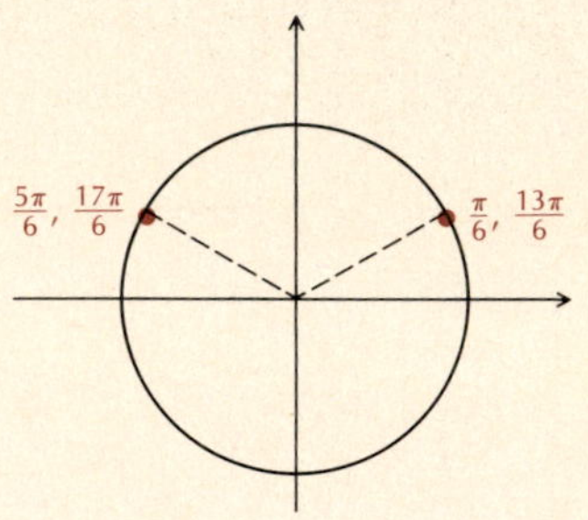

Using the unit circle, we find points x for which $\sin 2x = \frac{1}{2}$ and $0 \le 2x < 4\pi$. These values of $2x$ are $\pi/6$, $5\pi/6$, $13\pi/6$, and $17\pi/6$. Thus the desired values of x in $[0, 2\pi)$ are half of these. Therefore,

$$x = \frac{\pi}{12}, \frac{5\pi}{12}, \frac{13\pi}{12}, \frac{17\pi}{12}.$$

DO EXERCISE 3.

2 Using a Calculator

In solving some trigonometric equations, it is necessary to use a calculator. Answers can then be found in radians or degrees, depending on how the calculator is set. We will usually find answers in degrees.

EXAMPLE 4 Solve $2 + \sin x = 2.5299$ in $[0, 360°)$.

Solution We first solve for $\sin x$:

$$2 + \sin x = 2.5299$$
$$\sin x = 0.5299.$$

From a calculator, we find the reference angle, $x = 32°$. Since $\sin x$ is positive, the solutions are to be found in the first and second quadrants. The solutions are

$$32° \quad \text{or} \quad 148°.$$

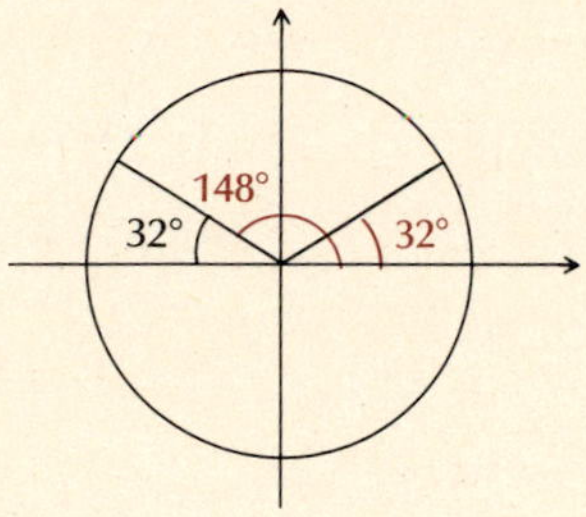

EXAMPLE 5 Solve $\sin x - 1 = -1.5299$ in $[0°, 360°)$.

Solution We first solve for $\sin x$:

$$\sin x - 1 = -1.5299$$
$$\sin x = -0.5299.$$

From a calculator, we find that $x = 32°$, so the reference angle is 32°. Since $\sin x$ is negative, the solutions are to be found in the third and fourth quadrants. The solutions are

$$212° \quad \text{or} \quad 328°.$$

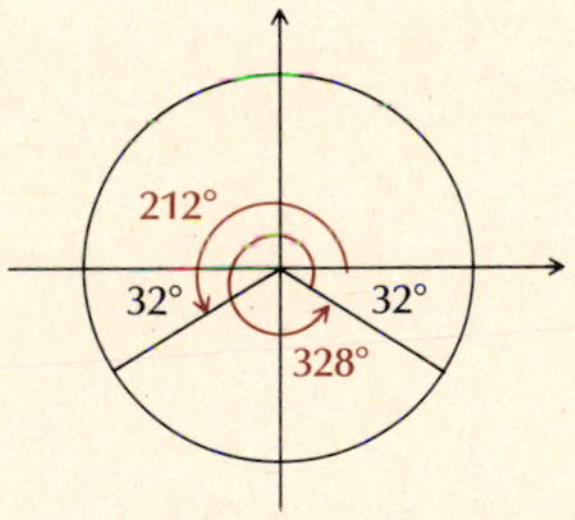

DO EXERCISES 4 AND 5.

3 Using Algebraic Techniques

In solving trigonometric equations, we can expect to apply some algebra before concerning ourselves with the trigonometric part. In the next examples, we begin by factoring. The equations are reducible to quadratic.

EXAMPLE 6 Solve $8 \cos^2 \theta - 2 \cos \theta = 1$ in $[0°, 360°)$.

Solution Since we will be using the principle of zero products, we first obtain a 0 on one side of the equation:

$$8 \cos^2 \theta - 2 \cos \theta - 1 = 0$$
$$(4 \cos \theta + 1)(2 \cos \theta - 1) = 0 \quad \text{Factoring}$$
$$4 \cos \theta + 1 = 0 \quad \text{or} \quad 2 \cos \theta - 1 = 0 \quad \text{Principle of zero products}$$
$$\cos \theta = -\tfrac{1}{4} = -0.25 \quad \text{or} \quad \cos \theta = \tfrac{1}{2}$$
$$\theta = 104°29', 255°31' \quad \text{or} \quad \theta = 60°, 300°.$$

The solutions in $[0°, 360°)$ are $104°29'$, $255°31'$, 60°, and 300°, or in decimal notation, 104.48°, 255.52°, 60°, and 300°.

Solve in $[0°, 360°)$.

4. $2 + \cos x = 2.7660$

5. $\cos x - 1 = -1.7660$

6. $8 \cos^2 \theta + 2 \cos \theta = 1$

7. $2 \cos^2 \phi + \cos \phi = 0$

8. Solve using the quadratic formula:

$10 \cos^2 x - 10 \cos x - 7 = 0.$

It may be helpful in accomplishing the algebraic part of solving trigonometric equations to make substitutions as in Chapter 2. If we use such an approach, the algebraic part of Example 6 would look like this:

$$8 \cos^2 \theta - 2 \cos \theta = 1.$$

We would then let $u = \cos \theta$:

$$\begin{aligned} 8u^2 - 2u &= 1 \\ (4u + 1)(2u - 1) &= 0 \\ 4u + 1 = 0 \quad \text{or} \quad 2u - 1 &= 0 \\ u = -\tfrac{1}{4} \quad \text{or} \quad u &= \tfrac{1}{2}. \end{aligned}$$

We then substitute $\cos \theta$ for u and solve for θ.

EXAMPLE 7 Solve $2 \sin^2 \phi + \sin \phi = 0$ in $[0, 2\pi)$.

Solution

$$\begin{aligned} 2 \sin^2 \phi + \sin \phi &= 0 \\ \sin \phi (2 \sin \phi + 1) &= 0 && \text{Factoring} \\ \sin \phi = 0 \quad \text{or} \quad 2 \sin \phi + 1 &= 0 && \text{Principle of zero products} \\ \sin \phi = 0 \quad \text{or} \quad \sin \phi &= -\frac{1}{2} \\ \phi = 0, \pi \quad \text{or} \quad \phi &= \frac{7\pi}{6}, \frac{11\pi}{6} \end{aligned}$$

The solutions in $[0, 2\pi)$ are 0, π, $7\pi/6$, and $11\pi/6$.

DO EXERCISES 6 AND 7.

If a trigonometric equation is quadratic but difficult or impossible to factor, we use the quadratic formula.

EXAMPLE 8 Solve $10 \sin^2 x - 12 \sin x - 7 = 0$ in $[0, 360°)$.

Solution It may help to make the substitution $u = \sin x$, in order to obtain the equation $10u^2 - 12u - 7 = 0$. We then use the quadratic formula to find u, or $\sin x$:

$$\begin{aligned} \sin x &= \frac{12 \pm \sqrt{144 + 280}}{20} && \text{Using the quadratic formula} \\ &= \frac{12 \pm \sqrt{424}}{20} = \frac{12 \pm 2\sqrt{106}}{20} = \frac{6 \pm \sqrt{106}}{10} \\ &= \frac{6 \pm 10.296}{10} \end{aligned}$$

$$\sin x = 1.6296 \quad \text{or} \quad \sin x = -0.4296.$$

Since sines are never greater than 1, the first of the equations has no solution. Using the other equation, we find the reference angle to be 25.44°, or 25°27′. Since $\sin x$ is negative, the solutions are to be found in the third and fourth quadrants. Thus the solutions are 205°27′ and 334°33′.

DO EXERCISE 8.

EXERCISE SET 7.8

1, **2** Solve, finding all solutions.

1. $\sin x = \frac{\sqrt{3}}{2}$
2. $\cos x = \frac{\sqrt{3}}{2}$
3. $\cos x = \frac{1}{\sqrt{2}}$
4. $\tan x = \sqrt{3}$
5. $\sin x = 0.3448$
6. $\cos x = 0.6406$

Solve, finding all solutions in $[0, 2\pi)$ or $[0°, 360°)$.

7. $\cos x = -0.5495$
8. $\sin x = -0.4279$
9. $2 \sin x + \sqrt{3} = 0$
10. $\sqrt{3} \tan x + 1 = 0$
11. $2 \tan x + 3 = 0$
12. $4 \sin x - 1 = 0$

3 Solve, finding all solutions in $[0, 2\pi)$ or $[0°, 360°)$.

13. $4 \sin^2 x - 1 = 0$
14. $2 \cos^2 x = 1$
15. $\cot^2 x - 3 = 0$
16. $\csc^2 x - 4 = 0$
17. $2 \sin^2 x + \sin x = 1$
18. $2 \cos^2 x + 3 \cos x = -1$
19. $\cos^2 x + 2 \cos x = 3$
20. $2 \sin^2 x - \sin x = 3$
21. $4 \sin^3 x - \sin x = 0$
22. $2 \cos^2 x - \sqrt{3} \cos x = 0$
23. $2 \sin^2 \theta + 7 \sin \theta = 4$
24. $2 \sin^2 \theta - 5 \sin \theta + 2 = 0$
25. $6 \cos^2 \phi + 5 \cos \phi + 1 = 0$
26. $2 \sin^2 \phi + \sin \phi - 1 = 0$
27. $2 \sin t \cos t + 2 \sin t - \cos t - 1 = 0$
28. $2 \sin t \tan t + \tan t - 2 \sin t - 1 = 0$
29. $\cos 2x \sin x + \sin x = 0$
30. $\sin 2x \cos x - \cos x = 0$
31. $5 \sin^2 x - 8 \sin x = 3$
32. $\cos^2 x + 6 \cos x + 4 = 0$
33. $2 \tan^2 x = 3 \tan x + 7$
34. $3 \sin^2 x = 3 \sin x + 2$
35. $7 = \cot^2 x + 4 \cot x$
36. $3 \tan^2 x + 2 \tan x = 7$

SYNTHESIS

Solve, restricting solutions to $[0, 2\pi)$ or $[0°, 360°)$ where sensible to do so.

37. $|\sin x| = \frac{\sqrt{3}}{2}$
38. $|\cos x| = \frac{1}{2}$
39. $\sqrt{\tan x} = \sqrt[4]{3}$
40. $12 \sin x - 7\sqrt{\sin x} + 1 = 0$
41. $16 \cos^4 x - 16 \cos^2 x + 3 = 0$
42. $\ln (\cos x) = 0$
43. $e^{\sin x} = 1$
44. $\sin (\ln x) = -1$
45. $e^{\ln (\sin x)} = 1$
46. A guy wire is attached to the top of a 50-ft pole and stretched to a point that is b ft from the bottom of the pole.
    a) Suppose that the angle of inclination of the wire to the top is θ. Show that
    $$\tan \theta = \frac{50}{b}.$$
    b) For what angle of inclination θ does $b = 40$ ft?
47. An airplane at an altitude of 2000 ft is flying toward an island. The straight-line distance from the airplane to the island is h ft.
    a) Suppose that the angle of depression is θ. Show that
    $$\sin \theta = \frac{2000}{h}.$$
    b) For what angle of depression θ does $h = 3000$ ft?
48. Solve graphically: $\sin x = \tan \frac{x}{2}$.

7.9 Identities in Solving Trigonometric Equations

OBJECTIVE

You should be able to:

1 Solve trigonometric equations requiring the use of identities.

1 When a trigonometric equation involves more than one function, we can use identities to put it in terms of a single function. This can usually be done in several ways. In the following three examples, we illustrate by solving the same equation using three different methods.

1. Solve $\sin x - \cos x = 1$, as in Example 1 (method 1).

EXAMPLE 1

Method 1. Solve the equation $\sin x + \cos x = 1$.

Solution To express $\cos x$ in terms of $\sin x$, we use the identity

$$\sin^2 x + \cos^2 x = 1.$$

From this, we obtain $\cos x = \pm\sqrt{1 - \sin^2 x}$. Now we substitute in the original equation:

$$\sin x \pm \sqrt{1 - \sin^2 x} = 1.$$

This is a radical equation. We get the radical alone on one side and then square both sides:

$$\begin{aligned}\pm\sqrt{1 - \sin^2 x} &= 1 - \sin x\\ (\pm\sqrt{1 - \sin^2 x})^2 &= (1 - \sin x)^2\\ 1 - \sin^2 x &= 1 - 2\sin x + \sin^2 x.\end{aligned}$$

This simplifies to

$$-2\sin^2 x + 2\sin x = 0.$$

Now we factor:

$$-2\sin x(\sin x - 1) = 0.$$

Using the principle of zero products gives us

$$\begin{aligned}-2\sin x = 0 \quad &\text{or} \quad \sin x - 1 = 0\\ \sin x = 0 \quad &\text{or} \quad \sin x = 1.\end{aligned}$$

The values of x in $[0, 2\pi)$ satisfying these are

$$x = 0, \qquad x = \pi, \qquad \text{or} \quad x = \frac{\pi}{2}.$$

Now we check these in the original equation. We find that π does not check but the other values do. Thus the solutions are 0 and $\pi/2$. The solutions, in degrees, are 0° and 90°.

CAUTION! It is important to check when solving trigonometric equations. Values are often obtained that are not solutions of the original equation.

DO EXERCISE 1.

EXAMPLE 1

Method 2. Solve $\sin x + \cos x = 1$.

Solution This time we will square both sides, obtaining

$$\sin^2 x + 2\sin x\cos x + \cos^2 x = 1.$$

Since $\sin^2 x + \cos^2 x = 1$, we can simplify to

$$2\sin x\cos x = 0.$$

Now we use the identity $2 \sin x \cos x = \sin 2x$, and obtain

$$\sin 2x = 0.$$

We are looking for solutions x to the equation for which

$$0 \leq x < 2\pi.$$

Multiplying by 2, we get

$$0 \leq 2x < 4\pi,$$

the interval we consider to solve $\sin 2x = 0$. These values of $2x$ are 0, π, 2π, and 3π. Thus the desired values of x in $[0, 2\pi)$ satisfying this equation are

$$x = 0, \quad x = \frac{\pi}{2}, \quad x = \pi, \quad \text{or} \quad x = \frac{3\pi}{2}.$$

Now we check these in the original equation. We find that π and $3\pi/2$ do not check but the other values do. Thus the solutions are 0 and $\pi/2$. The solutions, in degrees, are 0° and 90°.

DO EXERCISE 2.

EXAMPLE 1

Method 3. Solve $\sin x + \cos x = 1$.

Solution Let us recall the identity

$$c \sin ax + d \cos ax = A \sin (ax + b),$$

where $A = \sqrt{c^2 + d^2}$ and b is a number whose cosine is c/A and whose sine is d/A.

In this case, $c = d = a = 1$, so $A = \sqrt{2}$ and $b = \pi/4$. Thus our equation is equivalent to

$$\sqrt{2} \sin \left(x + \frac{\pi}{4}\right) = 1.$$

Then

$$\sin \left(x + \frac{\pi}{4}\right) = \frac{1}{\sqrt{2}} = \frac{\sqrt{2}}{2}$$

$$x + \frac{\pi}{4} = \frac{\pi}{4} \quad \text{or} \quad x + \frac{\pi}{4} = \frac{3\pi}{4},$$

$$x = 0 \quad \text{or} \quad x = \frac{\pi}{2}.$$

Both values check. The solutions in $[0, 2\pi)$ are 0 and $\pi/2$.

DO EXERCISE 3.

EXAMPLE 2 Solve: $2 \cos^2 x \tan x - \tan x = 0$.

Solution First, we factor:

$$2 \cos^2 x \tan x - \tan x = 0$$
$$\tan x \, (2 \cos^2 x - 1) = 0.$$

2. Solve $\sin x - \cos x = 1$, as in Example 1 (method 2).

3. Solve $\sin x - \cos x = 1$, as in Example 1 (method 3).

4. Solve: $\tan^2 x \cos x - \cos x = 0$.

5. Solve: $\sin 2x + \cos x = 0$.

6. Solve: $\sin 3\theta \cos \theta - \cos 3\theta \sin \theta = \frac{1}{2}$.

Now we use the identity $2\cos^2 x - 1 = \cos 2x$:

$$\tan x \cos 2x = 0$$

$$\tan x = 0 \quad \text{or} \quad \cos 2x = 0 \qquad \text{Principle of zero products}$$

$$x = 0, \pi \quad \text{or} \quad 2x = \frac{\pi}{2}, \frac{3\pi}{2}, \frac{5\pi}{2}, \frac{7\pi}{2}$$

$$x = 0, \pi \quad \text{or} \quad x = \frac{\pi}{4}, \frac{3\pi}{4}, \frac{5\pi}{4}, \frac{7\pi}{4}.$$

All values check. The solutions in $[0, 2\pi)$ are 0, π, $\pi/4$, $3\pi/4$, $5\pi/4$, and $7\pi/4$.

DO EXERCISE 4.

EXAMPLE 3 Solve: $\cos 2x + \sin x = 1$.

Solution We first use the identity $\cos 2x = 1 - 2\sin^2 x$, to get

$$1 - 2\sin^2 x + \sin x = 1$$

$$-2\sin^2 x + \sin x = 0$$

$$\sin x\,(1 - 2\sin x) = 0 \qquad \text{Factoring}$$

$$\sin x = 0 \quad \text{or} \quad 1 - 2\sin x = 0 \qquad \text{Principle of zero products}$$

$$\sin x = 0 \quad \text{or} \quad \sin x = \frac{1}{2}$$

$$x = 0, \pi \quad \text{or} \quad x = \frac{\pi}{6}, \frac{5\pi}{6}.$$

All values check. The solutions in $[0, 2\pi)$ are 0, π, $\pi/6$, and $5\pi/6$.

DO EXERCISE 5.

EXAMPLE 4 Solve: $\sin 5\theta \cos 2\theta - \cos 5\theta \sin 2\theta = \sqrt{2}/2$.

Solution We use the identity

$$\sin(\alpha - \beta) = \sin\alpha\cos\beta - \cos\alpha\sin\beta$$

to get $\sin 3\theta = \sqrt{2}/2$. We are looking for solutions to the equation for which

$$0 \le \theta < 2\pi.$$

Multiplying by 3, we get

$$0 \le 3\theta < 6\pi,$$

which is the interval we consider to solve $\sin 3\theta = \sqrt{2}/2$. These values of 3θ are

$$3\theta = \frac{\pi}{4}, \frac{3\pi}{4}, \frac{9\pi}{4}, \frac{11\pi}{4}, \frac{17\pi}{4}, \frac{19\pi}{4}.$$

Then

$$\theta = \frac{\pi}{12}, \frac{\pi}{4}, \frac{3\pi}{4}, \frac{11\pi}{12}, \frac{17\pi}{12}, \frac{19\pi}{12}.$$

Let us check $\pi/12$:

$$\sin\frac{5\pi}{12}\cos\frac{\pi}{6} - \cos\frac{5\pi}{12}\sin\frac{\pi}{6}.$$

We again use the difference identity to obtain $\sin(5\pi/12 - \pi/6)$. This is equal to $\sin(2\pi/12)$, or $\sin(\pi/4)$, and this is equal to $\sqrt{2}/2$. Thus $\pi/12$ checks. All of the above answers check. ■

DO EXERCISE 6.

EXAMPLE 5 Solve: $\tan^2 x + \sec x - 1 = 0$.

Solution We use the identity $1 + \tan^2 x = \sec^2 x$. Substituting, we get

$$\sec^2 x - 1 + \sec x - 1 = 0,$$

or

$$\sec^2 x + \sec x - 2 = 0$$

$$(\sec x + 2)(\sec x - 1) = 0 \qquad \text{Factoring}$$

$$\sec x = -2 \quad \text{or} \quad \sec x = 1 \qquad \text{Principle of zero products}$$

$$x = \frac{2}{3}\pi, \frac{4}{3}\pi \quad \text{or} \quad x = 0.$$

All of these values check. The solutions in $[0, 2\pi)$ are 0, $2\pi/3$, and $4\pi/3$. ■

EXAMPLE 6 Solve: $\sin x = \sin(2x - \pi)$.

Solution We can use either the identity $\sin(y - \pi) = -\sin y$ or the identity $\sin(\alpha - \beta) = \sin\alpha\cos\beta - \cos\alpha\sin\beta$ with the right-hand side. We get

$$\sin x = -\sin 2x.$$

Next, we use the identity $\sin 2x = 2\sin x\cos x$, and obtain

$$\sin x = -2\sin x\cos x, \quad \text{or} \quad 2\sin x\cos x + \sin x = 0$$

$$\sin x\,(2\cos x + 1) = 0 \qquad \text{Factoring}$$

$$\sin x = 0 \quad \text{or} \quad \cos x = -\frac{1}{2} \qquad \text{Principle of zero products}$$

$$x = 0, \pi \quad \text{or} \quad x = \frac{2\pi}{3}, \frac{4\pi}{3}.$$

All of these values check, so the solutions in $[0, 2\pi)$ are 0, π, $2\pi/3$, and $4\pi/3$. ■

DO EXERCISE 7.

7. Solve: $\cos x = \cos(\pi - 2x)$.

EXERCISE SET 7.9

1 Find all solutions of the following equations in $[0, 2\pi)$.

1. $\tan x \sin x - \tan x = 0$
2. $2\sin x\cos x + \sin x = 0$
3. $2\sec x\tan x + 2\sec x + \tan x + 1 = 0$
4. $2\csc x\cos x - 4\cos x - \csc x + 2 = 0$
5. $\sin 2x - \cos x = 0$
6. $\cos 2x - \sin x = 1$
7. $\sin 2x\sin x - \cos x = 0$
8. $\sin 2x\cos x - \sin x = 0$
9. $\sin 2x + 2\sin x\cos x = 0$
10. $\cos 2x\sin x + \sin x = 0$
11. $\cos 2x\cos x + \sin 2x\sin x = 1$
12. $\sin 2x\sin x - \cos 2x\cos x = -\cos x$
13. $\sin 4x - 2\sin 2x = 0$
14. $\sin 4x + 2\sin 2x = 0$
15. $\sin 2x + 2\sin x - \cos x - 1 = 0$
16. $\sin 2x + \sin x + 2\cos x + 1 = 0$
17. $\sec^2 x = 4\tan^2 x$
18. $\sec^2 x - 2\tan^2 x = 0$

19. $\sec^2 x + 3 \tan x - 11 = 0$

20. $\tan^2 x + 4 = 2 \sec^2 x + \tan x$

21. $\cot x = \tan(2x - 3\pi)$

22. $\tan x = \cot(2x + \pi)$

23. $\cos(\pi - x) + \sin\left(x - \frac{\pi}{2}\right) = 1$

24. $\sin(\pi - x) + \cos\left(\frac{\pi}{2} - x\right) = 1$

25. $\dfrac{\cos^2 x - 1}{\sin\left(\frac{\pi}{2} - x\right) - 1} = \dfrac{\sqrt{2}}{2} + 1$

26. $\dfrac{\sin^2 x - 1}{\cos\left(\frac{\pi}{2} - x\right) + 1} = \dfrac{\sqrt{2}}{2} - 1$

27. $2 \cos x + 2 \sin x = \sqrt{6}$

28. $2 \cos x + 2 \sin x = \sqrt{2}$

29. $\sqrt{3} \cos x - \sin x = 1$

30. $\sqrt{2} \cos x - \sqrt{2} \sin x = 2$

31. $\sec^2 x + 2 \tan x = 6$

32. $6 \tan^2 x = 5 \tan x + \sec^2 x$

33. $3 \cos 2x + \sin x = 1$

34. $5 \cos 2x + \sin x = 4$

SYNTHESIS

35. *Temperature during an illness.* The temperature T of a patient during a 12-day illness is given by

$$T(t) = 101.6^\circ + 3 \sin\left(\frac{\pi}{8} t\right).$$

Find the times t during the illness at which the patient's temperature was 103°.

36. *Satellite location.* A satellite circles the earth in such a manner that it is y miles from the equator (north or south, height from the surface not considered) t minutes after its launch, where

$$y = 5000\left[\cos \frac{\pi}{45}(t - 10)\right].$$

At what times t in the interval [0, 240], the first 4 hr, is the height 3000 mi north of the equator?

37. *Nautical mile.* (See Exercise 55, Exercise Set 7.4.) In Great Britain, the **nautical mile** is defined as the length of a minute of arc of the earth's radius. Since the earth is flattened at the poles, a British nautical mile varies with latitude. In fact, it is given, in feet, by the function

$$N(\phi) = 6066 - 31 \cos 2\phi,$$

where ϕ is the latitude in degrees. At what latitude north is the length of a British nautical mile found to be 6040 ft?

38. *Acceleration due to gravity.* (See Exercise 56, Exercise Set 7.4.) The acceleration due to gravity, often denoted by g in a formula such as $S = \frac{1}{2}gt^2$, means the distance that an object falls in time t. It has to do with the physics of motion near the earth's surface and is usually considered constant. In fact, however, g is not constant, but varies slightly with latitude. If ϕ stands for latitude, in degrees, g is given with good approximation by the formula

$$g = 9.78049(1 + 0.005288 \sin^2 \phi - 0.000006 \sin^2 2\phi),$$

where g is measured in m/sec^2 at sea level. At what latitude north does $g = 9.8$?

Solve.

39. $\arccos x = \arccos \frac{3}{5} - \arcsin \frac{4}{5}$

40. $\sin^{-1} x = \tan^{-1} \frac{1}{3} + \tan^{-1} \frac{1}{2}$

Solve graphically.

41. $\sin x - \cos x = \cot x$

42. $x \sin x = 1$

43. Suppose that $\sin x = 5 \cos x$. Find $\sin x \cos x$.

SUMMARY AND REVIEW: CHAPTER 7

TERMS TO KNOW

Identity
Sum–difference formulas
Cofunction identities
Reduction formulas
Double-angle identities
Half-angle identities
Basic identities
Pythagorean identities
Inverse trigonometric functions
Arcsine function
Arccosine function
Arctangent function
Trigonometric equations

REVIEW EXERCISES

1. Find an identity for $\cot(x - \pi)$.

Complete these Pythagorean identities.

2. $\sin^2 x + \cos^2 x =$ ________

3. $1 + \cot^2 x =$ ________

Complete these cofunction identities.

4. $\cos\left(x + \frac{\pi}{2}\right) =$ ________

5. $\cos\left(\frac{\pi}{2} - x\right) =$ ________

6. $\sin\left(x - \frac{\pi}{2}\right) =$ ________

7. Express $\tan x$ in terms of $\sec x$.

Simplify.

8. $\cos x(\tan x + \cot x)$

9. $\dfrac{\csc x(\sin^2 x + \cos^2 x \tan x)}{\sin x + \cos x}$

10. Rationalize the denominator: $\sqrt{\dfrac{\tan x}{\sec x}}$.

11. Rationalize the numerator: $\sqrt{\dfrac{\tan x}{\sec x}}$.

12. Given that $\sin\theta = 0.6820$, $\cos\theta = 0.7314$, $\tan\theta = 0.9325$, $\cot\theta = 1.0724$, $\sec\theta = 1.3673$, and $\csc\theta = 1.4663$, find the six function values for $90° - \theta$.

Use the sum and difference formulas to write equivalent expressions. You need not simplify.

13. $\cos\left(x + \frac{3\pi}{2}\right)$

14. $\tan(45° - 30°)$

15. Simplify: $\cos 27° \cos 16° + \sin 27° \sin 16°$.

16. Find $\cos 165°$ exactly.

17. Given that $\tan\alpha = \sqrt{3}$ and $\sin\beta = \sqrt{2}/2$ and that α and β are between 0 and $\pi/2$, evaluate $\tan(\alpha - \beta)$ exactly.

18. Find the angle from l_1 to l_2, given the equations

$$l_1 : y = 2x - 4 \quad \text{and} \quad l_2 : x - y = 2.$$

19. Find $\tan 2\theta$, $\cos 2\theta$, and $\sin 2\theta$ and the quadrant in which 2θ lies where $\cos\theta = -\frac{3}{5}$ and θ is in quadrant III.

20. Find $\sin(\pi/8)$ without using a table or a calculator.

21. Simplify: $\dfrac{\sin 2\theta}{\sin^2\theta}$.

Prove each of the following identities.

22. $\tan 2\theta = \dfrac{2\tan\theta}{1 - \tan^2\theta}$

23. $\dfrac{\sec x - \cos x}{\tan x} = \sin x$

24. Find an equivalent expression involving the sine function only: $6 \sin 3x + 2 \cos 3x$.

25. Find, in radians, $\sin^{-1} \frac{1}{2}$.

26. Use a calculator to find, in degrees, arccot 0.1584.

Find each of the following without the use of a table or a calculator.

27. $\sin^{-1}\left(-\frac{\sqrt{2}}{2}\right)$

28. $\text{arccot}(-\sqrt{3})$

Evaluate or simplify.

29. $\tan\left(\arctan \frac{7}{8}\right)$

30. $\cos^{-1}\left[\cos\left(-\frac{\pi}{3}\right)\right]$

31. $\cos^{-1}\left(\sin \frac{2\pi}{3}\right)$

32. $\sin\left(\arctan \frac{b}{5}\right)$

Solve, finding all solutions in $[0, 2\pi)$.

33. $\sin^2 x - 7 \sin x = 0$

34. $2\cos^2 x - 5\cos x + 2 = 0$

35. $\csc^2 x - 2\cot^2 x = 0$

36. $2\cot^2 x = 3\cot x + 1$

Solve, finding all solutions in $[0, 2\pi)$.

37. $|\sin x| = \frac{1}{2}$

38. $\cos(\ln 2x) = -1$

Graph.

39. $y + 1 = 2\cos^2 x$

40. $f(x) = 2\sin^{-1}\left(x + \frac{\pi}{2}\right)$

TEST: CHAPTER 7

Use the sum and difference formulas to write equivalent expressions. You need not simplify.

1. $\cos(\pi - x)$
2. $\tan(83° + 15°)$
3. Simplify: $\sin 40° \cos 5° - \cos 40° \sin 5°$.
4. Find $\cos 105°$ exactly.
5. Given that $\cos\alpha = \sqrt{2}/2$ and $\sin\beta = 1/2$ and that α and β are between 0 and $\pi/2$, evaluate $\tan(\alpha - \beta)$ exactly.
6. Find the angle from l_1 to l_2 given the equations l_1: $2x + y = 7$ and l_2: $6x - 3y = 4$.
7. Find an identity for $\sec(x - 450°)$.

Complete these identities.

8. $\sin^2 x + \cos^2 x =$ ________
9. $1 + \cot^2 x =$ ________
10. $\sin\left(x + \frac{\pi}{2}\right) =$ ________
11. $\cos\left(\frac{\pi}{2} - x\right) =$ ________
12. $\cos\left(x - \frac{\pi}{2}\right) =$ ________
13. Express $\csc x$ in terms of $\cot x$.

Simplify.

14. $\dfrac{\sqrt{\sec^2 x - 1}}{\sin x}$
15. $\dfrac{\tan^2 x \csc^2 x - 1}{\csc x \tan^2 x \sin x}$
16. Rationalize the denominator: $\sqrt{\dfrac{\sec x}{\csc x}}$.
17. Find $\sin 2\theta$, $\cos 2\theta$, and $\tan 2\theta$ and the quadrant in which 2θ lies if $\sin\theta = \frac{12}{13}$ and θ is in quadrant II.
18. Find $\tan\dfrac{5\pi}{12}$ exactly.
19. Simplify: $4\sin 2x \cos 2x$.
20. Prove the identity
$$\frac{1 - \cos 2\theta}{\sin 2\theta} = \tan\theta.$$
21. Find an equivalent expression involving the sine function only: $4\sin 2x + 3\cos 2x$.
22. Find, in radians, $\arcsin(-\sqrt{2}/2)$.
23. Use a calculator to find, in degrees, $\tan^{-1} 0.931304$.
24. Find $\cos^{-1}\dfrac{\sqrt{3}}{2}$.
25. Find $\text{arccot}(-1)$.
26. Find $\sin(\tan^{-1}\sqrt{3})$.
27. Find $\arccos\left(\cos\frac{\pi}{4}\right)$.

Find all solutions of the following equations in $[0, 2\pi)$.

28. $2\sin^2 x - 5\sin x = 3$
29. $2\cos^3 x - \cos x = 0$
30. $\tan^2 x + 3\tan x = 5$
31. Given that $\sin\theta = 0.4540$, $\cos\theta = 0.8910$, $\tan\theta = 0.5095$, $\cot\theta = 1.963$, $\sec\theta = 1.122$, and $\csc\theta = 2.203$, find the six function values for $90° - \theta$.

SYNTHESIS

32. Solve $\ln(\sin x) = 0$. Restrict solutions to $[0, 2\pi)$.

Triangles, Vectors, and Applications

3

In this chapter, we consider many applications of trigonometry. We first solve right triangles and related problems. We also use trigonometry to solve triangles that are not right (*oblique triangles*). Triangle trigonometry is important in many practical applications such as surveying and navigation.

The idea of a *vector* is related to the study of triangles. A vector is a quantity that has a direction. Vectors have many practical applications in the physical sciences.

We conclude the chapter with a continuation of the study of complex numbers begun in Chapter 2. Complex numbers have many applications in electricity and engineering.

FEATURE PROBLEM

The John Hancock Building is slanted on each side. It is 1107 ft tall, excluding the towers on top. The base of the building is a 165-ft–by–265-ft rectangle. The top of the building is a 100-ft–by–160-ft rectangle. Find the angle of inclination of each side of the building from the vertical.

THE MATHEMATICS

To find the angle of inclination of one of the sides, we find θ such that

$$\tan \theta = \frac{52.5}{1107}.$$

OBJECTIVES

You should be able to:

1 Solve right triangles.

2 Solve problems involving the use of right triangles and the trigonometric functions.

1. For the right triangle, using standard lettering, express each of the following using a, b, and c.
   a) $\sin B$
   b) $\cos B$
   c) $\tan B$
   d) $\csc B$
   e) $\sec B$
   f) $\cot B$

8.1 Applications of Right Triangles

As we look at the many applications of trigonometry, we begin with a study of right triangles. The basic circular, or trigonometric, functions can be considered to be ratios of sides of right triangles, placed in standard position, as developed in Chapter 6. For any triangle, whether or not it is in standard position, the ratios of sides are given by trigonometric functions.

It is customary to label the right angle C. The side opposite C, the hypotenuse, is labeled c. The acute angles are labeled A and B, and the lengths of the sides opposite them are labeled a and b, respectively. We refer to this as **standard lettering.**

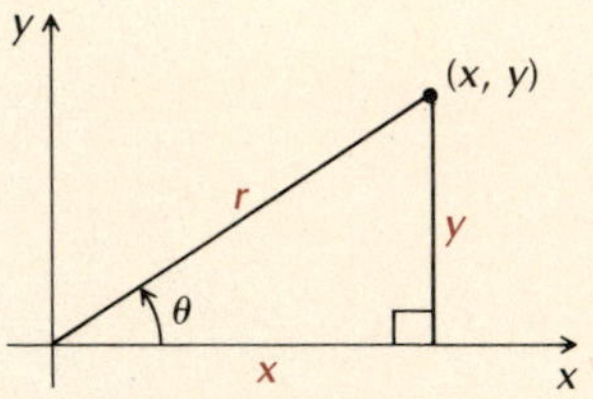

Standard position

$\sin \theta = \frac{y}{r}$

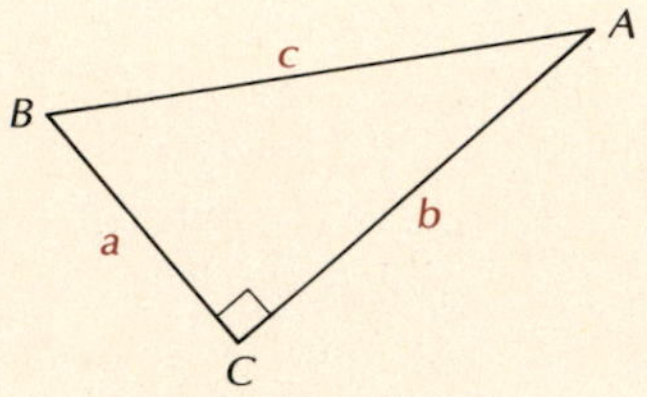

Standard lettering

$\sin A = \frac{a}{c}$

We know that:

$$\sin A = \frac{\text{opposite}}{\text{hypotenuse}} = \frac{a}{c}, \qquad \csc A = \frac{\text{hypotenuse}}{\text{opposite}} = \frac{c}{a},$$
$$\cos A = \frac{\text{adjacent}}{\text{hypotenuse}} = \frac{b}{c}, \qquad \sec A = \frac{\text{hypotenuse}}{\text{adjacent}} = \frac{c}{b},$$
$$\tan A = \frac{\text{opposite}}{\text{adjacent}} = \frac{a}{b}, \qquad \cot A = \frac{\text{adjacent}}{\text{opposite}} = \frac{b}{a}.$$

DO EXERCISE 1.

1 Solving Right Triangles

Now that we know how to find function values for any acute angle, we can begin to solve right triangles. Triangles that are not right triangles will be studied later. To **solve** a triangle means to find the lengths of all its sides and the measures of all its angles, provided they are not already known.

EXAMPLE 1 Find $m\angle B$.

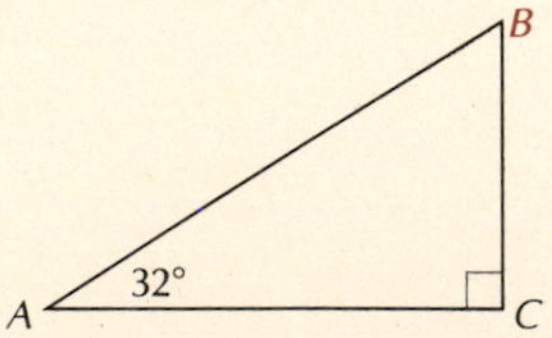

Solution Since the angle measures of any triangle add up to 180° and the right angle measures 90°, the acute angles must add up to 90°:

$$m\angle A + m\angle B = 90°.$$

Thus,

$$m\angle B = 90° - 32° = 58.$$

EXAMPLE 2 In this right triangle, find a and b.

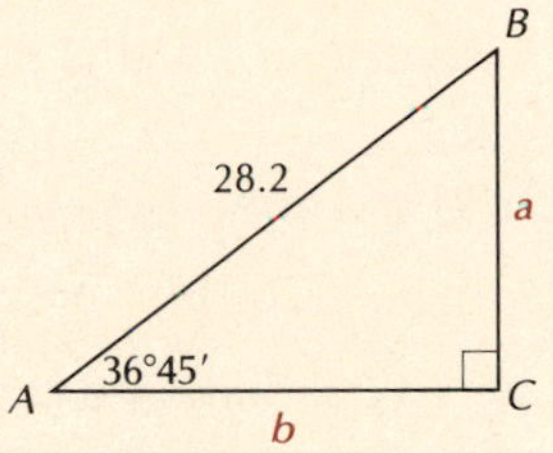

Solution By the definition of the sine ratio, we know that

$$\sin 36°45' = \frac{a}{28.2}.$$

To use a calculator, we convert 36°45′ to 36.75°. Substituting 36.75° for 36°45′ and solving for a, we have

$$a = 28.2 \sin 36.75°.$$

We find the sine of 36.75° and multiply:

$$a \approx 28.2 \times 0.5983 \quad \text{Using a calculator}$$
$$\approx 16.9.$$

To find b, we use the cosine ratio:

$$\cos 36.75° = \frac{b}{28.2}.$$

We solve for b:

$$b = 28.2 \cos 36.75°.$$

We find the cosine of 36.75° and multiply:

$$b \approx 28.2 \times 0.8013 \approx 22.6.$$

DO EXERCISES 2 AND 3.

Henceforth, for simplicity, we will use an equals sign, =, even when an approximation symbol, ≈, might be appropriate.

EXAMPLE 3 In this right triangle, find A and B.* Then find a. That is, *solve* the triangle.

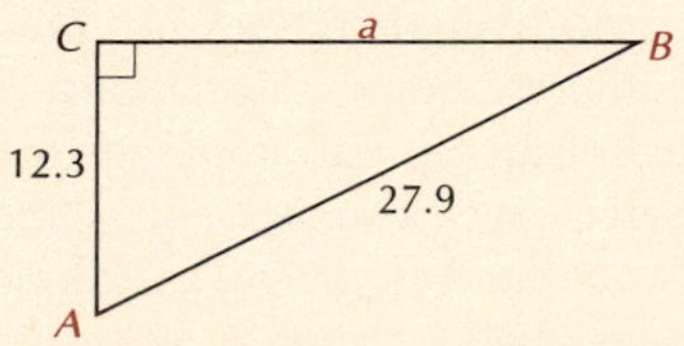

Solution We know the side adjacent to A and also the hypotenuse. That suggests the use of the cosine ratio:

$$\cos A = \frac{12.3}{27.9}$$
$$= 0.4409. \quad \text{Dividing}$$

*We will often shorten our writing or speaking, saying that we find B, rather than $m\angle B$.

2. Find $m\angle A$.

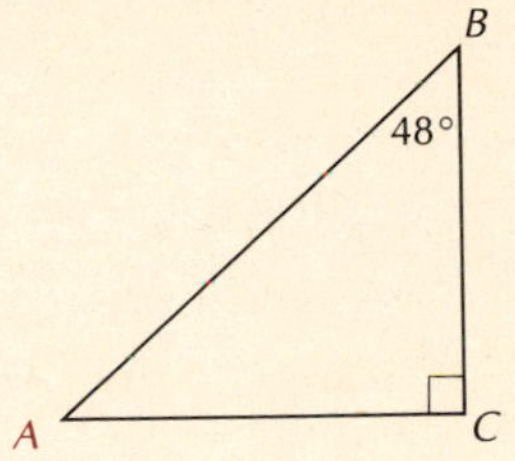

3. In this right triangle, find a and b.

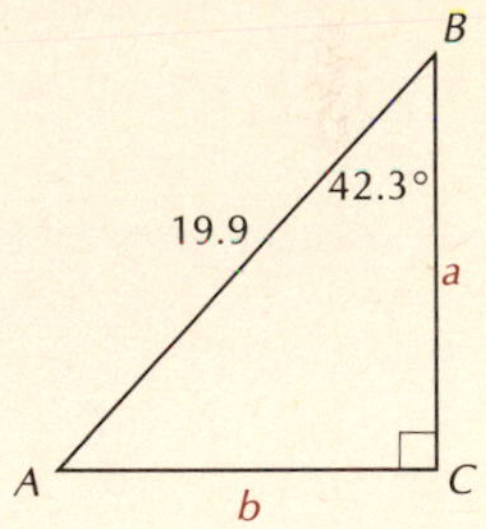

4. Solve this triangle.

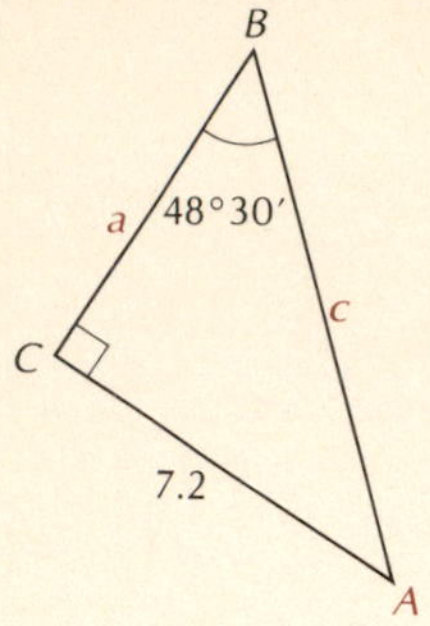

5. Solve this triangle. Standard lettering has been used.

$A = 43°20'$, $b = 8.62$

We now use a calculator to find the inverse cosine of 0.4409, which is A. We find that

$$A \text{ is about } 63.84°, \text{ or } 63°50'.$$

Now, since the measures of A and B must total 90°, we can find B by subtracting:

$$\begin{aligned} B &= 90° - 63.84° \\ &= 26.16°, \quad \text{or } 26°10'. \end{aligned}$$

To find a, there are several things we can do. We could use cos B, or we could use sin A. We could also use the tangent or cotangent ratios for either A or B. Let's use tan A:

$$\tan A = \frac{a}{12.3}, \quad \text{or}$$

$$\tan 63.84° = \frac{a}{12.3}.$$

Then

$$\begin{aligned} a &= 12.3 \tan 63.84° \qquad \text{Solving for } a \\ &= 12.3 \times 2.036 \\ &= 25.0. \end{aligned}$$

Precision

If Table 3 (at the back of the book) is used in solving triangles, distances are not precise to more than four digits and angles are not precise to less than 10′. If a calculator is used, the precision of the calculation can be much greater.

We should keep in mind that for most applications, three-digit accuracy is sufficient. In fact, many distances cannot even be measured to a precision greater than three digits. Thus in most examples, if we use more than three-digit precision, we are only fooling ourselves—those extra digits are actually meaningless. Thus values obtained from a calculator are too precise even to be realistic. Would it make much practical sense, for example, to calculate the distance from New York to Chicago to the nearest hundredth of an inch? In most cases, not.

The answers to the exercises in this text may be given with greater precision than is warranted. We have adopted the convention of rounding answers to three or four digits in most cases and not worrying about what is appropriate in an actual application. In a physics or engineering course, you will need to give attention to realistic rounding, but that is not the subject of this book.

You should realize, too, that there can be minor discrepancies due to different rounding procedures. Some calculations can be performed more conveniently on your calculator by not stopping to round. Answers will be found that way in the exercise sets. You may note some variance in the last one or two decimal places if you round as you go, as is done in the examples. Please do not be concerned about a small variation in an answer that might be due only to a difference in rounding.

DO EXERCISES 4 AND 5.

2 Applications of Right Triangles

Right triangles have many applications. To solve a problem, we locate a right triangle and then solve, or at least partially solve, it.

EXAMPLE 4 An observer stands on level ground, 200 m from the base of a television tower, and looks up at an angle of 26.5° to see the top of the tower.

a) How high is the tower above the observer's eye level?
b) How far is it from the observer's eye to the top of the tower?

Solution We draw a diagram and see that a right triangle is formed.

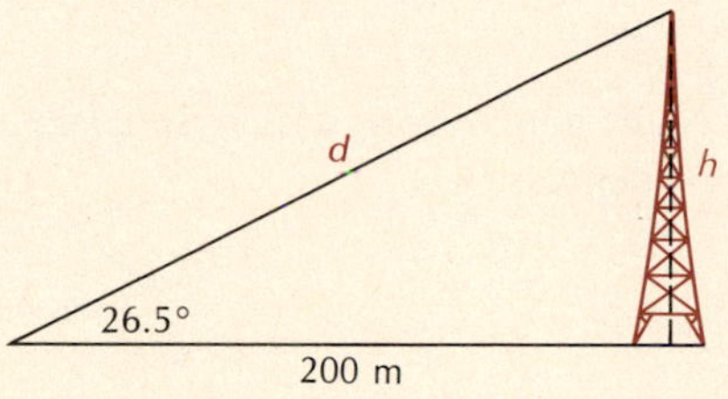

a) We use one of the trigonometric functions. The tangent function is most convenient. From the definition of the tangent function, we have

$$\frac{\text{opposite}}{\text{adjacent}} = \frac{h}{200} = \tan 26.5°.$$

Then $h = 200 \tan 26.5°$. Using a calculator, we find that $\tan 26.5° = 0.4986$, approximately. Thus, $h = 200 \times 0.4986 = 99.72$. The height of the tower is about 99.7 m.

b) To find the distance d from the observer's eye to the top of the tower, we use a different trigonometric function, one that involves the hypotenuse. We use the cosine function and get

$$\cos 26.5° = \frac{\text{adjacent}}{\text{hypotenuse}} = \frac{200}{d}.$$

Solving for d and using a calculator gives us

$$d = \frac{200}{\cos 26.5°} = \frac{200}{0.8949} = 223.5 \text{ ft}.$$

The distance from the observer's eye to the top of the tower is about 223.5 ft. ■

DO EXERCISE 6.

EXAMPLE 5 *Safe angle for ladders.* Suppose that a ladder has length L. It has been determined that the ladder is at its safest position on a wall when it is pulled out a distance D from the wall, where $L = 4D$. What angle does this safest position determine with the ground?

Solution We draw a diagram and then use the most convenient trigonometric function. From the definition of the cosine function, we have

$$\cos \theta = \frac{\text{adjacent}}{\text{hypotenuse}} = \frac{D}{L} = \frac{D}{4D} = \frac{1}{4}.$$

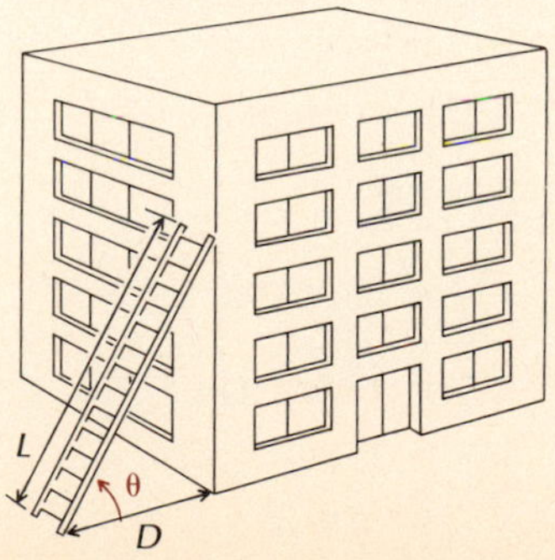

6. An observer stands 120 m from a tree and finds that the line of sight to the top of the tree is 32.3° above the horizontal.

a) Find the height of the tree above eye level.
b) Find the distance from the observer to the top of the tree.

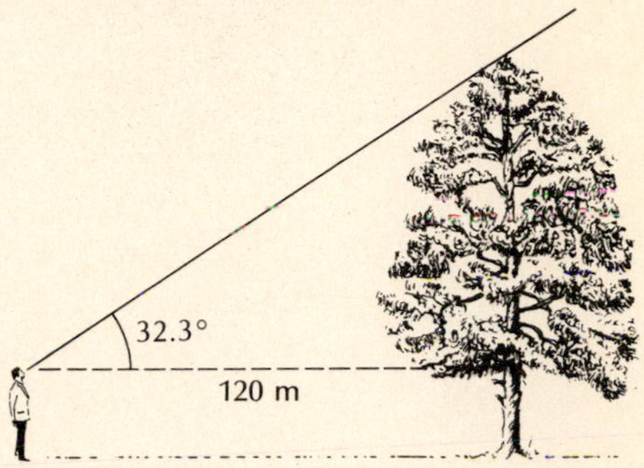

7. A guy wire is 13.6 m long and is fastened from the ground to a pole 6.5 m above the ground. What angle does the wire make with the ground?

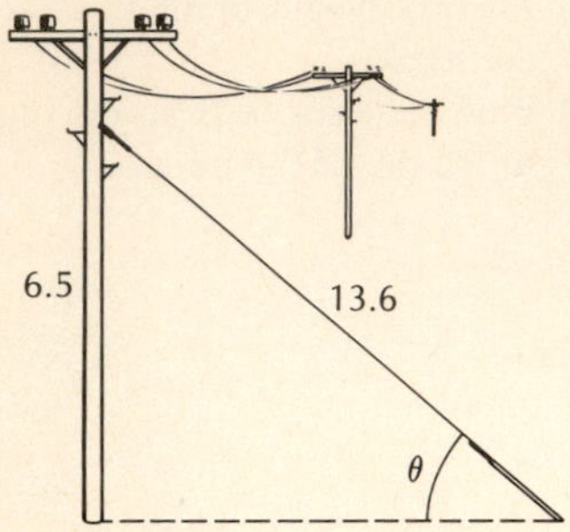

From a calculator, we find that θ is about $75.52°$, or $75°31'$. Thus the ladder is at its safest position when it makes an angle of about $75.52°$ with the ground.

DO EXERCISE 7.

Many applications with right triangles involve an angle of elevation or an angle of depression.

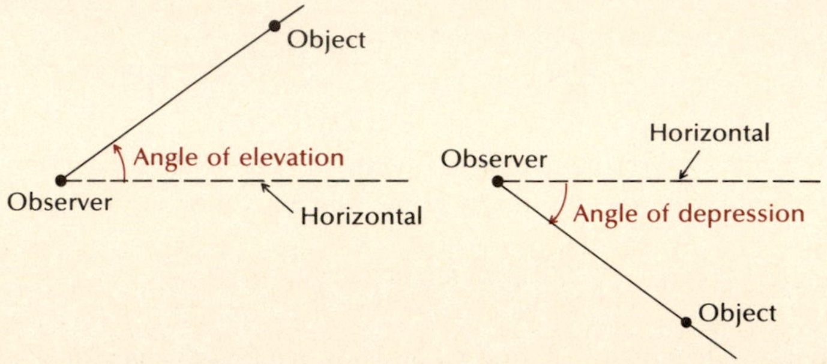

The angle between the horizontal and a line of sight *above* the horizontal is called an **angle of elevation.** The angle between the horizontal and a line of sight *below* the horizontal is called an **angle of depression.** For example, suppose you were looking straight ahead and then you moved your eyes upward toward an oncoming airplane. The angle your eyes pass through is an angle of elevation. If the pilot of the plane is looking forward and then moves his eyes down toward you, his eyes pass through an angle of depression.

EXAMPLE 6 *Finding a cloud height.* A device for measuring cloud height at night consists of a vertical beam of light, which makes a spot on the clouds. The spot is viewed from a point 135 ft away. Using a surveying device, the angle of elevation is found to be $67°40'$. Find the height of the clouds.

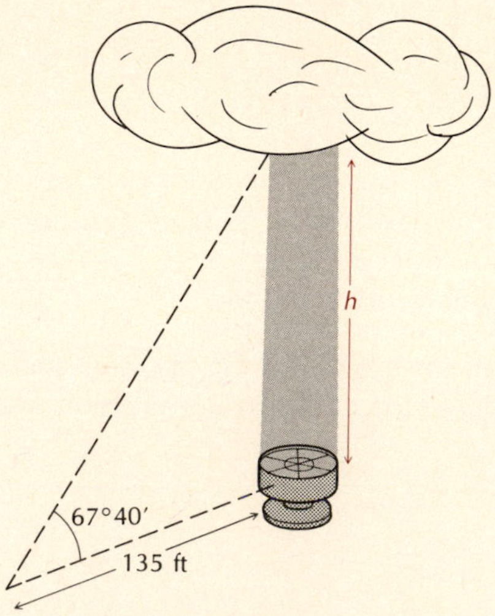

Solution From the figure, we have

$$\frac{h}{135} = \tan 67°40'$$

$$h = 135 \times \tan 67°40'$$

$$= 135 \times 2.434$$

$$= 329 \text{ ft.}$$

The following is a general procedure for solving problems involving triangles.

To solve a triangle problem:

1. **Draw a sketch of the problem situation.**
2. **Look for triangles and sketch them in.**
3. **Mark the known and unknown sides and angles.**
4. **Express the desired side or angle in terms of known trigonometric ratios. Then solve.**

DO EXERCISE 8.

EXAMPLE 7 *Surveying.* Horizontal distances must often be measured even though terrain is not level. One way of doing so is as follows. Distance down a slope is measured with a surveyor's tape, and the distance d is measured by making a level sighting from A to a pole held vertically at B, or the angle α is measured by an instrument placed at A. Suppose that a slope distance L is measured to be 121.3 ft and the angle α is measured to be $3°25'$. Find the horizontal distance H.

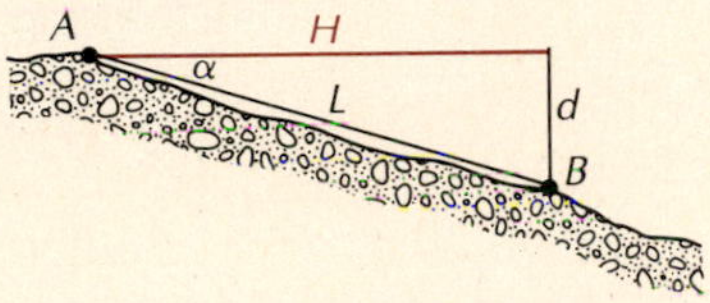

Solution From the figure, we see that $H/L = \cos \alpha$. Thus, $H = L \cos \alpha$, and in this case,

$$\begin{aligned} H &= 121.3 \times \cos 3°25' \\ &= 121.3 \times 0.9982 \\ &= 121.1 \text{ ft.} \end{aligned}$$

DO EXERCISE 9.

There are two applications of trigonometry that involve the idea of **direction,** or **bearing.** The first application is to aerial navigation and the second is to sea navigation, surveying, and other fields.

Bearing: First type. In aerial navigation, directions are given in degrees clockwise from north. Thus, east is 90°, south is 180°, and west is 270°. Several aerial directions or bearings are given below.

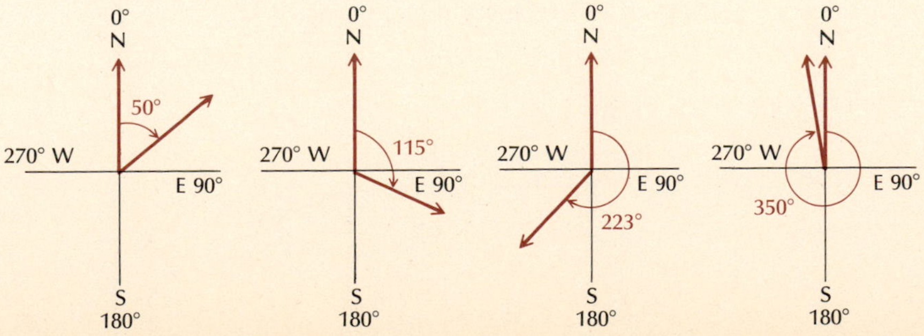

8. The length of a guy wire to a pole is 37.7 ft. It makes an angle of $71°20'$ with the ground, which is horizontal. How high above the ground is the guy wire attached to the pole?

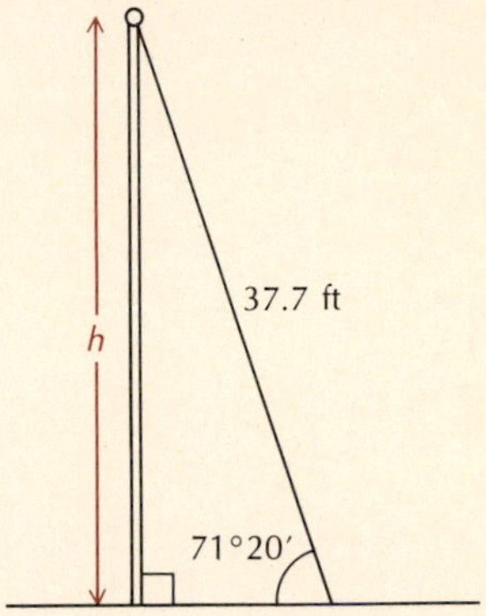

9. A downslope distance is measured to be 241.3 ft, and the angle of depression α is measured to be $5°15'$. Find the horizontal distance.

10. An airplane flies 150 km from an airport in a direction of 115°. How far east of the airport is the plane then? How far south?

EXAMPLE 8 An airplane leaves an airport and travels for 100 mi in a direction of 300°. How far north of the airport is the plane then? How far west?

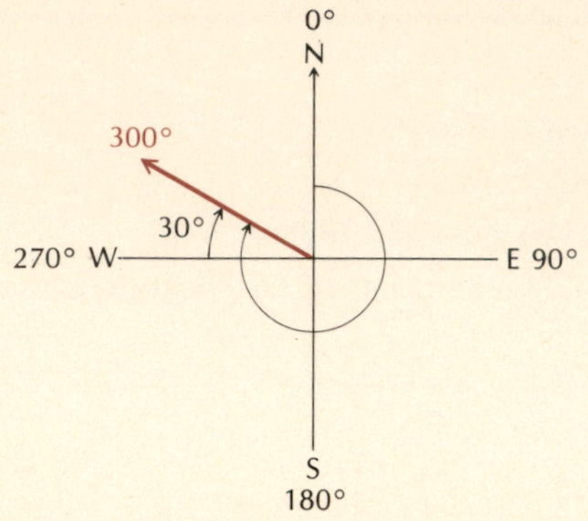

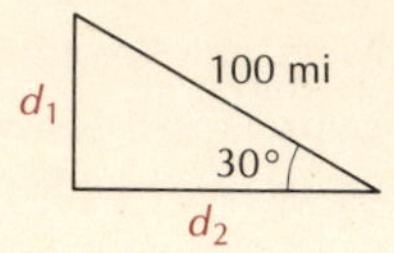

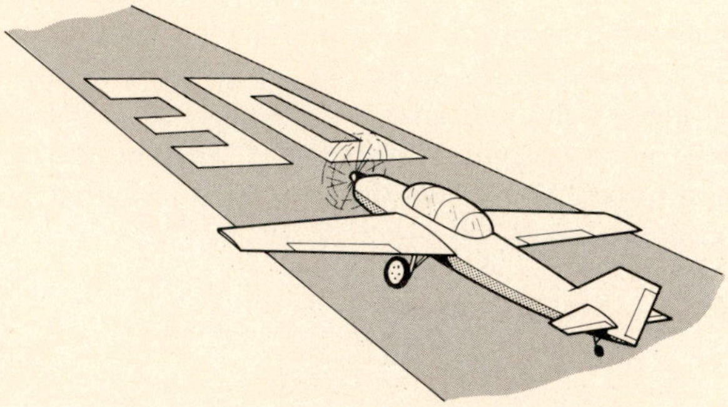

Solution The direction of flight is shown in the figure. In the triangle, d_1 is the northerly distance and d_2 is the westerly distance. Then

$$\frac{d_1}{100} = \sin 30° \quad \text{and} \quad \frac{d_2}{100} = \cos 30°.$$

Thus we have

$$d_1 = 100 \sin 30° = 100 \times 0.5$$
$$= 50 \text{ mi (to the nearest mile);}$$
$$d_2 = 100 \cos 30° = 100 \times 0.866$$
$$= 87 \text{ mi (to the nearest mile).}$$

The plane is 50 mi north and 87 mi west of the airport.

DO EXERCISE 10.

Bearing: Second type. The second way of giving direction, or bearing, involves reference to a north–south line using an acute angle. For example, N 43° W means 43° west of north and S 30° E means 30° east of south. Several bearings of this type are shown below.

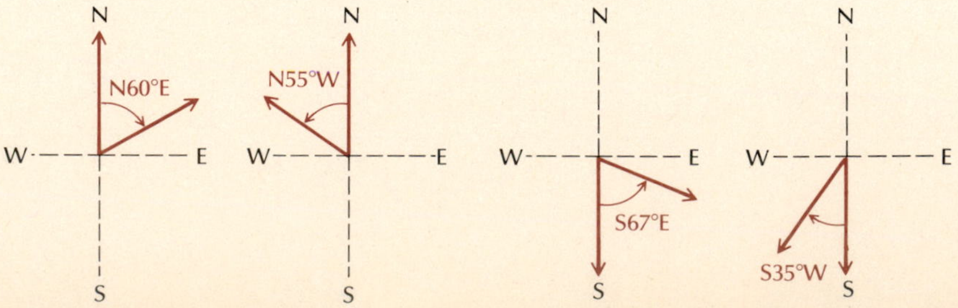

EXAMPLE 9 A forest ranger at point A sights a fire directly south. A second ranger at point B, 7 mi east, sights the same fire at a bearing of S 27°20′ W. How far from A is the fire?

$$\begin{aligned}\angle B &= 90° - 27°20' \\ &= 89°60' - 27°20' \\ &= 62°40'\end{aligned}$$

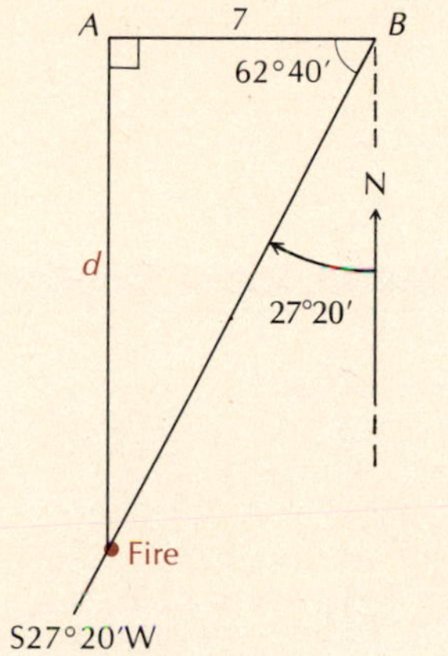

Solution From the figure, we see that the desired distance d is part of a right triangle, as shown. We have

$$\frac{d}{7} = \tan 62°40'$$

$$\begin{aligned}d &= 7 \tan 62°40' = 7 \times 1.935 \\ &= 13.5 \text{ mi.}\end{aligned}$$

DO EXERCISE 11.

EXAMPLE 10 From an observation tower, two markers are viewed on the ground. The markers and the base of the tower are on a line, and the observer's eye is 65.3 ft above the ground. The angles of depression to the markers are 53°10′ and 27°50′. How far is it from one marker to the other?

Solution We first make a sketch. We look for right triangles and sketch them in. Then we mark the known information on the sketch. The distance we seek is d, which is $d_1 - d_2$.

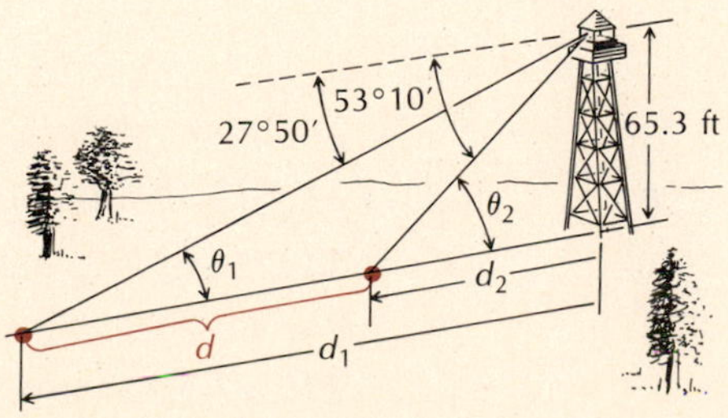

From the right angles in the figure, we have

$$\frac{d_1}{65.3} = \cot \theta_1 \quad \text{or} \quad \frac{d_2}{65.3} = \cot \theta_2.$$

When parallel lines are cut by a transversal, alternate interior angles are equal. Thus, $\theta_1 = 27°50'$ and $\theta_2 = 53°10'$. Then $d_1 = 65.3 \cot 27°50'$ and $d_2 = 65.3 \cot 53°10'$. We could calculate these and subtract, but the use of a

11. Directly east of a lookout station, there is a small forest fire. The bearing of this fire from a station 12 km south of the first is N 57°10′ E. How far is the fire from the southerly lookout station?

12. From an airplane flying 7500 ft above level ground, one can see two towns directly to the east. The angles of depression to the towns are 5°10′ and 77°30′. How far apart are the towns, to the nearest mile?

calculator is more efficient if we leave all calculations until the end.

$$\begin{aligned} d = d_1 - d_2 &= 65.3 \cot 27°50' - 65.3 \cot 53°10' \\ &= 65.3(\cot 27°50' - \cot 53°10') \\ &= 65.3(\cot 27.83° - \cot 53.17°) \quad \text{Converting to decimals} \\ &= 65.3(1.894 - 0.7490) \\ &= 74.8 \text{ ft.} \end{aligned}$$

DO EXERCISE 12.

EXERCISE SET 8.1

1 Solve each of the following right triangles. (Standard lettering has been used.)

1.

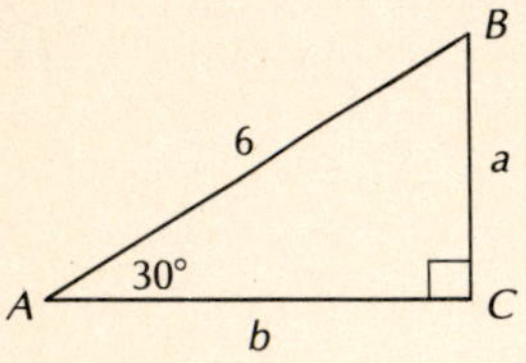

2.

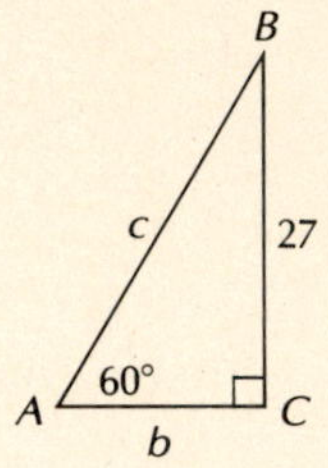

3.

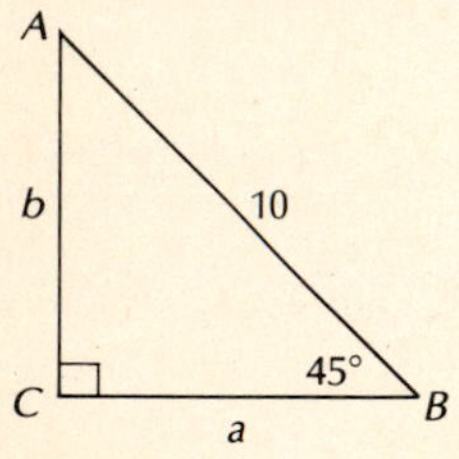

4.

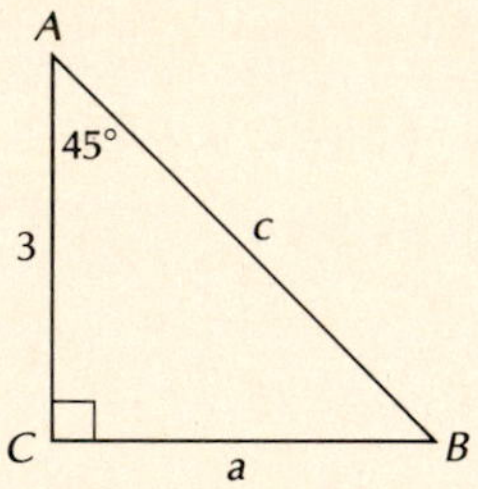

5.

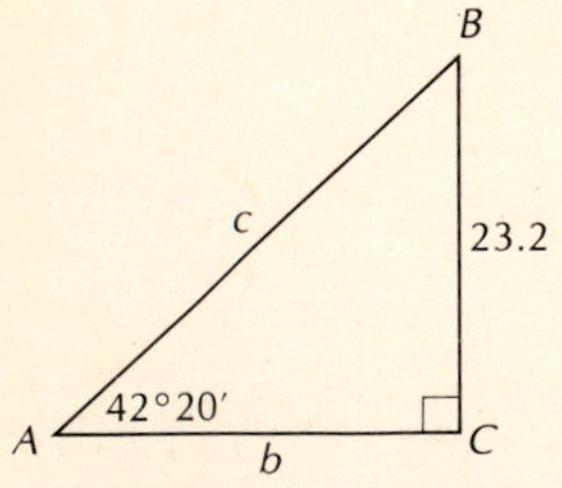

6.

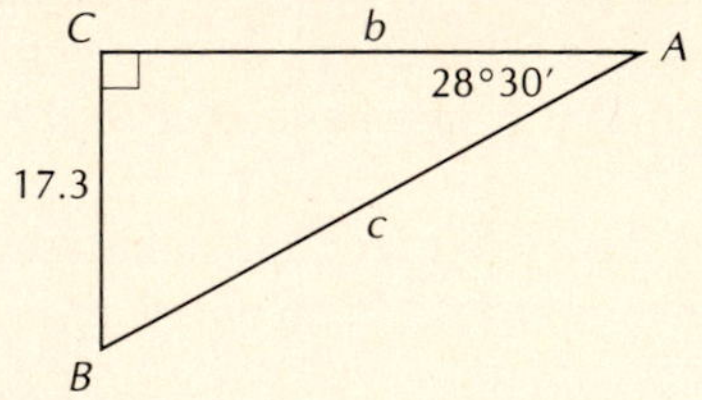

7. $A = 36°10'$, $a = 27.2$

8. $A = 87°40'$, $a = 9.73$

9. $B = 12°40'$, $b = 98.1$

10. $B = 69°50'$, $b = 127$

11. $A = 17°28'$, $b = 13.6$

12. $A = 78°42'$, $b = 1340$

13. $B = 23°12'$, $a = 350$

14. $B = 69°22'$, $a = 240$

15. $A = 47°35'$, $c = 48.3$

16. $A = 88°55'$, $c = 3950$

17. $B = 82°20'$, $c = 0.982$

18. $B = 56°30'$, $c = 0.0447$

19. $a = 12.5$, $b = 18.5$

20. $a = 10.2$, $b = 20.4$

21. $a = 16.0$, $c = 20.0$

22. $a = 15.0$, $c = 45.0$

23. $b = 1.86$, $c = 4.02$

24. $b = 100$, $c = 450$

2 Solve.

25. A guy wire to a pole makes an angle of 73°10′ with the level ground and is 14.5 ft from the pole at the ground. How far above the ground is the wire attached to the pole?

26. A guy wire to a pole makes an angle of 74°20′ with the level ground and is attached to the pole 34.2 ft above the ground. How far from the base of the pole is the wire attached to the ground?

27. A kite string makes an angle of 31°40′ with the (level) ground, and 455 ft of string is out. How high is the kite?

28. A kite string makes an angle of 41°40′ with the (level) ground when the kite is 114 ft high. How long is the string?

29. A road rises 3 m per 100 horizontal m. What angle does it make with the horizontal?

30. A kite is 120 ft high when 670 ft of string is out. What angle does the kite make with the ground?

31. What is the angle of elevation of the sun when a 6-ft man casts a 10.3-ft shadow?

32. What is the angle of elevation of the sun when a 35-ft mast casts a 20-ft shadow?

33. From a balloon 2500 ft high, a command post is seen with an angle of depression of 7°40′. How far is it from a point on the ground below the balloon to the command post?

34. From a lighthouse 55 ft above sea level, the angle of depression to a small boat is 11°20′. How far from the foot of the lighthouse is the boat?

35. An observer at a command post sights a balloon that is 2500 ft high at an angle of elevation of 8°20′. How far is it from the command post to a point directly under the balloon?

36. An observer sights the top of a building 173 ft higher than the eye, at an angle of elevation of 27°50′. How far is it from the observer to the building?

37. Ship *A* is due west of a lighthouse. Ship *B* is 12 km south of ship *A*. From ship *B*, the bearing to the lighthouse is N 63°20′ E. How far is ship *A* from the lighthouse?

38. Lookout station *A* is 15 km west of station *B*. The bearing from *A* to a fire directly south of *B* is S 37°50′ E. How far is the fire from *B*?

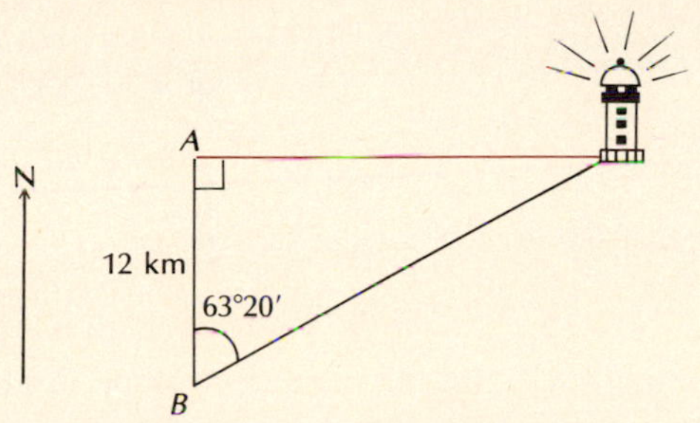

39. A regular pentagon has sides 30.5 cm long. Find the radius of the circumscribed circle.

40. A regular pentagon has sides 42.8 cm long. Find the radius of the inscribed circle.

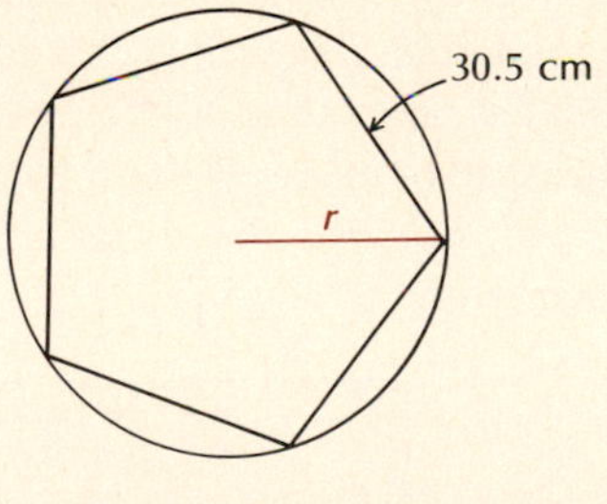

41. A regular hexagon has a perimeter of 50 cm and is inscribed in a circle. Find the radius of the circle.

42. A regular octagon is inscribed in a circle of radius 15.8 cm. Find the perimeter of the octagon.

43. A vertical antenna is mounted on top of a 50-ft pole. From a point on the level ground 75 ft from the base of the pole, the antenna subtends an angle of 10.5°. Find the length of the antenna.

44. An observer on a ladder looks at a building 100 ft away, noting that the angle of elevation of the top of the building is 18°40′ and the angle of depression of the bottom of the building is 6°20′. How tall is the building?

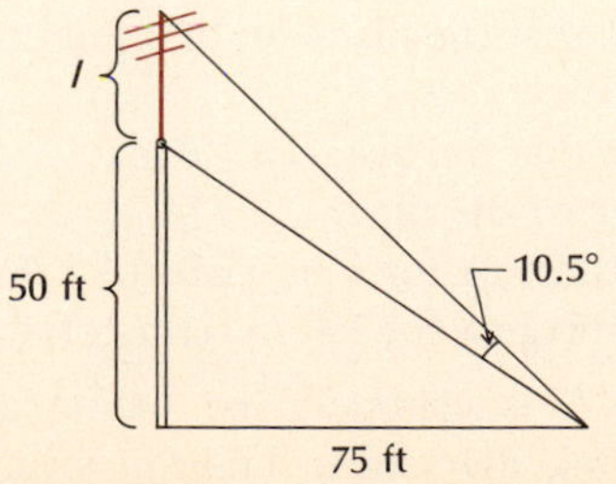

45. From a balloon 2 km high, the angles of depression to two towns, in line with the balloon, are 81°20′ and 13°40′. How far apart are the towns?

46. From a balloon 1000 m high, the angles of depression to two artillery posts, in line with the balloon, are 11°50′ and 84°10′. How far apart are the artillery posts?

47. A weather balloon is directly west of two observing stations that are 10 km apart. The angles of elevation of the balloon from the two stations are 17°50′ and 78°10′. How high is the balloon?

48. From two points south of a hill on level ground and 1000 ft apart, the angles of elevation of the hill are 12°20′ and 82°40′. How high is the hill?

49. An airplane travels at 120 km/h for 2 hr in a direction of 243° from Chicago. At the end of this time, how far south of Chicago is the plane?

50. An airplane travels at 150 km/h for 2 hr in a direction of 138° from Omaha. At the end of this time, how far east of Omaha is the plane?

SYNTHESIS

51. Find h.

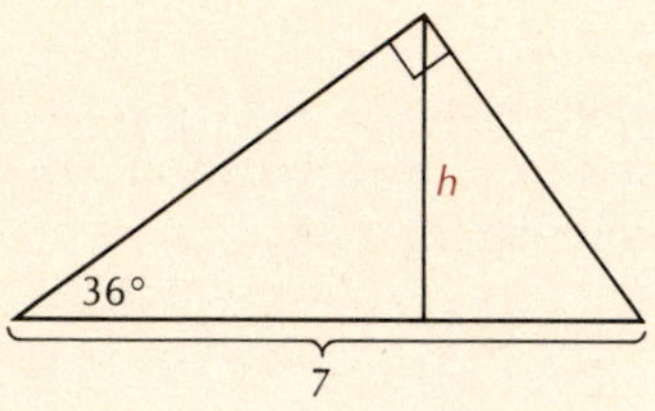

52. Find a.

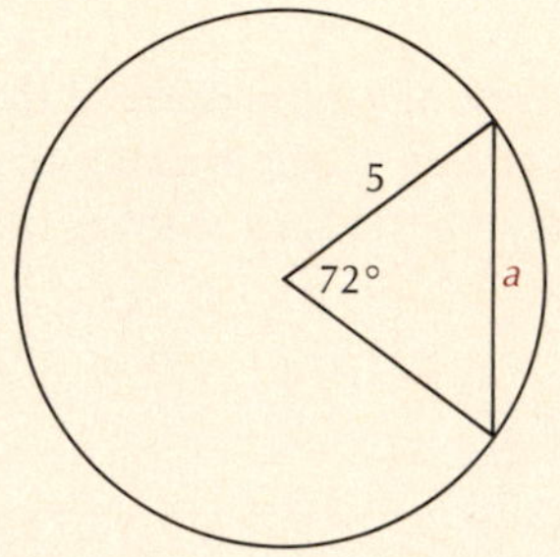

53. Show that the area of a right triangle is

$$\frac{1}{4}c^2 \sin 2A.$$

54. Show that the area of a right triangle is

$$\frac{1}{2}bc \sin A.$$

55. Find a formula for the distance to the horizon as a function of the height of the observer above the earth. Calculate the distance to the horizon from an airplane flying at an altitude of 1000 ft.

56. In finding horizontal distance from slope distance (see Example 7), we have $H = L - C$, where C is a correction. Show that a good approximation to C is $d^2/2L$.

57. *Carpentry.* A carpenter is constructing picnic pavilions in parks, as shown in the figure. The rafter ends are to be sawed in such a way that they will be vertical when in place. The front wall is 8 ft high, the back wall is $6\frac{1}{2}$ ft high, and the distance between walls is 8 ft. At what angle should the rafters be cut?

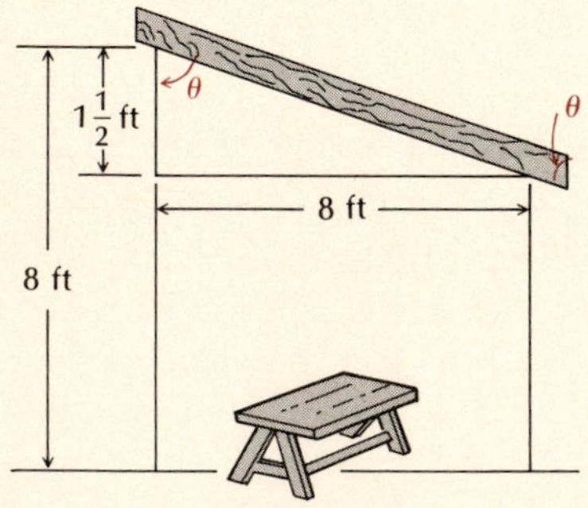

58. *Carpentry.* A V-gauge is used to find diameters of pipes. The advantage of such a device is that it is rugged, it is accurate, and it has no moving parts to break down. In the figure, the measure of angle AVB is 54°. A pipe is placed in the V-shaped slot and the distance VP is used to predict the diameter.

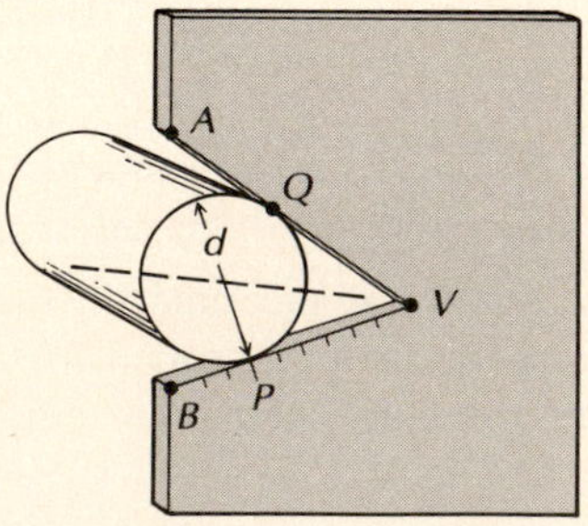

a) Suppose that the diameter of a pipe is 2 cm. What is the distance VP?

b) Suppose that the distance VP is 3.93 cm. What is the diameter of the pipe?

c) Find a formula for d in terms of VP.

d) Find a formula for VP in terms of d.

The line VP is calibrated by listing as its units the corresponding diameters. This, in effect, establishes a function between VP and d.

59. Show that the area of this (general) triangle is

$$\tfrac{1}{2}ab \sin \theta.$$

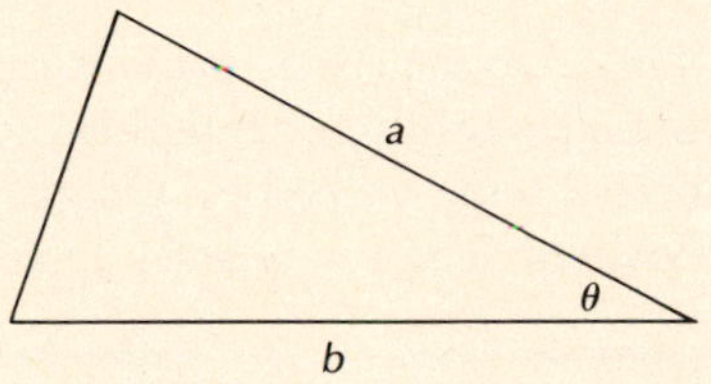

CHALLENGE

60. *Sound of an airplane.* It is a common experience to hear the sound of a low flying airplane, and look at the wrong place in the sky to see the plane. Suppose that a plane is traveling directly at you at a speed of 200 mph and an altitude of 3000 ft, and you hear the sound at what seems to be an angle of inclination of 20°. At what angle θ should you actually look in order to see the plane? Consider the speed of sound to be 1100 ft/sec.

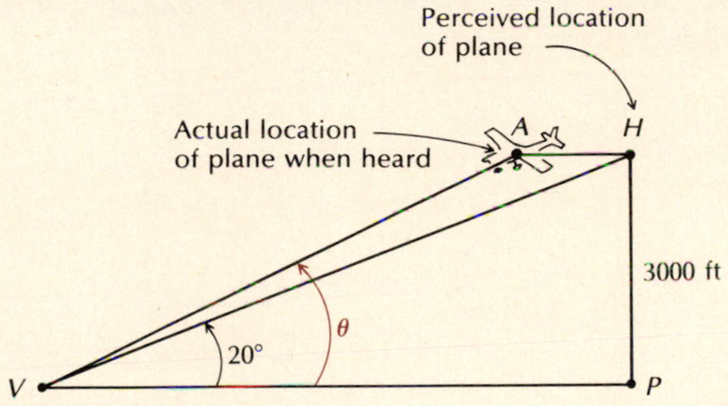

61. *Measuring the radius of the earth.* One way to measure the radius of the earth is to climb to the top of a mountain whose height above sea level is known and measure the angle between a vertical line to the center of the earth from the top of the mountain and a line drawn from the top of the mountain to the horizon, as shown in the figure. The height of Mt. Shasta in California is 14,162 ft. From the top of Mt. Shasta, one can see the horizon on the Pacific Ocean. The angle formed between a line to the horizon and the vertical is found to be 87°53′. Use this information to estimate the radius of the earth, in miles.

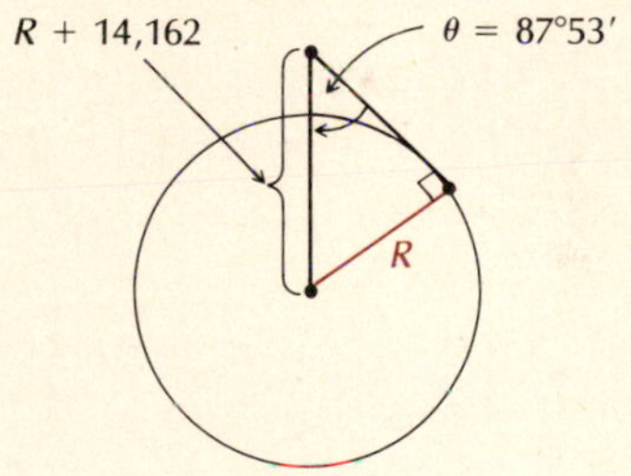

62. The John Hancock Building is slanted on each side. It is 1107 ft tall, excluding the towers on the top. The base of the building is a 165-ft–by–265-ft rectangle. The top of the building is a 100-ft–by–160-ft rectangle. Find the angle of inclination of each side of the building from the vertical.

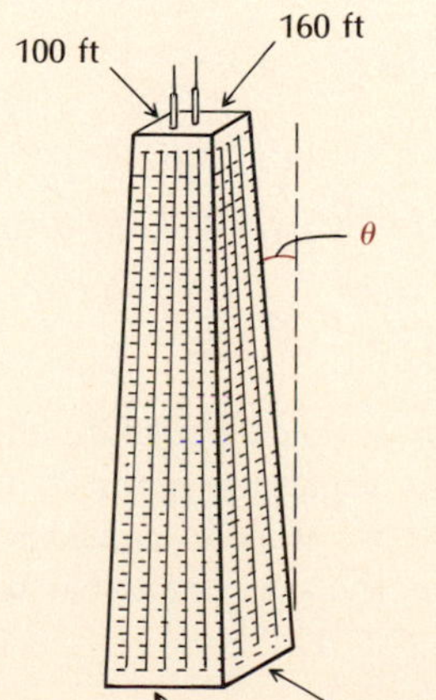

OBJECTIVES

You should be able to:

1 Use the law of sines to solve any triangle, given a side and two angles.

2 Use the law of sines to solve any triangle, given two sides and an angle opposite one of them, finding two solutions when they exist, and recognizing when a solution does not exist.

8.2 The Law of Sines

The trigonometric functions can be used to solve triangles that are not right triangles (**oblique triangles**). In order to solve oblique triangles, we need to derive some properties. One is the **law of sines** and the other is the **law of cosines.** The law of sines applies to the following situations.

The law of sines for solving oblique triangles applies to the following two cases:

1. **Two angles and any side of a triangle are known (AAS and ASA).**
2. **Two sides of a triangle and an angle opposite one of them are known (SSA).**

Keep in mind that there may be no solution, one solution, or two solutions. The latter is known as the *ambiguous case*.

We will consider any oblique triangle. It may or may not have an obtuse angle. We consider both cases, but the derivations are essentially the same. In the following figures, h is the same length.

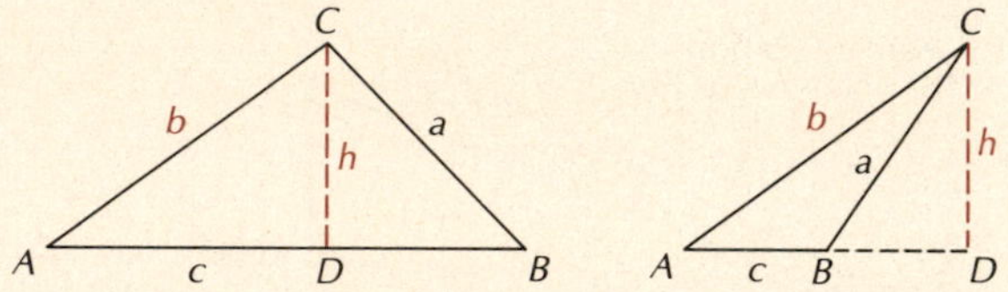

The triangles are lettered in the standard way, with angles A, B, and C and the sides opposite them a, b, and c, respectively. We have drawn an altitude from vertex C. It has length h. In either triangle we now have, from triangle ADC,

$$\frac{h}{b} = \sin A, \quad \text{or} \quad h = b \sin A.$$

From triangle DBC on the left above, we have $h/a = \sin B$, or $h = a \sin B$. On the right above, we have $h/a = \sin(180° - B) = \sin B$. So in either kind of triangle we now have

$$h = b \sin A \quad \text{and} \quad h = a \sin B.$$

It follows that

$$b \sin A = a \sin B.$$

We divide by $\sin A \sin B$ to obtain

$$\frac{a}{\sin A} = \frac{b}{\sin B}.$$

There is no danger of dividing by 0 here because we are dealing with triangles whose angles are never 0° or 180°.

If we were to consider an altitude from vertex A in the triangles shown above, the same argument would give us

$$\frac{b}{\sin B} = \frac{c}{\sin C}.$$

We combine these results to obtain the law of sines, which holds for right triangles as well as oblique triangles.

THEOREM 1 The Law of Sines

In any triangle ABC,

$$\frac{a}{\sin A} = \frac{b}{\sin B} = \frac{c}{\sin C}.$$

(The sides are proportional to the sines of the opposite angles.)

1 Solving Triangles (ASA and AAS)

When two angles and a side of any triangle are known, the law of sines can be used to solve the triangle.

EXAMPLE 1 In triangle ABC, $a = 4.56$, $A = 43°$, and $C = 57°$. Solve the triangle.

Solution

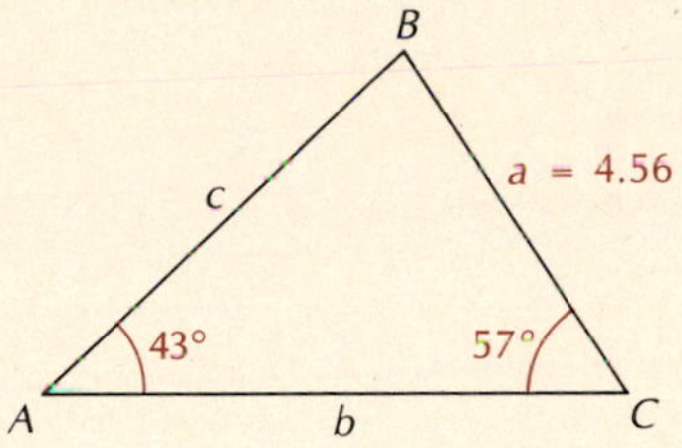

We first draw a sketch. We find B, as follows:

$$B = 180° - (43° + 57°) = 80°.$$

We can now find the other two sides, using the law of sines:

$$\frac{c}{\sin C} = \frac{a}{\sin A}$$

$$\frac{c}{\sin 57°} = \frac{4.56}{\sin 43°}$$

$$c = \frac{4.56 \sin 57°}{\sin 43°} \qquad \text{Solving for } c$$

$$c = \frac{4.56 \times 0.8387}{0.6820}$$

$$c = 5.61;$$

$$\frac{b}{\sin B} = \frac{a}{\sin A}$$

$$\frac{b}{\sin 80°} = \frac{4.56}{\sin 43°}$$

$$b = \frac{4.56 \sin 80°}{\sin 43°} \qquad \text{Solving for } b$$

$$b = \frac{4.56 \times 0.9848}{0.6820}$$

$$b = 6.58.$$

We have now found the unknown parts of the triangle: $B = 80°$, $c = 5.61$, and $b = 6.58$.

DO EXERCISE 1.

1. Solve this triangle.

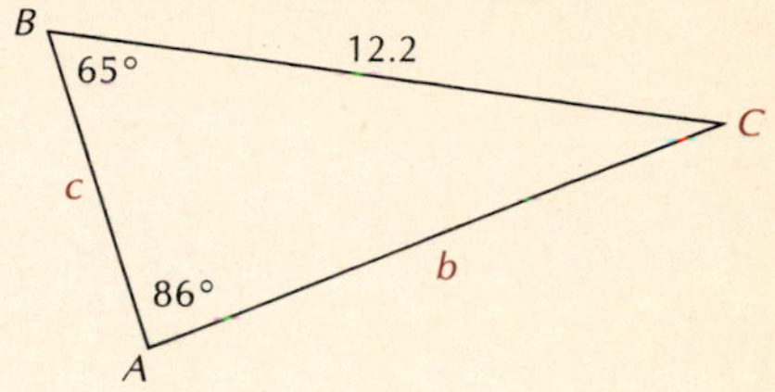

2. Solve this triangle.

   $a = 40, \quad b = 12, \quad B = 57^\circ$

2 The Ambiguous Case (SSA)

When two sides of a triangle and an angle opposite one of them are known, the law of sines can be used to solve the triangle. However, there may be no solution, one solution, or two solutions. The latter is known as the **ambiguous case.** Suppose a, b, and A are given. Then the various possibilities are as shown in the five cases below.

Case I

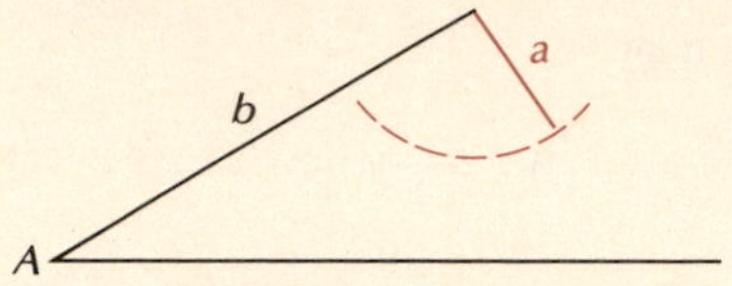

No solution, side a is too short to reach the base.

Case II

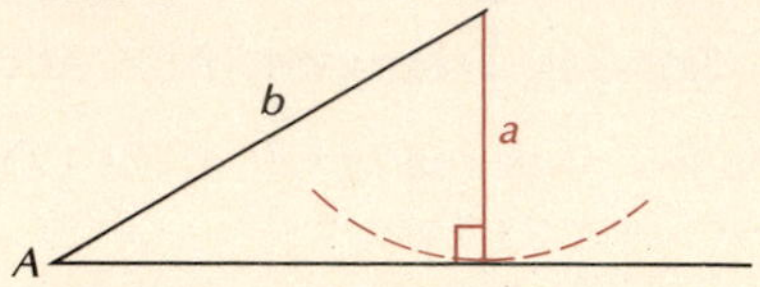

One solution, side a just reaches the base and is perpendicular to it.

Case III

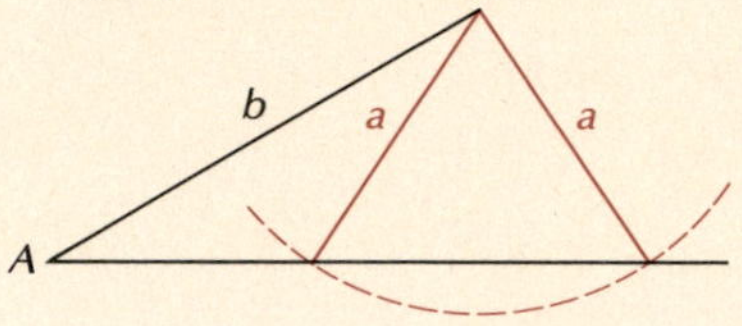

Two solutions, an arc of radius a meets the base at two points.

Case IV

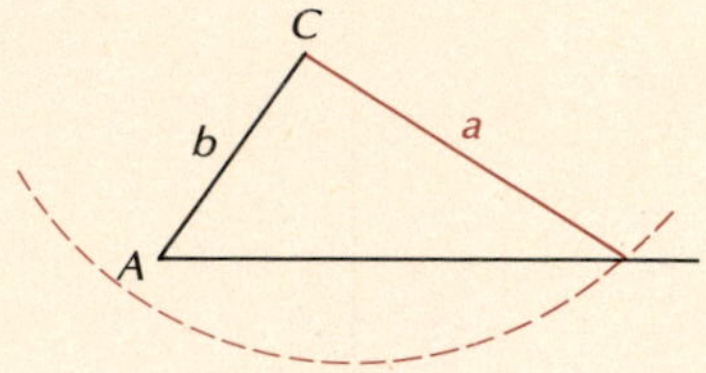

$a > b$. One solution, an arc of radius a meets the base at just one point.

Case V

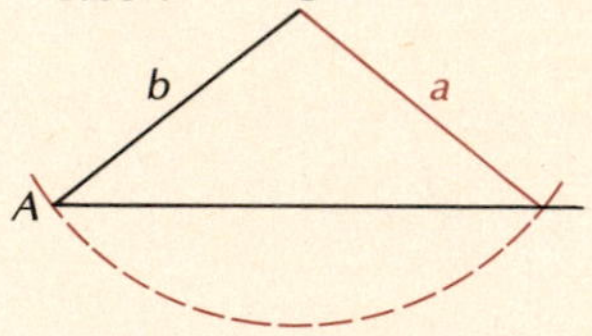

$a = b$. One solution, an arc of radius a meets the base at just one point, other than A.

The following examples correspond to the first four possibilities just described.

EXAMPLE 2 **Case I: No solution.** In triangle ABC, $a = 15$, $b = 25$, and $A = 47^\circ$. Solve the triangle.

Solution We look for B:

$$\frac{a}{\sin A} = \frac{b}{\sin B}$$

$$\frac{15}{\sin 47^\circ} = \frac{25}{\sin B}$$

$$\sin B = \frac{25 \sin 47^\circ}{15} \qquad \text{Solving for } \sin B$$

$$\sin B = \frac{25 \times 0.7314}{15}$$

$$\sin B = 1.219.$$

Since there is no angle having a sine greater than 1, there is *no* solution. ■

DO EXERCISE 2.

EXAMPLE 3 Case II: One solution. In triangle ABC, $a = 12$, $b = 5$, and $B = 24°38'$. Solve the triangle.

Solution We look for A:

$$\frac{a}{\sin A} = \frac{b}{\sin B}$$

$$\frac{12}{\sin A} = \frac{5}{\sin 24°38'}$$

$$\sin A = \frac{12 \sin 24°38'}{5}$$

$$\sin A = \frac{12 \times 0.4168}{5} = 1.00$$

and

$$A = 90°. \quad \text{We have a right triangle.}$$

Then $C = 180° - (90° + 24°38') = 65°22'$. We now find c:

$$\frac{c}{a} = \cos B$$

$$\frac{c}{12} = \cos 24°38'$$

$$c = 12 \cos 24°38' = 12 \times 0.9090$$

$$c = 10.9.$$

DO EXERCISE 3.

EXAMPLE 4 Case III: Two solutions. In triangle ABC, $a = 20$, $b = 15$, and $B = 30°$. Solve the triangle.

Solution We look for A:

$$\frac{a}{\sin A} = \frac{b}{\sin B}$$

$$\frac{20}{\sin A} = \frac{15}{\sin 30°}$$

$$\sin A = \frac{20 \sin 30°}{15} = \frac{20 \times 0.5}{15}$$

$$\sin A = 0.6667.$$

There are two angles less than 180° having a sine of 0.6667. They are 42° and 138°, to the nearest degree. This gives us two possible solutions.

Possible solution 1. If $A = 42°$, then

$$C = 180° - (30° + 42°) = 108°.$$

We now find c:

$$\frac{c}{\sin C} = \frac{b}{\sin B}$$

$$\frac{c}{\sin 108°} = \frac{15}{\sin 30°}$$

$$c = \frac{15 \sin 108°}{\sin 30°} = \frac{15 \times 0.9511}{0.5}$$

$$c = 28.5.$$

3. Solve this triangle.

$a = 3, \quad b = 4, \quad A = 48°35'$

4. In triangle ABC, $a = 25$, $b = 20$, and $B = 33°$. Solve the triangle.

These parts make a triangle, as shown; thus we have a solution.

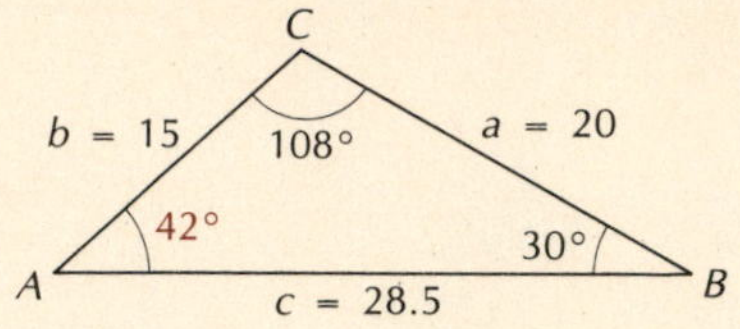

Possible solution 2. If $A = 138°$, then

$$C = 180° - (30° + 138°) = 12°.$$

We now find c:

$$\frac{c}{\sin C} = \frac{b}{\sin B}$$

$$\frac{c}{\sin 12°} = \frac{15}{\sin 30°}$$

$$c = \frac{15 \sin 12°}{\sin 30°}$$

$$c = \frac{15 \times 0.2079}{0.5}$$

$$c = 6.2.$$

These parts make a triangle; thus we have a second solution.

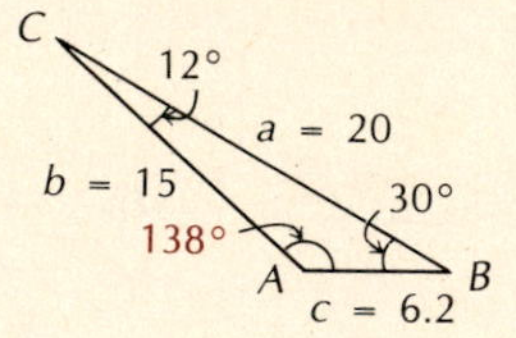

DO EXERCISE 4.

EXAMPLE 5 Case IV: One solution. In triangle ABC, $a = 25$, $b = 10$, and $A = 42°$. Solve the triangle.

Solution We look for B:

$$\frac{b}{\sin B} = \frac{a}{\sin A}$$

$$\frac{10}{\sin B} = \frac{25}{\sin 42°}$$

$$\sin B = \frac{10 \sin 42°}{25}$$

$$\sin B = \frac{10 \times 0.6691}{25}$$

$$\sin B = 0.2676.$$

Then $B = 15°31'$ or $B = 164°29'$. Since $a > b$, we know that there is only one solution. If we had not noticed this, we could tell it now. An angle of $164°29'$ cannot be an angle of this triangle because since it already has an angle of $42°$, these two would total more than $180°$.

$$C = 180° - (42° + 15°31') = 122°29',$$

$$\frac{c}{\sin C} = \frac{a}{\sin A}$$

$$\frac{c}{\sin 122°29'} = \frac{25}{\sin 42°}$$

$$c = \frac{25 \sin 122°29'}{\sin 42°} = \frac{25 \times 0.8435}{0.6691}$$

$$c = 31.5$$

DO EXERCISE 5.

5. In triangle ABC, $b = 20$, $c = 10$, and $B = 38°$. Solve the triangle.

EXERCISE SET 8.2

1 Solve each triangle, if possible.

1. $A = 133°$, $B = 30°$, $b = 18$
2. $B = 120°$, $C = 30°$, $a = 16$
3. $B = 38°$, $C = 21°$, $b = 24$
4. $A = 131°$, $C = 23°$, $b = 10$
5. $A = 68°30'$, $C = 42°40'$, $c = 23.5$ cm
6. $B = 118°20'$, $C = 45°40'$, $b = 42.1$ ft
7. $c = 3$ mi, $B = 37.48°$, $C = 32.16°$
8. $a = 200$ m, $A = 32.76°$, $C = 21.97°$
9. $c = 6019$ km, $A = 16°56'$, $B = 59°23'$
10. $A = 120°18'$, $C = 27°46'$, $a = 16{,}435$ mi
11. $A = 129°32'$, $C = 18°28'$, $b = 1204$ in.
12. $B = 37°51'$, $C = 19°47'$, $b = 240.123$ mm

2 Solve each triangle, if possible.

13. $A = 36°$, $a = 24$, $b = 34$
14. $C = 43°$, $c = 28$, $b = 27$
15. $A = 116°20'$, $a = 17.2$, $c = 13.5$
16. $A = 47°50'$, $a = 28.3$, $b = 18.2$
17. $C = 61°10'$, $c = 30.3$, $b = 24.2$
18. $B = 58°40'$, $a = 25.1$, $b = 32.6$
19. $a = 2345$ mi, $b = 2345$ mi, $A = 124.67°$
20. $b = 56.78$ yd, $c = 56.78$ yd, $C = 83°47'$
21. $a = 2$ cm, $b = 6$ cm, $A = 30°$
22. $A = 115°$, $c = 45.6$ yd, $a = 23.8$ yd
23. $a = 4000$ m, $b = 8000$ m, $A = 32°52'$
24. $a = 20.01$ cm, $b = 10.005$ cm, $A = 30°$
25. $b = 4.157$ km, $c = 3.446$ km, $C = 51°48'$
26. $b = 43.67$ mm, $a = 38.86$ mm, $C = 48.76°$
27. $A = 89°$, $a = 15.6$ in., $b = 18.4$ in.
28. $B = 36°$, $b = 8.0$ ft, $c = 10.0$ ft
29. $A = 41°50'$, $a = 90.0$ mi, $c = 110.0$ mi
30. $C = 46°30'$, $a = 56.2$ m, $c = 22.1$ m

1, **2** Solve. Keep in mind that the two types of bearing were considered in Section 8.1.

31. Points A and B are on opposite sides of a lunar crater. Point C is 50 m from A. The measure of $\angle BAC$ is determined to be 112° and the measure of $\angle ACB$ is determined to be 42°. What is the width of the crater?

32. A guy wire to a pole makes a 71° angle with level ground. At a point 25 ft farther from the pole than the guy wire, the angle of elevation of the top of the pole is 37°. How long is the guy wire?

33. A pole leans away from the sun at an angle of 7° to the vertical (see the figure at the right). When the angle of elevation of the sun is 51°, the pole casts a shadow 47 ft long on level ground. How long is the pole?

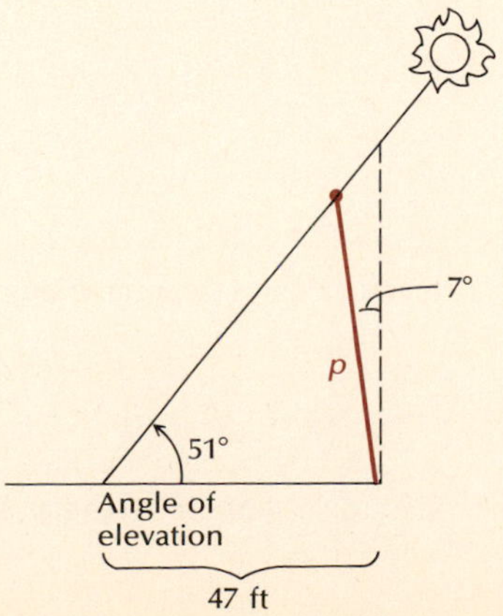

34. A vertical pole stands by a road that is inclined 10° to the horizontal. When the angle of elevation of the sun is 23°, the pole casts a shadow 38 ft long directly downhill along the road. How long is the pole?

35. A reconnaissance plane leaves its airport on the east coast of the United States and flies in a direction of 085°. Because of bad weather, it flies to another airport 230 km to the north of its home base. For the return trip, it flies in a direction of 283°. What was the total distance that it flew?

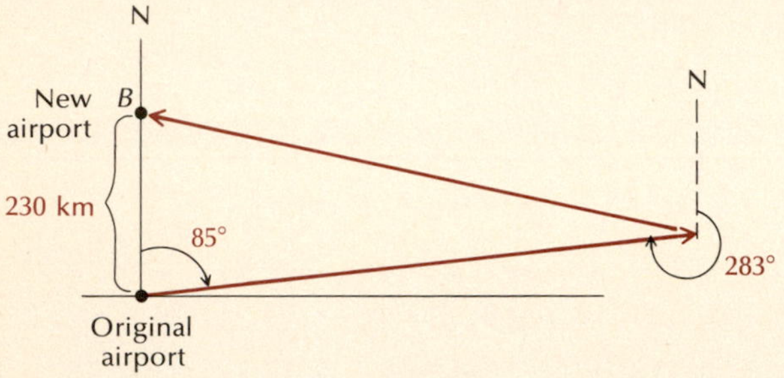

36. Lookout station B is 10.2 km east of station A. The bearing of a fire from A is S 10°40′ W. The bearing of the fire from B is S 31°20′ W. How far is the fire from A? from B?

37. A ranger in fire tower A spots a fire at a direction of 295°. A ranger at fire tower B, located 45 mi at a direction of 045° from tower A, spots the same fire at a direction of 255°. How far from tower A is the fire? from tower B?

38. An airplane leaves airport A and flies 200 km. At this time its direction from airport B, 250 km to the west of A, is 120°. How far is the airport from B?

39. A boat leaves a lighthouse A and sails 5.1 km. At this time it is sighted from lighthouse B, 7.2 km west of A. The bearing of the boat from B is N 65°10′ E. How far is the boat from B?

40. Mackinac Island is located 35 mi N 65°20′ W of Cheboygan, Michigan, where the Coast Guard cutter Mackinaw is stationed. A freighter in distress radios the Coast Guard cutter for help. It radios their position as N 25°40′ E of Mackinac Island and N 10°10′ W of Cheboygan. How far is the freighter from Cheboygan?

41. Three circles are arranged as shown in the figure below. Find the length PQ.

42. Three coins are arranged as shown in the figure below. Find the angle ϕ.

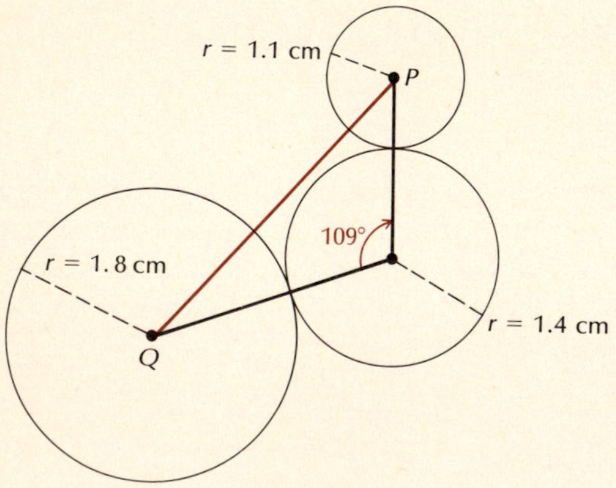

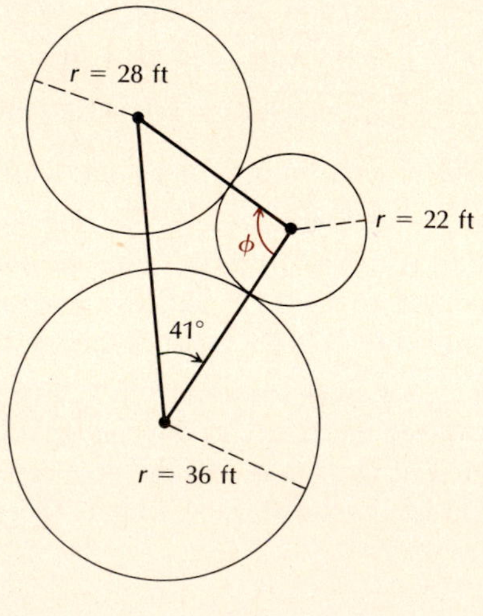

SYNTHESIS

Prove each of the following area formulas for a general triangle ABC with opposite sides of respective lengths a, b, and c and area represented by K.

43. $K = \frac{1}{2}bc \sin A = \frac{1}{2}ab \sin C = \frac{1}{2}ac \sin B$

44. $K = \dfrac{a^2 \sin B \sin C}{2 \sin A} = \dfrac{c^2 \sin A \sin B}{2 \sin C} = \dfrac{b^2 \sin C \sin A}{2 \sin B}$

Prove each of the following formulas for a general triangle ABC with opposite sides of respective lengths a, b, and c.

45. $\dfrac{a+b}{c} = \dfrac{\cos \frac{1}{2}(A - B)}{\sin \frac{1}{2}C}$

46. $\dfrac{a-b}{c} = \dfrac{\sin \frac{1}{2}(A - B)}{\cos \frac{1}{2}C}$

47. Prove that the area of a parallelogram is the product of two sides and the sine of the included angle.

48. Prove that the area of a quadrilateral is half the product of the lengths of its diagonals and the sine of the angle between the diagonals.

CHALLENGE

49. Consider the following triangle. Prove that

$$\frac{\sin \alpha}{z} + \frac{\sin \beta}{x} = \frac{\sin(\alpha + \beta)}{y}.$$

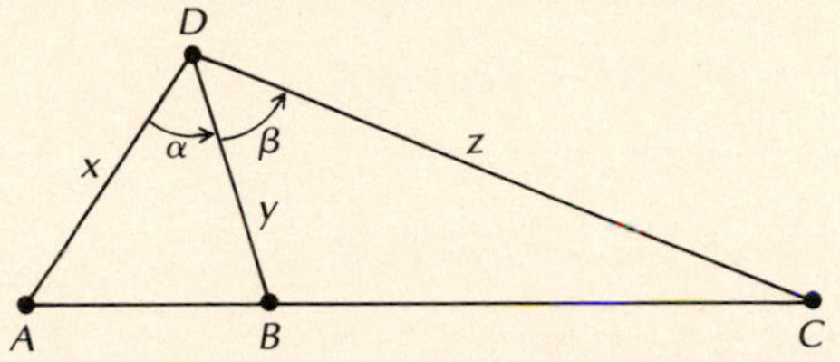

50. When two objects, such as ships, airplanes, or runners, move in straight-line paths, if the distance between them is decreasing and if the bearing from one of them to the other is constant, they will collide. ("Constant bearing means collision," as mariners put it.) Prove that this statement is true.

8.3 The Law of Cosines

OBJECTIVES

You should be able to:

1. Use the law of cosines, with the law of sines, to solve any triangle, given two sides and the included angle.
2. Use the law of cosines to solve any triangle, given three sides.
3. Determine whether the law of sines or the law of cosines should be applied to solve a triangle, and then solve the triangle.

A second property of triangles important in solving oblique triangles is called the **law of cosines.** The law of cosines applies to the following situations.

The law of cosines for solving oblique triangles applies to the following two cases:

1. **Two sides of a triangle and the included angle are known.**
2. **All three sides of the triangle are known.**

To derive this property, we consider any triangle ABC placed on a coordinate system. We place the origin at one of the vertices, say C, and the positive half of the x-axis along one of the sides, say CB. Let (x, y) be the coordinates of vertex A. Point B has coordinates $(a, 0)$, and point C has coordinates $(0, 0)$. Then

$$\cos C = \frac{x}{b} \quad \text{so} \quad x = b \cos C,$$

and

$$\sin C = \frac{y}{b} \quad \text{so} \quad y = b \sin C.$$

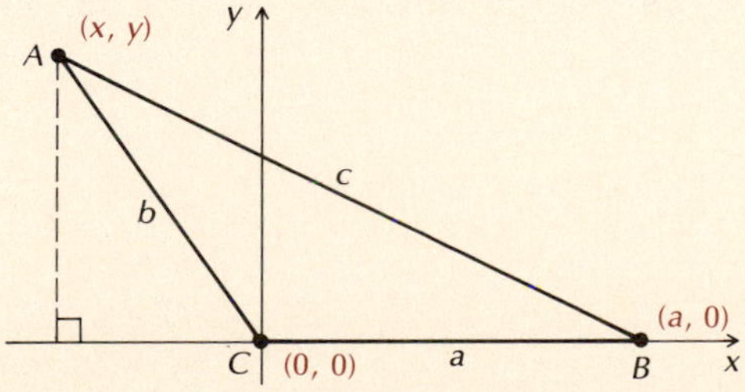

Thus,

$$(x, y) = (b \cos C, b \sin C).$$

So, point A has coordinates $(b \cos C, b \sin C)$.

Next, we use the distance formula to determine c^2:

$$c^2 = (b \cos C - a)^2 + (b \sin C - 0)^2.$$

Now we multiply and simplify:

$$c^2 = b^2 \cos^2 C - 2ab \cos C + a^2 + b^2 \sin^2 C$$
$$c^2 = a^2 + b^2(\sin^2 C + \cos^2 C) - 2ab \cos C$$
$$c^2 = a^2 + b^2 - 2ab \cos C.$$

Had we placed the origin at one of the other vertices, we would have obtained

$$a^2 = b^2 + c^2 - 2bc \cos A$$

or

$$b^2 = a^2 + c^2 - 2ac \cos B.$$

These results can be summarized as follows.

THEOREM 2 The Law of Cosines

In any triangle ABC,

$$a^2 = b^2 + c^2 - 2bc \cos A,$$
$$b^2 = a^2 + c^2 - 2ac \cos B,$$

and

$$c^2 = a^2 + b^2 - 2ab \cos C.$$

(In any triangle, the square of a side is the sum of the squares of the other two sides, minus twice the product of those sides and the cosine of the included angle.)

CAUTION! Should the law of sines be applied after the law of cosines, one should be alert for the possibility of the ambiguous case.

Only one of the above formulas need be memorized. The other two can be obtained by a change of letters.

1 Solving Triangles (SAS)

When two sides of a triangle and the included angle are known, we can use the law of cosines to find the third side. The law of cosines or the law of sines can then be used to finish solving the triangle.

EXAMPLE 1 In triangle ABC, $a = 24$, $c = 32$, and $B = 115°$. Solve the triangle.

Solution

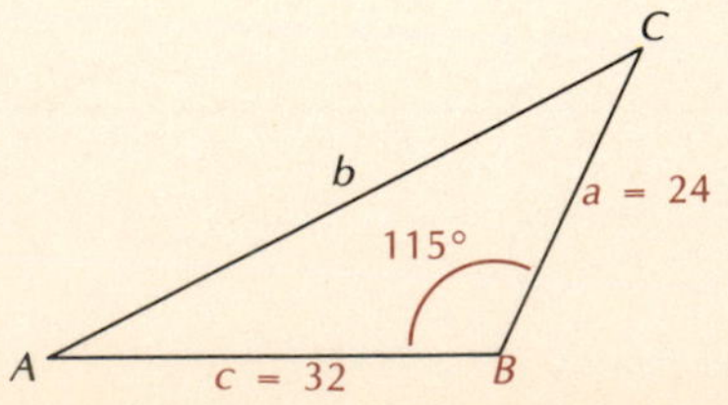

We first find the third side. From the law of cosines,

$$b^2 = a^2 + c^2 - 2ac \cos B$$
$$b^2 = 24^2 + 32^2 - 2 \cdot 24 \cdot 32(\cos 115°)$$
$$b^2 = 576 + 1024 - 1536(-0.4226)$$
$$b^2 = 2249.$$

Then $b = \sqrt{2249} = 47.4$. We now have $a = 24$, $b = 47.4$, and $c = 32$. We need to find the rest of the angle measures. At this point, we can find them in two ways. One way uses the law of sines. The ambiguous case may arise, however, and we would have to be alert to this possibility. The advantage of using the law of cosines again is that if we solve for the cosine and its value is negative, then we know that the angle is obtuse. If the value of the cosine is positive, then the angle is acute. Thus we use the law of cosines to find a second angle.

Let us find angle A. We select the formula from the law of cosines that contains $\cos A$ and substitute:

$$a^2 = b^2 + c^2 - 2bc \cos A$$
$$24^2 = 47.4^2 + 32^2 - 2(47.4)(32) \cos A$$
$$576 = 2246.76 + 1024 - 3033.6 \cos A$$
$$-2694.76 = -3033.6 \cos A$$
$$\cos A = 0.8883.$$

Then $A = 27°20'$. The third angle is now easy to find:

$$C = 180° - (115° + 27°20') = 37°40'.$$

DO EXERCISE 1.

1. In triangle ABC, $b = 18$, $c = 28$, and $A = 122°$. Solve the triangle.

2 Solving Triangles (SSS)

When all three sides of a triangle are known, the law of cosines can be used to solve the triangle.

EXAMPLE 2 In triangle ABC, $a = 18$, $b = 25$, and $c = 12$. Solve the triangle.

Solution

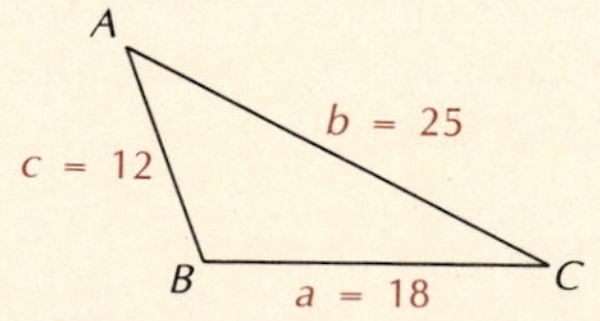

Let us find angle B. We select the formula from the law of cosines that contains $\cos B$; in other words, $b^2 = a^2 + c^2 - 2ac \cos B$. We substitute and solve for $\cos B$:

$$b^2 = a^2 + c^2 - 2ac \cos B$$
$$25^2 = 18^2 + 12^2 - 2(18)(12) \cos B$$
$$625 = 324 + 144 - 432 \cos B$$
$$157 = -432 \cos B$$
$$\cos B = -0.3634.$$

2. In triangle ABC, $a = 25$, $b = 10$, and $c = 20$. Solve the triangle.

Determine which law applies. Then solve the triangle.

3. $a = 50$, $b = 20$, $c = 40$

4. $b = 40$, $c = 20$, $B = 38°$

5. $b = 54$, $c = 84$, $A = 122°$

6. $a = 75$, $b = 60$, $B = 33°$

7. $A = 43°35'$, $a = 4.5$, $b = 6$

8. $b = 6$, $B = 57°$, $a = 20$

Then $B = 111°19'$. Similarly, we find angle A*:

$$\begin{aligned} a^2 &= b^2 + c^2 - 2bc \cos A \\ 18^2 &= 25^2 + 12^2 - 2(25)(12) \cos A \\ 324 &= 625 + 144 - 600 \cos A \\ -445 &= -600 \cos A \\ \cos A &= 0.7417. \end{aligned}$$

Thus, $A = 42°08'$. Then $C = 180° - (111°19' + 42°08') = 26°33'$. ■

DO EXERCISE 2.

3 Choosing the Appropriate Law

The following summarizes the situations in which to use the law of sines and the law of cosines.

> **To solve an oblique triangle:**
>
> **A) Use the *law of sines* when:**
>
> **1. Two angles and any side of a triangle are known (AAS and ASA).**
>
> **2. Two sides of a triangle and an angle opposite one of them are known (SSA). Keep in mind that there may be no solution, one solution, or two solutions. The latter is known as the *ambiguous case*.**
>
> **B) Use the *law of cosines* when:**
>
> **3. Two sides of a triangle and the included angle are known (SAS).**
>
> **4. All three sides of the triangle are known (SSS).**

Let us practice determining which law to use.

EXAMPLE 3 In triangle ABC, $a = 20$, $b = 15$, and $B = 30°$. Determine which law applies. Then solve the triangle.

Solution We are given two sides and an angle opposite one of them. In this case, the law of sines applies. Thus there may be no solution, one solution, or two solutions (the ambiguous case). This is indeed the ambiguous case. The triangle is solved in Example 4 of Section 8.2. ■

EXAMPLE 4 In triangle ABC, $a = 24$, $c = 32$, and $B = 115°$. Determine which law applies. Then solve the triangle.

Solution We are given two sides and the included angle. In this case, the law of cosines applies. The triangle is solved in Example 1 of this section. ■

DO EXERCISES 3–8.

*The law of sines could be used at this point, but be alert for the ambiguous case.

EXERCISE SET 8.3

1 Solve each triangle, if possible.

1. $A = 30°$, $b = 12$, $c = 24$
2. $C = 60°$, $a = 15$, $b = 12$
3. $A = 133°$, $b = 12$, $c = 15$
4. $B = 116°$, $a = 31$, $c = 25$
5. $B = 72°40'$, $c = 16$ m, $a = 78$ m
6. $A = 24°30'$, $b = 68$ ft, $c = 14$ ft
7. $C = 22.28°$, $a = 25.4$ cm, $b = 73.8$ cm
8. $B = 72.66°$, $a = 23.78$ km, $c = 25.74$ km
9. $A = 96°13'$, $b = 15.8$ yd, $c = 18.4$ yd
10. $C = 28°43'$, $a = 6$ mm, $b = 9$ mm
11. $a = 60.12$ mi, $b = 40.23$ mi, $C = 48.7°$
12. $b = 10.2$ in., $c = 17.3$ in., $A = 53.456°$

2 Solve each triangle, if possible.

13. $a = 12$, $b = 14$, $c = 20$
14. $a = 22$, $b = 22$, $c = 35$
15. $a = 16$ m, $b = 20$ m, $c = 32$ m
16. $a = 2.2$ cm, $b = 4.1$ cm, $c = 2.4$ cm
17. $a = 2$ ft, $b = 3$ ft, $c = 8$ ft
18. $a = 17$ yd, $b = 15.4$ yd, $c = 1.5$ yd
19. $a = 11.2$ cm, $b = 5.4$ cm, $c = 7$ cm
20. $a = 26.12$ km, $b = 21.34$ km, $c = 19.25$ km

3 Determine which law applies. Then solve each triangle.

21. $A = 70°$, $B = 12°$, $b = 21.4$
22. $a = 15$, $c = 7$, $B = 62°$
23. $a = 3.3$, $b = 2.7$, $c = 2.8$
24. $a = 1.5$, $b = 2.5$, $A = 58°$
25. $a = 60$, $b = 40$, $C = 48°$
26. $a = 3.6$, $b = 6.2$, $c = 4.1$
27. $B = 110°30'$, $C = 8°10'$, $c = 0.912$
28. $B = 52°$, $C = 15°$, $b = 60.4$

1, **2**

29. A ship leaves a harbor and sails 15 nautical mi east. It then sails 18 nautical mi in a direction of S 27° E. How far is it, then, from the harbor and in what direction?

30. An airplane leaves an airport and flies west 147 km. It then flies 200 km in a direction of 220°. How far is it, then, from the airport and in what direction?

31. Two ships leave harbor at the same time. The first sails N 15° W at 25 knots (a knot is one nautical mile per hour). The second sails N 32° E at 20 knots. After 2 hr, how far apart are the ships?

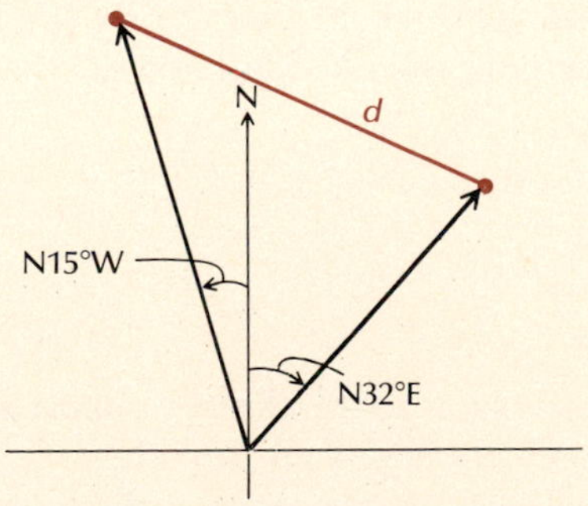

32. Two airplanes leave an airport at the same time. The first flies 150 km/h in a direction of 320°. The second flies 200 km/h in a direction of 200°. After 3 hr, how far apart are the planes?

33. A hill is inclined 5° to the horizontal. A 45-ft pole stands at the top of the hill. How long a rope will it take to reach from the top of the pole to a point 35 ft downhill from the base of the pole?

34. A hill is inclined 15° to the horizontal. A 40-ft pole stands at the top of the hill. How long a rope will it take to reach from the top of the pole to a point 68 ft downhill from the base of the pole?

35. A piece of wire 5.5 m long is bent into a triangular shape. One side is 1.5 m long and another is 2 m long. Find the angles of the triangle.

36. A triangular lot has sides of 120 ft, 150 ft, and 100 ft. Find the angles of the lot.

37. A slow-pitch softball diamond is a square 65 ft on a side. The pitcher's mound is 46 ft from home. How far is it from the pitcher's mound to first base?

38. A baseball diamond is a square 90 ft on a side. The pitcher's mound is 60.5 ft from home. How far does the pitcher have to run to cover first base?

39. The longer base of an isosceles trapezoid measures 14 ft. The nonparallel sides measure 10 ft, and the base angles measure 80°.

a) Find the length of a diagonal.
b) Find the area.

40. An isosceles triangle has a vertex angle of 38° and this angle is included by two sides, each measuring 20 ft. Find the area of the triangle.

41. A field in the shape of a parallelogram has sides that measure 50 yd and 70 yd. One angle of the field measures 78°. Find the area of the field.

42. An aircraft takes off to fly a 180-mi trip. After flying 75 mi, it is 10 mi off course. How much should the heading be corrected to then fly straight to the destination, assuming no wind correction?

3

43. A fence along one side of a triangular lot is 28 m long and makes angles of 25° and 58°20′ with the sides. How long are the other two sides?

44. A guy wire to a pole makes a 63°40′ angle with level ground. At a point 8.2 m farther from the pole than the guy wire, the angle of elevation of the top of the pole is 42°30′. How long is the guy wire?

45. A triangular swimming pool has sides of length 40 ft, 35.3 ft, and 26.6 ft. Find the angles of the triangle.

46. A triangular swimming pool measures 44 ft on one side and 32.8 ft on another side. These sides form an angle that measures 40°50′. How long is the other side?

47. A pole leans away from the sun at an angle of 8° to the vertical. When the angle of elevation of the sun is 53°, the pole casts a shadow 44 ft long on level ground. How long is the pole?

48. A hill is inclined 6° to the horizontal. A 42-ft pole stands at the top of the hill. How long a wire will it take to reach from the top of the pole to a point 35 ft downhill from the base of the pole?

SYNTHESIS

49. *Heron's formula.* If a, b, and c are the lengths of the sides of a triangle, then the area K of the triangle is given by

$$K = \sqrt{s(s-a)(s-b)(s-c)},$$

where $s = \frac{1}{2}(a+b+c)$. The number s is called the **semiperimeter**. Prove Heron's formula. (*Hint:* Use the area formula $K = \frac{1}{2}bc \sin A$ developed in Exercise 54 of Exercise Set 8.1.)

50. Use Heron's formula to find the area of the triangular swimming pool described in Exercise 45.

51. The measures of two sides of a parallelogram are 50 and 60 in., while one diagonal is 90 in. How long is the other diagonal?

52. The Port Huron–to–Mackinac sailboat race covers a distance of 259 mi from Port Huron, Michigan, to Mackinac Island. During the race, the course is measured at a heading of 035° from Port Huron to Cove Island for a distance of 181 mi. From Cove Island, the boats head 259° to Mackinac Island. With a north wind, one boat started on a course of 320° and sailed for 85 mi. The navigator then determined that the boat could "tack" (change heading) and sail directly for Cove Island. How far is the boat from Cove Island, and what must its new heading be in order to reach Cove Island?

53. From the top of a hill 20 ft above the surface of a lake, a tree is sighted across the lake. The angle of elevation is 11°. From the same position, the top of the tree is seen by reflection in the lake. The angle of depression is 14°. What is the height of the tree? (*Hint:* The angle of incidence, θ, equals the angle of reflection, α.)

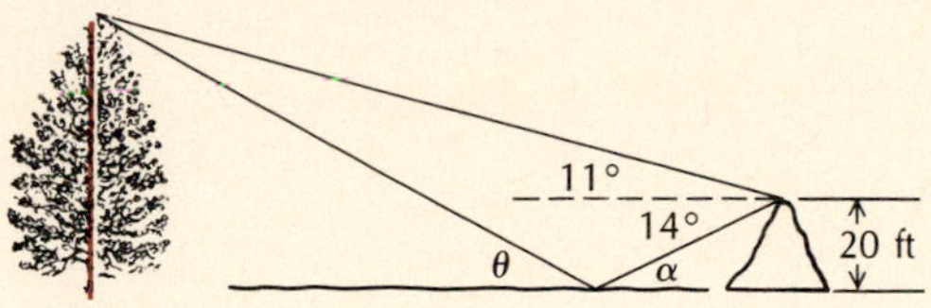

54. A bridge is being built across a canyon. The length of the bridge is 5045 ft. From the deepest point in the canyon, the angles of elevation of the ends of the bridge are 78° and 72°. How deep is the canyon?

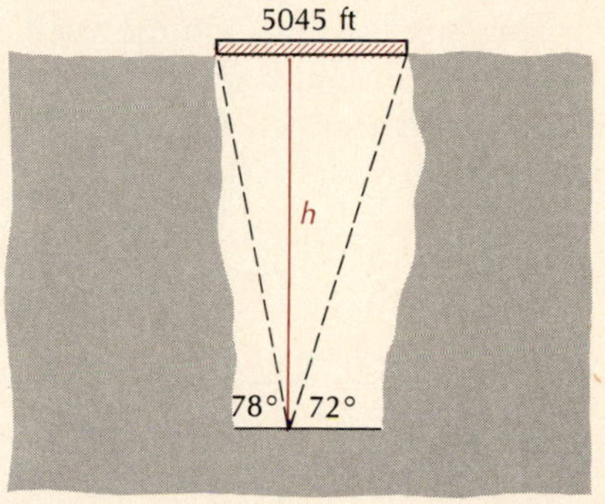

55. Find a formula for the area of an isosceles triangle in terms of the congruent sides and their included angle. Under what conditions will the area of a triangle with fixed congruent sides be maximum?

56. *Surveying.* In surveying, a series of bearings and distances is called a **transverse**. Measurements are taken as shown in the following figure.

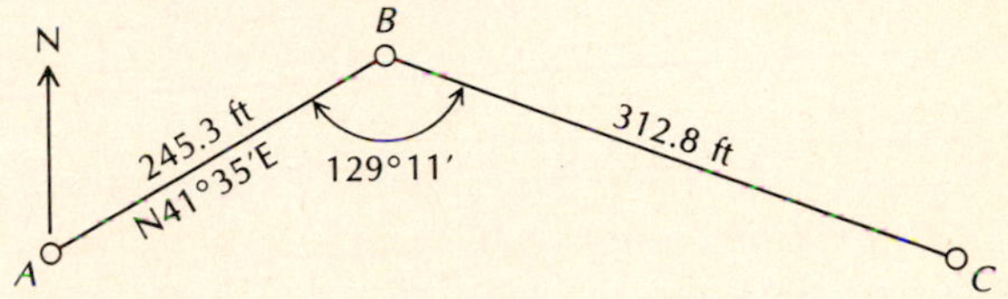

a) Compute the bearing BC.
b) Compute the distance and bearing AC.

CHALLENGE

57. Show that in any triangle ABC,

$$a^2 + b^2 + c^2 = 2(bc \cos A + ac \cos B + ab \cos C).$$

58. Show that in any triangle ABC,

$$\frac{\cos A}{a} + \frac{\cos B}{b} + \frac{\cos C}{c} = \frac{a^2 + b^2 + c^2}{2abc}.$$

59. A reconnaissance plane patrolling at 5000 ft sights a submarine at bearing 35° at an angle of depression of 25°. A carrier is at bearing 105° and at an angle of depression of 60°. How far is the submarine from the carrier?

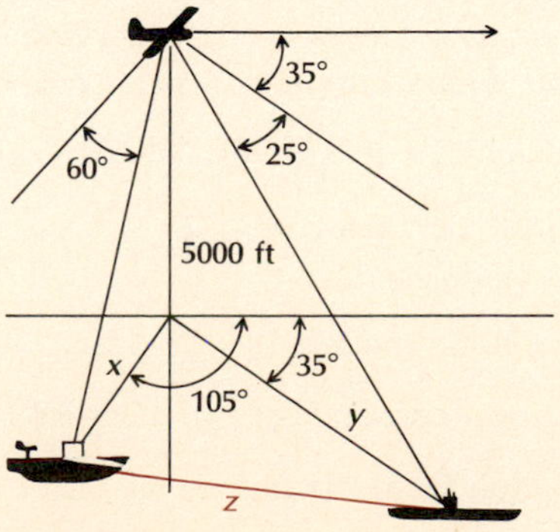

60. When two bubbles cling together in midair, their common surface is part of a sphere whose center C lies on the line passing through the centers of the bubbles. Also, the angles ADB and BDC each measure 60°. Use this information to find the radius CD of the common face in terms of the radii AD and BD of the two bubbles.

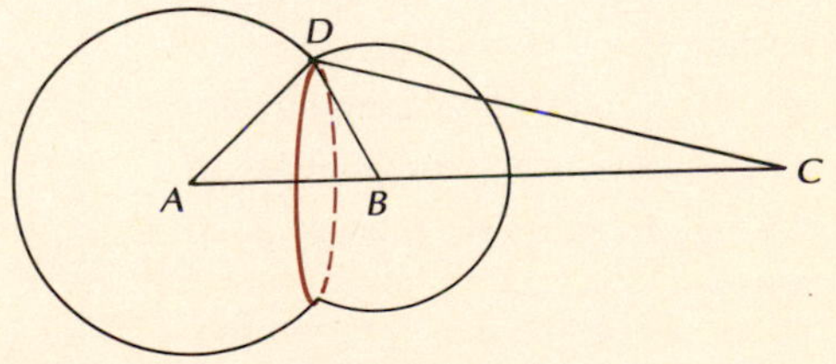

61. A satellite traveling in a circular orbit 1600 km above the earth is due to pass directly over a tracking station at noon. Assume that the satellite takes 2 hr to make an orbit and that the radius of the earth is 6400 km.

a) If the tracking antenna is aimed 30° above the horizon, at what time will the satellite pass through the beam of the antenna? (See the figure below.)

b) Find the distance between the satellite and the tracking station at 12:03 P.M.*

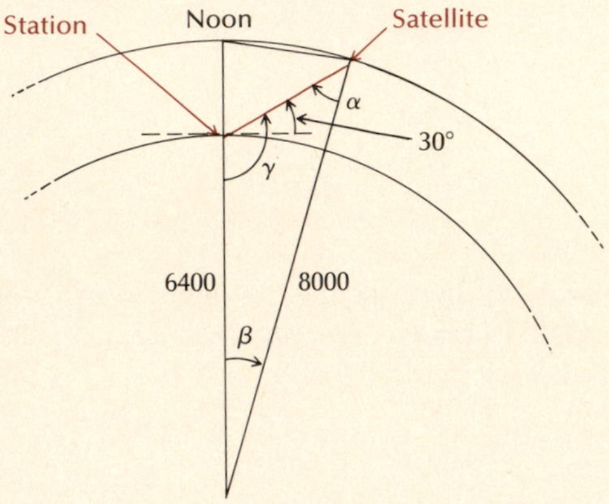

62. In reference to Exercise 61, at what angle above the horizon should the antenna be pointed so that its beam will intercept the satellite at 12:03 P.M.? (See the figure below.)†

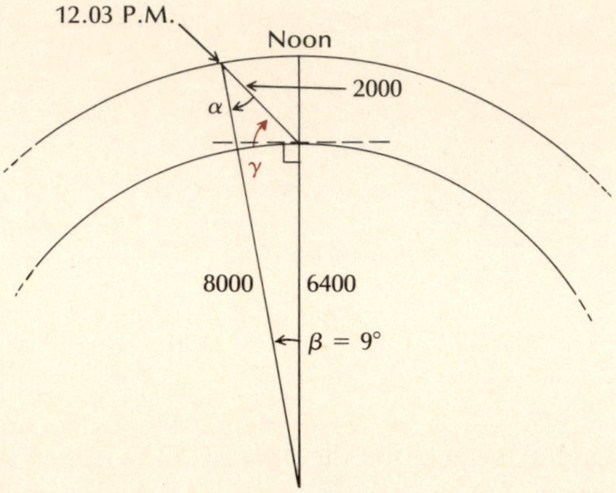

63. Two of NASA's tracking stations are located near the equator: One is in Ethiopia, at 40° east longitude; another near Quito, Ecuador, at 78° west longitude. Assume that both stations, represented by E and Q in the figure at the right, are on the equator and that the radius of the earth is 6380 km. A satellite in orbit over the equator is observed at the same instant from both tracking stations. The angles of elevation above the horizon are 5° from Quito and 10° from Ethiopia. Find the distance of the satellite from earth at the instant of observation.‡

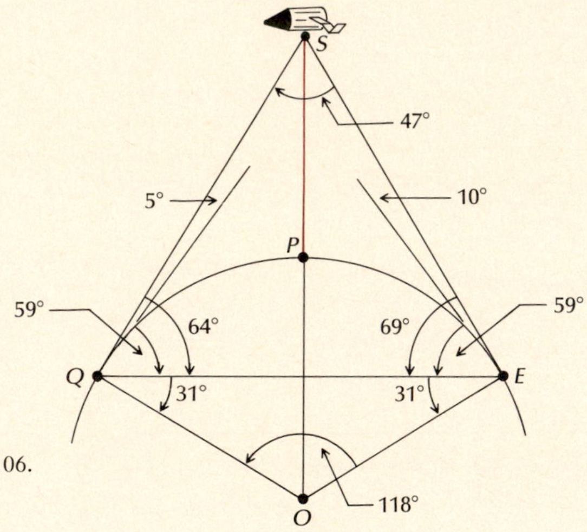

Source: Space Mathematics, National Aeronautics and Space Administration, pp. 104–106.
†*Ibid.*
‡*Ibid.*

8.4 Vectors and Applications

OBJECTIVES

You should be able to:

1. Given two vectors, find their sum, or resultant.
2. Solve applied problems involving finding sums of vectors.
3. Resolve vectors into components.
4. Add vectors using components.
5. Change from rectangular to polar notation for vectors, and from polar to rectangular notation for vectors.
6. Do simple manipulations in vector algebra.

In many applications, there occur certain quantities in which a direction is specified. Any such quantity having a *direction* and a *magnitude* is called a **vector quantity**, or **vector**. Here are some examples of vector quantities.

DISPLACEMENT. An object moves a certain distance in a certain direction.

A train travels 100 mi to the northeast.

A person takes 5 steps to the west.

A batter hits a ball 100 m along the left-field foul line.

VELOCITY. An object travels at a certain speed in a certain direction.

The wind is blowing 15 mph from the northwest.

An airplane is traveling 450 km/h in a direction of 243°.

FORCE. A push or pull is exerted on an object in a certain direction.

A 15-kg force* is exerted downward on the handle of a jack.

A 25-lb upward force is required to lift a box.

A wagon is being pulled up a 30° incline, requiring an effort of 200 kg.

We represent vectors abstractly by directed line segments, or arrows. The length is chosen, according to some scale, to represent the **magnitude of the vector,** and the direction of the arrow represents the **direction of the vector.** For example, if we let 1 cm represent 5 km/h, then a 15-km/h wind from the northwest would be represented by an arrow 3 cm long, as shown below.

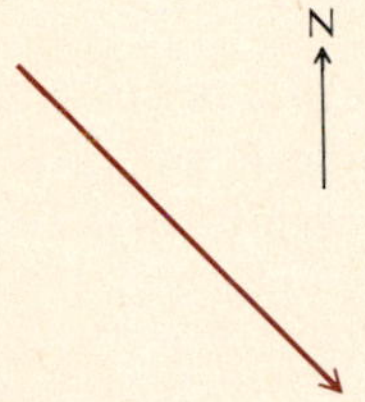

1 Vector Addition

To "add" vectors, we find a single vector that would have the same effect as the vectors combined. We will illustrate vector addition for displacements, but it is done the same way for any vector quantities. Suppose a person takes 4 steps east and then 3 steps north. He will then be 5 steps from the starting point in the direction shown. The **sum** of the two vectors is the vector 5 steps in magnitude and in the direction shown. The sum is also called the **resultant** of the two vectors.

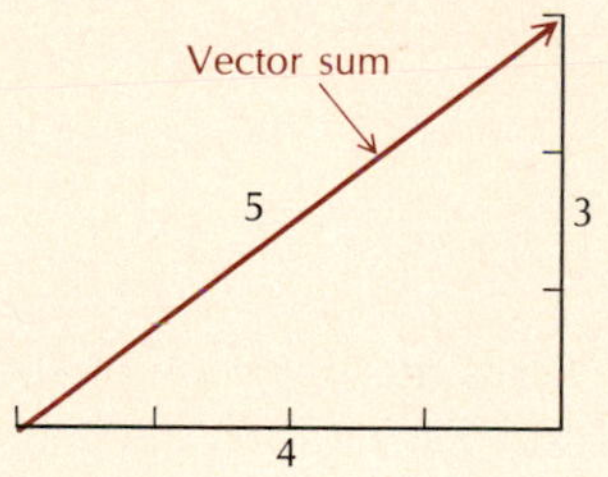

In general, if we have two vectors **a** and **b**, we can add them as in the above example.† That is, we place the tail of one arrow at the head of the other and then find the vector that forms the third side of a triangle. Or, we can place the tails of the arrows together, complete a parallelogram, and find the diagonal of the parallelogram.

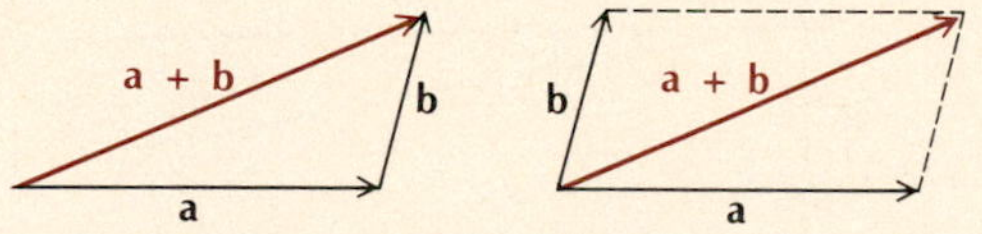

EXAMPLE 1 Forces of 15 kg and 25 kg act on an object at right angles to each other. Find their sum, or resultant, giving the angle that it makes with the larger force.

---

* Strictly speaking, *Force = Mass times Acceleration* and cannot be expressed as a mass. The standard metric unit of force is a *Newton*. The standard British unit of force is the *pound*. To simplify our discussion, we will allow either weight or mass to be a unit of force.

†It is general practice to denote vectors using boldface type; sometimes in handwriting, we write a bar over the letters.

1. Two forces of 5.0 kg and 14.0 kg act at right angles to each other. Find the resultant, specifying the angle that it makes with the larger force.

Solution

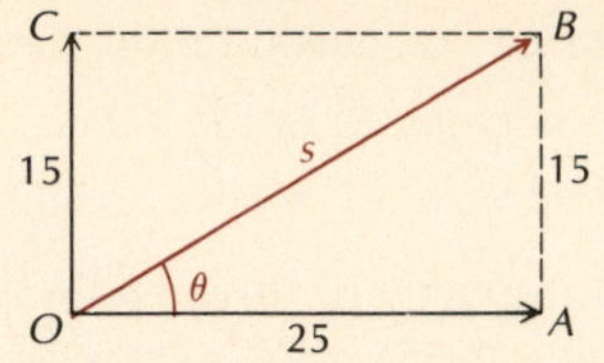

We make a drawing, this time a rectangle, using s for the length of vector **OB**. Since OAB is a right triangle, we have

$$\tan \theta = \frac{15}{25} = 0.6.$$

Thus θ, the angle that the resultant makes with the larger force, is 31° to the nearest degree. Now $s/15 = \csc \theta$, or $s = 15 \csc \theta = 15 \times 1.942 = 29.1$. Thus the resultant **OB** has a magnitude of 29.1 kg and makes an angle of 31° with the larger force. ■

DO EXERCISE 1.

Vector subtraction, $\mathbf{a} - \mathbf{b}$, is defined as $\mathbf{a} + (-\mathbf{b})$, where $-\mathbf{b}$ is the vector found from $\mathbf{b}$ by heading in the opposite direction. We can show $\mathbf{a} - \mathbf{b}$ as follows, though we will not give much attention to vector subtraction.

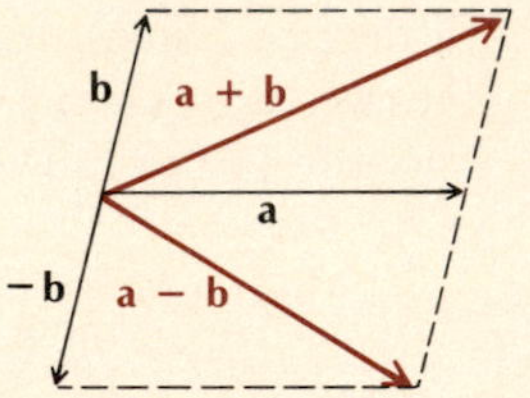

2 Applications

EXAMPLE 2 An airplane heads in a direction of 100° at a 180-km/h airspeed while a wind is blowing 40 km/h from 220°. Find the speed of the airplane over the ground and the direction of its track over the ground.

Solution

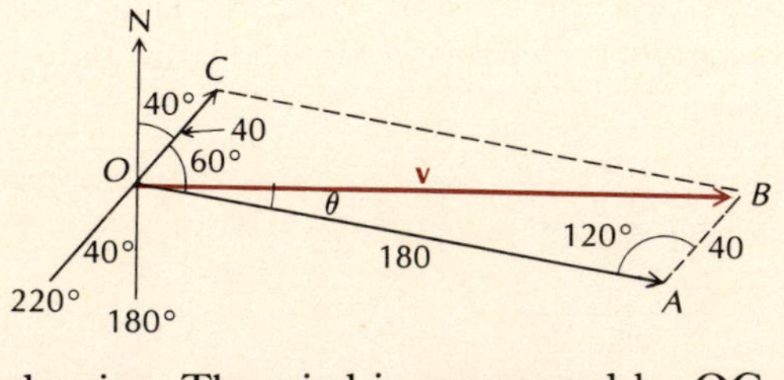

We first make a drawing. The wind is represented by **OC** and the velocity vector of the airplane by **OA**. The resultant velocity is **v**, the sum of the two vectors. We denote the length of **v** by $|\mathbf{v}|$. The measure of $\angle COA$ is 60°, so $\angle CBA = 60°$. Now since the sum of all the angles of the parallelogram is 360° and $\angle OCB$ and $\angle OAB$ have the same measure, each must be 120°. By the law of cosines in $\triangle OAB$, we have

$$|\mathbf{v}|^2 = 40^2 + 180^2 - 2 \cdot 40 \cdot 180 \cos 120°$$
$$|\mathbf{v}|^2 = 41{,}200.$$

Thus, $|\mathbf{v}|$ is 203 km/h. By the law of sines in the same triangle,

$$\frac{40}{\sin\theta} = \frac{203}{\sin 120°},$$

or

$$\sin\theta = \frac{40 \sin 120°}{203} = 0.1706.$$

Thus, $\theta = 10°$, to the nearest degree. Therefore, the ground speed of the airplane is 203 km/h, and its track is in the direction of $100° - 10°$, or $90°$.

DO EXERCISE 2.

3 Components of Vectors

Given a vector, it is often convenient to reverse the addition procedure, that is, to find two vectors whose sum is the given vector. Usually the two vectors we seek will be perpendicular. The two vectors we find are called **components** of the given vector.

EXAMPLE 3 A certain vector **a** has a magnitude of 130 and is inclined 40° with the horizontal. Resolve the vector into horizontal and vertical components.

Solution

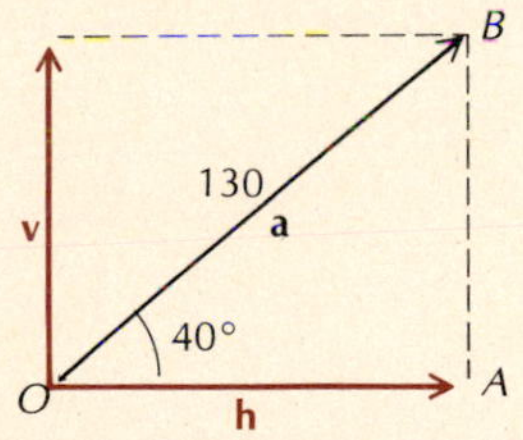

We first make a drawing showing horizontal and vertical vectors whose sum is the given vector **a**. From $\triangle OAB$, we see that

$$|\mathbf{h}| = 130 \cos 40° = 99.6$$

and

$$|\mathbf{v}| = 130 \sin 40° = 83.6.$$

These are the components we seek.

EXAMPLE 4 An airplane is flying at 200 km/h in a direction of 305°. Find the westerly and northerly components of its velocity.

Solution

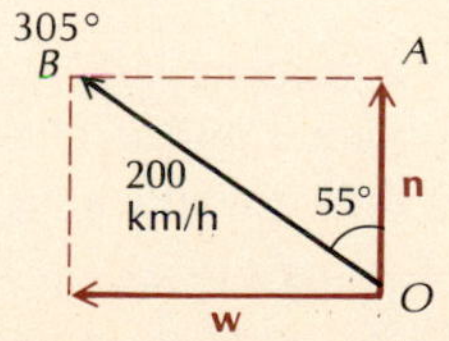

We first make a drawing showing westerly and northerly vectors whose sum is the given velocity. From $\triangle OAB$, we see that

$$|\mathbf{n}| = 200 \cos 55° = 115 \text{ km/h}$$

2. An airplane heads in a direction of 90° at a 140-km/h airspeed. The wind is 50 km/h from 210°. Find the direction and the speed of the airplane over the ground.

3. A vector of magnitude 100 points southeast. Resolve the vector into easterly and southerly components.

4. A wind has an easterly component (*from* the east) of 10 km/h and a southerly component (*from* the south) of 16 km/h. Find the magnitude and the direction of the wind.

and

$$|\mathbf{w}| = 200 \sin 55° = 164 \text{ km/h}.$$

DO EXERCISE 3.

4 Adding Vectors Using Components

Given a vector, we can resolve it into components. Given components of a vector, we can find the vector. We simply find the vector sum of the components.

EXAMPLE 5 A vector **v** has a westerly component of 12 and a southerly component of 16. Find the vector.

Solution

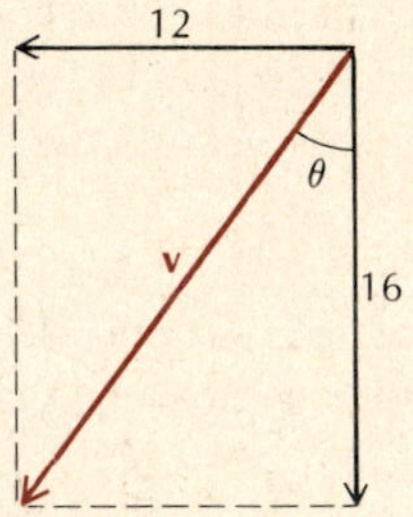

From the drawing, we see that $\tan \theta = 12/16 = 0.75$. Thus, $\theta = 36°52'$. Then $|\mathbf{v}| = 12 \csc 36°52' = 20$. Thus the vector has a magnitude of 20 and a direction of S 36°52′ W.

DO EXERCISE 4.

When we have two vectors resolved into components, we can add the two vectors by adding the components.

EXAMPLE 6 A vector **v** has a westerly component of 3 and a northerly component of 5. A second vector **w** has an easterly component of 8 and a northerly component of 4. Find the components of the sum. Find the sum.

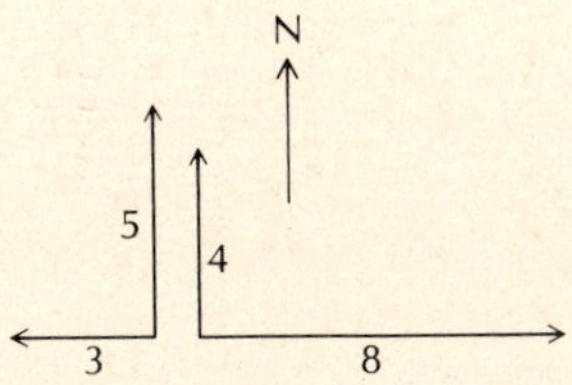

Solution Adding the east–west components, we obtain 5 east. Adding the north–south components, we obtain 9 north. The components of **v** + **w** are 5 east and 9 north. We now add these components:

$$\tan \theta = \tfrac{9}{5} = 1.8$$
$$\theta = 61°;$$

$$|\mathbf{v} + \mathbf{w}| = 9 \csc 61° = 10.3.$$

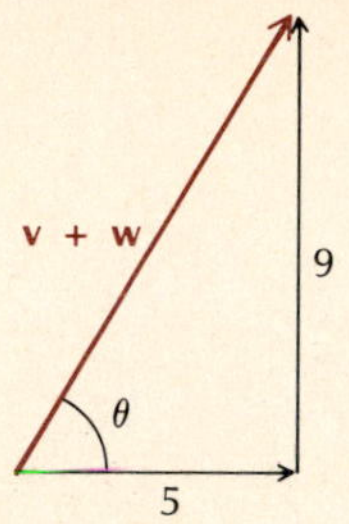

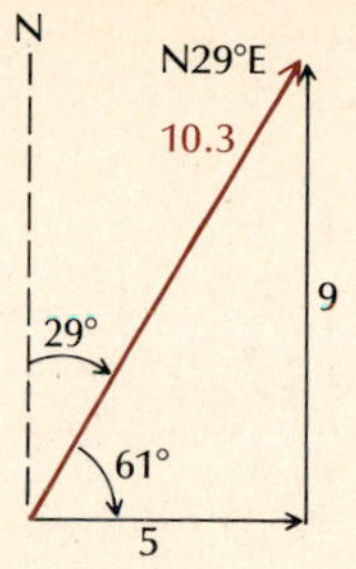

The sum $|\mathbf{v} + \mathbf{w}|$ has a magnitude of 10.3 and a direction of N29°E. ■

DO EXERCISE 5.

Analytic Representation of Vectors

If we place a coordinate system so that the origin is at the tail of an arrow representing a vector, we say that the vector is in **standard position**. Then if we know the coordinates of the other end of the vector, we know the vector. The coordinates will be the x-component and the y-component of the vector. Thus we can consider an ordered pair (a, b) to be a vector. To emphasize that we are thinking of a vector, we usually write $\langle a, b\rangle$. When vectors are given in this form, it is easy to add them. We simply add the respective components.

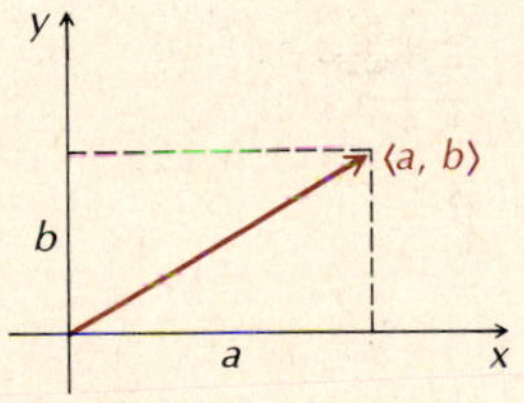

EXAMPLE 7 Find $\mathbf{u} + \mathbf{v}$, where $\mathbf{u} = \langle 3, -7\rangle$ and $\mathbf{v} = \langle 4, 2\rangle$.

Solution The sum is found by adding the x-components and then adding the y-components:

$$\mathbf{u} + \mathbf{v} = \langle 3, -7\rangle + \langle 4, 2\rangle = \langle 7, -5\rangle.$$ ■

DO EXERCISES 6 AND 7.

5 Polar and Rectangular Notation

Vectors can also be specified by giving their length and direction. When this is done, we say that we have **polar notation** for the vector. An example of polar notation is (15, 260°). The angle is measured from the positive half of the x-axis counterclockwise. In Section 8.6, we consider a coordinate system (called a **polar coordinate system**) in which polar notation is very natural.

EXAMPLE 8 Find polar notation for the vector $\mathbf{v}$, where $\mathbf{v} = \langle 7, -2\rangle$.

Solution We see the reference angle ϕ in the figure:

$$\tan \phi = \frac{-2}{7} = -0.2857$$

$$\phi = -16°;$$

$$|\mathbf{v}| = \sqrt{(-2)^2 + 7^2} = 7.28.$$

5. One vector has components of 7 up and 9 to the left. A second vector has components of 3 down and 12 to the left. Find the sum, expressing it:

   a) in terms of components;

   b) giving magnitude and direction.

Add the vectors, giving answers as ordered pairs.

6. $\langle 7, 2\rangle$ and $\langle 5, -10\rangle$

7. $\langle -2, 7\rangle$ and $\langle 14, -3\rangle$

8. Find polar notation for the vector $\langle -8, 3\rangle$.

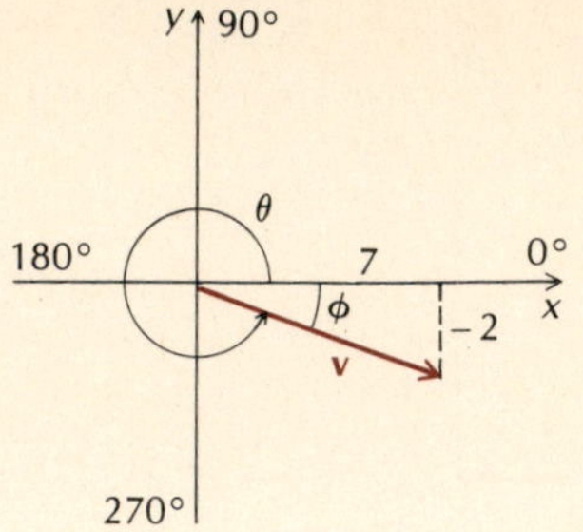

Thus polar notation for **v** is

$$(7.28, -16°), \quad \text{or} \quad (7.28, 344°).$$

DO EXERCISE 8.

Given polar notation for a vector, we can also find the ordered-pair notation (**rectangular notation**).

EXAMPLE 9 Find rectangular notation for the vector **w**, where **w** = (14, 155°).

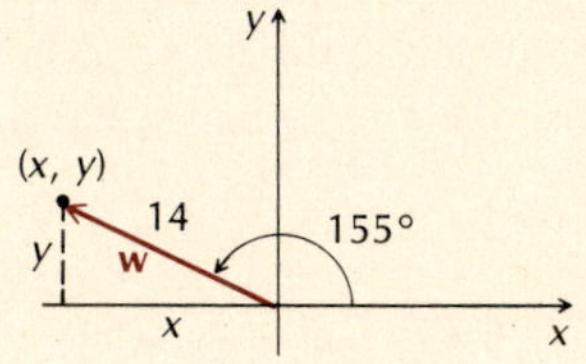

9. Find rectangular notation for the vector (15, 341°).

Solution From the figure, we have

$$x = 14 \cos 155° = -12.7,$$
$$y = 14 \sin 155° = 5.92.$$

Thus the rectangular notation for **w** is $\langle -12.7, 5.92\rangle$.

DO EXERCISE 9.

6 Properties of Vectors and Vector Algebra

With vectors represented in rectangular notation $\langle a, b\rangle$, it is easy to determine many of their properties. We consider the set of all ordered pairs of real numbers $\langle a, b\rangle$. This is the set of all vectors on a plane. Consider the sum of any two vectors:

$$\mathbf{u} = \langle a, b\rangle, \qquad \mathbf{v} = \langle c, d\rangle,$$
$$\mathbf{u} + \mathbf{v} = \langle a + c, b + d\rangle, \qquad \mathbf{v} + \mathbf{u} = \langle c + a, d + b\rangle.$$

We have just shown that addition of vectors is commutative. Other properties are also not difficult to show. The system of vectors is not a field, but it does have some of the familiar field properties.

Properties of Vectors

Commutative law	**Addition of vectors is commutative.**
Associative law	**Addition of vectors is associative.**
Identity	**There is an additive identity, the vector $\langle 0, 0\rangle$.**
Inverses	**Every vector v, given by $\mathbf{v} = \langle a, b\rangle$, has an additive inverse $-\mathbf{v}$, given by $-\mathbf{v} = \langle -a, -b\rangle$.**

Subtraction	We define $\mathbf{u} - \mathbf{v}$ to be the vector that when added to $\mathbf{v}$ gives $\mathbf{u}$. It follows that if $\mathbf{u} = \langle a, b\rangle$ and $\mathbf{v} = \langle c, d\rangle$, then $\mathbf{u} - \mathbf{v} = \langle a - c, b - d\rangle$. Thus subtraction is always possible. Also, $\mathbf{u} - \mathbf{v} = \mathbf{u} + (-\mathbf{v})$.
Length, or absolute value	If $\mathbf{u} = \langle a, b\rangle$, then it follows that $\lvert\mathbf{u}\rvert = \sqrt{a^2 + b^2}$.

A vector can be multiplied by a real number. A real number in this context is called a **scalar**. For example, $2\mathbf{v}$ is a vector in the same direction as $\mathbf{v}$ but twice as long. Analytically, we define scalar multiplication as follows.

DEFINITION Scalar Multiplication

If r is any real number and $\mathbf{v} = \langle a, b\rangle$, then $r\mathbf{v} = \langle ar, br\rangle$.

EXAMPLE 10 Do the following calculations, where $\mathbf{u} = \langle 4, 3\rangle$ and $\mathbf{v} = \langle -5, 8\rangle$: (a) $\mathbf{u} + \mathbf{v}$; (b) $\mathbf{u} - \mathbf{v}$; (c) $3\mathbf{u} - 4\mathbf{v}$; (d) $\lvert 3\mathbf{u} - 4\mathbf{v}\rvert$.

Solution

a) $\mathbf{u} + \mathbf{v} = \langle 4, 3\rangle + \langle -5, 8\rangle = \langle -1, 11\rangle$

b) $\mathbf{u} - \mathbf{v} = \mathbf{u} + (-\mathbf{v}) = \langle 4, 3\rangle + \langle 5, -8\rangle = \langle 9, -5\rangle$

c) $3\mathbf{u} - 4\mathbf{v} = 3\langle 4, 3\rangle - 4\langle -5, 8\rangle = \langle 12, 9\rangle - \langle -20, 32\rangle$
$= \langle 32, -23\rangle$

d) $\lvert 3\mathbf{u} - 4\mathbf{v}\rvert = \sqrt{32^2 + (-23)^2} = \sqrt{1553} \approx 39.41$

DO EXERCISE 10.

10. For $\mathbf{u} = \langle -3, 5\rangle$ and $\mathbf{v} = \langle 4, 2\rangle$, find each of the following.

a) $\mathbf{u} + \mathbf{v}$

b) $\mathbf{v} - \mathbf{u}$

c) $5\mathbf{u} - 2\mathbf{v}$

d) $\lvert 3\mathbf{u} + 2\mathbf{v}\rvert$

EXERCISE SET 8.4

1 Magnitudes of vectors **a** and **b** and the angle between the vectors θ are given. Find the resultant, giving the direction by specifying to the nearest degree the angle that it makes with the vector **a**.

1. $\lvert\mathbf{a}\rvert = 45$, $\lvert\mathbf{b}\rvert = 35$, $\theta = 90°$
2. $\lvert\mathbf{a}\rvert = 54$, $\lvert\mathbf{b}\rvert = 43$, $\theta = 90°$
3. $\lvert\mathbf{a}\rvert = 10$, $\lvert\mathbf{b}\rvert = 12$, $\theta = 67°$
4. $\lvert\mathbf{a}\rvert = 25$, $\lvert\mathbf{b}\rvert = 30$, $\theta = 75°$
5. $\lvert\mathbf{a}\rvert = 20$, $\lvert\mathbf{b}\rvert = 20$, $\theta = 117°$
6. $\lvert\mathbf{a}\rvert = 30$, $\lvert\mathbf{b}\rvert = 30$, $\theta = 123°$
7. $\lvert\mathbf{a}\rvert = 23$, $\lvert\mathbf{b}\rvert = 47$, $\theta = 27°$
8. $\lvert\mathbf{a}\rvert = 32$, $\lvert\mathbf{b}\rvert = 74$, $\theta = 72°$

2

9. Two forces of 5 kg and 12 kg act on an object at right angles. Find the magnitude of the resultant and the angle that it makes with the smaller force.
10. Two forces of 30 kg and 40 kg act on an object at right angles. Find the magnitude of the resultant and the angle that it makes with the smaller force.
11. Forces of 420 kg and 300 kg act on an object. The angle between the forces is 50°. Find the resultant, giving the angle that it makes with the larger force.
12. Forces of 410 kg and 600 kg act on an object. The angle between the forces is 47°. Find the resultant, giving the angle that it makes with the smaller force.
13. A balloon is rising 12 ft/sec while a wind is blowing 18 ft/sec. Find the speed of the balloon and the angle that it makes with the horizontal.
14. A balloon is rising 10 ft/sec while a wind is blowing 5 ft/sec. Find the speed of the balloon and the angle that it makes with the horizontal.
15. A boat heads 35°, propelled by a force of 750 lb. A wind from 320° exerts a force of 150 lb on the boat. How large is the resultant force, and in what direction is the boat moving?
16. A boat heads 220°, propelled by a force of 650 lb. A wind from 080° exerts a force of 100 lb on the boat. How large is the resultant force, and in what direction is the boat moving?

17. A ship sails N80°E for 120 nautical mi, then S20°W for 200 nautical mi. How far is it then from the starting point, and in what direction?

18. An airplane flies 032° for 210 km, then 280° for 170 km. How far is it, then, from the starting point, and in what direction?

19. A motorboat has a speed of 15 km/h. It crosses a river whose current has a speed of 3 km/h. In order to cross the river at right angles, in what direction should the boat be pointed?

20. The airplane has an airspeed of 150 km/h. It is to make a flight in a direction of 080° while there is a 25-km/h wind from 350°. What should the airplane's heading be?

1, **3**

21. A vector **u** with magnitude 150 is inclined to the right and upward 52° from the horizontal. Find the horizontal and vertical components of **u**.

22. A vector **u** with magnitude 170 is inclined to the right and downward 63° from the horizontal. Find the horizontal and vertical components of **u**.

23. An airplane is flying 220° at 250 km/h. Find the magnitude of the southerly and westerly components of its velocity **v**.

24. A wind is blowing from 310° at 25 mph. Find the magnitude of the southerly and easterly components of the wind velocity **v**.

2, **4**

25. A force **f** has a westerly component of 25 kg and a southerly component of 35 kg. Find the magnitude and the direction of **f**.

26. A force **f** has a component of 65 kg upward and a component of 90 kg to the left. Find the magnitude and the direction of **f**.

27. Vector **u** has a westerly component of 15 and a northerly component of 22. Vector **v** has a northerly component of 6 and an easterly component of 8. Find:
a) the components of **u** + **v**;
b) the magnitude and the direction of **u** + **v**.

28. Vector **u** has a component of 18 upward and a component of 12 to the left. Vector **v** has a component of 5 downward and a component of 35 to the right. Find:
a) the components of **u** + **v**;
b) the magnitude and the direction of **u** + **v**.

4 In Exercises 29-42, rectangular notation for vectors is given.

Add, writing rectangular notation for the answer.

29. $\langle 3, 7\rangle + \langle 2, 9\rangle$

30. $\langle 5, -7\rangle + \langle -3, 2\rangle$

31. $\langle 17, 7.6\rangle + \langle -12.2, 6.1\rangle$

32. $\langle -15.2, 37.1\rangle + \langle 7.9, -17.8\rangle$

33. $\langle -650, -750\rangle + \langle -12, 324\rangle$

34. $\langle -354, -973\rangle + \langle -75, 256\rangle$

3, **5** Find polar notation.

35. $\langle 3, 4\rangle$
36. $\langle 4, 3\rangle$
37. $\langle 10, -15\rangle$
38. $\langle 17, -10\rangle$
39. $\langle -3, -4\rangle$
40. $\langle -4, -3\rangle$
41. $\langle -10, 15\rangle$
42. $\langle -17, 10\rangle$

Find rectangular notation.

43. $(4, 30°)$
44. $(8, 60°)$
45. $(10, 235°)$
46. $(15, 210°)$
47. $(20, 330°)$
48. $(20, 200°)$
49. $(100, -45°)$
50. $(150, -60°)$

5, **6** Do the calculations for the following vectors: $\mathbf{u} = \langle 3, 4\rangle$, $\mathbf{v} = \langle 5, 12\rangle$, and $\mathbf{w} = \langle -6, 8\rangle$.

51. $3\mathbf{u} + 2\mathbf{v}$
52. $3\mathbf{u} - 2\mathbf{v}$
53. $(\mathbf{u} + \mathbf{v}) - \mathbf{w}$
54. $\mathbf{u} - (\mathbf{v} + \mathbf{w})$
55. $|\mathbf{u}| + |\mathbf{v}|$
56. $|\mathbf{u}| - |\mathbf{v}|$
57. $|\mathbf{u} + \mathbf{v}|$
58. $|\mathbf{u} - \mathbf{v}|$
59. $2|\mathbf{u} + \mathbf{v}|$
60. $2|\mathbf{u}| + 2|\mathbf{v}|$

SYNTHESIS

61. If **PQ** is any vector, what is **PQ** + **QP**?

62. The **inner product** of vectors $\mathbf{u} \cdot \mathbf{v}$ is a scalar, defined as follows: $\mathbf{u} \cdot \mathbf{v} = |\mathbf{u}||\mathbf{v}| \cos \theta$, where θ is the angle between the vectors. Show that vectors **u** and **v** are perpendicular if and only if $\mathbf{u} \cdot \mathbf{v} = 0$.

CHALLENGE

63. Show that scalar multiplication is distributive over vector addition.

64. Let $\mathbf{u} = \langle 3, 4 \rangle$. Find a vector that has the same direction as $\mathbf{u}$ but length 1.

65. Prove that for any vectors $\mathbf{u}$ and $\mathbf{v}$, $|\mathbf{u} + \mathbf{v}| \leq |\mathbf{u}| + |\mathbf{v}|$.

8.5 Forces in Equilibrium

OBJECTIVE

You should be able to:

1 **Solve applied problems in which several forces acting at a point are in equilibrium.**

1 When several forces act through the same point on an object, their vector sum must be **0** in order for a balance to occur. When a balance occurs, then the object is either stationary or moving in a straight line without acceleration. The fact that the vector sum must be **0** for a balance, and vice versa, allows us to solve many applied problems involving forces. In such problems, we usually resolve vectors into components. For a balance, the sum of the horizontal components must be **0**, and the sum of the vertical components must be 0.*

EXAMPLE 1 A 200-lb sign is hanging from the end of a horizontal hinged boom, supported by a cable from the end of the boom inclined 35° with the horizontal. Find the force (tension) in the cable and the force (compression) in the boom.

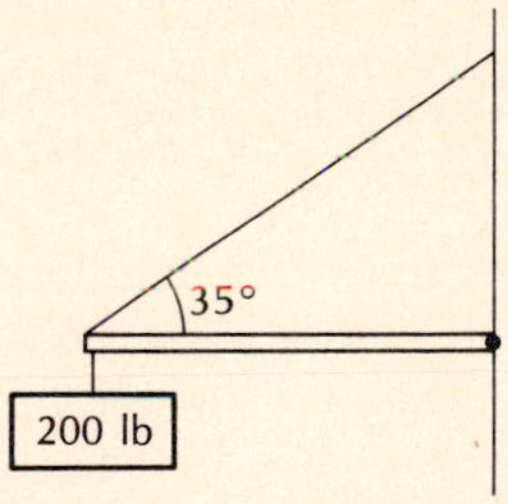

Solution At the end of the boom, we have three forces acting at a point: the weight of the sign acting down, the cable pulling up and to the right, and the boom pushing to the left. We draw a force diagram showing these.

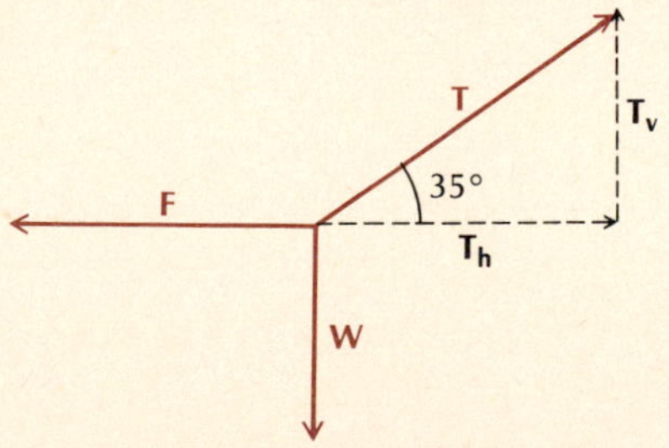

We have called the tension in the cable **T**, the compression in the boom **F**, and the weight **W**. We resolve the tension in the cable into horizontal and vertical components. We have a balance, so $\mathbf{F} + \mathbf{T} + \mathbf{W} = \mathbf{0}$. In fact, the sum of the horizontal components must be **0**, and the sum of the vertical components must also be **0**. Thus we know that $|\mathbf{T}_v| = 200$ lb and that $|\mathbf{F}| = |\mathbf{T}_h|$. Then we have

$$|\mathbf{T}_v| = |\mathbf{T}| \sin 35^\circ$$

$$|\mathbf{T}| = \frac{|\mathbf{T}_v|}{\sin 35^\circ} = \frac{200}{0.5736} = 349 \text{ lb.}$$

*For a balance, the rotational tendency of the forces must also be zero. This will always be the case if the forces are concurrent (act through the same point). In this section, we consider only applications in which forces are concurrent.

Then

$$|\mathbf{F}| = |\mathbf{T}_h| = 349 \cos 35° = 286 \text{ lb}.$$

The answer is that the tension in the cable is 349 lb and the compression in the boom is 286 lb. ■

DO EXERCISE 1.

EXAMPLE 2 A block weighing 100 kg rests on a 25° incline. Find the components of its weight perpendicular and parallel to the incline.

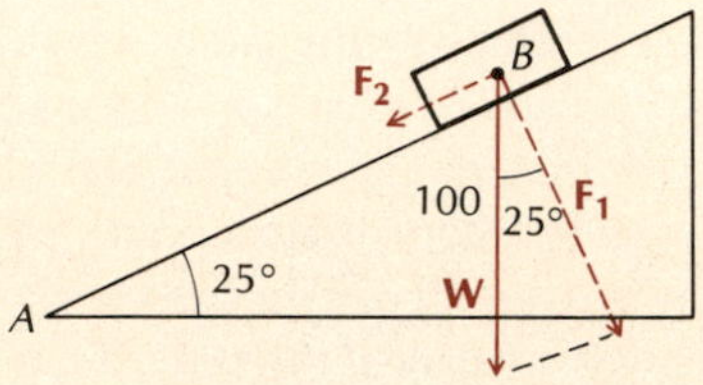

Solution The weight **W** is 100 kg acting downward, as shown. We draw $\mathbf{F}_1$ and $\mathbf{F}_2$, the components. The angle at B has the same measure as the angle at A because the sides of the two triangles are respectively perpendicular, making the triangles similar. Thus we have

$$|\mathbf{F}_1| = 100 \cos 25° = 90.6 \text{ kg} \quad \text{and} \quad |\mathbf{F}_2| = 100 \sin 25° = 42.3 \text{ kg}. \blacksquare$$

DO EXERCISE 2.

In a situation like that of Example 2, the force necessary to hold the block on the incline is a force the size of $\mathbf{F}_2$ but opposite in direction. The force with which the block pushes on the incline is $\mathbf{F}_1$.

EXAMPLE 3 A 500-kg block is suspended by two ropes, as shown. Find the tension in each rope.

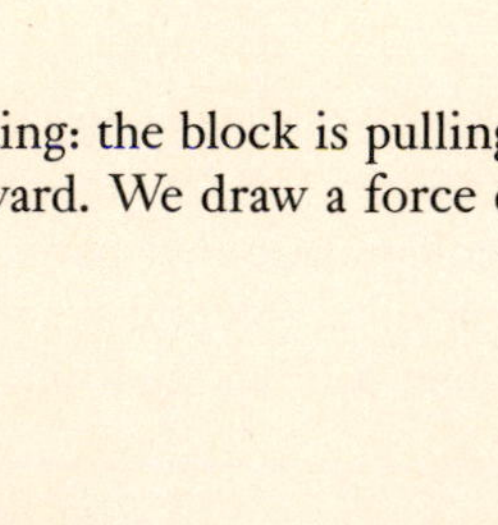

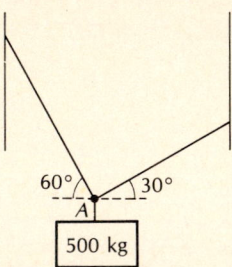

Solution At point A, there are three forces acting: the block is pulling down, and the two ropes are pulling upward and outward. We draw a force diagram showing these.

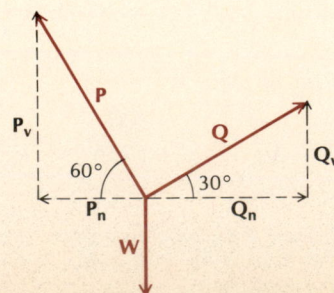

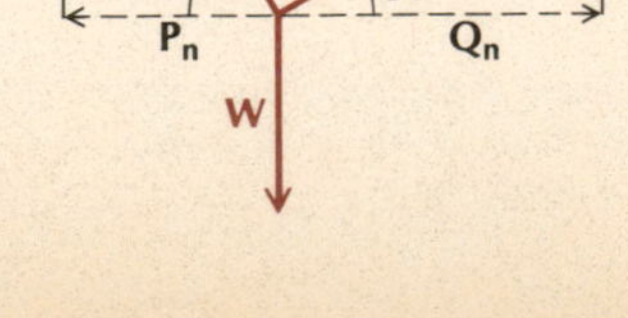

1. A weight of 300 lb is supported by two ropes at a point A. One rope is horizontal and the other makes an angle of 30° with the horizontal. Find the forces (of tension) in the two ropes.

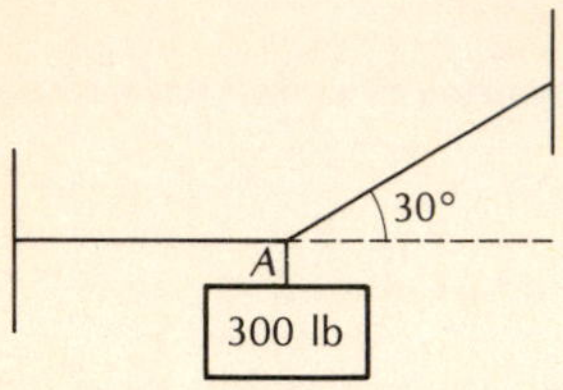

2. A block weighing 100 kg rests on a 30° incline. Find the components of its weight parallel and perpendicular to the incline.

We have called the forces exerted by the ropes **P** and **Q**. Their horizontal components have magnitudes $|\mathbf{P}|\cos 60°$ and $|\mathbf{Q}|\cos 30°$. Since there is a balance, these must be the same:

$$|\mathbf{P}|\cos 60° = |\mathbf{Q}|\cos 30°.$$

The vertical components in the ropes have magnitudes $|\mathbf{P}|\sin 60°$ and $|\mathbf{Q}|\sin 30°$. Since there is a balance, these must total 500 lb, the weight of the block:

$$|\mathbf{P}|\sin 60° + |\mathbf{Q}|\sin 30° = 500.$$

We now have two equations with unknowns $|\mathbf{P}|$ and $|\mathbf{Q}|$:

$$|\mathbf{P}|\cos 60° - |\mathbf{Q}|\cos 30° = 0, \qquad (1)$$

$$|\mathbf{P}|\sin 60° + |\mathbf{Q}|\sin 30° = 500. \qquad (2)$$

To solve, we begin by solving Eq. (1) for **P**:

$$|\mathbf{P}| = |\mathbf{Q}|\frac{\cos 30°}{\cos 60°}. \qquad (3)$$

Next, we substitute in Eq. (2):

$$|\mathbf{Q}|\frac{\cos 30° \sin 60°}{\cos 60°} + |\mathbf{Q}|\sin 30° = 500$$

$$|\mathbf{Q}|\frac{\cos 30° \sin 60°}{\cos 60°} + |\mathbf{Q}|\frac{\sin 30° \cos 60°}{\cos 60°} = 500$$

$$|\mathbf{Q}|(\cos 30° \sin 60° + \sin 30° \cos 60°) = 500 \cos 60°$$

$$|\mathbf{Q}|(\sin 90°) = 500 \cos 60° \qquad \text{Using an identity}$$

$$|\mathbf{Q}| = 500 \cdot \tfrac{1}{2} = 250 \text{ kg}.$$

Then substituting in Eq. (3), we obtain

$$|\mathbf{P}| = 250\,\frac{\cos 30°}{\cos 60°}$$

$$|\mathbf{P}| = 250\,\frac{\sqrt{3}/2}{1/2} = 250\sqrt{3}$$

$$|\mathbf{P}| = 433 \text{ kg}.$$

DO EXERCISE 3.

3. Two ropes support a 1000-kg weight, as shown. Find the tension in each rope.

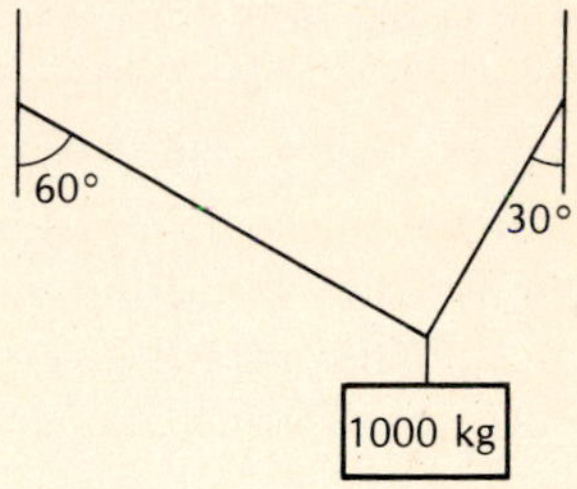

EXERCISE SET 8.5

1

1. A 150-lb sign is hanging from the end of a hinged boom, supported by a cable inclined 42° with the horizontal. Find the tension in the cable and the compression in the boom.

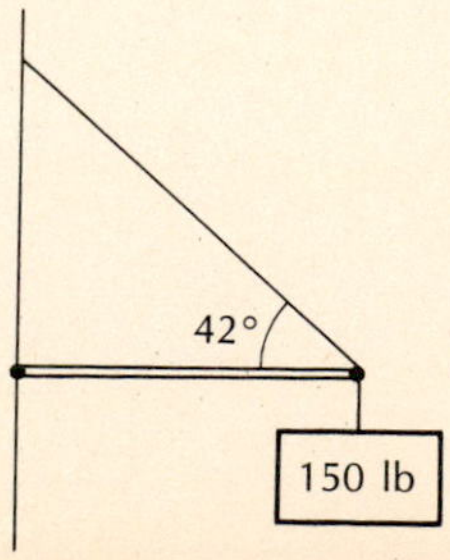

2. A 300-lb sign is hanging from the end of a hinged boom, supported by a cable inclined 51° from the horizontal. Find the tension in the cable and the compression in the boom (see the figure for Exercise 1).

3. A weight of 200 kg is supported by a frame made of two rods and hinged at A, B, and C (see the figure at right). Find the forces exerted by the two rods.

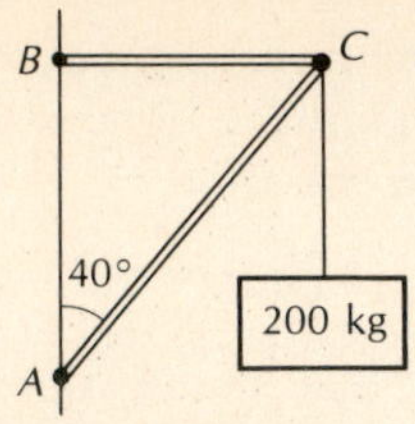

4. A weight of 300 kg is supported by the hinged frame of Exercise 3, but where the angle shown is 50°. Find the forces exerted by the two rods.

5. The force due to air movement on an airplane wing in flight is 2800 lb, acting at an angle of 28° with the vertical. What is the lift (vertical component) and what is the drag (horizontal component)?

6. The force due to air movement on an airplane wing in flight is 3500 lb, acting at an angle of 32° with the vertical. What is the lift (vertical component) and what is the drag (horizontal component)?

7. A 100-kg block of ice rests on a 37° incline. What force parallel to the incline is necessary to keep it from sliding down?

8. What force is necessary to pull a 6500-kg truck up a 7° incline?

9. The moon's gravity is one sixth that of the earth. To simulate the moon's gravity, an incline is used, as shown in the figure at right. The astronaut is suspended in a sling and is free to move horizontally. The component of her weight perpendicular to the incline should be one sixth of her weight. What must the angle θ be?

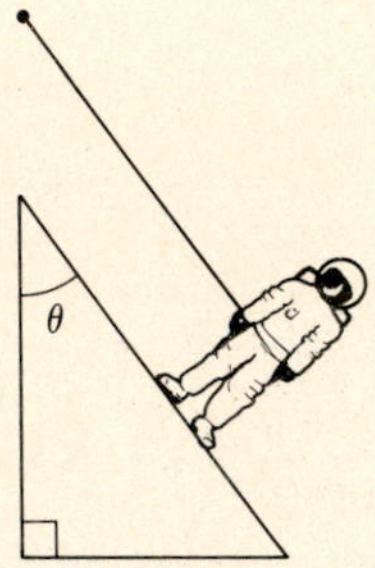

10. To simulate gravity on a planet having three eighths of the earth's gravity, what should the angle θ be? (See the figure and explanation for Exercise 9.)

11. A 400-kg block is suspended by two ropes, as shown. Find the tension in each rope.

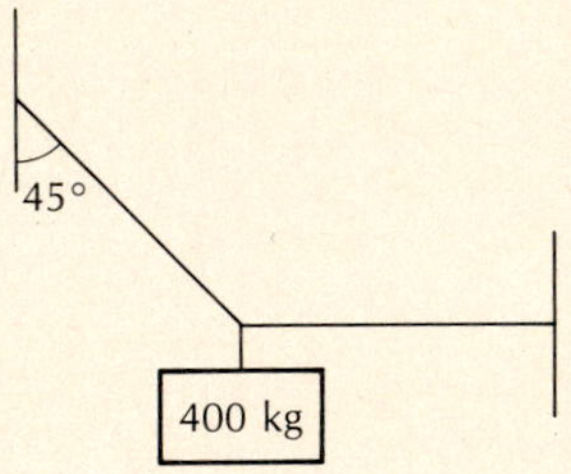

12. A 1000-lb block is suspended by two ropes, as shown. Find the tension in each rope.

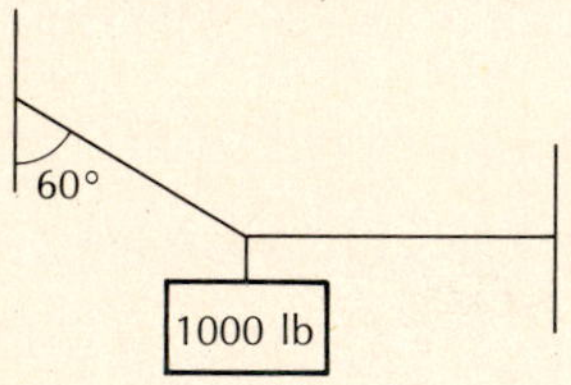

13. A 2000-kg block is suspended by two ropes, as shown. Find the tension in each rope.

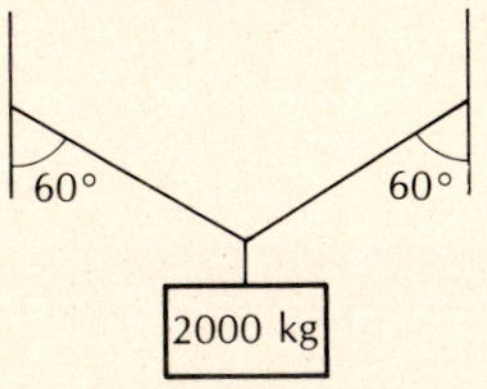

14. A 1500-lb block is suspended by two ropes, as shown. Find the tension in each rope.

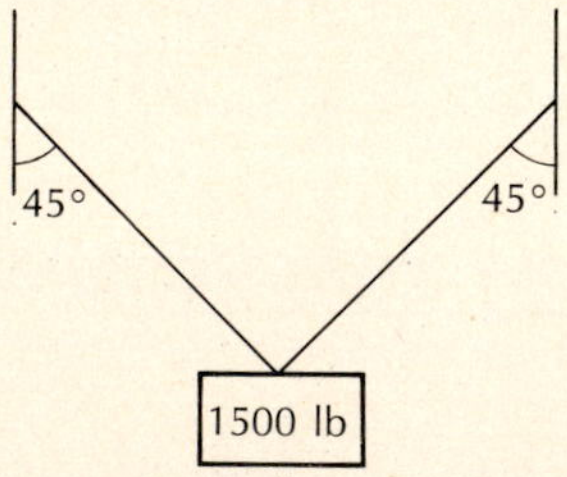

15. A 2500-kg block is suspended by two ropes, as shown. Find the tension in each rope.

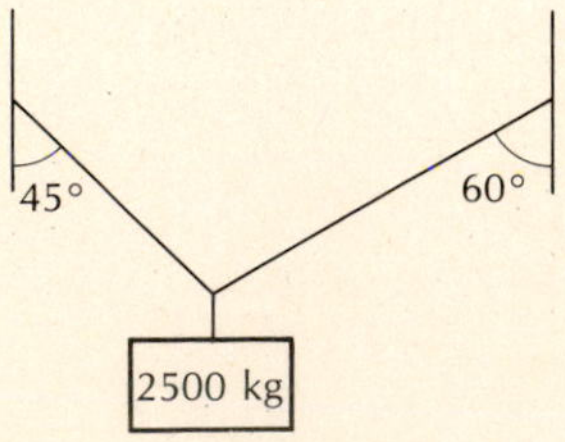

16. A 2000-kg block is suspended by two ropes, as shown. Find the tension in each rope.

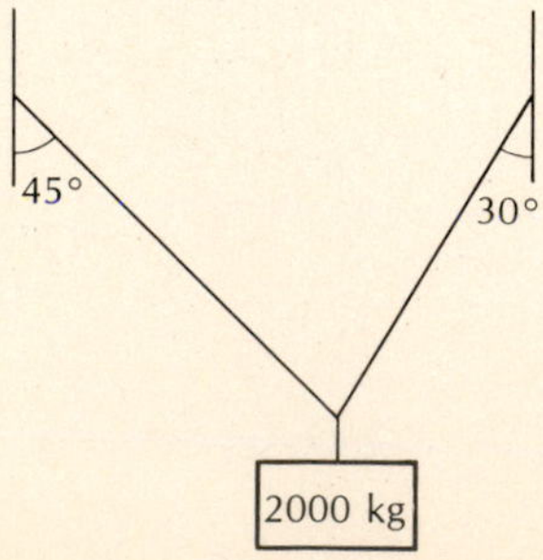

8.6 Polar Coordinates

OBJECTIVES

You should be able to:

1 Plot points, given their polar coordinates, and determine polar coordinates of points on a graph.

2 Convert from rectangular to polar equations and from polar to rectangular equations.

3 Graph polar equations.

1 In graphing, we locate a point with an ordered pair of numbers (a, b). We can consider any such ordered pair to be a vector. From our work with vectors, we know that we could also locate a point with a vector, given a length and a direction. When we use rectangular notation to locate points, we describe the coordinate system as **rectangular.*** When we use polar notation, we describe the coordinate system as **polar**. As this diagram shows, any point has **rectangular coordinates** (x, y) and **polar coordinates** (r, θ). On a polar graph, the origin is called the **pole** and the positive half of the x-axis is called the **polar axis.**

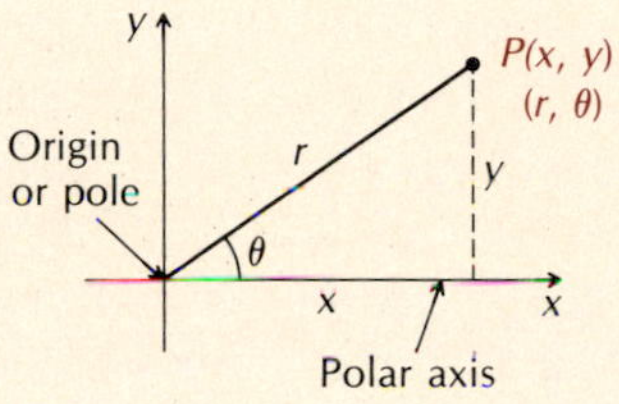

To plot points on a polar graph, we usually locate θ first, then move a distance r from the pole in that direction. If r is negative, we move in the opposite direction.

Polar graph paper, shown below, facilitates plotting. Point B illustrates that θ may be in radians. Point E illustrates that the polar coordinates of a point are not unique.

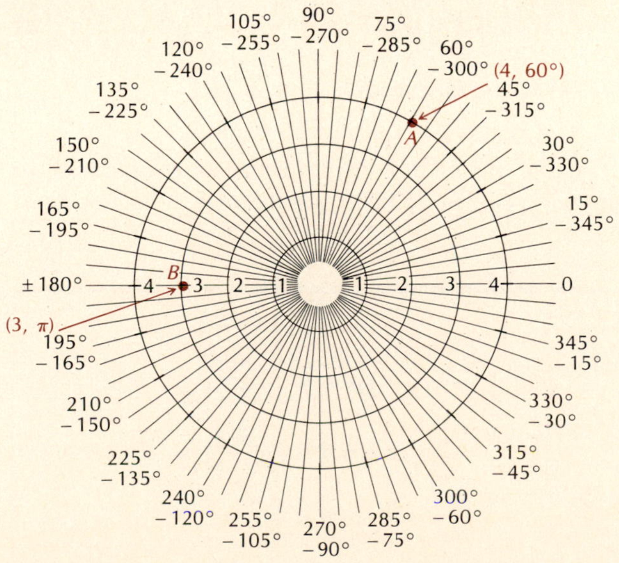

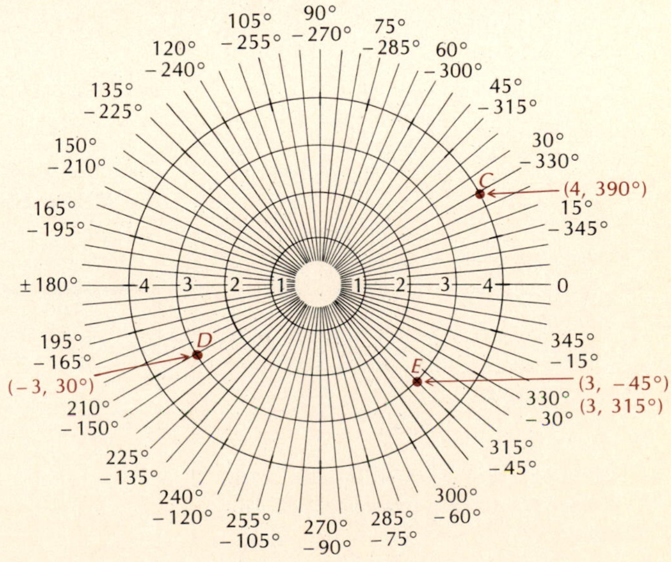

DO EXERCISES 1–4 ON THE FOLLOWING PAGE.

---

* Also called *Cartesian*, after the French mathematician René Descartes.

1. Plot each of the following points.
   a) $(3, 60^\circ)$ b) $(0, 10^\circ)$
   c) $(-5, 120^\circ)$ d) $(5, -60^\circ)$
   e) $(2, 3\pi/2)$

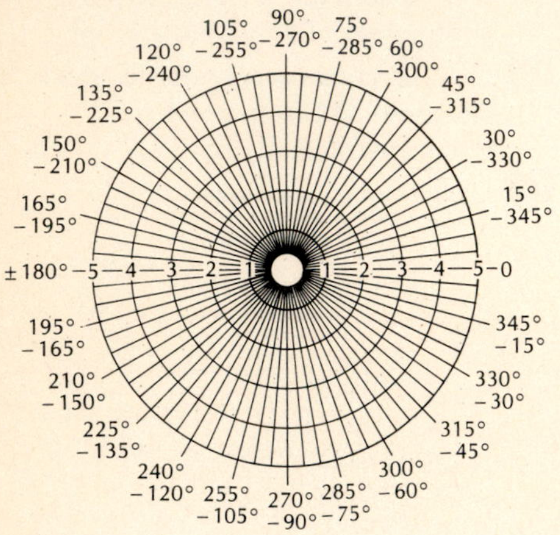

2. Find polar coordinates of each of these points. Give two answers for each.

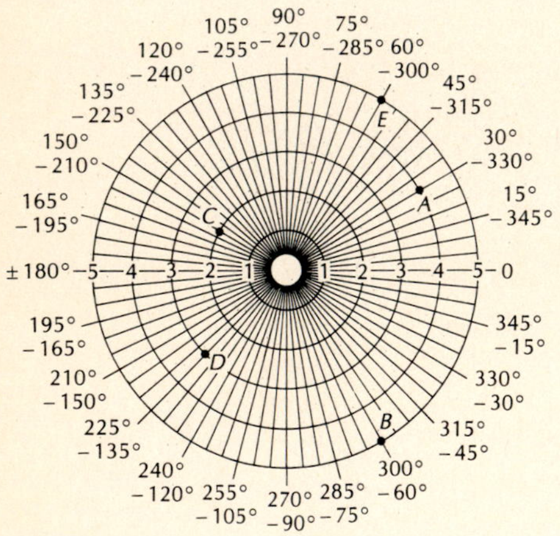

3. Find the polar coordinates of each of these points. (See Example 8 of Section 8.4.)
   a) $(3, 3)$ b) $(0, -4)$
   c) $(-3, 3\sqrt{3})$ d) $(2\sqrt{3}, -2)$

4. Find the rectangular coordinates of each of these points. (See Example 9 of Section 8.4.)
   a) $(5, 30^\circ)$ b) $(10, \pi/3)$
   c) $(-5, 45^\circ)$ d) $(-8, -5\pi/6)$

2 Polar and Rectangular Equations

Some curves have simpler equations in polar coordinates than in rectangular coordinates. For others, the reverse is true. Applying the definitions of sine and cosine, we obtain the following relationships between the two kinds of coordinates. These allow us to convert an equation from rectangular to polar coordinates:

$$x = r\cos\theta, \qquad y = r\sin\theta.$$

EXAMPLE 1 Convert to a polar equation: $x^2 + y^2 = 25$.

Solution

$$x^2 + y^2 = 25$$
$$(r\cos\theta)^2 + (r\sin\theta)^2 = 25 \qquad \text{Substituting for } x \text{ and } y$$
$$r^2\cos^2\theta + r^2\sin^2\theta = 25$$
$$r^2(\cos^2\theta + \sin^2\theta) = 25$$
$$r^2 = 25$$
$$r = \pm 5$$

Example 1 illustrates that the polar equation of a circle centered at the origin is much simpler than the rectangular equation.

EXAMPLE 2 Convert to a polar equation: $2x - y = 5$.

Solution

$$2x - y = 5$$
$$2(r\cos\theta) - (r\sin\theta) = 5$$
$$r(2\cos\theta - \sin\theta) = 5$$

DO EXERCISES 5 AND 6 ON THE FOLLOWING PAGE.

The relationships that we need to convert from polar equations to rectangular equations can easily be determined from the triangle. They are as follows:

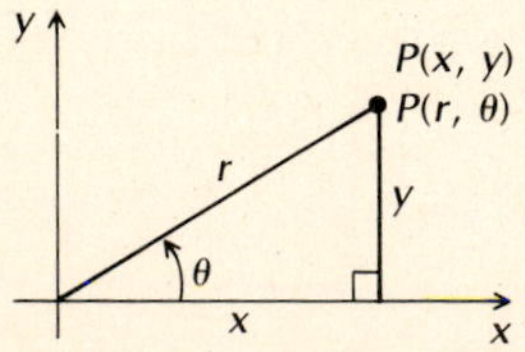

$$r = \pm\sqrt{x^2 + y^2},$$
$$\sin\theta = \frac{\pm y}{\sqrt{x^2 + y^2}},$$
$$\cos\theta = \frac{\pm x}{\sqrt{x^2 + y^2}},$$
$$\tan\theta = \frac{y}{x}$$

To convert to rectangular notation, we substitute as needed from either of the two lists, choosing the + or − signs as appropriate in each case.

EXAMPLE 3 Convert to a rectangular (Cartesian) equation: $r = 4$.

Solution

$$r = 4$$
$$+\sqrt{x^2 + y^2} = 4 \qquad \text{Substituting for } r$$
$$x^2 + y^2 = 16 \qquad \text{Squaring} \qquad \blacksquare$$

In squaring, as we did in Example 3, we must be careful not to introduce solutions of the equation that are not already present. We did not, because the graph is a circle of radius 4 centered at the origin, in both cases.

EXAMPLE 4 Convert to a rectangular equation: $r \cos \theta = 6$.

Solution From our first list, we know that $x = r \cos \theta$, so we have $x = 6$. $\blacksquare$

EXAMPLE 5 Convert to a rectangular equation: $r = 2 \cos \theta + 3 \sin \theta$.

Solution We first multiply on both sides by r:

$$r^2 = 2r \cos \theta + 3r \sin \theta$$
$$x^2 + y^2 = 2x + 3y. \qquad \text{Substituting} \qquad \blacksquare$$

DO EXERCISES 7–9.

3 Graphing Polar Equations

To graph a polar equation, we usually make a table of values, choosing values of θ and calculating corresponding values of r. We plot the points and then complete the graph, as in the rectangular case. A difference arises in the polar case, because as θ increases sufficiently, points may, in some cases, begin to repeat and the curve will be traced again and again. If such a point is reached, the curve is complete.

EXAMPLE 6 Graph: $r = \cos \theta$.

Solution We first make a table of values.

θ	0°	30°	45°	60°	90°	120°	135°	150°	180°
r	1	0.866	0.707	0.5	0	−0.5	−0.707	−0.866	−1

We plot these points and note that the last point is the same as the first.

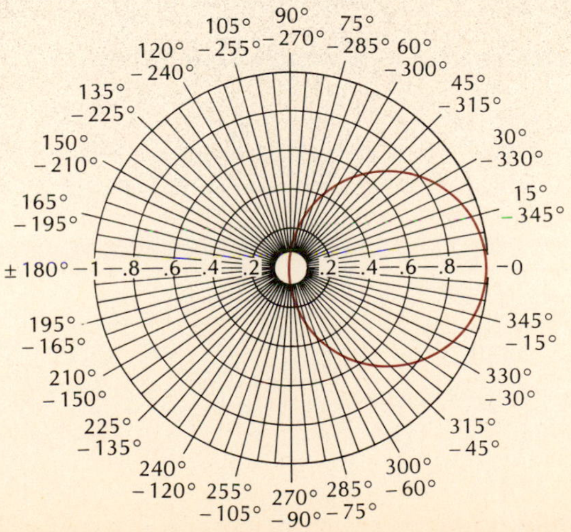

Convert to a polar equation.

5. $2x + 5y = 9$

6. $x^2 + y^2 + 8x = 0$

Convert to a Cartesian equation.

7. $r = 7$

8. $r \sin \theta = 5$

9. $r - 3 \cos \theta = 5 \sin \theta$

10. Graph: $r = 1 - \sin \theta$.

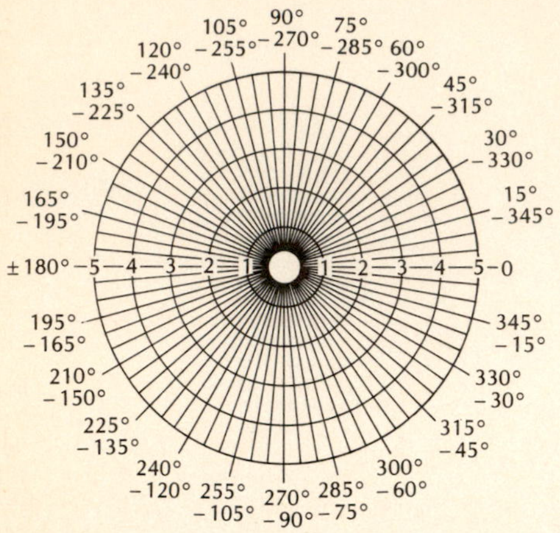

11. Graph: $r = -2 \sin 3\theta$.

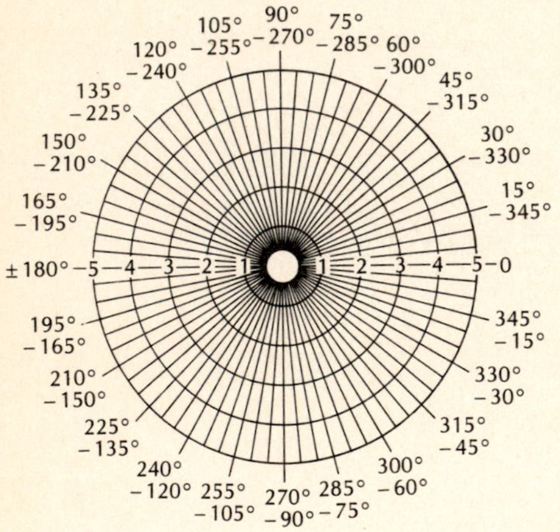

We try another point, $(-0.866, 210°)$, and find that it has already been plotted. Since the repeating has started, we plot no more points, but draw the graph, which is a circle as shown. ■

DO EXERCISE 10.

EXAMPLE 7 Graph: $r = 4 \sin 3\theta$.

Solution We first make a table of values.

θ	0°	15°	30°	45°	60°	75°	90°
r	0	2.83	4	2.83	0	−2.83	−4

θ	105°	120°	135°	150°	165°	180°
r	−2.83	0	2.83	4	2.83	0

We plot these points and sketch the curve. We then try another point, $(-2.83, 195°)$, and find that it has already been plotted. We plot no more points since the repeating has begun.

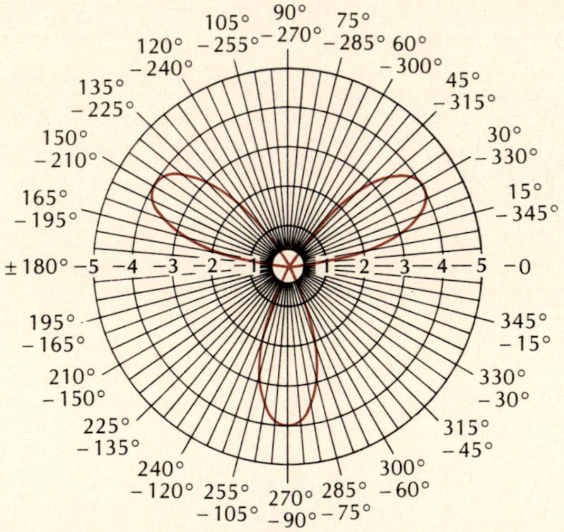

This curve is often referred to as a **three-leafed rose**. ■

DO EXERCISE 11.

EXERCISE SET 8.6

1 Using polar coordinates, graph each of the following points.

1. (4, 30°) **2.** (5, 45°) **3.** (0, 37°) **4.** (0, 48°)

5. (−6, 150°) **6.** (−5, 135°) **7.** (−8, 210°) **8.** (−5, 270°)

9. (3, −30°) **10.** (6, −45°) **11.** (7, −315°) **12.** (4, −270°)

13. (−3, −30°) **14.** (−6, −45°) **15.** (−3.2, 27°) **16.** (−6.8, 34°)

17. $\left(6, \frac{\pi}{4}\right)$ 18. $\left(5, \frac{\pi}{6}\right)$ 19. $\left(4, \frac{3\pi}{2}\right)$ 20. $\left(3, \frac{3\pi}{4}\right)$

21. $\left(-6, \frac{\pi}{4}\right)$ 22. $\left(-5, \frac{\pi}{6}\right)$ 23. $\left(-4, -\frac{3\pi}{2}\right)$ 24. $\left(-3, -\frac{3\pi}{4}\right)$

2 Find the polar coordinates of each of the following points.

25. $(4, 4)$ 26. $(5, 5)$ 27. $(0, 5)$ 28. $(0, -3)$
29. $(4, 0)$ 30. $(-5, 0)$ 31. $(3, 3\sqrt{3})$ 32. $(-3, -3\sqrt{3})$
33. $(\sqrt{3}, 1)$ 34. $(-\sqrt{3}, 1)$ 35. $(3\sqrt{3}, 3)$ 36. $(4\sqrt{3}, -4)$

Find the Cartesian coordinates of each of the following points.

37. $(4, 45°)$ 38. $(5, 60°)$ 39. $(0, 23°)$ 40. $(0, -34°)$
41. $(-3, 45°)$ 42. $(-5, 30°)$ 43. $(6, -60°)$ 44. $(3, -120°)$
45. $\left(10, \frac{\pi}{6}\right)$ 46. $\left(12, \frac{3\pi}{4}\right)$ 47. $\left(-5, \frac{5\pi}{6}\right)$ 48. $\left(-6, \frac{3\pi}{4}\right)$

Convert to a polar equation.

49. $3x + 4y = 5$ 50. $5x + 3y = 4$ 51. $x = 5$ 52. $y = 4$
53. $x^2 + y^2 = 36$ 54. $x^2 + y^2 = 16$ 55. $x^2 - 4y^2 = 4$ 56. $x^2 - 5y^2 = 5$

Convert to a rectangular equation.

57. $r = 5$ 58. $r = 8$ 59. $\theta = \frac{\pi}{4}$ 60. $\theta = \frac{3\pi}{4}$
61. $r \sin \theta = 2$ 62. $r \cos \theta = 5$ 63. $r = 4 \cos \theta$ 64. $r = -3 \sin \theta$
65. $r - r \sin \theta = 2$ 66. $r + r \cos \theta = 3$ 67. $r - 2 \cos \theta = 3 \sin \theta$ 68. $r + 5 \sin \theta = 7 \cos \theta$

3 Graph.

69. $r = 4 \cos \theta$
70. $r = 4 \sin \theta$
71. $r = 1 - \cos \theta$ (Cardioid)
72. $r = 1 + \sin \theta$ (Cardioid)
73. $r = \sin 2\theta$ (Four-leafed rose)
74. $r = 3 \cos 2\theta$ (Four-leafed rose)
75. $r = 2 \cos 3\theta$ (Three-leafed rose)
76. $r = \sin 3\theta$ (Three-leafed rose)
77. $r \cos \theta = 4$
78. $r \sin \theta = 6$
79. $r = \dfrac{5}{1 + \cos \theta}$ (Limaçon)
80. $r = \dfrac{3}{1 + \sin \theta}$ (Limaçon)
81. $r = \theta$ (Spiral of Archimedes)
82. $r = 3\theta$ (Spiral of Archimedes)
83. $r^2 = \sin 2\theta$ (Lemniscate)
84. $r^2 = 4 \cos 2\theta$ (Lemniscate)
85. $r = e^{\theta/10}$ or $\ln r = \theta/10$ (Logarithmic spiral)
86. $r = 10^{2\theta}$ or $\log r = 2\theta$ (Logarithmic spiral)
87. $r = \sin \theta \tan \theta$ (Cissoid)
88. $r = \cos 2\theta \sec \theta$ (Strophoid)

SYNTHESIS

Graph.

89. $r = 2 \cos 2\theta - 1$ (Bow tie)
90. $r = \cos 2\theta - 2$ (Peanut)
91. $r = \frac{1}{4} \tan^2 \theta \sec \theta$ (Semicubical parabola)
92. $r = \sin 2\theta + \cos \theta$ (Twisted sister)

CHALLENGE

93. Convert to a rectangular equation:
$$r = \sec^2 \frac{\theta}{2}.$$

94. The center of a regular hexagon is at the origin, and one vertex is the point $(4, 0°)$. Find the coordinates of the other vertices.

8.7 Complex Numbers

OBJECTIVES

You should be able to:

1 Graph a complex number $a + bi$, and graph the sum of two numbers.

2 Given rectangular, or binomial, notation for a complex number, find polar notation.

3 Given polar notation for a complex number, find binomial notation.

4 Use polar notation to multiply and divide complex numbers.

1. Graph each of the following complex numbers.
   a) $5 - 3i$
   b) $-3 + 4i$
   c) $-5 - 2i$
   d) $5 + 5i$

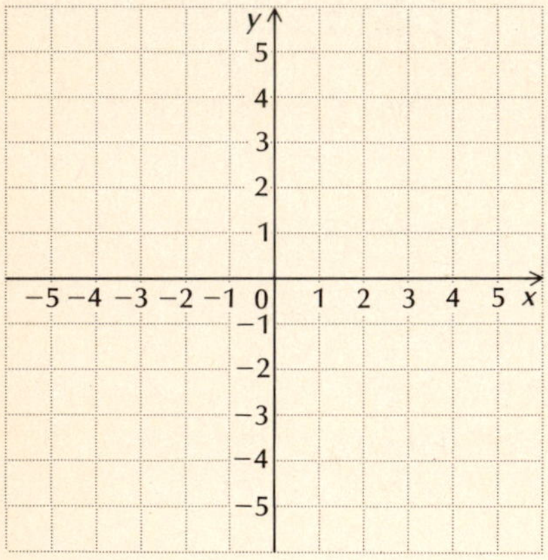

Graphical Representation and Polar Notation

1 Graphs

The real numbers can be graphed on a line. Complex numbers are graphed on a plane. We graph a complex number $a + bi$ in the same way that we graph an ordered pair of real numbers (a, b). However, in place of an x-axis, we have a **real axis**, and in place of a y-axis, we have an **imaginary axis.**

EXAMPLES Graph each of the following.

1. $3 + 2i$
2. $-4 + 5i$
3. $-5 - 4i$

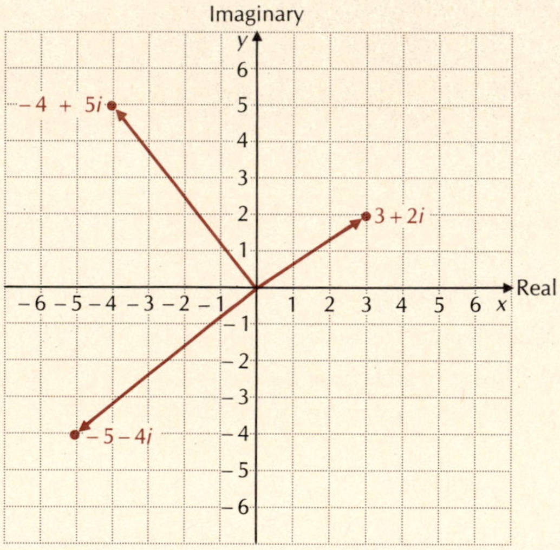

Horizontal distances correspond to the real part of a number. Vertical distances correspond to the imaginary part. Graphs of the numbers are shown above as vectors. The horizontal component of the vector for $a + bi$ is a and the vertical component is b. Complex numbers are sometimes used to study vectors.

DO EXERCISE 1.

Adding complex numbers is like adding vectors using components. For example, to add $3 + 2i$ and $5 + 4i$, we add the real parts and the imaginary parts to obtain $8 + 6i$. Graphically, then, the sum of two complex numbers looks like a vector sum. It is the diagonal of a parallelogram.

EXAMPLE 4 Show graphically $2 + 2i$ and $3 - i$. Show also their sum.

Solution

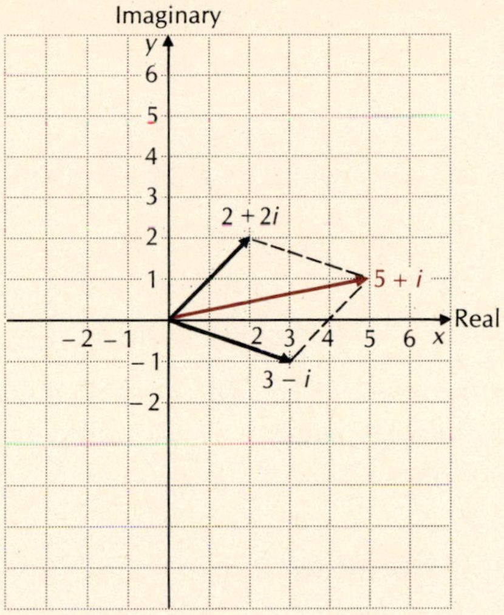

DO EXERCISE 2.

2 Polar Notation for Complex Numbers

We can locate endpoints of arrows as shown in the preceding examples or by giving the length of the arrow and the angle that it makes with the positive half of the real axis. We now develop **polar notation** using this idea. From the figure below, we see that the length of the vector is $\sqrt{a^2 + b^2}$. Note that this quantity is a real number. It is called the **absolute value** of $a + bi$.

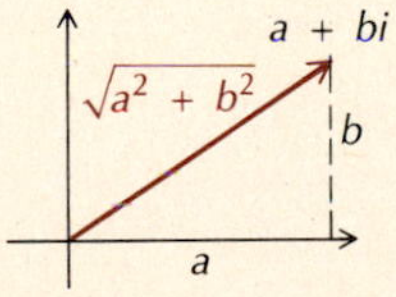

DEFINITION

The *absolute value* of a complex number $a + bi$ is denoted $|a + bi|$ and is defined to be $\sqrt{a^2 + b^2}$.

EXAMPLE 5 Find $|3 + 4i|$.

Solution

$$|3 + 4i| = \sqrt{3^2 + 4^2} = \sqrt{9 + 16} = 5$$

DO EXERCISE 3.

Now let us consider any complex number $a + bi$. Suppose that its absolute value is r. Let us also suppose that the angle that the vector makes with the real axis is θ. As this figure shows, we have $a/r = \cos\theta$ and $b/r = \sin\theta$, so

$$a = r\cos\theta \quad \text{and} \quad b = r\sin\theta.$$

2. Graph each pair of complex numbers and then graph their sum.

   a) $2 + 3i$, $1 - 5i$

   b) $-5 + 2i$, $-1 - 4i$

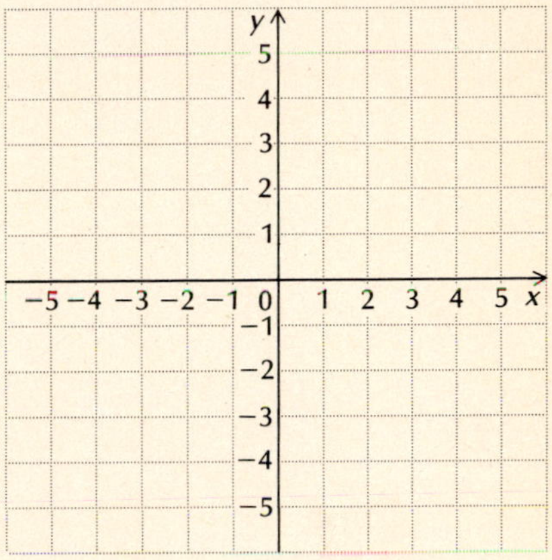

3. Find the absolute value.

   a) $|4 - 3i|$

   b) $|-12 - 5i|$

4. Write rectangular notation for

$\sqrt{2}(\cos 315^\circ + i \sin 315^\circ)$.

5. Write binomial notation for

$2 \text{ cis} \left(-\frac{\pi}{6}\right)$.

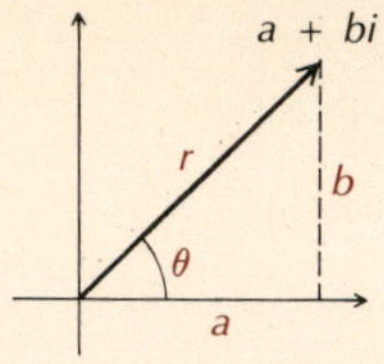

Thus,

$$a + bi = r\cos\theta + ir\sin\theta$$
$$= r(\cos\theta + i\sin\theta).$$

This is polar notation for $a + bi$. The angle θ is called the **argument**.

Polar notation for complex numbers is also called **trigonometric notation**, and is often abbreviated $r \text{ cis } \theta$.

2, 3 Change of Notation

To change from polar notation to **binomial**, or **rectangular**, notation $a + bi$, we proceed as in Section 8.6, recalling that $a = r\cos\theta$ and $b = r\sin\theta$.

EXAMPLE 6 Write binomial notation for $2(\cos 120^\circ + i \sin 120^\circ)$.

Solution

$$a = 2\cos 120^\circ = -1,$$
$$b = 2\sin 120^\circ = \sqrt{3}$$

Thus,

$$2(\cos 120^\circ + i\sin 120^\circ) = -1 + i\sqrt{3}.$$

EXAMPLE 7 Write binomial notation for $\sqrt{8} \text{ cis } \frac{7\pi}{4}$.

Solution

$$\sqrt{8} \text{ cis } \frac{7\pi}{4} = \sqrt{8}\left(\cos\frac{7\pi}{4} + \sin\frac{7\pi}{4}\right),$$
$$a = \sqrt{8}\cos\frac{7\pi}{4} = \sqrt{8}\cdot\frac{1}{\sqrt{2}} = 2,$$
$$b = \sqrt{8}\sin\frac{7\pi}{4} = \sqrt{8}\cdot\frac{-1}{\sqrt{2}} = -2$$

Thus,

$$\sqrt{8} \text{ cis } \frac{7\pi}{4} = 2 - 2i.$$

DO EXERCISES 4 AND 5.

To change from binomial notation to polar notation, we remember that $r = \sqrt{a^2 + b^2}$ and that θ is an angle for which $\sin\theta = b/r$ and $\cos\theta = a/r$.

EXAMPLE 8 Find polar notation for $1 + i$.

Solution We note that $a = 1$ and $b = 1$. Then

$$r = \sqrt{1^2 + 1^2} = \sqrt{2},$$

$$\sin\theta = \frac{1}{\sqrt{2}}, \text{ or } \frac{\sqrt{2}}{2}, \quad \text{and} \quad \cos\theta = \frac{1}{\sqrt{2}}, \text{ or } \frac{\sqrt{2}}{2}.$$

Thus, $\theta = \pi/4$, or 45°, and we have

$$1 + i = \sqrt{2}\text{ cis }\frac{\pi}{4}, \quad \text{or} \quad 1 + i = \sqrt{2}\text{ cis }45°.$$

EXAMPLE 9 Find polar notation for $\sqrt{3} - i$.

Solution We note that $a = \sqrt{3}$ and $b = -1$. Then

$$r = \sqrt{(\sqrt{3})^2 + (-1)^2} = 2,$$

$$\sin\theta = -\frac{1}{2}, \quad \text{and} \quad \cos\theta = \frac{\sqrt{3}}{2}.$$

Thus, $\theta = 11\pi/6$, or 330°, and we have

$$\sqrt{3} - i = 2\text{ cis }\frac{11\pi}{6}, \quad \text{or} \quad 2\text{ cis }330°.$$

In changing to polar notation, note that there are many angles satisfying the given conditions. We ordinarily choose the smallest positive angle.

DO EXERCISES 6 AND 7.

Write polar notation.

6. $1 - i$

7. $-3\sqrt{2} - 3\sqrt{2}i$

4 Multiplication and Division with Polar Notation

Multiplication of complex numbers is somewhat easier to do with polar notation than with rectangular notation. We simply multiply the absolute values and add the arguments. Let us state this more formally and prove it.

THEOREM 3

For any complex numbers r_1 cis θ_1 and r_2 cis θ_2,

$$(r_1\text{ cis }\theta_1)(r_2\text{ cis }\theta_2) = r_1 \cdot r_2\text{ cis }(\theta_1 + \theta_2).$$

Proof. Let us first multiply $a_1 + b_1 i$ by $a_2 + b_2 i$:

$$(a_1 + b_1 i)(a_2 + b_2 i) = (a_1 a_2 - b_1 b_2) + (a_2 b_1 + a_1 b_2)i.$$

Recall that

$$a_1 = r_1 \cos\theta_1, \qquad b_1 = r_1 \sin\theta_1,$$

and

$$a_2 = r_2 \cos\theta_2, \qquad b_2 = r_2 \sin\theta_2.$$

We substitute these in the product above, to obtain

$$\begin{aligned} r_1(\cos\theta_1 + i\sin\theta_1) \cdot r_2(\cos\theta_2 + i\sin\theta_2) \\ &= (r_1 r_2 \cos\theta_1 \cos\theta_2 - r_1 r_2 \sin\theta_1 \sin\theta_2) \\ &\quad + (r_1 r_2 \sin\theta_1 \cos\theta_2 + r_1 r_2 \cos\theta_1 \sin\theta_2)i. \end{aligned}$$

This simplifies to

$$r_1 r_2(\cos\theta_1 \cos\theta_2 - \sin\theta_1 \sin\theta_2) + r_1 r_2(\sin\theta_1 \cos\theta_2 + \cos\theta_1 \sin\theta_2)i.$$

Now, using identities for sums of angles, we simplify, obtaining

$$r_1r_2 \cos(\theta_1 + \theta_2) + r_1r_2 \sin(\theta_1 + \theta_2)i,$$

or

$$r_1r_2 \text{ cis } (\theta_1 + \theta_2),$$

which was to be shown.

To divide complex numbers, we do the reverse of the above. We state that fact, but omit the proof.

THEOREM 4

For any complex numbers r_1 cis θ_1 and r_2 cis θ_2 ($r_2 \neq 0$),

$$\frac{r_1 \text{ cis } \theta_1}{r_2 \text{ cis } \theta_2} = \frac{r_1}{r_2} \text{ cis } (\theta_1 - \theta_2).$$

EXAMPLE 10 Find the product of 3 cis 40° and 7 cis 20°.

Solution

$$3 \text{ cis } 40° \cdot 7 \text{ cis } 20° = 3 \cdot 7 \text{ cis } (40° + 20°)$$
$$= 21 \text{ cis } 60°.$$

EXAMPLE 11 Find the product of 2 cis π and 3 cis $(-\pi/2)$.

Solution

$$2 \text{ cis } \pi \cdot 3 \text{ cis } \left(-\frac{\pi}{2}\right) = 2 \cdot 3 \text{ cis } \left(\pi - \frac{\pi}{2}\right)$$
$$= 6 \text{ cis } \frac{\pi}{2}.$$

EXAMPLE 12 Convert to polar notation and multiply: $(1 + i)(\sqrt{3} - i)$.

Solution We first find polar notation (see Examples 8 and 9):

$$1 + i = \sqrt{2} \text{ cis } 45°, \qquad \sqrt{3} - i = 2 \text{ cis } 330°.$$

We now multiply, using Theorem 3:

$$(\sqrt{2} \text{ cis } 45°)(2 \text{ cis } 330°) = 2 \cdot \sqrt{2} \text{ cis } 375°, \quad \text{or} \quad 2\sqrt{2} \text{ cis } 15°.$$

EXAMPLE 13 Divide 2 cis π by 4 cis $\frac{\pi}{2}$.

Solution

$$\frac{2 \text{ cis } \pi}{4 \text{ cis } \frac{\pi}{2}} = \frac{2}{4} \text{ cis } \left(\pi - \frac{\pi}{2}\right)$$
$$= \frac{1}{2} \text{ cis } \frac{\pi}{2}$$

EXAMPLE 14 Convert to polar notation and divide: $(1 + i)/(1 - i)$.

Multiply.

8. 5 cis 25° · 4 cis 30°

9. $8 \text{ cis } \pi \cdot \frac{1}{2} \text{ cis } \frac{\pi}{4}$

10. Convert to polar notation and multiply:

$$(1 + i)(2 + 2i).$$

11. Divide:

$$10 \text{ cis } \frac{\pi}{2} \div 5 \text{ cis } \frac{\pi}{4}.$$

Solution We first convert to polar notation:

$$1 + i = \sqrt{2} \text{ cis } 45°,$$
$$1 - i = \sqrt{2} \text{ cis } 315°.$$

See Example 8.

We now divide, using Theorem 4:

$$\frac{1+i}{1-i} = \frac{\sqrt{2} \text{ cis } 45°}{\sqrt{2} \text{ cis } 315°} = 1 \cdot \text{cis } (45° - 315°)$$
$$= 1 \cdot \text{cis } (-270°), \quad \text{or} \quad \text{cis } 90°.$$

DO EXERCISES 8–12 (EXERCISES 8–11 ARE ON THE PRECEDING PAGE.)

12. Convert to polar notation and divide:

$$\frac{\sqrt{3} - i}{1 + i}.$$

EXERCISE SET 8.7

1 Graph each pair of complex numbers and then graph their sum.

1. $3 + 2i, 2 - 5i$
2. $4 + 3i, 3 - 4i$
3. $-5 + 3i, -2 - 3i$
4. $-4 + 2i, -3 - 4i$
5. $2 - 3i, -5 + 4i$
6. $3 - 2i, -5 + 5i$
7. $-2 - 5i, 5 + 3i$
8. $-3 - 4i, 6 + 3i$

2 Find rectangular notation.

9. $3(\cos 30° + i \sin 30°)$
10. $6(\cos 150° + i \sin 150°)$
11. $10 \text{ cis } 270°$
12. $5 \text{ cis } (-60°)$
13. $\sqrt{8}\left(\cos \frac{\pi}{4} + i \sin \frac{\pi}{4}\right)$
14. $5\left(\cos \frac{\pi}{3} + i \sin \frac{\pi}{3}\right)$
15. $\sqrt{8} \text{ cis } \frac{5\pi}{4}$
16. $\sqrt{8} \text{ cis } \left(-\frac{\pi}{4}\right)$

3 Find polar notation.

17. $1 - i$
18. $\sqrt{3} + i$
19. $10\sqrt{3} - 10i$
20. $-10\sqrt{3} + 10i$
21. -5
22. $-5i$

4 Convert to polar notation and then multiply or divide.

23. $(1 - i)(2 + 2i)$
24. $(1 + i\sqrt{3})(1 + i)$
25. $(2\sqrt{3} + 2i)(2i)$
26. $(3\sqrt{3} - 3i)(2i)$
27. $\dfrac{1 - i}{1 + i}$
28. $\dfrac{1 - i}{\sqrt{3} - i}$
29. $\dfrac{2\sqrt{3} - 2i}{1 + \sqrt{3}i}$
30. $\dfrac{3 - 3\sqrt{3}i}{\sqrt{3} - i}$

SYNTHESIS

31. Show that for any complex number z,
$$|z| = |-z|.$$
(*Hint:* Let $z = a + bi$.)

32. Show that for any complex number z,
$$|z| = |\bar{z}|.$$
(*Hint:* Let $z = a + bi$.)

33. Show that for any complex number z,
$$|z\bar{z}| = |z^2|.$$

34. Show that for any complex number z,
$$|z^2| = |z|^2.$$

35. Show that for any complex numbers z and w,
$$|z \cdot w| = |z| \cdot |w|.$$
(*Hint:* Let $z = r_1 \text{ cis } \theta_1$ and $w = r_2 \text{ cis } \theta_2$.)

36. Show that for any complex number z and any nonzero complex number w,
$$\left|\frac{z}{w}\right| = \frac{|z|}{|w|}.$$
(Use the hint for Exercise 35.)

37. On a complex plane, graph $|z| = 1$.

38. On a complex plane, graph $z + \bar{z} = 3$.

CHALLENGE

39. Find polar notation for $(\cos \theta + i \sin \theta)^{-1}$.

40. Compute $(5 \text{ cis } 20°)^3$.

8.8 Complex Numbers: DeMoivre's Theorem and nth Roots of Complex Numbers

OBJECTIVES

You should be able to:

1. Use DeMoivre's theorem to raise complex numbers to powers.
2. Find the nth roots of a complex number.

1. Find $(1 - i)^{10}$.

2. Find $(\sqrt{3} + i)^4$.

1 Powers of Complex Numbers

An important theorem about powers and roots of complex numbers is named for the French mathematician Abraham DeMoivre (1667–1754). Let us consider a number $r \text{ cis } \theta$ and its square:

$$(r \text{ cis } \theta)^2 = (r \text{ cis } \theta)(r \text{ cis } \theta) = r \cdot r \text{ cis } (\theta + \theta)$$
$$= r^2 \text{ cis } 2\theta.$$

Similarly, we see that

$$(r \text{ cis } \theta)^3 = r \cdot r \cdot r \text{ cis } (\theta + \theta + \theta)$$
$$= r^3 \text{ cis } 3\theta.$$

The generalization of this is DeMoivre's theorem.

THEOREM 5 DeMoivre's Theorem

For any complex number $r \text{ cis } \theta$ and any natural number n,

$$(r \text{ cis } \theta)^n = r^n \text{ cis } n\theta,$$

or

$$[r(\cos \theta + i \sin \theta)]^n = r^n[\cos n\theta + i \sin n\theta].$$

EXAMPLE 1 Find $(1 + i)^9$.

Solution We first find polar notation: $1 + i = \sqrt{2} \text{ cis } 45°$. Then

$$(1 + i)^9 = (\sqrt{2} \text{ cis } 45°)^9$$
$$= \sqrt{2}^9 \text{ cis } (9 \cdot 45°)$$
$$= 2^{9/2} \text{ cis } 405°$$
$$= 16\sqrt{2} \text{ cis } 45°$$

405° has the same terminal side as 45°.

$$= 16\sqrt{2} (\cos 45° + i \sin 45°)$$
$$= 16\sqrt{2}\left(\frac{\sqrt{2}}{2} + i\frac{\sqrt{2}}{2}\right)$$
$$= 16 + 16i.$$

EXAMPLE 2 Find $(\sqrt{3} - i)^{10}$.

Solution We first find polar notation: $\sqrt{3} - i = 2 \text{ cis } 330°$. Then

$$(\sqrt{3} - i)^{10} = (2 \text{ cis } 330°)^{10}$$
$$= 2^{10} \text{ cis } (10 \cdot 330°)$$
$$= 1024 \text{ cis } 3300°$$
$$= 1024 \text{ cis } 60°$$
$$= 1024[\cos 60° + i \sin 60°]$$
$$= 1024\left[\frac{1}{2} + i\frac{\sqrt{3}}{2}\right] = 512 + i512\sqrt{3}.$$

DO EXERCISES 1 AND 2.

2 Roots of Complex Numbers

As we will see, every nonzero complex number has two square roots. A number has three cube roots, four fourth roots, and so on. In general, a nonzero complex number has n different nth roots. They can be found by the formula that we now state and prove.

THEOREM 6

The nth roots of a complex number r cis θ are given by

$$r^{1/n} \text{ cis } \left(\frac{\theta}{n} + k \cdot \frac{360°}{n}\right),$$

where $k = 0, 1, 2, \ldots, n - 1$.

We show that this formula gives us n different roots, using DeMoivre's theorem. We take the expression for the nth roots and raise it to the nth power, to show that we get r cis θ:

$$\left[r^{1/n} \text{ cis } \left(\frac{\theta}{n} + k \cdot \frac{360°}{n}\right)\right]^n = (r^{1/n})^n \text{ cis } \left(\frac{\theta}{n} \cdot n + k \cdot n \cdot \frac{360°}{n}\right)$$
$$= r \text{ cis } (\theta + k \cdot 360°) = r \text{ cis } \theta.$$

Thus we know that the formula gives us nth roots for any natural number k. Next, we show that there are at least n different roots. To see this, consider substituting 0, 1, 2, and so on, for k. When $k = n$, the cycle begins to repeat, but from 0 to $n - 1$, the angles obtained and their sines and cosines are all different. There cannot be more than n different nth roots. That fact follows from the **fundamental theorem of algebra,** considered in the theory of polynomial functions.

EXAMPLE 3 Find the square roots of $2 + 2\sqrt{3}i$.

Solution We first find polar notation: $2 + 2\sqrt{3}i = 4$ cis 60°. Then

$$(4 \text{ cis } 60°)^{1/2} = 4^{1/2} \text{ cis } \left(\frac{60°}{2} + k \cdot \frac{360°}{2}\right), \quad k = 0, 1$$
$$= 2 \text{ cis } \left(30° + k \cdot \frac{360°}{2}\right), \quad k = 0, 1.$$

Thus the roots are 2 cis 30° and 2 cis 210°, or

$$\sqrt{3} + i \quad \text{and} \quad -\sqrt{3} - i.$$

DO EXERCISES 3 AND 4.

In Example 3, it should be noted that the two square roots of the number were additive inverses of each other. The same is true of the square roots of any complex number. To see this, let us find the square roots of any complex number r cis θ:

$$(r \text{ cis } \theta)^{1/2} = r^{1/2} \text{ cis } \left(\frac{\theta}{2} + k \cdot \frac{360°}{2}\right), \quad k = 0, 1$$
$$= r^{1/2} \text{ cis } \frac{\theta}{2} \quad \text{or} \quad r^{1/2} \text{ cis } \left(\frac{\theta}{2} + 180°\right).$$

Find the square roots.

3. $2i$

4. $10i$

5. Find and graph the cube roots of -1.

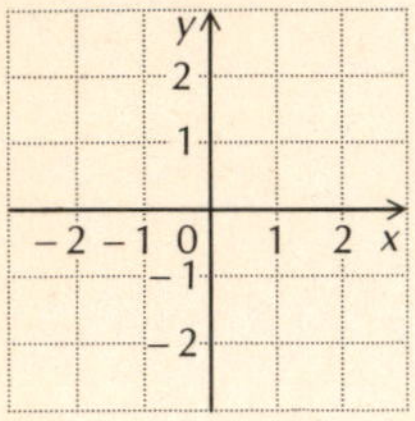

Now let us look at the two numbers on a graph. They lie on a line, so if one number has binomial notation $a + bi$, the other has binomial notation $-a - bi$. Thus their sum is 0 and they are additive inverses of each other.

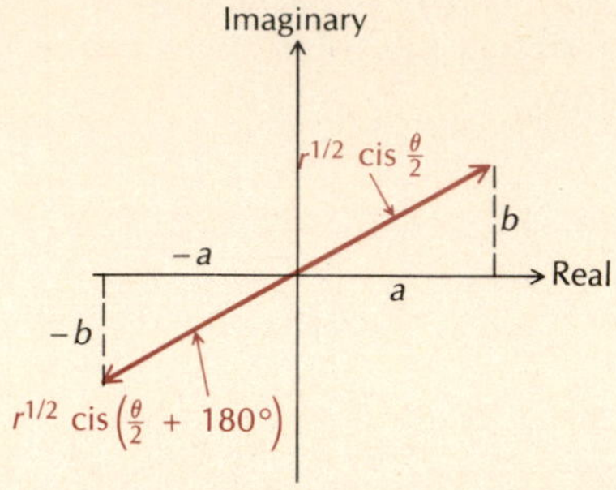

EXAMPLE 4 Find the cube roots of 1. Then locate them on a graph.

Solution

$$1 = 1 \text{ cis } 0°$$

$$(1 \text{ cis } 0°)^{1/3} = 1^{1/3} \text{ cis } \left(\frac{0°}{3} + k \cdot \frac{360°}{3}\right), \quad k = 0, 1, 2$$

The roots are 1 cis 0°, 1 cis 120°, and 1 cis 240°, or

$$1, \quad -\frac{1}{2} + \frac{\sqrt{3}}{2}i, \quad \text{and} \quad -\frac{1}{2} - \frac{\sqrt{3}}{2}i.$$

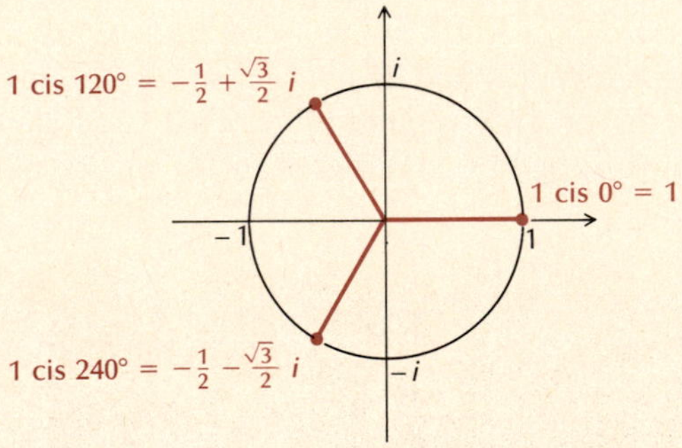

Note in Example 4 that the graphs of the cube roots lie equally spaced about a circle. The same is true of the nth roots of any complex number.

DO EXERCISE 5.

EXERCISE SET 8.8

1 Raise the number to the power and write polar notation for the answer.

1. $\left(2 \text{ cis } \frac{\pi}{3}\right)^3$
2. $\left(3 \text{ cis } \frac{\pi}{2}\right)^4$
3. $\left(2 \text{ cis } \frac{\pi}{6}\right)^6$
4. $\left(2 \text{ cis } \frac{\pi}{5}\right)^5$
5. $(1 + i)^6$
6. $(1 - i)^6$

Raise the number to the power and write rectangular notation for the answer.

7. $(2 \text{ cis } 240°)^4$
8. $(2 \text{ cis } 120°)^4$
9. $(1 + \sqrt{3}i)^4$
10. $(-\sqrt{3} + i)^6$
11. $\left(\frac{1}{\sqrt{2}} + \frac{1}{\sqrt{2}}i\right)^{10}$
12. $\left(\frac{1}{\sqrt{2}} - \frac{1}{\sqrt{2}}i\right)^{12}$
13. $\left(\frac{\sqrt{3}}{2} + \frac{1}{2}i\right)^{12}$
14. $\left(\frac{\sqrt{3}}{2} - \frac{1}{2}i\right)^{14}$

2 Solve, that is, find the square roots.

15. $x^2 = i$
16. $x^2 = -i$
17. $x^2 = 2\sqrt{2} - 2\sqrt{2}i$
18. $x^2 = 2\sqrt{2} + 2\sqrt{2}i$
19. $x^2 = -1 + \sqrt{3}i$
20. $x^2 = -\sqrt{3} - i$
21. $x^3 = i$
22. 🖩 $x^3 = 68.4321$
23. Find and graph the fourth roots of 16.
24. Find and graph the fourth roots of i.
25. Find and graph the fifth roots of -1.
26. Find and graph the sixth roots of 1.
27. Find the fourth roots of $-8 + i\,8\sqrt{3}$.
28. Find the cube roots of $-64i$.
29. Find the cube roots of $2\sqrt{3} - 2i$.
30. Find the cube roots of $1 - i\sqrt{3}$.

Find all the complex solutions of the equation.

31. $x^3 = 1$
32. $x^5 - 1 = 0$
33. $x^5 + 1 = 0$
34. $x^4 + i = 0$
35. $x^5 + \sqrt{3} + i = 0$
36. $x^3 + (4 - 4\sqrt{3}) = 0$
37. $x^4 + 81 = 0$
38. $x^6 + 1 = 0$

SYNTHESIS

Solve. Evaluate square roots.

39. $x^2 + (1 - i)x + i = 0$
40. $3x^2 + (1 + 2i)x + 1 - i = 0$

SUMMARY AND REVIEW: CHAPTER 8

TERMS TO KNOW

Angle of elevation
Angle of depression
Bearing
Right triangle
Oblique triangle
Law of sines
Law of cosines
Ambiguous case
Vector
Magnitude of a vector
Direction of a vector
Vector addition
Resultant
Components of vectors
Forces in equilibrium
Polar coordinates
Rectangular coordinates
Polar equation
Rectangular equation
Polar notation for complex numbers
Absolute value of a complex number
Rectangular notation for complex numbers
DeMoivre's theorem
Roots of complex numbers

REVIEW EXERCISES

Solve each of the following right triangles. Standard lettering has been used.

1. $a = 7.3$, $c = 8.6$
2. $a = 30.5$, $B = 51°10'$
3. One leg of a right triangle bears east. The hypotenuse is 734 cm long and bears N 57°20′ E. Find the perimeter of the triangle.
4. An observer's eye is 6 ft above the floor. A mural is being viewed. The bottom of the mural is at floor level. The observer looks downward 13° to see the bottom and upward 17° to see the top. How tall is the mural?
5. Solve triangle ABC, where $a = 25$, $b = 20$, and $B = 35°$.
6. Solve triangle ABC, where $B = 118°20'$, $C = 27°30'$, and $b = 0.974$.
7. In triangle ABC, $a = 3.7$, $c = 4.9$, and $B = 135°$. Find b.
8. In an isosceles triangle, the base angles each measure 52°20′ and the base is 513 cm long. Find the lengths of the other two sides.

9. A triangular flower garden has sides of lengths 11 ft, 9 ft, and 6 ft. Find the angles of the garden.

10. A parallelogram has sides of lengths 3.21 cm and 7.85 cm. One of its angles measures 147°. Find the area of the parallelogram.

11. A car is moving north at 45 mph. A ball is thrown from the car at 37 mph in a direction of N45°E. Find the speed and direction of the ball over the ground.

12. Vector **u** has components of 12 lb upward and 18 lb to the right. Vector **v** has components of 35 lb downward and 35 lb to the left. Find the components of **u** + **v** and the magnitude and the direction of **u** + **v**.

13. Find rectangular notation for the vector $(20, -200°)$.

14. Find polar notation for the vector $\langle -2, 3\rangle$.

15. Do the following calculations for $\mathbf{u} = \langle 4, 3\rangle$ and $\mathbf{v} = \langle -3, 4\rangle$: (a) $4\mathbf{u} - 3\mathbf{v}$; (b) $|\mathbf{u} + \mathbf{v}|$.

Find the polar coordinates of each of the following points.

16. $(0, -4)$

17. $(\sqrt{3}, -1)$

Find the Cartesian coordinates of each of the following points.

18. $(-4, 60°)$

19. $\left(4, \frac{2\pi}{3}\right)$

20. Convert to a polar equation:
$$x^2 + y^2 + 2x - 3y = 0.$$

21. Convert to a rectangular equation:
$$r = -4 \sin \theta.$$

Graph.

22. $r = \cos 2\theta$

23. $r = 2\theta$

24. $r = 3(1 - \sin \theta)$

25. A block weighing 150 lb rests on an inclined plane. The plane makes an angle of 45° with the ground. Find the components of the weight parallel and perpendicular to the plane.

26. A crane is supporting a 500-kg steel beam. The boom of the crane is inclined 30° from the vertical. Find the compression in the boom of the crane.

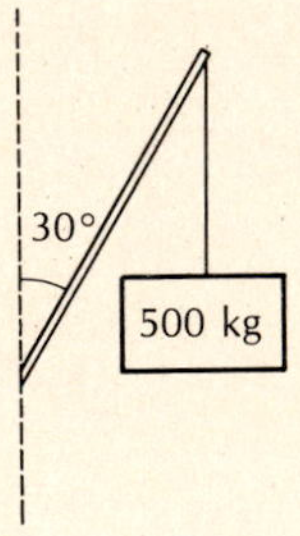

27. Graph the pair of complex numbers $-3 - 2i$ and $4 + 7i$ and their sum.

28. Find rectangular notation for $2(\cos 135° + i \sin 135°)$.

29. Find polar notation for $1 + i$.

30. Find the product of 7 cis 18° and 10 cis 32°.

31. Convert to polar notation and then divide:
$$\frac{1 + i}{\sqrt{3} + i}.$$

32. Find $(1 - i)^5$ and write polar notation for the answer.

33. Find $(3 \text{ cis } 120°)^4$ and write rectangular notation for the answer.

34. Find the cube roots of $1 + i$.

35. Solve $x^3 - 27 = 0$.

36. Find the square roots of $4i$.

SYNTHESIS

37. A parallelogram has sides of lengths 3.42 and 6.97. Its area is 18.4. Find the sizes of its angles.

38. Let $\mathbf{u} = \langle 12, 5\rangle$. Find a vector that has the same direction as **u** but length 3.

39. Convert to a rectangular equation:
$$r = \csc^2 \frac{\theta}{2}.$$

TEST: CHAPTER 8

Solve each of the following right triangles. Standard lettering has been used.

1. $A = 22°10'$, $b = 18.4$
2. $a = 11$, $c = 28$
3. A building 70 m high casts a shadow 100 m long. What is the angle of elevation of the sun?
4. Solve triangle ABC, where $B = 46°$, $C = 19°$, and $b = 70.2$.
5. In triangle ABC, $b = 11$, $c = 13$, and $A = 58°$. Find a.
6. A triangular flower bed measures 8 ft on one side, and that side makes angles of 38°40′ and 50° with the other sides. How long are the other two sides?
7. Two forces of 20 kg and 11 kg act on an object at right angles. Find the magnitude of the resultant and the angle that it makes with the smaller force.
8. Find rectangular notation for the vector $(25, 150°)$.
9. Find polar notation for the vector $\langle -10, 6 \rangle$.
10. Do the following calculations for $\mathbf{u} = \langle 2, -6 \rangle$ and $\mathbf{v} = \langle 3, 8 \rangle$: (a) $\mathbf{u} - \mathbf{v}$; (b) $|2\mathbf{u} + \mathbf{v}|$.
11. Vector **u** has an easterly component of 10 and a northerly component of 15. Vector **v** has a southerly component of 8 and a westerly component of 15. Find:
    a) the components of $\mathbf{u} + \mathbf{v}$;
    b) the magnitude and the direction of $\mathbf{u} + \mathbf{v}$.
12. Find Cartesian coordinates of the point $(-3, 30°)$.
13. Convert $r = 25$ to a rectangular equation.
14. A block weighing 120 kg is on a 24° incline. What force parallel to the incline is necessary to keep it from sliding down?
15. Let $\mathbf{u} = (-2, 7)$. Find a vector that has the same direction as **u** but has length 2.
16. Solve $x^2 = 8i$.
17. Graph the pair of complex numbers $2 - 7i$ and $-4 + 2i$ and their sum.
18. Find rectangular notation for $3 \text{ cis } (-30°)$.
19. Find polar notation for $-1 + \sqrt{3}i$.
20. Find the product of $2 \text{ cis } 50°$ and $5 \text{ cis } 40°$.
21. Divide $3 \text{ cis } 120°$ by $6 \text{ cis } 40°$.
22. Find $(1 + i)^6$. Write polar notation for the answer.
23. Find the cube roots of -8.

Graph.

24. $r = 4 \cos 3\theta$
25. $r = \dfrac{2}{1 + \sin \theta}$

SYNTHESIS

26. A parallelogram has sides of lengths 4.1 mm and 3.2 mm. One of its angles is 118°. Find the area of the parallelogram.

Systems of Linear Equations and Inequalities

9

The most difficult part of solving problems in algebra is almost always translating the problem situation to mathematical language. Once you get an equation, for example, the rest is straightforward. In this chapter, we study *systems of equations* and how to solve them using graphing, substitution, and elimination. One of the great advantages of using a system of equations is that many problem situations then become easier to translate to mathematical language.

Systems of equations have extensive application to many fields such as psychology, sociology, business, education, engineering, and science. Systems of inequalities are also useful in a branch of mathematics called *linear programming*. We include a brief introduction to linear programming in this chapter as well as a study of *matrices*, which can also be used to solve systems of equations.

FEATURE PROBLEM

A chemist has one solution of acid and water that is 25% acid and a second that is 65% acid. How many liters of each should be mixed together in order to get 8 L of a solution that is 40% acid?

THE MATHEMATICS

We let x = the amount of the 25% solution and y = the amount of the 65% solution and translate the information to a *system of equations*:

$$x + y = 8,$$
$$25\%x + 65\%y = 3.2.$$

To solve the problem, we solve the system for x and y.

OBJECTIVES

You should be able to:

1. Determine whether an ordered pair is a solution of a system of equations.
2. Solve a system of two linear equations in two variables by graphing.
3. Solve a system of two linear equations in two variables by the substitution method.
4. Solve a system of two linear equations in two variables using the elimination method.
5. Solve applied problems by translating them to a system of two linear equations and solving the system.

1. Determine whether $(-3, 2)$ is a solution of the conjunction

$2x - y = -8$ *and* $3x + 4y = -1$.

2. Determine whether $\left(0, \frac{1}{4}\right)$ is a solution of the conjunction

$5x + 12y = 13$ *and* $\sqrt{2}x + 9y = 10$.

3. Solve graphically:

$y - x = 1$ *and* $y + x = 3$.

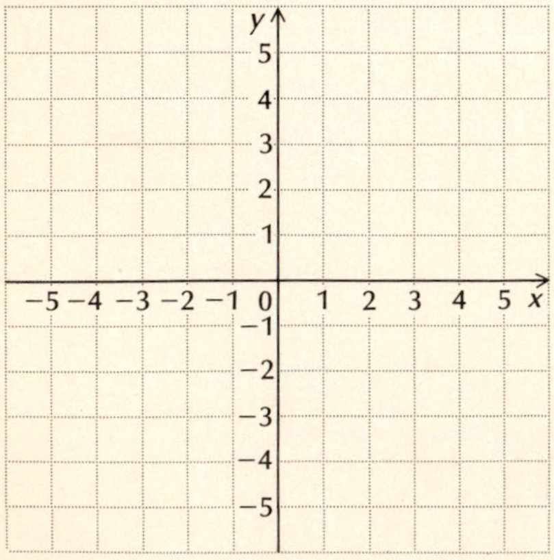

9.1 Systems of Equations in Two Variables

1 Identifying Solutions

Recall that a **conjunction** of sentences is formed by joining sentences with the word *and*. Here is an example:

$$x + y = 11 \quad \textit{and} \quad 3x - y = 5.$$

A **solution** of a sentence with two variables, such as $x + y = 11$, is an ordered pair. Some pairs in the solution set of $x + y = 11$ are

$$(5, 6), \quad (12, -1), \quad (4, 7), \quad (8, 3).$$

Some pairs in the solution set of $3x - y = 5$ are

$$(0, -5), \quad (4, 7), \quad (-2, -11), \quad (9, 22).$$

The solution set of the sentence

$$x + y = 11 \quad \textit{and} \quad 3x - y = 5$$

consists of all pairs that make *both* sentences true. That is, it is the *intersection* of the solution sets. Note that $(4, 7)$ is a solution of the conjunction; in fact, it is the only solution.

DO EXERCISES 1 AND 2.

2 Solving Systems of Equations Graphically

One way to find solutions of a conjunction of equations is by trial and error. Another way is to graph the equations and look for points of intersection. For example, the following graph shows the solution sets of $x + y = 11$ and $3x - y = 5$. Their intersection is the single ordered pair $(4, 7)$.

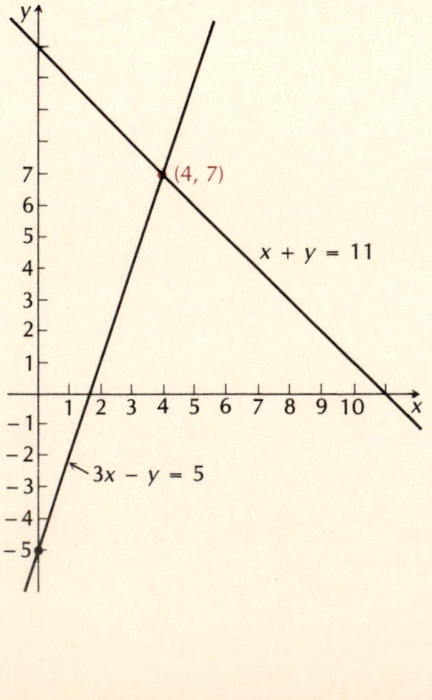

DO EXERCISE 3.

We often refer to a conjunction of equations as a **system of equations**. We usually omit the word *and* and very often write one equation under the other.

The graph of each equation in a system of two linear equations in two variables is a line.

Given the graphs of two lines, the following can happen:

a) The lines have no point in common—they are parallel. The system has no solution. (See (a) below.)

b) The lines have exactly one point in common. The system has exactly one solution. (See (b) below.)

c) The lines are the same—they have infinitely many points in common. The system has infinitely many solutions. (See (c) below.)

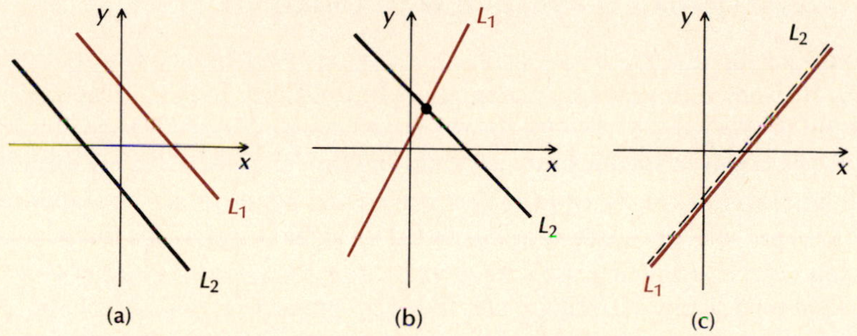

DO EXERCISES 4–6.

We now consider two algebraic methods for solving systems of linear equations.

3 The Substitution Method

To use the substitution method, we solve one equation for one of the variables. Then we substitute in the other equation and solve.

EXAMPLE 1 Solve the system

$$x + y = 11, \qquad (1)$$
$$3x - y = 5. \qquad (2)$$

Solution First we solve Eq. (1) for y. (We could just as well solve for x.)

$$y = 11 - x$$

Then we substitute $11 - x$ for y in Eq. (2). This gives an equation in one variable, which we know how to solve from earlier work:

$$3x - (11 - x) = 5$$
$$x = 4.$$

Now we substitute 4 for x in either Eq. (1) or (2) and solve for y. Let us use Eq. (1):

$$4 + y = 11$$
$$y = 7.$$

Solve the system graphically.

4. $2x - y = 1,$
$-6x + 3y = -3$

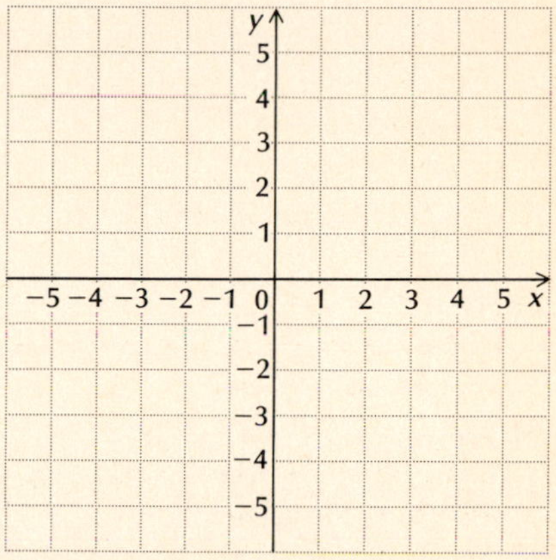

5. $x - 4y = -4,$
$-x + 4y = 8$

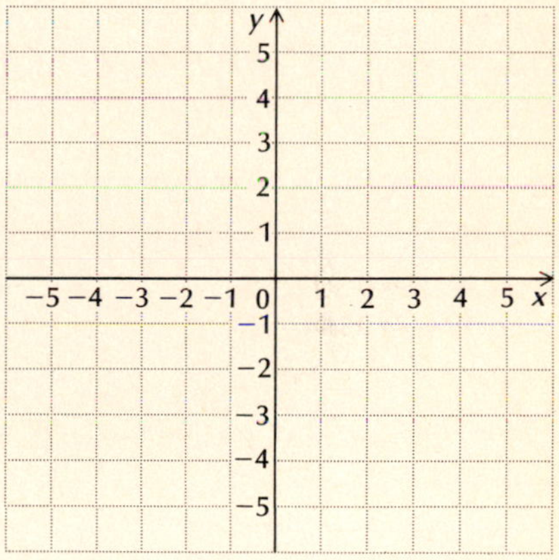

6. $y + x = 3,$
$y - x = 1$

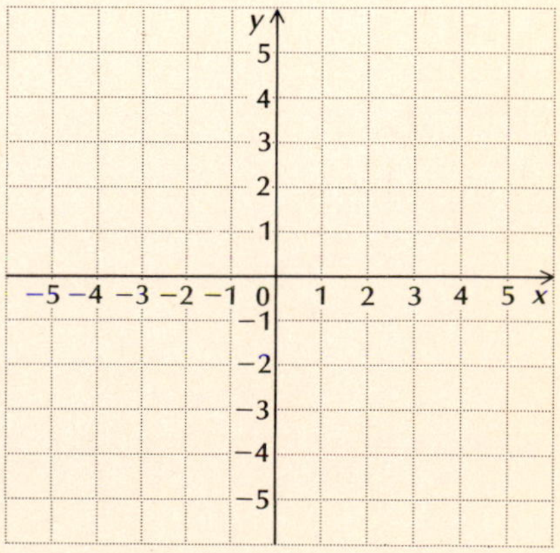

Solve, using the substitution method.

7. $2x + y = 6,$
$3x + 4y = 4$

8. $8x - 3y = -31,$
$2x + 6y = 26$

The solution is (4, 7). We list the coordinates of the solution in alphabetical order, 4 for x and 7 for y.

Check:

$x + y = 11$	
$4 + 7$	11
11	

$3x - y = 5$	
$3 \cdot 4 - 7$	5
$12 - 7$	
5	

DO EXERCISES 7 AND 8.

4 Gauss–Jordan Elimination with Equations

The next algebraic method to consider for solving systems of equations is called **Gauss–Jordan elimination**, or simply, **elimination.** It is an adaptation of methods developed by two German mathematicians, Carl Friedrich Gauss (1777–1855) and Wilhelm Jordan (1842–1899).

The **elimination method** makes use of the addition and multiplication principles for solving equations and is based on algebraic methods that you have probably learned in your earlier mathematics courses. Gauss–Jordan elimination is a special **algorithm,** or step-by-step procedure, which can be programmed on a computer. Our goal is to perform special operations on a system that will result in an "equivalent system" for which the solution is obvious. Two systems are **equivalent** if they have exactly the same solutions.

Suppose we want to solve the system

$$3x - 4y = -1,$$
$$-3x + 2y = 0.$$

Our goal is to carry out certain procedures to obtain an equivalent system of the type

$$Ax + By = C, \quad (1)$$
$$Dy = E. \quad (2)$$

When we obtain such a system, we can easily solve for the variables by multiplying on each side of Eq. (2) by the reciprocal of the coefficient of y. After we have solved for y, we substitute into Eq. (1) to find x.

Now considering the system given above, we add the left-hand sides, obtaining $-2y$, and then add the right-hand sides, obtaining -1. When we do this, we often say that we "added the two equations." In this way, we eliminate the x-term in the second equation to obtain a system of equations equivalent (having the same solutions) to the original:

$$3x - 4y = -1,$$
$$-3x + 2y = 0.$$

$$3x - 4y = -1 \quad (1)$$
$$-2y = -1 \quad (2)$$

$$\begin{array}{r} 3x - 4y = -1 \\ -3x + 2y = 0 \\ \hline -2y = -1 \end{array}$$ Adding

The system is now in the form

$$Ax + Cy = P,$$
$$By = Q.$$

We solve for y and then substitute into the first equation to find x:

$$-2y = -1 \qquad 3x - 4\left(\tfrac{1}{2}\right) = -1$$
$$y = \tfrac{1}{2}; \qquad 3x - 2 = -1$$
$$3x = 1$$
$$x = \tfrac{1}{3}.$$

This kind of substitution is often called *back substitution.*

We now know that the solution of the original system is $\left(\frac{1}{3}, \frac{1}{2}\right)$, because the solutions of this last system are obvious and we know that this system is equivalent to the original. We know that the last system is equivalent to the original, because the only computations we did consisted of using the addition principle, adding only expressions for which all replacements were meaningful, and multiplying by a nonzero constant. Each computation produces results equivalent to the original equations. For this reason, we do not need to check results, except to detect errors in computation.

EXAMPLE 2 Solve:

$$5x + 3y = 7,$$
$$3x - 5y = -23.$$

Solution We first multiply the second equation by 5 to make the x-coefficient a multiple of 5:

$$5x + 3y = 7,$$
$$15x - 25y = -115.$$

Now we multiply the first equation by -3 and add it to the second equation. This eliminates the x-term:

$$5x + 3y = 7, \qquad -15x - 9y = -21 \quad \text{Multiplying by } -3$$
$$-34y = -136. \leftarrow \qquad 15x - 25y = -115$$
$$-34y = -136 \quad \text{Adding}$$

Next, we solve the second equation for y. Then we substitute the result into the first equation to find x:

$$-34y = -136 \qquad 5x + 3(4) = 7$$
$$y = 4; \qquad 5x + 12 = 7$$
$$5x = -5$$
$$x = -1.$$

The solution is $(-1, 4)$. ■

There are some preliminary things that might be done to make the steps of the elimination method simpler. One is to first write the equations in the form $Ax + By = C$. Another is to interchange two equations before beginning. For example, if we have the system

$$5x + y = -2,$$
$$x + 7y = 3,$$

we may prefer to write the second equation first:

$$x + 7y = 3,$$
$$5x + y = -2.$$

This accomplishes two things. One is that the x-coefficient is 1 in the first equation. When we have found y later, this will make solving for x easier. The other is that this makes the x-coefficient in the second equation a multiple of the first.

Something else that might be done is to multiply one or more equations by a power of 10 before beginning in order to eliminate decimal points. For example, if we have the system

$$\begin{aligned} -0.3x + 0.5y &= 0.3, \\ 0.01x - 0.4y &= 1.2, \end{aligned}$$

we can multiply the first equation by 10 and the second by 100 to clear the decimals, to transform the system to

$$\begin{aligned} -3x + 5y &= 3, \\ x - 40y &= 120. \end{aligned}$$

We might also multiply in order to clear equations of fractions.

We transform the original system to an equivalent system of equations using any of the following operations, or transformations.

THEOREM 1 Transformations Producing Equivalent Systems

Each of the following will produce an equivalent system of equations:

a) Interchanging any two equations.

b) Multiplying each number or term of an equation by the same nonzero number.

c) Multiplying each number or term of one equation by the same nonzero number and adding the result to another equation.

EXAMPLE 3 Solve:

$$\begin{aligned} 5x + y &= -2, \\ x + 7y &= 3. \end{aligned}$$

Solution We first interchange the equations so that the x-coefficient of the second equation will be a multiple of the first:

$$\begin{aligned} x + 7y &= 3, \\ 5x + y &= -2. \end{aligned}$$

Next, we multiply the first equation by -5 and add the result to the second equation. This eliminates the x-term in the second equation:

$$\begin{aligned} x + 7y &= 3, \\ -34y &= -17. \end{aligned} \qquad \longleftarrow \qquad \begin{aligned} -5x - 35y &= -15 \\ 5x + y &= -2 \\ \hline -34y &= -17 \end{aligned} \quad \textbf{Adding}$$

Now we solve the second equation for y. Then we substitute the result in the first equation to find x.

$$\begin{aligned} -34y &= -17 \\ y &= \tfrac{1}{2} \end{aligned} \qquad\qquad \begin{aligned} x + 7\left(\tfrac{1}{2}\right) &= 3 \\ x + \tfrac{7}{2} &= 3 \\ x &= 3 - \tfrac{7}{2}, \quad \text{or} \quad -\tfrac{1}{2} \end{aligned}$$

The solution is $\left(-\frac{1}{2}, \frac{1}{2}\right)$. ■

DO EXERCISES 9–11.

5 Problem Solving

As you have already noticed, in the five-step problem-solving process, the most difficult and time-consuming part is the translation of a problem situation to mathematical language. Sometimes translating to an equation takes considerable thought and effort. In many cases, this task becomes easier if we translate to more than one equation in more than one variable.

EXAMPLE 4 An airplane flies the 3000-mi distance from Los Angeles to New York, with a tail wind, in 5 hr. On the return trip, against the wind, it makes the trip in 6 hr. Find the speed of the plane and the speed of the wind.

Solution

1. *Familiarize.* We first make a drawing.

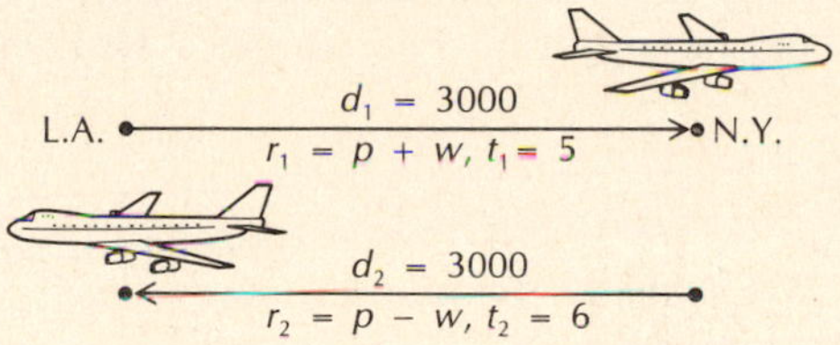

We let p represent the speed of the plane in still air and w represent the speed of the wind. Recall from Chapter 2 that distance, speed, and time are related by the motion formula, $d = rt$. We often need to use other formulas that we can derive from this one, namely, $r = d/t$ and $t = d/r$.

From the figure, we see that the distances are the same. When the plane flies east with the wind, its speed is $p + w$. When it flies west against the wind, its speed is $p - w$. We list the information in a table, the columns of the table coming from the formula $d = rt$.

	Distance	Speed	Time
East (with the wind)	3000	$p + w$	5
West (against the wind)	3000	$p - w$	6

2. *Translate.* Using $d = rt$ in each row of the table, we get an equation. Thus we have a system of equations:

$$3000 = (p + w)5 = 5p + 5w,$$
$$3000 = (p - w)6 = 6p - 6w.$$

3. *Carry out.* We solve the system

$$5p + 5w = 3000,$$
$$6p - 6w = 3000.$$

There is a common factor in each equation. We multiply by $\frac{1}{5}$ in the first equation and $\frac{1}{6}$ in the second equation to eliminate the common factors. Then we have

$$p + w = 600,$$
$$p - w = 500.$$

Next, we multiply the first equation by -1 and add it to the second

Solve.

9. $4x + 3y = -6,$
$-4x + 2y = 16$

10. $9x - 2y = -4,$
$3x + 4y = 1$

11. $0.2x + 0.3y = 0.1,$
$0.3x - 0.1y = 0.7$

12. The sum of two numbers is 10. The difference is 1. Find the numbers.

13. The sum of two numbers is 1. The difference is 10. Find the numbers.

14. It takes a boat 3 hr to travel 24 km upstream. It takes 2 hr to travel the 24 km downstream. Find the speed of the boat and the speed of the stream.

equation. This eliminates the p-term:

$$p + w = 600,$$
$$-2w = -100.$$

Then we solve the second equation for w and substitute into the first equation to find p:

$$-2w = -100 \qquad p + 50 = 600$$
$$w = 50; \qquad p = 550.$$

4. *Check.* We leave the check to the student.
5. *State.* The solution of the system of equations is (550, 50). That is, the speed of the plane is 550 mph and the speed of the wind is 50 mph.

DO EXERCISES 12–14.

EXAMPLE 5 Wine A is 5% alcohol and wine B is 15% alcohol. How many liters of each should be mixed in order to get a 10-L mixture that is 12% alcohol?

Solution

1. *Familiarize.* We organize the information in a table.

	Amount of solution	Percent of alcohol	Amount of alcohol in solution
A	x liters	5%	$5\%x$
B	y liters	15%	$15\%y$
Mixture	10 liters	12%	0.12×10, or 1.2 liters

Note that we have used x for the number of liters of A and y for the number of liters of B. To get the amount of alcohol, we multiply by the percentages, of course.

2. *Translate.* If we add x and y in the first column, we get 10, and this gives us one equation:

$$x + y = 10.$$

If we add the amounts of alcohol in the third column, we get 1.2, and this gives us another equation:

$$5\%x + 15\%y = 1.2.$$

After changing percents to decimals and clearing, we have this system:

$$x + y = 10,$$
$$5x + 15y = 120.$$

3. *Carry out.* Then solve the system. We leave this to the student. The solution is (3, 7). That is, 3 L of wine A and 7 L of wine B are possibilities for a solution to the original problem.
4. *Check.* We add the amounts of wine: 3 L + 7 L = 10 L. Thus the amount of wine checks. Next we check the amount of alcohol:

$$5\%(3) + 15\%(7) = 0.15 + 1.05, \quad \text{or } 1.2 \text{ L}.$$

Thus the amount of alcohol checks.

5. *State.* The solution of the problem is 3 L of wine A and 7 L of wine B.

DO EXERCISE 15.

15. Solution A is 25% acid. Solution B is 65% acid. How many liters of each should be mixed in order to get 8 L of a solution that is 40% acid?

EXERCISE SET 9.1

1

1. Determine whether $\left(\frac{1}{2}, 1\right)$ is a solution of the system

$$3x + y = \frac{5}{2}$$
$$2x - y = \frac{1}{4}.$$

2. Determine whether $\left(-2, \frac{1}{4}\right)$ is a solution of the system

$$x + 4y = -1,$$
$$2x + 8y = -2.$$

2 Solve graphically.

3. $x + y = 2,$ $3x + y = 0$

4. $x + y = 1,$ $3x + y = 7$

5. $y + 1 = 2x,$ $y - 1 = 2x$

6. $y + 1 = 2x,$ $3y = 6x - 3$

3 Solve using the substitution method.

7. $x - 5y = 4,$ $2x + y = 7$

8. $3x - y = 5,$ $x + y = \frac{1}{2}$

4 Solve using the elimination method.

9. $x - 3y = 2,$ $6x + 5y = -34$

10. $x + 3y = 0,$ $20x - 15y = 75$

11. $0.3x + 0.2y = -0.9,$ $0.2x - 0.3y = -0.6$

12. $0.2x - 0.3y = 0.3,$ $0.4x + 0.6y = -0.2$

13. $\frac{1}{5}x + \frac{1}{2}y = 6,$ $\frac{3}{5}x - \frac{1}{2}y = 2$

14. $\frac{2}{3}x + \frac{3}{5}y = -17,$ $\frac{1}{2}x - \frac{1}{3}y = -1$

15. $2a = 5 - 3b,$ $4a = 11 - 7b$

16. $7(a - b) = 14,$ $2a = b + 5$

5 **Problem Solving**

17. Find two numbers whose sum is -10 and whose difference is 1.

18. Find two numbers whose sum is -1 and whose difference is 10.

19. A boat travels 46 km downstream in 2 hr. It travels 51 km upstream in 3 hr. Find the speed of the boat and the speed of the stream.

20. An airplane travels 3000 km with a tail wind in 3 hr. It travels 3000 km with a head wind in 4 hr. Find the speed of the plane and the speed of the wind.

21. Antifreeze A is 18% alcohol and antifreeze B is 10% alcohol. How many liters of each should be mixed in order to get 20 L of a mixture that is 15% alcohol?

22. Beer A is 6% alcohol and beer B is 2% alcohol. How many liters of each should be mixed in order to get 50 L of a mixture that is 3.2% alcohol?

23. Two cars leave town traveling in opposite directions. One travels at a speed of 80 km/h and the other at 96 km/h. In how many hours will they be 528 km apart?

24. A train leaves a station and travels north at a speed of 75 km/h. Two hours later, a second train leaves on a parallel track and travels north at 125 km/h. How far from the station will they meet?

25. Two planes travel toward each other from cities that are 780 km apart at speeds of 190 and 200 km/h. They started at the same time. In how many hours will they meet?

26. Two motorcycles travel toward each other from Chicago and Indianapolis, which are about 350 km apart, at speeds of 110 and 90 km/h. They started at the same time. In how many hours will they meet?

27. One week, a business sold 40 scarves. White ones cost \$4.95 and printed ones cost \$7.95. In all, \$282 worth of scarves were sold. How many of each kind were sold?

28. One day, a store sold 30 sweatshirts. White ones cost \$9.95 and yellow ones cost \$10.50. In all, \$310.60 worth of sweatshirts were sold. How many of each color were sold?

29. Paula is 12 years older than her brother Bob. Four years from now, Bob will be $\frac{2}{3}$ as old as Paula. How old are they now?

30. Carlos is 8 years older than his sister Maria. Four years ago, Maria was $\frac{2}{3}$ as old as Carlos. How old are they now?

31. The perimeter of a lot is 190 m. The width is one-fourth the length. Find the dimensions.

32. The perimeter of a rectangular field is 628 m. The width of the field is 6 m less than the length. Find the dimensions.

33. The perimeter of a rectangle is 384 m. The length is 82 m greater than the width. Find the length and the width.

34. The perimeter of a rectangle is 86 cm. The length is 19 cm greater than the width. Find the area.

35. *Business.* Two investments are made that total \$15,000. For a certain year, these investments yield \$1432 in simple interest. Part of the \$15,000 is invested at 9% and part at 10%. Find the amount invested at each rate.

36. *Business.* For a certain year, \$3900 is received in interest from two investments. A certain amount is invested at 5%, and \$10,000 more than this is invested at 6%. Find the amount invested at each rate. (*Hint:* Express each equation in standard form $Ax + By = C$.)

SYNTHESIS

Solve.

37. $\dfrac{x+y}{4} - \dfrac{x-y}{3} = 1,$
$\dfrac{x-y}{2} + \dfrac{x+y}{4} = -9$

38. $\dfrac{x+y}{2} - \dfrac{y-x}{3} = 0,$
$\dfrac{x+y}{3} - \dfrac{x+y}{4} = 0$

Solve. Check by substituting.

39. $2.35x - 3.18y = 4.82,$
$1.92x + 6.77y = -3.87$

40. $0.0375x + 0.912y = -1.003,$
$463x - 801y = 946$

Problem Solving

41. Nancy jogs and walks to the university each day. She averages 4 km/h walking and 8 km/h jogging. The distance from home to the university is 6 km and she makes the trip in 1 hr. How far does she jog in a trip?

42. James and Joan are mathematics professors. They have a total of 46 years of teaching. Two years ago, James had taught 2.5 times as many years as Joan. How long has each taught?

43. A limited edition of a book published by a historical society was offered for sale to its membership. The cost was one book for \$12 or two books for \$20. The society sold 880 books, and the total amount of money taken in was \$9840. How many members ordered two books?

44. The ten's digit of a two-digit positive number is 2 more than three times the unit's digit. If the digits are interchanged, the new number is 13 less than half the given number. Find the given integer. (*Hint:* Let x = the ten's-place digit and y = the unit's-place digit; then $10x + y$ is the number.)

45. The numerator of a fraction is 12 more than the denominator. The sum of the numerator and the denominator is 5 more than three times the denominator. What is the reciprocal of the fraction?

46. The measure of one of two supplementary angles is 8° more than three times the measure of the other. Find the measure of the larger of the two angles.

47. A train leaves Union Station for Central Station, 216 km away, at 9 A.M. One hour later, a train leaves Central Station for Union Station. They meet at noon. If the second train had started at 9 A.M. and the first train at 10:30 A.M., they would still have met at noon. Find the speed of each train.

48. An automobile radiator contains 16 L of antifreeze and water. This mixture is 30% antifreeze. How much of this mixture should be drained and replaced with pure antifreeze so that there will be 50% antifreeze?

49. A stablehand agreed to work for one year. At the end of that time, he was to receive \$240 and one horse. After 7 months, the boy quit the job, but still received the horse and \$100. What is the value of the horse?

50. You are in line at a ticket window. There are two more people ahead of you in line than there are behind you. In the entire line, there are three times as many people as there are behind you. How many people are ahead of you in the line?

51. Phil and Phyllis are siblings. Phyllis has twice as many brothers as she has sisters. Phil has the same number of brothers and sisters. How many girls and how many boys are there in the family?

52. An automobile gets 18 miles per gallon (mpg) in city driving and 24 mpg in highway driving. The car is driven 465 mi on a full tank of 23 gal of gasoline. How many miles were driven in the city and how many were driven on the highway?

53. Two solutions of the equation $y = mx + b$ are $(-2, 3)$ and $(4, -5)$. Find m and b.

54. Two solutions of the equation $Ax + By = 1$ are $(3, -1)$ and $(-4, -2)$. Find A and B.

Business: Supply and demand. Find the equilibrium point (x_E, p_E) for the following supply and demand functions. (See Section 3.7.)

55. Demand: $x + 43p = 800$
Supply: $x - 16p = 210$

56. Demand: $x = 8800 - 30p$
Supply: $x = 7000 + 15p$

57. Demand: $x = 760 - 13p$
Supply: $x = 430 + 2p$

58. Demand: $x + 60p = 2000$
Supply: $x - 94p = 460$

Each of the following is a system of equations that is *not* linear. But each is *linear in form*, in that an appropriate substitution, say u for $1/x$ and v for $1/y$, yields a linear system. Solve for the new variable and then solve for the original variable.

59. $\frac{1}{x} - \frac{3}{y} = 2,$
$\frac{6}{x} + \frac{5}{y} = -34$

60. $2\sqrt[3]{x} + \sqrt{y} = 0,$
$5\sqrt[3]{x} + 2\sqrt{y} = -5$

61. $3|x| + 5|y| = 30,$
$5|x| + 3|y| = 34$

62. $15x^2 + 2y^3 = 6,$
$25x^2 - 2y^3 = -6$

CHALLENGE

63. A student, out hiking for the weekend, is standing on a railroad bridge, as shown in the figure below. A train is approaching from the direction shown by the arrow. If the student runs at a speed of 10 mph toward the train, she will reach point P on the bridge at the same moment that the train does. If she runs to point Q at the other end of the bridge at a speed of 10 mph, she will reach point Q also at the same moment that the train does. How fast, in miles per hour, is the train traveling?

9.2 Systems of Equations in Three or More Variables

1 Identifying Solutions

A **linear equation in three variables** is an equation equivalent to one of the type $Ax + By + Cz = D$. We now solve systems of these linear equations in three variables.

A **solution** of a system of equations in three variables is an ordered triple that makes all three equations true.

EXAMPLE 1 Determine whether $(2, -1, 0)$ is a solution of the system

$$4x + 2y + 5z = 6,$$
$$2x - y + z = 5,$$
$$x + 2y - z = 2.$$

OBJECTIVES

You should be able to:

1. Determine whether an ordered triple is a solution of a system of equations in three variables.
2. Solve systems of linear equations in three or more variables.
3. Solve applied problems by translating them to a system of three linear equations and solving the system.
4. Fit a quadratic function to data when three data points are given.

1. Consider the following system:

$$\begin{aligned} 4x - y + z &= 6, \\ 2x + y + 2z &= 3, \\ 3x - 2y + z &= 3. \end{aligned}$$

   a) Determine whether $\left(3, 0, \frac{1}{4}\right)$ is a solution.

   b) Determine whether $(2, 1, -1)$ is a solution.

Solution We substitute $(2, -1, 0)$ into each of the three equations:

$$\begin{array}{r|l} \multicolumn{2}{c}{4x + 2y + 5x = 6} \\ \hline 4(2) + 2(-1) + 5(0) & 6 \\ 8 - 2 + 0 & \\ 6 & \end{array} \qquad \begin{array}{r|l} \multicolumn{2}{c}{2x - y + z = 5} \\ \hline 2(2) - (-1) + 0 & 5 \\ 4 + 1 + 0 & \\ 5 & \end{array}$$

$$\begin{array}{r|l} \multicolumn{2}{c}{x + 2y - z = 2} \\ \hline 2 + 2(-1) - 0 & 2 \\ 2 - 2 - 0 & \\ 0 & \end{array}$$

Since $(2, -1, 0)$ is a solution of two of the equations but not *all* of the equations, it is not a solution of the system. ■

DO EXERCISE 1.

2 Solving Systems of Equations in Three or More Variables

Graphical methods of solving linear equations in three variables are unsatisfactory, because a three-dimensional coordinate system is required. The substitution method becomes cumbersome for most systems of more than two equations. Therefore, we will use the elimination method. It is essentially the same as for systems of two equations.

Our goal is to transform the original system to an equivalent one of the form

$$\begin{aligned} Ax + By + Cz &= D, \\ Ey + Fz &= G, \\ Hz &= K. \end{aligned}$$

Then we solve the third equation for z and back-substitute to find the other variables.

EXAMPLE 2 Solve

$$\begin{aligned} 2x - 4y + 6z &= 22, &\quad \text{(P1)} \\ 4x + 2y - 3z &= 4, &\quad \text{(P2)} \\ 3x + 3y - z &= 4, &\quad \text{(P3)} \end{aligned}$$

where (P1), (P2), and (P3) indicate the equation that is in the first, second, and third position, respectively. We will maintain this positional order throughout the solution, and refer to the equations by their positional number.

Solution We begin by multiplying (P3) by 2, to make each x-coefficient a multiple of the first.* Then we have the following:

$$\begin{aligned} 2x - 4y + 6z &= 22, &\quad \text{(P1)} \\ 4x + 2y - 3z &= 4, &\quad \text{(P2)} \\ 6x + 6y - 2z &= 8. &\quad \text{(P3)} \end{aligned}$$

* By proceeding in this manner, we avoid fractions. The method will work when fractions are allowed, but is more difficult that way.

Next, we multiply (P1) by −2 and add it to (P2). We also multiply (P1) by −3 and add it to (P3). Then we have the following:

$$\begin{aligned} 2x - 4y + 6z &= 22, & \text{(P1)}\\ 10y - 15z &= -40, & \text{(P2)}\\ 18y - 20z &= -58. & \text{(P3)} \end{aligned}$$

Now we multiply (P3) by −5 to make the *y*-coefficient a multiple of the *y*-coefficient in (P2):

$$\begin{aligned} 2x - 4y + 6z &= 22, & \text{(P1)}\\ 10y - 15z &= -40, & \text{(P2)}\\ -90y + 100z &= 290. & \text{(P3)} \end{aligned}$$

Next, we multiply (P2) by 9 and add it to (P3):

$$\begin{aligned} 2x - 4y + 6z &= 22, & \text{(P1)}\\ 10y - 15z &= -40, & \text{(P2)}\\ -35z &= -70. & \text{(P3)} \end{aligned}$$

Now we solve (P3) for z:

$$\begin{aligned} -35z &= -70\\ z &= 2. \end{aligned}$$

Next, we back-substitute 2 for z in (P2) and solve for y:

$$\begin{aligned} 10y - 15(2) &= -40\\ 10y - 30 &= -40\\ 10y &= -10\\ y &= -1. \end{aligned}$$

Finally, we back-substitute −1 for y and 2 for z in (P1) and solve for x:

$$\begin{aligned} 2x - 4(-1) + 6(2) &= 22\\ 2x + 4 + 12 &= 22\\ 2x + 16 &= 22\\ 2x &= 6\\ x &= 3. \end{aligned}$$

The solution is (3, −1, 2). To be sure that computational errors have not been made, one can check by substituting 3 for x, −1 for y, and 2 for z in all three original equations. If all are true, then the triple is a solution. ■

DO EXERCISE 2.

Although the solution of a system of three linear equations in three variables is difficult to do graphically, it is of interest to "see" what a solution might be. The graph of a linear equation in three variables is a plane. Thus the solution set of such a system is the intersection of three planes. Some possibilities are shown in the following figures.

2. Solve the system, using exactly the procedure used in the text.

$$\begin{aligned} x + 2y - z &= 5,\\ 2x - 4y + z &= 0,\\ 3x + 2y + 2z &= 3. \end{aligned}$$

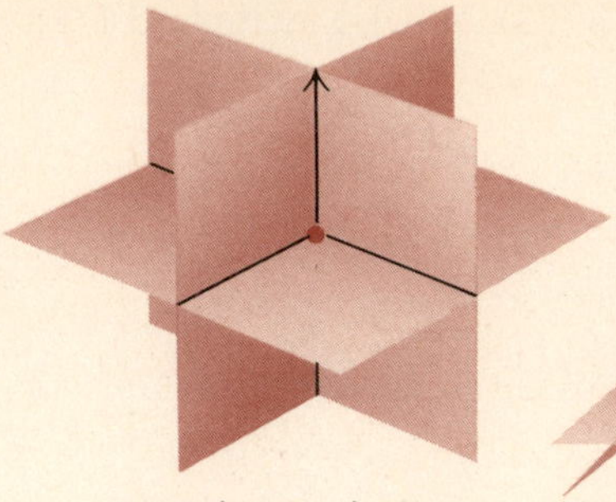

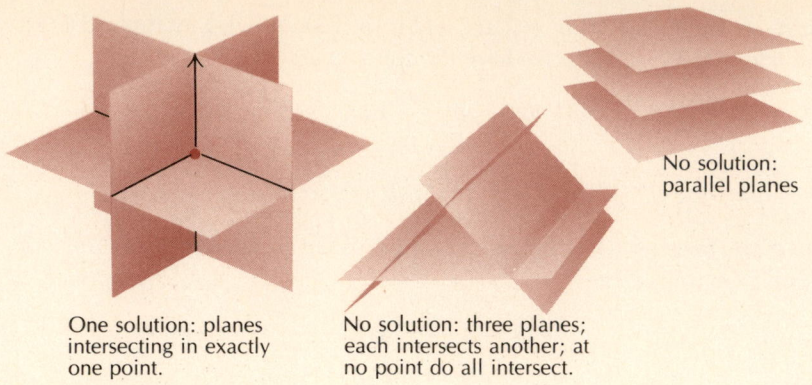

One solution: planes intersecting in exactly one point.

No solution: three planes; each intersects another; at no point do all intersect.

No solution: parallel planes

3 Problem Solving

Solving systems of three or more equations is important in many applications. Systems of equations arise very often in the use of statistics, for example, in such fields as the social sciences. They also occur in problems of business, science, and engineering.

EXAMPLE 3 In a triangle, the largest angle is 70° greater than the smallest angle. The largest angle is twice as large as the remaining angle. Find the measure of each angle.

Solution

1. *Familiarize.* The first thing to do with a problem like this is to make a drawing, or a sketch.

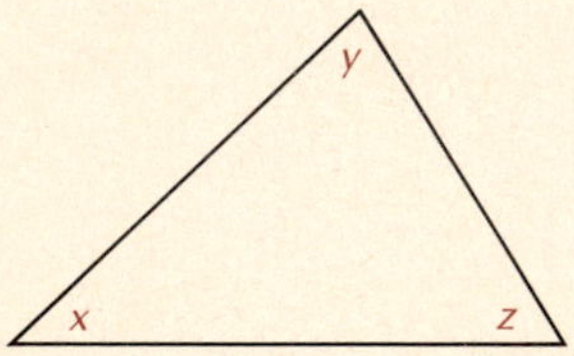

We don't know the size of any angle, so we have used x, y, and z for the measures of the angles. A geometric fact will be needed here—the fact that the measures of the angles of a triangle add up to 180°.

2. *Translate.* The geometric fact about triangles gives us one equation:

$$x + y + z = 180.$$

There are two statements in the problem that we can translate almost directly.

The largest angle	is	70°	greater than	the smallest angle.
z	$=$	70	$+$	x

The largest angle	is	twice as large as the remaining angle.
z	$=$	$2y$

We now have a system of three equations.

$$\begin{aligned} x + y + z &= 180, \\ x + 70 &= z, \\ 2y &= z; \end{aligned} \qquad \text{or} \qquad \begin{aligned} x + y + z &= 180, \\ x \qquad - z &= -70, \\ 2y - z &= 0. \end{aligned}$$

3. *Carry out.* We solve the system. The details are left to the student, but the solution is (30, 50, 100).

4. *Check.* The sum of the numbers is 180, so that checks. The largest angle measures 100° and the smallest measures 30°. The largest angle is thus 70° greater than the smallest. The remaining angle measures 50°. The largest angle measures 100°, so it is twice as large. We do have an answer to the problem.

5. *State.* The measures of the angles of the triangle are 30°, 50°, and 100°.

DO EXERCISE 3.

3. In a factory, there are three machines A, B, and C. When all three are running, they produce 222 suitcases per day. If A and B work but C does not, they produce 159 suitcases per day. If B and C work but A does not, they produce 147 suitcases per day. What is the daily production of each machine?

4 Mathematical Models and Problem Solving

In a situation in which a quadratic function will serve as a mathematical model, we may wish to find an equation, or formula, for the function. Recall that for a linear model, we can find an equation if we know two data points. For a quadratic function, we need three data points.

EXAMPLE 4 In a certain situation, it is believed that a quadratic function will be a good model. Find an equation of the function, given the data points $(1, -4)$, $(-1, -6)$, and $(2, -9)$.

Solution We want to find a quadratic function

$$f(x) = ax^2 + bx + c$$

containing the three given points, that is, a function for which the equation will be true when we substitute any of the ordered pairs of numbers into it. When we substitute, we get

for $(1, -4)$: $-4 = a \cdot 1^2 + b \cdot 1 + c$;

for $(-1, -6)$: $-6 = a(-1)^2 + b(-1) + c$;

for $(2, -9)$: $-9 = a \cdot 2^2 + b \cdot 2 + c$.

We now have a system of equations in the three unknowns a, b, and c:

$$\begin{aligned} a + b + c &= -4, \\ a - b + c &= -6, \\ 4a + 2b + c &= -9. \end{aligned}$$

We solve this system of equations, obtaining $(-2, 1, -3)$. Thus the function we are looking for is

$$f(x) = -2x^2 + x - 3.$$

DO EXERCISE 4.

4. Find a quadratic function that fits the data points $(1, 0)$, $(-1, 4)$, and $(2, 1)$.

EXAMPLE 5 *The cost of operating an automobile at various speeds.* Under certain conditions, it is found that the cost of operating an automobile as a function of speed is approximated by a quadratic function. Use the data shown below to find an equation of the function. Then use the equation to determine the cost of

5. The following table has values that will fit a quadratic function. Find the average number of accidents as a function of age. Then use the model to calculate the average number of accidents in which 16-year-olds are involved daily.

Age of driver	Average number of accidents per day
20	400
40	150
60	400

operating the automobile at 60 mph and at 80 mph.

Speed, in miles per hour	Operating cost per mile, in cents
10	22
20	20
50	20

Solution We use the three data points to obtain a, b, and c in the equation $f(x) = ax^2 + bx + c$:

$$\begin{aligned} 22 &= 100a + 10b + c, \\ 20 &= 400a + 20b + c, \quad \text{Substituting} \\ 20 &= 2500a + 50b + c. \end{aligned}$$

We solve this system of equations, obtaining $(0.005, -0.35, 25)$. Thus,

$$f(x) = 0.005x^2 - 0.35x + 25.$$

To find the cost of operating at 60 mph, we find $f(60)$:

$$f(60) = 0.005(60)^2 - 0.35(60) + 25 = 22¢.$$

We also find $f(80)$:

$$f(80) = 0.005(80)^2 - 0.35(80) + 25 = 29¢.$$

A graph of the cost function of Example 5 is as follows.

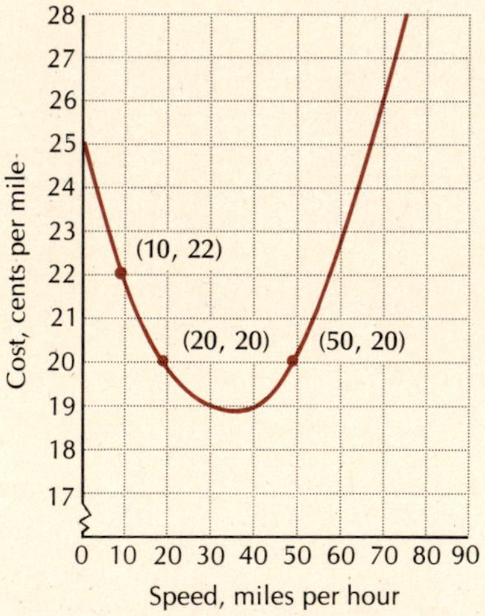

It should be noted that this cost function can give approximate results only within a certain interval. For example, $f(0) = 25$, meaning that it cost 25 cents per mile to stand still. This, of course, is absurd in the sense of mileage, although one does incur costs in owning a car whether one drives it or not. ■

DO EXERCISE 5.

EXERCISE SET 9.2

1 Consider the system

$$\begin{aligned} 2x + 3y - 5z &= 1, \\ 6x - 6y + 10z &= 3, \\ 4x - 9y + 5z &= 0. \end{aligned}$$

1. Determine whether $(-1, 1, 0)$ is a solution of the system.

2. Determine whether $\left(\frac{1}{2}, \frac{1}{3}, \frac{1}{5}\right)$ is a solution of the system.

2 Solve.

3. $$\begin{aligned} x + y + z &= 2, \\ 6x - 4y + 5z &= 31, \\ 5x + 2y + 2z &= 13 \end{aligned}$$

4. $$\begin{aligned} x + 6y + 3z &= 4, \\ 2x + y + 2z &= 3, \\ 3x - 2y + z &= 0 \end{aligned}$$

5. $$\begin{aligned} x - y + 2z &= -3, \\ x + 2y + 3z &= 4, \\ 2x + y + z &= -3 \end{aligned}$$

6. $$\begin{aligned} x + y + z &= 6, \\ 2x - y - z &= -3, \\ x - 2y + 3z &= 6 \end{aligned}$$

7. $$\begin{aligned} 4a + 9b &= 8, \\ 8a + 6c &= -1, \\ 6b + 6c &= -1 \end{aligned}$$

8. $$\begin{aligned} 3p + 2r &= 11, \\ q - 7r &= 4, \\ p - 6q &= 1 \end{aligned}$$

9. $$\begin{aligned} w + x + y + z &= 2, \\ w + 2x + 2y + 4z &= 1, \\ -w + x - y - z &= -6, \\ -w + 3x + y - z &= -2 \end{aligned}$$

10. $$\begin{aligned} w + x - y + z &= 0, \\ -w + 2x + 2y + z &= 5, \\ -w + 3x + y - z &= -4, \\ -2w + x + y - 3z &= -7 \end{aligned}$$

3 Problem Solving

11. The sum of three numbers is 26. Twice the first minus the second is 2 less than the third. The third is the second minus three times the first. Find the numbers.

12. The sum of three numbers is 5. The first number minus the second plus the third is 1. The first minus the third is 3 more than the second. Find the numbers.

13. In triangle ABC, the measure of angle B is three times the measure of angle A. The measure of angle C is 30° greater than the measure of angle A. Find the angle measures.

14. In triangle ABC, the measure of angle B is 2° more than three times the measure of angle A. The measure of angle C is 8° more than the measure of angle A. Find the angle measures.

15. A farmer picked strawberries on three days. She picked a total of 87 quarts. On Tuesday, she picked 15 quarts more than on Monday. On Wednesday, she picked 3 quarts fewer than on Tuesday. How many quarts did she pick each day?

16. Gina sells magazines part time. On Thursday, Friday, and Saturday, she sold \$66 worth. On Thursday, she sold \$3 more than on Friday. On Saturday, she sold \$6 more than on Thursday. How much did she take in each day?

17. Sawmills A, B, and C can produce 7400 board-feet of lumber per day. Mills A and B together can produce 4700 board-feet, while B and C together can produce 5200 board-feet. How many board-feet can each mill produce by itself?

18. In a factory there are three polishing machines, A, B, and C. When all three of them are working, 5700 lenses can be polished in one week. When only A and B are working, 3400 lenses can be polished in one week. When only B and C are working, 4200 lenses can be polished in one week. How many lenses can be polished in a week by each machine?

19. Three welders A, B, and C can weld 37 linear feet per hour when working together. If A and B together can weld 22 linear feet per hour, and A and C together can weld 25 linear feet per hour, how many linear feet per hour can each weld alone?

20. When three pumps, A, B, and C, are running together, they can pump 3700 gallons per hour. When only A and B are running, 2200 gallons per hour can be pumped. When only A and C are running, 2400 gallons per hour can be pumped. What is the pumping capacity of each pump?

21. On an 18-hole golf course, there are par-3 holes, par-4 holes, and par-5 holes. A golfer who shoots par on every hole has a total of 72. The sum of the number of par-3 holes and the number of par-5 holes is 8. How many of each type of hole are there on the golf course?

22. On an 18-hole golf course, there are par-3 holes, par-4 holes, and par-5 holes. A golfer who shoots par on every hole has a total of 70. There are twice as many par-4 holes as there are par-5 holes. How many of each type of hole are there on the golf course?

23. *Business.* A person receives \$212 per year in simple interest from three investments totaling \$2500. Part is invested at 7%, part at 8%, and part at 9%. There is \$1100 more invested at 9% than at 8%. Find the amount invested at each rate.

24. *Business.* A person receives \$341 per year in simple interest from three investments totaling \$3500. Part is invested at 8%, part at 9%, and part at 10%. There is \$2600 more invested at 10% than at 9%. Find the amount invested at each rate.

4 Solve.

25. *Curve fitting.* Find numbers a, b, and c such that a quadratic function $ax^2 + bx + c$ fits the data points $(1, 4)$, $(-1, -2)$, and $(2, 13)$. Write the equation for the function.

26. *Curve fitting.* Find numbers a, b, and c such that a quadratic function $ax^2 + bx + c$ fits the data points $(1, 4)$, $(-1, 6)$, and $(-2, 16)$. Write the equation for the function.

27. *Predicting earnings.* A business earns \$38 in the first week, \$66 in the second week, and \$86 in the third week. The manager graphs the points (1, 38), (2, 66), and (3, 86) and finds that a quadratic function might fit the data.

    a) Find a quadratic function that fits the data.
    b) Using the model, predict the earnings for the fourth week.

28. *Predicting earnings.* A business earns \$1000 in its first month, \$2000 in the second month, and \$8000 in the third month. The manager plots the points (1, 1000), (2, 2000), and (3, 8000) and finds that a quadratic function might fit the data.

    a) Find a quadratic function that fits the data.
    b) Using the model, predict the earnings for the fourth month.

29. *Biomedical: Death rate as a function of sleep.* (This problem is based on a study by Dr. Harold J. Morowitz.)

Average number of hours of sleep, x	Death rate per year per 100,000 males, y
5	1121
7	626
9	967

    a) Use the given data points to find a quadratic function $f(x) = ax^2 + bx + c$ that fits the data.
    b) Use the model to find the death rate of males who sleep 4 hr, 6 hr, and 10 hr.

30. *Counter reading on a VCR.* A person buys a video cassette recorder on which there is a revolution counter. There is also a booklet with a table that relates the time and the counter reading for which the tape has run.

Counter reading	Time of tape, in hours
000	0
300	1
500	2

    a) Find a quadratic function that fits the data.
    b) What amount of time has a tape run if the counter reading is 650?
    c) Use the function to find the counter reading after the tape has run for $1\frac{1}{2}$ hr.

31. *Shoe size and life expectancy.* In a recent study published in *Orthopedic Quarterly*, a team of Swedish orthopedists hypothesized a correlation between shoe size and life expectancy that closely fits a quadratic function. Data are given in the following table.

Shoe size (men)	Life expectancy, in years
8	72
11	82
14	69

    a) Find a quadratic function that fits the data.
    b) What is the life expectancy of a man with a shoe size of 10?
    c) A man's life expectancy is 79. What is his shoe size?

SYNTHESIS

Hint for Exercises 32 and 33: Let u represent $1/x$, v represent $1/y$, and w represent $1/z$. First solve for u, v, and w.

32. $\frac{2}{x} + \frac{2}{y} - \frac{3}{z} = 3,$
$\frac{1}{x} - \frac{2}{y} - \frac{3}{z} = 9,$
$\frac{7}{x} - \frac{2}{y} + \frac{9}{z} = -39$

33. $\frac{2}{x} - \frac{1}{y} - \frac{3}{z} = -1,$
$\frac{2}{x} - \frac{1}{y} + \frac{1}{z} = -9,$
$\frac{1}{x} + \frac{2}{y} - \frac{4}{z} = 17$

34. Pipes A, B, and C are connected to the same tank. When all three pipes are running, they can fill the tank in 3 hr. When pipes A and C are running, they can fill the tank in 4 hr. When pipes A and B are running, they can fill the tank in 8 hr. How long would it take each, running alone, to fill the tank?

35. When A, B, and C work together, they can do a job in 2 hr. When B and C work together, they can do the job in 4 hr. When A and B work together, they can do the job in $\frac{12}{5}$ hr. How long would it take each, working alone, to do the job?

36. Find the year in which the first U.S. transcontinental railroad was completed. The following are some facts about the number. The sum of the digits in the year is 24. The unit's digit is 1 more than the hundred's digit. Both the ten's and the unit's digits are multiples of three.

37. Find the sum of the angle measures at the tips of the star.

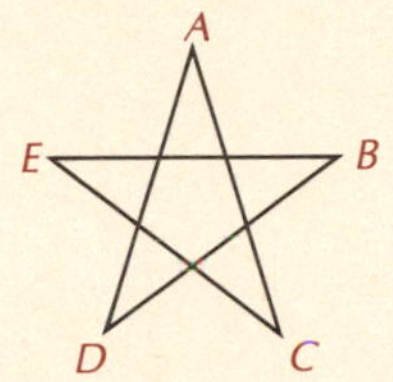

In each case, three solutions of an equation are given. Find the constants and write the equation. Use a system of equations.

38. $y = B - Mx - Nz$; $(1, 1, 2)$, $(3, 2, -6)$, and $\left(\frac{3}{2}, 1, 1\right)$

39. $Ax + By + Cz = 12$; $\left(1, \frac{3}{4}, 3\right)$, $\left(\frac{4}{3}, 1, 2\right)$, and $(2, 1, 1)$

CHALLENGE

40. A theater had 100 people in attendance. The audience consisted of men, women, and children. The ticket prices were $10 for men, $3 for women, and 50 cents for children. The total amount of money taken in was $100. How many men, women, and children were in attendance? Does there seem to be some information missing? Do some careful reasoning.

Solve.

41. $3.12x + 2.14y - 0.988z = 3.79,$
$3.84x - 3.53y + 1.96z = 7.80,$
$4.63x - 1.08y + 0.011z = 6.34$

42. $1.01x - 0.905y + 2.12z = -2.54,$
$1.32x + 2.05y + 2.97z = 3.97,$
$2.21x + 1.35y + 0.001z = -3.15$

43. Art, Bob, Carl, Denny, and Fred are on the same bowling team. They are all being truthful in the following comments regarding the last game they bowled.

Art: My score was a prime number. Fred finished third.

Bob: None of us bowled a score over 200.

Carl: Art beat me by exactly 23 pins. Denny's score was divisible by 10.

Denny: The sum of our five scores was exactly 885 pins. Bob's score was divisible by 8.

Fred: Art beat Bob by fewer than 10 pins. Denny beat Bob by exactly 14 pins.

Determine the score of each bowler in the game.

9.3 Special Cases

OBJECTIVE

You should be able to:

1 Solve systems of equations using the elimination method for special cases where systems may have no solution or infinitely many solutions. Then classify systems of equations as consistent or inconsistent, dependent or independent.

1 In Sections 9.1 and 9.2, each system had *exactly* one solution. Here we consider special cases where systems have no solution or infinitely many solutions.

Consistent and Inconsistent Systems

DEFINITION

A system of equations is *consistent* if and only if it has a solution.
A system of equations is *inconsistent* if and only if it has no solution.

Solve using the elimination method. Classify the system as consistent or inconsistent.

1. $4x - 2y = 2,$
   $2x - y = -8$

2. $4x - 2y = 2,$
   $2x + y = -3$

Let us consider a system that does not have a solution and see what happens when we apply the elimination method.

EXAMPLE 1 Solve using the elimination method. Classify the system as consistent or inconsistent.

$$x - 3y = 1,$$
$$-2x + 6y = 5$$

Solution Let us first look at what happens graphically. We graph each equation and find where they intersect. It turns out that the lines are parallel and have no point of intersection.

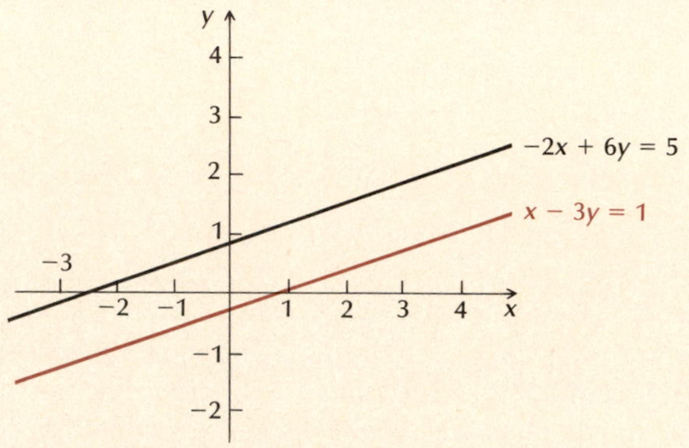

The solution set of the first equation is given by

$$S_1 = \{(x, y) | x - 3y = 1\}.$$

The solution set of the second equation is given by

$$S_2 = \{(x, y) | -2x + 6y = 5\}.$$

Now the solution set S of the system has no ordered pairs in it and is given by

$$S = S_1 \cap S_2 = \emptyset.$$

We see graphically that the system has no solution.

Let us see what happens if we apply the elimination method. We multiply the first equation by 2 and add the result to the second equation. This gives us

$$x - 3y = 1,$$
$$0 = 7.$$

The second equation says that $0 \cdot x + 0 \cdot y = 7$. There are no numbers x and y for which this is true.

Whenever we obtain a statement such as $0 = 7$, which is obviously false, we will know that the system we are trying to solve has no solutions. It is *inconsistent*. The solution set is $\emptyset$.

DO EXERCISES 1 AND 2.

Dependent and Independent Systems

DEFINITION

A system of linear equations is *dependent* if and only if removing one or more equations from the system results in a system that is equivalent to the original system. That is, if there exists a system of fewer equations with the same solutions, then the original system is dependent. Otherwise, the system is *independent*.

EXAMPLE 2 Solve the system using the elimination method. Classify it as consistent or inconsistent, dependent or independent.

$$2x + 3y = 6,$$
$$4x + 6y = 12.$$

Solution Let us look at what happens graphically. We graph each equation.

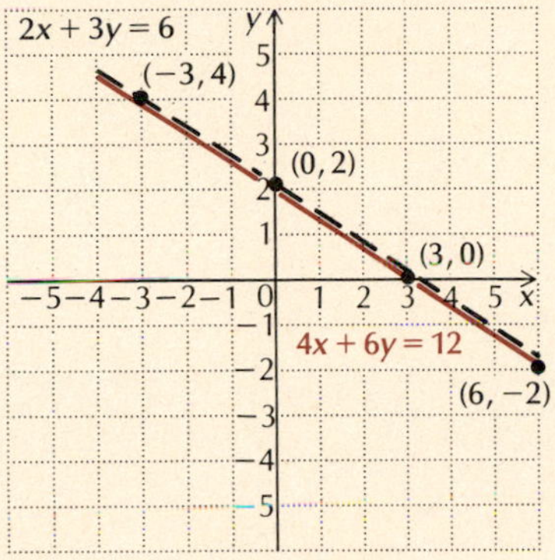

Now the solution set of the first equation is given by

$$S_1 = \{(x, y) | 2x + 3y = 6\}.$$

Some of the ordered pairs in S_1 are

$$(0, 2), (3, 0), (6, -2), \quad \text{and} \quad (-3, 4).$$

The solution set of the second equation is given by

$$S_2 = \{(x, y) | 4x + 6y = 12\}.$$

Some ordered pairs in S_2 are

$$(0, 2), (3, 0), (6, -2), \quad \text{and} \quad (-3, 4).$$

Indeed, the solution sets are the same:

$$S_1 = S_2.$$

Thus the solution sets of each equation are the same. If we remove one of the equations from the system, we still get the same solution set. Thus,

$$\begin{aligned} 2x + 3y &= 6, \\ 4x + 6y &= 12 \end{aligned} \quad \text{is equivalent to} \quad 2x + 3y = 6.$$

Thus the system is *dependent*. It is also *consistent*.

What happens when we apply the elimination method? We multiply the first equation by -2 and add. This gives us

$$2x + 3y = 6$$
$$0 = 0.$$

The equation $0 = 0$ is equivalent to $0x + 0y = 0$, which is true for any values of x and y. Thus it is true for any pair of numbers x and y that constitute a solution of the system. Therefore, the equation $0 = 0$ contributes nothing to the system and can be ignored. We then analyze the rest of the equations to see if the system they form has a solution. In this case, we know that the equation $2x + 3y = 6$ has infinitely many solutions, so the system is *dependent* and *consistent*. ■

EXAMPLE 3 Solve using the elimination method. Classify the system as consistent or inconsistent, dependent or independent.

$$x - 3y = 1,$$
$$x + y = 3,$$
$$5x - 7y = 9$$

Solution Let us look at what happens graphically.

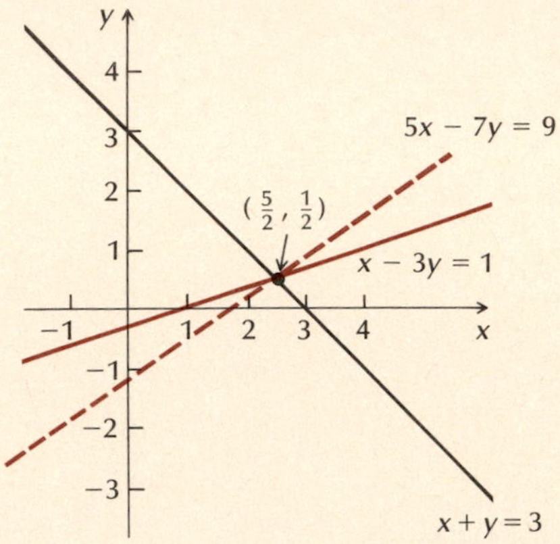

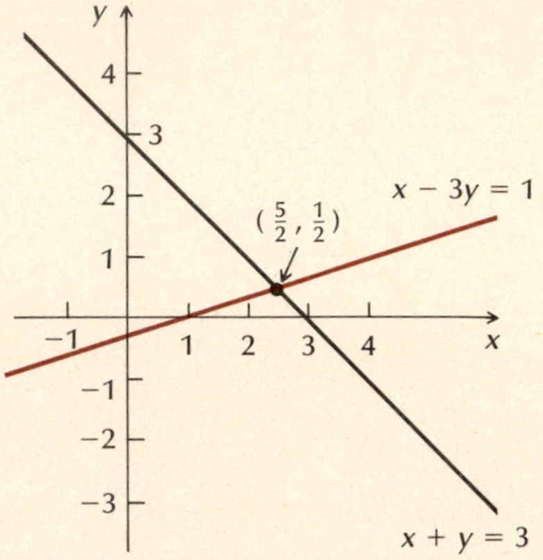

The solution set of the original system is $\{(\frac{5}{2}, \frac{1}{2})\}$. This is where the three graphs intersect. If we ignore the equation $5x - 7y = 9$, we obtain the graphs shown on the right. These still intersect at exactly one point, $(\frac{5}{2}, \frac{1}{2})$. Thus,

$$\begin{aligned} x - 3y &= 1, \\ x + y &= 3, \\ 5x - 7y &= 9 \end{aligned} \quad \text{is equivalent to} \quad \begin{aligned} x - 3y &= 1, \\ x + y &= 3, \end{aligned}$$

which means that the system is *dependent*.

Suppose we apply the elimination method to the original system. We start with

$$x - 3y = 1,$$
$$x + y = 3,$$
$$5x - 7y = 9.$$

We multiply the first equation by -1 and add the result to the second equation. We also multiply the first equation by -5 and add the result to the third equation:

$$x - 3y = 1,$$
$$4y = 2,$$
$$8y = 4.$$

Then we multiply the second equation by -2 and add the result to the third equation:

$$x - 3y = 1,$$
$$4y = 2,$$
$$0 = 0. \qquad \textbf{True for all } x \textbf{ and } y$$

We obtain the equation $0 = 0$, true for all values of x and y. This tells us that the system is *dependent*. To determine consistency, we consider the first two equations and find that the solution is $x = \frac{5}{2}$ and $y = \frac{1}{2}$. Thus the system is *consistent*.

DO EXERCISES 3 AND 4.

Solve using the elimination method. Classify the system as consistent or inconsistent, dependent or independent.

3. $4x - 2y = 6,$
 $-2x + y = -3$

4. $2x - y = 1,$
 $x + y = 5,$
 $x - 2y = -4$

In solving a system, how do we know that it is dependent?

If, at some stage, we find that two of the equations are identical, then we know that the system is *dependent*. If we obtain an obviously true statement, such as 0 = 0, then we know that the system is *dependent*. We cannot know whether such a system is consistent or inconsistent without further analysis.

A dependent and consistent system may have an infinite number of solutions. In such a case for systems of two variables, we can describe the solutions by expressing one variable in terms of the other.

EXAMPLE 4 Solve:

$$2x + 3y = 1,$$
$$4x + 6y = 2.$$

Solution We multiply the first equation by -2 and add. This gives us

$$2x + 3y = 1,$$
$$0 = 0.$$

Now we know that the system is dependent. The last equation contributes nothing, so we consider only the first one. Let us solve for x. We obtain

$$x = \frac{1 - 3y}{2}.$$

We can now describe the ordered pairs in the solution set, in terms of y only, as follows:

$$\left(\frac{1 - 3y}{2}, y\right).$$

Any value that we choose for y then gives us a value for x, and thus an ordered pair in the solution set. Some of these solutions are

$$(-4, 3), \quad \left(-\frac{5}{2}, 2\right), \quad \left(\frac{7}{2}, -2\right).$$

DO EXERCISES 5 AND 6.

When a system of three or more equations is dependent and consistent, we can describe its solutions by expressing one or more of the variables in terms of the others.

EXAMPLE 5 Solve:

$$x + 2y + 3z = 4, \quad \text{(P1)}$$
$$2x - y + z = 3, \quad \text{(P2)}$$
$$3x + y + 4z = 7, \quad \text{(P3)}$$

where (P1), (P2), and (P3) indicate the equations in the first, second, and third positions, respectively.

Solution Each x-coefficient is a multiple of the first. Thus we can begin solving by multiplying (P1) by -2 and adding it to (P2). We also multiply (P1)

Solve. Describe the solutions by expressing one variable in terms of the other. Give three of the ordered pairs in the solution set.

5. $-6x + 4y = 10,$
   $3x - 2y = -5$

6. $2x + 5y = 1,$
   $4x + 10y = 2$

7. Solve, giving the general form of the solutions. Then list three of the solutions.

$$\begin{aligned} 2x + y + z &= 9,\\ x + 2y - 4z &= -3,\\ x + y - z &= 2 \end{aligned}$$

by -3 and add it to (P3):

$$\begin{aligned} x + 2y + 3z &= 4, && \text{(P1)}\\ -5y - 5z &= -5, && \text{(P2)}\\ -5y - 5z &= -5. && \text{(P3)} \end{aligned}$$

Now (P2) and (P3) are identical. We no longer have a system of three equations, but a system of two. Thus we know that the system is dependent. If we were to multiply (P2) by -1 and add it to (P3), we would obtain $0 = 0$. We proceed by multiplying (P2) by $-\frac{1}{5}$ since the coefficients have the common factor -5:

$$\begin{aligned} x + 2y + 3z &= 4, && \text{(P1)}\\ y + z &= 1. && \text{(P2)} \end{aligned}$$

We will express two of the variables in terms of the other one. Let us choose z. Then we solve (P2) for y:

$$y = 1 - z.$$

We substitute this value of y in (P1), obtaining

$$x + 2(1 - z) + 3z = 4.$$

Solving for x, we get

$$x = 2 - z.$$

The solutions, then, are all of the form

$$(2 - z, 1 - z, z).$$

We obtain the solutions by choosing various values of z. If we let $z = 0$, we obtain the triple (2, 1, 0). If we let $z = 2$, we obtain (0, −1, 2), and so on. ■

DO EXERCISE 7.

Homogeneous Equations

When all the terms of a polynomial have the same degree, we say that the polynomial is **homogeneous**. Here are some examples:

$$3x^2 + 5y^2, \qquad 4x + 5y - 2z, \qquad 17x^3 - 4y^3 + 57z^3.$$

An equation formed by a homogeneous polynomial set equal to 0 is called a **homogeneous equation.** Let us now consider a system of homogeneous linear equations.

EXAMPLE 6 Solve:

$$\begin{aligned} 4x - 3y + z &= 0,\\ 2x \qquad - 3z &= 0,\\ -8x + 6y - 2z &= 0 \end{aligned}$$

Solution Any homogeneous system like this always has a solution (it can never be inconsistent), because (0, 0, 0) is a solution. This is called the **trivial solution.** There may or may not be other solutions. To find out, we proceed as in the case of nonhomogeneous equations.

First, we interchange the first two equations so that all the x-coefficients are multiples of the first:

$$\begin{aligned} 2x \qquad - 3z &= 0, && \text{(P1)} \\ 4x - 3y + z &= 0, && \text{(P2)} \\ -8x + 6y - 2z &= 0. && \text{(P3)} \end{aligned}$$

Now we multiply (P1) by -2 and add it to (P2). We also multiply (P1) by 4 and add it to (P3):

$$\begin{aligned} 2x \qquad - 3z &= 0, && \text{(P1)} \\ -3y + 7z &= 0, && \text{(P2)} \\ 6y - 14z &= 0. && \text{(P3)} \end{aligned}$$

Next, we multiply (P2) by 2 and add it to (P3):

$$\begin{aligned} 2x \qquad - 3z &= 0, && \text{(P1)} \\ -3y + 7z &= 0, && \text{(P2)} \\ 0 &= 0. && \text{(P3)} \end{aligned}$$

Now we know that the system is dependent, hence it has an infinite set of solutions. Next, we solve (P2) for y:

$$y = \tfrac{7}{3}z.$$

Typically, we would now substitute $\frac{7}{3}z$ for y in (P1), but since the y-term is missing, we need only solve for x:

$$x = \tfrac{3}{2}z.$$

We can now describe the members of the solution set as follows:

$$\left(\tfrac{3}{2}z, \tfrac{7}{3}z, z\right).$$

Some of the ordered pairs in the solution set are

$$\left(\tfrac{3}{2}, \tfrac{7}{3}, 1\right), \quad \left(3, \tfrac{14}{3}, 2\right), \quad \left(-\tfrac{3}{2}, -\tfrac{7}{3}, -1\right).$$

DO EXERCISES 8 AND 9.

Solve. If there is more than one solution, list three of them.

8. $x + y - z = 0,$
$x + 2y - 4z = 0,$
$2x + y + z = 0$

9. $x + y - z = 0,$
$x - y - z = 0,$
$x + y + 2z = 0$

EXERCISE SET 9.3

1 Solve. If a system has more than one solution, list three of them.

1. $9x - 3y = 15,$
$6x - 2y = 10$

2. $2s - 3t = 9,$
$4s - 6t = 9$

3. $5c + 2d = 24,$
$30c + 12d = 10$

4. $3x + 2y = 18,$
$9x + 6y = 5$

Solve.

5. $3x + 2y = 5,$
$4y = 10 - 6x$

6. $5x + 2y = 7y,$
$-14y + 4 = -10x$

7. $12y - 8x = 6,$
$4x + 3 = 6y$

8. $16x - 12y = 10,$
$6y + 5 = 8x$

Solve. If a system has more than one solution, list three of them.

9. $x + 2y - z = -8,$
$2x - y + z = 4,$
$8x + y + z = 2$

10. $x + 2y - z = 4,$
$4x - 3y + z = 8,$
$5x - y = 12$

11. $2x + y - 3z = 1,$
$x - 4y + z = 6,$
$4x + 16y + 4z = 24$

12. $4x + 12y + 16z = 4,$
$3x + 4y + 5z = 3,$
$x + 8y + 11z = 1$

13. $2x + y - 3z = 0,$
$x - 4y + z = 0,$
$4x - 16y + 4z = 0$

14. $4x + 12y + 16z = 0,$
$3x + 4y + 5z = 0,$
$x + 8y + 11z = 0$

15. $x + y - z = -3,$
$x + 2y + 2z = -1$

16. $x + y + 13z = 0,$
$x - y - 6z = 0$

17. $2x + y + z = 0,$
$x + y - z = 0,$
$x + 2y + 2z = 0$

18. $5x + 4y + z = 0,$
$10x + 8y - z = 0,$
$x - y - z = 0$

19. Classify each of the systems in the odd-numbered exercises 1–17 as consistent or inconsistent, dependent or independent.

20. Classify each of the systems in the even-numbered exercises 2–18 as consistent or inconsistent, dependent or independent.

SYNTHESIS

Solve.

21. $4.026x - 1.448y = 18.32,$
$0.724y = -9.16 + 2.013x$

22. $0.0284y = 1.052 - 8.114x,$
$0.0142y + 4.057x = 0.526$

23. a) Solve:

$$\begin{aligned} w + x + y + z &= 4,\\ w + x + y + z &= 3,\\ w + x + y + z &= 3. \end{aligned}$$

b) Classify the system as consistent or inconsistent.
c) Classify the system as dependent or independent.

24. a) Solve:

$$\begin{aligned} w - 8x + 3y + 2z &= 0,\\ -w + 5x + y - z &= 0,\\ -w + 2x + 5y &= 0,\\ 3x - 4y - z &= 0. \end{aligned}$$

b) Classify the system as consistent or inconsistent.
c) Classify the system as dependent or independent.

CHALLENGE

Determine the constant k such that each system is dependent.

25. $6x - 9y = -3,$
$-4x + 6y = k$

26. $8x - 16y = 20,$
$10x - 20y = k$

27. On an 18-hole golf course, there are par-3 holes, par-4 holes, and par-5 holes. A golfer who shoots par on every hole has a total of 72. There are the same number of par-3 holes as there are par-5 holes. There is at least one of each type of hole. How many of each type of hole are there on the golf course?

28. A two-digit number is such that the number is equal to four times the sum of the digits. Find the number.

9.4 Elimination Using Matrices

OBJECTIVE

You should be able to:

1 Solve systems of linear equations using matrices.

1 Solving Systems Using Matrices

In solving systems of equations, we perform computations with the constants. The variables play no important role in the process. We can simplify writing by omitting the variables. For example, the system

$$\begin{aligned} 3x + 4y &= 5,\\ x - 2y &= 1 \end{aligned}$$

simplifies to

$$\begin{matrix} 3 & 4 & 5\\ 1 & -2 & 1 \end{matrix}$$

if we leave off the variables and omit the operation and equals signs.

In the example above, we have written a rectangular array of numbers. Such an array is called a **matrix** (plural, **matrices**). We ordinarily write brackets around matrices, although this is not necessary when calculating with matrices. The following are matrices:

$$\begin{bmatrix} 4 & 1 & 3 & 5 \\ 1 & 0 & 1 & 2 \\ 6 & 3 & -2 & 0 \end{bmatrix}, \quad \begin{bmatrix} 6 & 2 & 1 & 4 & 7 \\ 1 & 2 & 1 & 3 & 1 \\ 4 & 0 & -2 & 0 & -3 \end{bmatrix}, \quad \begin{bmatrix} 1 & 2 \\ 145 & 0 \\ -7 & 9 \\ 8 & 1 \\ 0 & 0 \end{bmatrix}.$$

The **rows** of a matrix are horizontal, and the **columns** are vertical.

$$\begin{bmatrix} 5 & -2 & 2 \\ 1 & 0 & 1 \\ 0 & 1 & 2 \end{bmatrix} \begin{matrix} \leftarrow \text{row 1} \\ \leftarrow \text{row 2} \\ \leftarrow \text{row 3} \end{matrix}$$

$$\begin{matrix} \uparrow & \uparrow & \uparrow \\ \text{column 1} & \text{column 2} & \text{column 3} \end{matrix}$$

Let us now use matrices to solve systems of linear equations.

EXAMPLE 1 Solve:

$$\begin{aligned} 2x - y + 4z &= -3, \\ x - 4z &= 5, \\ 6x - y + 2z &= 10. \end{aligned}$$

Solution We first write a matrix, using only the constants. Note that where there are missing terms, we must write 0's:

$$\begin{bmatrix} 2 & -1 & 4 & -3 \\ 1 & 0 & -4 & 5 \\ 6 & -1 & 2 & 10 \end{bmatrix}.$$

We do exactly the same calculations using the matrix that we would do if we wrote the entire equations. The first step, if possible, is to interchange the rows so that each number in the first column below the first number is a multiple of that number. We do this by interchanging rows 1 and 2. This corresponds to interchanging the first two equations.

$$\begin{bmatrix} 1 & 0 & -4 & 5 \\ 2 & -1 & 4 & -3 \\ 6 & -1 & 2 & 10 \end{bmatrix}$$

Next, we multiply the first row by -2 and add it to the second row:

$$\begin{bmatrix} 1 & 0 & -4 & 5 \\ 0 & -1 & 12 & -13 \\ 6 & -1 & 2 & 10 \end{bmatrix}.$$

This corresponds to multiplying new equation (P1) by -2 and adding it to new equation (P2).*

Now we multiply the first row by -6 and add it to the third row:

$$\begin{bmatrix} 1 & 0 & -4 & 5 \\ 0 & -1 & 12 & -13 \\ 0 & -1 & 26 & -20 \end{bmatrix}.$$

This corresponds to multiplying equation (P1) by -6 and adding it to equation (P3).

---

* Recall that (P1), (P2), and (P3) indicate the equations that are in the first, second, and third position, respectively.

Solve using matrices.

1. $5x - 2y = -3,$
$2x + 5y = -24$

2. $x - 2y + 3z = 4,$
$2x - y + z = -1,$
$4x + y + z = 1$

Next we multiply row 2 by -1 and add it to the third row:

$$\begin{bmatrix} 1 & 0 & -4 & 5 \\ 0 & -1 & 12 & -13 \\ 0 & 0 & 14 & -7 \end{bmatrix}.$$

This corresponds to multiplying equation (P2) by -1 and adding it to equation (P3).

If we now put the variables back, we have

$$\begin{aligned} x \qquad - 4z &= 5, \\ -y + 12z &= -13, \\ 14z &= -7. \end{aligned}$$

Now we proceed as before. We solve (P3) for z and get $z = -\frac{1}{2}$. Next we back-substitute $-\frac{1}{2}$ for z in (P2) and solve for y: $-y + 12\left(-\frac{1}{2}\right) = -13$, so $y = 7$. Since there is no y-term in (P1), we need only substitute $-\frac{1}{2}$ for z in (P1) and solve for x: $x - 4\left(-\frac{1}{2}\right) = 5$, so $x = 3$. The solution is $\left(3, 7, -\frac{1}{2}\right)$.

Note in the preceding that our goal was to get the matrix in the form

$$\begin{bmatrix} a & b & c & d \\ 0 & e & f & g \\ 0 & 0 & h & k \end{bmatrix},$$

where there are just 0's below the **main diagonal**, formed by a, e, and h. Then we put the variables back and complete the solution.

All the operations used in the preceding example correspond to operations with the equations and they produce equivalent systems of equations. We call the matrices **row-equivalent** and the operations that produce them **row-equivalent operations.**

THEOREM 2

Each of the following row-equivalent operations produces an equivalent matrix:

a) Interchanging any two rows of a matrix.

b) Multiplying each element of a row by the same nonzero number.

c) Multiplying each element of a row by a nonzero number and adding the result to another row.

The best overall method for solving systems of equations is by row-equivalent matrices; even computers are programmed to use them.

DO EXERCISES 1 AND 2.

EXERCISE SET 9.4

1 Solve using matrices.

1. $4x + 2y = 11,$
$3x - y = 2$

2. $3x - 3y = 11,$
$9x - 2y = 5$

3. $x + 2y - 3z = 9,$
$2x - y + 2z = -8,$
$3x - y - 4z = 3$

4. $x - y + 2z = 0,$
$x - 2y + 3z = -1,$
$2x - 2y + z = -3$

5. $5x - 3y = -2,$
$4x + 2y = 5$

6. $3x + 4y = 7,$
$-5x + 2y = 10$

7. $4x - y - 3z = 1,$
$8x + y - z = 5,$
$2x + y + 2z = 5$

8. $3x + 2y + 2z = 3,$
$x + 2y - z = 5,$
$2x - 4y + z = 0$

9. $p + q + r = 1,$
$p + 2q + 3r = 4,$
$4p + 5q + 6r = 7$

10. $m + n + t = 9,$
$m - n - t = -15,$
$m + n + t = 3$

11. $-2w + 2x + 2y - 2z = -10,$
$w + x + y + z = -5,$
$3w + x - y + 4z = -2,$
$w + 3x - 2y + 2z = -6$

12. $-w + 2x - 3y + z = -8,$
$-w + x + y - z = -4,$
$w + x + y + z = 22,$
$-w + x - y - z = -14$

SYNTHESIS

13. A collection of 34 coins consists of dimes and nickels. The total value is \$1.90. How many dimes and how many nickels are there?

14. A collection of 43 coins consists of dimes and quarters. The total value is \$7.60. How many dimes and how many quarters are there?

15. A collection of 22 coins consists of nickels, dimes, and quarters. The total value is \$2.90. There are 6 more nickels than dimes. How many of each type of coin are there?

16. A collection of 18 coins consists of nickels, dimes, and quarters. The total value is \$2.55. There are 2 more quarters than dimes. How many of each type of coin are there?

17. A tobacco dealer has two kinds of tobacco. One is worth \$4.05 per pound and the other is worth \$2.70 per pound. The dealer wants to blend the two tobaccos to get a 15-lb mixture worth \$3.15 per pound. How much of each kind of tobacco should be used?

18. A grocer mixes candy worth \$0.80 per pound with nuts worth \$0.70 per pound to get a 20-lb mixture worth \$0.77 per pound. How many pounds of candy and how many pounds of nuts are used?

Recall the formula $I = prt$, for simple interest.

19. One year, \$8950 was received in interest from two investments. A certain amount was invested at $12\frac{1}{2}\%$, and \$10,000 more than this was invested at 13%. Find the amount of principal invested at each rate.

20. One year, a person had some money invested at $13\frac{1}{2}\%$ and another amount at $13\frac{3}{4}\%$. The income from the investments was \$3275. The income from the $13\frac{1}{2}\%$ investment was \$575 less than that from the $13\frac{3}{4}\%$ investment. How much was invested at each rate?

▦ Solve.

21. $4.83x + 9.06y = -39.42,$
$-1.35x + 6.67y = -33.99$

22. $3.11x - 2.04y = -24.39,$
$7.73x + 5.19y = -35.48$

23. $3.55x - 1.35y + 1.03z = 9.16,$
$-2.14x + 4.12y + 3.61z = -4.50,$
$5.48x - 2.44y - 5.86z = 0.813$

24. $4.12x - 1.35y - 18.2z = 601.3,$
$-3.41x + 68.9y + 38.7z = 1777,$
$0.955x - 0.813y - 6.53z = 160.2$

CHALLENGE

Solve.

25. $\sqrt{2}x + \pi y = 3,$
$\pi x - \sqrt{2}y = 1$

26. $ax + by = c,$
$dx + ey = f$

9.5 The Algebra of Matrices

OBJECTIVES

You should be able to:

1. **Add, subtract, and multiply matrices when possible.**
2. **Write a matrix equation equivalent to a system of equations.**

In Section 9.4, we used matrices to solve systems of equations. Here we study matrices as stand-alone entities, which can be added, subtracted, and multiplied.

Dimensions of a Matrix

A matrix of m rows and n columns is called a matrix with **dimensions** $m \times n$ (read "m by n").

Find the dimensions of the matrix.

1. $\begin{bmatrix} -3 & 5 \\ 4 & \frac{1}{4} \\ -\pi & 0 \end{bmatrix}$

2. $\begin{bmatrix} -3 & 0 \\ 0 & 3 \end{bmatrix}$

3. $\begin{bmatrix} 1 & 2 & 3 \\ 0 & 1 & 8 \\ 0 & 0 & 1 \end{bmatrix}$

4. $\begin{bmatrix} \pi & \sqrt{2} \end{bmatrix}$

5. $\begin{bmatrix} -5 \\ \pi \end{bmatrix}$

6. $[-3]$

7. Which of the above are square matrices?

8. Let

$$A = \begin{bmatrix} 4 & -1 \\ 6 & -3 \end{bmatrix} \text{ and } B = \begin{bmatrix} -6 & -5 \\ 7 & 3 \end{bmatrix}.$$

a) Find A + B.

b) Find B + A.

9. Add:

$$\begin{bmatrix} -3 & -4 & -5 \\ 0 & 1 & -1 \end{bmatrix} + \begin{bmatrix} 4 & 5 & -5 \\ 2 & 3 & -2 \end{bmatrix}.$$

10. Let

$$A = \begin{bmatrix} 4 & -3 \\ 5 & 8 \end{bmatrix} \text{ and } O = \begin{bmatrix} 0 & 0 \\ 0 & 0 \end{bmatrix}.$$

a) Find A + O.

b) Find O + A.

EXAMPLE 1 Find the dimensions of each matrix.

$$\begin{bmatrix} 2 & -3 & 4 \\ -1 & \frac{1}{2} & \pi \end{bmatrix}, \qquad \begin{bmatrix} -3 & 8 & 9 \\ \pi & -2 & 5 \\ -6 & 7 & 8 \end{bmatrix},$$

2 × 3 matrix   3 × 3 matrix

$$[-3 \quad 4], \qquad \begin{bmatrix} 10 \\ -7 \end{bmatrix}.$$

1 × 2 matrix   2 × 1 matrix

Square matrices have the same number of rows as columns. The 3 × 3 matrix is a square matrix.

DO EXERCISES 1–7.

1 Operations on Matrices

Matrix Addition

To add matrices, we add the corresponding elements or members. For this to be possible, the matrices must have the same dimensions.

EXAMPLES Add.

2. $$\begin{bmatrix} -5 & 0 \\ 4 & \frac{1}{2} \end{bmatrix} + \begin{bmatrix} 6 & -3 \\ 2 & 3 \end{bmatrix} = \begin{bmatrix} -5+6 & 0-3 \\ 4+2 & \frac{1}{2}+3 \end{bmatrix} = \begin{bmatrix} 1 & -3 \\ 6 & 3\frac{1}{2} \end{bmatrix}$$

3. $$\begin{bmatrix} 1 & 3 \\ -1 & 5 \\ 6 & 0 \end{bmatrix} + \begin{bmatrix} -1 & -2 \\ 1 & -2 \\ -3 & 1 \end{bmatrix} = \begin{bmatrix} 0 & 1 \\ 0 & 3 \\ 3 & 1 \end{bmatrix}$$

Addition of matrices is both commutative and associative.

DO EXERCISES 8 AND 9.

Zero Matrices

A matrix having zeros for all of its members is called a **zero matrix** and is often denoted by **O**. When a zero matrix is added to another matrix of the same dimensions, the original matrix is obtained. Thus a zero matrix is an **additive identity**.

EXAMPLE 4 Add.

$$\begin{bmatrix} 2 & -1 & 3 \\ 1 & 0 & -1 \end{bmatrix} + \begin{bmatrix} 0 & 0 & 0 \\ 0 & 0 & 0 \end{bmatrix} = \begin{bmatrix} 2 & -1 & 3 \\ 1 & 0 & -1 \end{bmatrix}$$

DO EXERCISE 10.

Additive Inverses and Subtraction

To subtract matrices, we subtract the corresponding members. Of course, the matrices must have the same dimensions for this to be possible.

EXAMPLE 5 Subtract.

$$\begin{bmatrix} 1 & 2 \\ -2 & 0 \\ -3 & -1 \end{bmatrix} - \begin{bmatrix} 1 & -1 \\ 1 & 3 \\ 2 & 3 \end{bmatrix} = \begin{bmatrix} 0 & 3 \\ -3 & -3 \\ -5 & -4 \end{bmatrix}$$

DO EXERCISES 11 AND 12.

The additive inverse, or opposite, of a matrix can be obtained by replacing each member by its additive inverse. Of course, when two matrices that are additive inverses or opposites of each other are added, a zero matrix is obtained.

EXAMPLE 6 Add.

$$\begin{bmatrix} 1 & 0 & 2 \\ 3 & -1 & 5 \end{bmatrix} + \begin{bmatrix} -1 & 0 & -2 \\ -3 & 1 & -5 \end{bmatrix} = \begin{bmatrix} 0 & 0 & 0 \\ 0 & 0 & 0 \end{bmatrix}$$

$$\mathbf{A} \quad + \quad (-\mathbf{A}) \quad = \quad \mathbf{O}$$

DO EXERCISES 13 AND 14.

With numbers, we can subtract by adding an inverse. This is also true of matrices. If we denote matrices by **A** and **B** and an additive inverse by $-\mathbf{B}$, this fact can be stated as follows:

$$\mathbf{A} - \mathbf{B} = \mathbf{A} + (-\mathbf{B}).$$

EXAMPLE 7

$$\begin{bmatrix} 3 & -1 \\ -2 & 4 \end{bmatrix} - \begin{bmatrix} 2 & 1 \\ 3 & -2 \end{bmatrix} = \begin{bmatrix} 1 & -2 \\ -5 & 6 \end{bmatrix}$$

$$\mathbf{A} \quad - \quad \mathbf{B}$$

$$\begin{bmatrix} 3 & -1 \\ -2 & 4 \end{bmatrix} + \begin{bmatrix} -2 & -1 \\ -3 & 2 \end{bmatrix} = \begin{bmatrix} 1 & -2 \\ -5 & 6 \end{bmatrix}$$

$$\mathbf{A} \quad + \quad (-\mathbf{B})$$

DO EXERCISE 15.

Multiplying Matrices and Numbers

We define a product of a matrix and a number.

> **DEFINITION**
>
> **The product of a number k and a matrix A is the matrix, denoted kA, obtained by multiplying each member in A by the number k.**

EXAMPLE 8 Let

$$\mathbf{A} = \begin{bmatrix} -3 & 0 \\ 4 & 5 \end{bmatrix}.$$

Find 3**A** and (-1)**A**.

Subtract.

11. $\begin{bmatrix} 1 & 3 & -2 \\ 4 & 0 & 5 \end{bmatrix} - \begin{bmatrix} 2 & -1 & 5 \\ 6 & 4 & -3 \end{bmatrix}$

12. $\begin{bmatrix} 1 & 2 \\ 4 & 1 \\ -5 & 4 \end{bmatrix} - \begin{bmatrix} 7 & -4 \\ 3 & 5 \\ 2 & -1 \end{bmatrix}$

13. Find the additive inverse:

$$\begin{bmatrix} 2 & -1 & 5 \\ 6 & 4 & -3 \end{bmatrix}.$$

14. Add:

$$\begin{bmatrix} 2 & -1 & 5 \\ 6 & 4 & -3 \end{bmatrix} + \begin{bmatrix} -2 & 1 & -5 \\ -6 & -4 & 3 \end{bmatrix}.$$

15. Add. Compare with Exercise 11.

$$\begin{bmatrix} 1 & 3 & -2 \\ 4 & 0 & 5 \end{bmatrix} + \begin{bmatrix} -2 & 1 & -5 \\ -6 & -4 & 3 \end{bmatrix}$$

Compute the product.

16. $5\begin{bmatrix} 1 & -2 & x \\ 4 & y & 1 \\ 0 & -5 & x^2 \end{bmatrix}$

17. $t\begin{bmatrix} 1 & -1 & 4 & x \\ y & 3 & -2 & y \\ 1 & 4 & -5 & y \end{bmatrix}$

18. Multiply:

$$[4 \quad -2 \quad 3]\begin{bmatrix} 2 \\ 3 \\ -5 \end{bmatrix}.$$

Solution

$$3\mathbf{A} = 3\begin{bmatrix} -3 & 0 \\ 4 & 5 \end{bmatrix} = \begin{bmatrix} -9 & 0 \\ 12 & 15 \end{bmatrix},$$

$$(-1)\mathbf{A} = -1\begin{bmatrix} -3 & 0 \\ 4 & 5 \end{bmatrix} = \begin{bmatrix} 3 & 0 \\ -4 & -5 \end{bmatrix}.$$

DO EXERCISES 16 AND 17.

Products of Matrices

We do not multiply two matrices by multiplying their corresponding members. The definition of matrix products comes from a need to convert a system of linear equations to a product of matrices.

Let us begin by considering one equation,

$$3x + 2y - 2z = 4.$$

We will write the coefficients on the left side in a 1×3 matrix (a **row matrix**) and the variables in a 3×1 matrix (a **column matrix**). The 4 on the right is written in a 1×1 matrix:

$$[3 \quad 2 \quad -2]\begin{bmatrix} x \\ y \\ z \end{bmatrix} = [4].$$

We can return to our original equation by multiplying the members of the row matrix by those of the column matrix, and adding:

$$[3 \quad 2 \quad -2]\begin{bmatrix} x \\ y \\ z \end{bmatrix} = [3x + 2y - 2z].$$

We define multiplication accordingly. In this special case, we have a *row matrix* **A** and a *column matrix* **B**. Their product **AB** is a 1×1 matrix, having the single member 4 (also called $3x + 2y - 2z$).

EXAMPLE 9 Find the product of these matrices.

$$[3 \quad 2 \quad -1]\begin{bmatrix} 1 \\ -2 \\ 3 \end{bmatrix} = [3 \cdot 1 + 2(-2) + (-1) \cdot 3] = [-4]$$

DO EXERCISE 18.

Let us continue by considering a system of equations:

$$\begin{aligned} 3x + 2y - 2z &= 4, \\ 2x - y + 5z &= 3, \\ -x + y + 4z &= 7. \end{aligned}$$

Consider the following matrices:

$$\underset{\mathbf{A}}{\begin{bmatrix} 3 & 2 & -2 \\ 2 & -1 & 5 \\ -1 & 1 & 4 \end{bmatrix}} \quad \underset{\mathbf{X}}{\begin{bmatrix} x \\ y \\ z \end{bmatrix}} \quad \underset{\mathbf{B}}{\begin{bmatrix} 4 \\ 3 \\ 7 \end{bmatrix}}$$

We call A the **coefficient matrix**. If we multiply the first row of A by the (only) column of X, as we did above, we get $3x + 2y - 2z$. If we multiply the second row of A by the column in X, in the same way, we get the following:

$$\begin{bmatrix} 2 & -1 & 5 \end{bmatrix} \begin{bmatrix} x \\ y \\ z \end{bmatrix} = 2x - y + 5z.$$

Note that the first members are multiplied, the second members are multiplied, the third members are multiplied, and the results are added, to get the single number $2x - y + 5z$. What do we get when we multiply the third row of A by the column in X?

$$\begin{bmatrix} -1 & 1 & 4 \end{bmatrix} \begin{bmatrix} x \\ y \\ z \end{bmatrix} = -x + y + 4z$$

We define the product AX to be the column matrix

$$\begin{bmatrix} 3x + 2y - 2z \\ 2x - y + 5z \\ -x + y + 4z \end{bmatrix}.$$

Now consider this matrix equation:

$$\begin{bmatrix} 3x + 2y - 2z \\ 2x - y + 5z \\ -x + y + 4z \end{bmatrix} = \begin{bmatrix} 4 \\ 3 \\ 7 \end{bmatrix}.$$

Equality for matrices is the same as for numbers, that is, a sentence such as $a = b$ says that a and b are two names for the same thing. Thus if the matrix equation above is true, the "two" matrices are really the same one. This means that $3x + 2y - 2z$ is 4, $2x - y + 5z$ is 3, and $-x + y + 4z$ is 7, or that

$$\begin{aligned} 3x + 2y - 2z &= 4, \\ 2x - y + 5z &= 3, \\ -x + y + 4z &= 7. \end{aligned}$$

Thus the matrix equation AX = B, or

$$\begin{bmatrix} 3 & 2 & -2 \\ 2 & -1 & 5 \\ -1 & 1 & 4 \end{bmatrix} \begin{bmatrix} x \\ y \\ z \end{bmatrix} = \begin{bmatrix} 4 \\ 3 \\ 7 \end{bmatrix},$$

is equivalent to the original system of equations.

EXAMPLE 10 Multiply.

$$\begin{bmatrix} 3 & 1 & -1 \\ 1 & 2 & 2 \\ -1 & 0 & 5 \\ 4 & 1 & 2 \end{bmatrix} \begin{bmatrix} 1 \\ 2 \\ 1 \end{bmatrix} = \begin{bmatrix} 3 \cdot 1 + 1 \cdot 2 - 1 \cdot 1 \\ 1 \cdot 1 + 2 \cdot 2 + 2 \cdot 1 \\ -1 \cdot 1 + 0 \cdot 2 + 5 \cdot 1 \\ 4 \cdot 1 + 1 \cdot 2 + 2 \cdot 1 \end{bmatrix} = \begin{bmatrix} 4 \\ 7 \\ 4 \\ 8 \end{bmatrix}$$

DO EXERCISE 19.

In all the examples discussed so far, the second matrix had only one column. If the second matrix has more than one column, we treat each of the columns in the same way when multiplying that we treated the single column. The product matrix will have as many columns as the second matrix.

19. Multiply:

$$\begin{bmatrix} 1 & 4 & 2 \\ -1 & 6 & 3 \\ 3 & 2 & -1 \\ 5 & 0 & 2 \end{bmatrix} \begin{bmatrix} 2 \\ 1 \\ 3 \end{bmatrix}.$$

20. Multiply:

$$\begin{bmatrix} 4 & 1 & 2 \\ -3 & 2 & 3 \\ 2 & 0 & 5 \\ 3 & 1 & 4 \end{bmatrix}\begin{bmatrix} 1 & 4 \\ 2 & 0 \\ -3 & 5 \end{bmatrix}.$$

EXAMPLE 11 Multiply (compare with Example 10).

$$\underset{A}{\begin{bmatrix} 3 & 1 & -1 \\ 1 & 2 & 2 \\ -1 & 0 & 5 \\ 4 & 1 & 2 \end{bmatrix}}\underset{B}{\begin{bmatrix} 1 & 0 \\ 2 & 1 \\ 1 & 3 \end{bmatrix}} = \begin{bmatrix} 4 & 3\cdot 0 + 1\cdot 1 + (-1)3 \\ 7 & 1\cdot 0 + 2\cdot 1 + \ 2\cdot 3 \\ 4 & -1\cdot 0 + 0\cdot 1 + \ 5\cdot 3 \\ 8 & 4\cdot 0 + 1\cdot 1 + \ 2\cdot 3 \end{bmatrix} = \begin{bmatrix} 4 & -2 \\ 7 & 8 \\ 4 & 15 \\ 8 & 7 \end{bmatrix}$$

First column: Same as in Example 10. Second column: The rows of A multiplied by the second column of B

DO EXERCISE 20.

EXAMPLE 12 Multiply.

$$\begin{bmatrix} 3 & 1 & -1 \\ 2 & 0 & 3 \end{bmatrix}\begin{bmatrix} 1 & 4 & 6 \\ 3 & -1 & 9 \\ 2 & 5 & 1 \end{bmatrix}$$

$$= \begin{bmatrix} 3\cdot 1 + 1\cdot 3 - 1\cdot 2 & 3\cdot 4 + 1\cdot(-1) - 1\cdot 5 & 3\cdot 6 + 1\cdot 9 - 1\cdot 1 \\ 2\cdot 1 + 0\cdot 3 + 3\cdot 2 & 2\cdot 4 + 0\cdot(-1) + 3\cdot 5 & 2\cdot 6 + 0\cdot 9 + 3\cdot 1 \end{bmatrix}$$

$$= \begin{bmatrix} 4 & 6 & 26 \\ 8 & 23 & 15 \end{bmatrix}$$

If matrix A has n columns and matrix B has n rows, then we can compute the product AB, regardless of other dimensions. The product will have as many rows as A and as many columns as B.

CAUTION! Given any two matrices A and B, you may or may not be able to add, subtract, or multiply them. **A + B** and **A − B** exist only when the dimensions are the same. **AB** exists only when the number of columns in **A** is the same as the number of rows in **B**.

For example, consider the matrices

$$A = \begin{bmatrix} 3 & 1 & -1 \\ 2 & 0 & 3 \end{bmatrix} \quad \text{and} \quad B = \begin{bmatrix} 1 & 4 & 6 \\ 3 & -1 & 9 \\ 2 & 5 & 1 \end{bmatrix}.$$

The dimensions of **A** are 2×3; the dimensions of **B** are 3×3. **A + B** and **A − B** do not exist because the dimensions of **A** and **B** are *not* the same. **AB** does exist because the number of columns in **A**, 3, is the same as the number of rows in **B**, 3. **AB** is given in Example 12.

$A_{2 \times 3}$ $B_{3 \times 3}$

AB does exist: $3 = 3$.

The dimensions of the product are 2×3.

But **BA** does *not* exist because the number of columns in **B**, 3, is not the same as the number of rows in A, 2.

$$B_{3\times 3} \qquad A_{2\times 3}$$

BA does not exist: $3 \neq 2$.

In this context, it can also be pointed out that matrix multiplication is not commutative: AB exists and BA does not, so $AB \neq BA$.

DO EXERCISES 21–24.

2 Equivalent Matrix Equations

For later purposes, it is important that we be able to write a matrix equation equivalent to a system of equations.

EXAMPLE 13 Write a matrix equation equivalent to this system of equations:

$$\begin{aligned} 4x + 2y - z &= 3, \\ 9x + z &= 5, \\ 4x + 5y - 2z &= 1, \\ x + y + z &= 0. \end{aligned}$$

Solution We write the coefficients on the left in a matrix. We write the product of that matrix and the column matrix containing the variables, and set the result equal to the column matrix containing the constants on the right:

$$\begin{bmatrix} 4 & 2 & -1 \\ 9 & 0 & 1 \\ 4 & 5 & -2 \\ 1 & 1 & 1 \end{bmatrix} \begin{bmatrix} x \\ y \\ z \end{bmatrix} = \begin{bmatrix} 3 \\ 5 \\ 1 \\ 0 \end{bmatrix}.$$

DO EXERCISE 25.

A Summary of Properties of Square Matrices

We now list a summary of some of the properties of square matrices of the same dimensions whose elements are real numbers. We now restrict our discussion to square matrices so all additions and multiplications are possible. Some of the proofs will be considered in the exercise set. Note that not all the field properties hold.

THEOREM 3

For any square matrices A, B, and C of the same dimensions, the following hold:

Commutativity: $A + B = B + A$.

Associativity: $A + (B + C) = (A + B) + C$, $A(BC) = (AB)C$.

Identity: There exists a unique matrix O, such that

$$A + 0 = 0 + A = A.$$

Inverses: There exists a unique matrix $-A$, such that.

$$A + (-A) = -A + A = 0.$$

(continued)

21. Find AB and BA if possible.

$$A = \begin{bmatrix} -2 & 4 & 0 \\ -3 & 0 & -8 \end{bmatrix}, \quad B = \begin{bmatrix} -1 & -2 & -3 \\ 0 & 1 & 0 \\ 4 & 5 & 2 \end{bmatrix}$$

22. Multiply:

$$\begin{bmatrix} 4 & 1 & 0 & 2 \end{bmatrix} \begin{bmatrix} 1 & 0 & 1 \\ 2 & -1 & 0 \\ 3 & 5 & 1 \\ 1 & 3 & 0 \end{bmatrix}.$$

23. Find AB and BA and compare.

$$A = \begin{bmatrix} -8 & 3 \\ -4 & 4 \end{bmatrix}, \quad B = \begin{bmatrix} 1 & -4 \\ 2 & 0 \end{bmatrix}$$

24. Find AI and IA. Comment.

$$A = \begin{bmatrix} 3 & 2 \\ -1 & 5 \end{bmatrix}, \quad I = \begin{bmatrix} 1 & 0 \\ 0 & 1 \end{bmatrix}$$

25. Write a matrix equation equivalent to this system of equations.

$$\begin{aligned} 3x + 4y - 2z &= 5, \\ 2x - 2y + 5z &= 3, \\ 6x + 7y - z &= 0 \end{aligned}$$

Distributivity: $A(B + C) = AB + AC, \quad (B + C)A = BA + CA.$

For any square matrices A and B, of the same dimensions, and any real numbers k and m,

$$k(A + B) = kA + kB,$$
$$(k + m)A = kA + mA,$$
$$(km)A = k(mA),$$

and

$$1A = A.$$

CAUTION! Note that, even with these restrictions, matrix multiplication is still *not* commutative. For example, let

$$A = \begin{bmatrix} 1 & 0 \\ 2 & 0 \end{bmatrix} \quad \text{and} \quad B = \begin{bmatrix} 3 & 4 \\ 0 & 0 \end{bmatrix}.$$

Then

$$AB = \begin{bmatrix} 3 & 4 \\ 6 & 8 \end{bmatrix} \quad \text{and} \quad BA = \begin{bmatrix} 11 & 0 \\ 0 & 0 \end{bmatrix},$$

so $AB \neq BA$.

EXERCISE SET 9.5

1 For Exercises 1–16, let

$$A = \begin{bmatrix} 1 & 2 \\ 4 & 3 \end{bmatrix}, \quad B = \begin{bmatrix} -3 & 5 \\ 2 & -1 \end{bmatrix}, \quad C = \begin{bmatrix} 1 & -1 \\ -1 & 1 \end{bmatrix}, \quad D = \begin{bmatrix} 1 & 1 \\ 1 & 1 \end{bmatrix},$$

$$E = \begin{bmatrix} 1 & 3 \\ 2 & 6 \end{bmatrix}, \quad F = \begin{bmatrix} 3 & 3 \\ -1 & -1 \end{bmatrix}, \quad O = \begin{bmatrix} 0 & 0 \\ 0 & 0 \end{bmatrix}, \quad \text{and} \quad I = \begin{bmatrix} 1 & 0 \\ 0 & 1 \end{bmatrix}.$$

Find.

1. A + B **2.** B + A **3.** E + O **4.** 2A

5. 3F **6.** (−1)D **7.** 3F + 2A **8.** A − B

9. B − A **10.** AB **11.** BA **12.** OF

13. CD **14.** EF **15.** AI **16.** IA

In Exercises 17–20, let

$$A = \begin{bmatrix} 1 & 0 & -2 \\ 0 & -1 & 3 \\ 3 & 2 & 4 \end{bmatrix}, \quad B = \begin{bmatrix} -1 & -2 & 5 \\ 1 & 0 & -1 \\ 2 & -3 & 1 \end{bmatrix}, \quad C = \begin{bmatrix} -2 & 9 & 6 \\ -3 & 3 & 4 \\ 2 & -2 & 1 \end{bmatrix}, \quad \text{and} \quad I = \begin{bmatrix} 1 & 0 & 0 \\ 0 & 1 & 0 \\ 0 & 0 & 1 \end{bmatrix}.$$

Find.

17. AB **18.** BA **19.** CI **20.** IC

Multiply.

21. $\begin{bmatrix} -3 & 2 \end{bmatrix}\begin{bmatrix} 4 \\ -2 \end{bmatrix}$

22. $\begin{bmatrix} -2 & 0 & 4 \end{bmatrix}\begin{bmatrix} 8 \\ -6 \\ \frac{1}{2} \end{bmatrix}$

23. $\begin{bmatrix} -5 & 1 & 2 \end{bmatrix}\begin{bmatrix} 1 & 3 \\ -1 & 0 \\ 4 & -2 \end{bmatrix}$

24. $\begin{bmatrix} -3 & 2 \\ 0 & 1 \\ -4 & 5 \end{bmatrix}\begin{bmatrix} 4 \\ 2 \end{bmatrix}$

2 Write a matrix equation equivalent to each of the following systems of equations.

25. $3x - 2y + 4z = 17,$
$2x + y - 5z = 13$

26. $3x + 2y + 5z = 9,$
$4x - 3y + 2z = 10$

27. $x - y + 2z - 4w = 12,$
$2x - y - z + w = 0,$
$x + 4y - 3z - w = 1,$
$3x + 5y - 7z + 2w = 9$

28. $2x + 4y - 5z + 12w = 2,$
$4x - y + 12z - w = 5,$
$-x + 4y + 2w = 13,$
$2x + 10y + z = 5$

Compute.

29. $\begin{bmatrix} 3.61 & -2.14 & 16.7 \\ -4.33 & 7.03 & 12.9 \\ 5.82 & -6.95 & 2.34 \end{bmatrix}\begin{bmatrix} 3.05 & 0.402 & -1.34 \\ 1.84 & -1.13 & 0.024 \\ -2.83 & 2.04 & 8.81 \end{bmatrix}$

30. $\begin{bmatrix} -1.23 & 4.51 & -17.4 \\ 61.2 & -8.81 & 0.123 \\ 14.14 & 6.92 & -14.4 \end{bmatrix}\begin{bmatrix} 4.24 & 16.1 & 41.3 \\ 0.146 & -6.06 & -18.9 \\ -8.43 & 1.12 & 0.0245 \end{bmatrix}$

SYNTHESIS

For Exercises 31–34, let

$$A = \begin{bmatrix} -1 & 0 \\ 2 & 1 \end{bmatrix} \text{ and } B = \begin{bmatrix} 1 & -1 \\ 0 & 2 \end{bmatrix}.$$

31. Show that
$$(A + B)(A - B) \neq A^2 - B^2,$$
where
$$A^2 = AA \text{ and } B^2 = BB.$$

32. Show that
$$(A + B)(A + B) \neq A^2 + 2AB + B^2.$$

33. Show that
$$(A + B)(A - B) = A^2 + BA - AB - B^2.$$

34. Show that
$$(A + B)(A + B) = A^2 + BA + AB + B^2.$$

35. Show that
$$\begin{bmatrix} \cos x & \sin x \\ -\sin x & \cos x \end{bmatrix}\begin{bmatrix} \cos y & \sin y \\ -\sin y & \cos y \end{bmatrix} = \begin{bmatrix} \cos(x+y) & \sin(x+y) \\ -\sin(x+y) & \cos(x+y) \end{bmatrix}.$$

CHALLENGE

Let

$$A = \begin{bmatrix} a_{11} & a_{12} \\ a_{21} & a_{22} \end{bmatrix}, \quad B = \begin{bmatrix} b_{11} & b_{12} \\ b_{21} & b_{22} \end{bmatrix}, \quad C = \begin{bmatrix} c_{11} & c_{12} \\ c_{21} & c_{22} \end{bmatrix}, \quad I = \begin{bmatrix} 1 & 0 \\ 0 & 1 \end{bmatrix}.$$

Prove each of the following.

36. $A + B = B + A$

37. $A + (B + C) = (A + B) + C$

38. $k(A + B) = kA + kB$

39. $(k + m)A = kA + mA$

40. $A(BC) = (AB)C$

41. $AI = IA = A$

9.6 Determinants and Cramer's Rule

1 Determinants of Two-By-Two Matrices

A matrix of m rows and n columns is called an $m \times n$ matrix (read "m by n"). If a matrix has the same number of rows and columns, it is called a **square matrix.** With every square matrix is associated a number called its **determinant,** defined as follows for 2×2 matrices.*

* The definition of the determinant of any square matrix is given in Section 9.7.

OBJECTIVES

You should be able to:

1. Evaluate determinants of 2×2 matrices.
2. Solve systems of two equations in two variables using Cramer's rule.
3. Evaluate determinants of 3×3 matrices.
4. Solve systems of three equations in three variables using Cramer's rule.

Evaluate.

1. $\begin{vmatrix} \sqrt{3} & -5 \\ -2 & -\sqrt{3} \end{vmatrix}$

2. $\begin{vmatrix} 1 & 2 \\ 3 & 4 \end{vmatrix}$

3. $\begin{vmatrix} -2 & -3 \\ 4 & x \end{vmatrix}$

DEFINITION

The determinant of the matrix $\begin{bmatrix} a & c \\ b & d \end{bmatrix}$ is denoted $\begin{vmatrix} a & c \\ b & d \end{vmatrix}$ and is defined as follows:

$$\begin{vmatrix} a & c \\ b & d \end{vmatrix} = ad - bc.$$

EXAMPLE 1 Evaluate: $\begin{vmatrix} \sqrt{2} & -3 \\ -4 & -\sqrt{2} \end{vmatrix}$.

Solution $\begin{vmatrix} \sqrt{2} & -3 \\ -4 & -\sqrt{2} \end{vmatrix}$ The arrows indicate the products involved.

$$= \sqrt{2}(-\sqrt{2}) - (-4)(-3) = -2 - 12 = -14$$

DO EXERCISES 1–3.

2 Cramer's Rule: 2 × 2 Systems

Determinants have many uses. One of these is in solving systems of linear equations, where the number of variables is the same as the number of equations, and where the constants are not all 0. Let us consider a system of two equations:

$$a_1x + b_1y = c_1,$$
$$a_2x + b_2y = c_2.$$

Using the methods of the preceding sections, we can solve. We obtain

$$x = \frac{c_1b_2 - c_2b_1}{a_1b_2 - a_2b_1}, \qquad y = \frac{a_1c_2 - a_2c_1}{a_1b_2 - a_2b_1}.$$

The numerators and denominators of the expressions for x and y can be written as determinants.

THEOREM 4 Cramer's Rule: 2 × 2 Systems

The solution of the system

$$a_1x + b_1y = c_1,$$
$$a_2x + b_2y = c_2,$$

if it is unique, is given by

$$x = \frac{\begin{vmatrix} c_1 & b_1 \\ c_2 & b_2 \end{vmatrix}}{\begin{vmatrix} a_1 & b_1 \\ a_2 & b_2 \end{vmatrix}}, \qquad y = \frac{\begin{vmatrix} a_1 & c_1 \\ a_2 & c_2 \end{vmatrix}}{\begin{vmatrix} a_1 & b_1 \\ a_2 & b_2 \end{vmatrix}}.$$

The equations above make sense only if the determinant in the denominator is not 0. If the denominator *is* 0, then one of two things happens.

1. If the denominator is 0 and the other two determinants in the numerators are also 0, then the system of equations is dependent.
2. If the denominator is 0 and at least one of the other determinants in the numerators is not 0, then the system is inconsistent.

To use this theorem, we compute the three determinants and compute x and y as shown above. Note that the denominator in both cases contains the coefficients of x and y, in the same position as in the original equations. For x, the numerator is obtained by replacing the x-coefficients (the a's) by the c's. For y, the numerator is obtained by replacing the y-coefficients (the b's) by the c's.

EXAMPLE 2 Solve using Cramer's rule:

$$2x + 5y = 7,$$
$$5x - 2y = -3.$$

Solution We have

$$x = \frac{\begin{vmatrix} 7 & 5 \\ -3 & -2 \end{vmatrix}}{\begin{vmatrix} 2 & 5 \\ 5 & -2 \end{vmatrix}} = \frac{7(-2) - (-3)5}{2(-2) - 5 \cdot 5} = -\frac{1}{29},$$

$$y = \frac{\begin{vmatrix} 2 & 7 \\ 5 & -3 \end{vmatrix}}{\begin{vmatrix} 2 & 5 \\ 5 & -2 \end{vmatrix}} = \frac{2(-3) - 5 \cdot 7}{-29} = \frac{41}{29}.$$

The solution is $\left(-\frac{1}{29}, \frac{41}{29}\right)$.

DO EXERCISES 4–6.

3 Determinants of Three-by-Three Matrices

DEFINITION

The *determinant* of a three-by-three matrix is defined as follows:

$$\begin{vmatrix} a_1 & b_1 & c_1 \\ a_2 & b_2 & c_2 \\ a_3 & b_3 & c_3 \end{vmatrix} = a_1 \cdot \begin{vmatrix} b_2 & c_2 \\ b_3 & c_3 \end{vmatrix} - a_2 \cdot \begin{vmatrix} b_1 & c_1 \\ b_3 & c_3 \end{vmatrix} + a_3 \cdot \begin{vmatrix} b_1 & c_1 \\ b_2 & c_2 \end{vmatrix}.$$

The two-by-two determinants on the right can be obtained from the three-by-three determinant by crossing out the row and the column in which the a-coefficients occur.

EXAMPLE 3 Evaluate.

$$\begin{vmatrix} -1 & 0 & 1 \\ -5 & 1 & -1 \\ 4 & 8 & 1 \end{vmatrix} = -1 \cdot \begin{vmatrix} 1 & -1 \\ 8 & 1 \end{vmatrix} - (-5) \cdot \begin{vmatrix} 0 & 1 \\ 8 & 1 \end{vmatrix} + 4 \cdot \begin{vmatrix} 0 & 1 \\ 1 & -1 \end{vmatrix}$$
$$= -1(1 + 8) + 5(-8) + 4(-1)$$
$$= -9 - 40 - 4 = -53$$

DO EXERCISES 7–9.

Solve using Cramer's rule.

4. $2x - y = 5,$
$x - 2y = 1$

5. $3x + 4y = -2,$
$5x - 7y = 1$

6. $\sqrt{2}x - \pi y = 3,$
$\pi x + \sqrt{2}y = 4$

Evaluate.

7. $\begin{vmatrix} 3 & 2 & 2 \\ -2 & 1 & 4 \\ 4 & -3 & 3 \end{vmatrix}$

8. $\begin{vmatrix} -5 & 0 & 0 \\ 4 & 2 & 0 \\ -3 & 5 & -6 \end{vmatrix}$

9. $\begin{vmatrix} x & 0 & x \\ 0 & x & 0 \\ 1 & 0 & x \end{vmatrix}$

4 Cramer's Rule: 3 × 3 Systems

THEOREM 5 Cramer's Rule: 3 × 3 Systems

The solution of the system

$$\begin{aligned} a_1x + b_1y + c_1z &= d_1, \\ a_2x + b_2y + c_2z &= d_2, \\ a_3x + b_3y + c_3z &= d_3 \end{aligned}$$

is found by considering the following determinants:

$$D = \begin{vmatrix} a_1 & b_1 & c_1 \\ a_2 & b_2 & c_2 \\ a_3 & b_3 & c_3 \end{vmatrix}, \qquad D_x = \begin{vmatrix} d_1 & b_1 & c_1 \\ d_2 & b_2 & c_2 \\ d_3 & b_3 & c_3 \end{vmatrix},$$

$$D_y = \begin{vmatrix} a_1 & d_1 & c_1 \\ a_2 & d_2 & c_2 \\ a_3 & d_3 & c_3 \end{vmatrix}, \qquad D_z = \begin{vmatrix} a_1 & b_1 & d_1 \\ a_2 & b_2 & d_2 \\ a_3 & b_3 & d_3 \end{vmatrix}.$$

The solution, if it is unique, is given by

$$x = \frac{D_x}{D}, \qquad y = \frac{D_y}{D}, \qquad z = \frac{D_z}{D}.$$

Note that we obtain the determinant D_x in the numerator for x from D by replacing the x-coefficients by d_1, d_2, and d_3. A similar thing happens with D_y and D_z. When $D = 0$, Cramer's rule cannot be used. If $D = 0$ and D_x, D_y, and D_z are 0, the system is dependent. If $D = 0$ and one of D_x, D_y, or D_z is not 0, then the system is inconsistent.

EXAMPLE 4 Solve using Cramer's rule:

$$\begin{aligned} x - 3y + 7z &= 13, \\ x + y + z &= 1, \\ x - 2y + 3z &= 4. \end{aligned}$$

Solution We have

$$D = \begin{vmatrix} 1 & -3 & 7 \\ 1 & 1 & 1 \\ 1 & -2 & 3 \end{vmatrix} = -10, \qquad D_x = \begin{vmatrix} 13 & -3 & 7 \\ 1 & 1 & 1 \\ 4 & 2 & 3 \end{vmatrix} = 20,$$

$$D_y = \begin{vmatrix} 1 & 13 & 7 \\ 1 & 1 & 1 \\ 1 & 4 & 3 \end{vmatrix} = -6, \qquad D_z = \begin{vmatrix} 1 & -3 & 13 \\ 1 & 1 & 1 \\ 1 & -2 & 4 \end{vmatrix} = -24.$$

Then

$$x = \frac{D_x}{D} = \frac{20}{-10} = -2,$$

$$y = \frac{D_y}{D} = \frac{-6}{-10} = \frac{3}{5},$$

$$z = \frac{D_z}{D} = \frac{-24}{-10} = \frac{12}{5}.$$

The solution is $(-2, \frac{3}{5}, \frac{12}{5})$. In practice, it is not necessary to evaluate D_z. When we have found values for x and y, we can substitute them into one of the equations and find z.

DO EXERCISE 10.

10. Solve using Cramer's rule:

$$\begin{aligned} x - 3y - 7z &= 6, \\ 2x + 3y + z &= 9, \\ 4x + y &= 7. \end{aligned}$$

EXERCISE SET 9.6

1 Evaluate.

1. $\begin{vmatrix} -2 & -\sqrt{5} \\ -\sqrt{5} & 3 \end{vmatrix}$

2. $\begin{vmatrix} \sqrt{5} & -3 \\ 4 & 2 \end{vmatrix}$

3. $\begin{vmatrix} x & 4 \\ x & x^2 \end{vmatrix}$

4. $\begin{vmatrix} y^2 & -2 \\ y & 3 \end{vmatrix}$

3 Evaluate.

5. $\begin{vmatrix} 3 & 1 & 2 \\ -2 & 3 & 1 \\ 3 & 4 & -6 \end{vmatrix}$

6. $\begin{vmatrix} 3 & -2 & 1 \\ 2 & 4 & 3 \\ -1 & 5 & 1 \end{vmatrix}$

7. $\begin{vmatrix} x & 0 & -1 \\ 2 & x & x^2 \\ -3 & x & 1 \end{vmatrix}$

8. $\begin{vmatrix} x & 1 & -1 \\ x^2 & x & x \\ 0 & x & 1 \end{vmatrix}$

2 Solve using Cramer's rule.

9. $-2x + 4y = 3,$ $3x - 7y = 1$

10. $5x - 4y = -3,$ $7x + 2y = 6$

11. $\sqrt{3}x + \pi y = -5,$ $\pi x - \sqrt{3}y = 4$

12. $\pi x - \sqrt{5}y = 2,$ $\sqrt{5}x + \pi y = -3$

4 Solve using Cramer's rule.

13. $3x + 2y - z = 4,$ $3x - 2y + z = 5,$ $4x - 5y - z = -1$

14. $3x - y + 2z = 1,$ $x - y + 2z = 3,$ $-2x + 3y + z = 1$

15. $6y + 6z = -1,$ $8x + 6z = -1,$ $4x + 9y = 8$

16. $3x + 5y = 2,$ $2x - 3z = 7,$ $4y + 2z = -1$

SYNTHESIS

Solve.

17. $\begin{vmatrix} x & 5 \\ -4 & x \end{vmatrix} = 24$

18. $\begin{vmatrix} y & 2 \\ 3 & y \end{vmatrix} = y$

19. $\begin{vmatrix} x & -3 \\ -1 & x \end{vmatrix} \geq 0$

20. $\begin{vmatrix} y & -5 \\ -2 & y \end{vmatrix} < 0$

21. $\begin{vmatrix} x+3 & 4 \\ x-3 & 5 \end{vmatrix} = -7$

22. $\begin{vmatrix} m+2 & -3 \\ m+5 & -4 \end{vmatrix} = 3m - 5$

23. $\begin{vmatrix} 2 & x & 1 \\ 1 & 2 & -1 \\ 3 & 4 & -2 \end{vmatrix} = -6$

24. $\begin{vmatrix} x & 2 & x \\ 3 & -1 & 1 \\ 1 & -2 & 2 \end{vmatrix} = -10$

Rewrite the expression using determinants. Answers may vary.

25. $2L + 2W$

26. $\pi r + \pi h$

27. $a^2 + b^2$

28. $\frac{1}{2}h(a + b)$

29. $2\pi r^2 + 2\pi rh$

30. $x^2y^2 - Q^2$

31. Show that

$$\begin{vmatrix} \cos x & \sin x \\ -\sin x & \cos x \end{vmatrix} = \begin{vmatrix} \cos x & -\sin x \\ \sin x & \cos x \end{vmatrix} = 1.$$

OBJECTIVES

You should be able to:

1. Find a specified element, a_{ij}, of a matrix, its minor, and its cofactor, and be able to expand its determinant across any row or down any column.
2. Use properties of determinants to simplify their evaluation.
3. Factor certain determinants.

1. For the matrix of Example 1, find a_{11}, a_{13}, a_{22}, a_{31}, and a_{32}.

9.7 Determinants of Higher Order

1 Evaluating Determinants of Higher Order

Further Notation

We will define the determinant function for square matrices of any dimension. To do this, we need some new notation. The members, or elements, of a matrix will now be denoted by lower-case letters with two subscripts, as follows:

$$A = \begin{bmatrix} a_{11} & a_{12} & a_{13} \\ a_{21} & a_{22} & a_{23} \\ a_{31} & a_{32} & a_{33} \end{bmatrix}.$$

The element in the ith row and jth column is denoted a_{ij}. We can also name the above matrix A as

$$[a_{ij}].$$

EXAMPLE 1 Consider

$$[a_{ij}] = \begin{bmatrix} -8 & 0 & 6 \\ 4 & -6 & 7 \\ -1 & -3 & 5 \end{bmatrix}.$$

Find a_{12}, a_{23}, and a_{33}.

Solution

$a_{12} = 0$ This is the intersection of the first row and second column.

$a_{23} = 7$ This is the intersection of the second row and third column.

$a_{33} = 5$ This is the intersection of the third row and third column. ■

DO EXERCISE 1.

Minors

We will restrict our attention to square matrices.

DEFINITION

In a matrix $[a_{ij}]$, the minor M_{ij} of an element a_{ij} is the determinant of the matrix found by deleting the ith row and the jth column.

Note that a minor is a certain determinant, hence is a number.

EXAMPLE 2 In the matrix given in Example 1, find M_{11} and M_{23}.

Solution To find M_{11}, we delete the first row and the first column:

$$\begin{bmatrix} -8 & 0 & 6 \\ 4 & -6 & 7 \\ -1 & -3 & 5 \end{bmatrix}.$$

We calculate the determinant of the matrix formed by the remaining elements:

$$M_{11} = \begin{vmatrix} -6 & 7 \\ -3 & 5 \end{vmatrix} = (-6) \cdot 5 - (-3) \cdot 7 = -30 - (-21)$$

$$= -30 + 21 = -9.$$

To find M_{23}, we delete the second row and the third column:

$$\begin{bmatrix} -8 & 0 & 6 \\ 4 & -6 & 7 \\ -1 & -3 & 5 \end{bmatrix}.$$

We calculate the determinant of the matrix formed by the remaining elements:

$$M_{23} = \begin{vmatrix} -8 & 0 \\ -1 & -3 \end{vmatrix} = -8(-3) - (-1)0 = 24.$$ ∎

DO EXERCISE 2.

2. For the matrix of Example 1, find the minors M_{22}, M_{32}, and M_{13}.

DEFINITION

In a matrix $[a_{ij}]$, the *cofactor* of an element a_{ij} is denoted A_{ij} and is given by

$$A_{ij} = (-1)^{i+j} M_{ij},$$

where M_{ij} is the minor of a_{ij}. In other words, to find the cofactor of an element, find its minor and multiply it by $(-1)^{i+j}$.

Note that $(-1)^{i+j}$ is 1 if $i + j$ is even and is -1 if $i + j$ is odd. Thus in calculating a cofactor, find the minor. Then add the number of the row and the number of the column. The sum is $i + j$. If this sum is odd, change the sign of the minor. If this sum is even, leave the minor as is.* Note also that the cofactor of an element is a number.

3. For the matrix of Example 1, find the cofactors A_{22}, A_{32}, and A_{13}.

EXAMPLE 3 In the matrix given in Example 1, find A_{11} and A_{23}.

Solution In Example 2, we found that $M_{11} = -9$. In A_{11}, the sum of the subscripts, $1 + 1 = 2$, is even, so we do not change the sign of the minor:

$$A_{11} = -9.$$

In Example 2, we found that $M_{23} = 24$. In A_{23}, the sum of the subscripts, $2 + 3 = 5$, is odd, so we do change the sign of the minor:

$$A_{23} = -24.$$ ∎

DO EXERCISE 3.

* $(-1)^{i+j}$ can also be found by counting through the matrix horizontally and/or vertically, starting with a_{11} and $(+)$, saying $+, -, +, -$, and so on, until you come to a_{ij}.

Start here $(+)$

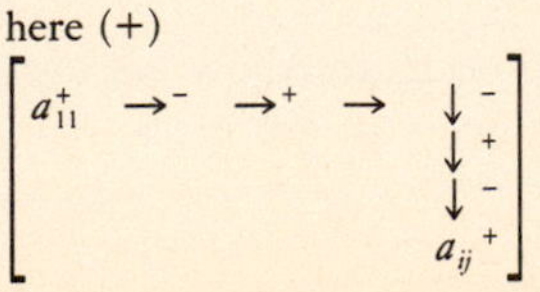

$$\begin{bmatrix} a_{11}^+ & \to^- & \to^+ & \to & \downarrow^- \\ & & & & \downarrow^+ \\ & & & & \downarrow^- \\ & & & & a_{ij}^+ \end{bmatrix}$$ The path does not matter.

4. Consider the matrix of Example 1. Find $|A|$ by expanding down the second column.

5. Consider the matrix of Example 1. Find $|A|$ by expanding down the third column.

Evaluating Determinants Using Cofactors

Consider the matrix **A** given by

$$\mathbf{A} = \begin{bmatrix} a_{11} & a_{12} & a_{13} \\ a_{21} & a_{22} & a_{23} \\ a_{31} & a_{32} & a_{33} \end{bmatrix}.$$

The determinant of the matrix, denoted $|\mathbf{A}|$, can be found as follows:

$$|\mathbf{A}| = a_{11}\mathbf{A}_{11} + a_{21}\mathbf{A}_{21} + a_{31}\mathbf{A}_{31}.$$

That is, multiply each element of the first column by its cofactor and add:

$$|\mathbf{A}| = a_{11} \cdot \begin{vmatrix} a_{22} & a_{23} \\ a_{32} & a_{33} \end{vmatrix} - a_{21} \cdot \begin{vmatrix} a_{12} & a_{13} \\ a_{32} & a_{33} \end{vmatrix} + a_{31} \cdot \begin{vmatrix} a_{12} & a_{13} \\ a_{22} & a_{23} \end{vmatrix}.$$

We have a minus sign with the second term since $2 + 1 = 3$, and 3 is odd. The last line is equivalent to the definition in Section 9.6. It can be shown that $|\mathbf{A}|$ can be found by picking *any* row or column, multiplying each element by its cofactor, and adding. This is called *expanding* across a row or down a column. We just expanded down the first column. We now define the determinant function for square matrices of any dimensions.

> **DEFINITION**
>
> **For any square matrix A of dimensions $n \times n$ ($n > 1$), we define the *determinant* of A, denoted $|A|$, as follows. Choose any row or column. Multiply each element in that row or column by its cofactor and add the results. The determinant of a 1×1 matrix is simply the element of the matrix.**

The value of a determinant will be the same no matter how it is evaluated.

EXAMPLE 4 Consider the matrix A of Example 1. Evaluate $|A|$ by expanding across the third row.

Solution

$$\begin{aligned}
|A| &= (-1)A_{31} + (-3)A_{32} + 5A_{33} \\
&= (-1)(-1)^{3+1} \cdot \begin{vmatrix} 0 & 6 \\ -6 & 7 \end{vmatrix} + (-3)(-1)^{3+2} \cdot \begin{vmatrix} -8 & 6 \\ 4 & 7 \end{vmatrix} \\
&\quad + 5(-1)^{3+3} \cdot \begin{vmatrix} -8 & 0 \\ 4 & -6 \end{vmatrix} \\
&= (-1) \cdot 1 \cdot [0 \cdot 7 - (-6)6] + (-3)(-1)[-8 \cdot 7 - 4 \cdot 6] \\
&\quad + 5 \cdot 1 \cdot [-8(-6) - 4 \cdot 0] \\
&= -[36] + 3[-80] + 5[48] \\
&= -36 - 240 + 240 \\
&= -36
\end{aligned}$$

The value of this determinant is -36 no matter how we evaluate it. That is, if we expand down the second column, we still get -36. ■

DO EXERCISES 4 AND 5.

2 Evaluation of Certain Determinants

We can simplify the evaluation of certain determinants using the following properties.

THEOREM 6

If a row (or column) of a matrix A has all elements 0, then $|\mathbf{A}| = 0$.

Proof. Just evaluate by expanding across a row (or down a column) that has all 0's.

EXAMPLES Evaluate.

5. $\begin{vmatrix} 0 & 6 \\ 0 & 7 \end{vmatrix} = 0$

6. $\begin{vmatrix} 4 & 5 & -7 \\ 0 & 0 & 0 \\ -3 & 9 & 6 \end{vmatrix} = 0$

DO EXERCISES 6 AND 7.

THEOREM 7

If two rows (or columns) of a matrix A are interchanged to obtain a new matrix B, then $|\mathbf{A}| = -|\mathbf{B}|$.

Proof. Choose one of the rows (or columns) to be interchanged and evaluate $|\mathbf{A}|$ by expanding across that row. Expand across the same row to evaluate $|\mathbf{B}|$. For that row, each $(-1)^{i+j}$ has changed signs, so $|\mathbf{A}| = -|\mathbf{B}|$.

EXAMPLES

7. $\begin{vmatrix} 6 & 7 & 8 \\ 4 & 1 & 2 \\ 2 & 9 & 0 \end{vmatrix} = -1 \begin{vmatrix} 6 & 8 & 7 \\ 4 & 2 & 1 \\ 2 & 0 & 9 \end{vmatrix}$

8. $\begin{vmatrix} -6 & 8 \\ 4 & -3 \end{vmatrix} = -1 \cdot \begin{vmatrix} 4 & -3 \\ -6 & 8 \end{vmatrix}$

DO EXERCISES 8 AND 9.

THEOREM 8

If two rows (or columns) of a matrix A are the same, then $|\mathbf{A}| = 0$.

Proof. Interchanging the rows (or columns) that are the same does not change A. Thus by Theorem 7, $|\mathbf{A}| = -|\mathbf{A}|$. This is possible only when $|\mathbf{A}| = 0$.

EXAMPLES Evaluate.

9. $\begin{vmatrix} 6 & 7 & 8 \\ -2 & 6 & 5 \\ -2 & 6 & 5 \end{vmatrix} = 0$

10. $\begin{vmatrix} -5 & 4 & -5 \\ 3 & 7 & 3 \\ 0 & 12 & 0 \end{vmatrix} = 0$

DO EXERCISE 10.

Evaluate.

6. $\begin{vmatrix} 2 & 3 \\ 0 & 0 \end{vmatrix}$

7. $\begin{vmatrix} 2 & 3 & 0 \\ -3 & 7 & 0 \\ 2 & 4 & 0 \end{vmatrix}$

8. Given

$\mathbf{A} = \begin{bmatrix} 5 & -9 \\ -2 & 0 \end{bmatrix}$ and

$\mathbf{B} = \begin{bmatrix} -2 & 0 \\ 5 & -9 \end{bmatrix}$.

a) Find $|\mathbf{A}|$ and $|\mathbf{B}|$.

b) Why does $|\mathbf{A}| = -|\mathbf{B}|$?

9. Given

$\mathbf{C} = \begin{vmatrix} -2 & 3 & 4 \\ 4 & 5 & 6 \\ -9 & -1 & 0 \end{vmatrix}$ and

$\mathbf{D} = \begin{vmatrix} 4 & 3 & -2 \\ 6 & 5 & 4 \\ 0 & -1 & -9 \end{vmatrix}$.

a) Find $|\mathbf{C}|$ and $|\mathbf{D}|$.

b) Why does $|\mathbf{C}| = -|\mathbf{D}|$?

10. Evaluate.

$\begin{vmatrix} -1 & 3 & -7 \\ -1 & 2 & -6 \\ -1 & 3 & -7 \end{vmatrix}$

Solve for x.

11. $\begin{vmatrix} -16 & 32 \\ -5 & -3 \end{vmatrix} = x \cdot \begin{vmatrix} 4 & -8 \\ -5 & -3 \end{vmatrix}$

12. $\begin{vmatrix} -3 & 12 & 2 \\ 5 & -6 & 3 \\ 0 & 18 & 5 \end{vmatrix} = x \cdot \begin{vmatrix} -3 & 2 & 2 \\ 5 & -1 & 3 \\ 0 & 3 & 5 \end{vmatrix}$

13. Without expanding, evaluate $|A|$.

$$A = \begin{bmatrix} 1 & 0 & -7 \\ -8 & 6 & -4 \\ 24 & -18 & 12 \end{bmatrix}$$

THEOREM 9

If all the elements of a row (or column) of a matrix A are multiplied by k, $|A|$ is multiplied by k. Or, if all the elements of a row (or column) of A have a common factor, we can factor it out of the determinant of A.

EXAMPLES

11. $\begin{vmatrix} 2 & 4 & 6 \\ -2 & 5 & 9 \\ 4 & -1 & -3 \end{vmatrix} = 3 \cdot \begin{vmatrix} 2 & 4 & 2 \\ -2 & 5 & 3 \\ 4 & -1 & -1 \end{vmatrix}$

12. $\begin{vmatrix} 10 & 25 \\ -4 & -7 \end{vmatrix} = 5 \cdot \begin{vmatrix} 2 & 5 \\ -4 & -7 \end{vmatrix}$

DO EXERCISES 11 AND 12.

Proof of Theorem 9. Evaluate the two determinants by expanding them across the same row (or down the same column) in question. The cofactors are the same and k can be factored out. Consider the case of a 3×3 matrix and the second column. Let

$$A = \begin{bmatrix} a_{11} & a_{12} & a_{13} \\ a_{21} & a_{22} & a_{23} \\ a_{31} & a_{32} & a_{33} \end{bmatrix}$$

and

$$B = \begin{bmatrix} a_{11} & ka_{12} & a_{13} \\ a_{21} & ka_{22} & a_{23} \\ a_{31} & ka_{32} & a_{33} \end{bmatrix}.$$

Then

$$\begin{aligned} |B| &= ka_{12}A_{12} + ka_{22}A_{22} + ka_{32}A_{32} \\ &= k(a_{12}A_{12} + a_{22}A_{22} + a_{32}A_{32}) \\ &= k|A|. \end{aligned}$$

EXAMPLE 13 Without expanding, find $|A|$.

$$A = \begin{bmatrix} -6 & 3 & 8 \\ 15 & -9 & -20 \\ -9 & -1 & 12 \end{bmatrix}$$

Solution

$$|A| = (-3) \cdot \begin{vmatrix} 2 & 3 & 8 \\ -5 & -9 & -20 \\ 3 & -1 & 12 \end{vmatrix} \qquad \text{Factoring } -3 \text{ out of the first column}$$

$$= (-3)(4) \cdot \begin{vmatrix} 2 & 3 & 2 \\ -5 & -9 & -5 \\ 3 & -1 & 3 \end{vmatrix} \qquad \text{Factoring 4 out of the third column}$$

$$= 0 \qquad \text{By Theorem 8}$$

DO EXERCISE 13.

THEOREM 10

If each element in a row (or column) is multiplied by a number k and each product is added to the corresponding element of another row (or column), we do not change the value of the determinant. That is, we can add a multiple of any row (or column) to any other row (or column) without changing the value of the determinant.

EXAMPLE 14 Find a determinant having the same value as the one on the left by adding three times the second column to the first column.

$$\begin{vmatrix} 0 & 1 & 2 \\ 4 & 5 & 6 \\ 7 & 8 & 9 \end{vmatrix} = \begin{vmatrix} 0+3(1) & 1 & 2 \\ 4+3(5) & 5 & 6 \\ 7+3(8) & 8 & 9 \end{vmatrix} = \begin{vmatrix} 3 & 1 & 2 \\ 19 & 5 & 6 \\ 31 & 8 & 9 \end{vmatrix}$$

EXAMPLE 15 Find a determinant having the same value as the one on the left by adding two times the third row to the first row.

$$\begin{vmatrix} 0 & 1 & 2 \\ 4 & 5 & 6 \\ 7 & 8 & 9 \end{vmatrix} = \begin{vmatrix} 0+2(7) & 1+2(8) & 2+2(9) \\ 4 & 5 & 6 \\ 7 & 8 & 9 \end{vmatrix} = \begin{vmatrix} 14 & 17 & 20 \\ 4 & 5 & 6 \\ 7 & 8 & 9 \end{vmatrix}$$

Proof of Theorem 10. We prove the theorem for the case of a 3×3 matrix where k times the first column has been added to the third column. Let

$$\mathbf{A} = \begin{bmatrix} a_{11} & a_{12} & a_{13} \\ a_{21} & a_{22} & a_{23} \\ a_{31} & a_{32} & a_{33} \end{bmatrix} \quad \text{and} \quad \mathbf{B} = \begin{bmatrix} a_{11} & a_{12} & ka_{11}+a_{13} \\ a_{21} & a_{22} & ka_{21}+a_{23} \\ a_{31} & a_{32} & ka_{31}+a_{33} \end{bmatrix}.$$

To show that $|\mathbf{A}| = |\mathbf{B}|$, we evaluate $|\mathbf{B}|$ by expanding down the third column:

$$\begin{aligned} |\mathbf{B}| &= (ka_{11}+a_{13})A_{13} + (ka_{21}+a_{23})A_{23} + (ka_{31}+a_{33})A_{33} \\ &= k(a_{11}A_{13} + a_{21}A_{23} + a_{31}A_{33}) + (a_{13}A_{13} + a_{23}A_{23} + a_{33}A_{33}) \\ &= k(a_{11}A_{13} + a_{21}A_{23} + a_{31}A_{33}) + |\mathbf{A}| \\ &= k \cdot \begin{vmatrix} a_{11} & a_{12} & a_{11} \\ a_{21} & a_{22} & a_{21} \\ a_{31} & a_{32} & a_{31} \end{vmatrix} + |\mathbf{A}| \\ &= k(0) + |\mathbf{A}| \qquad \text{By Theorem 8} \\ &= |\mathbf{A}|. \end{aligned}$$

DO EXERCISE 14.

We can use the properties of determinants to simplify their evaluation. We try to find another determinant where in some row or column, one element is 1 and the rest are 0.

EXAMPLE 16 Evaluate by first simplifying to a determinant where, in one row or column, one element is 1 and the rest are 0.

$$\begin{vmatrix} 6 & 2 & 3 \\ 6 & -1 & 5 \\ -2 & 3 & 1 \end{vmatrix}$$

14. Find a determinant equal to this one by adding twice the first row to the second row.

$$\begin{vmatrix} -2 & 3 & 4 \\ 1 & 4 & -3 \\ 0 & 9 & 7 \end{vmatrix}$$

Evaluate by first simplifying to a determinant where in one row or column, one element is 1 and the rest are 0.

15. $\begin{vmatrix} 3 & -1 & 1 \\ 2 & 2 & -4 \\ 2 & 4 & 1 \end{vmatrix}$

16. $\begin{vmatrix} 5 & -4 & 2 & -2 \\ 3 & -3 & -4 & 7 \\ -2 & 3 & 2 & 4 \\ -8 & 9 & 5 & -5 \end{vmatrix}$

17. Factor:

$$\begin{vmatrix} a^2 & b^2 & c^2 \\ a & b & c \\ 1 & 1 & 1 \end{vmatrix}.$$

Solution We will try to get two 0's and a 1 in the third row. It already has a 1; that is why we chose the third row. We first factor a 2 out of column 1:

$$2 \cdot \begin{vmatrix} 3 & 2 & 3 \\ 3 & -1 & 5 \\ -1 & 3 & 1 \end{vmatrix}. \quad \text{Theorem 9}$$

Now we multiply each element in column 3 by -3 and add the corresponding elements to column 2:

$$2 \cdot \begin{vmatrix} 3 & -7 & 3 \\ 3 & -16 & 5 \\ -1 & 0 & 1 \end{vmatrix}. \quad \text{Theorem 10}$$

Next, we add the elements in column 3 to the corresponding elements in column 1 (Theorem 10):

$$2 \cdot \begin{vmatrix} 6 & -7 & 3 \\ 8 & -16 & 5 \\ 0 & 0 & 1 \end{vmatrix}.$$

Finally, we evaluate the determinant by expanding across the last row:

$$2 \cdot \left(0 - 0 + 1 \cdot \begin{vmatrix} 6 & -7 \\ 8 & -16 \end{vmatrix}\right) = 2 \cdot [6(-16) - 8(-7)] = -80.$$

DO EXERCISES 15 AND 16.

3 Factoring Certain Determinants

EXAMPLE 17 Factor:

$$\begin{vmatrix} 1 & x & x^2 \\ 1 & y & y^2 \\ 1 & z & z^2 \end{vmatrix}.$$

Solution

$$\begin{vmatrix} 1 & x & x^2 \\ 1 & y & y^2 \\ 1 & z & z^2 \end{vmatrix} = \begin{vmatrix} 0 & x-y & x^2-y^2 \\ 1 & y & y^2 \\ 0 & z-y & z^2-y^2 \end{vmatrix}$$

By Theorem 10: Adding -1 times the second row to the first row, and -1 times the second row to the third row

$$= (x-y)(z-y) \cdot \begin{vmatrix} 0 & 1 & x+y \\ 1 & y & y^2 \\ 0 & 1 & z+y \end{vmatrix}$$

By Theorem 9: Factoring $x - y$ out of the first row and $z - y$ out of the third row

$$= (x-y)(z-y) \cdot \begin{vmatrix} 0 & 0 & x-z \\ 1 & y & y^2 \\ 0 & 1 & z+y \end{vmatrix}$$

By Theorem 10: adding -1 times the third row to the first row

$$= (x-y)(z-y)(x-z) \cdot \begin{vmatrix} 0 & 0 & 1 \\ 1 & y & y^2 \\ 0 & 1 & z+y \end{vmatrix}$$

By Theorem 9: Factoring $x - z$ out of the first row

$$= (x-y)(z-y)(x-z)$$

Expanding the determinant across the first row, we get 1.

DO EXERCISE 17.

EXERCISE SET 9.7

1 Use the following matrix for Exercises 1–10.

$$A = \begin{bmatrix} 7 & -4 & -6 \\ 2 & 0 & -3 \\ 1 & 2 & -5 \end{bmatrix}$$

1. Find a_{11}, a_{32}, and a_{22}.
2. Find a_{13}, a_{31}, and a_{23}.
3. Find M_{11}, M_{32}, and M_{22}.
4. Find M_{13}, M_{31}, and M_{23}.
5. Find A_{11}, A_{32}, and A_{22}.
6. Find A_{13}, A_{31}, and A_{23}.
7. Evaluate $|A|$ by expanding across the second row.
8. Evaluate $|A|$ by expanding down the second column.
9. Evaluate $|A|$ by expanding down the third column.
10. Evaluate $|A|$ by expanding across the first row.

Use the following matrix for Exercises 11–16.

$$A = \begin{bmatrix} 1 & 0 & 0 & -2 \\ 4 & 1 & 0 & 0 \\ 5 & 6 & 7 & 8 \\ -2 & -3 & -1 & 0 \end{bmatrix}$$

11. Find M_{41} and M_{33}.
12. Find M_{12} and M_{44}.
13. Find A_{24} and A_{43}.
14. Find A_{22} and A_{34}.
15. Evaluate $|A|$ by expanding across the first row.
16. Evaluate $|A|$ by expanding down the third column.

Evaluate.

17. $\begin{vmatrix} 5 & -4 & 2 & -2 \\ 3 & -3 & -4 & 7 \\ -2 & 3 & 2 & 4 \\ -8 & 9 & 5 & -5 \end{vmatrix}$

18. $\begin{vmatrix} x & p & q & r \\ 0 & y & s & t \\ 0 & 0 & z & u \\ 0 & 0 & 0 & w \end{vmatrix}$

2 Evaluate by first simplifying to a determinant where in one row or column, one element is 1 and the rest are 0.

19. $\begin{vmatrix} -4 & 5 \\ 6 & 10 \end{vmatrix}$

20. $\begin{vmatrix} 3 & -9 \\ -2 & 4 \end{vmatrix}$

21. $\begin{vmatrix} 2 & 1 & 1 \\ 2 & -3 & -1 \\ -4 & 5 & 2 \end{vmatrix}$

22. $\begin{vmatrix} 1 & 2 & 4 \\ 2 & 3 & 5 \\ 3 & 1 & 6 \end{vmatrix}$

23. $\begin{vmatrix} 11 & -15 & 20 \\ 16 & 24 & -8 \\ 6 & 9 & 15 \end{vmatrix}$

24. $\begin{vmatrix} 4 & -24 & 15 \\ -3 & 18 & -6 \\ 5 & -4 & 3 \end{vmatrix}$

25. $\begin{vmatrix} -3 & 0 & 2 & 6 \\ 2 & 4 & 0 & -1 \\ -1 & 0 & -5 & 2 \\ 0 & -1 & -2 & -3 \end{vmatrix}$

26. $\begin{vmatrix} -2 & 1 & 0 & 5 \\ 3 & 0 & -4 & -2 \\ 4 & -6 & -8 & -1 \\ 8 & 0 & -2 & -3 \end{vmatrix}$

Evaluate the determinant without expanding.

27. $\begin{vmatrix} x & y & z \\ 0 & 0 & 0 \\ p & q & r \end{vmatrix}$

28. $\begin{vmatrix} 5 & 5 & 5 \\ 3 & 3 & 3 \\ 2 & -7 & 8 \end{vmatrix}$

29. $\begin{vmatrix} 2a & t & -7a \\ 2b & u & -7b \\ 2c & v & -7c \end{vmatrix}$

30. $\begin{vmatrix} a & -1 & 4a \\ b & 2 & 4b \\ x & -3 & 4x \end{vmatrix}$

3 Factor.

31. $\begin{vmatrix} x^2 & x & 1 \\ y^2 & y & 1 \\ z^2 & z & 1 \end{vmatrix}$

32. $\begin{vmatrix} 1 & 1 & 1 \\ a & b & c \\ a^2 & b^2 & c^2 \end{vmatrix}$

33. $\begin{vmatrix} x & x^2 & x^3 \\ y & y^2 & y^3 \\ z & z^2 & z^3 \end{vmatrix}$

34. $\begin{vmatrix} 1 & 1 & 1 \\ a & b & c \\ a^3 & b^3 & c^3 \end{vmatrix}$

SYNTHESIS

35. If a line contains the points (x_1, y_1) and (x_2, y_2), an equation of the line can be written as follows:

$$\begin{vmatrix} x & y & 1 \\ x_1 & y_1 & 1 \\ x_2 & y_2 & 1 \end{vmatrix} = 0.$$

Prove this.

36. Show that the points (x_1, y_1), (x_2, y_2), and (x_3, y_3) are collinear (on the same straight line) if and only if

$$\begin{vmatrix} x_1 & y_1 & 1 \\ x_2 & y_2 & 1 \\ x_3 & y_3 & 1 \end{vmatrix} = 0.$$

CHALLENGE

37. Consider a triangle with vertices (x_1, y_1), (x_2, y_2), and (x_3, y_3). The area of this triangle is the absolute value of

$$\frac{1}{2} \cdot \begin{vmatrix} x_1 & y_1 & 1 \\ x_2 & y_2 & 1 \\ x_3 & y_3 & 1 \end{vmatrix}.$$

Prove this. (*Hint:* Look at the figure below. The area of triangle *ABC* is the area of trapezoid *ABDE* plus the area of trapezoid *AEFC* minus the area of trapezoid *BDFC*.)

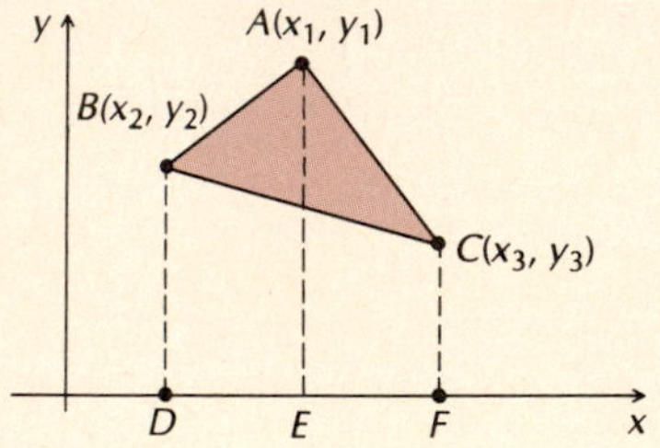

38. Prove that the lines $a_1x + b_1y = c_1$ and $a_2x + b_2y = c_2$ are parallel when

$$\begin{vmatrix} a_1 & b_1 \\ a_2 & b_2 \end{vmatrix} = 0$$

and either

$$\begin{vmatrix} c_1 & b_1 \\ c_2 & b_2 \end{vmatrix} \neq 0 \quad \text{or} \quad \begin{vmatrix} a_1 & c_1 \\ a_2 & c_2 \end{vmatrix} \neq 0.$$

OBJECTIVES

You should be able to:

1 Calculate the inverse of a square matrix, if it exists.

2 Use inverses of matrices to solve systems.

9.8 Inverses of Matrices

In this section, we learn the meaning of the *multiplicative inverse*, or simply *inverse*, of a square matrix. We learn how to compute such inverses and how to use them to solve equations.

Identity Matrices

The letter **I** is used to represent square matrices such as

$$\mathbf{I} = \begin{bmatrix} 1 & 0 \\ 0 & 1 \end{bmatrix} \quad \text{and} \quad \mathbf{I} = \begin{bmatrix} 1 & 0 & 0 \\ 0 & 1 & 0 \\ 0 & 0 & 1 \end{bmatrix}.$$

These square matrices have 1's extending from the upper left down to the lower right along what is called the **main diagonal**. The rest of the elements are 0.

DO EXERCISE 1.

In Margin Exercise 1, you have proved the following for the case $n = 2$.

THEOREM 11

For any square matrix A of dimensions $n \times n$,

$$AI = IA = A \qquad \text{(I is a multiplicative identity)},$$

where I is any square matrix, defined above, of dimensions $n \times n$. For any matrix A with exactly n rows,

$$IA = A,$$

where I is the square matrix of dimensions $n \times n$.

We usually say that I is an **identity matrix**. Suppose that for matrix A there is a matrix A^{-1} for which

$$A^{-1} \cdot A = I = A \cdot A^{-1}.$$

Then A^{-1} is the **multiplicative inverse**, or simply **inverse**, of A. For example, for the matrix

$$A = \begin{bmatrix} 5 & 3 \\ 3 & 2 \end{bmatrix},$$

we have

$$A^{-1} = \begin{bmatrix} 2 & -3 \\ -3 & 5 \end{bmatrix}.$$

We can check that

$$\begin{aligned} A \cdot A^{-1} &= \begin{bmatrix} 5 & 3 \\ 3 & 2 \end{bmatrix} \begin{bmatrix} 2 & -3 \\ -3 & 5 \end{bmatrix} \\ &= \begin{bmatrix} 1 & 0 \\ 0 & 1 \end{bmatrix} \\ &= I. \end{aligned}$$

We leave it to the student to verify that $A^{-1} \cdot A = I$.

DO EXERCISE 2.

1 Calculating Matrix Inverses: The Gauss–Jordan Reduction Method

We now learn a way to calculate the inverse of a square matrix, if it exists. Consider the matrix

$$A = \begin{bmatrix} 2 & -1 & 1 \\ 1 & -2 & 3 \\ 4 & 1 & 2 \end{bmatrix}.$$

First we form a new (**augmented**) matrix consisting, on the left, of the matrix

1. Consider

$$A = \begin{bmatrix} a & b \\ c & d \end{bmatrix}, \qquad X = \begin{bmatrix} x \\ y \end{bmatrix},$$

and

$$I = \begin{bmatrix} 1 & 0 \\ 0 & 1 \end{bmatrix}.$$

a) Find AI.
b) Find IA.
c) Compare AI and IA.
d) Find IX.
e) Find XI.
f) Compare IX and XI.

2. Let

$$A = \begin{bmatrix} 3 & 1 & 0 \\ 1 & -1 & 2 \\ 1 & 1 & 1 \end{bmatrix}$$

and

$$A^{-1} = \frac{1}{8}\begin{bmatrix} 3 & 1 & -2 \\ -1 & -3 & 6 \\ -2 & 2 & 4 \end{bmatrix}.$$

a) Find AA^{-1}.
b) Find $A^{-1}A$.
c) Compare AA^{-1} and $A^{-1}A$.

A and, on the right, of the corresponding identity matrix **I**:

This line is for separation and would not normally need to be written.

$$\left[\begin{array}{rrr|rrr} 2 & -1 & 1 & 1 & 0 & 0 \\ 1 & -2 & 3 & 0 & 1 & 0 \\ 4 & 1 & 2 & 0 & 0 & 1 \end{array}\right].$$

The matrix **A** — The identity matrix **I**

We now attempt to use the elimination method to transform **A** to an identity matrix, but whatever operations we perform, we do them on the entire augmented matrix. When we finish, we will get a matrix like the following:

$$\left[\begin{array}{ccc|ccc} 1 & 0 & 0 & a & b & c \\ 0 & 1 & 0 & d & e & f \\ 0 & 0 & 1 & g & h & i \end{array}\right].$$

The matrix on the right,

$$\begin{bmatrix} a & b & c \\ d & e & f \\ g & h & i \end{bmatrix},$$

will be $\mathbf{A}^{-1}$.

EXAMPLE 1 Find $\mathbf{A}^{-1}$.

$$\mathbf{A} = \begin{bmatrix} 2 & -1 & 1 \\ 1 & -2 & 3 \\ 4 & 1 & 2 \end{bmatrix}$$

Solution

a) We find the augmented matrix consisting of **A** and **I**:

$$\begin{bmatrix} 2 & -1 & 1 & 1 & 0 & 0 \\ 1 & -2 & 3 & 0 & 1 & 0 \\ 4 & 1 & 2 & 0 & 0 & 1 \end{bmatrix}.$$

Be sure to write the augmented matrix *before* doing any row operations. Otherwise, you may not obtain the inverse matrix.

b) We interchange the first and second rows so that the elements of the first column are multiples of the top number on the main diagonal:

$$\begin{bmatrix} 1 & -2 & 3 & 0 & 1 & 0 \\ 2 & -1 & 1 & 1 & 0 & 0 \\ 4 & 1 & 2 & 0 & 0 & 1 \end{bmatrix}.$$

c) Next we obtain 0's in the rest of the first column. We multiply the first row by -2 and add it to the second row. Then we multiply the first row by -4 and add it to the third row:

$$\begin{bmatrix} 1 & -2 & 3 & 0 & 1 & 0 \\ 0 & 3 & -5 & 1 & -2 & 0 \\ 0 & 9 & -10 & 0 & -4 & 1 \end{bmatrix}.$$

d) Next we move down the main diagonal to the number 3. We note that the number below it, 9, is a multiple of 3. We multiply the second row by -3 and add it to the third:

$$\begin{bmatrix} 1 & -2 & 3 & 0 & 1 & 0 \\ 0 & 3 & -5 & 1 & -2 & 0 \\ 0 & 0 & 5 & -3 & 2 & 1 \end{bmatrix}.$$

e) Now we move down the main diagonal to the number 5. We check to see if each number above 5 in the third column is a multiple of 5. Since this is not the case, we multiply the first row by -5:

$$\begin{bmatrix} -5 & 10 & -15 & 0 & -5 & 0 \\ 0 & 3 & -5 & 1 & -2 & 0 \\ 0 & 0 & 5 & -3 & 2 & 1 \end{bmatrix}.$$

f) Now we work back up. We add the third row to the second. We also multiply the third row by 3 and add it to the first:

$$\begin{bmatrix} -5 & 10 & 0 & -9 & 1 & 3 \\ 0 & 3 & 0 & -2 & 0 & 1 \\ 0 & 0 & 5 & -3 & 2 & 1 \end{bmatrix}.$$

g) We move back to the number 3 on the main diagonal. We multiply the first row by -3, so the element on the top of the second column is a multiple of 3:

$$\begin{bmatrix} 15 & -30 & 0 & 27 & -3 & -9 \\ 0 & 3 & 0 & -2 & 0 & 1 \\ 0 & 0 & 5 & -3 & 2 & 1 \end{bmatrix}.$$

h) We multiply the second row by 10 and add it to the first:

$$\begin{bmatrix} 15 & 0 & 0 & 7 & -3 & 1 \\ 0 & 3 & 0 & -2 & 0 & 1 \\ 0 & 0 & 5 & -3 & 2 & 1 \end{bmatrix}.$$

i) Finally, we get all 1's on the main diagonal. We multiply the first row by $\frac{1}{15}$, the second by $\frac{1}{3}$, and the third by $\frac{1}{5}$:

$$\begin{bmatrix} 1 & 0 & 0 & \frac{7}{15} & -\frac{1}{5} & \frac{1}{15} \\ 0 & 1 & 0 & -\frac{2}{3} & 0 & \frac{1}{3} \\ 0 & 0 & 1 & -\frac{3}{5} & \frac{2}{5} & \frac{1}{5} \end{bmatrix}.$$

We now have the matrix **I** on the left. Thus,

$$\mathbf{A}^{-1} = \begin{bmatrix} \frac{7}{15} & -\frac{1}{5} & \frac{1}{15} \\ -\frac{2}{3} & 0 & \frac{1}{3} \\ -\frac{3}{5} & \frac{2}{5} & \frac{1}{5} \end{bmatrix}.$$

The student can always check by doing the multiplication $\mathbf{A}^{-1}\mathbf{A}$ or $\mathbf{A}\mathbf{A}^{-1}$. If we cannot obtain the identity matrix on the left using the Gauss–Jordan reduction method, as would be the case when a system has no solution or infinitely many solutions, then $\mathbf{A}^{-1}$ does not exist. More specifically,

Find A^{-1}. Use the Gauss–Jordan reduction method.

3. $A = \begin{bmatrix} 1 & 0 & 1 \\ 2 & 1 & 0 \\ 1 & -1 & 1 \end{bmatrix}$

4. $A = \begin{bmatrix} 3 & 5 \\ 1 & -2 \end{bmatrix}$

5. Consider the system

$$4x - 2y = -1,$$
$$x + 5y = 1.$$

a) Write a matrix equation equivalent to the system.

b) Find the coefficient matrix A.

c) Find A^{-1}.

d) Use the inverse of the coefficient matrix to solve the system.

If we obtain a row of all 0's in either of the two matrices in the augmented matrix, then A^{-1} does not exist.

DO EXERCISES 3 AND 4.

2 Solving Systems Using Inverses

We can use matrix inverses to solve certain kinds of systems. Consider the system

$$3x + 5y = -1,$$
$$x - 2y = 4.$$

We write a matrix equation equivalent to this system:

$$\begin{bmatrix} 3 & 5 \\ 1 & -2 \end{bmatrix} \begin{bmatrix} x \\ y \end{bmatrix} = \begin{bmatrix} -1 \\ 4 \end{bmatrix}.$$

Now we let

$$\begin{bmatrix} 3 & 5 \\ 1 & -2 \end{bmatrix} = A, \quad \begin{bmatrix} x \\ y \end{bmatrix} = X, \quad \text{and} \quad \begin{bmatrix} -1 \\ 4 \end{bmatrix} = B.$$

Then we have the matrix equation

$$A \cdot X = B.$$

To solve a comparable equation $ax = b$ involving real numbers, we multiply on both sides by the real number a^{-1}, provided it exists. To solve the matrix equation $AX = B$, we multiply on both sides by the inverse, A^{-1}, of the coefficient matrix, A, provided the inverse exists. We found in Margin Exercise 4 that it does exist and is given by

$$A^{-1} = \begin{bmatrix} \frac{2}{11} & \frac{5}{11} \\ \frac{1}{11} & -\frac{3}{11} \end{bmatrix}, \quad \text{or} \quad \frac{1}{11}\begin{bmatrix} 2 & 5 \\ 1 & -3 \end{bmatrix}.$$

We solve the matrix equation $AX = B$ as follows:

$$A^{-1}(A \cdot X) = A^{-1} \cdot B \qquad \text{Multiplying by } A^{-1} \text{ on the left on each side}$$
$$(A^{-1} \cdot A) \cdot X = A^{-1} \cdot B \qquad \text{Associative law}$$
$$I \cdot X = A^{-1} \cdot B \qquad \text{Since } A^{-1} \cdot A = I$$
$$X = A^{-1} \cdot B. \qquad \text{Since } I \cdot X = I$$

Now we have, substituting,

$$X = A^{-1} \cdot B$$
$$\begin{bmatrix} x \\ y \end{bmatrix} = \frac{1}{11}\begin{bmatrix} 2 & 5 \\ 1 & -3 \end{bmatrix}\begin{bmatrix} -1 \\ 4 \end{bmatrix}$$
$$= \frac{1}{11}\begin{bmatrix} 18 \\ -13 \end{bmatrix}$$
$$= \begin{bmatrix} \frac{18}{11} \\ -\frac{13}{11} \end{bmatrix}.$$

The solution of the system of equations is $\left(\frac{18}{11}, -\frac{13}{11}\right)$, or $x = \frac{18}{11}$ and $y = -\frac{13}{11}$.

DO EXERCISE 5.

EXERCISE SET 9.8

1 Find A^{-1}, if it exists. Use the Gauss–Jordan reduction method. Check your answers by calculating AA^{-1} and $A^{-1}A$.

1. $A = \begin{bmatrix} 3 & 2 \\ 5 & 3 \end{bmatrix}$

2. $A = \begin{bmatrix} 3 & 5 \\ 1 & 2 \end{bmatrix}$

3. $A = \begin{bmatrix} 11 & 3 \\ 7 & 2 \end{bmatrix}$

4. $A = \begin{bmatrix} 8 & 5 \\ 5 & 3 \end{bmatrix}$

5. $A = \begin{bmatrix} 4 & -3 \\ 1 & 2 \end{bmatrix}$

6. $A = \begin{bmatrix} 0 & -1 \\ 1 & 0 \end{bmatrix}$

7. $A = \begin{bmatrix} 3 & 1 & 0 \\ 1 & 1 & 1 \\ 1 & -1 & 2 \end{bmatrix}$

8. $A = \begin{bmatrix} 1 & 0 & 1 \\ 2 & 1 & 0 \\ 1 & -1 & 1 \end{bmatrix}$

9. $A = \begin{bmatrix} 1 & -1 & 2 \\ 0 & 1 & 3 \\ 2 & 1 & -2 \end{bmatrix}$

10. $A = \begin{bmatrix} 1 & -1 & 2 \\ 0 & 1 & 2 \\ 1 & -3 & -4 \end{bmatrix}$

11. $A = \begin{bmatrix} 1 & -4 & 8 \\ 1 & -3 & 2 \\ 2 & -7 & 10 \end{bmatrix}$

12. $A = \begin{bmatrix} -2 & 5 & 3 \\ 4 & -1 & 3 \\ 7 & -2 & 5 \end{bmatrix}$

13. $A = \begin{bmatrix} 1 & 2 & 3 & 4 \\ 0 & 1 & 3 & -5 \\ 0 & 0 & 1 & -2 \\ 0 & 0 & 0 & -1 \end{bmatrix}$

14. $A = \begin{bmatrix} -2 & -3 & 4 & 1 \\ 0 & 1 & 1 & 0 \\ 0 & 4 & -6 & 1 \\ -2 & -2 & 5 & 1 \end{bmatrix}$

15. $A = \begin{bmatrix} -2 & 5 & 3 \\ 4 & -1 & 3 \\ 4 & -10 & -6 \end{bmatrix}$

16. $A = \begin{bmatrix} -2 & 6 \\ -1 & 3 \end{bmatrix}$

2 In Exercises 17–20, a system of equations is given, together with the inverse of the coefficient matrix. Use the matrix inverse to solve the system.

17. $11x + 3y = -4,$
$7x + 2y = 5;$
$A^{-1} = \begin{bmatrix} 2 & -3 \\ -7 & 11 \end{bmatrix}$

18. $8x + 5y = -6,$
$5x + 3y = 2;$
$A^{-1} = \begin{bmatrix} -3 & 5 \\ 5 & -8 \end{bmatrix}$

19. $3x + y = 2,$
$2x - y + 2z = -5,$
$x + y + z = 5;$
$A^{-1} = \frac{1}{9}\begin{bmatrix} 3 & 1 & -2 \\ 0 & -3 & 6 \\ -3 & 2 & 5 \end{bmatrix}$

20. $y - z = -4,$
$4x + y = -3,$
$3x - y + 3z = 1;$
$A^{-1} = -\frac{1}{2}\begin{bmatrix} 1 & -1 & -1 \\ -3 & 0 & 2 \\ -2 & 1 & 1 \end{bmatrix}$

Write a matrix equation equivalent to the system. Find the inverse of the coefficient matrix. Use the inverse of the coefficient matrix to solve each system. Show your work.

21. $4x - 3y = 2,$
$x + 2y = -1$

22. $3x + 5y = -4,$
$2x + 4y = -2$

23. $7x - 2y = -3,$
$9x + 3y = 4$

24. $5x + 3y = -2,$
$4x - y = 1$

25. $x + z = 1,$
$2x + y = 3,$
$x - y + z = 4$

26. $x + 2y + 3z = -1,$
$2x - 3y + 4z = 2,$
$-3x + 5y - 6z = 4$

27. $2w - 3x + 4y - 5z = 0,$
$3w - 2x + 7y - 3z = 2,$
$w + x - y + z = 1,$
$-w - 3x - 6y + 4z = 6$

28. $5w - 4x + 3y - 2z = -6,$
$w + 4x - 2y + 3z = -5,$
$2w - 3x + 6y - 9z = 14,$
$3w - 5x + 2y - 4z = -3$

SYNTHESIS

29. Let

$$A = \begin{bmatrix} a & b & c \\ d & e & f \\ g & h & i \end{bmatrix} \quad \text{and} \quad I = \begin{bmatrix} 1 & 0 & 0 \\ 0 & 1 & 0 \\ 0 & 0 & 1 \end{bmatrix}.$$

Show that $AI = IA = A$.

State the conditions under which A^{-1} exists. Then find a formula for A^{-1}.

30. $A = [x]$

31. $A = \begin{bmatrix} x & 0 \\ 0 & y \end{bmatrix}$

32. $A = \begin{bmatrix} 0 & 0 & x \\ 0 & y & 0 \\ z & 0 & 0 \end{bmatrix}$

33. $A = \begin{bmatrix} x & 1 & 1 & 1 \\ 0 & y & 0 & 0 \\ 0 & 0 & z & 0 \\ 0 & 0 & 0 & w \end{bmatrix}$

CHALLENGE

34. Consider

$$a_1 x + b_1 y = c_1,$$
$$a_2 x + b_2 y = c_2.$$

Find a general formula for the solution, and use it to prove Cramer's rule.

9.9 Systems of Inequalities and Linear Programming

OBJECTIVES

You should be able to:

1. Determine whether an ordered pair of numbers is a solution of an inequality in two variables.

2. Graph linear inequalities.

3. Graph systems of linear inequalities.

4. Solve linear programming problems.

A **graph** of an inequality is a drawing that represents its solutions. An inequality in one variable can be graphed on a number line. An inequality in two variables can be graphed on a coordinate plane.

1 Solutions of Inequalities in Two Variables

The solutions of inequalities in two variables are ordered pairs.

EXAMPLE 1 Determine whether $(-3, 2)$ is a solution of $5x - 4y \leq 13$.

Solution We use alphabetical order of variables. We replace x by -3 and y by 2.

$$\begin{array}{c|c} 5x - 4y \leq 13 & \\ \hline 5(-3) - 4 \cdot 2 & 13 \\ -15 - 8 & \\ -23 & \end{array}$$

Since $-23 \leq 13$ is true, $(-3, 2)$ is a solution. ■

EXAMPLE 2 Determine whether $(6, -7)$ is a solution of $5x - 4y \leq 13$.

Solution We use alphabetical order of variables. We replace x by 6 and y by -7.

$$\begin{array}{c|c} 5x - 4y \leq 13 & \\ \hline 5(6) - 4(-7) & 13 \\ 30 + 28 & \\ 58 & \end{array}$$

Since $58 \leq 13$ is false, $(6, -7)$ is not a solution. ■

DO EXERCISES 1 AND 2.

2 Graphs of Linear Inequalities in Two Variables

A **linear inequality** is an inequality that is equivalent to

$$Ax + By < C, \quad Ax + By \leq C, \quad Ax + By > C, \quad \text{or} \quad Ax + By \geq C.$$

That is, there is a first-degree polynomial on one side and a constant on the other. To graph linear inequalities, we use what we already know about graphing linear equations.

EXAMPLE 3 Graph: $y < x$.

Solution We first graph the line $y = x$ for comparison. Every solution of $y = x$ is an ordered pair like (4, 4). The first and second coordinates are the same. The graph of $y = x$ is shown at the left below.

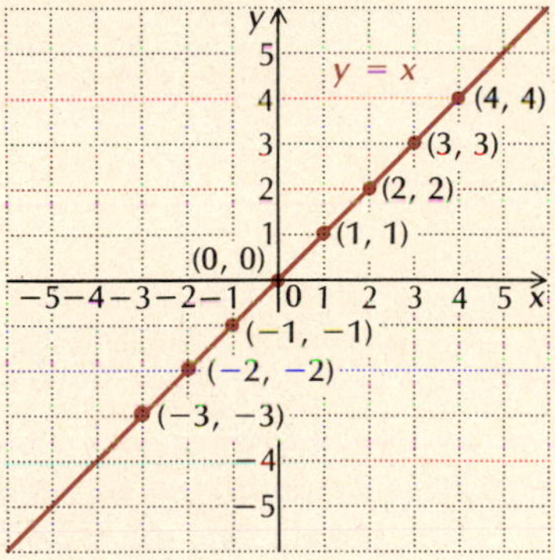

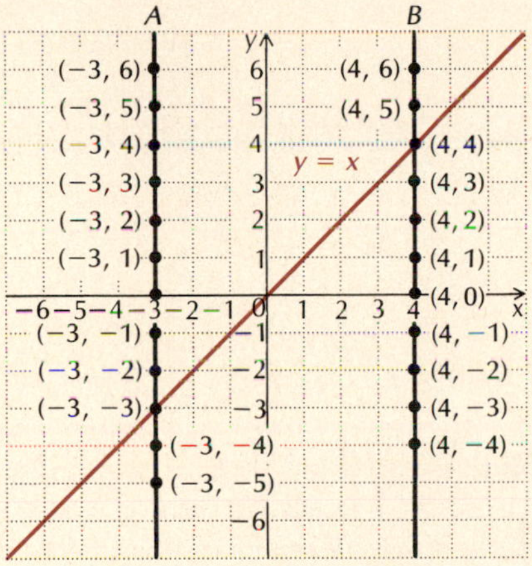

Now look at the graph to the right above. We consider a vertical line A and ordered pairs on it. For all points above $y = x$, the second coordinate is greater than the first, $y > x$. For all points below the line, $y < x$. The same thing happens for vertical line B and indeed for any vertical line. If we could draw all such bottom parts of the vertical lines, we would obtain the solutions of $y < x$. We see this on the graph at the left below. We generally use color shading to indicate these solutions, as shown on the graph at the right below. We shade the half-plane below $y = x$. This is the graph of $y < x$. Points on $y = x$ are not in the graph, so we draw it dashed.

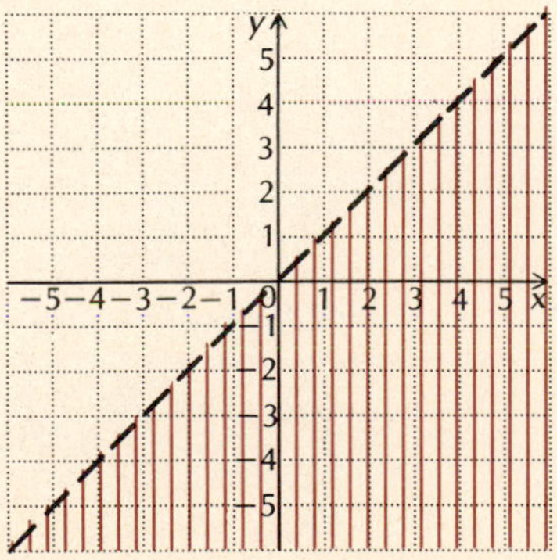

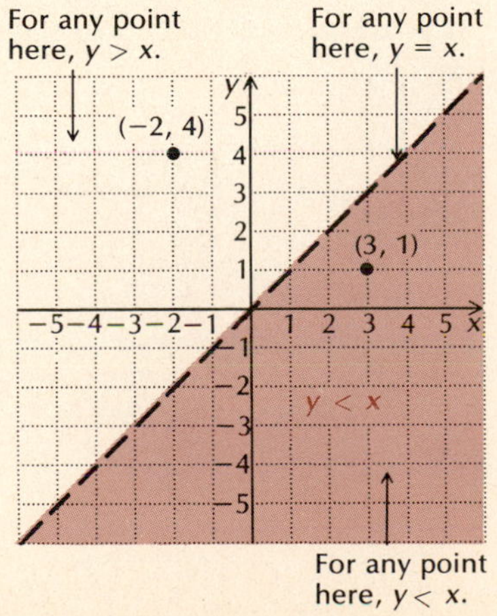

1. Determine whether (1, −4) is a solution of the inequality $4x - 5y \geq 12$.

2. Determine whether (4, −3) is a solution of the inequality $3y - 2x < -60$.

We can determine which half-plane to shade by considering a test point on either side of the line. We consider the pair (3, 1) as a test point. We normally do not know if it is a solution. We substitute to see if it is a solution:

$$\begin{array}{c|c} y & < x \\ \hline 1 & 3 \end{array}$$

We see that (3, 1) is a solution, and in fact any point on the same side of $y = x$ as (3, 1) is a solution. The points in that half-plane are solutions of $y > x$. ■

In a similar way, we can sketch the graph of $y > x$ as shown below.

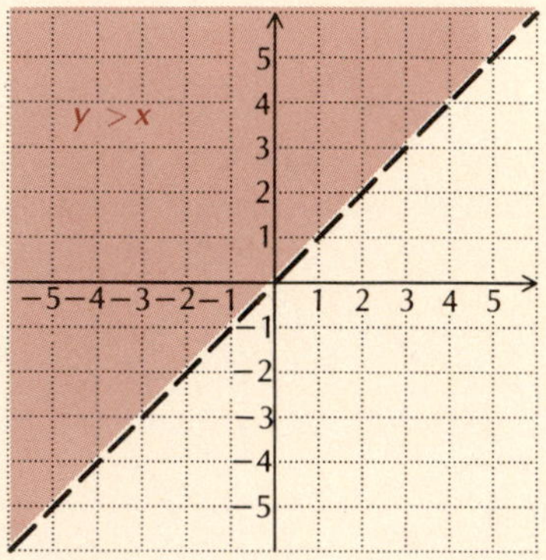

EXAMPLE 4 Graph: $3y - 2x \geq 1$.

Solution First we sketch the line $3y - 2x = 1$. Because points on the line $3y - 2x = 1$ are also in the graph of $3y - 2x \geq 1$, we draw the line solid. This indicates that all points on the line are solutions. The rest of the solutions are either in the half-plane above the line or in the half-plane below the line. To determine which, we select a point that is not on the line and determine whether it is a solution of $3y - 2x \geq 1$. We try $(-2, 4)$ as a test point:

$$\begin{array}{r|l} 3y - 2x & \geq 1 \\ \hline 3(4) - 2(-2) & 1 \\ 12 + 4 & \\ 16 & \end{array}$$

We see that $16 \geq 1$ is *true*. Thus, $(-2, 4)$ is a solution. All of the points in the half-plane containing $(-2, 4)$ are solutions, and the points in the opposite half-plane are *not* solutions. We shade the half-plane and obtain the graph as follows:

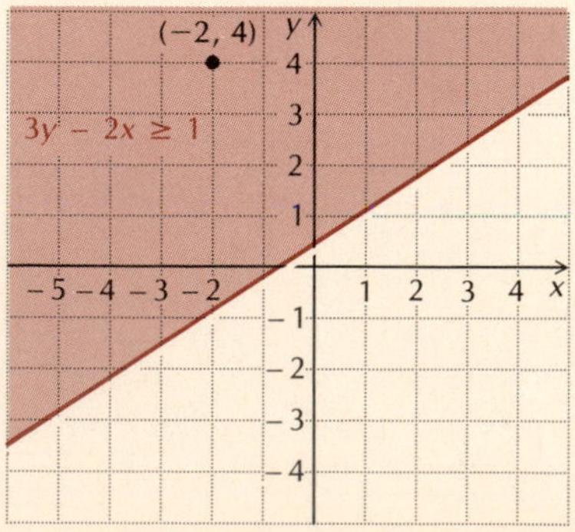

■

A **linear inequality** is one that we can get from a linear equation by changing the equals symbol to an inequality symbol. Every linear equation has a graph that is a straight line. The graph of a linear inequality is a half-plane, sometimes including the line along the edge. That line is called a **related equation.** We graph linear inequalities as follows.

To graph an inequality in two variables:

1. **Replace the inequality symbol with an equals sign and graph this related equation. If the inequality symbol is $<$ or $>$, draw the line dashed. If the inequality symbol is $\leq$ or $\geq$, draw the line solid.**
2. **The graph consists of a half-plane, either above or below or left or right of the line, and, if the line is solid, the line as well. To determine which half-plane to shade, pick a point not on the line as a test point. Substitute to find whether that point is a solution of the inequality. If so, shade the half-plane containing that point. If not, shade the opposite half-plane.**

EXAMPLE 5 Graph: $6x - 2y > 12$.

Solution We first graph the line $6x - 2y = 12$. The intercepts are $(0, -6)$ and $(2, 0)$. The point $(3, 3)$ is also on the graph. This line forms the boundary of the solutions of the inequality. In this case, points on the line are not solutions of the inequality, so we draw a dashed line. To determine which half-plane to shade, we consider a test point *not* on the line. We try $(0, 0)$ and substitute:

$6x - 2y$	≥ 12
$6(0) - 2(0)$	12
$0 + 0$	
0	

Since this inequality is *false*, the point $(0, 0)$ is not a solution; no point in the half-plane containing $(0, 0)$ is a solution. Thus the points in the opposite half-plane are solutions. The graph is shown below.

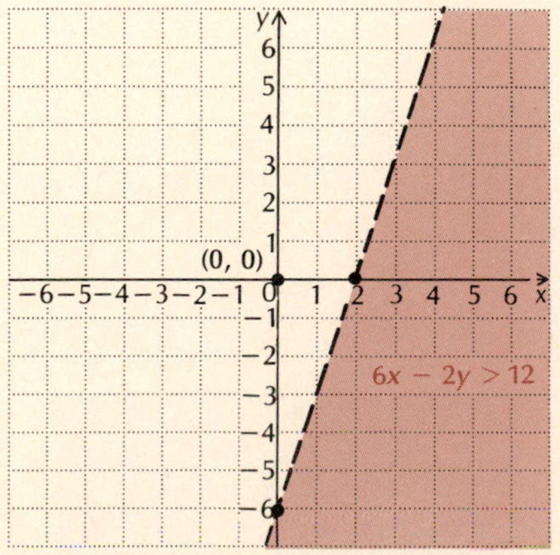

The point $(0, 0)$ is the easiest to use as a test point unless the line goes through the origin. If the related equation does go through the origin, we must test some other point not on the line. The point $(1, 1)$ is often another convenient point to try.

Graph.

3. $6x - 3y > 18$

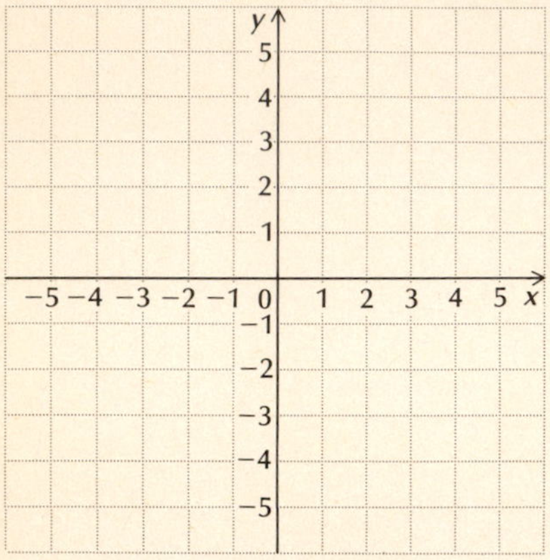

4. $4x + 3y \leq 12$

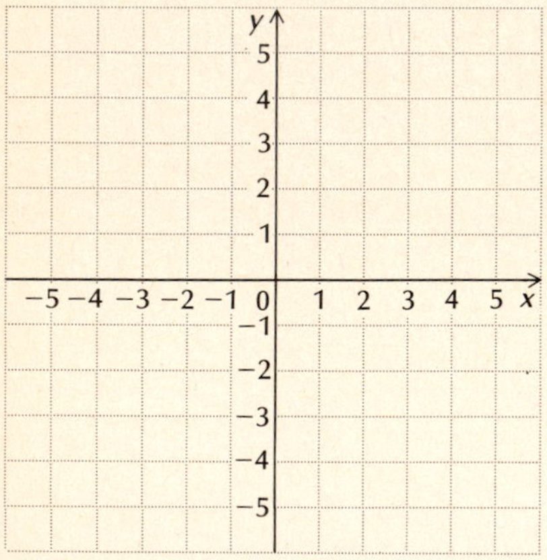

DO EXERCISES 3 AND 4.

EXAMPLE 6 Graph $x > -3$ on a plane.

Solution There is a missing variable in this inequality. If we graph the inequality on a line, we see that its graph is as follows:

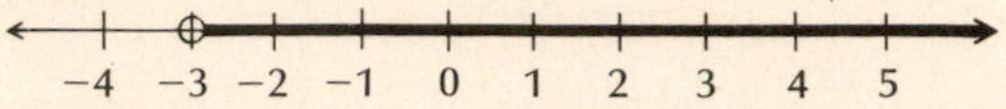

But we can also write this inequality as $x + 0y > -3$ and consider graphing it on the plane. We use the same technique that we have used with the other examples. We first graph the related equation $x = -3$ in the plane. We draw the graph with a dashed line.

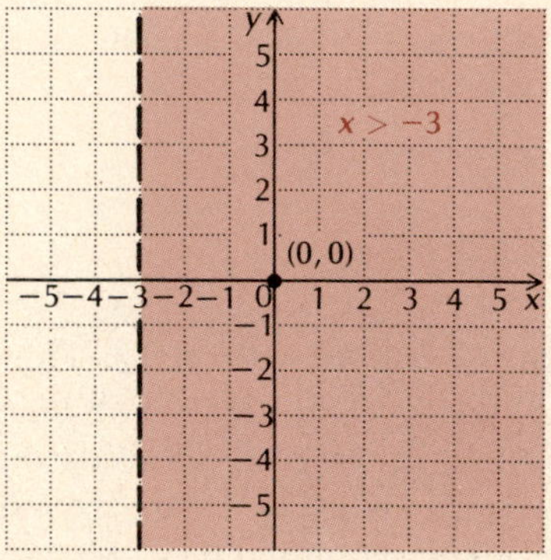

The rest of the graph is a half-plane to the right or left of the line $y = -3$. To determine which, we consider a test point (0, 0):

$x + 0y > -3$	
$0 + 0(0)$	-3
0	

We see that (0, 0) is a solution, so all the pairs in the half-plane containing (0, 0) are solutions. We shade that half-plane.

We see that the solutions of $x > -3$ are all those ordered pairs whose first coordinates are *greater than* -3. ■

EXAMPLE 7 Graph $y \leq 3$ on a plane.

Solution We first graph $y = 3$ using a solid line to indicate that all points on the line are solutions.

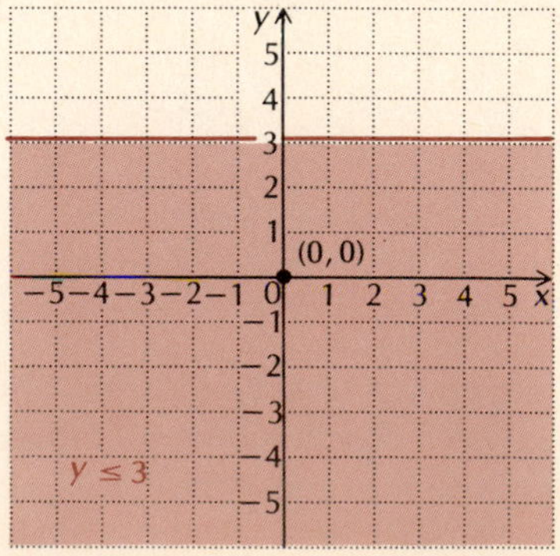

We then use (0, 0) as a test point and substitute:

$$\begin{array}{r|l} \multicolumn{2}{c}{0x + y \leq 3} \\ \hline 0(0) + 0 & 3 \\ 0 & \end{array}$$

We see that (0, 0) is a solution, so all points in the half-plane containing (0, 0) are solutions. Note that this half-plane consists of all ordered pairs whose second coordinate is *less than or equal to* 3.

DO EXERCISES 5 AND 6.

3 Systems of Linear Inequalities

A system of inequalities generally has a very large solution set. The most useful thing to do in problem situations involving such inequalities is to graph the solution set. To graph a system of equations, we graph the individual equations and then find the intersection of the individual graphs. We do the same thing for a system of inequalities, that is, we graph them and look for the intersection.

EXAMPLE 8 Graph the solutions of the system of inequalities

$$y - x \leq 4,$$
$$x + y \leq 4.$$

Solution First we graph the inequality $y - x \leq 4$. We graph $y - x = 4$ using a solid line. We consider (0, 0) as a test point and find that it is a solution, so we shade all points on that side of the line using color shading. The arrows at the ends of the line also indicate the half-plane that contains the solutions.

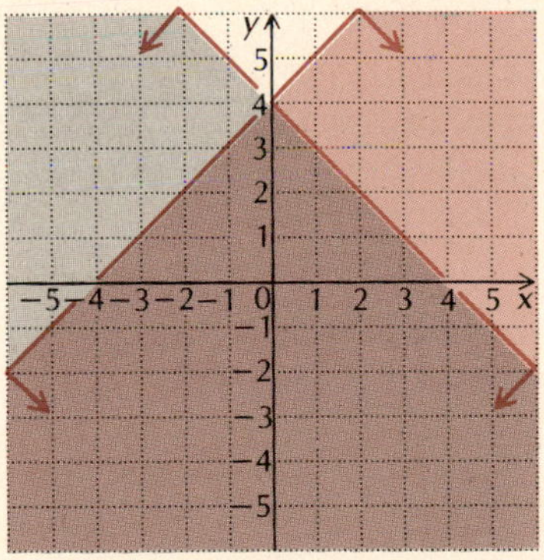

Next, we graph $x + y \leq 4$. We graph $x + y = 4$ using a solid line and consider (0, 0) as a test point. Again, since (0, 0) is a solution, we shade that side of the line using gray shading this time. The solution set of the system consists of the region that is shaded both color and gray and parts of the lines $y - x = 4$ and $x + y = 4$.

DO EXERCISE 7.

EXAMPLE 9 Graph: $-1 < y \leq 2$.

Solution This is a conjunction of two inequalities

$$-1 < y \quad and \quad y \leq 2.$$

Graph on a plane.

5. $x > 3$

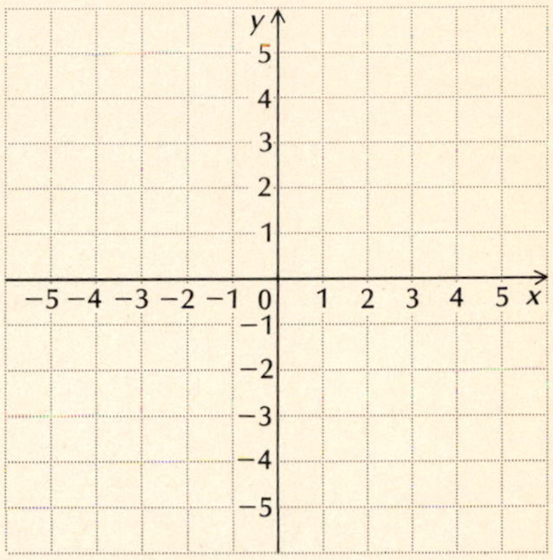

6. $y \geq -2$

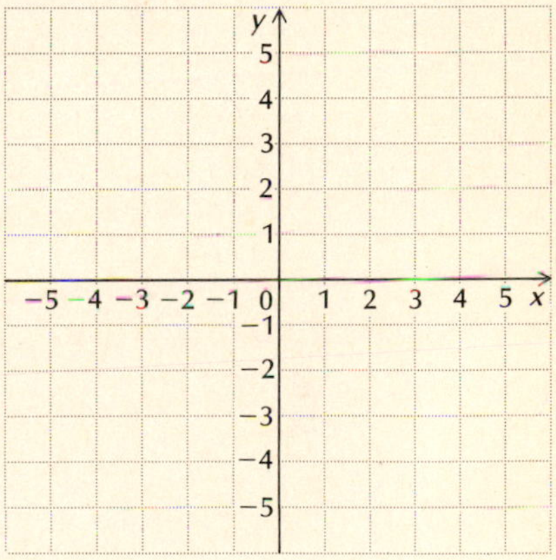

7. Graph:

$$x + y \geq 3,$$
$$y - x \geq 3.$$

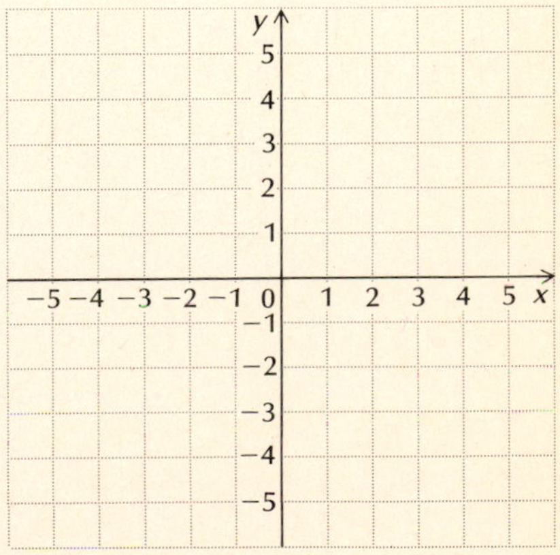

8. Graph:

$$-3 \le x < 4.$$

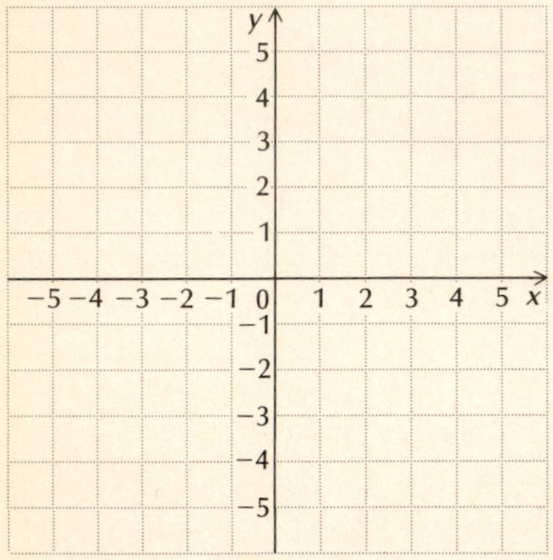

It will be true for any y that is both greater than -1 and less than or equal to 2. Here we show the graph of $y > -1$ in color and the graph of $y \le 2$ in gray. The arrows indicate which half-plane is to be shaded.

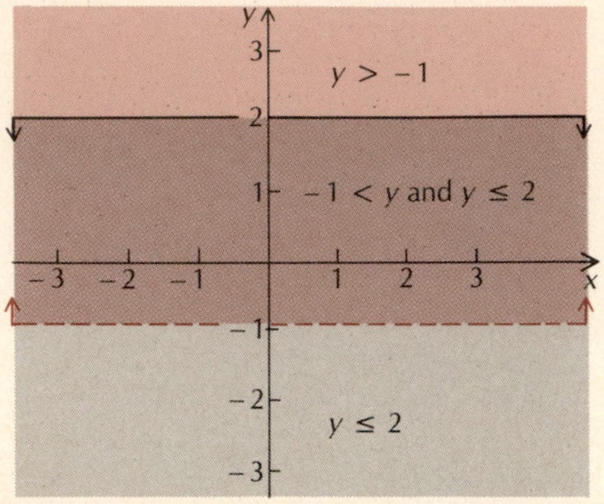

Since our inequality is a conjunction, the graph is the intersection of the graphs of the parts.

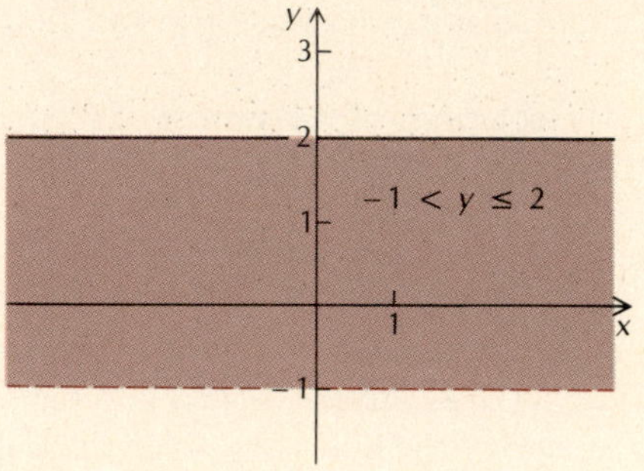

DO EXERCISE 8.

A system of inequalities may have a graph that consists of a polygon and its interior. It will be helpful later in the section if we are able to find the vertices of such a graph.

EXAMPLE 10 Graph the system of inequalities. Find the coordinates of any vertices formed.

$$\begin{aligned} 2x + y &\ge 2, \\ 4x + 3y &\le 12, \\ \tfrac{1}{2} \le x &\le 2, \\ y &\ge 0 \end{aligned}$$

Solution The separate graphs are shown on the left and the graph of the intersection, which is the graph of the system, is shown on the right at the top of the following page.

We find the vertex $\left(\frac{1}{2}, 1\right)$ by solving the system

$$\begin{aligned} 2x + y &= 2, \\ x &= \tfrac{1}{2}. \end{aligned}$$

We find the vertex $(1, 0)$ by solving the system

$$\begin{aligned} 2x + y &= 2, \\ y &= 0. \end{aligned}$$

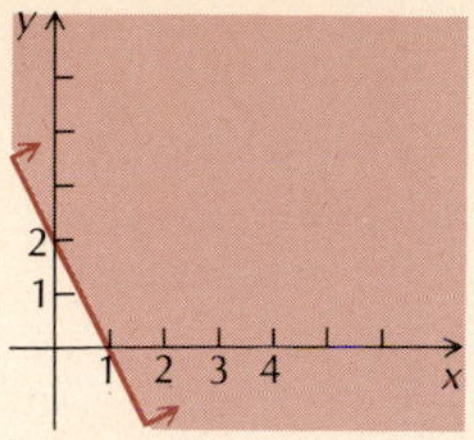

$2x + y \geq 2$

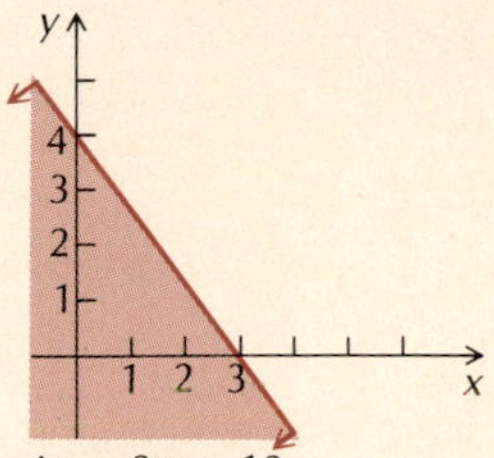

$4x + 3y \leq 12$

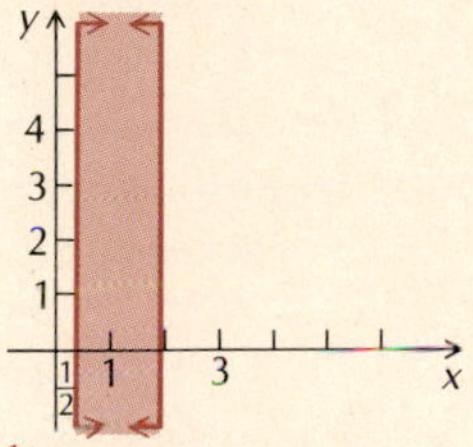

$\frac{1}{2} \leq x \leq 2$

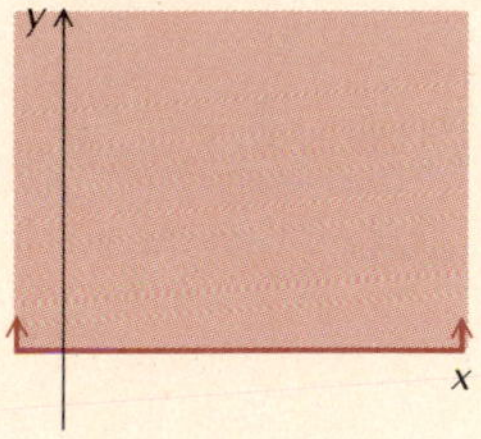

$y \geq 0$

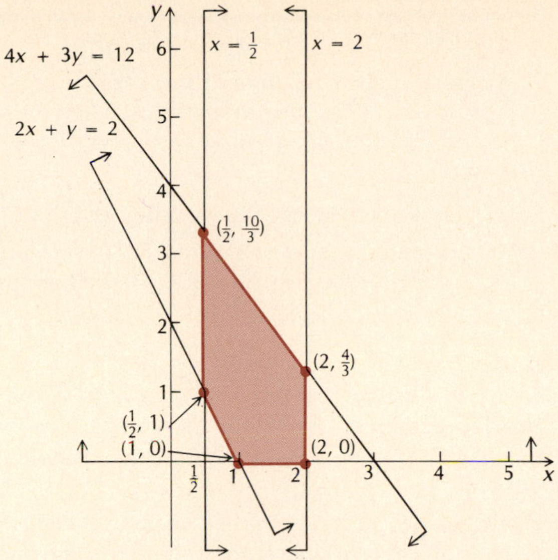

We find the vertex (2, 0) from the system

$$x = 2,$$
$$y = 0.$$

The vertices $\left(2, \frac{4}{3}\right)$ and $\left(\frac{1}{2}, \frac{10}{3}\right)$ were found by solving, respectively, the systems

$$\begin{aligned} x &= 2, \\ 4x + 3y &= 12; \end{aligned} \quad \text{and} \quad \begin{aligned} x &= \tfrac{1}{2}, \\ 4x + 3y &= 12. \end{aligned}$$

DO EXERCISES 9 AND 10.

4 An Application: Linear Programming

There are many problems in real life in which we want to find a greatest value (a maximum) or a least value (a minimum). For example, if you are in business, you would like to know how to make the *most* profit. Or you might like to know how to make your expenses the *least* possible. Many such problems can be solved using systems of inequalities and a branch of mathematics known as **linear programming**. The following is the basic theorem of linear programming.

9. Graph. If a polygon is formed, find the vertices.

$$\begin{aligned} 2x - 3y &\leq 6, \\ 2x + y &\geq 4, \\ 1 \leq y &\leq 3, \\ x &\geq 0. \end{aligned}$$

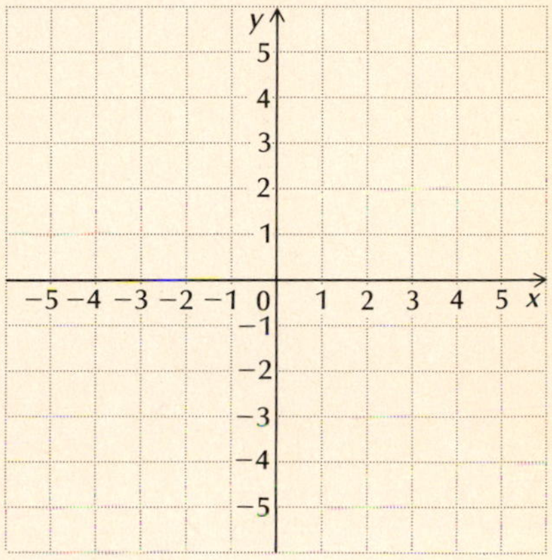

10. Graph. If a polygon is formed, find the vertices.

$$\begin{aligned} 5x + 6y &\leq 30, \\ 0 \leq y &\leq 3, \\ 0 \leq x &\leq 4 \end{aligned}$$

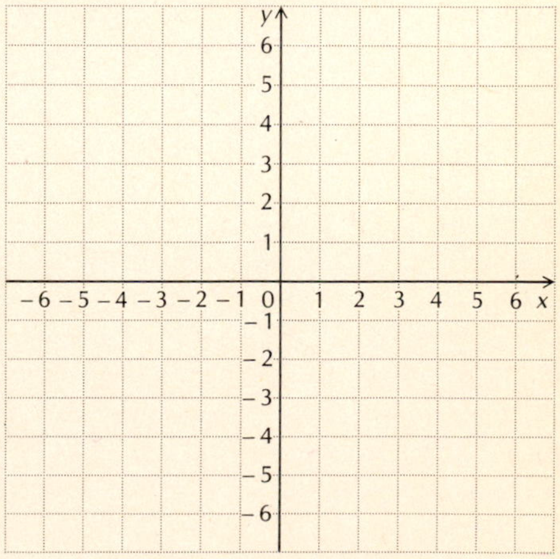

THEOREM 12 Linear Programming

Suppose a linear function of two variables $F = ax + by + c$ is defined over a system of linear inequalities. Then maximum or minimum values of the function will occur at certain vertices. To find the maximum or minimum:

1. **Graph the system of inequalities and find the vertices.**
2. **Compute the function values at the vertices. The largest and smallest of those values are the maximum and minimum of the function.**

This theorem was proven during World War II. Linear programming was developed then to deal with the complicated process of shipping men and supplies to Europe. Let us consider an example.

EXAMPLE 11 You are taking a test in which items of type A are worth 10 points and items of type B are worth 15 points. It takes 3 minutes for each item of type A and 6 minutes for each item of type B. The total time allowed is 60 minutes and you are not allowed to answer more than 16 questions. Assuming that all of your answers are correct, how many items of each type should you answer to get the best score?

Solution Let x = the number of items of type A and y = the number of items of type B. The total score T is a linear function of the two variables x and y:

$$T = 10x + 15y.$$

This function has a domain that is a set of ordered pairs of numbers (x, y). This domain is determined by the following inequalities, also called *constraints*:

Total number of questions allowed, not more than 16	$x + y \leq 16,$
Time, not more than 60 min	$3x + 6y \leq 60,$
Numbers of items answered will not be negative	$x \geq 0,$
	$y \geq 0.$

We now graph the domain of the function T. This is the graph of the system of inequalities above. We will determine the vertices, if any are formed.

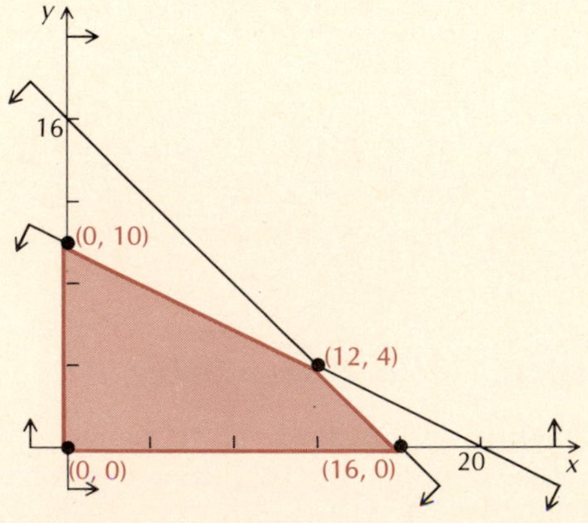

We know by Theorem 12 that the maximum and minimum values occur at certain vertices of the polygon. All we need do to find these values is to substitute the coordinates of the vertices in $T = 10x + 15y$.

Vertices (x, y)	Score $T = 10x + 15y$	
(0, 0)	$T = 10(0) + 15(0) = 0$	⟵ Minimum
(16, 0)	$T = 10(16) + 15(0) = 160$	
(12, 4)	$T = 10(12) + 15(4) = 180$	⟵ Maximum
(0, 10)	$T = 10(0) + 15(10) = 150$	

From the table, we see that the minimum value is 0 and the maximum is 180. To get this maximum, you must answer 12 items of type A and 4 items of type B.

EXAMPLE 12 Find the maximum and minimum values of $F = 9x + 40y$ subject to the constraints

$$y - x \geq 1,$$
$$y - x \leq 3,$$
$$2 \leq x \leq 5.$$

Solution We graph the system of inequalities, determine the vertices, and find the function values for those ordered pairs.

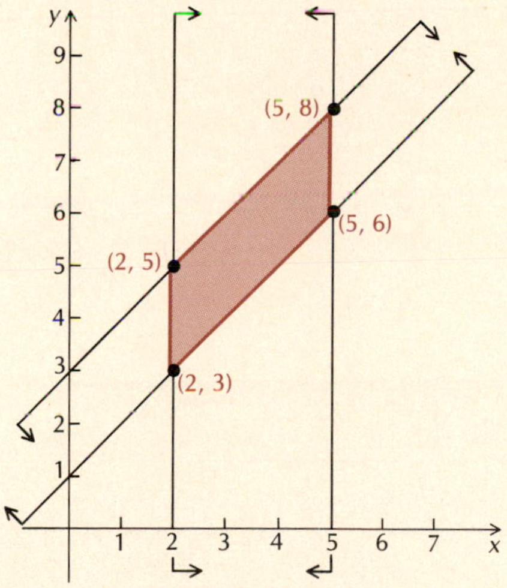

Vertices (x, y)	Score $F = 9x + 40y$	
(2, 3)	$F = 9(2) + 40(3) = 138$	⟵ Minimum
(2, 5)	$F = 9(2) + 40(5) = 218$	
(5, 6)	$F = 9(5) + 40(6) = 285$	
(5, 8)	$F = 9(5) + 40(8) = 365$	⟵ Maximum

The maximum value of F is 365 when $x = 5$ and $y = 8$. The minimum value of F is 138 when $x = 2$ and $y = 3$.

DO EXERCISES 11 AND 12.

EXAMPLE 13 A company manufactures motorcycles and bicycles. To stay in business, it must produce at least 10 motorcycles each month, but it does not have the facilities to produce more than 60 motorcycles. It also does not have the facilities to produce more than 120 bicycles. The total production of

11. Find the maximum and minimum values of

$$F = 34x + 6y$$

subject to

$$x + y \leq 6,$$
$$x + y \geq 1,$$
$$1 \leq x \leq 3.$$

12. Find the maximum and minimum values of

$$G = 3x - 5y + 27$$

subject to

$$x + 2y \leq 8,$$
$$0 \leq y \leq 3,$$
$$0 \leq x \leq 6.$$

13. A college snack bar cooks and sells hamburgers and hot dogs during the lunch hour. To stay in business, it must sell at least 10 hamburgers but cannot cook more than 40. It must also sell at least 30 hot dogs but cannot cook more than 70. It cannot cook more than 90 sandwiches altogether. The profit on a hamburger is \$0.33 and on a hot dog is \$0.21. How many of each kind of sandwich should they sell in order to make the maximum profit? What is the maximum profit?

motorcycles and bicycles cannot exceed 160. The profit on a motorcycle is \$134 and on a bicycle is \$20. Find the number of each that should be manufactured in order to maximize profit.

Solution Let x = the number of motorcycles to be produced and y = the number of bicycles to be produced. The profit P is given by

$$P = \$134x + \$20y,$$

subject to the constraints

$$10 \le x \le 60,$$
$$0 \le y \le 120,$$
$$x + y \le 160.$$

We graph the system of inequalities, determine the vertices, and find the function values for those ordered pairs.

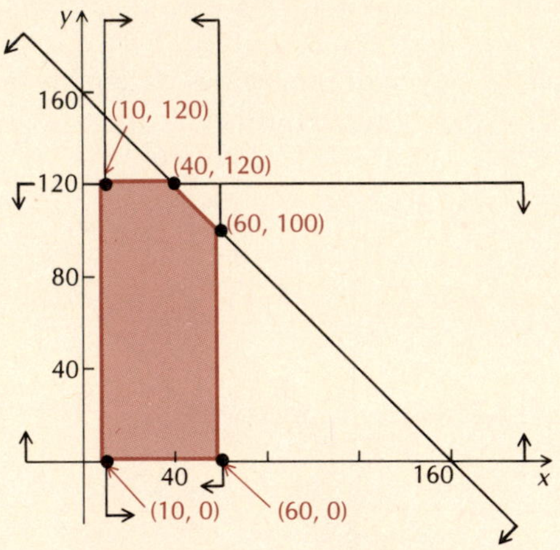

Vertices (x, y)	Profit $P = \$134x + \$20y$
(10, 0)	\$ 1,340
(60, 0)	\$ 8,040
(60, 100)	\$10,040
(40, 120)	\$ 7,760
(10, 120)	\$ 3,740

Thus the company will make a maximum profit of \$10,040 by producing 60 motorcycles and 100 bicycles.

DO EXERCISE 13.

EXERCISE SET 9.9

1 Determine whether the ordered pair is a solution of the inequality.

1. $(-4, 2)$; $2x + y > -5$
2. $(3, -6)$; $4x + 2y \le 0$
3. $(8, 14)$; $2y - 3x \ge 5$
4. $(7, 20)$; $3x - y < -1$

2 Graph the inequality on a plane, using graph paper.

5. $y > 2x$
6. $2y < x$
7. $y + x \ge 0$
8. $y + x < 0$
9. $y > x - 3$
10. $y \le x + 4$
11. $x + y < 4$
12. $x - y \ge 5$

13. $3x - 2y \leq 6$
14. $2x - 5y < 10$
15. $3y + 2x \geq 6$
16. $2y + x \leq 4$
17. $3x - 2 \leq 5x + y$
18. $2x - 6y \geq 8 + 2y$
19. $x < -4$
20. $y \geq 5$
21. $y > -3$
22. $x \leq 5$
23. $-4 < y < -1$
24. $-1 < y < 4$
25. $-4 \leq x \leq 4$
26. $-3 \leq x \leq 3$
27. $y \geq |x|$
28. $y \leq |x|$

3 Graph the system of inequalities. Find the coordinates of any vertices formed.

29. $y \leq x,$ $y \geq 3 - x$
30. $y \geq x,$ $y \leq x - 5$
31. $y \geq x,$ $y \leq x - 4$
32. $y \geq x,$ $y \leq 2 - x$
33. $y \geq -3,$ $x \geq 1$
34. $y \leq -2,$ $x \geq 2$
35. $x \leq 3,$ $y \geq 2 - 3x$
36. $x \geq -2,$ $y \leq 3 - 2x$
37. $x + y \leq 1,$ $x - y \leq 2$
38. $y + 3x \geq 0,$ $y + 3x \leq 2$
39. $2y - x \leq 2,$ $y + 3x \geq -1$
40. $y \leq 2x + 1,$ $y \geq -2x + 1,$ $x \leq 2$
41. $x - y \leq 2,$ $x + 2y \geq 8,$ $y \leq 4$
42. $x + 2y \leq 12,$ $2x + y \leq 12,$ $x \geq 0,$ $y \geq 0$
43. $4y - 3x \geq -12,$ $4y + 3x \geq -36,$ $y \leq 0,$ $x \leq 0$
44. $8x + 5y \leq 40,$ $x + 2y \leq 8,$ $x \geq 0,$ $y \geq 0$
45. $3x + 4y \geq 12,$ $5x + 6y \leq 30,$ $1 \leq x \leq 3$
46. $y - x \geq 1,$ $y - x \leq 3,$ $2 \leq x \leq 5$

4 Find the maximum and minimum values of the function and the values of x and y for which they occur.

47. $P = 17x - 3y + 60$, subject to
$$6x + 8y \leq 48,\quad 0 \leq y \leq 4,\quad 0 \leq x \leq 7.$$

48. $Q = 28x - 4y + 72$, subject to
$$5x + 4y \geq 20,\quad 0 \leq y \leq 4,\quad 0 \leq x \leq 3.$$

49. $F = 5x + 36y$, subject to
$$5x + 3y \leq 34,\quad 3x + 5y \leq 30,\quad x \geq 0,\quad y \geq 0.$$

50. $G = 16x + 14y$, subject to
$$3x + 2y \leq 12,\quad 7x + 5y \leq 29,\quad x \geq 0,\quad y \geq 0.$$

51. The Hockeypuck Biscuit Factory makes two types of biscuits, Biscuit Jumbos and Mitimite Biscuits. The oven can cook at most 200 biscuits per day. Jumbos each require 2 oz of flour, Mitimites require 1 oz of flour, and there is at most 300 oz of flour available. The income from Jumbos is \$0.10 and from Mitimites is \$0.08. How many of each type of biscuit should be made in order to maximize income? What is the maximum income?

52. A student owns a car and a moped. The student has at most 12 gal of gasoline to be used between the car and the moped. The car's tank holds at most 10 gal and the moped's 3 gal. The mileage for the car is 20 mpg and for the moped is 100 mpg. How many gallons of gasoline should each vehicle use if the student wants to travel as far as possible? What is the maximum number of miles?

53. You are about to take a test that contains questions of type A worth 10 points and questions of type B worth 25 points. You must do at least 3 questions of type A but time restricts doing more than 12. You must do at least 4 questions of type B but time restricts doing more than 15. You can do no more than 20 questions in total. How many of each type of question must you do to maximize your score? What is this maximum score?

54. You are about to take a test that contains questions of type A worth 4 points and questions of type B worth 7 points. You must do at least 5 questions of type A but time restricts doing more than 10. You must do at least 3 questions of type B but time restricts doing more than 10. You can do no more than 18 questions in total. How many of each type of question must you do to maximize your score? What is this maximum score?

55. A lumber company can convert logs into either lumber or plywood. In a given week, the mill can turn out 400 units of production, of which 100 units of lumber and 150 units of plywood are required by regular customers. The profit on a unit of lumber is \$20 and on a unit of plywood is \$30. How many units of each type should the mill produce in order to maximize the profit?

56. A farm consists of 240 acres of cropland. The farmer wishes to plant this acreage in corn or oats. Profit per acre in corn production is \$40 and in oats is \$30. An additional restriction is that the total number of hours of labor during the production period is 320. Each acre of land in corn production uses 2 hours of labor during the production period, while production of oats requires 1 hour per acre. Find how the land should be divided between corn and oats in order to give maximum profit.

57. A woman is planning to invest up to \$40,000 in corporate or municipal bonds, or both. The least she is allowed to invest in corporate bonds is \$6000, and she does not want to invest more than \$22,000 in corporate bonds. She also does not want to invest more than \$30,000 in municipal bonds. The interest on corporate bonds is 8% and on municipal bonds is $7\frac{1}{2}\%$. This is simple interest for one year. How much should she invest in each type of bond to maximize her income? What is the maximum income?

58. A man is planning to invest up to \$22,000 in bank X or bank Y, or both. He wants to invest at least \$2000 but no more than \$14,000 in bank X. Bank Y does not insure more than a \$15,000 investment, so he will invest no more than that in bank Y. The interest in bank X is 6% and in bank Y is $6\frac{1}{2}\%$. This is simple interest for one year. How much should he invest in each bank in order to maximize his income? What is the maximum income?

59. A pipe tobacco company has 3000 lb of English tobacco, 2000 lb of Virginia tobacco, and 500 lb of Latakia tobacco. To make one batch of Smello tobacco, it takes 12 lb of English tobacco and 4 lb of Latakia. To make one batch of Roppo tobacco, it takes 8 lb of English and 8 lb of Virginia tobacco. The profit is \$10.56 per batch for Smello and \$6.40 for Roppo. How many batches of each kind of tobacco should be made to yield maximum profit? What is the maximum profit? (*Hint:* Organize the information in a table.)

60. It takes a tailoring firm 2 hr of cutting and 4 hr of sewing to make a knit suit. To make a worsted suit, it takes 4 hr of cutting and 2 hr of sewing. At most 20 hr per day are available for cutting and at most 16 hr per day are available for sewing. The profit on a knit suit is \$34 and on a worsted suit is \$31. How many of each kind of suit should be made in order to maximize profit?

SYNTHESIS

Graph each system.

61. $y \geq x^2 - 2,$
$y \leq 2 - x^2$

62. $y < x + 1,$
$y \geq x^2$

63. *Hockey wins and losses.* A hockey team figures that it needs at least 60 points for the season to make the playoffs. A win w is worth 2 points and a tie t is worth 1 point. Find a system of inequalities that describes the situation. Graph the system.

64. *Elevators.* Many elevators have a capacity of 1 metric ton (1000 kg). An elevator contains c children, each weighing 35 kg, and a adults, each weighing 75 kg. Find a system of inequalities that asserts that the elevator is overloaded. Graph the system.

65. *Widths of a basketball floor.* Sizes of basketball floors vary due to building sizes and other constraints such as cost. The length L is to be at most 74 ft and the width is to be at most 50 ft. Find a system of inequalities that describes the perimeter of a basketball floor. Graph the system.

CHALLENGE

66. $|x| + |y| \leq 1$

67. $|x + y| \leq 1$

68. $|x - y| > 0$

69. $|x| > |y|$

70. *Allocation of resources in a manufacturing process.* A furniture manufacturer produces chairs and sofas. The chairs require 20 ft of wood, 1 lb of foam rubber, and 2 sq yd of material. The sofas require 100 ft of wood, 50 lb of foam rubber, and 20 sq yd of material. The manufacturer has in stock 1900 ft of wood, 500 lb of foam rubber, and 240 sq yd of material. The chairs can be sold for \$20 each and the sofas for \$300 each. How many of each should be produced in order to maximize the income?

SUMMARY AND REVIEW: CHAPTER 9

TERMS TO KNOW

Solution
System of equations
Substitution method
Gauss–Jordan elimination with equations
Consistent system
Inconsistent system
Dependent system
Independent system
Matrix
Elimination with matrices
Dimensions of a matrix
Square matrix
Zero matrix
Row matrix
Column matrix
Coefficient matrix
Determinant
Cramer's rule
Minor
Cofactor
Identity matrix
Inverse of a matrix
Gauss–Jordan method to find inverses
Linear inequality
System of inequalities
Linear programming

REVIEW EXERCISES

Solve.

1. $5x - 3y = -4,$
   $3x - y = -4$

2. $2x - 3y = 2,$
   $5x - y = -29$

3. $x + 5y = 12,$
   $5x + 25y = 12$

4. $2x - 4y + 3z = -3,$
   $-5x + 2y - z = 7,$
   $3x + 2y - 2z = 4$

5. $x + 5y + 3z = 0,$
   $3x - 2y + 4z = 0,$
   $2x + 3y - z = 0$

6. $x - y = 5,$
   $y - z = 6,$
   $z - w = 7,$
   $x + w = 8$

7. Classify each of the systems in Exercises 1–6 as consistent or inconsistent.

8. Classify each of the systems in Exercises 1–6 as dependent or independent.

Solve.

9. The value of 75 coins, consisting of nickels and dimes, is \$5.95. How many of each kind are there?

10. A family invested \$5000, part at 10% and the remainder at 10.5%. The annual income from both investments is \$517. What is the amount invested at each rate?

11. In triangle ABC, the measure of angle B is three times that of angle A. The measure of angle C is 20° more than that of angle A. Find the angular measurements.

12. A student has a total of 225 on three tests. The sum of the scores on the first and second tests exceeds the third score by 61. The first score exceeds the second by 6. Find the three scores.

Solve using matrices. If there is more than one solution, list three of them.

13. $x + 2y = 5,$
    $2x - 5y = -8$

14. $3x + 4y + 2z = 3,$
    $5x - 2y - 13z = 3,$
    $4x + 3y - 3z = 6$

15. $3x + 5y + z = 0,$
    $2x - 4y - 3z = 0,$
    $x + 3y + z = 0$

16. $w + x + y + z = -2,$
    $-3w - 2x + 3y + 2z = 10,$
    $2w + 3x + 2y - z = -12,$
    $2w + 4x - y + z = 1$

17. Find numbers a, b, and c such that the function $f(x) = ax^2 + bx + c$ fits the data points (0, 3), (1, 0), and (−1, 4). Then write the equations for the function.

Evaluate.

18. $\begin{vmatrix} 1 & -2 \\ 3 & 4 \end{vmatrix}$

19. $\begin{vmatrix} \sqrt{3} & -5 \\ -3 & -\sqrt{3} \end{vmatrix}$

20. $\begin{vmatrix} -2 & -3 \\ 4 & -x \end{vmatrix}$

21. $\begin{vmatrix} 2 & -1 & 1 \\ 1 & 2 & -1 \\ 3 & 4 & -3 \end{vmatrix}$

22. $\begin{vmatrix} -5.8 & 7.5 & 4.6 \\ 0 & 2.2 & 8.9 \\ 0 & 0 & 1.3 \end{vmatrix}$

23. $\begin{vmatrix} 3a & 3b & 3c \\ 5a & 5b & 5c \\ d & e & f \end{vmatrix}$

Solve for (x, y) using Cramer's rule.

24. $5x - 2y = 19,$
$7x + 3y = 15$

25. $ax - by = a^2,$
$bx + ay = ab$

26. Solve using Cramer's rule:

$$3x - 2y + z = 5,$$
$$4x - 5y - z = -1,$$
$$3x + 2y - z = 4.$$

For Exercises 27–34, let

$$A = \begin{bmatrix} 1 & -1 & 0 \\ 2 & 3 & -2 \\ -2 & 0 & 1 \end{bmatrix}, \quad B = \begin{bmatrix} -1 & 0 & 6 \\ 1 & -2 & 0 \\ 0 & 1 & -3 \end{bmatrix}, \quad \text{and} \quad C = \begin{bmatrix} -2 & 0 \\ 1 & 3 \end{bmatrix}.$$

Find each of the following, if possible.

27. A + B **28.** −3A **29.** −A **30.** AB

31. B + C **32.** A − B **33.** 2A − B **34.** A + 3B

Find A^{-1}, if it exists.

35. $A = \begin{bmatrix} -2 & 0 \\ 1 & 3 \end{bmatrix}$

36. $A = \begin{bmatrix} 0 & 0 & 3 \\ 0 & -2 & 0 \\ 4 & 0 & 0 \end{bmatrix}$

37. $A = \begin{bmatrix} 1 & 0 & 0 & 0 \\ 0 & 4 & -5 & 0 \\ 0 & 2 & 2 & 0 \\ 0 & 0 & 0 & 1 \end{bmatrix}$

38. Write a matrix equation equivalent to this system of equations.

$$3x - 2y + 4z = 13,$$
$$x + 5y - 3z = 7,$$
$$2x - 3y + 7z = -8$$

Evaluate.

39. $\begin{vmatrix} -4 & \sqrt{3} \\ \sqrt{3} & 7 \end{vmatrix}$

40. $\begin{vmatrix} 1 & -1 & 2 \\ -1 & 2 & 0 \\ -1 & 3 & 1 \end{vmatrix}$

41. $\begin{vmatrix} 0 & a & b \\ -a & 0 & c \\ -b & -c & 0 \end{vmatrix}$

42. $\begin{vmatrix} 4 & -7 & 6 & 7 \\ 0 & -3 & 9 & -8 \\ 0 & 0 & -2 & 6 \\ 0 & 0 & 0 & 5 \end{vmatrix}$

43. Without expanding, show that

$$\begin{vmatrix} 5a & 5b & 5c \\ 3a & 3b & 3c \\ d & e & f \end{vmatrix} = 0.$$

Factor.

44. $\begin{vmatrix} 1 & a & bc \\ 1 & b & ac \\ 1 & c & ab \end{vmatrix}$

45. $\begin{vmatrix} 1 & x^2 & x^3 \\ 1 & y^2 & y^3 \\ 1 & z^2 & z^3 \end{vmatrix}$

46. $\begin{vmatrix} 1 & a & a^2 & a^3 \\ 1 & b & b^2 & b^3 \\ 1 & c & c^2 & c^3 \\ 1 & d & d^2 & d^3 \end{vmatrix}$

47. Write a matrix equation equivalent to this system. Find the inverse of the coefficient matrix. Use the inverse of the coefficient matrix to solve each system. Show your work.

$$2x + 3y = 2,$$
$$5x - y = -29$$

48. Graph this system. Find the coordinates of any vertices formed.

$$2x + y \geq 9,$$
$$4x + 3y \geq 23,$$
$$x + 3y \geq 8,$$
$$x \geq 0,$$
$$y \geq 0$$

49. Maximize and minimize $T = 6x + 10y$ subject to

$$x + y \leq 10,$$
$$5x + 10y \geq 50,$$
$$x \geq 2,$$
$$y \geq 0.$$

50. You are about to take a test that contains questions of type A worth 7 points and questions of type B worth 12 points. The total number of questions worked must be at least 8. If you know that type A questions take 10 min and type B questions take 8 min and that the maximum time for the test is 80 min, how many of each type of question must you do in order to maximize your score? What is this maximum score?

SYNTHESIS

51. One year, a person invested a total of \$40,000, part at 12%, part at 13%, and the rest at $14\frac{1}{2}$%. The total interest received on the investments was \$5370. The interest received on the $14\frac{1}{2}$% investment was \$1050 more than the interest received on the 13% investment. How much was invested at each rate?

Solve.

52. $\frac{2}{3x} + \frac{4}{5y} = 8,$
$\frac{5}{4x} - \frac{3}{2y} = -6$

53. $\frac{3}{x} - \frac{4}{y} + \frac{1}{z} = -2,$
$\frac{5}{x} + \frac{1}{y} - \frac{2}{z} = 1,$
$\frac{7}{x} + \frac{3}{y} + \frac{2}{z} = 19$

Graph.

54. $|x| - |y| \leq 1$

55. $|xy| > 1$

56. On the basis of Exercise 42, conjecture and prove a theorem regarding determinants.

TEST: CHAPTER 9

Solve.

1. $0.2x - 0.5y = -0.1,$
$0.01x + 0.1y = 0.12$

2. A boat travels 36 km downstream in 2 hr. It travels 48 km upstream in 4 hr. Find the speed of the boat and the speed of the stream.

3. A chemist has one solution of acid and water that is 15% acid and a second that is 75% acid. Find how many gallons of each should be mixed together in order to get 20 gallons of a solution that is 39% acid.

4. A factory has three machines A, B, and C. With all three working, they make 500 toothbrushes per day. With A and B working, they make 284 toothbrushes per day. With A and C working, they make 260 toothbrushes per day. What is the daily production of each machine?

Solve using matrices. If there is more than one solution, list three of them.

5. $8x - 2y = 1,$
$12x + 8y = 7$

6. $12x - 6y - 19z = 0$
$3x - 2y - 6z = 0,$
$6x + 2y + 3z = 0$

7. $2x - 7y + z = 11,$
$3x + 2y - 4z = 15,$
$5x - 4y + 6z = 7$

Classify as consistent or inconsistent, dependent or independent.

8. $-x + y = -6,$
$-x - y = 6$

9. $x + 4y - 2z = 1,$
$2x + 6y - z = 0,$
$-3x - 10y + 3z = -1$

10. Find numbers a, b, and c such that the function $f(x) = ax^2 + bx + c$ fits the data points (2, 1), (0, 1), and (−1, −5). Then write the equation for the function.

Evaluate.

11. $\begin{vmatrix} -3 & 1 \\ -4 & \frac{2}{3} \end{vmatrix}$

12. $\begin{vmatrix} 1 & 1 & -2 \\ 0 & 2 & -6 \\ 4 & 0 & 3 \end{vmatrix}$

Solve using Cramer's rule.

13. $7x - 3y = 31,$
$4x + 2y = 14$

14. $x - 2y + 7z = 11,$
$2x + y - 3z = -5,$
$6x + z = 1$

For Exercises 15–24, let

$$\mathbf{A} = \begin{bmatrix} -1 & 3 \\ 0 & 4 \end{bmatrix}, \quad \mathbf{B} = \begin{bmatrix} -2 & 1 & 0 \\ 3 & 2 & 5 \end{bmatrix}, \quad \mathbf{C} = \begin{bmatrix} 4 & 0 \\ 1 & -1 \\ 2 & -3 \end{bmatrix},$$

$$\mathbf{D} = \begin{bmatrix} 0 & 2 & -1 \\ 3 & -1 & 1 \\ 0 & 4 & 3 \end{bmatrix}, \quad \mathbf{E} = \begin{bmatrix} 5 & -1 & 3 \\ 0 & 4 & 2 \\ 1 & 0 & -6 \end{bmatrix}, \quad \mathbf{F} = [2 \quad -1 \quad -3],$$

$$\mathbf{G} = \begin{bmatrix} -2 & -5 \\ 6 & -3 \end{bmatrix}, \quad \mathbf{O} = \begin{bmatrix} 0 & 0 \\ 0 & 0 \end{bmatrix}, \quad \mathbf{I} = \begin{bmatrix} 1 & 0 \\ 0 & 1 \end{bmatrix}.$$

Find each of the following, if possible.

15. **GC** 16. **O + A** 17. **A + I** 18. **D + E** 19. **A − G**

20. **BC** 21. **B + C** 22. **−F** 23. **2D − E** 24. **GI**

Find $\mathbf{A}^{-1}$, if it exists.

25. $\mathbf{A} = \begin{bmatrix} -3 & 1 \\ 2 & 0 \end{bmatrix}$

26. $\mathbf{A} = \begin{bmatrix} 3 & 4 & 3 \\ 1 & 0 & 1 \\ -2 & -5 & -2 \end{bmatrix}$

27. $\mathbf{A} = \begin{bmatrix} 2 & -1 & 0 \\ 3 & 0 & 1 \\ -2 & 4 & 0 \end{bmatrix}$

28. Let

$$\mathbf{A} = \begin{bmatrix} 3 & 1 & -2 \\ 2 & 0 & -1 \\ -5 & 0 & 4 \end{bmatrix}.$$

Find a_{12}, M_{12}, and A_{12}.

Evaluate.

29. $\begin{vmatrix} 1 & -1 & 3 \\ -2 & 4 & 2 \\ 1 & 2 & 1 \end{vmatrix}$

30. $\begin{vmatrix} -2 & 7 & -3 \\ 1 & 1 & 1 \\ 4 & -14 & 6 \end{vmatrix}$

31. $\begin{vmatrix} -2 & 3 & 4 & 6 \\ 1 & 0 & 0 & 0 \\ -1 & 1 & 2 & -3 \\ 0 & 5 & 1 & 3 \end{vmatrix}$

32. $\begin{vmatrix} 1 & -1 & -1 \\ 2 & -6 & 4 \\ -3 & 0 & 2 \end{vmatrix}$

33. Factor:

$$\begin{vmatrix} c^3 & b^3 & a^3 \\ c & b & a \\ 1 & 1 & 1 \end{vmatrix}.$$

34. Write a matrix equation equivalent to this system of equations and use the inverse of the coefficient matrix to solve the system. Show all your work.

$$2x - 3y = -9,$$
$$x + 4y = 1$$

35. Graph $4x - 3y \geq 12$.

36. Maximize and minimize $T = 20x + 60y$ subject to

$$x + 2y \leq 16,$$
$$2 \leq x \leq 5,$$
$$y \leq 0.$$

37. You are about to take a test that contains questions of type A worth 5 points and type B worth 12 points. You must complete the test in 72 minutes. Type A questions take 4 minutes; type B questions take 8 minutes. The total number of problems worked must not exceed 12. If you are told that you must work at least 2 questions of type B, how many of each type of question must you do in order to maximize your score? What is this maximum score?

SYNTHESIS

Solve.

38. $\frac{7}{x} - \frac{3}{y} = -16,$

$\frac{2}{x} + \frac{5}{y} = 13$

Conic Sections

10

The ellipse described here is a curve known as a *conic section*, meaning that the curve is formed as a cross section of a cone. In this chapter, we will study equations whose graphs are conic sections. We have actually studied three other conic sections—*lines, circles,* and *parabolas*—in some detail in Chapter 4. There are many applications of these equations of conic sections whose graphs have nothing to do with cones. We will consider many of them.

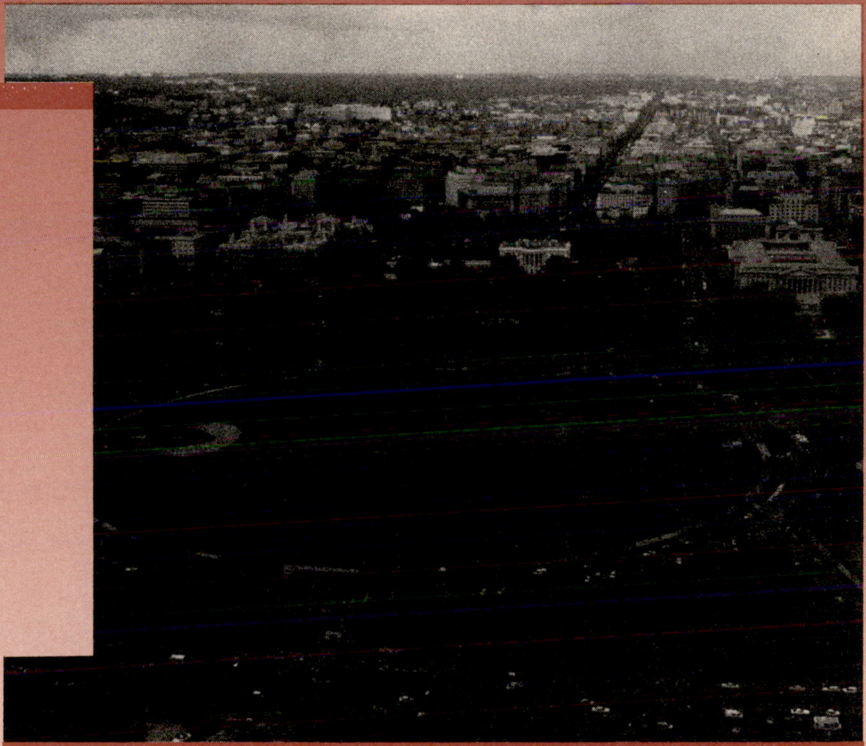

FEATURE PROBLEM

In Washington, D. C., there is a large grassy area south of the White House known as the *Ellipse*. It is actually an ellipse with major axis of length 1048 ft and minor axis of length 898 ft. Assuming a coordinate system is superimposed on the area in such a way that the center is at the origin and the major and minor axes are on the x- and y-axes of the coordinate system, find an equation of the ellipse.

THE MATHEMATICS

The equation has the general form

$$\frac{x^2}{a^2} + \frac{y^2}{b^2} = 1.$$

This is an equation of an *ellipse*.

10.1 Conic Sections: Lines and Ellipses

OBJECTIVES

You should be able to:

1. Graph an equation whose graph consists of a union of lines.
2. Graph an equation whose graph consists of a single point.
3. Given an equation of an ellipse centered at the origin, find the vertices and foci, and graph the ellipse.
4. Given an equation of an ellipse centered at other than the origin, complete the square if necessary and then find the center, vertices, and foci, and graph the ellipse.

In this chapter, we will study polynomial functions and relations that are graphs of second-degree equations of the type

$$Ax^2 + By^2 + Cx + Dy + E = 0.$$

Some of these equations are functions and some are relations that are not functions. Most of these equations have graphs that are **conic sections**, meaning that their graphs are geometric figures formed by cross sections of cones. Examples are shown below.

Conic sections in three dimensions

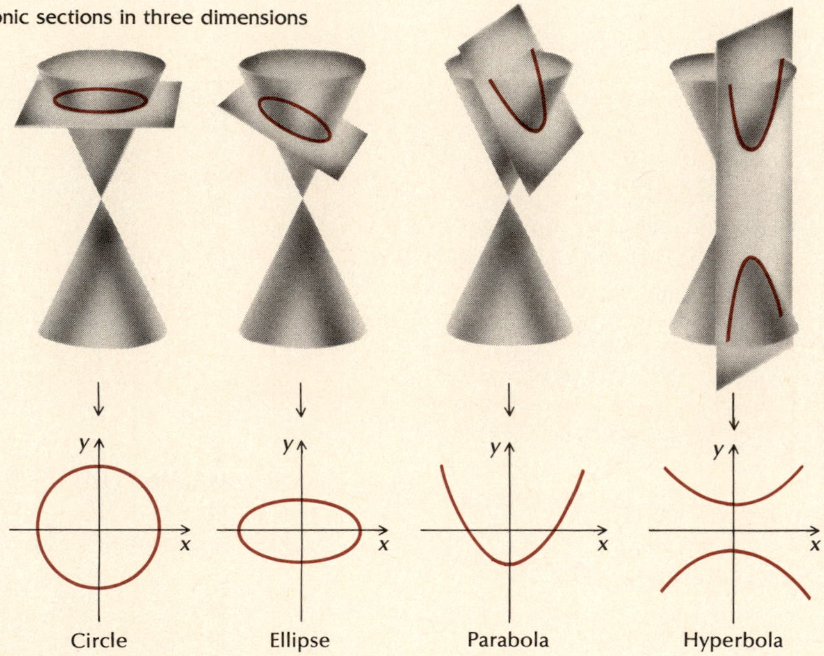

Conic sections graphed in a plane

Cones

Suppose that C is a circle with center O and P is a point not in the same plane as C such that the line $\overleftrightarrow{OP}$ is perpendicular to the plane of the circle C. The set of points on all lines through P and a point of the circle form a **right circular cone** (or **conical surface**). Any line contained in the surface is called a **surface**

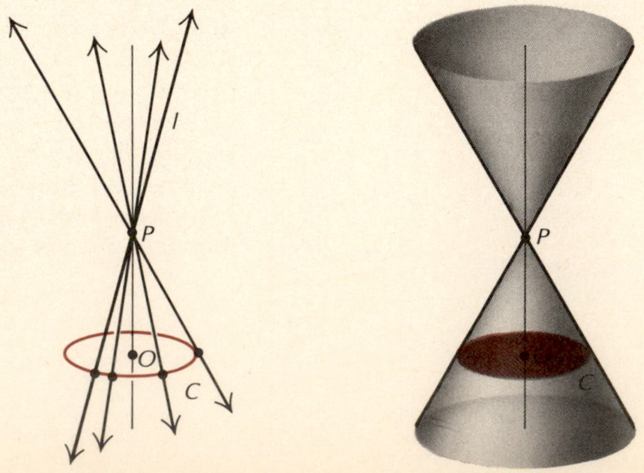

element. Note that there are two parts, or **nappes**, of a cone. Point P is called the **vertex** and line $\overleftrightarrow{OP}$ is called the **axis**.

Conic Sections

The intersection of any plane with a cone is nonempty. That intersection is called a **conic section**. Some other conic sections are shown below.

(a) Line (b) Intersecting lines (c) Single point

1 Lines

Some second-degree equations have graphs consisting of a line. Some have graphs consisting of two lines. The graphs are conic sections except when the two lines are parallel.

EXAMPLE 1 Graph: $3x^2 + 2xy - y^2 = 0$.

Solution We factor and use the principle of zero products:

$$3x^2 + 2xy - y^2 = 0$$
$$(3x - y)(x + y) = 0$$
$$3x - y = 0 \quad \text{or} \quad x + y = 0$$
$$y = 3x \quad \text{or} \quad y = -x.$$

The graph consists of two intersecting lines.

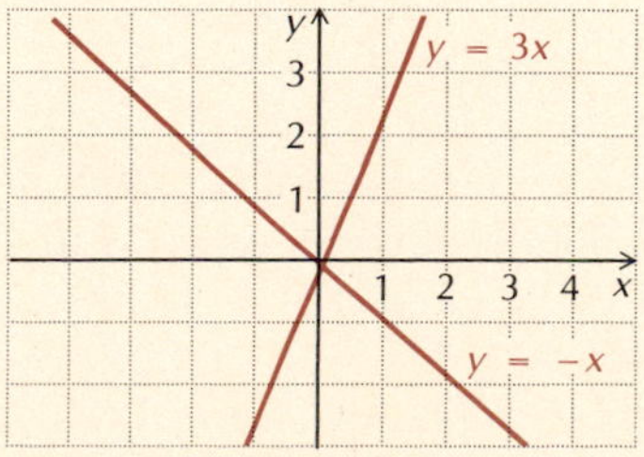

EXAMPLE 2 Graph: $(y - x)(y - x - 1) = 0$.

Solution

$$(y - x)(y - x - 1) = 0$$
$$y - x = 0 \quad \text{or} \quad y - x - 1 = 0 \qquad \text{Using the principle of zero products}$$
$$y = x \quad \text{or} \quad y = x + 1$$

Graph.

1. $x^2 - 4y^2 = 0$

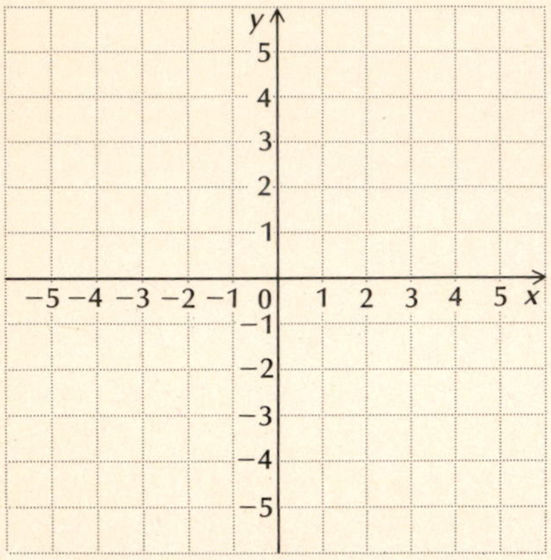

2. $y^2 = 4$

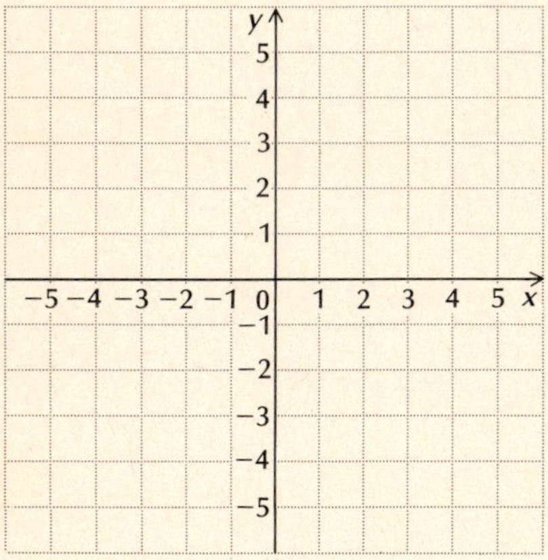

3. $y^2 + 9x^2 = 6xy$

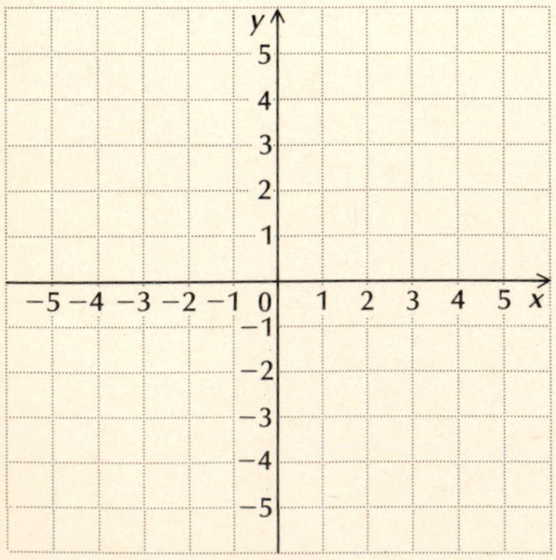

The graph consists of two parallel lines. This equation does *not* represent a conic section.

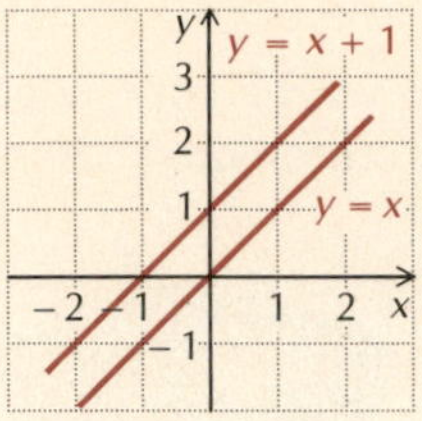

EXAMPLE 3 Graph: $y^2 + x^2 = 2xy$.

Solution

$$y^2 + x^2 = 2xy$$
$$y^2 - 2xy + x^2 = 0$$
$$(y - x)(y - x) = 0$$
$$y - x = 0 \quad \text{or} \quad y - x = 0$$
$$y = x \quad \text{or} \quad y = x$$

The graph consists of a single line.

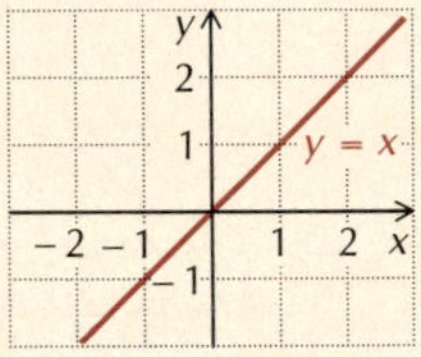

Generally, a second-degree equation with 0 on one side and a factorable expression on the other has a graph consisting of one or two lines. The equation $xy = 0$ has a graph that consists of the coordinate axes.

DO EXERCISES 1–3.

2 Single Points or No Points

When a plane intersects only the vertex of a cone, the result is a single point. The following is an equation for such a conic section.

EXAMPLE 4 Graph: $x^2 + 4y^2 = 0$.

Solution The expression $x^2 + 4y^2$ is not factorable in the real-number system. The only real-number solution of the equation is (0, 0).

EXAMPLE 5 Graph: $3x^2 + 7y^2 = -2$.

Solution Since squares of numbers are never negative, the left side of the equation can never be negative. The equation has no real-number solutions, hence there are no points on the graph. This is a type of equation whose graph is *not* a conic section. Every plane intersects every cone. Thus its solution set cannot be found by the intersection of a plane and a cone.

DO EXERCISES 4 AND 5 ON THE FOLLOWING PAGE.

3 Equations of Ellipses

Some equations of second degree have graphs that are circles. Circles are defined as follows.

DEFINITION

A *circle* is the locus or set of all points in a plane that are at a fixed distance from a fixed point in that plane.

When a plane intersects a cone perpendicular to the axis of the cone, as shown, a circle is formed.

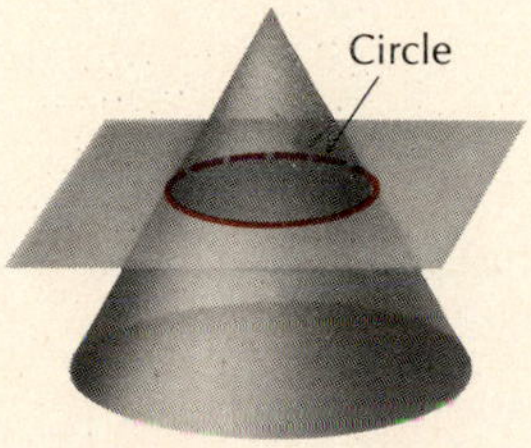

We studied equations of circles in Section 4.2. You may wish to review that section before continuing. The following is the standard equation of a circle with radius r and center at the point (h, k).

The equation, in standard form, of a circle with center (h, k) and radius r is

$$(x - h)^2 + (y - k)^2 = r^2.$$

Some equations of second degree have graphs that are *ellipses*. Ellipses are defined as follows.

DEFINITION

An *ellipse* is the locus or set of all points P in a plane such that the sum of the distances from P to two fixed points F_1 and F_2 in the plane is constant. F_1 and F_2 are called *foci* (singular, *focus*) of the ellipse. The *center* of the ellipse is the midpoint of the segment joining F_1 and F_2.

When a plane intersects a cone with the plane not perpendicular to the axis of the cone, as shown, an ellipse may be formed.

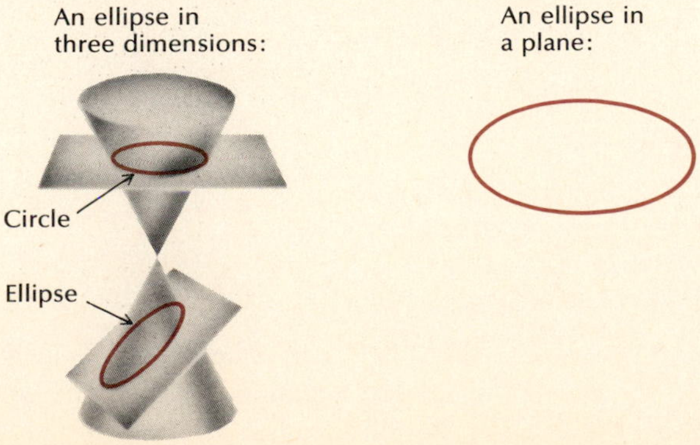

Graph.

4. $x^2 + 2y^2 = 0$

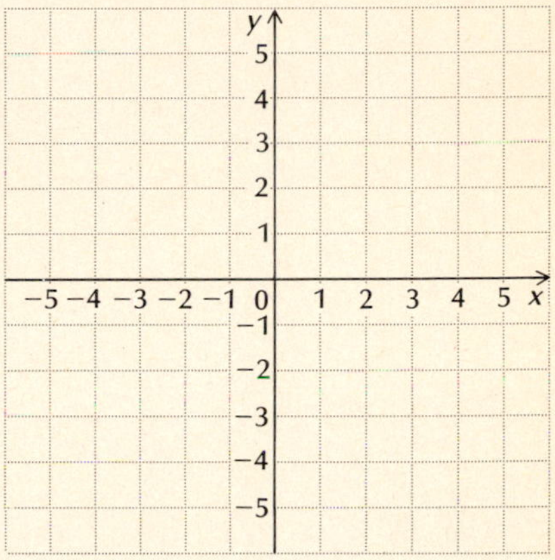

5. $x^2 + y^2 = -3$

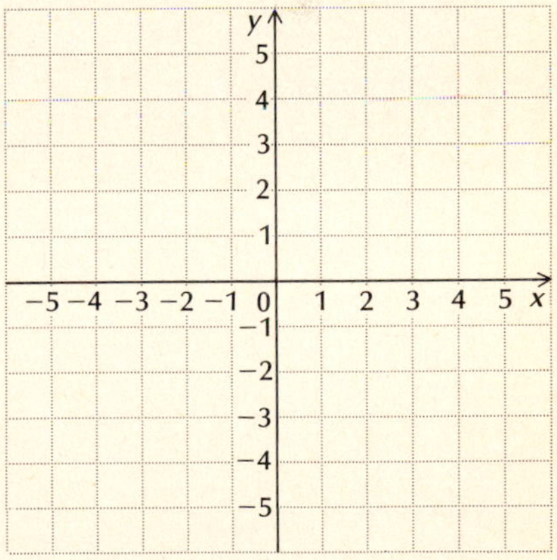

Here is a way to draw an ellipse. Stick two tacks in a piece of cardboard. These will be the foci F_1 and F_2. Attach a piece of string to the tacks. The length of the string will be the constant sum of the distances from the foci to any point on the ellipse. Take a pencil and pull the string tight. Now swing the pencil around, keeping the string tight.

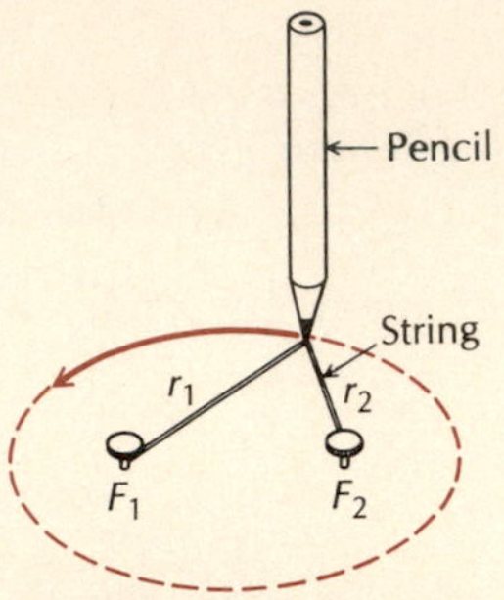

We first consider an equation of an ellipse whose center is at the origin and whose foci lie on one of the coordinate axes.

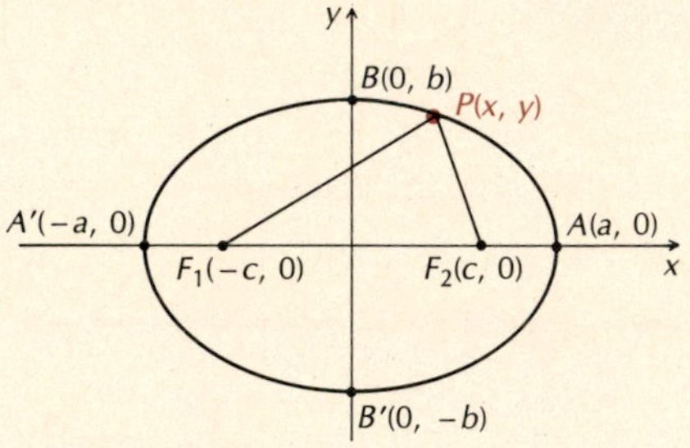

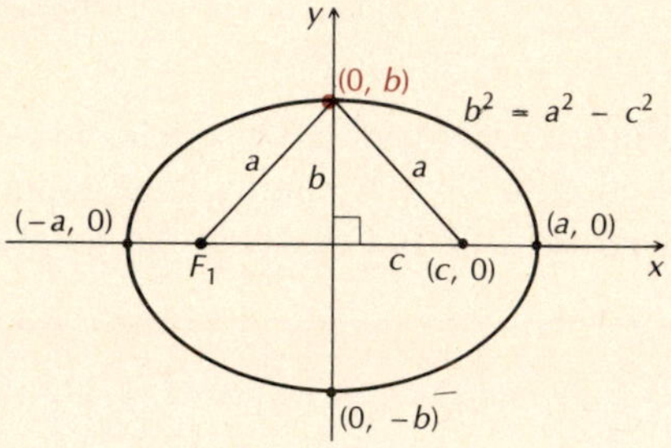

Suppose we have an ellipse with foci $F_1(-c, 0)$ and $F_2(c, 0)$. If $P(x, y)$ is a point on the ellipse, then F_1P is the distance from the first focus to the point $P(x, y)$ and F_2P is the distance from the second focus to the point $P(x, y)$. The sum $F_1P + F_2P$ is the given constant distance. We will call it $2a$:

$$F_1P + F_2P = 2a.$$

By the distance formula,

$$\sqrt{(x + c)^2 + y^2} + \sqrt{(x - c)^2 + y^2} = 2a,$$

or

$$\sqrt{(x + c)^2 + y^2} = 2a - \sqrt{(x - c)^2 + y^2}.$$

Squaring, we get

$$x^2 + 2cx + c^2 + y^2 = 4a^2 - 4a\sqrt{(x - c)^2 + y^2} + x^2 - 2cx + c^2 + y^2,$$

or

$$-4a^2 + 4cx = -4a\sqrt{(x - c)^2 + y^2}$$
$$-a^2 + cx = -a\sqrt{(x - c)^2 + y^2}.$$

Squaring again, we get

$$a^4 - 2a^2cx + c^2x^2 = a^2x^2 - 2cxa^2 + a^2c^2 + a^2y^2,$$

or

$$x^2(a^2 - c^2) + a^2y^2 = a^2(a^2 - c^2).$$

It follows from when P is at $(0, b)$ that $b^2 + c^2 = a^2$, or $b^2 = a^2 - c^2$. Substituting b^2 for $a^2 - c^2$ in the last equation, we have the equation of the ellipse $b^2x^2 + a^2y^2 = a^2b^2$, or the following.

> **The equation, in standard form, of an ellipse with center at the origin is**
>
> $$\frac{x^2}{a^2} + \frac{y^2}{b^2} = 1.$$

We have proved that if a point is on the ellipse, then its coordinates satisfy this equation. We also need to know the converse, that is, if the coordinates of a point satisfy this equation, then the point is on the ellipse. The proof of the latter will be omitted here. In the above, the longer axis of symmetry $\overline{A'A}$ is called the **major axis.** The foci are always on the major axis. The shorter axis of symmetry $\overline{B'B}$ is called the **minor axis.** The intersection of these axes is called the **center.** The points A, A', B, and B' are called **vertices.** If the center of an ellipse is at the origin, the vertices are also the intercepts.

Ellipses as Stretched Circles

A **unit circle** is a circle centered at the origin with radius 1:

$$x^2 + y^2 = 1.$$

If we replace x by x/a and y by y/b, we get an equation of an ellipse:

$$\left(\frac{x}{a}\right)^2 + \left(\frac{y}{b}\right)^2 = 1, \quad \text{or} \quad \frac{x^2}{a^2} + \frac{y^2}{b^2} = 1.$$

It follows that an ellipse is a circle transformed by a stretch or shrink in the x-direction and also in the y-direction. If $a = 2$, for example, the unit circle is stretched in the x-direction by a factor of 2. In any case, the x-intercepts become a and $-a$ and the y-intercepts become b and $-b$.

DO EXERCISE 6.

EXAMPLE 6 For the ellipse $x^2 + 16y^2 = 16$, find the vertices and the foci. Then graph the ellipse.

Solution

a) We first multiply by $\frac{1}{16}$ to find standard form:

$$\frac{x^2}{16} + \frac{y^2}{1} = 1, \quad \text{or} \quad \frac{x^2}{4^2} + \frac{y^2}{1^2} = 1.$$

Thus, $a = 4$ and $b = 1$. Two of the vertices are $(-4, 0)$ and $(4, 0)$. These are also x-intercepts. The other vertices are $(0, 1)$ and $(0, -1)$. These are also y-intercepts. Since we know that $c^2 = a^2 - b^2$, we have $c^2 = 16 - 1$, so $c = \sqrt{15}$ and the foci are $(-\sqrt{15}, 0)$ and $(\sqrt{15}, 0)$.

b) We plot the vertices found in (a) and connect them with a smooth, oval-shaped curve. To be accurate, it is often helpful to find some other points on the curve. We let $x = 2$ and solve for y:

$$\begin{aligned} x^2 + 16y^2 &= 16 \\ (2)^2 + 16y^2 &= 16 \\ 4 + 16y^2 &= 16 \\ 16y^2 &= 12 \\ y^2 &= \frac{12}{16} \\ y &= \pm\sqrt{\frac{12}{16}} = \pm\frac{\sqrt{3}}{2} \approx \pm 0.9. \end{aligned}$$

6. Suppose a circle has been distorted to get the ellipse shown below.

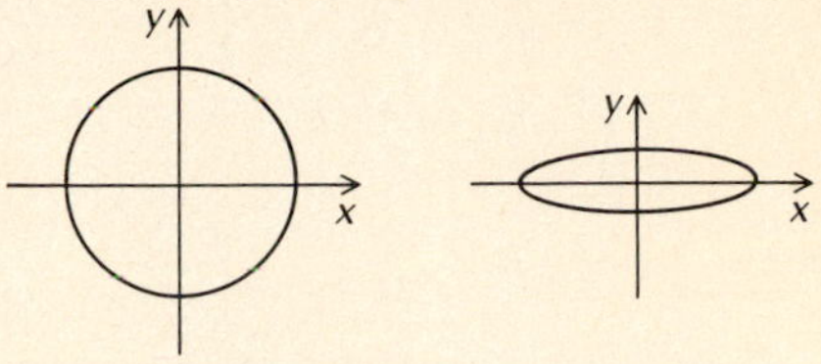

a) Has the circle been stretched or shrunk in the x-direction?

b) Has the circle been stretched or shrunk in the y-direction?

For each ellipse, find the vertices and the foci, and draw a graph.

7. $x^2 + 9y^2 = 9$

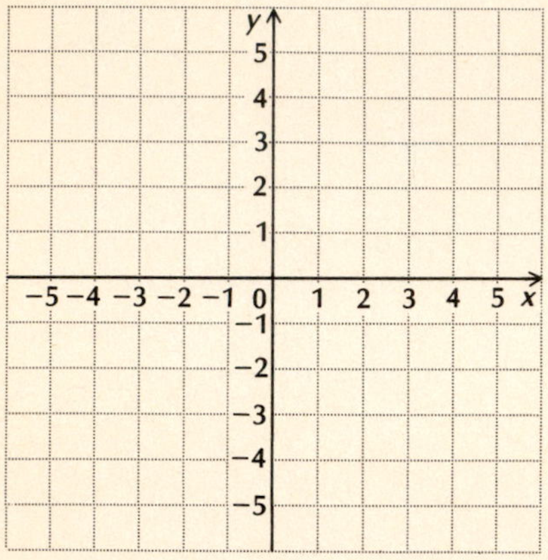

8. $9x^2 + 25y^2 = 225$

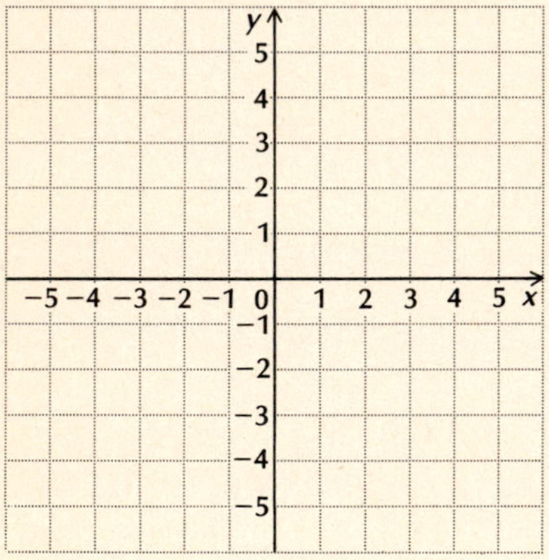

9. $2x^2 + 4y^2 = 8$

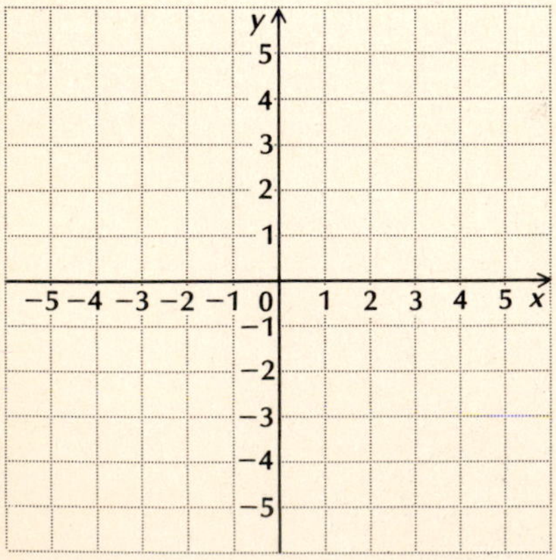

Thus, $(2, 0.9)$ and $(2, -0.9)$ can also be plotted and used to draw the graph. Similarly, the points $(-2, 0.9)$ and $(-2, -0.9)$ can be computed and plotted.

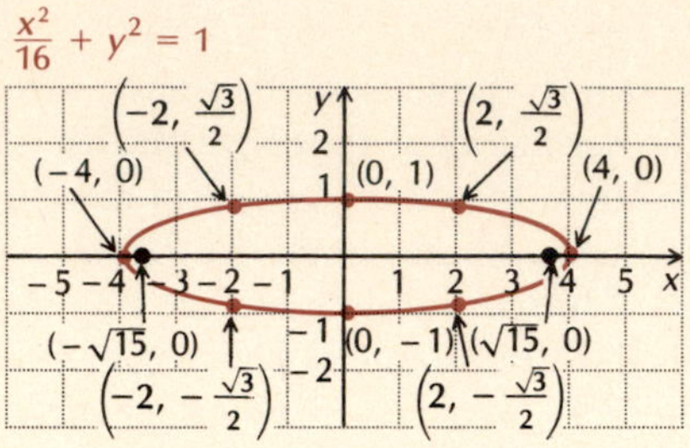

DO EXERCISES 7–9.

EXAMPLE 7 Graph this ellipse and its foci: $9x^2 + 2y^2 = 18$.

Solution

a) We first multiply by $\frac{1}{18}$:

$$\frac{x^2}{2} + \frac{y^2}{9} = 1, \quad \text{or} \quad \frac{x^2}{(\sqrt{2})^2} + \frac{y^2}{3^2} = 1.$$

Thus, $a = \sqrt{2}$ and $b = 3$, and the vertices are $(0, 3)$, $(0, -3)$, $(-\sqrt{2}, 0)$, and $(\sqrt{2}, 0)$.

b) Since $b > a$, the foci are on the y-axis and the major axis lies along the y-axis. To find c in this case, we proceed as follows:

$$c^2 = b^2 - a^2 = 9 - 2 = 7$$
$$c = \sqrt{7}.$$

The foci are $(0, \sqrt{7})$ and $(0, -\sqrt{7})$.

c) We plot the vertices and connect them with an oval-shaped curve as follows. Other points can be computed and plotted.

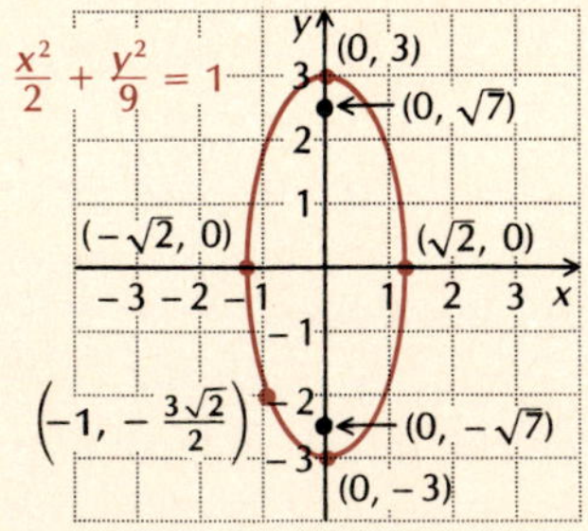

DO EXERCISES 10–12 ON THE FOLLOWING PAGE.

4 Standard Form by Completing the Square

If the center of an ellipse is not at the origin but at some point (h, k), then we can think of the ellipse

$$\frac{x^2}{a^2} + \frac{y^2}{b^2} = 1$$

being translated h units left or right and k units up or down. We know this by our work in Section 3.8 on translations. The standard form of the equation is then as follows.

> An equation, in standard form, of an ellipse with center at (h, k) is
>
> $$\frac{(x-h)^2}{a^2} + \frac{(y-k)^2}{b^2} = 1.$$
>
> If $a > b$, then the major axis and the foci are on the horizontal line $y = k$.
>
> If $b > a$, then the major axis and the foci are on the vertical line $x = h$.

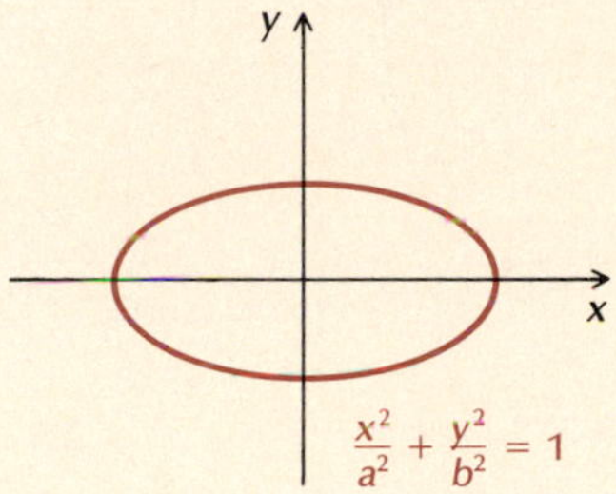

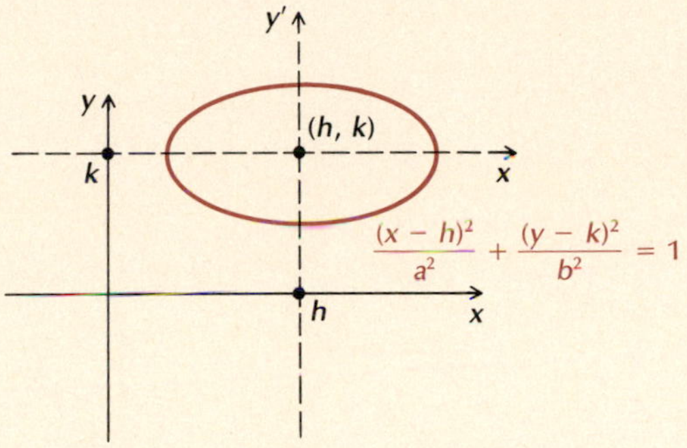

We can also think of such an ellipse under what is called **translation of axes** by considering the ellipse as being a graph of an equation

$$\frac{(x')^2}{a^2} + \frac{(y')^2}{b^2} = 1,$$

where the new axes are given by $x' = x - h$ and $y' = y - k$.

EXAMPLE 8 For the ellipse

$$16x^2 + 4y^2 + 96x - 8y + 84 = 0,$$

find the center, vertices, and foci. Then graph the ellipse.

Solution

a) We first complete the square to get standard form:

$$\begin{aligned} 16(x^2 + 6x + \quad) + 4(y^2 - 2y + \quad) &= -84 \\ 16(x^2 + 6x + 9 - 9) + 4(y^2 - 2y + 1 - 1) &= -84 \\ 16(x^2 + 6x + 9) + 4(y^2 - 2y + 1) &= -84 + 144 + 4 \\ 16(x+3)^2 + 4(y-1)^2 &= 64 \\ \frac{1}{64}[16(x+3)^2 + 4(y-1)^2] &= \frac{1}{64} \cdot 64 \\ \frac{(x+3)^2}{2^2} + \frac{(y-1)^2}{4^2} &= 1. \end{aligned}$$

The center is $(-3, 1)$, $a = 2$, and $b = 4$.

b) The vertices of the ellipse $x^2/2^2 + y^2/4^2 = 1$ are $(2, 0)$, $(-2, 0)$, $(0, 4)$, and $(0, -4)$. Now $c^2 = 16 - 4 = 12$, so $c = 2\sqrt{3}$. Since the denominator of the y^2-term is larger, the major axis is on the y-axis. Thus the foci are $(0, 2\sqrt{3})$ and $(0, -2\sqrt{3})$.

For each ellipse, find the center, the vertices, and the foci, and draw a graph.

10. $9x^2 + y^2 = 9$

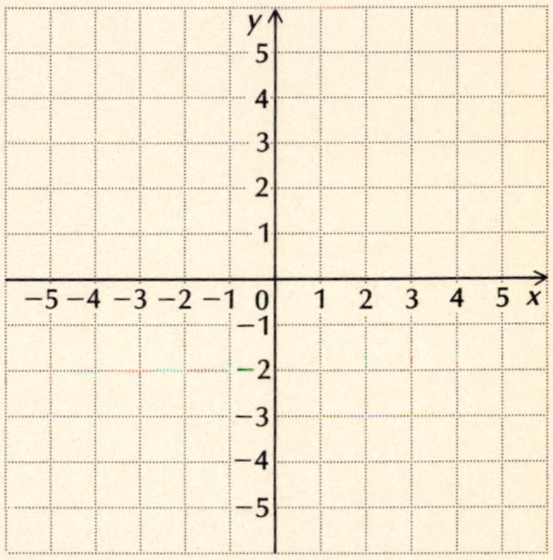

11. $25x^2 + 9y^2 = 225$

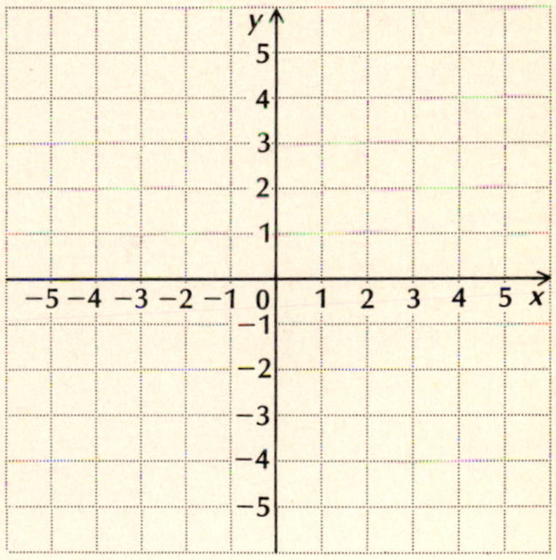

12. $4x^2 + 2y^2 = 8$

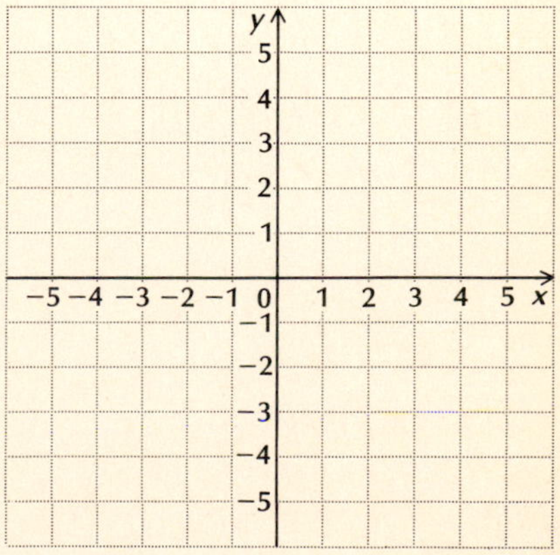

For each ellipse, find the center, the vertices, and the foci, and draw a graph.

13. $25x^2 + 9y^2 + 150x - 36y + 260 = 0$

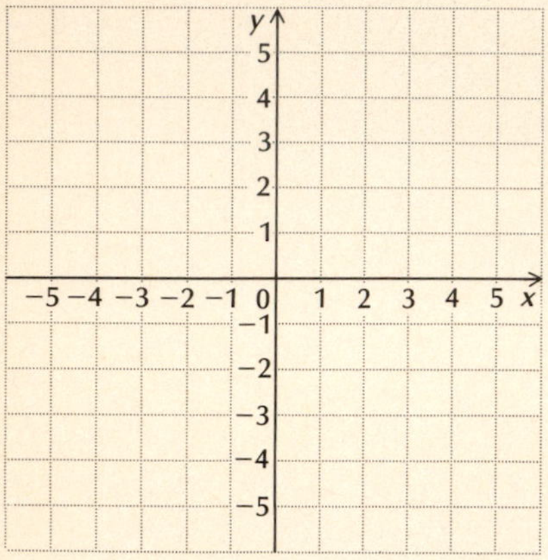

14. $9x^2 + 25y^2 - 36x + 150y + 260 = 0$

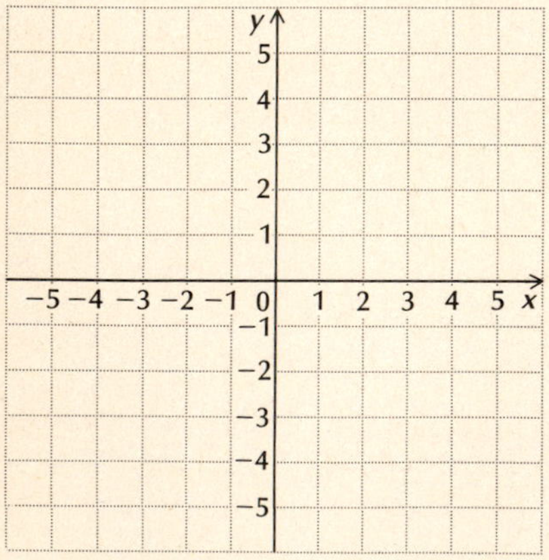

c) Then the vertices and foci of the translated ellipse are found by translation in the same way in which the center has been translated. Thus the vertices are

$$(-3 + 2, 1 + 0), \quad (-3 - 2, 1 + 0),$$
$$(-3 + 0, 1 + 4), \quad (-3 + 0, 1 - 4),$$

or

$$(-1, 1), \quad (-5, 1), \quad (-3, 5), \quad (-3, -3).$$

The foci are $(-3, 1 + 2\sqrt{3})$ and $(-3, 1 - 2\sqrt{3})$.

d) The graph is as follows:

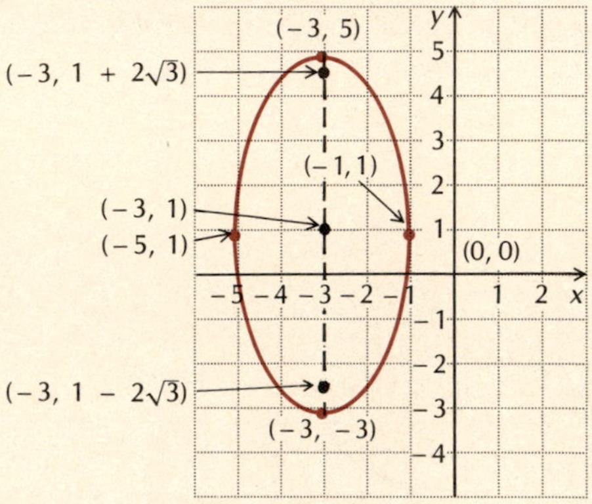

$16x^2 + 4y^2 + 96x - 8y + 84 = 0$

DO EXERCISES 13 AND 14.

Applications

Ellipses have many applications. For example, earth satellites travel in elliptical orbits. The planets travel around the sun in elliptical orbits with the sun at one focus.

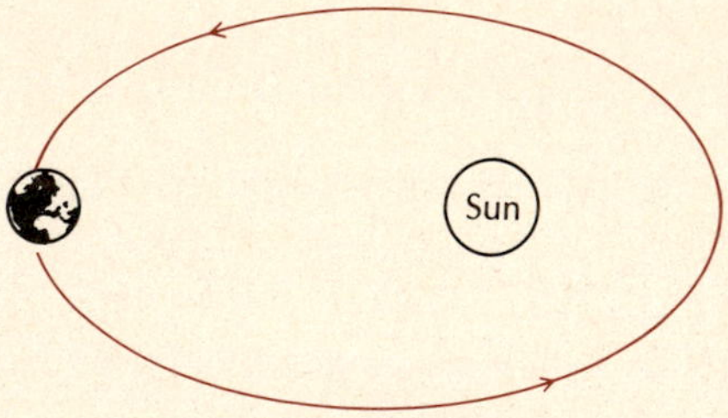

Planetary orbit

An interesting attraction found in museums is the *whispering gallery*. It is elliptical. Persons with their heads at the foci can whisper and hear each other clearly, while persons at other positions cannot hear them. This happens because the sound waves emanating from one focus are reflected to the other focus, being concentrated there.

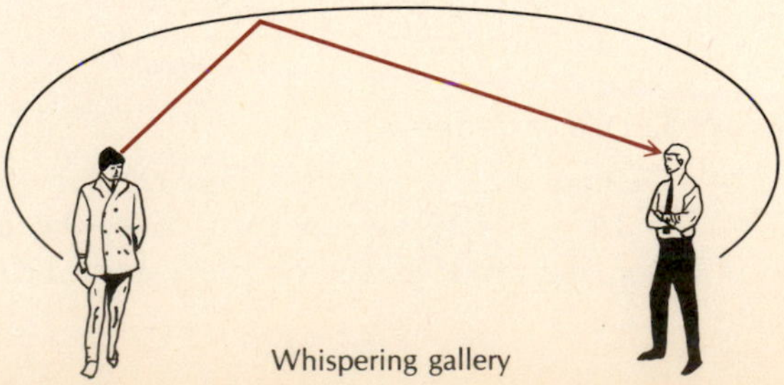

Whispering gallery

EXERCISE SET 10.1

1, 2 Graph.

1. $x^2 - y^2 = 0$
2. $x^2 - 9y^2 = 0$
3. $3x^2 + xy - 2y^2 = 0$
4. $x^2 - xy - 2y^2 = 0$
5. $2x^2 + y^2 = 0$
6. $5x^2 + y^2 = -3$

3 For each ellipse, find the vertices and the foci, and draw a graph.

7. $\frac{x^2}{4} + \frac{y^2}{1} = 1$
8. $\frac{x^2}{1} + \frac{y^2}{4} = 1$
9. $16x^2 + 9y^2 = 144$
10. $9x^2 + 16y^2 = 144$
11. $2x^2 + 3y^2 = 6$
12. $5x^2 + 7y^2 = 35$
13. $4x^2 + 9y^2 = 1$
14. $25x^2 + 16y^2 = 1$

4 Find the center, vertices, and foci, and draw a graph.

15. $\frac{(x-1)^2}{4} + \frac{(y-2)^2}{1} = 1$
16. $\frac{(x-1)^2}{1} + \frac{(y-2)^2}{4} = 1$
17. $\frac{(x+3)^2}{25} + \frac{(y-2)^2}{16} = 1$
18. $\frac{(x-2)^2}{25} + \frac{(y+3)^2}{16} = 1$
19. $3(x+2)^2 + 4(y-1)^2 = 192$
20. $4(x-5)^2 + 3(y-5)^2 = 192$
21. $4x^2 + 9y^2 - 16x + 18y - 11 = 0$
22. $x^2 + 2y^2 - 10x + 8y + 29 = 0$
23. $4x^2 + y^2 - 8x - 2y + 1 = 0$
24. $9x^2 + 4y^2 + 54x - 8y + 49 = 0$

For each ellipse, find the center and vertices.

25. ▦ $4x^2 + 9y^2 - 16.025x + 18.0927y - 11.346 = 0$
26. ▦ $9x^2 + 4y^2 + 54.063x - 8.016y + 49.872 = 0$

SYNTHESIS

Find the equation of the ellipse with the following vertices. (*Hint*: Graph the vertices.)

27. $(2, 0), (-2, 0), (0, 3), (0, -3)$
28. $(1, 0), (-1, 0), (0, 4), (0, -4)$
29. $(1, 1), (5, 1), (3, 6), (3, -4)$
30. $(-1, -1), (-1, 5), (-3, 2), (1, 2)$

Find the equation of the ellipse satisfying the given conditions.

31. Center at $(-2, 3)$ with major axis of length 4 and parallel to the y-axis, minor axis of length 1
32. Vertices $(3, 0)$ and $(-3, 0)$ and containing the point $(2, \frac{22}{3})$
33. a) Graph $9x^2 + y^2 = 9$. Is this relation a function?
    b) Solve $9x^2 + y^2 = 9$ for y.
    c) Graph $y = 3\sqrt{1 - x^2}$ and determine whether it is a function. Find the domain and the range.
    d) Graph $y = -3\sqrt{1 - x^2}$ and determine whether it is a function. Find the domain and the range.
34. Describe the graph of
$$\frac{x^2}{a^2} + \frac{y^2}{b^2} = 1$$
when $a^2 = b^2$.
35. ▦ Draw a large-scale precise graph of
$$\frac{x^2}{25} + \frac{y^2}{16} = 1$$
by calculating and plotting a large number of points.
36. The maximum distance of the earth from the sun is 9.3×10^7 miles. The minimum distance is 9.1×10^7 miles. The sun is at one focus of the elliptical orbit. Find the distance from the sun to the other focus.
37. The bridge support shown in this figure is the top half of an ellipse. Assuming that a coordinate system is superimposed on the drawing in such a way that the center of the ellipse is at point Q, find an equation of the ellipse.

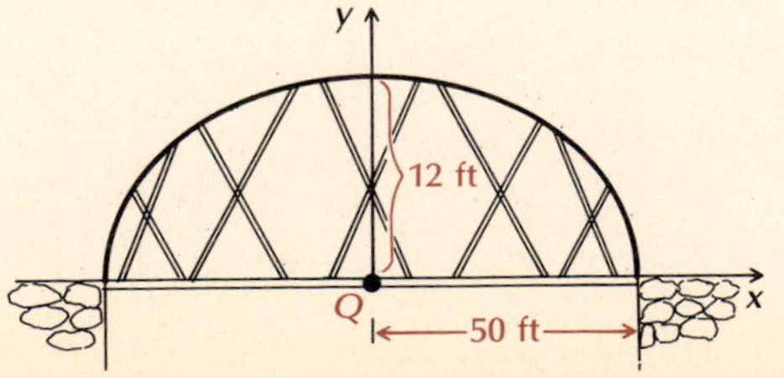

38. In Washington, D. C., there is a large grassy area south of the White House known as the **Ellipse**. It is actually an ellipse with major axis of length 1048 ft and minor axis of length 898 ft. Assuming that a coordinate system is superimposed on the area in such a way that the center is at the origin and the major and minor axes are on the x- and y-axes of the coordinate system, find an equation of the ellipse.

39. Consider the figures below. The circle on the left has a radius r. Suppose "each" radius shown is stretched or shrunk to form the ellipse on the right.

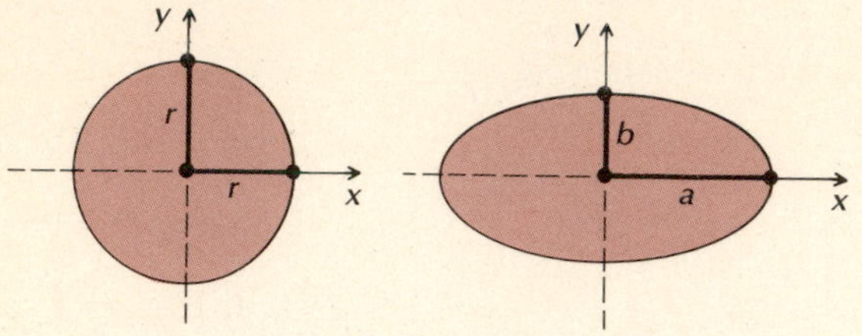

a) Conjecture a formula for the area of the ellipse:

$$\frac{x^2}{a^2} + \frac{y^2}{b^2} = 1.$$

(*Hint*: The area of the circle $x^2 + y^2 = r^2$ is $\pi \cdot r \cdot r$.)

b) Use the result of (a) to find the area of the ellipse

$$\frac{x^2}{16} + \frac{y^2}{25} = 1.$$

c) Use the result of (a) to find the area of the ellipse in Exercise 38.

40. The toy pictured here is called a "vacuum grinder." It consists of a rod hinged to two blocks A and B that slide in perpendicular grooves. One grasps the knob at C and grinds. Determine (and prove) whether or not the path of the handle C is an ellipse.

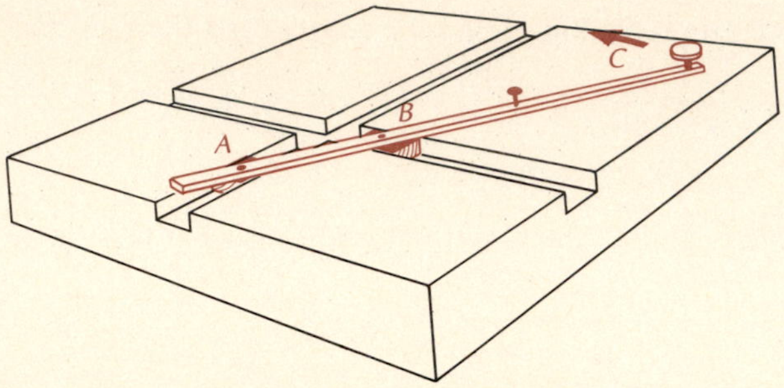

OBJECTIVES

You should be able to:

1 Given an equation of a hyperbola with center at the origin, find the vertices, foci, and asymptotes, and then graph.

2 Given an equation of a hyperbola with center not at the origin, complete the square if necessary, find the center, vertices, foci, and asymptotes, and then graph.

3 Graph hyperbolas having an equation $xy = k$.

10.2 Conic Sections: Hyperbolas

Some equations of second degree have graphs that are *hyperbolas*. Hyperbolas are defined as follows.

> DEFINITION
>
> **A *hyperbola* is a locus or set of all points P in a plane such that the absolute value of the difference of the distances from P to two fixed points F_1 and F_2 in the plane is constant. The points F_1 and F_2 are called *foci* (singular, *focus*) and the midpoint of the segment joining them is called the *center*.**

When a plane intersects a cone parallel to the axis of the cone as shown, a hyperbola is formed. Note that a hyperbola has two parts. These are called **branches**.

Hyperbola in three dimensions:

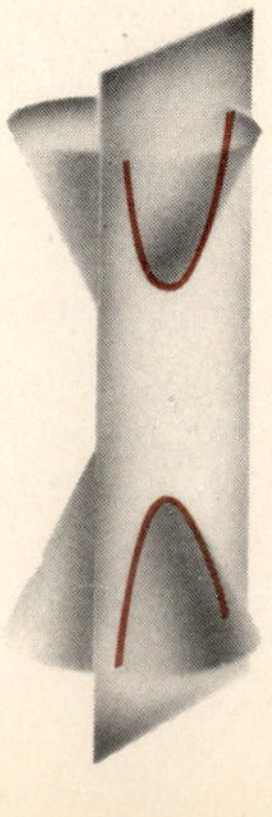

Hyperbola in a plane:

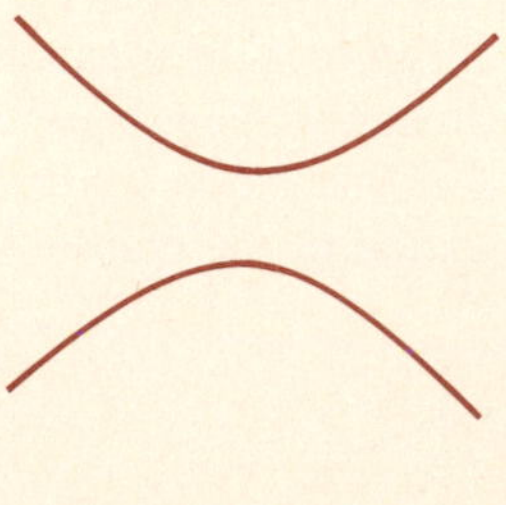

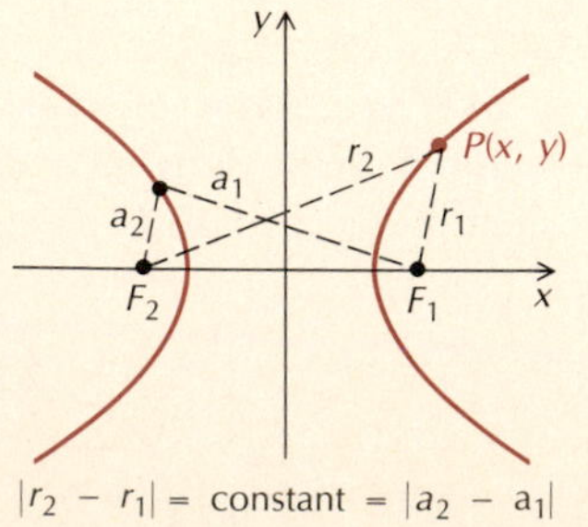

1 Equations of Hyperbolas

Now let us find equations for hyperbolas. We first consider an equation of a hyperbola whose center is at the origin and whose foci lie on one of the coordinate axes. Suppose we have a hyperbola, as shown, with foci $F_1(c, 0)$ and $F_2(-c, 0)$ on the x-axis. We consider a point $P(x, y)$ in the first quadrant. The proof for the other quadrants is similar to the proof that follows.

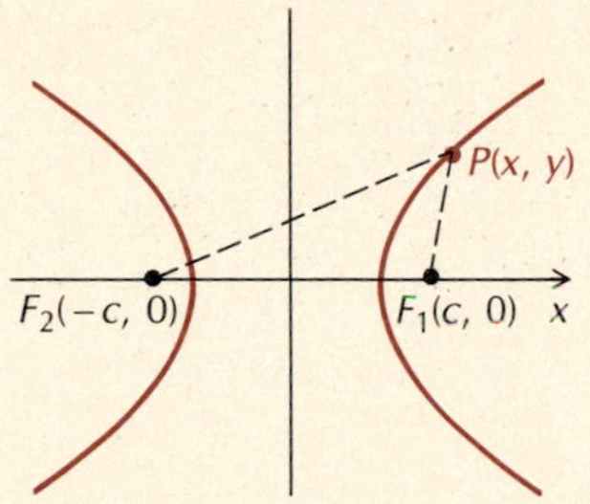

We know that $PF_2 > PF_1$, so $PF_2 - PF_1 > 0$ and $|PF_2 - PF_1| = PF_2 - PF_1$. Let the constant difference be $2a$. Then

$$|PF_2 - PF_1| = PF_2 - PF_1 = 2a.$$

In the triangle F_2PF_1, $F_1F_2 + PF_1 > PF_2$ by the triangle inequality, so $PF_2 - PF_1 < F_1F_2$, or $2a < 2c$; therefore, $a < c$. Using the distance formula, we have

$$\sqrt{(x + c)^2 + y^2} - \sqrt{(x - c)^2 + y^2} = 2a,$$

or

$$\sqrt{(x + c)^2 + y^2} = 2a + \sqrt{(x - c)^2 + y^2}.$$

Squaring, we get

$$x^2 + 2xc + c^2 + y^2 = 4a^2 + 4a\sqrt{(x - c)^2 + y^2} + x^2 - 2xc + c^2 + y^2,$$

which simplifies to

$$4cx - 4a^2 = 4a\sqrt{(x - c)^2 + y^2}$$

or

$$cx - a^2 = a\sqrt{(x - c)^2 + y^2}.$$

Squaring again, we get

$$c^2x^2 - 2a^2cx + a^4 = a^2x^2 - 2a^2cx + a^2c^2 + a^2y^2,$$

or

$$x^2(c^2 - a^2) - a^2y^2 = a^2(c^2 - a^2).$$

Since $c > a$, $c^2 > a^2$, so $c^2 - a^2$ is positive. We represent $c^2 - a^2$ by b^2. The previous equation then becomes

$$x^2b^2 - a^2y^2 = a^2b^2,$$

or the following:

The equation, in standard form, of a hyperbola with center at the origin and foci on the x-axis is

$$\frac{x^2}{a^2} - \frac{y^2}{b^2} = 1.$$

We have shown that if a point is on the hyperbola, it satisfies this equation. We also need to know the converse: If a point satisfies the equation, then it is on the hyperbola. We omit the proof.

The following figure is a hyperbola.

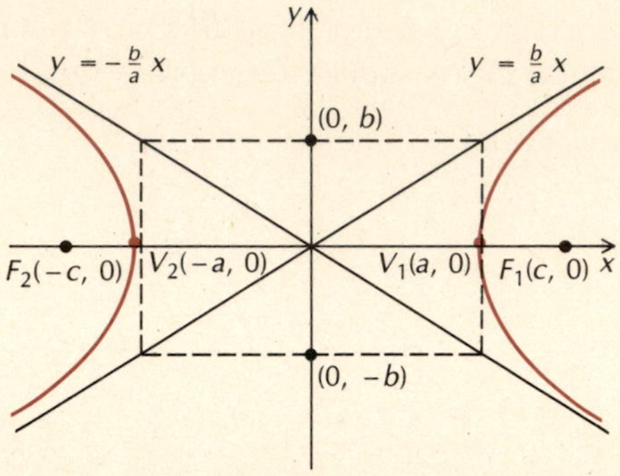

Points $V_1(a, 0)$ and $V_2(-a, 0)$ are called the **vertices**, and the line segment $\overline{V_1V_2}$ is called the **transverse axis**. The line segment from $(0, b)$ to $(0, -b)$ is called the **conjugate axis**. Note that when we replace either or both of x by $-x$ and/or y by $-y$, we get an equivalent equation. Thus the hyperbola is symmetric with respect to the origin, and the x- and y-axes are lines of symmetry.

The lines $y = (b/a)x$ and $y = -(b/a)x$ are *asymptotes*. They have slopes b/a and $-b/a$. **Asymptotes** are lines to which the graph gets closer and closer as x gets further away from 0.

EXAMPLE 1 For the hyperbola $9x^2 - 16y^2 = 144$, find the vertices, the foci, and the asymptotes. Then graph the hyperbola.

Solution

a) We first multiply by $\frac{1}{144}$ to find the standard form:

$$\frac{x^2}{16} - \frac{y^2}{9} = 1.$$

Thus, $a = 4$ and $b = 3$. Next, we find the x-intercepts, or *vertices*. Let $y = 0$, and we see that $x^2/4^2 = 1$, so $x = \pm 4$. The vertices are $(4, 0)$ and $(-4, 0)$. Hyperbolas have only two vertices. Suppose we try to find vertices on the y-axis: the y-intercepts. If we set $x = 0$, we get $y^2/9 = -1$, and this equation has no real-number solutions. The foci are always on the same axis as the vertices. Since

$$b^2 = c^2 - a^2,$$

$$c = \sqrt{a^2 + b^2} = \sqrt{4^2 + 3^2} = 5.$$

Thus the foci are $(5, 0)$ and $(-5, 0)$. The asymptotes are

$$y = \frac{b}{a}x = \frac{3}{4}x$$

and

$$y = -\frac{b}{a}x = -\frac{3}{4}x.$$

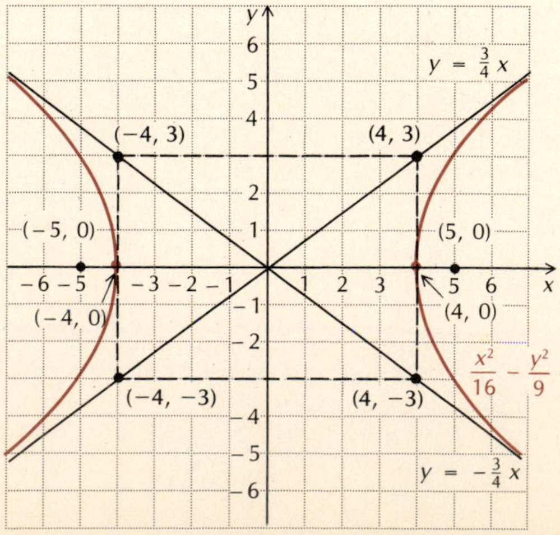

b) To graph the hyperbola, it is helpful to first graph the asymptotes. An easy way to do this is to draw the rectangle shown on the preceding page with vertices $(\pm a, \pm b)$, or $(\pm 4, \pm 3)$. Then draw the asymptotes out from the center to these vertices. The branches of the hyperbola are then drawn outward from the vertices toward the asymptotes. ■

Why are $y = (b/a)x$ and $y = -(b/a)x$ asymptotes? To answer this, we solve $x^2/16 - y^2/9 = 1$ for y^2:

$$9x^2 - 16y^2 = 144 \qquad \text{Multiplying by 144}$$
$$-16y^2 = 144 - 9x^2$$
$$16y^2 = 9x^2 - 144$$
$$y^2 = \frac{1}{16}(9x^2 - 144) = \frac{9x^2 - 144}{16}.$$

From this last equation, we see that as $|x|$ gets larger, the term -144 is very small compared to $9x^2$, so y^2 gets close to $9x^2/16$. That is, when $|x|$ is large,

$$y^2 \approx \frac{9x^2}{16},$$

so

$$y \approx \left|\frac{3}{4}x\right| \quad \text{or} \quad y \approx \pm\frac{3}{4}x.$$

Thus the lines $y = \frac{3}{4}x$ and $y = -\frac{3}{4}x$ are asymptotes.

DO EXERCISES 1 AND 2.

The foci of a hyperbola can be on the y-axis. In that case, the equation is as follows.

The equation, in standard form, of a hyperbola with center at the origin and foci on the y-axis is

$$\frac{y^2}{b^2} - \frac{x^2}{a^2} = 1.$$

In this case, the slopes of the asymptotes are still $\pm b/a$ and it is still true that $b^2 = c^2 - a^2$. There are now y-intercepts, and they are $(0, \pm b)$.

In summary,

Equation	Vertices	Foci
$\frac{x^2}{a^2} - \frac{y^2}{b^2} = 1$	$(-a, 0), (a, 0)$	$(-c, 0), (c, 0)$
$\frac{y^2}{b^2} - \frac{x^2}{a^2} = 1$	$(0, -b), (0, b)$	$(0, -c), (0, c)$
The asymptotes are $y = \pm\frac{b}{a}x$.		

EXAMPLE 2 For the hyperbola $25y^2 - 16x^2 = 400$, find the vertices, the foci, and the asymptotes. Then draw a graph.

Solution

a) We first multiply by $\frac{1}{400}$ to find the standard form:

$$\frac{y^2}{16} - \frac{x^2}{25} = 1.$$

For each hyperbola, find the vertices, the foci, and the asymptotes. Then draw a graph.

1. $4x^2 - 9y^2 = 36$

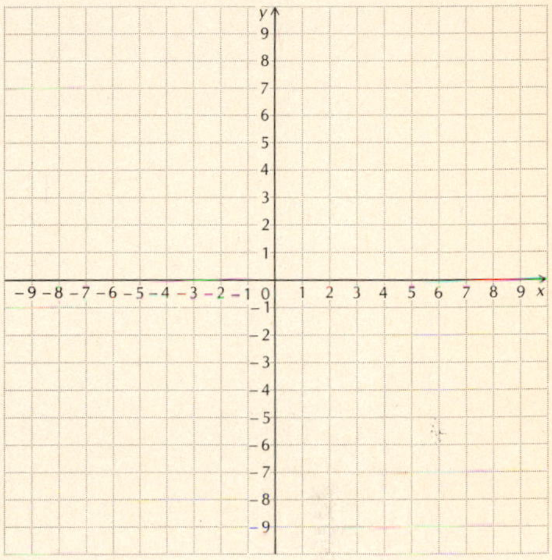

2. $x^2 - y^2 = 16$

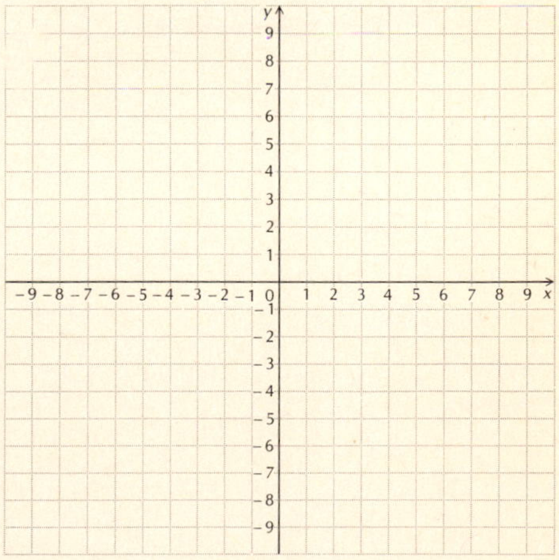

3. Find the vertices, the foci, and the asymptotes. Then draw a graph.

$$9y^2 - 25x^2 = 225$$

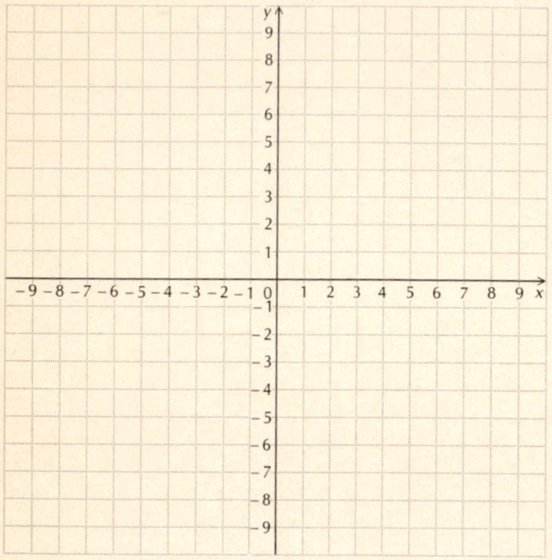

4. Find the vertices, the foci, and the asymptotes. Then draw a graph.

$$y^2 - x^2 = 25$$

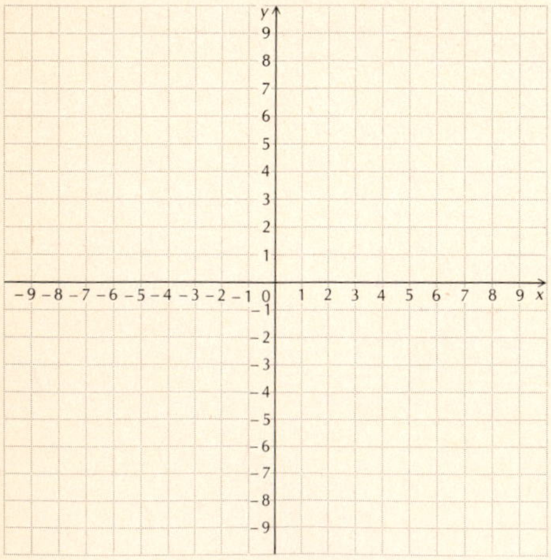

Thus, $a = 5$ and $b = 4$. The vertices are $(0, 4)$ and $(0, -4)$ since the x^2-term is negative. Since $c = \sqrt{4^2 + 5^2} = \sqrt{41}$, the foci are $(0, \sqrt{41})$ and $(0, -\sqrt{41})$. The asymptotes are $y = \frac{4}{5}x$ and $y = -\frac{4}{5}x$.

b) The graph is as shown.

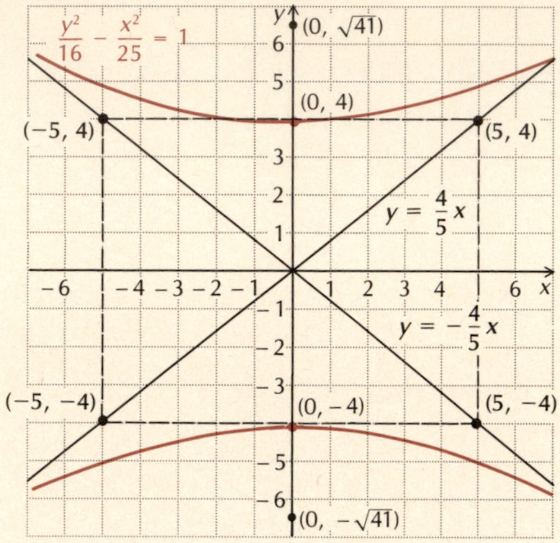

DO EXERCISES 3 AND 4.

2 Standard Form by Completing the Square

If the center of a hyperbola is not at the origin but at some point (h, k), then the standard equation is one of the following.

$\dfrac{(x-h)^2}{a^2} - \dfrac{(y-k)^2}{b^2} = 1$	**Hyperbola, transverse axis parallel to the x-axis**
$\dfrac{(y-k)^2}{b^2} - \dfrac{(x-h)^2}{a^2} = 1$	**Hyperbola, transverse axis parallel to the y-axis**

The asymptotes are $y - k = \pm\frac{b}{a}(x - h)$.

EXAMPLE 3 For the hyperbola

$$4x^2 - y^2 + 24x + 4y + 28 = 0,$$

find the center, the vertices, the foci, and the asymptotes. Then graph the hyperbola.

Solution

a) We complete the square to find standard form:

$$4(x^2 + 6x + \quad) - (y^2 - 4y + \quad) = -28$$
$$4(x^2 + 6x + 9 - 9) - (y^2 - 4y + 4 - 4) = -28$$
$$4(x^2 + 6x + 9) - (y^2 - 4y + 4) = -28 + 36 - 4$$
$$4(x + 3)^2 - (y - 2)^2 = 4$$
$$\frac{(x+3)^2}{1} - \frac{(y-2)^2}{4} = 1.$$

The center is $(-3, 2)$.

b) Consider $x^2/1 - y^2/4 = 1$. We have $a = 1$ and $b = 2$. The vertices of this hyperbola are $(1, 0)$ and $(-1, 0)$. Also, $c = \sqrt{1^2 + 2^2} = \sqrt{5}$, so the foci are $(\sqrt{5}, 0)$ and $(-\sqrt{5}, 0)$. The asymptotes are $y = 2x$ and $y = -2x$.

c) The vertices, foci, and asymptotes of the translated hyperbola are found in the same way in which the center has been translated. The vertices are $(-3 + 1, 2 + 0)$, $(-3 - 1, 2 + 0)$, or $(-2, 2)$, $(-4, 2)$. The foci are $(-3 + \sqrt{5}, 2)$ and $(-3 - \sqrt{5}, 2)$. The asymptotes are

$$y - 2 = 2(x + 3) \quad \text{and} \quad y - 2 = -2(x + 3).$$

d) The graph is as follows.

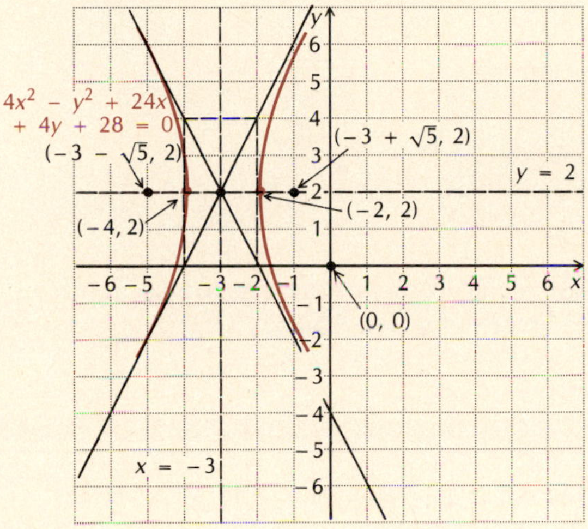

DO EXERCISES 5 AND 6.

3 Asymptotes on the Coordinate Axes

If a hyperbola has its center at the origin and its transverse axis at 45° to the coordinate axes, it has a simple equation. The coordinate axes are its asymptotes.

$xy = k$, k a nonzero constant	**Hyperbola, asymptotes the coordinate axes**

If k is positive, the branches of the hyperbola lie in the first and third quadrants. If k is negative, the branches lie in the second and fourth quadrants. In either case, the asymptotes are the x-axis and the y-axis. We graphed these equations in Chapter 3.

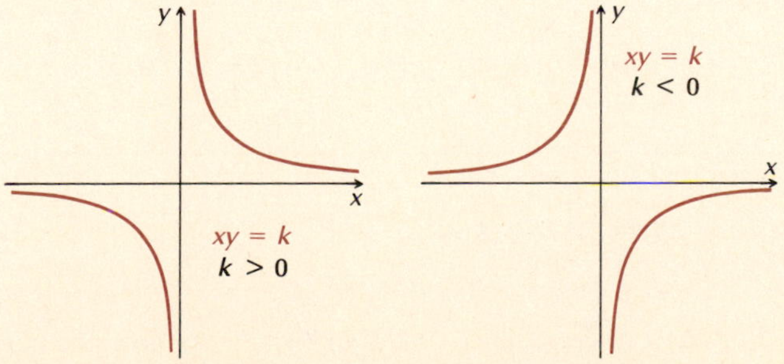

(a) (b)

DO EXERCISES 7 AND 8. (EXERCISE 8 IS ON THE FOLLOWING PAGE.)

For each hyperbola, find the center, the vertices, the foci, and the asymptotes. Then draw a graph.

5. $4x^2 - 25y^2 - 8x - 100y - 196 = 0$

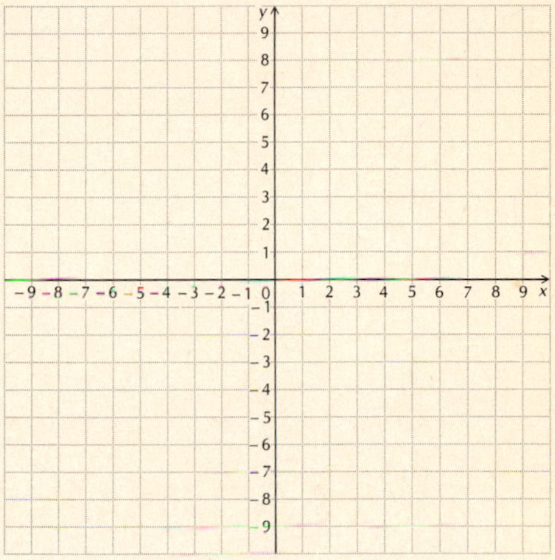

6. $16y^2 - 9x^2 - 64y - 18x - 89 = 0$

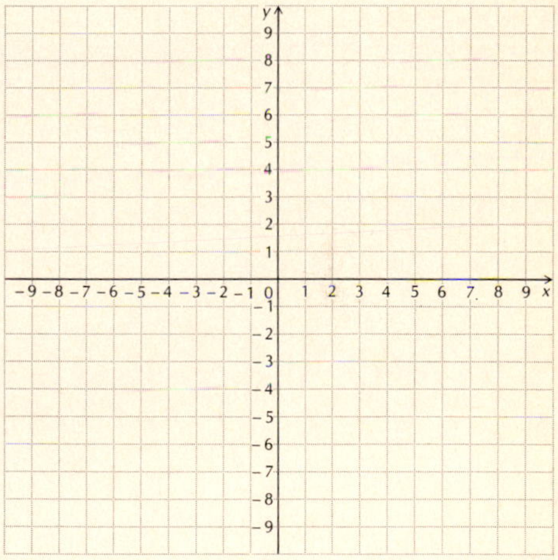

Graph.

7. $xy = 3$

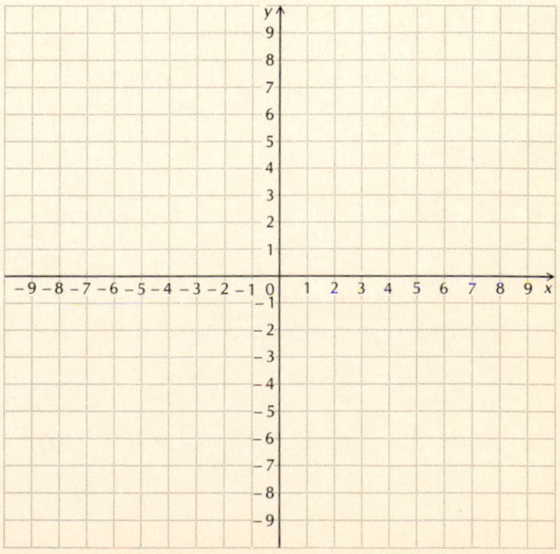

8. $xy = -12$

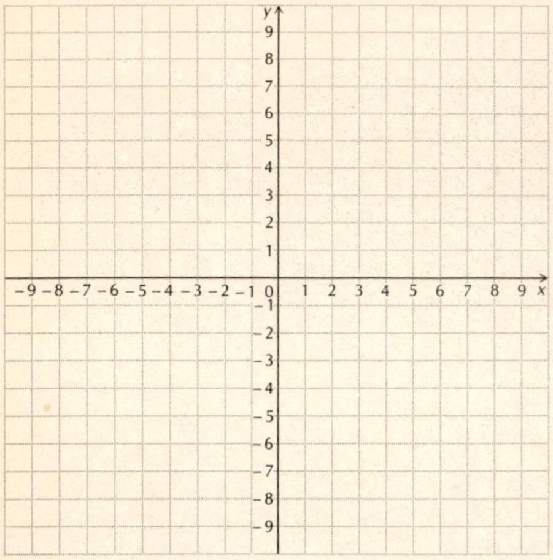

Applications

Hyperbolas also have many applications. For example, a jet breaking the sound barrier creates a sonic boom whose wave front has the shape of a cone. The cone intersects the ground in one branch of a hyperbola.

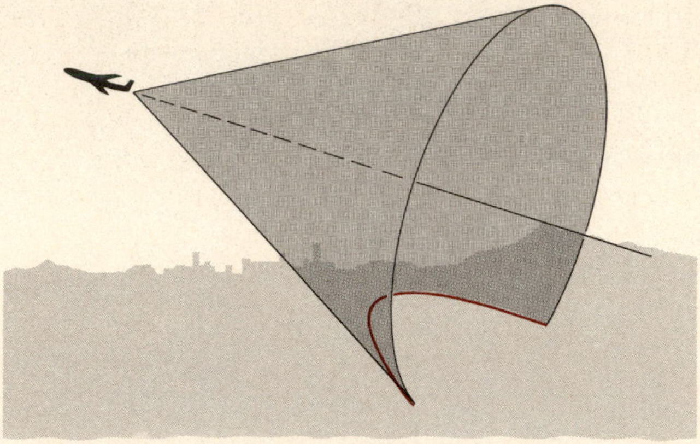

Some comets travel in hyperbolic orbits. A cross section of an amphitheater may be half of one branch of a hyperbola.

EXERCISE SET 10.2

1 For each hyperbola, find the center, the vertices, the foci, and the asymptotes, and graph the hyperbola.

1. $\frac{x^2}{9} - \frac{y^2}{1} = 1$

2. $\frac{x^2}{1} - \frac{y^2}{9} = 1$

3. $\frac{(x-2)^2}{9} - \frac{(y+5)^2}{1} = 1$

4. $\frac{(x-2)^2}{1} - \frac{(y+5)^2}{9} = 1$

5. $\frac{(y+3)^2}{4} - \frac{(x+1)^2}{16} = 1$

6. $\frac{(y+3)^2}{25} - \frac{(x+1)^2}{16} = 1$

7. $x^2 - 4y^2 = 4$

8. $4x^2 - y^2 = 4$

9. $4y^2 - x^2 = 4$

10. $y^2 - 4x^2 = 4$

11. $x^2 - y^2 = 2$

12. $x^2 - y^2 = 3$

13. $x^2 - y^2 = \frac{1}{4}$

14. $x^2 - y^2 = \frac{1}{9}$

2 For each hyperbola, find the center, the vertices, the foci, and the asymptotes, and graph the hyperbola.

15. $x^2 - y^2 - 2x - 4y - 4 = 0$

16. $4x^2 - y^2 + 8x - 4y - 4 = 0$

17. $36x^2 - y^2 - 24x + 6y - 41 = 0$

18. $9x^2 - 4y^2 + 54x + 8y + 45 = 0$

19. $9y^2 - 4x^2 - 18y + 24x - 63 = 0$

20. $x^2 - 25y^2 + 6x - 50y = 41$

21. $x^2 - y^2 - 2x - 4y = 4$

22. $9y^2 - 4x^2 - 54y - 8x + 41 = 0$

23. $y^2 - x^2 - 6x - 8y - 39 = 0$

24. $x^2 - y^2 = 8x - 2y - 13$

3 Graph.

25. $xy = 1$

26. $xy = -4$

27. $xy = -8$

28. $xy = 3$

SYNTHESIS

29. ▦ Find the center, the vertices, and the asymptotes:

$$x^2 - y^2 - 2.046x - 4.088y - 4.228 = 0.$$

Find an equation of a hyperbola having:

30. Vertices at $(1, 0)$ and $(-1, 0)$ and foci at $(2, 0)$ and $(-2, 0)$

31. Asymptotes $y = \frac{3}{2}x$ and $y = -\frac{3}{2}x$ and one vertex $(2, 0)$

32. Transverse axis parallel to the y-axis and of length 11, conjugate axis of length 6, and center $(3, -8)$

33. Find an equation of the hyperbola with vertices $(-9, 4)$ and $(-5, 4)$ and asymptotes $y = 3x + 25$ and $y = -3x - 17$.

34. Find an equation of the inverse of the relation of Exercise 33. Find the vertices and the center.

35. a) Graph $x^2 - 4y^2 = 4$. Is this relation a function?
b) Solve $x^2 - 4y^2 = 4$ for y.
c) Graph $y = \frac{1}{2}\sqrt{x^2 - 4}$ and determine whether it is a function. Find the domain and the range.
d) Graph $y = -\frac{1}{2}\sqrt{x^2 - 4}$ and determine whether it is a function. Find the domain and the range.

36. Show that the equation

$$\begin{vmatrix} \dfrac{x-h}{a} & \dfrac{y-k}{b} \\ \dfrac{y-k}{b} & \dfrac{x-h}{a} \end{vmatrix} = 1$$

is an equation of a hyperbola with center (h, k).

CHALLENGE

37. A rifle at A fires a bullet, which hits a target at B. A person at C hears the sound of the rifle shot and the sound of the bullet hitting the target simultaneously. Describe the set of all such points C.

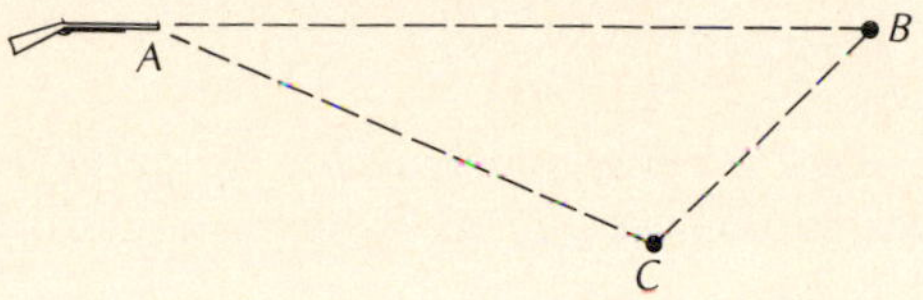

38. In a navigation system called Loran, a radio transmitter at M (the *master* station) sends out pulses. Each pulse triggers another transmitter at S (the *slave* station), which then also transmits a pulse. A ship or airplane at A receives pulses from both M and S and a device measures the difference in the time at which they arrive at A. Knowing this difference in time, the navigator can locate the vessel as being somewhere along a curve predrawn on a chart. What is the shape of that curve?

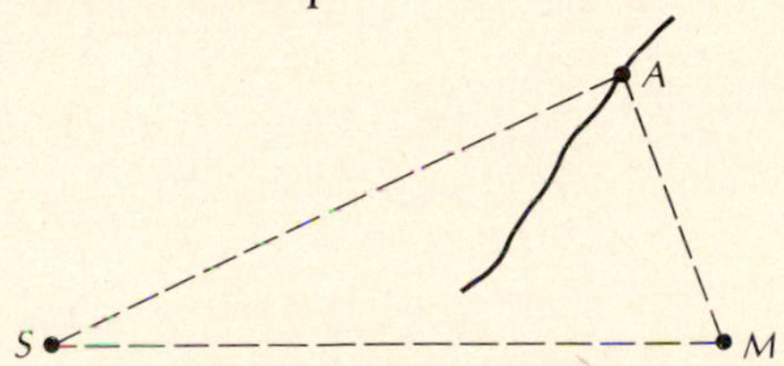

10.3 Conic Sections: Parabolas

OBJECTIVES

You should be able to:

1 Given an equation of a parabola, find the vertex, focus, and directrix, and graph the parabola.

2 Given the focus and directrix of a parabola, find an equation of the parabola.

3 Find the standard form of equations of parabolas by completing the square, if necessary, and graph the parabolas, having found the vertex, focus, and directrix.

4 Classify an equation as having a graph that is a circle, an ellipse, a hyperbola, or a parabola.

Some equations of second degree have graphs that are *parabolas*. Parabolas are defined as follows.

DEFINITION

A *parabola* is a locus or set of all points P in a plane equidistant from a fixed line and a fixed point in the plane. The fixed line is called the *directrix* and the fixed point is called the *focus*.

When a plane intersects a cone parallel to an element of the cone, a parabola is formed.

Parabola in three dimensions:

Parabola in a plane:

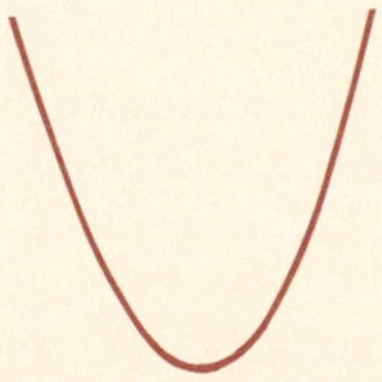

1 Equations of Parabolas

Now let us find equations for parabolas. Given the focus F and the directrix l, we place the coordinate axes as shown. The y-axis contains F and is perpendicular to l. The x-axis is halfway between F and l. We call the distance from F to the x-axis p. Then F has coordinates $(0, p)$ and l has the equation $y = -p$.

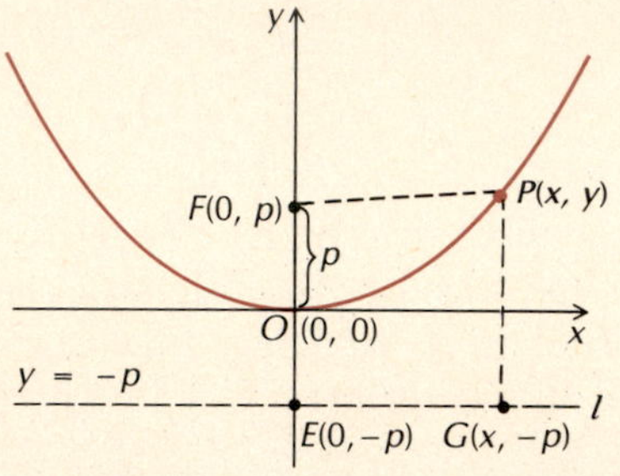

Let $P(x, y)$ be any point of the parabola and consider $\overline{PG}$ perpendicular to the line $y = -p$. The coordinates of G are $(x, -p)$. By definition of a parabola,

$$PF = PG.$$

Then using the distance formula, we have

$$\sqrt{(x-0)^2 + (y-p)^2} = \sqrt{(x-x)^2 + (y+p)^2}.$$

Squaring, we get

$$\begin{aligned} x^2 + y^2 - 2py + p^2 &= y^2 + 2py + p^2 \\ x^2 &= 4py. \end{aligned}$$

Thus we have the following.

$$x^2 = 4py$$

Standard equation of a parabola with focus at $(0, p)$ and directrix $y = -p$. The vertex is $(0, 0)$ and the y-axis is the only line of symmetry.

We have shown that if $P(x, y)$ is on the parabola, then its coordinates satisfy this equation. The converse is also true, but we omit the proof.

Note that if $p > 0$, as above, the graph opens upward. If $p < 0$, the graph opens downward and the focus and directrix exchange sides of the x-axis.

The inverse of the parabola above is described as follows.

$$y^2 = 4px$$

Standard equation of a parabola with focus at $(p, 0)$ and directrix $x = -p$. The vertex is $(0, 0)$ and the x-axis is the only line of symmetry.

EXAMPLE 1 For the parabola $y = x^2$, find the vertex, the focus, and the directrix, and draw a graph.

Solution We first write $y = x^2$ in the form $x^2 = 4py$. We do this by rewriting the coefficient of y, 1, as $4(\frac{1}{4})$:

$$x^2 = 4py = y$$
$$x^2 = 4(\tfrac{1}{4})y.$$

Vertex: $(0, 0)$
Focus: $(0, \tfrac{1}{4})$
Directrix: $y = -\tfrac{1}{4}$

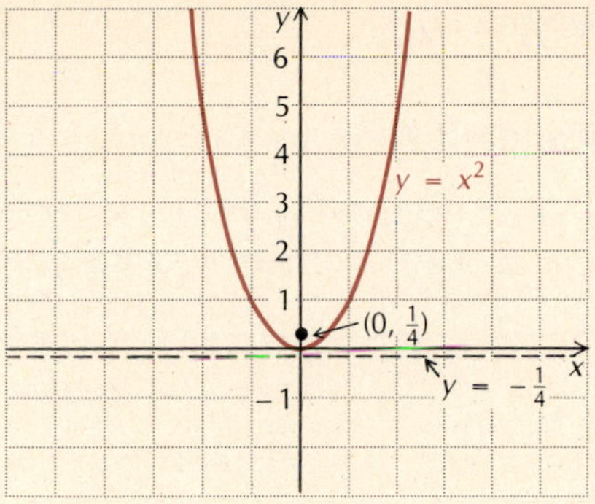

EXAMPLE 2 For the parabola $y^2 = -12x$, find the vertex, the focus, and the directrix, and draw a graph.

Solution We first write $y^2 = -12x$ in the form $y^2 = 4px$. We do this by rewriting the coefficient of x, -12, as $4(-3)$:

$$y^2 = 4px = -12x$$
$$y^2 = 4(-3)x.$$

Vertex: $(0, 0)$
Focus: $(-3, 0)$
Directrix: $x = -(-3)$
$= 3$

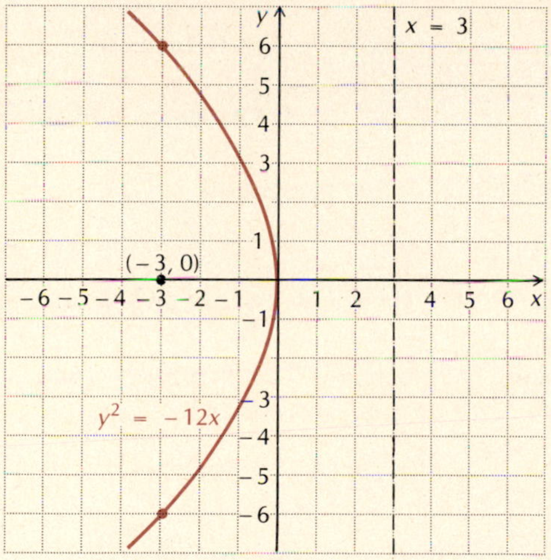

DO EXERCISES 1–3.

2 Focus and Directrix Known

EXAMPLE 3 Find an equation of a parabola with focus $(5, 0)$ and directrix $x = -5$.

Solution The focus is on the x-axis and $x = -5$ is the directrix, so the line of symmetry is the x-axis. Thus the equation is of the type

$$y^2 = 4px.$$

Since $p = 5$, the equation is $y^2 = 20x$.

EXAMPLE 4 Find an equation of a parabola with focus $(0, -7)$ and directrix $y = 7$.

Solution The focus is on the y-axis and $y = 7$ is the directrix, so the line of symmetry is the y-axis. Thus the equation is of the type

$$x^2 = 4py.$$

Since $p = -7$, we obtain $x^2 = -28y$.

For each parabola, find the vertex, the focus, and the directrix, and draw a graph.

1. $8y = x^2$

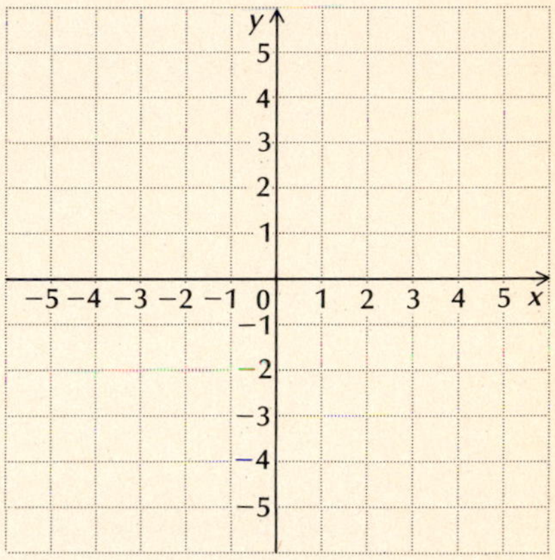

2. $y = 2x^2$

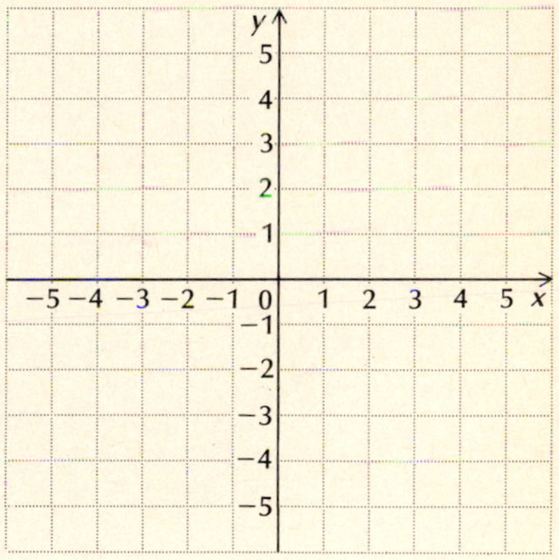

3. $y^2 = -6x$

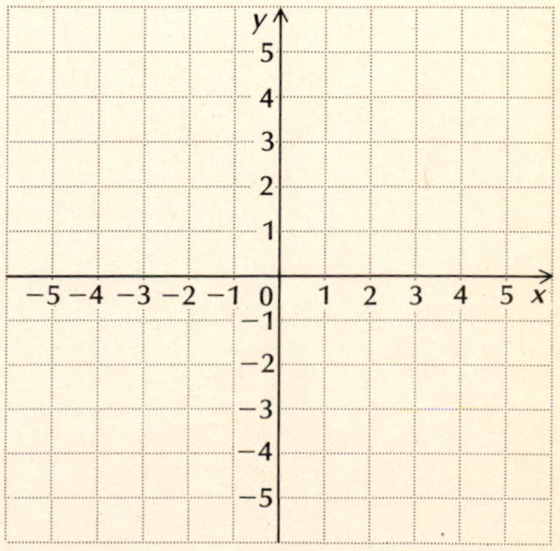

Find an equation of a parabola satisfying the given conditions.

4. Focus (3, 0), directrix $x = -3$

5. Focus $\left(0, \frac{1}{2}\right)$, directrix $y = -\frac{1}{2}$

6. Focus $(-6, 0)$, directrix $x = 6$

7. Focus $(0, -1)$, directrix $y = 1$

DO EXERCISES 4–7.

3 Standard Form by Completing the Square

If a parabola is translated so that its vertex is (h, k) and its axis of symmetry is parallel to the y-axis, it has an equation as follows:

$$(x - h)^2 = 4p(y - k),$$

where the vertex is (h, k), the focus is $(h, k + p)$, and the directrix is $y = k - p$.

If a parabola is translated so that its vertex is (h, k) and its axis of symmetry is parallel to the x-axis, it has an equation as follows:

$$(y - k)^2 = 4p(x - h),$$

where the vertex is (h, k), the focus is $(h + p, k)$, and the directrix is $x = h - p$.

EXAMPLE 5 For the parabola

$$x^2 + 6x + 4y + 5 = 0,$$

find the vertex, the focus, and the directrix, and graph the parabola.

Solution We complete the square:

$$\begin{aligned} x^2 + 6x \quad\quad\quad &= -4y - 5 \\ x^2 + 6x + 9 - 9 &= -4y - 5 \\ x^2 + 6x + 9 &= -4y + 4 \\ (x + 3)^2 &= -4(y - 1) = 4(-1)(y - 1). \end{aligned}$$

Vertex (h, k): $(-3, 1)$

Focus $(h, k + p)$: $(-3, 1 + (-1))$, or $(-3, 0)$

Directrix, $y = k - p$: $y = 1 - (-1)$, or $y = 2$

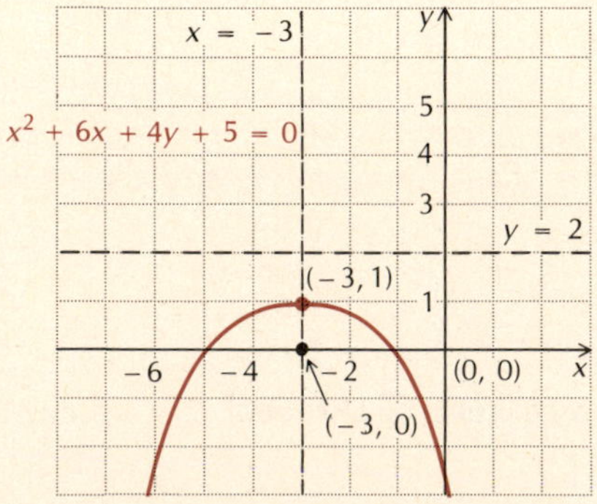

EXAMPLE 6 For the parabola

$$y^2 + 6y - 8x - 31 = 0,$$

find the vertex, the focus, and the directrix, and draw the graph.

Solution We complete the square:

$$
\begin{aligned}
y^2 + 6y &= 8x + 31 \\
y^2 + 6y + 9 - 9 &= 8x + 31 \\
y^2 + 6y + 9 &= 8x + 40 \\
(y + 3)^2 &= 8(x + 5) = 4(2)(x + 5).
\end{aligned}
$$

Vertex (h, k): $(-5, -3)$

Focus $(h + p, k)$: $(-5 + 2, -3)$ or $(-3, -3)$

Directrix, $x = h - p$: $x = -5 - 2$, or $x = -7$

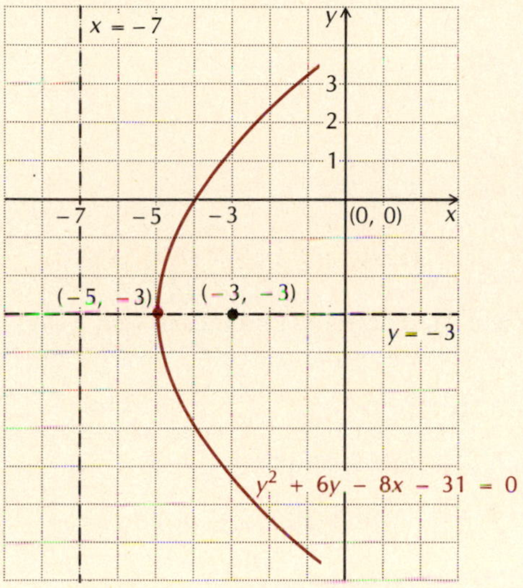

DO EXERCISES 8 AND 9.

Applications

Parabolas have many applications. For example, cross sections of headlights are parabolas. The bulb is located at the focus. All light from that point is reflected outward, parallel to the axis of symmetry.

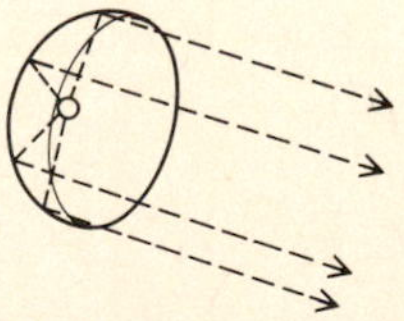

(a)

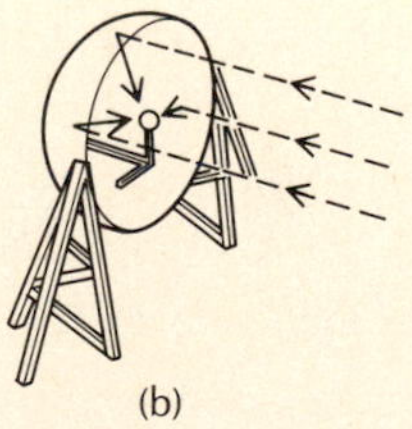

(b)

Radar and radio antennas may have cross sections that are parabolas. Incoming radio waves are reflected and concentrated at the focus. Cables hung between structures to form suspension bridges form parabolas. When a cable supports only its own weight, it does not form a parabola, but rather a curve called a *catenary*.

4 Classifying Equations

We said at the beginning of this chapter that certain equations of the type

$$Ax^2 + By^2 + Cx + Dy + E = 0$$

For each parabola, find the vertex, the focus, and the directrix, and graph the parabola.

8. $x^2 + 2x - 8y - 3 = 0$

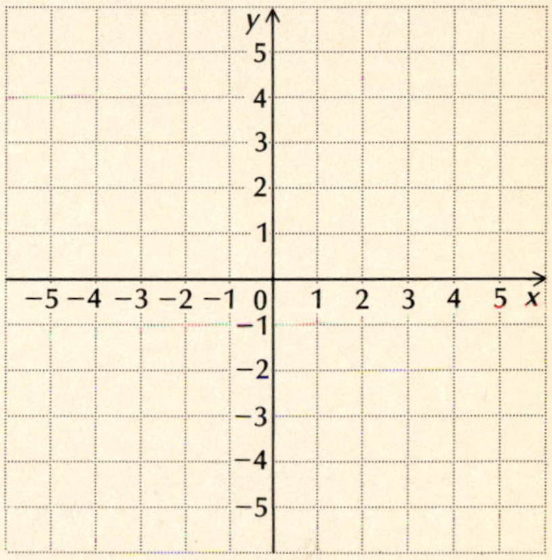

9. $y^2 + 2y + 4x - 7 = 0$

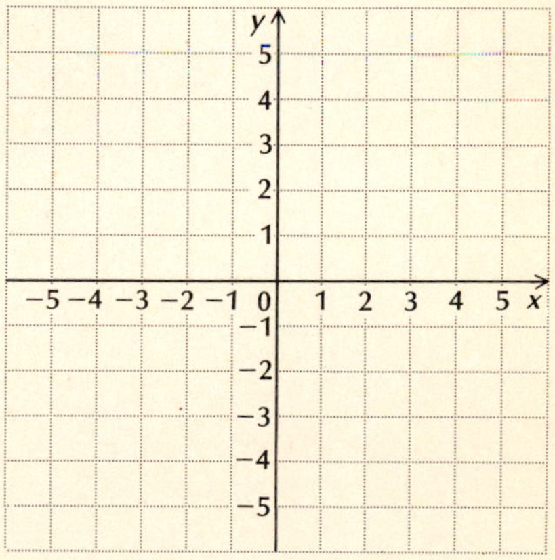

have graphs that are conic sections. For example, the equation $x^2 + y^2 = 4$ is a circle, but $x^2 + y^2 = -4$ has no real-number solutions, and it does *not* represent a conic section. How can we decide if the graph of an equation is a circle, an ellipse, a hyperbola, or a parabola? Certainly either A or B, or both, must be nonzero. Otherwise, we would not have a second-degree equation, and the graph is then probably a straight line, presuming either C or D is nonzero. The preceding shows that part of the answer rests in the value E. To know what the graph is, we would have to complete the square and try to find a standard form of one of the conic sections we have studied. Let us consider some other examples.

EXAMPLE 7 Classify each of the following equations as a circle, an ellipse, a hyperbola, or a parabola.

a) $6x^2 + 6y^2 = 24$
b) $y^2 + 6y = 8x + 31$
c) $4x^2 = 16y^2 + 64$
d) $16x^2 + 4y^2 + 96x = 8y - 84$

Solution

a) The fact that x and y are both squared tells us that we do not have a parabola. The fact that the squared terms both have positive coefficients tells us that we do not have a hyperbola. Do we have a circle? We need to get $x^2 + y^2$ by itself. Since both coefficients are the same, 6, we can factor 6 out of both terms on the left and then multiply by $\frac{1}{6}$:

$$6(x^2 + y^2) = 24$$
$$x^2 + y^2 = 4$$
$$x^2 + y^2 = 2^2.$$

The graph is a circle with center at the origin and radius 2.

b) We have only one of the variables squared. This tells us that the graph is not a circle, an ellipse, or a hyperbola. We find the following equivalent equation:

$$(y + 3)^2 = 4(2)(x + 5).$$

This tells us that we have a parabola that opens to the right and whose axis of symmetry is parallel to the x-axis.

c) Both variables are squared, so the graph is not a parabola. We can obtain the equivalent equation:

$$\frac{x^2}{16} - \frac{y^2}{4} = 1.$$

The minus sign then tells us that the graph is a hyperbola.

d) Both variables are squared, so the graph is not a parabola. We have a plus sign between the squared terms, so the graph is not a hyperbola. If the coefficients of the squared terms were the same, we might have the graph of a circle, as in part (a), but they are not. We find the equivalent equation:

$$\frac{(x + 3)^2}{4} + \frac{(y - 1)^2}{16} = 1.$$

Thus the graph is an ellipse.

We can also classify equations of the type

$$Ax^2 + By^2 + Cx + Dy + E = 0$$

by examining the coefficients. We can do this as follows:

Conic Section	Coefficients	Example
Ellipse	$A \neq B, AB > 0$	$16x^2 + 4y^2 = 64$
Circle	$A = B, A \neq 0, B \neq 0$	$x^2 + y^2 = 36$
Parabola	$A = 0$ *or* $B = 0$, but both cannot be 0.	$y^2 = 3(x - 7)$, $(x + 1)^2 = 6y + 11$
Hyperbola	$AB < 0$	$4x^2 = 16y^2 + 64$

DO EXERCISES 10–13.

Classify as a circle, an ellipse, a hyperbola, or a parabola.

10. $x^2 + 2x = 8y + 3$

11. $x^2 + y^2 + 4y = 14x + 11$

12. $4y^2 - 36 = 9x^2$

13. $36y = 25x^2 + 9y^2 + 150x + 260$

EXERCISE SET 10.3

1 For each parabola, find the vertex, the focus, and the directrix, and graph the parabola.

1. $x^2 = 8y$
2. $x^2 = 16y$
3. $y^2 = -6x$
4. $y^2 = -2x$
5. $x^2 - 4y = 0$
6. $y^2 + 4x = 0$
7. $y = 2x^2$
8. $y = \frac{1}{2}x^2$

2 Find an equation of a parabola satisfying the given conditions.

9. Focus $(4, 0)$, directrix $x = -4$
10. Focus $\left(0, \frac{1}{4}\right)$, directrix $y = -\frac{1}{4}$
11. Focus $(-\sqrt{2}, 0)$, directrix $x = \sqrt{2}$
12. Focus $(0, -\pi)$, directrix $y = \pi$
13. Focus $(3, 2)$, directrix $x = -4$
14. Focus $(-2, 3)$, directrix $y = -3$

3 Find the vertex, the focus, and the directrix, and graph.

15. $(x + 2)^2 = -6(y - 1)$
16. $(y - 3)^2 = -20(x + 2)$
17. $x^2 + 2x + 2y + 7 = 0$
18. $y^2 + 6y - x + 16 = 0$
19. $x^2 - y - 2 = 0$
20. $x^2 - 4x - 2y = 0$
21. $y = x^2 + 4x + 3$
22. $y = x^2 + 6x + 10$
23. $4y^2 - 4y - 4x + 24 = 0$
24. $4y^2 + 4y - 4x - 16 = 0$

For each parabola, find the vertex, the focus, and the directrix.

25. ▦ $x^2 = 8056.25y$
26. ▦ $y^2 = -7645.88x$

4 Classify as a circle, an ellipse, a hyperbola, or a parabola.

27. $x + 1 = 2y^2$
28. $10y + 40 = x^2 + y^2 + 8x$
29. $4y^2 + 25x^2 + 4 = 8y + 100x$
30. $9x^2 + 24y = 4y^2 + 36x + 36$
31. $2x + 13 + y^2 = 8y - x^2$
32. $x - \dfrac{5}{y} = 0$
33. $x = -16y + y^2 + 7$
34. $16x^2 + 5y^2 - 12x^2 + 8y^2 - 3x + 4y = 568$
35. $xy + 5x^2 = 9 + 7x^2 - 2x^2$
36. $56x^2 - 17y^2 = 234 - 13x^2 - 38y^2$

SYNTHESIS

37. Graph each of the following using the same set of axes.
$$x^2 - y^2 = 0, \quad x^2 - y^2 = 1,$$
$$x^2 + y^2 = 1, \quad y = x^2.$$
38. Graph each of the following using the same set of axes.
$$x^2 - 4y^2 = 0, \quad x^2 - 4y^2 = 1,$$
$$x^2 + 4y^2 = 1, \quad x = 4y^2.$$
39. Find an equation of the following parabola: Line of symmetry parallel to the y-axis, vertex $(-1, 2)$, and passing through $(-3, 1)$.
40. a) Graph $(y - 3)^2 = -20(x + 1)$. Is this relation a function?
 b) In general, is $(y - k)^2 = 4p(x - h)$ a function?

41. Find an equation of a parabola containing the point $(-3, 5)$, symmetric with respect to a horizontal line, and with vertex $(-2, 1)$.

42. The cables of a suspension bridge are 50 ft above the roadbed at the ends of the bridge and 10 ft above it in the center of the bridge. The roadbed is 200 ft long. Vertical cables are to be spaced every 20 ft along the bridge. Calculate the lengths of these vertical cables.

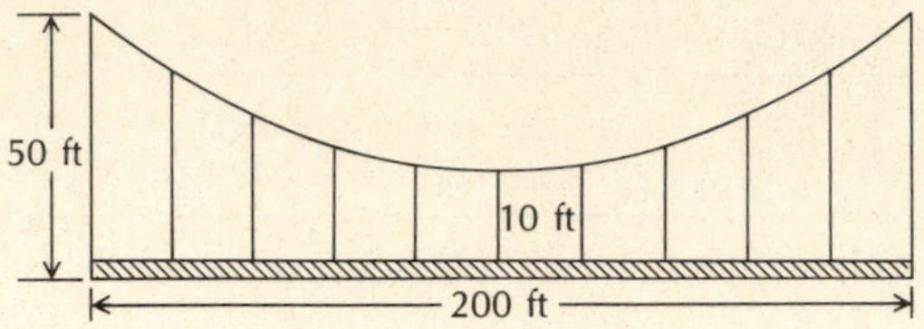

43. Show that the equation

$$\begin{vmatrix} y-k & x-h \\ 4p & y-k \end{vmatrix} = 0$$

is an equation of a parabola with vertex (h, k).

CHALLENGE

44. Prove that when a cable supports a load distributed uniformly horizontally, it hangs in the shape of a parabola. (*Hint:* Proceed as follows.)
   a) Place a coordinate system as shown here, with the origin at the lowest point of the cable.
   b) For a point $P(x, y)$ on the cable, write an equation of rotational equilibrium. The forces involved are the tensions in the cable, F_1 and F_2, and the weight supported, W (which is a function of x). The weight of the cable is essentially neglected. Use point P as the center of rotation.
   c) Solve for y.

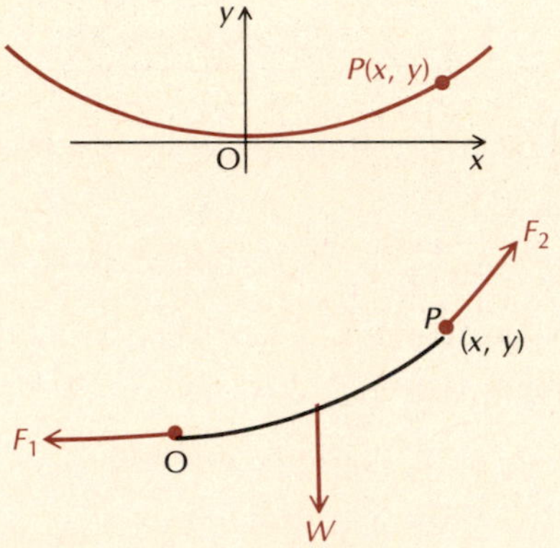

10.4 Systems of First-Degree and Second-Degree Equations

OBJECTIVES

You should be able to:

1 **Solve a system of one first-degree and one second-degree equation graphically or using the substitution method.**

2 **Solve applied problems involving the solution of one first-degree and one second-degree equation.**

When we studied systems of linear equations, we solved them both graphically and algebraically. Here we study systems in which one equation is of first degree and one is of second degree. We will use graphical and then algebraic methods of solving.

1 Solving Graphically and Algebraically

We consider a system of equations, an equation of a circle and an equation of a line. Let us think about the possible ways in which a circle and a line can intersect. The three possibilities are shown in the figure. For L_1 there is no point of intersection, hence the system of equations has no real solution. For L_2 there is one point of intersection, hence one real solution. For L_3 there are two points of intersection, hence two real solutions.

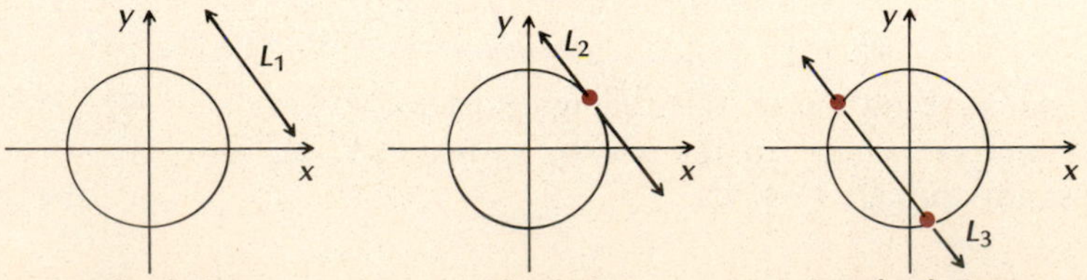

0 real solutions 1 real solution 2 real solutions

EXAMPLE 1 Solve this system graphically:

$$x^2 + y^2 = 25,$$
$$3x - 4y = 0.$$

Solution We graph the two equations using the same set of axes. The points of intersection have coordinates that must satisfy both equations. The solutions seem to be (4, 3) and (−4, −3).

We check.

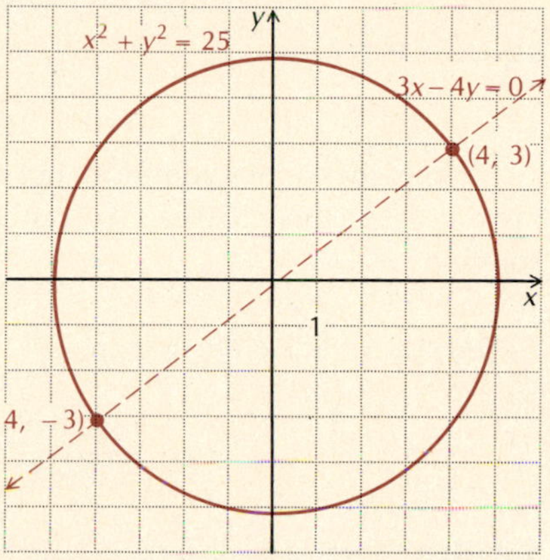

Check: For (−4, −3). We leave the check for (4, 3) to the student.

$3x - 4y = 0$	
$3(-4) - 4(-3)$	0
$-12 + 12$	
0	

$x^2 + y^2 = 25$	
$(-4)^2 + (-3)^2$	25
$16 + 9$	
25	

DO EXERCISES 1–3.

Remember that we used the *elimination* and *substitution* methods to solve systems of linear equations. In solving systems where one equation is of first degree and one is of second degree, it is preferable to use the *substitution* method.

EXAMPLE 2 Solve this system algebraically:

$$x^2 + y^2 = 25, \qquad (1)$$
$$3x - 4y = 0. \qquad (2)$$

Solution First we solve the linear equation (2) for x:

$$x = \tfrac{4}{3}y.$$

We then substitute $\frac{4}{3}y$ for x in Eq. (1) and solve for y:

$$x^2 + y^2 = 25$$
$$\left(\tfrac{4}{3}y\right)^2 + y^2 = 25$$
$$\tfrac{16}{9}y^2 + y^2 = 25$$
$$\tfrac{16}{9}y^2 + \tfrac{9}{9}y^2 = 25$$
$$\tfrac{25}{9}y^2 = 25$$
$$\tfrac{9}{25} \cdot \tfrac{25}{9}y^2 = \tfrac{9}{25} \cdot 25$$
$$y^2 = 9$$
$$y = \pm 3.$$

Solve the system graphically.

1. $x^2 + y^2 = 25,$
   $y - x = -1$

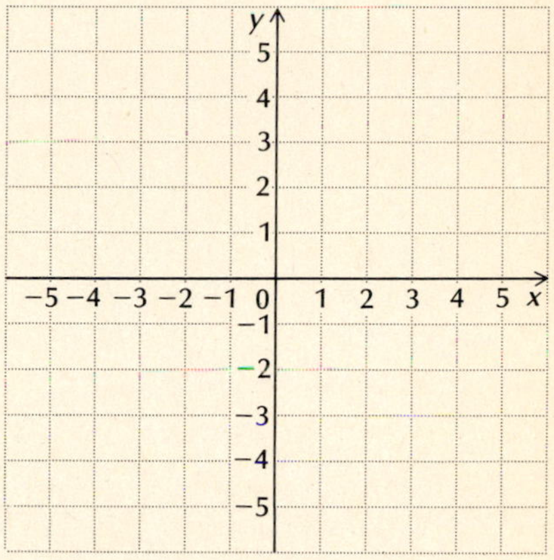

2. $y = x^2 - 2x - 1,$
   $y = x + 3$

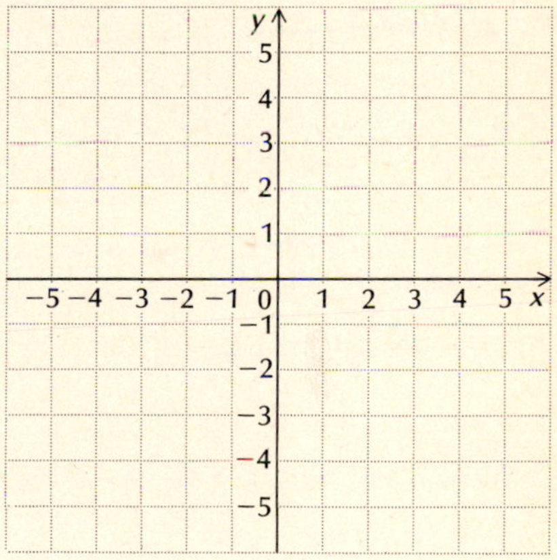

3. $y = \dfrac{x^2}{4},$

   $x + 2y = 4$

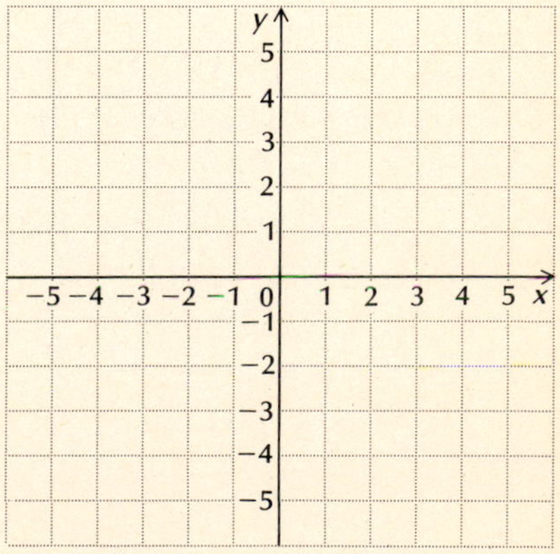

4. Solve the system in Exercise 1 algebraically.

5. Solve the system in Exercise 2 algebraically.

6. Solve the system in Exercise 3 algebraically.

7. Solve the system

$$y + 3x = 1,$$
$$x^2 - 2xy = 5.$$

8. Solve. Sketch the graphs to confirm your solution.

$$9x^2 - 4y^2 = 36,$$
$$5x + 2y = 0$$

Now we substitute these numbers into the linear equation $x = \frac{4}{3}y$ and solve for x:

$$x = \tfrac{4}{3}(3), \quad \text{or} \quad 4;$$
$$x = \tfrac{4}{3}(-3), \quad \text{or} \quad -4.$$

The pairs $(4, 3)$ and $(-4, -3)$ check, hence are solutions. ■

DO EXERCISES 4–6.

Sometimes an equation may not take on a familiar form like those studied previously in this chapter. We can still rely on the substitution method for solution.

EXAMPLE 3 Solve the system

$$y + 3 = 2x, \qquad \textbf{(1)}$$
$$x^2 + 2xy = -1. \qquad \textbf{(2)}$$

Solution First we solve the linear equation (1) for y:

$$y + 3 = 2x$$
$$y = 2x - 3.$$

We then substitute $2x - 3$ for y in Eq. (2) and solve for x:

$$x^2 + 2xy = -1$$
$$x^2 + 2x(2x - 3) = -1$$
$$x^2 + 4x^2 - 6x = -1$$
$$5x^2 - 6x + 1 = 0$$
$$(5x - 1)(x - 1) = 0$$
$$5x - 1 = 0 \quad \text{or} \quad x - 1 = 0$$
$$x = \tfrac{1}{5} \quad \text{or} \quad x = 1.$$

Now we substitute these numbers into the linear equation $y = 2x - 3$ and solve for y:

$$y = 2\left(\tfrac{1}{5}\right) - 3, \quad \text{or} \quad -\tfrac{13}{5};$$
$$y = 2(1) - 3, \quad \text{or} \quad -1.$$

The pairs $(\tfrac{1}{5}, -\tfrac{13}{5})$ and $(1, -1)$ check and are solutions. ■

DO EXERCISES 7 AND 8.

2 Problem Solving

We now consider solving problems in which the translation is a system of equations, one quadratic and one linear.

EXAMPLE 4 The perimeter of a rectangular field is 204 m, and the area is 2565 m². Find the dimensions of the field.

Solution

1. *Familiarize.* We draw a picture of the field, labeling the drawing. We let l = length and w = width.

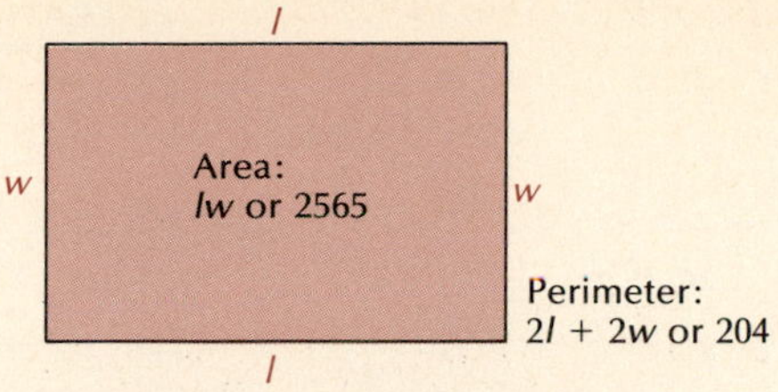

2. *Translate.* We then have the following translation:

$$\text{Perimeter: } 2l + 2w = 204;$$
$$\text{Area: } lw = 2565.$$

3. *Carry out.* We solve the system:

$$2l + 2w = 204, \quad \text{The graph is a line.}$$
$$lw = 2565. \quad \text{The graph is a hyperbola.}$$

We solve the second equation for l and get $l = 2565/w$. Then we substitute $2565/w$ for l in the first equation solve for w:

$$2l + 2w = 204$$
$$2\left(\frac{2565}{w}\right) + 2w = 204$$
$$2(2565) + 2w^2 = 204w \quad \text{Multiplying by } w$$
$$2w^2 - 204w + 2(2565) = 0 \quad \text{Standard form}$$
$$w^2 - 102w + 2565 = 0 \quad \text{Multiplying by } \tfrac{1}{2}$$
$$w = \frac{-(-102) \pm \sqrt{(-102)^2 - 4 \cdot (1) \cdot 2565}}{2(1)} \quad \text{Quadratic formula}$$
$$w = \frac{102 \pm 12}{2}$$
$$w = 57 \quad \text{or} \quad w = 45.$$

If $w = 57$, then

$$l = \frac{2565}{w} = \frac{2565}{57} = 45.$$

If $w = 45$, then

$$l = \frac{2565}{w} = \frac{2565}{45} = 57.$$

Since the length is usually considered to be longer than the width, we have the solution $l = 57$ and $w = 45$, or $(57, 45)$.

4. *Check.* If $l = 57$ and $w = 45$, then the perimeter is $2 \cdot 57 + 2 \cdot 45$, or 204. The area is $57 \cdot 45$, or 2565. The numbers check.

5. *State.* The answer is that the length is 57 m and the width is 45 m. ■

DO EXERCISES 9 AND 10.

9. The difference of two numbers is 4, and the difference of their squares is 72. What are the numbers?

10. The perimeter of a rectangular field is 34 ft, and the length of a diagonal is 13 ft. Find the dimensions of the field.

EXERCISE SET 10.4

1 Solve the system graphically. Then solve algebraically.

1. $x^2 + y^2 = 25,$ $y - x = 1$

2. $x^2 + y^2 = 100,$ $y - x = 2$

3. $y^2 - x^2 = 9,$ $2x - 3 = y$

4. $x + y = -6,$ $xy = -7$

5. $4x^2 + 9y^2 = 36,$ $3y + 2x = 6$

6. $9x^2 + 4y^2 = 36,$ $3x + 2y = 6$

7. $y^2 = x + 3,$ $2y = x + 4$

8. $y = x^2,$ $3x = y + 2$

Solve.

9. $x^2 + 4y^2 = 25,$ $x + 2y = 7$

10. $y^2 - x^2 = 16,$ $2x - y = 1$

11. $x^2 - xy + 3y^2 = 27,$ $x - y = 2$

12. $2y^2 + xy + x^2 = 7,$ $x - 2y = 5$

13. $3x + y = 7,$ $4x^2 + 5y = 56$

14. $2y^2 + xy = 5,$ $4y + x = 7$

15. $a + b = 7,$ $ab = 4$

16. $p + q = -6,$ $pq = -7$

17. $2a + b = 1,$ $b = 4 - a^2$

18. $4x^2 + 9y^2 = 36,$ $x + 3y = 3$

19. $a^2 + b^2 = 89,$ $a - b = 3$

20. $xy = 4,$ $x + y = 5$

21. $x^2 + y^2 = 5,$ $x - y = 8$

22. $4x^2 + 9y^2 = 36,$ $y - x = 8$

23. ▦ $x^2 + y^2 = 19{,}380{,}510.36,$ $27{,}942.25x - 6.125y = 0$

24. ▦ $2x + 2y = 1660,$ $xy = 35{,}325$

2 Solve.

25. The sum of two numbers is 12, and the sum of their squares is 90. What are the numbers?

26. The sum of two numbers is 15, and the difference of their squares is also 15. What are the numbers?

27. A rectangle has a perimeter of 28 cm, and the length of a diagonal is 10 cm. What are its dimensions?

28. A rectangle has a perimeter of 6 m, and the length of a diagonal is $\sqrt{5}$ m. What are its dimensions?

29. A rectangle has an area of 20 in^2 and a perimeter of 18 in. Find its dimensions.

30. A rectangle has an area of 2 yd^2 and a perimeter of 6 yd. Find its dimensions.

31. It will take 210 yd of fencing to enclose a rectangular field. The area of the field is 2250 yd^2. What are the dimensions?

32. The diagonal of a rectangle is 1 ft longer than the length of the rectangle and 3 ft longer than twice the width. Find the dimensions of the rectangle.

SYNTHESIS

33. Find two numbers whose product is 2 and the sum of whose reciprocals is $\frac{33}{8}$.

34. Find an equation of a circle that passes through $(-2, 3)$ and $(-4, 1)$ and whose center is on the line $5x + 8y = -2$.

35. A piece of wire 100 cm long is to be cut into two pieces and those pieces are each to be bent to make a square. The area of one square is to be 144 cm^2 greater than that of the other square. How should the wire be cut?

36. The sum of two numbers is 1, and their product is 1. Find the sum of their cubes. There is a method to solve this problem that is easier than solving a system of one first-degree equation and one second-degree equation. Can you discover it?

37. Find an equation of an ellipse centered at the origin that passes through the points $(1, \sqrt{3}/2)$ and $(\sqrt{3}, 1/2)$.

38. Find an equation of a hyperbola of the type
$$\frac{x^2}{a^2} - \frac{y^2}{b^2} = 1$$
that passes through the points $(-3, -3\sqrt{5}/2)$ and $(-3, 3\sqrt{5}/2)$.

Solve for x and y.

39. $x - y = a + 2b,$ $x^2 - y^2 = a^2 + 2ab + b^2$

40. $\dfrac{x}{a - b} + \dfrac{y}{a + b} = 1,$ $x^2 - y^2 = (a - b)^2$

41. Given the area A and the perimeter P of a rectangle, show that the length L and the width W are given by the formulas
$$L = \tfrac{1}{4}(P + \sqrt{P^2 - 16A}),$$
$$W = \tfrac{1}{4}(P - \sqrt{P^2 - 16A}).$$

42. Show that a hyperbola does not intersect its asymptotes. That is, solve the system
$$\frac{x^2}{a^2} - \frac{y^2}{b^2} = 1$$
$$y = \frac{b}{a}x \quad \left(\text{or } y = -\frac{b}{a}x\right).$$

43. Find an equation of a circle that passes through the points $(2, 4)$ and $(3, 3)$ and whose center is on the line $3x - y = 3$.

44. Find an equation of a circle that passes through the points $(7, 3)$ and $(5, 5)$ and whose center is on the line $y - 4x = 1$.

CHALLENGE

45. Solve:

$$x^3 + y^3 = 72,$$
$$x + y = 6.$$

46. Solve for h, k, and λ:

$$1 - 20\lambda k^2 = 0,$$
$$2 - 10\lambda hk = 0,$$
$$hk^2 - 640{,}000 = 0.$$

10.5 Systems of Second-Degree Equations

OBJECTIVES

You should be able to:

1. Solve systems of second-degree equations graphically.
2. Use the elimination method to solve systems of equations like
$$2x^2 + 5y^2 = 20,$$
$$3x^2 - y^2 = -1.$$
3. Use the substitution method to solve systems of equations like
$$x^2 + 4y^2 = 20,$$
$$xy = 4.$$
4. Solve applied problems involving systems of two second-degree equations.

We now consider systems of two second-degree equations. The following figure shows the ways in which a circle and a hyperbola can intersect.

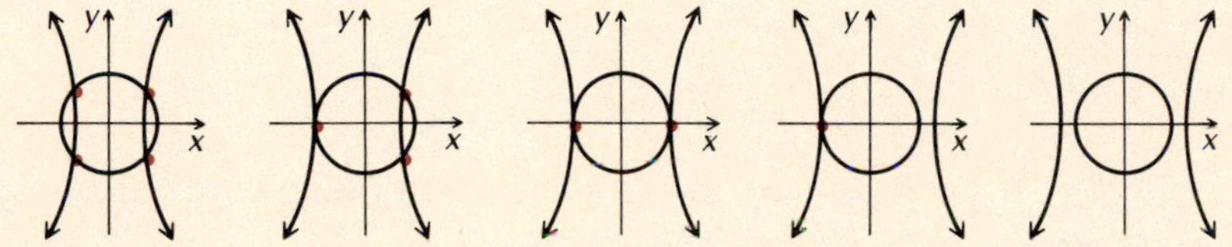

4 real solutions 3 real solutions 2 real solutions 1 real solution 0 real solutions

1 Solving Graphically

EXAMPLE 1 Solve this system graphically:

$$x^2 + y^2 = 25,$$
$$\frac{x^2}{25} - \frac{y^2}{25} = 1.$$

Solution We graph the two equations using the same set of axes. The points of intersection have coordinates that must satisfy both equations; the solutions seem to be (5, 0) and (−5, 0).

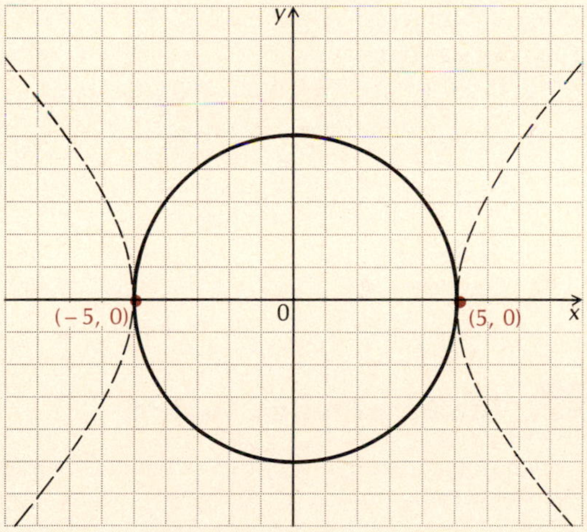

Check: Since $(5)^2 = 25$ and $(-5)^2 = 25$, we can do both checks at once.

$\frac{x^2}{25} - \frac{y^2}{25} = 1$	
$\frac{(\pm 5)^2}{25} - \frac{0^2}{25}$	1
$\frac{25}{25} - 0$	
1	

$x^2 + y^2 = 25$	
$(\pm 5)^2 + 0^2$	25
$25 + 0$	
25	

DO EXERCISES 1–3. (EXERCISES 2 AND 3 ARE ON THE FOLLOWING PAGE.)

Solve the system graphically.

1. $x^2 + y^2 = 4,$
$$\frac{x^2}{4} - \frac{y^2}{4} = 1$$

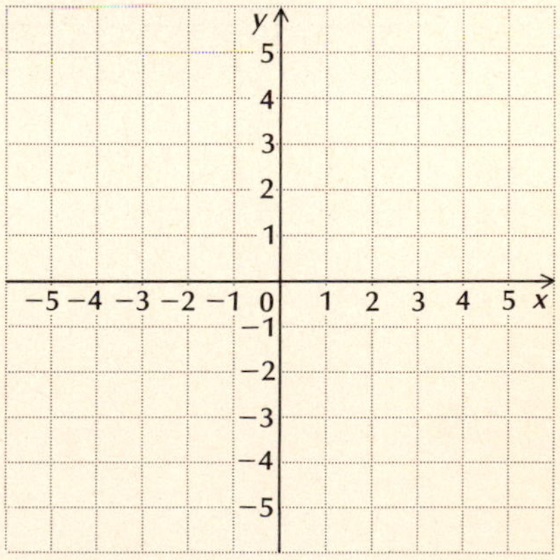

2. $x^2 + y^2 = 16,$

$$\frac{x^2}{16} + \frac{y^2}{9} = 1$$

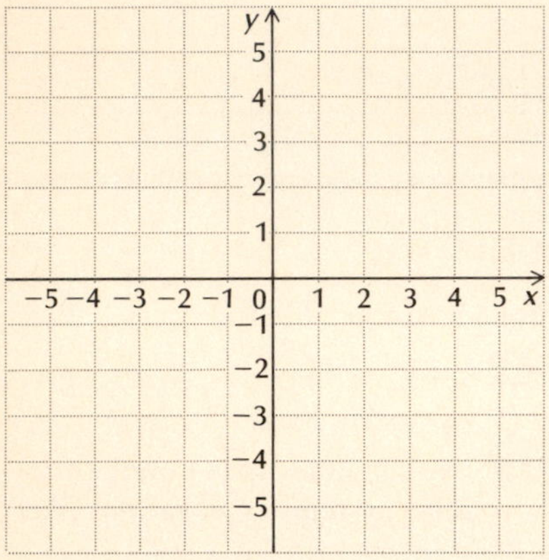

3. $x^2 + y^2 = 4,$

$$\frac{x^2}{25} + \frac{y^2}{4} = 1$$

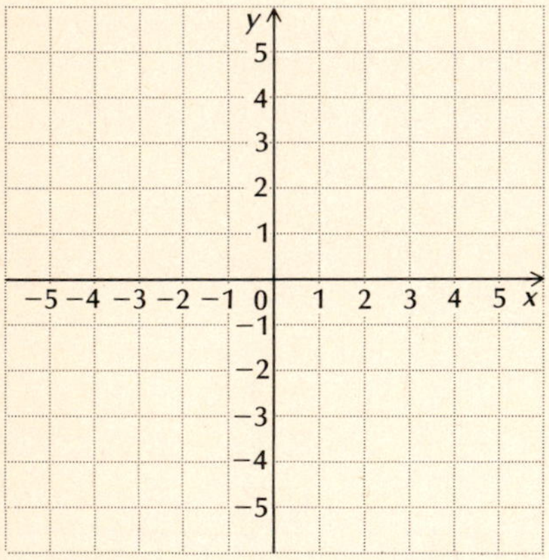

4. Solve this system:

$$2y^2 - 3x^2 = 6,$$
$$5y^2 + 2x^2 = 53.$$

2 The Elimination Method

To solve systems of two second-degree equations, we can use either the substitution or the elimination method. We begin with an example using the elimination method.

EXAMPLE 2 Solve this system:

$$2x^2 + 5y^2 = 22, \qquad (1)$$
$$3x^2 - y^2 = -1. \qquad (2)$$

Solution Here we multiply the second equation by 5 and use the elimination method:

$$\begin{aligned} 2x^2 + 5y^2 &= 22 \\ 15x^2 - 5y^2 &= -5 && \text{Multiplying by 5} \\ \hline 17x^2 \qquad &= 17 && \text{Adding} \\ x^2 &= 1 \\ x &= \pm 1. \end{aligned}$$

If $x = 1$, $x^2 = 1$, and if $x = -1$, $x^2 = 1$. Thus substituting 1 or -1 for x in Eq. (2), we have

$$\begin{aligned} 3 \cdot (\pm 1)^2 - y^2 &= -1 \\ 3 - y^2 &= -1 \\ -y^2 &= -4 \\ y^2 &= 4 \\ y &= \pm 2. \end{aligned}$$

Thus, if $x = 1$, $y = 2$, or $y = -2$, and if $x = -1$, $y = 2$, or $y = -2$. The possible solutions are $(1, 2)$, $(1, -2)$, $(-1, 2)$, and $(-1, -2)$.

Check: Since $(2)^2 = 4$, $(-2)^2 = 4$, $(1)^2 = 1$, and $(-1)^2 = 1$, we can check all four pairs at one time.

$$\begin{array}{r|l} \multicolumn{2}{c}{2x^2 + 5y^2 = 22} \\ \hline 2(\pm 1)^2 + 5(\pm 2)^2 & 22 \\ 2 + 20 & \\ 22 & \end{array} \qquad \begin{array}{r|l} \multicolumn{2}{c}{3x^2 - y^2 = -1} \\ \hline 3(\pm 1)^2 - (\pm 2)^2 & -1 \\ 3 - 4 & \\ -1 & \end{array}$$

The solutions are $(1, 2)$, $(1, -2)$, $(-1, 2)$, and $(-1, -2)$.

DO EXERCISE 4.

3 The Substitution Method

When one equation contains a product of variables and the other equation is of the form $Ax^2 + By^2 = C$, we often solve for one of the variables in the equation with the product and then substitute in the other.

EXAMPLE 3 Solve the system

$$x^2 + 4y^2 = 20, \qquad (1)$$
$$xy = 4. \qquad (2)$$

Solution Here we use the substitution method. First we solve Eq. (2) for y:

$$\begin{aligned} xy &= 4 \\ y &= \frac{4}{x}. \end{aligned}$$

Then we substitute $4/x$ for y in Eq. (1) and solve for x:

$$x^2 + 4y^2 = 20$$

$$x^2 + 4\left(\frac{4}{x}\right)^2 = 20$$

$$x^2 + \frac{64}{x^2} = 20$$

$$x^4 + 64 = 20x^2 \quad \text{Multiplying by } x^2$$

$$x^4 - 20x^2 + 64 = 0$$

$$u^2 - 20u + 64 = 0 \quad \text{Letting } u = x^2$$

$$(u - 16)(u - 4) = 0 \quad \text{Factoring}$$

$$u = 16 \quad \text{or} \quad u = 4. \quad \text{Principle of zero products}$$

Then $x^2 = 16$ or $x^2 = 4$, so $x = \pm 4$ or $x = \pm 2$. Since $y = 4/x$, if $x = 4$, $y = 1$; if $x = -4, y = -1$; if $x = 2, y = 2$; if $x = -2, y = -2$. The solutions are $(4, 1)$, $(-4, -1)$, $(2, 2)$, and $(-2, -2)$.

DO EXERCISE 5.

5. Solve this system:

$$x^2 + xy + y^2 = 19,$$
$$xy = 6.$$

4 Problem Solving

EXAMPLE 4 The area of a rectangle is 300 yd^2, and the length of a diagonal is 25 yd. Find the dimensions.

Solution

1. *Familiarize.* We draw a picture and label it. We note that there is a right triangle in the figure. We let l = the length of the rectangle and w = the width of the rectangle.

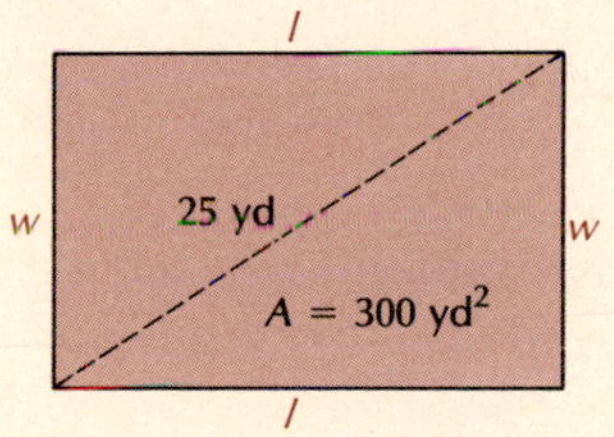

2. *Translate.* We translate to a system of equations.

From the Pythagorean theorem: $l^2 + w^2 = 25^2$; The graph is a circle.

From an area formula: $lw = 300$. The graph is a hyperbola.

3. *Carry out.* We solve the system

$$l^2 + w^2 = 625,$$
$$lw = 300.$$

We get $(20, 15)$ and $(-20, -15)$.

4. *Check.* Lengths of sides cannot be negative, so we need check only $(20, 15)$. In the right triangle, $20^2 + 15^2 = 400 + 225 = 625$, which is 25^2. The area is $20 \cdot 15 = 300$, so we have a solution.

5. *State.* The answer is that the length is 20 yd and the width is 15 yd.

DO EXERCISE 6.

6. The area of a rectangle is 2 ft^2, and the length of a diagonal is $\sqrt{5}$ ft. Find the dimensions of the rectangle.

EXERCISE SET 10.5

1, 2, 3 Solve the system graphically. Then solve algebraically.

1. $x^2 + y^2 = 25,$ $y^2 = x + 5$
2. $y = x^2,$ $x = y^2$
3. $x^2 + y^2 = 9,$ $x^2 - y^2 = 9$
4. $y^2 - 4x^2 = 4,$ $4x^2 + y^2 = 4$
5. $x^2 + y^2 = 25,$ $xy = 12$
6. $x^2 - y^2 = 16,$ $x + y^2 = 4$
7. $x^2 + y^2 = 4,$ $16x^2 + 9y^2 = 144$
8. $x^2 + y^2 = 25,$ $25x^2 + 16y^2 = 400$

2, **3** Solve.

9. $x^2 + y^2 = 16,$
$y^2 - 2x^2 = 10$

10. $x^2 + y^2 = 14,$
$x^2 - y^2 = 4$

11. $x^2 + y^2 = 5,$
$xy = 2$

12. $x^2 + y^2 = 20,$
$xy = 8$

13. $x^2 + y^2 = 13,$
$xy = 6$

14. $x^2 + 4y^2 = 20,$
$xy = 4$

15. $x^2 + y^2 + 6y + 5 = 0,$
$x^2 + y^2 - 2x - 8 = 0$

16. $2xy + 3y^2 = 7,$
$3xy - 2y^2 = 4$

17. $xy - y^2 = 2,$
$2xy - 3y^2 = 0$

18. $4a^2 - 25b^2 = 0,$
$2a^2 - 10b^2 = 3b + 4$

19. $m^2 - 3mn + n^2 + 1 = 0,$
$3m^2 - mn + 3n^2 = 13$

20. $ab - b^2 = -4,$
$ab - 2b^2 = -6$

21. $a^2 + b^2 = 14,$
$ab = 3\sqrt{5}$

22. $x^2 + xy = 5,$
$2x^2 + xy = 2$

23. $x^2 + y^2 = 25,$
$9x^2 + 4y^2 = 36$

24. $x^2 + y^2 = 1,$
$9x^2 - 16y^2 = 144$

25. ▦ $18.465x^2 + 788.723y^2 = 6408,$
$106.535x^2 - 788.723y^2 = 2692$

26. ▦ $0.319x^2 + 2688.7y^2 = 56{,}548,$
$0.306x^2 - 2688.7y^2 = 43{,}452$

4

27. Find two numbers whose product is 156 if the sum of their squares is 313.

28. Find two numbers whose product is 60 if the sum of their squares is 136.

29. The area of a rectangle is $\sqrt{3}$ m^2, and the length of a diagonal is 2 m. Find the dimensions.

30. The area of a rectangle is $\sqrt{2}$ m^2, and the length of a diagonal is $\sqrt{3}$ m. Find the dimensions.

31. A garden contains two square peanut beds. Find the length of each bed if the sum of their areas is 832 ft^2 and the difference of their areas is 320 ft^2.

32. A certain amount of money saved for 1 yr at a certain interest rate yielded \$7.50. If the principal had been \$25 more and the interest rate 1% less, the interest would have been the same. Find the principal and the rate.

SYNTHESIS

33. Find an equation of the circle that passes through the points (4, 6), (−6, 2), and (1, −3).

34. Find an equation of the circle that passes through the points (2, 3), (4, 5), and (0, −3).

35. The square of a certain number exceeds twice the square of another number by $\frac{1}{8}$. The sum of their squares is $\frac{5}{16}$. Find the numbers.

36. Four squares with sides 5 in. long are cut from the corners of a rectangular metal sheet that has an area of 340 in^2. The edges are bent up to form an open box with a volume of 350 in^3. Find the dimensions of the box.

Solve for x and y.

37. $x^2 + xy = a,$
$y^2 + xy = b$

38. $x^2 - y^2 = a^2 - b^2,$
$x - y = a - b$

Solve.

39. $p^2 + q^2 = 13,$
$\frac{1}{pq} = -\frac{1}{6}$

40. $a + b = \frac{5}{6},$
$\frac{a}{b} + \frac{b}{a} = \frac{13}{6}$

41. $x^2 + y^2 = 4,$
$(x - 1)^2 + y^2 = 4$

42. Solve for x and y.

$$a^2b^2 - 2b^2x^2 - a^2y^2 = 0,$$
$$a^2b^2 - b^2x^2 - 2a^2y^2 = 0$$

43. Find k such that the following equations have two roots in common.

$$(x - 2)^4 - (x - 2) = 0,$$
$$x^2 - kx + k = 0$$

Solve.

44. $10x^2 - xy + 4y^2 = 28,$
$2x^2 - 3xy - 2y^2 = 0$

45. $5^{x+y} = 100,$
$3^{2x-y} = 1000$

46. $e^x - e^{x+y} = 0,$
$e^y - e^{x-y} = 0$

47. The sum, the product, and the sum of the squares of two numbers are all the same. Find the numbers.

SUMMARY AND REVIEW: CHAPTER 10

TERMS TO KNOW

Conic Section
Ellipse
Foci
Vertex
Hyperbola
Asymptote
Parabola
Directrix

REVIEW EXERCISES

1. Graph: $2x^2 - 3xy - 2y^2 = 0$.

2. Find the center, vertices, and foci of the ellipse
$$16x^2 + 25y^2 - 64x + 50y - 311 = 0.$$
Then graph the ellipse.

3. Find an equation of the ellipse having vertices (3, 0) and (0, 4) and centered at the origin.

4. Find the center, vertices, foci, and asymptotes of the hyperbola
$$x^2 - 2y^2 + 4x + y - \tfrac{1}{8} = 0.$$

Graph.

5. $xy = -2$

6. $4y^2 - x^2 = 16$

7. Find an equation of the parabola with directrix $y = \frac{3}{2}$ and focus $(0, -\frac{3}{2})$.

8. Find the focus, vertex, and directrix of the parabola $y^2 = -12x$.

9. Find the vertex, focus, and directrix of the parabola $x^2 + 10x + 2y + 9 = 0$.

Classify the equation as a circle, an ellipse, a parabola, or a hyperbola.

10. $4x^2 + 4y^2 = 100 - 64x - 20y$

11. $9x^2 + 2y^2 = 18$

12. $y = -x^2 + 2x - 3 + 10y$

13. $\dfrac{y^2}{9} - \dfrac{x^2}{4} = 1$

14. $xy = 11$

15. $x = y^2 + 2y - 2 - 5x$

16. $xy = -3$

17. $x^2 + y^2 + 6x - 8y - 39 = 0$

Solve.

18. $x^2 - 16y = 0,$
$x^2 - y^2 = 64$

19. $4x^2 + 4y^2 = 65,$
$6x^2 - 4y^2 = 25$

20. $x^2 - y^2 = 33,$
$x + y = 11$

21. $x^2 - 2x + 2y^2 = 8,$
$2x + y = 6$

22. $x^2 - y = 3,$
$2x - y = 3$

23. $x^2 + y^2 = 25,$
$x^2 - y^2 = 7$

24. $x^2 - y^2 = 3,$
$y = x^2 - 3$

25. $x^2 + y^2 = 18,$
$2x + y = 3$

26. $x^2 + y^2 = 100,$
$2x^2 - 3y^2 = -120$

27. $x^2 + 2y^2 = 12,$
$xy = 4$

28. The sides of a triangle are 8, 10, and 14. Find the altitude to the longest side.

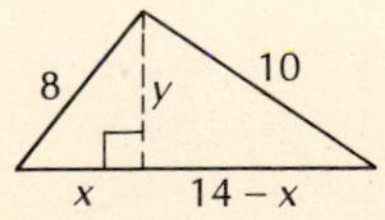

29. The sum of two numbers is 11 and the sum of their squares is 65. Find the numbers.

30. A rectangle has a perimeter of 38 m and an area of 84 m^2. What are its dimensions?

31. Find two positive integers whose sum is 12 and the sum of whose reciprocals is $\frac{3}{8}$.

32. The perimeter of a square is 12 cm more than the perimeter of another square. Its area exceeds the area of the other by 39 cm^2. Find the perimeter of each square.

33. The sum of the area of two circles is 130π ft^2. The difference of the areas is 112π ft^2. Find the radius of each circle.

SYNTHESIS

34. Find an equation of the ellipse that contains the point $(-1/2, 3\sqrt{3}/2)$ and two of whose vertices are $(-1, 0)$ and $(1, 0)$.

35. Find two numbers whose product is 4 and the sum of whose reciprocals is $\frac{65}{56}$.

36. Find an equation of the circle that passes through the points (10, 7), (−6, 7), and (−8, 1).

TEST: CHAPTER 10

1. Graph: $3x^2 - 5xy - 2y^2 = 0$.

2. Find the center, vertices, and foci of the ellipse
$$9x^2 + y^2 - 36x - 8y + 43 = 0.$$
Then graph the ellipse.

3. Find an equation of the ellipse with vertices (−7, 0), (7, 0), (0, −2), and (0, 2).

4. Find the center, vertices, foci, and asymptotes of the hyperbola
$$9y^2 - 25x^2 = 225.$$

5. Graph: $xy = 6$.

6. Find an equation of the parabola with directrix $y = 5$ and focus $(0, -5)$.

7. Find the vertex, focus, and directrix of the parabola $y^2 + 6y + 8x - 7 = 0$.

Solve.

8. $x^2 + y^2 = 100$,
$2y + x = 20$

9. $x^2 + y^2 = 5$,
$xy = 2$

10. The sum of two numbers is 4, and the difference of their squares is 32. What are the numbers?

11. The area of a rectangle is 10 yd^2, and the length of a diagonal is $\sqrt{101}$ yd. Find the dimensions.

12. In a fractional expression, the sum of the values of the numerator and the denominator is 23. The product of their values is 120. Find the values of the numerator and the denominator.

13. A rectangle with diagonal of length $5\sqrt{5}$ has an area of 22. Find the dimensions of the rectangle.

14. Two squares are such that the sum of their areas is 8 m^2 and the difference of their areas is 2 m^2. Find the length of a side of each square.

15. A rectangle has a diagonal of length 20 ft and a perimeter of 56 ft. Find the dimensions.

Classify the equation as a circle, an ellipse, a parabola, or a hyperbola.

16. $y = x^2 - 4x - 1 - 15y$

17. $x^2 + y^2 + 2x + 6y + 6 = 0$

18. $\frac{x^2}{9} - \frac{y^2}{4} = 1$

19. $16x^2 + 4y^2 = 64 - 128x$

20. $xy + 5 + 7x^2 = 7x^2$

21. $x = -y^2 + 8y + 16$

SYNTHESIS

22. Find an equation of the ellipse with vertices (−3, −4) and (−3, 2) and containing the point (−1, 1).

23. Find an equation of the circle through points (−3, 8), (4, 1), and (−3, −4).

24. Find the point on the y-axis that is equidistant from (−3, −5) and (4, −7).

25. The sum of two numbers is 36 and the product is 4. Find the sum of the reciprocals of the numbers.

Polynomial and Rational Functions

11

In this chapter, we extend a study of polynomial functions that began in Section 4.8. A *polynomial function* is a function that can be defined by a polynomial expression. A *rational function* is a function that can be defined as a quotient of two polynomials. We will study both kinds of functions. For polynomials, we will consider the finding of zeros or roots in greater depth. For rational functions, we will consider their graphs.

The polynomial and rational functions, together with the radical, trigonometric, logarithmic, and exponential functions, comprise the functions that we consider in elementary mathematics. These functions, which are considered in calculus, for example, are those most used in the applications of mathematics.

FEATURE PROBLEM

Weight above the earth. A person's weight above the earth at height h above sea level satisfies the equation

$$W = \left(\frac{r}{r + h}\right)^2 W_0,$$

where W_0 is a person's weight at sea level and r is the radius of the earth, in miles. Suppose a person weighs 150 lb at sea level. Express W as a function of h. Assume that the radius of the earth is 3965 mi.

THE MATHEMATICS

Substituting 150 for W_0 and 3965 for r, we obtain the function W described as follows:

$$W(h) = 150\left(\frac{3965}{3965 + h}\right)^2.$$

This is a *rational function*.

OBJECTIVES

You should be able to:

1 Determine the degree of a polynomial.

2 Determine whether a given number is a root or zero of a polynomial function.

3 By division, determine whether one polynomial is a factor of another.

4 Given a polynomial $P(x)$ and a divisor $d(x)$, determine the quotient and the remainder and express $P(x)$ as

$$P(x) = d(x) \cdot Q(x) + R(x).$$

Determine the degree of the polynomial.

1. $x^5 - x^3 + x^2 + 2i$

2. $3x - i$

3. $3x^2 - \sqrt{2}$

4. 0.5

5. 0

11.1 Division of Polynomials

In Section 4.8, we studied applications and graphs of polynomial functions. Most of the graphs we considered in that section had roots, or zeros, that were easy to find because the polynomials were either factored or fairly easy to factor. In general, finding roots, or zeros, is not easy or straightforward. In this chapter, we will delve more deeply into the problem of finding them. Let us first review some terminology.

1 Coefficients and Degrees

DEFINITION

A *polynomial function f* is given by

$$f(x) = a_nx^n + a_{n-1}x^{n-1} + \cdots + a_2x^2 + a_1x + a_0,$$

where n is a nonnegative integer and $a_n, a_{n-1}, \ldots, a_2, a_1, a_0$ are real or complex numbers, called the *coefficients* of the polynomial. The first nonzero coefficient is assumed to be a_n and is called the *leading coefficient*. The *degree* of the polynomial function is n.

Note that the definition of a polynomial function has been revised to allow the coefficients to be complex numbers. In certain cases, we will restrict the coefficients to be real numbers, rational numbers, or integers.

EXAMPLE 1 Determine the degree of each polynomial.

Polynomial	Type of Coefficient	Degree
$5x^3 - 3x^2 + (2 + 4i)x + i$	Complex	3
$-4x + \sqrt{2}$	Real	1
3.4	Rational	0
0	Integer	No degree

DO EXERCISES 1–5.

2 Zeros, or Roots, of Polynomials

We often refer to polynomials using function notation, such as $P(x)$. If a number a makes a polynomial 0, that is, $P(a) = 0$, we say that a is a **zero** of $P(x)$. Note that a is also a solution of the equation $P(x) = 0$, formed by setting $P(x)$ equal to 0. Thus, a is also the first coordinate of an x-intercept of $P(x) = 0$. We often say that a is a **root** of the equation $P(x) = 0$, or simply a root of the polynomial $P(x)$.*

EXAMPLE 2 Given $P(x) = x^3 + 2x^2 - 5x - 6$. Is 3 a *zero* of $P(x)$?

*Terminology varies from author to author. Traditionally, one referred to the *zeros* of a function and the *roots* of an equation. Subsequently, it was found to be convenient in exposition to speak also of the "roots" of a function, especially a polynomial function. In some books, the traditional distinction is still made today. We have found that the freer use of terminology is just as clear.

Solution We substitute 3 into the polynomial:

$$P(3) = (3)^3 + 2(3)^2 - 5(3) - 6 = 24.$$

Since $P(3) \neq 0$, 3 is *not* a zero of the polynomial.

EXAMPLE 3 Is -1 a root of $P(x)$ in Example 2?

Solution

$$P(-1) = (-1)^3 + 2(-1)^2 - 5(-1) - 6 = 0$$

Since $P(-1) = 0$, we know that -1 is a root of $P(x)$.

DO EXERCISES 6–8.

3 Determining Factors of Polynomials

When we divide one polynomial by another, we obtain a quotient and a remainder. If the remainder is 0, then the divisor is a **factor** of the dividend.

EXAMPLE 4 Divide to determine whether $x^2 + 9$ is a factor of $x^4 - 81$.

Solution

$$\begin{array}{r|l} & x^2 - 9 \\ \hline x^2 + 9 & x^4 \qquad\qquad -81 \\ & x^4 + 9x^2 \\ \hline & -9x^2 - 81 \\ & -9x^2 - 81 \\ \hline & 0 \end{array}$$

Note that spaces have been left for missing terms.

Since the reminder is 0, we know that $x^2 + 9$ is a factor of $x^4 - 81$.

EXAMPLE 5 Divide to determine whether $x^2 + 3x - 1$ is a factor of $x^4 - 81$.

Solution

$$\begin{array}{r|l} & x^2 - 3x + 10 \\ \hline x^2 + 3x - 1 & x^4 \qquad\qquad\qquad - 81 \\ & x^4 + 3x^3 - x^2 \\ \hline & -3x^3 + x^2 \\ & -3x^3 - 9x^2 + 3x \\ \hline & 10x^2 - 3x - 81 \\ & 10x^2 + 30x - 10 \\ \hline & -33x - 71 \end{array}$$

Since the remainder is not 0, we know that $x^2 + 3x - 1$ is not a factor of $x^4 - 81$.

DO EXERCISES 9 AND 10.

4 Dividends in Terms of Quotients, Divisors, and Remainders

In general, when we divide a polynomial $P(x)$ by a divisor $d(x)$, we obtain some polynomial $Q(x)$ for a quotient and some polynomial $R(x)$ for a remainder. The remainder must either be 0 or have degree less than that of $d(x)$. To check a division, we multiply the quotient by the divisor and add the remainder, to see if

6. Determine whether the following numbers are roots of the polynomial $x^2 - 4x - 21$.
   a) 7
   b) 3
   c) -7

7. Determine whether the following numbers are roots of the polynomial $x^4 - 16$.
   a) 2
   b) -2
   c) -1
   d) 0

8. Determine whether the following numbers are roots of the polynomial $x^2 + 1$.
   a) 1
   b) -1
   c) i
   d) $-i$

9. By division, determine whether the following polynomials are factors of the polynomial $x^4 - 16$.
   a) $x - 2$
   b) $x^2 + 3x - 1$

10. By division, determine whether the following polynomials are factors of the polynomial $x^3 + 2x^2 - 5x - 6$.
    a) $x + 1$
    b) $x - 3$
    c) $x^2 + 3x - 1$

11. Divide $x^3 + 2x^2 - 5x - 6$ by $x - 3$. Then express the dividend as

$$P(x) = d(x) \cdot Q(x) + R(x).$$

we get the dividend. Thus these polynomials are related as follows:

$$P(x) = d(x) \cdot Q(x) + R(x).$$

EXAMPLE 6 If $P(x) = x^4 - 81$ and $d(x) = x^2 + 9$, then $Q(x) = x^2 - 9$ and $R(x) = 0$, and

$$\underbrace{x^4 - 81}_{P(x)} = \underbrace{(x^2 + 9)}_{d(x)} \cdot \underbrace{(x^2 - 9)}_{Q(x)} + \underbrace{0}_{R(x)}.$$

EXAMPLE 7 If $P(x) = x^4 - 81$ and $d(x) = x^2 + 3x - 1$, then $Q(x) = x^2 - 3x + 10$ and $R(x) = -33x - 71$, and

$$\underbrace{x^4 - 81}_{P(x)} = \underbrace{(x^2 + 3x - 1)}_{d(x)} \cdot \underbrace{(x^2 - 3x + 10)}_{Q(x)} + \underbrace{(-33x - 71)}_{R(x)}.$$

DO EXERCISE 11.

EXERCISE SET 11.1

1 Determine the degree of the polynomial.

1. $x^4 - 3x^2 + 1$
2. $2x^5 - x^4 + \frac{1}{4}x - 7$
3. $-2x + 5$
4. $3x - \sqrt{\pi}$
5. $2x^2 - 3x + 4$
6. $\frac{1}{4}x^2 - 7$
7. 3
8. 0

2

9. Determine whether 2, 3, and -1 are roots, or zeros, of $P(x) = x^3 + 6x^2 - x - 30$.

10. Determine whether 2, 3, and -1 are roots, or zeros, of $P(x) = 2x^3 - 3x^2 + x - 1$.

3

11. For $P(x)$ in Exercise 9, which of the following are factors of $P(x)$?

    a) $x - 2$ b) $x - 3$ c) $x + 1$

12. For $P(x)$ in Exercise 10, which of the following are factors of $P(x)$?

    a) $x - 2$ b) $x - 3$ c) $x + 1$

4 In each of the following, a polynomial $P(x)$ and a divisor $d(x)$ are given. Find the quotient $Q(x)$ and the remainder $R(x)$ when $P(x)$ is divided by $d(x)$, and express $P(x)$ in the form $d(x) \cdot Q(x) + R(x)$.

13. $P(x) = x^3 + 6x^2 - x - 30$, $d(x) = x - 2$
14. $P(x) = 2x^3 - 3x^2 + x - 1$, $d(x) = x - 2$
15. $P(x) = x^3 + 6x^2 - x - 30$, $d(x) = x - 3$
16. $P(x) = 2x^3 - 3x^2 + x - 1$, $d(x) = x - 3$
17. $P(x) = x^3 - 8$, $d(x) = x + 2$
18. $P(x) = x^3 + 27$, $d(x) = x + 1$
19. $P(x) = x^4 + 9x^2 + 20$, $d(x) = x^2 + 4$
20. $P(x) = x^4 + x^2 + 2$, $d(x) = x^2 + x + 1$
21. $P(x) = 5x^7 - 3x^4 + 2x^2 - 3$, $d(x) = 2x^2 - x + 1$
22. $P(x) = 6x^5 + 4x^4 - 3x^2 + x - 2$, $d(x) = 3x^2 + 2x - 1$

SYNTHESIS

23. For $P(x) = x^5 - 64$:
    a) Find $P(2)$.
    b) Find the remainder when $P(x)$ is divided by $x - 2$, and compare your answer to (a).
    c) Find $P(-1)$.
    d) Find the remainder when $P(x)$ is divided by $x + 1$, and compare your answer to (c).

24. For $P(x) = x^3 + x^2$:
    a) Find $P(-1)$.
    b) Find the remainder when $P(x)$ is divided by $x + 1$, and compare your answer to (a).
    c) Find $P(2)$.
    d) Find the remainder when $P(x)$ is divided by $x - 2$, and compare your answer to (c).

25. If there are n teams in a sports league and each team plays each other once in a season, we can find the total number of games played by a polynomial function $f(n) = \frac{1}{2}(n^2 - n)$. Find the zeros of the function.

26. Medical dosage. The function
$$N(t) = -0.046t^3 + 2.08t + 2$$
gives the body concentration, in parts per million, of a certain dosage of medication after time t, in hours. Estimate the roots in the interval $[0, 8]$.

27. *Threshold weight.* In a medical study done by Alwin Shemesh in 1976, it was found that **threshold weight** W, defined as that weight above which the risk of mortality rises astronomically, is given as a function of height h by
$$W(x) = \left(\frac{h}{12.3}\right)^3,$$
where W is in pounds and h is in inches. Find the roots of the polynomial on the interval $[0, \infty)$.

28. *Beam deflection.* A beam rests at two points P and Q and has a concentrated load applied to the center of the beam, as shown in the figure. Let y denote the deflection of the beam at a distance of x units from point P to the left of the weight. The deflection depends on the elasticity of the board, the load, and other physical characteristics. Suppose under certain conditions that y is given by
$$y = \frac{1}{13}x^3 - \frac{1}{14}x.$$
Find the zeros of the polynomial in the interval $[0, 2]$.

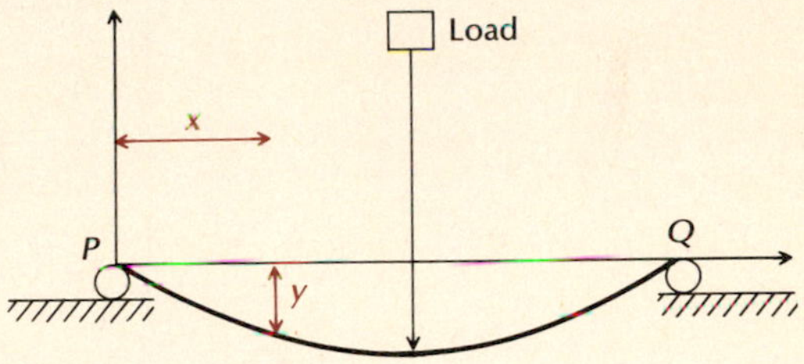

CHALLENGE

29. For $P(x) = 2x^2 - ix + 1$:
a) Find $P(-i)$.
b) Find the remainder when $P(x)$ is divided by $x + i$.

30. For $P(x) = 2x^2 + ix - i$:
a) Find $P(i)$.
b) Find the remainder when $P(x)$ is divided by $x - i$.

31. Find a rule for finding the degree of the product of two polynomials with real coefficients.

32. What can be said about the degree of the sum of two polynomials?

11.2 The Remainder and Factor Theorems

OBJECTIVES

You should be able to:

1. **Use synthetic division to find the quotient and the remainder when a polynomial is divided by $x - r$.**
2. **Use the remainder theorem to find a function value $P(r)$ when a polynomial $P(x)$ is divided by $x - r$.**
3. **Determine whether $x - r$ is a factor of $P(x)$ by determining whether $P(r) = 0$.**

Some of the exercises in the preceding exercise set illustrate the following theorem.

THEOREM 1 The Remainder Theorem

If a number r is substituted for x in the polynomial $P(x)$, then the result $P(r)$ is the remainder that would be obtained by dividing $P(x)$ by $x - r$. That is, if $P(x) = (x - r) \cdot Q(x) + R$, then $P(r) = R$.

Proof. The equation $P(x) = d(x) \cdot Q(x) + R(x)$, where $d(x) = x - r$, is the basis of this proof. If we divide $P(x)$ by $x - r$, we obtain a quotient $Q(x)$ and a remainder $R(x)$ related as follows:
$$P(x) = (x - r) \cdot Q(x) + R(x).$$
The remainder $R(x)$ must either be 0 or have degree less than $x - r$. Thus, $R(x)$ must be a constant. Let us call this constant R. In the expression above, we get a true sentence whenever we replace x by any number. Let us replace x by r. We

Use synthetic division to find the quotient and the remainder.

1. $(x^3 + 6x^2 - x - 30) \div (x - 2)$

2. $(x^3 - 2x^2 + 5x - 4) \div (x + 2)$

3. $(y^3 + 1) \div (y + 1)$

get

$$P(r) = (r - r) \cdot Q(r) + R$$
$$P(r) = 0 \cdot Q(r) + R$$
$$P(r) = R.$$

This tells us that the function value $P(r)$ is the remainder obtained when we divide $P(x)$ by $x - r$.

Theorem 1 motivates us to find a rapid way of dividing by $x - r$, in order to find function values.

1 Synthetic Division

To streamline division, we can arrange the work so that duplicate and unnecessary writing is avoided. Consider the following.

A.

$$
\begin{array}{r}
4x^2 + 5x + 11 \\
x - 2\overline{)4x^3 - 3x^2 + x + 7} \\
\underline{4x^3 - 8x^2} \\
5x^2 + x \\
\underline{5x^2 - 10x} \\
11x + 7 \\
\underline{11x - 22} \\
29
\end{array}
$$

B.

$$
\begin{array}{r}
4 \quad 5 \quad 11 \\
1 - 2\overline{)4 - 3 + 1 + 7} \\
\underline{4 - 8} \\
5 + 1 \\
\underline{5 - 10} \\
11 + 7 \\
\underline{11 - 22} \\
29
\end{array}
$$

The division in (B) is the same as that in (A), but we write only the coefficients. The color numerals are duplicated, so we look for an arrangement in which they are not duplicated. We can also simplify things by using the additive inverse of -2 and then adding instead of subtracting. When things are thus "collapsed," we have the algorithm known as **synthetic division**.

C. *Synthetic Division*

$$
\begin{array}{r|rrrr}
2 & 4 & -3 & 1 & 7 \\
 & & 8 & 10 & 22 \\
\hline
 & 4 & 5 & 11 & 29
\end{array}
$$

We "bring down" the 4. Then we multiply it by the 2 to get 8 and add to get 5. We then multiply 5 by 2 to get 10, add, and so on. The last number, 29, is the remainder. The others, 4, 5, and 11, are the coefficients of the quotient.

We write a 0 in the synthetic division for a missing term in the dividend.

EXAMPLE 1 Use synthetic division to find the quotient and remainder:

$$(2x^3 + 7x^2 - 5) \div (x + 3).$$

Solution First we note that $x + 3 = x - (-3)$.

$$
\begin{array}{r|rrrr}
-3 & 2 & 7 & 0 & -5 \\
 & & -6 & -3 & 9 \\
\hline
 & 2 & 1 & -3 & 4
\end{array}
$$

Note: We must write 0's for missing terms.

The quotient is $2x^2 + x - 3$. The remainder is 4. ■

DO EXERCISES 1–3.

2 Function Values for Polynomials

We can now apply synthetic division to find function values for polynomials.

EXAMPLE 2 Given that $P(x) = 2x^5 - 3x^4 + x^3 - 2x^2 + x - 8$, find $P(10)$.

Solution By Theorem 1, $P(10)$ is the remainder when $P(x)$ is divided by $x - 10$. We use synthetic division to find that remainder.*

10	2	−3	1	−2	1	−8
		20	170	1710	17,080	170,810
	2	17	171	1708	17,081	170,802

Thus, $P(10) = 170{,}802$.

DO EXERCISE 4.

EXAMPLE 3 Determine whether −4 is a zero, or root, of $P(x)$, where $P(x) = x^3 + 8x^2 + 8x - 32$.

Solution We use synthetic division and Theorem 1 to find $P(-4)$.

−4	1	8	8	−32
		−4	−16	32
	1	4	−8	0

Since $P(-4) = 0$, the number −4 is a root of $P(x)$.

DO EXERCISE 5.

3 Finding Factors of Polynomials

We now consider the following useful corollary of the remainder theorem.

THEOREM 2 The Factor Theorem

For a polynomial $P(x)$, if $P(r) = 0$, then $x - r$ is a factor of $P(x)$.

Proof. If we divide $P(x)$ by $x - r$, we obtain a quotient and a remainder, related as follows:

$$P(x) = (x - r) \cdot Q(x) + P(r).$$

Then if $P(r) = 0$, we have

$$P(x) = (x - r) \cdot Q(x),$$

so $x - r$ is a factor of $P(x)$.

This theorem is very useful in factoring polynomials, and hence in the solving of equations.

*Compare this with the work involved in a direct calculation! A calculator is most useful when finding polynomial function values by synthetic division. In Example 2, begin by entering 2. Then multiply by 10, add the result to −3, multiply that result by 10, and so on. The number 10 can be stored and recalled at each step where needed if the calculator has a memory. Since only the last result is needed, no intermediate values need be recorded.

4. Let $P(x) = x^5 - 2x^4 - 7x^3 + x^2 + 20$. Use synthetic division to find each of the following.
   a) $P(10)$
   b) $P(-8)$

5. Let $P(x) = x^3 + 6x^2 - x - 30$. Using synthetic division, determine whether the given numbers are roots of $P(x)$.
   a) 2
   b) 5
   c) −3

6. Determine whether $x - \frac{1}{2}$ is a factor of $4x^4 + 2x^3 + 8x - 1$.

7. Determine whether $x + 5$ is a factor of $x^3 + 625$.

8. a) Let $P(x) = x^3 + 6x^2 - x - 30$. Determine whether $x - 2$ is a factor of $P(x)$.
   b) Find another factor of $P(x)$.
   c) Find a complete factorization of $P(x)$.
   d) Solve $P(x) = 0$.

EXAMPLE 4 Let $P(x) = x^3 + 2x^2 - 5x - 6$. Factor $P(x)$ and solve the equation $P(x) = 0$.

Solution We look for linear factors of the form $x - r$. Let us try $x - 1$. We use synthetic division to see whether $P(1) = 0$.

$$\begin{array}{r|rrrr} 1 & 1 & 2 & -5 & -6 \\ & & 1 & 3 & -2 \\ \hline & 1 & 3 & -2 & -8 \end{array}$$

Since $P(1) \neq 0$, we know that $x - 1$ is not a factor of $P(x)$. We try $x + 1$ or $x - (-1)$ in the form $x - r$.

$$\begin{array}{r|rrrr} -1 & 1 & 2 & -5 & -6 \\ & & -1 & -1 & 6 \\ \hline & 1 & 1 & -6 & 0 \end{array}$$

Since $P(-1) = 0$, we know that $x + 1$ is one factor and the quotient, $x^2 + x - 6$, is another. Thus,

$$P(x) = (x + 1)(x^2 + x - 6).$$

The trinomial is easily factored, so we have

$$P(x) = (x + 1)(x + 3)(x - 2).$$

Our goal is to solve the equation $P(x) = 0$. To do so, we use the principle of zero products. The solutions are -1, -3, and 2. Thus the solution set is $\{-1, -3, 2\}$. ■

DO EXERCISES 6–8.

EXERCISE SET 11.2

1 Use synthetic division to find the quotient and the remainder.

1. $(2x^4 + 7x^3 + x - 12) \div (x + 3)$
2. $(x^3 - 7x^2 + 13x + 3) \div (x - 2)$
3. $(x^3 - 2x^2 - 8) \div (x + 2)$
4. $(x^3 - 3x + 10) \div (x - 2)$
5. $(x^4 - 1) \div (x - 1)$
6. $(x^5 + 32) \div (x + 2)$
7. $(2x^4 + 3x^2 - 1) \div \left(x - \frac{1}{2}\right)$
8. $(3x^4 - 2x^2 + 2) \div \left(x - \frac{1}{4}\right)$
9. $(x^4 - y^4) \div (x - y)$
10. $(x^3 + 3ix^2 - 4ix - 2) \div (x + i)$

2 Use synthetic division to find the function values.

11. $P(x) = x^3 - 6x^2 + 11x - 6$; find $P(1)$, $P(-2)$, and $P(3)$.
12. $P(x) = x^3 + 7x^2 - 12x - 3$; find $P(-3)$, $P(-2)$, and $P(1)$.
13. $P(x) = 2x^5 - 3x^4 + 2x^3 - x + 8$; find $P(20)$ and $P(-3)$.
14. $P(x) = x^5 - 10x^4 + 20x^3 - 5x - 100$; find $P(-10)$ and $P(5)$.
15. $P(x) = x^4 - 16$; find $P(2)$, $P(-2)$, and $P(3)$.
16. $P(x) = x^5 + 32$; find $P(2)$, $P(-2)$, and $P(3)$.

3 Using synthetic division, determine whether the numbers are roots of the polynomials.

17. $-3, 2$; $P(x) = 3x^3 + 5x^2 - 6x + 18$
18. $-4, 2$; $P(x) = 3x^3 + 11x^2 - 2x + 8$
19. $-3, \frac{1}{2}$; $P(x) = x^3 - \frac{7}{2}x^2 + x - \frac{3}{2}$
20. $i, -i, -2$; $P(x) = x^3 + 2x^2 + x + 2$

Factor the polynomial $P(x)$. Then solve the equation $P(x) = 0$.

21. $P(x) = x^3 + 4x^2 + x - 6$
22. $P(x) = x^3 + 5x^2 - 2x - 24$
23. $P(x) = x^3 - 6x^2 + 3x + 10$
24. $P(x) = x^3 + 2x^2 - 13x + 10$

25. $P(x) = x^3 - x^2 - 14x + 24$

26. $P(x) = x^3 - 3x^2 - 10x + 24$

27. $P(x) = x^4 - x^3 - 19x^2 + 49x - 30$

28. $P(x) = x^4 + 11x^3 + 41x^2 + 61x + 30$

SYNTHESIS

Solve.

29. $\dfrac{6x^2}{x^2 + 11} + \dfrac{60}{x^3 - 7x^2 + 11x - 77} = \dfrac{1}{x - 7}$

30. $\dfrac{2x^2}{x^2 - 1} + \dfrac{4}{x + 3} = \dfrac{32}{x^3 + 3x^2 - x - 3}$

31. $x^3 + 2x^2 - 13x + 10 > 0$

32. $x^4 - x^3 - 19x^2 + 49x - 30 < 0$

33. Find k so that $x + 2$ is a factor of
$$x^3 - kx^2 + 3x + 7k.$$

34. ▦ Given that
$$f(x) = 2.13x^5 - 42.1x^3 + 17.5x^2 + 0.953x - 1.98,$$
find $f(3.21)$ **(a)** by synthetic division; **(b)** by substitution.

35. For what values of k will the remainder be the same when $x^2 + kx + 4$ is divided by $x - 1$ or $x + 1$?

36. When $x^2 - 3x + 2k$ is divided by $x + 2$, the remainder is 7. Find the value of k.

CHALLENGE

37. Devise a way to use the method of synthetic division when the divisor is a polynomial such as $bx - r$.

38. Prove that $x - a$ is a factor of $x^n - a^n$, for any natural number n.

11.3 Theorems About Roots

OBJECTIVES

You should be able to:

1 **Factor polynomials and find their roots and their multiplicities.**

2 **Find a polynomial with specified roots.**

3 **In certain cases, given some of the roots of a polynomial, find such a polynomial and find the rest of its roots.**

The Fundamental Theorem of Algebra

A linear, or first-degree, polynomial $ax + b$ (where $a \neq 0$, of course) has just one root, $-b/a$. It can be shown that any quadratic polynomial with complex numbers for coefficients has at least one, and at most two, roots. The following theorem is a generalization. A proof is beyond this text.

> THEOREM 3 The Fundamental Theorem of Algebra
>
> **Every polynomial of degree greater than 0, with complex coefficients, has at least one root in the system of complex numbers.***

This is a very powerful theorem. Note that although it guarantees that a root exists, it does not tell how to find it. We now develop some theory that can help in finding roots. First, we prove a corollary of the fundamental theorem of algebra.

> THEOREM 4
>
> **Every polynomial of degree n, where $n > 0$, having complex coefficients, can be factored into n linear factors (not necessarily unique).**

Proof. Let us consider any polynomial of degree n, say $P(x)$. By the fundamental theorem, it has a root r_1. By the factor theorem, $x - r_1$ is a factor of

*It is wise to keep in mind that when we speak of "roots," or "zeros," of a polynomial $P(x)$, we are also speaking about "roots" or "solutions" of the polynomial equation $P(x) = 0$.

$P(x)$. Thus we know that

$$P(x) = (x - r_1) \cdot Q_1(x),$$

where $Q_1(x)$ is the quotient that would be obtained by dividing $P(x)$ by $x - r_1$. Let the leading coefficient of $P(x)$ be a_n. By considering the actual division process, we see that the leading coefficient of $Q_1(x)$ is also a_n and that the degree of $Q_1(x)$ is $n - 1$. Now if the degree of $Q_1(x)$ is greater than 0, then it has a root r_2, and we have

$$Q_1(x) = (x - r_2) \cdot Q_2(x),$$

where the degree of $Q_2(x)$ is $n - 2$ and the leading coefficient is a_n. Thus we have

$$P(x) = (x - r_1)(x - r_2) \cdot Q_2(x).$$

This process can be continued until a quotient $Q_n(x)$ having degree 0 is obtained. The leading coefficient will be a_n, so $Q_n(x)$ is actually the constant a_n. We now have the following:

$$P(x) = a_n(x - r_1)(x - r_2)(x - r_3) \cdots (x - r_n).$$

This completes the proof. We have actually shown a little more than what the theorem states. We see that $P(x)$ has been factored with one constant factor a_n and n linear factors having leading coefficient 1.

1 Finding Roots of Factored Polynomials

When a polynomial is factored into a product of linear factors, it is easy to find the roots by considering the principle of zero products.

EXAMPLE 1 Find the roots of

$$P(x) = (x - 3)(x + 4)(x + 1)(x - 1).$$

Solution To solve the equation $P(x) = 0$, we use the principle of zero products. The roots are 3, -4, -1, and 1. ■

EXAMPLE 2 Find the roots of

$$P(x) = 5(x - 2)(x - 2)(x - 2)(x + 1).$$

Solution To solve the equation $P(x) = 0$, we use the principle of zero products. The roots are 2 and -1. ■

In Example 2, the factor $x - 2$ occurs three times. In a case like this, we sometimes say that the root we obtain, 2, has a **multiplicity** of 3.

In Example 2, if we multiply out the right side, we obtain

$$P(x) = 5x^4 - 25x^3 + 30x^2 + 20x - 40.$$

Had we started with this expression, we might have had trouble finding the roots. We will be learning ways to do such factoring. Some polynomials can be factored using techniques we already know, such as factoring by grouping.

EXAMPLE 3 Find the roots of

$$P(x) = x^3 - 2x^2 - 9x + 18.$$

Solution We factor by grouping, as follows:

$$\begin{aligned} P(x) &= x^3 - 2x^2 - 9x + 18 \\ &= x^2(x-2) - 9(x-2) \\ &= (x^2 - 9)(x-2) \\ &= (x+3)(x-3)(x-2). \end{aligned}$$

Then by the principle of zero products, the equation $P(x) = 0$ has -3, 3, and 2 as roots. ■

Other factoring techniques can be used, as shown in Example 4.

EXAMPLE 4 Find the roots of

$$P(x) = x^4 + 4x^2 - 45.$$

Solution We factor as follows:

$$\begin{aligned} P(x) &= x^4 + 4x^2 - 45 \\ &= (x^2 - 5)(x^2 + 9) \\ &= (x - \sqrt{5})(x + \sqrt{5})(x - 3i)(x + 3i). \end{aligned}$$

Then by the principle of zero products, the equation has $\pm\sqrt{5}$ and $\pm 3i$ as roots. ■

From Theorem 4 and the examples above, we have the following.

> **THEOREM 5**
>
> **Every polynomial of degree n, where $n > 0$, has at least one root and at most n roots.***

DO EXERCISES 1–7.

Find the roots of the polynomial and state the multiplicity of each.

1. $P(x) = (x-5)(x-5)(x+6)$
2. $P(x) = 4(x+7)^2(x-3)$
3. $P(x) = (x+2)^3(x^2-9)$
4. $P(x) = (x^2 - 7x + 12)^2$
5. $P(x) = 5x^2 - 5$
6. $P(x) = x^4 + x^2 - 12$
7. $P(x) = 2x^3 + x^2 - 8x - 4$

2 Finding Polynomials with Given Roots

Given several numbers, we can find a polynomial having the given numbers as its roots.

EXAMPLE 5 Find a polynomial of degree 3, having the roots -2, 1, and $3i$.

Solution By Theorem 2, such a polynomial has factors $x + 2$, $x - 1$, and $x - 3i$, so we have

$$P(x) = a_n(x+2)(x-1)(x-3i).$$

The number a_n can be any nonzero number. The simplest polynomial will be obtained if we let it be 1. If we then multiply the factors, we obtain

$$P(x) = x^3 + (1 - 3i)x^2 + (-2 - 3i)x + 6i. \quad ■$$

*Theorem 5 is often stated as follows: "Every polynomial of degree n, where $n > 0$, has *exactly* n roots." This statement is not incompatible with Theorem 5, as it first seems, because to make sense of the statement just quoted, one must take multiplicities into account. Theorem 5, as stated here, is simpler and more straightforward.

EXAMPLE 6 Find a polynomial of degree 5 with -1 as a root of multiplicity 3, 4 as a root of multiplicity 1, and 0 as a root of multiplicity 1.

Solution Proceeding as in Example 5, letting $a_n = 1$, we obtain

$$(x + 1)^3(x - 4)(x - 0), \quad \text{or} \quad x^5 - x^4 - 9x^3 - 11x^2 - 4x.$$

DO EXERCISES 8–11.

8. Find a polynomial of degree 3 that has -1, 2, and 5 as roots.

9. Find a polynomial of degree 3 that has -1, 2, and $-5i$ as roots.

10. Find a polynomial of degree 5 with -2 as a root of multiplicity 3 and 0 as a root of multiplicity 2.

11. Find a polynomial of degree 4 with 1 as a root of multiplicity 3 and -5 as a root of multiplicity 1.

3 Roots of Polynomials with Real Coefficients

Consider the quadratic equation $x^2 - 2x + 2 = 0$, with real coefficients. Its roots are $1 + i$ and $1 - i$. Note that they are complex conjugates. This generalizes to any polynomial with real coefficients.

> THEOREM 6
>
> **If a complex number z is a root of a polynomial $P(x)$ with real coefficients, then its conjugate $\bar{z}$ is also a root. (Nonreal roots occur in conjugate pairs.)**

Proof. Let

$$P(x) = a_nx^n + a_{n-1}x^{n-1} + \cdots + a_1x + a_0,$$

where the coeffients are real numbers. Suppose z is a complex root of $P(x)$. Then $P(z) = 0$, or

$$a_nz^n + a_{n-1}z^{n-1} + \cdots + a_1z + a_0 = 0.$$

Now let us find the conjugate of each side of the equation. First note that $\bar{0} = 0$, since 0 is a real number. Then we have the following:

$$0 = \bar{0} = \overline{a_nz^n + a_{n-1}z^{n-1} + \cdots + a_1z + a_0}$$

$$= \overline{a_nz^n} + \overline{a_{n-1}z^{n-1}} + \cdots + \overline{a_1z} + \overline{a_0}$$

See Exercise 87 in Exercise Set 2.4.

$$= \overline{a_n} \cdot \overline{z^n} + \overline{a_{n-1}} \cdot \overline{z^{n-1}} + \cdots + \overline{a_1} \cdot \bar{z} + \overline{a_0}$$

See Exercise 88 in Exercise Set 2.4.

$$= a_n\overline{z^n} + a_{n-1}\overline{z^{n-1}} + \cdots + a_1\bar{z} + a_0$$

See Exercise 90 in Exercise Set 2.4.

$$= a_n\bar{z}^n + a_{n-1}\bar{z}^{n-1} + \cdots + a_1\bar{z} + a_0.$$

See Exercise 89 in Exercise Set 2.4.

Thus, $P(\bar{z}) = 0$, so $\bar{z}$ is a root of the polynomial.

For Theorem 6, it is essential that the coefficients be real numbers. We see this in Example 5, where the root $3i$ occurs, but its conjugate does not. This can happen because some of the coefficients of the polynomial are not real.

Rational Coefficients

When a polynomial has rational numbers for coefficients, certain irrational roots also occur in pairs, as described in the following theorem.

> THEOREM 7
>
> **Suppose $P(x)$ is a polynomial with rational coefficients. Then if either of the following is a root, so is the other: $a + c\sqrt{b}$, $a - c\sqrt{b}$, a and c rational, b not a square.**

This theorem can be proved in a manner analogous to Theorem 6, but we will not do so here. The theorem can be used to help in finding roots.

EXAMPLE 7 Suppose a polynomial of degree 6 with rational coefficients has $-2 + 5i$, $-2i$, and $1 - \sqrt{3}$ as some of its roots. Find the other roots.

Solution The other roots are $-2 - 5i$, $2i$, and $1 + \sqrt{3}$. There are no other roots since the degree is 6. ■

EXAMPLE 8 Find a polynomial of lowest degree with rational coefficients that has $1 - \sqrt{2}$ and $1 + 2i$ as some of its roots.

Solution The polynomial must also have the roots $1 + \sqrt{2}$ and $1 - 2i$. Thus the polynomial is

$$[x - (1 - \sqrt{2})][x - (1 + \sqrt{2})][x - (1 + 2i)][x - (1 - 2i)],$$

or

$$(x^2 - 2x - 1)(x^2 - 2x + 5),$$

or

$$x^4 - 4x^3 + 8x^2 - 8x - 5.$$ ■

DO EXERCISES 12–14.

EXAMPLE 9 Let $P(x) = x^4 - 5x^3 + 10x^2 - 20x + 24$. Find the other roots of $P(x)$, given that $2i$ is a root.

Solution Since $2i$ is a root, we know that $-2i$ is also a root. Thus,

$$P(x) = (x - 2i)(x + 2i) \cdot Q(x)$$

for some $Q(x)$. Since $(x - 2i)(x + 2i) = x^2 + 4$, we know that

$$P(x) = (x^2 + 4) \cdot Q(x).$$

We find, using division, that $Q(x) = x^2 - 5x + 6$, and since we can factor $x^2 - 5x + 6$, we get

$$P(x) = (x^2 + 4)(x - 2)(x - 3).$$

Thus the other roots are $-2i$, 2, and 3. ■

DO EXERCISE 15.

12. Suppose a polynomial of degree 5 with rational coefficients has -4, $7 - 2i$, and $3 + \sqrt{5}$ as roots. Find the other roots.

13. Find a polynomial of lowest degree with rational coefficients that has $2 + \sqrt{3}$ and $1 - i$ as some of its roots.

14. Find a polynomial of lowest degree with real coefficients that has $2i$ and 2 as some of its roots.

15. Find the other roots of

$$x^4 + x^3 - x^2 + x - 2,$$

given that i is a root.

EXERCISE SET 11.3

1 Find the roots of the polynomial or polynomial equation, and state the multiplicity of each.

1. $P(x) = (x + 3)^2(x - 1)$
2. $-8(x - 3)^2(x + 4)^3x^4 = 0$
3. $x^3(x - 1)^2(x + 4) = 0$
4. $P(x) = (x^2 - 5x + 6)^2$
5. $P(x) = x^4 - 4x^2 + 3$
6. $P(x) = x^4 - 10x^2 + 9$
7. $P(x) = x^3 + 3x^2 - x - 3$
8. $P(x) = x^3 - x^2 - 2x + 2$

2 Find a polynomial of degree 3 with the given numbers as roots.

9. $-2, 3, 5$
10. $2, i, -i$
11. $-3, 2i, -2i$
12. $1 + 4i, 1 - 4i, -1$
13. $\sqrt{2}, -\sqrt{2}, \sqrt{3}$

14. Find a polynomial equation of degree 4 with -2 as a root of multiplicity 1, 3 as a root of multiplicity 2, and -1 as a root of multiplicity 1.

3 Suppose a polynomial or polynomial equation of degree 5 with rational coefficients has the given numbers as roots. Find the other roots.

15. $6, -3 + 4i, 4 - \sqrt{5}$
16. $-2, 3, 4, 1 - i$

Find a polynomial of lowest degree with rational coefficients that has the given numbers as some of its roots.

17. $1 + i, 2$
18. $2 - i, -1$
19. $-4i, 5$
20. $2 - \sqrt{3}, 1 + i$
21. $\sqrt{5}, -3i$
22. $-\sqrt{2}, 4i$

Given that the polynomial or polynomial equation has the given root, find the other roots.

23. $x^4 - 5x^3 + 7x^2 - 5x + 6; -i$
24. $x^4 - 16 = 0; 2i$
25. $x^3 - 6x^2 + 13x - 20 = 0; 4$
26. $x^3 - 8; 2$

SYNTHESIS

Solve the equation. Use synthetic division, the quadratic formula, the theorems of this section, or whatever else you think might help.

27. $x^3 - 4x^2 + x - 4 = 0$
28. $x^3 - x^2 - 7x + 15 = 0$
29. $x^4 - 2x^3 - 2x - 1 = 0$
30. $x^4 + 9x^2 - 112 = 0$
31. The equation $x^2 + 2ax + b = 0$ has a double root. Find it.
32. Prove that a polynomial with positive coefficients cannot have a positive root.

CHALLENGE

33. Prove that every polynomial of odd degree, with real coefficients, has at least one real root.
34. Prove that every polynomial with real coefficients has a factorization into linear and quadratic factors (with real coefficients).
35. What does the fundamental theorem of algebra tell you about the following equation?

$$2\sin^5 x - 2\sin^3 x + \sin x = \frac{1}{8}$$

36. Prove that there is no polynomial P that defines the cosine function, $P(x) = \cos x$.

11.4 Rational Roots

OBJECTIVES

You should be able to:

1 **Given a polynomial with integer coefficients, find the rational roots and find the other roots, if possible.**

2 **Do the same for polynomials with rational coefficients.**

1 Integer Coefficients

It is not always easy to find the roots of a polynomial. However, if a polynomial has integer coefficients, there is a procedure that will yield all the rational roots.

THEOREM 8 The Rational Roots Theorem

Let

$$P(x) = a_nx^n + a_{n-1}x^{n-1} + \cdots + a_1x + a_0,$$

where all the coefficients are integers. Consider a rational number denoted by c/d, where c and d are relatively prime (having no common factor besides 1 and -1). If c/d is a root of $P(x)$, then c is a factor of a_0 and d is a factor of a_n.

Proof. Since c/d is a root of $P(x)$, we know that

$$a_n\left(\frac{c}{d}\right)^n + a_{n-1}\left(\frac{c}{d}\right)^{n-1} + \cdots + a_1\left(\frac{c}{d}\right)a_0 = 0. \qquad (1)$$

We multiply by d^n and get the equation

$$a_nc^n + a_{n-1}c^{n-1}d + \cdots + a_1cd^{n-1} + a_0d^n = 0. \qquad (2)$$

Then we have

$$a_nc^n = (-a_{n-1}c^{n-1} - \cdots - a_1cd^{n-2} - a_0d^{n-1})d.$$

Note that d is a factor of a_nc^n. Now d has no factor in common with c, other than 1 or -1, because c and d are relatively prime. Thus d has no factor in common with c^n. So d is a factor of a_n.

In a similar way, we can show from Eq. (2) that

$$a_0d^n = (-a_nc^{n-1} - a_{n-1}c^{n-2}d - \cdots - a_1d^{n-1})c.$$

Thus, c is a factor of a_0d^n. Again, c is not a factor of d^n, so it must be a factor of a_0.

EXAMPLE 1 Let $P(x) = 3x^4 - 11x^3 + 10x - 4$. Find the rational roots of $P(x)$. If possible, find the other roots.

Solution By the rational roots theorem, if c/d is a root of $P(x)$, then c must be a factor of -4 and d must be a factor of 3. Thus the possibilities for c and d are

$$c\colon\ 1, -1, 4, -4, 2, -2; \qquad d\colon\ 1, -1, 3, -3.$$

Then the resulting possibilities for c/d are

$$\frac{c}{d}\colon\ 1, -1, 4, -4, \frac{1}{3}, -\frac{1}{3}, \frac{4}{3}, -\frac{4}{3}, \frac{2}{3}, -\frac{2}{3}, 2, -2.$$

Of these 12 possibilities, we know that at most 4 of them could be roots because $P(x)$ is of degree 4. To find which are roots, we could use substitution, but synthetic division is usually more efficient. It is easiest to first consider the integers. Then we consider the fractions, if the integers do not produce all the roots.

We try 1:

$$\begin{array}{r|rrrr|r}
1 & 3 & -11 & 0 & 10 & -4 \\
 & & 3 & -8 & -8 & 2 \\
\hline
 & 3 & -8 & -8 & 2 & -2.
\end{array}$$

We try -1:

$$\begin{array}{r|rrrr|r}
-1 & 3 & -11 & 0 & 10 & -4 \\
 & & -3 & 14 & -14 & 4 \\
\hline
 & 3 & -14 & 14 & -4 & 0.
\end{array}$$

Since $P(1) = -2$, 1 is not a root; but $P(-1) = 0$, so -1 is a root. Using the results of the second synthetic division above, we can express $P(x)$ as follows:

$$P(x) = (x + 1)(3x^3 - 14x^2 + 14x - 4).$$

We now use $3x^3 - 14x^2 + 14x - 4$ and check the other possible roots. We

1. Let

   $P(x) = 2x^4 - 7x^3 - 35x^2 + 13x + 3.$

   If c/d is a rational root of $P(x)$, then:

   a) What are the possibilities for c?

   b) What are the possibilities for d?

   c) What are the possibilities for c/d?

   d) Find the rational roots.

   e) If possible, find the other roots.

use synthetic division again, to see whether -1 is a double root:

$$\begin{array}{r|rrrr} -1 & 3 & -14 & 14 & -4 \\ & & -3 & 17 & -31 \\ \hline & 3 & -17 & 31 & -35. \end{array}$$

It is not. There are no other roots that are integers, so we start checking the fractions. We try $\frac{2}{3}$:

$$\begin{array}{r|rrrr} \frac{2}{3} & 3 & -14 & 14 & -4 \\ & & 2 & -8 & 4 \\ \hline & 3 & -12 & 6 & 0. \end{array}$$

Since $P\left(\frac{2}{3}\right) = 0, \frac{2}{3}$ is a root. Again, using the results of the synthetic division, we can express $P(x)$ as

$$P(x) = (x + 1)\left(x - \tfrac{2}{3}\right)(3x^2 - 12x + 6).$$

Since the factor $3x^2 - 12x + 6$ is quadratic, we can use the quadratic formula to find that the other roots are $2 + \sqrt{2}$ and $2 - \sqrt{2}$. These are irrational numbers. Thus the rational roots are -1 and $\frac{2}{3}$. ■

DO EXERCISE 1.

EXAMPLE 2 Let $P(x) = x^3 + 6x^2 + x + 6$. Find the rational roots of $P(x)$. If possible, find the other roots.

Solution By the rational roots theorem, if c/d is a root of $P(x)$, then c must be a factor of 6 and d must be a factor of 1. Thus the possibilities for c and d are

$$c\text{:}\quad 1, -1, 2, -2, 3, -3, 6, -6; \qquad d\text{:}\quad 1, -1.$$

Then the resulting possibilities for c/d are

$$\frac{c}{d}\text{:}\quad 1, -1, 2, -2, 3, -3, 6, -6.$$

Note that these are the same as the possibilities for c. If the leading coefficient is 1, we need only check the factors of the last coefficient as possibilities for rational roots.

There is another aid in eliminating possibilities for rational roots. Note that all coefficients of $P(x)$ are positive. Thus when any positive number is substituted in $P(x)$, we get a positive value, never 0. Therefore, no positive number can be a root. Thus the only possibilities for roots are

$$-1, -2, -3, -6.$$

We try -6:

$$\begin{array}{r|rrrr} -6 & 1 & 6 & 1 & 6 \\ & & -6 & 0 & -6 \\ \hline & 1 & 0 & 1 & 0. \end{array}$$

Thus, $P(-6) = 0$, so -6 is a root. We could divide again to determine whether -6 is a double root, but since we can now factor $P(x)$ as a product of a linear and a quadratic polynomial, it is preferable to proceed that way. We have

$$P(x) = (x + 6)(x^2 + 1).$$

Now $x^2 + 1$ has the complex roots i and $-i$. Thus the only rational root of $P(x)$ is -6. ■

THEOREM 9

If all the coefficients of a polynomial are positive real numbers, then there are no positive real roots.

DO EXERCISES 2 AND 3.

EXAMPLE 3 Find the rational roots of $x^4 + 2x^3 + 2x^2 - 4x - 8$.

Solution Since the leading coefficient is 1, the only possibilities for rational roots are the factors of the last coefficient, -8:

$$1, -1, 2, -2, 4, -4, 8, -8.$$

But, using substitution or synthetic division, we find that none of the possibilities is a root. We leave it to the student to verify this. Thus there are no rational roots. ■

The polynomial given in Example 3 has no rational roots. We can approximate the irrational roots of such a polynomial by graphing it and determining the x-intercepts (see Section 11.5).

DO EXERCISES 4 AND 5.

2 Rational Coefficients

Suppose some (or all) of the coefficients of a polynomial are rational, but not integers. After multiplying on both sides by the LCM of the denominators, we can then find the rational roots.

EXAMPLE 4 Let $P(x) = \frac{1}{12}x^3 - \frac{1}{12}x^2 - \frac{2}{3}x + 1$. Find the rational roots of $P(x)$.

Solution The LCM of the denominators is 12. When we multiply on both sides by 12, we get

$$12P(x) = x^3 - x^2 - 8x + 12.$$

Note that all coefficients on the right are integers. Since the equation $P(x) = 0$ is equivalent to the equation $12P(x) = 0$, any root of $12P(x)$ is a root of $P(x)$. We leave it to the student to verify that 2 and -3 are the rational roots—in fact, the only roots—of $P(x)$. ■

DO EXERCISE 6.

2. Let $P(x) = x^3 + 6x^2 + 10x + 3$.
   If c/d is a rational root of $P(x)$, then:
   a) What are the possibilities for c?
   b) What are the possibilities for d?
   c) What are the possibilities for c/d?
   d) How can you tell without substitution or synthetic division that there are no positive roots?
   e) Find the rational roots of $P(x)$.
   f) Find the other roots if they exist.

3. Let $P(x) = x^3 + 5x^2 + 4x + 20$.
   If c/d is a rational root of $P(x)$, then:
   a) What are the possibilities for c?
   b) What are the possibilities for d?
   c) What are the possibilities for c/d?
   d) How can you tell without substitution or synthetic division that there are no positive roots?
   e) Find the rational roots of $P(x)$.
   f) Find the other roots if they exist.

4. Let $P(x) = x^4 + x^2 + 2x + 6$.
   a) How do you know at the outset that this polynomial has no positive roots?
   b) Find the rational roots of $P(x)$.

5. a) Find the rational roots of $x^2 + 3x + 3$.
   b) Can you find the other roots of this polynomial? What are they? Why can you find them?

6. Let $P(x) = x^4 - \frac{1}{6}x^3 - \frac{4}{3}x^2 + \frac{1}{6}x + \frac{1}{3}$.
   a) Which, if any, of the coefficients are *not* integers?
   b) What is the LCM of the denominators?
   c) Multiply by the LCM.
   d) Find the rational roots of the resulting polynomial.
   e) Are they rational roots of $P(x)$? Why?

EXERCISE SET 11.4

1 Use Theorem 8 to list all *possible* rational roots.

1. $x^5 - 3x^2 + 1$
2. $x^7 + 37x^5 - 6x^2 + 12$
3. $15x^6 + 47x^2 + 2$
4. $10x^{25} + 3x^{17} - 35x + 6$

Find the rational roots, if they exist, of each polynomial or equation. If possible, find the other roots. Then write the equation or polynomial in factored form.

5. $x^3 + 3x^2 - 2x - 6$
6. $x^3 - x^2 - 3x + 3 = 0$
7. $x^3 - 3x + 2 = 0$
8. $x^3 - 3x + 4$
9. $x^3 - 5x^2 + 11x - 19$
10. $x^3 - 7x^2 - 2x + 23$
11. $5x^4 - 4x^3 + 19x^2 - 16x - 4 = 0$
12. $3x^4 - 4x^3 + x^2 + 6x - 2$
13. $x^4 - 3x^3 - 20x^2 - 24x - 8$
14. $x^4 + 5x^3 - 27x^2 + 31x - 10$
15. $x^3 - 4x^2 + 2x + 4 = 0$
16. $x^3 - 8x^2 + 17x - 4$
17. $x^3 + 8$
18. $x^3 - 8 = 0$

2

19. $\frac{1}{3}x^3 - \frac{1}{2}x^2 - \frac{1}{6}x + \frac{1}{6}$
20. $\frac{2}{3}x^3 - \frac{1}{2}x^2 + \frac{2}{3}x - \frac{1}{2}$

Find only the rational roots.

21. $x^4 + 32$
22. $x^6 + 8 = 0$
23. $x^3 - x^2 - 4x + 3 = 0$
24. $2x^3 + 3x^2 + 2x + 3$
25. $x^4 + 2x^3 + 2x^2 - 4x - 8 = 0$
26. $x^4 + 6x^3 + 17x^2 + 36x + 66 = 0$
27. $x^5 - 5x^4 + 5x^3 + 15x^2 - 36x + 20$
28. $x^5 - 3x^4 - 3x^3 + 9x^2 - 4x + 12$

SYNTHESIS

29. The volume of a cube is 64 cm^3. Find the length of a side. (*Hint*: Solve $x^3 - 64 = 0$.)
30. The volume of a cube is 125 cm^3. Find the length of a side.
31. An open box of volume 48 cm^3 can be made from a piece of tin 10 cm on a side by cutting a square from each corner and folding up the edges. What is the length of a side of the squares?
32. An open box of volume 500 cm^3 can be made from a piece of tin 20 cm on a side by cutting a square from each corner and folding up the edges. What is the length of a side of the squares?

CHALLENGE

33. *Total profit.* The total profit P from the production and sale of x units of a product is given by the polynomial
$$P(x) = x^3 - 31x^2 + 230x - 200.$$
  a) Find the break-even values of the function.
  b) For what nonnegative values of x does the company make a profit? (See Section 4.7.)
  c) For what nonnegative values of x does the company have a loss (negative profit)?
34. The volume of a rectangular crate is $3\frac{3}{8}$ m^3. The length is 3 m longer than the width, and the height is 1 m less than the width. Find the dimensions of the crate.
35. Show that $\sqrt{5}$ is irrational by considering the equation $x^2 - 5 = 0$.
36. Generalize the result of Exercise 35 to find which positive integers have rational square roots.
37. Solve: $x^3 + 7 = 0$.
38. Solve: $x^3 - 5 = 0$.

11.5 Further Helps in Finding Roots

OBJECTIVES

You should be able to:

1 **Use Descartes' rule of signs to find information about the number of real roots of a polynomial with real coefficients.**

2 **Find upper and lower bounds of the roots of a polynomial with real coefficients, using synthetic division.**

(continued)

1 Descartes' Rule of Signs

The development of a rule that helps determine the number of positive real roots of a polynomial is credited to Descartes. To use the rule, we must have the polynomial arranged in descending or ascending order, with no zero terms written in and the leading coefficient positive. Then we determine the number of *variations of sign*, that is, the number of times, in going through the polynomial, that successive coefficients are of different sign.

EXAMPLE 1 Determine the number of variations of sign in the polynomial $2x^6 - 3x^2 + x + 4$.

Solution

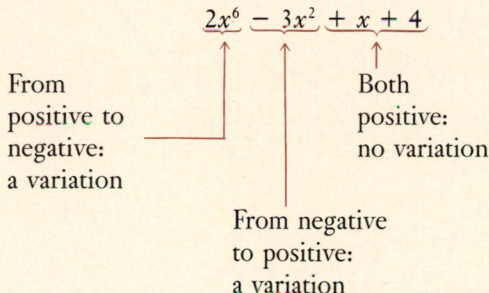

$$2x^6 - 3x^2 + x + 4$$

From positive to negative: a variation

From negative to positive: a variation

Both positive: no variation

The number of variations of sign is two.

DO EXERCISES 1 AND 2.

We now state Descartes' rule, without proof.

THEOREM 10 Descartes' Rule of Signs

The number of positive real roots of a polynomial with real coefficients is either:

1. **The same as the number of its variations of sign, or**
2. **Less than the number of its variations of sign by a positive even integer.**

A root of multiplicity m must be counted m times.

EXAMPLES In each case, what does Descartes' rule of signs tell you about the number of positive real roots?

2. $2x^5 - 5x^2 - 3x + 6$

   The number of variations of sign is two. Therefore, the number of positive real roots is either 2 or less than 2 by 2, 4, 6, etc. Thus the number of positive real roots is either 2 or 0, since a negative number of roots has no meaning.

3. $5x^4 - 3x^3 + 7x^2 - 12x + 4$

   There are four variations of sign. Thus the number of positive real roots is either

   $$4 \quad \text{or} \quad 4 - 2 \quad \text{or} \quad 4 - 4.$$

   That is, the number of positive real roots is 4, 2, or 0.

4. $6x^5 - 2x - 5 = 0$

   The number of variations of sign is one. Therefore, there is exactly one positive real root.

DO EXERCISES 3–5.

Negative Roots

Descartes' rule can also be used to help determine the number of negative real roots of a polynomial. Consider the following graphs of a polynomial equation

3 **Approximate zeros, or roots, of polynomial functions.**

Determine the number of variations of sign in the polynomial.

1. $3x^5 - 2x^3 - x^2 + x - 2$

2. $4p^7 + 6p^5 + 2p^3 - 5p^2 + 3$

In each case, what does Descartes' rule of signs tell you about the number of positive real roots?

3. $5x^3 - 4x - 5$

4. $6p^6 - 5p^4 + 3p^3 - 7p^2 + p - 2 = 0$

5. $3x^2 - 2x + 4$

$y = P(x)$ and its reflection across the y-axis, that is, $y = P(-x)$. The points at which the graphs cross the x-axis are the roots of the polynomials.

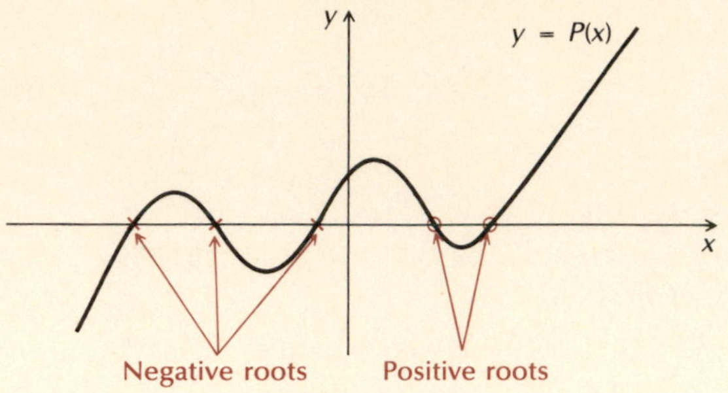

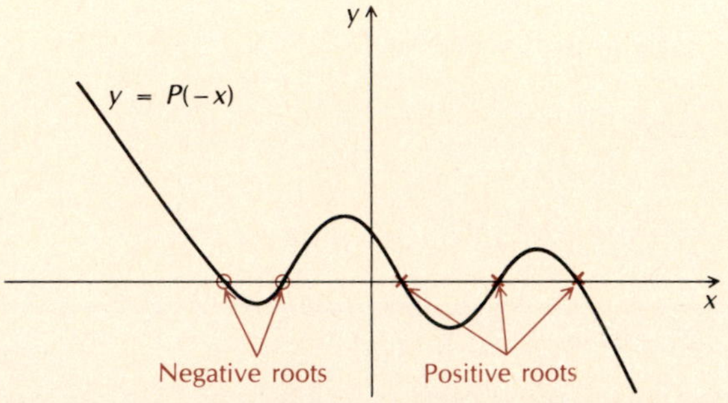

From the graphs, we see that the number of positive real roots of $P(-x)$ is the same as the number of negative real roots of $P(x)$.

THEOREM 11 Corollary to Descartes' Rule of Signs

The number of negative real roots of a polynomial $P(x)$ with real coefficients is either:

1. **The number of variations of sign of $P(-x)$,** or
2. **Less than the number of variations of sign of $P(-x)$ by a positive even integer.**

A root of multiplicity m must be counted m times.

To apply Theorem 11, we construct $P(-x)$ by replacing x by $-x$ whenever it occurs and then count the variations of sign.

EXAMPLES In each case, what does Descartes' rule of signs tell you about the number of negative real roots?

5. $5x^4 + 3x^3 + 7x^2 + 12x + 4$

$$P(-x) = 5(-x)^4 + 3(-x)^3 + 7(-x)^2 + 12(-x) + 4 \quad \text{Replacing } x \text{ by } -x$$
$$= 5x^4 - 3x^3 + 7x^2 - 12x + 4 \quad \text{Simplifying}$$

There are four variations of sign, so the number of negative real roots is either 4 or 2 or 0.*

*It is simple to construct $P(-x)$ mechanically. If $P(x)$ is in descending order with no missing terms, we change the sign of every second term beginning with the second term from the right. These terms have odd exponents.

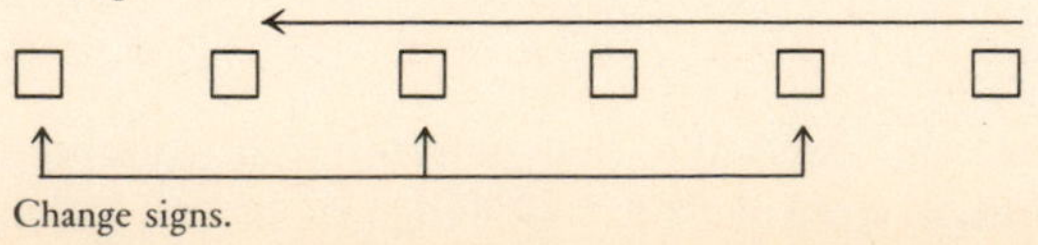

6. $2x^5 - 5x^2 + 3x + 6$

$$P(-x) = 2(-x)^5 - 5(-x)^2 + 3(-x) + 6 \quad \text{Replacing } x \text{ by } -x$$
$$= -2x^5 - 5x^2 - 3x + 6 \quad \text{Simplifying}$$

There is one variation of sign, so there is exactly one negative real root.

DO EXERCISES 6–8.

2 Positive-Integer Upper Bounds on Roots

Recall that when a polynomial $P(x)$ is divided by $x - a$, we obtain a quotient $Q(x)$ and a remainder R, related as follows:

$$P(x) = (x - a) \cdot Q(x) + R.$$

Let us consider the case in which a is positive, R is positive or zero, and all the coefficients of $Q(x)$ are nonnegative. Then $P(b)$ will be positive for any positive number b that is greater than a. We show this as follows.

$$P(b) = \underbrace{(b - a)}_{\text{positive}} \cdot \underbrace{Q(b)}_{\text{positive}} + \underbrace{R}_{\text{nonnegative}}$$

$b - a > 0$ because we assumed that $b > a$;

$Q(x) > 0$ because all coefficients are nonnegative and the number b that we are substituting is positive. (We know that $Q(x)$ is not the zero polynomial, so it has some nonzero coefficients.)

R is assumed to be nonnegative.

This shows that there can be no root of $P(x)$ greater than the positive number a. If there were such a root r, it would make $P(r) = 0$ and this cannot occur, because all values of $P(x)$ are positive when x is greater than a. In terms of synthetic division, this tells us the following.

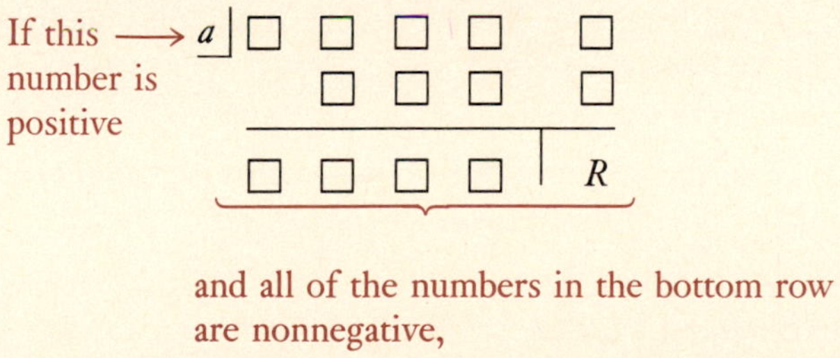
If this number is positive
a
R
and all of the numbers in the bottom row are nonnegative,

then there is no root greater than a. In other words, the number a is an **upper bound** to all the roots of $P(x)$. Of course, if the number R is 0, we know that a is actually a root, as well as an upper bound. We state this result as a theorem.

> THEOREM 12
>
> **If when a polynomial is divided by $x - a$, where a is positive, the remainder and all coefficients of the quotient are nonnegative, then the number a is an upper bound to the roots of the polynomial (all the roots are less than or equal to a).**

EXAMPLE 7 Find the smallest positive integer that is guaranteed by Theorem 12 to be an upper bound to the roots of

$$P(x) = 3x^4 - 11x^3 + 10x - 4.$$

In each case, what does Descartes' rule of signs tell you about the number of negative real roots?

6. $5x^3 - 4x - 5$

7. $6p^6 - 5p^4 + 3p^3 - 7p^2 + p - 2$

8. $3x^2 - 2x + 4$

In each case, determine the smallest positive integer that is guaranteed by Theorem 12 to be an upper bound to the roots.

9. $P(x) = 5x^4 - 18x^2 + 3x - 2$

10. $P(x) = x^3 - 75x^2 + 3$

11. Find the smallest positive integer that is guaranteed by Theorem 12 to be an upper bound to the roots of

$$P(x) = -4x^3 + 5x^2 - x + 2.$$

Solution We choose some *positive* number a and divide by $x - a$, using synthetic division. Let's try 1 for a.

$$\begin{array}{r|rrrr|r} 1 & 3 & -11 & 0 & 10 & -4 \\ & & 3 & -8 & -8 & 2 \\ \hline & 3 & -8 & -8 & 2 & -2 \end{array}$$

Since some of the coefficients are negative, the theorem does not guarantee that 1 is an upper bound. We choose a larger integer, this time 4.

$$\begin{array}{r|rrrr|r} 4 & 3 & -11 & 0 & 10 & -4 \\ & & 12 & 4 & 16 & 104 \\ \hline & 3 & 1 & 4 & 26 & 100 \end{array}$$

One reason for choosing 4 is that we know this number must be at least 11 in order for the addition to give a nonnegative result.

Since there are no negative numbers in the bottom row, 4 is an upper bound. The student can verify that neither 2 nor 3 satisfies the requirements of Theorem 12. Thus, 4 is the smallest positive integer that is guaranteed by Theorem 12 to be an upper bound to the roots of $P(x)$.*

CAUTION! In applying Theorem 12, remember that the number a must be a *positive* number.

DO EXERCISES 9 AND 10.

Note that when the leading coefficient of $P(x)$ is negative, it follows that the leading coefficient of the quotient (the first number in the bottom row in synthetic division) is also negative, regardless of our choice of a. When this situation arises, we make use of the fact that $P(x) = 0$ and $-P(x) = 0$ are equivalent equations and apply Theorem 12 to the polynomial $-P(x)$.

In determining bounds for roots, if the leading coefficient of $P(x)$ is negative, we simply convert to $-P(x)$ by changing the sign of every term. Then we proceed as before.

EXAMPLE 8 Find the smallest positive integer that is guaranteed by Theorem 12 to be an upper bound to the roots of $P(x) = -5x^3 - 12x^2 - 2x + 3$.

Solution Since the leading coefficient of $P(x)$ is negative, we change the sign of every term. We try 1 for a:

$$\begin{array}{r|rrr|r} 1 & 5 & 12 & 2 & -3 \\ & & 5 & 17 & 19 \\ \hline & 5 & 17 & 19 & 16 \end{array}$$

Since there are no negative numbers in the bottom row, 1 is an upper bound for the roots of $P(x)$.

DO EXERCISE 11.

*The converse of Theorem 12 is not true. That is, if the positive number a is an upper bound to the roots, it is not necessarily true that when we divide by $x - a$, all the coefficients of the quotient and the remainder will be nonnegative. Thus when we divide and find negative numbers in the bottom row, this does not mean that the number a is *not* an upper bound. We simply do not know whether it is or not. For example, while the roots of $P(x) = x^2 - 5x + 6$ are 2 and 3, Theorem 12 would not predict that 3 or 4 is an upper bound to the roots.

Negative-Integer Lower Bounds on Roots

A corollary of the theorem above can be used to find lower bounds of roots of a polynomial. Consider the following graphs.

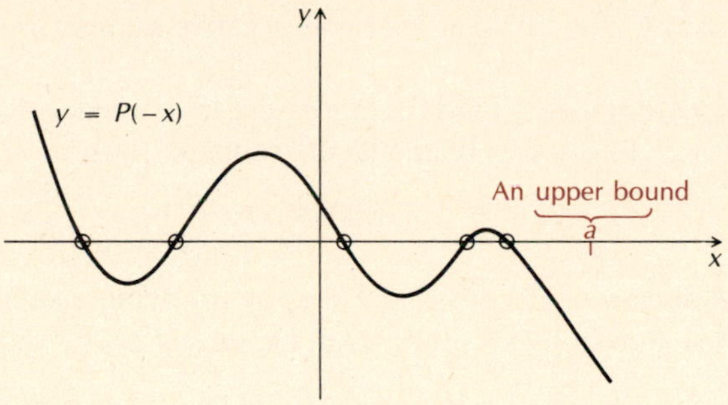

Here we have an upper bound, a, to the roots of the polynomial $P(-x)$. Now let us consider the reflection across the y-axis, $y = P(x)$.

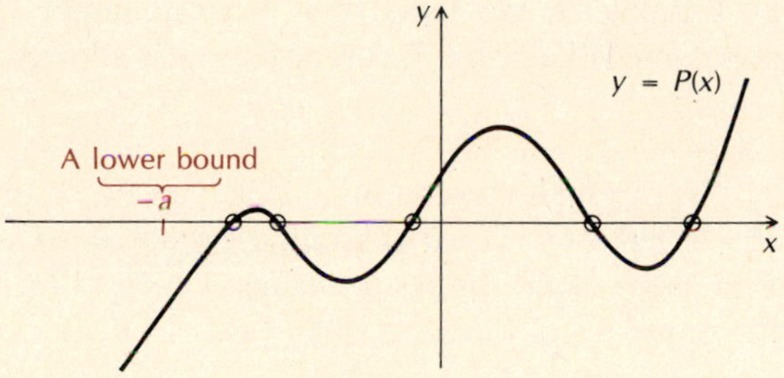

It is easy to see that the negative number $-a$ is a lower bound to the roots of $P(x)$.

> **THEOREM 13**
>
> **The number $-a$, where a is positive, is a lower bound to the roots of the polynomial $P(x)$, if a is an upper bound to the roots of $P(-x)$.**

EXAMPLE 9 Find the largest negative integer that is guaranteed by Theorem 13 to be a lower bound to the roots of

$$P(x) = 3x^4 + 11x^3 + 10x - 4.$$

Solution We first construct $P(-x)$ by replacing x with $-x$:

$$\begin{aligned} P(-x) &= 3(-x)^4 + 11(-x)^3 + 10(-x) - 4 \\ &= 3x^4 - 11x^3 - 10x - 4. \quad \text{Simplifying} \end{aligned}$$

Next, we use synthetic division, dividing by $x - 4$.*

$$\begin{array}{r|rrrrr} 4 & 3 & -11 & 0 & -10 & -4 \\ & & 12 & 4 & 16 & 24 \\ \hline & 3 & 1 & 4 & 6 & 20 \end{array}$$

*To use a shortcut, we can set up the synthetic division using $P(x)$.

$$\begin{array}{r|rrrrr} 4 & 3 & 11 & 0 & 10 & -4 \end{array}$$

Then we change the sign of every second number, beginning at the second from the right, and then divide.

$$\begin{array}{r|rrrrr} 4 & 3 & -11 & 0 & -10 & -4 \end{array}$$

↑ ↑
Change signs.

Determine the largest negative integer that is guaranteed by Theorem 13 to be a lower bound to the roots.

12. $5x^4 + 18x^3 + 3x - 2$

13. $4x^3 + 7x^2 + 3x + 5$

Since none of the numbers in the bottom row is negative, 4 is an upper bound to the roots of $P(-x)$. Thus, -4 is a lower bound to the roots of $3x^4 + 11x^3 + 10x - 4$, which is $P(x)$. If we use synthetic division, with any integer smaller than 4, we obtain at least one negative number in the bottom line. ■

EXAMPLE 10 Find the largest negative integer that is guaranteed by Theorem 13 to be a lower bound to the roots of

$$P(x) = 5x^3 - 12x^2 + 2x + 3.$$

Solution We construct $P(-x)$ by replacing x with $-x$ (or by simply changing the sign of every other term, beginning at the second from the right):

$$\begin{aligned} P(-x) &= 5(-x)^3 - 12(-x)^2 + 2(-x) + 3 \\ &= -5x^3 - 12x^2 - 2x + 3. \end{aligned}$$

In Example 8, we found that 1 is an upper bound for the roots of this polynomial. Thus, by Theorem 13, -1 is a lower bound for the roots of $P(x)$. ■

Be careful not to confuse $-P(x)$ with $P(-x)$ or $-P(-x)$. While the roots of $P(-x)$ and $-P(-x)$ are identical, the roots of $-P(x)$ are identical to the roots of $P(x)$ itself. As the figures preceding Theorem 13 indicate, the roots of $P(x)$ are the opposites of the roots of $P(-x)$.

DO EXERCISES 12 AND 13.

A Shortcut

On p. 682, you were cautioned to use only positive numbers when seeking upper bounds. Likewise, in looking for negative lower bounds, we have used only positive numbers, in accordance with Theorem 12. When we are looking for lower bounds, there is a way to use negative numbers in the synthetic division that may save a small amount of time. We illustrate, using the polynomial of Example 8. The following is what we have already done.

4	3	−11	0	−10	−4	← This is $P(-x)$.
		12	4	16	24	
	3	1	4	6	20	← This shows that −4 is a lower bound.

Now let us use -4 with $P(x)$ and compare.

−4	3	11	0	10	−4	← This is $P(x)$.
		−12	4	−16	24	
	3	−1	4	−6	20	

Note that the only difference is a change of sign in the indicated columns. This illustrates the following corollary to Theorem 13.

> THEOREM 14
>
> **When a polynomial is divided by $x - a$, where a is negative, a will be a lower bound to the roots if in the result of the synthetic division the first number is nonnegative, the second nonpositive, the third nonnegative, the fourth nonpositive, and so on.**

EXAMPLE 11 Find a largest negative integer that is a lower bound to the roots of $P(x)$, where

$$P(x) = 5x^3 - 12x^2 + 2x + 3.$$

Solution We try -1:

$$\begin{array}{r|rrr|r} -1 & 5 & -12 & 2 & 3 \\ & & -5 & 17 & -19 \\ \hline & 5 & -17 & 19 & -16 \end{array}$$

Since the odd-numbered coefficients (from left to right) are nonnegative and the even-numbered ones are nonpositive, -1 is a lower bound to the roots. ■

DO EXERCISE 14.

EXAMPLE 12 What does Descartes' rule of signs tell you about the roots of $P(x)$? Find upper and lower bounds to the real roots.

$$P(x) = 4x^4 + 3x^3 + x - 1$$

Solution

a) There is one variation of sign, so there is just one positive real root.

b) We have

$$\begin{aligned} P(-x) &= 4(-x)^4 + 3(-x)^3 + (-x) - 1 \\ &= 4x^4 - 3x^3 - x - 1. \end{aligned}$$

There is just one variation of sign, so there is just one negative real root.

c) Since the equation is of degree 4, it has four roots, considering multiplicities. Therefore, there are two nonreal roots.

d) We look for an upper bound.

$$\begin{array}{r|rrrr|r} 1 & 4 & 3 & 0 & 1 & -1 \\ & & 4 & 7 & 7 & 8 \\ \hline & 4 & 7 & 7 & 8 & 7 \end{array}$$

The number 1 is an upper bound. Let's try $\frac{1}{4}$.

$$\begin{array}{r|rrrr|r} \frac{1}{4} & 4 & 3 & 0 & 1 & -1 \\ & & 1 & 1 & \frac{1}{4} & \frac{5}{16} \\ \hline & 4 & 4 & 1 & \frac{5}{4} & -\frac{11}{16} \end{array}$$

We do not know whether or not $\frac{1}{4}$ is an upper bound, but we look no further.

e) We look for a lower bound, using Theorem 14.

$$\begin{array}{r|rrrr|r} -1 & 4 & 3 & 0 & 1 & -1 \\ & & -4 & 1 & -1 & 0 \\ \hline & 4 & -1 & 1 & 0 & -1 \end{array}$$

14. Find a negative integer that is a lower bound to the roots of

$$5x^4 + 18x^3 - 3x - 3.$$

In each case, what does Descartes' rule of signs tell you about the roots of the polynomial? Find upper and lower bounds to the real roots.

15. $x^4 - 6x^3 + 7x^2 + 6x - 2$

16. $x^4 + 2x^2 + x - 2$

We do not know whether or not -1 is a lower bound.

-2	4	3	0	1	-1
		-8	10	-20	38
	4	-5	10	-19	37

The number -2 is a lower bound and not a root (because $R \neq 0$).

To summarize, we have learned that $P(x)$ has two real roots, one between 0 and 1 and one between -2 and 0. It also has two nonreal roots.

From the information obtained in Example 12, we can actually say a little more. In part (d), we found that $P(1) = 7$ (a positive number) and $P(\frac{1}{4}) = -\frac{11}{16}$ (a negative number). Thus, $P(x)$ must be 0 for some number between $\frac{1}{4}$ and 1. One of the roots is therefore between $\frac{1}{4}$ and 1.

DO EXERCISES 15 AND 16.

3 The Method of Bisection: Approximating Solutions

Whenever we find the roots, or zeros, of a function $P(x)$, we have solved the equation $P(x) = 0$. How might a scientist proceed when the need arises to find roots of polynomials? One might use the rational roots theorem to check for rational roots, presuming the coefficients are integers. But, even if this is the case, only about 20% of the time will there be any rational roots. A more straightforward method of operating is to use a computer software package or a graphing calculator and estimate where the graph crosses the x-axis. Many such tools have "zoom-in" features that allow very good approximating, graphically, by considering the graph over smaller intervals. We could also look near this estimate for function values that differ in sign. Then we would know that there is a root somewhere between the input values.

In fact, most computer software packages also have provisions to go on and estimate the real roots, but let us operate as if that were not the case.

EXAMPLE 13 Approximate the irrational real roots of the polynomial $P(x) = 2x^3 - x + 2$ to the nearest hundredth.

Solution We use the rational roots theorem and find that there are no rational roots. If there were, we would factor the polynomial and consider the polynomial that results from factoring out the linear factors involving the rational roots. Next we use a computer software package and graph the polynomial. Below is a computer-generated graph of

$$P(x) = 2x^3 - x + 2.$$

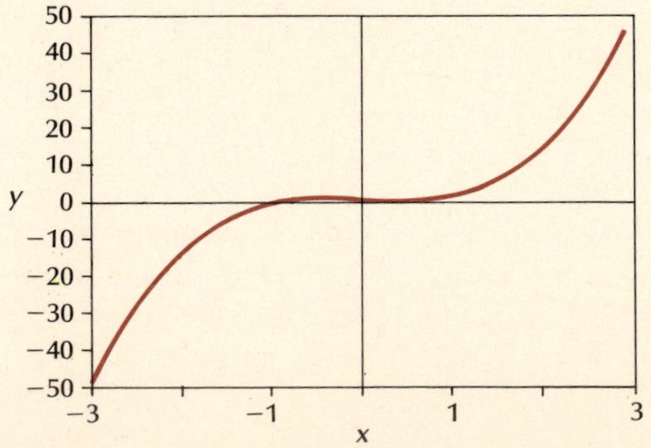

We see from the graph that there seems to be a root between -1 and -2. To be sure, we check function values and find that $P(-1) = 1$ and $P(-2) = -12$.

Since the signs differ, there must be a root somewhere between -1 and -2. To find a better approximation, we use computation by hand or a calculator or synthetic division.

x	$P(x)$	
-1	1	
-1.1	0.44	Since $P(-1.1) > 0$, there is no root between -1 and -1.1.
-1.2	-0.26	Since $P(-1.2) < 0$ and $P(-1.1) > 0$, there is a root somewhere between -1.1 and -1.2. We try -1.15, halfway between.
-1.15	0.11	Since $P(-1.2) < 0$ and $P(-1.15) > 0$, there is a root somewhere between -1.2 and -1.15. We try -1.17.
-1.17	-0.03	Since $P(-1.17)$ and $P(-1.15)$ have opposite signs, there is a root somewhere between -1.17 and -1.15. We try -1.16.
-1.16	0.04	Since $P(-1.17) < 0$ and $P(-1.16) > 0$, there is a root somewhere between -1.17 and -1.16.

We stop when we have the accuracy desired. We decide to approximate to the nearest hundredth. Since -0.03 is closer to zero than 0.04, we will accept -1.17 as a better approximation, to the nearest hundredth. There are no other irrational roots as evidenced by the graph of $P(x)$. ■

The method for approximating roots used in Example 13 is known as the **method of bisection.** There are many other methods of approximating used in mathematics. In fact, there is a branch of mathematics known as *numerical analysis*, which deals with approximation in great detail.

DO EXERCISE 17.

17. Approximate the irrational roots of $x^3 - 3x^2 + 1$ to the nearest hundredth.

EXERCISE SET 11.5

1 What does Descartes' rule of signs tell you about the number of positive real roots?

1. $3x^5 - 2x^2 + x - 1$
2. $5x^6 - 3x^3 + x^2 - x$
3. $6x^7 + 2x^2 + 5x + 4 = 0$
4. $-3x^5 - 7x^3 - 4x - 5 = 0$
5. $3p^{18} + 2p^4 - 5p^2 + p + 3$
6. $5t^{12} - 7t^4 + 3t^2 + t + 1$

What does Descartes' rule of signs tell you about the number of negative real roots?

7. $3x^5 - 2x^2 + x - 1$
8. $5x^6 - 3x^3 + x^2 - x$
9. $6x^7 + 2x^2 + 5x + 4 = 0$
10. $-3x^5 - 7x^3 - 4x - 5 = 0$
11. $3p^{18} + 2p^3 - 5p^2 + p + 3$
12. $5t^{11} - 7t^4 + 3t^2 + t + 1$

2 Find the smallest positive integer that is guaranteed by Theorem 12 to be an upper bound to the roots.

13. $3x^4 - 15x^2 + 2x - 3$
14. $4x^4 - 14x^2 + 4x - 2$
15. $6x^3 - 17x^2 - 3x - 1$
16. $5x^3 - 15x^2 + 5x - 4$

Find the largest negative integer that is guaranteed by Theorem 13 to be a lower bound to the roots.

17. $3x^4 - 15x^3 + 2x - 3$
18. $4x^4 - 17x^3 + 3x - 2$
19. $6x^3 + 15x^2 + 3x - 1$
20. $6x^3 + 12x^2 + 5x - 3$

1, **2** What does Descartes' rule of signs tell you about the roots of the polynomial or equation? Find upper and lower bounds to the real roots.

21. $P(x) = x^4 - 2x^2 + 12x - 8$
22. $P(x) = x^4 - 6x^2 + 20x - 24$
23. $P(x) = x^4 - 2x^2 - 8$
24. $P(x) = 3x^4 - 5x^2 - 4$
25. $P(x) = x^4 - 9x^2 - 6x + 4$
26. $P(x) = x^4 - 21x^2 + 4x + 6$
27. $P(x) = x^4 + 3x^2 + 2$
28. $P(x) = x^4 + 5x^2 + 6$

3 Approximate the irrational roots to the nearest tenth.

29. $x^3 - 3x - 2 = 0$
30. $x^3 - 3x^2 + 3 = 0$
31. $x^3 - 3x - 4 = 0$
32. $x^3 - 3x^2 + 5 = 0$
33. $x^4 + x^2 + 1 = 0$
34. $x^4 + 2x^2 + 2 = 0$
35. $x^4 - 6x^2 + 8 = 0$
36. $x^4 - 4x^2 + 2 = 0$
37. $x^5 + x^4 - x^3 - x^2 - 2x - 2 = 0$
38. $x^5 - 2x^4 - 2x^3 + 4x^2 - 3x + 6 = 0$

39. ▦ The following equation has a solution between 0 and 1. Approximate it, to the nearest hundredth.
$$2x^5 + 2x^3 - x^2 - 1 = 0$$

40. ▦ The following equation has a solution between 1 and 2. Approximate it, to the nearest hundredth.
$$x^4 - 2x^3 - 3x^2 + 4x + 2 = 0$$

Approximate the irrational roots, to the nearest hundredth.

41. ▦ $P(x) = x^3 - 2x^2 - x + 4$
42. ▦ $P(x) = x^3 - 4x^2 + x + 3$

SYNTHESIS

43. Prove that for n a positive even integer, $x^n - 1$ has only two real roots.

44. Prove that for n an odd positive integer, $x^n - 1$ has only one real root.

COMPUTER–CALCULATOR EXERCISES

There are many computer software packages and calculators that will graph polynomials and solve equations. Use a computer or a calculator to graph the following functions, approximate their roots, and solve the problems in Exercises 49 and 50.

45. $f(x) = x^3 - 9x^2 + 27x + 50$
46. $f(x) = x^3 - 3x + 1$
47. $f(x) = x^4 + 4x^3 - 36x^2 - 160x + 300$
48. $f(x) = x^6 + 4x^5 - 54x^4 - 160x^3 + 641x^2 + 828x - 1260$

49. See Exercise 26 in Exercise Set 11.1. After what time will the concentration be 0?

50. *Multiple investments.* At the beginning of a year, \$2000 is invested in a bank at a certain interest rate, compounded annually. At the beginning of the next year, \$1500 is invested in a second bank at the same interest rate. At the beginning of the third year, \$1800 is invested in a third bank at the same interest rate. At the beginning of the fourth year, there is a total of \$6213.02 in all three accounts. What is the interest rate?

CHALLENGE

51. Show that $x^4 + ax^2 + bx - c$, where a, b, and c are positive, has just two nonreal roots.

52. In the three-body problem that occurs in astronomy, the following equation occurs:
$$r^5 + (3 - k)r^4 + (3 - 2k)r^3 - kr^2 - 2kr - k = 0,$$
where $0 < k < 1$. Show that this equation has just one positive real solution.

11.6 Limits and Continuity

OBJECTIVES

You should be able to:

1 **Find $\lim_{x \to a} f(x)$, if it exists, using input–output tables, algebra, or graphs.**

(continued)

In this section, we give an intuitive (meaning "based on prior and present experience") treatment of two important concepts: limits and continuity. These concepts are relevant to the graphing of rational functions in the next section, and they provide a very important foundation for calculus as well.

2 Determine whether a graph is of a continuous function.

3 Determine whether a function is continuous at a given point a.

4 Find limits at infinity.

1 Limits

One aspect of a study of calculus is the analysis of how function values, or outputs, change when the inputs change. Basic to such a study is the notion of *limit*. Suppose the inputs get closer and closer to some number. If the corresponding outputs get closer and closer to a number, that number is called a **limit.**

Consider the function f given by

$$f(x) = 2x + 3.$$

Suppose we select input numbers x closer and closer to the number 4, and look at the output numbers $2x + 3$. Study the following input–output table and graph.

These inputs approach 4 from the left. — 4 — These inputs approach 4 from the right.

x	2	3.6	3.9	3.99	3.999	4.001	4.01	4.1	4.8	5
$f(x)$	7	10.2	10.8	10.98	10.998	11.002	11.02	11.2	12.6	13

These outputs approach 11. — 11 — These outputs approach 11.

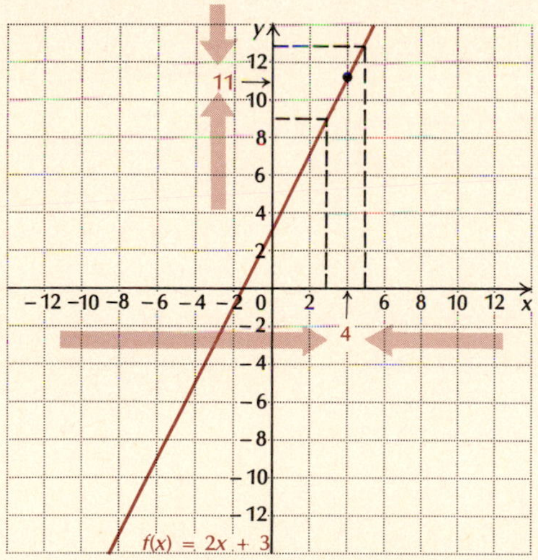

In the table and the graph, as input numbers approach 4 from the left, output numbers approach 11. As input numbers approach 4 from the right, output numbers approach 11. Thus we say:

$$\text{As } x \text{ } \textit{approaches } 4, \quad 2x + 3 \text{ } \textit{approaches } 11.$$

An arrow, $\to$, is often used for the word "approaches." Thus the above can be written

$$\text{As } x \to 4, \quad 2x + 3 \to 11.$$

The number 11 is said to be the *limit* of $2x + 3$ as x approaches 4. We can abbreviate this statement as follows:

$$\lim_{x \to 4} (2x + 3) = 11.$$

This is read, "The limit, as x approaches 4, of $2x + 3$ is 11."

1. Consider

$$f(x) = 3x - 1.$$

a) Complete this table. (calculator helpful, though not necessary)

x	$f(x)$
5	
5.8	
5.9	
5.99	
5.999	
6 ←	→ ?
6.001	
6.01	
6.1	
6.4	
7	

b) Find $\lim_{x \to 6} f(x)$.

c) Graph $f(x) = 3x - 1$.

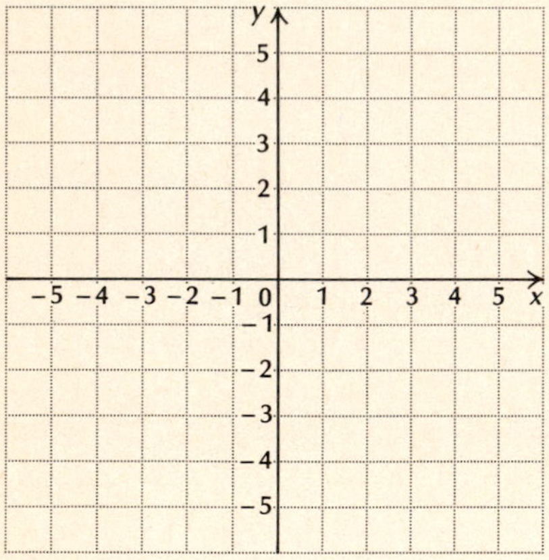

Use the graph to find each of the following limits.

d) $\lim_{x \to -1} f(x)$

e) $\lim_{x \to 2} f(x)$

f) $\lim_{x \to 0} f(x)$

DEFINITION

A function f has the *limit* L as x approaches a, written

$$\lim_{x \to a} f(x) = L,$$

if we can get $f(x)$ as close to L as we wish by restricting x to a sufficiently small interval about a but excluding a.

DO EXERCISE 1.

EXAMPLE 1 Consider the function

$$F(x) = \frac{1}{x}.$$

Find the following limits, if they exist.

a) $\lim_{x \to 3} F(x)$

b) $\lim_{x \to 0} F(x)$

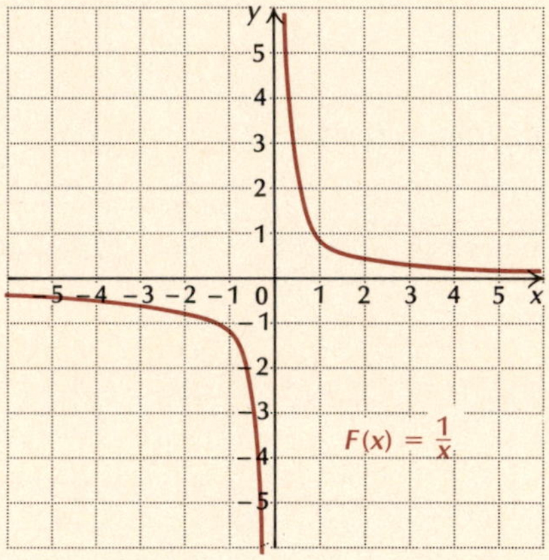

Solution

a) From the graph shown here, we see that as inputs x approach 3 from either the left or right, outputs $1/x$ approach $1/3$. We can also check this on an input–output table. Thus we have

$$\lim_{x \to 3} F(x) = \tfrac{1}{3}.$$

b) From the graph, or from an input–output table, we see that as inputs x approach 0 from the left, outputs get further and further negative without bound. Similarly, as inputs x approach 0 from the right, outputs get larger and larger without bound. We say that

$$\lim_{x \to 0} \frac{1}{x} \textit{ does not exist.}$$

DO EXERCISE 2 ON THE FOLLOWING PAGE.

The following fact is very important to keep in mind when determining whether a limit exists.

In order for a limit to exist, the limits from the left and the right must both exist and be the same.

We use the notation

$$\lim_{x \to a^+} f(x) \text{ to indicate the limit from the right}$$

and

$$\lim_{x \to a^-} f(x) \text{ to indicate the limit from the left.}$$

Then for a limit to exist, both of the limits above must exist and be the same.

EXAMPLE 2 Consider the function H defined as follows:

$$H(x) = \begin{cases} 2x + 2 & \text{for } x < 1, \\ 2x - 2 & \text{for } x \geq 1. \end{cases}$$

Find the following limits, if they exist.

a) $\lim_{x \to 1} H(x)$

b) $\lim_{x \to -3} H(x)$

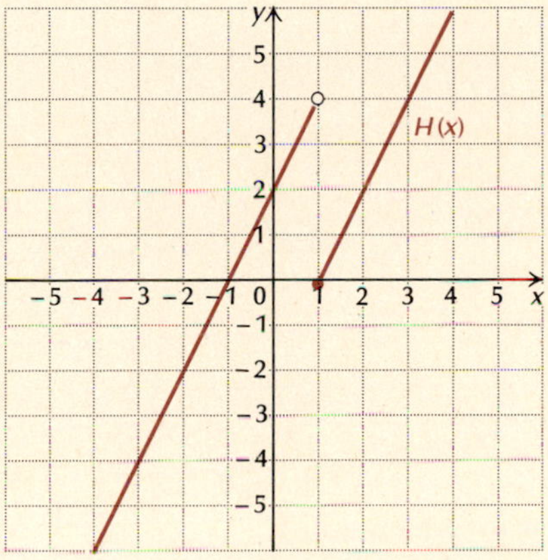

Solution

a) From the graph shown here, we see that as inputs x approach 1 from the left, outputs $H(x)$ approach 4. Thus the limit from the left is 4. That is,

$$\lim_{x \to 1^-} H(x) = 4.$$

But as inputs x approach 1 from the right, outputs $H(x)$ approach 0. Thus the limit from the right is 0. That is,

$$\lim_{x \to 1^+} H(x) = 0.$$

Since the limit from the left, 4, is not the same as the limit from the right, 0, we say that

$$\lim_{x \to 1} H(x) \textit{ does not exist.}$$

b) As inputs x approach -3 from the left, outputs $H(x)$ approach -4, so the limit from the left is -4. That is,

$$\lim_{x \to -3^-} H(x) = -4.$$

As inputs x approach -3 from the right, outputs $H(x)$ approach -4, so the limit from the right is -4. That is,

$$\lim_{x \to -3^+} H(x) = -4.$$

Since the limits from the left and from the right exist and are the same, we have

$$\lim_{x \to -3} H(x) = -4.$$

2. Consider the function

$$g(x) = \frac{1}{x - 3}.$$

a) Complete this table. (calculator helpful)

Inputs, x	Outputs, $g(x)$
1	
2	
2.9	
2.99	
2.999	
3 ←	→ ?
3.001	
3.01	
3.1	
3.8	
4	

b) Find $\lim_{x \to 3} g(x)$.

c) Graph $g(x) = \dfrac{1}{x - 3}$.

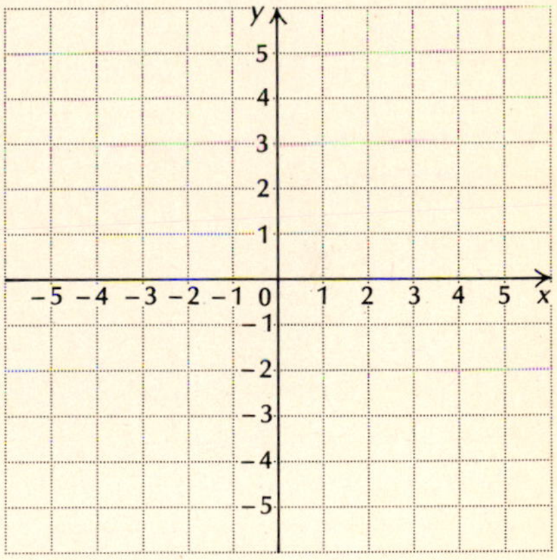

Use the graph to find each of these limits, if they exist.

d) $\lim_{x \to 3} g(x)$

e) $\lim_{x \to 1} g(x)$

f) $\lim_{x \to 4} g(x)$

3. Consider the following graph.

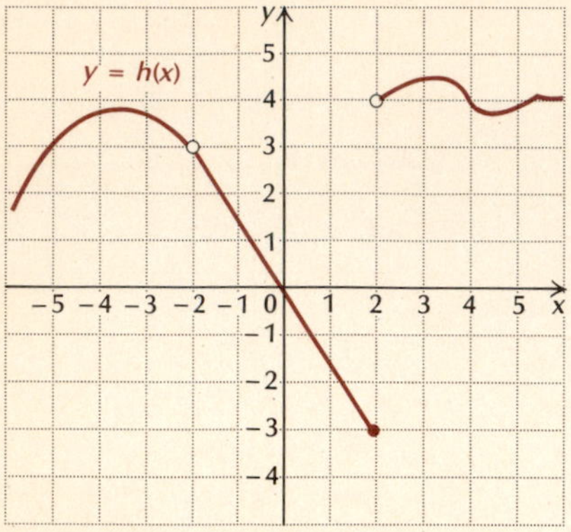

Find each of the following limits, if they exist.

a) $\lim_{x\to 2^-} h(x)$

b) $\lim_{x\to 2^+} h(x)$

c) $\lim_{x\to 2} h(x)$

d) $\lim_{x\to 0^-} h(x)$

e) $\lim_{x\to 0^+} h(x)$

f) $\lim_{x\to 0} h(x)$

g) $\lim_{x\to -2^-} h(x)$

h) $\lim_{x\to -2^+} h(x)$

i) $\lim_{x\to -2} h(x)$

4. Consider the following function:

$$f(x) = \begin{cases} 2 - x^2 & \text{for } x \geq 0, \\ x^2 - 2 & \text{for } x < 0. \end{cases}$$

a) Graph $y = f(x)$.

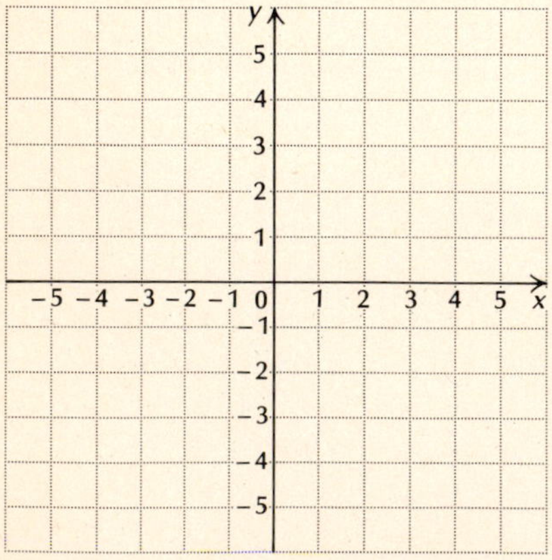

Find each of the following limits, if they exist.

b) $\lim_{x\to 0} f(x)$

c) $\lim_{x\to -2} f(x)$

DO EXERCISES 3 AND 4.

The following fact is also important in finding limits.

The limit at a point *a does not depend* on the function value at *a*, even if that function value, $f(a)$, exists. That is, whether or not a limit exists at *a* has nothing to do with the function value $f(a)$.

EXAMPLE 3 Consider the function G defined as follows:

$$G(x) = \begin{cases} 5 & \text{for } x = 1, \\ x + 1 & \text{for } x \neq 1. \end{cases}$$

Find each of the following limits, if they exist.

a) $\lim_{x\to 1} G(x)$

b) $\lim_{x\to 3} G(x)$

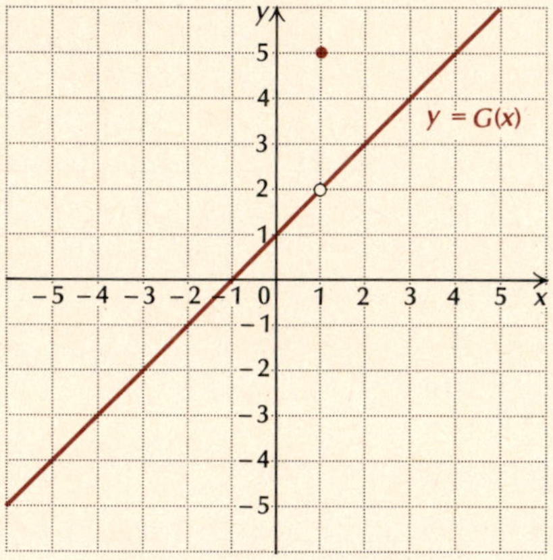

Solution

a) From the graph shown here, we see that as inputs x approach 1 from the left, outputs $G(x)$ approach 2, so the limit from the left is 2. As inputs x approach 1 from the right, outputs $G(x)$ approach 2, so the limit from the right is 2. Since the limit from the left, 2, is the same as the limit from the right, 2, we have

$$\lim_{x\to 1} G(x) = 2.$$

Note that the limit, 2, is not the same as the function value at 1, $G(1)$, which is 5.

b) We have $\lim_{x\to 3} G(x) = 4 = G(3)$. In this case, the function value and the limit are the same.

DO EXERCISES 5 AND 6 ON THE FOLLOWING PAGE.

2 Continuity

The situation in part (b) above, where the limit of a function is the same as its function value, satisfies a condition called "continuity at a point." We now consider the concept of continuity and connect it to our study of limits.

The following are graphs of functions that are *continuous* over the whole real line $(-\infty, \infty)$.

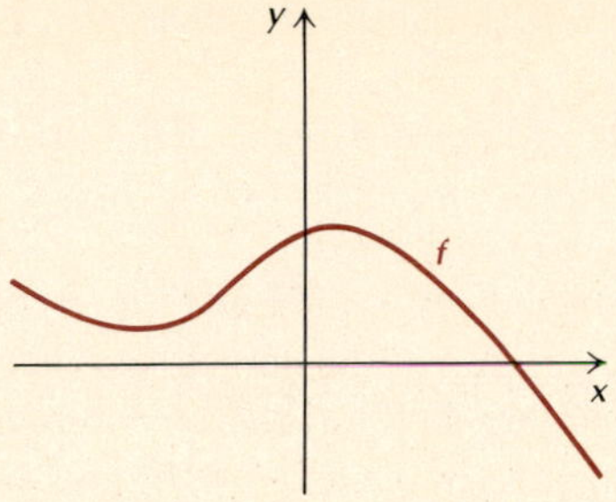

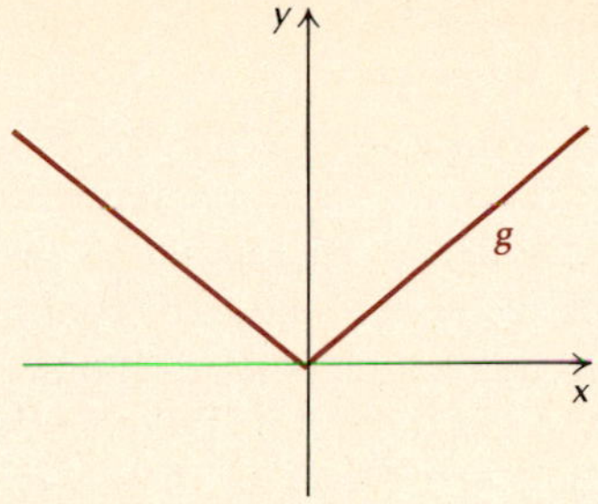

Note that there are no "jumps" or holes in the graphs. For now we will use a somewhat intuitive definition of continuity, which we will refine. We say that a function is **continuous** over, or on, some interval of the real line if its graph can be traced without lifting a pencil from the paper. The following are graphs of functions that are *not* continuous over the whole real line.

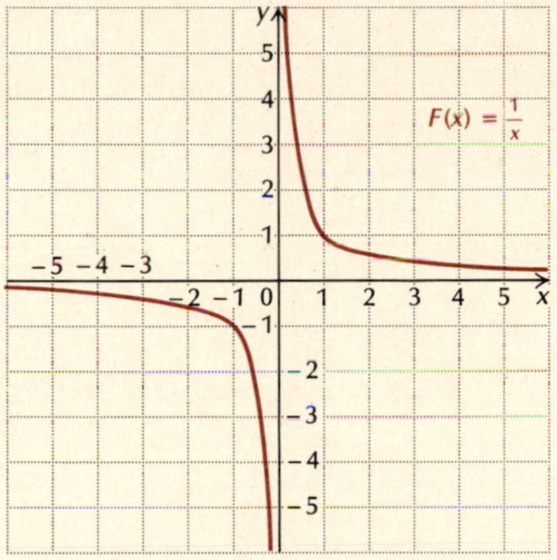

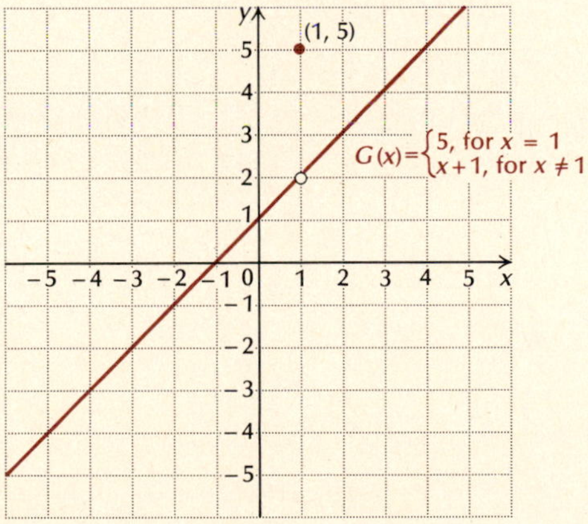

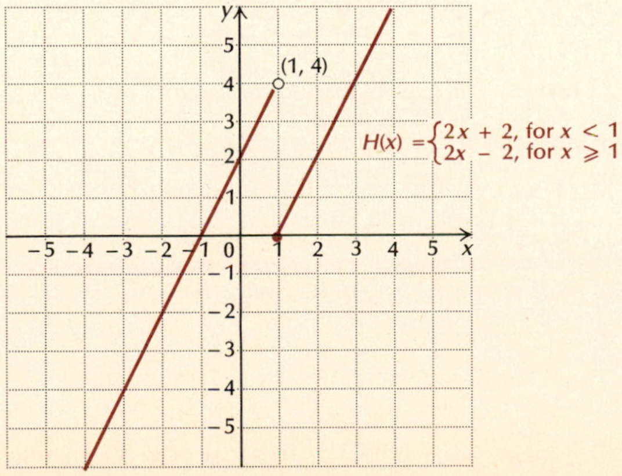

5. Consider the following function:

$$f(x) = \begin{cases} -3 & \text{for } x = -2, \\ x & \text{for } x \neq -2. \end{cases}$$

a) Graph $y = f(x)$.

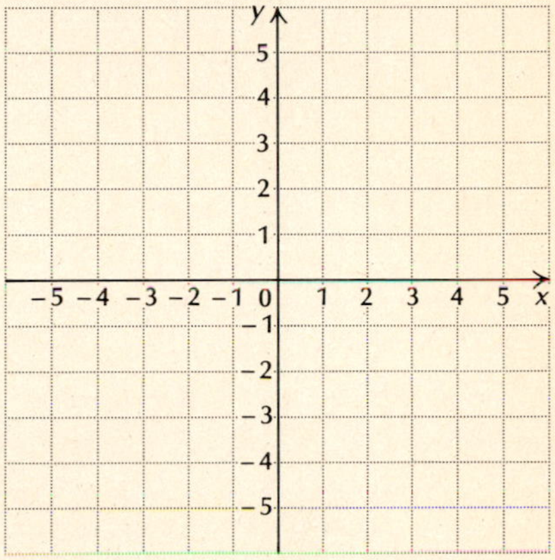

Find each of the following limits, if they exist.

b) $\lim_{x \to -2^-} f(x)$

c) $\lim_{x \to -2^+} f(x)$

d) $\lim_{x \to -2} f(x)$

e) Does $\lim_{x \to -2} f(x) = f(-2)$?

f) $\lim_{x \to 2^-} f(x)$

g) $\lim_{x \to 2^+} f(x)$

h) $\lim_{x \to 2} f(x)$

i) Does $\lim_{x \to 2} f(x) = f(2)$?

6. Consider the following graph.

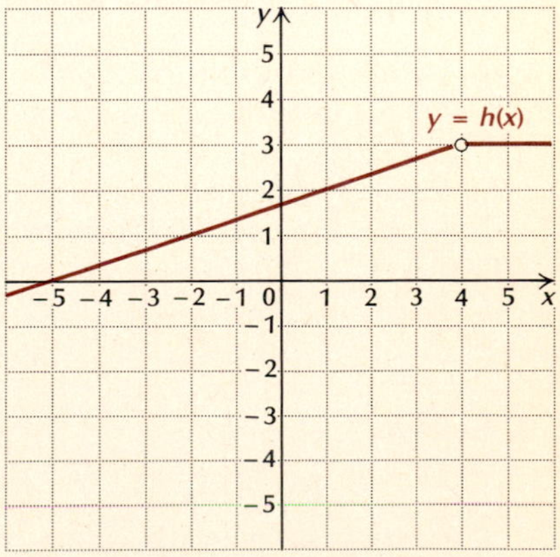

Find each of the following limits, if they exist.

a) $\lim_{h \to 1} h(x)$

b) $\lim_{h \to 4} h(x)$

c) Does $\lim_{h \to 1} h(x) = h(1)$?

d) Does $\lim_{h \to 4} h(x) = h(4)$?

7. Which of the following functions are continuous?

a)

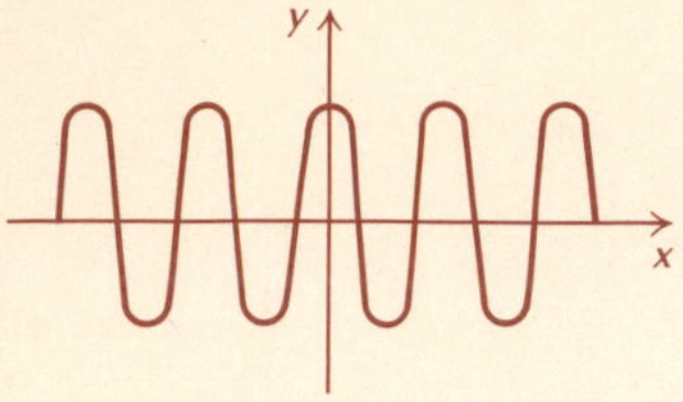

b)

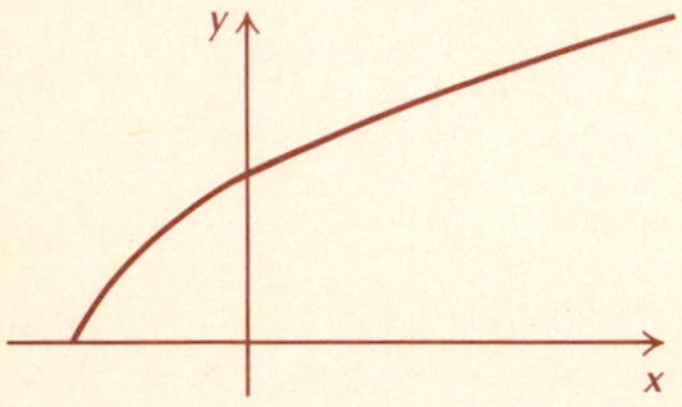

c)

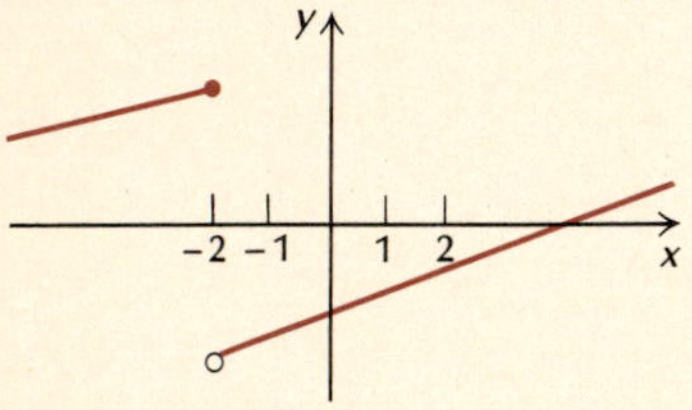

d)

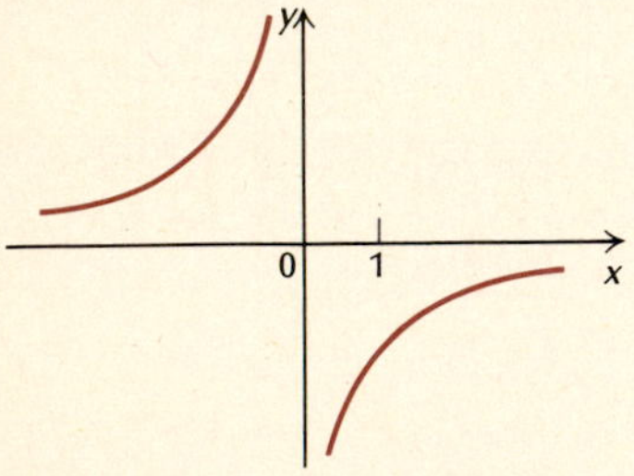

8. a) Decide whether the function in Margin Exercise 1(c) is continuous at -2; at 1.
   b) Decide whether the function in Margin Exercise 2(c) is continuous at 3; at 0.

9. Consider

$$f(x) = 3x - 1.$$

(See Margin Exercise 1.)

a) Does $f(6)$ exist? If so, what is it?
b) Does $\lim_{x\to 6} f(x)$ exist? If so, what is it?
c) Does $\lim_{x\to 6} f(x) = f(6)$?
d) Is f continuous at 6?

For functions G and H, the open circle indicates that the circled point is not part of the graph.

In each case, the graph *cannot* be traced without lifting the pencil from the paper. However, each case represents a different situation. Let us discuss why each case fails to be continuous over the whole real line.

The function F fails to be continuous over the *whole* real line $(-\infty, \infty)$. Since F is not defined at $x = 0$, the point $x = 0$ is not part of the domain, so $f(0)$ does not exist and there is no point $(0, f(0))$ on the graph. Thus there is no point to trace at $x = 0$. However, F is continuous on the intervals $(-\infty, 0)$ and $(0, \infty)$.

The function G is not continuous over the whole real line because it is not continuous at $x = 1$. Let us trace the graph of G to the left of $x = 1$. As x approaches 1, $G(x)$ seems to approach 2; but at $x = 1$, $G(x)$ *jumps* up to 5, while to the right of $x = 1$, $G(x)$ *jumps* back to some value close to 2. Thus G is discontinuous at $x = 1$.

The function H is not continuous over the whole real line because it is not continuous at $x = 1$. Let us trace the graph of H starting to the left of $x = 1$. As x approaches 1, $H(x)$ seems to approach 4; but at $x = 1$, $H(x)$ *jumps* down to 0, while just to the right of $x = 1$, $H(x)$ is close to 0. Thus, $H(x)$ is discontinuous at $x = 1$.

DO EXERCISES 7 AND 8.

3 Limits and Continuity

After working through Example 3, we might ask, "When can we substitute to find a limit?" The answer lies in the following definition, which also happens to provide a more formal definition of continuity.

DEFINITION

A function f is *continuous* at $x = a$ if:

1. **$f(a)$ exists,**
2. **$\lim_{x\to a} f(x)$ exists, and**
3. **$\lim_{x\to a} f(x) = f(a)$.**

A function is *continuous over an interval I* if it is continuous at each point in *I*.

EXAMPLE 4 Determine whether the function given by

$$f(x) = 2x + 3$$

is continuous at $x = 4$.

Solution This function is continuous at 4 because

1. $f(4)$ exists, $f(4) = 11$;
2. $\lim_{x\to 4} f(x)$ exists, $\lim_{x\to 4} f(x) = 11$ (as shown earlier);
3. $\lim_{x\to 4} f(x) = f(4)$.

In fact, $f(x) = 2x + 3$ is continuous at any point on the real line. ■

DO EXERCISE 9.

EXAMPLE 5 Determine whether the function in Example 1 is continuous at $x = 0$.

Solution The function is *not* continuous at $x = 0$ because $f(0)$ does not exist. ■

EXAMPLE 6 Determine whether the function in Example 2 is continuous at $x = 1$.

Solution The function is *not* continuous at $x = 1$ because $\lim_{x \to 1} H(x)$ does not exist. ■

EXAMPLE 7 Determine whether the function in Example 3 is continuous at $x = 1$.

Solution The function is *not* continuous at $x = 1$ because $G(1) = 5$, but $\lim_{x \to 1} G(x) = 2$. ■

DO EXERCISES 10–12.

Continuity Principles

The following continuity principles, which we will not prove, allow us to build up continuous functions.

i) Any constant function is continuous (such a function never varies).

ii) For any positive integer n, x^n and $\sqrt[n]{x}$ are continuous. When n is even, the inputs of $\sqrt[n]{x}$ are restricted to $[0, \infty)$.

iii) If $f(x)$ and $g(x)$ are continuous, then so are

$$f(x) + g(x), \quad f(x) - g(x), \quad \text{and} \quad f(x) \cdot g(x).$$

iv) If $f(x)$ is continuous, so is $1/f(x)$, so long as the inputs x do not yield outputs $f(x) = 0$.

EXAMPLE 8 Provide an argument to show that

$$f(x) = x^2 - 3x + 2$$

is continuous.

Solution Now x^2 is continuous, by (ii). The constant function 3 is continuous by (i), and the function x is continuous by (ii), so the product $3x$ is continuous by (iii). Thus, $x^2 - 3x$ is continuous by (iii), and since the constant 2 is continuous, we can apply (iii) again to show that $x^2 - 3x + 2$ is continuous. ■

In similar fashion, we can show that any polynomial such as

$$f(x) = x^4 - 5x^3 + x^2 - 7$$

is continuous. A rational function is a quotient of two polynomials:

$$r(x) = \frac{f(x)}{g(x)}.$$

Thus by (iv), a rational function is continuous so long as the inputs x are such that $g(x) \neq 0$.

DO EXERCISE 13.

10. Determine whether the function in Margin Exercise 2 is continuous at
    a) $x = 3$;
    b) $x = -2$.

11. Determine whether the function in Margin Exercise 4 is continuous at $x = 0$.

12. Determine whether the function in Margin Exercise 5 is continuous at $x = -2$.

13. Provide an argument to show that the function given by

$$f(x) = \frac{\sqrt[3]{x} - 7x^2}{x - 2}$$

is continuous so long as $x \neq 2$.

Find each limit, if it exists.

14. $\lim_{x \to -2} (x^4 - 5x^3 + x^2 - 7)$

15. $\lim_{x \to 1} \sqrt{x^2 + 3x + 4}$

Using Limit Principles

If a function is continuous at a, we can substitute to find the limit.

EXAMPLE 9 Find $\lim_{x \to 2} (x^4 - 5x^3 + x^2 - 7)$.

Solution It follows from the continuity principles that $x^4 - 5x^3 + x^2 - 7$ is continuous. Thus the limit can be found by substitution:

$$\lim_{x \to 2} (x^4 - 5x^3 + x^2 - 7) = 2^4 - 5 \cdot 2^3 + 2^2 - 7$$
$$= 16 - 40 + 4 - 7 = -27.$$

EXAMPLE 10 Find $\lim_{x \to 0} \sqrt{x^2 - 3x + 2}$.

Solution By using the continuity principles, we have shown that $x^2 - 3x + 2$ is continuous; and so long as x is restricted to values for which $x^2 - 3x + 2$ is nonnegative, it follows from principle (ii) that $\sqrt{x^2 - 3x + 2}$ is continuous. Thus we can substitute to find the limit:

$$\lim_{x \to 0} \sqrt{x^2 - 3x + 2} = \sqrt{0^2 - 3 \cdot 0 + 2} = \sqrt{2}.$$

DO EXERCISES 14 AND 15.

There are limit principles that correspond to the continuity principles. We can use them to find limits when we are uncertain of the continuity of a function at a given point.

Limit Principles

If $\lim_{x \to a} f(x) = L$ and $\lim_{x \to a} g(x) = M$, then we have the following:

L1. $\lim_{x \to a} c = c$.

(The limit of a constant is the constant.)

L2. $\lim_{x \to a} x^n = a^n$, $\lim_{x \to a} \sqrt[n]{x} = \sqrt[n]{a}$, **for any positive integer n.**

(When n is even, the inputs of $\sqrt[n]{x}$ must be restricted to $[0, \infty)$.)

L3. $\lim_{x \to a} [f(x) \pm g(x)] = \lim_{x \to a} f(x) \pm \lim_{x \to a} g(x) = L \pm M$.

(The limit of a sum or a difference is the sum or the difference of the limits.)

$$\lim_{x \to a} [f(x) \cdot g(x)] = \left[\lim_{x \to a} f(x)\right] \cdot \left[\lim_{x \to a} g(x)\right] = L \cdot M.$$

(The limit of a product is the product of the limits.)

L4. $\lim_{x \to a} \frac{1}{f(x)} = \frac{1}{\lim_{x \to a} f(x)} = \frac{1}{L}$, **provided $L \neq 0$.**

(The limit of a reciprocal is the reciprocal of the limit.)

L5. $\lim_{x \to a} cf(x) = c \cdot \lim_{x \to a} f(x) = c \cdot L$.

(The limit of a constant times a function is the constant times the limit of the function.)

L5 can actually be proved using L1 and L3, but we state it for emphasis.

EXAMPLE 11 Find

$$\lim_{x \to -3} \frac{x^2 - 9}{x + 3}.$$

Solution The function $(x^2 - 9)/(x + 3)$ is not continuous at $x = -3$. We use some algebraic simplification and then some limit principles.

$$\lim_{x \to -3} \frac{x^2 - 9}{x + 3} = \lim_{x \to -3} \frac{(x + 3)(x - 3)}{x + 3}$$

$$= \lim_{x \to -3} (x - 3) \quad \text{Simplifying, assuming } x \neq -3$$

$$= \lim_{x \to -3} x - \lim_{x \to -3} 3 \quad \text{By L3}$$

$$= -3 - 3 = -6$$

DO EXERCISES 16 AND 17.

4 Limits and Infinity

We discussed in Example 1 that

$$\lim_{x \to 0} \frac{1}{x} \textit{ does not exist.}$$

Go back and look at the graph. As x approaches 0 from the right, the outputs get larger and larger without bound. These numbers do not approach any real number, though it might be said that the limit from the right is ∞ (infinity). This is written

$$\lim_{x \to 0^+} F(x) = \infty.$$

Similarly, as we approach 0 from the left, the function values get smaller and smaller without bound. We say that the limit as x approaches 0 from the left is $-\infty$ (read "negative infinity") and this is written

$$\lim_{x \to 0^-} F(x) = -\infty.$$

CAUTION! Keep in mind that ∞ and $-\infty$ are *not* symbols for real numbers. We associate ∞ with numbers increasing without bound in a positive direction, and $-\infty$ with numbers decreasing without bound in a negative direction.

DO EXERCISE 18.

Limits at Infinity

We sometimes need to determine limits when the inputs get larger and larger without bound, that is, when they approach infinity. In such cases, we are finding *limits at infinity*. Such a limit would be expressed as

$$\lim_{x \to \infty} f(x).$$

EXAMPLE 12 Find

$$\lim_{x \to \infty} \left(\frac{3x - 1}{x}\right).$$

Find each limit, if it exists.

16. $\lim_{x \to -4} \frac{x^2 - 16}{x + 4}$

17. $\lim_{x \to 3} \frac{x - 3}{x^2 - 9}$

18. *Earned-run average.* A pitcher's earned-run average (the average number of runs given up every 9 innings or 1 game) is given by

$$E = 9 \cdot \frac{n}{i},$$

where n = the number of earned runs allowed and i = the number of innings pitched. Suppose we fix the number of earned runs allowed at 4 and let i vary. We get a function given by

$$E(i) = 9 \cdot \frac{4}{i}.$$

a) Complete the following table, rounding to two decimal places.

Innings pitched (i)	Earned-run average (E)
9	
8	
7	
6	
5	
4	
3	
2	
1	
$\frac{2}{3}$ (2 outs)	
$\frac{1}{3}$ (1 outs)	

b) Find $\lim_{i \to 0} E(i)$.

c) On the basis of (a) and (b), what might a pitcher's earned-run average be if 4 runs were allowed and there were 0 outs?

19. Consider $f(x) = \frac{2x + 5}{x}$.

a) Complete this table. (calculator helpful)

Inputs, x	Outputs, $\frac{2x+5}{x}$
4	
20	
80	
200	
1,000	
10,000	

b) Find $\lim_{x\to\infty} \frac{2x + 5}{x}$.

20. Find $\lim_{x\to\infty} \frac{2x^2 + x - 7}{3x^2 - 4x + 1}$.

Solution One way to find such a limit is to use an input–output table, as follows.

Inputs, x	1	10	50	100	2000
Outputs, $\frac{3x-1}{x}$	2.0	2.9	2.98	2.99	2.9995

As the inputs get larger and larger without bound, the outputs get closer and closer to 3. Thus,

$$\lim_{x\to\infty} \left(\frac{3x - 1}{x}\right) = 3.$$

Another way to do this is to use some algebra, and the fact that as $x \to \infty$ or $x \to -\infty$, $b/ax^n \to 0$, for any positive integer n.

We multiply by 1, using $(1/x) \div (1/x)$. This amounts to dividing both numerator and denominator by x:

$$\begin{aligned}
\lim_{x\to\infty} \frac{3x - 1}{x} &= \lim_{x\to\infty} \frac{3x - 1}{x} \cdot \frac{(1/x)}{(1/x)} \\
&= \lim_{x\to\infty} \frac{(3x - 1)\frac{1}{x}}{x \cdot \frac{1}{x}} \\
&= \lim_{x\to\infty} \frac{3x \cdot \frac{1}{x} - 1 \cdot \frac{1}{x}}{1} \\
&= \lim_{x\to\infty} \left(3 - \frac{1}{x}\right) \\
&= \lim_{x\to\infty} 3 - \lim_{x\to\infty} \frac{1}{x} \\
&= 3 - 0 \\
&= 3.
\end{aligned}$$

DO EXERCISE 19.

EXAMPLE 13 Find

$$\lim_{x\to\infty} \frac{3x^2 - 7x + 2}{7x^2 + 5x + 1}.$$

Solution The highest power of x in the denominator is x^2. We divide both numerator and denominator by x^2:

$$\begin{aligned}
\lim_{x\to\infty} \frac{3x^2 - 7x + 2}{7x^2 + 5x + 1} &= \lim_{x\to\infty} \frac{3 - \frac{7}{x} + \frac{2}{x^2}}{7 + \frac{5}{x} + \frac{1}{x^2}} \\
&= \frac{3 - 0 + 0}{7 + 0 + 0} \\
&= \frac{3}{7}.
\end{aligned}$$

DO EXERCISE 20.

EXERCISE SET 11.6

2 Determine whether each of the following is continuous.

1.

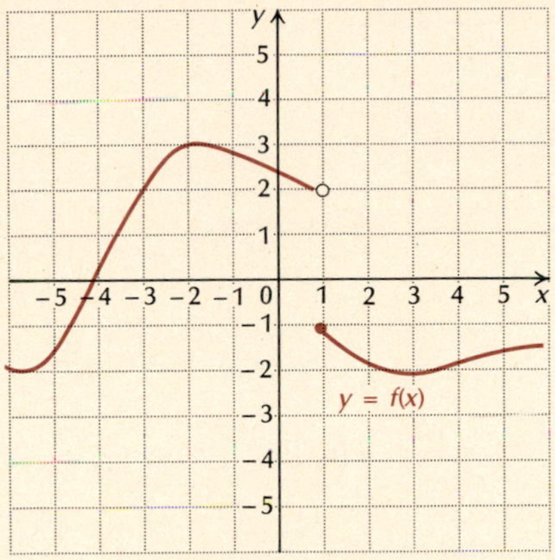

2.

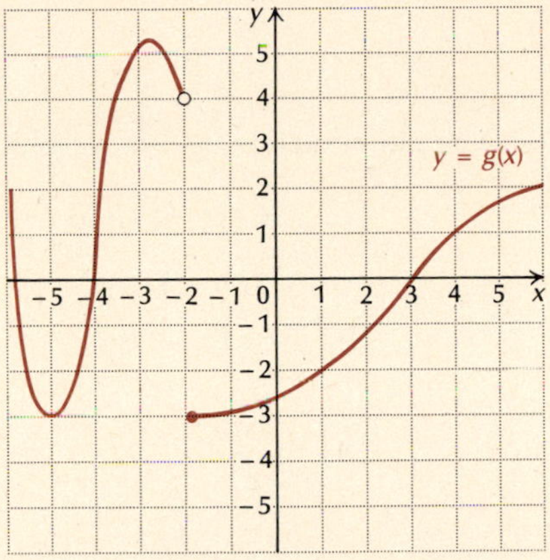

3.

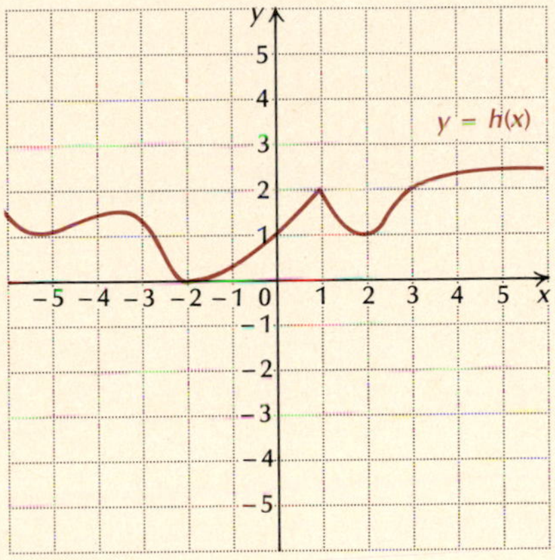

4.

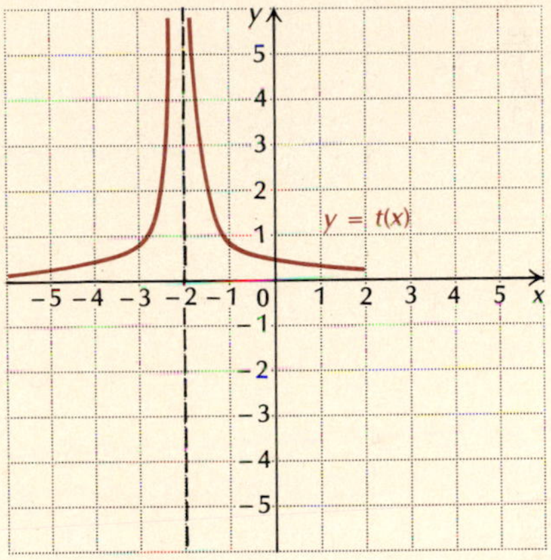

1, **3** Use the graphs and functions in Exercises 1–4 to answer the following.

5. a) Find $\lim_{x\to 1^-} f(x)$, $\lim_{x\to 1^+} f(x)$, and $\lim_{x\to 1} f(x)$.
   b) Find $f(1)$.
   c) Is f continuous at $x = 1$?
   d) Find $\lim_{x\to -2} f(x)$.
   e) Find $f(-2)$.
   f) Is f continuous at $x = -2$?

6. a) Find $\lim_{x\to 1^-} g(x)$, $\lim_{x\to 1^+} g(x)$, and $\lim_{x\to 1} g(x)$.
   b) Find $g(1)$.
   c) Is g continuous at $x = 1$?
   d) Find $\lim_{x\to -2} g(x)$.
   e) Find $g(-2)$.
   f) Is g continuous at $x = -2$?

7. a) Find $\lim_{x\to 1} h(x)$.
   b) Find $h(1)$.
   c) Is h continuous at $x = 1$?
   d) Find $\lim_{x\to -2} h(x)$.
   e) Find $h(-2)$.
   f) Is h continuous at $x = -2$?

8. a) Find $\lim_{x\to 1} t(x)$.
   b) Find $t(1)$.
   c) Is t continuous at $x = 1$?
   d) Find $\lim_{x\to -2} t(x)$.
   e) Find $t(-2)$.
   f) Is t continuous at $x = -2$?

The postage function. Postal rates are as follows: 25¢ for the first ounce and 20¢ for each additional ounce or fraction thereof. Formally, if x is the weight of a letter in ounces, then $p(x)$ is the cost of mailing the letter, where

$$p(x) = 25¢, \quad \text{if } 0 < x \le 1,$$
$$p(x) = 45¢, \quad \text{if } 1 < x \le 2,$$
$$p(x) = 65¢, \quad \text{if } 2 < x \le 3,$$

and so on, up to 12 ounces (at which point postal cost also depends on distance). The graph of p is shown at the right.

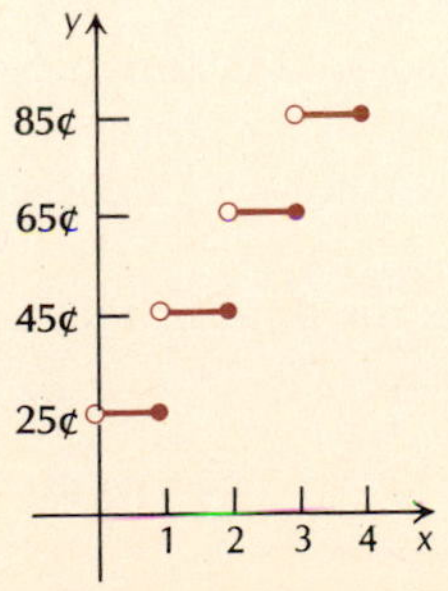

9. Is p continuous at 1? at $1\frac{1}{2}$? at 2? at 2.53?

10. Is p continuous at 3? at $3\frac{1}{4}$? at 4? at 3.98?

Using the graph above, find the limit, if it exists.

11. $\lim_{x\to 1^-} p(x)$, $\lim_{x\to 1^+} p(x)$, $\lim_{x\to 1} p(x)$

12. $\lim_{x\to 2^-} p(x)$, $\lim_{x\to 2^+} p(x)$, $\lim_{x\to 2} p(x)$

13. $\lim_{x\to 2.3} p(x)$

14. $\lim_{x\to 1/2} p(x)$

1, **4** Find the limit. Use any method: algebra, graphs, or input–output tables.

15. $\lim_{x\to 1} (x^2 - 3)$

16. $\lim_{x\to 1} (x^2 + 4)$

17. $\lim_{x\to 0} \dfrac{3}{x}$

18. $\lim_{x\to 0} \dfrac{-4}{x}$

19. $\lim_{x\to 3} (2x + 5)$

20. $\lim_{x\to 4} (5 - 3x)$

21. $\lim_{x\to -5} \dfrac{x^2 - 25}{x + 5}$

22. $\lim_{x\to -4} \dfrac{x^2 - 16}{x + 4}$

23. $\lim_{x\to -2} \dfrac{5}{x}$

24. $\lim_{x\to -5} \dfrac{-2}{x}$

25. $\lim_{x\to 2} \dfrac{x^2 + x - 6}{x - 2}$

26. $\lim_{x\to -4} \dfrac{x^2 - x - 20}{x + 4}$

27. $\lim_{x\to 5} \sqrt[3]{x^2 - 17}$

28. $\lim_{x\to 2} \sqrt{x^2 + 5}$

29. $\lim_{x\to 1} (x^4 - x^3 + x^2 + x + 1)$

30. $\lim_{x\to 2} (2x^5 - 3x^4 + x^3 - 2x^2 + x + 1)$

31. $\lim_{x\to 2} \dfrac{1}{x - 2}$

32. $\lim_{x\to 1} \dfrac{1}{(x - 1)^2}$

33. $\lim_{x\to 2} \dfrac{3x^2 - 4x + 2}{7x^2 - 5x + 3}$

34. $\lim_{x\to -1} \dfrac{4x^2 + 5x - 7}{3x^2 - 2x + 1}$

35. $\lim_{x\to 2} \dfrac{x^2 + x - 6}{x^2 - 4}$

36. $\lim_{x\to 4} \dfrac{x^2 - 16}{x^2 - x - 12}$

37. ▦ $\lim_{x\to 1} \dfrac{1 - \sqrt{x}}{1 - x}$

38. ▦ $\lim_{x\to 4} \dfrac{\sqrt{x} - 2}{x - 4}$

39. $\lim_{x\to \infty} \dfrac{2x - 4}{5x}$

40. $\lim_{x\to \infty} \dfrac{3x + 1}{4x}$

41. $\lim_{x\to \infty} \left(5 - \dfrac{2}{x}\right)$

42. $\lim_{x\to \infty} \left(7 + \dfrac{3}{x}\right)$

43. $\lim_{x\to \infty} \dfrac{2x - 5}{4x + 3}$

44. $\lim_{x\to \infty} \dfrac{6x + 1}{5x - 2}$

45. $\lim_{x\to \infty} \dfrac{2x^2 - 5}{3x^2 - x + 7}$

46. $\lim_{x\to \infty} \dfrac{4 - 3x - 12x^2}{1 + 5x + 3x^2}$

47. Consider

$$f(x) = \begin{cases} 1 & \text{for } x \neq 2, \\ -1 & \text{for } x = 2. \end{cases}$$

Find each of the following.

a) $\lim_{x\to 0} f(x)$

b) $\lim_{x\to 2^-} f(x)$

c) $\lim_{x\to 2^+} f(x)$

d) $\lim_{x\to 2} f(x)$

e) Is f continuous at 0? at 2?

48. Consider

$$g(x) = \begin{cases} -4 & \text{for } x = 3, \\ 2x + 5 & \text{for } x \neq 3. \end{cases}$$

Find each of the following.

a) $\lim_{x\to 3^-} g(x)$

b) $\lim_{x\to 3^+} g(x)$

c) $\lim_{x\to 3} g(x)$

d) $\lim_{x\to 2} g(x)$

e) Is g continuous at 3? at 2?

49. *Business: Depreciation.* A new conveyor system costs \$10,000. In any year, it depreciates 8% of its value at the beginning of that year.

a) What is the annual depreciation in each of the first five years?

b) What is the total depreciation at the end of ten years?

c) What is the limit of the sum of the annual depreciation costs?

50. *Business: Depreciation.* A new car costs \$6000. In any year, it depreciates 30% of its value at the beginning of that year.

a) What is the annual depreciation in each of the first five years?

b) What is the total depreciation at the end of ten years?

c) What is the limit of the sum of the annual depreciation costs?

51. Inside its own 5-yd line, a defensive football team is penalized half the distance to the goal. Suppose a defensive team keeps getting such a penalty. What is the limit of the distance of the offensive team from the goal? Can the offensive team ever score a touchdown in this manner?

SYNTHESIS

Find the limit, if it exists.

52. $\lim_{x \to 0} \frac{|x|}{x}$

53. 🖩 $\lim_{x \to 1} \frac{2 - \sqrt{x + 3}}{x - 1}$

54. $\lim_{x \to 1} \frac{x^3 - 1}{x^2 - 1}$

55. $\lim_{x \to \infty} \frac{4 - 3x}{5 - 2x^2}$

56. $\lim_{x \to \infty} \frac{6x^5 - x^4}{4x^2 - 3x^3}$

11.7 Rational Functions

OBJECTIVE

You should be able to:

1 Sketch a graph of a rational function.

1 A **rational function** is a function definable as a quotient of two polynomials. Here are some examples:

$$f(x) = \frac{x^2 + 3x - 5}{x - 4}, \qquad f(x) = \frac{8}{x^2 + 1}, \qquad f(x) = \frac{6x^5 - 7x + 11}{4}.$$

DEFINITION

A *rational function* is a function *f* that can be described by

$$f(x) = \frac{P(x)}{Q(x)},$$

where $P(x)$ and $Q(x)$ are polynomials having no common factor other than 1 and −1, and with $Q(x)$ not the zero polynomial. The domain of f consists of all inputs x for which $Q(x) \neq 0$.

Polynomial functions are themselves special kinds of rational functions, since $Q(x)$ can be the polynomial 1. Here we are interested in graphing rational functions in which the denominator is not a constant. We begin with the simplest such functions.

Graphs of Transformations of Equations of the Type $f(x) = a/x^n$

EXAMPLE 1 Graph: $f(x) = \frac{1}{x}$.

Solution We have considered this graph before. We consider it again from the viewpoint of setting up the general graphing of rational functions.

a) The domain of this function consists of all real numbers except 0.
b) For nonzero x or y, the equation above is equivalent to $xy = 1$. Now it is easy to see that the graph is symmetric with respect to the line $y = x$ (because interchanging x and y produces an equivalent equation).
c) The function is odd. Thus the graph is also symmetric with respect to the origin (because replacing x with $-x$ and y with $-y$ produces an equivalent equation).
d) Now we find some values, keeping in mind the two symmetries.

x	1	2	3	4	5
$f(x)$	1	$\frac{1}{2}$	$\frac{1}{3}$	$\frac{1}{4}$	$\frac{1}{5}$

e) We plot these points. We use one symmetry to get the points in the first quadrant. We use the other symmetry to get the points in the third quadrant.

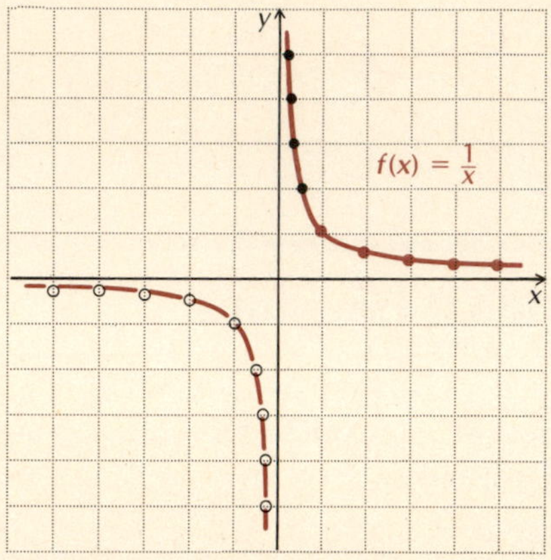

The points indicated by ● are obtained from the table. Those marked ● are obtained by reflection across the line $y = x$. Following this, the points marked ○ are obtained by reflection across the origin.

Asymptotes

Note that the curve in Example 1 does not touch either axis, but comes very close. As $|x|$ becomes very large, the curve comes very near to the x-axis. In fact, we can find points as close to the x-axis as we please by choosing x large enough or x small enough. The following limit statements convey this information:

$$\lim_{x \to -\infty} \frac{1}{x} = 0 \quad \text{and} \quad \lim_{x \to \infty} \frac{1}{x} = 0.$$

From these limit facts, we say that the curve approaches the x-axis *asymptotically*, and we say that the x-axis is a **horizontal asymptote** to the curve.

DEFINITION

The line $y = b$ is a *horizontal asymptote* if either or both of the following limit statements are true:

$$\lim_{x \to -\infty} f(x) = b \quad \text{or} \quad \lim_{x \to \infty} f(x) = b.$$

The following figures illustrate horizontal asymptotes.

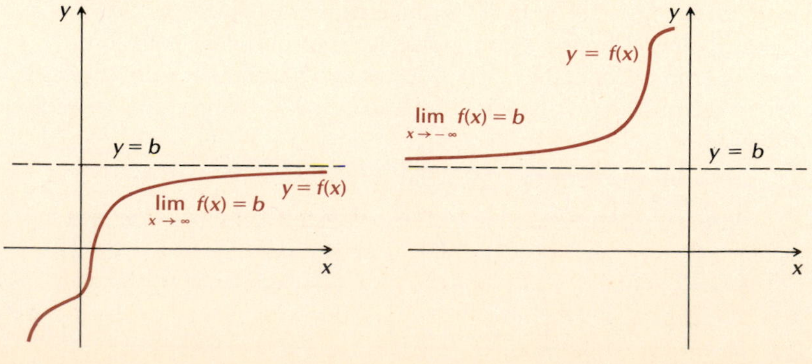

We also note in Example 1 that as x values get closer and closer to 0 from the left, the function values (y-values) get more and more negative. In fact, by selecting x close enough to 0, we can get the function values as small negatively as we wish. Similarly, as we approach 0 from the right, the function values get as large as we wish. The following limit statements convey this information:

$$\lim_{x\to 0^-} \frac{1}{x} = -\infty \quad \text{and} \quad \lim_{x\to 0^+} \frac{1}{x} = \infty.$$

We say that the y-axis is a **vertical asymptote** to this curve.

DEFINITION

The line $x = a$ is a *vertical asymptote* if any of the following limit statements are true:

$$\lim_{x\to a^-} f(x) = \infty \quad \text{or} \quad \lim_{x\to a^-} f(x) = -\infty \quad \text{or}$$
$$\lim_{x\to a^+} f(x) = \infty \quad \text{or} \quad \lim_{x\to a^+} f(x) = -\infty.$$

The following figure shows the four ways in which a vertical asymptote can occur.

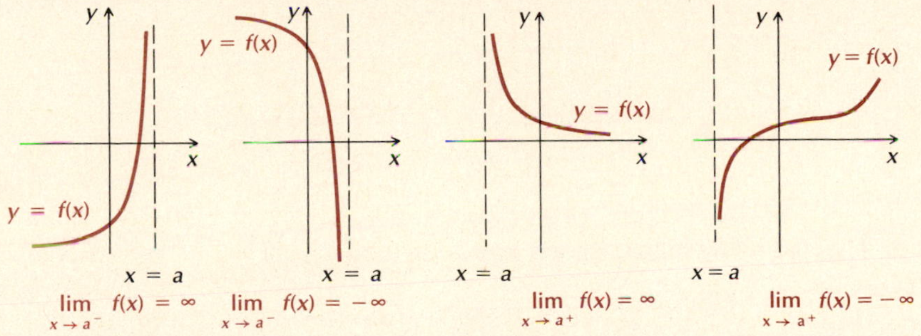

The following fact is true regarding horizontal and vertical asymptotes.

THEOREM 15

The graph of a rational function may or may not cross a horizontal asymptote. The graph of a rational function never crosses a vertical asymptote. If a is a zero of the denominator of the rational function, then the line $x = a$ is a vertical asymptote. (Remember that this theorem applies to rational functions, defined such that the numerator and denominator have no common factor other than 1 or -1.)

Note that if a is a zero of the denominator, then the function is not defined at $x = a$. Thus it cannot have a function value at a, so the graph cannot cross the line $x = a$, which happens to be a vertical asymptote.

EXAMPLE 2 Graph: $f(x) = \dfrac{1}{x^2}$.

Solution

a) Note that this function is defined for all values of x except 0. Note also that for all other values of x, $x^2 > 0$, so $1/x^2 > 0$. Therefore, the entire graph is above the x-axis.

b) Since $\lim_{x\to 0^-} (1/x^2) = \infty$, we know that the line $x = 0$, the y-axis, is a vertical asymptote. We also know this since $\lim_{x\to 0^+} (1/x^2) = \infty$. As $|x|$ gets very close to 0, $f(x)$ gets very large.

Sketch a graph of the rational function. Use graph paper.

1. $f(x) = \frac{4}{x}$

2. $f(x) = \frac{32}{x^4}$

3. $f(x) = \frac{-1}{x^2}$

4. $f(x) = \frac{-1}{x^3}$

c) Since $\lim_{x\to\infty} (1/x^2) = 0$ and $\lim_{x\to-\infty} (1/x^2) = 0$, it follows that the line $y = 0$, the x-axis, is a horizontal asymptote. That is, as $|x|$ gets very large, $f(x)$ approaches 0.

d) This function is even. Therefore, the graph is symmetric with respect to the y-axis (because replacing x with $-x$ produces an equivalent equation).

e) We compute some function values and using those points and the preceding information sketch the following graph.

x	1	2	3	4	$\frac{1}{2}$	$\frac{1}{3}$
$f(x)$	1	$\frac{1}{4}$	$\frac{1}{9}$	$\frac{1}{16}$	4	9

The graph is as follows. Points marked ● are obtained from the table. Points marked ● are obtained by reflection across the y-axis.

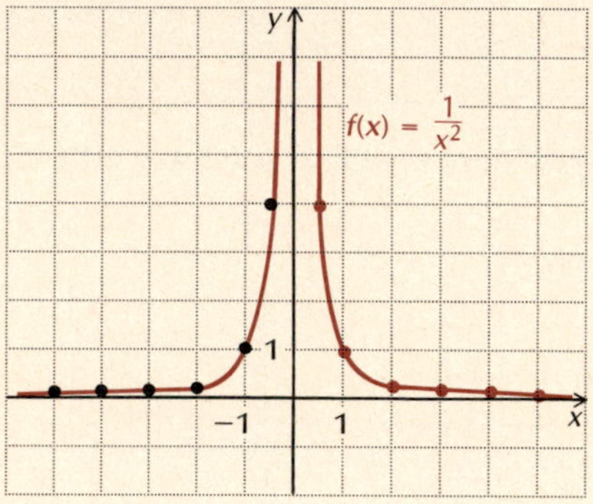

The following figure shows the four general curves of the graphs of $f(x) = a/x^n$.

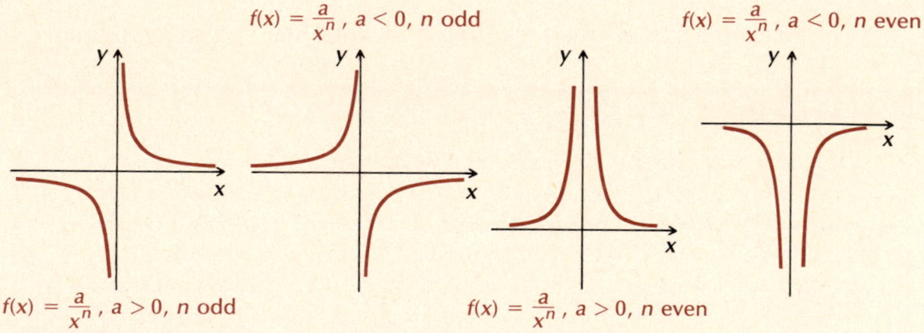

DO EXERCISES 1–4.

Using the idea of transformation (Section 3.8), we can easily graph variations of the preceding curves.

EXAMPLE 3 Graph: $f(x) = \dfrac{1}{x-2}$.

Solution The graph of $f(x) = 1/(x - 2)$ is a translation of the graph of $f(x) = 1/x$ to the right 2 units. The x-axis is a horizontal asymptote to the curve, and the line $x = 2$ is a vertical asymptote.

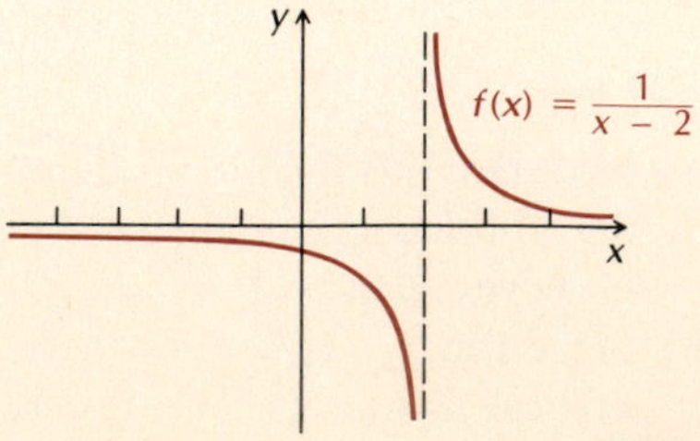

DO EXERCISES 5–10.

Occurrences of Asymptotes

It is important in graphing rational functions to determine where the asymptotes, if any, occur. Vertical asymptotes are easy to locate when a denominator is factored. The x-values that make a denominator 0, but *do not* make the numerator 0, are those of the vertical asymptotes.

EXAMPLES Determine the vertical asymptotes.

4. $f(x) = \dfrac{3x - 2}{x(x - 5)(x + 3)}$

The vertical asymptotes are the lines $x = 0$, $x = 5$, and $x = -3$.

5. $f(x) = \dfrac{x - 2}{x^3 - x}$

We factor the denominator:

$$x^3 - x = x(x + 1)(x - 1).$$

The vertical asymptotes are $x = 0$, $x = -1$, and $x = 1$.

DO EXERCISES 11 AND 12.

Horizontal asymptotes occur when the degree of the numerator is the same as or less than that of the denominator. Let us first consider a function for which the degree of the numerator is less than the degree of the denominator:

$$f(x) = \frac{2x + 3}{x^3 - 2x^2 + 4}.$$

We multiply by $(1/x^3)/(1/x^3)$ to obtain

$$f(x) = \frac{\dfrac{2}{x^2} + \dfrac{3}{x^3}}{1 - \dfrac{2}{x} + \dfrac{4}{x^3}}.$$

Let us now consider what happens to the function values as $|x|$ becomes very large. Each expression with x in its denominator takes on smaller and smaller values, approaching 0. Thus the numerator approaches 0 and the denominator approaches 1; hence the entire expression takes on values closer and closer to 0. Thus,

$$\lim_{x\to-\infty} f(x) = 0 \quad \text{and} \quad \lim_{x\to\infty} f(x) = 0,$$

so the x-axis is a horizontal asymptote.

Next we consider a function for which the numerator and the denominator have the same degree:

$$f(x) = \frac{3x^2 + 2x - 4}{2x^2 - x + 1} = \frac{3x^2 + 2x - 4}{2x^2 - x + 1} \cdot \frac{\frac{1}{x^2}}{\frac{1}{x^2}}$$

$$= \frac{3 + \dfrac{2}{x} - \dfrac{4}{x^2}}{2 - \dfrac{1}{x} + \dfrac{1}{x^2}}.$$

Sketch a graph of the rational function. Use graph paper.

5. $f(x) = \dfrac{1}{x + 5}$

6. $f(x) = \dfrac{3x + 1}{x}$

7. $f(x) = 3 - \dfrac{2}{x}$

8. $f(x) = \dfrac{1}{3x + 15}$

[*Hint*: $3x + 15 = 3(x + 5)$.] Compare with Exercise 5.

9. $f(x) = \dfrac{1}{(x - 3)^2}$

10. $f(x) = \dfrac{-3x^2 + 1}{x^2}$

(*Hint*: Divide the numerator by the denominator.)

Find the vertical asymptotes.

11. $f(x) = \dfrac{x + 3}{x^3 - x^2 - 6x}$

12. $f(x) = \dfrac{x^2 + 5}{2x^3 + x^2 - 8x - 4}$

13. For which of the following is the x-axis an asymptote?

a) $f(x) = \dfrac{3x^4 - x^2 + 4}{18x^4 - x^3 + 44}$

b) $f(x) = \dfrac{17x^2 + 14x - 5}{x^3 - 2x - 1}$

c) $f(x) = \dfrac{135x^5 - x^2}{x^7}$

Find the horizontal asymptotes.

14. $f(x) = \dfrac{3x^3 + 4x - 9}{6x^3 - 7x^2 + 3}$

15. $f(x) = \dfrac{9x^4 - 7x^2 - 9}{3x^4 + 7x^2 + 9}$

As $|x|$ gets very large, the numerator approaches 3 and the denominator approaches 2. Therefore, the function gets very close to $\frac{3}{2}$. Thus,

$$\lim_{x\to\infty} f(x) = \tfrac{3}{2} \quad \text{and} \quad \lim_{x\to-\infty} f(x) = \tfrac{3}{2}.$$

Therefore, the line $y = \frac{3}{2}$ is a horizontal asymptote. From this example, we can see that when the degrees of the numerator and the denominator are the same, the asymptote can be determined by dividing the leading coefficients of the two polynomials.

THEOREM 16

Whenever the degree of the numerator is less than the degree of the denominator, the x-axis, or $y = 0$, is a horizontal asymptote.

Whenever the degree of the numerator equals the degree of the denominator, the line $y = a/b$ is a horizontal asymptote, where a is the leading coefficient of the numerator and b is the leading coefficient of the denominator.

EXAMPLES Determine the horizontal asymptotes.

6. $f(x) = \dfrac{5x^3 - x^2 + 7}{3x^5 + x - 10}$

The line $y = 0$ is a horizontal asymptote.

7. $f(x) = \dfrac{-7x^4 - 10x^2 + 1}{11x^4 + x - 2}$

The line $y = -\frac{7}{11}$ is a horizontal asymptote.

DO EXERCISES 13–15.

Oblique Asymptotes

There are asymptotes that are neither vertical nor horizontal. Lines that are nonvertical and nonhorizontal can be asymptotes. They are called **oblique asymptotes.** Such a situation is shown in the following graph.

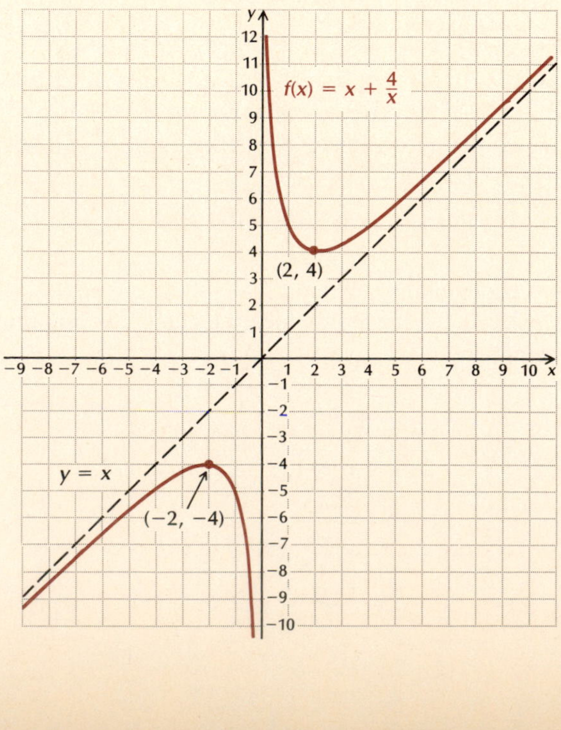

DEFINITION

The line $y = mx + b$ is an *oblique asymptote* of the rational function $f(x) = P(x)/Q(x)$ if

$$\lim_{x\to -\infty} f(x) = mx + b \quad \text{or} \quad \lim_{x\to\infty} f(x) = mx + b.$$

Oblique asymptotes occur when the degree of the numerator is 1 more than the degree of the denominator. One way to find such an asymptote is by dividing the numerator by the denominator. Consider

$$f(x) = \frac{2x^2 - 7x - 7}{x - 5}.$$

When we divide the numerator by the denominator, we obtain a quotient of $2x + 3$ and a remainder of 8. Thus,

$$f(x) = (2x + 3) + \frac{8}{x - 5}.$$

Now we can see that when $|x|$ gets very large, $8/(x - 5)$ approaches 0, and the y-values approach $2x + 3$. Thus,

$$\lim_{x\to\infty} f(x) = 2x + 3 \quad \text{and} \quad \lim_{x\to -\infty} f(x) = 2x + 3.$$

For very large and very small values of x, $f(x)$ approaches the line $y = 2x + 3$ as a limit. Thus, $y = 2x + 3$ is an oblique asymptote.

DO EXERCISES 16 AND 17.

The following summarizes the conditions under which asymptotes occur.

Asymptotes of a rational function can occur as follows.

1. ***Vertical asymptotes* occur at the x-values that make the denominator *zero*, but the numerator nonzero.**
2. **The *x-axis is an asymptote* when the degree of the numerator is less than the degree of the denominator.**
3. ***Horizontal asymptotes other than the x-axis* occur when the numerator and the denominator have the same degree.**
4. ***Oblique asymptotes* occur when the degree of the numerator is 1 more than the degree of the denominator.**

Zeros, or Roots

Zeros, or roots, of a rational function occur when the numerator is 0 but the denominator is not. The zeros of a function occur at points where the graph crosses the x-axis. Therefore, knowing the zeros helps in making a graph. If the numerator can be factored, the zeros are easy to determine.

EXAMPLE 8 Find the zeros of the rational function

$$f(x) = \frac{x^3 - x^2 - 6x}{x^2 - 3x + 2}.$$

Solution We factor the numerator and the denominator:

$$f(x) = \frac{x(x + 2)(x - 3)}{(x - 1)(x - 2)}.$$

Find the oblique asymptotes.

16. $f(x) = \dfrac{3x^2 - 7x + 2}{x - 2}$

17. $f(x) = \dfrac{5x^3 + 2x + 1}{x^2 - 4}$

Find the zeros.

18. $f(x) = \dfrac{x(x-3)(x+5)}{(x+2)(x-4)}$

19. $f(x) = \dfrac{x^3 + 2x^2 - 3x}{x^2 + 5}$

The x-values making the numerator zero are 0, -2, and 3. Since none of these makes the denominator zero, they are the zeros of the function. The graph will cross the x-axis at $(0, 0)$, $(-2, 0)$, and $(3, 0)$. ■

DO EXERCISES 18 AND 19.

The following is an outline of the procedure to be followed in graphing rational functions.

To graph a rational function:

a) Determine any symmetries.
b) Determine any horizontal or oblique asymptotes and sketch them.
c) Factor the numerator and the denominator.
  i) Determine any vertical asymptotes and sketch them.
  ii) Determine the zeros, if possible, and plot them.
d) Determine where the function values are positive or negative.
e) Make a table of values and plot them.
f) Sketch the curve.

EXAMPLE 9 Graph: $g(x) = \dfrac{1}{x^2 - 3}$.

Solution We follow the outline above.

a) The function is even, so the graph is symmetric about the y-axis.

b) Since the degree of the numerator is less than the degree of the denominator, the x-axis is a horizontal asymptote. There are no oblique asymptotes.

c) We can factor the denominator, but not the numerator. We have

$$g(x) = \frac{1}{(x - \sqrt{3})(x + \sqrt{3})}.$$

The zeros of the denominator are $\pm\sqrt{3}$. Thus, $x = \sqrt{3}$ and $x = -\sqrt{3}$ are vertical asymptotes of the graph of g. We draw these asymptotes with dashed lines. The function g has no zeros.

d) The vertical asymptotes divide the x-axis into intervals as follows. To find where the function is positive or negative, we use the procedure of Section 4.7. We try a test point in each interval.

Interval	$(-\infty, -\sqrt{3})$	$(-\sqrt{3}, \sqrt{3})$	$(\sqrt{3}, \infty)$
Test value	$g(-4) = \frac{1}{13}$	$g(0) = \frac{1}{3}$	$g(4) = \frac{1}{13}$
Sign of $g(x)$	Positive	Negative	Positive
Location of points on graphs	Above x-axis	Below x-axis	Above x-axis

e) We use the preceding information, compute some other function values, and complete the sketch of the graph of g.

x	$g(x)$, approximately
−5	0.05
−3	0.2
−2	1
−1	−0.5
0	−0.3
1	−0.5
2	1
3	0.2
5	0.05

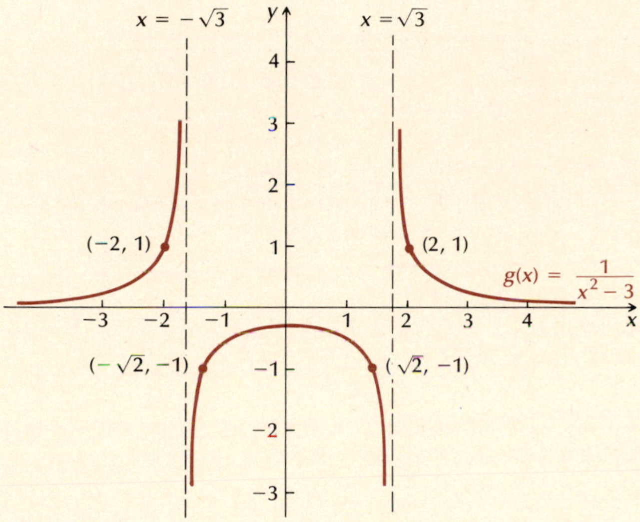

DO EXERCISES 20 AND 21.

EXAMPLE 10 Graph: $f(x) = \dfrac{x^3 - x^2 - 6x}{x^2 - 3x + 2}$.

Solution We follow the outline above.

a) The function is neither even nor odd and no symmetries are apparent.

b) The degree of the numerator is 1 more than the degree of the denominator, so we have an oblique asymptote. Dividing numerator by denominator will show that the line $y = x + 2$ is an oblique asymptote. We sketch the graph of $y = x + 2$ with a dashed line.

c) The numerator and denominator are easily factorable. We have

$$f(x) = \frac{x(x + 2)(x + 3)}{(x - 1)(x - 2)}.$$

The zeros of the denominator are 1 and 2. Thus the vertical asymptotes are $x = 1$ and $x = 2$. We sketch the vertical asymptotes using dashed lines. The zeros of the function are 0, −2, and 3. We plot the points (0, 0), (−2, 0), and (3, 0).

d) The zeros and vertical asymptotes divide the x-axis into intervals as follows. To find where the function is positive or negative, we use the procedure of Section 4.7. We try a test point in each interval to see if each satisfies the original inequality.

Sketch a graph of the rational function. Use graph paper.

20. $f(x) = \dfrac{1}{x^2 - 1}$

21. $f(x) = \dfrac{1}{x^2 + 4}$

Interval	$(-\infty, -2)$	$(-2, 0)$	$(0, 1)$	$(1, 2)$	$(2, 3)$	$(3, \infty)$
Test value	$f(-3) = -0.9$	$f(-1) = 0.7$	$f(0.5) = -4.2$	$f(1.5) = 31.5$	$f(2.5) = -7.5$	$f(4) = 4$
Sign of $f(x)$	Negative	Positive	Negative	Positive	Negative	Positive
Location of points on graph	Below x-axis	Above	Below	Above	Below	Above

e) Next we make a table of function values in addition to those found as test values.

x	$f(x)$, approximately
-10	-7.9
-8	-5.9
-6	-3.9
-5	-2.9
5	5.8
6	7.2
8	9.5
10	11.7

f) Using all available information, we draw the graph.

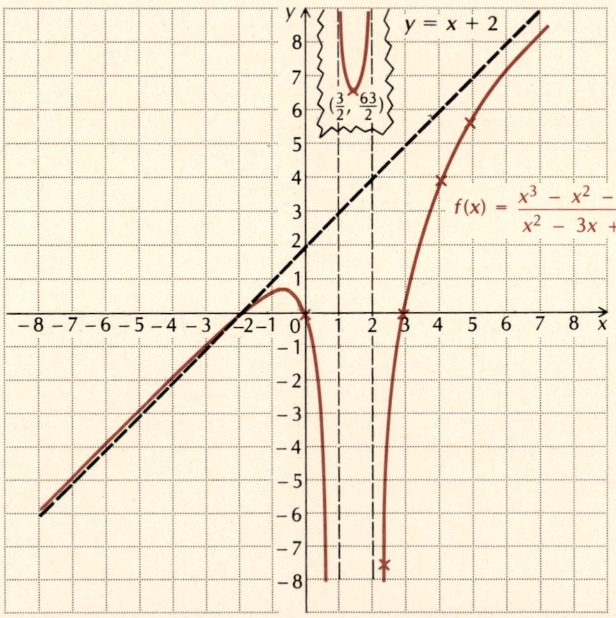

The lower left part of the graph is as shown below. The curve crosses the oblique asymptote at $(-2, 0)$, and then, moving to the left, comes back close to the asymptote from above. Note that a small "hump" occurs in the graph between -6 and -5. Are there other points of intersection of the graph with the line $y = x + 2$? Setting $f(x) = x + 2$ and solving will show that the only solution is -2, so there are no other points of intersection.

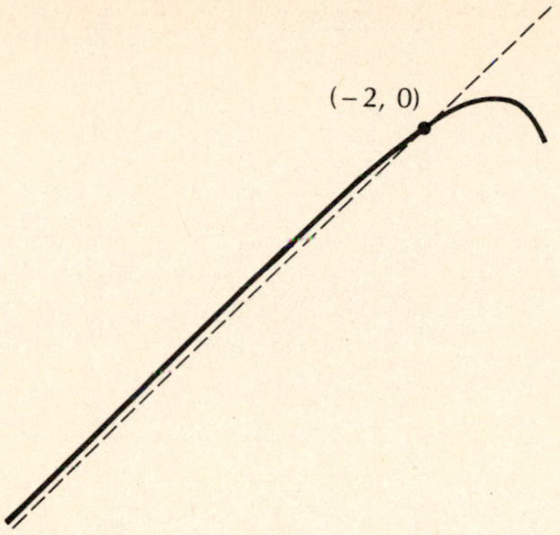

It is sometimes written that graphs do not cross their asymptotes. The curve in Example 10 and the one in Example 11, which follows, illustrate that this is indeed not the case for horizontal or oblique asymptotes. But it is true that graphs of functions do not cross vertical asymptotes.

EXAMPLE 11 Graph:

$$f(x) = \frac{x^2 - 1}{x^2 + x - 6}.$$

Solution Following the outline of steps, we obtain the following graph. Note that the graph crosses its horizontal asymptote and then approaches it from below as x gets very large.

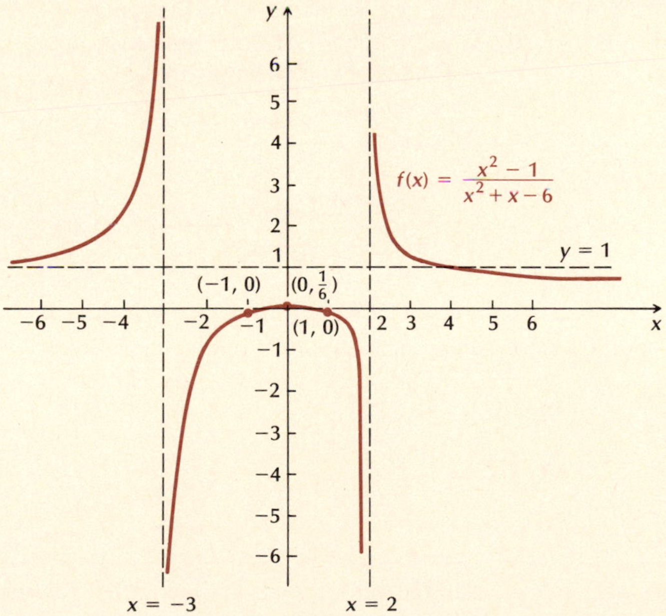

DO EXERCISES 22 AND 23.

Sketch the graph of the function.

22. $f(x) = \dfrac{x(x-3)(x+5)}{(x+2)(x-4)}$

23. $f(x) = \dfrac{x^2 - 4}{x^2 - x - 12}$

EXERCISE SET 11.7

1 Sketch the graph of the rational function. Be sure to sketch all the asymptotes.

1. $f(x) = \dfrac{1}{x - 3}$

2. $f(x) = \dfrac{1}{x - 5}$

3. $f(x) = \dfrac{-2}{x - 5}$

4. $f(x) = \dfrac{-3}{x - 3}$

5. $f(x) = \dfrac{2x + 1}{x}$

6. $f(x) = \dfrac{3x - 1}{x}$

7. $f(x) = \dfrac{1}{(x - 2)^2}$

8. $f(x) = \dfrac{-2}{(x - 3)^2}$

9. $f(x) = \dfrac{2}{x^2}$

10. $f(x) = \dfrac{1}{3x^2}$

11. $f(x) = \dfrac{1}{x^2 + 3}$

12. $f(x) = \dfrac{-1}{x^2 + 2}$

13. $f(x) = \dfrac{x - 1}{x + 2}$

14. $f(x) = \dfrac{x - 2}{x + 1}$

15. $f(x) = \dfrac{3x}{x^2 + 5x + 4}$

16. $f(x) = \dfrac{x + 3}{2x^2 - 5x - 3}$

17. $f(x) = \dfrac{x^2 - 4}{x - 1}$

18. $f(x) = \dfrac{x^2 - 9}{x + 1}$

19. $f(x) = \dfrac{x^2 + x - 2}{2x^2 + 1}$

20. $f(x) = \dfrac{x^2 - 2x - 3}{3x^2 + 2}$

21. $f(x) = \dfrac{x - 1}{x^2 - 2x - 3}$

22. $f(x) = \dfrac{x + 2}{x^2 + 2x - 15}$

23. $f(x) = \dfrac{x + 2}{(x - 1)^3}$

24. $f(x) = \dfrac{x - 3}{(x + 1)^3}$

25. $f(x) = \dfrac{x^3 + 1}{x}$

26. $f(x) = \dfrac{x^3 - 1}{x}$

27. $f(x) = \dfrac{x^3 + 2x^2 - 15x}{x^2 - 5x - 14}$

28. $f(x) = \dfrac{x^3 + 2x^2 - 3x}{x^2 - 25}$

29. $f(x) = \dfrac{5x^4}{x^4 + 1}$

30. $f(x) = \dfrac{x + 1}{x^2 + x - 6}$

31. $f(x) = \dfrac{x^2 - x - 2}{x + 2}$

32. $f(x) = \dfrac{x^2}{x^2 - x - 2}$

33. *Distance as a function of time*. The distance from Kansas City to Indianapolis is 500 miles. A car makes the trip in time t, in hours, at various speeds r, in miles per hour.
 a) Express the time t as a rational function of the speed r.
 b) Sketch a graph of the function over the interval $(0, \infty)$.

34. *Electrical resistance*. The total resistance in a circuit with two resistors in parallel is given by

$$R = \frac{R_1 R_2}{R_1 + R_2},$$

where R_1 and R_2 are positive.
 a) A circuit has a 10-ohm resistor and a variable resistor in parallel. Express the total resistance of the circuit as a function of the resistance of the variable resistor, r.
 b) Graph the function.
 c) What are the domain and the range?
 d) What number does the resistance approach as the resistance gets larger and larger?

SYNTHESIS

Sketch a graph of the function.

35. $f(x) = \dfrac{x^3 + 4x^2 + x - 6}{x^2 - x - 2}$

36. $f(x) = \dfrac{2x^3 + x^2 - 8x - 4}{x^3 + x^2 - 9x - 9}$

COMPUTER-CALCULATOR EXERCISES

Sketch a graph of each of the following using a computer software graphing package or a graphing calculator.

37. $f(x) = \dfrac{x^4 + 3x^3 + 21x^2 - 50x + 80}{x^4 + 8x^3 - x^2 + 20x - 10}$

38. $f(x) = \dfrac{x^3 + x^2 + x}{x^2 - 1}$

39. $f(x) = \dfrac{x^3}{x^2 - 1}$

40. $f(x) = \dfrac{x^3 + 4}{x}$

CHALLENGE

41. *Time of flight*. A pilot can make a round-trip flight between two cities of the same latitude in time t_0 flying at velocity v, if there is no wind. If there is a wind of velocity w due east or west, the time of flight T is given by

$$T = \frac{v^2 t_0}{v^2 - w^2}.$$

 a) Express the time of flight T as a function of the wind speed w, given that the cities are 300 km apart and the plane flies between them in still air at a speed of 200 km/h.
 b) What is the time of flight if the wind speed is 5 km/h? 10 km/h? 20 km/h?
 c) What are the domain and the range of the function? Use the constraints of the problem in your determination.
 d) Graph the function.

42. *Weight above the earth.* A person's weight at height h above sea level satisfies the equation

$$W = \left(\frac{r}{r + h}\right)^2 W_0,$$

where W_0 is a person's weight at sea level and r is the radius of the earth, in miles.

a) Suppose a person weighs 150 lb at sea level. Express W as a function of h. Assume that the radius of the earth is about 3965 miles.

b) Sketch a graph of this function over the interval $[0, 20{,}000]$.

c) At what height would the person weigh 50 lb?

43. *Surface area.* A container firm is designing an open-top rectangular box, with a square base, that will hold 108 cubic centimeters (cc).

a) Express the surface area S as a function of the length x of a side of the base.

b) Sketch a graph of the function on the interval $(0, \infty)$.

c) Estimate the minimum surface area and the value of x that will yield the minimum surface area.

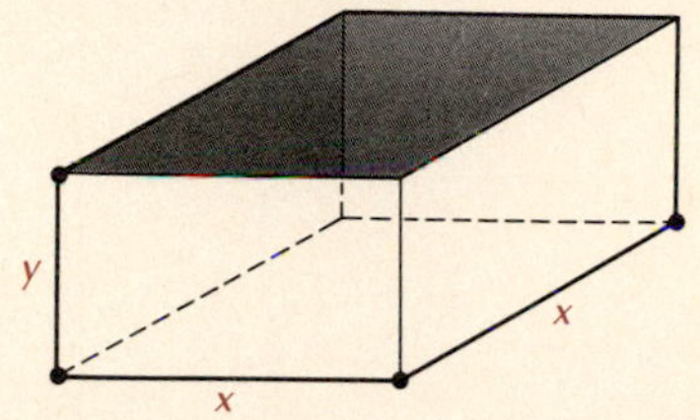

11.8 Partial Fractions

OBJECTIVE

You should be able to:

1 **Decompose fractional expressions into partial fractions.**

1 There are situations in which it is helpful to do the reverse of adding fractional expressions, that is, to **decompose** a fractional expression into a sum of several fractional expressions. The following illustrates:

$$\frac{3}{x + 2} - \frac{2}{2x - 3} = \frac{4x - 13}{2x^2 + x - 6}.$$

Adding is straightforward. The present problem is to take the sum and find the fractions that were added to get it. That procedure turns out to be rather straightforward also, and it is based on the following theorem, whose proof we omit.

Theorem 17

Any rational expression $P(x)/Q(x)$, with no common factor other than 1 or -1, where the degree of the numerator is less than that of the denominator, can be decomposed into partial fractions.

To each linear factor $ax + b$ of $Q(x)$, there corresponds a fraction $A/(ax + b)$. To each linear factor $ax + b$ occurring twice in $Q(x)$, there correspond two fractions: $B_1/(ax + b)$ and $B_2/(ax + b)^2$. If $ax + b$ occurs three times, there corresponds an additional fraction, $B_3/(ax + b)^3$, and so on.

To each quadratic factor $ax^2 + bx + c$ (presuming it is *irreducible*, meaning that it cannot be factored into linear factors with real coefficients) of $Q(x)$, there corresponds a fraction

$$(Ax + B)/(ax^2 + bx + c).$$

If the quadratic factor occurs twice, there is an additional fraction, $(Cx + D)/(ax^2 + bx + c)^2$, and so on.* The expressions in the numerators $A, B, \ldots,$ are constants, hence they do not depend on x.

*This theorem covers all situations, because any polynomial $Q(x)$ with real coefficients has a factorization into linear and quadratic factors.

1. Decompose into partial fractions:

$$\frac{13x + 5}{3x^2 - 7x - 6}.$$

Theorem 17 assumes that $P(x)/Q(x)$ is simplified. For our first example, we use the fractional expression in the opening paragraph.

EXAMPLE 1 Decompose into partial fractions:

$$\frac{4x - 13}{2x^2 + x - 6}.$$

Solution We begin by factoring the denominator: $(x + 2)(2x - 3)$. By Theorem 17, we know that there are constants A and B such that

$$\frac{4x - 13}{(x + 2)(2x - 3)} = \frac{A}{x + 2} + \frac{B}{2x - 3}.$$

To determine A and B, we add on the right:

$$\frac{4x - 13}{(x + 2)(2x - 3)} = \frac{A(2x - 3) + B(x + 2)}{(x + 2)(2x - 3)}.$$

Next, we equate the numerators:

$$4x - 13 = A(2x - 3) + B(x + 2).$$

Since the latter equation containing A and B is true for all x, we can substitute any value of x whatever and still have a true equation. We let $2x - 3 = 0$, or $x = \frac{3}{2}$. Then we get

$$4\left(\tfrac{3}{2}\right) - 13 = A\left(2 \cdot \tfrac{3}{2} - 3\right) + B\left(\tfrac{3}{2} + 2\right)$$
$$-7 = 0 + \tfrac{7}{2}B.$$

Solving, we obtain $B = -2$. Next we let $x + 2 = 0$, or $x = -2$. Substituting, we get

$$4(-2) - 13 = A[2(-2) - 3] + B(-2 + 2).$$

Solving, we obtain $A = 3$.

The decomposition is as follows:

$$\frac{3}{x + 2} - \frac{2}{2x - 3}.$$

To check, we would add, to see if we get the original expression.

The values A and B can also be determined with a system of equations resulting from equating the corresponding coefficients. The equation $4x - 13 = A(2x - 3) + B(x + 2)$ can be written as the equation $4x - 13 = (2A + B)x + (2B - 3A)$. Using this form, we can equate the coefficients and solve the resulting system, $4 = 2A + B$ and $-13 = 2B - 3A$. We will consider this in more depth later.

DO EXERCISE 1.

EXAMPLE 2 Decompose into partial fractions:

$$\frac{7x^2 - 29x + 24}{(2x - 1)(x - 2)^2}.$$

Solution By Theorem 17, the decomposition looks like the following:

$$\frac{A}{2x - 1} + \frac{B}{x - 2} + \frac{C}{(x - 2)^2}.$$

As in Example 1, we add and equate numerators. This gives us

$$7x^2 - 29x + 24 = A(x-2)^2 + B(2x-1)(x-2) + C(2x-1). \quad (1)$$

Since the equation containing A, B, and C is true for all x, we can substitute any value of x whatever and still have a true equation. We let x be such that $2x - 1 = 0$, or $x = \frac{1}{2}$. Then we get

$$7\left(\frac{1}{2}\right)^2 - 29 \cdot \frac{1}{2} + 24 = A\left(\frac{1}{2} - 2\right)^2 + 0.$$

Solving, we obtain $A = 5$. Next, we let $x - 2 = 0$, or $x = 2$. Substituting gives us

$$7(2)^2 - 29(2) + 24 = 0 + C(2 \cdot 2 - 1).$$

Solving, we obtain $C = -2$.

To find B, we first simplify Eq. (1):

$$\begin{aligned} 7x^2 - 29x + 24 &= A(x^2 - 4x + 4) + B(2x^2 - 5x + 2) + C(2x - 1) \\ &= Ax^2 - 4Ax + 4A + 2Bx^2 - 5Bx + 2B + 2Cx - C \\ &= (A + 2B)x^2 + (-4A - 5B + 2C)x + (4A + 2B - C). \end{aligned}$$

Then we equate the coefficients of x^2:

$$7 = A + 2B.$$

Substituting 5 for A and solving for B gives us $B = 1$. The decomposition is as follows:

$$\frac{5}{2x-1} + \frac{1}{x-2} - \frac{2}{(x-2)^2}.$$

DO EXERCISE 2.

EXAMPLE 3 Decompose into partial fractions:

$$\frac{x^2 - 17x + 35}{(x^2+1)(x-4)}.$$

Solution By Theorem 17, the decomposition looks like the following:

$$\frac{Ax + B}{x^2 + 1} + \frac{C}{x - 4}.$$

Adding and equating numerators, we get

$$x^2 - 17x + 35 = (Ax + B)(x - 4) + C(x^2 + 1). \quad (2)$$

Letting $x = 4$, we get

$$4^2 - 17 \cdot 4 + 35 = 0 + C(4^2 + 1).$$

Solving, we obtain $C = -1$.

To find A and B, we first simplify Eq. (2):

$$\begin{aligned} x^2 - 17x + 35 &= Ax^2 + Bx - 4Ax - 4B + Cx^2 + C \\ &= (A + C)x^2 + (B - 4A)x + (-4B + C). \end{aligned}$$

Equating the coefficients of x^2, we get $1 = A + C$. Since $C = -1$, we know that $A = 2$. Equating the constant terms, we get $35 = -4B + C$. This gives us $B = -9$. The decomposition is as follows:

$$\frac{2x - 9}{x^2 + 1} - \frac{1}{x - 4}.$$

2. Decompose into partial fractions:

$$\frac{3x^2 - 3x - 2}{(x+1)(x-1)^2}.$$

Theorem 17 refers only to rational expressions $P(x)/Q(x)$ in which the degree of the numerator is less than the degree of the denominator. As we show in the following example, this is not much of a restriction. For an expression not satisfying the condition, we can divide numerator by denominator. The result will be some polynomial plus a fraction $R(x)/Q(x)$, where $R(x)$ is the remainder. The latter fraction meets the conditions of Theorem 17, hence it can be decomposed.

DO EXERCISE 3.

EXAMPLE 4 Decompose into partial fractions:

$$\frac{6x^3 + 5x^2 - 7}{3x^2 - 2x - 1}.$$

Solution Since the degree of the numerator is not less than that of the denominator, we divide:

$$\begin{array}{r} 2x + 3 \\ 3x^2 - 2x - 1 \overline{\smash{)}\, 6x^3 + 5x^2 - 7} \\ \underline{6x^3 - 4x^2 - 2x} \\ 9x^2 + 2x - 7 \\ \underline{9x^2 - 6x - 3} \\ 8x - 4 \end{array}$$

The original expression is thus equivalent to the following:

$$2x + 3 + \frac{8x - 4}{3x^2 - 2x - 1}.$$

We decompose the fraction as follows:

$$\frac{8x - 4}{(3x + 1)(x - 1)} = \frac{5}{3x + 1} + \frac{1}{x - 1}.$$

The final result is the following:

$$2x + 3 + \frac{5}{3x + 1} + \frac{1}{x - 1}.$$

DO EXERCISE 4.

We can also use systems of equations to decompose fractional expressions into a sum of fractional expressions. Let us reconsider Example 2.

EXAMPLE 5 Decompose into partial fractions:

$$\frac{7x^2 - 29x + 24}{(2x - 1)(x - 2)^2}.$$

Solution By Theorem 17, the decomposition looks like the following:

$$\frac{A}{2x - 1} + \frac{B}{x - 2} + \frac{C}{(x - 2)^2}.$$

We first add and equate the numerators:

$$\begin{aligned} 7x^2 - 29x + 24 &= A(x - 2)^2 + B(2x - 1)(x - 2) + C(2x - 1) \\ &= A(x^2 - 4x + 4) + B(2x^2 - 5x + 2) + C(2x - 1), \end{aligned}$$

or

$$7x^2 - 29x + 24 = (A + 2B)x^2 + (-4A - 5B + 2C)x + (4A + 2B - C).$$

3. Decompose into partial fractions:

$$\frac{2x^2 + 4x + 5}{(x^2 + 1)(x + 2)}.$$

4. Decompose into partial fractions:

$$\frac{6x^3 + 29x^2 - 8x + 18}{2x^2 + 9x - 5}.$$

Then we equate corresponding coefficients:

$$7 = A + 2B, \quad \text{The coefficients of the } x^2\text{-terms}$$
$$-29 = -4A - 5B + 2C, \quad \text{The coefficients of the } x\text{-terms}$$
$$24 = 4A + 2B - C. \quad \text{The constant terms}$$

We now have a system of three equations. We solve the system to obtain

$$A = 5, \quad B = 1, \quad \text{and} \quad C = -2.$$

The decomposition is as follows:

$$\frac{5}{2x-1} + \frac{1}{x-2} - \frac{2}{(x-2)^2}.$$

DO EXERCISE 5.

EXAMPLE 6 Decompose into partial fractions:

$$\frac{11x^2 - 8x - 7}{(2x^2 - 1)(x - 3)}.$$

Solution By Theorem 17, the decomposition looks like the following:

$$\frac{Ax + B}{2x^2 - 1} + \frac{C}{x - 3}.$$

Adding and equating the numerators, we get

$$11x^2 - 8x - 7 = (Ax + B)(x - 3) + C(2x^2 - 1)$$
$$= Ax^2 - 3Ax + Bx - 3B + 2Cx^2 - C,$$

or $$11x^2 - 8x - 7 = (A + 2C)x^2 + (-3A + B)x + (-3B - C).$$

We then equate corresponding coefficients:

$$11 = A + 2C, \quad \text{The coefficients of the } x^2\text{-terms}$$
$$-8 = -3A + B, \quad \text{The coefficients of the } x\text{-terms}$$
$$-7 = -3B - C. \quad \text{The constant terms}$$

We solve this system of three equations and obtain

$$A = 3, \quad B = 1, \quad \text{and} \quad C = 4.$$

The decomposition is as follows:

$$\frac{3x + 1}{2x^2 - 1} + \frac{4}{x - 3}.$$

DO EXERCISE 6.

5. Decompose into partial fractions:

$$\frac{x^2 + 13x + 7}{(x - 4)(x + 1)^2}.$$

6. Decompose into partial fractions:

$$\frac{5x^2 - 7x + 7}{(3x^2 + 2)(x - 1)}.$$

EXERCISE SET 11.8

I Decompose into partial fractions.

1. $\dfrac{x + 7}{(x - 3)(x + 2)}$
2. $\dfrac{2x}{(x + 1)(x - 1)}$
3. $\dfrac{7x - 1}{6x^2 - 5x + 1}$
4. $\dfrac{13x + 46}{12x^2 - 11x - 15}$
5. $\dfrac{3x^2 - 11x - 26}{(x^2 - 4)(x + 1)}$
6. $\dfrac{5x^2 + 9x - 56}{(x - 4)(x - 2)(x + 1)}$
7. $\dfrac{9}{(x + 2)^2(x - 1)}$
8. $\dfrac{x^2 - x - 4}{(x - 2)^3}$
9. $\dfrac{2x^2 + 3x + 1}{(x^2 - 1)(2x - 1)}$
10. $\dfrac{x^2 - 10x + 13}{(x^2 - 5x + 6)(x - 1)}$
11. $\dfrac{x^4 - 3x^3 - 3x^2 + 10}{(x + 1)^2(x - 3)}$
12. $\dfrac{10x^3 - 15x^2 - 35x}{x^2 - x - 6}$

13. $\dfrac{-x^2 + 2x - 13}{(x^2 + 2)(x - 1)}$

14. $\dfrac{26x^2 + 208x}{(x^2 + 1)(x + 5)}$

15. $\dfrac{6 + 26x - x^2}{(2x - 1)(x + 2)^2}$

16. $\dfrac{5x^3 + 6x^2 + 5x}{(x^2 - 1)(x + 1)^3}$

17. $\dfrac{6x^3 + 5x^2 + 6x - 2}{2x^2 + x - 1}$

18. $\dfrac{2x^3 + 3x^2 - 11x - 10}{x^2 + 2x - 3}$

19. $\dfrac{2x^2 - 11x + 5}{(x - 3)(x^2 + 2x - 5)}$

20. $\dfrac{3x^2 - 3x - 8}{(x - 5)(x^2 + x - 4)}$

Decompose into partial fractions using a system of equations.

21. $\dfrac{-4x^2 - 2x + 10}{(3x + 5)(x + 1)^2}$

22. $\dfrac{26x^2 - 36x + 22}{(x - 4)(2x - 1)^2}$

23. $\dfrac{36x + 1}{12x^2 - 7x - 10}$

24. $\dfrac{-17x + 61}{6x^2 + 39x - 21}$

25. $\dfrac{-4x^2 - 9x + 8}{(3x^2 + 1)(x - 2)}$

26. $\dfrac{11x^2 - 39x + 16}{(x^2 + 4)(x - 8)}$

SYNTHESIS

27. Decompose into partial fractions:
$$\frac{x}{x^4 - a^4}.$$

28. Decompose into partial fractions:
$$\frac{9x^3 - 24x^2 + 48x}{(x - 2)^4(x + 1)}.$$
[*Hint*: Let the expression equal
$$\frac{A}{x + 1} + \frac{P(x)}{(x - 2)^4}$$
and find $P(x)$.]

Decompose into partial fractions and then graph by addition of ordinates (adding respective fraction values).

29. $f(x) = \dfrac{3x}{x^2 + 5x + 4}$

30. $f(x) = \dfrac{x - 1}{x^2 - 2x - 3}$

CHALLENGE

Decompose into partial fractions.

31. $\dfrac{1 + \ln x^2}{(\ln x + 2)(\ln x - 3)^2}$

32. $\dfrac{1}{e^{-x} + 3 + 2e^x}$

SUMMARY AND REVIEW: CHAPTER 11

TERMS TO KNOW

Polynomial function
Zero
Root
Remainder theorem
Synthetic division
Factor theorem
Fundamental theorem of algebra
Rational roots theorem
Descartes' rule of signs
Upper bound on roots
Lower bound on roots
Limit
Continuity
Rational function
Asymptote
Horizontal asymptote
Vertical asymptote
Oblique asymptote
Partial fraction

REVIEW EXERCISES

1. Find the remainder when
$$x^4 + 3x^3 + 3x^2 + 3x + 2$$
is divided by $x + 2$.

2. Use synthetic division to find the quotient and the remainder:
$$(2x^4 - 6x^3 + 7x^2 - 5x + 1) \div (x + 2).$$

3. Use synthetic division to find $P(3)$:
$$P(x) = 2x^4 - 3x^3 + x^2 - 3x + 7.$$

4. Factor the polynomial $P(x)$. Then solve the equation $P(x) = 0$.
$$P(x) = x^3 + 7x^2 + 7x - 15$$

5. Determine whether $x + 1$ is a factor of
$$x^3 + 6x^2 + x + 30.$$

6. Find the roots of
$$P(x) = x^2(x^2 + x - 12)^2(x^2 - 16)$$
and state the multiplicity of each.

7. Find a polynomial of degree 3 with roots 0, 1, and 2.

8. Find a polynomial of lowest degree having roots 1 and -1, and having 2 as a root of multiplicity 2 and -3 as a root of multiplicity 3.

9. The equation $x^4 - 81 = 0$ has $3i$ for a root. Find the other roots.

10. A polynomial of degree 4 with rational coefficients has roots $-8 - 7i$ and $10 + \sqrt{5}$. Find the other roots.

11. List all possible rational roots of
$$2x^6 - 12x^4 + 17x^2 + 12.$$

12. Let $P(x) = x^4 - x^3 - 3x^2 - 9x - 108$. Find the rational roots of $P(x)$. If possible, find the other roots.

13. What does Descartes' rule of signs tell you about the number of positive real roots of
$$3x^{12} + 3x^4 - 7x^2 + x + 5?$$

14. What does Descartes' rule of signs tell you about the number of negative real roots of
$$6x^8 - 12x^4 + 5x^2 + x + 2?$$

15. Find the smallest positive integer that is guaranteed by Theorem 12 to be an upper bound to the roots of
$$2x^4 - 7x^2 + 2x - 1.$$

16. Find the largest negative integer that is guaranteed by Theorem 13 to be a lower bound to the roots of
$$12x^3 + 24x^2 + 10x - 6.$$

17. The equation $x^5 + x^4 - x^3 - x^2 - 2x - 2$ has a root between 1 and 2. Approximate this root to hundredths.

18. Approximate the irrational roots of
$$P(x) = x^3 - 3x^2 + 3.$$

19. Graph: $f(x) = \dfrac{x^2 + x - 6}{x^2 - x - 20}$.

20. Decompose into partial fractions: $\dfrac{5}{(x + 2)^2(x + 1)}$.

Find the limit, if it exists.

21. $\lim\limits_{x \to -2} \dfrac{8}{x}$

22. $\lim\limits_{x \to 1} (4x^3 - x^2 + 7x)$

23. $\lim\limits_{x \to -7} \dfrac{x^2 + 4x - 21}{x + 7}$

24. $\lim\limits_{x \to \infty} \dfrac{5x + 6}{x}$

Determine whether the graph is continuous.

25.

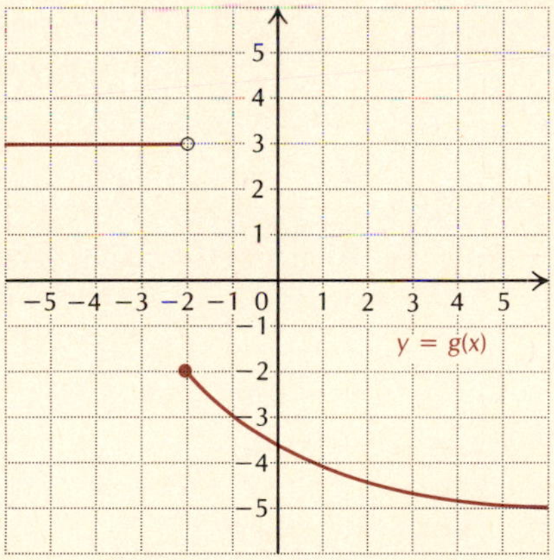

26.

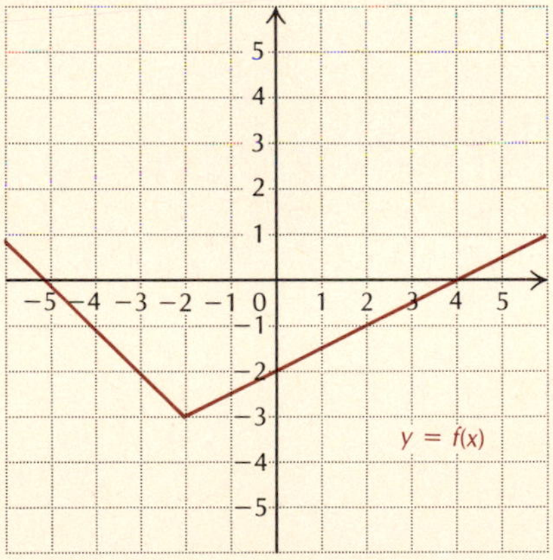

For the function in Exercise 25, answer the following.

27. Find $\lim_{x \to 1} g(x)$.

28. Find $g(1)$.

29. Is g continuous at 1?

30. Find $\lim_{x \to -2} g(x)$.

31. Find $g(-2)$.

32. Is g continuous at -2?

SYNTHESIS

33. Find a complete factorization of $x^3 - 1$.

34. Find k such that $x + 3$ is a factor of
$$x^3 + kx^2 + kx - 15.$$

35. The equation $x^2 - 8x + c = 0$ has a double root. Find it.

36. When $x^2 - 4x + 3k$ is divided by $x + 5$, the remainder is 33. Find the value of k.

37. Graph: $y = 1 - \dfrac{1}{x^2 + 4}$.

TEST: CHAPTER 11

1. Find the remainder when $3x^4 - 6x^3 + x - 2$ is divided by $x - 3$.

2. Determine whether $x + i$ is a factor of $x^3 + x$.

3. Use the synthetic division to find the quotient and the remainder. Show all your work.

$$(5x^3 - x^2 + 4x - 3) \div (x + 1)$$

4. Use synthetic division to find $P(-5)$:

$$P(x) = 2x^4 - 7x^3 + 8x^2 - 10.$$

5. Factor the polynomial $P(x)$. Then solve the equation $P(x) = 0$.

$$P(x) = x^3 - 4x^2 + x + 6$$

6. Find a polynomial of degree 4 with roots 2, -2, $2 + i$, and $2 - i$.

7. Find a polynomial of lowest degree having roots 0 and -1, and having 2 as a root of multiplicity 2 and 1 as a root of multiplicity 3.

8. Find a polynomial of lowest degree with rational coefficients that has $3 - i$ and 2 as two of its roots.

9. List all possible rational roots of

$$3x^5 - 4x^3 - 2x + 6.$$

10. What does Descartes' rule of signs tell you about the number of negative real roots of

$$5x^7 - 2x^6 + x^4 + 2x - 6?$$

11. Find the smallest positive integer that is guaranteed by Theorem 12 to be an upper bound to the roots of

$$3x^4 - 2x^3 - 15x - 10.$$

12. Approximate the irrational roots of

$$P(x) = x^4 - x - 2.$$

13. Graph:

$$f(x) = \frac{x - 2}{x^2 - 2x - 15}.$$

14. Decompose into partial fractions:

$$\frac{-8x + 23}{2x^2 + 5x - 12}.$$

Determine whether the function is continuous.

15.

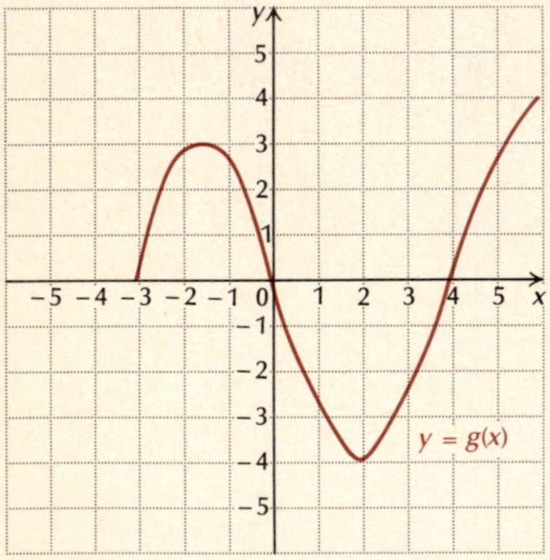

16.

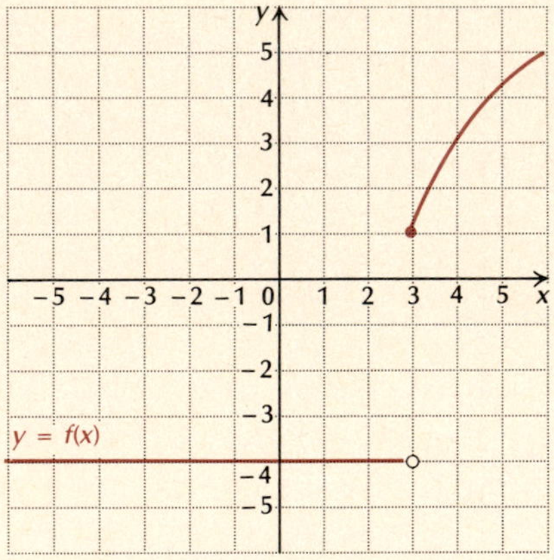

For the function in Question 16, answer each of the following.

17. Find $\lim_{x\to 3} f(x)$.

18. Find $f(3)$.

19. Is f continuous at 3?

20. Find $\lim_{x\to 4} f(x)$.

21. Find $f(4)$.

22. Is f continuous at 4?

Find the limit, if it exists.

23. $\lim_{x\to 1} 3x^4 - 2x^2 + 5$

24. $\lim_{x\to 1} \dfrac{x - 1}{x^2 - 1}$

25. $\lim_{x\to 0} \dfrac{7}{x}$

26. $\lim_{x\to \infty} \dfrac{4x - 3}{x}$

SYNTHESIS

27. Graph: $y = \left|2 - \dfrac{1}{3x}\right|$.

28. Solve: $x^4 - 2x^3 + 3x^2 - 2x + 2 = 0$.

Sequences, Series, and Combinatorics

12

The problem here illustrates a counting, or combinatoric, problem basic to a study of theoretical probability. A desire to calculate odds in games of chance gave rise to the *theory of probability*. Today, the theory of probability has many applications to business, medicine, sociology, psychology, and science.

The first part of this chapter is devoted to *sequences* and *series*. The idea of a sequence is a familiar one. For example, when a manager makes out a batting order, a sequence is being formed. Each batter is associated with a natural number; that is, a function is defined. When the members of a sequence are numbers, we can think of adding them. Such a sum is called a *series*. We also study a method of proof known as *mathematical induction*, which enables us to prove many important formulas, as well as other mathematical results.

FEATURE PROBLEM

The state of Michigan runs a 6-out-of-44-number lotto twice a week that pays at least \$1.5 million. You purchase a card for \$1 and pick any 6 numbers from 1 to 44. If your 6 numbers match those that the state draws, you win. How many 6-number combinations are there for the drawing?

THE MATHEMATICS

The number of combinations is

$$
\begin{aligned}
{}_{44}C_6 &= \binom{44}{6} \\
&= \frac{44 \cdot 43 \cdot 42 \cdot 41 \cdot 40 \cdot 39}{6 \cdot 5 \cdot 4 \cdot 3 \cdot 2 \cdot 1} \\
&= 7{,}059{,}052.
\end{aligned}
$$

OBJECTIVES

You should be able to:

1. Given a formula for the nth term (general term) of a sequence, find any term in the sequence, given a value for n.
2. Given a sequence, look for a pattern and try to guess a rule or formula for the general term.
3. Find and evaluate a series.
4. Convert between sigma (Σ) notation and other notation for a series.
5. Given a recursively defined sequence, construct its terms.

12.1 Sequences and Series

In this section, we discuss sets of numbers, considered in order, and their sums.

1 Sequences

Suppose \$1000 is invested at 8%, compounded annually. The amounts to which the account will grow after 1 year, 2 years, 3 years, 4 years, and so on, are as follows:

①	②	③	④
↓	↓	↓	↓
\$1080.00	\$1166.40	\$1259.71	\$1360.49, ...

Note that we can think of this as a function that maps 1 to the number \$1080.00, 2 to the number \$1166.40, 3 to the number \$1259.71, 4 to the number \$1360.49, and so on. A **sequence** is thus a *function*, where the domain is a set of consecutive positive integers.

If we keep computing the amounts in the account forever, we obtain an **infinite sequence**:

\$1080.00, \$1166.40, \$1259.71, \$1360.49, \$1469.33, \$1586.87, ...

The three dots at the end indicate that the sequence goes on without stopping. If we stop after a certain number of years, we obtain a **finite sequence**:

\$1080.00, \$1166.40, \$1259.71, \$1360.49.

> **DEFINITION Sequence**
>
> **An *infinite sequence* is a function having for its domain the set of positive integers: $\{1, 2, 3, 4, 5, \ldots\}$.**
>
> **A *finite sequence* is a function having for its domain the set of positive integers $\{1, 2, 3, 4, 5, \ldots, n\}$, for some positive integer n.**

As another example, consider the sequence given by

$$a(n) = 2^n, \quad \text{or} \quad a_n = 2^n.$$

The notation a_n means the same as $a(n)$ but is more commonly used with sequences. Some of the function values (also known as **terms** of the sequence) are as follows:

$$a_1 = 2^1 = 2,$$
$$a_2 = 2^2 = 4,$$
$$a_3 = 2^3 = 8,$$
$$a_6 = 2^6 = 64.$$

The first term of the sequence is a_1, the fifth term is a_5, and the nth term, or **general** term, is a_n. This sequence can also be denoted in the following ways:

a) $2, 4, 8, \ldots;$

b) $2, 4, 8, \ldots, 2^n, \ldots.$

EXAMPLE 1 Find the first 4 terms and the 57th term of the sequence whose general term is given by $a_n = (-1)^n/(n+1)$.

Solution

$$a_1 = \frac{(-1)^1}{1+1} = -\frac{1}{2},$$
$$a_2 = \frac{(-1)^2}{2+1} = \frac{1}{3},$$
$$a_3 = \frac{(-1)^3}{3+1} = \frac{1}{4},$$
$$a_4 = \frac{(-1)^4}{4+1} = \frac{1}{5},$$
$$a_{57} = \frac{(-1)^{57}}{57+1} = -\frac{1}{58}.$$

Note in Example 1 that the power $(-1)^n$ causes the signs of the terms to alternate between positive and negative, depending on whether n is even or odd.

DO EXERCISES 1 AND 2.

2 Finding the General Term

When a sequence is described merely by naming the first few terms, we do not know for sure what the general term is, but the reader is expected to make a guess by looking for a pattern.

EXAMPLES For each sequence, make a guess at the general term.

2. 1, 4, 9, 16, 25, . . .

These are squares of numbers, so the general term may be n^2.

3. $\sqrt{1}, \sqrt{2}, \sqrt{3}, \sqrt{4}, \ldots$

These are square roots of numbers, so the general term may be $\sqrt{n}$.

4. $-1, 2, -4, 8, -16, \ldots$

These are powers of 2 with alternating signs, so the general term may be $(-1)^n\,[2^{n-1}]$.

5. 2, 4, 8, . . .

If we see the pattern of powers of 2, we will see 16 as the next term and guess 2^n for the general term. We would then write the sequence with more terms as

$$2, 4, 8, 16, 32, 64, 128, \ldots .$$

If we see that we can get the second term by adding 2, the third term by adding 4, and the next term by adding 6, and so on, we will see 14 as the next term. A general term for the sequence is then $n^2 - n + 2$, and we would then write the sequence with more terms as

$$2, 4, 8, 14, 22, 32, 44, 58, \ldots .$$

Example 5 illustrates that the fewer the given number of terms of a sequence, the more difficult it is to be certain about the nth term.

DO EXERCISES 3–7.

1. A sequence is given by $a(n) = 2n - 1$.

a) Find the first 3 terms.

b) Find the 34th term.

2. A sequence is given by $a_n = \dfrac{(-1)^n}{n-1}$, $n \geq 2$.

a) Find the first 4 terms.

b) Find the 48th term.

For each sequence, try to find a rule for finding the general term, or the nth term. Answers may vary.

3. 2, 4, 6, 8, 10, . . .

4. 1, 2, 3, 4, 5, 6, . . .

5. 1, 8, 27, 64, 125, . . .

6. $x, \frac{x^2}{2}, \frac{x^3}{3}, \frac{x^4}{4}, \frac{x^5}{5}, \ldots$

7. 1, 2, 4, 8, 16, 32, . . .

8. For the sequence 1, −1, 2, −2, 3, −3, 4, −4, find each of the following.

a) S_4

b) S_7

3 Sums and Series

DEFINITION Series

Given the infinite sequence

$$a_1, a_2, a_3, a_4, \ldots, a_n, \ldots,$$

the sum of the terms

$$a_1 + a_2 + a_3 + \cdots + a_n + \cdots$$

is called an *infinite series*. A *partial sum* is the sum of the first n terms:

$$a_1 + a_2 + a_3 + \cdots + a_n.$$

A partial sum is also called a *finite series*, and is denoted S_n.

Consider the sequence

$$3, 5, 7, 9, \ldots, 2n + 1.$$

We construct some partial sums:

$S_1 = 3,$ This is the first term of the given sequence.

$S_2 = 3 + 5 = 8,$ The sum of the first two terms

$S_3 = 3 + 5 + 7 = 15,$ The sum of the first three terms

$S_4 = 3 + 5 + 7 + 9 = 24.$ The sum of the first four terms

Note that if we write these partial sums in order, we create a new sequence:

$$3, 8, 15, 24, \ldots.$$

EXAMPLE 6 For the sequence −2, 4, −6, 8, −10, 12, −14, find (a) S_3 and (b) S_5.

Solution

a) $S_3 = -2 + 4 + (-6) = -4$

b) $S_5 = -2 + 4 + (-6) + 8 + (-10) = -6$

DO EXERCISE 8.

4 Sigma Notation

The Greek letter Σ (sigma) can be used to simplify notation when a series has a formula for the general term.

The sum of the first four terms of the sequence 3, 5, 7, 9, . . . , $2k + 1$ can be named as follows, using what is called **sigma notation**, or **summation notation:**

$$\sum_{k=1}^{4} (2k + 1).$$

This is read "the sum as k goes from 1 to 4 of $(2k + 1)$." The letter k is called the **index of summation**. Sometimes the index of summation starts at a number other than 1.

EXAMPLES Rename and evaluate each sum.

7. $\displaystyle\sum_{k=1}^{5} k^2 = 1^2 + 2^2 + 3^2 + 4^2 + 5^2$

$= 1 + 4 + 9 + 16 + 25 = 55$

8. $\sum_{k=1}^{4} (-1)^k(2k) = (-1)^1(2 \cdot 1) + (-1)^2(2 \cdot 2) + (-1)^3(2 \cdot 3) + (-1)^4(2 \cdot 4)$
$= -2 + 4 - 6 + 8 = 4$

9. $\sum_{k=0}^{3} (2^k + 5) = (2^0 + 5) + (2^1 + 5) + (2^2 + 5) + (2^3 + 5)$
$= 6 + 7 + 9 + 13 = 35$ ∎

DO EXERCISES 9–11.

EXAMPLES Write sigma notation for each sum.

10. $-1 + 3 - 5 + 7$

These are odd integers with alternating signs. Therefore, the general term is $(-1)^k(2k - 1)$, beginning with $k = 1$. Sigma notation is

$$\sum_{k=1}^{4} (-1)^k(2k - 1).$$

11. $3 + 9 + 27 + 81 + \cdots$

This is a sum of powers of 3, and it is also an infinite series. We use the symbol ∞ to represent infinity and name the infinite series with sigma notation, as follows:

$$\sum_{k=1}^{\infty} 3^k.$$

12. $x + x^2 + x^3 + x^4 + x^5 + x^6 = \sum_{k=1}^{6} x^k$

13. $\sin\frac{\pi}{2} - \sin\frac{3\pi}{2} + \sin\frac{5\pi}{2} - \sin\frac{7\pi}{2} + \cdots$
$= \sum_{k=1}^{\infty} (-1)^{k+1} \sin\left(\frac{2k-1}{2}\right)\pi$ ∎

DO EXERCISES 12–15.

5 Recursive Definitions

A sequence may be defined by a **recursive definition**. Such a definition lists the first term, or the first few terms, and then tells how to get the rest of the terms from the given terms.

EXAMPLE 14 Find the first 5 terms of the sequence defined by

$$a_1 = 5,$$
$$a_{k+1} = 2a_k - 3, \quad \text{for } k \geq 1.$$

Solution We have

$$a_1 = 5,$$
$$a_2 = 2a_1 - 3 = 2 \cdot 5 - 3 = 7,$$
$$a_3 = 2a_2 - 3 = 2 \cdot 7 - 3 = 11,$$
$$a_4 = 2a_3 - 3 = 2 \cdot 11 - 3 = 19,$$
$$a_5 = 2a_4 - 3 = 2 \cdot 19 - 3 = 35.$$ ∎

DO EXERCISE 16.

Rename and evaluate the sum.

9. $\sum_{k=1}^{3} \left(2 + \frac{1}{k}\right)$

10. $\sum_{k=0}^{4} 5^k$

11. $\sum_{k=8}^{11} k^3$

Name the series with sigma notation.

12. $2 + 4 + 6 + 8 + 10$

13. $1 + 8 + 27 + 64$

14. $x + \frac{x^2}{2} + \frac{x^3}{3} + \frac{x^4}{4} + \frac{x^5}{5} + \frac{x^6}{6} + \cdots$

15. $4 + 9 + 16 + 25 + 36 + \cdots$

16. Find the first 5 terms of this recursively defined sequence:

$$a_1 = -3,$$
$$a_{k+1} = (-1) \cdot a_k^2, \quad \text{for } k \geq 1.$$

17. Find the first 8 terms of the recursively defined sequence

$$a_1 = 5,$$
$$a_2 = 5,$$
$$a_{k+1} = a_k + a_{k-1}, \quad \text{for } k \geq 2.$$

EXAMPLE 15 *The Fibonacci sequence.* One of the most famous recursively defined sequences is the *Fibonacci sequence*. So much mathematics has been derived from it that there is a journal, called the *Fibonacci Quarterly*, devoted to publishing the results. The Fibonacci sequence is defined as follows:

$$a_1 = 1,$$
$$a_2 = 1,$$
$$a_{k+1} = a_k + a_{k-1}, \quad \text{for } k \geq 2.$$

Find the first 7 terms of the Fibonacci sequence.

Solution

$$a_1 = 1,$$
$$a_2 = 1,$$
$$a_3 = a_2 + a_1 = 1 + 1 = 2,$$
$$a_4 = a_3 + a_2 = 2 + 1 = 3,$$
$$a_5 = a_4 + a_3 = 3 + 2 = 5,$$
$$a_6 = a_5 + a_4 = 5 + 3 = 8,$$
$$a_7 = a_6 + a_5 = 8 + 5 = 13.$$

DO EXERCISE 17.

EXERCISE SET 12.1

1 In each of the following, the nth term of a sequence is given. In each case, find the first 4 terms, the 10th term, a_{10}, and the 15th term, a_{15}.

1. $a_n = 4n - 1$
2. $a_n = (n-1)(n-2)(n-3)$
3. $a_n = \dfrac{n}{n-1}, n \geq 2$
4. $a_n = n^2 - 1$
5. $a_n = n^2 + 2n$
6. $a_n = \dfrac{n^2 - 1}{n^2 + 1}$
7. $a_n = n + \dfrac{1}{n}$
8. $a_n = \left(-\dfrac{1}{2}\right)^{n-1}$
9. $a_n = (-1)^n n^2$
10. $a_n = (-1)^n(n + 3)$
11. $a_n = (-1)^{n+1}(3n - 5)$
12. $a_n = (-1)^n(n^3 - 1)$
13. $a_n = \dfrac{n+2}{n+5}$
14. $a_n = \dfrac{2n - 1}{3n - 4}$

Find the indicated term of the given sequence.

15. $a_n = 5n - 6$; a_8
16. $a_n = 3n + 10$; a_9
17. $a_n = (3n - 4)(2n + 5)$; a_7
18. $a_n = (2n - 3)^2$; a_6
19. $a_n = (-1)^{n-1}(4.6n - 18.3)$; a_{12}
20. $a_n = (-2)^{n-2}(54.76 - 1.3n)$; a_{23}
21. $a_n = 5n^2(4n - 100)$; a_{11}
22. $a_n = 4n^2(11n + 31)$; a_{22}
23. $a_n = \left(1 + \dfrac{1}{n}\right)^2$; a_{20}
24. $a_n = \left(1 - \dfrac{1}{n}\right)^3$; a_{15}
25. $a_n = \log 10^n$; a_{43}
26. $a_n = \ln e^n$; a_{67}
27. $a_n = 1 + \dfrac{1}{n^2}$; a_{38}
28. $a_n = 2 - \dfrac{1000}{n}$; a_{100}

2 For each sequence, find the general term, or nth term, a_n, or a rule for finding a_n. Answers may vary.

29. 1, 3, 5, 7, 9, . . .
30. 3, 9, 27, 81, 243, . . .
31. −2, 6, −18, 54, . . .
32. −2, 3, 8, 13, 18, . . .

33. $\frac{2}{3}, \frac{3}{4}, \frac{4}{5}, \frac{5}{6}, \frac{6}{7}, \ldots$

34. $\sqrt{2}, \sqrt{4}, \sqrt{6}, \sqrt{8}, \sqrt{10}, \ldots$

35. $\sqrt{3}, 3, 3\sqrt{3}, 9, 9\sqrt{3}, \ldots$

36. $1 \cdot 2, 2 \cdot 3, 3 \cdot 4, 4 \cdot 5, \ldots$

37. $-1, -4, -7, -10, -13, \ldots$

38. $\log 1, \log 10, \log 100, \log 1000, \ldots$

3 For each sequence, find the indicated partial sum.

39. $1, 2, 3, 4, 5, 6, 7, \ldots; S_7$

40. $1, -3, 5, -7, 9, -11, \ldots; S_8$

41. $2, 4, 6, 8, \ldots; S_5$

42. $1, \frac{1}{4}, \frac{1}{9}, \frac{1}{16}, \frac{1}{25}, \ldots; S_5$

4 Rename without using sigma notation.

43. $\sum_{k=1}^{5} \frac{1}{2k}$

44. $\sum_{k=1}^{6} \frac{1}{2k+1}$

45. $\sum_{k=0}^{5} 2^k$

46. $\sum_{k=4}^{7} \sqrt{2k-1}$

47. $\sum_{k=7}^{10} \log k$

48. $\sum_{k=0}^{4} \pi k$

49. $\sum_{k=1}^{8} \frac{k}{k+1}$

50. $\sum_{k=1}^{4} \frac{k-1}{k+3}$

51. $\sum_{k=1}^{5} (-1)^k$

52. $\sum_{k=1}^{5} (-1)^{k+1}$

53. $\sum_{k=1}^{8} (-1)^{k+1} 3k$

54. $\sum_{k=1}^{7} (-1)^k 4^{k+1}$

55. $\sum_{k=1}^{6} \frac{2}{k^2+1}$

56. $\sum_{k=1}^{10} k(k+1)$

57. $\sum_{k=0}^{5} (k^2 - 2k + 3)$

58. $\sum_{k=0}^{5} (k^2 - 3k + 4)$

59. $\sum_{k=1}^{10} \frac{1}{k(k+1)}$

60. $\sum_{k=1}^{10} \frac{2^k}{2^k+1}$

Write sigma notation.

61. $\frac{1}{2} + \frac{2}{3} + \frac{3}{4} + \frac{4}{5} + \frac{5}{6} + \frac{6}{7}$

62. $3 + 6 + 9 + 12 + 15$

63. $-2 + 4 - 8 + 16 - 32 + 64$

64. $\frac{1}{1^2} + \frac{1}{2^2} + \frac{1}{3^2} + \frac{1}{4^2} + \frac{1}{5^2}$

65. $4 - 9 + 16 - 25 + \cdots + (-1)^n n^2$

66. $9 - 16 + 25 + \cdots + (-1)^{n+1} n^2$

67. $5 + 10 + 15 + 20 + 25 + \cdots$

68. $7 + 14 + 21 + 28 + 35 + \cdots$

69. $\frac{1}{1 \cdot 2} + \frac{1}{2 \cdot 3} + \frac{1}{3 \cdot 4} + \frac{1}{4 \cdot 5} + \cdots$

70. $\frac{1}{1 \cdot 2^2} + \frac{1}{2 \cdot 3^2} + \frac{1}{3 \cdot 4^2} + \frac{1}{4 \cdot 5^2} + \cdots$

5 Find the first 4 terms of each recursively defined sequence.

71. $a_1 = 4, a_{k+1} = 1 + \frac{1}{a_k}$

72. $a_1 = 256, a_{k+1} = \sqrt{a_k}$

73. $a_1 = 6561, a_{k+1} = (-1)^k \sqrt{a_k}$

74. $a_1 = e^Q, a_{k+1} = \ln a_k$

75. $a_1 = 2, a_2 = 3, a_{k+1} = a_k + a_{k-1}$

76. $a_1 = -10, a_2 = 8, a_{k+1} = a_k - a_{k-1}$

SYNTHESIS

Find the first 5 terms of the sequence, and then find S_5.

77. $a_n = \frac{1}{2^n} \log 1000^n$

78. $a_n = i^n, i = \sqrt{-1}$

79. $a_n = \ln (1 \cdot 2 \cdot 3 \cdots n)$

80. $a_n = \sin \frac{n\pi}{2}$

81. $a_n = \cos^{-1} (-1)^n$

82. $a_n = |\cos (x + \pi n)|$

83. a) Find the first few terms of the sequence $a_n = n^2 - n + 41$.
b) What pattern do you observe?
c) Find the 41st term. Does the pattern you found in (b) still hold?

Find decimal notation, rounded to six decimal places, for the first 6 terms of each sequence.

84. ▦ $a_n = \left(1 + \frac{1}{n}\right)^n$

85. ▦ $a_n = \sqrt{n+1} - \sqrt{n}$

86. ▦ $a_1 = 2, a_{k+1} = \sqrt{1 + \sqrt{a_k}}$

87. ▦ $a_1 = 2, a_{k+1} = \frac{1}{2}\left(a_k + \frac{2}{a_k}\right)$

88. A single cell of bacteria divides into two every 15 min. Suppose that the same rate of division is maintained for 4 hr. Give a sequence that lists the number of cells after successive 15-min periods.

89. The value of an office machine is \$5200. Its scrap value each year is 75% of its value the year before. Give a sequence that lists the scrap value of the machine for each year of a 10-year period.

90. A student gets \$4.20 for working in a warehouse for a publishing company. Each year, the student gets a \$0.15 hourly raise. Give a sequence that lists the hourly salary of the student over a 10-year period.

CHALLENGE

For each sequence, find a formula for S_n.

91. $a_n = \ln n$

92. $a_n = \frac{1}{n} - \frac{1}{n+1}$

12.2 Arithmetic Sequences and Series

OBJECTIVES

You should be able to:

1 **Use Theorem 1 to find, for any arithmetic sequence,**
- **a) the *n*th term when *n* is given;**
- **b) *n*, when an *n*th term is given.**
- **c) Given two terms of a sequence, find the common difference and construct the sequence.**

2 **Use Theorems 2 and 3 to find the sum of the first *n* terms of an arithmetic sequence.**

3 **Solve problems using arithmetic sequences.**

4 **Insert arithmetic means between two numbers.**

In this section, we concentrate on what is called an *arithmetic sequence.* If we start with a particular first term and then add the same number successively, we obtain an **arithmetic sequence.** We will also study *arithmetic series.*

1 Arithmetic Sequences

Consider this sequence:

$$2, 5, 8, 11, 14, 17, \ldots.$$

Note that adding 3 to any term produces the following term. In other words, the difference between any term and the preceding one is 3. This is an example of an arithmetic sequence.

DEFINITION Arithmetic Sequence

A sequence is *arithmetic* if there exists a number d, called the *common difference,* such that $a_n = a_{n-1} + d$, for any $n \geq 2$.

Arithmetic sequences are also called **arithmetic progressions.**

EXAMPLES The following are arithmetic sequences. Identify the first term a_1 and the common difference d.

Sequence	First term, a_1	Common difference, d
1. 4, 9, 14, 19, 24, ...	4	5
2. 34, 27, 20, 13, 6, −1, −8, ...	34	−7
3. $2, 2\frac{1}{2}, 3, 3\frac{1}{2}, 4, 4\frac{1}{2}, \ldots$	2	$\frac{1}{2}$

We obtain d for the first sequence by picking any term beyond the first—say, the second term, 9—and subtract the preceding term from it:

$9 - 4 = 5$. Then we can check by adding 5 to each term to see if we obtain the next:

$$a_1 = 4,$$
$$a_2 = 4 + 5 = 9,$$
$$a_3 = 9 + 5 = 14,$$
$$a_4 = 14 + 5 = 19, \quad \text{and so on.}$$

We now find a formula for the general, or nth, term of any arithmetic sequence. Let us denote the difference between successive terms (called the **common difference**) by d, and write out the first few terms:

$$\begin{aligned} &a_1, \\ &a_2 = a_1 + d, \\ &a_3 = a_2 + d = (a_1 + d) + d = a_1 + 2d, \\ &a_4 = a_3 + d = (a_1 + 2d) + d = a_1 + 3d. \end{aligned}$$

Note that the coefficient of d in each case is 1 less than the number of the term, n.

DO EXERCISE 1.

Generalizing, we obtain the following.

THEOREM 1

The nth term of an arithmetic sequence is given by

$$a_n = a_1 + (n - 1)d, \quad \text{for any } n \geq 1.$$

EXAMPLE 4 Find the 14th term of the arithmetic sequence 4, 7, 10, 13, ...

Solution First note that $a_1 = 4$, $d = 3$, and $n = 14$. Then using the formula

$$a_n = a_1 + (n - 1)d,$$

we obtain

$$a_{14} = 4 + (14 - 1) \cdot 3 = 4 + 13 \cdot 3 = 4 + 39 = 43.$$

The 14th term is 43. ∎

DO EXERCISE 2.

EXAMPLE 5 In the sequence in Example 4, which term is 301? That is, what is n if $a_n = 301$?

Solution We substitute into the formula of Theorem 1 and solve for n:

$$\begin{aligned} a_n &= a_1 + (n - 1)d \\ 301 &= 4 + (n - 1) \cdot 3 \\ 301 &= 4 + 3n - 3 \\ 301 &= 3n + 1 \\ 300 &= 3n \\ 100 &= n. \end{aligned}$$

The 100th term is 301. ∎

1. In the following arithmetic sequence, identify the first term a_1 and the common difference d.

   3.1, 3.9, 4.7, 5.5, 6.3

2. Find the 13th term of the sequence 2, 6, 10, 14, ...

3. In the sequence given in Margin Exercise 2, which term is 298? That is, what is n if $a_n = 298$?

DO EXERCISE 3.

Given two terms and their places in an arithmetic sequence, we can construct the sequence.

EXAMPLE 6 The 3rd term of an arithmetic sequence is 8, and the 16th term is 47. Find a_1 and d and construct the sequence.

Solution We know that $a_3 = 8$ and $a_{16} = 47$. Thus we would have to add d thirteen times to get from 8 to 47. That is,

$$8 + 13d = 47.$$

Solving, we obtain

$$\begin{aligned} 13d &= 39 \\ d &= 3. \end{aligned}$$

Since $a_3 = 8$, we subtract d twice to get to a_1. Thus,

$$a_1 = 8 - 2 \cdot 3 = 2.$$

The sequence is 2, 5, 8, 11, ...

DO EXERCISE 4.

4. The 7th term of an arithmetic sequence is 79, and the 13th term is 151. Find a_1 and d. Construct the sequence.

2 Sum of the First *n* Terms of an Arithmetic Sequence

Suppose we add the first 4 terms of the sequence

$$3, 5, 7, 9, 11, \ldots.$$

We get what is called an **arithmetic series:**

$$3 + 5 + 7 + 9, \quad \text{or} \quad 24.$$

The sum of the first n terms of a sequence is denoted S_n. Thus, for the preceding sequence, $S_4 = 24$. We want to find a formula for S_n when the sequence is arithmetic. We can denote an arithmetic sequence as

This is the next-to-last term. If you add d to this term, the result is a_n.

$$a_1, \quad (a_1 + d), \quad (a_1 + 2d), \quad \ldots, \quad (a_n - 2d), \quad (a_n - d), \quad a_n.$$

This term is two terms back from the last. If you add d to this term, you get the next-to-last term, $a_n - d$.

Then S_n is given by

$$S_n = a_1 + (a_1 + d) + (a_1 + 2d) + \cdots + (a_n - 2d) + (a_n - d) + a_n. \quad \textbf{(1)}$$

If we reverse the order of addition, we get

$$S_n = a_n + (a_n - d) + (a_n - 2d) + \cdots + (a_1 + 2d) + (a_1 + d) + a_1. \quad \textbf{(2)}$$

Suppose we add corresponding terms of each side of Eqs. (1) and (2). Then we get

$$\begin{aligned}2S_n = [a_1 + a_n] + [(a_1 + d) + (a_n - d)] \\ + [(a_1 + 2d) + (a_n - 2d)] + \cdots + [(a_n - 2d) + (a_1 + 2d)] \\ + [(a_n - d) + (a_1 + d)] + [a_n + a_1].\end{aligned}$$

This simplifies to

$$2S_n = (a_1 + a_n) + (a_1 + a_n) + (a_1 + a_n) + \cdots + (a_1 + a_n).$$

Since there are n binomials $(a_1 + a_n)$ being added, it follows that

$$2S_n = n(a_1 + a_n),$$

from which we get the following formula.

THEOREM 2

The sum of the first n terms of an arithmetic sequence is given by

$$S_n = \frac{n}{2}(a_1 + a_n).$$

EXAMPLE 7 Find the sum of the first 100 natural numbers.

Solution The sum is

$$1 + 2 + 3 + \cdots + 99 + 100.$$

This is the sum of the first 100 terms of the arithmetic sequence for which

$$a_1 = 1, \quad a_n = 100, \quad \text{and} \quad n = 100.$$

Then substituting into the formula

$$S_n = \frac{n}{2}(a_1 + a_n),$$

we get

$$S_{100} = \frac{100}{2}(1 + 100) = 50(101) = 5050.$$

DO EXERCISE 5.

The preceding formula is useful when we know both a_1 and a_n, the first and last terms, but it often happens that we do not know a_n. We thus need a formula in terms of a_1, n, and d. We substitute the expression for a_n, given in Theorem 1, $a_n = a_1 + (n - 1)d$, into the formula of Theorem 2:

$$S_n = \frac{n}{2}(a_1 + [a_1 + (n - 1)d]).$$

This gives us the following.

THEOREM 3

The sum of the first n terms of an arithmetic sequence is given by

$$S_n = \frac{n}{2}[2a_1 + (n - 1)d].$$

EXAMPLE 8 Find the sum of the first 15 terms of the arithmetic sequence 4, 7, 10, 13, ...

5. Find the sum of the first 200 natural numbers.

6. Find the sum of the first 15 terms of the arithmetic sequence

$$1, 3, 5, 7, 9, \ldots .$$

7. Find the sum:

$$\sum_{k=1}^{10} (9k - 4).$$

Solution Note that

$$a_1 = 4, \quad d = 3, \quad \text{and} \quad n = 15.$$

Here we use the formula of Theorem 3 since we do not know a_n:

$$S_n = \frac{n}{2}[2a_1 + (n - 1)d].$$

We get

$$S_{15} = \frac{15}{2}[2 \cdot 4 + (15 - 1)3] = \frac{15}{2}[8 + 14 \cdot 3] = \frac{15}{2}[8 + 42]$$
$$= \frac{15}{2}[50] = 375.$$

DO EXERCISE 6.

EXAMPLE 9 Find the sum: $\sum_{k=1}^{13} (4k + 5)$.

Solution It is helpful to write out a few terms:

$$9 + 13 + 17 + \cdots .$$

We see that the sum is an arithmetic series coming from an arithmetic sequence with $a_1 = 9$, $d = 4$, and $n = 13$. We use the formula of Theorem 3:

$$S_n = \frac{n}{2}[2a_1 + (n - 1)d]$$
$$S_{13} = \frac{13}{2}[2 \cdot 9 + (13 - 1)4] = \frac{13}{2}[18 + 12 \cdot 4] = \frac{13}{2} \cdot 66 = 429.$$

DO EXERCISE 7.

3 Problem Solving

For some problem situations, the translations may involve sequences or series. We look at some examples.

EXAMPLE 10 You take a job starting with an hourly rate of \$14.25. You are promised a raise of 15¢ per hour every 2 months for 5 years. At the end of 5 years, what will be your hourly wage?

Solution One thing to do is write down your hourly wage for several two-month time periods. What appears is a *sequence of numbers*: 14.25, 14.40, 14.55, ... Is it an arithmetic sequence? Yes, because we add 0.15 each time in order to get the next term.

We ask ourselves what we know about arithmetic sequences. We recall, or if necessary look up, the pertinent formula(s). There are three:

$$a_n = a_1 + (n - 1)d, \quad S_n = \frac{n}{2}(a_1 + a_n), \quad \text{and} \quad S_n = \frac{n}{2}[2a_1 + (n - 1)d].$$

In this case, we are not looking for a *sum*, so the first formula will give us our answer. We want to know the last term in a sequence. We will need to know a_1, n, and d. From our list above, we see that

$$a_1 = 14.25 \quad \text{and} \quad d = 0.15.$$

What is n? That is, how many terms are in the sequence? Each year, there are 6 raises, since you get a raise every 2 months. There are 5 years, so the total number of raises will be 5×6, or 30. There will be 31 terms: the original wage and 30 increased rates.

We want to find the 31st term of an arithmetic sequence, with $a_1 = 14.25$, $d = 0.15$, and $n = 31$. We substitute into the formula:

$$a_{31} = 14.25 + (31 - 1) \times 0.15 = \$18.75.$$

At the end of 5 years, your hourly wage will be \$18.75. ■

Example 10 is one in which the calculations or the translation could be done in a number of ways. Here is a point to remember: There is often a variety of ways in which a problem can be solved. In this chapter, however, we will concentrate on the use of sequences and series and their related formulas in problem solving.

DO EXERCISE 8.

EXAMPLE 11 A stack of telephone poles has 30 poles in the bottom row. There are 29 poles in the second row, 28 in the next row, and so on. How many poles are in the stack?

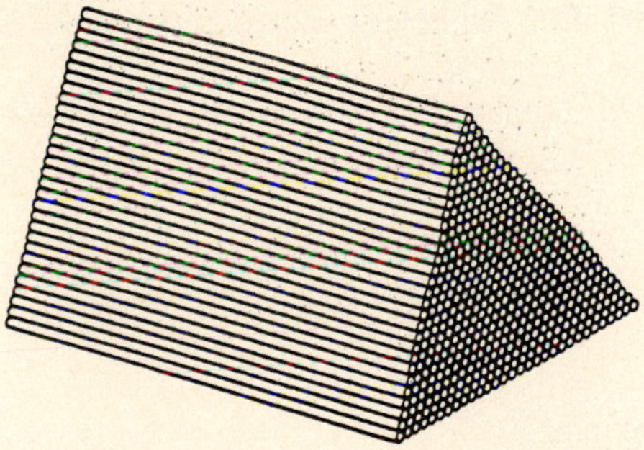

Solution A picture will help in this case. The following figure shows the ends of the poles and the way in which they stack. There are 30 poles on the bottom, and we see that there will be one fewer in each succeeding row. How many rows will there be?

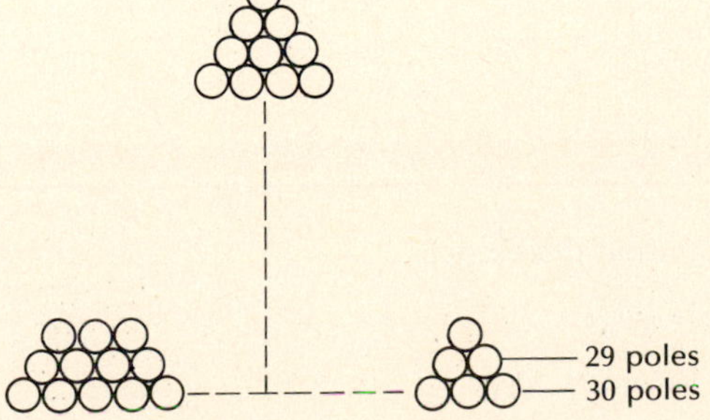

We go from 30 poles in a row, down to one pole in the top row, so there must be 30 rows.

We want the sum

$$30 + 29 + 28 + \cdots + 1.$$

Thus we want the sum of an arithmetic sequence. We recall the formula for the

8. You take a job starting with an hourly rate of \$16.75. You are promised a raise of 25¢ per hour every 2 months for 7 years. At the end of 7 years, what will be your hourly wage?

9. A stack of iron rods has 112 rods in the bottom row, 111 rods in the next row, and so on. The top row has 48 rods. How many rods are in the stack?

10. Insert three arithmetic means between 4 and 16.

sum of an arithmetic sequence:

$$S_n = \frac{n}{2}(a_1 + a_n).$$

We want to find the sum of an arithmetic sequence, where $a_1 = 30$ and, since there are 30 terms, $n = 30$. There is just one pole on top, so $a_{30} = 1$. Substituting into the formula, we have

$$S_{30} = \frac{30}{2}(30 + 1) = 465.$$

The answer is that there are 465 poles in the stack. ■

DO EXERCISE 9.

4 Arithmetic Means

If p, m, and q form an arithmetic sequence, it can be shown (see Exercise 60) that $m = (p + q)/2$. We call m the **arithmetic mean** of p and q. Given two numbers p and q, if we find k other numbers

$$p, m_1, m_2, \ldots, m_k, q,$$

we say that we have "inserted k arithmetic means between 4 and 13."

EXAMPLE 12 Insert three arithmetic means between 4 and 13.

Solution We look for numbers m_1, m_2, and m_3 such that 4, m_1, m_2, m_3, 13 is an arithmetic sequence. In this case, $a_1 = 4$, $n = 5$, and $a_5 = 13$. We use the formula of Theorem 1:

$$a_n = a_1 + (n - 1)d$$
$$13 = 4 + (5 - 1)d.$$

Then $d = 2\frac{1}{4}$, so we have

$$m_1 = a_1 + d = 4 + 2\tfrac{1}{4} = 6\tfrac{1}{4},$$
$$m_2 = m_1 + d = 6\tfrac{1}{4} + 2\tfrac{1}{4} = 8\tfrac{1}{2},$$
$$m_3 = m_2 + d = 8\tfrac{1}{2} + 2\tfrac{1}{4} = 10\tfrac{3}{4}.$$ ■

DO EXERCISE 10.

EXERCISE SET 12.2

1 For the arithmetic sequence, find the first term and the common difference.

1. 3, 8, 13, 18, . . .
2. 1.08, 1.16, 1.24, 1.32, . . .
3. 9, 5, 1, −3, . . .
4. −8, −5, −2, 1, 4, . . .
5. $\frac{3}{2}, \frac{9}{4}, 3, \frac{15}{4}, \ldots$
6. $\frac{3}{5}, \frac{1}{10}, -\frac{2}{5}, \ldots$
7. \$1.07, \$1.14, \$1.21, \$1.28, . . .
8. \$316, \$313, \$310, \$307, . . .
9. Find the 12th term of the arithmetic sequence
   2, 6, 10, ...
10. Find the 11th term of the arithmetic sequence
    0.07, 0.12, 0.17, ...
11. Find the 17th term of the arithmetic sequence
    7, 4, 1, ...
12. Find the 14th term of the arithmetic sequence
    $3, \frac{7}{3}, \frac{5}{3}, \ldots$.

13. Find the 13th term of the arithmetic sequence
$$\$1200, \$964.32, \$728.64, \ldots.$$

14. Find the 10th term of the arithmetic sequence
$$\$2345.78, \$2967.54, \$3589.30, \ldots.$$

15. In the sequence of Exercise 9, what term is 106?

16. In the sequence of Exercise 10, what term is 1.67?

17. In the sequence of Exercise 11, what term is -296?

18. In the sequence of Exercise 12, what term is -27?

19. Find a_{17} when $a_1 = 5$ and $d = 6$.

20. Find a_{20} when $a_1 = 14$ and $d = -3$.

21. Find a_1 when $d = 4$ and $a_8 = 33$.

22. Find d when $a_1 = 8$ and $a_{11} = 26$.

23. Find n when $a_1 = 5$, $d = -3$, and $a_n = -76$.

24. Find n when $a_1 = 25$, $d = -14$, and $a_n = -507$.

25. In an arithmetic sequence, $a_{17} = -40$ and $a_{28} = -73$. Find a_1 and d. Write the first 5 terms of the sequence.

26. In an arithmetic sequence, $a_{17} = \frac{25}{3}$ and $a_{32} = \frac{95}{6}$. Find a_1 and d. Write the first 5 terms of the sequence.

2

27. Find the sum of the first 20 terms of the series
$$5 + 8 + 11 + 14 + \cdots.$$

28. Find the sum of the first 14 terms of the series
$$11 + 7 + 3 + \cdots.$$

29. Find the sum of the first 300 natural numbers.

30. Find the sum of the first 400 natural numbers.

31. Find the sum of the even numbers from 2 to 100, inclusive.

32. Find the sum of the odd numbers from 1 to 99, inclusive.

33. Find the sum of the multiples of 7, from 7 to 98 inclusive.

34. Find the sum of all multiples of 4 that are between 14 and 523.

35. If an arithmetic series has $a_1 = 2$, $d = 5$, and $n = 20$, what is S_n?

36. If an arithmetic series has $a_1 = 7$, $d = -3$, and $n = 32$, what is S_n?

3

37. A gardener is making a triangular planting, with 35 plants in the front row, 31 in the second row, 27 in the third row, and so on. If the pattern is consistent, how many plants will there be in the last row?

38. A formation of a marching band has 14 marchers in the front row, 16 in the second row, 18 in the third row, and so on, for 25 rows. How many marchers are in the last row? How many marchers are there altogether?

39. How many poles will be in a pile of telephone poles if there are 50 in the first layer, 49 in the second, and so on, until there is 1 in the last layer?

40. If 10¢ is saved on October 1, 20¢ on October 2, 30¢ on October 3, and so on, how much is saved during October? (October has 31 days.)

41. A family saves money in an arithmetic sequence. They save \$600 the first year, \$700 the second, and so on, for 20 years. How much do they save in all (disregarding interest)?

42. A student saves \$30 on August 1, \$50 on August 2, \$70 on August 3, and so on. How much would be saved in August?

43. Theaters are often built with more seats per row as the rows move toward the back. Suppose that the main floor of a theater has 28 seats in the first row, 32 in the second, 36 in the third, and so on, for 50 rows. How many seats are on the main floor?

44. A person sets up an investment such that it will return \$5000 the first year, \$6125 the second year, \$7250 the third year, for 25 years. How much in all is received from the investment?

4

45. Insert four arithmetic means between 4 and 13.

46. Insert three arithmetic means between -3 and 5.

SYNTHESIS

47. Find a formula for the sum of the first n odd natural numbers:
$$1 + 3 + 5 + \cdots + (2n - 1).$$

48. Find a formula for the sum of the first n natural numbers:
$$1 + 2 + 3 + \cdots + n.$$

49. Find three numbers in an arithmetic sequence such that the sum of the first and third is 10 and the product of the first and second is 15.

50. Find the first term and the common difference for the arithmetic sequence where
$$a_2 = 40 - 3q \quad \text{and} \quad a_4 = 10p + q.$$

51. ▦ Find the first 10 terms of the arithmetic sequence for which $a_1 = \$8760$ and $d = -\$798.23$.

52. ▦ Find the sum of the first 10 terms of the sequence given in Exercise 51.

53. The zeros of this polynomial function form an arithmetic sequence. Find them.

$$f(x) = x^4 + 4x^3 - 84x^2 - 176x + 640$$

54. Insert enough arithmetic means between 1 and 50 so that the sum of the resulting series will be 459.

55. Suppose that the lengths of the sides of a right triangle form an arithmetic sequence. Prove that the triangle is similar to a right triangle whose sides have lengths 3, 4, and 5.

56. *Business: Straight-line depreciation.* A company buys an office machine for \$5200 on January 1 of a given year. The machine is expected to last for 8 years, at the end of which time its **trade-in,** or **salvage, value** will be \$1100. If the company figures the decline in value to be the same each year, then the **book values,** or **salvage values,** after t years, $0 \le t \le 8$, form an arithmetic sequence given by

$$a_t = C - t\left(\frac{C - S}{N}\right),$$

where C = the original cost of the item (\$5200), N = the years of expected life (8), and S = the salvage value (\$1100).

a) Find the formula for a_t for the straight-line depreciation of the office machine.
b) Find the salvage value after 0 years, 1 year, 2 years, 3 years, 4 years, 7 years, 8 years.

CHALLENGE

57. Prove that an expression for the general term of an arithmetic sequence defines a linear function.

58. Prove that if p, m, and q form an arithmetic sequence, then

$$m = \frac{p + q}{2}.$$

12.3 Geometric Sequences and Series

OBJECTIVES

You should be able to:

1. Identify the common ratio of a geometric sequence. Use Theorem 4 for geometric sequences to find a given term.
2. Use Theorem 5 to find the sum of the first n terms of a geometric sequence.
3. Use Theorem 6 to find the sum of an infinite geometric series, if it exists.
4. Use geometric sequences to solve problems.

For arithmetic sequences, we added a certain number to each term to get the next term. With the kind of sequence we consider now, we multiply each term by a certain number to get the next term. These are **geometric sequences.** We also consider geometric series.

1 Geometric Sequences

Consider the sequence

$$2, 6, 18, 54, 162, \ldots.$$

If we multiply each term by 3, we get the next term. Sequences in which each term can be multiplied by a certain number to get the next term are called **geometric.** We usually denote this number r. We refer to it as the **common ratio,** because we can get r by dividing any term by the preceding term.

DEFINITION Geometric Sequence

A sequence is *geometric* if there exists a number r, called the *common ratio*, such that

$$\frac{a_{n+1}}{a_n} = r, \quad \text{or} \quad a_{n+1} = a_n r \quad \text{for any } n \ge 1.$$

A geometric sequence is also called a **geometric progression.**

EXAMPLES The following are geometric sequences. Identify the common ratio.

Sequence	Common Ratio	
1. 3, 6, 12, 24, 48, 96, . . .	2	$6/3 = 2$, $12/6 = 2$, and so on
2. 3, −6, 12, −24, 48, −96, . . .	−2	$-6/3 = -2$, $12/-6 = -2$, and so on
3. $5200, $3900, $2925, $2193.75, . . .	0.75	$\$3900/\$5200 = 0.75$, $\$2925/\$3900 = 0.75$
4. $1000, $1080, $1166.40, . . .	1.08	$\$1080/\$1000 = 1.08$
5. $1, \frac{1}{2}, \frac{1}{4}, \frac{1}{8}, \ldots$	$\frac{1}{2}$	$\frac{1}{2}/1 = \frac{1}{2}$, $\frac{1}{4}/\frac{1}{2} = \frac{1}{2}$ ■

DO EXERCISES 1–5.

We now find a formula for the general, or nth, term of any geometric sequence. Let a_1 be the 1st term, and let r be the common ratio. We write out the first few terms as follows:

$$a_1,$$
$$a_2 = a_1 r,$$
$$a_3 = a_2 r = (a_1 r)r = a_1 r^2,$$
$$a_4 = a_3 r = (a_1 r^2)r = a_1 r^3.$$

→ Note that the exponent is 1 less than the number of the term.

Generalizing, we obtain the following.

THEOREM 4

The nth term of a geometric sequence is given by
$$a_n = a_1 r^{n-1} \quad \text{for any } n \geq 1.$$

EXAMPLE 6 Find the 7th term of the geometric sequence 4, 20, 100,

Solution First note that
$$a_1 = 4 \quad \text{and} \quad n = 7.$$

To find the common ratio, we can divide any term by its predecessor, provided it has one. Since the second term is 20 and the first is 4, we get
$$r = \frac{20}{4}, \quad \text{or } 5.$$

Then using the formula
$$a_n = a_1 r^{n-1},$$
we have
$$a_7 = 4 \cdot 5^{7-1} = 4 \cdot 5^6 = 4 \cdot 15{,}625 = 62{,}500. \quad ■$$

DO EXERCISE 6.

EXAMPLE 7 Find the 10th term of the geometric sequence
$$64, -32, 16, -8, \ldots .$$

Solution First note that
$$a_1 = 64, \quad n = 10, \quad \text{and} \quad r = \frac{-32}{64}, \quad \text{or } -\frac{1}{2}.$$

The following are geometric sequences. Identify the common ratio.

1. 1, 5, 25, 125,

2. 3, −9, 27, −81, . . .

3. $6000, $5100, $4335, $3684.75, . . .

4. $100, $109, $118.81, . . .

5. $1, \frac{1}{2}, \frac{1}{4}, \frac{1}{8}, \ldots$

6. Find the 9th term of the geometric sequence
$$2, 4, 8, 16, \ldots .$$

7. Find the 6th term of the geometric sequence
$$-3, 1, -\tfrac{1}{3}, \tfrac{1}{9}, \ldots.$$

Then using the formula
$$a_n = a_1 r^{n-1},$$
we have
$$a_{10} = 64 \cdot \left(-\frac{1}{2}\right)^{10-1} = 64 \cdot \left(-\frac{1}{2}\right)^9 = 2^6 \cdot \left(-\frac{1}{2^9}\right) = -\frac{1}{2^3} = -\frac{1}{8}.$$

DO EXERCISE 7.

2 Sum of the First n Terms of a Geometric Sequence

We want to find a formula for the sum S_n of the first n terms of a geometric sequence
$$a_1, a_1r, a_1r^2, a_1r^3, \ldots, a_1r^{n-1}, \ldots.$$
The **geometric series** is given by
$$S_n = a_1 + a_1r + a_1r^2 + \cdots + a_1r^{n-2} + a_1r^{n-1}. \qquad (1)$$
We want to develop a formula that allows us to find this sum without a great amount of adding. If we multiply on both sides of Eq. (1) by r, we have
$$rS_n = a_1r + a_1r^2 + a_2r^3 + \cdots + a_1r^{n-1} + a_1r^n. \qquad (2)$$
When we multiply on both sides of Eq. (1) by -1, we get
$$-S_n = -a_1 - a_1r - a_1r^2 - \cdots - a_1r^{n-2} - a_1r^{n-1}. \qquad (3)$$
Then, when we add corresponding sides of Eqs. (2) and (3), certain terms are additive inverses of each other and have a sum that is 0, so we get
$$rS_n - S_n = a_1r^n - a_1,$$
or
$$(r - 1)S_n = a_1(r^n - 1),$$
from which we get the following formula.

THEOREM 5

The sum of the first n terms of a geometric sequence is given by
$$S_n = \frac{a_1(r^n - 1)}{r - 1}, \quad \text{for any } r \neq 1.$$

EXAMPLE 8 Find the sum of the first 7 terms of the geometric sequence
$$3, 15, 75, 375, \ldots.$$

Solution First note that
$$a_1 = 3, \quad n = 7, \quad \text{and} \quad r = \tfrac{15}{3}, \quad \text{or } 5.$$
Then using the formula
$$S_n = \frac{a_1(r^n - 1)}{r - 1},$$
we have
$$S_7 = \frac{3(5^7 - 1)}{5 - 1} = \frac{3(78{,}125 - 1)}{4} = \frac{3(78{,}124)}{4} = 58{,}593.$$

DO EXERCISES 8 AND 9.

EXAMPLE 9 Find the sum: $\sum_{k=1}^{11} (0.3)^k$.

Solution This is a geometric series. The first term is 0.3, $r = 0.3$, and $n = 11$. Then

$$S_{11} = \frac{0.3[(0.3)^{11} - 1]}{0.3 - 1} \approx 0.42857\ldots.$$

DO EXERCISE 10.

3 Infinite Geometric Series

Suppose we consider the sum of the terms of an infinite geometric sequence, such as 2, 4, 8, 16, 32, ... We get what is called an **infinite geometric series:**

$$2 + 4 + 8 + 16 + 32 + \cdots.$$

As n grows larger and larger, the sum of the first n terms, S_n, becomes larger and larger without bound. There are infinite series that get closer and closer to some specific number. Here is an example:

$$\frac{1}{2} + \frac{1}{4} + \frac{1}{8} + \frac{1}{16} + \cdots + \frac{1}{2^n} + \cdots.$$

Let's consider the partial sums S_n for some values of n:

$$S_1 = \tfrac{1}{2} = \tfrac{1}{2} = 0.5,$$
$$S_2 = \tfrac{1}{2} + \tfrac{1}{4} = \tfrac{3}{4} = 0.75,$$
$$S_3 = \tfrac{1}{2} + \tfrac{1}{4} + \tfrac{1}{8} = \tfrac{7}{8} = 0.875,$$
$$S_4 = \tfrac{1}{2} + \tfrac{1}{4} + \tfrac{1}{8} + \tfrac{1}{16} = \tfrac{15}{16} = 0.9375,$$
$$S_5 = \tfrac{1}{2} + \tfrac{1}{4} + \tfrac{1}{8} + \tfrac{1}{16} + \tfrac{1}{32} = \tfrac{31}{32} = 0.96875.$$

The denominator of each term is a power of 2, 2^n, and the numerator is 1 less than the denominator. Thus we can describe S_n as

$$S_n = \frac{2^n - 1}{2^n}.$$

Note that the numerator is less than the denominator for all values of n, but as n gets larger and larger, the values of S_n get closer and closer to 1. We say that 1 is the **limit** of S_n and that 1 is the **sum** of the **infinite geometric series**. The sum of the infinite series, if it exists, is denoted S_∞. It can be shown (but we will not do it here) that the sum of the terms of an infinite geometric series exists if and only if $|r| < 1$ (that is, the absolute value of the common ratio is less than 1).

We want to find a formula for the sum of an infinite geometric series. We first consider the sum of the first n terms:

$$S_n = \frac{a_1(r^n - 1)}{r - 1} = \frac{a_1 - a_1 r^n}{1 - r}.$$

For $|r| < 1$, it follows that values of r^n get closer and closer to 0 as n gets large. (Choose a number between -1 and 1 and check this by finding larger and larger powers on your calculator.) As r^n gets closer and closer to 0, so does $a_1 r^n$. Thus, S_n gets closer and closer to $a_1/(1 - r)$.

8. Find the sum of the first 8 terms of the geometric sequence

$$4, 12, 36, 108, \ldots.$$

9. Find the sum of the first 10 terms of the geometric sequence

$$2, -1, \tfrac{1}{2}, -\tfrac{1}{4}, \ldots.$$

10. Find the sum:

$$\sum_{k=1}^{5} 3^k.$$

Determine whether the infinite geometric series has a sum. If so, find it.

11. $1 + 7 + 49 + 343 + \cdots$

12. $1 + (-1) + 1 + (-1) + \cdots$

13. $\frac{1}{2} + \frac{1}{4} + \frac{1}{8} + \frac{1}{16} + \frac{1}{32} + \cdots$

14. $625 + 250 + 100 + 40 + \cdots$

Find fractional notation.

15. $0.222\overline{2}$

16. $0.13\overline{13}$

THEOREM 6

When $|r| < 1$, the sum of an infinite geometric series is given by

$$S_\infty = \frac{a_1}{1 - r}.$$

EXAMPLE 10 Determine whether this infinite geometric series has a sum. If so, find it.

$$1 + 3 + 9 + 27 + \cdots$$

Solution We have $|r| = |3| = 3$, and since $|r| > 1$, the series does *not* have a sum. ■

EXAMPLE 11 Determine whether this infinite geometric series has a sum. If so, find it.

$$1 - \frac{1}{2} + \frac{1}{4} - \frac{1}{8} + \frac{1}{16} - \cdots$$

Solution

a) $|r| = \left|-\frac{1}{2}\right| = \frac{1}{2}$, and since $|r| < 1$, the series does have a sum.

b) The sum is given by

$$S_\infty = \frac{1}{1 - \left(-\frac{1}{2}\right)} = \frac{1}{\frac{3}{2}} = \frac{2}{3}.$$ ■

DO EXERCISES 11–14.

EXAMPLE 12 Find fractional notation for 0.63636363 . . . , or $0.\overline{63}$.

Solution We can express this as

$$0.63 + 0.0063 + 0.000063 + \cdots.$$

This is an infinite geometric series, where $a_1 = 0.63$ and $r = 0.01$. Since $|r| < 1$, this series has a sum:

$$S_\infty = \frac{a_1}{1 - r} = \frac{0.63}{1 - 0.01} = \frac{0.63}{0.99} = \frac{63}{99}, \quad \text{or } \frac{7}{11}.$$

Thus fractional notation for 0.63636363 . . . is $\frac{7}{11}$. ■

DO EXERCISES 15 AND 16.

4 Problem Solving

For some problem-solving situations, the translation may involve geometric sequences or series.

EXAMPLE 13 Suppose someone offered you a job for the month of September (30 days) under the following conditions. You will be paid \$0.01 for the first day, \$0.02 for the second, \$0.04 for the third, and so on, doubling your previous day's salary each day. How much would you earn? (Would you take the job? Make a guess before reading further.)

Solution You earn \$0.01 the first day, \$0.01(2) the second day, \$0.01(2)(2) the third day, and so on. The amounts form a geometric sequence with $a_1 = \$0.01$, $r = 2$, and $n = 30$.

The amount earned is the geometric series

$$\$0.01 + \$0.01(2) + \$0.01(2^2) + \$0.01(2^3) + \cdots + \$0.01(2^{29}),$$

where

$$a_1 = \$0.01, \quad n = 30, \quad \text{and} \quad r = 2.$$

Then using the formula

$$S_n = \frac{a_1(r^n - 1)}{r - 1},$$

we have

$$\begin{aligned} S_{30} &= \frac{\$0.01(2^{30} - 1)}{2 - 1} \\ &\approx \$0.01(1{,}074{,}000{,}000 - 1) && \text{Using a calculator to approximate } 2^{30} \\ &\approx \$0.01(1{,}074{,}000{,}000) && 1{,}074{,}000{,}000 - 1 \approx 1{,}074{,}000{,}000 \\ &= \$10{,}740{,}000. \end{aligned}$$

Since the salary for September is more than \$10 million, most people would take the job.

DO EXERCISE 17.

EXAMPLE 14 *The economic multiplier.* The NCAA finals have a tremendous effect on the economy of the host city. Recently, the finals were held in Dallas, Texas. Suppose that 20,000 people visited the city and spent \$400 each while there. Then assume that 80% of that money is spent again in the city, and then 80% of that money is spent again, and so on. Find the total effect of this money on the economy. This is known as the **economic multiplier effect.**

Solution According to certain economic theory, the money that this effectively puts into the economy can be calculated as the sum of an infinite geometric sequence as follows. The "initial effect" is 20,000 × \$400, or \$8,000,000. The total effect is

$$\$8{,}000{,}000 + \$8{,}000{,}000(0.80) + \$8{,}000{,}000(0.80)^2 + \$8{,}000{,}000(0.80)^3 + \cdots.$$

Using the formula of Theorem 6, we find this amount to be

$$S_\infty = \frac{\$8{,}000{,}000}{1 - 0.80} = \$40{,}000{,}000.$$

Do you see why cities work so hard to be chosen to host the NCAA finals?

DO EXERCISE 18.

17. Under the conditions of Example 13, how much would you make in October, which has 31 days?

18. In Example 14, suppose that 95% of the money will be spent again, and so on. What is the economic multiplier effect?

EXERCISE SET 12.3

1 Find the common ratio.

1. 2, 4, 8, 16, . . .
2. 18, −6, 2, $-\frac{2}{3}$, . . .
3. −1, 1, −1, 1, . . .
4. −8, −0.8, −0.08, −0.008, . . .
5. $\frac{1}{2}, -\frac{1}{4}, \frac{1}{8}, -\frac{1}{16}, \ldots$
6. $\frac{2}{3}, -\frac{4}{3}, \frac{8}{3}, -\frac{16}{3}, \ldots$

7. $75, 15, 3, \frac{3}{5}, \ldots$

8. $6.275, 0.6275, 0.06275, \ldots$

9. $\frac{1}{x}, \frac{1}{x^2}, \frac{1}{x^3}, \ldots$

10. $5, \frac{5m}{2}, \frac{5m^2}{4}, \frac{5m^3}{8}, \ldots$

11. \$780, \$858, \$943.80, \$1038.18, . . .

12. \$5600, \$5320, \$5054, \$4801.30, . . .

Find the indicated term.

13. $2, 4, 8, 16, \ldots$; the 6th term

14. $2, -10, 50, -250, \ldots$; the 9th term

15. $2, 2\sqrt{3}, 6, \ldots$; the 9th term

16. $1, -1, 1, -1, \ldots$; the 57th term

17. $\frac{8}{243}, \frac{8}{81}, \frac{8}{27}, \ldots$; the 10th term

18. $\frac{7}{625}, \frac{-7}{25}, 7, \ldots$; the 23rd term

19. \$1000, \$1080, \$1166.40, . . . ; the 5th term

20. \$1000, \$1070, \$1144.90, . . . ; the 6th term

Find the *n*th, or general, term.

21. $1, 3, 9, \ldots$

22. $25, 5, 1, \ldots$

23. $1, -1, 1, -1, \ldots$

24. $2, 4, 8, \ldots$

25. $\frac{1}{x}, \frac{1}{x^2}, \frac{1}{x^3}, \ldots$

26. $5, \frac{5m}{2}, \frac{5m^2}{4}, \ldots$

2

27. Find the sum of the first 7 terms of the geometric series

$$6 + 12 + 24 + \cdots.$$

28. Find the sum of the first 6 terms of the geometric series

$$16 - 8 + 4 - \cdots.$$

29. Find the sum of the first 7 terms of the geometric series

$$\tfrac{1}{18} - \tfrac{1}{6} + \tfrac{1}{2} - \cdots.$$

30. Find the sum of the geometric series

$$-8 + 4 + (-2) + \cdots + \left(-\tfrac{1}{32}\right).$$

31. Find the sum of the first 8 terms of the series

$$1 + x + x^2 + x^3 + \cdots.$$

32. Find the sum of the first 10 terms of the series

$$1 + x^2 + x^4 + x^6 + \cdots.$$

33. Find the sum of the first 16 terms of the geometric sequence

$$\$200, \$200(1.06), \$200(1.06)^2, \ldots.$$

34. Find the sum of the first 23 terms of the geometric sequence

$$\$1000, \$1000(1.08), \$1000(1.08)^2, \ldots.$$

35. Find the sum:

$$\sum_{k=1}^{\infty} \left(\tfrac{1}{2}\right)^{k-1}.$$

36. Find the sum:

$$\sum_{k=1}^{\infty} 2^k.$$

3 Determine whether each of the following infinite geometric series has a sum. If so, find it.

37. $4 + 2 + 1 + \cdots$

38. $7 + 3 + \frac{9}{7} + \cdots$

39. $25 + 20 + 16 + \cdots$

40. $12 + 9 + \frac{27}{4} + \cdots$

41. $100 - 10 + 1 - \frac{1}{10} + \cdots$

42. $-6 + 18 - 54 + 162 - \cdots$

43. $8 + 40 + 200 + \cdots$

44. $-6 + 3 - \frac{3}{2} + \frac{3}{4} - \cdots$

45. $0.6 + 0.06 + 0.006 + \cdots$

46. $0.37 + 0.0037 + 0.000037 + \cdots$

47. $\$500(1.11)^{-1} + \$500(1.11)^{-2} + \$500(1.11)^{-3} + \cdots$

48. $\$1000(1.08)^{-1} + 1000(1.08)^{-2} + \$1000(1.08)^{-3} + \cdots$

49. $\sum_{k=1}^{\infty} 16(0.1)^{k-1}$

50. $\sum_{k=1}^{\infty} 4(0.6)^{k-1}$

51. $\sum_{k=1}^{\infty} \frac{1}{2^{k-1}}$

52. $\sum_{k=1}^{\infty} \frac{8}{3}\left(\frac{1}{2}\right)^{k-1}$

Find fractional notation for each of the following infinite sums. (These are geometric series.)

53. $0.777\overline{7}$

54. $8.999\overline{9}$

55. $0.533\overline{3}$

56. $0.644\overline{4}$

57. $5.1515\overline{15}$

58. $0.4125\overline{125}$

4 Problem solving

59. A ping-pong ball is dropped from a height of 16 ft and always rebounds $\frac{1}{4}$ of the distance fallen. How high does it rebound the 6th time?

60. Approximate the total amount of the rebound heights of the ball in Exercise 59.

61. Gaintown has a population of 100,000 now, and the population is increasing by 3% each year. What will the population be in 15 years?

62. How long will it take for the population of Gaintown to double? (See Exercise 61.)

63. A student borrows \$1200. The loan is to be repaid in 13 years at 12% interest, compounded annually. How much will be repaid at the end of 13 years?

64. A piece of paper is 0.01 in. thick. It is folded repeatedly in such a way that its thickness is doubled each time for 20 times. How thick is the result?

65. A superball dropped from the top of the Washington Monument (556 ft high) always rebounds $\frac{3}{4}$ of the distance fallen. How far (up and down) will the ball have traveled when it hits the ground for the 6th time?

66. Approximate the total distance that the ball of Exercise 65 will have traveled when it comes to rest.

67. Suppose someone offered you a job for the month of February (28 days) under the following conditions. You will be paid \$0.01 the 1st day, \$0.02 the 2nd, \$0.04 the 3rd, and so on, doubling your previous day's salary each day. How much would you earn altogether?

68. *The amount of an annuity*. A person decides to save money in a savings account for retirement. At the beginning of each year, \$1000 is invested at 11%, compounded annually. How much will be in the retirement fund at the end of 40 years?

69. *The economic multiplier*. The government is making a \$13,000,000,000 expenditure for a new type of aircraft. If 85% of this gets spent again, and 85% of this gets spent again, and so on, what is the effect on the economy?

70. Repeat Exercise 69, for \$9,400,000,000 and 99%.

71. *Advertising effect*. A company is marketing a new product in a city of 5,000,000 people. They plan an advertising campaign that they think will induce 40% of the people to buy the product. They estimate that if those people like the product, they will induce 40% (of the 40% of 5,000,000) more to buy the product, and those will induce 40%, and so on. In all, how many people will buy the product as a result of the advertising campaign? What percentage of the population is this?

72. Repeat Exercise 71, for 3,000,000 and 55%.

SYNTHESIS

73. Prove that $\sqrt{3} - \sqrt{2}, 4 - \sqrt{6}$, and $6\sqrt{3} - 2\sqrt{2}$ form a geometric sequence.

74. Consider the sequence $x + 3, x + 7, 4x - 2, \ldots$.
    a) If the sequence is arithmetic, find x and then determine each of the three terms and the 4th term.
    b) If the sequence is geometric, find x and then determine each of the three terms and the 4th term.

75. Find the sum of the first n terms of
$$1 + x + x^2 + \cdots.$$

76. Find the sum of the first n terms of
$$x^2 - x^3 + x^4 - x^5 + \cdots.$$

For Exercises 77 and 78, assume that $a_1, a_2, a_3, \ldots$, is a geometric sequence.

77. Prove that $a_1^2, a_2^2, a_3^2, \ldots$, is a geometric sequence.

78. Prove that $\ln a_1, \ln a_2, \ln a_3, \ldots$, is an arithmetic sequence.

79. Prove that $5^{a_1}, 5^{a_2}, 5^{a_3}, \ldots$, is a geometric sequence, if $a_1, a_2, a_3, \ldots$, is an arithmetic sequence.

80. ▦ The function $f(x) = e^x$ is given by the infinite series
$$e^x = 1 + x + \frac{x^2}{2!} + \frac{x^3}{3!} + \cdots$$
($n!$ is read "n-factorial," and is defined to be the product $1 \cdot 2 \cdot 3 \cdots n$). Approximate e to three decimal places using the first six terms of the series.

81. An infinite sequence is defined recursively by

$$a_1 = 2,\ a_{k+1} = \frac{1}{2}\left(a_k + \frac{2}{a_k}\right).$$

Find decimal notation, rounded to six decimal places, for each of the first seven terms of the sequence. Make a conjecture about the limit of the sequence.

82. The sides of a square are 16 cm long. A second square is inscribed by joining the midpoints of the sides, successively. In the second square, we repeat the process, inscribing a third square. If this process is continued indefinitely, what is the sum of all the areas of all the squares? (*Hint*: Use an infinite geometric series.)

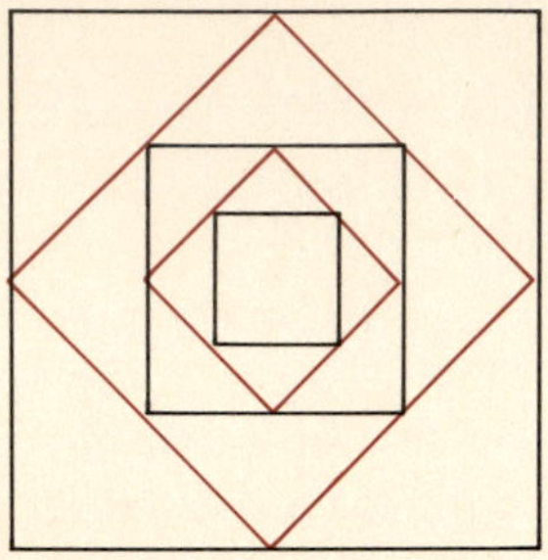

83. The infinite series

$$S = 2 + \frac{1}{2} + \frac{1}{2\cdot 3} + \frac{1}{2\cdot 3\cdot 4} + \frac{1}{2\cdot 3\cdot 4\cdot 5} + \frac{1}{2\cdot 3\cdot 4\cdot 5\cdot 6} + \cdots$$

is not geometric, but does have a sum. Find values of S_1, S_2, S_3, S_4, S_5, and S_6. Make a conjecture about the value of S.

OBJECTIVES

You should be able to:

1. List the statements of an infinite sequence that is defined by a formula.
2. Do proofs by mathematical induction.

12.4 Mathematical Induction

1 Sequences of Statements

Infinite sequences of statements occur often in mathematics. In an infinite sequence of statements, there is, of course, a statement for each natural number. For example, consider the sentence

For x between 0 and 1, $0 < x^n < 1$.

Let us think of this as $S(n)$, or S_n. Substituting natural numbers for n gives a sequence of statements. We list a few of them.*

Statement 1 (S_1): For x between 0 and 1, $0 < x^1 < 1$.
Statement 2 (S_2): For x between 0 and 1, $0 < x^2 < 1$.
S_3: For x between 0 and 1, $0 < x^3 < 1$.
S_4: For x between 0 and 1, $0 < x^4 < 1$.

EXAMPLE 1 List the first few statements in the sequence obtainable from $\log n < n$.

Solution This time, S_n is "$\log n < n$."

S_1: $\log 1 < 1$
S_2: $\log 2 < 2$
S_3: $\log 3 < 3$

*Note that S_1, S_2, and so on do *not* represent sums in this context.

Many sequences of statements concern sums.

EXAMPLE 2 List the first few statements in the sequence obtainable from

$$1 + 3 + 5 + \cdots + (2n - 1) = n^2.$$

Solution This time, the entire equation is the statement S_n.

S_1: $1 = 1^2$

S_2: $1 + 3 = 2^2$

S_3: $1 + 3 + 5 = 3^2$

S_4: $1 + 3 + 5 + 7 = 4^2$ ■

DO EXERCISES 1 AND 2.

1. List the first 5 statements in the sequence obtainable from
$$n^2 + 1 > n + 1.$$

2. List the first 5 statements in the sequence obtainable from
$$1 + 2 + \cdots + n = \frac{n(n + 1)}{2}.$$

2 Proving Infinite Sequences of Statements

The method of proof of this section, called **mathematical induction**, allows us to prove infinite sequences of statements. They are usually verbalized somewhat as follows:

For all natural numbers n, S_n,

where S_n is some sentence such as those in the preceding examples. Of course, we cannot prove each statement of an infinite sequence individually. Instead, we try to show that whenever S_k holds, then S_{k+1} must hold. We abbreviate this as $S_k \to S_{k+1}$. (This is read "S_k *implies* S_{k+1}.") If we can establish that this holds for all natural numbers k, we then have the following:

$S_1 \to S_2$ meaning whenever S_1 holds, S_2 must hold:

$S_2 \to S_3$ meaning whenever S_2 holds, S_3 must hold;

$S_3 \to S_4$ meaning whenever S_3 holds, S_4 must hold;

and so on, indefinitely.

At this stage, we do not yet know whether there is *any* k for which S_k holds. All we know is that if S_k holds, then S_{k+1} must hold. So we now show that S_k holds for some k, usually $k = 1$. We then have the following.

S_1 is true.	We have verified, or proved, this.
$S_1 \to S_2$ Therefore, S_2 is true.	This means that whenever S_1 holds, S_2 must hold.
$S_2 \to S_3$ Therefore, S_3 is true.	This means that whenever S_2 holds, S_3 must hold.
And so on.	

We conclude that S_n is true for all natural numbers n.

We now state the principle of mathematical induction.

Principle of Mathematical Induction

We can prove an infinite sequence of statements S_n by showing the following.

a) S_1 is true. (This is called the *basis step*.)

b) For all natural numbers k, $S_k \to S_{k+1}$. (This is called the *induction step*.)

3. Prove that if x is a number greater than 1, then for any natural number n,

$$x \leq x^n.$$

The situation with mathematical induction is analogous to lining up a sequence of dominoes. The induction step tells us that if any one domino is knocked over, then the one next to it will be hit and knocked over. The basis step tells us that the first domino can indeed be knocked over. Note that for all the dominoes to fall, *both* conditions must be satisfied.

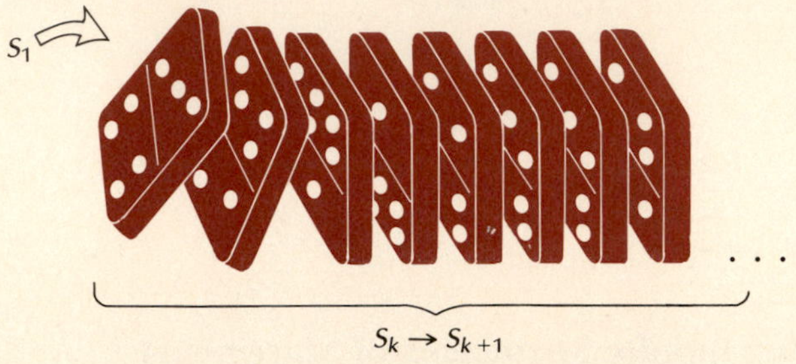

When you are learning to do proofs by mathematical induction, it is helpful to first write out S_n, S_1, S_k, and S_{k+1}. This helps to identify what is to be assumed and what is to be deduced.

EXAMPLE 3 Prove: For every natural number n, $n < 2^n$.

Proof. We first list S_n, S_1, S_k, and S_{k+1}.

S_n: $n < 2^n$

S_1: $1 < 2^1$

S_k: $k < 2^k$

S_{k+1}: $k + 1 < 2^{k+1}$

a) *Basis step.* S_1, as listed, is obviously true.

b) *Induction step.* We assume S_k as the hypothesis and try to show that it implies S_{k+1}. For any k,

$k < 2^k$ **By hypothesis (this is S_k)**

$2k < 2 \cdot 2^k$ **Multiplying on both sides by 2**

$2k < 2^{k+1}$. **Adding exponents on the right**

Now, since k is any natural number,

$1 \leq k$

$k + 1 \leq k + k$ **Adding k on both sides**

$k + 1 \leq 2k.$

Thus we have

$$k + 1 \leq 2k < 2^{k+1}$$

and

$$k + 1 < 2^{k+1}. \quad \textbf{This is } S_{k+1}.$$

We have now shown that $S_k \rightarrow S_{k+1}$ for any natural number k. Thus the induction step and the proof are complete. We have proved that for every natural number n, $n < 2^n$. ■

DO EXERCISE 3.

EXAMPLE 4 Prove: For every natural number n,

$$1 + 3 + 5 + 7 + \cdots + (2n - 1) = n^2.$$

Proof. We first list S_n, S_1, S_k, and S_{k+1}.

S_n: $1 + 3 + 5 + \cdots + (2n - 1) = n^2$

S_1: $1 = 1^2$

S_k: $1 + 3 + 5 + \cdots + (2k - 1) = k^2$

S_{k+1}: $1 + 3 + 5 + \cdots + (2k - 1) + [2(k + 1) - 1] = (k + 1)^2$

a) *Basis step.* S_1, as listed, is obviously true.

b) *Induction step.* We assume S_k as hypothesis and try to show that it implies S_{k+1}:

$$1 + 3 + 5 + \cdots + (2k - 1) = k^2 \text{ for any natural number } k.$$

This is the hypothesis (it is S_k). Starting with the left side of S_{k+1} and substituting k^2 for $1 + 3 + \cdots + (2k - 1)$, we have

$$\begin{aligned} &\underbrace{1 + 3 + \cdots + (2k - 1)} + [2(k + 1) - 1] \\ &= k^2 + [2(k + 1) - 1] \\ &= k^2 + 2k + 1 \\ &= (k + 1)^2. \end{aligned}$$

We have derived S_{k+1}. Thus we have shown that for all natural numbers k, $S_k \to S_{k+1}$. This completes the induction step and the proof is complete. We have proved that the equation is true for all natural numbers n. ∎

DO EXERCISE 4.

4. Consider

$$2 + 4 + 6 + \cdots + 2n = n(n + 1).$$

a) List S_1 and S_2.

b) List S_k.

c) List S_{k+1}.

d) Complete the basis step; that is, verify that S_1 is true.

e) Complete the proof that the formula holds for all n, by proving that S_k implies S_{k+1} for all natural numbers k.

EXAMPLE 5 Prove

$$\sum_{p=1}^{n} \frac{1}{2^p} = \frac{2^n - 1}{2^n},$$

for all natural numbers n. In other words, prove that for all natural numbers n,

$$\frac{1}{2} + \frac{1}{4} + \frac{1}{8} + \cdots + \frac{1}{2^n} = \frac{2^n - 1}{2^n}.$$

Proof. We first list S_n, S_1, S_k, and S_{k+1}.

S_n: $\displaystyle\sum_{p=1}^{n} \frac{1}{2^p} = \frac{1}{2} + \frac{1}{4} + \frac{1}{8} + \cdots + \frac{1}{2^n} = \frac{2^n - 1}{2^n}$

S_1: $\displaystyle\frac{1}{2} = \frac{2^1 - 1}{2^1}$

S_k: $\displaystyle\sum_{p=1}^{k} \frac{1}{2^p} = \frac{1}{2} + \frac{1}{4} + \cdots + \frac{1}{2^k} = \frac{2^k - 1}{2^k}$

S_{k+1}: $\displaystyle\sum_{p=1}^{k+1} \frac{1}{2^p} = \frac{1}{2} + \frac{1}{4} + \frac{1}{8} + \cdots + \frac{1}{2^k} + \frac{1}{2^{k+1}} = \frac{2^{k+1} - 1}{2^{k+1}}$

a) *Basis step.* Since

$$\frac{2^1 - 1}{2^1} = \frac{2 - 1}{2} = \frac{1}{2},$$

S_1 is true.

5. Prove that

$$\sum_{p=1}^{n} (3p - 1) = \frac{n(3n + 1)}{2}$$

for all natural numbers n.

b) *Induction step*. Using S_k as the hypothesis, we have, for all natural numbers k,

$$\frac{1}{2} + \frac{1}{4} + \frac{1}{8} + \cdots + \frac{1}{2^k} = \frac{2^k - 1}{2^k}.$$

Starting with the left side of S_{k+1} and substituting

$$\frac{2^k - 1}{2^k}$$

for

$$\frac{1}{2} + \frac{1}{4} + \cdots + \frac{1}{2^k},$$

we have

$$\underbrace{\frac{1}{2} + \frac{1}{4} + \cdots + \frac{1}{2^k}} + \frac{1}{2^{k+1}}$$

$$= \frac{2^k - 1}{2^k} + \frac{1}{2^{k+1}} = \frac{2^k - 1}{2^k} \cdot \frac{2}{2} + \frac{1}{2^{k+1}}$$

$$= \frac{(2^k - 1) \cdot 2 + 1}{2^{k+1}}$$

$$= \frac{2^{k+1} - 1}{2^{k+1}}.$$

We have arrived at S_{k+1}. Thus we have shown that for all natural numbers k, $S_k \to S_{k+1}$. This completes the induction and the proof is complete. We have proved that the equation is true for all natural numbers n. ■

DO EXERCISE 5.

EXERCISE SET 12.4

1 List the first 5 statements in the sequence obtainable from each of the following.

1. $n^2 < n^3$
2. $n^2 - n + 41$ is prime.
3. A polygon of n sides has $[n(n - 3)]/2$ diagonals.
4. The sum of the angles of a polygon of n sides is $(n - 2) \cdot 180°$.

2 Use mathematical induction to prove each of the following, for every natural number n.

5. $1 + 2 + 3 + \cdots + n = \frac{n(n + 1)}{2}$
6. $4 + 8 + 12 + \cdots + 4n = 2n(n + 1)$
7. $1 + 5 + 9 + \cdots + (4n - 3) = n(2n - 1)$
8. $3 + 6 + 9 + \cdots + 3n = \frac{3n(n + 1)}{2}$
9. $\frac{1}{1 \cdot 2} + \frac{1}{2 \cdot 3} + \cdots + \frac{1}{n(n + 1)} = \frac{n}{n + 1}$
10. $2 + 4 + 8 + \cdots + 2^n = 2(2^n - 1)$
11. $n < n + 1$
12. $2 \leq 2^n$
13. $3^n < 3^{n+1}$
14. $2n \leq 2^n$
15. $1^3 + 2^3 + 3^3 + \cdots + n^3 = \frac{n^2(n + 1)^2}{4}$
16. $\frac{1}{1 \cdot 2 \cdot 3} + \frac{1}{2 \cdot 3 \cdot 4} + \frac{1}{3 \cdot 4 \cdot 5} + \cdots + \frac{1}{n(n + 1)(n + 2)} = \frac{n(n + 3)}{4(n + 1)(n + 2)}$

SYNTHESIS

17. $\left(1 + \frac{1}{1}\right)\left(1 + \frac{1}{2}\right)\left(1 + \frac{1}{3}\right) \cdots \left(1 + \frac{1}{n}\right) = n + 1$

18. $a_1 + (a_1 + d) + (a_2 + d) + \cdots + [a_1 + (n - 1)d] = \frac{n}{2}[2a_1 + (n - 1)d]$

(This is a formula for the sum of the first n terms of an *arithmetic* sequence (Theorem 3).)

19. $a_1 + a_1r + a_1r^2 + \cdots + a_1r^{n-1} = \dfrac{a_1 - a_1r^n}{1 - r}$

(This is a formula for the sum of the first n terms of a *geometric* sequence (Theorem 5).)

20. $\cos n\pi = (-1)^n$

(*Hint*: Use the identity for $\cos(\alpha + \beta)$.)

21. *DeMoivre's theorem*. For any angle θ and any nonnegative number r, and $i^2 = -1$,

$$[r(\cos\theta + i\sin\theta)]^n = r^n(\cos n\theta + i\sin n\theta).$$

22. $|\sin(nx)| \le n\,|\sin x|$

Prove using mathematical induction.

23. For every natural number $n \ge 2$,

$$\cos x \cdot \cos 2x \cdot \cos 4x \cdot \cdots \cdot \cos 2^{n-1}x = \frac{\sin 2^n x}{2^n \sin x}$$

24. For every natural number $n \ge 2$,

$$\log_a(b_1b_2 \ldots b_n) = \log_a b_1 + \log_a b_2 + \cdots + \log_a b_n$$

25. For every natural number $n \ge 2$,

$$\left(1 - \frac{1}{2^2}\right)\left(1 - \frac{1}{3^2}\right) \cdots \left(1 - \frac{1}{n^2}\right) = \frac{n + 1}{2n}.$$

Prove each of the following for any complex numbers $z_1, \ldots, z_n$, where $i^2 = -1$ and $\bar{z}$ is the conjugate of z (see Section 2.4).

26. $\overline{z^n} = \bar{z}^n$

27. $\overline{z_1 + z_2 + \cdots + z_n} = \bar{z}_1 + \bar{z}_2 + \cdots + \bar{z}_n$

28. $\overline{z_1 \cdot z_2 \cdot \cdots \cdot z_n} = \bar{z}_1 \cdot \bar{z}_2 \cdot \cdots \cdot \bar{z}_n$

29. i^n is either 1, -1, i, or $-i$.

For any integers a and b, b is a factor of a if there exists an integer c such that $a = bc$. Prove each of the following for any natural number n.

30. 3 is a factor of $n^3 + 2n$.

31. 2 is a factor of $n^2 + n$.

32. 5 is a factor of $n^5 - n$.

33. 3 is a factor of $n(n + 1)(n + 2)$.

CHALLENGE

34. Use mathematical induction to prove that for every natural number $n \ge 2$,

$$\frac{1}{\sqrt{1}} + \frac{1}{\sqrt{2}} + \frac{1}{\sqrt{3}} + \cdots + \frac{1}{\sqrt{n}} > \sqrt{n}.$$

35. *The Tower of Hanoi problem*. There are three pegs on a board. On one peg are n disks, each smaller than the one on which it rests. The problem is to move this pile of disks to another peg. The final order must be the same, but you can move only one disk at a time and you can never place a larger disk on a smaller one.

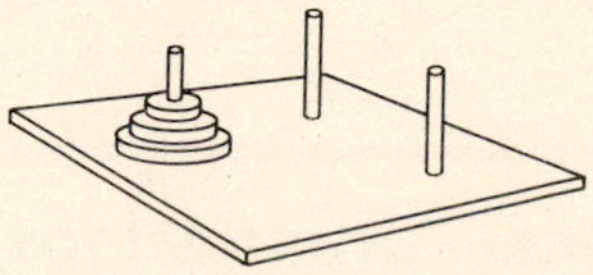

a) What is the *least* number of moves it takes to move 3 disks?
b) What is the *least* number of moves it takes to move 4 disks?
c) What is the *least* number of moves it takes to move 2 disks?
d) What is the *least* number of moves it takes to move 1 disk?
e) Conjecture a formula for the *least* number of moves it takes to move n disks. Prove it by mathematical induction.

36. Consider the statement: For every natural number n, $n = n + 1$.

a) Can you prove the basis step? If so, do it.
b) Can you prove the induction step? If so, do it.
c) Is the statement true? This illustrates the need for the basis step in an induction proof.

37. Find the error in this proof.
Statement. Everyone is of the same sex.

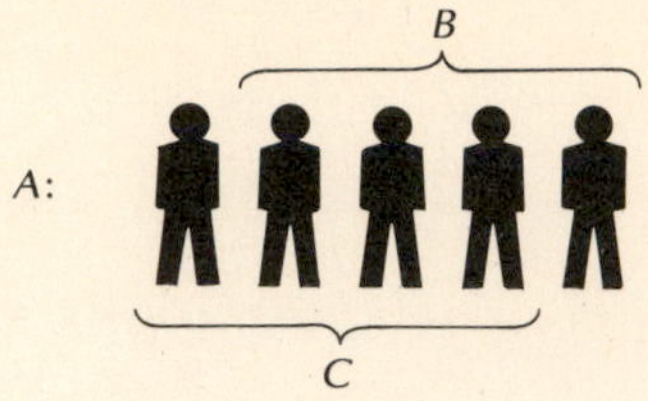

Proof. Let S_n be the statement: If A is a set of n people, then all the people are of the same sex. Clearly, S_1 is true. Assume S_k. Let A be a set of $k + 1$ people. Then A is the union of two overlapping sets B and C, each containing k people. (Consider the illustration for $k = 5$.) By S_k, all the people in B are of the same sex. Since B and C overlap, all the people in A are of the same sex, and $S_k \to S_{k+1}$.

38. *Fractals*. A **fractal**, simply stated, is an infinite recurring geometric figure. Fractals have application to the field of computer graphs and, subsequently, to physics and other fields of mathematics, such as topology and complex variables. Normally, if you make a drawing using a computer and zoom in on the drawing, you get to the place where you see the drawing made up of small squares called *pixels* (which means picture element). To avoid this happening, one uses a fractal technique, which in effect gives infinite resolution. Here is how a fractal might be constructed. You start with a basic figure, and at each step of an infinite sequence of steps, you build on the preceding figure.

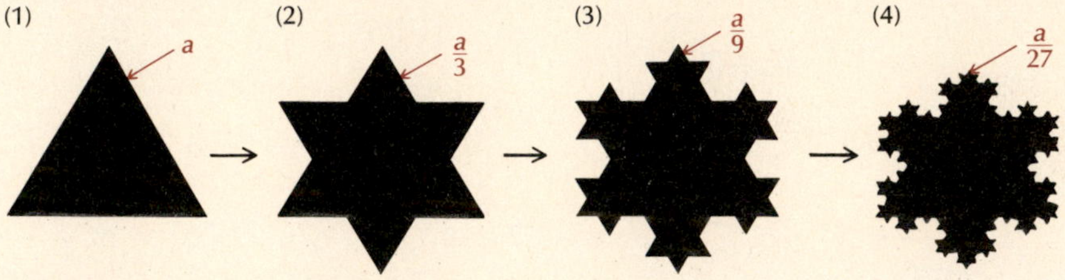

Suppose we start with an equilateral triangle with side of length a, as in Figure (1). Then you add to each side another equilateral triangle, with side of length $a/3$, as in Figure (2). Then to each side of Figure (2) you add another equilateral triangle with side of length $a/9$, and so on, obtaining a design that looks very much like a snowflake.

a) Complete the following table and look for patterns.

Figure	Number of sides	Perimeter	Area
(1)	3	$3a$	$\frac{a^2}{4}\sqrt{3}$
(2)			
(3)			
(4)			

b) Conjecture a formula for the number of sides of the nth figure. Prove it by mathematical induction.
c) Conjecture a formula for the perimeter of the nth figure. Prove it by mathematical induction.
d) Conjecture a formula for the area of the nth figure. Prove it by mathematical induction.
e) Find the limit of these areas.

12.5 Combinatorics: Permutations

OBJECTIVE

You should be able to:

1 **Evaluate factorial and permutation notation and solve related counting problems.**

In order to study probability, it is first necessary to study the theory of counting, called **combinatorics**. Such a study concerns itself with determining the number of ways in which a set can be arranged or combined, certain objects can be chosen, or a succession of events can occur.

1 The part of combinatorics that we will consider here is the study of *permutations*.

> **The study of *permutations* involves *order* and *arrangement*.**

EXAMPLE 1 How many 3-letter code symbols can be formed with the letters A, B, C, *without* repetition?

Solution Examples of such symbols are ABC, CBA, ACB, and so on. Consider placing the letters in these frames.

We can select any of the 3 letters for the first letter in the symbol. Once this letter has been selected, the second can be selected from the 2 remaining letters. The third letter is already determined, since only 1 possibility is left. The possibilities can be arrived at with a **tree diagram**.

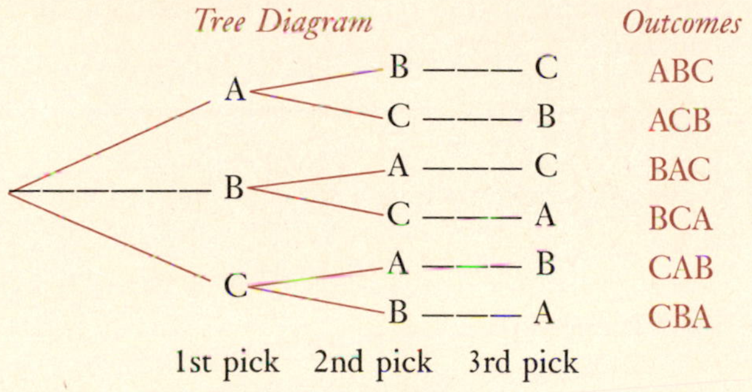

There are $3 \cdot 2 \cdot 1$, or 6, possibilities. The set of all of them is as follows:

$$\{ABC, ACB, BAC, BCA, CAB, CBA\}.$$

Suppose we perform an experiment such as selecting letters (as in the preceding example), flipping a coin, or drawing a card. The results are called **outcomes**. An **event** is a set of outcomes. The following theorem concerns events that occur together, or are combined.

> **Fundamental Counting Principle**
>
> **Given a combined action, or event, in which the first action can be performed in n_1 ways, the second action can be performed in n_2 ways, and so on. Then the total number of ways in which the combined action can be performed is the product**
>
> $$n_1 \cdot n_2 \cdot n_3 \cdot \cdots \cdot n_k.$$

EXAMPLE 2 How many 3-letter code symbols can be formed with the letters A, B, and C, *with* repetition?

Solution There are 3 choices for the first letter and, since we allow repetition, 3 choices for the second and 3 for the third. Thus by the fundamental counting principle, there are $3 \cdot 3 \cdot 3$, or 27, choices.

DO EXERCISES 1–3.

1. How many 3-digit numbers can be named using all the digits 5, 6, 7 without repetition? with repetition?

2. *Zip codes in Canada.* A zip code for Montreal, Quebec, in Canada, is H2N 1M5. It consists of a letter in the first, third, and fifth places, and a number from 0 to 9 in the second, fourth, and sixth places.
   a) How many such zip codes are there?
   b) There are 25 million people in Canada. Can each person have his or her own unique zip code?

3. In how many ways can 5 different cars be parked in a row in a parking lot?

4. How many permutations are there in a set of 5 objects? Consider a set {A, B, C, D, E}.

Compute.

5. ${}_3P_3$

6. ${}_5P_5$

7. ${}_6P_6$

8. In how many different ways can 4 horses be lined up for a race?

9. In how many different ways can 6 people line up at a ticket window?

10. In how many different ways can the 9-person batting order of a baseball team be made up, if you assume that the pitcher bats last?

DEFINITION

A *permutation* of a set of n objects is an ordered arrangement of all n objects.

Consider, for example, a set of 4 objects:

$$\{A, B, C, D\}.$$

To find the number of ordered arrangements of the set, we select a first letter: There are 4 choices. Then we select a second letter: There are 3 choices. Then we select a third letter: There are 2 choices. Finally, there is 1 choice for the last selection. Thus by the fundamental counting principle, there are $4 \cdot 3 \cdot 2 \cdot 1$, or 24, permutations of a set of 4 objects.

DO EXERCISE 4.

We can find a formula for the total number of permutations of all objects in a set of n objects. We have n choices for the first selection, $n - 1$ for the second, $n - 2$ for the third, and so on. For the nth selection, there is only 1 choice.

THEOREM 7

The total number of permutations of a set of n objects, denoted ${}_nP_n$, is given by

$${}_nP_n = n(n-1)(n-2)\cdots(3)(2)(1).$$

EXAMPLE 3 Find (a) ${}_4P_4$ and (b) ${}_7P_7$.

Solution

a) ${}_4P_4 = 4 \cdot 3 \cdot 2 \cdot 1 = 24$

b) ${}_7P_7 = 7 \cdot 6 \cdot 5 \cdot 4 \cdot 3 \cdot 2 \cdot 1 = 5040$

EXAMPLE 4 In how many different ways can 9 different letters be placed in 9 mailboxes, one letter to a box?

Solution

$${}_9P_9 = 9 \cdot 8 \cdot 7 \cdot 6 \cdot 5 \cdot 4 \cdot 3 \cdot 2 \cdot 1 = 362{,}880$$

DO EXERCISES 5–10.

Factorial Notation

Products of successive natural numbers, such as $7 \cdot 6 \cdot 5 \cdot 4 \cdot 3 \cdot 2 \cdot 1$ and $9 \cdot 8 \cdot 7 \cdot 6 \cdot 5 \cdot 4 \cdot 3 \cdot 2 \cdot 1$, are used so often that it is convenient to adopt a notation for them.

For the product $7 \cdot 6 \cdot 5 \cdot 4 \cdot 3 \cdot 2 \cdot 1$, we write 7!, read "7-factorial."

DEFINITION

$$n! = n(n-1)(n-2)\cdots(3)(2)(1)$$

Here are some examples.

$$\begin{aligned}
7! &= 7 \cdot 6 \cdot 5 \cdot 4 \cdot 3 \cdot 2 \cdot 1 = 5040\\
6! &= 6 \cdot 5 \cdot 4 \cdot 3 \cdot 2 \cdot 1 = 720\\
5! &= 5 \cdot 4 \cdot 3 \cdot 2 \cdot 1 = 120\\
4! &= 4 \cdot 3 \cdot 2 \cdot 1 = 24\\
3! &= 3 \cdot 2 \cdot 1 = 6\\
2! &= 2 \cdot 1 = 2\\
1! &= 1 = 1
\end{aligned}$$

DO EXERCISES 11 AND 12.

We also define 0! to be 1. We do this so that certain formulas and theorems can be stated concisely and with a consistent pattern.

We can now simplify the formula of Theorem 7 as follows:

$$_nP_n = n!.$$

DO EXERCISE 13.

Note that $8! = 8 \cdot 7!$. We can see this as follows. By definition of factorial notation,

$$\begin{aligned}
8! &= 8 \cdot 7 \cdot 6 \cdot 5 \cdot 4 \cdot 3 \cdot 2 \cdot 1\\
&= 8 \cdot (7 \cdot 6 \cdot 5 \cdot 4 \cdot 3 \cdot 2 \cdot 1)\\
&= 8 \cdot 7!.
\end{aligned}$$

Generalizing, we get the following.

For any natural number n, $n! = n(n-1)!$.

By using this result repeatedly, we can further manipulate factorial notation.

EXAMPLE 5 Rewrite 7! with a factor of 5!.

Solution

$$7! = 7 \cdot 6 \cdot 5!$$

DO EXERCISE 14.

Permutations of n Objects Taken r at a Time

Consider a set of 6 objects, say {A, B, C, D, E, F}. How many ordered arrangements are there having 3 members without repetition? We can select the first object in 6 ways. There are then 5 choices for the second and then 4 choices for the third. By the fundamental counting principle, there are then $6 \cdot 5 \cdot 4$ ways to construct the subset. In other words, there are $6 \cdot 5 \cdot 4$ permutations of a set of 6 objects taken 3 at a time. Note that

$$6 \cdot 5 \cdot 4 = \frac{6 \cdot 5 \cdot 4 \cdot 3 \cdot 2 \cdot 1}{3 \cdot 2 \cdot 1}, \quad \text{or} \quad \frac{6!}{3!}.$$

11. Find 8!.

12. Find 9!.

13. Using factorial notation only, represent the number of permutations of 18 objects.

14. a) Rewrite 10! with a factor of 9!.
    b) Rewrite 20! with a factor of 15!.

15. Compute $_7P_3$.

16. Compute each of the following.

a) $_{10}P_4$

b) $_8P_2$

c) $_{11}P_5$

d) $_nP_1$

e) $_nP_2$

f) $_nP_0$

DEFINITION

A *permutation* of a set of *n* objects taken *r* at a time is an ordered arrangement of *r* objects taken from the set.

Consider a set of n objects and the selecting of an ordered arrangement of r objects. The first object can be selected in n ways. The second can be selected in $n - 1$ ways, and so on. The rth can be selected in $n - (r - 1)$ ways. By the fundamental counting principle, the total number of permutations is

$$n(n-1)(n-2)\cdots[n-(r-1)].$$

We now multiply by 1:

$$n(n-1)(n-2)\cdots[n-(r-1)]\frac{(n-r)!}{(n-r)!}$$
$$= \frac{n(n-1)(n-2)(n-3)\cdots[n-(r-1)](n-r)!}{(n-r)!}.$$

The numerator is now the product of all natural numbers from n to 1, hence is $n!$. Thus the total number of permutations is

$$\frac{n!}{(n-r)!}.$$

This gives us the following theorem.

Theorem 8

The number of permutations of a set of *n* objects taken *r* at a time, denoted $_nP_r$, is given by

$$_nP_r = n(n-1)(n-2)\cdots[n-(r-1)] \qquad (1)$$
$$= \frac{n!}{(n-r)!}. \qquad (2)$$

Formula (1) is most useful in application, but formula (2) will be important in a later development.

EXAMPLE 6 Compute $_6P_4$ using both formulas of Theorem 8.

Solution Using formula (1), we have

$$_6P_4 = 6\cdot5\cdot4\cdot3$$
$$= 360.$$

Note that the 6 in $_6P_4$ shows where to start and the 4 in $_6P_4$ shows how many factors there are.

Using formula (2) of Theorem 2, we have

$$_6P_4 = \frac{6!}{(6-4)!} = \frac{6!}{2!} = \frac{6\cdot5\cdot4\cdot3\cdot2\cdot1}{2\cdot1} = 6\cdot5\cdot4\cdot3 = 360.$$ ■

DO EXERCISES 15 AND 16.

EXAMPLE 7 In how many ways can the letters of the set {A, B, C, D, E, F, G} be arranged without repetition to form code words of (a) 7 letters? (b) 5 letters? (c) 4 letters? (d) 2 letters?

Solution

a) ${}_7P_7 = 7 \cdot 6 \cdot 5 \cdot 4 \cdot 3 \cdot 2 \cdot 1 = 5040$

b) ${}_7P_5 = 7 \cdot 6 \cdot 5 \cdot 4 \cdot 3 = 2520$

c) ${}_7P_4 = 7 \cdot 6 \cdot 5 \cdot 4 = 840$

d) ${}_7P_2 = 7 \cdot 6 = 42$

EXAMPLE 8 A baseball manager arranges the batting order as follows: The 4 infielders will bat first, then the outfielders, catcher, and pitcher will follow, not necessarily in that order. How many different batting orders are possible?

Solution The infielders can bat in 4! different ways; the rest in 5! different ways. Then by the fundamental counting principle, we have ${}_4P_4 \cdot {}_5P_5 = 4! \cdot 5!$, or 2880, possible batting orders.

DO EXERCISES 17–19.

Permutations of Sets with Nondistinguishable Objects

Consider a set of 7 marbles, 4 of which are red and 3 of which are black. Although the marbles are all different, when they are lined up, one black marble will look just like any other black marble. In this sense, we say that the red marbles are nondistinguishable and the black marbles are nondistinguishable.

We know that there are 7! permutations of this set. Many of them will look alike, however. We develop a formula for finding the number of distinguishable permutations.

Consider a set of n objects in which n_1 are of one kind, n_2 are of a second kind, . . . , n_k are of the kth kind. By Theorem 7, the total number of permutations of the set is $n!$. Let P be the number of distinguishable permutations. For each of these P permutations, there are $n_1!$ actual permutations, obtained by permuting the objects of the first kind. For each of these $P \cdot n_1!$ permutations, there are $n_2!$ actual permutations, obtained by permuting the objects of the second kind. And so on. By the fundamental counting principle, the total number of actual permutations is

$$P \cdot n_1! \cdot n_2! \cdot \cdots \cdot n_k!.$$

Then we have $P \cdot n_1! \cdot n_2! \cdot \cdots \cdot n_k! = n!$. Solving for P, we obtain

$$P = \frac{n!}{n_1! n_2! \cdots n_k!}.$$

This proves the following theorem.

THEOREM 9

For a set of n objects in which n_1 are of one kind, n_2 are of another kind, . . . , n_k are of a kth kind, the number of distinguishable permutations is

$$\frac{n!}{n_1! \cdot n_2! \cdot \cdots \cdot n_k!}.$$

EXAMPLE 9 In how many distinguishable ways can the letters of the word CINCINNATI be arranged?

17. In how many ways can a 5-woman starting unit be selected from a 12-woman basketball squad and arranged in a straight line?

18. Many nations use flags consisting of three vertical stripes similar to the one shown here. For example, the flag of Ireland has its 1st stripe green, 2nd white, and 3rd gold. Suppose the following 9 colors are available: *black*, *yellow*, *red*, *blue*, *white*, *gold*, *orange*, *pink*, *purple*. How many different flags can be made up without repetition of colors? This assumes that the order in which a color appears as a stripe is considered.

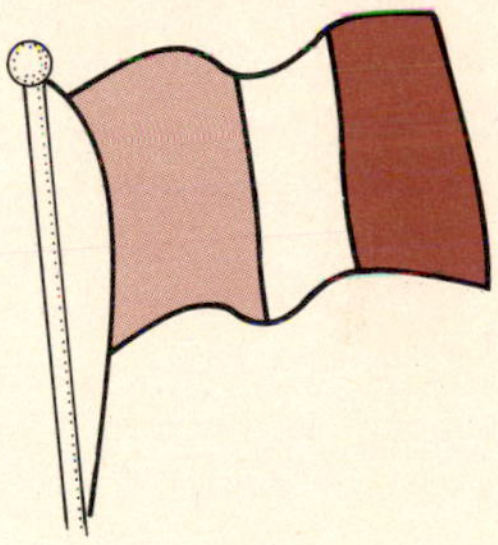

19. How many 7-digit numbers can be named, without repetition, using the digits 2, 3, 4, 5, 6, 7, and 8, if the even digits come first?

20. In how many distinguishable ways can the letters of the word MISSISSIPPI be arranged?

21. How many 6-digit numbers can be named with all the digits 3, 3, 3, 4, 4, and 5?

22. How many vertical signal-flag arrangements can be formed with 3 solid red, 3 solid green, and 2 solid yellow flags?

23. In how many ways can 7 men be arranged around a round table?

24. In how many ways can the numbers on a clock face be arranged?

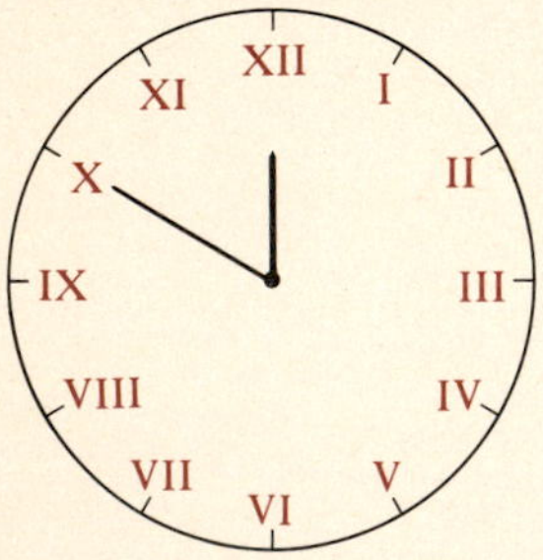

25. How many 5-letter code symbols can be formed by repeated use of the letters of the alphabet?

Solution *Note*: There are 2 C's, 3 I's, 3 N's, 1 A, and 1 T, for a total of 10. Thus,

$$P = \frac{10!}{2! \cdot 3! \cdot 3! \cdot 1! \cdot 1!}, \quad \text{or} \quad 50{,}400. \quad \blacksquare$$

DO EXERCISES 20–22.

Circular Arrangements

Suppose we arrange the 4 letters A, B, C, and D in a circular arrangement.

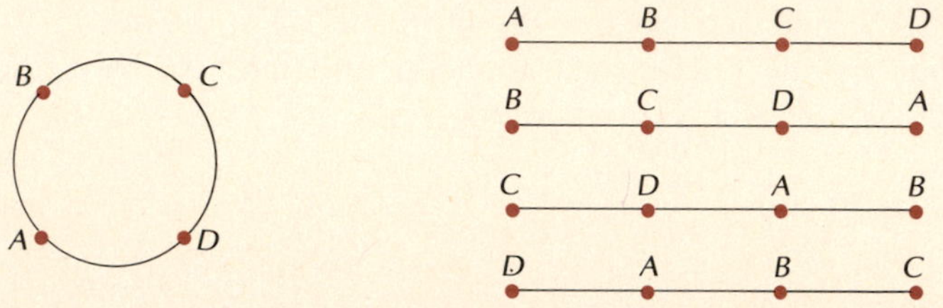

Note that the circular arrangements ABCD, BCDA, CDAB, and DABC are not distinguishable, because they are on a line. For each circular arrangement, there are 4 distinguishable arrangements on a line. Thus if there are P circular arrangements, these yield $4 \cdot P$ arrangements on a line, which we know is $_4P_4$, or 4!. Thus, $4 \cdot P = 4!$, so $P = 4!/4$, or 3!. To generalize:

> **THEOREM 10**
>
> The number of distinct (different) circular arrangements of n objects is $(n - 1)!$.

EXAMPLE 10 In how many ways can 9 different foods be arranged around a lazy Susan?

Solution

$$(9 - 1)! = 8! = 8 \cdot 7 \cdot 6 \cdot 5 \cdot 4 \cdot 3 \cdot 2 \cdot 1 = 40{,}320 \quad \blacksquare$$

DO EXERCISES 23 AND 24.

Repeated Use of the Same Object

EXAMPLE 11 How many 5-letter code symbols can be formed with the letters A, B, C, and D if we allow a letter to occur more than once?

Solution We have five spaces:

We can select the first letter in 4 ways, the second in 4 ways, and so on. Thus there are 4^5, or 1024, arrangements. $\quad \blacksquare$

Generalizing, we have the following.

> **THEOREM 11**
>
> The number of distinct arrangements of n objects taken r at a time, allowing repetition, is n^r.

DO EXERCISE 25.

EXERCISE SET 12.5

1 Evaluate.

1. ${}_4P_3$
2. ${}_7P_5$
3. ${}_{10}P_7$
4. ${}_{10}P_3$

5. How many 5-digit numbers can be named using the digits 5, 6, 7, 8, and 9 without repetition? with repetition?

6. How many 4-digit numbers can be named using the digits 2, 3, 4, and 5 without repetition? with repetition?

7. In how many ways can 5 students be arranged in a straight line? in a circle?

8. In how many ways can 7 athletes be arranged in a straight line? in a circle?

9. In how many distinguishable ways can the letters of the word DIGIT be arranged?

10. In how many distinguishable ways can the letters of the word RABBIT be arranged?

11. How many 7-digit phone numbers can be formed with the digits 0, 1, 2, 3, 4, 5, 6, 7, 8, and 9, assuming that no digit is used more than once and the first digit is not 0?

12. A program is planned to have 5 rock numbers and 4 speeches. In how many ways can this be done if a rock number and a speech are to alternate and the rock numbers come first?

13. Suppose the expression $a^2b^3c^4$ is rewritten without exponents. In how many ways can this be done?

14. Suppose the expression a^3bc^2 is rewritten without exponents. In how many ways can this be done?

15. In how many ways could King Arthur and his 12 knights sit at his Round Table?

16. In how many ways can 4 people be seated at a bridge table?

17. A penny, nickel, dime, quarter, and half-dollar (if you have one) are arranged in a straight line.
    a) Considering just the coins, in how many ways can they be lined up?
    b) Considering the coins and heads and tails, in how many ways can they be lined up?

18. A penny, nickel, dime, and quarter are arranged in a straight line.
    a) Considering just the coins, in how many ways can they be lined up?
    b) Considering the coins and heads and tails, in how many ways can they be lined up?

19. ▦ Compute ${}_{52}P_4$.

20. ▦ Compute ${}_{50}P_5$.

21. A professor is going to grade her 24 students on a curve. She will give 3 A's, 5 B's, 9 C's, 4 D's, and 3 F's. In how many ways can she do this?

22. A professor is planning to grade his 20 students on a curve. He will give 2 A's, 5 B's, 8 C's, 3 D's, and 2 F's. In how many ways can he do this?

23. ▦ How many distinguishable code symbols can be formed from the letters of the word MATH? BUSINESS? PHILOSOPHICAL?

24. ▦ How many distinguishable code symbols can be formed from the letters of the word ORANGE? BIOLOGY? MATHEMATICS?

25. ▦ A state forms its license plates by first listing a number that corresponds to the county in which the car owner lives (the names of the counties are alphabetized and the number is its location in that order). Then the plate lists a letter of the alphabet, and this is followed by a number from 1 to 9999. How many such plates are possible if there are 80 counties?

26. How many code symbols can be formed using 4 out of 5 letters of A, B, C, D, E if the letters:
    a) are not repeated?
    b) can be repeated?
    c) are not repeated but must begin with D?
    d) are not repeated but must end with DE?

27. *Zip codes.* A zip code in Dallas, Texas, is 75247. A zip code in Cambridge, Massachusetts, is 02142.
    a) How many zip codes are possible if any of the digits 0 to 9 can be used?
    b) If each post office has its own zip code, how many possible post offices can there be?

28. *Zip codes.* Zip codes are sometimes given using a 9-digit number like 75247-5456, where the last 4 digits represent a post office box number.
    a) How many 9-digit zip codes are possible?
    b) There are 243 million people in the United States. If each person has a zip code and there were enough post office boxes, are there enough zip codes?

29. *Social security numbers*. A social security number is a 9-digit number like 293-36-0391.

a) How many social security numbers can there be?

b) There are 243 million people in the United States. Can each person have a social security number?

30. ▦ How "long" is 15!? You own 15 different books and decide to actually make up all possible arrangements of the books on a shelf. About how long, in years, would it take if you can make one arrangement per second?

SYNTHESIS

Solve for n.

31. ${}_nP_5 = 7 \cdot {}_nP_4$

32. ${}_nP_4 = 8 \cdot {}_{n-1}P_3$

33. ${}_nP_5 = 9 \cdot {}_{n-1}P_4$

34. ${}_nP_4 = 8 \cdot {}_nP_3$

CHALLENGE

35. In a single-elimination sports tournament consisting of n teams, a team is eliminated when it loses one game. How many games are required to complete the tournament?

36. In a double-elimination softball tournament consisting of n teams, a team is eliminated when it loses two games. At most, how many games are required to complete the tournament?

37. Find a formula for the total number of placements of n keys onto a key ring.

12.6 Combinatorics: Combinations

OBJECTIVE

You should be able to:

1 Evaluate combination notation and solve related problems.

1. Consider the set {A, B, C, D, E}. How many combinations are there taken:

a) 5 at a time?

b) 4 at a time?

c) 2 at a time?

d) 1 at a time?

e) 0 at a time?

1 If you play cards, you know that in most situations the *order* in which you hold cards *is not important*! It is just the contents of the hand, or set, of cards. We may sometimes make selections from a set *without regard to order*. Such selections are called **combinations.**

[A] → [B] → [C] [C] [A] [B]

Permutation: Order considered! *Combination:* Order *not* considered!

EXAMPLE 1 Find all the combinations of taking 3 elements from the set of 5 elements {A, B, C, D, E}. How many are there?

Solution The combinations are

{A, B, C}, {A, B, D}, {A, B, E}, {A, C, D}, {A, C, E},
{A, D, E}, {B, C, D}, {B, C, E}, {B, D, E}, {C, D, E}.

There are 10 combinations of 5 objects taken 3 at a time. ■

When we find all the combinations of 5 objects taken 3 at a time, we are finding all the 3-element subsets. When we are naming a set, the order of the listing is *not* considered. Thus,

{A, C, B} names the same set as {A, B, C}.

DO EXERCISE 1.

DEFINITION

The set A is a *subset* of B, denoted $A \subseteq B$, if every element of A is an element of B.

$A \subseteq B$

A is a subset of B.

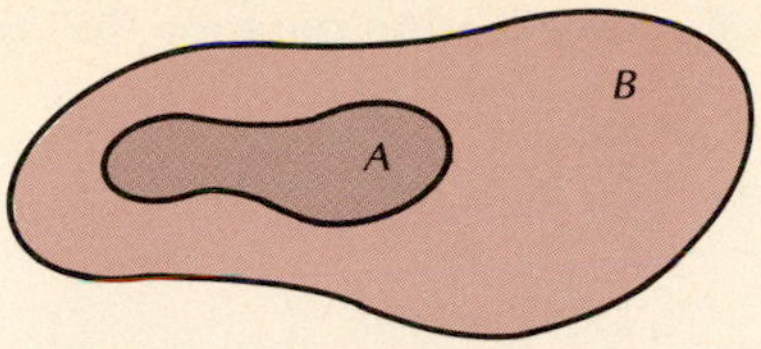

> **DEFINITION**
>
> **A *combination* containing r objects is a subset of a set that has n objects, $r \leq n$.**

The elements of a subset are not ordered.

> **When thinking of *combinations*, do *not* think about order!**

EXAMPLE 2 Find all the subsets of the set {A, B, C}. Identify these as combinations. How many subsets are there in all?

Solution

a) The empty set has 0 elements in it. It is denoted $\varnothing$. The empty set is a subset of every set. In this case, it is the combination of 3 objects taken 0 at a time. There is 1 such combination, $\varnothing$.

b) The following are all the one-element subsets of {A, B, C}:

{A}, {B}, {C}.

These are the combinations of 3 objects taken 1 at a time. There are 3 such combinations.

c) The following are all the two-element subsets of {A, B, C}:

{A, B}, {A, C}, {B, C}.

These are the combinations of 3 objects taken 2 at a time. There are 3 such combinations.

d) The following are all the three-element subsets of {A, B, C}:

{A, B, C}.

These are the combinations of 3 objects taken 3 at a time. There is only 1 such combination. A set is always a subset of itself.

The total number of subsets is 1 + 3 + 3 + 1, or 8. ■

DO EXERCISES 2 AND 3.

We want to develop a formula for computing the number of combinations of n objects taken r at a time without actually listing the combinations, or subsets.

> **DEFINITION**
>
> **The notation for the number of combinations taken r at a time from a set of n objects is denoted ${}_nC_r$.**

2. Consider the set {A, B}.
   a) List all the subsets with 0 elements. How many such subsets are there?
   b) List all the subsets with 1 element. How many such subsets are there?
   c) List all the subsets with 2 elements. How many such subsets are there?
   d) How many subsets of the set {A, B} are there in all?

3. Consider the set {A}.
   a) List all the subsets with 0 elements. How many such subsets are there?
   b) List all the subsets with 1 element. How many such subsets are there?
   c) How many subsets of the set {A} are there in all?

We call ${}_nC_r$ **combination notation.** In Example 1 and Margin Exercise 1, we see that

$${}_5C_5 = 1, \quad {}_5C_4 = 5, \quad {}_5C_3 = 10, \quad {}_5C_2 = 10, \quad {}_5C_1 = 5, \quad {}_5C_0 = 1$$

and that

The total number of subsets of a set of 5 objects

$$= {}_5C_5 + {}_5C_4 + {}_5C_3 + {}_5C_2 + {}_5C_1 + {}_5C_0$$
$$= 1 + 5 + 10 + 10 + 5 + 1 = 32.$$

We can derive some general results here. First, it is always true that ${}_nC_n = 1$, because a set with n objects has only 1 subset with n objects, the set itself. Second, ${}_nC_1 = n$ because a set with n objects has n subsets with 1 element each. Finally, ${}_nC_0 = 1$ because a set with n objects has only one subset with 0 elements, namely, the empty set $\varnothing$.

We want to derive a general formula for ${}_nC_r$, for any $r \leq n$. Let us return to Example 1 and compare the number of combinations with the number of permutations.

Combinations		*Permutations*					
{A, B, C}	⟶	ABC	BCA	CAB	CBA	BAC	ACB
{A, B, D}	⟶	ABD	BDA	DAB	DBA	BAD	ADB
{A, B, E}	⟶	ABE	BEA	EAB	EBA	BAE	AEB
{A, C, D}	⟶	ACD	CDA	DAC	DCA	CAD	ADC
{A, C, E}	⟶	ACE	CEA	EAC	ECA	CAE	AEC
{A, D, E}	⟶	ADE	DEA	EAD	EDA	DAE	AED
{B, C, D}	⟶	BCD	CDB	DBC	DCB	CBD	BDC
{B, C, E}	⟶	BCE	CEB	EBC	ECB	CBE	BEC
{B, D, E}	⟶	BDE	DEB	EBD	EDB	DBE	BED
{C, D, E}	⟶	CDE	DEC	ECD	EDC	DCE	CED

Note that each combination of 3 objects, say {A, C, E}, yields 3!, or 6, permutations, as shown above. It follows that

$$3! \cdot {}_5C_3 = 60 = {}_5P_3 = 5 \cdot 4 \cdot 3,$$

so

$${}_5C_3 = \frac{{}_5P_3}{3!} = \frac{5 \cdot 4 \cdot 3}{3 \cdot 2 \cdot 1} = 10.$$

In general, the number of combinations of n objects taken r at a time, ${}_nC_r$, times the number of permutations of these r objects, $r!$, must equal the number of permutations of n objects taken r at a time:

$$r! \cdot {}_nC_r = {}_nP_r$$

$${}_nC_r = \frac{{}_nP_r}{r!} = \frac{1}{r!} \cdot {}_nP_r = \frac{1}{r!} \cdot \frac{n!}{(n-r)!} = \frac{n!}{r!(n-r)!}.$$

This now gives us two formulas for computing ${}_nC_r$.

THEOREM 12

The total number of combinations of n objects taken r at a time, denoted ${}_nC_r$, is given by

$${}_nC_r = \frac{n!}{r!(n-r)!}, \tag{1}$$

or

$${}_nC_r = \frac{{}_nP_r}{r!} = \frac{n(n-1)(n-2)\cdots[n-(r-1)]}{r!}. \tag{2}$$

There is another kind of notation that is also used for ${}_nC_r$. It is called **binomial coefficient notation.** The reason for such terminology will be seen later.

DEFINITION

$$\binom{n}{r} = {}_nC_r$$

You should be able to use either notation and either formula.

EXAMPLE 3 Evaluate $\binom{7}{5}$, using formulas (1) and (2).

Solution

a) By formula (1),

$$\binom{7}{5} = \frac{7!}{5!2!} = \frac{7 \cdot 6 \cdot 5 \cdot 4 \cdot 3 \cdot 2 \cdot 1}{5 \cdot 4 \cdot 3 \cdot 2 \cdot 1 \cdot 2 \cdot 1} = \frac{7 \cdot 6 \cdot 5 \cdot 4 \cdot 3}{5 \cdot 4 \cdot 3 \cdot 2 \cdot 1} = \frac{7 \cdot 6}{2 \cdot 1} = 21.$$

b) By formula (2),

The 7 tells where to start.

$$\binom{7}{5} = \frac{7 \cdot 6 \cdot 5 \cdot 4 \cdot 3}{5 \cdot 4 \cdot 3 \cdot 2 \cdot 1} = \frac{7 \cdot 6}{2 \cdot 1} = 21$$

The 5 tells us how many factors there are in both numerator and denominator and where to start the denominator.

CAUTION!

$\binom{n}{r}$ **does not mean** $n \div r$ **or** $\frac{n}{r}$.

DO EXERCISES 4–7.

The method in Example 3(b), using formula (2), is easiest to carry out, but in some situations formula (1) does become useful.

EXAMPLE 4 Evaluate $\binom{n}{0}$ and $\binom{n}{2}$.

Solution We use formula (1) for the first expression and formula (2) for the second. Then

$$\binom{n}{0} = \frac{n!}{0!(n-0)!} = \frac{n!}{1 \cdot n!} = 1,$$

using formula (1), and

$$\binom{n}{2} = \frac{n(n-1)}{2!} = \frac{n(n-1)}{2}, \quad \text{or} \quad \frac{n^2 - n}{2},$$

using formula (2).

DO EXERCISES 8 AND 9.

Evaluate.

4. $\binom{10}{3}$

5. $\binom{10}{7}$

6. ${}_9C_4$

7. ${}_9C_5$

Evaluate.

8. $\binom{n}{1}$

9. $\binom{n}{3}$

10. a) Evaluate $\binom{8}{5}$ and $\binom{8}{3}$.

b) Which seemed easier to compute?

11. Evaluate $\binom{100}{97}$.

Note that

$$\binom{7}{2} = \frac{7 \cdot 6}{2 \cdot 1} = 21,$$

so that from Example 3,

$$\binom{7}{5} = \binom{7}{2}.$$

This says that the number of 5-element subsets of a set of 7 objects is the same as the number of 2-element subsets of a set of 7 objects. When 5 elements are chosen from a set, one also chooses *not* to include 2 elements. To see this, consider such a set:

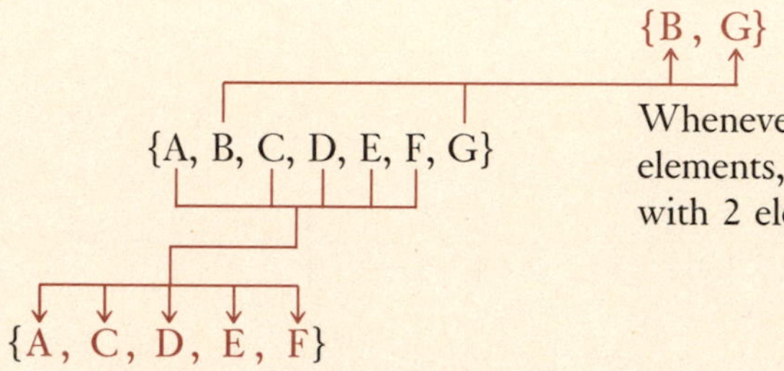

Whenever we form a subset with 5 elements, we leave behind a subset with 2 elements, and vice versa.

Thus the numbers of each type of subset are the same. In general:

THEOREM 13

$$\binom{n}{r} = \binom{n}{n-r} \quad \text{and} \quad {}_nC_r = {}_nC_{n-r}$$

The number of subsets of size r of a set with n objects is the same as the number of subsets of size $n - r$. The number of combinations of n objects taken r at a time is the same as the number of combinations of n objects taken $n - r$ at a time.

Theorem 13 provides an alternative way to compute. For example, instead of computing ${}_{52}C_{48}$, it is a lot easier to compute ${}_{52}C_4$.

DO EXERCISES 10 AND 11.

We now solve problems involving combinations.

EXAMPLE 5 *Michigan lotto.* The state of Michigan runs a 6-out-of-44-number lotto twice a week that pays at least $1.5 million. You purchase a card for $1 and pick any 6 numbers from 1 to 44. If your 6 numbers match those that the state draws, you win.

a) How many possible 6-number combinations are there for drawing?
b) Suppose it takes 10 minutes to pick your numbers and buy a ticket. How many tickets can you buy in 4 days?
c) How many people would you have to hire to buy all the tickets and ensure that you win?

Solution

a) No order is implied here. You pick any 6 numbers from 1 to 44. Thus the number of combinations is

$$ {}_{44}C_6 = \binom{44}{6} = \frac{44 \cdot 43 \cdot 42 \cdot 41 \cdot 40 \cdot 39}{6 \cdot 5 \cdot 4 \cdot 3 \cdot 2 \cdot 1} = 7{,}059{,}052.$$

b) In four days, there are $4 \cdot 24 \cdot 60$, or 5760, minutes, so you could buy 5760/10, or 576, tickets in that entire time period.
c) You would need to hire 7,059,052/576, or about 12,256, people to buy all the tickets and ensure a win. (This presumes lottery tickets can be bought 24 hours a day, which is questionable.) ■

DO EXERCISE 12.

EXAMPLE 6 How many committees can be formed from a group of 5 governors and 7 senators if each committee contains 3 governors and 4 senators?

Solution The 3 governors can be selected in ${}_5C_3$ ways and the 4 senators can be selected in ${}_7C_4$ ways. If we use the fundamental counting principle, it follows that the number of possible committees is

$$ {}_5C_3 \cdot {}_7C_4 = 10 \cdot 35 = 350. $$ ■

DO EXERCISE 13.

12. An examination consists of 10 questions. A student is required to answer 8 of them. In how many different ways can the student choose 8 questions to answer? (*Hint:* Is the *order* in which the student answers the questions important, assuming that the answers themselves are numbered?)

13. A committee is to be formed from a group of 12 men and 8 women and is to consist of 3 men and 2 women. How many committees can be formed?

EXERCISE SET 12.6

1 Evaluate.

1. ${}_{13}C_2$
2. ${}_9C_6$
3. $\binom{13}{11}$
4. $\binom{9}{3}$
5. $\binom{7}{1}$
6. $\binom{8}{8}$
7. $\frac{{}_5P_3}{3!}$
8. $\frac{{}_{10}P_5}{5!}$
9. $\binom{6}{0}$
10. $\binom{6}{1}$
11. $\binom{6}{2}$
12. $\binom{6}{3}$
13. ${}_{12}C_{11}$
14. ${}_{12}C_{10}$
15. ${}_{12}C_9$
16. ${}_{12}C_8$
17. $\binom{m}{2}$
18. $\binom{t}{4}$
19. $\binom{p}{3}$
20. $\binom{m}{m}$

Find the sum.

21. $\binom{7}{0} + \binom{7}{1} + \binom{7}{2} + \binom{7}{3} + \binom{7}{4} + \binom{7}{5} + \binom{7}{6} + \binom{7}{7}$
22. $\binom{6}{0} + \binom{6}{1} + \binom{6}{2} + \binom{6}{3} + \binom{6}{4} + \binom{6}{5} + \binom{6}{6}$
23. ${}_{100}C_0 + {}_{100}C_1 + \cdots + {}_{100}C_{100}$
24. ${}_nC_0 + {}_nC_1 + \cdots + {}_nC_n$

In each of the following exercises, give an expression for the answer in terms of permutation notation, combination notation, factorial notation, or other products. Then evaluate.

25. There are 23 students in a fraternity. How many sets of 4 officers can be selected?
26. How many basketball games can be played in a 9-team league if each team plays all other teams once? twice?
27. On a test, a student is to select 6 out of 10 questions. In how many ways can he do this?
28. On a test, a student is to select 7 out of 11 questions. In how many ways can she do this?
29. How many lines are determined by 8 points, no 3 of which are collinear? How many triangles are determined by the same points?
30. How many lines are determined by 7 points, no 3 of which are collinear? How many triangles are determined by the same points?

31. Of the first 10 questions on a test, a student must answer 7. On the second 5 questions, she must answer 3. In how many ways can this be done?

32. Of the first 8 questions on a test, a student must answer 6. On the second 4 questions, he must answer 3. In how many ways can this be done?

33. Suppose the Senate of the United States consists of 58 Democrats and 42 Republicans. How many committees made up of 6 Democrats and 4 Republicans can be formed? You need not simplify the expression.

34. Suppose the Senate of the United States consists of 63 Republicans and 37 Democrats. How many committees made up of 8 Republicans and 12 Democrats can be formed? You need not simplify the expression.

35. How many 5-card poker hands consisting of 3 aces and 2 cards that are not aces are possible with a 52-card deck? (See Section 12.8 for a description of a 52-card deck.)

36. How many 5-card poker hands consisting of 2 kings and 3 cards that are not kings are possible with a 52-card deck?

37. Bresler's Ice Cream, a national firm, sells ice cream in 33 flavors.

a) How many 3-dip cones are possible if order of flavors is to be considered and no flavor is repeated?
b) How many 3-dip cones are possible if order is to be considered and a flavor can be repeated?
c) How many 3-dip cones are possible if order is not considered and no flavor is repeated?

38. Baskin-Robbins Ice Cream, a national firm, sells ice cream in 31 flavors.

a) How many 2-dip cones are possible if order of flavors is to be considered and no flavor is repeated?
b) How many 2-dip cones are possible if order is to be considered and a flavor can be repeated?
c) How many 2-dip cones are possible if order is not considered and no flavor is repeated?

39. Pizza Hut, a national pizza firm, has the following toppings for pizzas:

cheese, pepperoni, sausage, mushroom, onion, green pepper, beef, Italian sausage, black olives, jalapeno peppers, ham, anchovies.

How many different kinds of pizzas can Pizza Hut serve excluding size and thickness of pizzas?

40. Pizza Hut serves round pizzas in three sizes—9-inch, 13-inch, and 15-inch—and two thicknesses—thin-and-crispy and pan. Including all the toppings listed in Exercise 39, and considering sizes and thicknesses, how many different kinds of pizzas can Pizza Hut serve?

SYNTHESIS

41. ▦ How many 5-card poker hands are possible with a 52-card deck?

42. ▦ How many 13-card bridge hands are possible with a 52-card deck?

43. There are 8 points on a circle. How many triangles can be inscribed with these points as vertices?

44. There are n points on a circle. How many quadrilaterals can be inscribed with these points as vertices?

45. A set of 5 parallel lines crosses another set of 8 parallel lines at angles that are not right angles. How many parallelograms are formed?

46. Prove: For any natural numbers n and $r \leq n$,
$$\binom{n}{r} = \binom{n}{n-r}.$$

47. How many games are played in a league with 8 teams if each team plays each other team once? twice?

48. How many games are played in a league with n teams if each team plays each other team once? twice?

Solve for n.

49. $\binom{n+1}{3} = 2 \cdot \binom{n}{2}$

50. $\binom{n}{n-2} = 6$

51. $\binom{n+2}{4} = 6 \cdot \binom{n}{2}$

52. $\binom{n}{3} = 2 \cdot \binom{n-1}{2}$

CHALLENGE

53. How many line segments are determined by the 5 vertices of a pentagon? Of these, how many are diagonals?

54. How many line segments are determined by the 6 vertices of a hexagon? Of these, how many are diagonals?

55. How many line segments are determined by the n vertices of a n-gon? Of these, how many are diagonals? Use mathematical induction to prove the result for the diagonals.

56. Prove: For any natural numbers n and $r \leq n$,
$$\binom{n}{r-1} + \binom{n}{r} = \binom{n+1}{r}.$$

12.7 The Binomial Theorem

OBJECTIVES

You should be able to:

1. Expand a power of a binomial $(a + b)^n$ using Pascal's triangle.
2. Expand a power of a binomial $(a + b)^n$ using the preceding coefficient.
3. Expand a power of a binomial using factorial notation.
4. Find a specific term of a binomial expansion.
5. Find the total number of subsets of a set of n objects.

1 Binomial Expansions Using Pascal's Triangle

Consider the following expanded powers of $(a + b)^n$, where $a + b$ is any binomial and n is a whole number. Look for patterns.

$$\begin{aligned}
(a + b)^0 &= 1\\
(a + b)^1 &= a + b\\
(a + b)^2 &= a^2 + 2ab + b^2\\
(a + b)^3 &= a^3 + 3a^2b + 3ab^2 + b^3\\
(a + b)^4 &= a^4 + 4a^3b + 6a^2b^2 + 4ab^3 + b^4\\
(a + b)^5 &= a^5 + 5a^4b + 10a^3b^2 + 10a^2b^3 + 5ab^4 + b^5
\end{aligned}$$

Each expansion is a polynomial. There are some patterns to be noted in the expansions.

1. In each term, the sum of the exponents is n.
2. The exponents of a start with n and decrease to 0. The last term has no factor of a. The first term has no factor of b. The exponents of b start in the second term with 1 and increase to n, or we can think of them starting in the first term with 0 and increasing to n.
3. There is one more term than the power n. That is, there are $n + 1$ terms in the expansion of $(a + b)^n$.
4. Now we consider the coefficients. The first and last coefficients are 1, and the coefficients have a symmetry to them. They start at 1 and increase through certain values about "half"-way and then decrease through these same values back to 1. Let us explore this further.

Suppose we wanted to find an expansion of $(a + b)^8$. If the patterns we have noticed were to continue, then we know that there are 9 terms in the expansion, which would be in the following form:

$$a^8 + c_1a^7b + c_2a^6b^2 + c_3a^5b^3 + c_4a^4b^4 + c_3a^3b^5 + c_2a^2b^6 + c_1ab^7 + b^8.$$

How can we determine these coefficients? We can answer this question in three different ways. Your instructor may direct you regarding which to learn. The first method seems to be the easiest, but is not always. It involves writing down the coefficients in a triangular array as follows. We get what is known as **Pascal's triangle:**

$(a+b)^0$:						1					
$(a+b)^1$:					1		1				
$(a+b)^2$:				1		2		1			
$(a+b)^3$:			1		3		3		1		
$(a+b)^4$:		1		4		6		4		1	
$(a+b)^5$:	1		5		10		10		5		1

There are many patterns in the triangle. Find as many as you can.

DO EXERCISE 1.

1. Find as many patterns as you can in Pascal's triangle. See if you can use the patterns to write three more rows of the triangle.

Perhaps you discovered a way to write the next row of numbers, given the numbers in the row above it. There are always 1's on the outside. Each

2. Write one more row of Pascal's triangle and use it to expand $(a + b)^9$.

remaining number is found by adding the two numbers above. This is shown as follows:

1
1 1
1 2 1
1 3 3 1
1 4 6 4 1
1 5 10 10 5 1
1 6 15 20 15 6 1

We see that

the 1st number is 1;
the 2nd number is 1 + 5, or 6;
the 3rd number is 5 + 10, or 15;
the 4th number is 10 + 10, or 20;
the 5th number is 10 + 5, or 15; and
the 6th number is 5 + 1, or 6.

Thus the expansion of $(a + b)^6$ is

$$(a + b)^6 = a^6 + 6a^5b + 15a^4b^2 + 20a^3b^3 + 15a^2b^4 + 6ab^5 + b^6.$$

To find the expansion for $(a + b)^8$, we complete two more rows of Pascal's triangle:

1
1 1
1 2 1
1 3 3 1
1 4 6 4 1
1 5 10 10 5 1
1 6 15 20 15 6 1
1 7 21 35 35 21 7 1
1 8 28 56 70 56 28 8 1

Thus the expansion of $(a + b)^8$ is

$$(a + b)^8 = a^8 + 8a^7b + 28a^6b^2 + 56a^5b^3 + 70a^4b^4 + 56a^3b^5 + 28a^2b^6 + 8ab^7 + b^8.$$

We can generalize our results as follows:

THEOREM 14 The Binomial Theorem

For any binomial $a + b$ and any natural number n,

$$(a + b)^n = c_0a^nb^0 + c_1a^{n-1}b + c_2a^{n-2}b^2 + \cdots + c_{n-1}a^1b^{n-1} + c_na^0b^n,$$

where the numbers $c_0, c_1, c_2, \ldots, c_n$ are from the $(n + 1)$st row of Pascal's triangle, and $c_0 = c_n = 1$.

DO EXERCISE 2.

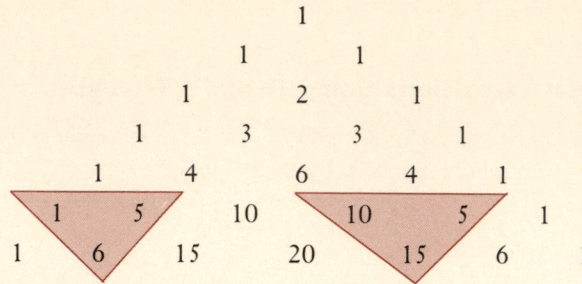

EXAMPLE 1 Expand: $(u - v)^5$.

Solution Note that $a = u$, $b = -v$, and $n = 5$. We use the 6th row of Pascal's triangle:

$$1 \quad 5 \quad 10 \quad 10 \quad 5 \quad 1.$$

Then we have

$$\begin{aligned}(u - v)^5 &= 1(u)^5 + 5(u)^4(-v)^1 + 10(u)^3(-v)^2 \\ &\quad + 10(u)^2(-v)^3 + 5(u)(-v)^4 + 1(-v)^5 \\ &= u^5 - 5u^4v + 10u^3v^2 - 10u^2v^3 + 5uv^4 - v^5.\end{aligned}$$

Note that the signs of the terms alternate between + and −. When the power of $-v$ is odd, the sign is −. ■

DO EXERCISE 3.

3. Expand: $(x - y)^4$.

EXAMPLE 2 Expand: $\left(2t + \frac{3}{t}\right)^6$.

Solution Note that $a = 2t$, $b = 3/t$, and $n = 6$. We use the 7th row of Pascal's triangle:

$$1 \quad 6 \quad 15 \quad 20 \quad 15 \quad 6 \quad 1.$$

Then we have

$$\begin{aligned}\left(2t + \frac{3}{t}\right)^6 &= (2t)^6 + 6(2t)^5\left(\frac{3}{t}\right)^1 + 15(2t)^4\left(\frac{3}{t}\right)^2 + 20(2t)^3\left(\frac{3}{t}\right)^3 \\ &\quad + 15(2t)^2\left(\frac{3}{t}\right)^4 + 6(2t)^1\left(\frac{3}{t}\right)^5 + \left(\frac{3}{t}\right)^6 \\ &= 64t^6 + 6(32t^5)\left(\frac{3}{t}\right) + 15(16t^4)\left(\frac{9}{t^2}\right) + 20(8t^3)\left(\frac{27}{t^3}\right) \\ &\quad + 15(4t^2)\left(\frac{81}{t^4}\right) + 6(2t)\left(\frac{243}{t^5}\right) + \frac{729}{t^6} \\ &= 64t^6 + 576t^4 + 2160t^2 + 4320 + 4860t^{-2} + 2916t^{-4} \\ &\quad + 729t^{-6}.\end{aligned}$$

■

DO EXERCISE 4.

4. Expand: $\left(2t + \frac{1}{t}\right)^7$.

2 Binomial Expansion Using the Preceding Coefficient

The disadvantage in using Pascal's triangle is that one must compute all the preceding rows in the table in order to obtain the row needed for the expansion. The following two methods avoid this difficulty. To understand one of these methods, let us look at the expansion of $(a + b)^5$.

	a^5	$+ 5a^4b$	$+ 10a^3b^2$	$+ 10a^2b^3$	$+ 5ab^4$	$+ b^5$
Term:	1st	2nd	3rd	4th	5th	6th

This method of finding a coefficient involves considering the preceding coefficient.

The *first-term coefficient* is 1.

The *second-term coefficient* is the preceding coefficient, 1, times the exponent of a in the preceding term, 5, divided by 1, which is

2 − 1, the number of the term minus 1: $\dfrac{1 \cdot 5}{1} = 5$.

5. Expand: $(x - y)^9$.

The *third-term coefficient* is the preceding coefficient, 5, times the exponent of a in the preceding term, 4, divided by 2, which is

$3 - 1$, the number of the term minus 1: $\frac{5 \cdot 4}{2} = 10$.

The *fourth-term coefficient* is the preceding coefficient, 10, times the exponent of a in the preceding term, 3, divided by 3, which is

$4 - 1$, the number of the term minus 1: $\frac{10 \cdot 3}{3} = 10$.

The *fifth-term coefficient* is the preceding coefficient, 10, times the exponent of a in the preceding term, 2, divided by 4, which is

$5 - 1$, the number of the term minus 1: $\frac{10 \cdot 2}{4} = 5$.

The *sixth-term coefficient* is the preceding coefficient, 5, times the exponent of a in the preceding term, 1, divided by 5, which is

$6 - 1$, the number of the term minus 1: $\frac{5 \cdot 1}{5} = 1$.

In general:

The *n*th coefficient of the binomial expansion is:

1. **1 if $n = 1$,**
2. **AB/C, where A is the preceding coefficient, B is the exponent of a in the preceding term, and C is $n - 1$.**

EXAMPLE 3 Expand: $(m + n)^{10}$.

Solution We find each term in sequence. There are 11 terms in the expansion. When we get to the 6th coefficient, we can use symmetry to obtain the others.

The 1st term is m^{10}.

The 2nd term is $\frac{1 \cdot 10}{1} m^9 n^1 = 10m^9 n$.

The 3rd term is $\frac{10 \cdot 9}{2} m^8 n^2 = 45m^8 n^2$.

The 4th term is $\frac{45 \cdot 8}{3} m^7 n^3 = 120m^7 n^3$.

The 5th term is $\frac{120 \cdot 7}{4} m^6 n^4 = 210m^6 n^4$.

The 6th term is $\frac{210 \cdot 6}{5} m^5 n^5 = 252m^5 n^5$.

The rest of the coefficients are 210, 120, 45, 10, and 1. The complete expansion is

$$(m + n)^{10} = m^{10} + 10m^9 n + 45m^8 n^2 + 120m^7 n^3 + 210m^6 n^4 + 252m^5 n^5 + 210m^4 n^6 + 120m^3 n^7 + 45m^2 n^8 + 10mn^9 + n^{10}.$$

DO EXERCISE 5.

3 Binomial Expansion Using Factorial Notation

Suppose all we wanted was the 8th term in the expansion $(a + b)^{11}$. The preceding methods would yield this term, but the computations would be lengthy. The last method we use will allow us to find the rth term of an expansion without computing any others. This last method is also useful in more advanced courses such as finite mathematics and calculus, and uses, for obvious reasons, the **binomial coefficient notation** $\binom{n}{r}$ developed in Section 12.6.

We can restate the binomial theorem as follows.

THEOREM 15 The Binomial Theorem

For any binomial $a + b$ and any natural number n,

$$(a + b)^n = \binom{n}{0}a^n + \binom{n}{1}a^{n-1}b + \binom{n}{2}a^{n-2}b^2 + \cdots + \binom{n}{n}b^n.$$

The binomial theorem can be proved by mathematical induction, but we will not do that here.

Sigma notation for a binomial series is as follows:

$$(a + b)^n = \sum_{r=0}^{n} \binom{n}{r}a^{n-r}b^r.$$

Because of Theorem 15, $\binom{n}{r}$ is called a **binomial coefficient**. It should now be apparent why $0!$ is defined to be 1. In the binomial expansion, we want $\binom{n}{0}$ to equal 1 and we also want the definition

$$\binom{n}{r} = \frac{n!}{r!(n - r)!}$$

to hold for all whole numbers n and r. Thus we must have

$$\binom{n}{0} = \frac{n!}{0!(n - 0)!} = \frac{n!}{0!n!} = 1.$$

This will be satisfied if $0!$ is defined to be 1.

EXAMPLE 4 Expand: $(x^2 - 2y)^5$.

Solution Note that $a = x^2$, $b = -2y$, and $n = 5$. Then, using the binomial theorem, we have

$$\begin{aligned}(x^2 - 2y)^5 &= \binom{5}{0}(x^2)^5 + \binom{5}{1}(x^2)^4(-2y) + \binom{5}{2}(x^2)^3(-2y)^2 \\ &\quad + \binom{5}{3}(x^2)^2(-2y)^3 + \binom{5}{4}x^2(-2y)^4 + \binom{5}{5}(-2y)^5 \\ &= \frac{5!}{0!5!}x^{10} + \frac{5!}{1!4!}x^8(-2y) + \frac{5!}{2!3!}x^6(-2y)^2 + \frac{5!}{3!2!}x^4(-2y)^3 \\ &\quad + \frac{5!}{4!1!}x^2(-2y)^4 + \frac{5!}{5!0!}(-2y)^5 \\ &= x^{10} - 10x^8y + 40x^6y^2 - 80x^4y^3 + 80x^2y^4 - 32y^5.\end{aligned}$$

DO EXERCISES 6 AND 7.

EXAMPLE 5 Expand: $(2/x + 3\sqrt{x})^4$.

Solution Note that $a = 2/x$, $b = 3\sqrt{x}$, and $n = 4$. Then, using the binomial

6. Expand: $(x + 5b)^5$.

7. Expand: $(x^2 - 1)^5$.

8. Expand: $\left(2x + \frac{1}{y}\right)^4$.

9. Expand: $(x - \sqrt{2})^6$.

10. Find the 4th term of $(3a + 4b)^7$.

11. Find the 9th term of $(3x - 2)^{10}$.

theorem, we have

$$\left(\frac{2}{x} + 3\sqrt{x}\right)^4 = \binom{4}{0}\cdot\left(\frac{2}{x}\right)^4 + \binom{4}{1}\cdot\left(\frac{2}{x}\right)^3(3\sqrt{x}) + \binom{4}{2}\cdot\left(\frac{2}{x}\right)^2(3\sqrt{x})^2$$
$$+ \binom{4}{3}\left(\frac{2}{x}\right)(3\sqrt{x})^3 + \binom{4}{4}(3\sqrt{x})^4$$
$$= \frac{4!}{0!4!}\cdot\frac{16}{x^4} + \frac{4!}{1!3!}\cdot\frac{8}{x^3}3\sqrt{x} + \frac{4!}{2!2!}\cdot\frac{4}{x^2}\cdot 9x$$
$$+ \frac{4!}{3!1!}\cdot\frac{2}{x}\cdot 27x^{3/2} + \frac{4!}{4!0!}\cdot 81x^2$$
$$= \frac{16}{x^4} + \frac{96}{x^{5/2}} + \frac{216}{x} + 216\sqrt{x} + 81x^2.$$

DO EXERCISES 8 AND 9.

4 Finding a Specific Term

Suppose we wanted to determine just a particular term of an expansion. The method we have developed will allow us to find such a term without computing all the rows of Pascal's triangle or all the preceding coefficients.

THEOREM 16

The $(r + 1)$st term of $(a + b)^n$ is

$$\binom{n}{r}a^{n-r}b^r.$$

EXAMPLE 6 Find the 5th term in the expansion of $(2x - 5y)^6$.

Solution First, we note that $5 = 4 + 1$. Thus, $r = 4$, $a = 2x$, $b = -5y$, and $n = 6$. Then the 5th term of the expansion is

$$\binom{6}{4}(2x)^{6-4}(-5y)^4, \quad\text{or}\quad \frac{6!}{4!2!}(2x)^2(-5y)^4, \quad\text{or}\quad 37{,}500x^2y^4.$$

EXAMPLE 7 Find the 8th term in the expansion of $(3x - 2)^{10}$.

Solution First, we note that $8 = 7 + 1$. Thus, $r = 7$, $a = 3x$, $b = -2$, and $n = 10$. Then the 8th term of the expansion is

$$\binom{10}{7}(3x)^{10-7}(-2)^7, \quad\text{or}\quad \frac{10!}{7!3!}(3x)^3(-128), \quad\text{or}\quad -414{,}720x^3.$$

DO EXERCISES 10 AND 11.

5 Subsets

Suppose a set has n objects. The number of subsets containing r members is $\binom{n}{r}$, by Theorem 12. The total number of subsets of a set is the number with 0 elements, plus the number with 1 element, plus the number with 2 elements, and so on. The total number of subsets of a set with n members is

$$\binom{n}{0} + \binom{n}{1} + \binom{n}{2} + \cdots + \binom{n}{n}.$$

Now let us expand $(1 + 1)^n$:

$$(1 + 1)^n = \binom{n}{0} + \binom{n}{1} + \binom{n}{2} + \cdots + \binom{n}{n}.$$

Thus the total number of subsets is $(1 + 1)^n$, or 2^n. We have proved the following theorem.

THEOREM 17

The total number of subsets of a set with n members is 2^n.

EXAMPLE 8 The set {A, B, C, D, E} has how many subsets?

Solution The set has 5 members, so the number of subsets is 2^5, or 32.

EXAMPLE 9 Wendy's, a fast-food restaurant, advertised at one time that it made and sold hamburgers in 256 ways, using combinations of 8 seasonings. Show why.

Solution The total number of combinations is

$$\binom{8}{0} + \binom{8}{1} + \cdots + \binom{8}{8} = 2^8 = 256.$$

DO EXERCISES 12 AND 13.

12. How many subsets are there of the set {a, b, c, d, e, f}?

13. How many subsets are there of the set of all states of the United States?

EXERCISE SET 12.7

1, **2**, **3** Expand.

1. $(m + n)^5$
2. $(a - b)^4$
3. $(x - y)^6$
4. $(p + q)^7$
5. $(x^2 - 3y)^5$
6. $(3c - d)^7$
7. $(3c - d)^6$
8. $(t^{-2} + 2)^6$
9. $(x - y)^3$
10. $(x - y)^5$
11. $\left(\frac{1}{x} + y\right)^7$
12. $(2s - 3t^2)^3$
13. $\left(a - \frac{2}{a}\right)^9$
14. $\left(2x + \frac{1}{x}\right)^9$
15. $(1 - 1)^n$
16. $(1 + 3)^n$
17. $(\sqrt{3} - t)^4$
18. $(\sqrt{5} + t)^6$
19. $(\sqrt{2} + 1)^6 - (\sqrt{2} - 1)^6$
20. $(1 - \sqrt{2})^4 + (1 + \sqrt{2})^4$
21. $(x^{-2} + x^2)^4$
22. $\left(\frac{1}{\sqrt{x}} - \sqrt{x}\right)^6$

4 Find the indicated term of the binomial expression.

23. 3rd, $(a + b)^6$
24. 6th, $(x + y)^7$
25. 12th, $(a - 2)^{14}$
26. 11th, $(x - 3)^{12}$
27. 5th, $\left(2x^3 - \sqrt{y}\right)^8$
28. 4th, $\left(\frac{1}{b^2} + \frac{b}{3}\right)^7$
29. Middle, $(2u - 3v^2)^{10}$
30. Middle two, $\left(\sqrt{x} + \sqrt{3}\right)^5$

5 Determine the number of subsets of each of the following.

31. A set of 7 members
32. A set of 6 members
33. ▦ The set of letters of the English alphabet, which contains 26 letters
34. ▦ The set of letters of the Greek alphabet, which contains 24 letters

SYNTHESIS

Expand.

35. $(\sqrt{2} - i)^4$, where $i^2 = -1$
36. $(1 + i)^6$, where $i^2 = -1$
37. $(\sin t - \csc t)^7$
38. $(\tan\theta + \cot\theta)^{11}$

39. Find a formula for
$$(a-b)^n.$$
Use sigma notation.

40. Expand and simplify:
$$\frac{(x+h)^n - x^n}{h}.$$
Use sigma notation.

Solve for x.

41. $\sum_{r=0}^{8} \binom{8}{r} x^{8-r} 3^r = 0$

42. $\sum_{r=0}^{4} \binom{4}{r} 5^{4-r} x^r = 64$

43. $\sum_{r=0}^{5} \binom{5}{r} (-1)^r x^{5-r} 3^r = 32$

44. $\sum_{r=0}^{9} \binom{9}{r} \sin^r x = 0$

45. ▦ At one point in a recent season, Darryl Strawberry of the New York Mets had a batting average of 0.313. Suppose he came to bat 5 times in a game. The probability of his getting exactly 3 hits is the 3rd term of the binomial expansion of $(0.313 + 0.687)^5$. Find that term and use your calculator to estimate the probability.

46. ▦ The probability that a woman will be either widowed or divorced is 85%. Suppose 8 women are interviewed. The probability that exactly 5 of them will be either widowed or divorced in her lifetime is the 6th term of the binomial expansion of $(0.15 + 0.85)^8$. Find that term and use your calculator to estimate the probability.

47. ▦ In reference to Exercise 45, the probability that Strawberry will get at most 3 hits is found by adding the last 4 terms of the binomial expansion of $(0.313 + 0.687)^5$. Find these terms and use your calculator to estimate the probability.

48. ▦ In reference to Exercise 46, the probability that at least 6 of them will be widowed or divorced is found by adding the last 3 terms of the binomial expansion of $(0.15 + 0.85)^8$. Find these terms and use your calculator to estimate the probability.

49. Find the middle term of the expansion of $(8u + 3v^2)^{10}$.

50. Find the two middle terms of the expansion of $(\sqrt{x} - \sqrt{3})^5$.

51. Find the term of
$$\left(\frac{3x^2}{2} - \frac{1}{3x}\right)^{12}$$
that does not contain x.

52. Find the middle term of $(x^2 - 6y^{3/2})^8$.

53. Find the ratio of the 4th term of $\left(p^2 - \frac{1}{2}p\sqrt[3]{q}\right)^5$ to the third term.

54. Find the term of $(\sqrt[3]{x} - 1/\sqrt{x})^7$ containing $1/x^{1/6}$.

55. What is the degree of $(x^5 + 3)^4$?

56. A money clip contains one each of the following bills: \$1, \$2, \$5, \$10, \$20, \$50, and \$100. How many different sums of money can be formed using the bills?

57. Find four consecutive integers such that the sum of the cubes of the three smallest of these is the cube of the fourth.

CHALLENGE

Simplify.

58. $\sum_{r=0}^{23} \binom{23}{r} (\log_a x)^{23-r} (\log_a t)^r$

59. $\ln \left[\sum_{r=0}^{15} \binom{15}{r} (\cos t)^{30-2r} (\sin t)^{2r} \right]$

60. Use mathematical induction and the property
$$\binom{n}{r-1} + \binom{n}{r} = \binom{n+1}{r}$$
to prove the binomial theorem.

12.8 Probability

OBJECTIVE

You should be able to:

1 **Compute the probability of a simple event.**

We say that when a coin is tossed, the chances that it will fall heads are 1 out of 2, or the **probability** that it will fall heads is $\frac{1}{2}$. Of course this does not mean that if a coin is tossed ten times, it will necessarily fall heads exactly five times. If the coin is tossed a great number of times, however, it will fall heads very nearly half of them.

Experimental and Theoretical Probability

If we toss a coin a great number of times, say 1000, and count the number of heads, we can determine the probability of getting a head. If there are 503 heads, we would calculate the probability of getting a head to be

$$\frac{503}{1000}, \quad \text{or} \quad 0.503.$$

This is an **experimental** determination of probability. Such a determination of probability is quite common. Here, for example, are some probabilities that have been determined *experimentally*:

1. If you kiss someone who has a cold, the probability of your catching a cold is 0.07.
2. A person just released from prison has an 80% probability of returning.

If we consider a coin and reason that it is just as likely to fall heads as tails, we would calculate the probability to be $\frac{1}{2}$. This is a **theoretical** determination of probability. Here, for example, are some probabilities that have been determined *theoretically*:

1. If there are 30 people in a room, the probability that two of them have the same birthday (excluding year of birth) is 0.706.
2. You are on a trip. You meet someone, and after a period of conversation, you discover that you have a common acquaintance. The typical reaction "It's a small world!" is actually not appropriate, because the probability of such an occurrence is quite high, just over 22%.

It is results like these that lend credence to the value of a study of probability. You might ask, "What is the *true* probability?" In fact, there is none. Experimentally, we can determine probabilities within certain limits. These may or may not agree with what we obtain theoretically.

1 Computing Probabilities

Experimental Probabilities

We first consider experimental determination of probability. The basic principle we use in computing such probabilities is as follows.

> Principle P (Experimental)
>
> **An experiment is performed in which n observations are made. If a situation E, or event, occurs m times out of the n observations, then we say that the *experimental probability* of that event is given by**
>
> $$P(E) = \frac{m}{n}.$$

EXAMPLE 1 *Sociological survey.* An actual experiment was conducted to determine the number of people who are left-handed, right-handed, or both. The results are shown in the graph.

a) Determine the probability that a person is left-handed.

b) Determine the probability that a person is ambidextrous (uses both hands equally well).

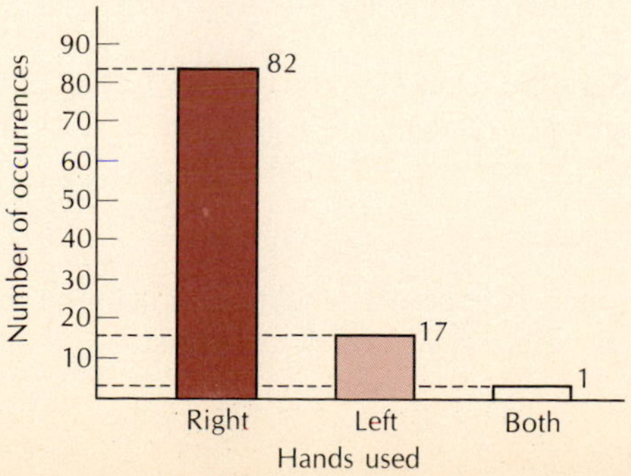

1. In reference to Example 1, what is the probability that a person is right-handed?

Solution

a) The number of people who are right-handed was 82, the number who are left-handed was 17, and there was 1 person who was ambidextrous. The total number of observations was 82 + 17 + 1, or 100. Thus the probability that a person is left-handed is P, where

$$P = \frac{17}{100}.$$

b) The probability that a person is ambidextrous is P, where

$$P = \frac{1}{100}.$$

DO EXERCISE 1.

EXAMPLE 2 *Quality control.* It is very important to a manufacturer to maintain the quality of its products. The goal is to produce as few defective products as possible. But can a company afford to check every product to see if it is defective if it is producing thousands of them every day? Most often, it is not profitable. To find out about the quality of its production, that is, what percentage of its products are defective, the company checks a smaller sample.

The U.S. Department of Agriculture requires that 80% of the seeds that a company produces must sprout. To find out about the quality of the seeds it has produced, a company takes 500 seeds from those it has produced and plants them. It finds that 417 of the seeds sprout.

a) What is the probability that a seed will sprout?
b) Did the seeds pass government standards?

2. With another large batch of seeds, the company of Example 2 plants 500 seeds and 367 of them sprout. What is the probability that a seed will sprout? Did this batch of seeds pass government inspection?

Solution

a) We know that 500 seeds were planted and 417 sprouted. The probability of a seed sprouting is P, where

$$P = \frac{417}{500} = 0.834, \quad \text{or} \quad 83.4\%.$$

b) Since the percentage of seeds exceeded the 80% requirement, the company deduces that it is producing quality seeds.

DO EXERCISE 2.

EXAMPLE 3 *Estimating wildlife populations.* To determine the number of fish in a lake, a conservationist catches 225 fish, tags them, and puts them back into the lake. Later, 108 fish are caught, and 15 of them are found to be tagged. Estimate how many fish are in the lake.

Solution Let F = the number of fish in the lake. If there are F fish in the lake and 225 of them are tagged, the probability that a fish is tagged is the ratio

$$\frac{225}{F}.$$

Later, 108 fish are caught and it is found that 15 of them are tagged. The ratio of the fish tagged to fish caught is

$$\frac{15}{108}.$$

This can also be thought of as an estimate, at least, of the probability that a fish is tagged. Thus we assume that the two ratios are the same. This gives us a proportion:

$$\frac{225}{F} = \frac{15}{108}.$$

We solve the proportion for F. We multiply by the LCM, which is $108F$:

$$108F \cdot \frac{225}{F} = 108F \cdot \frac{15}{108}$$
$$108 \cdot 225 = F \cdot 15$$
$$\frac{108 \cdot 225}{15} = F$$
$$1620 = F.$$

Thus we estimate that there are about 1620 fish in the lake. ■

DO EXERCISE 3.

3. To determine the number of deer in a forest, a conservationist catches 612 deer, tags them, and lets them loose. Later, 244 deer are caught, and 72 of them are tagged. Estimate how many deer are in the forest.

EXAMPLE 4 *TV ratings*. The major television networks and others such as cable TV are always concerned about the percentages of homes that have TVs and are watching their programs. It is too costly and unmanageable to contact every home in the country so a sample, or portion, of the homes are contacted. This is done by an electronic device attached to the TVs of about 1400 homes across the country. Viewing information is then fed into a computer. The following are the results of a recent survey.

Network	CBS	ABC	NBC	Other or not watching
Number of homes watching	258	231	206	705

What is the probability that a home was tuned to CBS during the time period considered? to ABC?

Solution The probability that a home was tuned to CBS is P, where

$$P = \frac{258}{1400} \approx 0.184 = 18.4\%.$$

The probability that a home was tuned to ABC is P, where

$$P = \frac{231}{1400} \approx 0.165 = 16.5\%.$$ ■

DO EXERCISE 4.

4. In Example 4, what is the probability that a home was tuned to NBC? What is the probability that a home was tuned to a network other than CBS, ABC, or NBC, or was not tuned in at all?

The numbers that we found in Example 4 and in Margin Exercise 4 (18.4 for CBS, 16.5 for ABC, and 14.7 for NBC) are called the *ratings*.

Theoretical Probabilities

We need some terminology before we can continue. Suppose we perform an experiment such as flipping a coin, throwing a dart, drawing a card from a deck, or checking an item off an assembly line for quality. The results of an experiment are called **outcomes**. The set of all possible outcomes is called the **sample space**. An **event** is a set of outcomes, that is, a subset of the sample space. For example, for the experiment "throwing a dart," suppose the dartboard is as follows.

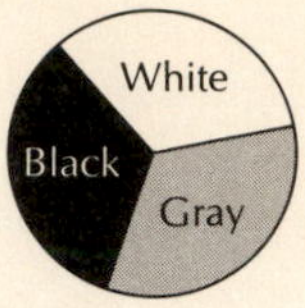

Then one event is

$$\{\text{black}\}, \qquad \text{(the outcome is "hitting black")}$$

which is a subset of the sample space

$$\{\text{black, white, gray}\}, \qquad \text{(sample space)}$$

assuming that the dart must hit the target somewhere.

We denote the probability that an event E occurs as $P(E)$. For example, "getting a head" may be denoted by H. Then $P(H)$ represents the probability of getting a head. When all the outcomes of an experiment have the same probability of occurring, we say that they are *equally likely*. To see the distinction between events that are equally likely and those that are not, consider the dartboards shown below.

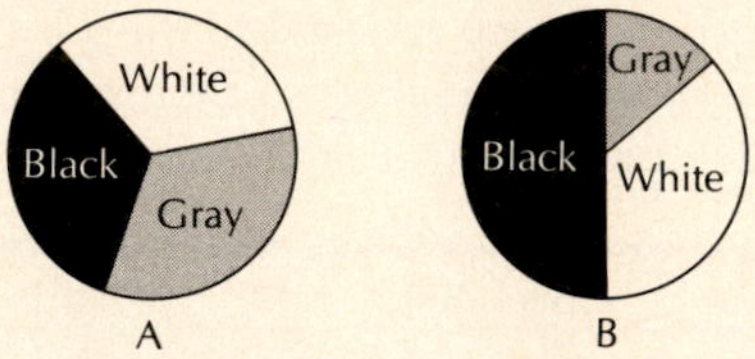

For dartboard A, the events hitting *black*, *white*, and *gray* are equally likely, but for board B they are not. A sample space that can be expressed as a union of equally likely events can allow us to calculate probabilities of other events.

> **Principle *P* (Theoretical)**
>
> **If an event E can occur m ways out of n possible equally likely outcomes of a sample space S, then the *theoretical probability* of that event is given by**
>
> $$P(E) = \frac{m}{n}.$$

A die (pl., dice) is a cube, with six faces, each containing a number of dots from 1 to 6.

EXAMPLE 5 What is the probability of rolling a 3 on a die?

Solution On a fair die, there are 6 equally likely outcomes and there is 1 way to get a 3. By Principle P, $P(3) = \frac{1}{6}$.

EXAMPLE 6 What is the probability of rolling an even number on a die?

Solution The event is getting an *even* number. It can occur in 3 ways (getting 2, 4, or 6). The number of equally likely outcomes is 6. By Principle P, $P(\text{even}) = \frac{3}{6}$, or $\frac{1}{2}$.

DO EXERCISE 5.

We now use a number of examples related to a standard bridge deck of 52 cards. Such a deck is made up as shown in the following figure.

A DECK OF 52 CARDS:

EXAMPLE 7 What is the probability of drawing an ace from a well-shuffled deck of 52 cards?

Solution Since there are 52 outcomes (cards in the deck) and they are equally likely (from a well-shuffled deck) and there are 4 ways to obtain an ace, by Principle P we have

$$P(\text{drawing an ace}) = \frac{4}{52}, \quad \text{or} \quad \frac{1}{13}.$$

EXAMPLE 8 Suppose we select, without looking, one marble from a bag containing 3 red marbles and 4 green marbles. What is the probability of selecting a red marble?

Solution There are 7 equally likely ways of selecting any marble, and since the number of ways of getting a red marble is 3,

$$P(\text{selecting a red marble}) = \tfrac{3}{7}.$$

DO EXERCISES 6 AND 7.

The following are some results that follow from Principle P.

THEOREM 18

If an event E cannot occur, then $P(E) = 0$.

5. What is the probability of rolling a prime number on a die?

6. Suppose we draw a card from a well-shuffled deck of 52 cards.
   a) What is the probability of drawing a king?
   b) What is the probability of drawing a spade?
   c) What is the probability of drawing a black card?
   d) What is the probability of drawing a jack or a queen?

7. Suppose we select, without looking, one marble from a bag containing 5 red marbles and 6 green marbles. What is the probability of selecting a green marble?

8. On a single roll of a die, what is the probability of getting a 7?

9. On a single roll of a die, what is the probability of getting a 1, 2, 3, 4, 5, or 6?

10. Suppose 3 cards are drawn from a well-shuffled deck of 52 cards. What is the probability that all 3 of them are spades?

11. Suppose 2 people are selected at random from a group that consists of 8 men and 5 women. What is the probability that both of them are women?

12. Suppose 3 people are selected at random from a group that consists of 8 men and 6 women. What is the probability that 2 men and 1 woman are selected?

For example, in coin tossing, the event that a coin would land on its edge has probability 0.

THEOREM 19

If an event E is certain to occur (that is, every trial is a success), then $P(E) = 1$.

For example, in coin tossing, the event that a coin falls either heads or tails has probability 1.

In general:

THEOREM 20

The probability that an event E will occur is a number from 0 to 1:

$$0 \le P(E) \le 1.$$

DO EXERCISES 8 AND 9.

In the following examples, we use the combinations that we studied in Sections 12.5 and 12.6 to calculate theoretical probabilities.

EXAMPLE 9 Suppose 2 cards are drawn from a well-shuffled deck of 52 cards. What is the probability that both of them are spades?

Solution The number of ways n of drawing 2 cards from a deck of 52 is ${}_{52}C_2$. Now 13 of the 52 cards are spades, so the number of ways m of drawing 2 spades is ${}_{13}C_2$. Thus,

$$P(\text{getting 2 spades}) = \frac{m}{n} = \frac{{}_{13}C_2}{{}_{52}C_2} = \frac{78}{1326} = \frac{1}{17}.$$

EXAMPLE 10 Suppose 2 people are selected at random from a group that consists of 6 men and 4 women. What is the probability that both of them are women?

Solution The number of ways of selecting 2 people from a group of 10 is ${}_{10}C_2$. The number of ways of selecting 2 women from a group of 4 is ${}_4C_2$. Thus the probability of selecting 2 women from the group of 10 is P, where

$$P = \frac{{}_4C_2}{{}_{10}C_2} = \frac{6}{45} = \frac{2}{15}.$$

EXAMPLE 11 Suppose 3 people are selected at random from a group that consists of 6 men and 4 women. What is the probability that 1 man and 2 women are selected?

Solution The number of ways of selecting 3 people from a group of 10 is ${}_{10}C_3$. One man can be selected in ${}_6C_1$ ways, and 2 women can be selected in ${}_4C_2$ ways. By the fundamental counting principle, the number of ways of selecting 1 man and 2 women is ${}_6C_1 \cdot {}_4C_2$. Thus the probability is

$$P = \frac{{}_6C_1 \cdot {}_4C_2}{{}_{10}C_3}, \quad \text{or} \quad \frac{3}{10}.$$

DO EXERCISES 10–12.

EXAMPLE 12 What is the probability of getting a total of 8 on a roll of a pair of dice? (Assume that the dice are different, say one red and one black.)

Solution On each die, there are 6 possible outcomes. The outcomes are paired so there are $6 \cdot 6$, or 36, possible ways in which the two can fall.

Red die						
6	(1, 6)	(2, 6)	(3, 6)	(4, 6)	(5, 6)	(6, 6)
5	(1, 5)	(2, 5)	(3, 5)	(4, 5)	(5, 5)	(6, 5)
4	(1, 4)	(2, 4)	(3, 4)	(4, 4)	(5, 4)	(6, 4)
3	(1, 3)	(2, 3)	(3, 3)	(4, 3)	(5, 3)	(6, 3)
2	(1, 2)	(2, 2)	(3, 2)	(4, 2)	(5, 2)	(6, 2)
1	(1, 1)	(2, 1)	(3, 1)	(4, 1)	(5, 1)	(6, 1)
	1	2	3	4	5	6 Black die

The pairs that total 8 are as shown. Thus there are 5 possible ways of getting a total of 8, so the probability is $\frac{5}{36}$. ■

DO EXERCISE 13.

13. What is the probability of getting a total of 7 on a roll of a pair of dice?

Origin and Use of Probability

A desire to calculate odds in games of chance gave rise to the theory of probability. Today the theory of probability and its closely related field, mathematical statistics, have many applications, most of them not related to games of chance. Opinion polls, with such uses as predicting elections, are a familiar example. Quality control, in which a prediction about the percentage of faulty items manufactured is made without testing them all, is an important application, among many, in business. Still other applications are in the areas of genetics, medicine, and the kinetic theory of gases.

EXERCISE SET 12.8

1

1. In an actual survey, 100 people were polled to determine the probability of a person wearing either glasses or contact lenses. Of those polled, 57 wore either glasses or contacts. What is the probability that a person wears either glasses or contacts? What is the probability that a person wears neither?

2. In another survey, 100 people were polled and asked to select a number from 1 to 5. The results are shown in the following table.

Number choices	1	2	3	4	5
Number of people who selected that number	18	24	23	23	12

What is the probability that the number selected is 1? 2? 3? 4? 5? What general conclusion might a psychologist make from this experiment?

Linguistics. An experiment was conducted to determine the relative occurrence of various letters of the English alphabet. A paragraph from a newspaper, one from a textbook, and one from a magazine were considered. In all, there was a total of 1044 letters. The number of occurrences of each letter of the alphabet is listed in the following table.

Letter	A	B	C	D	E	F	G	H	I	J	K	L	M
Number of occurrences	78	22	33	33	140	24	22	63	60	2	9	35	30

Letter	N	O	P	Q	R	S	T	U	V	W	X	Y	Z
Number of occurrences	74	74	27	4	67	67	95	31	10	22	8	13	1

Round answers to Exercises 3–6 to three decimal places.

3. What is the probability of the occurrence of the letter A? E? I? O? U?

4. What is the probability of a vowel occurring?

5. What is the probability of a consonant occurring?

6. What letter has the least probability of occurring? What is the probability of this letter not occurring?

7. To determine the number of deer in a game preserve, a conservationist catches 318 deer, tags them, and lets them loose. Later, 168 deer are caught; 56 of them are tagged. How many deer are in the preserve?

8. To determine the number of trout in a lake, a conservationist catches 112 trout, tags them, and throws them back into the lake. Later, 82 trout are caught; 32 of them are tagged. How many trout are in the lake?

Suppose we draw a card from a well-shuffled deck of 52 cards.

9. How many equally likely outcomes are there?

10. What is the probability of drawing a queen?

11. What is the probability of drawing a heart?

12. What is the probability of drawing a club?

13. What is the probability of drawing a 4?

14. What is the probability of drawing a red card?

15. What is the probability of drawing a black card?

16. What is the probability of drawing an ace or a deuce?

17. What is the probability of drawing a 9 or a king?

Suppose we select, without looking, one marble from a bag containing 4 red marbles and 10 green marbles.

18. What is the probability of selecting a red marble?

19. What is the probability of selecting a green marble?

20. What is the probability of selecting a purple marble?

21. What is the probability of selecting a white marble?

Suppose 4 cards are drawn from a well-shuffled deck of 52 cards.

22. What is the probability that all 4 are spades?

23. What is the probability that all 4 are hearts?

24. If 4 marbles are drawn at random all at once from a bag containing 8 white marbles and 6 black marbles, what is the probability that 2 will be white and 2 will be black?

25. From a group of 8 men and 7 women, a committee of 4 is chosen. What is the probability that 2 men and 2 women will be chosen?

26. What is the probability of getting a total of 6 on a roll of a pair of dice?

27. What is the probability of getting a total of 3 on a roll of a pair of dice?

28. What is the probability of getting snake eyes (a total of 2) on a roll of a pair of dice?

29. What is the probability of getting box-cars (a total of 12) on a roll of a pair of dice?

30. From a bag containing 5 nickels, 8 dimes, and 7 quarters, 5 coins are drawn at random, all at once. What is the probability of getting 2 nickels, 2 dimes, and 1 quarter?

31. From a bag containing 6 nickels, 10 dimes, and 4 quarters, 6 coins are drawn at random, all at once. What is the probability of getting 3 nickels, 2 dimes, and 1 quarter?

Roulette. A roulette wheel contains slots numbered 00, 0, 1, 2, 3, . . . , 35, 36. Eighteen of the slots numbered 1 through 36 are colored red and eighteen are colored black. The 00 and 0 slots are uncolored. The wheel is spun, and a ball is rolled around the rim until it falls into a slot. What is the probability that the ball falls in:

32. a black slot?

33. a red slot?

34. a red or black slot?

35. the 00 slot?

36. the 0 slot?

37. either the 00 or 0 slot? (Here the house always wins.)

38. an odd-numbered slot?

SYNTHESIS

Five-card poker hands and probabilities. In part (a) of each problem, give a reasoned expression as well as the answer. Read all the problems before beginning.

39. ▦ How many 5-card poker hands can be dealt from a standard 52-card deck?

40. ▦ A *royal flush* consists of a 5-card hand with A-K-Q-J-10 of the same suit.

a) How many royal flushes are there?

b) What is the probability of getting a royal flush?

41. A *straight flush* consists of 5 cards in sequence in the same suit, but excludes royal flushes. An ace can be used low, before a two.

a) How many straight flushes are there?
b) What is the probability of getting a straight flush?

42. *Four of a kind* is a 5-card hand in which 4 of the cards are of the same denomination, such as J-J-J-J-6, 7-7-7-7-A, or 2-2-2-2-5.

a) How many are there?
b) What is the probability of getting four of a kind?

CHALLENGE

43. A *full house* consists of a pair and three of a kind, such as Q-Q-Q-4-4.

a) How many are there?
b) What is the probability of getting a full house?

44. A *pair* is a 5-card hand in which just 2 of the cards are of the same denomination, such as Q-Q-8-A-3.

a) How many are there?
b) What is the probability of getting a pair?

45. *Three of a kind* is a 5-card hand in which exactly 3 of the cards are of the same denomination and the other 2 are *not* of the same denomination, such as Q-Q-Q-10-7.

a) How many are there?
b) What is the probability of getting three of a kind?

46. A *flush* is a 5-card hand in which all the cards are of the same suit, but not all in sequence (not a straight flush or royal flush).

a) How many are there?
b) What is the probability of getting a flush?

47. *Two pairs* is a hand like Q-Q-3-3-A.

a) How many are there?
b) What is the probability of getting two pairs?

48. A *straight* is any 5 cards in sequence, but not of the same suit—for example, 4 of spades, 5 of spades, 6 of diamonds, 7 of hearts, and 8 of clubs.

a) How many are there?
b) What is the probability of getting a straight?

49. Suppose a dartboard is a mosaic made up of circles and squares like those in the figure shown here. Also assume that each time you throw the dart you hit the board.

a) Find the probability p of hitting inside a circle.
b) Find the probability of hitting inside a square but outside a circle.
c) Solve the answer to part (a) for π. Suppose you throw a dart at the board 100 times and you hit the circle 78 times. Explain how you can use the result to find an estimate of π.

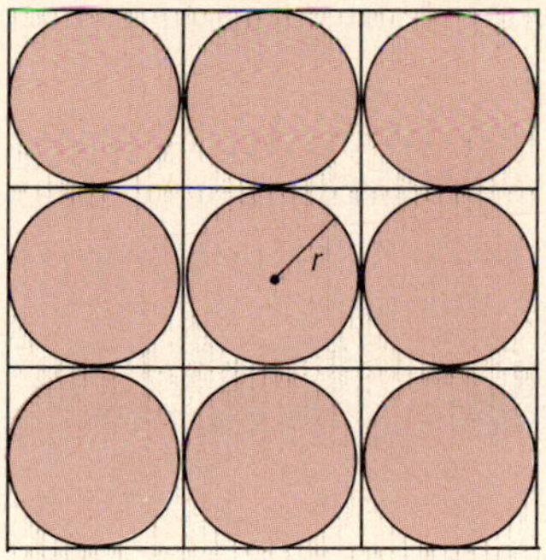

SUMMARY AND REVIEW: CHAPTER 12

TERMS TO KNOW

Sequence
Infinite sequence
Finite sequence
Series
Infinite series
Partial sum
Finite series
Sigma notation
Recursive definition
Arithmetic sequence
Common difference
Arithmetic mean
Geometric sequence
Common ratio
Infinite geometric series
Limit or sum of an infinite geometric series
Mathematical induction
Fundamental counting principle
Permutation
Factorial notation
Combination
Subset
Binomial theorem
Probability

REVIEW EXERCISES

1. Find the 10th term in the arithmetic sequence $\frac{3}{4}$, $\frac{13}{12}$, $\frac{17}{12}$, ...

2. Find the 6th term in the arithmetic sequence $a - b$, a, $a + b$, ...

3. Find the sum of the first 18 terms of the arithmetic sequence 4, 7, 10, ...

4. Find the sum of the first 30 positive integers.

5. The first term of an arithmetic sequence is 5. The 17th term is 53. Find the 3rd term.

6. The common difference in an arithmetic sequence is 3. The 10th term is 23. Find the first term.

7. For a geometric sequence, $a_1 = -2$, $r = 2$, and $a_n = -64$. Find n and S_n.

8. For a geometric sequence, $r = \frac{1}{2}$, $n = 5$, and $S_n = \frac{31}{2}$. Find a_1 and a_n.

9. Determine whether this geometric sequence has a sum.

$$25, 27.5, 30.25, 33.275, \ldots$$

10. Determine whether this geometric sequence has a sum.

$$0.27, 0.0027, 0.000027, \ldots$$

11. Find this infinite sum. The series is geometric.

$$\frac{1}{2} - \frac{1}{6} + \frac{1}{18} - \cdots$$

12. Find fractional notation for $2.\overline{13}$.

13. Insert four arithmetic means between 5 and 9.

14. A golf ball is dropped from a height of 30 ft to the pavement, and the rebound is one fourth of the distance that it drops. If, after each descent, it continues to rebound one fourth of the distance dropped, what is the total distance that the ball has traveled when it reaches the pavement on its 10th descent?

15. You receive 10¢ on the first day of the year, 12¢ on the 2nd day, 14¢ on the 3rd day, and so on. How much will you receive on the 365th day? What is the sum of all these 365 gifts?

16. The present population of a city is 30,000. Its population is supposed to double every 10 yr. What will its population be at the end of 80 yr?

17. The sides of a square are each 16 in. long. A second square is inscribed by joining the midpoints of the sides, successively. In the second square, we repeat the process, inscribing a third square. If this process is continued indefinitely, what is the sum of the perimeters of all of the squares? (*Hint*: Use an infinite geometric series.)

18. A pendulum is moving back and forth in such a way that it traverses an arc 10 cm in length, and thereafter arcs are $\frac{4}{7}$ the length of the previous arc. What is the sum of the arc lengths that the pendulum traverses?

Use mathematical induction.

19. Prove: For all natural numbers n,

$$1 + 4 + 7 + \cdots + (3n - 2) = \frac{n(3n - 1)}{2}.$$

20. Prove: For all natural numbers n,

$$1 + 3 + 3^2 + \cdots + 3^{n-1} = \frac{3^n - 1}{2}.$$

21. Prove: For every natural number $n \geq 2$,

$$\left(1 - \frac{1}{2}\right)\left(1 - \frac{1}{3}\right)\cdots\left(1 - \frac{1}{n}\right) = \frac{1}{n}.$$

22. Find the first 4 terms of this recursively defined sequence.

$$a_1 = 5, \quad a_{k+1} = 2a_k^2 + 1$$

23. Write Σ notation for this sequence.

$$0 + 3 + 8 + 15 + 24 + 35 + 48$$

24. In how many different ways can 6 books be arranged on a shelf?

25. If 9 different signal flags are available, how many different displays are possible using 4 flags in a row?

26. The winner of a contest can choose any 8 of 15 prizes. How many different selections can be made?

27. The Greek alphabet contains 24 letters. How many fraternity or sorority names can be formed using 3 different letters?

28. In how many distinguishable ways can the letters of the word TENNESSEE be arranged?

29. A manufacturer of houses has one floor plan but achieves variety by having 3 different colored roofs, 4 different ways of attaching the garage, and 3 different types of entrance. Find the number of different houses that can be produced.

30. How many code symbols can be formed using 5 out of 6 of the letters of G, H, I, J, K, L if the letters:

a) cannot be repeated?
b) can be repeated?
c) cannot be repeated but must begin with K?
d) cannot be repeated but must end with IGH?

31. In how many different ways can 7 people be seated at a round table?

32. Find the 4th term of $(a + x)^{12}$.

33. Find the 12th term of $(a + x)^{18}$. Do not multiply out the factorials.

Expand.

34. $(m + n)^7$

35. $(x^2 + 3y)^4$

36. $(5i + 1)^6$, where $i^2 = -1$

37. $(\cos t + \sec t)^8$

38. Before an election, a poll was conducted to see which candidate was favored. Three people were running for a particular office. During the polling, 86 favored A, 97 favored B, and 23 favored C. Assuming that the poll is a valid indicator of the election, what is the probability that the election will be won by A? B? C?

39. What is the probability of rolling a 10 on a roll of a pair of dice? on a roll of one die?

40. From a deck of 52 cards, 1 card is drawn. What is the probability that it is a club?

41. From a deck of 52 cards, 3 are drawn at random without replacement. What is the probability that 2 are aces and 1 is a king?

SYNTHESIS

42. Explain why the following cannot be proved by mathematical induction: For every natural number n:

a) $3 + 5 + \cdots + (2n + 1) = (n + 1)^2$;
b) $1 + 3 + \cdots + (2n - 1) = n^2 + 3$.

43. Suppose a and b are geometric sequences. Prove that c is a geometric sequence, where $c_n = a_n b_n$.

44. Suppose a is an arithmetic sequence. Prove that c is a geometric sequence, where $c_n = b^{a_n}$, for some positive number b.

45. Suppose a is an arithmetic sequence. Under what conditions is b an arithmetic sequence where:

a) $b_n = |a_n|$?
b) $b_n = a_n + 8$?
c) $b_n = 7a_n$?
d) $b_n = \dfrac{1}{a_n}$?
e) $b_n = \log a_n$?
f) $b_n = a_n^3$?

46. The zeros of this polynomial form an arithmetic sequence. Find them.

$$x^4 - 4x^3 - 4x^2 + 16x$$

47. Write the first 3 terms of the infinite geometric sequence with $S_\infty = \frac{3}{11}$ and $r = 0.01$.

48. Write the first 3 terms of the infinite geometric sequence with $r = -\frac{1}{3}$ and $S_\infty = \frac{3}{8}$.

49. Simplify:

$$\sum_{r=0}^{10} (-1)^r \binom{10}{r} (\log x)^{10-r} (\log y)^r.$$

Solve for n.

50. $\binom{n}{n-1} = 36$

51. $26 \cdot \binom{n}{1} = \binom{n}{3}$

TEST: CHAPTER 12

1. Find the 18th term of the arithmetic sequence $\frac{1}{4}, 1, \frac{7}{4}, \frac{5}{2}, \ldots$.

2. The 2nd term of an arithmetic sequence is 9, and the 9th term is 37. Find the common difference.

3. Which term of the arithmetic sequence $1, \frac{3}{2}, 2, \frac{5}{2}, \ldots$ is $\frac{31}{2}$?

4. Insert three arithmetic means between 3 and 14.

5. Find the 8th term of the geometric sequence

$$0.2, 0.6, 1.8, \ldots.$$

6. Find the sum

$$\sum_{k=1}^{5} \left(\frac{1}{2}\right)^{k+1}.$$

7. Which of the following infinite geometric sequences have sums?

   a) 2, 0.2, 0.02, 0.002, . . .
   b) 3, −6, 12, −24, 48, . . .
   c) $\frac{1}{20}, \frac{1}{10}, \frac{1}{5}, \frac{2}{5}, \ldots$

8. Find the sum of the infinite geometric sequence

$$25, -5, 1, -\tfrac{1}{5}, \ldots.$$

9. A student made deposits in a savings account as follows: \$20.50 the first month, \$26 the second month, \$31.50 the third month, and so on, for 2 years. What was the sum of the deposits?

10. A publishing company prints only $\frac{3}{5}$ as many books with each new printing of a book. If 100,000 copies of a book are printed originally, how many will be printed in the 5th printing?

11. Find fractional notation for $0.12\overline{8}$.

12. Use mathematical induction. Prove that for every natural number n,

$$5 + 10 + 50 + \cdots + 5n = \frac{5n(n+1)}{2}.$$

13. Find the first 4 terms of this recursively defined sequence.

$$a_1 = 5, \quad a_{k+1} = 4a_k + 3$$

14. How many code symbols can be formed using 4 out of 6 of the letters of D, E, F, G, H, I if the letters:

   a) can be repeated?
   b) cannot be repeated?
   c) cannot be repeated but must begin with FH?

15. In how many different ways can 6 people be seated at a round table?

16. On a test, a student must answer 4 out of 7 questions. In how many ways can this be done?

17. From a group of 20 seniors and 14 juniors, how many committees consisting of 3 seniors and 2 juniors are possible?

18. In how many distinguishable ways can the letters of the word ARKANSAS be arranged?

19. Determine the number of subsets of a set of 8 members.

20. Find the 3rd term of $(2a + b)^7$.

21. Expand $(x - \sqrt{2})^5$.

22. What is the probability of getting a total of 6 on a roll of a pair of dice?

23. From a deck of 52 cards, 1 card is drawn. What is the probability of drawing a 3 or a queen?

24. If 3 marbles are drawn at random all at once from a bag containing 5 green marbles, 7 red marbles, and 4 white marbles, what is the probability that 2 will be green and 1 will be white?

SYNTHESIS

25. Find 4 numbers in an arithmetic sequence such that twice the second minus the fourth is 1 and the sum of the first and third is 14.

26. How many diagonals does a dodecagon (12-sided polygon) have?

27. Solve for n:

$$\binom{n}{6} = 3 \cdot \binom{n-1}{5}.$$

28. Solve for a:

$$\sum_{r=0}^{5} 9^{5-r} a^r = 0.$$

Tables

TABLE 1 Powers, Roots and Reciprocals

n	n^2	n^3	$\sqrt{n}$	$\sqrt[3]{n}$	$\sqrt{10n}$	$\frac{1}{n}$	n	n^2	n^3	$\sqrt{n}$	$\sqrt[3]{n}$	$\sqrt{10n}$	$\frac{1}{n}$
1	1	1	1.000	1.000	3.162	1.0000	**51**	2,601	132,651	7.141	3.708	22.583	.0196
2	4	8	1.414	1.260	4.472	.5000	**52**	2,704	140,608	7.211	3.733	22.804	.0192
3	9	27	1.732	1.442	5.477	.3333	**53**	2,809	148,877	7.280	3.756	23.022	.0189
4	16	64	2.000	1.587	6.325	.2500	**54**	2,916	157,464	7.348	3.780	23.238	.0185
5	25	125	2.236	1.710	7.071	.2000	**55**	3,025	166,375	7.416	3.803	23.452	.0182
6	36	216	2.449	1.817	7.746	.1667	**56**	3,136	175,616	7.483	3.826	23.664	.0179
7	49	343	2.646	1.913	8.367	.1429	**57**	3,249	185,193	7.550	3.849	23.875	.0175
8	64	512	2.828	2.000	8.944	.1250	**58**	3,364	195,112	7.616	3.871	24.083	.0172
9	81	729	3.000	2.080	9.487	.1111	**59**	3,481	205,379	7.681	3.893	24.290	.0169
10	100	1,000	3.162	2.154	10.000	.1000	**60**	3,600	216,000	7.746	3.915	24.495	.0167
11	121	1,331	3.317	2.224	10.488	.0909	**61**	3,721	226,981	7.810	3.936	24.698	.0164
12	144	1,728	3.464	2.289	10.954	.0833	**62**	3,844	238,328	7.874	3.958	24.900	.0161
13	169	2,197	3.606	2.351	11.402	.0769	**63**	3,969	250,047	7.937	3.979	25.100	.0159
14	196	2,744	3.742	2.410	11.832	.0714	**64**	4,096	262,144	8.000	4.000	25.298	.0156
15	225	3,375	3.873	2.466	12.247	.0667	**65**	4,225	274,625	8.062	4.021	25.495	.0154
16	256	4,096	4.000	2.520	12.648	.0625	**66**	4,356	287,496	8.124	4.041	25.690	.0152
17	289	4,913	4.123	2.571	13.038	.0588	**67**	4,489	300,763	8.185	4.062	25.884	.0149
18	324	5,832	4.243	2.621	13.416	.0556	**68**	4,624	314,432	8.246	4.082	26.077	.0147
19	361	6,859	4.359	2.668	13.784	.0526	**69**	4,761	328,509	8.307	4.102	26.268	.0145
20	400	8,000	4.472	2.714	14.142	.0500	**70**	4,900	343,000	8.367	4.121	26.458	.0143
21	441	9,261	4.583	2.759	14.491	.0476	**71**	5,041	357,911	8.426	4.141	26.646	.0141
22	484	10,648	4.690	2.802	14.832	.0455	**72**	5,184	373,248	8.485	4.160	26.833	.0139
23	529	12,167	4.796	2.844	15.166	.0435	**73**	5,329	389,017	8.544	4.179	27.019	.0137
24	576	13,824	4.899	2.884	15.492	.0417	**74**	5,476	405,224	8.602	4.198	27.203	.0135
25	625	15,625	5.000	2.924	15.811	.0400	**75**	5,625	421,875	8.660	4.217	27.386	.0133
26	676	17,576	5.099	2.962	16.125	.0385	**76**	5,776	438,976	8.718	4.236	27.568	.0132
27	729	19,683	5.196	3.000	16.432	.0370	**77**	5,929	456,533	8.775	4.254	27.749	.0130
28	784	21,952	5.292	3.037	16.733	.0357	**78**	6,084	474,552	8.832	4.273	27.928	.0128
29	841	24,389	5.385	3.072	17.029	.0345	**79**	6,241	493,039	8.888	4.291	28.107	.0127
30	900	27,000	5.477	3.107	17.321	.0333	**80**	6,400	512,000	8.944	4.309	28.284	.0125
31	961	29,791	5.568	3.141	17.607	.0323	**81**	6,561	531,441	9.000	4.327	28.460	.0123
32	1,024	32,768	5.657	3.175	17.889	.0312	**82**	6,724	551,368	9.055	4.344	28.636	.0122
33	1,089	35,937	5.745	3.208	18.166	.0303	**83**	6,889	571,787	9.110	4.362	28.810	.0120
34	1,156	39,304	5.831	3.240	18.439	.0294	**84**	7.056	592,704	9.165	4.380	28.983	.0119
35	1,225	42,875	5.916	3.271	18.708	.0286	**85**	7,225	614,125	9.220	4.397	29.155	.0118
36	1,296	46,656	6.000	3.302	18.974	.0278	**86**	7,396	636,056	9.274	4.414	29.326	.0116
37	1,369	50,653	6.083	3.332	19.235	.0270	**87**	7,569	658,503	9.327	4.431	29.496	.0115
38	1,444	54,872	6.164	3.362	19.494	.0263	**88**	7,744	681,472	9.381	4.448	29.665	.0114
39	1,521	59,319	6.245	3.391	19.748	.0256	**89**	7,921	704,969	9.434	4.465	29.833	.0112
40	1,600	64,000	6.325	3.420	20.000	.0250	**90**	8,100	729,000	9.487	4.481	30.000	.0111
41	1,681	68,921	6.403	3.448	20.248	.0244	**91**	8,281	753,571	9.539	4.498	30.166	.0110
42	1,764	74,088	6.481	3.476	20.494	.0238	**92**	8,464	778,688	9.592	4.514	30.332	.0109
43	1,849	79,507	6.557	3.503	20.736	.0233	**93**	8,649	804,357	9.644	4.531	30.496	.0108
44	1,936	85,184	6.633	3.530	20.976	.0227	**94**	8,836	830,584	9.695	4.547	30.659	.0106
45	2,025	91,125	6.708	3.557	21.213	.0222	**95**	9,025	857,375	9.747	4.563	30.822	.0105
46	2,116	97,336	6.782	3.583	21.448	.0217	**96**	9,216	884,736	9.798	4.579	30.984	.0104
47	2,209	103,823	6.856	3.609	21.679	.0213	**97**	9,409	912,673	9.849	4.595	31.145	.0103
48	2,304	110,592	6.928	3.634	21.909	.0208	**98**	9,604	941,192	9.899	4.610	31.305	.0102
49	2,401	117,649	7.000	3.659	22.136	.0204	**99**	9,801	970,299	9.950	4.626	31.464	.0101
50	2,500	125,000	7.071	3.684	22.361	.0200	**100**	10,000	1,000,000	10.000	4.642	31.623	.0100

TABLE 2 Common Logarithms

x	0	1	2	3	4	5	6	7	8	9
1.0	.0000	.0043	.0086	.0128	.0170	.0212	.0253	.0294	.0334	.0374
1.1	.0414	.0453	.0492	.0531	.0569	.0607	.0645	.0682	.0719	.0755
1.2	.0792	.0828	.0864	.0899	.0934	.0969	.1004	.1038	.1072	.1106
1.3	.1139	.1173	.1206	.1239	.1271	.1303	.1335	.1367	.1399	.1430
1.4	.1461	.1492	.1523	.1553	.1584	.1614	.1644	.1673	.1703	.1732
1.5	.1761	.1790	.1818	.1847	.1875	.1903	.1931	.1959	.1987	.2014
1.6	.2041	.2068	.2095	.2122	.2148	.2175	.2201	.2227	.2253	.2279
1.7	.2304	.2330	.2355	.2380	.2405	.2430	.2455	.2480	.2504	.2529
1.8	.2553	.2577	.2601	.2625	.2648	.2672	.2695	.2718	.2742	.2765
1.9	.2788	.2810	.2833	.2856	.2878	.2900	.2923	.2945	.2967	.2989
2.0	.3010	.3032	.3054	.3075	.3096	.3118	.3139	.3160	.3181	.3201
2.1	.3222	.3243	.3263	.3284	.3304	.3324	.3345	.3365	.3385	.3404
2.2	.3424	.3444	.3464	.3483	.3502	.3522	.3541	.3560	.3579	.3598
2.3	.3617	.3636	.3655	.3674	.3692	.3711	.3729	.3747	.3766	.3784
2.4	.3802	.3820	.3838	.3856	.3874	.3892	.3909	.3927	.3945	.3962
2.5	.3979	.3997	.4014	.4031	.4048	.4065	.4082	.4099	.4116	.4133
2.6	.4150	.4166	.4183	.4200	.4216	.4232	.4249	.4265	.4281	.4298
2.7	.4314	.4330	.4346	.4362	.4378	.4393	.4409	.4425	.4440	.4456
2.8	.4472	.4487	.4502	.4518	.4533	.4548	.4564	.4579	.4594	.4609
2.9	.4624	.4639	.4654	.4669	.4683	.4698	.4713	.4728	.4742	.4757
3.0	.4771	.4786	.4800	.4814	.4829	.4843	.4857	.4871	.4886	.4900
3.1	.4914	.4928	.4942	.4955	.4969	.4983	.4997	.5011	.5024	.5038
3.2	.5051	.5065	.5079	.5092	.5105	.5119	.5132	.5145	.5159	.5172
3.3	.5185	.5198	.5211	.5224	.5237	.5250	.5263	.5276	.5289	.5307
3.4	.5315	.5328	.5340	.5353	.5366	.5378	.5391	.5403	.5416	.5428
3.5	.5441	.5453	.5465	.5478	.5490	.5502	.5514	.5527	.5539	.5551
3.6	.5563	.5575	.5587	.5599	.5611	.5623	.5635	.5647	.5658	.5670
3.7	.5682	.5694	.5705	.5717	.5729	.5740	.5752	.5763	.5775	.5786
3.8	.5798	.5809	.5821	.5832	.5843	.5855	.5866	.5877	.5888	.5899
3.9	.5911	.5922	.5933	.5944	.5955	.5966	.5977	.5988	.5999	.6010
4.0	.6021	.6031	.6042	.6053	.6064	.6075	.6085	.6096	.6107	.6117
4.1	.6128	.6138	.6149	.6160	.6170	.6180	.6191	.6201	.6212	.6222
4.2	.6232	.6243	.6253	.6263	.6274	.6284	.6294	.6304	.6314	.6325
4.3	.6335	.6345	.6355	.6365	.6375	.6385	.6395	.6405	.6415	.6425
4.4	.6435	.6444	.6454	.6464	.6474	.6484	.6493	.6503	.6513	.6522
4.5	.6532	.6542	.6551	.6561	.6571	.6580	.6590	.6599	.6609	.6618
4.6	.6628	.6637	.6646	.6656	.6665	.6675	.6684	.6693	.6702	.6712
4.7	.6721	.6730	.6739	.6749	.6758	.6767	.6776	.6785	.6794	.6803
4.8	.6812	.6821	.6830	.6839	.6848	.6857	.6866	.6875	.6884	.6893
4.9	.6902	.6911	.6920	.6928	.6937	.6946	.6955	.6964	.6972	.6981
5.0	.6990	.6998	.7007	.7016	.7024	.7033	.7042	.7050	.7059	.7067
5.1	.7076	.7084	.7093	.7101	.7110	.7118	.7126	.7135	.7143	.7152
5.2	.7160	.7168	.7177	.7185	.7193	.7202	.7210	.7218	.7226	.7235
5.3	.7243	.7251	.7259	.7267	.7275	.7284	.7292	.7300	.7308	.7316
5.4	.7324	.7332	.7340	.7348	.7356	.7364	.7372	.7380	.7388	.7396

TABLE 2 (*continued*)

x	0	1	2	3	4	5	6	7	8	9
5.5	.7404	.7412	.7419	.7427	.7435	.7443	.7451	.7459	.7466	.7474
5.6	.7482	.7490	.7497	.7505	.7513	.7520	.7528	.7536	.7543	.7551
5.7	.7559	.7566	.7574	.7582	.7589	.7597	.7604	.7612	.7619	.7627
5.8	.7634	.7642	.7649	.7657	.7664	.7672	.7679	.7686	.7694	.7701
5.9	.7709	.7716	.7723	.7731	.7738	.7745	.7752	.7760	.7767	.7774
6.0	.7782	.7789	.7796	.7803	.7810	.7818	.7825	.7832	.7839	.7846
6.1	.7853	.7860	.7868	.7875	.7882	.7889	.7896	.7903	.7910	.7917
6.2	.7924	.7931	.7938	.7945	.7952	.7959	.7966	.7973	.7980	.7987
6.3	.7993	.8000	.8007	.8014	.8021	.8028	.8035	.8041	.8048	.8055
6.4	.8062	.8069	.8075	.8082	.8089	.8096	.8102	.8109	.8116	.8122
6.5	.8129	.8136	.8142	.8149	.8156	.8162	.8169	.8176	.8182	.8189
6.6	.8195	.8202	.8209	.8215	.8222	.8228	.8235	.8241	.8248	.8254
6.7	.8261	.8267	.8274	.8280	.8287	.8293	.8299	.8306	.8312	.8319
6.8	.8325	.8331	.8338	.8344	.8351	.8357	.8363	.8370	.8376	.8382
6.9	.8388	.8395	.8401	.8407	.8414	.8420	.8426	.8432	.8439	.8445
7.0	.8451	.8457	.8463	.8470	.8476	.8482	.8488	.8494	.8500	.8506
7.1	.8513	.8519	.8525	.8531	.8537	.8543	.8549	.8555	.8561	.8567
7.2	.8573	.8579	.8585	.8591	.8597	.8603	.8609	.8615	.8621	.8627
7.3	.8633	.8639	.8645	.8651	.8657	.8663	.8669	.8675	.8681	.8686
7.4	.8692	.8698	.8704	.8710	.8716	.8722	.8727	.8733	.8739	.8745
7.5	.8751	.8756	.8762	.8768	.8774	.8779	.8785	.8791	.8797	.8802
7.6	.8808	.8814	.8820	.8825	.8831	.8837	.8842	.8848	.8854	.8859
7.7	.8865	.8871	.8876	.8882	.8887	.8893	.8899	.8904	.8910	.8915
7.8	.8921	.8927	.8932	.8938	.8943	.8949	.8954	.8960	.8965	.8971
7.9	.8976	.8982	.8987	.8993	.8998	.9004	.9009	.9015	.9020	.9025
8.0	.9031	.9036	.9042	.9047	.9053	.9058	.9063	.9069	.9074	.9079
8.1	.9085	.9090	.9096	.9101	.9106	.9112	.9117	.9122	.9128	.9133
8.2	.9138	.9143	.9149	.9154	.9159	.9165	.9170	.9175	.9180	.9186
8.3	.9191	.9196	.9201	.9206	.9212	.9217	.9222	.9227	.9232	.9238
8.4	.9243	.9248	.9253	.9258	.9263	.9269	.9274	.9279	.9284	.9289
8.5	.9294	.9299	.9304	.9309	.9315	.9320	.9325	.9330	.9335	.9340
8.6	.9345	.9350	.9355	.9360	.9365	.9370	.9375	.9380	.9385	.9390
8.7	.9395	.9400	.9405	.9410	.9415	.9420	.9425	.9430	.9435	.9440
8.8	.9445	.9450	.9455	.9460	.9465	.9469	.9474	.9479	.9484	.9489
8.9	.9494	.9499	.9504	.9509	.9513	.9518	.9523	.9528	.9533	.9538
9.0	.9542	.9547	.9552	.9557	.9562	.9566	.9571	.9576	.9581	.9586
9.1	.9590	.9595	.9600	.9605	.9609	.9614	.9619	.9624	.9628	.9633
9.2	.9638	.9643	.9647	.9652	.9657	.9661	.9666	.9671	.9675	.9680
9.3	.9685	.9689	.9694	.9699	.9703	.9708	.9713	.9717	.9722	.9727
9.4	.9731	.9736	.9741	.9745	.9750	.9754	.9759	.9763	.9768	.9773
9.5	.9777	.9782	.9786	.9791	.9795	.9800	.9805	.9809	.9814	.9818
9.6	.9823	.9827	.9832	.9836	.9841	.9845	.9850	.9854	.9859	.9863
9.7	.9868	.9872	.9877	.9881	.9886	.9890	.9894	.9899	.9903	.9908
9.8	.9912	.9917	.9921	.9926	.9930	.9934	.9939	.9943	.9948	.9952
9.9	.9956	.9961	.9965	.9969	.9974	.9978	.9983	.9987	.9991	.9996
x	0	1	2	3	4	5	6	7	8	9

TABLE 3 Values of Trigonometric Functions

Degrees	Radians	Sin	Cos	Tan	Cot	Sec	Csc		
0° 00′	.0000	.0000	1.0000	.0000	——	1.000	——	1.5708	**90° 00′**
10	029	029	000	029	343.8	000	343.8	679	50
20	058	058	000	058	171.9	000	171.9	650	40
30	.0087	.0087	1.0000	.0087	114.6	1.000	114.6	1.5621	30
40	116	116	.9999	116	85.94	000	85.95	592	20
50	145	145	999	145	68.75	000	68.76	563	10
1° 00′	.0175	.0175	.9998	.0175	57.29	1.000	57.30	1.5533	**89° 00′**
10	204	204	998	204	49.10	000	49.11	504	50
20	233	233	997	233	42.96	000	42.98	475	40
30	.0262	.0262	.9997	.0262	38.19	1.000	38.20	1.5446	30
40	291	291	996	291	34.37	000	34.38	417	20
50	320	320	995	320	31.24	001	31.26	388	10
2° 00′	.0349	.0349	.9994	.0349	28.64	1.001	28.65	1.5359	**88° 00′**
10	378	378	993	378	26.43	001	26.45	330	50
20	407	407	992	407	24.54	001	24.56	301	40
30	.0436	.0436	.9990	.0437	22.90	1.001	22.93	1.5272	30
40	465	465	989	466	21.47	001	21.49	243	20
50	495	494	988	495	20.21	001	20.23	213	10
3° 00′	.0524	.0523	.9986	.0524	19.08	1.001	19.11	1.5184	**87° 00′**
10	553	552	985	553	18.07	002	18.10	155	50
20	582	581	983	582	17.17	002	17.20	126	40
30	.0611	.0610	.9981	.0612	16.35	1.002	16.38	1.5097	30
40	640	640	980	641	15.60	002	15.64	068	20
50	669	669	978	670	14.92	002	14.96	039	10
4° 00′	.0698	.0698	.9976	.0699	14.30	1.002	14.34	1.5010	**86° 00′**
10	727	727	974	729	13.73	003	13.76	981	50
20	756	756	971	758	13.20	003	13.23	952	40
30	.0785	.0785	.9969	.0787	12.71	1.003	12.75	1.4923	30
40	814	814	967	816	12.25	003	12.29	893	20
50	844	843	964	846	11.83	004	11.87	864	10
5° 00′	.0873	.0872	.9962	.0875	11.43	1.004	11.47	1.4835	**85° 00′**
10	902	901	959	904	11.06	004	11.10	806	50
20	931	929	957	934	10.71	004	10.76	777	40
30	.0960	.0958	.9954	.0963	10.39	1.005	10.43	1.4748	30
40	989	987	951	992	10.08	005	10.13	719	20
50	.1018	.1016	948	.1022	9.788	005	9.839	690	10
6° 00′	.1047	.1045	.9945	.1051	9.514	1.006	9.567	1.4661	**84° 00′**
10	076	074	942	080	9.255	006	9.309	632	50
20	105	103	939	110	9.010	006	9.065	603	40
30	.1134	.1132	.9936	.1139	8.777	1.006	8.834	1.4573	30
40	164	161	932	169	8.556	007	8.614	544	20
50	193	190	929	198	8.345	007	8.405	515	10
7° 00′	.1222	.1219	.9925	.1228	8.144	1.008	8.206	1.4486	**83° 00′**
10	251	248	922	257	7.953	008	8.016	457	50
20	280	276	918	287	7.770	008	7.834	428	40
30	.1309	.1305	.9914	.1317	7.596	1.009	7.661	1.4399	30
40	338	334	911	346	7.429	009	7.496	370	20
50	367	363	907	376	7.269	009	7.337	341	10
8° 00′	.1396	.1392	.9903	.1405	7.115	1.010	7.185	1.4312	**82° 00′**
10	425	421	899	435	6.968	010	7.040	283	50
20	454	449	894	465	6.827	011	6.900	254	40
30	.1484	.1478	.9890	.1495	6.691	1.011	6.765	1.4224	30
40	513	507	886	524	6.561	012	6.636	195	20
50	542	536	881	554	6.435	012	6.512	166	10
9° 00′	.1571	.1564	.9877	.1584	6.314	1.012	6.392	1.4137	**81° 00′**
		Cos	Sin	Cot	Tan	Csc	Sec	Radians	Degrees

TABLE 3 *(continued)*

Degrees	Radians	Sin	Cos	Tan	Cot	Sec	Csc		
9° 00′	.1571	.1564	.9877	.1584	6.314	1.012	6.392	1.4137	**81° 00′**
10	600	593	872	614	197	013	277	108	50
20	629	622	868	644	084	013	166	079	40
30	.1658	.1650	.9863	.1673	5.976	1.014	6.059	1.4050	30
40	687	679	858	703	871	014	5.955	1.4021	20
50	716	708	853	733	769	015	855	992	10
10° 00′	.1745	.1736	.9848	.1763	5.671	1.015	5.759	1.3963	**80° 00′**
10	774	765	843	793	576	016	665	934	50
20	804	794	838	823	485	016	575	904	40
30	.1833	.1822	.9833	.1853	5.396	1.017	5.487	1.3875	30
40	862	851	827	883	309	018	403	846	20
50	891	880	822	914	226	018	320	817	10
11° 00′	.1920	.1908	.9816	.1944	5.145	1.019	5.241	1.3788	**79° 00′**
10	949	937	811	974	066	019	164	759	50
20	978	965	805	.2004	4.989	020	089	730	40
30	.2007	.1994	.9799	.2035	4.915	1.020	5.016	1.3701	30
40	036	.2022	793	065	843	021	4.945	672	20
50	065	051	787	095	773	022	876	643	10
12° 00′	.2094	.2079	.9781	.2126	4.705	1.022	4.810	1.3614	**78° 00′**
10	123	108	775	156	638	023	745	584	50
20	153	136	769	186	574	024	682	555	40
30	.2182	.2164	.9763	.2217	4.511	1.024	4.620	1.3526	30
40	211	193	757	247	449	025	560	497	20
50	240	221	750	278	390	026	502	468	10
13° 00′	.2269	.2250	.9744	.2309	4.331	1.026	4.445	1.3439	**77° 00′**
10	298	278	737	339	275	027	390	410	50
20	327	306	730	370	219	028	336	381	40
30	.2356	.2334	.9724	.2401	4.165	1.028	4.284	1.3352	30
40	385	363	717	432	113	029	232	323	20
50	414	391	710	462	061	030	182	294	10
14° 00′	.2443	.2419	.9703	.2493	4.011	1.031	4.134	1.3265	**76° 00′**
10	473	447	696	524	3.962	031	086	235	50
20	502	476	689	555	914	032	039	206	40
30	.2531	.2504	.9681	.2586	3.867	1.033	3.994	1.3177	30
40	560	532	674	617	821	034	950	148	20
50	589	560	667	648	776	034	906	119	10
15° 00′	.2618	.2588	.9659	.2679	3.732	1.035	3.864	1.3090	**75° 00′**
10	647	616	652	711	689	036	822	061	50
20	676	644	644	742	647	037	782	032	40
30	.2705	.2672	.9636	.2773	3.606	1.038	3.742	1.3003	30
40	734	700	628	805	566	039	703	974	20
50	763	728	621	836	526	039	665	945	10
16° 00′	.2793	.2756	.9613	.2867	3.487	1.040	3.628	1.2915	**74° 00′**
10	822	784	605	899	450	041	592	886	50
20	851	812	596	931	412	042	556	857	40
30	.2880	.2840	.9588	.2962	3.376	1.043	3.521	1.2828	30
40	909	868	580	994	340	044	487	799	20
50	938	896	572	.3026	305	045	453	770	10
17° 00′	.2967	.2924	.9563	.3057	3.271	1.046	3.420	1.2741	**73° 00′**
10	996	952	555	089	237	047	388	712	50
20	.3025	979	546	121	204	048	356	683	40
30	.3054	.3007	.9537	.3153	3.172	1.049	3.326	1.2654	30
40	083	035	528	185	140	049	295	625	20
50	113	062	520	217	108	050	265	595	10
18° 00′	.3142	.3090	.9511	.3249	3.078	1.051	3.236	1.2566	**72° 00′**
		Cos	Sin	Cot	Tan	Csc	Sec	Radians	Degrees

(Continued)

TABLE 3 (*continued*)

Degrees	Radians	Sin	Cos	Tan	Cot	Sec	Csc		
18° 00′	.3142	.3090	.9511	.3249	3.078	1.051	3.236	1.2566	**72° 00′**
10	171	118	502	281	047	052	207	537	50
20	200	145	492	314	018	053	179	508	40
30	.3229	.3173	.9483	.3346	2.989	1.054	3.152	1.2479	30
40	258	201	474	378	960	056	124	450	20
50	287	228	465	411	932	057	098	421	10
19° 00′	.3316	.3256	.9455	.3443	2.904	1.058	3.072	1.2392	**71° 00′**
10	345	283	446	476	877	059	046	363	50
20	374	311	436	508	850	060	021	334	40
30	.3403	.3338	.9426	.3541	2.824	1.061	2.996	1.2305	30
40	432	365	417	574	798	062	971	275	20
50	462	393	407	607	773	063	947	246	10
20° 00′	.3491	.3420	.9397	.3640	2.747	1.064	2.924	1.2217	**70° 00′**
10	520	448	387	673	723	065	901	188	50
20	549	475	377	706	699	066	878	159	40
30	.3578	.3502	.9367	.3739	2.675	1.068	2.855	1.2130	30
40	607	529	356	772	651	069	833	101	20
50	636	557	346	805	628	070	812	072	10
21° 00′	.3665	.3584	.9336	.3839	2.605	1.071	2.790	1.2043	**69° 00′**
10	694	611	325	872	583	072	769	1.2014	50
20	723	638	315	906	560	074	749	985	40
30	.3752	.3665	.9304	.3939	2.539	1.075	2.729	1.1956	30
40	782	692	293	973	517	076	709	926	20
50	811	719	283	.4006	496	077	689	897	10
22° 00′	.3840	.3746	.9272	.4040	2.475	1.079	2.669	1.1868	**68° 00′**
10	869	773	261	074	455	080	650	839	50
20	898	800	250	108	434	081	632	810	40
30	.3927	.3827	.9239	.4142	2.414	1.082	2.613	1.1781	30
40	956	854	228	176	394	084	595	752	20
50	985	881	216	210	375	085	577	723	10
23° 00′	.4014	.3907	.9205	.4245	2.356	1.086	2.559	1.1694	**67° 00′**
10	043	934	194	279	337	088	542	665	50
20	072	961	182	314	318	089	525	636	40
30	.4102	.3987	.9171	.4348	2.300	1.090	2.508	1.1606	30
40	131	.4014	159	383	282	092	491	577	20
50	160	041	147	417	264	093	475	548	10
24° 00′	.4189	.4067	.9135	.4452	2.246	1.095	2.459	1.1519	**66° 00′**
10	218	094	124	487	229	096	443	490	50
20	247	120	112	522	211	097	427	461	40
30	.4276	.4147	.9100	.4557	2.194	1.099	2.411	1.1432	30
40	305	173	088	592	177	100	396	403	20
50	334	200	075	628	161	102	381	374	10
25° 00′	.4363	.4226	.9063	.4663	2.145	1.103	2.366	1.1345	**65° 00′**
10	392	253	051	699	128	105	352	316	50
20	422	279	038	734	112	106	337	286	40
30	.4451	.4305	.9026	.4770	2.097	1.108	2.323	1.1257	30
40	480	331	013	806	081	109	309	228	20
50	509	358	001	841	066	111	295	199	10
26° 00′	.4538	.4384	.8988	.4877	2.050	1.113	2.281	1.1170	**64° 00′**
10	567	410	975	913	035	114	268	141	50
20	596	436	962	950	020	116	254	112	40
30	.4625	.4462	.8949	.4986	2.006	1.117	2.241	1.1083	30
40	654	488	936	.5022	1.991	119	228	054	20
50	683	514	923	059	977	121	215	1.1025	10
27° 00′	.4712	.4540	.8910	.5095	1.963	1.122	2.203	1.0996	**63° 00′**
		Cos	Sin	Cot	Tan	Csc	Sec	Radians	Degrees

TABLE 3 (*continued*)

Degrees	Radians	Sin	Cos	Tan	Cot	Sec	Csc		
27° 00′	.4712	.4540	.8910	.5095	1.963	1.122	2.203	1.0996	**63° 00′**
10	741	566	897	132	949	124	190	966	50
20	771	592	884	169	935	126	178	937	40
30	.4800	.4617	.8870	.5206	1.921	1.127	2.166	1.0908	30
40	829	643	857	243	907	129	154	879	20
50	858	669	843	280	894	131	142	850	10
28° 00′	.4887	.4695	.8829	.5317	1.881	1.133	2.130	1.0821	**62° 00′**
10	916	720	816	354	868	134	118	792	50
20	945	746	802	392	855	136	107	763	40
30	.4974	.4772	.8788	.5430	1.842	1.138	2.096	1.0734	30
40	.5003	797	774	467	829	140	085	705	20
50	032	823	760	505	816	142	074	676	10
29° 00′	.5061	.4848	.8746	.5543	1.804	1.143	2.063	1.0647	**61° 00′**
10	091	874	732	581	792	145	052	617	50
20	120	899	718	619	780	147	041	588	40
30	.5149	.4924	.8704	.5658	1.767	1.149	2.031	1.0559	30
40	178	950	689	696	756	151	020	530	20
50	207	975	675	735	744	153	010	501	10
30° 00′	.5236	.5000	.8660	.5774	1.732	1.155	2.000	1.0472	**60° 00′**
10	265	025	646	812	720	157	1.990	443	50
20	294	050	631	851	709	159	980	414	40
30	.5323	.5075	.8616	.5890	1.698	1.161	1.970	1.0385	30
40	352	100	601	930	686	163	961	356	20
50	381	125	587	969	675	165	951	327	10
31° 00′	.5411	.5150	.8572	.6009	1.664	1.167	1.942	1.0297	**59° 00′**
10	440	175	557	048	653	169	932	268	50
20	469	200	542	088	643	171	923	239	40
30	.5498	.5225	.8526	.6128	1.632	1.173	1.914	1.0210	30
40	527	250	511	168	621	175	905	181	20
50	556	275	496	208	611	177	896	152	10
32° 00′	.5585	.5299	.8480	.6249	1.600	1.179	1.887	1.0123	**58° 00′**
10	614	324	465	289	590	181	878	094	50
20	643	348	450	330	580	184	870	065	40
30	.5672	.5373	.8434	.6371	1.570	1.186	1.861	1.0036	30
40	701	398	418	412	560	188	853	1.0007	20
50	730	422	403	453	550	190	844	977	10
33° 00′	.5760	.5446	.8387	.6494	1.540	1.192	1.836	.9948	**57° 00′**
10	789	471	371	536	530	195	828	919	50
20	818	495	355	577	520	197	820	890	40
30	.5847	.5519	.8339	.6619	1.511	1.199	1.812	.9861	30
40	876	544	323	661	501	202	804	832	20
50	905	568	307	703	1.492	204	796	803	10
34° 00′	.5934	.5592	.8290	.6745	1.483	1.206	1.788	.9774	**56° 00′**
10	963	616	274	787	473	209	781	745	50
20	992	640	258	830	464	211	773	716	40
30	.6021	.5664	.8241	.6873	1.455	1.213	1.766	.9687	30
40	050	688	225	916	446	216	758	657	20
50	080	712	208	959	437	218	751	628	10
35° 00′	.6109	.5736	.8192	.7002	1.428	1.221	1.743	.9599	**55° 00′**
10	138	760	175	046	419	223	736	570	50
20	167	783	158	089	411	226	729	541	40
30	.6196	.5807	.8141	.7133	1.402	1.228	1.722	.9512	30
40	225	831	124	177	393	231	715	483	20
50	254	854	107	221	385	233	708	454	10
36° 00′	.6283	.5878	.8090	.7265	1.376	1.236	1.701	.9425	**54° 00′**
		Cos	Sin	Cot	Tan	Csc	Sec	Radians	Degrees

(Continued)

TABLE 3 (*continued*)

Degrees	Radians	Sin	Cos	Tan	Cot	Sec	Csc		
36° 00′	.6283	.5878	.8090	.7265	1.376	1.236	1.701	.9425	**54° 00′**
10	312	901	073	310	368	239	695	396	50
20	341	925	056	355	360	241	688	367	40
30	.6370	.5948	.8039	.7400	1.351	1.244	1.681	.9338	30
40	400	972	021	445	343	247	675	308	20
50	429	995	004	490	335	249	668	279	10
37° 00′	.6458	.6018	.7986	.7536	1.327	1.252	1.662	.9250	**53° 00′**
10	487	041	969	581	319	255	655	221	50
20	516	065	951	627	311	258	649	192	40
30	.6545	.6088	.7934	.7673	1.303	1.260	1.643	.9163	30
40	574	111	916	720	295	263	636	134	20
50	603	134	898	766	288	266	630	105	10
38° 00′	.6632	.6157	.7880	.7813	1.280	1.269	1.624	.9076	**52° 00′**
10	661	180	862	860	272	272	618	047	50
20	690	202	844	907	265	275	612	.9018	40
30	.6720	.6225	.7826	.7954	1.257	1.278	1.606	.8988	30
40	749	248	808	.8002	250	281	601	959	20
50	778	271	790	050	242	284	595	930	10
39° 00′	.6807	.6293	.7771	.8098	1.235	1.287	1.589	.8901	**51° 00′**
10	836	316	753	146	228	290	583	872	50
20	865	338	735	195	220	293	578	843	40
30	.6894	.6361	.7716	.8243	1.213	1.296	1.572	.8814	30
40	923	383	698	292	206	299	567	785	20
50	952	406	679	342	199	302	561	756	10
40° 00′	.6981	.6428	.7660	.8391	1.192	1.305	1.556	.8727	**50° 00′**
10	.7010	450	642	441	185	309	550	698	50
20	039	472	623	491	178	312	545	668	40
30	.7069	.6494	.7604	.8541	1.171	1.315	1.540	.8639	30
40	098	517	585	591	164	318	535	610	20
50	127	539	566	642	157	322	529	581	10
41° 00′	.7156	.6561	.7547	.8693	1.150	1.325	1.524	.8552	**49° 00′**
10	185	583	528	744	144	328	519	523	50
20	214	604	509	796	137	332	514	494	40
30	.7243	.6626	.7490	.8847	1.130	1.335	1.509	.8465	30
40	272	648	470	899	124	339	504	436	20
50	301	670	451	952	117	342	499	407	10
42° 00′	.7330	.6691	.7431	.9004	1.111	1.346	1.494	.8378	**48° 00′**
10	359	713	412	057	104	349	490	348	50
20	389	734	392	110	098	353	485	319	40
30	.7418	.6756	.7373	.9163	1.091	1.356	1.480	.8290	30
40	447	777	353	217	085	360	476	261	20
50	476	799	333	271	079	364	471	232	10
43° 00′	.7505	.6820	.7314	.9325	1.072	1.367	1.466	.8203	**47° 00′**
10	534	841	294	380	066	371	462	174	50
20	563	862	274	435	060	375	457	145	40
30	.7592	.6884	.7254	.9490	1.054	1.379	1.453	.8116	30
40	621	905	234	545	048	382	448	087	20
50	650	926	214	601	042	386	444	058	10
44° 00′	.7679	.6947	.7193	.9657	1.036	1.390	1.440	.8029	**46° 00′**
10	709	967	173	713	030	394	435	999	50
20	738	988	153	770	024	398	431	970	40
30	.7767	.7009	.7133	.9827	1.018	1.402	1.427	.7941	30
40	796	030	112	884	012	406	423	912	20
50	825	050	092	942	006	410	418	883	10
45° 00′	.7854	.7071	.7071	1.000	1.000	1.414	1.414	.7854	**45° 00′**
		Cos	Sin	Cot	Tan	Csc	Sec	Radians	Degrees

TABLE 4 Exponential Functions

x	e^x	e^{-x}	x	e^x	e^{-x}	x	e^x	e^{-x}
0.00	1.0000	1.0000	0.55	1.7333	0.5769	3.6	36.598	0.0273
0.01	1.0101	0.9900	0.60	1.8221	0.5488	3.7	40.447	0.0247
0.02	1.0202	0.9802	0.65	1.9155	0.5220	3.8	44.701	0.0224
0.03	1.0305	0.9704	0.70	2.0138	0.4966	3.9	49.402	0.0202
0.04	1.0408	0.9608	0.75	2.1170	0.4724	4.0	54.598	0.0183
0.05	1.0513	0.9512	0.80	2.2255	0.4493	4.1	60.340	0.0166
0.06	1.0618	0.9418	0.85	2.3396	0.4274	4.2	66.686	0.0150
0.07	1.0725	0.9324	0.90	2.4596	0.4066	4.3	73.700	0.0136
0.08	1.0833	0.9231	0.95	2.5857	0.3867	4.4	81.451	0.0123
0.09	1.0942	0.9139	1.0	2.7183	0.3679	4.5	90.017	0.0111
0.10	1.1052	0.9048	1.1	3.0042	0.3329	4.6	99.484	0.0101
0.11	1.1163	0.8958	1.2	3.3201	0.3012	4.7	109.95	0.0091
0.12	1.1275	0.8869	1.3	3.6693	0.2725	4.8	121.51	0.0082
0.13	1.1388	0.8781	1.4	4.0552	0.2466	4.9	134.29	0.0074
0.14	1.1503	0.8694	1.5	4.4817	0.2231	5	148.41	0.0067
0.15	1.1618	0.8607	1.6	4.9530	0.2019	6	403.43	0.0025
0.16	1.1735	0.8521	1.7	5.4739	0.1827	7	1096.6	0.0009
0.17	1.1853	0.8437	1.8	6.0496	0.1653	8	2981.0	0.0003
0.18	1.1972	0.8353	1.9	6.6859	0.1496	9	8103.1	0.0001
0.19	1.2092	0.8270	2.0	7.3891	0.1353	10	22026	0.00005
0.20	1.2214	0.8187	2.1	8.1662	0.1225	11	59874	0.00002
0.21	1.2337	0.8106	2.2	9.0250	0.1108	12	162,754	0.000006
0.22	1.2461	0.8025	2.3	9.9742	0.1003	13	442,413	0.000002
0.23	1.2586	0.7945	2.4	11.023	0.0907	14	1,202,604	0.0000008
0.24	1.2712	0.7866	2.5	12.182	0.0821	15	3,269,017	0.0000003
0.25	1.2840	0.7788	2.6	13.464	0.0743			
0.26	1.2969	0.7711	2.7	14.880	0.0672			
0.27	1.3100	0.7634	2.8	16.445	0.0608			
0.28	1.3231	0.7558	2.9	18.174	0.0550			
0.29	1.3364	0.7483	3.0	20.086	0.0498			
0.30	1.3499	0.7408	3.1	22.198	0.0450			
0.35	1.4191	0.7047	3.2	24.533	0.0408			
0.40	1.4918	0.6703	3.3	27.113	0.0369			
0.45	1.5683	0.6376	3.4	29.964	0.0334			
0.50	1.6487	0.6065	3.5	33.115	0.0302			

TABLE 5 Factorials and Large Powers of 2

n	$n!$	2^n
0	1	1
1	1	2
2	2	4
3	6	8
4	24	16
5	120	32
6	720	64
7	5040	128
8	40,320	256
9	362,880	512
10	3,628,800	1024
11	39,916,800	2048
12	479,001,600	4096
13	6,227,020,800	8192
14	87,178,291,200	16,384
15	1,307,674,368,000	32,768
16	20,922,789,888,000	65,536
17	355,687,428,096,000	131,072
18	6,402,373,705,728,000	262,144
19	121,645,100,408,832,000	524,288
20	2,432,902,008,176,640,000	1,048,576

TABLE 6 Tables of Measures

LENGTH

1 *kilo*meter (km) = 1000 meters (m)
1 *hecto*meter (hm) = 100 meters
1 *deka*meter (dam) = 10 meters
1 *deci*meter (dm) = 0.1 meter
1 *centi*meter (cm) = 0.01 meter
1 *milli*meter (mm) = 0.001 meter

MASS OR WEIGHT

1 *kilo*gram (kg) = 1000 grams (g)
1 *hecto*gram (hg) = 100 grams
1 *deka*gram (dag) = 10 grams
1 *deci*gram (dg) = 0.1 gram
1 *centi*gram (dg) = 0.01 gram
1 metric ton (MT or t) = 1000 kilograms

AREA

1 *hectare* (ha) = 100 are (a), or 10,000 sq m (m^2)
1 are (a) = 100 sq m (m^2)
1 *centare* (ca) = 0.01 are, or 1 m^2
The word "are" is pronounced "AIR."

VOLUME

1000 cubic centimeters (cc or cm^3) = 1 liter (L)
1 cubic centimeter (cc) = 1 milliliter (mL)
1 mL of water weighs 1 g
1 stere = 1 cubic meter

Metric-American Conversions (Approximate)

LENGTH

1 m = 39.37 in. = 3.3 ft
1 in. = 2.54 cm
1 km = 0.62 mi
1 mi = 1.6 km
1 cm = 3/8 in.

MASS OR WEIGHT

1 kg = 2.2 lb
1 MT = 1.1 tons
1 lb = 454 g
1 oz = 28 g

AREA

1 hectare = 2.47 acres
1 are = 120 sq yd

VOLUME

1 liter = 1.057 qt = 2.1 pt
1 cup = 240 mL
1 ounce (liquid) = 30 mL
1 gallon = 3.78 liters
1 tablespoon = 15 mL
1 teaspoon = 5 mL

Answers

CHAPTER 1

Margin Exercises, Section 1.1

1. 1, 19 **2.** 0, 1, 19 **3.** $-6, 0, 1, 19$ **4.** All of them **5.** Rational **6.** Rational **7.** Rational **8.** Rational **9.** Irrational **10.** Irrational **11.** -12 **12.** -4.7 **13.** $-\frac{4}{5}$ **14.** -0.2 **15.** 5 **16.** 0 **17.** $-6, -6$ **18.** 8, 8 **19.** 3.4, 3.4 **20.** $2, -4$ **21.** -24 **22.** $\frac{21}{25}$ **23.** 144 **24.** 1.3 **25.** 17 **26.** $-\frac{11}{5}$ **27.** -13 **28.** 4 **29.** -3 **30.** $-\frac{8}{3}$ **31.** 2 **32.** No **33.** Yes **34.** Yes **35.** No

Exercise Set 1.1, pp. 10–11

1. 3, 14 **3.** $\sqrt{3}, -\sqrt{7}, \sqrt[3]{2}$ **5.** $-6, 0, 3, -2, 14$ **7.** Rational **9.** Rational **11.** Rational **13.** Irrational **15.** Irrational **17.** Irrational **19.** Rational **21.** Irrational **23.** 7, 7 **25.** $-57, -57$ **27.** -87 **29.** -16 **31.** -10.3 **33.** $\frac{39}{10}$ **35.** 28 **37.** -49.2 **39.** 210 **41.** $-\frac{833}{5}$ **43.** 5 **45.** $-\frac{1}{7}$ **47.** $-\frac{3}{49}$ **49.** 25 **51.** -4 **53.** 18 **55.** -11.6 **57.** $-\frac{35}{8}$ **59.** **(a)** 1.96, 1.9881, 1.999396, 1.999962, 1.999990; **(b)** $\sqrt{2}$ **61.** Identity (+) **63.** Distributive **65.** Commutative (+) **67.** Associative (×) **69.** Inverse (×) **71.** Commutative (+) **73.** Commutative (+) **75.** $7 - 5 \neq 5 - 7$; $7 - 5 = 2$, $5 - 7 = -2$ **77.** $16 \div (4 \div 2) \neq (16 \div 4) \div 2$; $16 \div (4 \div 2) = 8$, $(16 \div 4) \div 2 = 2$ **79.** $\frac{3927}{1250}$ **81.** 1 **83.** $\frac{183{,}062}{9990}$ **85.** $(b + c)a = a(b + c) = ab + ac = ba + ca$

Margin Exercises, Section 1.2

1. 8^4 **2.** x^3 **3.** $(4y)^4$ **4.** $3 \cdot 3 \cdot 3 \cdot 3$, or 81 **5.** $5x \cdot 5x \cdot 5x \cdot 5x$, or $625x^4$ **6.** $(-5)(-5)(-5)(-5)$, or 625 **7.** $-[5 \cdot 5 \cdot 5 \cdot 5]$, or -625 **8.** 1 **9.** $25y^2$ **10.** $-8x^3$ **11.** 4^{-3} **12.** $\frac{1}{10^4}$, or $\frac{1}{10 \cdot 10 \cdot 10 \cdot 10}$, or $\frac{1}{10{,}000}$ **13.** $\frac{1}{4^3}$, $\frac{1}{4 \cdot 4 \cdot 4}$, $\frac{1}{64}$ **14.** 8^4 **15.** y^5 **16.** $-18x^{11}$ **17.** $-\frac{75}{x^{14}}$ **18.** $-\frac{10y^2}{x^{12}}$ **19.** $60y$ **20.** 4^3 **21.** 5^6 **22.** $\frac{1}{10^{14}}$ **23.** $\frac{1}{9^6}$ **24.** y^{11} **25.** $\frac{5}{y}$ **26.** $-\frac{2y^9}{x^3}$ **27.** 3^{49} **28.** $\frac{1}{8^{14}}$ **29.** $\frac{1}{y^{28}}$ **30.** $8x^3y^3$ **31.** $\frac{16y^{14}}{x^4}$ **32.** $\frac{z^{15}}{27x^{12}y^6}$ **33.** $\frac{8y^{45}z^3}{x^{30}}$ **34.** 4.65×10^5 **35.** 3.789×10^3 **36.** 1.45×10^{-4} **37.** 6.7×10^{-10} **38.** 0.0000467 **39.** 7,894,000,000,000 **40.** 8,166,000,000 **41.** 0.000001103 **42.** 3.1536×10^7 **43.** **(a)** 79; **(b)** 87 **44.** 27 **45.** **(a)** -13; **(b)** -13 **46.** 2 **47.** $\sqrt{3}$ **48.** 11.3 **49.** $\frac{3}{4}$ **50.** 20 **51.** 20 **52.** 4 **53.** 4 **54.** $6|a||b|$ **55.** x^8 **56.** $10m^2n^2|n|$ **57.** $\frac{2x^2|x|}{y^2}$

Exercise Set 1.2, pp. 19–20

1. $\frac{1}{2}$ **3.** 1 **5.** 4^3 **7.** $6x^5$ **9.** $\frac{15b^5}{a}$ **11.** $72x^5$ **13.** $-18x^7yz$ **15.** b^3 **17.** $\frac{x^3}{y^3}$ **19.** 1 **21.** $3ab^2$ **23.** $\frac{4xy}{7z^5}$

25. $8a^3b^6$ 27. $16x^{12}$ 29. $-16x^{12}$ 31. $36a^4b^6c^2$
33. $\frac{1}{25}c^2d^4$ 35. 1 37. 32 39. $\frac{27a^8c^{18}}{4b^{10}}$ 41. $\frac{3}{4}xy$
43. $\frac{32{,}768a^{20}c^{10}}{b^{25}}$ 45. $-25, 25$ 47. $-1.1664, 1.1664$
49. 5.8×10^7 51. 3.65×10^5 53. 2.7×10^{-6}
55. 2.7×10^{-2} 57. 9.11×10^{-28} 59. 3.664×10^9
61. 400,000 63. 0.0062 65. 7,690,000,000,000
67. 0.000000567 69. 9,460,000,000,000
71. 0.0000000769 73. 256,700,000 75. 1.512×10^{10}
77. 10^{-9} sec 79. 1.46×10^{12} mi 81. 10 83. 3
85. 2048 87. $\frac{243}{8}$ 89. 12 91. 47 93. $7|a|$ 95. $8x^6$
97. $9|x||y|$ 99. $3a^2|b|$ 101. x^{8t} 103. t^{8x} 105. $(xy)^{ac+bc}$
107. $9x^{2a}y^{2b}$ 109. \$750.43 111. In $(x^3)^2$, exponents were added instead of multiplied; x^{10} 113. In 2^3, the base 2 and the exponent 3 were multiplied. In $(x^{-4})^3$, the exponents were added. In $(y^6)^3$, the exponents were subtracted. In $(z^3)^3$, the exponents were added; $8x^{-12}y^{18}z^9$

Margin Exercises, Section 1.3

1. 8, 6, 4, 9, 0; 9 2. 4, 4, 5, 6, 0; 6 3. $9x^3y^2 - 2x^2y^3$
4. $7xy^2 - 2x^2y$ 5. $3x^4\sqrt{y} + 2$ 6. $-4x^3 + 2x^2 - 4x - \frac{3}{2}$
7. $5p^2q^4 + p^2q^2 - 6pq^2 - 3q + 5$
8. $-(5x^2t^2 - 4xy^2t - 3xt + 6x - 5)$, $-5x^2t^2 + 4xy^2t + 3xt - 6x + 5$
9. $-(-3x^2y + 5xy - 7x + 4y + 2)$, $3x^2y - 5xy + 7x - 4y - 2$
10. $8xy^4 - 9xy^2 + 4x^2 + 2y - 7$
11. $3x^2y - 9x^3y^2 + 5x^2y^3 - x^2y^2 + 9y$

Exercise Set 1.3, p. 24

1. 4, 3, 2, 1, 0; 4 3. 3, 6, 6, 0; 6 5. 5, 6, 2, 1, 0; 6
7. $3x^2y - 5xy^2 + 7xy + 2$
9. $-10pq^2 - 5p^2q + 7pq - 4p + 2q + 3$
11. $3x + 2y - 2z - 3$ 13. $5x\sqrt{y} - 4y\sqrt{x} - \frac{2}{5}$
15. $-(5x^3 - 7x^2 + 3x - 6)$, $-5x^3 + 7x^2 - 3x + 6$
17. $-2x^2 + 6x - 2$ 19. $6a - 5b - 2c + 4d$
21. $x^4 - 3x^3 - 4x^2 + 9x - 3$ 23. $9x\sqrt{y} - 3y\sqrt{x} + 9.1$
25. $-1.047p^2q - 2.479pq^2 + 8.879pq - 104.144$

Margin Exercises, Section 1.4

1. $3x^3y^2 + 4x^2y^2 - xy^2 + 6y^2$
2. $2p^4q^2 + 3p^3q^2 + 3p^2q^2 + 2q^2$ 3. $2x^3y - 4xy + 3x^3 - 6x$
4. $15x^2 - xy - 6y^2$ 5. $6xy - 2\sqrt{2}x + 3\sqrt{2}y - 2$
6. $16x^2 - 40xy + 25y^2$ 7. $4y^4 + 24x^2y^3 + 36x^4y^2$
8. $16x^2 - 49$ 9. $25x^4y^2 - 4y^2$ 10. $16y^4 - 3$
11. $4x^2 + 12x + 9 - 25y^2$ 12. $25t^2 - 4x^6y^4$
13. $x^3 + 3x^2 + 3x + 1$ 14. $x^3 - 3x^2 + 3x - 1$
15. $t^6 - 9t^4b + 27t^2b^2 - 27b^3$
16. $8a^9 - 60a^6b^2 + 150a^3b^4 - 125b^6$

Exercise Set 1.4, pp. 27–28

1. $6x^3 + 4x^2 + 32x - 64$
3. $4a^3b^2 - 10a^2b^2 + 3ab^3 + 4ab^2 - 6b^3 + 4a^2b - 2ab + 3b^2$
5. $a^3 - b^3$ 7. $4x^2 + 8xy + 3y^2$ 9. $12x^3 + x^2y - \frac{3}{2}xy - \frac{1}{8}y^2$
11. $10p^3q^4 - 4p^2q^3r - 5r^2pq + 2r^3$ 13. $4x^2 + 12xy + 9y^2$
15. $4x^4 - 12x^2y + 9y^2$ 17. $4x^6 + 12x^3y^2 + 9y^4$
19. $\frac{1}{4}x^4 - \frac{3}{5}x^2y + \frac{9}{25}y^2$ 21. $0.25x^2 + 0.70xy^2 + 0.49y^4$
23. $9x^2 - 4y^2$ 25. $x^4 - y^2z^2$ 27. $9x^4 - 2$
29. $4x^2 + 12xy + 9y^2 - 16$ 31. $x^4 + 6x^2y + 9y^2 - y^4$
33. $x^4 - 1$ 35. $16x^4 - y^4$
37. $0.002601x^2 + 0.00408xy + 0.0016y^2$
39. $2462.0358x^2 - 945.0214x - 38.908$
41. $y^3 + 15y^2 + 75y + 125$
43. $m^6 - 6m^4n + 12m^2n^2 - 8n^3$
45. $2x^3 - 2\sqrt{2}x^2y - \sqrt{2}xy^2 + 2y^3$ 47. $a^{2n} - b^{2n}$
49. $x^{3m} - 3x^{2m}t^n + 3x^mt^{2n} - t^{3n}$ 51. $x^6 - 1$
53. $16x^4 - 32x^3 + 16x^2$ 55. $x^{a^2-b^2}$
57. $a^2 + b^2 + c^2 + 2ab + 2ac + 2bc$
59. $a^4 + 4a^3b + 6a^2b^2 + 4ab^3 + b^4$ 61. $m^5 + t^5$ 63. A term is missing—found by calculating twice the product of the terms; the first term is $9a^2$, not $3a^2$, where the 3 was not squared; $9a^2 + 6ab + b^2$ 65. In step (1), $3x$ should be $6x$; the 2 and 3 were not multiplied. Similarly, -3 should be -12; the 4 and -3 were not multiplied. In step (2), $3x - 3$ is not x; $6x^2 + 6x - 12$. 67. $x^n - y^n$

Margin Exercises, Section 1.5

1. $4x^2y(5x + 3)$ 2. $(p + q)(2x + y + 2)$
3. $(4x^2 - 3)(x + 5)$ 4. $(x - 4)(x + 4)$
5. $(5y^2 + 4x)(5y^2 - 4x)$ 6. $2(y^2 + 4x^2)(y - 2x)(y + 2x)$
7. $(x - \sqrt{3})(x + \sqrt{3})$ 8. $(x + 5)(x + 1)$ 9. $(x + 7)(x - 2)$
10. $(w^2 - 5)(w^2 - 2)$ 11. $(3x + 2)(x + 1)$
12. $3(2x^2y^3 + 5)(x^2y^3 - 4)$ 13. Not factorable
14. $(3y - 5)^2$ 15. $(4x + 9y)^2$ 16. $-3y^2(2x^2 - 5y^3)^2$
17. $(x - 2)(x^2 + 2x + 4)$ 18. $(4 - t)(16 + 4t + t^2)$
19. $(3x + y)(9x^2 - 3xy + y^2)$
20. $(2m + 5t)(4m^2 - 10mt + 25t^2)$
21. $2y(4y^2 - 5x^2)(16y^4 + 20x^2y^2 + 25x^4)$
22. $(p + 2)(p - 2)(p^2 - 2p + 4)(p^2 + 2p + 4)$

Exercise Set 1.5, pp. 33–34

1. $(p + 4)(p + 2)$ 3. $(n + 8)(n - 7)$ 5. $(y^2 - 7)(y^2 + 3)$
7. $3ab(6a - 5b)$ 9. $(a + c)(b - 2)$ 11. $(x^2 + 6)(x + 3)$
13. $(y + 2)(y - 2)(y - 3)$ 15. $(3x - 5)(3x + 5)$
17. $4x(y^2 - z)(y^2 + z)$ 19. $(y - 3)^2$ 21. $(1 - 4x)^2$
23. $(2x - \sqrt{5})(2x + \sqrt{5})$ 25. $(xy - 7)^2$
27. $4a(x + 7)(x - 2)$ 29. $(a + b + c)(a + b - c)$
31. $(x + y - a - b)(x + y + a + b)$
33. $5(y^2 + 4x^2)(y - 2x)(y + 2x)$ 35. $(x + 2)(x^2 - 2x + 4)$
37. $3\left(x - \frac{1}{2}\right)\left(x^2 + \frac{1}{2}x + \frac{1}{4}\right)$ 39. $(x + 0.1)(x^2 - 0.1x + 0.01)$
41. $3(z - 2)(z^2 + 2z + 4)$
43. $(a - t)(a + t)(a^2 - at + t^2)(a^2 + at + t^2)$
45. $2ab(2a^2 + 3b^2)(4a^4 - 6a^2b^2 + 9b^4)$
47. $(x + 4.19524)(x - 4.19524)$
49. $37(x + 0.626y)(x - 0.626y)$ 51. $h(3x^2 + 3xh + h^2)$
53. $(y^2 + 12)(y^2 - 7)$ 55. $\left(y + \frac{4}{7}\right)\left(y - \frac{2}{7}\right)$
57. $(t + 0.9)(t - 0.3)$ 59. $(x^n + 8)(x^n - 3)$
61. $(x + a)(x + b)$ 63. $\left(\frac{1}{2}t - \frac{2}{5}\right)^2$
65. $(5y^m - x^n + 1)(5y^m + x^n - 1)$
67. $3(x^n - 2y^m)(x^{2n} + 2x^ny^m + 4y^{2m})$ 69. $y(y - 1)^2(y - 2)$
71. $5\left(x - \frac{9}{5}\right)$ 73. (a) $x^5 + x - 1$; (b) $(x^2 - x + 1)(x^3 + x^2 - 1)$

Margin Exercises, Section 1.6

1. All real numbers except 3 2. All real numbers except -3 and -4 3. $\frac{(x+y)(x+y)}{(2x^2-1)(7x)}$ 4. $\frac{(x-2)(x+4)}{(x+2)(x+2)}$ 5. $\frac{x^2+5x+6}{x^2-2x-15}$; all real numbers except 5; all real numbers except 5 and -3
6. $\frac{3x+2}{x+2}$; all real numbers except 0 and -2; all real numbers except -2 7. $\frac{y+2}{y-1}$; all real numbers except 1 and -1; all real numbers except 1 8. $\frac{3x-3y}{x+y}$ 9. $\frac{2a^2b+2ab^2}{a-b}$
10. $\frac{3x^2+4x+2}{x-5}$ 11. $\frac{2x^2+11}{x-5}$ 12. $\frac{4x^2-xy+4y^2}{2(2x-y)(x-y)}$
13. $\frac{x-6}{(x+4)(x+6)}$ 14. $\frac{1}{a-x}$ 15. $\frac{a^2b^2}{b^2-ab+a^2}$

Exercise Set 1.6, pp. 41–43

1. All numbers except 0 and 1 3. All numbers except -5, -2, and 2 5. $\frac{3}{x}$; all numbers except 0 7. $\frac{7}{(x+2)(x+5)}$; all numbers except -5 and -2 9. $\frac{5}{2}x$; all numbers 11. $\frac{x-2}{x+2}$; all numbers except -2 13. $\frac{1}{x-y}$ 15. $\frac{(x+5)(2x+3)}{7x}$
17. $\frac{a+2}{a-5}$ 19. $m+n$ 21. $\frac{3(x-4)}{2(x+4)}$ 23. $\frac{1}{x+y}$
25. $\frac{x-y-z}{x+y+z}$ 27. 1 29. $\frac{y-2}{y-1}$ 31. $\frac{x+y}{2x-3y}$
33. $\frac{3x-4}{x^2-4}$ 35. $\frac{3y-10}{(y-5)(y+4)}$ 37. $\frac{4x-8y}{x^2-y^2}$
39. $\frac{3x-4}{(x-2)(x-1)}$ 41. $\frac{5a^2+10ab-4b^2}{(a-b)(a+b)}$
43. $\frac{11x^2-18x+8}{(2+x)(2-x)^2}$ 45. 0 47. $\frac{x+y}{x}$ 49. $\frac{a^2-1}{a^2+1}$
51. $\frac{c^2-2c+4}{c}$ 53. $\frac{xy}{x-y}$ 55. $x-y$ 57. $\frac{x^2-y^2}{xy}$
59. $\frac{1+a}{1-a}$ 61. $\frac{b+a}{b-a}$ 63. $2x+h$ 65. $3x^2+3xh+h^2$
67. x^5 69. Step (1) uses the wrong reciprocal. The reciprocal of a sum is not the sum of the reciprocals. Step (2) would be correct if step (1) had been. Step (3) would be correct if steps (1) and (2) had been. Step (4) has an improper simplification of the b in the denominator; $\frac{12a}{b(4a+3b)}$. 71. $\frac{(n+1)(n+2)(n+3)}{2\cdot 3}$

Margin Exercises, Section 1.7

1. Yes, no 2. No, yes 3. Yes, yes 4. Yes, yes
5. $|x+2|$ 6. $|x|\cdot|y-2|$, or $|x(y-2)|$ 7. $|x+2|$
8. $|x+4|$ 9. $-4xy$ 10. $\sqrt{133}$ 11. $\sqrt{x^2-4y^2}$ 12. $\sqrt[4]{81}$, or 3 13. $10\sqrt{3}$ 14. $6|y|$ 15. $|x+1|\sqrt{2}$ 16. $2\cdot\sqrt[3]{2}$
17. $(a+b)\cdot\sqrt[3]{a+b}$ 18. $\frac{7}{8}$ 19. $\frac{5}{|y|}$ 20. $\frac{2}{3}$ 21. $\frac{\sqrt[3]{7}}{5}$
22. 5 23. $\frac{|x|}{5}$ 24. $\frac{2x}{y}$ 25. 3^{10} 26. 3^4 27. $-6\sqrt{5}$
28. $(10y+7)\sqrt[3]{2y}$ 29. $-4-9\sqrt{6}$ 30. 37.42 mph
31. 15.65 m 32. $\frac{\sqrt{3}+\sqrt{5}}{-2}$ 33. $\frac{x-7\sqrt{x}+10}{x-4}$
34. $\frac{1}{\sqrt{a+2}+\sqrt{a}}$ 35. $\frac{x-5}{x+2\sqrt{5x}+5}$

Exercise Set 1.7, pp. 49–51

1. No, yes 3. Yes, no 5. Yes, no 7. Yes, yes 9. 11
11. $4|x|$ 13. $|b+1|$ 15. $-3x$ 17. $|x-2|$ 19. 2
21. $6\sqrt{5}$ 23. $3\sqrt[3]{2}$ 25. $8\sqrt{2}|c|d^2$ 27. $3\sqrt{2}$
29. $2x^2y\sqrt{6}$ 31. $3x\sqrt[3]{4y}$ 33. $2(x+4)\sqrt[3]{(x+4)^2}$
35. $\sqrt{7b}$ 37. 2 39. $\frac{1}{2x}$ 41. $\sqrt{a+b}$ 43. $\frac{3a\sqrt{2b}}{4b}$
45. $\frac{y\cdot\sqrt[3]{20x^2z^2}}{5x^2}$ 47. $8x^2\sqrt[3]{2}$ 49. $ab^2x^2y\sqrt{a}$ 51. $51\sqrt{2}$
53. $-12\sqrt{5}-2\sqrt{2}$ 55. $19\sqrt[3]{x^2}-3x$ 57. $4y\sqrt{3}-2y\sqrt{6}$
59. 1 61. $4+2\sqrt{3}$ 63. $t-2x\sqrt{t}+x^2$ 65. $10\sqrt{7}$
67. x 69. About 42.43 mph 71. 1.57 sec, 3.14 sec, 8.88 sec, 11.1 sec 73. About 13,709.5 ft 75. $h=\frac{a}{2}\sqrt{3}$ 77. $\sqrt{2s}$
79. 8 81. $\frac{3(3-\sqrt{5})}{2}$ 83. $\frac{2\sqrt[3]{6}}{3}$
85. $\frac{8x-20\sqrt{xy}-6x\sqrt{y}+15y\sqrt{x}}{4x-25y}$ 87. $\frac{2-5a}{6(\sqrt{2}-\sqrt{5a})}$
89. $\frac{x}{x+2-2\sqrt{x+1}}$ 91. $\frac{a}{3(\sqrt{a+3}+\sqrt{3})}$
93. $10.124x^2y$ 95. $\frac{0.5933a\sqrt{b}}{b}$ 97. $\frac{(2+x^2)\sqrt{1+x^2}}{1+x^2}$
99. Let $a=16$ and $b=9$. Then $\sqrt{a+b}=5$ and $\sqrt{a}+\sqrt{b}=7$. 101. 0.0188 m 103. (a) 6.95 mi/sec; (b) 3.1 mi/sec

Margin Exercises, Section 1.8

1. $n\sqrt{n}$ 2. $\frac{1}{\sqrt[7]{y^6}}$, or $\frac{\sqrt[7]{y}}{y}$ 3. 16 4. $\frac{1}{16}$ 5. $(5ab)^{4/3}$, or $(5ab)\sqrt[3]{5ab}$ 6. 8 7. $a^{2/3}$, or $\sqrt[3]{a^2}$ 8. $2^{1/3}$, or $\sqrt[3]{2}$ 9. $5^{11/6}$, or $5\sqrt[6]{5^5}$ 10. $\sqrt[4]{a^5}$, or $a\sqrt[4]{a}$ 11. $\frac{1}{\sqrt[5]{x^6}}$, or $\frac{1}{x\sqrt[5]{x}}$, or $\frac{\sqrt[5]{x^4}}{x^2}$
12. $\sqrt[4]{2^3}+\frac{1}{\sqrt[4]{2}}$, or $3\frac{\sqrt[4]{2^3}}{2}$ 13. $\sqrt[6]{200}$
14. $\sqrt[6]{x^4y^3z^5}$ 15. $\sqrt[4]{(x+y)}$ 16. 5.98 ft
17. $\frac{p^{14}-q^{12}}{p^6q^5}$ 18. $\frac{5x+4y^{3/4}}{x^{1/3}y^{1/4}}$ 19. $\frac{5x^2(x+3)}{(2x+5)^{3/2}}$

Exercise Set 1.8, pp. 55–56

1. $\sqrt[4]{x^3}$ 3. 8 5. $\frac{1}{5}$ 7. $\frac{a}{b}\sqrt[4]{ab}$ 9. $20^{2/3}$ 11. $13^{5/4}$, or $13\sqrt[4]{13}$ 13. $11^{1/6}$ 15. $5^{5/6}$ 17. 4 19. $2y^2$
21. $(a^2+b^2)^{1/3}$ 23. $3ab^3$ 25. $\frac{m^2n^4}{2}$ 27. $8a^{4/2}$, or $8a^2$
29. $\frac{x^{-3}}{3^{-1}b^2}$, or $\frac{3}{x^3b^2}$ 31. $xy^{1/3}$, or $x\sqrt[3]{y}$ 33. $\sqrt[6]{288}$
35. $\sqrt[12]{x^{11}y^7}$ 37. $a\sqrt[6]{a^5}$ 39. $(a+x)\sqrt[12]{(a+x)^{11}}$ 41. 24.685
43. 43.138 45. 32.942 47. 5.56 ft 49. 7.07 ft 51. 34 hr, 16.3 hr, 14.5 hr, 13.2 hr, 8.2 hr, 6.2 hr 53. $\frac{b^{10}-a^5}{a^2b^5}$
55. $\frac{5a+2b}{a^{1/3}b^{1/2}}$ 57. $\frac{y-x}{x^{1/3}y^{1/4}}$ 59. $\frac{(3x-2)(x+1)^{1/4}}{(2x-3)^3}$
61. $\frac{-2(14x+19)}{(x+1)^{1/2}(3x+4)^{3/4}}$ 63. $3(x^2+1)^2(7x^2-10x+1)$
65. $\frac{-x^2-3}{x^4}$ 67. $\frac{-x^2-2}{x^3(x^2+1)^{1/2}}$ 69. $a^{a/2}$

Margin Exercises, Section 1.9

1. $18{,}600\ \frac{\text{m}}{\text{sec}}$ **2.** $0.5\ \frac{\text{m}}{\text{sec}}$ **3.** 62 ft **4.** $\frac{23}{20}$ kg **5.** $105\ \frac{\text{cm}}{\text{sec}}$ **6.** 12 yd **7.** 80 oz **8.** $\frac{7}{10}$ **9.** $11.25\ \frac{\text{in.-lb}}{\text{hr}^2}$ **10.** $4\ \frac{\text{lb}^2}{\text{m}^2}$ **11.** 1224 in. **12.** 58,080 ft **13.** 18,000 sec **14.** 20 yd **15.** 36.96 km **16.** 100 hr **17.** $176\ \frac{\text{ft}}{\text{sec}}$ **18.** $0.36\ \text{m}^2$ **19.** $50\ \frac{\text{g}}{\text{cm}^3}$ **20.** $300\ \frac{¢}{\text{hr}}$

Exercise Set 1.9, p. 60

1. 12 yd **3.** 48 hr **5.** 3 g **7.** 8 m **9.** $12\ \text{ft}^3$ **11.** $\frac{7\ \text{kg}^2}{10\ \text{m}^2}$ **13.** $720\ \frac{\text{lb-mi}^2}{\text{hr}^2\text{-ft}}$ **15.** $\frac{15}{2}\ \frac{\text{cm}^5\text{-kg}}{\text{sec}^3}$ **17.** 6 ft **19.** 172,800 sec **21.** $600\ \frac{\text{g}}{\text{cm}}$ **23.** $2{,}160{,}000\ \text{cm}^2$ **25.** $150\ \frac{¢}{\text{hr}}$ **27.** $6.228\ \frac{\text{L}}{\text{hr}}$ **29.** $5{,}865{,}696{,}000{,}000\ \frac{\text{mi}}{\text{yr}}$ **31.** $1621.8\ \frac{\text{m}}{\text{min}}$ **33.** $1638.4\ \text{km}^2$ **35.** 7.5 g, 1250 g **37.** 1600 g **39.** 15 mol **41.** $4.4937 \times 10^{20}\ \frac{\text{g m}^2}{\text{sec}^2}$ **43.** $29.979\ \frac{\text{cm}}{\text{nanosecond}}$

Review Exercises: Chapter 1, pp. 61–62

1. [1.1] 12, −3, −1, −19, 31, 0 **2.** [1.1] 12, 31 **3.** [1.1] All except $\sqrt{7}$, $\sqrt[3]{10}$ **4.** [1.1] All **5.** [1.1] $\sqrt{7}$, $\sqrt[3]{10}$ **6.** [1.1] 0, 12, 31 **7.** [1.1] −4 **8.** [1.1] −8 **9.** [1.1] −5 **10.** [1.1] 30 **11.** [1.1] −6 **12.** [1.1] 153 **13.** [1.1] −3000 **14.** [1.1] $-\frac{3}{16}$ **15.** [1.1] $\frac{31}{24}$ **16.** [1.1] 117 **17.** [1.1] −10 **18.** [1.2] 3,261,000 **19.** [1.2] 0.00041 **20.** [1.2] 277,000,000 **21.** [1.2] 0.0001009 **22.** [1.2] 1.432×10^{-2} **23.** [1.2] 4.321×10^4 **24.** [1.2] $-14a^{-2}b^7$, or $-\frac{14b^7}{a^2}$ **25.** [1.2] $6x^9y^{-6}z^6$, or $\frac{6x^9z^6}{y^6}$ **26.** [1.7] 3 **27.** [1.7] −2 **28.** [1.6] $\frac{b}{a}$ **29.** [1.6] $\frac{x+y}{xy}$ **30.** [1.7] −4 **31.** [1.7] $25x^4 - 10x^2\sqrt{2} + 2$ **32.** [1.7] $13\sqrt{5}$ **33.** [1.4] $x^3 + t^3$ **34.** [1.4] $125a^3 + 300a^2b + 240ab^2 + 64b^3$ **35.** [1.3] $8xy^4 - 9xy^2 + 4x^2 + 2y - 7$ **36.** [1.5] $(x^2 - 3)(x + 2)$ **37.** [1.5] $3a(2a - 3b^2)(2a + 3b^2)$ **38.** [1.5] $(x + 12)^2$ **39.** [1.5] $x(9x - 1)(x + 4)$ **40.** [1.5] $(2x - 1)(4x^2 + 2x + 1)$ **41.** [1.5] $(3x^2 + 5y^2)(9x^4 - 15x^2y^2 + 25y^4)$ **42.** [1.8] $y^3 \cdot \sqrt[6]{y}$ **43.** [1.8] $\sqrt[3]{(a+b)^2}$ **44.** [1.8] $\sqrt[5]{b^7}$, $b\sqrt[5]{b^2}$ **45.** [1.8] $\frac{m^4n^2}{3}$ **46.** [1.6] 3 **47.** [1.6] $\frac{x-5}{(x+3)(x+5)}$ **48.** [1.7] $\frac{x-y}{x+2\sqrt{xy}+y}$ **49.** [1.7] $\frac{x-2\sqrt{xy}+y}{x-y}$ **50.** [1.8] $\frac{2x-3y}{x^{1/2}y^{3/4}}$ **51.** [1.8] $\frac{(4x+3)(3x+5)^{3/2}}{(x-2)^{3/4}}$ **52.** [1.7] 18.8 ft **53.** [1.9] $\frac{500}{3}$, or $166\frac{2}{3}\ \frac{\text{m}}{\text{min}}$ **54.** [1.1] Inverse (+) **55.** [1.1] Distributive **56.** [1.1] Associative (×) **57.** [1.1] Commutative (×) **58.** [1.4] $x^{2n} + 6x^n - 40$ **59.** [1.4] $t^{2a} + 2 + t^{-2a}$ **60.** [1.4] $y^{2b} - z^{2c}$ **61.** [1.4] $a^{3n} - 3a^{2n}b^m + 3a^nb^{2m} - b^{3m}$ **62.** [1.5] $(y^n + 8)^2$ **63.** [1.5] $(x^t - 7)(x^t + 4)$ **64.** [1.5] $m^{3n}(m^n - 1)(m^{2n} + m^n + 1)$ **65.** [1.6] $\frac{2xn^5}{(n+1)^5}$ **66.** [1.6] $\frac{(n-1)(n-2)(n-3)(n-4)}{-24}$

Test: Chapter 1, pp. 62–63

1. [1.1] 0, 233 **2.** [1.1] $\sqrt{8}$, $-\sqrt[3]{11}$ **3.** [1.1] All **4.** [1.1] All except $\sqrt{8}$, $-\sqrt[3]{11}$ **5.** [1.1] 233 **6.** [1.1] −14, −5, 0, 233 **7.** [1.1] 0 **8.** [1.1] 2 **9.** [1.1] 12 **10.** [1.1] 8 **11.** [1.1] −3 **12.** [1.1] $\frac{726}{5}$ **13.** [1.2] 0.002834 **14.** [1.2] 470 **15.** [1.2] 4,450,000,000 **16.** [1.2] 0.0000445 **17.** [1.2] 8.16×10^{-4} **18.** [1.2] 4.8057×10^2 **19.** [1.2] $-12x^9y^{-6}$ **20.** [1.2] $\frac{2}{3}p^4q^{14}r^{-9}$ **21.** [1.7] −3 **22.** [1.7] 5 **23.** [1.7] 6 **24.** [1.6] $\frac{x+y}{x^2y}$ **25.** [1.4] $25a^4 - 20a^2b + 4b^2$ **26.** [1.3] $8x^2y - 5xy + y^2 + y - 12$ **27.** [1.4] $64y^3 - 144y^2 + 108y - 27$ **28.** [1.8] $\sqrt[20]{(c+d)^9}$ **29.** [1.8] $\sqrt[7]{t^2}$ **30.** [1.5] $4(2x+3)^2$ **31.** [1.5] $(t - 7)(t^2 + 7t + 49)$ **32.** [1.5] $m^3(m - 3n)(m + 3n)$ **33.** [1.5] $3(4p^2 - 5)(p^2 + 2)$ **34.** [1.6] $x - 5$ **35.** [1.6] $\frac{x+3}{(x+6)(x+2)}$ **36.** [1.7] $\frac{49 - 14\sqrt{x} + x}{49 - x}$ **37.** [1.4] $x^{3t} + 3x^t + 3x^{-t} + x^{-3t}$ **38.** [1.7] $\frac{49 - x}{49 + 14\sqrt{x} + x}$ **39.** [1.8] $\frac{b^{3/5}(b - a^2)}{a^{2/3}}$ **40.** [1.7] 90.8 ft **41.** [1.8] $\frac{x^2(5x+9)}{(2x+3)^{3/2}}$ **42.** [1.9] $72\ \frac{\text{km}}{\text{h}}$ **43.** [1.5] $(x^8 + 4)(x^4 + 2)(x^4 - 2)$ **44.** [1.2] $\frac{1}{2}\left|\frac{n(x-2)}{n+1}\right|$

CHAPTER 2

Margin Exercises, Section 2.1

1. {9} **2.** {0, −1} **3.** $\left\{\frac{4}{3}\right\}$ **4.** $\left\{\frac{17}{2}\right\}$ **5.** $\left\{-\frac{19}{8}\right\}$ **6.** ∅ **7.** All real numbers **8.** $\left\{7, -\frac{3}{2}\right\}$ **9.** {5, −4} **10.** {0, 5} **11.** $\left\{-\frac{1}{5}\right\}$ **12.** $\left\{0, -\frac{1}{3}, 4\right\}$ **13.** $\left\{1, -1, -\frac{1}{5}\right\}$ **14.** $\left\{x \mid x > \frac{3}{2}\right\}$ **15.** $\left\{y \mid \frac{22}{13} \le y\right\}$ **16.** $\{x \mid x < 5\}$ **17.** $\left\{x \mid x > \frac{5}{2}\right\}$ **18.** $\{y \mid y \ge -7\}$ **19.** $\{x \mid x^2 = 5\}$ **20.** $\{x \mid x \ge 2\}$ **21.** $\{x \mid x \ge -3\}$ **22.** $\left\{x \mid \frac{11}{2} \ge x\right\}$

Exercise Set 2.1, pp. 70–71

1. {12} **3.** {−6} **5.** {8} **7.** $\left\{\frac{4}{5}\right\}$ **9.** {2} **11.** $\left\{-\frac{3}{2}\right\}$ **13.** {−2} **15.** $\left\{\frac{3}{2}, \frac{2}{3}\right\}$ **17.** {0, 1, −2} **19.** $\left\{\frac{2}{3}, -1\right\}$ **21.** {4, 1} **23.** {−1, −2} **25.** $\left\{-\frac{5}{3}, 4, \frac{5}{2}\right\}$ **27.** $\left\{0, \frac{1}{4}, -\frac{1}{4}\right\}$ **29.** {0, 3} **31.** $\left\{0, -\frac{1}{3}, 2\right\}$ **33.** $\left\{\frac{3}{2}, -\frac{2}{3}, 1\right\}$ **35.** $\left\{\frac{1}{2}, 0, -3\right\}$ **37.** ∅ **39.** $\left\{-1, -\frac{1}{7}, 1\right\}$ **41.** {−2, −1, 1} **43.** All real numbers **45.** $\{x \mid x > 3\}$ **47.** $\left\{x \mid x \ge -\frac{5}{12}\right\}$ **49.** $\left\{y \mid y \ge \frac{22}{13}\right\}$ **51.** $\left\{x \mid x \le \frac{15}{34}\right\}$ **53.** $\{x \mid x < 1\}$ **55.** $\{x \mid x > 2.5\}$ **57.** $\{t \mid t^2 = 5\}$ **59.** $\{x \mid x \ge 3\}$ **61.** $\left\{x \mid \frac{3}{4} \ge x\right\}$ **63.** {0.7892} **65.** {0, 2.1522} **67.** $\{x \mid x < -0.7848\}$ **69.** {−2, 2} **71.** {−5, −4, 5}

Margin Exercises, Section 2.2

1. Yes **2.** Yes **3.** No **4.** Yes **5.** No **6.** No **7.** Add $5x^2$. **8.** Add $-5x^2$. **9.** ∅ **10.** {4} **11.** {−6, 6} **12.** $\left\{\frac{16}{5}\right\}$ **13.** {10}

Exercise Set 2.2, pp. 76–77

1. Yes **3.** No **5.** No **7.** $\{\frac{20}{9}\}$ **9.** $\emptyset$ **11.** $\{286\}$ **13.** $\{3, 2\}$ **15.** $\{-2\}$ **17.** $\{6\}$ **19.** $\emptyset$ **21.** $\emptyset$ **23.** $\{8, -5\}$ **25.** $\{\frac{5}{3}\}$ **27.** $\emptyset$ **29.** $\{0.94656\}$ **31.** $\{-5\}$ **33.** $\{2\}$ **35.** $\{-\frac{19}{5}\}$ **37.** All real numbers **39.** $\{x \mid x \neq 3\}$ **41.** $\{x \mid x \neq -2\}$ **43.** (1) equivalent to (2); (2) not equivalent to (3); (3) equivalent to (4) **45.** Identity **47.** Identity **49.** Not an identity

Margin Exercises, Section 2.3

1. $F = \frac{9}{5}C + 32$ **2.** $r_2 = \dfrac{Rr_1}{r_1 - R}$ **3.** 18% **4.** 57 **5.** 27 **6.** \$2600 **7.** \$2565.78 **8.** \$2637.93 **9.** 20 ft **10.** 36 km/h **11.** 375 km **12.** 50 km/h, 60 km/h **13.** $2\frac{2}{9}$ hr **14.** Helen: 12 hr; Fran: 6 hr

Exercise Set 2.3, pp. 87–90

1. $w = \dfrac{P - 2l}{2}$ **3.** $b = \dfrac{2A}{h}$ **5.** $r = \dfrac{d}{t}$ **7.** $I = \dfrac{E}{R}$ **9.** $T_1 = \dfrac{T_2P_1V_1}{P_2V_2}$ **11.** $v_1 = \dfrac{H}{Sm} + v_2$ **13.** $p = \dfrac{Fm}{m - F}$ **15.** $x = \dfrac{5 + ab}{a - b}$ **17.** $x = -\dfrac{a}{9}$ **19.** 44% **21.** 6% **23.** \$14,500; \$16,095 **25.** \$650 **27.** 26°, 130°, 24° **29.** 68 m, 93 m **31.** 91% **33.** 2 cm **35.** 810,000 **37.** 12 km/h **39.** A: 46 mph; B: 58 mph **41.** 98.3 mi **43.** $1\frac{34}{71}$ hr **45.** 6.21 hr **47.** (a) \$1137.50; (b) \$1142.23; (c) \$1144.75; (d) \$1147.37; (e) \$1147.40 **49.** 32 mph **51.** $53\frac{6}{23}$ mph **53.** $51\frac{3}{7}$ mph **55.** $10{:}38\frac{2}{11}$ **57.** \$44,926 **59.** (a) 1201.2 mi; (b) less time to return to Los Angeles **61.** $k = \frac{100}{3}$ **63.** \$16

Margin Exercises, Section 2.4

1. $i\sqrt{6}$, or $\sqrt{6}i$ **2.** $-i\sqrt{10}$, or $-\sqrt{10}i$ **3.** $2i$ **4.** $-5i$ **5.** $-\sqrt{10}$ **6.** $\sqrt{11}$ **7.** $i\sqrt{7}$ **8.** $7i$ **9.** $3i$ **10.** $(\sqrt{17} + 3)i$ **11.** i **12.** -1 **13.** $-i$ **14.** $12 + i$ **15.** $5 - i$ **16.** $2 + 14i$ **17.** 8 **18.** $-6 + 8i$ **19.** $3i$ **20.** $(x + 2i)(x - 2i)$ **21.** $(3 + yi)(3 - yi)$ **22.** Yes **23.** $x = -1, y = 2$ **24.** $7 - 2i$ **25.** $6 + 4i$ **26.** $5i$ **27.** $-3i$ **28.** -3 **29.** 8 **30.** $\dfrac{9}{13} + \dfrac{7}{13}i$ **31.** $\dfrac{4}{13} + \dfrac{7}{13}i$ **32.** $\dfrac{1}{3 + 4i}, \dfrac{3}{25} - \dfrac{4}{25}i$ **33.** $2 + 5i$

Exercise Set 2.4, pp. 96–97

1. $i\sqrt{15}$ **3.** $9i$ **5.** $-2i\sqrt{3}$ **7.** $9i$ **9.** $i(\sqrt{7} - \sqrt{10})$ **11.** $-\sqrt{55}$ **13.** $2\sqrt{5}$ **15.** $\sqrt{\dfrac{5}{2}}i$ **17.** $-\dfrac{3}{2}i$ **19.** -2 **21.** -1 **23.** $-i$ **25.** $-i$ **27.** -1 **29.** $6 + 5i$ **31.** 8 **33.** $2 + 4i$ **35.** $-4 - i$ **37.** $-5 + 5i$ **39.** $7 - i$ **41.** $-6 + 12i$ **43.** $-5 + 12i$ **45.** $(2x + 5yi)(2x - 5yi)$ **47.** Yes **49.** $x = -\dfrac{3}{2}, y = 7$ **51.** $\dfrac{1}{2} + \dfrac{7}{2}i$ **53.** $\dfrac{1}{3} + \dfrac{2\sqrt{2}}{3}i$ **55.** $2 - 3i$ **57.** $\dfrac{1}{5} + \dfrac{2}{5}i$ **59.** $-\dfrac{1}{2} - \dfrac{1}{2}i$ **61.** $\dfrac{28}{65} - \dfrac{29}{65}i$ **63.** $-\dfrac{1}{2} + \dfrac{3}{2}i$ **65.** $\dfrac{5}{2} + \dfrac{13}{2}i$ **67.** $\dfrac{4}{25} - \dfrac{3}{25}i$ **69.** $\dfrac{5}{29} + \dfrac{2}{29}i$ **71.** $-i$ **73.** $\dfrac{1}{4}i$ **75.** $\dfrac{2}{5} + \dfrac{6}{5}i$ **77.** $\dfrac{8}{5} - \dfrac{9}{5}i$ **79.** $2 - i$ **81.** $\dfrac{11}{25} + \dfrac{2}{25}i$ **83.** For example, $\sqrt{-1}\sqrt{-1} = i^2 = -1$, but $\sqrt{(-1)(-1)} = \sqrt{1} = 1$. **85.** Let $z = a + bi$. Then $z \cdot \bar{z} = (a + bi)(a - bi) = a^2 - b^2i^2 = a^2 + b^2$. Since a and b are real numbers, so is $a^2 + b^2$. Thus, $z \cdot \bar{z}$ is real. **87.** Let $z = a + bi$ and $w = c + di$. Then $z + w = (a + bi) + (c + di) = (a + c) + (b + d)i$, by adding. We now take the conjugate and obtain $(a + c) - (b + d)i$. Now $\bar{z} + \bar{w} = \overline{(a + bi)} + \overline{(c + di)} = (a - bi) + (c - di)$, taking the conjugates. We will now add to obtain $(a + c) - (b + d)i$, the same result as before. Thus, $\overline{z + w} = \bar{z} + \bar{w}$. **89.** By the definition of exponents, the conjugate of z^n is the conjugate of the product of n factors of z. Using the result of Exercise 88, we see that the conjugate of n factors of z is the product of n factors of $\bar{z}$. Thus, $\overline{z^n} = \bar{z}^n$. **91.** $3\bar{z}^5 - 4\bar{z}^2 + 3\bar{z} - 5$ **93.** $7 + \dfrac{8}{9}i$ **95.** $-bi$ **97.** $\sqrt{2} + i\sqrt{2}, -\sqrt{2} - i\sqrt{2}$ **99.** $\dfrac{ac + bd}{a^2 + b^2} + \dfrac{ad - bc}{a^2 + b^2}i$

Margin Exercises, Section 2.5

1. $\{\pm\sqrt{3}\}$ **2.** $\{0\}$ **3.** $\left\{\pm\sqrt{\dfrac{\pi}{3}}\right\}$ **4.** $\left\{\pm\sqrt{\dfrac{n}{m}}\right\}$ **5.** $\left\{\pm\dfrac{\sqrt{2}}{2}i\right\}$ **6.** $\{-4 \pm \sqrt{7}\}$ **7.** $\{5 \pm \sqrt{3}\}$ **8.** $\{-3, -7\}$ **9.** $4, (x + 2)^2$ **10.** $9, (x - 3)^2$ **11.** $\dfrac{25}{4}, \left(x + \dfrac{5}{2}\right)^2$ **12.** $\dfrac{49}{4}, \left(x - \dfrac{7}{2}\right)^2$ **13.** $\dfrac{9}{64}, \left(x + \dfrac{3}{8}\right)^2$ **14.** $\dfrac{1}{4}, \left(x - \dfrac{1}{2}\right)^2$ **15.** $\{-2 \pm \sqrt{7}\}$ **16.** $\{4, 2\}$ **17.** $\{2, 3\}$ **18.** $\left\{\dfrac{-1 \pm \sqrt{7}}{2}\right\}$ **19.** $\left\{-1, \dfrac{1}{4}\right\}$ **20.** $\left\{\dfrac{1}{2}, -4\right\}$ **21.** $\left\{\dfrac{4 \pm \sqrt{31}}{5}\right\}$ **22.** $\left\{\dfrac{1 \pm i\sqrt{7}}{2}\right\}$ **23.** Two real **24.** One real **25.** Two nonreal **26.** $3x^2 + 7x - 20 = 0$ **27.** $x^2 + \sqrt{2}x - 4 = 0$ **28.** $x^2 + 25 = 0$

Exercise Set 2.5, pp. 104–105

1. $\{\pm 3\}$ **3.** $\{\pm i\}$ **5.** $\{\pm\sqrt{5}\}$ **7.** $\{0\}$ **9.** $\left\{\pm\dfrac{\sqrt{6}}{2}\right\}$ **11.** $\{\pm i\sqrt{7}\}$ **13.** $\left\{\pm\sqrt{\dfrac{b}{a}}\right\}$ **15.** $\{7 \pm \sqrt{5}\}$ **17.** $\left\{\pm\dfrac{3}{2}\right\}$ **19.** $\{h \pm \sqrt{a + 1}\}$ **21.** $\{-3 \pm \sqrt{5}\}$ **23.** $\{-10, 3\}$ **25.** $\left\{\dfrac{2 \pm \sqrt{14}}{5}\right\}$ **27.** $\left\{-5, \dfrac{3}{2}\right\}$ **29.** $\{-5, 1\}$ **31.** $\left\{-\dfrac{1}{2}, 2\right\}$ **33.** $\left\{\dfrac{-4 \pm \sqrt{7}}{3}\right\}$ **35.** $\{6 \pm \sqrt{33}\}$ **37.** $\left\{\dfrac{1 \pm i\sqrt{3}}{2}\right\}$ **39.** $\{2 \pm 3i\}$ **41.** $\left\{\dfrac{13 \pm \sqrt{509}}{10}\right\}$ **43.** $\{0.04 \pm 0.02\sqrt{79}\}$ **45.** $\left\{\dfrac{-19 \pm \sqrt{445}}{14}\right\}$ **47.** $\left\{\dfrac{-1 \pm i\sqrt{15}}{2}\right\}$ **49.** One real **51.** Two nonreal **53.** Two real **55.** One real **57.** Two nonreal **59.** Two real **61.** Two real **63.** One real **65.** $x^2 + 2x - 99 = 0$ **67.** $x^2 - 14x + 49 = 0$ **69.** $x^2 - \dfrac{4}{5}x - \dfrac{12}{25} = 0$, or $25x^2 - 20x - 12 = 0$ **71.** $x^2 - \left(\dfrac{c + d}{2}\right)x + \dfrac{cd}{4} = 0$ **73.** $x^2 - 4\sqrt{2}x + 6 = 0$

75. $x^2 + 9 = 0$ **77.** $\{1.1754, -0.4254\}$ **79.** $\left\{\frac{3}{2}, \frac{2}{3}\right\}$

81. $\{-0.1 \pm \sqrt{0.31}\}$ **83.** $\left\{\frac{-1 \pm \sqrt{1 + 4\sqrt{2}}}{2}\right\}$

85. $\left\{\frac{-\sqrt{5} \pm \sqrt{5 + 4\sqrt{3}}}{2}\right\}$ **87.** $\left\{\frac{\sqrt{6} \pm \sqrt{6 + 8\sqrt{10}}}{4}\right\}$

89. $\left\{-2, \frac{3}{4}\right\}$ **91.** $\left\{\frac{1 \pm \sqrt{113}}{2}\right\}$ **93.** $\{3 \pm \sqrt{5}\}$

95. **(a)** $\frac{-b + \sqrt{b^2 - 4ac}}{2a} + \frac{-b - \sqrt{b^2 - 4ac}}{2a} = \frac{-2b}{2a} = -\frac{b}{a}$; **(b)** $\frac{-b + \sqrt{b^2 - 4ac}}{2a} \cdot \frac{-b - \sqrt{b^2 - 4ac}}{2a} =$
$\frac{b^2 - (b^2 - 4ac)}{4a^2} = \frac{4ac}{4a^2} = \frac{c}{a}$ **97.** **(a)** $k = -\frac{3}{5}$; **(b)** $-\frac{1}{3}$

99. **(a)** $k = 9 + 9i$; **(b)** $3 + 3i$ **101.** $x^2 - \sqrt{3}x + 8 = 0$
103. $h = -36, k = 15$ **105.** 2

Margin Exercises, Section 2.6

1. $r = \sqrt{\frac{3V}{\pi h}}$ **2.** $t = \frac{-v_0 + \sqrt{v_0^2 + 64S}}{32}$ **3.** 18.75%
4. $12 - 2\sqrt{22} \approx 2.619$ ft **5.** **(a)** 4.33 sec; **(b)** 1.87 sec; **(c)** 44.9 m

Exercise Set 2.6, pp. 109–111

1. $d = \sqrt{\frac{kM_1M_2}{F}}$ **3.** $t = \sqrt{\frac{2S}{a}}$ **5.** $t = \frac{v_0 \pm \sqrt{v_0^2 - 64S}}{32}$
7. $n = \frac{3 + \sqrt{9 + 8d}}{2}$ **9.** $i = -1 + \sqrt{\frac{A}{P}}$ **11.** 18.75%
13. 11% **15.** 9 **17.** $24 - 4\sqrt{34} \approx 0.676$ ft **19.** 2 ft
21. 4.685 cm **23.** A: 15 mph; B: 20 mph **25.** **(a)** 3.91 sec; **(b)** 1.906 sec; **(c)** 79.6 m **27.** 3.237 cm **29.** $35 - 5\sqrt{33}$, or 6.28 ft **31.** $\frac{15 - \sqrt{115}}{2}$, or 2.14 cm **33.** First part: $\frac{75 + 5\sqrt{185}}{2}$, or 71.5 mph; second part: $\frac{65 + 5\sqrt{185}}{2}$, or 66.5 mph

35. $6 + 3\sqrt{5}$, or 12.7 mph **37.** 7 **39.** 12 **41.** $2, -\frac{3}{k}$

43. $\frac{1}{m + n}, \frac{-2}{m + n}$ **45.** 11.7% **47.** $a_3 = \sqrt{a_1^2 + a_2^2}$
49. $x = 4$ cm, $y = 3$ cm

Margin Exercises, Section 2.7

1. $\emptyset$ **2.** $\{4\}$ **3.** $\left\{\frac{17}{3}\right\}$ **4.** $\{9\}$ **5.** $\{5\}$ **6.** $m = \sqrt{\frac{1}{1 - P^2}}$, or $\frac{\sqrt{1 - P^2}}{1 - P^2}$

Exercise Set 2.7, pp. 114–115

1. $\left\{\frac{5}{3}\right\}$ **3.** $\{\pm\sqrt{2}\}$ **5.** $\emptyset$ **7.** $\{4\}$ **9.** $\emptyset$ **11.** $\{-6\}$

13. $\{3, -1\}$ **15.** $\left\{\frac{80}{9}\right\}$ **17.** $\{1\}$ **19.** $\{7, 3\}$

21. $\{62.4459\}$ **23.** $\{-8\}$ **25.** $\{81\}$ **27.** $\left\{\frac{1}{64}\right\}$

29. $\{-125\}$ **31.** $L = \frac{gT^2}{4\pi^2}, g = \frac{4L\pi^2}{T^2}$ **33.** 208 mi

35. 14,400 ft **37.** $\{5 \pm 2\sqrt{2}\}$ **39.** $\left\{-\frac{8}{9}\right\}$ **41.** $\{2\}$

43. $\left\{\frac{-5 + \sqrt{61}}{18}\right\}$ **45.** $\{9\}$ **47.** $\{8\}$ **49.** $\{10\}$

Margin Exercises, Section 2.8

1. **(a)** 9; **(b)** $\sqrt{x} = 12 - x$; $(12 - x)^2 = x$; $x = 144 - 24x + x^2$; $0 = x^2 - 25x + 144$; $0 = (x - 9)(x - 16)$. The procedure in (a) was probably easier, since the factoring was easier.

2. $\pm\sqrt{\frac{5 + \sqrt{3}}{2}}, \pm\sqrt{\frac{5 - \sqrt{3}}{2}}$ **3.** $\pm\sqrt{3}, 0$ **4.** $125, -8$
5. $\pm 3i, \pm 2$ **6.** 350.63 ft

Exercise Set 2.8, pp. 119–120

1. $\{1, 81\}$ **3.** $\{\pm\sqrt{5}\}$ **5.** $\{-27, 8\}$ **7.** $\{16\}$
9. $\{7, 5, -1, 1\}$ **11.** $\left\{1, 4, \frac{5 \pm \sqrt{37}}{2}\right\}$

13. $\{\pm\sqrt{2 + \sqrt{6}}, \pm\sqrt{2 - \sqrt{6}}\}$
15. $\left\{-\frac{1}{2}, \frac{1}{3}\right\}$ **17.** $\{-1, 2\}$ **19.** $\{\pm 5, \pm i\}$

21. $\left\{-1 \pm \sqrt{3}, \frac{9 \pm \sqrt{89}}{2}\right\}$ **23.** $\left\{\frac{100}{99}\right\}$ **25.** $\left\{-\frac{6}{7}\right\}$

27. $\left\{1 \pm \sqrt{2}, \frac{-1 \pm \sqrt{5}}{2}\right\}$ **29.** 132.66 ft **31.** 7%

33. $\{2.0485\}$ **35.** $\{1, 4\}$ **37.** $\{19\}$
39. $\left\{1, 3, \frac{-1 \pm i\sqrt{3}}{2}, \frac{-3 \pm 3i\sqrt{3}}{2}\right\}$ **41.** $\left\{-1, 6, \frac{5 \pm \sqrt{37}}{2}\right\}$
43. $\left\{-2 \pm \sqrt{2}, \frac{1 \pm i\sqrt{7}}{2}\right\}$ **45.** $\left\{\frac{-51 + 7\sqrt{61}}{194}\right\}$

Margin Exercises, Section 2.9

1. $y = 160x$ **2.** 4.5 kg **3.** 50 volts **4.** 176,250 tons
5. $y = \frac{6.4}{x}$ **6.** 7.5 hr **7.** $y = 3x^2$ **8.** $y = \frac{9}{x^2}$ **9.** $y = 7xz$

10. $y = 7\frac{xz}{w^2}$ **11.** 2 sec **12.** **(a)** 128 lb; **(b)** 4000 mi

Exercise Set 2.9, pp. 125–126

1. $y = \frac{3}{2}x$ **3.** $y = \frac{4000}{x}$ **5.** $y = 5.375x$ **7.** $y = \frac{0.0015}{x^2}$

9. $y = \frac{xz}{w}$ **11.** $y = \frac{5}{4} \cdot \frac{xz}{w^2}$ **13.** y is doubled. **15.** y is multiplied by $\frac{1}{n^2}$. **17.** 532,500 tons **19.** L is multiplied by 16. **21.** 68.56 m **23.** 624.24 m² **25.** 97 **27.** If p varies directly as q, then $p = kq$. Thus, $q = \frac{1}{k}p$, so q varies directly as p. **29.** $\frac{\pi}{4}$ **31.** **(a)** $N = 0.001\frac{P_1P_2}{d^2}$; **(b)** 5658; **(c)** 1173 km; **(d)** Division by 0 is not defined.

Review Exercises: Chapter 2, pp. 127–128

1. [2.4] $-2\sqrt{10}i$ **2.** [2.4] $-4\sqrt{15}$ **3.** [2.4] $14 + 2i$
4. [2.4] $1 - 4i$ **5.** [2.4] $2 - i$ **6.** [2.4] $\frac{11}{10} + \frac{3}{10}i$ **7.** [2.4]

No 8. [2.4] $\frac{6}{85} + \frac{7}{85}i$ 9. [2.4] $x = 2, y = -4$ 10. [2.4] $\left\{-\frac{7}{15} + \frac{3}{5}i\right\}$ 11. [2.2] $\{-1\}$ 12. [2.1] $\left\{3, -\frac{2}{3}, -2\right\}$ 13. [2.1] $\left\{\frac{4}{3}, -2\right\}$ 14. [2.5] $\{1 \pm 3i\}$ 15. [2.5] $\{-5, -2, 2\}$ 16. [2.2] $\left\{\frac{27}{7}\right\}$ 17. [2.8] $\left\{\pm\sqrt{\frac{3+\sqrt{5}}{2}}, \pm\sqrt{\frac{3-\sqrt{5}}{2}}\right\}$ 18. [2.8] $\{1\}$ 19. [2.8] $\{\pm\sqrt{3}, 0\}$ 20. [2.8] $\{-8, 125\}$ 21. [2.7] $\{5\}$ 22. [2.7] $\{0, 3\}$ 23. [2.5] $\{8, -2\}$ 24. [2.1] $\{6, -3\}$ 25. [2.1] $\{-20\}$ 26. [2.1] $\{-5, 3\}$ 27. [2.1] $\{-2, 1\}$ 28. [2.1] $\{y|y > -2\}$ 29. [2.1] $\{x|x \geq 5\}$ 30. [2.1] $\{x|x \leq 4\}$ 31. [2.5] Two nonreal solutions 32. [2.5] Two real solutions 33. [2.5] $x^2 + \frac{5}{2}x - \frac{3}{2} = 0$, or $2x^2 + 5x - 3 = 0$ 34. [2.5] $x^2 - 2x + 5 = 0$ 35. [2.7] $h = \frac{v^2}{2g}$ 36. [2.3] $t = \frac{ab}{a+b}$ 37. [2.3] 94% 38. [2.3] $1\frac{1}{3}$ hr 39. [2.3] $1\frac{1}{2}$ hr 40. [2.3] 60 41. [2.3] 4.5 42. [2.6] 80 km/h 43. [2.6] 8, 15, 17 44. [2.6] $2 + 2\sqrt{2} \approx 4.8$ km/h 45. [2.9] $y = \frac{0.5}{x^2}$ 46. [2.9] $T = \frac{1}{180} \cdot \frac{x^2}{p}$ 47. [2.9] \$2.27 per share 48. [2.9] $s = 16t^2$; $7\frac{1}{2}$ sec 49. [2.2] No 50. [2.2] No 51. [2.5] $-(a + c)$ 52. [2.2] Yes 53. [2.9] $A = \frac{1}{4\pi} \cdot C^2$; $\frac{1}{4\pi}$ 54. [2.7] $\{256\}$ 55. [2.2] No 56. [2.4] $x = 2 - i$, $y = -1 - 3i$

Test: Chapter 2, p. 129

1. [2.4] $-i$ 2. [2.4] $4\sqrt{3}$ 3. [2.4] $18 + 26i$ 4. [2.4] $4 + 14i$ 5. [2.4] $\frac{16}{13} + \frac{11}{13}i$ 6. [2.4] 2 7. [2.4] $\frac{1}{5} - \frac{1}{10}i$ 8. [2.4] $x = 2, y = -3$ 9. [2.1] $\left\{\frac{9}{2}, -4, 5\right\}$ 10. [2.8] $\{1, 36\}$ 11. [2.1] $\left\{\frac{7}{2}, -3\right\}$ 12. [2.5] $\left\{\frac{3 \pm \sqrt{89}}{8}\right\}$ 13. [2.2] $\{11\}$ 14. [2.5] $\left\{\frac{2 \pm 2i\sqrt{14}}{5}\right\}$ 15. [2.7] $\{15\}$ 16. [2.2] $\{0\}$ 17. [2.5] $\{8, -4\}$ 18. [2.1] $\{y|y > -2\}$ 19. [2.1] $\{6, -7\}$ 20. [2.1] $\{0\}$ 21. [2.3] 200% 22. [2.3] 24 hr 23. [2.3] 12 mph 24. [2.5] $\{2 \pm \sqrt{6}\}$ 25. [2.5] Two real 26. [2.5] $x^2 + 25 = 0$ 27. [2.3] $T_2 = \frac{S_1T_1W_2}{W_1S_2}$ 28. [2.6] 30 ft, 40 ft 29. [2.9] $y = \frac{6xw}{z^2}$ 30. [2.9] 16 cm 31. [2.2] Yes 32. [2.7] $\{x|2 \geq x\}$ 33. [2.2, 2.5] $\left\{\frac{-1 \pm \sqrt{5}}{2}\right\}$ 34. [2.7] $\{10 + 2\sqrt{21}\}$

CHAPTER 3

Margin Exercises, Section 3.1

1. {(1983, \$35,883), (1984, \$38,828), (1985, \$40,715), (1987, \$48,440)} 2. {(DeConcini, Arizona), (McCain, Arizona), (Mack, Florida), (Graham, Florida), (Dixon, Illinois), (Simon, Illinois), (Hatfield, Oregon), (Packwood, Oregon)} 3. (a) $\{(d, 1), (d, 2), (e, 1), (e, 2), (f, 1), (f, 2)\}$; (b) $\{(1, d), (1, e), (1, f), (2, d), (2, e), (2, f)\}$ 4. {(1, 1), (1, 2), (1, 3), (1, 4), (2, 1), (2, 2), (2, 3), (2, 4), (3, 1), (3, 2), (3, 3), (3, 4), (4, 1), (4, 2), (4, 3), (4, 4)} 5. {(−2, −2), (−2, −1), (−2, 0), (−2, 5), (−2, 7), (−1, −2), (−1, −1), (−1, 0), (−1, 5), (−1, 7), (0, −2), (0, −1), (0, 0), (0, 5), (0, 7), (5, −2), (5, −1), (5, 0), (5, 5), (5, 7), (7, −2), (7, −1), (7, 0), (7, 5), (7, 7)} 6. {(1, 1), (2, 2), (3, 3), (4, 4)} 7. {(2, 1), (3, 1), (3, 2), (4, 1), (4, 2), (4, 3)} 8. $\{(1, a), (2, a), (3, a)\}$, answers may vary. 9. Domain: {1983, 1984, 1985, 1987}; range: {\$35,883, \$38,828, \$40,715, \$48,440} 10. Domain: {DeConcini, McCain, Mack, Graham, Dixon, Simon, Hatfield, Packwood}; range: {Arizona, Florida, Illinois, Oregon} 11. Domain: {2, 3, 4}; range: {1, 2, 3} 12. Domain: {2, −4, −6}; range: {2, 3, 5, 7}

Exercise Set 3.1, pp. 135–136

1. {(New York, Mets), (New York, Giants), (Atlanta, Braves), (Atlanta, Falcons), (Houston, Astros), (Houston, Oilers), (San Diego, Padres), (San Diego, Chargers)} 3. {(−3, 3), (3, 3), (−2, 2), (2, 2), (−1, 1), (1, 1), (0, 0)} 5. $\{(0, a), (0, b), (0, c), (2, a), (2, b), (2, c), (3, a), (3, b), (3, c), (4, a), (4, b), (4, c), (5, a), (5, b), (5, c)\}$ 7. $\{(x, 1), (x, 2), (y, 1), (y, 2), (z, 1), (z, 2)\}$ 9. {(5, 5), (5, 6), (5, 7), (5, 8), (6, 5), (6, 6), (6, 7), (6, 8), (7, 5), (7, 6), (7, 7), (7, 8), (8, 5), (8, 6), (8, 7), (8, 8)} 11. (a) $\{(0, a), (0, b), (0, c), (2, a), (2, b), (2, c)\}$; (b) $\{(a, 0), (a, 2), (b, 0), (b, 2), (c, 0), (c, 2)\}$; (c) {(0, 0), (0, 2), (2, 0), (2, 2)}; (d) $\{(a, a), (a, b), (a, c), (b, a), (b, b), (b, c), (c, a), (c, b), (c, c)\}$ 13. {(−7, −3), (−7, 1), (−7, 2), (−7, 5), (−3, 1), (−3, 2), (−3, 5), (1, 2), (1, 5), (2, 5)} 15. {(−7, −7), (−7, −3), (−7, 1), (−7, 2), (−7, 5), (−3, −3), (−3, 1), (−3, 2), (−3, 5), (1, 1), (1, 2), (1, 5), (2, 2), (2, 5), (5, 5)} 17. {(−7, −7), (−3, −3), (1, 1), (2, 2), (5, 5)} 19. Answers will vary. 21. Domain: {5, 6, 8}; range: {2, 4, 6} 23. Domain: {6, 7, 8, −4}; range: {0, 5, −7} 25. Domain: {8, −8, 5, −3}; range: {1} 27. Domain: {5}; range: {−6} 29. Domain: {New York, Atlanta, Houston, San Diego}; range: {Mets, Giants, Braves, Falcons, Astros, Oilers, Padres, Chargers} 31. Domain: {−3, 3, −2, 2, −1, 1, 0}; range: {3, 2, 1, 0} 33. Domain: {−7, −3, 1, 2}; range: {−3, 1, 2, 5} 35. Domain: {−7, −3, 1, 2, 5}; range: {−7, −3, 1, 2, 5} 37. (a) {(−1, −1), (−1, 0), (−1, 1), (−1, 2), (0, −1), (0, 0), (0, 1), (0, 2), (1, −1), (1, 0), (1, 1), (1, 2), (2, −1), (2, 0), (2, 1), (2, 2)}; (c) Domain: {0, 1}, range: {0, 1, 2} 39. {(2, 4), (2, 5), (3, 5)}

Margin Exercises, Section 3.2

1. (a)

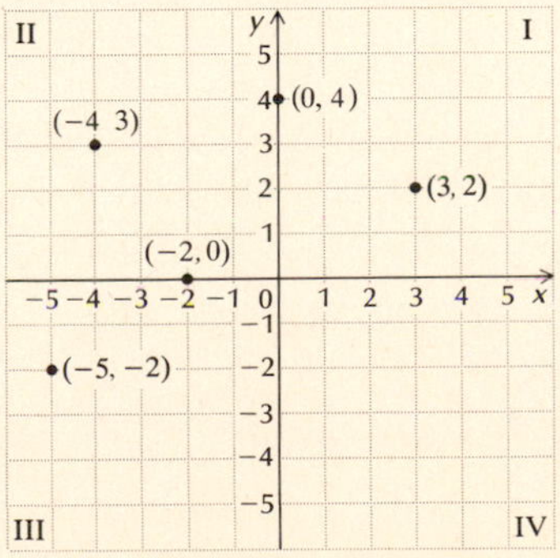

(b) Domain: {3, −5, −4, −2, 0}; range: {2, −2, 3, 0, 4} 2. Yes 3. No 4. No 5. Yes

6.

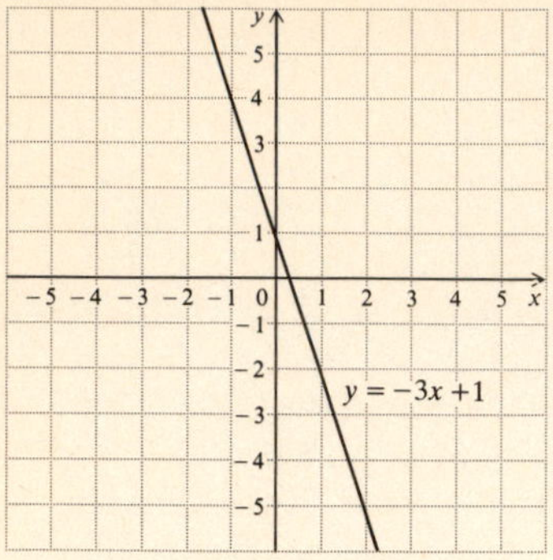

7.

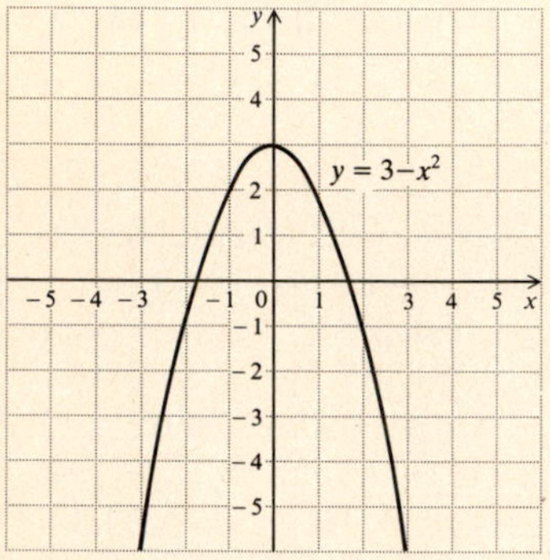

8. The shapes are the same, but this curve opens to the right instead of up.

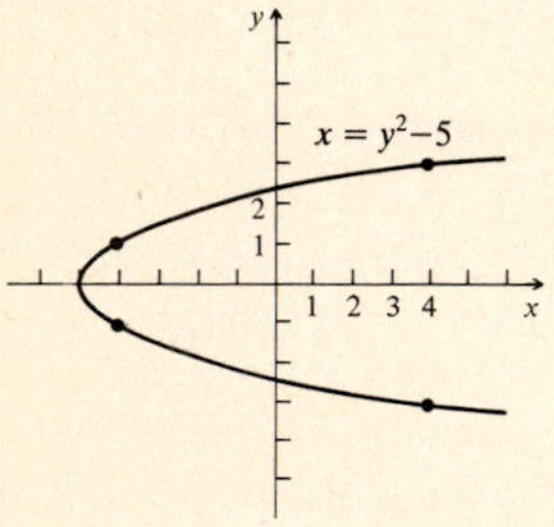

9.

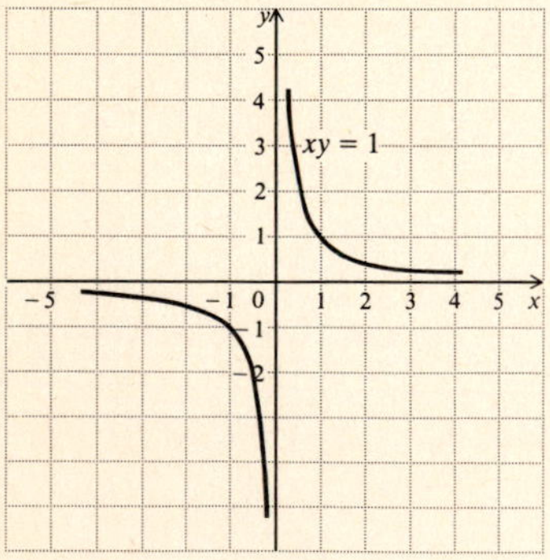

10. The shapes are the same, but this graph opens to the right instead of up.

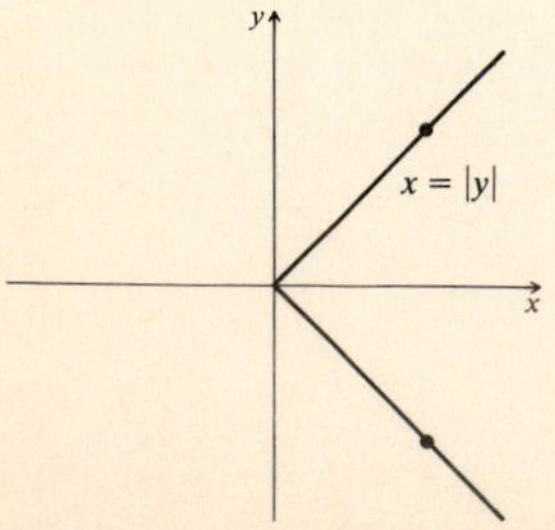

11.

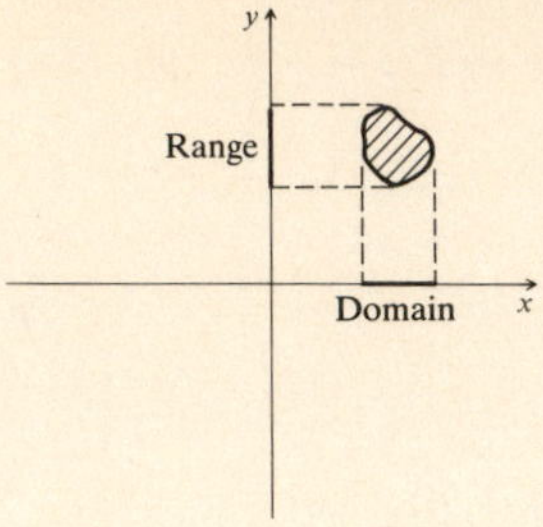

12.

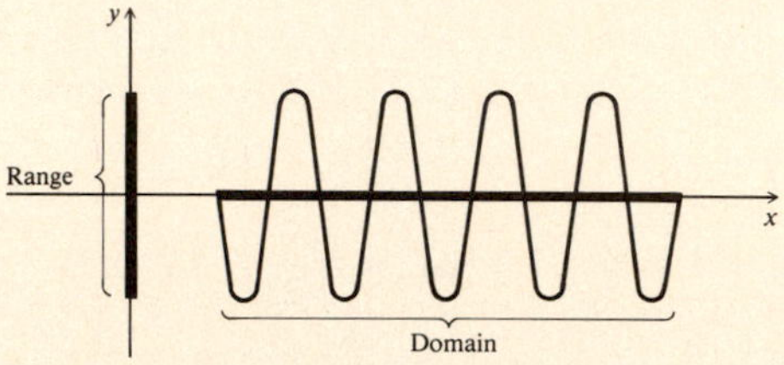

13.

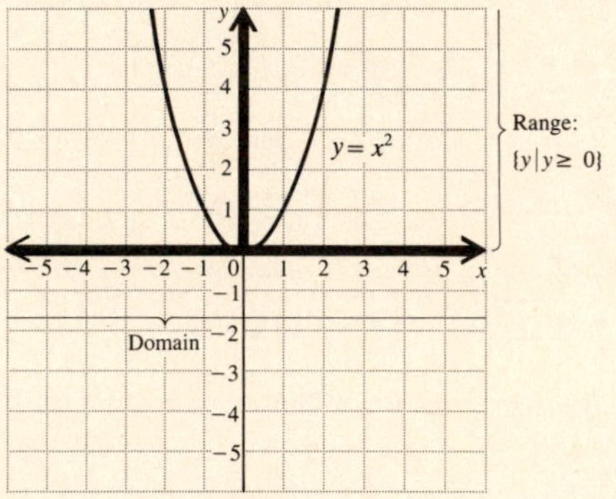

Exercise Set 3.2, pp. 143–144

1.

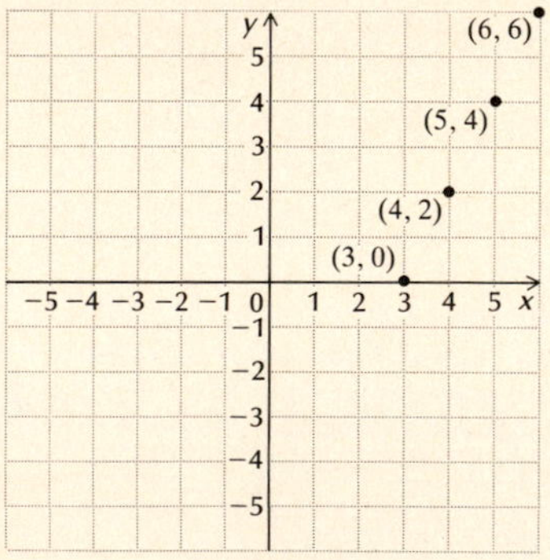

3.

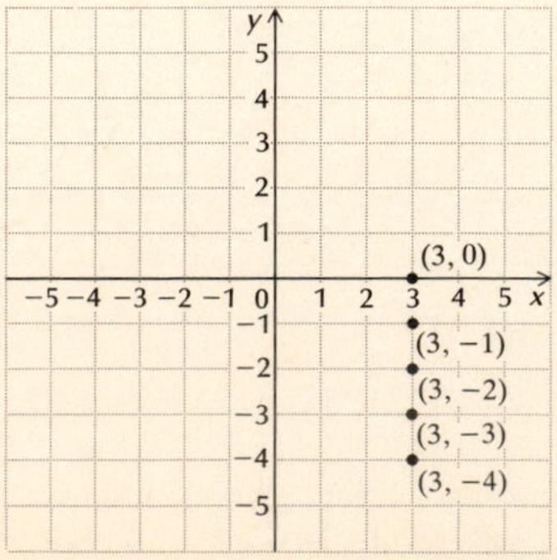

5.

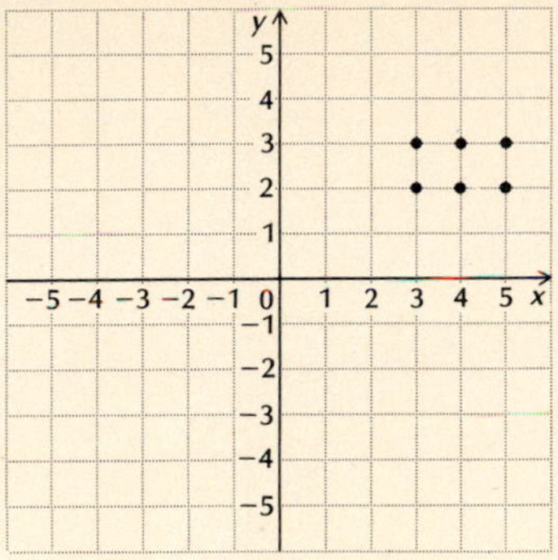

7.

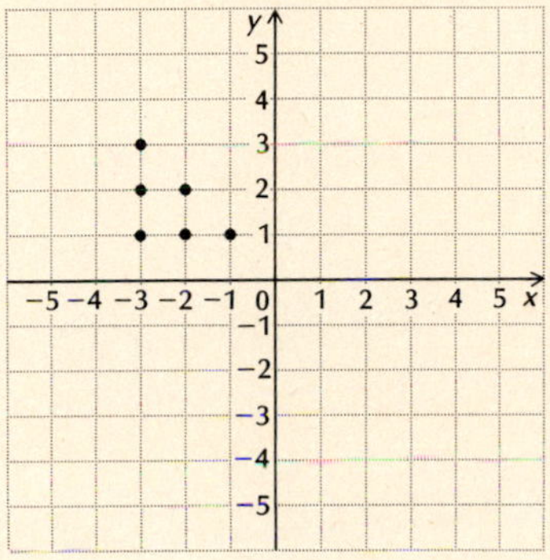

9. Yes, no **11.** No, no
13. Yes, yes **15.** Yes, no
17. Yes, no **19.** No, yes

21.

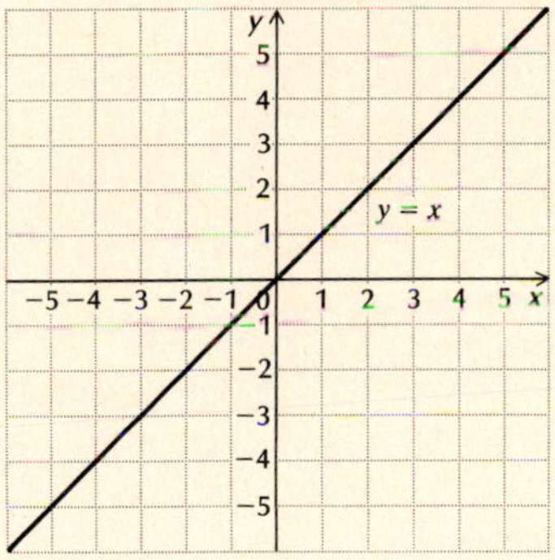

23.

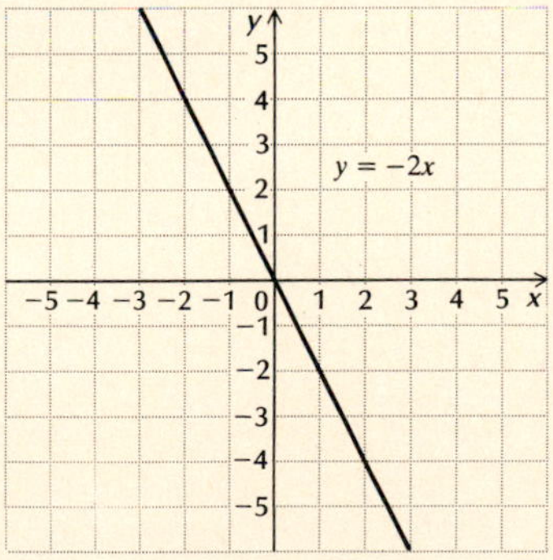

25.

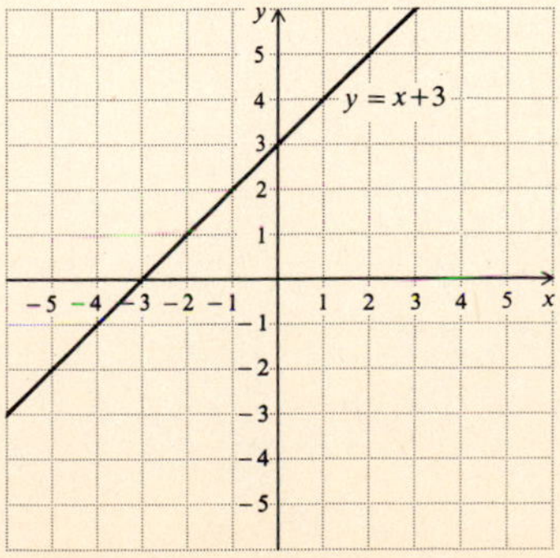

27.

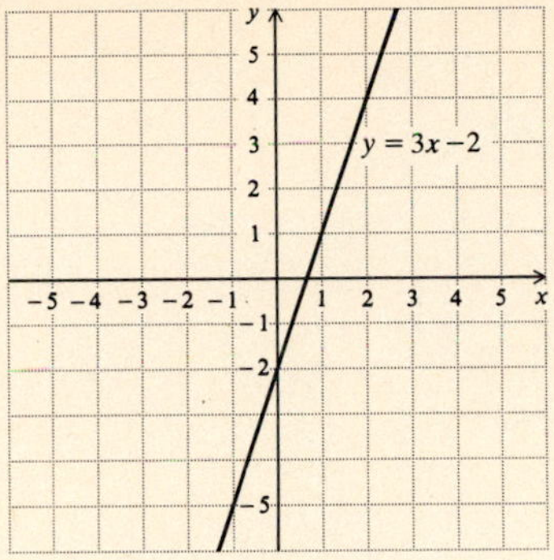

29.

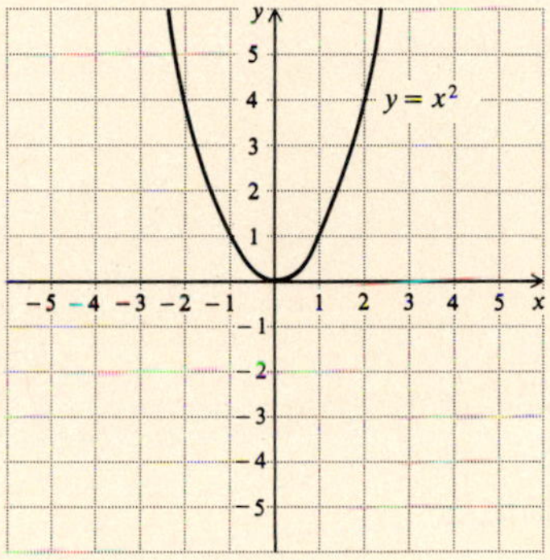

31.

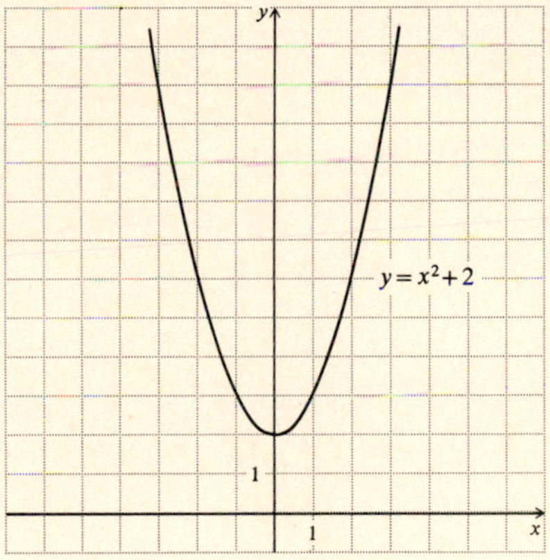

33.

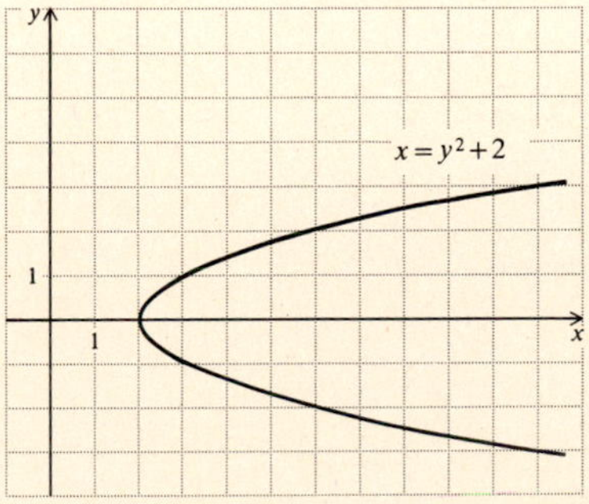

35.

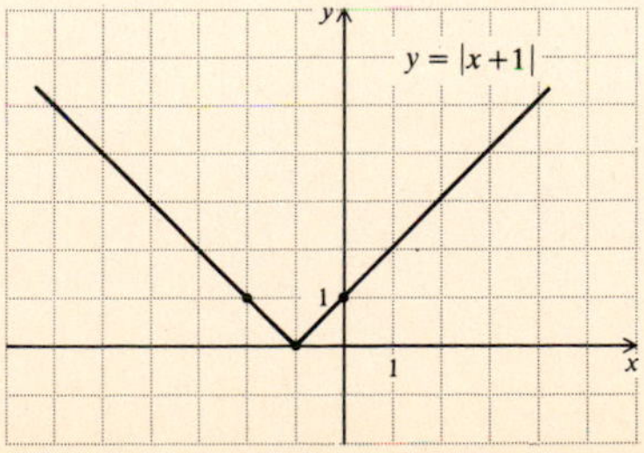

37.

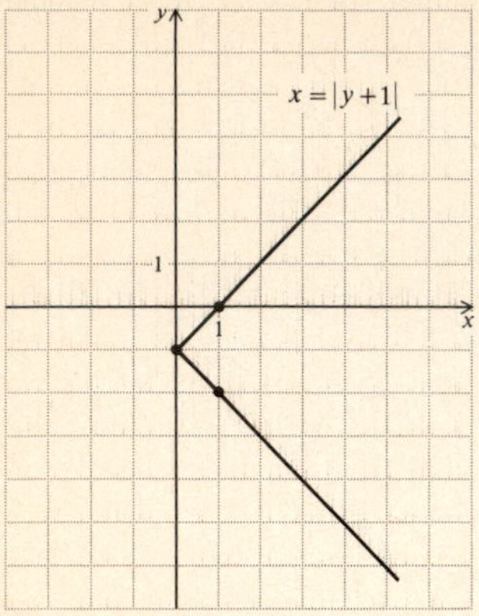

39.

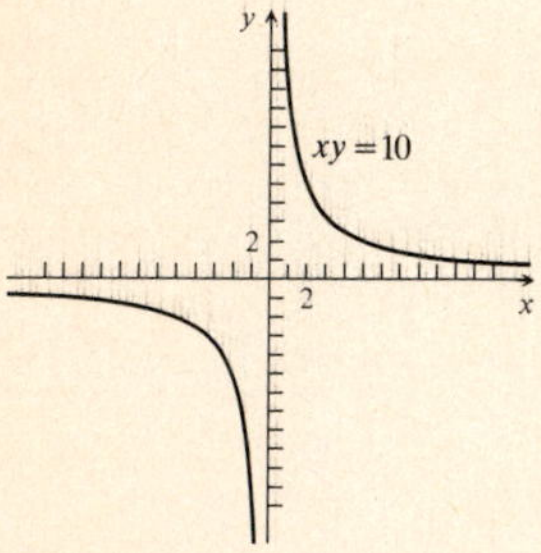

41.

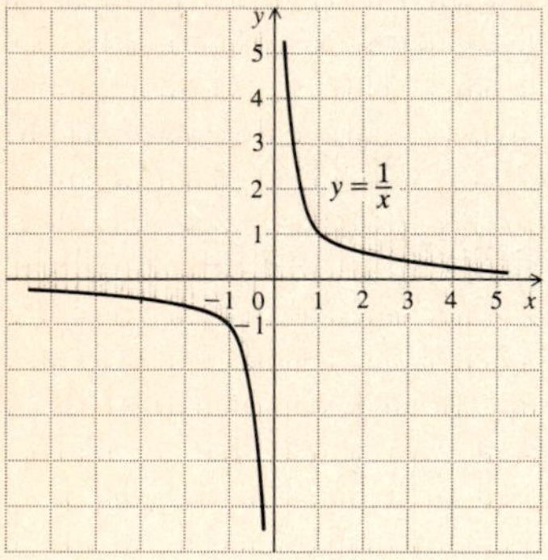

43.

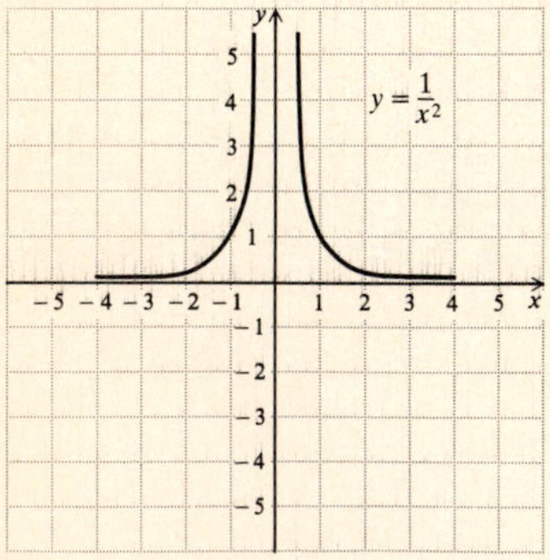

45.

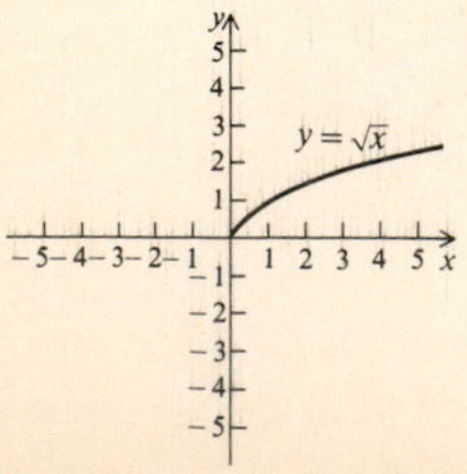

47.

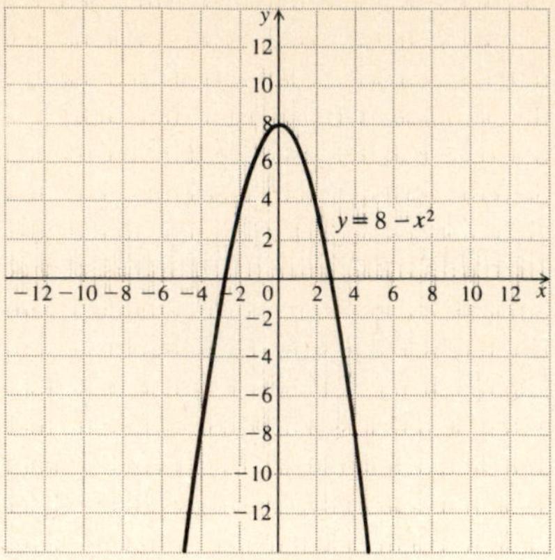

49. Same graphs.

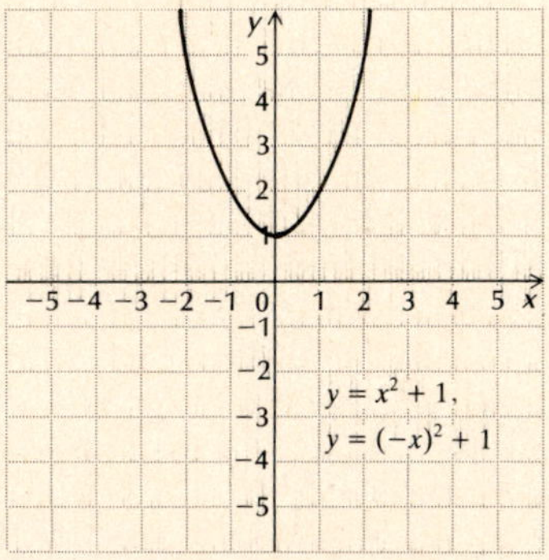

51. Domain: $\{x|2 \leq x \leq 6\}$; range: $\{y|1 \leq y \leq 5\}$
53. Domain: the set of real numbers; range: $\{y|y \leq 0\}$
55. Domain: $\{x|x \geq 0\}$; range: the set of real numbers
57. Domain: $\{x|x \geq 0\}$; range: $\{y|y \geq 0\}$ **59.** Domain: the set of real numbers; range: $\{y|y \leq 8\}$ **61.** Horizontal line through (0, 3) **63.** Line through (0, 1) and (−1, 0) **65.** Line through (0, 0) and (1, 2) **67.** See Exercise 29.

69.

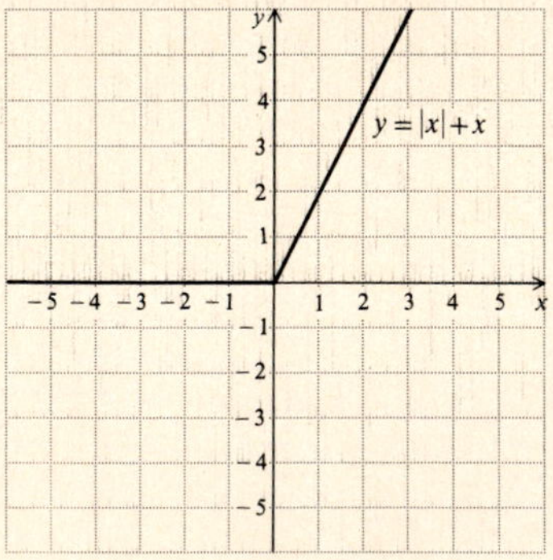

71.

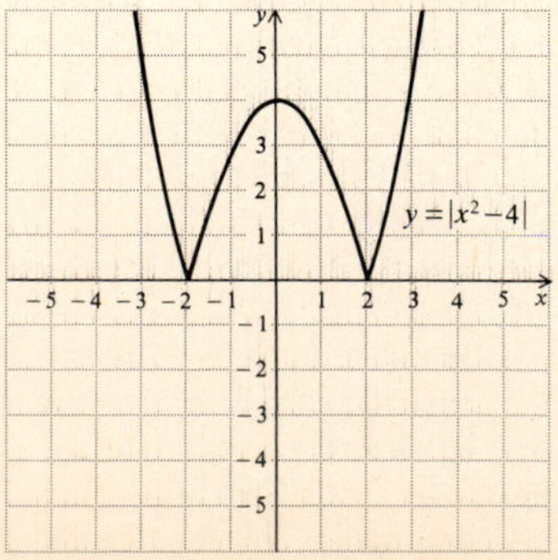

73.

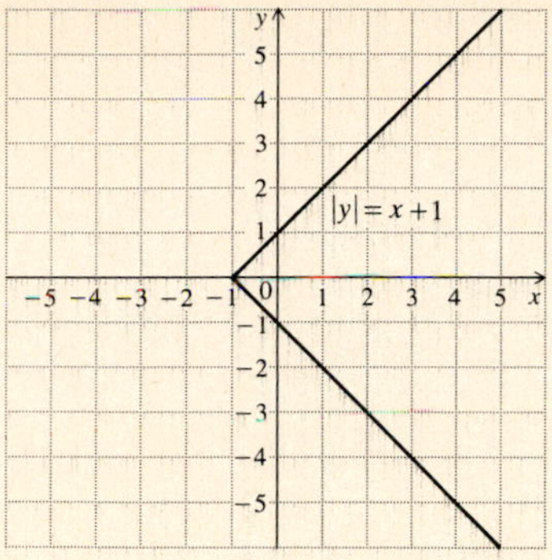

75.

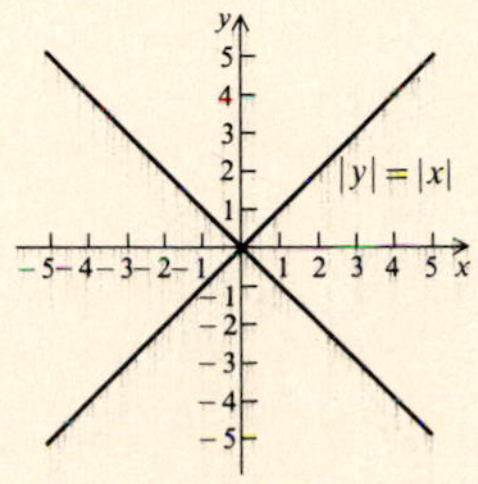

77.

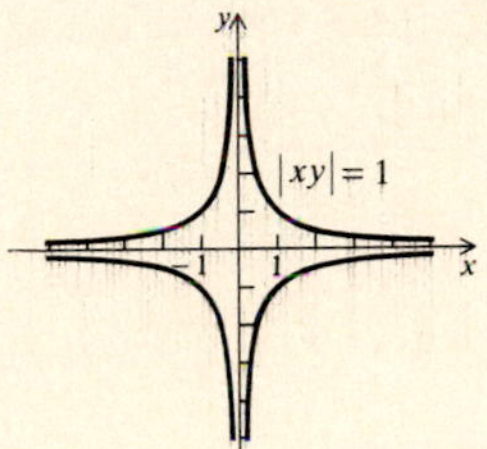

79.

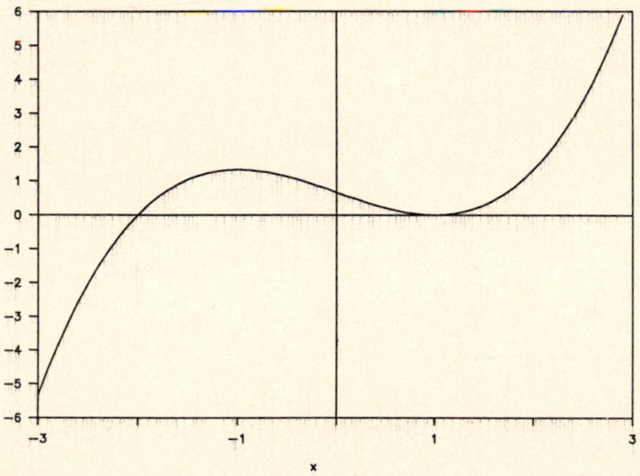

81.

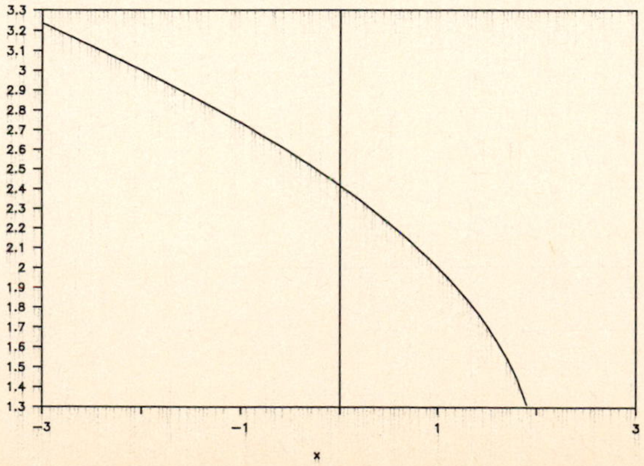

83.

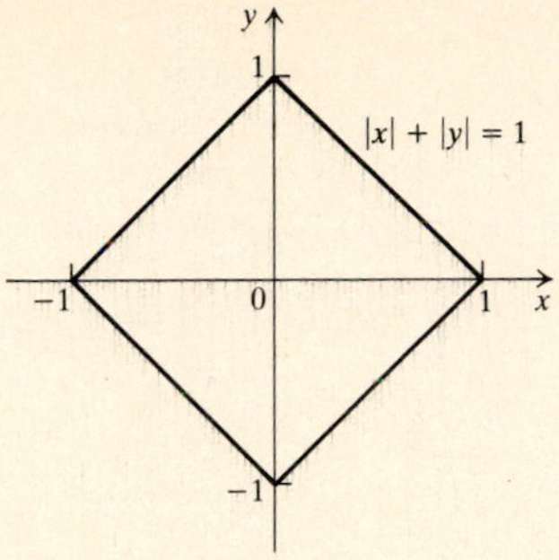

Margin Exercises, Section 3.3

1. Yes **2.** Yes **3.** No **4.** Yes **5.** B, C, D, E **6.** (a), (b), (c) **7.** $f(1) = -5, f(2) = 4, f(3) = 3, f(4) = 4$; domain $= \{1, 2, 3, 4\}$; range $= \{-5, 3, 4\}$ **8.** $f(-3) = 4.5, f(0) = 3, f(2) = 1, f(4) = 2.75$ **9.** $f(-1) = -5, f(0) = -3, f(5.6) = 8.2, f(10) = 17$ **10.** **(a)** -1; **(b)** 5; **(c)** -5; **(d)** $4a^2 + 10a - 1$; **(e)** $a^2 + 7a + 5$ **11.** $f(16) = 4, f(3) = \sqrt{3}, f(-4)$ is not possible **12.** $G(0) = 7, G(-3) = 7, G(\frac{1}{2}) = 7$; range: $\{7\}$ **13.** $6a + 3h$ **14.** $2a - 1 + h$ **15.** $\left\{x \mid x \neq -\frac{4}{3} \text{ and } x \neq -2\right\}$ **16.** $\{x \mid x \geq -2.5\}$ **17.** All real numbers

Exercise Set 3.3, pp. 152–154

1. No **3.** Yes **5.** Yes **7.** No **9.** Yes **11.** No **13.** Yes **15.** No **17.** Yes **19.** **(a)** 0; **(b)** 1; **(c)** 57; **(d)** $5t^2 + 4t$; **(e)** $5t^2 - 6t + 1$; **(f)** $10a + 5h + 4$ **21.** **(a)** 5; **(b)** -2; **(c)** -4; **(d)** $4|y| + 6y$; **(e)** $2|a + h| + 3a + 3h$; **(f)** $\dfrac{2|a + h| + 3h - 2|a|}{h}$ **23.** **(a)** 3.14977; **(b)** 55.73147; **(c)** 3178.20675; **(d)** 1116.70323 **25.** **(a)** $\frac{2}{3}$; **(b)** $\frac{10}{9}$; **(c)** 0; **(d)** not possible; **(e)** $\dfrac{h^2 - 3h}{2h^2 - 3h - 5}$; **(f)** $\dfrac{a^2 + 2ab + b^2 - a - b - 2}{2a^2 + 4ab + 2b^2 - 5a - 5b - 3}$ **27.** $2a + h$ **29.** $f(0)$ does not exist as a real number, $f(2) = 2 + \sqrt{3}, f(10) = 10 + 3\sqrt{11}$ **31.** $f(-1) = 2, f(7) = 9, f(5) = -6, f(-3) = 4$; domain: $\{-1, -3, 5, 7\}$; range: $\{2, 4, -6, 9\}$ **33.** $g(-2) = 0, g(-3) = -3, g(0) = 3, g(2) = 2$ **35.** All real numbers **37.** $\{x \mid x \neq 0\}$ **39.** $\left\{x \mid x \geq -\frac{4}{7}\right\}$ **41.** $\{x \mid x \neq 2, -2\}$ **43.** $\left\{x \mid x \neq -\frac{3}{4}, 2\right\}$ **45.** $\{x \mid x \neq 0, -2, 1\}$ **47.** $-4 + 3i$ **49.** $\dfrac{-1}{x(x + h)}$ **51.** $\dfrac{1}{\sqrt{x + h} + \sqrt{x}}$ **53.** $\{x \mid x \neq 2, -1 \text{ and } x \geq -3\}$ **55.** All real numbers

Margin Exercises, Section 3.4

1.

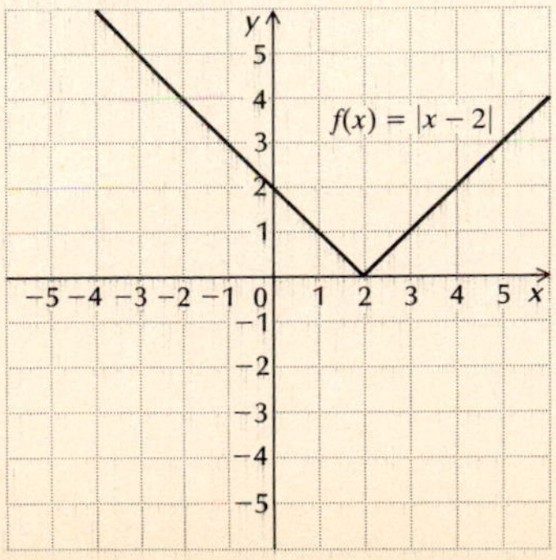

2.

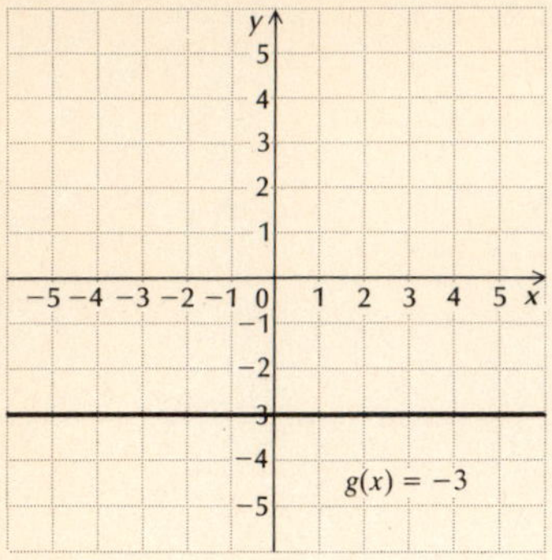

3. 209.08 in^2

4. 1076.6 ft/sec, 1147.0 ft/sec, 1183.1 ft/sec

5. (a) $A(x) = x(40 - x) = 40x - x^2$;

(b)

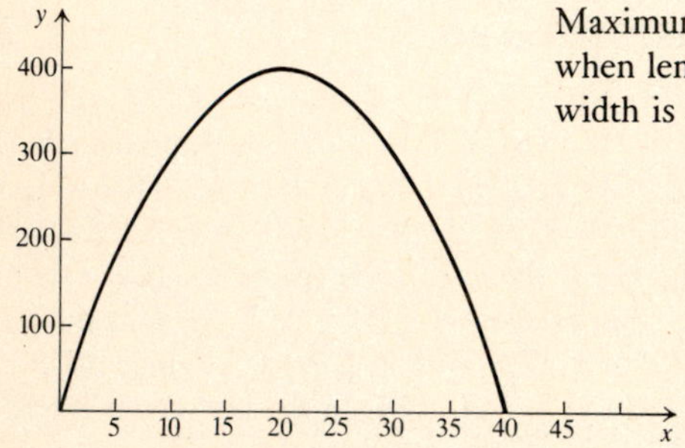

Maximum area is 400 ft^2 when length is 20 ft and width is 20 ft.

6. $d(t) = 40\sqrt{9t^2 + 100}$

Exercise Set 3.4, pp. 158–161

1.

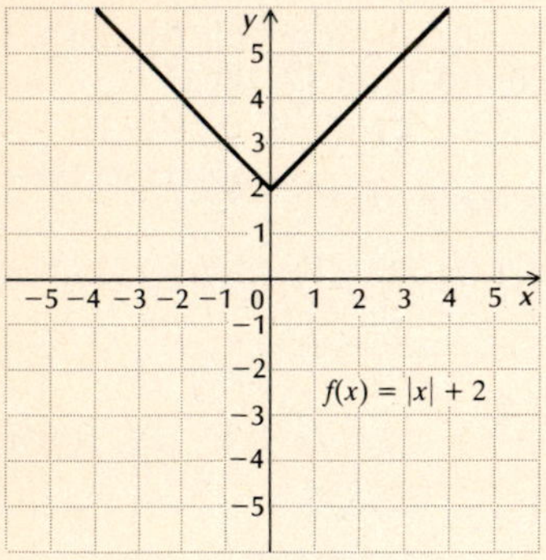

3.

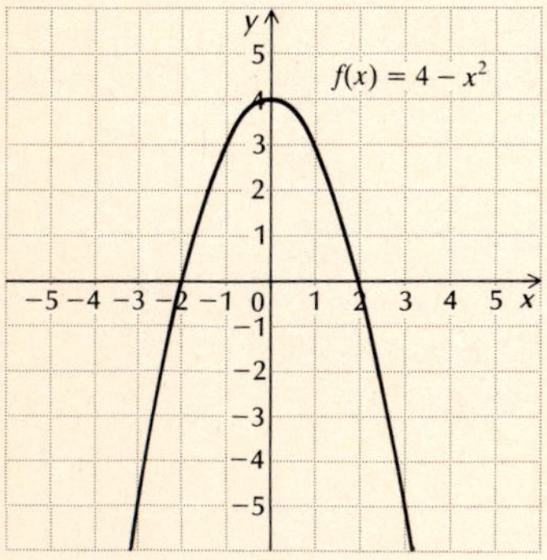

5.

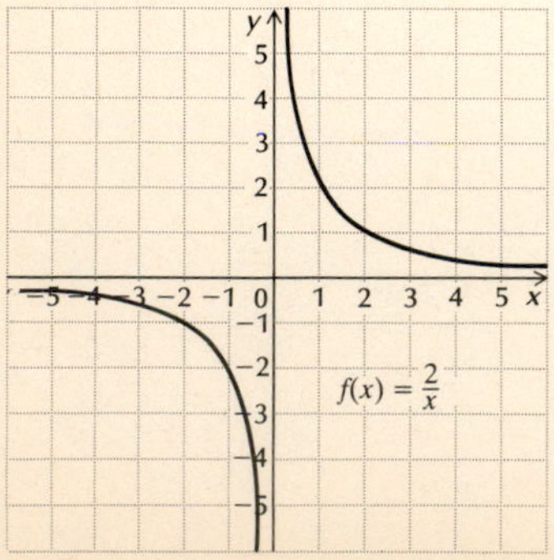

7.

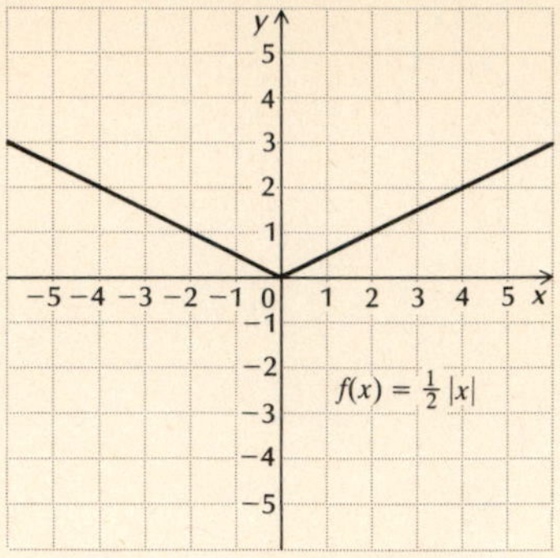

9.

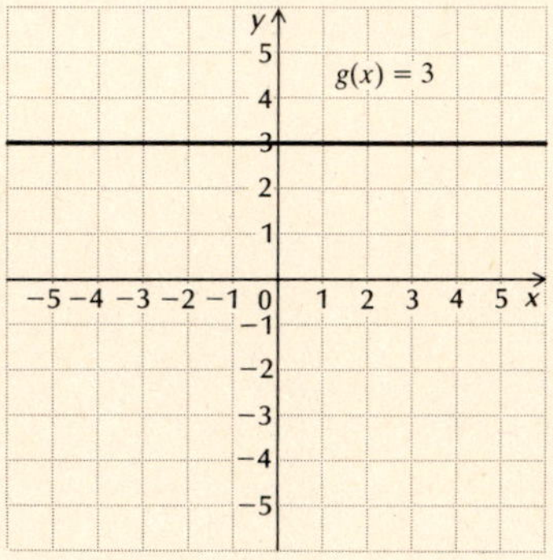

11.

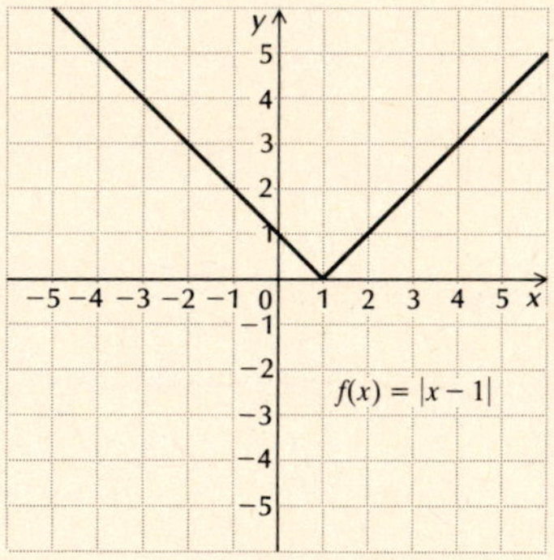

13.

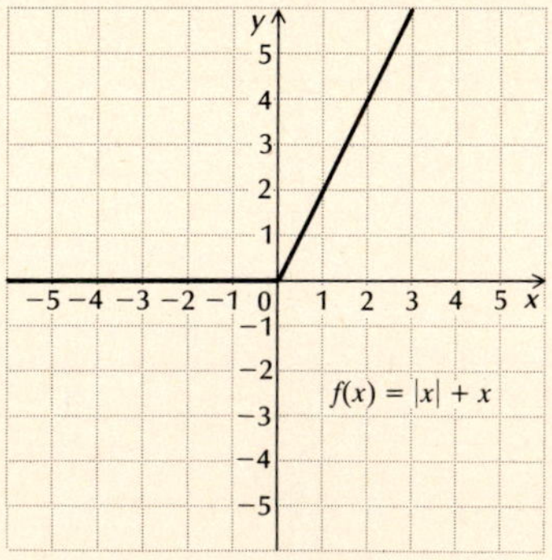

15.

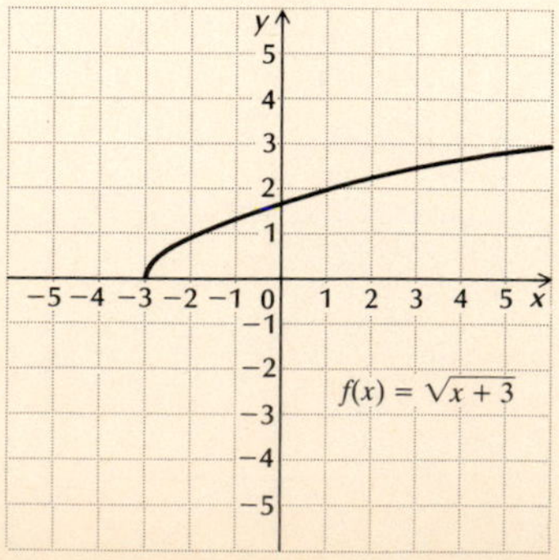

17. **(a)** $A(0) = 19.7$, $A(1) = 19.78$, $A(10) = 20.5$, $A(30) = 22.1$, $A(40) = 22.9$; **(b)** 23.38 **19.** Pentagon: 5; octagon: 20; decagon: 35; dodecagon: 54 **21.** **(a)** 645 m above sea level; **(b)** at sea level **23.** **(a)** $A(L) = L(L - 4)$, or $L^2 - 4L$; **(b)** $A(W) = W(W + 4)$, or $W^2 + 4W$ **25.** $A(x) = x(17 - x)$, or $17x - x^2$ **27.** $A(x) = x\sqrt{256 - x^2}$ **29.** $A(i) = 1000(1 + i)^4$ **31.** **(a)** $V(x) = 5x^2 - \frac{1}{2}x^3$;
(b)

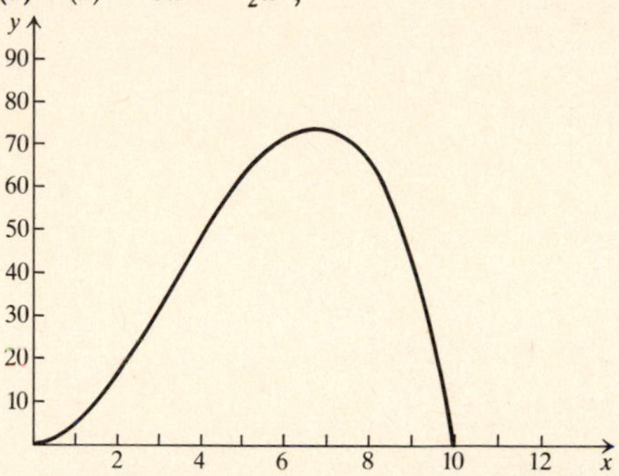

Using the graph, we see that the maximum value seems to be about 72 in³ when x is about 6 in. The exact maximum is $74\frac{2}{27}$ in³ when x is $6\frac{2}{3}$ in. **33.** $SA(x) = x^2 + \dfrac{432}{x}$ **35.** $d(s) = \dfrac{14}{s}$ **37.** $h(d) = \sqrt{d^2 - 13{,}690{,}000}$ **39.** $A(a) = \dfrac{\sqrt{3}}{4}a^2$ **41.** $S(x) = \dfrac{x^2}{4\pi} + \left(\dfrac{24 - x}{4}\right)^2$ **43.** **(a)** $h(r) = \dfrac{30 - 5r}{3}$; **(b)** $V(r) = \pi r^2\left(\dfrac{30 - 5r}{3}\right)$; **(c)** $V(h) = \pi\left(\dfrac{30 - 3h}{5}\right)^2 h$

Margin Exercises, Section 3.5

1. **(a)** $(-3, 2)$; **(b)** $(4, -5)$

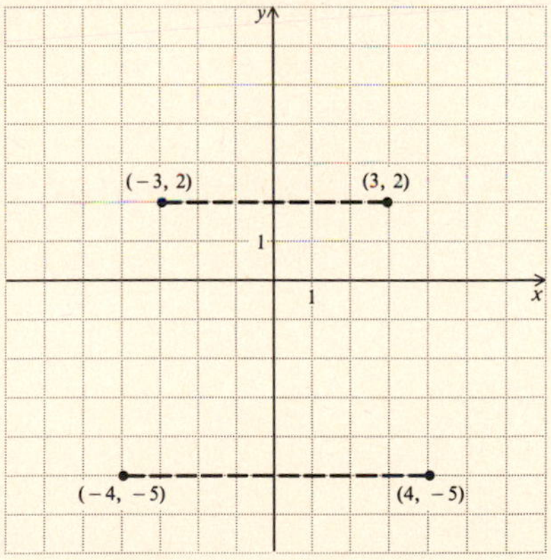

2. **(a)** $(4, -3)$; **(b)** $(3, 5)$

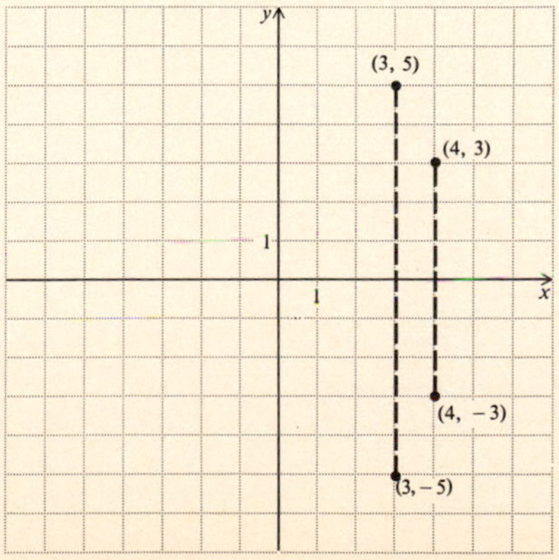

3. x-axis: no; y-axis: yes **4.** x-axis: yes; y-axis: no **5.** x-axis: yes; y-axis: yes **6.** x-axis: yes; y-axis: yes **7.** a-axis: no; b-axis: no **8.** p-axis: no; q-axis: no **9.** **(a)** $(-3, -2)$; **(b)** $(4, -3)$; **(c)** $(5, 7)$

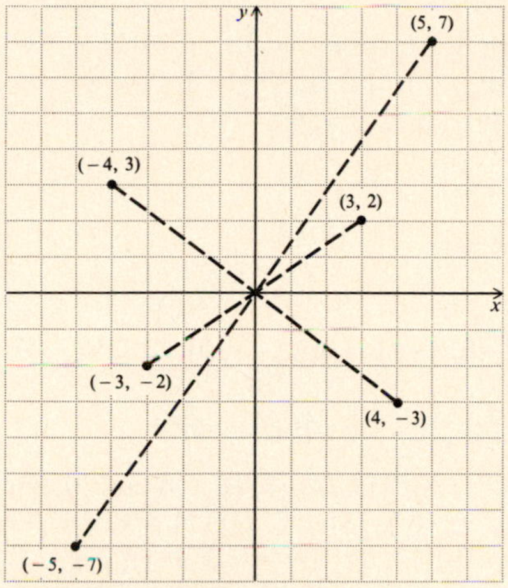

10. Yes **11.** Yes **12.** Yes **13.** Yes **14.** Yes **15.** No **16.** **(a)** Yes; **(b)** yes; **(c)** no; **(d)** yes **17.** **(a)** No; **(b)** yes; **(c)** yes; **(d)** yes **18.** **(a)** Odd; **(b)** odd; **(c)** even; **(d)** odd **19.** **(a)** Neither; **(b)** even; **(c)** odd; **(d)** neither; **(e)** neither

Exercise Set 3.5, pp. 170–171

1. x-axis, no; y-axis, yes; origin, no **3.** x-axis, no; y-axis, yes; origin, no **5.** All yes **7.** All yes **9.** All no **11.** All no **13.** Yes **15.** Yes **17.** Yes **19.** Yes **21.** No **23.** No **25.** Yes **27.** Yes **29.** **(a)** Even; **(b)** even; **(c)** odd; **(d)** neither **31.** Neither **33.** Even **35.** Neither **37.** Even **39.** Odd **41.** Neither **43.** Odd **45.** Even and odd **47.** Even

49.

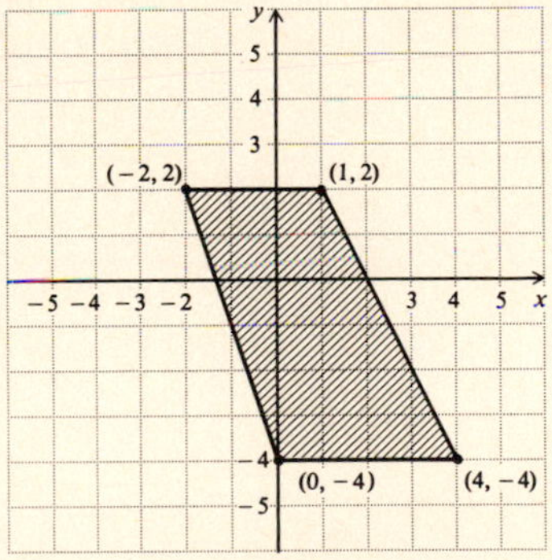

51.

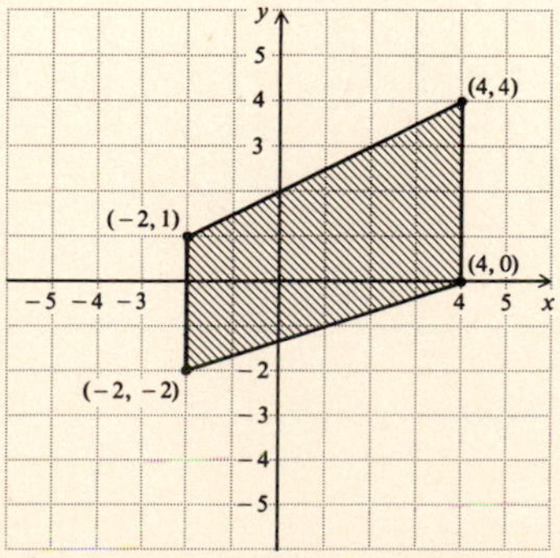

53. Let E $[x, y]$ represent an equation in x and y; let $\sim$ mean "is equivalent to"; and let w.r.t. mean "with respect to."

If the graph is symmetric w.r.t. the x-axis and origin, i.e., if E $[x, -y] \sim$ E $[x, y]$ and E $[-x, -y] \sim$ E $[x, y]$, then E $[x, -y] \sim$ E $[-x, -y]$ which becomes, by symmetry w.r.t. the origin, E $[-x, y] \sim$ E $[x, y]$; hence the graph is symmetric w.r.t. the y-axis.

If the graph is symmetric w.r.t. the y-axis and origin, i.e., if E $[-x, y] \sim$ E $[x, y]$ and E $[-x, -y] \sim$ E $[x, y]$, then E $[-x, y] \sim$ E $[-x, -y]$ which becomes, by symmetry w.r.t. the origin, E $[x, -y] \sim$ E $[x, y]$; hence the graph is symmetric w.r.t. the x-axis.

If the graph is symmetric w.r.t. both axes, i.e., if E $[x, -y] \sim$ E $[x, y]$ and E $[-x, y] \sim$ E $[x, y]$, then E $[x, -y] \sim$ E $[-x, y]$; hence the graph is symmetric w.r.t. the origin.

Margin Exercises, Section 3.6

1. **(a)** $(-1, 3)$; **(b)** $(1, 4)$ **2.** **(a)** $(-2, 3)$; **(b)** $(0, 1)$; **(c)** $\left(-\frac{1}{4}, \sqrt{2}\right)$ **3.** **(a)** $[-1, 4]$; **(b)** $(-1, 4]$; **(c)** $[-1, 0)$; **(d)** $(-1, 4)$ **4.** **(a)** $\left[4, 5\frac{1}{2}\right]$; **(b)** $(-3, 0]$; **(c)** $\left[-\frac{1}{2}, \frac{1}{2}\right)$; **(d)** $(-\pi, \pi)$ **5.** **(a)** $(-\infty, 5]$; **(b)** $(4, \infty)$; **(c)** $(-\infty, 4.8)$; **(d)** $[3, \infty)$ **6.** **(a)** $[8, \infty)$; **(b)** $(-\infty, -7)$; **(c)** $(10, \infty)$; **(d)** $(-\infty, 0.78]$ **7.** $f(x) = f(x + 4)$, $f(x) = f(x + 6)$ **8.** **(a)** Yes; **(b)** 3 **9.** $t(x) = t(x + 1)$; t is periodic. It does not have a period, because there is no smallest positive p for which $t(x + p) = t(x)$. **10.** **(a)** No; **(b)** yes; **(c)** no; **(d)** yes; **(e)** no **11.** Where $x = -3, 0, 2$ **12.** **(a)** Increasing; **(b)** increasing; **(c)** decreasing; **(d)** neither **13.** Increasing: $[-3, 0]$; decreasing: $[0, 3]$; there are many answers **14.** $(2, \$9)$

15.

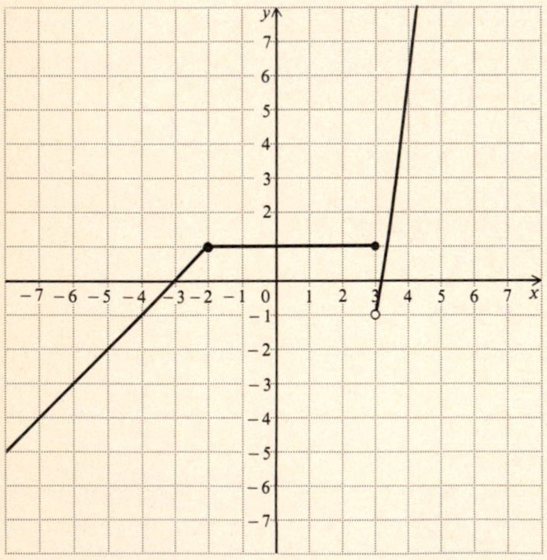

16.

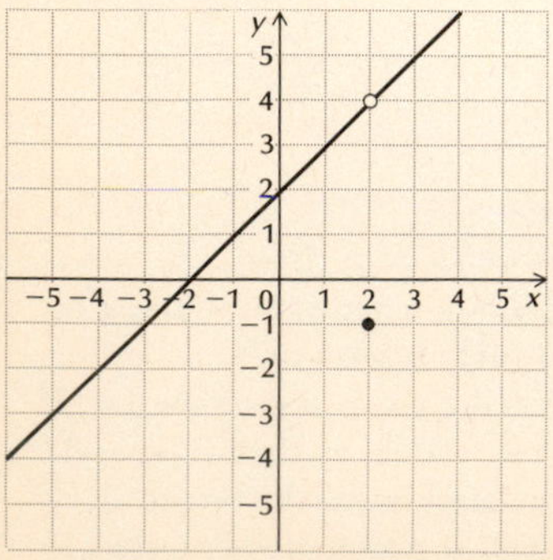

17. **(a)** $7, -2, 1, 3, -3, -1$;
(b)

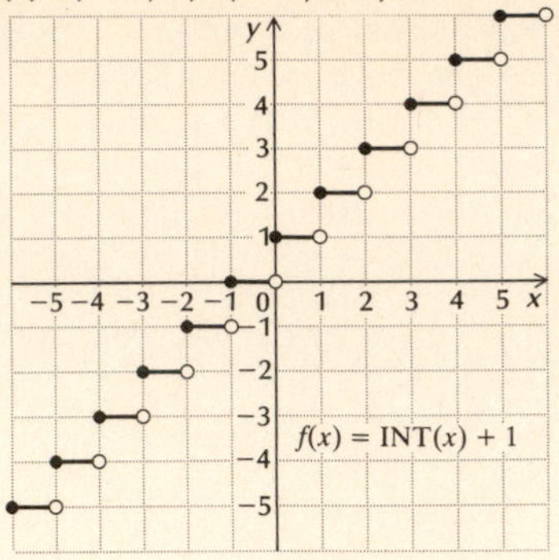

(c)

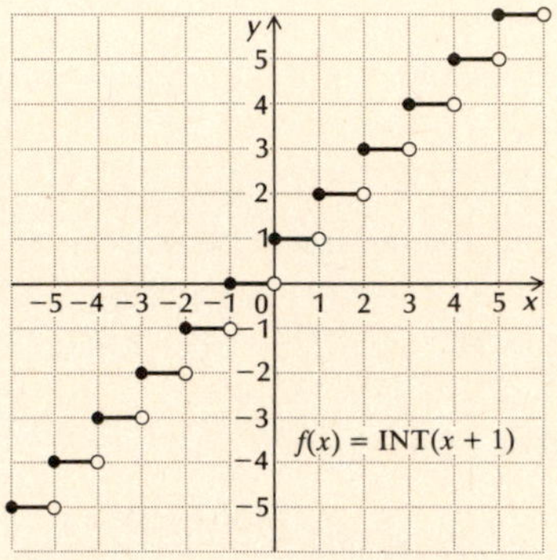

Exercise Set 3.6, pp. 178–181

1. $(0, 5)$ **3.** $[-9, -4)$ **5.** $[x, x + h]$ **7.** (p, ∞) **9.** $[-3, 3]$ **11.** $[-14, -11)$ **13.** $(-\infty, -4]$ **15.** $(-\infty, 3.8)$ **17.** **(a)** No; **(b)** yes; **(c)** yes; **(d)** no **19.** 4 **21.** **(a)** Yes; **(b)** yes; **(c)** no; **(d)** yes; **(e)** yes **23.** Where $x = -3$ and $x = 2$ **25.** **(a)** Increasing; **(b)** neither; **(c)** decreasing; **(d)** neither **27.** $(2, \$4)$ **29.** $(1, \$4)$ **31.** $(2, \$4)$

33.

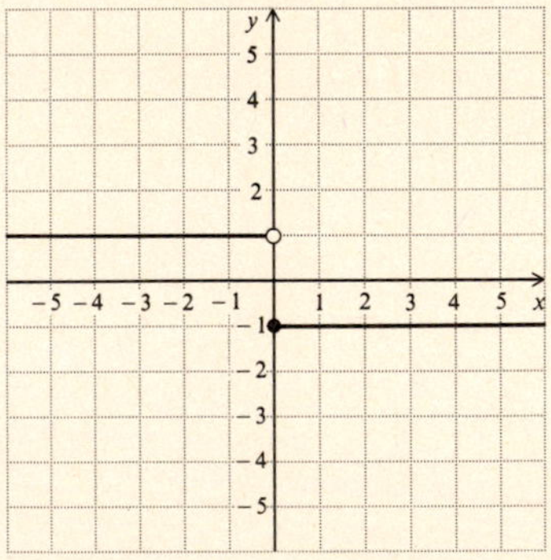

35.

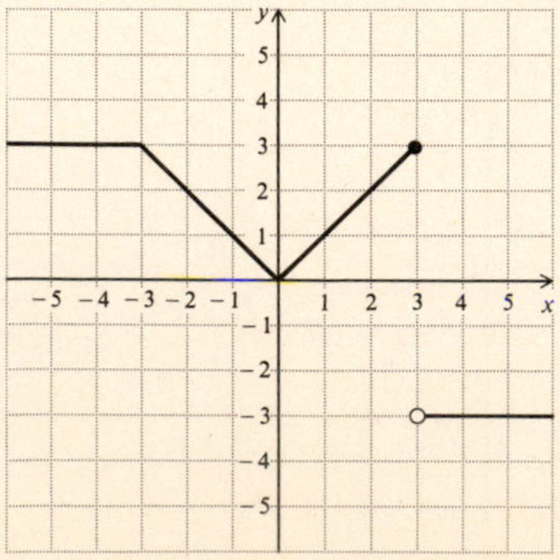

37.

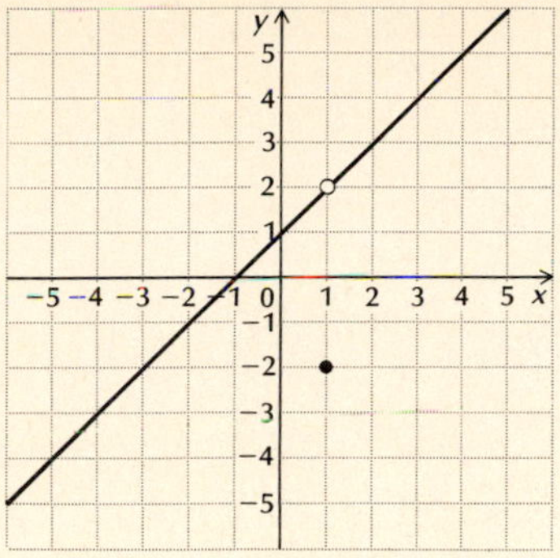

39.

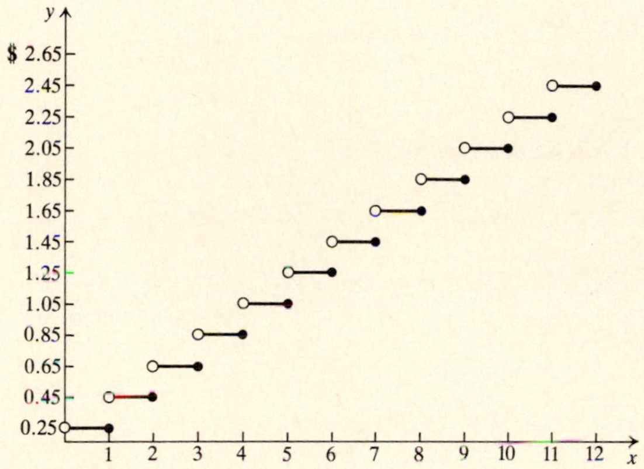

41.

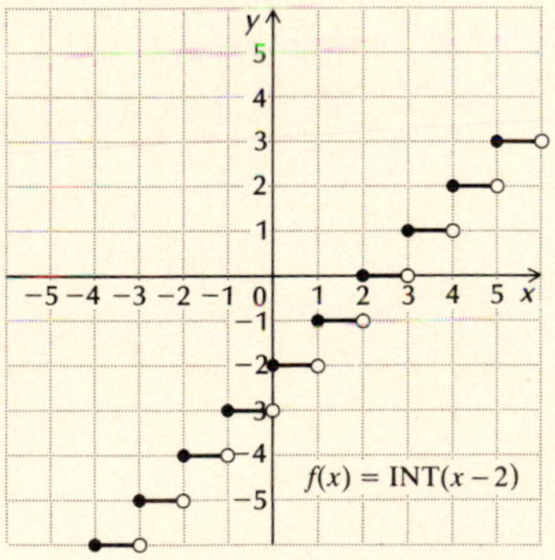

43.

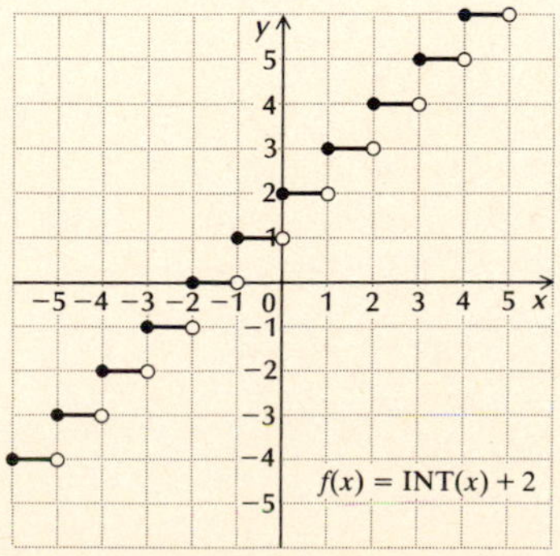

45. Increasing: [0, 1]; decreasing: [−1, 0]; many possible answers 47. Increasing: (a), (e); decreasing: (b); neither: (c), (d), (f)

49.

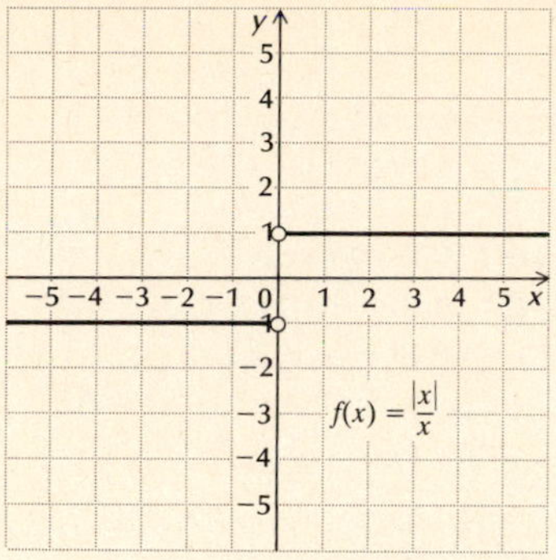

51. $\{x \mid 4 \le x < 5\}$
53. (a) 182; (b) 171

55.

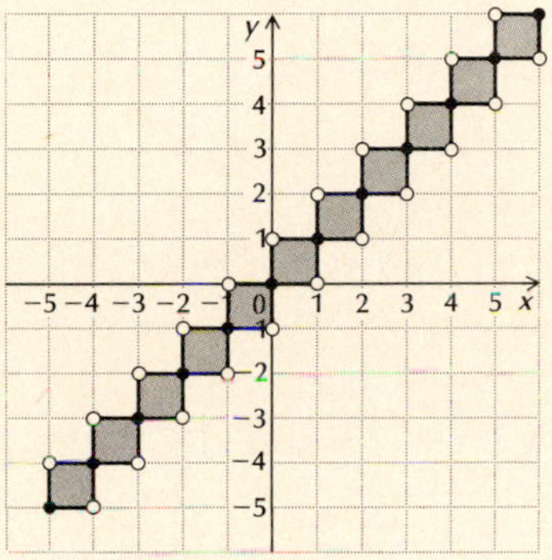

No

Margin Exercises, Section 3.7

1. (a) $(f+g)(5) = \frac{1}{3}$, $(f+g)(2)$ does not exist; **(b)** $(f-g)(7) = -\frac{119}{5}$, $(f-g)(-3) = \frac{79}{5}$; **(c)** $fg(-3) = \frac{16}{5}$, $gg(1) = 576$; **(d)** $(f/g)(1) = \frac{1}{24}$, $(g/f)(5) = 0$, $(f/g)(5)$ does not exist, $(f/g)(-5)$ does not exist; **(e)** all reals except 2, all reals, all reals except 2, all reals except 2, all reals except 2, all reals except 2, 5, and -5 **2. (a)** $2x^2 - 6$; **(b)** 12; **(c)** $x^4 - 3x^2 - 27$; **(d)** $\frac{x^2+3}{x^2-9}$; **(e)** $x^4 + 6x^2 + 9$; **(f)** all reals, all reals, all reals, all reals except 3 and -3 **3. (a)** $P(x) = -0.5x^2 + 40x - 3$; **(b)** $R(40) = 1200$, $C(40) = 403$, $P(40) = 797$ **4. (a)** $f \circ g(-3) = 13$, $g \circ f(-3) = 3$; **(b)** $f \circ g(x) = x^2 + 4$, $g \circ f(x) = x^2 + 10x + 24$ **5. (a)** $f \circ g(x) = \sqrt{x^2+3}$, $g \circ f(x) = x + 3$; **(b)** Domain of f: all reals greater than or equal to -3; domain of g: all reals; domain of $f \circ g$: all reals **6. (a)** $K \circ C(F) = \frac{5}{9}(F - 32) + 273$. This function converts Fahrenheit F temperature to degrees in Kelvin units; **(b)** 248° **7.** Answers may vary. **(a)** $f(x) = \sqrt[3]{x+1}$, $g(x) = x^2$; **(b)** $f(x) = \frac{1}{x^4}$, $g(x) = x + 5$

Exercise Set 3.7, pp. 187–189

1. (a) $(f+g)(x) = 2x + 1$, $(f-g)(x) = -7$, $fg(x) = x^2 + x - 12$, $ff(x) = x^2 - 6x + 9$, $(f/g)(x) = \frac{x-3}{x+4}$, $(g/f)(x) = \frac{x+4}{x-3}$, $f \circ g(x) = x + 1$, $g \circ f(x) = x + 1$; **(b)** all reals, all reals, all reals, all reals, all reals, all reals except -4, all reals except 3, all reals, all reals **3. (a)** $(f+g)(x) = x^3 + 2x^2 + 9x - 3$, $(f-g)(x) = x^3 - 2x^2 - 9x + 3$, $fg(x) = 2x^5 + 9x^4 - 3x^3$, $ff(x) = x^6$, $(f/g)(x) = \frac{x^3}{2x^2 + 9x - 3}$, $(g/f)(x) = \frac{2x^2 + 9x - 3}{x^3}$, $f \circ g(x) = (2x^2 + 9x - 3)^3$, $g \circ f(x) = 2x^6 + 9x^3 - 3$; **(b)** all reals, all reals, all reals, all reals, all reals, all reals, all reals except

$\frac{-9 \pm \sqrt{105}}{4}$, all reals except 0, all reals, all reals **5.** -6
7. $x^2 - 2x - 9$ **9.** 55 **11.** Does not exist
13. $2x^3 + 5x^2 - 8x - 20$ **15.** $\frac{2x + 5}{x^2 - 4}$
17. $4x^2 + 20x + 21$ **19.** $4x + 15$ **21. (a)** $P(x) = -0.4x^2 + 57x - 13$; **(b)** $R(20) = 1040$, $C(20) = 73$, $P(20) = 967$ **23.** $f \circ g(x) = x$, $g \circ f(x) = x$ **25.** $f \circ g(x) = x$, $g \circ f(x) = x$ **27.** $f \circ g(x) = x$, $g \circ f(x) = x$ **29.** $f \circ g(x) = |x|$, $g \circ f(x) = x$ **31.** $f \circ g(x) = x$, $g \circ f(x) = x$ **33.** $f \circ g(x) = -6$, $g \circ f(x) = 12$ **35.** $f(x) = x^5$, $g(x) = 4 - 3x$ **37.** $f(x) = \frac{1}{x^4}$, $g(x) = x - 1$ **39.** $f(x) = \frac{x - 1}{x + 1}$, $g(x) = x^3$ **41.** $f(x) = x^6$, $g(x) = \frac{2 + x^3}{2 - x^3}$ **43.** $f(x) = \sqrt{x}$, $g(x) = \frac{x - 5}{x + 2}$
45. $f(x) = x^5 + x^4 + x^3 - x^2 + 4x$, $g(x) = x + 3$
47. (a) $a(t) = 250t$; **(b)** $P(a) = 300 + a$; **(c)** $P \circ a(t) = 300 + 250t$

49.

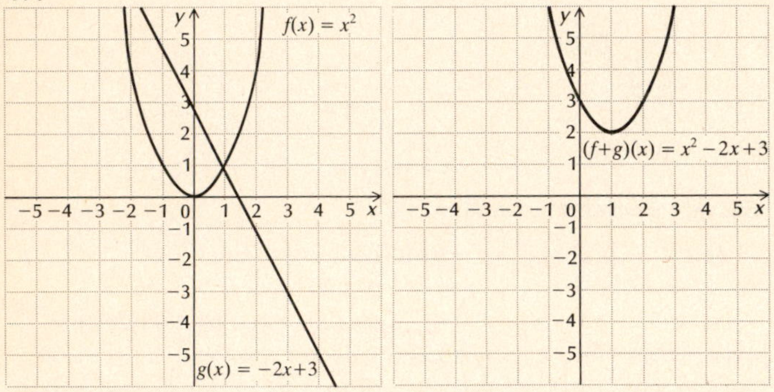

51.

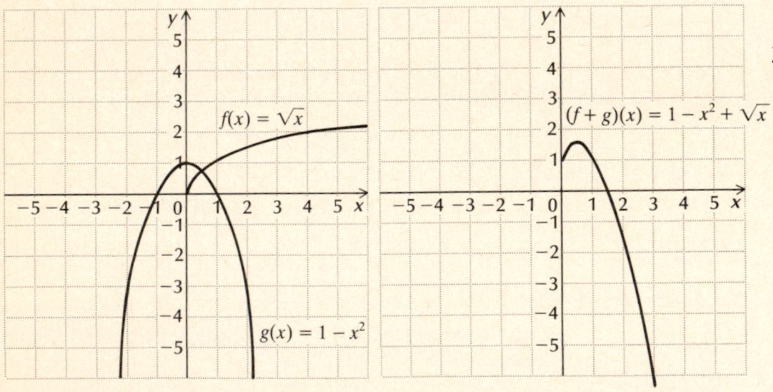

53. $b = -2$ **55.** $f \circ f(x) = \frac{x - 1}{x}$, $f \circ f \circ f(x) = x$ **57.** Let $f(x)$ and $g(x)$ be even functions. Then by definition, $f(x) = f(-x)$ and $g(x) = g(-x)$. Thus,
$(f + g)(x) = f(x) + g(x) = f(-x) + g(-x) = (f + g)(-x)$ and $f + g$ is even. **59.** Let $f(x)$ and $g(x)$ be odd functions. Then by definition, $-f(x) = f(-x)$, or $f(x) = -f(-x)$, and $-g(x) = g(-x)$, or $g(x) = -g(-x)$. Thus, $f \circ g(x) = f(g(x)) = f(-g(-x)) = -f(g(-x)) = -f(-g(x))$ and $f \circ g$ is odd.
61. Let f and g be increasing functions, and thus by definition for all a and b in the domains of f and g, if $a < b$, then $f(a) < f(b)$ and $g(a) < g(b)$. Since f and g are each increasing, $g(a) < g(b)$ and $f(g(a)) < f(g(b))$, or $f \circ g(a) < f \circ g(b)$. Thus, $f \circ g$ is increasing. Using the addition property of inequalities, we get $f(a) + g(a) < f(b) + g(b)$, or $(f + g)(a) < (f + g)(b)$ and $f + g$ is increasing. **63.** $O(-x) = \frac{f(-x) - f(-(-x))}{2} = \frac{f(-x) - f(x)}{2}$, $-O(x) = -\frac{f(x) - f(-x)}{2} = \frac{f(-x) - f(x)}{2}$. Thus, $O(-x) = -O(x)$ and O is odd.

65.

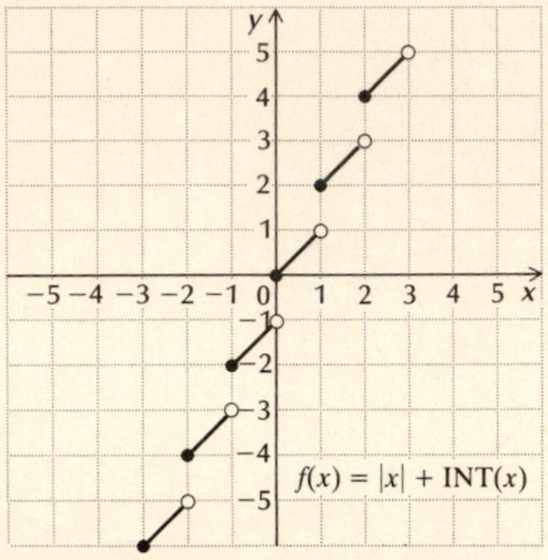

Margin Exercises, Section 3.8

1.

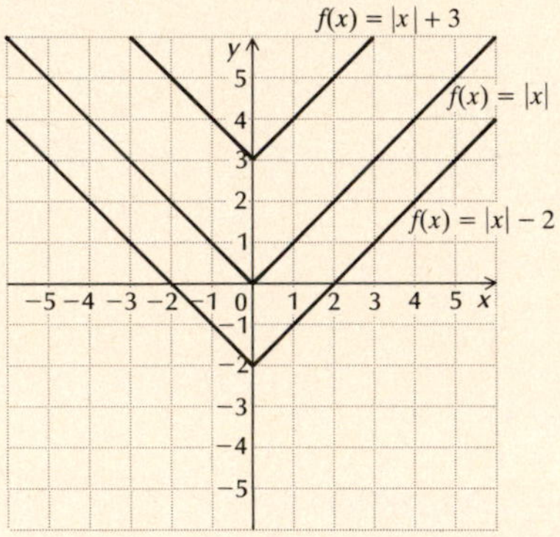

2.

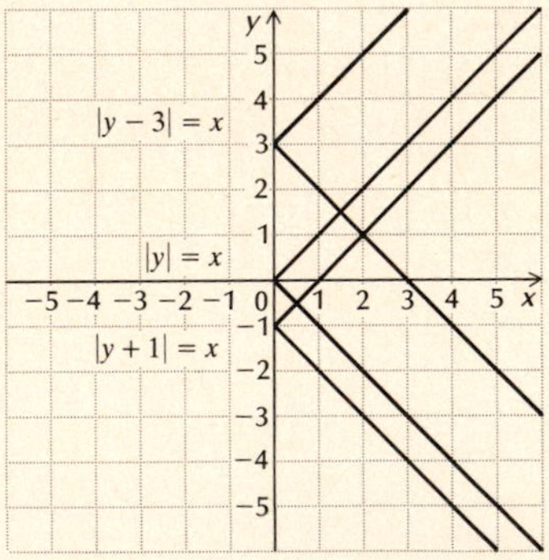

3.

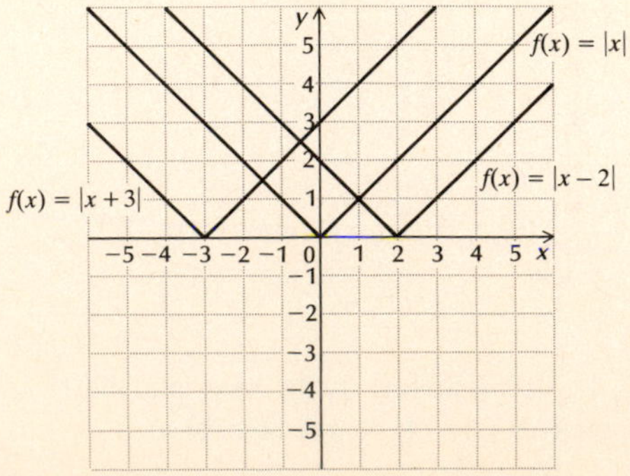

4.

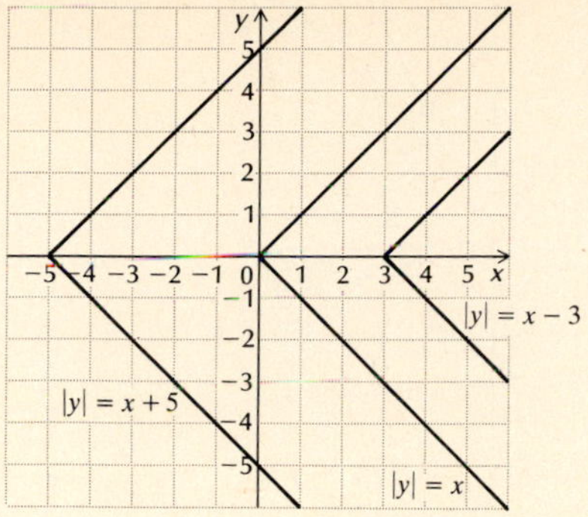
$|y| = x + 5$
$|y| = x$
$|y| = x - 3$

5.

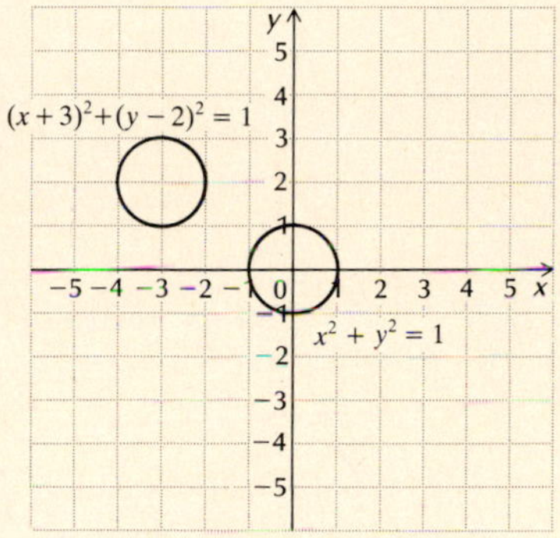
$(x+3)^2+(y-2)^2 = 1$
$x^2 + y^2 = 1$

6.

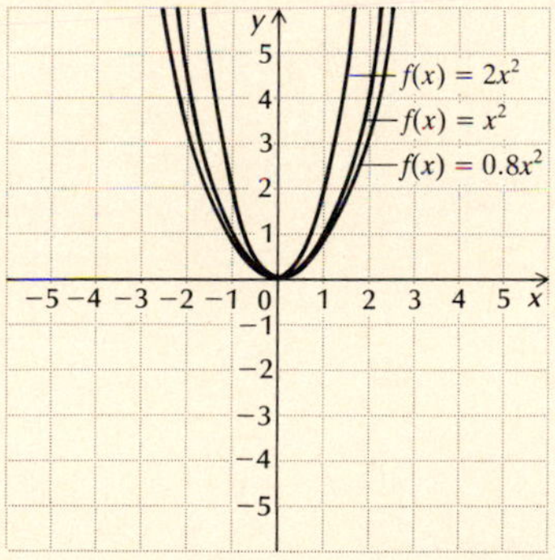
$f(x) = 2x^2$
$f(x) = x^2$
$f(x) = 0.8x^2$

7.

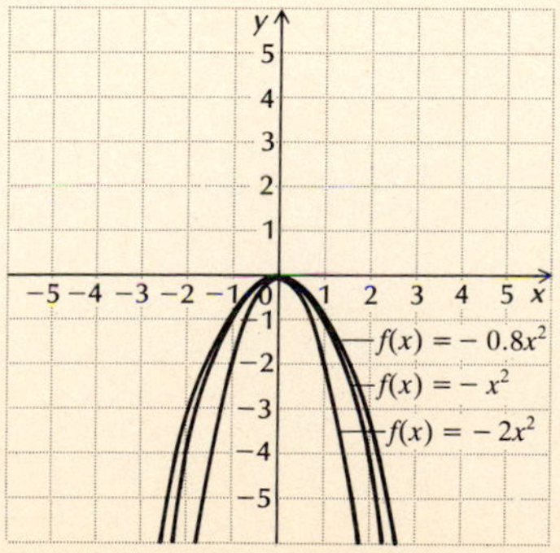
$f(x) = -0.8x^2$
$f(x) = -x^2$
$f(x) = -2x^2$

8.

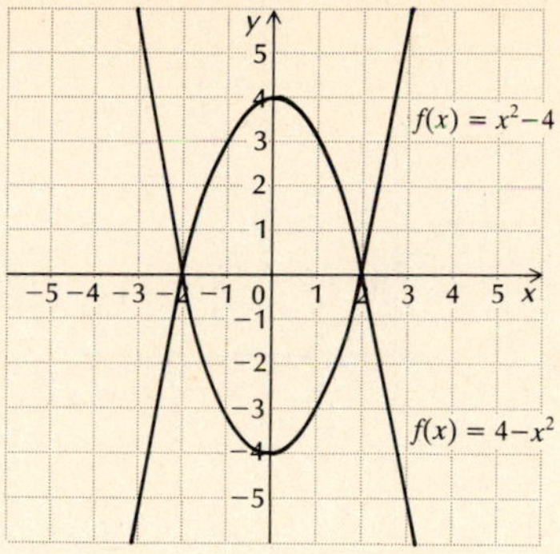
$f(x) = x^2-4$
$f(x) = 4-x^2$

9.

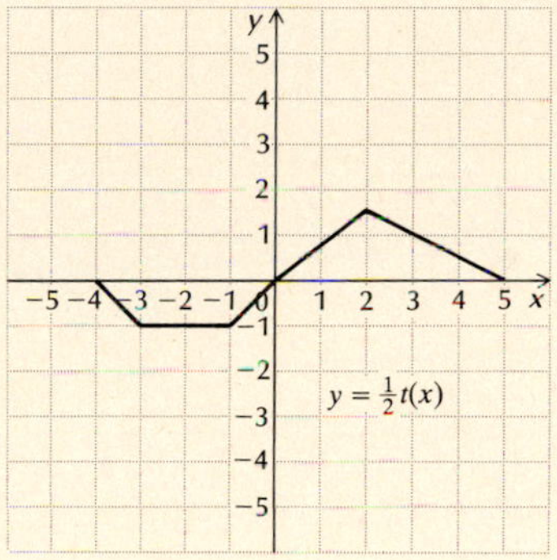
$y = \frac{1}{2}t(x)$

10.

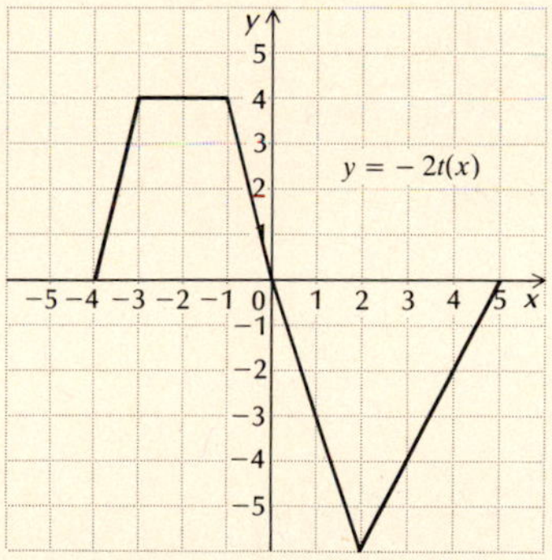
$y = -2t(x)$

11.

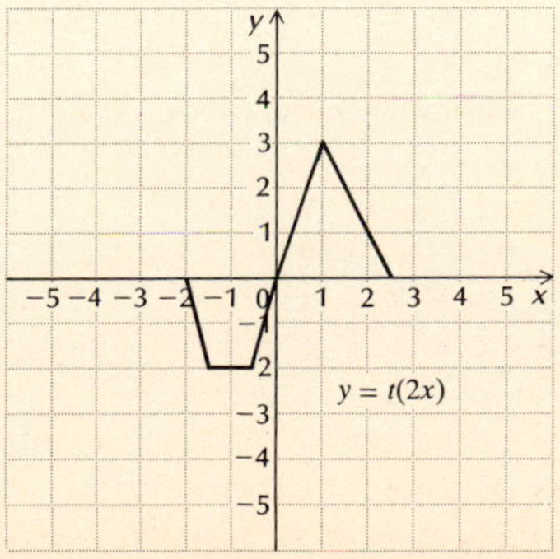
$y = t(2x)$

12.

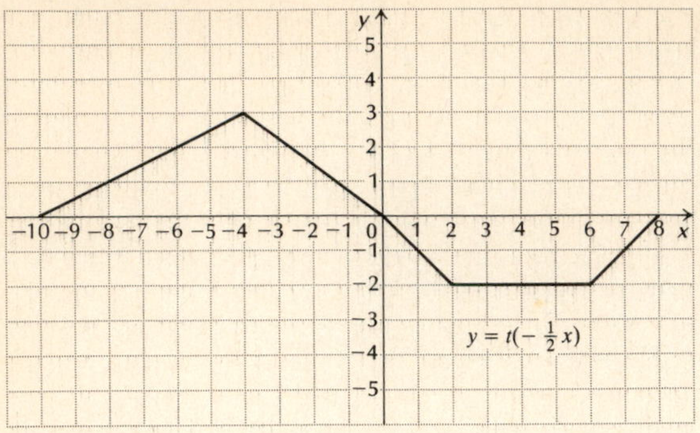

13.

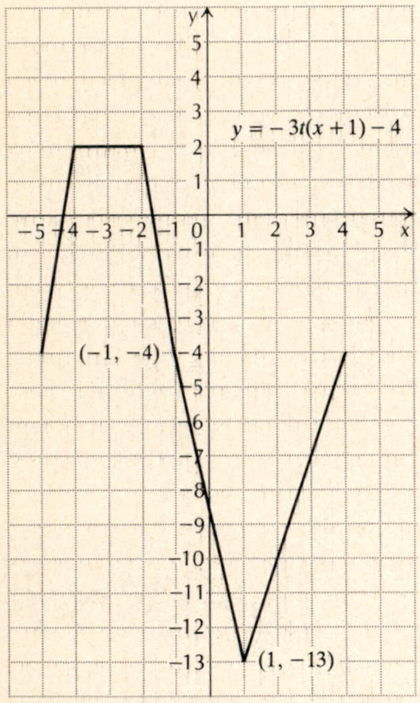

Exercise Set 3.8, pp. 197–199

1. and 3.

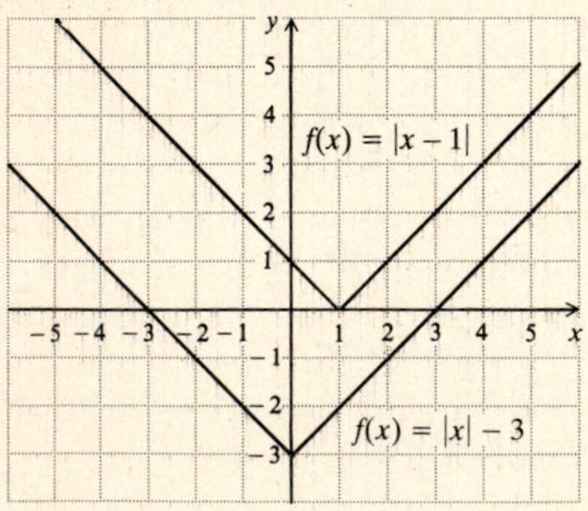

5 and 7.

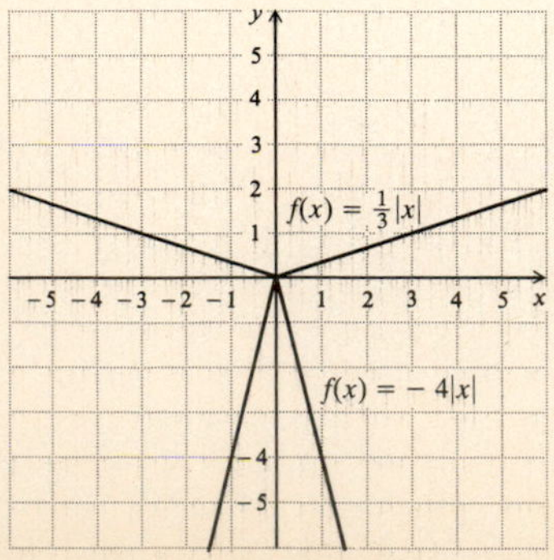

9. and 11.

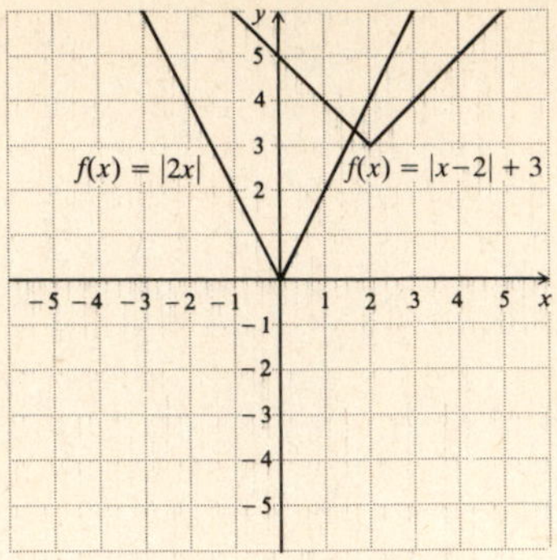

13.

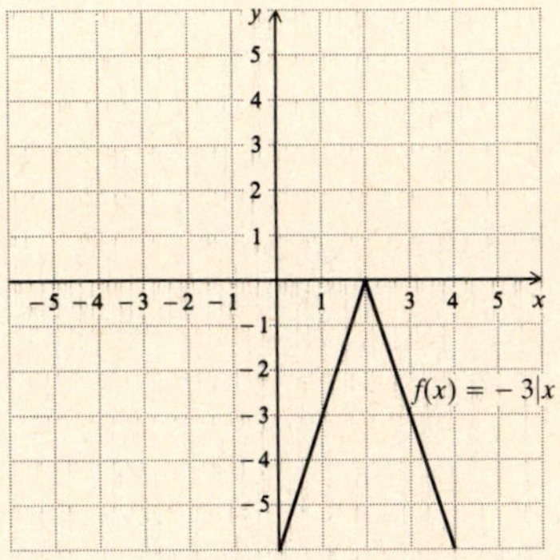

15. and 17.

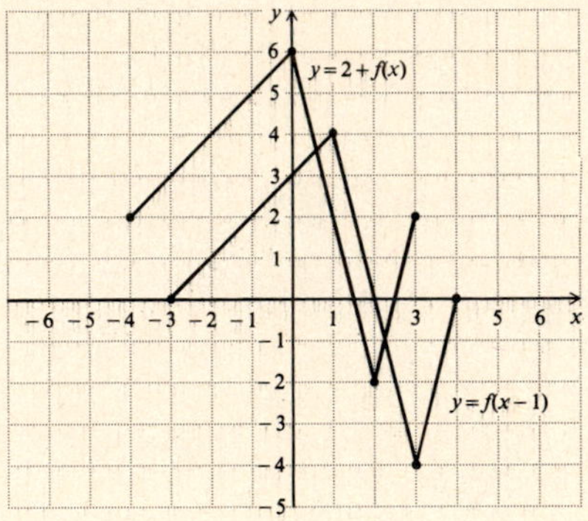

19.

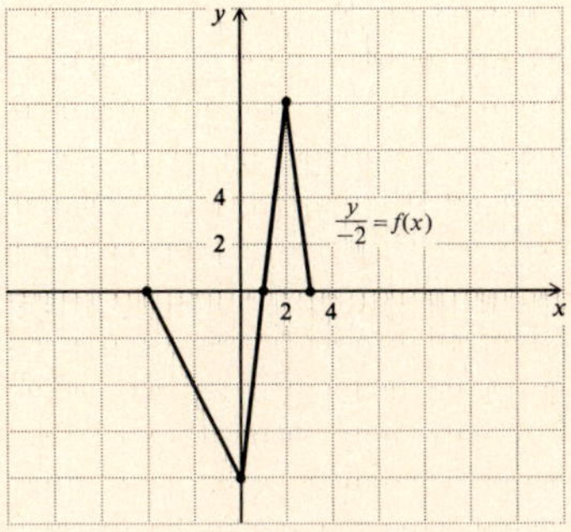

21.

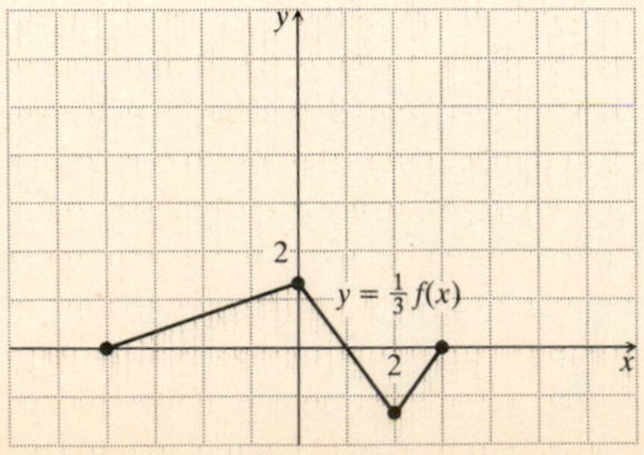

23.

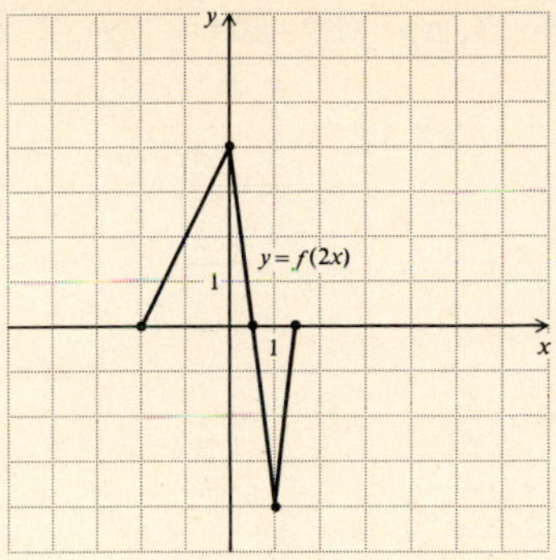

25.

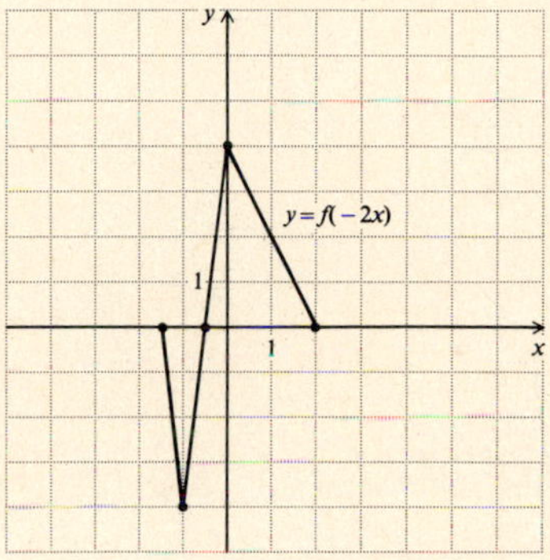

27.

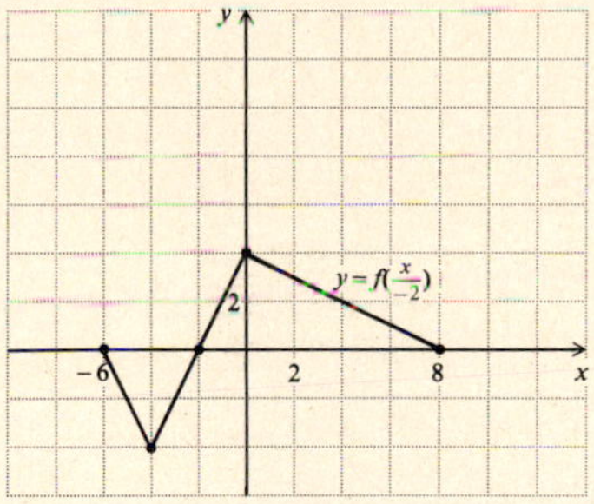

29.

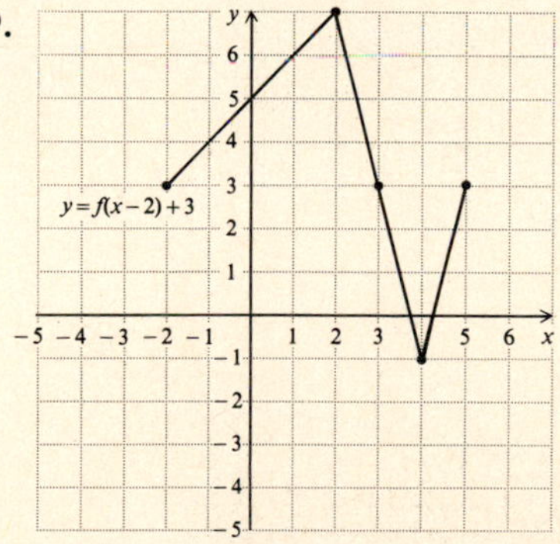

31.

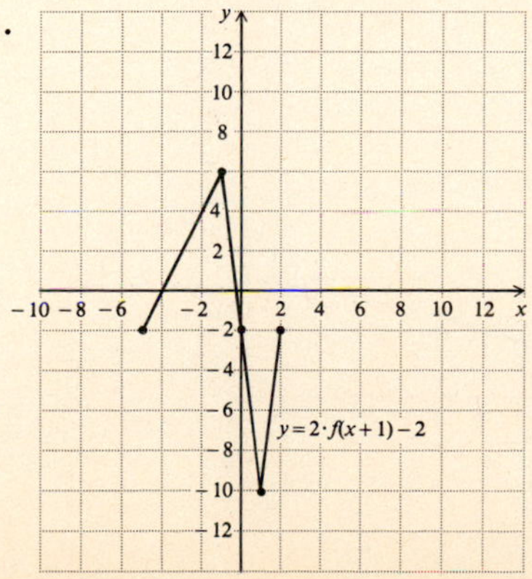

33.

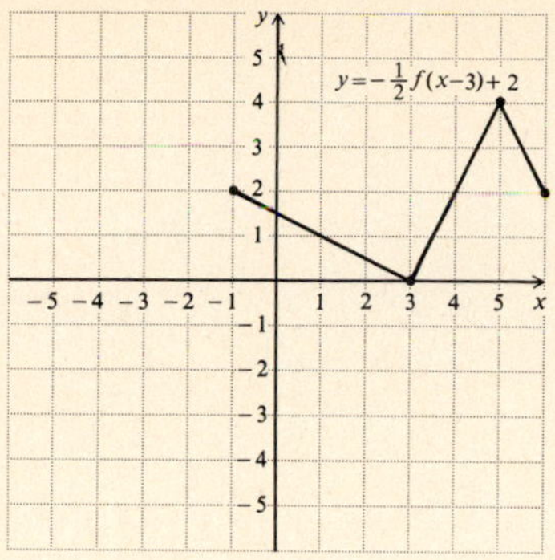

35.

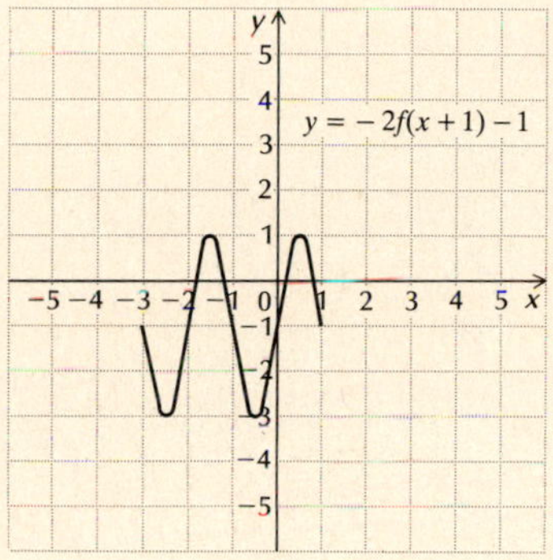

37.

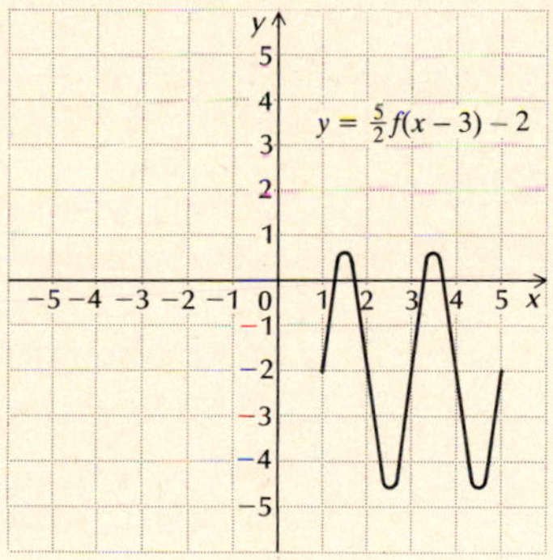

39.

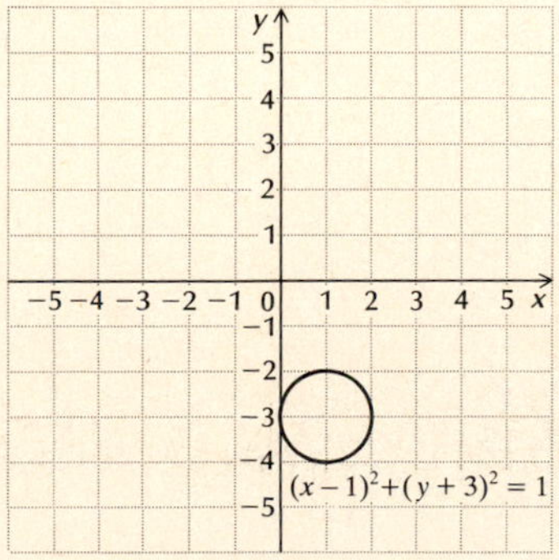

41.

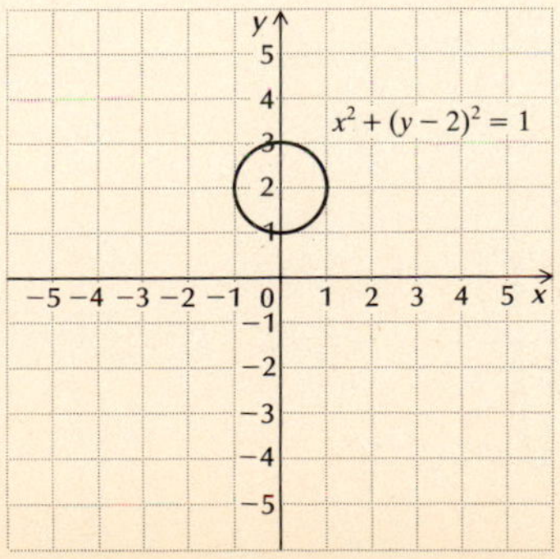

43.

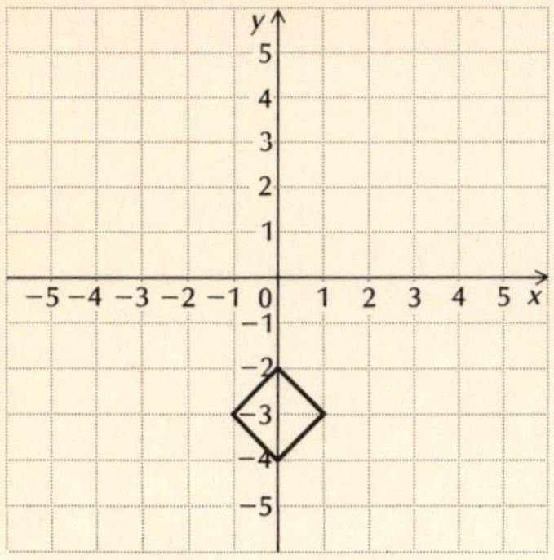

45.

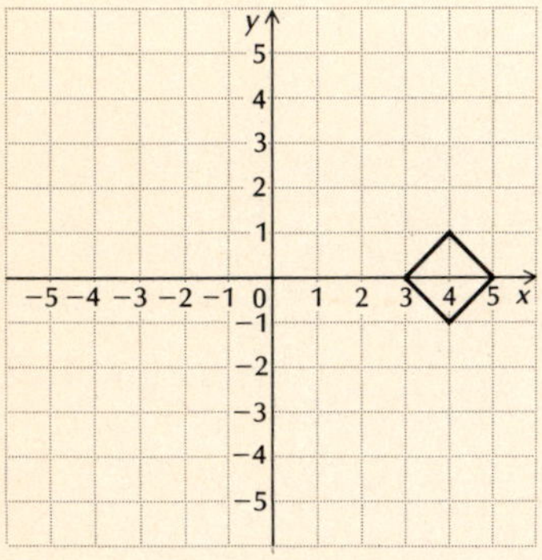

47. The graph is translated 1.8 units to the left, stretched vertically by a factor of $\sqrt{2}$, and reflected across the x-axis.

49. – – – – – – (1) $y = f(x)$ is the given function.

—— —— (2) $\frac{y}{3} = f(2x)$ is the result of shrinking (1) horizontally by a factor of $\frac{1}{2}$ and stretching vertically by a factor of 3

———— (3) $\frac{y}{3} = f\left[2\left(x + \frac{1}{4}\right)\right]$, the required graph, is a translation of (2) to the left through $\frac{1}{4}$ unit.

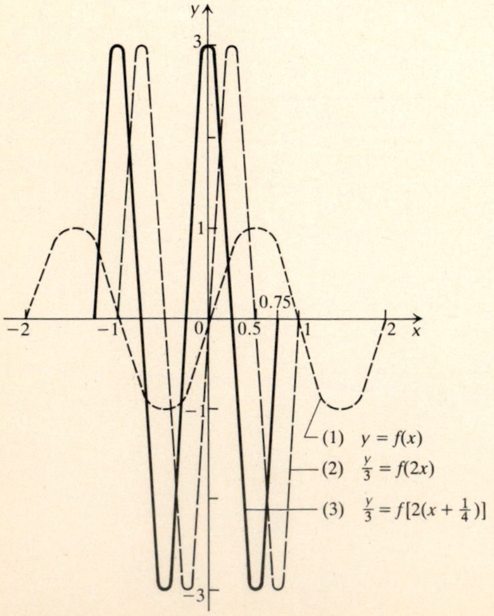

Review Exercises: Chapter 3, pp. 199–201

1. [3.1] {(1, 1), (1, 3), (1, 5), (1, 7), (3, 1), (3, 3), (3, 5), (3, 7), (5, 1), (5, 3), (5, 5), (5, 7), (7, 1), (7, 3), (7, 5), (7, 7)}
2. [3.1] No 3. [3.1] {3, 5, 7} 4. [3.1] {1, 3, 5, 7}

5. [3.2]

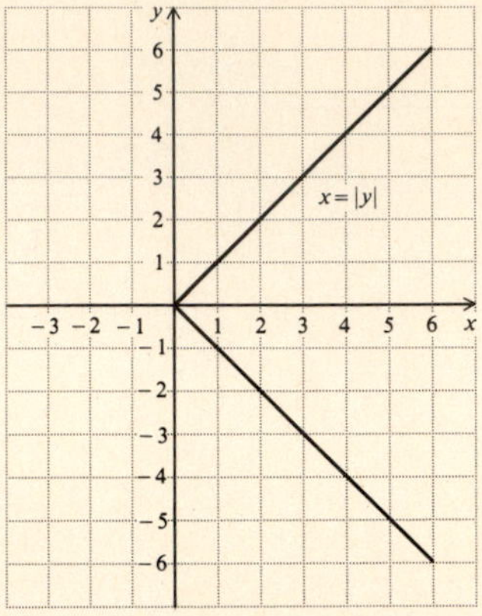

6. [3.2]

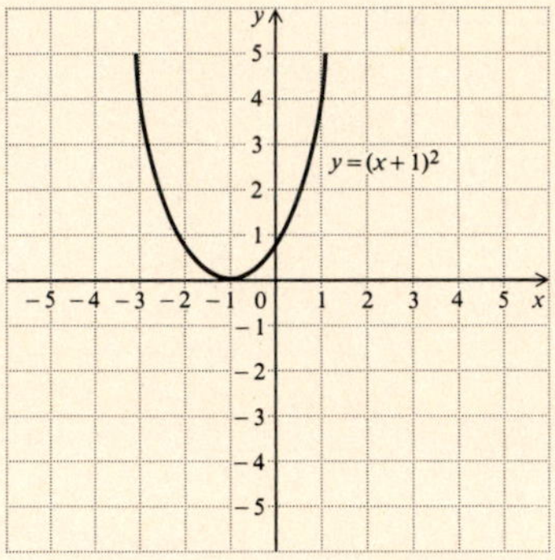

7. [3.4]

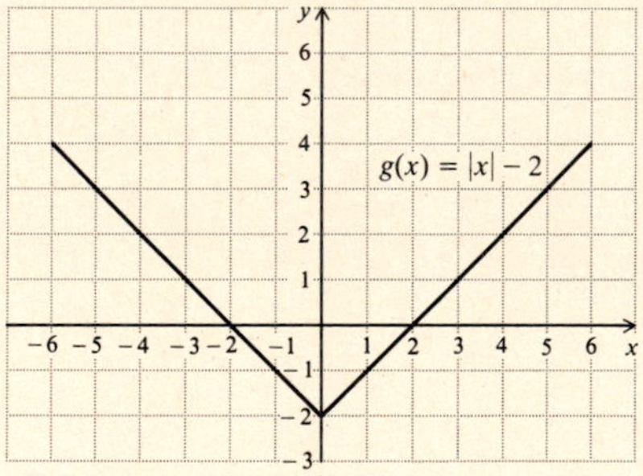

8. [3.4]

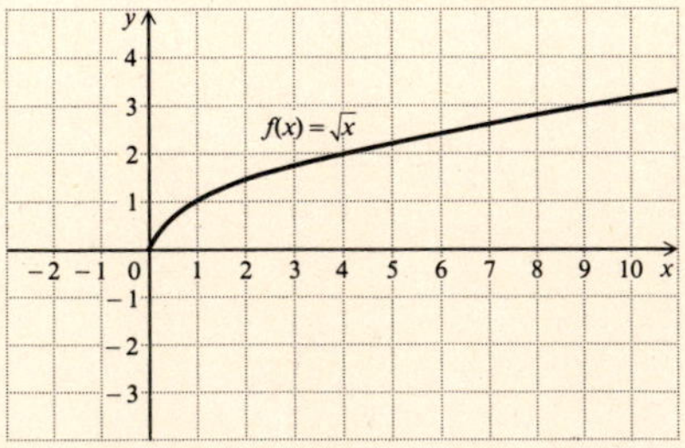

9. [3.4]

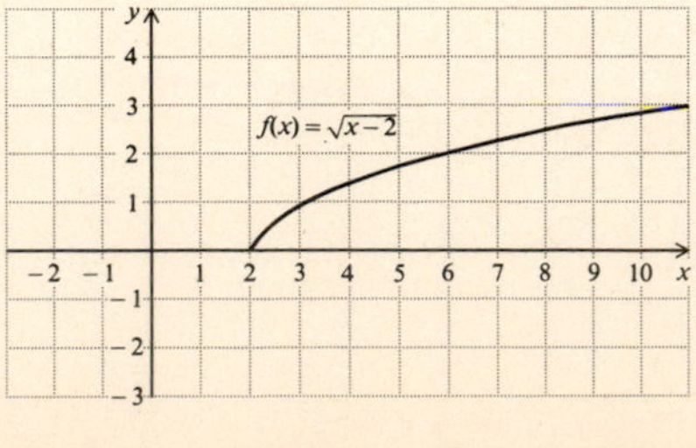

10. [3.4]

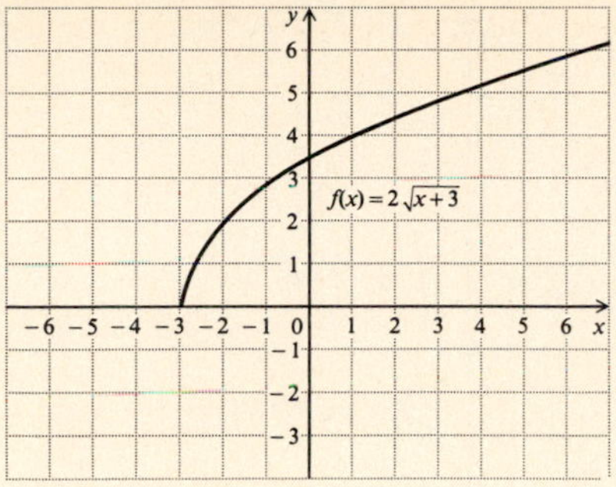

11. [3.4]

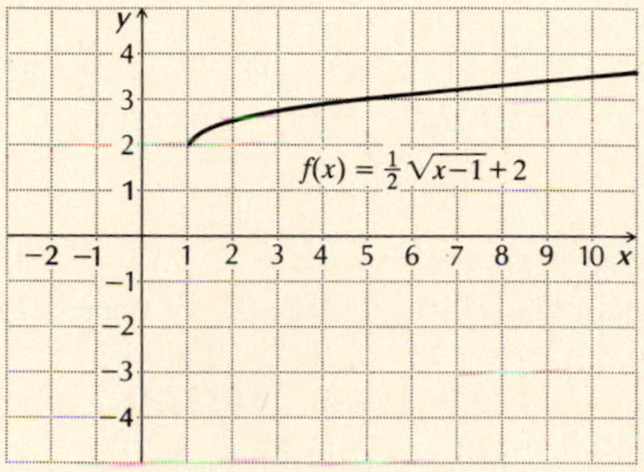

12. [3.6]

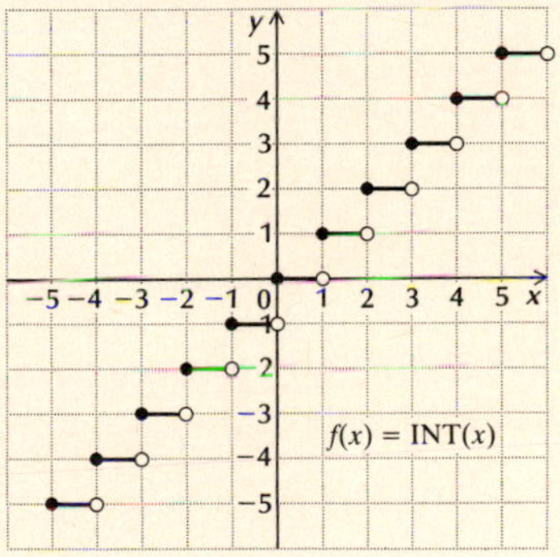

13. [3.6]

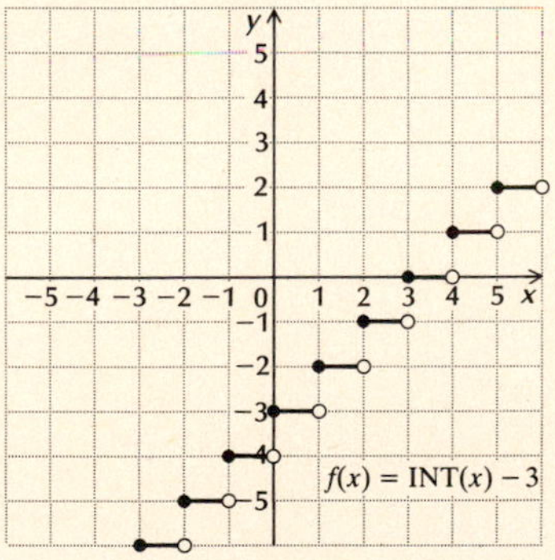

14. [3.6]

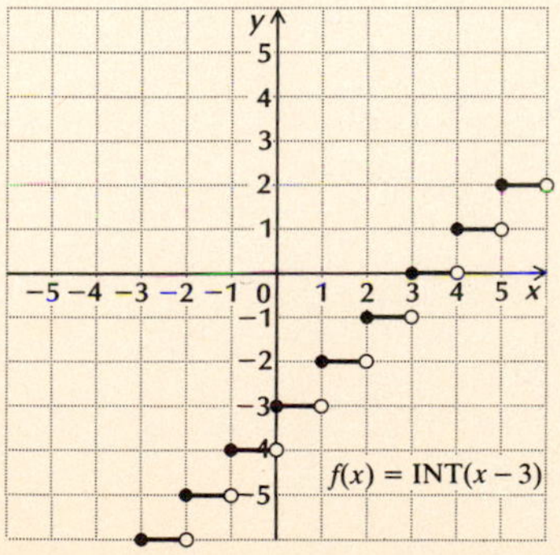

15. [3.5] (b), (d), (f) **16.** [3.5] (a), (b), (d), (g) **17.** [3.5] (b), (c), (d), (h) **18.** [3.7] $P(x) = -0.5x^2 + 105x - 6$ **19.** [3.6] (3, \$16) **20.** [3.3] (b) **21.** [3.3] -3 **22.** [3.3] 9 **23.** [3.3] $2a + h - 1$ **24.** [3.3] 0 **25.** [3.3] 4 **26.** [3.3] $2\sqrt{a+1}$ **27.** [3.3] $\left\{x \middle| x \leq \frac{7}{3}\right\}$ **28.** [3.3] $\{x | x \neq 1, 5\}$ **29.** [3.7] **(a)** $(f+g)(x) = 3 - 2x + \frac{4}{x^2}$, $(f-g)(x) = \frac{4}{x^2} - 3 + 2x$, $fg(x) = \frac{12}{x^2} - \frac{8}{x}$, $(f/g)(x) = \frac{4}{3x^2 - 2x^3}$, $f \circ g(x) = \frac{4}{(3-2x)^2}$, $g \circ f(x) = 3 - \frac{8}{x^2}$; **(b)** all reals except 0, all reals, all reals except 0, all reals except 0, all reals except 0, all reals except 0 and $\frac{3}{2}$, all reals except $\frac{3}{2}$, all reals except 0 **30.** [3.7] **(a)** $(f+g)(x) = 3x^2 + 6x - 1$, $(f-g)(x) = 3x^2 + 2x + 1$, $fg(x) = 6x^3 + 5x^2 - 4x$, $(f/g)(x) = \frac{3x^2 + 4x}{2x - 1}$, $f \circ g(x) = 12x^2 - 4x - 1$, $g \circ f(x) = 6x^2 + 8x - 1$; **(b)** all reals, all reals, all reals, all reals, all reals, all reals except $\frac{1}{2}$, all reals, all reals

31. [3.8] $y = 1 + f(x)$

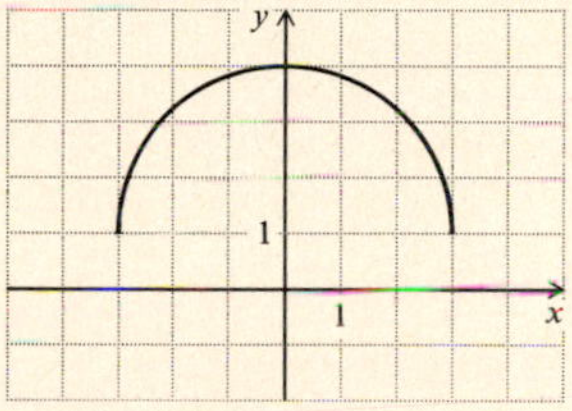

32. [3.8] $y = \frac{1}{2}f(x)$

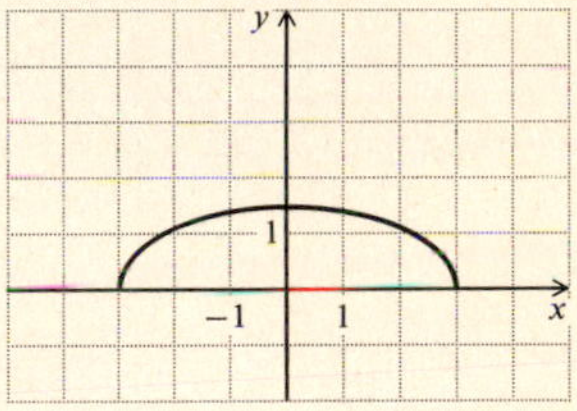

33. [3.8] $y = f(x + 1)$

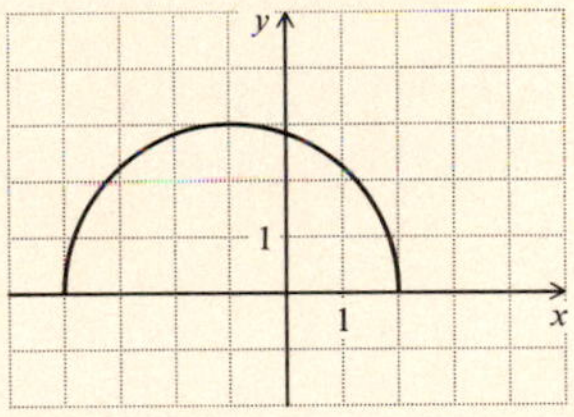

34. [3.5] (a), (b), (c) **35.** [3.5] (e), (f) **36.** [3.5] (d) **37.** [3.6] (a), (c) **38.** [3.6] 2 **39.** [3.6] **(a)** Yes; **(b)** no **40.** [3.6] (c) **41.** [3.6] (a) **42.** [3.6] (b) **43.** [3.6] $[-\pi, 2\pi]$ **44.** [3.6] (0, 1] **45.** [3.6] $(-\infty, 14)$ **46.** [3.6]

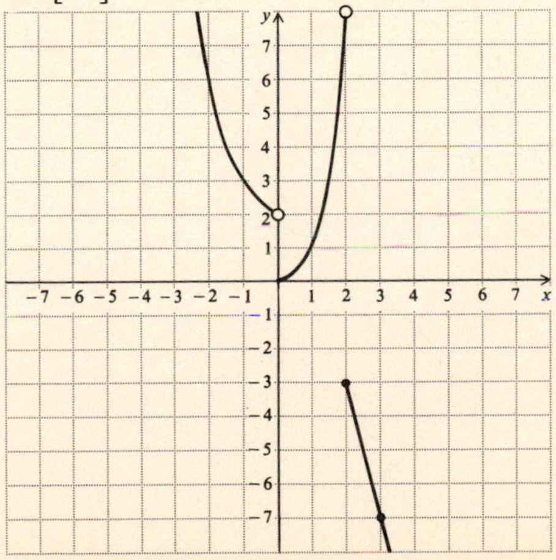

47. [3.6]

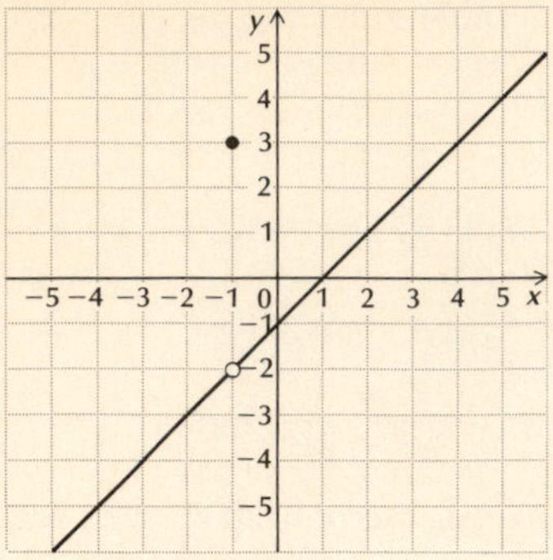

48. [3.4] $d(t) = \sqrt{(55t)^2 + (50t)^2} = t\sqrt{5525} = 5t\sqrt{221}$
49. [3.4] $V(a) = 8\pi a^2 - 128$ **50.** [3.7] **(a)** $f(x) = \sqrt{x}$, $g(x) = 5x + 2$; **(b)** $f(x) = \dfrac{x+1}{x-1}$, $g(x) = x^3$ **51.** [3.3] $\{x | x \neq 0, 3, -3\}$ **52.** [3.3] $\{x | x < 0\}$ **53.** [3.8] Graph $y = f(x)$. Then reflect that portion that lies below the x-axis, across the x-axis.

54. [3.2]

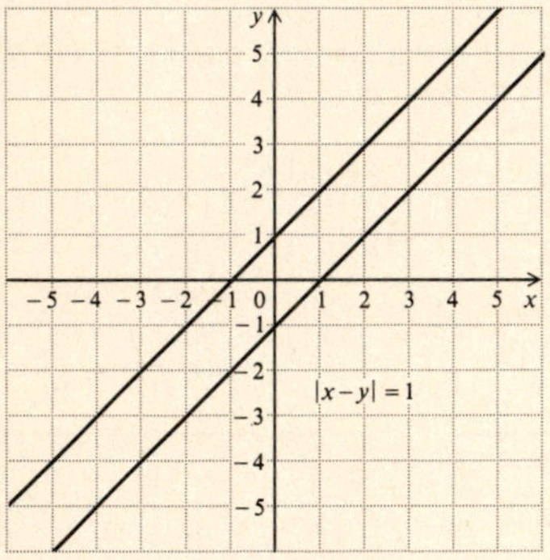

Test: Chapter 3, pp. 201–204

1. [3.1] $\{(1, 3), (1, 6), (3, 3), (3, 6), (7, 3), (7, 6)\}$ **2.** [3.1] Yes
3. [3.1] $\{-2, 2, -7, 7\}$ **4.** [3.1] $\{-2, 0, 2, 7\}$

5. [3.4]

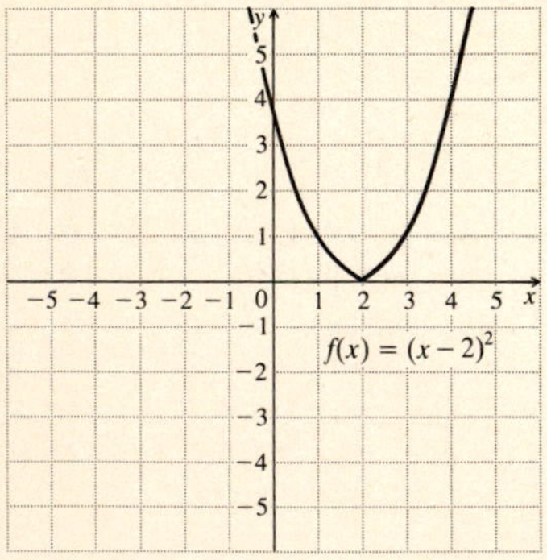

6. [3.2]

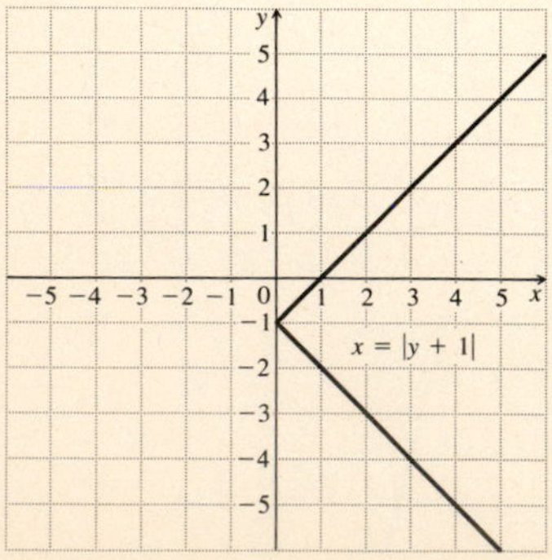

7. [3.6]

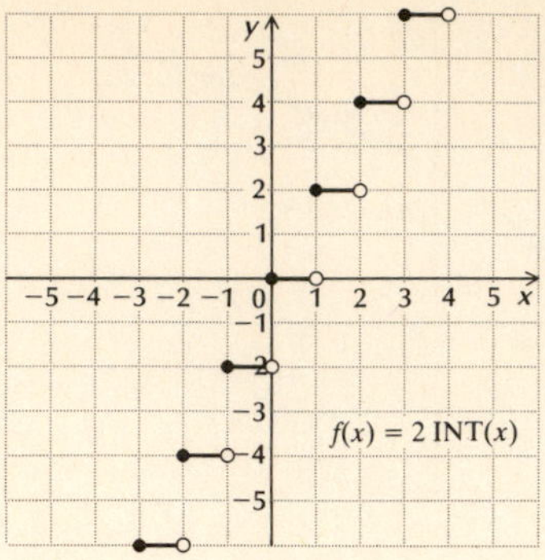

8. [3.5] (a), (d), (e) **9.** [3.5] (b), (e) **10.** [3.3] $\{x | x \neq -4, 4\}$
11. [3.7] $(f + g)(x) = x^2 + 2x + 5$, $(f - g)(x) = -x^2 + 2x - 7$, $fg(x) = 2x^3 - x^2 + 12x - 6$, $(f/g)(x) = \dfrac{2x - 1}{x^2 + 6}$, $f \circ g(x) = 2x^2 + 11$, $g \circ f(x) = 4x^2 - 4x + 7$ **12.** [3.7] All reals, all reals, all reals, all reals, all reals, all reals, all reals, all reals
13. [3.3] (b) **14.** [3.3] 4 **15.** [3.3] 2 **16.** [3.3] $a^2 - 3a + 4$ **17.** [3.3] $2a - 1 + h$ **18.** [3.7] $p(x) = -0.1x^2 + 100x - 20$ **19.** [3.6] (3, $27)

20. [3.8] **(a)**

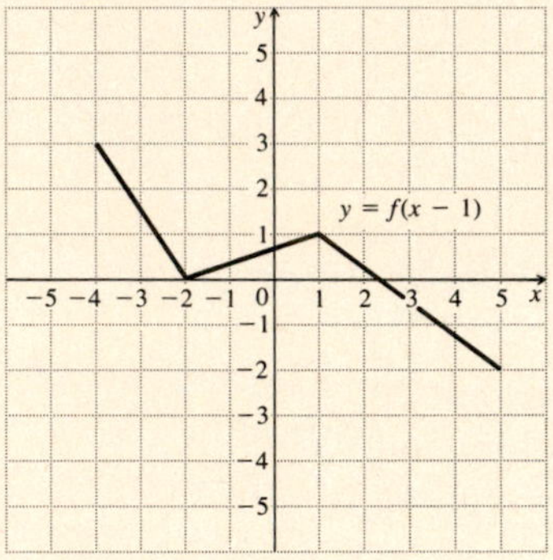

(b)

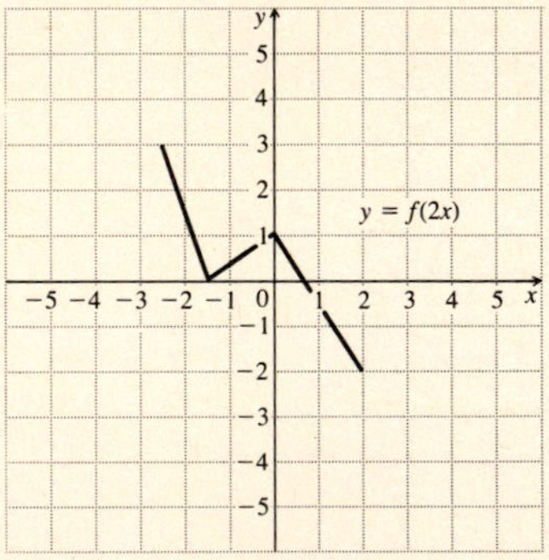

(c)

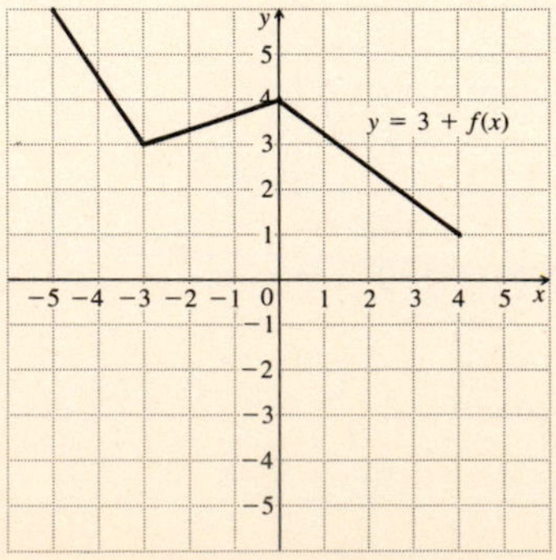

21. [3.5] (a), (c), (e) **22.** [3.5] (d) **23.** [3.6] $(-7, 2)$
24. [3.6] $[-2, \infty)$ **25.** [3.6] (c) **26.** [3.6] 2 **27.** [3.6] **(a)** No; **(b)** yes **28.** [3.6] (b) **29.** [3.6] (c)

30. [3.6]

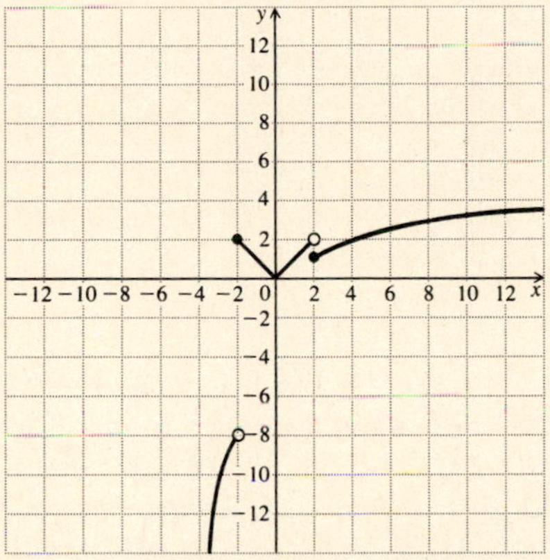

31. [3.4] $A(x) = 2x\sqrt{4 - x^2}$ **32.** [3.7] **(a)** $f(x) = \dfrac{1}{\sqrt{x}}$, $g(x) = 7x + 2$; **(b)** $f(x) = 4x^2 + 9$, $g(x) = 5x - 1$ **33.** [3.7] Let $f(x)$ and $g(x)$ be odd functions. Then by definition, $f(-x) = -f(x)$, or $f(x) = -f(-x)$, and $g(-x) = -g(x)$, or $g(x) = -g(-x)$. Thus, $(f + g)(x) = f(x) + g(x) = -f(-x) + [-g(-x)] = -[f(-x) + g(-x)] = -(f + g)(-x)$ and $f + g$ is odd.

CHAPTER 4

Margin Exercises, Section 4.1

1. (a) Yes; **(b)** yes; **(c)** no; **(d)** no

2.

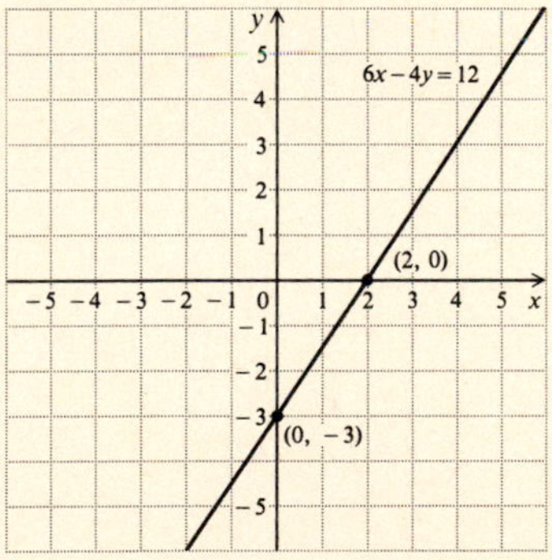

3.

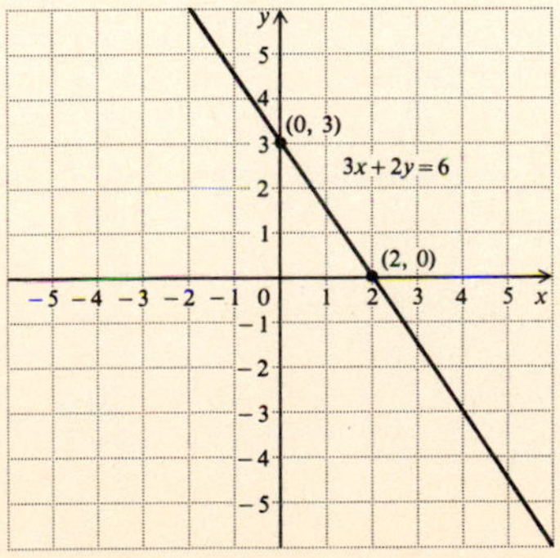

4.

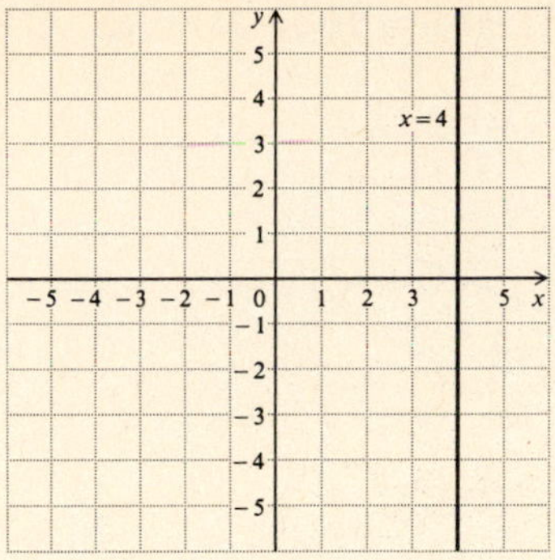

5.

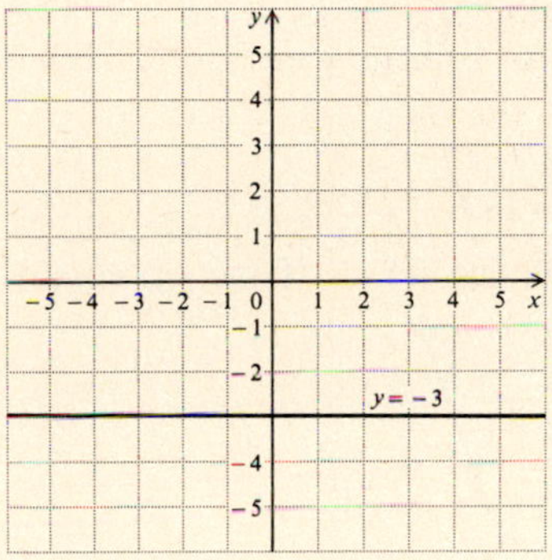

6. $m = 2$ **7.** $m = -2$ **8.** $m = 2$ **9.** $m = 6$ **10.** $m = -1$ **11.** $-\frac{1}{3}$ **12.** 0 **13.** m is not defined **14.** 0 **15.** $-\frac{12}{41}$ **16.** $y = -3x - \frac{23}{4}$ **17.** $y = \dfrac{x}{4} - 9$ **18.** $y = -\dfrac{x}{2} + \dfrac{5}{2}$ **19.** $y = -3x + 7$ **20.** $y = -\dfrac{10x}{3} + 4$ **21.** $m = -7$; $b = 11$ **22.** $m = 0$; $b = -4$ **23. (a)** $y = \frac{2}{3}x + 2$; **(b)** $m = \frac{2}{3}$; $b = 2$

24.

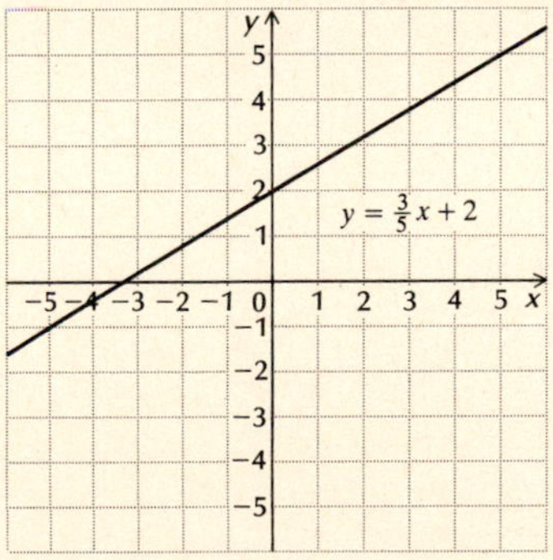

25.

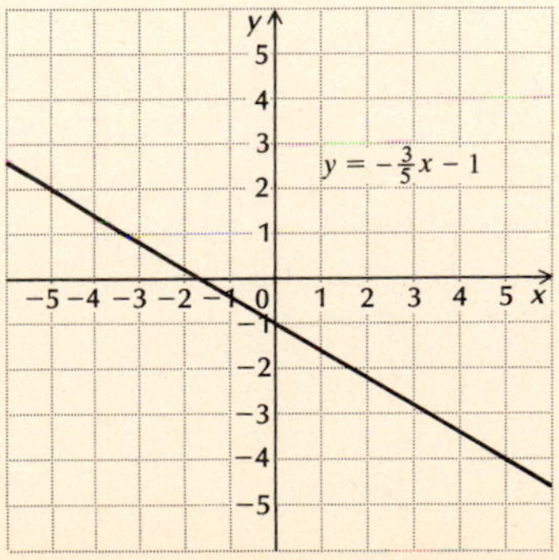

26. **(a)** $P(0) = 1$, $P(5) = 1\frac{5}{33}$, $P(10) = 1\frac{10}{33}$, $P(33) = 2$, $P(200) = 7\frac{2}{33}$;

(b)

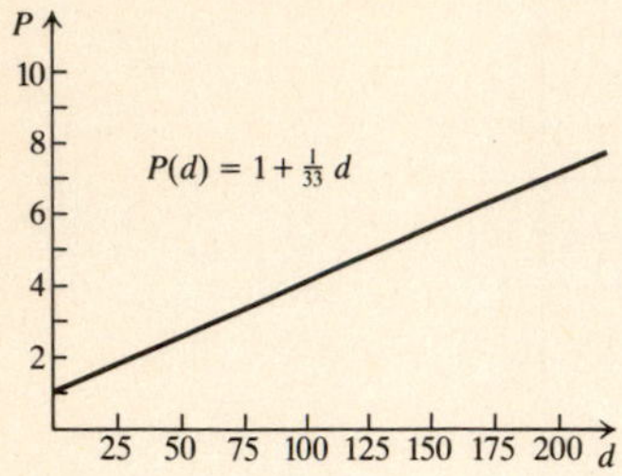

(c) Pressure can only be nonnegative. Thus the function is meaningful only for values of x for which $1 + \frac{1}{33}d \geq 0$. Thus, $d \geq -33$. Since d also can only be nonnegative, the domain is the interval $[0, \infty)$.

Exercise Set 4.1, pp. 215–217

1. **(a)** Yes; **(b)** yes; **(c)** no; **(d)** yes; **(e)** no; **(f)** no; **(g)** no; **(h)** yes

3.

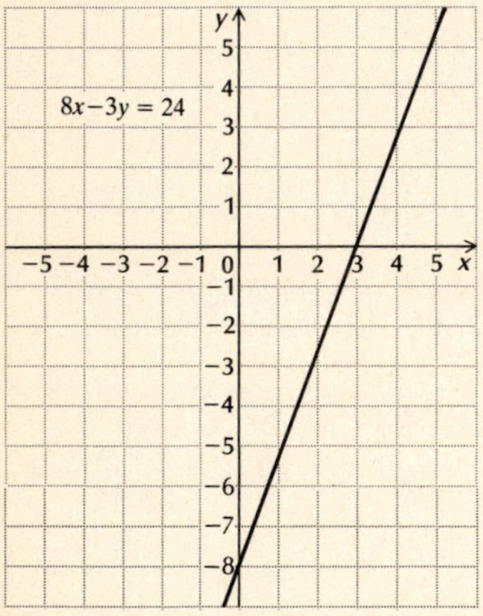

5.

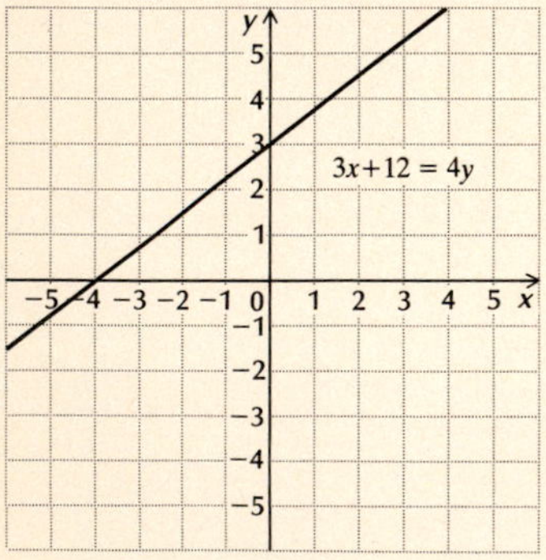

7.

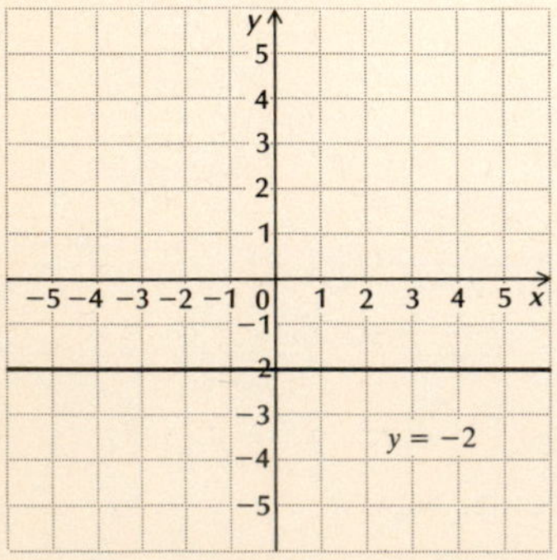

9.

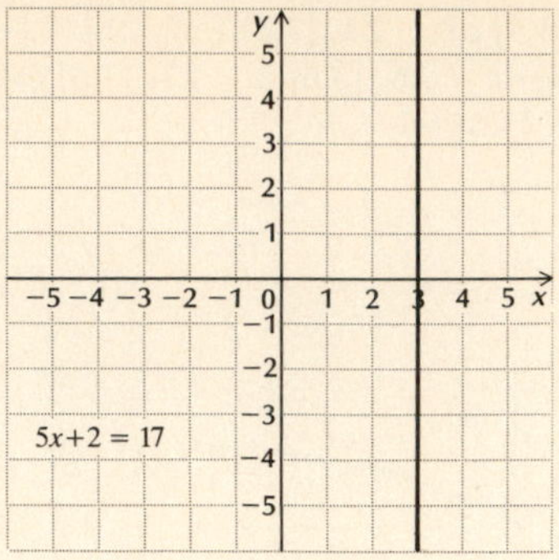

11. $\frac{1}{8}$ **13.** $\frac{1}{2}$

15. Not defined **17.** 0 **19.** Grade is 6.7%; $y = 0.067x$ **21.** Grade is -3% **23.** Approximately $\frac{2}{5}$ ft, or 4.8 in. **25.** $y = 4x - 10$ **27.** $y = 2x - 5$ **29.** $y = -\frac{2}{3}x + \frac{13}{3}$ **31.** $y = -8$ **33.** $y = \frac{1}{2}x + \frac{7}{2}$ **35.** $y = 6x + 17$ **37.** $y = 6$ **39.** $m = 2$; $b = 3$ **41.** $m = -3$; $b = 5$ **43.** $m = \frac{3}{4}$; $b = -3$ **45.** $m = 0$; $b = -\frac{10}{3}$

47.

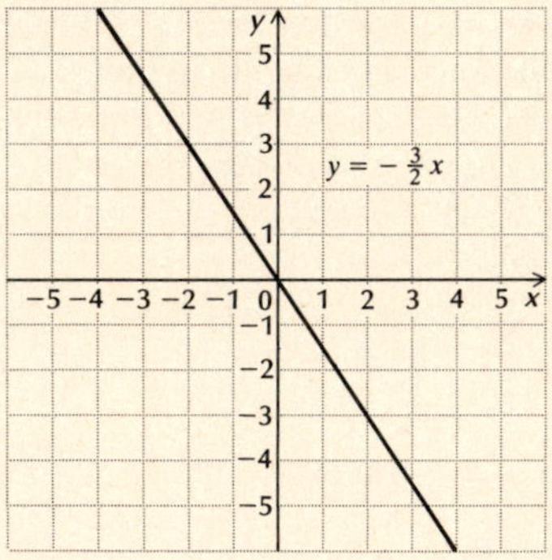

49.

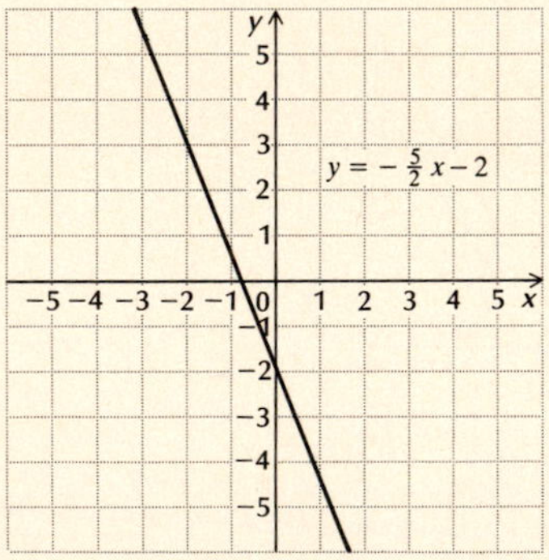

51.

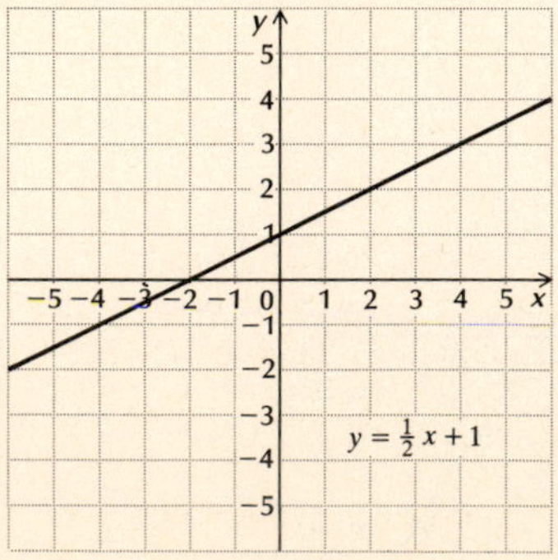

53.

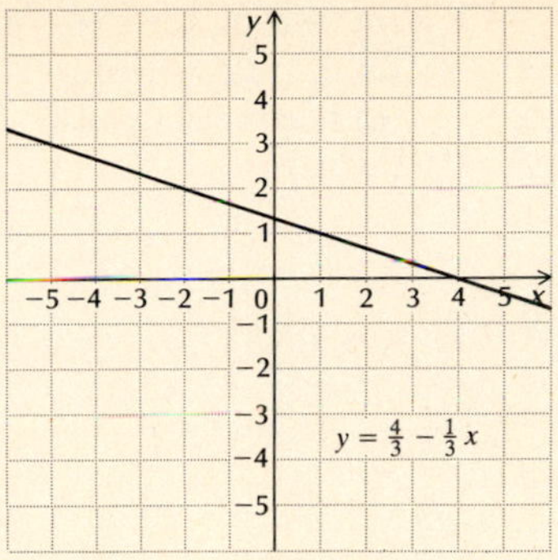

55. $y = 3.516x - 13.1602$ 57. $y = 1.2222x + 1.0949$

59. (a) $T(5) = 70$, $T(20) = 220$, $T(1000) = 10{,}020$;

(b)

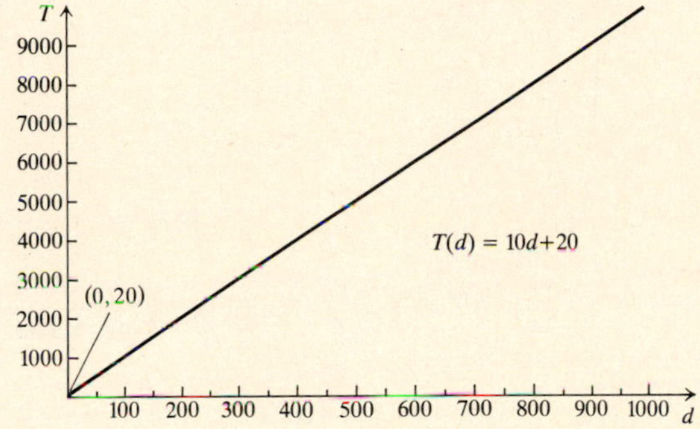

(c) The domain is the interval $[0, 5600]$.

61. (a) $T(0) = -1.18$, $T(50) = 5.97$, $T(80) = 10.26$;

(b)

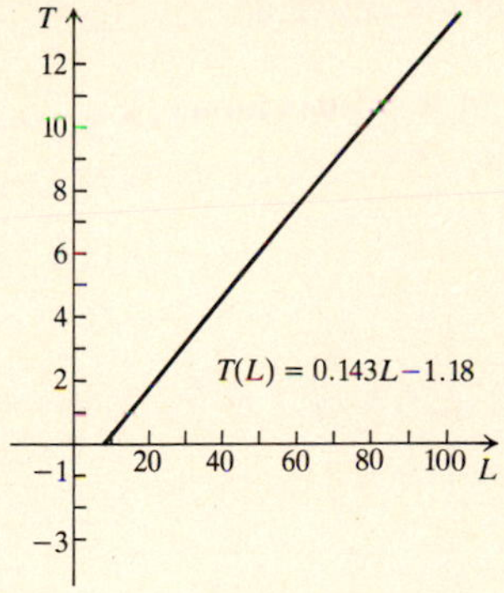

(c) The domain is the interval $[8.25, \infty)$.

63. (a) $A(0) = 2$, $A(1) = 3.1$, $A(4) = 6.4$, $A(10) = 13$;

(b)

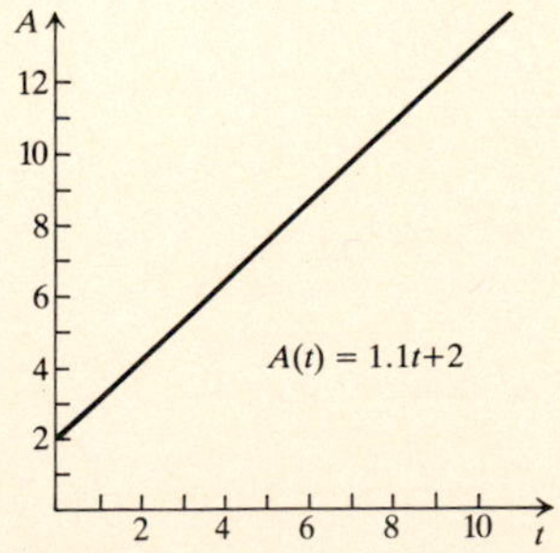

(c) Since the area is given to be greater than or equal to 2 and any value for t less than 0 results in an area less than 2, the domain must be the interval $[0, \infty)$. 65. $f(x) = mx$ 67. $f(x) = x + b$ 69. False 71. False 73. True 75. Yes 77. $\overline{AB}$, $\overline{DC}$: same slope; $\overline{BC}$, $\overline{AD}$: same slope 79. $F = \frac{9}{5}C + 32$

81. $P = mQ + b$, $m \neq 0$. Then we can solve for Q: $Q = \dfrac{P}{m} - \dfrac{b}{m}$.

Margin Exercises, Section 4.2

1. No 2. Yes 3. Perpendicular 4. Neither 5. Parallel 6. Parallel: $y - 4 = -2(x - 3)$, or $y = -2x + 10$; perpendicular: $y - 4 = \frac{1}{2}(x - 3)$, or $y = \frac{1}{2}x + \frac{5}{2}$ 7. Parallel: $x = 5$; perpendicular: $y = -4$ 8. Parallel: $y = \frac{11}{2}$; perpendicular: $x = -6$ 9. $\sqrt{149}$ 10. $6\sqrt{2}$ 11. 16 12. 8 13. Yes 14. No 15. $\left(\frac{3}{2}, -\frac{5}{2}\right)$ 16. $(9, -5)$ 17. $x^2 + y^2 = 20$ 18. $(x + 3)^2 + (y - 7)^2 = 25$ 19. $(-1, 3)$, 2 20. $(7, -2)$, 8 21. $(x + 1)^2 + (y - 4)^2 = 41$

Exercise Set 4.2, pp. 224–226

1. Neither 3. Perpendicular 5. $y = 3x + 3$, $y = -\frac{1}{3}x + 3$ 7. $x = 3$, $y = 8$ 9. $y = -3$, $x = -2$ 11. $y = \frac{5}{2}x + \frac{5}{2}$, $y = -\frac{2}{5}x - \frac{31}{5}$ 13. $x = 0$, $y = 3$ 15. $y = -7$, $x = -3$ 17. $y = 0.6114x + 3.4094$ 19. 5 21. $3\sqrt{2}$ 23. $\sqrt{a^2 + 64}$ 25. $\sqrt{a^2 + b^2}$ 27. $2\sqrt{a}$ 29. 18.8061 31. Yes 33. $(-\frac{1}{2}, -1)$ 35. $(a, 0)$ 37. $(-0.4485, -0.2733)$

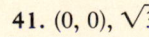

39. $(0, 0)$, 6

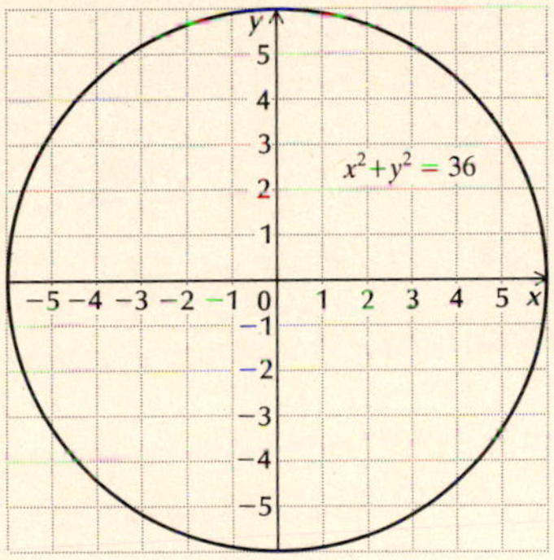

41. $(0, 0)$, $\sqrt{3}$

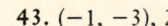

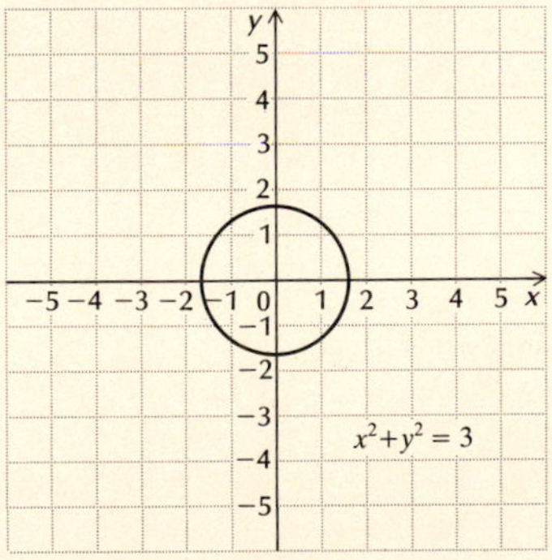

43. $(-1, -3)$, 2

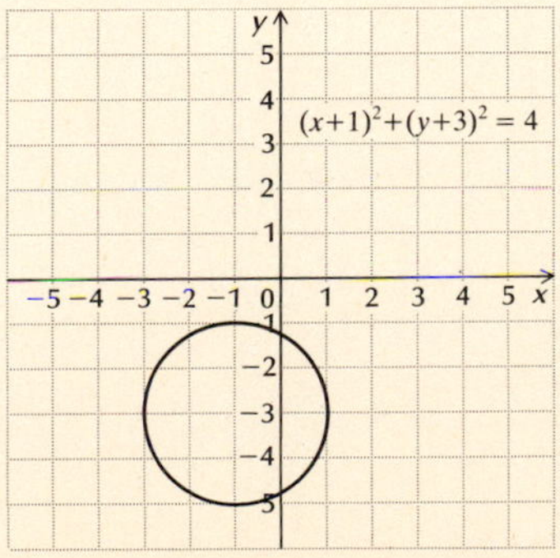

45. $(8, -3)$, $2\sqrt{10}$

47. $(3, 0)$, $\dfrac{1}{5}$

49. $(-4, 3)$, $2\sqrt{10}$

51. $(-3, 0)$, 3

53. $(-4, 0)$, 10

55. $\left(-\frac{21}{2}, -\frac{33}{2}\right)$, $\dfrac{\sqrt{1462}}{2}$

57. $(-4.123, 3.174)$, 10.071

59. $(0, 0)$, $\frac{1}{3}$

61. $x^2 + y^2 = 25$

63. $(x + 4)^2 + (y - 1)^2 = 20$ 65. $y = -\frac{7}{3}x + \frac{22}{3}$, $y = \frac{3}{7}x - \frac{26}{7}$ 67. $(5, 0)$ 69. $k = -\frac{31}{4}$ 71. $y = \frac{5}{4}x + \frac{75}{8}$
73. $(x - 2)^2 + (y - 4)^2 = 16$
75. $(x - 1)^2 + (y - 2)^2 = 41$
77. $(x + 8)^2 + (y - 5)^2 = 25$ 79. $(x - 8)^2 + (y - 9)^2 = \frac{25}{2}$
81. $x^2 + y^2 = 70.56$ 83. $(x - h)^2 + (y - k)^2 = r^2$
85. Yes 87. Yes

89. We have the points $O(0, 0)$, $B(b, 0)$, $H(0, h)$, and $P\left(\frac{b}{2}, \frac{h}{2}\right)$.
Then each of the three distances PO, PB, PH is $\dfrac{\sqrt{b^2 + h^2}}{2}$.

91. $P_1P_2 = \sqrt{(x - x)^2 + (y_2 - y_1)^2} = |y_2 - y_1|$;
$P_1P_2 = \sqrt{(x_2 - x_1)^2 + (y - y)^2} = |x_2 - x_1|$

Margin Exercises, Section 4.3

1. (a)

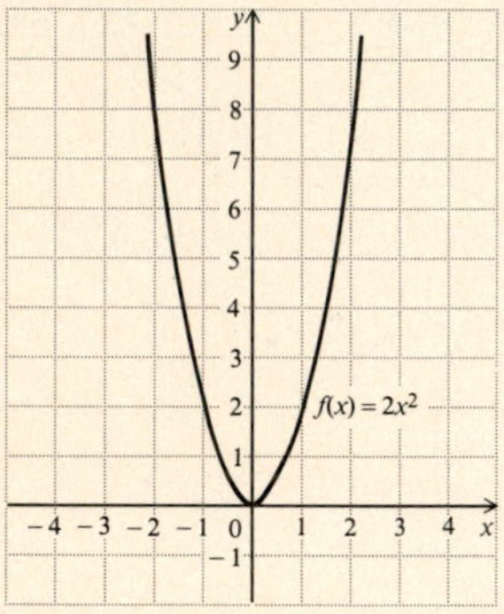

(b) upward; (c) y-axis, $x = 0$; (d) 0; (e) $(0, 0)$

2. (a)

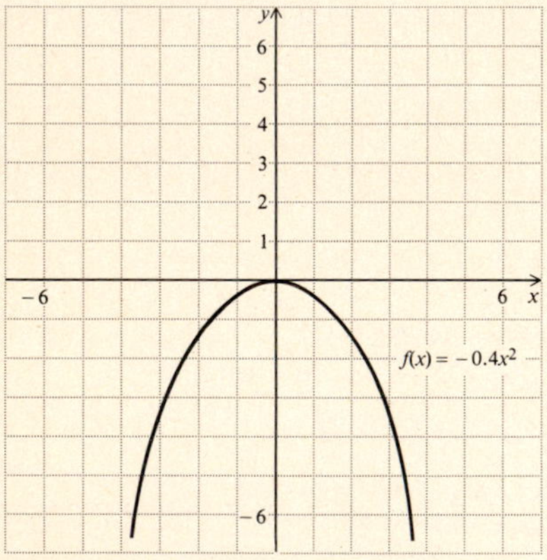

(b) downward; (c) y-axis, $x = 0$; (d) 0; (e) $(0, 0)$

3. (a) and (b)

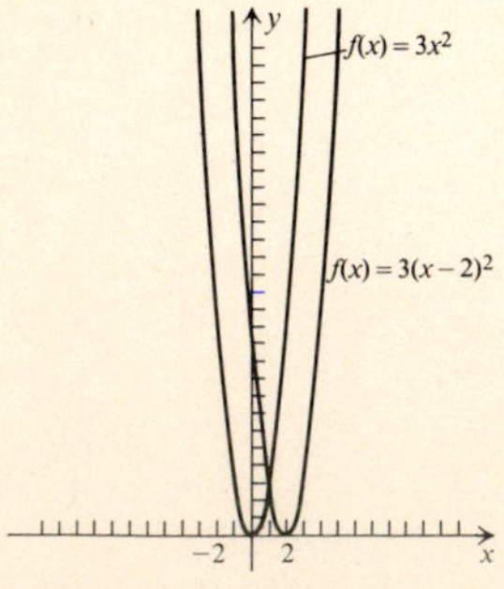

(c) $(2, 0)$; (d) $x = 2$; (e) 0; (f) upward; (g) horizontal translation to the right

4. (a) and (b)

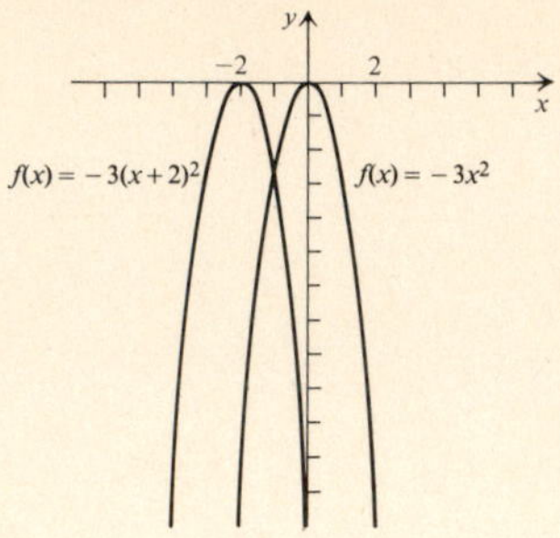

(c) $(-2, 0)$; (d) $x = -2$; (e) 0; (f) downward; (g) horizontal translation to the left

5. (a) $(2, 4)$; (b) $x = 2$; (c) no; (d) yes, 4

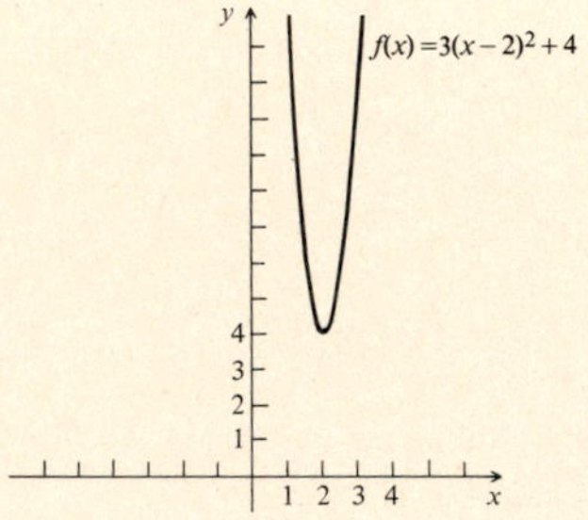

6. (a) $(-2, -1)$; (b) $x = -2$; (c) yes, -1; (d) no

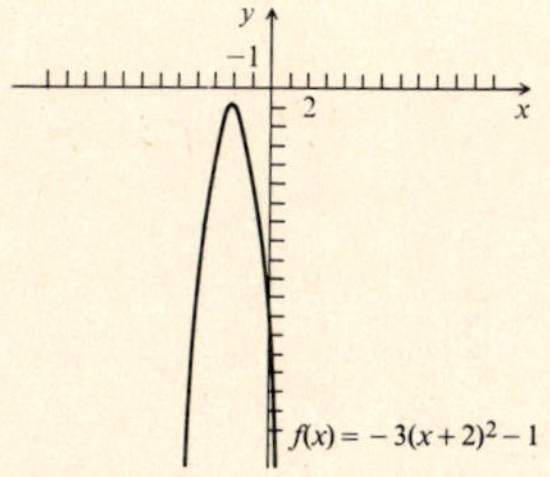

7. (a) $(5, \pi)$; (b) $x = 5$; (c) no; (d) yes, π
8. (a) $(5, 0)$; (b) $x = 5$; (c) yes, 0; (d) no
9. (a) $(-\frac{1}{4}, -6)$; (b) $x = -\frac{1}{4}$; (c) no; (d) yes, -6
10. (a) $(-9, 3)$; (b) $x = -9$; (c) yes, 3; (d) no
11. $f(x) = (x - 2)^2 + 3$ 12. $f(x) = 3(x + 4)^2 - 38$
13. (a) $(\frac{3}{2}, -14)$, $x = \frac{3}{2}$; (b) -14 is a minimum
14. Vertex: $(-3, 34)$; maximum: 34

15.

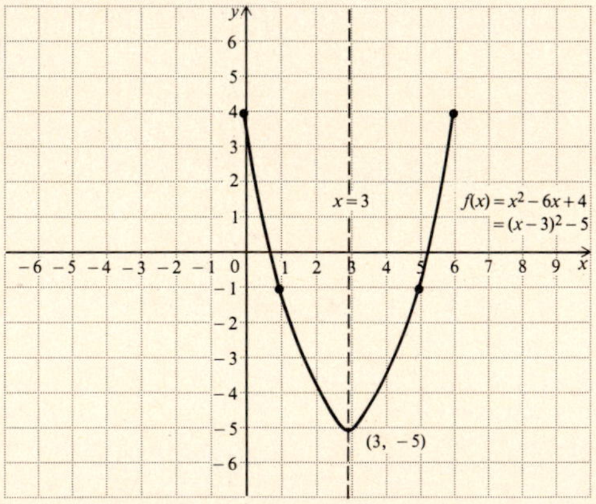

16.

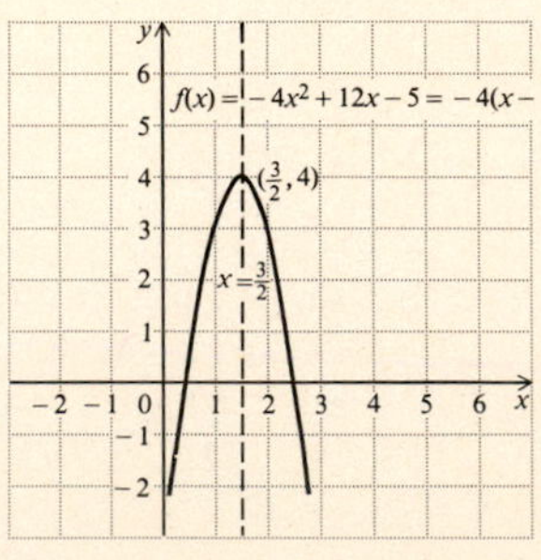

17. $(1 - \sqrt{6}, 0)$, $(1 + \sqrt{6}, 0)$ 18. $(-1, 0)$, $(3, 0)$
19. $(-4, 0)$ 20. None

Exercise Set 4.3, p. 234

1. (a) $(0, 0)$; (b) $x = 0$; (c) 0 is a minimum **3.** (a) $(9, 0)$; (b) $x = 9$; (c) 0 is a maximum **5.** (a) $(1, -4)$; (b) $x = 1$; (c) -4 is a minimum **7.** (a) $f(x) = -(x - 1)^2 + 4$; (b) $(1, 4)$; (c) 4 is a maximum **9.** (a) $f(x) = (x + \frac{3}{2})^2 - \frac{9}{4}$; (b) $(-\frac{3}{2}, -\frac{9}{4})$; (c) $-\frac{9}{4}$ is a minimum **11.** (a) $f(x) = -\frac{3}{4}(x - 4)^2 + 12$; (b) $(4, 12)$; (c) 12 is a maximum **13.** (a) $f(x) = 3(x + \frac{1}{6})^2 - \frac{49}{12}$; (b) $(-\frac{1}{6}, -\frac{49}{12})$; (c) $-\frac{49}{12}$ is a minimum **15.** Maximum: 0 **17.** Minimum: 5

19. Maximum: $-\dfrac{207}{16}$ **21.** Minimum: $-\dfrac{3}{4}$ **23.** Maximum: $\$\dfrac{435{,}625}{3}$

25. $f(x) = -x^2 + 2x + 3$
$= -(x - 1)^2 + 4$

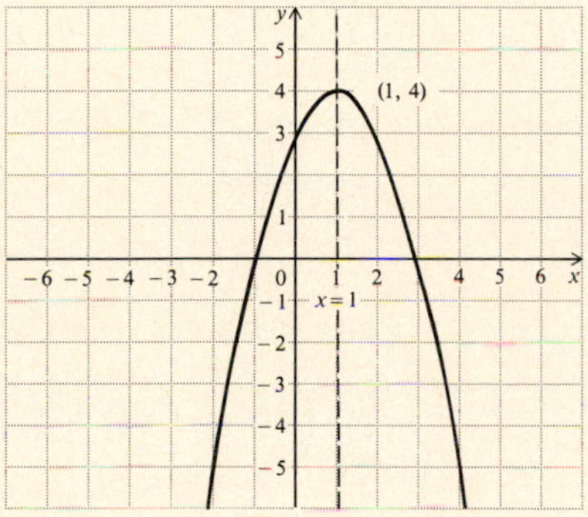

27. $f(x) = x^2 - 8x + 19 = (x - 4)^2 + 3$

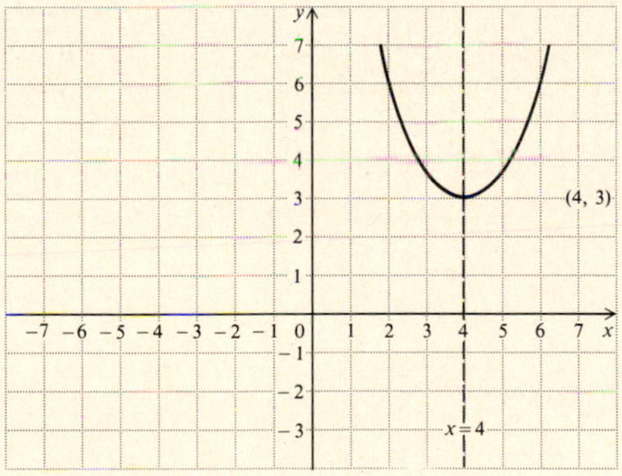

29. $f(x) = -\frac{1}{2}x^2 - 3x + \frac{1}{2}$
$= -\frac{1}{2}(x + 3)^2 + 5$

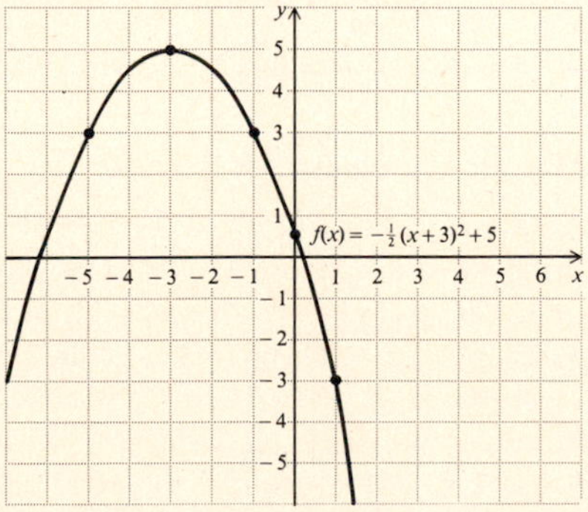

31. $f(x) = 3x^2 - 24x + 50$
$= 3(x - 4)^2 + 2$

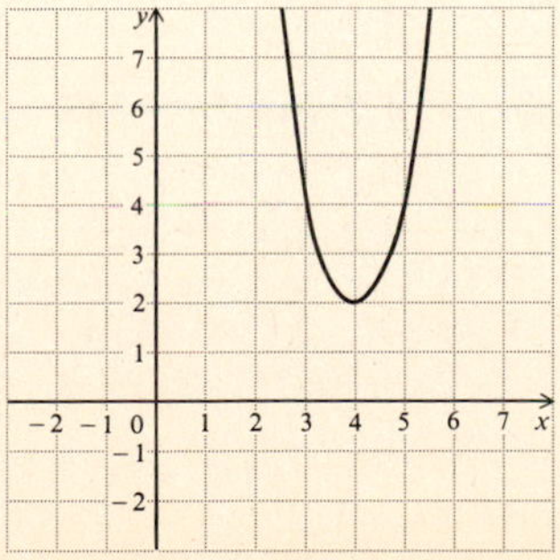

33. $(3, 0), (-1, 0)$ **35.** $(4 \pm \sqrt{11}, 0)$ **37.** None

39. $f(x) = a\left[x - \left(-\dfrac{b}{2a}\right)\right]^2 + \dfrac{4ac - b^2}{4a}$

41.

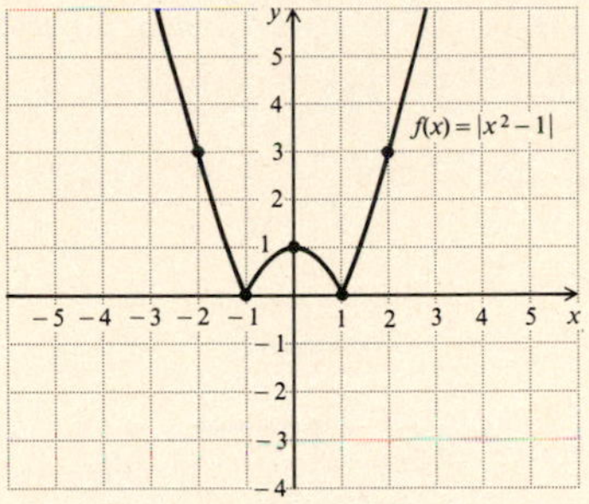

43. Minimum, -6.95 **45.** $a = \dfrac{3}{4}$ **47.** $c = -\dfrac{945}{4}$ **49.** $\left(\dfrac{1}{q}, \dfrac{q^2 - 1}{q}\right)$ **51.** $c = 9$

Margin Exercises, Section 4.4

1. (a) $S = \frac{14}{15}d + 9\frac{2}{3}$; (b) 85

2. (a)

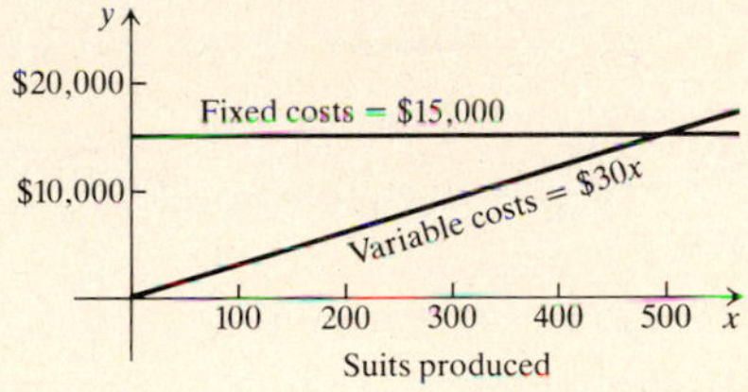

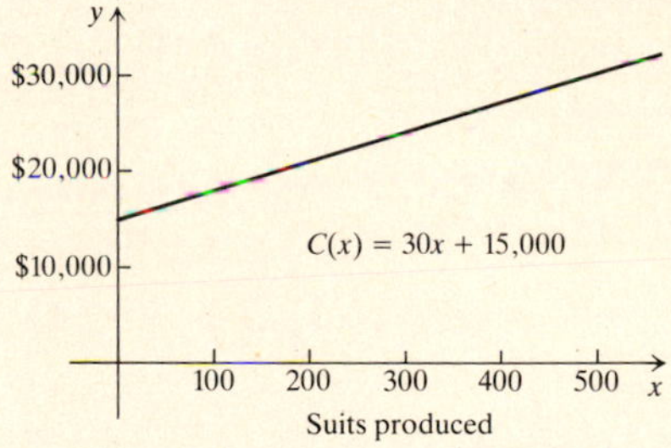

(b) $C(100) = \$18{,}000$, $C(400) = \$27{,}000$; (c) \$9000

3. (a) and (b)

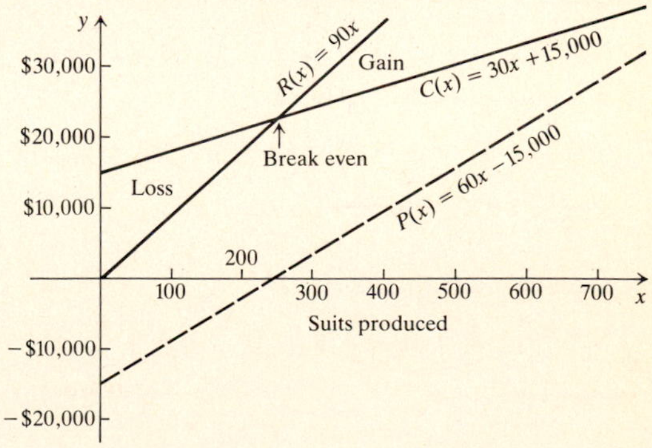

(c) 250

4. Maximum height 12.4 m, at $t = \dfrac{2}{7}$. Reaches ground in 1.88 sec. **5.** Maximum area: 400 ft^2; length: 20 ft; width: 20 ft

Exercise Set 4.4, pp. 242–245

1. (a) $E = 0.15t + 72$; (b) 78.45, 79.05
3. (a) $R = -0.01t + 10.43$; (b) 9.67, 9.63; (c) 2063
5. (a) $P = 0.005375t + 0.016$; (b) 15.575%, 22.5625%
7. (a) $C(x) = 40x + 22{,}500$; (b) $R(x) = 85x$; (c) $P(x) = 45x - 22{,}500$; (d) Profit: \$112,500; (e) 500 **9.** 25 and 25; maximum product = 625 **11.** Base: 10 cm; height: 10 cm

13. 30 yd by 60 yd; maximum area = 1800 yd^2 **15.** 13.5 ft by 13.5 ft; 182.25 ft^2 **17.** 46 units; maximum profit = \$1048 **19.** 70 units; maximum profit = \$19 **21.** **(a)** $R(x) = 150x - 0.5x^2$; **(b)** $P(x) = -0.75x^2 + 150x - 4000$; **(c)** 100; **(d)** \$3500; **(e)** \$100 **23.** **(a)** 1662.5 m after 15 sec; **(b)** after 33.4 sec. **25.** $x = y = \dfrac{24}{4 + \pi}$ **27.** \$26 **29.** \$6.50 **31.** 6 cm by 4.5 cm; 27 cm^2

Margin Exercises, Section 4.5

1. $\{-3, 4\}$ **2.** $\{2, d, e\}$ **3.** 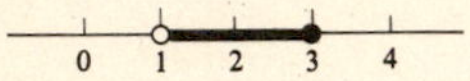

4. 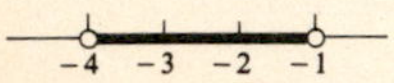 **5.** 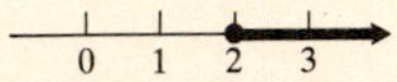

6. $\emptyset$; nothing to graph **7.** $\{1, 2, 3, 4, 5\}$
8. $\{-3, -4, 1, 2, 3, 4, 8, 9, 11\}$ **9.** $\{1, 2, a, b, c, d, e\}$

10. 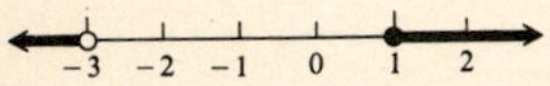 **11.** 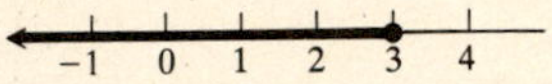

12. 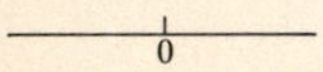 **13.** 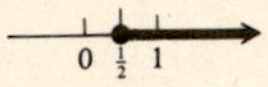 **14.** $4 < x < 8$

15. $-3 \leq x < 0$ **16.** $-5 \leq x \leq -2$ **17.** $-1 < x \leq -\frac{1}{4}$
18. $-\frac{1}{2} < x$ and $x < 1$ **19.** $-\frac{17}{3} \leq x$ and $x < -2$
20. $\frac{19}{4} \leq x$ and $x \leq \frac{37}{6}$
21. $\{x \mid -\frac{2}{3} < x \leq 4\} = \{x \mid -\frac{2}{3} < x\} \cap \{x \mid x \leq 4\}$

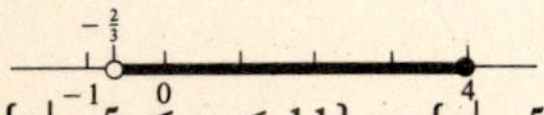

22. $\{x \mid -5 < x < 11\} = \{x \mid -5 < x\} \cap \{x \mid x < 11\}$

23. $\{x \mid -\frac{1}{3} \leq x \leq \frac{1}{6}\} = \{x \mid -\frac{1}{3} \leq x\} \cap \{x \mid x \leq \frac{1}{6}\}$

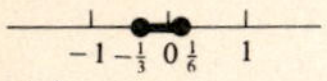

24. $\{x \mid x < -7 \text{ or } x > -1\} = \{x \mid x < -7\} \cup \{x \mid x > -1\}$

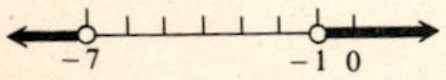

25. $\{x \mid x \leq -1 \text{ or } x > 4\} = \{x \mid x \leq -1\} \cup \{x \mid x > 4\}$

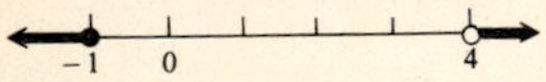

26. $\{x \mid x \geq \frac{5}{3} \text{ or } x \leq 1\} = \{x \mid x \geq \frac{5}{3}\} \cup \{x \mid x \leq 1\}$

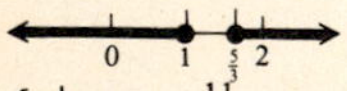

27. $\{x \mid x < -\frac{11}{4} \text{ or } x \geq \frac{1}{4}\} = \{x \mid x < -\frac{11}{4}\} \cup \{x \mid x \geq \frac{1}{4}\}$

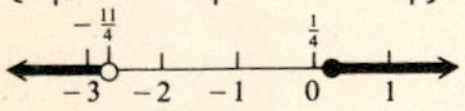

Exercise Set 4.5, pp. 249–251

1. $\{3, 4, 5\}$ **3.** $\{0, 2, 4, 6, 8, 9\}$ **5.** $\{c\}$

7. 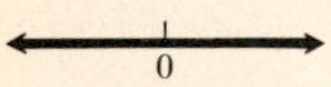 **9.** 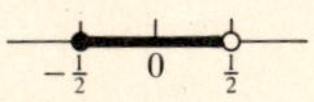

11. 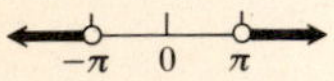 **13.** 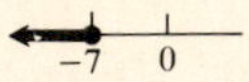

15. $\emptyset$ **17.** $\{x \mid -3 \leq x < 3\}$ **19.** $\{x \mid 8 \leq x \leq 10\}$
21. $\{-7\}$ **23.** $\{x \mid -\frac{3}{2} < x < 2\}$ **25.** $\{x \mid 1 < x \leq 5\}$
27. $\{x \mid -\frac{11}{3} < x \leq \frac{13}{3}\}$ **29.** $\{x \mid x \leq -2 \text{ or } x > 1\}$
31. $\{x \mid x \leq -\frac{7}{2} \text{ or } x \geq \frac{1}{2}\}$ **33.** $\{x \mid x < 9.6 \text{ or } x > 10.4\}$
35. $\{x \mid x \leq -\frac{57}{4} \text{ or } x \geq -\frac{55}{4}\}$ **37.** $\{w \mid 4.885 \text{ cm} < w < 53.67 \text{ cm}\}$
39. $\{S \mid 97\% \leq S \leq 100\%\}$; yes **41.** $\{F \mid 1981\frac{2}{5}° \leq F < 4676°\}$
43. $\{d \mid 62\frac{2}{9} < d < 136\frac{8}{27}\}$ **45.** $\{1\}$ **47.** $\{x \mid x > -\frac{1}{5}\}$
49. $\{x \mid x \geq -\frac{3}{2}\}$ **51.** $\{x \mid x > 2\}$

Margin Exercises, Section 4.6

1. $\{-5, 5\}$ 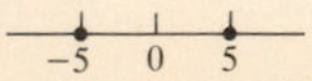 **2.** $\{-\frac{1}{4}, \frac{1}{4}\}$ 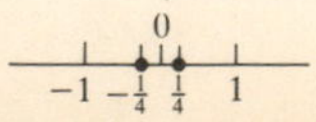

3. $\{x \mid -5 < x < 5\}$, or $(-5, 5)$ 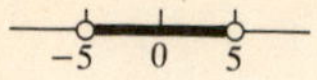

4. $\{x \mid -\frac{1}{4} \leq x \leq \frac{1}{4}\}$, or $[-\frac{1}{4}, \frac{1}{4}]$ 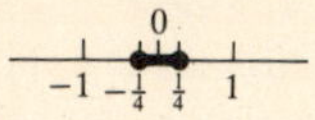

5. $\{x \mid x \leq -5 \text{ or } x \geq 5\}$, or $(-\infty, -5] \cup [5, \infty)$ 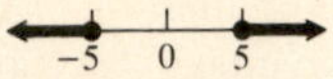

6. $\{x \mid x < -\frac{1}{4} \text{ or } x > \frac{1}{4}\}$, or $(-\infty, -\frac{1}{4}) \cup (\frac{1}{4}, \infty)$ 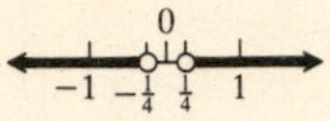

7. $\{-5, 3\}$ **8.** $\{x \mid -9 < x < -5\}$, or $(-9, -5)$
9. $\{x \mid x < 1 \text{ or } x > 5\}$, or $(-\infty, 1) \cup (5, \infty)$ **10.** $\{-1, \frac{11}{3}\}$
11. $\{x \mid -\frac{3}{10} < x < \frac{1}{10}\}$, or $(-\frac{3}{10}, \frac{1}{10})$
12. $\{x \mid x < -1 \text{ or } x > \frac{11}{3}\}$, or $(-\infty, -1) \cup (\frac{11}{3}, \infty)$
13. $\{-\frac{2}{3}, 8\}$ **14.** $\{-\frac{3}{2}\}$ **15.** $\{x \mid x \geq -\frac{15}{13}\}$, or $(-\frac{15}{13}, \infty)$
16. $\{x \mid x \leq -\frac{5}{13}\}$, or $(-\infty, -\frac{5}{13})$

Exercise Set 4.6, pp. 255–256

1. $\{-7, 7\}$; 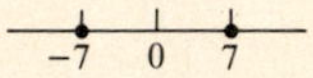

3. $\{x \mid -7 < x < 7\}$, or $(-7, 7)$; 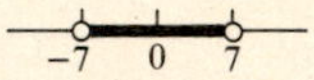

5. $\{x \mid x \leq -\pi \text{ or } x \geq \pi\}$, $(-\infty, -\pi] \cup [\pi, \infty)$; 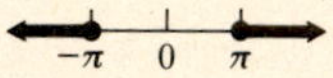

7. $\{-3, 5\}$ **9.** $(-17, 1)$ **11.** $(-\infty, -17] \cup [1, \infty)$
13. $(-\frac{1}{4}, \frac{3}{4})$ **15.** $\{-\frac{1}{3}, \frac{1}{3}\}$ **17.** $\{-1, -\frac{1}{3}\}$ **19.** $(-\frac{1}{3}, \frac{1}{3})$
21. $[-6, 3]$ **23.** $(-\infty, 4.9) \cup (5.1, \infty)$ **25.** $[-\frac{7}{3}, 1]$
27. $[-\frac{1}{2}, \frac{7}{2}]$ **29.** $(-\infty, -8) \cup (7, \infty)$ **31.** $(-\infty, -\frac{7}{4}) \cup (-\frac{3}{2}, \infty)$
33. $[\frac{3}{8}, \frac{9}{8}]$ **35.** $\emptyset$ **37.** $\emptyset$ **39.** $(-\infty, -22.2182) \cup (-12.2158, \infty)$
41. $(-\infty, -0.5746] \cup [7.9277, \infty)$ **43.** $\{\frac{5}{3}, 11\}$ **45.** $\{0, \frac{24}{7}\}$
47. $[2, \infty)$ **49.** $\{-\frac{3}{8}, \frac{7}{6}\}$ **51.** $[\frac{2}{5}, 5]$ **53.** $(-\infty, \frac{3}{5}] \cup [5, \infty)$
55. $\{\frac{4}{5}, 2\}$ **57.** $\{-4, 4\}$ **59.** $(-\infty, \frac{3}{2}]$ **61.** $(-\frac{9}{2}, \frac{11}{2})$
63. $(-\infty, -\frac{8}{3}) \cup (-2, \infty)$ **65.** $\{-\frac{3}{2}, \frac{5}{4}\}$ **67.** If $a = 0$, then $-|0| \leq 0 \leq |0|$. If $a < 0$, then $-|a| = a \leq a \leq |a|$, or $-|a| \leq a \leq |a|$. If $a > 0$, then $-|a| = -a \leq a \leq |a|$, or $-|a| \leq a \leq |a|$.

69. By the triangle inequality, $|a + b| < |a| + |b| < \dfrac{e}{2} + \dfrac{e}{2}$; hence $|a + b| < e$. **71.** 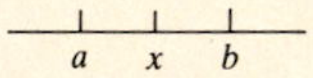 Assume: $a \neq b$. Since x is the midpoint, $|x - a| = |x - b|$, or $x - a = \pm(x - b)$. The + sign gives $a = b$, not admissable; the − sign gives $x = \dfrac{a + b}{2}$.

Margin Exercises, Section 4.7

1. $\{x \mid x < -3 \text{ or } x > 1\}$, or $(-\infty, -3) \cup (1, \infty)$
2. $\{x \mid -3 < x < 1\}$, or $(-3, 1)$ **3.** $\{x \mid -3 \leq x \leq 1\}$, or $[-3, 1]$ **4.** $\{x \mid x < -4 \text{ or } x > 1\}$, or $(-\infty, -4) \cup (1, \infty)$
5. $\{x \mid -4 \leq x \leq 1\}$, or $[-4, 1]$ **6.** $\{x \mid x < -4 \text{ or } x > 1\}$, or $(-\infty, -4) \cup (1, \infty)$ **7.** $\{x \mid -4 \leq x \leq 1\}$, or $[-4, 1]$
8. $\{x \mid x < -1 \text{ or } 0 < x < 1\}$, or $(-\infty, -1) \cup (0, 1)$
9. $\{x \mid 2 < x \leq \frac{7}{2}\}$, or $(2, \frac{7}{2}]$ **10.** $\{x \mid x < 5 \text{ or } x > 10\}$, or $(-\infty, 5) \cup (10, \infty)$

Exercise Set 4.7, pp. 261–263

1. $(-\infty, -5) \cup (3, \infty)$ **3.** $[-2, 1]$ **5.** $(-2, 1)$ **7.** $(-\infty, -1] \cup [1, \infty)$ **9.** $(-\infty, -3] \cup [3, \infty)$ **11.** $(-\infty, \infty)$ **13.** $(2, 4)$
15. $\left(-3, \dfrac{5}{4}\right)$ **17.** $\left(-\infty, \dfrac{-1 - \sqrt{41}}{4}\right) \cup \left(\dfrac{-1 + \sqrt{41}}{4}, \infty\right)$
19. $(-\infty, -2) \cup (0, 2)$ **21.** $(-3, -1) \cup (2, \infty)$
23. $(-\infty, -3) \cup (-2, 1)$ **25.** $(4, \infty)$ **27.** $\left(0, \dfrac{1}{3}\right)$
29. $\left(-\infty, -\dfrac{2}{3}\right) \cup (3, \infty)$ **31.** $[-2, 0)$ **33.** $(\frac{3}{2}, 4]$

35. $(-\infty, -\frac{5}{2}] \cup (-2, \infty)$ **37.** $(-\infty, 0)$ **39.** $(-\infty, -\frac{11}{7})$
41. $(1, \infty)$ **43.** $(0, 2) \cup (2, \infty)$
45. $(-\infty, 0) \cup [1, \infty)$ **47.** $(-\infty, -3) \cup (-2, -1) \cup (2, 0)$
49. $[-\sqrt{2}, \sqrt{2}]$ **51.** $(-\infty, \frac{5}{3}) \cup (11, \infty)$ **53.** $\emptyset$
55. $(-\infty, -\frac{1}{4}) \cup (\frac{1}{2}, \infty)$ **57.** $(-\infty, -5] \cup [5, \infty)$
59. $(-\infty, 0) \cup (0, \infty)$ **61.** $(-4, -2) \cup (-1, 1)$
63. $\{h \mid h > -2 + 2\sqrt{6} \text{ cm}\}$ **65. (a)** $\{x \mid 10 < x < 200\}$;
(b) $\{x \mid 0 \leq x < 10 \text{ or } x > 200\}$ **67.** $\{n \mid 13 \leq n \leq 50\}$
69. (a) 10, 35; **(b)** $\{x \mid 10 < x < 35\}$; **(c)** $\{x \mid x < 10 \text{ or } x > 35\}$
71. (a) $\{k \mid k > 2 \text{ or } k < -2\}$; **(b)** $\{k \mid -2 < k < 2\}$
73. $\{x \mid -1 \leq x \leq 1\}$ **75.** $\{x \mid x \leq -3 \text{ or } x \geq 1\}$
77. Roots: $-2, 1, 3$; $f(x) < 0$: $\{x \mid x < -2 \text{ or } 1 < x < 3\}$;
$f(x) > 0$: $\{x \mid -2 < x < 1 \text{ or } x > 3\}$
79. No roots; $f(x) < 0$: $\{x \mid x < 0\}$; $f(x) > 0$: $\{x \mid x > 0\}$
81. Roots: $-2, 1, 2, 3$; $f(x) < 0$: $\{x \mid -2 < x < 1 \text{ or } 2 < x < 3\}$;
$f(x) > 0$: $\{x \mid x < -2 \text{ or } 1 < x < 2 \text{ or } x > 3\}$

Margin Exercises, Section 4.8

1.

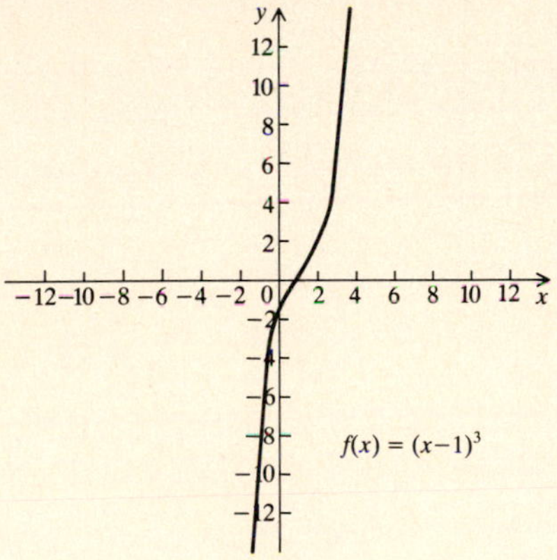

2.

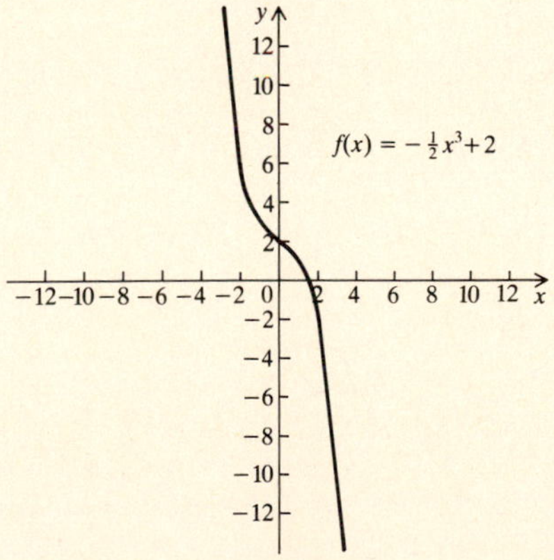

3.

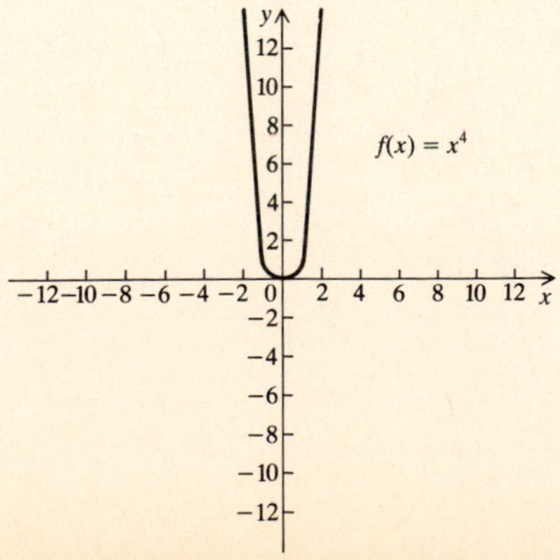

4.

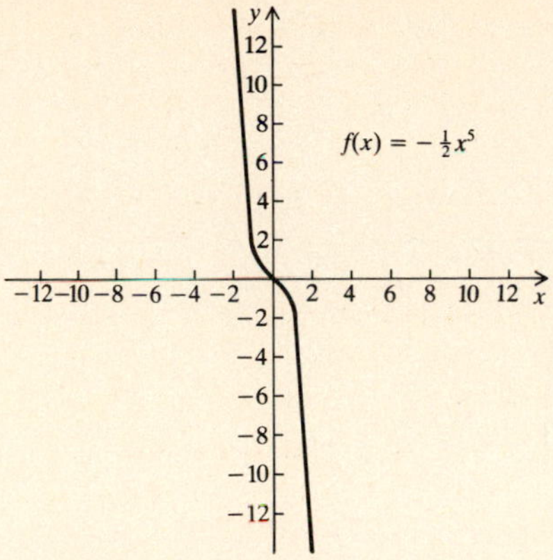

5.

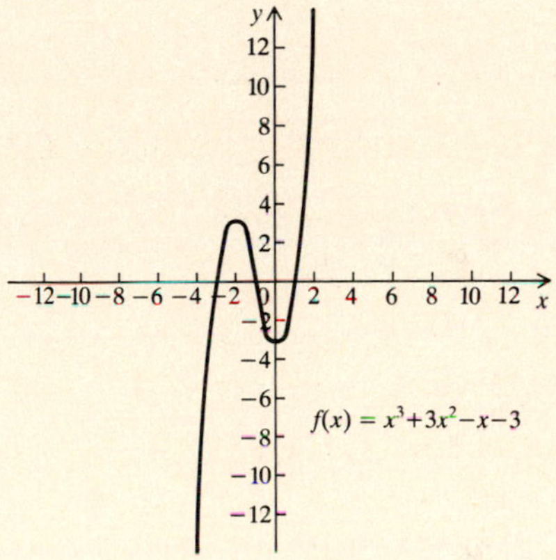

6.

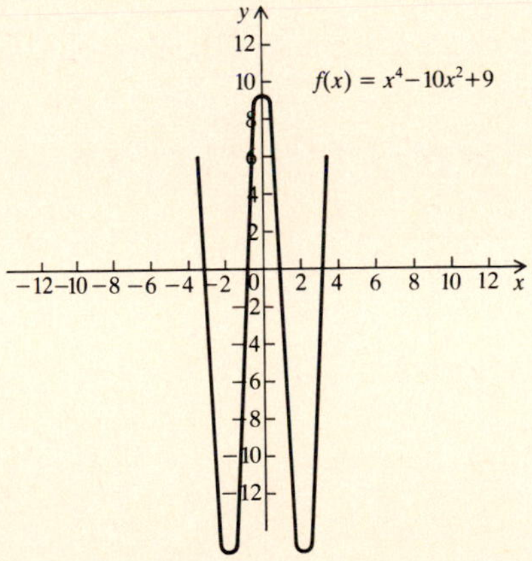

7. 455; 2925

Exercise Set 4.8, pp. 209–271

1.

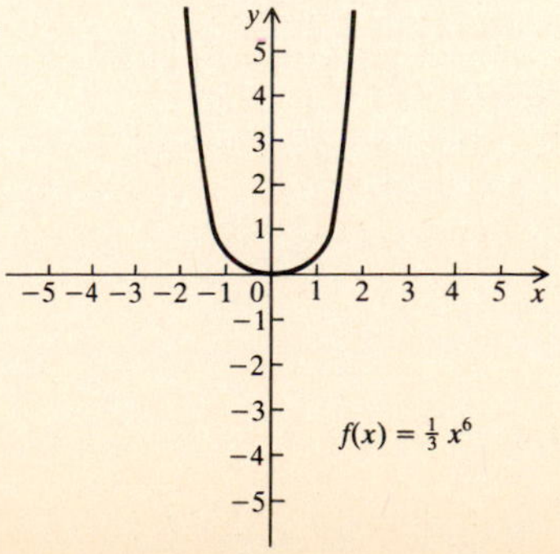

3.

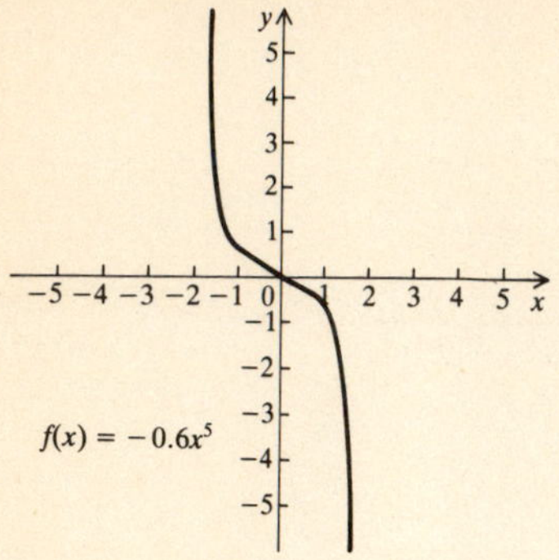
$f(x) = -0.6x^5$

5.

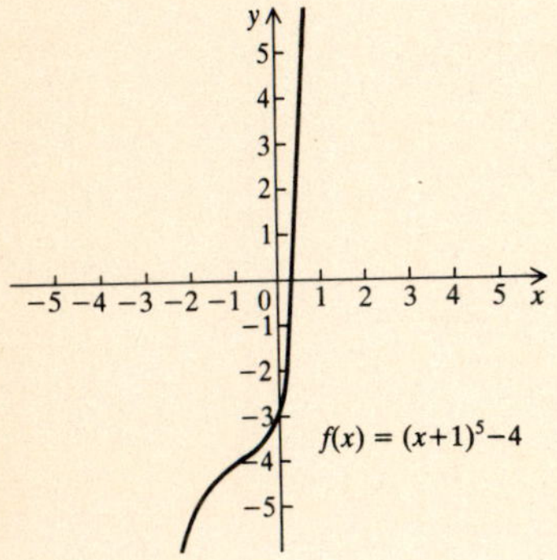
$f(x) = (x+1)^5-4$

7.

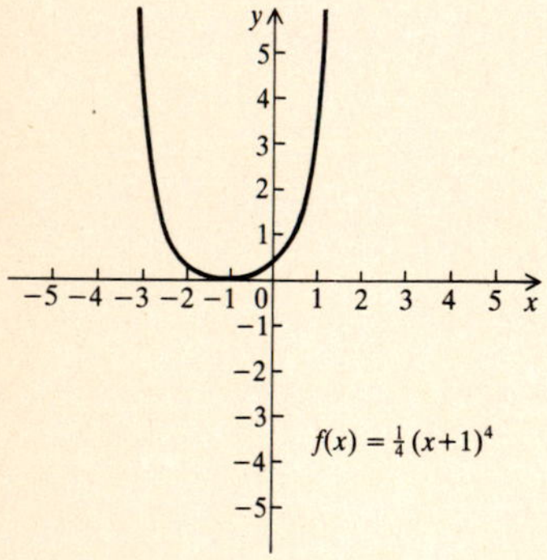
$f(x) = \frac{1}{4}(x+1)^4$

9.

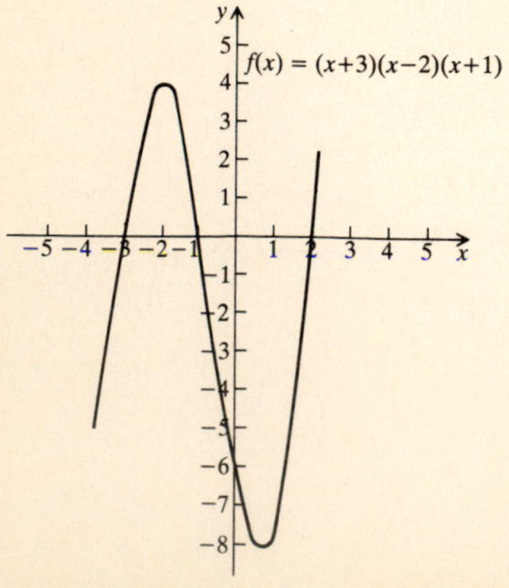
$f(x) = (x+3)(x-2)(x+1)$

11.

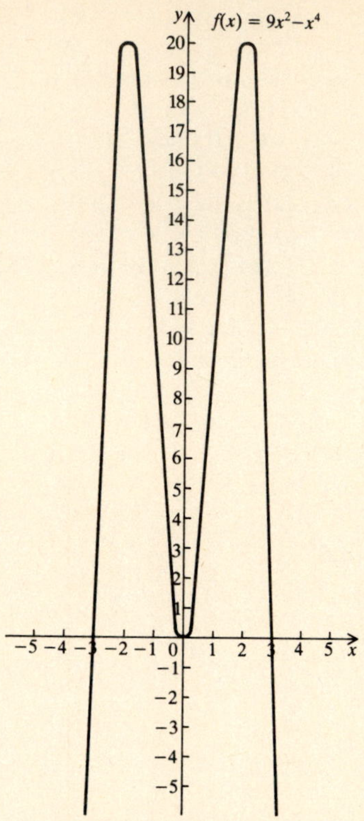
$f(x) = 9x^2-x^4$

13.

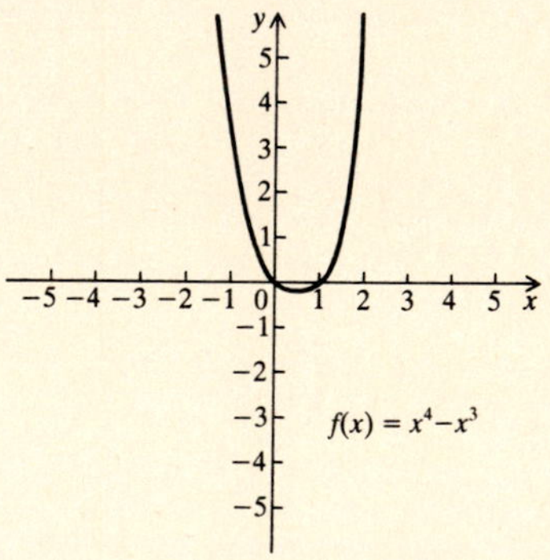
$f(x) = x^4-x^3$

15.

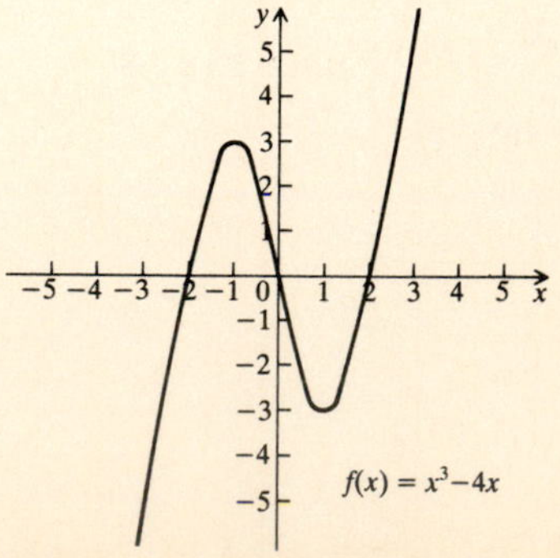
$f(x) = x^3-4x$

17.

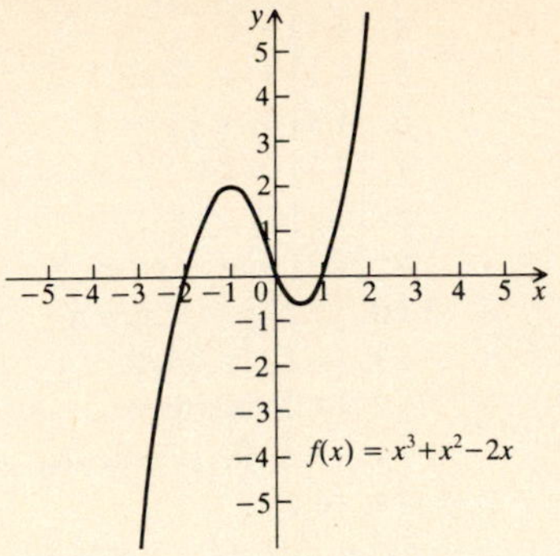

19.

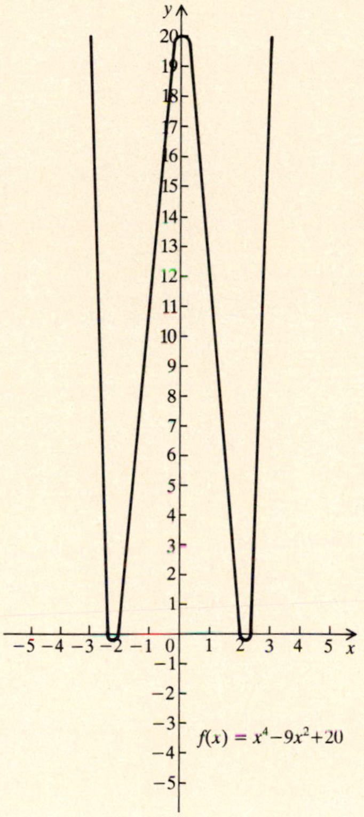

21.

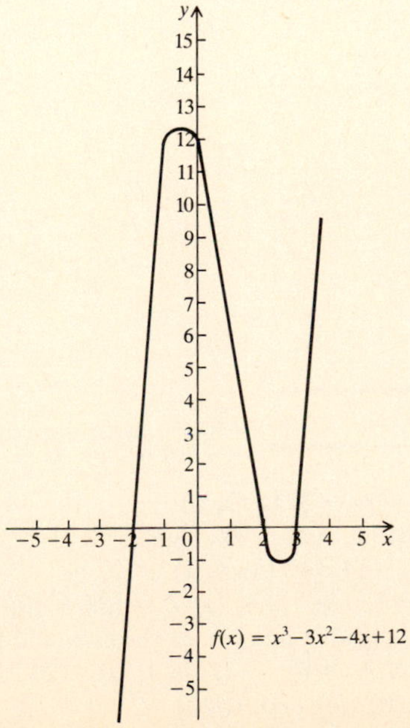

23.

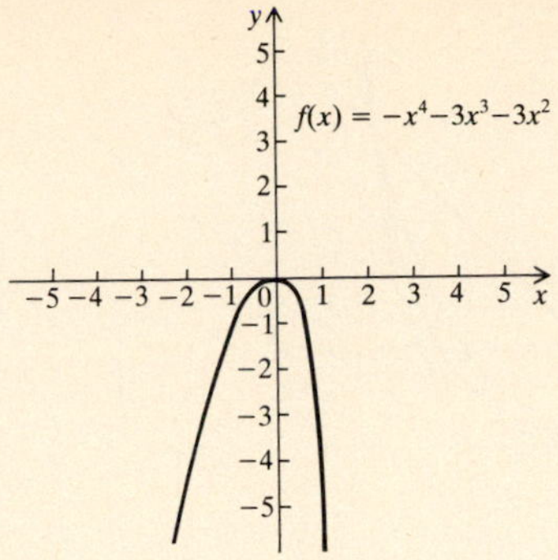

25.

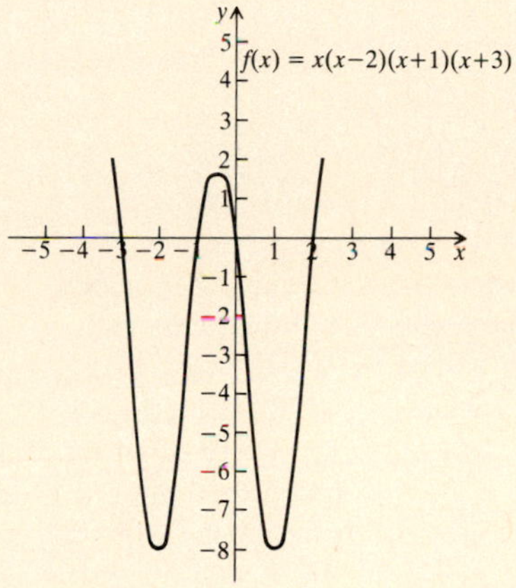

27. **(a)** 0.0055, 0.4725, 1.8626, 16.1868, 38.8132, 55.4341, 76.2088; **(b)**

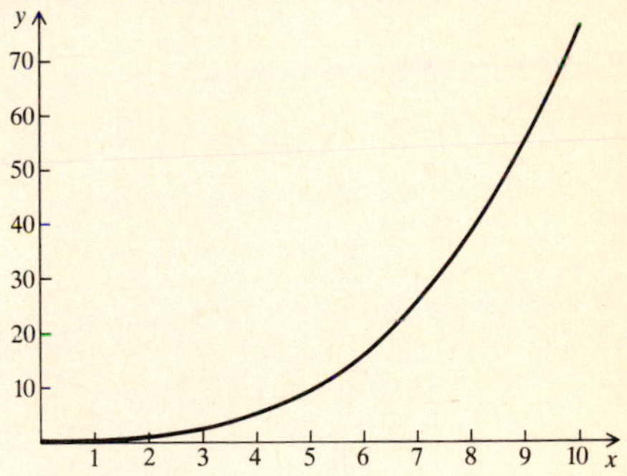

29. **(a)** 161.6 lb, 184.3 lb; **(b)** no, $W = 209.1$ lb

31. **(a)** $V(x) = x^2\left(\frac{10 - x}{2}\right)$, or $V(x) = 5x^2 - \frac{1}{2}x^3$; **(b)**

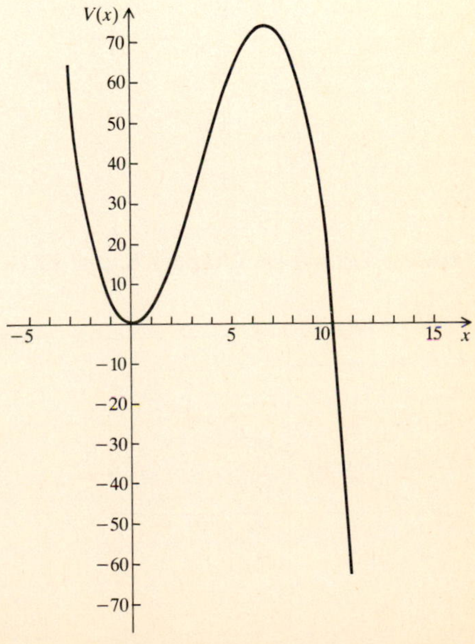

(c) $(-\infty, 0)$ and $(0, 10)$

33. **(a)**

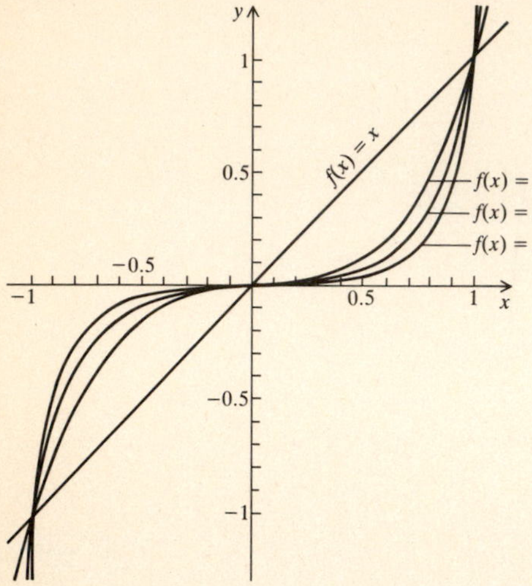

(b) All functions of the form $f(x) = x^k$, where k is an odd positive integer, contain the points $(-1, -1)$, $(0, 0)$, and $(1, 1)$.

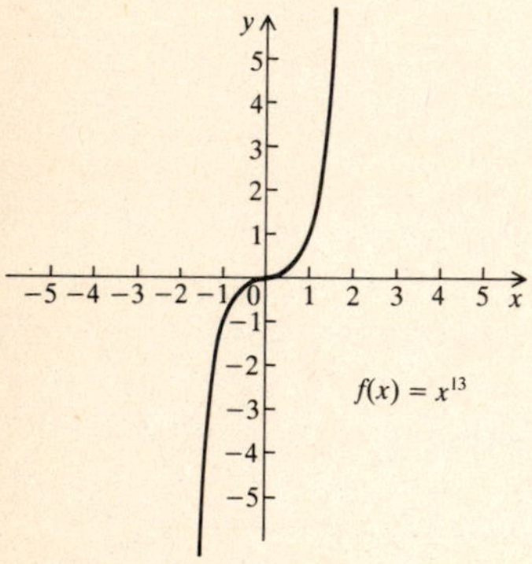

35. 1, 4, 11, 12, 14, 19, 20 **37.** Every exponent of the polynomial must be even or zero or $f(x) = 0$. **39.** Roots: $-3, -2, 1$ **41.** Roots: -1.41421, -1, 1.41421, 2

43. **(a)**

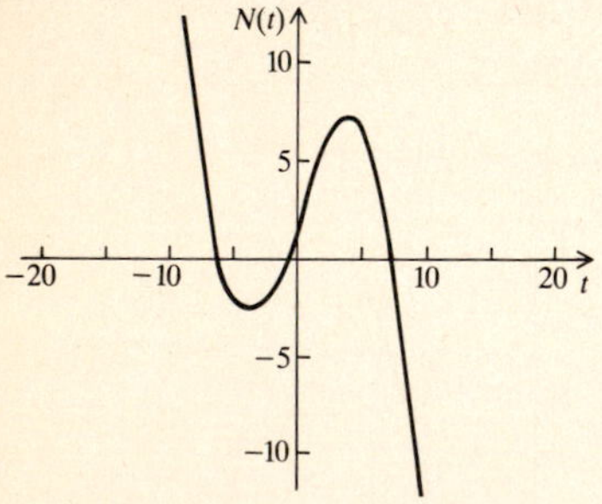

(b) Roots: -6.17908, -0.98251, 7.16159; **(c)** approximately 7.2 hr

Review Exercises: Chapter 4, pp. 271–273

1. [4.1]

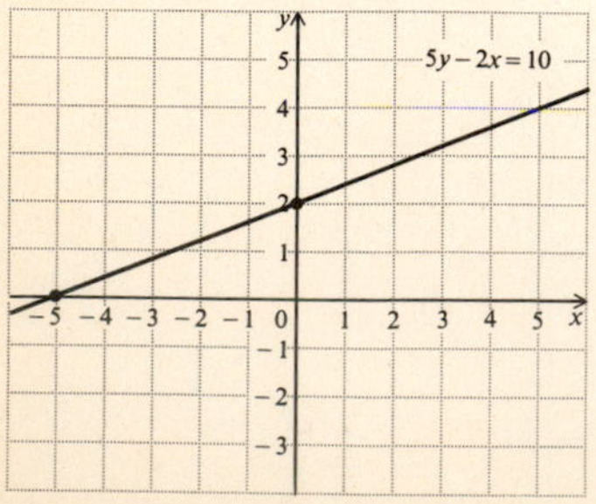

2. [4.1] $m = -2$; y-intercept: $(0, -7)$

3. [4.1]

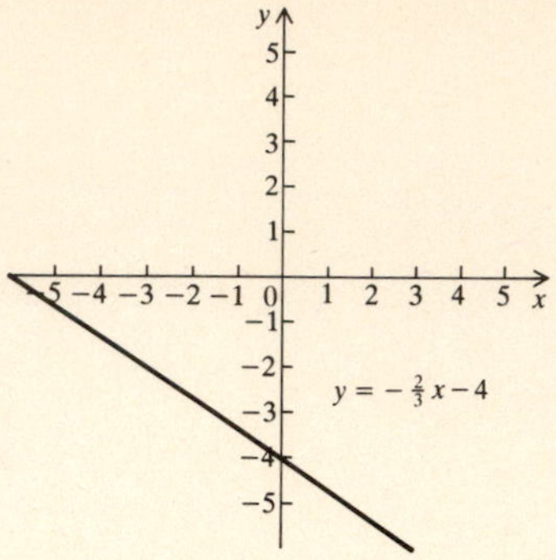

4. [4.1] -1
5. [4.1] $y = 3x + 5$
6. [4.1] $y = \frac{1}{3}x - \frac{1}{3}$
7. [4.2] $\sqrt{34}$
8. [4.2] $(\frac{1}{2}, \frac{11}{2})$
9. [4.2] $y = -\frac{2}{3}x - \frac{1}{3}$
10. [4.2] $y = \frac{3}{2}x - \frac{5}{2}$
11. [4.2] Parallel
12. [4.2] Neither
13. [4.2] Perpendicular
14. [4.2] $(x + 2)^2 + (y - 6)^2 = 13$

15. [4.2] Center: $(-1, 3)$; radius: $\frac{3}{2}$;

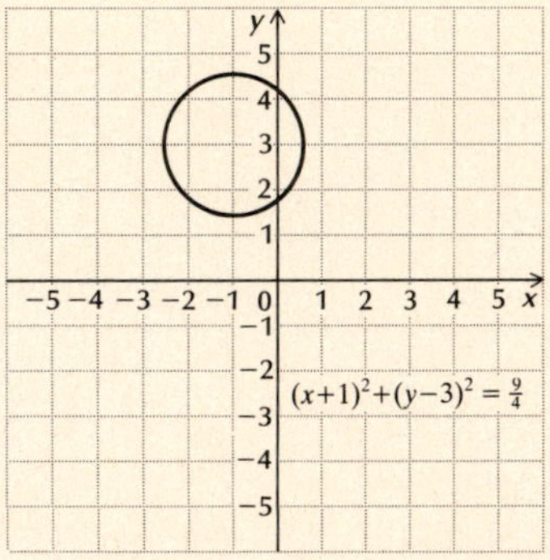

16. [4.2] Center: $(5, -2)$; radius: $2\sqrt{2}$ **17.** [4.2] $(x - 3)^2 + (y - 4)^2 = 25$ **18.** [4.2] $(x - 2)^2 + (y - 4)^2 = 26$ **19.** [4.3] **(a)** $f(x) = 3(x + 1)^2 - 2$; **(b)** $(-1, -2)$; **(c)** $x = -1$; **(d)** minimum: -2 **20.** [4.3] **(a)** $f(x) = -2(x + \frac{3}{4})^2 + \frac{57}{8}$; **(b)** $(-\frac{3}{4}, \frac{57}{8})$; **(c)** $x = -\frac{3}{4}$; **(d)** maximum: $\frac{57}{8}$
21. [4.3]

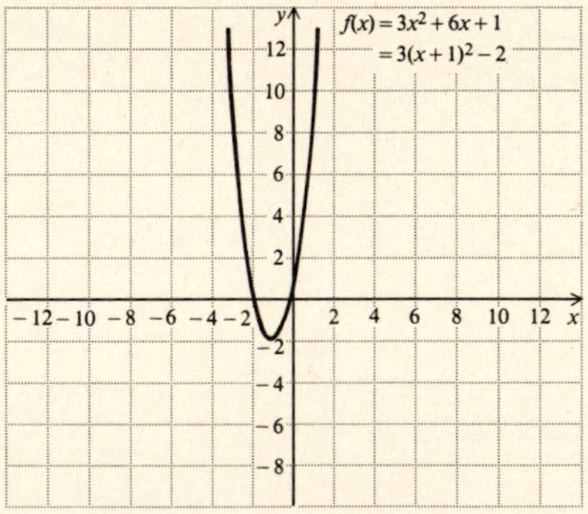

22. [4.3] None **23.** 4.5 {5}
24. [4.5] {3, 4, 5, 7, 8, 9, 11, 12, 13}

25. [4.5]

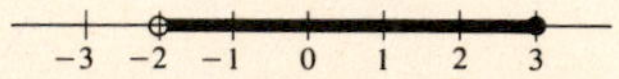

26. [4.5]

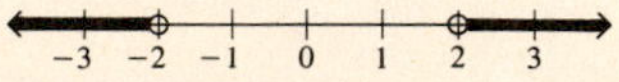

27. [4.5] [2, 4]

28. [4.6] (1, 11) **29.** [4.6] $(-\infty, -\frac{4}{3}) \cup (0, \infty)$ **30.** [4.6] $[-20, 4]$ **31.** [4.6] $\{2, -7\}$ **32.** [4.7] $(-3, 3)$ **33.** [4.7] $(-\infty, -\frac{1}{2}) \cup (2, \infty)$ **34.** [4.7] $(-4, 1) \cup (2, \infty)$ **35.** [4.7] $(-\infty, -\frac{14}{3}) \cup (-3, \infty)$ **36.** [4.4] **(a)** $T = \frac{1}{4}N + 40$; **(b)** 59°F, 65°F; **(c)** 40°F; since a negative number of chirps would be meaningless, the domain is $\{N | N \geq 0\}$. **37.** [4.4] 20 × 20 **38.** [4.4] **(a)** 20; **(b)** $\{x | x > 20\}$; **(c)** $\{x | x < 20\}$

39. [4.8]

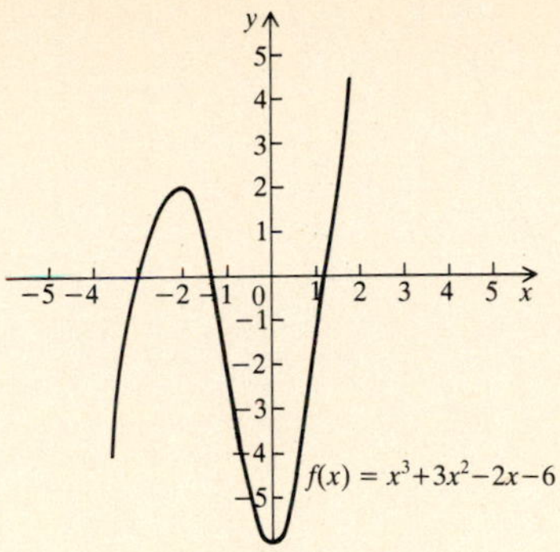

40. [4.8]

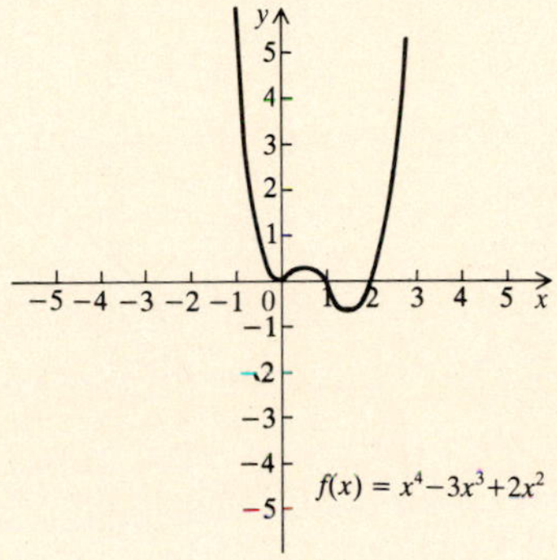

41. [4.5] $[-2, \frac{1}{3})$ **42.** [4.6] $(-\infty, -\frac{1}{2}) \cup (\frac{1}{2}, \infty)$ **43.** [4.7] $(-\infty, 2)$ **44.** [4.1], [4.3] $m \neq 0$ **45.** [4.6] $\{x|\frac{1}{3} \leq x \leq 1\}$ **46.** [4.6] $\{x|-1 < x < \frac{3}{7}\}$ **47.** [4.8] $\{n|n \geq 15\}$

Test: Chapter 4, pp. 273–274

1. [4.1] $m = -\frac{3}{7}$, y-intercept is $\frac{10}{7}$ **2.** [4.1] $y = 5x + 12$ **3.** [4.1] $y = 2x - 7$ **4.** [4.1] $\sqrt{13}$ **5.** [4.2] $(\frac{11}{2}, -\frac{3}{2})$ **6.** [4.2] Perpendicular

7. [4.2]

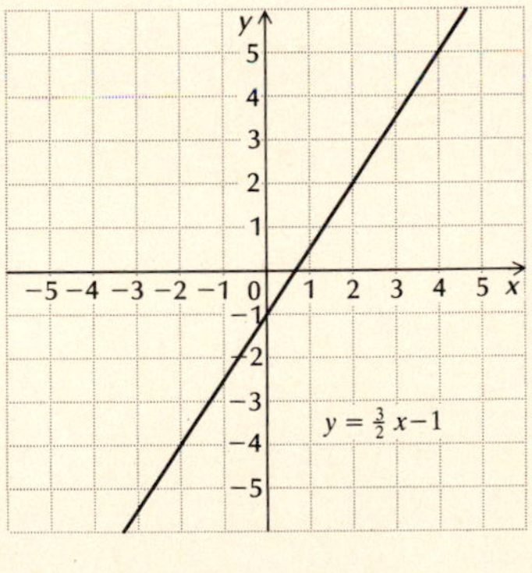

8. [4.2] $y = \frac{1}{3}x - \frac{14}{3}$
9. [4.2] $(x + 2)^2 + (y - 4)^2 = 16$
10. [4.2] Center: $(1, -3)$; radius: $\sqrt{5}$
11. [4.3] **(a)** $f(x) = 5(x - 1)^2 - 2$; **(b)** $(1, -2)$; **(c)** minimum: -2
12. [4.3] **(a)** $f(x) = -4(x - \frac{3}{8})^2 - \frac{7}{16}$; **(b)** $(\frac{3}{8}, -\frac{7}{16})$; **(c)** maximum: $-\frac{7}{16}$

13. [4.3]

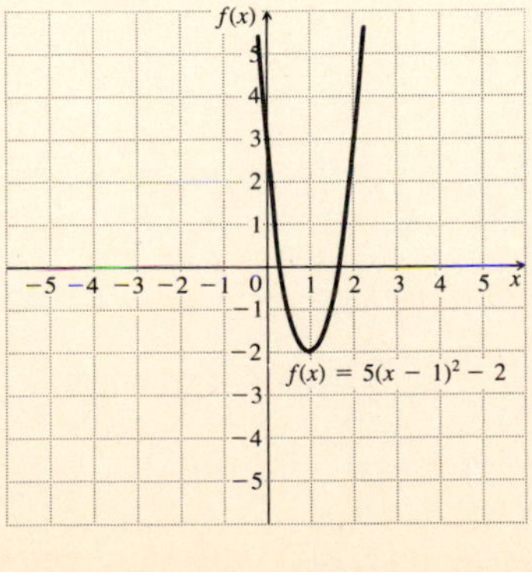

14. [4.3] $\left(\frac{1 - \sqrt{13}}{6}, 0\right)$, $\left(\frac{1 + \sqrt{13}}{6}, 0\right)$
15. [4.5] $\{2, 3, 4, 6, 7, 8, 10, 11, 12\}$

16. [4.5]

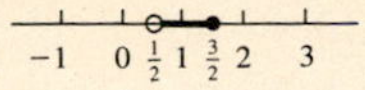

17. [4.5] $(-5, -3]$
18. [4.6] $(-\infty, -\frac{3}{2}] \cup [3, \infty)$ **19.** [4.6] $(1, 5)$ **20.** [4.6] $\{-2, 3\}$ **21.** [4.7] $(-\infty, 2] \cup (6, \infty)$ **22.** [4.7] $(-\frac{3}{2}, \frac{1}{4})$ **23.** [4.7] $(-\infty, -7) \cup (-\frac{3}{2}, \infty)$ **24.** [4.4] $-10, -10$ **25.** [4.4] **(a)** 10, 30; **(b)** $\{x|10 < x < 30\}$; **(c)** $\{x|x < 10 \text{ or } x > 30\}$ **26.** [4.4] **(a)** $f(x) = 1148x + 12{,}877$; **(b)** 28,949, 35,837; **(c)** $\{x|x \geq 0\}$

27. [4.8]

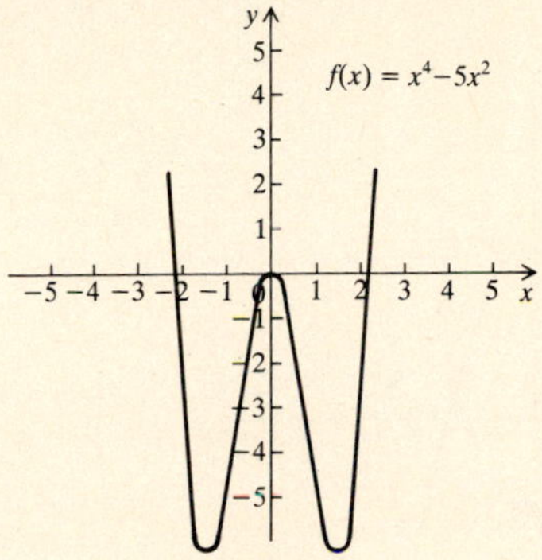

28. [4.8]

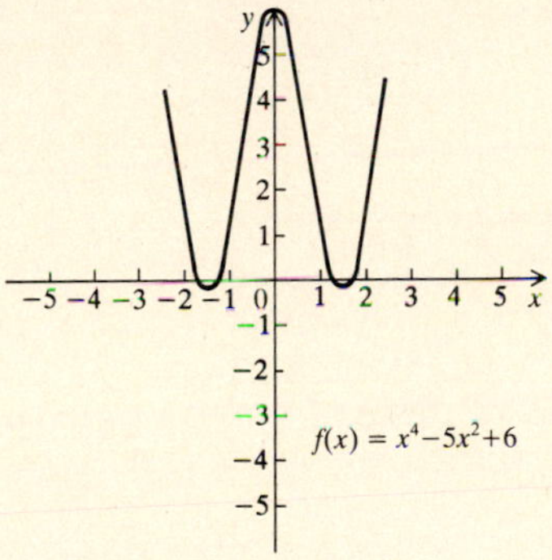

29. [4.7] **(a)** After $2\frac{1}{2}$ seconds, maximum height of 324 ft is attained. **(b)** 7 seconds; **(c)** between 2 and 3 seconds **30.** [4.6] $(-\infty, -\frac{1}{8}) \cup (\frac{1}{4}, \infty)$ **31.** [4.6] $\{x|x \leq -5 \text{ or } x \geq 2\}$

CHAPTER 5

Margin Exercises, Section 5.1

1. $\{(4, -1), (5, 2), (-3, 0), (1, 5)\}$ **2. (a)** $x = 3y + 2$; **(b)** $x = y$; **(c)** $y^2 + 3x^2 = 4$; **(d)** $x = 5y^2 + 2$; **(e)** $x^2 = 4y - 5$; **(f)** $yx = 5$
3. (a), (b)

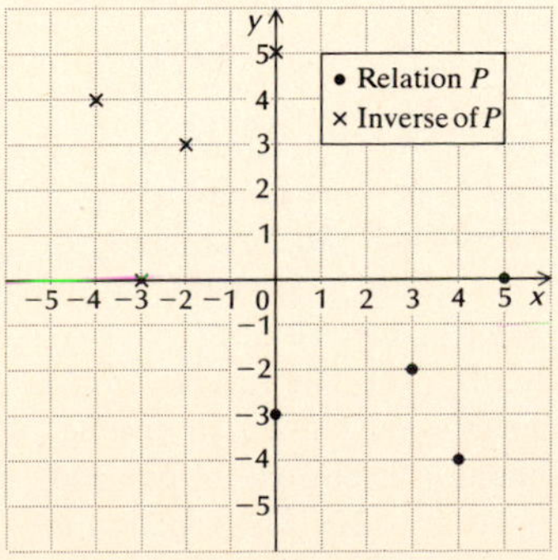

inverse of $P = \{(0, 5), (-2, 3), (-3, 0), (-4, 4)\}$
4. (d) They are reflections across the line $y = x$.

5. (a)

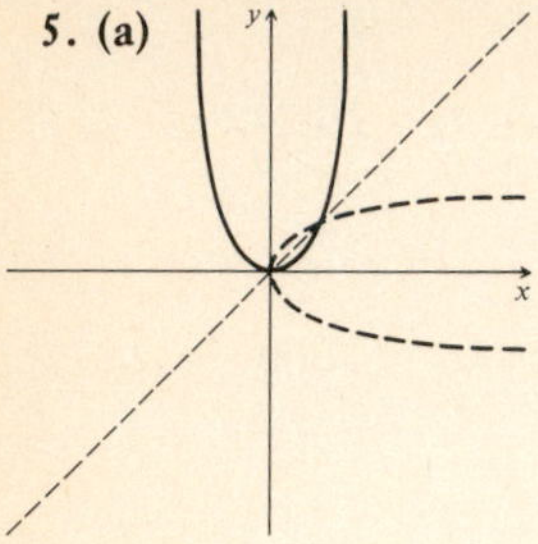

(b)

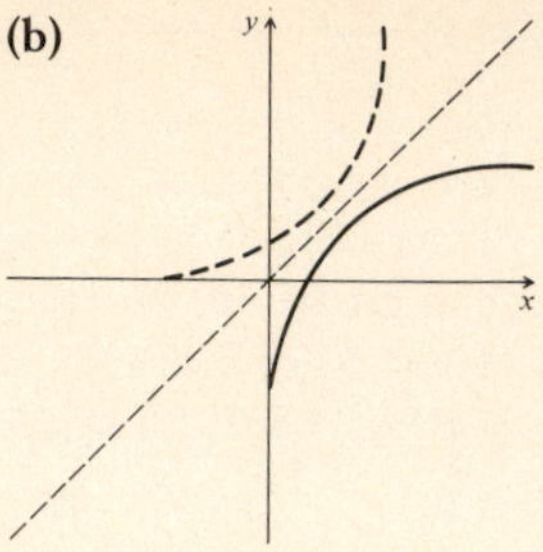

(c)

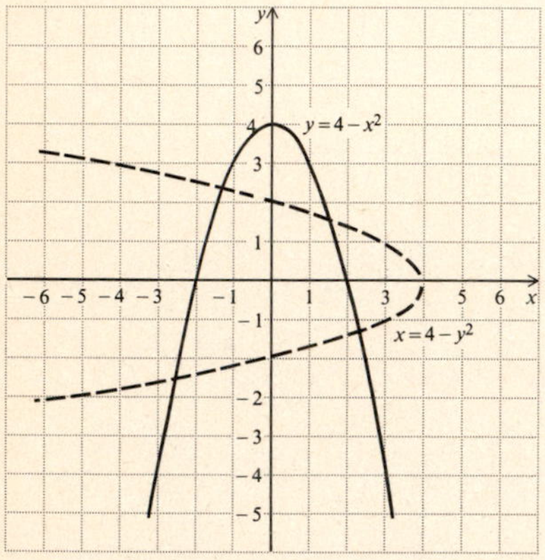

6. Yes 7. Yes 8. Yes 9. No 10. Yes 11. Yes 12. No 13. No

14.

Women's Dress Sizes	
Domain (United States)	**Range (France)**
6 ←	38
8 ←	40
10 ←	42
12 ←	44
14 ←	46
16 ←	48
18 ←	50

; yes

15.

Sports Teams	
Domain	**Range**
Lakers →	Los Angeles
Dodgers →	Los Angeles
Rams →	Los Angeles
Knickerbockers →	New York
Yankees →	New York
Giants →	New York

; no

16. (a) G is a function; (b) inverse of $G = \{(-4, 5), (2, 9), (-6, 8), (-4, 10)\}$; (c) The inverse of G is not a function because there are two ordered pairs, namely, $(-4, 5)$ and $(-4, 10)$ that have the same first coordinates but different second coordinates. 17. (a) H is a function; (b) inverse of $H = \{(3, 2), (-1, 9), (4, 7), (-3, 4), (12, 10)\}$; (c) The inverse of H is a function. 18. The function is one-to-one and has an inverse that is also a function. 19. The function is not one-to-one and does not have an inverse that is a function 20. The function is not one-to-one and does not have an inverse that is a function. 21. (a) and (d) 22. (a) Yes; (b) $f^{-1}(x) = 3 - x$ 23. (a) Yes; (b) $g^{-1}(x) = \dfrac{x+2}{3}$

24.

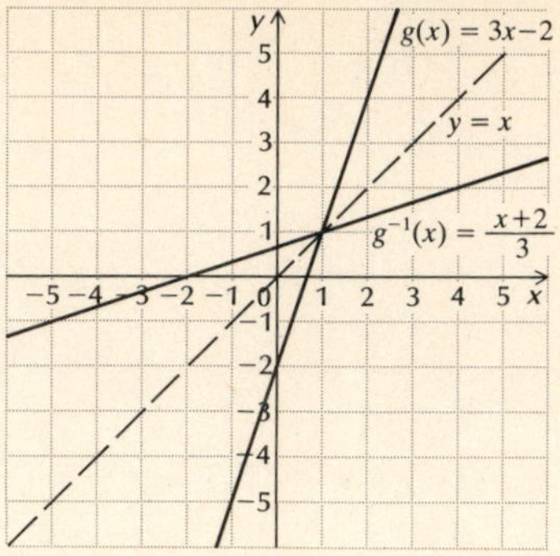

25. (a) The graph of $f(x) = x^3 + 1$ passes the horizontal-line test and thus has an inverse. (b) $f^{-1}(x) = \sqrt[3]{x - 1}$; (c)

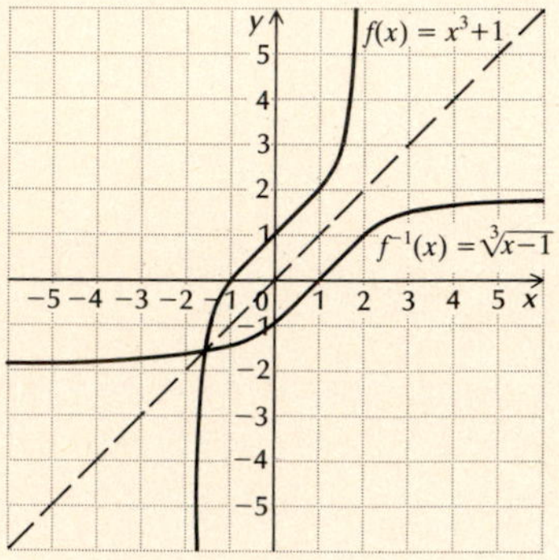

26. $f^{-1}(x) = \sqrt{x + 4}$, $x \geqslant -4$

27. $f^{-1} \circ f(x) = f^{-1}\left(\dfrac{4}{x} - 3\right) = \dfrac{4}{\left(\dfrac{4}{x} - 3\right) + 3} = \dfrac{4}{\dfrac{4}{x}} = x$,

$f \circ f^{-1}(x) = f\left(\dfrac{4}{x+3}\right) = \dfrac{4}{\dfrac{4}{x+3}} - 3 = x + 3 - 3 = x$

28. $f(f^{-1}(1992)) = 1992$, $f^{-1}(f(-23{,}456)) = -23{,}456$

Exercise Set 5.1, pp. 286–288

1. $\{(1, 0), (6, 5), (-4, -2)\}$
3. $\{(8, 7), (8, -2), (-4, 3), (-8, 8)\}$ 5. $x = 4y - 5$
7. $y^2 - 3x^2 = 3$ 9. $x = 3y^2 + 2$ 11. $yx = 7$

13.

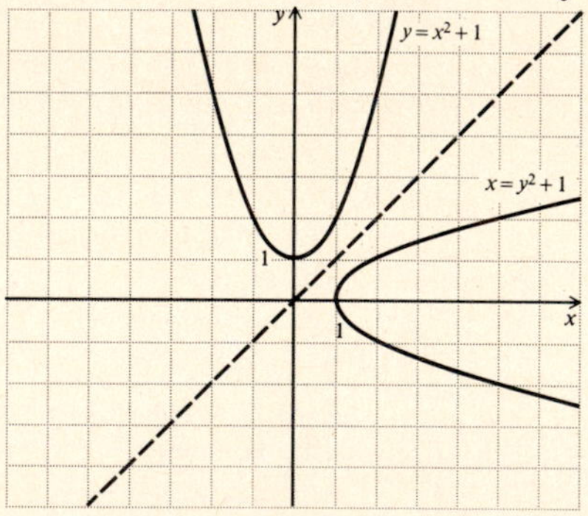

15.

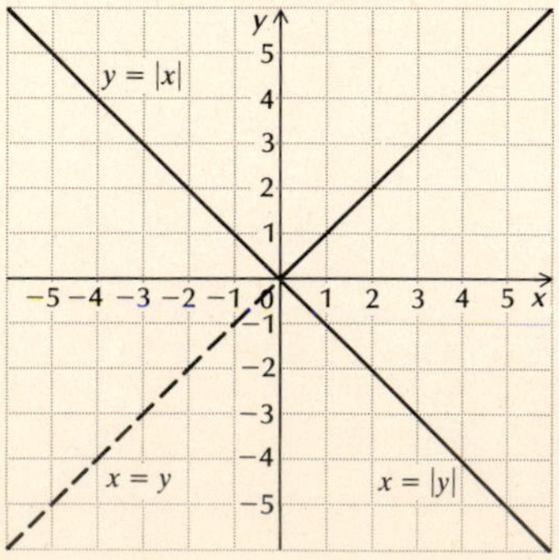

17. No 19. Yes 21. Yes 23. Yes 25. No 27. Yes 29. Yes 31. No 33. No 35. No 37. No 39. Yes

41. (a) Yes; (b) $f^{-1}(x) = x - 4$ **43.** (a) Yes; (b) $f^{-1}(x) = 5 - x$ **45.** (a) Yes; (b) $f^{-1}(x) = x + 3$ **47.** (a) Yes; (b) $f^{-1}(x) = \frac{1}{2}x$ **49.** (a) Yes; (b) $f^{-1}(x) = \frac{x - 5}{2}$ **51.** (a) Yes; (b) $f^{-1}(x) = \frac{4}{x} - 7$ **53.** (a) Yes; (b) $f^{-1}(x) = \frac{1}{x}$ **55.** (a) Yes; (b) $f^{-1}(x) = \frac{4x - 3}{2}$ **57.** (a) Yes; (b) $f^{-1}(x) = \frac{4 + 3x}{x - 1}$ **59.** (a) Yes; (b) $f^{-1}(x) = \sqrt[3]{x + 1}$ **61.** (a) Yes; (b) $f^{-1}(x) = \sqrt[3]{x} + 4$ **63.** (a) Yes; (b) $f^{-1}(x) = x^3$ **65.** (a) Yes; (b) $f^{-1}(x) = \frac{\sqrt{x - 3}}{2}$, $x > 39$ **67.** (a) Yes; (b) $f^{-1}(x) = x^2 - 1$, $x \geqslant 0$ **69.** (a), (c)

71.

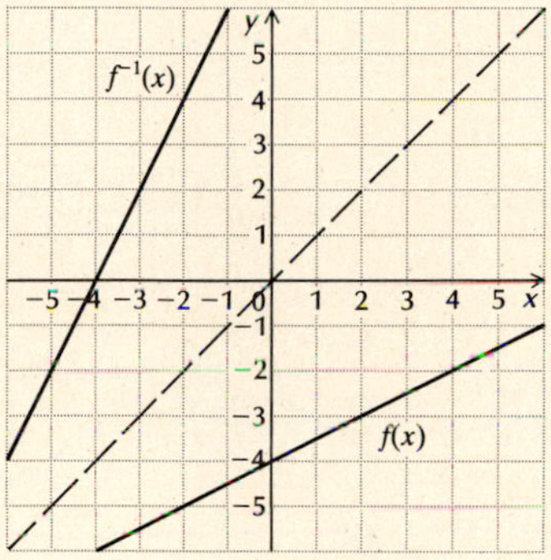

73.

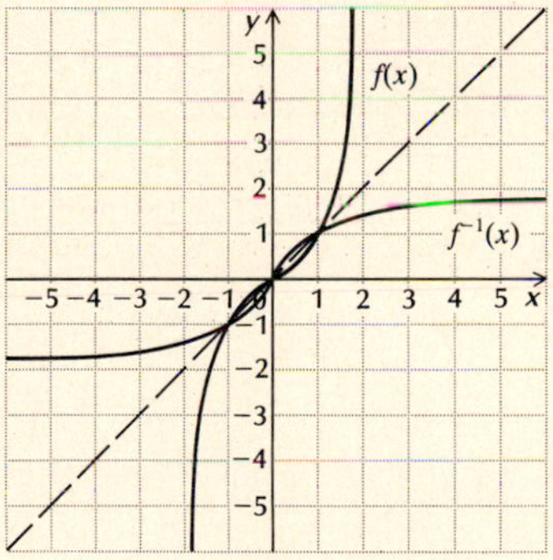

75.

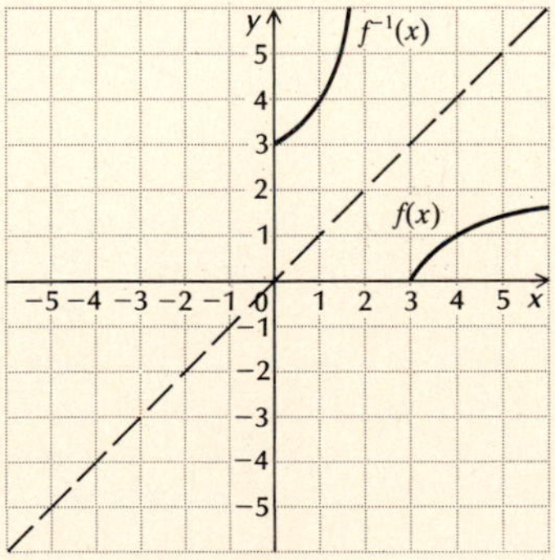

77.

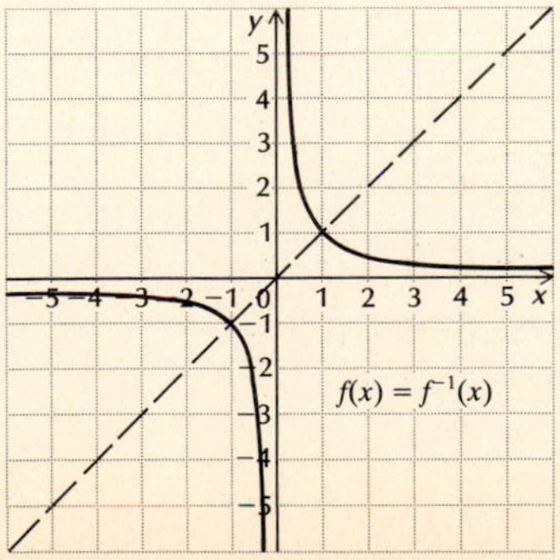

79.

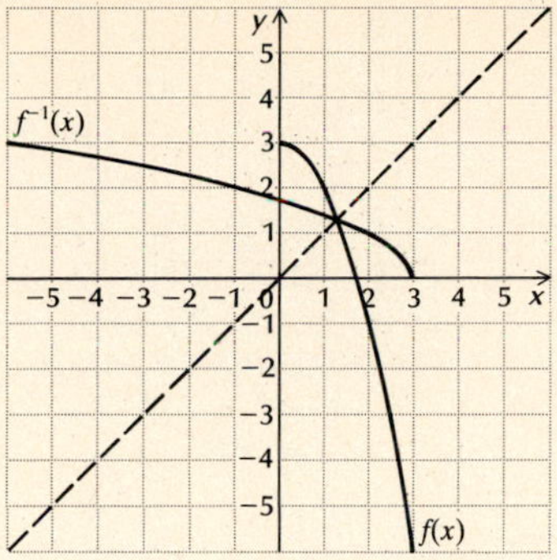

81. $f^{-1} \circ f(x) = f^{-1}(f(x)) = f^{-1}\left(\frac{7}{8}x\right) = \frac{8}{7}\left(\frac{7}{8}x\right) = x$, $f \circ f^{-1}(x) = f(f^{-1}(x)) = f\left(\frac{8}{7}x\right) = \frac{7}{8}\left(\frac{8}{7}x\right) = x$

83. $f^{-1} \circ f(x) = f^{-1}(f(x)) = f^{-1}\left(\frac{1 - x}{x}\right) = \frac{1}{\frac{1 - x}{x} + 1} = \frac{1}{\frac{1}{x}} = x,$

$$f \circ f^{-1}(x) = f(f^{-1}(x)) = f\left(\frac{1}{x + 1}\right) = \frac{1 - \frac{1}{x + 1}}{\frac{1}{x + 1}} = \frac{\frac{x}{x + 1}}{\frac{1}{x + 1}} = x$$

85. 3; -125 **87.** 12,053; $-17{,}243$ **89.** (a) 40, 42, 46, 50; (b) yes, $f^{-1}(x) = x - 32$; (c) 8, 10, 14, 18 **91.** No; the function $f(x) = 5$ does not pass the horizontal-line test. **93.** No **95.** No **97.** Answers may vary. $f(x) = 2/x$, $f(x) = 4 - x$, $f(x) = x$

99.

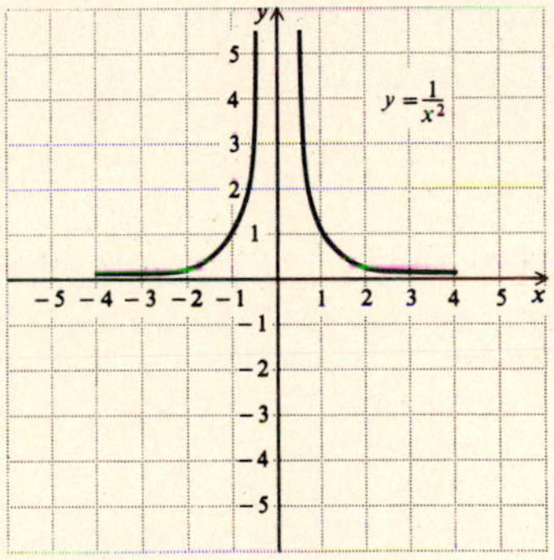

x-axis: no; y-axis: yes; origin: no; $y = x$: no **101.** The inverse of the given relation $|x| - |y| = 1$ is $|y| - |x| = 1$. Each of the two graphs is symmetric with respect to the x-axis, the y-axis, and the origin. Neither is symmetric with respect to the line $y = x$.

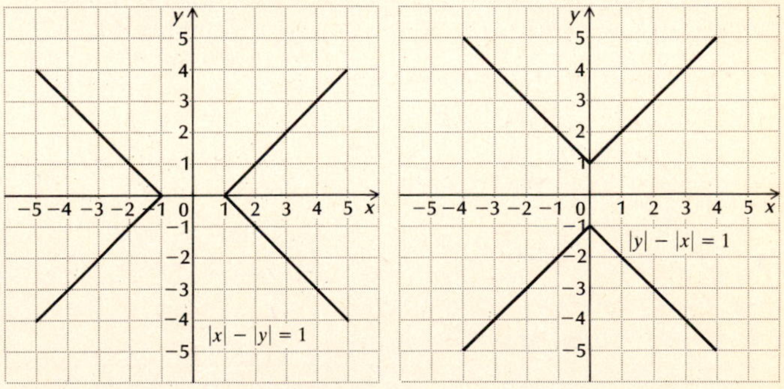

Margin Exercises, Section 5.2

1. (a) 1, 3, 9, 27, $\frac{1}{3}$, $\frac{1}{9}$, $\frac{1}{27}$; (b)

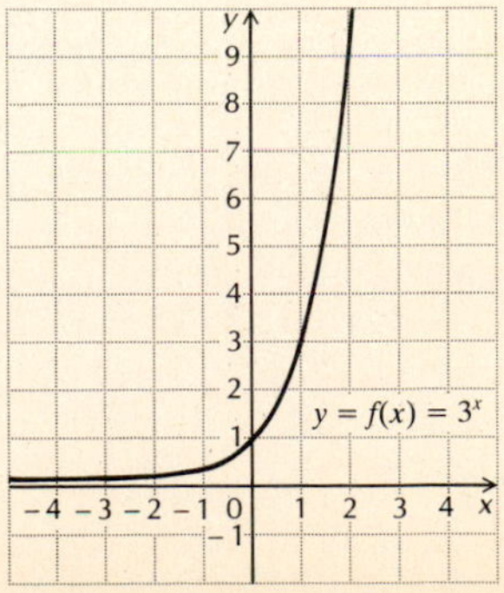

2. (a) $1, \frac{1}{3}, \frac{1}{9}, \frac{1}{27}, 3, 9, 27$;

(b)

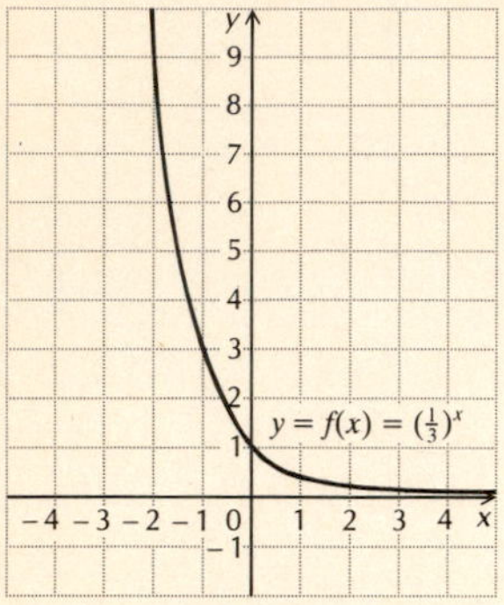

3.

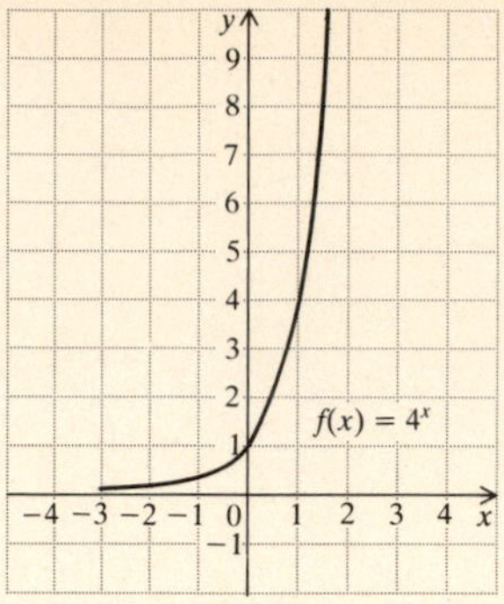

5.

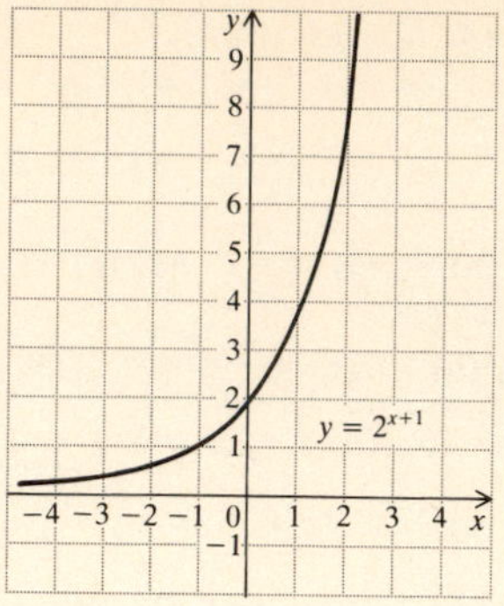

7.

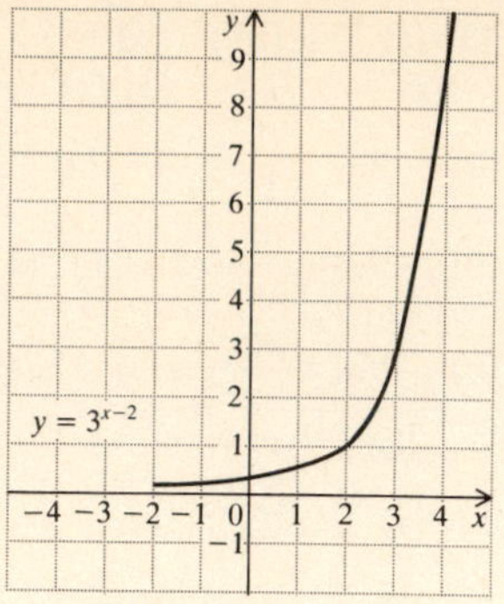

4.

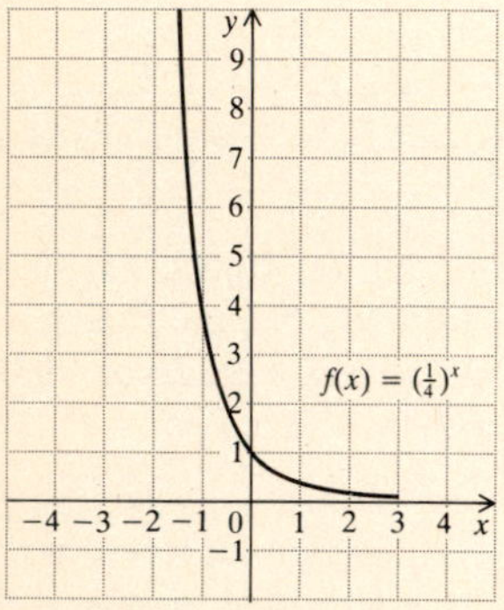

5.

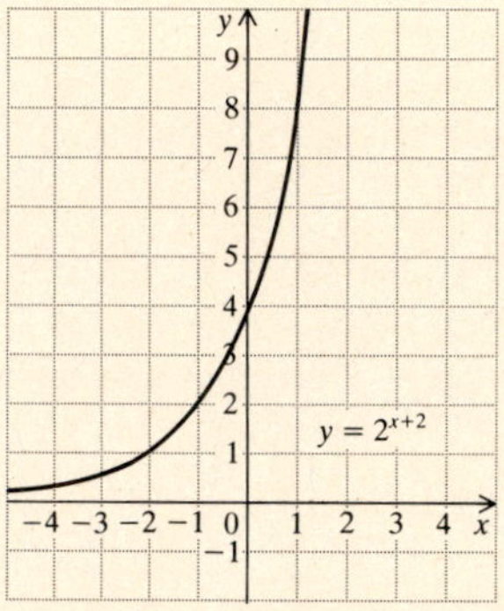

9.

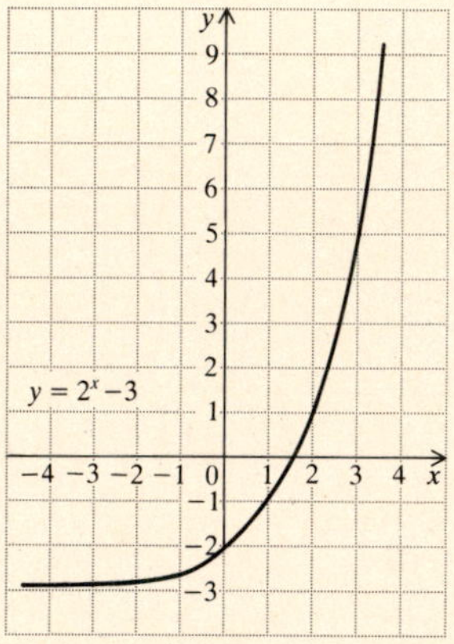

11.

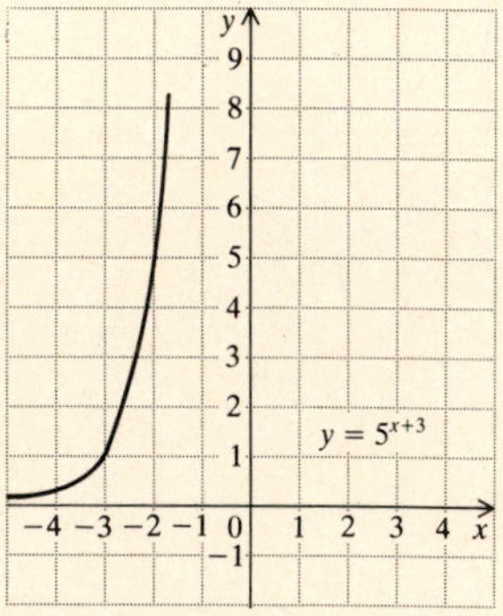

6.

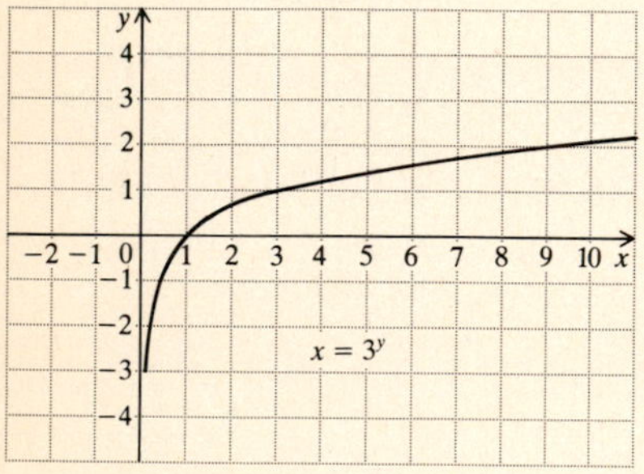

13.

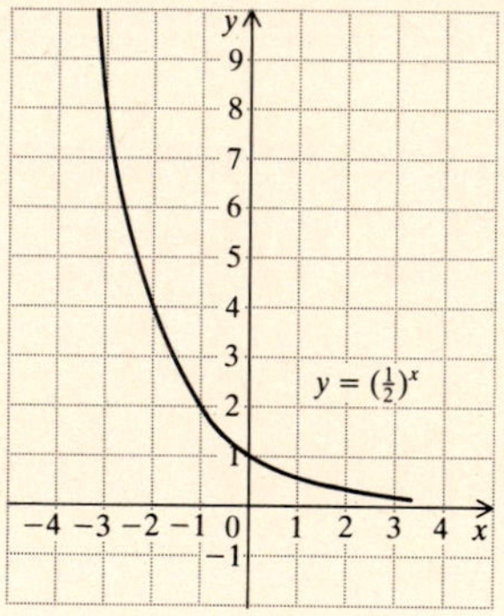

15.

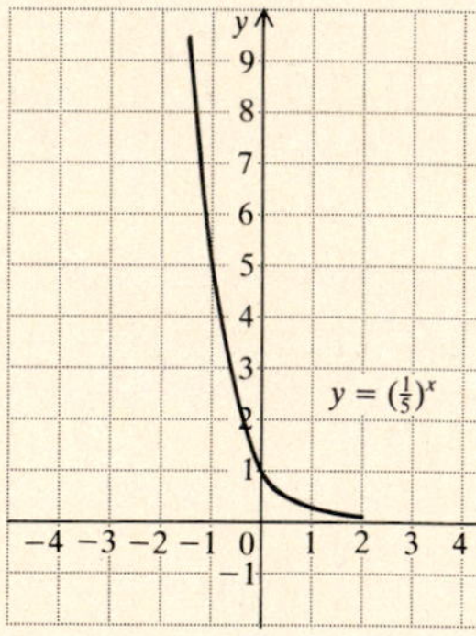

7. (a) $A(t) = 80{,}000(1.08)^t$; (b) \$80,000, \$108,839.12, \$148,074.42, \$172,714;

(c)

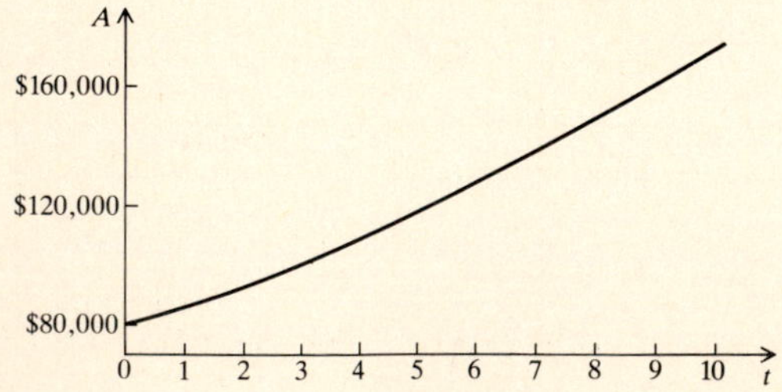

17.

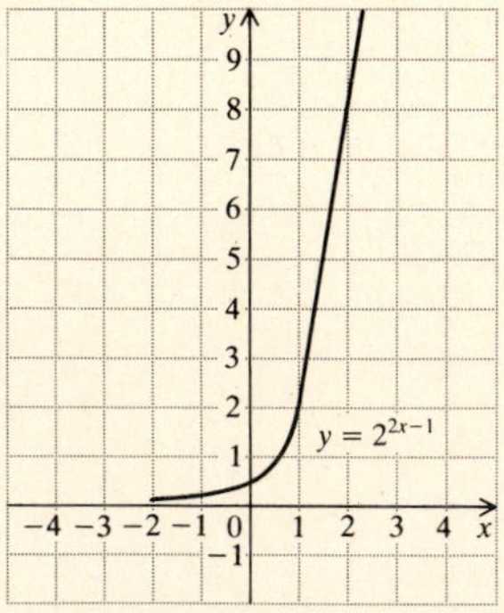

19.

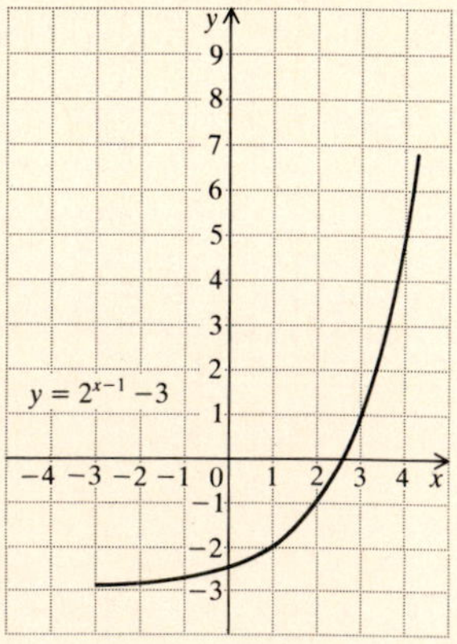

Exercise Set 5.2, pp. 294–296

1.

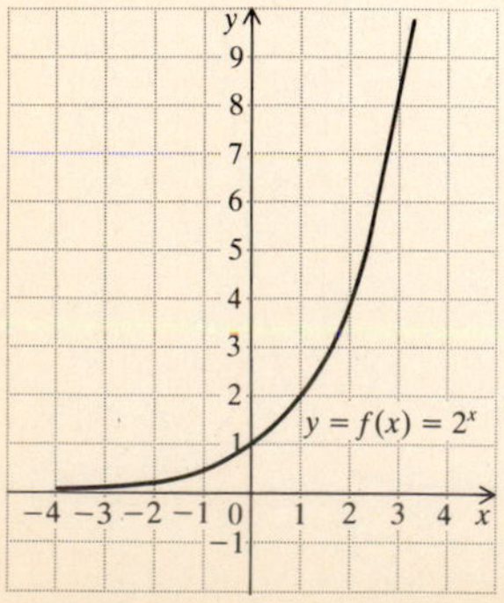

3.

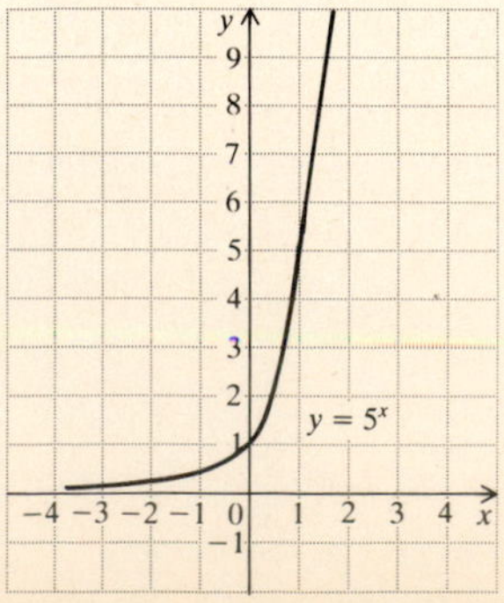

21.

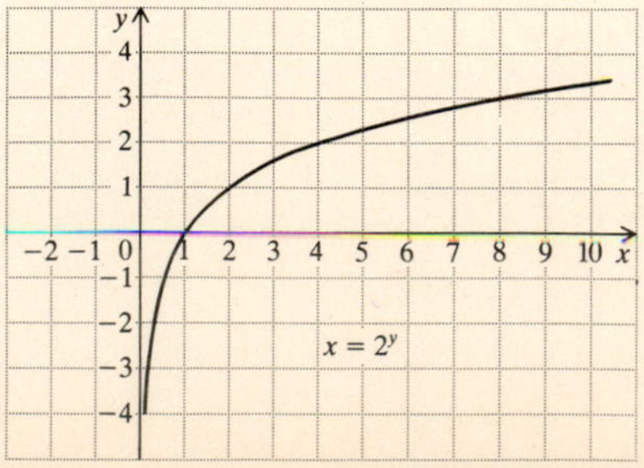

23.

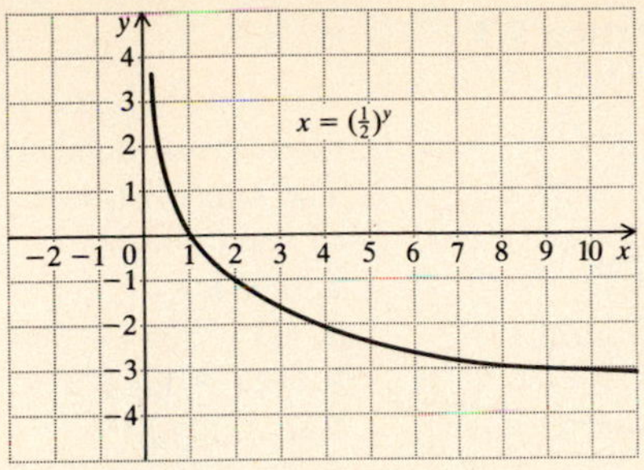

25.

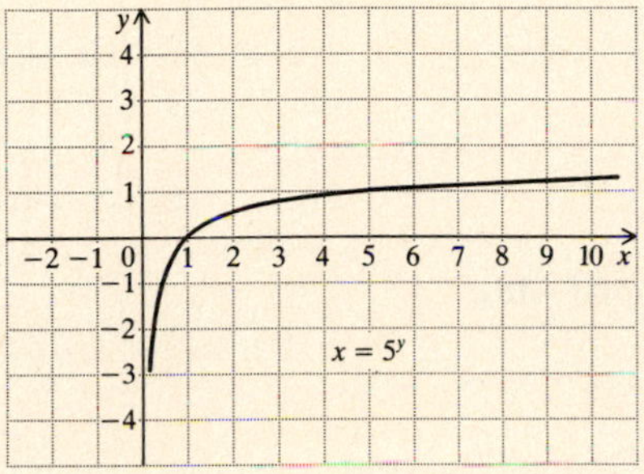

27.

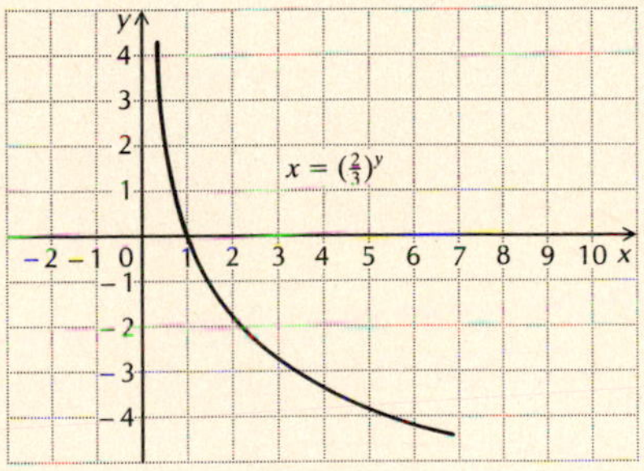

29.

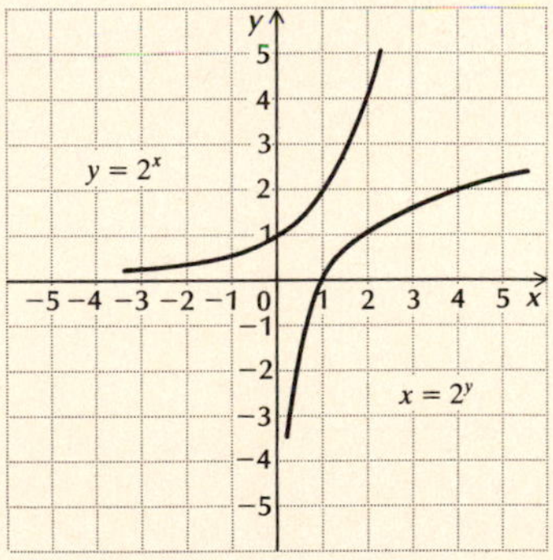

31.

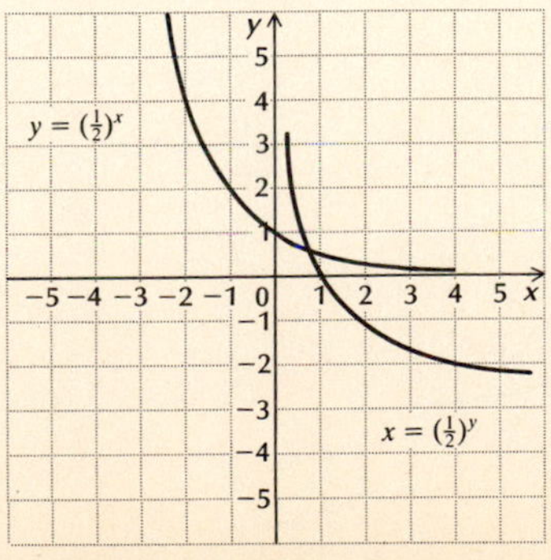

33. **(a)** 18.4 million, 45 million, 661.4 million, 58,320 million, 5,142,752.7 million; **(b)**

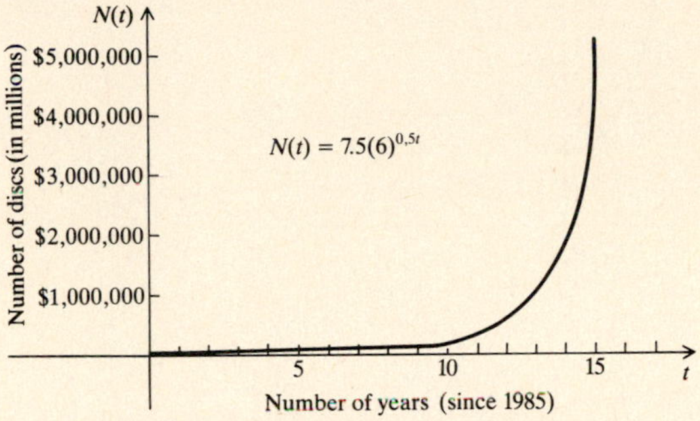

35. **(a)** $A(t) =$ \$50,000$(1.09)^t$; **(b)** \$50,000, \$70,579.08, \$99,628.13, \$118,368.18; **(c)**

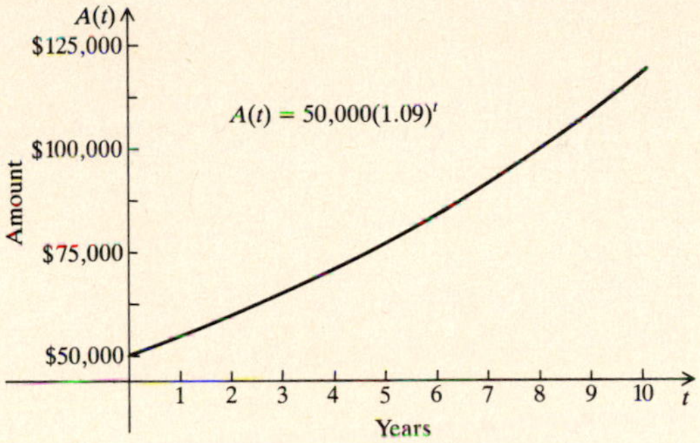

37. **(a)** \$5200, \$3900, \$2925, \$1233.98, \$292.83; **(b)**

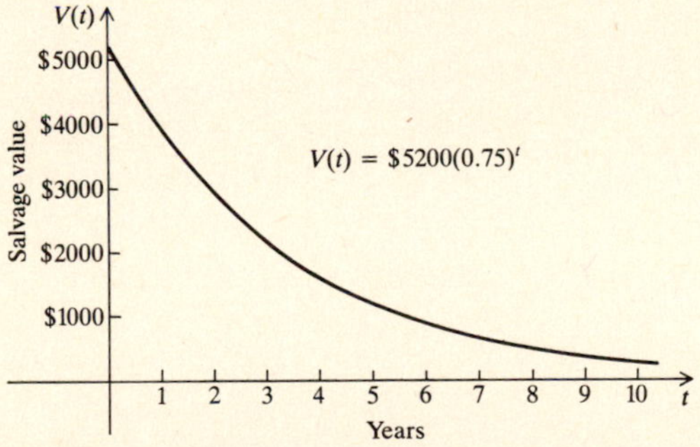

39. **(a)** 343; **(b)** 416.681217; **(c)** 450.409815; **(d)** 451.287125; **(e)** 451.726421; **(f)** 451.805540 **41.** $\pi^{3.2}$

43.

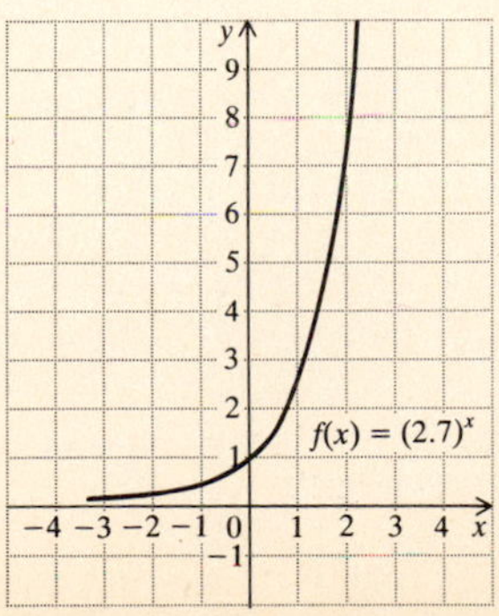

45.

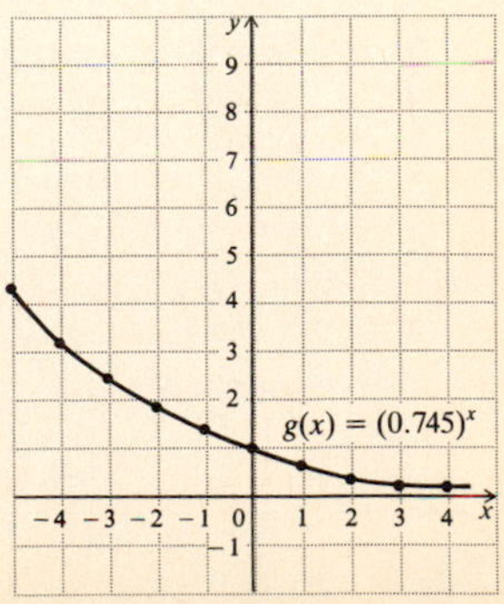

47.

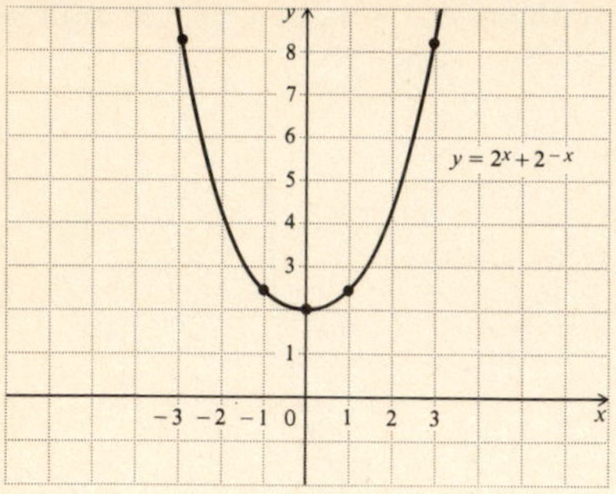

49.

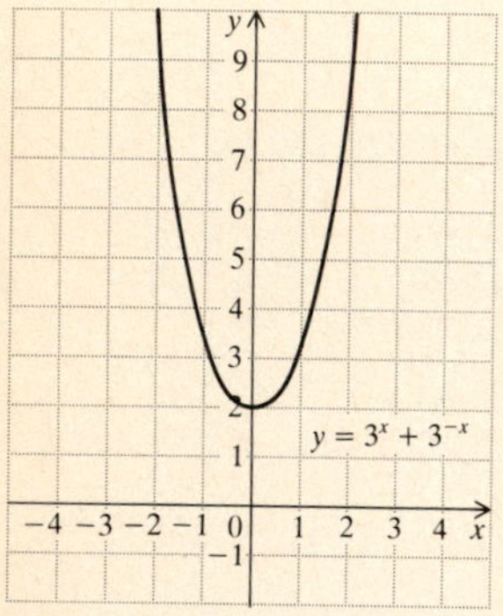

51.

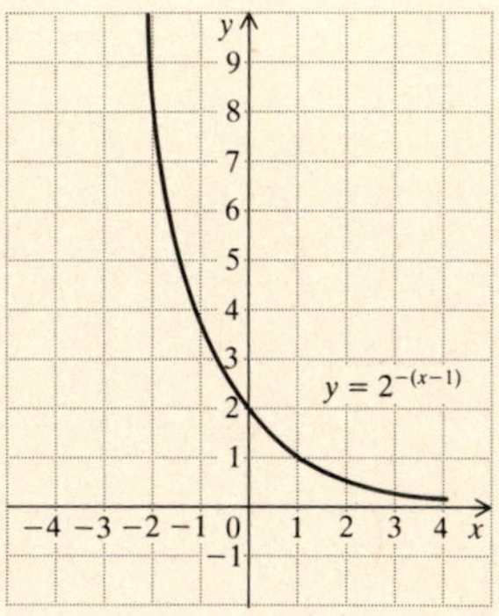

53.

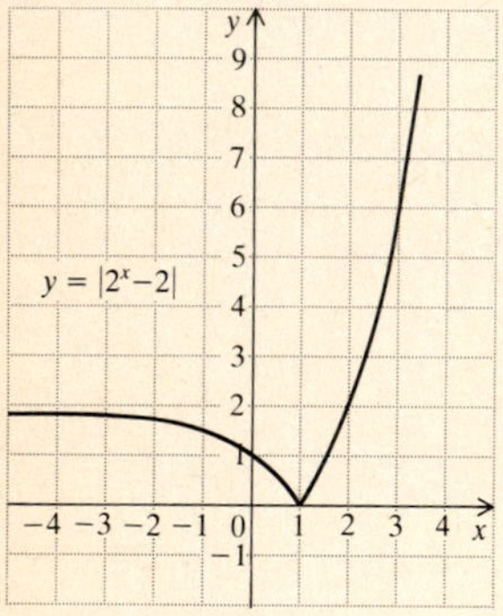

55.

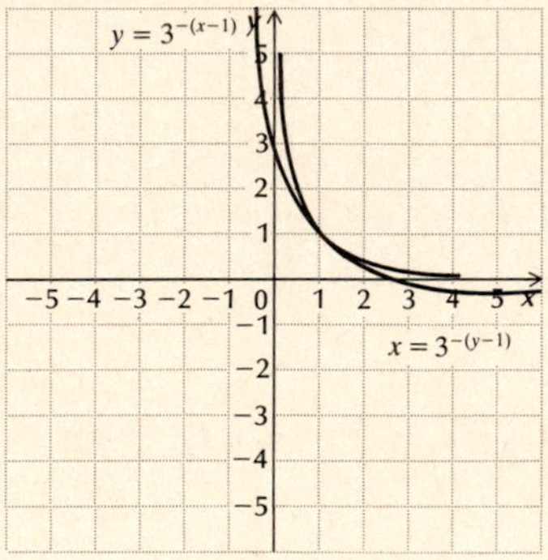

57. $\{x|x > 0\}$ 59. **(a)** 155.7 wph, 190.2 wph, 199.5 wph, 199.9 wph;
(b)

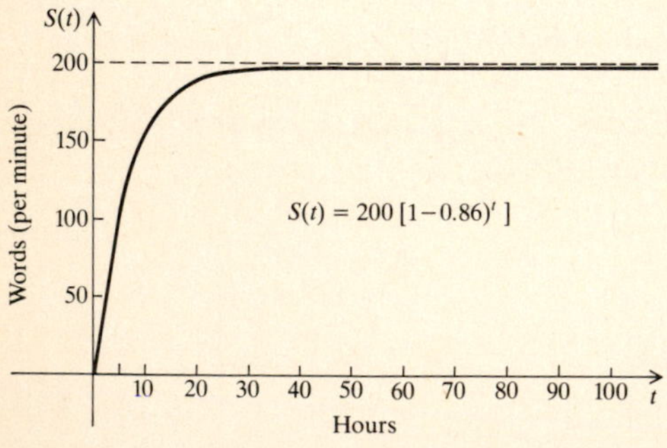

61.

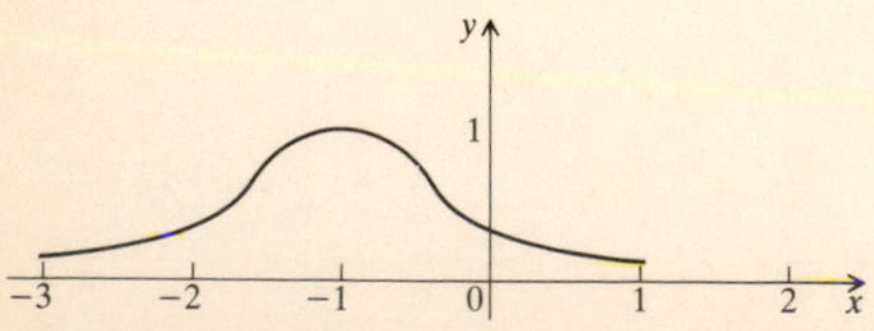

Margin Exercises, Section 5.3

1.

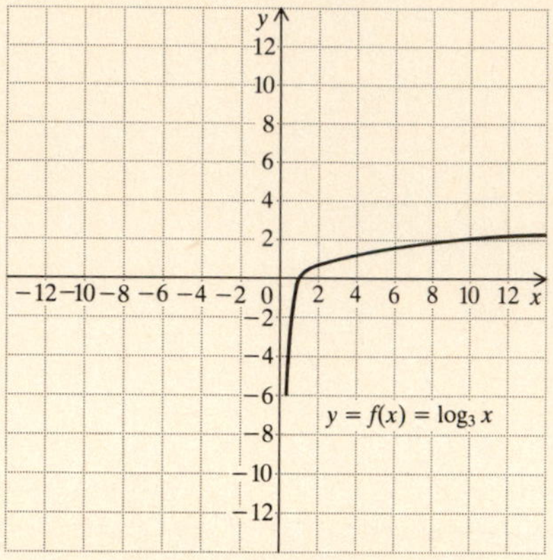

The domain is the set of positive real numbers. The range is the set of all real numbers.
2. $0 = \log_6 1$
3. $-3 = \log_{10} 0.001$
4. $0.25 = \log_{16} 2$
5. $T = \log_m P$
6. $2^5 = 32$
7. $10^3 = 1000$
8. $a^7 = Q$
9. $t^x = M$ **10.** 10,000
11. 3 **12.** $\frac{1}{4}$ **13.** 5
14. -4 **15.** 0 **16.** 0
17. 1 **18.** 1 **19.** 0

Exercise Set 5.3, pp. 301–302

1.

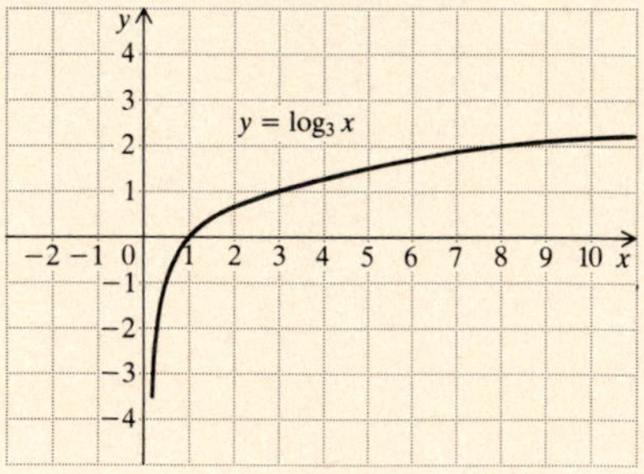

3.

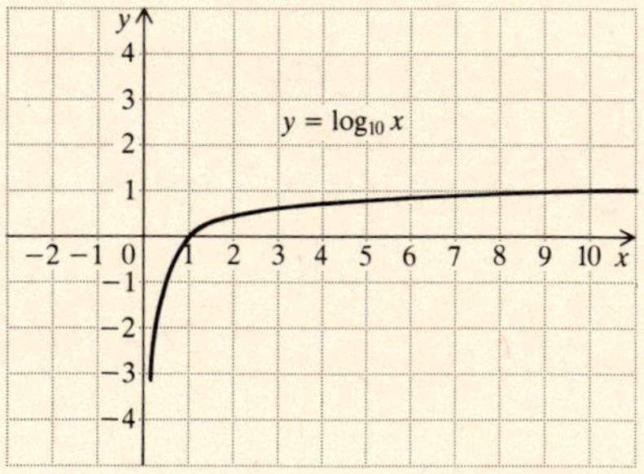

5.

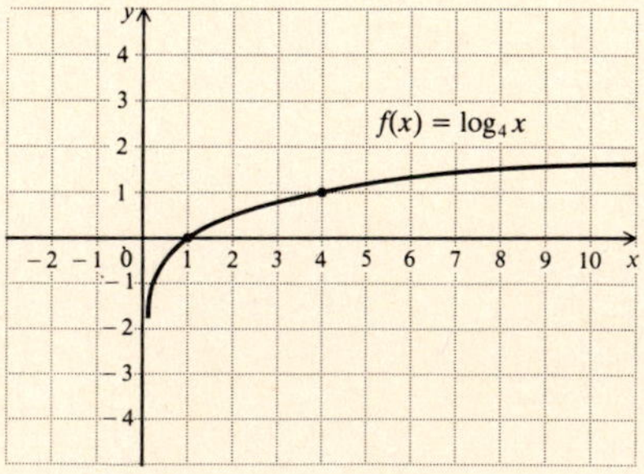

7.

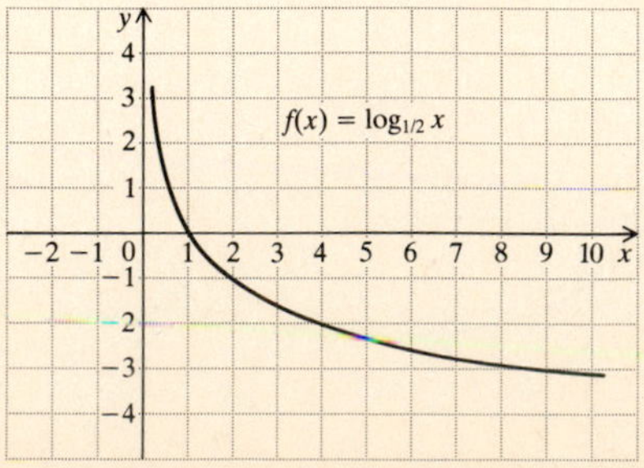

9. 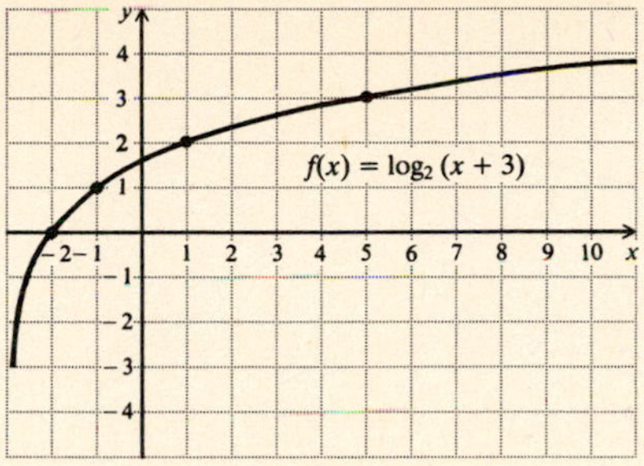

11. 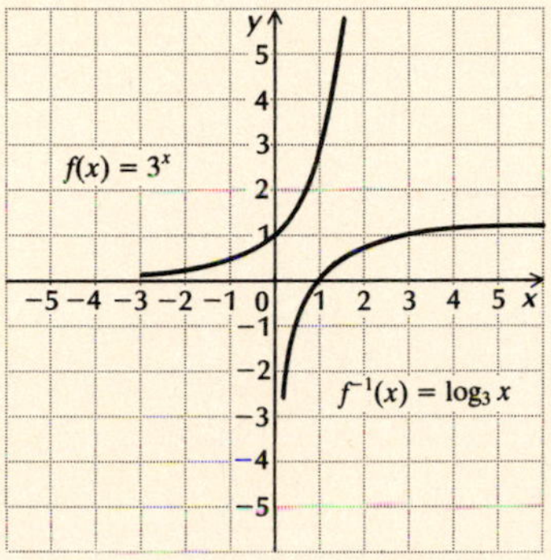

13. $3 = \log_{10} 1000$ **15.** $-3 = \log_5 \frac{1}{125}$ **17.** $\frac{1}{3} = \log_8 2$ **19.** $0.3010 = \log_{10} 2$ **21.** $3 = \log_e t$ **23.** $t = \log_Q x$ **25.** $3 = \log_e 20.0855$ **27.** $-1 = \log_e 0.3679$ **29.** $4^t = 7$ **31.** $2^5 = 32$ **33.** $10^{-1} = 0.1$ **35.** $10^{0.845} = 7$ **37.** $e^{3.4012} = 30$ **39.** $t^k = Q$ **41.** $e^{-0.9676} = 0.38$ **43.** $r^{-x} = M$ **45.** 1000 **47.** 1 **49.** 9 **51.** 4 **53.** $\frac{1}{2}$ **55.** 2 **57.** 3 **59.** -1 **61.** 0 **63.** 4 **65.** -2 **67.** 0 **69.** 0 **71.** $\frac{1}{2}$ **73.** 5 **75.** m

77. 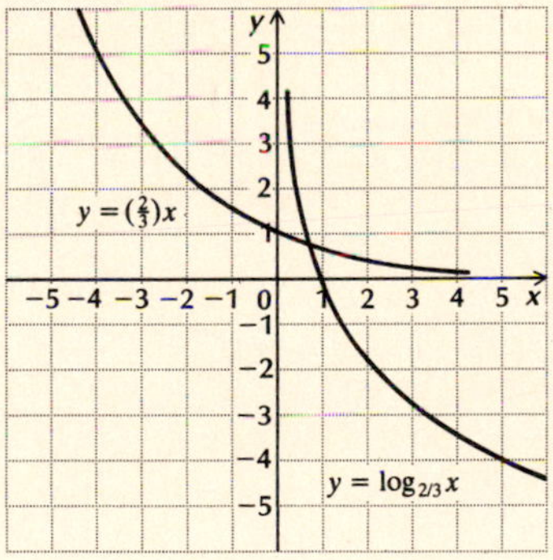

79. 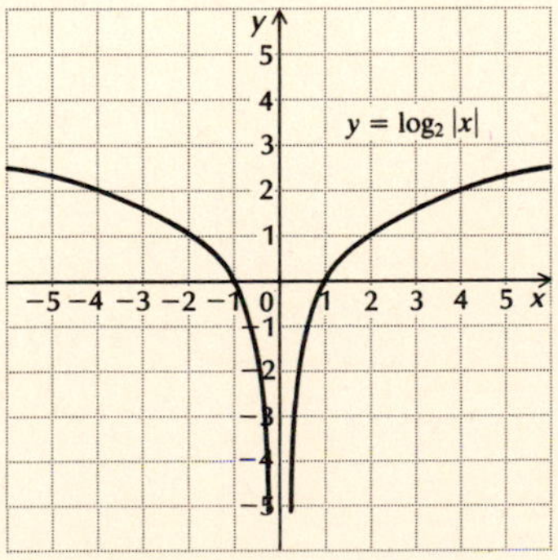

81. 25 **83.** $(\sqrt{5})^{-3}$, or $5^{-3/2}$ **85.** 6 **87.** 1296 **89.** 3 **91.** 0 **93.** -8 **95.** All real numbers **97.** $\{x|x \neq 0\}$ **99.** $\{x|x > \frac{4}{3}\}$ **101.** $\{x|x < -3 \text{ or } x > 3\}$ **103.** $\{x|x \geqslant 16\}$

Margin Exercises, Section 5.4

1. $\log_5 25 + \log_5 5$ **2.** $\log_b P + \log_b Q$ **3.** $\log_3 35$ **4.** $\log_a CABIN$ **5.** $5 \log_7 4$ **6.** $\frac{1}{2} \log_a 5$ **7.** $\log_b P - \log_b Q$ **8.** $\log_2 \dfrac{x}{25}$ **9.** $\frac{1}{2}(3 \log_a z - \log_a x - \log_a y)$ **10.** $2 \log_a x - 3 \log_a y + 1$ **11.** $3 \log_m a + 4 \log_m b - 5 - 9 \log_m n$ **12.** $\log_a \dfrac{x^5\sqrt[4]{z}}{y}$ **13.** $-\frac{3}{2} \log_a b$ **14.** 0.602 **15.** 1 **16.** -0.398 **17.** 0.398 **18.** -0.699 **19.** $\frac{3}{2}$ **20.** 1.699 **21.** 1.204 **22.** 8 **23.** 4.3 **24.** 23 **25.** 3 **26.** x **27.** 42

Exercise Set 5.4, pp. 308–309

1. $\log_2 64 + \log_2 8$ **3.** $\log_4 32 + \log_4 64$ **5.** $\log_c Q + \log_c P$ **7.** $\log_b 720$ **9.** $\log_c PQ$ **11.** $4 \log_a x$ **13.** $5 \log_c y$ **15.** $-6 \log_b Q$ **17.** $\log_a 76 - \log_a 13$ **19.** $\log_b 5 - \log_b 4$ **21.** $\log_a \frac{18}{5}$ **23.** $3 \log_a x + 2 \log_a y + \log_a z$ **25.** $2 \log_b x + \log_b y - 3$ **27.** $\frac{1}{3}(4 \log_c x - 3 \log_c y - 2 \log_c z)$ **29.** $\frac{1}{4}(8 \log_a m + 12 \log_a n - 3 - 5 \log_a b)$ **31.** $\log_a \dfrac{x^{2/5}}{y^{1/3}}$ **33.** $\log_a \dfrac{2x^4}{y^3}$ **35.** $\frac{1}{2} - \log_a x$ **37.** 1.3023 **39.** -0.2457 **41.** -0.5283 **43.** $\frac{3}{2}$ **45.** 1.5283 **47.** 1.548 **49.** 11 **51.** $|x - 4|$ **53.** $4x$ **55.** Q **57.** 15 **59.** -7 **61.** False **63.** True **65.** False **67.** False **69.** True **71.** $\log_a (x^6 - x^4y^2 + x^2y^4 - y^6)$ **73.** $\frac{1}{2}[\log_a (2 - x) + \log_a (2 + x)]$ **75.** $\frac{10}{3}$ **77.** $\{x|x > 0\}$ **79.** $\frac{9}{8}$ **81.** 3 **83.** $\{x|x > -2\}$ **85.** -2 **87.** $\log_a \dfrac{x + \sqrt{x^2 - 5}}{5} \cdot \dfrac{x - \sqrt{x^2 - 5}}{x - \sqrt{x^2 - 5}} = \log_a \dfrac{5}{5(x - \sqrt{x^2 - 5})} = -\log_a (x - \sqrt{x^2 - 5})$ **89.** Let $\log_{a^{1/n}} x = M$. Then $(a^{1/n})^M = x$, or $(a^M)^{1/n} = x$, and $((a^M)^{1/n})^n = x^n$, or $a^M = x^n$, or $\log_a x^n = M$. Thus, $\log_{a^{1/n}} x = \log_a x^n$.

Margin Exercises, Section 5.5

1. 7.9913 **2.** -3.2097 **3.** 1.7251 **4.** -1.2711 **5.** 1,274,383.09 **6.** 1.0029 **7.** 0.00003 **8.** 0.7169 **9.** 11.3518 **10.** -6.7882 **11.** -0.9416 **12.** 7.3427 **13.** 1272.1962 **14.** 1.0124 **15.** 0.3296 **16.** 0.0010 **17.** 1.1606 **18.** 5.7369

Exercise Set 5.5, pp. 314–315

1. 0.4771 **3.** 0.9031 **5.** 0.3692 **7.** 1.8129 **9.** 1.7952 **11.** 2.7259 **13.** 4.1271 **15.** -0.2441 **17.** -1.2840 **19.** -2.0084 **21.** 1000 **23.** 501.1872 **25.** 3.0001 **27.** 0.2841 **29.** 0.0011 **31.** 5.7498 **33.** Does not exist **35.** 1.0088 **37.** 1.0986 **39.** 2.0794 **41.** 4.4067 **43.** 9.0318 **45.** -5.1328 **47.** 36.7890 **49.** 0.0023 **51.** 1.0057 **53.** 5.8346×10^{14} **55.** 66.5569 **57.** -30.6182 **59.** 1637.9488 **61.** 2.8044 **63.** 0.0000027 **65.** 3.3219 **67.** 3.5850 **69.** -0.26144 **71.** -2.8074 **73.** -1.0959 **75.** 4.0229 **77.** $\ln x = \dfrac{\log x}{\log e} = \dfrac{1}{\log e} \cdot \log x = 2.3026 \log x$

79. 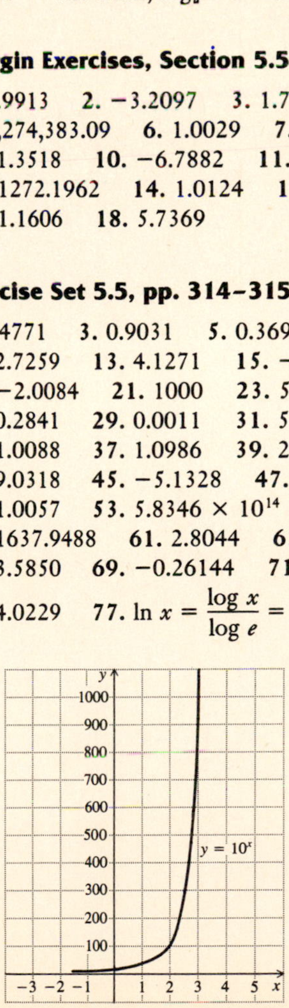

81.

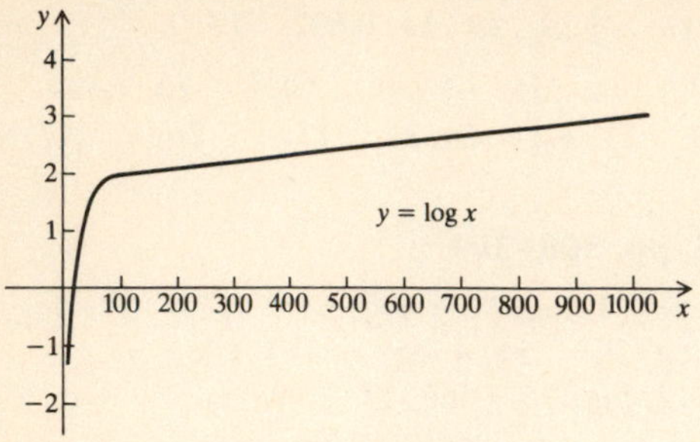

83. 3 **85.** 253.3282 **87.** 2.9617

89. Let $a = e$, $b = 10$, and $M = e$. Substitute in the change-of-base formula:

$$\log e = \frac{\ln e}{\ln 10} = \frac{1}{\ln 10}.$$

91. Let $a = M$, $b = b$, and $M = M$. Substitute in the change-of-base formula:

$$\log_b M = \frac{\log_M M}{\log_M b} = \frac{1}{\log_M b}.$$

93. $\log_a(\log_a x) = \log_a\left(\frac{\log_b x}{\log_b a}\right) = \log_a(\log_b x) - \log_a(\log_b a)$

95. 4, 2.8680, 2.7320, 2.7196, 2.7184 **97.** $e^{\sqrt{\pi}}$

Margin Exercises, Section 5.6

1.

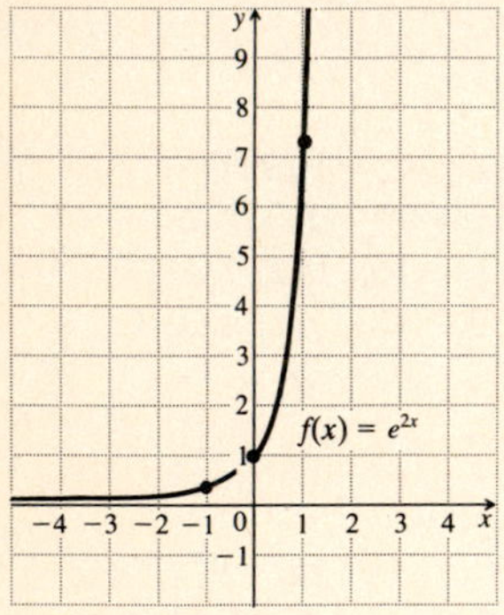

2.

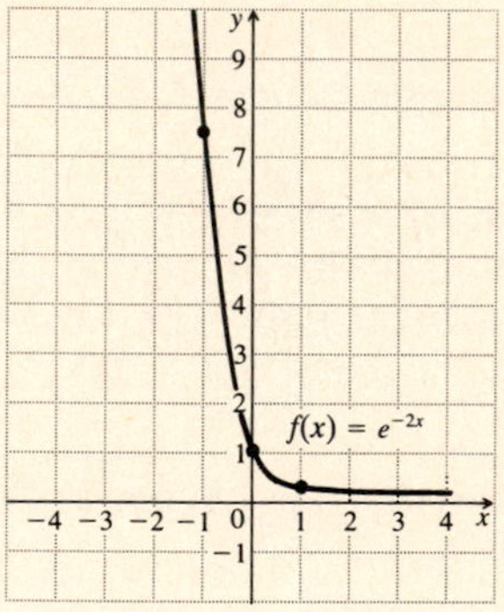

3.

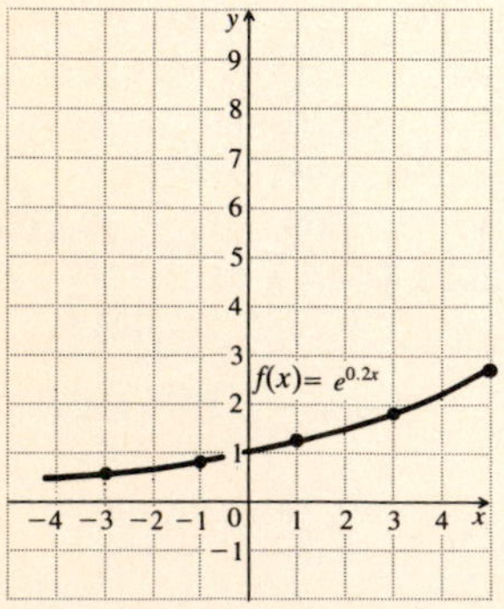

4.

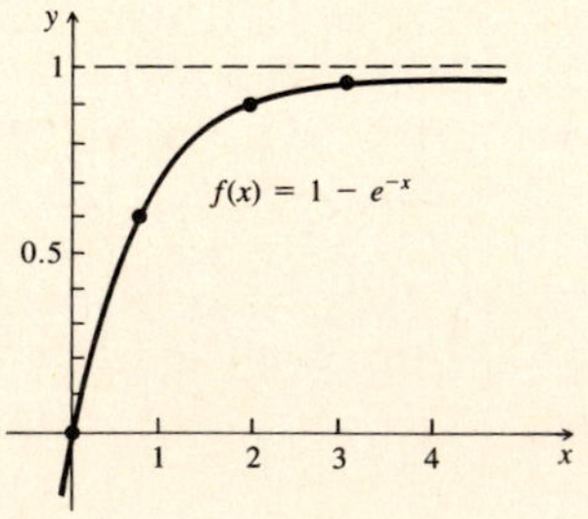

5.

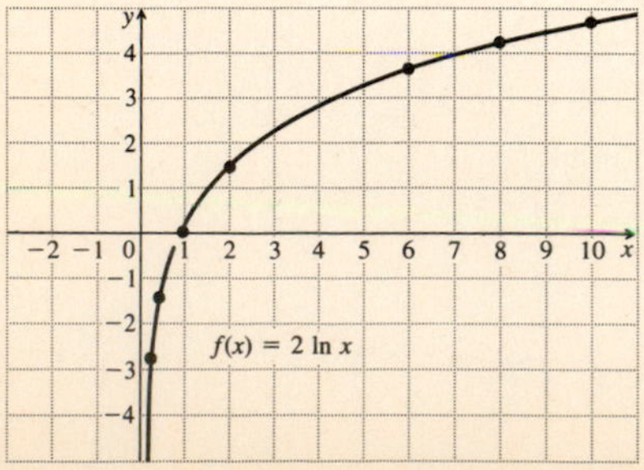

6.

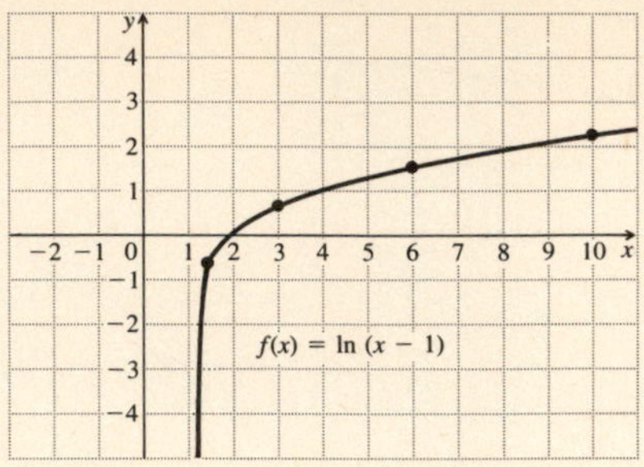

7. (a) \$59.63;
(b)

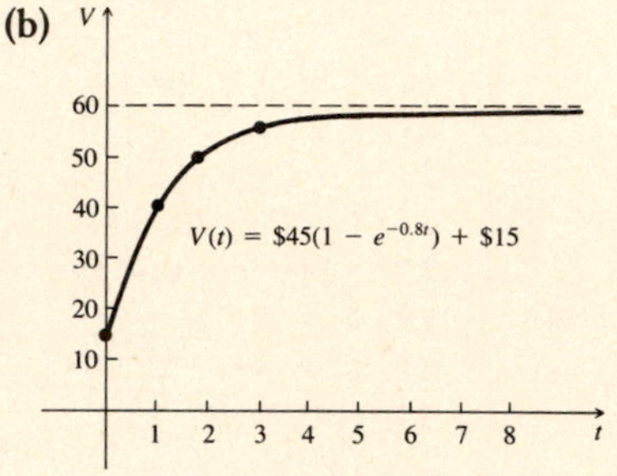

Exercise Set 5.6, pp. 318–319

1.

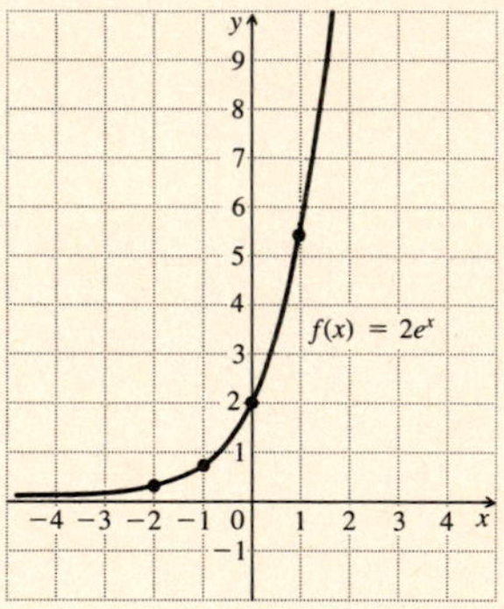

3.

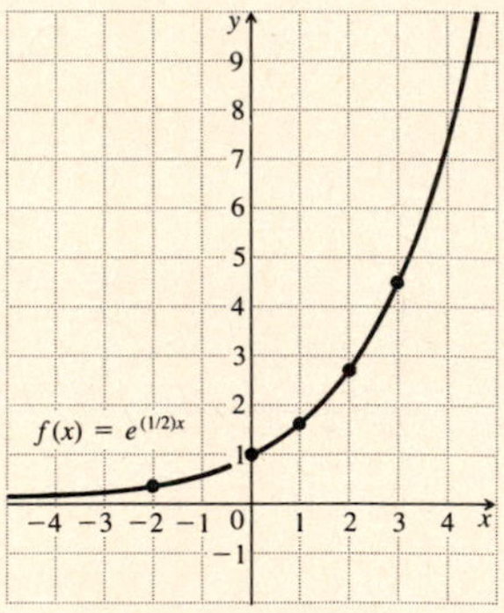

5.

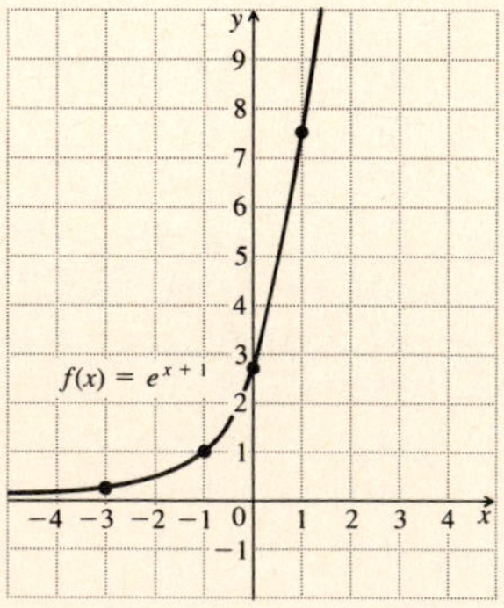

7.

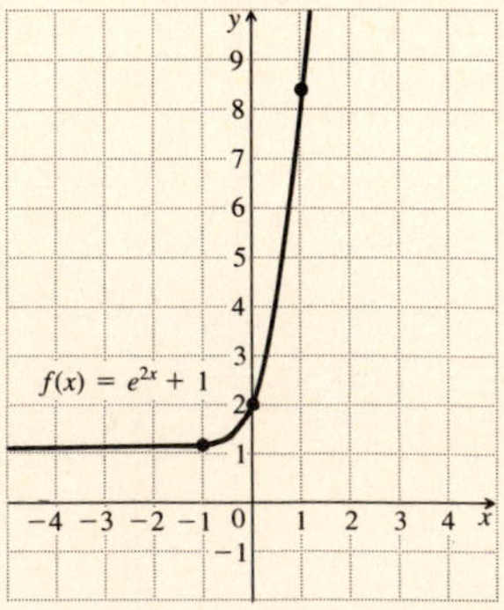

9.

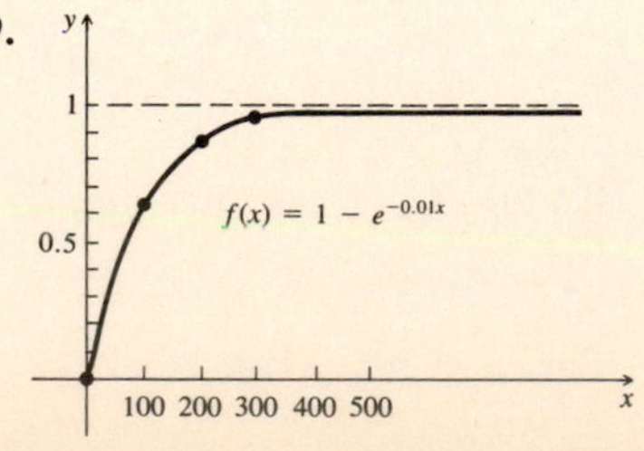

11.

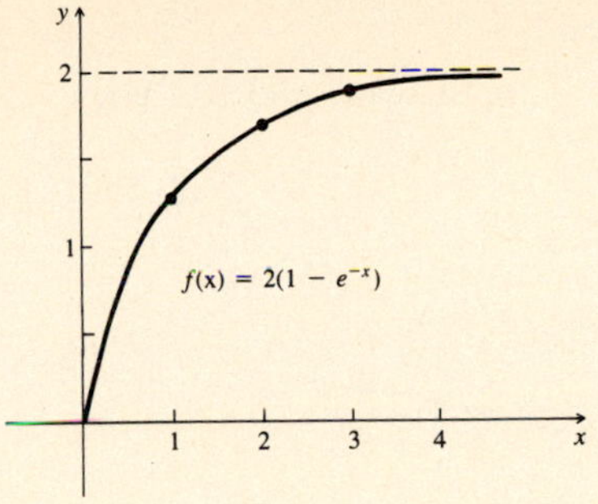

13.

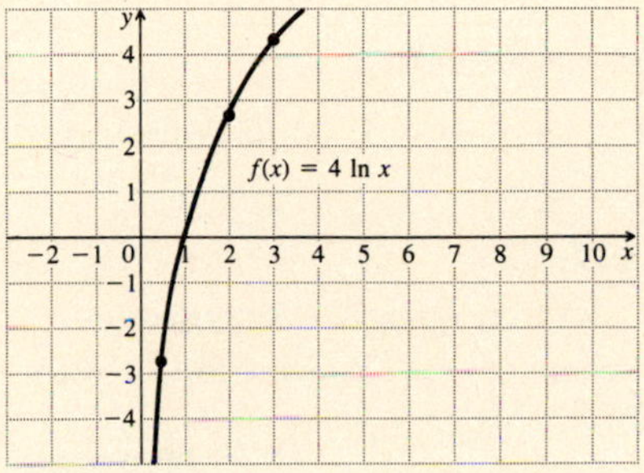

15.

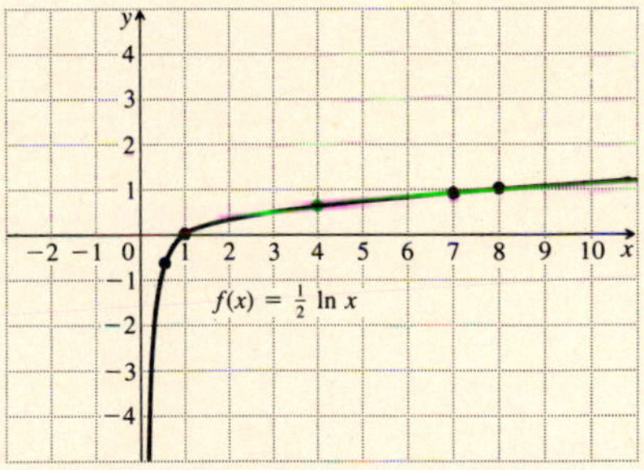

17.

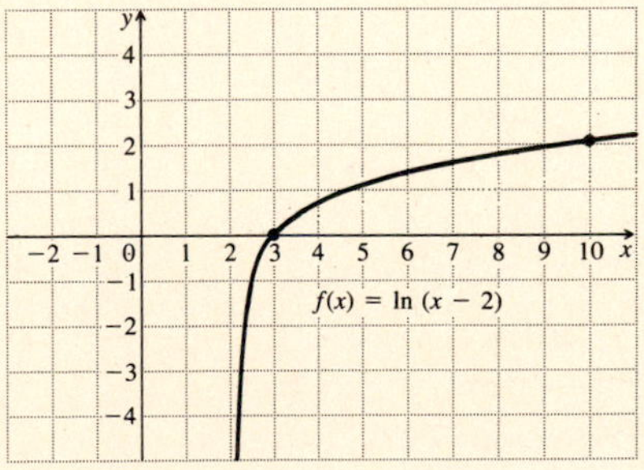

19.

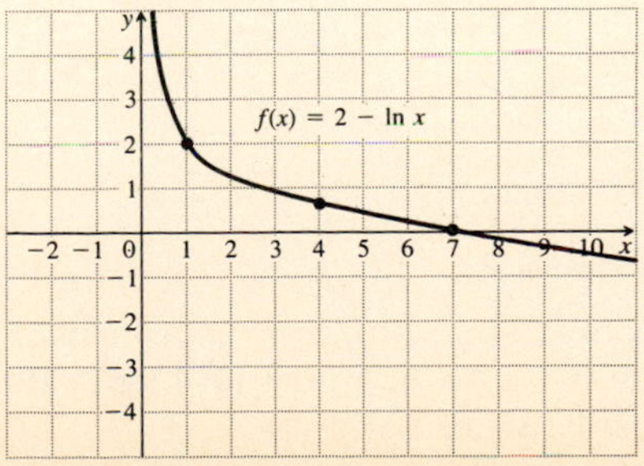

21. **(a)** 18.1%, 55.1%, 69.9%, 90.9%; **(b)**

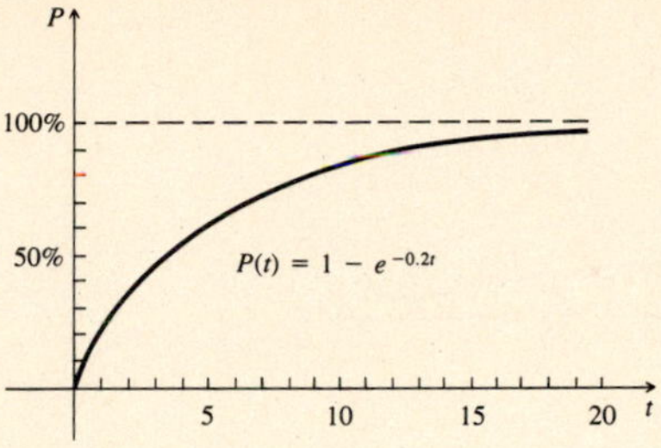

23. **(a)** \$58.69, \$71.57, \$77.29, \$77.92, \$78.00; **(b)**

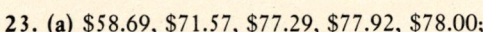

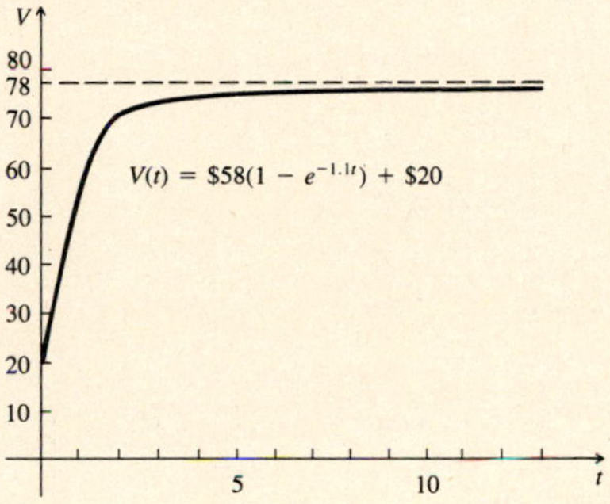

25.

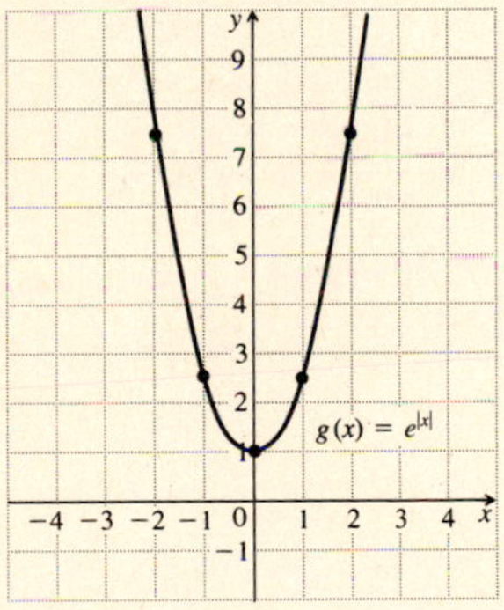

27.

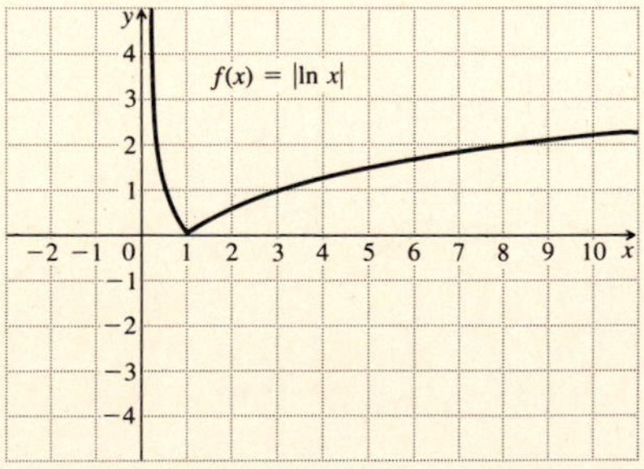

29.

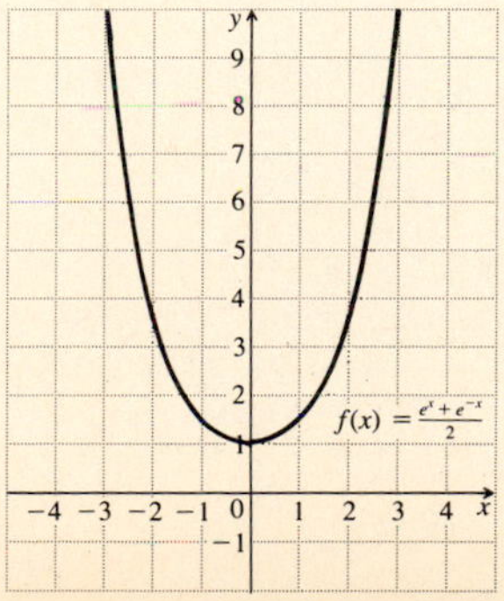

31. **(a)** 20, 42, 234, 602; **(b)**

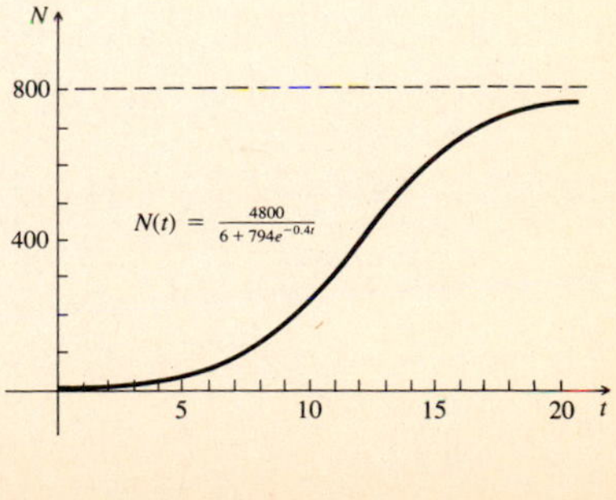

33. (a)

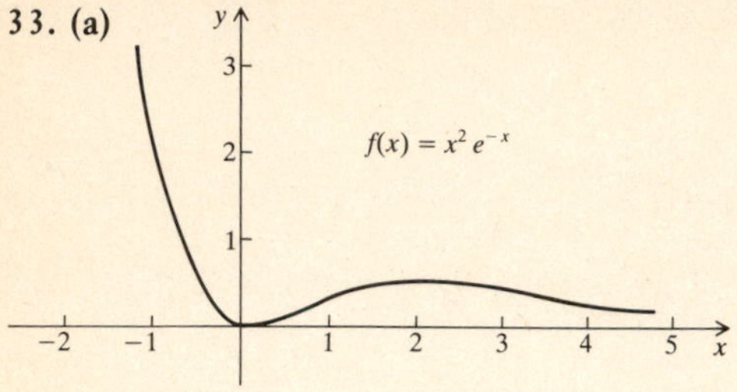

(b) 0; (c) minimum: 0

35. (a)

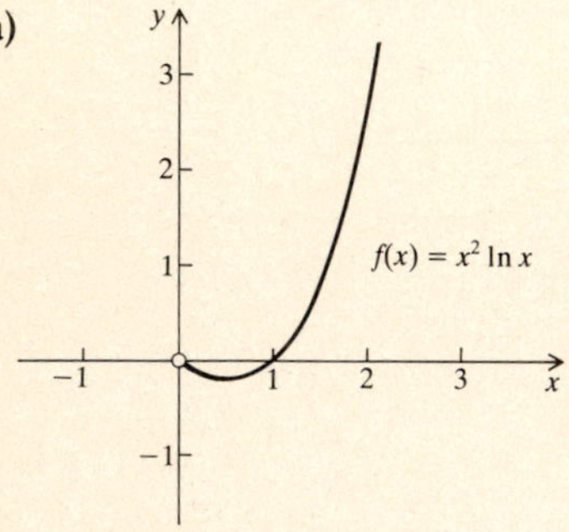

(b) 0, 1; (c) minimum: −0.2

Margin Exercises, Section 5.7

1. 1 2. $\frac{3}{2}$ 3. $\frac{\log 20}{\log 7} \approx 1.5395$ 4. $\frac{\ln 80}{0.3} \approx 14.6068$

5. $x = \ln(t + \sqrt{t^2 + 1})$ 6. 8 7. $\frac{35}{4}$ 8. 5

Exercise Set 5.7, pp. 323–325

1. 5 3. 4 5. $\frac{3}{2}$ 7. $\frac{3}{7}$ 9. $\frac{\log 33}{\log 2} \approx 5.0444$

11. $\frac{\log 40}{\log 2} \approx 5.3219$ 13. $\frac{5}{2}$ 15. −3, −1

17. $\frac{\log 70}{\log 84} \approx 0.9589$ 19. $\ln 1000 \approx 6.9078$

21. $-\ln 0.3 \approx 1.2040$ 23. $\frac{\ln 0.08}{-0.03} \approx 84.1910$

25. $\frac{\log 2}{\log 2 - \log 3} \approx -1.7095$ 27. $\frac{\log 48}{\log 3.9} \approx 2.8444$

29. $\frac{\log 250}{\log 1.87} \approx 8.8211$ 31. $\frac{12}{5}$

33. $x = \ln\left(\frac{5}{2}t + \sqrt{\frac{25t^2}{4} + 1}\right)$ 35. $x = \frac{1}{2}\ln\frac{t+1}{t-1}$

37. 625 39. $\frac{1}{125}$ 41. 100 43. 0.001 45. e 47. e^{-2}

49. $-\frac{117}{7}$ 51. 10 53. $\frac{1}{3}$ 55. 3 57. $\frac{1}{3}$ 59. 5

61. 1, 10^6 63. $\pm\sqrt{24}$ 65. 1, 100 67. 4 69. $\emptyset$

71. $\pm\sqrt{58}$ 73. 10^{1000} 75. −9, 9 77. $\frac{1}{100}$, 100

79. 3, −8 81. $\frac{1}{10}, \frac{1}{1000}$ 83. $-\frac{8}{3}$ 85. 1 87. $-\frac{1}{2}$

89. $t = \frac{\ln P - \ln P_0}{k}$ 91. $t = -\frac{1}{k}\ln\frac{T - T_0}{T_1 - T_0}$ 93. $Q = a^b\sqrt[3]{y}$

95. $\log_3 x = -3$ 97. If $a > 1$, $x \geqslant 1$. If $0 < a < 1$, $0 < x \leqslant 1$.

99. $x > \frac{\log 0.8}{\log 0.5} \approx 0.3219$ 101. 1, 4 103. $a = \frac{2}{3}b$

Margin Exercises, Section 5.8

1. (a) $A(t) = \$80{,}000(1.07)^t$; (b) 16.2 years; (c) 10.2 years
2. (a) 5000; (b) 5169;
(c)

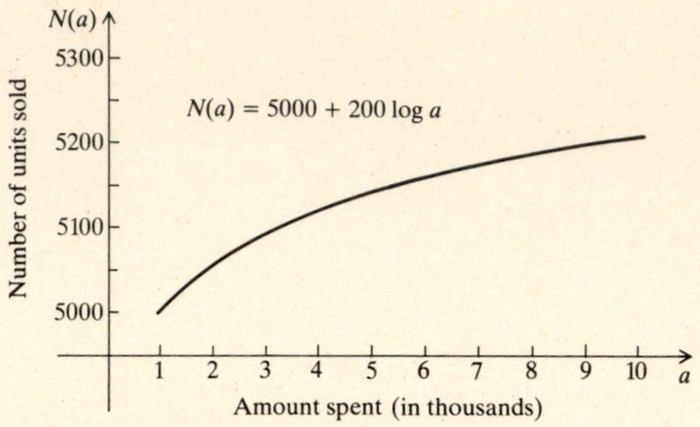

(d) \$5011.87

3. (a) 5.6; (b) 1.6×10^{-6} 4. 7.85 5. (a) 65 decibels; (b) 140 decibels

Exercise Set 5.8, pp. 329–331

1. (a) 5.5 years; (b) 0.8 year 3. (a) $A(t) = \$50{,}000(1.09)^t$; (b) 25.5 years; (c) 8.04 years 5. (a) 6.6 years; (b) 3.1 years
7. (a) 68%; (b) 54%, 40%;
(c)

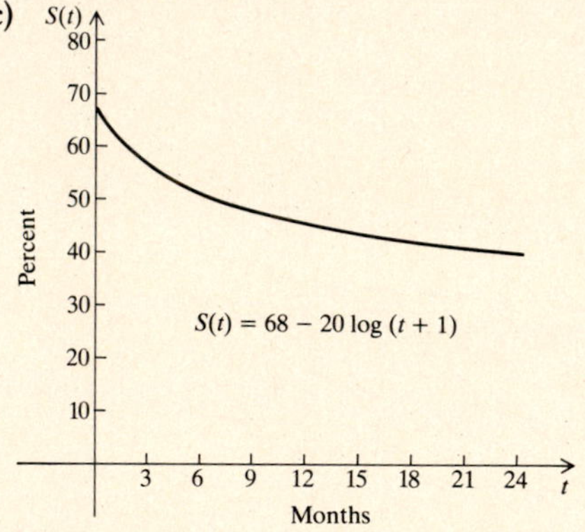

(d) 6.9 yr

9. (a) 1000; (b) 1140;
(c)

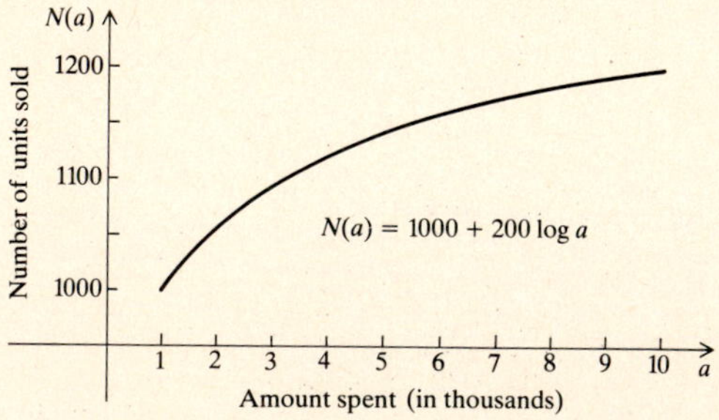

(d) \$23,988.33
11. 3.8
13. 4.2
15. 3.9×10^{-6}
17. 1.6×10^{-5}
19. 8.25
21. 34 decibels
23. 60 decibels

25. (a) 4.6 hr; (b) 9.2 hr 27. $I = 10^R I_0$
29. $[H^+] = 10^{-pH}$ 31. (a) 3.0; (b) 436,515.8; (c) 100 times
33. (a)
$$y = bx^k$$
$$\log y = \log(bx^k)$$
$$\log y = \log b + k \log x$$
$$Y = kX + \log b;$$
(b) $Y = 0.287X + 1.47$;
(c) $y = 10^{1.47}x^{0.287}$; (d) 119.6 mm

Margin Exercises, Section 5.9

1. (a) 3.0 ft/sec; (b) 1.3 ft/sec; (c) 2.4 ft/sec 2. 260 million, 310 million 3. (a) 33¢; (b) in 110.5 yr from 1932, or in 2043
4. 6,010,389 5. (a) $k = 0.085$, or 8.5%;

(b) $P(t) = \$10{,}000e^{0.085t}$; **(c)** \$23,396.47; **(d)** 8.2 yr
6. 16.9 yr **7.** 5.0% **8.** 10.2% **9.** 13,412

Exercise Set 5.9, pp. 339–342

1. 3.4 ft/sec **3.** 1.5 ft/sec **5. (a)** $C(t) = 5e^{0.097t}$; **(b)** \$0.76, \$1.99; **(c)** in 47.5 years from 1962;
(d)

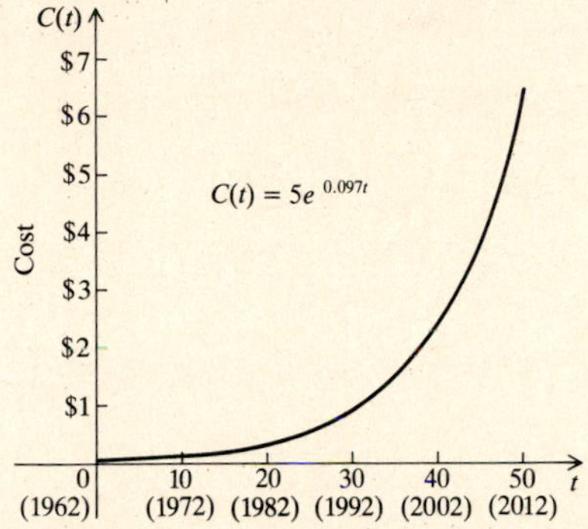

(e) 7.1 yr

7. (a) $P(t) = P_0e^{0.06t}$; **(b)** \$536.56; **(c)** \$724.27;
(d)

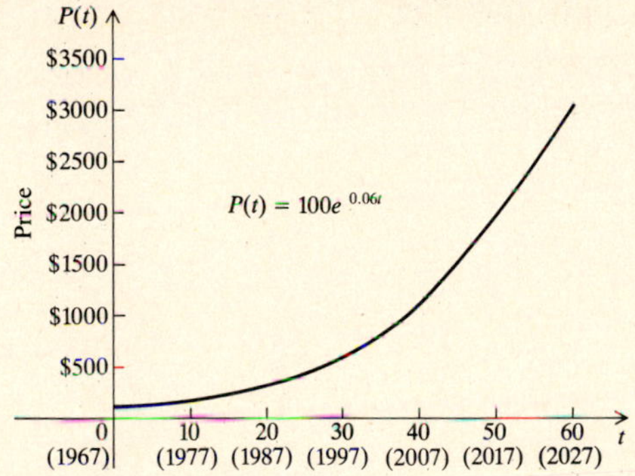

9. (a)

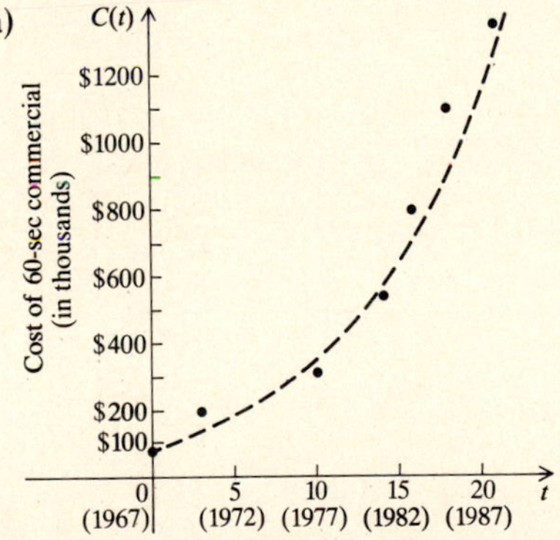

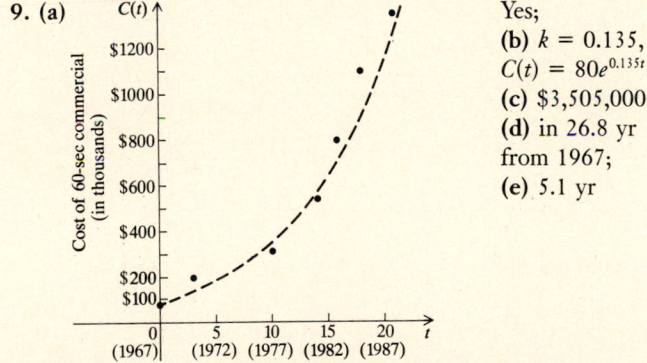

Yes; **(b)** $k = 0.135$, $C(t) = 80e^{0.135t}$; **(c)** \$3,505,000; **(d)** in 26.8 yr from 1967; **(e)** 5.1 yr

11. (a) $P(t) = P_0e^{0.09t}$; **(b)** \$5470.87, \$5986.09; **(c)** 7.7 yr
13. 19.8 yr **15.** 9.9% **17. (a)** $k = 0.16$; $P(t) = 84{,}000e^{0.16t}$; **(b)** \$1,240,241,652; **(c)** 4.3 yr; **(d)** 58.7 yr
19. In 17.3 yr from 1990
21. (a) $k = 0.007$; $P(t) = 2{,}812{,}000e^{0.007t}$; **(b)** 3,373,315
23. 5135 **25.** 23.1% per minute **27.** 7.2 days **29. (a)** 0.7%; **(b)** 81.1% **31. (a)** 13.3 lb/in^2; **(b)** 7.24 lb/in^2; **(c)** 46,052 ft; **(d)** 72,589 ft **33. (a)** 24.7%, 1.5%, 0.09%, (3.98×10^{-31})%; **(b)** 0.00008% **35.** $R = e^{v/c}$ **37.** (6, \$404) **39.** 51.8° F

Review Exercises: Chapter 5, pp. 342–344

1. [5.1] {(5, −4), (−3, 2), (7, 1), (8, 8), (−4, 5)} **2.** [5.1] $x = 3y^2 + 2y - 1$ **3.** [5.1] $x = \sqrt{y + 2}$ **4.** [5.1] (b), (e) **5.** [5.1] (d) **6.** [5.1] $f^{-1}(x) = (2x - 4)^2$ **7.** [5.1] $f^{-1}(x) = \sqrt[3]{x - 8}$ **8.** [5.1] a **9.** [5.1] t

10. [5.3]

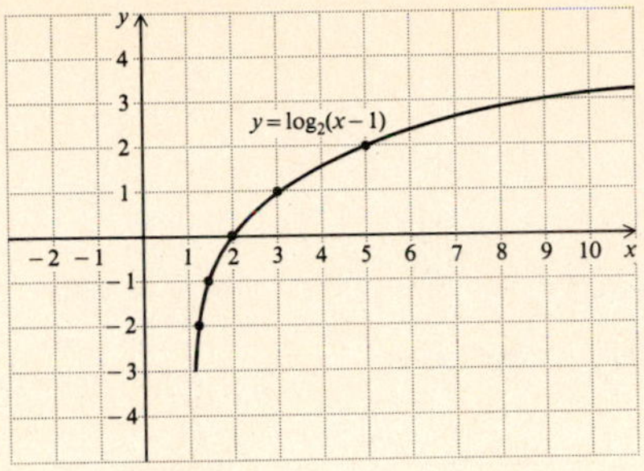

11. [5.2]

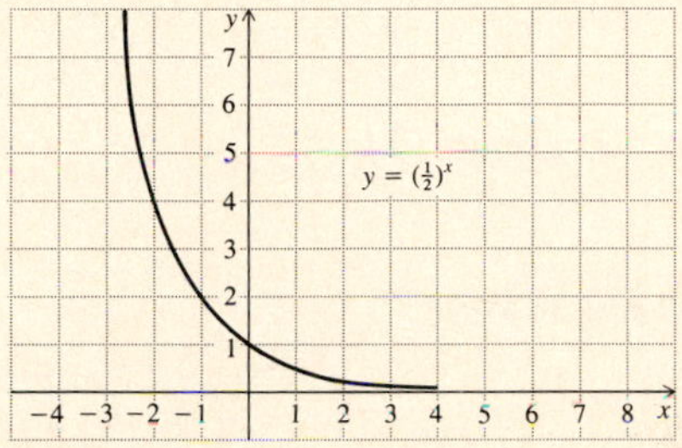

12. [5.6]

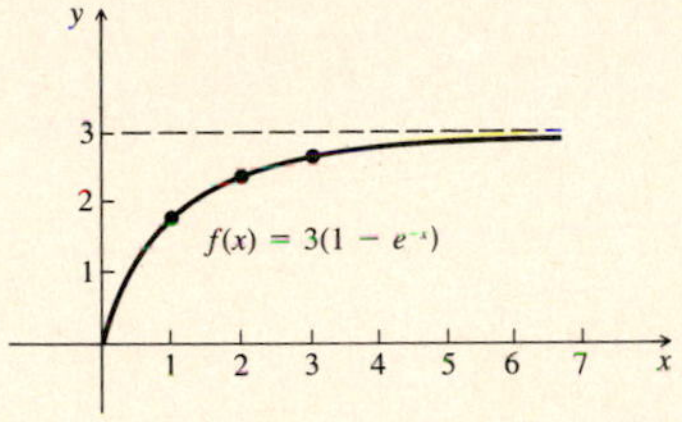

13. [5.6]

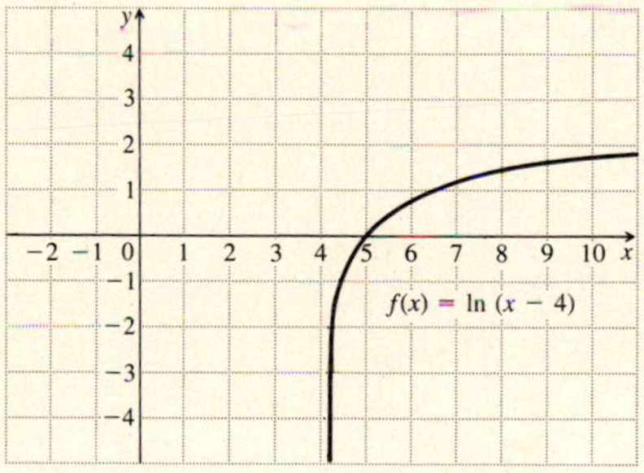

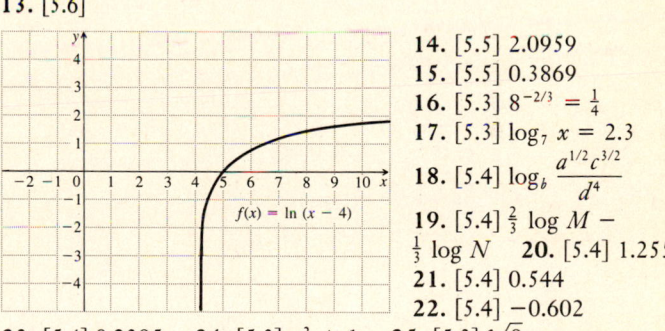

14. [5.5] 2.0959
15. [5.5] 0.3869
16. [5.3] $8^{-2/3} = \frac{1}{4}$
17. [5.3] $\log_7 x = 2.3$
18. [5.4] $\log_b \dfrac{a^{1/2}c^{3/2}}{d^4}$
19. [5.4] $\frac{2}{3}\log M - \frac{1}{3}\log N$ **20.** [5.4] 1.255
21. [5.4] 0.544
22. [5.4] −0.602
23. [5.4] 0.2385 **24.** [5.3] $x^2 + 1$ **25.** [5.3] $\sqrt{9}$ **26.** [5.3] 4 **27.** [5.3] $\frac{1}{2}$ **28.** [5.3] 3 **29.** [5.7] $\frac{1}{5}$ **30.** [5.7] 4.3820 **31.** [5.7] 1 **32.** [5.7] 9 **33.** [5.7] 3 **34.** [5.7] 3 **35.** [5.8] 5.7 yr **36.** [5.8] 30 decibels **37.** [5.9] 6.25 g **38.** [5.8] **(a)** 82%; **(b)** 50%; **(c)** 4.5 months **39.** [5.9] **(a)** $k = 0.05$, $C(t) = 4.65e^{0.05t}$; **(b)** \$51.26; **(c)** in 29.2 years since 1962; **(d)** 13.9 yr **40.** [5.9] 3.9% **41.** [5.9] 8.1 yr **42.** [5.9] 2623 **43.** [5.8] 6.4 **44.** [5.8] 8 **45.** [5.5] −2.6655 **46.** [5.5] 6.1278 **47.** [5.5] −4.6910 **48.** [5.5] 11.3780 **49.** [5.5] −10.4919 **50.** [5.5] 24.5004 **51.** [5.5] 1068.32 **52.** [5.5] 0.000003442 **53.** [5.5] 1.0019 **54.** [5.5] 712.799 **55.** [5.5] 113,210.02 **56.** [5.5] 0.0168 **57.** [5.7] 64, $\frac{1}{64}$ **58.** [5.7] 1 **59.** [5.3]

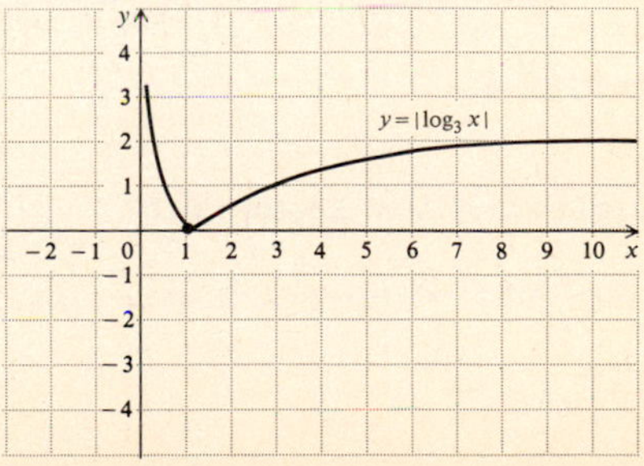

60. [5.6]

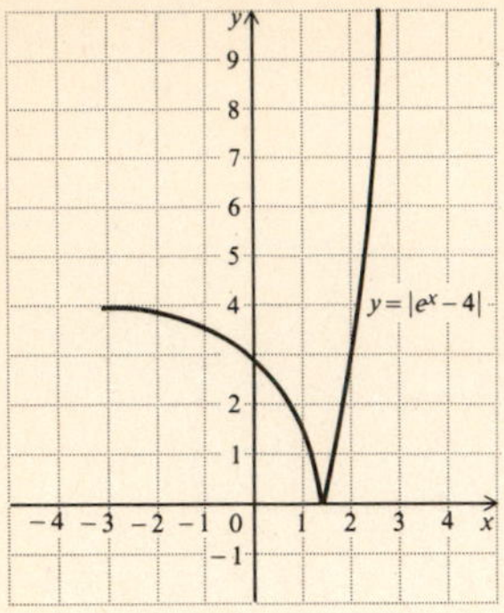

61. [5.6] $\{x|x > e^{6/5}\}$ **62.** [5.6] $\{x|x \neq \frac{1}{4} \ln 10\}$

Test: Chapter 5, pp. 344–345

1. [5.1] $\{(-5, 3), (-3, 6), (-3, 8), (3, -5), (-2.7, 1.3)\}$
2. [5.1] $x = |y|$ **3.** [5.1] (a), (c) **4.** [5.1] (a), (b), (d)
5. [5.1] $f^{-1}(x) = x^2 + 6$ **6.** [5.1] 3
7. [5.3]

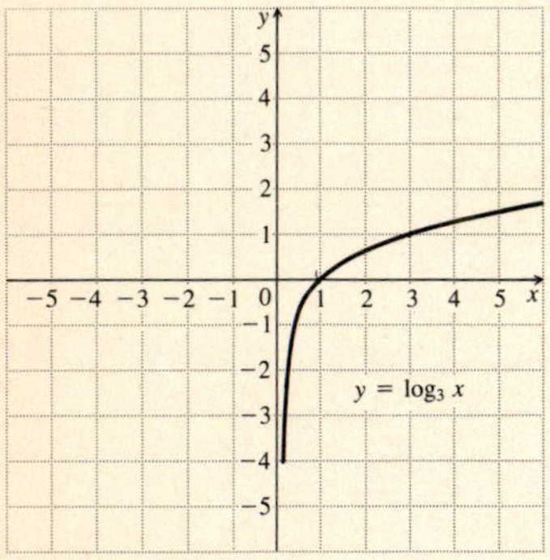

8. [5.6]

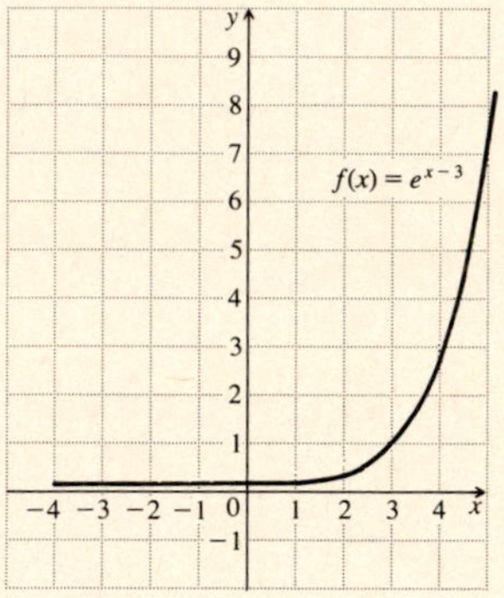

9. [5.5] 1.9482 **10.** [5.3] $\sqrt{3}^4 = 9$
11. [5.3] $\log_x 0.03125 = 5$ **12.** [5.3] $3x$
13. [5.4] $\log_c \dfrac{x^3\sqrt{z}}{y^4}$ **14.** [5.4] $\dfrac{1}{4}\log w + \dfrac{3}{4}\log r$ **15.** [5.4] 0.477
16. [5.4] 1.699 **17.** [5.4] 0.233 **18.** [5.4] $0, \frac{1}{2}$ **19.** [5.3] 16
20. [5.7] 2 **21.** [5.7] 4 **22.** [5.7] 1.6094
23. [5.5] -2.2804 **24.** [5.5] 1.3507 **25.** [5.5] 118.9871
26. [5.5] 7.6676 **27.** [5.5] 4.9405 **28.** [5.5] 0.0099
29. [5.5] -15.0534 **30.** [5.5] Does not exist **31.** [5.8] 6.1 yr
32. [5.8] 47 decibels **33.** [5.9] 156,183 **34.** [5.8] **(a)** 2.5 ft/sec; **(b)** 437,502 **35.** [5.9] **(a)** $P(t) = 209e^{0.01t}$; **(b)** 308.7 million, 384.7 million; **(c)** in 36.1 yr from 1959 **36.** [5.9] 2.3%
37. [5.9] 8.1 yr **38.** [5.9] 3984 yr **39.** [5.8] 24 decibels
40. [5.8] 8.7 **41.** [5.7] False **42.** [5.3, 5.6] $\{x|x > 1\}$
43. [5.7] 16

CHAPTER 6

Margin Exercises, Section 6.1

1. (a) $\dfrac{\pi}{2}$; **(b)** $\dfrac{3\pi}{4}$; **(c)** $\dfrac{3\pi}{2}$ **2. (a)** $\dfrac{\pi}{4}$; **(b)** $\dfrac{7\pi}{4}$; **(c)** $\dfrac{5\pi}{4}$ **3. (a)** $\dfrac{\pi}{6}$; **(b)** $\dfrac{4\pi}{3}$; **(c)** $\dfrac{11\pi}{6}$

4.

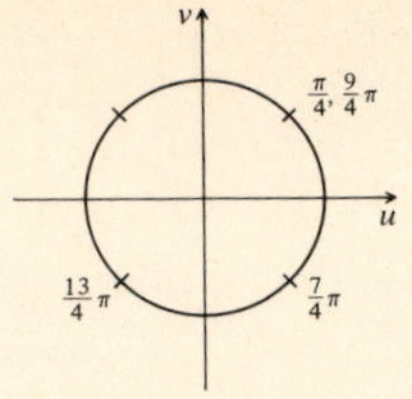

5.

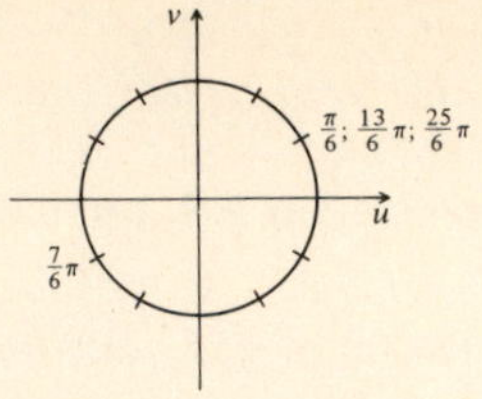

6.

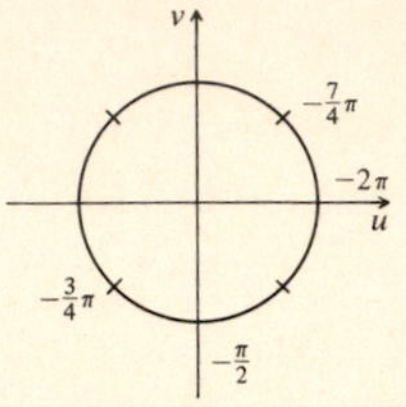

7. (a) $\left(\dfrac{3}{5}, \dfrac{4}{5}\right)$; **(b)** $\left(-\dfrac{3}{5}, -\dfrac{4}{5}\right)$; **(c)** $\left(-\dfrac{3}{5}, \dfrac{4}{5}\right)$; **(d)** yes, by symmetry
8. (a) $\left(-\dfrac{\sqrt{35}}{6}, \dfrac{1}{6}\right)$; **(b)** $\left(\dfrac{\sqrt{35}}{6}, -\dfrac{1}{6}\right)$; **(c)** $\left(\dfrac{\sqrt{35}}{6}, \dfrac{1}{6}\right)$; **(d)** yes, by symmetry
9. $N\left(\dfrac{\sqrt{3}}{2}, -\dfrac{1}{2}\right)$, $P\left(-\dfrac{\sqrt{3}}{2}, -\dfrac{1}{2}\right)$, $R\left(-\dfrac{\sqrt{3}}{2}, \dfrac{1}{2}\right)$

Exercise Set 6.1, pp. 353–354

1.

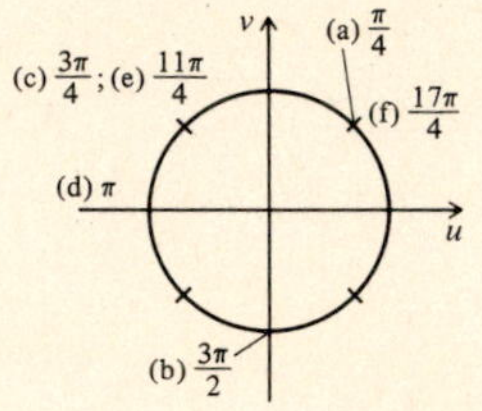

3.

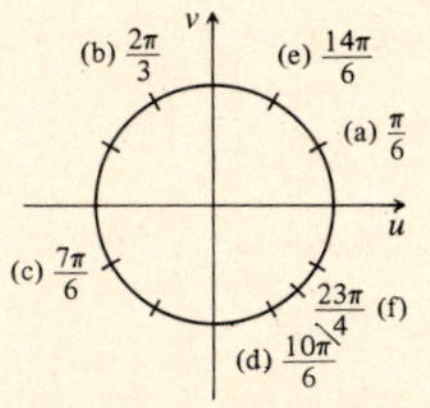

5.

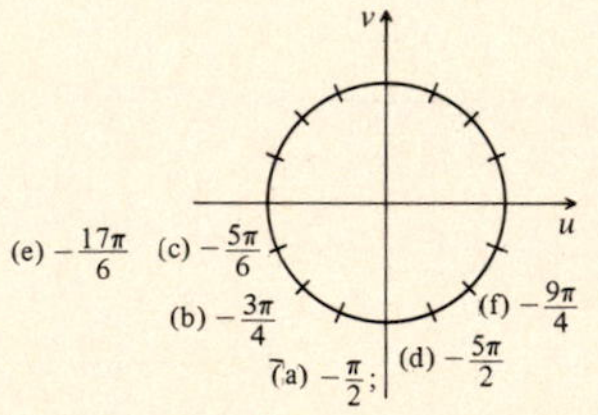

7. M: $\dfrac{2}{3}\pi, -\dfrac{4}{3}\pi$; N: $\dfrac{5}{6}\pi, -\dfrac{7}{6}\pi$; P: $\dfrac{5}{4}\pi, -\dfrac{3}{4}\pi$; Q: $\dfrac{11}{6}\pi, -\dfrac{\pi}{6}$
9. (a) $(-0.375, -0.927)$; **(b)** $(0.375, 0.927)$; **(c)** $(0.375, -0.927)$; **(d)** yes, by symmetry
11. (a) $\left(-\dfrac{\sqrt{2}}{2}, -\dfrac{\sqrt{2}}{2}\right)$; **(b)** $\left(\dfrac{\sqrt{2}}{2}, \dfrac{\sqrt{2}}{2}\right)$; **(c)** $\left(\dfrac{\sqrt{2}}{2}, -\dfrac{\sqrt{2}}{2}\right)$
13. (a) $\left(\dfrac{2}{3}, -\dfrac{\sqrt{5}}{3}\right)$; **(b)** $\left(-\dfrac{2}{3}, \dfrac{\sqrt{5}}{3}\right)$; **(c)** $\left(-\dfrac{2}{3}, -\dfrac{\sqrt{5}}{3}\right)$
15. $M\ (0, 1)$; $N\left(-\dfrac{\sqrt{2}}{2}, \dfrac{\sqrt{2}}{2}\right)$; $P\ (0, -1)$; $Q\left(\dfrac{\sqrt{2}}{2}, -\dfrac{\sqrt{2}}{2}\right)$
17. $\left(\dfrac{\sqrt{3}}{2}, -\dfrac{1}{2}\right)$ **19.** $\left(\dfrac{3}{4}, \dfrac{\sqrt{7}}{4}\right)$ **21.** M: $\dfrac{8\pi}{3}$; N: $\dfrac{17\pi}{6}$; P: $\dfrac{13\pi}{4}$; Q: $\dfrac{23\pi}{6}$ **23.** $\pm\dfrac{2\sqrt{2}}{3}$ **25.** ± 0.96649

Margin Exercises, Section 6.2

1. (a) 0; **(b)** -1; **(c)** $\dfrac{\sqrt{2}}{2}$; **(d)** $-\dfrac{\sqrt{2}}{2}$; **(e)** $\dfrac{\sqrt{3}}{2}$; **(f)** $-\dfrac{1}{2}$ **2. (a)** -1; **(b)** 0; **(c)** $\dfrac{\sqrt{2}}{2}$; **(d)** $\dfrac{\sqrt{2}}{2}$; **(e)** $-\dfrac{\sqrt{3}}{2}$; **(f)** $\dfrac{1}{2}$

3.

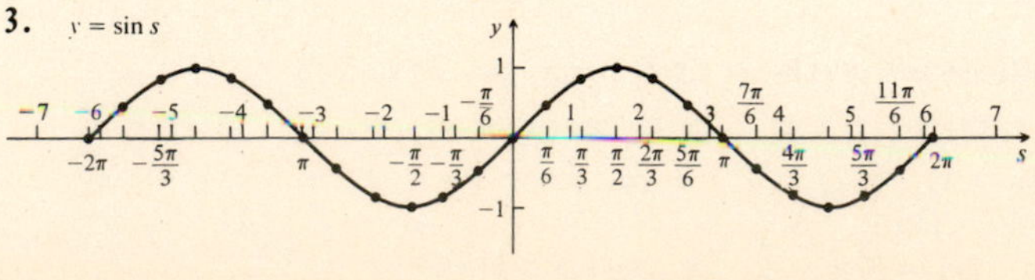

4. Yes, 2π **5.** Odd **6.** Yes **7.** The set of all real numbers **8.** The set of real numbers from -1 to 1, inclusive
9. (a) $(-a, -b)$; **(b)** $(-a, -b)$

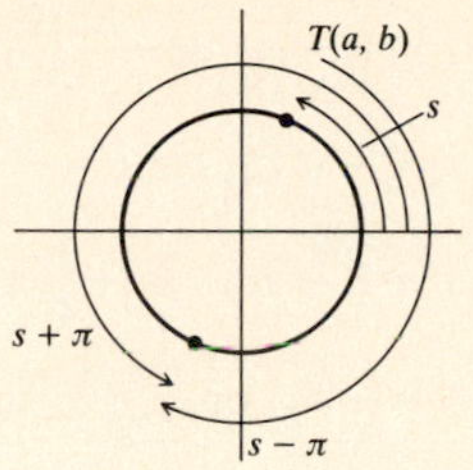

10.

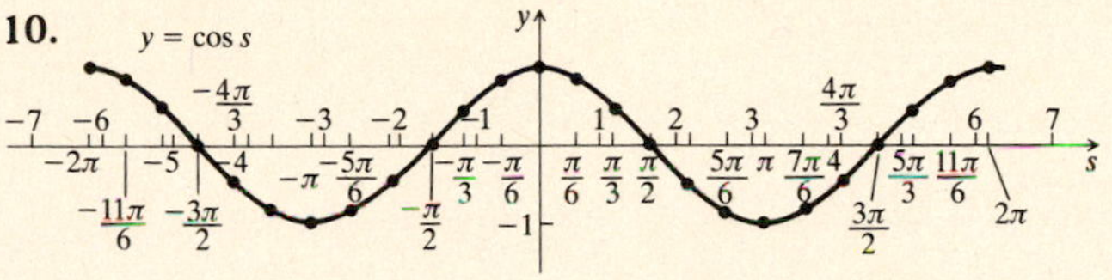

11. Yes, 2π **12.** Even **13.** Yes **14.** The set of all real numbers **15.** The set of all real numbers from -1 to 1, inclusive **16.** $\cos y = d$, $\sin y = e$

Exercise Set 6.2, pp. 363–364

1. $\frac{\sqrt{2}}{2}$ **3.** $\frac{1}{2}$ **5.** -1 **7.** $\frac{\sqrt{3}}{2}$ **9.** $-\frac{\sqrt{2}}{2}$ **11.** $\frac{1}{2}$ **13.** 0 **15.** $-\frac{\sqrt{2}}{2}$ **17.** $\frac{\sqrt{2}}{2}$ **19.** $-\frac{\sqrt{3}}{2}$ **21.** $-\frac{\sqrt{3}}{2}$ **23.** $\frac{\sqrt{2}}{2}$ **25.** $-\frac{\sqrt{2}}{2}$ **27.** 0 **29.** $-\frac{\sqrt{2}}{2}$ **31.** 0 **33.** $\frac{\sqrt{2}}{2}$ **35.** 1 **37.** $-\frac{\sqrt{2}}{2}$ **39.** 0 **41.** $\frac{\sqrt{3}}{2}$ **43.** $-\frac{1}{2}$ **45.** $\cos x$ **47.** $-\sin x$ **49.** $-\cos x$ **51.** $\cos x$ **53.** $-\cos x$
55. (a) See Margin Exercise 3;
(b)

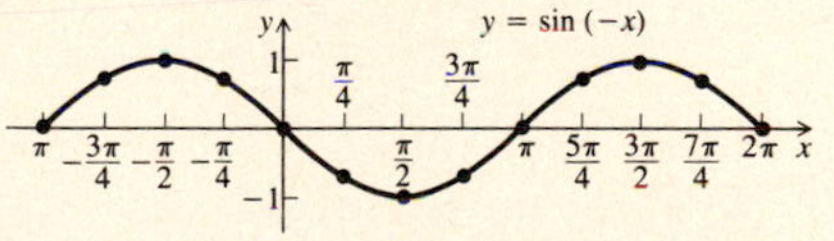

(c) same as (b); **(d)** same as (b)

57. (a) See Margin Exercise 3;
(b)

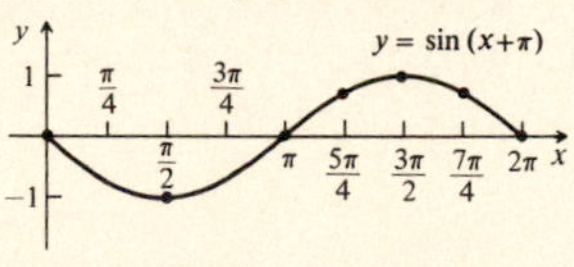

(c) same as (b); **(d)** the same

59. (a) See Margin Exercise 10;
(b)

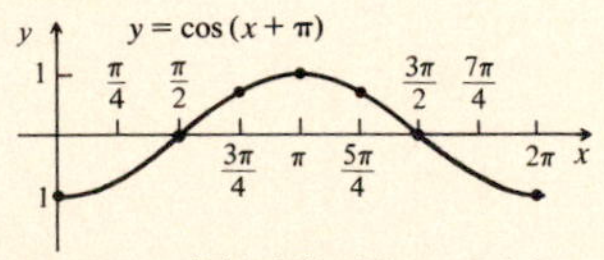

(c) same as (b): **(d)** (b) and (c) have identical graphs

61. (a) $\frac{\pi}{2} + 2k\pi$, k any integer; **(b)** $\frac{3\pi}{2} + 2k\pi$, k any integer
63. $\theta = k\pi$, k any integer **65.** $f \circ g(x) = \cos^2 x + 2\cos x$, $g \circ f(x) = \cos(x^2 + 2x)$ **67. (a)** 0.8660; **(b)** 0.7071
69.

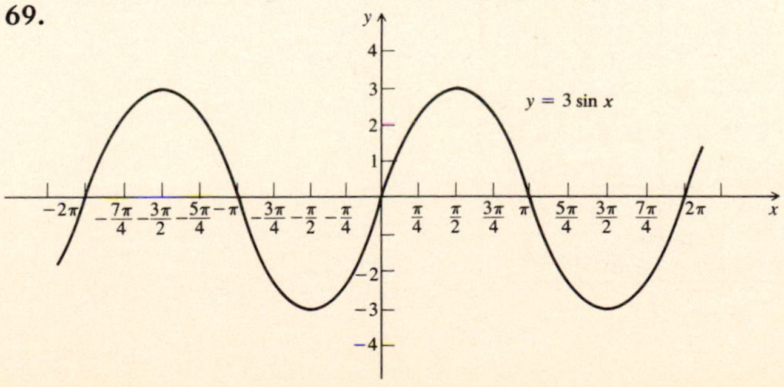

71. The domain is the set of all real numbers. The range is the set of all real numbers from 0 to 1 inclusive. The period is π. The amplitude is $\frac{1}{2}$. **73.** The domain consists of the intervals $\left[-\frac{\pi}{2} + 2k\pi, \frac{\pi}{2} + 2k\pi\right]$, k any integer. **75.** The domain is the set of all real numbers except $\frac{\pi}{2} + k\pi$ for any integer k.
77. The limit of $(\sin\theta)/\theta$ as θ approaches 0 is 1.

Margin Exercises, Section 6.3

1. 0, not defined, -1, not defined **2.** $-1, -1, -\sqrt{2}, \sqrt{2}$
3. $\frac{\sqrt{3}}{3}, \sqrt{3}, \frac{2\sqrt{3}}{3}, 2$ **4.** $-\sqrt{3}, -\frac{\sqrt{3}}{3}, 2, -\frac{2\sqrt{3}}{3}$ **5.** Not defined, not defined, not defined **6.** $\frac{\pi}{2}, -\frac{\pi}{2}, \frac{3\pi}{2}, -\frac{3\pi}{2}, \frac{5\pi}{2}, -\frac{5\pi}{2}$, etc. **7.**

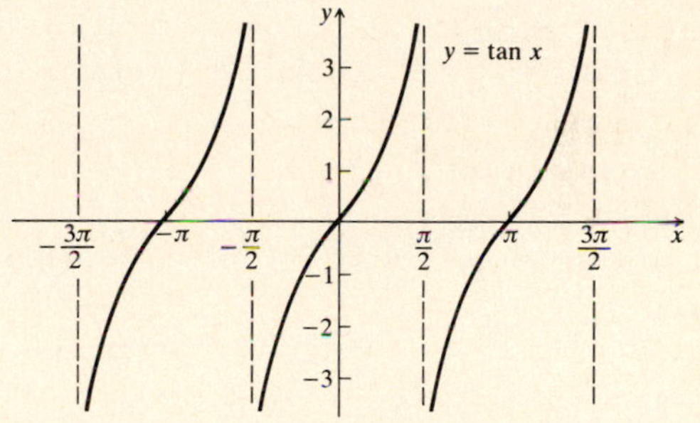

8. π **9.** The set of all real numbers except $\frac{\pi}{2} + k\pi$, k any integer
10. The set of all real numbers **11.** Odd **12.** $\pi, -\pi, 2\pi, -2\pi$, etc. **13.**

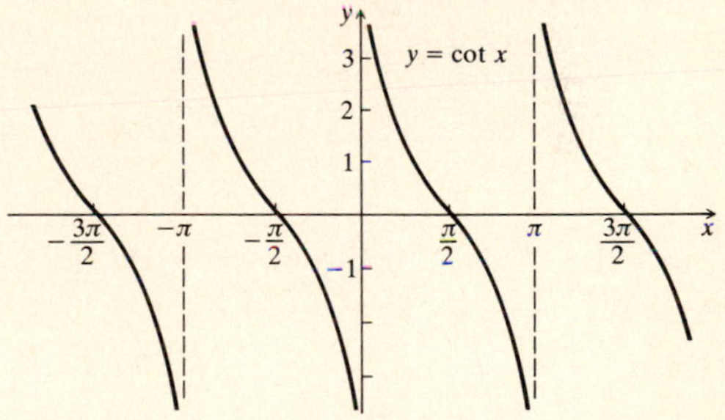

14. π **15.** The set of all real numbers except $k\pi$, k any integer
16. The set of all real numbers **17.** Positive: I, III; negative: II, IV **18.** Odd **19. (a)** $\frac{\pi}{2}, -\frac{\pi}{2}, \frac{3\pi}{2}, -\frac{3\pi}{2}$, etc.; **(b)** $0, \pi, -\pi, 2\pi, -2\pi$, etc. **20.**

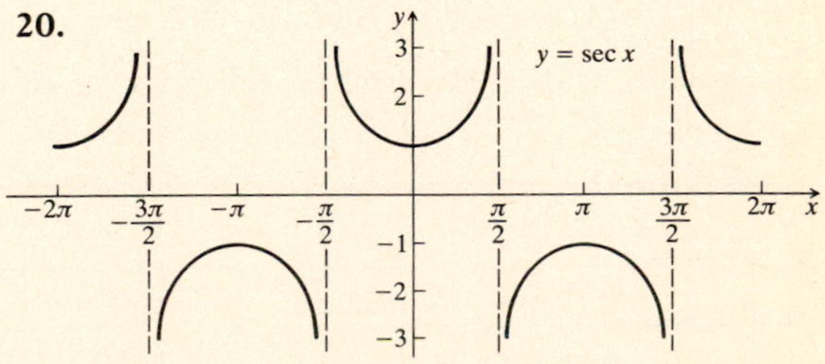

21.

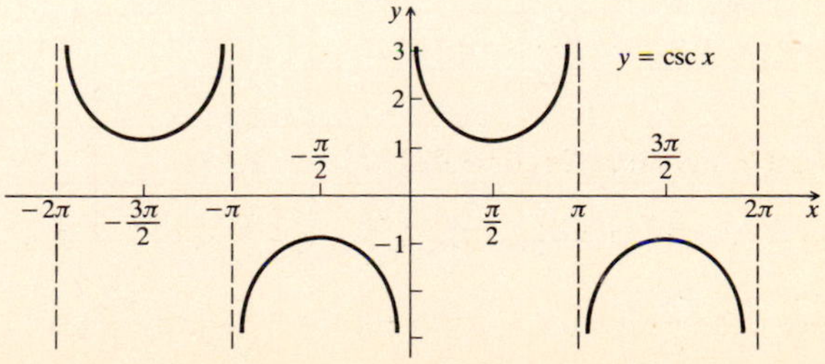

22. 2π; 2π **23.** The set of real numbers except $\frac{\pi}{2} + k\pi$, k any integer; the set of real numbers except $k\pi$, k any integer **24.** Its range consists of all real numbers 1 and greater, in addition to all real numbers -1 and less; its range is the same as that of the secant function. **25.** Positive: I, IV; negative: II, III **26.** Positive: I, II; negative: III, IV **27.** Even **28.** Odd
29.

Function	I	II	III	IV
sin, csc	+	+	−	−
cos, sec	+	−	−	+
tan, cot	+	−	+	−

30. $\tan s = \frac{\sin s}{\cos s} = \frac{1}{\frac{\cos s}{\sin s}} = \frac{1}{\cot s}$ **31.** $\cos s = \frac{1}{\frac{1}{\cos s}} = \frac{1}{\sec s}$

32. $\sin^2 s + \cos^2 s = 1$

$$\frac{\sin^2 s}{\cos^2 s} + \frac{\cos^2 s}{\cos^2 s} = \frac{1}{\cos^2 s}$$

$$\tan^2 s + 1 = \sec^2 s$$

33. $\sin^2 x = 1 - \cos^2 x$

34. $\sin x = \pm\sqrt{1 - \cos^2 x}$ **35.** $\cos s = -\frac{\sqrt{55}}{8}$; $\tan s = \frac{3}{\sqrt{55}}$, or $\frac{3\sqrt{55}}{55}$; $\cot s = \frac{\sqrt{55}}{3}$; $\sec s = -\frac{8}{\sqrt{55}}$, or $-\frac{8\sqrt{55}}{55}$; $\csc s = -\frac{8}{3}$ **36.** $\sin s = \frac{3}{\sqrt{10}}$, or $\frac{3\sqrt{10}}{10}$; $\cos s = -\frac{1}{\sqrt{10}}$, or $-\frac{\sqrt{10}}{10}$; $\cot s = -\frac{1}{3}$; $\sec s = -\sqrt{10}$; $\csc s = \frac{\sqrt{10}}{3}$

37. $\cot(-x) = \frac{\cos(-x)}{\sin(-x)} = \frac{\cos x}{-\sin x} = -\cot x$; $\therefore \cot(-x) = -\cot x$

38. $\tan(x - \pi) = \frac{\sin(x - \pi)}{\cos(x - \pi)} = \frac{-\sin x}{-\cos x} = \frac{\sin x}{\cos x} = \tan x$; $\therefore \tan(x - \pi) = \tan x$ **39.** $\csc(\pi - x) = \frac{1}{\sin(\pi - x)} = \frac{1}{\sin x} = \csc x$; $\therefore \csc(\pi - x) = \csc x$

Exercise Set 6.3, pp. 374–375

1. 1 **3.** $\frac{\sqrt{3}}{3}$ **5.** $\sqrt{2}$ **7.** Undefined **9.** $-\sqrt{3}$ **11.** $\sqrt{2}$ **13.** $-\frac{\sqrt{3}}{3}$ **15.** cos, sec **17.** sin, cos, sec, csc **19.** Positive: I, III; negative: II, IV **21.** Positive: I, IV; negative: II, III

23.

	$\pi/16$	$\pi/8$	$\pi/6$	$\pi/4$	$3\pi/8$	$7\pi/16$
sin	0.19509	0.38268	0.50000	0.70711	0.92388	0.98079
cos	0.98079	0.92388	0.86603	0.70711	0.38268	0.19509
tan	0.19891	0.41421	0.57735	1.00000	2.41424	5.02737
cot	5.02737	2.41424	1.73206	1.00000	0.41421	0.19891
sec	1.01959	1.08239	1.15469	1.41421	2.61315	5.12584
csc	5.12584	2.61315	2.00000	1.41421	1.08239	1.01959

25. $\sin s = \frac{2\sqrt{2}}{3}$; $\tan s = 2\sqrt{2}$; $\cot s = \frac{1}{2\sqrt{2}}$, or $\frac{\sqrt{2}}{4}$; $\sec s = 3$; $\csc s = \frac{3}{2\sqrt{2}}$, or $\frac{3\sqrt{2}}{4}$ **27.** $\sin s = -\frac{3}{\sqrt{10}}$, or $-\frac{3\sqrt{10}}{10}$; $\cos s = -\frac{1}{\sqrt{10}}$, or $-\frac{\sqrt{10}}{10}$; $\cot s = \frac{1}{3}$; $\sec s = -\sqrt{10}$; $\csc s = -\frac{\sqrt{10}}{3}$ **29.** $\sin s = \frac{4}{5}$; $\cos s = -\frac{3}{5}$; $\tan s = -\frac{4}{3}$; $\cot s = -\frac{3}{4}$; $\csc s = \frac{5}{4}$ **31.** $\cos s = -\frac{\sqrt{21}}{5}$; $\tan s = \frac{2}{\sqrt{21}}$, or $\frac{2\sqrt{21}}{21}$; $\cot s = \frac{\sqrt{21}}{2}$; $\sec s = -\frac{5}{\sqrt{21}}$, or $-\frac{5\sqrt{21}}{21}$; $\csc s = -\frac{5}{2}$ **33.** $\sec(-x) = \frac{1}{\cos(-x)} = \frac{1}{\cos x} = \sec x$

35. $\cot(x + \pi) = \frac{\cos(x + \pi)}{\sin(x + \pi)} = \frac{-\cos x}{-\sin x} = \cot x$

37. $\sec(x + \pi) = \frac{1}{\cos(x + \pi)} = \frac{1}{-\cos x} = -\sec x$

39. The graph of sec $(x - \pi)$ is like that of sec x, moved π units to the right. The graph of $-\sec x$ is that of sec x reflected across the x-axis. The graphs are identical.
41. If the graph of tan x were reflected across the y-axis and then translated to the right a distance of $\pi/2$, the graph of cot x would be obtained. There are other ways to describe the relation.
43. The sine and tangent functions; the cosine and cotangent functions
45.

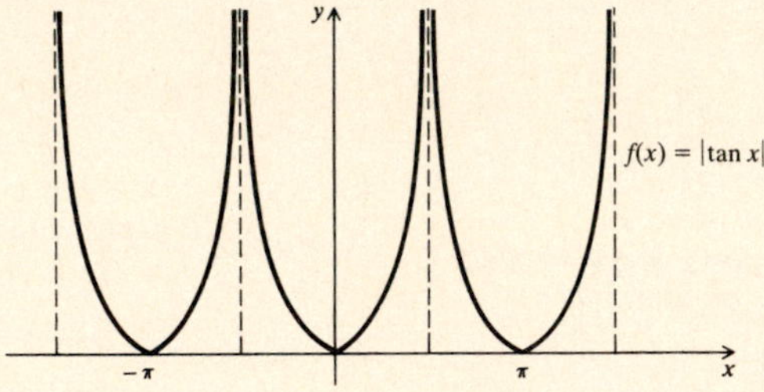

47.

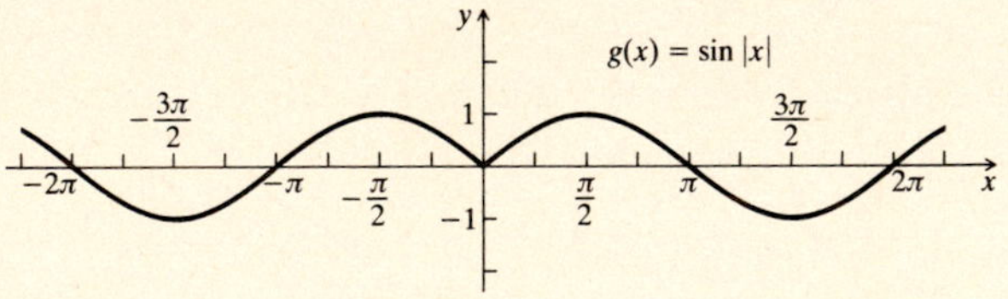

49. It is true that $\cos\theta \le \sec\theta$ for all θ in the intervals $\left(-\frac{\pi}{2} + k \cdot 2\pi, \frac{\pi}{2} + k \cdot 2\pi\right)$, k any integer.

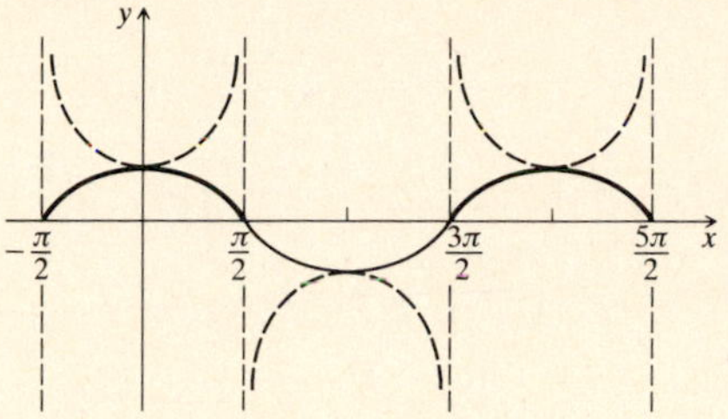

Margin Exercises, Section 6.4

1. (a)

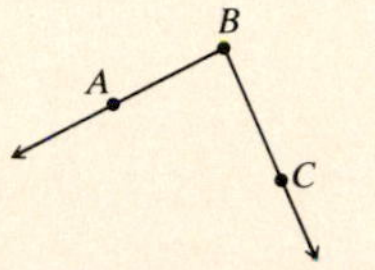

(b)

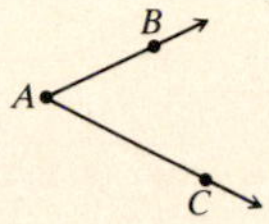

2. (a) I; (b) III; (c) IV; (d) III; (e) I; (f) IV; (g) I **3.** (a) 360°; (b) 180°; (c) 90°; (d) 45°; (e) 60°; (f) 30° **4.** Positive: 405°, 765°; negative: −315°, −675°. Answers may vary. **5.** (a) Obtuse; (b) right; (c) acute; (d) straight **6.** (a) $\frac{1}{60}$; (b) $\frac{1}{3600}$; (c) 56, 23, 11 **7.** Complement: 4°47′22″; supplement: 94°47′22″ **8.** Complement: 55.11°; supplement: 145.11° **9.** 67.2211° **10.** 37°27′ **11.** (a) $\frac{5}{4}\pi$; (b) $\frac{7}{4}\pi$; (c) -4π **12.** (a) 1.26; (b) 5.23; (c) −5.50 **13.** (a) 240°; (b) 450°; (c) −144° **14.** (a) III; (b) I; (c) IV; (d) III **15.** 57.6 cm **16.** 6 **17.** $\frac{3}{2}$ **18.** 377 cm/sec **19.** r in cm, ω in radians/sec **20.** km/yr **21.** 3.6 radians/sec **22.** 70.4 radians, or 11.21 revolutions

Exercise Set 6.4, pp. 385–388

1. I **3.** III **5.** I **7.** II **9.** Positive: 418°, 778°; negative: −302°, −662°. Answers may vary. **11.** Positive: 240°, 600°; negative: −480°, −840°. Answers may vary. **13.** 32°37′; 122°37′ **15.** 16°14′49″; 106°14′49″ **17.** 22.69°; 112.69° **19.** 78.7656°; 168.7656° **21.** 8°36′ **23.** 72°15′ **25.** 46°23′ **27.** 67°50′ **29.** 87°20′44″ **31.** 48°1′30″ **33.** 9.75° **35.** 35.83° **37.** 80.55° **39.** 3.03° **41.** 19.7897° **43.** 31.9653° **45.** $\frac{\pi}{6}$ **47.** $\frac{\pi}{3}$ **49.** $\frac{5\pi}{12}$ **51.** 0.2095π **53.** 1.1922π **55.** 2.093 **57.** 5.582 **59.** 3.489 **61.** 2.0550 **63.** 0.0236 **65.** 57.32° **67.** 1440° **69.** 135° **71.** 74.69° **73.** 135.6° **75.** $30° = \frac{\pi}{6}$, $60° = \frac{\pi}{3}$, $135° = \frac{3\pi}{4}$, $180° = \pi$, $225° = \frac{5\pi}{4}$, $270° = \frac{3\pi}{2}$, $315° = \frac{7\pi}{4}$ **77.** 1.1 radians, 63° **79.** 5.233 radians **81.** 16 m **83.** 3150 cm/min **85.** 52.3 cm/sec **87.** 1047 mph **89.** 68.8 radians/sec **91.** 10 mph **93.** 75.4 ft **95.** 377 radians **97.** (a) 53.33; (b) 170; (c) 25; (d) 142.86 **99.** 111.7 km, 69.81 mi **101.** $\frac{1}{30}$ radian **103.** 2.093 radians/sec **105.** (a) 395.35 rpm/sec, angular acc $= \frac{\Delta v}{t}$; (b) 41.4 radians/sec^2 **107.** 37.5° **109.** 1.15 statute miles

Margin Exercises, Section 6.5

1. 0 **2.** 0 **3.** $\frac{\sqrt{2}}{2}$ **4.** 1 **5.** $-\frac{\sqrt{2}}{2}$ **6.** $\sin\theta = -\frac{3}{5}$; $\cos\theta = -\frac{4}{5}$; $\tan\theta = \frac{3}{4}$; $\cot\theta = \frac{4}{3}$; $\sec\theta = -\frac{5}{4}$; $\csc\theta = -\frac{5}{3}$ **7.** $\sin\theta = \frac{\sqrt{3}}{2}$; $\cos\theta = -\frac{1}{2}$; $\tan\theta = -\sqrt{3}$; $\cot\theta = -\frac{1}{\sqrt{3}}$, or $-\frac{\sqrt{3}}{3}$; $\sec\theta = -2$; $\csc\theta = \frac{2}{\sqrt{3}}$, or $\frac{2\sqrt{3}}{3}$ **8.** $\sin\theta = -\frac{2}{\sqrt{5}}$, or $-\frac{2\sqrt{5}}{5}$; $\cos\theta = -\frac{1}{\sqrt{5}}$, or $-\frac{\sqrt{5}}{5}$; $\tan\theta = 2$ **9.** $\sin\theta$ and $\csc\theta$: +, +, −, −; $\cos\theta$ and $\sec\theta$: +, −, −, +; $\tan\theta$ and $\cot\theta$: +, −, +, − **10.** The cosine and secant are positive; the other four function values are negative. **11.** The sine and cosecant are positive; the other four function values are negative. **12.** The cosine and secant are positive; the other four function values are negative. **13.** $\sin 270° = -1$; $\cos 270° = 0$; $\tan 270°$: undefined; $\cot 270° = 0$; $\sec 270°$: undefined; $\csc 270° = -1$ **14.** (a) $\sin\theta = \frac{12}{13}$, $\cos\theta = \frac{5}{13}$, $\tan\theta = \frac{12}{5}$, $\cot\theta = \frac{5}{12}$, $\sec\theta = \frac{13}{5}$, $\csc\theta = \frac{13}{12}$; (b) $\sin\theta = \frac{5}{13}$, $\cos\theta = \frac{12}{13}$, $\tan\theta = \frac{5}{12}$, $\cot\theta = \frac{12}{5}$, $\sec\theta = \frac{13}{12}$, $\csc\theta = \frac{13}{5}$ **15.** 127.3 ft **16.** $\sin\theta = -\frac{\sqrt{7}}{4}$; $\tan\theta = -\frac{\sqrt{7}}{3}$; $\cot\theta = -\frac{3}{\sqrt{7}}$, or $-\frac{3\sqrt{7}}{7}$; $\sec\theta = \frac{4}{3}$; $\csc\theta = -\frac{4}{\sqrt{7}}$, or $-\frac{4\sqrt{7}}{7}$ **17.** $\sin\theta = \frac{1}{\sqrt{10}}$, or $\frac{\sqrt{10}}{10}$; $\cos\theta = -\frac{3}{\sqrt{10}}$, or $-\frac{3\sqrt{10}}{10}$; $\tan\theta = -\frac{1}{3}$; $\sec\theta = -\frac{\sqrt{10}}{3}$; $\csc\theta = \sqrt{10}$

Exercise Set 6.5, pp. 399–401

1. −1 **3.** $\frac{\sqrt{2}}{2}$ **5.** $-\frac{\sqrt{2}}{2}$ **7.** $-\frac{1}{\sqrt{3}}$, or $-\frac{\sqrt{3}}{3}$ **9.** $\sin\theta = \frac{5}{13}$; $\cos\theta = -\frac{12}{13}$; $\tan\theta = -\frac{5}{12}$; $\cot\theta = -\frac{12}{5}$; $\sec\theta = -\frac{13}{12}$; $\csc\theta = \frac{13}{5}$ **11.** $\sin\theta = -\frac{3}{4}$; $\cos\theta = -\frac{\sqrt{7}}{4}$; $\tan\theta = \frac{3}{\sqrt{7}}$, or $\frac{3\sqrt{7}}{7}$; $\cot\theta = \frac{\sqrt{7}}{3}$; $\sec\theta = -\frac{4}{\sqrt{7}}$, or $-\frac{4\sqrt{7}}{7}$; $\csc\theta = -\frac{4}{3}$ **13.** $\sin\theta = -\frac{2}{\sqrt{13}}$, or $-\frac{2\sqrt{13}}{13}$; $\cos\theta = \frac{3}{\sqrt{13}}$, or $\frac{3\sqrt{13}}{13}$; $\tan\theta = -\frac{2}{3}$ **15.** $\sin\theta = \frac{5}{\sqrt{41}}$, or $\frac{5\sqrt{41}}{41}$; $\cos\theta = \frac{4}{\sqrt{41}}$, or $\frac{4\sqrt{41}}{41}$; $\tan\theta = \frac{5}{4}$ **17.** The cosine and secant are positive; the other four function values are negative. **19.** The sine and cosecant are positive; the other four function values are negative. **21.** The sine and cosecant are positive; the other four function values are negative. **23.** The sine and cosecant are positive; the other four function values are negative. **25.** $\sin 90° = 1$; $\cos 90° = 0$; $\tan 90°$ is undefined; $\cot 90° = 0$; $\sec 90°$ is undefined; $\csc 90° = 1$ **27.** $\sin(-180°) = 0$; $\cos(-180°) = -1$; $\tan(-180°) = 0$; $\cot(-180°)$ is undefined; $\sec(-180°) = -1$; $\csc(-180°)$ is undefined. **29.** $\sin\theta = \frac{8}{17}$; $\cos\theta = \frac{15}{7}$;

$\tan\theta = \frac{8}{15}$; $\cot\theta = \frac{15}{8}$; $\sec\theta = \frac{17}{15}$; $\csc\theta = \frac{17}{8}$ **31.** $\sin\theta = \frac{3}{h}$; $\cos\theta = \frac{7}{h}$; $\tan\theta = \frac{3}{7}$; $\cos\theta = \frac{7}{3}$; $\sec\theta = \frac{h}{7}$; $\csc\theta = \frac{h}{3}$

33. $\sin\theta = 0.8788$; $\cos\theta = 0.4771$; $\tan\theta = 1.8419$; $\cot\theta = 0.5429$; $\sec\theta = 2.0958$; $\csc\theta = 1.1379$

35. $\sin 37.5° = \frac{28}{l}$; $\cos 37.5° = \frac{d}{l}$; $\tan 37.5° = \frac{28}{d}$; $\cot 37.5° = \frac{d}{28}$; $\sec 37.5° = \frac{l}{d}$; $\csc 37.5° = \frac{l}{28}$

37.

θ	sin	cos	tan	cot	sec	csc
45°	$\sqrt{2}/2$	$\sqrt{2}/2$	1	1	$2/\sqrt{2}$	$2/\sqrt{2}$
30°	1/2	$\sqrt{3}/2$	$1/\sqrt{3}$	$\sqrt{3}$	$2/\sqrt{3}$	2
60°	$\sqrt{3}/2$	1/2	$\sqrt{3}$	$1/\sqrt{3}$	2	$2/\sqrt{3}$

39. 62.4 m **41.** $\cos\theta = -\frac{2\sqrt{2}}{3}$; $\tan\theta = 2\sqrt{2}$; $\cot\theta = \frac{1}{2\sqrt{2}}$, or $\frac{\sqrt{2}}{4}$; $\sec\theta = -\frac{3}{2\sqrt{2}}$, or $-\frac{3\sqrt{2}}{4}$; $\csc\theta = -3$

43. $\sin\theta = -\frac{4}{5}$; $\tan\theta = -\frac{4}{3}$; $\cot\theta = -\frac{3}{4}$; $\sec\theta = \frac{5}{3}$; $\csc\theta = -\frac{5}{4}$ **45.** $\sin\theta = -\frac{1}{\sqrt{5}}$, or $-\frac{\sqrt{5}}{5}$; $\cos\theta = \frac{2}{\sqrt{5}}$, or $\frac{2\sqrt{5}}{5}$; $\tan\theta = -\frac{1}{2}$; $\sec\theta = \frac{\sqrt{5}}{2}$; $\csc\theta = -\sqrt{5}$

47. $\cos\theta = -\frac{2\sqrt{2}}{3}$; $\tan\theta = -\frac{1}{2\sqrt{2}}$, or $-\frac{\sqrt{2}}{4}$; $\cot\theta = -2\sqrt{2}$; $\sec\theta = -\frac{3}{2\sqrt{2}}$, or $-\frac{3\sqrt{2}}{4}$; $\csc\theta = 3$ **49.** $\cos\theta = \frac{7}{25}$; $\tan\theta = \frac{24}{7}$; $\cot\theta = \frac{7}{24}$; $\sec\theta = \frac{25}{7}$; $\csc\theta = \frac{25}{24}$

51. $\sin\phi = \frac{2}{\sqrt{5}}$, or $\frac{2\sqrt{5}}{5}$; $\cos\phi = \frac{1}{\sqrt{5}}$, or $\frac{\sqrt{5}}{5}$; $\cot\phi = \frac{1}{2}$; $\sec\phi = \sqrt{5}$; $\csc\phi = \frac{\sqrt{5}}{2}$

53.

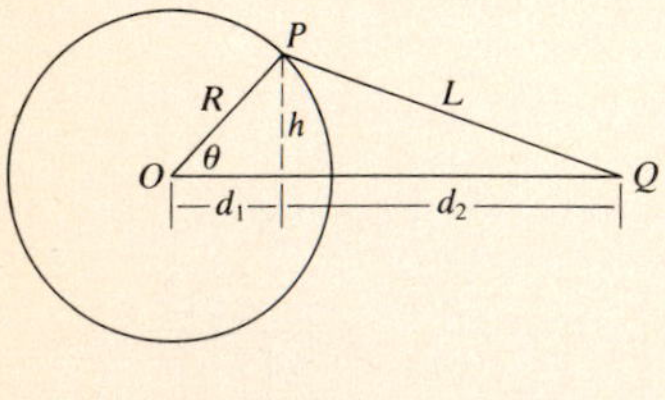

$x = OQ = d_1 + d_2 =$
$R\cos\theta + \sqrt{L^2 - h^2} =$
$R\cos\theta + \sqrt{L^2 - (R\sin\theta)^2} =$
$R\cos\omega t + \sqrt{L^2 - R^2\sin^2\omega t}$

Margin Exercises, Section 6.6

1. $\sin 120° = \frac{\sqrt{3}}{2}$; $\cos 120° = -\frac{1}{2}$; $\tan 120° = -\sqrt{3}$; $\cot 120° = -\frac{1}{\sqrt{3}}$, or $-\frac{\sqrt{3}}{3}$; $\sec 120° = -2$; $\csc 120° = \frac{2}{\sqrt{3}}$, or $\frac{2\sqrt{3}}{3}$ **2.** $\sin 570° = -\frac{1}{2}$; $\cos 570° = -\frac{\sqrt{3}}{2}$; $\tan 570° = \frac{1}{\sqrt{3}}$, or $\frac{\sqrt{3}}{3}$; $\cot 570° = \sqrt{3}$; $\sec 570° = -\frac{2}{\sqrt{3}}$, or $-\frac{2\sqrt{3}}{3}$; $\csc 570° = -2$ **3.** $\sin(-945°) = \frac{\sqrt{2}}{2}$; $\cos(-945°) = -\frac{\sqrt{2}}{2}$; $\tan(-945°) = -1$; $\cot(-945°) = -1$; $\sec(-945°) = -\frac{2}{\sqrt{2}}$, or $-\sqrt{2}$; $\csc(-945°) = \frac{2}{\sqrt{2}}$, or $\sqrt{2}$

4. $\sin\frac{17\pi}{6} = \frac{1}{2}$; $\cos\frac{17\pi}{6} = -\frac{\sqrt{3}}{2}$; $\tan\frac{17\pi}{6} = -\frac{1}{\sqrt{3}}$, or $-\frac{\sqrt{3}}{3}$; $\cot\frac{17\pi}{6} = -\sqrt{3}$; $\sec\frac{17\pi}{6} = -\frac{2}{\sqrt{3}}$, or $-\frac{2\sqrt{3}}{3}$; $\csc\frac{17\pi}{6} = 2$

5. $\sin\left(-\frac{22\pi}{3}\right) = \frac{\sqrt{3}}{2}$; $\cos\left(-\frac{22\pi}{3}\right) = -\frac{1}{2}$; $\tan\left(-\frac{22\pi}{3}\right) = -\sqrt{3}$; $\cot\left(-\frac{22\pi}{3}\right) = -\frac{1}{\sqrt{3}}$, or $-\frac{\sqrt{3}}{3}$; $\sec\left(-\frac{22\pi}{3}\right) = -2$; $\csc\left(-\frac{22\pi}{3}\right) = \frac{2}{\sqrt{3}}$, or $\frac{2\sqrt{3}}{3}$ **6.** $\sin\frac{29\pi}{4} = -\frac{\sqrt{2}}{2}$; $\cos\frac{29\pi}{4} = -\frac{\sqrt{2}}{2}$; $\tan\frac{29\pi}{4} = 1$; $\cot\frac{29\pi}{4} = 1$; $\sec\frac{29\pi}{4} = -\frac{2}{\sqrt{2}}$, or $-\sqrt{2}$; $\csc\frac{29\pi}{4} = -\frac{2}{\sqrt{2}}$, or $-\sqrt{2}$ **7.** −0.754710 **8.** 0.416260 **9.** 0.292372 **10.** 1.683835 **11.** −0.379265 **12.** −1.143983 **13.–15.** The calculator was set in degree mode; the angle, or rotation, measure was in radians. **16.** 0.4710 **17.** −0.2301 **18.** 1.7535 **19.** −0.88546 **20.** 0.36397 **21.** 0.83666 **22.** −0.27676 **23.** 0.65486 **24.** −5.84248 **25.** 24°48′ **26.** 50° **27.** 45°30′ **28.** −13.67° **29.** 2.1788 **30.** 124°50′ **31.** 1.5628 **32.** 304.83°, or 304°50′ **33.** 111.98°, or 111°59′ **34.** 193.67°, or 193°40′ **35.** 213.7°, or 213°42′

Exercise Set 6.6, pp. 409–411

1. $\frac{2}{\sqrt{2}}$, or $\sqrt{2}$ **3.** $\frac{1}{2}$ **5.** $\sqrt{3}$ **7.** −1 **9.** −1 **11.** $-\frac{1}{2}$ **13.** 1 **15.** $\frac{2}{\sqrt{2}}$, or $\sqrt{2}$ **17.** $\frac{\sqrt{2}}{2}$ **19.** −1 **21.** $-\frac{\sqrt{2}}{2}$ **23.** $-\sqrt{3}$ **25.** $\sin\frac{23\pi}{6} = -\frac{1}{2}$; $\cos\frac{23\pi}{6} = \frac{\sqrt{3}}{2}$; $\tan\frac{23\pi}{6} = -\frac{1}{\sqrt{3}}$, or $-\frac{\sqrt{3}}{3}$; $\cot\frac{23\pi}{6} = -\sqrt{3}$; $\sec\frac{23\pi}{6} = \frac{2}{\sqrt{3}}$, or $\frac{2\sqrt{3}}{3}$; $\csc\frac{23\pi}{6} = -2$ **27.** $\sin\left(-\frac{19\pi}{3}\right) = -\frac{\sqrt{3}}{2}$; $\cos\left(-\frac{19\pi}{3}\right) = \frac{1}{2}$; $\tan\left(-\frac{19\pi}{3}\right) = -\sqrt{3}$; $\cot\left(-\frac{19\pi}{3}\right) = -\frac{1}{\sqrt{3}}$, or $-\frac{\sqrt{3}}{3}$; $\sec\left(-\frac{19\pi}{3}\right) = 2$; $\csc\left(-\frac{19\pi}{3}\right) = -\frac{2}{\sqrt{3}}$, or $-\frac{2\sqrt{3}}{3}$

29. $\sin\frac{31\pi}{3} = \frac{\sqrt{3}}{2}$; $\cos\frac{31\pi}{3} = \frac{1}{2}$; $\tan\frac{31\pi}{3} = \sqrt{3}$; $\cot\frac{31\pi}{3} = \frac{1}{\sqrt{3}}$, or $\frac{\sqrt{3}}{3}$; $\sec\frac{31\pi}{3} = 2$; $\csc\frac{31\pi}{3} = \frac{2}{\sqrt{3}}$, or $\frac{2\sqrt{3}}{3}$

31. $\sin(-750°) = -\frac{1}{2}$; $\cos(-750°) = \frac{\sqrt{3}}{2}$; $\tan(-750°) = -\frac{1}{\sqrt{3}}$, or $-\frac{\sqrt{3}}{3}$; $\cot(-750°) = -\sqrt{3}$; $\sec(-750°) = \frac{2}{\sqrt{3}}$, or $\frac{2\sqrt{3}}{3}$; $\csc(-750°) = -2$ **33.** 0.951057 **35.** 0.045410 **37.** 0.885664 **39.** 0.963241 **41.** 2.050197 **43.** 1.017124 **45.** −0.362184 **47.** −249.1107 **49.** −2.121903

51. −1.193718 **53.** 0.593653 **55.** 0.188238
57. −0.643538 **59.** −0.839072 **61.** 0 **63.** 0.680122
65. 1.236068 **67.** −4.493959 **69.** −1 **71.** −0.747022
73. 30.83°, 0.5381 **75.** 49.37°, 0.8616 **77.** 82.33°, 1.4369
79. 9.17°, 0.1600 **81.** 275.38° **83.** 200.15° **85.** 193.82°
87. 162.51° **89.** 2.5270 **91.** 1.7839 **93.** 4.7648
95. 0.089321 **97.** 38.25 inches **99.** 2°30′, or 0.0436 radians
101. 25°42′51″ **103.** 0.47943 **105.** 865,000
107. **(a)** $t = \dfrac{v \sin\theta + \sqrt{v^2 \sin^2\theta + 2gh}}{g}$;
(b) $x = v\cos\theta \dfrac{v\sin\theta + \sqrt{v^2\sin^2\theta + 2gh}}{g}$; **(c)** 14.026 m/sec;
(d) x is larger when h is larger; thus a taller shotputter can put the shot a greater distance; **(e)** $t \approx 2.57$ sec, $v \approx 124$ ft/sec

Margin Exercises, Section 6.7

1.

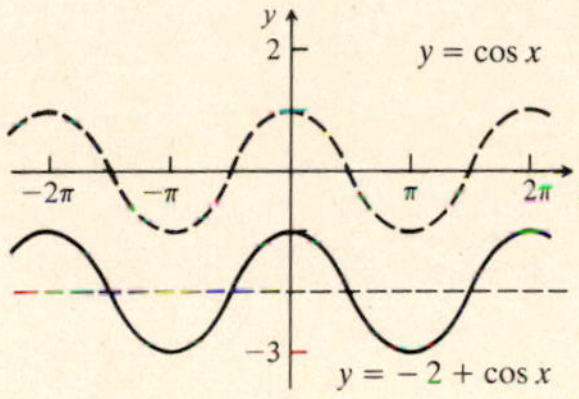

2. The amplitude is 2.

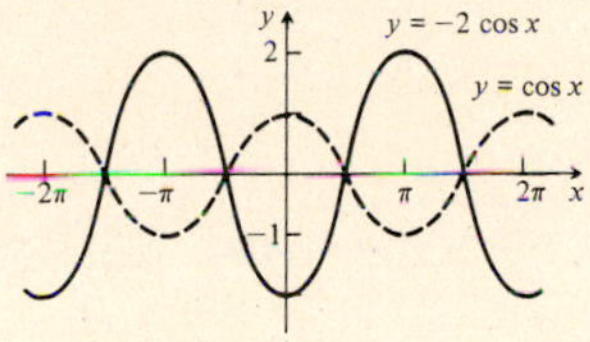

3. The amplitude is $\frac{2}{3}$.

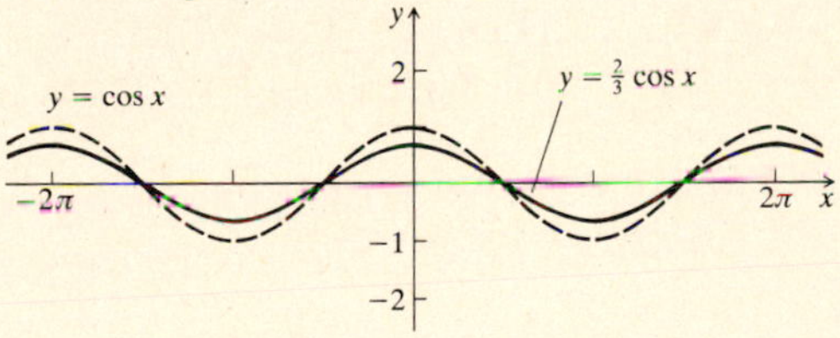

4.

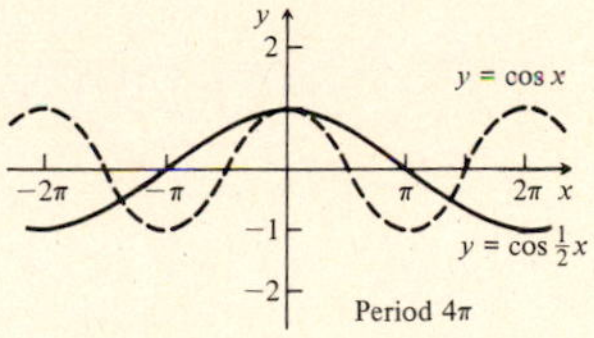

5.

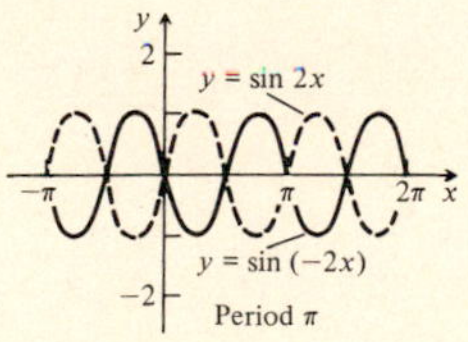

6.

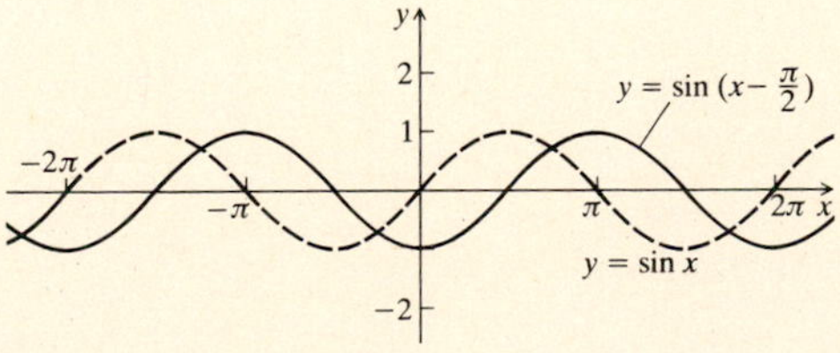

7.

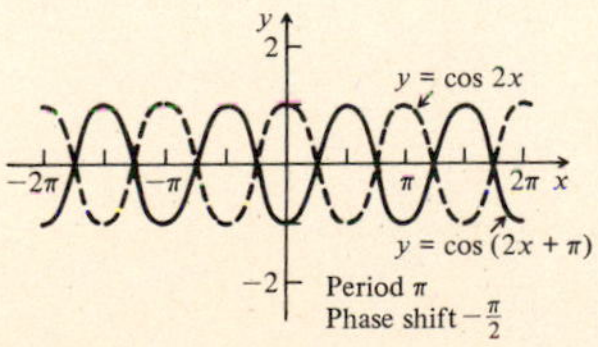

8.

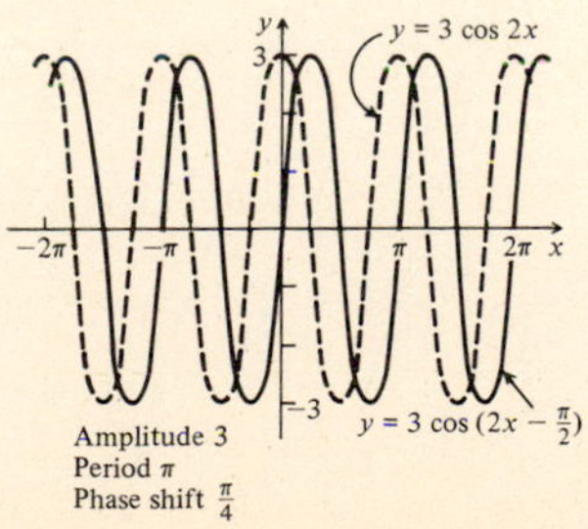

9.

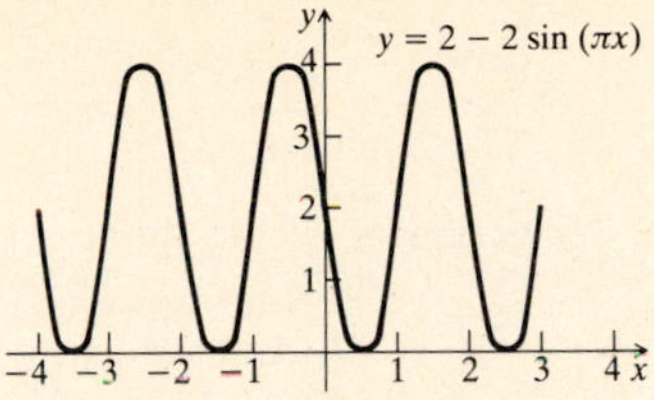

10.

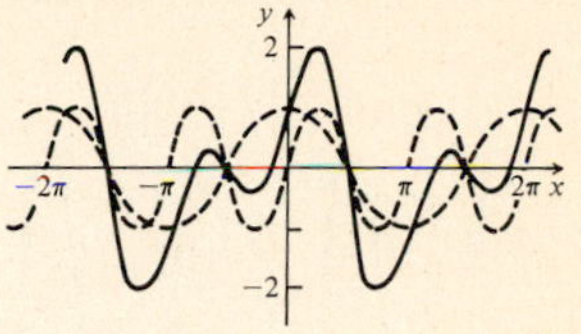

Exercise Set 6.7, pp. 420–422

1.

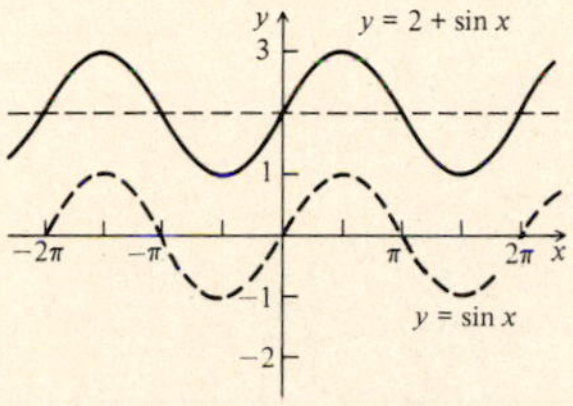

3.

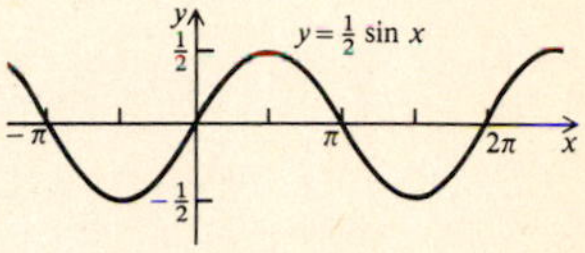

5.

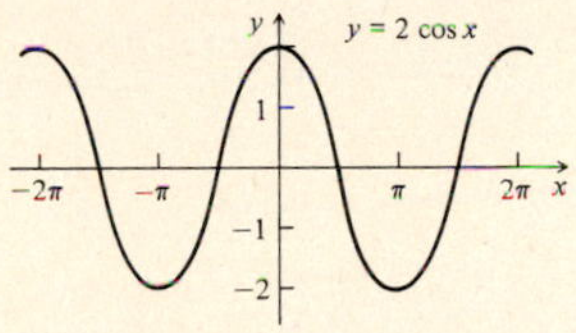

7.

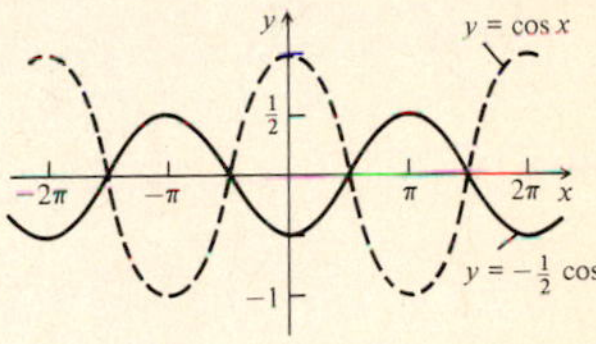

9. Amplitude, 2; period, 2; phase shift, 0

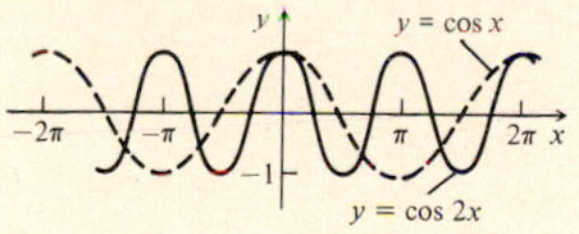

11.

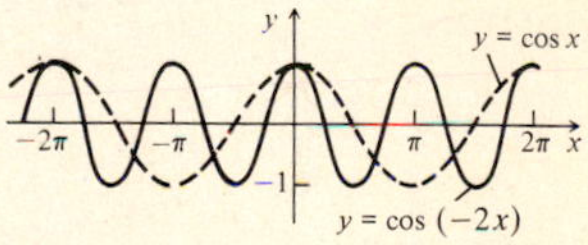

13.

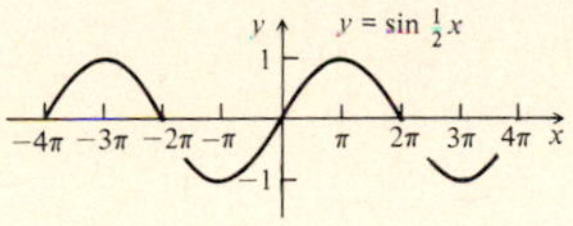

15.

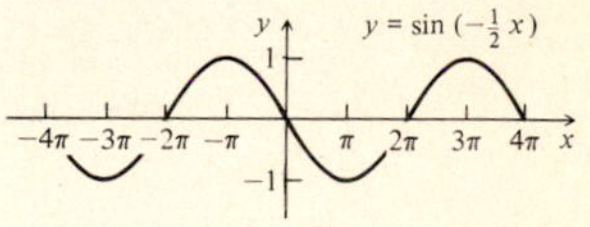

17.

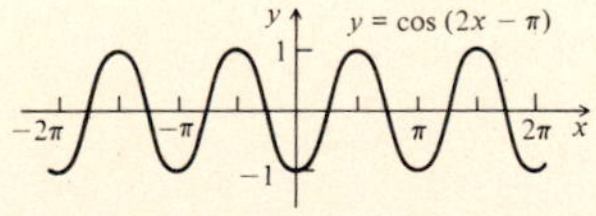

19.

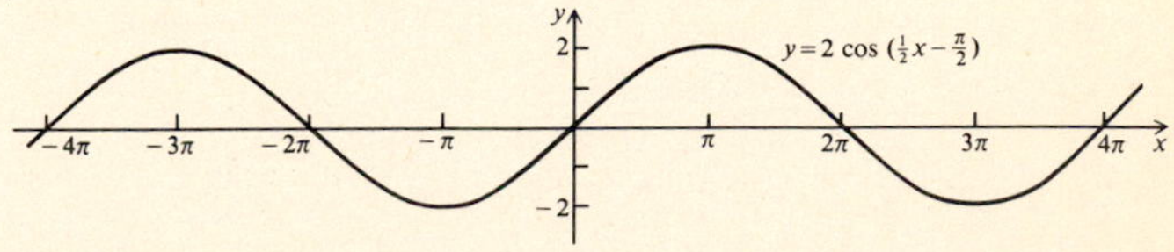

21.

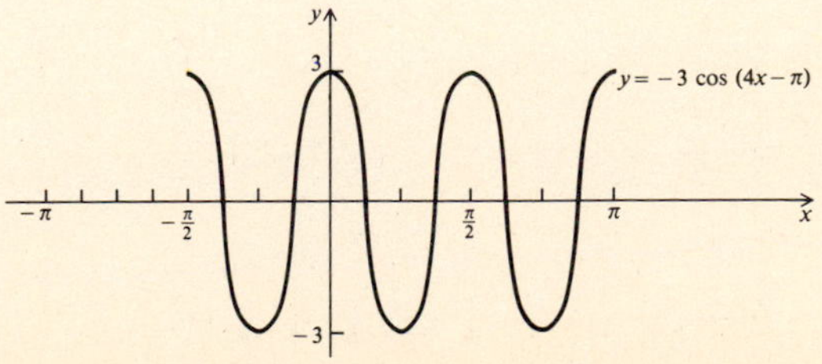

23.

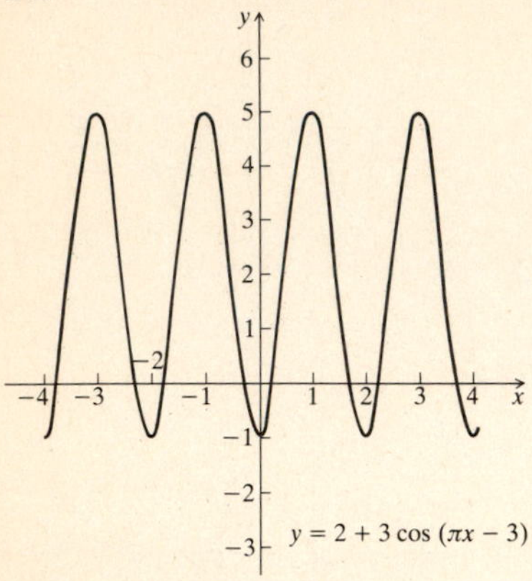

25. Amplitude, 3; period, $\frac{2\pi}{3}$; phase shift, $\frac{\pi}{6}$
27. Amplitude, 5; period, $\frac{\pi}{2}$; phase shift, $-\frac{\pi}{12}$
29. Amplitude, $\frac{1}{2}$; period, 1; phase shift, $-\frac{1}{2}$

31.

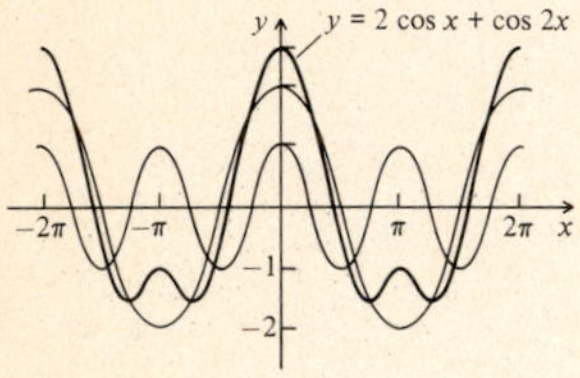

33.

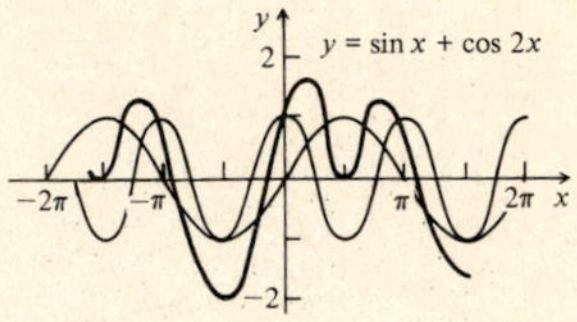

35.

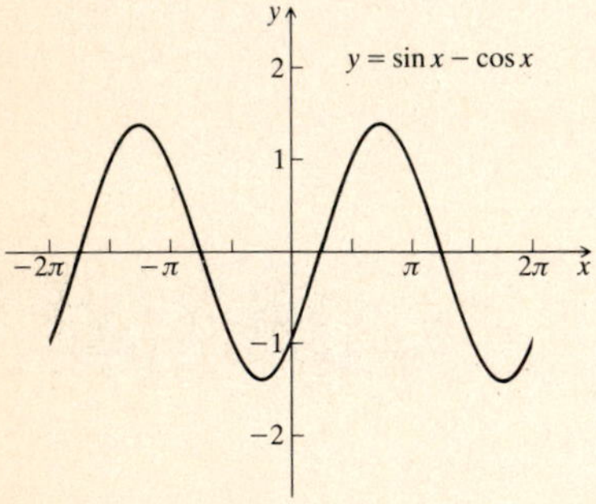

37.

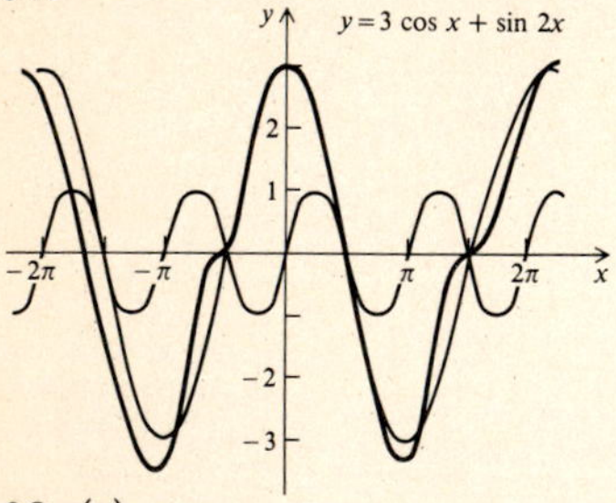

39. (a)

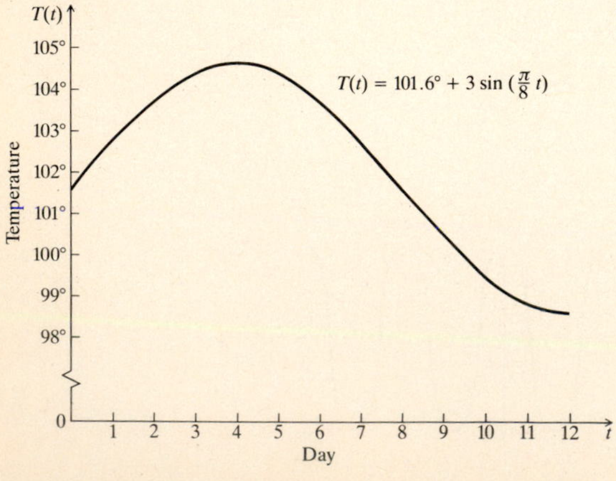

(b) 104.6°, 98.6°

41. (a)

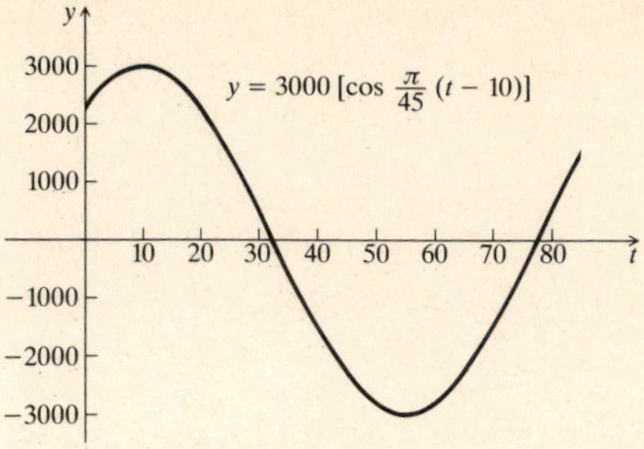

(b) Amplitude: 3000; period: 90; phase shift: 10

43. $y = 3 - 8.6\sin\left[\frac{2\pi}{5}(x + 11)\right]$; answers may vary.

45. $y = 2 + 16\cos\left[6\left(x + \frac{2}{\pi}\right)\right]$

47.

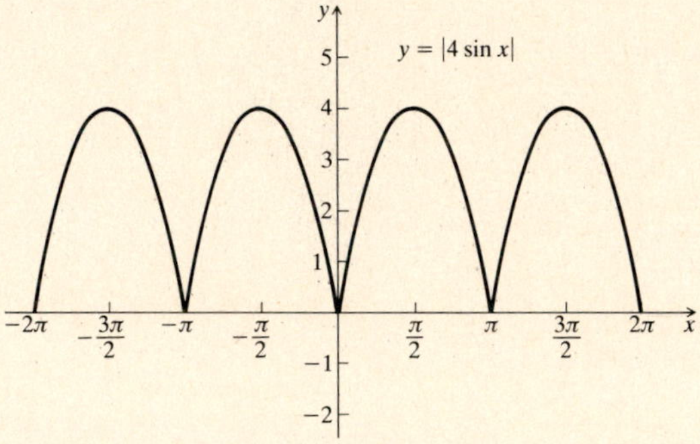

49.

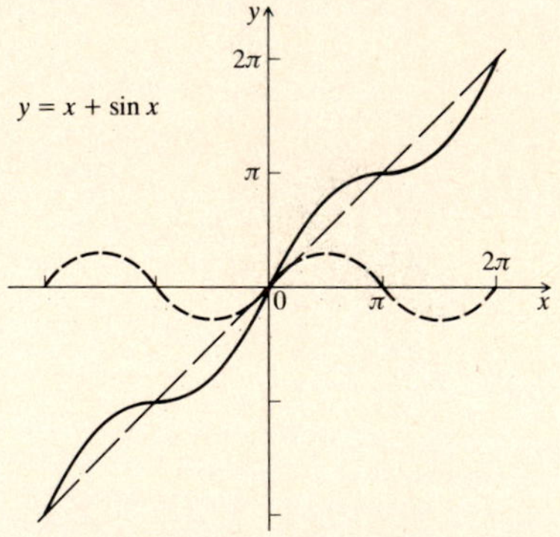

Margin Exercises, Section 6.8

1.

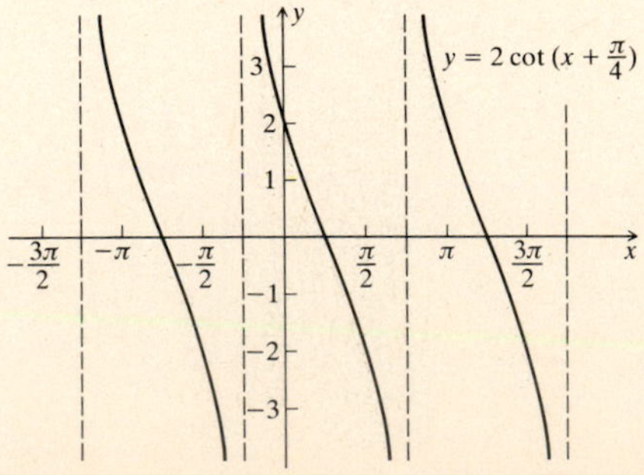

2.

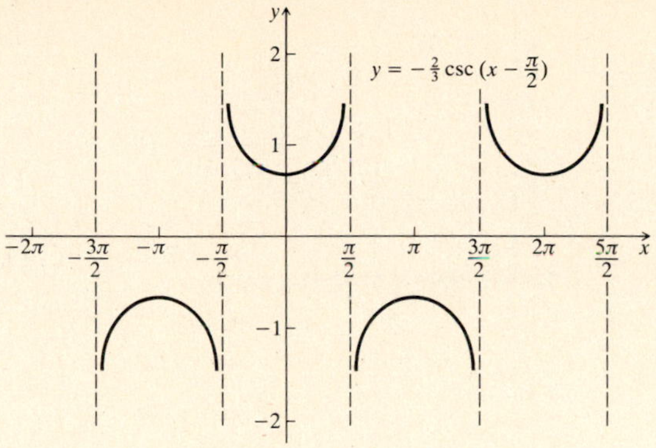

3.

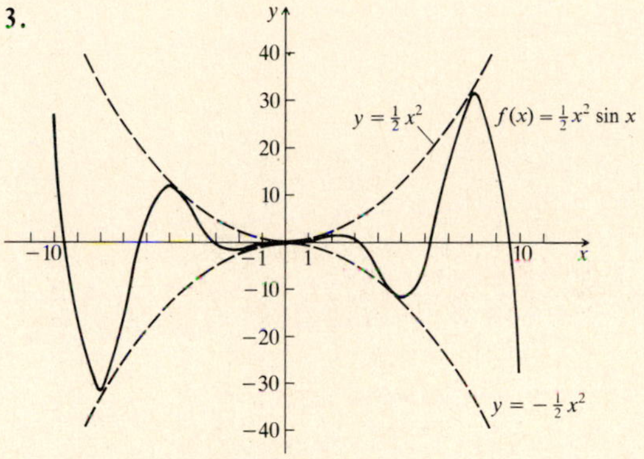

Exercise Set 6.8, pp. 426–427

1.

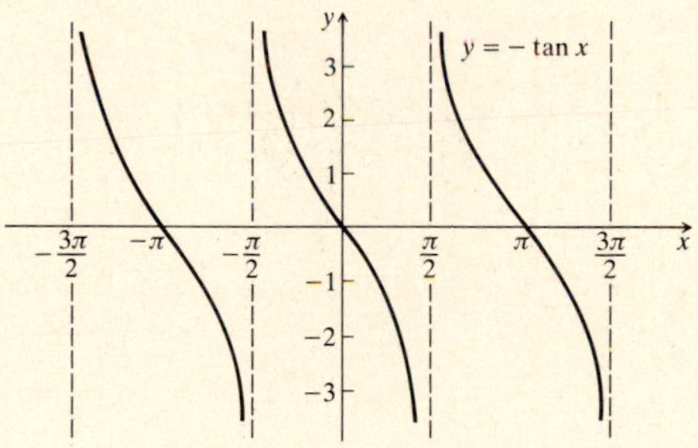

3.

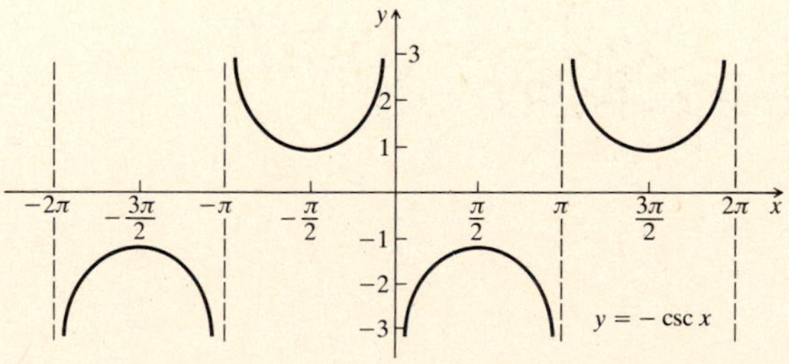

5.

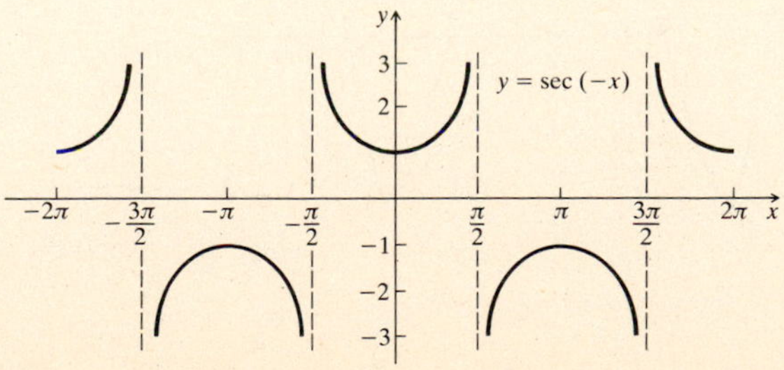

7.

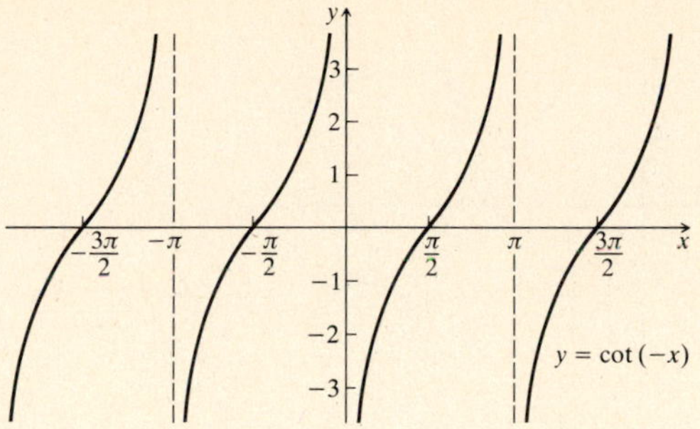

9.

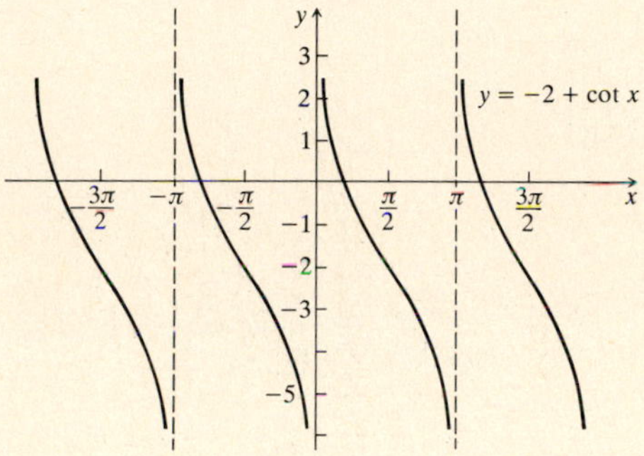

11.

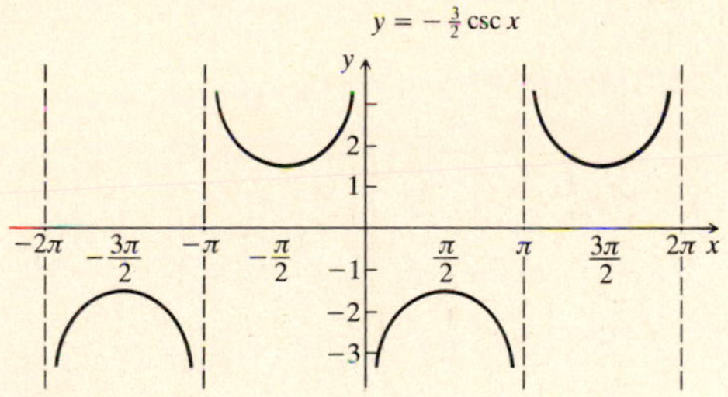

13.

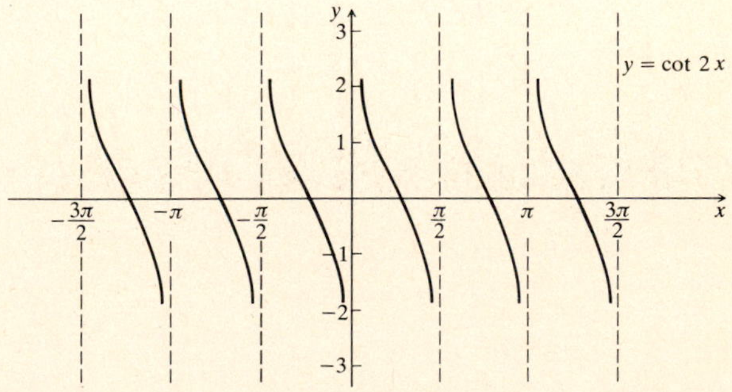

15.

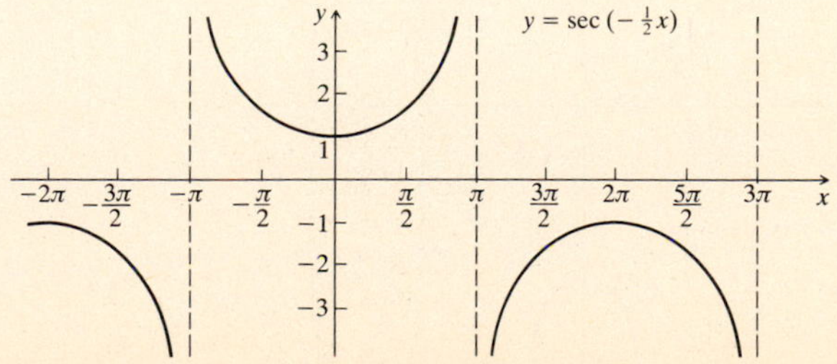

17.

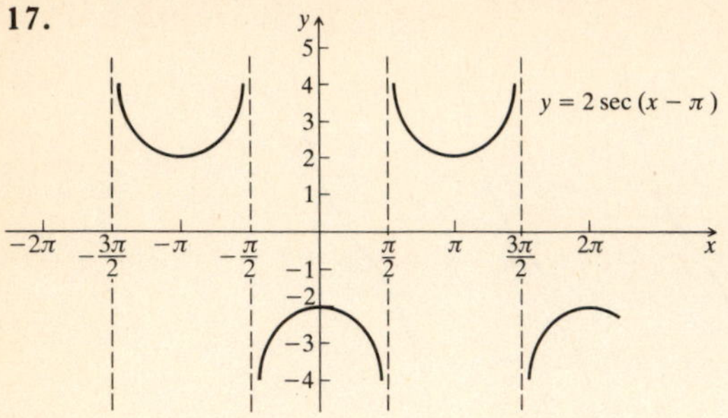

19.

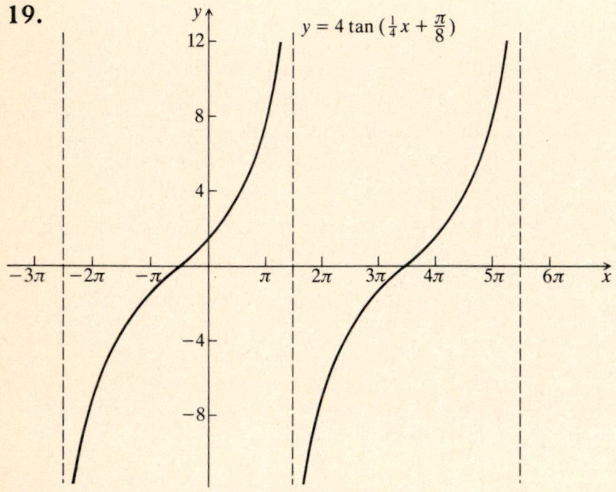

21.

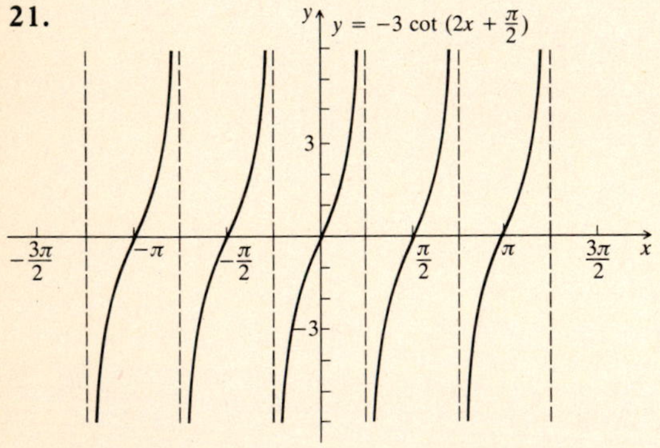

23.

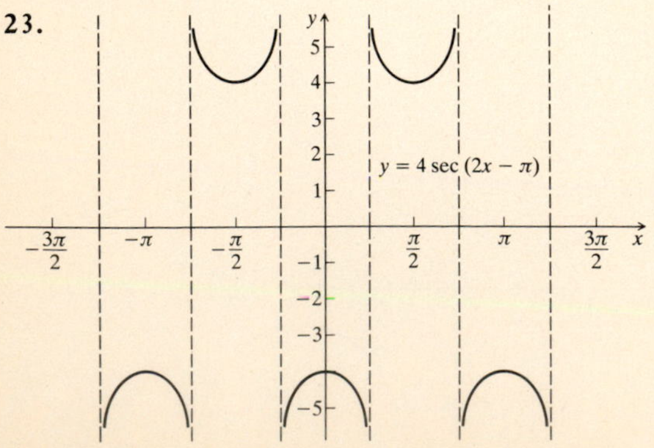

25.

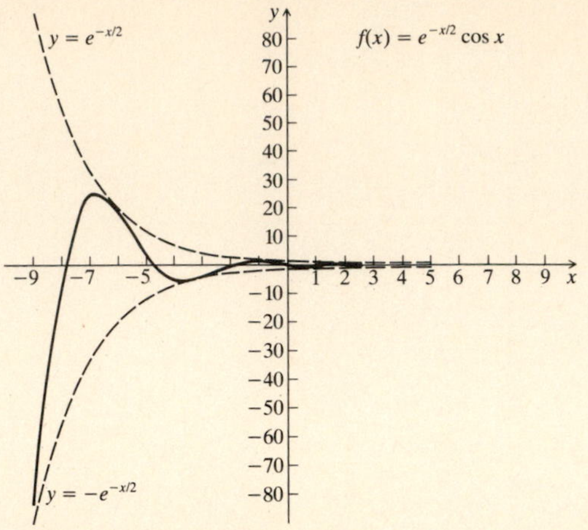

27.

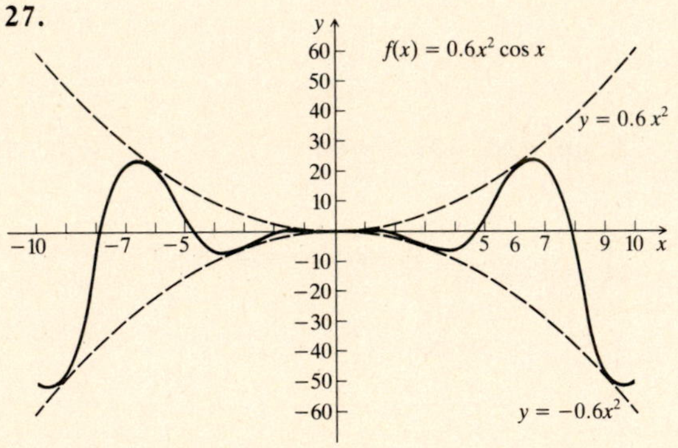

29.

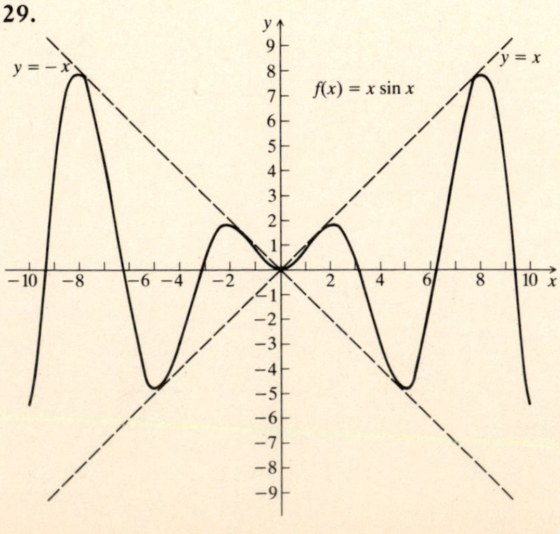

31.

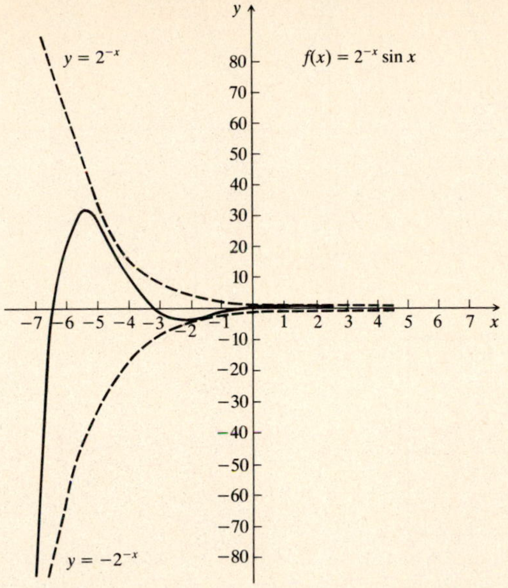

33.

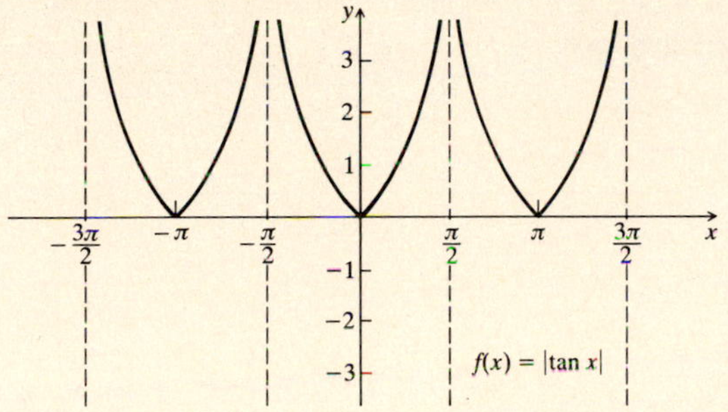

35.

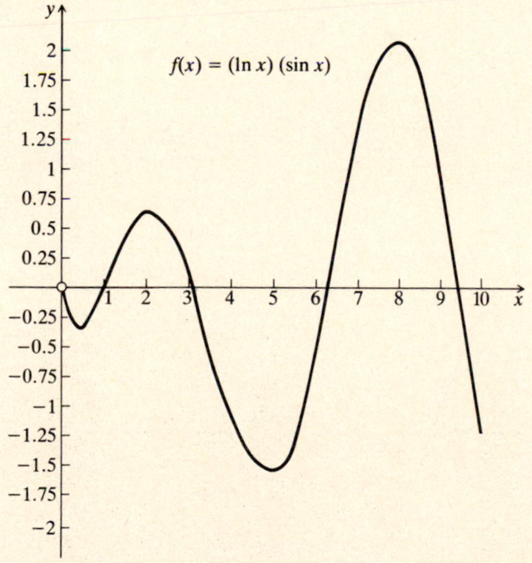

37.

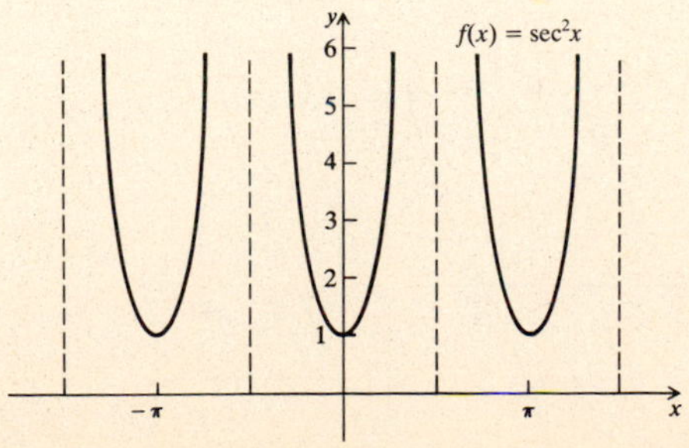

39. (a)

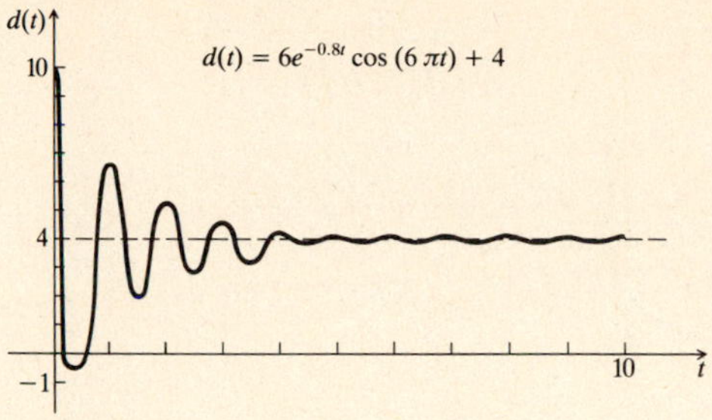

(b) 4 inches

41. (a)

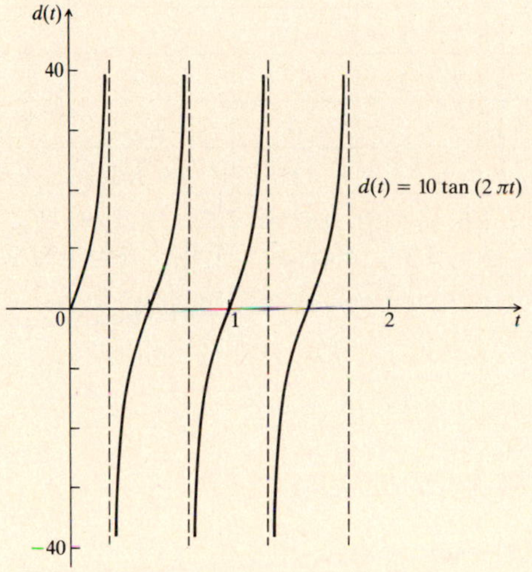

(b) The function is undefined when the beam is parallel to the wall.

43.

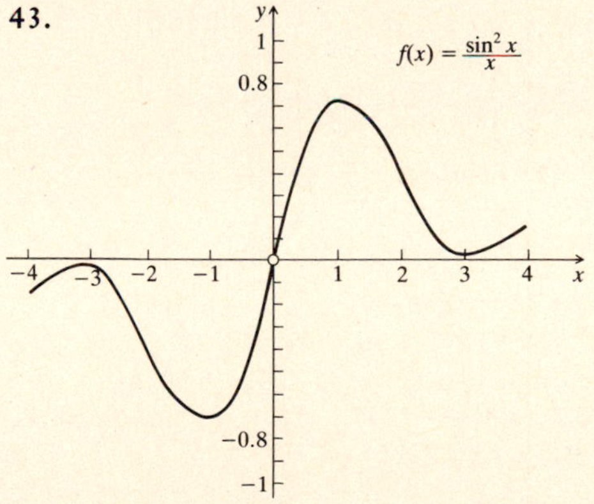

45.

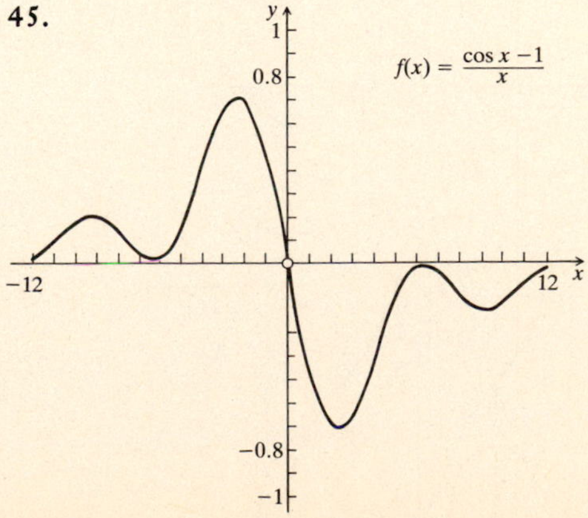

Review Exercises: Chapter 6, pp. 427–428

1. [6.1] u-axis: $\left(\frac{3}{5}, \frac{4}{5}\right)$; v-axis: $\left(-\frac{3}{5}, -\frac{4}{5}\right)$; origin: $\left(-\frac{3}{5}, \frac{4}{5}\right)$

2. [6.1]

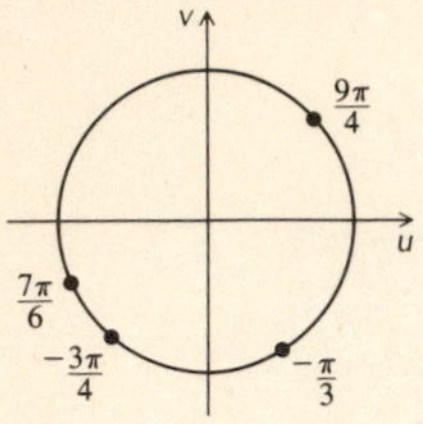

3. [6.3] I, IV

4. [6.2, 6.3]

	$\pi/6$	$\pi/4$	$\pi/2$	$2\pi/3$	$-\pi$	$-5\pi/4$
$\sin x$	$1/2$	$\sqrt{2}/2$	1	$\sqrt{3}/2$	0	$\sqrt{2}/2$
$\cos x$	$\sqrt{3}/2$	$\sqrt{2}/2$	0	$-1/2$	-1	$-\sqrt{2}/2$
$\cot x$	$\sqrt{3}$	1	0	$-\sqrt{3}/3$	Undefined	-1
$\csc x$	2	$\sqrt{2}$	1	$2\sqrt{3}/3$	Undefined	$\sqrt{2}$

5. [6.4] I, 0.483π, 1.52 **6.** [6.4] II, 0.806π, 2.53 **7.** [6.4] IV, -0.167π, -0.524 **8.** [6.4] 270° **9.** [6.4] 171.89° **10.** [6.4] Positive: 425°, 785°; negative: −295°, −655° **11.** [6.4] $\frac{7\pi}{4}$, or 5.5 cm **12.** [6.4] 2.25, 129° **13.** [6.4] 1131 cm/min **14.** [6.4] 146,215 radians/hr **15.** [6.5] $\sin\theta = \frac{3}{\sqrt{13}}$, or $\frac{3\sqrt{13}}{13}$; $\cos\theta = -\frac{2}{\sqrt{13}}$, or $-\frac{2\sqrt{13}}{13}$; $\tan\theta = -\frac{3}{2}$; $\cot\theta = -\frac{2}{3}$; $\sec\theta = -\frac{\sqrt{13}}{2}$; $\csc\theta = \frac{\sqrt{13}}{3}$ **16.** [6.5] $\sin\theta = \frac{3}{\sqrt{73}}$, or $\frac{3\sqrt{73}}{73}$; $\cos\theta = \frac{8}{\sqrt{73}}$, or $\frac{8\sqrt{73}}{73}$; $\tan\theta = \frac{3}{8}$; $\cot\theta = \frac{8}{3}$; $\sec\theta = \frac{\sqrt{73}}{8}$; $\csc\theta = \frac{\sqrt{73}}{3}$ **17.** [6.6] $\frac{\sqrt{2}}{2}$ **18.** [6.6] 1 **19.** [6.6] $\sqrt{3}$ **20.** [6.6] $-\frac{\sqrt{3}}{2}$ **21.** [6.6] Undefined **22.** [6.6] $-\frac{2\sqrt{3}}{3}$ **23.** [6.5] $\sin\theta = -\frac{2}{3}$; $\cos\theta = -\frac{\sqrt{5}}{3}$; $\cot\theta = \frac{\sqrt{5}}{2}$; $\sec\theta = -\frac{3\sqrt{5}}{5}$; $\csc\theta = -\frac{3}{2}$ **24.** [6.4] 22°12′ **25.** [6.4] 47.5575° **26.** [6.6] 0.185095 **27.** [6.6] 0.993577 **28.** [6.6] 8.5796 **29.** [6.6] 187° **30.** [6.6] 105°30′ **31.** [6.6] 0.2792 **32.** [6.6] 1.4108 **33.** [6.6] −0.9056 **34.** [6.6] −1.7013 **35.** [6.6] −1.7744 **36.** [6.6] Undefined **37.** [6.6] −6.1685 **38.** [6.6] 0.2225

39. [6.2]

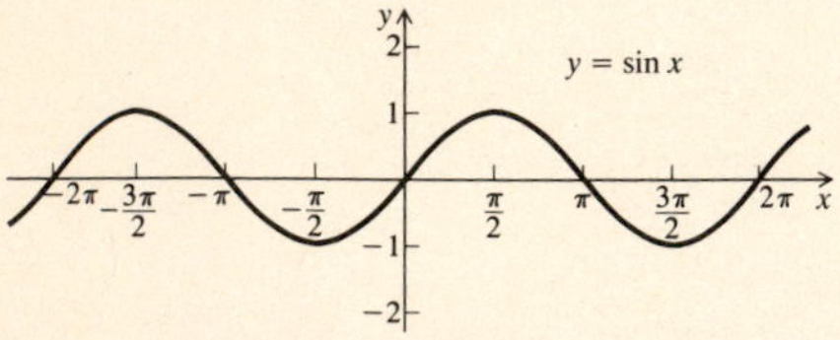

40. [6.3]

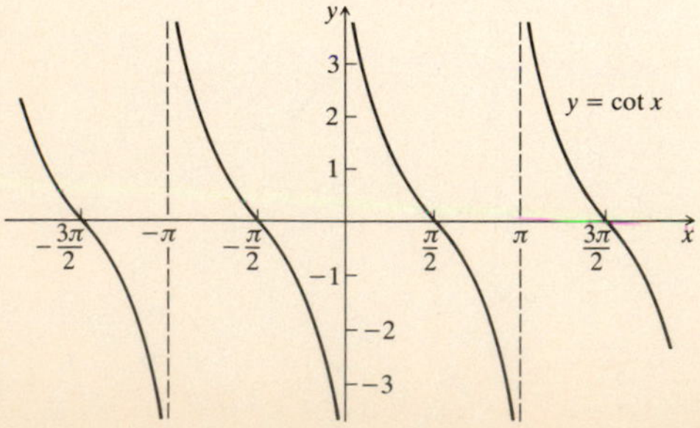

41. [6.3]

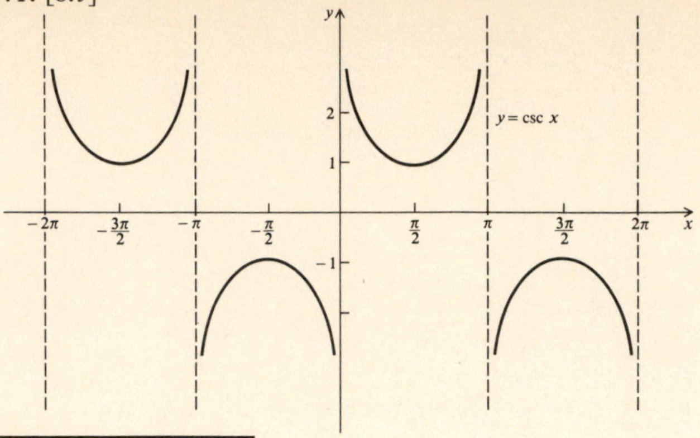

42. [6.2] Period: 2π; range: all reals from −1 to 1, inclusive **43.** [6.2] Amplitude: 1; domain: all reals **44.** [6.3] Period: 2π; range: all real numbers 1 and greater, in addition to all real numbers −1 and less

45. [6.7]

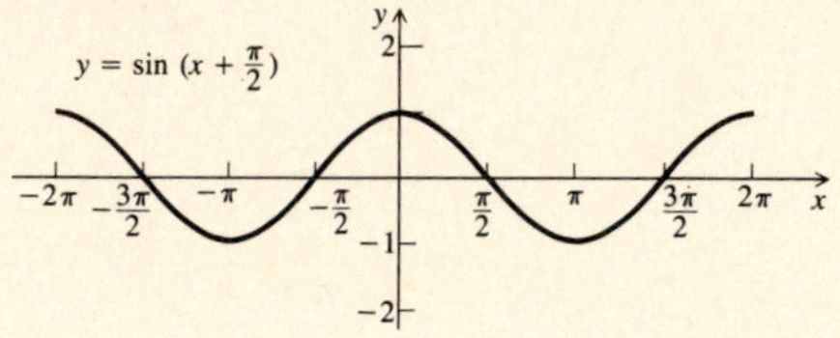

46. [6.7]

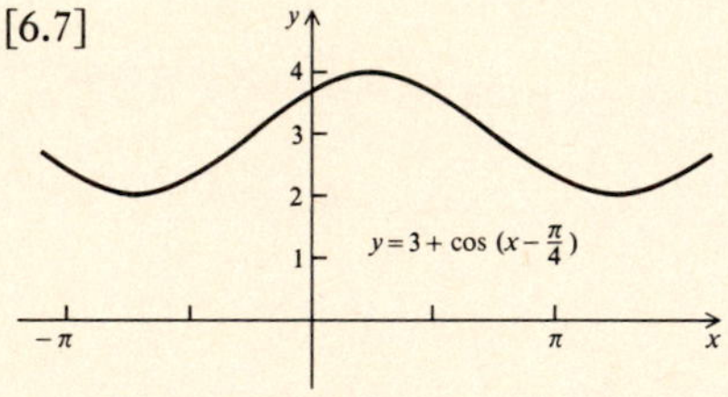

47. [6.7] $\frac{\pi}{4}$ **48.** [6.7] 2π

49. [6.7]

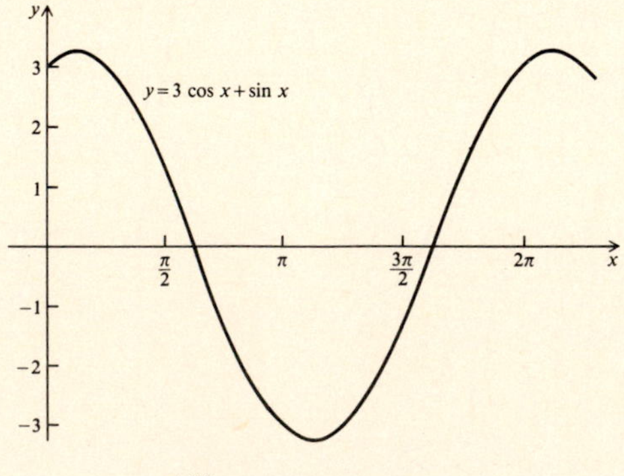

50. [6.8]

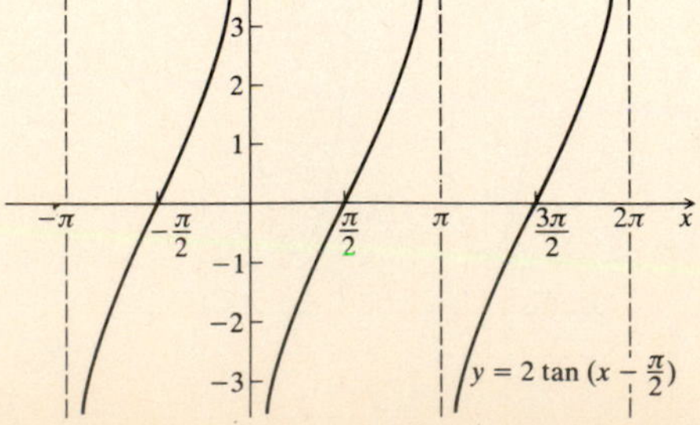

51. [6.8]

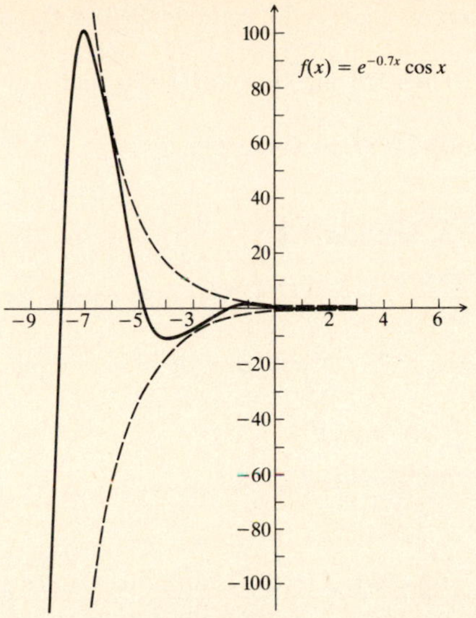

52. No, $\sin x = \frac{7}{5}$, but sines are never greater than 1. **53.** All values **54.** Domain: set of all reals; range: $[-3, 3]$; period: 4π

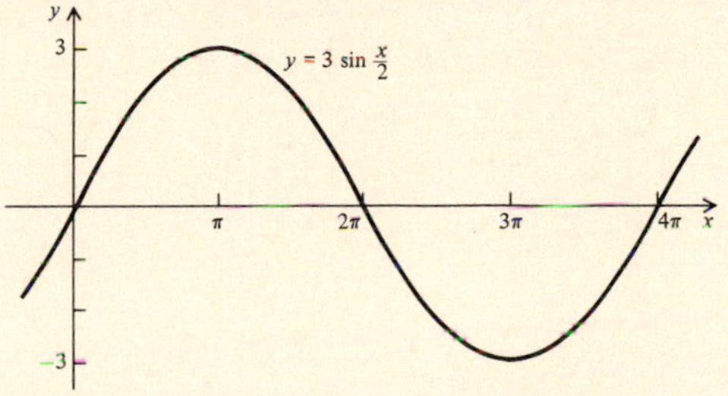

55. The domain consists of the intervals $\left(-\frac{\pi}{2} + 2k\pi, \frac{\pi}{2} + 2k\pi\right)$, k any integer **56.** $\cos x = -0.7890$; $\tan x = -0.7787$; $\cot x = -1.2842$; $\sec x = -1.2674$; $\csc x = 1.6276$

Test: Chapter 6, pp. 429–430

1. [6.1] Origin: $\left(\frac{4}{5}, \frac{2}{5}\right)$; x-axis: $\left(-\frac{4}{5}, \frac{2}{5}\right)$; y-axis: $\left(\frac{4}{5}, -\frac{2}{5}\right)$
2. [6.1]

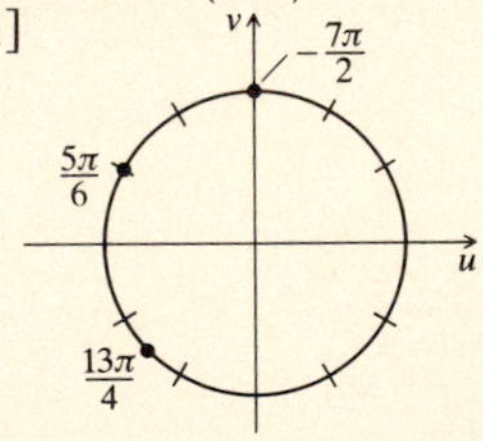

3. [6.2]

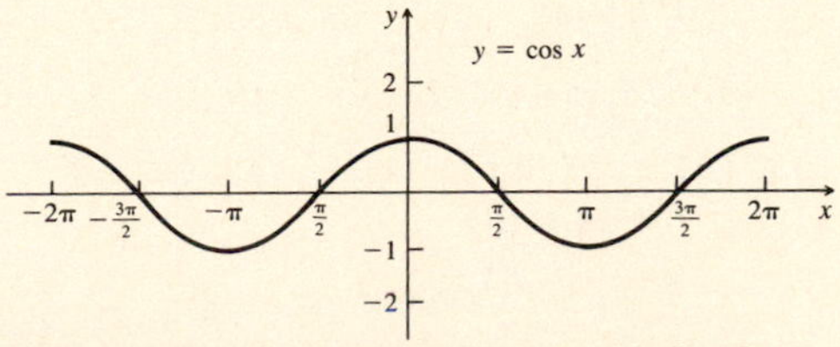

4. [6.2] All reals from -1 to 1 inclusive **5.** [6.2] 2π
6. [6.2] All reals **7.** [6.2] 1
8. [6.2]

	$\pi/2$	$-4\pi/2$	$7\pi/2$
$\sin x$	1	0	−1
$\cos x$	0	1	0

9. [6.3]

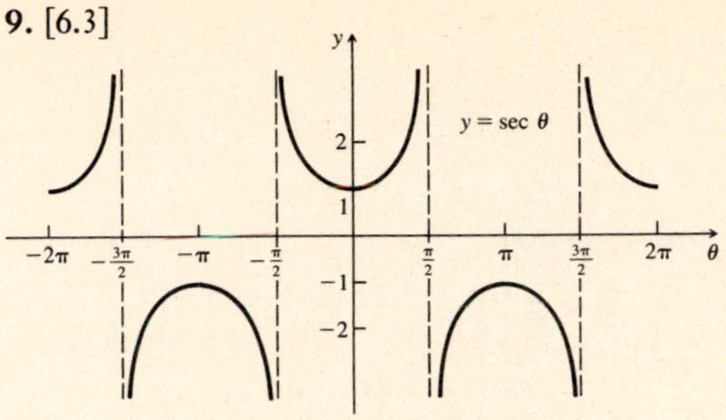

10. [6.3] 2π

11. [6.3] All reals except $\frac{\pi}{2} + k\pi$ **12.** [6.2, 6.3] II, IV

13. [6.4] III, $-\frac{5\pi}{6}$, -2.62 **14.** [6.4] II, $\frac{13\pi}{20}$, 2.04 **15.** [6.4] III, $\frac{5\pi}{4}$, 3.93 **16.** [6.4] 315° **17.** [6.4] 114°35′ **18.** [6.4] 10.47 cm **19.** [6.4] $\frac{1}{2}$, 29° **20.** [6.4] 2010 **21.** [6.4] 1250

22. [6.5] $\sin \theta = \frac{5}{\sqrt{34}}$, or $\frac{5\sqrt{34}}{34}$; $\cos \theta = -\frac{3}{\sqrt{34}}$, or $-\frac{3\sqrt{34}}{34}$; $\tan \theta = -\frac{5}{3}$; $\cot \theta = -\frac{3}{5}$; $\sec \theta = -\frac{\sqrt{34}}{3}$; $\csc \theta = \frac{\sqrt{34}}{5}$

23. [6.6] -1 **24.** [6.6] $-\frac{1}{2}$ **25.** [6.6] $\frac{1}{2}$ **26.** [6.5] $\cos \theta = -\frac{\sqrt{15}}{4}$; $\tan \theta = -\frac{\sqrt{15}}{15}$; $\cot \theta = -\sqrt{15}$; $\sec \theta = -\frac{4\sqrt{15}}{15}$; $\csc \theta = 4$ **27.** [6.4] 19°15′ **28.** [6.4] 72.8° **29.** [6.6] -0.830949 **30.** [6.6] 2.80201 **31.** [6.6] 0.614213 **32.** [6.6] 1.27396 **33.** [6.6] -3.73281 **34.** [6.6] 0.781286 **35.** [6.6] 60.62°, 1.0580 **36.** [6.6] 7.13°, 0.1244 **37.** [6.6] 154.50°
38. [6.6] -1.4834
39. [6.7]

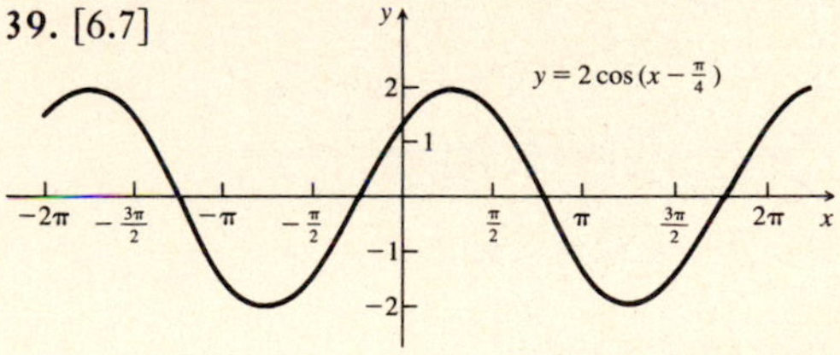

40. [6.7] 2π
41. [6.7] $\frac{\pi}{4}$

42. [6.7]

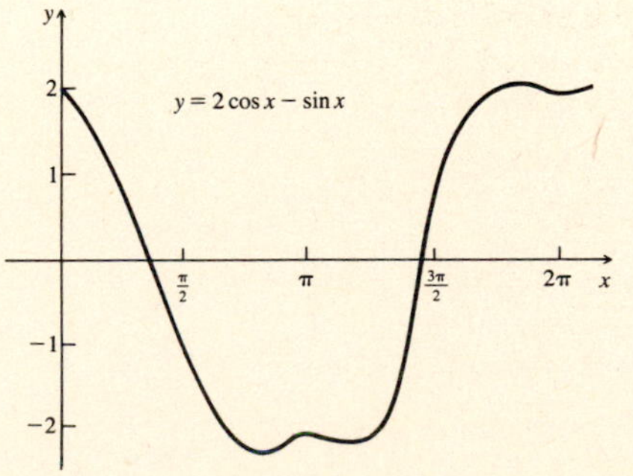

43. (a) 0.99518; (b) 0.99518
44. 132.5°

CHAPTER 7

Margin Exercises, Section 7.1

1. $\cos x + 1$ **2.** $\sin x$ **3.** $\sec \theta$ **4.** $\frac{1 + \sin x}{1 - \cot x}$

5. $\frac{5 \sin x - \cos x}{\sin^2 x - \cos^2 x}$ **6.** $\frac{2}{\cos^2 x\,(\cot x - 2)}$

7. $\tan x \sin x \sqrt{\sin x}$ 8. $\dfrac{\cos x}{\sqrt{3 \cos x}}$ 9. $\cos\theta = \pm\dfrac{1}{\sqrt{1 + \tan^2\theta}}$

10. $\sqrt{4 - x^2} = 2\cos\theta$; $\cos\theta = \dfrac{\sqrt{4 - x^2}}{2}$; $\tan\theta = \dfrac{x}{\sqrt{4 - x^2}}$

Exercise Set 7.1, pp. 436–437

1. $\sin^2 x - \cos^2 x$ 3. $\sin x - \sec x$ 5. $\sin y + \cos y$
7. $\cot x - \tan x$ 9. $1 - 2\sin y \cos y$ 11. $2\tan x + \sec^2 x$
13. $\sin^2 y + \csc^2 y - 2$ 15. $\cos^3 x - \sec^3 x$
17. $\cot^3 x - \tan^3 x$ 19. $\cos^2 x$ 21. $\cos x(\sin x + \cos x)$
23. $(\sin\theta - \cos\theta)(\sin\theta + \cos\theta)$ 25. $\sin x(\sec x + 1)$
27. $(\sin x + \cos x)(\sin x - \cos x)$ 29. $3(\cot y + 1)^2$
31. $(\csc^2 x + 5)(\csc x + 1)(\csc x - 1)$ 33. $(\sin y + 3) \times (\sin^2 y - 3\sin y + 9)$ 35. $(\sin v - \csc v)(\sin^2 v + 1 + \csc^2 v)$
37. $\tan x$ 39. $\dfrac{2\cos^2 x}{9\sin x}$ 41. $\cos x - 1$
43. $\cos\alpha + 1$ 45. $\dfrac{2\tan x + 1}{3\sin x + 1}$ 47. $\sec\theta$ 49. 1
51. $\dfrac{1}{2\cos x}$, or $\dfrac{1}{2}\sec x$ 53. $\dfrac{3\tan x}{\cos x - \sin x}$ 55. $\dfrac{1}{3}\cot\gamma$
57. $\dfrac{1 - 2\sin y + 2\cos y}{\sin^2 y - \cos^2 y}$ 59. -1 61. $\dfrac{5(\sin x - 3)}{3}$
63. $\sin x \cos x$ 65. $\sqrt{\sin y}(\sin y + \cos y)$ 67. $\sin x + \cos x$
69. $1 - \sin y$ 71. $\sin x(\sqrt{2} + \sqrt{\cos x})$ 73. $\dfrac{\sqrt{\sin x \cos x}}{\cos x}$
75. $\dfrac{\sqrt{\cos x}}{\cot x}$ 77. $\dfrac{\sqrt{2}\cot x}{2}$ 79. $\dfrac{\cos x}{1 - \sin x}$
81. $\dfrac{\sin x}{\sqrt{\cos x \sin x}}$ 83. $\dfrac{\sin x}{\sqrt{\cos x}}$ 85. $\dfrac{\cot x}{\sqrt{2}}$ 87. $\dfrac{1 + \sin x}{\cos x}$
89. $\cos\theta = \pm\sqrt{1 - \sin^2\theta}$; $\tan\theta = \pm\dfrac{\sin\theta}{\sqrt{1 - \sin^2\theta}}$; $\cot\theta = \pm\dfrac{\sqrt{1 - \sin^2\theta}}{\sin\theta}$; $\sec\theta = \pm\dfrac{1}{\sqrt{1 - \sin^2\theta}}$; $\csc\theta = \dfrac{1}{\sin\theta}$
91. $\sin\theta = \pm\dfrac{\tan\theta}{\sqrt{1 + \tan^2\theta}}$; $\cos\theta = \pm\dfrac{1}{\sqrt{1 + \tan^2\theta}}$; $\cot\theta = \dfrac{1}{\tan\theta}$; $\sec\theta = \pm\sqrt{1 + \tan^2\theta}$; $\csc\theta = \pm\dfrac{\sqrt{1 + \tan^2\theta}}{\tan\theta}$
93. $\sin\theta = \pm\dfrac{\sqrt{\sec^2\theta - 1}}{\sec\theta}$; $\cos\theta = \dfrac{1}{\sec\theta}$; $\tan\theta = \pm\sqrt{\sec^2\theta - 1}$; $\cot\theta = \pm\dfrac{1}{\sqrt{\sec^2\theta - 1}}$; $\csc\theta = \pm\dfrac{\sec\theta}{\sqrt{\sec^2\theta - 1}}$
95. $\cos\theta = \dfrac{\sqrt{a^2 - x^2}}{a}$; $\tan\theta = \dfrac{x}{\sqrt{a^2 - x^2}}$ 97. $\sin\theta = \dfrac{\sqrt{x^2 - 9}}{x}$; $\cos\theta = \dfrac{3}{x}$
99. $\sin\theta\tan\theta$ 101. Let $x = \dfrac{\pi}{4}$. Then $\left(\sin\dfrac{\pi}{4} + \cos\dfrac{\pi}{4}\right)^2 = \left(\dfrac{\sqrt{2}}{2} + \dfrac{\sqrt{2}}{2}\right)^2 = (\sqrt{2})^2 = 2$, and $\sin^2\dfrac{\pi}{4} + \cos^2\dfrac{\pi}{4} = \left(\dfrac{\sqrt{2}}{2}\right)^2 + \left(\dfrac{\sqrt{2}}{2}\right)^2 = \dfrac{2}{4} + \dfrac{2}{4} = 1.$ 103. Let $x = \dfrac{\pi}{6}$ and $y = \dfrac{\pi}{4}$. Then $\dfrac{\sin\frac{\pi}{6}}{\sin\frac{\pi}{4}} = \dfrac{\frac{1}{2}}{\frac{\sqrt{2}}{2}} = \dfrac{1}{\sqrt{2}}$, and $\dfrac{x}{y} = \dfrac{\frac{\pi}{6}}{\frac{\pi}{4}} = \dfrac{2}{3}$. 105. Let $\alpha = \dfrac{\pi}{3}$ and $\beta = \dfrac{2\pi}{3}$. Then $\cos\left(\dfrac{\pi}{3} + \dfrac{2\pi}{3}\right) = \cos\pi = -1$, and $\cos\dfrac{\pi}{3} + \cos\dfrac{2\pi}{3} = \dfrac{1}{2} + \left(-\dfrac{1}{2}\right) = 0.$ 107. Let $\theta = \dfrac{\pi}{2}$. Then $\cos\left(2 \cdot \dfrac{\pi}{2}\right) = \cos\pi = -1$, and $2\cos\dfrac{\pi}{2} = 2 \cdot 0 = 0.$
109. Let $\theta = \dfrac{\pi}{2}$. Then $\ln\left(\sin\dfrac{\pi}{2}\right) = \ln(1) = 0$, and $\sin\left(\ln\dfrac{\pi}{2}\right) = \sin(0.451583) = 0.436390.$

Margin Exercises, Section 7.2

1. $\dfrac{1}{2}$ 2. $\dfrac{-\sqrt{6} + \sqrt{2}}{4}$, or $\dfrac{\sqrt{2}}{4}(1 - \sqrt{3})$ 3. $\cos\dfrac{7\pi}{12}$, or $\cos\left(-\dfrac{7\pi}{12}\right)$ 4. $\cos 25°$ 5. $\cos(\alpha + \beta)$ 6. $\dfrac{\sqrt{2} + \sqrt{6}}{4}$, or $\dfrac{\sqrt{2}}{4}(1 + \sqrt{3})$ 7. $\sin(\alpha - \beta)$

8. $$\tan(\alpha - \beta) = \frac{\sin(\alpha - \beta)}{\cos(\alpha - \beta)} = \frac{\sin\alpha\cos\beta - \cos\alpha\sin\beta}{\cos\alpha\cos\beta + \sin\alpha\sin\beta} \cdot \frac{\frac{1}{\cos\alpha\cos\beta}}{\frac{1}{\cos\alpha\cos\beta}} = \frac{\frac{\sin\alpha\cos\beta}{\cos\alpha\cos\beta} - \frac{\cos\alpha\sin\beta}{\cos\alpha\cos\beta}}{\frac{\cos\alpha\cos\beta}{\cos\alpha\cos\beta} + \frac{\sin\alpha\sin\beta}{\cos\alpha\cos\beta}} = \frac{\tan\alpha - \tan\beta}{1 + \tan\alpha\tan\beta}$$

9. $\dfrac{3 + \sqrt{3}}{3 - \sqrt{3}}$ 10. $\sin\dfrac{\pi}{6}$, or $\dfrac{1}{2}$ 11. $\dfrac{3 + \sqrt{105}}{16}$

12. 30°

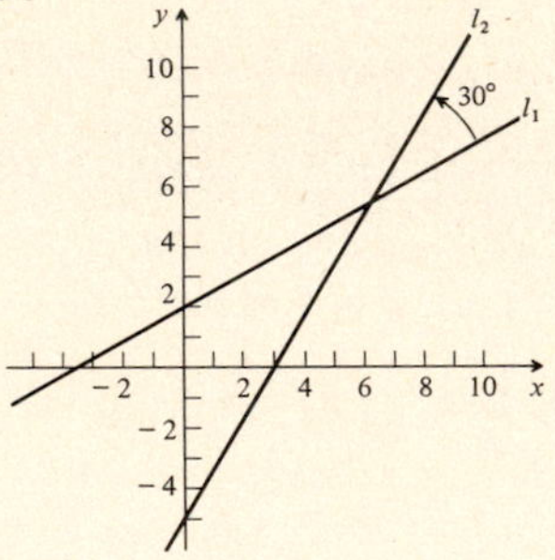

13. 150°

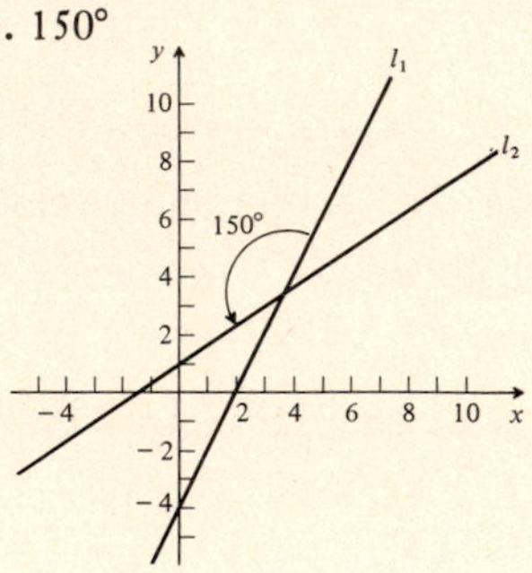

14. 46.4°, or 46°24′

Exercise Set 7.2, pp. 444–446

1. $\dfrac{\sqrt{6} + \sqrt{2}}{4}$ 3. $\dfrac{\sqrt{6} - \sqrt{2}}{4}$ 5. $\dfrac{\sqrt{6} + \sqrt{2}}{4}$ 7. $\dfrac{3 + \sqrt{3}}{3 - \sqrt{3}}$
9. sin 59°, or 0.8572 11. tan 52°, or 1.280 13. 1 15. 0
17. Undefined 19. 0.8448 21. $2\sin\alpha\cos\beta$
23. $2\cos\alpha\cos\beta$ 25. $\cos u$ 27. $\dfrac{2\pi}{3}$ 29. 0° 31. 86.63°, or 86°38′ 33. 135° 35. 90°21′ 37. $2\sin\theta\cos\theta$

39. $$\cot(\alpha + \beta) = \frac{\cos(\alpha + \beta)}{\sin(\alpha + \beta)} = \frac{\cos\alpha\cos\beta - \sin\alpha\sin\beta}{\sin\alpha\cos\beta + \cos\alpha\sin\beta} \cdot \frac{\frac{1}{\sin\alpha\sin\beta}}{\frac{1}{\sin\alpha\sin\beta}} = \frac{\frac{\cos\alpha\cos\beta}{\sin\alpha\sin\beta} - 1}{\frac{\sin\alpha\cos\beta}{\sin\alpha\sin\beta} + \frac{\cos\alpha\sin\beta}{\sin\alpha\sin\beta}} = \frac{\cot\alpha\cot\beta - 1}{\cot\alpha + \cot\beta}$$

41. $\sin\left(x+\frac{3\pi}{2}\right)=\sin x\cos\frac{3\pi}{2}+\cos x\sin\frac{3\pi}{2}=(\sin x)(0)+(\cos x)(-1)=-\cos x$ **43.** $\cos\left(x+\frac{3\pi}{2}\right)=\cos x\cos\frac{3\pi}{2}-\sin x\sin\frac{3\pi}{2}=(\cos x)(0)-(\sin x)(-1)=\sin x$

45. $\tan\left(x+\frac{\pi}{4}\right)=\dfrac{\tan x+\tan\frac{\pi}{4}}{1-\tan x\tan\frac{\pi}{4}}=\dfrac{1+\tan x}{1-\tan x}$

47. $$\frac{\tan\alpha+\tan\beta}{1+\tan\alpha\tan\beta}=\frac{\dfrac{\sin\alpha}{\cos\alpha}+\dfrac{\sin\beta}{\cos\beta}}{1+\dfrac{\sin\alpha}{\cos\alpha}\cdot\dfrac{\sin\beta}{\cos\beta}}=\frac{\dfrac{\sin\alpha\cos\beta+\cos\alpha\sin\beta}{\cos\alpha\cos\beta}}{\dfrac{\cos\alpha\cos\beta+\sin\alpha\sin\beta}{\cos\alpha\cos\beta}}=\frac{\sin\alpha\cos\beta+\cos\alpha\sin\beta}{\cos\alpha\cos\beta+\sin\alpha\sin\beta}=\frac{\sin(\alpha+\beta)}{\cos(\alpha-\beta)}$$

49. $\sin(\alpha+\beta)+\sin(\alpha-\beta)=\sin\alpha\cos\beta+\cos\alpha\sin\beta+\sin\alpha\cos\beta-\cos\alpha\sin\beta=2\sin\alpha\cos\beta$ **51.** $\frac{1}{7}$

53. $\frac{1}{6}$ **55.** 73°, approximately **57.** 22.83°, or 22°50′

59. $\dfrac{\cos(x+h)-\cos x}{h}=\dfrac{\cos x\cos h-\sin x\sin h-\cos x}{h}=\dfrac{\cos x\cos h-\cos x}{h}-\dfrac{\sin x\sin h}{h}=\cos x\left(\dfrac{\cos h-1}{h}\right)-\sin x\left(\dfrac{\sin h}{h}\right)$ **61.** $\cos(\alpha+\beta)=\cos\alpha\cos\beta-\sin\alpha\sin\beta=\cos\alpha\cos\beta-\cos\left(\frac{\pi}{2}-\alpha\right)\cos\left(\frac{\pi}{2}-\beta\right)$

Margin Exercises, Section 7.3

1. $\cot\left(\frac{\pi}{2}-\theta\right)=\dfrac{\cos\left(\frac{\pi}{2}-\theta\right)}{\sin\left(\frac{\pi}{2}-\theta\right)}=\dfrac{\sin\theta}{\cos\theta}=\tan\theta$

2. $\sec\left(\frac{\pi}{2}-\theta\right)=\dfrac{1}{\cos\left(\frac{\pi}{2}-\theta\right)}=\dfrac{1}{\sin\theta}=\csc\theta$

3. $\csc\left(\frac{\pi}{2}-\theta\right)=\dfrac{1}{\sin\left(\frac{\pi}{2}-\theta\right)}=\dfrac{1}{\cos\theta}=\sec\theta$

4. $\sin 15°=0.2588$; $\cos 15°=0.9659$; $\tan 15°=0.2679$; $\cot 15°=3.732$; $\sec 15°=1.035$; $\csc 15°=3.864$

5. (a) and **(b)**

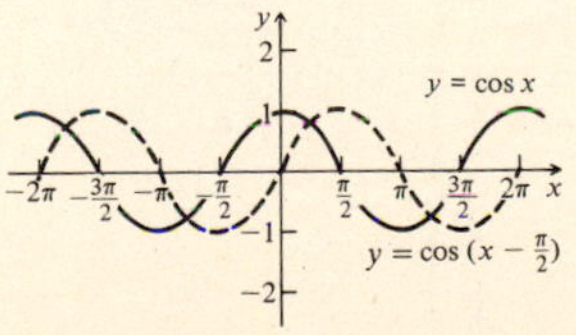

(c)

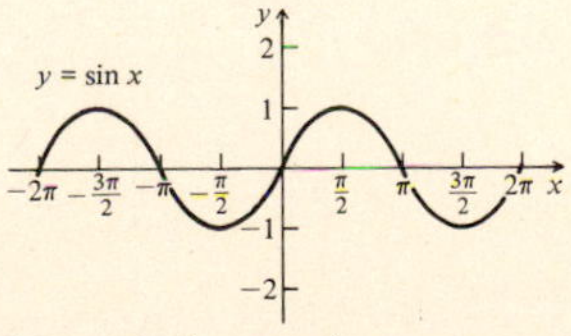

(d) the graphs of b and c are the same; **(e)** $\cos\left(x-\frac{\pi}{2}\right)=\sin x$

6. $\cos\left(x-\frac{\pi}{2}\right)=\cos x\cos\frac{\pi}{2}+\sin x\sin\frac{\pi}{2}=\cos x\cdot 0+\sin x\cdot 1=\sin x$ **7.** $\sin\left(x-\frac{\pi}{2}\right)=\sin x\cos\frac{\pi}{2}-\cos x\sin\frac{\pi}{2}=\sin x\cdot 0-\cos x\cdot 1=-\cos x$ **8. (a)** Translate the graph of $y=\cos x$ to obtain the graph of $y=\cos\left(x+\frac{\pi}{2}\right)$. Then graph $y=\sin x$ and reflect the graph to obtain $y=-\sin x$.

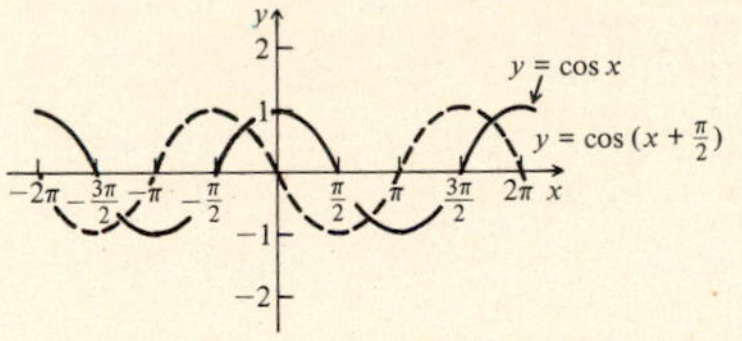

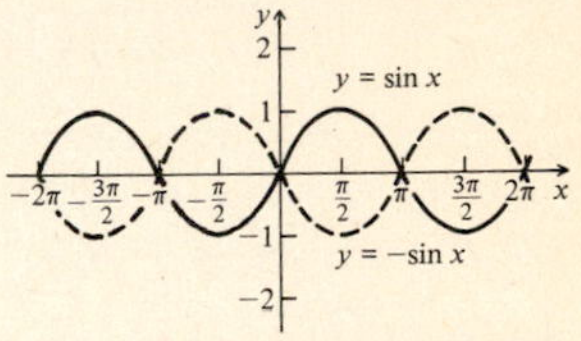

The graphs of $y=\cos\left(x+\frac{\pi}{2}\right)$ and $y=-\sin x$ are the same. Thus, $\cos\left(x+\frac{\pi}{2}\right)=-\sin x$. **(b)** $\cos\left(x+\frac{\pi}{2}\right)=\cos x\cos\frac{\pi}{2}-\sin x\sin\frac{\pi}{2}=\cos x\cdot 0-\sin x\cdot 1=-\sin x$

9. $\cot\left(x+\frac{\pi}{2}\right)=\dfrac{\cos\left(x+\frac{\pi}{2}\right)}{\sin\left(x+\frac{\pi}{2}\right)}=\dfrac{-\sin x}{\cos x}=-\tan x$

10. $\csc\left(x+\frac{\pi}{2}\right)=\dfrac{1}{\sin\left(x-\frac{\pi}{2}\right)}=\dfrac{1}{-\cos x}=-\sec x$

11. $\sin(x+\pi)=\sin x\cos\pi+\cos x\sin\pi=\sin x\cdot(-1)+\cos x\cdot 0=-\sin x$ **12.** $\tan(\theta+270°)=\dfrac{\sin(\theta+270°)}{\cos(\theta+270°)}=\dfrac{\sin\theta\cos 270°+\cos\theta\sin 270°}{\cos\theta\cos 270°-\sin\theta\sin 270°}=\dfrac{\sin\theta\cdot 0+\cos\theta\cdot(-1)}{\cos\theta\cdot 0-\sin\theta\cdot(-1)}=\dfrac{-\cos\theta}{\sin\theta}=-\cot\theta$ **13.** $\tan(\theta+180°)=\tan(90°\cdot 2+\theta)=\tan\theta$
14. $\sin(270°-\theta)=\sin(90°\cdot 3-\theta)=-\cos\theta$
15. $\cos(3\pi-x)=\cos\left(\frac{\pi}{2}\cdot 6-x\right)=-\cos x$

16. $\sqrt{2}\sin\left(2x+\frac{\pi}{4}\right)$ **17.** $13\sin(x+b)$, where $\cos b=\frac{12}{13}$ and $\sin b=-\frac{5}{13}$ **18.** $\sqrt{3}\sin x+\cos x=2\sin\left(x+\frac{\pi}{6}\right)$

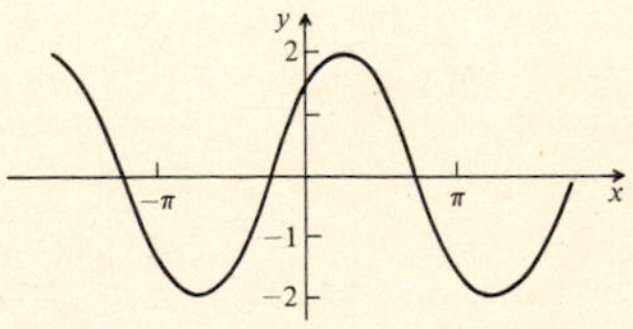

Exercise Set 7.3, pp. 453–454

1. $\sin 25°=0.4226$; $\cos 25°=0.9063$; $\tan 25°=0.4663$; $\cot 25°=2.145$; $\sec 25°=1.103$; $\csc 25°=2.366$
3. (a) $\tan 22.5°=\dfrac{\sqrt{2-\sqrt{2}}}{\sqrt{2+\sqrt{2}}}$, $\cot 22.5°=\dfrac{\sqrt{2+\sqrt{2}}}{\sqrt{2-\sqrt{2}}}$, $\sec 22.5°=\dfrac{2}{\sqrt{2+\sqrt{2}}}$, $\csc 22.5°=\dfrac{2}{\sqrt{2-\sqrt{2}}}$;
(b) $\sin 67.5°=\dfrac{\sqrt{2+\sqrt{2}}}{2}$, $\cos 67.5°=\dfrac{\sqrt{2-\sqrt{2}}}{2}$, $\tan 67.5°=\dfrac{\sqrt{2+\sqrt{2}}}{\sqrt{2-\sqrt{2}}}$, $\cot 67.5°=\dfrac{\sqrt{2-\sqrt{2}}}{\sqrt{2+\sqrt{2}}}$,

$\sec 67.5° = \dfrac{2}{\sqrt{2-\sqrt{2}}}$, $\csc 67.5° = \dfrac{2}{\sqrt{2+\sqrt{2}}}$

5. $\tan\left(x - \dfrac{\pi}{2}\right) = \dfrac{\sin\left(x-\dfrac{\pi}{2}\right)}{\cos\left(x-\dfrac{\pi}{2}\right)} = \dfrac{-\cos x}{\sin x} = -\cot x$

7. $\csc\left(x + \dfrac{\pi}{2}\right) = \dfrac{1}{\sin\left(x+\dfrac{\pi}{2}\right)} = \dfrac{1}{\cos x} = \sec x$

9. $\sin(x - \pi) = \sin x \cos \pi - \cos x \sin \pi = (\sin x)(-1) - (\cos x)(0) = -\sin x$ **11.** $\tan(\theta - 270°) = -\cot\theta$ **13.** $\sin(\theta + 450°) = \cos\theta$ **15.** $\sec(\pi - \theta) = -\sec\theta$ **17.** $\tan(x - 4\pi) = \tan x$

19. $\cos\left(\dfrac{9\pi}{2} + x\right) = -\sin x$ **21.** $\csc(540° - \theta) = \csc\theta$

23. $\cot\left(x - \dfrac{3\pi}{2}\right) = -\tan x$ **25.** $2\sin\left(2x + \dfrac{\pi}{3}\right)$

27. $5\sin(x + b)$, where $\cos b = \dfrac{4}{5}$, $\sin b = \dfrac{3}{5}$

29. $7.89 \sin(0.374x + 31.2°)$

31. $y = \sqrt{2}\sin\left(2x - \dfrac{\pi}{4}\right)$

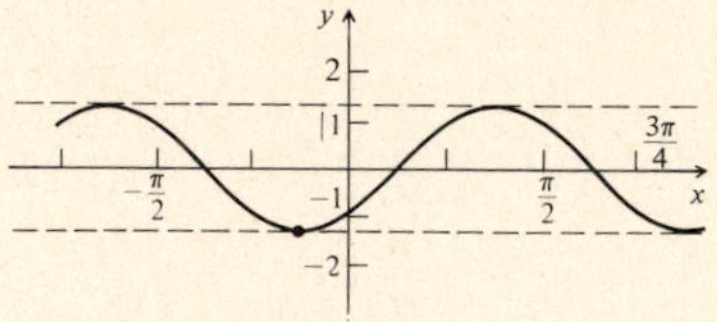

33. **(a)** $\cos\theta = -\dfrac{2\sqrt{2}}{3}$; $\tan\theta = -\dfrac{1}{2\sqrt{2}}$, or $-\dfrac{\sqrt{2}}{4}$; $\cot\theta = -2\sqrt{2}$; $\sec\theta = -\dfrac{3}{2\sqrt{2}}$, or $-\dfrac{3\sqrt{2}}{4}$; $\csc\theta = 3$; **(b)** $\sin\left(\dfrac{\pi}{2} - \theta\right) = -\dfrac{2\sqrt{2}}{3}$, $\cos\left(\dfrac{\pi}{2} - \theta\right) = \dfrac{1}{3}$, $\tan\left(\dfrac{\pi}{2} - \theta\right) = -2\sqrt{2}$, $\cot\left(\dfrac{\pi}{2} - \theta\right) = -\dfrac{\sqrt{2}}{4}$, $\sec\left(\dfrac{\pi}{2} - \theta\right) = 3$, $\csc\left(\dfrac{\pi}{2} - \theta\right) = -\dfrac{3\sqrt{2}}{4}$; **(c)** $\sin(\pi + \theta) = -\dfrac{1}{3}$, $\cos(\pi + \theta) = \dfrac{2\sqrt{2}}{3}$, $\tan(\pi + \theta) = -\dfrac{\sqrt{2}}{4}$, $\cot(\pi + \theta) = -2\sqrt{2}$, $\sec(\pi + \theta) = \dfrac{3\sqrt{2}}{4}$, $\csc(\pi + \theta) = -3$; **(d)** $\sin(\pi - \theta) = \dfrac{1}{3}$, $\cos(\pi - \theta) = \dfrac{2\sqrt{2}}{3}$, $\tan(\pi - \theta) = \dfrac{\sqrt{2}}{4}$, $\cot(\pi - \theta) = 2\sqrt{2}$, $\sec(\pi - \theta) = \dfrac{3\sqrt{2}}{4}$, $\csc(\pi - \theta) = 3$; **(e)** $\sin(2\pi - \theta) = -\dfrac{1}{3}$, $\cos(2\pi - \theta) = -\dfrac{2\sqrt{2}}{3}$, $\tan(2\pi - \theta) = \dfrac{\sqrt{2}}{4}$, $\cot(2\pi - \theta) = 2\sqrt{2}$, $\sec(2\pi - \theta) = -\dfrac{3\sqrt{2}}{4}$, $\csc(2\pi - \theta) = -3$ **35.** **(a)** $\cos 27° = 0.89101$, $\tan 27° = 0.50952$, $\cot 27° = 1.9626$, $\sec 27° = 1.1223$, $\csc 27° = 2.2027$; **(b)** $\sin 63° = 0.89101$, $\cos 63° = 0.45399$, $\tan 63° = 1.9626$, $\cot 63° = 0.50952$, $\sec 63° = 2.2027$, $\csc 63° = 1.1223$

37. $\sin 128° = 0.78801$, $\cos 128° = -0.61566$, $\tan 128° = -1.2799$, $\cot 128° = -0.78129$, $\sec 128° = -1.6243$, $\csc 128° = 1.2690$

39. $\sin^2 x$ **41.** $\sin x(\cos y + \tan y)$ **43.** $-\cos x(1 + \cot x)$

45. $\cos x + 1$ **47.** $\dfrac{\cos x - 1}{1 + \tan x}$ **49.** $\cos x - 1$ **51.** $-\tan^2 y$

Margin Exercises, Section 7.4

1. 24/25 **2.** II, $\cos 2\theta = -119/169$, $\tan 2\theta = -120/119$, $\sin 2\theta = 120/169$ **3.** $\cos\theta - 4\sin^2\theta\cos\theta$ **4.** $\sin^3 x = \sin x \dfrac{1 - \cos 2x}{2}$ **5.** $\dfrac{\sqrt{2+\sqrt{3}}}{2}$ **6.** $\dfrac{1}{2+\sqrt{3}}$, or $2 - \sqrt{3}$ **7.** $\tan 2x$ **8.** $\dfrac{\frac{1}{2}(\cos^2 x - \sin^2 x)}{\sin x \cos x} = \cot 2x$

Exercise Set 7.4, pp. 459–460

1. $\sin 2\theta = \dfrac{24}{25}$, $\cos 2\theta = -\dfrac{7}{25}$, $\tan 2\theta = -\dfrac{24}{7}$, II

3. $\sin 2\theta = \dfrac{24}{25}$, $\cos 2\theta = \dfrac{7}{25}$, $\tan 2\theta = \dfrac{24}{7}$, I **5.** $\sin 2\theta = \dfrac{24}{25}$, $\cos 2\theta = -\dfrac{7}{25}$, $\tan 2\theta = -\dfrac{24}{7}$, II **7.** $8\sin\theta\cos^3\theta - 4\sin\theta\cdot\cos\theta$, or $4\sin\theta\cos^3\theta - 4\sin^3\theta\cos\theta$, or $4\sin\theta\cos\theta - 8\sin^3\theta\cos\theta$ **9.** $\dfrac{3 - 4\cos 2\theta + \cos 4\theta}{8}$ **11.** $\dfrac{\sqrt{2+\sqrt{3}}}{2}$

13. $2 + \sqrt{3}$ **15.** $\dfrac{\sqrt{2+\sqrt{2}}}{2}$ **17.** $\dfrac{\sqrt{2+\sqrt{2}}}{2}$ **19.** $\dfrac{\sqrt{2+\sqrt{2}}}{2}$

21. $\dfrac{\sqrt{2-\sqrt{2}}}{2}$ **23.** 0.6421 **25.** 0.9844 **27.** 0.1734

29. $\cos x$ **31.** $\cos x$ **33.** $\sin x$ **35.** $\cos x$ **37.** 1

39. 1 **41.** 8 **43.** $\sin 2x$ **45.** $(\sin x + \cos x)^2 = 1 + \sin 2x$

47. $\dfrac{2\cot x}{\cot^2 x - 1} = \tan 2x$ **49.** $2\sin^2 2x + \cos 4x = 1$

51. $\sin\theta = \sqrt{\dfrac{1}{2} + \dfrac{\sqrt{6}}{5}}$; $\cos\theta = \sqrt{\dfrac{1}{2} - \dfrac{\sqrt{6}}{5}}$; $\tan\theta = \sqrt{\dfrac{5 + 2\sqrt{6}}{5 - 2\sqrt{6}}}$ **53.** $\sin\theta = -\dfrac{8}{17}$; $\cos\theta = \dfrac{15}{17}$; $\tan\theta = -\dfrac{8}{15}$

55. **(a)** 6062.76 ft; **(b)** 6097 ft; $N(\phi) = 6066 - 31(2\cos^2\phi - 1)$

57. Since $\cos^2 x - \sin^2 x = \cos 2x$, the graph of $f(x) = \cos^2 x - \sin^2 x$ is the same as the graph of $f(x) = \cos 2x$.

Margin Exercises, Section 7.5

1. *Method 1:*

Left side	Right side
$\dfrac{\csc t - 1}{t\csc t}$	$\dfrac{1 - \sin t}{t}$
$\dfrac{\frac{1}{\sin t} - 1}{t\cdot\frac{1}{\sin t}}$	
$\dfrac{\frac{1}{\sin t} - 1}{t\cdot\frac{1}{\sin t}}\cdot\dfrac{\sin t}{\sin t}$	
$\dfrac{1 - \sin t}{t}$	

2. *Method 1:*

Left side	Right side
$\sec\theta\csc\theta$	$\tan\theta + \cot\theta$
$\dfrac{1}{\cos\theta}\cdot\dfrac{1}{\sin\theta}$	
$\dfrac{1}{\cos\theta\sin\theta}\cdot\dfrac{\sin^2\theta + \cos^2\theta}{1}$	
$\dfrac{\sin\theta}{\cos\theta} + \dfrac{\cos\theta}{\sin\theta}$	
$\tan\theta + \cot\theta$	

3. *Method 1:*

Left side	Right side
$(\sin\theta - \cos\theta)^2$	$1 - \sin 2\theta$
	$1 - 2\sin\theta\cos\theta$
	$(\sin^2\theta + \cos^2\theta) - 2\sin\theta\cos\theta$
	$\sin^2\theta - 2\sin\theta\cos\theta + \cos^2\theta$
	$(\sin\theta - \cos\theta)^2$

4. *Method 1:*

Left side	Right side
$(\sec u - \tan u)(1 + \sin u)$	$\cos u$
$\left(\dfrac{1}{\cos u} - \dfrac{\sin u}{\cos u}\right)(1 + \sin u)$	
$\dfrac{1}{\cos u} + \dfrac{\sin u}{\cos u} - \dfrac{\sin u}{\cos u} - \dfrac{\sin^2 u}{\cos u}$	
$\dfrac{1 - \sin^2 u}{\cos u}$	
$\dfrac{\cos^2 u}{\cos u}$	
$\cos u$	

5. *Method 1:*

Left side	Right side
$\dfrac{\sin t}{1 + \cos t}$	$\dfrac{1 - \cos t}{\sin t}$
$\dfrac{\sin t}{1 + \cos t} \cdot \dfrac{1 - \cos t}{1 - \cos t}$	
$\dfrac{(\sin t)(1 - \cos t)}{1 - \cos^2 t}$	
$\dfrac{(\sin t)(1 - \cos t)}{\sin^2 t}$	
$\dfrac{1 - \cos t}{\sin t}$	

6. *Method 2:*

Left side	Right side
$\cot^2 x - \cos^2 x$	$\cos^2 x \cot^2 x$
$\dfrac{\cos^2 x}{\sin^2 x} - \cos^2 x$	$\cos^2 x \cdot \dfrac{\cos^2 x}{\sin^2 x}$
$\dfrac{\cos^2 x - \sin^2 x \cos^2 x}{\sin^2 x}$	$\dfrac{\cos^4 x}{\sin^2 x}$
$\dfrac{\cos^2 x\,(1 - \sin^2 x)}{\sin^2 x}$	
$\dfrac{\cos^4 x}{\sin^2 x}$	

7. *Method 2:*

Left side	Right side
$\dfrac{\sin 2\theta + \sin\theta}{\cos 2\theta + \cos\theta + 1}$	$\tan\theta$
$\dfrac{2\sin\theta\cos\theta + \sin\theta}{(2\cos^2\theta - 1) + \cos\theta + 1}$	$\dfrac{\sin\theta}{\cos\theta}$
$\dfrac{\sin\theta\,(2\cos\theta + 1)}{2\cos^2\theta + \cos\theta}$	
$\dfrac{\sin\theta\,(2\cos\theta + 1)}{\cos\theta\,(2\cos\theta + 1)}$	
$\dfrac{\sin\theta}{\cos\theta}$	

8.

$\sin u \cdot \cos v$	$\frac{1}{2}[\sin(u + v) + \sin(u - v)]$
	$\frac{1}{2}[\sin u\cos v + \cos u\sin v + \sin u\cos v - \cos u\sin v]$
	$\frac{1}{2}(2\sin u\cos v)$
	$\sin u \cdot \cos v$

Exercise Set 7.5, pp. 466–468

1.

$\csc x - \cos x\cot x$	$\sin x$
$\dfrac{1}{\sin x} - \cos x\dfrac{\cos x}{\sin x}$	
$\dfrac{1 - \cos^2 x}{\sin x}$	
$\dfrac{\sin^2 x}{\sin x}$	
$\sin x$	

3.

$\dfrac{1 + \cos\theta}{\sin\theta} + \dfrac{\sin\theta}{\cos\theta}$	$\dfrac{\cos\theta + 1}{\sin\theta\cos\theta}$
$\dfrac{1 + \cos\theta}{\sin\theta} \cdot \dfrac{\cos\theta}{\cos\theta} + \dfrac{\sin\theta \cdot \sin\theta}{\cos\theta\sin\theta}$	$\dfrac{1 + \cos\theta}{\sin\theta\cos\theta}$
$\dfrac{\cos\theta + \cos^2\theta + \sin^2\theta}{\sin\theta\cos\theta}$	
$\dfrac{1 + \cos\theta}{\sin\theta\cos\theta}$	

5.

$\dfrac{1 - \sin x}{\cos x}$	$\dfrac{\cos x}{1 + \sin x}$
$\dfrac{1 - \sin x}{\cos x} \cdot \dfrac{\cos x}{\cos x}$	$\dfrac{\cos x}{1 + \sin x} \cdot \dfrac{1 - \sin x}{1 - \sin x}$
$\dfrac{\cos x - \sin x\cos x}{\cos^2 x}$	$\dfrac{\cos x - \sin x\cos x}{1 - \sin^2 x}$
	$\dfrac{\cos x - \sin x\cos x}{\cos^2 x}$

7.

$\dfrac{1 + \tan \theta}{1 + \cot \theta}$	$\dfrac{\sec \theta}{\csc \theta}$
$\dfrac{1 + \dfrac{\sin \theta}{\cos \theta}}{1 + \dfrac{\cos \theta}{\sin \theta}}$	$\dfrac{\sin \theta}{\cos \theta}$
$\dfrac{\dfrac{\cos \theta + \sin \theta}{\cos \theta}}{\dfrac{\sin \theta + \cos \theta}{\sin \theta}}$	$\tan \theta$
$\dfrac{\cos \theta + \sin \theta}{\cos \theta} \cdot \dfrac{\sin \theta}{\sin \theta + \cos \theta}$	
$\dfrac{\sin \theta}{\cos \theta}$	
$\tan \theta$	

9.

$\dfrac{\sin x + \cos x}{\sec x + \csc x}$	$\dfrac{\sin x}{\sec x}$
$\dfrac{\sin x + \cos x}{\dfrac{1}{\cos x} + \dfrac{1}{\sin x}}$	
$\dfrac{\sin x + \cos x}{\dfrac{\sin x + \cos x}{\sin x \cos x}}$	
$(\sin x + \cos x) \cdot \dfrac{\sin x \cos x}{\sin x + \cos x}$	
$\sin x \cos x$	
$\dfrac{\sin x}{\sec x}$	

11.

$\dfrac{1 + \tan \theta}{1 - \tan \theta} + \dfrac{1 + \cot \theta}{1 - \cot \theta}$	0
$\dfrac{1 + \dfrac{\sin \theta}{\cos \theta}}{1 - \dfrac{\sin \theta}{\cos \theta}} + \dfrac{1 + \dfrac{\cos \theta}{\sin \theta}}{1 - \dfrac{\cos \theta}{\sin \theta}}$	
$\dfrac{\dfrac{\cos \theta + \sin \theta}{\cos \theta}}{\dfrac{\cos \theta - \sin \theta}{\cos \theta}} + \dfrac{\dfrac{\sin \theta + \cos \theta}{\sin \theta}}{\dfrac{\sin \theta - \cos \theta}{\sin \theta}}$	
$\dfrac{\cos \theta + \sin \theta}{\cos \theta} \cdot \dfrac{\cos \theta}{\cos \theta - \sin \theta} + \dfrac{\sin \theta + \cos \theta}{\sin \theta} \cdot \dfrac{\sin \theta}{\sin \theta - \cos \theta}$	
$\dfrac{\cos \theta + \sin \theta}{\cos \theta - \sin \theta} + \dfrac{\sin \theta + \cos \theta}{\sin \theta - \cos \theta}$	
$\dfrac{\cos \theta + \sin \theta}{\cos \theta - \sin \theta} - \dfrac{\cos \theta + \sin \theta}{\cos \theta - \sin \theta}$	
0	

13.

$\dfrac{1 + \cos 2\theta}{\sin 2\theta}$	$\cot \theta$
$\dfrac{1 + 2\cos^2 \theta - 1}{2 \sin \theta \cos \theta}$	$\dfrac{\cos \theta}{\sin \theta}$
$\dfrac{\cos \theta}{\sin \theta}$	

15.

$\sec 2\theta$	$\dfrac{\sec^2 \theta}{2 - \sec^2 \theta}$
$\dfrac{1}{\cos 2\theta}$	$\dfrac{1 + \tan^2 \theta}{2 - 1 - \tan^2 \theta}$
$\dfrac{1}{\cos^2 \theta - \sin^2 \theta}$	$\dfrac{1 + \tan^2 \theta}{1 - \tan^2 \theta}$
	$\dfrac{1 + \dfrac{\sin^2 \theta}{\cos^2 \theta}}{1 - \dfrac{\sin^2 \theta}{\cos^2 \theta}}$
	$\dfrac{\cos^2 \theta + \sin^2 \theta}{\cos^2 \theta - \sin^2 \theta}$
	$\dfrac{1}{\cos^2 \theta - \sin^2 \theta}$

17.

$\dfrac{\sin (\alpha + \beta)}{\cos \alpha \cos \beta}$	$\tan \alpha + \tan \beta$
$\dfrac{\sin \alpha \cos \beta + \cos \alpha \sin \beta}{\cos \alpha \cos \beta}$	$\dfrac{\sin \alpha}{\cos \alpha} + \dfrac{\sin \beta}{\cos \beta}$
	$\dfrac{\sin \alpha \cos \beta + \cos \alpha \sin \beta}{\cos \alpha \cos \beta}$

19.

$1 - \cos 5\theta \cos 3\theta - \sin 5\theta \cos 3\theta$	$2 \sin^2 \theta$
$1 - [\cos 5\theta \cos 3\theta + \sin 5\theta \sin 3\theta]$	$1 - \cos 2\theta$
$1 - \cos (5\theta - 3\theta)$	
$1 - \cos 2\theta$	

21.

$\dfrac{\tan \theta + \sin \theta}{2 \tan \theta}$	$\cos^2 \dfrac{\theta}{2}$
$\dfrac{1}{2}\left[\dfrac{\dfrac{\sin \theta + \sin \theta \cos \theta}{\cos \theta}}{\dfrac{\sin \theta}{\cos \theta}}\right]$	$\dfrac{1 + \cos \theta}{2}$
$\dfrac{1}{2}\left[\dfrac{\sin \theta\,(1 + \cos \theta)}{\cos \theta} \cdot \dfrac{\cos \theta}{\sin \theta}\right]$	
$\dfrac{1 + \cos \theta}{2}$	

23.

$\cos^4 x - \sin^4 x$	$\cos 2x$
$(\cos^2 x - \sin^2 x)(\cos^2 x + \sin^2 x)$	$\cos^2 x - \sin^2 x$
$\cos^2 x - \sin^2 x$	

25.

$\frac{\tan 3\theta - \tan \theta}{1 + \tan 3\theta \tan \theta}$	$\frac{2 \tan \theta}{1 - \tan^2 \theta}$
$\tan (3\theta - \theta)$	$\tan 2\theta$
$\tan 2\theta$	

27.

$\frac{\cos^3 x - \sin^3 x}{\cos x - \sin x}$	$\frac{2 + \sin 2x}{2}$
$\frac{(\cos x - \sin x)(\cos^2 x + \cos x \sin x + \sin^2 x)}{\cos x - \sin x}$	$\frac{2 + 2 \sin x \cos x}{2}$
$1 + \cos x \sin x$	$1 + \sin x \cos x$

29.

$\sin (\alpha + \beta) \sin (\alpha - \beta)$	$\sin^2 \alpha - \sin^2 \beta$
$(\sin \alpha \cos \beta + \cos \alpha \sin \beta)(\sin \alpha \cos \beta - \cos \alpha \sin \beta)$	$1 - \cos^2 \alpha - (1 - \cos^2 \beta)$
$\sin^2 \alpha \cos^2 \beta - \cos^2 \alpha \sin^2 \beta$	$\cos^2 \beta - \cos^2 \alpha$
$\cos^2 \beta (1 - \cos^2 \alpha) - \cos^2 \alpha (1 - \cos^2 \beta)$	
$\cos^2 \beta - \cos^2 \alpha \cos^2 \beta - \cos^2 \alpha + \cos^2 \alpha \cos^2 \beta$	
$\cos^2 \beta - \cos^2 \alpha$	

31.

$\cos (\alpha + \beta) + \cos (\alpha - \beta)$	$2 \cos \alpha \cos \beta$
$\cos \alpha \cos \beta - \sin \alpha \sin \beta + \cos \alpha \cos \beta + \sin \alpha \sin \beta$	
$2 \cos \alpha \cos \beta$	

33.

$\sin^2 x - \cos^2 x$	$1 - 2 \cos^2 x$
$-(\cos^2 x - \sin^2 x)$	$-(2 \cos^2 x - 1)$
$-\cos 2x$	$-\cos 2x$

35.

$\sin^2 \theta$	$\cos^2 \theta (\sec^2 \theta - 1)$
	$\cos^2 \theta \cdot \tan^2 \theta$
	$\cos^2 \theta \cdot \frac{\sin^2 \theta}{\cos^2 \theta}$
	$\sin^2 \theta$

37.

$\frac{\tan x}{\sec x - \cos x}$	$\frac{\sec x}{\tan x}$
$\frac{\frac{\sin x}{\cos x}}{\frac{1}{\cos x} - \cos x}$	$\frac{\frac{1}{\cos x}}{\frac{\sin x}{\cos x}}$
$\frac{\frac{\sin x}{\cos x}}{\frac{1 - \cos^2 x}{\cos x}}$	$\frac{1}{\cos x} \cdot \frac{\cos x}{\sin x}$
$\frac{\sin x}{1 - \cos^2 x}$	$\frac{1}{\sin x}$
$\frac{\sin x}{\sin^2 x}$	
$\frac{1}{\sin x}$	

39.

$\frac{\cos \theta + \sin \theta}{\cos \theta}$	$1 + \tan \theta$
$\frac{\cos \theta}{\cos \theta} + \frac{\sin \theta}{\cos \theta}$	
$1 + \tan \theta$	

41.

$\frac{\tan^2 x}{1 + \tan^2 x}$	$\sin^2 x$
$\frac{\tan^2 x}{\sec^2 x}$	
$\frac{\frac{\sin^2 x}{\cos^2 x}}{\frac{1}{\cos^2 x}}$	
$\sin^2 x$	

43.

$\frac{\cot x + \tan x}{\csc x}$	$\sec x$
$\frac{\frac{\cos x}{\sin x} + \frac{\sin x}{\cos x}}{\frac{1}{\sin x}}$	$\frac{1}{\cos x}$
$\frac{\cos^2 x + \sin^2 x}{\sin x \cos x} \cdot \frac{\sin x}{1}$	
$\frac{1}{\cos x}$	

45.

$\frac{\tan \theta + \cot \theta}{\csc \theta}$	$\sec \theta$
$\frac{\frac{\sin \theta}{\cos \theta} + \frac{\cos \theta}{\sin \theta}}{\frac{1}{\sin \theta}}$	$\frac{1}{\cos \theta}$
$\frac{\sin^2 \theta + \cos^2 \theta}{\cos \theta \sin \theta} \cdot \frac{\sin \theta}{1}$	
$\frac{1}{\cos \theta}$	

47.

$\frac{\sin x}{1 + \cos x} + \frac{1 + \cos x}{\sin x}$	$2 \csc x$
$\frac{\sin^2 x + (1 + \cos x)^2}{(1 + \cos x) \sin x}$	$\frac{2}{\sin x}$
$\frac{\sin^2 x + 1 + 2 \cos x + \cos^2 x}{(1 + \cos x) \sin x}$	
$\frac{2 + 2 \cos x}{(1 + \cos x) \sin x}$	
$\frac{2(1 + \cos x)}{(1 + \cos x) \sin x}$	
$\frac{2}{\sin x}$	

49.

$\cos \theta (1 + \csc \theta) - \cot \theta$	$\cos \theta$
$\cos \theta \left(1 + \frac{1}{\sin \theta}\right) - \frac{\cos \theta}{\sin \theta}$	
$\cos \theta + \frac{\cos \theta}{\sin \theta} - \frac{\cos \theta}{\sin \theta}$	
$\cos \theta$	

51.

$\dfrac{\tan x + \cot x}{\sec x + \csc x}$	$\dfrac{1}{\cos x + \sin x}$
$\dfrac{\dfrac{\sin x}{\cos x} + \dfrac{\cos x}{\sin x}}{\dfrac{1}{\cos x} + \dfrac{1}{\sin x}}$	
$\dfrac{\dfrac{\sin^2 x + \cos^2 x}{\cos x \sin x}}{\dfrac{\sin x + \cos x}{\cos x \sin x}}$	
$\dfrac{\sin^2 x + \cos^2 x}{\sin x + \cos x}$	
$\dfrac{1}{\cos x + \sin x}$	

53.

$(\sec\theta - \tan\theta)(1 + \csc\theta)$	$\cot\theta$
$\left(\dfrac{1}{\cos\theta} - \dfrac{\sin\theta}{\cos\theta}\right)\left(1 + \dfrac{1}{\sin\theta}\right)$	$\dfrac{\cos\theta}{\sin\theta}$
$\left(\dfrac{1 - \sin\theta}{\cos\theta}\right)\left(\dfrac{\sin\theta + 1}{\sin\theta}\right)$	
$\dfrac{1 - \sin^2\theta}{\cos\theta\sin\theta}$	
$\dfrac{\cos^2\theta}{\cos\theta\sin\theta}$	
$\dfrac{\cos\theta}{\sin\theta}$	

55.

$(\sec x + \tan x)(1 - \sin x)$	$\cos x$
$\left(\dfrac{1}{\cos x} + \dfrac{\sin x}{\cos x}\right)(1 - \sin x)$	
$\left(\dfrac{1 + \sin x}{\cos x}\right)(1 - \sin x)$	
$\dfrac{1 - \sin^2 x}{\cos x}$	
$\dfrac{\cos^2 x}{\cos x}$	
$\cos x$	

57.

$\cos^2\theta - \sin^2\theta$	$\cos^4\theta - \sin^4\theta$
	$(\cos^2\theta + \sin^2\theta)\cdot(\cos^2\theta - \sin^2\theta)$
	$\cos^2\theta - \sin^2\theta$

59.

$\dfrac{\cos\theta}{1 - \cos\theta}$	$\dfrac{1 + \sec\theta}{\tan^2\theta}$
	$\dfrac{1 + \sec\theta}{\sec^2\theta - 1}$
	$\dfrac{1}{\sec\theta - 1}$
	$\dfrac{1}{\dfrac{1}{\cos\theta} - 1}$
	$\dfrac{1}{\dfrac{1 - \cos\theta}{\cos\theta}}$
	$\dfrac{\cos\theta}{1 - \cos\theta}$

61.

$\dfrac{\sin x - \cos x}{\cos^2 x}$	$\dfrac{\tan^2 x - 1}{\sin x + \cos x}$
	$\dfrac{\dfrac{\sin^2 x}{\cos^2 x} - 1}{\sin x + \cos x}$
	$\dfrac{\sin^2 x - \cos^2 x}{\cos^2 x}\cdot\dfrac{1}{\sin x + \cos x}$
	$\dfrac{\sin x - \cos x}{\cos^2 x}$

63.

$1 + \sec^4\theta$	$\tan^4\theta + 2\sec^2\theta$
$1 + (1 + \tan^2\theta)^2$	$\tan^4\theta + 2(1 + \tan^2\theta)$
$1 + 1 + 2\tan^2\theta + \tan^4\theta$	$\tan^4\theta + 2\tan^2\theta + 2$
$\tan^4\theta + 2\tan^2\theta + 2$	

65.

$\dfrac{1 + \tan x}{1 - \tan x}$	$\dfrac{\cot x + 1}{\cot x - 1}$
$\dfrac{1 + \dfrac{1}{\cot x}}{1 - \dfrac{1}{\cot x}}$	
$\dfrac{\dfrac{\cot x + 1}{\cot x}}{\dfrac{\cot x - 1}{\cot x}}$	
$\dfrac{\cot x + 1}{\cot x - 1}$	

67.

$\dfrac{\sin\theta\tan\theta}{\sin\theta + \tan\theta}$	$\dfrac{\tan\theta - \sin\theta}{\tan\theta\sin\theta}$
$\dfrac{\sin\theta\cdot\dfrac{\sin\theta}{\cos\theta}}{\sin\theta + \dfrac{\sin\theta}{\cos\theta}}$	$\dfrac{\dfrac{\sin\theta}{\cos\theta} - \sin\theta}{\dfrac{\sin\theta}{\cos\theta}\cdot\sin\theta}$
$\dfrac{\dfrac{\sin^2\theta}{\cos\theta}}{\dfrac{\sin\theta\cos\theta + \sin\theta}{\cos\theta}}$	$\dfrac{\dfrac{\sin\theta - \sin\theta\cos\theta}{\cos\theta}}{\dfrac{\sin^2\theta}{\cos\theta}}$
$\dfrac{\sin^2\theta}{\sin\theta\cos\theta + \sin\theta}$	$\dfrac{\sin\theta - \sin\theta\cos\theta}{\sin^2\theta}$
$\dfrac{\sin\theta}{\cos\theta + 1}$	$\dfrac{1 - \cos\theta}{\sin\theta}$
$\dfrac{\sin\theta}{\cos\theta + 1}\cdot\dfrac{1 - \cos\theta}{1 - \cos\theta}$	
$\dfrac{\sin\theta\,(1 - \cos\theta)}{1 - \cos^2\theta}$	
$\dfrac{\sin\theta\,(1 - \cos\theta)}{\sin^2\theta}$	
$\dfrac{1 - \cos\theta}{\sin\theta}$	

69.

$\dfrac{\cos x + 1}{\sin x} + \dfrac{\sin x}{\cos x + 1}$	$\dfrac{2}{\sin x}$
$\dfrac{(\cos x + 1)^2 + \sin^2 x}{\sin x\,(\cos x + 1)}$	
$\dfrac{\cos^2 x + 2\cos x + 1 + \sin^2 x}{\sin x\,(\cos x + 1)}$	
$\dfrac{2\cos x + 2}{\sin x\,(\cos x + 1)}$	
$\dfrac{2(\cos x + 1)}{\sin x\,(\cos x + 1)}$	
$\dfrac{2}{\sin x}$	

71.

$\sec\theta\csc\theta + \dfrac{\cot\theta}{\tan\theta - 1}$	$\dfrac{\tan\theta}{1 - \cot\theta} - 1$
$\dfrac{1}{\cos\theta}\cdot\dfrac{1}{\sin\theta} + \dfrac{\frac{\cos\theta}{\sin\theta}}{\frac{\sin\theta}{\cos\theta} - 1}$	$\dfrac{\frac{\sin\theta}{\cos\theta}}{1 - \frac{\cos\theta}{\sin\theta}} - 1$
$\dfrac{1}{\cos\theta\sin\theta} + \dfrac{\cos\theta}{\sin\theta}\cdot\dfrac{\cos\theta}{\sin\theta - \cos\theta}$	$\dfrac{\sin\theta}{\cos\theta}\cdot\dfrac{\sin\theta}{\sin\theta - \cos\theta} - 1$
$\dfrac{1}{\cos\theta\sin\theta} + \dfrac{\cos^2\theta}{\sin\theta\,(\sin\theta - \cos\theta)}$	$\dfrac{\sin^2\theta - \cos\theta\,(\sin\theta - \cos\theta)}{\cos\theta\,(\sin\theta - \cos\theta)}$
$\dfrac{\sin\theta - \cos\theta + \cos^3\theta}{\cos\theta\sin\theta\,(\sin\theta - \cos\theta)}$	$\dfrac{\sin^2\theta - \cos\theta\sin\theta + \cos^2\theta}{\cos\theta\,(\sin\theta - \cos\theta)}$
$\dfrac{\sin\theta - \cos\theta\,(1 - \cos^2\theta)}{\cos\theta\sin\theta\,(\sin\theta - \cos\theta)}$	$\dfrac{1 - \cos\theta\sin\theta}{\cos\theta\,(\sin\theta - \cos\theta)}$
$\dfrac{\sin\theta - \cos\theta\cdot\sin^2\theta}{\cos\theta\sin\theta\,(\sin\theta - \cos\theta)}$	
$\dfrac{1 - \cos\theta\sin\theta}{\cos\theta\,(\sin\theta - \cos\theta)}$	

73.

$\cot x + \csc x$	$\dfrac{\sin x}{1 - \cos x}$
$\dfrac{\cos x}{\sin x} + \dfrac{1}{\sin x}$	$\dfrac{\sin x}{1 - \cos x}\cdot\dfrac{1 + \cos x}{1 + \cos x}$
$\dfrac{\cos x + 1}{\sin x}$	$\dfrac{\sin x\,(1 + \cos x)}{1 - \cos^2 x}$
	$\dfrac{\sin x\,(1 + \cos x)}{\sin^2 x}$
	$\dfrac{1 + \cos x}{\sin x}$

75.

$\tan\theta + \cot\theta$	$\dfrac{1}{\cot\theta\sin^2\theta}$
$\dfrac{\sin\theta}{\cos\theta} + \dfrac{\cos\theta}{\sin\theta}$	$\dfrac{1}{\frac{\cos\theta}{\sin\theta}\cdot\sin^2\theta}$
$\dfrac{\sin^2\theta + \cos^2\theta}{\cos\theta\sin\theta}$	$\dfrac{1}{\cos\theta\sin\theta}$
$\dfrac{1}{\cos\theta\sin\theta}$	

77.

$2\cos^2 x - 1$	$\cos^4 x - \sin^4 x$
$2\cos^2 x - (\sin^2 x + \cos^2 x)$	$(\cos^2 x + \sin^2 x)(\cos^2 x - \sin^2 x)$
$\cos^2 x - \sin^2 x$	$\cos^2 x - \sin^2 x$

79.

$(\cos x + \sin x)(1 - \sin x\cos x)$	$\cos^3 x + \sin^3 x$
	$(\cos x + \sin x)(\cos^2 x - \cos x\sin x + \sin^2 x)$
	$(\cos x + \sin x)(1 - \sin x\cos x)$

81.

$\cos u\cdot\sin v$	$\frac{1}{2}[\sin(u + v) - \sin(u - v)]$
	$\frac{1}{2}[\sin u\cos v + \cos u\sin v - \sin u\cos v + \cos u\sin v]$
	$\frac{1}{2}(2\cos u\sin v)$
	$\cos u\cdot\sin v$

The other formulas follow similarly.

83. $\sin 3\theta - \sin 5\theta = 2\cos\dfrac{8\theta}{2}\sin\dfrac{-2\theta}{2} = -2\cos 4\theta\sin\theta$

85. $\sin 8\theta + \sin 5\theta = 2\sin\dfrac{13\theta}{2}\cos\dfrac{3\theta}{2}$

87. $\sin 7u\sin 5u = \frac{1}{2}(\cos 2u - \cos 12u)$

89. $7\cos\theta\sin 7\theta = \frac{7}{2}[\sin 8\theta - \sin(-6\theta)] = \frac{7}{2}(\sin 8\theta + \sin 6\theta)$

91. $\cos 55^\circ\sin 25^\circ = \frac{1}{2}(\sin 80^\circ - \sin 30^\circ) = \frac{1}{2}\sin 80^\circ - \frac{1}{4}$

93.

$\sin 4\theta + \sin 6\theta$	$\cot\theta\,(\cos 4\theta - \cos 6\theta)$
$2\sin\dfrac{10\theta}{2}\cos\dfrac{-2\theta}{2}$	$\dfrac{\cos\theta}{\sin\theta}\left(2\sin\dfrac{10\theta}{2}\sin\dfrac{2\theta}{2}\right)$
$2\sin 5\theta\cos(-\theta)$	$\dfrac{\cos\theta}{\sin\theta}(2\sin 5\theta\sin\theta)$
$2\sin 5\theta\cos\theta$	$2\sin 5\theta\cos\theta$

95.

$\cot 4x\,(\sin x + \sin 4x + \sin 7x)$	$\cos x + \cos 4x + \cos 7x$
$\dfrac{\cos 4x}{\sin 4x}\left(\sin 4x + 2\sin\dfrac{8x}{2}\cos\dfrac{-6x}{2}\right)$	$\cos 4x + 2\cos\dfrac{8x}{2}\cdot\cos\dfrac{6x}{2}$
$\dfrac{\cos 4x}{\sin 4x}(\sin 4x + 2\sin 4x\cos 3x)$	$\cos 4x + 2\cos 4x\cdot\cos 3x$
$\cos 4x\,(1 + 2\cos 3x)$	$\cos 4x\,(1 + 2\cos 3x)$

97.

$\cot\dfrac{x + y}{2}$	$\dfrac{\sin y - \sin x}{\cos x - \cos y}$
$\dfrac{\cos\frac{x + y}{2}}{\sin\frac{x + y}{2}}$	$\dfrac{2\cos\frac{x + y}{2}\sin\frac{y - x}{2}}{2\sin\frac{x + y}{2}\sin\frac{y - x}{2}}$
	$\dfrac{\cos\frac{x + y}{2}}{\sin\frac{x + y}{2}}$

99.

$\tan\frac{\theta+\phi}{2}(\sin\theta-\sin\phi)$	$\tan\frac{\theta-\phi}{2}(\sin\theta+\sin\phi)$
$\frac{\sin\frac{\theta+\phi}{2}}{\cos\frac{\theta+\phi}{2}}\left(2\cos\frac{\theta+\phi}{2}\sin\frac{\theta-\phi}{2}\right)$	$\frac{\sin\frac{\theta-\phi}{2}}{\cos\frac{\theta-\phi}{2}}\left(2\sin\frac{\theta+\phi}{2}\cos\frac{\theta-\phi}{2}\right)$
$2\sin\frac{\theta+\phi}{2}\cdot\sin\frac{\theta-\phi}{2}$	$2\sin\frac{\theta+\phi}{2}\cdot\sin\frac{\theta-\phi}{2}$

101. $\ln|\sec x| = \ln\left|\frac{1}{\cos x}\right| = \ln\frac{1}{|\cos x|} = \ln(|\cos x|)^{-1} = -\ln|\cos x|$ **103.** $\ln|\tan x| = \ln\left|\frac{\sin x}{\cos x}\right| = \ln\frac{|\sin x|}{|\cos x|} = \ln|\sin x| - \ln|\cos x|$ **105.** $\ln e^{\sin t} = (\sin t)(\ln e) = \sin t\cdot 1 = \sin t$

107.

$\ln\lvert\csc\theta-\cot\theta\rvert$	$-\ln\lvert\csc\theta+\cot\theta\rvert$
$\ln\left\lvert\frac{1}{\sin\theta}-\frac{\cos\theta}{\sin\theta}\right\rvert$	$-\ln\left\lvert\frac{1}{\sin\theta}+\frac{\cos\theta}{\sin\theta}\right\rvert$
$\ln\left\lvert\frac{1-\cos\theta}{\sin\theta}\right\rvert$	$-\ln\left\lvert\frac{1+\cos\theta}{\sin\theta}\right\rvert$
$\ln\left\lvert\frac{1-\cos\theta}{\sin\theta}\cdot\frac{1+\cos\theta}{1+\cos\theta}\right\rvert$	$\ln\left\lvert\frac{1+\cos\theta}{\sin\theta}\right\rvert^{-1}$
$\ln\left\lvert\frac{1-\cos^2\theta}{\sin\theta(1+\cos\theta)}\right\rvert$	$\ln\left\lvert\frac{\sin\theta}{1+\cos\theta}\right\rvert$
$\ln\left\lvert\frac{\sin^2\theta}{\sin\theta(1+\cos\theta)}\right\rvert$	
$\ln\left\lvert\frac{\sin\theta}{1+\cos\theta}\right\rvert$	

109.
$$\begin{aligned}\log(\cos x-\sin x)+\log(\cos x+\sin x) &= \log[(\cos x-\sin x)(\cos x+\sin x)]\\ &= \log(\cos^2 x-\sin^2 x) = \log\cos 2x\end{aligned}$$

111.
$$\frac{1}{\omega C(\tan\theta+\tan\phi)} = \frac{1}{\omega C\left(\frac{\sin\theta}{\cos\theta}+\frac{\sin\phi}{\cos\phi}\right)} = \frac{1}{\omega C\left(\frac{\sin\theta\cos\phi+\sin\phi\cos\theta}{\cos\theta\cos\phi}\right)} = \frac{\cos\theta\cos\phi}{\omega C\sin(\theta+\phi)}$$

Margin Exercises, Section 7.6

1.

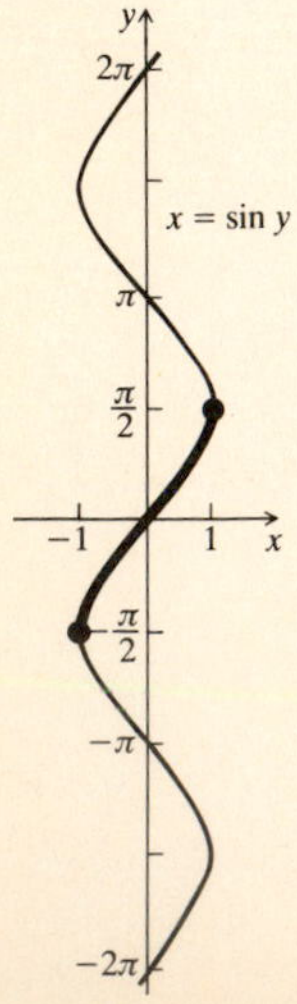

2.

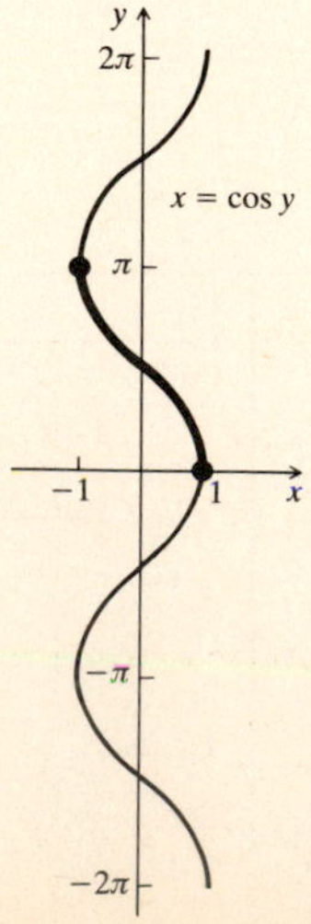

3.

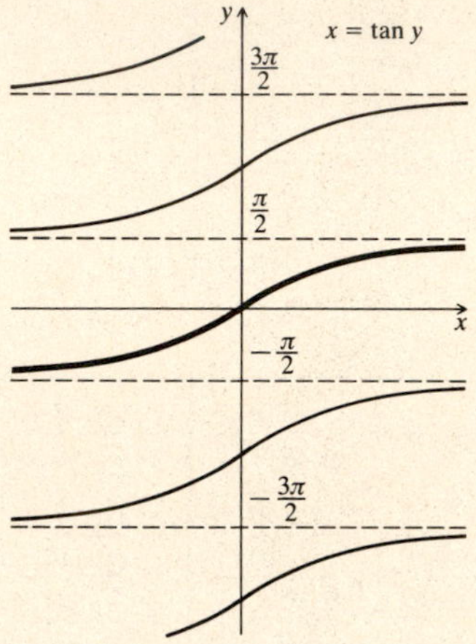

4.

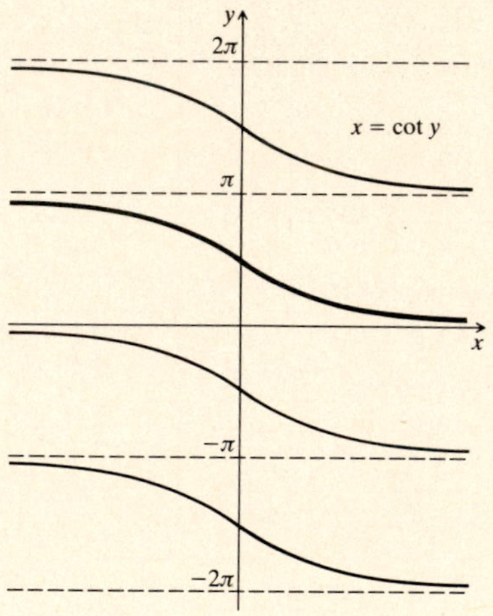

5. $\frac{\pi}{3}$, or 60° **6.** $\frac{3\pi}{4}$, or 135° **7.** $\frac{3\pi}{4}$, or 135° **8.** $-\frac{\pi}{4}$, or −45° **9.** 50.76°, or 50°46′ **10.** −5.86°, or −5°51′ **11.** 53.38°, or 53°23′ **12.** −87.45°, or −87°27′

Exercise Set 7.6, pp. 474–475

1. $\frac{\pi}{4}$, or 45° **3.** $\frac{\pi}{4}$, or 45° **5.** $-\frac{\pi}{4}$, or −45° **7.** $\frac{3\pi}{4}$, or 135° **9.** $\frac{\pi}{3}$, or 60° **11.** $\frac{\pi}{4}$, or 45° **13.** $-\frac{\pi}{6}$, or −30° **15.** $\frac{3\pi}{4}$, or 135° **17.** 0, or 0° **19.** $\frac{\pi}{2}$, or 90° **21.** $-\frac{\pi}{4}$, or −45° **23.** $\frac{\pi}{3}$, or 60° **25.** $-\frac{\pi}{3}$, or −60° **27.** $\frac{3\pi}{4}$, or 135° **29.** $-\frac{\pi}{6}$, or −30° **31.** $\frac{2\pi}{3}$, or 120° **33.** 23° **35.** 38.31°, or 38°18′ **37.** 36.97°, or 36°58′ **39.** 168.82°, or 168°49′ **41.** 20.17°, or 20°10′ **43.** −38.33°, or −38°20′ **45.** 31.09°, or 31°5′ **47.** −9.17°, or −9°10′ **49.** 13.50°, or 13°30′ **51.** −39.50°, or −39°30′ **53.** −62.70°, or −62°42′ **55.** −22.23°, or −22°14′ **57.** −9.56°, or −9°34′ **59.** −2.49°, or −2°29′

61.

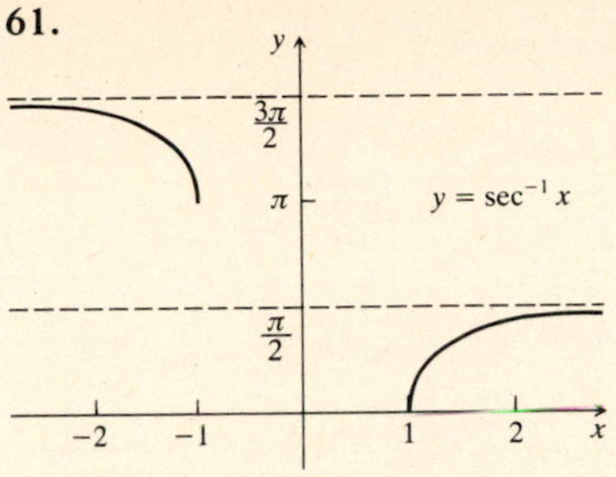

63. Let $x = \frac{\sqrt{2}}{2}$. Then $\sin^{-1}\frac{\sqrt{2}}{2} = \frac{\pi}{4} \approx 0.7854$, and $\left(\sin\frac{\sqrt{2}}{2}\right)^{-1} =$ $(0.6496)^{-1} = 1.5393$. 65. Let $x = 1$. Then $\tan^{-1} 1 = \frac{\pi}{4} \approx 0.7854$, and $(\tan 1)^{-1} \approx (1.5574)^{-1} \approx 0.6421$. 67. Let $x = \frac{\sqrt{2}}{2}$. Then $\tan^{-1}\frac{\sqrt{2}}{2} \approx 0.6155$ and $\dfrac{\sin^{-1}\frac{\sqrt{2}}{2}}{\cos^{-1}\frac{\sqrt{2}}{2}} = \frac{\pi/4}{\pi/4} = 1$.

69. $\tan\theta = \frac{50}{b}$, or $\theta = \tan^{-1}\frac{50}{b}$ 71. (a) 3.1415927; (b) π

Margin Exercises, Section 7.7

1. $\frac{1}{2}$ 2. 1 3. $\frac{\sqrt{2}}{2}$ 4. The expression cannot be evaluated. 5. $\frac{2\pi}{3}$ 6. $-\frac{\pi}{4}$ 7. $\frac{\pi}{6}$ 8. $\frac{2\pi}{3}$ 9. $\frac{\sqrt{2}}{2}$ 10. $\frac{\sqrt{3}}{2}$ 11. $\frac{\pi}{3}$ 12. $\frac{\pi}{2}$ 13. 1 14. $\frac{3}{\sqrt{b^2+9}}$ 15. $\frac{2}{\sqrt{5}}$ 16. $\frac{t}{\sqrt{1-t^2}}$ 17. 1 18. $\frac{1}{3}$

Exercise Set 7.7, pp. 479–480

1. 0.3 3. −4.2 5. $\frac{\pi}{3}$ 7. $-\frac{\pi}{4}$ 9. $\frac{\pi}{5}$ 11. $-\frac{\pi}{3}$ 13. $\frac{\sqrt{3}}{2}$ 15. $\frac{1}{2}$ 17. 1 19. $\frac{\pi}{6}$ 21. $\frac{\pi}{3}$ 23. $\frac{\pi}{2}$ 25. $\frac{x}{\sqrt{x^2+4}}$ 27. $\frac{\sqrt{x^2-9}}{3}, x > 0$ 29. $\frac{\sqrt{b^2-a^2}}{a}, b > 0$ 31. $\frac{3}{\sqrt{11}}$ 33. $\frac{1}{3\sqrt{11}}$ 35. $-\frac{1}{2\sqrt{6}}$ 37. $\frac{1}{\sqrt{1+y^2}}$ 39. $\frac{1}{\sqrt{1+t^2}}$ 41. $\frac{\sqrt{1-y^2}}{y}$ 43. $\sqrt{1-x^2}$ 45. $\frac{1}{2}$ 47. $\frac{\sqrt{2}+\sqrt{3}}{2}$ 49. $\frac{24}{25}$ 51. $\frac{119}{169}$ 53. $\frac{3+4\sqrt{3}}{10}$ 55. $-\frac{\sqrt{2}}{10}$ 57. $xy + \sqrt{(1-x^2)(1-y^2)}$ 59. $y\sqrt{1-x^2} - x\sqrt{1-y^2}$ 61. 0.9861 63. $\frac{p}{3}$ 65. $\sin\theta = x$; $\cos\theta = \sqrt{1-x^2}$; $\tan\theta = \frac{x}{\sqrt{1-x^2}}$; $\cot\theta = \frac{\sqrt{1-x^2}}{x}$; $\sec\theta = \frac{1}{\sqrt{1-x^2}}$; $\csc\theta = \frac{1}{x}$ 67. $\sin\theta = \frac{x}{\sqrt{1+x^2}}$; $\cos\theta = \frac{1}{\sqrt{1+x^2}}$; $\tan\theta = x$; $\cot\theta = \frac{1}{x}$; $\sec\theta = \sqrt{1+x^2}$; $\csc\theta = \frac{\sqrt{1+x^2}}{x}$

69.

$\sin^{-1}x + \cos^{-1}x$	$\frac{\pi}{2}$
$\sin(\sin^{-1}x + \cos^{-1}x)$	$\sin\frac{\pi}{2}$
$[\sin(\sin^{-1}x)][\cos(\cos^{-1}x)] + [\cos(\sin^{-1}x)][\sin(\cos^{-1}x)]$	1
$x \cdot x + \sqrt{1-x^2} \cdot \sqrt{1-x^2}$	
$x^2 + 1 - x^2$	
1	

71.

$\sin^{-1}x$	$\tan^{-1}\frac{x}{\sqrt{1-x^2}}$
$\sin(\sin^{-1}x)$	$\sin\left(\tan^{-1}\frac{x}{\sqrt{1-x^2}}\right)$
x	x

73.

$\arcsin x$	$\arccos\sqrt{1-x^2}$
$\sin(\arcsin x)$	$\sin(\arccos\sqrt{1-x^2})$
x	x

75. $\theta = \arctan\frac{y+h}{x} - \arctan\frac{y}{x}$

Margin Exercises, Section 7.8

1. $60° + 360°k$, $300° + 360°k$; $\frac{\pi}{3} + 2k\pi$, $\frac{5\pi}{3} + 2k\pi$ 2. $\frac{\pi}{6}, \frac{5\pi}{6}, \frac{7\pi}{6}, \frac{11\pi}{6}$; 30°, 150°, 210°, 330° 3. $\frac{\pi}{6}, \frac{5\pi}{6}, \frac{7\pi}{6}, \frac{11\pi}{6}$ 4. 40°, 320° 5. 140°, 220° 6. 120°, 240°, 75.52° (or 75°31′), 284.48° (or 284°29′) 7. 120°, 240°, 90°, 270° 8. 241.66° (or 241°40′), 118.34° (or 118°20′)

Exercise Set 7.8, p. 485

1. $\frac{\pi}{3} + 2k\pi$, $\frac{2\pi}{3} + 2k\pi$ 3. $\frac{\pi}{4} + 2k\pi$, $\frac{-\pi}{4} + 2k\pi$ 5. 20.17° (or 20°10′) + $k \cdot 360°$, 159.83° (or 159°50′) + $k \cdot 360°$ 7. 236.67° (or 236°40′), 123.33° (or 123°20′) 9. $\frac{4\pi}{3}, \frac{5\pi}{3}$ 11. 123.69° (or 123°41′), 303.69° (or 303°41′) 13. $\frac{\pi}{6}, \frac{5\pi}{6}, \frac{7\pi}{6}, \frac{11\pi}{6}$ 15. $\frac{\pi}{6}, \frac{5\pi}{6}, \frac{7\pi}{6}, \frac{11\pi}{6}$ 17. $\frac{\pi}{6}, \frac{5\pi}{6}, \frac{3\pi}{2}$ 19. 0 21. 0, $\frac{\pi}{6}, \frac{5\pi}{6}$; π, $\frac{7\pi}{6}, \frac{11\pi}{6}$ 23. $\frac{\pi}{6}, \frac{5\pi}{6}$ 25. 109.47° (or 109°28′), 250.53° (or 250°32′), 120°, 240° 27. $\frac{\pi}{6}, \frac{5\pi}{6}, \pi$ 29. 0, π, $\frac{\pi}{2}, \frac{3\pi}{2}$ 31. 198.27° (or 198°16′), 341.73° (or 341°44′) 33. 70.12° (or 70°7′), 128.31° (or 128°19′), 250.12° (or 250°7′), 308.31° (or 308°19′) 35. 37.22° (or 37°13′), 169.35° (or 169°21′), 217.22° (or 217°13′), 349.35° (or 349°21′) 37. 60°, 120°, 240°, 300° 39. 60°, 240° 41. 30°, 60°, 120°, 150°, 210°, 240°, 300°, 330° 43. 0, π 45. $\frac{\pi}{2}$

47. (a)

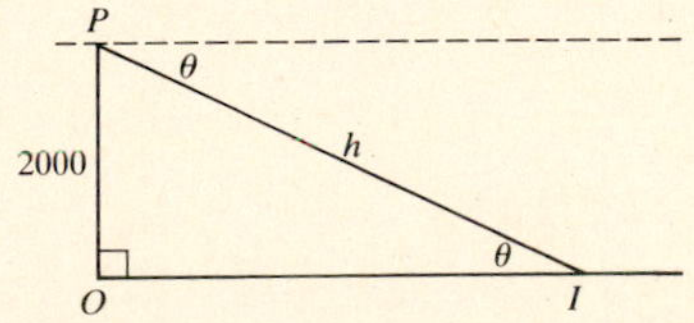

The angle of elevation is equal to the angle of depression. Thus, $\sin\theta = \frac{2000}{h}$. (b) 41.81° (or 41°49′)

Margin Exercises, Section 7.9

1. $\frac{\pi}{2}, \pi$ 2. $\frac{\pi}{2}, \pi$ 3. $\frac{\pi}{2}, \pi$ 4. $\frac{\pi}{4}, \frac{3\pi}{4}, \frac{5\pi}{4}, \frac{7\pi}{4}$ 5. $\frac{\pi}{2}, \frac{3\pi}{2}, \frac{7\pi}{6}, \frac{11\pi}{6}$ 6. $\frac{\pi}{12}, \frac{5\pi}{12}, \frac{13\pi}{12}, \frac{17\pi}{12}$ 7. $\frac{\pi}{3}, \pi, \frac{5\pi}{3}$

Exercise Set 7.9, pp. 489–490

1. $0, \pi$ **3.** $\frac{3\pi}{4}, \frac{7\pi}{4}$ **5.** $\frac{\pi}{2}, \frac{3\pi}{2}, \frac{\pi}{6}, \frac{5\pi}{6}$ **7.** $\frac{\pi}{2}, \frac{3\pi}{2}, \frac{\pi}{4}, \frac{3\pi}{4}, \frac{5\pi}{4}, \frac{7\pi}{4}$ **9.** $0, \frac{\pi}{2}, \pi, \frac{3\pi}{2}$ **11.** 0 **13.** $0, \frac{\pi}{2}, \pi, \frac{3\pi}{2}$ **15.** $\frac{\pi}{6}, \frac{5\pi}{6}, \pi$ **17.** $\frac{\pi}{6}, \frac{5\pi}{6}, \frac{7\pi}{6}, \frac{11\pi}{6}$ **19.** 63.43° (or 63°26′), 243.43° (or 243°26′), 101.31° (or 101°19′), 281.31° (or 281°19′) **21.** $\frac{\pi}{6}, \frac{\pi}{2}, \frac{5\pi}{6}, \frac{7\pi}{6}, \frac{3\pi}{2}, \frac{11\pi}{6}$ **23.** $\frac{2\pi}{3}, \frac{4\pi}{3}$ **25.** $\frac{\pi}{4}, \frac{7\pi}{4}$ **27.** $\frac{\pi}{12}, \frac{5\pi}{12}$ **29.** $\frac{\pi}{6}, \frac{3\pi}{2}$ **31.** 0.9669, 1.853, 4.108, 4.995 **33.** 0.7297, 2.412, 3.665, 5.760 **35.** $t \approx 1.24, 6.76$ **37.** 16.5° **39.** 1 **41.** 1.15, 5.65, −0.63, −0.516, etc. **43.** 0.1923

Review Exercises: Chapter 7, pp. 491–492

1. [7.3] $\cot(x - \pi) = \cot x$ **2.** [7.5] 1 **3.** [7.5] $\csc^2 x$ **4.** [7.3] $-\sin x$ **5.** [7.3] $\sin x$ **6.** [7.3] $-\cos x$ **7.** [7.1] $\tan x = \pm\sqrt{\sec^2 x - 1}$ **8.** [7.1] $\csc x$ **9.** [7.1] 1 **10.** [7.1] $\frac{\sqrt{\tan x \sec x}}{\sec x}$ **11.** [7.1] $\frac{\tan x}{\sqrt{\sec x \tan x}}$ **12.** [7.3] $\sin(90° - \theta) = 0.7314$; $\cos(90° - \theta) = 0.6820$; $\tan(90° - \theta) = 1.0724$; $\cot(90° - \theta) = 0.9325$; $\sec(90° - \theta) = 1.4663$; $\csc(90° - \theta) = 1.3673$ **13.** [7.2] $\cos x \cos \frac{3\pi}{2} - \sin x \sin \frac{3\pi}{2}$ **14.** [7.2] $\frac{\tan 45° - \tan 30°}{1 + \tan 45° \tan 30°}$ **15.** [7.2] $\cos(27° - 16°)$ or $\cos 11°$ **16.** [7.2] $\frac{-\sqrt{6} - \sqrt{2}}{4}$ **17.** [7.2] $2 - \sqrt{3}$ **18.** [7.2] 161.57° (or 161°34′) **19.** [7.4] $\sin 2\theta = \frac{24}{25}$, $\cos 2\theta = -\frac{7}{25}$, $\tan 2\theta = -\frac{24}{7}$, 2θ is in quadrant II **20.** [7.4] $\frac{1}{2}\sqrt{2 - \sqrt{2}}$ **21.** [7.4] $2 \cot \theta$

22. [7.5]

$\tan 2\theta$	$\frac{2 \tan \theta}{1 - \tan^2 \theta}$
$\frac{\sin 2\theta}{\cos 2\theta}$	$\frac{2\frac{\sin \theta}{\cos \theta}}{\frac{\cos^2 \theta}{\cos^2 \theta} - \frac{\sin^2 \theta}{\cos^2 \theta}}$
$\frac{2 \sin \theta \cos \theta}{\cos^2 \theta - \sin^2 \theta}$	$\frac{2 \sin \theta}{\cos \theta} \cdot \frac{\cos^2 \theta}{\cos^2 \theta - \sin^2 \theta}$
	$\frac{2 \sin \theta \cos \theta}{\cos^2 \theta - \sin^2 \theta}$

23. [7.5]

$\frac{\sec x - \cos x}{\tan x}$	$\sin x$
$\frac{\frac{1}{\cos x} - \cos x}{\frac{\sin x}{\cos x}}$	
$\frac{1 - \cos^2 x}{\cos x} \cdot \frac{\cos x}{\sin x}$	
$\frac{1 - \cos^2 x}{\sin x}$	
$\frac{\sin^2 x}{\sin x}$	

24. [7.3] $\sqrt{40} \sin(3x + b)$, where $\cos b = \frac{3\sqrt{10}}{10}$ and $\sin b = \frac{\sqrt{10}}{10}$ **25.** [7.6] $\frac{\pi}{6}$ **26.** [7.6] 81° **27.** [7.6] −45°, or $-\frac{\pi}{4}$ **28.** [7.6] 150°, or $\frac{5\pi}{6}$ **29.** [7.7] $\frac{7}{8}$ **30.** [7.7] $\frac{\pi}{3}$ **31.** [7.7] $\frac{\pi}{6}$ **32.** [7.7] $\frac{b}{\sqrt{b^2 + 25}}$ **33.** [7.8] $0, \pi$ **34.** [7.8] $\frac{\pi}{3}, \frac{5\pi}{3}$ **35.** [7.9] $\frac{\pi}{4}, \frac{3\pi}{4}, \frac{5\pi}{4}, \frac{7\pi}{4}$ **36.** [7.9] 0.5117, 1.845, 3.6533, 4.9861 **37.** [7.8] $\frac{\pi}{6}, \frac{5\pi}{6}, \frac{7\pi}{6}, \frac{11\pi}{6}$ **38.** [7.8] No solution in $[0, 2\pi)$.

39. [7.4]

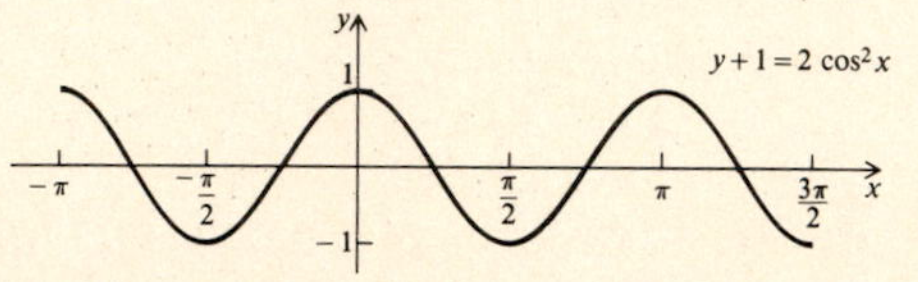

40. [7.6]

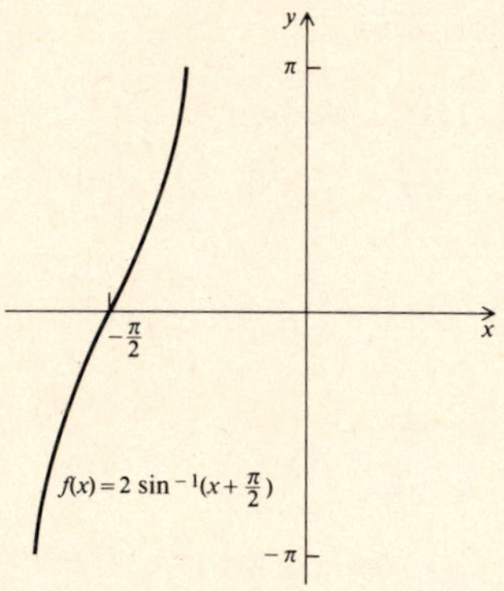

Test: Chapter 7, p. 492

1. [7.2] $\cos \pi \cos x + \sin \pi \sin x$ **2.** [7.2] $\frac{\tan 83° + \tan 15°}{1 - \tan 83° \tan 15°}$ **3.** [7.2] $\sin 35°$ **4.** [7.2] $\frac{\sqrt{2} - \sqrt{6}}{4}$ **5.** [7.2] $2 - \sqrt{3}$ **6.** [7.2] 126.87°, or 126°52′ **7.** [7.3] $\csc x$ **8.** [7.5] 1 **9.** [7.5] $\csc^2 x$ **10.** [7.3] $\cos x$ **11.** [7.3] $\sin x$ **12.** [7.3] $\sin x$ **13.** [7.1] $\csc x = \pm\sqrt{1 + \cot^2 x}$ **14.** [7.1] $\sec x$ **15.** [7.1] 1 **16.** [7.1] $\frac{\sqrt{\sec x \csc x}}{\csc x}$ **17.** [7.4] $\sin 2\theta = -\frac{120}{169}$; $\cos 2\theta = -\frac{119}{169}$; $\tan 2\theta = \frac{120}{119}$; quadrant III **18.** [7.4] $2 + \sqrt{3}$ **19.** [7.4] $2 \sin 4x$

20. [7.5]

$\frac{1 - \cos 2\theta}{\sin 2\theta}$	$\tan \theta$
$\frac{1 - (1 - 2\sin^2 \theta)}{2 \sin \theta \cos \theta}$	$\frac{\sin \theta}{\cos \theta}$
$\frac{2 \sin^2 \theta}{2 \sin \theta \cos \theta}$	
$\frac{\sin \theta}{\cos \theta}$	

21. [7.3] $5 \sin(2x + b)$, where $\cos b = \frac{4}{5}$ and $\sin b = \frac{3}{5}$ **22.** [7.6] $-\frac{\pi}{4}$ **23.** [7.6] 42°58′ **24.** [7.6] $\frac{\pi}{6}$ **25.** [7.7] $\frac{3\pi}{4}$ **26.** [7.7] $\frac{\sqrt{3}}{2}$ **27.** [7.7] $\frac{\pi}{4}$ **28.** [7.8] $\frac{7\pi}{6}, \frac{11\pi}{6}$ **29.** [7.9] $\frac{\pi}{2}$,

$\frac{3\pi}{2}, \frac{\pi}{4}, \frac{3\pi}{4}, \frac{5\pi}{4}, \frac{7\pi}{4}$ **30.** [7.8] 0.8730, 1.805, 4.015, 4.947 **31.** [7.3] $\sin(90° - \theta) = 0.8910$, $\cos(90° - \theta) = 0.4540$, $\tan(90° - \theta) = 1.963$, $\cot(90° - \theta) = 0.5095$, $\sec(90° - \theta) = 2.203$, $\csc(90° - \theta) = 1.122$ **32.** [7.8] $\frac{\pi}{2}$

CHAPTER 8

Margin Exercises, Section 8.1

1. (a) $\frac{b}{c}$; (b) $\frac{a}{c}$; (c) $\frac{b}{a}$; (d) $\frac{c}{b}$; (e) $\frac{c}{a}$; (f) $\frac{a}{b}$ **2.** 42° **3.** $a = 14.7$, $b = 13.4$ **4.** $A = 41°30'$, $a = 6.37$, $c = 9.613$ **5.** $B = 46°40'$, $a = 8.13$, $c = 11.9$ **6.** (a) 75.86 m; (b) 142.0 m **7.** 28.55°, or 28°33′ **8.** 35.7 ft **9.** 240.3 ft **10.** East: 136 km; south: 63 km **11.** 22.1 km **12.** 15 mi

Exercise Set 8.1, pp. 502–505

1. $B = 60°$, $a = 3$, $b = 3\sqrt{3}$, or 5.20 **3.** $A = 45°$, $a = b = 5\sqrt{2}$, or 7.07 **5.** $B = 47°40'$, $b = 25.5$, $c = 34.5$ **7.** $B = 53°50'$, $b = 37.2$, $c = 46.1$ **9.** $A = 77°20'$, $a = 436.5$, $c = 447.4$ **11.** $B = 72°32'$, $a = 4.3$, $c = 14.3$ **13.** $A = 66°48'$, $b = 150$, $c = 381$ **15.** $B = 42°25'$, $a = 35.7$, $b = 32.6$ **17.** $A = 7°40'$, $a = 0.131$, $b = 0.973$ **19.** $A = 34.05°$, or 34°03′; $B = 55.95°$, or 55°57′; $c = 22.3$ **21.** $A = 53.13°$, or 53°08′; $B = 36.87°$, or 36°52′; $b = 12.0$ **23.** $A = 62.44°$, or 62°26′; $B = 27.56°$, or 27°34′; $a = 3.56$ **25.** 47.9 ft **27.** 239 ft **29.** 1.72°, or 1°43′ **31.** 30.22°, or 30°13′ **33.** 3.52 mi **35.** 17,067 ft **37.** 23.9 km **39.** 25.9 cm **41.** $\frac{25}{3}$ cm **43.** 22.9 ft **45.** 7.92 km **47.** 3.45 km **49.** 109 km **51.** 3.3287 **53.** Area $= \frac{1}{2} \cdot b \cdot a = \frac{1}{2}(c \cos A)(c \sin A) = \frac{1}{2}c^2(\cos A \sin A) = \frac{1}{2}c^2\left(\frac{1}{2}\sin 2A\right) = \frac{1}{4}c^2 \sin 2A$ **55.** $d = \sqrt{8000h + h^2}$, where d and h are in miles; 38.9 miles **57.** Cut so that $\theta = 79.38°$, or 79°23′ **59.** $A = \frac{1}{2}bh$, where $h = a \sin \theta$. Thus $A = \frac{1}{2}b(a \sin \theta)$, or $\frac{1}{2}ab \sin \theta$ **61.** 3928 mi

Margin Exercises, Section 8.2

1. $C = 29°$, $c = 5.93$, $b = 11.1$ **2.** $\sin A = 2.8$; impossible **3.** $B = 90°$, $C = 41°25'$, $c = 5$ **4.** (a) $A = 42°54'$, $C = 104°06'$, $c = 35.6$; (b) $A = 137°06'$, $C = 9°54'$, $c = 6.31$ **5.** $C = 17°56'$, $A = 124°04'$, $a = 26.9$

Exercise Set 8.2, pp. 511–513

1. $C = 17°$, $a = 26.3$, $c = 10.5$ **3.** $A = 121°$, $a = 33.4$, $c = 14.0$ **5.** $B = 68°50'$, $a = 32.3$ cm, $b = 32.3$ cm **7.** $A = 110.36°$, $a = 5.28$ mi, $b = 3.43$ mi **9.** $C = 103°41'$, $a = 1804$ km, $b = 5331$ km **11.** $B = 32°$, $a = 1752$ in., $c = 720$ in. **13.** $B = 56°23'$, $C = 87°37'$, $c = 40.8$, and $B = 123°37'$, $C = 20°23'$, $c = 14.2$ **15.** $B = 18°58'$, $C = 44°42'$, $b = 6.24$ **17.** $A = 74°26'$, $B = 44°24'$, $a = 33.3$ **19.** Not possible **21.** Not possible **23.** Not possible **25.** $B = 71°26'$, $A = 56°46'$, $a = 3.668$; or $B = 108°34'$, $A = 19°38'$, $a = 1.473$ **27.** Not possible **29.** $C = 54°36'$; $B = 83°34'$, $b = 134.1$ mi; or $C = 125°23'$, $B = 12°47'$, $b = 29.9$ mi **31.** 76.3 m **33.** 50.8 ft **35.** 1470 km **37.** From A: 35 mi; from B: 65.8 mi **39.** 10.6 km or 2.43 km **41.** 4.7 cm **43.**

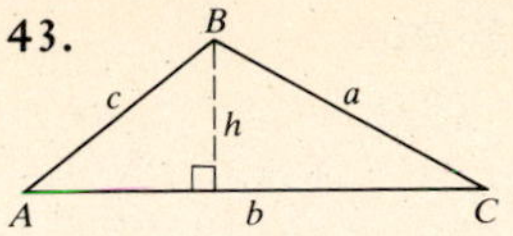

$K = \frac{1}{2}bh = \frac{1}{2}bc \sin A$ where $h = c \sin A$.

45. See Student's Solution Manual or Instructor's Solution Manual.

47.

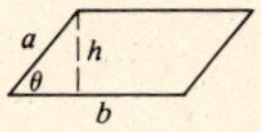

$A = bh$, $h = a \sin \theta$, so $A = ab \sin \theta$

49. Area $\triangle ADB$ + Area $\triangle BDC$ = Area $\triangle ADC$. Using the result of Exercise 47, we have

$$xy \sin \alpha + yz \sin \beta = xz \sin (\alpha + \beta).$$

Multiplying both sides by $\frac{1}{x \cdot y \cdot z}$ gives

$$\frac{\sin \alpha}{z} + \frac{\sin \beta}{x} = \frac{\sin(\alpha + \beta)}{y}.$$

Margin Exercises, Section 8.3

1. $a = 40.5$, $B = 22°09'$, $C = 35°51'$ **2.** $A = 108°13'$, $B = 22°20'$, $C = 49°27'$ **3.** Law of cosines; $A = 108°13'$, $B = 22°20'$, $C = 49°27'$ **4.** Law of sines; $C = 17°56'$, $A = 124°04'$, $a = 53.8$ **5.** Law of cosines; $a = 122$, $B = 22°07'$, $C = 35°53'$ **6.** Law of sines; $A = 42°54'$, $C = 104°06'$, $c = 106.8$; or $A = 137°06'$, $C = 9°54'$, $c = 18.9$ **7.** Law of sines; $B = 66°49'$; $C = 69°36'$, $c = 6.1$; or $B = 113°11'$, $C = 23°14'$, $c = 2.6$ **8.** Law of sines; not possible

Exercise Set 8.3, pp. 517–520

1. $a = 14.9$, $B = 23°45'$, $C = 126°15'$ **3.** $a = 24.8$, $B = 20°43'$, $C = 26°17'$ **5.** $b = 74.8$ m, $A = 95°33'$, $C = 11°47'$ **7.** $c = 51.2$ cm, $A = 10.84°$, $B = 146.88°$ **9.** $a = 25.5$ yd, $C = 45°47'$, $B = 38°$ **11.** $c = 45.17$ mi, $B = 42°$, $C = 89.3°$ **13.** $A = 36°11'$, $B = 43°32'$, $c = 100°17'$ **15.** $A = 24°09'$, $B = 30°45'$, $C = 125°06'$ **17.** Not possible **19.** $A = 128.7°$, $B = 22.1°$, $C = 29.2°$ **21.** Law of sines; $C = 98°$, $a = 96.7$, $c = 101.9$ **23.** Law of cosines; $A = 73.7°$, $B = 51.8°$, $C = 54.5°$ **25.** Law of cosines; $c = 44.6$, $B = 42°$, $A = 90°$ **27.** Law of sines; $A = 61°20'$, $a = 5.633$, $b = 6.014$ **29.** 28.2 nautical mi, S 55°20′ E **31.** 37 nautical mi **33.** 59.4 ft **35.** 68°, 68°, 44° **37.** 45.96 ft **39.** (a) 15.73 ft; (b) 120.8 ft² **41.** 3424 yd² **43.** 11.9 m, 24.0 m **45.** 79.1°, 40.8°, 60.1° **47.** 49.7 ft **49.** See Student's Solution Manual or Instructor's Solution Manual. **51.** 64 in. **53.** 161.5 ft **55.** $A = \frac{1}{2}a^2 \sin \theta$, when $\theta = 90°$ **57.** Let $S^2 = a^2 + b^2 + c^2$. Then

$$\begin{aligned} a^2 &= b^2 + c^2 - 2bc \cos A \\ b^2 &= c^2 + a^2 - 2ca \cos B \\ (+)\quad c^2 &= a^2 + b^2 - 2ab \cos C \\ \hline S^2 &= S^2 + S^2 - 2(bc \cos A + ac \cos B + ab \cos C). \\ \therefore\ S^2 &= a^2 + b^2 + c^2 = 2(bc \cos A + ac \cos B + ab \cos C) \end{aligned}$$

59. 10,106 ft **61.** (a) 11:15 A.M.; (b) 1955 km **63.** 9468 km

Margin Exercises, Section 8.4

1. 14.9 kg, 19°40′ **2.** 75°, 171 km/h **3.** E: $50\sqrt{2}$, S: $50\sqrt{2}$ **4.** 18.9 km/h from S 32° E **5.** (a) 4 up, 21 left; (b) 21.4, 10°47′ with horizontal **6.** $\langle 12, -8\rangle$ **7.** $\langle 12, 4\rangle$ **8.** $(8.54, 159°27')$ **9.** $\langle 14.2, -4.88\rangle$ **10.** (a) $\langle 1, 7\rangle$; (b) $\langle 7, -3\rangle$; (c) $\langle -23, 21\rangle$; (d) 19.03

Exercise Set 8.4, pp. 527–529

1. 57, 38° **3.** 18.4, 37° **5.** 20.9, 58° **7.** 68.3, 18° **9.** 13 kg, 67° **11.** 655 kg, 21° **13.** 21.6 ft/sec, 34° **15.** 726 lb, 47° **17.** 174 nautical mi, S 15° E **19.** An angle of 12° upstream **21.** Vertical 118.2, horizontal 92.3 **23.** S. 192 km/h, W. 161 km/h **25.** 43 kg, S 35°32′ W **27.** **(a)** N. 28, W. 7; **(b)** 28.9, N 14°02′ W **29.** $\langle 5, 16\rangle$ **31.** $\langle 4.8, 13.7\rangle$ **33.** $\langle -662, -426\rangle$ **35.** (5, 53°08′) **37.** (18, 303°41′) **39.** (5, 233°08′) **41.** (18, 123°41′) **43.** $\langle 3.46, 2\rangle$ **45.** $\langle -5.74, -8.19\rangle$ **47.** $\langle 17.3, -10\rangle$ **49.** $\langle 70.7, -70.7\rangle$ **51.** $\langle 19, 36\rangle$ **53.** $\langle 14, 8\rangle$ **55.** 18 **57.** 17.89 **59.** 35.78 **61.** $\langle 0, 0\rangle$ **63.** To show: If k is a scalar, then $k(\mathbf{u}+\mathbf{v}) = k\mathbf{u} + k\mathbf{v}$. Proof: Let $\mathbf{u} = \langle a, b\rangle$ and $\mathbf{v} = \langle c, d\rangle$. Then

$$\begin{aligned} k(\mathbf{u}+\mathbf{v}) &= k(\langle a, b\rangle + \langle c, d\rangle) \\ &= k\langle a+c, b+d\rangle \\ &= \langle k(a+c), k(b+d)\rangle \\ &= \langle ka+kc, kb+kd\rangle \\ &= \langle ka, kb\rangle + \langle kc, kd\rangle \\ &= k\langle a, b\rangle + k\langle c, d\rangle \\ &= k\mathbf{u} + k\mathbf{v} \end{aligned}$$

65.

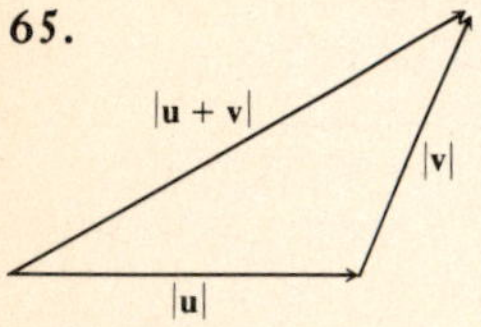

If the vectors are *not* collinear (or parallel), they form a triangle as shown. From plane geometry (Euclid), we know that (any) one side of a triangle is *less than* the sum of the other two sides.

If the vectors *are* collinear, then either

$$|\mathbf{u}+\mathbf{v}| = |\mathbf{u}| + |\mathbf{v}|$$

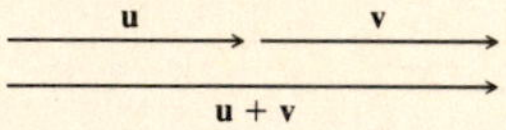

or

$$|\mathbf{u}+\mathbf{v}| < |\mathbf{u}| + |\mathbf{v}|.$$

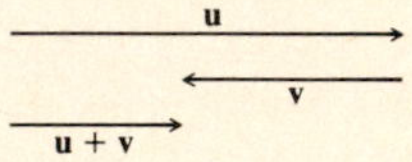

Margin Exercises, Section 8.5

1. Horizontal rope $300\sqrt{3}$ lb; other rope 600 lb **2.** Parallel to incline 50 kg; perpendicular to incline $50\sqrt{3}$ or 86.6 kg **3.** 500 kg on left, 866 kg on right

Exercise Set 8.5, pp. 531–532

1. Cable 224-lb tension, boom 166-lb compression **3.** Horizontal rod 168-kg tension, other rod 261-kg compression **5.** Lift 2472 lb, drag 1315 lb **7.** 60 kg **9.** 9°36′ **11.** Horizontal rope 400 kg, other rope 566 kg **13.** 2000 kg in each **15.** 2241 kg on left, 1830 kg on right

Margin Exercises, Section 8.6

1.

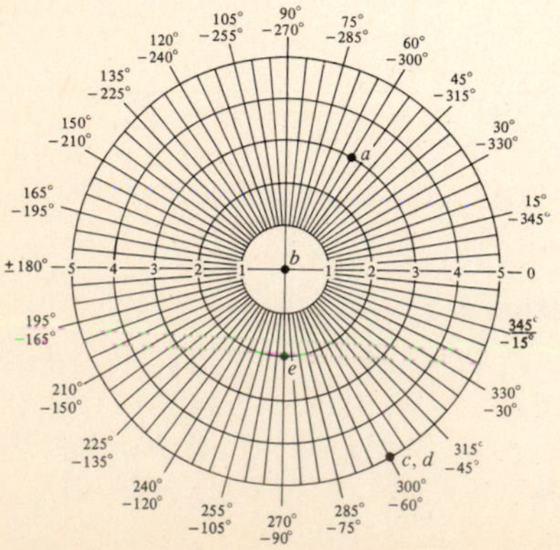

2. Many answers possible. A: (4, 30°), (4, 390°), (−4, 210°), etc.; B: (5, −60°), (5, 300°); C: (2, 150°), (2, −210°); D: (3, 225°), (3, −135°); E: (5, 60°), (−5, −120°) **3.** **(a)** $(3\sqrt{2}, 45°)$; **(b)** (4, 270°); **(c)** (6, 120°); **(d)** (4, 330°) **4.** **(a)** $\left(\frac{5}{2}\sqrt{3}, \frac{5}{2}\right)$; **(b)** $(5, 5\sqrt{3})$; **(c)** $\left(-\frac{5}{\sqrt{2}}, -\frac{5}{\sqrt{2}}\right)$; **(d)** $(4\sqrt{3}, 4)$ **5.** $2r\cos\theta + 5r\sin\theta = 9$ **6.** $r^2 + 8r\cos\theta = 0$ **7.** $x^2 + y^2 = 49$ **8.** $y = 5$ **9.** $x^2 + y^2 - 3x = 5y$

10.

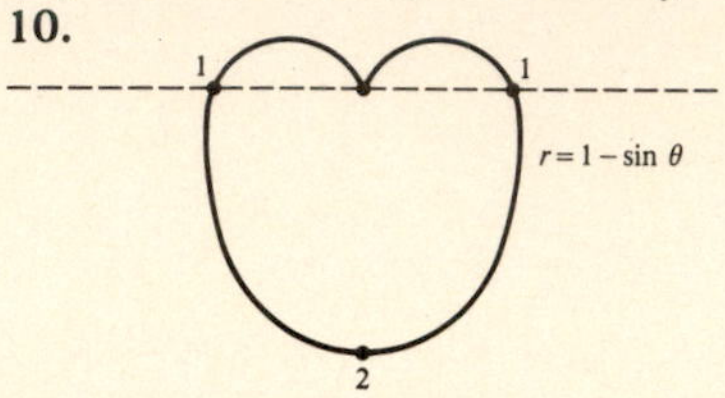

11.

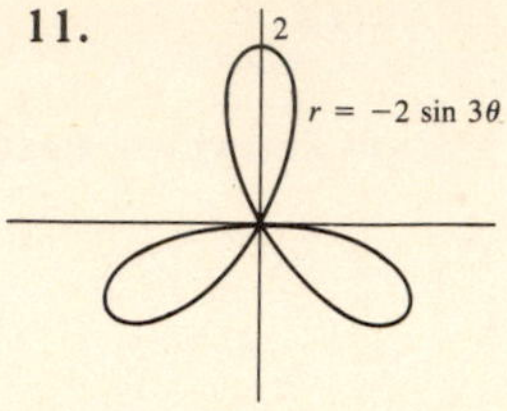

Exercise Set 8.6, pp. 536–537

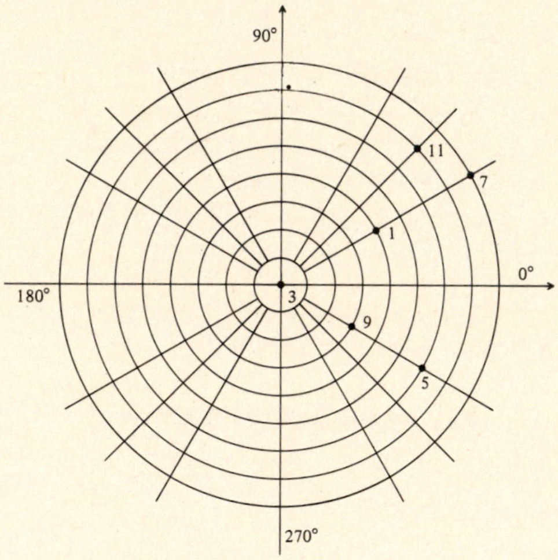

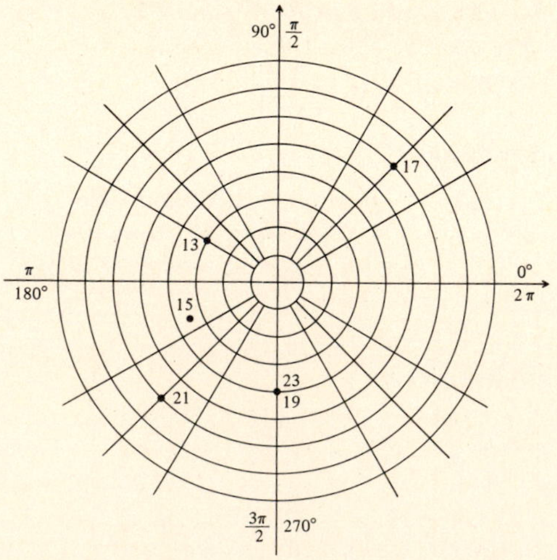

25. $(4\sqrt{2}, 45°)$ **27.** (5, 90°) **29.** (4, 0°) **31.** (6, 60°) **33.** (2, 30°) **35.** (6, 30°) **37.** $\left(\frac{4}{\sqrt{2}}, \frac{4}{\sqrt{2}}\right)$ or (2.83, 2.83) **39.** (0, 0) **41.** $\left(\frac{-3}{\sqrt{2}}, \frac{-3}{\sqrt{2}}\right)$ **43.** $(3, -3\sqrt{3})$ **45.** $(5\sqrt{3}, 5)$ **47.** (4.33, −2.5) **49.** $3r\cos\theta + 4r\sin\theta = 5$

51. $r\cos\theta = 5$ **53.** $r^2 = 36$ **55.** $r^2(\cos^2\theta - 4\sin^2\theta) = 4$
57. $x^2 + y^2 = 25$ **59.** $y = x$ **61.** $y = 2$ **63.** $x^2 + y^2 = 4x$
65. $x^2 - 4y = 4$ **67.** $x^2 - 2x + y^2 - 3y = 0$

69.

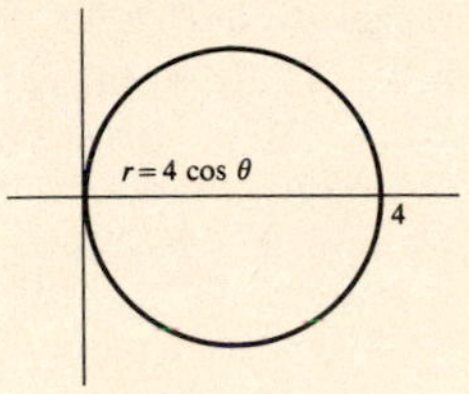

71.

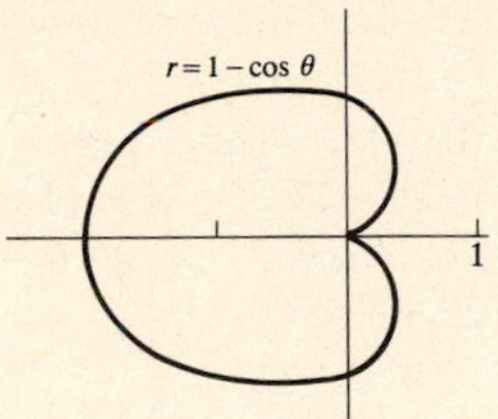

73.

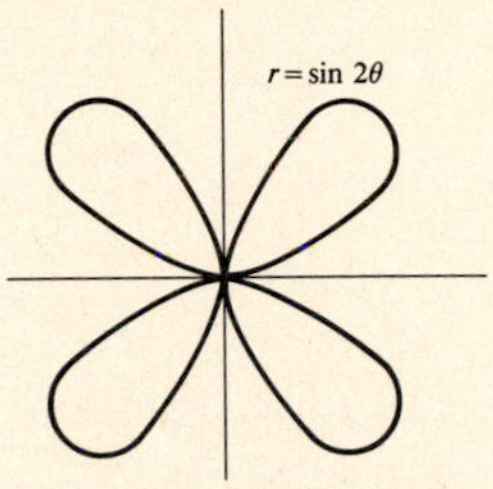

75.

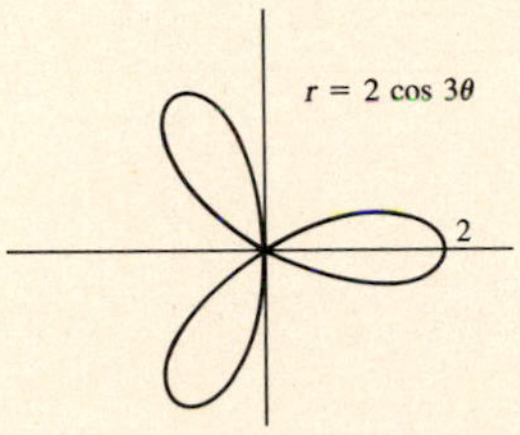

77.

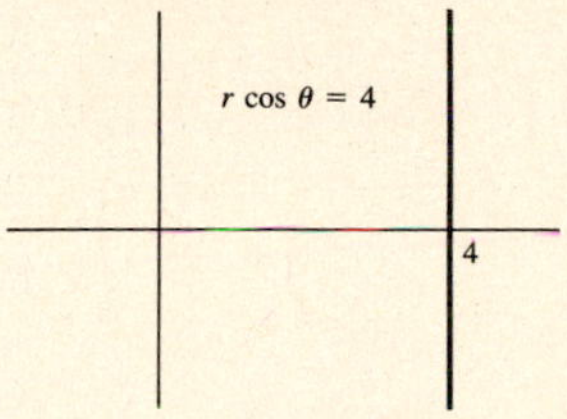

79.

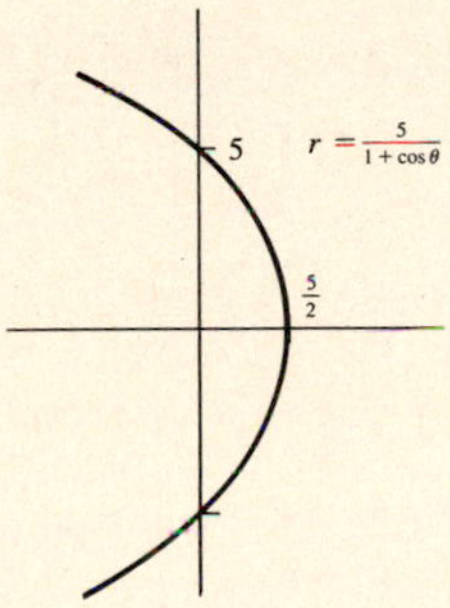

81.

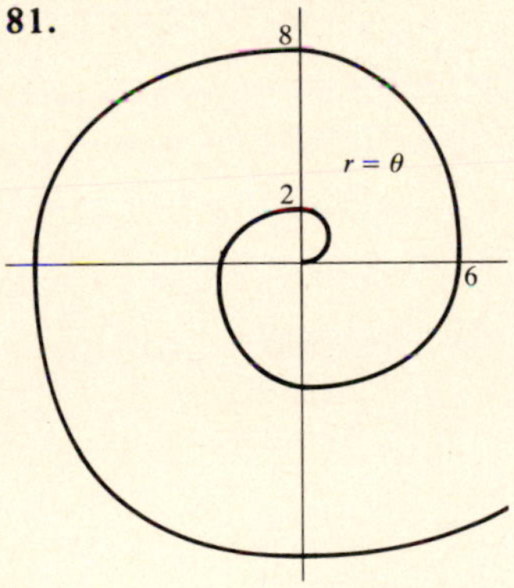

83.

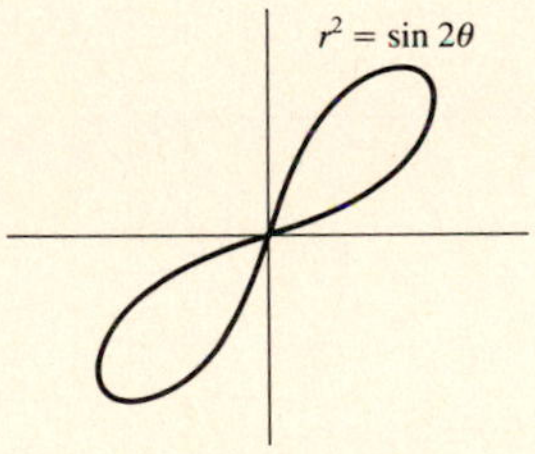

85.

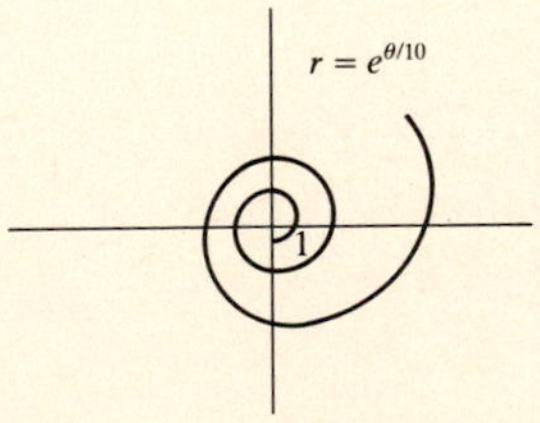

87.

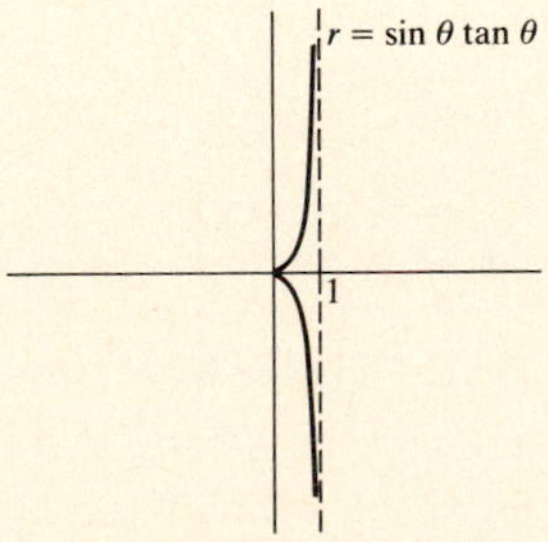

89.

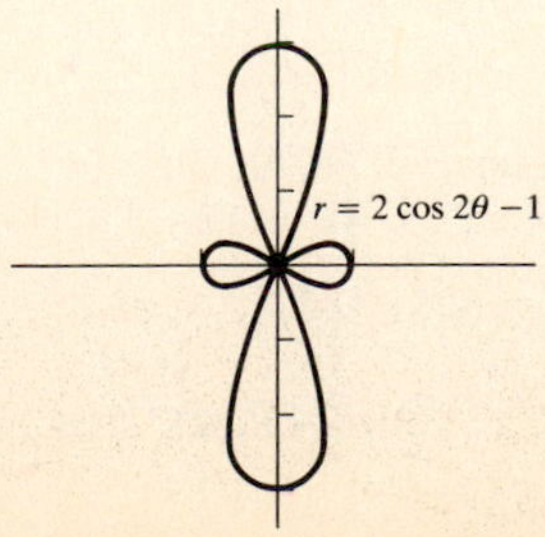

91.

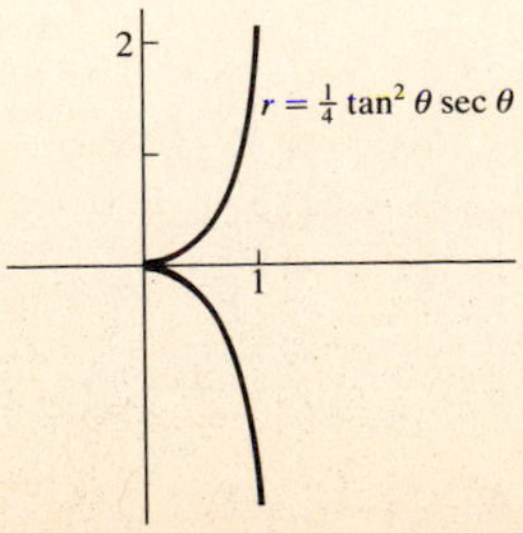

93. $y^2 = -4x + 4$

Margin Exercises, Section 8.7

1.

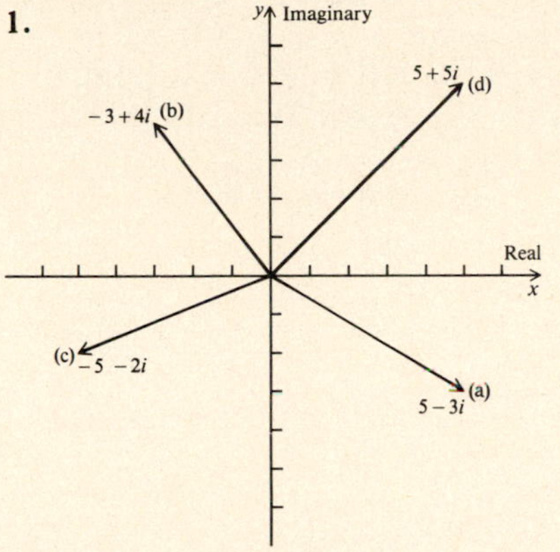

2.

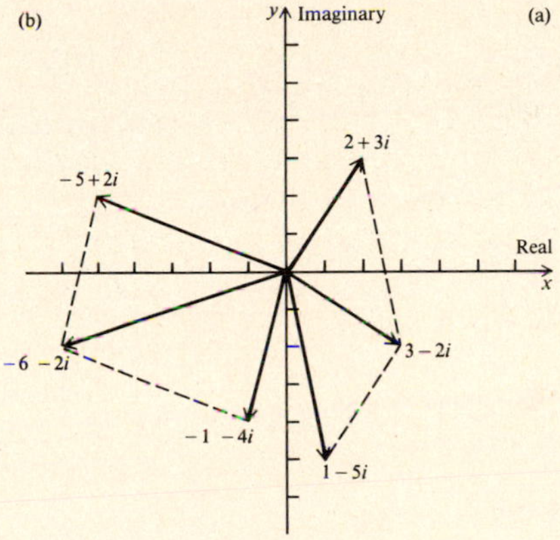

3. (a) 5; (b) 13
4. $1 - i$
5. $\sqrt{3} - i$
6. $\sqrt{2}$ cis 315°
7. 6 cis 225°
8. 20 cis 55°
9. 4 cis $\frac{5\pi}{4}$
10. 4 cis 90°
11. 2 cis $\frac{\pi}{4}$
12. $\sqrt{2}$ cis 285°

Exercise Set 8.7, p. 543

1.

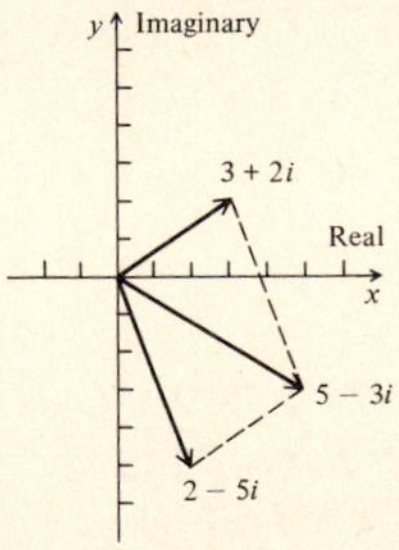

3.

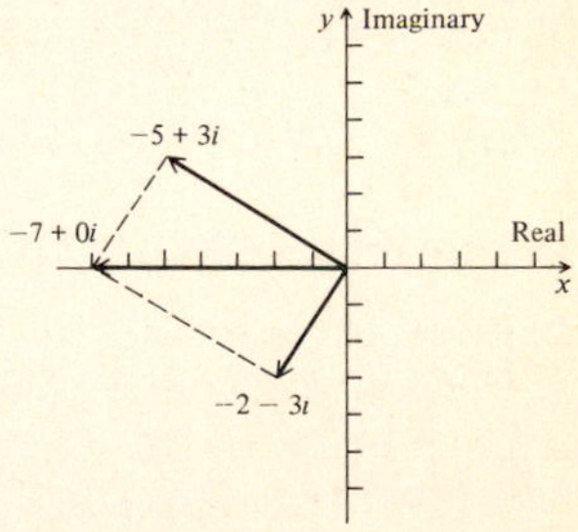

5.

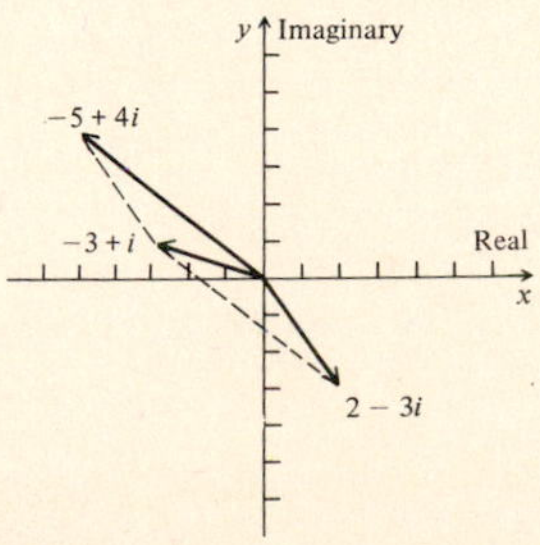

7.

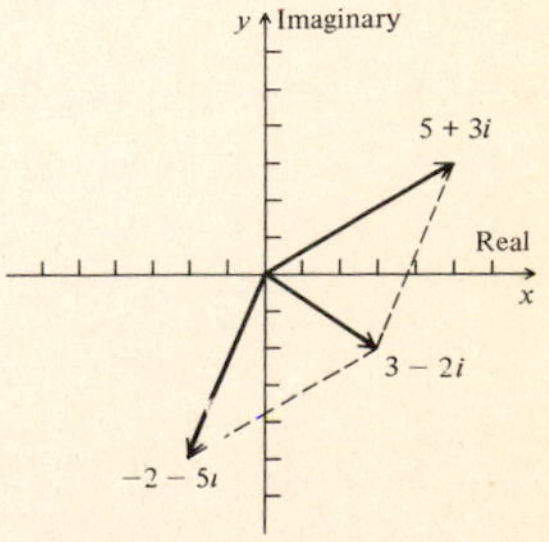

9. $\frac{3\sqrt{3}}{2} + \frac{3}{2}i$ **11.** $-10i$ **13.** $2 + 2i$ **15.** $-2 - 2i$
17. $\sqrt{2}$ cis 315° **19.** 20 cis 330° **21.** 5 cis 180° **23.** 4 cis 0°, or 4 **25.** 8 cis 120° **27.** cis 270°, or $-i$ **29.** 2 cis 270°, or $-2i$ **31.** $z = a + bi$, $|z| = \sqrt{a^2 + b^2}$; $-z = -a - bi$, $|-z| = \sqrt{(-a)^2 + (-b)^2} = \sqrt{a^2 + b^2}$, $\therefore |z| = |-z|$

33. $|(a + bi)(a - bi)| = |a^2 + b^2| = a^2 + b^2$; $|(a + bi)^2| = |a^2 + 2abi - b^2| = |a^2 - b^2 - 2abi| = \sqrt{(a^2 - b^2)^2 + (2ab)^2} = \sqrt{a^4 + 2a^2b^2 + b^4} = a^2 + b^2$
35. $z \cdot w = (r_1 \text{ cis } \theta_1)(r_2 \text{ cis } \theta_2) = r_1 r_2 \text{ cis } (\theta_1 + \theta_2)$, $|z \cdot w| = \sqrt{[r_1 r_2 \cos(\theta_1 + \theta_2)]^2 + [r_1 r_2 \sin(\theta_1 + \theta_2)]^2} = \sqrt{(r_1 r_2)^2} = |r_1 r_2|$, $|z| = \sqrt{(r_1 \cos \theta_1)^2 + (r_1 \sin \theta_1)^2} = \sqrt{r_1^2} = |r_1|$, $|w| = \sqrt{(r_2 \cos \theta_2)^2 + (r_2 \sin \theta_2)^2} = \sqrt{r_2^2} = |r_2|$. Then $|z| \cdot |w| = |r_1| \cdot |r_2| = |r_1 r_2| = |z \cdot w|$
37.

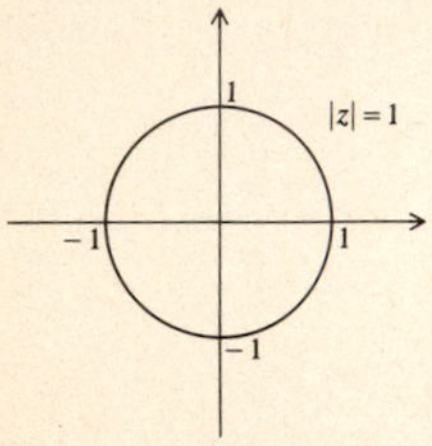

39. $\cos \theta - i \sin \theta$

Margin Exercises, Section 8.8

1. 32 cis 270°, or $-32i$ **2.** 16 cis 120°, or $-8 + 8i\sqrt{3}$
3. $1 + i$, $-1 - i$ **4.** $\sqrt{5} + i\sqrt{5}$, $-\sqrt{5} - i\sqrt{5}$ **5.** 1 cis 60°, 1 cis 180°, 1 cis 300°; or $\frac{1}{2} + \frac{\sqrt{3}}{2}i$, -1, $\frac{1}{2} - \frac{\sqrt{3}}{2}i$

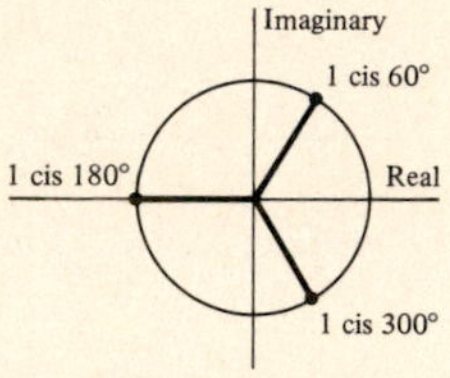

Exercise Set 8.8, pp. 546–547

1. 8 cis π **3.** 64 cis π **5.** 8 cis 270° **7.** $-8 - 8\sqrt{3}i$
9. $-8 - 8\sqrt{3}i$ **11.** i **13.** 1 **15.** cis 45°, cis 225°; or $\frac{\sqrt{2}}{2} + i\frac{\sqrt{2}}{2}$, $-\frac{\sqrt{2}}{2} - i\frac{\sqrt{2}}{2}$ **17.** 2 cis 157.5°, 2 cis 337.5°
19. $\sqrt{2}$ cis 60°, $\sqrt{2}$ cis 240°; or $\frac{\sqrt{6}}{2} + \frac{\sqrt{2}}{2}i$, $-\frac{\sqrt{6}}{2} - \frac{\sqrt{2}}{2}i$
21. cis 30°, cis 150°, cis 270°; or $\frac{\sqrt{3}}{2} + \frac{1}{2}i$, $\frac{-\sqrt{3}}{2} + \frac{1}{2}i$, $-i$
23. 2 cis 0°, 2 cis 90°, 2 cis 180°, 2 cis 270°; or 2, $2i$, -2, $-2i$

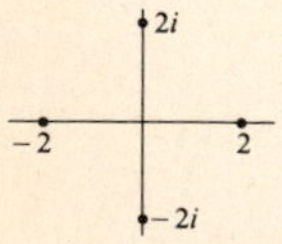

25. cis 36°, cis 108°, cis 180°, cis 252°, cis 324°
27. 2 cis 30°, 2 cis 120°, 2 cis 210°, 2 cis 300°; or $\sqrt{3} + i$, $-1 + \sqrt{3}i$, $-\sqrt{3} - i$, $1 - \sqrt{3}i$
29. $\sqrt[3]{4}$ cis 110°, $\sqrt[3]{4}$ cis 230°, $\sqrt[3]{4}$ cis 350°
31. cis 0°, cis 120°, cis 240°; or 1, $-\frac{1}{2} + \frac{\sqrt{3}}{2}i$, $-\frac{1}{2} - \frac{\sqrt{3}}{2}i$
33. cis 36°, cis 108°, cis 180°, cis 252°, cis 324° **35.** $\sqrt[5]{2}$ cis 42°, $\sqrt[5]{2}$ cis 114°, $\sqrt[5]{2}$ cis 186°, $\sqrt[5]{2}$ cis 258°, $\sqrt[5]{2}$ cis 330°
37. 3 cis 45°, 3 cis 135°, 3 cis 225°, 3 cis 315°; or $\frac{3\sqrt{2}}{2} + \frac{3\sqrt{2}}{2}i$, $-\frac{3\sqrt{2}}{2} + \frac{3\sqrt{2}}{2}i$, $-\frac{3\sqrt{2}}{2} - \frac{3\sqrt{2}}{2}i$, $\frac{3\sqrt{2}}{2} - \frac{3\sqrt{2}}{2}i$ **39.** $-1.366 + 1.366i$, $0.366 - 0.366i$

Review Exercises: Chapter 8, pp. 547–548

1. [8.1] $A = 58.09°$, or $58°05'$; $B = 31.91°$, or $31°55'$; $b = 4.5$
2. [8.1] $A = 38°50'$; $b = 37.9$; $c = 48.6$ **3.** [8.1] 1748 cm
4. [8.1] 13.95 ft **5.** [8.2] $A = 45°48'$, $C = 99°12'$, $c = 34.4$; or $A = 134°12'$, $C = 10°48'$, $c = 6.53$ **6.** [8.2] $A = 34°10'$, $a = 0.621$, $c = 0.511$ **7.** [8.4] 7.96 **8.** [8.2] 420 cm **9.** [8.3] $A = 92.1°$, $B = 33.0°$, $C = 54.8°$ **10.** [8.1] 13.72 cm²
11. [8.4] 75.8 mph; N 20°11′ E **12.** [8.4] 23 lb down, 17 lb left; 28.6 lb, 53.5° downward from horizontal (to left) **13.** [8.4] $\langle -18.79, 6.84 \rangle$ **14.** [8.4] $(\sqrt{13}, 123°41')$ **15.** [8.4] **(a)** $\langle 25, 0 \rangle$; **(b)** $5\sqrt{2}$ **16.** [8.6] (4, 270°) **17.** [8.6] (2, 330°) **18.** [8.6] $(-2, -2\sqrt{3})$ **19.** [8.6] $(-2, 2\sqrt{3})$ **20.** [8.6] $r + 2\cos\theta - 3\sin\theta = 0$ **21.** [8.6] $x^2 + y^2 = -4y$
22. [8.6]

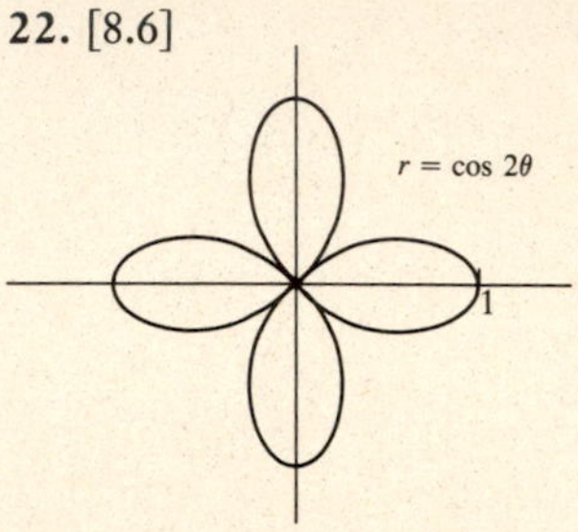

23. [8.6]

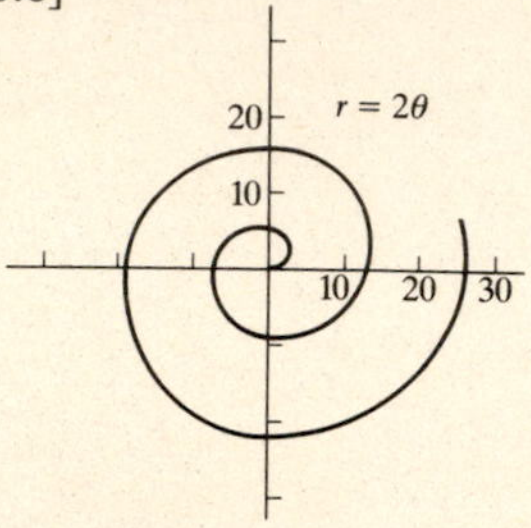

24. [8.6]

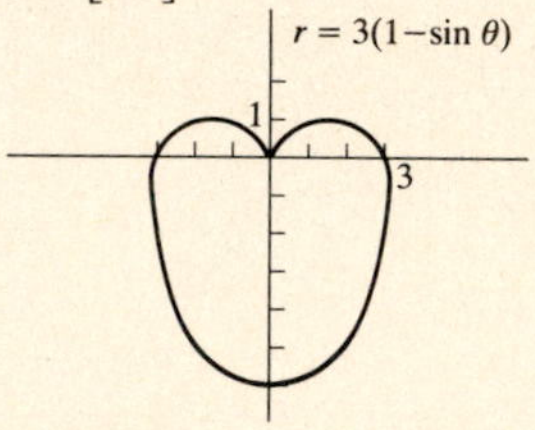

25. [8.5] Parallel: 106 lb; perpendicular: 106 lb
26. [8.5] 577 kg

27. [8.7]

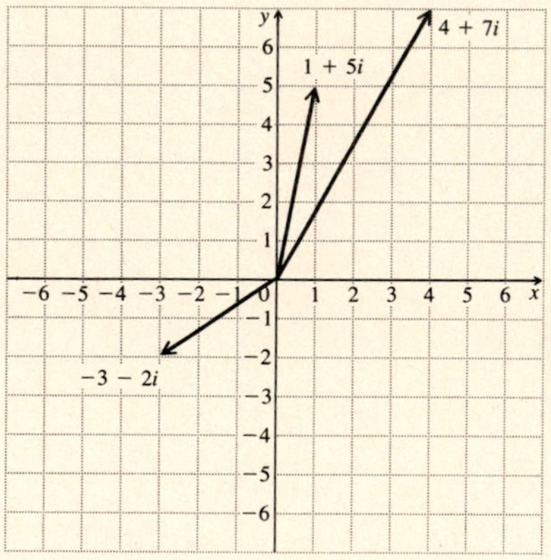

28. [8.7] $-\sqrt{2} + \sqrt{2}i$
29. [8.7] $\sqrt{2}$ cis 45°
30. [8.7] 70 cis 50°
31. [8.7] $\frac{\sqrt{2}}{2}$ cis 15°
32. [8.8] $4\sqrt{2}$ cis 135°
33. [8.8] $-\frac{81}{2} + \frac{81\sqrt{3}}{2}i$
34. [8.8] $\sqrt[6]{2}$ cis 15°; $\sqrt[6]{2}$ cis 135°; $\sqrt[6]{2}$ cis 255°
35. [8.8] 3, $\frac{3}{2}(-1 \pm \sqrt{3}i)$ **36.** [8.8] 2 cis 45°, 2 cis 225°; or $\sqrt{2} + \sqrt{2}i$, $-\sqrt{2} - \sqrt{2}i$ **37.** [8.2] 50.52°, 129.48°
38. [8.4] $\langle \frac{36}{13}, \frac{15}{13} \rangle$ **39.** [8.6] $y^2 = 4x + 4$

Test: Chapter 8, p. 549

1. [8.1] $B = 67°50'$, $a = 7.5$, $c = 19.9$ **2.** [8.1] $B = 66.87°$, or $66°52'$; $A = 23.13°$, or $23°08'$; $b = 25.7$ **3.** [8.1] 35° **4.** [8.2] $A = 115°$, $a = 88.4$, $c = 31.8$ **5.** [8.3] 11.8 **6.** [8.2] 6.13 ft, 5.00 ft **7.** [8.4] 22.8 kg, 61°11′ **8.** [8.4] $\left(-\frac{25\sqrt{3}}{2}, \frac{25}{2}\right)$
9. [8.4] (11.7, 149°02′) **10.** [8.4] **(a)** $\langle -1, -14 \rangle$; **(b)** 8.062
11. [8.4] **(a)** N. 7, W. 5; **(b)** 8.6, N 35°32′ W **12.** [8.6] $\left(-\frac{3\sqrt{3}}{2}, -\frac{3}{2}\right)$ **13.** [8.6] $x^2 + y^2 = 625$ **14.** [8.5] 48.8 lb
15. [8.6] $\left(-\frac{4}{\sqrt{53}}, \frac{14}{\sqrt{53}}\right)$ **16.** [8.8] $2\sqrt{2}$ cis 45°, $2\sqrt{2}$ cis 225°; or $2 + 2i$, $-2 - 2i$

17. [8.7]

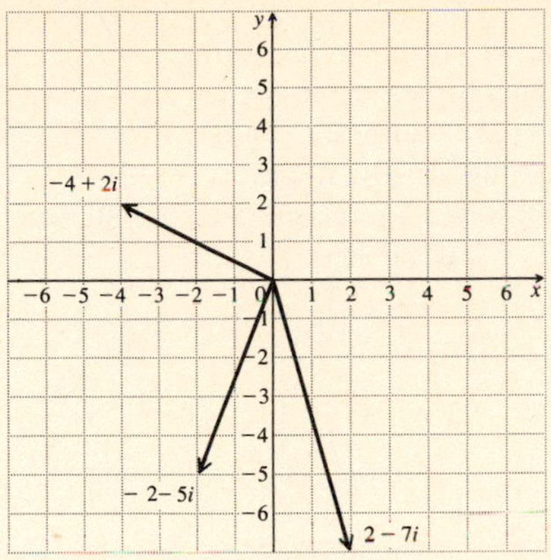

18. [8.7] $\frac{3\sqrt{3}}{2} - \frac{3}{2}i$
19. [8.7] 2 cis 120°
20. [8.7] 10 cis 90°
21. [8.7] $\frac{1}{2}$ cis 80°
22. [8.8] 8 cis 270°
23. [8.8] 2 cis 60°, 2 cis 180°, 2 cis 300°; or $1 + \sqrt{3}i$, -2, $1 - \sqrt{3}i$

24. [8.6]

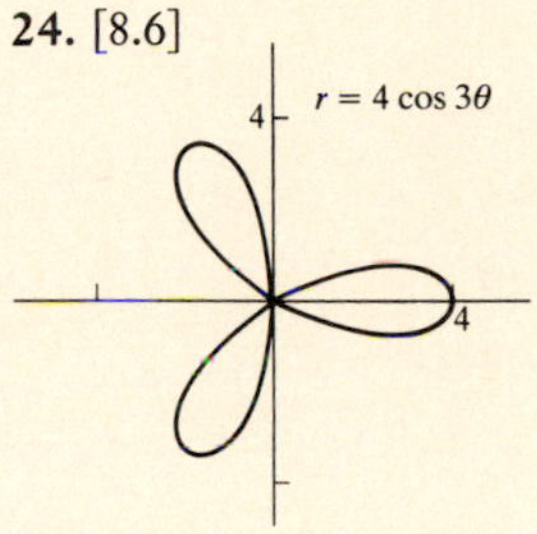

25. [8.6]

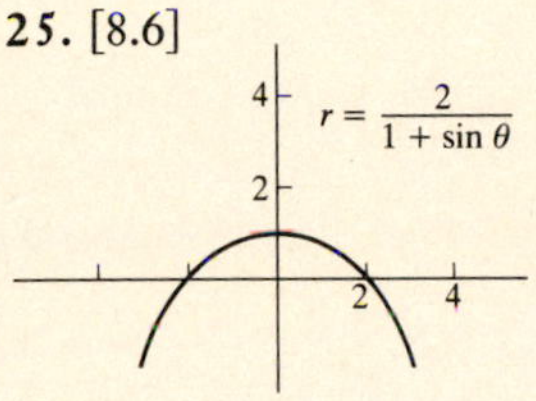

26. [8.2] 11.58 mm^2

CHAPTER 9

Margin Exercises, Section 9.1

1. Yes **2.** No **3.** (1, 2) **4.** Infinitely many solutions **5.** No solution **6.** (1, 2) **7.** (4, −2) **8.** (−2, 5) **9.** (−3, 2) **10.** $\left(-\frac{1}{3}, \frac{1}{2}\right)$ **11.** (2, −1) **12.** $\frac{11}{2}, \frac{9}{2}$ **13.** $\frac{11}{2}, -\frac{9}{2}$ **14.** 10 km/h, 2 km/h **15.** 5 L, 3 L

Exercise Set 9.1, pp. 559–561

1. No **3.** (−1, 3) **5.** No solution **7.** $\left(\frac{39}{11}, -\frac{1}{11}\right)$ **9.** (−4, −2) **11.** (−3, 0) **13.** (10, 8) **15.** (1, 1) **17.** $-\frac{11}{2}, -\frac{9}{2}$ **19.** 20 km/h, 3 km/h **21.** 12.5 L, 7.5 L **23.** 3 hr **25.** 2 hr **27.** 12 white, 28 printed **29.** Paula is 32, Bob is 20 **31.** 76 m, 19 m **33.** 137 m, 55 m **35.** \$6800 at 9%, \$8200 at 10% **37.** (−12, 0) **39.** (0.924, −0.833) **41.** 4 km **43.** 180 **45.** $\frac{7}{19}$ **47.** First train: 36 km/h; second train: 54 km/h **49.** \$96 **51.** 4 boys, 3 girls **53.** $m = -\frac{4}{3}$, $b = \frac{1}{3}$ **55.** (370, \$10) **57.** (474, \$22) **59.** $\left(-\frac{1}{4}, -\frac{1}{2}\right)$ **61.** {(5, 3), (−5, 3), (5, −3), (−5, −3)} **63.** 40 mph

Margin Exercises, Section 9.2

1. (a) No; (b) yes **2.** $\left(2, \frac{1}{2}, -2\right)$ **3.** A: 75; B: 84; C: 63 **4.** $f(x) = x^2 - 2x + 1$ **5.** $f(x) = \frac{5}{8}x^2 - 50x + 1150$; 510 accidents

Exercise Set 9.2, pp. 566–569

1. No **3.** (3, −2, 1) **5.** (−3, 2, 1) **7.** $\left(\frac{1}{2}, \frac{2}{3}, -\frac{5}{6}\right)$ **9.** (1, −2, 4, −1) **11.** 8, 21, −3 **13.** A = 30°, B = 90°, C = 60° **15.** 20 on Mon., 35 on Tues., 32 on Wed. **17.** A: 2200; B: 2500; C: 2700 **19.** A: 10; B: 12; C: 15 **21.** Par-3: 4; par-4: 10; par-5: 4 **23.** 7%: \$400; 8%: \$500; 9%: \$1600 **25.** $y = 2x^2 + 3x - 1$ **27.** (a) $E = -4t^2 + 40t + 2$; (b) \$98 **29.** (a) $f(x) = 104.5x^2 - 1501.5x + 6016$; (b) 1682, 769, 1451 **31.** (a) $f(x) = -\frac{23}{18}x^2 + \frac{497}{18}x - \frac{1208}{18}$; (b) 81.2 years; (c) 12.3 **33.** $\left(-1, \frac{1}{5}, -\frac{1}{2}\right)$ **35.** A: 4 hr; B: 6 hr; C: 12 hr **37.** 180° **39.** $3x + 4y + 2z = 12$ **41.** (1.49, 0.54, 2.03) **43.** Art: 181; Bob: 176; Carl: 158; Denny: 190; Fred: 180

Margin Exercises, Section 9.3

1. No solution; inconsistent **2.** $\left(-\frac{1}{2}, -2\right)$; consistent **3.** Infinitely many solutions; consistent; dependent **4.** (2, 3); consistent; dependent **5.** $\left(\frac{2y-5}{3}, y\right)$ or $\left(x, \frac{3x+5}{2}\right)$, (1, 4), (3, 7), $\left(0, \frac{5}{2}\right)$, etc. **6.** $\left(x, \frac{1-2x}{5}\right)$ or $\left(\frac{1-5y}{2}, y\right)$, (−7, 3), (3, −1), $\left(0, \frac{1}{5}\right)$, etc. **7.** $(-2z + 7, 3z - 5, z)$; (7, −5, 0), (5, −2, 1), (3, 1, 2), etc. **8.** $(-2z, 3z, z)$; (−2, 3, 1), (2, −3, −1), (−4, 6, 2), etc. **9.** (0, 0, 0), only solution

Exercise Set 9.3, pp. 575–576

1. $\left(\frac{y+5}{3}, y\right)$ or $(x, 3x - 5)$; (0, −5), (1, −2), (−1, −8), etc. **3.** ∅ **5.** $\left(\frac{5-2y}{3}, y\right)$ or $\left(x, \frac{5-3x}{2}\right)$; (3, −2), etc. **7.** $\left(\frac{6y-3}{4}, y\right)$ or $\left(x, \frac{4x+3}{6}\right)$; $\left(1, \frac{7}{6}\right)$, etc. **9.** ∅ **11.** $\left(\frac{10+11z}{9}, \frac{-11+5z}{9}, z\right)$; $\left(\frac{10}{9}, -\frac{11}{9}, 0\right)$, etc. **13.** $\left(\frac{11}{9}z, \frac{5}{9}z, z\right)$; $\left(\frac{11}{9}, \frac{5}{9}, 1\right)$, etc. **15.** $(4z - 5, -3z + 2, z)$; (−1, −1, 1), etc. **17.** (0, 0, 0) **19.** Consistent: 1, 5, 7, 11, 13, 15, 17, the others are inconsistent; dependent: 1, 5, 7, 11, 13, 15, the others are independent **21.** $\left(\frac{724y + 9160}{2013}, y\right)$ or $\left(x, \frac{2013x - 9160}{724}\right)$ **23.** (a) ∅; (b) inconsistent; (c) dependent **25.** $k = 2$ **27.** $(x, 18 - 2x, x)$ where $1 \le x \le 8$, x is an integer

Margin Exercises, Section 9.4

1. $\left(-\frac{63}{29}, -\frac{114}{29}\right)$ **2.** (−1, 2, 3)

Exercise Set 9.4, pp. 578–579

1. $\left(\frac{3}{2}, \frac{5}{2}\right)$ **3.** (−1, 2, −2) **5.** $\left(\frac{1}{2}, \frac{3}{2}\right)$ **7.** $\left(\frac{3}{2}, -4, 3\right)$ **9.** $(r - 2, 3 - 2r, r)$ **11.** (1, −3, −2, −1) **13.** 4 dimes, 30 nickels **15.** 10 nickels, 4 dimes, 8 quarters **17.** 5 lb of \$4.05; 10 lb of \$2.70 **19.** \$30,000 at $12\frac{1}{2}$%; \$40,000 at 13% **21.** (1.0128, −4.8909) **23.** (1.23, −2.11, 1.89) **25.** $x = \frac{3\sqrt{2} + \pi}{2 + \pi^2}$ and $y = \frac{3\pi - \sqrt{2}}{\pi^2 + 2}$

Margin Exercises, Section 9.5

1. 3 × 2 **2.** 2 × 2 **3.** 3 × 3 **4.** 1 × 2 **5.** 2 × 1 **6.** 1 × 1 **7.** 2, 3, 6 **8.** $\mathbf{A} + \mathbf{B} = \begin{bmatrix} -2 & -6 \\ 13 & 0 \end{bmatrix} = \mathbf{B} + \mathbf{A}$

9. $\begin{bmatrix} 1 & 1 & -10 \\ 2 & 4 & -3 \end{bmatrix}$ 10. $A + 0 = \begin{bmatrix} 4 & -3 \\ 5 & 8 \end{bmatrix} = 0 + A = A$

11. $\begin{bmatrix} -1 & 4 & -7 \\ -2 & -4 & 8 \end{bmatrix}$ 12. $\begin{bmatrix} -6 & 6 \\ 1 & -4 \\ -7 & 5 \end{bmatrix}$

13. $\begin{bmatrix} -2 & 1 & -5 \\ -6 & -4 & 3 \end{bmatrix}$ 14. $\begin{bmatrix} 0 & 0 & 0 \\ 0 & 0 & 0 \end{bmatrix}$

15. $\begin{bmatrix} -1 & 4 & -7 \\ -2 & -4 & 8 \end{bmatrix}$ 16. $\begin{bmatrix} 5 & -10 & 5x \\ 20 & 5y & 5 \\ 0 & -25 & 5x^2 \end{bmatrix}$

17. $\begin{bmatrix} t & -t & 4t & tx \\ ty & 3t & -2t & ty \\ t & 4t & -5t & ty \end{bmatrix}$ 18. $[-13]$ 19. $\begin{bmatrix} 12 \\ 13 \\ 5 \\ 16 \end{bmatrix}$

20. $\begin{bmatrix} 0 & 26 \\ -8 & 3 \\ -13 & 33 \\ -7 & 32 \end{bmatrix}$

21. $AB = \begin{bmatrix} 2 & 8 & 6 \\ -29 & -34 & -7 \end{bmatrix}$; BA not possible

22. $[8 \quad 5 \quad 4]$ 23. $AB = \begin{bmatrix} -2 & 32 \\ 4 & 16 \end{bmatrix}$, $BA = \begin{bmatrix} 8 & -13 \\ -16 & 6 \end{bmatrix}$

24. $AI = \begin{bmatrix} 3 & 2 \\ -1 & 5 \end{bmatrix} = IA = A$

25. $\begin{bmatrix} 3 & 4 & -2 \\ 2 & -2 & 5 \\ 6 & 7 & -1 \end{bmatrix} \begin{bmatrix} x \\ y \\ z \end{bmatrix} = \begin{bmatrix} 5 \\ 3 \\ 0 \end{bmatrix}$

Exercise Set 9.5, pp. 586–587

1. $\begin{bmatrix} -2 & 7 \\ 6 & 2 \end{bmatrix}$ 3. $\begin{bmatrix} 1 & 3 \\ 2 & 6 \end{bmatrix}$ 5. $\begin{bmatrix} 9 & 9 \\ -3 & -3 \end{bmatrix}$ 7. $\begin{bmatrix} 11 & 13 \\ 5 & 3 \end{bmatrix}$

9. $\begin{bmatrix} -4 & 3 \\ -2 & -4 \end{bmatrix}$ 11. $\begin{bmatrix} 17 & 9 \\ -2 & 1 \end{bmatrix}$ 13. $\begin{bmatrix} 0 & 0 \\ 0 & 0 \end{bmatrix}$

15. $\begin{bmatrix} 1 & 2 \\ 4 & 3 \end{bmatrix}$ or A 17. $\begin{bmatrix} -5 & 4 & 3 \\ 5 & -9 & 4 \\ 7 & -18 & 17 \end{bmatrix}$

19. $\begin{bmatrix} -2 & 9 & 6 \\ -3 & 3 & 4 \\ 2 & -2 & 1 \end{bmatrix}$ or C 21. $[-16]$ 23. $[2 \quad -19]$

25. $\begin{bmatrix} 3 & -2 & 4 \\ 2 & 1 & -5 \end{bmatrix} \begin{bmatrix} x \\ y \\ z \end{bmatrix} = \begin{bmatrix} 17 \\ 13 \end{bmatrix}$

27. $\begin{bmatrix} 1 & -1 & 2 & -4 \\ 2 & -1 & -1 & 1 \\ 1 & 4 & -3 & -1 \\ 3 & 5 & -7 & 2 \end{bmatrix} \begin{bmatrix} x \\ y \\ z \\ w \end{bmatrix} = \begin{bmatrix} 12 \\ 0 \\ 1 \\ 9 \end{bmatrix}$

29. $\begin{bmatrix} -40.19 & 37.94 & 142.24 \\ -36.78 & 16.63 & 119.62 \\ -1.66 & 14.97 & 12.65 \end{bmatrix}$

31. $(A + B)(A - B) = \begin{bmatrix} -2 & 1 \\ 2 & -1 \end{bmatrix}$, $A^2 - B^2 = \begin{bmatrix} 0 & 3 \\ 0 & -3 \end{bmatrix}$

33. $(A + B)(A - B) = \begin{bmatrix} -2 & 1 \\ 2 & -1 \end{bmatrix} = A^2 + BA - AB - B^2$

35. By matrix multiplication, we obtain
$$\begin{bmatrix} \cos x \cos y - \sin x \sin y & \cos x \sin y + \sin x \cos y \\ -\sin x \cos y - \cos x \sin y & -\sin x \sin y + \cos x \cos y \end{bmatrix}.$$
The use of identities produces the desired result.

37. $A + (B + C) = (A + B) + C$, is similar to Exercise 36, but uses associativity of addition of real numbers.

39. $$(k + m)A = \begin{bmatrix} (k + m)a_{11} & (k + m)a_{12} \\ (k + m)a_{21} & (k + m)a_{22} \end{bmatrix}$$
$$= \begin{bmatrix} ka_{11} + ma_{11} & ka_{12} + ma_{12} \\ ka_{21} + ma_{21} & ka_{22} + ma_{22} \end{bmatrix}$$
$$= \begin{bmatrix} ka_{11} & ka_{12} \\ ka_{21} & ka_{22} \end{bmatrix} + \begin{bmatrix} ma_{11} & ma_{12} \\ ma_{21} & ma_{22} \end{bmatrix}$$
$$= kA + mA$$

41. Find AI, IA, and compare with A.

Margin Exercises, Section 9.6

1. -13 2. -2 3. $-2x + 12$ 4. $(3, 1)$ 5. $\left(-\frac{10}{41}, -\frac{13}{41}\right)$

6. $\left(\frac{3\sqrt{2} + 4\pi}{2 + \pi^2}, \frac{4\sqrt{2} - 3\pi}{2 + \pi^2}\right)$ 7. 93 8. 60 9. $x^3 - x^2$

10. $(1, 3, -2)$

Exercise Set 9.6, p. 591

1 -11 3. $x^3 - 4x$ 5. -109 7. $-x^4 + x^2 - 5x$

9. $\left(-\frac{25}{2}, -\frac{11}{2}\right)$ 11. $\left(\frac{4\pi - 5\sqrt{3}}{3 + \pi^2}, \frac{4\sqrt{3} + 5\pi}{-3 - \pi^2}\right)$

13. $\left(\frac{3}{2}, \frac{13}{14}, \frac{33}{14}\right)$ 15. $\left(\frac{1}{2}, \frac{2}{3}, -\frac{5}{6}\right)$ 17. 2, -2

19. $\{x \mid x \leq -\sqrt{3} \text{ or } x \geq \sqrt{3}\}$ 21. -34 23. 4

25. $\begin{vmatrix} L & -W \\ 2 & 2 \end{vmatrix}$ 27. $\begin{vmatrix} a & b \\ -b & a \end{vmatrix}$ 29. $\begin{vmatrix} 2\pi r & 2\pi r \\ -h & r \end{vmatrix}$

31. Evaluating each determinant gives $\cos^2 x + \sin^2 x = 1$.

Margin Exercises, Section 9.7

1. $a_{11} = -8$, $a_{13} = 6$, $a_{22} = -6$, $a_{31} = -1$, $a_{32} = -3$

2. $M_{22} = \begin{vmatrix} -8 & 6 \\ -1 & 5 \end{vmatrix} = -34$, $M_{32} = \begin{vmatrix} -8 & 6 \\ 4 & 7 \end{vmatrix} = -80$,
$M_{13} = \begin{vmatrix} 4 & -6 \\ -1 & -3 \end{vmatrix} = -18$

3. $A_{22} = -34$, $A_{32} = 80$, $A_{13} = -18$

4. $|A| = 0 \cdot A_{12} + (-6)A_{22} + (-3)A_{32} = -36$

5. $|A| = 6 \cdot A_{13} + 7A_{23} + 5A_{33} = -36$

6. 0 7. 0 8. (a) $|A| = -18$, $|B| = 18$;
(b) The rows are interchanged. 9. (a) $|C| = -10$, $|D| = 10$;
(b) The first and third columns are interchanged.

10. 0 11. $x = -4$ 12. $x = 6$ 13. 0 14. $\begin{vmatrix} -2 & 3 & 4 \\ -3 & 10 & 5 \\ 0 & 9 & 7 \end{vmatrix}$

15. 68 16. -195 17. $(b - a)(c - a)(b - c)$

Exercise Set 9.7, pp. 599–600

1. $a_{11} = 7$, $a_{32} = 2$, $a_{22} = 0$
3. $M_{11} = 6$, $M_{32} = -9$, $M_{22} = -29$

5. $A_{11} = 6, A_{32} = 9, A_{22} = -29$ 7. $|A| = -10$
9. $|A| = -10$ 11. $M_{41} = -14, M_{33} = 20$
13. $A_{24} = 15, A_{43} = 30$ 15. $|A| = 110$ 17. -195
19. -70 21. -4 23. 9072 25. -153 27. 0 29. 0
31. $(x - y)(y - z)(x - z)$ 33. $xyz(x - y)(y - z)(z - x)$
35. Evaluate the determinant and compare with the two-point equation of a line.

37. $\frac{1}{2} \cdot \begin{vmatrix} x_1 & y_1 & 1 \\ x_2 & y_2 & 1 \\ x_3 & y_3 & 1 \end{vmatrix}$

$$= \tfrac{1}{2}\left(x_1 \cdot \begin{vmatrix} y_2 & 1 \\ y_3 & 1 \end{vmatrix} - y_1 \cdot \begin{vmatrix} x_2 & 1 \\ x_3 & 1 \end{vmatrix} + 1 \cdot \begin{vmatrix} x_2 & y_2 \\ x_3 & y_3 \end{vmatrix}\right)$$
$$= \tfrac{1}{2}[x_1(y_2 - y_3) - y_1(x_2 - x_3) + (x_2y_3 - x_3y_2)]$$
$$= \tfrac{1}{2}[x_1y_2 - x_1y_3 - x_2y_1 + x_3y_1 + x_2y_3 - x_3y_2];$$

Area of triangle ABC

$$= \tfrac{1}{2} \cdot (x_1 - x_2)(y_1 + y_2) + \tfrac{1}{2} \cdot (x_3 - x_1)(y_1 + y_3) - \tfrac{1}{2} \cdot (x_3 - x_2)(y_2 + y_3)$$
$$= \tfrac{1}{2}(x_1y_1 + x_1y_2 - x_2y_1 - x_2y_2 + x_3y_1 + x_3y_3 - x_1y_1 - x_1y_3 - x_3y_2 - x_3y_3 + x_2y_2 + x_2y_3)$$
$$= \tfrac{1}{2}(x_1y_2 - x_2y_1 + x_3y_1 - x_1y_3 - x_3y_2 + x_2y_3)$$

Margin Exercises, Section 9.8

1. (a) A; (b) A; (c) both equal A; (d) X; (e) X; (f) both equal to X

2. (a) $\begin{bmatrix} 1 & 0 & 0 \\ 0 & 1 & 0 \\ 0 & 0 & 1 \end{bmatrix} = \mathbf{I}$; (b) I; (c) both equal I

3. $A^{-1} = \begin{bmatrix} -\frac{1}{2} & \frac{1}{2} & \frac{1}{2} \\ 1 & 0 & -1 \\ \frac{3}{2} & -\frac{1}{2} & -\frac{1}{2} \end{bmatrix}$ 4. $A^{-1} = \frac{1}{11}\begin{bmatrix} 2 & 5 \\ 1 & -3 \end{bmatrix}$

5. (a) $\begin{bmatrix} 4 & -2 \\ 1 & 5 \end{bmatrix}\begin{bmatrix} x_1 \\ x_2 \end{bmatrix} = \begin{bmatrix} -1 \\ 1 \end{bmatrix}$; (b) $A = \begin{bmatrix} 4 & -2 \\ 1 & 5 \end{bmatrix}$;
(c) $A^{-1} = \frac{1}{22}\begin{bmatrix} 5 & 2 \\ -1 & 4 \end{bmatrix}$; (d) $x_1 = -\frac{3}{22}, x_2 = \frac{5}{22}$

Exercise Set 9.8, pp. 605–606

1. $A^{-1} = \begin{bmatrix} -3 & 2 \\ 5 & -3 \end{bmatrix}$ 3. $A^{-1} = \begin{bmatrix} 2 & -3 \\ -7 & 11 \end{bmatrix}$

5. $A^{-1} = \begin{bmatrix} \frac{2}{11} & \frac{3}{11} \\ -\frac{1}{11} & \frac{4}{11} \end{bmatrix}$ 7. $A^{-1} = \begin{bmatrix} \frac{3}{8} & -\frac{1}{4} & \frac{1}{8} \\ -\frac{1}{8} & \frac{3}{4} & -\frac{3}{8} \\ -\frac{1}{4} & \frac{1}{2} & \frac{1}{4} \end{bmatrix}$

9. $A^{-1} = \begin{bmatrix} \frac{1}{3} & 0 & \frac{1}{3} \\ -\frac{2}{5} & \frac{2}{5} & \frac{1}{5} \\ \frac{2}{15} & \frac{1}{5} & -\frac{1}{15} \end{bmatrix}$ 11. A^{-1} does not exist.

13. $A^{-1} = \begin{bmatrix} 1 & -2 & 3 & 8 \\ 0 & 1 & -3 & 1 \\ 0 & 0 & 1 & -2 \\ 0 & 0 & 0 & -1 \end{bmatrix}$ 15. A^{-1} does not exist.

17. $(-23, 83)$ 19. $(-1, 5, 1)$

21. $\begin{bmatrix} 4 & -3 \\ 1 & 2 \end{bmatrix}\begin{bmatrix} x \\ y \end{bmatrix} = \begin{bmatrix} 2 \\ -1 \end{bmatrix}$, $A^{-1} = \frac{1}{11}\begin{bmatrix} 2 & 3 \\ -1 & 4 \end{bmatrix}$, $\left(\frac{1}{11}, -\frac{6}{11}\right)$

23. $\begin{bmatrix} 7 & -2 \\ 9 & 3 \end{bmatrix}\begin{bmatrix} x \\ y \end{bmatrix} = \begin{bmatrix} -3 \\ 4 \end{bmatrix}$, $A^{-1} = \frac{1}{39}\begin{bmatrix} 3 & 2 \\ -9 & 7 \end{bmatrix}$, $\left(-\frac{1}{39}, \frac{55}{39}\right)$

25. $\begin{bmatrix} 1 & 0 & 1 \\ 2 & 1 & 0 \\ 1 & -1 & 1 \end{bmatrix}\begin{bmatrix} x \\ y \\ z \end{bmatrix} = \begin{bmatrix} 1 \\ 3 \\ 4 \end{bmatrix}$; $A^{-1} = \frac{1}{2}\begin{bmatrix} -1 & 1 & 1 \\ 2 & 0 & -2 \\ 3 & -1 & -1 \end{bmatrix}$, $(3, -3, -2)$

27. $(1, -1, 0, 1)$ 29. Find AI and IA and compare with A.

31. A^{-1} exists if and only if $xy \neq 0$. $A^{-1} = \begin{bmatrix} x^{-1} & 0 \\ 0 & y^{-1} \end{bmatrix}$.

33. A^{-1} exists if and only if $xyzw \neq 0$.

$$A^{-1} = \begin{bmatrix} \frac{1}{x} & -\frac{1}{xy} & -\frac{1}{xz} & -\frac{1}{xw} \\ 0 & \frac{1}{y} & 0 & 0 \\ 0 & 0 & \frac{1}{z} & 0 \\ 0 & 0 & 0 & \frac{1}{w} \end{bmatrix}$$

Margin Exercises, Section 9.9

1. Yes 2. No

3.

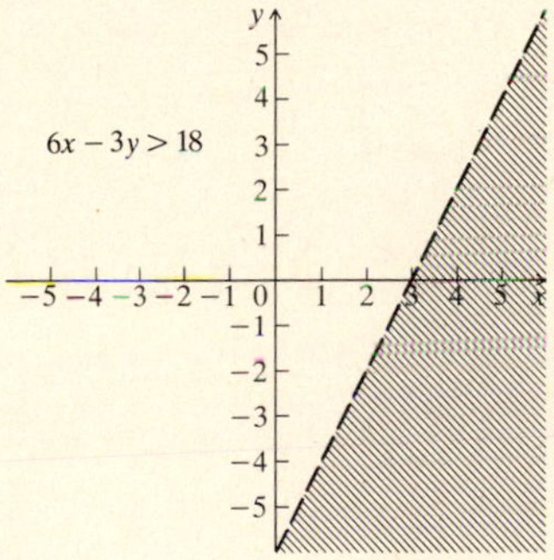

4.

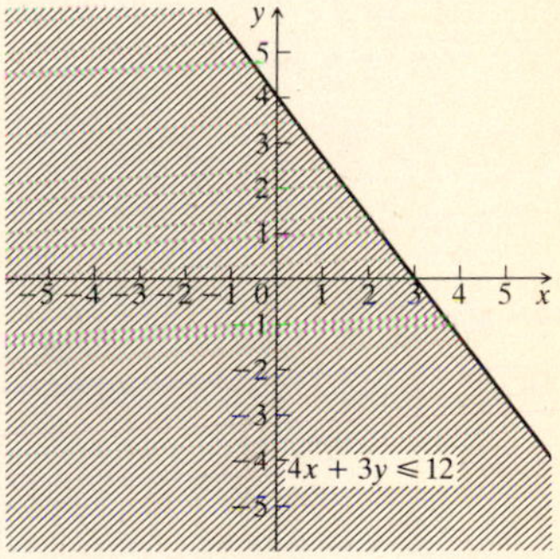

5.

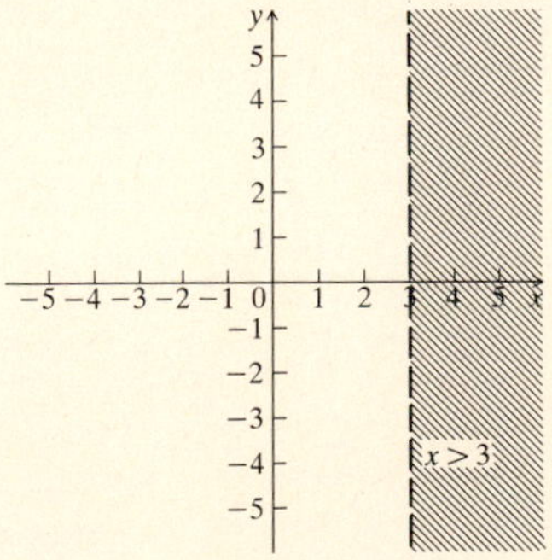

6.

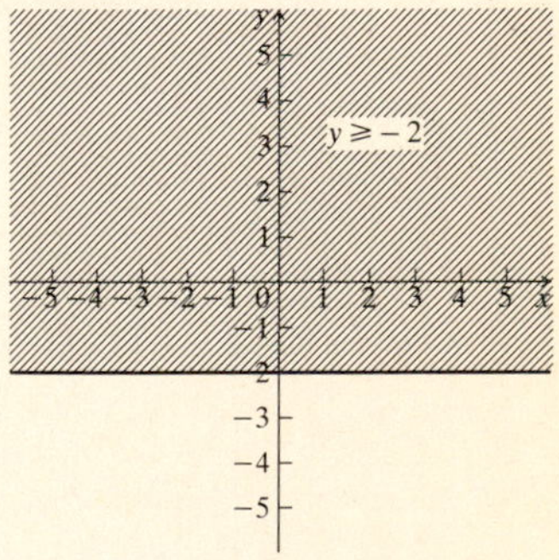

7.

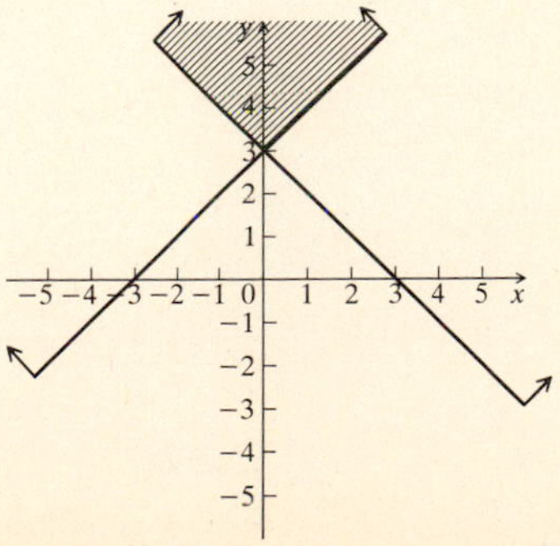

8.

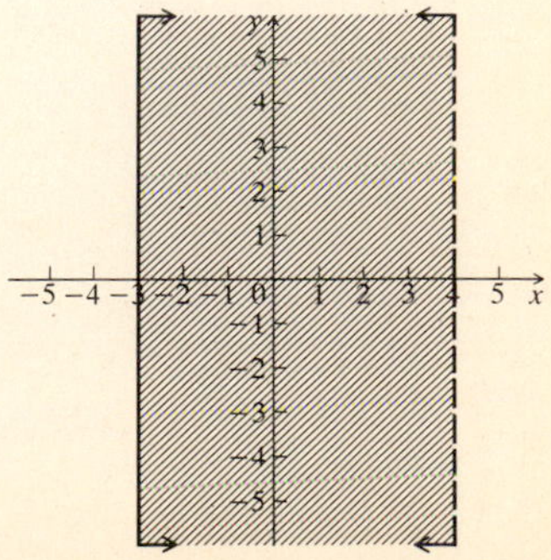

9. Vertices: $\left(\frac{3}{2}, 1\right), \left(\frac{1}{2}, 3\right), \left(\frac{15}{2}, 3\right), \left(\frac{9}{2}, 1\right)$

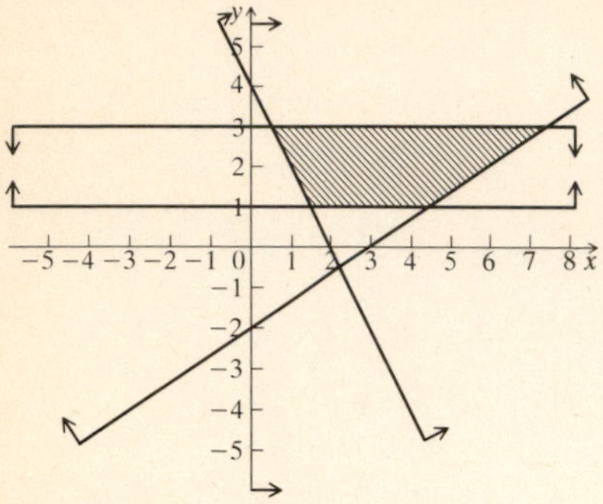

10. Vertices: $(0, 0), (4, 0), \left(4, \frac{5}{3}\right), (0, 3), \left(\frac{12}{5}, 3\right)$

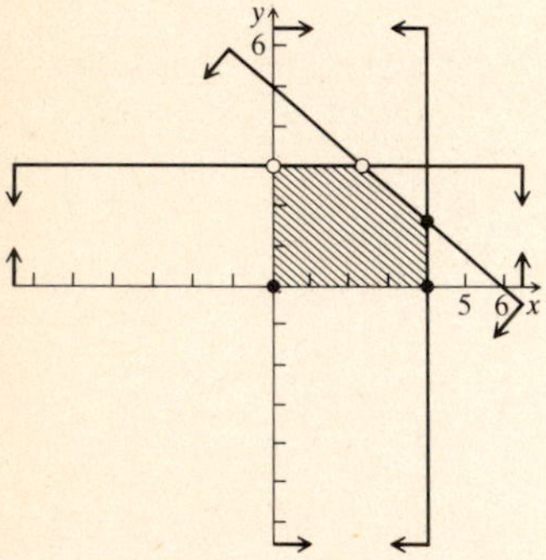

11. Maximum 120, when $x = 3, y = 3$; minimum 34, when $x = 1, y = 0$ **12.** Maximum 45, when $x = 6, y = 0$; minimum 12, when $x = 0, y = 3$
13. Maximum \$23.70, by selling 50 hot dogs and 40 hamburgers

Exercise Set 9.9, pp. 616–618

1. No **3.** No

5.

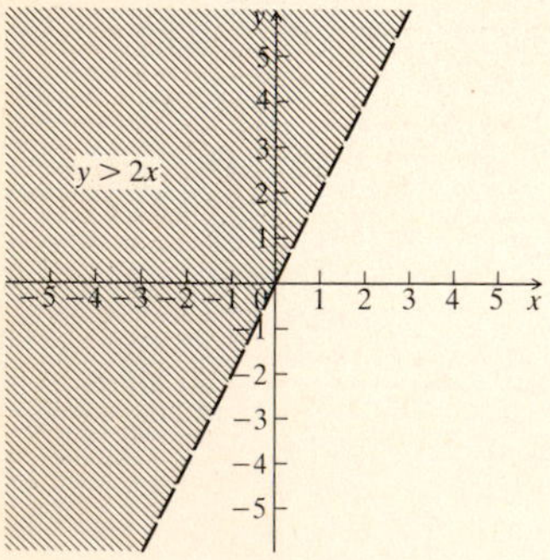

7.

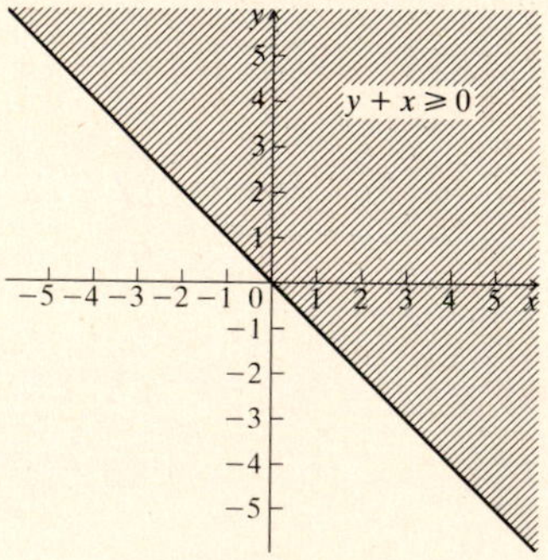

9.

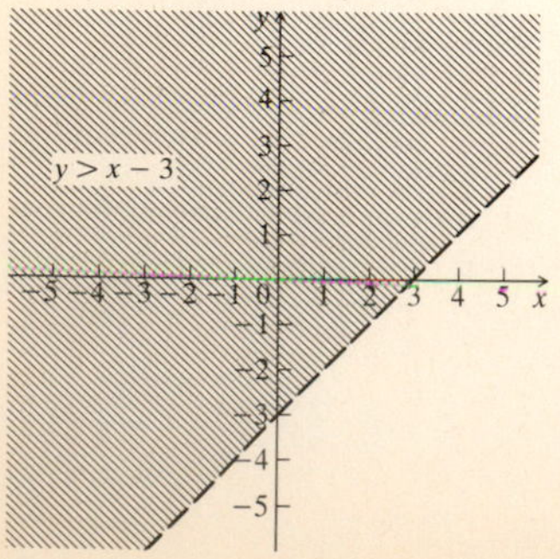

11.

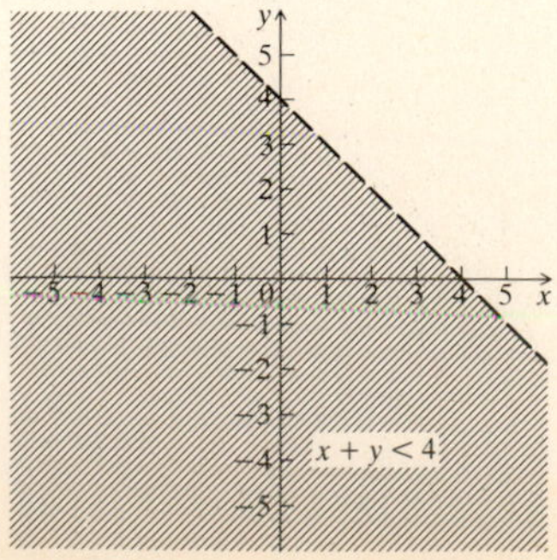

13.

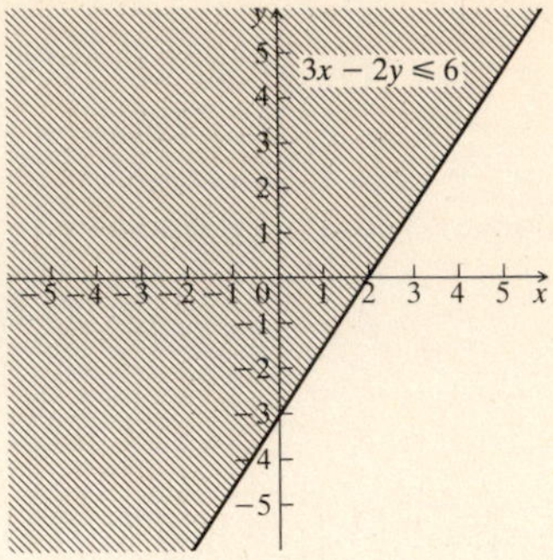

15.

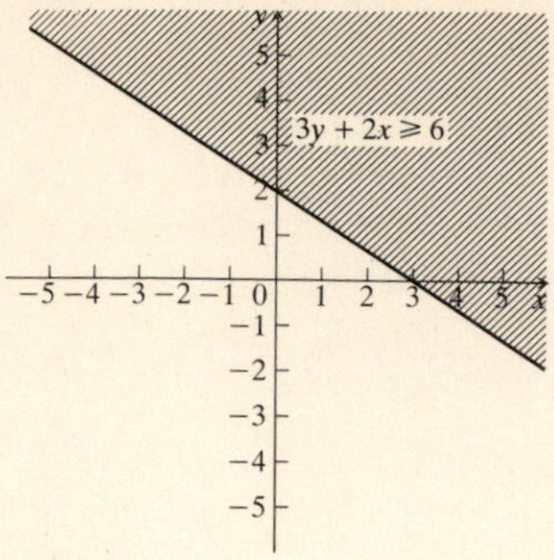

17.

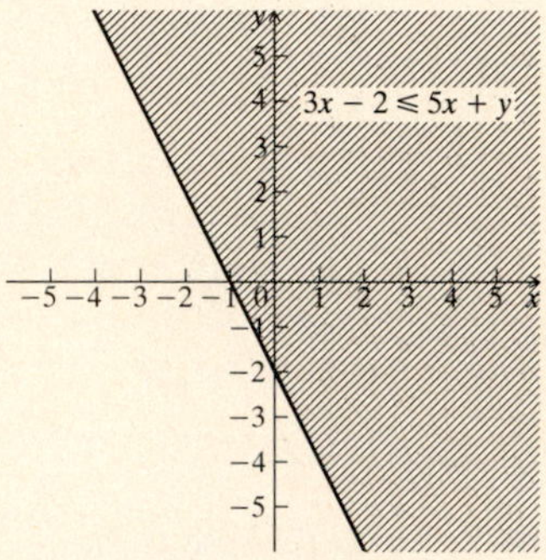

19.

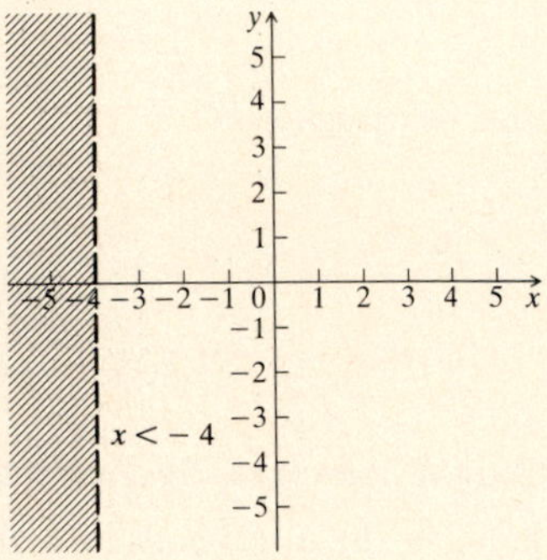

21.

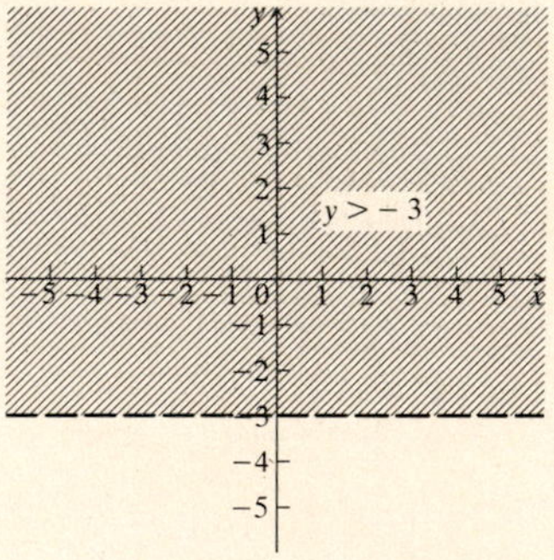

23.

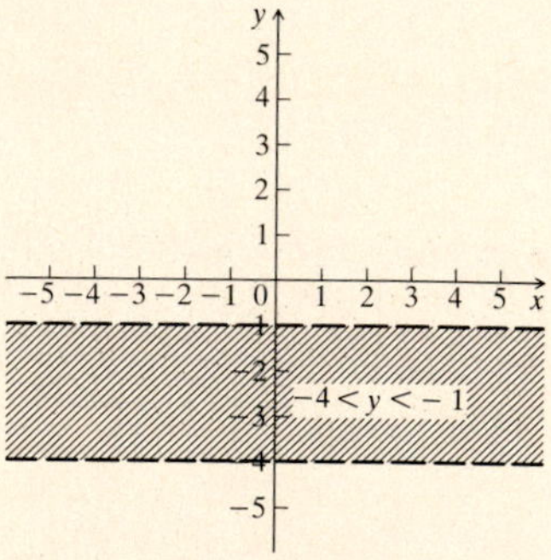

25.

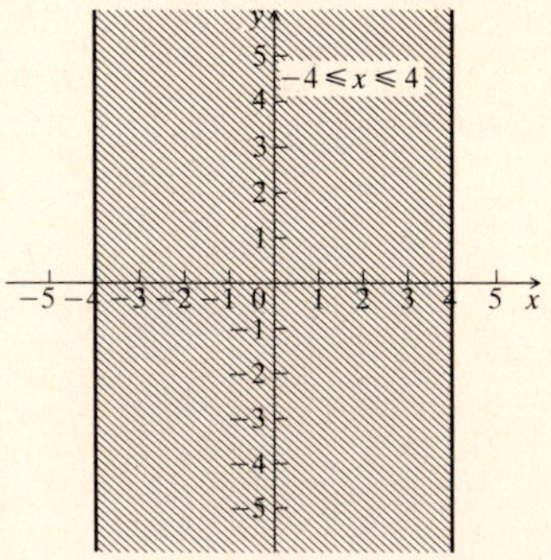

27.

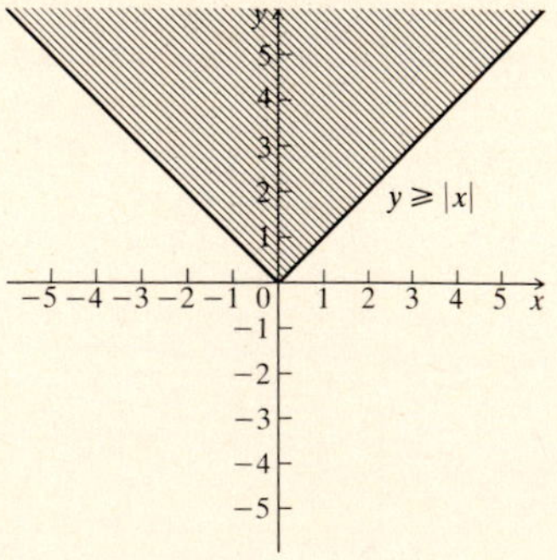

29.

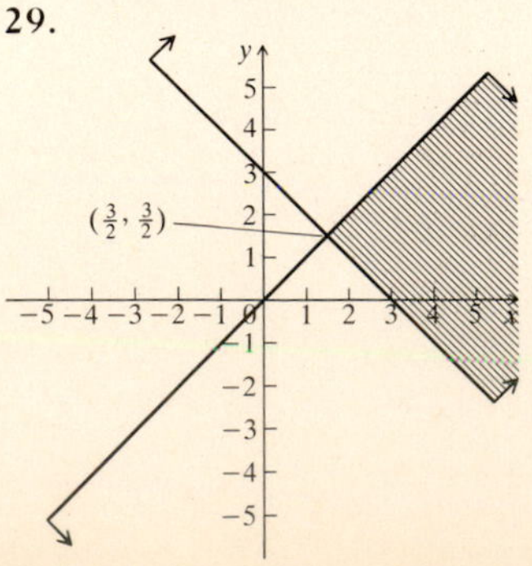

31.

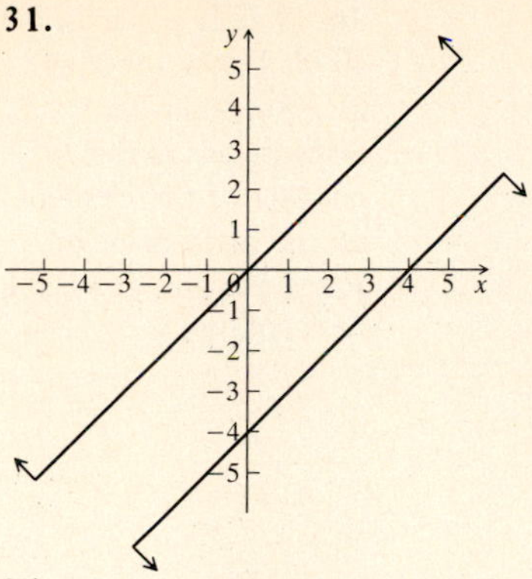

33.

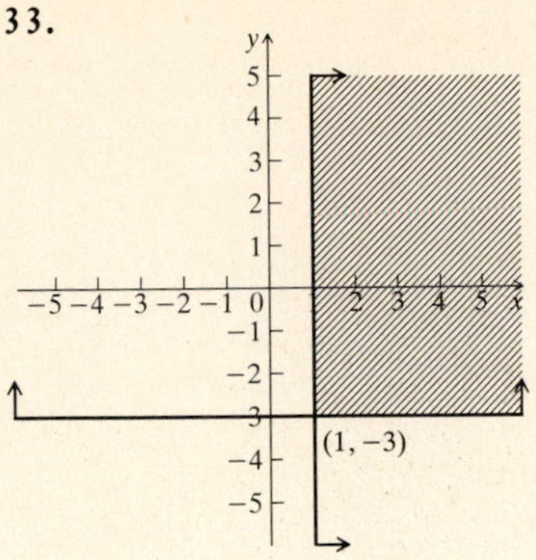

45.

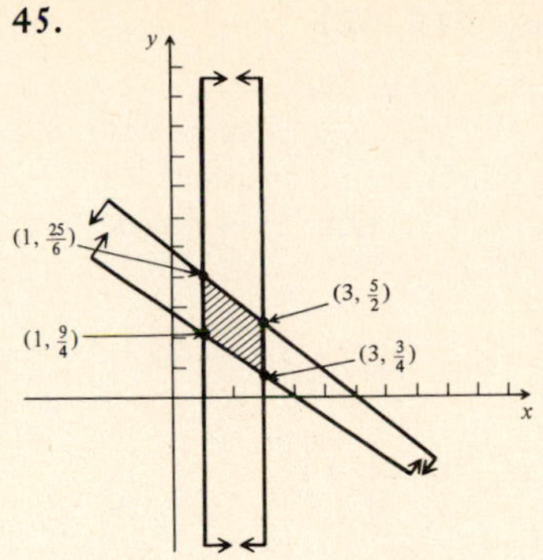

47. The maximum value of P is 179 when $x = 7$ and $y = 0$. The minimum value of P is 48 when $x = 0$ and $y = 4$. **49.** The maximum value of F is 216 when $x = 0$ and $y = 6$. The minimum value of F is 0 when $x = 0$ and $y = 0$. **51.** Maximum income of \$18 when 100 of each type of biscuit is made. **53.** Maximum score of 425 when 5 questions of type A and 15 of type B are answered. **55.** Maximum profit of \$11,000 is achieved by producing 100 units of lumber and 300 units of plywood. **57.** Maximum income of \$3110 is achieved when \$22,000 is invested in corporate bonds and \$18,000 is invested in municipal bonds. **59.** Maximum profit of \$2520 when 125 batches of Smello and 187.5 batches of Roppo are made.

35.

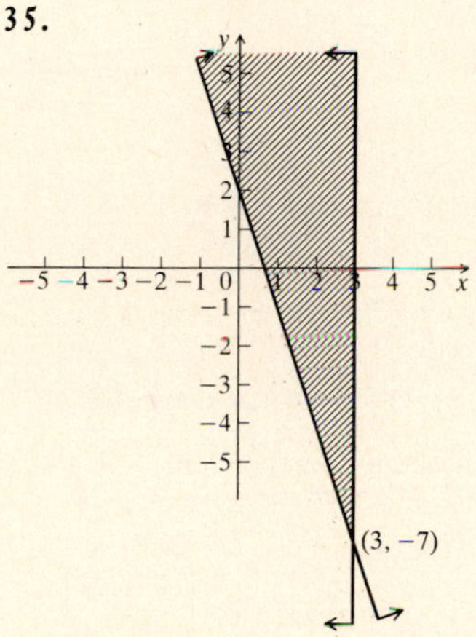

37.

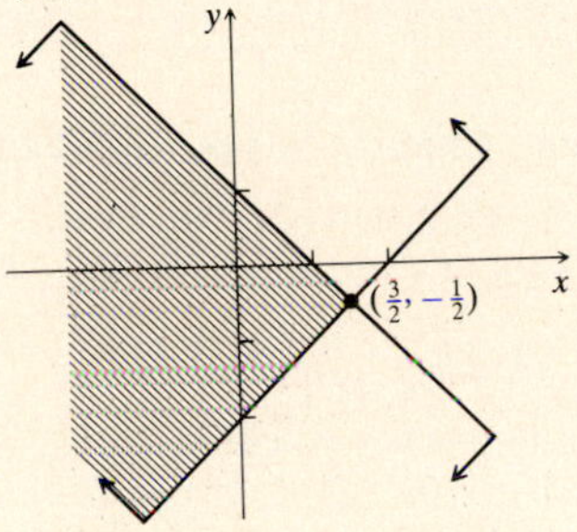

61.

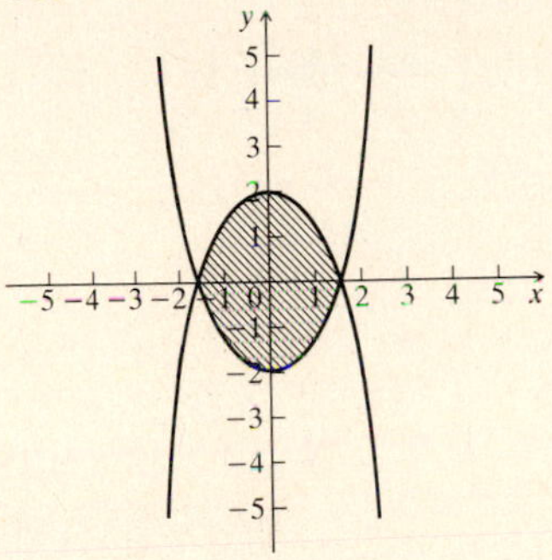

39.

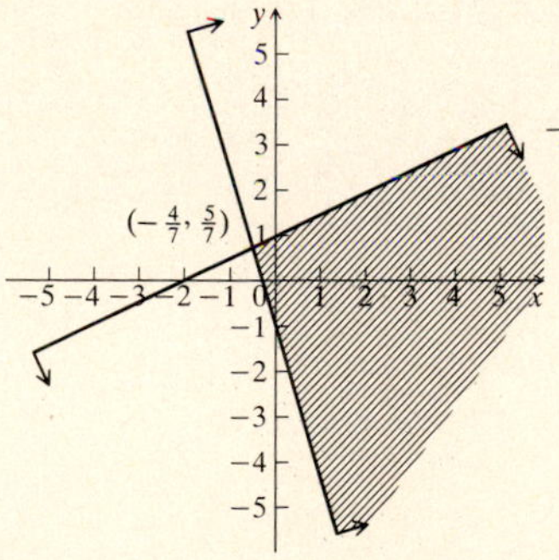

41.

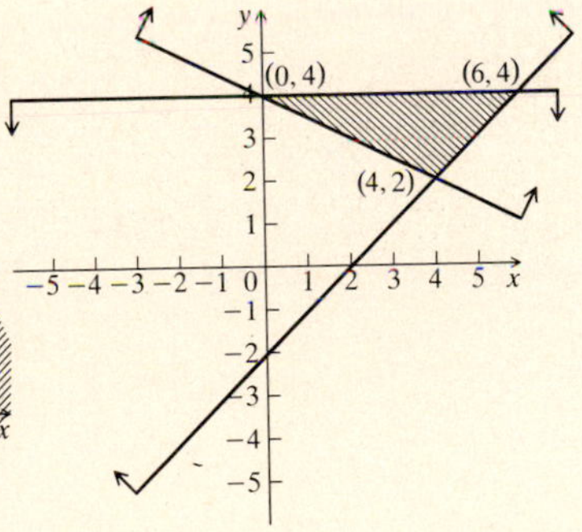

63. $2w + t \geq 60,$
$w \geq 0,$
$t \geq 0$

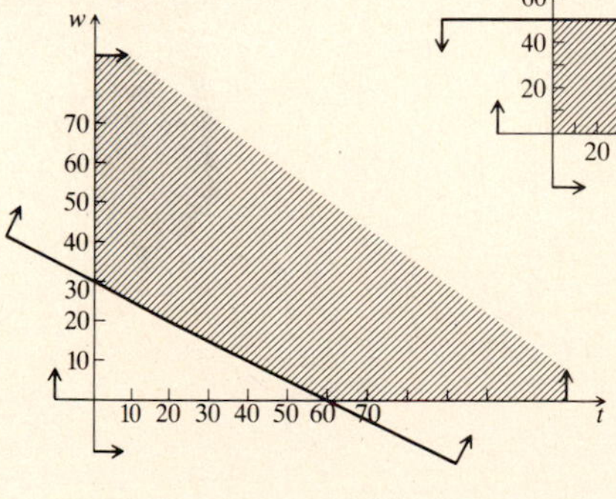

65.

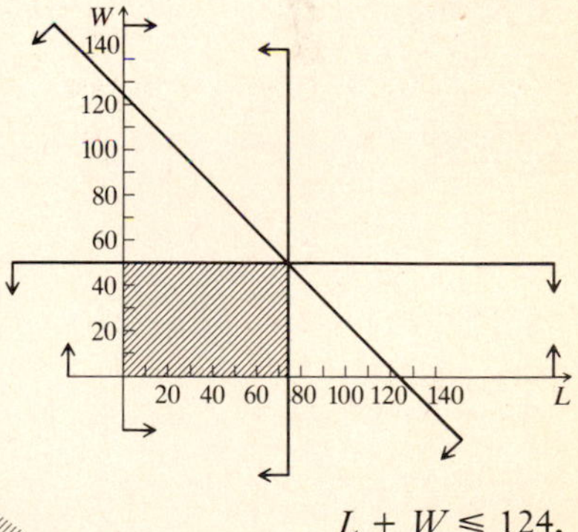

$L + W \leq 124,$
$0 \leq L \leq 74,$
$0 \leq W \leq 50$

43.

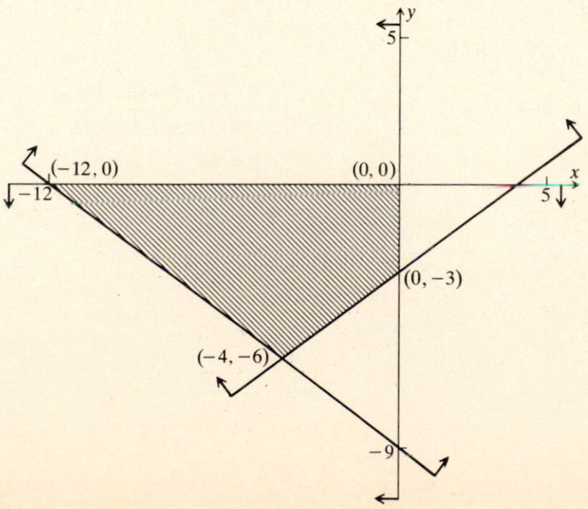

67.

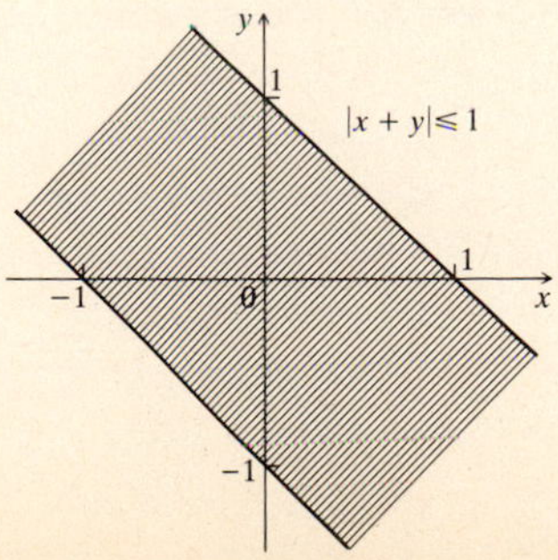

69.

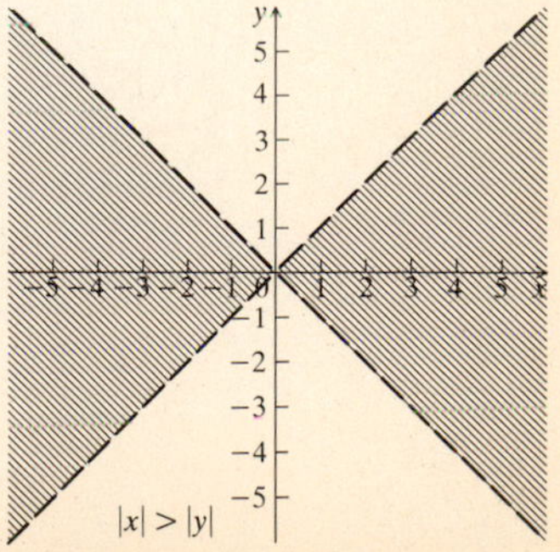

Review Exercises: Chapter 9, pp. 619–621

1. [9.1] $(-2, -2)$ **2.** [9.1] $(-5, 4)$ **3.** [9.3] $\emptyset$
4. [9.3] $\emptyset$ **5.** [9.2] $(0, 0, 0)$ **6.** [9.2] $(w, x, y, z) = (-5, 13, 8, 2)$
7. [9.3] Consistent: 1, 2, 5, 6; the others are inconsistent
8. [9.3] All are independent. **9.** [9.1] 31 nickels, 44 dimes
10. [9.1] \$1600 at 10%, \$3400 at 10.5%
11. [9.2] A: 32°; B: 96°; C: 52°
12. [9.2] A: 74.5; B: 68.5; C: 82 **13.** [9.4] $(1, 2)$
14. [9.4] $(-3, 4, -2)$
15. [9.3], [9.4] $\left(\frac{z}{2}, -\frac{z}{2}, z\right)$; $(0, 0, 0)$, $\left(\frac{1}{2}, -\frac{1}{2}, 1\right)$, $(1, -1, 2)$, etc. **16.** [9.4] $(-4, 1, -2, 3)$
17. [9.2] $y = -x^2 - 2x + 3$ **18.** [9.6] 10
19. [9.6] -18 **20.** [9.6] $2x + 12$ **21.** [9.6] -6
22. [9.6] -16.588 **23.** [9.6] 0 **24.** [9.6] $(3, -2)$
25. [9.6] $(a, 0)$ **26.** [9.6] $\left(\frac{3}{2}, \frac{13}{14}, \frac{33}{14}\right)$

27. [9.5] $\begin{bmatrix} 0 & -1 & 6 \\ 3 & 1 & -2 \\ -2 & 1 & -2 \end{bmatrix}$ **28.** [9.5] $\begin{bmatrix} -3 & 3 & 0 \\ -6 & -9 & 6 \\ 6 & 0 & -3 \end{bmatrix}$

29. [9.5] $\begin{bmatrix} -1 & 1 & 0 \\ -2 & -3 & 2 \\ 2 & 0 & -1 \end{bmatrix}$ **30.** [9.5] $\begin{bmatrix} -2 & 2 & 6 \\ 1 & -8 & 18 \\ 2 & 1 & -15 \end{bmatrix}$

31. [9.5] Not possible **32.** [9.5] $\begin{bmatrix} 2 & -1 & -6 \\ 1 & 5 & -2 \\ -2 & -1 & 4 \end{bmatrix}$

33. [9.5] $\begin{bmatrix} 3 & -2 & -6 \\ 3 & 8 & -4 \\ -4 & -1 & 5 \end{bmatrix}$ **34.** [9.5] $\begin{bmatrix} -2 & -1 & 18 \\ 5 & -3 & -2 \\ -2 & 3 & -8 \end{bmatrix}$

35. [9.8] $\begin{bmatrix} -\frac{1}{2} & 0 \\ \frac{1}{6} & \frac{1}{3} \end{bmatrix}$ **36.** [9.8] $\begin{bmatrix} 0 & 0 & \frac{1}{4} \\ 0 & -\frac{1}{2} & 0 \\ \frac{1}{3} & 0 & 0 \end{bmatrix}$

37. [9.8] $\begin{bmatrix} 1 & 0 & 0 & 0 \\ 0 & \frac{1}{9} & \frac{5}{18} & 0 \\ 0 & -\frac{1}{9} & \frac{2}{9} & 0 \\ 0 & 0 & 0 & 1 \end{bmatrix}$

38. [9.5] $\begin{bmatrix} 3 & -2 & 4 \\ 1 & 5 & -3 \\ 2 & -3 & 7 \end{bmatrix} \begin{bmatrix} x \\ y \\ z \end{bmatrix} = \begin{bmatrix} 13 \\ 7 \\ -8 \end{bmatrix}$ **39.** [9.7] -31

40. [9.7] -1 **41.** [9.7] 0 **42.** [9.7] 120

43. [9.7] $\begin{vmatrix} 5a & 5b & 5c \\ 3a & 3b & 3c \\ d & e & f \end{vmatrix} = 5(3) \begin{vmatrix} a & b & c \\ a & b & c \\ d & e & f \end{vmatrix} = 0$, since the first two rows are the same. **44.** [9.7] $(a - b)(b - c)(c - a)$
45. [9.7] $(x - y)(y - z)(z - x)(xy + yz + zx)$
46. [9.7] $(b - a)(c - a)(d - a)(c - b)(d - b)(d - c)$
47. [9.8] $(-5, 4)$ **48.** [9.9] $(0, 9)$, $(2, 5)$, $(5, 1)$, $(8, 0)$
49. [9.9] Minimum = 52 at $(2, 4)$; maximum = 92 at $(2, 8)$
50. [9.9] Type A: 0; type B: 10; maximum score = 120 pts
51. [9.2] \$10,000 at 12%, \$12,000 at 13%, \$18,000 at $14\frac{1}{2}$%
52. [9.1] $\left(\frac{5}{18}, \frac{1}{7}\right)$ **53.** [9.2] $\left(1, \frac{1}{2}, \frac{1}{3}\right)$ **54.** [9.9]

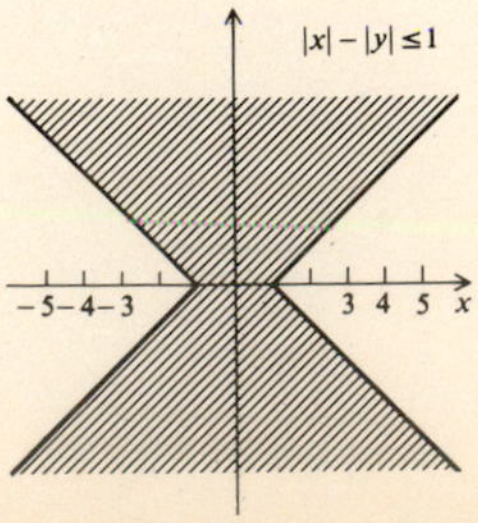

55. [9.9]

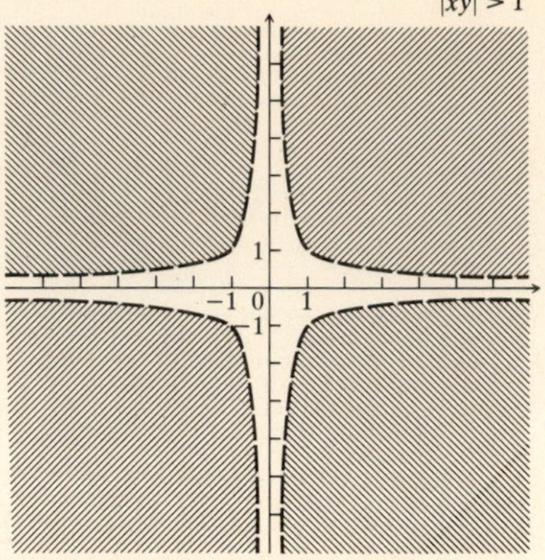

56. [9.7] If a matrix has all 0's below the main diagonal, then its determinant is the product of the elements on the main diagonal. *Proof:* Expand down the first column.

Test: Chapter 9, pp. 621–623

1. [9.1] $(2, 1)$ **2.** [9.1] Boat: 15 km/h; stream: 3 km/h **3.** [9.1] 12 gal of 15%, 8 gal of 75% **4.** [9.2] A: 44; B: 240; C: 216
5. [9.4] $\left(\frac{1}{4}, \frac{1}{2}\right)$ **6.** [9.4] $\left(\frac{z}{3}, -\frac{5z}{2}, z\right)$: $(0, 0, 0)$, $\left(\frac{1}{3}, -\frac{5}{2}, 1\right)$, $\left(\frac{2}{3}, -5, 2\right)$, $\left(1, -\frac{15}{2}, 3\right)$, $(2, -15, 6)$, etc. Answers may vary.
7. [9.4] $(3, -1, -2)$ **8.** [9.3] Consistent, independent **9.** [9.3] Consistent, dependent **10.** [9.2] $y = -2x^2 + 4x + 1$
11. [9.6] 2 **12.** [9.7] -2 **13.** [9.6] $(4, -1)$ **14.** [9.6] $(0, -2, 1)$ **15.** [9.5] Not possible **16.** [9.5] A or $\begin{bmatrix} -1 & 3 \\ 0 & 4 \end{bmatrix}$

17. [9.5] $\begin{bmatrix} 0 & 3 \\ 0 & 5 \end{bmatrix}$ **18.** [9.5] $\begin{bmatrix} 5 & 1 & 2 \\ 3 & 3 & 3 \\ 1 & 4 & -3 \end{bmatrix}$ **19.** [9.5] $\begin{bmatrix} 1 & 8 \\ -6 & 7 \end{bmatrix}$ **20.** [9.5] $\begin{bmatrix} -7 & -1 \\ 24 & -17 \end{bmatrix}$ **21.** [9.5] Not possible

22. [9.5] $[-2 \;\; 1 \;\; 3]$ **23.** [9.5] $\begin{bmatrix} -5 & 5 & -5 \\ 6 & -6 & 0 \\ -1 & 8 & 12 \end{bmatrix}$ **24.** [9.5] $\begin{bmatrix} -2 & -5 \\ 6 & -3 \end{bmatrix}$ **25.** [9.8] $\begin{bmatrix} 0 & \frac{1}{2} \\ 1 & \frac{3}{2} \end{bmatrix}$ **26.** [9.8] Does not exist

27. [9.8] $\begin{bmatrix} \frac{2}{3} & 0 & \frac{1}{6} \\ \frac{1}{3} & 0 & \frac{1}{3} \\ -2 & 1 & -\frac{1}{2} \end{bmatrix}$ **28.** [9.7] $a_{12} = 1$, $M_{12} = \begin{vmatrix} 2 & -1 \\ -5 & 4 \end{vmatrix} = 3$, $A_{12} = -3$ **29.** [9.7] -28
30. [9.7] 0 **31.** [9.7] 99 **32.** [9.7] 22
33. [9.7] $(a - b)(b - c)(c - a)(a + b + c)$
34. [9.8] $\begin{bmatrix} 2 & -3 \\ 1 & 4 \end{bmatrix} \begin{bmatrix} x \\ y \end{bmatrix} = \begin{bmatrix} -9 \\ 1 \end{bmatrix}$, $(-3, 1)$
35. [9.9]

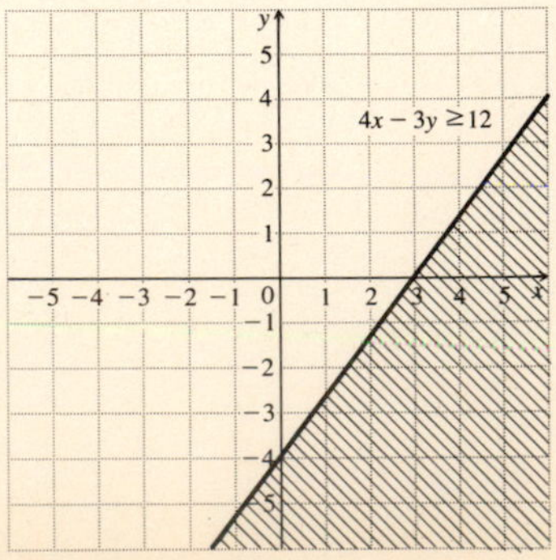

36. [9.9] Minimum 40 at $(2, 0)$, maximum 460 at $(2, 7)$
37. [9.9] Type A: 0; type B: 9; maximum score: 108
38. [9.1] $\left(-1, \frac{1}{3}\right)$

CHAPTER 10

Margin Exercises, Section 10.1

1. The union of the graphs of $y = -\frac{1}{2}x$ and $y = \frac{1}{2}x$
2. The union of the graphs of $y = 2$ and $y = -2$
3. Graph consists of a single line, $y = 3x$
4. Graph consists just of $(0, 0)$
5. There is no real-number solution, hence no graph.
6. **(a)** Stretched; **(b)** shrunk
7. V: $(-3, 0)$, $(3, 0)$, $(0, -1)$, $(0, 1)$; F: $(-2\sqrt{2}, 0)$, $(2\sqrt{2}, 0)$

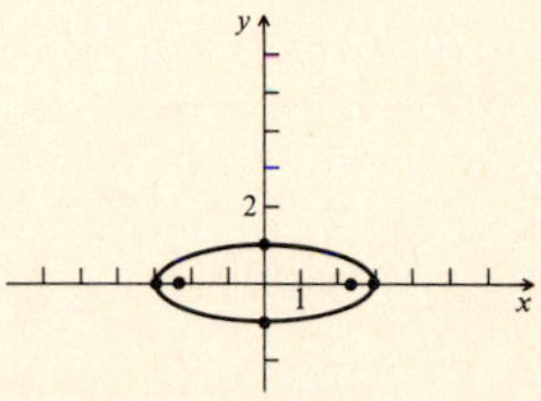

8. V: $(-5, 0)$, $(5, 0)$, $(0, -3)$, $(0, 3)$; F: $(-4, 0)$, $(4, 0)$

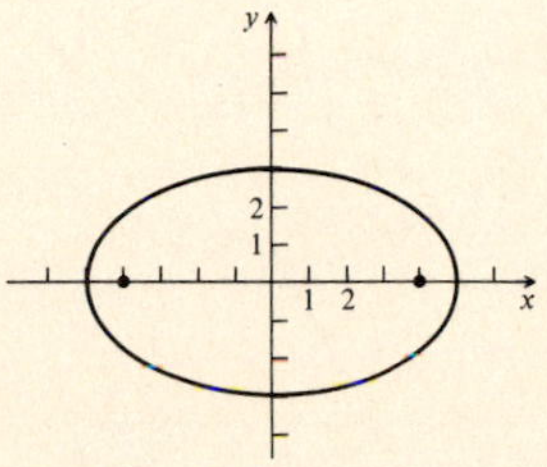

9. V: $(-2, 0)$, $(2, 0)$, $(0, \sqrt{2})$ $(0, -\sqrt{2})$; F: $(-\sqrt{2}, 0)$, $(\sqrt{2}, 0)$

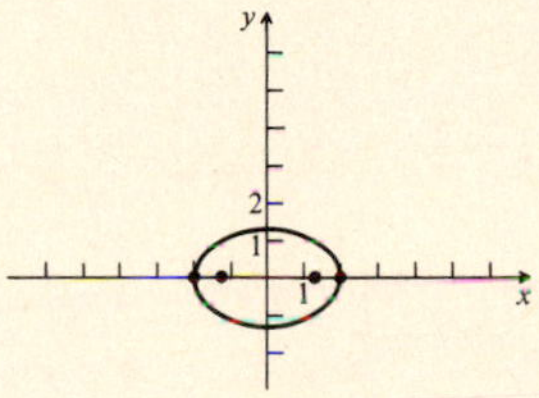

10. C: $(0, 0)$; V: $(-1, 0)$, $(1, 0)$, $(0, 3)$, $(0, -3)$; F: $(0, 2\sqrt{2})$, $(0, -2\sqrt{2})$

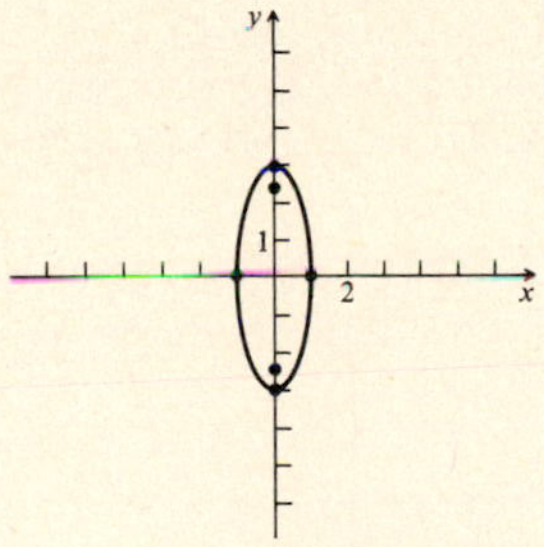

11. C: $(0, 0)$; V: $(3, 0)$, $(-3, 0)$, $(0, 5)$, $(0, -5)$; F: $(0, 4)$, $(0, -4)$

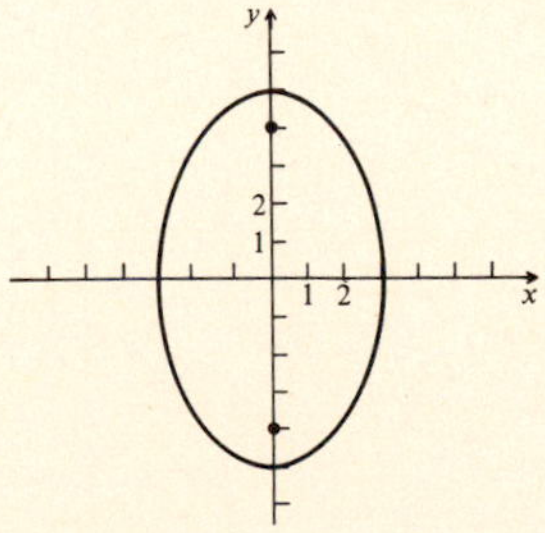

12. C: $(0, 0)$; V: $(\sqrt{2}, 0)$, $(-\sqrt{2}, 0)$, $(0, 2)$, $(0, -2)$; F: $(0, \sqrt{2})$, $(0, -\sqrt{2})$

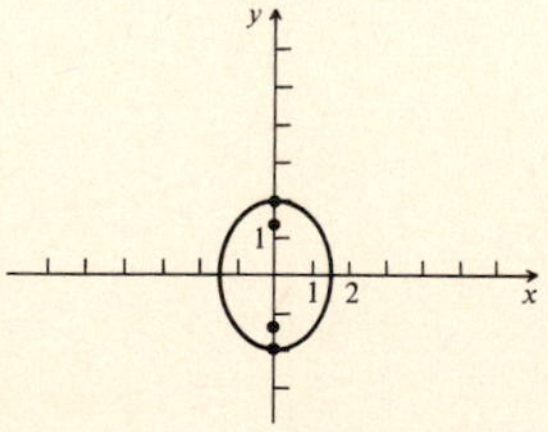

13. C: $(-3, 2)$; V: $\left(-2\frac{4}{5}, 2\right)$, $\left(-3\frac{1}{5}, 2\right)$, $\left(-3, 2\frac{1}{3}\right)$, $\left(-3, 1\frac{2}{3}\right)$; F: $\left(-3, 2\frac{4}{15}\right)$, $\left(-3, 1\frac{11}{15}\right)$

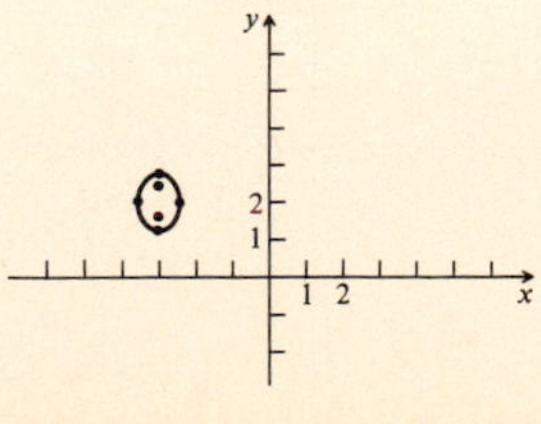

14. C: $(2, -3)$; V: $\left(2\frac{1}{3}, -3\right)$, $\left(1\frac{2}{3}, -3\right)$, $\left(2, -2\frac{4}{5}\right)$, $\left(2, -3\frac{1}{5}\right)$; F: $\left(2\frac{4}{15}, -3\right)$, $\left(1\frac{11}{15}, -3\right)$

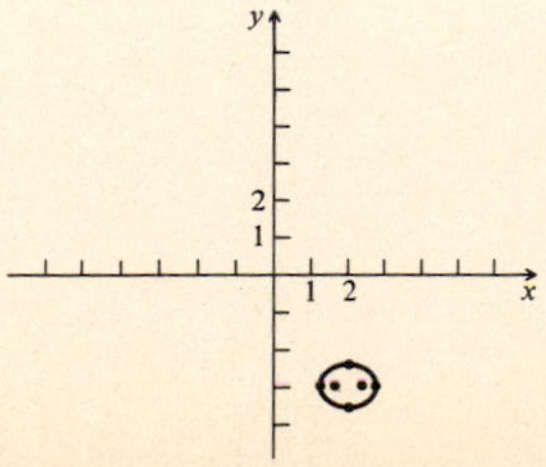

Exercise Set 10.1, pp. 635–636

1. The union of the graphs of $y = x$ and $y = -x$
3. The union of the graphs of $y = -x$ and $y = \frac{3}{2}x$
5. The point $(0, 0)$
7. V: $(2, 0)$, $(-2, 0)$, $(0, 1)$, $(0, -1)$; F: $(\sqrt{3}, 0)$, $(-\sqrt{3}, 0)$

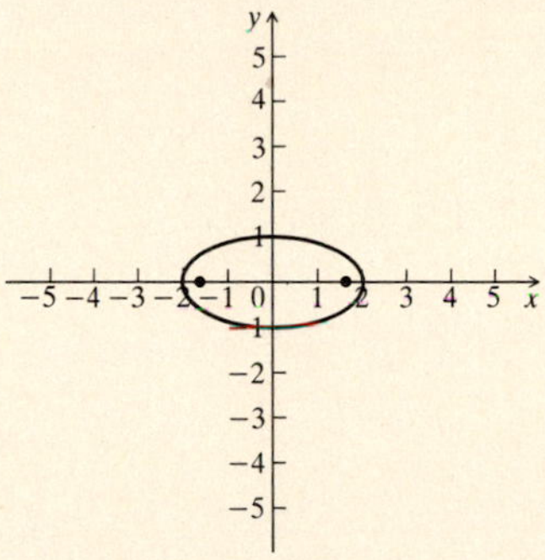

9. V: $(-3, 0)$, $(3, 0)$, $(0, 4)$, $(0, -4)$; F: $(0, \sqrt{7})$, $(0, -\sqrt{7})$

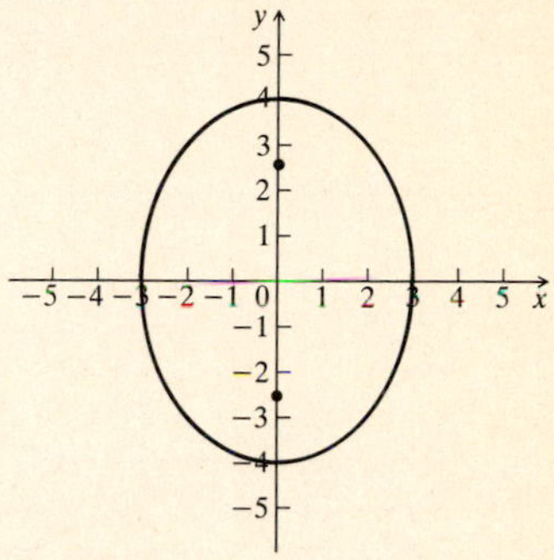

11. V: $(-\sqrt{3}, 0)$, $(\sqrt{3}, 0)$, $(0, \sqrt{2})$, $(0, -\sqrt{2})$; F: $(-1, 0)$, $(1, 0)$

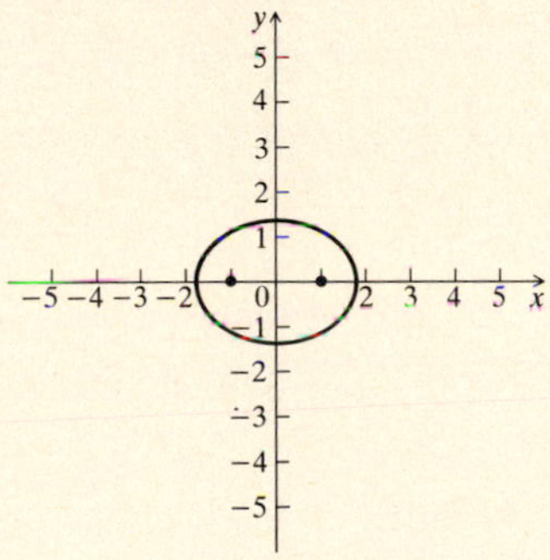

13. V: $\left(\pm\frac{1}{2}, 0\right)$, $\left(0, \pm\frac{1}{3}\right)$; F: $\left(\frac{\pm\sqrt{5}}{6}, 0\right)$

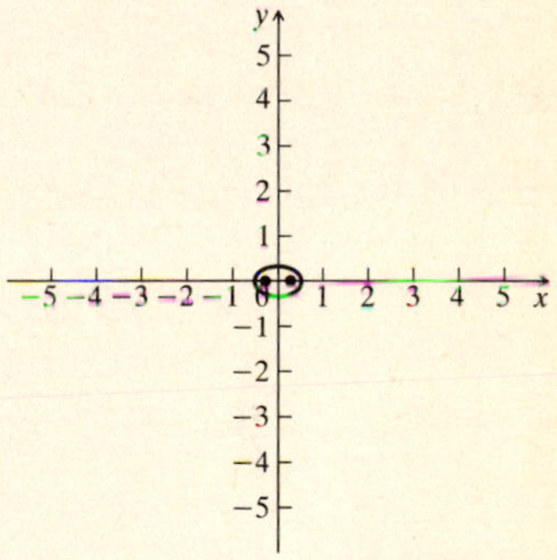

15. C: $(1, 2)$; V: $(3, 2)$, $(-1, 2)$, $(1, 3)$, $(1, 1)$; F: $(1 \pm \sqrt{3}, 2)$

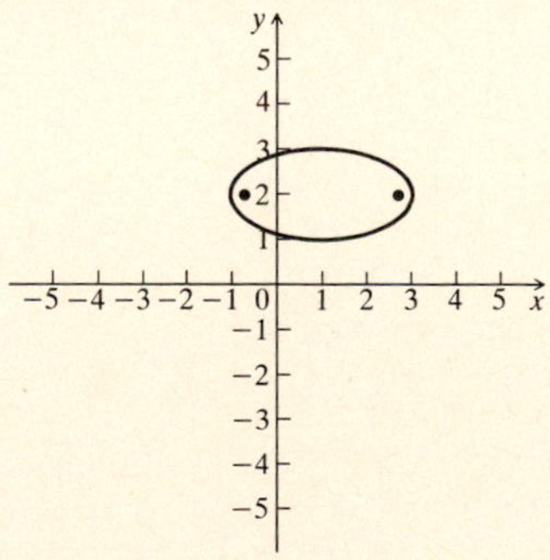

17. C: $(-3, 2)$; V: $(2, 2)$, $(-8, 2)$, $(-3, 6)$, $(-3, -2)$; F: $(0, 2)$, $(-6, 2)$

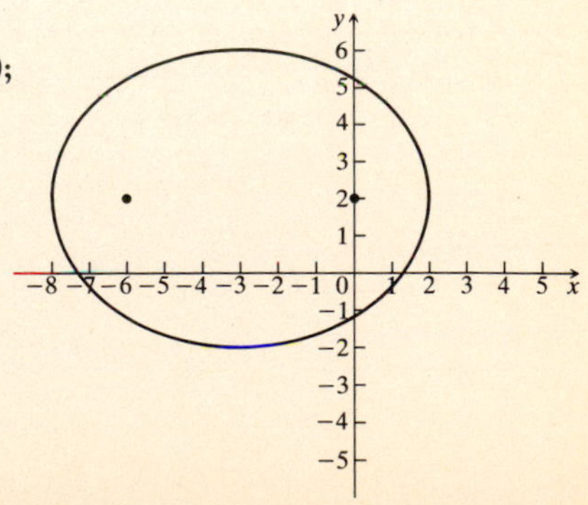

19. C: $(-2, 1)$; V: $(-10, 1)$, $(6, 1)$, $(-2, 1 \pm 4\sqrt{3})$;
F: $(-6, 1)$, $(2, 1)$

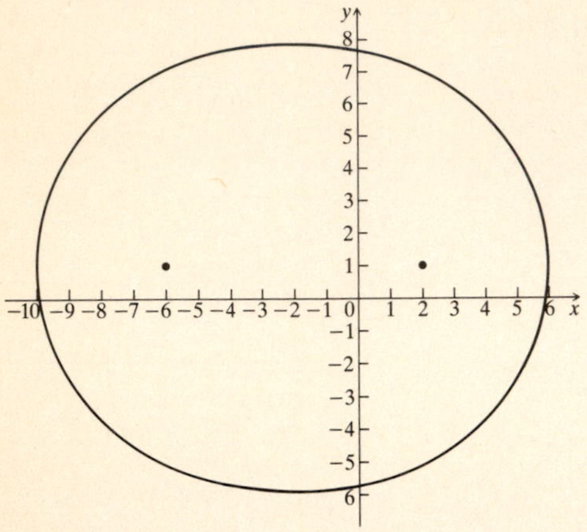

21. C: $(2, -1)$; V: $(-1, -1)$, $(5, -1)$, $(2, 1)$, $(2, -3)$; F: $(2 \pm \sqrt{5}, -1)$

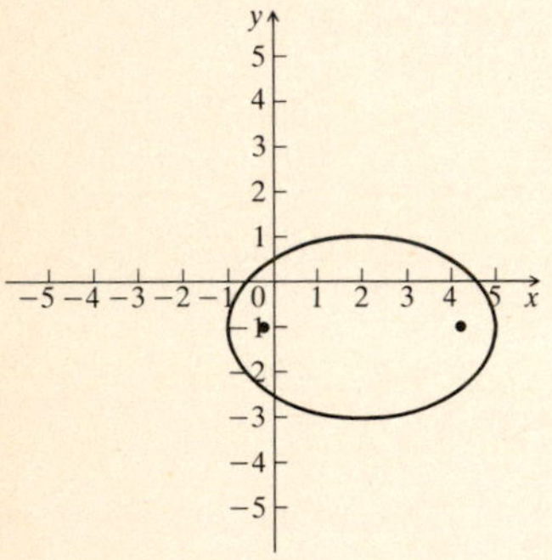

23. C: $(1, 1)$; V: $(0, 1)$, $(2, 1)$, $(1, 3)$, $(1, -1)$; F: $(1, 1 \pm \sqrt{3})$

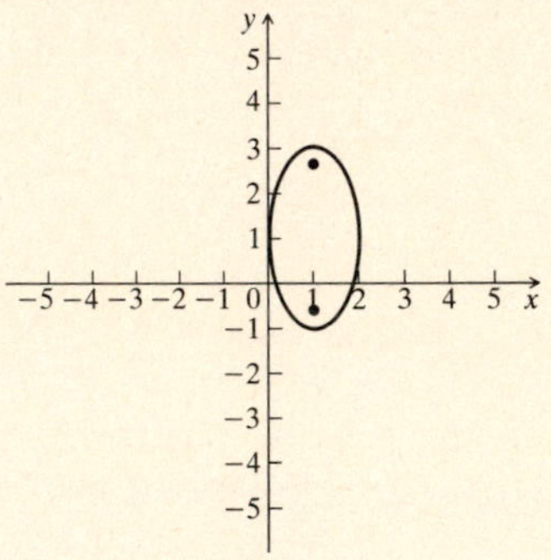

25. C: $(2.003125, -1.00515)$; V: $(5.0234304, -1.00515)$, $(-1.0171804, -1.00515)$, $(2.003125, -3.0186869)$, $(2.003125, 1.0083869)$

27. $\dfrac{x^2}{4} + \dfrac{y^2}{9} = 1$

29. $\dfrac{(x-3)^2}{4} + \dfrac{(y-1)^2}{25} = 1$

31. $\dfrac{(x+2)^2}{\frac{1}{4}} + \dfrac{(y-3)^2}{4} = 1$

33. **(a)** No; **(b)** $y = \pm 3\sqrt{1 - x^2}$;
(c) yes; domain $\{x \mid -1 \leq x \leq 1\}$, range $\{y \mid 0 \leq y \leq 3\}$
(d) yes; domain $\{x \mid -1 \leq x \leq 1\}$, range $\{y \mid -3 \leq y \leq 0\}$

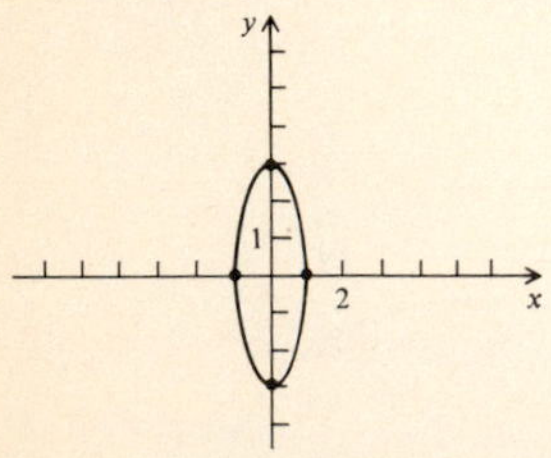

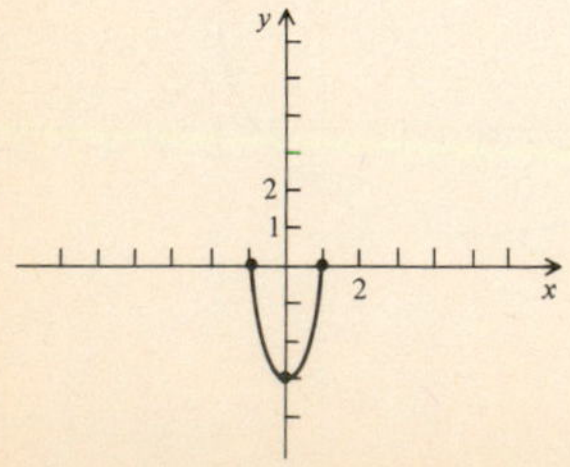

37. $\dfrac{x^2}{50^2} + \dfrac{y^2}{12^2} = 1$, or $\dfrac{x^2}{2500} + \dfrac{y^2}{144} = 1$ **39.** **(a)** $A = \pi \cdot a \cdot b$; **(b)** 20π; **(c)** 739,141.4 ft^2

Margin Exercises, Section 10.2

1. V: $(3, 0)$, $(-3, 0)$; F: $(\sqrt{13}, 0)$, $(-\sqrt{13}, 0)$; A: $y = \frac{2}{3}x$, $y = -\frac{2}{3}x$

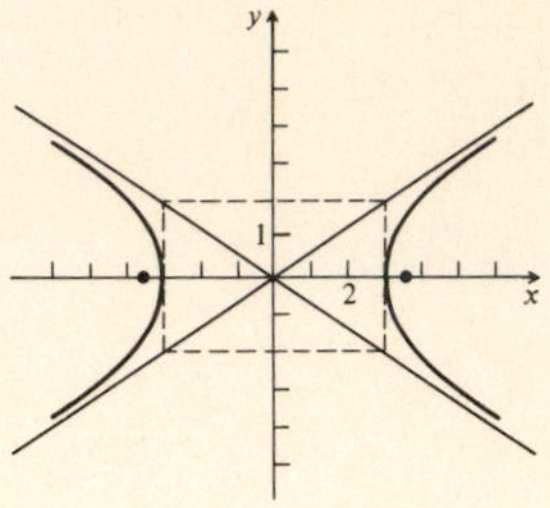

2. V: $(4,0)$, $(-4,0)$; F: $(4\sqrt{2},0)$, $(-4\sqrt{2}, 0)$; A: $y = x$, $y = -x$

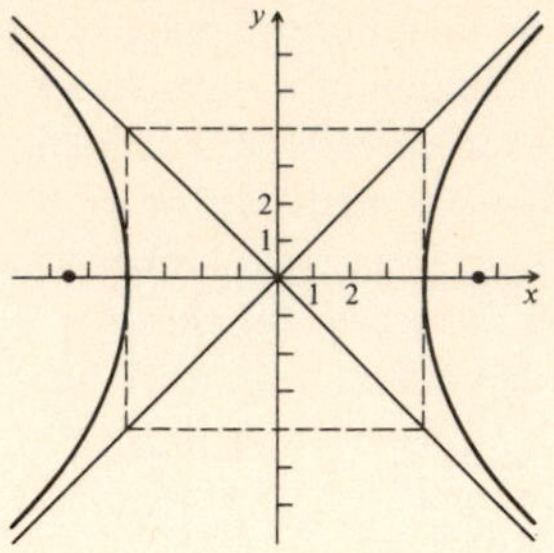

3. V: $(0,5)$, $(0,-5)$; F: $(0,\sqrt{34})$, $(0, -\sqrt{34})$; A: $y = \frac{5}{3}x$, $y = -\frac{5}{3}x$

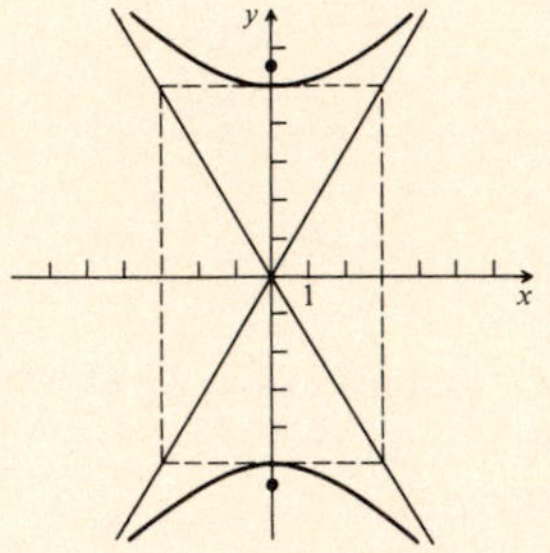

4. V: $(0,5)$, $(0,-5)$; F: $(0,5\sqrt{2})$, $(0, -5\sqrt{2})$; A: $y = x$, $y = -x$

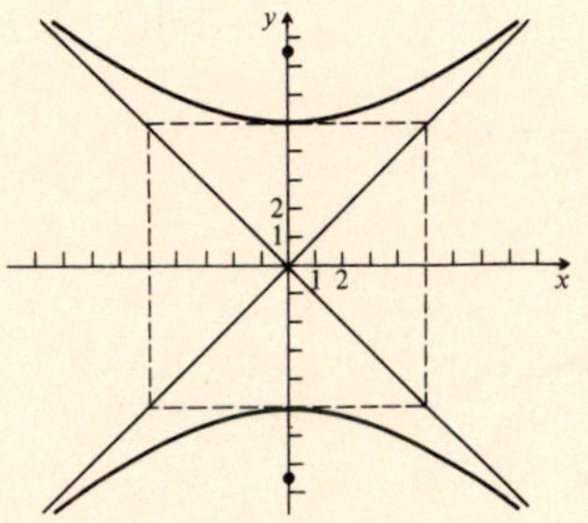

5. C: $(1,-2)$; V: $(6,-2)$, $(-4,-2)$; F: $(1+\sqrt{29},-2)$, $(1-\sqrt{29},-2)$; A: $y = \frac{2}{5}x - \frac{12}{5}$, $y = -\frac{2}{5}x - \frac{8}{5}$

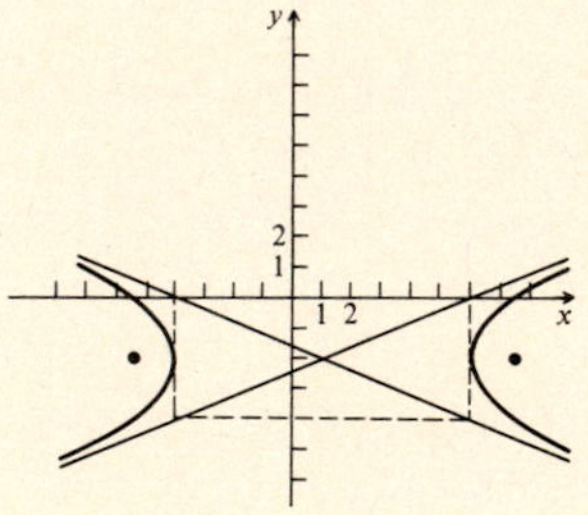

6. C: $(-1, 2)$; V: $(-1, 5)$, $(-1, -1)$; F: $(-1, 7)$, $(-1, -3)$; A: $y = \frac{3}{4}x + \frac{11}{4}$, $y = -\frac{3}{4}x + \frac{5}{4}$

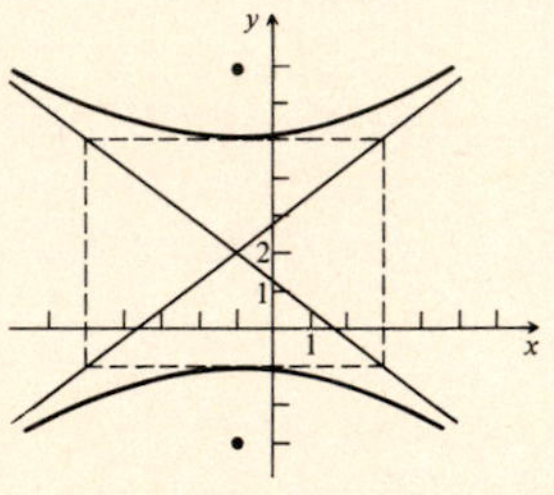

7.

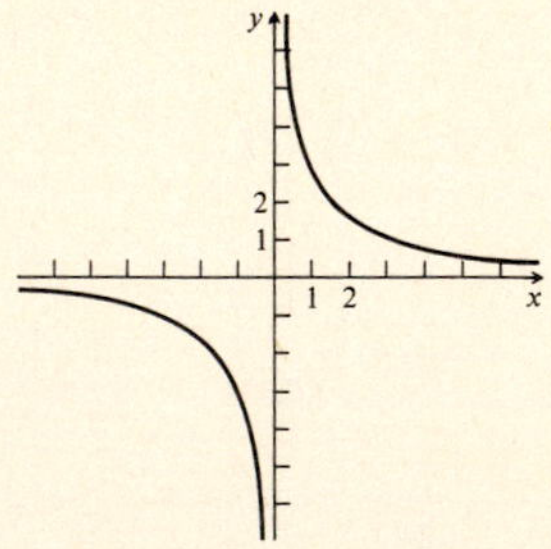

8.

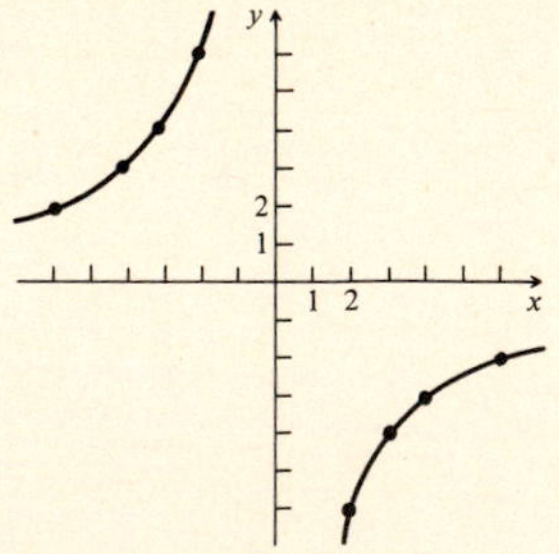

Exercise Set 10.2, pp. 642–643

1. C: $(0, 0)$; V: $(-3, 0)$, $(3, 0)$; F: $(-\sqrt{10}, 0)$, $(\sqrt{10}, 0)$; A: $y = \frac{1}{3}x$, $y = -\frac{1}{3}x$

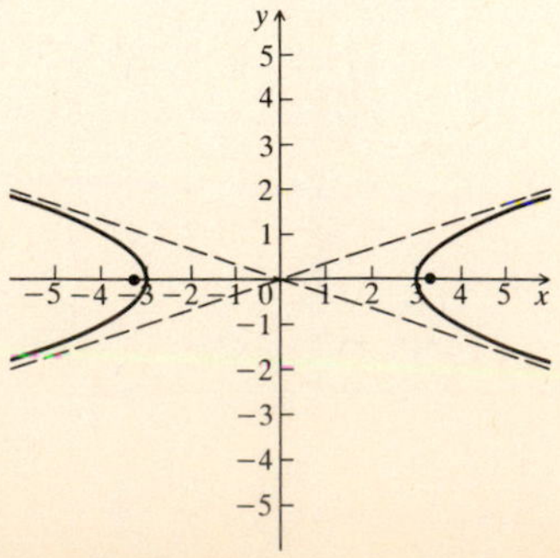

3. C: $(2, -5)$; V: $(-1, -5)$, $(5, -5)$; F: $(2 - \sqrt{10}, -5)$, $(2 + \sqrt{10}, -5)$; A: $y = -\dfrac{x}{3} - \dfrac{13}{3}$, $y = \dfrac{x}{3} - \dfrac{17}{3}$

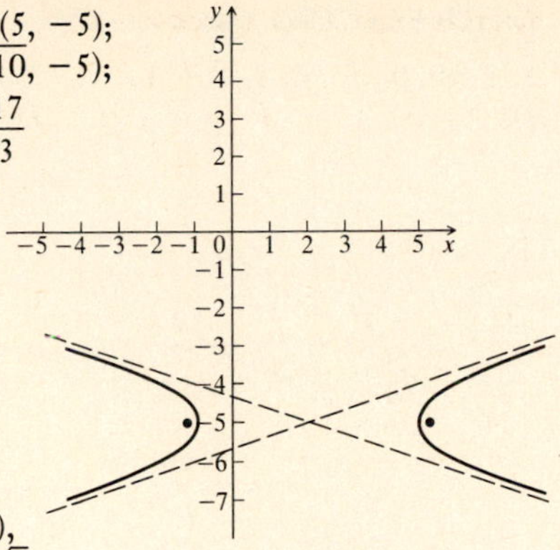

5. C: $(-1, -3)$; V: $(-1, -1)$, $(-1, -5)$; F: $(-1, -3 + 2\sqrt{5})$, $(-1, -3 - 2\sqrt{5})$; A: $y = \frac{1}{2}x - \frac{5}{2}$, $y = -\frac{1}{2}x - \frac{7}{2}$

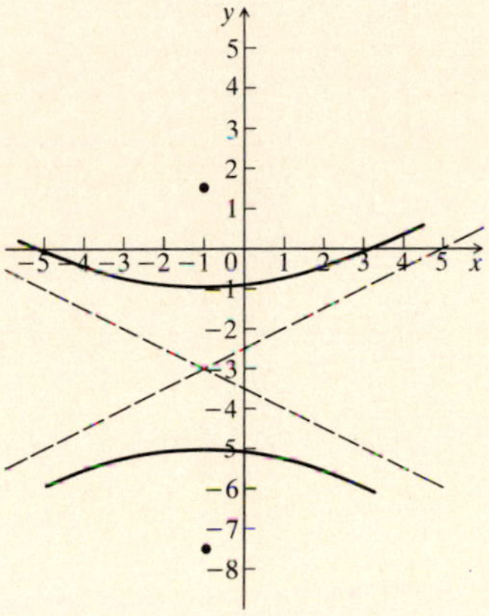

7. C: $(0, 0)$; V: $(-2, 0)$, $(2, 0)$; F: $(-\sqrt{5}, 0)$, $(\sqrt{5}, 0)$; A: $y = -\frac{1}{2}x$, $y = \frac{1}{2}x$

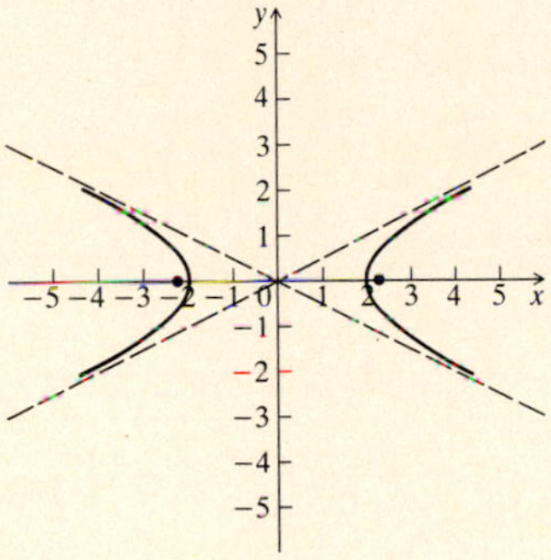

9. C: $(0, 0)$; V: $(0, 1)$, $(0, -1)$; F: $(0, \sqrt{5})$, $(0, -\sqrt{5})$; A: $y = -\frac{1}{2}x$, $y = \frac{1}{2}x$

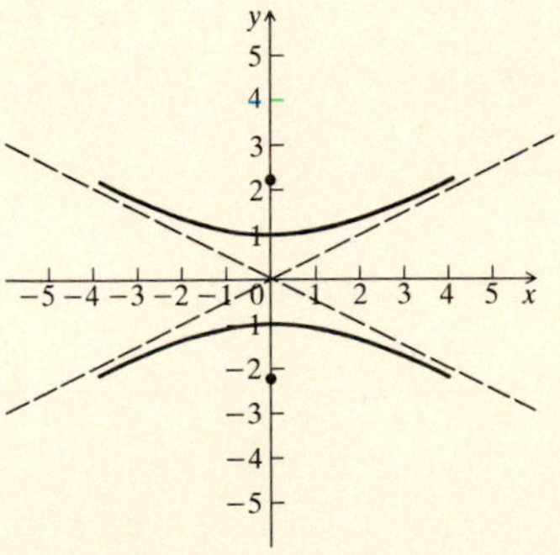

11. C: $(0, 0)$; V: $(-\sqrt{2}, 0)$, $(\sqrt{2}, 0)$; F: $(-2, 0)$, $(2, 0)$; A: $y = \pm x$

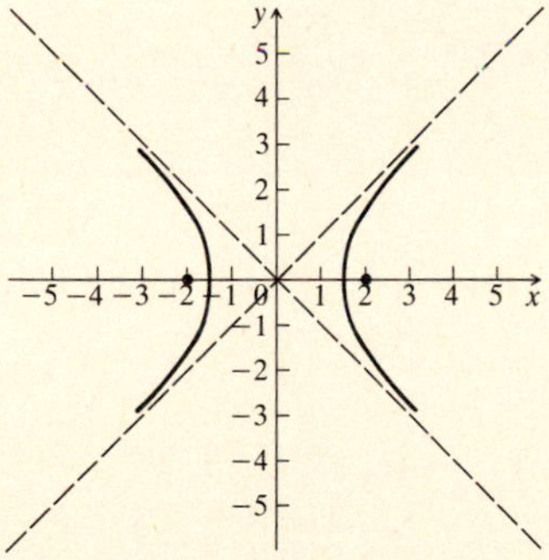

13. C: $(0, 0)$; V: $\left(-\frac{1}{2}, 0\right)$, $\left(\frac{1}{2}, 0\right)$; F: $\left(-\dfrac{\sqrt{2}}{2}, 0\right)$, $\left(\dfrac{\sqrt{2}}{2}, 0\right)$; A: $y = \pm x$

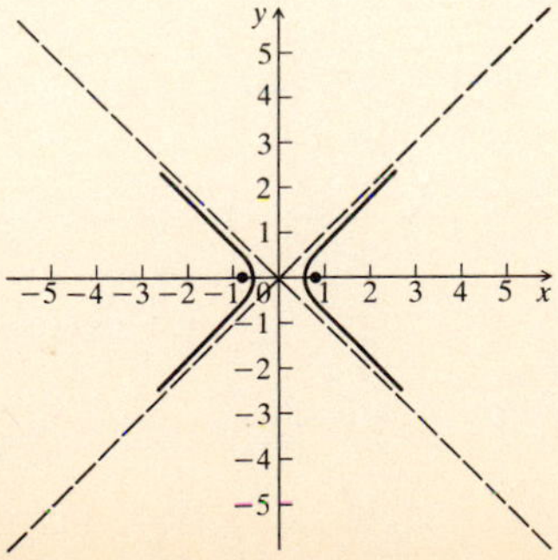

15. C: $(1, -2)$; V: $(0, -2)$, $(2, -2)$; F: $(1 - \sqrt{2}, -2)$, $(1 + \sqrt{2}, -2)$; A: $y = -x - 1$, $y = x - 3$

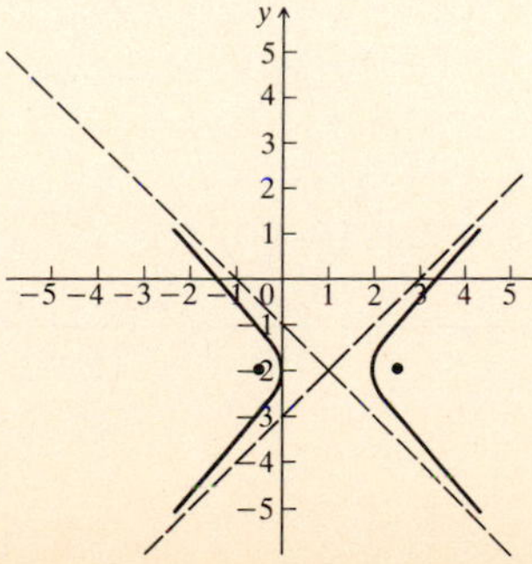

17. C: $\left(\frac{1}{3}, 3\right)$; V: $\left(-\frac{2}{3}, 3\right)$; $\left(\frac{4}{3}, 3\right)$; F: $\left(\frac{1}{3} - \sqrt{37}, 3\right)$, $\left(\frac{1}{3} + \sqrt{37}, 3\right)$; A: $y = 6x + 1$, $y = -6x + 5$

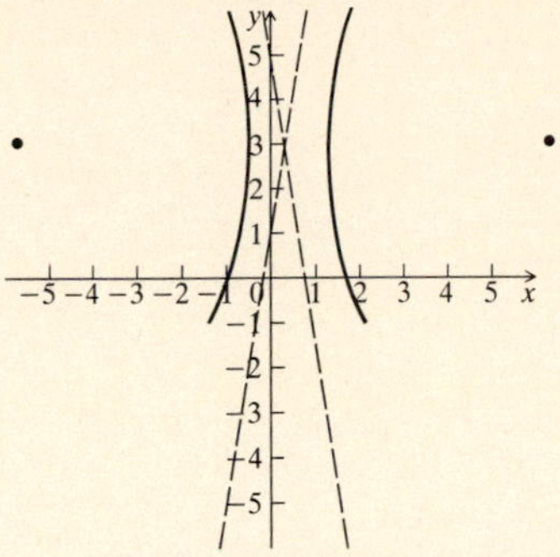

19. C: $(3, 1)$; V: $(3, 3)$, $(3, -1)$; F: $(3, 1 + \sqrt{13})$, $(3, 1 - \sqrt{13})$; A: $y = \frac{2}{3}x - 1$, $y = -\frac{2}{3}x + 3$

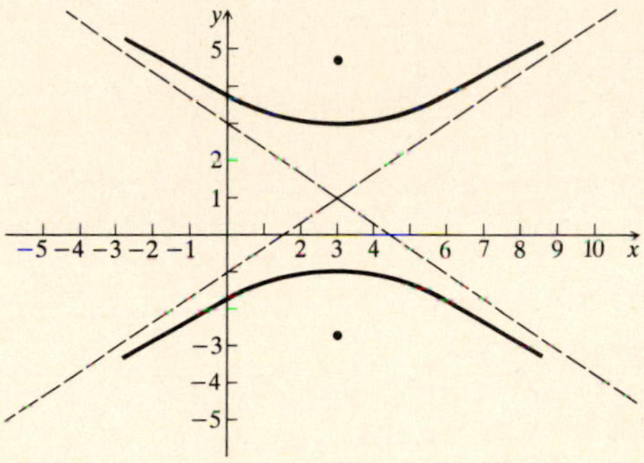

21. C: $(1, -2)$; V: $(2, -2)$, $(0, -2)$; F: $(1 + \sqrt{2}, -2)$, $(1 - \sqrt{2}, -2)$; A: $y = x - 3$, $y = -x - 1$

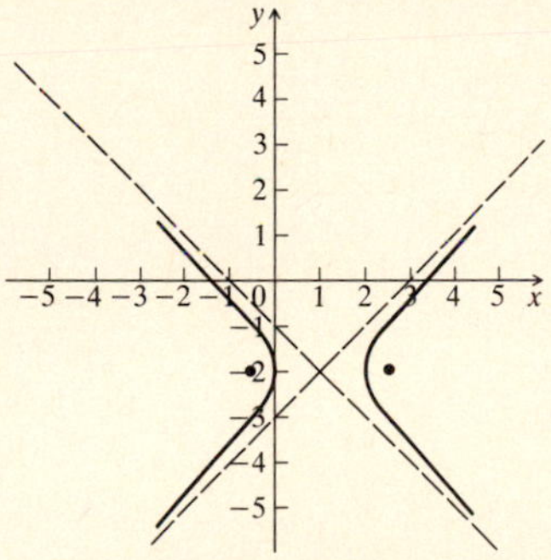

23. C: $(-3, 4)$; V: $(-3, 10)$, $(-3, -2)$; F: $(-3, 4 + 6\sqrt{2})$, $(-3, 4 - 6\sqrt{2})$; A: $y = x + 7$, $y = -x + 1$

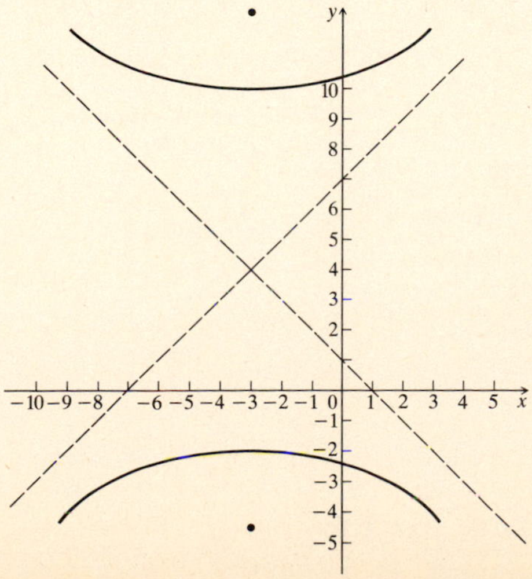

25.

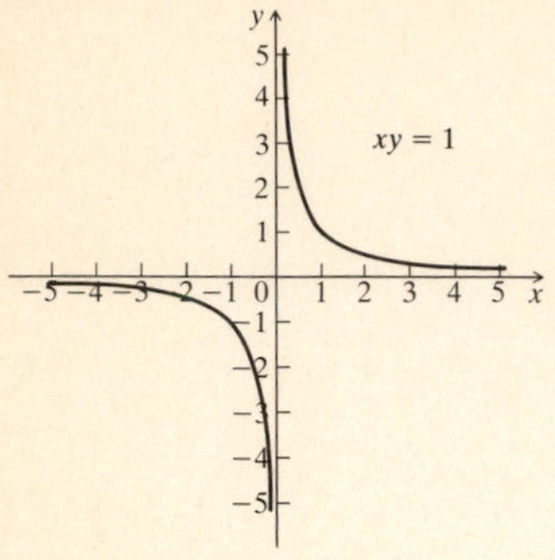

27.

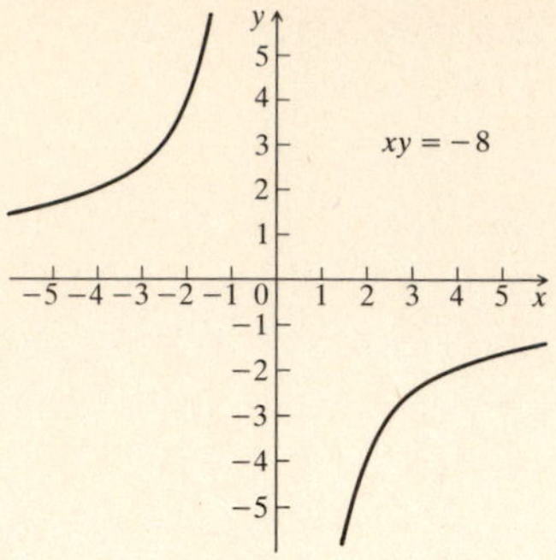

29. C: (1.023, −2.044); V: (2.07, −2.044), (−0.024, −2.044); A: $y = x - 3.067$, $y = -x - 1.021$

31. $\frac{x^2}{4} - \frac{y^2}{9} = 1$ 33. $\frac{(x+7)^2}{4} - \frac{(y-4)^2}{36} = 1$

35. (a) No; (b) $y = \pm\frac{1}{2}\sqrt{x^2 - 4}$;
(c) Yes, domain: $\{x | x \leq -2 \text{ or } x \geq 2\}$, range: $\{y | y \geq 0\}$;
(d) Yes, domain: $\{x | x \leq -2 \text{ or } x \geq 2\}$, range: $\{y | y \leq 0\}$

37. The (absolute value of the) difference of the distances from C to two fixed points A and B is a constant. By definition, such a locus is a hyperbola.

Margin Exercises, Section 10.3

1. V: (0, 0); F: (0, 2); D: $y = -2$

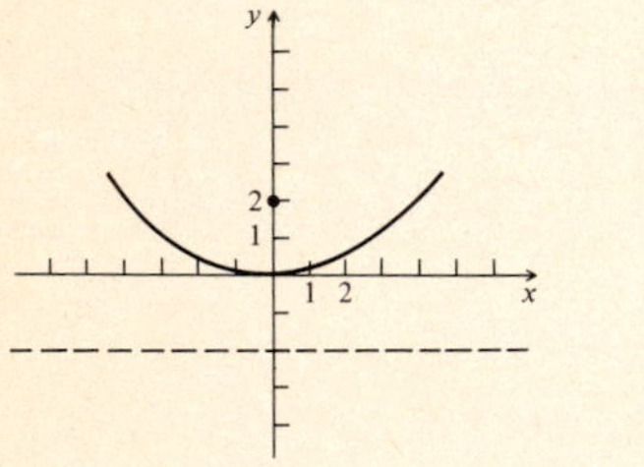

2. V: (0, 0); F: $\left(0, \frac{1}{8}\right)$; D: $y = -\frac{1}{8}$

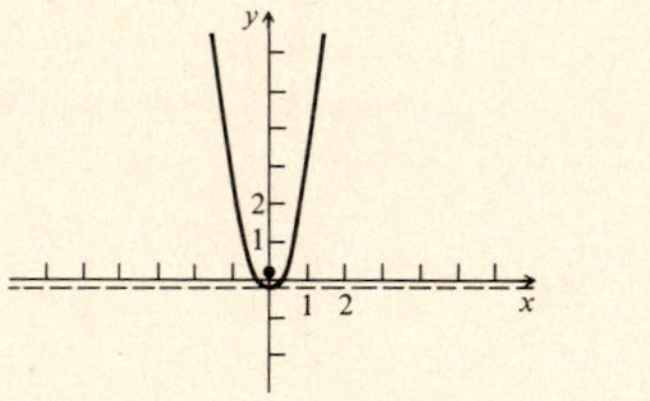

3. V: (0, 0); F: $\left(-\frac{3}{2}, 0\right)$; D: $x = \frac{3}{2}$

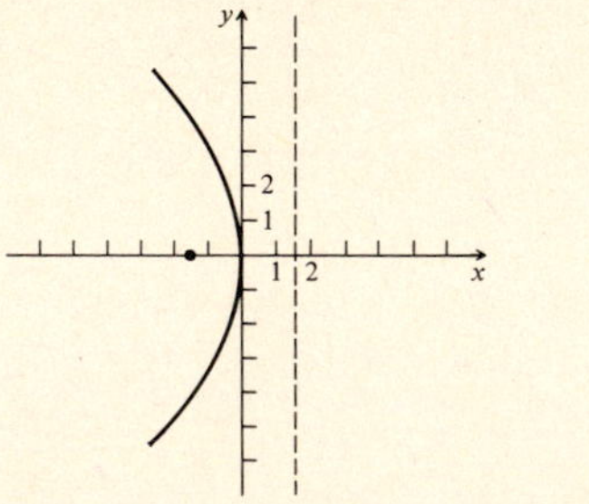

4. $y^2 = 12x$ 5. $x^2 = 2y$
6. $y^2 = -24x$
7. $x^2 = -4y$

8. V: $\left(-1, -\frac{1}{2}\right)$; F: $\left(-1, \frac{3}{2}\right)$; D: $y = -\frac{5}{2}$

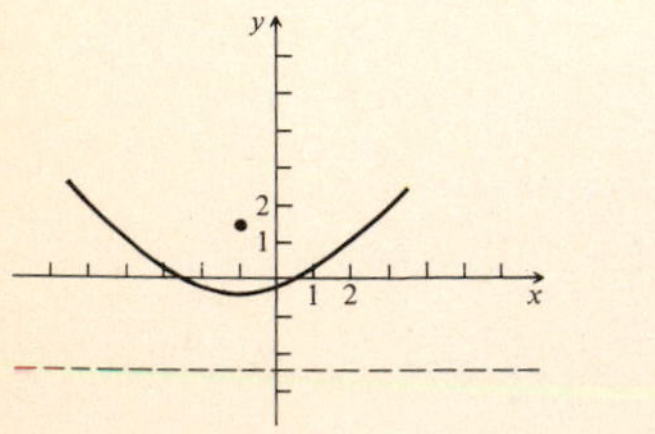

9. V: (2, −1); F: (1, −1); D: $x = 3$

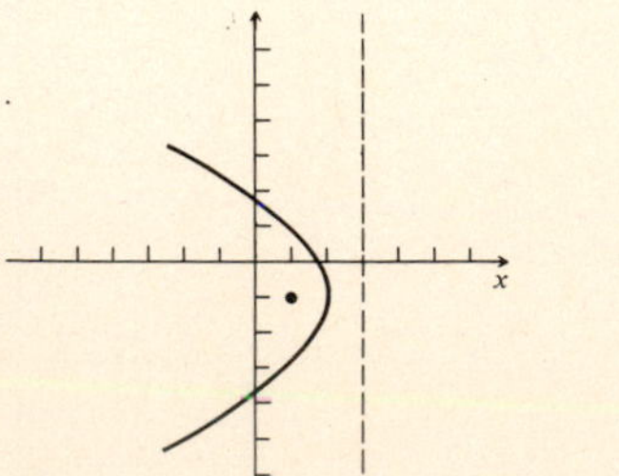

10. Parabola 11. Circle 12. Hyperbola 13. Ellipse

Exercise Set 10.3, pp. 649–650

1. V: (0, 0); F: (0, 2); D: $y = -2$

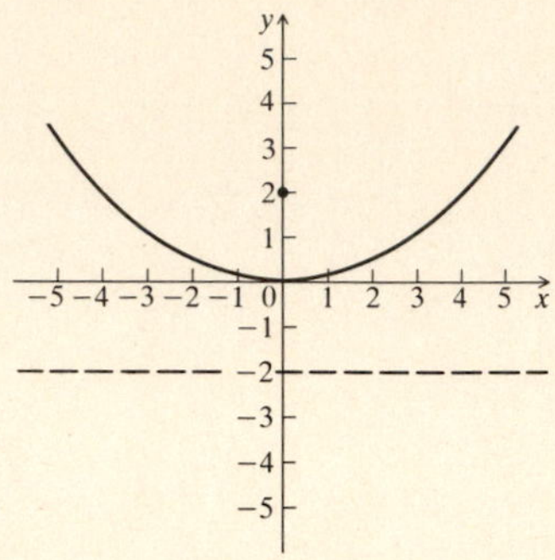

3. V: (0, 0); F: $\left(-\frac{3}{2}, 0\right)$; D: $x = \frac{3}{2}$

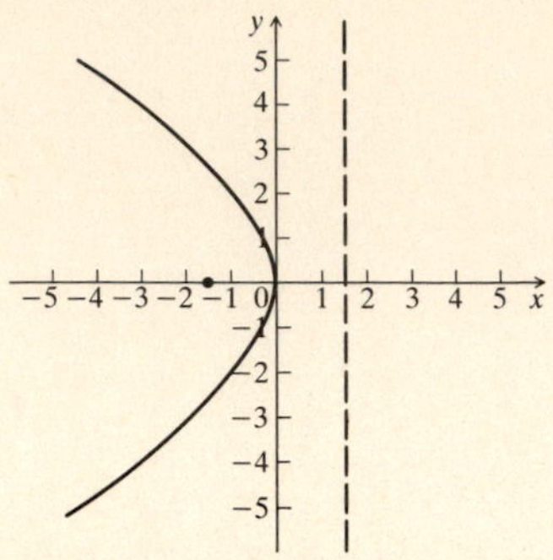

5. V: (0, 0); F: (0, 1); D: $y = -1$

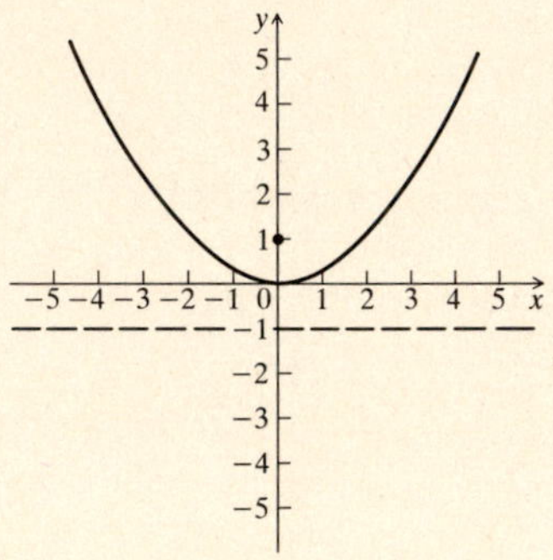

7. V: (0, 0); F: $\left(0, \frac{1}{8}\right)$; D: $y = -\frac{1}{8}$

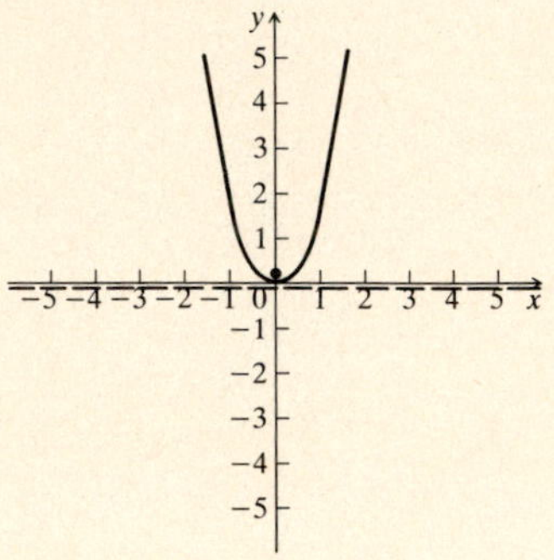

9. $y^2 = 16x$ 11. $y^2 = -4\sqrt{2}x$ 13. $(y-2)^2 = 14\left(x + \frac{1}{2}\right)$

15. V: (−2, 1); F: $\left(-2, -\frac{1}{2}\right)$; D: $y = \frac{5}{2}$

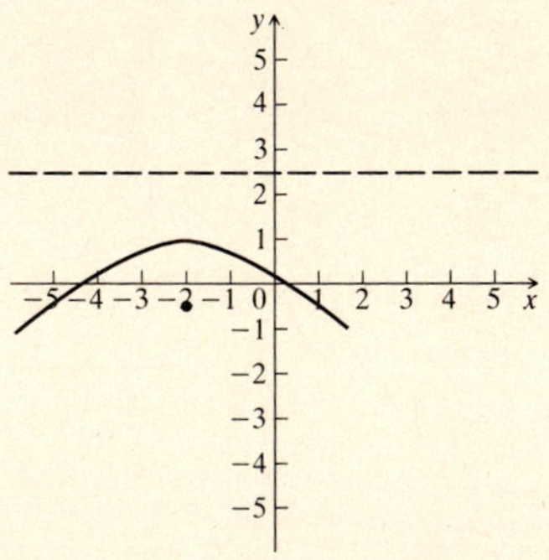

17. V: (−1, −3); F: $\left(-1, -\frac{7}{2}\right)$; D: $y = -\frac{5}{2}$

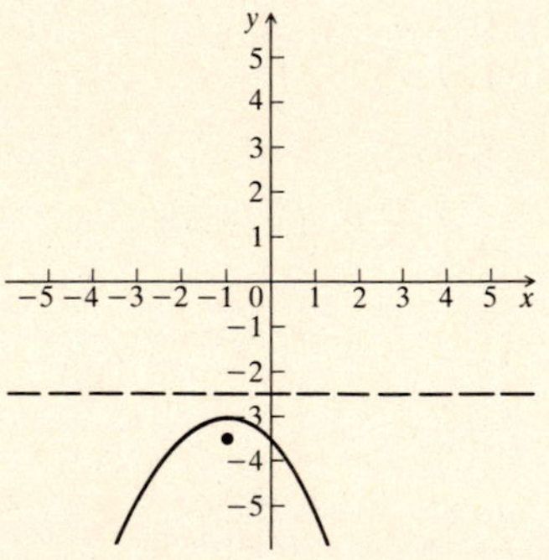

19. V: (0, −2); F: $\left(0, -1\frac{3}{4}\right)$; D: $y = -2\frac{1}{4}$

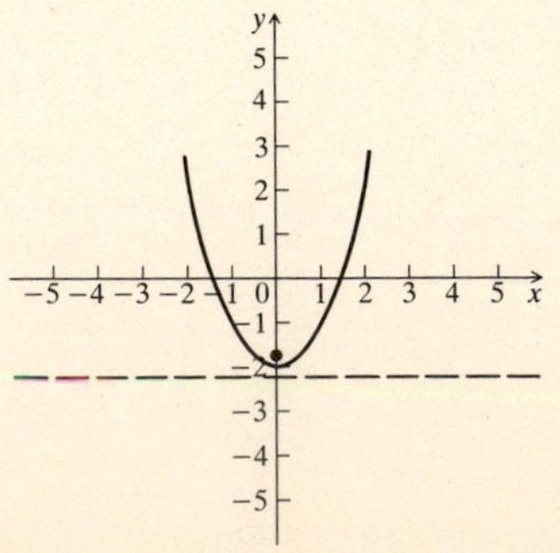

21. V: (−2, −1); F: $\left(-2, -\frac{3}{4}\right)$; D: $y = -1\frac{1}{4}$

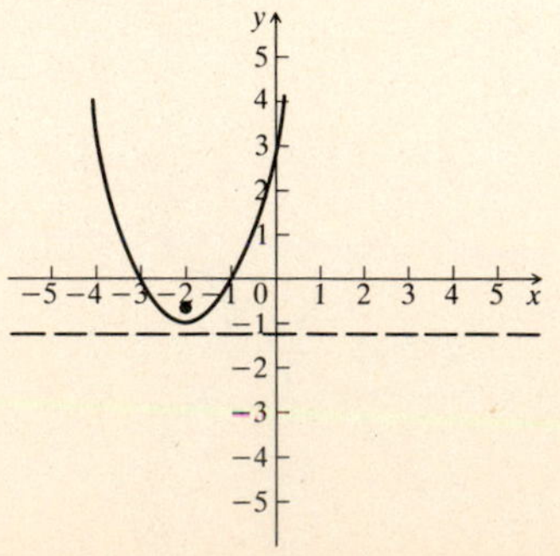

23. V: $\left(5\frac{3}{4}, \frac{1}{2}\right)$; F: $\left(6, \frac{1}{2}\right)$; D: $x = 5\frac{1}{2}$

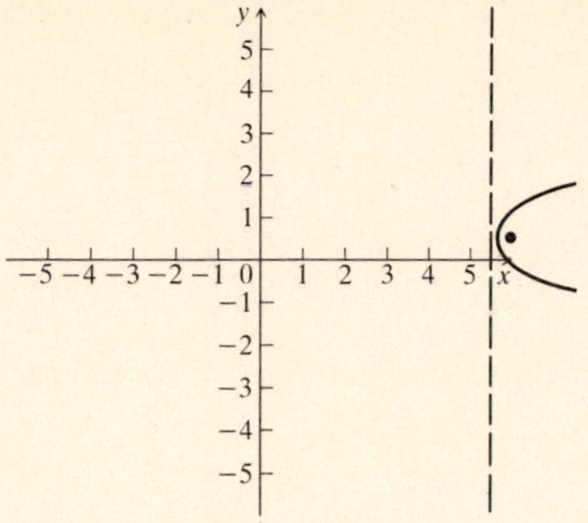

25. V: (0, 0); F: (0, 2014.0625); D: $y = -2014.0625$ 27. Parabola 29. Elllipse 31. Circle 33. Parabola 35. Hyperbola 37. The graph of $x^2 - y^2 = 0$ is two lines; the others are, respectively, a hyperbola, a circle, and a parabola. 39. $(x + 1)^2 = -4(y - 2)$ 41. $(x + 2)^2 = \frac{1}{4}(y - 1)$

43. $(y - k)^2 - (x - h)(4p) = 0$; $(y - k)^2 = 4p(x - h)$

Margin Exercises, Section 10.4

1. (4, 3), (−3, −4) 2. (4, 7), (−1, 2) 3. (−4, 4), (2, 1) 4. (4, 3), (−3, −4) 5. (4, 7), (−1, 2) 6. (−4, 4), (2, 1) 7. $\left(-\frac{5}{7}, \frac{22}{7}\right)$, (1, −2) 8. 7, 11 9. 5 ft, 12 ft

Exercise Set 10.4, pp. 653–655

1. (−4, −3), (3, 4) 3. (0, −3), (4, 5) 5. (3, 0), (0, 2) 7. (−2, 1) 9. (3, 2), $\left(4, \frac{3}{2}\right)$

11. $\left(\frac{5 + \sqrt{70}}{3}, \frac{-1 + \sqrt{70}}{3}\right), \left(\frac{5 - \sqrt{70}}{3}, \frac{-1 - \sqrt{70}}{3}\right)$

13. $\left(\frac{15 + \sqrt{561}}{8}, \frac{11 - 3\sqrt{561}}{8}\right), \left(\frac{15 - \sqrt{561}}{8}, \frac{11 + 3\sqrt{561}}{8}\right)$

15. $\left(\frac{7 - \sqrt{33}}{2}, \frac{7 + \sqrt{33}}{2}\right), \left(\frac{7 - \sqrt{33}}{2}, \frac{7 + \sqrt{33}}{2}\right)$

17. (3, −5), (−1, 3) 19. (8, 5), (−5, −8)

21. $\left(\frac{8 + 3i\sqrt{6}}{2}, \frac{-8 + 3i\sqrt{6}}{2}\right), \left(\frac{8 - 3i\sqrt{6}}{2}, \frac{-8 - 3i\sqrt{6}}{2}\right)$

23. (0.965, 4402.33), (−0.965, −4402.33) 25. 3, 9 27. 6 cm, 8 cm 29. 4 in. by 5 in. 31. 75 yd by 30 yd 33. 8, $\frac{1}{4}$

35. 61.52 cm, 38.48 cm 37. $\frac{x^2}{4} + y^2 = 1$

39. $\left(\frac{2a^2 + 6ab + 5b^2}{2(a + 2b)}, \frac{-2ab - 3b^2}{2(a + 2b)}\right)$

41. $2(L + W) = P$, $L + W = \frac{P}{2}$, $LW = A$, $L = \frac{P}{2} - W$,

$W\left(\frac{P}{2} - W\right) = A$, $W^2 - \frac{WP}{2} + A = 0$,

$$W = \frac{\frac{P}{2} \pm \sqrt{\left(\frac{P}{2}\right)^2 - 4A}}{2} = \frac{P}{4} \pm \frac{\sqrt{P^2 - 16A}}{4}$$

$$= \frac{1}{4}(P \pm \sqrt{P^2 - 16A})$$

43. $(x - 2)^2 + (y - 3)^2 = 1$ 45. (2, 4), (4, 2)

Margin Exercises, Section 10.5

1. (2, 0), (−2, 0) 2. (4, 0), (−4, 0) 3. (0, 2), (0, −2) 4. (2, 3), (2, −3), (−2, 3), (−2, −3) 5. (3, 2), (−3, −2), (2, 3), (−2, −3) 6. 1 ft, 2 ft

Exercise Set 10.5, pp. 657–658

1. (−5, 0), (4, 3), (4, −3) 3. (3, 0), (−3, 0) 5. (4, 3), (−4, −3), (3, 4), (−3, −4) 7. No solution

9. $(\sqrt{2}, \sqrt{14}), (-\sqrt{2}, \sqrt{14}), (\sqrt{2}, -\sqrt{14}), (-\sqrt{2}, -\sqrt{14})$
11. (1, 2), (−1, −2), (2, 1), (−2, −1)
13. (3, 2), (−3, −2), (2, 3), (−2, −3)

15. $\left(\frac{5 - 9\sqrt{15}}{20}, \frac{-45 + 3\sqrt{15}}{20}\right), \left(\frac{5 + 9\sqrt{15}}{20}, \frac{-45 - 3\sqrt{15}}{20}\right)$

17. (3, 2), (−3, −2) 19. (2, 1), (−2, −1), (1, 2), (−1, −2)

21. $(3, \sqrt{5}), (-3, -\sqrt{5}), (\sqrt{5}, 3), (-\sqrt{5}, -3)$

23. $\left(\frac{8\sqrt{5}}{5}i, \frac{3\sqrt{105}}{5}\right), \left(\frac{8\sqrt{5}}{5}i, -\frac{3\sqrt{105}}{5}\right), \left(-\frac{8\sqrt{5}}{5}i, \frac{3\sqrt{105}}{5}\right), \left(-\frac{8\sqrt{5}}{5}i, -\frac{3\sqrt{105}}{5}\right)$

25. (8.53, 2.53), (8.53, −2.53), (−8.53, 2.53), (−8.53, −2.53)
27. 13, 12 and −13, −12 29. 1 m by $\sqrt{3}$ m

31. 16 ft, 24 ft 33. $\left(x + \frac{5}{13}\right)^2 + \left(y - \frac{32}{13}\right)^2 = \frac{5365}{169}$

35. $\left(\frac{1}{2}, \frac{1}{4}\right), \left(\frac{1}{2}, -\frac{1}{4}\right), \left(-\frac{1}{2}, \frac{1}{4}\right), \left(-\frac{1}{2}, -\frac{1}{4}\right)$

37. $\left(\frac{a}{\sqrt{a + b}}, \frac{b}{\sqrt{a + b}}\right), \left(\frac{-a}{\sqrt{a + b}}, \frac{-b}{\sqrt{a + b}}\right)$

39. (3, −2), (−3, 2), (2, −3), (−2, 3)

41. $\left(\frac{1}{2}, \sqrt{4 - x^2}\right), \left(\frac{1}{2}, -\sqrt{4 - x^2}\right)$ 43. $k = 3$

45. $\left(\frac{2 \log 3 + 3 \log 5}{3(\log 3 \cdot \log 5)}, \frac{4 \log 3 - 3 \log 5}{3(\log 3 \cdot \log 5)}\right)$ 47. (0, 0), $\left(\frac{3 - i\sqrt{3}}{2}, \frac{3 + i\sqrt{3}}{2}\right), \left(\frac{3 + i\sqrt{3}}{2}, \frac{3 - i\sqrt{3}}{2}\right)$

Review Exercises: Chapter 10, pp. 659–660

1. [10.1]

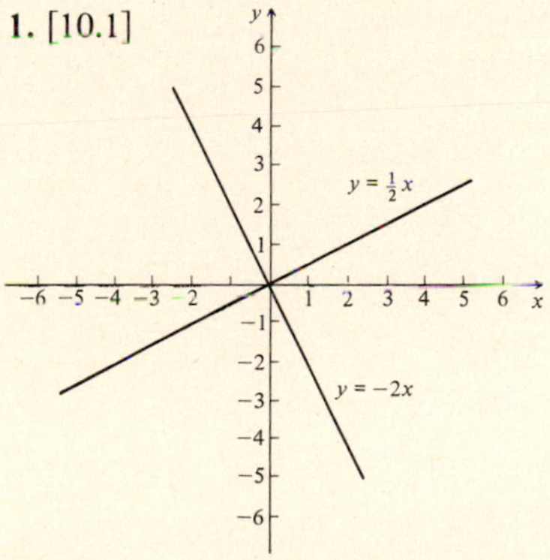

2. [10.1]

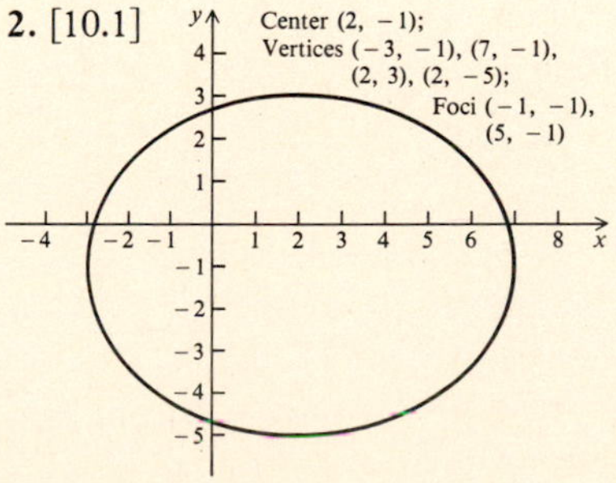

C: (2, −1); V: (−3, −1), (7, −1), (2, 3), (2, −5); F: (−1, −1), (5, −1) 3. [10.1] $\frac{x^2}{9} + \frac{y^2}{16} = 1$ 4. [10.2] C: $\left(-2, \frac{1}{4}\right)$; V: $\left(0, \frac{1}{4}\right), \left(-4, \frac{1}{4}\right)$; F: $\left(-2 + \sqrt{6}, \frac{1}{4}\right), \left(-2 - \sqrt{6}, \frac{1}{4}\right)$; A: $y - \frac{1}{4} = \pm\frac{\sqrt{2}}{2}(x + 2)$

5. [10.2]

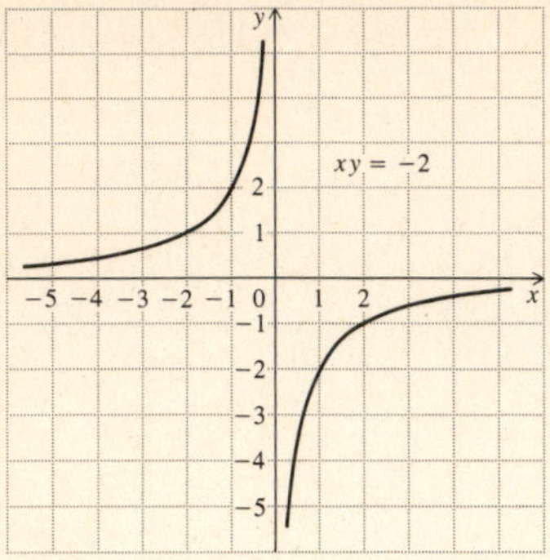

6. [10.2]

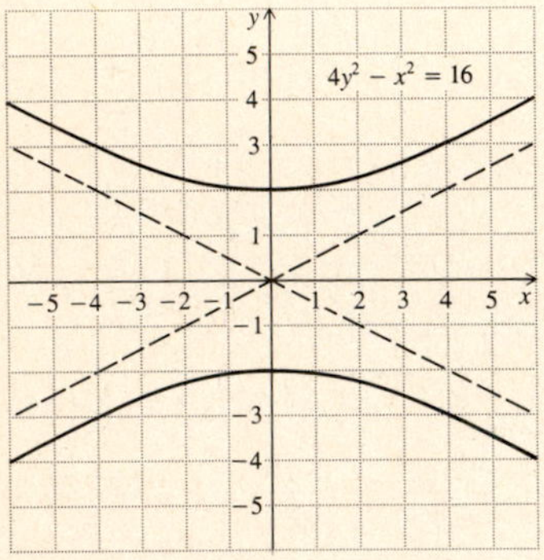

7. [10.3] $x^2 = -6y$ **8.** [10.3] F: $(-3, 0)$; V: $(0, 0)$; D: $x = 3$
9. [10.3] V: $(-5, 8)$; F: $\left(-5, \frac{15}{2}\right)$; D: $y = \frac{17}{2}$ **10.** [10.3] Circle
11. [10.3] Ellipse **12.** [10.3] Parabola **13.** [10.3] Hyperbola
14. [10.3] Hyperbola **15.** [10.3] Parabola
16. [10.3] Hyperbola **17.** [10.3] Circle **18.** [10.5] $(-8\sqrt{2}, 8)$, $(8\sqrt{2}, 8)$ **19.** [10.5] $\left(3, \frac{\sqrt{29}}{2}\right)$, $\left(-3, \frac{\sqrt{29}}{2}\right)$, $\left(3, -\frac{\sqrt{29}}{2}\right)$, $\left(-3, -\frac{\sqrt{29}}{2}\right)$ **20.** [10.4] $(7, 4)$ **21.** [10.4] $(2, 2)$, $\left(\frac{32}{9}, -\frac{10}{9}\right)$
22. [10.4] $(0, -3)$, $(2, 1)$ **23.** [10.5] $(4, 3)$, $(4, -3)$, $(-4, 3)$, $(-4, -3)$ **24.** [10.5] $(-\sqrt{3}, 0)$, $(\sqrt{3}, 0)$, $(-2, 1)$, $(2, 1)$
25. [10.4] $\left(-\frac{3}{5}, \frac{21}{5}\right)$, $(3, -3)$ **26.** [10.5] $(6, 8)$, $(6, -8)$, $(-6, 8)$, $(-6, -8)$ **27.** [10.5] $(2, 2)$, $(-2, -2)$, $(2\sqrt{2}, \sqrt{2})$, $(-2\sqrt{2}, -\sqrt{2})$ **28.** [10.5] $\frac{16\sqrt{6}}{7} \approx 5.6$ **29.** [10.4] 7, 4
30. [10.4] 12 by 7 **31.** [10.4] 4, 8 **32.** [10.4] 32 cm, 20 cm
33. [10.5] 11 ft, 3 ft **34.** [10.1] $x^2 + \frac{y^2}{9} = 1$ **35.** [10.5] $\frac{8}{7}, \frac{7}{2}$
36. [10.1] $(x - 2)^2 + (y - 1)^2 = 100$

Test: Chapter 10, p. 660

1. [10.1]

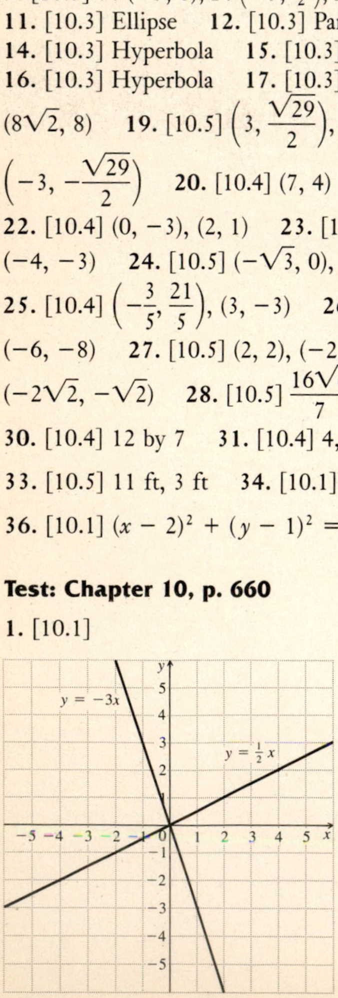

2. [10.1] C: $(2, 4)$; V: $(1, 4)$, $(3, 4)$, $(2, 1)$, $(2, 7)$; F: $(2, 4 + 2\sqrt{2})$, $(2, 4 - 2\sqrt{2})$

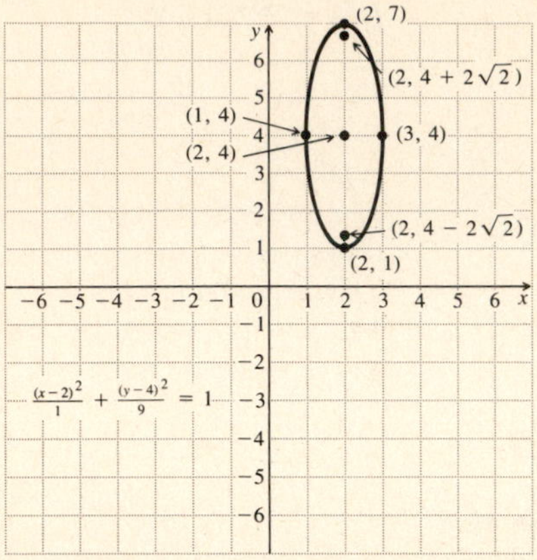

3. [10.1] $\frac{x^2}{49} + \frac{y^2}{4} = 1$
4. [10.2] C: $(0, 0)$; V: $(0, 5)$, $(0, -5)$; F: $(0, \sqrt{34})$, $(0, -\sqrt{34})$; A: $y = \frac{5}{3}x$, $y = -\frac{5}{3}x$

5. [10.2]

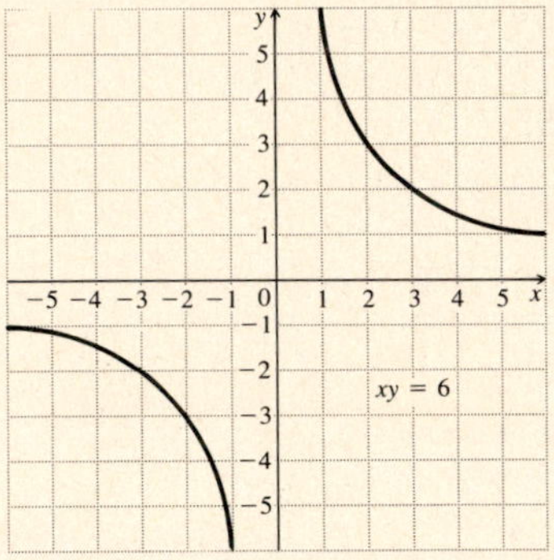

6. [10.3] $x^2 = -20y$
7. [10.3] V: $(2, -3)$; F: $(0, -3)$; D: $x = 4$
8. [10.4] $(8, 6)$, $(0, 10)$
9. [10.5] $(1, 2)$, $(2, 1)$, $(-2, -1)$, $(-1, -2)$
10. [10.4] $-2, 6$
11. [10.5] 10 yd by 1 yd
12. [10.4] Numerator 15, denominator 8; or numerator 8, denominator 15
13. [10.5] 11 by 2
14. [10.5] $\sqrt{5}$ m, $\sqrt{3}$ m
15. [10.4] 16 ft by 12 ft **16.** [10.3] Parabola **17.** [10.3] Circle **18.** [10.3] Hyperbola **19.** [10.3] Ellipse **20.** [10.3] Hyperbola **21.** [10.3] Parabola **22.** [10.1] $\frac{5(x + 3)^2}{36} + \frac{(y + 1)^2}{9} = 1$ **23.** [10.1] $(x + 2)^2 + (y - 2)^2 = 37$ **24.** [10.1] $\left(0, -\frac{31}{4}\right)$ **25.** [10.4] 9

CHAPTER 11

Margin Exercises, Section 11.1

1. 5 **2.** 1 **3.** 2 **4.** 0 **5.** No degree **6.** 7 **7.** 2, −2
8. $i, -i$ **9.** (a) Yes; (b) no **10.** (a) Yes; (b) No; (c) no
11. $Q(x) = x^2 + 5x + 10$, $R(x) = 24$, $P(x) = (x - 3)(x^2 + 5x + 10) + 24$

Exercise Set 11.1, pp. 664–665

1. 4 **3.** 1 **5.** 2 **7.** 0
9. 2 yes; 3 no, −1 no **11.** (a) Yes; (b) no; (c) no
13. $Q(x) = x^2 + 8x + 15$, $R(x) = 0$, $P(x) = (x - 2)(x^2 + 8x + 15) + 0$
15. $Q(x) = x^2 + 9x + 26$, $R(x) = 48$, $P(x) = (x - 3)(x^2 + 9x + 26) + 48$
17. $Q(x) = x^2 - 2x + 4$, $R(x) = -16$, $P(x) = (x + 2)(x^2 - 2x + 4) - 16$
19. $Q(x) = x^2 + 5$, $R(x) = 0$, $P(x) = (x^2 + 4)(x^2 + 5) + 0$
21. $P(x) = (2x^2 - x + 1) \cdot \left(\frac{5}{2}x^5 + \frac{5}{4}x^4 - \frac{5}{8}x^3 - \frac{39}{16}x^2 - \frac{29}{32}x + \frac{113}{64}\right) + \frac{171x - 305}{64}$

23. (a) -32; (b) -32; (c) -65; (d) -65 **25.** 0, 1 **27.** 0 **29.** (a) -2; (b) -2 **31.** The degree of the product is the sum of the degrees of the factors.

Margin Exercises, Section 11.2

1. $Q(x) = x^2 + 8x + 15$, $R(x) = 0$ **2.** $Q(x) = x^2 - 4x + 13$, $R(x) = -30$ **3.** $Q(x) = x^2 - x + 1$, $R(x) = 0$
4. (a) $P(10) = 73{,}120$; (b) $P(-8) = -37{,}292$
5. (a) Yes; (b) no; (c) yes **6.** No **7.** No
8. (a) Yes; (b) $x^2 + 8x + 15$; (c) $(x - 2)(x + 5)(x + 3)$; (d) 2, -5, -3

Exercise Set 11.2, pp. 668–669

1. $Q(x) = 2x^3 + x^2 - 3x + 10$, $R(x) = -42$
3. $Q(x) = x^2 - 4x + 8$, $R(x) = -24$
5. $Q(x) = x^3 + x^2 + x + 1$, $R(x) = 0$
7. $Q(x) = 2x^3 + x^2 + \frac{7}{2}x + \frac{7}{4}$, $R(x) = -\frac{1}{8}$
9. $Q(x) = x^3 + x^2y + xy^2 + y^3$, $R(x) = 0$
11. $P(1) = 0$, $P(-2) = -60$, $P(3) = 0$
13. $P(20) = 5{,}935{,}988$, $P(-3) = -772$
15. $P(2) = 0$, $P(-2) = 0$, $P(3) = 65$
17. -3 yes, 2 no **19.** -3 no, $\frac{1}{2}$ no
21. $P(x) = (x - 1)(x + 2)(x + 3)$; 1, -2, -3
23. $P(x) = (x - 2)(x - 5)(x + 1)$; 2, 5, -1
25. $P(x) = (x - 2)(x - 3)(x + 4)$; 2, 3, -4
27. $P(x) = (x - 1)(x - 2)(x - 3)(x + 5)$; 1, 2, 3, -5
29. $-1, \frac{7}{6}$ **31.** $-5 < x < 1$ or $x > 2$ **33.** $\frac{14}{3}$ **35.** $k = 0$
37. $P(x) = (bx - r)Q(x) + R$ or $P(x) = \left(x - \frac{r}{b}\right) \cdot bQ(x) + R$ or $P(x) = \left(x - \frac{r}{b}\right)Q_1(x) + R$. Thus we would divide by $x - \frac{r}{b}$ and multiply the resulting quotient $Q_1(x)$ by $\frac{1}{b}$ to obtain $Q(x)$. R is unchanged.

Margin Exercises, Section 11.3

1. 5, mult. 2; -6, mult. 1 **2.** -7, mult. 2; 3, mult. 1
3. -2, mult. 3; 3, mult. 1; -3, mult. 1 **4.** 4, mult. 2; 3, mult. 2
5. 1, -1, each has mult. 1 **6.** $\pm 2i$, $\pm\sqrt{3}$, each has mult. 1
7. 2, -2, $-\frac{1}{2}$, each has mult. 1 **8.** $x^3 - 6x^2 + 3x + 10$
9. $x^3 + (-1 + 5i)x^2 + (-2 - 5i)x - 10i$
10. $x^5 + 6x^4 + 12x^3 + 8x^2$ **11.** $x^4 + 2x^3 - 12x^2 + 14x - 5$
12. $7 + 2i$, $3 - \sqrt{5}$ **13.** $x^4 - 6x^3 + 11x^2 - 10x + 2$
14. $x^3 - 2x^2 + 4x - 8$ **15.** $-i$, -2, 1

Exercise Set 11.3, pp. 673–674

1. -3, mult. 2; 1, mult. 1 **3.** 0, mult. 3; 1, mult. 2; -4, mult. 1
5. $\pm\sqrt{3}$, ± 1; each has mult. 1 **7.** -3, -1, 1; each has mult. 1
9. $x^3 - 6x^2 - x + 30$ **11.** $x^3 + 3x^2 + 4x + 12$
13. $x^3 - \sqrt{3}x^2 - 2x + 2\sqrt{3}$; no **15.** $-3 - 4i$, $4 + \sqrt{5}$
17. $x^3 - 4x^2 + 6x - 4$ **19.** $x^3 - 5x^2 + 16x - 80$
21. $x^4 + 4x^2 - 45$ **23.** i, 2, 3 **25.** $1 + 2i$, $1 - 2i$
27. 4, i, $-i$ **29.** i, $-i$, $1 + \sqrt{2}$, $1 - \sqrt{2}$ **31.** $-a$
33. Since $P(x)$ of odd degree n has n linear factors, of which an even number corresponds to all the pairs of conjugate nonreal roots, there is at least one other factor $(x - a)$ that gives a real root a.
35. There is at least one complex number $a + bi$ such that $\sin x = a + bi$. But the equation $\sin x = a + bi$ may have no solution or infinitely many solutions, depending on the value of $a + bi$.

Margin Exercises, Section 11.4

1. (a) 3, -3, 1, -1; (b) 2, -2, 1, -1; (c) $\frac{3}{2}$, $-\frac{3}{2}$, 3, -3, $\frac{1}{2}$, $-\frac{1}{2}$, 1, -1; (d) $\frac{1}{2}$, -3 (e) $3 + \sqrt{10}$, $3 - \sqrt{10}$
2. (a) 3, -3, 1, -1; (b) 1, -1; (c) same as for c (see part a); (d) all coefficients positive; (e) -3; (f) $\frac{-3 \pm \sqrt{5}}{2}$
3. (a) 20, -20, 10, -10, 5, -5, 4, -4, 2, -2, 1, -1; (b) 1, -1; (c) same as for c; (d) all coefficients positive; (e) -5; (f) $2i$, $-2i$ **4.** (a) All coefficients positive; (b) none
5. (a) None; (b) yes, $\frac{-3 \pm i\sqrt{3}}{2}$, quadratic formula
6. (a) $-\frac{1}{6}$, $-\frac{4}{3}$, $\frac{1}{6}$, $\frac{1}{3}$; (b) 6; (c) $6P(x) = 6x^4 - x^3 - 8x^2 + x + 2$; (d) 1, -1, $-\frac{1}{2}$, $\frac{2}{3}$; (e) yes, $P(x) = 0$ and $6P(x) = 0$ are equivalent.

Exercise Set 11.4, pp. 677–678

1. 1, -1 **3.** $\pm\left(1, \frac{1}{3}, \frac{1}{5}, \frac{1}{15}, 2, \frac{2}{3}, \frac{2}{5}, \frac{2}{15}\right)$
5. -3, $\sqrt{2}$, $-\sqrt{2}$; $(x + 3)(x - \sqrt{2})(x + \sqrt{2})$
7. -2, 1; $(x + 2)(x - 1)^2 = 0$ **9.** No rational roots
11. $-\frac{1}{5}$, 1, $2i$, $-2i$; $\left(x + \frac{1}{5}\right)(x - 1)(x - 2i)(x + 2i) = 0$
13. -1, -2, $3 + \sqrt{13}$, $3 - \sqrt{13}$; $(x + 1)(x + 2)(x - 3 - \sqrt{13})(x - 3 + \sqrt{13})$
15. 2, $1 \pm \sqrt{3}$; $(x - 2)(x - 1 - \sqrt{3})(x - 1 + \sqrt{3}) = 0$
17. -2, $1 \pm i\sqrt{3}$; $(x + 2)(x - 1 - i\sqrt{3})(x - 1 + i\sqrt{3})$
19. $\frac{1}{2}, \frac{1 \pm \sqrt{5}}{2}$; $\left(x - \frac{1}{2}\right)\left(x - \frac{1 + \sqrt{5}}{2}\right)\left(x - \frac{1 - \sqrt{5}}{2}\right)$
21. None **23.** None **25.** None **27.** -2, 1, 2
29. 4 cm **31.** 3 cm, $\frac{7 - \sqrt{33}}{2}$ cm
33. (a) 1, 10, 20; (b) $\{x \mid 1 < x < 10 \text{ or } x > 20\}$; (c) $\{x \mid 0 < x < 1 \text{ or } 10 < x < 20\}$
35. $\sqrt{5}$ is a root of $x^2 - 5 = 0$, which has no rational roots (since ± 1, ± 5 are not roots). Thus $\sqrt{5}$ must be irrational.
37. $-\sqrt[3]{7}$, $\sqrt[3]{7}\left(\frac{1}{2} + \frac{\sqrt{3}}{2}i\right)$, $\sqrt[3]{7}\left(\frac{1}{2} - \frac{\sqrt{3}}{2}i\right)$

Margin Exercises, Section 11.5

1. 3 **2.** 2 **3.** Just 1 **4.** 5, 3, or 1 **5.** 2 or 0
6. 2 or 0 **7.** 1 **8.** 0 **9.** 2
10. 75 **11.** 2 **12.** -4 **13.** -2 **14.** -4
15. Positive roots, 3 or 1; negative roots, 1; upper bound 6 (answer may vary); lower bound -1 (answer may vary).
16. Positive roots, 1; negative roots, 1. Thus 2 nonreal roots; upper bound 1 (answer may vary); lower bound -1 (in fact, -1 is a root—answer may vary). **17.** -0.53, 0.65, 2.88

Exercise Set 11.5, pp. 687–688

1. 3 or 1 **3.** 0 **5.** 2 or 0 **7.** 0 **9.** 3 or 1 **11.** 2 or 0
13. 3 **15.** 4 **17.** -1 **19.** -3
21. 3 or 1 positive; 1 negative; upper bound, 2; lower bound, -3 **23.** 1 positive; 1 negative; 2 nonreal; upper bound, 2; lower bound, -2 **25.** 2 or 0 positive; 2 or 0 negative; upper bound, 4; lower bound, -3 **27.** 0 positive, 0 negative
29. All roots are rational. **31.** 2.2 **33.** No real roots
35. -1.4, 1.4 **37.** -1.4, 1.4 **39.** 0.79 **41.** -1.27

43. Let $P(x) = x^n - 1$. There is one variation of sign, so there is just one positive root. Since n is even, $P(-x) = P(x)$. Hence $P(-x)$ has just one variation of sign, and there is just one negative root. Zero is not a root, so the total number of real roots is two.
45. -1.3 47. 1.5, 5.7 49. 7.16 hr
51. Since $a, b, c > 0$, we have $f(x)$: $+ + + -$ and $f(-x)$: $+ + - -$. Thus there are one positive root and one negative root; and 0 is not a root. Then $4 - (1 + 1) = 2$ nonreal roots.

Margin Exercises, Section 11.6

1. (a) 14, 16.4, 16.7, 16.97, 16.997; 17.003, 17.03, 17.3, 18.2, 20; (b) 17;
(c)

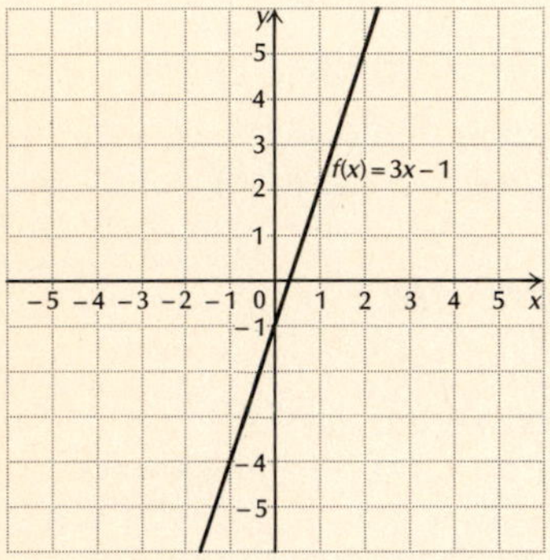

(d) -4; (e) 5; (f) -1
2. (a) -0.5, -1, -10, -100, -1000; 1000, 100, 10, 1.25, 1; (b) does not exist;
(c)

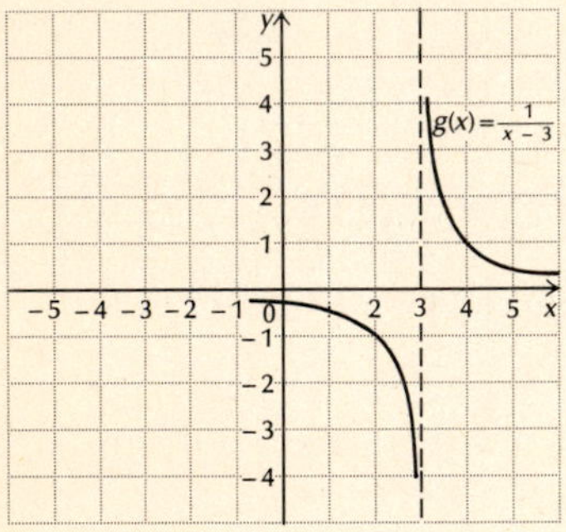

(d) does not exist; (e) -0.5; (f) 1
3. (a) -3; (b) 4; (c) does not exist; (d) 0; (e) 0; (f) 0; (g) 3; (h) 3; (i) 3
4. (a)

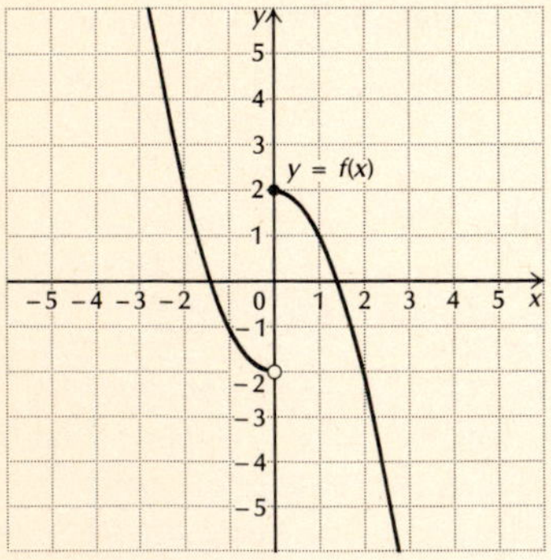

(b) does not exist; (c) 2
5. (a)

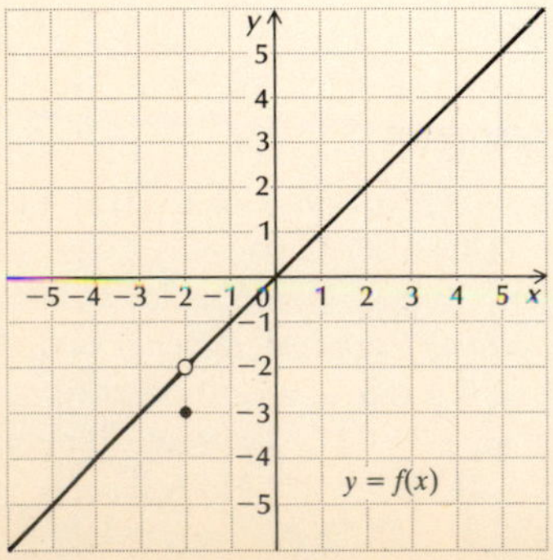

(b) -2; (c) -2; (d) -2; (e) no; (f) 2; (g) 2; (h) 2; (i) yes
6. (a) 2; (b) 3; (c) yes; (d) no
7. (a), (b) 8. (a) Yes, yes; (b) no, yes
9. (a) Yes, 17; (b) yes, 17; (c) yes; (d) yes
10. (a) No; (b) yes
11. No 12. No
13. $\sqrt[3]{x}$ is continuous by (ii); 7 is continuous by (i) and x^2 is continuous by (ii), so $7x^2$ is continuous by (iii). Then $\sqrt[3]{x} - 7x^2$ is continuous by (iii). Now x is continuous by (ii) and 2 is continuous by (i), so $x - 2$ is continuous by (iii) and $\dfrac{1}{x-2}$ is continuous by (iv). Thus $\dfrac{\sqrt[3]{x} - 7x^2}{x-2}$ is continuous by (iii).
14. 53 15. $\sqrt{8}$ 16. -8 17. $\frac{1}{6}$
18. (a) 4, 4.5, 5.14, 6, 7.2, 9, 12, 18, 36, 54, 108; (b) $+\infty$; (c) $+\infty$
19. (a) 3.25, 2.25, 2.0625, 2.025, 2.005, 2.0005; (b) 2 20. $\frac{2}{3}$

Exercise Set 11.6, pp. 699–701

1. No 3. Yes 5. (a) 2, -1, does not exist; (b) -1; (c) no; (d) 3; (e) 3; (f) yes 7. (a) 2; (b) 2; (c) yes; (d) 0; (e) 0; (f) yes 9. No, yes, no, yes 11. 25, 45, does not exist 13. 65¢ 15. -2 17. Does not exist 19. 11 21. -10 23. $-\frac{5}{2}$ 25. 5 27. 2 29. 3 31. Does not exist 33. $\frac{2}{7}$ 35. $\frac{5}{4}$ 37. $\frac{1}{2}$ 39. $\frac{2}{5}$ 41. 5 43. $\frac{1}{2}$ 45. $\frac{2}{3}$ 47. (a) 1; (b) 1; (c) 1; (d) 1; (e) yes; no 49. (a) \$800, \$736, \$677.12, \$622.95, \$573.11; (b) \$5656.12; (c) \$10,000 51. 0, no 53. -0.25 55. 0

Margin Exercises, Section 11.7

1.

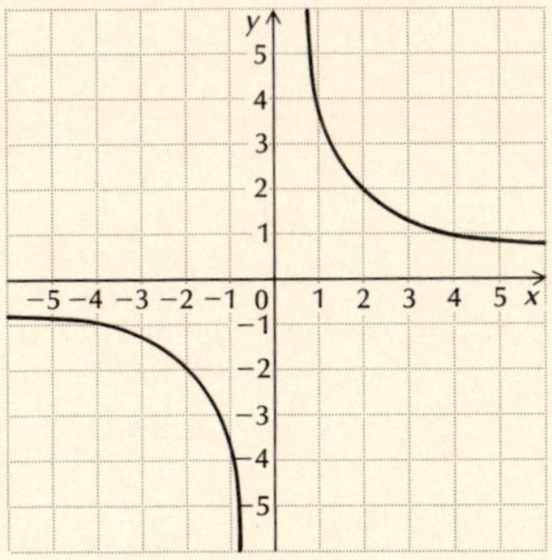

2.

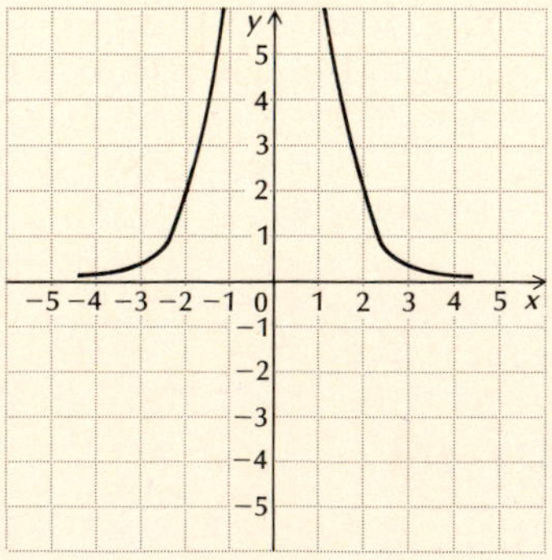

3.

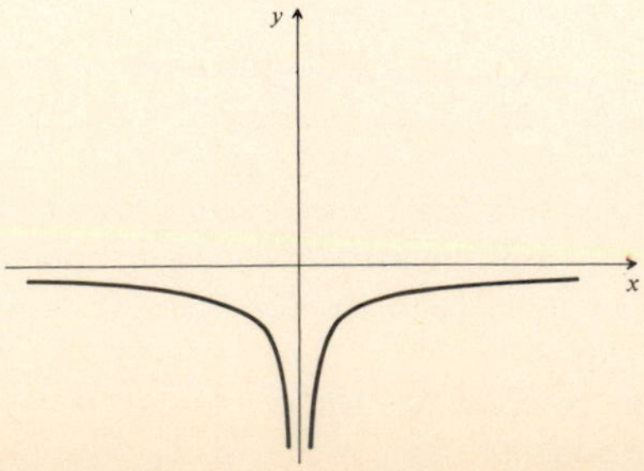

4. 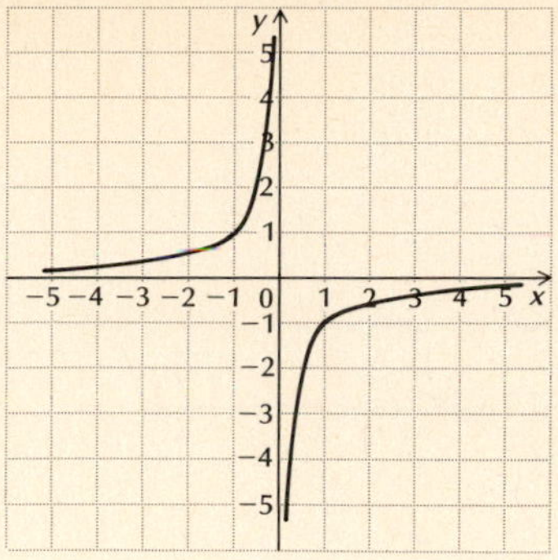

5. 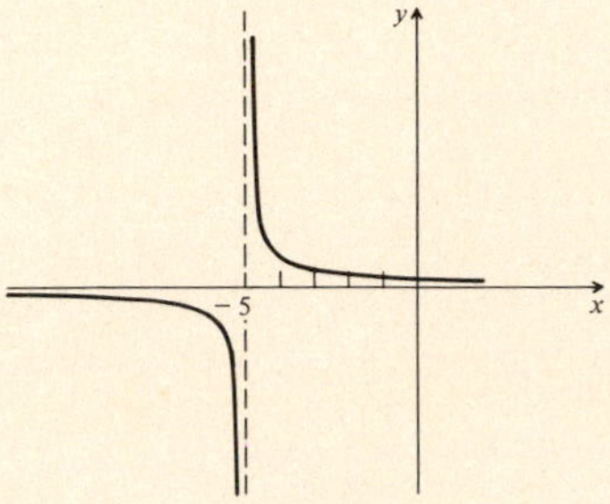

6. 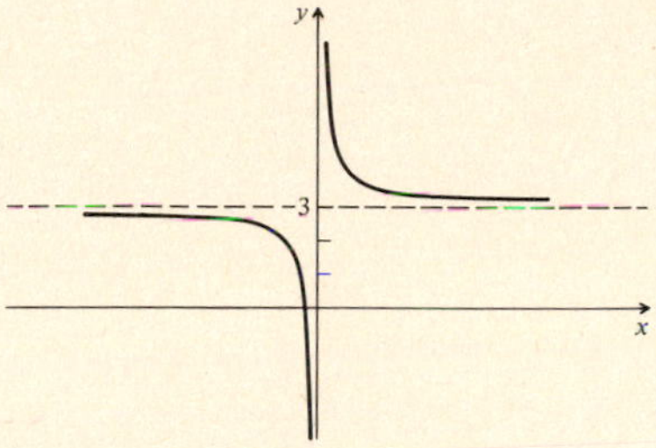

7. 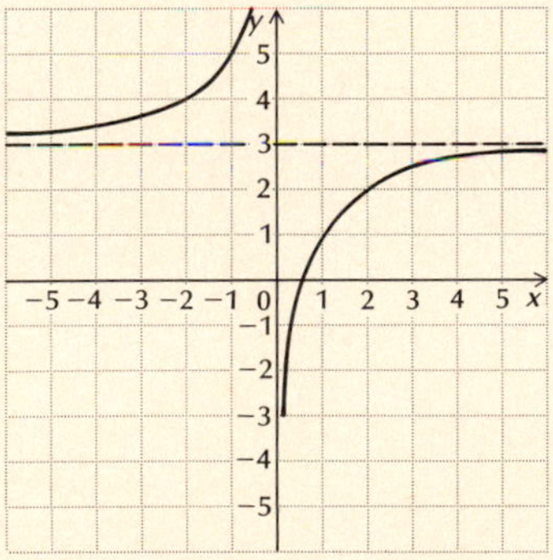

8. 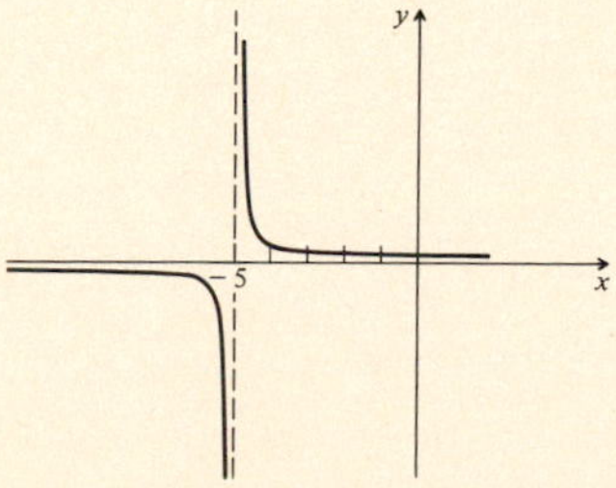

9. 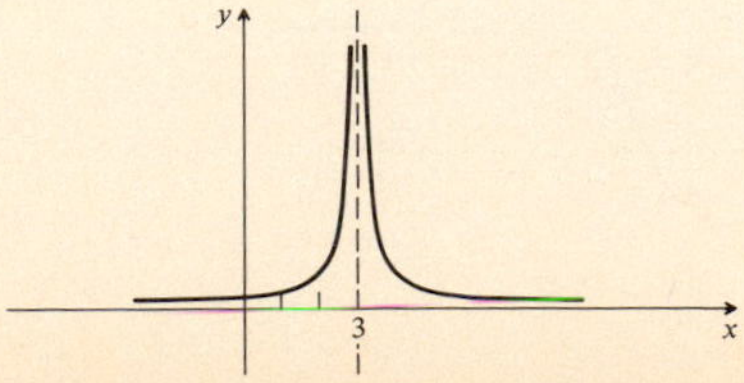

10. 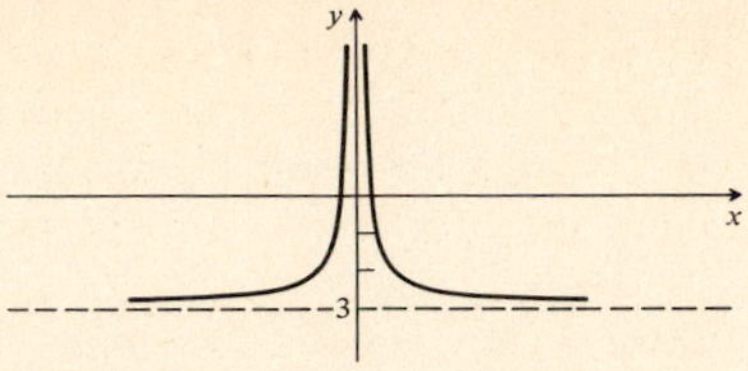

11. $x = 0, x = -2, x = 3$ 12. $x = 2, x = -2, x = -\frac{1}{2}$
13. (b) and (c) 14. $y = \frac{1}{2}$ 15. $y = 3$ 16. $y = 3x - 1$
17. $y = 5x$ 18. $0, 3, -5$ 19. $0, 1, -3$

20. 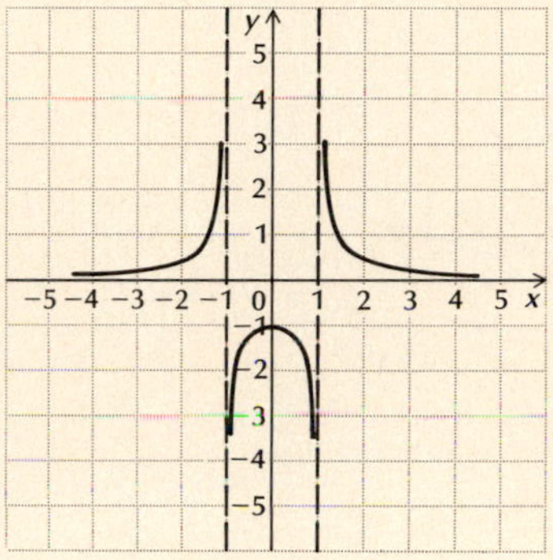

21. 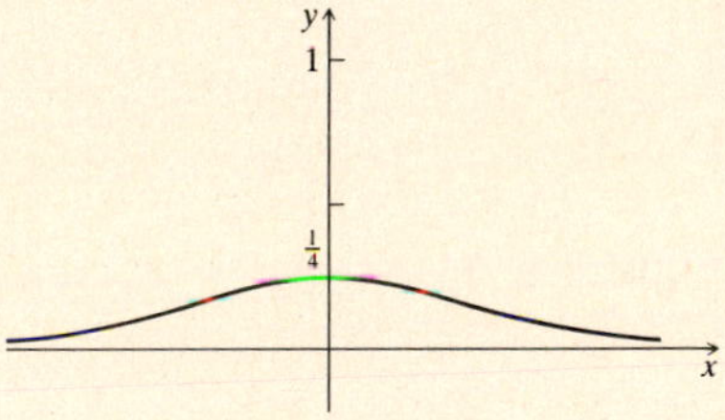

22. 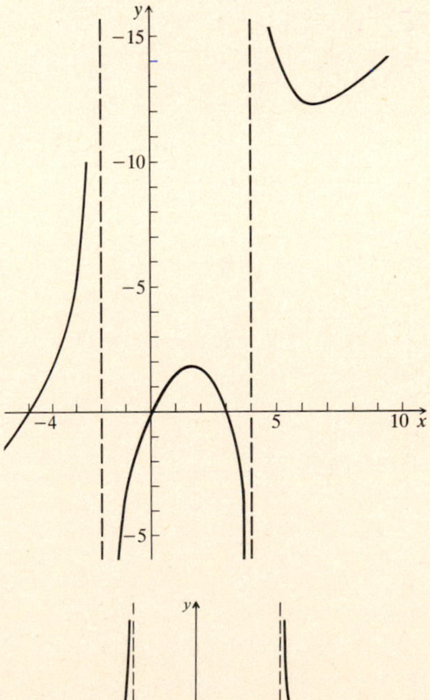

23. 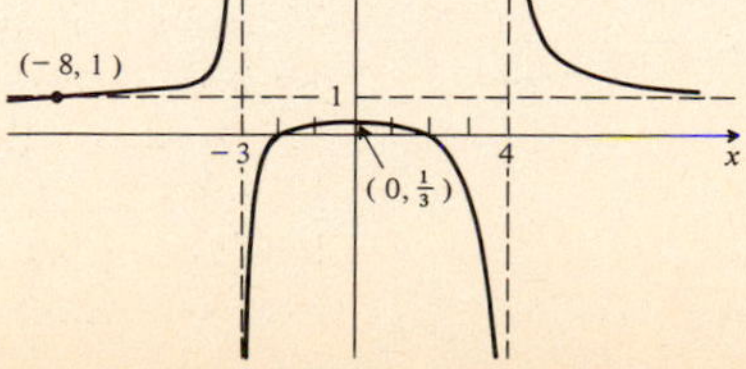

Exercise Set 11.7, pp. 711–713

1.

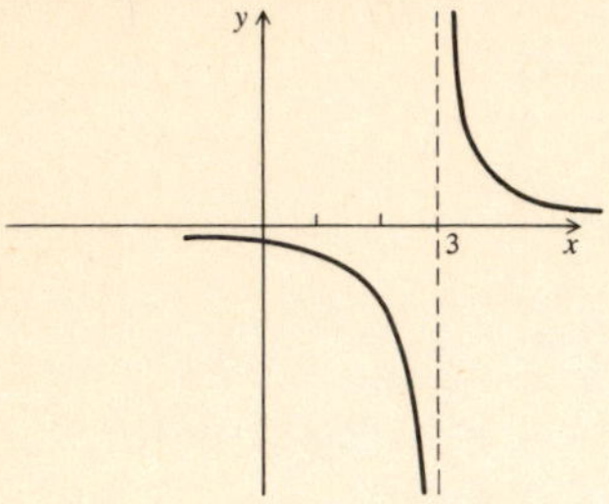

3.

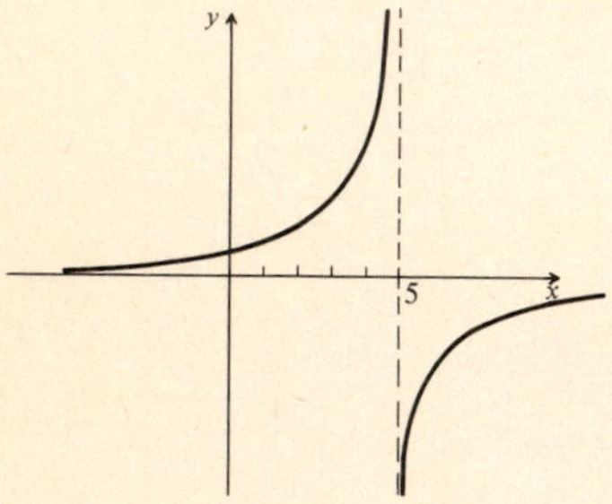

5.

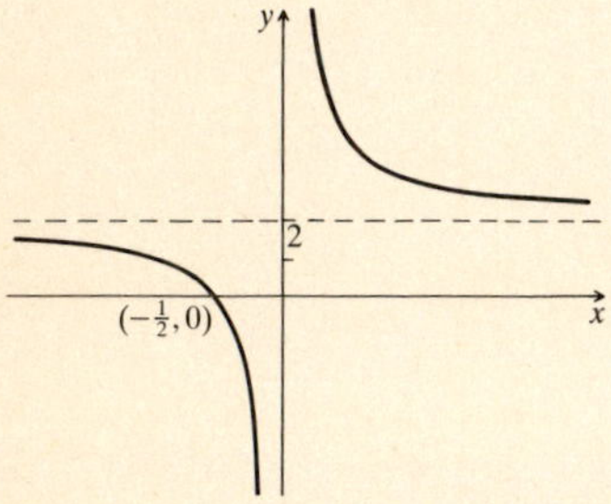

7.

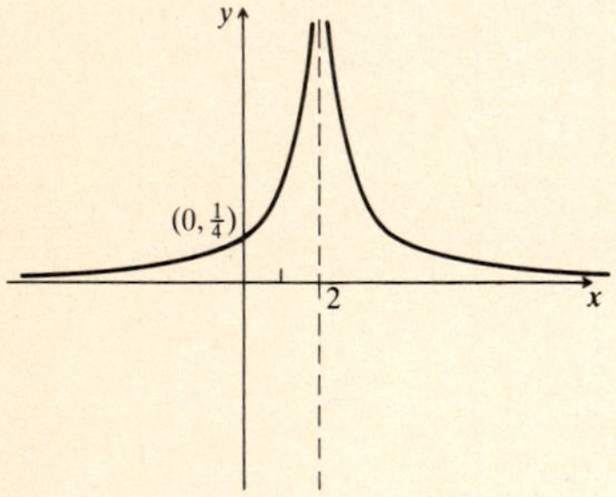

9.

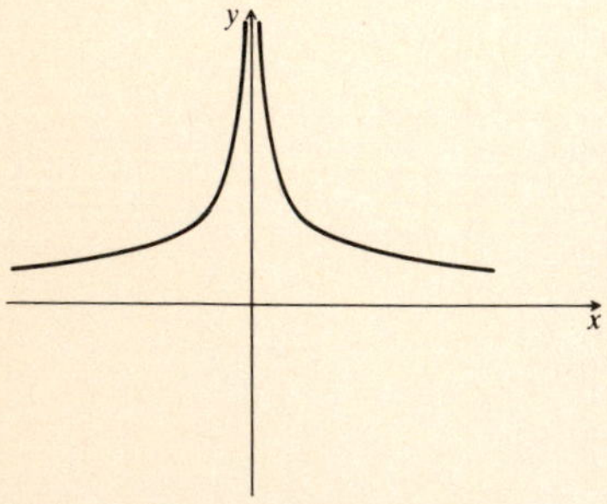

11.

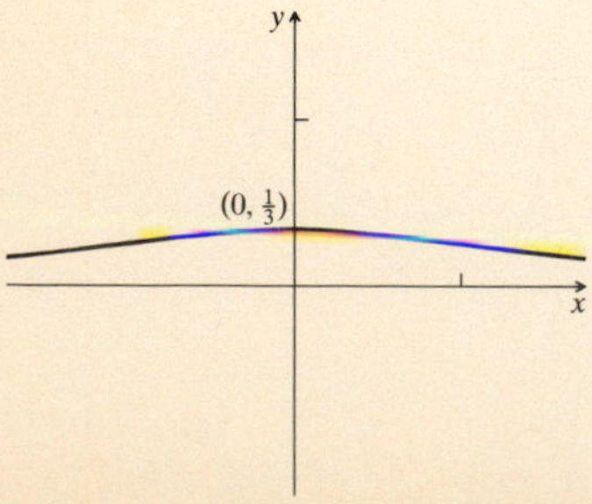

13.

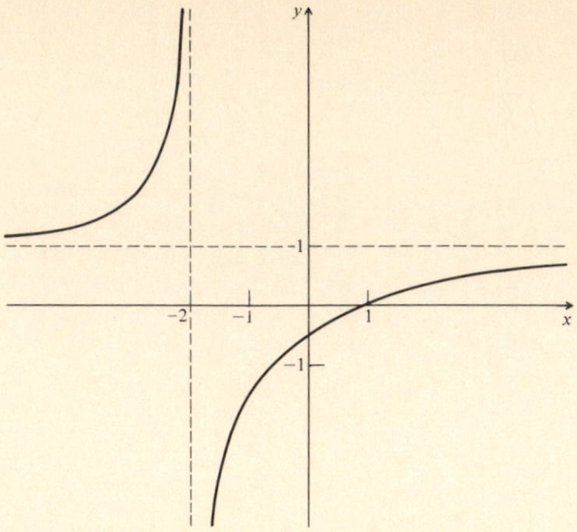

15.

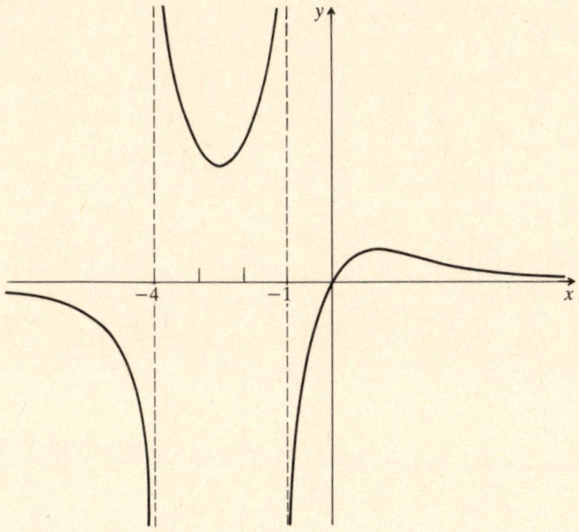

17.

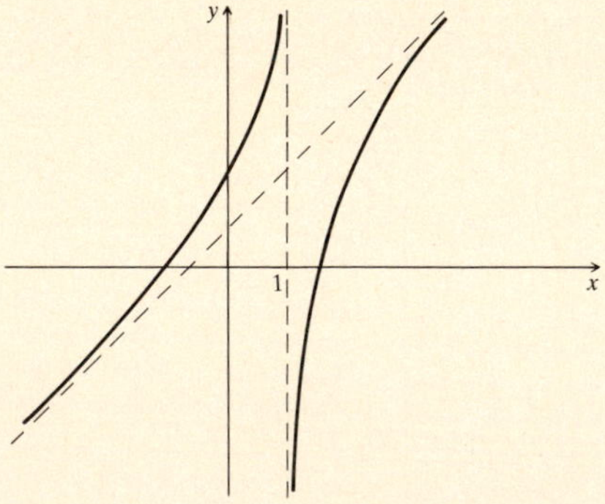

19.

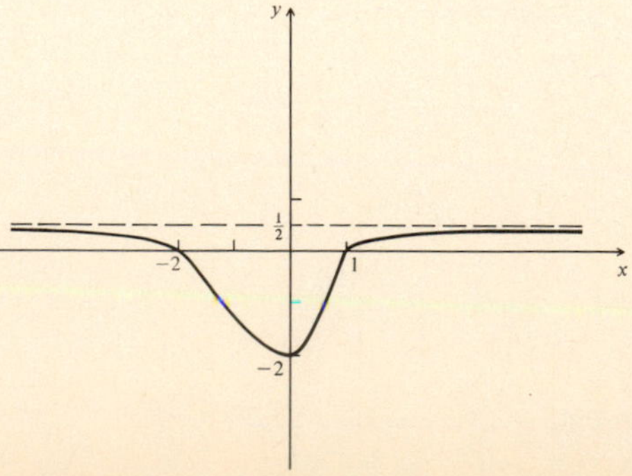

21.

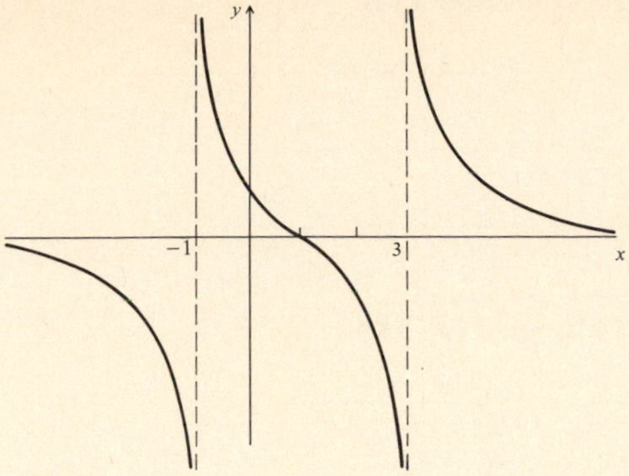

23.

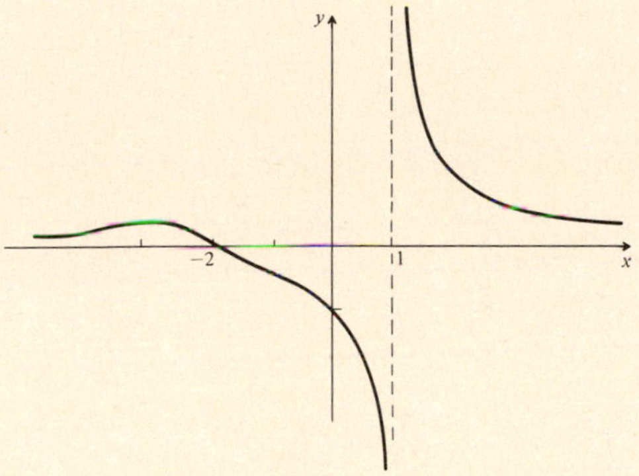

25.

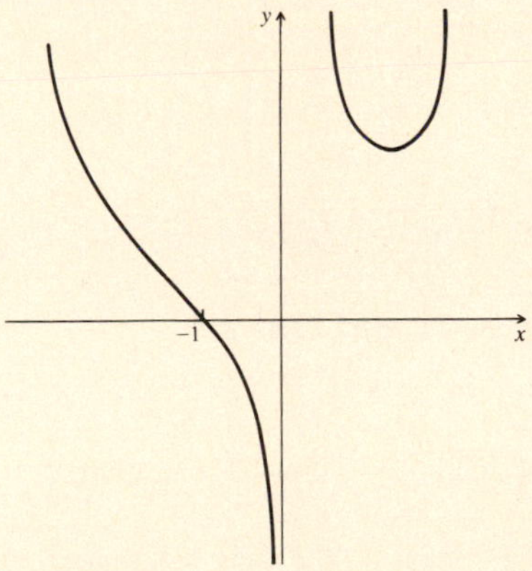

27.

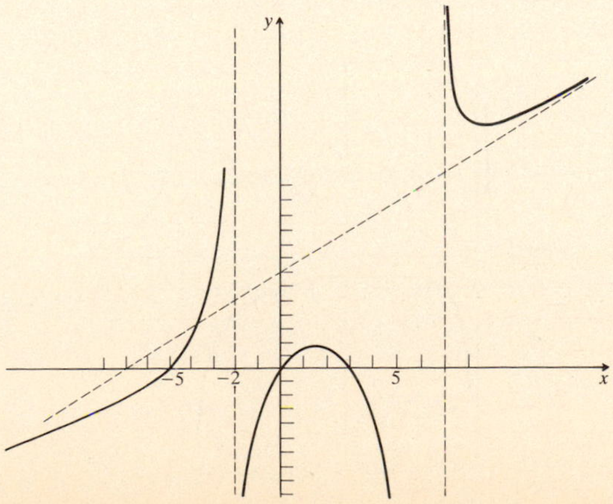

29.

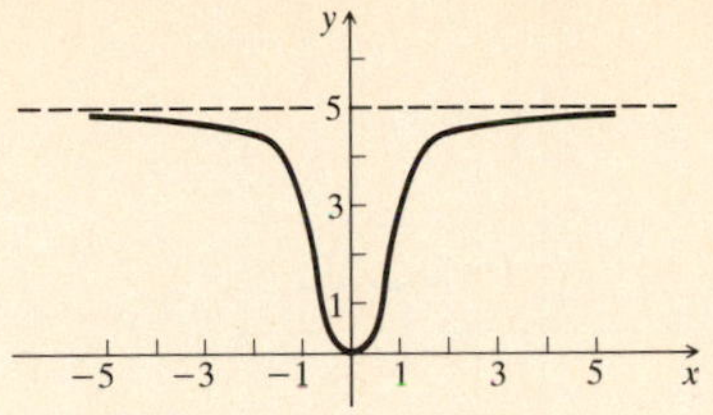

31.

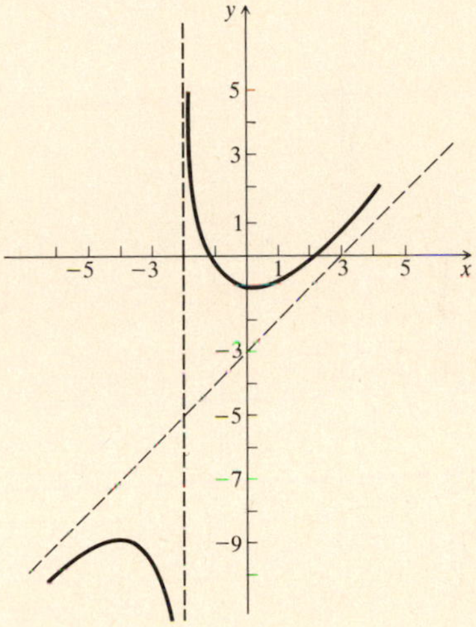

33. (a) $t = \dfrac{500}{r}$;

(b)

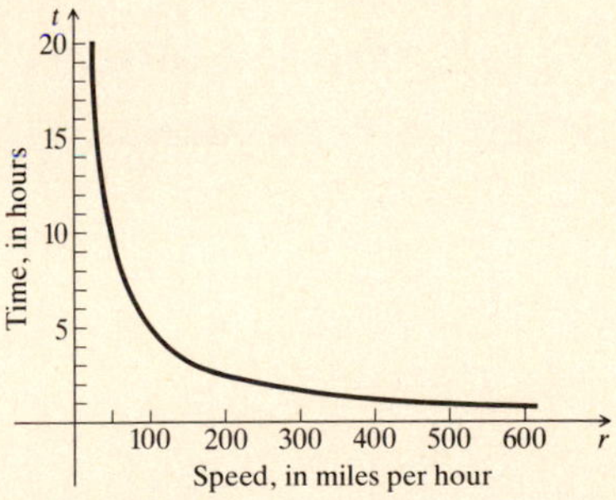

35.

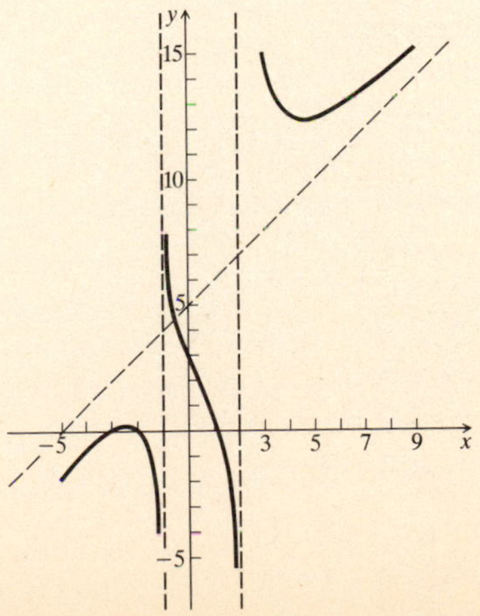

37.

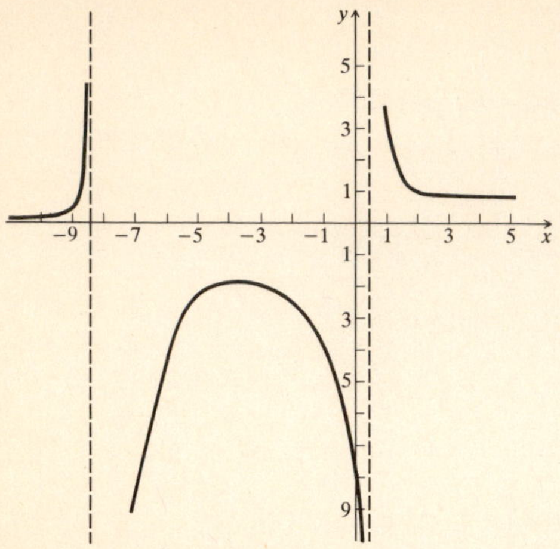

39.

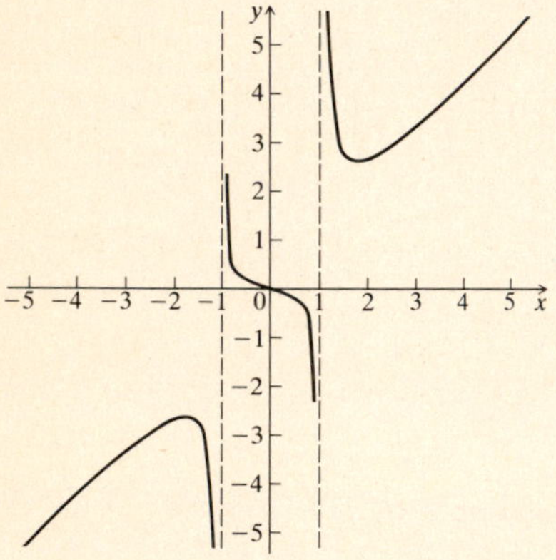

41. **(a)** $T(w) = \dfrac{120{,}000}{40{,}000 - w^2}$; **(b)** 3.001876 hr, 3.007519 hr, 3.030303 hr; **(c)** Domain $= \{w \mid 0 \leqslant w < 200\}$, range $= \{T \mid T \geqslant 3\}$;

(d)

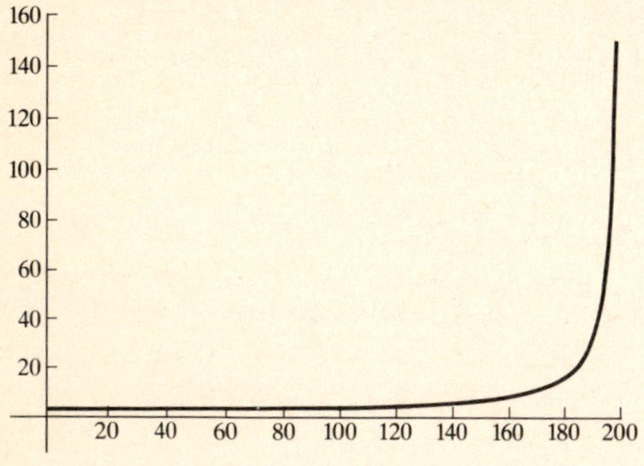

48. **(a)** $S(x) = x^2 + \dfrac{432}{x}$;

(b)

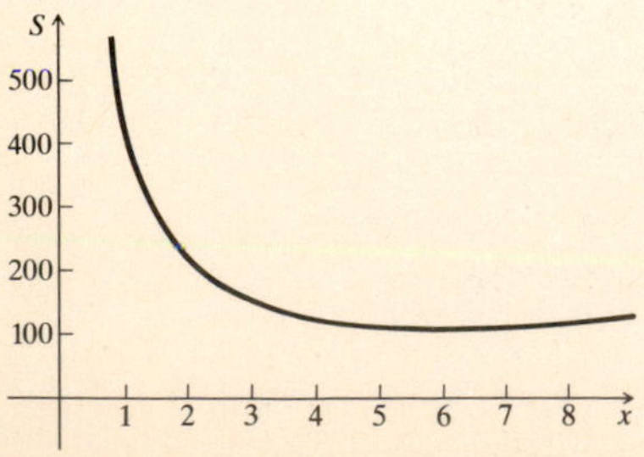

Margin Exercises, Section 11.8

1. $\dfrac{1}{3x+2} + \dfrac{4}{x-3}$ **2.** $\dfrac{1}{x+1} + \dfrac{2}{x-1} - \dfrac{1}{(x-1)^2}$

3. $\dfrac{x+2}{x^2+1} + \dfrac{1}{x+2}$ **4.** $3x + 1 + \dfrac{4}{2x-1} - \dfrac{3}{x+5}$

5. $\dfrac{3}{x-4} - \dfrac{2}{x+1} + \dfrac{1}{(x+1)^2}$ **6.** $\dfrac{2x-5}{3x^2+2} + \dfrac{1}{x-1}$

Exercise Set 11.8, pp. 717–718

1. $\dfrac{2}{x-3} + \dfrac{-1}{x+2}$ **3.** $\dfrac{-4}{3x-1} + \dfrac{5}{2x-1}$

5. $\dfrac{-3}{x-2} + \dfrac{2}{x+2} + \dfrac{4}{x+1}$ **7.** $\dfrac{-3}{(x+2)^2} + \dfrac{-1}{x+2} + \dfrac{1}{x-1}$

9. $\dfrac{3}{x-1} + \dfrac{-4}{2x-1}$ **11.** $x - 2 + \dfrac{-\frac{11}{4}}{(x+1)^2} + \dfrac{\frac{17}{16}}{x+1} + \dfrac{-\frac{17}{16}}{x-3}$

13. $\dfrac{3x+5}{x^2+2} + \dfrac{-4}{x-1}$ **15.** $\dfrac{-2}{x+2} + \dfrac{10}{(x+2)^2} + \dfrac{3}{2x-1}$

17. $3x + 1 + \dfrac{2}{2x-1} + \dfrac{3}{x+1}$ **19.** $\dfrac{-1}{x-3} + \dfrac{3x}{x^2+2x-5}$

21. $\dfrac{5}{3x+5} - \dfrac{3}{x+1} + \dfrac{4}{(x+1)^2}$ **23.** $\dfrac{8}{4x-5} + \dfrac{3}{3x+2}$

25. $\dfrac{2x-5}{3x^2+1} - \dfrac{2}{x-2}$ **27.** $-\dfrac{\frac{1}{2a^2}x}{x^2+a^2} + \dfrac{\frac{1}{4a^2}}{x-a} + \dfrac{\frac{1}{4a^2}}{x+a}$

29. $\dfrac{-1}{x+1} + \dfrac{4}{x+4}$

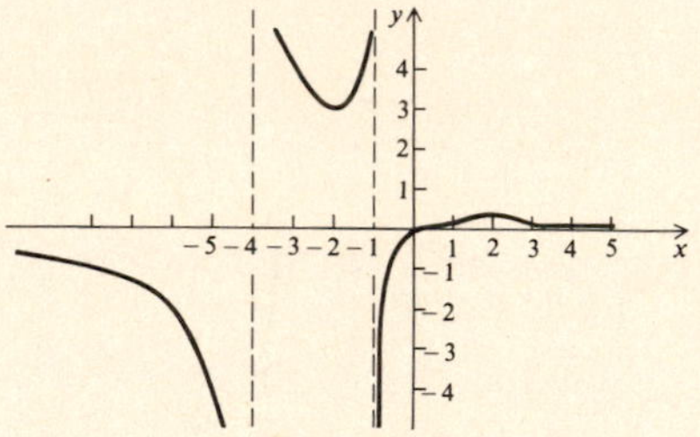

31. $-\dfrac{3}{25(\ln x + 2)} + \dfrac{3}{25(\ln x - 3)} + \dfrac{7}{5(\ln x - 3)^2}$

Review Exercises: Chapter 11, pp. 718–719

1. [11.2] 0 **2.** [11.2] $Q = 2x^3 - 10x^2 + 27x - 59$, $R = 119$
3. [11.2] 88 **4.** [11.2] $(x-1)(x+3)(x+5)$; 1, −3, −5
5. [11.2] No **6.** [11.3] 0, mult. 2; 3, mult. 2; −4, mult. 3; 4, mult. 1 **7.** [11.3] $x^3 - 3x^2 + 2x$
8. [11.3] $(x^2-1)(x-2)^2(x+3)^3$ **9.** [11.3] $\pm 3, -3i$
10. [11.3] $-8 + 7i$, $10 - \sqrt{5}$
11. [11.4] $\pm(1, 2, 3, 4, 6, 12, \frac{1}{2}, \frac{3}{2})$ **12.** [11.4] $-3, 4, \pm 3i$
13. [11.5] 2 or none **14.** [11.5] 4, 2, or none **15.** [11.5] 2
16. [11.5] −2 **17.** [11.5] 1.41 **18.** [11.5] −0.9, 1.3, 2.5
19. [11.7]

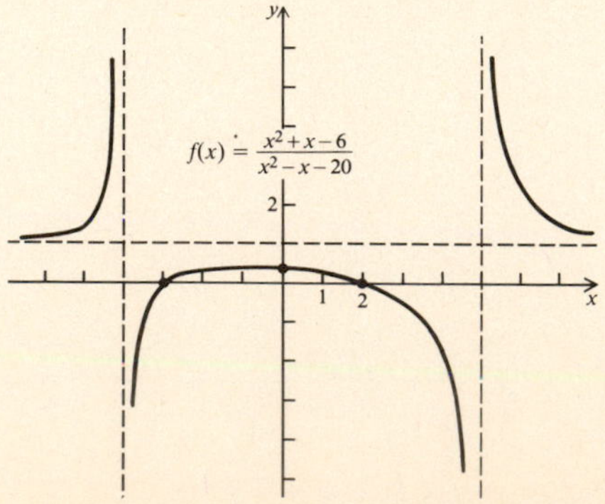

20. [11.8] $\dfrac{5}{x+1} - \dfrac{5}{x+2} - \dfrac{5}{(x+2)^2}$
21. [11.6] -4 22. [11.6] 10 23. [11.6] -10
24. [11.6] 5 25. [11.6] No 26. [11.6] Yes
27. [11.6] -4 28. [11.6] -4 29. [11.6] Yes
30. [11.6] Does not exist 31. [11.6] -2 32. [11.6] No
33. [11.3] $(x-1)\left(x+\dfrac{1}{2}+i\dfrac{\sqrt{3}}{2}\right)\left(x+\dfrac{1}{2}-i\dfrac{\sqrt{3}}{2}\right)$
34. [11.2] 7 35. [11.3] 4 36. [11.2] -4
37. [11.8]

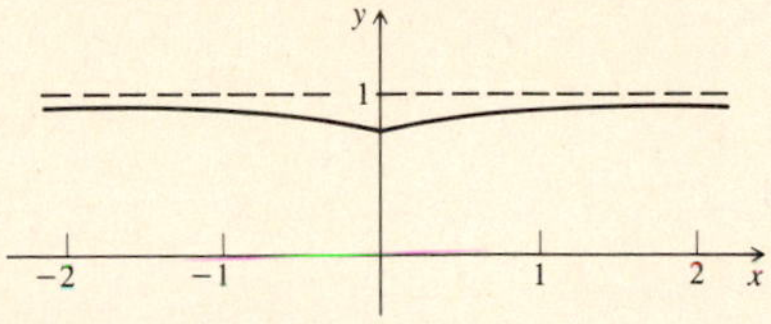

Test: Chapter 11, p. 720

1. [11.2] 82 2. [11.2] Yes 3. [11.2] The quotient is $5x^2 - 6x + 10$. The remainder is -13. 4. [11.2] 2315
5. [11.2] $(x-2)(x+1)(x-3)$; $-1, 2, 3$
6. [11.3] $x^4 - 4x^3 + x^2 + 16x - 20$
7. [11.3] $x(x+1)(x-2)^2(x-1)^3$
8. [11.3] $x^3 - 8x^2 + 22x - 20$
9. [11.4] $\pm\left(\frac{1}{3}, \frac{2}{3}, 1, 2, 3, 6\right)$ 10. [11.5] 2 or none
11. [11.5] 3 12. [11.5] 1.4
13. [11.7]

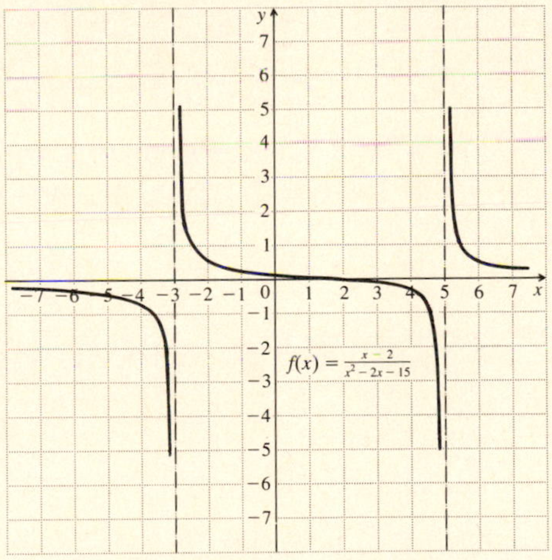

14. [11.8] $\dfrac{2}{2x-3} - \dfrac{5}{x+4}$ 15. [11.6] Yes 16. [11.6] No
17. [11.6] Does not exist 18. [11.6] 1 19. [11.6] No
20. [11.6] 3 21. [11.6] 3 22. [11.6] Yes 23. [11.6] 6
24. [11.6] $\frac{1}{2}$ 25. [11.6] Does not exist 26. [11.6] 4
27. [11.7]

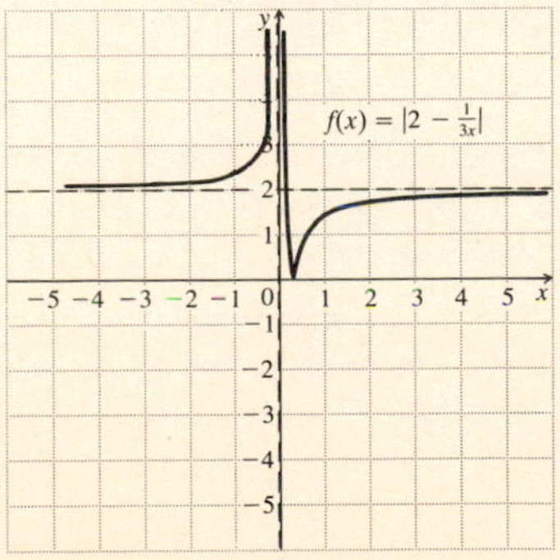

28. [11.4], [11.5] $i, -i, 1-i, 1+i$

CHAPTER 12

Margin Exercises, Section 12.1

1. (a) 1, 3, 5; (b) 67 2. (a) $1, -\frac{1}{2}, \frac{1}{3}, -\frac{1}{4}$; (b) $\frac{1}{47}$
3. $2n$ 4. n 5. n^3 6. $\dfrac{x^n}{n}$ 7. 2^{n-1}
8. (a) 0; (b) 4 9. $3 + 2\frac{1}{2} + 2\frac{1}{3} = 7\frac{5}{6}$
10. $5^0 + 5^1 + 5^2 + 5^3 + 5^4 = 781$
11. $8^3 + 9^3 + 10^3 + 11^3 = 3572$
12. $\sum_{k=1}^{5} 2k$ 13. $\sum_{k=1}^{4} k^3$ 14. $\sum_{k=1}^{\infty} \dfrac{x^k}{k}$
15. $\sum_{k=2}^{\infty} k^2$ 16. $-3, -9, -81, -6561, -43{,}046{,}721$
17. 5, 5, 10, 15, 25, 40, 65, 105

Exercise Set 12.1, pp. 726–728

1. 3, 7, 11, 15; 39, 59 3. $2, \frac{3}{2}, \frac{4}{3}, \frac{5}{4}; \frac{10}{9}; \frac{15}{14}$
5. 3, 8, 15, 24; 120; 255 7. $2, 2\frac{1}{2}, 3\frac{1}{3}, 4\frac{1}{4}; 10\frac{1}{10}; 15\frac{1}{15}$
9. $-1, 4, -9, 16; 100; -225$ 11. $-2, -1, 4, -7; -25; 40$
13. $\frac{1}{2}, \frac{4}{7}, \frac{5}{8}, \frac{2}{3}; \frac{4}{5}; \frac{17}{20}$ 15. 34 17. 323 19. -36.9
21. $-33{,}880$ 23. $\dfrac{441}{400}$ 25. 43 27. $1\dfrac{1}{1444}$ 29. $2n - 1$
31. $(-1)^n \cdot 2 \cdot 3^{n-1}$ 33. $\dfrac{n+1}{n+2}$ 35. $(\sqrt{3})^n$
37. $(-1)(3n-2)$ 39. 28 41. 30
43. $\frac{1}{2} + \frac{1}{4} + \frac{1}{6} + \frac{1}{8} + \frac{1}{10}$
45. $2^0 + 2^1 + 2^2 + 2^3 + 2^4 + 2^5 = 63$
47. $\log 7 + \log 8 + \log 9 + \log 10$
49. $\frac{1}{2} + \frac{2}{3} + \frac{3}{4} + \frac{4}{5} + \frac{5}{6} + \frac{6}{7} + \frac{7}{8} + \frac{8}{9}$
51. $-1 + 1 - 1 + 1 - 1 = -1$
53. $3 - 6 + 9 - 12 + 15 - 18 + 21 - 24 = -12$
55. $1 + \frac{2}{5} + \frac{1}{5} + \frac{2}{17} + \frac{1}{13} + \frac{2}{37}$
57. $3 + 2 + 3 + 6 + 11 + 18 = 43$
59. $\frac{1}{2} + \frac{1}{6} + \frac{1}{12} + \frac{1}{20} + \frac{1}{30} + \frac{1}{42} + \frac{1}{56} + \frac{1}{72} + \frac{1}{90} + \frac{1}{110}$
61. $\sum_{k=1}^{6} \dfrac{k}{k+1}$ 63. $\sum_{k=1}^{6} (-2)^k$ 65. $\sum_{k=2}^{n} (-1)^k k^2$
67. $\sum_{k=1}^{\infty} 5k$ 69. $\sum_{k=1}^{\infty} \dfrac{1}{k(k+1)}$ 71. $4, 1\frac{1}{4}, 1\frac{4}{5}, 1\frac{5}{9}$
73. $6561, -81, 9i, -3\sqrt{i}$ 75. 2, 3, 5, 8
77. $\frac{3}{2}, \frac{3}{2}, \frac{9}{8}, \frac{3}{4}, \frac{15}{32}; \frac{171}{32}$ 79. 0, 0.693, 1.792, 3.178, 4.787; 10.450 81. $\pi, 0, \pi, 0, \pi; 3\pi$ 83. (a) 41, 43, 47, 53, 61, 71; (b) all numbers are prime; (c) 1681; no 85. 0.414214, 0.317837, 0.267949, 0.236068, 0.213422, 0.196262
87. 2, 1.5, 1.416667, 1.414216, 1.414214, 1.414214 89. \$3900, \$2925, \$2193.75, \$1645.31, \$1233.98, \$925.49, \$694.12, \$520.59, \$390.44, \$292.83 91. $\ln n!$

Margin Exercises, Section 12.2

1. $a_1 = 3.1, d = 0.8$ 2. 50 3. 75 4. $a_1 = 7, d = 12$; 7, 19, 31, 43, . . . 5. 20,100 6. 225 7. 455 8. \$27.25
9. 5200 10. 7, 10, 13

Exercise Set 12.2, pp. 734–736

1. $a_1 = 3, d = 5$ 3. $a_1 = 9, d = -4$ 5. $a_1 = \frac{3}{2}, d = \frac{3}{4}$
7. $a_1 = \$1.07, d = \$0.07$ 9. $a_{12} = 46$ 11. $a_{17} = -41$
13. $a_{13} = -\$1628.16$ 15. 27th 17. 102nd 19. $a_{17} = 101$

21. $a_1 = 5$ **23.** $n = 28$ **25.** $a_1 = 8$; $d = -3$; 8, 5, 2, -1, -4
27. 670 **29.** 45,150 **31.** 2550 **33.** 735 **35.** 990
37. 3 **39.** 1275 **41.** \$31,000 **43.** 6300
45. $5\frac{4}{5}, 7\frac{3}{5}, 9\frac{2}{5}, 11\frac{1}{5}$ **47.** n^2 **49.** 3, 5, 7 **51.** \$8760, \$7961.77, \$7163.54, \$6365.31, \$5567.08, \$4768.85, \$3970.62, \$3172.39, \$2374.16, \$1575.93 **53.** -10, -4, 2, 8 **55.** Sides are a, $a + d$, $a + 2d$; and $a^2 + (a + d)^2 = (a + 2d)^2$. Solving, we get $d = \frac{a}{3}$. Thus the sides are a, $\frac{4a}{3}$, and $\frac{5a}{3}$ in the ratio 3:4:5.
57. $a_n = a_1 + (n - 1)d$; $a_n = f(n) = d \cdot n + (a_1 - d)$, where $f(n)$ is a linear function of n with slope d and y-intercept $(a_1 - d)$.

Margin Exercises, Section 12.3

1. 5 **2.** -3 **3.** 0.85 **4.** 1.09 **5.** $\frac{1}{2}$ **6.** $a_9 = 512$
7. $a_6 = \frac{1}{81}$ **8.** $S_8 = 13{,}120$ **9.** $S_{10} = \frac{341}{256}$ **10.** 363
11. No **12.** No **13.** Yes, 1 **14.** Yes, $\frac{3125}{3}$ **15.** $\frac{2}{9}$
16. $\frac{13}{99}$ **17.** \$21,474,836.47 **18.** \$160,000,000

Exercise Set 12.3, pp. 741–744

1. 2 **3.** -1 **5.** $-\frac{1}{2}$ **7.** $\frac{1}{5}$ **9.** $\frac{1}{x}$ **11.** 1.1 **13.** 64
15. 162 **17.** 648 **19.** \$1360.49 **21.** 3^{n-1} **23.** $(-1)^{n-1}$
25. $\frac{1}{x^n}$ **27.** 762 **29.** $\frac{547}{18}$ **31.** $\frac{x^8 - 1}{x - 1}$ **33.** \$5134.51
35. 2 **37.** Yes, 8 **39.** Yes, 125 **41.** Yes, $\frac{1000}{11}$ **43.** No
45. Yes, $\frac{2}{3}$ **47.** Yes, \$4545.45 **49.** Yes, $\frac{160}{9}$ **51.** Yes, 2
53. $\frac{7}{9}$ **55.** $\frac{8}{15}$ **57.** $\frac{510}{99}$, or $\frac{170}{33}$ **59.** $\frac{1}{256}$ ft **61.** 155,797
63. \$5236.19 **65.** 3100 ft **67.** \$2,684,355
69. \$86,666,666,667 **71.** 3.33×10^6, $66\frac{2}{3}\%$
73. $(4 - \sqrt{6})/(\sqrt{3} - \sqrt{2}) = 2\sqrt{3} + \sqrt{2}$, $(6\sqrt{3} - 2\sqrt{2})/(4 - \sqrt{6}) = 2\sqrt{3} + \sqrt{2}$; there exists a common ratio, $2\sqrt{3} + \sqrt{2}$; thus the sequence is geometric.
75. $S_n = \frac{x^n - 1}{x - 1}$ **77.** $\frac{a_{n+1}}{a_n} = r$, so $\frac{(a_{n+1})^2}{(a_n)^2} = r^2$; thus $a_1^2, a_2^2, \ldots,$ is geometric, with ratio r^2.
79. Let the arithmetic sequence have common difference $d = a_{n+1} - a_n$. Then the sequence $5^{a_1}, 5^{a_2}, 5^{a_3}, \ldots,$ has $\frac{5^{a_{n+1}}}{5^{a_n}} = 5^d$ for a common ratio and is therefore geometric.
81. 2, 1.5, 1.416667, 1.414216, 1.414214, 1.414214, 1.414214; the limit is $\sqrt{2}$.
83. 2, 2.5, 2.666667, 2.708333, 2.716667, 2.718056; $e \approx 2.7182818$

Margin Exercises, Section 12.4

1. $1^2 + 1 > 1 + 1$ or $2 > 2$; $2^2 + 1 > 2 + 1$ or $5 > 3$; $3^2 + 1 > 3 + 1$ or $10 > 4$; $4^2 + 1 > 4 + 1$ or $17 > 5$; $5^2 + 1 > 5 + 1$ or $26 > 6$ **2.** $1 = \frac{1(1 + 1)}{2}$ or $1 = 1$; $1 + 2 = \frac{2(2 + 1)}{2}$, or $1 + 2 = 3$; $1 + 2 + 3 = 6$; $1 + 2 + 3 + 4 = 10$; $1 + 2 + 3 + 4 + 5 = 15$
3. S_1: $x \leq x$; S_2: $x \leq x^2$. Both obviously true if $x > 1$. Thus basis step is complete. S_k: $x \leq x^k$; S_{k+1}: $x \leq x^{k+1}$. Assume S_k: $x \leq x^k$. We know by hypothesis that $1 < x$. Multiply the inequalities to get $x \cdot 1 \leq x^k \cdot x$, or $x < x^{k+1}$. We have arrived at S_{k+1}, hence have shown that $S_k \to S_{k+1}$ for all natural numbers k. We can now conclude that $x \leq x^k$ for all natural numbers n.
4. **(a)** $2 \cdot 1 = 1(1 + 1)$, $2 + 4 = 2(2 + 1)$;
(b) $2 + 4 + \cdots + 2k = k(k + 1)$;
(c) $2 + 4 + \cdots + 2(k + 1) = (k + 1)[(k + 1) + 1]$;
(d) $2 \cdot 1 \stackrel{?}{=} 1(1 + 1)$, $2 = 1 \cdot 2 = 2$;
(e) Assume S_k as hypothesis:
$$2 + 4 + \cdots + 2k = k(k + 1).$$
Add $2(k + 1)$ on both sides:
$$2 + 4 + \cdots + 2k + 2(k + 1) = k(k + 1) + 2(k + 1) = (k + 1)(k + 2) \quad \text{(Simplifying)}$$
We now have
$$2 + 4 + \cdots + 2(k + 1) = (k + 1)(k + 2)$$
or
$$(k + 1)[(k + 1) + 1].$$
This is S_{k+1}. Hence $S_k \to S_{k+1}$ for all k. Finally, we conclude that $2 + 4 + \cdots + 2n = n(n + 1)$ for all natural numbers n.
5. S_1: $2 = \frac{1(3 + 1)}{2}$. True. S_k: $\sum_{p=1}^{k} (3p - 1) = \frac{k(3k + 1)}{2}$, or $2 + 5 + 8 + \cdots + 3k - 1 = \frac{k(3k + 1)}{2}$.
S_{k+1}: $\sum_{p=1}^{k+1} (3p - 1) = \frac{(k + 1)[3(k + 1) + 1]}{2}$, or $2 + 5 + \cdots + 3k - 1 + 3(k + 1) - 1 = \frac{(k + 1)[3(k + 1) + 1]}{2}$. Assume S_k. Then add $[3(k + 1) - 1]$ on both sides:
$$\begin{aligned} 2 + 5 + \cdots + [3(k + 1) - 1] &= \frac{k(3k + 1)}{2} + [3(k + 1) - 1] \\ &= \frac{k(3k + 1) + 2[3(k + 1) - 1]}{2} \\ &= \frac{3k^2 + 7k + 4}{2} = \frac{(k + 1)(3k + 4)}{2} \\ &= \frac{(k + 1)[3(k + 1) + 1]}{2} \end{aligned}$$
We have arrived at
$$\sum_{p=1}^{k+1} (3p - 1) = \frac{(k + 1)[3(k + 1) - 1]}{2}.$$
This is S_{k+1}. So $S_k \to S_{k+1}$ for all k. We conclude that
$$\sum_{p=1}^{n} (3p - 1) = \frac{n(3n + 1)}{2}$$
for *all* natural numbers n.

Exercise Set 12.4, pp. 748–750

1. $1^2 < 1^3$, $2^2 < 2^3$, $3^2 < 3^3$, etc.
3. A polygon of 3 sides has $\frac{3(3 - 3)}{2}$ diagonals. A polygon of 4 sides has $\frac{4(4 - 3)}{2}$ diagonals, etc.
5. S_n: $1 + 2 + 3 + \cdots + n = \frac{n(n + 1)}{2}$
S_1: $1 = \frac{1(1 + 1)}{2}$
S_k: $1 + 2 + 3 + \cdots + k = \frac{k(k + 1)}{2}$
S_{k+1}: $1 + 2 + 3 + \cdots + k + (k + 1) = \frac{(k + 1)(k + 2)}{2}$

1. Basis step: S_1 true by substitution.
2. Induction step: Assume S_k. Deduce S_{k+1}.
Starting with the left side of S_{k+1}, we have

$$\underbrace{1 + 2 + 3 + \cdots + k}+ (k + 1)$$
$$= \frac{k(k+1)}{2} + (k+1) \quad (\text{by } S_k)$$
$$= \frac{k(k+1) + 2(k+1)}{2} \quad (\text{adding})$$
$$= \frac{(k+1)(k+2)}{2} \quad (\text{distributive law})$$

7. S_n: $1 + 5 + 9 + \cdots + (4n - 3) = n(2n - 1)$
S_1: $1 = 1(2 \cdot 1 - 1)$
S_k: $1 + 5 + 9 + \cdots + (4k - 3) = k(2k - 1)$
S_{k+1}: $1 + 5 + 9 + \cdots + (4k - 3) + [4(k + 1) - 3]$
$$= (k+1)[2(k+1) - 1]$$
$$= (k+1)(2k+1)$$

1. Basic step: S_1 true by substitution.
2. Induction step: Assume S_k. Deduce S_{k+1}.
Starting with the left side of S_{k+1}, we have

$$\underbrace{1 + 5 + 9 + \cdots + (4k - 3)} + [4(k+1) - 3]$$
$$= k(2k - 1) + [4(k+1) - 3] \quad (\text{by } S_k)$$
$$= 2k^2 - k + 4k + 4 - 3$$
$$= (k+1)(2k+1)$$

9. S_n: $\frac{1}{1 \cdot 2} + \frac{1}{2 \cdot 3} + \cdots + \frac{1}{n(n+1)} = \frac{n}{n+1}$

S_1: $\frac{1}{1 \cdot 2} = \frac{1}{1+1}$

S_k: $\frac{1}{1 \cdot 2} + \frac{1}{2 \cdot 3} + \cdots + \frac{1}{k(k+1)} = \frac{k}{k+1}$

S_{k+1}: $\frac{1}{1 \cdot 2} + \frac{1}{2 \cdot 3} + \cdots + \frac{1}{k(k+1)} + \frac{1}{(k+1)(k+2)}$
$$= \frac{k+1}{k+2}$$

2. Induction step: Assume S_k. Deduce S_{k+1}. Add
$$\frac{1}{(k+1)(k+2)}$$
to both sides and simplify the right side.

11. *2. Induction step:* Assume S_k. Deduce S_{k+1}. Now
$$k < k + 1 \quad (\text{by } S_k)$$
$$k + 1 < k + 1 + 1 \quad (\text{adding } 1)$$
$$\therefore k + 1 < k + 2$$

13. S_1: $3^1 < 3^{1+1}$
S_k: $3^k < 3^{k+1}$
$3^k \cdot 3 < 3^{k+1} \cdot 3$
$3^{k+1} < 3^{(k+1)+1}$

15. S_1: $1^3 = \frac{1^2(1+1)^2}{4} = 1$

S_k: $1^3 + 2^3 + \cdots + k^3 = \frac{k^2(k+1)^2}{4}$

$$1^3 + 2^3 + \cdots + (k+1)^3 = \frac{k^2(k+1)^2}{4} + (k+1)^3$$
$$= \frac{(k+1)^2}{4}[k^2 + 4(k+1)]$$
$$= \frac{(k+1)^2(k+2)^2}{4}$$

17. S_1: $1 + \frac{1}{1} = 1 + 1$

S_k: $\left(1 + \frac{1}{1}\right) \cdots \left(1 + \frac{1}{k}\right) = k + 1$

Multiply by $\left(1 + \frac{1}{k+1}\right)$:
$$\left(1 + \frac{1}{1}\right) \cdots \left(1 + \frac{1}{k+1}\right) = (k+1)\left(1 + \frac{1}{k+1}\right)$$
$$= (k+1)\left(\frac{k+1+1}{k+1}\right)$$
$$= (k+1) + 1$$

19. S_1: $a_1 = \frac{a_1 - a_1 r}{1 - r} = \frac{a_1(1 - r)}{1 - r} = a_1$

S_k: $a_1 + \cdots + a_1 r^{k-1} = \frac{a_1 - a_1 r^k}{1 - r}$

Add $a_1 r^k$.
$$a_1 + \cdots + a_1 r^k = \frac{a_1 - a_1 r^k}{1 - r} + a_1 r^k \frac{1 - r}{1 - r}$$
$$= \frac{a_1 - a_1 r^k + a_1 r^k - a_1 r^{k+1}}{1 - r}$$
$$= \frac{a_1 - a_1 r^{k+1}}{1 - r}$$

21. *2. Induction step:* Assume S_k. Deduce S_{k+1}.
Starting with the left side of S_{k+1} we have:
$$[r(\cos\theta + i\sin\theta)]^{k+1}$$
$$= [r(\cos\theta + i\sin\theta)]^k[r(\cos\theta + i\sin\theta)]$$
$$= r^k[\cos k\theta + i\sin k\theta][r(\cos\theta + i\sin\theta)]$$
$$= r^{k+1}(\cos k\theta + i\sin k\theta)(\cos\theta + i\sin\theta)$$
$$= r^{k+1}[\cos k\theta\cos\theta + i\sin k\theta\cos\theta + i\cos k\theta\sin\theta - \sin k\theta\sin\theta]$$
$$= r^{k+1}[(\cos k\theta\cos\theta - \sin k\theta\sin\theta) + i(\sin k\theta\cos\theta + \cos k\theta\sin\theta)]$$
$$= r^{k+1}[\cos(k\theta + \theta) + i\sin(k\theta + \theta)]$$
$$= r^{k+1}[\cos(k+1)\theta + i\sin(k+1)\theta]$$

23. *2. Induction step:* Assume S_k. Deduce S_{k+1}.
We start with the left side of S_{k+1}.
$$\underbrace{\cos x \cdot \cos 2x \cdots \cos 2^{k-1}x} \cdot \cos 2^k x$$
$$= \frac{\sin 2^k x}{2^k \sin x} \cdot \cos 2^k x$$
$$= \frac{\sin 2^k x \cos 2^k x}{2^k \sin x}$$
$$= \frac{\frac{1}{2}\sin 2(2^k x)}{2^k \sin x} \quad (\text{since } \sin 2\theta = 2\sin\theta\cos\theta)$$
$$= \frac{\sin 2^{k+1}x}{2^{k+1}\sin x}$$

25. S_n: $\left(1 - \frac{1}{2^2}\right)\left(1 - \frac{1}{3^2}\right) \cdots \left(1 - \frac{1}{n^2}\right) = \frac{n+1}{2n}$

S_2: $1 - \frac{1}{2^2} = \frac{2+1}{2 \cdot 2}$

S_k: $\left(1 - \frac{1}{2^2}\right)\left(1 - \frac{1}{3^2}\right) \cdots \left(1 - \frac{1}{k^2}\right) = \frac{k+1}{2k}$

S_{k+1}: $\left(1 - \frac{1}{2^2}\right)\left(1 - \frac{1}{3^2}\right) \cdots \left(1 - \frac{1}{k^2}\right)\left(1 - \frac{1}{(k+1)^2}\right) = \frac{k+2}{2(k+1)}$

1. Basis step: S_2 is true by substitution.
2. Induction step: Assume S_k. Deduce S_{k+1}.
Starting with the left side of S_{k+1} we have
$$\underbrace{\left(1 - \frac{1}{2^2}\right)\left(1 - \frac{1}{3^2}\right) \cdots \left(1 - \frac{1}{k^2}\right)}\left(1 - \frac{1}{(k+1)^2}\right)$$
$$= \frac{k+1}{2k}\left(1 - \frac{1}{(k+1)^2}\right) = \frac{k+1}{2k} - \frac{1}{2k(k+1)}$$
$$= \frac{(k+1)(k+1) - 1}{2k(k+1)} = \frac{k^2 + 2k + 1 - 1}{2k(k+1)}$$
$$= \frac{k^2 + 2k}{2k(k+1)} = \frac{k(k+2)}{2k(k+1)} = \frac{k+2}{2(k+1)}$$

27. S_2: $\overline{z_1 + z_2} = \overline{z_1} + \overline{z_2}$:

$\overline{(a + bi) + (c + di)} = \overline{(a + c) + (b + d)i}$
$= (a + c) - (b + d)i$
$\overline{(a + bi)} + \overline{(c + di)} = a - bi + c - di$
$= (a + c) - (b + d)i.$

S_k: $\overline{z_1 + z_2 + \cdots + z_k} = \overline{z_1} + \overline{z_2} + \cdots + \overline{z_k}$.

$\overline{(z_1 + z_2 + \cdots + z_k) + z_{k+1}} = \overline{(z_1 + z_2 + \cdots + z_k)} + \overline{z_{k+1}}$ (by S_2).
$= \overline{z_1} + \overline{z_2} + \cdots + \overline{z_k} + \overline{z_{k+1}}$ (by S_k).

29. S_1: i is either i or -1 or $-i$ or 1.
S_k: i^k is either i or -1 or $-i$ or 1.
$i^{k+1} = i^k \cdot i$ is then $i \cdot i = -1$ or $-1 \cdot i = -i$ or $-i \cdot i = 1$ or $1 \cdot i = i$.

31. S_1: 2 is a factor of $1^2 + 1$.
S_k: 2 is a factor of $k^2 + k$.
$(k + 1)^2 + (k + 1) = k^2 + 2k + 1 + k + 1$
$= k^2 + k + 2(k + 1).$
By S_k, 2 is a factor of $k^2 + k$; hence 2 is a factor of the right-hand side, so is a factor of $(k + 1)^2 + (k + 1)$.

33. S_1: 3 is a factor of $1(1 + 1)(1 + 2)$
S_k: 3 is a factor of $k(k + 1)(k + 2)$, or $k(k^2 + 3k + 2)$.
$(k + 1)(k + 1 + 1)(k + 1 + 2) = (k + 1)(k + 2)(k + 3)$
$= (k^2 + 3k + 2)(k + 3)$
$= k(k^2 + 3k + 2) + 3(k^2 + 3k + 2)$
By S_k, 3 is a factor of $k(k^2 + 3k + 2)$; hence 3 is a factor of the right-hand side, so 3 is a factor of $(k + 1)(k + 2)(k + 3)$.

35. The least number of moves for

1 disk(s) is $1 = 2^1 - 1$,
2 disk(s) is $3 = 2^2 - 1$,
3 disk(s) is $7 = 2^3 - 1$,
4 disk(s) is $15 = 2^4 - 1$; etc.

Let P_n be the least number of moves for n disks.
We conjecture and must show:
S_n: $P_n = 2^n - 1$.
1. Basis step: S_1 true by substitution.
2. Induction step: Assume S_k for k disks: $P_k = 2^k - 1$. Show: $P_{k+1} = 2^{k+1} - 1$. Now suppose there are $k + 1$ disks on one peg. Move k of them to another peg in $2^k - 1$ moves (by S_k) and move the remaining disk to the free peg (1 move). Then move the k disks onto it in (another) $2^k - 1$ moves. Thus the total moves P_{k+1} is $2(2^k - 1) + 1 = 2^{k+1} - 1$: $P_{k+1} = 2^{k+1} - 1$.

37. B is a set of k people of the same sex (by S_k), and C is a set of k people of the same sex (by S_k). But B might be all men and C might be all women; and then the argument fails.

Margin Exercises, Section 12.5

1. $3 \cdot 2 \cdot 1$, or 6; $3 \cdot 3 \cdot 3$, or 27
2. (a) $26 \cdot 10 \cdot 26 \cdot 10 \cdot 26 \cdot 10$, or 17,576,000; (b) no
3. $5 \cdot 4 \cdot 3 \cdot 2 \cdot 1$, or 120 **4.** $5 \cdot 4 \cdot 3 \cdot 2 \cdot 1$, or 120
5. $3 \cdot 2 \cdot 1$, or 6 **6.** $5 \cdot 4 \cdot 3 \cdot 2 \cdot 1$, or 120
7. $6 \cdot 5 \cdot 4 \cdot 3 \cdot 2 \cdot 1$, or 720
8. $4 \cdot 3 \cdot 2 \cdot 1$, or 24
9. $6 \cdot 5 \cdot 4 \cdot 3 \cdot 2 \cdot 1$, or 720
10. $8 \cdot 7 \cdot 6 \cdot 5 \cdot 4 \cdot 3 \cdot 2 \cdot 1$, or 40,320 **11.** 40,320
12. 362,880 **13.** 18! **14.** (a) $10! = 10 \cdot 9!$; (b) $20! = 20 \cdot 19 \cdot 18 \cdot 17 \cdot 16 \cdot 15!$
15. ${}_7P_3 = 7 \cdot 6 \cdot 5 = 210$; ${}_7P_3 = \frac{7!}{4!} = \frac{7 \cdot 6 \cdot 5 \cdot 4 \cdot 3 \cdot 2 \cdot 1}{4 \cdot 3 \cdot 2 \cdot 1} = 7 \cdot 6 \cdot 5 = 210$ **16.** (a) ${}_{10}P_4 = 10 \cdot 9 \cdot 8 \cdot 7 = 5040$; (b) ${}_8P_2 = 8 \cdot 7 = 56$; (c) ${}_{11}P_5 = 11 \cdot 10 \cdot 9 \cdot 8 \cdot 7 = 55{,}440$; (d) ${}_nP_1 = n$; (e) ${}_nP_2 = n(n - 1) = n^2 - n$; (f) 1
17. ${}_{12}P_5 = 12 \cdot 11 \cdot 10 \cdot 9 \cdot 8 = 95{,}040$
18. ${}_9P_3 = 9 \cdot 8 \cdot 7 = 504$
19. ${}_4P_4 \cdot {}_3P_3 = 4! \cdot 3! = 144$ **20.** $\frac{11!}{1!4!4!2!} = 34{,}650$
21. $\frac{6!}{3!2!1!} = 60$ **22.** $\frac{8!}{3!3!2!} = 560$ **23.** $6! = 720$
24. $11! = 39{,}916{,}800$ **25.** $26^5 = 11{,}881{,}376$

Exercise Set 12.5, pp. 757–758

1. $4 \cdot 3 \cdot 2$, or 24 **3.** ${}_{10}P_7 = 10 \cdot 9 \cdot 8 \cdot 7 \cdot 6 \cdot 5 \cdot 4$, or 604,800
5. 120; 3125 **7.** 120; 24 **9.** $\frac{5!}{2!1!1!1!} = 5 \cdot 4 \cdot 3 = 60$
11. $9 \cdot 9 \cdot 8 \cdot 7 \cdot 6 \cdot 5 \cdot 4$, or 544,320 **13.** $\frac{9!}{2!3!4!} = 1260$
15. 12!, or 479,001,600 **17.** (a) 120; (b) 3840
19. $52 \cdot 51 \cdot 50 \cdot 49 = 6{,}497{,}400$
21. $\frac{24!}{3!5!9!4!3!} = 16{,}491{,}024{,}950{,}400$
23. $4! = 24$, $8! \div 3! = 6720$, $\frac{13!}{2!2!2!2!2!} = 194{,}594{,}400$
25. $80 \cdot 26 \cdot 9999 = 20{,}797{,}920$
27. (a) 10^5, or 100,000; (b) 100,000
29. (a) 10^9, or 1,000,000,000; (b) yes
31. 11 **33.** 9 **35.** $n - 1$ **37.** Depending on the interpretation of the question, there can be either $(n - 1)!$ or $\frac{(n - 1)!}{2}$ placements. See the Student's Solution Manual or the Instructor's Solution Manual for a detailed description of the interpretations.

Margin Exercises, Section 12.6

1. (a) 1; (b) 5; (c) 10; (d) 5; (e) 1
2. (a) Ø, 1; (b) {A}, {B}, 2; (c) {A, B}, 1; (d) 4
3. (a) Ø, 1; (b) {A}, 1; (c) 2
4. 120 **5.** 120 **6.** 126 **7.** 126 **8.** n
9. $\frac{n(n - 1)(n - 2)}{6}$ **10.** (a) 56, 56; (b) $\binom{8}{3}$
11. 161,700 **12.** 45 **13.** 6160

Exercise Set 12.6, pp. 763–764

1. 78 **3.** 78 **5.** 7 **7.** 10 **9.** 1 **11.** 15 **13.** 12
15. 220 **17.** $\frac{m!}{2!(m - 2)!}$ **19.** $\frac{p!}{3!(p - 3)!}$ **21.** 2^7, or 128
23. 2^{100} **25.** $\binom{23}{4} = 8855$ **27.** $\binom{10}{6} = 210$ **29.** $\binom{8}{2} = 28$, $\binom{8}{3} = 56$ **31.** $\binom{10}{7} \cdot \binom{5}{3} = 1200$ **33.** $\binom{58}{6} \cdot \binom{42}{4}$
35. $\binom{4}{3} \cdot \binom{48}{2} = 4512$ **37.** (a) ${}_{33}P_3 = 32{,}736$; (b) $33^3 = 35{,}937$; (c) ${}_{33}C_3 = 5456$ **39.** $2^{12} = 4096$

41. $\binom{52}{5} = 2{,}598{,}960$ **43.** $\binom{8}{3} = 56$ **45.** $\binom{5}{2}\binom{8}{2} = 280$
47. 28, 56 **49.** 5 **51.** 7 **53.** ${}_5C_2 = 10$; ${}_5C_2 - 5 = 5$
55. (a) ${}_nC_2 = \dfrac{n(n-1)}{2}$;

(b) ${}_nC_2 - n = \dfrac{n(n-1)}{2} - \dfrac{2n}{2} = \dfrac{n(n-3)}{2}$, where $n = 4, 5, 6, \ldots$;

(c) Let D_n be the number of diagonals of an n-gon. We must prove S_n (below) using mathematical induction. We have

$$S_n\colon D_n = \frac{n(n-3)}{2}, \text{ for } n = 4, 5, 6, \ldots$$

$$S_4\colon D_4 = \frac{4 \cdot 1}{2}$$

$$S_k\colon D_k = \frac{k(k-3)}{2}$$

$$S_{k+1}\colon D_{k+1} = \frac{(k+1)(k-2)}{2}$$

1. Basis step: S_4 is true (a quadrilateral has 2 diagonals).

2. Induction step: Assume S_k. Observe that when an additional vertex V_{k+1} is added to the k-gon, we gain k segments, 2 of which are sides [of the $(k+1)$-gon], and a former side $\overline{V_1V_k}$ becomes a diagonal. Thus the additonal number of diagonals is $k - 2 + 1$, or $k - 1$. Then the new total of diagonals is $D_k + (k-1)$, or

$$\begin{aligned} D_{k+1} &= D_k + (k-1) \\ &= \frac{k(k-3)}{2} + (k-1) \qquad (\text{by } S_k) \\ &= \frac{(k+1)(k-2)}{2}. \end{aligned}$$

Margin Exercises, Section 12.7

1. 1 6 15 20 15 6 1; 1 7 21 35 35 21 7 1; 1 8 28 56 70 56 28 8 1 **2.** $a^9 + 9a^8b + 36a^7b^2 + 84a^6b^3 + 126a^5b^4 + 126a^4b^5 + 84a^3b^6 + 36a^2b^7 + 9ab^8 + b^9$
3. $x^4 - 4x^3y + 6x^2y^2 - 4xy^3 + y^4$ **4.** $128t^7 + 448t^5 + 672t^3 + 560t + 280t^{-1} + 84t^{-3} + 14t^{-5} + t^{-7}$ **5.** $x^9 - 9x^8y + 36x^7y^2 - 84x^6y^3 + 126x^5y^4 - 126x^4y^5 + 84x^3y^6 - 36x^2y^7 + 9xy^8 - y^9$
6. $x^5 + 25x^4b + 250x^3b^2 + 1250x^2b^3 + 3125xb^4 + 3125b^5$
7. $x^{10} - 5x^8 + 10x^6 - 10x^4 + 5x^2 - 1$
8. $16x^4 + 32x^3y^{-1} + 24x^2y^{-2} + 8xy^{-3} + y^{-4}$
9. $x^6 - 6\sqrt{2}x^5 + 30x^4 - 40\sqrt{2}x^3 + 60x^2 - 24\sqrt{2}x + 8$
10. $181{,}440a^4b^3$ **11.** $103{,}680x^2$ **12.** 2^6, or 64 **13.** 2^{50}

Exercise Set 12.7, pp. 771–772

1. $m^5 + 5m^4n + 10m^3n^2 + 10m^2n^3 + 5mn^4 + n^5$
3. $x^6 - 6x^5y + 15x^4y^2 - 20x^3y^3 + 15x^2y^4 - 6xy^5 + y^6$
5. $x^{10} - 15x^8y + 90x^6y^2 - 270x^4y^3 + 405x^2y^4 - 243y^5$
7. $729c^6 - 1458c^5d + 1215c^4d^2 - 540c^3d^3 + 135c^2d^4 - 18cd^5 + d^6$ **9.** $x^3 - 3x^2y + 3xy^2 - y^3$
11. $x^{-7} + 7x^{-6}y + 21x^{-5}y^2 + 35x^{-4}y^3 + 35x^{-3}y^4 + 21x^{-2}y^5 + 7x^{-1}y^6 + y^7$
13. $a^9 - 18a^7 + 144a^5 - 672a^3 + 2016a - 4032a^{-1} + 5376a^{-3} - 4608a^{-5} + 2304a^{-7} - 512a^{-9}$
15. $1 - n + \binom{n}{2} - \binom{n}{3} + \cdots + \binom{n}{n}(-1)^n$
17. $9 - 12\sqrt{3}t + 18t^2 - 4\sqrt{3}t^3 + t^4$ **19.** $140\sqrt{2}$
21. $x^{-8} + 4x^{-4} + 6 + 4x^4 + x^8$ **23.** $15a^4b^2$
25. $-745{,}472a^3$ **27.** $1120x^{12}y^2$ **29.** $-1{,}959{,}552u^5v^{10}$
31. 2^7, or 128 **33.** 2^{26}, or 67,108,864 **35.** $-7 - 4\sqrt{2}i$
37. $\sin^7 t - 7\sin^5 t + 21\sin^3 t - 35\sin t + 35\csc t - 21\csc^3 t + 7\csc^5 t - \csc^7 t$ **39.** $\displaystyle\sum_{r=0}^{n} \binom{n}{r}(-1)^r a^{n-r} b^r$
41. -3 **43.** 5 **45.** $\binom{5}{2}(0.313)^3(0.687)^2 \approx 0.14473$
47. 0.96403 **49.** $2{,}006{,}581{,}248u^5v^{10}$ **51.** $\dfrac{55}{144}$ **53.** $-\dfrac{\sqrt[3]{q}}{2p}$
55. 20 **57.** 3, 4, 5, 6 **59.** 0

Margin Exercises, Section 12.8

1. $\frac{82}{100}$ **2.** 73.4%; no **3.** 2074 **4.** 14.7%, 50.4% **5.** $\frac{1}{2}$
6. (a) $\frac{1}{13}$; (b) $\frac{1}{4}$; (c) $\frac{1}{2}$; (d) $\frac{2}{13}$ **7.** $\frac{6}{11}$ **8.** 0 **9.** 1
10. $\frac{11}{850}$ **11.** $\frac{5}{39}$ **12.** $\frac{6}{13}$ **13.** $\frac{1}{6}$

Exercise Set 12.8, pp. 779–781

1. 0.57, 0.43 **3.** 0.075, 0.134, 0.057, 0.071, 0.030
5. 0.633 **7.** 954 **9.** 52 **11.** $\frac{1}{4}$ **13.** $\frac{1}{13}$ **15.** $\frac{1}{2}$
17. $\frac{2}{13}$ **19.** $\frac{5}{7}$ **21.** 0 **23.** $\frac{11}{4165}$ **25.** $\frac{28}{65}$ **27.** $\frac{1}{18}$
29. $\frac{1}{36}$ **31.** $\frac{30}{323}$ **33.** $\frac{9}{19}$ **35.** $\frac{1}{38}$ **37.** $\frac{1}{19}$ **39.** 2,598,960
41. (a) 36; (b) 1.39×10^{-5}
43. (a) $(13 \cdot {}_4C_3) \cdot (12 \cdot {}_4C_2) = 3744$;
(b) $\dfrac{3744}{{}_{52}C_5} = \dfrac{3744}{2{,}598{,}960} \approx 0.00144$
45. (a) $13 \cdot \binom{4}{3} \cdot \binom{48}{2} - 3744 = 54{,}912$;
(b) $\dfrac{54{,}912}{{}_{52}C_5} = \dfrac{54{,}912}{2{,}598{,}960} \approx 0.0211$
47. (a) $\binom{13}{2}\binom{4}{2}\binom{4}{2}\binom{44}{1} = 123{,}552$;
(b) $\dfrac{123{,}552}{{}_{52}C_5} = \dfrac{123{,}552}{2{,}598{,}960} \approx 0.0475$
49. (a) $\dfrac{\pi}{4}$; (b) $\dfrac{4 - \pi}{4}$; (c) $\pi = 4p$, since $p \approx \dfrac{78}{100}$, $\pi \approx 4(0.78) = 3.12$

Review Exercises: Chapter 12, pp. 782–783

1. [12.2] $3\frac{3}{4}$ **2.** [12.2] $a + 4b$ **3.** [12.2] 531
4. [12.2] 465 **5.** [12.2] 11 **6.** [12.2] -4
7. [12.3] $n = 6$, $S_n = -126$ **8.** [12.3] $a_1 = 8$, $a_5 = \frac{1}{2}$
9. [12.3] No **10.** [12.3] Yes **11.** [12.3] $\frac{3}{8}$ **12.** [12.3] $\frac{211}{99}$
13. [12.2] $5\frac{4}{5}$, $6\frac{3}{5}$, $7\frac{2}{5}$, $8\frac{1}{5}$ **14.** [12.3] ≈ 50 ft
15. [12.2] \$7.38, \$1365.10 **16.** [12.3] 7,680,000
17. [12.3] $\dfrac{64\sqrt{2}}{\sqrt{2} - 1}$ in., or $128 + 64\sqrt{2}$ in. **18.** [12.3] $23\frac{1}{3}$ cm
19. [12.4] $S_n\colon 1 + 4 + 7 + \cdots + (3n - 2) = \dfrac{n(3n-1)}{2}$

$$S_1\colon 1 = \frac{1(3-1)}{2}$$

$$S_k\colon 1 + 4 + 7 + \cdots + (3k - 2) = \frac{k(3k-1)}{2}$$

$$\begin{aligned} S_{k+1}\colon & 1 + 4 + 7 + \cdots + [3(k+1) - 2] \\ &= 1 + 4 + 7 + \cdots + (3k - 2) + (3k + 1) \\ &= \frac{(k+1)(3k+2)}{2} \end{aligned}$$

1. Basis step: $1 = \frac{2}{2} = \frac{1(3-1)}{2}$ is true.

2. Induction step: Assume S_k. Add $(3k + 1)$ to both sides.

$$1 + 4 + 7 + \cdots + (3k-2) + (3k+1) = \frac{k(3k-1)}{2} + (3k+1) = \frac{k(3k-1)}{2} + \frac{2(3k+1)}{2} = \frac{3k^2 - k + 6k + 2}{2} = \frac{3k^2 + 5k + 2}{2} = \frac{(k+1)(3k+2)}{2}$$

20. [12.4] S_1: $1 = \frac{3^1 - 1}{2}$; S_2: $1 + 3 = \frac{3^2 - 1}{2}$

S_k: $1 + 3 + 3^2 + \cdots + 3^{k-1} = \frac{3^k - 1}{2}$

2. Induction step: Assume S_k. Add 3^k on both sides.

$$1 + 3 + \cdots + 3^{k-1} + 3^k = \frac{3^k - 1}{2} + 3^k = \frac{3^k - 1}{2} + 3^k \cdot \frac{2}{2} = \frac{3 \cdot 3^k - 1}{2} = \frac{3^{k+1} - 1}{2}$$

21. [12.4]

S_n: $\left(1 - \frac{1}{2}\right)\left(1 - \frac{1}{3}\right)\cdots\left(1 - \frac{1}{n}\right) = \frac{1}{n}$

S_2: $\left(1 - \frac{1}{2}\right) = \frac{1}{2}$

S_k: $\left(1 - \frac{1}{2}\right)\left(1 - \frac{1}{3}\right)\cdots\left(1 - \frac{1}{k}\right) = \frac{1}{k}$

S_{k+1}: $\left(1 - \frac{1}{2}\right)\left(1 - \frac{1}{3}\right)\cdots\left(1 - \frac{1}{k}\right)\left(1 - \frac{1}{k+1}\right) = \frac{1}{k+1}$

1. Basis step: S_2 is true by substitution.

2. Induction step: Assume S_k. Deduce S_{k+1}. Starting with the left side of S_{k+1} we have:

$$\underbrace{\left(1 - \frac{1}{2}\right)\left(1 - \frac{1}{3}\right)\cdots\left(1 - \frac{1}{k}\right)}\left(1 - \frac{1}{k+1}\right)$$

$$= \frac{1}{k} \cdot \left(1 - \frac{1}{k+1}\right) \quad \text{(by } S_k\text{)}$$

$$= \frac{1}{k} \cdot \left(\frac{k+1-1}{k+1}\right)$$

$$= \frac{1}{k} \cdot \frac{k}{k+1}$$

$$= \frac{1}{k+1} \quad \text{(simplifying)}$$

22. [12.1] 5; 51; 5203; 54,142,419

23. [12.1] $\sum_{n=1}^{7} (n^2 - 1)$ or $\sum_{n=0}^{6} n(n+2)$

24. [12.5] $6! = 720$ **25.** [12.5] $9 \cdot 8 \cdot 7 \cdot 6 = 3024$

26. [12.6] $\binom{15}{8} = 6435$ **27.** [12.5] $24 \cdot 23 \cdot 22 = 12{,}144$

28. [12.5] $\frac{9!}{1!4!2!2!} = 3780$ **29.** [12.5] 36 **30.** [12.5], [12.6] **(a)** ${}_6P_5 = 720$; **(b)** $6^5 = 7776$; **(c)** ${}_5P_4 = 120$; **(d)** ${}_3P_2 = 6$

31. [12.5] $6! = 720$ **32.** [12.7] $220a^9x^3$ **33.** [12.7] $\binom{18}{11}a^7x^{11}$

34. [12.7] $m^7 + 7m^6n + 21m^5n^2 + 35m^4n^3 + 35m^3n^4 + 21m^2n^5 + 7mn^6 + n^7$

35. [12.7] $x^8 + 12x^6y + 54x^4y^2 + 108x^2y^3 + 81y^4$

36. [12.7] $-6624 + 16{,}280i$

37. [12.7] $\cos^8 t + 8\cos^6 t + 28\cos^4 t + 56\cos^2 t + 70 + 56\sec^2 t + 28\sec^4 t + 8\sec^6 t + \sec^8 t$

38. [12.8] $\frac{86}{206} \approx 0.42$, $\frac{97}{206} \approx 0.47$, $\frac{23}{206} \approx 0.11$

39. [12.8] $\frac{1}{12}$, 0 **40.** [12.8] $\frac{1}{4}$ **41.** [12.8] $\frac{6}{5525}$

42. [12.4] S_1 fails.

43. [12.3] $\frac{a_{k+1}}{a_k} = r_1$, $\frac{b_{k+1}}{b_k} = r_2$, so $\frac{a_{k+1}b_{k+1}}{a_kb_k} = r_1r_2$ (constant)

44. [12.2] $a_{k+1} - a_k = d$, so $\frac{c_{k+1}}{c_k} = \frac{b^{a_{k+1}}}{b^{a_k}} = b^{a_{k+1}-a_k} = b^d$ (constant)

45. [12.2] **(a)** a_n is all positive or all negative; **(b)** always; **(c)** always; **(d)** $a_n = k$, k a constant; **(e)** $a_n = k$, k a constant; **(f)** $a_n = k$, k a constant

46. [12.2] -2, 0, 2, 4 **47.** [12.3] 0.27, 0.0027, 0.000027

48. [12.3] $\frac{1}{2}$, $-\frac{1}{6}$, $\frac{1}{18}$

49. [12.6] $\left(\log \frac{x}{y}\right)^{10}$ **50.** [12.6] 36 **51.** [12.6] 14

Test: Chapter 12, pp. 783–784

1. [12.2] 13 **2.** [12.2] 4 **3.** [12.2] 30th

4. [12.2] $\frac{23}{4}$, $\frac{17}{2}$, $\frac{45}{4}$ **5.** [12.3] 437.4 **6.** [12.3] $\frac{31}{64}$

7. [12.3] a **8.** [12.3] $\frac{125}{6}$ **9.** [12.2] \$2010

10. [12.3] 12,960 **11.** [12.3] $\frac{29}{225}$

12. [12.4] S_n: $5 + 10 + 15 + \cdots + 5n = \frac{5n(n+1)}{2}$

S_1: $5 = \frac{5 \cdot 1(1+1)}{2}$

S_k: $5 + 10 + 15 + \cdots + 5k = \frac{5k(k+1)}{2}$

S_{k+1}: $5 + 10 + 15 + \cdots + 5k + 5(k+1) = \frac{5(k+1)(k+2)}{2}$

1. Basis step: $\frac{5 \cdot 1(1+1)}{2} = 5$, so S_1 is true.

2. Induction step: Assume S_k. Then add $5(k+1)$ on both sides.

$$5 + 10 + 15 + \cdots + 5k + 5(k+1) = \frac{5k(k+1)}{2} + 5(k+1) = \frac{5k(k+1)}{2} + \frac{10(k+1)}{2} = \frac{5k^2 + 15k + 10}{2} = \frac{5(k^2 + 3k + 2)}{2} = \frac{5(k+1)(k+2)}{2}$$

13. [12.1] 5, 23, 95, 383

14. [12.5] **(a)** 6^4, or 1296; **(b)** 360; **(c)** 12

15. [12.5] 5!, or 120 **16.** [12.6] 35 **17.** [12.6] 103,740

18. [12.5] 3360 **19.** [12.6] 2^8, or 256

20. [12.7] $672a^5b^2$

21. [12.7] $x^5 - 5\sqrt{2}x^4 + 20x^3 - 20\sqrt{2}x^2 + 20x - 4\sqrt{2}$

22. [12.8] $\frac{5}{36}$ **23.** [12.8] $\frac{2}{13}$ **24.** [12.8] $\frac{1}{14}$

25. [12.2] 4, 7, 10, 13 **26.** [12.6] 54 **27.** [12.6] 18

28. [12.3] 9 cis 60°; 9 cis 120°; 9 cis 180°, or -9; 9 cis 240°; 9 cis 300°

Index

Miscellaneous formulas of general interest

A

B

C

D

Q

R

GEOMETRIC FORMULAS

Plane Geometry

Rectangle
Area: $A = lw$
Perimeter: $P = 2l + 2w$

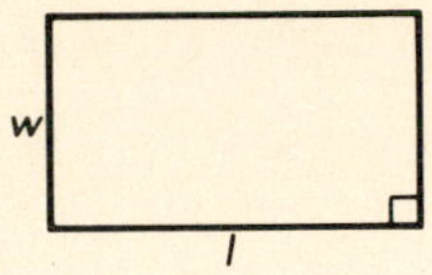

Square
Area: $A = s^2$
Perimeter: $P = 4s$

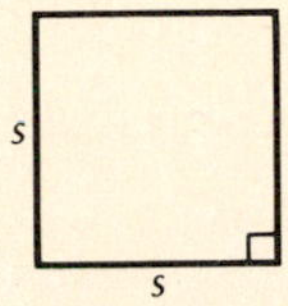

Triangle
Area: $A = \frac{1}{2}bh$

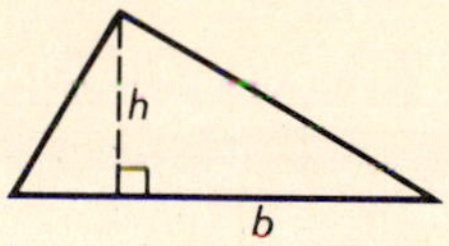

Sum of Angle Measures:
$A + B + C = 180°$

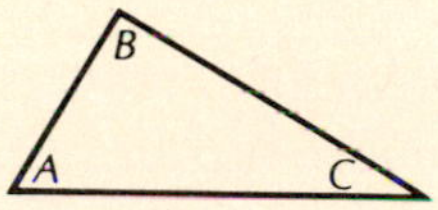

Right Triangle
Pythagorean Theorem
(Equation):
$a^2 + b^2 = c^2$

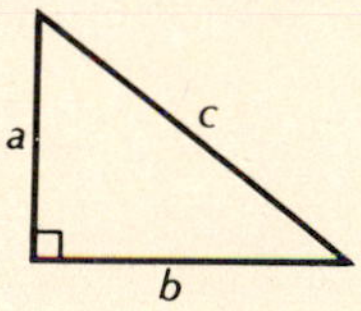

Parallelogram
Area: $A = bh$

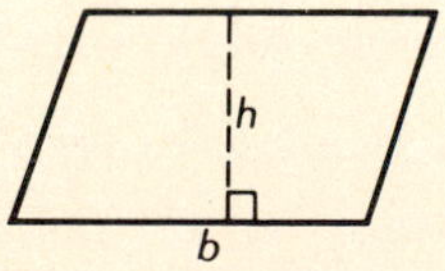

Trapezoid
Area: $A = \frac{1}{2}h(a + b)$

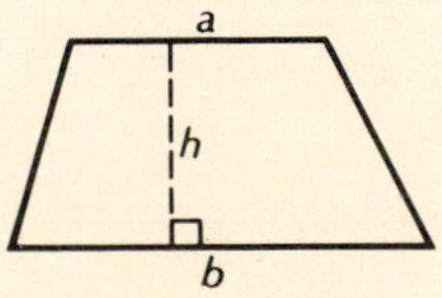

Circle
Area: $A = \pi r^2$
Circumference:
$C = \pi D = 2\pi r$
($\frac{22}{7}$ and 3.14 are different approximations for π)

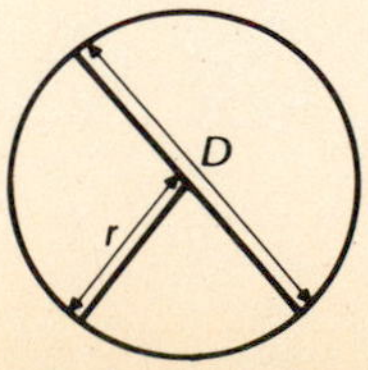

Solid Geometry

Rectangular Solid
Volume: $V = lwh$

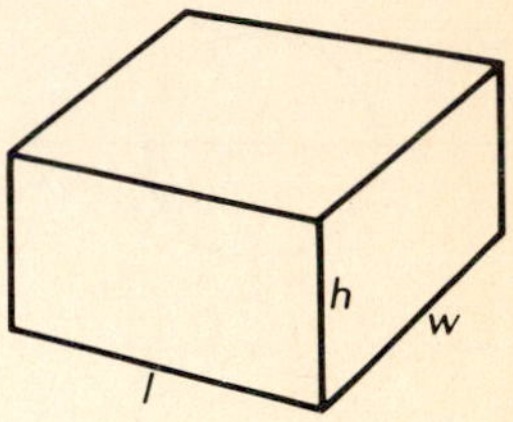

Cube
Volume: $V = s^3$

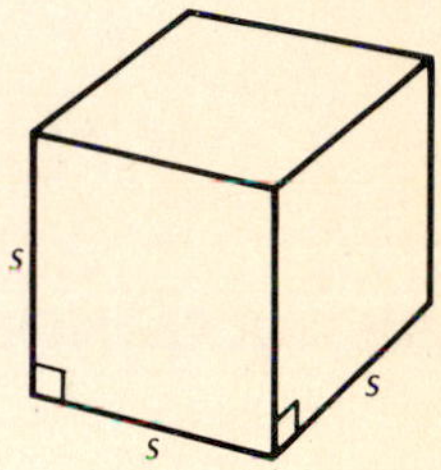

Right Circular Cylinder
Volume: $V = \pi r^2 h$
Lateral Surface Area:
$L = 2\pi rh$
Total Surface Area:
$S = 2\pi rh + 2\pi r^2$

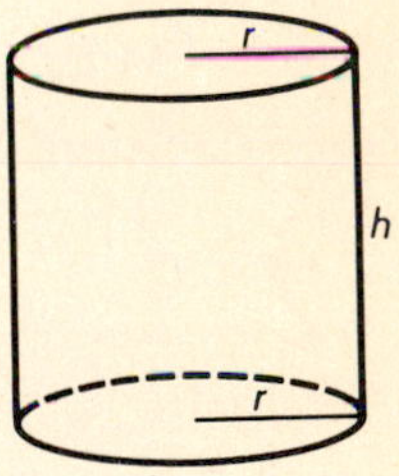

Right Circular Cone
Volume: $V = \frac{1}{3}\pi r^2 h$
Lateral Surface Area:
$L = \pi rs$
Total Surface Area:
$S = \pi r^2 + \pi rs$
Slant Height:
$s = \sqrt{r^2 + h^2}$

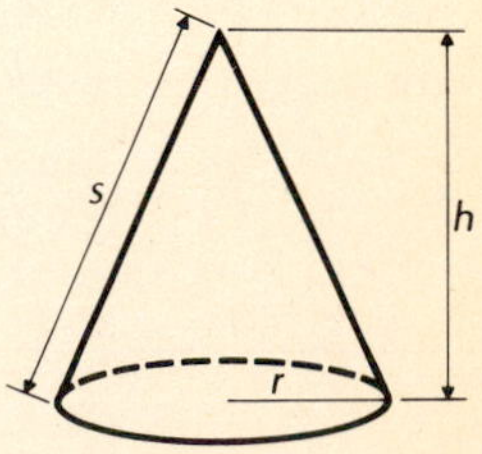

Sphere
Volume: $V = \frac{4}{3}\pi r^3$
Surface Area: $S = 4\pi r^2$

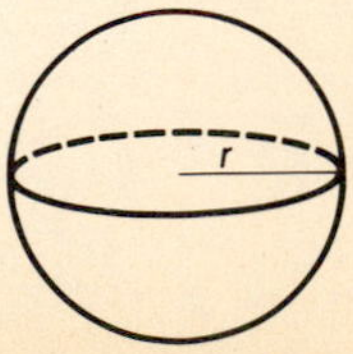

TRIGONOMETRIC FUNCTIONS

Real Numbers

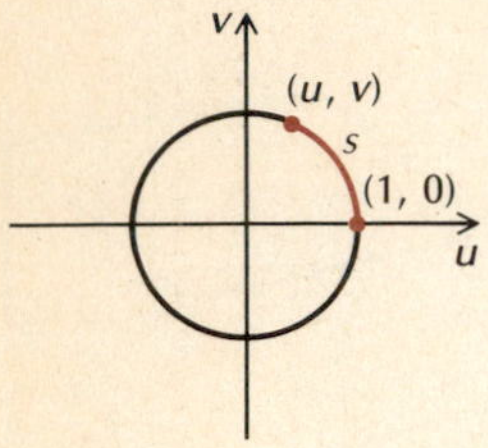

$\sin\theta = v, \quad \csc\theta = \frac{1}{v},$

$\cos\theta = u, \quad \sec\theta = \frac{1}{u},$

$\tan\theta = \frac{v}{u}, \quad \cot\theta = \frac{u}{v}$

Any Angle

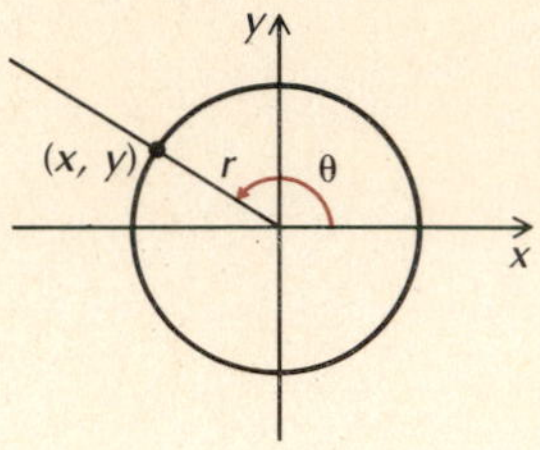

$\sin\theta = \frac{y}{r}, \quad \csc\theta = \frac{r}{y},$

$\cos\theta = \frac{x}{r}, \quad \sec\theta = \frac{r}{x},$

$\tan\theta = \frac{y}{x}, \quad \cot\theta = \frac{x}{y}$

Acute Angles

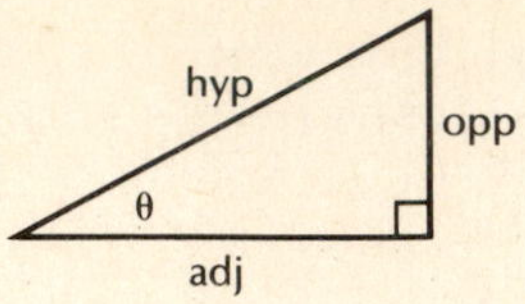

$\sin\theta = \frac{\text{opp}}{\text{hyp}}, \quad \csc\theta = \frac{\text{hyp}}{\text{opp}},$

$\cos\theta = \frac{\text{adj}}{\text{hyp}}, \quad \sec\theta = \frac{\text{hyp}}{\text{adj}},$

$\tan\theta = \frac{\text{opp}}{\text{adj}}, \quad \cot\theta = \frac{\text{adj}}{\text{opp}}$

BASIC TRIGONOMETRIC IDENTITIES

$\sin(-x) = -\sin x,$

$\cos(-x) = \cos x,$

$\tan(-x) = -\tan x,$

$\tan x = \frac{\sin x}{\cos x},$

$\cot x = \frac{\cos x}{\sin x},$

$\csc x = \frac{1}{\sin x},$

$\sec x = \frac{1}{\cos x},$

$\cot x = \frac{1}{\tan x}$

Pythagorean Identities

$\sin^2 x + \cos^2 x = 1,$

$1 + \tan^2 x = \sec^2 x,$

$1 + \cot^2 x = \csc^2 x$

Cofunction Identities

$\sin\left(x \pm \frac{\pi}{2}\right) = \pm\cos x,$

$\cos\left(x \pm \frac{\pi}{2}\right) = \mp\sin x,$

$\tan\left(x \pm \frac{\pi}{2}\right) = -\cot x$

Sum and Difference Identities

$\sin(\alpha \pm \beta) = \sin\alpha\cos\beta \pm \cos\alpha\sin\beta,$

$\cos(\alpha \pm \beta) = \cos\alpha\cos\beta \mp \sin\alpha\sin\beta,$

$\tan(\alpha \pm \beta) = \frac{\tan\alpha \pm \tan\beta}{1 \mp \tan\alpha\tan\beta}$

Double-Angle Identities

$\sin 2x = 2\sin x\cos x,$

$$\cos 2x = \cos^2 x - \sin^2 x = 1 - 2\sin^2 x = 2\cos^2 x - 1,$$

$\tan 2x = \frac{2\tan x}{1 - \tan^2 x},$

$\sin^2 x = \frac{1 - \cos 2x}{2},$

$\cos^2 x = \frac{1 + \cos 2x}{2}$

Half-Angle Identities

$\sin\frac{x}{2} = \pm\sqrt{\frac{1 - \cos x}{2}},$

$\cos\frac{x}{2} = \pm\sqrt{\frac{1 + \cos x}{2}},$

$\tan\frac{x}{2} = \pm\sqrt{\frac{1 - \cos x}{1 + \cos x}} = \frac{\sin x}{1 + \cos x} = \frac{1 - \cos x}{\sin x}$